U.S. Customary and SI unit systems.

Base Dimension	System of Units: U.S. Customary	SI
force	pound (lb)	newton[a] (N) $\equiv$ kg·m/s^2
mass	slug[a] $\equiv$ lb·s^2/ft	kilogram (kg)
length	foot (ft)	meter (m)
time	second (s)	second (s)

[a] Derived unit.

Conversion factors between U.S. Customary and SI unit systems.

	U.S. Customary		SI
length	1 in.	=	0.0254 m (2.54 cm, 25.4 mm)[a]
	1 ft (12 in.)	=	0.3048 m[a]
	1 mi (5280 ft)	=	1.609 km
force	1 lb	=	4.448 N
	1 kip (1000 lb)	=	4.448 kN
mass	1 slug (1 lb·s^2/ft)	=	14.59 kg

[a] Exact.

Common prefixes used in the SI unit systems.

Multiplication Factor		Prefix	Symbol
1 000 000 000 000 000 000 000 000	10^{24}	yotta	Y
1 000 000 000 000 000 000 000	10^{21}	zetta	Z
1 000 000 000 000 000 000	10^{18}	exa	E
1 000 000 000 000 000	10^{15}	peta	P
1 000 000 000 000	10^{12}	tera	T
1 000 000 000	10^{9}	giga	G
1 000 000	10^{6}	mega	M
1 000	10^{3}	kilo	k
100	10^{2}	hecto	h
10	10^{1}	deka	da
0.1	10^{-1}	deci	d
0.01	10^{-2}	centi	c
0.001	10^{-3}	milli	m
0.000 001	10^{-6}	micro	μ
0.000 000 001	10^{-9}	nano	n
0.000 000 000 001	10^{-12}	pico	p
0.000 000 000 000 001	10^{-15}	femto	f
0.000 000 000 000 000 001	10^{-18}	atto	a
0.000 000 000 000 000 000 001	10^{-21}	zepto	z
0.000 000 000 000 000 000 000 001	10^{-24}	yocto	y

Engineering Mechanics
DYNAMICS

Engineering Mechanics
DYNAMICS
SECOND EDITION

Gary L. Gray
Department of Engineering Science and Mechanics
Penn State University

Francesco Costanzo
Department of Engineering Science and Mechanics
Penn State University

Michael E. Plesha
Department of Engineering Physics
University of Wisconsin—Madison

The McGraw-Hill Companies

ENGINEERING MECHANICS: DYNAMICS, SECOND EDITION

Published by McGraw-Hill, a business unit of The McGraw-Hill Companies, Inc., 1221 Avenue of the Americas, New York, NY 10020.

This book is printed on acid-free paper.

1 2 3 4 5 6 7 8 9 0 DOW/DOW 1 0 9 8 7 6 5 4 3 2

ISBN 978–0–07–338030–8
MHID 0–07–338030–X

Vice President, Editor-in-Chief: *Marty Lange*
Vice President, EDP: *Kimberly Meriwether David*
Senior Director of Development: *Kristine Tibbetts*
Global Publisher: *Raghothaman Srinivasan*
Editorial Director: *Michael Lange*
Executive Editor: *Bill Stenquist*
Developmental Editor: *Darlene M. Schueller*
Senior Marketing Manager: *Curt Reynolds*
Lead Project Manager: *Sheila M. Frank*
Senior Buyer: *Sherry L. Kane*
Lead Media Project Manager: *Judi David*
Senior Designer: *David W. Hash*
Cover Designer: *Asylum Studios*
Cover Image: *The bright sun greets the International Space Station in this Nov. 22, 2009 scene from the Russian section of the orbital outpost, photographed by one of the STS-129 crew members. NASA*
Lead Photo Research Coordinator: *Carrie K. Burger*
Photo Research: *Danny Meldung/Photo Affairs, Inc*
Compositor: *Aptara, Inc.*
Typeface: *10/13 Times LT Std*
Printer: *R. R. Donnelley*

Library of Congress Cataloging-in-Publication Data

Gray, Gary L.
Engineering mechanics. Dynamics / Gary L. Gray, Francesco Costanzo, Michael E. Plesha. – 2nd ed.
p. cm.
Includes index.
ISBN 978–0–07–338030–8 — ISBN 0–07–338030–X (alk. paper)
1. Dynamics. 2. Mechanics, Applied. I. Costanzo, Francesco, 1964- II. Plesha, Michael E. III. Title.
TA352.G724 2013
620.1'04–dc23
2011049531

www.mhhe.com

ABOUT THE AUTHORS

Gary L. Gray is an Associate Professor of Engineering Science and Mechanics in the Department of Engineering Science and Mechanics at Penn State in University Park, PA. He received a B.S. in Mechanical Engineering (cum laude) from Washington University in St. Louis, MO, an S.M. in Engineering Science from Harvard University, and M.S. and Ph.D. degrees in Engineering Mechanics from the University of Wisconsin-Madison. His primary research interests are in dynamical systems, dynamics of mechanical systems, and mechanics education. For his contributions to mechanics education, he has been awarded the Outstanding and Premier Teaching Awards from the Penn State Engineering Society, the Outstanding New Mechanics Educator Award from the American Society for Engineering Education, the Learning Excellence Award from General Electric, and the Collaborative and Curricular Innovations Special Recognition Award from the Provost of Penn State. In addition to dynamics, he also teaches mechanics of materials, mechanical vibrations, numerical methods, advanced dynamics, and engineering mathematics.

Francesco Costanzo is a Professor of Engineering Science and Mechanics in the Engineering Science and Mechanics Department at Penn State. He received the Laurea in Ingegneria Aeronautica from the Politecnico di Milano, Milan, Italy. After coming to the U.S. as a Fulbright scholar he received his Ph.D. in aerospace engineering from Texas A&M University. His primary research interest is the mathematical and numerical modeling of material behavior. His specific research interests include the theoretical and numerical characterization of dynamic fracture in materials subject to thermo-mechanical loading, the development of multi-scale methods for predicting continuum-level material properties from molecular calculations, and modeling and computational problems in bio-medical applications. In addition to scientific research, he has contributed to various projects for the advancement of mechanics education under the sponsorship of several organizations, including the National Science Foundation. For his contributions, he has received various awards, including the 1998 and the 2003 GE Learning Excellence Awards, and the 1999 ASEE Outstanding New Mechanics Educator Award. In addition to teaching dynamics, he also teaches statics, mechanics of materials, continuum mechanics, and mathematical theory of elasticity.

Michael E. Plesha is a Professor of Engineering Mechanics in the Department of Engineering Physics at the University of Wisconsin-Madison. Professor Plesha received his B.S. from the University of Illinois-Chicago in structural engineering and materials, and his M.S. and Ph.D. from Northwestern University in structural engineering and applied mechanics. His primary research areas are computational mechanics, focusing on the development of finite element and discrete element methods for solving static and dynamic nonlinear problems, and the development of constitutive models for characterizing behavior of materials. Much of his work focuses on problems featuring contact, friction, and material interfaces. Applications include nanotribology, high temperature rheology of ceramic composite materials, modeling geomaterials including rock and soil, penetration mechanics, and modeling crack growth in structures. He is co-author of the book *Concepts and Applications of Finite Element Analysis* (with R. D. Cook, D. S. Malkus, and R. J. Witt). He teaches courses in statics, basic and advanced mechanics of materials, mechanical vibrations, and finite element methods.

The authors thank their families for their patience, understanding, and, most importantly, encouragement during the long years it took to bring these books to completion. Without their support, none of this would have been possible.

BRIEF CONTENTS

Dynamics

1 Introduction to Dynamics 1

2 Particle Kinematics 29

3 Force and Acceleration Methods for Particles 169

4 Energy Methods for Particles 241

5 Momentum Methods for Particles 313

6 Planar Rigid Body Kinematics 433

7 Newton-Euler Equations for Planar Rigid Body Motion 521

8 Energy and Momentum Methods for Rigid Bodies 583

9 Mechanical Vibrations 663

10 Three-Dimensional Dynamics of Rigid Bodies 721

Appendices

A Mass Moments of Inertia A-1

B Angular Momentum of a Rigid Body A-11

C Answers to Even-Numbered Problems A-15

Credits C-1

Index I-1

TABLE OF CONTENTS

Preface xiii

1 Introduction to Dynamics 1

1.1 The Newtonian Equations 1

1.2 Fundamental Concepts in Dynamics 5

Space and time **5**

Force, mass, and inertia **5**

Particle and rigid body **6**

Vectors and their Cartesian representation **6**

Useful vector “tips and tricks” **9**

Units **10**

1.3 Dynamics and Engineering Design 26

System modeling **27**

2 Particle Kinematics 29

2.1 Position, Velocity, Acceleration, and Cartesian Coordinates . . . 29

Position vector **30**

Trajectory **30**

Velocity vector and speed **30**

Acceleration vector **32**

Cartesian coordinates **32**

2.2 Elementary Motions 47

Rectilinear motion relations **47**

Circular motion and angular velocity **49**

2.3 Projectile Motion 68

2.4 The Time Derivative of a Vector 80

Time derivative of a unit vector **80**

Time derivative of an arbitrary vector **81**

2.5 Planar Motion: Normal-Tangential Components 92

Normal-tangential components **92**

2.6 Planar Motion: Polar Coordinates 105

Polar coordinates and position, velocity, and acceleration **105**

2.7 Relative Motion Analysis and Differentiation of Geometrical Constraints 121

Relative motion **121**

Differentiation of geometrical constraints **122**

2.8 Motion in Three Dimensions 142

Cylindrical coordinates **142**
Spherical coordinates **143**
Cartesian coordinates **144**

Chapter Review **155**

3 Force and Acceleration Methods for Particles 169

3.1 Rectilinear Motion **169**
Applying Newton's second law **169**
Force laws **171**
Equation(s) of motion **173**
Inertial reference frames **173**
Degrees of freedom **174**

3.2 Curvilinear Motion **192**
Newton's second law in 2D and 3D component systems **192**

3.3 Systems of Particles **217**
Engineering materials one atom at a time **217**
Newton's second law for systems of particles **217**

Chapter Review **234**

4 Energy Methods for Particles 241

4.1 Work-Energy Principle for a Particle **241**
Work-energy principle and its relation with $\vec{F} = m\vec{a}$ **241**
Work of a force **243**

4.2 Conservative Forces and Potential Energy **257**
Work done by the constant force of gravity **257**
Work of a central force **257**
Conservative forces and potential energy **258**
Work-energy principle for any type of force **260**
When is a force conservative? **260**

4.3 Work-Energy Principle for Systems of Particles **281**
Internal work and work-energy principle for a system **281**
Kinetic energy for a system of particles **282**

4.4 Power and Efficiency **298**
Power developed by a force **298**
Efficiency **298**

Chapter Review **305**

5 Momentum Methods for Particles 313

5.1 Momentum and Impulse **313**
Impulse-momentum principle **313**
Conservation of linear momentum **316**

5.2 **Impact** **335**
Impacts are short, dramatic events **335**
Definition of impact and notation **335**
Line of impact and contact force between impacting objects **335**
Impulsive forces and impact-relevant FBDs **336**
Coefficient of restitution **336**
Unconstrained direct central impact **338**
Unconstrained oblique central impact **338**
Impact and energy **339**

5.3 **Angular Momentum** **361**
Moment-angular momentum relation for a particle **361**
Angular impulse-momentum for a system of particles **362**
Euler's first and second laws of motion **365**

5.4 **Orbital Mechanics** **385**
Determination of the orbit **385**
Energy considerations **391**

5.5 **Mass Flows** **401**
Steady flows **401**
Variable mass flows and propulsion **404**

Chapter Review **422**

6 **Planar Rigid Body Kinematics** **433**

6.1 **Fundamental Equations, Translation, and Rotation About a Fixed Axis** **433**
Crank, connecting rod, and piston motion **433**
Qualitative description of rigid body motion **434**
General motion of a rigid body **435**
Elementary rigid body motions: translations **437**
Elementary rigid body motions: rotation about a fixed axis **438**
Planar motion in practice **439**

6.2 **Planar Motion: Velocity Analysis** **452**
Vector approach **452**
Differentiation of constraints **453**
Instantaneous center of rotation **453**

6.3 **Planar Motion: Acceleration Analysis** **474**
Vector approach **474**
Differentiation of constraints **474**
Rolling without slip: acceleration analysis **475**

6.4 **Rotating Reference Frames** **494**
The general kinematic equations for the motion of a point relative to a rotating reference frame **494**
Coriolis component of acceleration **498**

Chapter Review 512

7 Newton-Euler Equations for Planar Rigid Body Motion. 521

7.1 Newton-Euler Equations for Bodies Symmetric with Respect to the Plane of Motion. 521

Linear momentum: translational equations 521

Angular momentum: rotational equations 522

Graphical interpretation of the equations of motion 526

7.2 Newton-Euler Equations: Translation. 529

7.3 Newton-Euler Equations: Rotation About a Fixed Axis 539

7.4 Newton-Euler Equations: General Plane Motion 553

Newton-Euler equations for general plane motion 553

Chapter Review 575

8 Energy and Momentum Methods for Rigid Bodies 583

8.1 Work-Energy Principle for Rigid Bodies. 583

Kinetic energy of rigid bodies in planar motion 583

Work-energy principle for a rigid body 585

Work done on rigid bodies 585

Potential energy and conservation of energy 586

Work-energy principle for systems 588

Power 588

8.2 Momentum Methods for Rigid Bodies. 618

Impulse-momentum principle for a rigid body 618

Angular impulse-momentum principle for a rigid body 619

8.3 Impact of Rigid Bodies. 639

Rigid body impact: basic nomenclature and assumptions 640

Classification of impacts 640

Central impact 640

Eccentric impact 642

Constrained eccentric impact 643

Chapter Review 656

9 Mechanical Vibrations. 663

9.1 Undamped Free Vibration 663

Oscillation of a railcar after coupling 663

Standard form of the harmonic oscillator 665

Linearizing nonlinear systems 666

Energy method 667

9.2 Undamped Forced Vibration 681

Standard form of the forced harmonic oscillator 681

9.3 **Viscously Damped Vibration** 695
Viscously damped free vibration **695**
Viscously damped forced vibration **698**
Chapter Review 714

10 Three-Dimensional Dynamics of Rigid Bodies 721
10.1 **Three-Dimensional Kinematics of Rigid Bodies** 721
Computation of angular accelerations **722**
Summing angular velocities **722**
10.2 **Three-Dimensional Kinetics of Rigid Bodies** 738
Newton-Euler equations for three-dimensional motion **738**
Kinetic energy of a rigid body in three-dimensional motion **743**
Chapter Review 759

A Mass Moments of Inertia A-1
Definition of mass moments and products of inertia **A-1**
How are mass moments of inertia used? **A-3**
Radius of gyration **A-4**
Parallel axis theorem **A-4**
Principal moments of inertia **A-6**
Moment of inertia about an arbitrary axis **A-9**
Evaluation of moments of inertia using composite shapes **A-10**

B Angular Momentum of a Rigid Body A-11
Angular momentum of a rigid body undergoing three-dimensional motion **A-11**
Angular momentum of a rigid body in planar motion **A-14**

C Answers to Even-Numbered Problems A-15

Credits C-1

Index I-1

PREFACE

Dynamics is the science that relates motion to the forces that cause and are caused by that motion. Dynamics is at the heart of the design and analysis of mechanical systems whose operating principles rely on motion or are meant to control motion. The engineering applications of dynamics are many and varied. Traditional applications include the design of mechanisms, engines, turbines, and airplanes. Other (perhaps less known) applications include the kinesiology of the human body, the analysis of cell motion, and the design of some micro- and nano-size devices, including both sensors and actuators. All of these applications stem from the combination of kinematics, which describes the geometry of motion, with a few basic principles anchored in Newton's laws of motion, such as the work-energy and the impulse-momentum principles.

With this book we hope to provide a teaching and learning experience that is not only effective but also motivates the study and application of dynamics. We have structured the book to achieve four main objectives. First, we provide a rigorous introduction to the fundamental principles of particle and rigid body dynamics. In a constantly changing technological landscape, it is by relying on fundamentals that we can find new ways of applying what we know. Second, we incorporate those pedagogical principles that recent research in math, science, and engineering education has identified as essential for improving student learning. While it is commonly accepted that a good conceptual understanding is important to improve problem-solving skills, it has been discovered that problem-solving skills and concepts need to be taught in different ways. Third, we have made *modeling* the underlying theme of our approach to problem solving. We believe that modeling, understood as the making of sensible assumptions to reduce a real complex problem to a simpler but solvable problem, is also something that must be taught and discussed alongside the basic principles. Fourth, we emphasize a systematic approach to solving every problem, an integral part of which is creating the aforementioned model. The four objectives that animate this textbook have been incorporated in a series of clearly identifiable features that are used consistently throughout the book. We believe these features make the book new and unique, and we hope that they will improve both the teaching and the learning experience.

This book is the second volume of a Statics and Dynamics series. Let's see in detail what makes these books different.

Why Another Statics and Dynamics Series?

These books provide thorough coverage of all the pertinent topics traditionally associated with statics and dynamics. Indeed, many of the currently available texts also provide this. However, the new books by Gray/Costanzo/Plesha offer several major innovations that enhance the learning objectives and outcomes in these subjects.

What Then Are the Major Differences between Gray/Costanzo/Plesha and Other Engineering Mechanics Texts?

A Consistent and Systematic Approach to Problem Solving One of the main objectives of this text is to foster the habit of solving problems using a systematic approach. Therefore, the example problems in Gray/Costanzo/Plesha follow a structured four-step problem-solving methodology that will help you develop your problem-solving skills not only in statics and dynamics, but also in all other mechanics subjects that follow. This structured problem-solving approach consists of the following steps: Road Map & Modeling, Governing Equations, Computation, and Discussion & Verification. The Road Map provides some of the general objectives of the problem and develops a strategy for how the solution will be developed. Modeling is next, where a real-life problem is idealized by a model. This step results in the creation of a free body diagram and the selection of the balance laws needed to solve the problem. The Governing Equations step is devoted to writing all the equations needed to solve the problem. These equations typically include the Equilibrium Equations, and, depending upon the particular problem, Force Laws (e.g. spring laws or frictional laws) and Kinematic Equations. In the Computation step, the governing equations are solved. In the final step, Discussion & Verification, the solution is interrogated to ensure that it is meaningful and accurate. This four-step problem-solving methodology is followed for all examples that involve a balance principle such as Newton's second law or the work-energy principle. Some problems (e.g., kinematics problems) do not involve balance principles, and for these the Modeling step is not needed.

Contemporary Examples, Problems, and Applications The examples, homework problems, and design problems were carefully constructed to help show you how the various topics of statics and dynamics are used in engineering practice. Statics and dynamics are immensely important subjects in modern engineering and science, and one of our goals is to excite you about these subjects and the career that lies ahead of you.

A Focus on Design A major difference between Gray/Costanzo/Plesha and other books is the systematic incorporation of design and modeling of real-life problems throughout. In statics, topics include important discussions on design, ethics, and professional responsibility. In dynamics, the emphasis is on parametric analysis and motion over ranges of time and space. These books show you that meaningful engineering design is possible using the concepts of statics and dynamics. Not only is the ability to develop a design very satisfying, but it also helps you develop a greater understanding of basic concepts and helps sharpen your ability to apply these concepts. Because the main focus of statics and dynamics textbooks should be the establishment of a firm understanding of basic concepts and correct problem-solving techniques, design topics do not have an overbearing presence in the books. Rather, design topics are included where they are most appropriate. While some of the discussions on design could be described as "common sense," such a characterization trivializes the importance and necessity for discussing pertinent issues such as safety, uncertainty in determining loads, the designer's responsibility to anticipate uses, even unintended uses, communications, ethics, and uncertainty in workmanship. Perhaps the most important feature of our inclusion of design and modeling topics is that you get a glimpse of what engineering is about and where your career in engineering is headed. The book is structured so that design topics and design problems are offered in a variety of places, and it is possible to pick when and where the coverage of design is most effective.

Computational Tools Some examples and problems are appropriate for solution using computer software. The use of computers extends the types of problems that can be solved while alleviating the burden of solving equations. Such examples and problems give you insight into the power of computer tools and further insight into how statics and dynamics are used in engineering practice.

Modern Pedagogy Numerous modern pedagogical elements have been included. These elements are designed to reinforce concepts and they provide additional information to help you make meaningful connections with real-world applications. Marginal notes (i.e., Helpful Information, Common Pitfalls, Interesting facts, and Concept Alerts) help you place topics, ideas, and examples in a larger context. These notes will help you study (e.g., Helpful Information and Common Pitfalls), will provide real-world examples of how different aspects of statics and dynamics are used (e.g., Interesting Facts), and will drive home important concepts or help dispel misconceptions (e.g., Concepts Alerts and Common Pitfalls). Mini Examples are used throughout the text to immediately and quickly illustrate a point or concept without having to wait for the worked-out examples at the end of the section.

Answers to Problems Answers to most even-numbered problems are posted as a freely downloadable PDF file at www.mhhe.com/pgc2e. Providing answers in this manner allows for the inclusion of more complex information than would otherwise be possible. In addition to final numerical and/or symbolic answers, plots for Computer Problems are included.

Changes to the Second Edition

The second edition of Engineering Mechanics: Dynamics retains all of the major pedagogical innovations of the first edition, including a consistent approach to problem-solving in all example problems, contemporary engineering applications in the examples and homework exercises, an emphasis on engineering design and its implications for problem-solving, and application of computational tools where applicable. In addition, the outstanding accuracy of the first edition has been preserved by again directly publishing the author's source files and through the use of an accuracy checking process unrivaled in mechanics textbooks. The second edition adds over 350 new homework problems, a substantial number of which are also included in the McGraw-Hill Connect® online homework system, which is an algorithmic homework and course management tool. These added homework problems bring the total number of problems to 1416, not counting the design problems. Another major enhancement is the adoption of a simplified and more direct presentation style of the theory sections of the textbook, with a coordinated addition of over a dozen introductory-level examples. Instructors will also find a substantial enhancement of the solutions manual, in which the solution of every kinetics problem now follows the structured problem-solving approach presented in the examples of the textbook.

The following individuals have been instrumental in ensuring the highest standard of content and accuracy. We are deeply indebted to them for their tireless efforts.

Second Edition

Reviewers

George G. Adams
Northeastern University
Stephen Bechtel
The Ohio State University
J. A. M. Boulet
University of Tennessee-Knoxville
Janet Brelin-Fornari
Kettering University
Suren Chen
Colorado State University
Nicola Ferrier
University of Wisconsin-Madison
Michael W. Keller
The University of Tulsa
Yohannes Ketema
University of Minnesota
Dragomir C. Marinkovich
Milwaukee School of Engineering
Tom Mase
California Polytechnic State University, San Luis Obispo
Richard McNitt
Penn State University
William R. Murray
California Polytechnic State University
Chris Passerello
Michigan Technological University
Gordon R. Pennock
Purdue University
Vincent C. Prantil
Milwaukee School of Engineering
Bidhan C. Roy
University of Wisconsin-Platteville
David A. Rubenstein
Oklahoma State University
John Schmitt
Oregon State University
Larry M. Silverberg
North Carolina State University
Richard E. Stanley
Kettering University
T. W. Wu
University of Kentucky
Jack Xin
Kansas State University

Focus Group Participants

Brock E. Barry
United States Military Academy
Daniel Dickrell, III
University of Florida
Ali Gordon
University of Central Florida-Orlando
Stephanie Magleby
Brigham Young University
Tom Mase
California Polytechnic State University, San Luis Obispo
Gregory Miller
University of Washington
Carisa H. Ramming
Oklahoma State University
Robert J. Witt
University of Wisconsin-Madison

First Edition

Board of Advisors

Janet Brelin-Fornari
Kettering University
Manoj Chopra
University of Central Florida
Pasquale Cinnella
Mississippi State University
Ralph E. Flori
Missouri University of Science and Technology
Christine B. Masters
Penn State University
Mark Nagurka
Marquette University
David W. Parish
North Carolina State University
Gordon R. Pennock
Purdue University
Michael T. Shelton
California State Polytechnic University-Pomona
Joseph C. Slater
Wright State University
Arun R. Srinivasa
Texas A&M University
Carl R. Vilmann
Michigan Technological University
Ronald W. Welch
The University of Texas at Tyler
Robert J. Witt
University of Wisconsin-Madison

Reviewers

Makola M. Abdullah
Florida Agricultural and Mechanical University
Murad Abu-Farsakh
Louisiana State University
George G. Adams
Northeastern University
Farid Amirouche
University of Illinois at Chicago
Stephen Bechtel
Ohio State University
Kenneth Belanus
Oklahoma State University
Glenn Beltz
University of California-Santa Barbara
Haym Benaroya
Rutgers University
Sherrill B. Biggers
Clemson University
James Blanchard
University of Wisconsin-Madison
Janet Brelin-Fornari
Kettering University
Pasquale Cinnella
Mississippi State University
Ted A. Conway
University of Central Florida
Joseph Cusumano
Penn State University
Bogdan I. Epureanu
University of Michigan
Ralph E. Flori
Missouri University of Science and Technology
Barry Goodno
Georgia Institute of Technology
Kurt Gramoll
University of Oklahoma

Hartley T. Grandin, Jr.
Professor Emeritus, Worcester Polytechnic Institute
Roy J. Hartfield, Jr.
Auburn University
Paul R. Heyliger
Colorado State University
James D. Jones
Purdue University
Yohannes Ketema
University of Minnesota
Carl R. Knospe
University of Virginia
Sang-Joon John Lee
San Jose State University
Jia Lu
The University of Iowa
Ron McClendon
University of Georgia
Paul Mitiguy
Consulting Professor, Stanford University
William R. Murray
California Polytechnic State University, San Luis Obispo
Mark Nagurka
Marquette University
Robert G. Oakberg
Montana State University
James J. Olsen
Wright State University
Chris Passerello
Michigan Technological University
Gary A. Pertmer
University of Maryland
David Richardson
University of Cincinnati

William C. Schneider
Texas A&M University
Sorin Siegler
Drexel University
Joseph C. Slater
Wright State University
Ahmad Sleiti
University of Central Florida
Arun R. Srinivasa
Texas A&M University
Josef S. Torok
Rochester Institute of Technology
John J. Uicker
Professor Emeritus, University of Wisconsin-Madison
David G. Ullman
Professor Emeritus, Oregon State University
Carl R. Vilmann
Michigan Technological University
Claudia M. D. Wilson
Florida State University
C. Ray Wimberly
University of Texas at Arlington
Robert J. Witt
University of Wisconsin-Madison
T. W. Wu
University of Kentucky
X. J. Xin
Kansas State University
Henry Xue
California State Polytechnic University, Pomona
Joseph R. Zaworski
Oregon State University
M. A. Zikry
North Carolina State University

Symposium Attendees

Farid Amirouche
University of Illinois at Chicago
Subhash C. Anand
Clemson University
Manohar L. Arora
Colorado School of Mines
Stephen Bechtel
Ohio State University
Sherrill B. Biggers
Clemson University

J. A. M. Boulet
University of Tennessee
Janet Brelin-Fornari
Kettering University
Louis M. Brock
University of Kentucky
Amir Chaghajerdi
Colorado School of Mines
Manoj Chopra
University of Central Florida

Pasquale Cinnella
Mississippi State University
Adel ElSafty
University of North Florida
Ralph E. Flori
Missouri University of Science and Technology
Walter Haisler
Texas A&M University

Kimberly Hill
University of Minnesota
James D. Jones
Purdue University
Yohannes Ketema
University of Minnesota
Charles Krousgrill
Purdue University
Jia Lu
The University of Iowa
Mohammad Mahinfalah
Milwaukee School of Engineering
Tom Mase
California Polytechnic State University, San Luis Obispo
Christine B. Masters
Penn State University
Daniel A. Mendelsohn
The Ohio State University
Faissal A. Moslehy
University of Central Florida
LTC Mark Orwat
United States Military Academy at West Point
David W. Parish
North Carolina State University
Arthur E. Peterson
Professor Emeritus, University of Alberta
W. Tad Pfeffer
University of Colorado at Boulder
David G. Pollock
Washington State University
Robert L. Rankin
Professor Emeritus, Arizona State University
Mario Rivera-Borrero
University of Puerto Rico at Mayaguez
Hani Salim
University of Missouri
Brian P. Self
California Polytechnic State University, San Luis Obispo
Michael T. Shelton
California State Polytechnic University-Pomona
Lorenz Sigurdson
University of Alberta
Larry Silverberg
North Carolina State University
Joseph C. Slater
Wright State University
Arun R. Srinivasa
Texas A&M University
David G. Ullman
Professor Emeritus, Oregon State University
Carl R. Vilmann
Michigan Technological University
Anthony J. Vizzini
Mississippi State University
Andrew J. Walters
Mississippi State University
Ronald W. Welch
The University of Texas at Tyler
Robert J. Witt
University of Wisconsin-Madison
T. W. Wu
University of Kentucky
Musharraf Zaman
University of Oklahoma-Norman
Joseph R. Zaworski
Oregon State University

Focus Group Attendees

Janet Brelin-Fornari
Kettering University
Yohannes Ketema
University of Minnesota
Mark Nagurka
Marquette University
C. Ray Wimberly
University of Texas at Arlington
M. A. Zikry
North Carolina State University

Accuracy Checkers

Walter Haisler
Texas A&M University
Richard McNitt
Penn State University
Mark Nagurka
Marquette University

ACKNOWLEDGMENTS

The authors are grateful to the many instructors who have provided valuable feedback on the first edition of this book. We are especially thankful to Professor Robert J. Witt of the University of Wisconsin-Madison, Professor Christine B. Masters of Penn State University, Professor Carl R. Vilmann of Michigan Technological University, and Professor Mark L. Nagurka of Marquette University for the guidance they have provided over the years these books have been in development.

Special thanks go to Andrew Miller for infrastructure he created to keep the authors, manuscript, and solutions manual in sync. His knowledge of programming, scripting, subversion, and many other computer technologies made a gargantuan task feel just a little more manageable.

Force and Acceleration Methods for Particles

3

An aerobatic maneuver performed by the Red Arrows, the aerobatic team of the British Royal Air Force. Using Newton's second law we can relate the acceleration of these airplanes to the forces acting on them.

Newton's second law is an *axiom*, that is, a statement we accept as true and not derivable from other principles. Therefore, this chapter will emphasize *how* $\vec{F} = m\vec{a}$ is applied, and it begins the study of *kinetics*, which is the study of the forces that cause and are caused by motion. By the time we complete this chapter, we will be able to use the kinematics discussed in Chapter 2, along with Newton's second law, to either (1) predict the motion of a particle system caused by given forces or (2) determine the forces needed for a particle system to move in a prescribed way.

3.1 Rectilinear Motion

In this chapter we show how Newton's second law, which we stated in Eq. (1.2) as

$$\vec{F} = m\vec{a}, \tag{3.1}$$

is applied to study the motion of bodies that are modeled as particles. We begin by looking at straight line or rectilinear motion (see Fig. 3.1).

Figure 3.1
A man pushing a crate over a flat horizontal surface. The crate is in rectilinear motion.

Applying Newton's second law

Newton's second law has associated with it a rich terminology and a well-established sequence of steps to formulate a problem and find a solution. We now present the procedure and terminology used when applying Newton's second law to mechanical systems.

Step 1. Road Map & Modeling. ***Review the given information, identify the system, state assumptions, sketch the FBD, and identify a problem-solving strategy.*** The ultimate goal of this first step is the creation of a model and associated free body diagram (FBD) for the system under consideration. An FBD is a diagram of

Chapter Introduction

Each chapter begins with an introductory statement of the purpose and goals of the chapter.

A Closer Look We can now finish the parametric analysis begun in Example 6.6 on p. 460. We plot the angular acceleration of the CR α_{BC}, as well as the acceleration of the piston a_{Cy}, for the same conditions considered in Example 6.11. What is remarkable is the magnitude of the accelerations of the CR. Referring to Figs. 3 and 4, we

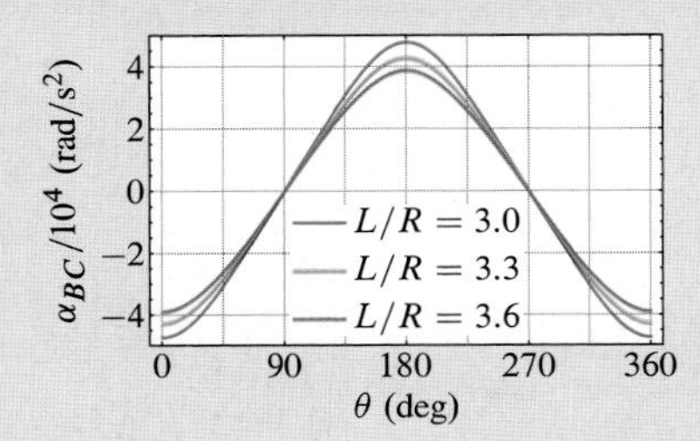

Figure 3. The angular acceleration of the CR for $\dot{\theta} = 3500$ rpm, $L = 150$ mm, and three values of L/R.

Computer Solutions

We make use of computer solutions in some problems and it is important that you be able to easily identify when this is the case. Therefore, anywhere a computer is used for a solution, you will see the symbol. If this occurs within part of an example problem or within the discussion, then the part requiring the use of a computer will be enclosed in the following symbols. If one of the exercises requires a computer for its solution, then the computer symbol and its mirror image will appear on either side of the problem heading.

Mini-Examples

Mini-examples are used throughout the text to immediately and quickly illustrate a point or concept without having to wait for the worked-out examples at the end of the section.

force of attraction between two bodies. The gravitational force on a mass m_1 due to a mass m_2 a distance r away from m_1 is

$$\vec{F}_{12} = \frac{Gm_1 m_2}{r^2}\hat{u}, \tag{1.5}$$

where $\hat{u}$ is a unit vector pointing from m_1 to m_2 and G is the *universal gravitational constant** (sometimes called the *constant of gravitation* or *constant of universal gravitation*). The following example demonstrates the application of this law.

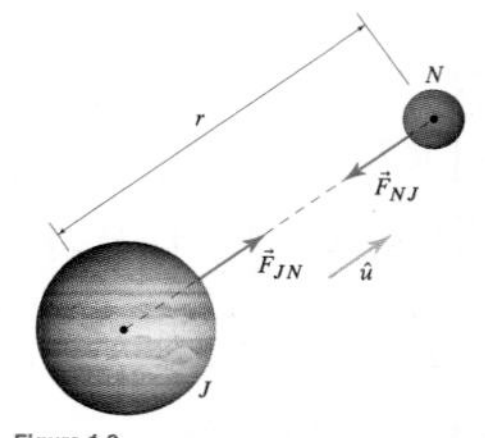

Figure 1.2
The gravitational force between the planets Jupiter J and Neptune N. The relative sizes of the planets are accurate, but their separation distance is not.

Mini-Example
Using the planets Jupiter and Neptune as an example, the force on Jupiter due to the gravitational attraction of Neptune, $\vec{F}_{JN}$, is given by (see Fig. 1.2)

$$\vec{F}_{JN} = \frac{Gm_J m_N}{r^2}\hat{u}, \tag{1.6}$$

where r is the distance between the two bodies, m_J is the mass of Jupiter, m_N is the mass of Neptune, and $\hat{u}$ is a unit vector pointing from the center of Jupiter to the center of Neptune. The mass of Jupiter is 1.9×10^{27} kg, and that of Neptune is 1.02×10^{26} kg. Since the mean radius of Jupiter's orbit is 778,300,000 km and that of Neptune is 4,505,000,000 km, we assume that their closest approach to one another is approximately 3,727,000,000 km. Thus, at their closest approach, the magnitude of the force between these two planets is

$$|\vec{F}_{JN}| = \left(6.674\times10^{-11}\,\frac{\text{m}^3}{\text{kg}\cdot\text{s}^2}\right)\frac{(1.9\times10^{27}\,\text{kg})(1.02\times10^{26}\,\text{kg})}{(3.727\times10^{12}\,\text{m})^2} = 9.312\times10^{17}\,\text{N}. \tag{1.7}$$

It is interesting to compare this force with the force of gravitation between Jupiter and the Sun. The Sun's mass is 1.989×10^{30} kg, and we have already stated that the mean radius of Jupiter's orbit is 778,300,000 km. Applying Eq. (1.6) between Jupiter and the Sun gives 4.164×10^{23} N, which is almost 450,000 times larger.

Acceleration due to gravity. Equation (1.5) allows us to determine the force of Earth's gravity on an object of mass m on the surface of the Earth. This is done by noting that the radius of the Earth is 6371.0 km (see the marginal note) and the mass of the Earth is 5.9736×10^{24} kg and then applying Eq. (1.5):

Interesting Fact
The radius of the Earth. The Earth is not a perfect sphere. Therefore, there are different notions of "radius of the Earth." The given

EXAMPLE 3.6 *Tension in a Wrecking Ball Cable*

The wrecking ball A shown in Fig. 1 is released from rest when $\theta = \theta_0 = 30°$, and it swings freely about the fixed point at O. Assuming that the weight of the ball is W = 2500 lb and L = 30 ft, determine the tension in the cable to which the ball is attached when the ball reaches $\theta = 0°$.

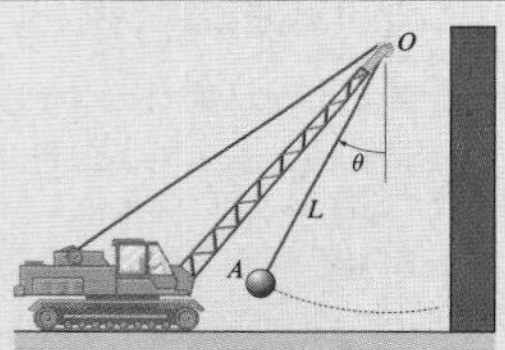

Figure 1

SOLUTION

Road Map & Modeling Modeling the wrecking ball as a particle and neglecting all forces except the weight force W and the cable tension T, the FBD is as shown in Fig. 2. Applying Newton's second law in the polar component system shown should allow us to find the tension in the cable as a function of its swing angle and thus, find its tension when $\theta = 0°$.

Governing Equations

Balance Principles Referring to the FBD in Fig. 2 and applying Newton's second law, we obtain

$$\sum F_\theta: \quad -W\sin\theta = ma_\theta, \tag{1}$$

$$\sum F_r: \quad W\cos\theta - T = ma_r, \tag{2}$$

where $m = W/g$.

Force Laws All forces are accounted for on the FBD.

Kinematic Equations Writing a_θ and a_r in polar components gives

$$a_\theta = r\ddot{\theta} + 2\dot{r}\dot{\theta} = L\ddot{\theta} \quad \text{and} \quad a_r = \ddot{r} - r\dot{\theta}^2 = -L\dot{\theta}^2, \tag{3}$$

where we have replaced r with the constant length L.

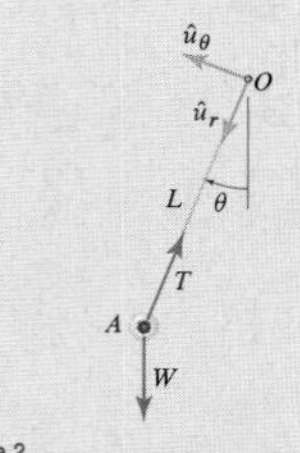

Figure 2
FBD of the wrecking ball as it swings downward.

Computation Substituting Eqs. (3) into Eqs. (1) and (2), we obtain

$$-W\sin\theta = mL\ddot{\theta} \quad \text{and} \quad W\cos\theta - T = -mL\dot{\theta}^2$$

$$\Rightarrow \quad \ddot{\theta} = -\frac{g}{L}\sin\theta \quad \text{and} \quad T = W\cos\theta + mL\dot{\theta}^2. \tag{4}$$

Notice that the tension is a function of $\dot{\theta}$, so we need to integrate $\ddot{\theta}(\theta)$ to find $\dot{\theta}(\theta)$ using the chain rule, that is,

$$\ddot{\theta} = \dot{\theta}\frac{d\dot{\theta}}{d\theta} = -\frac{g}{L}\sin\theta \quad \Rightarrow \quad \int_0^{\dot{\theta}} \dot{\theta}\,d\dot{\theta} = -\frac{g}{L}\int_{\theta_0}^{\theta}\sin\theta\,d\theta$$

$$\Rightarrow \quad \dot{\theta}^2 = 2\frac{g}{L}(\cos\theta - \cos\theta_0). \tag{5}$$

Substituting Eq. (5) into the expression for T in Eq. (4) gives $T(\theta)$ as

$$T = W(3\cos\theta - 2\cos\theta_0) \quad \Rightarrow \quad \boxed{T(\theta = 0) = W(3 - 2\cos\theta_0) = 3170\,\text{lb},} \tag{6}$$

where we have used W = 2500 lb and $\theta_0 = 30°$ to obtain the final numerical result.

Discussion & Verification The final result in Eq. (6) is dimensionally correct, and the magnitude of the tension seems reasonable. Interestingly, the tension does not depend on the length of the supporting cable. That is, if the initial angle is 30° and the wrecking ball is released from rest, the tension in the cable will always be 3170 lb, regardless of the length of the suspending cable.

Examples

Consistent Problem-Solving Methodology

Every problem in the text employs a carefully defined problem-solving methodology to encourage systematic problem formulation, while reinforcing the steps needed to arrive at correct and realistic solutions.

Each example problem contains these four steps:

- **Road Map & Modeling**
- **Governing Equations**
- **Computation**
- **Discussion & Verification**

Some examples include a Closer Look (noted with a magnifying glass icon) that offers additional insight into the example.

Concept Alerts and Concept Problems

Two additional features are the Concept Alert and the Concept Problems. These have been included because research has shown (and it has been our experience) that even though you may do quite well in a science or engineering course, your conceptual understanding may be lacking. **Concept Alerts** are marginal notes and are used to drive home important concepts (or help dispel misconceptions) that are related to the material being developed at that point in the text. **Concept Problems** are mixed in with the problems that appear at the end of each section. These are questions designed to get you thinking about the application of a concept or idea presented within that section. They should never require calculation and should require answers of no more than a few sentences.

Concept Alert

Direction of velocity vectors. One of the most important concepts in kinematics is that the velocity of a particle is always tangent to the particle's path.

Problem 3.2

An object is lowered very slowly onto a conveyor belt that is moving to the right. What is the direction of the friction force acting on the object at the instant the object touches the belt?

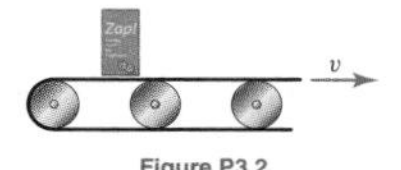

Figure P3.2

Problem 3.3

A person is trying to move a heavy crate by pushing on it. While the person is pushing, what is the resultant force acting on the crate if the crate does not move?

Common Pitfall

Newton's second law and inertial frames. Since the application of Newton's second law requires the use of an inertial reference frame, the component system shown in Fig. 2 must be understood as originating from an xy coordinate ... ground—this is the ine... It would be a mistake... nate system moving w... the truck is deceleratin... ground and, therefore, i... of reference.

Interesting Fact

Cyclic loading and fatigue. The fact that, under the given conditions, the rotor bearing experiences a cyclic load 1000 times per second means that it will quickly experience a large number of load cycles. It turns out that even a rather low stress can cause ... after millions of load ... stress, the smaller the ...uired. This mechanism ...*gue*. Since the number ...e rotor bearing of a ...kly, even a small im...ilure due to fatigue. To ...ue, see W. D. Callister, ...*and Engineering: An* ... John Wiley & Sons,

Helpful Information

The right-hand rule. In three dimensions, a Cartesian coordinate system uses three orthogonal reference directions. These are the x, y, and z directions shown below.

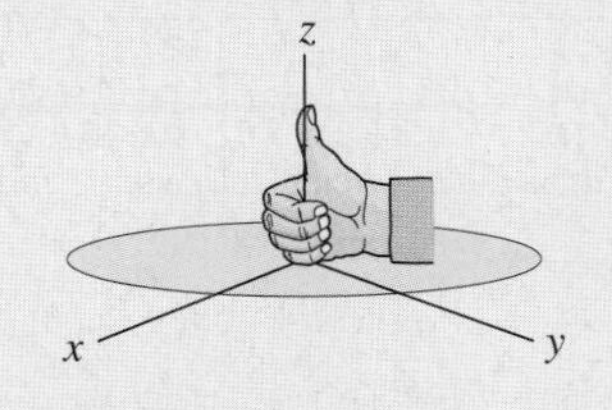

Proper interpretation of many vector operations, such as the cross product, requires that the x, y, and z directions be arranged in a consistent manner. The convention in mechanics and vector mathematics in general is that if the axes are arranged as shown, then, according to the *right-hand rule*, rotating the x direction into the y direction yields the z direction. The result is called a *right-handed coordinate system*.

Marginal Notes

Marginal notes have been implemented that will help place topics, ideas, and examples in a larger context. This feature will help students study (using **Helpful Information** and **Common Pitfalls**) and will provide real-world examples of how different aspects of dynamics are used (using **Interesting Facts**).

Sections and End of Section Summary

Each chapter is organized into several sections. There is a wealth of information and features within each section, including examples, problems, marginal notes, and other pedagogical aids. Each section concludes with an end of section summary that succinctly summarizes that section. In many cases, cross-referenced important equations are presented again for review and reinforcement before the student proceeds to the examples and homework problems.

9.2 Undamped Forced Vibration

Many systems are *forced* to vibrate by an external excitation. This section is devoted to the forced vibration of mechanical systems.

Standard form of the forced harmonic oscillator

A standard forced harmonic oscillator is shown in Fig. 9.11, in which the block of mass m is attached to a fixed support by a linear spring of constant k and is also being driven by the time-dependent force $P(t) = F_0 \sin \omega_0 t$. Modeling the block as a particle, its FBD is as shown in Fig. 9.12, where F_s is the spring force acting on the block. Summing forces in the x direction, we obtain

$$\sum F_x: \quad P(t) - F_s = ma_x, \tag{9.29}$$

where the force law is given by $F_s = kx$ and the kinematic equation is $a_x = \ddot{x}$. Substituting these relations as well as $P(t)$ into Eq. (9.29), we obtain

$$F_0 \sin \omega_0 t - kx = m\ddot{x} \quad \Rightarrow \quad \ddot{x} + \frac{k}{m}x = \frac{F_0}{m} \sin \omega_0 t. \tag{9.30}$$

Noting that $\omega_n^2 = k/m$, this last equation becomes

$$\ddot{x} + \omega_n^2 x = \frac{F_0}{m} \sin \omega_0 t, \tag{9.31}$$

which is the *standard form of the forced harmonic oscillator equation*. It is a *nonhomogeneous* version of Eq. (9.12) on p. 665 as a result of the term $(F_0/m) \sin \omega_0 t$. The term on the right-hand side of Eq. (9.31) is a function of *only* the independent variable t. It is often called a *forcing function* because it forces the system to vibrate. This particular type of forcing is harmonic because it is a harmonic function of time.

The theory of differential equations tells us that the *general solution* of Eq. (9.31) is the sum of the *complementary solution* $x_c(t)$ and a *particular solution* $x_p(t)$. The *complementary solution** is the solution of the associated homogeneous equation

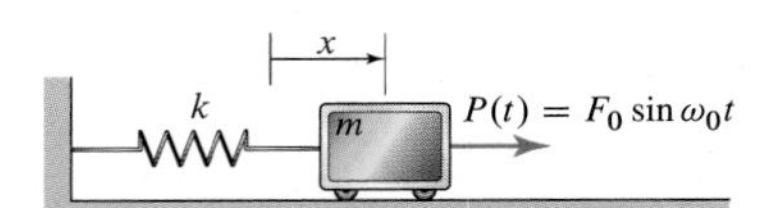

Figure 9.11
A forced harmonic oscillator whose equation of motion is given by Eq. (9.31) with $\omega_n = \sqrt{k/m}$. The position x is measured from the equilibrium position of the system when $F_0 = 0$.

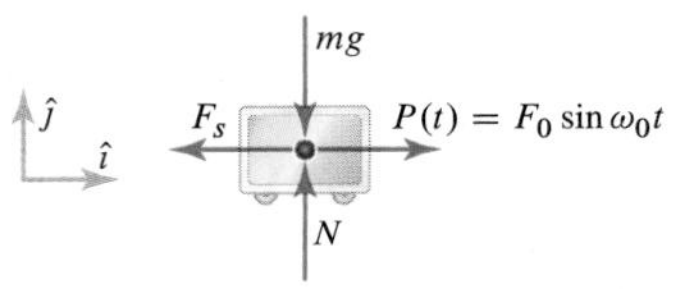

Figure 9.12
FBD of the forced harmonic oscillator in Fig. 9.11.

Interesting Fact

How practical is harmonic forcing? The answer to this question lies in an amazing result due to Jean Baptiste Joseph Fourier (1768–1830) and later contributors. It says ...ise smooth function ...an infinite series of ...lled *Fourier series* ...*his means that any* ...*egarded as the sum* ...h addition, given the ...Eq. (9.31) (i.e., it is ...he overall particular ...monic forcing terms ...particular solutions ...monic forcing term. ...ther allow engineers ...utions to problems ...g as a sum of sim- ...oscillator solutions. ...ng is ubiquitous in ...s is one of the most ...ed mathematics.

End of Section Summary

When a harmonic oscillator is subject to harmonic forcing, the standard form of the equation of motion is

Eq. (9.31), p. 681
$$\ddot{x} + \omega_n^2 x = \frac{F_0}{m} \sin \omega_0 t,$$

where F_0 is the amplitude of the forcing and ω_0 is its frequency (see Fig. 9.16). The general solution to this equation consists of the sum of the complementary solution and a particular solution. The *complementary solution* x_c is the solution of the associated homogeneous equation, which is given by, for example, Eq. (9.13). For $\omega_0 \neq \omega_n$, a particular solution was found to be

Eq. (9.35), p. 682
$$x_p = \frac{F_0/k}{1 - (\omega_0/\omega_n)^2} \sin \omega_0 t,$$

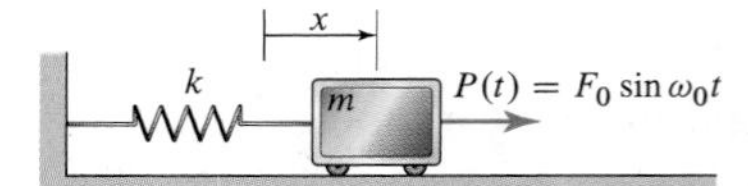

Figure 9.16
A forced harmonic oscillator whose equation of motion is given by Eq. (9.31) with $\omega_n = \sqrt{k/m}$. The position x is measured from the equilibrium position of the block.

Problems

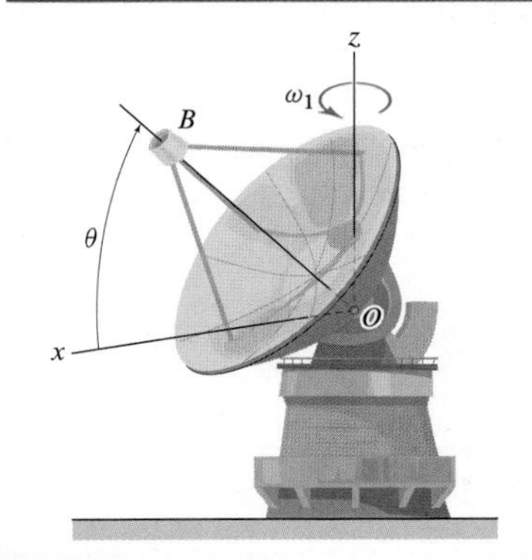

Figure P10.1 and P10.2

Problems 10.1 and 10.2

The radar dish can rotate about the vertical z axis at rate ω_1 and about the horizontal y axis (not shown in the figure) at rate $\dot{\theta}$. The distance between the center of rotation at O and the subreflector at B is ℓ.

Problem 10.1 If ω_1 and $\dot{\theta}$ are both constant, determine the velocity and acceleration of the subreflector B in terms of the elevation angle θ.

Problem 10.2 If $\omega_1(t)$ and $\dot{\theta}(t)$ are known functions of time, determine the velocity and acceleration of the subreflector B in terms of the elevation angle θ.

Problems 10.3 and 10.4

The truncated cone rolls without slipping on the xy plane. At the instant shown, the angular speed about the z axis is ω_1, and it is changing at $\dot{\omega}_1$.

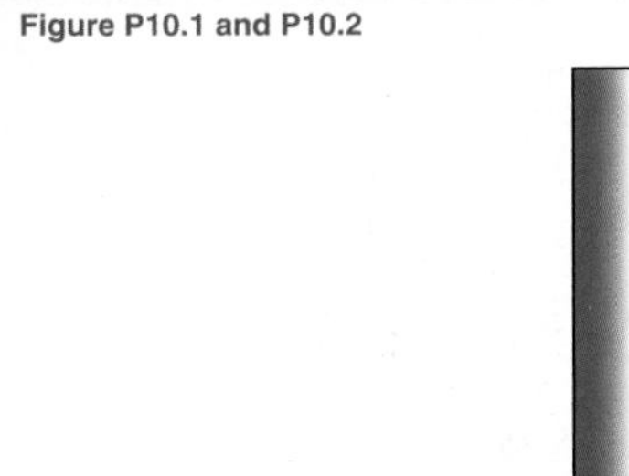

Figure P2.132 and P2.133

Problem 2.134

A micro spiral pump consists of a spiral channel attached to a stationary plate. This plate has two ports, one for fluid inlet and the other for outlet, the outlet being farther from the center of the plate than the inlet. The system is capped by a rotating disk. The fluid trapped between the rotating disk and stationary plate is put in motion by the rotation of the top disk, which pulls the fluid through the spiral channel. With this in mind, consider a channel with geometry given by the equation $r = \eta\theta + r_0$, where $\eta = 12\,\mu\text{m}$ is called the polar slope, $r_0 = 146\,\mu\text{m}$ is the radius at the inlet, r is the distance from the spin axis, and θ, measured in radians, is the angular position of a point in the spiral channel. If the top disk rotates with a constant angular speed $\omega = 30{,}000$ rpm, and assuming that the fluid particles in contact with the rotating disk are essentially stuck to it, determine the velocity and acceleration of one such fluid particle when it is at $r = 170\,\mu\text{m}$.* Express the answer using the component system shown (which rotates with the top disk).

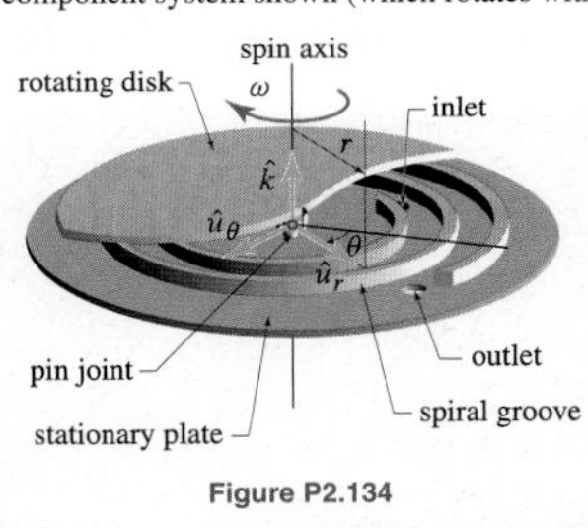

Figure P2.134

Modern Problems

Problems of varying difficulty follow each section. These problems allow students to develop their ability to apply concepts of dynamics on their own. The most common question asked by students is "How do I set this problem up?" What is really meant by this question is "How do I develop a good mathematical model for this problem?" The only way to develop this ability is by practicing numerous problems. Answers to most even-numbered problems are posted as a freely downloadable PDF file at www.mhhe.com/pgc2e. Providing answers in this manner allows for more complex information than would otherwise be possible. Each problem in the book is accompanied by a thermometer icon that indicates the approximate level of difficulty. Those considered to be "introductory" are indicated with the symbol 🌡. Problems considered to be "representative" are indicated with the symbol 🌡, and problems that are considered to be "challenging" are indicated with the symbol 🌡.

Engineering Design and Design Problems

Several design problems are presented where appropriate throughout the book. These problems can be tackled with the knowledge and skill set that are typical of introductory-level courses, although the use of mathematical software is strongly recommended. These problems are open ended and their solution requires the definition of a parameter space in which the dynamics of the system must be analyzed. In dynamics we have chosen to emphasize the role played by *parametric analyses* in the overall design process, as opposed to cost-benefit analyses or the choice of specific materials and/or components.

Design Problems

Design Problem 7.1

Revisit the calculations done at the beginning of the chapter concerning the determination of the maximum acceleration that can be achieved by a motorcycle without causing the front wheel to lift off the ground. Specifically, construct a new model of the motorcycle by selecting a real-life motorcycle and researching its geometry and inertia properties, including the inertia properties of the wheels. Then analyze your model to determine how the maximum acceleration in question depends on the horizontal and vertical positions of the center of mass with respect to the points of contact between the ground and the wheels. Include in your analysis a comparison of results that account for the inertia of the front wheel with results that neglect the inertia of the front wheel.

Figure DP7.1

Chapter Review

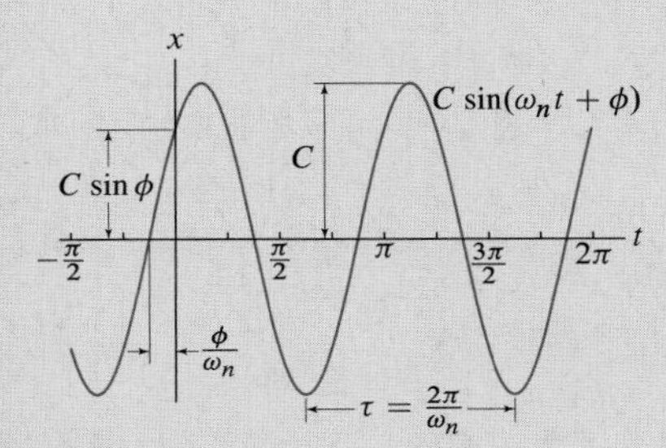

Figure 9.30
Plot of Eq. (9.3) showing the amplitude C, phase angle ϕ, and period τ of a harmonic oscillator.

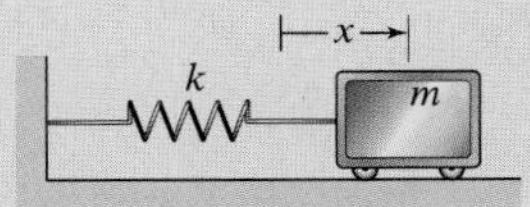

Figure 9.31
A simple spring-mass harmonic oscillator whose equation of motion is given by $m\ddot{x}+kx=0$ and for which $\omega_n=\sqrt{k/m}$.

In this chapter, we studied the vibration or oscillation of mechanical systems about their equilibrium position. We considered only harmonic oscillators, although we did study the effects of viscous damping and harmonic forcing on the response of a harmonic oscillator.

Undamped free vibration

Any one DOF system whose equation of motion is of the form

Eq. (9.12), p. 665

$$\ddot{x}+\omega_n^2 x=0,$$

is called a *harmonic oscillator*, and the above expression is referred to as the *standard form* of the harmonic oscillator equation. The solution of this equation can be written as (see Fig. 9.30)

Eq. (9.3), p. 664

$$x(t)=C\sin(\omega_n t+\phi),$$

where ω_n is the *natural frequency*, C is the *amplitude*, and ϕ is the *phase angle* of vibration.

A simple example of a harmonic oscillator is a system consisting of a mass m attached at the free end of a spring with constant k and with the other end fixed (see Fig. 9.31). The natural frequency of such a system is given by

Eq. (9.4), p. 664

$$\omega_n=\sqrt{\frac{k}{m}}.$$

In addition, the amplitude C and the phase angle ϕ are given by, respectively,

Eqs. (9.5) and (9.6), p. 664

$$C=\sqrt{\frac{v_i^2}{\omega_n^2}+x_i^2} \quad \text{and} \quad \tan\phi=\frac{x_i\omega_n}{v_i},$$

where we let $t=0$ be the initial time, $x_i=x(0)$ (i.e., x_i is the initial position), and $v_i=\dot{x}(0)$ (i.e., v_i is the initial velocity). If $v_i=0$, then ϕ can be chosen equal to $-\pi/2$ or $\pi/2$ rad for $x_i<0$ and $x_i>0$, respectively. An alternative form of the solution to Eq. (9.12) is given by

Eq. (9.15), p. 666

$$x(t)=x_i\cos\omega_n t+\frac{v_i}{\omega_n}\sin\omega_n t.$$

The *period* of the oscillation is given by

Eq. (9.7), p. 664

$$\text{Period}=\tau=\frac{2\pi}{\omega_n},$$

and the *frequency* of vibration is

Eq. (9.8), p. 664

$$\text{Frequency}=f=\frac{1}{\tau}=\frac{\omega_n}{2\pi}.$$

End-of-Chapter Review and Problems

Every chapter concludes with a succinct, yet comprehensive chapter review and a wealth of review problems.

Review Problems

Problem 2.280

The velocity and acceleration of point P expressed relative to frame A at some time t are

$$\vec{v}_{P/A} = (12.5\,\hat{\imath}_A + 7.34\,\hat{\jmath}_A)\text{ m/s} \quad\text{and}\quad \vec{a}_{P/A} = (7.23\,\hat{\imath}_A - 3.24\,\hat{\jmath}_A)\text{ m/s}^2.$$

Knowing that frame B does not move relative to frame A, determine the expressions for the velocity and acceleration of P with respect to frame B. Verify that the speed of P and the magnitude of P's acceleration are the same in the two frames.

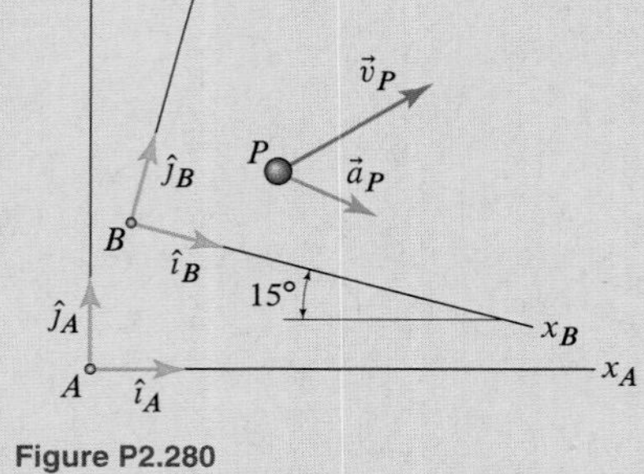

Figure P2.280

Problem 2.281

The motion of a point P with respect to a Cartesian coordinate system is described by $\vec{r} = \{2\sqrt{t}\,\hat{\imath} + [4\ln(t+1) + 2t^2]\,\hat{\jmath}\}$ ft, where t is time expressed in seconds. Determine the average velocity between $t_1 = 4$ s and $t_2 = 6$ s. Then find the time $\bar{t}$ for which the x component of P's velocity is *exactly* equal to the x component of P's average velocity between times t_1 and t_2. Is it possible to find a time at which P's velocity and P's average velocity are exactly equal? Explain why. *Hint:* Velocity is a vector.

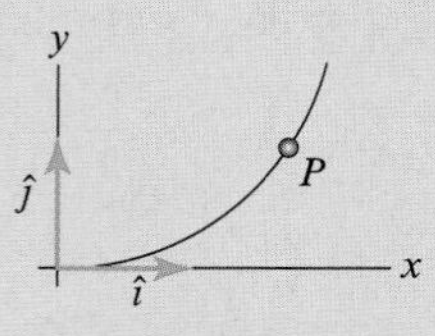

Figure P2.281

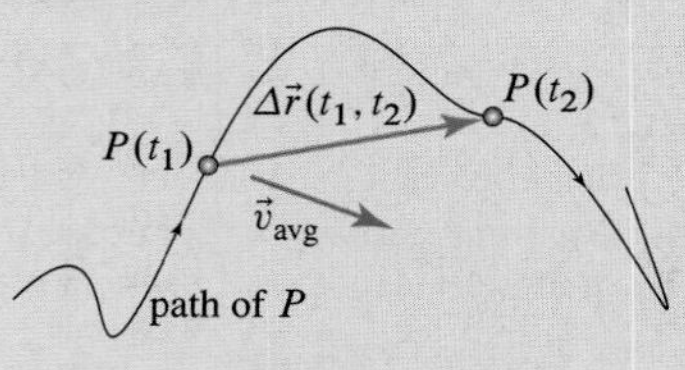

Figure P2.282

Problem 2.282

The figure shows the displacement vector of a point P between two time instants t_1 and t_2. Is it possible for the vector $\vec{v}_{\text{avg}}$ shown to be the average velocity of P over the time interval $[t_1, t_2]$?

Problems 2.283 and 2.284

A dynamic fracture model proposed to explain the behavior of cracks propagating at high velocity views the crack path as a *wavy path.** In this model, a crack tip appearing to travel along a straight path actually travels at roughly the speed of sound along a wavy path. Let the wavy path of the crack tip be described by the function $y = h\,\sin(2\pi x/\lambda)$, where h is the amplitude of the crack tip fluctuations in the direction perpendicular to the crack plane and λ is the corresponding period. Assume that the crack tip travels along the wavy path at a constant speed v_s (e.g., the speed of sound).

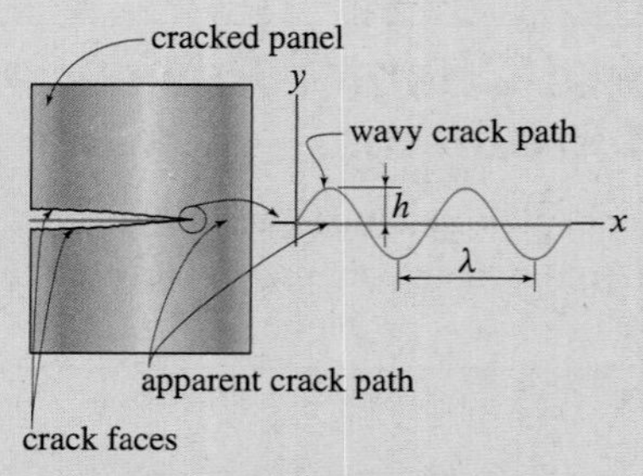

Figure P2.283 and P2.284

Problem 2.283 Find the expression for the x component of the crack tip velocity as a function of v_s, λ, h, and x.

Problem 2.284 Denote the *apparent* crack tip velocity by v_a, and define it as the average value of the x component of the crack velocity, that is,

$$v_a = \frac{1}{\lambda}\int_0^{\lambda} v_x\,dx.$$

* In dynamic fracture, the structural failure of a material occurs at speeds close to the speed of sound (in that material). This field of study is very important in the design of impact- and/or blast-resistant structures. The model mentioned in these problems is due to H. Gao, "Surface Roughening and Branching Instabilities in Dynamic Fracture," *Journal of the Mechanics and Physics of Solids*, **41**(3), pp. 457–486, 1993.

What Resources Support This Textbook?

McGraw-Hill offers various tools and technology products to support ***Engineering Mechanics: Statics*** and ***Engineering Mechanics: Dynamics***. Instructors can obtain teaching aids by calling the McGraw-Hill Customer Service Department at 1-800-338-3987, visiting our online catalog at www.mhhe.com, or contacting their local McGraw-Hill sales representative.

McGraw-Hill Connect Engineering. McGraw-Hill Connect Engineering is a web-based assignment and assessment platform that gives students the means to better connect with their coursework, with their instructors, and with the important concepts that they will need to know for success now and in the future. With Connect Engineering, instructors can deliver assignments, quizzes and tests easily online. Connect Engineering is available at www.mhhe.com/pgc2e.

Problem Answers. For the use of students, answers to most even-numbered problems are posted as a freely downloadable PDF file at www.mhhe.com/pgc2e.

Solutions Manual. The Solutions Manual that accompanies the second edition features detailed typeset solutions to the homework problems. These solutions use the same consistent problem solving methodology used throughout the text. The Solutions Manual is available on the password-protected instructor's website at www.mhhe.com/pgc2e.

PowerPoint Slides. A full set of PowerPoint slides containing the theory and examples associated with each section of the text are located on the instructor's website at www.mhhe.com/pgc2e.

CourseSmart. This text is offered through CourseSmart for both instructors and students. CourseSmart is an online browser where students can purchase access to this and other McGraw-Hill textbooks in a digital format. Through their browser, students can access the complete text online at almost half the cost of a traditional text. Purchasing the eTextbook also allows students to take advantage of CourseSmart's web tools for learning, which include full text search, notes and highlighting, and e-mail tools for sharing notes among classmates. To learn more about CourseSmart options, contact your sales representative or visit www.coursesmart.com.

Hands-on Mechanics. Hands-on Mechanics is a website designed for instructors who are interested in incorporating three dimensional, hands-on teaching aids into their lectures. Developed through a partnership between the McGraw-Hill Engineering Team and the Department of Civil and Mechanical Engineering at the United States Military Academy at West Point, this website not only provides detailed instructions on how to build 3-D teaching tools using materials found in any lab or local hardware store but also provides a community where educators can share ideas, trade best practices, and submit their own demonstrations for posting on the site. Visit www.handsonmechanics.com.

McGraw-Hill Create™ Craft your teaching resources to match the way you teach! With McGraw-Hill Create, you can easily rearrange chapters, combine material from other content sources, and quickly upload content you have written like your course syllabus or teaching notes. Find the content you need in Create by searching through thousands of leading McGraw-Hill textbooks. Arrange your book to fit your teaching style. Create even allows you to personalize your book's appearance by selecting the cover and adding your name, school, and course information. Order a Create book and you'll receive a complimentary print review copy in 3–5 business days or a complimentary electronic review copy (eComp) via email in minutes. Go to www.mcgrawhillcreate.com today and register to experience how McGraw-Hill Create empowers you to teach *your* students *your* way.

Engineering Mechanics
DYNAMICS

Introduction to Dynamics

In Section 1.1, we introduce Isaac Newton's (1643–1727) laws of motion and his universal law of gravitation. In Section 1.2, we review those elements of physics and vector algebra needed to develop the material in the remainder of the book. In Section 1.3, we touch upon the role of dynamics in engineering design.

Raphael's *School of Athens* depicts ancient Greek philosophers, such as Aristotle, Plato, Euclid, and Pythagoras. This fresco celebrates the kinship that the renaissance humanists felt with the great minds from antiquity as they explored new ways of thinking about the arts, sciences, and engineering.

1.1 The Newtonian Equations

The *dynamics* we study in this book is the part of mechanics concerned with the motion of bodies, the forces causing their motion, and/or the forces caused by their motion. Dynamics builds upon statics in that the ability to draw free body diagrams and to write the corresponding balance equations for particles and rigid bodies are fundamental to dynamics. Dynamics also complements mechanics of materials in that it develops your ability to find forces due to the acceleration of objects, forces that can be used to find stresses using mechanics of materials.

Since the middle of the 20th century, dynamics has also included the study and analysis of any time-varying process, be it mechanical, electrical, chemical, or biological. While we focus on mechanical processes, much of what we study is also applicable to other time-varying phenomena. Our goal is to provide an introduction to the science, skill, and art involved in modeling mechanical systems to predict their motion.

Newton's laws of motion

Newton's three laws of motion are

First law *A particle remains at rest, or moves in a straight line with a constant speed, as long as the total force acting on the particle is zero.*

Second law *The time rate of change of momentum of a particle is equal to the resultant force acting on that particle.*

Third law *The forces of action and reaction between interacting particles are equal in magnitude, opposite in direction, and collinear.*

The second law, stated mathematically, is

$$\boxed{\vec{F} = \frac{d\vec{p}}{dt} = \frac{d(m\vec{v})}{dt},} \tag{1.1}$$

where $\vec{F}$ is the net force acting on the particle, $\vec{p}$ is the momentum of the particle, m is the mass of the particle, and $\vec{v}$ is the velocity of the particle. We have used the definition of momentum, which is $\vec{p} = m\vec{v}$. Throughout this book, we will denote vectors by using either a superposed arrow ($\vec{}$) or a superposed caret or hat ($\hat{}$) if the vector is a *unit vector*. Newton's second law is often written as

$$\boxed{\vec{F} = m\vec{a},} \tag{1.2}$$

which explicitly accounts for the fact that a particle is generally understood to have constant mass.* We will learn in Chapter 3 that the first law is simply a special case of the second. The second and third laws, along with the ideas developed by Leonhard Euler (1707–1783) for rigid body dynamics, are all that is needed to solve a broad spectrum of problems involving particles and rigid bodies.

The third law, stated mathematically, is

$$\vec{F}_{ij} = -\vec{F}_{ji}, \tag{1.3}$$

$$\vec{F}_{ij} \times (\vec{r}_i - \vec{r}_j) = \vec{0}, \tag{1.4}$$

where, for any interacting particles i and j, $\vec{F}_{ij}$ is the force on particle i due to particle j and $\vec{r}_i$ is the position of the ith particle (Fig. 1.1). Some people refer to Newton's third law as just Eq. (1.3), while others require both Eqs. (1.3) and (1.4). Requiring both equations is sometimes referred to as the *strong form of Newton's third law*.

Interesting Fact

Newton's Third Law in modern mechanics. Modern mechanics generally discards Newton's third law and replaces it with a much more general result based on the concept of *angular momentum*. Since the 1950s, it has been proposed that an even more general notion called *the principle of material frame indifference* could be used to replace Newton's third law. This latter principle states that the properties of materials and the actions of bodies on one another are the same for all observers.

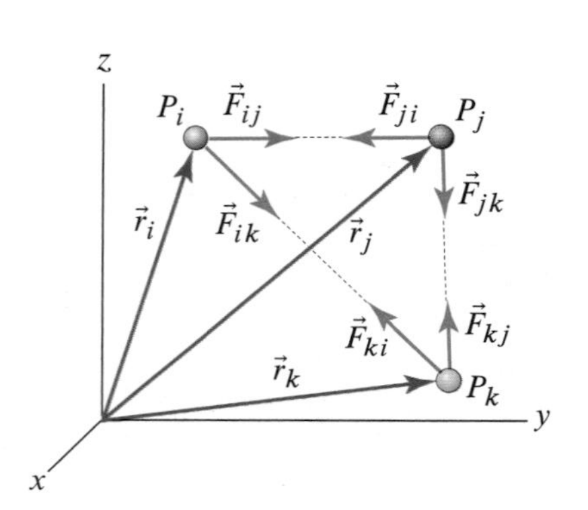

Figure 1.1
A system of particles interacting with one another.

Newton's universal law of gravitation

Newton used his laws of dynamics along with the laws postulated by Johannes Kepler (1571–1630) to deduce the *universal law of gravitation*, which describes the

* The application of Eq. (1.1) to variable mass systems will be considered in Section 5.5.

force of attraction between two bodies. The gravitational force on a mass m_1 due to a mass m_2 a distance r away from m_1 is

$$\boxed{\vec{F}_{12} = \frac{G m_1 m_2}{r^2}\,\hat{u},} \tag{1.5}$$

where $\hat{u}$ is a unit vector pointing from m_1 to m_2 and G is the *universal gravitational constant** (sometimes called the *constant of gravitation* or *constant of universal gravitation*). The following example demonstrates the application of this law.

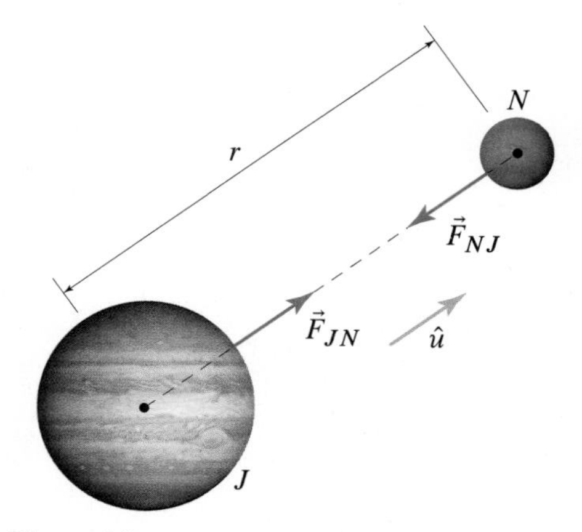

Figure 1.2
The gravitational force between the planets Jupiter J and Neptune N. The relative sizes of the planets are accurate, but their separation distance is not.

Mini-Example

Using the planets Jupiter and Neptune as an example, the force on Jupiter due to the gravitational attraction of Neptune, $\vec{F}_{JN}$, is given by (see Fig. 1.2)

$$\vec{F}_{JN} = \frac{G m_J m_N}{r^2}\,\hat{u}, \tag{1.6}$$

where r is the distance between the two bodies, m_J is the mass of Jupiter, m_N is the mass of Neptune, and $\hat{u}$ is a unit vector pointing from the center of Jupiter to the center of Neptune. The mass of Jupiter is 1.9×10^{27} kg, and that of Neptune is 1.02×10^{26} kg. Since the mean radius of Jupiter's orbit is 778,300,000 km and that of Neptune is 4,505,000,000 km, we assume that their closest approach to one another is approximately 3,727,000,000 km. Thus, at their closest approach, the magnitude of the force between these two planets is

$$\begin{aligned}|\vec{F}_{JN}| &= \left(6.674\times10^{-11}\,\frac{\text{m}^3}{\text{kg}\cdot\text{s}^2}\right)\frac{(1.9\times10^{27}\,\text{kg})(1.02\times10^{26}\,\text{kg})}{(3.727\times10^{12}\,\text{m})^2}\\ &= 9.312\times10^{17}\,\text{N}.\end{aligned} \tag{1.7}$$

It is interesting to compare this force with the force of gravitation between Jupiter and the Sun. The Sun's mass is 1.989×10^{30} kg, and we have already stated that the mean radius of Jupiter's orbit is 778,300,000 km. Applying Eq. (1.6) between Jupiter and the Sun gives 4.164×10^{23} N, which is almost 450,000 times larger.

Acceleration due to gravity. Equation (1.5) allows us to determine the force of Earth's gravity on an object of mass m on the surface of the Earth. This is done by noting that the radius of the Earth is 6371.0 km (see the marginal note) and the mass of the Earth is 5.9736×10^{24} kg and then applying Eq. (1.5):

$$\begin{aligned}F_s &= \left(6.674\times10^{-11}\,\frac{\text{m}^3}{\text{kg}\cdot\text{s}^2}\right)\frac{(5.9736\times10^{24}\,\text{kg})m}{(6371.0\times10^3\,\text{m})^2}\\ &= (9.8222\,\text{m/s}^2)m.\end{aligned} \tag{1.8}$$

This result† tells us that the force of gravity (in N) on an object on the Earth's surface is about 9.8 times the object's mass (in kg). This factor of 9.8 is so prevalent in engineering that it is given the label g, and it is called the *acceleration due to*

Interesting Fact

The radius of the Earth. The Earth is not a perfect sphere. Therefore, there are different notions of "radius of the Earth." The given value of 6371.0 km is the *volumetric radius* when rounded to 5 significant digits. The Earth's volumetric radius is the radius of a perfect sphere with volume equal to that of the Earth. Other measures of the Earth's radius, rounded to 5 significant digits, are the *quadratic mean radius*, the *authalic mean radius*, and the *meridional Earth radius*, which are equal to 6372.8, 6371.0, and 6367.4 km, respectively.

---

* Henry Cavendish (1731–1810) was the first to measure G and did so in 1798. The generally accepted value is $G = 6.674\times10^{-11}\,\text{m}^3/(\text{kg}\cdot\text{s}^2) = 3.439\times10^{-8}\,\text{ft}^3/(\text{slug}\cdot\text{s}^2)$.

† We will normally round the result of all calculations to 4 significant digits. Here we are using 5 significant figures because the data used in this particular calculation is known to that degree of accuracy.

gravity because it has units of acceleration and its value is the acceleration of objects in free fall near the surface of the Earth. We will take the value of g to be $9.81\,\mathrm{m/s^2}$ in SI units and $32.2\,\mathrm{ft/s^2}$ in U.S. Customary units. Notice that the value of g obtained in Eq. (1.8) is slightly greater than the $9.81\,\mathrm{m/s^2}$ that we will use in this book. The difference between these values has several causes, including that the Earth is not perfectly spherical, does not have uniform mass distribution, and is rotating. Because of these factors, the actual acceleration due to gravity is about 0.27% lower at the equator, and 0.26% higher at the poles, relative to the standard value of $g = 9.81\,\mathrm{m/s^2}$, which is for a north or south latitude of 45° at sea level. There may also be small local variations in gravity due to geological formations. Nonetheless, throughout this book we will use the standard value of g stated above.

Change in acceleration due to altitude. There is a formula that allows us to find how the acceleration due to gravity changes with altitude. To find it, we begin by equating Eqs. (1.2) and (1.5) to determine the acceleration a at a height h above the surface of the Earth

$$a = \frac{Gm_e}{(r_e + h)^2}, \tag{1.9}$$

where r_e is the radius of the Earth, m_e is the mass of the Earth, and we have canceled the mass of the object on both sides of the equation. Now, at the surface of the Earth, we know that $a = g$ and $h = 0$, so Eq. (1.9) becomes

$$g = Gm_e/r_e^2 \quad \Rightarrow \quad Gm_e = gr_e^2. \tag{1.10}$$

Substituting Eq. (1.10) into Eq. (1.9), we see that a is given by

$$\boxed{a = g\frac{r_e^2}{(r_e + h)^2},} \tag{1.11}$$

where g is the acceleration due to gravity at the surface of the Earth. Equation (1.11) is very handy because it requires knowledge of only the radius of the Earth to get the acceleration due to gravity rather than having to know both the radius of the Earth *and* the universal gravitational constant G.

1.2 Fundamental Concepts in Dynamics

Space and time

Space

Space is the environment in which objects move, and we consider it to be a collection of locations or *points*. The position of a point is indicated by the point's coordinates in a chosen coordinate system. Figure 1.3 shows a three-dimensional *Cartesian coordinate system* with *origin* at O and mutually orthogonal axes x, y, and z. The *Cartesian coordinates* of the point P are x_P, y_P, and z_P, which are scalars obtained by measuring the distance between O and the perpendicular projections of the point P onto the axes x, y, and z, respectively. Note that x_P, y_P, and z_P have a positive or negative sign depending on whether, in going from O to the projections of P along each axis, one moves in the positive or negative direction of these axes. In Chapter 2, we will introduce additional coordinate systems.

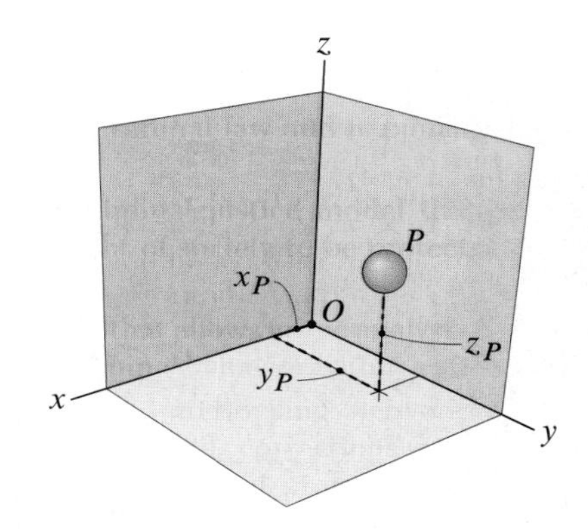

Figure 1.3
A point in a three-dimensional space.

Time

Time is a scalar variable that allows us to specify the order of a sequence of events. In classical mechanics and in this book, the most important assumption about time is that it is *absolute*. We assume that the duration of an event is independent of the motion of the observer making time measurements. Einstein's theory of relativity rejects this assumption.

Helpful Information

The right-hand rule. In three dimensions, a Cartesian coordinate system uses three orthogonal reference directions. These are the x, y, and z directions shown below.

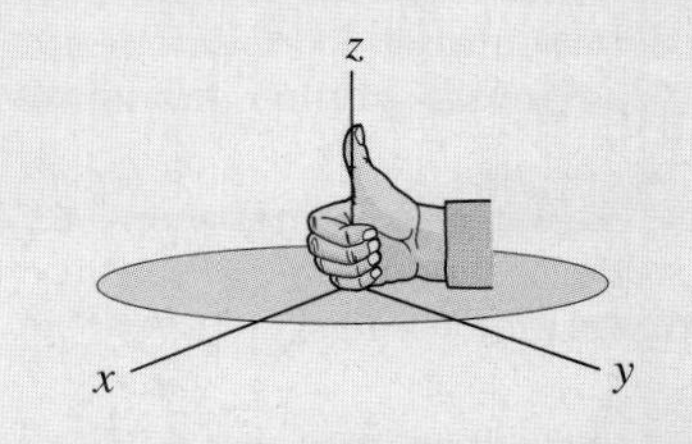

Proper interpretation of many vector operations, such as the cross product, requires that the x, y, and z directions be arranged in a consistent manner. The convention in mechanics and vector mathematics in general is that if the axes are arranged as shown, then, according to the *right-hand rule*, rotating the x direction into the y direction yields the z direction. The result is called a *right-handed coordinate system*.

Force, mass, and inertia

Force

The *force* acting on an object is the interaction between that object and its environment. A more precise description of this interaction requires that we know something about the interaction in question. For example, if two objects collide or slide against one another, we say that they interact via *contact* forces. Regardless of the type, the characteristics of a force are its magnitude, its line of action, and its orientation or direction. This is why we use *vectors* to represent forces.

Mass

The *mass* of an object is a measure of the amount of matter in the object. Along with the concept of force, the concept of mass is considered a *primitive concept*, i.e., not explainable via more elementary ideas. Newton's second law postulates that the force acting on a body is *proportional* to the body's acceleration — the constant of proportionality is the *mass* of the body.

Inertia

Inertia is commonly understood as a body's resistance to changing its state of motion in response to the application of a force system. In this book, we use *inertia* as an umbrella term encompassing both the idea of mass and that of mass distribution over a region of space. The *inertia properties* of an object are its mass and a quantitative description of the mass distribution.

Particle and rigid body

Particle

A *particle* is an object whose mass is concentrated at a point; therefore, it is also called a point mass. The inertia properties of a particle consist only of its mass. A particle is generally understood to have zero volume. It is meaningless to talk about the rotation of a particle whose position is held fixed, although we do say that a particle can "rotate about a point," meaning that a particle can move along a path around a point. Regardless of its volume, when we choose to model an object as a particle, we neglect the possibility that the object might rotate in the sense of "change its orientation" relative to some chosen reference.

Rigid body

A *rigid body* is an object whose mass is (1) distributed over a region of space and (2) such that the distance between any two points on it never changes. Since its mass is not concentrated at a point, the *rigid body* is the simplest model for the study of motions that include the possibility of rotation, i.e., a change of orientation relative to a chosen reference. We model objects as rigid bodies when we want to account for the possibility of rotation while neglecting the effects of deformation. Finally, the mass distribution of a rigid body does not change relative to an observer moving with the body. This fact makes it possible to describe the inertia properties of a three-dimensional rigid body with seven pieces of information consisting of the body's mass and six mass moments of inertia.*

Vectors and their Cartesian representation

Notation

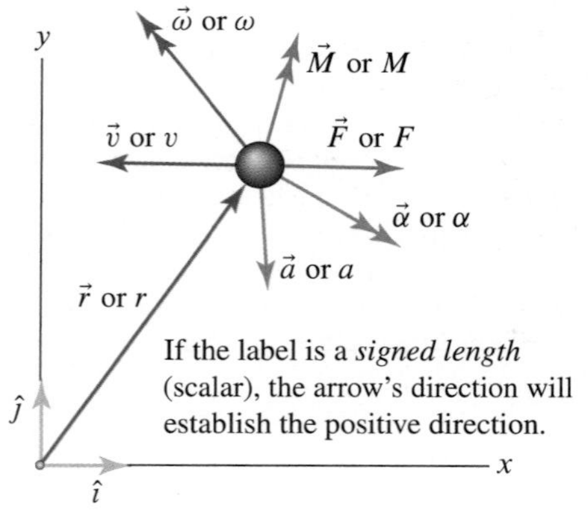

Scalars. By *scalar* we mean a *real number*. Scalars will be denoted by italic roman characters (e.g., a, h, or W) or by Greek letters (e.g., α, ω, or δ).

Vectors. We will *always* denote vectors by placing arrows or carets (in the case of unit vectors) over letters, such as $\vec{F}$ or $\hat{\imath}$. The conventions we use to depict vectors in figures are shown in Fig. 1.4. The color scheme used in the figure is defined in the caption. Depending on what we want or need to emphasize in a figure, a vector will be labeled with a letter that has an arrow placed above it (e.g., $\vec{a}$ or $\vec{\omega}$) or with just a letter (e.g., a or ω) according to the following conventions:

- In figures, a vector will be labeled with arrows over letters when it is important to emphasize the arbitrary directional nature of the vector (e.g., a velocity) or the vectorial nature of the quantity (e.g., a unit vector).

- Base vectors in Cartesian components will be designated using the unit vectors $\hat{\imath}$, $\hat{\jmath}$, and $\hat{k}$. A *unit vector* is a vector with magnitude equal to 1. In any other context, e.g., in other component systems, unit vectors will be designated using a caret over the letter u, that is, $\hat{u}$, often accompanied by a subscript indicating the direction of the vector, such as $\hat{u}_r$.

- In figures, the label of a vector with known direction will generally not be a letter with an arrow over it. A vector with known direction will usually be

Figure 1.4
Notation and colors for commonly used vectors. Position vectors will always be blue ■ ($\vec{r}$), velocity (linear and angular) vectors purple ■ ($\vec{v}$ and $\vec{\omega}$), and acceleration (linear and angular) vectors green ■ ($\vec{a}$ and $\vec{\alpha}$). Forces and moments will always be red ■ ($\vec{F}$ and $\vec{M}$) and unit vectors orange ■ ($\hat{\imath}$ and $\hat{\jmath}$). Vectors with no particular physical significance will be black ■, magenta ■, or gray ■.

* For a definition of the mass moments of inertia of a rigid body, see Section 10.3 on p. 553 of M. E. Plesha, G. L. Gray, and F. Costanzo, *Engineering Mechanics: Statics*, McGraw-Hill, Dubuque, IA, 2010.

labeled as a *signed length* (i.e., a scalar component) whose positive direction is that of the arrow in the figure.

- Double-headed arrows will designate vectors associated with "rotational" quantities, i.e., moments, angular velocities, and angular accelerations (angular velocities and accelerations will be discussed in Chapter 2).

Cartesian vector representation

We now review those aspects of vectors in two dimensions that are most important for our applications. This presentation is easily extended to three dimensions.

Figure 1.5 shows the position of a point P with respect to the origin O of a

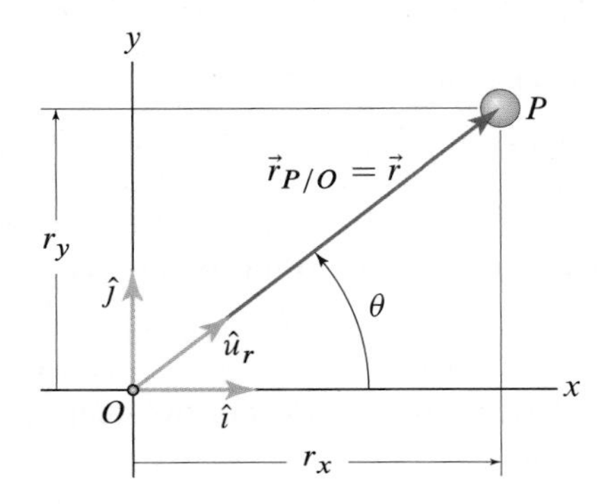

Figure 1.5. Description of the position of a point P. The curved arrows indicating an angle with a single arrowhead designate an angle's positive direction.

rectangular coordinate system. The position of P is represented by the arrow that starts at O and ends at P, which we call the vector $\vec{r}_{P/O}$. The subscript $_{P/O}$ is read "P relative to O," or "P as seen by O," or "P with respect to O." When only one point is being discussed, we typically drop the $_{P/O}$ part of the notation and simply indicate position as $\vec{r}$.

The *Cartesian representation* of $\vec{r}$ is

$$\vec{r} = r_x\,\hat{\imath} + r_y\,\hat{\jmath}, \tag{1.12}$$

where $\hat{\imath}$ and $\hat{\jmath}$ are unit vectors in the x and y directions, respectively. The quantities r_x and r_y are the *(scalar) Cartesian components* of $\vec{r}$. Using trigonometry, we have

$$r_x = |\vec{r}|\cos\theta \quad \text{and} \quad r_y = |\vec{r}|\sin\theta, \tag{1.13}$$

where θ is the orientation of the segment $\overline{OP}$ (the bar over the letters O and P designates the line segment connecting the points O and P) relative to the x axis and $|\vec{r}|$, called the *magnitude of* $\vec{r}$ or *length of* $\vec{r}$, is the length of $\overline{OP}$. Equation (1.12) could be written as $\vec{r} = \vec{r}_x + \vec{r}_y$, where the vectors $\vec{r}_x = r_x\,\hat{\imath}$ and $\vec{r}_y = r_y\,\hat{\jmath}$ are called the x and y *vector* components of $\vec{r}$, respectively. In this book, *component* will always mean *scalar component*. When talking about *vector components*, we will explicitly say *vector components*.

Generalizing what we said about $\vec{r}_{P/O}$, given points A and B with coordinates (x_A, y_A) and (x_B, y_B), respectively, the vector

$$\vec{r}_{A/B} = (x_A - x_B)\,\hat{\imath} + (y_A - y_B)\,\hat{\jmath} \tag{1.14}$$

will be called *the position of A with respect to B*, or position of A relative to B (Fig. 1.6).

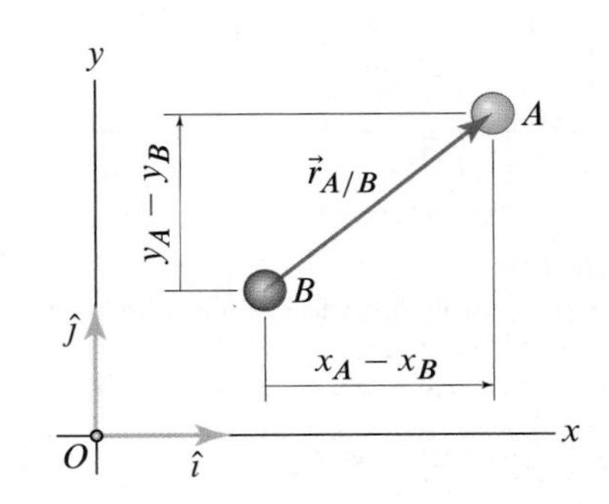

Figure 1.6 Vector representation of the position of A relative to B.

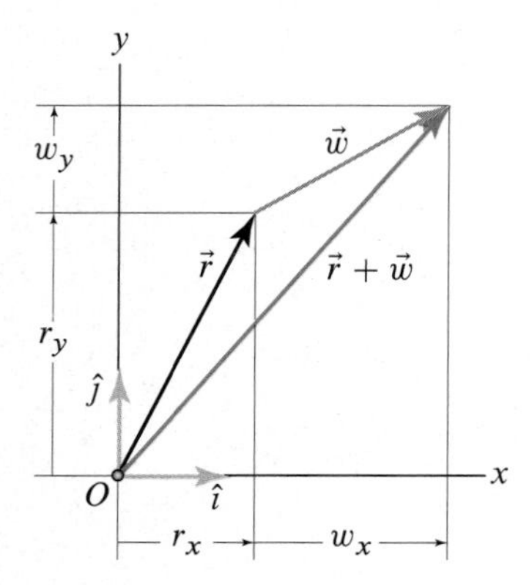

Figure 1.7
Graphical representation of the vector addition of $\vec{r}$ and $\vec{w}$ showing the "triangle law."

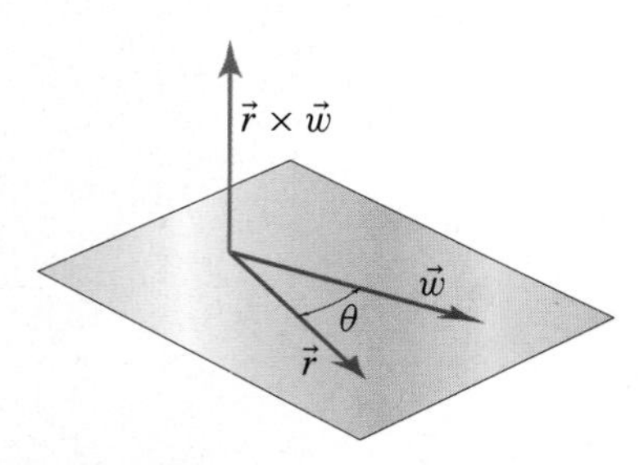

Figure 1.8
Graphical representation of the vector cross product of $\vec{r}$ and $\vec{w}$.

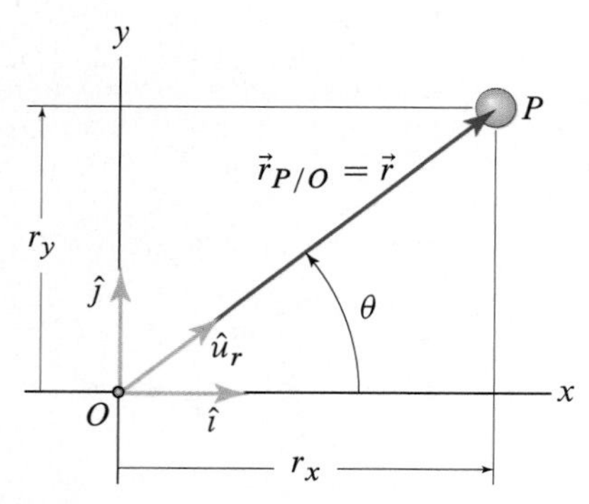

Figure 1.9
Description of the position of a particle P.

Vector operations

Here are the vector operations we will use:

1. A vector $\vec{r}$ can be multiplied by a scalar a in the following way:
$$a\vec{r} = ar_x\,\hat{\imath} + ar_y\,\hat{\jmath}. \tag{1.15}$$
This *scales* the magnitude of $\vec{r}$ by the factor $|a|$. The object $a\vec{r}$ is a *vector* (not a scalar) with the same line of action as $\vec{r}$; the direction of $a\vec{r}$ is the same as that of $\vec{r}$ if $a > 0$, whereas it is opposite to $\vec{r}$ if $a < 0$.

2. Two vectors can be summed to obtain another vector as follows:
$$\vec{r} + \vec{w} = (r_x + w_x)\,\hat{\imath} + (r_y + w_y)\,\hat{\jmath}, \tag{1.16}$$
which conforms to the triangle law of *vector addition* (see Fig. 1.7).

3. The operation of summing a scalar with a vector is not defined.

4. Referring to Fig. 1.8, the *dot* or *scalar product* of two vectors $\vec{r}$ and $\vec{w}$ is denoted by $\vec{r}\cdot\vec{w}$ and yields the following *scalar* quantity:
$$\vec{r}\cdot\vec{w} = |\vec{r}||\vec{w}|\cos\theta. \tag{1.17}$$

5. Referring to Fig. 1.8, the *cross product* of two vectors $\vec{r}$ and $\vec{w}$ is the *vector* denoted by $\vec{r}\times\vec{w}$ with
   (a) magnitude
   $$|\vec{r}\times\vec{w}| = |\vec{r}||\vec{w}|\sin\theta, \tag{1.18}$$
   (b) line of action perpendicular to the plane containing $\vec{r}$ and $\vec{w}$, and
   (c) direction determined by the right-hand rule.

In Eqs. (1.17) and (1.18), θ is the smallest angle that will rotate one of the vectors into the other. For the cross product, this choice of θ ensures that Eq. (1.18) always yields a nonnegative value. For the dot product, since $\cos\theta = \cos(2\pi - \theta)$, θ can be replaced by $2\pi - \theta$. Finally, the definition of cross product implies that the cross product is *anticommutative*, that is,
$$\vec{r}\times\vec{w} = -\vec{w}\times\vec{r}. \tag{1.19}$$

Referring to Fig. 1.9, we recall that $\vec{r}$ represents the length and orientation of the segment $\overline{OP}$. Using the Pythagorean theorem and trigonometry, and expressing angles in radians, we have
$$\text{length of } \vec{r} = |\vec{r}| = \sqrt{r_x^2 + r_y^2}, \tag{1.20}$$
and
$$\text{direction of } \vec{r} = \theta = \tan^{-1}\left(\frac{r_y}{r_x}\right) \pm n\pi, \quad n = 0, 1, 2, \ldots \tag{1.21}$$
where n is found by identifying the quadrant containing P (for the case in Fig. 1.9, $n = 0$). Finally, Eqs. (1.12) and (1.13) allow us to rewrite $\vec{r}$ as
$$\vec{r} = |\vec{r}|\,\hat{u}_r, \quad \text{where} \quad \hat{u}_r = \cos\theta\,\hat{\imath} + \sin\theta\,\hat{\jmath}. \tag{1.22}$$
Since $\hat{u}_r$ is a unit vector in the direction of $\vec{r}$, Eq. (1.22) implies that

> *The information carried by any vector can be written as the product of its magnitude and a unit vector pointing in the direction of that vector.*

Useful vector "tips and tricks"

Components of a vector

We now review how to find the components of a vector, since this operation occurs often in dynamics.

Figure 1.10 shows two perpendicular and oriented lines ℓ_1 and ℓ_2, where by *oriented* we mean that they have a positive and a negative direction. The lines are oriented using the unit vector $\hat{u}_1$ for ℓ_1 and $\hat{u}_2$ for ℓ_2. We also have a vector $\vec{q}$ oriented arbitrarily relative to ℓ_1 and ℓ_2. Our goal is to find the scalar components of $\vec{q}$ along ℓ_1 and ℓ_2.

If we apply Eq. (1.17) to Fig. 1.10 and let $\vec{r}$ be $\vec{q}$, $\vec{w}$ be $\hat{u}_1$, and θ be θ_1, we see that the dot product gives us q_1 directly, that is,

$$q_1 = \vec{q} \cdot \hat{u}_1 = |\vec{q}||\hat{u}_1| \cos\theta_1 = |\vec{q}| \cos\theta_1. \tag{1.23}$$

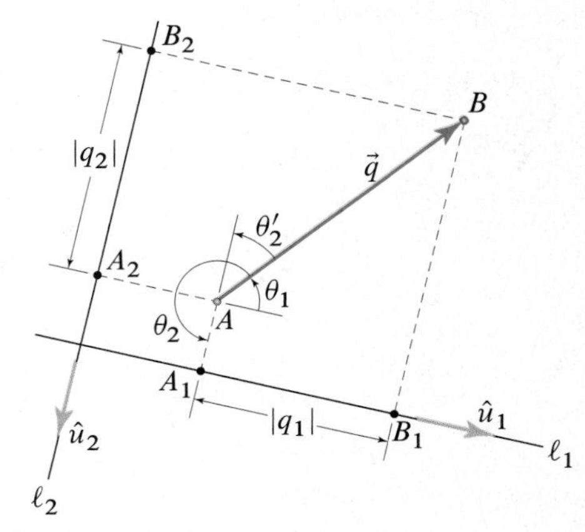

Figure 1.10
Diagram showing the components of $\vec{q}$ in the directions of $\hat{u}_1$ and $\hat{u}_2$.

The quantity q_1 in Eq. (1.23) is what we were looking for because, according to the definition of scalar component of a vector and Fig. 1.10,

1. $|q_1|$ is the distance between A_1 and B_1.

2. The sign of $\vec{q} \cdot \hat{u}_1$ is determined by the sign of $\cos\theta_1$, which is positive if $0° \le \theta_1 < 90°$ and negative if $90° < \theta_1 \le 180°$ (if $\theta_1 = 90°$, A_1 and B_1 coincide so that $q_1 = 0$).

In summary,

$$\boxed{\text{component of } \vec{q} \text{ along } \ell_1 = q_1 = \vec{q} \cdot \hat{u}_1.} \tag{1.24}$$

Repeating the foregoing discussion in the case of q_2 we have that the

$$\text{component of } \vec{q} \text{ along } \ell_2 = q_2 = \vec{q} \cdot \hat{u}_2 = |\vec{q}| \cos\theta_2 = -|\vec{q}| \cos\theta_2', \tag{1.25}$$

since $\theta_2 = \theta_2' + \pi$ and $\cos(\theta_2' + \pi) = -\cos\theta_2'$.

In dynamics we often face the situation depicted in Fig. 1.11, in which we need to calculate the Cartesian components of two mutually orthogonal vectors $\vec{q}$ and $\vec{r}$. If the angle θ defining the orientation of $\vec{q}$ relative to the y axis is given, to express $\vec{q}$ and $\vec{r}$ in components, we can write $\vec{q}$ as

$$\begin{aligned} \vec{q} = (\vec{q} \cdot \hat{\imath})\,\hat{\imath} + (\vec{q} \cdot \hat{\jmath})\,\hat{\jmath} &= |\vec{q}| \cos(\theta + \tfrac{\pi}{2})\,\hat{\imath} + |\vec{q}| \cos\theta\,\hat{\jmath} \\ &= -|\vec{q}| \sin\theta\,\hat{\imath} + |\vec{q}| \cos\theta\,\hat{\jmath} \\ &= |\vec{q}|\,\underbrace{(-\sin\theta\,\hat{\imath} + \cos\theta\,\hat{\jmath})}_{\text{unit vector}}, \end{aligned} \tag{1.26}$$

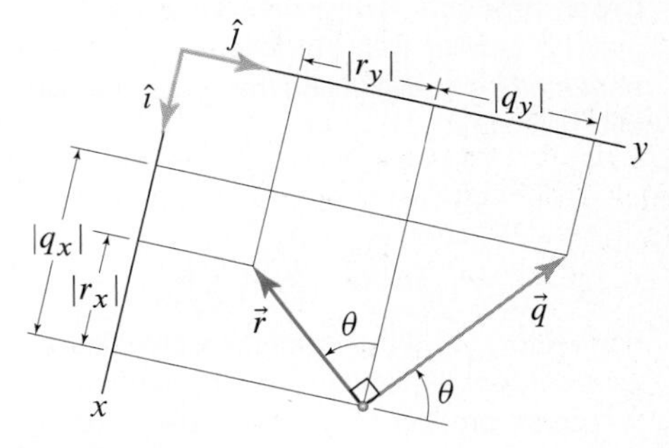

Figure 1.11
Diagram showing the Cartesian components of the vector $\vec{q}$ as well as the vector $\vec{r}$ that is orthogonal to $\vec{q}$.

and then we can write $\vec{r}$ as

$$\begin{aligned} \vec{r} = (\vec{r} \cdot \hat{\imath})\,\hat{\imath} + (\vec{r} \cdot \hat{\jmath})\,\hat{\jmath} &= |\vec{r}| \cos(\pi - \theta)\,\hat{\imath} + |\vec{r}| \cos(\theta + \tfrac{\pi}{2})\,\hat{\jmath} \\ &= -|\vec{r}| \cos\theta\,\hat{\imath} - |\vec{r}| \sin\theta\,\hat{\jmath} \\ &= |\vec{r}|\,\underbrace{(-\cos\theta\,\hat{\imath} - \sin\theta\,\hat{\jmath})}_{\text{unit vector}}. \end{aligned} \tag{1.27}$$

Equations (1.26) and (1.27) demonstrate that the two equations have the following structure:

- Each vector is equal to its magnitude times a unit vector with one sine and one cosine term.

- The argument of the sine and cosine terms is the angle orienting one of the vectors with respect to one of the component directions.

- In the two equations there will *always* be three positive terms and one negative term or three negative terms and one positive term.

If, in planar problems, we take the components of two orthogonal vectors in two orthogonal directions and get anything other than this structure, then we should immediately recognize that we have made a mistake.

Cross products

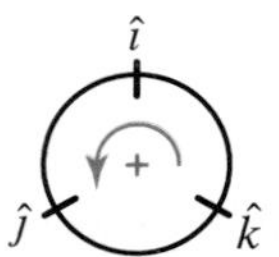

Figure 1.12
A little "trick" to help remember the cross products between Cartesian unit vectors

Since we will often encounter cross products in dynamics, here is a useful procedure to help us evaluate them. Consider three unit vectors $\hat{\imath}$, $\hat{\jmath}$, and $\hat{k}$ such that, according to the right-hand rule, $\hat{\imath} \times \hat{\jmath} = \hat{k}$. Arrange the three vectors as shown in Fig. 1.12. To calculate the product of, say, $\hat{\jmath} \times \hat{k}$, just move around the circle, starting from $\hat{\jmath}$ and going toward $\hat{k}$. Now notice that (1) the next vector on the circle is $\hat{\imath}$ and (2) in going from $\hat{\jmath}$ to $\hat{k}$ we move with the arrow (counterclockwise). Hence, $\hat{\jmath} \times \hat{k} = +\hat{\imath}$. Now consider $\hat{k} \times \hat{\jmath}$ and notice that in going from $\hat{k}$ toward $\hat{\jmath}$, the next vector along the circle is $\hat{\imath}$, and we move opposite to the arrow. Therefore, the result is negative, and we have $\hat{k} \times \hat{\jmath} = -\hat{\imath}$.

Helpful Information

Cross products using determinants. You may be familiar with the following *determinant method* of evaluating the cross product of two vectors:

$$\vec{a} \times \vec{b} = \begin{vmatrix} \hat{\imath} & \hat{\jmath} & \hat{k} \\ a_x & a_y & a_z \\ b_x & b_y & b_z \end{vmatrix}.$$

For vectors in 3D, this method provides a very efficient evaluation. As an alternative, the cross product may be evaluated on a term-by-term basis by expanding the following product:

$$\vec{a} \times \vec{b} = (a_x \hat{\imath} + a_y \hat{\jmath} + a_z \hat{k}) \times (b_x \hat{\imath} + b_y \hat{\jmath} + b_z \hat{k}).$$

When expanded, nine terms such as $a_x \hat{\imath} \times b_x \hat{\imath}$ and $a_x \hat{\imath} \times b_y \hat{\jmath}$ must be evaluated, and this is accomplished quickly using Fig. 1.12. In this book, we primarily do cross products of vectors in 2D, and we will likely find the term-by-term evaluation to be quicker.

Units

Units are essential to any quantifiable measure. Newton's second law in scalar form, $F = ma$, provides for the formulation of a consistent and unambiguous system of units. We will use both U.S. Customary units and SI units (International System*) as shown in Table 1.1. Each system has three *base dimensions* and a fourth *derived dimension*. In the U.S. Customary system, the base dimensions are force, length, and time, whose corresponding *base units* are lb (pounds), ft (feet), and s (seconds), respectively. The corresponding derived dimension is mass, which is obtained from the equation $m = F/a$. This gives the mass unit as lb·s^2/ft. This unit of mass is often called the *slug*.

Table 1.1. U.S. Customary and SI unit systems.

Base dimension	System of units	
	U.S. Customary	*SI*
force	pound (lb)	newton[a] (N) $\equiv$ kg·m/s^2
mass	slug[a] $\equiv$ lb·s^2/ft	kilogram (kg)
length	foot (ft)	meter (m)
time	second (s)	second (s)

[a] derived unit

In the SI system, the base dimensions are mass, length, and time, whose corresponding base units are kg (kilogram), m (meter), and s (second), respectively. The corresponding derived dimension is force, the unit of which is obtained from the

* SI has been adopted as the abbreviation for the French *Le Système International d'Unités*.

equation $F = ma$, which gives the force unit as $\text{kg} \cdot \text{m/s}^2$. This unit of force is referred to as a *newton*, and its abbreviation is N.

Because of the difference in base dimensions between the U.S. Customary system and the SI system, when using the U.S. Customary system, we normally specify the weight of an object (typically in lb) instead of its mass; and, conversely, when using the SI system, we normally specify the mass of an object (typically in kg) instead of its weight.

For both systems, we may occasionally use different, but consistent, units for some dimensions. For example, we may use minutes rather than seconds, inches instead of feet, grams instead of kilograms.

Plane angles are dimensionless quantities (they are defined as the ratio of two lengths). In both the U.S. Customary and SI systems, angles are expressed in radians, abbreviated rad. Another commonly used unit to express angle measurements is the degree, indicated by the symbol °. Angle measurements in degrees and in radians are related as follows:

$$\boxed{180^\circ = \pi \text{ rad.}} \tag{1.28}$$

Common Pitfall

Weight and mass. Unfortunately, it is common to refer to weight using mass units. For example, the person who says, "I weigh 70 kg" really means "My mass is 70 kg." In science and engineering it is essential that accurate nomenclature be used. Weights and forces must be reported using appropriate force units, and masses must be reported using appropriate mass units.

Dimensional homogeneity and unit conversions

Equations must be dimensionally homogeneous. This means that the quantities on the two sides of the equal sign must have the same dimensions. Our strong recommendation is that appropriate units always be used in all equations during a calculation to make sure that the results are dimensionally correct. Such practice helps avoid catastrophic blunders and provides a useful check on a solution, for if an equation is found to be dimensionally inconsistent, then an error has certainly been made. In September 1999, NASA (National Aeronautics and Space Administration) lost a $125 million Mars orbiter because the climate orbiter spacecraft team at the contractor who built the spacecraft used U.S. Customary units when computing rocket thrust, while the mission navigation team at NASA used metric units for this key spacecraft operation. This units error, which came into play when the spacecraft was to be inserted into orbit around Mars, caused the spacecraft to approach Mars at too low an altitude, thus causing it to burn up in Mars' atmosphere.

Unit conversions are often needed and are easily done using conversion factors, such as those shown in Table 1.2, and rules of algebra. The basic idea is to multiply either or both sides of an equation by dimensionless factors of unity, where each factor of unity embodies an appropriate unit conversion. This procedure is illustrated in the examples at the end of this section.

Table 1.2
Conversion factors between U.S. Customary and SI unit systems.

	U.S. Customary	SI
length	1 in.	0.0254 m (25.4 mm)
	1 ft (12 in.)	0.3048 m
	1 mi (5280 ft)	1.609 km
force	1 lb	4.448 N
	1 kip (1000 lb)	4.448 kN
mass	1 slug ($1\ \text{lb}\cdot\text{s}^2/\text{ft}$)	14.59 kg

Prefixes

Prefixes are a useful alternative to scientific notation for representing numbers that are very large or very small. Common prefixes and a summary of rules for their use are given in Table 1.3.

Here is a list of common rules for correct prefix use:

1. With few exceptions, prefixes should be used only in the numerator of unit combinations; e.g., use the unit km/s (kilometer per second) and avoid the unit m/ms (meter per millisecond). One common exception to this rule is kg, which may appear in numerator or denominator; e.g., use the unit kW/kg (kilowatt per kilogram) and avoid the unit W/g (watt per gram).

Table 1.3. Common prefixes used in the SI unit systems.

Multiplication factor		Prefix	Symbol
1 000 000 000 000 000 000 000 000	10^{24}	yotta	Y
1 000 000 000 000 000 000 000	10^{21}	zetta	Z
1 000 000 000 000 000 000	10^{18}	exa	E
1 000 000 000 000 000	10^{15}	peta	P
1 000 000 000 000	10^{12}	tera	T
1 000 000 000	10^{9}	giga	G
1 000 000	10^{6}	mega	M
1 000	10^{3}	kilo	k
100	10^{2}	hecto	h
10	10^{1}	deka	da
0.1	10^{-1}	deci	d
0.01	10^{-2}	centi	c
0.001	10^{-3}	milli	m
0.000 001	10^{-6}	micro	μ
0.000 000 001	10^{-9}	nano	n
0.000 000 000 001	10^{-12}	pico	p
0.000 000 000 000 001	10^{-15}	femto	f
0.000 000 000 000 000 001	10^{-18}	atto	a
0.000 000 000 000 000 000 001	10^{-21}	zepto	z
0.000 000 000 000 000 000 000 001	10^{-24}	yocto	y

2. Double prefixes must be avoided; e.g., use the unit GHz (gigahertz) and avoid the unit kMHz (kilo-megahertz).

3. Use a center dot or dash to denote multiplication of units, e.g., N·m or N-m. In this book, we denote multiplication of units by a dot, e.g., N·m.

4. Exponentiation applies to both the unit and prefix, e.g., $\text{mm}^2 = (\text{mm})^2$.

5. If the number of digits on either side of a decimal point exceeds 4, it is common to group the digits into groups of 3, with the groups separated by commas or thin spaces. Since many countries use a comma to represent a decimal point, the thin space is sometimes preferable; e.g., 1234.0 could be written as is, but by contrast, 12345.0 should be written as 12,345.0 or as 12 345.0.

While prefixes can often be incorporated into an expression by inspection, the rules for doing this are identical to those for performing unit transformations.

Accuracy of numbers in calculations

Throughout this book, we will generally assume that the data given for problems is accurate to three significant digits. When calculations are performed, such as in example problems, all intermediate results are stored in the memory of a calculator or computer using the full precision these machines offer. However, when these intermediate results are reported in this book, they are rounded to four significant digits. Final answers are usually reported with three significant digits. If you verify the calculations described in this book using the rounded numbers that are reported, you may occasionally calculate results that are slightly different from those shown.

End of Section Summary

Review of vector operations. Referring to Fig. 1.13, the Cartesian representation of a two-dimensional vector $\vec{r}$ takes the form

Eq. (1.12), p. 7

$$\vec{r} = r_x\,\hat{\imath} + r_y\,\hat{\jmath},$$

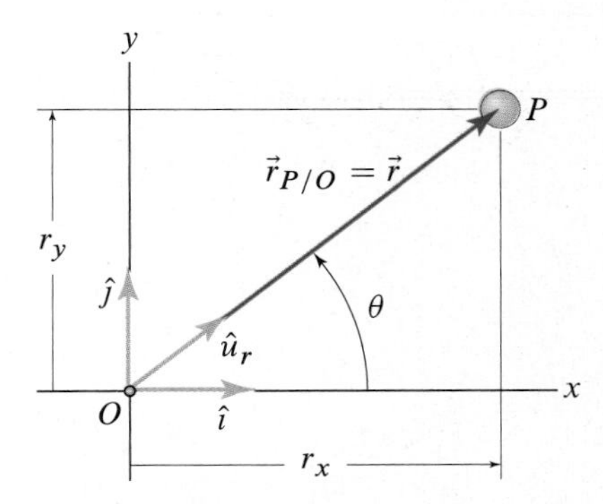

Figure 1.13
Description of the position of a particle P.

where $\hat{\imath}$ and $\hat{\jmath}$ are unit vectors in the positive x and y directions, respectively, and where r_x and r_y are the x and y (*scalar*) *components* of $\vec{r}$, respectively. Using trigonometry, r_x and r_y are given by

Eqs. (1.13), p. 7

$$r_x = |\vec{r}|\cos\theta \quad \text{and} \quad r_y = |\vec{r}|\sin\theta,$$

where θ is the orientation of the segment $\overline{OP}$ relative to the x axis and $|\vec{r}|$, called the *magnitude of* $\vec{r}$ or *length of* $\vec{r}$, is the length of $\overline{OP}$.

The *dot* or *scalar product* of the vectors $\vec{r}$ and $\vec{w}$ gives the *scalar*

Eq. (1.17), p. 8

$$\vec{r}\cdot\vec{w} = |\vec{r}||\vec{w}|\cos\theta,$$

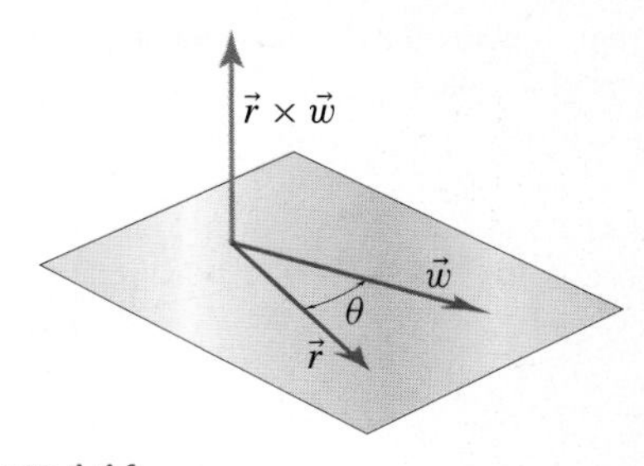

Figure 1.14
Graphical representation of the vector cross product of $\vec{r}$ and $\vec{w}$.

with θ being the smallest angle that will rotate one of the vectors into the other.

Referring to Fig. 1.14, the *cross product* of two vectors $\vec{r}$ and $\vec{w}$ is denoted by $\vec{r}\times\vec{w}$ and gives a *vector* with

1. Magnitude equal to

   Eq. (1.18), p. 8

   $$|\vec{r}\times\vec{w}| = |\vec{r}||\vec{w}|\sin\theta,$$

   where θ is the smallest angle that will rotate one of the vectors into the other (to ensure that $|\vec{r}||\vec{w}|\sin\theta \geq 0$).

2. Whose line of action is perpendicular to the plane containing $\vec{r}$ and $\vec{w}$ and whose direction is determined by the right-hand rule.

3. For which the cross product is anticommutative; i.e., $\vec{r}\times\vec{w} = -\vec{w}\times\vec{r}$.

Referring to Fig 1.13, given r_x and r_y, we compute $|\vec{r}|$ and θ as follows:

Eqs. (1.20) and (1.21), p. 8

$$\text{length of } \vec{r} = |\vec{r}| = \sqrt{r_x^2 + r_y^2},$$

$$\text{direction of } \vec{r} = \theta = \tan^{-1}\left(\frac{r_y}{r_x}\right).$$

Another useful representation of a vector $\vec{r}$ is as follows:

Eq. (1.22), p. 8

$$\vec{r} = |\vec{r}|\,\hat{u}_r, \quad \text{where} \quad \hat{u}_r = \cos\theta\,\hat{\imath} + \sin\theta\,\hat{\jmath},$$

where $\hat{u}_r$ is the unit vector in the direction of $\vec{r}$.

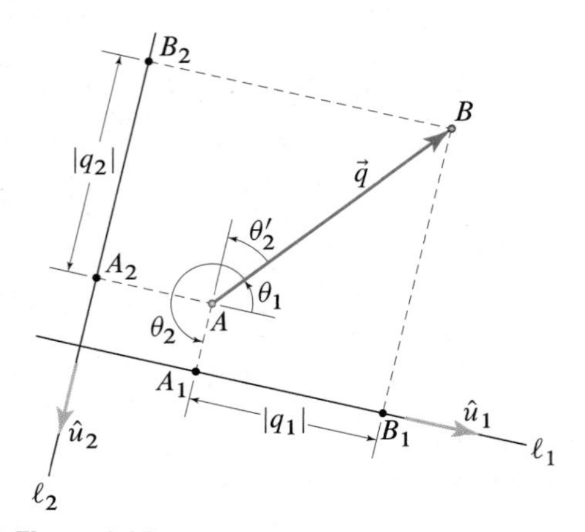

Figure 1.15
Diagram showing the components of $\vec{q}$ in the directions of $\hat{u}_1$ and $\hat{u}_2$.

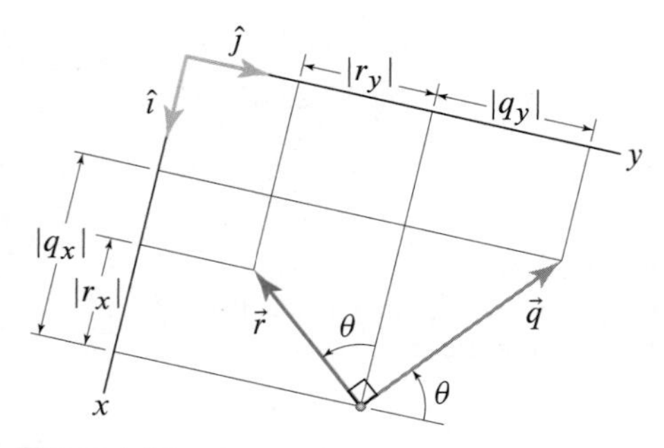

Figure 1.16
Diagram showing the Cartesian components of the vector $\vec{q}$ as well as the vector $\vec{r}$ that is orthogonal to $\vec{q}$.

Useful vector "tips and tricks." Referring to Fig. 1.15, the component of a vector in a given direction can be computed using the dot product. For example,

Eq. (1.23), p. 9

$$q_1 = \vec{q} \cdot \hat{u}_1 = |\vec{q}||\hat{u}_1| \cos\theta_1 = |\vec{q}| \cos\theta_1,$$

that is,

Eq. (1.24), p. 9

$$\text{component of } \vec{q} \text{ along } \ell_1 = q_1 = \vec{q} \cdot \hat{u}_1.$$

Referring to Fig. 1.16, for two mutually orthogonal vectors $\vec{r}$ and $\vec{q}$ the following relations hold:

Eqs. (1.26) and (1.27), p. 9

$$\begin{aligned}
\vec{q} = (\vec{q} \cdot \hat{\imath})\,\hat{\imath} + (\vec{q} \cdot \hat{\jmath})\,\hat{\jmath} &= |\vec{q}| \cos(\theta + \tfrac{\pi}{2})\,\hat{\imath} + |\vec{q}| \cos\theta\,\hat{\jmath} \\
&= -|\vec{q}| \sin\theta\,\hat{\imath} + |\vec{q}| \cos\theta\,\hat{\jmath} \\
&= |\vec{q}|(-\sin\theta\,\hat{\imath} + \cos\theta\,\hat{\jmath}), \\
\vec{r} = (\vec{r} \cdot \hat{\imath})\,\hat{\imath} + (\vec{r} \cdot \hat{\jmath})\,\hat{\jmath} &= |\vec{r}| \cos(\pi - \theta)\,\hat{\imath} + |\vec{r}| \cos(\theta + \tfrac{\pi}{2})\,\hat{\jmath} \\
&= -|\vec{r}| \cos\theta\,\hat{\imath} - |\vec{r}| \sin\theta\,\hat{\jmath} \\
&= |\vec{r}|(-\cos\theta\,\hat{\imath} - \sin\theta\,\hat{\jmath}).
\end{aligned}$$

EXAMPLE 1.1 *Components of Vectors*

At the instant shown, the acceleration of the airplane in Fig. 1 is the vector

$$\vec{a} = (5.63\,\hat{u}_t + 37.2\,\hat{u}_n)\,\text{m/s}^2, \tag{1}$$

where the mutually perpendicular unit vectors $\hat{u}_t$ and $\hat{u}_n$ are tangent and perpendicular to the airplane's path, respectively. The angle between $\hat{u}_t$ and the horizontal direction is $\theta = 26°$. Determine ϕ, the angle between $\vec{a}$ and $\hat{u}_t$, and the expression of $\vec{a}$ relative to the unit vectors $\hat{\imath}$ and $\hat{\jmath}$, which are horizontal and vertical, respectively.

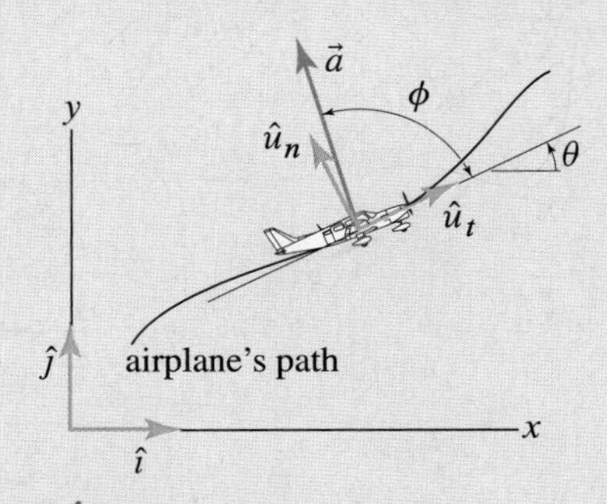

Figure 1
An airplane performing a maneuver.

SOLUTION

Road Map Letting a_t and a_n be the components of $\vec{a}$ in the $(\hat{u}_n, \hat{u}_t)$ component system, Eq. (1) implies that $a_t = 5.63\,\text{m/s}^2$ and $a_n = 37.2\,\text{m/s}^2$. Since a_t and a_n are known, ϕ can be found via Eq. (1.21) on p. 8 after replacing the quantities r_x and r_y in that equation with a_t and a_n, respectively. To express $\vec{a}$ using $\hat{\imath}$ and $\hat{\jmath}$, we need to determine the components of $\vec{a}$ in the $\hat{\imath}$ and $\hat{\jmath}$ directions. This can be done via Eq. (1.26) on p. 9.

Determination of ϕ

Computation Replacing r_x and r_y in Eq. (1.21) on p. 8 with a_t and a_n, respectively, we have

$$\phi = \tan^{-1}\left(\frac{a_n}{a_t}\right) \pm n\pi, \quad n = 0, 1, 2, \ldots. \tag{2}$$

Since the components a_n and a_t are both positive, the vector $\vec{a}$ lies in the first quadrant of the $(\hat{u}_t, \hat{u}_n)$ system. Therefore, we can choose $n = 0$ in Eq. (2). Recalling that $a_t = 5.63\,\text{m/s}^2$ and $a_n = 37.2\,\text{m/s}^2$, Eq. (2) can be evaluated to obtain

$$\boxed{\phi = 81.39°.} \tag{3}$$

Expression of $\vec{a}$ via $\hat{\imath}$ and $\hat{\jmath}$

Computation Replacing the vector $\vec{q}$ with $\vec{a} = a_t\,\hat{u}_t + a_n\,\hat{u}_n$ in the first equality in Eq. (1.26) on p. 9, we have

$$\begin{aligned}\vec{a} &= \left[(a_t\,\hat{u}_t + a_n\,\hat{u}_n)\cdot\hat{\imath}\right]\hat{\imath} + \left[(a_t\,\hat{u}_t + a_n\,\hat{u}_n)\cdot\hat{\imath}\right]\hat{\jmath} \\ &= (a_t\hat{u}_t\cdot\hat{\imath} + a_n\hat{u}_n\cdot\hat{\imath})\,\hat{\imath} + (a_t\hat{u}_t\cdot\hat{\jmath} + a_n\hat{u}_n\cdot\hat{\jmath})\,\hat{\jmath}.\end{aligned} \tag{4}$$

Referring to Fig. 2, the angles between $\hat{u}_t$ and the unit vectors $\hat{\imath}$ and $\hat{\jmath}$ are θ and $90° - \theta$, respectively. Similarly, the angles between $\hat{u}_n$ and the unit vectors $\hat{\imath}$ and $\hat{\jmath}$ are $\theta + 90°$ and θ, respectively. Therefore, we have

$$\hat{u}_t\cdot\hat{\imath} = \cos\theta, \qquad \hat{u}_t\cdot\hat{\jmath} = \cos(90° - \theta) = \sin\theta, \tag{5}$$

$$\hat{u}_n\cdot\hat{\imath} = \cos(\theta + 90°) = -\sin\theta, \quad \hat{u}_n\cdot\hat{\jmath} = \cos\theta. \tag{6}$$

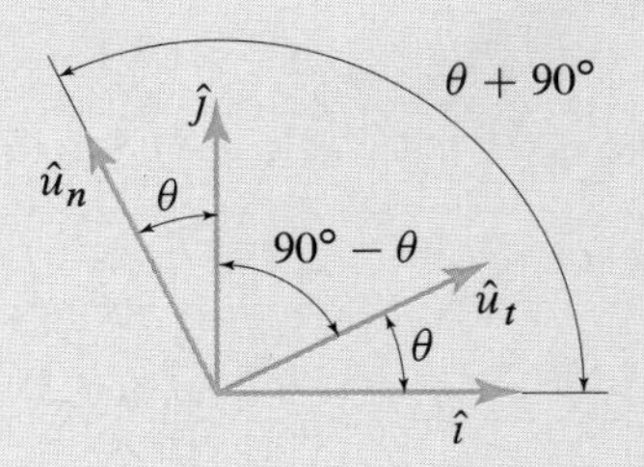

Figure 2
Orientation of the unit vectors $\hat{u}_t$ and $\hat{u}_n$ relative to the unit vectors $\hat{\imath}$ and $\hat{\jmath}$.

Using Eqs. (5) and (6), Eq. (4) can be simplified to

$$\vec{a} = (a_t\cos\theta - a_n\sin\theta)\,\hat{\imath} + (a_t\sin\theta + a_n\cos\theta)\,\hat{\jmath}. \tag{7}$$

Recalling that $a_t = 5.63\,\text{m/s}^2$, $a_n = 37.2\,\text{m/s}^2$, and $\theta = 26°$, Eq. (7) gives

$$\boxed{\vec{a} = (-11.25\,\hat{\imath} + 35.90\,\hat{\jmath})\,\text{m/s}^2.} \tag{8}$$

Discussion & Verification The value of ϕ seems reasonable given how much larger a_n is relative to a_t. To verify the result in Eq. (8), we can calculate the magnitude of $\vec{a}$ using its components relative to both the $(\hat{\imath}, \hat{\jmath})$ and the $(\hat{u}_t, \hat{u}_n)$ systems and checking that we obtain the same value. In the $(\hat{\imath}, \hat{\jmath})$ system, we have $|\vec{a}| = (-11.25^2 + 35.90^2)^{1/2}\,\text{m/s}^2 = 37.62\,\text{m/s}^2$. In the $(\hat{u}_t, \hat{u}_n)$ system we have $|\vec{a}| = (5.63^2 + 37.2^2)^{1/2}\,\text{m/s}^2 = 37.62\,\text{m/s}^2$. Since the result is as expected, we can say that the result in Eq. (8) appears to be correct.

EXAMPLE 1.2 *Position Vectors, Relative Position Vectors, and Components*

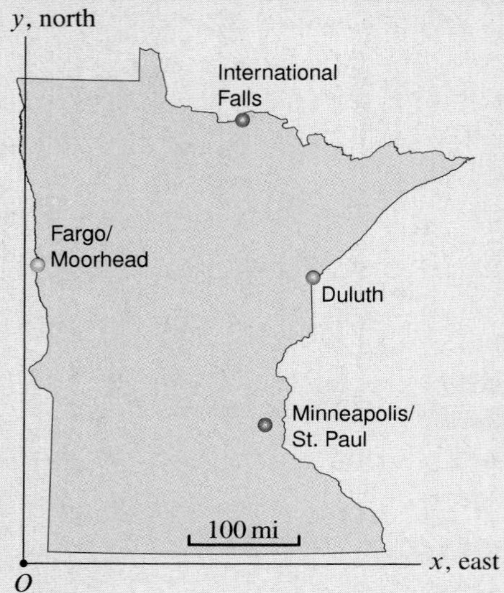

Figure 1
A map of the state of Minnesota with the locations of four of its cities. An east-north or xy coordinate frame with origin at O is also defined.

The map of the state of Minnesota in Fig. 1 shows four of its cities and defines a coordinate system whose origin is at O. The coordinates of the four cities relative to O are given in Table 1. Assuming the Earth is flat and ignoring errors due to the map projection used, the x and y directions can be considered to be east and north, respectively. Using the information in Table 1, determine

(a) The position of Duluth (D) relative to Minneapolis/St. Paul (M), $\vec{r}_{D/M}$.

(b) The orientation, relative to north, of the position of International Falls (I) relative to Fargo/Moorhead (F).

(c) The east and north (scalar) components of the position of Fargo/Moorhead relative to Minneapolis/St. Paul.

(d) The position of the point H halfway between Minneapolis/St. Paul and International Falls.

Table 1. Coordinates of the four cities in the state of Minnesota shown in Fig. 1. All coordinates are relative to the origin O.

City	x, east (mi)	y, north (mi)
Minneapolis/St. Paul (M)	216	130
Duluth (D)	259	267
Fargo/Moorhead (F)	12	278
International Falls (I)	195	413

SOLUTION

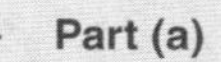

Part (a)

Road Map Since the coordinates of points D and M are available, the components of $\vec{r}_{D/M}$ can be found by using Eq. (1.14) on p. 7 to compute the difference between the coordinates of D and M.

Computation Referring to Fig. 2 and Table 1 and then taking the difference between $\vec{r}_D$ and $\vec{r}_M$, we obtain

$$\begin{aligned}\vec{r}_{D/M} &= \vec{r}_D - \vec{r}_M = (x_D - x_M)\,\hat{\imath} + (y_D - y_M)\,\hat{\jmath} \\ &= \left[(259 - 216)\,\hat{\imath} + (267 - 130)\,\hat{\jmath}\right]\text{mi} \\ &= (43.00\,\hat{\imath} + 137.0\,\hat{\jmath})\,\text{mi} = 143.6\,\text{mi} \ @\ 72.57^\circ \measuredangle,\end{aligned} \tag{1}$$

where $\theta_{D/M} = 72.57°$.

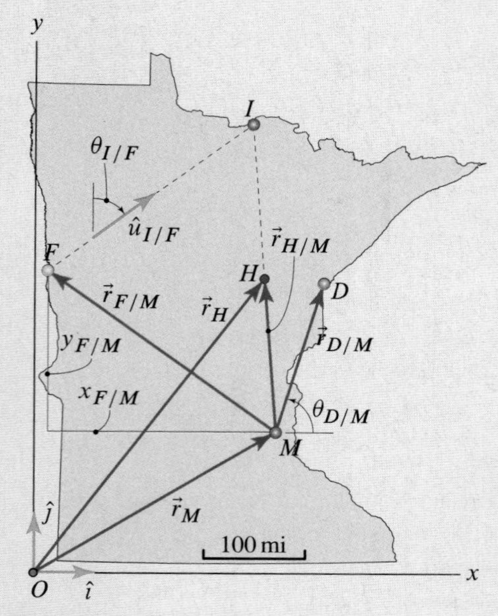

Figure 2
Vectors, projections, and angles needed to compute the quantities of interest.

Discussion & Verification Referring to Fig. 2 and taking advantage of the scale indicated on the map, we can graphically verify that our answers are correct. We note that in this problem, the answer given as a length (143.6 mi) and a direction ($\theta_{D/M} = 72.57°$) is probably more straightforward to verify than the answer given in terms of Cartesian components, since we could directly measure the distance from Minneapolis/St. Paul to Duluth on the map itself.

Part (b)

Road Map To determine the orientation of I relative to F, finding either the unit vector $\hat{u}_{I/F}$ or the angle $\theta_{I/F}$ will suffice. To find $\hat{u}_{I/F}$, we can find $\vec{r}_{I/F}$ and then divide by its magnitude (Eq. (1.20) on page 8).

Computation Again referring to Fig. 2 and Table 1 and using Eq. (1.14), the vector $\hat{u}_{I/F}$ is given by

$$\hat{u}_{I/F} = \frac{\vec{r}_{I/F}}{|\vec{r}_{I/F}|} = \frac{(195-12)\,\hat{\imath} + (413-278)\,\hat{\jmath}}{\sqrt{(195-12)^2 + (413-278)^2}} = 0.8047\,\hat{\imath} + 0.5936\,\hat{\jmath}. \tag{2}$$

Now, we can find the angle $\theta_{I/F}$ by applying Eq. (1.21), which gives

$$\boxed{\theta_{I/F} = \tan^{-1}\left(\frac{x_{I/F}}{y_{I/F}}\right) = \tan^{-1}\left(\frac{183}{135}\right) = 53.58°,} \tag{3}$$

which would be approximately northeast.

Discussion & Verification As with Part (a), we have expressed the answer in two different ways, and the version given in terms of an angle (i.e., Eq. (3)) probably allows for an easier verification that the solution is reasonable.

Part (c)

Road Map To find the east (x) and north (y) components of $\vec{r}_{F/M}$, we can use Eq. (1.24) on p. 9.

Computation Referring to Fig. 2 and using Eq. (1.24), we find that

$$\boxed{\begin{aligned} x_{F/M} &= \vec{r}_{F/M} \cdot \hat{\imath} = \big[(12-216)\,\hat{\imath} + (278-130)\,\hat{\jmath}\big] \cdot \hat{\imath}\ \text{mi} = -204.0\ \text{mi}, \\ y_{F/M} &= \vec{r}_{F/M} \cdot \hat{\jmath} = \big[(12-216)\,\hat{\imath} + (278-130)\,\hat{\jmath}\big] \cdot \hat{\jmath}\ \text{mi} = 148.0\ \text{mi}, \end{aligned}} \quad \begin{aligned} (4) \\ (5) \end{aligned}$$

where we have used the fact that $\hat{\imath} \cdot \hat{\imath} = \hat{\jmath} \cdot \hat{\jmath} = 1$ and $\hat{\imath} \cdot \hat{\jmath} = \hat{\jmath} \cdot \hat{\imath} = 0$.

Discussion & Verification We calculated that Fargo/Moorhead is 148.0 mi north of Minneapolis/St. Paul, and it is 204.0 mi *west* (i.e., −204 mi east), which certainly seem reasonable given the figure.

Part (d)

Road Map As we can see from Fig. 2, the position of the point H, which is halfway between Minneapolis/St. Paul and International Falls, is given by the vector $\vec{r}_H$. The key to the solution is then seeing that we can write this position as $\vec{r}_H = \vec{r}_M + \vec{r}_{H/M}$.

Computation We begin with the decomposition of $\vec{r}_H$, which is given by

$$\vec{r}_H = \vec{r}_M + \vec{r}_{H/M}. \tag{6}$$

We can write $\vec{r}_{H/M}$ as

$$\vec{r}_{H/M} = \tfrac{1}{2}\vec{r}_{I/M} = \tfrac{1}{2}\big[(195-216)\,\hat{\imath} + (413-130)\,\hat{\jmath}\big]\ \text{mi} = (-10.50\,\hat{\imath} + 141.5\,\hat{\jmath})\ \text{mi}, \tag{7}$$

which, when substituted into Eq. (6), gives

$$\boxed{\vec{r}_H = \big[(216\,\hat{\imath} + 130\,\hat{\jmath}) + (-10.50\,\hat{\imath} + 141.5\,\hat{\jmath})\big]\ \text{mi} = (205.5\,\hat{\imath} + 271.5\,\hat{\jmath})\ \text{mi},} \tag{8}$$

where we have used $\vec{r}_M = (216\,\hat{\imath} + 130\,\hat{\jmath})$ mi from Table 1.

Discussion & Verification Again referring to Fig. 2, the result in Eq. (8) looks reasonable. In addition, the power of vectors starts to come into focus in this part of the example. That is, once we knew the position of International Falls relative to Minneapolis/St. Paul, computing the position at any fraction of the distance in between was trivial. Once that was computed, vector addition allowed us to easily find the position of the halfway point relative to the origin at O.

EXAMPLE 1.3 *Dimensional Analysis and Unit Usage*

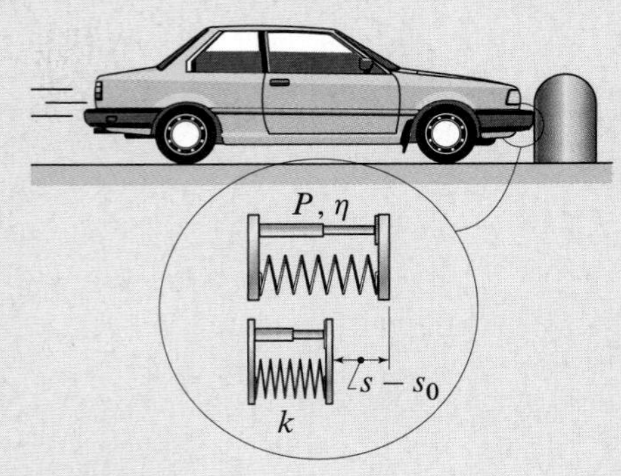

Figure 1
A car about to collide with a concrete block.

In a collision, the bumper of a car can undergo both elastic (i.e., reversible) and permanent deformation. A model for the force F transmitted by the bumper to the car is given by $F = P + k(s - s_0) + \eta\, ds/dt$, where $s - s_0$ is the compression experienced by the bumper and has the dimension of length, s_0 is a constant with dimension of length, and ds/dt is the time rate of change of s, which has the dimension of length over time. The quantity P is the force needed to permanently deform the bumper, k is the stiffness of the system, and η is a constant that relates the overall force to the speed of deformation. Determine

(a) The dimensions of P, k, and η, and

(b) The units that these quantities would have in the SI and the U.S. Customary systems.

SOLUTION

Part (a)

Road Map The first step in dimensional analysis is the identification of a basic relation, such as a law of nature, containing the quantities to analyze and for which the dimensions are known. Since we are dealing with the expression of a force, we can use Eq. (1.2) on p. 2 as the basic relation. Notice that the given force law consists of the sum of three terms. For this sum to be meaningful, each of the terms in question must have dimensions of force. We call this property *dimensional homogeneity*, and it is the key to solving the given problem.

Computation Let L, M, and T denote length, mass, and time, respectively. Writing "[*something*]" to mean the "dimensions of *something*," for Eq. (1.2), we have

$$[F] = [ma] = [m]\,[a] = M\frac{L}{T^2}. \tag{1}$$

Considering the given force law, we have

$$[F] = \left[P + k(s - s_0) + \eta\frac{ds}{dt}\right] = [P] + [k(s - s_0)] + \left[\eta\frac{ds}{dt}\right]. \tag{2}$$

Comparing Eq. (1) with Eq. (2) and enforcing dimensional homogeneity between them, we see that $[P]$, $[k(s - s_0)]$, and $[\eta\, ds/dt]$ must each be ML/T^2. Therefore, the quantity P must have the dimensions of force:

$$\boxed{[P] = M\frac{L}{T^2}.} \tag{3}$$

For the term k, we have

$$[k(s - s_0)] = [k]\,[s - s_0] = [k]L = M\frac{L}{T^2}, \tag{4}$$

where we have used the fact that the dimension of s and s_0 is L. Simplifying the last equality in Eq. (4), we have

$$\boxed{[k] = \frac{M}{T^2}.} \tag{5}$$

Next, considering the term with η in Eq. (2), we have

$$\left[\eta\frac{ds}{dt}\right] = [\eta]\left[\frac{ds}{dt}\right] = [\eta]\frac{L}{T} = M\frac{L}{T^2}, \tag{6}$$

where we have used the fact that the dimensions of ds/dt are L/T. Simplifying the last equality in Eq. (6), we have

$$\boxed{[\eta] = \frac{M}{T}.} \tag{7}$$

Concept Alert

Dimensional homogeneity. In writing Eq. (2), we have used a property stating that "[something + something else] = [something] + [something else]." This property expresses the requirement that the sum of two physical quantities makes sense only when these quantities have the same dimensions. Using a more formal language, we say that the quantities in question must satisfy *dimensional homogeneity* or, equivalently, must be *dimensionally homogeneous*.

Discussion & Verification The correctness of our dimensional analysis in the case of P is apparent, since P must have the dimensions of a force. In the case of k and η, we can verify the correctness of our solution by substituting these quantities back into the expression for F. Doing so confirms that the dimensions of k and η are correct.

Part (b)

Road Map To solve Part (b) of the problem, we need to match the dimensions obtained in Part (a) with their corresponding units according to the conventions established by the SI and U.S. Customary systems. To do this, we need only look at Table 1.1 on p. 10.

Computation Since P has the same dimensions as a force, its SI units can be simply taken to be N (newtons) or, using the SI system base units, $kg{\cdot}m/s^2$. In the U.S. Customary system, P is measured in lb (pounds).

Since the dimensions of k are mass over time squared, the corresponding units are kg/s^2 in the SI system and lb/ft in the U.S. Customary system. Notice that in the SI system, the units of k can also be expressed as N/m.

Recalling that η has dimensions of mass over time, the units of η are kg/s in the SI system and slug/s in the U.S. Customary system. Using the base units of the U.S. Customary system, the units of η are lb·s/ft.

All of these results are summarized in Table 1.

Table 1. Summary of the solution to the second part of the problem.

Quantity	SI units	U.S. Customary units
P	N or $kg{\cdot}m/s^2$	lb
k	kg/s^2 or N/m	lb/ft
η	kg/s	lb·s/ft

Discussion & Verification The correctness of our results can be verified by replacing the units with the dimensions they correspond to. For example, for P we have that the dimensions corresponding to the unit of pound are those of a force, i.e., ML/T^2, which are the dimensions of P obtained in Part (a) of the problem solution. Repeating this process for the other quantities and for both the SI system and the U.S. Customary system, we see that our results are correct.

EXAMPLE 1.4 *Dimensional Analysis and Unit Conversion*

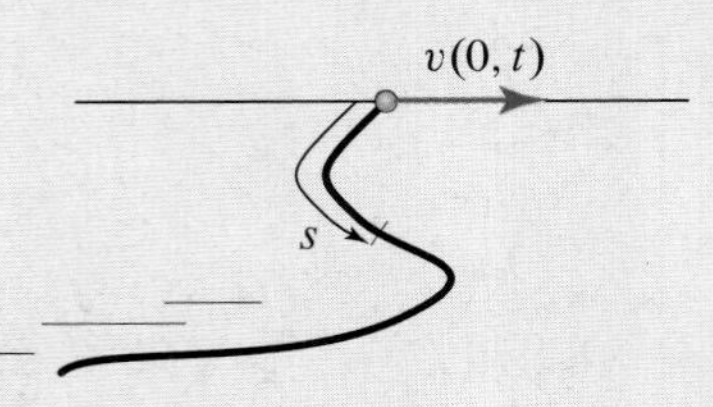

Figure 1
A moving string with one end being dragged.

In studying the motion of a string, it is determined that the speed at various points along the string is given by the function

$$v(s,t) = \alpha + \beta t^2 - \gamma s + \delta \frac{s}{t}, \tag{1}$$

where s is the coordinate of points along the string, t is time, and α, β, γ, and δ are constants.

(a) What are the dimensions of α, β, γ, and δ?

(b) If α, β, γ, and δ are all equal to 1 in SI units, what are they in U.S. Customary units?

Interesting Fact

Are engineers really interested in the motion of strings? The answer is "Actually, yes!" Highly detailed models of real physical objects tend to be quite complicated, difficult to tackle from a numerical standpoint, and often difficult to interpret. Therefore, before delving into the complexity of highly detailed models, both physicists and engineers tend to "simplify things" and model real systems as simple objects, such as strings (or beams). Simple models can be very effective in capturing the essential physical behavior of real physical objects and, in that way, give us useful insight into the physical world.

SOLUTION

Part (a)

Road Map Since the dimensions of speed are L/T (with corresponding units of m/s in SI and ft/s in U.S. Customary), the dimensions of every term on the right-hand side of Eq. (1) must also be L/T.

Computation Begin with $[\alpha]$, which must have the same dimensions as those of the speed v, so they are simply L/T. As for $[\beta]$, we know that

$$\left[\beta t^2\right] = [\beta]T^2 = L/T \quad \Rightarrow \quad \boxed{[\beta] = L/T^3.} \tag{2}$$

To get $[\gamma]$, we proceed as in Eq. (2) to obtain

$$[\gamma s] = [\gamma]L = L/T \quad \Rightarrow \quad \boxed{[\gamma] = 1/T.} \tag{3}$$

Finally, $[\delta]$ is obtained similarly as

$$\left[\delta \frac{s}{t}\right] = [\delta]L/T = L/T \quad \Rightarrow \quad \boxed{[\delta] \text{ is dimensionless.}} \tag{4}$$

Discussion & Verification The verification of the results for α is immediate since no calculations were performed to obtain them. In the case of β, γ, and δ, substituting the results in Eqs. (2)–(4), we see that our results are correct.

Part (b)

Road Map After expressing α, δ, and γ in SI units, we need to convert them into U.S. Customary units. Note that no conversion is needed for δ since it is dimensionless.

Computation Based on Part (a), the SI units of α, β, and γ are m/s, m/s^3, and s^{-1}, respectively. Converting unit values for α, β, and γ to U.S. Customary units gives

$$\alpha = 1\,\text{m/s} = 1\frac{\cancel{\text{m}}}{\text{s}}\left(\frac{\text{ft}}{0.3048\,\cancel{\text{m}}}\right) = 3.281\,\text{ft/s}, \tag{5}$$

$$\beta = 1\,\text{m/s}^3 = 1\frac{\cancel{\text{m}}}{\text{s}^3}\left(\frac{\text{ft}}{0.3048\,\cancel{\text{m}}}\right) = 3.281\,\text{ft/s}^3, \tag{6}$$

$$\gamma = 1\,\text{s}^{-1}, \tag{7}$$

where no conversion is needed for γ since 1 s^{-1} is the same in either SI or U.S. Customary units.

Discussion & Verification The only results to verify are those for α and β, which have dimensions of L/T and L/T^3, respectively. Therefore, since the base unit for time is the same in both the SI and U.S. Customary systems, the unit conversion was expected to affect only the L dimension and yield the same value (namely, 3.28) for both α and β. Finally, the value in question was expected to be close to 3 since a meter is a little over 3 ft. Thus, our results appear to be correct.

Problems

Problem 1.1

Determine $(r_{B/A})_x$ and $(r_{B/A})_y$, the x and y components of the vector $\vec{r}_{B/A}$, so as to be able to write $\vec{r}_{B/A} = (r_{B/A})_x\,\hat{\imath} + (r_{B/A})_y\,\hat{\jmath}$.

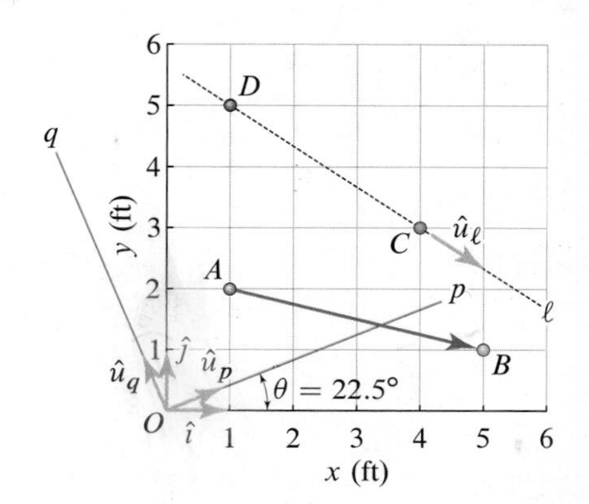

Figure P1.1–P1.5

Problem 1.2

If the positive direction of line ℓ is from D to C, find the component of the vector $\vec{r}_{B/A}$ along ℓ.

Problem 1.3

Find the components of $\vec{r}_{B/A}$ along the p and q axes.

Problem 1.4

Determine expressions for the vector $\vec{r}_{B/A}$ using both the xy and the pq coordinate systems. Next, determine $|\vec{r}_{B/A}|$, the magnitude of $\vec{r}_{B/A}$, using both the xy and the pq representations and establish whether or not the two values for $|\vec{r}_{B/A}|$ are equal to each other.

Problem 1.5

Suppose that you were to compute the quantities $|\vec{r}_{B/A}|_{xy}$ and $|\vec{r}_{B/A}|_{pq}$, that is, the magnitude of the vector $\vec{r}_{B/A}$ computed using the xy and pq frames, respectively. Do you expect these two scalar values to be the same or different? Why?

Problem 1.6

The magnitude of the velocity vector of the car is $|\vec{v}| = 80\,\text{ft/s}$. If the vector $\vec{v}$ forms an angle $\theta = 0.09\,\text{rad}$ with the horizontal direction, determine the Cartesian representation of $\vec{v}$ relative to the $(\hat{\imath}, \hat{\jmath})$ component system.

Figure P1.6 and P1.7

Problem 1.7

The velocity of the car has the following representation: $\vec{v} = (8.30\,\hat{\imath} + 0.726\,\hat{\jmath})\,\text{m/s}$. Determine the magnitude of $\vec{v}$. In addition, knowing that the angle θ describes the orientation of $\vec{v}$, determine θ and express its value in degrees.

Problem 1.8

The acceleration of the car has the following representation: $\vec{a} = -(3.53\,\hat{\imath} + 0.309\,\hat{\jmath})\,\text{m/s}^2$. Knowing that $\vec{a}$ is parallel to the incline, determine the angles θ and ϕ and express their value in radians.

Figure P1.8

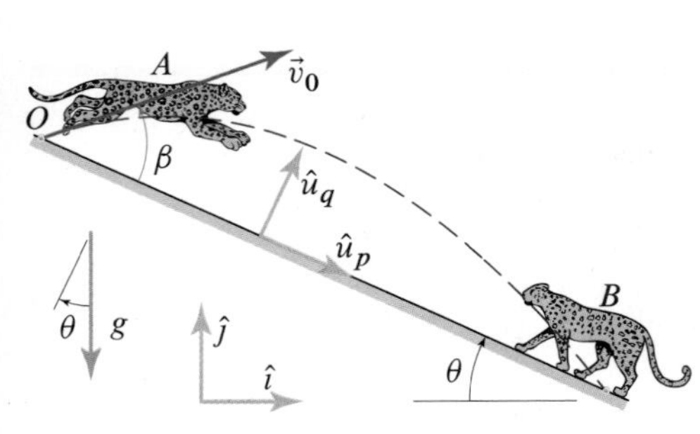

Figure P1.9

Problem 1.9

A jaguar A leaps from O with a velocity $\vec{v}_0$ to try and intercept a panther B. The unit vectors $\hat{u}_p$ and $\hat{u}_q$ are parallel and perpendicular to the incline, respectively. The unit vectors $\hat{\imath}$ and $\hat{\jmath}$ are horizontal and vertical, respectively. While airborne, the jaguar is subject to a constant acceleration with magnitude g and direction opposite to $\hat{\jmath}$. Denoting the magnitude of $\vec{v}_0$ by v_0 and denoting the (vector) acceleration of the jaguar by $\vec{a}_A$, provide the expression of $\vec{v}_0$ in the $(\hat{\imath}, \hat{\jmath})$ component system and the expression of $\vec{a}_A$ in the $(\hat{u}_p, \hat{u}_q)$ component system. Treat the angles β and θ as known.

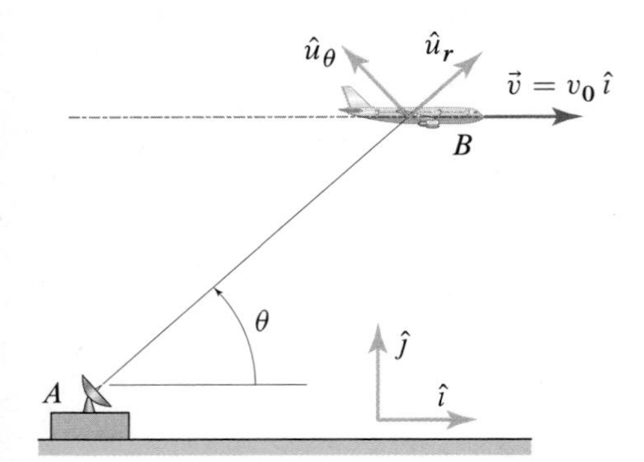

Figure P1.10

Problem 1.10

The velocity vector of the airplane is $\vec{v} = v_0\,\hat{\imath}$, with $v_0 = 420$ mph. Determine the components of the vector $\vec{v}$ in the $\hat{u}_r$ and $\hat{u}_\theta$ directions for $\theta = 35°$. Express the result in feet per second.

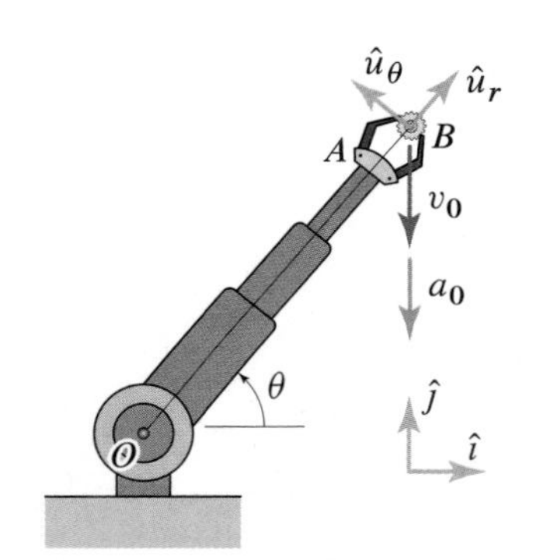

Figure P1.11

Problem 1.11

The motion of the telescopic arm is such that the velocity and acceleration vectors of the gear B are $\vec{v} = -v_0\,\hat{\jmath}$ and $\vec{a} = -a_0\,\hat{\jmath}$, respectively, with $v_0 = 8$ ft/s and $a_0 = 0.5$ ft/s^2. Determine the components of $\vec{v}$ and $\vec{a}$ in the direction of the unit vectors $\hat{u}_r$ and $\hat{u}_\theta$ for $\theta = 32°$.

Problem 1.12

At the instant shown, the velocity and acceleration vectors of the airplane have the following expressions:

$$\vec{v} = (215\,\hat{\imath} + 332\,\hat{\jmath})\text{ ft/s} \quad \text{and} \quad \vec{a} = (-190\,\hat{\imath} + 76.0\,\hat{\jmath})\text{ ft/s}^2.$$

Use Eq. (1.17) on p. 8 to determine the angle ϕ, the smaller of the two angles formed by $\vec{v}$ and $\vec{a}$. Express the result in degrees.

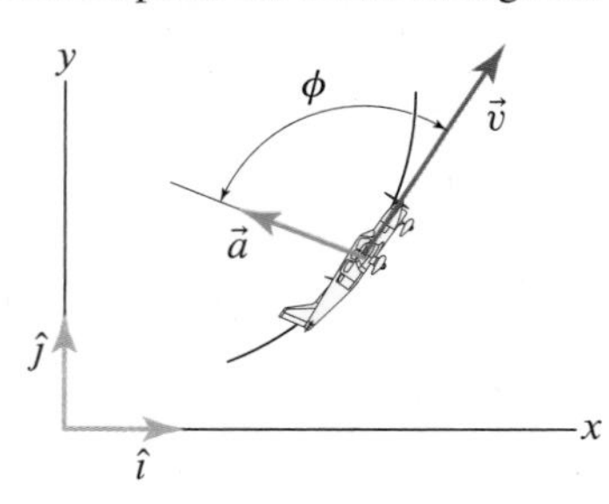

Figure P1.12

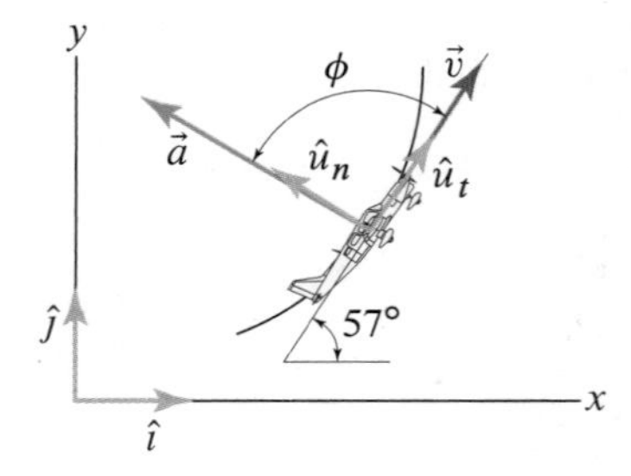

Figure P1.13

Problem 1.13

At the instant shown, when expressed via the $(\hat{u}_t, \hat{u}_n)$ component system, the airplane's velocity and acceleration are

$$\vec{v} = 135\,\hat{u}_t\text{ m/s} \quad \text{and} \quad \vec{a} = (-7.25\,\hat{u}_t + 182\,\hat{u}_n)\text{ m/s}^2.$$

Determine the angle ϕ between the velocity and acceleration vectors. In addition, treating the $(\hat{u}_t, \hat{u}_n)$ and $(\hat{\imath}, \hat{\jmath})$ component systems as stationary relative to one another, express the airplane's velocity and acceleration in the $(\hat{\imath}, \hat{\jmath})$ component system.

Problem 1.14

The components of the position vector $\vec{r}$ of point P relative to the $(\hat{\imath}_1, \hat{\jmath}_1)$ component system are $r_{x1} = 2$ ft and $r_{y1} = 5$ ft. If $\theta = 30°$, determine coordinates of P relative to the (x_2, y_2) coordinate system.

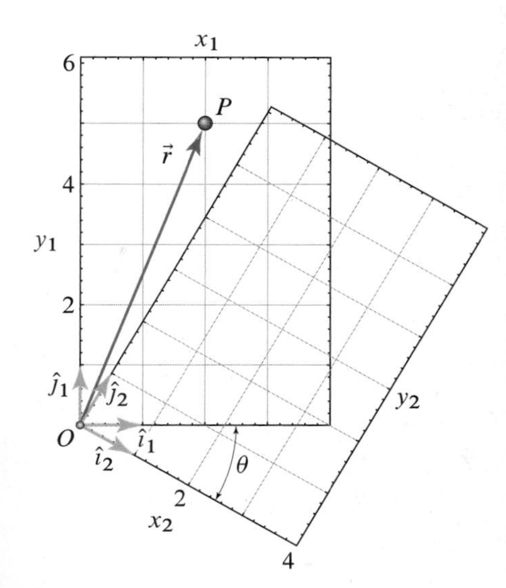

Figure P1.14

Problem 1.15

The velocity of point P relative to frame A is $\vec{v}_{P/A} = (-14.9\,\hat{\imath}_A + 19.4\,\hat{\jmath}_A)$ ft/s, and the acceleration of P relative to frame B is $\vec{a}_{P/B} = (3.97\,\hat{\imath}_B + 4.79\,\hat{\jmath}_B)$ ft/s^2. Frames A and B do not move relative to one another. Determine the expressions for the velocity of P in frame B and the acceleration of P in frame A.

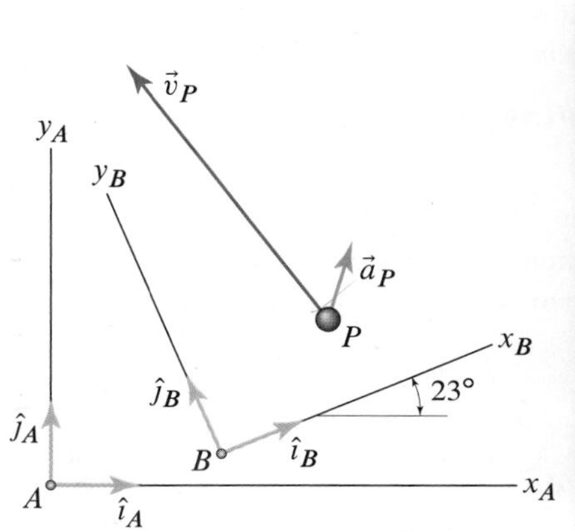

Figure P1.15

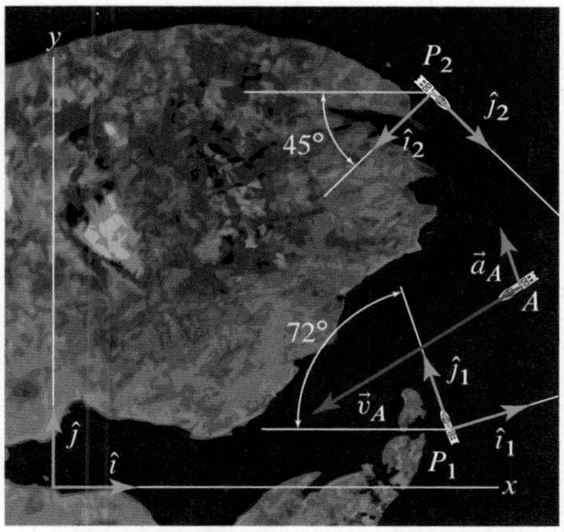

Figure P1.16

Problem 1.16

Two Coast Guard patrol boats P_1 and P_2 are stationary while monitoring the motion of a surface vessel A. The velocity of A with respect to P_1 is expressed by

$$\vec{v}_A = (-23\,\hat{\imath}_1 - 6\,\hat{\jmath}_1)\ \text{ft/s},$$

whereas the acceleration of A, expressed relative to P_2, is given by

$$\vec{a}_A = (-2\,\hat{\imath}_2 - 4\hat{\jmath}_2)\ \text{ft/s}^2.$$

Determine the velocity and the acceleration of A expressed with respect to the land-based component system $(\hat{\imath}, \hat{\jmath})$.

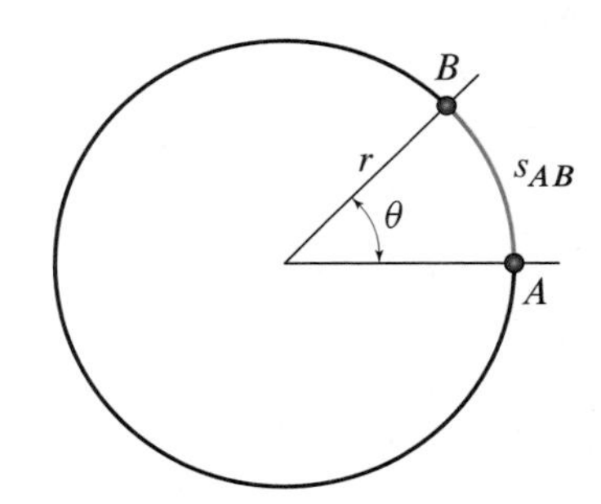

Figure P1.17

Problem 1.17

The measure of angles in radians is defined according to the following relation: $r\theta = s_{AB}$, where r is the radius of the circle and s_{AB} denotes the length of the circular arc. Determine the dimensions of the angle θ.

Problem 1.18

Letting C denote the circumference of a circle, a 1° angle is, by definition, an angle that subtends an arc of length ℓ such that $C/\ell = 360$. Apply the definition of degree and determine the radius of the circle shown knowing that the length s of the arc subtended by the 4° angle in the figure is 1.84 mm.

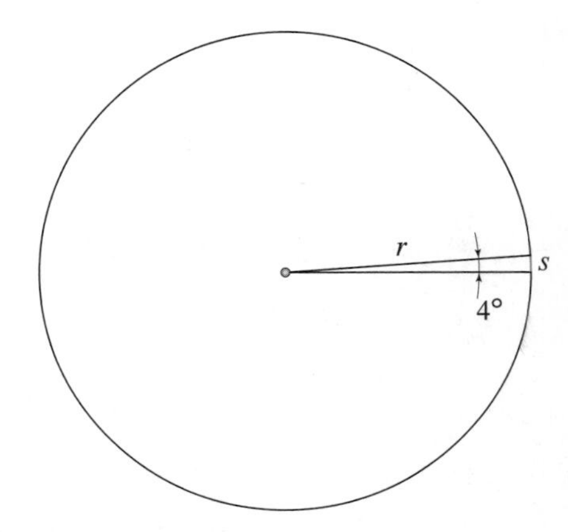

Figure P1.18

Problem 1.19

A simple oscillator consists of a linear spring fixed at one end and a mass attached at the other end, which is free to move. Suppose that the periodic motion of a simple oscillator is described by the relation $y = Y_0 \sin(2\pi\omega_0 t)$, where y has units of length and denotes the vertical position of the oscillator, Y_0 is the oscillation amplitude, ω_0 is the oscillation frequency, and t is time. Recalling that the argument of a trigonometric function is an angle, determine the dimensions of Y_0 and ω_0, as well as their units in both the SI and the U.S. Customary systems.

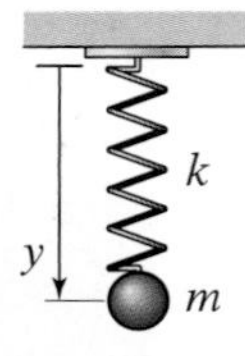

Figure P1.19

Problem 1.20

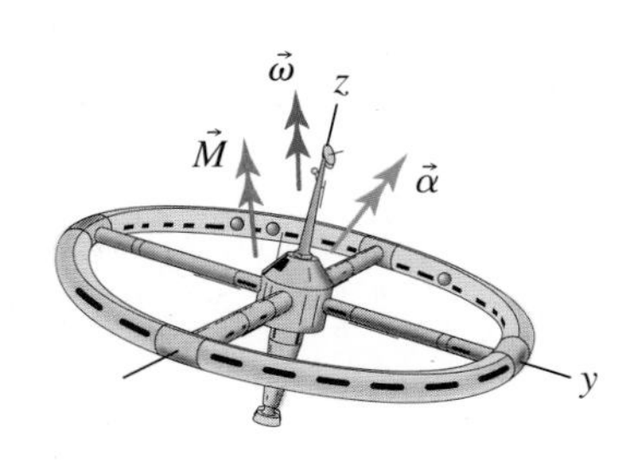

Figure P1.20

To study the motion of a space station, the station can be modeled as a rigid body and the equations describing its motion can be chosen to be Euler's equations, which read

$$M_x = I_{xx}\,\alpha_x - (I_{yy} - I_{zz})\omega_y\,\omega_z,$$
$$M_y = I_{yy}\,\alpha_y - (I_{zz} - I_{xx})\omega_x\,\omega_z,$$
$$M_z = I_{zz}\,\alpha_z - (I_{xx} - I_{yy})\omega_x\,\omega_y.$$

In the previous equations, M_x, M_y, and M_z denote the x, y, and z components of the moment applied to the body;* ω_x, ω_y, and ω_z denote the corresponding components of the angular velocity of the body, where angular velocity is defined as the time rate of change of an angle; α_x, α_y, and α_z denote the corresponding components of the angular acceleration of the body, where angular acceleration is defined as the time rate of change of an angular velocity. The quantities I_{xx}, I_{yy}, and I_{zz} are called the *principal mass moments of inertia* of the body. Determine the dimensions of I_{xx}, I_{yy}, and I_{zz} and determine their units in SI, as well as in the U.S. Customary system.

Problem 1.21

The lift force F_L generated by the airflow moving over a wing is often expressed as follows:

$$F_L = \tfrac{1}{2}\rho v^2 C_L(\theta) A, \qquad (1)$$

where ρ, v, and A denote the mass density of air, the airspeed (relative to the wing), and the wing's nominal surface area, respectively. The quantity C_L is called the *lift coefficient*, and it is a function of the wing's angle of attack θ. Find the dimensions of C_L and determine its units in the SI system.

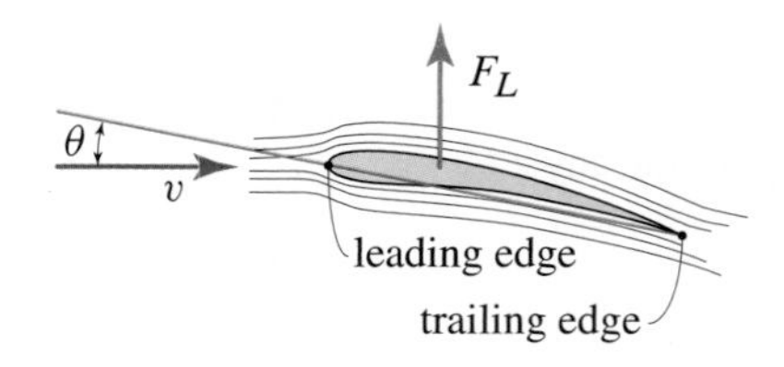

Figure P1.21

* The xyz frame used in Euler's equations is a special frame that moves with the body, but for the purpose of the problem solution, it will suffice to say that it consists of three mutually orthogonal axes.

Problem 1.22

Are the words *units* and *dimensions* synonyms?

Problem 1.23

A rock is released from rest into water. The magnitude F_d of the drag force acting on the rock due to its motion through water can be modeled as $F_d = C_d v$, where v is the speed of the rock and C_d is a constant drag coefficient. Determine the units used to measure C_d in the U.S. Customary system.

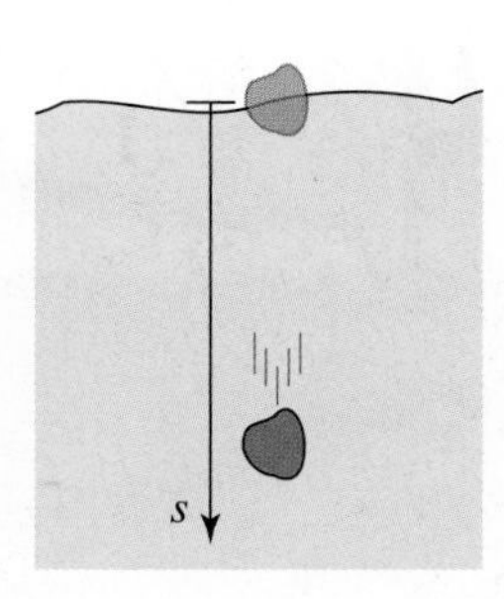

Figure P1.23

Problem 1.24

In elementary beam theory, for a uniform beam supported as shown, the relation between the force P applied at the end of a beam and the corresponding end deflection δ is $P = (3EI/L^3)\delta$, where E is a constant called the modulus of elasticity, I is a constant called the centroidal area moment of inertia, and L is the length of the beam. If the dimensions of I are length to the power four, determine the SI units used to measure the constant E.

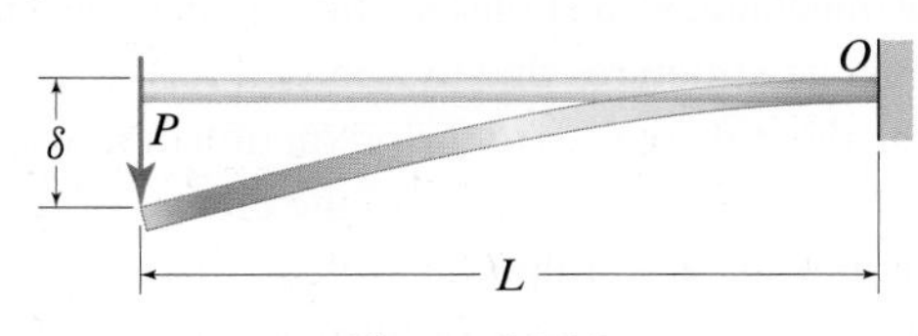

Figure P1.24

1.3 Dynamics and Engineering Design*

Figure 1.17
Dynamic loading due to the interaction of the structure with the wind played a crucial role in the collapse of the Tacoma Narrows bridge on November 7, 1940.

Design is the goal of all engineering endeavors. While there is no single accepted definition of engineering design, we can define it as the process that results in the specification of how a *system* can be produced to meet the *needs* and *requirements* identified in the design process. The "system" to which we refer might be as simple as a bolt or as complex as the Apollo missions. The need might be to hold two steel plates together or to put a human on the moon. The requirements can be technical, environmental, political, or financial, among others.

One of the hard things about engineering design is that neither the process used to reach the final design nor the final result of the design is unique (this is in contrast to the typical homework or exam problem you have been given, where there is only *one* correct answer). There are generally many (sometimes infinitely many) designs that will satisfy the requirements of most design specifications.

In the design process, mechanicians† are generally concerned with the determination of a variety of mechanical responses, such as deflection, stress, strain, and temperature in machines or structures due to imposed forces. Mechanicians are also concerned with relating these responses to the prediction of a structure's or machine's fatigue life, durability, and safety, as well as to the choice of materials and fabrication procedures. Dynamics plays an important role in this determination since accelerations are an important cause of forces in mechanical systems (via the equations developed by Newton and Euler).

There are several methods to structure the design process. Regardless of the method used, the design process should be done iteratively by identifying needs, prioritizing, making value decisions, and exploiting physical laws to develop a sound solution that optimizes the objectives while satisfying the constraints that have been identified. As you continue your education, you will learn about structured and standard procedures for design, including information about performance standards, safety standards, and design codes. For now, we will focus on that part of the design process that can be introduced based on your knowledge of calculus, statics, and dynamics.

Objectives of design

At a minimum, a mechanical design product must

- Accomplish the design goals.
- Not fail during normal use.
- Minimize hazards.
- Attempt to anticipate and account for all foreseeable uses.
- Be thoroughly documented and archived.

In addition, a design should also take into consideration

- Cost of manufacture, purchase, and ownership.
- Ease of manufacture and maintenance.

* See D. G. Ullman, *The Mechanical Design Process*, 3rd edition, McGraw-Hill, 2003, for a thorough treatment of design.

† *Mechanicians* are people who study mechanics. *Mechanics* are people who fix cars or other machinery.

- Energy efficiency in manufacture and use.
- Impact on the environment in its manufacture, use, and retirement.

Our goal is not to teach you engineering design, but to introduce you to some aspects of it. The aspects that we will focus on most heavily are *system modeling* and *parametric studies*. It can be argued that dynamics is about modeling a mechanical system and then studying its behavior as parameters of the system are varied.

System modeling

In dynamics, we will consider *modeling* to be the process of translating real life into mathematical equations to predict the behavior of the model. Once you learn to model mechanical systems, you will have a very powerful tool at your disposal. Unfortunately, that modeling process does not come easily for many students, and it is only mastered through *a lot* of practice.

When creating models, we have to remember that models are *not* exact representations of reality, but they should give us enough information to tell us something meaningful about real physical systems. For example, the wings of a commercial airliner undergo significant bending and torsion during flight. However, in building a model of an airplane's performance, we typically start by assuming that the wings are perfectly rigid. To assume that the wings are rigid allows us to simplify the equations of our model and more directly predict the airplane's behavior for a range of flight conditions. Based on these predictions, we can then refine the model to study the airplane's behavior in those circumstances where we need to account for the wings' deformation.

In general, once a model is created, you should compare the predictions of that model with data obtained from the real system. If the behavior predicted by the model agrees with the real behavior of the system, then, within the assumptions used to create the model, you can use that model to make further predictions about the system's behavior. If the predictions do not agree, then you must determine how your model can be improved.

It is not always possible to compare the results of a model with the results of a laboratory test. For example, when engineers are creating a new type of aircraft, they cannot, for financial reasons and/or time constraints, build a prototype for every design they create to test it. They must use experience garnered from previous models and designs and extrapolate, using their knowledge of engineering and physics to create a new system. In fact, when new commercial or military aircraft are designed, the first prototypes built are generally those that are flight-tested—there are no intermediate steps.

Particle Kinematics

2

Jim Wooding of the U.S. performing a long jump during the 1984 Olympic Games in Los Angeles, California. Projectile motion can be used to model the long jump.

Kinematics studies the *geometry of motion* without reference to the causes of motion. It is rooted in vector algebra and calculus, and to many "it looks and feels like math." Kinematics is essential for the application of Newton's second law, $\vec{F} = m\vec{a}$, since it allows us to describe the $\vec{a}$ in $\vec{F} = m\vec{a}$. Unfortunately, just writing $\vec{F} = m\vec{a}$ doesn't tell us how something moves; it only describes the relationship between force and acceleration. In general, we also want to know a particle's *motion*, where by *motion* we mean "all the positions occupied by an object over time." It is kinematics that allows us to translate $\vec{a}$ into the *motion* of a point by using calculus. This is why in this chapter we study the concepts of position, velocity, and acceleration, and how these quantities relate to one another. In addition, we will learn how to write position, velocity, and acceleration in the various component systems most commonly used in dynamics.

2.1 Position, Velocity, Acceleration, and Cartesian Coordinates

Kinematics describes motion without reference to the causes and the effects of motion. Kinematics is a core component of any theory relating motion to forces. It is often a vital aspect of many branches of technology, such as tracking aircraft so as to make air traffic control possible, the design of mechanisms, or relating the shape of a roller coaster track and the speed of a roller coaster car to the accelerations experienced during the ride. In this section, we develop those concepts that allow us to characterize the motion of a body as a function of time. Vectors play a fundamental role in this development. In fact, the three main fundamental concepts we cover, namely, position, velocity, and acceleration, are all vector quantities.

A notation for time derivatives

In kinematics we write derivatives with respect to time so often that it is convenient to have a shorthand notation. If $f(t)$ is a function of time, we place a dot over it to mean $df(t)/dt$. Furthermore, the *number* of dots over a quantity indicates the order of the derivative; that is,

$$\dot{f}(t) = \frac{df(t)}{dt}; \quad \ddot{f}(t) = \frac{d^2 f(t)}{dt^2}; \quad \dddot{f}(t) = \frac{d^3 f(t)}{dt^3}; \quad \text{etc.} \tag{2.1}$$

Position vector

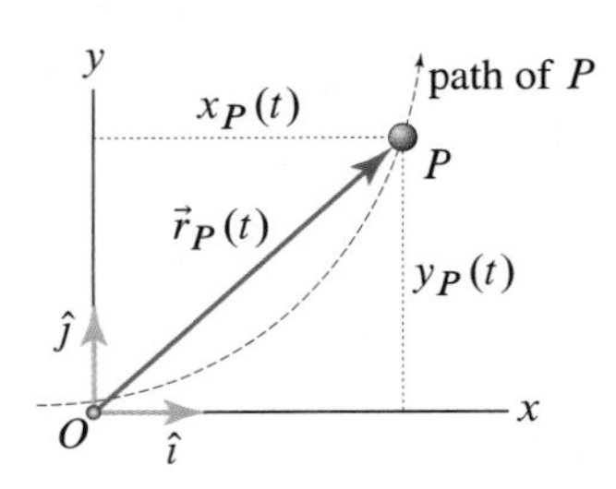

Figure 2.1
A point P and its position vector relative to O.

Referring to Fig. 2.1, the *position vector* of the point P at time t relative to the origin O is the *vector* $\vec{r}_P(t)$ going from O to P at time t. In two dimensions (Section 2.8 on p. 142 covers the three-dimensional case) and using Cartesian components,

$$\vec{r}_P(t) = x_P(t)\,\hat{\imath} + y_P(t)\,\hat{\jmath}. \tag{2.2}$$

When studying the motion of a single point, we can drop the subscript indicating the point and simply write $\vec{r} = x(t)\,\hat{\imath} + y(t)\,\hat{\jmath}$ instead of $\vec{r}_P(t) = x_P(t)\,\hat{\imath} + y_P(t)\,\hat{\jmath}$.

The notion of position is meaningless without the specification of a reference point relative to which position is defined. If the reference point is not the origin of a coordinate system, or if several coordinate systems are used concurrently, then we use the notation introduced in Section 1.2 (see Eq. (1.14) on p. 7) to indicate the position of a point *relative* to another.

Trajectory

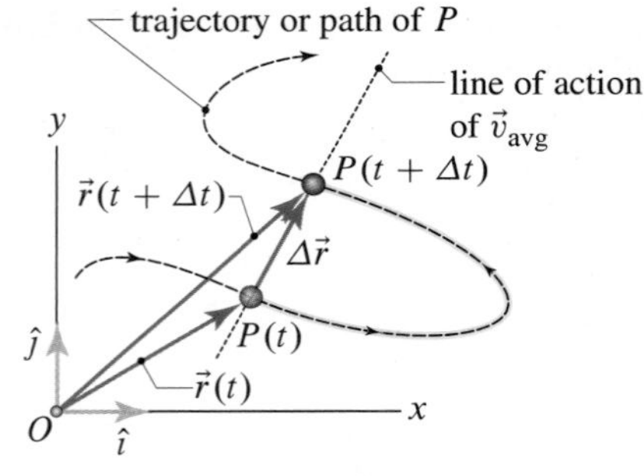

Figure 2.2
Displacement of point P between times t and $t + \Delta t$, with $\Delta t > 0$. The distance traveled by P between t and $t + \Delta t$ is highlighted in yellow.

The *trajectory* of a point is the line traced through space by the point during its motion (Fig. 2.2). Synonyms for *trajectory* are *path* and *space curve*. The trajectory of a point can be an open line, a closed loop, or some other kind of self-intersecting line. An object's trajectory by itself cannot tell us anything about how fast an object moves or how many times a point goes past a specific location.

Velocity vector and speed

Displacement vector. Referring to Fig. 2.2, we define the *displacement (or position change)* of P between t and $t + \Delta t$ (with $\Delta t > 0$) as the vector $\Delta\vec{r} = \vec{r}(t + \Delta t) - \vec{r}(t)$. In general, the length of $\Delta\vec{r}$ *does not* measure the distance traveled by P between t and $t + \Delta t$ (highlighted in yellow in Fig. 2.2). This is so because the distance traveled between t and $t + \Delta t$ depends on the geometry of the path of P, whereas $\Delta\vec{r}$ depends only on $\vec{r}(t)$ to $\vec{r}(t + \Delta t)$ without reference to *how* P moved from one position to the other.

Average velocity vector. Referring again to Fig. 2.2, we define the *average velocity vector* of P between t and $t + \Delta t$ as

$$\vec{v}_{avg} = \underbrace{\frac{1}{(t + \Delta t) - t}}_{\text{scalar}} \underbrace{\left[\vec{r}(t + \Delta t) - \vec{r}(t)\right]}_{\text{vector}} = \frac{\Delta\vec{r}}{\Delta t}. \tag{2.3}$$

The vectors $\vec{v}_{avg}$ and $\Delta\vec{r}$ have the same direction since the term $1/\Delta t$ in Eq. (2.3) is a positive scalar.

Velocity vector. Given $\vec{v}_{avg}$ between t and $t + \Delta t$, the *velocity vector* at time t is

$$\vec{v}(t) = \lim_{\Delta t \to 0} \vec{v}_{avg} = \lim_{\Delta t \to 0} \frac{\Delta\vec{r}}{\Delta t}. \tag{2.4}$$

The second limit in Eq. (2.4) is the time derivative of $\vec{r}(t)$, so that we write

$$\boxed{\vec{v}(t) = \frac{d\vec{r}(t)}{dt} = \dot{\vec{r}}(t).} \tag{2.5}$$

The velocity vector is the time rate of change of the position vector.

Speed. The *speed* of a point is the *magnitude of its velocity*:

$$\boxed{v(t) = \left|\vec{v}(t)\right|.} \tag{2.6}$$

By definition, the *speed is never negative.*

The velocity vector is *always* tangent to the path. The velocity has an important property: *The velocity vector at a point along the trajectory is tangent to the trajectory at that point!* To see this we recall that $\vec{v}_{\text{avg}}$ between t and $t + \Delta t$ has the same direction as the displacement $\Delta\vec{r}$ between the same time instants. Referring to Fig. 2.3, let's consider the sequence of vectors $\Delta\vec{r}_n = \vec{r}(t + \Delta t/n) - \vec{r}(t)$ between t and $t + \Delta t/n$ ($n = 1, 2, \ldots$). As $n \to \infty$, $\Delta\vec{r}_n$ and the associated average velocity become tangent to the path, thus implying that $\vec{v}(t)$ is tangent to the path at $P(t)$.

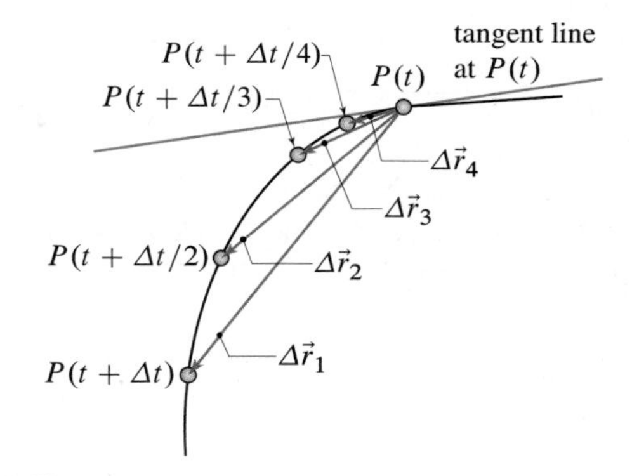

Figure 2.3
Displacement vectors $\Delta\vec{r}_n$ between time t and subsequent times $t + \Delta t/n$ ($n = 1, 2, \ldots$).

Distance traveled and speed. Let $s(t)$ be the distance traveled as a function of time. This quantity only increases with time, and calculus tells us that $ds/dt \geq 0$. We now show that the quantity ds/dt coincides with the speed. Referring to Fig. 2.4, we consider the infinitesimal displacement $d\vec{r}$ between time instants t and $t + dt$, with $dt > 0$. Recalling that $\vec{v} = d\vec{r}/dt$, we can write

$$d\vec{r} = \vec{v}\,dt. \tag{2.7}$$

Since $d\vec{r}$ is infinitesimal, it is tangent to the path and has magnitude equal to ds:

$$\left|d\vec{r}\right| = ds = \left|\vec{v}\,dt\right| \quad \Rightarrow \quad ds = \left|\vec{v}\right| dt \quad \Rightarrow \quad ds = v\,dt, \tag{2.8}$$

where we have used Eq. (2.6). The last of Eqs. (2.8) implies that

$$\boxed{\frac{ds}{dt} = v.} \tag{2.9}$$

The speed is therefore both the magnitude of the velocity and the time rate of change of the distance traveled. Another consequence of the last of Eqs. (2.8) is obtained by integrating this equation between two time instants t_1 and t_2, with $t_1 < t_2$:

$$\int_{s_1}^{s_2} ds = \int_{t_1}^{t_2} v\,dt \quad \Rightarrow \quad s_2 - s_1 = \int_{t_1}^{t_2} v\,dt. \tag{2.10}$$

Recalling that the average speed between t_1 and t_2 is $v_{\text{avg}} = \left(\int_{t_1}^{t_2} v\,dt\right)/(t_1 - t_2)$, dividing both sides of the last of Eqs. (2.10) by $t_2 - t_1$, we have

$$\boxed{v_{\text{avg}} = \frac{s_2 - s_1}{t_2 - t_1}.} \tag{2.11}$$

The ratio of the distance traveled between two time instants t_1 and t_2 and the time interval size $t_2 - t_1$ measures the *average speed* between t_1 and t_2.

Concept Alert

Direction of velocity vectors. One of the most important concepts in kinematics is that the velocity of a particle is always tangent to the particle's path.

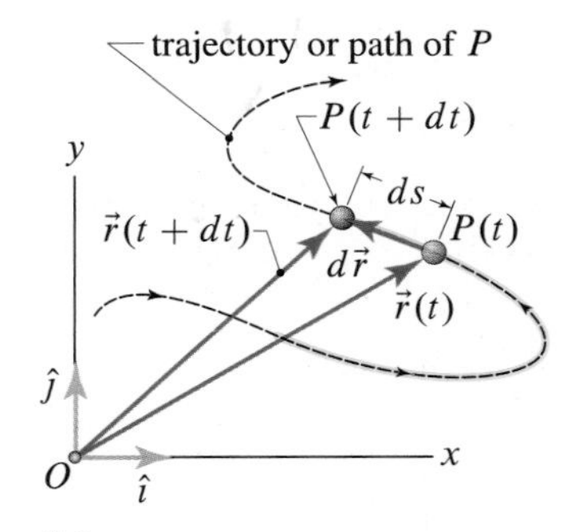

Figure 2.4
A point P and its path. The distance traveled (indicated in yellow) between time instants t and $t + dt$ is ds, which, as $dt \to 0$, represents the magnitude of the vector $d\vec{r}$.

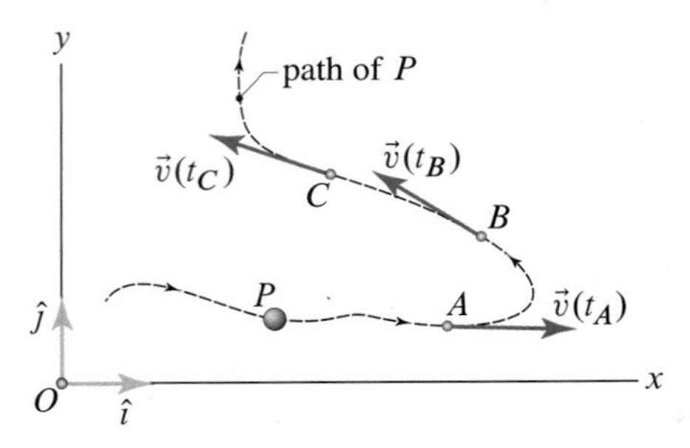

Figure 2.5
Velocity vectors of a particle P at three instants in time t_A, t_B, and t_C. The particle is moving with constant speed.

Acceleration vector

The *acceleration vector* is the time rate of change of the velocity vector, i.e.,

$$\vec{a}(t) = \frac{d\vec{v}(t)}{dt} = \frac{d^2\vec{r}(t)}{dt^2} = \dot{\vec{v}}(t) = \ddot{\vec{r}}(t). \tag{2.12}$$

The acceleration is generally not tangent to the path. Contrary to what we discovered about the velocity, the acceleration vector is generally not tangent to the path. To see why, consider a point P moving at constant speed as shown in Fig. 2.5. Although the speed is constant, the acceleration $\vec{a}(t)$ of P, which measures the changes in both the magnitude and the *direction* of $\vec{v}(t)$, is not zero because $\vec{v}(t)$ changes direction so as to remain tangent to the *curved* path. Note that the velocity change, i.e., $\vec{a}(t)$, is more pronounced between points A and B than it is between B and C since the path's curvature is larger between A and B than between B and C.* Furthermore, even if the speed v is constant, $\vec{a}(t)$ depends on v because v determines how quickly P moves along its path and therefore, how fast $\vec{v}(t)$ changes direction. Hence, $\vec{a}(t)$ along a curved path depends on both the speed and the path's curvature. Finally, to get a sense of the *direction* of $\vec{a}(t)$, consider the velocity between times t and $t + \Delta t$, as shown in Fig. 2.6. Recalling that

$$\vec{a}(t) = \dot{\vec{v}}(t) = \lim_{\Delta t \to 0} \frac{\vec{v}(t + \Delta t) - \vec{v}(t)}{\Delta t}, \tag{2.13}$$

the direction of $\vec{a}(t)\,\Delta t$, which is the same as that of $\vec{a}(t)$, can be approximated by the direction of $\vec{v}(t + \Delta t) - \vec{v}(t)$ if Δt is small enough. Referring to Fig. 2.6, the vector $\vec{v}(t + \Delta t) - \vec{v}(t)$, and therefore $\vec{a}(t)$, points toward the concave side of

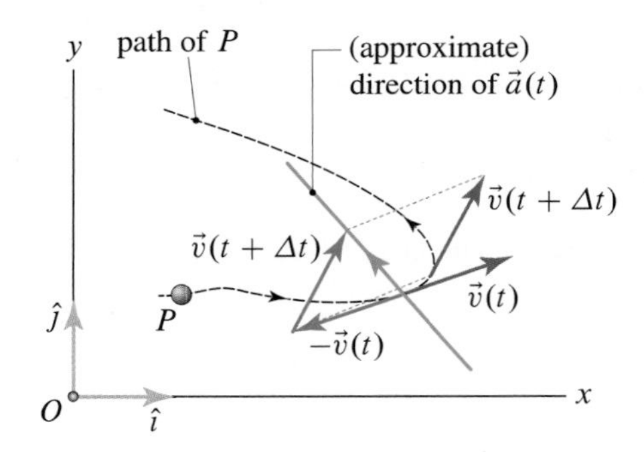

Figure 2.6. Velocity of P at times t and $t + \Delta t$ showing the approximate direction of $\vec{a}(t)$.

the trajectory instead of being tangent to it. While the arguments just provided are qualitative, in Section 2.5 we will quantitatively show that, for a curved path, not only does the acceleration have a component pointing toward the concave side of the trajectory, but also this component is proportional to the square of the speed and inversely proportional to the path's radius of curvature.

Cartesian coordinates

We now discuss the relation between the Cartesian *coordinates* of a point and the *components* of the point's position, velocity, and acceleration vectors.

* We will mathematically define *curvature* in Section 2.5.

Coordinates and position. Referring to Fig. 2.7, the Cartesian coordinates of P are $x(t)$ and $y(t)$ (for the three-dimensional case see Section 2.8). Cartesian coordinate systems have the distinctive property that the coordinates of a point are also the components of the position vector of the point relative to the system's origin. That is, if r_x and r_y are the Cartesian components of $\vec{r}$ in the x and y directions, respectively, then $r_x = x(t)$ and $r_y = y(t)$. Therefore, we express the *position vector $\vec{r}(t)$ in a Cartesian component system* as

$$\boxed{\vec{r}(t) = x(t)\,\hat{\imath} + y(t)\,\hat{\jmath}.} \tag{2.14}$$

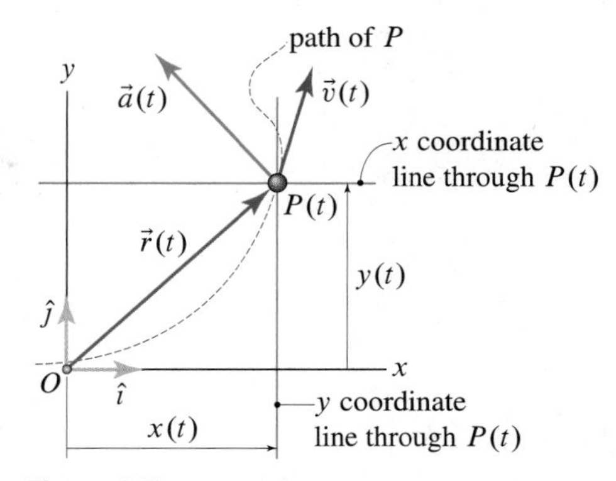

Figure 2.7
Position, velocity, and acceleration of a moving point P along with a Cartesian coordinate system.

Velocity and acceleration components in terms of time derivatives of the coordinates. The velocity $\vec{v}(t)$ and the acceleration $\vec{a}(t)$ of P are vectors based at $P(t)$, as shown in Fig. 2.8. The components of $\vec{v}(t)$ and $\vec{a}(t)$ are understood to be

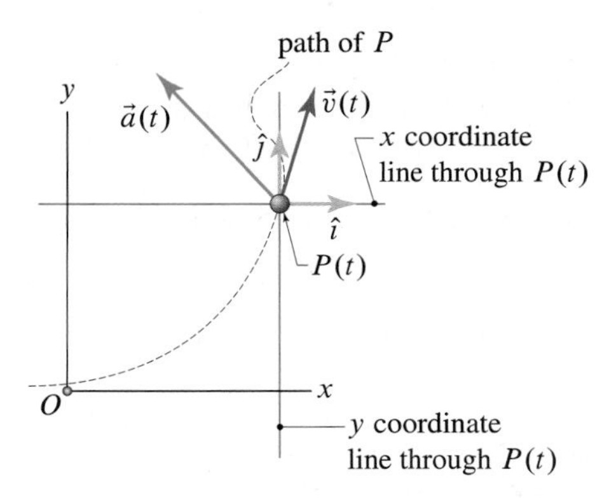

Figure 2.8. A moving point P at various times with its corresponding companion base vectors.

relative to the *vectors $\hat{\imath}$ and $\hat{\jmath}$ that are also based at $P(t)$ and tangent to the coordinate lines*. Computing $\vec{v}$ as the time derivative of $\vec{r}$ in Eq. (2.14), we have

$$\vec{v} = \dot{x}(t)\,\hat{\imath} + x(t)\,\frac{d\hat{\imath}}{dt} + \dot{y}(t)\,\hat{\jmath} + y(t)\,\frac{d\hat{\jmath}}{dt}, \tag{2.15}$$

where the terms $d\hat{\imath}/dt$ and $d\hat{\jmath}/dt$ are due to the fact that, since $\hat{\imath}$ and $\hat{\jmath}$ are based at the moving point $P(t)$, $\hat{\imath}$ and $\hat{\jmath}$ need to be regarded as implicit functions of time. However, Cartesian coordinate systems have the property that their base vectors do not change direction from point to point, i.e., $\hat{\imath}$ and $\hat{\jmath}$ are constants so that

$$\frac{d\hat{\imath}}{dt} = \vec{0} \quad \text{and} \quad \frac{d\hat{\jmath}}{dt} = \vec{0}, \tag{2.16}$$

and the *velocity vector $\vec{v}(t)$ in a Cartesian component system* becomes

$$\boxed{\vec{v} = \dot{x}(t)\,\hat{\imath} + \dot{y}(t)\,\hat{\jmath} = v_x(t)\,\hat{\imath} + v_y(t)\,\hat{\jmath},} \tag{2.17}$$

where the velocity components are obtained by direct time differentiation of the coordinates, i.e., $v_x = \dot{x}$ and $v_y = \dot{y}$. Similarly, the *acceleration vector $\vec{a}(t)$ in a Cartesian component system* is

$$\boxed{\vec{a} = \ddot{x}(t)\,\hat{\imath} + \ddot{y}(t)\,\hat{\jmath} = \dot{v}_x(t)\,\hat{\imath} + \dot{v}_y(t)\,\hat{\jmath} = a_x(t)\,\hat{\imath} + a_y(t)\,\hat{\jmath},} \tag{2.18}$$

where $a_x = \dot{v}_x = \ddot{x}$ and $a_y = \dot{v}_y = \ddot{y}$. In later sections, we will discover that in non-Cartesian coordinate systems the *components* of $\vec{v}$ and $\vec{a}$ are *not* obtained by the direct time differentiation of the *coordinates*.

End of Section Summary

Position. The position of a point is a *vector* going from the origin of the chosen frame of reference to the point in question.

Trajectory. The trajectory of a moving point is the line traced by the point during its motion. Another name for trajectory is *path*.

Displacement. The displacement between positions A and B is the vector going from A to B. In general, the magnitude of the displacement between two positions is not the distance traveled along the path between these positions.

Velocity. The velocity vector is the time rate of change of the position vector. The velocity is *always* tangent to the path.

Speed. The speed is the magnitude of the velocity and is a nonnegative scalar quantity. The speed measures the time rate of change of the distance traveled along the path.

Acceleration. The acceleration vector is the time rate of change of the velocity vector. Contrary to what happens for the velocity, the acceleration vector is, in general, not tangent to the trajectory.

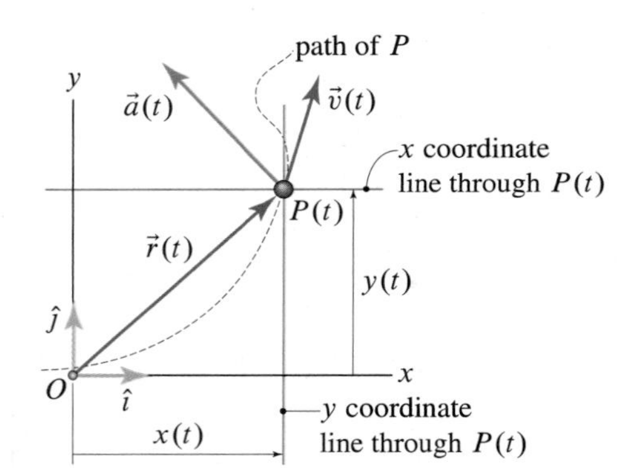

Figure 2.9
The position vector $\vec{r}(t)$ of the point P in Cartesian coordinates.

Cartesian coordinates. The Cartesian coordinates of a point P moving along some path are shown in Fig. 2.9. The position vector is given by

Eq. (2.14), p. 33

$$\vec{r}(t) = x(t)\,\hat{\imath} + y(t)\,\hat{\jmath}.$$

In Cartesian components, the velocity and acceleration vectors are given by

Eqs. (2.17) and (2.18), p. 33

$$\vec{v}(t) = \dot{x}(t)\,\hat{\imath} + \dot{y}(t)\,\hat{\jmath} = v_x(t)\,\hat{\imath} + v_y(t)\,\hat{\jmath},$$
$$\vec{a}(t) = \ddot{x}(t)\,\hat{\imath} + \ddot{y}(t)\,\hat{\jmath} = \dot{v}_x(t)\,\hat{\imath} + \dot{v}_y(t)\,\hat{\jmath} = a_x(t)\,\hat{\imath} + a_y(t)\,\hat{\jmath},$$

where $v_x = \dot{x}$, $v_y = \dot{y}$, $a_x = \dot{v}_x = \ddot{x}$, and $a_y = \dot{v}_y = \ddot{y}$.

An important note regarding example problems. In Chapter 2, all examples will begin to employ *part* of the problem-solving framework that is formally introduced in Chapter 3. That is, each example problem will include a *Road Map* step, a *Computation* step, and a *Discussion & Verification* step.* The Road Map step will lay out the path to the solution by identifying given information and unknowns and then proposing a problem-solving strategy. The Computation step will set up the appropriate equations and will solve them. Finally, the Discussion & Verification step will check whether or not the solution is reasonable.

* The problem-solving framework introduced in Chapter 3 includes additional steps since it is applied to kinetics problems, which are generally more involved than kinematics problems.

EXAMPLE 2.1 *How Do You Get to Carnegie Hall? . . . Practice!*

A cab picks up a passenger outside Radio City Music Hall on the corner of the Avenue of the Americas and E 51st St. (point A) and drops her off in front of Carnegie Hall on 7th Ave. (point D) after 5 min, following the route shown. Find the cab's displacement, average velocity, distance traveled, and average speed in going from A to D. The distance from A to B is 2200 ft, from B to C is 906 ft, and from C to D is 700 ft.

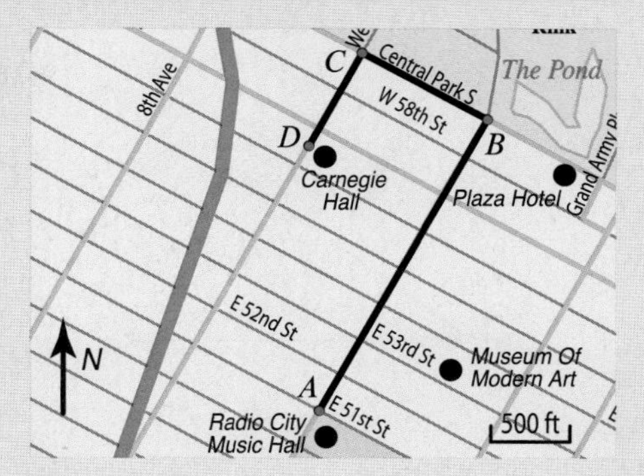

Figure 1
Cab route.

SOLUTION

Road Map To solve the problem, we need to set up a coordinate system and identify the coordinates of points A, B, C, and D that define the cab's path. Then the problem's questions can be answered by applying the definitions of displacement, average velocity, distanced traveled, and speed.

Computation We select a Cartesian coordinate system with origin at A and aligned with the city grid (Fig. 2). Using the given information, the coordinates of A, B, C, and D are given in Table 1. The displacement from A to D, which we will denote by $\Delta\vec{r}_{AD}$, is the difference between $\vec{r}_D$ and $\vec{r}_A$, namely, the position vectors of A and D:

$$\Delta\vec{r}_{AD} = \vec{r}_D - \vec{r}_A = (1500\,\hat{\imath} + 906.0\,\hat{\jmath})\text{ ft}. \tag{1}$$

Applying the definition of average velocity in Eq. (2.3) between t_A and t_D we have

$$\vec{v}_{\text{avg}} = \frac{\Delta\vec{r}_{AD}}{t_D - t_A} = (5.000\,\hat{\imath} + 3.020\,\hat{\jmath})\text{ ft/s}, \tag{2}$$

where $t_D - t_A = 5\text{ min} = 300\text{ s}$. Next, the distance traveled by the cab, which we will denote by d, is given by the sum of the lengths of the segments $\overline{AB}$, $\overline{BC}$, and $\overline{CD}$, i.e,

$$d = (x_B - x_A) + (y_C - y_B) + (x_C - x_D) = 3806\text{ ft}. \tag{3}$$

As shown in Eq. (2.11) on p. 31, the average speed between t_A and t_D is the ratio between the distance traveled between these time instants and the time interval duration $t_D - t_A$. Hence, we have

$$v_{\text{avg}} = \frac{d}{t_D - t_A} = 12.69\text{ ft/s}. \tag{4}$$

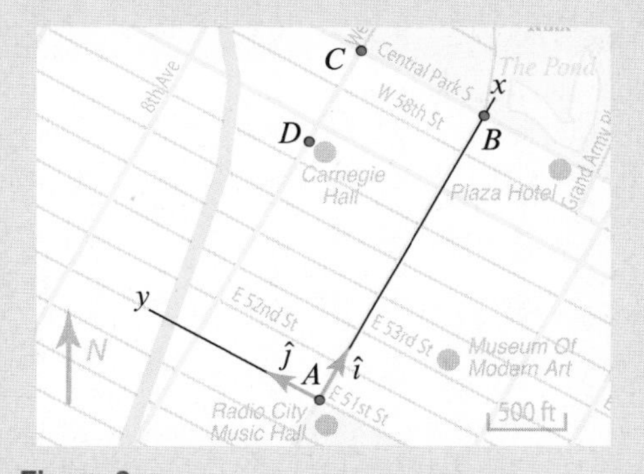

Figure 2
Cartesian coordinate system with origin at the pickup point A. The city grid is such that the line through B and C is parallel to the y axis and the line through C and D is parallel to the x axis.

Table 1
Coordinates of the points defining the cab's route.

Point	x (ft)	y (ft)
A	0	0
B	2200	0
C	2200	906
D	1500	906

Discussion & Verification The results obtained are dimensionally correct. We observe that v_{avg} in miles per hour is 8.650 mph. Hence, we can consider the result acceptable since such a speed is typical of midtown Manhattan.

A Closer Look Observing that $|\Delta\vec{r}_{AD}| = 1752\text{ ft}$, and that $|\vec{v}_{\text{avg}}| = 5.841\text{ ft/s}$, we see that this example reinforces the idea that we should never confuse the magnitude of the displacement for the distance traveled or the magnitude of the average velocity for the average speed.

EXAMPLE 2.2 *Trajectory, Velocity, and Acceleration*

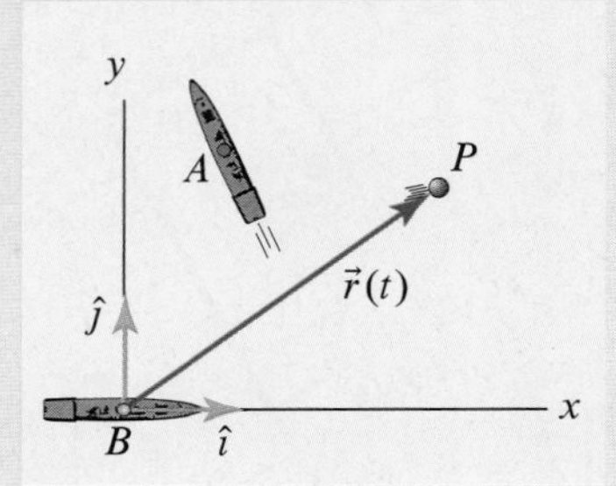

Figure 1
Point P represents an unmanned aerial vehicle launched from ship A.

Helpful Information

Trajectory and time. The trajectory (or path) is what the motion looks like once time is removed from the motion's description.

Ship B tracks an unmanned aerial vehicle P launched from A and flying parallel to the water. Relative to the Cartesian frame shown in Fig. 1 and for the first 8 seconds of flight, the recorded motion of P is

$$\vec{r}(t) = \left[\left(225 + 2.13t^3\right)\hat{\imath} + \left(225 + 0.993t^3\right)\hat{\jmath}\right]\text{m}, \tag{1}$$

where t is in seconds. Determine

(a) The trajectory of P for $0 \le t \le 5$ s.
(b) The velocity and the speed of P.
(c) The acceleration of P and its orientation relative to the path.

SOLUTION

Part (a): Path of P

Road Map We have the motion of P as a function of time. Since the trajectory of P is the line traced by P in space, we can find the path by eliminating time from the motion.

Computation Letting $x(t)$ and $y(t)$ denote the coordinates of P relative to the coordinate system in Fig. 1, $\vec{r}(t)$ can be expressed as $\vec{r}(t) = x(t)\,\hat{\imath} + y(t)\,\hat{\jmath}$, where, by comparison with Eq. (1),

$$x(t) = \left(225 + 2.13t^3\right)\text{m} \quad \text{and} \quad y(t) = \left(225 + 0.993t^3\right)\text{m}. \tag{2}$$

To eliminate time, we solve the first of Eqs. (2) for t^3 and then substitute the result in the second of Eqs. (2). This yields

$$y = \left[225 + \frac{0.993}{2.13}(x - 225)\right]\text{m} \quad \Rightarrow \quad y = \left(120.1 + 0.4662x\right)\text{m}. \tag{3}$$

The last of Eqs. (3) is the equation of the straight line $\mathcal{L}$ depicted in Fig. 2. The answer to Part (a) of the problem is the segment of $\mathcal{L}$ traveled by P for $0 \le t \le 5$ s. To find this segment, we first determine the positions of P at $t = 0$ and $t = 5$ s:

$$\vec{r}(0) = (225.0\,\hat{\imath} + 225.0\,\hat{\jmath})\,\text{m} \quad \text{and} \quad \vec{r}(5\,\text{s}) = (491.2\,\hat{\imath} + 349.1\,\hat{\jmath})\,\text{m}. \tag{4}$$

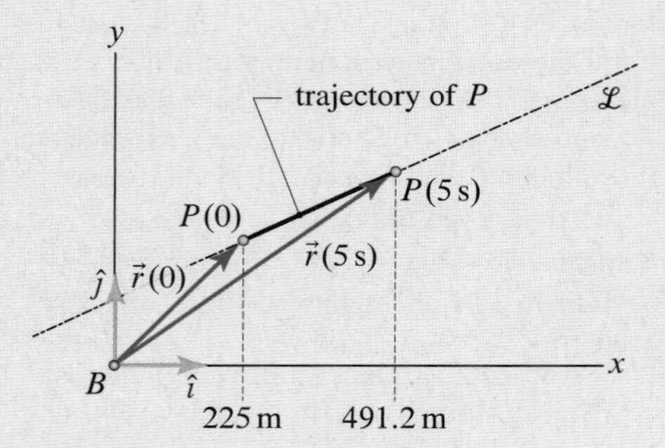

Figure 2
Path of P as seen by frames A and B.

We observe that $x(t)$ and $y(t)$ in Eq. (2) only increase with time so that, for $0 < t < 5$ s, P is on $\mathcal{L}$ between the points identified by $\vec{r}(0)$ and $\vec{r}(5\,\text{s})$. Therefore, referring to Eqs. (4) and Fig. 2, the trajectory of P for $0 \le t \le 5$ s is the segment of $\mathcal{L}$ for $225.0\,\text{m} \le x \le 491.2\,\text{m}$, corresponding to $225.0\,\text{m} \le y \le 349.1\,\text{m}$:

$$\boxed{y = \left(120.1 + 0.4662x\right)\text{m} \quad \text{for} \quad 225.0\,\text{m} \le x \le 491.2\,\text{m}.} \tag{5}$$

Discussion & Verification The fact that the trajectory P is a straight-line segment is reasonable since both x and y depend on time in the same way, namely, via the term t^3. Also, the trajectory was expected to be some finite segment since the time interval indicated by the problem was finite.

Part (b): Velocity and Speed of P

Road Map Since $\vec{r}(t)$ is given, we can compute the velocity $\vec{v}(t)$ by differentiating $\vec{r}(t)$ with respect to time. The speed is then found by computing the magnitude of $\vec{v}(t)$.

Computation Differentiating Eq. (1) with respect to time and simplifying, we have

$$\boxed{\vec{v}(t) = \left(6.390t^2\,\hat{\imath} + 2.979t^2\,\hat{\jmath}\right)\text{m/s}.} \tag{6}$$

Using Eq. (6) and the Pythagorean theorem, the magnitude of $\vec{v}(t)$ is

$$\boxed{v(t) = \left(7.050t^2\right)\text{m/s}.} \tag{7}$$

Discussion & Verification Since the position is a cubic polynomial in t, we expect the velocity to be a quadratic polynomial in t, as is in fact the case. As far as $v(t)$ is concerned, we observe that it is greater or equal to zero for any possible value of time, as it should be.

Part (c): Acceleration of P & Its Orientation

Road Map Since $\vec{v}(t)$ is known, we can find the acceleration $\vec{a}(t)$ by differentiating $\vec{v}(t)$ with respect to time. To determine the orientation of $\vec{a}(t)$ relative to the path, we recall that $\vec{v}(t)$ is tangent to the path. Hence, the orientation of $\vec{a}(t)$ relative to the path is the same as that of $\vec{a}(t)$ relative to $\vec{v}(t)$. In turn, this orientation is described by the angle between $\vec{a}(t)$ and $\vec{v}(t)$.

Computation Differentiating Eq. (6) with respect to time, we have

$$\boxed{\vec{a}(t) = \big(12.78t\,\hat{\imath} + 5.958t\,\hat{\jmath}\big)\,\text{m/s}^2.} \tag{8}$$

Denoting by $\theta_{\vec{v}\vec{a}}$ the angle between $\vec{v}(t)$ and $\vec{a}(t)$, we recall that the dot product of $\vec{v}(t)$ and $\vec{a}(t)$ has the form

$$\vec{v}(t)\cdot\vec{a}(t) = \big|\vec{v}(t)\big|\big|\vec{a}(t)\big|\cos\theta_{\vec{v}\vec{a}} \quad \Rightarrow \quad \cos\theta_{\vec{v}\vec{a}} = \frac{\vec{v}(t)\cdot\vec{a}(t)}{\big|\vec{v}(t)\big|\big|\vec{a}(t)\big|}. \tag{9}$$

To determine the numerator of the right-hand side of the last of Eqs. (9), we use the component expressions for $\vec{v}(t)$ and $\vec{a}(t)$ in Eqs. (6) and (8), respectively:

$$\begin{aligned}\vec{v}(t)\cdot\vec{a}(t) &= \big[\big(6.390t^2\,\hat{\imath} + 2.979t^2\,\hat{\jmath}\big)\,\text{m/s}\big]\cdot\big[\big(12.78t\,\hat{\imath} + 5.958t\,\hat{\jmath}\big)\,\text{m/s}^2\big]\\ &= \big(99.41t^3\big)\,\text{m}^2/\text{s}^3.\end{aligned} \tag{10}$$

As for the denominator of the right-hand side of the last of Eqs. (9), we observe that $\big|\vec{v}(t)\big|$ is given in Eq. (7) and $\big|\vec{a}(t)\big|$ can be obtained from Eq. (8) via the Pythagorean theorem:

$$\big|\vec{a}(t)\big| = \big(14.10t\big)\,\text{m/s}^2. \tag{11}$$

Substituting the results in Eqs. (7), (10), and (11) into the last of Eqs. (9) we have

$$\cos\theta_{\vec{v}\vec{a}} = 1, \tag{12}$$

which can be solved for $\theta_{\vec{v}\vec{a}}$ to obtain

$$\boxed{\theta_{\vec{v}\vec{a}} = 0.} \tag{13}$$

Equation (13) implies that $\vec{a}(t)$ is tangent to the trajectory at every time instant.

Discussion & Verification The result for the acceleration appears to be reasonable since it is a first-order polynomial in t, which is consistent with the fact that the velocity is a second-order polynomial in t. Also, the fact that $\vec{a}(t)$ is tangent to the path is not surprising since the trajectory computed earlier in the problem turned out to be a straight-line segment.

Common Pitfall

Does Eq. (13) contradict what we said earlier about the acceleration not being tangent to the trajectory? Earlier in the section we stated that *in general* the acceleration is not tangent to the path. We also indicated that if the trajectory is curved, then we must expect the acceleration not to be parallel to the path. Therefore, we can conclude that (1) there can be points along a path at which the acceleration is tangent to the path and that (2) at these points the path's curvature must be equal to zero. This is consistent with what we found in our example since the path in this example is a straight line, that is, *a line with no curvature*.

EXAMPLE 2.3 *Relating the Path and the Speed to Velocity and Acceleration*

A particle P moves with constant speed v_0 along the parabola $y^2 = 4ax$ with $\dot{y} > 0$. Determine the

(a) Cartesian components of the velocity vector as a function of y.

(b) Cartesian components of the acceleration vector as a function of y.

(c) Angle between the velocity and acceleration vectors as a function of y.

In addition, plot the Cartesian components of the velocity and acceleration vectors as functions of y for $a = 0.2\,\text{m}$ and $v_0 = 3\,\text{m/s}$.

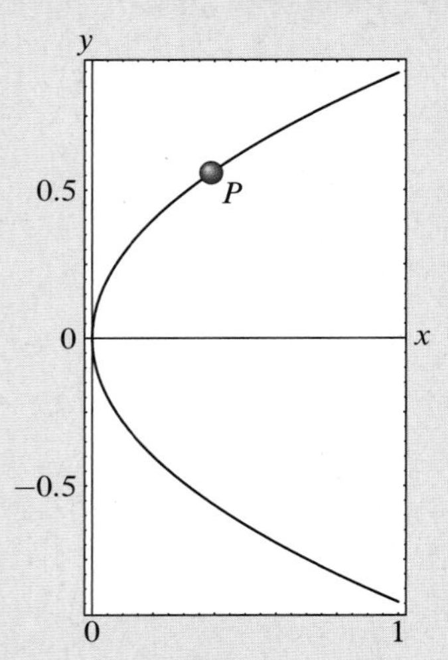

Figure 1
Parabolic path of the particle P for $a = 0.2\,\text{m}$.

Helpful Information

Why find components as a function of y instead of x? We chose to find the velocity and acceleration components as a function of y rather than x because of the fact that for any given value of y, there is one value of x. The converse is not true; that is, for a given value of x, there are two values of y, and this makes the analysis more complicated.

SOLUTION

Road Map We are given the path of the particle and its speed along the path, but we do not know how x and y vary with time, so we cannot immediately apply Eqs. (2.17) and (2.18). On the other hand, we can differentiate the equation of the curve with respect to time and make use of the fact that the speed is given by $\sqrt{\dot{x}^2 + \dot{y}^2}$.

Computation We start by differentiating the particle's path with respect to time to obtain

$$\frac{d}{dt}(y^2 = 4ax) \quad \Rightarrow \quad \frac{d(y^2)}{dy}\frac{dy}{dt} = \frac{d(4ax)}{dt} \quad \Rightarrow \quad 2y\dot{y} = 4a\dot{x}, \tag{1}$$

where, in differentiating with respect to time, we have used the chain rule. Solving the last of Eqs. (1) for $\dot{y}$, we obtain

$$\dot{y} = \frac{2a\dot{x}}{y}. \tag{2}$$

Now, we also know that the speed is related to the components of the velocity via

$$v_0^2 = \dot{x}^2 + \dot{y}^2 \quad \Rightarrow \quad v_0^2 = \dot{x}^2 + \left(\frac{2a\dot{x}}{y}\right)^2, \tag{3}$$

where we have substituted in Eq. (2). We can now solve Eq. (3) for $\dot{x}$ to obtain

$$\boxed{\dot{x} = \frac{v_0 y}{\sqrt{y^2 + 4a^2}},} \tag{4}$$

where we have chosen the plus sign when taking the square root to be consistent with Eq. (2), which, given that $\dot{y} > 0$, tells us that for $y < 0$ we must have $\dot{x} < 0$ and for $y > 0$ we must have $\dot{x} > 0$. Now we can find $\dot{y}(y)$ by substituting Eq. (4) into Eq. (2), which gives

$$\boxed{\dot{y} = \frac{2v_0 a}{\sqrt{y^2 + 4a^2}},} \tag{5}$$

where $\dot{y} > 0$ as expected. Figure 2 shows $\dot{x}$ and $\dot{y}$ as functions of y.

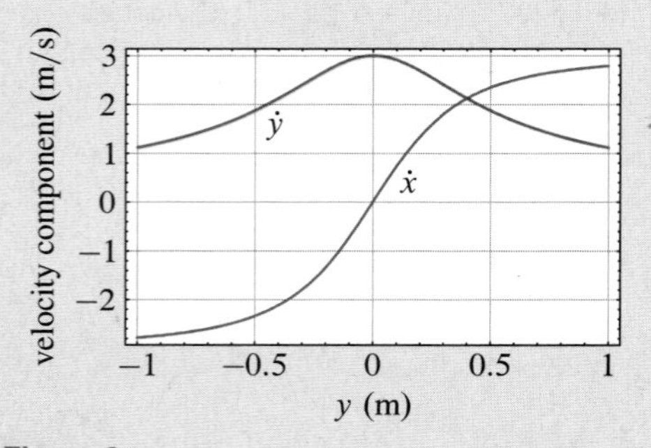

Figure 2
The x and y components of the velocity as a function of y.

There are several ways to obtain the x and y components of the acceleration. We will differentiate the last of Eqs. (1) with respect to time, and after simplifying, we obtain

$$\dot{y}^2 + y\ddot{y} = 2a\ddot{x} \quad \Rightarrow \quad \ddot{x} = \frac{1}{2a}(\dot{y}^2 + y\ddot{y}). \tag{6}$$

We see that we need $\ddot{y}$ to know $\ddot{x}$. Hence, differentiating Eq. (5) with respect to time, we obtain

$$\ddot{y} = 2v_0 a\left[\frac{-y\dot{y}}{(y^2 + 4a^2)^{3/2}}\right], \tag{7}$$

which, after substituting $\dot{y}$ from Eq. (5), becomes

$$\boxed{\ddot{y} = \frac{-4v_0^2 a^2 y}{(y^2 + 4a^2)^2}.} \tag{8}$$

We can now get the final version of $\ddot{x}$ by substituting Eqs. (5) and (8) into the second of Eqs. (6):

$$\ddot{x} = \frac{1}{2a}\left[\frac{4v_0^2 a^2}{y^2 + 4a^2} - y\,\frac{4v_0^2 a^2 y}{(y^2 + 4a^2)^2}\right] \quad\Rightarrow\quad \boxed{\ddot{x} = \frac{8v_0^2 a^3}{(y^2 + 4a^2)^2}.} \tag{9}$$

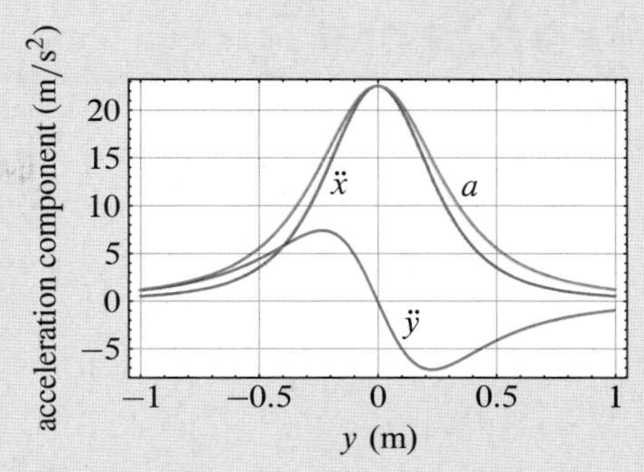

Figure 3
The x component (blue), y component (red), and the total acceleration (green) of the particle as a function of y.

Figure 3 shows $\ddot{x}$, $\ddot{y}$, and $a = \sqrt{\ddot{x}^2 + \ddot{y}^2}$ as a function of y.

Finally, to find the angle between the velocity and acceleration vectors, we first need to form those vectors as

$$\vec{v} = \dot{x}\,\hat{\imath} + \dot{y}\,\hat{\jmath} \quad\text{and}\quad \vec{a} = \ddot{x}\,\hat{\imath} + \ddot{y}\,\hat{\jmath}. \tag{10}$$

Then, using Eq. (1.17) on p. 8, the angle ϕ between these two vectors is

$$\cos\phi = \frac{\vec{v}\cdot\vec{a}}{|\vec{v}||\vec{a}|} = \frac{(\dot{x}\,\hat{\imath} + \dot{y}\,\hat{\jmath})\cdot(\ddot{x}\,\hat{\imath} + \ddot{y}\,\hat{\jmath})}{\sqrt{\dot{x}^2 + \dot{y}^2}\sqrt{\ddot{x}^2 + \ddot{y}^2}} = \frac{\dot{x}\ddot{x} + \dot{y}\ddot{y}}{\sqrt{\dot{x}^2 + \dot{y}^2}\sqrt{\ddot{x}^2 + \ddot{y}^2}}. \tag{11}$$

Although it is tedious, we can substitute Eqs. (4), (5), (8), and (9) into the numerator of the last expression in Eq. (11) to find

$$\dot{x}\ddot{x} + \dot{y}\ddot{y} = 0 \quad\Rightarrow\quad \cos\phi = 0 \quad\Rightarrow\quad \boxed{\phi = 90^\circ.} \tag{12}$$

Since $\phi = 90^\circ$ for all values of y, $\vec{v}$ is *always* perpendicular to $\vec{a}$.

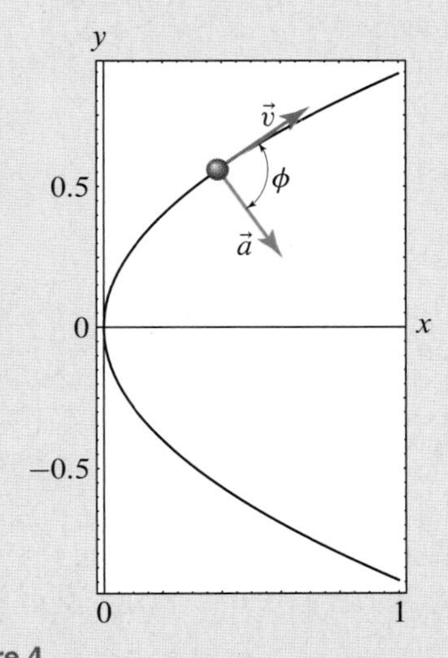

Figure 4
The velocity vector, acceleration vector, and the angle ϕ for one position of the particle as it moves along the parabola.

Discussion & Verification Recall that $\dot{y} > 0$. This agrees with the plot in Fig. 2, as it should, since we chose the plus sign when taking the square root to get Eq. (5). We also see that $\dot{y}$ is largest at $y = 0$. This also makes sense due to the fact that the particle has constant speed. Recalling that the velocity is always tangent to the path, note that at $y = 0$, the velocity must be completely in the y direction. Therefore, at $y = 0$, $\dot{x}$ must be zero, and so $\dot{y}$ achieves a maximum equal to $v_0 = 3\,\text{m/s}$, as can be seen in Fig. 2. As for the x component of the velocity, it is negative for $y < 0$ and positive for $y > 0$. This is what we should expect, given that x is decreasing for $y < 0$ and is increasing for $y > 0$ (this is why we chose the + sign in Eq. (4)). In addition, notice that $\dot{x} = 0$ when $y = 0$, which is also to be expected from the parabola shown in Fig. 1.

To verify our acceleration results, recall that the particle moves at a constant speed along a *parabola*. Since the particle's path is curved, we expect the components of the acceleration vector to be different from zero, as can be seen in Fig. 3. In addition, we expect $\ddot{y} = 0$ at $y = 0$ because, as argued earlier, $\dot{y}$ is maximum at $y = 0$. In addition, note that $\ddot{x}$ is largest at $y = 0$ even though $\dot{x} = 0$ there. While perhaps counterintuitive, this result is consistent with the fact that part of the acceleration is proportional to curvature of the path on which the particle moves. Since the curvature of the parabola is largest at $y = 0$, the magnitude of the acceleration, given by $a = \sqrt{\ddot{x}^2 + \ddot{y}^2}$, achieves a maximum at $y = 0$.*

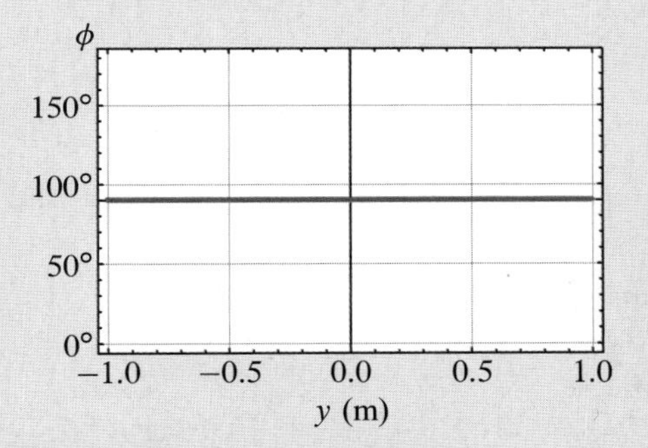

Figure 5
The heavy red line shows the angle $\phi(y)$ between the velocity vector and the acceleration vector.

As for the angle ϕ between $\vec{v}$ and $\vec{a}$, we discovered that it is always equal to 90°. Since the speed is constant, we know that $\vec{v}$ *cannot* be changing along its line of action, and so any change in $\vec{v}$ (that is, any $\vec{a}$) must be perpendicular to $\vec{v}$. We will discuss this topic in detail in Sections 2.4 and 2.5.

* This topic is discussed in detail in Section 2.5.

Problems

Problem 2.1

If $\vec{v}_{\text{avg}}$ is the average velocity of a point P over a given time interval, is $|\vec{v}_{\text{avg}}|$, the magnitude of the average velocity, equal to the average speed of P over the time interval in question?

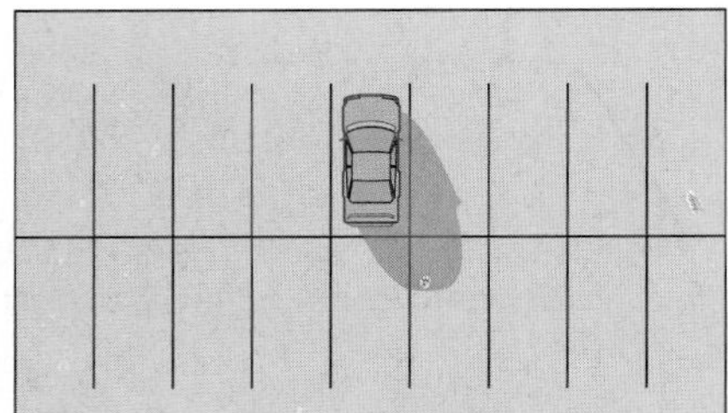

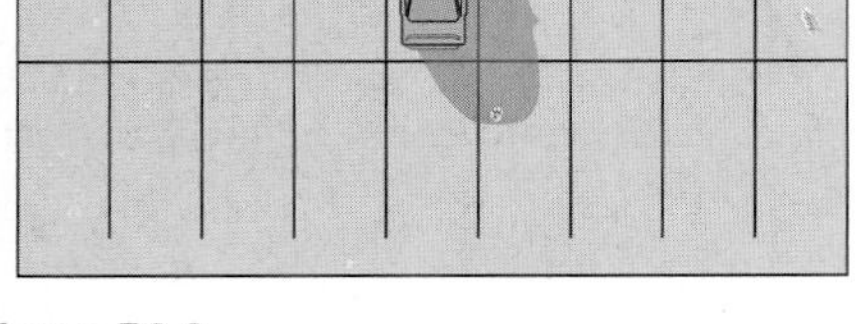

Figure P2.2

Problem 2.2

A car is seen parked in a given parking space at 8:00 A.M. on a Monday morning and is then seen parked in the same spot the next morning at the same time. What is the displacement of the car between the two observations? What is the distance traveled by the car during the two observations?

Problem 2.3

Is it possible for the vector $\vec{v}$ shown to represent the velocity of the point P?

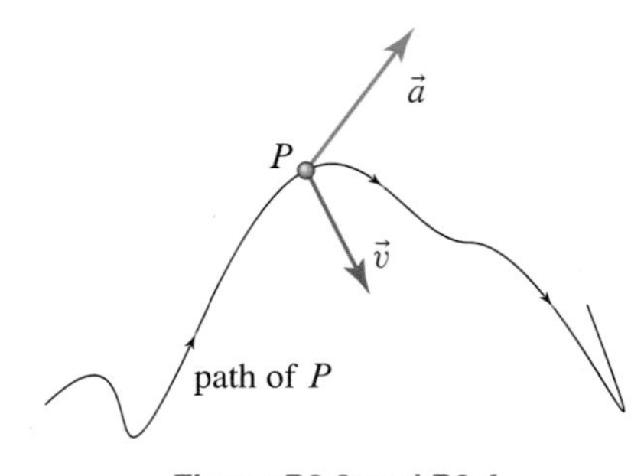

Figure P2.3 and P2.4

Problem 2.4

Is it possible for the vector $\vec{a}$ shown to be the acceleration of the point P?

Problem 2.5

Two points P and Q happen to go by the same location in space (though at different times).

(a) What must the paths of P and Q have in common if, at the location in question, P and Q have identical speeds?

(b) What must the paths of P and Q have in common if, at the location in question, P and Q have identical velocities?

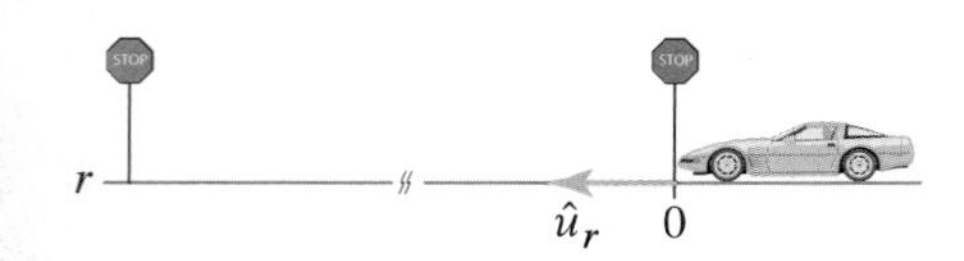

Figure P2.6

Problem 2.6

The position of a car traveling between two stop signs along a straight city block is given by $r = [9t - (45/2)\sin(2t/5)]$ m, where t denotes time (in seconds), and where the argument of the sine function is measured in radians. Compute the displacement of the car between 2.1 and 3.7 s, as well as between 11.1 and 12.7 s. For each of these time intervals compute the average velocity.

Figure P2.7

Problem 2.7

A city bus covers a 15 km route in 45 min. If the initial departure and final arrival points coincide, determine the average velocity and the average speed of the bus over the entire duration of the ride. Express the answers in m/s.

Problem 2.8

An airplane A is performing a loop with constant radius ρ. When $\theta = 120°$, the speed of the airplane is $v_0 = 210\,\text{mph}$. Modeling the airplane as a point, find the velocity of the airplane at this instant using the component system shown. Express your answer in ft/s.

Problem 2.9

An airplane A is performing a loop with constant radius $\rho = 300\,\text{m}$. From elementary physics, we know that the acceleration of a point in uniform circular motion (i.e., circular motion at constant speed) is directed toward the center of the circle and has magnitude equal to v^2/ρ, where v is the speed. Assuming that A can maintain its speed constant and using the component system shown, provide the expressions of the velocity and acceleration of A when $\theta = 40°$ and $|\vec{a}| = 3g$, where $\vec{a}$ is the acceleration of A and g is the acceleration due to gravity.

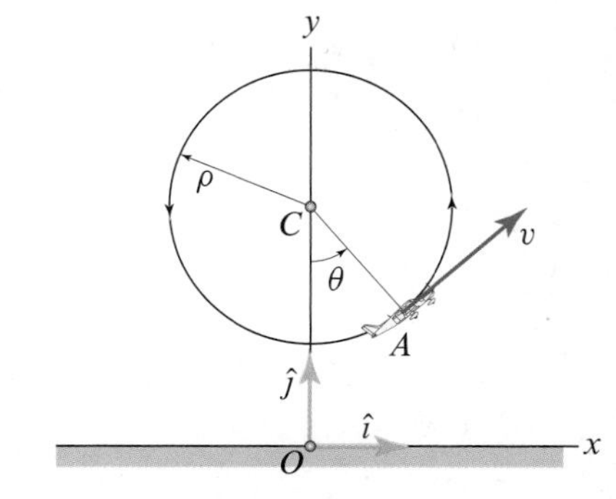

Figure P2.8 and P2.9

Problem 2.10

An airplane takes off as shown following a trajectory described by equation $y = \kappa x^2$, where $\kappa = 2 \times 10^{-4}\,\text{ft}^{-1}$. When $x = 1200\,\text{ft}$, the speed of the plane is $v_0 = 110\,\text{mph}$. Using the component system shown, provide the expression for the velocity of the airplane when $x = 1200\,\text{ft}$. Express your answer in ft/s.

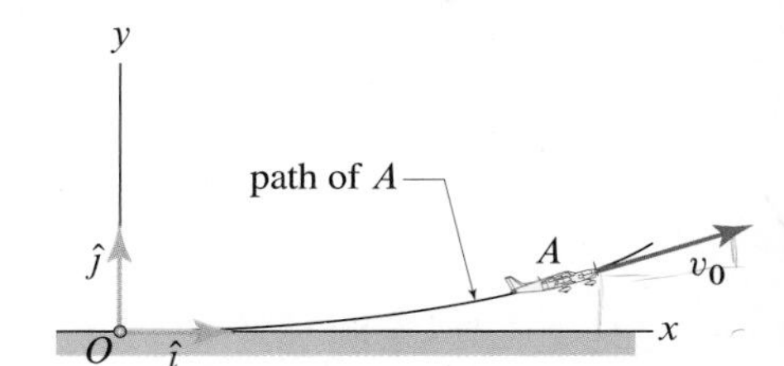

Figure P2.10

Problem 2.11

The position of a car as a function of time t, with $t > 0$ and expressed in seconds, is

$$\vec{r}(t) = \left[\left(5.98t^2 + 0.139t^3 - 0.0149t^4\right)\hat{\imath} + \left(0.523t^2 + 0.0122t^3 - 0.00131t^4\right)\hat{\jmath}\right]\text{ft}.$$

Determine the velocity, speed, and acceleration of the car for $t = 15\,\text{s}$.

Figure P2.11 and P2.12

Problem 2.12

The position of a car as a function of time t, with $t > 0$ and expressed in seconds, is

$$\vec{r}(t) = \left[12.3\left(t + 1.54e^{-0.65t}\right)\hat{\imath} + 2.17\left(t + 1.54e^{-0.65t}\right)\hat{\jmath}\right]\text{m}.$$

Find the difference between the average velocity over the time interval $0 \le t \le 2\,\text{s}$ and the true velocity computed at the midpoint of the interval, i.e., at $t = 1\,\text{s}$. Repeat the calculation for the time interval $8\,\text{s} \le t \le 10\,\text{s}$. Explain why the difference between the average velocity and the true velocity over the time interval $0 \le t \le 2\,\text{s}$ is not equal to that over $8\,\text{s} \le t \le 10\,\text{s}$.

Problem 2.13

The position of a car as a function of time t, with $t > 0$ and expressed in seconds, is

$$\vec{r}(t) = \left[(66t - 120)\,\hat{\imath} + \left(1.2 + 31.7t - 8.71t^2\right)\hat{\jmath}\right]\text{ft}.$$

If the speed limit is 55 mph, determine the time at which the car will exceed this limit.

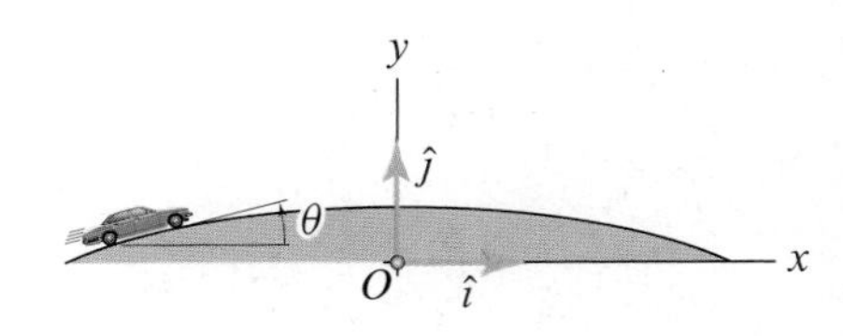

Figure P2.13 and P2.14

Problem 2.14

The position of a car as a function of time t, with $t > 0$ and expressed in seconds, is

$$\vec{r}(t) = \left[(66t - 120)\,\hat{\imath} + \left(1.2 + 31.7t - 8.71t^2\right)\hat{\jmath}\right]\text{ft}.$$

Determine the slope θ of the trajectory of the car for $t_1 = 1$ s and $t_2 = 3$ s. In addition, find the angle ϕ between velocity and acceleration for $t_1 = 1$ s and $t_2 = 3$ s. Based on the values of ϕ at t_1 and t_2, argue whether the speed of the car is increasing or decreasing at t_1 and t_2.

Problem 2.15

Let $\vec{r} = [t\,\hat{\imath} + (2 + 3t + 2t^2)\,\hat{\jmath}]$ m describe the motion of the point P relative to the Cartesian frame of reference shown. Determine an analytic expression of the type $y = y(x)$ for the trajectory of P for $0 \le t \le 5$ s.

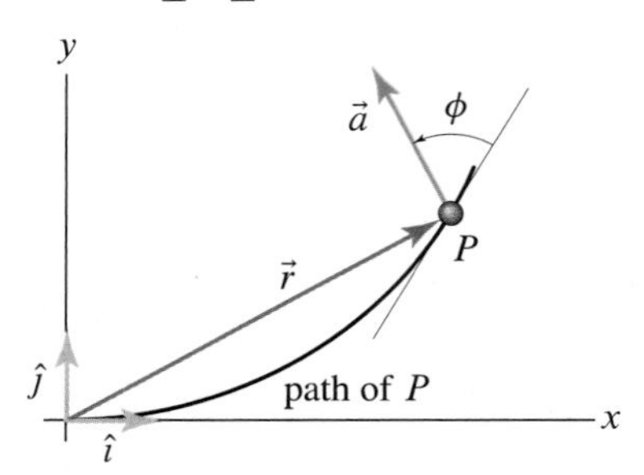

Figure P2.15 and P2.16

Problem 2.16

Let $\vec{r} = [t\,\hat{\imath} + (2 + 3t + 2t^2)\,\hat{\jmath}]$ ft describe the motion of a point P relative to the Cartesian frame of reference shown. Recalling that for any two vectors $\vec{p}$ and $\vec{q}$ we have that $\vec{p} \cdot \vec{q} = |\vec{p}|\,|\vec{q}| \cos\beta$, where β is the angle formed by $\vec{p}$ and $\vec{q}$, and recalling that the velocity vector is *always* tangent to the trajectory, determine the function $\phi(x)$ describing the angle between the acceleration vector and the tangent to the path of P.

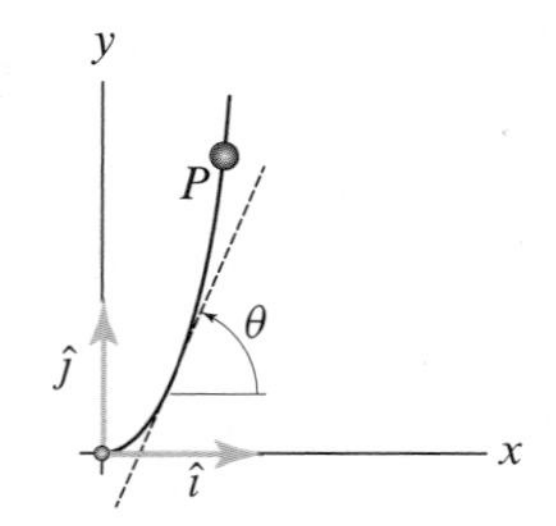

Figure P2.17 and P2.18

Problems 2.17 and 2.18

The motion of a point P with respect to a Cartesian coordinate system is described by $\vec{r} = \left[2\sqrt{t}\,\hat{\imath} + \left(4\ln(t+1) + 2t^2\right)\hat{\jmath}\right]$ ft, where t denotes time, $t > 0$, and is expressed in seconds.

Problem 2.17 Determine the angle θ formed by the tangent to the path and the horizontal direction at $t = 3$ s.

Problem 2.18 Determine the average acceleration of P between times $t_1 = 4$ s and $t_2 = 6$ s and find the difference between it and the true acceleration of P at $t = 5$ s.

Problem 2.19

The motion of a stone thrown into a pond is described by

$$\vec{r}(t) = \left[\left(1.5 - 0.3e^{-13.6t}\right)\hat{\imath} + \left(0.094e^{-13.6t} - 0.094 - 0.72t\right)\hat{\jmath}\right]\text{m},$$

where t is time expressed in seconds, and $t = 0$ s is the time when the stone first hits the water. Determine the stone's velocity and acceleration. In addition, find the initial angle of impact θ of the stone with the water, i.e., the angle formed by the stone's trajectory and the horizontal direction at $t = 0$.

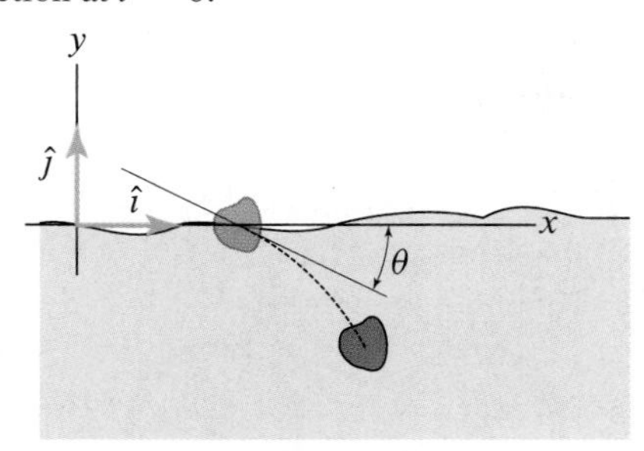

Figure P2.19

Problem 2.20

As part of a mechanism, a peg P is made to slide within a rectilinear guide with the following prescribed motion:

$$\vec{r}(t) = x_0\big[\sin(2\pi\omega t) - 3\sin(\pi\omega t)\big]\,\hat{\imath},$$

where t denotes time in seconds, $x_0 = 1.2$ in., and $\omega = 0.5$ rad/s. Determine the displacement and the distance traveled over the time interval $0 \le t \le 4$ s. In addition, determine the corresponding average velocity and average speed. Express displacement and distance traveled in ft, and express velocity and speed in ft/s. You may find useful the following trigonometric identity: $\cos(2\beta) = 2\cos^2\beta - 1$.

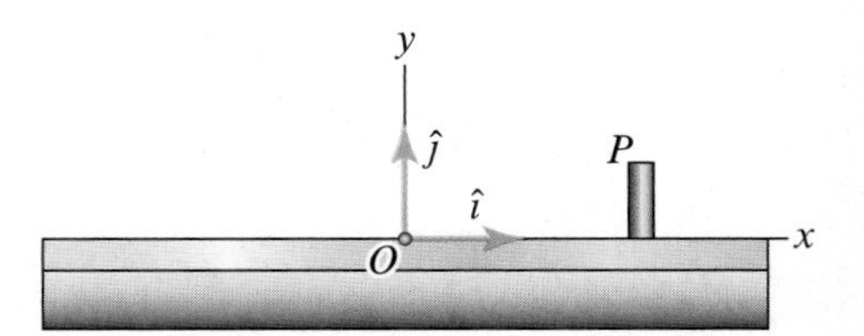

Figure P2.20

Problems 2.21 and 2.22

The position of point P as a function of time t, $t \ge 0$ and expressed in seconds, is

$$\vec{r}(t) = 2.0\,[0.5 + \sin(\omega t)]\,\hat{\imath} + \Big[9.5 + 10.5\sin(\omega t) + 4.0\sin^2(\omega t)\Big]\,\hat{\jmath},$$

where $\omega = 1.3$ rad/s and the position is measured in meters.

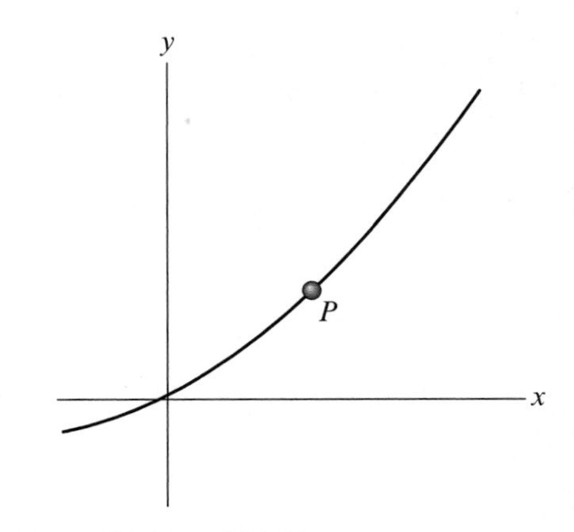

Figure P2.21 and P2.22

Problem 2.21 Find the trajectory of P in Cartesian components and then, using the x component of $\vec{r}(t)$, find the maximum and minimum values of x reached by P. The equation for the trajectory is valid for all values of x, yet the maximum and minimum values of x as given by the x component of $\vec{r}(t)$ are finite. What is the origin of this discrepancy?

Problem 2.22

(a) Plot the trajectory of P for $0 \le t \le 0.6$ s, $0 \le t \le 1.4$ s, $0 \le t \le 2.3$ s, and $0 \le t \le 5$ s.

(b) Plot the $y(x)$ trajectory for $-10\text{ m} \le x \le 10\text{ m}$.

(c) You will notice that the trajectory found in (b) does not agree with any of those found in (a). Explain this discrepancy by analytically determining the minimum and maximum values of x reached by P.

As you look at this sequence of plots, why does the trajectory change between some times and not others?

Problems 2.23 through 2.25

A bicycle is moving to the right at a speed $v_0 = 20$ mph on a horizontal and straight road. The radius of the bicycle's wheels is $R = 1.15$ ft. Let P be a point on the periphery of the front wheel. One can show that the x and y coordinates of P are described by the following functions of time:

$$x(t) = v_0 t + R\sin(v_0 t/R) \quad \text{and} \quad y(t) = R\big[1 + \cos(v_0 t/R)\big].$$

Figure P2.23–P2.25

Problem 2.23 Determine the expressions for the velocity, speed, and acceleration of P as functions of time.

Problem 2.24 Determine the maximum and minimum speed achieved by P, as well as the y coordinate of P when the maximum and minimum speeds are achieved. Finally, compute the acceleration of P when P achieves its maximum and minimum speeds.

Problem 2.25 Plot the trajectory of P for $0 \le t \le 1$ s. For the same time interval, plot the speed as a function of time, as well as the components of the velocity and acceleration of P.

Problem 2.26

Find the x and y components of the acceleration in Example 2.3 (except for the plots) by simply differentiating Eqs. (4) and (5) with respect to time. Verify that you get the results given in Example 2.3.

Problem 2.27

Find the x and y components of the acceleration in Example 2.3 (except for the plots) by differentiating the first of Eqs. (3) and the last of Eqs. (1) with respect to time and then solving the resulting two equations for $\ddot{x}$ and $\ddot{y}$. Verify that you get the results given in Example 2.3.

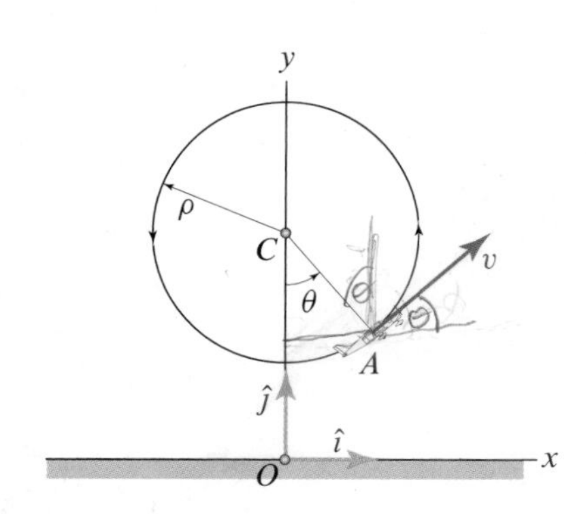

Figure P2.28

Problem 2.28

Airplane A is performing a loop with constant radius $\rho = 1000$ ft. The equation describing the loop is as follows:

$$(x - x_C)^2 + (y - y_C)^2 = \rho^2,$$

where $x_C = 0$ and $y_C = 1500$ ft are the coordinates of the center of the loop. If the plane were capable of maintaining its speed constant and equal to $v_0 = 160$ mph, determine the velocity and acceleration of the plane for $\theta = 30°$.

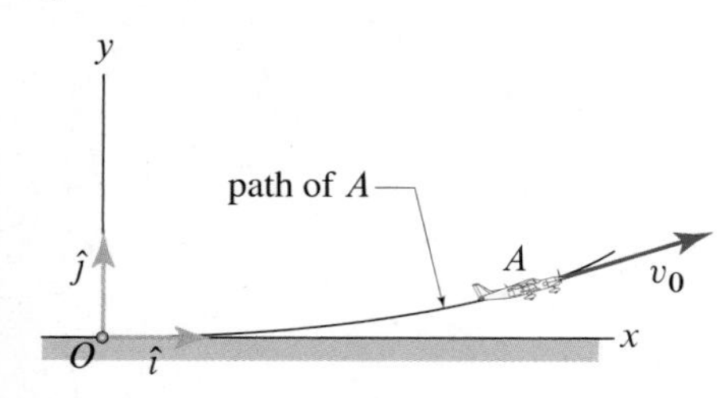

Figure P2.29

Problem 2.29

An airplane A takes off as shown with a constant speed equal to $v_0 = 160$ km/h. The path of the airplane is described by the equation $y = \kappa x^2$, where $\kappa = 6 \times 10^{-4}\,\text{m}^{-1}$. Using the component system shown, provide the expression for the velocity and acceleration of the airplane when $x = 400$ m. Express the velocity in m/s and the acceleration in m/s^2.

Problem 2.30

A test track for automobiles has a portion with a specific profile described by:

$$y = h\big[1 - \sin(x/w)\big],$$

where $h = 0.5$ ft and $w = 8$ ft, and where the argument of the sine function is understood to be in radians. A car travels in the positive x direction such that the horizontal component of velocity remains constant and equal to 55 mph. Modeling the car as a point moving along the given profile, determine the maximum speed of the car. Express your answer in ft/s.

Figure P2.30–P2.32

Problem 2.31

A test track for automobiles has a portion with a specific profile described by:

$$y = h\big[1 - \cos(x/w)\big],$$

where $h = 0.20$ m and $w = 2$ m, and where the argument of the cosine function is understood to be in radians. A car travels in the positive x direction with a constant x component of velocity equal to 100 km/h. Modeling the car as a point moving along the given profile, determine the velocity and acceleration (expressed in m/s and m/s^2, respectively) of the car for $x = 24$ m.

Problem 2.32

A test track for automobiles has a portion with a specific profile described by:

$$y = h\big[1 - \cos(x/w)\big],$$

where $h = 0.75$ ft and $w = 10$ ft, and where the argument of the cosine function is understood to be in radians. A car drives at a constant speed $v_0 = 35$ mph. Modeling the car as a point moving along the given profile, find the velocity and acceleration of the car for $x = 97$ ft. Express velocity in ft/s and acceleration in ft/s^2.

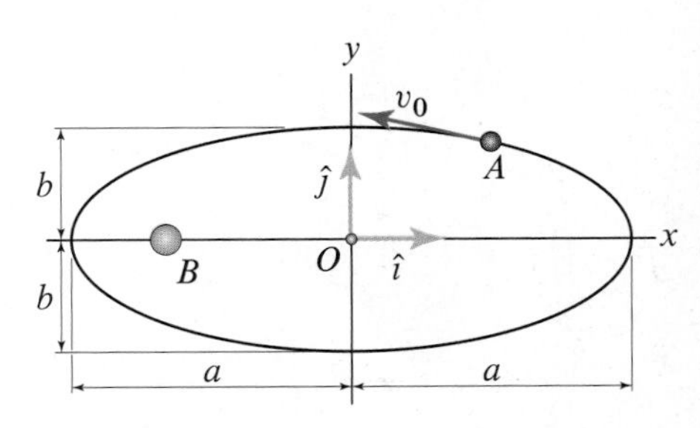

Figure P2.33

Problem 2.33

The orbit of a satellite A around planet B is the ellipse shown and is described by the equation $(x/a)^2 + (y/b)^2 = 1$, where a and b are the semimajor and semiminor axes of the ellipse, respectively. When $x = a/2$ and $y > 0$, the satellite is moving with a speed v_0 as shown. Determine the expression for the satellite's velocity $\vec{v}$ in terms of v_0, a, and b for $x = a/2$ and $y > 0$.

Problems 2.34 and 2.35

In the mechanism shown, block B is fixed and has a profile described by the following relation:

$$y = h\left[1 + \frac{1}{2}\left(\frac{x}{d}\right)^2 - \frac{1}{4}\left(\frac{x}{d}\right)^4\right].$$

The follower moves with the shuttle A, and the tip C of the follower remains in contact with B.

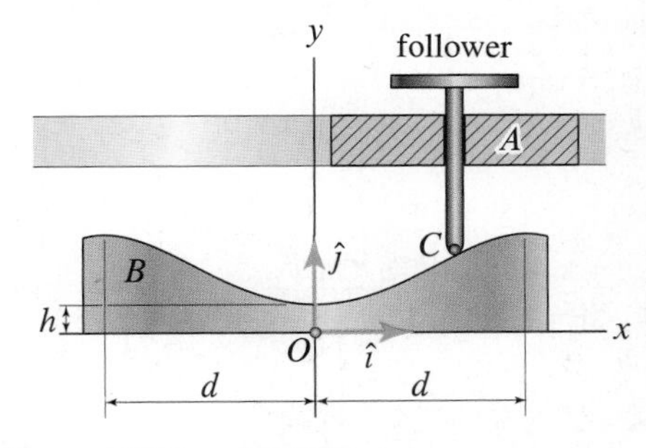

Figure P2.34 and P2.35

Problem 2.34 Assume that $h = 0.25$ in., $d = 1$ in., and the horizontal position of C is $x = d\sin(\omega t)$, where $\omega = 2\pi$ rad/s, and t is time in seconds. Determine an analytical expression for the speed of C as a function of x and the parameters d, h, and ω. Then, evaluate the speed of C for $x = 0$, $x = 0.5$ in., and $x = 1$ in. Express your answers in ft/s.

Problem 2.35 Assume that $h = 2$ mm, $d = 20$ mm, and A is made to move from $x = -d$ to $x = d$ with a constant speed $v_0 = 0.1$ m/s. Determine the acceleration of C for $x = 15$ mm. Express your answer in m/s^2.

Problem 2.36

The Center for Gravitational Biology Research at NASA's Ames Research Center runs a large centrifuge capable of $20g$ of acceleration, where g is the acceleration due to gravity ($12.5g$ is the maximum for human subjects). The distance from the axis of rotation to the cab at either A or B is $R = 25$ ft. The trajectory of A is described by $y_A = \sqrt{R^2 - x_A^2}$ for $y_A \geq 0$ and by $y_A = -\sqrt{R^2 - x_A^2}$ for $y_A < 0$. If A moves at the constant speed $v_A = 120$ ft/s, determine the velocity and acceleration of A when $x_A = -20$ ft and $y_A > 0$.

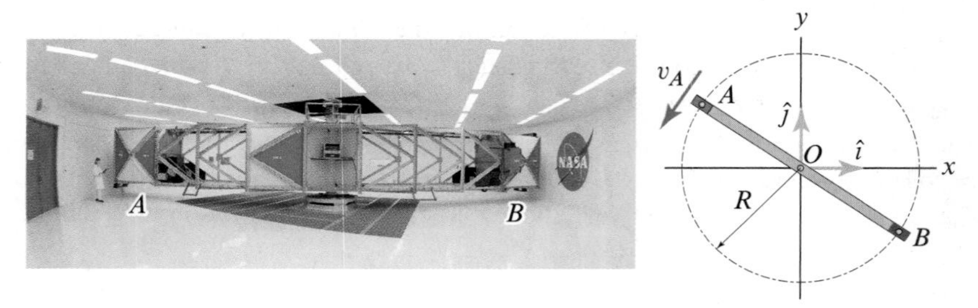

Figure P2.36

Problems 2.37 and 2.38

Point C is a point on the connecting rod of a mechanism called a *slider-crank*. The x and y coordinates of C can be expressed as follows: $x_C = R\cos\theta + \frac{1}{2}\sqrt{L^2 - R^2\sin^2\theta}$ and $y_C = (R/2)\sin\theta$, where θ describes the position of the crank. The crank rotates at a constant rate such that $\theta = \omega t$, where t is time.

Problem 2.37 Find expressions for the velocity, speed, and acceleration of C as functions of the angle θ and the parameters, R, L, and ω.

Problem 2.38 Let t be expressed in seconds, $R = 0.1$ m, $L = 0.25$ m, and $\omega = 250$ rad/s. Plot the trajectory of point C for $0 \leq t \leq 0.025$ s. For the same interval of time, plot the speed as a function of time, as well as the components of the velocity and acceleration of C.

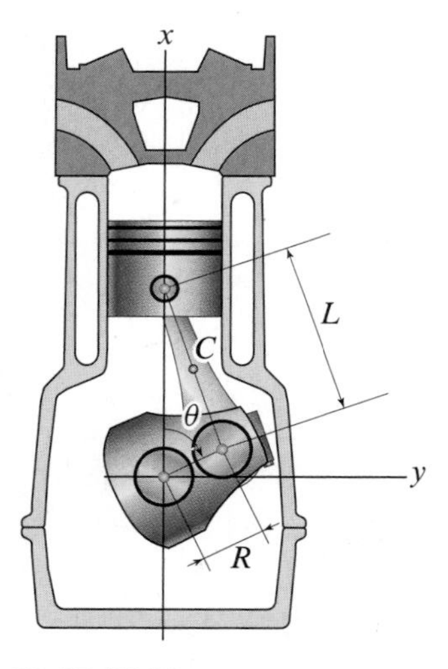

Figure P2.37–P2.38

2.2 Elementary Motions

We now examine how to relate acceleration to position and velocity in a variety of applications. To focus on how these relations are built, here we avoid dealing with vector quantities and we examine only one-dimensional motions.

Figure 2.10
Usain Bolt winning the gold medal in the 100 m sprint at the 2008 Beijing Olympic Games. In this race, each runner is in rectilinear motion.

Rectilinear motion relations

Rectilinear motion is motion that occurs along a straight line, though the relations that govern it are useful for describing other types of one-dimensional motions, even when the trajectory is not a straight line (see circular motion on p. 49). Referring to Fig. 2.11, we denote by $s(t)$ the rectilinear position coordinate of a point P. The velocity and acceleration of P will be denoted by $v(t) = \dot{s}(t)$ and $a(t) = \ddot{s}(t)$, respectively. Since physical measurements and Newton's second law usually provide us with data of the form $a(t)$, $a(v)$, or $a(s)$, we now investigate how to relate acceleration information of these types to velocity and position for rectilinear motions.

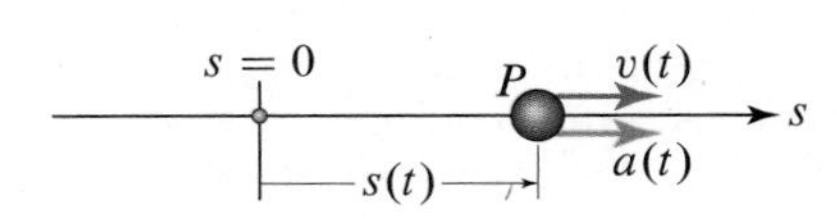

Figure 2.11
The position coordinate $s(t)$ for rectilinear motion.

If $a(t)$ is known

If we know the acceleration as a function of time $a(t)$, then we can determine $v(t)$ and $s(t)$ by time integration. We rewrite $a(t) = dv/dt$ as $dv = a(t)\,dt$, and letting $v = v_0$ for $t = t_0$, we have

$$\int_{v_0}^{v} dv = \int_{t_0}^{t} a(t)\,dt, \tag{2.19}$$

or

$$\boxed{v(t) = v_0 + \int_{t_0}^{t} a(t)\,dt.} \tag{2.20}$$

Rewriting $v(t) = ds/dt$ as $ds = v(t)\,dt$ and letting $s = s_0$ for $t = t_0$, we can determine $s(t)$ from Eq. (2.20) as follows:

$$\int_{s_0}^{s} ds = \int_{t_0}^{t} v(t)\,dt = \int_{t_0}^{t} \left[v_0 + \int_{t_0}^{t} a(t)\,dt \right] dt, \tag{2.21}$$

or

$$\boxed{s(t) = s_0 + v_0(t - t_0) + \int_{t_0}^{t} \left[\int_{t_0}^{t} a(t)\,dt \right] dt.} \tag{2.22}$$

If $a(v)$ is known

If we know the acceleration as a function of velocity $a(v)$, then finding v and s by integration is slightly more complicated. Starting from $a = dv/dt$, when $a(v)$ is given, we can separate the v and t variables by writing

$$dt = \frac{dv}{a(v)}. \tag{2.23}$$

If $a(v) \neq 0$ during the time interval considered, we can integrate Eq. (2.23) as follows:

$$\int_{t_0}^{t} dt = \int_{v_0}^{v} \frac{1}{a(v)}\,dv \quad \Rightarrow \quad \boxed{t(v) = t_0 + \int_{v_0}^{v} \frac{1}{a(v)}\,dv,} \tag{2.24}$$

where v_0 is the value of v for $t = t_0$. While the last of Eqs. (2.24) gives time as a function of velocity, sometimes we can invert $t(v)$ to find $v(t)$.

To compute the position we can try to solve the last of Eqs. (2.24) for $v(t)$. Then we can try to integrate $v(t)$ with respect to t to obtain $s(t)$ as was done in Eq. (2.22). Unfortunately, this is often difficult or impossible to do. Alternatively, we can obtain $s(v)$ instead of $s(t)$, using the chain rule of calculus, i.e.,

$$a = \frac{dv}{dt} = \frac{dv}{ds}\frac{ds}{dt} = v\,\frac{dv}{ds}. \tag{2.25}$$

Then, recalling that we have $a(v)$, we can separate the variables s and v as follows:

$$ds = \frac{v}{a(v)}\,dv. \tag{2.26}$$

Setting $s = s_0$ when $v = v_0$, Eq. (2.26) can be integrated to obtain

$$\int_{s_0}^{s} ds = \int_{v_0}^{v} \frac{v}{a(v)}\,dv \quad \Rightarrow \quad s(v) = s_0 + \int_{v_0}^{v} \frac{v}{a(v)}\,dv. \tag{2.27}$$

Helpful Information

The chain rule. Since it is often used, let's look more closely at the chain rule. Taking a bit of liberty, the chain rule can be presented as follows:

$$\frac{d(\text{Groucho})}{d(\text{Harpo})} = \frac{d(\text{Groucho})}{d(\text{Zeppo})}\frac{d(\text{Zeppo})}{d(\text{Harpo})}.$$

Although Zeppo did not appear on the left-hand side of the equation above, we were able to force him to appear on the right-hand side. This chain-rule-based "trick" will come in handy. The reason for using the Marx Brothers in this example is that the chain rule works only if all its terms are related to one another (mathematically, they must be functions of one another). Now, the connection with Eq. (2.25) is that we needed to make the variable s come into the picture even though, at first, it was not present in $a = dv/dt$. Therefore, we made v pose as our Groucho and t as our Harpo. Letting s take on the role of Zeppo, we were able to accomplish what we wanted.

If $a(s)$ is known

When the acceleration is known as a function of position, i.e., $a = a(s)$, we can again start from $a = v\,dv/ds$ given in Eq. (2.25). However, since we have $a(s)$, this time the correct separation of variables gives us $v\,dv = a(s)\,ds$. Then, letting $v = v_0$ for $s = s_0$, we can obtain the velocity as a function of position $v(s)$ as follows:

$$\int_{v_0}^{v} v\,dv = \int_{s_0}^{s} a(s)\,ds \quad \Rightarrow \quad \tfrac{1}{2}v^2 - \tfrac{1}{2}v_0^2 = \int_{s_0}^{s} a(s)\,ds, \tag{2.28}$$

or

$$v^2(s) = v_0^2 + 2\int_{s_0}^{s} a(s)\,ds. \tag{2.29}$$

Finally, once $v(s)$ is known through Eq. (2.29), we can obtain time as a function of position $t(s)$ starting from $v = ds/dt$. Then, recalling that we have $v(s)$, we can separate the variables t and s as:

$$dt = \frac{ds}{v(s)} \quad \Rightarrow \quad \int_{t_0}^{t} dt = \int_{s_0}^{s} \frac{ds}{v(s)}, \tag{2.30}$$

where we have again let $s = s_0$ when $t = t_0$. Completing the integration of the left-hand side, we obtain

$$t(s) = t_0 + \int_{s_0}^{s} \frac{ds}{v(s)}. \tag{2.31}$$

What if a is constant?

If the acceleration is a constant, then the equations we have derived simplify substantially. The constant acceleration relations are important because there are many problems in dynamics in which the acceleration is constant. For example, in studying the motion of a projectile, we generally assume that the projectile's acceleration is constant.

If the *acceleration is a constant* a_c, Eq. (2.20) becomes

$$v = v_0 + a_c(t - t_0) \quad \text{(constant acceleration)}, \tag{2.32}$$

Eq. (2.22) becomes

$$s = s_0 + v_0(t - t_0) + \tfrac{1}{2}a_c(t - t_0)^2 \quad \text{(constant acceleration)}, \tag{2.33}$$

and Eq. (2.29) becomes

$$v^2 = v_0^2 + 2a_c(s - s_0) \quad \text{(constant acceleration)}. \tag{2.34}$$

Circular motion and angular velocity

The relationships for rectilinear motion are applicable to any one-dimensional motion. To demonstrate this idea, we will now apply them to a common one-dimensional curvilinear motion: *circular motion*.

In Fig. 2.12, a particle A is moving in a circle of radius r and center O. Since r is constant, the position of A can be described with a single coordinate, such as the oriented arc length s or the angle θ. If the line OA rotates through the angle $\Delta\theta$ in the time Δt, then we can define an average time rate of change of the angle θ as $\omega_{\text{avg}} = \Delta\theta/\Delta t$. Letting $\Delta t \to 0$, we obtain the instantaneous time rate of change of θ, i.e., $\dot{\theta}$, called the *angular velocity*, as

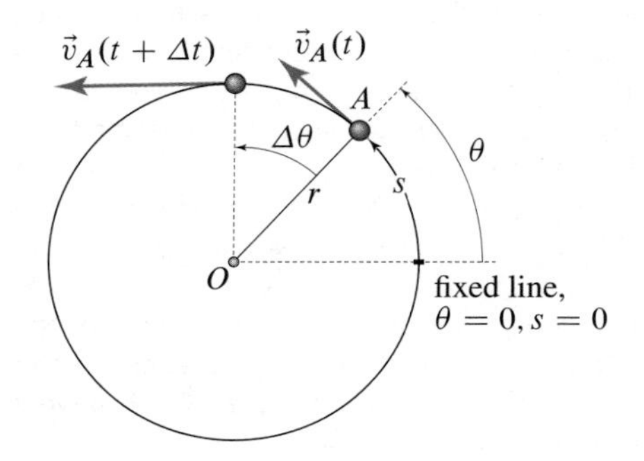

Figure 2.12
Particle A with speed $v_A = |\vec{v}_A|$ moving in a circle of radius r centered at O.

$$\omega(t) = \lim_{\Delta t \to 0} \frac{\Delta\theta}{\Delta t} = \frac{d\theta(t)}{dt} = \dot{\theta}(t). \tag{2.35}$$

We can then define *angular acceleration* α by differentiating Eq. (2.35) with respect to time, i.e.,

$$\alpha(t) = \frac{d\omega(t)}{dt} = \dot{\omega}(t) = \ddot{\theta}(t). \tag{2.36}$$

When using the coordinate s, since $s = r\theta$ and r is constant, we can write

$$\dot{s} = r\dot{\theta} = \omega r \quad \text{and} \quad \ddot{s} = r\ddot{\theta} = \alpha r. \tag{2.37}$$

Circular motion relations

All of the relationships we developed for rectilinear motion apply equally well to circular motion, except that we need to replace the rectilinear variables with their circular counterparts. For example, Eq. (2.20) becomes

$$\omega(t) = \omega_0 + \int_{t_0}^{t} \alpha(t)\,dt. \tag{2.38}$$

Table 2.1 lists each kinematic variable in rectilinear motion and the corresponding kinematic variable for circular motion. Replacing each rectilinear motion variable with its circular motion counterpart in Eqs. (2.20)–(2.34), we obtain the corresponding circular motion equations. Finally, if the angular acceleration is constant, we can use the constant acceleration relations with a_c replaced by α_c.

Table 2.1
Correspondence of kinematic variables between rectilinear and circular motion.

Kinematic variable	Rectilinear motion	Circular motion[a]
time	t	t
position	s	θ
velocity	v	ω
acceleration	a	α

[a] Except for time, each of these should have the word *angular* in front of its kinematic variable name.

End of Section Summary

In this section we have developed relationships that link a single coordinate and its time derivatives. Note that, except for the constant acceleration relations, they are rarely applied directly, and the acceleration is integrated as was done in the development of these equations.

1. If the acceleration is provided as a function of time, i.e., $a = a(t)$, for velocity and position, we have

Eqs. (2.20) and (2.22), p. 47

$$v(t) = v_0 + \int_{t_0}^{t} a(t)\,dt,$$

$$s(t) = s_0 + v_0(t - t_0) + \int_{t_0}^{t} \left[\int_{t_0}^{t} a(t)\,dt \right] dt.$$

2. If the acceleration is provided as a function of velocity, i.e., $a = a(v)$, for time and position, we have

Eq. (2.24), p. 47, and Eq. (2.27), p. 48

$$t(v) = t_0 + \int_{v_0}^{v} \frac{1}{a(v)}\,dv,$$

$$s(v) = s_0 + \int_{v_0}^{v} \frac{v}{a(v)}\,dv.$$

3. If the acceleration is provided as a function of position, i.e., $a = a(s)$, for velocity and time, we have

Eqs. (2.29) and (2.31), p. 48

$$v^2(s) = v_0^2 + 2\int_{s_0}^{s} a(s)\,ds,$$

$$t(s) = t_0 + \int_{s_0}^{s} \frac{ds}{v(s)}.$$

4. If the acceleration is a constant a_c, for velocity and position, we have

Eqs. (2.32)–(2.34), p. 49

$$v = v_0 + a_c(t - t_0),$$

$$s = s_0 + v_0(t - t_0) + \tfrac{1}{2}a_c(t - t_0)^2,$$

$$v^2 = v_0^2 + 2a_c(s - s_0).$$

Circular motion. For circular motion, the equations summarized in items 1–4 above hold as long as we use the replacement rules

$$s \to \theta, \quad v \to \omega, \quad a \to \alpha,$$

where $\omega = \dot{\theta}$ and $\alpha = \ddot{\theta}$ are the *angular velocity* and *angular acceleration*, respectively.

EXAMPLE 2.4 *Acceleration as a Function of Time*

A rocket sled accelerates from rest along a straight rail (Fig. 1). The sled's acceleration is $a = \beta t^2$, where β is a constant and t is time in seconds. The sled achieves a final speed $v_f = 180$ mph after traveling a distance $d = 300$ ft. Determine β and the time required to achieve v_f.

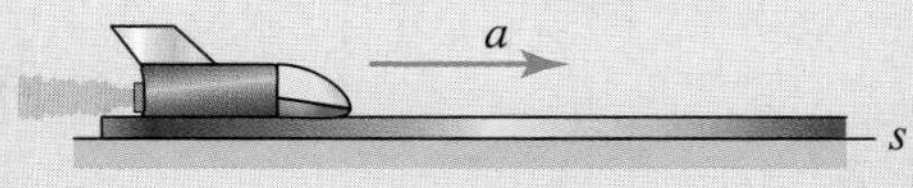

Figure 1
Rocket sled on rectilinear rail.

SOLUTION

Road Map The acceleration a is given as a function of time and needs to be related to speed and position. The speed is obtained from the velocity, which is obtained by integrating a with respect to time. Once the velocity is known as a function of time, the position is obtained by integrating the velocity with respect to time.

Computation Since $a = dv/dt$, where v denotes the velocity, and since the acceleration is a known function of time, we can separate the variables v and t as follows: $dv = a(t)\,dt$. Using the given expression for $a(t)$, we can then write

$$dv = \beta t^2\,dt \quad \Rightarrow \quad \int_0^v dv = \int_0^t \beta t^2\,dt \quad \Rightarrow \quad v = \tfrac{1}{3}\beta t^3, \tag{1}$$

where the limits of integration reflect that $v = 0$ for $t = 0$ (the sled starts from rest). Next we recall that $v = ds/dt$, where, referring to Fig. 1, s denotes the position. Hence, we can write $ds = v\,dt$, which, using the last of Eqs. (1) gives

$$ds = \tfrac{1}{3}\beta t^3\,dt. \tag{2}$$

Letting s_0 denote the position at time $t = 0$, we can integrate Eq. (2) as follows:

$$\int_{s_0}^{s} ds = \int_0^t \tfrac{1}{3}\beta t^3\,dt \quad \Rightarrow \quad s - s_0 = \tfrac{1}{12}\beta t^4. \tag{3}$$

Let t_f denote the time at which $v = v_f$ and $s - s_0 = d$. Then, for $t = t_f$, the last of Eqs. (1) and the last of Eqs. (3) become

$$v_f = \tfrac{1}{3}\beta t_f^3 \quad \text{and} \quad d = \tfrac{1}{12}\beta t_f^4. \tag{4}$$

Equations (4) form a system of two equations in the two unknowns β and t_f, whose solution is

$$\beta = \frac{3v_f^4}{64d^3} \quad \text{and} \quad t_f = 4\frac{d}{v_f}. \tag{5}$$

Recalling that $d = 300$ ft and $v_f = 180$ mph $= 264.0$ ft/s, we can evaluate the expressions in Eqs. (5) to obtain

$$\boxed{\beta = 8.433\ \text{ft/s}^4 \quad \text{and} \quad t_f = 4.545\ \text{s}.} \tag{6}$$

Discussion & Verification Since $a = \beta t^2$, β has dimensions of length over time to the fourth power. Also, t_f has dimensions of time. Hence, our results have the correct dimensions and units. To check whether or not the numerical results are reasonable, we can find the time t_f under constant acceleration conditions and then compare this result to what we have obtained. Under constant acceleration, $t_f = 2d/v_f$, i.e., half of what we obtained. This is consistent with the fact that under constant acceleration the velocity increases linearly with time, whereas in our problem the velocity increases cubically. Therefore, our result can be considered acceptable. In turn, assuming that we have not made any algebra mistakes, we can say that the corresponding result for β is also acceptable.

EXAMPLE 2.5 *Acceleration as a Function of Velocity*

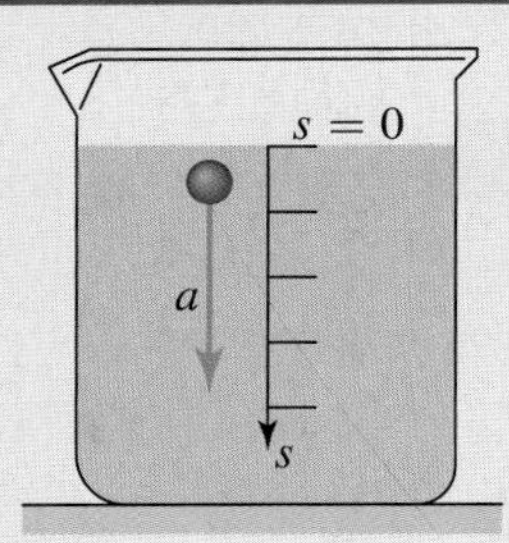

Figure 1
A sphere sinking in a beaker filled with a polymer fluid.

A sphere is dropped from rest at $s = 0$ in a thick polymer fluid (say, shampoo). A typical model for the motion of a sphere in such a fluid tells us that the acceleration of the sphere has the form $a = g - \eta v$, where g is the acceleration due to gravity, v is the velocity, and η is a constant coefficient with dimensions of one over time. If $\eta = 30\,\text{s}^{-1}$, determine the time it takes for the velocity of the sphere to become equal to 1 ft/s. Also, determine the corresponding position of the sphere.

SOLUTION

Road Map To relate a change in velocity to a corresponding time interval, we will apply the definition of acceleration, since, in such a definition, velocity and time appear directly. To relate a change in position to a corresponding change in velocity, we rewrite the acceleration in such a way that the expression for the acceleration is in terms of velocity and position. This can be done by applying the chain rule.

Computation Recalling that $a = dv/dt$, since the acceleration is a given function of velocity, we can separate the variable v and t as follows: $dt = dv/a(v)$. Then, using the given expression for the acceleration, we have

$$dt = \frac{dv}{g - \eta v}. \tag{1}$$

Letting t_f be the time at which $v = v_f = 1$ ft/s, and recalling that $v = 0$ for $t = 0$, we can integrate Eq. (1) as follows:

$$\int_0^{t_f} dt = \int_0^{v_f} \frac{dv}{g - \eta v} \quad \Rightarrow \quad t_f = -\frac{1}{\eta}\big[\ln(g - \eta v_f) - \ln g\big]. \tag{2}$$

Recalling that $\eta = 30\,\text{s}^{-1}$, $g = 32.2\,\text{ft/s}^2$, and $v_f = 1$ ft/s (the subscript f stands for final), we can evaluate the last of Eqs. (2) to obtain

$$\boxed{t_f = 0.08945\,\text{s}.} \tag{3}$$

To determine the position of the sphere corresponding to $v = v_f$, we use the chain rule to express the acceleration as $a = v\,dv/ds$, which allows us to write

$$g - \eta v = v\frac{dv}{ds} \quad \Rightarrow \quad ds = \frac{v}{g - \eta v}\,dv \quad \Rightarrow \quad ds = \left(-\frac{1}{\eta} + \frac{g/\eta}{g - \eta v}\right)dv, \tag{4}$$

where in the second of Eqs. (4) we have separated the variables s and v, and in the last of Eqs. (4) we have expressed the function $v/(g - \eta v)$ in a way that is convenient for integration. Recalling that $v = 0$ for $s = 0$, and letting s_f denote the position for $v = v_f$, we can integrate the last of Eqs. (4) as follows:

$$\int_0^{s_f} ds = \int_0^{v_f}\left(-\frac{1}{\eta} + \frac{g/\eta}{g - \eta v}\right)dv \quad \Rightarrow \quad s_f = -\frac{v_f}{\eta} - \frac{g}{\eta^2}\big[\ln(g - \eta v_f) - \ln g\big]. \tag{5}$$

Recalling again that $\eta = 30\,\text{s}^{-1}$, $g = 32.2\,\text{ft/s}^2$, and $v_f = 1$ ft/s, we can evaluate the last of Eqs. (5) to obtain

$$\boxed{s_f = 0.06268\,\text{ft}.} \tag{6}$$

Discussion & Verification Our results have the correct dimensions and units. As for their numerical values, we expect that to achieve the same velocity $v_f = 1$ ft/s, a sphere sinking in a thick fluid will take longer and will need to sink deeper than a sphere falling freely due to gravity. In this latter case, the acceleration of the sphere would be equal to g and t_f and s_f would be $v_f/g = 0.03106$ s and $v_f^2/(2g) = 0.01553$ ft, respectively, which are indeed smaller than the results we have obtained in Eqs. (3) and (6), respectively.

EXAMPLE 2.6 *Acceleration as a Function of Position*

In a braking test, a car traveling at a speed $v_0 = 80\,\text{km/h}$ is made to apply the brakes so as to lock the wheels and slide. The pavement has a roughness that increases as the car moves to the right. Letting $s = 0$ characterize the location at which sliding begins, the pavement's roughness is such that the acceleration of the car is $a = -g\mu_0[1 + (s/\lambda)]$, where g is the acceleration due to gravity, $\mu_0 = 0.6$, and λ is a constant with dimensions of length. Determine λ knowing that the car stops after a distance $d = 30\,\text{m}$.

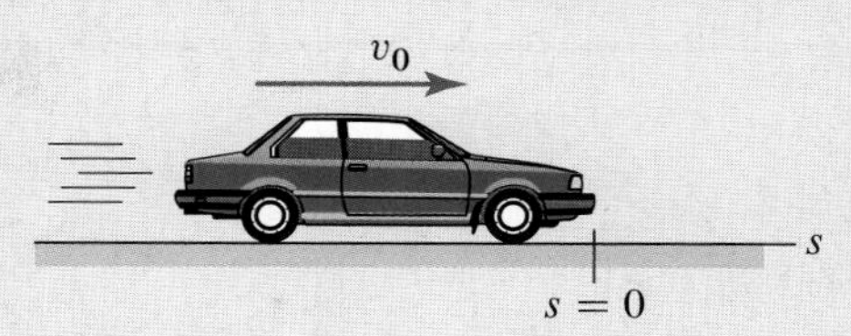

Figure 1

SOLUTION

Road Map We can determine λ by relating the acceleration to the braking distance. Since the acceleration is given as a function of position, we first apply the chain rule to rewrite the acceleration in terms of velocity and position. Then we will be able to relate the change in velocity to the corresponding change in position, thus obtaining an equation we can solve for λ.

Computation By definition, $a = dv/dt$, where v is the velocity. Hence, making use of the chain rule, we have

$$a = \frac{dv}{dt} = \frac{dv}{ds}\frac{ds}{dt} = v\frac{dv}{ds} \quad \Rightarrow \quad -g\mu_0\left[1 + \frac{s}{\lambda}\right] = v\frac{dv}{ds}, \tag{1}$$

where we have used the given expression for a and the fact that $v = ds/dt$. The last of Eqs. (1) can be rewritten to separate the variables s and v, i.e.,

$$-\left[g\mu_0 + \frac{g\mu_0}{\lambda}s\right]ds = v\,dv, \tag{2}$$

where we have expanded the product on the left-hand side of the last of Eqs. (1). We now observe that $s = 0$ when $v = v_0$ and that $s = d$ when $v = 0$. Hence, the expression in Eq. (2) can be integrated as follows:

$$-\int_0^d \left[g\mu_0 + \frac{g\mu_0}{\lambda}s\right]ds = \int_{v_0}^0 v\,dv \quad \Rightarrow \quad -g\mu_0 d - \frac{g\mu_0 d^2}{2\lambda} = -\tfrac{1}{2}v_0^2. \tag{3}$$

The last of Eqs. (3) can be solved for λ to obtain

$$\lambda = \frac{g\mu_0 d^2}{v_0^2 - 2g\mu_0 d}. \tag{4}$$

Recalling that $g = 9.81\,\text{m/s}^2$, $\mu_0 = 0.6$, $d = 30\,\text{m}$, and $v_0 = 80\,\text{km/h} = 22.22\,\text{m/s}$, we can evaluate the expression in Eq. (4) to obtain

$$\boxed{\lambda = 37.66\,\text{m}.} \tag{5}$$

Discussion & Verification The expression on the right-hand side of Eq. (4) has dimensions of length, as it should. To check whether or not the result is acceptable, we observe that Eq. (4) implies that $\lambda \to 0$ for $d \to 0$ and $\lambda \to \infty$ for $d \to v_0^2/(2g\mu_0)$, where $v_0^2/(2g\mu_0) = 41.95\,\text{m}$ can be shown to be the stopping distance if the acceleration of the car were constant and equal to $-g\mu_0$. Recalling that $a = -g\mu_0[1 + (s/\lambda)]$, this behavior is consistent with the fact that if $\lambda \to 0$, a tends to negative infinity and for $\lambda \to \infty$, a becomes equal to $-g\mu_0$. Therefore, overall, we can say that the value of λ we found is certainly within the range of admissible values for the parameter in question.

EXAMPLE 2.7 Measuring the Depth of a Well by Relating Time, Velocity, and Acceleration

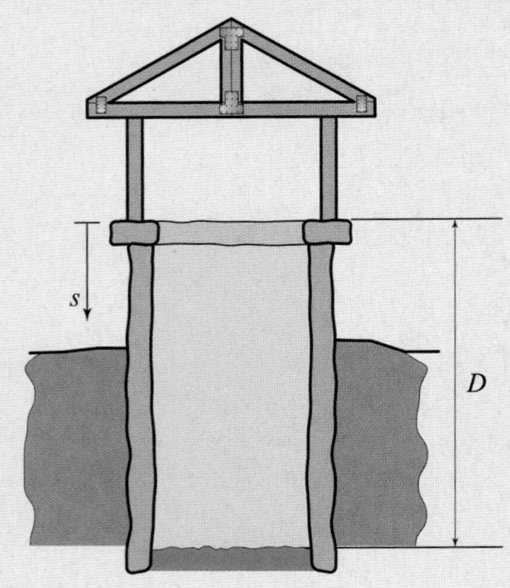

Figure 1
A well of depth D showing the positive direction of the coordinate s.

We can estimate the depth of a well by measuring the time it takes for a rock dropped from the top of the well to reach the water below. Assuming that gravity is the only force acting on the rock, estimate a well's depth under two different assumptions: the speed of sound is (a) finite and equal to $v_s = 340\,\text{m/s}$ and (b) infinite. Also, compare the two estimates to provide a "rule of thumb" as to when we can assume that the speed of sound is infinite.

SOLUTION

Road Map In this problem our time measure is the sum of two parts: (1) the time taken by the rock to go from the top to the bottom of the well and (2) the time taken by sound to go from the bottom to the top of the well. The well's depth can be related to the first time by assuming that the rock travels at a constant acceleration, namely, $g = 9.81\,\text{m/s}^2$. The well's depth can also be related to the second time by assuming that sound travels at a constant speed, which will be assumed to be finite in Part (a) and infinite in Part (b). By requiring that the two depth estimates be identical, we will be able to find a relation between the well's depth and the overall measured time.

Part (a): Finite Sound Speed

Computation Figure 1 shows a well of unknown depth D. Let t_m, t_i, and t_s be the (total) measured time, the time taken by the rock to fall the distance D and hit the water, and the time it takes sound to go back up, respectively, so that

$$t_m = t_i + t_s. \tag{1}$$

The motion of the rock falling the distance D is a rectilinear motion with constant acceleration $g = 9.81\,\text{m/s}^2$. Hence, by choosing a coordinate axis pointing from the top to the bottom of the well, noting that the rock starts at $s_0 = 0\,\text{m}$, and assuming that the rock is released with initial velocity v_0 equal to zero, Eq. (2.33) tells us that

$$D = \tfrac{1}{2} g t_i^2 \quad \Rightarrow \quad t_i = \sqrt{\frac{2D}{g}}. \tag{2}$$

As soon as the rock hits the water, a sound wave traveling with a constant velocity $v_s = 340\,\text{m/s}$, and therefore, with constant acceleration $a_s = 0\,\text{m/s}^2$, goes from the bottom of the well up to the observer's ear at $s = 0$. Hence, the time taken by the sound is

$$t_s = \frac{D}{v_s}. \tag{3}$$

Next, using the expressions for t_i and t_s in Eqs. (2) and (3), respectively, Eq. (1) becomes

$$t_m = \sqrt{\frac{2D}{g}} + \frac{D}{v_s}. \tag{4}$$

This equation can be solved for D to obtain (see the Helpful Information note in the margin for details)

$$\boxed{D = v_s t_m - \frac{v_s^2}{g}\left(\sqrt{1 + \frac{2 t_m g}{v_s}} - 1\right).} \tag{5}$$

> **Helpful Information**
>
> **Solving Eq. (4) for D.** To solve Eq. (4) for D, we first rewrite it to isolate the square root term, i.e.,
>
> $$D - v_s t_m = -v_s \sqrt{2D/g}.$$
>
> We then square each side to obtain
>
> $$D^2 - 2D v_s t_m + v_s^2 t_m^2 = v_s^2 \frac{2D}{g},$$
>
> which can be rearranged to read
>
> $$D^2 - 2v_s\left(t_m + \frac{v_s}{g}\right) D + v_s^2 t_m^2 = 0.$$
>
> This is a quadratic equation in D with the following two roots:
>
> $$D = v_s t_m + \frac{v_s^2}{g}\left(1 \pm \sqrt{1 + \frac{2 t_m g}{v_s}}\right).$$
>
> Only one of these roots is physically meaningful. The solution with the plus sign in front of the square root term yields a nonzero value for D when $t_m = 0$. This result contradicts Eq. (4), so the only acceptable solution is the one with the minus sign.

Part (b): Infinite Sound Speed

Computation If the speed of sound were infinite, Eq. (3) would imply that $t_s = 0$. Hence, from Eq. (1), we see that $t_m = t_i$ so that the first of Eqs. (2) yields

$$\boxed{D = \tfrac{1}{2} g t_m^2.} \tag{6}$$

Discussion & Verification The result in Eq. (5) is dimensionally correct. Since the given data t_m, g, and v_s have dimensions of time (T), length over time squared (L/T^2), and length over time (L/T), respectively, then note that the argument of the square root term in Eq. (5) is nondimensional, i.e.,

$$\left[\frac{2t_m g}{v_s}\right] = [t_m][g]\frac{1}{[v_s]} = T\frac{L}{T^2}\frac{T}{L} = 1. \tag{7}$$

Consequently, the dimensions of D in Eq. (5) are

$$[D] = \left[v_s t_m + \frac{v_s^2}{g}\right] = [v_s][t_m] + [v_s]^2\frac{1}{[g]} = \frac{L}{T}T + \frac{L^2}{T^2}\frac{T^2}{L} = L, \tag{8}$$

as expected. The result in Eq. (6) can be shown to be dimensionally correct in a similar manner.

A Closer Look We now compare the solutions with finite and infinite sound speed to understand under what conditions it is important to account for the finiteness of the speed of sound. Consider Fig. 2, which presents three curves derived under three different sets of assumptions (the curves have D on the horizontal axis to allow us to more easily make comments based on the well's depth). Figure 2 not only shows the functions in Eqs. (5) and (6), but also the solution we would obtain by taking into account both the finiteness of the speed of sound *and air resistance*. This latter curve was obtained by assuming that the rock used for the measurement (1) is spherical with a radius $r = 1$ cm and (2) is made out of granite, with a density of $2.75\,\mathrm{g/cm^3}$; and it is subject to an aerodynamic drag force given by $F_D = C_D \rho A v^2/2$, where the dimensionless drag coefficient C_D was chosen to be equal to 0.3, ρ is the density of air at ground level, and A is the frontal area of the spherical rock.* Treating the curve obtained by this drag model as the "true" relation between D and t_m, observe that the red curve, representing Eq. (5), and the black curve, corresponding to Eq. (6), diverge from the true curve as the depth of the well increases. However, the three curves essentially coincide near the origin of the plot. So estimating the depth of a well while disregarding air resistance and the finiteness of the speed of sound is not that bad for shallow wells, where one could define *shallow* to mean, say, less than about 30 m. The second conclusion we can draw is that, by accounting for the finiteness of the speed of sound, our formula can now be applied for depths all the way up to 80 m without having to resort to complex theories of rock-air interaction. Finally, notice that the curves provided here allow us to get a quantitative appreciation for the error we would make in estimating the depth of the well depending on the curve used. For example, consider the case in which we measure $t_m = 4$ s. How deep is the well? Well, according to the red curve, the depth is roughly 70.5 m, whereas the black line indicates a depth of 78.5 m. Therefore, we could go with the estimate given by the black line (since it is easier to compute) with the knowledge that the error is of the order of 12%.

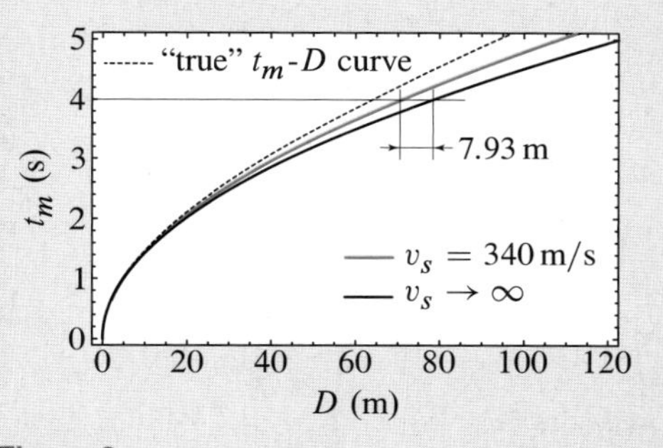

Figure 2
Measured time vs. well depth curves for various sets of assumptions.

* This formula for the aerodynamic drag is often discussed in fluid mechanics courses.

EXAMPLE 2.8 *Acceleration Function of Velocity: Descent of a Skydiver*

Figure 1
A skydiver descending.

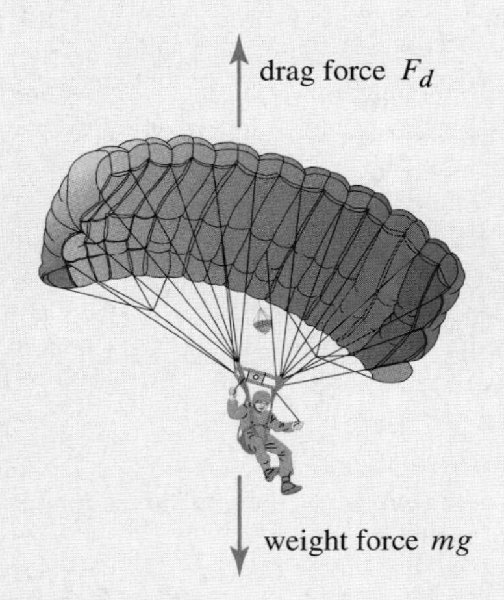

Figure 2
Skydiver with drag and weight forces depicted.

The skydiver shown in Figs. 1 and 2 has deployed his parachute after free-falling at $v_0 = 44.5\,\text{m/s}$. As we will learn in Chapter 3, if we model the drag force F_d due to air resistance as being proportional to the square of the skydiver's velocity, Newton's second law tells us that the skydiver's acceleration is $a = g - C_d v^2/m$, where g is the acceleration due to gravity, m is the skydiver's mass, and C_d is a constant drag coefficient.* Letting $C_d = 43.2\,\text{kg/m}$, $m = 110\,\text{kg}$, and $g = 9.81\,\text{m/s}^2$, determine the skydiver's velocity as a function of time. In addition, determine the skydiver's terminal velocity.

SOLUTION

Part (a): From Acceleration to Velocity

Road Map Since $a = dv/dt$ and the acceleration is of the form $a = a(v)$, we obtain an expression of the type $dt = dv/a(v)$. Therefore, in this problem we first obtain time as a function of velocity, i.e., $t = t(v)$, and then we will try to invert this relationship to obtain $v = v(t)$. This is the strategy followed in developing Eq. (2.24) for the case when $a = a(v)$.

Computation As discussed in the Road Map, starting with $dt = dv/a(v)$ and using the given expression for $a(v)$, we have

$$dt = \frac{dv}{g - \frac{C_d}{m}v^2} \quad \Rightarrow \quad \int_0^t dt = \int_{v_0}^{v} \frac{dv}{g - \frac{C_d}{m}v^2} \quad \Rightarrow \quad t = \int_{v_0}^{v} \frac{dv}{g - \frac{C_d}{m}v^2}, \tag{1}$$

where we have used the fact that $v = v_0$ for $t = 0$. To carry out the integral on the right-hand side of the last of Eqs. (1), we can factor out the term C_d/m at the denominator and rewrite the integrand as follows:

$$\frac{1}{g - \frac{C_d}{m}v^2} = \frac{m}{C_d}\frac{1}{\frac{mg}{C_d} - v^2}. \tag{2}$$

We observe that the denominator of the last fraction in Eq. (2) contains the term mg/C_d, which has the same dimensions as a velocity squared. Hence, we define the quantity

$$\tilde{v} = \sqrt{\frac{mg}{C_d}}, \tag{3}$$

with the same dimensions as a velocity, and rewrite Eq. (2) as follows:

$$\frac{1}{g - \frac{C_d}{m}v^2} = \frac{m}{C_d}\frac{1}{\tilde{v}^2 - v^2} = \frac{m}{2\tilde{v}C_d}\left(\frac{1}{\tilde{v} - v} + \frac{1}{\tilde{v} + v}\right). \tag{4}$$

Substituting the last expression in Eq. (4) into the last of Eqs. (1) and integrating, we have

$$t = -\frac{m}{2\tilde{v}C_d}\left[\ln(\tilde{v} - v)\Big|_{v_0}^{v} - \ln(\tilde{v} + v)\Big|_{v_0}^{v}\right]. \tag{5}$$

Recalling that $\ln a - \ln b = \ln(a/b)$, Eq. (5) can be simplified to read

$$t = -\frac{m}{2\tilde{v}C_d}\ln\left(\frac{\tilde{v} - v}{\tilde{v} + v}\frac{\tilde{v} + v_0}{\tilde{v} - v_0}\right). \tag{6}$$

As anticipated in the Road Map, we now have an expression for time as a function of velocity. To obtain the desired result, namely, velocity as function of time, we need to solve

* The coefficient C_d is related to the coefficient C_D in Example 2.7 as follows: $C_d = \frac{1}{2}C_D\rho A$.

Eq. (6) for v. To do so, we first isolate the logarithmic term and then take the exponential of both sides of the equation, i.e.,

$$-\frac{2\tilde{v}C_d}{m}t = \ln\left(\frac{\tilde{v}-v}{\tilde{v}+v}\frac{\tilde{v}+v_0}{\tilde{v}-v_0}\right) \quad \Rightarrow \quad e^{-(2\tilde{v}C_d/m)t} = \frac{\tilde{v}-v}{\tilde{v}+v}\frac{\tilde{v}+v_0}{\tilde{v}-v_0}. \tag{7}$$

The last of Eqs. (7) can be solved for v to obtain

$$\boxed{v = \tilde{v}\frac{\tilde{v}+v_0-(\tilde{v}-v_0)e^{-(2\tilde{v}C_d/m)t}}{\tilde{v}+v_0+(\tilde{v}-v_0)e^{-(2\tilde{v}C_d/m)t}}.} \tag{8}$$

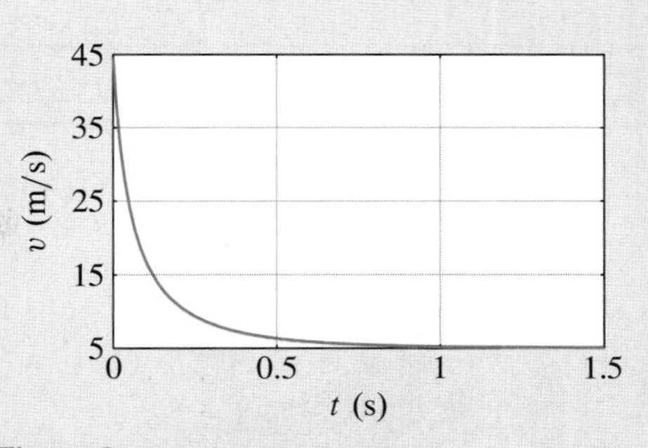

Figure 3
Velocity of the skydiver as he descends with his parachute deployed.

The expression in Eq. (8) has been plotted in Fig. 3 for the parameters given. Notice that the skydiver starts out at 44.5 m/s at $t = 0$ s and quickly (in about one second) slows down to approximately 5 m/s.

Part (b): Terminal Velocity

Road Map The *terminal velocity* is defined as the velocity reached after an infinite amount of time. Therefore, we can find the terminal velocity by examining the behavior of Eq. (8) as t grows larger and larger.

Computation As t becomes larger and larger, the two exponential terms in Eq. (8) go to zero, and the expression for v simplifies to the following expression: $v = \tilde{v}$. From Eq. (3), we recall that $\tilde{v} = \sqrt{mg/C_d}$, and therefore, we can conclude that

$$\boxed{v_{\text{term}} = \sqrt{\frac{mg}{C_d}} = 4.998\,\text{m/s},} \tag{9}$$

where we have used the fact that $m = 110$ kg, $g = 9.81\,\text{m/s}^2$, and $C_d = 43.2$ kg/m. This agrees with the plot in Fig. 3, which shows that for larger time values the velocity tends to become constant and equal to the value 5 m/s.

Helpful Information

Another commonly accepted and consistent definition of *terminal velocity*. When a body is in free fall in a medium such as air (or water), the body experiences an aerodynamic (or fluid dynamic) resistance, called *drag*, that opposes gravity and increases with speed. If the body falls for enough time, the aerodynamic drag will end up equilibrating the force of gravity and the body will stop accelerating. The *terminal velocity* can be defined as the value of the velocity at which the acceleration becomes equal to zero.

Discussion & Verification In defining $\tilde{v}$ we had observed that this quantity has the same dimensions as velocity. Using this fact, we can then readily verify that Eq. (8) has the correct dimensions since the fraction on the right-hand side is nondimensional. Also, the fact that a finite terminal velocity exists does match experience, since we observe that bodies free falling in air stop accelerating after some time (provided they can fall for long enough time).

EXAMPLE 2.9 *Constant Angular Acceleration: Propeller and Supersonic Effects*

Figure 1
V-22 Osprey aircraft.

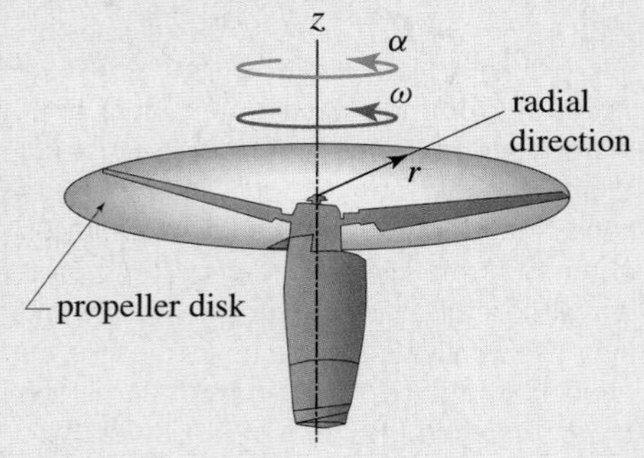

Figure 2
View of one of the engines and its companion propeller of the V-22 Osprey. The radius of the V-22 propellers is 19 ft.

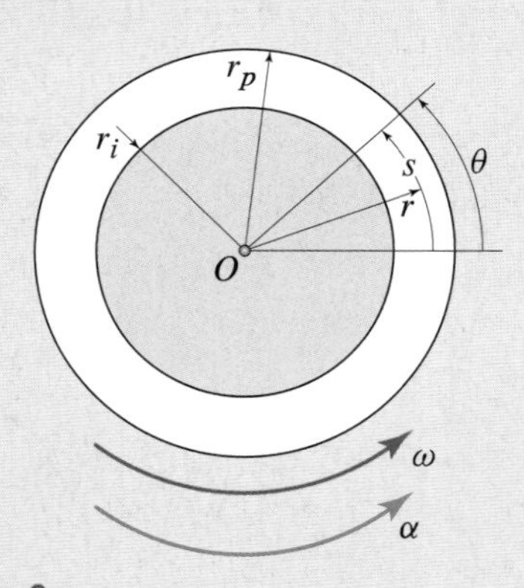

Figure 3
Definition of the coordinates s of a point a distance r from the center of rotation O. Using the angular coordinate θ, we have that $s = r\theta$. The shaded circle has an area equal to 50% of the total area of the propeller disk.

In Fig. 2, the propeller shown has radius $r_p = 19\,\text{ft}$, and it rotates about its axis while keeping the propeller disk stationary.* Suppose that the propeller starts from rest with a constant angular acceleration $\alpha = 50\,\text{rad/s}^2$.† Knowing that the speed of sound at sea level under standard conditions is $v_s = 1130\,\text{ft/s}$, find

(a) ω_s, the angular speed of the propeller, expressed in rpm, at which 50% of the area of the propeller disk operates in the supersonic regime.

(b) t_s, the time it takes to achieve ω_s.

(c) θ_s, the number of revolutions experienced by the propeller in going from rest to ω_s.

SOLUTION

Road Map As the propeller spins up, different points along a propeller's blade experience different speeds. This is so because the motion of each point on the propeller is circular, and while the angular velocity and acceleration for this motion are the same for all points, the distance from the axis of rotation is not. The overall motion is a constant *angular* acceleration motion, and we can use the formulas derived for this case.

Part (a): Calculation of ω_s

Computation Referring to Fig. 3 and using Eq. (2.37) to describe the velocity of points in circular motion, we have that the speed $|\dot{s}|$ of a point at a distance r from the spin axis is

$$|\dot{s}| = r|\omega|, \tag{1}$$

which shows that $|\dot{s}|$ is proportional to the distance from the axis of rotation. Hence, if a point interior to the propeller disk moves at supersonic speeds, all of the points between it and the propeller's periphery will also move at supersonic speeds. Letting r_i be the radius of the inner disk of the propeller whose area is 50% of the total disk area (see Fig. 3) yields

$$\pi r_i^2 = \frac{\pi r_p^2}{2} \quad \Rightarrow \quad r_i = \frac{r_p}{\sqrt{2}} = 13.44\,\text{ft}. \tag{2}$$

If points on the circle of radius r_i have achieved the speed of sound, then combining Eq. (1) and the second of Eqs. (2), we have

$$v_s = r_i \omega_s \quad \Rightarrow \quad \boxed{\omega_s = \frac{v_s\sqrt{2}}{r_p} = 84.11\,\text{rad/s} = 803.2\,\text{rpm},} \tag{3}$$

where $\omega_s > 0$ since it is an angular *speed*.

Part (b): Calculation of t_s

Computation To establish how long it takes to achieve ω_s, recall that the angular acceleration α is the time derivative of the angular velocity ω, i.e., $\alpha = \dot{\omega} = d\omega/dt$, so that we can write $d\omega = \alpha\,dt$. Therefore, observing that α is constant and letting $\omega_0 = 0$ be the initial angular velocity (the propeller starts from rest), we have

$$\int_{\omega_0}^{\omega} d\omega = \int_0^{t_s} \alpha\,dt \quad \Rightarrow \quad \omega_s = \omega_0 + \int_0^{t_s} \alpha\,dt \quad \Rightarrow \quad \omega_s = \alpha t_s. \tag{4}$$

Combining the results in Eqs. (3) and (4), we have

$$\boxed{t_s = \frac{v_s\sqrt{2}}{\alpha r_p} = 1.682\,\text{s}.} \tag{5}$$

* The propeller disk is the disk spanned by the propeller blades as they rotate.

† This is roughly what it takes to go from 0 to 1430 rpm in 3 s and is therefore, very easy to achieve even with an average small-car engine.

Part (c): Calculation of θ_s

Since a *revolution* is an angular displacement equal to 2π rad, we can calculate the number of revolutions experienced by the propeller by computing the difference in the propeller's angular position θ between $t = 0$ and $t = t_s$. Since the angular acceleration is constant, we can use Eq. (2.32) along with Table 2.1 to arrive at the following relationship between ω_s, α, and θ_s:

$$\omega_s^2 = \omega_0^2 + 2\alpha(\theta_s - \theta_0), \tag{6}$$

where θ_0 is the angular position of a point on the propeller disk at $t = 0$. Since all points on the propeller disk experience the same angular displacement, by choosing a point with $\theta_0 = 0$ and recalling $\omega_0 = 0$, Eq. (2.1) can be solved to obtain

$$\boxed{\theta_s = \frac{\omega_s^2}{2\alpha} = 70.74\,\text{rad} = 11.26\,\text{rev}.} \tag{7}$$

Discussion & Verification Let L and T denote dimensions of length and time, respectively. Then, referring to Eq. (3), we see that the dimensions of ω_s are given by

$$[\omega_s] = \left[\frac{v_s}{r_p}\right] = [v_s]\frac{1}{[r_p]} = \frac{L}{T}\frac{1}{L} = \frac{1}{T}, \tag{8}$$

as expected. In addition, recalling that a radian is nondimensional, we see that ω_s has the right dimensions and is expressed in appropriate units. Next, referring to Eq. (5), the dimensions of t_s are given by

$$[t_s] = \left[\frac{v_s}{\alpha r_p}\right] = [v_s]\frac{1}{[\alpha]}\frac{1}{[r_p]} = \frac{L}{T}\frac{1}{T^{-2}}\frac{1}{L} = T, \tag{9}$$

as expected. Considering Eq. (5) again, we see that t_s has been expressed in appropriate units. Finally, referring to Eq. (7), the dimensions of θ_s are given by

$$[\theta_s] = \left[\frac{\omega_s^2}{\alpha}\right] = [\omega_s^2]\frac{1}{[\alpha]} = \frac{1}{T^2}\frac{1}{T^{-2}} = 1, \tag{10}$$

as expected. Considering Eq. (7) again and recalling that a radian is nondimensional, θ_s has been expressed in appropriate units.

As far as the numerical values of our results are concerned, because the propeller is accelerating from rest to the angular speed ω_s during the time interval $0 \le t \le t_s$, the angular displacement θ_s we computed must be smaller than the value $t_s\omega_s$, which represents the angular displacement the propeller would have experienced if it had been rotating at the *constant* angular speed ω_s for $0 \le t \le t_s$. Since $t_s\omega_s = 141.5\,\text{rad}$, our expectation is met.

Interesting Fact

Propellers in high-performance planes. Propeller propulsion was extremely popular until the early 1950s. However, especially for military applications, jet propulsion quickly replaced propeller propulsion starting at the end of World War II. By the end of World War II, the German Luftwaffe had already put in service four different jet planes, while the U.S. Army Air Forces (USAAF), the British RAF, and the Japanese Imperial Navy were testing their first prototypes. Propeller propulsion was abandoned because of an intrinsic aerodynamic problem: propellers do not perform well at supersonic speeds. This same problem contributes to keeping helicopters in the realm of "slow" flying machines. To understand this issue, we need to realize that, aerodynamically, each blade on a propeller is a *wing*. As a blade rotates, it deflects air from one side to the other side of the propeller disk. Due to Newton's third law, the *backward* motion imparted to the air by the propeller results in a corresponding *forward* motion of the propeller and of the plane to which the propeller is attached. Therefore the air velocity as seen by a propeller's blade is the sum of the velocity due to rotatory motion and the velocity due to the airplane's forward motion. Consequently, the propeller blades experience supersonic speeds much sooner than the plane does, and unless special design strategies are implemented, this quickly decreases the propeller's efficiency.

Problems

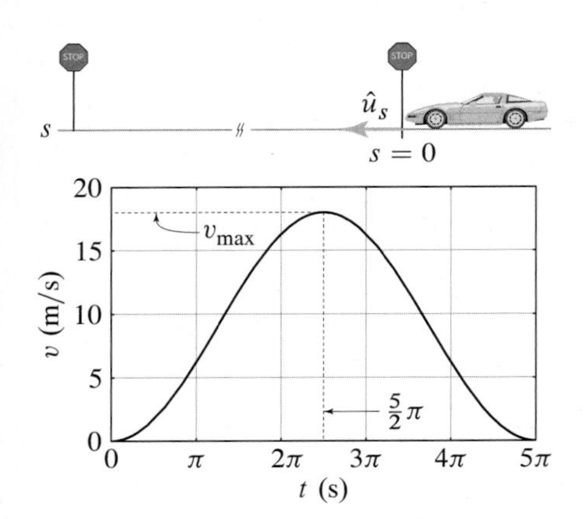

Figure P2.39–P2.42

Problems 2.39 through 2.42

The following four problems refer to a car traveling between two stop signs, in which the car's velocity is assumed to be given by $v(t) = [9 - 9\cos(2t/5)]$ m/s for $0 \le t \le 5\pi$ s.

Problem 2.39 Determine v_{max}, the maximum velocity reached by the car. Furthermore, determine the position $s_{v_{max}}$ and the time $t_{v_{max}}$ at which v_{max} occurs.

Problem 2.40 Determine the time at which the brakes are applied and the car starts to slow down.

Problem 2.41 Determine the average velocity of the car between the two stop signs.

Problem 2.42 Determine $|a|_{max}$, the maximum of the magnitude of the acceleration reached by the car, and determine the position(s) at which $|a|_{max}$ occurs.

Problem 2.43

The acceleration of a sled is prescribed to have the following form: $a = \beta\sqrt{t}$, where t is time expressed in seconds, and β is a constant. The sled starts from rest at $t = 0$. Determine β in such a way that the distance traveled after 1 s is 25 ft.

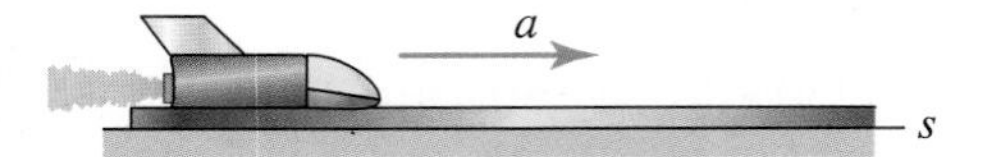

Figure P2.43 and P2.44

Problem 2.44

The acceleration of a sled can be prescribed to have one of the following forms: $a = \beta_1\sqrt{t}$, $a = \beta_2 t$, and $a = \beta_3 t^2$, where t is time expressed in seconds, $\beta_1 = 1\,\text{m/s}^{5/2}$, $\beta_2 = 1\,\text{m/s}^3$, and $\beta_3 = 1\,\text{m/s}^4$. The sled starts from rest at $t = 0$. Determine which of the three cases allows the sled to cover the largest distance in 1 s. In addition, determine the distance covered for the case in question.

Problems 2.45 and 2.46

A peg is constrained to move in a rectilinear guide and is given the following acceleration: $a = a_0 \sin\omega t$, where $a_0 = 20\,\text{ft/s}^2$, $\omega = 250$ rad/s, and t is time expressed in seconds.

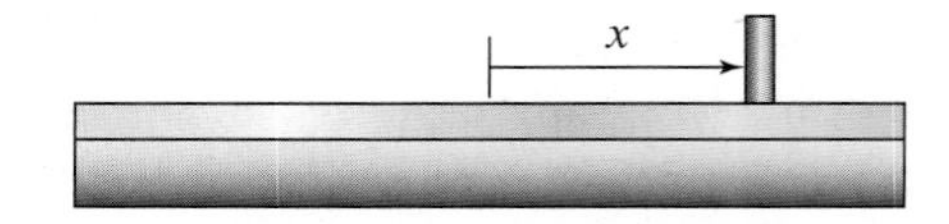

Figure P2.45 and P2.46

Problem 2.45 If $x = 0$ and $v = 0$ for $t = 0$, determine the position of the peg at $t = 4$ s.

Problem 2.46 Determine the value of the velocity of the peg at $t = 0$ so that $x(t)$ is periodic.

Problem 2.47

A ring is thrown straight upward from a height $h = 2.5\,\mathrm{m}$ off the ground and with an initial velocity $v_0 = 3.45\,\mathrm{m/s}$. Gravity causes the ring to have a constant downward acceleration $g = 9.81\,\mathrm{m/s^2}$. Determine h_{max}, the maximum height reached by the ring.

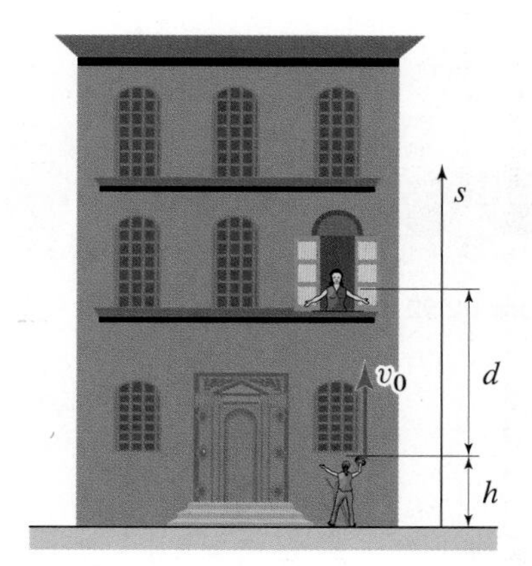

Figure P2.47 and P2.48

Problem 2.48

A ring is thrown straight upward from a height $h = 2.5\,\mathrm{m}$ off the ground. Gravity causes the ring to have a constant downward acceleration $g = 9.81\,\mathrm{m/s^2}$. Letting $d = 5.2\,\mathrm{m}$, if the person at the window is to receive the ring in the gentlest possible manner, determine the initial velocity v_0 the ring must be given when first released.

Problem 2.49

A hot air balloon is climbing with a velocity of $7\,\mathrm{m/s}$ when a sandbag (used as ballast) is released at an altitude of $305\,\mathrm{m}$. Assuming that the sandbag is subject only to gravity and that therefore its acceleration is given by $\ddot{y} = -g$, g being the acceleration due to gravity, determine how long the sandbag takes to hit the ground and its impact velocity.

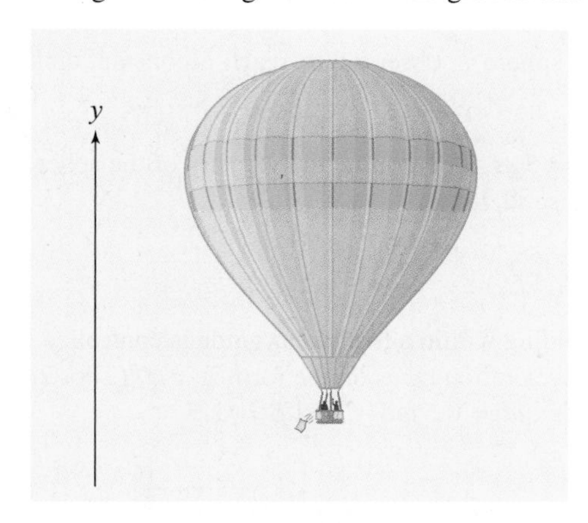

Figure P2.49

Problem 2.50

Approximately 1 h 15 min into the movie *King Kong* (the one directed by Peter Jackson), there is a scene in which Kong is holding Ann Darrow (played by the actress Naomi Watts) in his hand while swinging his arm in anger. A quick analysis of the movie indicates that at a particular moment Kong displaces Ann from rest by roughly 10 ft in a span of four frames. Knowing that the DVD plays at 24 frames per second and assuming that Kong subjects Ann to a constant acceleration, determine the acceleration Ann experiences in the scene in question. Express your answer in terms of the acceleration due to gravity g. Comment on what would happen to a person *really* subjected to this acceleration.

Figure P2.50

Problem 2.51

A car travels on a rectilinear stretch of road at a constant speed $v_0 = 65\,\mathrm{mph}$. At $s = 0$ the driver applies the brakes hard enough to cause the car to skid. Assume that the car keeps sliding until it stops, and assume that throughout this process the car's acceleration is given by $\ddot{s} = -\mu_k g$, where $\mu_k = 0.76$ is the kinetic friction coefficient and g is the acceleration of gravity. Compute the car's stopping distance and time.

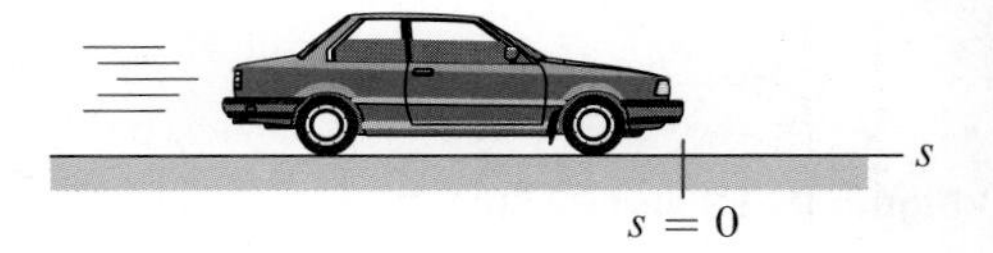

Figure P2.51

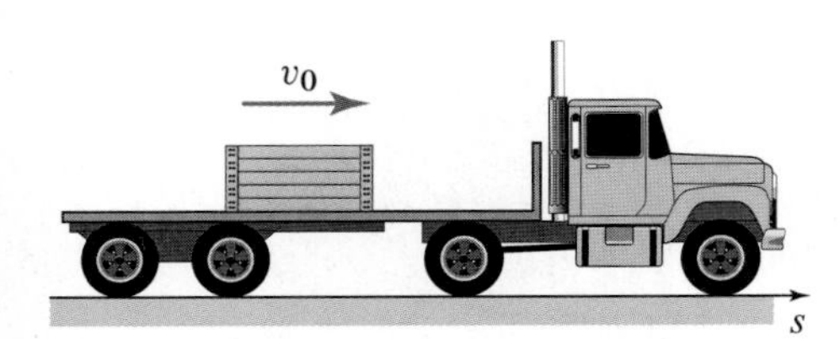

Figure P2.52 and P2.53

2-66

Problems 2.52 and 2.53

If the truck brakes and the crate slides to the right relative to the truck, the horizontal acceleration of the crate is given by $\ddot{s} = -g\mu_k$, where g is the acceleration of gravity, $\mu_k = 0.87$ is the kinetic friction coefficient, and s is the position of the crate relative to a coordinate system attached to the ground (rather than the truck).

Problem 2.52 Assuming that the crate slides without hitting the right end of the truck bed, determine the time it takes to stop if its velocity at the start of the sliding motion is $v_0 = 55\,\text{mph}$.

Problem 2.53 Assuming that the crate slides without hitting the right end of the truck bed, determine the distance it takes to stop if its velocity at the start of the sliding motion is $v_0 = 75\,\text{km/h}$.

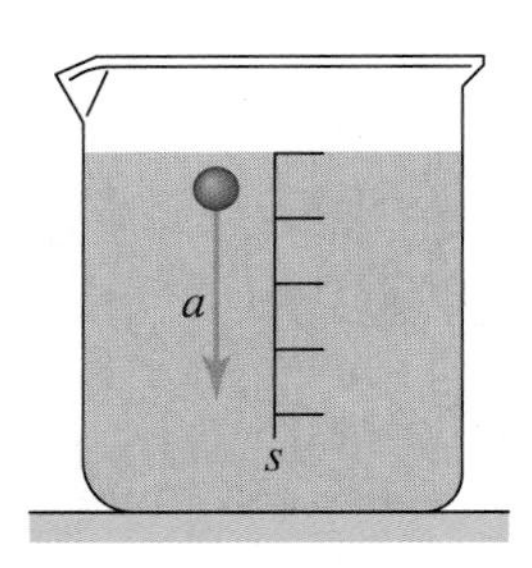

Figure P2.54 and P2.55

Problems 2.54 and 2.55

A sphere is dropped from rest at the free surface of a thick polymer fluid. The acceleration of the sphere has the form $a = g - \eta v$, where g is the acceleration due to gravity, η is a constant, and v is the sphere's velocity.

Problem 2.54 The sphere is observed to reach a constant sinking velocity equal to $0.1\,\text{m/s}$. Determine η.

Problem 2.55 If $\eta = 50\,\text{s}^{-1}$ determine the velocity of the sphere after $0.02\,\text{s}$. Express the result in feet per second.

Problems 2.56 and 2.57

The motion of a peg sliding within a rectilinear guide is controlled by an actuator in such a way that the peg's acceleration takes on the form $\ddot{x} = a_0(2\cos 2\omega t - \beta \sin \omega t)$, where t is time, $a_0 = 3.5\,\text{m/s}^2$, $\omega = 0.5\,\text{rad/s}$, and $\beta = 1.5$.

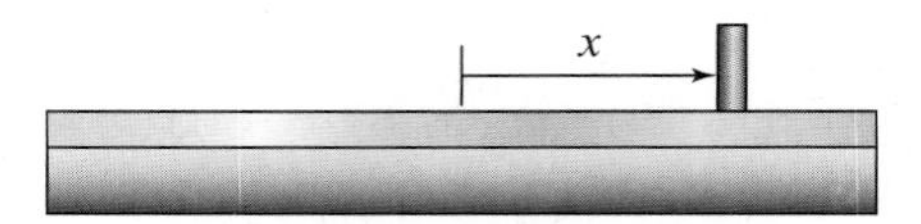

Figure P2.56 and P2.57

Problem 2.56 Determine the expressions for the velocity and the position of the peg as functions of time if $\dot{x}(0) = 0\,\text{m/s}$ and $x(0) = 0\,\text{m}$.

Problem 2.57 Determine the total distance traveled by the peg during the time interval $0\,\text{s} \le t \le 5\,\text{s}$ if $\dot{x}(0) = a_0\beta/\omega$.

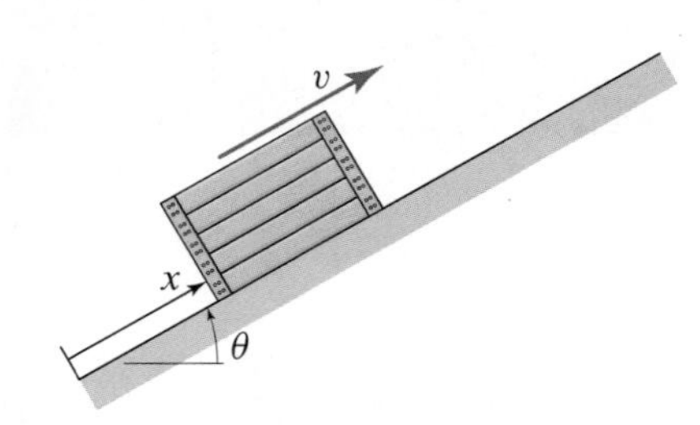

Figure P2.58 and P2.59

Problems 2.58 and 2.59

A package is pushed up an incline at $x = 0$ with an initial speed v_0. The incline is coated with a thin viscous layer so that the acceleration of the package is given by $a = -(g\sin\theta + \eta v)$, where g is the acceleration due to gravity, η is a constant, and v is the velocity of the package.

Problem 2.58 If $\theta = 30°$, $v_0 = 10\,\text{ft/s}$, and $\eta = 8\,\text{s}^{-1}$, determine the time it takes for the package to come to a stop.

Problem 2.59 If $\theta = 25°$, $v_0 = 7\,\text{m/s}$, and $\eta = 8\,\text{s}^{-1}$, determine the distance d traveled by the package before it comes to a stop.

Problem 2.60

Referring to Example 2.8 on p. 56, and defining *terminal velocity* as the velocity at which a falling object stops accelerating, determine the skydiver's terminal velocity without performing any integrations.

Figure P2.60 and P2.61

Problem 2.61

Referring to Example 2.8 on p. 56, determine the distance d traveled by the skydiver from the instant the parachute is deployed until the difference between the velocity and the terminal velocity is 10% of the terminal velocity.

Problems 2.62 and 2.63

In a physics experiment, a sphere with a given electric charge is constrained to move along a rectilinear guide with the following acceleration: $a = a_0 \sin(2\pi s/\lambda)$, where $a_0 = 8\,\mathrm{m/s^2}$, π is measured in radians, s is the position of the sphere measured in meters, $-\lambda \le s \le \lambda$, and $\lambda = 0.25\,\mathrm{m}$.

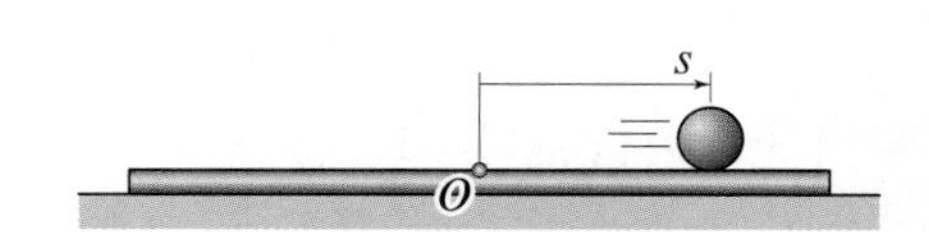

Figure P2.62 and P2.63

Problem 2.62 If the sphere is placed at rest at $s = 0$ and then gently nudged away from this position, what is the maximum speed that the sphere could achieve, and where would this maximum occur?

Problem 2.63 Suppose that the velocity of the sphere is equal to zero for $s = \lambda/4$. Determine the range of motion of the sphere, that is, the interval along the s axis within which the sphere moves. *Hint:* Determine the speed of the sphere and the interval along the s axis within which the speed has admissible values.

Problems 2.64 and 2.65

The acceleration of an object in rectilinear free fall while immersed in a linear viscous fluid is $a = g - C_d v/m$, where g is the acceleration of gravity, C_d is a constant drag coefficient, v is the object's velocity, and m is the object's mass.

Problem 2.64 Letting $t_0 = 0$ and $v_0 = 0$, find the velocity as a function of time and find the terminal velocity.

Problem 2.65 Letting $s_0 = 0$ and $v_0 = 0$, find the position as a function of velocity.

Problem 2.66

A 1.5 kg rock is released from rest at the surface of a calm lake. If the resistance offered by the water as the rock falls is directly proportional to the rock's velocity, the rock's acceleration is $a = g - C_d v/m$, where g is the acceleration of gravity, C_d is a constant drag coefficient, v is the rock's velocity, and m is the rock's mass. Letting $C_d = 4.1\,\mathrm{kg/s}$, determine the rock's velocity after 1.8 s.

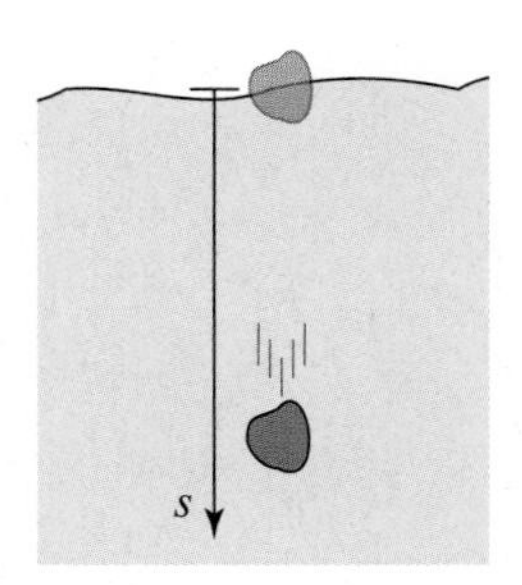

Figure P2.64–P2.68

Problems 2.67 and 2.68

A 3.1 lb rock is released from rest at the surface of a calm lake, and its acceleration is $a = g - C_d v/m$, where g is the acceleration of gravity, $C_d = 0.27\,\mathrm{lb{\cdot}s/ft}$ is a constant drag coefficient, v is the rock's velocity, and m is the rock's mass.

Problem 2.67 Determine the depth to which the rock will have sunk when the rock achieves 99% of its terminal velocity.

Problem 2.68 Determine the rock's velocity after it drops 5 ft.

Problem 2.69

Suppose that the acceleration of an object of mass m along a straight line is $a = g - C_d v/m$, where the constants g and C_d are given and v is the object's velocity. If $v(t)$ is unknown and $v(0)$ is given, can you determine the object's velocity with the following integral?

$$v(t) = v(0) + \int_0^t \left(g - \frac{C_d}{m} v\right) dt.$$

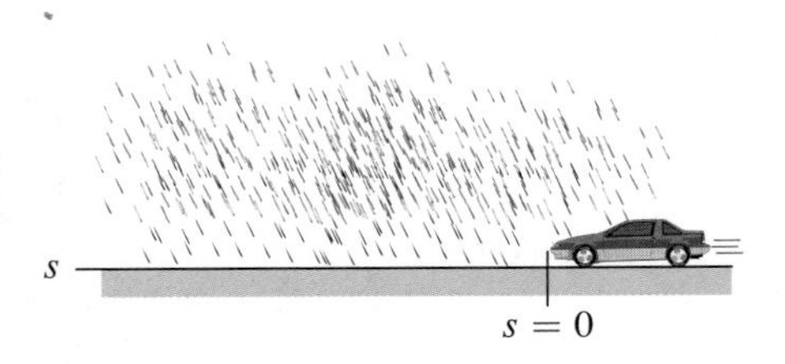

Figure P2.70

Problem 2.70

Heavy rains cause a particular stretch of road to have a coefficient of friction that changes as a function of location. Specifically, measurements indicate that the friction coefficient has a 3% decrease per meter. Under these conditions the acceleration of a car skidding while trying to stop can be approximated by $\ddot{s} = -(\mu_k - cs)g$ (the 3% decrease in friction was used in deriving this equation for acceleration), where μ_k is the friction coefficient under dry conditions, g is the acceleration of gravity, and c, with units of m^{-1}, describes the rate of friction decrement. Let $\mu_k = 0.5$, $c = 0.015\,\mathrm{m}^{-1}$, and $v_0 = 45\,\mathrm{km/h}$, where v_0 is the initial velocity of the car. Determine the distance it will take the car to stop and the percentage of increase in stopping distance with respect to dry conditions, i.e., when $c = 0$.

Problem 2.71

A car stops 4 s after the application of the brakes while covering a rectilinear stretch 337 ft long. If the motion occurred with a constant acceleration a_c, determine the initial speed v_0 of the car and the acceleration a_c. Express v_0 in mph and a_c in terms of g, the acceleration of gravity.

Figure P2.71

Problems 2.72 through 2.75

As you will learn in Chapter 3, the angular acceleration of a simple pendulum is given by $\ddot{\theta} = -(g/L)\sin\theta$, where g is the acceleration of gravity and L is the length of the pendulum cord.

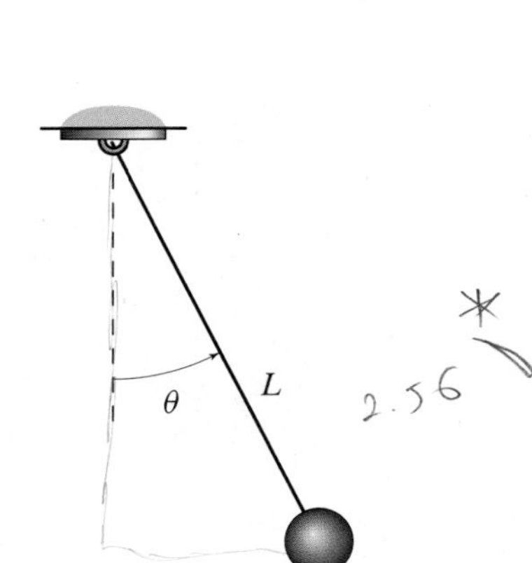

Figure P2.72–P2.75

Problem 2.72 Derive the expression of the angular velocity $\dot{\theta}$ as a function of the angular coordinate θ. The initial conditions are $\theta(0) = \theta_0$ and $\dot{\theta}(0) = \dot{\theta}_0$.

Problem 2.73 Let the length of the pendulum cord be $L = 1.5\,\mathrm{m}$. If $\dot{\theta} = 3.7\,\mathrm{rad/s}$ when $\theta = 14°$, determine the maximum value of θ achieved by the pendulum.

Problem 2.74 The given angular acceleration remains valid even if the pendulum cord is replaced by a massless rigid bar. For this case, let $L = 5.3\,\mathrm{ft}$ and assume that the pendulum is placed in motion at $\theta = 0°$. What is the minimum angular velocity at this position for the pendulum to swing through a full circle?

Problem 2.75 Let $L = 3.5\,\mathrm{ft}$ and suppose that at $t = 0\,\mathrm{s}$ the pendulum's position is $\theta(0) = 32°$ with $\dot{\theta}(0) = 0\,\mathrm{rad/s}$. Determine the pendulum's period of oscillation, i.e., from its initial position back to this position.

Problems 2.76 through 2.78

As we will see in Chapter 3, the acceleration of a particle of mass m suspended by a linear spring with spring constant k and unstretched length L_0 (when the spring length is equal to L_0, the spring exerts no force on the particle) is given by $\ddot{x} = g - (k/m)(x - L_0)$.

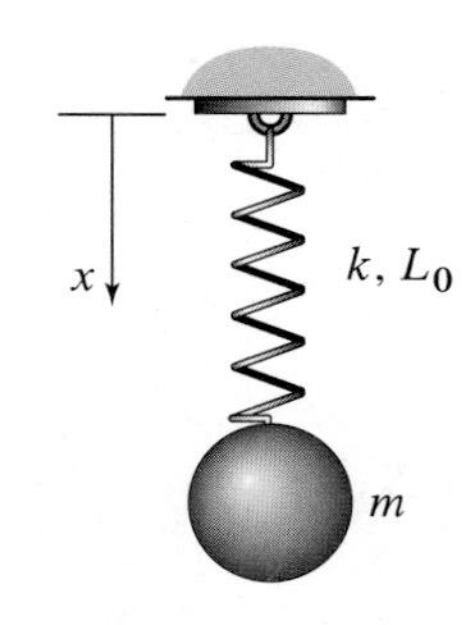

Figure P2.76–P2.78

Problem 2.76 Derive the expression for the particle's velocity $\dot{x}$ as a function of position x. Assume that at $t = 0$, the particle's velocity is v_0 and its position is x_0.

Problem 2.77 Let $k = 100\,\text{N/m}$, $m = 0.7\,\text{kg}$, and $L_0 = 0.75\,\text{m}$. If the particle is released from rest at $x = 0\,\text{m}$, determine the maximum length achieved by the spring.

Problem 2.78 Let $k = 8\,\text{lb/ft}$, $m = 0.048\,\text{slug}$, and $L_0 = 2.5\,\text{ft}$. If the particle is released from rest at $x = 0\,\text{ft}$, determine how long it takes for the spring to achieve its maximum length. *Hint:* A good table of integrals will come in handy.

Problem 2.79

A weight A with mass $m = 18\,\text{kg}$ is attached to the free end of a nonlinear spring such that the acceleration of A is $a = g - (\gamma/m)(y - L_0)^3$, where g is the acceleration due to gravity, γ is a constant, and $L_0 = 0.5\,\text{m}$. Determine γ such that A does not fall below $y = 1\,\text{m}$ when released from rest at $y = L_0$.

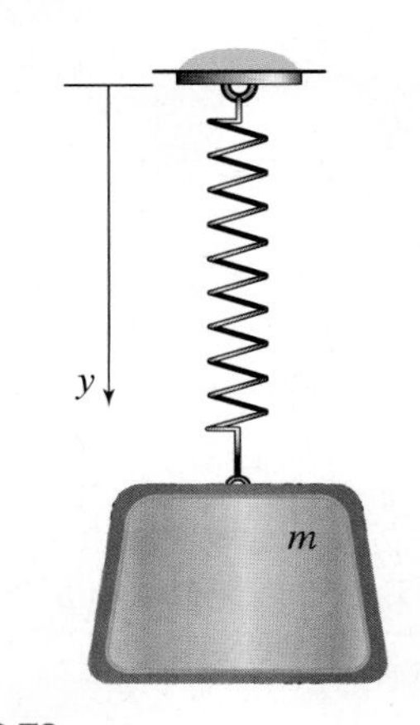

Figure P2.79

Problems 2.80 and 2.81

Two masses m_A and m_B are placed at a distance r_0 from one another. Because of their mutual gravitational attraction, the acceleration of sphere B as seen from sphere A is given by

$$\ddot{r} = -G\left(\frac{m_A + m_B}{r^2}\right),$$

where $G = 6.674 \times 10^{-11}\,\text{m}^3/(\text{kg}\cdot\text{s}^2) = 3.439 \times 10^{-8}\,\text{ft}^3/(\text{slug}\cdot\text{s}^2)$ is the universal gravitational constant.

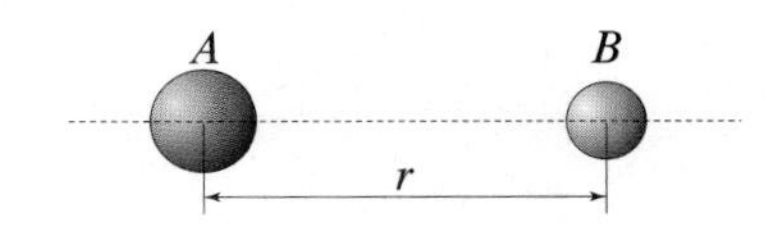

Figure P2.80 and P2.81

Problem 2.80 If the spheres are released from rest, determine

(a) The velocity of B (as seen by A) as a function of the distance r.

(b) The velocity of B (as seen by A) at impact if $r_0 = 7\,\text{ft}$, the weight of A is 2.1 lb, the weight of B is 0.7 lb, and

(i) The diameters of A and B are $d_A = 1.5\,\text{ft}$ and $d_B = 1.2\,\text{ft}$, respectively.

(ii) The diameters of A and B are infinitesimally small.

Problem 2.81 Assume that the particles are released from rest at $r = r_0$.

(a) Determine the expression relating their relative position r and time. *Hint:*

$$\int \sqrt{x/(1-x)}\,dx = \sin^{-1}(\sqrt{x}) - \sqrt{x(1-x)}.$$

(b) Determine the time it takes for the objects to come into contact if $r_0 = 3\,\text{m}$, A and B have masses of 1.1 and 2.3 kg, respectively, and

(i) The diameters of A and B are $d_A = 22\,\text{cm}$ and $d_B = 15\,\text{cm}$, respectively.

(ii) The diameters of A and B are infinitesimally small.

Problem 2.82

Suppose that the acceleration $\ddot{r}$ of an object moving along a straight line takes on the form

$$\ddot{r} = -G\left(\frac{m_A + m_B}{r^2}\right),$$

where the constants G, m_A, and m_B are known. If $\dot{r}(0)$ is given, under what conditions can you determine $\dot{r}(t)$ via the following integral?

$$\dot{r}(t) = \dot{r}(0) - \int_0^t G\,\frac{m_A + m_B}{r^2}\,dt.$$

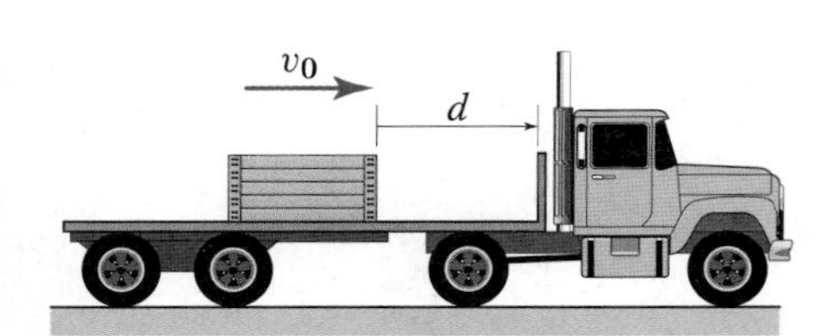

Figure P2.83

Problem 2.83

If the truck brakes hard enough that the crate slides to the right relative to the truck, the distance d between the crate and the front of the trailer changes according to the relation

$$\ddot{d} = \begin{cases} \mu_k g + a_T & \text{for } t < t_s, \\ \mu_k g & \text{for } t > t_s, \end{cases}$$

where t_s is the time it takes the truck to stop, a_T is the acceleration of the truck, g is the acceleration of gravity, and μ_k is the kinetic friction coefficient between the truck and the crate. Suppose that the truck and the crate are initially traveling to the right at $v_0 = 60$ mph and the brakes are applied so that $a_T = -10.0\,\text{ft/s}^2$. Determine the minimum value of μ_k so that the crate does not hit the right end of the truck bed if the initial distance d is 12 ft. *Hint:* The truck stops *before* the crate stops.

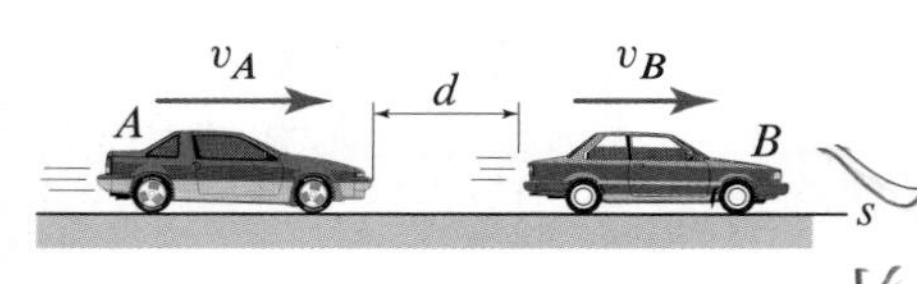

Figure P2.84

Problem 2.84

Cars A and B are traveling at $v_A = 72$ mph and $v_B = 67$ mph, respectively, when the driver of car B applies the brakes abruptly, causing the car to slide to a stop. The driver of car A takes 1.5 s to react to the situation and applies the brakes in turn, causing car A to slide as well. If A and B slide with equal accelerations, i.e., $\ddot{s}_A = \ddot{s}_B = -\mu_k g$, where $\mu_k = 0.83$ is the kinetic friction coefficient and g is the acceleration of gravity, compute the minimum distance d between A and B at the time B starts sliding to avoid a collision.

Problems 2.85 through 2.87

The spool of paper used in a printing process is unrolled with velocity v_p and acceleration a_p. The thickness of the paper is h, and the outer radius of the spool at any instant is r.

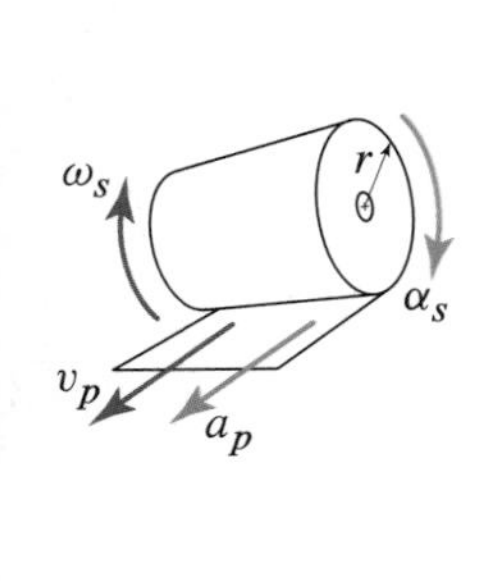

Figure P2.85–P2.87

Problem 2.85 If the velocity at which the paper is unrolled is *constant*, determine the angular acceleration α_s of the spool as a function of r, h, and v_p. Evaluate your answer for $h = 0.0048$ in., for $v_p = 1000$ ft/min, and two values of r, that is, $r_1 = 25$ in. and $r_2 = 10$ in.

Problem 2.86 If the velocity at which the paper is unrolled is *not constant*, determine the angular acceleration α_s of the spool as a function of r, h, v_p, and a_p. Evaluate your answer for $h = 0.0048$ in., $v_p = 1000$ ft/min, $a_p = 3\,\text{ft/s}^2$, and two values of r, that is, $r_1 = 25$ in. and $r_2 = 10$ in.

Problem 2.87 If the velocity at which the paper is unrolled is *constant*, determine the angular acceleration α_s of the spool as a function of r, h, and v_p. Plot your answer for $h = 0.0048$ in. and $v_p = 1000$ ft/min as a function of r for 1 in. $\leq r \leq 25$ in. Over what range does α_s vary?

Problem 2.88

Derive the constant acceleration relation in Eq. (2.32), starting from Eq. (2.24). State what assumption you need to make about the acceleration a to complete the derivation. Finally, use Eq. (2.27), along with the result of your derivation, to derive Eq. (2.33). Be careful to do the integral in Eq. (2.27) before substituting your result for $v(t)$ (try it without doing so, to see what happens). After completing this problem, notice that Eqs. (2.32) and (2.33) are *not* subject to the same assumption you needed to make to solve both parts of this problem.

2.3 Projectile Motion

Figure 2.13
A person shooting a basketball. The trajectory of the basketball is accurately described using the definition of projectile motion given in Eq. (2.39).

In this section we present a simple model to study the motion of projectiles. We will generally model projectile motion as a constant acceleration motion, but there are a few homework problems in which atmospheric drag is considered.

We define *projectile motion* to be a motion with *constant* acceleration given by

$$\boxed{a_{\text{horiz}} = 0 \quad \text{and} \quad a_{\text{vert}} = -g,} \tag{2.39}$$

where, referring to Fig. 2.14, a_{horiz} and a_{vert} are the components of the acceleration

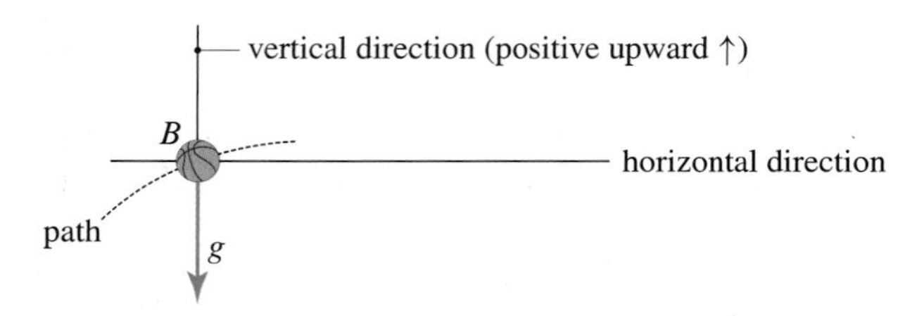

Figure 2.14. Acceleration of a basketball B in projectile motion.

in the horizontal and vertical directions, respectively, and where we have chosen the vertical direction to be positive *upward*. Since these accelerations are both *constant*, we can use Eqs. (2.32)–(2.34) to analyze the motion in the horizontal and vertical directions independently of one another.

This definition of projectile motion comes from the assumption that the only force acting on the projectile during flight is the *constant* force of gravity. As we will formally study in Chapter 3 and as you probably already know, the free body diagram of a projectile in flight looks like that in Fig. 2.15, where $W = mg$ is the weight force, m is the mass of the basketball, and we have assumed no air resistance. There are no forces in the x or horizontal direction, which means there is no acceleration in that direction. In the vertical direction, applying Newton's second law immediately leads to the second of Eqs. (2.39).

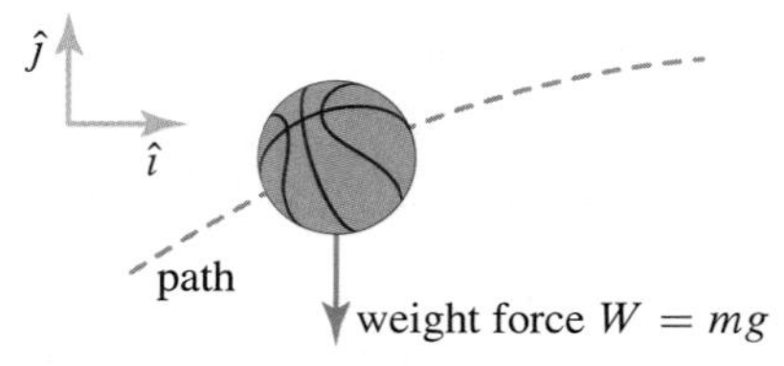

Figure 2.15
Free body diagram of a basketball in flight.

This definition of projectile motion is a very simplified description of true projectile motion because it neglects air resistance and changes in gravitational attraction with changes in height, but it is very accurate for low-speed projectiles moving over short distances.

Note: To describe the velocity, position, and trajectory of a projectile, we typically use a Cartesian coordinate system with axes parallel and perpendicular to the direction of gravity, as is done in Fig. 2.15. However, other choices are possible and, in some cases, more convenient.

EXAMPLE 2.10 *Time of Impact and Range of a Projectile*

A golfer chips the ball from the rough at the edge of the green. If the ball leaves the rough with a speed $v_0 = 15\,\text{ft/s}$ and an angle $\beta = 45°$, determine the flight time of the ball and the corresponding distance d covered.

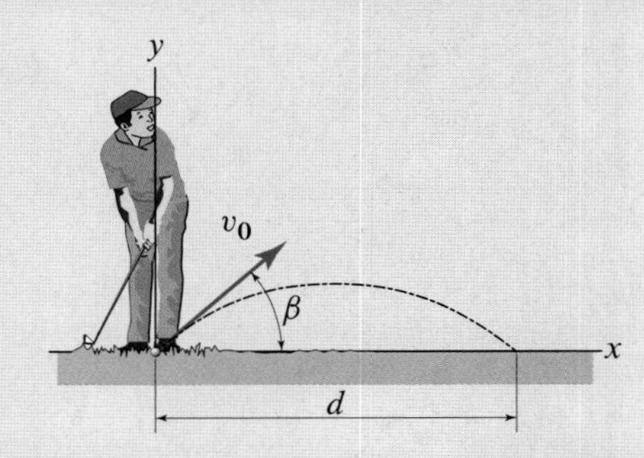

Figure 1

SOLUTION

Road Map & Modeling We model the motion of the golf ball as a projectile motion. This will allow us to use constant acceleration equations to determine the x and y coordinates of the ball as a function of time. The time of flight is the value of time for which the ball comes back to the ground, i.e., to $y = 0$. The distance d is then the corresponding value of the coordinate x.

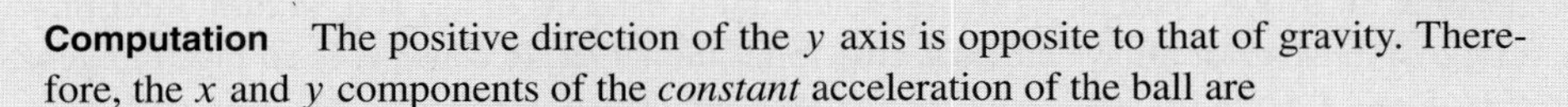

Computation The positive direction of the y axis is opposite to that of gravity. Therefore, the x and y components of the *constant* acceleration of the ball are

$$\ddot{x} = 0 \quad \text{and} \quad \ddot{y} = -g. \tag{1}$$

Letting t denote time, Eqs. (1), along with Eq. (2.33) on p. 49, tell us that the x and y coordinates of the ball as functions of time are

$$x(t) = x(0) + \dot{x}(0)t \quad \text{and} \quad y(t) = y(0) + \dot{y}(0)t - \tfrac{1}{2}gt^2, \tag{2}$$

where $\dot{x}(t)$ and $\dot{y}(t)$ are the x and y components of the golf ball's velocity. Since the ball starts at the origin of the xy coordinate system, and using the quantities v_0 and β, we have

$$x(0) = 0, \quad y(0) = 0, \quad \dot{x}(0) = v_0 \cos\beta, \quad \text{and} \quad \dot{y}(0) = v_0 \sin\beta. \tag{3}$$

Substituting Eqs. (3) into Eqs. (2), $x(t)$ and $y(t)$ become:

$$x(t) = v_0 t \cos\beta \quad \text{and} \quad y(t) = v_0 t \sin\beta - \tfrac{1}{2}gt^2. \tag{4}$$

Letting t_f be the time at which the ball comes back to the green, we have

$$y(t_f) = 0 \quad \Rightarrow \quad v_0 t_f \sin\beta - \tfrac{1}{2}gt_f^2 = 0 \quad \Rightarrow \quad t_f = \frac{2v_0 \sin\beta}{g}. \tag{5}$$

Recalling that $v_0 = 15\,\text{ft/s}$, $\beta = 45°$, and $g = 32.2\,\text{ft/s}^2$, the last of Eqs. (5) yields

$$\boxed{t_f = 0.6588\,\text{s}.} \tag{6}$$

The distance d is equal to $x(t_f)$. Substituting the last of Eqs. (5) into the first of Eqs. (4), we have

$$d = \frac{2v_0^2 \sin\beta \cos\beta}{g}, \tag{7}$$

which can be evaluated to obtain

$$\boxed{d = 6.988\,\text{ft}.} \tag{8}$$

Discussion & Verification The expressions for t_f and d have correct dimensions, and the corresponding numerical values are expressed in appropriate units. To check whether or not our results are reasonable, we can say that the value of d is correct if t_f is correct, due to the simplicity of the calculation. To check whether or not the value of t_f is reasonable, we observe that $t_f/2$ is the time that the golf ball takes to drop from its maximum height above ground. This indicates that the maximum elevation reached by the golf ball is $h = g(t_f/2)^2/2 = 1.747\,\text{ft}$, which is quite reasonable given the low initial speed of the ball and the value of the initial angle β.

Helpful Information

Determination of t_f. The time t_f is the solution of the second of Eqs. (5). The equation in question is a second-order algebraic equation in t_f, and therefore we expect two roots. One of these roots is that reported in the last of Eqs. (5). The second is simply $t_f = 0$. This latter result simply indicates that $y = 0$ for $t = 0$, which is true since the ball starts out on the ground. However, from the viewpoint of the calculation of the time of flight, $t_f = 0$ makes no sense, and it has therefore been disregarded.

EXAMPLE 2.11 Range of Elevation Angles of a Projectile

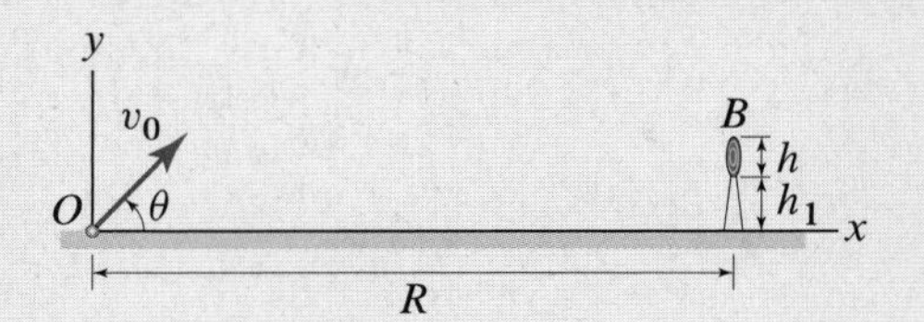

Figure 1
A projectile launched from O in an attempt to hit the target at B. Not drawn to scale.

A projectile is launched from O at speed $v_0 = 1100\,\text{ft/s}$ to hit a point B on a target that is $R = 1000\,\text{ft}$ away. The bottom of the target is $h_1 = 4\,\text{ft}$ above the ground, and the target is $h = 3\,\text{ft}$ high. Determine the range of angles at which the projectile can be fired in order to hit the target, and compare this with the angle subtended by the target as seen from O.

SOLUTION

Road Map This is a projectile motion with given initial and final positions, as well as initial speed v_0. Referring to Fig. 1, the components of the projectile's acceleration are $a_x = 0$ and $a_y = -g$. We will relate the projectile's time of flight t_f to y_B, the vertical position of B. We can then write the launch angle in terms of y_B and, in turn, infer the range of angles that allows the projectile to hit the target. Because R is so much larger than h and h_1, we should expect the range of angles we will find to be small (i.e., our aim will need to be very accurate). Therefore, in carrying out our calculations we will use more significant digits than we normally do. We will discuss the practical implications of this choice in the Discussion & Verification section.

Computation The initial and final x and y positions of B can be related to t_f by applying Eq. (2.33) (on p. 49) in the x and y directions

$$x_B = x_0 + v_{0x} t_f \quad \text{and} \quad y_B = y_0 + v_{0y} t_f - \tfrac{1}{2} g t_f^2. \tag{1}$$

Considering Fig. 1, we have $x_0 = y_0 = 0$, $v_{0x} = v_0 \cos\theta$, $v_{0y} = v_0 \sin\theta$, and $x_B = R$. We will set $y_B = h_1 = 4\,\text{ft}$ to find $\theta = \theta_{\text{min}}$ and $y_B = h_1 + h = 7\,\text{ft}$ to find $\theta = \theta_{\text{max}}$. Therefore, treating y_B as a known quantity, we have

$$R = v_0 t_f \cos\theta \quad \text{and} \quad y_B = v_0 t_f \sin\theta - \tfrac{1}{2} g t_f^2. \tag{2}$$

Equations (2) are two equations in the unknowns t_f and θ. Solving the first of Eqs. (2) for t_f and substituting the result into the second of Eqs. (2), we obtain

$$y_B = v_0 \left(\frac{R}{v_0 \cos\theta} \right) \sin\theta - \tfrac{1}{2} g \left(\frac{R}{v_0 \cos\theta} \right)^2. \tag{3}$$

Since $\sin\theta / \cos\theta = \tan\theta$ and $1/\cos^2\theta = \sec^2\theta$, we have

$$y_B = R\tan\theta - \left(\frac{gR^2}{2v_0^2} \right) \sec^2\theta. \tag{4}$$

Finally, by noting that $\sec^2\theta = 1 + \tan^2\theta$ and then rearranging, Eq. (4) becomes

$$\left(\frac{gR^2}{2v_0^2} \right) \tan^2\theta - R\tan\theta + \left(y_B + \frac{gR^2}{2v_0^2} \right) = 0, \tag{5}$$

which is a quadratic equation in $\tan\theta$. Dividing through by the coefficient of the $\tan^2\theta$ term, we obtain

$$\tan^2\theta - \left(\frac{2v_0^2}{gR} \right) \tan\theta + \left(\frac{2 y_B v_0^2}{gR^2} + 1 \right) = 0. \tag{6}$$

Equation (6) can be solved for $\tan\theta$ to obtain the two solutions

$$\tan\theta = \frac{v_0^2 \pm \sqrt{v_0^4 - g\left(gR^2 + 2 y_B v_0^2\right)}}{gR}, \tag{7}$$

which means that for each value of y_B, there are two possible values of the firing angle θ: θ_1 and θ_2. Substituting in values for all constants, including the two different values for y_B, we obtain

$$y_B = 4\,\text{ft} \quad \Rightarrow \quad \begin{cases} \theta_1 = 0.991678°, \\ \theta_2 = 89.2375041°, \end{cases} \tag{8}$$

$$y_B = 7\,\text{ft} \quad \Rightarrow \quad \begin{cases} \theta_1 = 1.163590°, \\ \theta_2 = 89.2374736°. \end{cases} \tag{9}$$

Helpful Information

Number of digits in calculations. In the numerical calculations shown in this example, the differences in some of the numbers are *so* small that we are keeping many more digits than we normally would. If we did not, the differences would not be apparent.

Equations (8) and (9) give us the values of θ needed to hit the bottom and top of the sign, respectively. Referring to Eq. (8), it is probably intuitive that if we chose $\theta_1 < \theta < \theta_2$, we would overshoot the bottom of the target, whereas we would undershoot it for $\theta < \theta_1$ and $\theta > \theta_2$. For example, substituting $\theta = 45°$ (i.e., a value of θ between those in Eq. (8)) into Eq. (4) gives $y_B = 973.4$ ft, which, as expected, is larger than 4 ft. Extending the discussion to Eq. (9), we can then say that if $\theta_1 < \theta < \theta_2$ in Eq. (9), we would overshoot the top of the sign, whereas we would undershoot it for $\theta < \theta_1$ and $\theta > \theta_2$. We can therefore conclude that there are two ranges of firing angles, such that our projectile will hit the target and that these ranges are given by

$$0.991678° \le \theta \le 1.163590° \tag{10}$$

and

$$89.23747360° \le \theta \le 89.23750406°. \tag{11}$$

Discussion & Verification In obtaining the angle ranges in Eqs. (10) and (11), we went through a simple verification step by computing the answer for $\theta = 45°$, in which we saw that our results were as expected. To extend our discussion, let's now consider the size of the ranges in Eqs. (10) and (11):

$$\Delta\theta_1 = 1.163590° - 0.991678° = 0.171912°, \tag{12}$$

$$\Delta\theta_2 = 89.23747360° - 89.23750406° = -0.00003046°. \tag{13}$$

Equations (10) and (12) tell us that to hit our target, we need to elevate our launcher about 1° with an accuracy of 0.17°. Equations (11) and (13) tell us that we can also hit the target if we elevate our launcher to approximately 89.2°, but this time we have an *extremely small* margin of error. In fact, we need to be accurate to within three one-hundred thousandths of a degree! In addition, we also need to keep in mind that our model does not account for aerodynamic effects, which place an additional accuracy burden on our aim. These considerations tell us that, in practical applications, we need to rely on an *active guidance system* rather than the accuracy of the launch angles.

Let's complete this example by comparing the angle $\Delta\theta_1$ with the angle subtended by the target at a distance of 1000 ft. Referring to Fig. 2, we can see that the angle subtended is given by

$$\beta = \gamma - \phi, \tag{14}$$

where

$$\gamma = \tan^{-1}\left(\frac{7}{1000}\right) = 0.401064° \quad \text{and} \quad \phi = \tan^{-1}\left(\frac{4}{1000}\right) = 0.229182°, \tag{15}$$

so that $\beta = 0.171882°$. Since β is very close to the $\Delta\theta_1$ in Eq. (12) and since computing β is simpler than computing $\Delta\theta_1$, we might think that we could have computed β to approximate $\Delta\theta_1$. However, in general, the elevation angle ranges and angle subtended by the target can be substantially different.

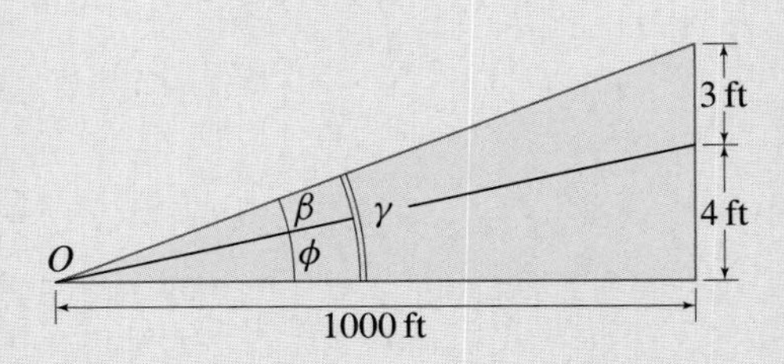

Figure 2
Not drawn to scale—the vertical dimension has been greatly exaggerated so that the angles can be easily seen.

EXAMPLE 2.12 *Initial Speed and Elevation Angle of a Projectile*

A baseball batter makes contact with a ball about 4 ft above the ground and hits it hard enough that it *just* clears the center field wall, which is 400 ft away and is 9 ft high. How fast must the ball be moving and at what angle must it be hit so that it just clears the center field wall as shown in Fig. 1?

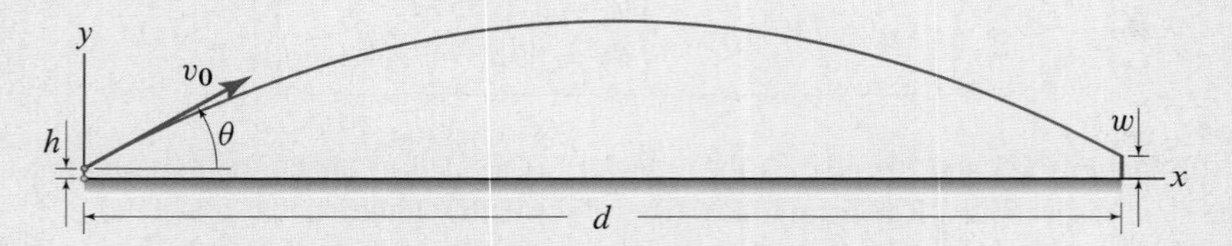

Figure 1. Side view, drawn to scale, of the given baseball field with all parameters defined. In the trajectory shown, the baseball was hit 4 ft off the ground, at 123.2 ft/s, and at a 30° angle so that it *just* clears a 9 ft fence that is 400 ft away.

SOLUTION

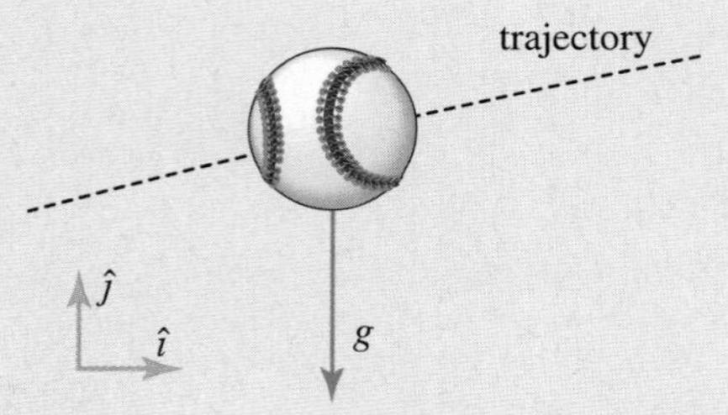

Figure 2
The only nonzero component of acceleration of the baseball.

Road Map Referring to Fig. 2, we model the ball as a projectile with acceleration given by $a_x = 0$ and $a_y = -g$. We know the starting and ending locations of the projectile, and we wish to determine the v_0 and θ required to get it from start to finish. Therefore, we can proceed as in Example 2.11, i.e., by writing the projectile's x and y positions as a function of time and then, eliminating time, we will obtain an expression for v_0 in terms of θ.

Computation Since both components of acceleration are constant, we can apply the constant acceleration equation, Eq. (2.33) (on p. 49), in both the x and y directions to obtain

$$x = x_0 + v_{0x}t \quad \Rightarrow \quad d = 0 + v_0 t \cos\theta, \tag{1}$$

$$y = y_0 + v_{0y}t + \tfrac{1}{2}a_y t^2 \quad \Rightarrow \quad w = h + v_0 t \sin\theta - \tfrac{1}{2}gt^2. \tag{2}$$

Equations (1) and (2) are two equations for the three unknowns v_0, θ, and t. Since we are interested in v_0 and θ, we can eliminate t from these two equations and then solve for v_0 as a function of θ to obtain

$$\boxed{v_0 = d\sqrt{\frac{g}{2\cos\theta[(h-w)\cos\theta + d\sin\theta]}}.} \tag{3}$$

This result tells us that there are infinitely many combinations of v_0 and θ that will *just* get the baseball over the center field fence.

Discussion & Verification The solution to the problem is an expression rather than a specific quantitative answer. To verify that Eq. (3) is correct, we first check that its dimensions are correct. Letting L and T denote dimensions of length and time, we have $[g] = L/T^2$ and $[h] = [w] = [d] = L$. Therefore, the dimensions of the argument of the square root in Eq. (3) are T^{-2}, so that the overall dimensions of the right-hand side of Eq. (3) are L/T, as expected. Further verification requires that we study the behavior of the expression in Eq. (3), as is done next.

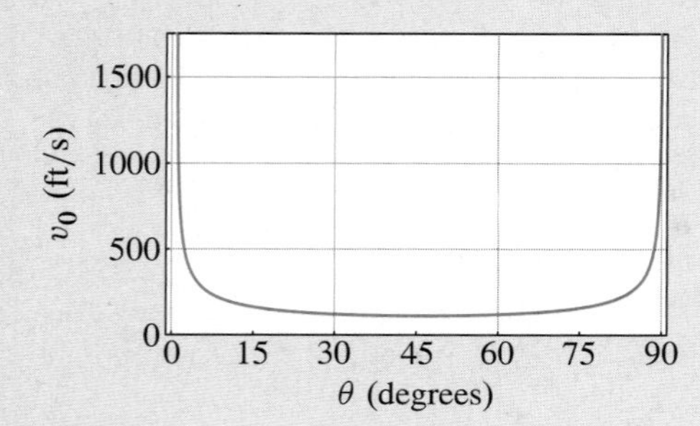

Figure 3
Plot of Eq. (3), that is, v_0 as a function of θ.

A Closer Look For $d = 400$ ft, $h = 4$ ft, $w = 9$ ft, and $g = 32.2\,\text{ft/s}^2$, the plot of the required v_0 for $0° \le \theta \le 90°$ is shown in Fig. 3. By careful inspection of the left side of the curve, we see that the curve approaches an asymptote value of θ other than 0°.

This is so because the point at which the batter makes contact with the ball is lower than the height of the fence, so that even if the batter hits the ball infinitely hard, there is an angle below which the ball will not clear the fence (we leave it to the reader to show that this angle is 0.7162°). On the right side of the curve, we see that the required speed again approaches infinity, but now it is as the angle approaches 90°. Finally, it is also apparent that the optimal angle to hit the ball is near 45°, where by *optimal angle* we mean the angle corresponding to the smallest possible v_0. As it turns out, the optimal angle is not exactly at 45° since we are trying to get the maximum distance for the minimum speed between two points of *unequal* height. To find the optimal angle, we could differentiate Eq. (3) with respect to θ, set the result equal to zero, and then solve for the optimal angle. We can also square both sides and differentiate that. Doing so and setting the result equal to zero, we have

$$\left.\frac{d(v_0^2)}{d\theta}\right|_{\theta=\theta_0} = \frac{gd^2}{2}\left\{\frac{-d\cos^2\theta_0 + 2(h-w)\cos\theta_0\sin\theta_0 + d\sin^2\theta_0}{\cos^2\theta_0[(h-w)\cos\theta_0 + d\sin\theta_0]^2}\right\} = 0, \quad (4)$$

where we have replaced θ with θ_0, which is the optimal value of θ. Since the denominator and numerator do not go to zero at the same values of θ (if they did, we would have to apply l'Hopital's rule), we can solve Eq. (4) by setting the numerator of the fraction within curly braces to zero. Doing this and using some trigonometric identities,* we obtain

$$-d\cos(2\theta_0) + (h-w)\sin(2\theta_0) = 0, \quad (5)$$

or

$$\tan(2\theta_0) = \frac{d}{h-w} = -80. \quad (6)$$

Equation (6) has infinitely many solutions given by

$$2\theta_0 = 90.72° \pm n180°, \qquad n = 0, 1, \ldots, \infty, \quad (7)$$

but the only meaningful solution in this context is $2\theta_0 = 90.72°$, or $\theta_0 = 45.36°$. This means that the optimal angle to hit the ball when trying to hit it over an object that is higher than the initial position of the ball is greater than 45°. From Eq. (5) we can also see that the optimal angle is less than 45° when we try to hit the ball over a lower object, and it is exactly equal to 45° only if $h = w$, i.e., if we are trying to hit the ball over an object whose height is the same as the ball's initial position.

This discussion leads to the conclusion that the solution we derived does indeed have a behavior that matches our expectations and physical intuition.

---

* $\sin^2 x - \cos^2 x = -\cos(2x)$ and $2\sin x\cos x = \sin(2x)$.

Problems

Problem 2.89

The discussion in Example 2.12 revealed that the angle θ had to be greater than $\theta_{\min} = 0.716°$. Find an analytical expression for $\theta_{\min}$ in terms of h, w, and d.

Problem 2.90

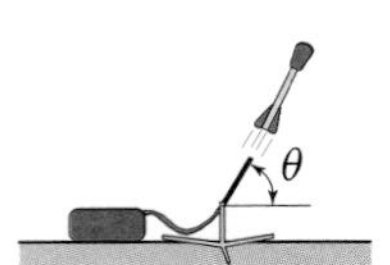

Figure P2.90

A stomp rocket is a toy consisting of a hose connected to a "blast pad" (i.e., an air bladder) at one end and to a short pipe mounted on a tripod at the other end. A rocket with a hollow body is mounted onto the pipe and is propelled into the air by "stomping" on the blast pad. Some manufacturers claim that one can shoot a rocket over 200 ft in the air. Neglecting air resistance, determine the rocket's minimum initial speed such that it reaches a maximum flight height of 200 ft.

Problem 2.91

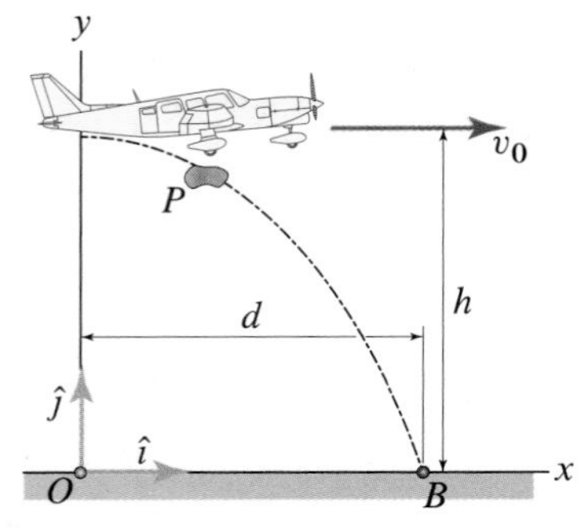

Figure P2.91 and P2.92

An airplane flying horizontally at elevation $h = 150$ ft and at a constant speed $v_0 = 80$ mph drops a package P when passing over point O. Determine the horizontal distance d between the drop point and point B at which the package hits the ground.

Problem 2.92

An airplane flying horizontally at elevation $h = 60$ m and at a constant speed $v_0 = 120$ km/h drops a package P when passing over point O. Determine the time it takes for the package to hit the ground at point B. In addition, determine the velocity of the package at B.

Problem 2.93

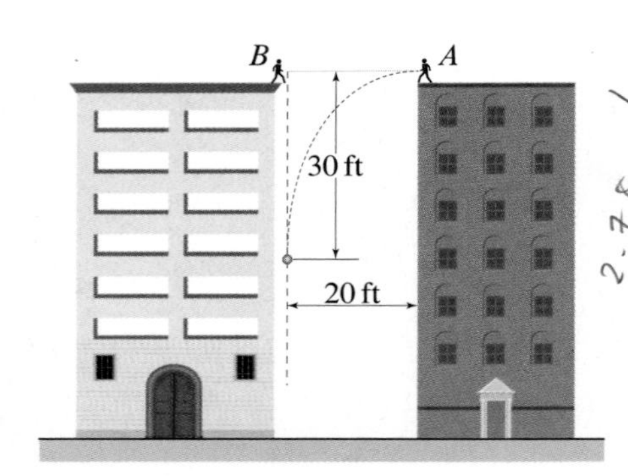

Figure P2.93

Stuntmen A and B are shooting a movie scene in which A needs to pass a gun to B. Stuntman B is supposed to start falling vertically precisely when A throws the gun to B. Treating the gun and the stuntman B as particles, find the velocity of the gun as it leaves A's hand so that B will catch it after falling 30 ft.

Problem 2.94

The jaguar A leaps from O at speed $v_0 = 6$ m/s and angle $\beta = 35°$ relative to the incline to try to intercept the panther B at C. Determine the distance R that the jaguar jumps from O to C (i.e., R is the distance between the two points of the trajectory that intersect the incline), given that the angle of the incline is $\theta = 25°$.

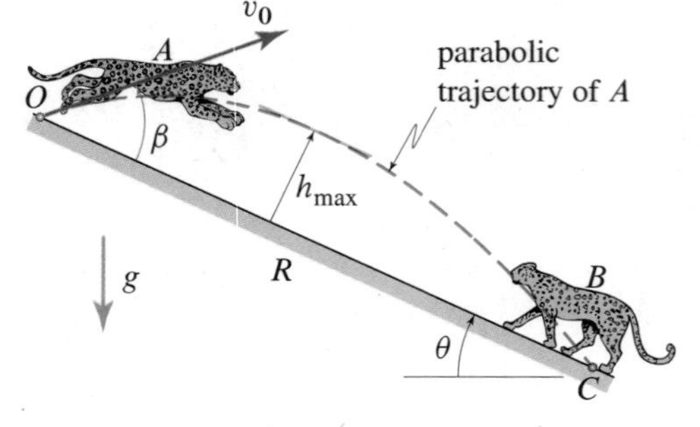

Figure P2.94

Problem 2.95

If the projectile is released at A with initial speed v_0 and angle β, derive the projectile's trajectory, using the coordinate system shown. Neglect air resistance.

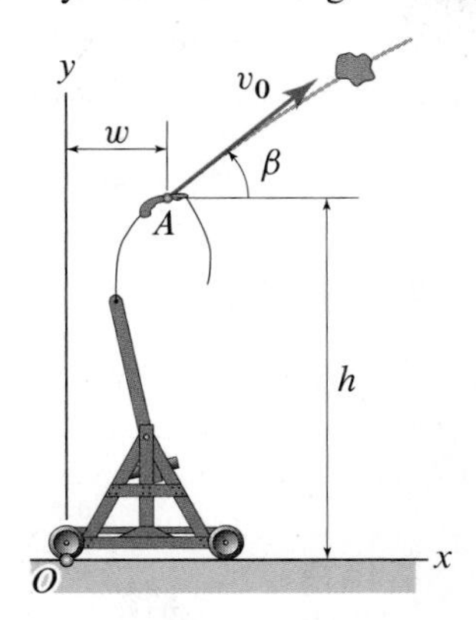

Figure P2.95

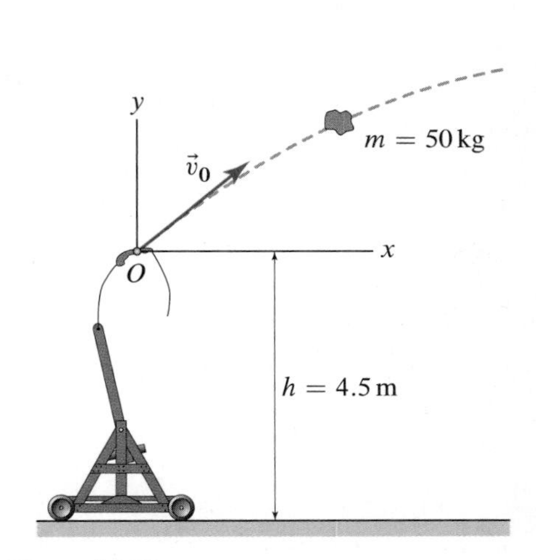

Figure P2.96

Problem 2.96

A trebuchet releases a rock with mass $m = 50\,\text{kg}$ at point O. The initial velocity of the projectile is $\vec{v}_0 = (45\,\hat{\imath} + 30\,\hat{\jmath})\,\text{m/s}$. Neglecting aerodynamic effects, determine where the rock will land and its time of flight.

Problem 2.97

A golfer chips the ball into the hole on the fly from the rough at the edge of the green. Letting $\alpha = 4°$ and $d = 2.4\,\text{m}$, verify that the golfer will place the ball within 10 mm of the center of the hole if the ball leaves the rough with a speed $v_0 = 5.03\,\text{m/s}$ and an angle $\beta = 41°$.

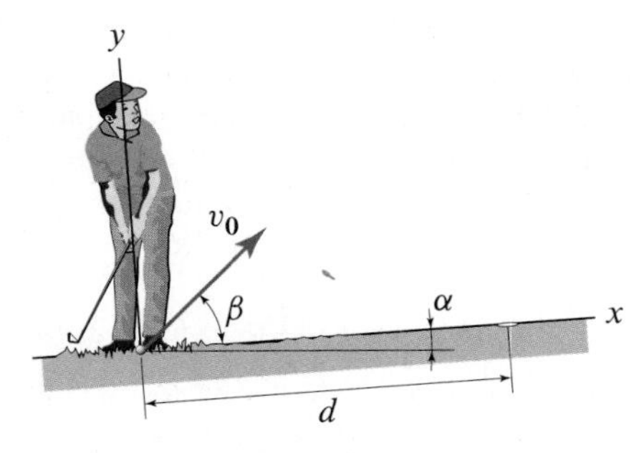

Figure P2.97

Problems 2.98 and 2.99

In a movie scene involving a car chase, a car goes over the top of a ramp at A and lands at B below.

Problem 2.98 If $\alpha = 20°$ and $\beta = 23°$, determine the distance d covered by the car if the car's speed at A is 45 km/h. Neglect aerodynamic effects.

Problem 2.99 Determine the speed of the car at A if the car is to cover distance $d = 150\,\text{ft}$ for $\alpha = 20°$ and $\beta = 27°$. Neglect aerodynamic effects.

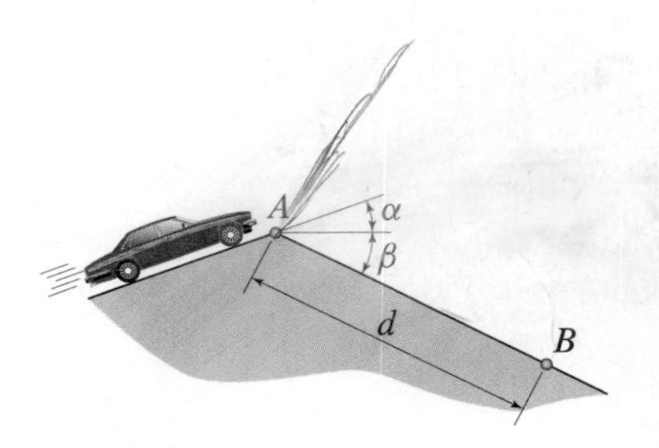

Figure P2.98 and P2.99

Problem 2.100

The M777 lightweight 155 mm howitzer is a piece of artillery whose rounds are ejected from the gun with a speed of 829 m/s. Assuming that the gun is fired over a flat battlefield and ignoring aerodynamic effects, determine (*a*) the elevation angle needed to achieve the maximum range, (*b*) the maximum possible range of the gun, and (*c*) the time it would take a projectile to cover the maximum range. Express the result for the range as a percentage of the actual maximum range of this weapon, which is 30 km for unassisted ammunition.

Figure P2.100

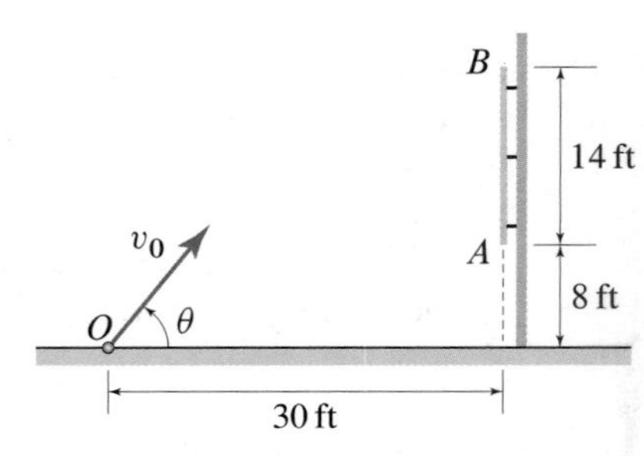

Figure P2.101

Problem 2.101

You want to throw a rock from point O to hit the vertical advertising sign AB, which is $R = 30$ ft away. You can throw a rock at the speed $v_0 = 45$ ft/s. The bottom of the sign is 8 ft off the ground and the sign is 14 ft tall. Determine the range of angles at which the projectile can be thrown in order to hit the target, and compare this with the angle subtended by the target as seen from an observer at point O. Compare your results with those found in Example 2.11.

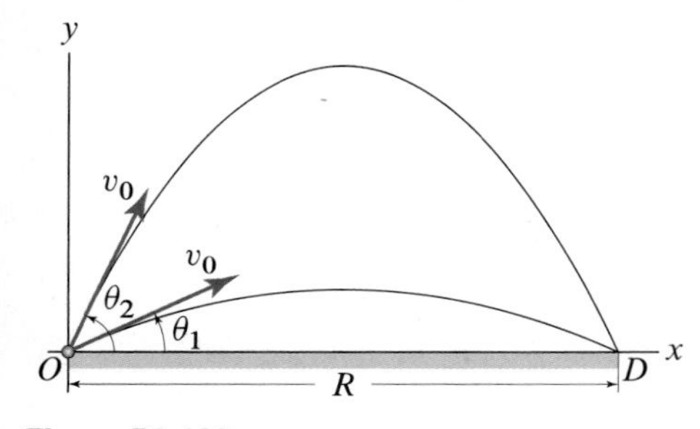

Figure P2.102

Problem 2.102

Suppose that you can throw a projectile at a large enough v_0 so that it can hit a target a distance R downrange. Given that you know v_0 and R, determine the general expressions for the *two* distinct launch angles θ_1 and θ_2 that will allow the projectile to hit D. For $v_0 = 30$ m/s and $R = 70$ m, determine numerical values for θ_1 and θ_2.

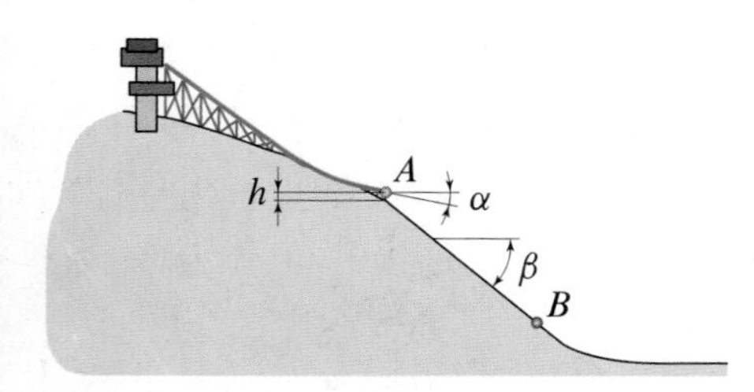

Figure P2.103

Problem 2.103

An alpine ski jumper can fly distances in excess of 100 m* by using his or her body and skis as a "wing" and therefore, taking advantage of aerodynamic effects. With this in mind and assuming that a ski jumper could survive the jump, determine the distance the jumper could "fly" without aerodynamic effects, i.e., if the jumper were in free fall after clearing the ramp. For the purpose of your calculation, use the following typical data: $\alpha = 11°$ (slope of ramp at takeoff point A), $\beta = 36°$ (average slope of the hill),† $v_0 = 86$ km/h (speed at A), $h = 3$ m (height of takeoff point with respect to the hill). Finally, for simplicity, let the jump distance be the distance between the takeoff point A and the landing point B.

* On March 20, 2005, using the very large ski ramp at Planica, Slovenia, Bjørn Einar Romøren of Norway set the world record by flying a distance of 239 m.

† While the given average slope of the landing hill is accurate, you should know that, according to regulations, the landing hill must have a curved profile. Here, we have chosen to use a landing hill with a *constant* slope of 36° to simplify the problem.

Problems 2.104 and 2.105

A soccer player practices kicking a ball from A directly into the goal (i.e., the ball does not bounce first) while clearing a 6 ft tall fixed barrier.

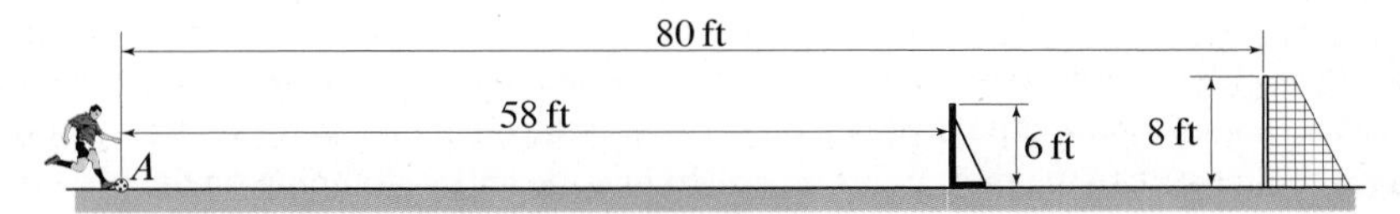

Figure P2.104 and P2.105

Problem 2.104 Determine the minimum speed that the player needs to give the ball to accomplish the task. *Hint:* Consider the equation for the projectile's trajectory of the form $y = C_0 + C_1 x + C_2 x^2$, with the y axis parallel to the direction of gravity, for the case in which the ball reaches the goal at its base. Solve this equation for the initial speed v_0 as a function of the initial angle θ, and finally find $(v_0)_{\min}$ as you learned in calculus. Don't forget to check whether or not the ball clears the barrier.

Problem 2.105 Find the initial speed and angle that allow the ball to barely clear the barrier while barely reaching the goal at its base. *Hint:* A projectile's trajectory can be given the form $y = C_1 x - C_2 x^2$, where the coefficients C_1 and C_2 can be found by forcing the parabola to go through two given points.

Problems 2.106 and 2.107

In a circus act a tiger is required to jump from point A to point C so that it goes through the ring of fire at B. *Hint:* A projectile's trajectory can be given the form $y = C_1 x - C_2 x^2$, where the coefficients C_1 and C_2 can be found by forcing the parabola to go through two given points.

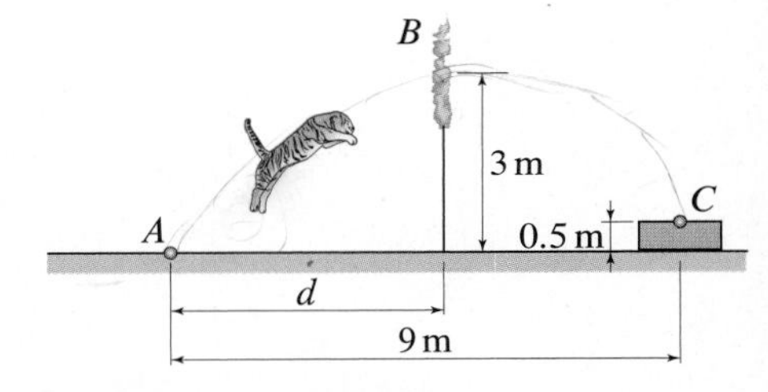

Figure P2.106 and P2.107

Problem 2.106 Determine the tiger's initial velocity if the ring of fire is placed at a distance $d = 5.5$ m from A. Furthermore, determine the slope of the tiger's trajectory as the tiger goes through the ring of fire.

Problem 2.107 Determine the tiger's initial velocity, as well as the distance d so that the slope of the tiger's trajectory as the tiger goes through the ring of fire is completely horizontal.

Problem 2.108

A jaguar A leaps from O at speed v_0 and angle β relative to the incline to attack a panther B at C. Determine an expression for the maximum *perpendicular* height $h_{\max}$ above the incline achieved by the leaping jaguar, given that the angle of the incline is θ.

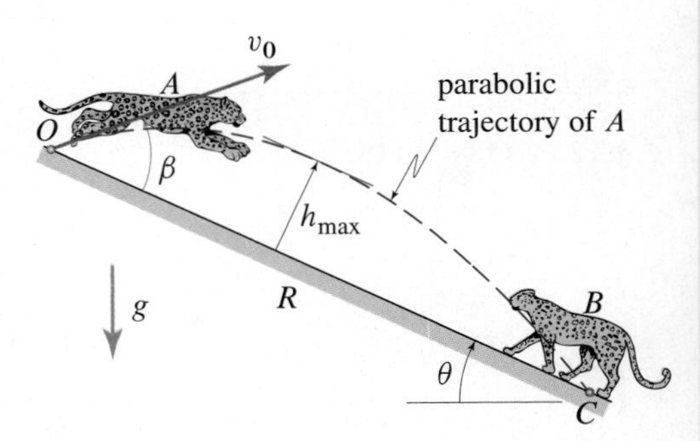

Figure P2.108–P2.110

Problems 2.109 and 2.110

The jaguar A leaps from O at speed v_0 and angle β relative to the incline to intercept the panther B at C. The distance along the incline from O to C is R, and the angle of the incline with respect to the horizontal is θ.

Problem 2.109 Determine an expression for v_0 as a function of β for A to be able to get from O to C.

Problem 2.110 Derive v_0 as a function of β to leap a given distance R along with the optimal value of launch angle β, i.e., the value of β necessary to leap a given distance R with the minimum v_0. Then plot v_0 as a function of β for $g = 9.81\,\text{m/s}^2$,

$R = 7$ m, and $\theta = 25°$, and find a numerical value of the optimal β and the corresponding value of v_0 for the given set of parameters.

Figure P2.111 and P2.112

Problems 2.111 and 2.112

A stomp rocket is a toy consisting of a hose connected to a blast pad (i.e., an air bladder) at one end and to a short pipe mounted on a tripod at the other end. A rocket with a hollow body is mounted onto the pipe and is propelled into the air by stomping on the blast pad.

Problem 2.111 If the rocket can be imparted an initial speed $v_0 = 120\,\text{ft/s}$, and if the rocket's landing spot at B is at the same elevation as the launch point, i.e., $h = 0\,\text{ft}$, neglect air resistance and determine the rocket's launch angle θ such that the rocket achieves the maximum possible range. In addition, compute R, the rocket's maximum range, and t_f, the corresponding flight time.

Problem 2.112 Assuming the rocket can be given an initial speed $v_0 = 120\,\text{ft/s}$, the rocket's landing spot at B is 10 ft higher than the launch point, i.e., $h = 10\,\text{ft}$, and neglecting air resistance, find the rocket's launch angle θ such that the rocket achieves the maximum possible range. In addition, as part of the solution, compute the corresponding maximum range and flight time. To do this:

(a) Determine the range R as a function of time.

(b) Take the expression for R found in (a), square it, and then differentiate it with respect to time to find the flight time that corresponds to the maximum range, and then find that maximum range.

(c) Use the time found in (b) to then find the angle required to achieve the maximum range.

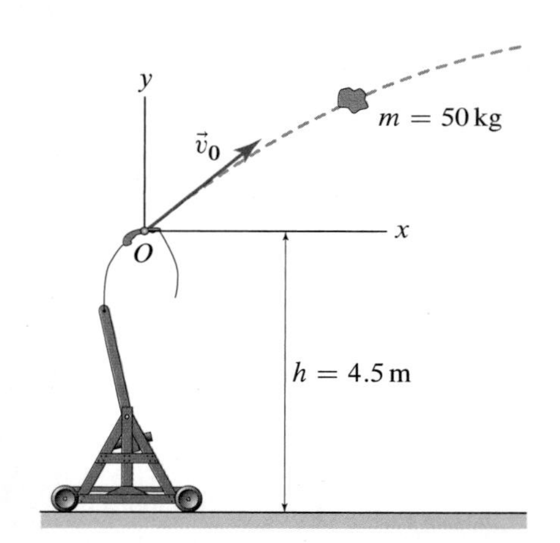

Figure P2.113–P2.116

Problem 2.113

A trebuchet releases a rock with mass $m = 50\,\text{kg}$ at the point O. The initial velocity of the projectile is $\vec{v}_0 = (45\,\hat{\imath} + 30\,\hat{\jmath})\,\text{m/s}$. If one were to model the effects of air resistance via a drag force directly proportional to the projectile's velocity, the resulting accelerations in the x and y directions would be $\ddot{x} = -(\eta/m)\dot{x}$ and $\ddot{y} = -g - (\eta/m)\dot{y}$, respectively, where g is the acceleration of gravity and $\eta = 0.64\,\text{kg/s}$ is a viscous drag coefficient. Find an expression for the trajectory of the projectile.

Problem 2.114

Continue Prob. 2.113 and, for the case where $\eta = 0.64\,\text{kg/s}$, determine the maximum height from the ground reached by the projectile and the time it takes to achieve it. Compare the result with what you would obtain in the absence of air resistance.

Problem 2.115

Continue Prob. 2.113 and, for the case where $\eta = 0.64\,\text{kg/s}$, determine t_I and x_I, the value of t, and the x position corresponding to the projectile's impact with the ground.

Problem 2.116

With reference to Probs. 2.113 and 2.115, assume that an experiment is conducted so that the measured value of x_I is 10% smaller than what is predicted in the absence of viscous drag. Find the value of η that would be required for the theory in Prob. 2.113 to match the experiment.

Design Problems

Design Problem 2.1

In the manufacture of steel balls of the type used for ball bearings, it is important that their material properties be sufficiently uniform. One way to detect gross differences in their material properties is to observe how a ball rebounds when dropped on a hard strike plate. The rebound characteristics of a ball can be assessed via a quantity called the *coefficient of restitution* (COR).* Specifically, if $(v_n)_{\text{strike}}$ is the component of a ball's impact velocity normal to the strike plate, then the COR is given by

$$\text{COR} = \frac{|(v_n)_{\text{rebound}}|}{|(v_n)_{\text{strike}}|}, \qquad \text{(COR equation)}$$

where $(v_n)_{\text{rebound}}$ is the component of the rebound velocity normal to the strike plate.

Given that each ball has a radius $R = 0.2$ in., design a sorting device to select the balls with $0.800 < \text{COR} < 0.825$. The device consists of an incline defined by the angle θ and length L. The strike plate is placed at the bottom of a well with depth h and width w. Finally, at a distance ℓ from the incline, there is a thin vertical barrier with a gap of size d placed at a height b from the bottom of the well. Releasing a ball from rest at the top of the incline and assuming that the ball rolls without slip, we know the ball will reach the bottom of the incline with a speed†

$$v_0 = \sqrt{\tfrac{10}{7} g L \sin\theta},$$

where g is the acceleration due to gravity. After rolling off the incline, each ball will rebound off the strike plate such that the horizontal component of velocity is unaffected by the impact while the vertical component will behave as described in the COR equation. After rebounding, the balls to be isolated will pass through the vertical gap, whereas the rest of the balls will not go through the gap. In your design, choose appropriate values of L, $\theta < 45°$, h, w, d, and ℓ to accomplish the desired task while ensuring that the overall dimensions of the device do not exceed 4 ft in both the horizontal and vertical directions.

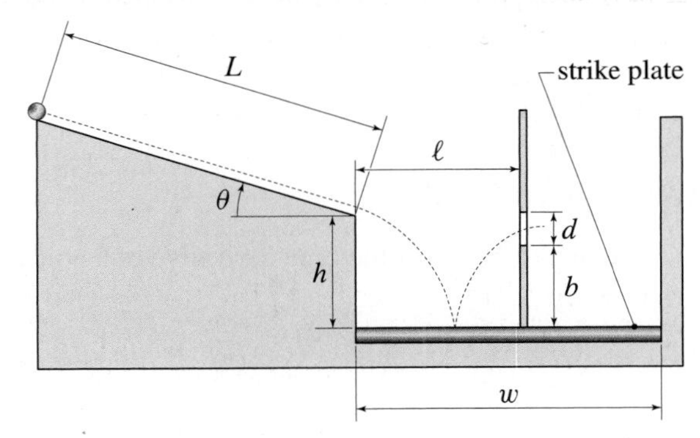

Figure DP2.1

* We will study the COR in detail in Section 5.2.

† We will see how to derive this formula in Chapters 7 and 8.

2.4 The Time Derivative of a Vector

Time derivatives of vector quantities are ubiquitous in dynamics. We have already seen the time derivative of position and velocity vectors. In coming chapters, we will also see the time derivatives of quantities, such as angular velocity, momentum, and angular momentum, all of which are vectors. It is therefore worthwhile to devote a little time to the time derivative of a vector so as to strengthen our understanding of this operation.

Vectors can change with time in two ways:

1. They can change in magnitude.
2. They can change in direction due to *rotation*.

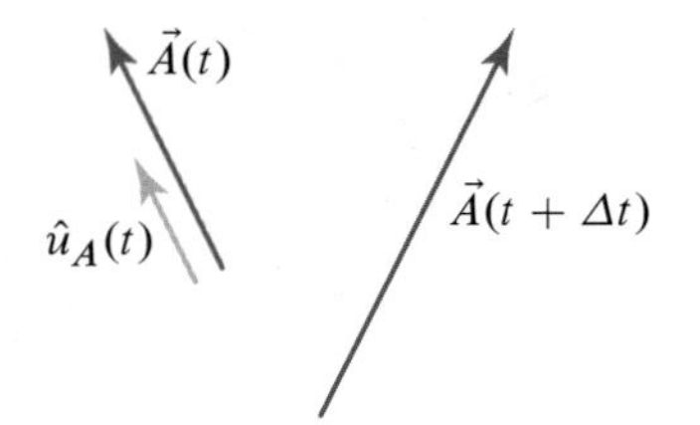

Figure 2.16
A vector $\vec{A}$ changing from $\vec{A}(t)$ to $\vec{A}(t+\Delta t)$.

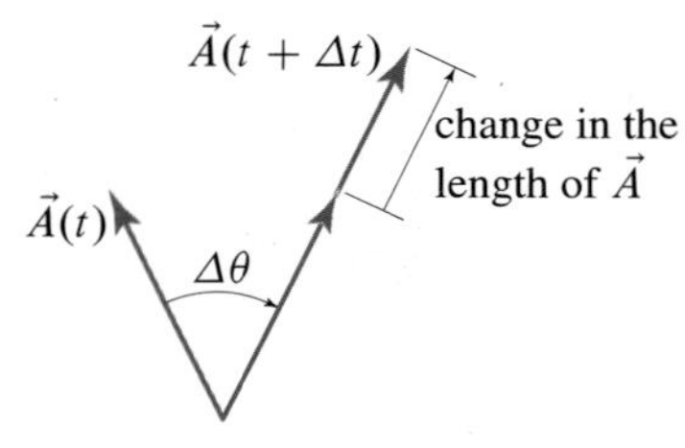

Figure 2.17
Figure 2.16 with vectors $\vec{A}(t)$ and $\vec{A}(t+\Delta t)$ drawn with their tail points coinciding.

Consider a vector $\vec{A}$ in the plane of the page at time t and time $t+\Delta t$, as shown in Fig. 2.16. The vector $\vec{A}$ is changing in length and direction, as well as in the position of its "tail." We now consider the change in $\vec{A}$, as depicted in Fig. 2.17, in which the tail points at t and $t+\Delta t$ are made to coincide. Since we can express $\vec{A}$ as a magnitude times a unit vector $\hat{u}_A$ in the direction of $\vec{A}$, differentiating it with respect to time, we obtain

$$\dot{\vec{A}}(t) = \frac{d}{dt}\big[A(t)\,\hat{u}_A(t)\big] = \frac{dA}{dt}\,\hat{u}_A + A\,\frac{d\hat{u}_A}{dt} = \dot{A}\,\hat{u}_A + A\,\dot{\hat{u}}_A, \tag{2.40}$$

where $A = |\vec{A}|$ is the magnitude of $\vec{A}$, and we have used the product rule of differentiation. Equation (2.40) shows that the time derivative of $\vec{A}$ consists of two parts:

1. The term $\dot{A}\,\hat{u}_A$, which is a vector in the direction of $\vec{A}$ measuring the time rate of change of the *magnitude* of $\vec{A}$, and
2. The term $A\,\dot{\hat{u}}_A$, which is the magnitude of $\vec{A}$ multiplied by the time derivative of the unit vector $\hat{u}_A$ measuring the time rate of change of the *direction* of $\vec{A}$.

To understand $\dot{\vec{A}}$, we now need to understand the vector $\dot{\hat{u}}_A$.

Time derivative of a unit vector

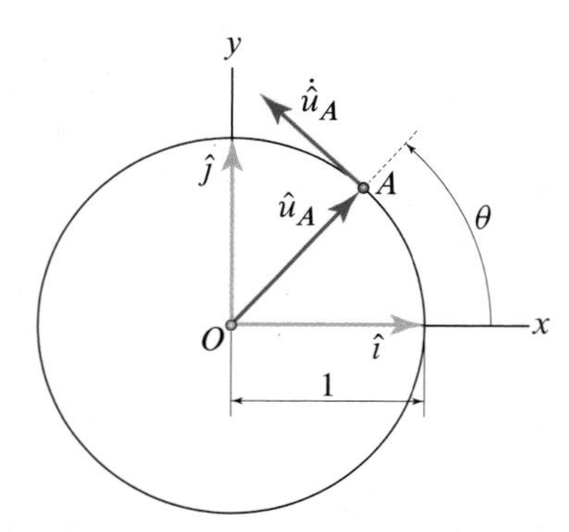

Figure 2.18
A rotating unit vector $\hat{u}_A$.

To understand and obtain a formula for the time derivative of a unit vector, we begin by expressing it as (see Fig. 2.18)

$$\hat{u}_A = \cos\theta\,\hat{\imath} + \sin\theta\,\hat{\jmath}, \tag{2.41}$$

so that

$$\dot{\hat{u}}_A = \dot{\theta}(-\sin\theta\,\hat{\imath} + \cos\theta\,\hat{\jmath}). \tag{2.42}$$

The term $-\sin\theta\,\hat{\imath} + \cos\theta\,\hat{\jmath}$ in Eq. (2.42) is a unit vector perpendicular to $\hat{u}_A$. This implies that $\dot{\hat{u}}_A$ is perpendicular to $\hat{u}_A$ (this is shown in Fig. 2.18). The term $\dot{\theta}$ in Eq. (2.42) is the rotation rate of $\hat{u}_A$ since θ measures the orientation of $\hat{u}_A$. Equation (2.42) gives us $\dot{\hat{u}}_A$ in this case, but can we find a more general expression? We can if we define the angular velocity of $\hat{u}_A$ as a vector. In particular, we define the following:

axis of rotation is the "hinge line" about which an object rotates. For the door in Fig. 2.19, the indicated hinge line AB is the axis of rotation of the door.

direction of rotation orients the axis of rotation via the right-hand rule, with the thumb pointing in the direction of rotation (see Fig. 2.20). This direction can be indicated by a unit vector.

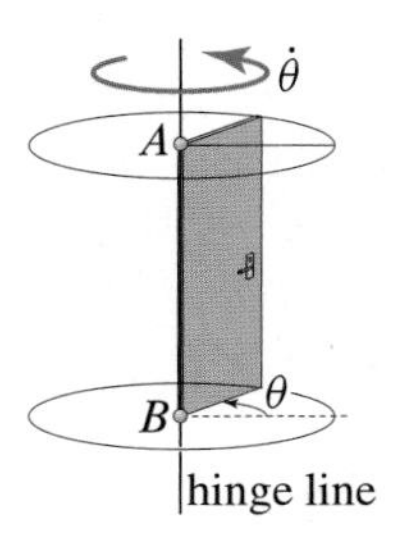

Figure 2.19
The hinge line of a swinging door.

With these definitions in mind, we can now define the angular velocity vector $\vec{\omega}_A$ of the unit vector $\hat{u}_A$ in Fig. 2.18 as

$$\vec{\omega}_A = \dot{\theta}\,\hat{k}, \tag{2.43}$$

where we note that since $\hat{u}_A$ is rotating in the xy plane, its angular velocity must be perpendicular to it. Equation (2.43) implies that the vectors $\hat{u}_A$, $\dot{\hat{u}}_A$, and $\vec{\omega}_A$ are mutually perpendicular and that the vector $\vec{\omega}_A \times \hat{u}_A$ is parallel to $\dot{\hat{u}}_A$, that is,

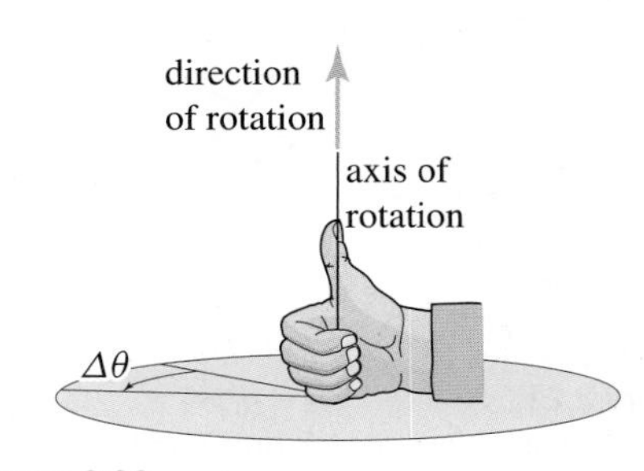

Figure 2.20
Direction of rotation defined using the right-hand rule.

$$\begin{aligned}\vec{\omega}_A \times \hat{u}_A &= \dot{\theta}\,\hat{k} \times (\cos\theta\,\hat{\imath} + \sin\theta\,\hat{\jmath}) \\ &= \dot{\theta}(\cos\theta\,\hat{k} \times \hat{\imath} + \sin\theta\,\hat{k} \times \hat{\jmath}) \\ &= \dot{\theta}(\cos\theta\,\hat{\jmath} - \sin\theta\,\hat{\imath}).\end{aligned} \tag{2.44}$$

Equation (2.44) is a remarkable result because, by comparing it with Eq. (2.42), it tells us that $\vec{\omega} \times \hat{u}_A$ is not just parallel to $\dot{\hat{u}}_A$, it is $\dot{\hat{u}}_A$! That is,

$$\dot{\hat{u}}_A = \vec{\omega}_A \times \hat{u}_A. \tag{2.45}$$

It turns out that the result in Eq. (2.45) is *universal*: whether in 2D or 3D, the time rate of change of a unit vector can always be represented as the cross product of the angular velocity of that unit vector with the unit vector in question.

We now have a general and physically motivated way to express the *time derivative of any unit vector* $\vec{u}$ as

$$\boxed{\dot{\hat{u}} = \vec{\omega}_u \times \hat{u},} \tag{2.46}$$

where $\vec{\omega}_u$ is the angular velocity of $\vec{u}$. Equation (2.46) is best remembered as

> *The time derivative of a unit vector is the angular velocity of the vector crossed with the vector itself.*

Time derivative of an arbitrary vector

Using Eq. (2.46), the interpretation of the term $A\,\dot{\hat{u}}_A$ in Eq. (2.40) is now clear: it is the magnitude of $\vec{A}$ times $\vec{\omega}_A \times \hat{u}_A$, so Eq. (2.40) becomes

$$\dot{\vec{A}}(t) = \dot{A}\,\hat{u}_A + A\,\vec{\omega}_A \times \hat{u}_A = \dot{A}\,\hat{u}_A + \vec{\omega}_A \times A\,\hat{u}_A, \tag{2.47}$$

or

$$\boxed{\dot{\vec{A}}(t) = \dot{A}\,\hat{u}_A + \vec{\omega}_A \times \vec{A}.} \tag{2.48}$$

We can now interpret the time derivative of any vector as the time rate of change of the magnitude of the vector plus a time rate of change of direction of that vector

$$\underbrace{\dot{\vec{A}}(t)}_{\substack{\text{change}\\ \text{in } \vec{A}}} = \underbrace{\dot{A}\,\hat{u}_A}_{\substack{\text{change in}\\ \text{magnitude}}} + \underbrace{\vec{\omega}_A \times \vec{A}}_{\substack{\text{change in}\\ \text{direction}}}, \tag{2.49}$$

where the time rate of change of direction is given by the cross product of the vector's angular velocity and the vector itself. This relationship applies to *any* vector, and we will use it *extensively*.

End of Section Summary

The time derivative of a vector consists of a part due to the vector's change in length and a part due to the vector's change in direction. Since a unit vector never changes its length, its time derivative only consists of a contribution from a change in direction:

Eq. (2.46), p. 81

$$\underbrace{\dot{\hat{u}}(t)}_{\text{change in } \hat{u}} = \underbrace{\vec{\omega}_u \times \hat{u}}_{\text{change in direction}}.$$

For an arbitrary vector $\vec{A}$, the following relationship provides its time derivative:

Eq. (2.48), p. 81

$$\underbrace{\dot{\vec{A}}(t)}_{\text{change in } \vec{A}} = \underbrace{\dot{A}\,\hat{u}_A}_{\text{change in magnitude}} + \underbrace{\vec{\omega}_A \times \vec{A}}_{\text{change in direction}},$$

where, referring to Fig. 2.21, $\hat{u}_A$ is a unit vector in the direction of $\vec{A}$, $\dot{A} = d|\vec{A}|/dt$, and $\vec{\omega}_A$ is the angular velocity of the vector $\vec{A}$.

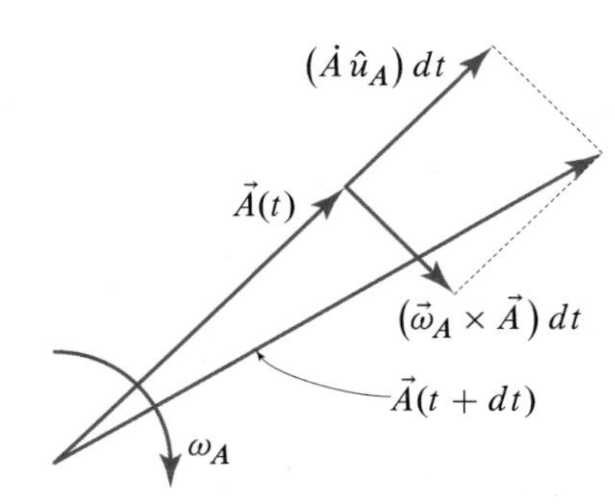

Figure 2.21
Depiction of the vector $\vec{A}$ at time t and at time $t+dt$ showing its change in magnitude and change in direction.

EXAMPLE 2.13 *Time Derivative of a Vector and Circular Motion*

The point Q is moving on a circular path of radius r, which is centered at the fixed point O (see Fig. 1). Use the notion of the time derivative of a vector as developed in this section to determine expressions for the velocity and acceleration vectors of a particle whose motion is circular.

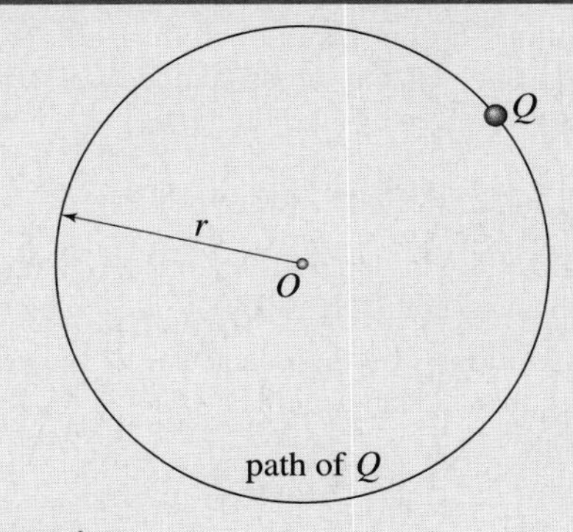

Figure 1
Point Q moving on a circle centered at O.

SOLUTION

Road Map As with the previous example, we start by defining a vector to differentiate with respect to time. In this case, we choose the vector defining the position of Q relative to O. We will then determine the time derivatives of this vector, which will, in turn, give us the velocity and acceleration of the point Q as it moves in a circle.

Computation Referring to Fig. 2, the position of Q relative to O is

$$\vec{r}_Q = r\,\hat{u}_Q, \tag{1}$$

where r is the radius of the circle and $\hat{u}_Q$ is the unit vector pointing from O to Q. Using Eq. (2.48) to compute the velocity of Q, we obtain

$$\vec{v}_Q = \dot{\vec{r}}_Q = \dot{r}\,\hat{u}_Q + \vec{\omega}_Q \times \vec{r}_Q = r\,\vec{\omega}_Q \times \hat{u}_Q, \tag{2}$$

where we have used the fact that r is constant and we have used Eq. (1) to obtain the last equality.

To determine $\vec{\omega}_Q$, which is the angular velocity of the unit vector $\hat{u}_Q$, observe that $\hat{u}_Q$ always lies in the xy plane, so that its axis of rotation must be parallel to the z axis. In addition, it is clear that the unit vector $\hat{u}_Q$ rotates at the rate $\dot{\beta}$, and so we must have

$$\vec{\omega}_Q = \omega_Q\,\hat{k} \quad \text{and} \quad \omega_Q = \dot{\beta}. \tag{3}$$

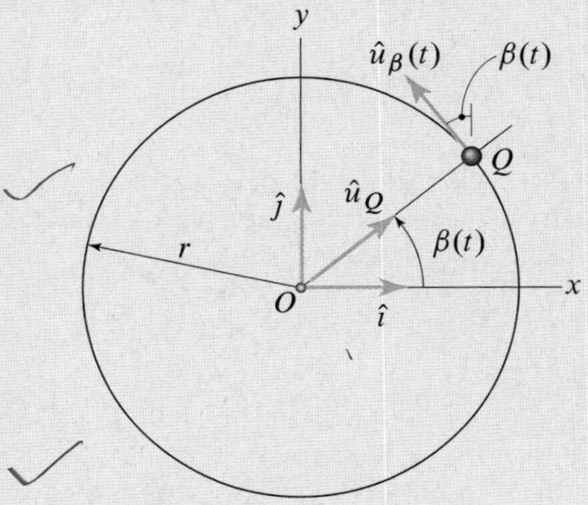

Figure 2
Particle Q moving along a circle centered at O.

Substituting Eqs. (3) into Eq. (2), we obtain

$$\vec{v}_Q = r\dot{\beta}\,\hat{k} \times \hat{u}_Q, \tag{4}$$

which we will now evaluate in a couple of ways.

For the first way, referring to Fig. 2, we can write $\hat{u}_Q$ in terms of its Cartesian components as

$$\hat{u}_Q = \cos\beta\,\hat{\imath} + \sin\beta\,\hat{\jmath}, \tag{5}$$

which, when substituted into Eq. (4), gives

$$\vec{v}_Q = r\dot{\beta}\,\hat{k} \times (\cos\beta\,\hat{\imath} + \sin\beta\,\hat{\jmath}) = r\dot{\beta}(-\sin\beta\,\hat{\imath} + \cos\beta\,\hat{\jmath}). \tag{6}$$

Again referring to Fig. 2, notice that the term $-\sin\beta\,\hat{\imath} + \cos\beta\,\hat{\jmath}$ in Eq. (6) is equal to the unit vector $\hat{u}_\beta$, which is tangent to the circle at Q and pointing in the direction of increasing β. Therefore, $\vec{v}_Q$ can be given the following compact form

$$\boxed{\vec{v}_Q = r\dot{\beta}\,\hat{u}_\beta,} \tag{7}$$

which reminds us that the velocity vector is *always* tangent to the path and is directly proportional to both the angular velocity and the radius of the circular path.

We can also arrive at Eq. (7) by noting that the right-hand rule tells us that $\hat{k} \times \hat{u}_Q$ must be perpendicular to both $\hat{k}$ and $\hat{u}_Q$, and so points in the direction of $\hat{u}_\beta$. In addition, since $\hat{k}$ and $\hat{u}_Q$ are perpendicular to one another, the sine of the angle between them is equal to 1, and so

$$\hat{k} \times \hat{u}_Q = \hat{u}_\beta, \tag{8}$$

which, when substituted into Eq. (4), gives Eq. (7).

Helpful Information

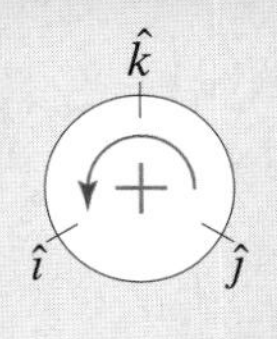

Cross products. Since this example requires the computation of several cross products, here is a reminder of a visual aid from Chapter 1 that helps with cross products. Arrange the three vectors $\hat{\imath}$, $\hat{\jmath}$, and $\hat{k}$ as shown; to calculate $\hat{\jmath} \times \hat{k}$, just move around the circle, starting from $\hat{\jmath}$ and going toward $\hat{k}$. Now notice that the next vector on the circle is $\hat{\imath}$, and in going from $\hat{\jmath}$ to $\hat{k}$ we have moved in the direction of the arrow. Hence, $\hat{\jmath} \times \hat{k} = +\hat{\imath}$. If instead we want to determine the outcome of $\hat{\imath} \times \hat{k}$, we need to notice that, moving along the circle starting from $\hat{\imath}$ and going toward $\hat{k}$, the subsequent vector is $\hat{\jmath}$, and we move opposite to the arrow. Therefore, we have that $\hat{\imath} \times \hat{k} = -\hat{\jmath}$.

Helpful Information

Another cross product reminder. The magnitude of the cross product of two vectors is given by (see Eq. (1.18))

$$|\hat{k} \times \hat{u}_Q| = |\hat{k}||\hat{u}_Q| \sin\theta,$$

where θ is the angle between the two vectors. Since both $\hat{k}$ and $\hat{u}_Q$ are unit vectors and $\hat{k}$ and $\hat{u}_Q$ are perpendicular to one another, this relation becomes

$$|\hat{k} \times \hat{u}_Q| = 1,$$

which is exactly what we have in Eq. (8).

Next, we will obtain the acceleration by taking the time derivative of the velocity as given by Eq. (2):

$$\vec{a}_Q = r\left(\dot{\vec{\omega}}_Q \times \hat{u}_Q + \vec{\omega}_Q \times \dot{\hat{u}}_Q\right). \tag{9}$$

Equation (2.46) tells us that $\dot{\hat{u}}_Q = \vec{\omega}_Q \times \hat{u}_Q$. Hence, Eq. (9) becomes

$$\vec{a}_Q = r\left[\dot{\vec{\omega}}_Q \times \hat{u}_Q + \vec{\omega}_Q \times (\vec{\omega}_Q \times \hat{u}_Q)\right]. \tag{10}$$

Differentiating Eq. (3) with respect to time, we obtain

$$\dot{\vec{\omega}}_Q = \ddot{\beta}\,\hat{k} + \dot{\beta}\,\dot{\hat{k}} = \ddot{\beta}\,\hat{k}, \tag{11}$$

where we have used the fact that $\dot{\hat{k}} = \vec{0}$. Substituting Eqs. (3), (5), and (11) into Eq. (10) and carrying out all the cross products, we obtain the following expression for the acceleration of Q:

$$\vec{a}_Q = -r(\ddot{\beta}\sin\beta - \dot{\beta}^2\cos\beta)\,\hat{\imath} + r(\ddot{\beta}\cos\beta - \dot{\beta}^2\sin\beta)\,\hat{\jmath}. \tag{12}$$

Rearranging the acceleration terms in the following way

$$\vec{a}_Q = -\dot{\beta}^2\, r(\cos\beta\,\hat{\imath} + \sin\beta\,\hat{\jmath}) + \ddot{\beta}\, r(-\sin\beta\,\hat{\imath} + \cos\beta\,\hat{\jmath}) \tag{13}$$

allows us to write $\vec{a}_Q$ as

$$\boxed{\vec{a}_Q = -r\dot{\beta}^2\,\hat{u}_Q + r\ddot{\beta}\,\hat{u}_\beta.} \tag{14}$$

Note that we can also compute $\vec{a}_Q$ by differentiating Eq. (7) with respect to time to obtain

$$\vec{a}_Q = \dot{r}\dot{\beta}\,\hat{u}_\beta + r\ddot{\beta}\,\hat{u}_\beta + r\dot{\beta}\dot{\hat{u}}_\beta = r\ddot{\beta}\,\hat{u}_\beta + r\dot{\beta}\dot{\hat{u}}_\beta, \tag{15}$$

where we have used the fact that r is constant. Now, since $\dot{\hat{u}}_\beta = \vec{\omega}_\beta \times \hat{u}_\beta$, $\vec{\omega}_\beta = \vec{\omega}_Q = \dot{\beta}\,\hat{k}$, and $\dot{\beta}\,\hat{k} \times \hat{u}_\beta = -\dot{\beta}\,\hat{u}_Q$, Eq. (15) becomes

$$\vec{a}_Q = r\ddot{\beta}\,\hat{u}_\beta + (r\dot{\beta})(-\dot{\beta}\,\hat{u}_Q) = r\ddot{\beta}\,\hat{u}_\beta - r\dot{\beta}^2\,\hat{u}_Q, \tag{16}$$

which is identical to Eq. (14).

Discussion & Verification Equation (14) tells us that, in a circular motion, the acceleration vector has two components:

1. A *tangential component*, which is *tangent* to the circular path and is proportional to the angular acceleration $\ddot{\beta}$.
2. A *radial component*, which is
   (a) Proportional to the *square* of the angular velocity.
   (b) *Always* directed *radially* inward toward the center of the circular trajectory.

Finally, observe that both the tangential and radial components of the acceleration are *directly proportional* to the radius of the circular trajectory.

EXAMPLE 2.14 *Time Derivative of a Vector Applied to a Tracking Problem*

An airplane B is flying at a constant speed v_0 and at a constant altitude h (see Fig. 1). The radar station at A tracks the plane by measuring the distance r between it and the plane, the rate at which r is changing, the antenna orientation θ, and the angular velocity of the antenna. Determine the relationships between those quantities that can be found by the tracking station and the speed and height of the airplane.

Figure 1
A radar station tracking a plane in flight.

SOLUTION

Road Map We will see that the height of the plane is found by using a simple geometric relationship. To obtain the relationship for speed that we seek, we will make use of an idea that we will call on many times: we can equate a *specific* expression for something (i.e., the velocity of the plane is horizontal) to a *general* expression for that something (i.e., the velocity of the plane using Eq. (2.48)).

Computation Referring to Fig. 2, given that we know θ and r, and assuming that h_A is also known, finding the altitude of the plane h in terms of θ and r is an easy matter since

$$\boxed{h = r\sin\theta + h_A.} \tag{1}$$

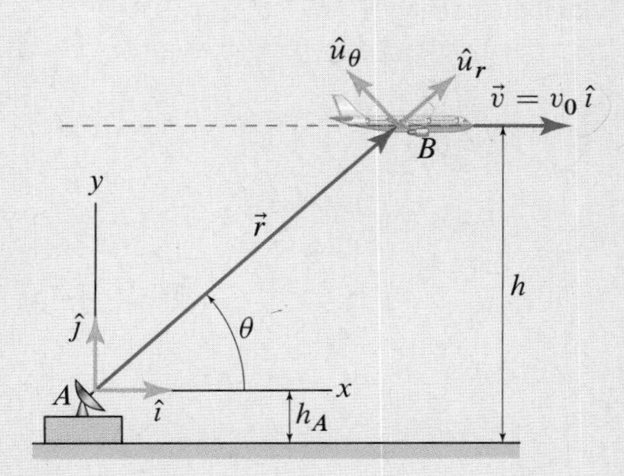

Figure 2
The given and defined dimensions and coordinate directions.

To find the speed of the plane v_0, we need to first write its position and then differentiate it. Again referring to Fig. 2, the natural position vector to use is the position of the plane relative to the radar station, which can be represented as its magnitude times a unit vector parallel to $\vec{r}$ and pointing from A to B

$$\vec{r} = r\,\hat{u}_r, \tag{2}$$

where $r = |\vec{r}|$. Using this expression for $\vec{r}$, we can employ Eq. (2.48) to write the velocity of the plane as

$$\dot{\vec{r}} = \dot{r}\,\hat{u}_r + \vec{\omega}_r \times \vec{r} = \dot{r}\,\hat{u}_r + \vec{\omega}_r \times r\hat{u}_r. \tag{3}$$

To interpret $\vec{\omega}_r$, observe that the position vector $\vec{r}$ remains in the xy plane at all times, and its rotation rate is measured by $\dot{\theta}$. Using the right-hand rule tells us that $\vec{\omega}_r$ is given by

$$\vec{\omega}_r = \dot{\theta}\,\hat{k} = \omega_r\,\hat{k}, \tag{4}$$

where $\omega_r = \dot{\theta}$ will be negative if θ turns out to be decreasing. Substituting Eq. (4) into Eq. (3), we obtain

$$\dot{\vec{r}} = \dot{r}\,\hat{u}_r + \omega_r\,\hat{k} \times r\hat{u}_r = \dot{r}\,\hat{u}_r + r\omega_r\,\hat{u}_\theta, \tag{5}$$

where, referring to Fig. 2, we have used the fact that $\hat{k} \times \hat{u}_r = \hat{u}_\theta$ and we have simply defined $\hat{u}_\theta$ to be a unit vector perpendicular to $\hat{u}_r$ and pointing in the direction of increasing θ.*

Now, here is where we equate the general with the specific as mentioned in the Road Map. We *know* that the velocity of the plane is straight and horizontal, so it can be written as

$$\vec{v} = v_0\,\hat{\imath}. \tag{6}$$

This *specific* relation for the velocity *must* equal the general relation for the velocity given in Eq. (5). Therefore, it must be true that

$$v_0\,\hat{\imath} = \dot{r}\,\hat{u}_r + r\omega_r\,\hat{u}_\theta. \tag{7}$$

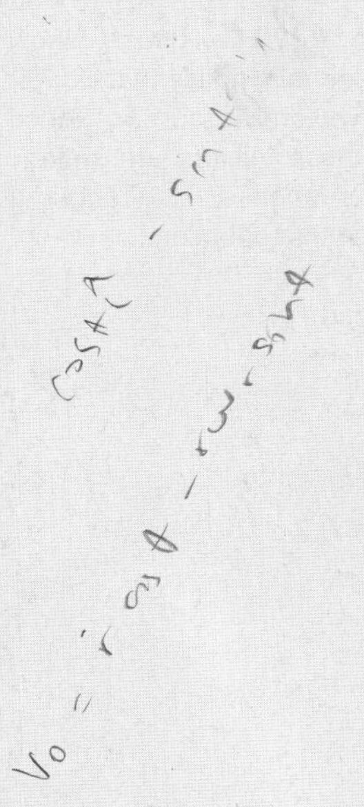

* We will see *a lot* more of these two unit vectors in Section 2.6.

Helpful Information

Equating the general with the specific. This is a technique we will use again and again. In this case, we have a *general* expression for $\dot{\vec{r}}$ given in Eq. (5) that is true for *any* motion of $\vec{r}$. We also have another expression for $\dot{\vec{r}}$, namely, $\dot{\vec{r}} = v_0\,\hat{\imath}$, that is true for this example. Even though the two expressions for $\dot{\vec{r}}$ are different, they must both remain true. A useful idea to retain is that there is a lot to be gained by *forcing* the two relations to take on the same form. In more abstract terms, there is a lot to be gained by forcing a *general* relation for a given quantity to match a *specific* condition for that quantity.

The problem with Eq. (7) is that it is written in two different component systems. For Eq. (7) to be useful, we need to write it using a single component system. Expressing $\hat{\imath}$ in terms of $\hat{u}_r$ and $\hat{u}_\theta$, we obtain

$$v_0(\cos\theta\,\hat{u}_r - \sin\theta\,\hat{u}_\theta) = \dot{r}\,\hat{u}_r + r\omega_r\,\hat{u}_\theta. \tag{8}$$

Now, we equate the coefficients of $\hat{u}_r$ and the coefficients of $\hat{u}_\theta$ to obtain the following two equations:

$$v_0\cos\theta = \dot{r}, \tag{9}$$

$$-v_0\sin\theta = r\omega_r. \tag{10}$$

Since we know r, θ, $\dot{r}$, and ω_r, Eqs. (9) and (10) are two equations for one unknown: v_0. We can solve both equations and find that

$$\boxed{v_0 = \frac{\dot{r}}{\cos\theta} = -\frac{r\omega_r}{\sin\theta},} \tag{11}$$

where either expression would give v_0 from the radar measurements.

Discussion & Verification Recalling that r has dimensions of length, θ is nondimensional, and ω_r has dimensions of 1 over time, Eq. (11) is dimensionally correct since it has the dimensions of length over time, as expected. In addition, note that for the situation depicted in Fig. 2, we expect $\omega_r = \dot{\theta}$ to be negative, and this is consistent with Eq. (11). By solving Eq. (11) for ω_r, we obtain

$$\omega_r = -\frac{v_0\sin\theta}{r}, \tag{12}$$

which implies that $\omega_r < 0$ since $\sin\theta$ is positive, r is positive, and v_0 is positive because the plane is moving to the right. This result is also consistent with the right-hand rule since, for an airplane moving to the right, $\vec{r}$ rotates in the *negative* z direction.

Helpful Information

Where is θ in Eq. (5)? How does θ show up in Eq. (5)? To see this, notice that Eq. (5) contains the unit vectors $\hat{u}_r$ and $\hat{u}_\theta$. Since the directions of these vectors are entirely determined by the angle θ, the angle θ implicitly appears in that equation.

A Closer Look We stated that Eqs. (9) and (10) form a system of two equations in the one unknown v_0. Is there something wrong with this? The answer is no, and to understand this we need to go back and look more carefully at Eq. (5). This equation allows us to determine the velocity of the plane from radar measurements of r, θ, $\dot{r}$, and ω_r *no matter how the plane is moving* (see also marginal note entitled "Where is θ in Eq. (5)?"). However, in solving the problem we took advantage of the fact that the plane is flying in a specific *known* direction, and this resulted in having more information than is actually needed to solve the problem. So we have two equations to find one unknown. This situation is acceptable so long as the two equations we have do not contradict each other. Our solution is acceptable because Eqs. (9) and (10) are consistent with one another as they both imply that the plane is flying at a constant altitude.

Problems

Problem 2.117

Consider the vectors $\vec{a} = 2\,\hat{\imath} + 1\,\hat{\jmath} + 7\,\hat{k}$ and $\vec{b} = 1\,\hat{\imath} + 2\,\hat{\jmath} + 3\,\hat{k}$. Compute the following quantities.

(a) $\vec{a} \times \vec{b}$

(b) $\vec{b} \times \vec{a}$

(c) $\vec{a} \times \vec{b} + \vec{b} \times \vec{a}$

(d) $\vec{a} \times \vec{a}$

(e) $(\vec{a} \times \vec{a}) \times \vec{b}$

(f) $\vec{a} \times (\vec{a} \times \vec{b})$

Parts (a)–(d) of this problem are meant to be a reminder that the cross product is an *anticommutative* operation, while Parts (e) and (f) are meant to be a reminder that the cross product is an operation that is *not associative*.

Problem 2.118

Consider two vectors $\vec{a} = 1\,\hat{\imath} + 2\,\hat{\jmath} + 3\,\hat{k}$ and $\vec{b} = -6\,\hat{\imath} + 3\,\hat{\jmath}$.

(a) Verify that $\vec{a}$ and $\vec{b}$ are perpendicular to one another.

(b) Compute the vector triple product $\vec{a} \times (\vec{a} \times \vec{b})$.

(c) Compare the result from calculating $\vec{a} \times (\vec{a} \times \vec{b})$ with the vector $-|\vec{a}|^2\,\vec{b}$.

The purpose of this exercise is to show that as long as $\vec{a}$ and $\vec{b}$ are perpendicular to one another, you can always write $\vec{a} \times (\vec{a} \times \vec{b}) = -|\vec{a}|^2\,\vec{b}$. This identity turns out to be very useful in the study of the planar motion of rigid bodies.

Problem 2.119

Let $\vec{r}$ be the position vector of a point P with respect to a Cartesian coordinate system with axes x, y, and z. Let the motion of P be confined to the xy plane, so that $\vec{r} = r_x\,\hat{\imath} + r_y\,\hat{\jmath}$ (i.e., $\vec{r} \cdot \hat{k} = 0$). Also, let $\vec{\omega}_r = \omega_r\,\hat{k}$ be the angular velocity vector of the vector $\vec{r}$. Compute the outcome of the products $\vec{\omega}_r \times (\vec{\omega}_r \times \vec{r})$ and $\vec{\omega}_r \times (\vec{r} \times \vec{\omega}_r)$.

Problem 2.120

The three propellers shown are all rotating with the same *angular speed* of 1000 rpm about different coordinate axes.

(a) Provide the proper vector expressions for the *angular velocity* of each of the three propellers.

(b) Suppose that an identical propeller rotates at 1000 rpm about the axis ℓ oriented by the unit vector $\hat{u}_\ell$. Let any point P on ℓ have coordinates such that $x_P = y_P = z_P$. Find the vector representation of the angular velocity of this fourth propeller.

Express the answers using units of radians per second.

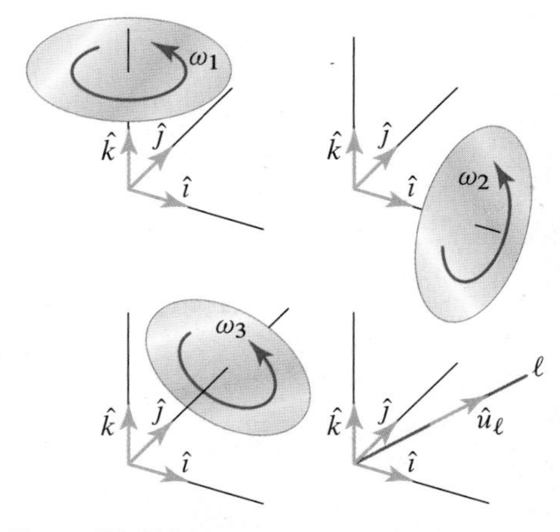

Figure P2.120

Problem 2.121

Point P is constrained to move along a straight line ℓ whose positive orientation is described by the unit vector $\hat{u}_\ell$. Point A is a fixed reference point on ℓ. Let the vector $\vec{r}_{P/A}$ denote the position of P relative to A and let $\hat{u}_{P/A}$ be a unit vector pointing from A to P. Use the concept of time derivative of a vector to describe the velocity and acceleration of P. In addition, comment on what happens to the description of the velocity and acceleration when P happens to coincide with the fixed point A.

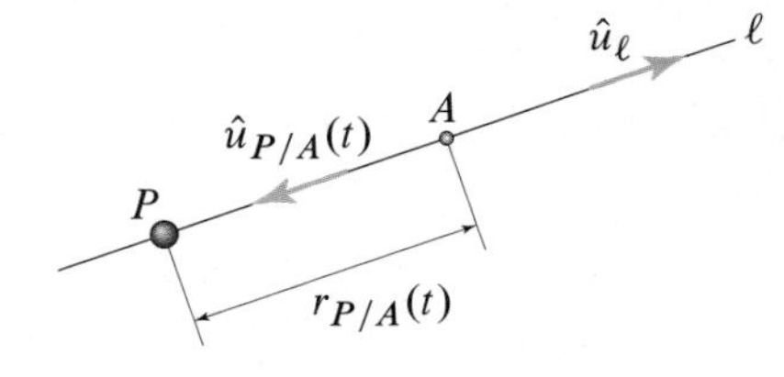

Figure P2.121

Problem 2.122

Starting with Eq. (2.48), show that the second derivative with respect to time of an arbitrary vector $\vec{A}$ is given by

$$\ddot{\vec{A}} = \ddot{A}\,\hat{u}_A + 2\vec{\omega}_A \times \dot{A}\,\hat{u}_A + \dot{\vec{\omega}}_A \times \vec{A} + \vec{\omega}_A \times (\vec{\omega}_A \times \vec{A}).$$

Keep the answer in pure vector form, and do not resort to using components in any component system.

Problem 2.123

The propeller shown has a diameter of 38 ft and is rotating with a constant angular speed of 400 rpm. At a given instant, a point P on the propeller is at $\vec{r}_P = (12.5\,\hat{\imath} + 14.3\,\hat{\jmath})$ ft. Use Eq. (2.48) and the equation derived in Prob. 2.122 to compute the velocity and acceleration of P, respectively.

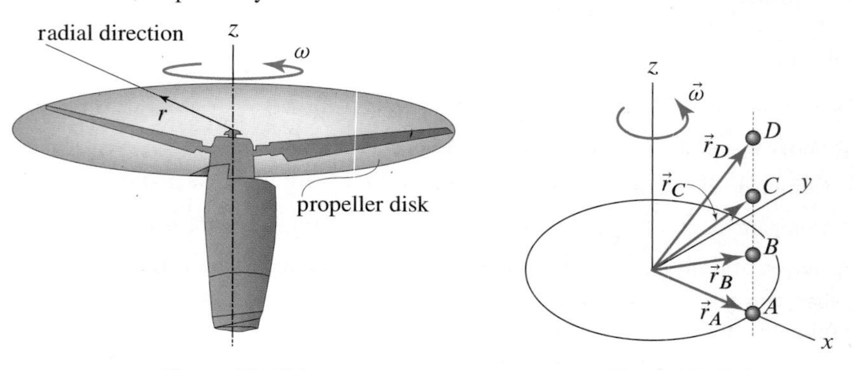

Figure P2.123

Figure P2.124

Problem 2.124

Consider the four points whose positions are given by the vectors $\vec{r}_A = (2\,\hat{\imath} + 0\,\hat{k})$ m, $\vec{r}_B = (2\,\hat{\imath} + 1\,\hat{k})$ m, $\vec{r}_C = (2\,\hat{\imath} + 2\,\hat{k})$ m, and $\vec{r}_D = (2\,\hat{\imath} + 3\,\hat{k})$ m. Knowing that the magnitude of these vectors is constant and that the angular velocity of these vectors at a given instant is $\vec{\omega} = 5\,\hat{k}$ rad/s, apply Eq. (2.48) to find the velocities $\vec{v}_A$, $\vec{v}_B$, $\vec{v}_C$, and $\vec{v}_D$. Explain why all the velocity vectors are the same even though the position vectors are not.

Problem 2.125

A child on a merry-go-round is moving radially outward at a constant rate of 4 ft/s. If the merry-go-round is spinning at 30 rpm, determine the velocity and acceleration of point P on the child when the child is 0.5 and 2.3 ft from the spin axis. Express the answers using the component system shown.

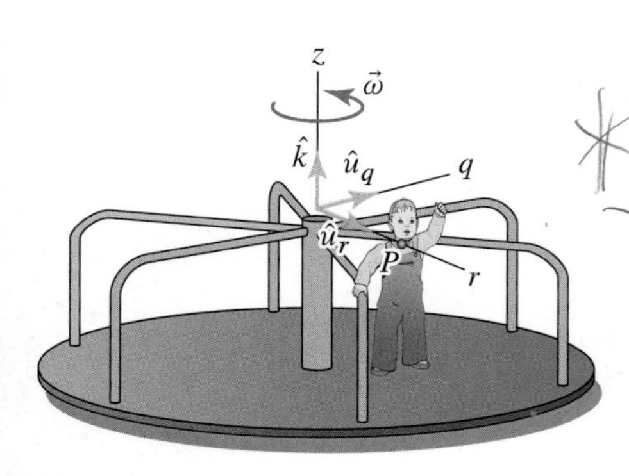

Figure P2.125

Problem 2.126

When a wheel rolls without slipping on a stationary surface, the point O on the wheel that is in contact with the rolling surface has zero velocity. With this in mind, consider a nondeformable wheel rolling without slip on a flat stationary surface. The center of the wheel P is traveling to the right with a constant speed $v_0 = 23$ m/s. Letting $R = 0.35$ m, determine the angular velocity of the wheel, using the stationary component system shown.

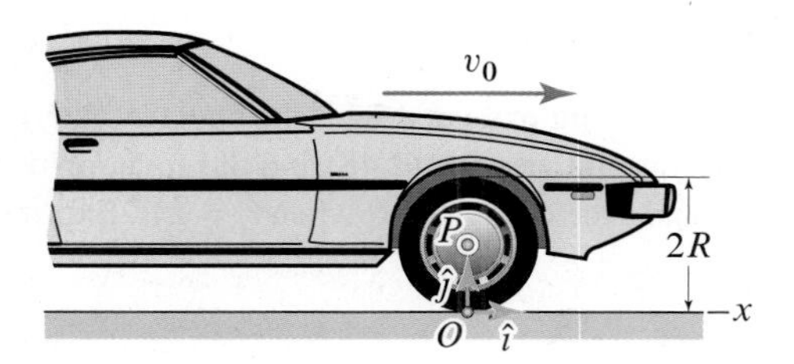

Figure P2.126

Problem 2.127

The radar station at O is tracking the meteor P as it moves through the atmosphere. At the instant shown, the station measures the following data for the motion of the meteor: $r = 21{,}000\,\text{ft}$, $\theta = 40°$, $\dot{r} = -22{,}440\,\text{ft/s}$, and $\dot{\theta} = -2.935\,\text{rad/s}$. Use Eq. (2.48) to determine the magnitude and direction (relative to the xy coordinate system shown) of the velocity vector at this instant.

Problem 2.128

The radar station at O is tracking the meteor P as it moves through the atmosphere. At the instant shown, the station measures the following data for the motion of the meteor: $r = 21{,}000\,\text{ft}$, $\theta = 40°$, $\dot{r} = -22{,}440\,\text{ft/s}$, $\dot{\theta} = -2.935\,\text{rad/s}$, $\ddot{r} = 187{,}500\,\text{ft/s}^2$, and $\ddot{\theta} = -5.409\,\text{rad/s}^2$. Use the equation derived in Prob. 2.122 to determine the magnitude and direction (relative to the xy coordinate system shown) of the acceleration vector at this instant.

Figure P2.127 and P2.128

Problem 2.129

A plane B is approaching a runway along the trajectory shown while the radar antenna A is monitoring the distance r between A and B, as well as the angle θ. If the plane has a constant approach speed v_0 as shown, use Eq. (2.48) to determine the expressions for $\dot{r}$ and $\dot{\theta}$ in terms of r, θ, v_0, and ϕ.

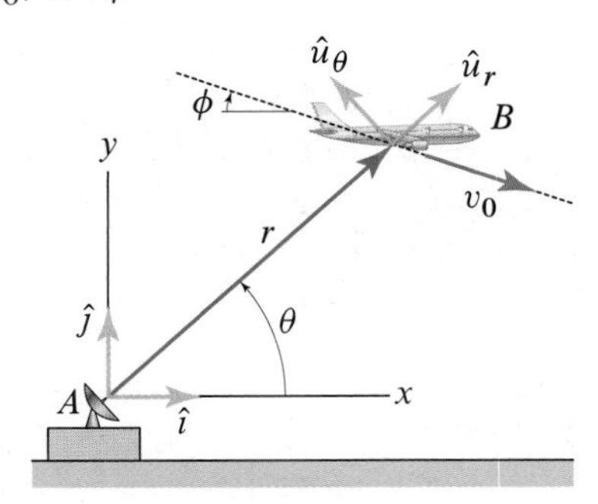

Figure P2.129 and P2.130

Problem 2.130

A plane B is approaching a runway along the trajectory shown with $\phi = 15°$, while the radar antenna A is monitoring the distance r between A and B, as well as the angle θ. The plane has a constant approach speed v_0. In addition, when $\theta = 20°$, it is known that $\dot{r} = 216\,\text{ft/s}$ and $\dot{\theta} = -0.022\,\text{rad/s}$. Use Eq. (2.48) to determine the corresponding values of v_0 and of the distance between the plane and the radar antenna.

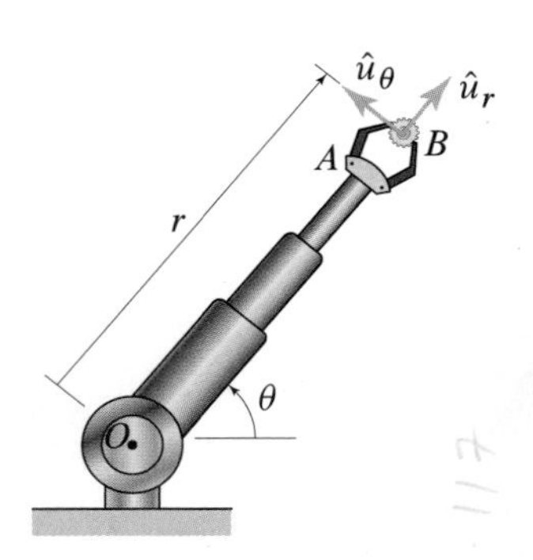

Figure P2.131

Problem 2.131

The end B of a robot arm is being extended with the constant rate $\dot{r} = 4\,\text{ft/s}$. Knowing that $\dot{\theta} = 0.4\,\text{rad/s}$ and is constant, use Eq. (2.48) and the equation derived in Prob. 2.122 to determine the velocity and acceleration of B when $r = 2\,\text{ft}$. Express your answer using the component system shown.

Problem 2.132

The end B of a robot arm is moving vertically down with a constant speed $v_0 = 2\,\text{m/s}$. Letting $d = 1.5\,\text{m}$, apply Eq. (2.48) to determine the rate at which r and θ are changing when $\theta = 37°$.

Problem 2.133

The end B of a robot arm is moving vertically down with a constant speed $v_0 = 6\,\text{ft/s}$. Letting $d = 4\,\text{ft}$, use Eq. (2.48) and the equation derived in Prob. 2.122 to determine $\dot{r}$, $\dot{\theta}$, $\ddot{r}$, and $\ddot{\theta}$ when $\theta = 0°$.

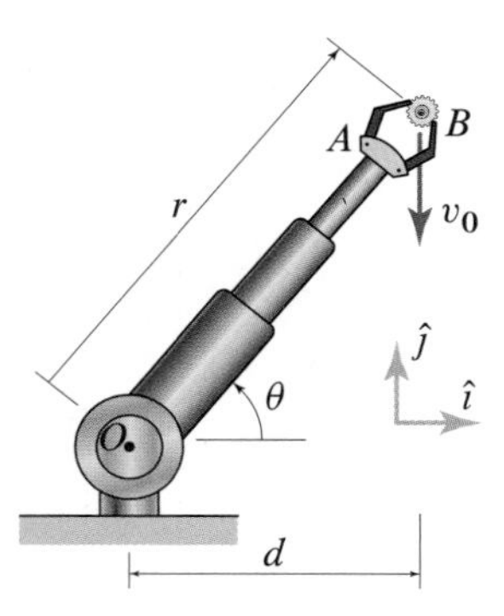

Figure P2.132 and P2.133

Problem 2.134

A micro spiral pump consists of a spiral channel attached to a stationary plate. This plate has two ports, one for fluid inlet and the other for outlet, the outlet being farther from the center of the plate than the inlet. The system is capped by a rotating disk. The fluid trapped between the rotating disk and stationary plate is put in motion by the rotation of the top disk, which pulls the fluid through the spiral channel. With this in mind, consider a channel with geometry given by the equation $r = \eta\theta + r_0$, where $\eta = 12\,\mu\text{m}$ is called the polar slope, $r_0 = 146\,\mu\text{m}$ is the radius at the inlet, r is the distance from the spin axis, and θ, measured in radians, is the angular position of a point in the spiral channel. If the top disk rotates with a constant angular speed $\omega = 30{,}000\,\text{rpm}$, and assuming that the fluid particles in contact with the rotating disk are essentially stuck to it, determine the velocity and acceleration of one such fluid particle when it is at $r = 170\,\mu\text{m}$.* Express the answer using the component system shown (which rotates with the top disk).

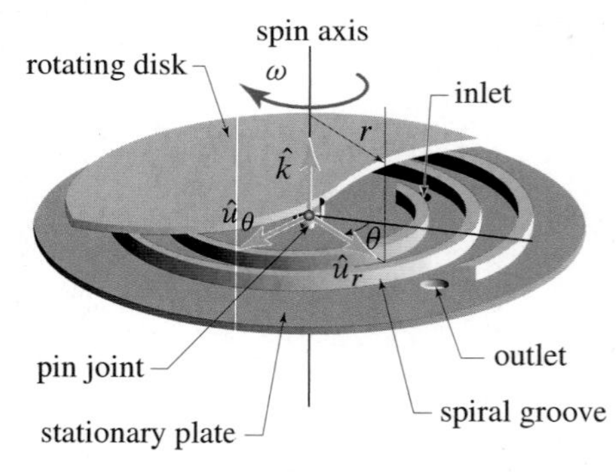

Figure P2.134

Problem 2.135

A disk rotates about its center, which is the fixed point O. The disk has a straight channel whose centerline passes by O and within which a collar A is allowed to slide. If, when A passes by O, the speed of A relative to the channel is $v = 14\,\text{m/s}$ and is increasing

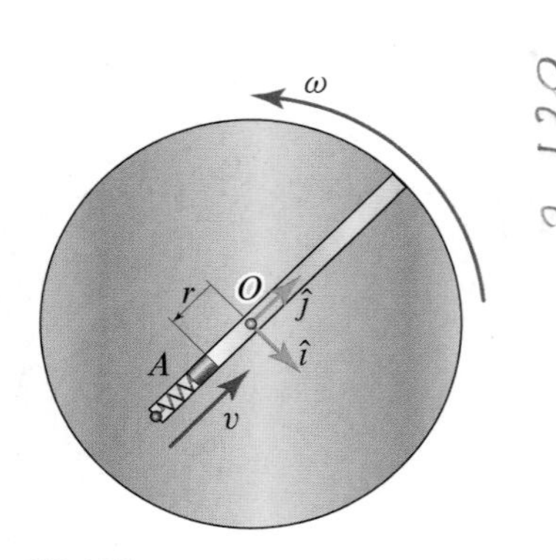

Figure P2.135

* The spiral pump was originally invented in 1746 by H. A. Wirtz, a Swiss pewterer from Zurich. Recently, the spiral pump concept has seen a comeback in microdevice design. The data used in this example is taken from M. I. Kilani, P. C. Galambos, Y. S. Haik, and C.-J. Chen, "Design and Analysis of a Surface Micromachined Spiral-Channel Viscous Pump," *Journal of Fluids Engineering*, **125**, pp. 339–344, 2003.

in the direction shown with a rate of $5\,\mathrm{m/s^2}$, determine the acceleration of A given that $\omega = 4\,\mathrm{rad/s}$ and is constant. Express the answer using the component system shown, which rotates with the disk. *Hint:* Apply the equation derived in Prob. 2.122 to the vector describing the position of A relative to O and then let $r = 0$.

Problem 2.136

At the instant shown, the angular velocity and acceleration of the merry-go-round are as indicated in the figure. The distance of the child from the spin axis is r_P, so his acceleration is $\vec{a}_P = \ddot{r}_P\,\hat{u}_r + \dot{r}_P\,\dot{\hat{u}}_r + \dot{\vec{\omega}} \times r_P\,\hat{u}_r + \vec{\omega} \times \dot{r}_P\,\hat{u}_r + \vec{\omega} \times r_P\,\dot{\hat{u}}_r$. Assuming that the child is walking along a radial line, should the child walk outward or inward to make sure that he does not experience any sideways acceleration (i.e., in the direction of $\hat{u}_q$)?

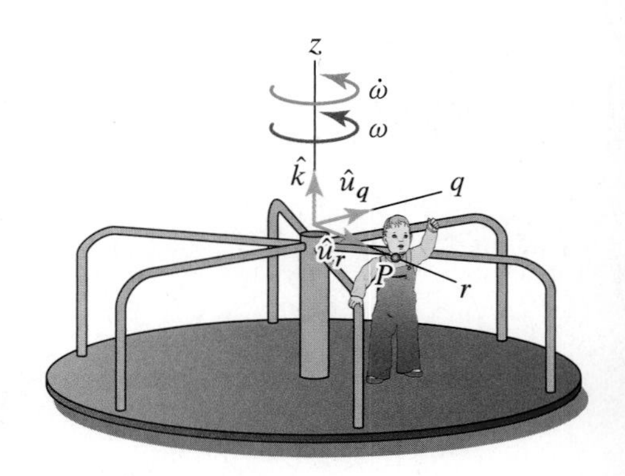

Figure P2.136 and P2.137

Problem 2.137

Assuming that the child shown is moving on the merry-go-round along a radial line, use the equation derived in Prob. 2.122 to determine the relation that ω, $\dot{\omega}$, r, and $\dot{r}$ must satisfy so that the child will not experience any sideways acceleration.

Problem 2.138

The mechanism shown is called a *swinging block* slider crank. First used in various steam locomotive engines in the 1800s, this mechanism is often found in door-closing systems. If the disk is rotating with a constant angular velocity $\dot{\theta} = 60\,\mathrm{rpm}$, $H = 4\,\mathrm{ft}$, $R = 1.5\,\mathrm{ft}$, and r is the distance between B and O, compute $\dot{r}$ and $\dot{\phi}$ when $\theta = 90°$. *Hint:* Apply Eq. (2.48) to the vector describing the position of B relative to O.

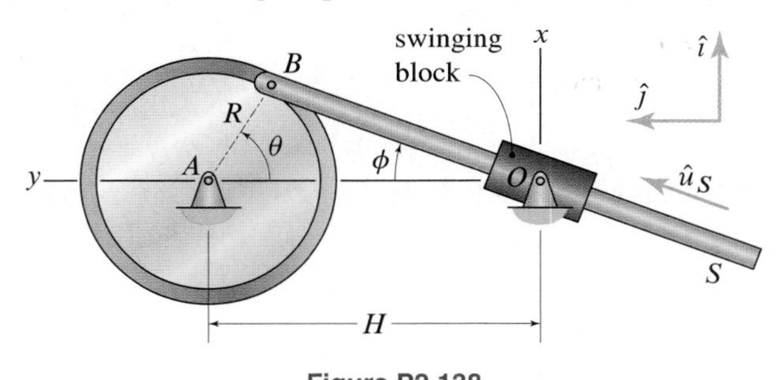

Figure P2.138

Problems 2.139 and 2.140

A sprinkler essentially consists of a pipe AB mounted on a hollow shaft. The water comes in the pipe at O and goes out the nozzles at A and B, causing the pipe to rotate. Assume that the particles of water move through the pipe at a constant rate *relative to the pipe* of $5\,\mathrm{ft/s}$ and that the pipe AB is rotating at a constant angular velocity of $250\,\mathrm{rpm}$. In all cases, express the answers using the right-handed and orthogonal component system shown.

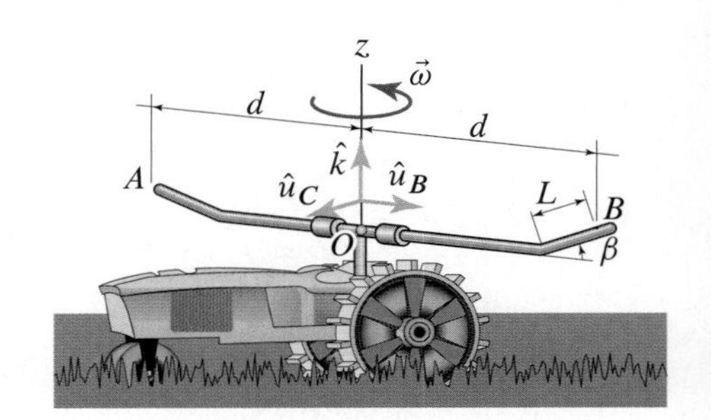

Figure P2.139 and P2.140

Problem 2.139 Determine the acceleration of the water particles when they are at $d/2$ from O (still within the horizontal portion of the pipe). Let $d = 7\,\mathrm{in.}$

Problem 2.140 Determine the acceleration of the water particles right before they are expelled at B. Let $d = 7\,\mathrm{in.}$, $\beta = 15°$, and $L = 2\,\mathrm{in.}$ *Hint:* In this case, the vector describing the position of a water particle at B goes from O to B and is best written as $\vec{r} = r_B\,\hat{u}_B + r_z\,\hat{k}$.

2.5 Planar Motion: Normal-Tangential Components

It is not always convenient to study motion using a Cartesian coordinate system. In this section and the next, we will learn about two additional ways to describe motion. Here we introduce a way to describe the velocity and acceleration vectors of a particle that is based *entirely* on the path of the particle.

Normal-tangential components

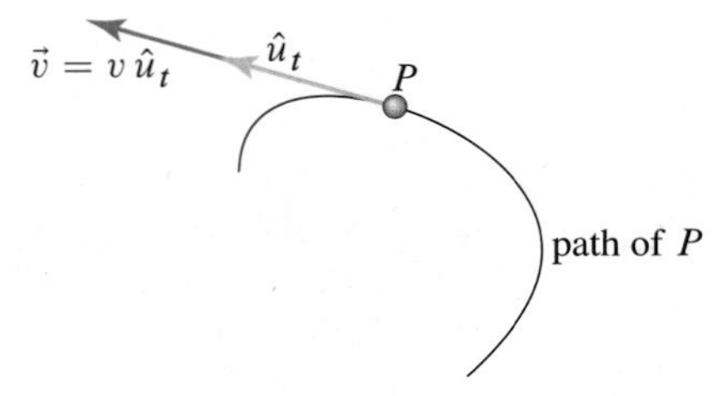

Figure 2.22
The velocity vector $\vec{v}$ defined in terms of the unit tangent vector $\hat{u}_t$.

Referring to Fig. 2.22, the particle P is moving along an arbitrary path. We denote by $\hat{u}_t$ the unit vector tangent to the path at P and pointing in the direction of motion. Since the velocity vector is always tangent to the path, we can write it using $\hat{u}_t$ as

$$\boxed{\vec{v} = v\,\hat{u}_t,} \tag{2.50}$$

where v is the speed of the particle. To obtain the acceleration of P, we differentiate Eq. (2.50) with respect to time to obtain

$$\vec{a} = \dot{\vec{v}} = \dot{v}\,\hat{u}_t + v\,\dot{\hat{u}}_t. \tag{2.51}$$

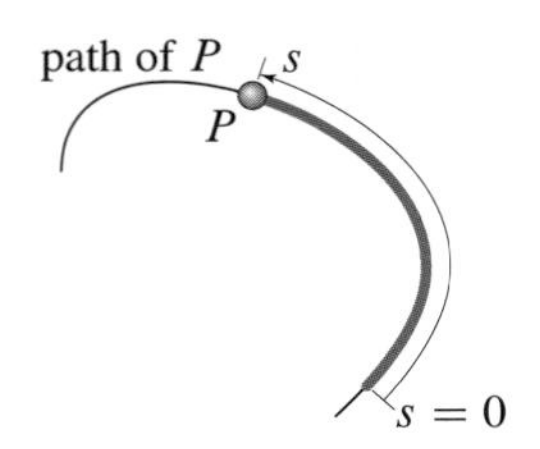

Figure 2.23
A point P moving along a path showing the arc length s.

To determine a convenient expression for $\dot{\hat{u}}_t$, we begin by introducing the *arc length* $s(t)$, defined as the *distance traveled by P along its path from $t = 0$ to the current time t* (Fig 2.23). Since s is the distance traveled, $s \geq 0$ and, no matter which direction P moves along the path, s continues to increase. Now, since $\vec{v}$ changes direction as P moves along the path, we can view the unit vector $\hat{u}_t$ as a function of $s(t)$ (unless the path is straight). We can use the rate of change of $\hat{u}_t(s)$ with respect to s, i.e., the vector $d\hat{u}_t(s)/ds$, to describe how "bendy" the path is, that is, its curvature. The *curvature* of the path, traditionally denoted by the Greek letter κ (kappa), is defined as

$$\kappa(s) = \left|\frac{d\hat{u}_t(s)}{ds}\right|, \tag{2.52}$$

which has dimensions of 1 over length. If the path is straight, $\hat{u}_t(s) =$ constant and $\kappa(s) = 0$. Writing $\dot{\hat{u}}_t$ using the chain rule, we see that

$$\dot{\hat{u}}_t = \frac{d\hat{u}_t}{ds}\frac{ds}{dt} \quad\Rightarrow\quad \vec{a} = \dot{v}\,\hat{u}_t + v^2\,\frac{d\hat{u}_t}{ds}, \tag{2.53}$$

where we have used Eq. (2.51) and $v = ds/dt$ from Eq. (2.9) on p. 31. We now find an expression for $d\hat{u}_t/ds$ by recalling that $\hat{u}_t(s)$ is a unit vector, which implies that $\hat{u}_t(s) \cdot \hat{u}_t(s) = 1 =$ constant. Therefore, we have

$$\frac{d}{ds}(\hat{u}_t \cdot \hat{u}_t) = \frac{d\hat{u}_t}{ds} \cdot \hat{u}_t + \hat{u}_t \cdot \frac{d\hat{u}_t}{ds} = 2\hat{u}_t \cdot \frac{d\hat{u}_t}{ds} = 0, \tag{2.54}$$

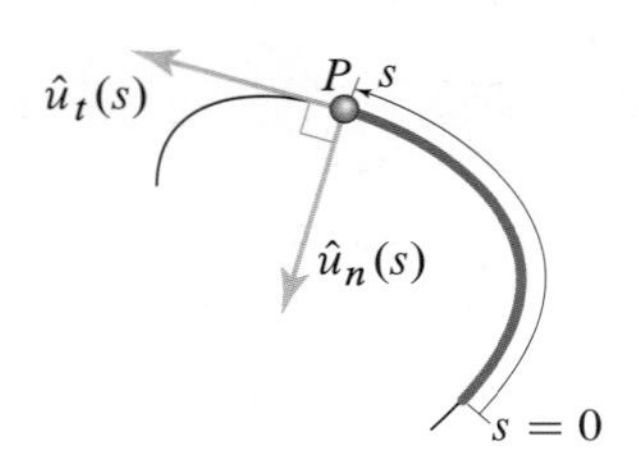

Figure 2.24
Principal unit normal to the path at P.

which means that $\hat{u}_t$ and $d\hat{u}_t/ds$ are orthogonal to one another. Consequently, when $\kappa \neq 0$, we can define a curvature-related unit normal to the path as:

$$\hat{u}_n(s) = \frac{d\hat{u}_t(s)/ds}{|d\hat{u}_t(s)/ds|} = \frac{1}{\kappa(s)}\frac{d\hat{u}_t(s)}{ds}, \tag{2.55}$$

where the subscript n stands for *normal*. Referring to Fig. 2.24, the unit vector $\hat{u}_n(s)$ is called the *principal unit normal* to the path, and it *always points toward the concave side of the curve*. Because $\hat{u}_n(s)$ is only defined when $\kappa \neq 0$, we cannot use $\hat{u}_n(s)$ when dealing with straight lines or at inflection points of curved lines.

If at P the curvature $\kappa \neq 0$, then Eq. (2.55) can be rewritten as

$$\frac{d\hat{u}_t}{ds} = \frac{1}{\rho}\hat{u}_n, \quad \text{where} \quad \rho(s) = \frac{1}{\kappa(s)} \tag{2.56}$$

is called the *radius of curvature* of the path and, in general, it changes along the path. If $\rho(s)$ is constant, then the path is a circle. Going back to the expression for the acceleration, substituting the first of Eqs. (2.56) into the second of Eqs. (2.53), we obtain

$$\vec{a} = \dot{v}\,\hat{u}_t + \frac{v^2}{\rho}\hat{u}_n = a_t\,\hat{u}_t + a_n\,\hat{u}_n, \tag{2.57}$$

where $a_t = \dot{v}$ and $a_n = v^2/\rho$ are the tangential and normal components of the acceleration, respectively (see Fig. 2.25).

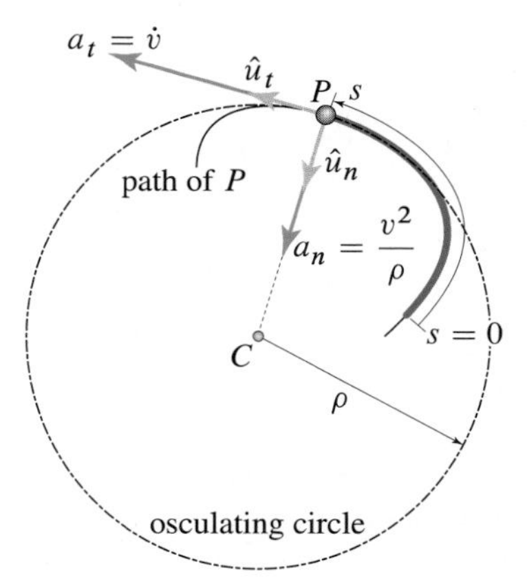

Figure 2.25
Acceleration in normal-tangential components. The *osculating circle* is the circle tangent to the path at P with radius ρ and center C on the concave side of the path. Even for three-dimensional motions, the acceleration vector lies always in the same plane as the osculating circle.

In a three-dimensional context, the direction perpendicular to the plane defined by $\hat{u}_t$ and $\hat{u}_n$ is identified by the unit vector

$$\hat{u}_b(s) = \hat{u}_t(s) \times \hat{u}_n(s), \tag{2.58}$$

which is called the *binormal* unit vector.

Radius of curvature in Cartesian coordinates. In Cartesian coordinates, the planar path of a particle is usually given the form $y = y(x)$. In this case, geometry tells us that the radius of curvature at any position x is given by

$$\rho(x) = \frac{\left[1 + (dy/dx)^2\right]^{3/2}}{\left|d^2y/dx^2\right|}. \tag{2.59}$$

As an example, Fig. 2.26 shows a plot of the path $y(x) = (1 - x)\sin x$ for $0 \leq x \leq 2\pi$. We have used Eq. (2.59) to compute the radii of curvature of $y(x)$ at three different points. The first point, the local minimum at $x = 2.24$, has $\rho_1 = 0.450$. The second, the local maximum at $x = 4.96$, has $\rho_2 = 0.231$. Finally, at $x = 5.60$, $\rho_3 = 6.70$. Figure 2.26 demonstrates the intuitive notion that the "tighter the bend" in the path, the smaller its radius of curvature and the "straighter the path," the larger its radius of curvature.

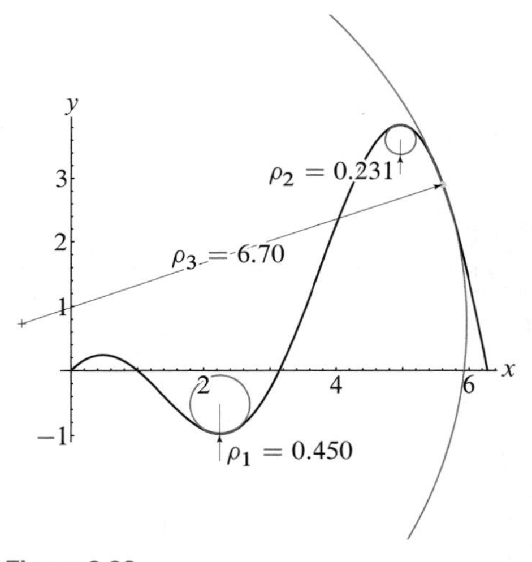

Figure 2.26
A 2D path showing its radius of curvature at three different points.

Connection with the time derivative of a vector. We conclude our discussion by going back to Eq. (2.53) to obtain $\vec{a}$, using the ideas from Section 2.4. Doing so, i.e., applying Eq. (2.48) on p. 81, gives

$$\vec{a} = \dot{\vec{v}} = \dot{v}\,\hat{u}_t + \vec{\omega}_v \times \vec{v} = \dot{v}\,\hat{u}_t + v\vec{\omega}_v \times \hat{u}_t. \tag{2.60}$$

Comparing Eq. (2.60) with Eq. (2.57), we have

$$v\vec{\omega}_v \times \hat{u}_t = \frac{v^2}{\rho}\hat{u}_n. \tag{2.61}$$

Canceling v, noting from Eq. (2.58) that $\hat{u}_n = \hat{u}_b \times \hat{u}_t$, and assuming that $\vec{\omega}_v$ is perpendicular to the plane containing $\hat{u}_t$ and $\hat{u}_n$, we obtain

$$\vec{\omega}_v \times \hat{u}_t = \frac{v}{\rho}\hat{u}_b \times \hat{u}_t \quad \Rightarrow \quad \vec{\omega}_v = \frac{v}{\rho}\hat{u}_b. \tag{2.62}$$

That is, the rate of rotation of the velocity vector is directly proportional to the speed of the particle, as well as the curvature (i.e., $1/\rho$) of the path.

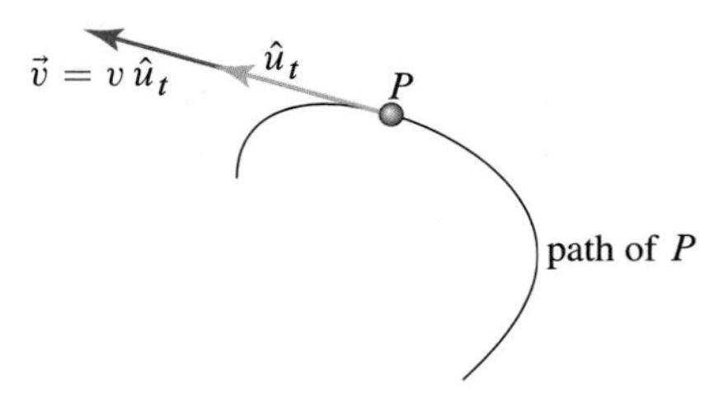

Figure 2.27
The velocity vector $\vec{v}$ defined in terms of the unit tangent vector $\hat{u}_t$.

End of Section Summary

In this section we have derived expressions for the velocity and acceleration of a point using the normal-tangential component system. Referring to Fig. 2.22, the velocity vector has the form

Eq. (2.50), p. 92

$$\vec{v} = v\,\hat{u}_t,$$

where v is the speed and $\hat{u}_t$ is the tangent unit vector at the point P.

Referring to Fig. 2.28, the acceleration vector in normal-tangential components

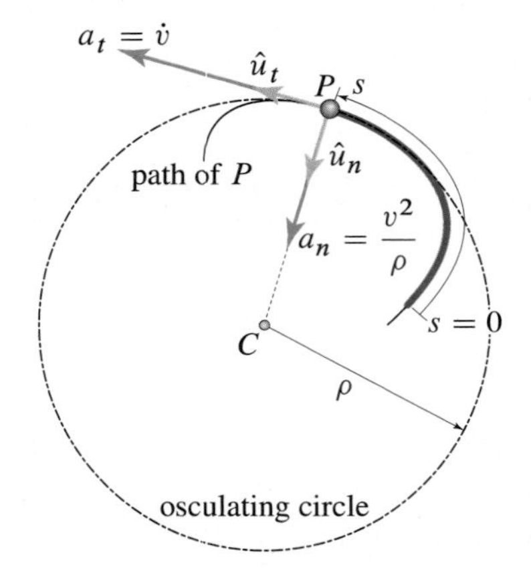

Figure 2.28. Acceleration in normal-tangential components.

has the form

Eq. (2.57), p. 93

$$\vec{a} = \dot{v}\,\hat{u}_t + \frac{v^2}{\rho}\,\hat{u}_n = a_t\,\hat{u}_t + a_n\,\hat{u}_n,$$

where $\dot{v} = a_t$ is the tangential component of acceleration and $v^2/\rho = a_n$ is the normal component of acceleration.

When we are using a Cartesian coordinate system, for a path expressed by a relation, such as $y = y(x)$, the path's radius of curvature is given by ρ:

Eq. (2.59), p. 93

$$\rho(x) = \frac{\left[1 + (dy/dx)^2\right]^{3/2}}{\left|d^2y/dx^2\right|}.$$

EXAMPLE 2.15 *Relating the Shape of the Path to Acceleration*

Let's assume that by *lateral G-force* the Federation Internationale de l'Automobile (FiA) that compiled the map in Fig. 1 really meant to provide a measurement of the acceleration normal to the path of the racing cars expressed in "units of g," where g is the acceleration due to gravity. Use this information along with the reported speed to estimate the radius of curvature of the Monaco Formula 1 track (1) at the stretch preceding the Rascasse and (2) at the end of the Tunnel.

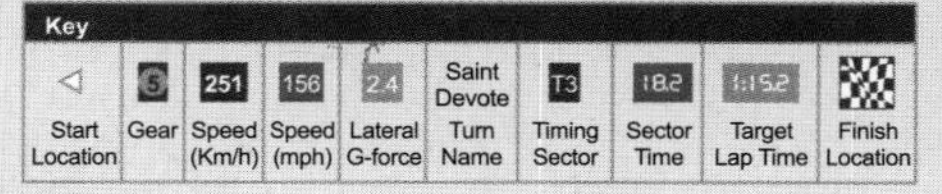

Figure 1
Formula 1 track at Monaco. The locations of the Tunnel and the Rascasse are indicated in gold. Additional information is provided including typical speed, acceleration, and gear information at various important points along the track.

SOLUTION

Road Map If we assume that a Formula 1 car can be modeled as a particle, then the solution to this problem is obtained by using the relation linking the speed to the component of the acceleration normal to the path, that is, $a_n = v^2/\rho$.

Computation As indicated in the problem statement, if one then assumes that the lateral G-force is the component of the acceleration normal to the path, then at the end of the stretch preceding the Rascasse (see Fig. 2) we have

$$v = 141\,\text{km/h} = 39.17\,\text{m/s} \quad \text{and} \quad a_n = 1.5g = 14.72\,\text{m/s}^2, \tag{1}$$

so that

$$\boxed{\rho = \frac{v^2}{a_n} = 104.2\,\text{m}.} \tag{2}$$

Proceeding in a similar way for the Tunnel (see Fig. 3), we have

$$v = 264\,\text{km/h} = 73.33\,\text{m/s} \quad \text{and} \quad a_n = 2.6g = 25.51\,\text{m/s}^2, \tag{3}$$

so that

$$\boxed{\rho = \frac{v^2}{a_n} = 210.8\,\text{m}.} \tag{4}$$

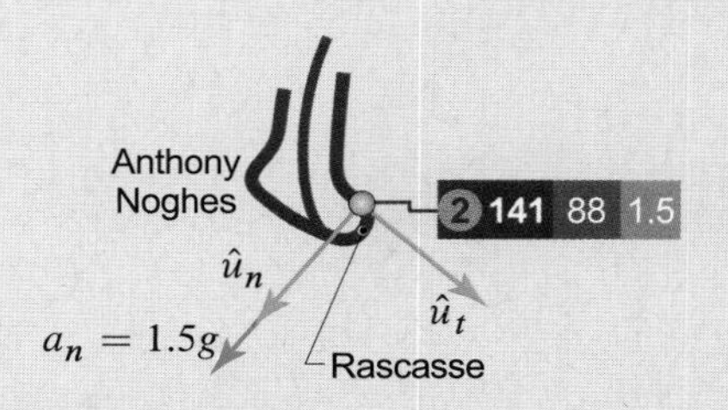

Figure 2
Normal tangential component system right before the Rascasse. The normal component of the acceleration is also shown. Refer to Fig. 1 for the key to the numbers shown in the blue boxes.

Discussion & Verification The solution was obtained as a direct application of the formula for the normal component of the acceleration, and is therefore, elementary to verify that the dimensions are correct and proper units have been used.

A Closer Look A basic question to ask is, How good are these estimates? After some research, the authors were able to obtain a map of the circuit from which the radii of curvature of most turns could be directly measured, although without taking into account changes in elevation along the curves. From this map, the radius of curvature in the first calculation was measured to be roughly 90 m, whereas the radius of curvature midway through the tunnel was found to be roughly 200 m. Hence, we can conclude that the gross estimates we have derived from the map in Fig. 1 and those that were obtained by direct (although still approximate) measurement are in reasonable agreement.

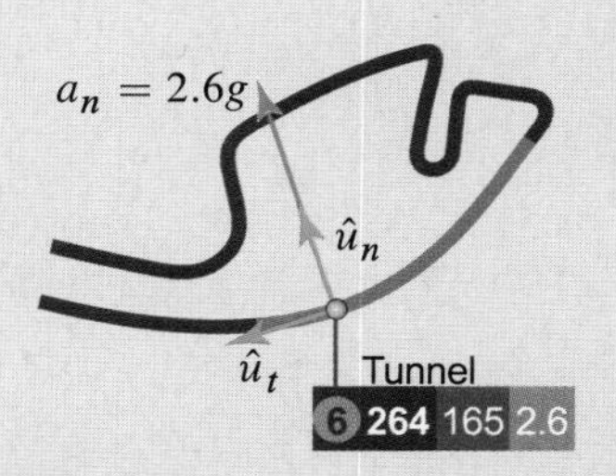

Figure 3
Normal tangential component system at the end of the Tunnel. The normal component of the acceleration is also shown. Refer to Fig. 1 for the key to the numbers shown in the blue boxes.

EXAMPLE 2.16 Curvature and Projectile Motion

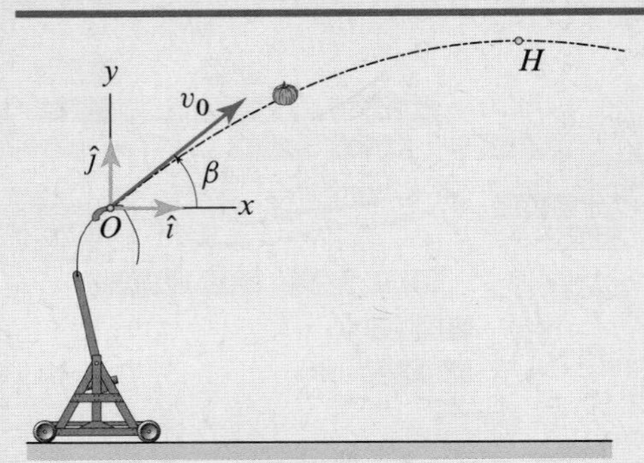

Figure 1
A pumpkin launched by a trebuchet.

A pumpkin has been launched from a trebuchet at the annual World Championship "Punkin Chunkin" competition in Millsboro, Delaware. The pumpkin, shown in Fig. 1, is assumed to have been released at O with an initial speed v_0 and an elevation angle β. Determine the time rate of change of the speed at the time of release and the radius of curvature of the pumpkin's trajectory at its highest point.

SOLUTION

Road Map By modeling the pumpkin's motion as a projectile motion, the pumpkin's acceleration is the acceleration of gravity g in the negative y direction. The key to the solution is then to compute the tangential and normal components of the acceleration, since the rate of change of the speed is $\dot{v} = a_t$ and the radius of curvature ρ is such that $a_n = v^2/\rho$.

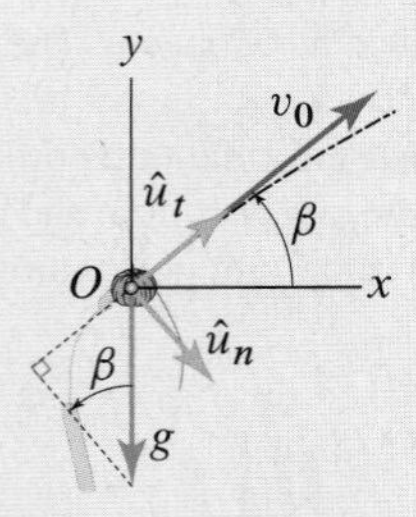

Figure 2
Normal-tangential component system at O.

Computation Referring to Fig. 2, recall that the velocity is always tangent to the path. Therefore, the tangent to the path at O is oriented at an angle β with respect to the x axis. Therefore, we have

$$\boxed{\dot{v}_O = a_{Ot} = -g \sin\beta,} \tag{1}$$

where v_O and a_{Ot} denote the speed at O and the tangential component of the acceleration at O, respectively.

Considering the situation at the highest point on the trajectory, referring to Fig. 3, observe that the tangent to the trajectory at this point is completely horizontal. Therefore, calling H the highest point on the trajectory, since the pumpkin's acceleration is completely in the y direction, we have

$$a_{Ht} = 0 \quad \text{and} \quad a_{Hn} = g. \tag{2}$$

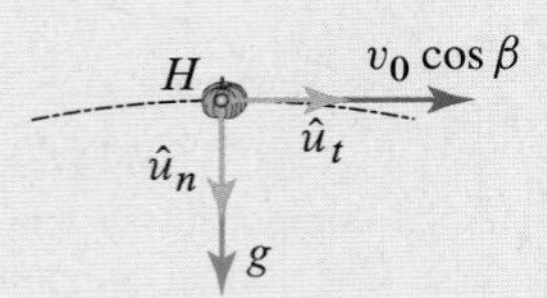

Figure 3
Normal-tangential component system at H.

Furthermore, the velocity at H is completely in the x direction. From projectile motion, we know that the horizontal component of the velocity remains constant throughout the motion. Therefore, we have

$$\vec{v}_H = v_{Hx}\,\hat{\imath} = v_0 \cos\beta\,\hat{\imath} \quad \Rightarrow \quad v_H = v_0 \cos\beta. \tag{3}$$

Using the normal component of Eq. (2.57), we have $a_n = v^2/\rho$. Therefore, using Eqs. (2) and (3), we have

$$\boxed{\rho_H = \frac{v_H^2}{a_{Hn}} = \frac{v_0^2 \cos^2\beta}{g}.} \tag{4}$$

Discussion & Verification Both the results in Eqs. (1) and (4) are dimensionally correct. Referring to Eq. (1), observe that $\sin\beta$ is nondimensional, and therefore, $\dot{v}_O$ has the same dimensions of g, i.e., the dimensions of acceleration, as expected. As far as Eq. (4) is concerned, recall that the dimensions of v_0 are length over time and those of g are length over time squared. Therefore, ρ has dimensions of length, as expected. From Eq. (1), observe that $\dot{v}_O$ is negative. This is to be expected since, up until the projectile reaches point H, the speed of the projectile is expected to decrease. Finally, from Eq. (4), observe that ρ increases with the horizontal component of the initial velocity. This is to be expected since the "flatness" of the overall trajectory is governed by the horizontal component of the velocity. Hence, the solution we have obtained appears to be correct.

EXAMPLE 2.17 *Accelerations in Circular Motion*

The Center for Gravitational Biology Research at NASA's Ames Research Center runs a large centrifuge capable of simulating $20g$ of acceleration ($12.5g$ is the maximum for human subjects). The radius of the centrifuge is 29 ft, and the distance from the axis of rotation to the cab at either A or B is about 25 ft. Assuming that the centrifuge accelerates uniformly to reach its final speed and that it takes 12.5 s to do so, determine

(a) The angular velocity of the centrifuge ω_f required to maintain a final acceleration of $20g$ in cab A.

(b) The magnitude of the acceleration of cab A as a function of time from the moment the centrifuge starts spinning until its final speed is achieved.

Figure 1
The $20g$ centrifuge at the Center for Gravitational Biology Research, which is part of NASA's Ames Research Center in Moffett Field, California.

SOLUTION

Road Map Since we know the magnitude of the final acceleration of A ($20g$), we can use the acceleration relationships developed in this section to determine the final value of the angular velocity of the centrifuge. Because the centrifuge accelerates uniformly during spin-up, we can use the constant acceleration relations from Section 2.2 to relate the centrifuge's final angular velocity to the needed angular acceleration during spin-up. Once the centrifuge's acceleration is known, we will be able to derive an expression for the acceleration of A as a function of time during spin-up.

Computation The key element of the solution of this problem is the relation between the motion of A and the motion of the centrifuge as a whole. To establish this relation, consider Fig. 2 and observe that the path of A is a circle with center C on the spin axis of the centrifuge. Since $\hat{u}_n$ remains pointing toward C, its angular velocity is the angular velocity of the centrifuge, namely, $\omega\,\hat{u}_b$. Observing that $\hat{u}_t$ must remain perpendicular to $\hat{u}_n$, we see $\hat{u}_t$ must also rotate with angular velocity $\omega\,\hat{u}_b$. This angular velocity is the angular velocity $\vec{\omega}_v$ that appears in Eqs. (2.60)–(2.62). Consequently, using Eq. (2.62), whether during spin-up or not, the angular velocity of the centrifuge and the motion of A are related by the relation

$$\omega = \frac{v}{\rho}, \tag{1}$$

where v is the speed of A and ρ is the radius of the path of A.

When the magnitude of the acceleration of A reaches its final value $a_f = 20g$, the speed of A will become constant, i.e., $\dot{v}_f = 0$, and a_f will coincide with *just* the normal component of Eq. (2.57) so that

$$a_f = \frac{v_f^2}{\rho} = \rho\omega_f^2 \quad \Rightarrow \quad 20g = 20(32.2\,\text{ft/s}^2) = \rho\omega_f^2, \tag{2}$$

where we have used Eq. (1), ω_f is the final value of ω, and we have used $g = 32.2\,\text{ft/s}^2$. Since $\rho = 25$ ft, solving for ω_f, we obtain

$$\boxed{\omega_f = 5.075\,\text{rad/s} = 48.47\,\text{rpm}.} \tag{3}$$

We can now use Eq. (2.32), along with Table 2.1, to obtain the angular acceleration of the centrifuge during spin-up as

$$\omega_f = \omega_0 + \alpha t_f \quad \Rightarrow \quad \alpha = \omega_f / t_f \quad \Rightarrow \quad \alpha = 0.4060\,\text{rad/s}^2, \tag{4}$$

where we have used $\omega_0 = 0$ and $t_f = 12.5$ s.

To relate α to the acceleration of A, we rewrite Eq. (1) as $v = \rho\omega$, and differentiating with respect to time, we obtain

$$\dot{v} = \rho\dot{\omega} = \rho\alpha, \tag{5}$$

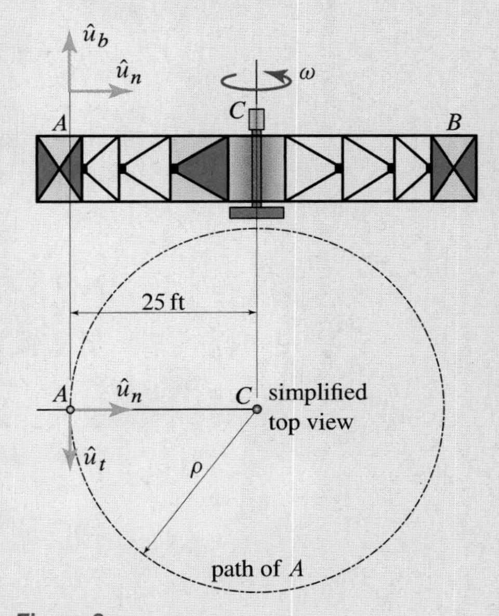

Figure 2
The $20g$ centrifuge showing the axis of rotation and the path of A as well as the normal-tangential component system at A.

where we have used the fact that ρ is constant. The quantity $\dot{v}$ is the tangential component of the acceleration of A. The normal component of the acceleration of A is given by

$$a_n = \frac{v^2}{\rho} = \rho\omega^2 = \rho\alpha^2 t^2, \tag{6}$$

where we have used $\omega = \alpha t$ from Eq. (4), which is the value of ω at an arbitrary time t during spin-up. Therefore, the magnitude of the acceleration during spin-up is given by

$$a = \sqrt{a_t^2 + a_n^2} = \sqrt{\rho^2\alpha^2 + \rho^2\alpha^4 t^4} = \rho\alpha\sqrt{1 + \alpha^2 t^4}. \tag{7}$$

Recalling that $\rho = 25\,\text{ft}$ and using the result in Eq. (4), we have

$$\boxed{a = \left(10.15\,\text{ft/s}^2\right)\sqrt{1 + (0.1649\,\text{s}^{-4})t^4}.} \tag{8}$$

Discussion & Verification Equations (2) and (7) present our results in symbolic form and allow us to easily verify that our results are dimensionally correct. In addition, the numerical form of the results presented in Eqs. (3) and (8) is expressed using appropriate and consistent units. Overall, our solution indicates that the speed of A increases uniformly, and this is consistent with the given piece of information that the spin-up of the centrifuge occurs at a constant rate. Hence, our solution appears to be correct.

A Closer Look The magnitude of the acceleration of A, as given in Eq. (8), is an increasing function of time. Hence, a is largest at $t = 12.5\,\text{s}$ (the end of the spin-up) and its value is given by

$$a_{\text{max}} = 644.1\,\text{ft/s}^2 = 20.00g, \tag{9}$$

which appears to be the same as the value of a_f. The result in Eq. (9) is somewhat unexpected because the value of a_f is based only on the normal acceleration of A at the end of spin-up, whereas a_{max} includes the contributions of both the normal and tangential components of the acceleration of A. This indicates that the tangential component of acceleration contributes an insignificant amount to the total acceleration. When Eq. (7) is evaluated at the end of spin-up ($t = 12.5\,\text{s}$), $a_t = 10.15\,\text{ft/s}^2$ (see the leading term on the right-hand side of Eq. (8)), and $a_n = 644.0\,\text{ft/s}^2$, so we can see that a_n is over 63 times larger than a_t.

We conclude by observing that our solution assumes that the angular acceleration is constant until $t = 12.5\,\text{s}$, at which time it becomes zero, so the angular velocity becomes constant. This assumption is not entirely realistic since such abrupt changes in acceleration are not typically found in applications.

Problems

Problem 2.141

A particle P is moving along the curve C, whose equation is given by

$$(y^2 - x^2)(x - 1)(2x - 3) = 4(x^2 + y^2 - 2x)^2,$$

at a *constant* speed v_c. For any position on the curve C for which the radius of curvature is defined (i.e., *not* equal to infinity), what *must* be the angle ϕ between the velocity vector $\vec{v}$ and the acceleration vector $\vec{a}$?

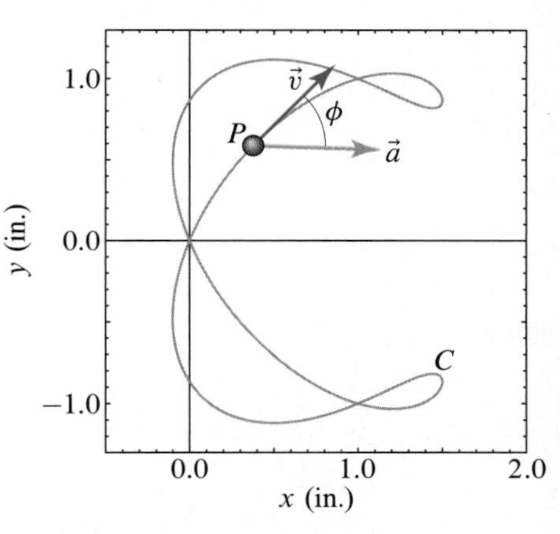

Figure P2.141

Problem 2.142

A particle P is moving along a path with the velocity shown. Is the sketch of the normal-tangential component system at P correct?

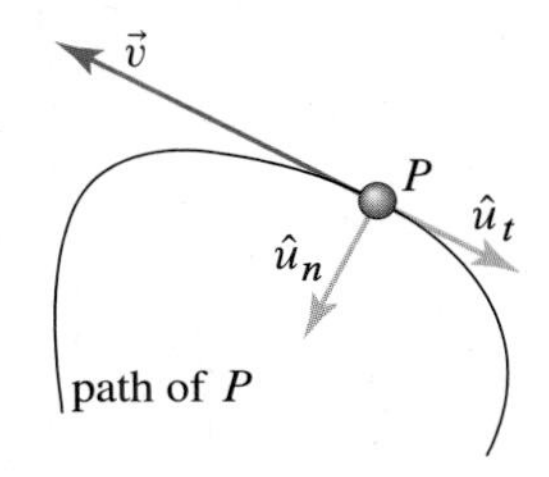

Figure P2.142

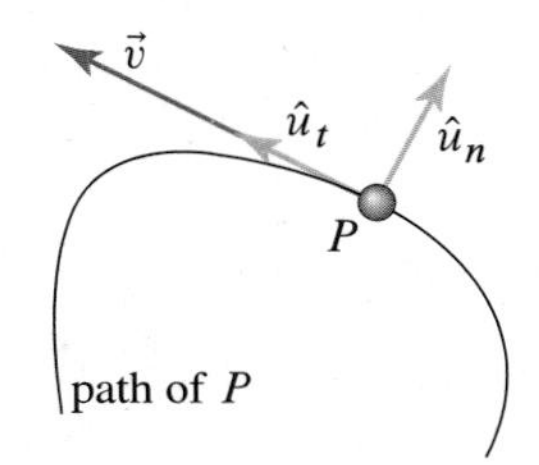

Figure P2.143

Problem 2.143

A particle P is moving along a path with the velocity shown. Is the sketch of the normal-tangential component system at P correct?

Problem 2.144

A particle P is moving along a straight line with the velocity and acceleration shown. What is wrong with the unit vectors shown in the figure?

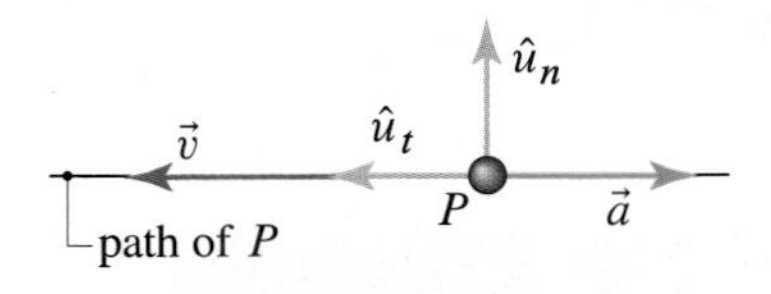

Figure P2.144

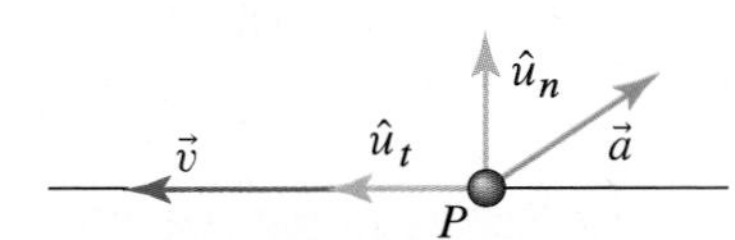

Figure P2.145

Problem 2.145

A particle P is moving along some path with the velocity and acceleration shown. Can the path of P be the straight line shown?

Problem 2.146

The water jet of a fountain is let out at a speed $v_0 = 80\,\text{ft/s}$ and at an angle $\beta = 60°$. Determine the radius of curvature of the jet at its highest point.

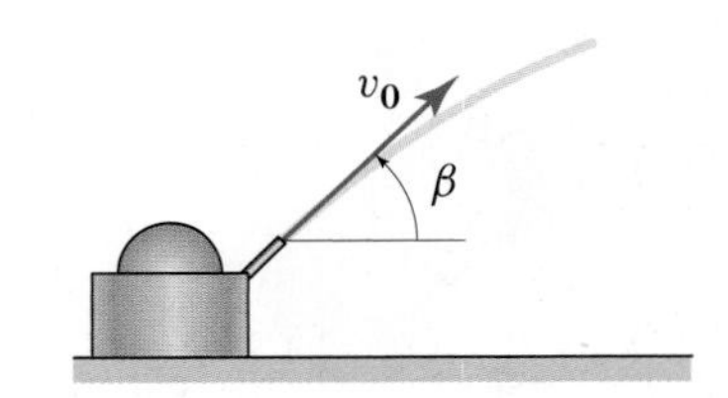

Figure P2.146

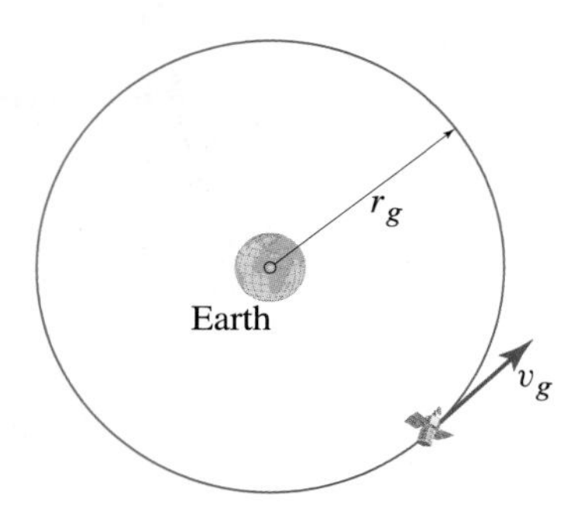

Figure P2.147

Problem 2.147

A telecommunications satellite is made to orbit the Earth in such a way as to appear to hover in the same point in the sky as seen by a person standing on the surface of the Earth. Assuming that the satellite's orbit is circular with radius $r_g = 1.385 \times 10^8$ ft and knowing that the speed of the satellite is constant and equal to $v_g = 1.008 \times 10^4$ ft/s, determine the magnitude of the acceleration of the satellite.

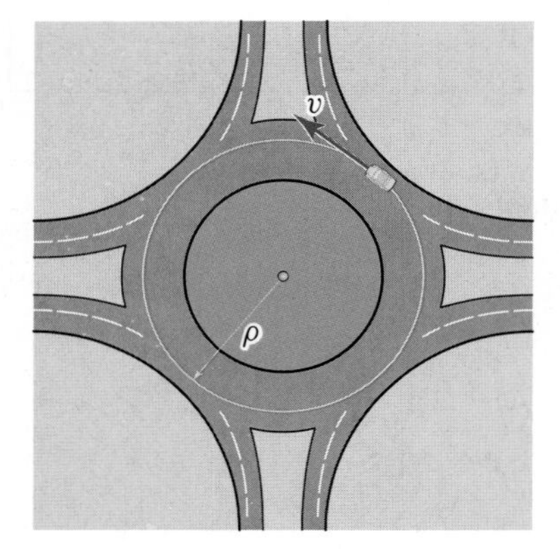

Figure P2.148 and P2.149

Problem 2.148

A car travels along a city roundabout with radius $\rho = 30$ m. At the instant shown, the speed of the car is $v = 35$ km/h and the magnitude of the acceleration of the car is $4.5\,\text{m/s}^2$. If the car is increasing its speed, determine the time rate of change of the speed of the car at the instant shown.

Problem 2.149

A car travels along a city roundabout with radius $\rho = 100$ ft. At the instant shown, the speed of the car is $v = 25$ mph and the speed is decreasing at the rate $8\,\text{ft/s}^2$. Determine the magnitude of the acceleration of the car at the instant shown.

Problem 2.150

Making the same assumptions stated in Example 2.15, consider the map of the Formula 1 circuit at Hockenheim in Germany and estimate the radius of curvature of the curves Südkurve and Nordkurve (at the locations indicated in gold).

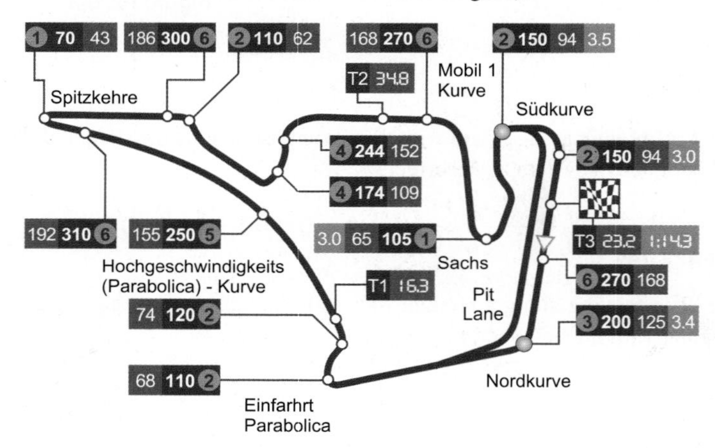

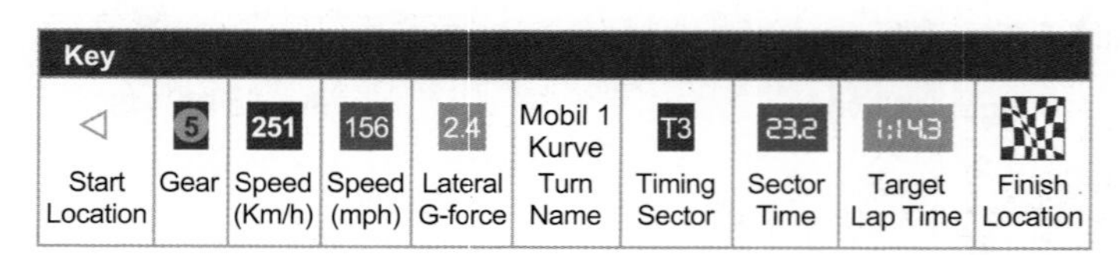

Figure P2.150

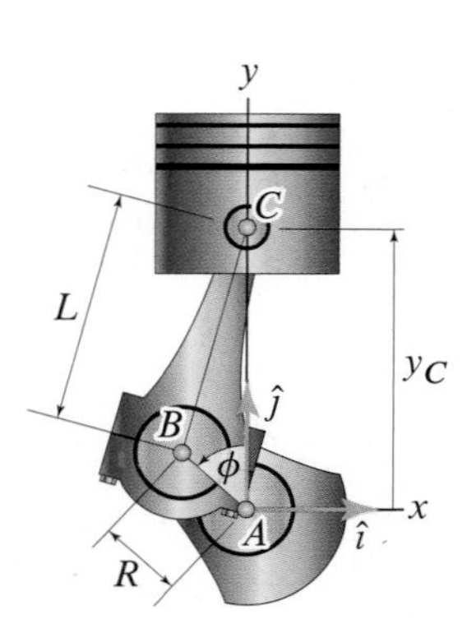

Figure P2.151

Problem 2.151

The position of the piston C, as a function of the crank angle ϕ and the lengths of the crank AB and connecting rod BC, is given by $y_C = R\cos\phi + L\sqrt{1 - (R\sin\phi/L)^2}$ and $x_C = 0$. Using the component system shown, express $\hat{u}_t$, the unit vector tangent to the trajectory of C, as a function of the crank angle ϕ for $0 \le \phi \le 2\pi$ rad.

Problem 2.152

An aerobatics plane initiates the basic loop maneuver such that, at the bottom of the loop, the plane is going 140 mph, while subjecting the plane to approximately $4g$ of acceleration. Estimate the corresponding radius of the loop.

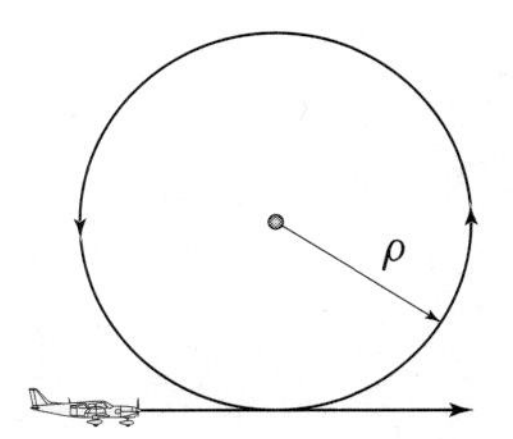

Figure P2.152

Problems 2.153 through 2.155

The portion of a race track between points A (corresponding to $x = 0$) and B is part of a parabolic curve described by the equation $y = \kappa x^2$, where κ is a constant. Let g denote the acceleration due to gravity.

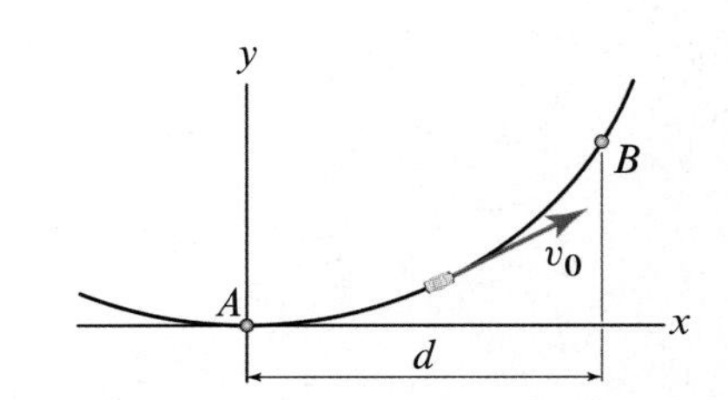

Figure P2.153–P2.155

Problem 2.153 Determine κ such that a car driving at constant speed $v_0 = 180$ mph experiences at A an acceleration with magnitude equal to $1.5g$.

Problem 2.154 If $\kappa = 0.4 \times 10^{-3}\,\text{ft}^{-1}$, determine d such that a car driving at constant speed $v_0 = 180$ mph experiences at B an acceleration with magnitude equal to g.

Problem 2.155 Suppose a car travels from A to B with a constant speed $v_0 = 180$ mph. Let $|\vec{a}|_{\text{min}}$ and $|\vec{a}|_{\text{max}}$ denote the minimum and maximum values of the magnitude of the acceleration, respectively. Determine $|\vec{a}|_{\text{min}}$ if $d = 1200$ ft and $|\vec{a}|_{\text{max}} = 1.5g$.

Problems 2.156 through 2.158

An airplane is flying straight and level at a constant speed v_0 when it starts climbing along a path described by the equation $y = h + \beta x^3$, where h and β are constants. Let g denote the acceleration due to gravity.

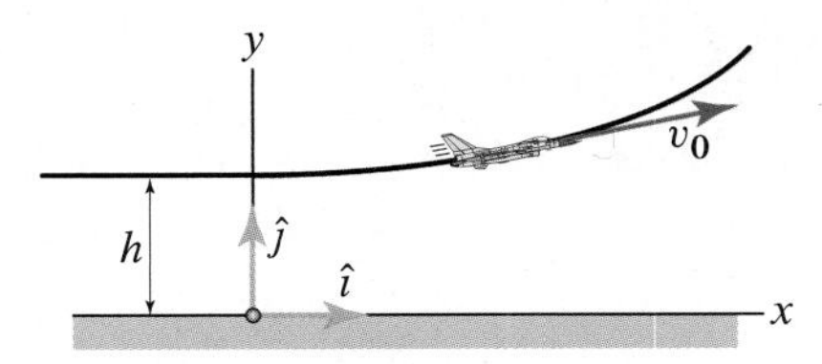

Figure P2.156–P2.158

Problem 2.156 Determine the acceleration of the airplane at $x = 0$.

Problem 2.157 If $\beta = 0.05 \times 10^{-3}\,\text{m}^{-2}$ and v_0 remains constant, find v_0 such that the magnitude of the acceleration of the airplane is equal to $3g$ for $x = 300$ m.

Problem 2.158 If $v_0 = 600$ km/h and $\beta = 0.025 \times 10^{-4}\,\text{m}^{-2}$, determine the acceleration of the airplane for $x = 350$ m and express it in the Cartesian component system shown.

Problem 2.159

A jet is flying at a constant speed $v_0 = 750$ mph while performing a constant speed circular turn. If the magnitude of the acceleration needs to remain constant and equal to $9g$, where g is the acceleration due to gravity, determine the radius of curvature of the turn.

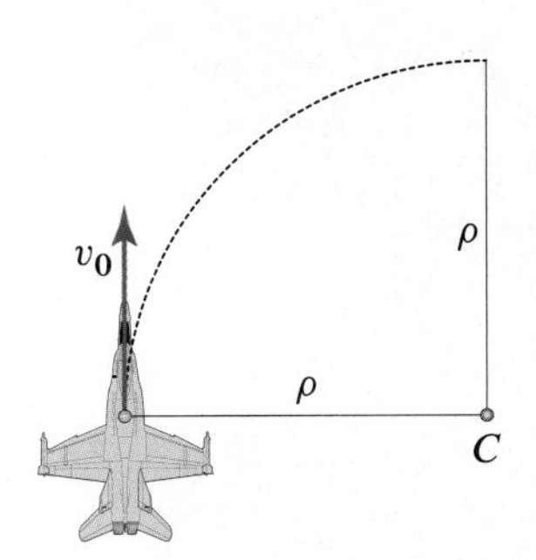

Figure P2.159

Problem 2.160

Particles A and B are moving in the plane with the same constant speed v, and their paths are tangent at P. Do these particles have zero acceleration at P? If not, do these particles have the same acceleration at P?

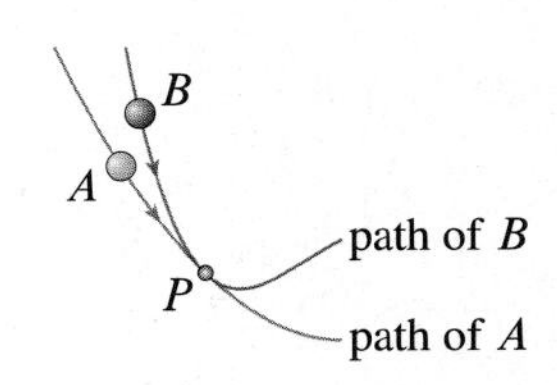

Figure P2.160

Problem 2.161

Uranium is used in light water reactors to produce a controlled nuclear reaction for the generation of power. When first mined, uranium comes out as the oxide U_3O_8, 0.7% of which is the isotope U-235 and 99.3% the isotope U-238.* For it to be used in a nuclear reactor, the concentration of U-235 must be in the 3–5% range.† The process of increasing the percentage of U-235 is called *enrichment*, and it is done in a number of ways. One method uses centrifuges, which spin at very high rates to create artificial gravity. In these centrifuges, the heavy U-238 atoms concentrate on the outside of the cylinder (where the acceleration is largest), and the lighter U-235 atoms concentrate near the spin axis. Before centrifuging, the uranium is processed into gaseous uranium hexafluoride or UF_6, which is then injected into the centrifuge. Assuming that the radius of the centrifuge is 20 cm and that it spins at 70,000 rpm, determine

(a) The velocity of the outer surface of the centrifuge.

(b) The acceleration in g experienced by an atom of uranium that is on the inside of the outer wall of the centrifuge.

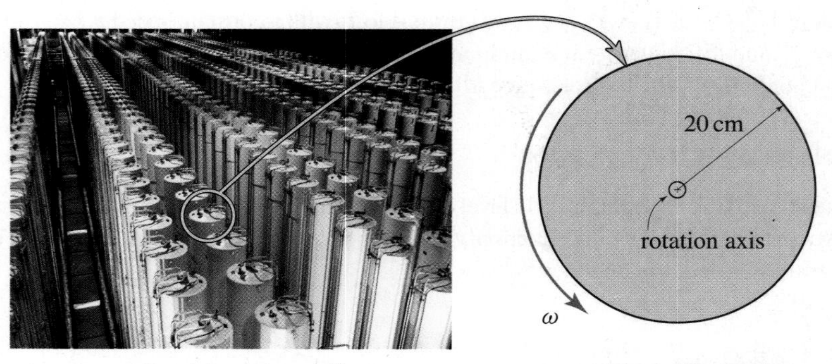

array of enrichment centrifuges centrifuge cross section

Figure P2.161

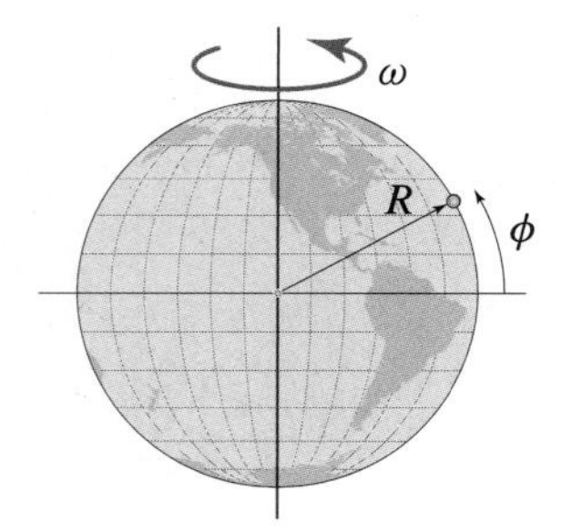

Figure P2.162

Problem 2.162

Treating the center of the Earth as a fixed point, determine the magnitude of the acceleration of points on the surface of the earth as a function of the angle ϕ shown. Use $R = 6371$ km as the radius of the Earth.

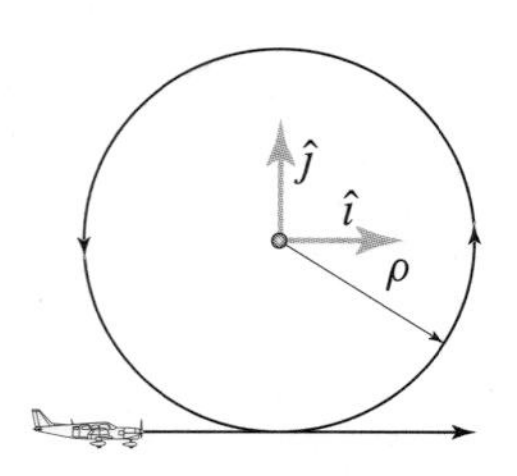

Figure P2.163–P2.165

Problems 2.163 through 2.165

An airplane is flying straight and level at a speed $v_0 = 150$ mph and with a constant time rate of increase of speed $\dot{v} = 20\,\text{ft/s}^2$, when it starts to climb along a circular path with a radius of curvature $\rho = 2000$ ft. The airplane maintains $\dot{v}$ constant for about 30 s.

Problem 2.163 Determine the acceleration of the airplane right at the start of the climb and express the result in the Cartesian component system shown.

Problem 2.164 Determine the acceleration of the airplane 25 s after the start of the climb and express the result in the Cartesian component system shown.

Problem 2.165 Determine the acceleration of the airplane after it has traveled 150 ft along the path and express the result in the Cartesian component system shown.

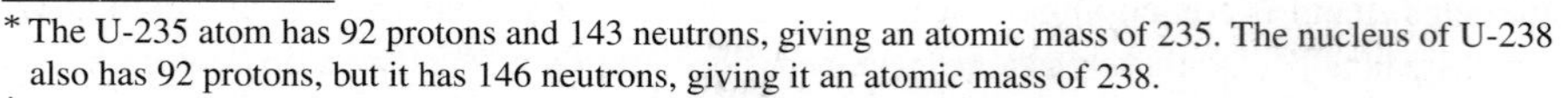

* The U-235 atom has 92 protons and 143 neutrons, giving an atomic mass of 235. The nucleus of U-238 also has 92 protons, but it has 146 neutrons, giving it an atomic mass of 238.

† For nuclear weapons, the concentration of U-235 must be about 90%.

Problem 2.166

Suppose that a highway exit ramp is designed to be a circular segment of radius $\rho = 130\,\text{ft}$. A car begins to exit the highway at A while traveling at a speed of 65 mph and goes by point B with a speed of 25 mph. Compute the acceleration vector of the car as a function of the arc length s, assuming that the tangential component of the acceleration is constant between points A and B.

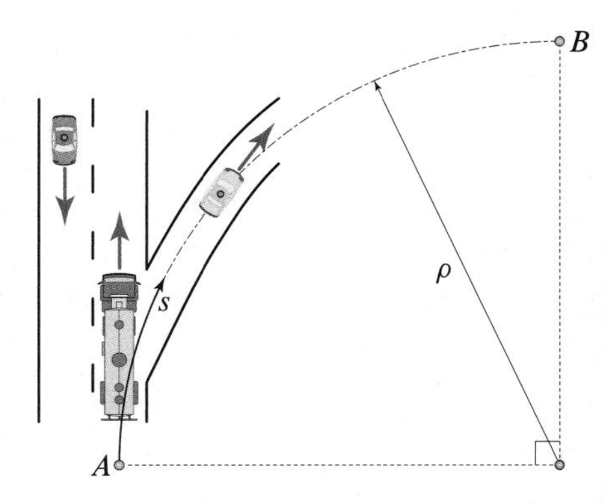

Figure P2.166 and P2.167

Problem 2.167

Suppose that a highway exit ramp is designed to be a circular segment of radius $\rho = 130\,\text{ft}$. A car begins to exit the highway at A while traveling at a speed of 65 mph and goes by point B with a speed of 25 mph. Compute the acceleration vector of the car as a function of the arc length s, assuming that between A and B the speed was controlled so as to maintain constant the rate dv/ds.

Problem 2.168

A water jet is ejected from the nozzle of a fountain with a speed $v_0 = 12\,\text{m/s}$. Letting $\beta = 33°$, determine the rate of change of the speed of the water particles as soon as these are ejected as well as the corresponding radius of curvature of the water path.

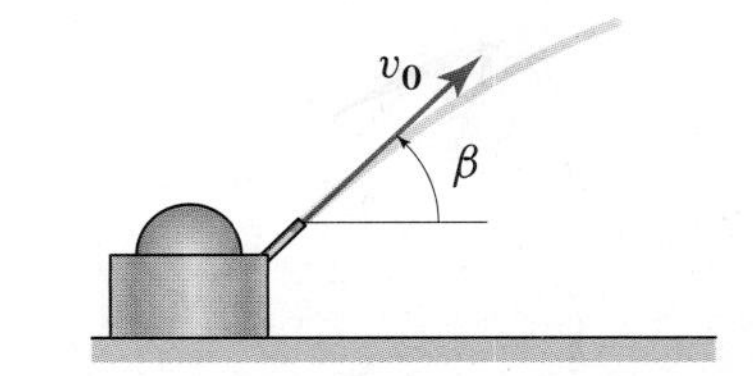

Figure P2.168 and P2.169

Problem 2.169

A water jet is ejected from the nozzle of a fountain with a speed v_0. Letting $\beta = 21°$, determine v_0 so that the radius of curvature at the highest point on the water arch is 10 ft.

Problem 2.170

A car traveling with a speed $v_0 = 65\,\text{mph}$ almost loses contact with the ground when it reaches the top of the hill. Determine the radius of curvature of the hill at its top.

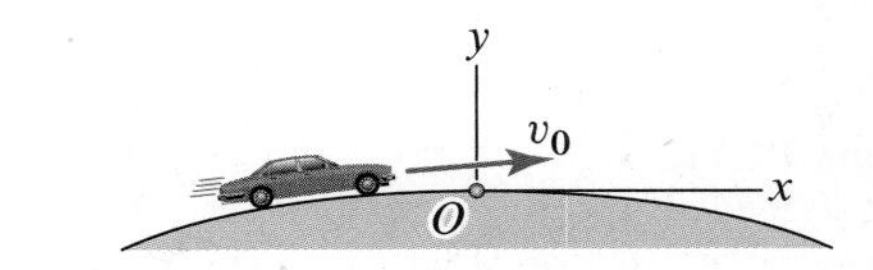

Figure P2.170 and P2.171

Problem 2.171

A car is traveling at a constant speed over a hill. If, using a Cartesian coordinate system with origin O at the top of the hill, the hill's profile is described by the function $y = -(0.003\,\text{m}^{-1})x^2$, where x and y are in meters, determine the minimum speed at which the car would lose contact with the ground at the top of the hill. Express the answer in km/h.

Problem 2.172

A race boat is traveling at a constant speed $v_0 = 130\,\text{mph}$ when it performs a turn with constant radius ρ to change its course by 90° as shown. The turn is performed while losing speed uniformly in time so that the boat's speed at the end of the turn is $v_f = 125\,\text{mph}$. If the maximum allowed normal acceleration is equal to $2g$, where g is the acceleration due to gravity, determine the tightest radius of curvature possible and the time needed to complete the turn.

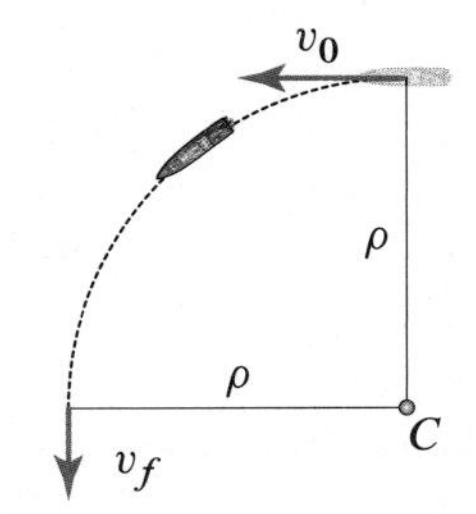

Figure P2.172 and P2.173

Problem 2.173

A race boat is traveling at a constant speed $v_0 = 130\,\text{mph}$ when it performs a turn with constant radius ρ to change its course by 90° as shown. The turn is performed while losing speed uniformly in time so that the boat's speed at the end of the turn is $v_f = 116\,\text{mph}$. If the magnitude of the acceleration is not allowed to exceed $2g$, where g is the acceleration due to gravity, determine the tightest radius of curvature possible and the time needed to complete the turn.

Problem 2.174

A truck enters an exit ramp with an initial speed v_0. The ramp is a circular arc with radius ρ. Derive an expression for the magnitude of the acceleration of the truck as a function of the path coordinate s (and the parameters v_0 and ρ) if the truck stops at B and travels from A to B with a constant rate of change of the speed with respect to s.

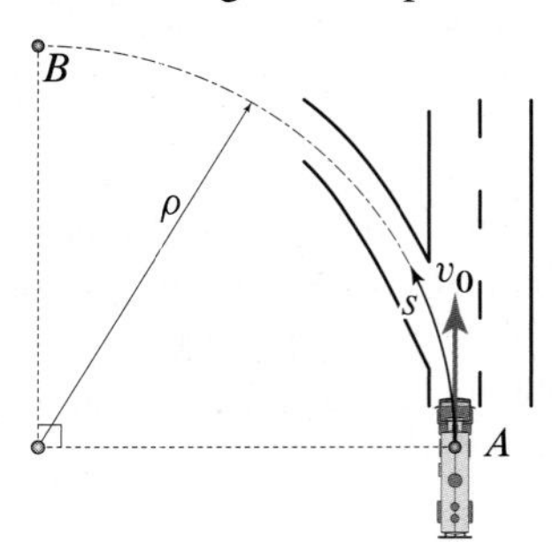

Figure P2.174

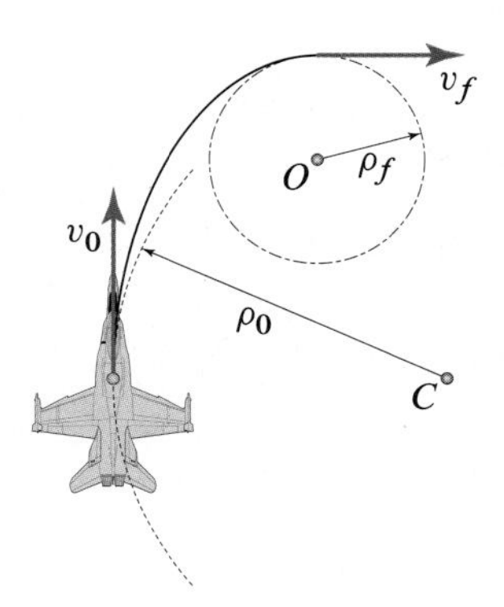

Figure P2.175

Problem 2.175

A jet is flying straight and level at a speed $v_0 = 1100\,\text{km/h}$ when it turns to change its course by 90° as shown. In an attempt to progressively tighten the turn, the speed of the plane is uniformly decreased in time while keeping the normal acceleration constant and equal to $8g$, where g is the acceleration due to gravity. At the end of the turn, the speed of the plane is $v_f = 800\,\text{km/h}$. Determine the radius of curvature ρ_f at the end of the turn and the time t_f that the plane takes to complete its change in course.

Problem 2.176

A car is traveling over a hill with a constant speed $v_0 = 70\,\text{mph}$. Using the Cartesian coordinate system shown, the hill's profile is given by the function $y = -(0.0005\,\text{ft}^{-1})x^2$, where x and y are measured in feet. At $x = -300\,\text{ft}$, the driver applies the brakes, causing a constant time rate of change of speed $\dot{v} = -3\,\text{ft/s}^2$ until the car arrives at O. Determine the distance traveled while applying the brakes along with the time to cover this distance. *Hint:* To compute the distance traveled by the car along the car's path, observe that $ds = \sqrt{dx^2 + dy^2} = \sqrt{1 + (dy/dx)^2}\,dx$, and that

$$\int \sqrt{1 + C^2x^2}\,dx = \frac{x}{2}\sqrt{1 + C^2x^2} + \frac{1}{2C}\ln\left(Cx + \sqrt{1 + C^2x^2}\right).$$

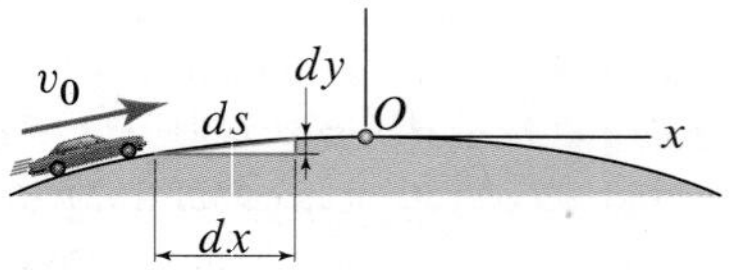

Figure P2.176

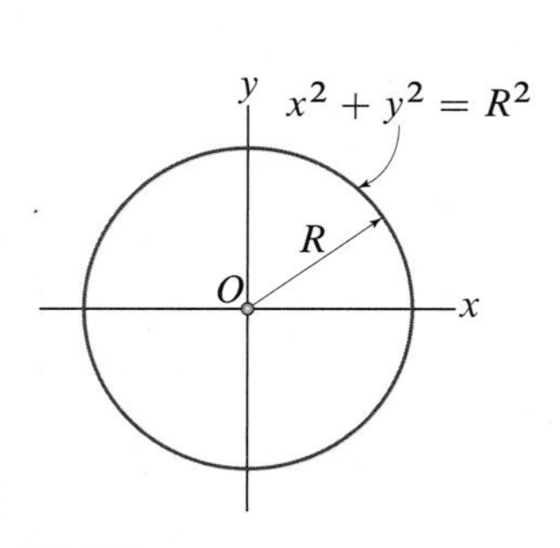

Figure P2.177

Problem 2.177

Recalling that a circle of radius R and center at the origin O of a Cartesian coordinate system with axes x and y can be expressed by the formula $x^2 + y^2 = R^2$, use Eq. (2.59) to verify that the radius of curvature of this circle is equal to R.

2.6 Planar Motion: Polar Coordinates

In this section we describe the position, velocity, and acceleration of a point moving in a plane when the point's coordinates are given relative to a polar coordinate system.

Polar coordinates and position, velocity, and acceleration

Figure 2.29 shows a particle P moving along a path in the xy plane. The position of P relative to the origin at O is given by $\vec{r}$. The distance r between O and P and the angle θ, measured with respect to the x axis, identify the position of P in the plane of motion. The quantities r and θ are called the *polar coordinates** of P relative to the origin O and the reference line coinciding with the x axis. The position vector of P is then

$$\vec{r} = r\,\hat{u}_r, \tag{2.63}$$

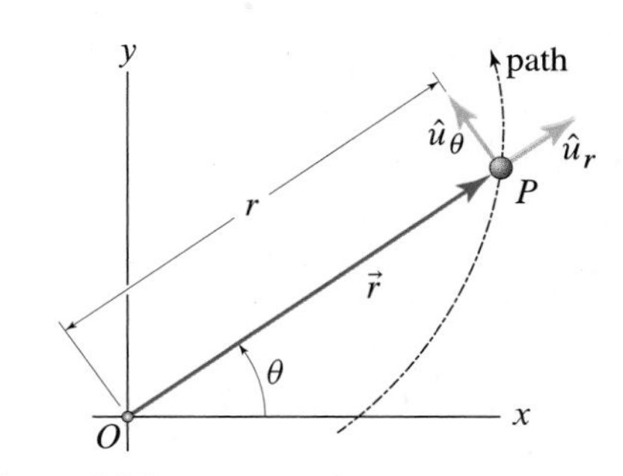

Figure 2.29
The polar coordinate system defining the position of the plane at B.

where $\hat{u}_r$ is the unit vector pointing from O to P. Although θ does not *explicitly* appear in Eq. (2.63), $\vec{r}$ depends on θ because θ defines the direction of $\hat{u}_r$.

Taking the time derivative of Eq. (2.63) to find the velocity in polar coordinates, we obtain

$$\vec{v} = \dot{r}\,\hat{u}_r + r\,\dot{\hat{u}}_r. \tag{2.64}$$

The time derivative of the unit vector $\hat{u}_r$ can be evaluated using Eq. (2.46) on p. 81, which allows us to rewrite Eq. (2.64) as

$$\vec{v} = \dot{r}\,\hat{u}_r + r\,\vec{\omega}_r \times \hat{u}_r, \tag{2.65}$$

where $\vec{\omega}_r$ is the angular velocity of $\hat{u}_r$. Since θ increases in the counterclockwise direction, using the right-hand rule, we have $\vec{\omega}_r = \dot{\theta}\,\hat{k}$. Using $\vec{\omega}_r = \dot{\theta}\,\hat{k}$, the last term in Eq. (2.65) becomes $r\dot{\theta}\,\hat{k} \times \hat{u}_r = r\dot{\theta}\,\hat{u}_\theta$, where $\hat{u}_\theta$ is given by $\hat{k} \times \hat{u}_r = \hat{u}_\theta$. Equation (2.65) then becomes

$$\vec{v} = \dot{r}\,\hat{u}_r + r\dot{\theta}\,\hat{u}_\theta = v_r\,\hat{u}_r + v_\theta\,\hat{u}_\theta, \tag{2.66}$$

where

$$v_r = \dot{r} \quad \text{and} \quad v_\theta = r\dot{\theta} \tag{2.67}$$

are the *radial* and *transverse components of the velocity*, respectively. The time derivative of Eq. (2.66) yields

$$\vec{a} = \ddot{r}\,\hat{u}_r + \dot{r}\,\dot{\hat{u}}_r + \dot{r}\dot{\theta}\,\hat{u}_\theta + r\ddot{\theta}\,\hat{u}_\theta + r\dot{\theta}\,\dot{\hat{u}}_\theta. \tag{2.68}$$

Since $\dot{\hat{u}}_r = \vec{\omega}_r \times \hat{u}_r$, $\dot{\hat{u}}_\theta = \vec{\omega}_\theta \times \hat{u}_\theta$, and $\vec{\omega}_r = \vec{\omega}_\theta = \dot{\theta}\,\hat{k}$, Eq. (2.68) becomes

$$\vec{a} = \ddot{r}\,\hat{u}_r + \dot{r}\dot{\theta}\,\hat{k} \times \hat{u}_r + \dot{r}\dot{\theta}\,\hat{u}_\theta + r\ddot{\theta}\,\hat{u}_\theta + r\dot{\theta}^2\,\hat{k} \times \hat{u}_\theta. \tag{2.69}$$

Noting that $\hat{k} \times \hat{u}_\theta = -\hat{u}_r$ and combining the coefficients of $\hat{u}_r$ and $\hat{u}_\theta$, we can rewrite Eq. (2.69) as

$$\vec{a} = (\ddot{r} - r\dot{\theta}^2)\hat{u}_r + (r\ddot{\theta} + 2\dot{r}\dot{\theta})\hat{u}_\theta = a_r\,\hat{u}_r + a_\theta\,\hat{u}_\theta, \tag{2.70}$$

where

$$a_r = \ddot{r} - r\dot{\theta}^2 \quad \text{and} \quad a_\theta = r\ddot{\theta} + 2\dot{r}\dot{\theta} \tag{2.71}$$

are the *radial* and *transverse components of the acceleration*, respectively.

Concept Alert

Direction of $\hat{u}_r$ and $\hat{u}_\theta$. The unit vector $\hat{u}_r$ always points away from the origin. The unit vector $\hat{u}_\theta$ always points in the direction of increasing θ.

* Polar coordinates are also called *radial-transverse coordinates*, in which *radial* refers to the r direction and *transverse* refers to the θ direction. In Section 2.8, we will see the cylindrical coordinate system, which is the three-dimensional generalization of the polar coordinate system.

EXAMPLE 2.18 *Constant Velocity Motion in Polar Coordinates*

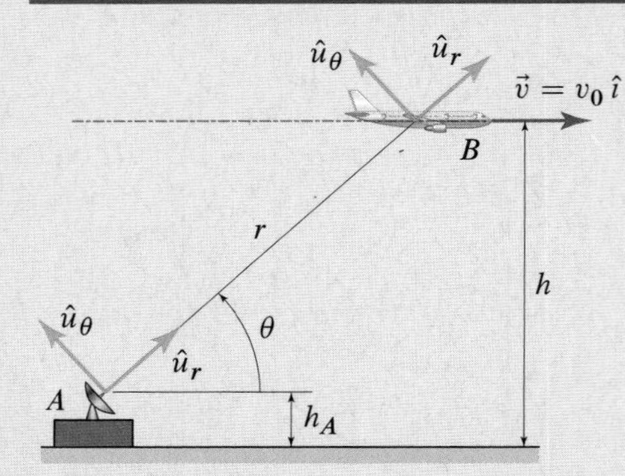

Figure 1
Radar station at A tracking a plane at B.

What relationships do the radar readings obtained by the station at A need to satisfy to conclude that the plane at B shown in Fig. 1 is flying straight and level at altitude h and at a constant speed v_0?

SOLUTION

Road Map It is not hard to imagine that a radar station, such as the one at A, would record *discrete* readings of $r(t)$ and $\theta(t)$ for any object it is tracking. Given a tabulated list of r and θ as a function of time, the question is, How can we use these readings to determine whether or not a plane is flying (1) straight and level and (2) at a constant speed?

The solution strategy used in this example will be used often. We will use expressions for positions, velocities, and accelerations in a chosen coordinate system to give form to specific requirements (e.g., that the altitude be constant or that the speed be maintained). Once a requirement is obtained, it can be manipulated further (e.g., differentiated with respect to time) to obtain new relationships that will be consistent with the requirements' initial statements.

Computation To assess the plane's altitude, first we need to express the altitude using the given data, and then we need to *enforce* the requirement that the altitude be maintained constant. Hence, from Fig. 1, we have

$$h = r\sin\theta + h_A. \tag{1}$$

So h will remain constant as long as

$$\boxed{r(t)\sin\theta(t) = \text{constant.}} \tag{2}$$

Since the speed is obtained as the magnitude of the velocity vector, to find what relationship must be satisfied in order for the plane to maintain the given constant speed, let's look at the expression for the plane's velocity. The velocity vector in polar coordinates is given by Eq. (2.66), that is, $\vec{v} = \dot{r}\,\hat{u}_r + r\dot{\theta}\,\hat{u}_\theta$, and so its magnitude is given by the square root of the sum of the squares of its components, i.e.,

$$|\vec{v}| = v = \sqrt{\dot{r}^2 + r^2\dot{\theta}^2}. \tag{3}$$

Therefore, for the speed to be a constant equal to v_0 we must have

$$\boxed{v_0 = \sqrt{\dot{r}^2 + r^2\dot{\theta}^2} = \text{constant.}} \tag{4}$$

Discussion & Verification Since r has dimensions of length and θ is nondimensional, then $\dot{r}$ and $r\dot{\theta}$ both have dimensions of length over time, i.e., of velocity. Hence, we can conclude that Eq. (4) is dimensionally correct.

A Closer Look While Eq. (4) gives the relationship we wanted, it is often desirable to have relations that require as few mathematical operations as possible. For example, it turns out that there is no need to compute a square root to verify whether or not the speed is a constant—if it is true that v_0 is a constant, then we necessarily must have that v_0^2 is also a constant. Therefore, as long as

$$\boxed{\dot{r}^2(t) + r^2(t)\dot{\theta}^2(t) = \text{constant,}} \tag{5}$$

we can conclude that the plane is flying at a constant speed.

EXAMPLE 2.19 *Acceleration During Orbital Motion*

For a satellite orbiting a planet, empirical observations (see Kepler's laws in Chapter 1) tell us that in the polar coordinate system shown in Fig. 1, the quantity $r^2\dot{\theta}$ remains constant throughout the satellite's motion. Show that for such a motion, the satellite's acceleration is purely in the radial direction; i.e., the transverse component of the satellite's acceleration is equal to zero.

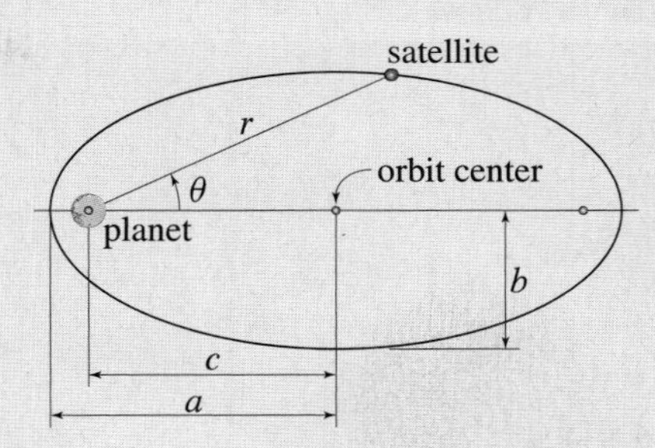

Figure 1
A satellite orbiting a planet. The orbit is an ellipse with major and minor semiaxes equal to a and b, respectively. The length $c = \sqrt{a^2 - b^2}$ denotes the distance between either focus (one of which is occupied by the planet) and the orbit center.

SOLUTION

Road Map We need to show that $a_\theta = 0$. To do this, we can determine how the expression for a_θ given by Eq. (2.70) relates to the law of Kepler stating that $r^2\dot{\theta}$ is constant.

Computation Equation (2.70) tells us that in polar coordinates the r and θ components of the satellite's acceleration are given by

$$a_r = \ddot{r} - r\dot{\theta}^2 \quad \text{and} \quad a_\theta = r\ddot{\theta} + 2\dot{r}\dot{\theta}. \tag{1}$$

Kepler's observations indicate that

$$r^2\dot{\theta} = K, \tag{2}$$

where K is a constant whose value depends on the particular orbit followed by the satellite. Now, if we differentiate Eq. (2) with respect to time, we obtain

$$2r\dot{r}\dot{\theta} + r^2\ddot{\theta} = 0, \tag{3}$$

where we have used the fact that K is constant so that $\dot{K} = 0$. Equation (3) tells us that

$$2r\dot{r}\dot{\theta} + r^2\ddot{\theta} = r\left(r\ddot{\theta} + 2\dot{r}\dot{\theta}\right) = 0 \quad \Rightarrow \quad r\ddot{\theta} + 2\dot{r}\dot{\theta} = 0, \tag{4}$$

where we have used the fact that r is never zero to obtain the final expression. Comparing Eq. (4) with Eq. (1), we see that

$$a_\theta = 0, \tag{5}$$

which is what we set out to show.

Discussion & Verification The derivation of our result is elementary, and it is correct because we have correctly applied the chain and product rules of calculus in taking the derivative of Eq. (2).

A Closer Look The result we have obtained is based on astronomical observations that predate the work of Newton. From a historical viewpoint this is important because, in formulating his second law of motion and his law of universal gravitation, Newton needed to formulate a theory consistent with Kepler's observations. Therefore, it is not by chance that Newton's law of gravitation demands that the force of gravity between two particles be directed along the line connecting the particles. This requirement, along with Newton's second law, $\vec{F} = m\vec{a}$, causes the acceleration of the planet in Fig. 1 to be *completely* along the radial line connecting the satellite and the planet (i.e., $a_\theta = 0$) so that, overall, both the universal law of gravitation and Newton's second law are consistent with Kepler's observations.

We will come across Eq. (2) again in Section 5.3 because Eq. (2) is also the mathematical expression of the fact that the angular momentum of the satellite, computed relative to the planet, is conserved. More generally, Eq. (2) expresses the conservation of angular momentum for a particle moving under the action of a central force (provided that the origin of the polar coordinate system used is the center of the force).

EXAMPLE 2.20 *Rectilinear Motion & Polar Coordinates*

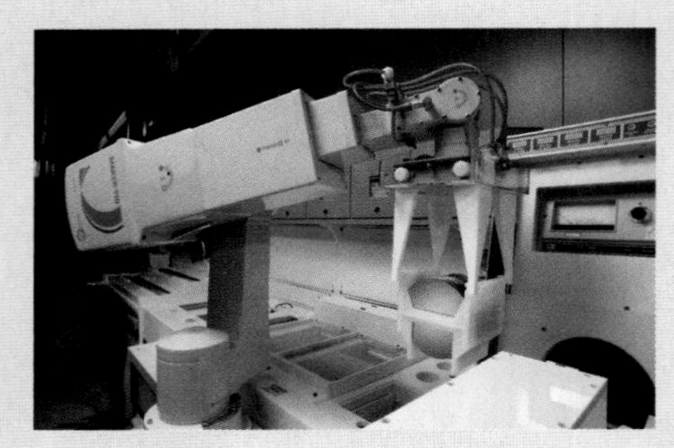

Figure 1
A robotic arm.

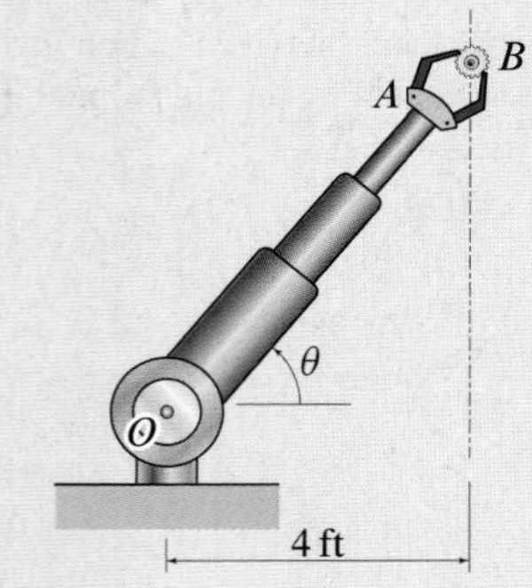

Figure 2
Schematic of robot arm shown in Fig. 1.

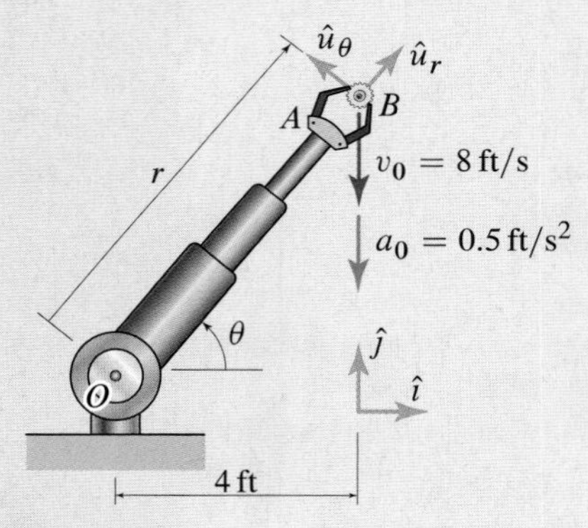

Figure 3
Robotic arm showing the polar and Cartesian coordinate systems we will use, as well as the velocity and acceleration of the end effector.

As a part of an assembly process, the end effector A on the robotic arm in Fig. 2 needs to move the gear B along the vertical line shown in a specified fashion. Arm OA can vary its length by telescoping via internal actuators. A motor at O allows the arm to pivot in the vertical plane. When $\theta = 50°$, B is moving downward with a speed $v_0 = 8\,\text{ft/s}$ and a downward acceleration with magnitude $a_0 = 0.5\,\text{ft/s}^2$. At this instant, determine the required length of the arm, the rate at which the arm is extending, and its rotation rate $\dot{\theta}$. In addition, determine the second time derivatives of both the arm's length and the angle θ.

SOLUTION

Road Map We know the gear's path, velocity, and acceleration as well as the angle θ at the instant shown. Therefore, we can determine the length r at this instant by simple trigonometry. As far as determining $\dot{r}$ and $\dot{\theta}$ is concerned, observe that the known velocity of B is easily written in terms of the unit vector $\hat{\jmath}$ shown in Fig. 3. Once this is done, we can use trigonometry to rewrite the velocity of B via $\hat{u}_r$ and $\hat{u}_\theta$. Then $\dot{r}$ and $\dot{\theta}$ are found by equating this *specific* expression for the velocity of B to the general expression of the velocity vector in polar coordinates. Finally, $\ddot{r}$ and $\ddot{\theta}$ are found by applying to the acceleration the same strategy just described for the velocity.

Computation We know that $\theta = 50°$ at this instant, so the geometry in Fig. 3 tells us that

$$\boxed{r = \frac{4\,\text{ft}}{\cos\theta} = 6.223\,\text{ft}.} \tag{1}$$

Since B moves vertically downward, the velocity of the end effector is

$$\vec{v} = -v_0\,\hat{\jmath} = -(8\,\text{ft/s})\,\hat{\jmath}. \tag{2}$$

Equating Eq. (2) with the expression of the velocity in polar coordinates, we have

$$-(8\,\text{ft/s})\,\hat{\jmath} = \dot{r}\,\hat{u}_r + r\dot{\theta}\,\hat{u}_\theta. \tag{3}$$

Now, we can write $\hat{\jmath}$ in terms of the polar unit vectors as

$$\hat{\jmath} = \sin\theta\,\hat{u}_r + \cos\theta\,\hat{u}_\theta = \sin 50°\,\hat{u}_r + \cos 50°\,\hat{u}_\theta. \tag{4}$$

Substituting Eq. (4) into Eq. (3) and equating components, we have

$$\dot{r} = -8\sin 50°\,\text{ft/s} \quad\text{and}\quad r\dot{\theta} = -8\cos 50°\,\text{ft/s}, \tag{5}$$

which, by using the result in Eq. (1), can then be solved for $\dot{r}$ and $\dot{\theta}$ to obtain

$$\boxed{\dot{r} = -6.128\,\text{ft/s} \quad\text{and}\quad \dot{\theta} = -0.8264\,\text{rad/s}.} \tag{6}$$

Dealing with the acceleration of B as we have done for the velocity, we have

$$\vec{a} = -a_0\,\hat{\jmath} = -\big(0.5\,\text{ft/s}^2\big)\,\hat{\jmath}. \tag{7}$$

Equating Eq. (7) with the expression for acceleration in polar coordinates, we have

$$-(0.5\,\text{ft/s}^2)\,\hat{\jmath} = (\ddot{r} - r\dot{\theta}^2)\,\hat{u}_r + (r\ddot{\theta} + 2\dot{r}\dot{\theta})\,\hat{u}_\theta. \tag{8}$$

Using Eq. (4) again and equating components, we have

$$\ddot{r} - r\dot{\theta}^2 = -0.5\sin 50°\,\text{ft/s}^2 \quad\text{and}\quad r\ddot{\theta} + 2\dot{r}\dot{\theta} = -0.5\cos 50°\,\text{ft/s}^2. \tag{9}$$

Solving Eqs. (9) for $\ddot{r}$ and $\ddot{\theta}$, we have

$$\ddot{r} = -0.5 \sin 50^\circ \text{ ft/s}^2 + r\dot{\theta}^2 \quad \text{and} \quad \ddot{\theta} = -\frac{1}{r}(0.5 \cos 50^\circ \text{ ft/s}^2 + 2\dot{r}\dot{\theta}). \tag{10}$$

Substituting in Eqs. (10) the results we have already obtained for r, $\dot{r}$, and $\dot{\theta}$, we have

$$\boxed{\ddot{r} = 3.866 \text{ ft/s}^2 \quad \text{and} \quad \ddot{\theta} = -1.679 \text{ rad/s}^2.} \tag{11}$$

Discussion & Verification Our results are dimensionally correct, and appropriate units have been used. To understand whether or not the signs of our results are correct, we begin by noticing that in Eqs. (6) $\dot{r} < 0$ and $\dot{\theta} < 0$, which means that the arm is actually getting shorter while rotating clockwise. Intuition tells us that if the end effector is moving straight down, then the arm does need to get shorter and rotate clockwise. Fortunately, both of these observations are consistent with the results in Eq. (6).

As for the sign of $\ddot{r}$ in Eqs. (11), the fact that $\ddot{r} > 0$ indicates that while the arm is getting shorter (i.e., $\dot{r} < 0$), the rate at which this happens is decreasing. That is, since $\ddot{r}$ has a sign opposite to $\dot{r}$, we should expect that as B keeps moving down, the length of the arm will stop shortening. This result is correct because the length of the arm will stop shortening when the arm becomes horizontal, and then it will start increasing for negative values of θ. As far as the sign of $\ddot{\theta}$ is concerned, our result indicates that the rate of clockwise rotation is increasing. This result is to be expected even if the acceleration of B were equal to zero. In this case, i.e., if $a_\theta = 0 = r\ddot{\theta} + 2\dot{r}\dot{\theta}$, the sign of $\ddot{\theta}$ is opposite to the sign of the product $\dot{r}\dot{\theta}$. For us, such a product is positive because we found that both $\dot{r}$ and $\dot{\theta}$ were negative. In our case $a_\theta \neq 0$, but B is accelerating downward and therefore, provides no contribution to $\ddot{\theta}$ in the counterclockwise direction.

EXAMPLE 2.21 *Projectile Motion in Polar Coordinates*

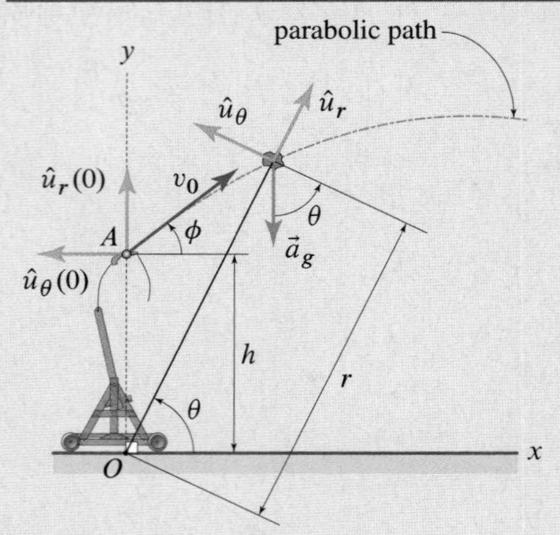

Figure 1
Projectile motion in polar coordinates. The point O is the origin of the coordinate system. The point A is the point at which the object was released, and it is taken to be the initial position of the projectile.

In Section 2.3, we saw that the equations describing the motion of a projectile in Cartesian coordinates took the *simple* form $\ddot{x} = 0$ and $\ddot{y} = -g$. Referring to Fig. 1, revisit the projectile problem and derive the equations describing the motion of a projectile released at A, using polar coordinates. In addition, derive the expressions for the initial conditions of the motion in the same polar coordinate system, given that the projectile is launched from point A at speed v_0 in the direction shown in Fig. 1.

SOLUTION

Road Map What we want to show is that any problem can be formulated using any coordinate system we choose. However, this freedom of choice comes with the consequence that if we do not choose wisely, the mathematical complexity of the problem can be substantial.

Since we know the acceleration of the projectile at every point along its path, the idea is to relate that known acceleration to the direction of a_r and a_θ at every point. A similar idea will apply to finding the initial conditions; that is, we know the initial conditions relative to point A, and we just need to translate those to the polar coordinate system whose origin is at point O.

Computation We begin by finding the polar components of the projectile's acceleration due to gravity, which we denote by $\vec{a}_g$. The components of this vector along the radial and transverse directions are

$$a_{gr} = \vec{a}_g \cdot \hat{u}_r = -g \sin\theta, \tag{1}$$

$$a_{g\theta} = \vec{a}_g \cdot \hat{u}_\theta = -g \cos\theta, \tag{2}$$

where g is the acceleration due to gravity. Equating the *general* expressions for the components of acceleration in polar coordinates given by Eq. (2.71) to the corresponding components in Eqs.(1) and (2) gives

$$\boxed{\ddot{r} - r\dot{\theta}^2 = -g \sin\theta,} \tag{3}$$

$$\boxed{r\ddot{\theta} + 2\dot{r}\dot{\theta} = -g \cos\theta.} \tag{4}$$

Equations (3) and (4) form a system of *coupled, nonlinear* ordinary differential equations which, mathematically, are far more complex than the equations we obtained in Cartesian coordinates.

To integrate Eqs. (3) and (4) so as to determine the motion of the projectile, we would need to complement these differential equations with corresponding *initial conditions*. These conditions consist of the position and velocity of A at time $t = 0$ expressed in polar coordinates. Referring to Fig. 1, we see that at time $t = 0$ the projectile is at A, and therefore, we have

$$r(0) = h \quad \text{and} \quad \theta(0) = \frac{\pi}{2}. \tag{5}$$

For the velocity at $t = 0$ we have

$$\vec{v}(0) = v_0 \sin\phi \, \hat{u}_r(0) - v_0 \cos\phi \, \hat{u}_\theta(0), \tag{6}$$

where the quantities v_0 and ϕ are known. Equating Eq. (6) to the general expression for the velocity vector in polar coordinates given by Eq. (2.66), we have

$$\vec{v}(0) = \dot{r}(0)\, \hat{u}_r(0) + r(0)\dot{\theta}(0)\, \hat{u}_\theta(0) = v_0 \sin\phi \, \hat{u}_r(0) - v_0 \cos\phi \, \hat{u}_\theta(0), \tag{7}$$

that is,

$$\dot{r}(0) = v_0 \sin\phi \quad \text{and} \quad r(0)\dot{\theta}(0) = -v_0 \cos\phi. \tag{8}$$

Concept Alert

Unit vectors in path and polar component systems are functions of time. When we express vectors using normal-tangential components or polar components, it is important to keep in mind that the unit vectors of these component systems are themselves functions of time. This is why, in expressing the velocity at time $t = 0$, we had to use the unit vectors $\hat{u}_r$ and $\hat{u}_\theta$ at $t = 0$.

In summary, the initial conditions for this problem take on the form

$$r(0) = h, \qquad \theta(0) = \frac{\pi}{2}, \tag{9}$$

$$\dot{r}(0) = v_0 \sin\phi, \qquad \dot{\theta}(0) = -\frac{v_0}{h}\cos\phi, \tag{10}$$

where we have used the fact that $r(0) = h$ from Eq. (5).

Discussion & Verification Recalling that r has dimensions of length and that θ is nondimensional, we can verify that the left-hand sides of Eqs. (3) and (4) have dimensions of length over time squared, as expected since the dimensions of the right-hand sides of these equations are those of the acceleration of gravity g. We can verify the dimensional correctness of Eqs. (9) and (10) in a similar manner.

A Closer Look Equations (3), (4), (9), and (10) are much more complicated-looking than the corresponding equations in Cartesian coordinates (i.e., $\ddot{x} = 0$ and $\ddot{y} = -g$, along with the initial conditions $x(0) = 0$ and $y(0) = h$). However, the trajectory obtained by solving Eqs. (3) and (4) is identical to that obtained by solving the corresponding equations in Cartesian coordinates (i.e., $\ddot{x} = 0$ and $\ddot{y} = -g$); that is, we would obtain exactly the same parabola in either case (provided, of course, that the same initial position and velocity are used).

A question often asked by students is, Can I solve this problem using this or that coordinate system? The answer to this question is, in general, yes. However, whether you are solving the problem analytically or numerically, the coordinate system chosen *does* make a difference in how involved the problem's solution becomes. In this example, we have derived (although not solved) the equations governing the motion of a projectile. Clearly, the choice of coordinate system did not change the underlying physics of the problem. However, the resulting system of equations cannot be easily solved by hand because they are nonlinear and, above all, coupled; that is, the equation containing $\ddot{r}$ also contains θ and $\dot{\theta}$, and the equation containing $\ddot{\theta}$ also contains r and $\dot{r}$. Furthermore, even the derivation of the initial conditions required several steps to complete.

Another lesson to be learned concerns the polar coordinate system in particular. We should be aware of the fact that this problem becomes *ill-posed* if point A, the initial position of the projectile, is chosen to coincide with point O, i.e., the origin of the chosen coordinate system. In this case, although we could still write Eqs. (3) and (4), we can no longer write meaningful initial conditions because the value of θ for a point at the origin is arbitrary and (as h would be equal to zero) $\dot{\theta}$ becomes undefined. This implies that when using the polar coordinate system, we should choose the origin of the coordinate system so as *not* to coincide with points where initial conditions are specified.

Problems

Problem 2.178

A particle P is moving along a path with the velocity shown. Discuss in detail whether or not there are incorrect elements in the sketch of the polar component system at P.

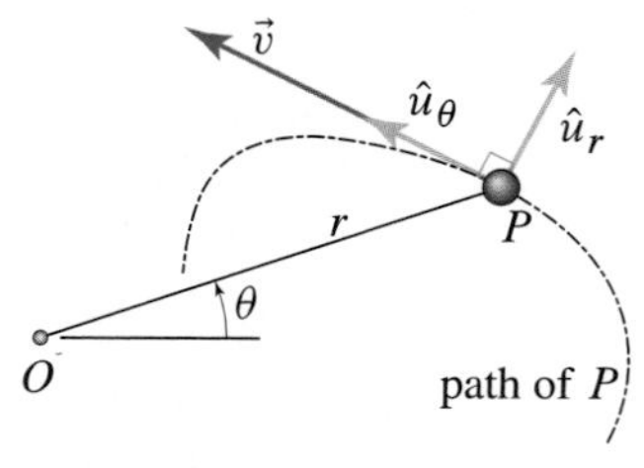

Figure P2.178

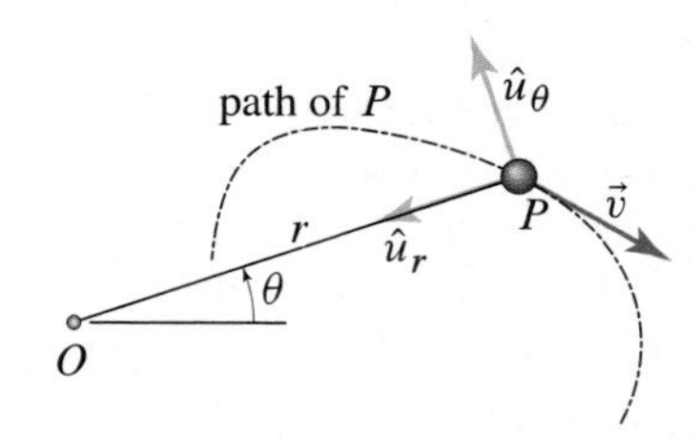

Figure P2.179

Problem 2.179

A particle P is moving along a path with the velocity shown. Discuss in detail whether or not there are incorrect elements in the sketch of the polar component system at P.

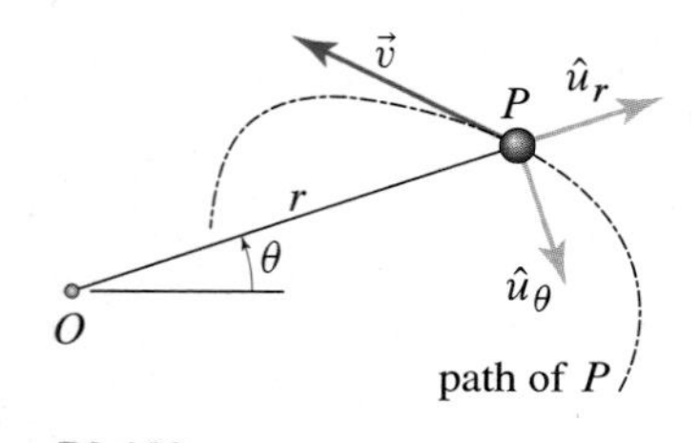

Figure P2.180

Problem 2.180

A particle P is moving along a path with the velocity shown. Discuss in detail whether or not there are incorrect elements in the sketch of the polar component system at P.

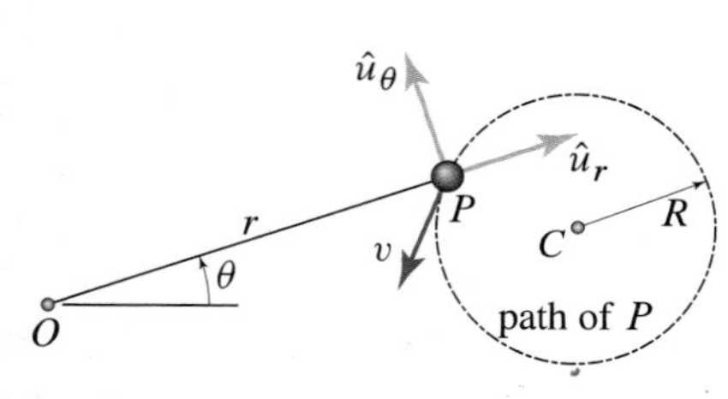

Figure P2.181

Problem 2.181

A particle P is moving along a circle with center C and radius R in the direction shown. Letting O be the origin of a polar coordinate system with the coordinates r and θ shown, discuss in detail whether or not there are incorrect elements in the sketch of the polar component system at P.

Problem 2.182

A radar station is tracking a plane flying at a constant altitude with a constant speed $v_0 = 550$ mph. If at a given instant $r = 7$ mi and $\theta = 32°$, determine the corresponding values of $\dot{r}$, $\dot{\theta}$, $\ddot{r}$, and $\ddot{\theta}$.

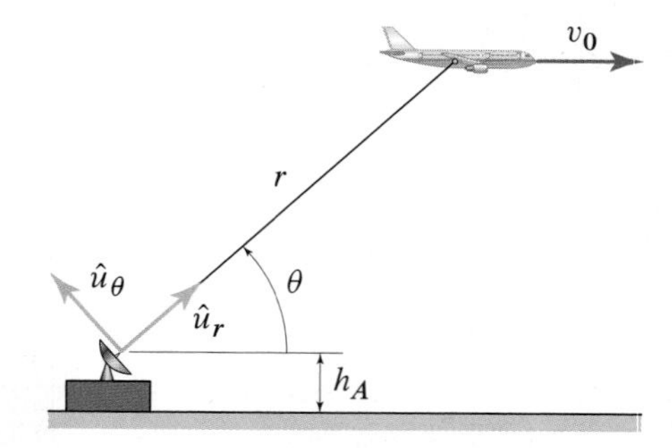

Figure P2.182

Problem 2.183

A basketball moves along the trajectory shown. Modeling the motion of the ball as a projectile motion, determine the radial and transverse components of the acceleration when $\theta = 65°$. Express your answer in SI units.

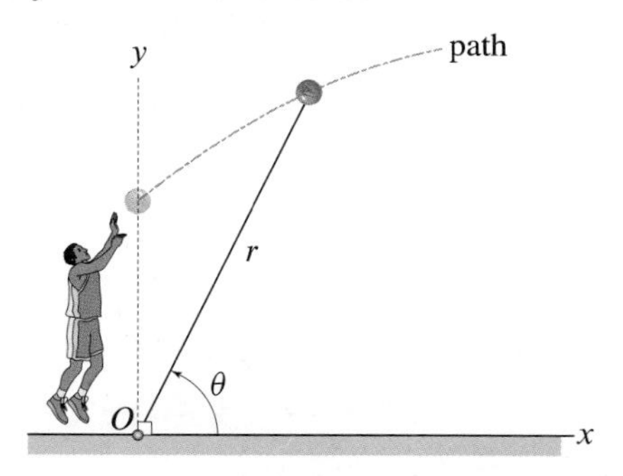

Figure P2.183

Problem 2.184

At a given instant, the merry-go-round is rotating with an angular velocity $\omega = 20$ rpm while the child is moving radially outward at a constant rate of 0.7 m/s. Assuming that the angular velocity of the merry-go-round remains constant, i.e., $\alpha = 0$, determine the magnitudes of the speed and of the acceleration of the child when he is 0.8 m away from the spin axis.

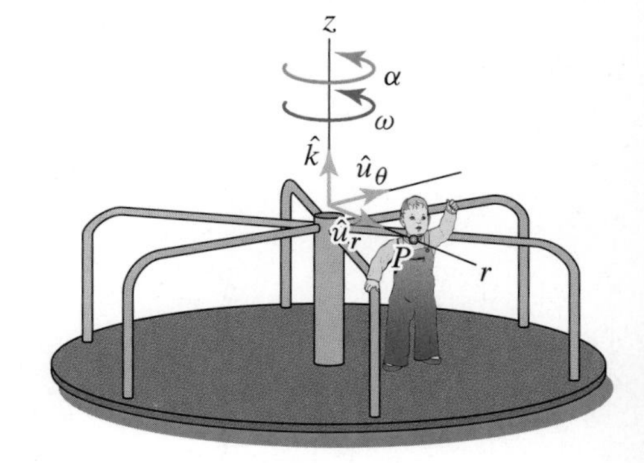

Figure P2.184–P2.186

Problem 2.185

At a given instant, the merry-go-round is rotating with an angular velocity $\omega = 18$ rpm, and it is slowing down at a rate of 0.4 rad/s^2. When the child is 2.5 ft away from the spin axis, determine the time rate of change of the child's distance from the spin axis so that the child experiences no transverse acceleration while moving along a radial line.

Problem 2.186

At a given instant, the merry-go-round is rotating with an angular velocity $\omega = 18$ rpm. When the child is 0.45 m away from the spin axis, determine the second derivative with respect to time of the child's distance from the spin axis so that the child experiences no radial acceleration.

Problem 2.187

A ball is dropped from rest from a height $h = 5$ ft. If the distance $d = 3$ ft, determine the radial and transverse components of the acceleration and the velocity of the ball when the ball has traveled a distance $h/2$ from its release position.

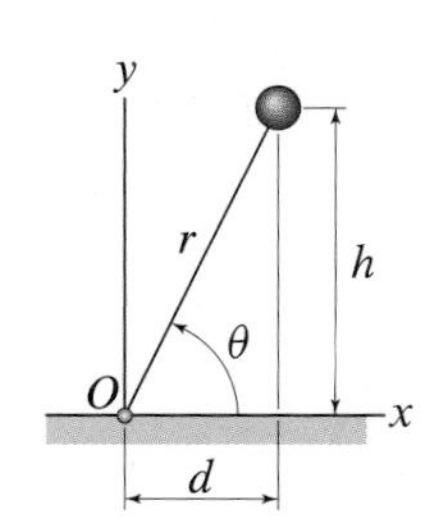

Figure P2.187

Problem 2.188

The polar coordinates of a particle are the following functions of time:

$$r = r_0 \sin(t^3/\tau^3) \quad \text{and} \quad \theta = \theta_0 \cos(t/\tau),$$

where r_0 and θ_0 are constants, $\tau = 1$ s, and where t is time in seconds. Determine r_0 and θ_0 such that the velocity of the particle is completely in the radial direction for $t = 15$ s and the corresponding speed is equal to 6 m/s.

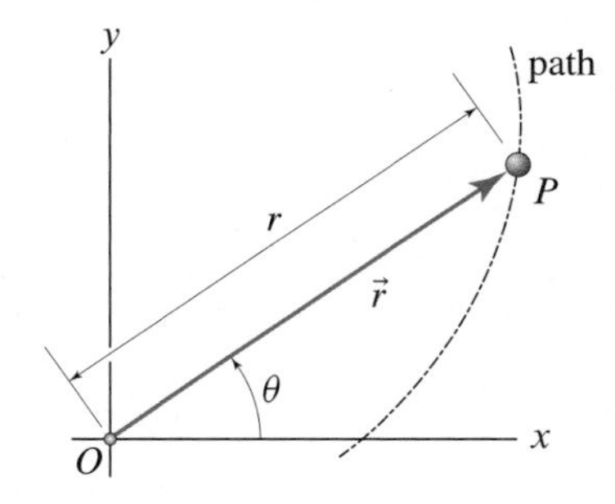

Figure P2.188

Problem 2.189

A space station is rotating in the direction shown at a constant rate of 0.22 rad/s. A crew member travels from the periphery to the center of the station through one of the radial shafts at a constant rate of 1.3 m/s (relative to the shaft) while holding onto a handrail in the shaft. Taking $t = 0$ to be the instant at which travel through the shaft begins and knowing that the radius of the station is 200 m, determine the velocity and acceleration of the crew member as a function of *time*. Express your answer using a polar coordinate system with origin at the center of the station.

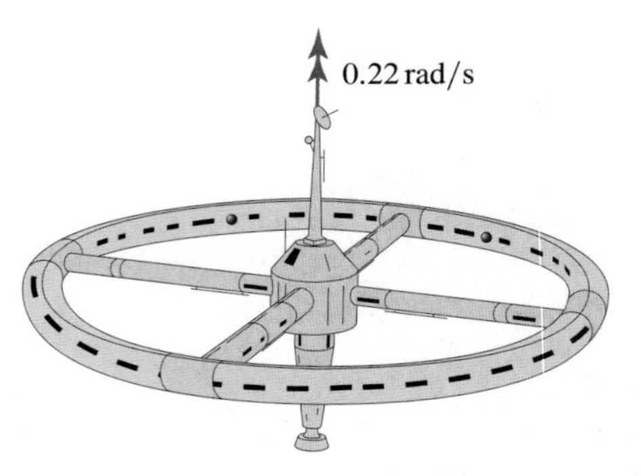

Figure P2.189 and P2.190

Problem 2.190

Solve Prob. 2.189 and express your answers as a function of *position* along the shaft traveled by the astronaut.

Problem 2.191

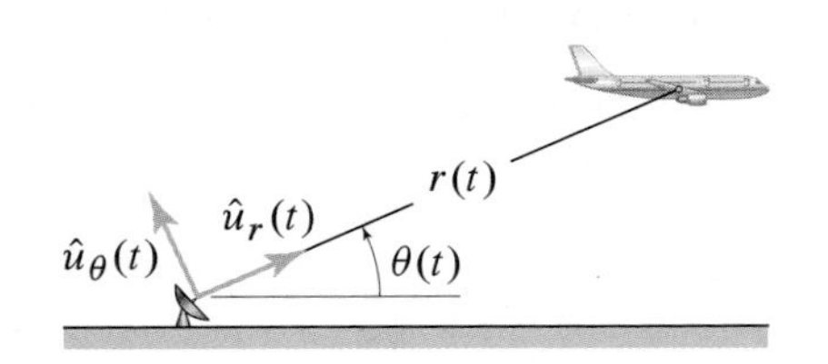

Figure P2.191

During a given time interval, a radar station tracking an airplane records the readings

$$\dot{r}(t) = [449.8\cos\theta(t) + 11.78\sin\theta(t)]\,\text{mph},$$
$$r(t)\dot{\theta}(t) = [11.78\cos\theta(t) - 449.8\sin\theta(t)]\,\text{mph},$$

where t denotes time. Determine the speed of the plane. Furthermore, determine whether the plane being tracked is ascending or descending and the corresponding climbing rate (i.e., the rate of change of the plane's altitude) expressed in ft/s.

Problems 2.192 and 2.193

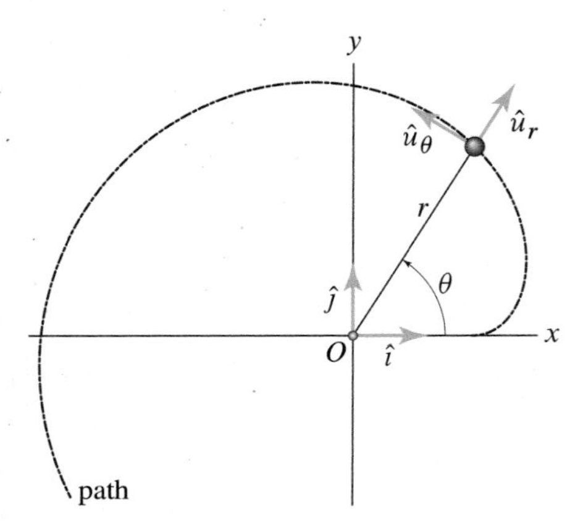

Figure P2.192 and P2.193

The polar coordinates of a particle are the following functions of time:

$$r = r_0\left(1 + \frac{t}{\tau}\right) \quad \text{and} \quad \theta = \theta_0 \frac{t^2}{\tau^2},$$

where $r_0 = 3$ ft, $\theta_0 = 1.2$ rad, $\tau = 20$ s, and t is time in seconds.

Problem 2.192 Determine the velocity and the acceleration of the particle for $t = 35$ s and express the result using the polar component system formed by the unit vectors $\hat{u}_r$ and $\hat{u}_\theta$ at $t = 35$ s.

Problem 2.193 Determine the velocity and the acceleration of the particle for $t = 35$ s and express the result using the Cartesian component system formed by the unit vectors $\hat{\imath}$ and $\hat{\jmath}$.

Problems 2.194 and 2.195

A particle is moving such that the time rate of change of its polar coordinates are

$$\dot{r} = \text{constant} = 3\,\text{ft/s} \quad \text{and} \quad \dot{\theta} = \text{constant} = 0.25\,\text{rad/s}.$$

Problem 2.194 Knowing that at time $t = 0$, a particle has polar coordinates $r_0 = 0.2\,\text{ft}$ and $\theta_0 = 15°$, determine the position, velocity, and acceleration of the particle for $t = 10\,\text{s}$. Express your answers in the polar component system formed by the unit vectors $\hat{u}_r$ and $\hat{u}_\theta$ at $t = 10\,\text{s}$.

Problem 2.195 Knowing that at time $t = 0$, a particle has polar coordinates $r_0 = 0.2\,\text{ft}$ and $\theta_0 = 15°$, determine the position, velocity, and acceleration of the particle for $t = 10\,\text{s}$. Express your answers in the Cartesian component system formed by the unit vectors $\hat{\imath}$ and $\hat{\jmath}$.

Figure P2.194 and P2.195

Problems 2.196 and 2.197

A micro spiral pump* consists of a spiral channel attached to a stationary plate. This plate has two ports, one for fluid inlet and another for outlet, the outlet being farther from the center of the plate than the inlet. The system is capped by a rotating disk. The fluid trapped between the rotating disk and the stationary plate is put in motion by the rotation of the top disk, which pulls the fluid through the spiral channel.

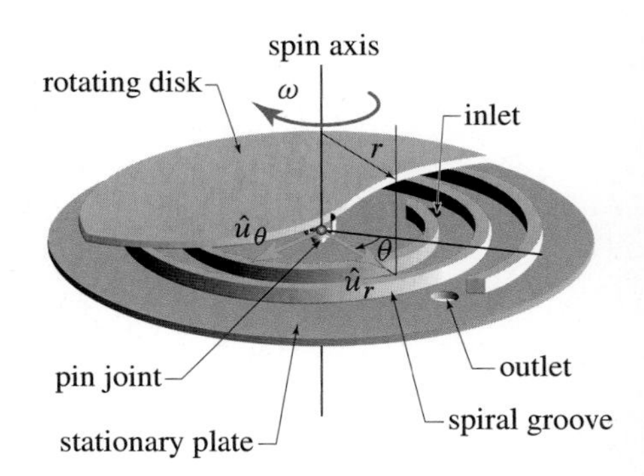

Figure P2.196 and P2.197

Problem 2.196 Consider a spiral channel with the geometry given by the equation $r = \eta\theta + r_0$, where $r_0 = 146\,\mu\text{m}$ is the starting radius, r is the distance from the spin axis, and θ, measured in radians, is the angular position of a point in the spiral channel. Assume that the radius at the outlet is $r_{\text{out}} = 190\,\mu\text{m}$, that the top disk rotates with a constant angular speed ω, and that the fluid particles in contact with the rotating disk are essentially stuck to it. Determine the constant η and the value of ω (in rpm) such that after 1.25 rev of the top disk, the speed of the particles in contact with this disk is $v = 0.5\,\text{m/s}$ at the outlet.

Problem 2.197 Consider a spiral channel with the geometry given by the equation $r = \eta\theta + r_0$, where $\eta = 12\,\mu\text{m}$ is called the polar slope, $r_0 = 146\,\mu\text{m}$ is the starting radius, r is the distance from the spin axis, and θ, measured in radians, is the angular position of a point in the spiral channel. If the top disk rotates with a constant angular speed $\omega = 30{,}000\,\text{rpm}$, and assuming that the fluid particles in contact with the rotating disk are essentially stuck to it, use the polar coordinate system shown and determine the velocity and acceleration of one fluid particle when it is at $r = 170\,\mu\text{m}$.

Problem 2.198

The cutaway of the gun barrel shows a projectile that, upon exit, moves with a speed $v_s = 5490\,\text{ft/s}$ relative to the gun barrel. The length of the gun barrel is $L = 15\,\text{ft}$. Assuming that the angle θ is increasing at a constant rate of $0.15\,\text{rad/s}$, determine the speed of the projectile right when it leaves the barrel. In addition, assuming that the projectile acceleration along the barrel is constant and that the projectile starts from rest, determine the magnitude of the acceleration upon exit.

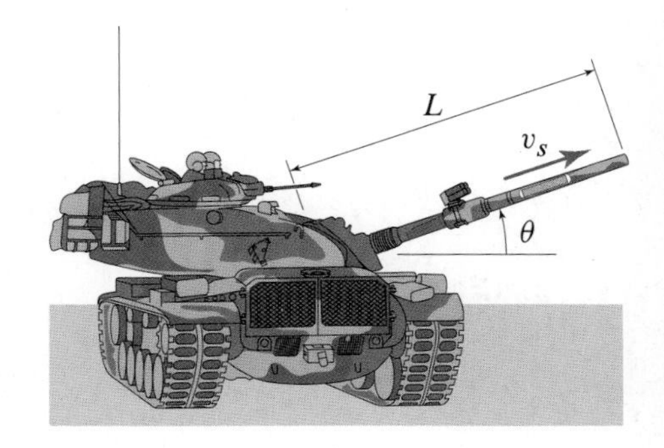

Figure P2.198

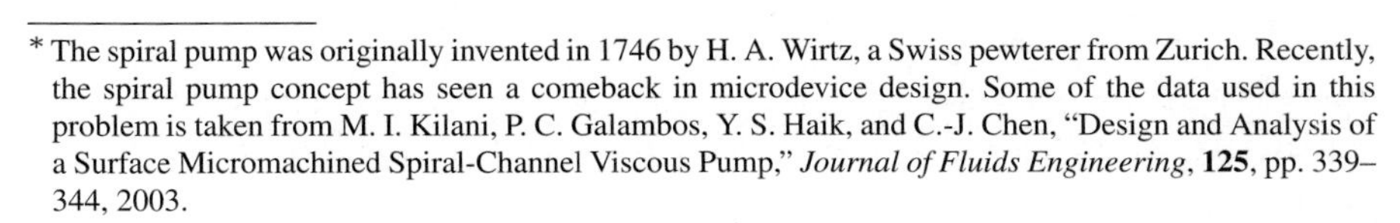

* The spiral pump was originally invented in 1746 by H. A. Wirtz, a Swiss pewterer from Zurich. Recently, the spiral pump concept has seen a comeback in microdevice design. Some of the data used in this problem is taken from M. I. Kilani, P. C. Galambos, Y. S. Haik, and C.-J. Chen, "Design and Analysis of a Surface Micromachined Spiral-Channel Viscous Pump," *Journal of Fluids Engineering*, **125**, pp. 339–344, 2003.

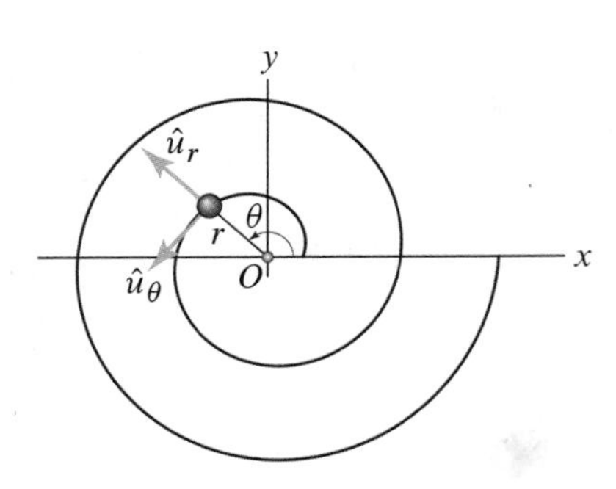

Figure P2.199 and P2.200

Problem 2.199

A particle moves along a spiral described by the equation $r = r_0 + \kappa\theta$, where r_0 and κ are constants, and where θ is in radians. Assume that $\dot{\theta} = \alpha t$, where $\alpha = 0.15\,\text{rad/s}^2$ and t is time expressed in seconds. If $r = 0.25\,\text{m}$ and $\theta = 0$ for $t = 0$, determine κ such that, for $t = 10\,\text{s}$, the acceleration is completely in the radial direction. In addition, determine the value of the polar coordinates of the point for $t = 10\,\text{s}$.

Problem 2.200

A point is moving counterclockwise at constant speed v_0 along a spiral described by the equation $r = r_0 + \kappa\theta$, where r_0 and κ are constants with dimensions of length. Determine the expressions of the velocity and the acceleration of the particle as a function of θ expressed in the polar component system shown.

Problem 2.201

A person driving along a rectilinear stretch of road is fined for speeding, having been clocked at 75 mph when the radar gun was pointing as shown. The driver claims that, because the radar gun is off to the side of the road instead of directly in front of his car, the radar gun overestimates his speed. Is he right or wrong and why?

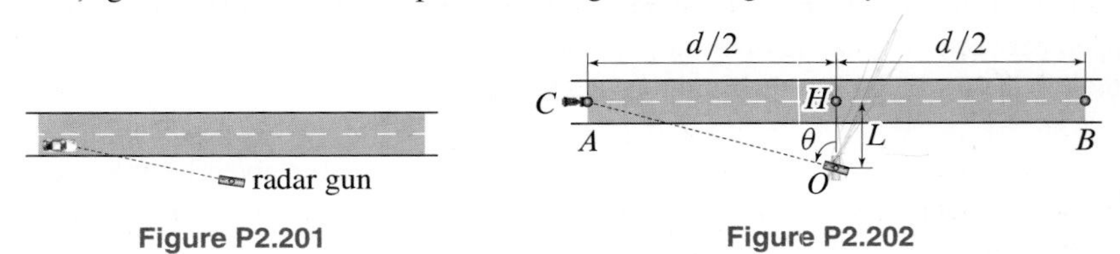

Figure P2.201 Figure P2.202

Problem 2.202

A motion tracking camera is placed along a rectilinear stretch of a racetrack (the figure is not to scale). A car C enters the stretch at A with a speed $v_A = 110\,\text{mph}$ and accelerates uniformly in time so that at B it has a speed $v_B = 175\,\text{mph}$, where $d = 1\,\text{mi}$. Letting the distance $L = 50\,\text{ft}$, if the camera is to track C, determine the camera's angular velocity and the time rate of change of the angular velocity when the car is at A and at H.

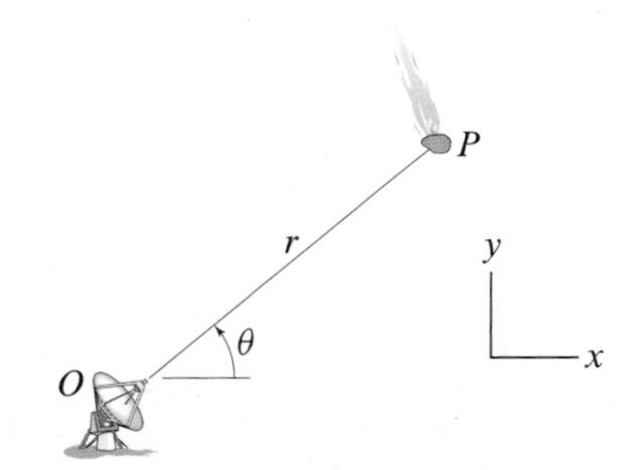

Figure P2.203

Problem 2.203

The radar station at O is tracking a meteor P as it moves through the atmosphere. At the instant shown, the station measures the following data for the motion of the meteor: $r = 21{,}000\,\text{ft}$, $\theta = 40°$, $\dot{r} = -22{,}440\,\text{ft/s}$, $\dot{\theta} = -2.935\,\text{rad/s}$, $\ddot{r} = 187{,}500\,\text{ft/s}^2$, and $\ddot{\theta} = -5.409\,\text{rad/s}^2$.

(a) Determine the magnitude and direction (relative to the xy coordinate system shown) of the velocity vector at this instant.

(b) Determine the magnitude and direction (relative to the xy coordinate system shown) of the acceleration vector at this instant.

Problem 2.204

The time derivative of the acceleration, i.e., $\dot{\vec{a}}$, is usually referred to as the *jerk*. * Starting from Eq. (2.70), compute the jerk in polar coordinates.

* For passengers riding in vehicles, high values of jerk usually make for an uncomfortable ride. Elevator manufacturers are very interested in jerk since they want to move passengers quickly from one floor to another without large changes in acceleration.

Problem 2.205

The *reciprocating rectilinear motion* mechanism shown consists of a disk pinned at its center at A that rotates with a constant angular velocity ω_{AB}, a slotted arm CD that is pinned at C, and a bar that can oscillate within the guides at E and F. As the disk rotates, the peg at B moves within the slotted arm, causing it to rock back and forth. As the arm rocks, it provides a slow advance and a quick return to the reciprocating bar due to the change in distance between C and B. Letting $\theta = 30°$, $\omega_{AB} = 50$ rpm, $R = 0.3$ ft, and $h = 0.6$ ft, determine $\dot{\phi}$ and $\ddot{\phi}$, i.e., the angular velocity and angular acceleration of the slotted arm CD, respectively.

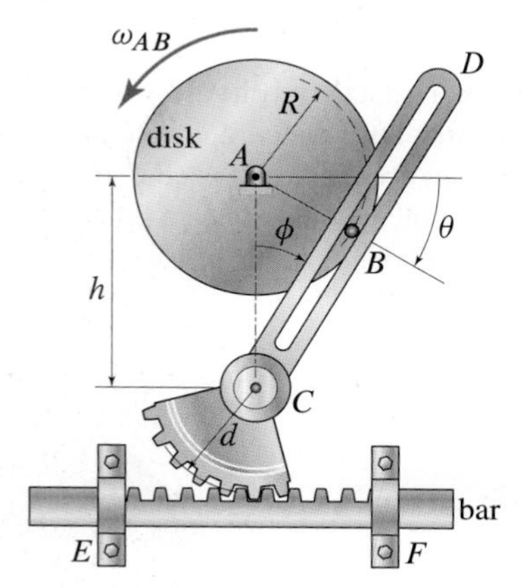

Figure P2.205

Problems 2.206 and 2.207

As a part of an assembly process, the end effector at A on the robotic arm needs to move the gear at B along the vertical line shown with some known velocity v_0 and acceleration a_0. Arm OA can vary its length by telescoping via internal actuators, and a motor at O allows it to pivot in the vertical plane.

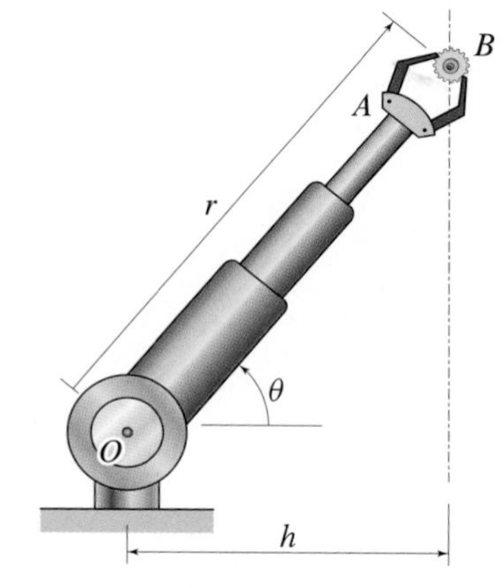

Figure P2.206 and P2.207

Problem 2.206 When $\theta = 50°$, it is required that $v_0 = 8$ ft/s (down) and that it be slowing down at $a_0 = 2$ ft/s^2. Using $h = 4$ ft, determine, at this instant, the values for $\ddot{r}$ (the extensional acceleration) and $\ddot{\theta}$ (the angular acceleration). 2.166

Problem 2.207 Letting v_0 and a_0 be positive if the gear moves and accelerates upward, determine expressions for r, $\dot{r}$, $\ddot{r}$, $\dot{\theta}$, and $\ddot{\theta}$ that are valid for any value of θ.

Problems 2.208 and 2.209

In the cutting of sheet metal, the robotic arm OA needs to move the cutting tool at C counterclockwise at a constant speed v_0 along a circular path of radius ρ. The center of the circle is located in the position shown relative to the base of the robotic arm at O.

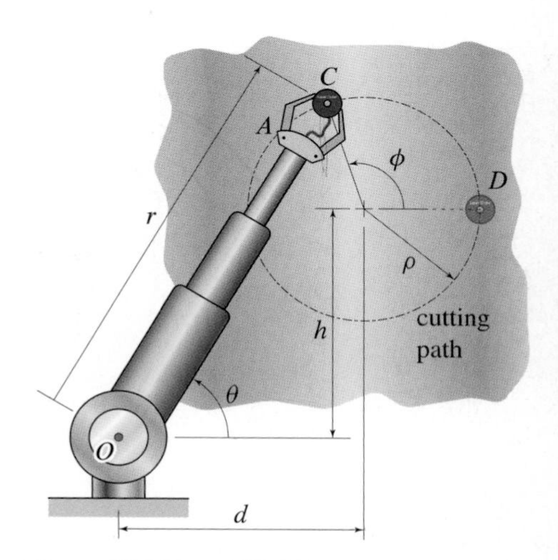

Figure P2.208 and P2.209

Problem 2.208 When the cutting tool is at D ($\phi = 0$), determine r, $\dot{r}$, $\dot{\theta}$, $\ddot{r}$, and $\ddot{\theta}$ as functions of the given quantities (i.e., d, h, ρ, v_0).

Problem 2.209 For all positions along the circular cut (i.e., for any value of ϕ), determine r, $\dot{r}$, $\dot{\theta}$, $\ddot{r}$, and $\ddot{\theta}$ as functions of the given quantities (i.e., d, h, ρ, v_0). These quantities can be found "by hand," but it is tedious, so you might consider using symbolic algebra software, such as Mathematica or Maple.

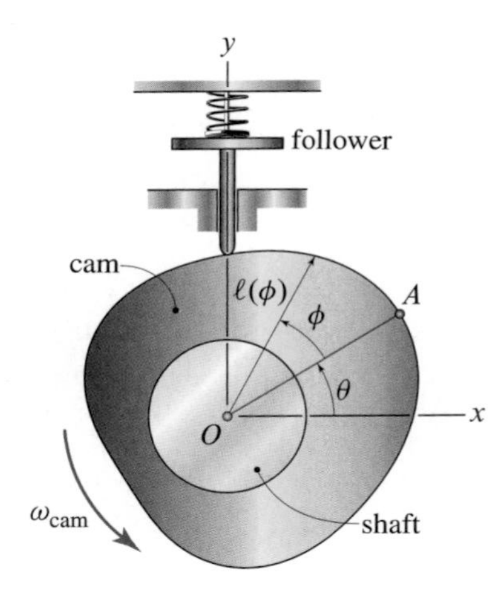

Figure P2.210

Problem 2.210

The cam is mounted on a shaft that rotates about O with constant angular velocity ω_{cam}. The profile of the cam is described by the function $\ell(\phi) = R_0(1 + 0.25\cos^3\phi)$, where the angle ϕ is measured relative to the segment OA, which rotates with the cam. Letting $\omega_{cam} = 3000$ rpm and $R_0 = 3$ cm, determine the velocity and acceleration of the follower when $\theta = 33°$. Express the acceleration of the follower in terms of g, the acceleration due to gravity.

Problem 2.211

The collar is mounted on the horizontal arm shown, which is originally rotating with the angular velocity ω_0. Assume that after the cord is cut, the collar slides along the arm in such a way that the collar's total acceleration is equal to zero. Determine an expression of the radial component of the collar's velocity as a function of r, the distance from the spin axis. *Hint:* Using polar coordinates, observe that $d(r^2\dot{\theta})/dt = ra_\theta$.

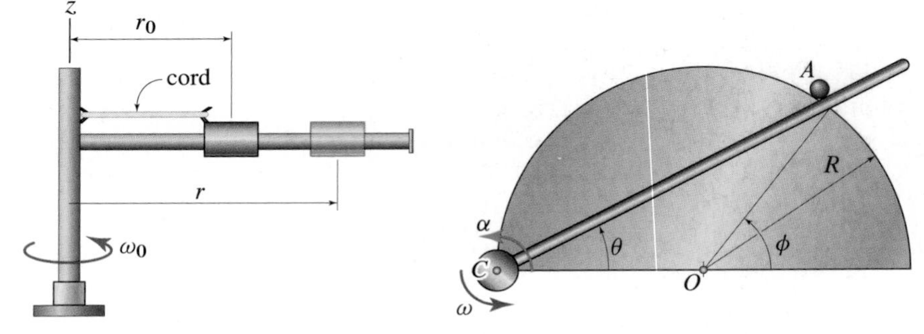

Figure P2.211

Figure P2.212

Problem 2.212

Particle A slides over the semicylinder while pushed by the arm pinned at C. The motion of the arm is controlled such that it starts from rest at $\theta = 0$, ω increases uniformly as a function of θ, and $\omega = 0.5$ rad/s for $\theta = 45°$. Letting $R = 4$ in., determine the speed and the magnitude of the acceleration of A when $\phi = 32°$.

Problem 2.213

The mechanism shown is called a *swinging block* slider crank. First used in various steam locomotive engines in the 1800s, this mechanism is often found in door-closing systems. If the disk is rotating with a constant angular velocity $\dot{\theta} = 60$ rpm, $H = 4$ ft, $R = 1.5$ ft, and r denotes the distance between B and O, compute $\dot{r}$, $\dot{\phi}$, $\ddot{r}$, and $\ddot{\phi}$ when $\theta = 90°$.

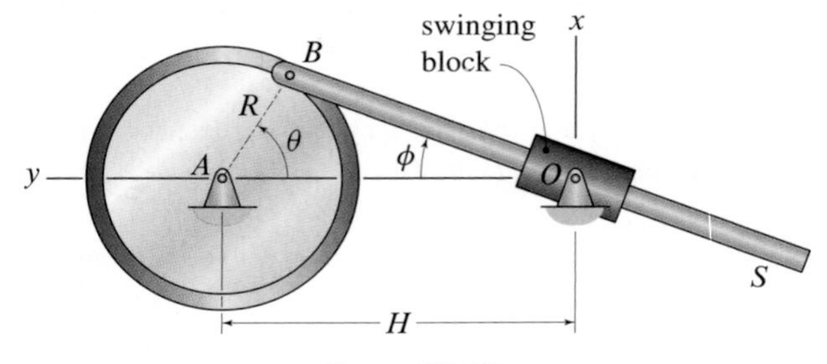

Figure P2.213

Problem 2.214

A satellite is moving along the elliptical orbit shown. Using the polar coordinate system in the figure, the satellite's orbit is described by the equation

$$r(\theta) = 2b^2 \frac{a + \sqrt{a^2 - b^2}\cos\theta}{a^2 + b^2 - (a^2 - b^2)\cos(2\theta)},$$

which implies the following identity

$$\frac{rr'' - 2(r')^2 - r^2}{r^3} = -\frac{a}{b^2},$$

where the prime indicates differentiation with respect to θ. Using this identity and knowing that the satellite moves so that $K = r^2\dot{\theta}$ with K constant (i.e., according to Kepler's laws), show that the radial component of acceleration is proportional to $-1/r^2$, which is in agreement with Newton's universal law of gravitation.

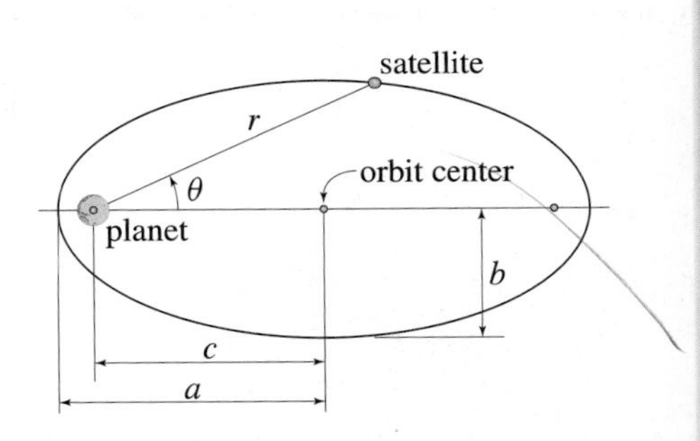

Figure P2.214

Problem 2.215

At a given instant, an airplane flying at an altitude $h_0 = 10{,}000$ ft begins its descent in preparation for landing when it is $r(0) = 20$ mi from the radar station at the destination's airport. At that instant, the aiplane's speed is $v_0 = 300$ mph, the climb rate is constant and equal to -5 ft/s, and the horizontal component of velocity is decreasing steadily at a rate of $15\,\text{ft/s}^2$. Determine the $\dot{r}$, $\dot{\theta}$, $\ddot{r}$, and $\ddot{\theta}$ that would be observed by the radar station.

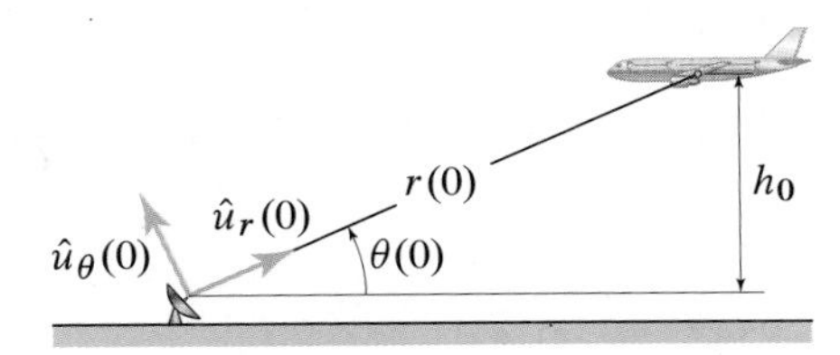

Figure P2.215

Problem 2.216

Considering the system analyzed in Example 2.21, let $h = 15$ ft, $v_0 = 55$ mph, and $\phi = 25°$. Plot the trajectory of the projectile in two different ways: (1) by solving the projectile motion problem using Cartesian coordinates and plotting y versus x and (2) by using a computer to solve Eqs. (3), (4), (9), and (10) in Example 2.21. You should, of course, get the same trajectory regardless of the coordinate system used.

Design Problems

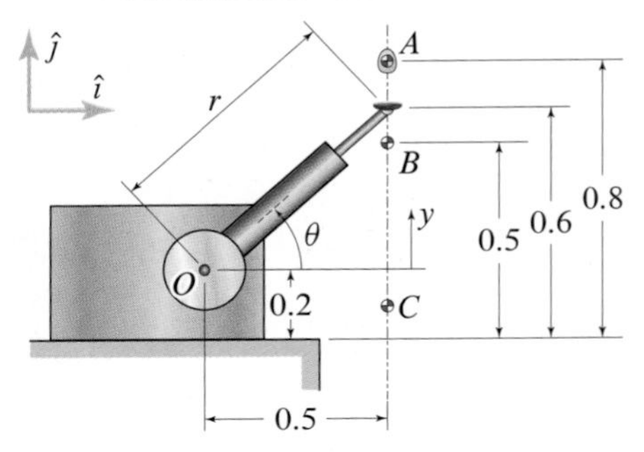

Figure DP2.2 and DP2.3

Design Problem 2.2

As a part of a robotics competition, a robotic arm is to be designed so as to catch an egg without breaking it. The egg is released at point A from rest while the arm is initially also at rest in the position shown. The arm starts moving when the egg is released, and it catches the egg at B with the same speed and acceleration that the egg has at that instant (matching the speed avoids impact and matching the acceleration keeps them together after impact).

Assume that the arm will start slowing the egg down at B and will bring the egg to a complete stop at C. Also assume that the vertical rate of change of deceleration (the time derivative of the acceleration) felt by the egg remains constant and that the egg arrives at C with zero acceleration. For the motion between B and C, determine

(a) The rate of change of the deceleration.

(b) The function $y(t)$ of the vertical motion.

(c) The time it takes for the arm to bring the egg to a stop.

(d) The vertical position C at which they come to a stop.

(e) Then, using the fact that $\dot{\vec{a}} \cdot \hat{\imath}$ is 0 and $\dot{\vec{a}} \cdot \hat{\jmath}$ is the time rate of change of the vertical acceleration due to the motion of the arm along a vertical line, plot the functions $r(t)$ and $\theta(t)$ required to achieve the given motion from B to C.

(f) Finally, use the geometrical constraints $0.5 \tan\theta = y$ and $r^2 = y^2 + (0.5)^2$ to determine analytical expressions for $r(t)$ and $\theta(t)$, and compare the plots of these analytical expressions with the plots found in Part (e). They should, of course, be the same.

Design Problem 2.3

As a part of a robotics competition, a robotic arm is to be designed so as to catch an egg without breaking it. The egg is released at point A from rest while the arm is initially also at rest in the position shown. The arm starts moving when the egg is released, and it is to catch the egg at point B in such a way as to avoid any impact between the egg and the robot hand. See Design Problem 2.2 for how the robot arm needs to be moving for it to *catch* the egg. With this in mind, find an acceleration profile (i.e., $\ddot{r}(t)$ and $\ddot{\theta}(t)$) of the arm as it moves from its initial position to B that satisfies this condition.

2.7 Relative Motion Analysis and Differentiation of Geometrical Constraints

In this section we discuss relative motion and differentiation of constraints. These concepts are used to solve problems with multiple moving objects and are important in the development of rigid body kinematics. We will study relative motion using frames of reference that only translate relative to one another. The general case, which includes frames that also rotate relative to one another, is presented in Section 6.4.

Relative motion

Consider two points A and B moving on separate paths in a plane (Fig 2.30). The XY Cartesian frame of reference has base vectors $\hat{I}$ and $\hat{J}$ and is fixed — this is the *stationary frame*. The xy Cartesian frame of reference has base vectors $\hat{\imath}$ and $\hat{\jmath}$, it is attached to A, and it only translates relative to the XY frame — this is the *moving frame*. Using Eq. (1.14) on p. 7, the position of B relative to A in the XY frame is

$$\vec{r}_{B/A} = (X_B - X_A)\,\hat{I} + (Y_B - Y_A)\,\hat{J}, \tag{2.72}$$

where (X_A, X_B) and (X_B, Y_B) are the coordinates of A and B relative to the XY frame, respectively, and where we recall that the subscript B/A is read "B relative to A" (see discussion of Eq. (1.14) on p. 7).

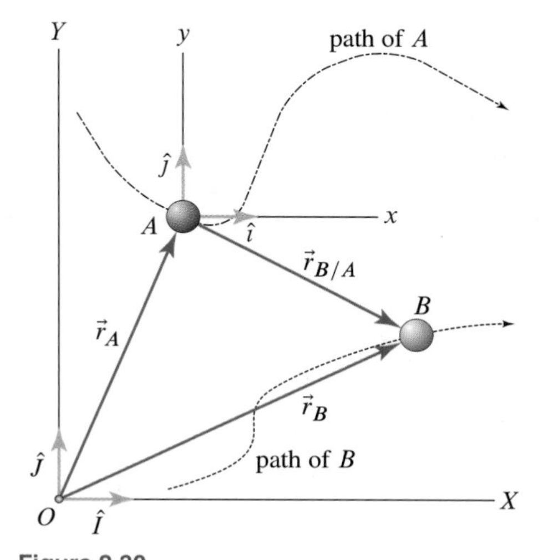

Figure 2.30
Two particles A and B and the definition of their relative position vector $\vec{r}_{B/A}$. The xy frame only translates relative to the XY frame.

The position of B relative to A as viewed by the xy frame is

$$\vec{r}_{B/A} = x_B\,\hat{\imath} + y_B\,\hat{\jmath}, \tag{2.73}$$

where x_B and y_B are the coordinates of B relative to the xy frame. Since the xy frame is Cartesian, when the moving observer differentiates the vector $\vec{r}_{B/A}$ with respect to time, this observer obtains

$$\big(\dot{\vec{r}}_{B/A}\big)_{xy\text{ frame}} = \dot{x}_B\,\hat{\imath} + \dot{y}_B\,\hat{\jmath}. \tag{2.74}$$

By contrast, when the stationary observer computes the same time derivative, the result is

$$\begin{aligned}\big(\dot{\vec{r}}_{B/A}\big)_{XY\text{ frame}} &= (\dot{X}_B - \dot{X}_A)\,\hat{I} + (\dot{Y}_B - \dot{Y}_A)\,\hat{J}\\ &= \dot{x}_B\,\hat{\imath} + x_B\,\dot{\hat{\imath}} + \dot{y}_B\,\hat{\jmath} + y_B\,\dot{\hat{\jmath}} = \dot{x}_B\,\hat{\imath} + \dot{y}_B\,\hat{\jmath},\end{aligned} \tag{2.75}$$

where the second line in Eq. (2.75) is obtained by letting the stationary observer compute the time derivative of Eq. (2.73) and we have used the fact that $\dot{\hat{\imath}} = \dot{\hat{\jmath}} = 0$ since xy is not rotating. Comparing Eqs. (2.74) and (2.75) we see that

$$\big(\dot{\vec{r}}_{B/A}\big)_{xy\text{ frame}} = \big(\dot{\vec{r}}_{B/A}\big)_{XY\text{ frame}}, \tag{2.76}$$

which states that the time rate of change of the vector $\vec{r}_{B/A}$ is the same for the stationary and moving observers when these observers only translate relative to one another.

To relate position, velocity, and acceleration between the xy and XY reference frames, we now consider the vector triangle OAB, for which we have (see Fig. 2.30)

$$\boxed{\vec{r}_B = \vec{r}_A + \vec{r}_{B/A},} \tag{2.77}$$

where $\vec{r}_A$ and $\vec{r}_B$ are the position vectors of A and B relative to the XY reference frame, respectively. The time derivative of Eq. (2.77) gives

$$\boxed{\vec{v}_B = \vec{v}_A + \vec{v}_{B/A},} \tag{2.78}$$

where $\vec{v}_{B/A} = d\vec{r}_{B/A}/dt$ is the relative velocity of B with respect to A. Differentiating Eq. (2.78) with respect to time, we have

$$\boxed{\vec{a}_B = \vec{a}_A + \vec{a}_{B/A},} \tag{2.79}$$

where $\vec{a}_{B/A} = d^2\vec{r}_{B/A}/dt^2$ is the relative acceleration of B with respect to A. Because of Eq. (2.76), $\vec{v}_{B/A}$ and $\vec{a}_{B/A}$ are identical in the xy and XY frames. Equation (2.78) says that *the velocity of B, as seen by the stationary observer, is equal to the velocity of A, as seen by the stationary observer, plus the relative velocity of B with respect to A, which is the velocity of B as seen by the moving observer*. By replacing *velocity* with *acceleration*, Eq. (2.79) can be read in a similar way.

Differentiation of geometrical constraints

In addition to the vector-based methods we have studied, sometimes we can find the motion of points by writing geometrical constraint equations involving the position of these points and then differentiating the constraint equations with respect to time. We will show how this works with two short examples.

A rigid system sliding down a wall

Referring to Fig. 2.31(a), the particles A and B are connected by a rigid bar of length L. The system is sliding down the wall at A and across the floor at B. By writing constraint equations involving the positions of A and B and the angle θ, we can derive equations relating the time rate of change of these quantities. Since the bar is rigid,

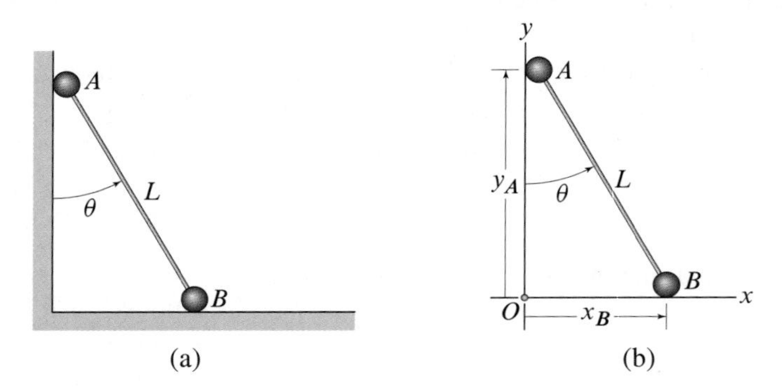

Figure 2.31. (a) A bar sliding down a wall. (b)

points A and B move under the constraints that A only moves vertically, B only moves horizontally, and the distance L between A and B remains constant. Therefore, referring to Fig. 2.31(b), neglecting the size of A and B, and using the Cartesian coordinate system shown, the positions of A and B are completely characterized in terms of the θ and L as follows:

$$x_B = L\sin\theta \quad \text{and} \quad y_A = L\cos\theta. \tag{2.80}$$

Differentiating Eqs. (2.80) with respect to time, we obtain the velocity and acceleration of A and B:

$$\dot{x}_B = v_B = L\dot{\theta}\cos\theta, \qquad \dot{y}_A = v_A = -L\dot{\theta}\sin\theta, \tag{2.81}$$

$$\ddot{x}_B = a_B = L\ddot{\theta}\cos\theta - L\dot{\theta}^2\sin\theta, \quad \ddot{y}_A = a_A = -L\ddot{\theta}\sin\theta - L\dot{\theta}^2\cos\theta. \tag{2.82}$$

Equations (2.81) and (2.82) can be used in several ways. For example, if we are given the position of B as a function of time, we could solve the six expressions in Eqs. (2.80) and (2.82) to find the $\theta(t)$, $\dot{\theta}(t)$, $\ddot{\theta}(t)$, $y_A(t)$, $\dot{y}_A(t)$, and $\ddot{y}_A(t)$.

A simple pulley system

As an another example of a constrained system, we consider the pulley systems in Fig. 2.32, for which we want to determine how the velocity and acceleration of block P are related to the velocity and acceleration of block Q under the constraint that the cords in the system are *inextensible*.

The key to the analysis of any pulley system is the notion of cord (or cable or rope) length and its first and second time derivatives. In Fig. 2.32, there are three cords. However, cords GI and JH simply keep block P attached to pulley G and pulley H attached to the fixed ceiling at J, respectively. Therefore,

$$v_P = v_G \quad \Rightarrow \quad a_P = a_G, \tag{2.83}$$

$$v_H = v_J = 0 \quad \Rightarrow \quad a_H = a_J = 0, \tag{2.84}$$

where Eqs. (2.84) hold because J is fixed. We now look at the cord $ABCDEF$.

Let the length of cord $ABCDEF$ be L, which is a constant since all cords are assumed inextensible. We now express L in terms of the quantities shown in Fig. 2.32, i.e.,

$$L = \overline{AB} + \widetilde{BC} + \overline{CD} + \widetilde{DE} + \overline{EF}, \tag{2.85}$$

where $\widetilde{BC}$ and $\widetilde{DE}$ are the lengths of the *curved* segments of cord that wrap around pulleys G and H, respectively, and the letters with overbars represent the lengths of the corresponding *straight* line segments. The lengths of $\widetilde{BC}$ and $\widetilde{DE}$ are both constant, the sum of which we will call K. The lengths of the straight segments can be *written in terms of the coordinates of blocks P and Q*, which we do as follows:

$$\overline{AB} = y_P - \overline{GI}, \qquad \overline{CD} = y_P - \overline{GI} - \overline{JH}, \qquad \overline{EF} = y_Q - \overline{JH}. \tag{2.86}$$

Then Eq. (2.85) becomes

$$L = 2y_P + y_Q - 2\overline{GI} - 2\overline{JH} + K. \tag{2.87}$$

We now differentiate Eq. (2.87) with respect to time and recognize that $d(\overline{GI})/dt = 0$ and $d(\overline{JH})/dt = 0$, due to Eqs. (2.83) and (2.84), respectively. Therefore, we obtain

$$\dot{L} = 2\dot{y}_P + \dot{y}_Q. \tag{2.88}$$

Recalling that L is constant because the cord is inextensible, $\dot{L} = 0$ and Eq. (2.88) yields the relation between the velocities of blocks P and Q, i.e.,

$$2\dot{y}_P + \dot{y}_Q = 0 \quad \text{or} \quad 2v_P + v_Q = 0. \tag{2.89}$$

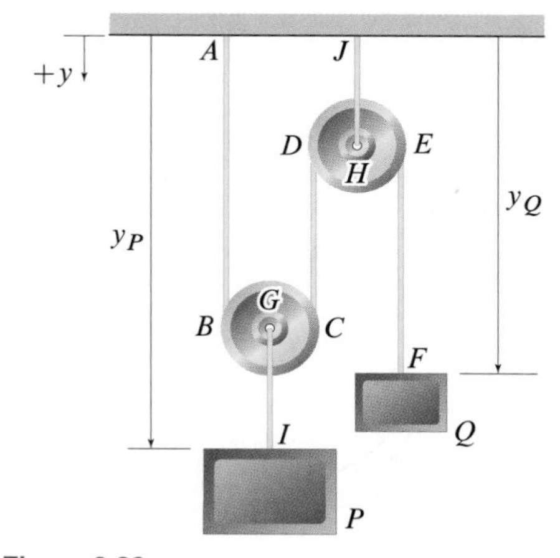

Figure 2.32
Simple pulley system used to demonstrate the principles behind the differentiation of geometric constraints.

Helpful Information

What if the cord length is not constant? In some problems the cord length is not constant, such as when a motor or winch is pulling in or letting out cord. In these cases, we modify the kinematics by setting the time derivative of the appropriate cord length equal to the rate at which the cord is becoming longer or shorter, i.e.,

$$\dot{L} = \begin{cases} +\text{ rate of increase,} \\ \text{or} \\ -\text{ rate of decrease.} \end{cases}$$

We now note the following about Eq. (2.89):

- Equation (2.89) says that if P is moving 4 m/s *down*, then Q must be moving 8 m/s *up*. This is so because $v_Q = -2v_P$ and we have defined *both* y_P and y_Q to be positive downward.
- We can differentiate Eq. (2.89) with respect to time to obtain a relationship between the accelerations of blocks P and Q

$$2a_P + a_Q = 0. \tag{2.90}$$

- We did not need to know the length of each cord — we only needed to know that each length was constant. This will often be the case.

End of Section Summary

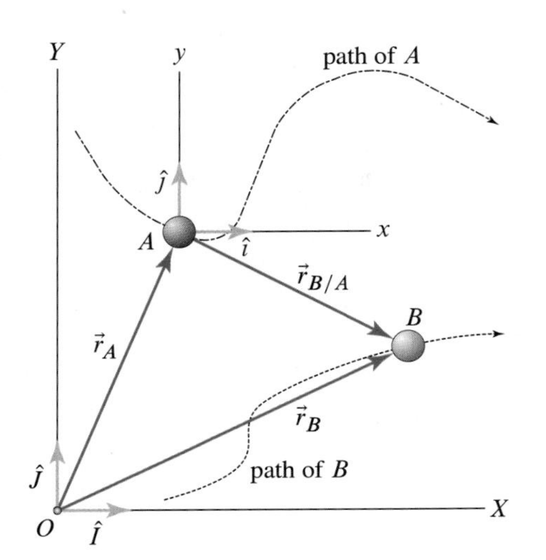

Figure 2.33
Two particles A and B and the definition of their relative position vector $\vec{r}_{B/A}$.

Relative motion. Referring to Fig. 2.33, consider the planar motion of points A and B. The positions of A and B relative to the XY frame are $\vec{r}_A$ and $\vec{r}_B$, respectively. Attached to A there is a frame xy that translates but does not rotate relative to frame XY. In either frame, the position of B relative to A is given by $\vec{r}_{B/A}$. Using vector addition, the vectors $\vec{r}_A$, $\vec{r}_B$, and $\vec{r}_{B/A}$ are related as follows:

Eq. (2.77), p. 121

$$\vec{r}_B = \vec{r}_A + \vec{r}_{B/A}.$$

The first and second time derivatives of the above equation are, respectively,

Eqs. (2.78) and (2.79), p. 122

$$\vec{v}_B = \vec{v}_A + \vec{v}_{B/A},$$
$$\vec{a}_B = \vec{a}_A + \vec{a}_{B/A},$$

where $\vec{v}_{B/A} = \dot{\vec{r}}_{B/A}$ and $\vec{a}_{B/A} = \ddot{\vec{r}}_{B/A}$ are the relative velocity and acceleration of B with respect to A, respectively. In general, the vectors $\vec{v}_{B/A}$ and $\vec{a}_{B/A}$ computed by the xy observer are different from the vectors $\vec{v}_{B/A}$ and $\vec{a}_{B/A}$ computed by the XY observer. However, the xy and XY observers compute the same $\vec{v}_{B/A}$ and the same $\vec{a}_{B/A}$ if these observers do not rotate relative to one another.

Constrained motion. There are very few dynamics problems in which the motion is *not* constrained in some way. In certain classes of systems, constraints are described by geometrical relations between points in the system. We analyzed a pulley system whose motion was constrained by the inextensibility of the cords in the system. The key to constrained motion analysis is the awareness that we can differentiate the equations describing the geometrical constraints to obtain velocities and accelerations of points of interest.

EXAMPLE 2.22 *Relative Speed and Acceleration*

The driver of car B sees a police car P and applies his brakes, causing the car to decelerate at a constant rate of $25\,\text{ft/s}^2$. At the same time, the police car is traveling at a constant speed $v_P = 35$ mph, and using a radar gun, the police officer sees B coming toward her at 65 mph when $\theta = 22°$. At the instant that the radar gun measurement was taken, determine the corresponding true speed of B and the magnitude of the relative acceleration of B with respect to P.

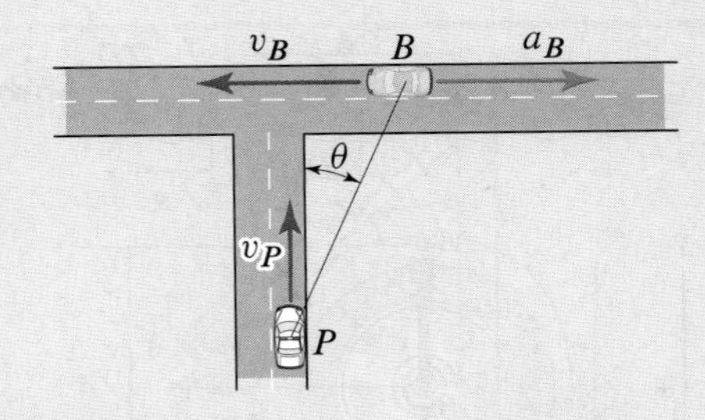

Figure 1

SOLUTION

Road Map We need to find the speed of B with respect to the road, which we choose as our stationary frame. The speed measured by the radar gun is relative to the moving observer P and is the component of the velocity of B relative to P along the line connecting B and P. Hence, we will find the velocity of B relative to P and then consider the component of this velocity along the line PB. As for the relative acceleration of B with respect to P, we can calculate it as a direct application of Eq. (2.79).

Computation Letting v_B be the speed of B, referring to Fig. 2, the velocity of B relative to P is

$$\vec{v}_{B/P} = \vec{v}_B - \vec{v}_P = -v_B\,\hat{\imath} - v_P\,\hat{\jmath}. \tag{1}$$

We denote the radar gun speed reading by v_r, and we observe that v_r is the magnitude of the component of $\vec{v}_{B/P}$ along the line connecting points P and B. The sign of the component in question is negative because the police officer sees B coming *toward* her. Therefore, denoting by $\hat{u}_{B/P}$ the unit vector pointing from P to B and observing that $\hat{u}_{B/P} = \sin\theta\,\hat{\imath} + \cos\theta\,\hat{\jmath}$, we have

$$\vec{v}_{B/P}\cdot\hat{u}_{B/P} = -v_r \quad\Rightarrow\quad -v_B\sin\theta - v_P\cos\theta = -v_r, \tag{2}$$

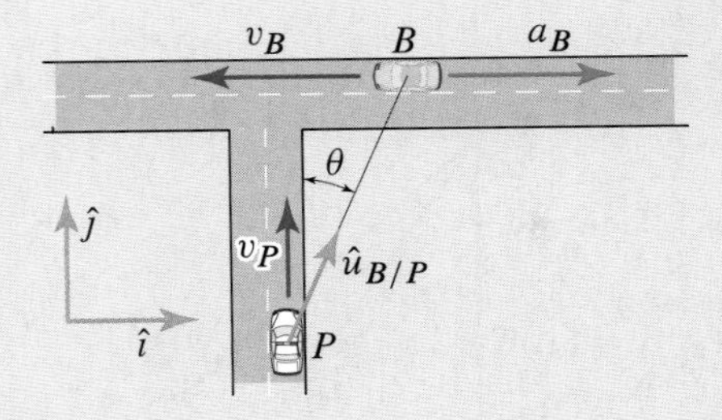

Figure 2
Stationary component system.

where we have used the expression for $\vec{v}_{B/P}$ in Eq. (1). Solving the second of Eqs. (2) for v_B, we have

$$\boxed{v_B = \frac{v_r - v_P\cos\theta}{\sin\theta} = 86.89\,\text{mph}.} \tag{3}$$

Applying Eq. (2.79), the acceleration of B relative to P is

$$\vec{a}_{B/P} = \vec{a}_B - \vec{a}_P = (25.00\,\text{ft/s}^2)\,\hat{\imath}, \tag{4}$$

since the police car is traveling at a constant velocity. Therefore, we have

$$\boxed{\left|\vec{a}_{B/P}\right| = 25.00\,\text{ft/s}^2.} \tag{5}$$

Discussion & Verification Since the terms $\cos\theta$ and $\sin\theta$ are nondimensional, the result in Eq. (3) is dimensionally correct. The result in Eq. (5) is also dimensionally correct since it was derived as a direct application of a dimensionally correct formula for the relative acceleration.

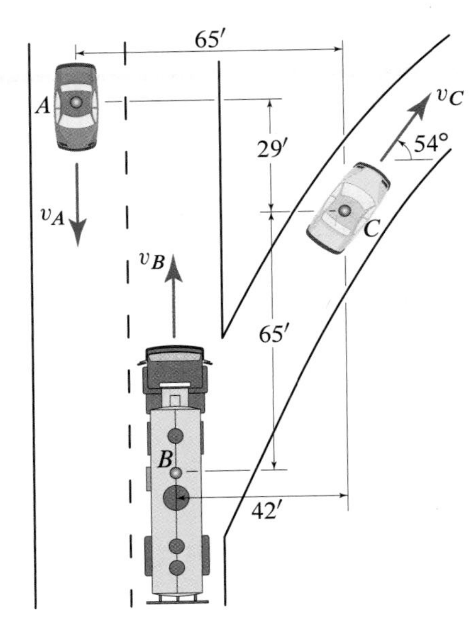

Figure P2.223 and P2.224

Problem 2.223

Three vehicles A, B, and C are in the positions shown and are moving with the indicated directions. We define the *rate of separation* (ROS) of two particles P_1 and P_2 as the component of the relative velocity of, say, P_2 with respect to P_1 in the direction of the relative position vector of P_2 with respect to P_1, which is along the line that connects the two particles. At the given instant, determine the rates of separation ROS_{AB} and ROS_{CB}, that is, the rate of separation between A and B and between C and B. Let $v_A = 60\,\mathrm{mph}$, $v_B = 55\,\mathrm{mph}$, and $v_C = 35\,\mathrm{mph}$. Furthermore, treat the vehicles as particles and use the dimensions shown in the figure.

Problem 2.224

Car A is moving at a constant speed $v_A = 75\,\mathrm{km/h}$, while car C is moving at a constant speed $v_C = 42\,\mathrm{km/h}$ on a circular exit ramp with radius $\rho = 80\,\mathrm{m}$. Determine the velocity and acceleration of C relative to A.

Problems 2.225 and 2.226

During practice, a player P punts a ball B with a speed $v_0 = 25\,\mathrm{ft/s}$, at an angle $\theta = 60°$, and at a height h from the ground. Then the player sprints along a straight line and catches the ball at the same height from the ground at which the ball was initially kicked. The length d denotes the horizontal distance between the player's position at the start of the sprint and the ball's position when the ball leaves the player's foot. Also, let Δt denote the time interval between the instant at which the ball leaves the player's foot and the instant at which the player starts sprinting.

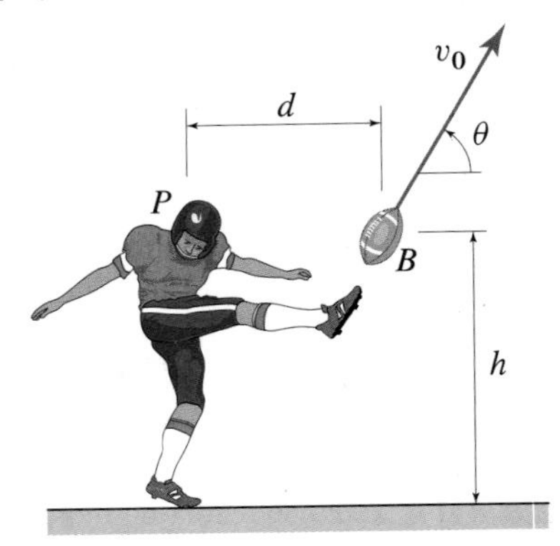

Figure P2.225 and P2.226

Problem 2.225 Assume that $d = 0$ and $\Delta t = 0$, and determine the average speed of the player so that he catches the ball.

Problem 2.226 Assume that $d = 3\,\mathrm{ft}$ and $\Delta t = 0.2\,\mathrm{s}$, and determine the average speed of the player so that he catches the ball.

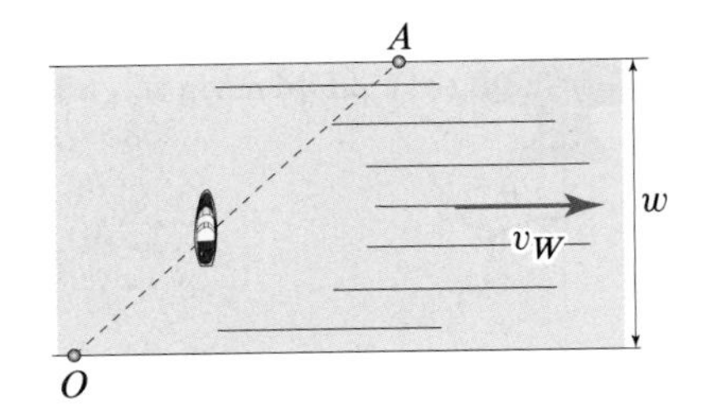

Figure P2.227

Problem 2.227

A remote controlled boat, capable of a maximum speed of $10\,\mathrm{ft/s}$ in still water, is made to cross a stream with a width $w = 35\,\mathrm{ft}$ that is flowing with a speed $v_W = 7\,\mathrm{ft/s}$. If the boat starts from point O and keeps its orientation parallel to the cross-stream direction, find the location of point A at which the boat reaches the other bank while moving at its maximum speed. Furthermore, determine how much time the crossing requires.

Problem 2.228

A remote controlled boat, capable of a maximum speed of 10 ft/s in still water, is made to cross a stream of width $w = 35$ ft that is flowing with a speed $v_W = 7$ ft/s. The boat is placed in the water at O, and it is *intended* to arrive at A by using a homing device that makes the boat always point toward A. Determine the time the boat takes to get to A and the path it follows. Also, consider a case in which the maximum speed of the boat is equal to the speed of the current. In such a case, does the boat ever make it to point A? *Hint:* To solve the problem, write $\vec{v}_{B/W} = v_{B/W}\,\hat{u}_{A/B}$, where the unit vector $\hat{u}_{A/B}$ always points from the boat to point A and is therefore, a function of time.

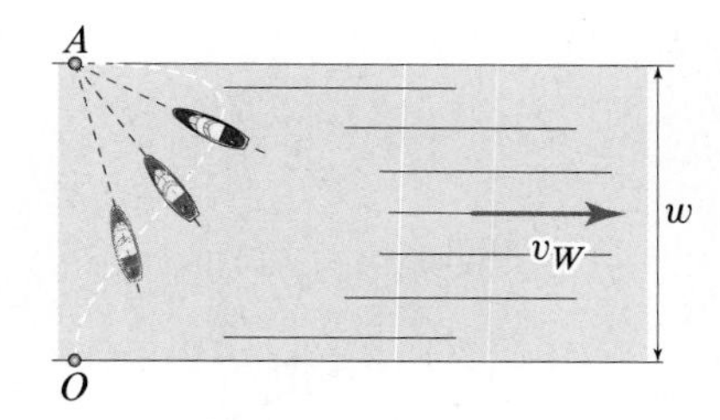

Figure P2.228

Problem 2.229

An airplane flying horizontally with a speed $v_p = 110$ km/h relative to the water drops a crate onto a carrier when vertically over the back end of the ship, which is traveling at a speed $v_s = 26$ km/h relative to the water. If the airplane drops the crate from a height $h = 20$ m, at what distance from the back of the ship will the crate first land on the deck of the ship?

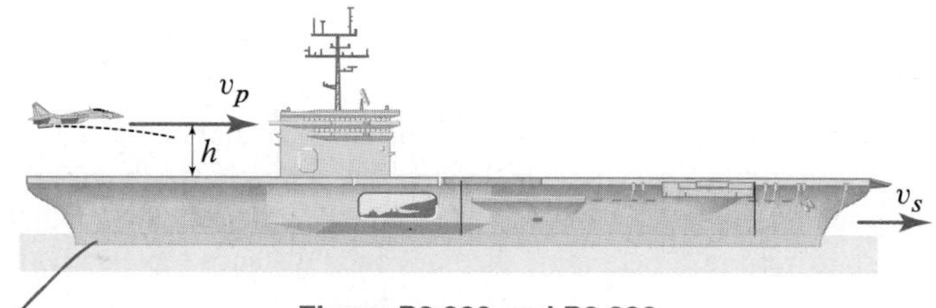

Figure P2.229 and P2.230

Problem 2.230

An airplane flying horizontally with a speed v_p relative to the water drops a crate onto a carrier when vertically over the back end of the ship, which is traveling at a speed $v_s = 32$ mph relative to the water. The length of the carrier's deck is $\ell = 1000$ ft, and the drop height is $h = 50$ ft. Determine the maximum value of v_p so that the crate will first impact within the rear half of the deck.

Problem 2.231

An airplane is initially flying north with a speed $v_0 = 430$ mph relative to the ground, while the wind has a constant speed $v_W = 12$ mph, forming an angle $\theta = 23°$ with the north-south direction. The airplane performs a course change of $\beta = 75°$ eastward while maintaining a constant reading of the airspeed indicator. Letting $\vec{v}_{P/A}$ be the velocity of the airplane relative to the air and assuming that the airspeed indicator measures the magnitude of the component of $\vec{v}_{P/A}$ in the direction of motion of the airplane, determine the speed of the airplane relative to the ground after the course correction.

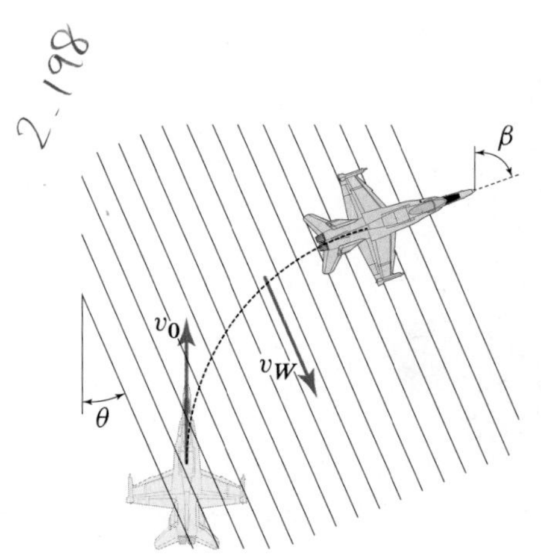

Figure P2.231

Problem 2.232

At the instant shown, block B is sliding over the ground with a velocity $\vec{v}_B$ while block A is sliding over block B and has an absolute velocity $\vec{v}_A = -(4\,\hat{\imath} + 4\,\hat{\jmath})$ ft/s. Determine $\vec{v}_B$ if $\theta = 30°$.

Problem 2.233

At the instant shown, $\vec{v}_B = 5\,\hat{\imath}$ m/s. If $\theta = 25°$, determine the speed of A relative to B in order for A to travel only in the vertical direction while sliding over B.

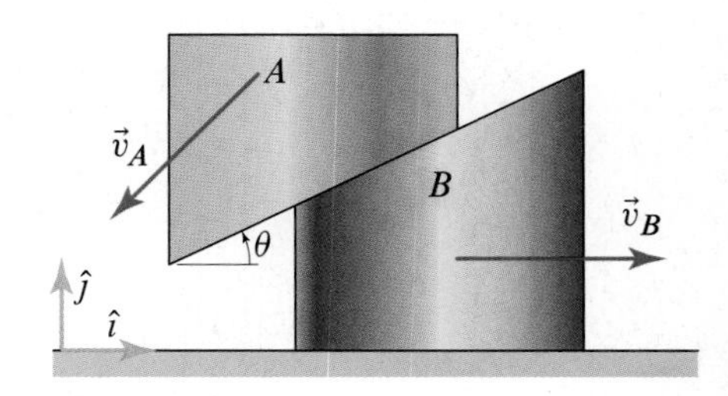

Figure P2.232 and P2.233

Problem 2.234

An interesting application of the relative motion equations is the experimental determination of the speed at which rain falls. Say you perform an experiment in your car in which you park your car in the rain and measure the angle the falling rain makes on your side window. Let this angle be $\theta_{\text{rest}} = 20°$. Next you drive forward at 25 mph and measure the new angle $\theta_{\text{motion}} = 70°$ that the rain makes with the vertical. Determine the speed of the falling rain.

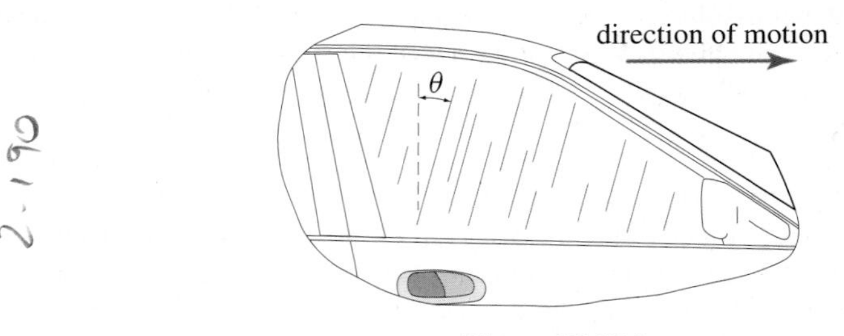

Figure P2.234

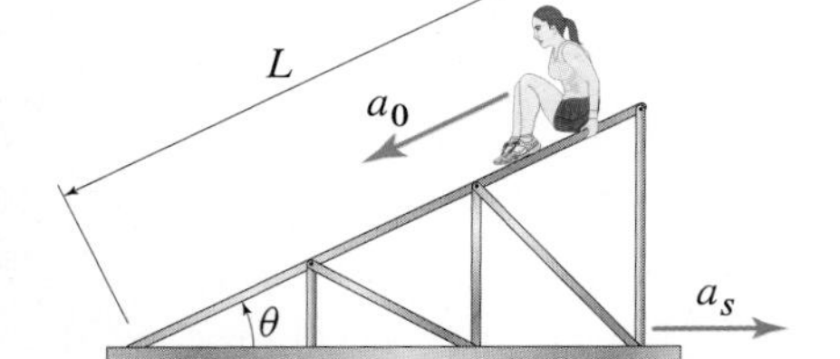

Figure P2.235

Problem 2.235

A woman is sliding down an incline with a constant acceleration of $a_0 = 2.3\,\text{m/s}^2$ relative to the incline. At the same time the incline is accelerating to the right at $1.2\,\text{m/s}^2$ relative to the ground. Letting $\theta = 34°$ and $L = 4\,\text{m}$ and assuming that both the woman and the incline start from rest, determine the horizontal distance traveled by the woman with respect to the ground when she reaches the bottom of the slide.

Problem 2.236

The pendulum bob A swings about O, which is a fixed point, while bob B swings about A. Express the components of the acceleration of B relative to the Cartesian component system shown with origin at the fixed point O in terms of L_1, L_2, θ, ϕ, and the necessary time derivatives of ϕ and θ.

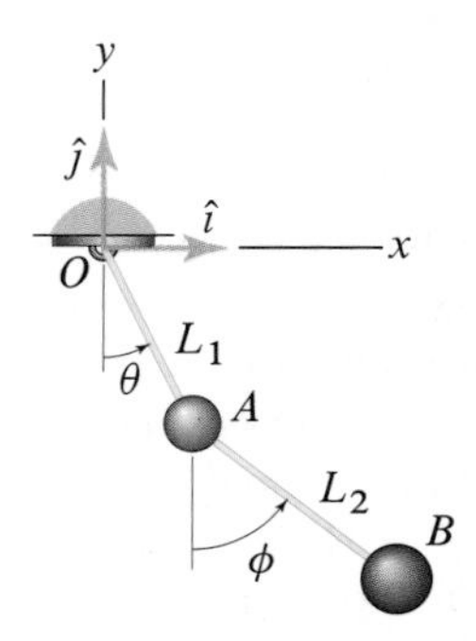

Figure P2.236

Problem 2.237

Revisit Example 2.24 in which the movie's hero is traveling on train car A with constant speed $v_A = 18\,\text{m/s}$ while the target B is moving at a constant speed $v_B = 40\,\text{m/s}$ (so that $a_B = 0$). Recall that 4 s before an otherwise inevitable collision between A and B, a projectile P traveling at a speed of 300 m/s relative to A is shot toward B. Take advantage of the solution in Example 2.24 and determine the time it takes the projectile P to reach B and the projectile's distance traveled.

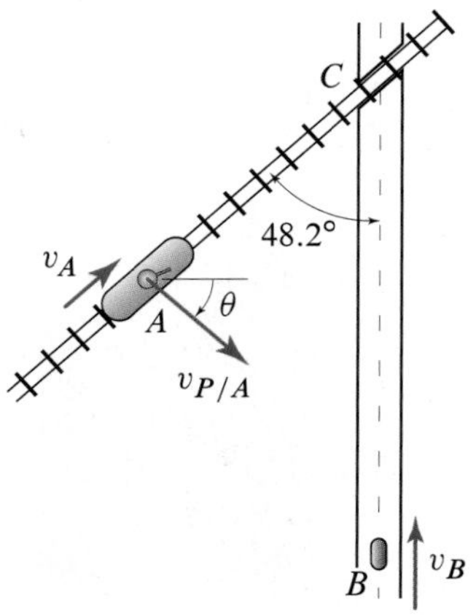

Figure P2.237

Problem 2.238

Consider the following variation of the problem in Example 2.24 in which a movie hero needs to destroy a mobile robot B, except this time they are not going to collide at C. Assume that the hero is traveling on the train car A with constant speed $v_A = 18\,\mathrm{m/s}$, while the robot B travels at a constant speed $v_B = 50\,\mathrm{m/s}$. In addition, assume that at time $t = 0\,\mathrm{s}$ the train car A and the robot B are 72 and 160 m away from C, respectively. To prevent B from reaching its intended target, at $t = 0\,\mathrm{s}$ the hero fires a projectile P at B. If P can travel at a constant speed of 300 m/s relative to the gun, determine the orientation θ that must be given to the gun to hit B. *Hint:* An equation of the type $\sin\beta \pm A\cos\beta = C$ has the solution $\beta = \mp\gamma + \sin^{-1}(C\cos\gamma)$, if $|C\cos\gamma| \le 1$, where $\gamma = \tan^{-1} A$.

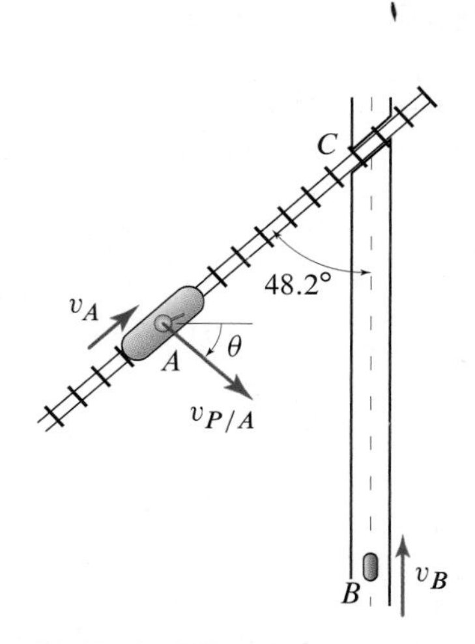

Figure P2.238 and P2.239

Problem 2.239

Consider the following variation of the problem in Example 2.24 in which a movie hero needs to destroy a mobile robot B. As was done in that problem, assume that the movie hero is traveling on the train car A with constant speed $v_A = 18\,\mathrm{m/s}$ and that, 4 s before an otherwise inevitable collision at C, the hero fires a projectile P traveling at 300 m/s relative to A. Unlike Example 2.24, assume here that the robot B travels with a constant acceleration $a_B = 10\,\mathrm{m/s^2}$ and that $v_B(0) = 20\,\mathrm{m/s}$, where $t = 0$ is the time of firing. Determine the orientation θ of the gun fired by the hero so that B can be destroyed before the collision at C.

Problem 2.240

A park ranger R is aiming a rifle armed with a tranquilizer dart at a bear (the figure is not to scale). The bear is moving in the direction shown at a constant speed $v_B = 25\,\mathrm{mph}$. The ranger fires the rifle when the bear is at C at a distance of 150 ft. Knowing that $\alpha = 10°$, $\beta = 108°$, the dart travels with a constant speed of 425 ft/s, and the dart and the bear are moving in a horizontal plane, determine the orientation θ of the rifle so that the ranger can hit the bear. *Hint:* An equation of the type $\sin\beta \pm A\cos\beta = C$ has the solution $\beta = \mp\gamma + \sin^{-1}(C\cos\gamma)$, if $|C\cos\gamma| \le 1$, where $\gamma = \tan^{-1} A$.

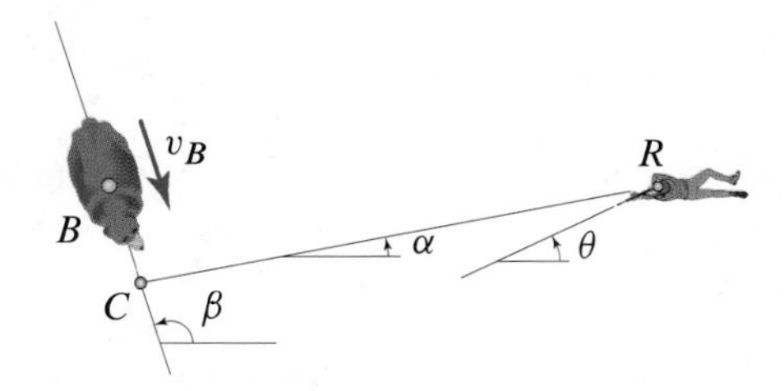

Figure P2.240

Problem 2.241

The object in the figure is called a *gun tackle*, and it used to be very common on sailboats to help in the operation of front-loaded guns. If the end at A is pulled down at a speed of 1.5 m/s, determine the velocity of B. Neglect the fact that some portions of the rope are not vertically aligned.

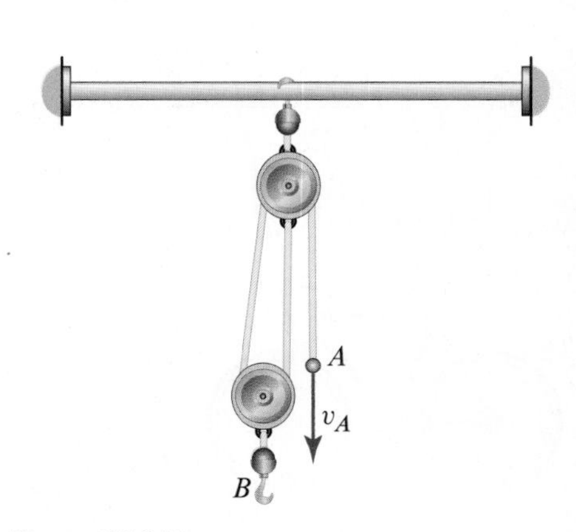

Figure P2.241

Problem 2.242

The gun tackle shown is operated with the help of a horse. If the horse moves to the right at a constant speed of 7 ft/s, determine the velocity and acceleration of B when the horizontal distance from B to A is 15 ft. Except for the part of the rope between C and A, neglect the fact that some portions are not vertically aligned. Also neglect the change in the amount of rope wrapped around pulley C as the horse moves to the right.

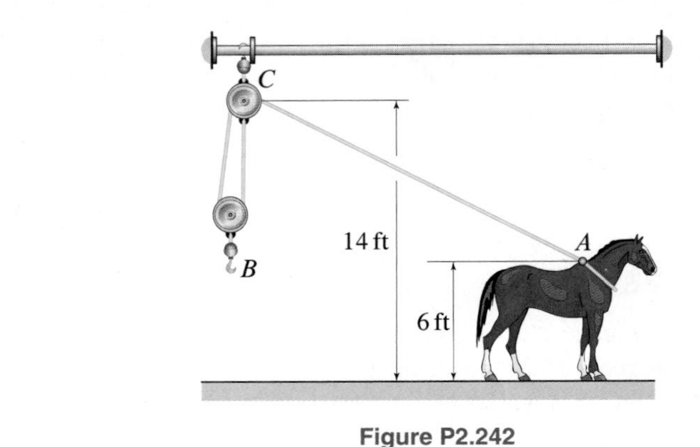

Figure P2.242

Problem 2.243

The figure shows an inverted gun tackle with snatch block, which used to be common on sailboats. If the end at A is pulled at a speed of 1.5 m/s, determine the velocity of B. Neglect the fact that some portions of the rope are not vertically aligned.

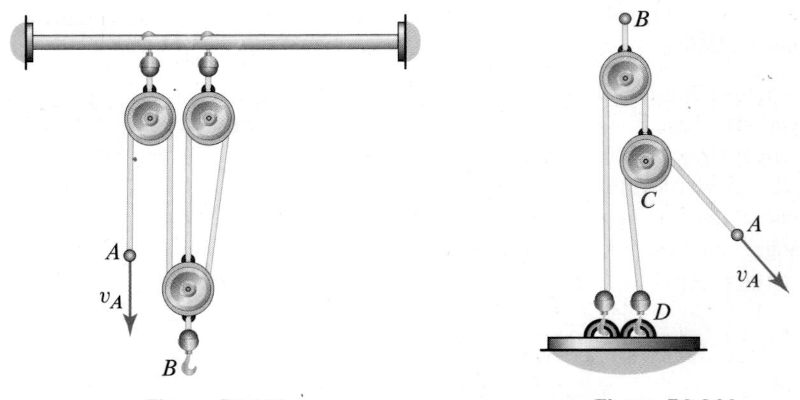

Figure P2.243 Figure P2.244

Problem 2.244

In maritime speak, the system in the figure is often called a *whip-upon-whip purchase* and is used for controlling certain types of sails on small cutters (by attaching point B to the sail to be unfurled). If the end of the rope at A is pulled with a speed of 4 m/s, determine the velocity of B. Neglect the fact that the segment of the rope between C and D is not vertically aligned, and assume that the slope of segment AC is constant.

Problem 2.245

The pulley system shown is used to store a bicycle in a garage. If the bicycle is hoisted by a winch that winds the rope at a rate $v_0 = 5$ in./s, determine the vertical speed of the bicycle.

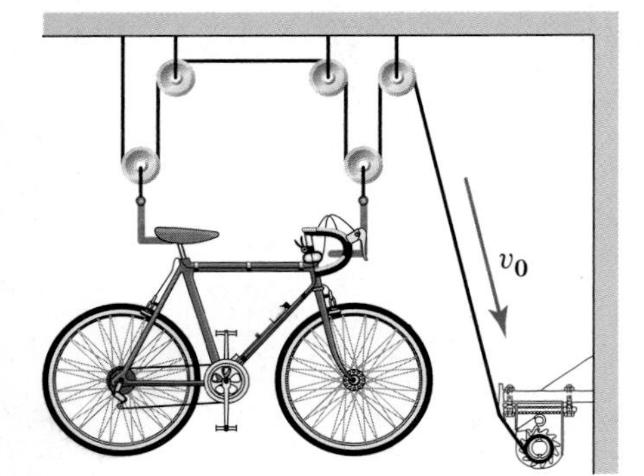

Figure P2.245

Problem 2.246

Letting $\theta = 50°$, determine the vertical component of the velocity of A if B is moving downward with a speed $v_B = 3\,\text{ft/s}$.

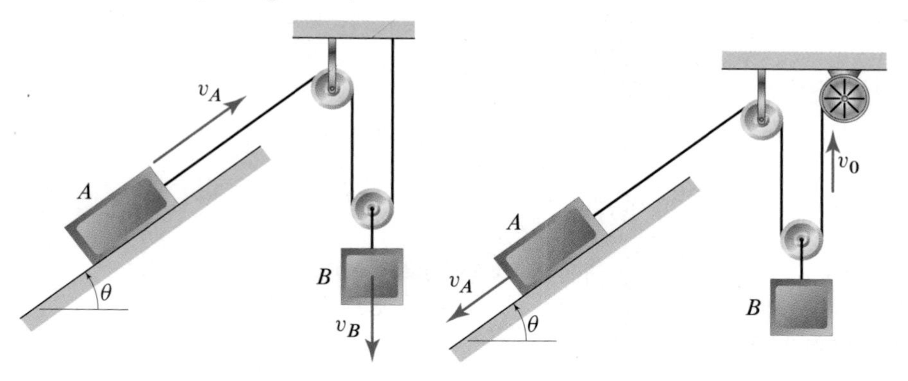

Figure P2.246 Figure P2.247

Problem 2.247

Determine the speed of block B if block A is sliding down the incline with a speed $v_A = 1.5\,\text{m/s}$ while the cord is retracted by a winch at a constant rate $v_0 = 2.5\,\text{m/s}$.

Problem 2.248

Block A is released from rest and starts sliding down the incline with an acceleration $a_0 = 3.7\,\text{m/s}^2$. Determine the acceleration of block B relative to the incline. Also, determine the time needed for B to move a distance $d = 0.2\,\text{m}$ relative to A.

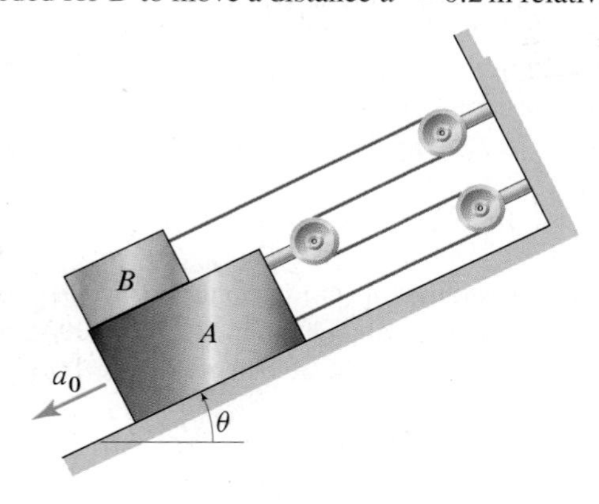

Figure P2.248

Problems 2.249 and 2.250

In the pulley system shown, the segment AD and the motion of A are not impeded by the load G. Assume all ropes are vertically aligned.

Problem 2.249 Determine the velocity and acceleration of the load G if $v_0 = 3\,\text{ft/s}$ and $a_0 = 1\,\text{ft/s}^2$.

Problem 2.250 The load G is initially at rest when the end A of the rope is pulled with the constant acceleration a_0. Determine a_0 so that G is lifted 2 ft in 4.3 s.

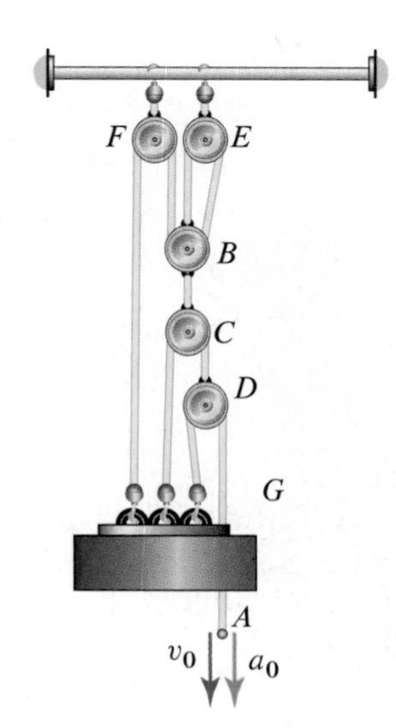

Figure P2.249 and P2.250

Problem 2.251

A crate A is being pulled up an inclined ramp by a winch retracting the cord at a constant rate $v_0 = 2\,\text{ft/s}$. Letting $h = 1.5\,\text{ft}$, determine the speed of the crate when $d = 4\,\text{ft}$.

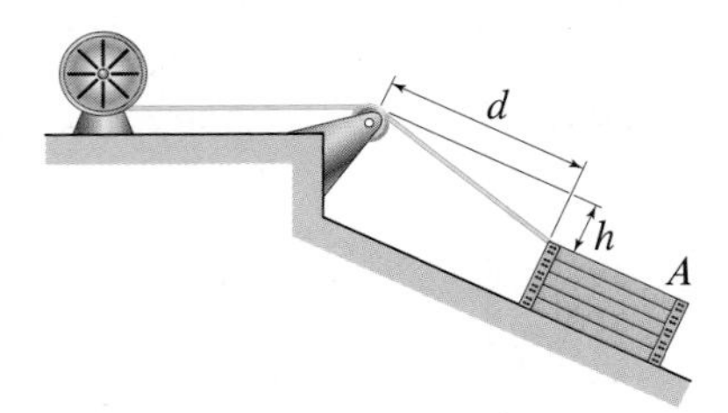

Figure P2.251 and P2.252

Problem 2.252

A crate A is being pulled up an inclined ramp by a winch. The rate of winding of the cord is controlled so as to hoist the crate up the incline with a constant speed v_0. Letting $\dot{\ell}$ denote the length of cord retracted by the winch per unit time, determine an expression for $\dot{\ell}$ in terms of v_0, h, and d.

Problem 2.253

The piston head at C is constrained to move along the y axis. Let the crank AB be rotating counterclockwise at a constant angular speed $\dot{\theta} = 2000\,\text{rpm}$, $R = 3.5\,\text{in.}$, and $L = 5.3\,\text{in.}$ Determine the velocity of C when $\theta = 35°$.

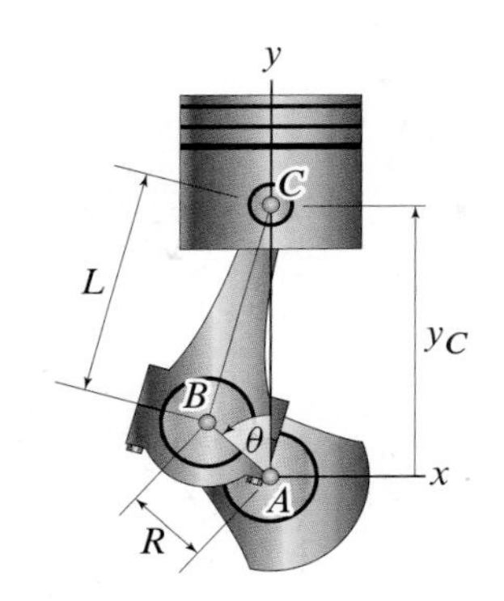

Figure P2.253–P2.255

Problem 2.254

Let $\vec{\omega}_{BC}$ denote the angular velocity of the relative position vector $\vec{r}_{C/B}$. As such, $\vec{\omega}_{BC}$ is also the angular velocity of the connecting rod BC. Using the concept of time derivative of a vector given in Section 2.4 on p. 80, determine the component of the relative velocity of C with respect to B along the direction of the connecting rod BC.

Problem 2.255

The piston head at C is constrained to move along the y axis. Let the crank AB be rotating counterclockwise at a constant angular speed $\dot{\theta} = 2000\,\text{rpm}$, $R = 3.5\,\text{in.}$, and $L = 5.3\,\text{in.}$ Determine expressions for the velocity and acceleration of C as a function of θ and the given parameters.

Problems 2.256 and 2.257

In the cutting of sheet metal, the robotic arm OA needs to move the cutting tool at C counterclockwise at a constant speed v_0 along a circular path of radius ρ. The center of the circle is located in the position shown relative to the base of the robotic arm at O.

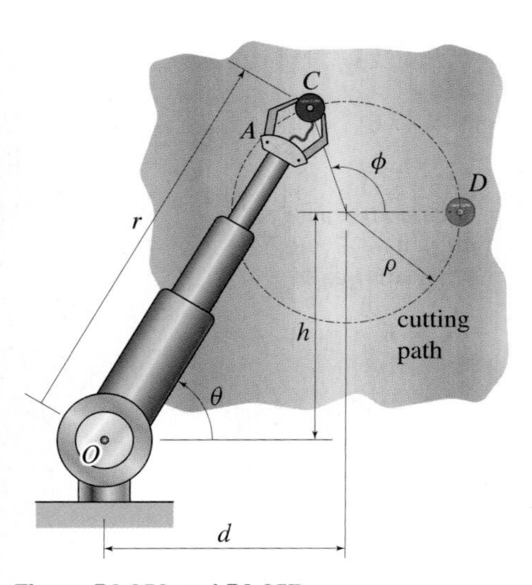

Figure P2.256 and P2.257

Problem 2.256 For all positions along the circular cut (i.e., for any value of ϕ), determine r, $\dot{r}$, and $\dot{\theta}$ as functions of the given quantities (i.e., d, h, ρ, v_0). Use one or more geometric constraints and their derivatives to do this. These quantities can be found "by hand," but it is tedious, so you might consider using symbolic algebra software, such as Mathematica or Maple.

Problem 2.257 For all positions along the circular cut (i.e., for any value of ϕ), determine $\ddot{r}$ and $\ddot{\theta}$ as functions of the given quantities (i.e., d, h, ρ, v_0). These quantities can be found by hand, but it is very tedious, so you might consider using symbolic algebra software, such as Mathematica or Maple.

Problem 2.258

At the instant shown, block A is moving at a constant speed $v_0 = 3\,\text{m/s}$ to the left and $w = 2.3\,\text{m}$. Using $h = 2.7\,\text{m}$, determine how much time is needed to lower B 0.75 m from this position.

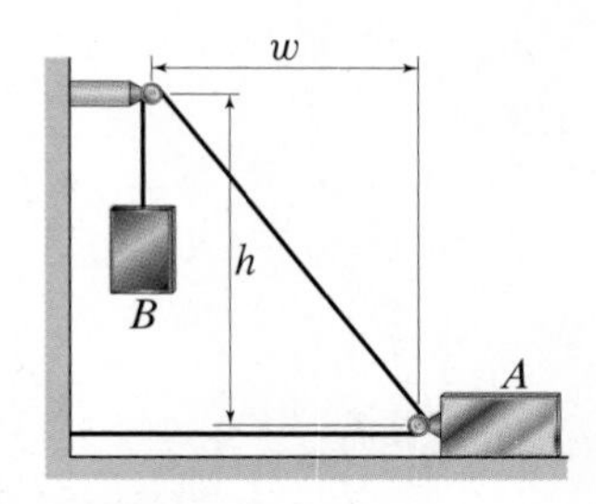

Figure P2.258 and P2.259

Problem 2.259

At the instant shown, $h = 10\,\text{ft}$, $w = 8\,\text{ft}$, and block B is moving with a speed $v_0 = 5\,\text{ft/s}$ and an acceleration $a_0 = 1\,\text{ft/s}^2$, both downward. Determine the velocity and acceleration of block A.

Problem 2.260

As a part of a robotics competition, a robotic arm with a rigid open hand at C is to be designed so that the hand catches an egg without breaking it. The egg is released from rest at $t = 0$ from point A. The arm, initially at rest in the position shown, starts moving when the egg is released. The hand must catch the egg without any impact with the egg. This can be done by specifying that the hand and the egg must be at the same position at the same time with identical velocities. A student proposes to do this using a constant value of $\ddot{\theta}$ for which (after a fair bit of work) it is found that the arm catches the egg at $t = 0.4391\,\text{s}$ for $\ddot{\theta} = -13.27\,\text{rad/s}^2$. Using these values of t and $\ddot{\theta}$, determine the acceleration of both the hand and the egg at the time of catch. Then, explain whether or not using a constant value of $\ddot{\theta}$, as has been proposed, is an acceptable strategy.

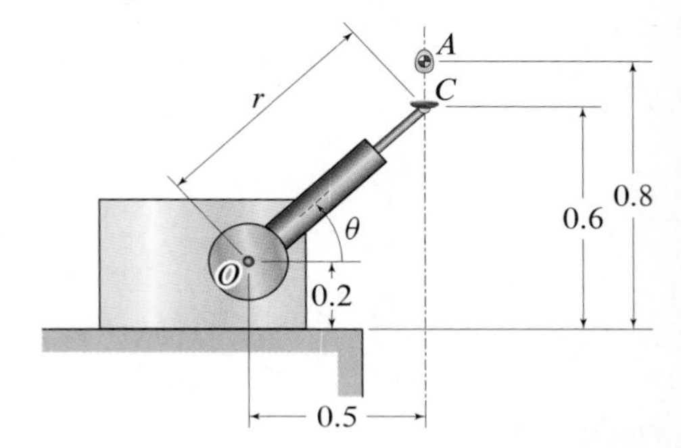

Figure P2.260 and P2.261

Problem 2.261

Referring to the problem of a robot arm catching an egg (Prob. 2.260), the strategy is that the arm and the egg must have the same velocity and the same position at the same time for the arm to gently catch the egg. In addition, what should be true about the accelerations of the arm and the egg for the catch to be successful *after* they rendezvous with the same velocity at the same position and time? Describe what happens if the accelerations of the arm and egg do not match.

Figure 2.34
A skysurfer inverted and spinning. The motion of every point on the surfer and his board is three-dimensional.

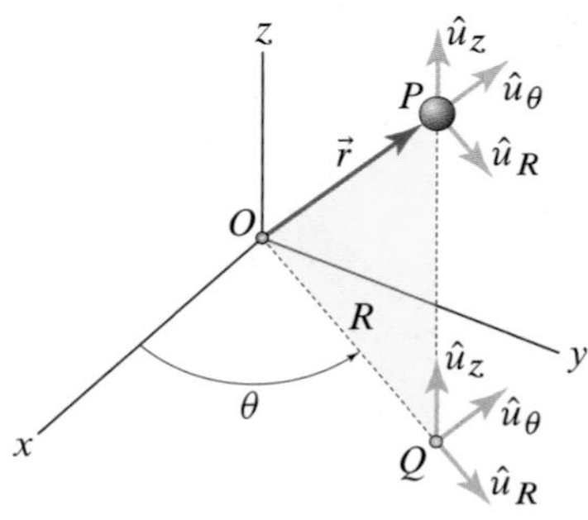

Figure 2.35
Coordinate directions in the cylindrical coordinate system. As with polar coordinates, defined in Section 2.6, the θ coordinate is implicitly contained in $\hat{u}_R$.

2.8 Motion in Three Dimensions

So far in Chapter 2, we have focused on planar motion, which, while occurring in a three-dimensional environment, is such that we can describe the position of moving points using only two coordinates. In this section, we consider the motion of points that are not constrained to move in a plane and therefore, require that we use three coordinates when describing their position.

Cylindrical coordinates

The three-dimensional extension of the polar coordinate system is called the *cylindrical coordinate system*. Figure 2.35 shows the coordinate directions for a cylindrical coordinate system with origin at O and with the coordinate θ defined to be positive counterclockwise as viewed down the positive z axis. If θ is defined to be positive in the clockwise direction, the results that follow will change. Referring to Fig. 2.35, the cylindrical coordinates of a point P are

$$R,\ \theta,\ \text{and}\ z, \tag{2.91}$$

where R and θ are the polar coordinates of the projection of point P onto the $R\theta$ plane, namely, point Q, and the coordinate z is the distance between P and Q taken positive or negative, depending on whether or not P is reached from Q by moving in the positive z direction.

To describe vectors, we introduce the following set of three mutually orthogonal base vectors:

$$\hat{u}_R,\ \hat{u}_\theta,\ \text{and}\ \hat{u}_z\ \text{with}\ \hat{u}_z = \hat{u}_R \times \hat{u}_\theta. \tag{2.92}$$

Here, $\hat{u}_R$ and $\hat{u}_\theta$ are determined as in the case of the polar coordinate system and are therefore parallel to the $R\theta$ plane, whereas $\hat{u}_z$ is a unit vector parallel to the z axis and pointing in the positive z direction. It follows that the position *vector* of P relative to the origin O is given by

$$\boxed{\vec{r} = R\,\hat{u}_R + z\,\hat{u}_z.} \tag{2.93}$$

Equation (2.93) expresses the fact that we can locate a point by tracking the point's projection on the $R\theta$ plane, given by the term $R\,\hat{u}_R$, and then adding to this the point's elevation z with respect to the $R\theta$ plane.

Differentiating Eq. (2.93) with respect to time, we obtain the following result for the velocity in cylindrical coordinates

$$\boxed{\vec{v} = \dot{R}\,\hat{u}_R + R\dot{\theta}\,\hat{u}_\theta + \dot{z}\,\hat{u}_z = v_R\,\hat{u}_R + v_\theta\,\hat{u}_\theta + v_z\,\hat{u}_z,} \tag{2.94}$$

where

$$\boxed{v_R = \dot{R}, \quad v_\theta = R\dot{\theta}, \quad \text{and} \quad v_z = \dot{z}} \tag{2.95}$$

are the cylindrical components of the velocity vector. In writing Eq. (2.94), we have used Eq. (2.66) on p. 105 and the fact that $\hat{u}_z$ is constant because the z axis is fixed. We obtain the acceleration in cylindrical coordinates by differentiating Eq. (2.94) with respect to time, i.e.,

$$\boxed{\begin{aligned}\vec{a} &= \big(\ddot{R} - R\dot{\theta}^2\big)\,\hat{u}_R + \big(R\ddot{\theta} + 2\dot{R}\dot{\theta}\big)\,\hat{u}_\theta + \ddot{z}\,\hat{u}_z \\ &= a_R\,\hat{u}_R + a_\theta\,\hat{u}_\theta + a_z\,\hat{u}_z,\end{aligned}} \tag{2.96}$$

where

$$a_R = \ddot{R} - R\dot{\theta}^2, \quad a_\theta = R\ddot{\theta} + 2\dot{R}\dot{\theta}, \quad \text{and} \quad a_z = \ddot{z} \tag{2.97}$$

are the cylindrical components of the acceleration vector. In writing Eq. (2.96), we have used Eq. (2.70) on p. 105 and the constancy of $\hat{u}_z$.

Spherical coordinates

Another coordinate system that is often used in the study of three-dimensional motion, such as in problems involving navigation and motion relative to the Earth, is the *spherical coordinate system*.

The position of a point P in a three-dimensional space can be described by r, ϕ, and θ, which are called the *spherical coordinates* of P (Fig. 2.36). The angles ϕ and θ give orientation to the line going from the origin O to P, while r is the distance between P and O. In Fig. 2.36, we have also indicated three mutually orthogonal unit vectors $\hat{u}_r$, $\hat{u}_\phi$, and $\hat{u}_\theta$. These unit vectors are oriented such that $\hat{u}_r$ points from O to P, while $\hat{u}_\phi$ and $\hat{u}_\theta$ point in the direction of increasing ϕ and θ, respectively, with $\hat{u}_\phi \times \hat{u}_\theta = \hat{u}_r$. If ϕ and θ, as well as the unit vectors $\hat{u}_r$, $\hat{u}_\phi$, and $\hat{u}_\theta$, are not defined as given here, the results that follow may not be applicable.

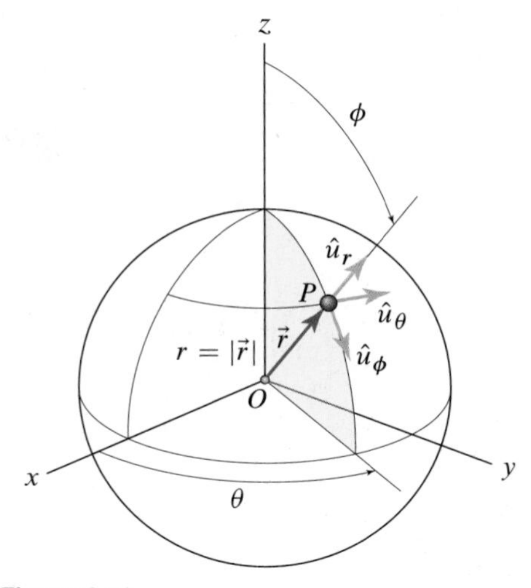

Figure 2.36
Definition of the spherical coordinates and corresponding orthogonal base vectors whose length is one, which are often called *orthonormal base vectors*.

In spherical coordinates, the position vector of P relative to the origin O is

$$\vec{r} = r\,\hat{u}_r. \tag{2.98}$$

Differentiating $\vec{r}$ with respect to time, we obtain

$$\vec{v} = \dot{\vec{r}} = \dot{r}\,\hat{u}_r + r\,\dot{\hat{u}}_r, \tag{2.99}$$

which implies that to describe the velocity vector we need to express $\dot{\hat{u}}_r$ in terms of the base vectors $\hat{u}_r$, $\hat{u}_\theta$, and $\hat{u}_\phi$. To find $\dot{\hat{u}}_r$, we need to find $\dot{\hat{u}}_r = \vec{\omega}_{\text{bv}} \times \hat{u}_r$, where $\vec{\omega}_{\text{bv}}$ is the angular velocity of the unit base vectors. To do this, we note that if the coordinates θ and ϕ undergo infinitesimal changes, the angular velocity corresponding to these infinitesimal changes is

$$\vec{\omega}_{\text{bv}} = \dot{\phi}\,\hat{u}_\theta + \dot{\theta}\,\hat{u}_z = \dot{\theta}\cos\phi\,\hat{u}_r - \dot{\theta}\sin\phi\,\hat{u}_\phi + \dot{\phi}\,\hat{u}_\theta, \tag{2.100}$$

where the last equation is obtained by observing that $\hat{u}_z = \cos\phi\,\hat{u}_r - \sin\phi\,\hat{u}_\phi$.

Using the result in Eq. (2.100) along with Eq. (2.46) on p. 81, we can find the time derivative of each of the unit vectors $\hat{u}_r$, $\hat{u}_\theta$, and $\hat{u}_\phi$ as

$$\dot{\hat{u}}_r = \vec{\omega}_{\text{bv}} \times \hat{u}_r = \dot{\phi}\,\hat{u}_\phi + \dot{\theta}\sin\phi\,\hat{u}_\theta, \tag{2.101}$$

$$\dot{\hat{u}}_\phi = \vec{\omega}_{\text{bv}} \times \hat{u}_\phi = -\dot{\phi}\,\hat{u}_r + \dot{\theta}\cos\phi\,\hat{u}_\theta, \tag{2.102}$$

$$\dot{\hat{u}}_\theta = \vec{\omega}_{\text{bv}} \times \hat{u}_\theta = -\dot{\theta}\sin\phi\,\hat{u}_r - \dot{\theta}\cos\phi\,\hat{u}_\phi. \tag{2.103}$$

Substituting Eq. (2.101) into Eq. (2.99), we obtain the velocity in spherical coordinates as

$$\vec{v} = \dot{r}\,\hat{u}_r + r\dot{\phi}\,\hat{u}_\phi + r\dot{\theta}\sin\phi\,\hat{u}_\theta = v_r\,\hat{u}_r + v_\phi\,\hat{u}_\phi + v_\theta\,\hat{u}_\theta, \tag{2.104}$$

where

$$v_r = \dot{r}, \quad v_\phi = r\dot{\phi}, \quad \text{and} \quad v_\theta = r\dot{\theta}\sin\phi \tag{2.105}$$

Helpful Information

Where are ϕ and θ in the position vector? We have seen this before, that is, the position vector *seems* not to depend on all of the "coordinates". In this case, the position vector $\vec{r} = r\,\hat{u}_r$ seems not to be a function of either θ or ϕ. However, the angles θ and ϕ are *implicitly* contained in the definition of the direction of $\hat{u}_r$.

are the spherical components of the velocity vector. Differentiating Eq. (2.104) with respect to time and using Eqs. (2.101)–(2.103) give the acceleration as

$$\begin{aligned}\vec{a} &= \left(\ddot{r} - r\dot{\phi}^2 - r\dot{\theta}^2 \sin^2\phi\right)\hat{u}_r \\ &\quad + \left(r\ddot{\phi} + 2\dot{r}\dot{\phi} - r\dot{\theta}^2 \sin\phi\cos\phi\right)\hat{u}_\phi \\ &\quad + \left(r\ddot{\theta}\sin\phi + 2\dot{r}\dot{\theta}\sin\phi + 2r\dot{\phi}\dot{\theta}\cos\phi\right)\hat{u}_\theta \\ &= a_r\,\hat{u}_r + a_\phi\,\hat{u}_\phi + a_\theta\,\hat{u}_\theta,\end{aligned} \tag{2.106}$$

where

$$\begin{aligned}a_r &= \ddot{r} - r\dot{\phi}^2 - r\dot{\theta}^2\sin^2\phi, \\ a_\phi &= r\ddot{\phi} + 2\dot{r}\dot{\phi} - r\dot{\theta}^2\sin\phi\cos\phi, \\ a_\theta &= r\ddot{\theta}\sin\phi + 2\dot{r}\dot{\theta}\sin\phi + 2r\dot{\phi}\dot{\theta}\cos\phi\end{aligned} \tag{2.107}$$

are the spherical components of the acceleration vector.

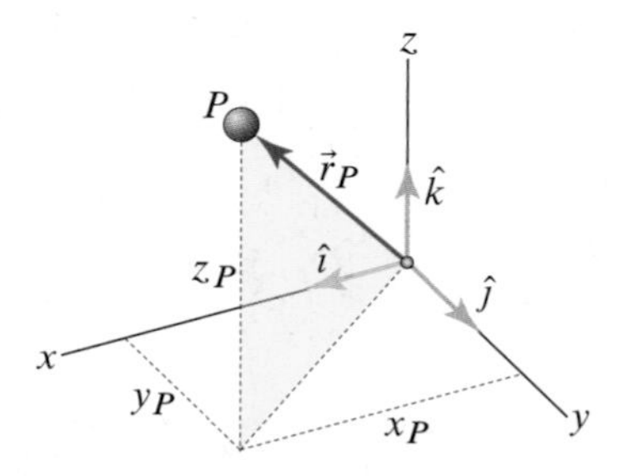

Figure 2.37
The Cartesian coordinate system and its unit vectors in three dimensions.

Cartesian coordinates

The Cartesian coordinate system and its unit vectors in three dimensions are defined as shown in Fig. 2.37. Because the directions of its base vectors do not change, we obtain the following relations for position, velocity, and acceleration, respectively, in Cartesian coordinates:

$$\vec{r}(t) = x(t)\,\hat{\imath} + y(t)\,\hat{\jmath} + z(t)\,\hat{k}, \tag{2.108}$$

$$\vec{v}(t) = \dot{x}\,\hat{\imath} + \dot{y}\,\hat{\jmath} + \dot{z}\,\hat{k} = v_x\,\hat{\imath} + v_y\,\hat{\jmath} + v_z\,\hat{k}, \tag{2.109}$$

$$\vec{a}(t) = \ddot{x}\,\hat{\imath} + \ddot{y}\,\hat{\jmath} + \ddot{z}\,\hat{k} = a_x\,\hat{\imath} + a_y\,\hat{\jmath} + a_z\,\hat{k}. \tag{2.110}$$

End of Section Summary

Cylindrical coordinates. Referring to Fig. 2.38, the position of a point P in three dimensions can be described by the three quantities R, θ, and z that are the point's cylindrical coordinates. In addition, we see that cylindrical coordinates involve the orthogonal triad of unit vectors $\hat{u}_R$, $\hat{u}_\theta$, and $\hat{u}_z$. Using this triad, the position in cylindrical coordinates is given by

Eq. (2.93), p. 142

$$\vec{r} = R\,\hat{u}_R + z\,\hat{u}_z.$$

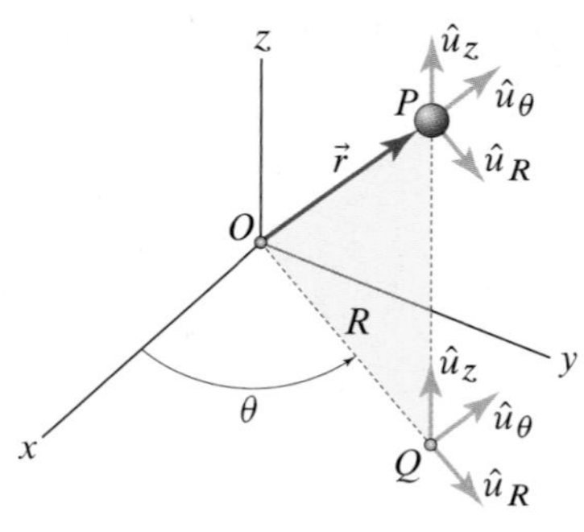

Figure 2.38
Coordinate directions in the cylindrical coordinate system. As with polar coordinates defined in Section 2.6, the θ coordinate is implicitly contained in $\hat{u}_r$.

The velocity vector in cylindrical coordinates is given by

Eq. (2.94), p. 142

$$\vec{v} = \dot{R}\,\hat{u}_R + R\dot{\theta}\,\hat{u}_\theta + \dot{z}\,\hat{u}_z = v_R\,\hat{u}_R + v_\theta\,\hat{u}_\theta + v_z\,\hat{u}_z,$$

where

Eqs. (2.95), p. 142

$$v_R = \dot{R}, \quad v_\theta = R\dot{\theta}, \quad \text{and} \quad v_z = \dot{z}$$

are the components of the velocity vector in the $\hat{u}_R$, $\hat{u}_\theta$, and $\hat{u}_z$ directions, respectively. Finally, the acceleration vector in cylindrical coordinates is given by

Eq. (2.96), p. 142

$$\vec{a} = (\ddot{R} - R\dot{\theta}^2)\,\hat{u}_R + (R\ddot{\theta} + 2\dot{R}\dot{\theta})\,\hat{u}_\theta + \ddot{z}\,\hat{u}_z = a_R\,\hat{u}_R + a_\theta\,\hat{u}_\theta + a_z\,\hat{u}_z,$$

where

Eqs. (2.97), p. 143

$$a_R = \ddot{R} - R\dot{\theta}^2, \quad a_\theta = R\ddot{\theta} + 2\dot{R}\dot{\theta}, \quad \text{and} \quad a_z = \ddot{z}$$

are the components of the acceleration vector in the $\hat{u}_R$, $\hat{u}_\theta$, and $\hat{u}_z$ directions, respectively.

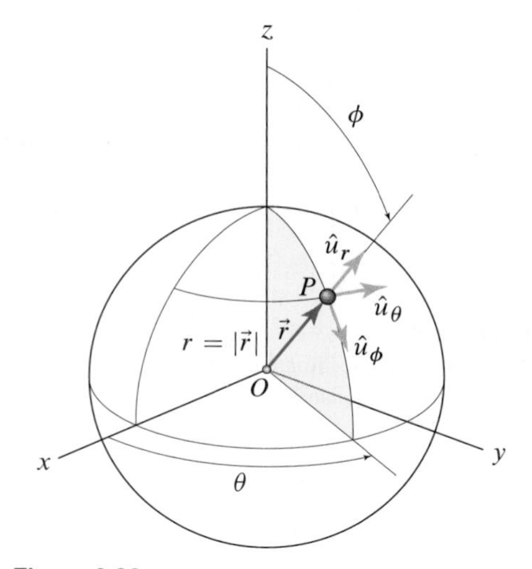

Figure 2.39
Definition of the unit vectors in spherical coordinates.

Spherical coordinates. Referring to Fig. 2.39, the position of a point P in three dimensions can be described by the three quantities r, θ, and ϕ that are the point's spherical coordinates. In addition, spherical coordinates involve the orthogonal triad of unit vectors $\hat{u}_r$, $\hat{u}_\phi$, and $\hat{u}_\theta$, with $\hat{u}_\phi \times \hat{u}_\theta = \hat{u}_r$. Using this triad, the position vector in spherical coordinates is given by

Eq. (2.98), p. 143

$$\vec{r} = r\,\hat{u}_r.$$

The velocity vector in spherical coordinates is given by

Eq. (2.104), p. 143

$$\vec{v} = \dot{r}\,\hat{u}_r + r\dot{\phi}\,\hat{u}_\phi + r\dot{\theta}\sin\phi\,\hat{u}_\theta = v_r\,\hat{u}_r + v_\phi\,\hat{u}_\phi + v_\theta\,\hat{u}_\theta,$$

where

Eqs. (2.105), p. 143

$$v_r = \dot{r}, \quad v_\phi = r\dot{\phi}, \quad \text{and} \quad v_\theta = r\dot{\theta}\sin\phi$$

are the components of the velocity vector in the $\hat{u}_r$, $\hat{u}_\phi$, and $\hat{u}_\theta$ directions, respectively. Finally, the acceleration vector in spherical coordinates is given by

Eq. (2.106), p. 144

$$\begin{aligned}\vec{a} = {} & (\ddot{r} - r\dot{\phi}^2 - r\dot{\theta}^2\sin^2\phi)\,\hat{u}_r \\ & + (r\ddot{\phi} + 2\dot{r}\dot{\phi} - r\dot{\theta}^2\sin\phi\cos\phi)\,\hat{u}_\phi \\ & + (r\ddot{\theta}\sin\phi + 2\dot{r}\dot{\theta}\sin\phi + 2r\dot{\phi}\dot{\theta}\cos\phi)\,\hat{u}_\theta \\ = {} & a_r\,\hat{u}_r + a_\phi\,\hat{u}_\phi + a_\theta\,\hat{u}_\theta,\end{aligned}$$

where

Eqs. (2.107), p. 144

$$\begin{aligned}a_r &= \ddot{r} - r\dot{\phi}^2 - r\dot{\theta}^2\sin^2\phi, \\ a_\phi &= r\ddot{\phi} + 2\dot{r}\dot{\phi} - r\dot{\theta}^2\sin\phi\cos\phi, \\ a_\theta &= r\ddot{\theta}\sin\phi + 2\dot{r}\dot{\theta}\sin\phi + 2r\dot{\phi}\dot{\theta}\cos\phi\end{aligned}$$

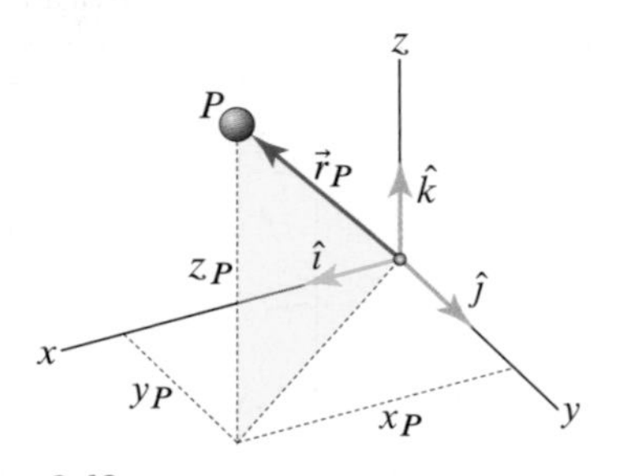

Figure 2.40
The Cartesian coordinate system and its unit vectors in three dimensions.

are the components of the acceleration vector in the $\hat{u}_r$, $\hat{u}_\phi$, and $\hat{u}_\theta$ directions, respectively.

Cartesian coordinates. Referring to Fig. 2.40, in the three-dimensional Cartesian coordinate system, the position, velocity, and acceleration, respectively, are given in component form as

Eqs. (2.108)–(2.110), p. 144

$$\vec{r}(t) = x(t)\,\hat{\imath} + y(t)\,\hat{\jmath} + z(t)\,\hat{k},$$
$$\vec{v}(t) = \dot{x}\,\hat{\imath} + \dot{y}\,\hat{\jmath} + \dot{z}\,\hat{k} = v_x\,\hat{\imath} + v_y\,\hat{\jmath} + v_z\,\hat{k},$$
$$\vec{a}(t) = \ddot{x}\,\hat{\imath} + \ddot{y}\,\hat{\jmath} + \ddot{z}\,\hat{k} = a_x\,\hat{\imath} + a_y\,\hat{\jmath} + a_z\,\hat{k}.$$

EXAMPLE 2.27 *Application of Cylindrical Coordinates*

A top-slewing crane (also called a tower crane) is lifting an object C at a constant rate of 1.5 m/s while rotating at a constant rate of 0.15 rad/s in the direction shown in Fig. 1. If the distance between the object and the axis of rotation of the crane's boom is 45 m, find the velocity and acceleration of C, assuming that the swinging motion of C can be neglected.

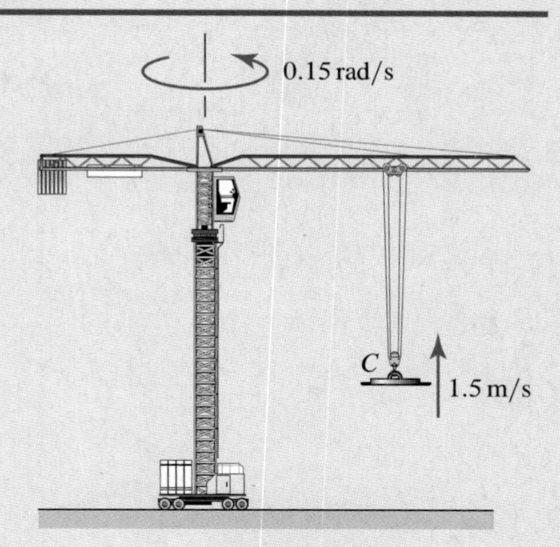

Figure 1
A top-slewing crane. This kind of crane is very common on big construction sites.

SOLUTION

Road Map This problem is readily solved using cylindrical coordinates. We are provided with quantities that are naturally defined in a cylindrical coordinate system with its origin placed at the intersection of the $\hat{u}_R$ and $\hat{u}_z$ vectors at the base of the crane (see Fig. 2).

Computation According to our choice of cylindrical coordinates, the velocity and acceleration vectors of C are given by Eqs. (2.94) and (2.96), respectively, which are

$$\vec{v}_C = \dot{R}\,\hat{u}_R + R\dot{\theta}\,\hat{u}_\theta + \dot{z}\,\hat{u}_z, \tag{1}$$

$$\vec{a}_C = \left(\ddot{R} - R\dot{\theta}^2\right)\hat{u}_R + \left(R\ddot{\theta} + 2\dot{R}\dot{\theta}\right)\hat{u}_\theta + \ddot{z}\,\hat{u}_z. \tag{2}$$

Recalling that $R = 45\,\text{m}$, $\dot{R} = 0$, $\ddot{R} = 0$, $\dot{z} = 1.5\,\text{m/s}$, $\ddot{z} = 0$, $\dot{\theta} = 0.15\,\text{rad/s}$, and $\ddot{\theta} = 0$, Eqs. (1) and (2) become

$$\vec{v}_C = (6.750\,\hat{u}_\theta + 1.500\,\hat{u}_z)\,\text{m/s}, \tag{3}$$

$$\vec{a}_C = -1.012\,\hat{u}_R\,\text{m/s}^2. \tag{4}$$

Discussion & Verification The solution seems reasonable, since the motion of C is the composition of a uniform circular motion with angular velocity equal to 0.15 rad/s and a uniform rectilinear motion in the positive z direction. We were expecting the velocity not to have a radial component, while the acceleration was expected to have only a radial component, which is what Eqs. (3) and (4) reflect.

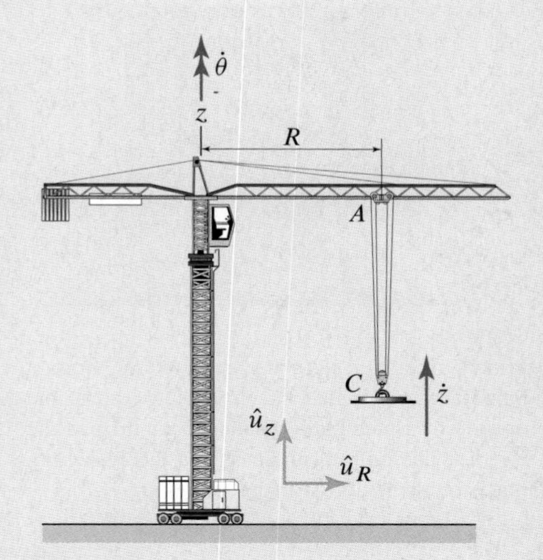

Figure 2
The top-slewing crane with a cylindrical coordinate system defined.

A Closer Look This example demonstrates how easy it can be to find velocities and accelerations when the appropriate component system is used. Even if:

- The trolley at A were moving inward or outward with *any* known $R(t)$ (so that R, $\dot{R}$, and $\ddot{R}$ were known).
- The payload C were moving up or down at *any* known $z(t)$ (so that $\dot{z}$ and $\ddot{z}$ were known).
- The tower crane were rotating with *any* known $\theta(t)$ (so that $\dot{\theta}$ and $\ddot{\theta}$ were known),

we could still easily determine the values of all the quantities in Eqs. (1) and (2) to find the velocity and acceleration of the payload at C.

EXAMPLE 2.28 *Application of Cartesian Coordinates*

Figure 1
Photos of an electrostatic precipitator from a manufacturing plant. The top photo shows the outside of a precipitator with the air inlet in the middle. The bottom photo shows the collecting plates on the inside of the precipitator.

Electrostatic precipitators are used to "scrub" or clean the emissions from coal-fired power plants (see Fig. 1). They work by sending the particulate-laden flue gas through a large structure that slows down the particles so that they can be efficiently collected. The gas enters at 50–60 ft/s and slows down to 3–6 ft/s as it expands. Inside the main structure are alternating rows of collection and discharge electrodes. Applying a high voltage (about 45,000–70,000 V for large precipitators) to the discharge electrodes located between the collection plates causes them to emit electrons into the gas, thus ionizing the immediate area. When the flue gas passes through this "corona," the particles become negatively charged and are then attracted to the positively charged collecting plates from which they can be collected and removed.

Using the geometry defined in Fig. 2, determine where the particle P will land on the collecting plate given that $\theta_1 = 30°$, $\theta_2 = 55°$, $d = 1.5$ ft, $|\vec{v}_P| = v_P = 5$ ft/s, and that gravity acts in the $-z$ direction. Assume that the collecting plate imposes a constant y component of acceleration on the particle of $a_{\text{ep}} = 250\,\text{ft/s}^2$.

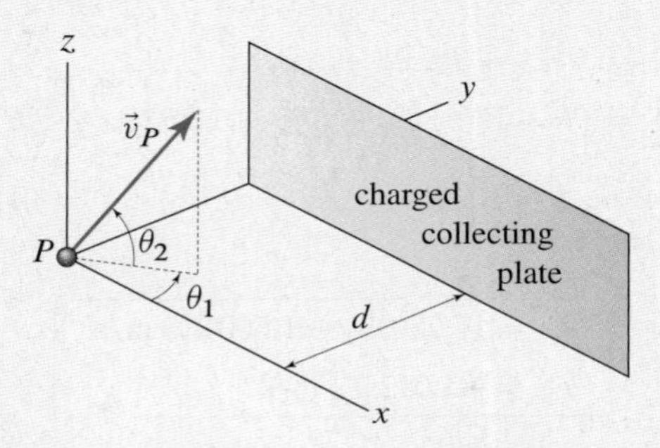

Figure 2. Schematic of a particle moving near a positively charged collecting plate in an electrostatic precipitator. Collection efficiencies of modern precipitators range from 99.5–99.9%.

SOLUTION

Road Map The key to the solution is to first determine the time it takes the particle to get to the plate. We can find that time since we know the acceleration in the y direction, and we can find the initial velocity in the y direction. Once that time is known, we can use constant acceleration kinematics to find the position at which the particle hits the plate.

Computation We begin by finding the components of the initial velocity of P in all three Cartesian coordinate directions. Referring to Fig. 3, we can see that the initial component of $\vec{v}_P$ in the xy plane is $v_P \cos\theta_2$ and that the initial z component of $\vec{v}_P$ is given by

$$v_{Pz} = v_P \sin\theta_2. \tag{1}$$

Now that we have the component of v_P in the xy plane, it is easier to see that the initial x and y components are given by

$$v_{Px} = (v_P \cos\theta_2)\cos\theta_1, \tag{2}$$

$$v_{Py} = (v_P \cos\theta_2)\sin\theta_1. \tag{3}$$

Next, we are told that the acceleration of the particle is

$$\vec{a}_P = a_{\text{ep}}\,\hat{\jmath} - g\,\hat{k} = \big(250\,\hat{\jmath} - 32.2\,\hat{k}\big)\,\text{ft/s}^2, \tag{4}$$

which is constant in all component directions.

Now that we have all accelerations and the initial velocities, we can use the y component to determine the time it takes for the particle to hit the plate. Using Eq. (2.33), we have

$$d = v_{Py} t_c + \tfrac{1}{2} a_{\text{ep}} t_c^2, \tag{5}$$

where t_c is the time for P to reach the collector plate. Equation (5) can be solved for t_c to obtain

$$t_c = -\frac{v_P}{a_{\text{ep}}} \cos\theta_2 \sin\theta_1 \pm \frac{1}{a_{\text{ep}}} \sqrt{2 d a_{\text{ep}} + v_P^2 \cos^2\theta_2 \sin^2\theta_1}, \tag{6}$$

where Eq. (3) has been used. Substituting in the given parameters, we find that t_c equals either 0.1040 s or −0.1154 s, with the physically appropriate answer being the positive one

$$t_c = 0.1040\,\text{s}. \tag{7}$$

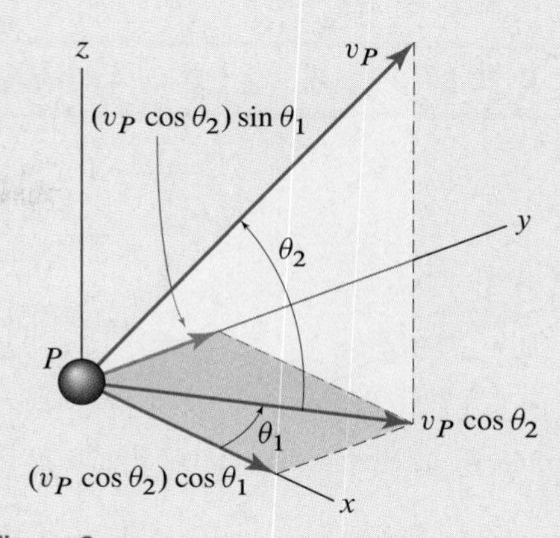

Figure 3
Graphical depiction of the initial velocity of the particle in the precipitator.

Knowing the time to reach the collector, we can now determine the distance traveled in the x and z directions. Again applying Eq. (2.33), but now in the x and z directions, we obtain

$$x_c = v_{Px} t_c = 0.2582\,\text{ft}, \tag{8}$$

$$z_c = v_{Pz} t_c - \tfrac{1}{2} g t_c^2 = 0.2518\,\text{ft}. \tag{9}$$

Figure 4 shows the trajectory of the particle that has been attracted to the collecting plate.

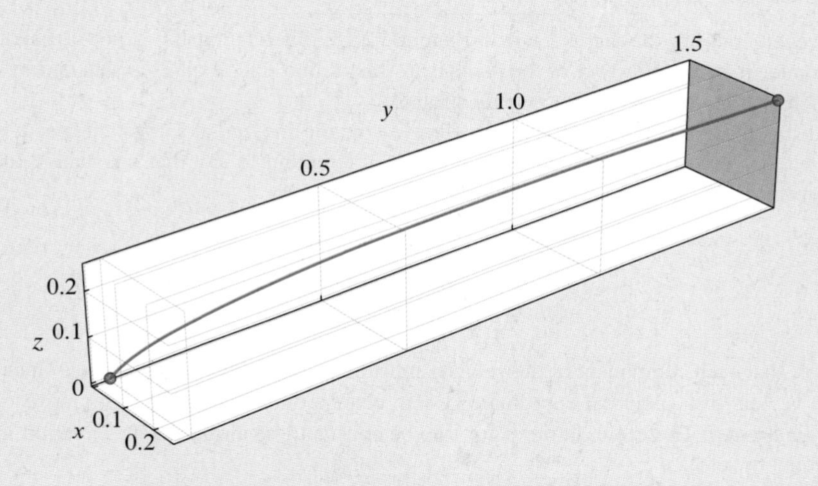

Figure 4. Trajectory of the particle P. All dimensions are in feet.

Discussion & Verification Figure 4 is very helpful in assessing whether or not our solution is reasonable. Given that the initial speed of the particle is 5 ft/s and that it is accelerating toward the collecting plate at 250 ft/s², it seems reasonable that it would not move very far down the precipitator (i.e., in the x direction) before hitting the collecting plate.

Interesting Fact

Is 250 ft/s² a reasonable acceleration? We stated that the particle is accelerating at 250 ft/s² toward the collector plate. This is reasonable when we consider that the particles collected on these plates are *very* small. For example, in some electrostatic precipitators, the particles range from smaller than 1 μm to up to 3 μm in diameter. A 1 μm particle of sodium would have a mass of about 10^{-15} kg, and so to generate an acceleration of 250 ft/s², which is equal to 76.20 m/s², requires a force of only about 8×10^{-14} N.

EXAMPLE 2.29 Application of Spherical Coordinates

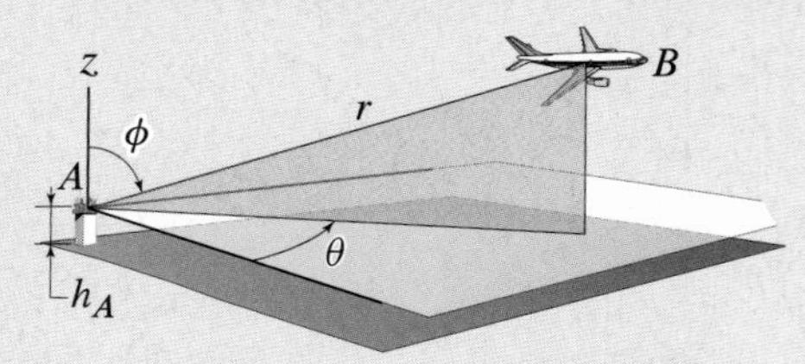

Figure 1
Airliner flying straight and level at altitude h.

Revisit Example 2.18 and determine, using spherical coordinates, what relationships the radar readings obtained by the station at A need to satisfy for you to conclude that the jet at B shown in Fig. 1 is flying

(a) At a level altitude.

(b) In a straight line at constant speed v_0.

SOLUTION

Road Map As with Example 2.18, the idea is to write the imposed constraints in terms of the chosen coordinate system. For this system, that means that the altitude is constant and that the velocity vector is constant.

Computation Given the coordinate system indicated in the figure, the plane's altitude h is given by

$$h = r\cos\phi + h_A, \tag{1}$$

where h_A is the elevation of the radar antenna above ground. Since h_A is constant, the plane can be said to be flying at a constant altitude as long as

$$\boxed{r(t)\cos\phi(t) = \text{constant.}} \tag{2}$$

However, contrary to what we saw in Example 2.18, this relationship is not sufficient to guarantee that the trajectory of the plane is a straight line since Eq. (2) is satisfied by *any* trajectory lying on a plane h above the ground.

If the plane being tracked is flying along a straight line *and* at a constant speed, then its velocity vector must be constant. While it is tempting to say that constant velocity means that

$$\dot{r} = \text{constant}, \tag{3}$$

$$r\dot{\phi} = \text{constant}, \tag{4}$$

$$r\dot{\theta}\sin\phi = \text{constant}, \tag{5}$$

that is, that each component of the velocity must be constant, this is not the case since the base vectors of a spherical coordinate system change direction as the plane moves (see marginal note). Therefore, in order for $\vec{v}$ to be constant, the airplane's acceleration must be equal to zero, i.e.,

$$\boxed{\begin{aligned} \ddot{r} - r\dot{\phi}^2 - r\dot{\theta}^2\sin^2\phi &= 0, \\ r\ddot{\phi} + 2\dot{r}\dot{\phi} - r\dot{\theta}^2\sin\phi\cos\phi &= 0, \\ r\ddot{\theta}\sin\phi + 2\dot{r}\dot{\theta}\sin\phi + 2r\dot{\phi}\dot{\theta}\cos\phi &= 0. \end{aligned}} \tag{6}$$

Finally, to measure the airplane's speed v_0, we need to compute the magnitude of $\vec{v}$, which is given by the square root of the sum of the squares of the components of the velocity vector, that is,

$$\boxed{v_0 = \sqrt{\dot{r}^2 + (r\dot{\phi})^2 + (r\dot{\theta}\sin\phi)^2}.} \tag{7}$$

Discussion & Verification Our results are correct since they were obtained in symbolic form by a direct application of Eqs. (2.107) and (2.105), respectively, i.e., without additional manipulations that could have introduced errors.

A Closer Look Verifying that the speed is constant does not allow us to say that the plane's trajectory is straight or level.

Helpful Information

What does *constant velocity* mean? We know that mathematically, constant velocity means that $\dot{\vec{v}} = \vec{0}$. We know that this means that the velocity vector doesn't change magnitude or direction. In spherical coordinates, this relationship takes the form

$$\dot{\vec{v}} = \frac{d}{dt}\Big(\dot{r}\,\hat{u}_r + r\dot{\phi}\,\hat{u}_\phi + r\dot{\theta}\sin\phi\,\hat{u}_\theta\Big).$$

When this derivative is expanded, we must differentiate the scalar coefficients of the unit vectors, *as well as the unit vectors themselves.* So, as we have already seen, constant scalar components of a vector do not imply that the vector is constant since the unit vectors can change direction. This is why Eqs. (3)–(5) are not sufficient to say that $\vec{v}$ is constant. In addition, this is why it *is* sufficient in Cartesian coordinates to say that the constancy of the scalar coefficients is sufficient to say that a vector is constant.

Problems

Problem 2.262

Although point P is moving on a sphere, its motion is being studied with the *cylindrical* coordinate system shown. Discuss in detail whether or not there are incorrect elements in the sketch of the cylindrical component system at P.

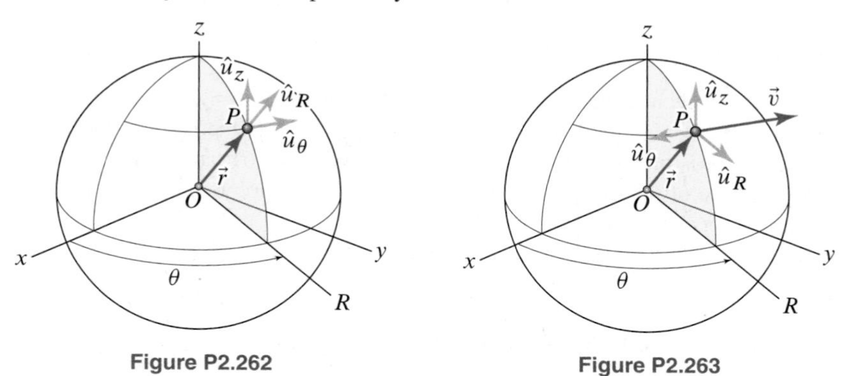

Figure P2.262 Figure P2.263

Problem 2.263

Although point P is moving on a sphere, its motion is being studied with the *cylindrical* coordinate system shown. Discuss in detail whether or not there are incorrect elements in the sketch of the cylindrical component system at P.

Problem 2.264

Discuss in detail whether or not (a) there are incorrect elements in the sketch of the spherical component system at P and (b) the formulas for the velocity and acceleration components derived in the section can be used with the coordinate system shown.

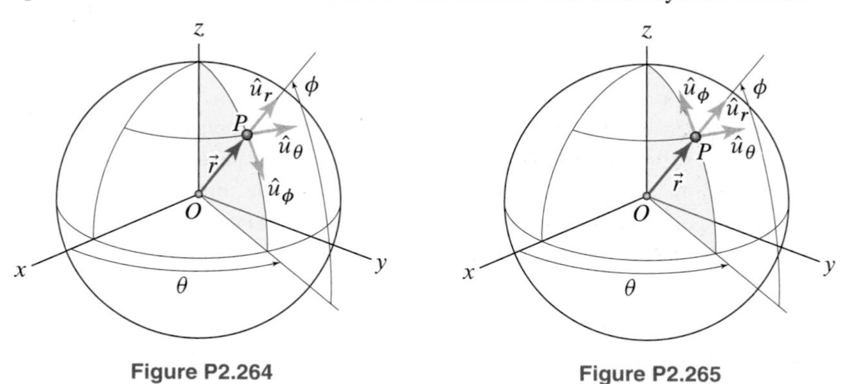

Figure P2.264 Figure P2.265

Problem 2.265

Discuss in detail whether or not (a) there are incorrect elements in the sketch of the spherical component system at P and (b) the formulas for the velocity and acceleration components derived in the section can be used with the coordinate system shown.

Problem 2.266

A glider is descending with a constant speed $v_0 = 30\,\text{m/s}$ and a constant descent rate of $1\,\text{m/s}$ along a helical path with a constant radius $R = 400\,\text{m}$. Determine the time the glider takes to complete a full 360° turn about the axis of the helix (the z axis).

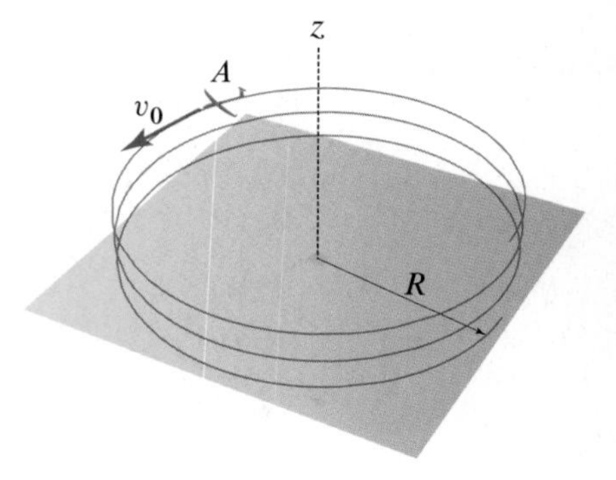

Figure P2.266

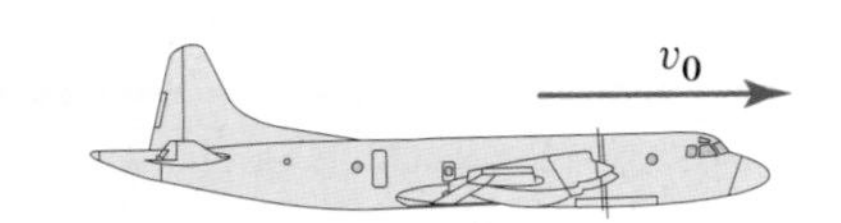

Figure P2.267

Problem 2.267

An airplane is flying horizontally at a constant speed $v_0 = 320$ mph while its propellers rotate at a constant angular speed $\omega = 1500$ rpm. If the propellers have a diameter $d = 14$ ft, determine the magnitude of the acceleration of a point on the periphery of the propeller blades.

Problem 2.268

A top-slewing crane is lifting an object C at a constant rate of $\dot{z} = 5.3$ ft/s while rotating at a constant rate $\omega = 0.12$ rad/s about the vertical axis. If the distance between the object and the axis of rotation of the crane's boom is $r = 46$ ft and it is being reduced at a constant rate of 6.5 ft/s, find the velocity and acceleration of C, assuming that the swinging motion of C can be neglected.

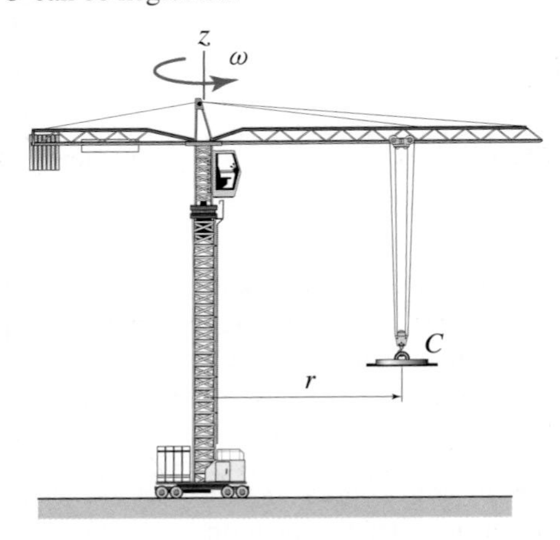

Figure P2.268

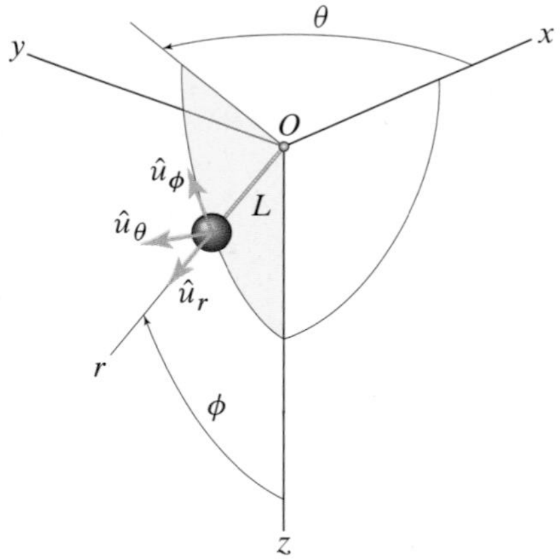

Figure P2.269

Problem 2.269

The system depicted in the figure is called a *spherical pendulum*. The fixed end of the pendulum is at O. Point O behaves as a spherical joint; i.e., the location of O is fixed while the pendulum's cord can swing in any direction in the three-dimensional space. Assume that the pendulum's cord has a constant length L, and use the coordinate system depicted in the figure to derive the expression for the acceleration of the pendulum.

Problem 2.270

Revisit Example 2.29, and assuming that the plane is accelerating, determine the relation(s) that the radar readings obtained by the station at A need to satisfy for you to conclude that the jet is flying along a straight line whether at constant altitude or not.

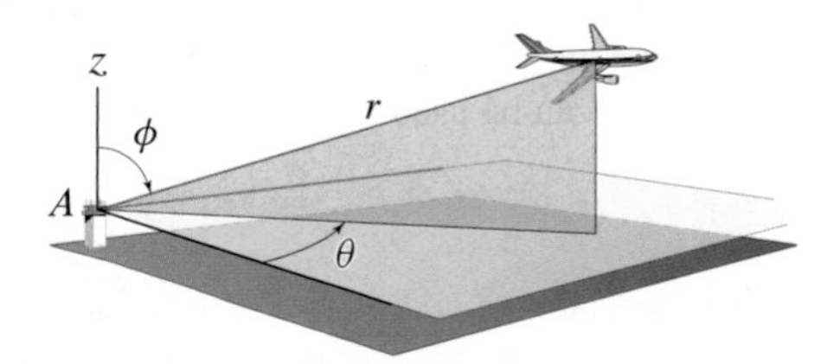

Figure P2.270

Problem 2.271

A golfer chips the ball on a flat, level part of a golf course as shown. Letting $\alpha = 23°$, $\beta = 41°$, and the initial speed be $v_0 = 6\,\mathrm{m/s}$, determine the x and y coordinates of the place where the ball will land.

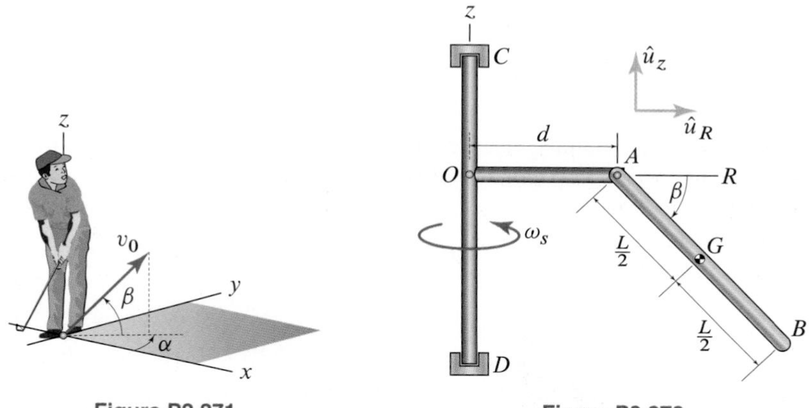

Figure P2.271

Figure P2.272

Problem 2.272

Relative to the cylindrical coordinate system shown, with origin at O, the radial and z coordinates of point G are $R = d + (L/2)\cos\beta$ and $z = -(L/2)\sin\beta$, respectively, where $d = 0.5\,\mathrm{m}$ and $L = 0.6\,\mathrm{m}$. The shaft CD rotates as shown with a constant angular velocity $\omega_s = 10\,\mathrm{rad/s}$, and the angle β varies with time as follows: $\beta = \beta_0 \sin(2\omega t)$, where $\beta_0 = 0.3\,\mathrm{rad}$, $\omega = 2\,\mathrm{rad/s}$, and t is time in seconds. Determine the velocity and the acceleration of G for $t = 3\,\mathrm{s}$ (express the result in the cylindrical component system $(\hat{u}_R, \hat{u}_\theta, \hat{u}_z)$, with $\hat{u}_\theta = \hat{u}_z \times \hat{u}_R$).

Problem 2.273

An airplane is traveling at a constant altitude of 10,000 ft, with a constant speed of 450 mph, within the plane whose equation is given by $x + y = 10\,\mathrm{mi}$ and in the direction of increasing x. Find the expressions for $\dot{r}$, $\dot{\theta}$, $\dot{\phi}$, $\ddot{r}$, $\ddot{\theta}$, and $\ddot{\phi}$ that would be measured when the airplane is closest to the radar station.

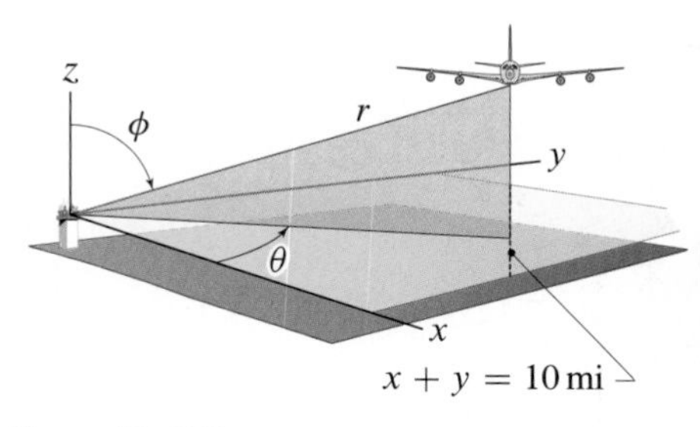

Figure P2.273

Problem 2.274

A carnival ride called the *octopus* consists of eight arms that rotate about the z axis at the constant angular velocity $\dot{\theta} = 6\,\mathrm{rpm}$. The arms have a length $L = 22\,\mathrm{ft}$ and form an angle ϕ with the z axis. Assuming that ϕ varies with time as $\phi(t) = \phi_0 + \phi_1 \sin\omega t$ with $\phi_0 = 70.5°$, $\phi_1 = 25.5°$, and $\omega = 1\,\mathrm{rad/s}$, determine the magnitude of the acceleration of the outer end of an arm when ϕ achieves its maximum value.

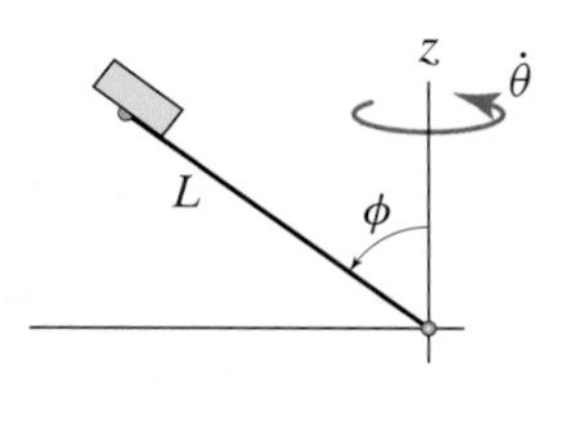

Figure P2.274

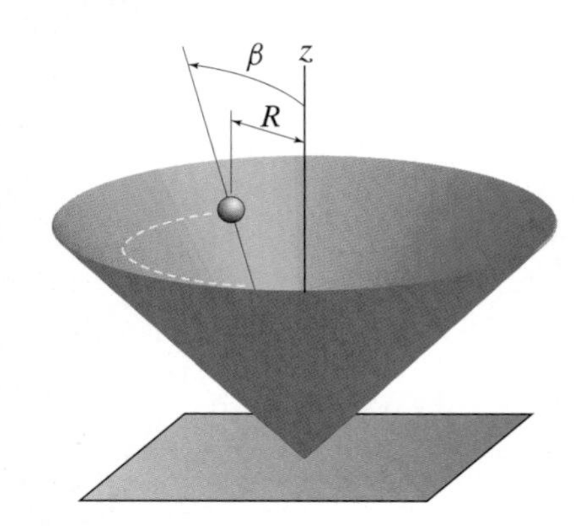

Figure P2.275

Problem 2.275

A particle is moving over the surface of a right cone with angle β and under the constraint that $R^2\dot{\theta} = K$, where K is a constant. The equation describing the cone is $R = z \tan \beta$. Determine the expressions for the velocity and the acceleration of the particle in terms of K, β, z, and the time derivatives of z.

Problem 2.276

Solve Prob. 2.275 for general surfaces of revolution; that is, R is no longer equal to $z \tan \beta$, but is now an arbitrary function of z, that is, $R = f(z)$. The expressions you need to find will contain K, $f(z)$, derivatives of $f(z)$ with respect to z, and derivatives of z with respect to time.

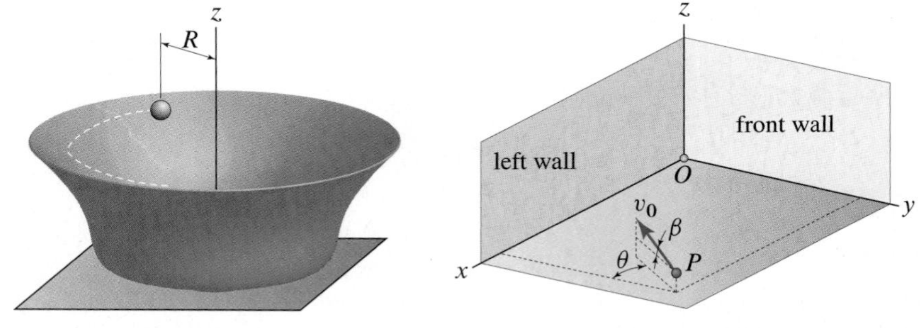

Figure P2.276 Figure P2.277

Problem 2.277

In a racquetball court, at point P with coordinates $x_P = 35\,\text{ft}$, $y_P = 16\,\text{ft}$, and $z_P = 1\,\text{ft}$, a ball is imparted a speed $v_0 = 90\,\text{mph}$ and a direction defined by the angles $\theta = 63^\circ$ and $\beta = 8^\circ$ (β is the angle formed by the initial velocity vector and the xy plane). The ball bounces off the left vertical wall to then hit the front wall of the court. Assume that the rebound off the left vertical wall occurs such that (1) the component of the ball's velocity tangent to the wall before and after rebound is the same and (2) the component of velocity normal to the wall right after impact is equal in magnitude and opposite in direction to the same component of velocity right before impact. Accounting for the effect of gravity, determine the coordinates of the point on the front wall that will be hit by the ball after rebounding off the left wall.

Problem 2.278

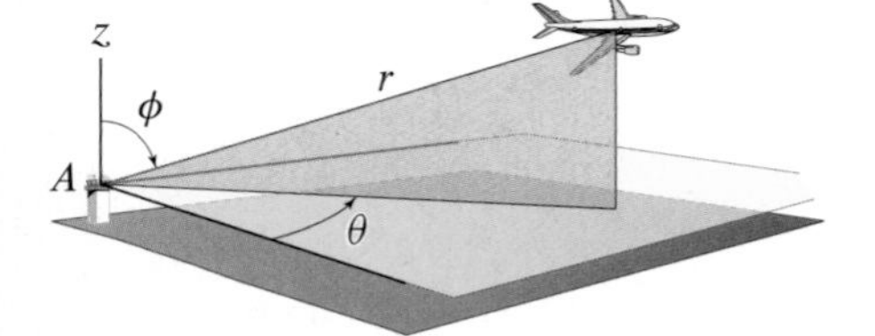

Figure P2.278 and P2.279

An airplane is being tracked by a radar station at A. At the instant $t = 0$, the following data is recorded: $r = 15\,\text{km}$, $\phi = 80^\circ$, $\theta = 15^\circ$, $\dot{r} = 350\,\text{km/h}$, $\dot{\phi} = -0.002\,\text{rad/s}$, $\dot{\theta} = 0.003\,\text{rad/s}$. If the airplane is flying to keep each of the spherical velocity components constant for a few minutes, determine the spherical components of the airplane's acceleration when $t = 30\,\text{s}$.

Problem 2.279

An airplane is being tracked by a radar station at A. At the instant $t = 0$, the following data is recorded: $r = 15\,\text{km}$, $\phi = 80^\circ$, $\theta = 15^\circ$, $\dot{r} = 350\,\text{km/h}$, $\dot{\phi} = -0.002\,\text{rad/s}$, $\dot{\theta} = 0.003\,\text{rad/s}$. If the airplane is flying to keep each of the spherical velocity components constant, plot the trajectory of the airplane for $0 < t < 150\,\text{s}$.

Chapter Review

In this chapter we presented some basic definitions needed to study the motion of objects. We also developed some basic tools for the analysis of motion in both two and three dimensions.

Position, velocity, acceleration, and Cartesian coordinates

Position. The position of a point is a *vector* going from the origin of the chosen frame of reference to the point in question.

Trajectory. The trajectory of a moving point is the line traced by the point during its motion. Another name for trajectory is *path*.

Displacement. The displacement between positions A and B is the vector going from A to B. In general, the magnitude of the displacement between two positions is not the distance traveled along the path between these positions.

Velocity. The velocity vector is the time rate of change of the position vector. The velocity is *always* tangent to the path.

Speed. The speed is the magnitude of the velocity and is a nonnegative scalar quantity. The speed measures the time rate of change of the distance traveled along the path.

Acceleration. The acceleration vector is the time rate of change of the velocity vector. Contrary to what happens for the velocity, the acceleration vector is, in general, not tangent to the trajectory.

Cartesian coordinates. The Cartesian coordinates of a particle P moving along some path are shown in Fig. 2.41. The position vector is given by

Eq. (2.14), p. 33

$$\vec{r}(t) = x(t)\,\hat{\imath} + y(t)\,\hat{\jmath}.$$

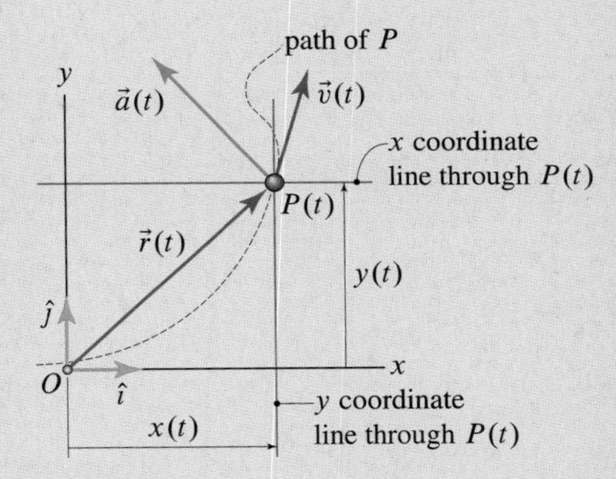

Figure 2.41
The position vector $\vec{r}(t)$ of the point P in Cartesian coordinates.

In Cartesian components, the velocity and acceleration vectors are given by

Eqs. (2.17) and (2.18), p. 33

$$\vec{v}(t) = \dot{x}(t)\,\hat{\imath} + \dot{y}(t)\,\hat{\jmath} = v_x(t)\,\hat{\imath} + v_y(t)\,\hat{\jmath},$$
$$\vec{a}(t) = \ddot{x}(t)\,\hat{\imath} + \ddot{y}(t)\,\hat{\jmath} = \dot{v}_x(t)\,\hat{\imath} + \dot{v}_y(t)\,\hat{\jmath} = a_x(t)\,\hat{\imath} + a_y(t)\,\hat{\jmath},$$

where $v_x = \dot{x}$, $v_y = \dot{y}$, $a_x = \dot{v}_x = \ddot{x}$, and $a_y = \dot{v}_y = \ddot{y}$.

Elementary motions

In applications, the acceleration of a point can be a function of time, position, or sometimes velocity. Except for the constant acceleration relations, these equations are rarely applied directly, and the acceleration is integrated as was done in their development.

1. If the acceleration is provided as a function of time, i.e., $a = a(t)$, for velocity and position, we have

Eqs. (2.20) and (2.22), p. 47

$$v(t) = v_0 + \int_{t_0}^{t} a(t)\,dt,$$

$$s(t) = s_0 + v_0(t - t_0) + \int_{t_0}^{t}\left[\int_{t_0}^{t} a(t)\,dt\right]dt.$$

2. If the acceleration is provided as a function of velocity, i.e., $a = a(v)$, for time and position, we have

Eq. (2.24), p. 47, and Eq. (2.27), p. 48

$$t(v) = t_0 + \int_{v_0}^{v} \frac{1}{a(v)}\, dv,$$
$$s(v) = s_0 + \int_{v_0}^{v} \frac{v}{a(v)}\, dv.$$

3. If the acceleration is provided as a function of position, i.e., $a = a(s)$, for velocity and time, we have

Eqs. (2.29) and (2.31), p. 48

$$v^2(s) = v_0^2 + 2\int_{s_0}^{s} a(s)\, ds,$$
$$t(s) = t_0 + \int_{s_0}^{s} \frac{ds}{v(s)}.$$

4. If the acceleration is a constant a_c, for velocity and position, we have

Eqs. (2.32)–(2.34), p. 49

$$v = v_0 + a_c(t - t_0),$$
$$s = s_0 + v_0(t - t_0) + \tfrac{1}{2}a_c(t - t_0)^2,$$
$$v^2 = v_0^2 + 2a_c(s - s_0).$$

Circular motion. For circular motion, the equations summarized in items (1)–(4) above hold as long as we use the replacement rules

$$s \to \theta, \quad v \to \omega, \quad a \to \alpha,$$

where $\omega = \dot{\theta}$ and $\alpha = \ddot{\theta}$ are the *angular velocity* and *angular acceleration*, respectively.

Projectile motion

A common application of the constant acceleration equations summarized above is the study of projectile motion. We defined *projectile motion* as the motion of a particle in free flight, neglecting the forces due to air drag and neglecting changes in gravitational attraction with changes in height. In this case, referring to Fig. 2.42, the only force on the particle is the *constant* gravitational force, and the equations describing the motion are

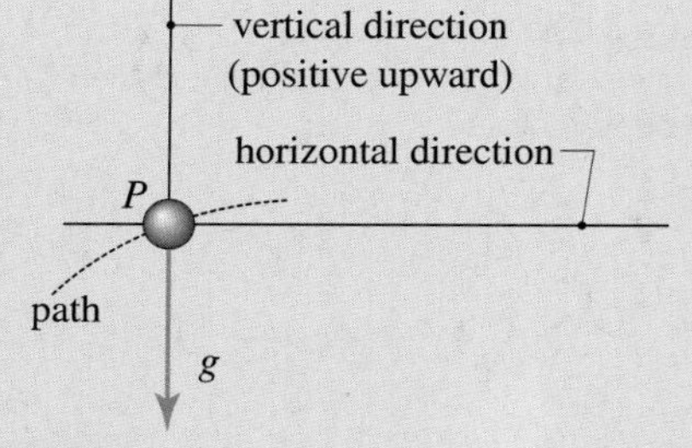

Figure 2.42
Acceleration of a point P in projectile motion.

Eqs. (2.39), p. 68

$$a_{\text{horiz}} = 0 \quad \text{and} \quad a_{\text{vert}} = -g.$$

Time derivative of a vector

A very important skill in the study of motion is the ability to compute the time derivative of vector quantities. Vectors are characterized by their magnitude and direction, and therefore the time derivative of a vector consists of a change in length and a change in direction. In the case of a unit vector, its time derivative consists of only a change in direction:

Eq. (2.46), p. 81

$$\underbrace{\dot{\hat{u}}(t)}_{\substack{\text{change} \\ \text{in } \vec{u}}} = \underbrace{\vec{\omega}_u \times \hat{u}}_{\substack{\text{change in} \\ \text{direction}}}.$$

In the case of an arbitrary vector $\vec{A}$ we have

Eq. (2.48), p. 81

$$\underbrace{\dot{\vec{A}}(t)}_{\text{change in } \vec{A}} = \underbrace{\dot{A}\,\hat{u}_A}_{\text{change in magnitude}} + \underbrace{\vec{\omega}_A \times \vec{A}}_{\text{change in direction}},$$

where, referring to Fig. 2.43, $\hat{u}_A$ is a unit vector in the direction of $\vec{A}$, $\dot{A} = d|\vec{A}|/dt$, and $\vec{\omega}_A$ is the angular velocity of the vector $\vec{A}$.

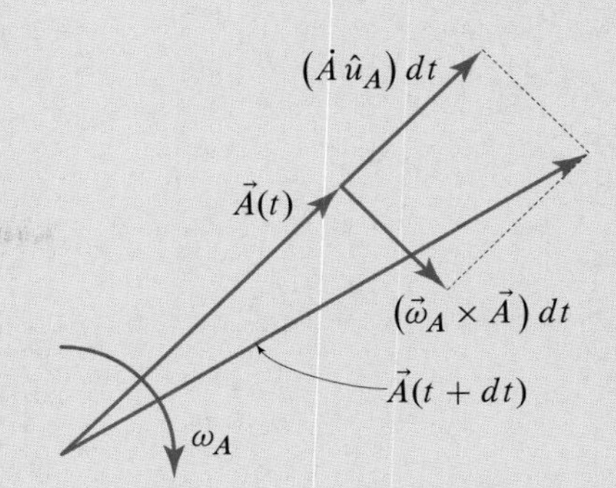

Figure 2.43
Depiction of the vector $\vec{A}$ at time t and at time $t + dt$ showing its change in magnitude and change in direction.

Normal-tangential components and polar coordinates

While we can always represent vectors using Cartesian components, it is sometimes convenient to represent vector quantities in other component systems.

Normal-tangential component system. Referring to Fig. 2.44, the velocity vector has the form

Eq. (2.50), p. 92

$$\vec{v} = v\,\hat{u}_t,$$

where v is the speed and $\hat{u}_t$ is the tangent unit vector at the point P. Referring to

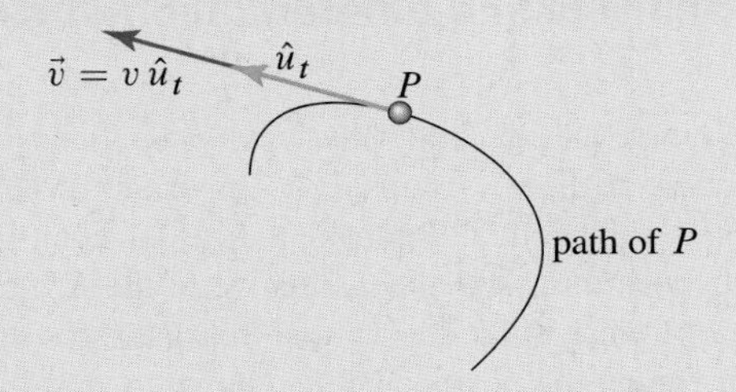

Figure 2.44. Representation of the velocity in normal-tangential components.

Fig. 2.45, the acceleration vector in normal-tangential components has the form

Eq. (2.57), p. 93

$$\vec{a} = \dot{v}\,\hat{u}_t + \frac{v^2}{\rho}\,\hat{u}_n = a_t\,\hat{u}_t + a_n\,\hat{u}_n,$$

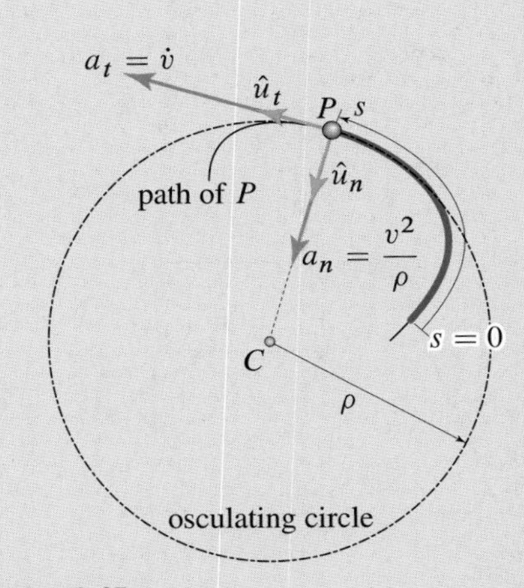

Figure 2.45
Acceleration in normal-tangential components.

where $\dot{v} = a_t$ is the tangential component of acceleration and $v^2/\rho = a_n$ is the normal component of acceleration. When we are using a Cartesian coordinate system, for a path expressed by a relation, such as $y = y(x)$, the path's radius of curvature is given by ρ:

Eq. (2.59), p. 93

$$\rho(x) = \frac{\left[1 + (dy/dx)^2\right]^{3/2}}{\left|d^2y/dx^2\right|}.$$

Polar coordinates. Referring to Fig. 2.46, r and θ are the polar coordinates of point P. The coordinate θ is chosen positive in the counterclockwise direction as viewed down the positive z axis. Using polar coordinates, the position vector of P is

Eq. (2.63), p. 105

$$\vec{r} = r\,\hat{u}_r.$$

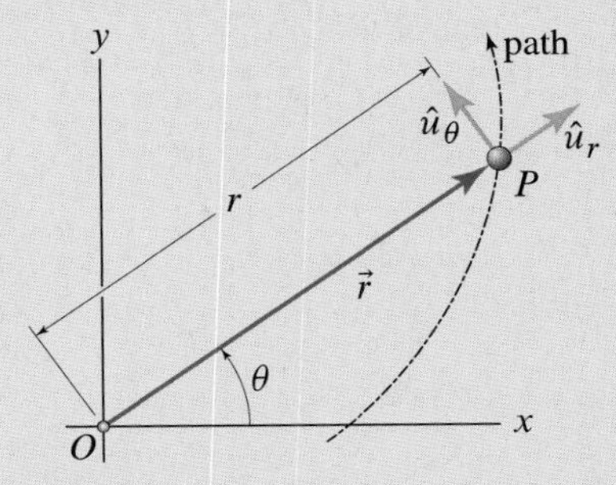

Figure 2.46
The position $\vec{r}$ of a particle defined using the polar coordinates r and θ.

Differentiating the position with respect to time, we obtain the velocity vector in polar coordinates as

Eq. (2.66), p. 105

$$\vec{v} = \dot{r}\,\hat{u}_r + r\dot{\theta}\,\hat{u}_\theta = v_r\,\hat{u}_r + v_\theta\,\hat{u}_\theta,$$

where

Eqs. (2.67), p. 105

$$v_r = \dot{r} \quad \text{and} \quad v_\theta = r\dot{\theta}$$

are the radial and transverse components of the velocity, respectively. Differentiating the velocity with respect to time, the acceleration vector in polar coordinates is

Eq. (2.70), p. 105

$$\vec{a} = \left(\ddot{r} - r\dot{\theta}^2\right)\hat{u}_r + \left(r\ddot{\theta} + 2\dot{r}\dot{\theta}\right)\hat{u}_\theta = a_r\,\hat{u}_r + a_\theta\,\hat{u}_\theta,$$

where

Eqs. (2.71), p. 105

$$a_r = \ddot{r} - r\dot{\theta}^2 \quad \text{and} \quad a_\theta = r\ddot{\theta} + 2\dot{r}\dot{\theta}$$

are the radial and transverse components of the acceleration, respectively.

Relative motion analysis and differentiation of geometrical constraints

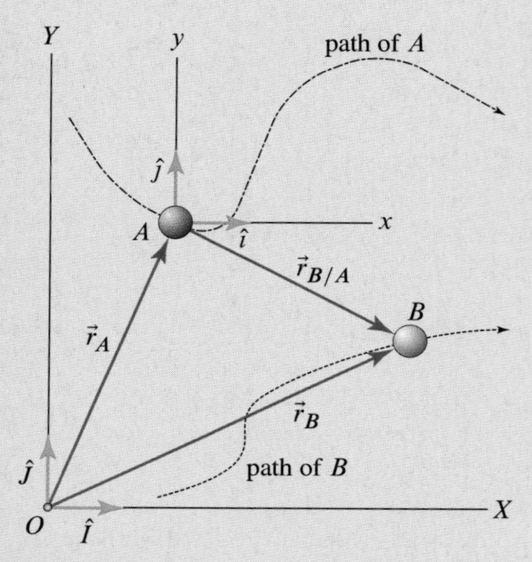

Figure 2.47
Two particles A and B and the definition of their relative position vector $\vec{r}_{B/A}$.

Relative motion analysis. In general, physical systems consist of several moving parts. To study the motion of these systems, it is important to be able to describe the motion of one object relative to another. Referring to Fig. 2.47, consider the planar motion of points A and B. The positions of A and B relative to the XY frame are $\vec{r}_A$ and $\vec{r}_B$, respectively. Attached to A there is a frame xy that translates, but does not rotate relative to frame XY. In either frame, the position of B relative to A is given by $\vec{r}_{B/A}$. Using vector addition, the vectors $\vec{r}_A$, $\vec{r}_B$, and $\vec{r}_{B/A}$ are related as follows:

Eq. (2.77), p. 121

$$\vec{r}_B = \vec{r}_A + \vec{r}_{B/A}.$$

The first and second time derivatives of the above equation are, respectively,

Eqs. (2.78) and (2.79), p. 122

$$\vec{v}_B = \vec{v}_A + \vec{v}_{B/A},$$
$$\vec{a}_B = \vec{a}_A + \vec{a}_{B/A},$$

where $\vec{v}_{B/A} = \dot{\vec{r}}_{B/A}$ and $\vec{a}_{B/A} = \ddot{\vec{r}}_{B/A}$ are the relative velocity and acceleration of B with respect to A, respectively. In general, the vectors $\vec{v}_{B/A}$ and $\vec{a}_{B/A}$ computed by the xy observer are different from the vectors $\vec{v}_{B/A}$ and $\vec{a}_{B/A}$ computed by the XY observer. However, the xy and XY observers compute the same $\vec{v}_{B/A}$ and the same $\vec{a}_{B/A}$ if these observers do not rotate relative to one another.

Differentiation of geometrical constraints. There are very few dynamics problems in which the motion is *not* constrained in some way. In certain classes of systems, constraints are described by geometrical relations between points in the system. We analyzed a pulley system whose motion was constrained by the inextensibility of the cords in the system. The key to constrained motion analysis is knowing that we can differentiate the equations describing the geometrical constraints to obtain velocities and accelerations of points of interest.

Motion in three dimensions

When the motion of points in a system is not constrained to be planar, we need to use component systems that are three-dimensional. We studied three such component systems.

Cylindrical coordinates. The position of a point P in three dimensions can be described by the three quantities R, θ, and z that are the point's cylindrical coordinates (Fig 2.48). In addition, we see that cylindrical coordinates involve the orthogonal triad of unit vectors $\hat{u}_R$, $\hat{u}_\theta$, and $\hat{u}_z$. Using this triad, the position in cylindrical coordinates is given by

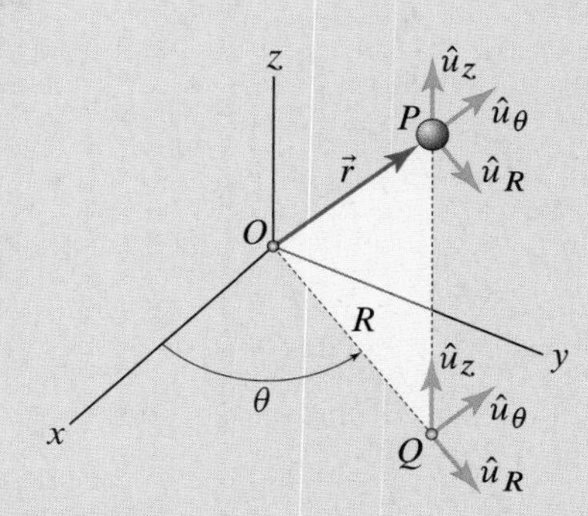

Figure 2.48
Coordinate directions in the cylindrical coordinate system. As with polar coordinates defined in Section 2.6, the θ coordinate is implicitly contained in $\hat{u}_r$.

Eq. (2.93), p. 142

$$\vec{r} = R\,\hat{u}_R + z\,\hat{u}_z.$$

The velocity vector in cylindrical coordinates is given by

Eq. (2.94), p. 142

$$\vec{v} = \dot{R}\,\hat{u}_R + R\dot{\theta}\,\hat{u}_\theta + \dot{z}\,\hat{u}_z = v_R\,\hat{u}_R + v_\theta\,\hat{u}_\theta + v_z\,\hat{u}_z,$$

where

Eqs. (2.95), p. 142

$$v_R = \dot{R}, \quad v_\theta = R\dot{\theta}, \quad \text{and} \quad v_z = \dot{z}$$

are the components of the velocity vector in the $\hat{u}_R$, $\hat{u}_\theta$, and $\hat{u}_z$ directions, respectively. Finally, the acceleration vector in cylindrical coordinates is given by

Eq. (2.96), p. 142

$$\begin{aligned}\vec{a} &= \left(\ddot{R} - R\dot{\theta}^2\right)\hat{u}_R + \left(R\ddot{\theta} + 2\dot{R}\dot{\theta}\right)\hat{u}_\theta + \ddot{z}\,\hat{u}_z \\ &= a_R\,\hat{u}_R + a_\theta\,\hat{u}_\theta + a_z\,\hat{u}_z,\end{aligned}$$

where

Eqs. (2.97), p. 143

$$a_R = \ddot{R} - R\dot{\theta}^2, \quad a_\theta = R\ddot{\theta} + 2\dot{R}\dot{\theta}, \quad \text{and} \quad a_z = \ddot{z}$$

are the components of the acceleration vector in the $\hat{u}_R$, $\hat{u}_\theta$, and $\hat{u}_z$ directions, respectively.

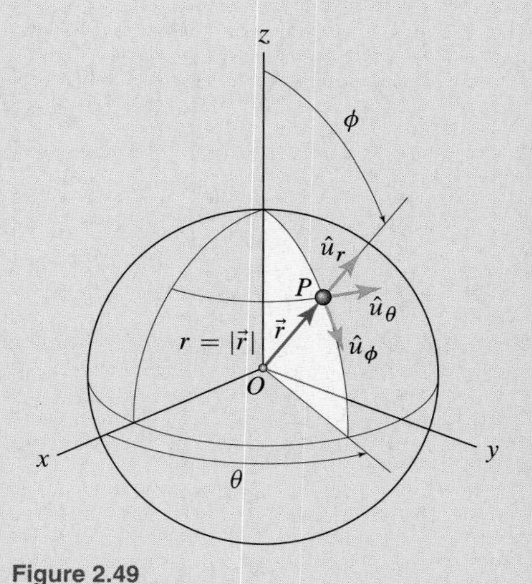

Figure 2.49
Definition of the unit vectors in spherical coordinates.

Spherical coordinates. The position of a point P in three dimensions can be described by the three quantities r, θ, and ϕ that are the point's spherical coordinates (Fig. 2.49). In addition, spherical coordinates involve the orthogonal triad of unit vectors $\hat{u}_r$, $\hat{u}_\phi$, and $\hat{u}_\theta$, with $\hat{u}_\phi \times \hat{u}_\theta = \hat{u}_r$. Using this triad, the position vector in spherical coordinates is given by

Eq. (2.98), p. 143

$$\vec{r} = r\,\hat{u}_r.$$

The velocity vector in spherical coordinates is given by

Eq. (2.104), p. 143

$$\vec{v} = \dot{r}\,\hat{u}_r + r\dot{\phi}\,\hat{u}_\phi + r\dot{\theta}\sin\phi\,\hat{u}_\theta = v_r\,\hat{u}_r + v_\phi\,\hat{u}_\phi + v_\theta\,\hat{u}_\theta,$$

where

Eqs. (2.105), p. 143

$$v_r = \dot{r}, \quad v_\phi = r\dot{\phi}, \quad \text{and} \quad v_\theta = r\dot{\theta}\sin\phi$$

are the components of the velocity vector in the $\hat{u}_r$, $\hat{u}_\phi$, and $\hat{u}_\theta$ directions, respectively. Finally, the acceleration vector in spherical coordinates is given by

Eq. (2.106), p. 144

$$\begin{aligned}\vec{a} &= \left(\ddot{r} - r\dot{\phi}^2 - r\dot{\theta}^2\sin^2\phi\right)\hat{u}_r \\ &\quad + \left(r\ddot{\phi} + 2\dot{r}\dot{\phi} - r\dot{\theta}^2\sin\phi\cos\phi\right)\hat{u}_\phi \\ &\quad + \left(r\ddot{\theta}\sin\phi + 2\dot{r}\dot{\theta}\sin\phi + 2r\dot{\phi}\dot{\theta}\cos\phi\right)\hat{u}_\theta \\ &= a_r\,\hat{u}_r + a_\phi\,\hat{u}_\phi + a_\theta\,\hat{u}_\theta,\end{aligned}$$

where

Eqs. (2.107), p. 144

$$\begin{aligned}a_r &= \ddot{r} - r\dot{\phi}^2 - r\dot{\theta}^2\sin^2\phi, \\ a_\phi &= r\ddot{\phi} + 2\dot{r}\dot{\phi} - r\dot{\theta}^2\sin\phi\cos\phi, \\ a_\theta &= r\ddot{\theta}\sin\phi + 2\dot{r}\dot{\theta}\sin\phi + 2r\dot{\phi}\dot{\theta}\cos\phi\end{aligned}$$

are the components of the acceleration vector in the $\hat{u}_r$, $\hat{u}_\phi$, and $\hat{u}_\theta$ directions, respectively.

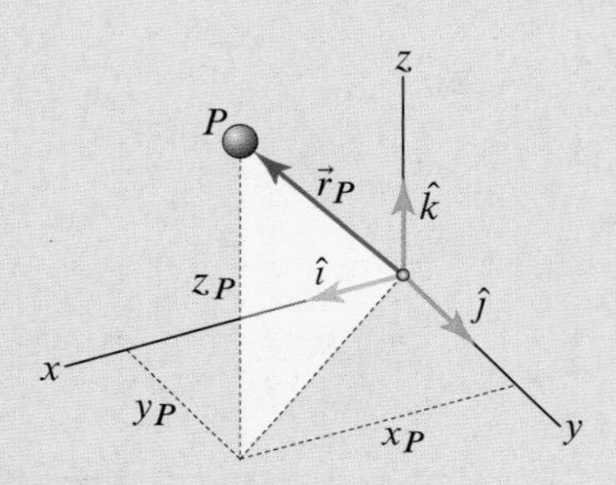

Figure 2.50
The Cartesian coordinate system and its unit vectors in three dimensions.

Cartesian coordinates. Referring to Fig. 2.50, in the three-dimensional Cartesian coordinate system, the position, velocity, and acceleration, respectively, are given in component form as

Eqs. (2.108)–(2.110), p. 144

$$\begin{aligned}\vec{r}(t) &= x(t)\,\hat{\imath} + y(t)\,\hat{\jmath} + z(t)\,\hat{k}, \\ \vec{v}(t) &= \dot{x}\,\hat{\imath} + \dot{y}\,\hat{\jmath} + \dot{z}\,\hat{k} = v_x\,\hat{\imath} + v_y\,\hat{\jmath} + v_z\,\hat{k}, \\ \vec{a}(t) &= \ddot{x}\,\hat{\imath} + \ddot{y}\,\hat{\jmath} + \ddot{z}\,\hat{k} = a_x\,\hat{\imath} + a_y\,\hat{\jmath} + a_z\,\hat{k}.\end{aligned}$$

Review Problems

Problem 2.280

The velocity and acceleration of point P expressed relative to frame A at some time t are

$$\vec{v}_{P/A} = (12.5\,\hat{\imath}_A + 7.34\,\hat{\jmath}_A)\ \text{m/s} \quad \text{and} \quad \vec{a}_{P/A} = (7.23\,\hat{\imath}_A - 3.24\,\hat{\jmath}_A)\ \text{m/s}^2.$$

Knowing that frame B does not move relative to frame A, determine the expressions for the velocity and acceleration of P with respect to frame B. Verify that the speed of P and the magnitude of P's acceleration are the same in the two frames.

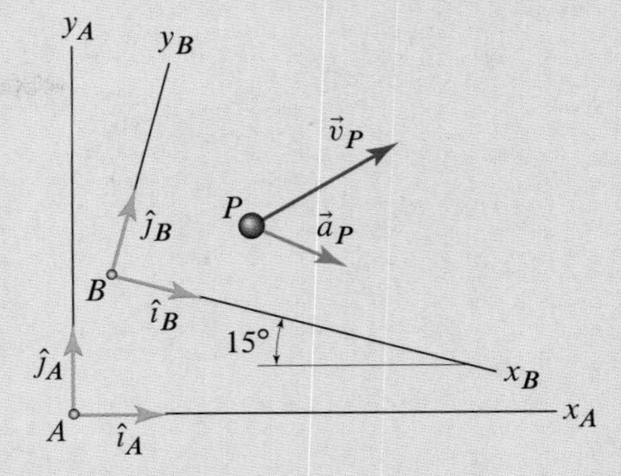

Figure P2.280

Problem 2.281

The motion of a point P with respect to a Cartesian coordinate system is described by $\vec{r} = \{2\sqrt{t}\,\hat{\imath} + [4\ln(t+1) + 2t^2]\,\hat{\jmath}\}$ ft, where t is time expressed in seconds. Determine the average velocity between $t_1 = 4$ s and $t_2 = 6$ s. Then find the time $\bar{t}$ for which the x component of P's velocity is *exactly* equal to the x component of P's average velocity between times t_1 and t_2. Is it possible to find a time at which P's velocity and P's average velocity are exactly equal? Explain why. *Hint:* Velocity is a vector.

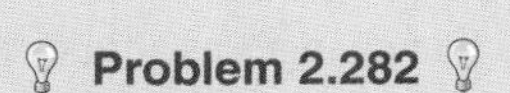

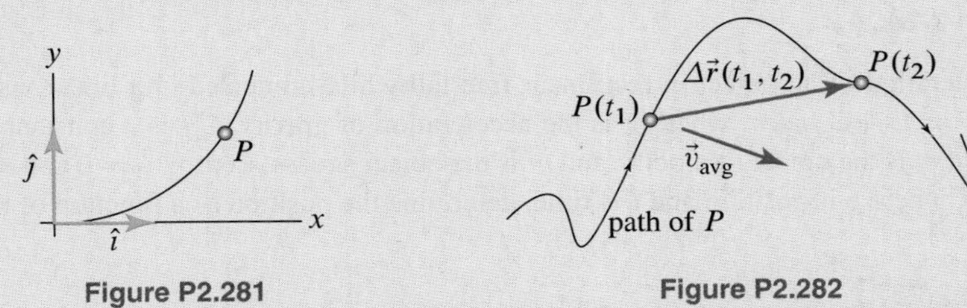

Figure P2.281

Figure P2.282

Problem 2.282

The figure shows the displacement vector of a point P between two time instants t_1 and t_2. Is it possible for the vector $\vec{v}_{avg}$ shown to be the average velocity of P over the time interval $[t_1, t_2]$?

Problems 2.283 and 2.284

A dynamic fracture model proposed to explain the behavior of cracks propagating at high velocity views the crack path as a *wavy path*.* In this model, a crack tip appearing to travel along a straight path actually travels at roughly the speed of sound along a wavy path. Let the wavy path of the crack tip be described by the function $y = h\sin(2\pi x/\lambda)$, where h is the amplitude of the crack tip fluctuations in the direction perpendicular to the crack plane and λ is the corresponding period. Assume that the crack tip travels along the wavy path at a constant speed v_s (e.g., the speed of sound).

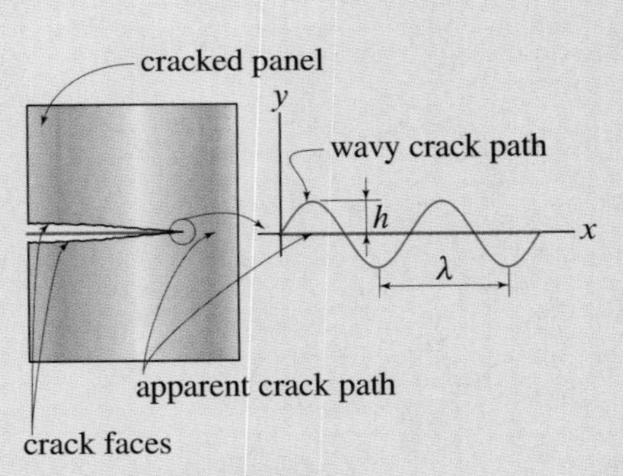

Figure P2.283 and P2.284

Problem 2.283 Find the expression for the x component of the crack tip velocity as a function of v_s, λ, h, and x.

Problem 2.284 Denote the *apparent* crack tip velocity by v_a, and define it as the average value of the x component of the crack velocity, that is,

$$v_a = \frac{1}{\lambda}\int_0^{\lambda} v_x\,dx.$$

* In dynamic fracture, the structural failure of a material occurs at speeds close to the speed of sound (in that material). This field of study is very important in the design of impact- and/or blast-resistant structures. The model mentioned in these problems is due to H. Gao, "Surface Roughening and Branching Instabilities in Dynamic Fracture," *Journal of the Mechanics and Physics of Solids*, **41**(3), pp. 457–486, 1993.

The second equation tells us that $N = mg$, which can then be substituted into the first to obtain

$$\ddot{x} = \frac{P}{m} - \mu_k g \quad \Rightarrow \quad v_f^2 = 2\left(\frac{P}{m} - \mu_k g\right)d \quad \Rightarrow \quad d = \frac{m v_f^2}{2(P - \mu_k m g)}, \tag{3.12}$$

where we have used the fact that $\ddot{x}$ is constant and Eq. (2.34) on p. 49 to obtain the second of Eqs. (3.12).

For the verification step, the final result for d has the dimension of length, as it should. In addition, we see that the distance it takes the crate to reach speed v_f increases as v_f increases and decreases as P increases, both of which agree with our intuition.

Springs

Springs, which are often thought of as having the coiled shape shown in Fig. 3.4, come in many shapes, sizes, and materials. These include bungee cords, beams made of metal or other materials, metal plates, and torsional rods. Many objects that can deform generally do so with some elasticity and can, therefore, be treated as springs.

Figure 3.4
Simple coil springs from a pen.

Geometrically, we describe a spring in terms of its length (Fig. 3.5). The *unstretched length of the spring* is the length of the spring when no force is applied to it. Letting L and L_0 denote the current and unstretched lengths of a spring, respectively, we define the *stretch*, denoted by the Greek letter δ, to be the quantity

$$\boxed{\delta = L - L_0.} \tag{3.13}$$

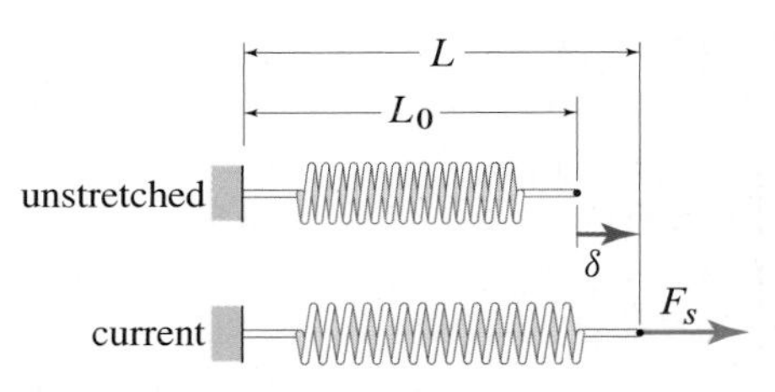

Figure 3.5
Stretch δ of a spring due to force F_s.

When $\delta > 0$ the spring is stretched and when $\delta < 0$ it is compressed.

We will always assume that springs are massless, which means that the force internal to a spring is equal to the external force applied to the spring.

Linear elastic springs. A spring is said to be *linear elastic* if the internal force in the spring is linearly related to the amount the spring is stretched or compressed. Referring to Fig. 3.6, the force F_s required to stretch a linear elastic spring by an amount δ is given by

$$\boxed{F_s = k\delta = k(L - L_0),} \tag{3.14}$$

where k is the *spring constant* and has dimensions of force over length. Unless we indicate otherwise, we will always assume that springs are linear elastic.

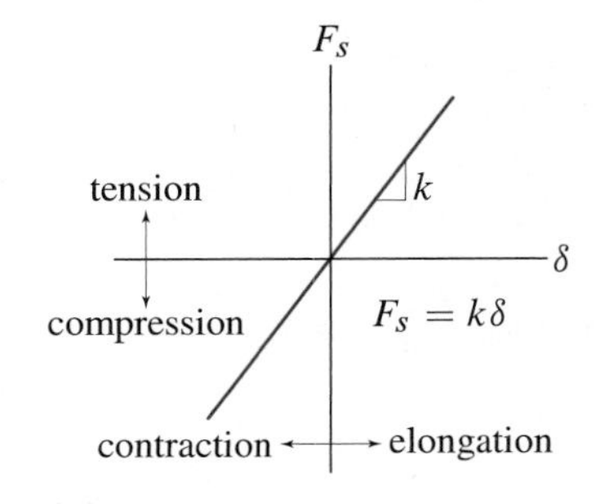

Figure 3.6
Spring law for a linear elastic spring.

Equation(s) of motion

The equation(s) whose solution allows us to determine the motion of a particle or rigid body as a function of time is (are) the *equation(s) of motion* for that particle or rigid body. The first of Eqs (3.12) is the equation of motion of the crate since it provides the constant acceleration of the crate, which can be solved for the position of the crate x as a function of time t. In general, equation(s) of motion for a system involve the system's position variables, their time derivatives, and time.

Inertial reference frames

The use of Newton's second law *requires* that the acceleration $\vec{a}$ be measured with respect to an *inertial frame of reference*. An *inertial reference frame* is one in which

Newton's first and second laws are valid, at least to the level of accuracy desired. In addition, if we have found a frame that *acceptably* satisfies this definition (i.e., it is inertial), then any frame that is not accelerating relative to such an inertial frame is also inertial. For all but a small class of engineering problems (e.g., those that involve relativistic effects or orbital mechanics), a frame attached to the surface of the Earth can be considered inertial.

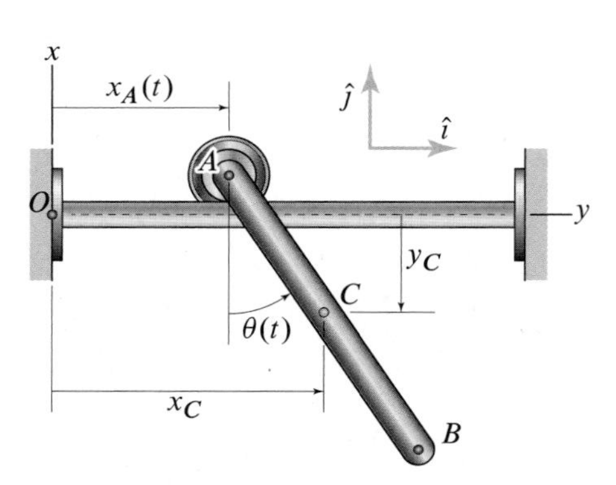

Figure 3.7
An illustration of *degrees of freedom* using a rigid bar with a roller at one end. The length of the bar is L.

Degrees of freedom

A system's *degrees of freedom* are the independent coordinates needed to uniquely specify the position of that system. A more intuitive way of thinking about degrees of freedom is to think of the *number of degrees of freedom* as the number of different coordinates in a system that must be fixed in order to keep the system from moving. For example, the bar shown in Fig. 3.7 can pivot about the hinge at A and can move horizontally along the guide bar. This bar has two degrees of freedom since we would need to fix the two coordinates x_A and θ to prevent the system from moving. It is useful to identify the degrees of freedom of a system because *the number of equations of motion of a system is equal to its number of degrees of freedom.* We saw this with the sliding crate — it has one degree of freedom and we ended up with one equation of motion.

End of Section Summary

Helpful Information

We will structure the solution of kinetics problems using the following four steps:

1. *Road Map & Modeling:* Identify data and unknowns, identify the system, state assumptions, sketch the FBD, and identify a problem-solving strategy.
2. *Governing Equations:* Write the balance principles, force laws, and kinematic equations.
3. *Computation:* Solve the assembled system of equations.
4. *Discussion & Verification:* Study the solution and perform a "sanity check."

Applying Newton's second law. We developed a four-step problem-solving procedure for applying Newton's second law to mechanical systems. The central element of this procedure is the derivation of the governing equations, which originate from the balance principles, force laws, and kinematic equations. In this chapter, the balance principle is Newton's second law, which we write as

Eq. (3.1), p. 169

$$\vec{F} = m\vec{a},$$

and which we apply in component form as

Eqs. (3.2), p. 170

$$\sum F_a = ma_a, \quad \sum F_b = ma_b, \quad \text{and} \quad \sum F_c = ma_c,$$

where a, b, and c are the orthogonal directions of the chosen component system and where we usually need only two directions for planar problems. The four-step procedure we presented for applying Newton's second law, outlined in the margin note, is given as a guide to the *order* in which things should be done, and we will use it consistently in each example we present.

Governing equations and equations of motion. The *governing equations* for a system consist of the (1) balance principles, (2) force laws, and (3) kinematic equations. The *equations of motion* are the equations derived from the governing equations that allow for the determination of the motion.

Friction. We will include friction via the Coulomb friction model. According to this model, in the absence of slip, the magnitude of the friction force F satisfies the following inequality:

Eq. (3.3), p. 171

$$|F| \le \mu_s |N|,$$

where μ_s is the *coefficient of static friction* and N is the force normal to the contact surface. The relation

Eq. (3.4), p. 172

$$F = \mu_s N$$

defines the case of impending slip.

If A and B are two objects sliding with respect to one another, the magnitude of the friction force exerted by B onto A is given by

Eq. (3.6), p. 172

$$F = \mu_k N,$$

where μ_k is the *coefficient of kinetic friction* and where the direction of the friction force must be consistent with the fact that friction opposes the relative motion of A and B.

Springs. A spring is said to be *linear elastic* if the internal force in the spring is linearly related to the amount the spring is stretched or compressed. The force F_s required to stretch a linear elastic spring by an amount δ is given by

Eq. (3.14), p. 173

$$F_s = k\delta = k(L - L_0),$$

where k is the *spring constant* and where L and L_0 are the current and unstretched lengths of the spring, respectively. When $\delta > 0$, the spring is said to be stretched, and when $\delta < 0$, the spring is said to be compressed.

EXAMPLE 3.1 A Chameleon Capturing an Insect

Figure 1
A chameleon capturing an insect.

When a chameleon propels its long tongue to snatch an insect for a meal (see Fig. 1), the process occurs so quickly that a high-speed video camera is needed to capture the event. Video data tells us that it takes 0.15 s for the chameleon to completely retrieve the insect. We will assume that the initial speed of the insect is zero and that the final speed of the insect must be zero since it ends up in the chameleon's mouth for ingestion. The distance traveled is experimentally found to be 0.3 m along a straight line, so we will assume the velocity profile $v(t) = 2[1 - \cos(41.89t)]$ m/s (t is time in seconds), a plot of which is shown in Fig. 2. Given that the insect is a cockroach whose mass is 6 g, determine the "stickiness" force of the chameleon's tongue required to retrieve the insect.

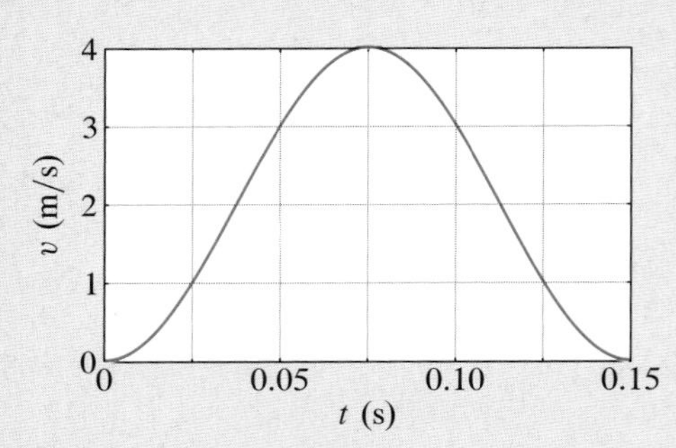

Figure 2
Velocity versus time profile for the tip of the chameleon's tongue. The area under this curve is equal to the 0.3 m distance traveled by the insect.

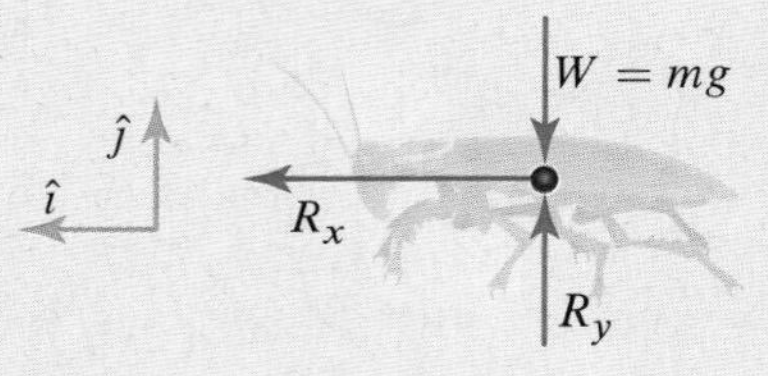

Figure 3
FBD of the insect retrieved by the chameleon.

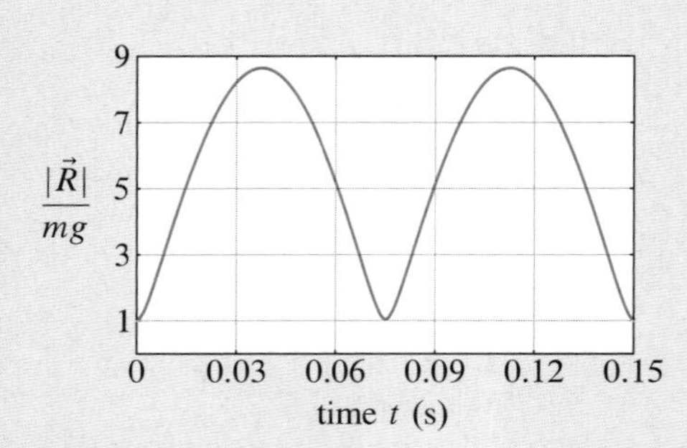

Figure 4
Required force on the insect as it is pulled into the mouth of the chameleon.

SOLUTION

Road Map & Modeling The FBD of the insect as it is being pulled in by the chameleon is shown in Fig. 3, where the x axis is aligned with the direction of motion. Knowing $v(t) = \dot{x}(t)$ for the insect, we can determine acceleration and then apply Newton's second law to determine the required force.

Governing Equations

Balance Principles Applying Newton's second law to the FBD in Fig. 3, obtain

$$\sum F_x: \quad R_x = ma_x, \tag{1}$$

$$\sum F_y: \quad R_y - mg = ma_y. \tag{2}$$

Force Laws All forces are accounted for on the FBD.

Kinematic Equations We are told that the insect moves along a straight line, so $a_y = 0$. To determine a_x, we differentiate $v(t)$ with respect to time to obtain

$$a_x = dv/dt = (83.78\,\text{m/s}^2)\sin(41.89t). \tag{3}$$

Computation Substituting $a_y = 0$ and Eq. (3) into Eqs. (1) and (2), we obtain

$$R_x = [(83.78\,\text{m/s}^2)\sin(41.89t)]m \quad \text{and} \quad R_y = mg, \tag{4}$$

which gives a total required force of

$$\boxed{|\vec{R}| = \sqrt{R_x^2 + R_y^2} = m\sqrt{[(83.78\,\text{m/s}^2)\sin(41.89t)]^2 + g^2}.} \tag{5}$$

The value of $|\vec{R}|$ given by Eq. (5), divided by the insect's weight, has been plotted in Fig. 4.

Discussion & Verification Figure 4 shows that the maximum force required is almost $9mg$, and this occurs at two different points in time. Referring to Eqs. (1) and (3), the first time the maximum force is achieved corresponds to a positive value of R_x so that the force is directed toward the chameleon. The second maximum corresponds to a negative value of R_x, which is needed to slow down the insect before it comes to a stop in the chameleon's mouth. For a cockroach with mass $m = 6$ g, the maximum value of $|\vec{R}|$ can be calculated by differentiating Eq. (5) with respect to time and setting the result equal to zero, which tells us that $R_{\text{max}} = 0.5061$ N. This value occurs at $t_1 = 0.03750$ s and $t_2 = 0.1125$ s.

EXAMPLE 3.2 *Spring Stopping a Moving Crate on an Incline*

The crate of mass $m = 45$ kg shown in Fig. 1 is moving down the incline with speed $v_0 = 8$ m/s at the instant that the spring with elastic constant $k = 100$ N/m is unstretched. Modeling the crate as a particle, determine the distance d the crate moves from the given position before it momentarily comes to a stop. Assume that friction between the crate and the incline is negligible, that the unstretched length of the spring is $L_0 = 1.2$ m, and that $\theta = 25°$.

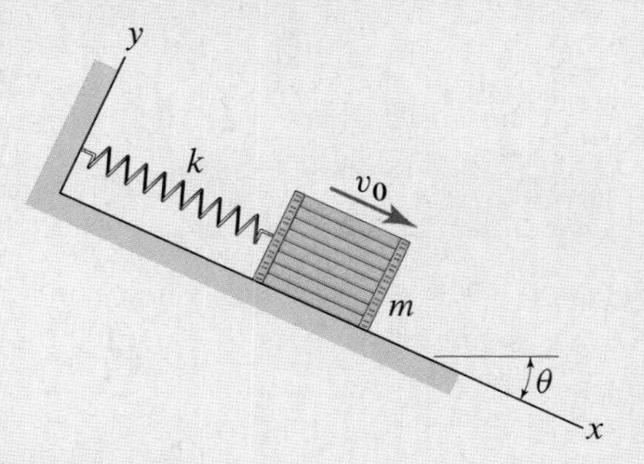

Figure 1
A crate moving with speed v_0 on an incline.

SOLUTION

Road Map & Modeling Our goal is to use Newton's second law to determine the acceleration of the crate. The acceleration can then be manipulated to obtain the position of the crate. Since the crate is sliding on the incline, the FBD is as shown in Fig. 2, where F_s is the force exerted on the crate by the spring, N is the normal force between the crate and the incline, and mg is the force of gravity. The position x is measured from the wall to which the spring is attached. As such, x measures the current length of the spring.

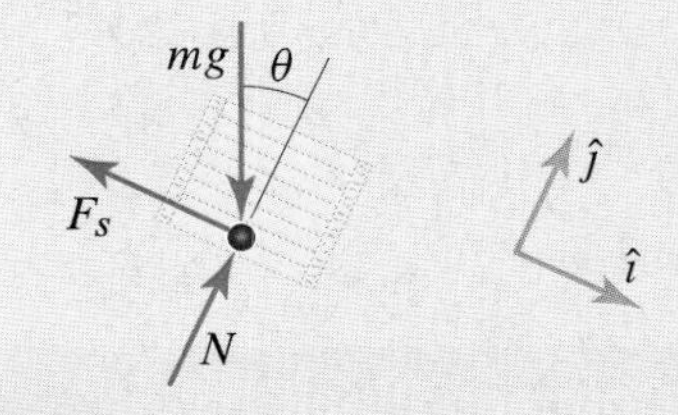

Figure 2
The FBD of the crate as it slides down the incline.

Governing Equations

Balance Principles Referring to the FBD in Fig. 2, Newton's second law gives

$$\sum F_x: \quad -F_s + mg\sin\theta = ma_x, \tag{1}$$

$$\sum F_y: \quad N - mg\cos\theta = ma_y. \tag{2}$$

Force Laws Since x measures the current length of the spring, the expression for F_s for any position x is

$$F_s = k(x - L_0). \tag{3}$$

Kinematic Equations Since the motion is constrained to just the x direction, the kinematic equations are

$$a_x = \ddot{x} \quad \text{and} \quad a_y = 0. \tag{4}$$

Computation Substituting Eqs. (3) and (4) into Eqs. (1) and (2), we obtain

$$-k(x - L_0) + mg\sin\theta = m\ddot{x} \quad \text{and} \quad N = mg\cos\theta. \tag{5}$$

Dividing through by m, applying the chain rule to the first of Eqs. (5), and then integrating from the initial position to the point where the crate first comes to a stop, we obtain

$$v\frac{dv}{dx} = g\sin\theta - \frac{k}{m}(x - L_0)$$

$$\Rightarrow \quad \int_{v_0}^{0} v\,dv = \int_{L_0}^{L_0+d}\left[g\sin\theta + \frac{k}{m}(L_0 - x)\right]dx$$

$$\Rightarrow \quad -\tfrac{1}{2}v_0^2 = gd\sin\theta - \frac{kd^2}{2m}, \tag{6}$$

which is a quadratic equation for the distance d. Solving and then substituting in given data, we obtain

$$d = \frac{mg}{k}\sin\theta \pm \frac{1}{k}\sqrt{m\left(kv_0^2 + mg^2\sin^2\theta\right)}$$

$$\Rightarrow \quad d = -3.816\,\text{m} \quad \text{and} \quad d = 7.547\,\text{m} \quad \Rightarrow \quad \boxed{d = 7.547\,\text{m},} \tag{7}$$

where we have chosen the physically meaningful result as the final answer.

Discussion & Verification The dimensions in Eqs. (6) and (7) are both correct, and the stopping distance seems physically reasonable given the initial speed and lack of friction.

EXAMPLE 3.3 *Friction and Impending Slip*

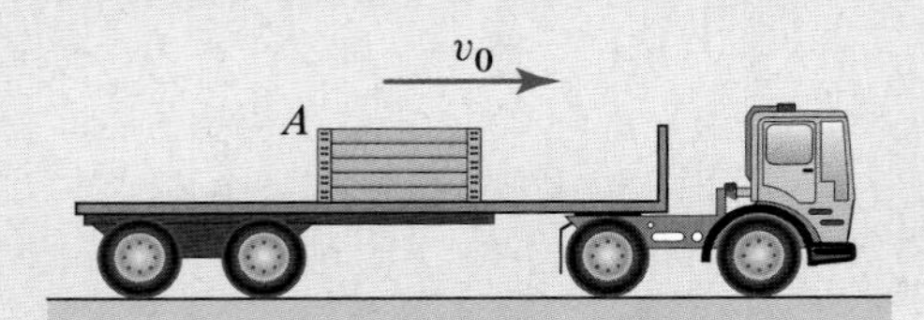

Figure 1
A truck hauling a large crate.

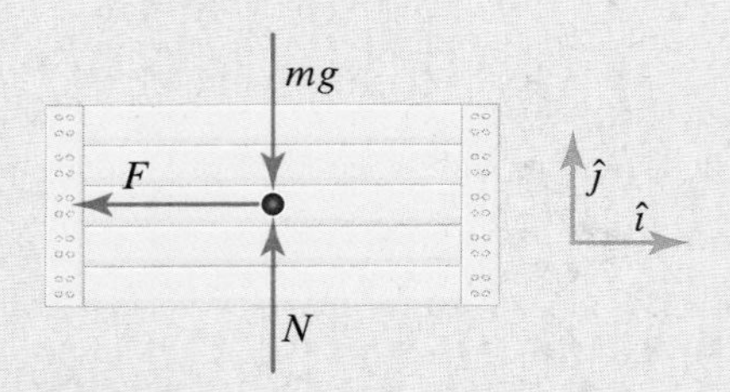

Figure 2
FBD of the crate.

> **Common Pitfall**
>
> **Newton's second law and inertial frames.** Since the application of Newton's second law requires the use of an inertial reference frame, the component system shown in Fig. 2 must be understood as originating from an xy coordinate system *fixed* with the ground—this is the inertial reference frame. It would be a mistake to choose a coordinate system moving with the truck because the truck is decelerating with respect to the ground and, therefore, is not an inertial frame of reference.

The truck shown in Fig. 1 is traveling at $v_0 = 100\,\text{km/h}$ when the driver slams on the brakes and comes to a stop as quickly as possible. If the coefficient of static friction between the crate A and the bed of the truck is 0.35, determine the minimum stopping distance d_{min} and the minimum stopping time t_{min} of the truck for which the crate does not slide forward on the truck.

SOLUTION

Road Map & Modeling The truck must stop as fast as possible without causing the *crate* to slide. Therefore, the system to analyze is the crate, which we model as a particle. In the FBD shown in Fig. 2, we have assumed that the only relevant forces are the crate's weight mg, the normal force N between the crate and the truck, and the friction force F between the crate and the truck. The force F points left because if the crate were to slip, it would slip to the right relative to the truck, and F must oppose this motion. Since the motion is rectilinear, we have chosen a Cartesian coordinate system with the x axis parallel to the direction of motion. The maximum deceleration without sliding corresponds to the maximum possible friction force acting on the crate, which is the force corresponding to *impending slip*. Our solution strategy will be to relate the maximum friction force to the acceleration via Newton's second law. Once we have an expression for the maximum acceleration, we will apply our knowledge of kinematics to compute d_{min} and t_{min}.

Governing Equations

Balance Principles Referring to Fig. 2, Newton's second law yields

$$\sum F_x: \quad -F = ma_x, \tag{1}$$

$$\sum F_y: \quad N - mg = ma_y. \tag{2}$$

Force Laws The friction law for *impending slip* is

$$F = \mu_s N. \tag{3}$$

Kinematic Equations The kinematic equations are

$$a_x = a_{\text{max}} \quad \text{and} \quad a_y = 0, \tag{4}$$

which express the fact that while a_x is still unknown, we are seeking the maximum value of acceleration, and also expresses the fact that the crate is not moving vertically.

Computation Equations (1)–(4) are four equations in the unknowns N, a_y, a_{max}, and F. Equation (2) and the second of Eqs. (4) tell us that $N = mg$. Substituting this result into Eq. (3) and, in turn, substituting that into Eq. (1), we obtain

$$-\mu_s mg = ma_{\text{max}} \quad \Rightarrow \quad a_{\text{max}} = -\mu_s g, \tag{5}$$

where the negative sign indicates that the crate (and truck) is decelerating when the driver applies the brakes.

Since the maximum deceleration is constant (both μ_s and g are constants), we can determine the stopping distance by using Eq. (2.34) on p. 49:

$$v^2 = v_0^2 + 2a_c(x - x_0) \quad \Rightarrow \quad 0 = v_0^2 - 2\mu_s g d_{\text{min}}, \tag{6}$$

so that

$$\boxed{d_{\text{min}} = \frac{v_0^2}{2\mu_s g} = 112.4\,\text{m},} \tag{7}$$

where we have used the conversion 100 km/h = 27.78 m/s. To determine the stopping time, we apply Eq. (2.32) on p. 49:

$$v = v_0 + a_c t \quad \Rightarrow \quad 0 = v_0 - \mu_s g t_{\min}, \tag{8}$$

so that

$$\boxed{t_{\min} = \frac{v_0}{\mu_s g} = 8.090\,\mathrm{s}.} \tag{9}$$

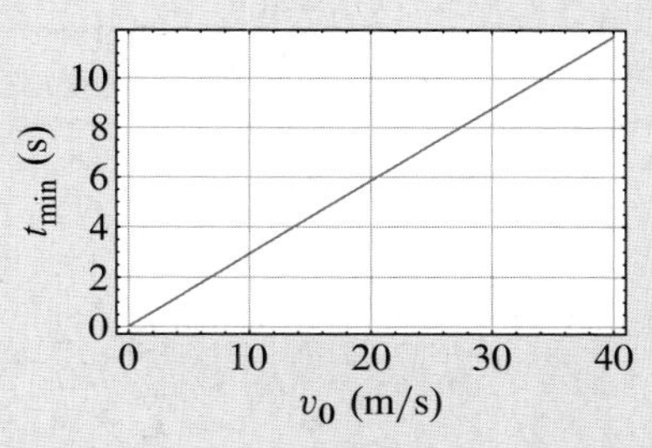

Figure 3
The minimum stopping time of the crate as a function of the initial speed of the truck.

Discussion & Verification The symbolic forms of the results in Eqs. (7) and (9) have the correct dimensions for $d_{\min}$ and $t_{\min}$. Also, the final numerical results have been expressed using appropriate units. As far as the values we have obtained for $d_{\min}$ and $t_{\min}$ are concerned, these are certainly reasonable since, in absolute value, the maximum deceleration possible under the stated assumptions is roughly one-third of the acceleration of gravity.

A Closer Look The model used here, which is based on the Coulomb friction model, is such that the deceleration we calculated is completely independent of the crate's mass. Therefore, the crate could have been twice as heavy, and we would have obtained the same answer. This is a basic property of this friction model. Another important observation to make is that, again because of the friction model used, the minimum time required to stop is linearly proportional to the initial speed v_0, and the minimum distance required to stop is proportional to the initial speed squared v_0^2. This is shown in Figs. 3 and 4, respectively.

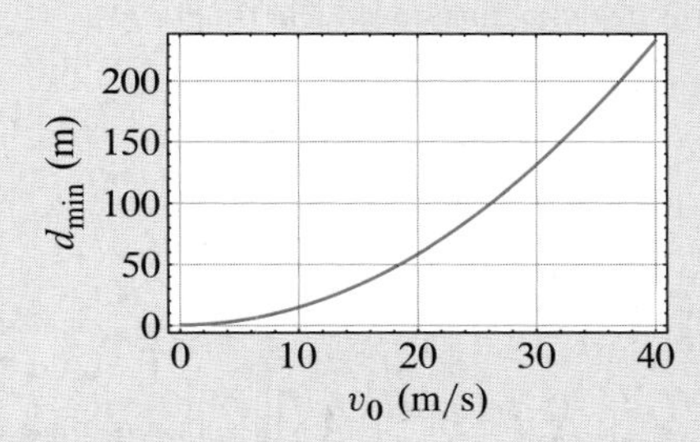

Figure 4
The minimum stopping distance of the crate as a function of the initial speed of the truck.

EXAMPLE 3.4 *Transition from Static to Kinetic Friction*

Figure 1
A person pushing a crate of mass m over a rough surface with a time-dependent force $P(t)$.

A simple problem of practical importance that illustrates the nature of friction is the study of the initiation of motion of an object on a rough surface. Let's look at what happens when a person pushes a crate of mass m on a rough floor, as shown in Fig. 1. Letting μ_s and μ_k, with $\mu_k < \mu_s$, be the static and kinetic friction coefficients between the crate and the ground, respectively, consider a case in which the force $P(t)$ exerted by the person increases linearly with time; i.e., $P(t) = P_0 t$. For this case, determine the time at which the motion begins and the friction force as a function of time.

SOLUTION

Road Map & Modeling We are given one of the forces applied to the crate as a function of time. To determine the time at which the motion starts, we need to determine when the horizontal component of the applied force P overcomes the maximum friction force allowed by the *static* friction coefficient, which is done by applying Newton's second law. Referring to Fig. 2, we model the crate as a particle under the action of gravity mg, the friction force F, the normal force N between the crate and the ground, and the force $P(t)$ applied by the person. The force $P(t)$ has been plotted as a function of time in Fig. 3. For the sake of generality, the direction of the applied force $P(t)$ is assumed to form a generic angle θ with the horizontal direction.

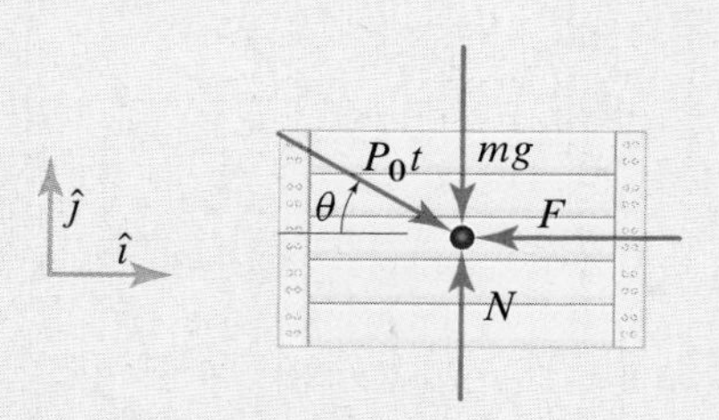

Figure 2
FBD of the crate.

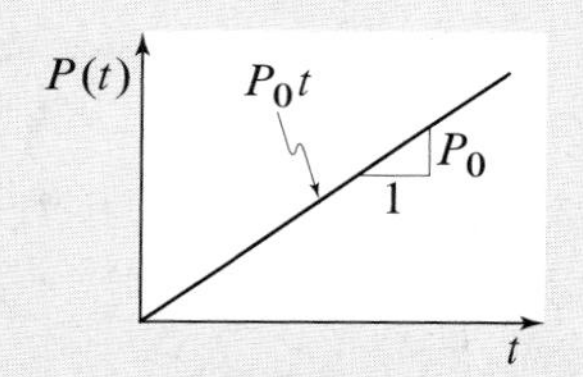

Figure 3
Plot of $P(t)$ versus time.

Governing Equations

Balance Principles Based on the FBD in Fig. 2, Newton's second law gives

$$\sum F_x: \quad P_0\, t \cos\theta - F = m a_x, \tag{1}$$

$$\sum F_y: \quad N - P_0\, t \sin\theta - mg = m a_y. \tag{2}$$

Force Laws The force law for friction will depend on the value of F. If there were no friction, the crate would move to the right, and so the force F must point to the left, as indicated on the FBD in Fig. 2. In addition, $F \leq \mu_s N$ when the crate is not moving, and $F = \mu_k N$ after the crate starts moving. As long as the component of $P(t)$ propelling the crate forward is less than the maximum possible static friction, the crate will not move, that is, as long as $P(t)\cos\theta \leq \mu_s N$. Once $P(t)\cos\theta > \mu_s N$, then the crate starts to move, kinetic friction comes into play, and the friction force *immediately* drops to $\mu_k N$ (since $\mu_k < \mu_s$). Therefore, to compute the time at which the motion starts, the force law to use is

$$F = \mu_s N. \tag{3}$$

Once the crate starts moving, this equation will need to be replaced by

$$F = \mu_k N. \tag{4}$$

The force law describing the weight of the crate is elementary and has been indicated directly on the FBD as mg. We do not have an explicit force law for N since its value is determined by the fact that the crate cannot move in the y direction.

Kinematic Equations Since the crate does not move in the y direction, we have

$$a_y = 0. \tag{5}$$

In the x direction, before the motion starts, we have

$$a_x = 0. \tag{6}$$

Once the motion starts, a_x becomes an unknown.

Computation Substituting Eq. (5) into Eq. (2) and solving for N, we obtain

$$N = P_0 t \sin\theta + mg. \tag{7}$$

Recalling that $a_x = 0$ until the motion begins, we solve Eq. (1) for F and set the result equal to $\mu_s N$, where N comes from Eq. (7). This yields

$$\mu_s(P_0 t_s \sin\theta + mg) = P_0 t_s \cos\theta, \tag{8}$$

where t_s is the time at which motion begins. Solving for t_s in Eq. (8), we obtain

$$t_s = \frac{mg}{P_0}\left(\frac{\mu_s}{\cos\theta - \mu_s \sin\theta}\right). \tag{9}$$

Once $t = t_s$, the crate starts to move, and as indicated in Eq. (4), F immediately drops from $\mu_s N$ to $\mu_k N$. Also, for $t > t_s$, a_x is no longer known. Therefore, for $t > t_s$, we have four equations—Eqs. (1), (2), (4), and (5)—to solve for the four unknowns a_x, a_y, F, and N. Solving, we obtain

$$\left.\begin{aligned} a_x &= \frac{P_0 t}{m}(\cos\theta - \mu_k \sin\theta) - \mu_k g,\\ a_y &= 0,\\ F &= \mu_k(mg + P_0 t \sin\theta),\\ N &= mg + P_0 t \sin\theta. \end{aligned}\right\} \text{for } t > t_s. \tag{10}$$

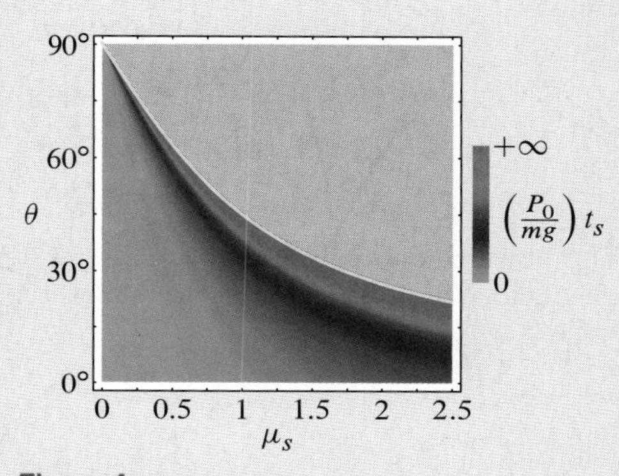

Figure 4
A contour plot of the term in parentheses in Eq. (9). Note that combinations of θ and μ_s lying in the gray area above the yellow curve correspond to values for which the crate will never move.

Discussion & Verification Observing that the quantity P_0 has dimensions of force per unit time, we can easily verify that the dimensions of the results in Eqs. (9) and (10) are correct.

A Closer Look This example demonstrates several features of the Coulomb friction model. First, Eq. (9) tells us that the crate will never move if $\cos\theta \leq \mu_s \sin\theta$ (since the denominator becomes negative). Also, the closer $\cos\theta$ is to $\mu_s \sin\theta$, the closer the denominator in Eq. (9) is to zero and the longer it takes to get the crate moving.

An interesting representation of Eq. (9) can be found in Fig. 4, which shows t_s as a function of θ and μ_s. The yellow curve at the edge of the red region represents the θ and μ_s values for which the denominator of Eq. (9) is zero. Since the crate only moves for positive values of t_s, no matter how long we wait or how hard we push, the crate will never move for any θ and μ_s values above and to the right of the yellow curve. As we approach the yellow curve from below, it takes longer and longer to get the crate moving. For example, as θ approaches 90°, it becomes impossible to move the crate no matter how small μ_s is, unless $\mu_s = 0$. Since friction is the key ingredient in this problem, it is illustrative to plot the friction force F as a function of time. Until the crate starts to move, Eq. (1) gives F as

$$F = P_0 t \cos\theta, \quad \text{for } t < t_s. \tag{11}$$

As we have just seen, Eq. (10) gives us F for $t > t_s$. A plot of F vs. t for some particular values of the system parameters is shown in Fig. 5. Notice the discontinuity in the friction force when the crate starts to slip at $t = t_s$. This is a general feature of Coulomb friction and will always happen as long as $\mu_s \neq \mu_k$. Finally, note that F keeps increasing with time after the crate starts moving because the value of N increases to balance the corresponding increasing value of the vertical component of $P(t)$.

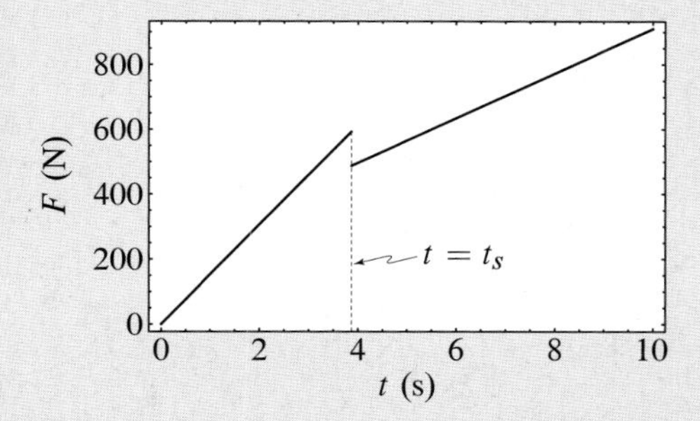

Figure 5
Friction F vs. time t. The parameters used were $m = 50\,\text{kg}$, $g = 9.81\,\text{m/s}^2$, $P_0 = 200\,\text{N/s}$, $\theta = 40°$, $\mu_s = 0.6$, and $\mu_k = 0.45$.

EXAMPLE 3.5 Motion under the Action of Spring Forces

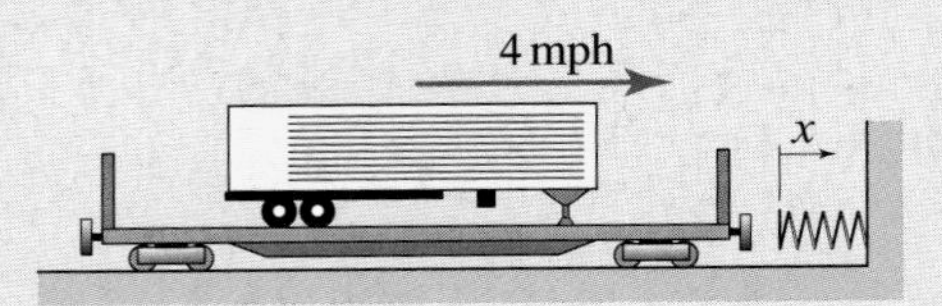

Figure 1
Railcar running into a large spring.

A 60 ton railcar and its cargo, a 27 ton trailer, are moving to the right at 4 mph, as shown in Fig. 1, when they encounter a large linear spring that has been designed to stop a 60 ton railcar moving at 5 mph in a distance of 3 ft when it is initially uncompressed. If the trailer does not slip relative to the railcar and if the spring is initially uncompressed, determine (a) how much the spring compresses in stopping the 87 ton loaded railcar and (b) how long it takes for the spring to stop the railcar.

SOLUTION

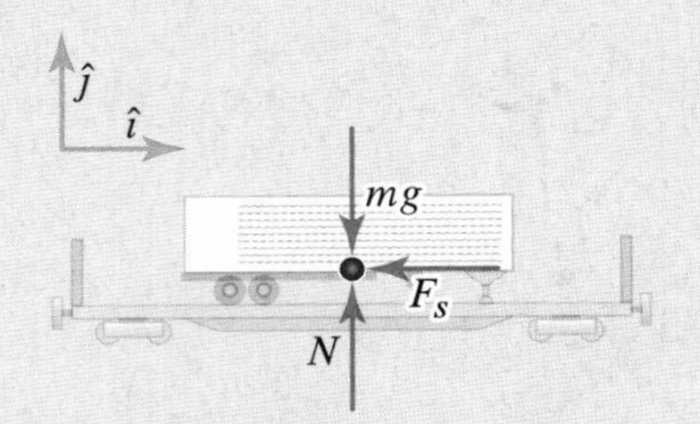

Figure 2
FBD of the railcar shown in Fig. 1, where m is either the mass of the railcar or the railcar and trailer.

Road Map & Modeling The given information consists of (1) the conditions under which the railcar and its cargo impact the spring and (2) the spring's design criterion. We need to compute the stopping distance and time of the railcar and its cargo, and so "railcar and cargo" is the system to analyze. We will model this system as a particle, and its FBD is shown in Fig. 2. We have assumed that the only relevant forces are gravity, the reaction with the rails, and the spring force. Note that we are not given direct information about the spring constant or its unstretched length. However, using the stated design criteria in conjunction with Newton's second law, we will be able to first determine the spring's force law and then use the result to compute the system's acceleration. Once we have the system's acceleration, we will need to apply kinematics to obtain the corresponding stopping time and distance information.

Governing Equations

Balance Principles Based on the FBD in Fig. 2, Newton's second law gives

$$\sum F_x: \quad -F_s = ma_x, \tag{1}$$

$$\sum F_y: \quad N - mg = ma_y. \tag{2}$$

Force Laws While the spring constant k is unknown, the spring's force law can still be given the form

$$F_s = kx, \tag{3}$$

where x is measured from the uncompressed end position of the spring.

Kinematic Equations The kinematic relations are

$$a_x = \ddot{x} \quad \text{and} \quad a_y = 0, \tag{4}$$

where we have chosen to represent a_x as $\ddot{x}$ because the problem is asking us to relate acceleration to position, and where the second of Eqs. (4) states that there is no motion in the vertical direction.

Computation Since there is no motion in the y direction, we will disregard the equations in the vertical direction. Next, substituting the first of Eqs. (4) and Eq. (3) into Eq. (1) and rearranging, we obtain

$$\ddot{x} + \frac{k}{m}x = 0. \tag{5}$$

Rearranging Eq. (5) and making use of the chain rule, we obtain

$$\ddot{x} = \dot{x}\,\frac{d\dot{x}}{dx} = -\frac{k}{m}x \quad \Rightarrow \quad \dot{x}\,d\dot{x} = -\frac{k}{m}x\,dx, \tag{6}$$

which can be integrated as

$$\int_{v_i}^{v} \dot{x}\,d\dot{x} = -\int_{x_i}^{x} \frac{k}{m}x\,dx \quad \Rightarrow \quad v^2 - v_i^2 = -\frac{k}{m}\left(x^2 - x_i^2\right), \tag{7}$$

where v is the speed of the railcar at the location x and where the subscript i stands for *initial*. Letting v_f be the speed corresponding to the final position x_f, solving Eq. (7) for k, and substituting in numbers corresponding to the spring's design criteria, we obtain

$$k = \frac{m_r\left(v_f^2 - v_i^2\right)}{x_i^2 - x_f^2} = 22{,}270\,\mathrm{lb/ft}, \tag{8}$$

where we have used the unit conversions 5 mph = 7.333 ft/s and 1 ton = 2000 lb, m_r is the mass of the railcar, $v_f = 0$, $x_i = 0$, and $x_f = 3$ ft.

Part (a)

Now that we have k, we can determine how far the spring compresses with the loaded railcar by applying Eq. (7) again. Letting $v = v_f$ for $x = x_f$, solving Eq. (7) for x_f, and substituting in the numbers for the loaded railcar, we obtain

$$x_f = \sqrt{x_i^2 - \frac{m_t}{k}\left(v_f^2 - v_i^2\right)} = 2.890\,\mathrm{ft}, \tag{9}$$

where we have used the unit conversion 4 mph = 5.867 ft/s and m_t is the total mass of the railcar and the trailer.

Part (b)

To determine the stopping time t_f, we go back to Eq. (7), solve it for v, and rearrange the result to integrate it with respect to time, i.e.,

$$v = \frac{dx}{dt} = \sqrt{v_i^2 - \frac{k}{m_t}\left(x^2 - x_i^2\right)} \quad \Rightarrow \quad \int_{x_i}^{x_f} \frac{dx}{\sqrt{v_i^2 - \frac{k}{m_t}\left(x^2 - x_i^2\right)}} = \int_{t_i}^{t_f} dt. \tag{10}$$

Using a table of integrals, we obtain

$$t_f - t_i = \sqrt{\frac{m_t}{k}}\left[\sin^{-1}\left(\frac{\sqrt{\frac{k}{m_t}}\,x_f}{\sqrt{v_i^2 + \frac{k}{m_t}x_i^2}}\right) - \sin^{-1}\left(\frac{\sqrt{\frac{k}{m_t}}\,x_i}{\sqrt{v_i^2 + \frac{k}{m_t}x_i^2}}\right)\right] \quad \Rightarrow \quad t_f = 0.7714\,\mathrm{s}, \tag{11}$$

where we substituted in $x_f = 2.890$ ft, $v_i = 5.867$ ft/s, $k = 22{,}270$ lb/ft, $m_t = (174{,}000/32.2)$ slug, $t_i = 0$ s, and $x_i = 0$ ft.

Discussion & Verification The result in Eq. (9) tells us that a 174,000 lb loaded railcar traveling at 4 mph compresses the spring a little less than 3 ft. This result seems reasonable. Although the loaded car is 45% heavier than when it is empty, the spring compression should have not been expected to necessarily increase because the loaded car is moving with a speed that is only 80% of the design speed for an impact between an unloaded railcar and the bumper. Clearly the increase in weight is larger than the decrease in speed; however, the spring compression depends on the *square* of the speed, i.e., in this problem, speed is a more significant factor than weight.

Helpful Information

The integral in Eq. (10). In computing the integral in Eq. (10), it might help to notice that if we let

$$a = \sqrt{v_i^2 + \frac{k}{m_t}x_i^2} \quad \text{and} \quad u = \sqrt{\frac{k}{m_t}}\,x,$$

then Eq. (10), with lower limits set equal to zero, becomes

$$\int_0^{t_f} dt = \sqrt{\frac{m_t}{k}} \int_0^{\sqrt{\frac{k}{m_t}}x_f} \frac{du}{\sqrt{a^2 - u^2}},$$

or

$$t_f = \sqrt{\frac{m_t}{k}} \sin^{-1}\left(\frac{u}{a}\right)\Big|_0^{\sqrt{\frac{k}{m_t}}x_f} = \sqrt{\frac{m_t}{k}} \sin^{-1}\left(\frac{\sqrt{\frac{k}{m_t}}\,x_f}{\sqrt{v_i^2 + \frac{k}{m_t}x_i^2}}\right).$$

Problems

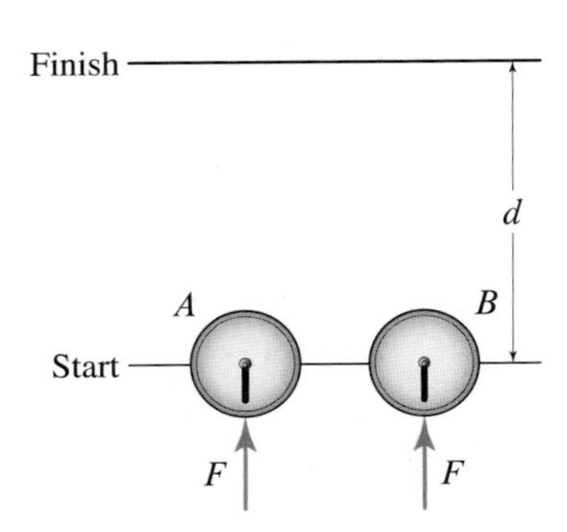

Figure P3.1

Problem 3.1

Two curling stones A and B, with masses m and $4m$, respectively, and initially at rest on the start line, are pushed by two identical forces F over the distance d. Which stone arrives first to the finish line?

Problem 3.2

An object is lowered very slowly onto a conveyor belt that is moving to the right. What is the direction of the friction force acting on the object at the instant the object touches the belt?

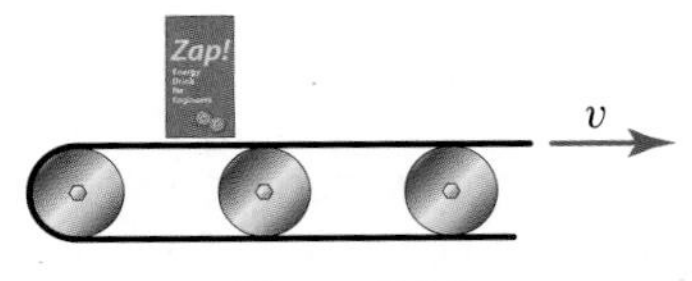

Figure P3.2

Figure P3.3

Problem 3.3

A person is trying to move a heavy crate by pushing on it. While the person is pushing, what is the resultant force acting on the crate if the crate does not move?

Problem 3.4

A person is lifting a 75 lb crate A by applying a constant force $P = 40$ lb to the pulley system shown. Neglecting friction and the inertia of the pulleys, determine the acceleration of the crate. Treat all rope segments as purely vertical.

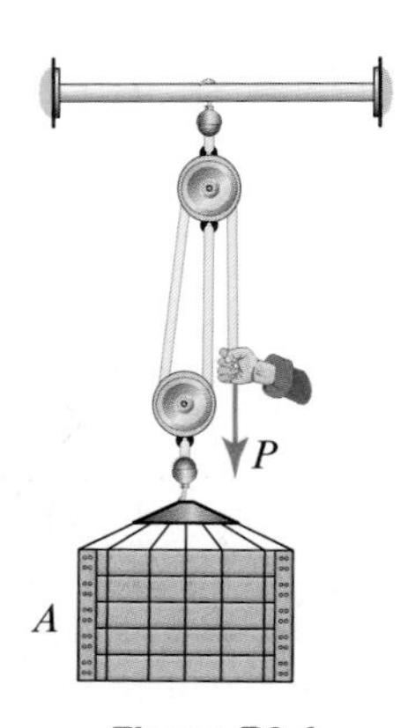

Figure P3.4

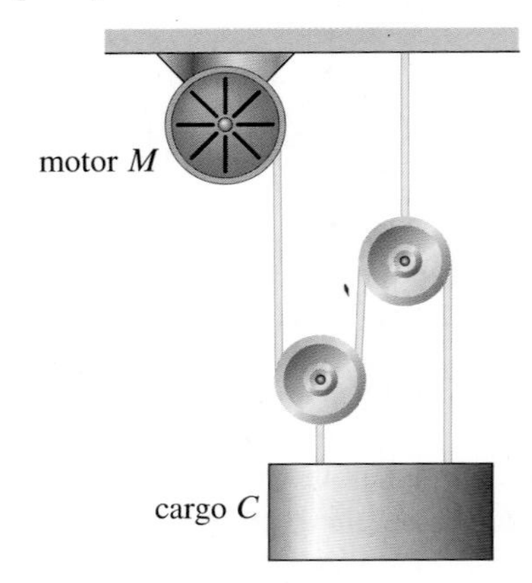

Figure P3.5

Problem 3.5

The motor M is at rest when someone flips a switch and it starts pulling in the rope. The acceleration of the rope is uniform and is such that it takes 1 s to achieve a retraction rate of 4 ft/s. After 1 s the retraction rate becomes constant. Determine the tension in the rope during and after the initial 1 s interval. The cargo C weighs 130 lb, the weight of the ropes and pulleys is negligible, and friction in the pulleys is negligible.

Problems 3.6 and 3.7

A crate of weight $W = 550$ lb has been attached to a pickup truck by a rope whose tensile strength is $T_{max} = 350$ lb. If the truck and crate start at rest with $\theta = 30°$, determine the maximum acceleration of the truck such that the rope does not break.

Figure P3.6 and P3.7

Problem 3.6 Determine the solution for the case in which friction between the crate and the ground is negligible.

Problem 3.7 Determine the solution for the case in which friction between the crate and the ground is *not* negligible and $\mu_s = 0.4$ and $\mu_k = 0.25$.

Problem 3.8

The crate A of mass m and the wedge B on which it rests are both initially at rest. The wedge, whose face is inclined at $\theta = 30°$ with the horizontal, is given an acceleration a_B to the left as shown. Given that the coefficient of static friction between the crate and the wedge is $\mu_s = 0.6$, determine the maximum value of a_B such that the crate does not slip on the wedge.

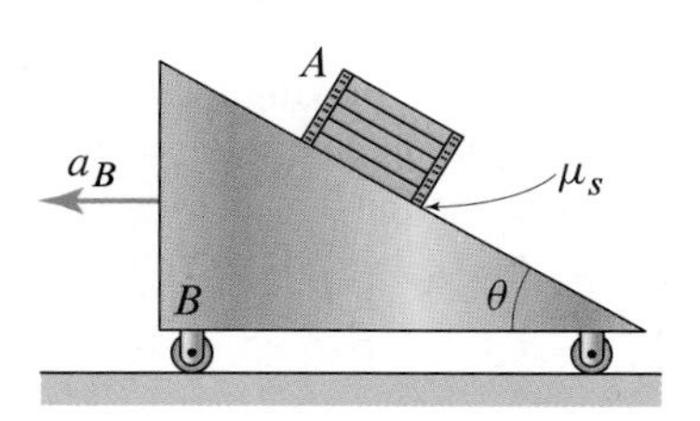

Figure P3.8

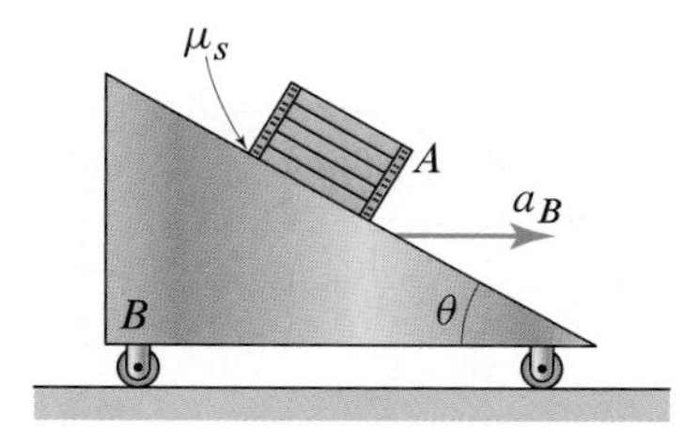

Figure P3.9

Problem 3.9

The crate A of mass m and the wedge B on which it rests are both initially at rest. The wedge, whose face is inclined at $\theta = 30°$ with the horizontal, is given an acceleration a_B to the right as shown. Given that the coefficient of static friction between the crate and the wedge is $\mu_s = 0.6$, determine the maximum value of a_B such that the crate does not slip on the wedge.

Problem 3.10

A hammer hits a mass m on the end of a metal bar. In Chapter 5, we will see that this imparts an instantaneous initial velocity v_0 at $x = 0$ to the mass. Treating the bar as a massless spring, determine the equation of motion of the mass m. The equivalent spring constant of a bar in compression is given by $k_{eq} = EA/L$, where E is Young's modulus of the bar, A is the cross-sectional area of the bar, and L is the length of the bar.*

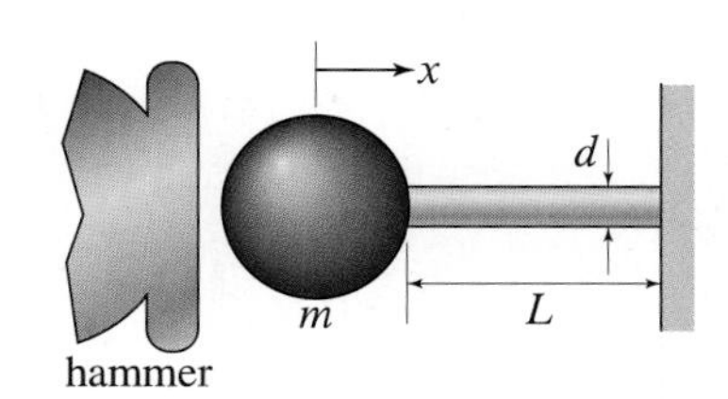

Figure P3.10 and P3.11

Problem 3.11

For the mass described in Prob. 3.10:

(a) Integrate the equation of motion to determine the speed of the mass $v(x)$ as a function of x.

(b) Use the result found in Part (a) to obtain the position of the mass as a function of time $x(t)$ from the initial time up until the mass stops for the first time. *Hint:* $\int (a^2 - bx^2)^{-1/2}\,dx = \sin^{-1}(\sqrt{b}x/a)/\sqrt{b}$.

* From strength of materials, $\sigma = F/A = E\epsilon = E\Delta L/L$. Therefore, $F = (EA/L)\Delta L = (EA/L)x$. Since $F = k_{eq}x$ for this system, we see that $k_{eq} = EA/L$.

Problem 3.12

The crate A of mass m and the wedge B on which it rests are moving together down the incline with the acceleration a_B as shown. The angle of the incline is $\theta = 30°$ with respect to the horizontal. Given that the coefficient of static friction between the crate and the wedge is $\mu_s = 0.6$, determine the maximum value of a_B before the crate starts to slip on the wedge.

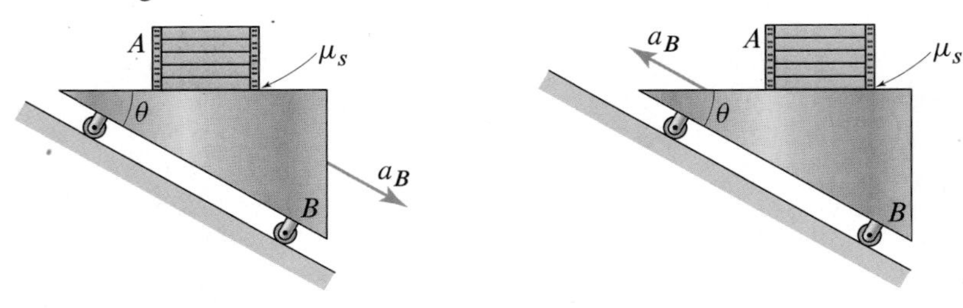

Figure P3.12 Figure P3.13

Problem 3.13

The crate A of mass m and the wedge B on which it rests are moving together up the incline with the acceleration a_B as shown. The angle of the incline is $\theta = 30°$ with respect to the horizontal. Given that the coefficient of static friction between the crate and the wedge is $\mu_s = 0.6$, determine the maximum value of a_B before the crate starts to slip on the wedge.

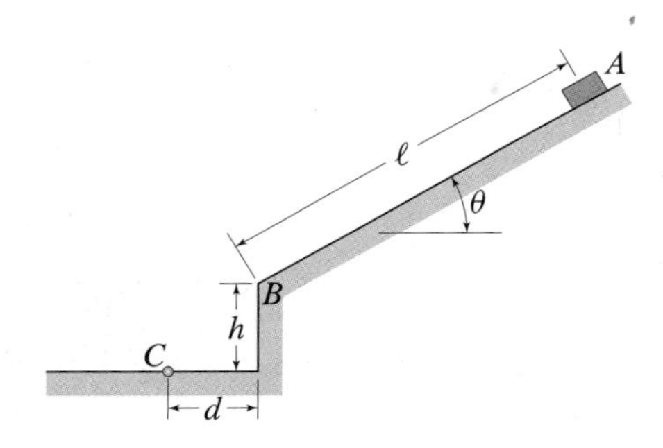

Figure P3.14 and P3.15

Problems 3.14 and 3.15

A suitcase is released from rest at A on the $\theta = 30°$ ramp. It slides a distance $\ell = 25$ ft and then goes over the edge at B and drops a height $h = 5$ ft. Determine the horizontal distance d to the landing spot at C.

Problem 3.14 Assume that friction on the incline between A and B is negligible.

Problem 3.15 Assume that the coefficient of static friction is insufficient to prevent slipping and that the coefficient of kinetic friction on the incline between A and B is $\mu_k = 0.3$.

Problems 3.16 and 3.17

A vehicle is stuck on the railroad tracks as a 430,000 lb locomotive is approaching with a speed of 75 mph. As soon as the problem is detected, the locomotive's emergency brakes are activated, locking the wheels and causing the locomotive to slide.

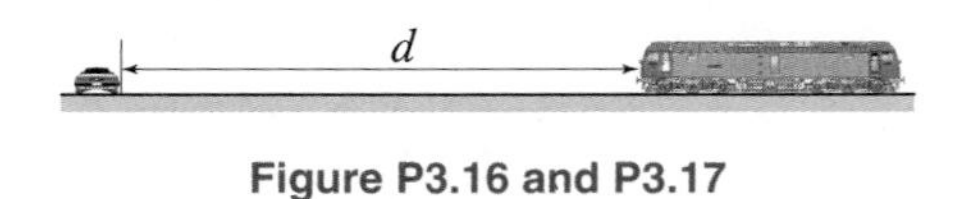

Figure P3.16 and P3.17

Problem 3.16 If the coefficient of kinetic friction between the locomotive and the track is 0.45, what is the minimum distance d at which the brakes must be applied to avoid a collision? What would that distance be if instead of a locomotive, there was a 30×10^6 lb train? Treat the locomotive and the train as particles, assume that the railroad tracks are rectilinear and horizontal, and note that only the locomotive's brakes are applied.

Problem 3.17 Continue Prob. 3.16 and determine the time required to stop the locomotive.

Problem 3.18

As the skydiver moves downward with a speed v, the air drag exerted by the parachute on the skydiver has a magnitude $F_d = C_d v^2$ (C_d is a drag coefficient) and a direction opposite to the direction of motion. Determine the expression of the skydiver's acceleration in terms of C_d, v, the mass of the skydiver m, and the acceleration due to gravity.

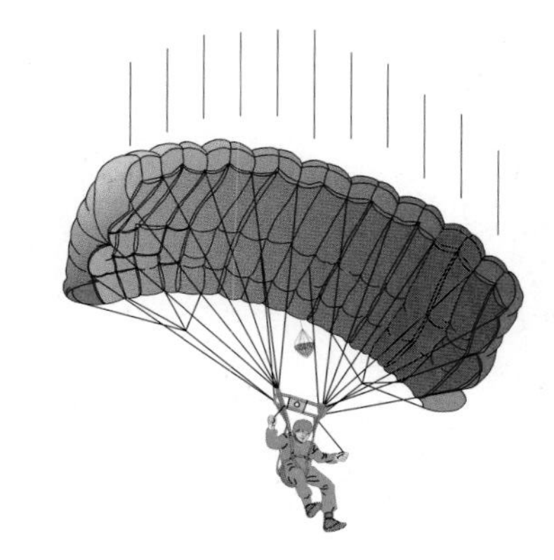

Figure P3.18

Problem 3.19

A car is driving down a 23° rough incline at 55 km/h when its brakes are applied. Treating the car as a particle and neglecting all forces except gravity and friction, determine the stopping distance if

(a) The tires slide and the coefficient of kinetic friction between the tires and the road is 0.7.

(b) The car is equipped with antilock brakes and the tires do not slide. Use 0.9 for the coefficient of static friction between the tires and the road.

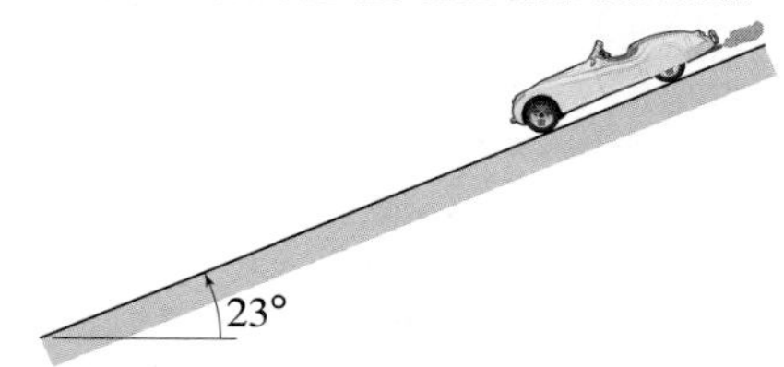

Figure P3.19

Problems 3.20 through 3.22

The truck shown is traveling at $v_0 = 60$ mph when the driver applies the brakes to come to a stop. The deceleration of the truck is constant, and the truck comes to a complete stop after braking for a distance of 350 ft. Treat the crate as a particle so that tipping can be neglected.

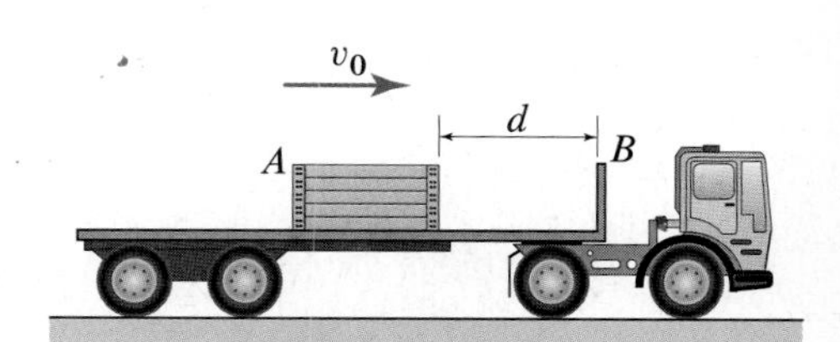

Figure P3.20–P3.22

Problem 3.20 Determine the minimum coefficient of static friction between the crate A and the truck so that the crate does *not* slide relative to the truck.

Problem 3.21 If the coefficient of kinetic friction between the crate A and the bed of the truck is 0.3 and static friction is not sufficient to prevent slip, determine the minimum distance d between the crate and the truck B so that the crate never hits the truck at B.

Problem 3.22 If the coefficient of kinetic friction between the crate A and the bed of the truck is 0.3, static friction is not sufficient to prevent slip, and the distance d from the front of the crate to the truck at B is 10 ft, determine the speed *relative to the truck* with which the crate strikes the truck at B.

Problem 3.23

A metal ball with mass $m = 0.15$ kg is dropped from rest in a fluid. The magnitude of the resistance due to the fluid is given by $C_d v$, where C_d is a drag coefficient and v is the ball's speed. If $C_d = 2.1$ kg/s, determine the ball's speed 4 s after release.

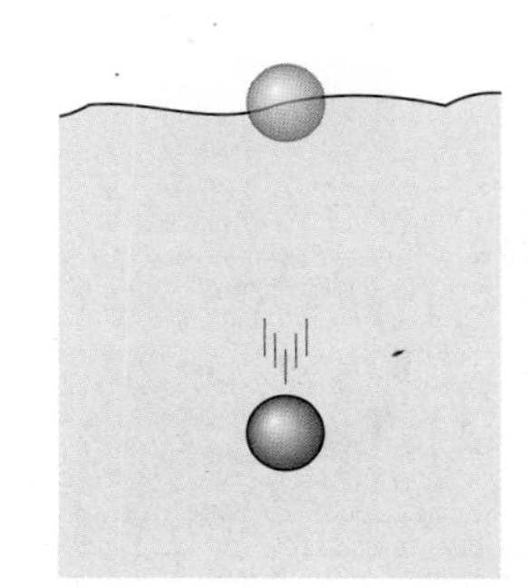

Figure P3.23 and P3.24

Problem 3.24

A metal ball weighing 0.35 lb is dropped from rest in a fluid. The magnitude of the resistance due to the fluid is given by $C_d v$, where C_d is a drag coefficient and v is the ball's speed. It is observed that 2 s after release, the speed of the ball is 25 ft/s. Determine the value of C_d.

Problem 3.25

A horse is lifting a 500 lb crate by moving to the right at a constant speed $v_0 = 3\text{ ft/s}$. Observing that B is fixed and letting $h = 6\text{ ft}$ and $\ell = 14\text{ ft}$, determine the tension in the rope when the horizontal distance d between B and point A on the horse is 10 ft. Treat all rope segments between B and C as vertical, and ignore the change in the amount of rope that wraps around the pulley at B.

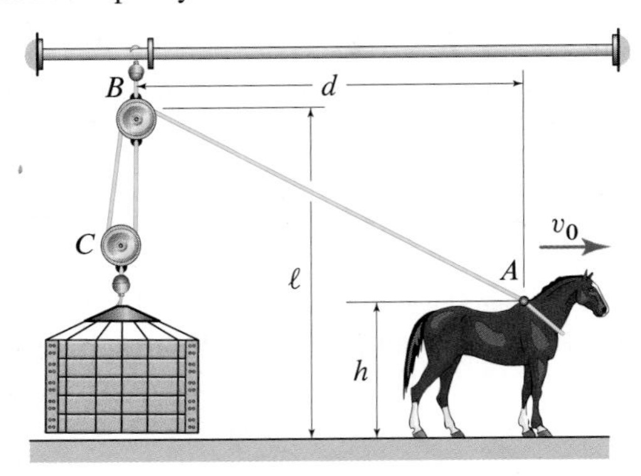

Figure P3.25

Problem 3.26

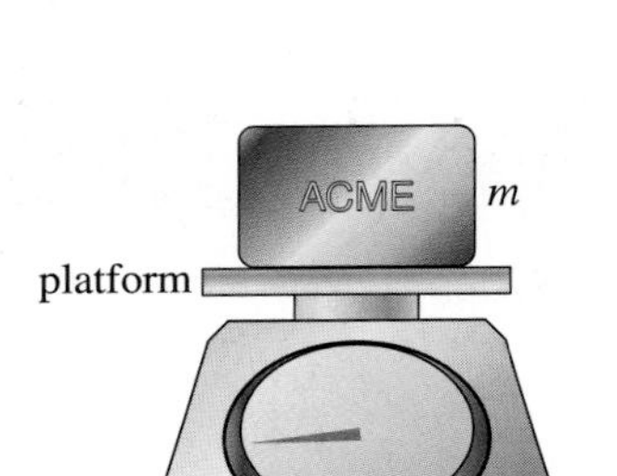

Figure P3.26

The centers of two spheres A and B with masses $m_A = 1\text{ kg}$ and $m_B = 2\text{ kg}$ are a distance $r_0 = 1\text{ m}$ apart. B is fixed in space, and A is initially at rest. Using Eq. (1.5) on p. 3, which is Newton's universal law of gravitation, determine the speed with which A impacts B if the radii of the two spheres are $r_A = 0.05\text{ m}$ and $r_B = 0.15\text{ m}$. Assume that the two masses are infinitely far from any other mass so that they are only influenced by their mutual attraction.

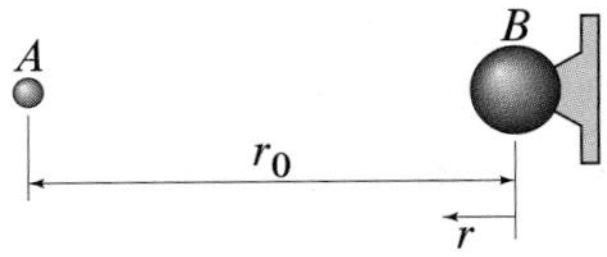

Problems 3.27 and 3.28

Figure P3.27 and P3.28

Spring scales work by measuring the displacement of a spring that supports both the platform and the object, of mass m, whose weight is being measured. Neglect the mass of the platform on which the mass sits and assume that the spring is uncompressed before the mass is placed on the platform. In addition, assume that the spring is linear elastic with spring constant k.

Problem 3.27 If the mass m is gently placed on the spring scale (i.e., it is released from zero height above the scale), determine the *maximum reading* on the scale after the mass is released.

Problem 3.28 If the mass m is gently placed on the spring scale (i.e., it is released from zero height above the scale), determine the *maximum speed* attained by the mass m as the spring compresses.

Problem 3.29

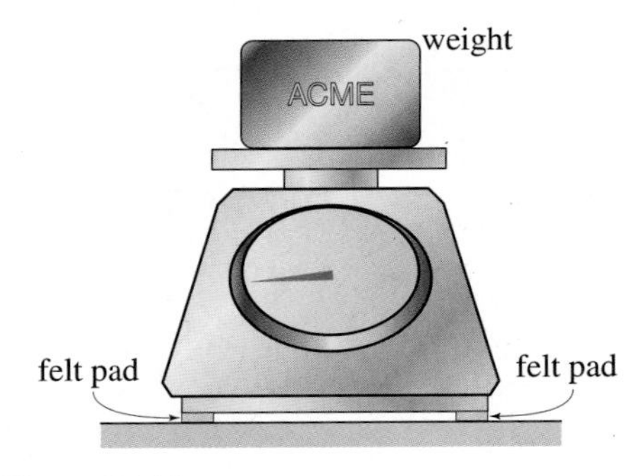

Figure P3.29

A scale is to be used on a wood countertop made from some very nice Brazilian cherry. To protect the countertop, the owner attaches self-sticking felt pads to the feet of the scale. When the weight was placed on the scale before the felt pads were applied, the scale read a certain value. Will the value be higher, lower, or the same when the same weight is placed on the scale but with the felt pads between the scale and the countertop? Ignore the transient dynamic effects that occur immediately after the weight is placed on the scale.

Problems 3.30 through 3.32

Car bumpers are designed to limit the extent of damage to the car in the case of low-velocity collisions. Consider a 3300 lb passenger car impacting a concrete barrier while traveling at a speed of 4.0 mph. Model the car as a particle, and consider two types of bumpers: (1) a simple linear spring with constant k and (2) a linear spring of constant k in parallel with a shock-absorbing unit generating a nearly constant force of 700 lb over 0.25 ft.

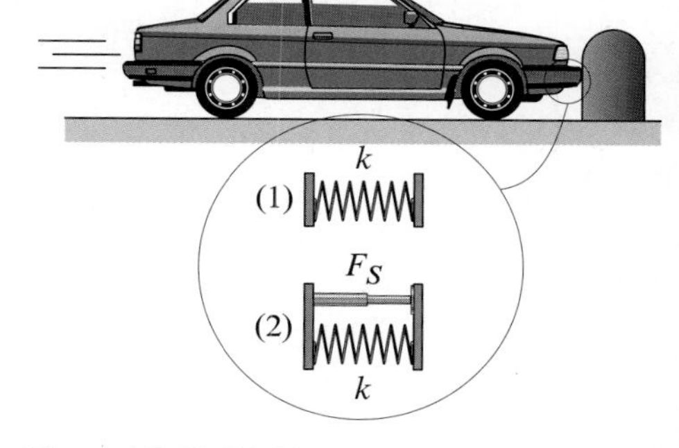

Figure P3.30–P3.32

Problem 3.30 If the bumper is of type 1 and if $k = 6500\,\text{lb/ft}$, find the spring compression necessary to stop the car.

Problem 3.31 If the bumper is of type 1, find the value of k necessary to stop the car when the bumper is compressed 0.25 ft.

Problem 3.32 If the bumper is of type 2, find the value of k necessary to stop the car when the bumper is compressed 0.25 ft.

Problem 3.33

What would it mean for the static or kinetic friction coefficients to be negative? Is this possible? Can either the static or kinetic friction coefficients be greater than 1? If yes, explain and give an example.

Problem 3.34

Packages for transporting delicate items (e.g., a laptop or glass) are designed to "absorb" some of the energy of the impact in order to protect the contents. These energy absorbers can get pretty complicated to model (e.g., the mechanics of styrofoam peanuts is not easy), but we can begin to understand how they work by modeling them as a linear elastic spring of constant k that is placed between the contents (an expensive vase) of mass m and the package P. Assuming that $m = 3\,\text{kg}$ and that the box is dropped from a height of 1.5 m, determine the magnitude of the maximum displacement of the vase relative to the box, as well as the magnitude of the maximum force exerted on the vase by the packaging if $k = 3500\,\text{N/m}$. Treat the vase as a particle, and neglect all forces except for gravity and the spring force. Assume that the spring relaxes after the box is dropped and that it does not oscillate.

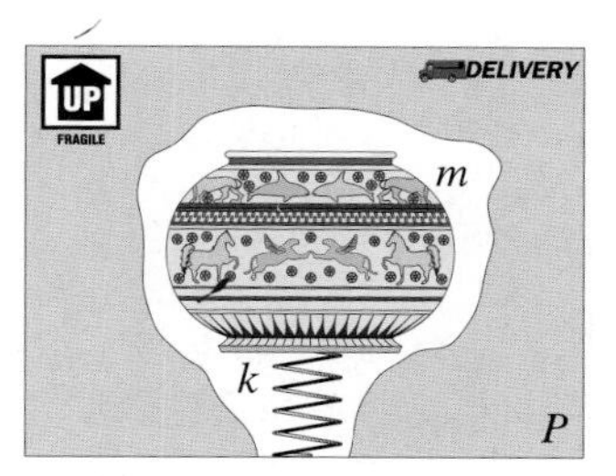

Figure P3.34

Problems 3.35 and 3.36

Car bumpers are designed to limit the extent of damage to the car in the case of low-velocity collisions. Consider a passenger car impacting a concrete barrier while traveling at a speed of 4.0 mph. Model the car as a particle of mass m, and assume that the bumper has a spring element in parallel with a shock absorber so that the overall force exerted by the bumper is $F_B = k\delta + \eta\dot{\delta}$, where k, δ, and η denote the spring constant, the spring compression, and the bumper damping coefficient, respectively.

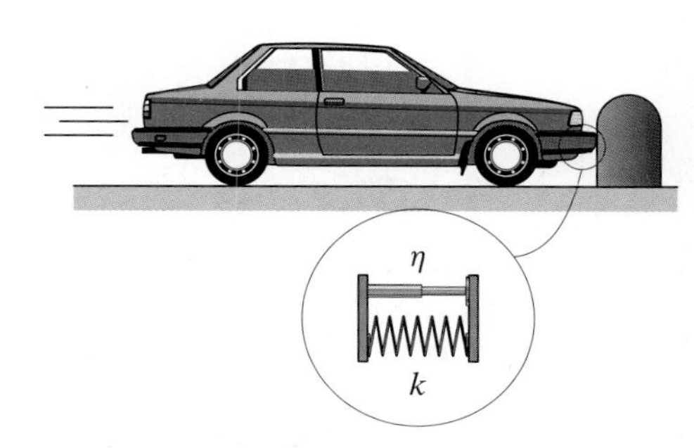

Figure P3.35 and P3.36

Problem 3.35 Derive the equations of motion for the car during the collision.

Problem 3.36 Let the weight of the car be 3300 lb, $k = 6500\,\text{lb/ft}$, and $\eta = 300\,\text{lb·s/ft}$, and let the car be traveling at 4.0 mph at impact. Determine the maximum compression of the bumper necessary to bring the car to a stop. Also determine the time required to stop the car.

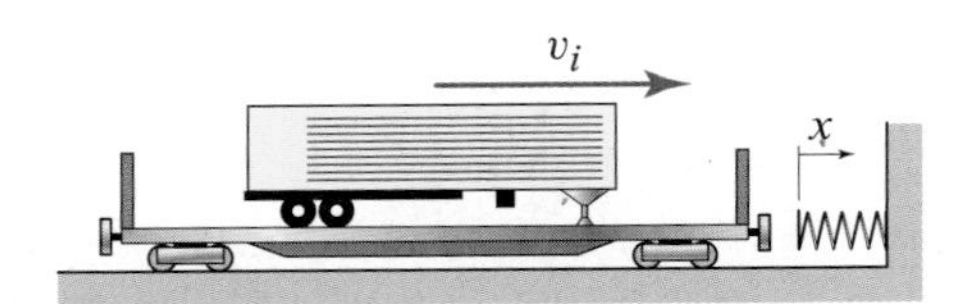

Figure P3.37–P3.39

Problems 3.37 through 3.39

A railcar with an overall mass of 75,000 kg traveling with a speed v_i is approaching a barrier equipped with a bumper consisting of a nonlinear spring whose force vs. compression law is given by $F_s = \beta x^3$, where $\beta = 640 \times 10^6\,\mathrm{N/m^3}$ and x is the compression of the bumper.

Problem 3.37 Treating the system as a particle and assuming that the contact between the railcar and rails is frictionless, determine the maximum value of v_i so that the compression of the bumper is limited to 20 cm.

Problem 3.38 Treating the system as a particle, assuming that the contact between railcar and rails is frictionless, and letting $v_i = 6\,\mathrm{km/h}$, determine the bumper compression necessary to bring the railcar to a stop.

Problem 3.39 Treating the system as a particle, assuming that the contact between the railcar and rails is frictionless, and letting $v_i = 6\,\mathrm{km/h}$, determine how long it takes for the bumper to bring the railcar to a stop.

Problem 3.40

A 6 lb collar is constrained to travel along a rectilinear and frictionless bar of length $L = 5\,\mathrm{ft}$. The springs attached to the collar are identical and are unstretched when the collar is at B. Treating the collar as a particle, neglecting air resistance, and knowing that at A the collar is moving to the right with a speed of 11 ft/s, determine the linear spring constant k so that the collar reaches D with zero speed. Points E and F are fixed.

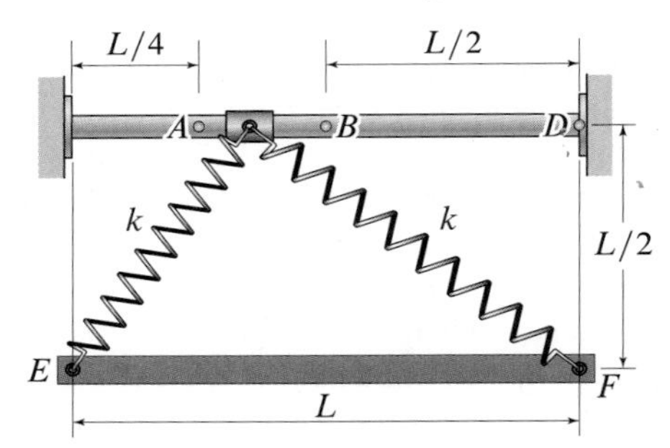

Figure P3.40 and P3.41

Problem 3.41

A 10 lb collar is constrained to travel along a rectilinear and frictionless bar of length $L = 5\,\mathrm{ft}$. The springs attached to the collar are identical, they have a spring constant $k = 4\,\mathrm{lb/ft}$, and they are unstretched when the collar is at B. Treating the collar as a particle, neglecting air resistance, and knowing that at A the collar is moving to the right with a speed of 14 ft/s, determine the speed with which the collar arrives at D. Points E and F are fixed.

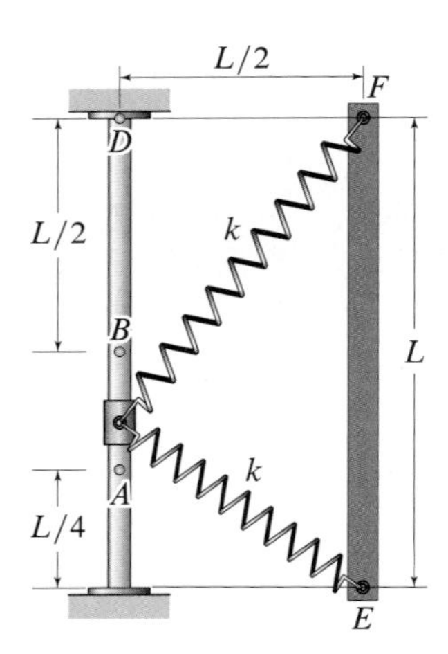

Figure P3.42

Problem 3.42

An 11 kg collar is constrained to travel along a rectilinear and frictionless bar of length $L = 2\,\mathrm{m}$. The springs attached to the collar are identical, and they are unstretched when the collar is at B. Treating the collar as a particle, neglecting air resistance, and knowing that at A the collar is moving upward with a speed of 23 m/s, determine the linear spring constant k so that the collar reaches D with zero speed. Points E and F are fixed.

Problem 3.43

Derive the equation of motion of the mass m released from rest at $x = x_0$ from the slingshot-like device (the mass m is attached to the elastic cords). Assume that the cords connecting the mass to the device are linear springs with spring constant k and unstretched length L_0. In addition, assume that the mass is equidistant from the two supports and that the mass and both springs lie in the xy plane. Ignore gravity and assume $L > L_0$.

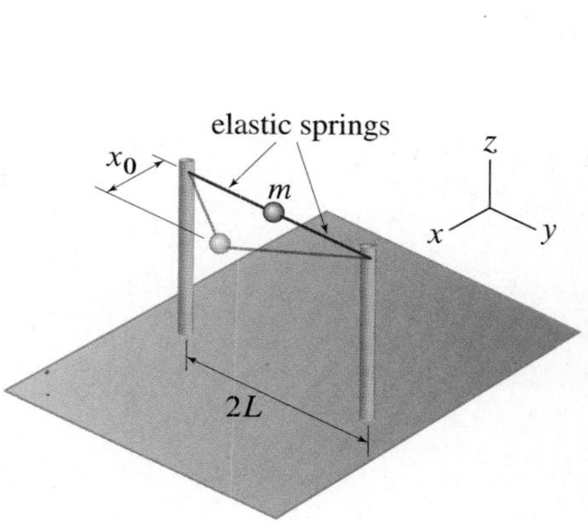

Figure P3.43–P3.46

Problem 3.44

Determine the speed of the mass m when it reaches $x = 0$ if it is released from rest at $x = x_0$ from the device in Prob. 3.43. Assume that the cords connecting the mass to the device are linear springs with spring constant k and unstretched length L_0. Ignore gravity and assume $L > L_0$.

Problem 3.45

Given the approximation

$$\frac{1}{\sqrt{(x/L)^2 + 1}} \approx 1 - \frac{(x/L)^2}{2}, \tag{1}$$

show that the equation of motion for the mass m when it is released from rest at $x = x_0$ from the device in Prob. 3.43 can be written as

$$\ddot{x} + \omega_0^2 x(1 + \Lambda x^2) = 0, \tag{2}$$

where

$$\omega_0^2 = \frac{2k}{m}\left(\frac{L - L_0}{L}\right) \quad \text{and} \quad \Lambda = \frac{L_0}{2L^2(L - L_0)}. \tag{3}$$

Assume that the cords connecting the mass to the device are linear springs with spring constant k, length L, and unstretched length L_0. Ignore gravity and assume $L > L_0$. Equation (2) is a famous equation in mechanics called *Duffing's equation*.

Problem 3.46

The force on the mass for the device in Prob. 3.43 and the force on the mass in the Duffing equation obtained from the device equation of motion (and defined by Eqs. (2) and (3)) can be plotted as a function of x. The nature of that force depends on whether or not the springs are initially stretched ($L > L_0$) or initially compressed ($L < L_0$). The figure shows the elastic restoring force on the mass m as a function of the displacement x for four different cases:

- Force for device with $L = 2$ and $L_0 = 1$.
- Force for device with $L = 1$ and $L_0 = 2$.
- Force for the Duffing equation with $L = 2$ and $L_0 = 1$.
- Force for the Duffing equation with $L = 1$ and $L_0 = 2$.

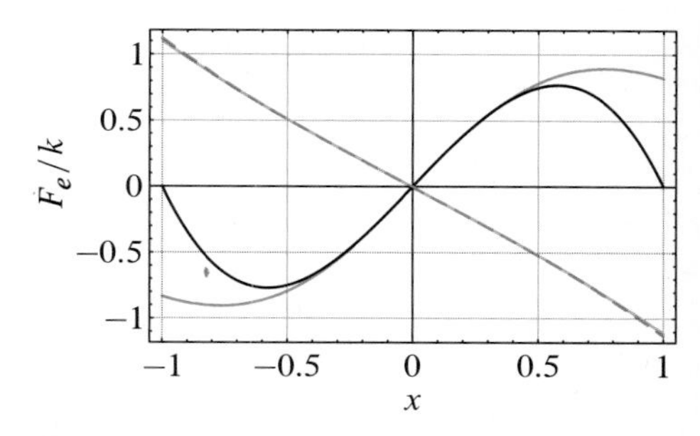

Figure P3.46

For small x, the force given by the Duffing equation is a good approximation to the force in the device. Explain which of the curves corresponds to a "hardening" spring (a spring that gets stiffer as you pull it) and which corresponds to a "softening" spring (a spring that gets less stiff as you pull it), and explain physically why we see this behavior.

3.2 Curvilinear Motion

Figure 3.8
Marco Andretti at the 2007 Indianapolis 500. When going around a turn, the Formula One cars are approximately in planar motion, but are not traveling in a straight line.

This section builds on what we learned in Section 3.1 by applying Newton's second law to systems undergoing curvilinear motion. No new material is introduced, thus, allowing us to strengthen our modeling and problem-solving skills. The application of Newton's second law to curvilinear problems is formally identical to what we saw in Section 3.1 with the same four-step problem-solving procedure.

Newton's second law in 2D and 3D component systems

When applying $\vec{F} = m\vec{a}$, it is done in component form, using Eq. (3.2) on p. 170, after choosing a convenient component system. For systems in curvilinear motion, we will choose one or more component systems from those described in Chapter 2. We now present Newton's second law as it appears in each component system studied in Chapter 2.

Two-dimensional component systems

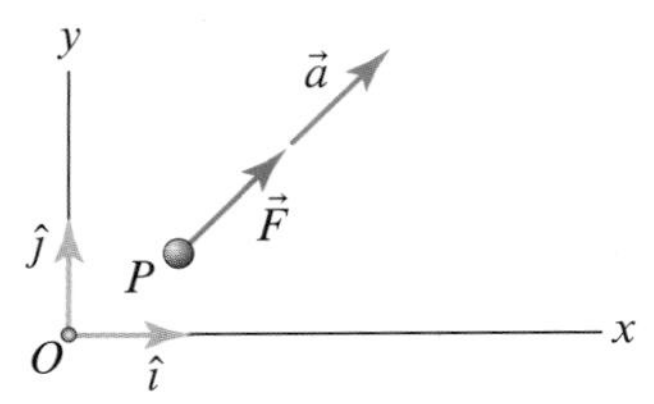

Figure 3.9
Force, acceleration, and a Cartesian component system.

Cartesian components. Referring to Fig. 3.9, Newton's second law in planar Cartesian components is

$$\sum F_x = ma_x \quad \text{and} \quad \sum F_y = ma_y. \tag{3.15}$$

The associated kinematic equations are given by Eqs. (2.17), which imply

$$a_x = \ddot{x} \quad \text{and} \quad a_y = \ddot{y}. \tag{3.16}$$

Path components. Referring to Fig. 3.10, Newton's second law in planar path components is

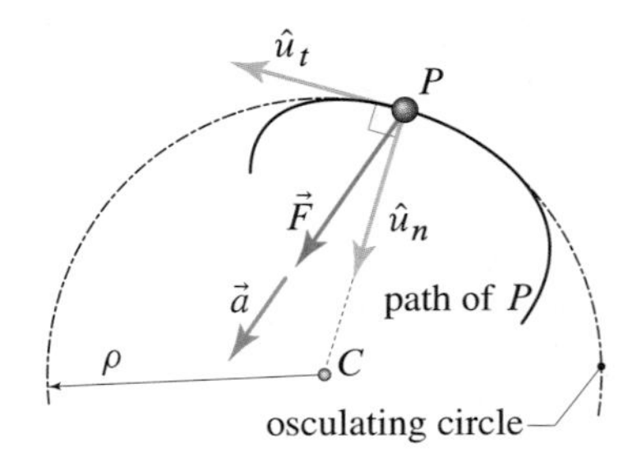

Figure 3.10. Force, acceleration, and a path component system.

$$\sum F_n = ma_n \quad \text{and} \quad \sum F_t = ma_t. \tag{3.17}$$

The associated kinematic equations are given by Eqs. (2.78), which imply

$$a_n = \frac{v^2}{\rho} \quad \text{and} \quad a_t = \dot{v}. \tag{3.18}$$

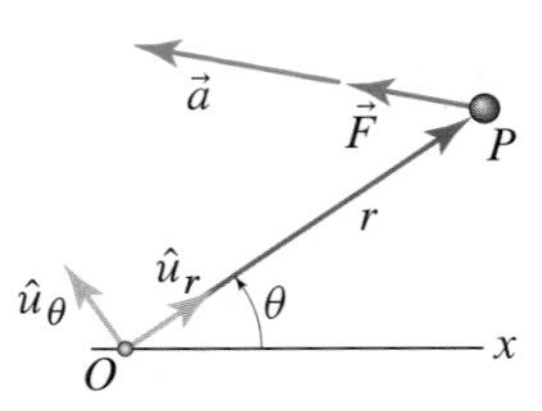

Figure 3.11
Force, acceleration, and a polar component system.

Polar components. Referring to Fig. 3.11, Newton's second law in planar polar components is

$$\sum F_r = ma_r \quad \text{and} \quad \sum F_\theta = ma_\theta. \tag{3.19}$$

The associated kinematic equations are given by Eqs. (2.90), which are

$$a_r = \ddot{r} - r\dot{\theta}^2 \quad \text{and} \quad a_\theta = r\ddot{\theta} + 2\dot{r}\dot{\theta}. \tag{3.20}$$

Three-dimensional component systems

Cartesian components. Referring to Fig. 3.12, Newton's second law in three-

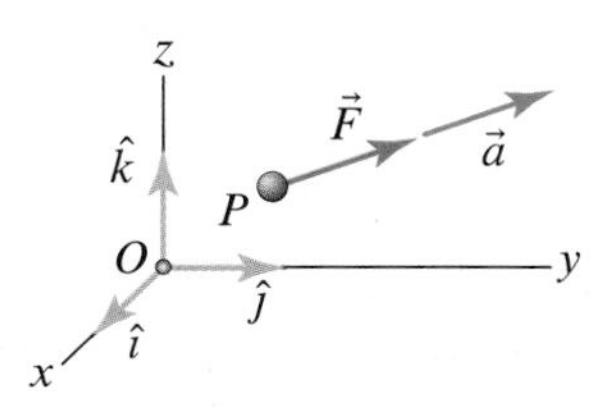

Figure 3.12. Force, acceleration, and a 3D Cartesian component system.

dimensional Cartesian components is

$$\sum F_x = ma_x, \quad \sum F_y = ma_y, \quad \text{and} \quad \sum F_z = ma_z. \tag{3.21}$$

The associated kinematic equations are given by Eqs. (2.143), which imply

$$a_x = \ddot{x}, \quad a_y = \ddot{y}, \quad \text{and} \quad a_z = \ddot{z}. \tag{3.22}$$

Path components. Referring to Fig. 3.13, Newton's second law in three-dimensional path components is

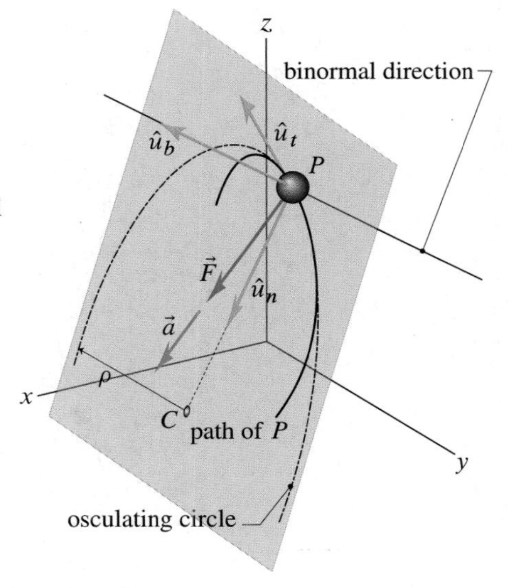

Figure 3.13
Force, acceleration, and a 3D path component system.

$$\sum F_n = ma_n, \quad \sum F_t = ma_t, \quad \text{and} \quad \sum F_b = ma_b. \tag{3.23}$$

The associated kinematic equations are given by Eqs. (2.78), which imply

$$a_n = \frac{v^2}{\rho}, \quad a_t = \dot{v}, \quad \text{and} \quad a_b = 0, \tag{3.24}$$

where $a_b = 0$ because, by definition, $\hat{u}_b$ is perpendicular to the plane containing the velocity *and* the acceleration vectors (see Section 2.5).

Cylindrical components. Referring to Fig. 3.14, Newton's second law in three-

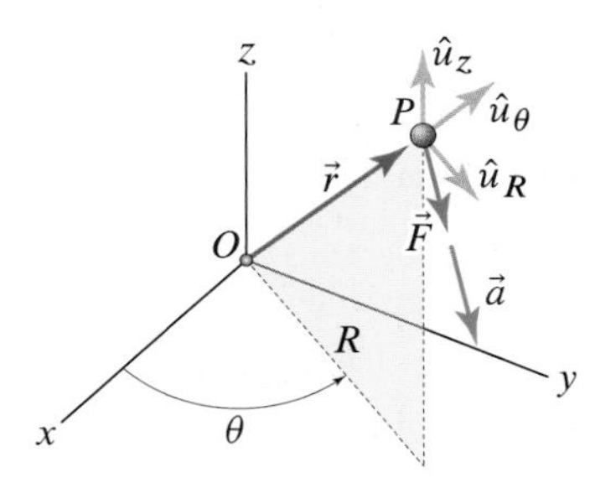

Figure 3.14. Force, acceleration, and a cylindrical component system.

dimensional cylindrical components is

$$\sum F_R = ma_R, \quad \sum F_\theta = ma_\theta, \quad \text{and} \quad \sum F_z = ma_z. \tag{3.25}$$

The associated kinematic equations are given by Eqs. (2.124), which are

$$a_R = \ddot{R} - R\dot{\theta}^2, \quad a_\theta = R\ddot{\theta} + 2\dot{R}\dot{\theta}, \quad \text{and} \quad a_z = \ddot{z}. \tag{3.26}$$

Spherical components. Referring to Fig. 3.15, Newton's second law in three-

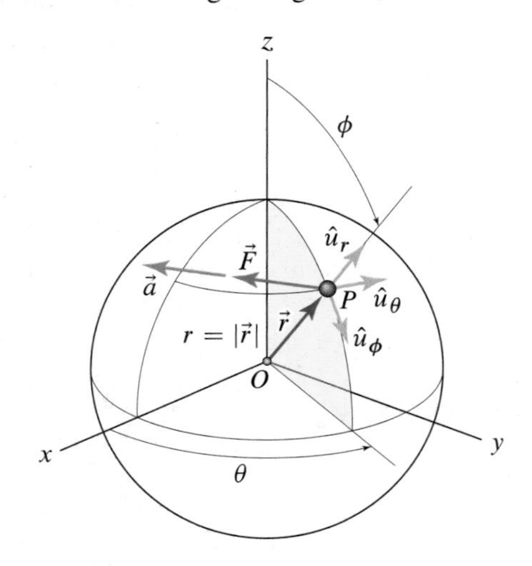

Figure 3.15. Force, acceleration, and a spherical component system.

dimensional spherical components is

$$\sum F_r = ma_r, \quad \sum F_\phi = ma_\phi, \quad \text{and} \quad \sum F_\theta = ma_\theta. \tag{3.27}$$

The associated kinematic equations are given by Eqs. (2.140), which are

$$\begin{aligned} a_r &= \ddot{r} - r\dot{\phi}^2 - r\dot{\theta}^2 \sin^2\phi, \\ a_\phi &= r\ddot{\phi} + 2\dot{r}\dot{\phi} - r\dot{\theta}^2 \sin\phi\cos\phi, \\ a_\theta &= r\ddot{\theta}\sin\phi + 2\dot{r}\dot{\theta}\sin\phi + 2r\dot{\phi}\dot{\theta}\cos\phi. \end{aligned} \tag{3.28}$$

EXAMPLE 3.6 Tension in a Wrecking Ball Cable

The wrecking ball A shown in Fig. 1 is released from rest when $\theta = \theta_0 = 30°$, and it swings freely about the fixed point at O. Assuming that the weight of the ball is $W = 2500$ lb and $L = 30$ ft, determine the tension in the cable to which the ball is attached when the ball reaches $\theta = 0°$.

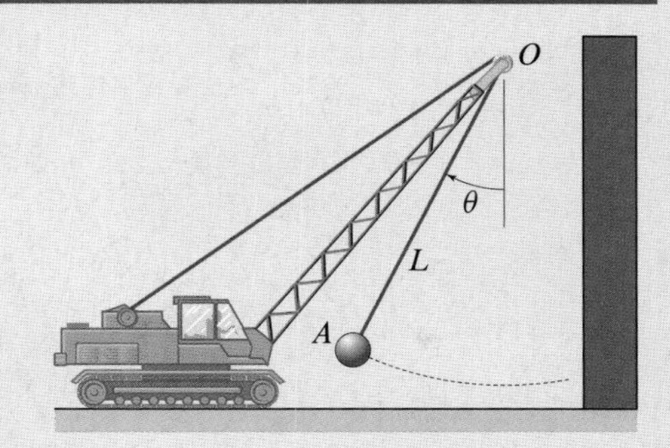

Figure 1

SOLUTION

Road Map & Modeling Modeling the wrecking ball as a particle and neglecting all forces except the weight force W and the cable tension T, the FBD is as shown in Fig. 2. Applying Newton's second law in the polar component system shown should allow us to find the tension in the cable as a function of its swing angle and thus, find its tension when $\theta = 0°$.

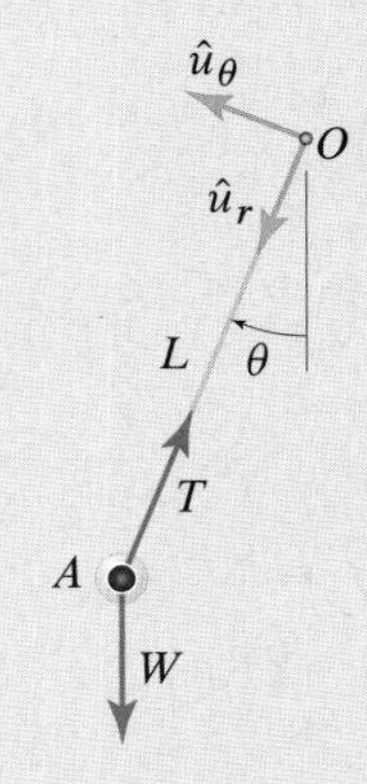

Figure 2
FBD of the wrecking ball as it swings downward.

Governing Equations

Balance Principles Referring to the FBD in Fig. 2 and applying Newton's second law, we obtain

$$\sum F_\theta: \quad -W \sin\theta = m a_\theta, \tag{1}$$

$$\sum F_r: \quad W\cos\theta - T = m a_r, \tag{2}$$

where $m = W/g$.

Force Laws All forces are accounted for on the FBD.

Kinematic Equations Writing a_θ and a_r in polar components gives

$$a_\theta = r\ddot{\theta} + 2\dot{r}\dot{\theta} = L\ddot{\theta} \quad \text{and} \quad a_r = \ddot{r} - r\dot{\theta}^2 = -L\dot{\theta}^2, \tag{3}$$

where we have replaced r with the constant length L.

Computation Substituting Eqs. (3) into Eqs. (1) and (2), we obtain

$$-W\sin\theta = mL\ddot{\theta} \quad \text{and} \quad W\cos\theta - T = -mL\dot{\theta}^2$$

$$\Rightarrow \quad \ddot{\theta} = -\frac{g}{L}\sin\theta \quad \text{and} \quad T = W\cos\theta + mL\dot{\theta}^2. \tag{4}$$

Notice that the tension is a function of $\dot{\theta}$, so we need to integrate $\ddot{\theta}(\theta)$ to find $\dot{\theta}(\theta)$ using the chain rule, that is,

$$\ddot{\theta} = \dot{\theta}\frac{d\dot{\theta}}{d\theta} = -\frac{g}{L}\sin\theta \quad \Rightarrow \quad \int_0^{\dot{\theta}} \dot{\theta}\, d\dot{\theta} = -\frac{g}{L}\int_{\theta_0}^{\theta} \sin\theta\, d\theta$$

$$\Rightarrow \quad \dot{\theta}^2 = 2\frac{g}{L}(\cos\theta - \cos\theta_0). \tag{5}$$

Substituting Eq. (5) into the expression for T in Eq. (4) gives $T(\theta)$ as

$$T = W(3\cos\theta - 2\cos\theta_0) \quad \Rightarrow \quad \boxed{T(\theta = 0) = W(3 - 2\cos\theta_0) = 3170\,\text{lb},} \tag{6}$$

where we have used $W = 2500$ lb and $\theta_0 = 30°$ to obtain the final numerical result.

Discussion & Verification The final result in Eq. (6) is dimensionally correct, and the magnitude of the tension seems reasonable. Interestingly, the tension does not depend on the length of the supporting cable. That is, if the initial angle is 30° and the wrecking ball is released from rest, the tension in the cable will always be 3170 lb, regardless of the length of the suspending cable.

EXAMPLE 3.7 *Projectile Motion with Drag*

Figure 1
Rory McIlroy of Northern Ireland hits a shot in the 17th fairway during the second round of the 111th U.S. Open.

The projectile motion model presented in Section 2.3, in which a projectile is subject only to constant gravity and the trajectory is a parabola, is not adequate for studying the trajectory of objects like golf balls. In general, the trajectory of a golf ball is not a parabola and is affected by many factors, such as the dimple pattern on the ball, the spin imparted to the ball, and the local density of the air. Here we can consider a simple improvement on the model in Section 2.3 by lumping the effects just mentioned into an aerodynamic drag, which we assume to be proportional to the square of the ball's speed and directed opposite to the ball's velocity. Use this model to derive the equations of motion of the golf ball.

SOLUTION

Road Map & Modeling We model the golf ball as a particle subject to its own weight mg and a drag force F_d (Fig. 2). We choose a Cartesian component system with the y axis parallel to gravity because it simplifies the representation of the weight force and it allows for an easy comparison between the current model and that in Section 2.3. The direction of the drag force is opposite to the ball's velocity, which is in the $\hat{u}_t$ direction. Although the orientation of $\hat{u}_t$ is currently unknown, for convenience we have oriented it using the time-dependent angle θ. As we saw in Section 3.1, the ball's equations of motion are derived by combining Newton's second law with the force laws and the kinematic equations.

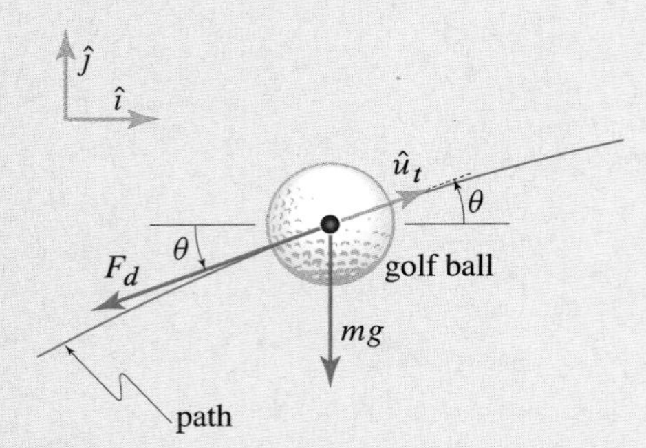

Figure 2
FBD of a golf ball in flight, modeled as a particle and subject to gravity and the aerodynamic drag force F_d. The unit vector $\hat{u}_t$ is tangent to the ball's path and indicates the direction of the ball's velocity.

Governing Equations

Balance Principles Using the FBD in Fig. 2, Newton's second law, in component form, gives

$$\sum F_x: \quad -F_d \cos\theta = m a_x, \tag{1}$$

$$\sum F_y: \quad -F_d \sin\theta - mg = m a_y. \tag{2}$$

Force Laws Because F_d is proportional to the square of the speed, we have

$$F_d = C_d v^2, \tag{3}$$

where C_d is a *drag coefficient** and v is the ball's speed.

Kinematic Equations In the chosen component system, we have

$$a_x = \ddot{x} \quad \text{and} \quad a_y = \ddot{y}. \tag{4}$$

In addition, because the ball's velocity can be written as $\vec{v} = \dot{x}\,\hat{\imath} + \dot{y}\,\hat{\jmath}$ and the speed is the magnitude of $\vec{v}$, we have

$$v = \sqrt{\dot{x}^2 + \dot{y}^2}. \tag{5}$$

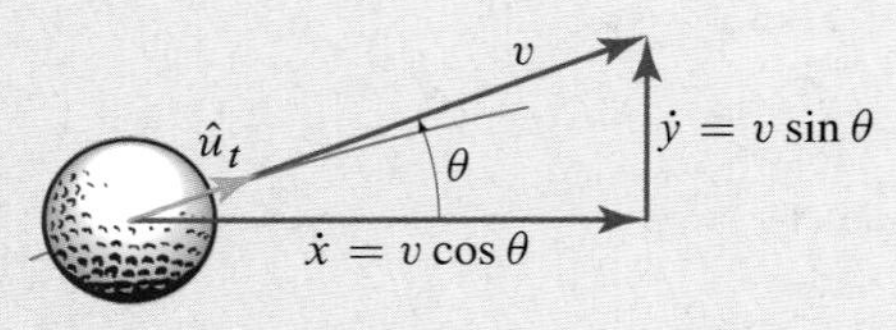

Figure 3
Components of the velocity vector of the golf ball.

Finally, given that θ is the orientation of $\vec{v}$ relative to the x direction, we have

$$\cos\theta = \frac{\dot{x}}{v} = \frac{\dot{x}}{\sqrt{\dot{x}^2 + \dot{y}^2}} \quad \text{and} \quad \sin\theta = \frac{\dot{y}}{v} = \frac{\dot{y}}{\sqrt{\dot{x}^2 + \dot{y}^2}}, \tag{6}$$

where the components of $\vec{v}$ are depicted in Fig. 3.

* The coefficient C_d in Eq. (3) has dimensions of mass over length and should not be confused with the nondimensional drag coefficient normally used in aerodynamics.

Computation Substituting Eqs. (3)–(6) into Eqs. (1) and (2), we obtain

$$-C_d \dot{x} \sqrt{\dot{x}^2 + \dot{y}^2} = m\ddot{x}, \tag{7}$$

$$-C_d \dot{y} \sqrt{\dot{x}^2 + \dot{y}^2} - mg = m\ddot{y}. \tag{8}$$

Equations (7) and (8) are the equations of motion for this problem.

Discussion & Verification Since the dimensions of C_d are mass over length, Eqs. (7) and (8) are dimensionally correct. In addition, referring to Eq. (7), note that for a positive $\dot{x}$, the left-hand side is negative, thus, indicating that the effect of the drag is that of slowing the particle down, as expected. A similar argument can be made for Eq. (8). Finally, note that if we set $C_d = 0$ in Eqs. (7) and (8), we recover the equations of motion of a projectile according to the model in Section 2.3, namely, $\ddot{x} = 0$ and $\ddot{y} = -g$. Therefore, we can conclude that, overall, the equations of motion we have derived appear to be correct.

A Closer Look When $C_d \neq 0$, Eqs. (7) and (8) cannot be solved analytically. However, they can easily be solved numerically using mathematical software as long as we provide values for m, g, and C_d, as well as the initial position and velocity of the ball. A typical golf ball weighs 1.61 oz, corresponding to a mass of 3.125×10^{-3} slug. To estimate C_d, the authors have calibrated the current model, using experimental data pertaining to the swing of Tiger Woods who, using a driver, imparted to a ball an initial speed of 186 mph and an upward direction of 11.2° with respect to the horizontal, and caused the ball to travel on the fly a distance of roughly 270 yd, or 810 ft. Using this information and with the help of a computer, we estimated that $C_d = 4.71 \times 10^{-7}\,\mathrm{lb{\cdot}s^2/ft^2}$. Figure 4 shows a comparison between the trajectory corresponding to the stated values of C_d and the initial conditions and a trajectory with the same initial conditions but with $C_d = 0$. The reduction of the range of the ball due to drag is about 8%.

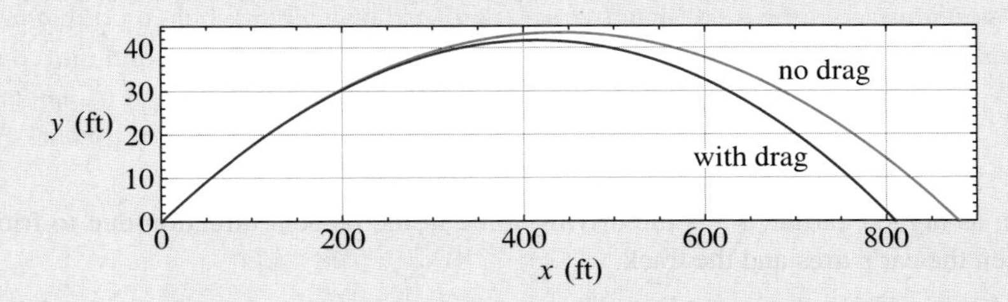

Figure 4. Trajectories of a golf ball with and without drag computed using the parameters and initial conditions stated in the text. To allow the two trajectories to be easily distinguished, the x and y axes have been given different scales.

EXAMPLE 3.11 *Anomalous Acceleration on a Merry-Go-Round*

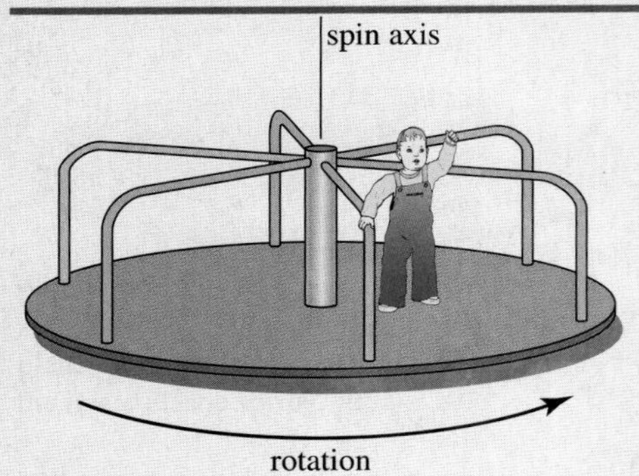

Figure 1
A small child walking radially on a spinning merry-go-round.

Figure 1 shows a typical playground merry-go-round. In walking over the platform while spinning, not only does the child feel like he is being thrown radially outward, but also he feels like he is being thrown sideways. Investigate this motion, and determine the forces required to walk radially at a constant rate v_0 on a platform of radius ρ, while spinning at constant angular rate ω_0.

SOLUTION

Road Map & Modeling We are told how a child moves, and we are asked to find what forces are necessary for the motion to occur as described, which means that we are given the $\vec{a}$ in $\vec{F} = m\vec{a}$. Referring to Fig. 2, we model the child as a particle and describe his

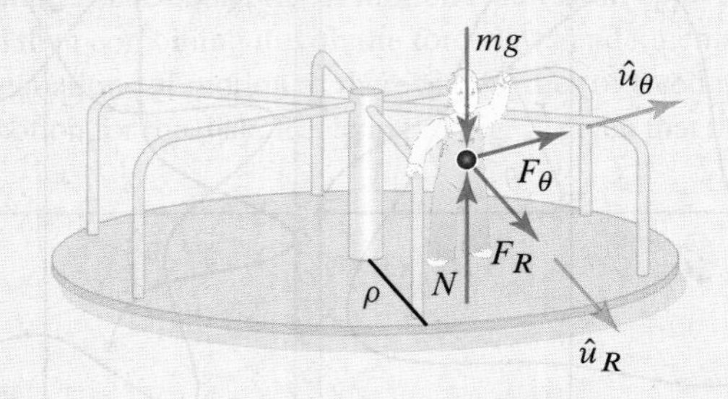

Figure 2. FBD of the child walking radially outward on the merry-go-round.

motion using a cylindrical coordinate system with origin on the spin axis of the merry-go-round. We assume that the child is subject to gravity, namely, mg, the normal reaction N at the floor, and any other forces that might be required for the motion to be as described. This is why the FBD includes the forces F_R and F_θ, in the radial and transverse directions, respectively. Generally, F_R and F_θ will be due to friction between the child and the spinning platform, as well as the fact that the child can hold onto the railings mounted on the merry-go-round.

Governing Equations

Balance Principles Using the FBD in Fig. 2, Newton's second law gives

$$\sum F_R: \quad F_R = ma_R, \tag{1}$$

$$\sum F_\theta: \quad F_\theta = ma_\theta, \tag{2}$$

$$\sum F_z: \quad N - mg = ma_z. \tag{3}$$

Force Laws All known forces are accounted for on the FBD. The forces N, F_R, and F_θ are unknowns of the problem.

Kinematic Equations Using cylindrical components, since $\dot{\theta} = \omega_0$ and $\dot{r} = v_0$ are both constant, we have

$$a_R = \ddot{R} - R\dot{\theta}^2 = -R\omega_0^2, \tag{4}$$

$$a_\theta = R\ddot{\theta} + 2\dot{R}\dot{\theta} = 2v_0\omega_0, \tag{5}$$

$$a_z = \ddot{z} = 0. \tag{6}$$

Computation Equations (3) and (6) simply tell us that $N = mg$. Equations (1), (2), (4), and (5) tell us that

$$F_R = -mR\omega_0^2, \tag{7}$$

$$F_\theta = 2mv_0\omega_0. \tag{8}$$

which are the forces required to keep the child moving in the radial direction at a constant speed as the merry-go-round spins at a constant angular velocity.

Discussion & Verification Equation (7) says that F_R is independent of how fast the child walks and that it always acts inward given that m, r, and ω_0^2 are positive. As intuition might suggest, this corresponds to a physical sensation of being "thrown" outward.* By contrast, F_θ in Eq. (8) can be positive or negative because v_0 and ω_0 can each be positive or negative, thus leading to four possible cases. In each of the cases, when we use right and left to indicate direction, they will be relative to the child, who we assume is facing in the direction of motion.

Case 1 In this case, let $v_0 > 0$ and $\omega_0 > 0$, so that the child is walking *outward* and the merry-go-round is spinning as shown in Fig. 1, i.e., counterclockwise (ccw) as viewed from above. This situation is shown in Fig. 3(a). For this case, Eq. (8) tells us that $F_\theta > 0$ so that the force *applied to the child* while walking *outward* will be to his *left*. The child will then feel as though he is thrown to the *right*.

Case 2 In this case, $v_0 < 0$ and $\omega_0 > 0$, so that the child is walking *inward* and the merry-go-round is spinning counterclockwise (ccw). This situation is shown in Fig. 3(b). For this case, Eq. (8) tells us that $F_\theta < 0$ so that the force *applied to the child* while walking *inward* will again be to his *left*. Once again, the child will then feel as though he is thrown to the *right*.

Case 3 In this case, $v_0 > 0$ and $\omega_0 < 0$, so that the child is walking *outward* and the merry-go-round is now spinning clockwise (cw). This situation is shown in Fig. 3(c). For this case, Eq. (8) tells us that $F_\theta < 0$ so that the force *applied to the child* while walking *outward* will now be to his *right*. This time, the child will then feel as though he is thrown to the *left*.

Case 4 In this case, $v_0 < 0$ and $\omega_0 < 0$, so that the child is walking *inward* and the merry-go-round is again spinning clockwise (cw). This situation is shown in Fig. 3(d). For this case, Eq. (8) tells us that $F_\theta > 0$ so that the force *applied to the child* while walking *inward* will now be to his *right*, and the child will again feel as though he is thrown to the *left*.

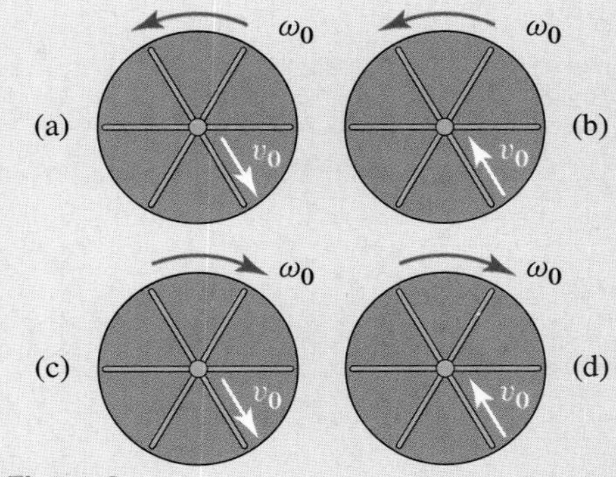

Figure 3
(a) Case 1: $v_0 > 0$, $\omega_0 > 0$; (b) Case 2: $v_0 < 0$, $\omega_0 > 0$; (c) Case 3: $v_0 > 0$, $\omega_0 < 0$; (d) Case 4: $v_0 < 0$, $\omega_0 < 0$.

The analysis of these four cases tells us that whether the child walks inward or outward, as long as the merry-go-round spins counterclockwise (ccw), the child is always thrown to the right. Similarly, whether or not the child walks inward or outward, as long as the merry-go-round spins clockwise (cw), the child is always thrown to the left. It turns out that we can generalize these results for a person walking in *any* direction to state the following:

> *No matter which direction a person walks on a rotating reference frame, if the frame is rotating counterclockwise (ccw), then the person is thrown to his or her right, whereas the person is thrown to her or his left if the reference frame is rotating clockwise (cw).*

This result will be easy to demonstrate once we learn about the kinematics of motion relative to rotating reference frames, which we will cover in Section 6.4.

---

* For example, the force *on* you as you round a curve in a car is in the same direction as your acceleration, i.e., toward the center of the curve. However, the physical sensation you experience is that of being thrown *away* from the center of the curve.

Problems

Problem 3.47

A train is traveling with constant speed v_0 along the level track OAB, which lies in the horizontal plane. The section between points O and A is straight, and the section between points A and B is circular with radius of curvature ρ. The train starts at point O at time $t = 0$, reaches point A at time t_A, and reaches point B at time t_B. For this motion, sketch the magnitude of the acceleration vector as a function of time. Would you want to design a train track with this shape?

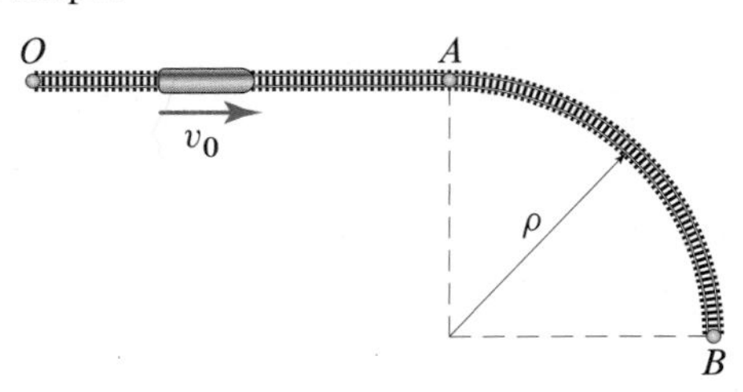

Figure P3.47

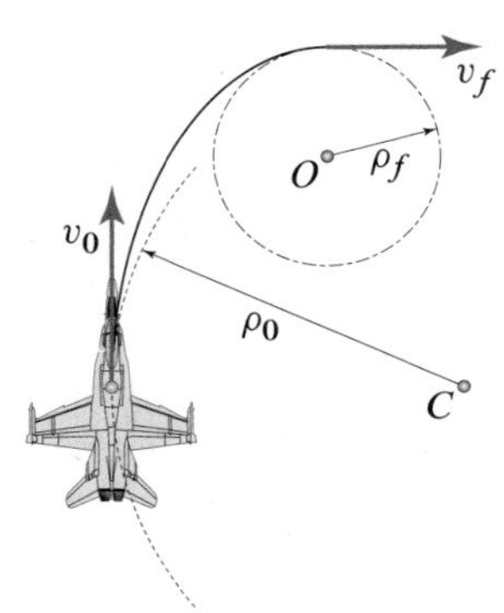

Figure P3.48

Problem 3.48

A plane is turning along a horizontal path at constant speed. What is the component of force in the direction of the path acting on the plane?

Problem 3.49

Body D is in equilibrium when cable AC suddenly breaks. If B is fixed, does the tension in the inextensible cable BC increase or decrease at the instant of release?

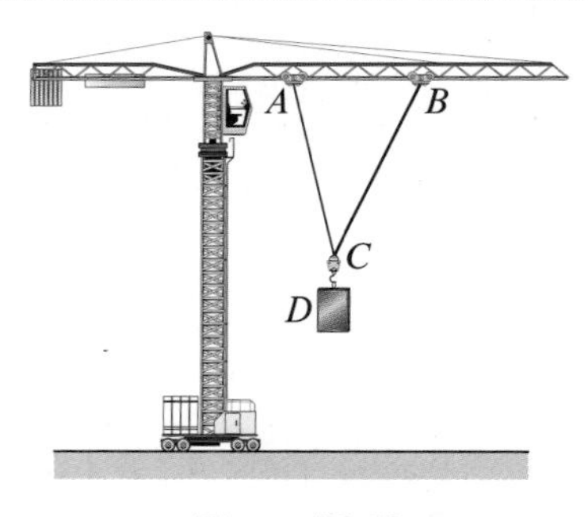

Figure P3.49

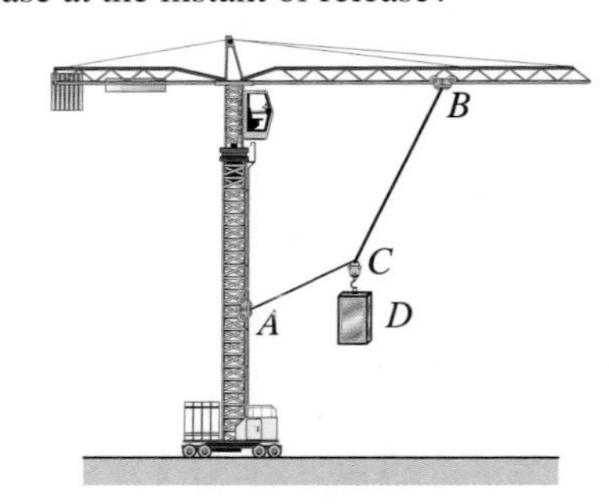

Figure P3.50

Problem 3.50

Body D is in equilibrium when cable AC suddenly breaks. If B is fixed, does the tension in the inextensible cable BC increase or decrease at the instant of release?

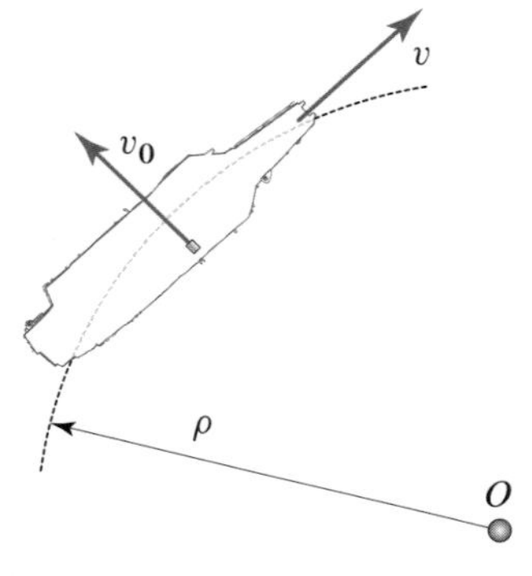

Figure P3.51

Problem 3.51

An aircraft carrier is turning with a constant speed v along a circular path with radius ρ and center O. During the maneuver, a forklift is being driven across the deck of the ship at a constant speed v_0, relative to the deck. Does the friction force between the forklift and the deck have a component perpendicular to the relative velocity of the forklift with respect to the deck?

Problem 3.52

A jet is coming out of a dive, and a sensor in the pilot's seat measures a force of 800 lb for a pilot whose weight is 180 lb. If the jet's instruments indicate that the plane is traveling at 850 mph, determine the radius of curvature of the plane's path at this instant.

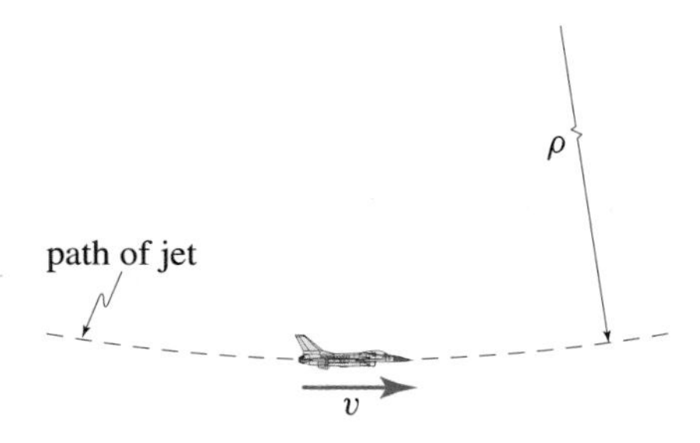

Figure P3.52

Problems 3.53 and 3.54

A partial cross section of an amusement park ride is shown. While the ride spins up to the angular speed ω_c, there is a small platform at F on which the person P stands. Once the ride reaches the desired angular speed, the platform falls away and only friction keeps the person from sliding to the floor of the ride. The wall, against which the person lies, is inclined at the angle $\theta = 15°$ with respect to the vertical. Model the person as a particle that is a distance $d = 20$ ft from the spin axis AB and let the coefficient of static friction between the person and the wall be $\mu_s = 0.7$.

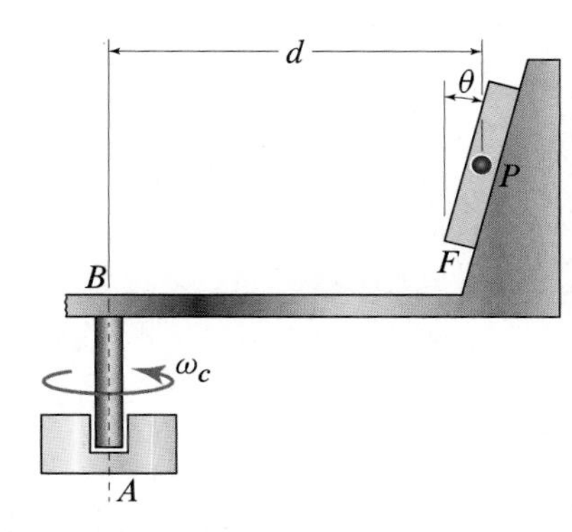

Figure P3.53 and P3.54

Problem 3.53 Determine the minimum value of ω_c at which the platform at F can be withdrawn.

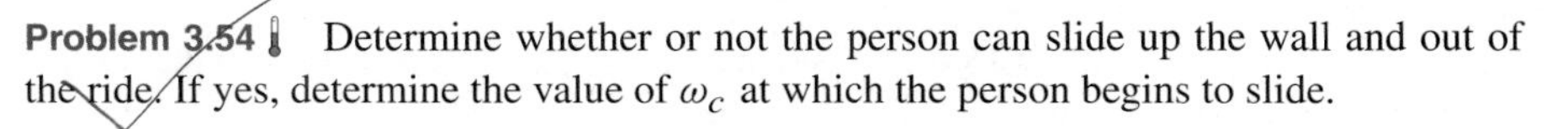

Problem 3.54 Determine whether or not the person can slide up the wall and out of the ride. If yes, determine the value of ω_c at which the person begins to slide.

Problem 3.55

A person is swinging a ball of mass m above their head in a horizontal plane. When the ball is moving at a constant speed, the string forms an angle θ with respect to the horizontal. Assuming that the motion of point O is negligible and that the distance between the point O and the mass m is L, determine the speed of the ball.

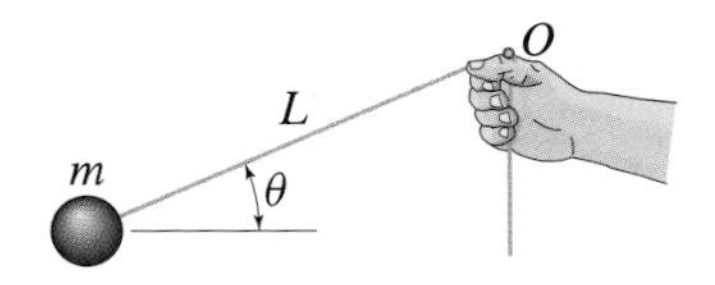

Figure P3.55

Problem 3.56

Initially, the wrecking ball A of mass m is held stationary by the horizontal cable AB. The cable AB is then released so that the wrecking ball A starts swinging about the fixed point O. Determine the tension in the cable OA before the cable AB is released and immediately after it is released. What is the percent change in cable tension if $\theta = 30°$?

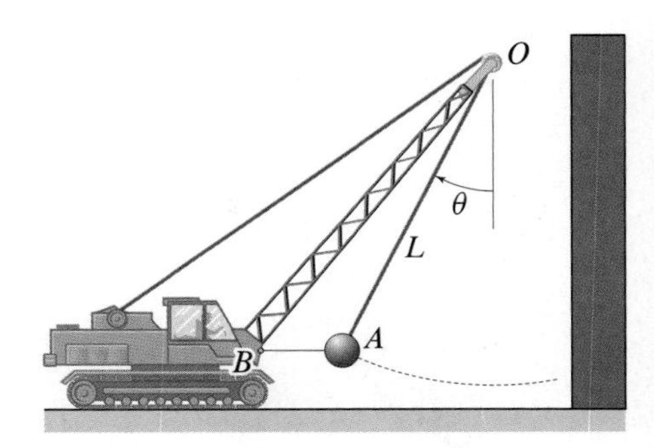

Figure P3.56

Problem 3.57

A car traveling over a hill starts to lose contact with the ground at the top of the hill at O. If the radius of curvature of the hill is 282 ft, determine the speed of the car at O.

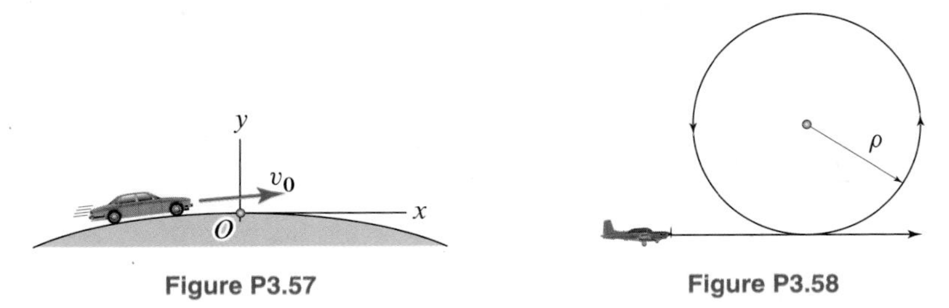

Figure P3.57 Figure P3.58

Problem 3.58

A 950 kg aerobatics plane initiates the basic loop maneuver at the bottom of a loop with radius $\rho = 110$ m and a constant speed of 225 km/h. At this instant, determine the magnitude of the plane's acceleration, expressed in terms of g, the acceleration due to gravity, and the magnitude of the lift provided by the wings.

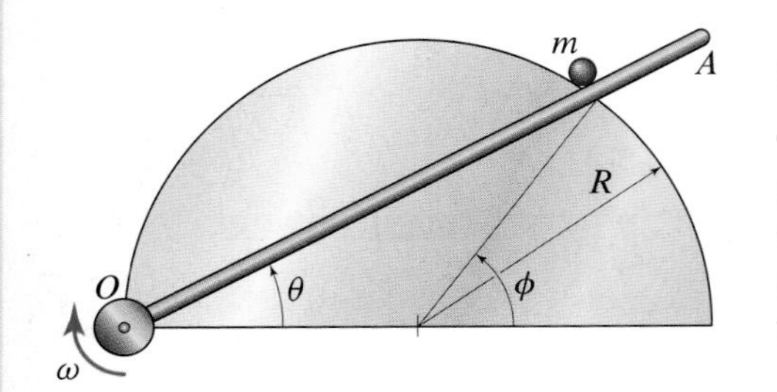

Figure P3.59

Problem 3.59

The ball of mass m is guided along the vertical circular path of radius $R = 1$ m using the arm OA. If the arm starts from $\phi = 90°$ and rotates clockwise with a constant angular velocity $\omega = 0.87$ rad/s, determine the angle ϕ at which the particle starts to leave the surface of the semicylinder. Neglect all friction forces acting on the ball, neglect the thickness of the arm OA, and treat the ball as a particle.

Problem 3.60

Referring to Example 3.10, instead of using polar coordinates as was done in that example, work the problem using a Cartesian coordinate system with origin at O (cf. Fig. 1 of Example 3.10) and derive the problem's equations of motion.

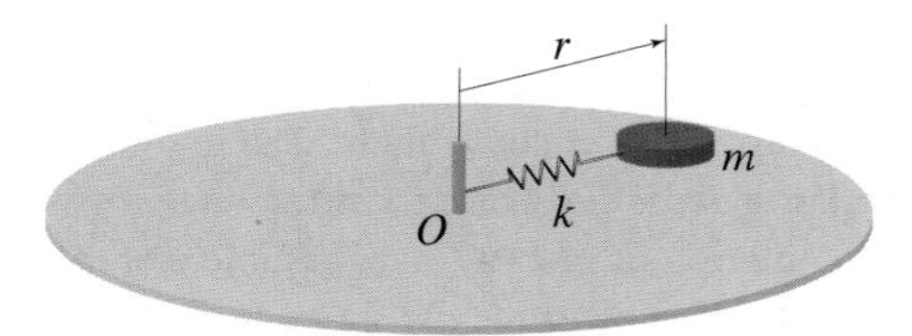

Figure P3.60 and P3.61

Problem 3.61

Continue Prob. 3.60 and using mathematical software, numerically solve the equations of motion. Use the same parameters and initial conditions that were used in Example 3.10, and compare your results with those presented in that example.

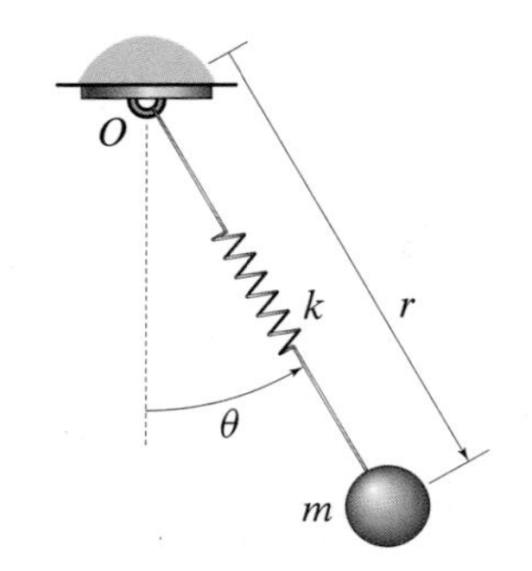

Figure P3.62 and P3.63

Problem 3.62

Derive the equations of motion for the pendulum supported by a linear spring of constant k and unstretched length r_u. Neglect friction at the pivot O, the mass of the spring, and air resistance. Treat the pendulum bob as a particle of mass m, and use polar coordinates.

Problem 3.63

Using mathematical software, solve the equations of motion for the pendulum supported by a linear spring of constant k (derived in Prob. 3.62). Plot the trajectory of the mass m in the vertical plane for a number of different values of k/m. The unstretched length of the spring is 0.25 m, and the mass is released when the pendulum is vertical, the spring is stretched 0.75 m, and the mass is moving to the right at 1 m/s.

Problems 3.64 and 3.65

The system shown is initially at rest when the bent bar starts to rotate about the vertical axis AB with constant angular acceleration $\alpha_0 = 3\,\text{rad/s}^2$. The coefficient of static friction between the collar of mass $m = 2\,\text{kg}$ and the bent bar is $\mu_s = 0.35$, and the collar is initially $d = 70\,\text{cm}$ from the spin axis AB.

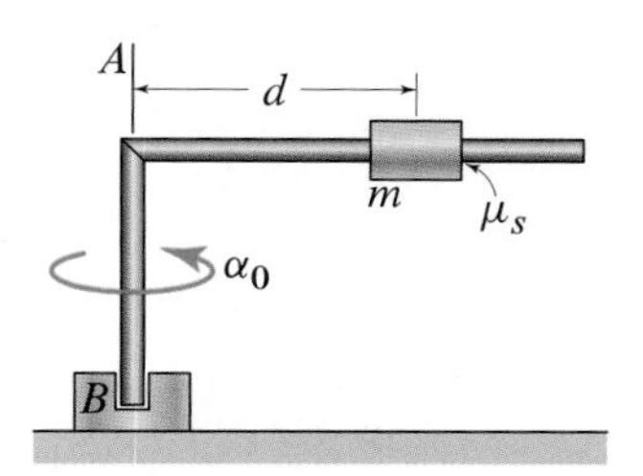

Figure P3.64 and P3.65

Problem 3.64 Assuming the motion starts at $t = 0$, determine the time at which the collar starts to slip relative to the bent bar.

Problem 3.65 Determine the number of rotations undergone by the bent bar when the collar starts to slip relative to it.

Problem 3.66

Revisiting Example 3.7, assume that the drag force in the x direction is proportional to the square of the x component of velocity, but that the ball's trajectory is shallow enough to neglect the drag force in the y direction. Using this assumption, letting O be the initial position of the ball, and letting the ball's initial velocity have a magnitude v_0 and letting it form an angle θ_0 with the x axis, determine an expression for the trajectory of the ball of the form $y = y(x)$.

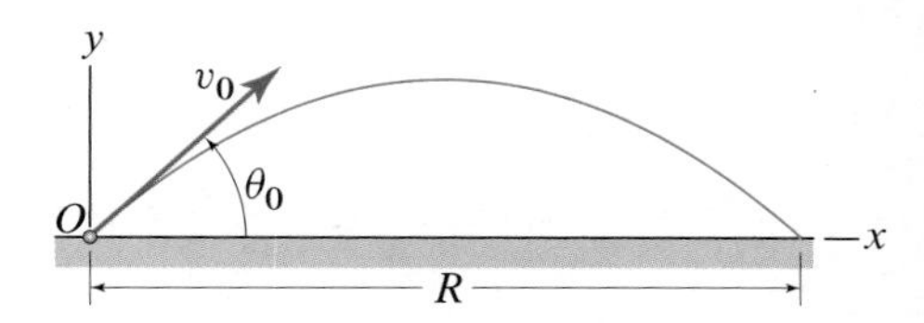

Figure P3.66 and P3.67

Problem 3.67

Revisit Example 3.7 and assume that the trajectory of the ball is shallow enough that the effects of the drag force in the y direction can be neglected and that the component of the drag force in the x direction is proportional to the square of the x component of velocity. Using this assumption and the same drag coefficient discussed in the example ($C_d = 4.71\times10^{-7}\,\text{lb}\cdot\text{s}^2/\text{ft}^2$), compute the horizontal distance R traveled by a 1.61 oz golf ball subject to the same initial conditions given in the example, i.e., the initial velocity has a magnitude $v_0 = 187\,\text{mph}$ and an initial orientation $\theta_0 = 11.2°$.

Problem 3.68

Using a cylindrical component system whose origin is at the center of curvature of the path of the car, derive the governing equations for Example 3.8 on p. 198. Solve the equations and verify that you get the same solution.

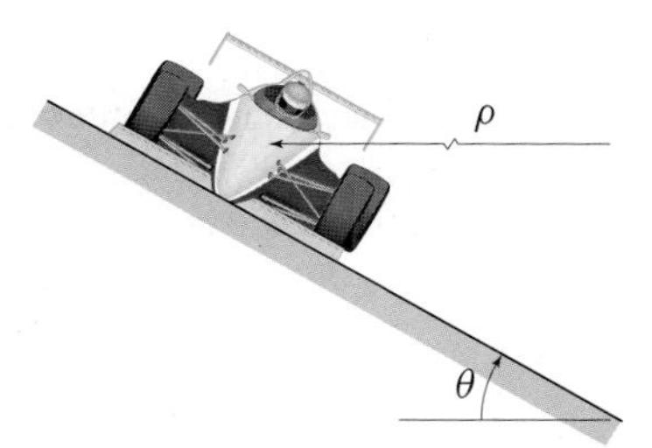

Figure P3.68–P3.70

Problem 3.69

A race car is traveling at a constant speed over a circular banked turn. Oil on the track has caused the static friction coefficient between the tires and the track to be $\mu_s = 0.2$. If the radius of the car's trajectory is $\rho = 320$ m and the bank angle is $\theta = 33°$, determine the range of speeds within which the car must travel not to slide sideways.

Problem 3.70

A race car is traveling at a constant speed $v = 200$ mph over a circular banked turn. Let the weight of the car be $W = 4000$ lb, the radius of the car's trajectory be $\rho = 1100$ ft, the bank angle be $\theta = 33°$, and the coefficient of static friction between the car and the track be $\mu_s = 1.7$. Determine the component of the friction force perpendicular to the direction of motion.

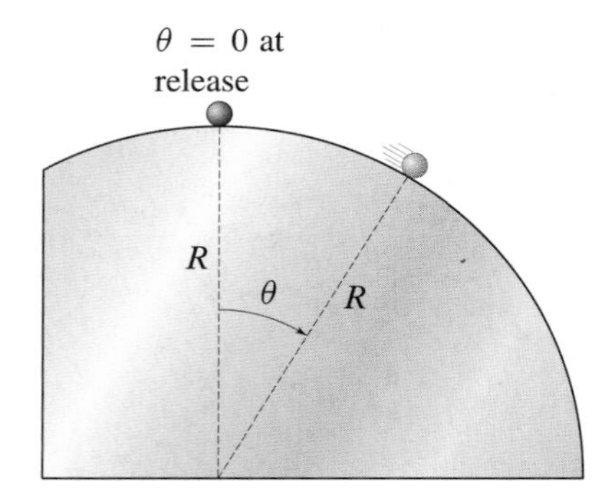

Figure P3.71

Problem 3.71

Revisit Example 3.9 on p. 200 by letting the sphere be released at $\theta = 0$ with a speed $v_0 = 0.5$ m/s. Neglecting friction, compute the angle at which the sphere separates from the cylinder if $R = 1.35$ m.

Problem 3.72

The small object of mass m is placed on the rotating conical surface at the radius shown. If the coefficient of static friction between the object and the rotating surface is 0.8, calculate the maximum angular velocity ω_c of the cone about the vertical axis for which the object will not slip. Assume the ω_c is increased very gradually so that the angular acceleration of the cone can be ignored.

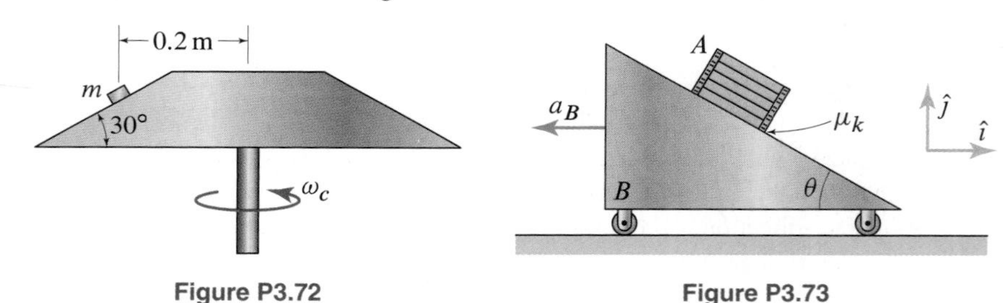

Figure P3.72 Figure P3.73

Problem 3.73

The wedge-shaped cart B is moving to the left with acceleration a_B. The coefficient of static friction between the crate and the cart is insufficient to prevent slipping between the two. If the mass of A is m and the coefficient of kinetic friction between the crate and the cart is μ_k, determine the acceleration of the crate in the component system shown.

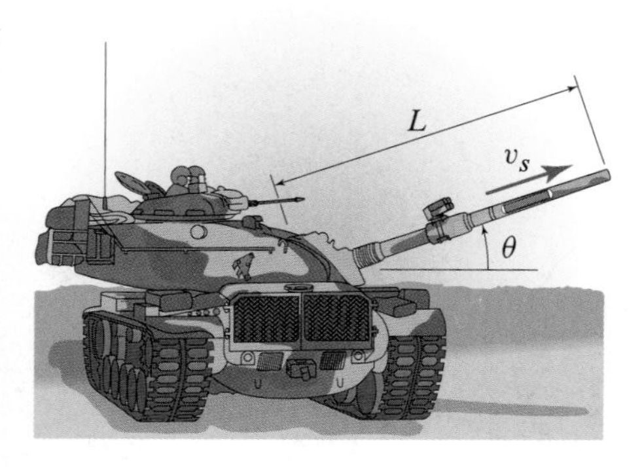

Figure P3.74

Problem 3.74

The cutaway of the gun barrel shows a projectile moving through the barrel. If the projectile's exit speed is $v_s = 1675$ m/s (relative to the barrel), the projectile's mass is 18.5 kg, the length of the barrel is $L = 4.4$ m, the acceleration of the projectile down the gun barrel is constant, and θ is increasing at a constant rate of 0.18 rad/s, determine

(a) The acceleration of the projectile.

(b) The pressure force acting on the back of the projectile.

(c) The normal force on the gun barrel due to the projectile.

as the projectile leaves the gun, but while it is still in the barrel. Assume that the projectile exits the barrel when $\theta = 20°$, and ignore friction between the projectile and the barrel.

Problems 3.75 and 3.76

A simple sling can be built by placing a projectile in a tube and then spinning it. Consider a simple model in which the tube is pinned as shown and is rotated about the pin in the *horizontal* plane at constant angular velocity ω. Assume that there is no friction between the projectile and the inside of the tube and that the projectile is initially kept fixed at a distance d from the open end of the tube.

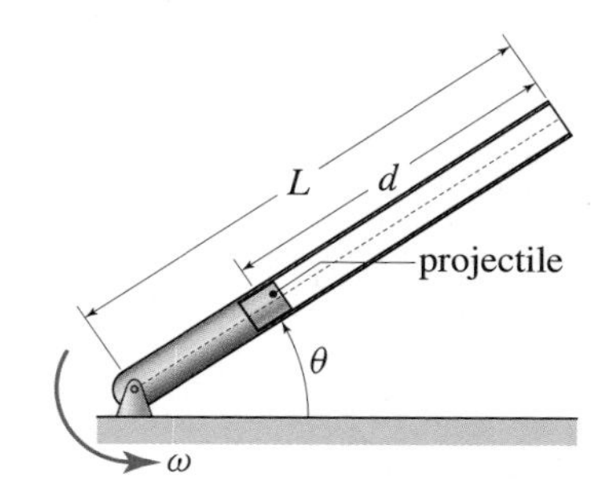

Figure P3.75 and P3.76

Problem 3.75 After the projectile is released, compute the normal force exerted by the inside of the tube on the projectile as a function of position of the projectile along the tube.

Problem 3.76 Letting $d = 3$ ft and $L = 7$ ft, determine the value of the tube's angular velocity ω if, after release, the projectile exits the tube with a speed of 90 ft/s.

Problem 3.77

The wedge-shaped cart B is moving up and to the left with acceleration a_B. The coefficient of static friction between the crate and the cart is insufficient to prevent slipping between the two. If the mass of A is m and the coefficient of kinetic friction between the crate and the cart is μ_k, determine the acceleration of the crate in the component system shown.

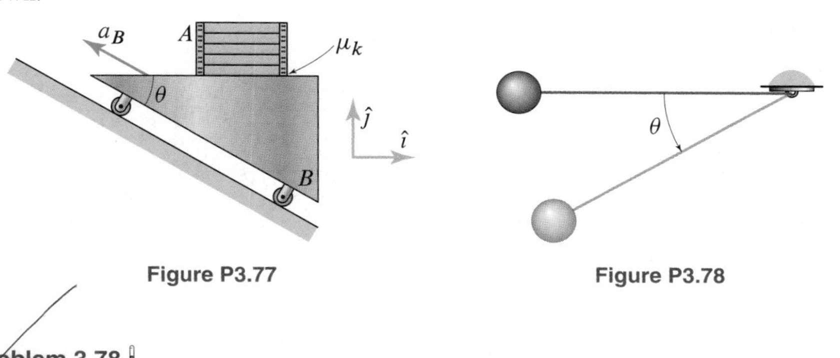

Figure P3.77

Figure P3.78

Problem 3.78

The pendulum is released from rest when $\theta = 0°$. If the string holding the pendulum bob breaks when the tension is twice the weight of the bob, at what angle does the string break? Treat the pendulum as a particle, ignore air resistance, and let the string be inextensible and massless.

Problem 3.79

The trolley T moves along rails on the horizontal truss of the tower crane. The trolley and the load of mass m are both initially at rest (with $\theta = 0$) when the trolley starts moving to the right with constant acceleration $a_T = g$. Determine

(a) The maximum angle θ_{max} achieved by the load m.

(b) The tension in the supporting cable as a function of θ.

Treat the load m as a particle, and ignore the mass of the supporting cable.

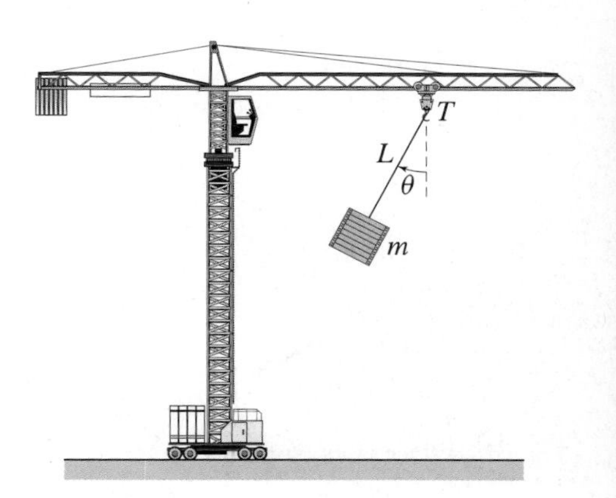

Figure P3.79

Problem 3.80

If the particle is constrained to only move back and forth in the plane of the center of the bowl, how many degrees of freedom must it have? Recall that this will also be the number of equations of motion of the particle.

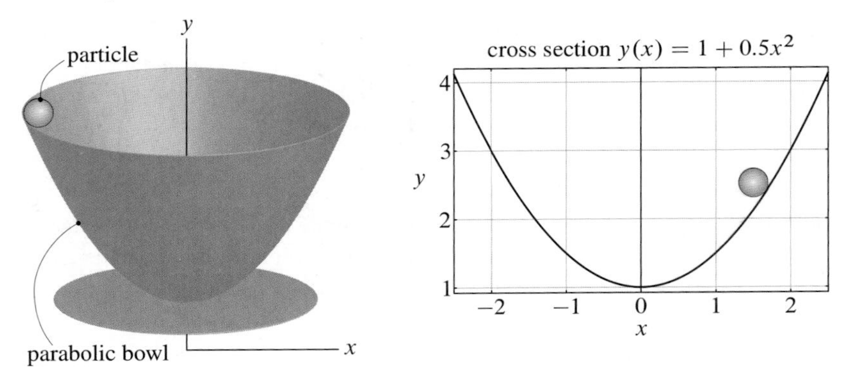

Figure P3.80 and P3.81

Problem 3.81

Derive the equation(s) of motion for a particle moving on the inner surface of the smooth parabolic bowl shown. Assume that the particle only moves in the vertical xy plane that goes through the center of the bowl, that the equation of the parabola is $y(x)=1+0.5x^2$, and that gravity is acting in the $-y$ direction. The bowl's cross section is shown on the right side of the figure.

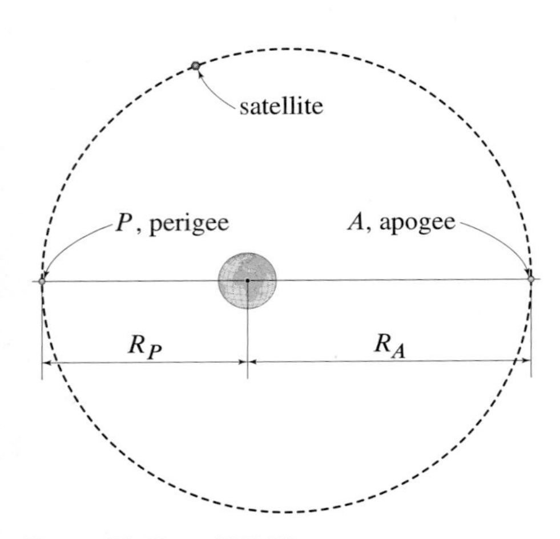

Figure P3.82 and P3.83

Problems 3.82 and 3.83

A satellite orbits the Earth as shown. Model the satellite as a particle, and assume that the center of the Earth can be chosen as the origin of an inertial frame of reference.

Problem 3.82 Using a polar coordinate system and letting m_e be the mass of the Earth, determine the equations of motion of the satellite.

Problem 3.83 The minimum and maximum distances from the center of the Earth are $R_P = 4.5\times10^7$ m and $R_A = 6.163\times10^7$ m, respectively, where the subscripts P and A stand for *perigee* (the point on the orbit closest to Earth) and *apogee* (the point on the orbit farthest from Earth), respectively. If the satellite's speed at P is $v_P = 3.2\times 10^3$ m/s, determine the satellite's speed at A.

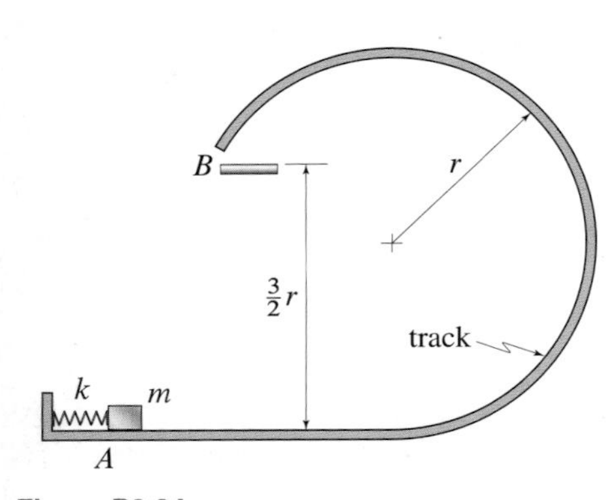

Figure P3.84

Problem 3.84

The package handling system is designed to launch the small package of mass m from A, using a compressed linear spring of constant k. After launch, the package slides along the track until it lands on the conveyor belt at B. The track has small, well-oiled rollers, making any friction between the packages and the track negligible. Modeling the package as a particle, determine the minimum initial compression of the spring so that the package gets to B without separating from the track, and determine the corresponding speed with which the package reaches the conveyor at B.

Problem 3.85

A particle moves over the inner surface of an inverted cone. Assuming that the cone's surface is frictionless and using the cylindrical coordinate system shown, show that the particle's equations of motion are

$$\ddot{R}(1 + \cot^2 \phi) - R\dot{\theta}^2 = -g \cot \phi,$$
$$R\ddot{\theta} + 2\dot{R}\dot{\theta} = 0.$$

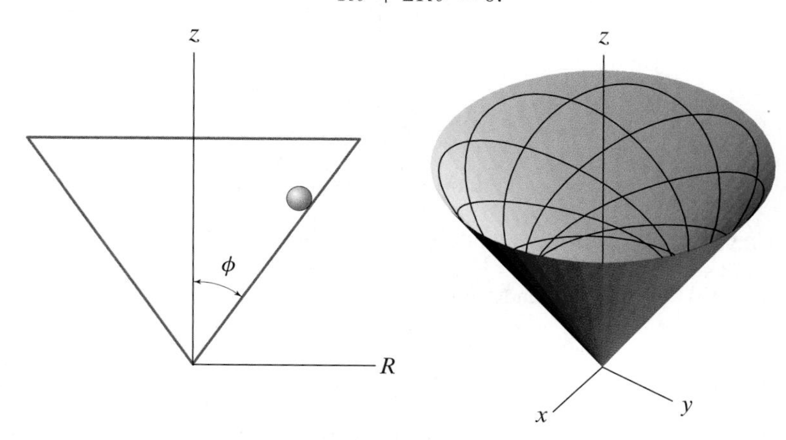

Figure P3.85 and P3.86

Problem 3.86

Continue Prob. 3.85 by integrating the particle's equations of motion and plotting its trajectory for $0 \le t \le 25$ s. Use the following parameter values and initial conditions: $g = 9.81\,\text{m/s}^2$, $\phi = 30°$, $\theta(0) = 0°$, $\dot{\theta}(0) = 1.00\,\text{rad/s}$, $R(0) = 5\,\text{m}$, $\dot{R}(0) = 0\,\text{m/s}$.

Problem 3.87

The disk shown, weighing 3 lb, rotates about O by sliding without friction over the horizontal surface shown. The spring is linear with constant k and unstretched length $L_0 = 0.75$ ft. The maximum distance achieved by the disk from point O is $d_{max} = 1.85$ ft while traveling at a speed $v_0 = 20$ ft/s. Determine the value of k such that the minimum distance between the disk and O is $d_{min} = d_{max}/2$.

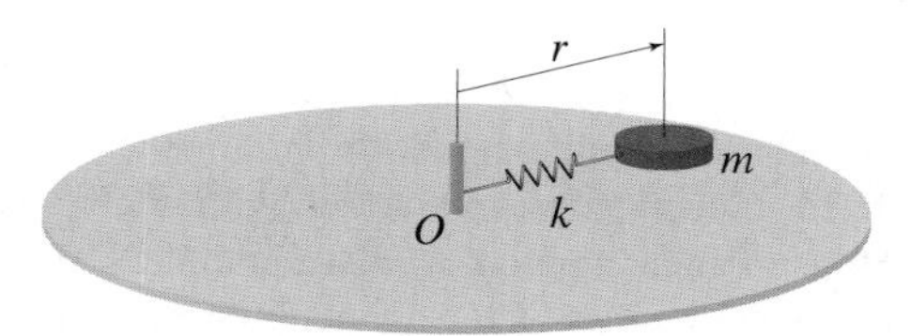

Figure P3.87

Problems 3.88 and 3.89

A pendulum with cord length $L = 6$ ft and a bob weighing 3 lb is released from rest at an angle θ_i. Once the pendulum has swung to the vertical position (i.e., $\theta = 0$), its cord runs into a small fixed obstacle. In solving this problem, neglect the size of the obstacle; model the pendulum's bob as a particle and the pendulum's cord as massless and inextensible; and let gravity and the tension in the cord be the only relevant forces.

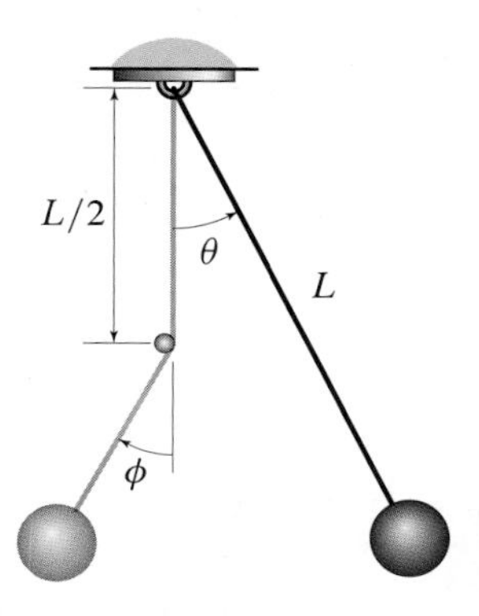

Figure P3.88 and P3.89

Problem 3.88 What is the maximum height reached by the pendulum, measured from its lowest point, if $\theta_i = 20°$?

Problem 3.89 If the bob is released from rest at $\theta_i = 90°$, at what angle ϕ does the cord go slack?

Problems 3.90 through 3.93

A spherical pendulum is suspended from point O and put in motion.

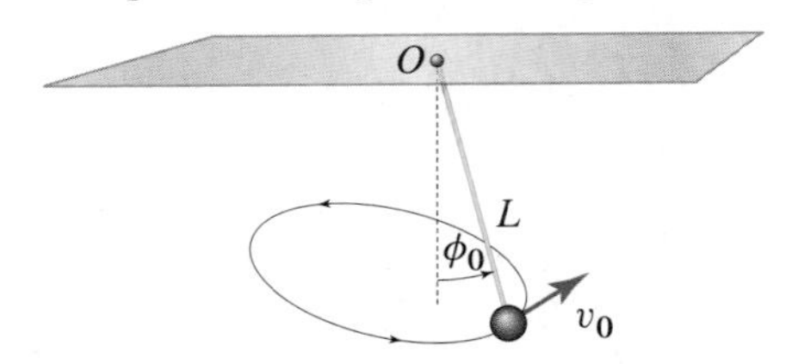

Figure P3.90–P3.93

Problem 3.90 Derive the pendulum's equations of motion using Cartesian coordinates.

Problem 3.91 Derive the pendulum's equations of motion using cylindrical coordinates.

Problem 3.92 Derive the pendulum's equations of motion using spherical coordinates.

Problem 3.93 At the lowest point on its trajectory, the pendulum in the figure has a speed $v_0 = 2.5\,\text{ft/s}$ while $\phi_0 = 15°$. Letting $L = 2\,\text{ft}$, plot the trajectory of the pendulum bob for 5 s.

Problems 3.94 through 3.97

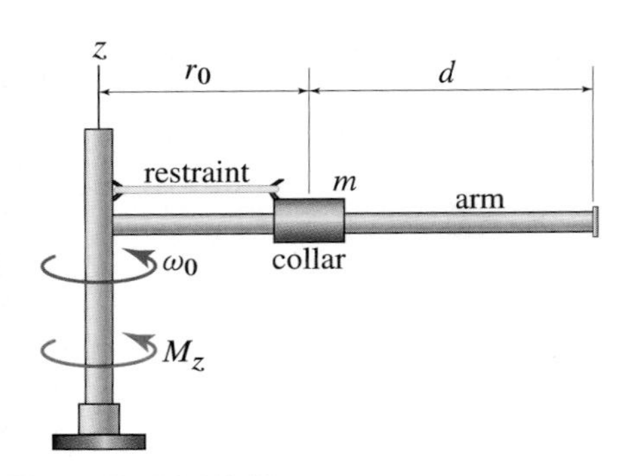

Figure P3.94–P3.97

Consider a collar with mass m that is free to slide with no friction along a rotating arm, which has negligible mass. The system is initially rotating with an angular velocity ω_0 while the collar is kept a distance r_0 away from the z axis. At some point, the restraint keeping the collar in place is removed so that the collar is allowed to slide.

Problem 3.94 Derive the collar's equations of motion when $M_z = 0$.

Problem 3.95 If no external forces and moments are applied to the system, what are the radial speed and the total speed of the collar when it reaches the end of the arm? Use $m = 2\,\text{kg}$, $\omega_0 = 1\,\text{rad/s}$, $r_0 = 0.5\,\text{m}$, and $d = 1\,\text{m}$.

Problem 3.96 Compute the moment M_z that you would need to apply to the arm, as a function of r, to keep the arm rotating at a *constant* angular velocity ω_0. In addition, determine the radial speed, as well as the total speed with which the collar would reach the end of the arm with this moment applied. *Hint:* The moment M_z applied to the arm of negligible mass is equivalent to a force M_z/r applied to the collar in the plane of motion and perpendicular to the arm.

Problem 3.97 Compute and plot the collar's trajectory from the moment of release until the collar reaches the end of the arm. Use the parameters and initial conditions given in Prob. 3.95.

Problem 3.98

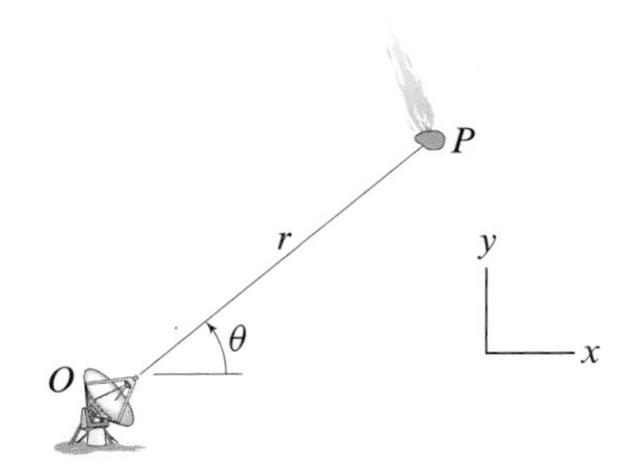

Figure P3.98

The radar station at O is tracking the meteor P as it moves through the atmosphere. The data measured by the radar station indicates that the acceleration vector of P is almost exactly in the opposite direction of the velocity vector. Explain why this is what we should expect.

Problem 3.99

The particle P is placed on the turntable, and both are initially at rest. The turntable is then turned on so that the disk starts spinning. Assuming that there is no friction between the turntable disk and the particle, what will be the motion of the particle after the disk starts spinning? Explain.

Figure P3.99

Problem 3.100

The conveyor belt moves parts, each with mass m, at a constant speed v_0. When the parts get to A, they begin moving over a circular path of radius ρ. If the coefficient of static friction between the belt and the parts is μ_s, determine the angle θ at which the parts will start to slide on the belt. Neglect the size of the parts. After determining θ in terms of μ_s, v_0, ρ, and g, evaluate θ for $\mu_s = 0.6$, $v_0 = 3\,\text{mph}$, and $\rho = 14\,\text{in.}$

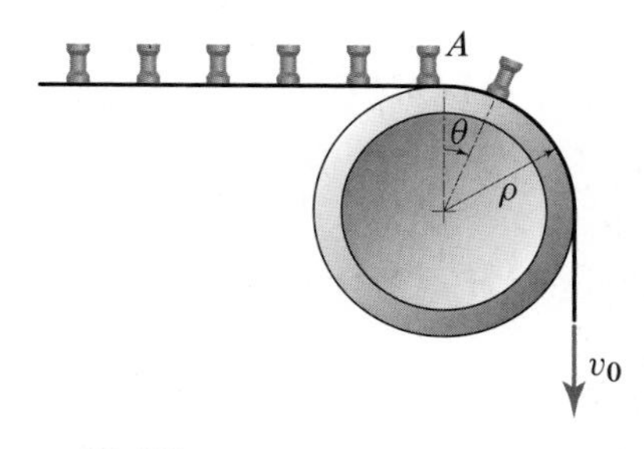

Figure P3.100

Design Problems

Design Problem 3.1

The mechanism in the figure needs to be designed so as to capture parcels at A and to deliver them at B. The impact speed at B must not exceed 1.5 m/s. The parcels have masses ranging between 5 and 10 kg incoming at speeds between 3 and 6 m/s. Determine the acceptable ranges of the spring constant, the free length of the spring, and the kinetic friction coefficient between the sliding surface and the cradle to achieve the desired result.

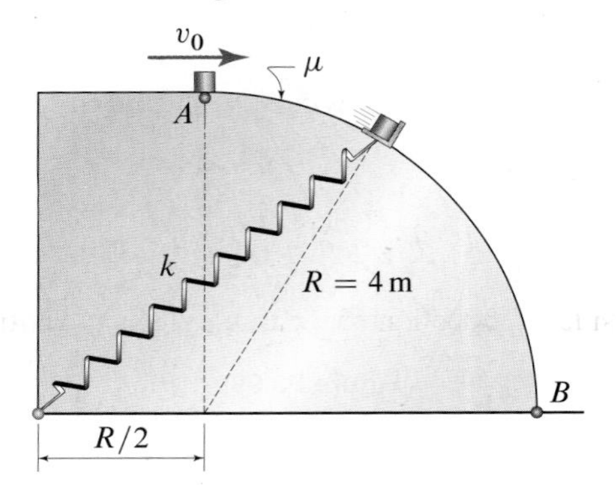

Figure DP3.1

Design Problem 3.2

Consider the cam-follower system in the figure. The cam is on a shaft that rotates with speeds up to 3000 rpm. The profile of the cam is described by the function $\ell(\phi) = R_0(1 + 0.25\cos^3\phi)$, where the angle ϕ is measured relative to the segment OA, which rotates with the cam, and where $R_0 = 3$ cm. The follower has a mass of 90 g. Assuming that the contact between cam and follower is frictionless, choose appropriate values of the spring constant k and the spring's unstretched length L_0 such that the follower always remains in contact with the cam while minimizing the value of the contact force between the follower and the cam.

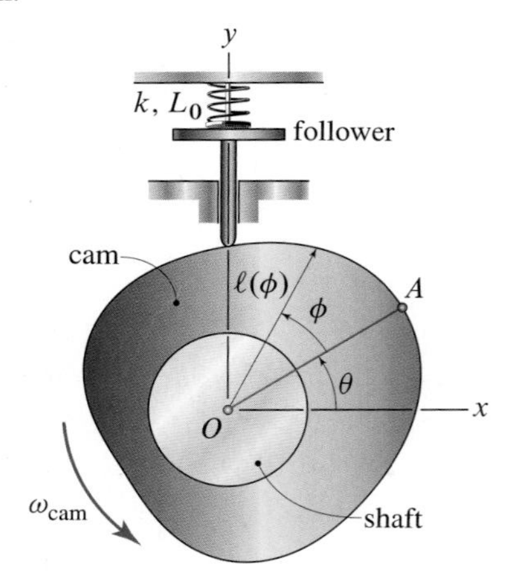

Figure DP3.2

3.3 Systems of Particles

Engineering materials one atom at a time

When applied to large collections of atoms, the particle dynamics we are studying plays an important role in the burgeoning field of nanotechnology. The dynamics of systems of particles is a core component of a nanoscale modeling approach called *molecular dynamics* (MD), in which the force laws describing atomic interactions are used together with Newton's second law to study the motion of a system of atoms. Using information from these simulations, we can estimate and predict a variety of physical properties (see Fig. 3.16). We will now develop the equations that allow us to apply Newton's second law to a system of particles.

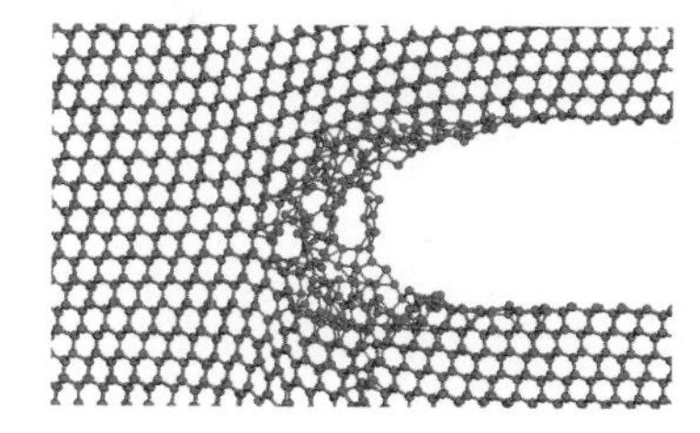

Figure 3.16
Snapshot from an MD simulation of fracture in silicon used to estimate the fracture toughness of the material under various conditions.

Newton's second law for systems of particles

In Fig. 3.17, we consider a collection of n particles. The quantities m_i and $\vec{r}_i$ denote

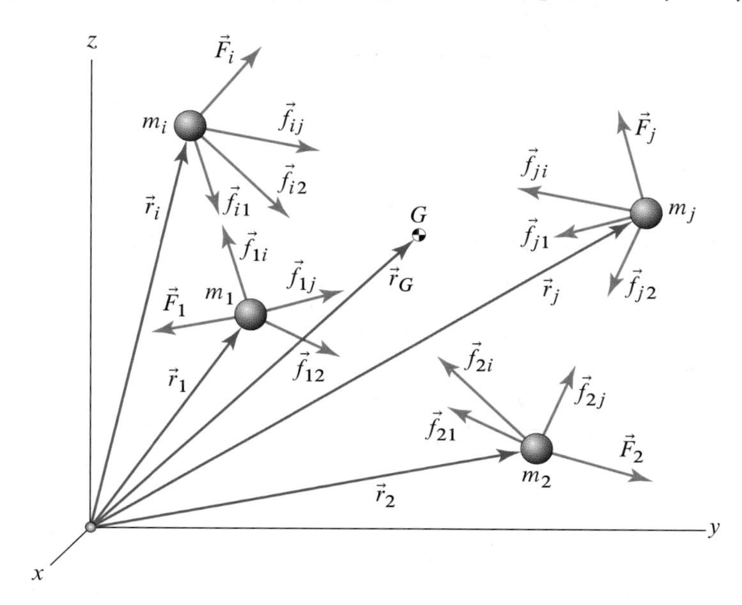

Figure 3.17. System of n particles showing position vectors $\vec{r}_i$, external forces $\vec{F}_i$, and internal forces $\vec{f}_{ij}$.

the mass and the position of particle i, respectively. A force acting on a particle of the system is called *internal* if it arises from the interaction between that particle and other particles within the system; otherwise the force is called *external*. The force denoted by $\vec{F}_i$ is the *sum* of all external forces acting on particle i (e.g., gravity, air drag, etc.), and it would exist even if the rest of the particles were removed from the system. With $\vec{f}_{ij}$ we denote the force on particle i due to particle j. Therefore, the forces $\vec{f}_{i1}, \vec{f}_{i2}, \ldots, \vec{f}_{in}$ are the *internal* forces acting on particle i.

Applying Newton's second law to each particle within the system, we have

$$\vec{F}_1 + \sum_{\substack{j=1 \\ j\neq 1}}^{n} \vec{f}_{1j} = m_1 \vec{a}_1, \quad \vec{F}_2 + \sum_{\substack{j=1 \\ j\neq 2}}^{n} \vec{f}_{2j} = m_2 \vec{a}_2, \quad \ldots \quad \vec{F}_n + \sum_{\substack{j=1 \\ j\neq n}}^{n} \vec{f}_{nj} = m_n \vec{a}_n, \tag{3.29}$$

where we have noted that in equation i, we cannot have $j = i$ because we assume that a particle does not exert a force on itself.

To understand the effect of Newton's third law on the system as a whole, we sum all n equations in Eqs. (3.29) to obtain

$$\sum_{i=1}^{n} \vec{F}_i + \sum_{i=1}^{n} \sum_{\substack{j=1 \\ j \neq i}}^{n} \vec{f}_{ij} = \sum_{i=1}^{n} m_i \vec{a}_i. \tag{3.30}$$

The first term in Eq. (3.30), $\sum_{i=1}^{n} \vec{F}_i$, is the sum of all external forces acting on the system of particles, which we will denote by $\vec{F}$, i.e.,

$$\vec{F} = \sum_{i=1}^{n} \vec{F}_i. \tag{3.31}$$

The second term in Eq. (3.30) is the sum of all internal forces. Referring to Fig. 3.17, observe that for every $\vec{f}_{ij}$ in the system there will always be an $\vec{f}_{ji}$. By Newton's third law, these forces are equal and opposite and they will *always* cancel one another. For example, if $n = 3$, the second term in Eq. (3.30) yields

$$\sum_{i=1}^{3} \sum_{\substack{j=1 \\ j \neq i}}^{3} \vec{f}_{ij} = \vec{f}_{12} + \vec{f}_{13} + \vec{f}_{21} + \vec{f}_{23} + \vec{f}_{31} + \vec{f}_{32} = \vec{0}, \tag{3.32}$$

given that $\vec{f}_{12} = -\vec{f}_{21}$, $\vec{f}_{13} = -\vec{f}_{31}$, and $\vec{f}_{23} = -\vec{f}_{32}$. Therefore, by Newton's third law, the second sum on the left-hand side of Eq. (3.30) *always* vanishes, i.e.,

$$\sum_{i=1}^{n} \sum_{\substack{j=1 \\ j \neq i}}^{n} \vec{f}_{ij} = \vec{0}. \tag{3.33}$$

Finally, the last term in Eq. (3.30), $\sum_{i=1}^{n} m_i \vec{a}_i$, can be written as

$$\sum_{i=1}^{n} m_i \vec{a}_i = \sum_{i=1}^{n} m_i \frac{d^2 \vec{r}_i}{dt^2} = \frac{d^2}{dt^2} \sum_{i=1}^{n} m_i \vec{r}_i, \tag{3.34}$$

where, in moving d^2/dt^2 outside the summation, we used the fact that the mass of each particle is constant. We now recall that the term $\sum_{i=1}^{n} m_i \vec{r}_i$ defines the position of the system's center of mass through the relation

$$m \vec{r}_G = \sum_{i=1}^{n} m_i \vec{r}_i, \tag{3.35}$$

where $m = \sum_{i=1}^{n} m_i$ is the total mass of the system and $\vec{r}_G$ is the position of the mass center G (see Fig. 3.17). Using Eq. (3.35), Eq. (3.34) becomes

$$\sum_{i=1}^{n} m_i \vec{a}_i = \frac{d^2}{dt^2}(m \vec{r}_G) = m \frac{d^2 \vec{r}_G}{dt^2} = m \vec{a}_G, \tag{3.36}$$

where we have used the fact that, because each mass m_i is constant, the total mass m must also be constant. Substituting Eqs. (3.31), (3.33), and (3.36) into Eq. (3.30), we obtain

$$\boxed{\vec{F} = \sum_{i=1}^{n} m_i \vec{a}_i = m \vec{a}_G.} \tag{3.37}$$

Equation (3.37) says that the mass center of a system of particles moves as if it were a particle of mass m subjected to *only* the total external force $\vec{F}$. Consequently,

- If $\vec{F} = \vec{0}$, then the mass center moves with constant velocity. Therefore, if $\vec{F} = \vec{0}$ *and* the mass center is initially at rest, the mass center will remain at rest.

- If $\vec{F} = \vec{0}$, then $m\vec{a}_G = \vec{0}$, and its time integral must be constant, i.e.,

$$\int m\vec{a}_G \, dt = m\vec{v}_G = \text{constant}, \tag{3.38}$$

  which is a statement of the *conservation of linear momentum* for a system of particles. We will discuss this conservation law in Section 5.1.

EXAMPLE 3.12 Motion with Conservation of Linear Momentum

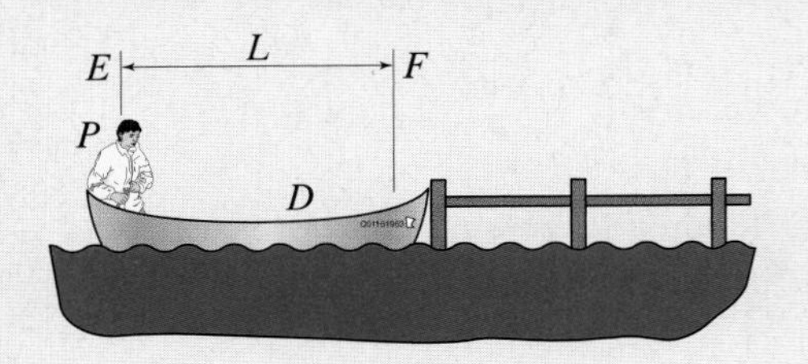

Figure 1
A canoeist who has just reached the dock and is about to walk the length of the canoe in order to exit onto the dock.

The canoeist P has paddled her canoe D to the dock, as shown in Fig. 1. After the front end of her canoe reaches the end of the dock, she decides to walk the distance L from the back of the canoe to the front and exit onto the dock. Assuming that the canoe can slide in the water with negligible resistance, determine the distance between the person and the dock when she reaches the front end of the canoe to determine whether or not she will be able to exit onto the dock without getting wet. Assume that the person weighs 185 lb, the canoe weighs 70 lb, and $L = 10$ ft.

SOLUTION

Road Map & Modeling Viewing the person and canoe as a two-particle system, Fig. 2 shows the FBD of the system *as a whole*, which includes only forces external to the

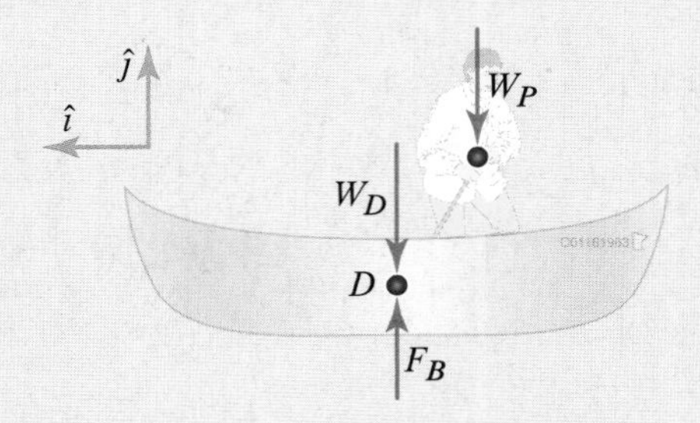

Figure 2. FBD of the canoe and canoeist. We have modeled the canoe and the canoeist as particles.

system. Our model includes the weight of both the person and the canoe, as well as the buoyancy force F_B. We have neglected any horizontal resistance offered by the water and/or air. Because the total horizontal force is equal to zero and the system is initially at rest, the position of the mass center will not change even as the person moves relative to the canoe. We will write Newton's second law for the system *as a whole* and integrate it with respect to time to determine the position of each part of the system.

Governing Equations

Balance Principles Referring to the FBD in Fig. 2, applying the first of Eqs. (3.37) in the x direction, we have

$$\sum F_x: \quad 0 = m_P a_{Px} + m_D a_{Dx}, \tag{1}$$

where a_{Px} and a_{Dx} are the x components of the acceleration of P and D, respectively.

Force Laws All forces are accounted for on the FBD.

Kinematic Equations Because our analysis is limited to the x direction, the kinematic relations we need are

$$a_{Px} = \dot{v}_{Px}, \quad v_{Px} = \dot{x}_P, \quad a_{Dx} = \dot{v}_{Dx}, \quad \text{and} \quad v_{Dx} = \dot{x}_D, \tag{2}$$

where, referring to Fig. 3, the coordinate x is measured from the end of the dock.

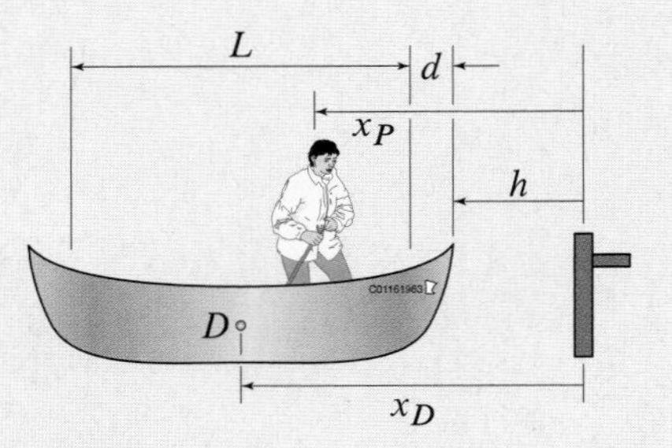

Figure 3
Definitions of distances and coordinate directions for the problem's kinematics.

Computation Integrating Eq. (1) with respect to time yields

$$m_P v_{Px} + m_D v_{Dx} = C_1, \tag{3}$$

where C_1 is a constant of integration. This equation says that *for any arbitrary position of P and D*, the sum of the product of their masses and the x component of their velocities is constant. Hence, we can evaluate C_1 by recalling that both the canoe D and the canoeist P have zero velocity after they reach the dock and before the canoeist starts to walk the length of the canoe, that is,

$$m_P \cdot 0 + m_D \cdot 0 = C_1 \quad \Rightarrow \quad C_1 = 0. \tag{4}$$

Next, substituting $C_1 = 0$ into Eq. (3) and integrating with respect to time, we have

$$m_P x_P + m_D x_D = C_2, \tag{5}$$

where C_2 is a second constant of integration. Referring to Fig. 3, to find C_2 we evaluate Eq. (5) when $h = 0$, i.e., when the canoe is touching the dock and the canoeist is at the left end of the canoe. This yields

$$m_P(d + L) + m_D(d + L/2) = C_2, \tag{6}$$

where d is the constant distance from the end of the canoe to the point in the canoe at which the canoeist would step out. Substituting the expression for C_2 given by Eq. (6) into Eq. (5), we have that the positions of the canoeist and the canoe are related as follows:

$$m_P x_P + m_D x_D = m_P(d + L) + m_D(d + L/2). \tag{7}$$

We can now determine where the canoe ends up when the canoeist walks to the front end. When the canoeist is at the front end of the canoe, we must have $x_{D/P} = x_D - x_P = L/2$. Therefore, $x_D = x_P + L/2$, and Eq. (7) becomes

$$m_P x_P + m_D(x_P + L/2) = m_P(d + L) + m_D(d + L/2). \tag{8}$$

Solving for x_P, we have

$$x_P = \frac{m_P(d + L) + m_D d}{m_P + m_D}. \tag{9}$$

Finally, since $h = x_P - d$ when the canoeist is at the front end of the canoe, for this situation we obtain

$$\boxed{h = \frac{m_P}{m_P + m_D} L = 7.255\ \text{ft},} \tag{10}$$

where we have used the given data to obtain the final numerical answer. The result in Eq. (10) indicates that the poor canoeist will not be able to step from the canoe onto the dock without getting wet!

Discussion & Verification The result in Eq. (10) has the dimension of length, as it should. The distance h is smaller than the length of the canoe and is therefore within reason.

A Closer Look One solution to the problem of the canoe moving away from the dock would be for someone on the dock to tie the canoe to the dock after the canoe arrives.

An interesting feature of Eq. (10) is that the speed with which the person walks along the canoe from one end to the other does not appear in the solution. In addition, it would be reasonable to expect that if the canoe were tied to the dock with a rope and then the canoeist walked the length of the canoe, there would be a tension in the rope. It turns out that it is not the speed that determines the tension in the rope — it is the acceleration of the person that determines the rope's tension. We will see an interesting variation of this problem in Chapter 5 when we study momentum methods for systems of particles (see Example 5.4).

EXAMPLE 3.13 *Pulley System Analysis*

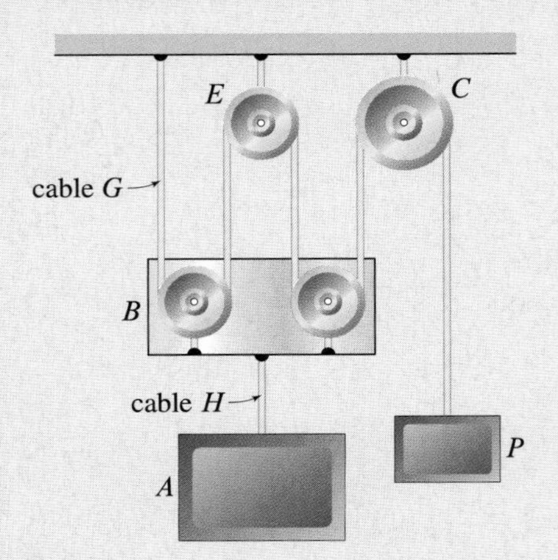

Figure 1
A pulley system designed for lifting large loads.

The pulley system in Fig. 1 is designed to lift a heavy load at A by attaching a mass at P. Assuming that the payload A has mass m_A and the pulley housing B (which includes the pulleys) has mass m_B, determine the accelerations of A and P if a mass m_P is attached at P. Neglect friction in, and rotational inertia of, the pulleys.

SOLUTION

Road Map & Modeling We want to determine the acceleration of specific elements of the system, so we isolate these elements and sketch an FBD for each (as opposed to sketching an FBD for the whole system, as was done in Example 3.12). Modeling each moving element as a particle, ignoring the masses of the cables and assuming they are inextensible, we obtain the FBDs shown in Fig. 2. Since the motion of A, B, and P is

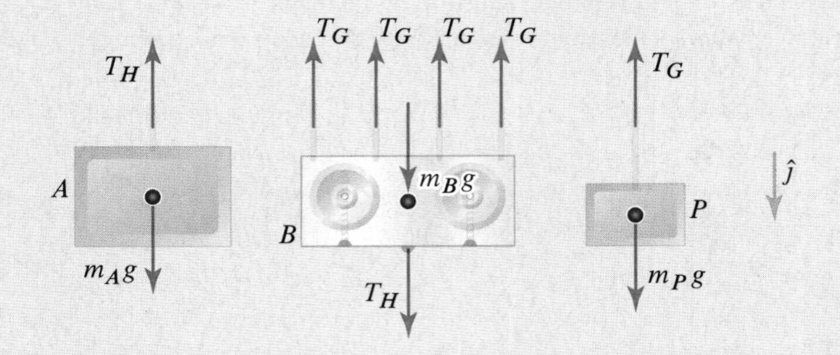

Figure 2. FBD of mass A, mass P, and pulley housing B.

only in the vertical direction, the problem is one-dimensional, and our component system consists of only one unit vector ($\hat{\jmath}$), as shown in Fig. 2.

Governing Equations

Balance Principles Using the FBDs in Fig. 2, application of Newton's second law to A, B, and P yields

$$\left(\sum F_y\right)_A: \quad m_A g - T_H = m_A a_A, \tag{1}$$

$$\left(\sum F_y\right)_B: \quad m_B g + T_H - 4T_G = m_B a_B, \tag{2}$$

$$\left(\sum F_y\right)_P: \quad m_P g - T_G = m_P a_P. \tag{3}$$

Force Laws The weights of A, B, and P are already accounted for in Fig. 2. As far as the cables are concerned, their inextensibility is enforced by kinematic constraints instead of force laws.

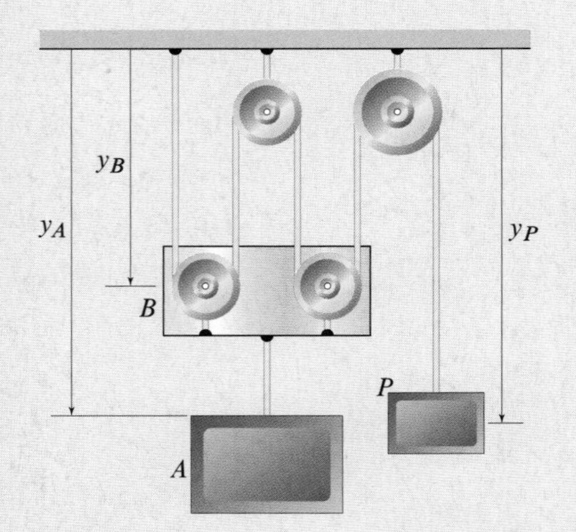

Figure 3
The positions of A, B, and P.

Kinematic Equations Since the cables are inextensible, from Fig. 3 we have

$$L_G = 4y_B + y_P \quad \Rightarrow \quad 4a_B + a_P = 0, \tag{4}$$

where L_G denotes the length of cable G. In addition, the inextensibility of the cable connecting A and B demands that

$$a_A = a_B. \tag{5}$$

Computation Equations (1)–(5) provide a system of five equations in the five unknowns a_A, a_B, a_P, T_G, and T_H that can be solved to obtain

$$T_G = \frac{5m_P(m_A + m_B)}{16m_P + (m_A + m_B)}g, \tag{6}$$

$$T_H = \frac{20m_A m_P}{16m_P + (m_A + m_B)}g. \tag{7}$$

and

$$a_A = a_B = -\frac{4m_P - (m_A + m_B)}{16m_P + (m_A + m_B)}g, \tag{8}$$

$$a_P = \frac{4[4m_P - (m_A + m_B)]}{16m_P + (m_A + m_B)}g, \tag{9}$$

Discussion & Verification The fractions in Eqs. (8) and (9) are dimensionless, so the results in these equations have the dimensions of g, i.e., acceleration, as they should. The fractions in Eqs. (6) and (7) have dimensions of mass, so the results in these equations have dimensions of force, as they should. Equations (8) and (9) tell us that the accelerations of A and P are in the opposite direction, again as expected. Finally, the cable tension T_G is *always* positive, so the cable never goes slack, as it should. Hence, our solution seems to be correct.

A Closer Look Focusing on either Eq. (8) or (9), we see that the "bang for the buck" that we get with the mass at P is 4 times its mass since the numerator of either of these equations contains the term $4m_P - (m_A + m_B)$. This term is positive if $m_P > \frac{1}{4}(m_A + m_B)$; and furthermore, if $m_P > \frac{1}{4}(m_A + m_B)$, then $a_P > 0$ (otherwise, $a_P < 0$). Thus, we see that it is not m_P that must be greater than the combined mass of A and B for P to pull A up: m_P need only be greater than $\frac{1}{4}(m_A + m_B)$!

Helpful Information

Sign conventions. The positive direction of the y axis is downward. Hence, when a_{Ay} and a_{Py} have opposite signs, it means that they are accelerating in opposite directions. Indeed, A and P move in opposite directions, as can be seen by differentiating the first of Eqs. (4) with respect to time.

EXAMPLE 3.14 *Sliding with Friction*

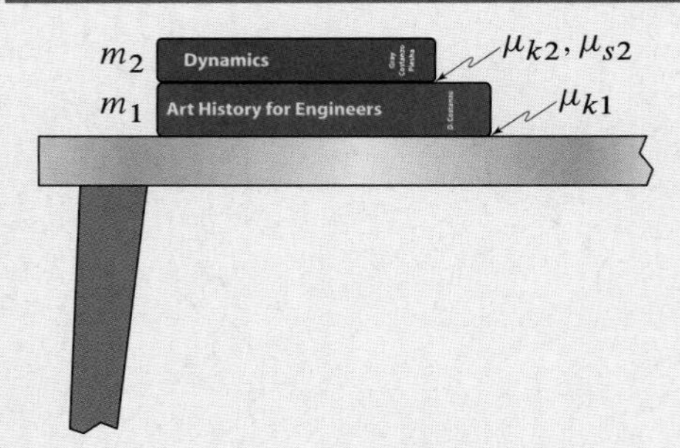

Figure 1
A pair of stacked books thrown horizontally on a rough table.

A pair of stacked books with masses $m_1 = 1.5\,\text{kg}$ and $m_2 = 1\,\text{kg}$ is thrown on a table (Fig. 1). The books strike the table with essentially zero vertical speed, and their common horizontal speed is $v_0 = 0.75\,\text{m/s}$. Letting $\mu_{k1} = 0.45$ be the coefficient of kinetic friction between the bottom book and the table, and letting $\mu_{s2} = 0.4$ and $\mu_{k2} = 0.3$ be the coefficients of static and kinetic friction between the two books, respectively, determine the books' final positions relative to where they hit the table and their position relative to one another. Model both books as particles with the same initial horizontal position.

SOLUTION

Road Map & Modeling When the books strike the table, the bottom book *must* slide on the table because it would otherwise experience an infinite deceleration in going from v_0 to zero. Since there is no similar argument telling us that the top book must slide relative to the bottom book, we begin by *assuming* that m_2 *does not* slip on m_1. After we obtain the solution using this assumption, we will check whether or not the results are consistent with it. If not, we will conclude that the books slide relative to one another, and we will have to compute a new solution. Since we need to determine their individual positions, we draw an FBD of each book as shown in Fig. 2. We have chosen the x axis to be parallel to the trajectory of the books, with the origin taken to be the point at which the books first impact the table.

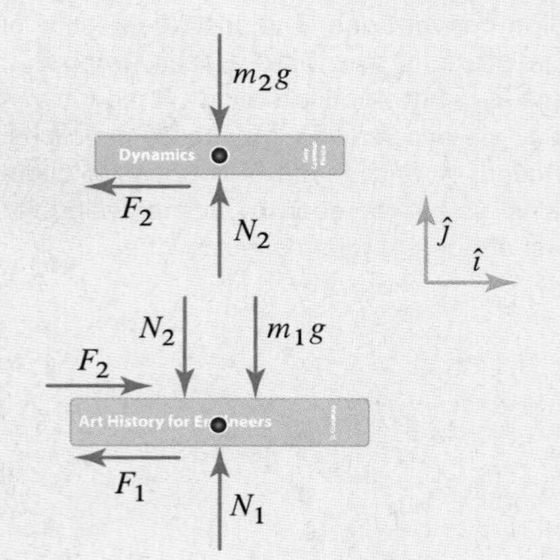

Figure 2
FBDs for the two books modeled as particles.

Governing Equations

Balance Principles Applying Newton's second law to the FBDs in Fig. 2, we obtain

$$\left(\sum F_x\right)_2: \quad -F_2 = m_2 a_{2x}, \tag{1}$$

$$\left(\sum F_y\right)_2: \quad N_2 - m_2 g = m_2 a_{2y}, \tag{2}$$

$$\left(\sum F_x\right)_1: \quad F_2 - F_1 = m_1 a_{1x}, \tag{3}$$

$$\left(\sum F_y\right)_1: \quad N_1 - N_2 - m_1 g = m_1 a_{1y}. \tag{4}$$

Force Laws Using the no-slip assumption between books 1 and 2 and the sliding assumption between book 1 and the table, the friction laws are

$$F_1 = \mu_{k1} N_1 \quad \text{and} \quad |F_2/N_2| < \mu_{s2}. \tag{5}$$

Kinematic Equations Letting a be the common acceleration of the books, we have

$$a_{2x} = a, \quad a_{2y} = 0, \quad a_{1x} = a, \quad \text{and} \quad a_{1y} = 0. \tag{6}$$

Computation The first of Eqs. (5), along with the equations resulting from substituting Eqs. (6) into Eqs. (1)–(4), forms a system of five equations in the five unknowns F_1, F_2, a, N_1, and N_2. Since our first objective is to verify whether or not the no-slip assumption is correct, we begin by solving for just F_2 and N_2:

$$F_2 = \mu_{k1} m_2 g = 4.414\,\text{N} \quad \text{and} \quad N_2 = m_2 g = 9.810\,\text{N}. \tag{7}$$

Discussion & Verification Substituting the results of Eqs. (7) into the inequality in Eq. (5), we have

$$|F_2/N_2| = 0.45 \not< \mu_{s2} = 0.4, \tag{8}$$

which means that the no-slip assumption is incorrect and that m_2 *does* slip over m_1.

m_2 slides over m_1

We now rework the problem, assuming that the books slide relative to one another. The FBDs in Fig. 2 still apply, so Eqs. (1)–(4) are still valid. However, the force laws and corresponding kinematic equations need to reflect the new working assumption.

Force Laws The force laws are now

$$F_1 = \mu_{k1} N_1 \quad \text{and} \quad F_2 = \mu_{k2} N_2. \tag{9}$$

Kinematic Equations The x components of acceleration of m_1 and m_2 are now different, so we have

$$a_{2x} = a_2, \quad a_{2y} = 0, \quad a_{1x} = a_1, \quad \text{and} \quad a_{1y} = 0. \tag{10}$$

Computation The relations obtained by substituting Eqs. (10) into Eqs. (1)–(4), along with Eqs. (9), form a system of six equations in the six unknowns a_1, a_2, F_1, F_2, N_1, and N_2. Solving this system, we have

$$a_1 = -(g/m_1)[\mu_{k1}(m_1 + m_2) - \mu_{k2} m_2] = -5.396\,\text{m/s}^2, \tag{11}$$
$$a_2 = -\mu_{k2} g = -2.943\,\text{m/s}^2, \tag{12}$$
$$F_1 = \mu_{k1} g(m_1 + m_2) = 11.04\,\text{N}, \tag{13}$$
$$F_2 = \mu_{k2} m_2 g = 2.943\,\text{N}, \tag{14}$$
$$N_1 = g(m_1 + m_2) = 24.52\,\text{N}, \tag{15}$$
$$N_2 = g m_2 = 9.810\,\text{N}. \tag{16}$$

Now that we know a_1 and a_2, and recalling that v_0 is the initial speed of both books, we can compute how far each of them slides, using constant acceleration kinematics from Section 2.2. This yields

$$0 = v_0^2 + 2a_1 x_1 = v_0^2 - \frac{2g}{m_1}[\mu_{k1}(m_1 + m_2) - \mu_{k2} m_2] x_1, \tag{17}$$
$$0 = v_0^2 + 2a_2 x_2 = v_0^2 - 2\mu_{k2} g x_2. \tag{18}$$

Solving for x_1 and x_2 and using the given parameters, we have

$$x_1 = \frac{m_1 v_0^2}{2g[\mu_{k1}(m_1 + m_2) - \mu_{k2} m_2]} = 0.05213\,\text{m}, \tag{19}$$
$$x_2 = \frac{v_0^2}{2g\mu_{k2}} = 0.09557\,\text{m}. \tag{20}$$

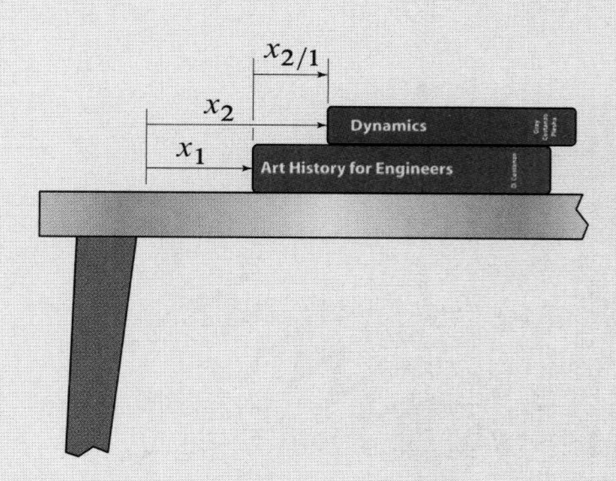

Figure 3
Relative position kinematics of the two books.

Referring to Fig. 3, our solution implies that book 2, as viewed by an inertial observer, i.e., someone sitting on the table, slides 9.6 cm. However, *relative* to the bottom book, it slides

$$x_{\text{top/bottom}} = x_{2/1} = x_2 - x_1 = 0.04344\,\text{m}. \tag{21}$$

Discussion & Verification The signs of the results in Eqs. (19)–(21) are what we expect given that both books move to the right, and book 2 stops to the right of book 1. Also, the distances slid by the books are on the order of centimeters, which seems reasonable. Because their accelerations are constant and the two books start their motion from the same position with identical speeds, the fact that $x_1 < x_2$ implies that book 1 stops first. This tells us that the friction laws used in the present solution remained valid throughout the motion of both books. Had the situation been reversed, we would have had to compute a new solution, although such a scenario is not achievable using the Coulomb friction model. In conclusion, the current solution appears to be correct.

Problems

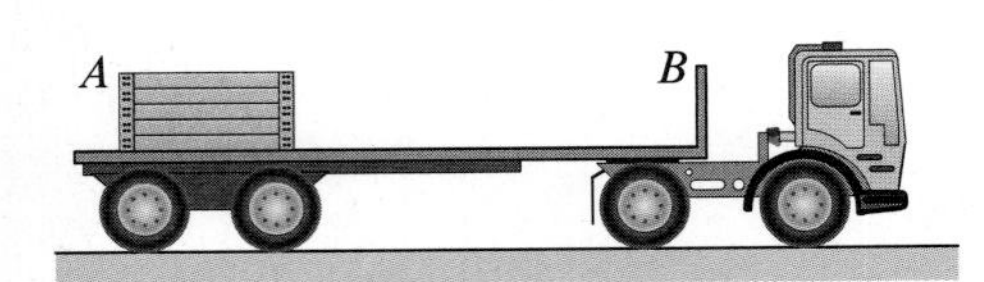

Figure P3.101

Problem 3.101

The driver of the truck suddenly applies the brakes, and the truck comes to a stop. During braking, either the crate slides or it does not. Considering the forces acting on the truck during braking, will the truck stop in a shorter distance (or time) if the crate slides, or will the distance (or time) be shorter if it does not? Justify your answer.

Problem 3.102

A car is being pulled to the right in the two ways shown. Neglecting the inertia of the pulleys and rope, as well as any friction in the pulleys, if the car is allowed to roll freely, will the acceleration of the car in (a) be smaller, equal to, or larger than the acceleration of the car in (b)?

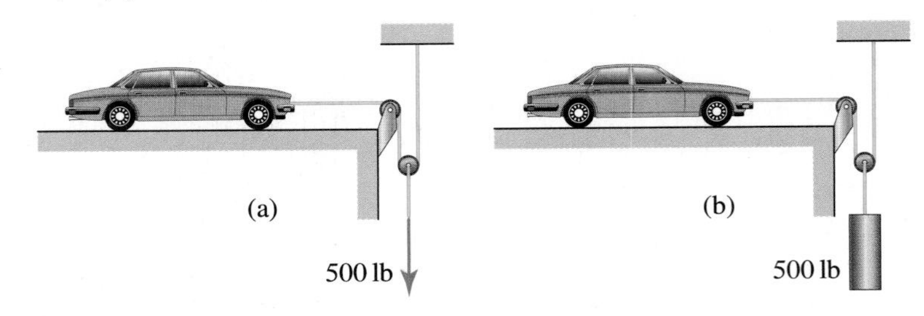

Figure P3.102

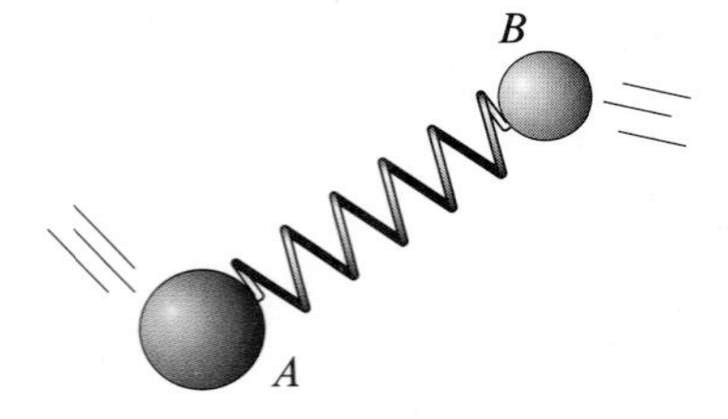

Figure P3.103

Problem 3.103

Particles A and B, which are connected with a massless linear spring, have been thrown up in the air and are moving under the action of the spring force and their own weight. Assuming that no other forces are affecting the motion of the particles, what will be the acceleration of their center of mass?

Problem 3.104

Two particles A and B with masses m_A and m_B, respectively, are placed at a distance r_0 from one another. Assuming that the only force acting on the masses is their mutual gravitational attraction, determine the acceleration of particle B as seen from particle A.

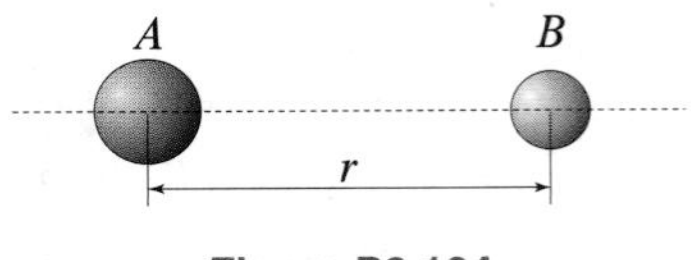

Figure P3.104

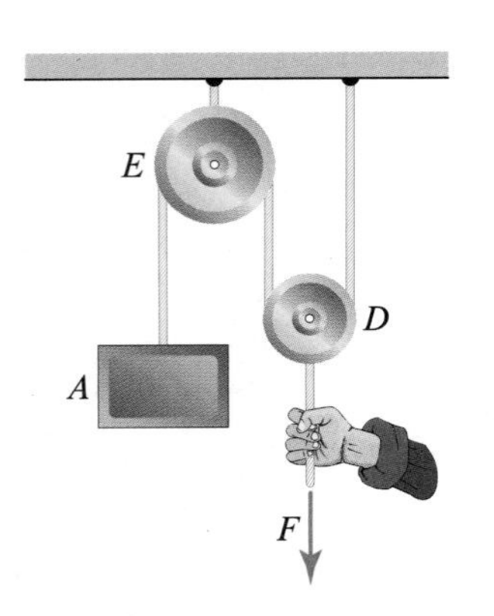

Figure P3.105–P3.107

Problem 3.105

A person lifts the 80 kg load A by pulling down on the rope with a constant force F as shown. Neglecting any source of friction as well as the inertia of the ropes and the pulleys, determine F if A accelerates upward at $0.5\,\text{m/s}^2$.

Problem 3.106

The load A weighs 185 lb. Neglecting any source of friction as well as the inertia of the ropes and the pulleys, determine the acceleration of A if a person pulls down on the rope with a constant force $F = 185\,\text{lb}$ as shown.

Problem 3.107

A person lifts the 80 kg load A by pulling down on the rope with a constant force F as shown. Neglecting friction, the inertia of the ropes, and the rotational inertia of the pulleys, but accounting for the fact that pulley D has a mass $m_D = 8\,\text{kg}$, determine F if A accelerates upward at $2.5\,\text{m/s}^2$.

Problem 3.108

Revisit Example 3.13 and determine the expression for the acceleration of A if the load at P is replaced by a force with magnitude equal to the load's weight, i.e., $F = m_P g$.

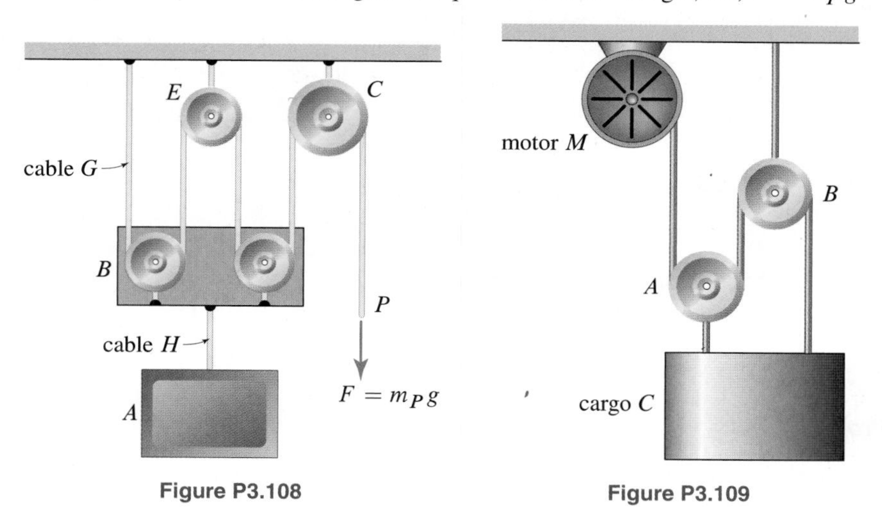

Figure P3.108

Figure P3.109

Problem 3.109

The motor M is at rest when someone flips a switch and it starts pulling in the rope. The acceleration of the rope is uniform, and it takes 1 s to achieve a retraction rate of 4 ft/s. After 1 s the retraction rate becomes constant. Determine the tension in the cable during and after the initial 1 s interval. The cargo C weighs 130 lb, pulleys A and B each weigh 12 lb, and the weight of the ropes is negligible. Neglect friction in the pulleys and the rotational inertia of the pulleys.

Problem 3.110

Revisit Example 3.14 and assume that the static coefficient of friction between the two books is $\mu_{s2} = 0.55$, while all other parameters stay as specified in the example. Determine the acceleration of each of the books.

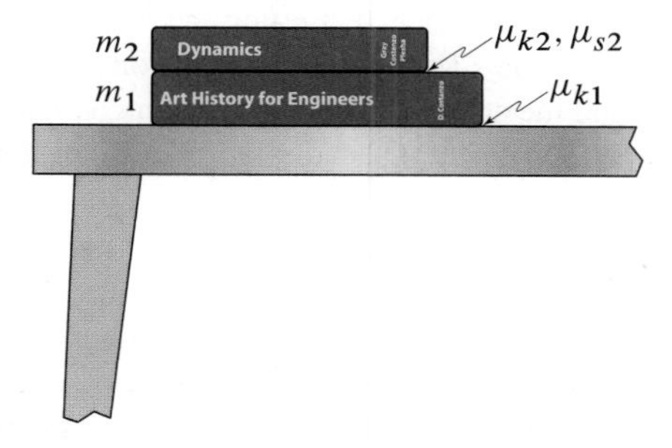

Figure P3.110

Problem 3.111

As seen in Fig. P3.111(a), a window washing platform is controlled by the two pulley systems at AB and CD. The workers E and F can raise and lower the platform P by pulling on the cables H and I, respectively. The weight of each of the workers is 185 lb, and the platform P weighs 200 lb. A schematic representation of the pulley system is shown in Fig. P3.111(b). If the workers start from rest and, in 1.5 s, uniformly start pulling the cable in at 2.5 ft/s, determine the force each worker must exert on the cables H and I during that 1.5 s. Neglect the mass of the pulleys, friction in the pulleys, and the mass of the cable. Assume each worker pulls with the same force, and ignore the departure from vertical of segments H and I.

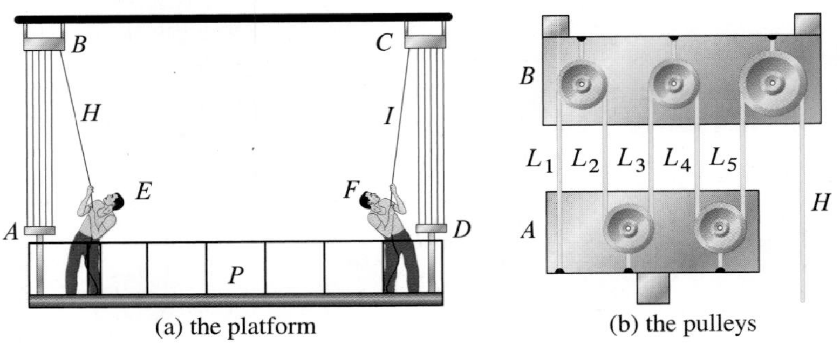

Figure P3.111

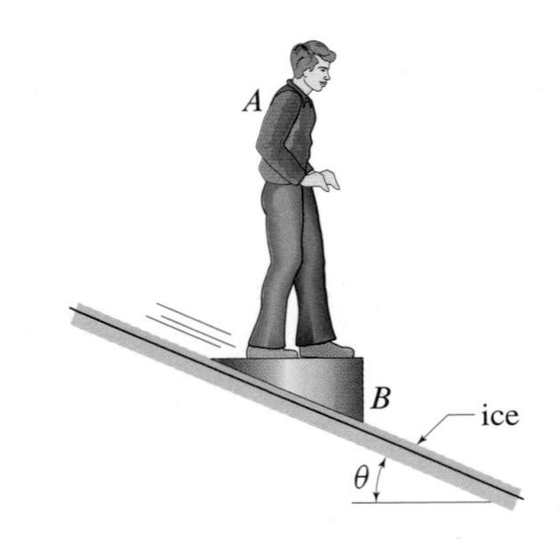

Figure P3.112 and P3.113

Problem 3.112

A man A is trying to keep his balance while on a metal wedge B that is sliding down an icy incline. Letting $m_A = 78$ kg and $m_B = 25$ kg be the masses of A and B, respectively, and assuming that there is enough friction between A and B for A not to slide with respect to B, determine the value of the normal reaction force between A and B, as well as the magnitude of their acceleration if $\theta = 20°$. Friction between the wedge and the incline is negligible.

Problem 3.113

A man A is trying to keep his balance while on a metal wedge B that is sliding down an icy incline. The weights of A and B are $W_A = 181$ lb and $W_B = 50$ lb, respectively. Determine the minimum coefficient of static friction μ_s between A and B required for A not to slide with respect to B if $\theta = 23°$. Friction between the wedge and the incline is negligible.

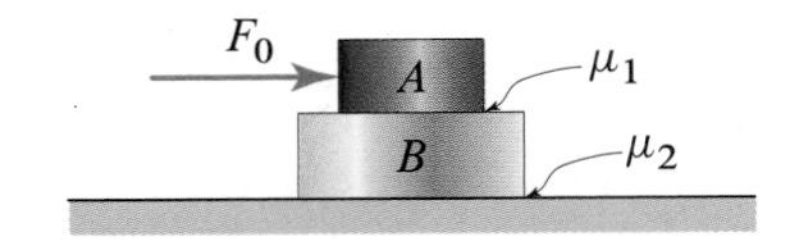

Figure P3.114

Problem 3.114

A force F_0 of 400 lb is applied to block A. Letting the weights of A and B be 55 and 73 lb, respectively, and letting the static *and* kinetic friction coefficients between blocks A and B be $\mu_1 = 0.25$, and the static *and* kinetic friction coefficients between block B and the ground be $\mu_2 = 0.45$, determine the accelerations of both blocks.

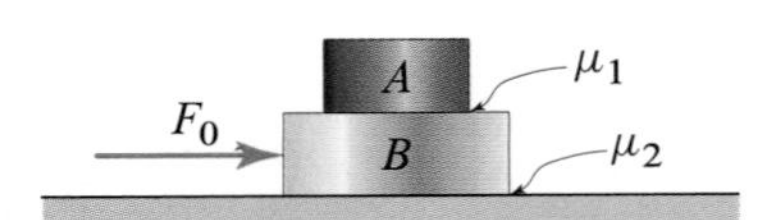

Figure P3.115

Problem 3.115

A force F_0 of 400 lb is applied to block B. Letting the weights of A and B be 55 and 73 lb, respectively, and letting the static *and* kinetic friction coefficients between blocks A and B be $\mu_1 = 0.25$, and the static *and* kinetic friction coefficients between block B and the ground be $\mu_2 = 0.45$, determine the accelerations of both blocks.

Problem 3.116

Blocks A and B are connected by a pulley system. The coefficient of kinetic friction between the block A and the incline is $\mu_k = 0.25$, and static friction is insufficient to prevent slipping. The mass of A is $m_A = 7\,\text{kg}$, the mass of B is $m_B = 20\,\text{kg}$, and the angle between the incline and the horizontal is $\theta = 30°$. Determine the acceleration of A, the acceleration of B, and the tension in the rope after the system is released.

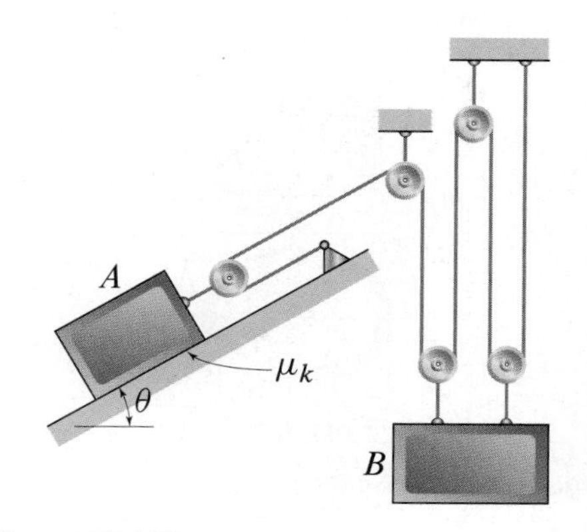

Figure P3.116

Problem 3.117

Two identical balls, each of mass m, are connected by a string of negligible mass and length $2l$. A short string is attached to the middle of the string connecting the two balls and is pulled vertically with a constant force P. If the system starts from rest at $\theta = \theta_0$ and assuming that the balls only move in the horizontal direction, determine the expression for the speed of the two balls as θ approaches 90°. Neglect the size of the balls as well as friction between the balls and the surface on which they slide.

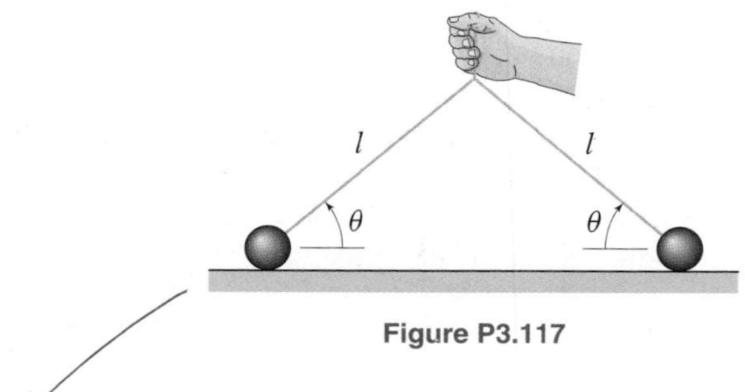

Figure P3.117

Problem 3.118

The two sliders A and B of mass $m_A = 4\,\text{kg}$ and $m_B = 3\,\text{kg}$, respectively, move with negligible friction in the slots shown, which lie in the vertical plane. They are connected by a rigid bar of negligible mass and length $L = 0.5\,\text{m}$. If, for $y_B = 0.3\,\text{m}$, the system is initially at rest, determine the acceleration of each slider and the force in the bar immediately after release. *Hint:* The force that the bar exerts on each slider has the same direction as the bar itself.

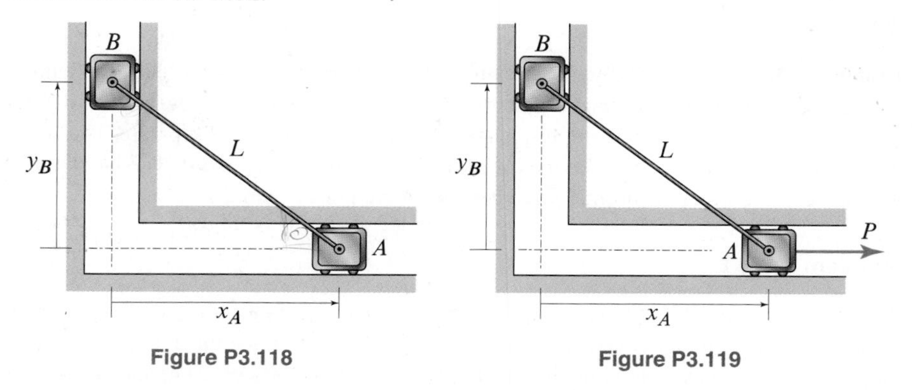

Figure P3.118 Figure P3.119

Problem 3.119

The two sliders A and B of weight $W_A = 8\,\text{lb}$ and $W_B = 6\,\text{lb}$, respectively, move with negligible friction in the slots shown, which lie in the vertical plane. They are connected by a rigid bar of negligible weight and length $L = 1.75\,\text{ft}$. Slider A is also subject to the force $P = 15\,\text{lb}$. If, for $y_B = 1.0\,\text{ft}$, the system is initially at rest, determine the acceleration of each slider and the force in the bar immediately after release. *Hint:* The force that the bar exerts on each slider has the same direction as the bar itself.

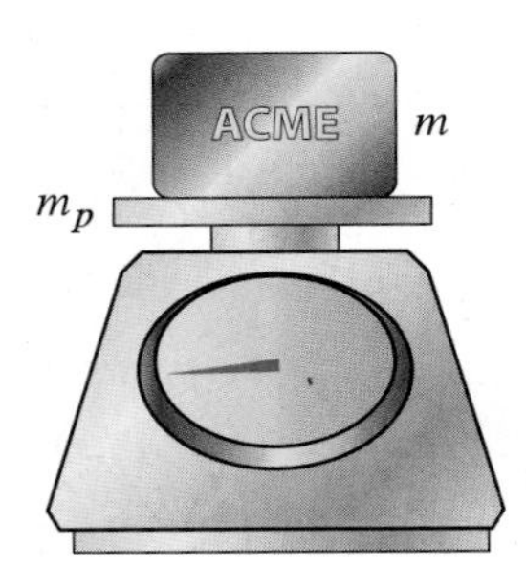

Figure P3.120 and P3.121

Problems 3.120 and 3.121

Spring scales work by measuring the displacement of a spring that supports both the platform of mass m_p and the object, of mass m, whose weight is being measured. Most scales read zero when no mass m has been placed on them; that is, they are calibrated so that the weight reading accounts for the mass of the platform m_p. Assume that the spring is linear elastic with spring constant k.

Problem 3.120 If the mass m is gently placed on the spring scale (i.e., it is released from zero height above the scale), determine the *maximum reading* on the scale after the mass is released.

Problem 3.121 If the mass m is gently placed on the spring scale (i.e., it is released from zero height above the scale), determine the expression for *maximum speed* attained by the mass m as the spring compresses.

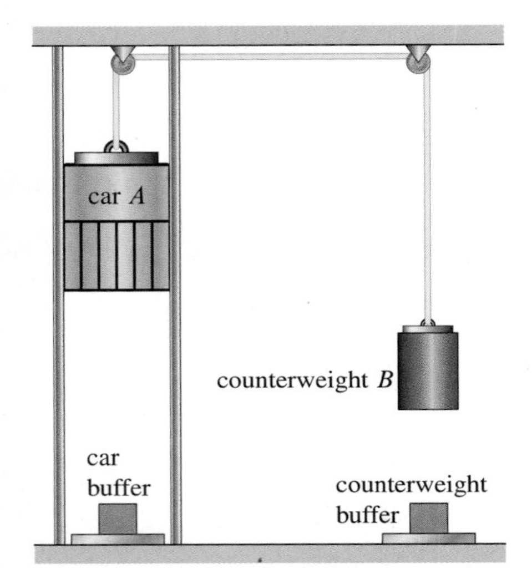

Figure P3.122

Problem 3.122

A simple elevator consists of a 15,000 kg car A connected to a 12,000 kg counterweight B. Suppose that a failure occurs when the car is at rest and 50 m above its buffer, causing the elevator car to fall. Model the car and the counterweight as particles and the cord as massless and inextensible; and model the action of the emergency brakes using a Coulomb friction model with kinetic friction coefficient $\mu_k = 0.5$ and a normal force equal to 35% of the car's weight. Determine the speed with which the car impacts the buffer.

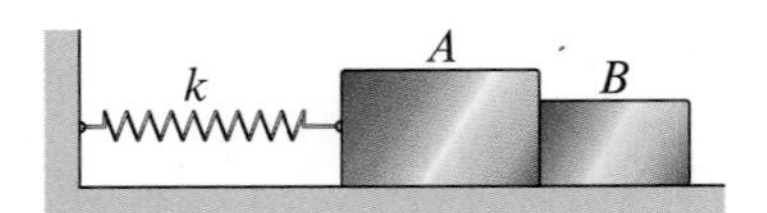

Figure P3.123 and P3.124

Problems 3.123 and 3.124

The linear elastic spring with stiffness k and unstretched length ℓ_u is attached to both the vertical wall and the metal block A of mass m_A. The metal block B is pushed into A so that the spring compresses a distance d. The block B is then released from rest.

Problem 3.123 Assuming that friction between the blocks and the horizontal surface is negligible, determine the distance the blocks slide before they separate and their speed at separation. *Hint:* The blocks will start to separate when the normal force between them goes to zero.

Problem 3.124 Assuming that friction between the blocks and the surface is non-negligible and that the coefficient of static friction is insufficient to prevent motion, determine the condition on the compression distance d for the blocks to separate. The coefficient of kinetic friction between the blocks and the surface is μ_k. *Hint:* The blocks will start to separate when the normal force between them goes to zero.

Problem 3.125

Two particles A and B with masses m_A and m_B, respectively, are a distance r_0 apart, and both masses are initially at rest. Using Eq. (1.5) on p. 3, determine the amount of time it takes for the two masses to come into contact if $m_A = 1$ kg, $m_B = 2$ kg, and $r_0 = 1$ m. Assume that the two masses are only influenced by their mutual gravitational attraction. *Hint:* $\int_0^{r_0} \sqrt{r_0 r/(r_0 - r)}\, dr = \frac{1}{2}\pi r_0^{3/2}$.

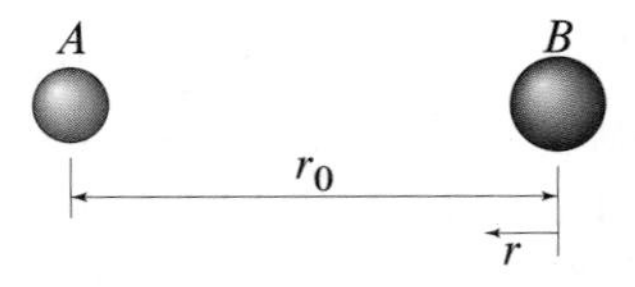

Figure P3.125

Problem 3.126

Energy storage devices that use spinning flywheels to store energy are becoming available. To maximize energy storage, the flywheel must spin as fast as possible. Unfortunately, if it spins too fast, internal stresses in the flywheel cause it to come apart catastrophically. Therefore, it is important to keep the speed at the periphery of the flywheel below about 1000 m/s. It is also critical that the flywheel be well balanced to avoid the damaging vibrations that would otherwise result. With this in mind, let the flywheel D with diameter 0.3 m rotate at $\omega = 60{,}000$ rpm. In addition, assume that the cart B is constrained to move rectilinearly along the guide tracks. Given that the flywheel is not perfectly balanced, that the unbalanced weight A has mass m_A, and that the total mass of the flywheel D, cart B, and electronics package E is m_B, determine the constraint force between the wheels of the cart and the guide tracks as a function of θ, the masses, the diameter, and the angular speed of the flywheel. What is the *maximum* constraint force between the wheels of the cart and the guide tracks? Finally, evaluate your answers for $m_A = 1$ g (about the mass of a paper clip) and $m_B = 70$ kg. Assume that the unbalanced mass is at the periphery of the flywheel.

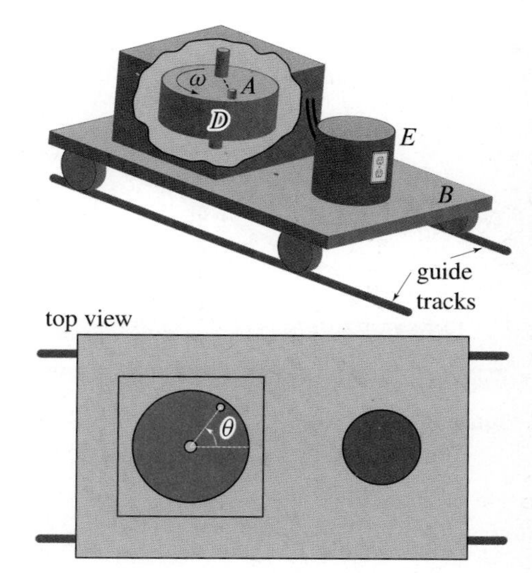

Figure P3.126

Problem 3.127

The double pendulum shown consists of two particles with masses $m_1 = 7.5$ kg and $m_2 = 12$ kg connected by two inextensible cords of length $L_1 = 1.4$ m and $L_2 = 2$ m and negligible mass. If the system is released from rest when $\theta = 10°$ and $\phi = 20°$, determine the tension in the two cords at the instant of release.

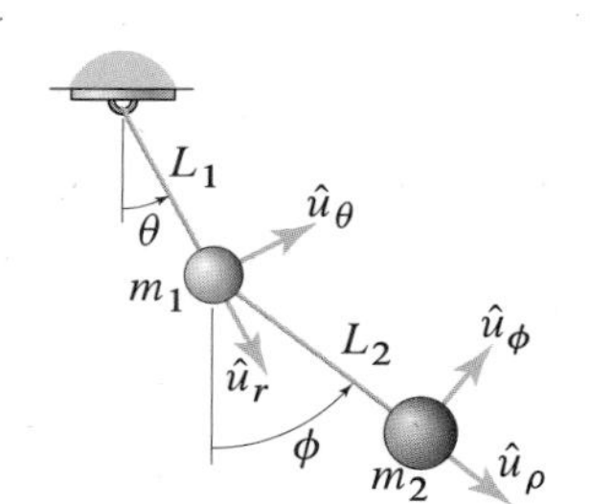

Figure P3.127

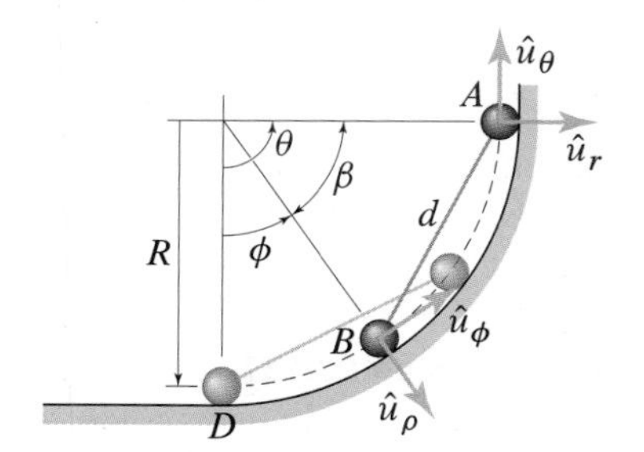

Figure P3.128 and P3.129

Problems 3.128 and 3.129

Two small spheres A and B, each of mass m, are attached at either end of a rod of length d. The system is released from rest in the position shown. Neglect friction, treat the spheres as particles (assume that their diameter is negligible), neglect the mass of the rod, assume the rod is rigid, and assume that $d < R$. *Hint:* The force that the rod exerts on either ball has the same direction as the rod itself.

Problem 3.128 Using the angle θ as the dependent variable, derive the equation of motion for the particle system from the moment of release until particle B reaches point D.

Problem 3.129 Determine the expression for the speed of the spheres immediately before B reaches point D.

Problem 3.130

A 62 kg woman A sits atop the 60 kg cart B, both of which are initially at rest. The cart is rigidly attached to a wall by the rope CD. If the woman slides down the frictionless incline of length $L = 3.5$ m, determine the tension in the rope CD as she slides down the incline. Ignore the mass of the wheels on which the cart can roll. The angle $\theta = 26°$.

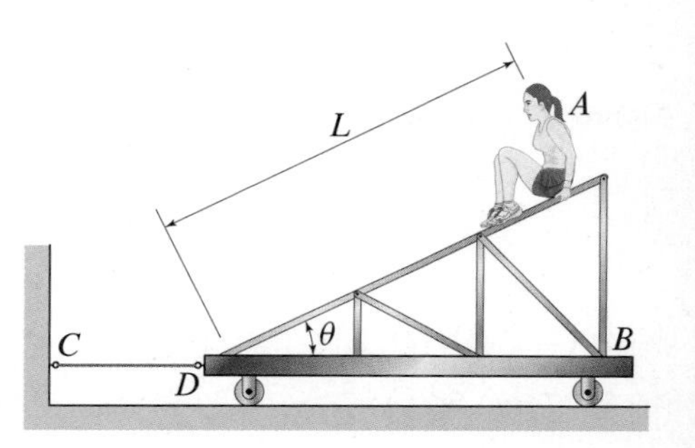

Figure P3.130

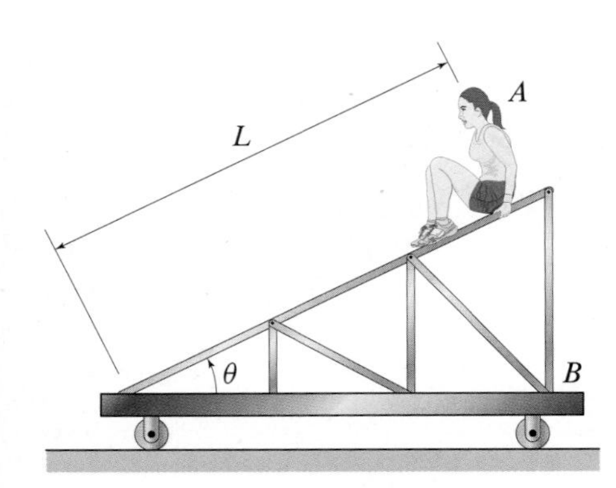

Figure P3.131

Problem 3.131

A 62 kg woman A sits atop the 60 kg cart B, both of which are initially at rest. If $\theta = 26°$ and the woman slides down the incline of length $L = 3.5$ m, determine the velocity of both the woman and the cart when she reaches the bottom of the incline. Ignore the mass of the wheels on which the cart rolls and friction in their bearings, and neglect friction between the woman and the incline.

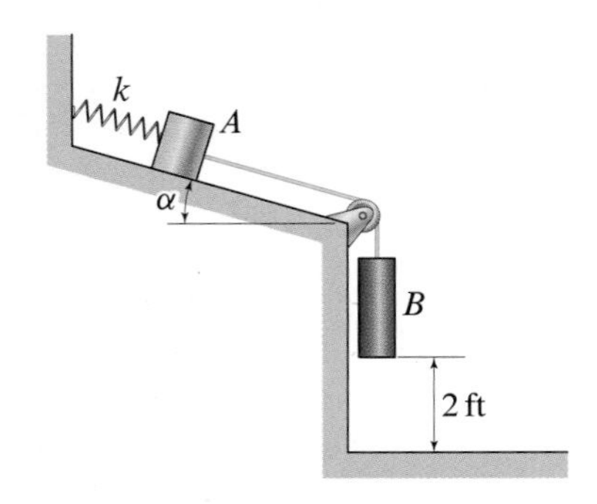

Figure P3.132–P3.135

Problems 3.132 through 3.135

Two blocks A and B weighing 123 and 234 lb, respectively, are released from rest as shown. At the moment of release the spring is unstretched. In solving these problems, model A and B as particles, neglect air resistance, and assume that the cord is inextensible. *Hint:* If B hits the ground, then its maximum displacement is equal to the distance between the initial position of B and the ground.

Problem 3.132 Determine the maximum displacement and the maximum speed of block B if $\alpha = 0°$, the contact between A and the surface is frictionless, and the spring constant is $k = 30$ lb/ft.

Problem 3.133 Determine the maximum displacement and the maximum speed of block B if $\alpha = 20°$, the contact between A and the incline is frictionless, and the spring constant is $k = 30$ lb/ft.

Problem 3.134 Determine the maximum displacement and the maximum speed of block B if $\alpha = 20°$, the contact between A and the incline is frictionless, and the spring constant is $k = 300$ lb/ft.

Problem 3.135 Determine the maximum displacement and the maximum speed of block B if $\alpha = 35°$, the static and kinetic friction coefficients are $\mu_s = 0.25$ and $\mu_k = 0.2$, respectively, and the spring constant is $k = 25$ lb/ft.

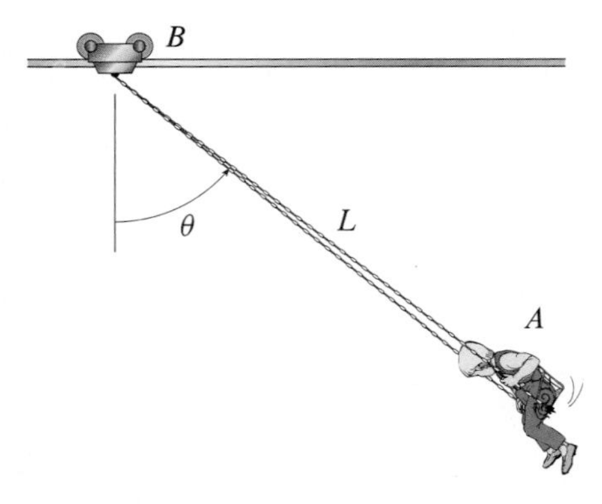

Figure P3.136 and P3.137

Problems 3.136 and 3.137

In the ride shown, a person A sits in a seat that is attached by a cable of length L to a freely moving trolley B of mass m_B. The total mass of the person and the seat is m_A. The trolley is constrained by the beam to move in only the horizontal direction. The system is released from rest at the angle $\theta = \theta_0$ and is allowed to swing in the vertical plane. Neglect the mass of the cable, and treat the person and the seat as a single particle.

Problem 3.136 Derive the system's equations of motion, using the position of the trolley and the angle θ as dependent variables.

Problem 3.137 Derive the system's equations of motion, using the position of the trolley and the angle θ as dependent variables, and then use a computer to solve these equations for one full period/cycle of the motion. Plot the speed of the trolley and the speed of the person vs. the angle θ for $m_A = 45$ kg, $m_B = 10$ kg, $L = 3$ m, and $\theta_0 = 70°$.

Design Problems

Design Problem 3.3

In pulley problems, we have assumed that the ropes are inextensible and massless. In some engineering problems, such as in the design of fast elevators, this may not be a reasonable assumption. Here we will confront one aspect of any design process concerning the *quality* of the information provided by models of different complexity. In this problem, we will consider the effects of the rope's deformation. However, to avoid excessive complexity, we will model the rope's elasticity as being "lumped" at one end.

Let the system be released from rest and the velocity of A be controlled so that it accelerates uniformly downward to a speed of $3\,\mathrm{m/s}$ in $1.2\,\mathrm{s}$. Over this time interval, determine and plot, for different values of k, the position of B as a function of time, as well as the tension in the rope as a function of time. Keep in mind that the inextensible rope model can be viewed as a special case of the deformable rope model with an infinite stiffness k. Consider various values of k, starting with extremely large values (to simulate infinity) and gradually decreasing the value of k to understand at what point the two models provide significantly different answers (you will need to decide what *significant* means in this context).

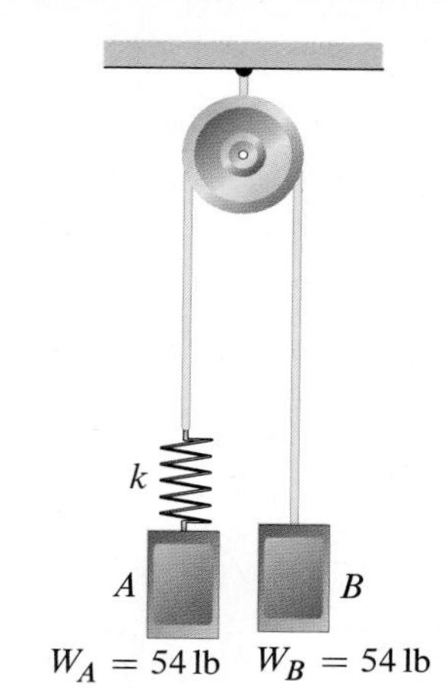

Figure DP3.3

Design Problem 3.4

In this problem we will explore the assumption that we can neglect the mass of cables or ropes in pulley systems.

Consider the simple pendulum in the left part of the figure. Let the pendulum bob be set in motion in the position shown with a speed $v_0 = 3\,\mathrm{m/s}$. Determine the motion of the pendulum bob and, in particular, its maximum displacement from its starting position. Next consider a double pendulum with an equally long cord, but with a cord of mass m_C. To gain an appreciation for the effect of the mass of the cord on the pendulum's motion, let the entire mass of the cord be lumped at its midpoint (we then neglect the mass of the two cords connecting O to m_C and m_C to m_B). Next let this double pendulum be set in motion in the same way as in the previous case. Again, determine the motion of the pendulum bob m_B, and determine the value of m_C necessary to cause a 10% difference in the maximum displacement from its starting position.

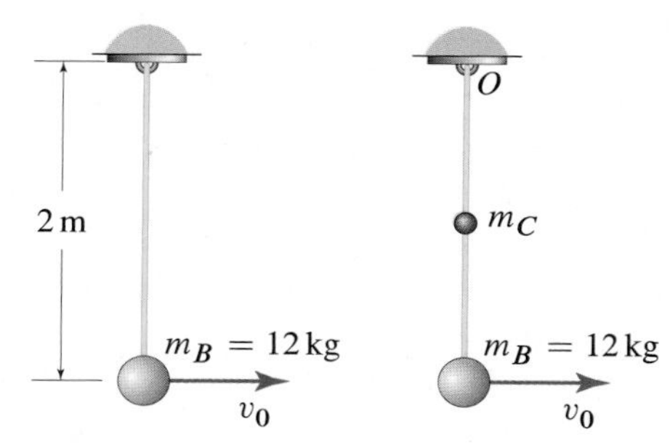

Figure DP3.4

Design Problem 3.5

In this problem we will explore the assumption that we can neglect the mass of cables or ropes in pulley systems.

Consider the simple pendulum in the left part of the figure. Let the pendulum bob be set in motion in the position shown with a speed $v_0 = 3\,\mathrm{m/s}$. Determine the motion of the pendulum bob and, in particular, its maximum displacement from its starting position. Next consider a pendulum with an equally long cord but with a cord of mass m_C. To gain an appreciation for the effect of the mass of the cord on the pendulum's motion, let the entire mass of the cord be lumped at its midpoint (we then neglect the mass of the single rigid rod connecting O to m_C to m_B). Next let this rigid pendulum be set in motion in the same way as in the previous case. Again, determine the motion of the pendulum bob m_B, and determine the value of m_C necessary to cause a 10% difference in the maximum displacement from its starting position. *Hint:* Assume that the rigid bar can provide only forces that are parallel to the bar itself.

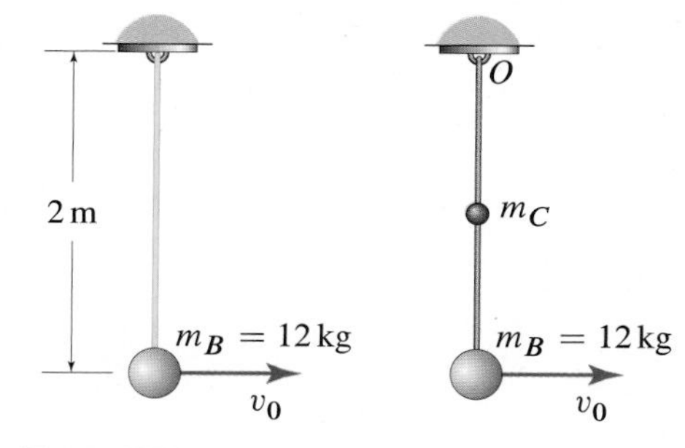

Figure DP3.5

Chapter Review

In this chapter, we have presented a general approach to the solution of kinetics problems involving the motion of a single particle or a system of particles. We now present a concise summary.

Helpful Information

We will structure the solution of kinetics problems using the following four steps:

1. *Road Map & Modeling:* Identify data and unknowns, identify the system, state assumptions, sketch the FBD, and identify a problem-solving strategy.
2. *Governing Equations:* Write the balance principles, force laws, and kinematic equations.
3. *Computation:* Solve the assembled system of equations.
4. *Discussion & Verification:* Study the solution and perform a "sanity check."

Applying Newton's second law. We developed a four-step problem-solving procedure for applying Newton's second law to mechanical systems. The central element of this procedure is the derivation of the governing equations, which originate from the

1. Balance principles.
2. Force laws.
3. Kinematic equations.

In this chapter, the balance principle was Newton's second law, which we write as

Eq. (3.1), p. 169

$$\vec{F} = m\vec{a}.$$

We applied Newton's second law in component form as

Eq. (3.2), p. 170

$$\sum F_a = ma_a, \quad \sum F_b = ma_b, \quad \text{and} \quad \sum F_c = ma_c,$$

where a, b, and c are the orthogonal directions of the chosen component system, and we usually need only two directions for planar problems. The four-step "recipe" we presented for applying Newton's second law, outlined in the margin note, is given as a guide to the *order* in which things should be done, and we will use it consistently in each example we present.

Governing equations and equations of motion. The *governing equations* for a system consist of the (1) balance principles, (2) force laws, and (3) kinematic equations. The *equations of motion* are the equations derived from the governing equations that allow for the determination of the motion.

Degrees of freedom. A system's *degrees of freedom* are the independent coordinates needed to completely describe a system's position. The number of degrees of freedom is equal to the number of different coordinates in a system that must be fixed in order to keep the system from moving. The required number of equations of motion is equal to the number of degrees of freedom for a system.

Friction. We will include friction via the Coulomb friction model. According to this model, in the absence of slip, the magnitude of the friction force F satisfies the inequality

Eq. (3.3), p. 171

$$|F| \leq \mu_s |N|,$$

where μ_s is the *coefficient of static friction* and N is the force normal to the contact surface. The relation

Eq. (3.4), p. 172

$$F = \mu_s N$$

defines the case of impending slip.

If A and B are two objects sliding with respect to one another, the magnitude of the friction force exerted by B onto A is given by

Eq. (3.6), p. 172

$$F = \mu_k N,$$

where μ_k is the *coefficient of kinetic friction* and where the direction of the friction force must be consistent with the fact that friction opposes the relative motion of A and B.

Springs. A spring is said to be *linear elastic* if the internal force in the spring is linearly related to the amount the spring is stretched or compressed. The force F_s required to stretch or compress a linear elastic spring by an amount δ is given by

Eq. (3.14), p. 173

$$F_s = k\delta = k(L - L_0),$$

where k is the *spring constant* and where L and L_0 are the current and unstretched lengths of the spring, respectively. When $\delta > 0$, the spring is said to be stretched; and when $\delta < 0$, the spring is said to be compressed.

Systems of particles

In practice, applying Newton's second law to a system of particles is essentially identical to applying it to a single particle — we write $(\vec{F}_i)_{\text{tot}} = m_i \vec{a}_i$, $i = 1, \ldots, n$, for each of the n particles in the system, where $(\vec{F}_i)_{\text{tot}}$ is the total force acting on particle i, thus including both the external and internal forces acting on the particle. In addition to Newton's second law, we need to enforce Newton's third law, which is crucial to account for the effect of the internal forces.

Motion of the mass center. The motion of the mass center of a system of particles is governed by the relation

Eq. (3.37), p. 218

$$\vec{F} = \sum_{i=1}^{n} m_i \vec{a}_i = m\vec{a}_G,$$

where $\vec{F}$ is the total external force on the system, m is the total mass of the system, and $\vec{a}_G$ is the acceleration of the center of mass of the system. This equation has some important consequences:

- If $\vec{F} = \vec{0}$, then the mass center moves with constant velocity; and if the mass center is initially at rest, it will remain at rest.
- If $\vec{F} = \vec{0}$, then we can conclude that $m\vec{a}_G$ must be zero and its time integral must be a constant.

Review Problems

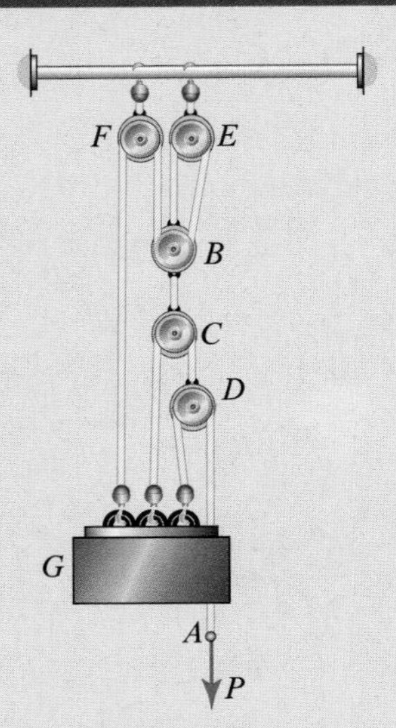

Figure P3.138 and P3.139

Problem 3.138

A constant force P is applied at A to the rope running behind the load G, which has a mass of 300 kg. Assuming that any source of friction and the inertia of the pulleys can be neglected, determine P such that G has an upward acceleration of $1\ \mathrm{m/s^2}$.

Problem 3.139

A constant force $P = 300$ lb is applied at A to the rope running behind the load G, which weighs 1000 lb. If each of the pulleys weighs 7 lb, and assuming that any source of friction and the rotational inertia of the pulleys can be neglected, determine the acceleration of G and the tension in the rope connecting pulleys B and C.

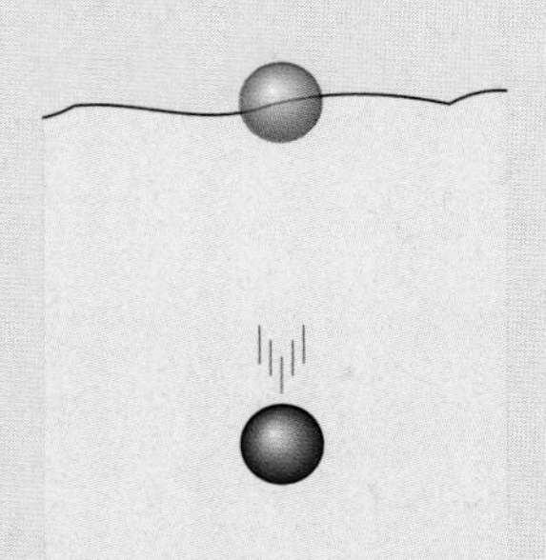

Figure P3.140 and P3.141

Problem 3.140

A metal ball weighing 0.2 lb is dropped from rest in a fluid. If the magnitude of the resistance due to the fluid is given by $C_d v$, where $C_d = 0.5\ \mathrm{lb{\cdot}s/ft}$ is a drag coefficient and v is the ball's speed, determine the depth at which the ball will have sunk when the ball achieves a speed of 0.3 ft/s.

Problem 3.141

A metal ball weighing 0.2 lb is dropped from rest in a fluid. After falling 1 ft, the ball has a speed of 2.25 ft/s. If the magnitude of the resistance due to the fluid is given by $C_d v$, where C_d is a drag coefficient and v is the ball's speed, determine the value of C_d.

Problem 3.142

Two particles A and B, with masses m_A and m_B, respectively, are a distance r_0 apart. Particle B is fixed in space, and A is initially at rest. Using Eq. (1.5) on p. 3 and assuming that the diameters of the masses are negligible, determine the time it takes for the two particles to come into contact if $m_A = 1$ kg, $m_B = 2$ kg, and $r_0 = 1$ m. Assume that the two masses are infinitely far from any other mass. *Hint:* $\int_0^{r_0} \sqrt{r_0 r/(r_0 - r)}\, dr = \frac{1}{2}\pi r_0^{3/2}$.

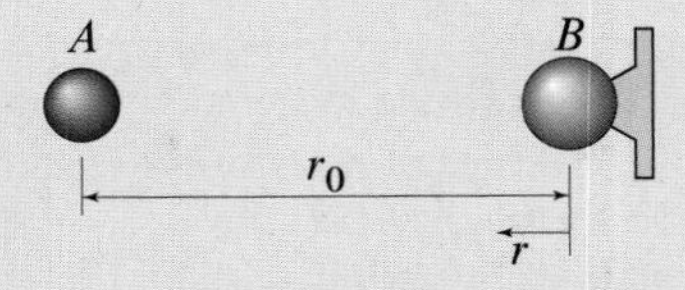

Figure P3.142

Problems 3.143 and 3.144

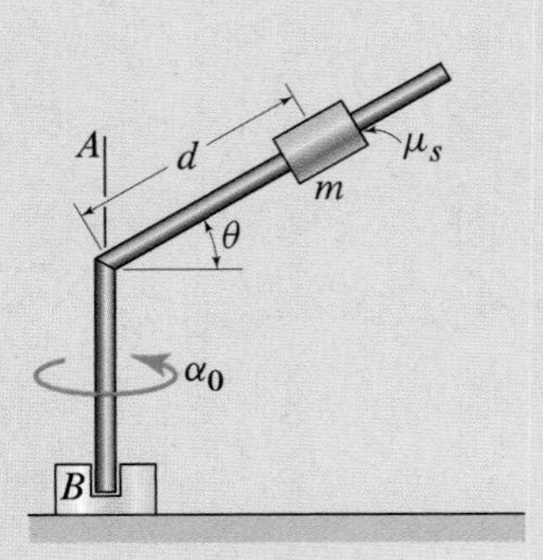

Figure P3.143 and P3.144

The system shown is initially at rest when the bent bar starts to rotate about the vertical axis AB with constant angular acceleration $\alpha_0 = 3\ \mathrm{rad/s^2}$. The coefficient of static friction between the collar of mass $m = 2$ kg and the bent bar is $\mu_s = 0.6$, the angle of the bend in the bar is $\theta = 30°$, and the collar is initially at $d = 70$ cm from the spin axis AB.

Problem 3.143 Assuming the motion starts at $t = 0$, determine the time at which the collar starts to slip relative to the bent bar.

Problem 3.144 Determine the number of rotations undergone by the bent bar when the collar starts to slip relative to it.

Problem 3.145

The centers of two spheres A and B, with weights $W_A = 3$ lb and $W_B = 7$ lb, respectively, are a distance $r_0 = 5$ ft apart when they are released from rest. Using Eq. (1.5) on p. 3, determine the speed with which they collide if the diameters of spheres A and B are $d_A = 2.5$ in. and $d_B = 4$ in., respectively. Assume that the two masses are only influenced by their mutual gravitational attraction.

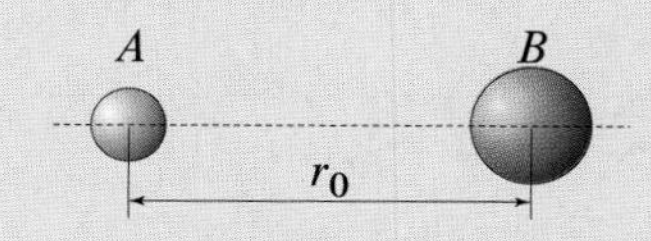

Figure P3.145

Problem 3.146

A roller coaster goes over the top A of the track shown with a speed $v = 135$ km/h. If the radius of curvature at A is $\rho = 60$ m, determine the minimum force that a restraint must apply to a person with a mass of 85 kg to keep the person on his or her seat.

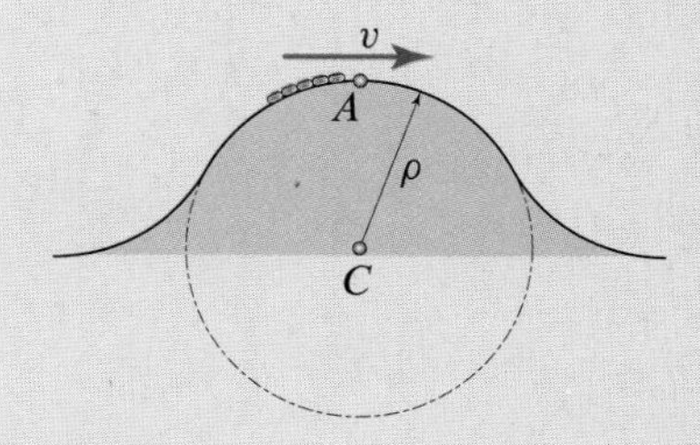

Figure P3.146

Problem 3.147

A 50,000 lb aircraft is flying along a rectilinear path at a constant altitude with a speed $v = 720$ mph when the pilot initiates a turn by banking the plane 20° to the right. Assuming that the initial rate of change of speed is negligible, determine the components of the acceleration of the aircraft right at the beginning of the turn if the pilot does not adjust the attitude of the aircraft so that the magnitude of the lift remains the same as when the plane is flying straight and the aerodynamic drag remains in the horizontal plane. Also determine the radius of curvature at the beginning of the turn.

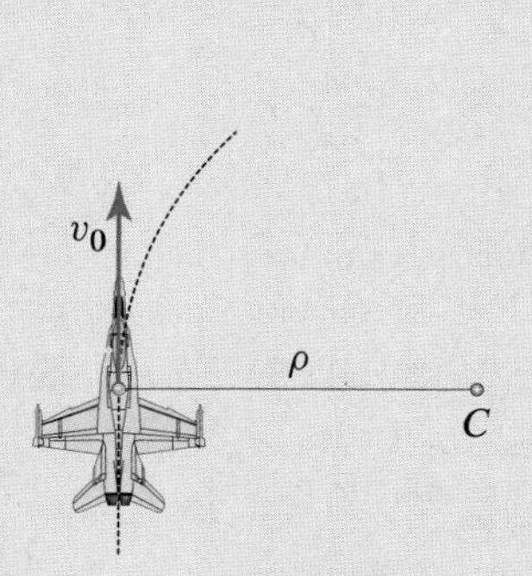

Figure P3.147

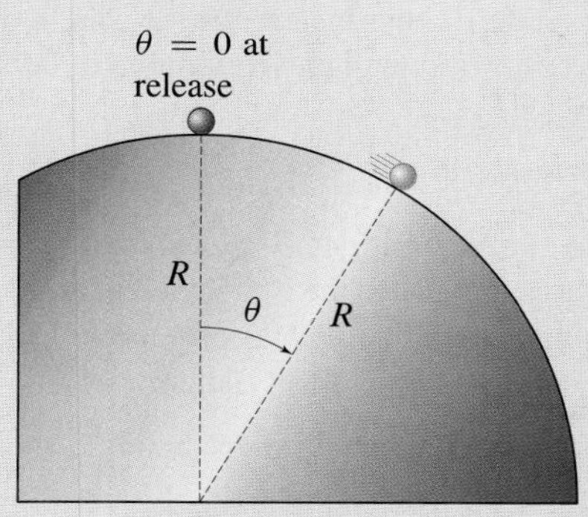

Figure P3.148

Problem 3.148

Referring to Example 3.9 on p. 200, let $R = 1.25$ ft and let the angle at which the sphere separates from the cylinder be $\theta_s = 34°$. If the sphere were placed in motion at the very top of the cylinder, determine the sphere's initial speed.

Problem 3.149

Referring to Example 3.8 on p. 198, show that, for $\theta = 33°$ and under the assumption that $\mu_s > 1/\tan\theta$, the no-slip solution in Eqs. (15) and (16) satisfies the no-slip condition $|F| \leq \mu_s |N|$ for any value of the car's speed.

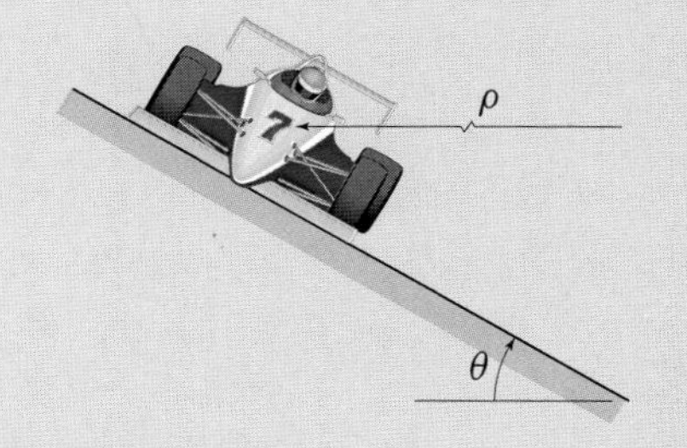

Figure P3.149

Problem 3.150

Revisit Example 3.7 and assume that the drag force acting on the ball has the form $\vec{F}_d = -\eta\vec{v}$, where $\vec{v}$ is the velocity of the ball and η is a drag coefficient. Determine the trajectory of the ball, expressing it in the form $y = y(x)$.

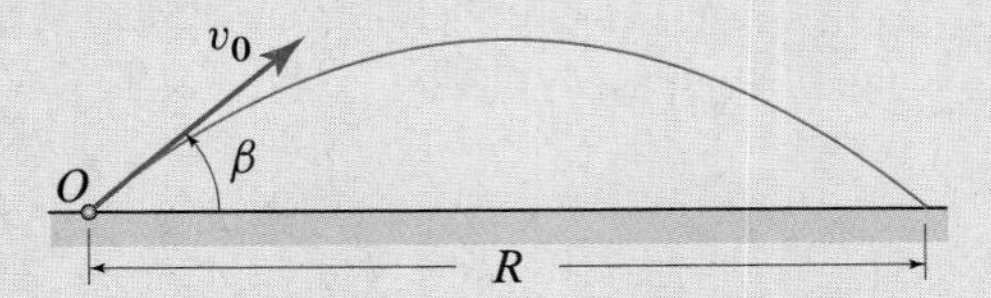

Figure P3.150 and P3.151

Problem 3.151

Revisit Example 3.7 and assume that the drag force acting on the ball has the form $\vec{F}_d = -\eta\vec{v}$, where $\vec{v}$ is the velocity of the ball and η is a drag coefficient. Determine the value of η such that a 1.61 oz ball has a range $R = 270$ yd when put in motion with an initial velocity of magnitude $v_0 = 186$ mph and initial direction $\beta = 11.2°$.

Problem 3.152

The load B has a mass $m_B = 250$ kg, and the load A has a mass $m_A = 120$ kg. Let the system be released from rest, and neglecting any source of friction, as well as the inertia of the ropes and the pulleys, determine the acceleration of A and the tension in the cord to which A is attached.

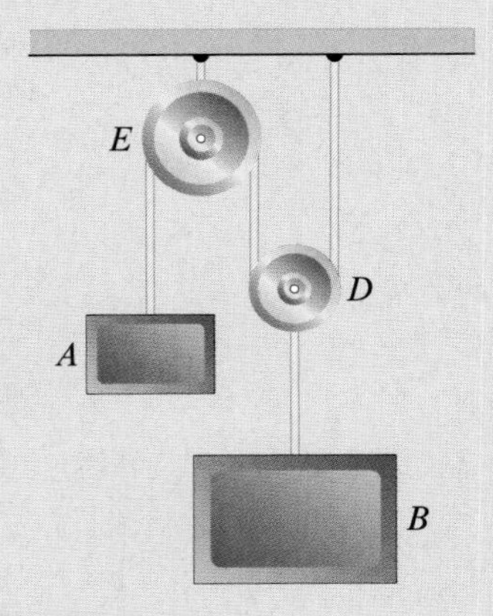

Figure P3.152 and P3.153

Problem 3.153

The load B weighs 300 lb. Neglecting any source of friction, as well as the inertia of the ropes and the pulleys, determine the weight of A if, after the system is released from rest, B moves upward with an acceleration of $0.75\ \text{ft/s}^2$.

Problem 3.154

A crate of mass m is gently placed with zero initial velocity on an inextensible conveyor belt that is moving to the right at a constant speed v_0. Treating the crate as a particle and assuming that the coefficients of static and kinetic friction between the crate and conveyor are μ_s and μ_k, respectively, determine:

(a) the distance the crate slides before it stops slipping relative to the belt, and

(b) the time it takes for the crate to stop sliding.

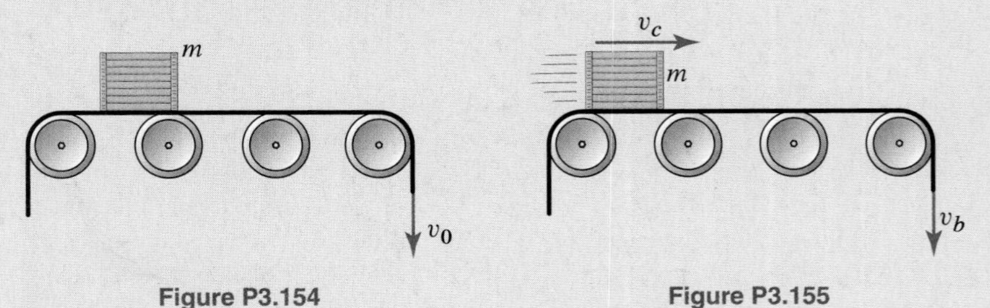

Figure P3.154 Figure P3.155

Problem 3.155

A crate of mass m is thrown horizontally with speed v_c onto an inextensible conveyor belt that is moving to the right at a constant speed v_b. Treating the crate as a particle, knowing that $v_b > v_c$, and assuming that the coefficients of static and kinetic friction between the crate and conveyor are μ_s and μ_k, respectively, determine:

(a) the distance the crate slides before it stops slipping relative to the belt,

(b) the time it takes for the crate to stop sliding, and

(c) the distance the crate moves relative to the belt.

Problem 3.156

A man A is trying to keep his balance while on a metal wedge B that is sliding down an icy incline. Let $m_A = 78$ kg and $m_B = 25$ kg be the masses of A and B, respectively. In addition, let the static and kinetic friction coefficients between A and B be $\mu_s = 0.4$ and $\mu_k = 0.35$, respectively. Determine the acceleration of A if $\theta = 23°$. Friction between the wedge and the incline is negligible.

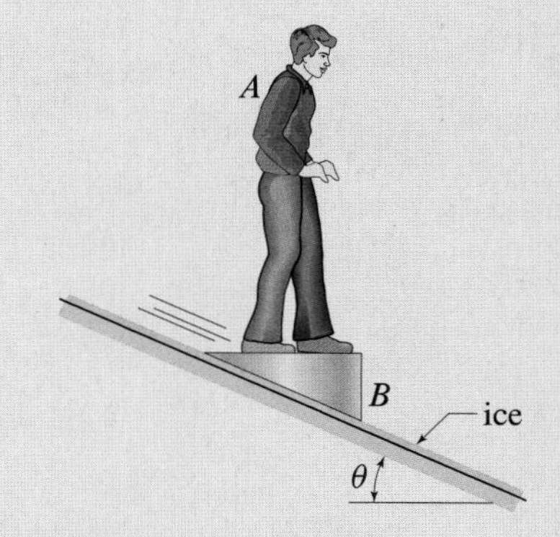

Figure P3.156

Problem 3.157

Derive the equations of motion for the double pendulum shown.

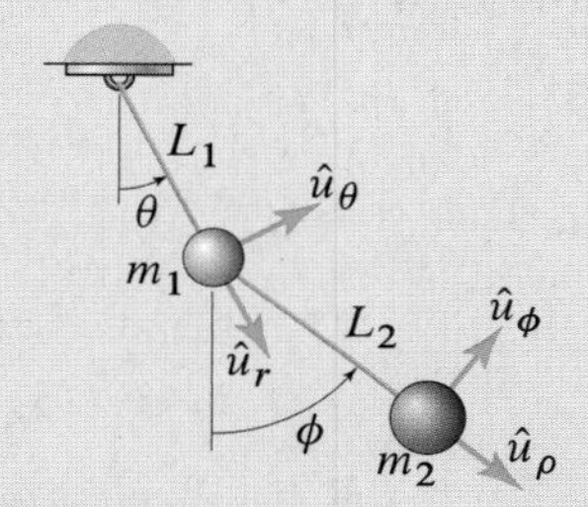

Figure P3.157 and P3.158

Problem 3.158

Derive the equations of motion for the double pendulum shown. After doing so, let $L_1 = 1.4$ m, $L_2 = 2$ m, $m_1 = 7.5$ kg, and $m_2 = 12$ kg, and release the pendulum from rest with $\theta(0) = 25°$ and $\phi(0) = -37°$. Integrate the equations of motion, and plot the trajectory of each of the particles for at least 5 s.

Problem 3.159

Blocks A and B are connected by the pulley system shown. Friction between the block A and the incline is negligible. The weight of A is $W_A = 12$ lb, the weight of B is $W_B = 30$ lb, and the angle between the incline and the horizontal is $\theta = 30°$. Determine the acceleration of A, the acceleration of B, and the tension in the rope after the system is released.

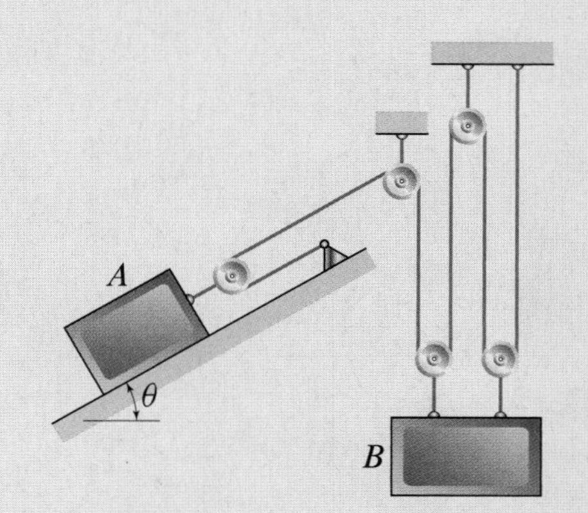

Figure P3.159

Energy Methods for Particles

This chapter presents the concepts of *work of a force* and *kinetic energy of a particle*. These two quantities play a crucial role in a balance law called the work-energy principle, which is intimately related to Newton's second law. We will see that the work-energy principle can be derived from $\vec{F} = m\vec{a}$ by integrating $\vec{F} = m\vec{a}$ with respect to position. We sometimes refer to the work-energy principle as the "pre-integrated" form of Newton's second law to remind us of the connection between these two fundamental laws of physics. The chapter will conclude with a presentation of the concepts of *power* and *efficiency*, which are important in measuring the performance of engines and machines.

Blanka Vlašić performing a high jump at the 2008 Indoor World Championship in Valencia, Spain. Using the work-energy principle we can see how speed is converted to height.

4.1 Work-Energy Principle for a Particle

Various problems in Chapter 3 required us to find a change in speed from an acceleration that was a function of position (e.g., Examples 3.2 and 3.5). The solution was based on the application of the chain rule followed by a corresponding integration. In this chapter, we will learn a more direct solution method based on the application of a balance law that *pre-integrates* Newton's second law with respect to position.

Work-energy principle and its relation with $\vec{F} = m\vec{a}$

Consider a particle of mass m moving along a path $\mathcal{L}_{1\text{-}2}$ between points P_1 and P_2 under the action of a force $\vec{F}$ (Fig. 4.1). Newton's second law says that $\vec{F} = m\vec{a}$ for the particle. Dotting both sides of $\vec{F} = m\vec{a}$ with the particle's infinitesimal displacement $d\vec{r}$, we have

$$\vec{F} \cdot d\vec{r} = m\vec{a} \cdot d\vec{r}. \tag{4.1}$$

Recalling that $d\vec{r} = \vec{v}\,dt$ and $\vec{a} = d\vec{v}/dt$, we can rewrite Eq. (4.1) as

$$\vec{F} \cdot d\vec{r} = m\,\frac{d\vec{v}}{dt} \cdot \vec{v}\,dt = m\vec{v} \cdot d\vec{v}, \tag{4.2}$$

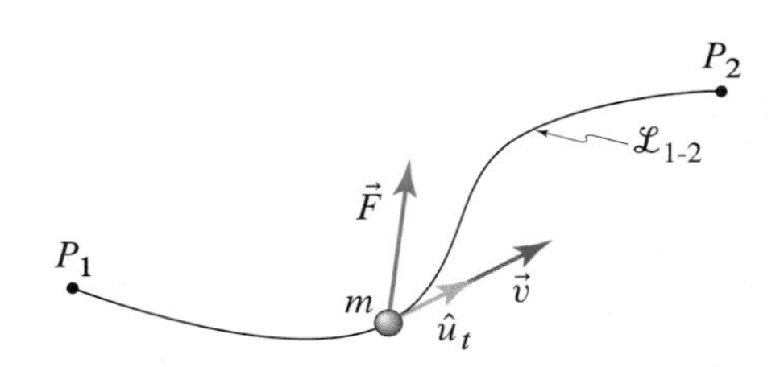

Figure 4.1
Particle of mass m moving along the path $\mathcal{L}_{1\text{-}2}$ while being subjected to the force $\vec{F}$.

where the last expression was obtained by canceling dt. We now observe that the differential of $\frac{1}{2}m\vec{v}\cdot\vec{v}$ is equal to the last term in Eq. (4.2), that is,

$$d\left(\tfrac{1}{2}m\vec{v}\cdot\vec{v}\right) = \tfrac{1}{2}m\left(d\vec{v}\cdot\vec{v} + \vec{v}\cdot d\vec{v}\right) = \tfrac{1}{2}m\left(2\vec{v}\cdot d\vec{v}\right) = m\vec{v}\cdot d\vec{v}, \tag{4.3}$$

where $d()$ is the differential of the quantity in parentheses. Substituting Eq. (4.3) into Eq. (4.2) and integrating along the path of the particle from the initial point P_1 to the final point P_2, we obtain

$$\int_{\mathcal{L}_{1\text{-}2}} \vec{F}\cdot d\vec{r} = \int_{\mathcal{L}_{1\text{-}2}} d\left(\tfrac{1}{2}m\vec{v}\cdot\vec{v}\right). \tag{4.4}$$

Helpful Information

Line integral of an exact differential. All introductory calculus textbooks discuss line integrals (see, e.g., G. B. Thomas, Jr., and R. L. Finney, *Calculus and Analytic Geometry*, 9th ed., Addison-Wesley, Boston, 1996). A theorem pertaining to line integrals states that if $\varphi = \varphi(\vec{r})$, then

$$\int_{\mathcal{L}_{1\text{-}2}} d\varphi = \varphi(\vec{r}_2) - \varphi(\vec{r}_1),$$

where $\vec{r}_1$ is the initial point of $\mathcal{L}_{1\text{-}2}$ and $\vec{r}_2$ is the endpoint of $\mathcal{L}_{1\text{-}2}$. The only restrictions on this result are that $\varphi(\vec{r})$ must be continuous and single-valued in the region containing $\mathcal{L}_{1\text{-}2}$ and that $\mathcal{L}_{1\text{-}2}$ must be smooth. It is this theorem that we apply in Eq. (4.5).

Both integrals in Eq. (4.4) are *path* or *line* integrals, which you have probably studied in your calculus courses. The integral on the right-hand side of Eq. (4.4) is the line integral of an *exact differential* (see the Helpful Information marginal note) and can be written as

$$\int_{\mathcal{L}_{1\text{-}2}} d\left(\tfrac{1}{2}m\vec{v}\cdot\vec{v}\right) = \tfrac{1}{2}m\vec{v}_2\cdot\vec{v}_2 - \tfrac{1}{2}m\vec{v}_1\cdot\vec{v}_1, \tag{4.5}$$

where $\vec{v}_1$ and $\vec{v}_2$ are the velocities of the particle at points P_1 and P_2, respectively. Since $\vec{v}\cdot\vec{v} = v^2$, substituting Eq. (4.5) into Eq. (4.4) yields

$$\boxed{\int_{\mathcal{L}_{1\text{-}2}} \vec{F}\cdot d\vec{r} = \tfrac{1}{2}mv_2^2 - \tfrac{1}{2}mv_1^2.} \tag{4.6}$$

We now introduce the following definitions:

$$U_{1\text{-}2} = \int_{\mathcal{L}_{1\text{-}2}} \vec{F}\cdot d\vec{r} = \text{the } \textit{work} \text{ done by } \vec{F} \text{ on the particle in moving from } P_1 \text{ to } P_2 \text{ along the path } \mathcal{L}_{1\text{-}2}, \tag{4.7}$$

$$T = \tfrac{1}{2}mv^2 = \text{the } \textit{kinetic energy} \text{ of the particle.} \tag{4.8}$$

Concept Alert

Work equals change in kinetic energy. The work done on an object of mass m equals the object's change in kinetic energy *regardless of the object's mass.* That is, the same force F acting through the same distance always results in the same amount of work done, whether the mass of a particle is, say, m, $4m$, or $100m$. This is not to say that the speed of particles of different mass will be the same, but they will have *exactly the same kinetic energy.*

By using these definitions, it can be seen why we refer to Eq. (4.6) as the *work-energy principle*. In words, we interpret the work-energy principle as saying that *the change in kinetic energy of a particle is equal to the work done on that particle*. Work and kinetic energy are *scalar* quantities and, by definition, the kinetic energy is *never* negative. Using the definitions in Eqs. (4.7) and (4.8), the work-energy principle can be given the form

$$\boxed{T_1 + U_{1\text{-}2} = T_2.} \tag{4.9}$$

Units for work and kinetic energy

The dimension of both work and kinetic energy are (force) × (length) or, equivalently, (mass) × (length)^2/(time)^2. Table 4.1 gives the units of energy we will use in both

Table 4.1. Units of work, energy, and moment.

Quantity	U.S. Customary	SI
energy or work	ft·lb	J (joule)
moment	ft·lb	N·m

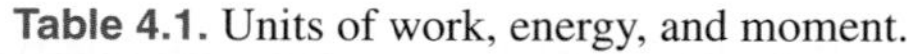

the U.S. Customary system and the SI system. The choice for the U.S. Customary system is somewhat arbitrary since ft·lb is the same as lb·ft. We have also included the units of a moment in Table 4.1. From a dimensional viewpoint, energy, work, and moment are equivalent. This is not a problem in the SI system since, when referring to energy, the unit N·m is called a *joule* and is written with the symbol J. By contrast, in the U.S. Customary system the unit ft·lb is used for moment, work, and energy.

Work of a force

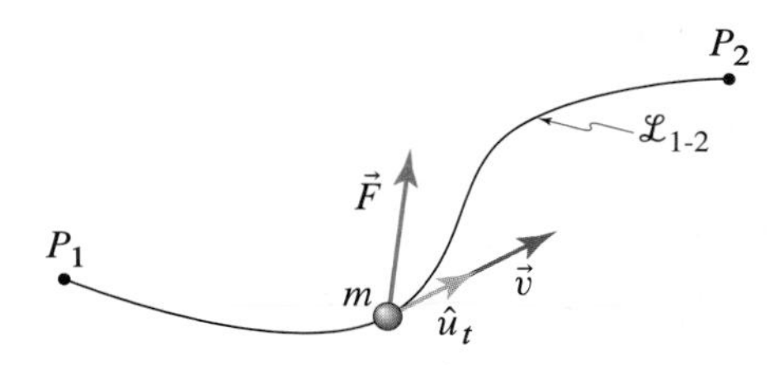

Figure 4.2
Particle of mass m moving along the path $\mathcal{L}_{1\text{-}2}$ while being subjected to the force $\vec{F}$.

We now want to more closely examine what it means for a force to do work. Referring to Fig. 4.2, the work of a force $\vec{F}$ can be written in all of the following ways:

$$U_{1\text{-}2} = \int_{\mathcal{L}_{1\text{-}2}} \vec{F}\cdot d\vec{r} = \int_{t_1}^{t_2} \vec{F}\cdot\vec{v}\,dt = \int_{t_1}^{t_2} \vec{F}\cdot v\,\hat{u}_t\,dt$$

$$= \int_{t_1}^{t_2} \vec{F}\cdot\frac{ds}{dt}\,\hat{u}_t\,dt = \int_{s_1}^{s_2} \vec{F}\cdot\hat{u}_t\,ds, \tag{4.10}$$

where t_1 and t_2 are the times at which the particle is at P_1 and P_2, respectively, s_1 and s_2 are the arc lengths at P_1 and P_2, respectively, and we have used $d\vec{r} = \vec{v}\,dt$.

The last expression in Eqs. (4.10) clearly shows that the work done by the force $\vec{F}$ in moving along the path $\mathcal{L}_{1\text{-}2}$ *depends only on the component of* $\vec{F}$ *in the direction of motion*, i.e., $\vec{F}\cdot\hat{u}_t$. Equations (4.10) also tell us that

- The sign of the work is determined by the sign of $\vec{F}\cdot\hat{u}_t$: it is positive if $\vec{F}$ promotes the motion, whereas it is negative if $\vec{F}$ hinders the motion (see Fig. 4.3).

- Since the second integral in Eqs. (4.10) is with respect to time, $\vec{F}\cdot\vec{v}$ can be interpreted as the *time rate of work* done by the force $\vec{F}$. We will explore this idea in Section 4.4 where we study power and efficiency.

- Forces of constraint, such as normal forces, *never* contribute to $U_{1\text{-}2}$ because they are always perpendicular to the path.*

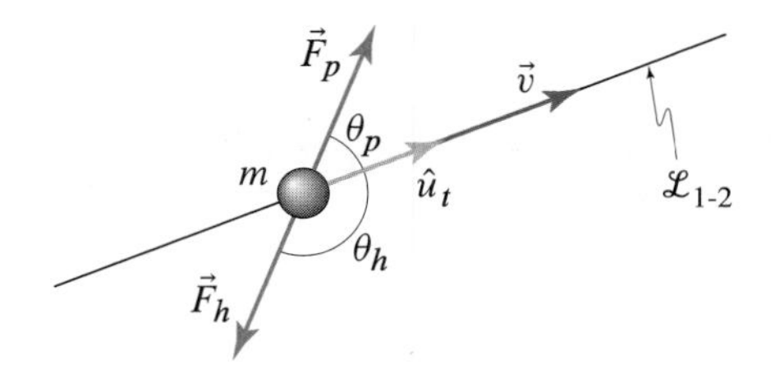

Figure 4.3. The angle θ_p between the force $\vec{F}_p$ and the velocity vector $\vec{v}$ is acute. Hence, $\cos\theta_p$ is positive and the work of $\vec{F}_p$ is positive. By contrast, the angle θ_h between the force $\vec{F}_h$ and the velocity vector $\vec{v}$ is obtuse. Hence, $\cos\theta_h$ is negative and the work of $\vec{F}_h$ is negative.

Work of a constant force

Here we consider a simple two-dimensional example in which we compute the work done by a constant force. The generalization to three dimensions is done similarly.

---

* If the constraint is *moving* and it has a component of velocity normal to the constraint surface, then Eqs. (4.10) tells us that the constraint force *will* do work. We will not consider such cases in this book, as this is a topic for advanced dynamics courses.

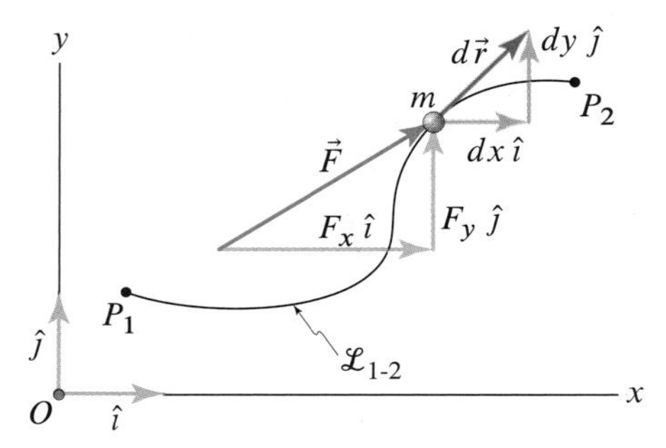

Figure 4.4
A constant force $\vec{F}$ acting on a particle.

We will consider more complex examples in Section 4.2, where we will compute the work of forces, such as spring forces and gravitation as given by Newton's universal law of gravitation.

We can express a force $\vec{F}$ in two dimensions in Cartesian components as $\vec{F} = F_x\,\hat{\imath} + F_y\,\hat{\jmath}$ (Fig. 4.4). We will assume that $\vec{F}$ is *constant*. Letting $d\vec{r}$ denote the infinitesimal displacement of the point of application of $\vec{F}$, in Cartesian components we can write $d\vec{r}$ as $d\vec{r} = dx\,\hat{\imath} + dy\,\hat{\jmath}$. We then have the following expression for the work done by $\vec{F}$:

$$\begin{aligned} U_{1\text{-}2} &= \int_{\mathcal{L}_{1\text{-}2}} \vec{F}\cdot d\vec{r} = \int_{\mathcal{L}_{1\text{-}2}} (F_x\,\hat{\imath} + F_y\,\hat{\jmath})\cdot(dx\,\hat{\imath} + dy\,\hat{\jmath}) \\ &= \int_{x_1}^{x_2} F_x\,dx + \int_{y_1}^{y_2} F_y\,dy = F_x\int_{x_1}^{x_2} dx + F_y\int_{y_1}^{y_2} dy \\ &= F_x(x_2 - x_1) + F_y(y_2 - y_1) = \vec{F}\cdot(\vec{r}_2 - \vec{r}_1). \end{aligned} \tag{4.11}$$

Equation (4.11) tells us that the work of a constant force only depends on the end-points of the path over which the force acts and not the path itself.

End of Section Summary

The work-energy principle is expressed by the following equation:

Eq. (4.9), p. 242

$$T_1 + U_{1\text{-}2} = T_2,$$

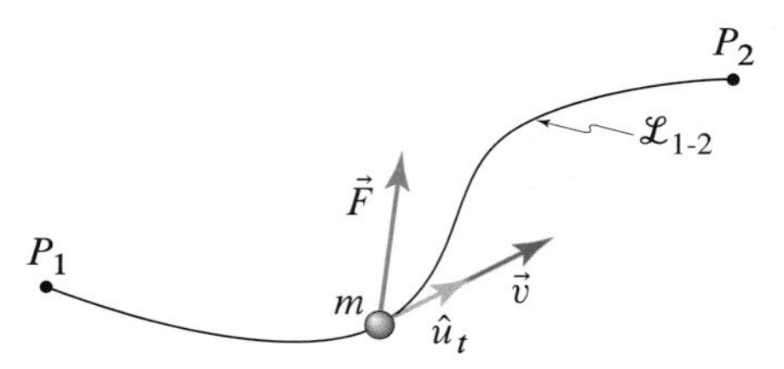

Figure 4.5
Particle of mass m moving along the path $\mathcal{L}_{1\text{-}2}$ while being subjected to the force $\vec{F}$.

where $U_{1\text{-}2}$ is the *work* and is defined as (see Fig. 4.5)

Eq. (4.7), p. 242

$$U_{1\text{-}2} = \int_{\mathcal{L}_{1\text{-}2}} \vec{F}\cdot d\vec{r}$$

and T is the *kinetic energy* and is defined as

Eq. (4.8), p. 242

$$T = \tfrac{1}{2}mv^2.$$

Work and kinetic energy have the following basic properties:

- Work and kinetic energy are *scalar* quantities.
- The work depends on only the component of $\vec{F}$ in the direction of motion, that is, on $\vec{F}\cdot\hat{u}_t$, and its sign is determined by the sign of $\vec{F}\cdot\hat{u}_t$.
- The kinetic energy is *never* negative.

Common Pitfall

Kinetic energy is never negative. If you compute a kinetic energy using a velocity component that is negative, *don't* be tempted to write $T = -\frac{1}{2}mv^2$; since v is squared, regardless of the sign of the velocity component, the kinetic energy can *never* be negative.

EXAMPLE 4.1 *Ball Sliding Down a Cylindrical Surface*

As shown in Fig. 1, a small ball is given a slight nudge from rest at the top of a semicylinder and slides downward. If the mass of the ball is m, the radius of the semicylinder is R, and friction between the ball and the semicylinder is negligible, determine the angle θ at which the ball separates from the semicylinder. We solved this problem using Newton's second law in Example 3.9 on p. 200.

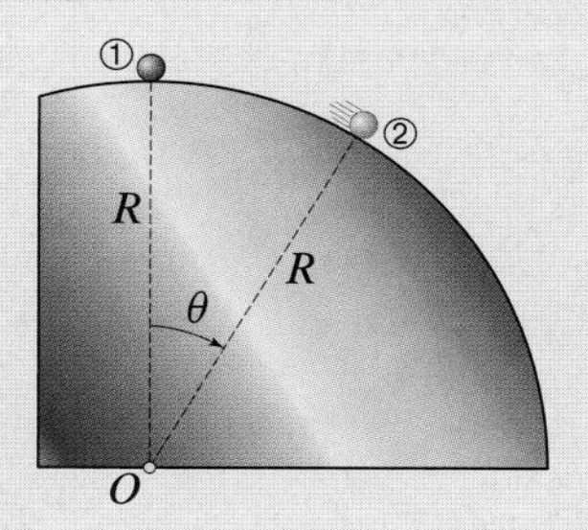

Figure 1
A particle sliding on a frictionless semicylinder.

SOLUTION

Road Map & Modeling The FBD of the ball as it slides down the semicylinder is shown in Fig. 2, where we have used polar components to describe the motion. As in Example 3.9, the key is to find the normal force N as a function of the angle θ and then determine where N becomes zero. We can apply Newton's second law to find $N(\theta, \dot{\theta})$ and then the work-energy principle to find $\dot{\theta}(\theta)$.

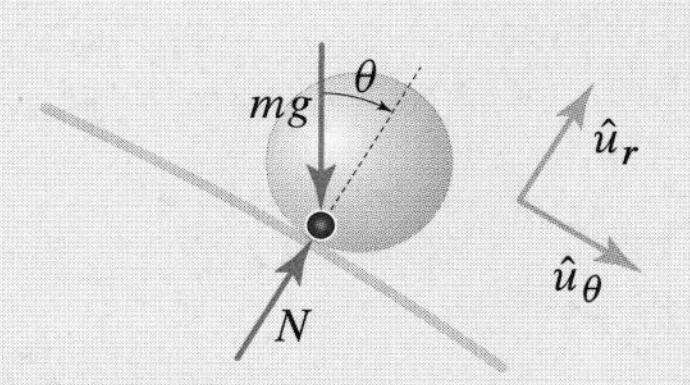

Figure 2
FBD at an arbitrary θ of the ball as it slides down the semicylinder.

Governing Equations

Application of Newton's Second Law

Balance Principles Applying Newton's second law in the radial direction, we obtain

$$\sum F_r: \quad N - mg\cos\theta = ma_r. \tag{1}$$

Force Laws All forces are accounted for on the FBD.

Kinematic Equations Since the motion of the ball is circular, we have that $a_r = -R\dot{\theta}^2$.

Computation Substituting the expression for a_r into Eq. (1), we obtain

$$N = mg\cos\theta - mR\dot{\theta}^2, \tag{2}$$

where we see that we still need to find $\dot{\theta}(\theta)$.

Application of the Work-Energy Principle

Balance Principles Applying the work-energy principle between ① and ②, we obtain

$$T_1 + U_{1\text{-}2} = T_2, \tag{3}$$

where the kinetic energies at ① and ② are, respectively,

$$T_1 = 0 \quad \text{and} \quad T_2 = \tfrac{1}{2}mv_2^2, \tag{4}$$

where v_2 is the speed of the ball in ②.

Force Laws Only the weight force does work on the ball, so we compute its work using the last of Eqs. (4.10) as

$$U_{1\text{-}2} = \int_{s_1}^{s_2} mg(\sin\theta\,\hat{u}_\theta - \cos\theta\,\hat{u}_r)\cdot\hat{u}_t\,ds = mgR\int_0^\theta \sin\theta\,d\theta = mgR(1-\cos\theta), \tag{5}$$

where we have used $\hat{u}_\theta = \hat{u}_t$ (see Fig. 3) and $ds = R\,d\theta$ to obtain the second equation.

Kinematic Equations Using polar components, we can write $v_2 = R\dot{\theta}$.

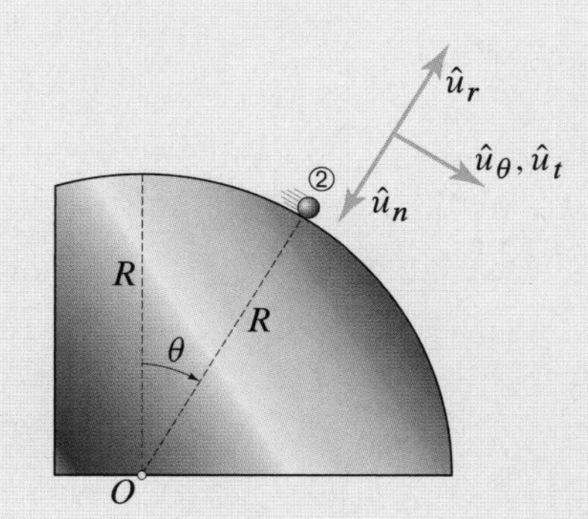

Figure 3
Equality of $\hat{u}_\theta$ and $\hat{u}_t$ for circular motion.

Computation Substituting Eqs. (4), (5), and $v_2 = R\dot{\theta}$ into Eq. (3), we obtain

$$mgR(1-\cos\theta) = \tfrac{1}{2}m(R\dot{\theta})^2 \quad \Rightarrow \quad \dot{\theta}^2 = \frac{2g}{R}(1-\cos\theta)$$

$$\Rightarrow \quad N = mg\cos\theta - 2mg(1-\cos\theta) = mg(3\cos\theta - 2), \tag{6}$$

which, for $N = 0$, implies that $\cos\theta = \frac{2}{3}$ or $\boxed{\theta = 48.19°.}$

Discussion & Verification This is the same angle we obtained in Example 3.9.

EXAMPLE 4.2 *Relating Kinetic Energy to Average Force*

Figure 1
An F/A-18 Hornet taking off from an aircraft carrier.

An F/A-18 Hornet (see Fig. 1) takes off from an aircraft carrier using two separate propulsion systems: its two jet engines and a steam-powered catapult. During launch, a fully loaded Hornet weighing 50,000 lb goes from 0 (relative to the carrier) to 165 mph (relative to the surface of the Earth) in a distance of 300 ft (relative to the carrier) while each of its two engines is at full power, generating about 22,000 lb of thrust, and the catapult is engaged. For a *stationary* aircraft carrier, neglecting aerodynamic forces, determine:

(a) The total work done on the aircraft during launch.

(b) The work done by the catapult on the aircraft during launch.

(c) The average force exerted by the catapult on the aircraft.

SOLUTION

Road Map & Modeling Referring to the FBD in Fig. 2, we will model the airplane as a particle under the action of gravity, the engines' thrust F_T, the force of the catapult F_C, and the normal reaction between the aircraft and the deck. Based on this model, the work-energy principle tells us that the work done on the airplane is equal to the airplane's change in kinetic energy. We can calculate the kinetic energy because we are given the airplane's weight and change in speed. Since the engines' thrust is constant and we know the takeoff distance, we can compute the work done by the engines by a direct application of the definition of work. Subtracting the work of the engines from the total work will allow us to find the work of the catapult.

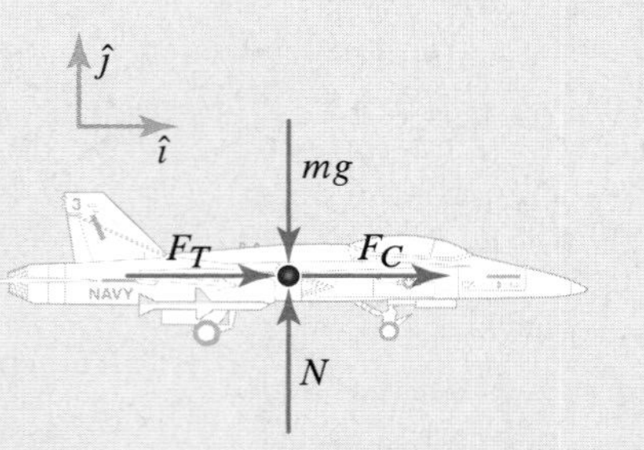

Figure 2
FBD of the aircraft as it is propelled by the jet engines and by the catapult.

Governing Equations

Balance Principles The work-energy principle gives

$$T_1 + U_{1\text{-}2} = T_2, \tag{1}$$

where ① is at the start of the launch, ② is right before takeoff, $U_{1\text{-}2}$ is the total work done on the aircraft between ① and ②, and T_1 and T_2 are the aircraft's kinetic energy at ① and ②, respectively, which are given by

$$T_1 = \tfrac{1}{2}mv_1^2 \quad \text{and} \quad T_2 = \tfrac{1}{2}mv_2^2, \tag{2}$$

where v_1 and v_2 are the speed of the aircraft at ① and ②, respectively.

Force Laws Both gravity and the engines' thrust are modeled as constant forces. Because we are only interested in the average value of the catapult force, we will model F_C as a constant. Since the work is computed using forces and force laws, in work-energy problems, we will always determine expressions for the work in the Force Laws portion of our solution procedure. Hence, the expressions for the work done by the engines and catapult are

$$(U_{1\text{-}2})_{\text{engines}} = \int_{\mathcal{L}_{1\text{-}2}} \vec{F}_T \cdot d\vec{r} = \int_{0\,\text{ft}}^{300\,\text{ft}} (44{,}000\,\text{lb})\,dx = 1.320\times 10^7\,\text{ft·lb}, \tag{3}$$

$$(U_{1\text{-}2})_{\text{catapult}} = \int_{\mathcal{L}_{1\text{-}2}} \vec{F}_C \cdot d\vec{r} = \int_{0\,\text{ft}}^{300\,\text{ft}} F_C\,dx = (300.0\,\text{ft})F_C, \tag{4}$$

where $\mathcal{L}_{1\text{-}2}$ is the 300 ft rectilinear stretch traveled by the plane between ① and ②. Note that the aircraft's weight and the normal reaction N do no work since the motion of the airplane is assumed to be horizontal and therefore, perpendicular to these forces.

Kinematic Equations Recalling the definition of ① and ②, we have

$$v_1 = 0 \quad \text{and} \quad v_2 = 165\,\text{mph} = 242.0\,\text{ft/s}. \tag{5}$$

Computation Combining Eqs. (2) and (5) with Eq. (1) and solving for $U_{1\text{-}2}$, we find the total work done on the aircraft during launch to be

$$\boxed{U_{1\text{-}2} = T_2 - T_1 = 4.547 \times 10^7\ \text{ft}\cdot\text{lb}.} \tag{6}$$

Therefore, the work done by the catapult is the total work done minus the work done by the engines, that is,

$$(U_{1\text{-}2})_{\text{catapult}} = U_{1\text{-}2} - (U_{1\text{-}2})_{\text{engines}}. \tag{7}$$

Substituting Eqs. (3) and (6) into Eq. (7), we have

$$\boxed{(U_{1\text{-}2})_{\text{catapult}} = 3.227 \times 10^7\ \text{ft}\cdot\text{lb}.} \tag{8}$$

Finally, solving Eq. (4) for F_C and using the result in Eq. (8), we have

$$\boxed{F_C = \frac{(U_{1\text{-}2})_{\text{catapult}}}{300\ \text{ft}} = 107{,}600\ \text{lb}.} \tag{9}$$

Discussion & Verification The calculation of the total work done on the aircraft and the work done by the catapult was done by a direct application of the work-energy principle, and therefore, we need only verify that proper units were used to express our result, which is indeed the case. As far as the result in Eq. (9) is concerned, since the dimensions of work are force times length, we know that the dimensions of our result are correct and proper units were used to express it.

A Closer Look This problem emphasizes an important point: *no matter how many or what kinds of forces are acting on a particle, the work done by all forces is equal to the change in kinetic energy of the body.*

The amount of work done on the F/A-18 is equivalent to the work required to push a 70 lb wooden crate over level concrete ($\mu_k \approx 0.6$) at a constant speed for more than 205 miles! Also, notice that the force due to the catapult is more than twice that provided by the aircraft's engines. This tells us that an airplane such as an F/A-18 could not take off unassisted from an aircraft carrier.

EXAMPLE 4.3 *Relating Speed to Position*

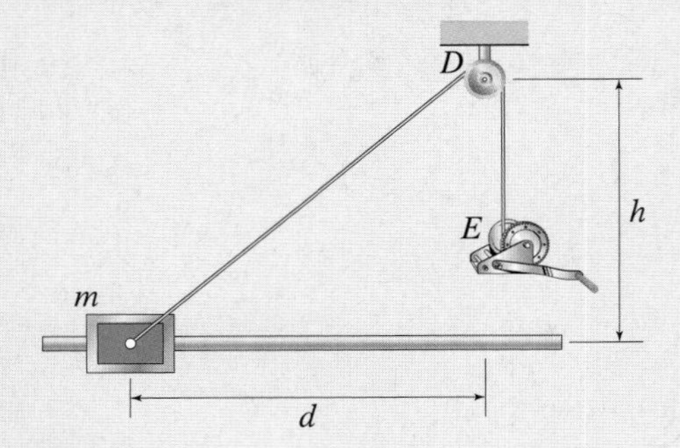

Figure 1
Winch and block system geometry.

The block of mass $m = 20\,\text{kg}$ is connected by a pulley at D to the winch at E by an inextensible cord. The winch, which is mounted at a fixed location, can exert a constant force $P = 130\,\text{N}$ on the cord. The friction between the block and the horizontal bar on which the block slides is negligible. Letting $h = 0.8\,\text{m}$, if the block starts from rest in the position shown, determine the speed of the block when it has moved the distance $d = 1.15\,\text{m}$ and is directly under the pulley.

SOLUTION

Road Map & Modeling Since we are interested in relating speed to position, we will solve this problem by using the work-energy principle. Referring to the FBD in Fig. 2, we will model the block as a particle subject to its own weight, the normal reaction between

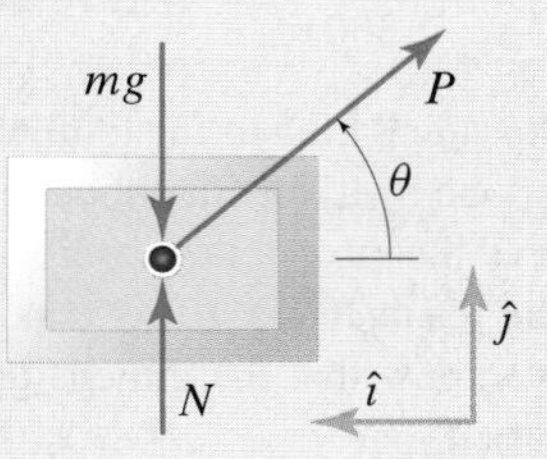

Figure 2. FBD of the block in a generic position between its initial and final positions.

the block and the horizontal guide, and the force in the cord. Notice that neither the weight nor the force N does any work, given that the block's motion is in the horizontal direction.

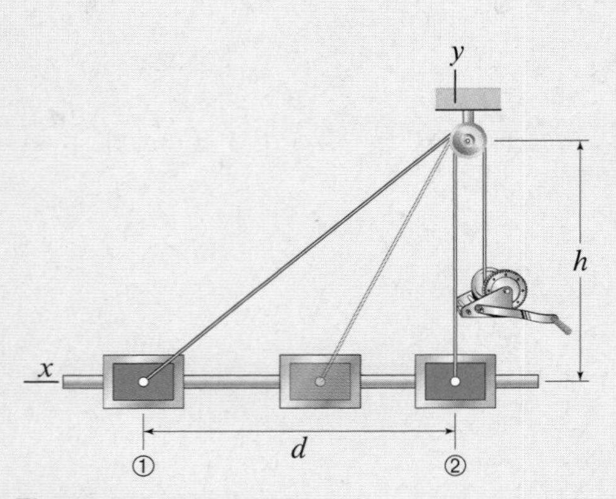

Figure 3
Winch and block system showing ① and ② and the Cartesian coordinate system used.

Governing Equations

Balance Principles Referring to Fig. 3 and applying the work-energy principle between ① and ②, we obtain

$$T_1 + U_{1\text{-}2} = T_2. \tag{1}$$

The kinetic energies are given by

$$T_1 = \tfrac{1}{2} m v_1^2 \quad \text{and} \quad T_2 = \tfrac{1}{2} m v_2^2, \tag{2}$$

where v_1 and v_2 are the block's speed at ① and ②, respectively.

Force Laws The only force doing work is the force P, whose magnitude is constant and given. As mentioned in Example 4.2, when using the work-energy principle, we devote the Force Laws step of our solution procedure to the computation of the work done by the forces appearing in the FBD. Hence, we have

$$U_{1\text{-}2} = \int_{\mathcal{L}_{1\text{-}2}} \vec{F} \cdot d\vec{r} = \int_d^0 -P\cos\theta\, dx, \tag{3}$$

where we have used $\vec{F} = -P\cos\theta\,\hat{\imath} + P\sin\theta\,\hat{\jmath}$ and $d\vec{r} = dx\,\hat{\imath}$.

Kinematic Equations Recalling that the block is released from rest and that the speed at ② is the unknown of this problem, we have

$$v_1 = 0. \tag{4}$$

Observe also that the integral in Eq. (3) contains the variable θ in the integrand and uses the variable x as the variable of integration. To compute this integral, we need to express either the variable θ as a function of x or the variable x as a function of θ. We will choose the latter strategy (the final result must be the same no matter which strategy we use) and will use differentiation of constraints (see Section 2.7 on p. 121) to obtain the relation we need. Noticing that $x \tan\theta = h$ and then taking the differential of this relation, we have

$$\tan\theta\, dx + x \sec^2\theta\, d\theta = \tan\theta\, dx + \frac{h}{\tan\theta}\sec^2\theta\, d\theta = 0, \tag{5}$$

where we have used the fact that $x = h/\tan\theta$ and that $dh = 0$ since h is a constant. Solving for dx, we find that

$$dx = \frac{-h}{\sin^2\theta}\, d\theta. \tag{6}$$

Computation Substituting Eq. (6) into Eq. (3), we have

$$U_{1\text{-}2} = \int_{\tan^{-1}(h/d)}^{\pi/2} P\left(\frac{h\cos\theta}{\sin^2\theta}\right) d\theta, \tag{7}$$

where we have used the fact that $\theta = \tan^{-1}(h/d)$ when $x = d$ and $\theta = \pi/2$ rad when $x = 0$. Evaluating the integral in Eq. (7), we obtain

$$U_{1\text{-}2} = -P\frac{h}{\sin\theta}\bigg|_{\tan^{-1}(h/d)}^{\pi/2} = -P\left(h - \sqrt{d^2 + h^2}\right), \tag{8}$$

where we have used the identity $\sin(\tan^{-1} x) = x/\sqrt{1 + x^2}$ (see Fig. 4).

Combining Eqs. (2), (4), and (8) with Eq. (1), we obtain

$$P\left(\sqrt{d^2 + h^2} - h\right) = \tfrac{1}{2}mv_2^2, \tag{9}$$

which, upon solving for v_2, gives

$$\boxed{v_2 = \sqrt{\frac{2P}{m}\left(\sqrt{d^2 + h^2} - h\right)} = 2.795\,\text{m/s}.} \tag{10}$$

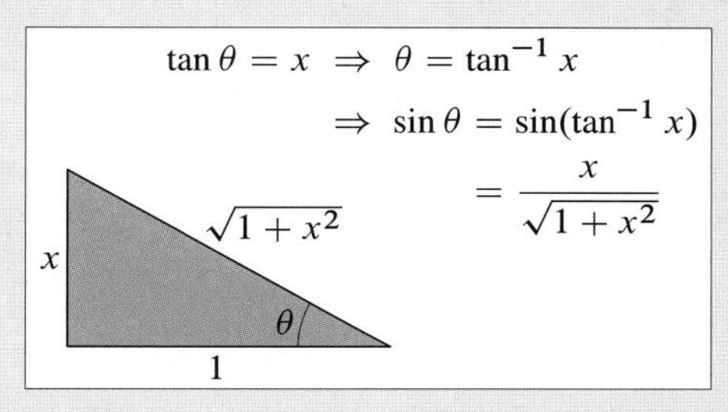

Figure 4
Demonstration of the trigonometric identity used in Eq. (8).

Discussion & Verification Observing that the term P/m has dimensions of acceleration and that the term $\sqrt{d^2 + h^2} - h$ has dimensions of length, we see that v_2 has dimensions of length over time, as it should. Furthermore, the final numerical result was expressed with appropriate units. Also, observe that the argument of the square root in Eq. (10) contains the term $P(\sqrt{d^2 + h^2} - h)$, which can be interpreted as the work of a *constant* force (i.e., constant in both magnitude and direction) with magnitude P along the distance $\Delta = \sqrt{d^2 + h^2} - h$. What makes Δ interesting is that it corresponds to the amount of cord that is wound onto the winch as the block goes from ① to ②. In fact, this is the work done by the tension in the vertical branch of the cord, i.e., the portion of the cord that goes from the pulley to the winch. So overall our result appears to be correct.

A Closer Look We should consider whether or not we can obtain the result in Eq. (10) without resorting to the rather involved integration in Eqs. (7) and (8), instead computing $U_{1\text{-}2}$ in some more physically based way. We will explore this question in Example 4.4.

EXAMPLE 4.4 Choosing a Convenient FBD to Relate Speed to Position

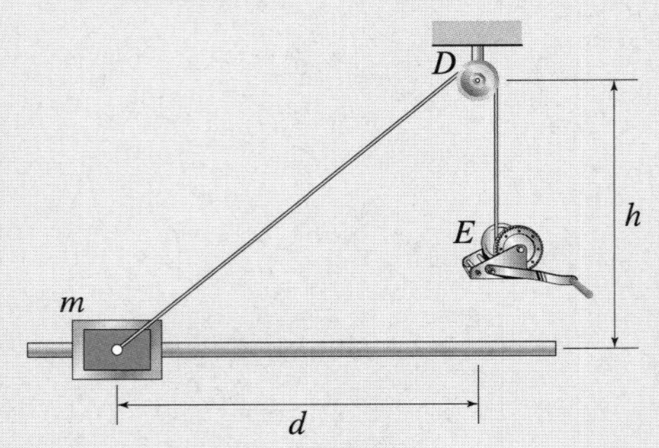

Figure 1
Winch and block system geometry.

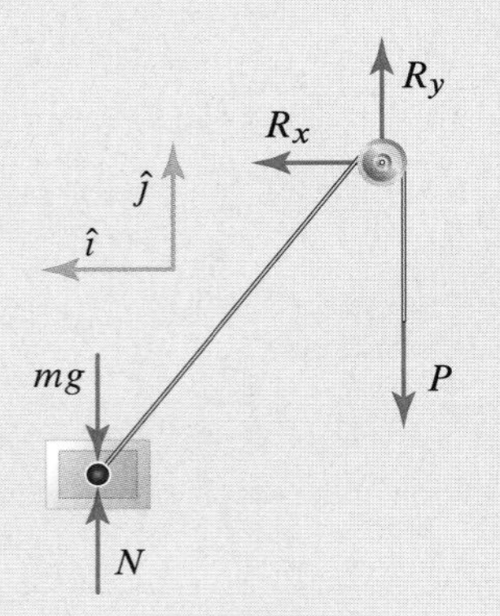

Figure 2
FBD of the block, cord, and pulley as the block moves from ① to ②.

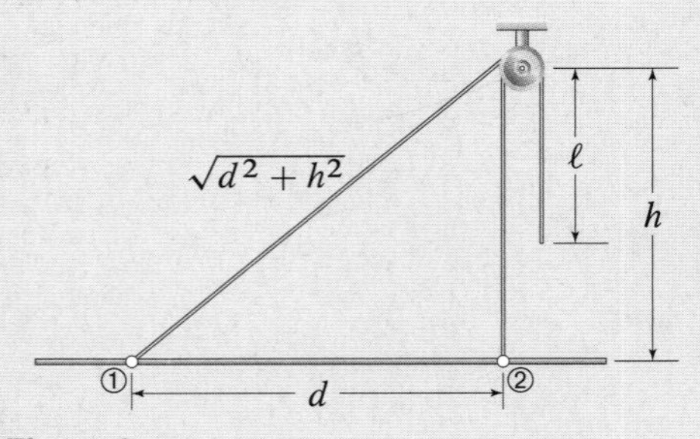

Figure 3
The length of the cord at ① and ②. The length ℓ is the constant vertical distance between the pulley and winch.

Let's revisit Example 4.3 and again determine the speed of the block when it has moved the distance d and is directly under the pulley. However, instead of focusing on the FBD of only the block, solve the problem by using an FBD that includes the pulley at D.

SOLUTION

Road Map & Modeling As in Example 4.3, we will apply the work-energy principle, but we will use the FBD suggested in the problem statement, which is shown in Fig. 2. Again, the only force doing work on the block is P because the point of application of the reactions R_x and R_y is fixed (hence, R_x and R_y do no work) and the forces mg and N are perpendicular to the direction of motion of the block.

Governing Equations

Balance Principles Letting ① be the position of the block at the instant of release and ② be the position of the block when the block is directly under the pulley, applying the work-energy principle between ① and ②, we obtain

$$T_1 + U_{1\text{-}2} = T_2. \tag{1}$$

The block's kinetic energies are given by

$$T_1 = \tfrac{1}{2}mv_1^2 \quad \text{and} \quad T_2 = \tfrac{1}{2}mv_2^2, \tag{2}$$

where v_1 and v_2 are the block's speed at ① and ②, respectively.

Force Laws This time $U_{1\text{-}2}$ takes on the following simple form:

$$U_{1\text{-}2} = P\Delta, \tag{3}$$

where Δ is the amount of cord wound up on the winch as the block moves from ① to ② and is therefore given by (see Fig. 3)

$$\Delta = L_{①} - L_{②} = \left(\sqrt{d^2+h^2} + \ell\right) - (h + \ell) = \sqrt{d^2+h^2} - h, \tag{4}$$

where $L_{①}$ and $L_{②}$ are the length of the cord at ① and ②, respectively.

Kinematic Equations Recalling that m starts from rest and that v_2 is the unknown of this problem, we have

$$v_1 = 0. \tag{5}$$

Computation Substituting Eqs. (2)–(5) into Eq. (1), we obtain

$$P\left(\sqrt{d^2+h^2} - h\right) = \tfrac{1}{2}mv_2^2, \tag{6}$$

which, upon solving for v_2, gives

$$\boxed{v_2 = \sqrt{\frac{2P}{m}\left(\sqrt{d^2+h^2} - h\right)}.} \tag{7}$$

Discussion & Verification As expected, we obtained the same result found in Example 4.3.

A Closer Look The approach followed in this example is much more straightforward than that in Example 4.3 even though the only difference between the two examples is how we computed $U_{1\text{-}2}$. The lesson here is that judiciously selecting the system to analyze can save significant time and effort.

Problems

Problem 4.1

A rocket lifts off with an acceleration a. During liftoff, in terms of absolute values, is the work done on an astronaut by gravity larger than, equal to, or smaller than the work done by the normal reaction between the astronaut and her or his seat?

Figure P4.1

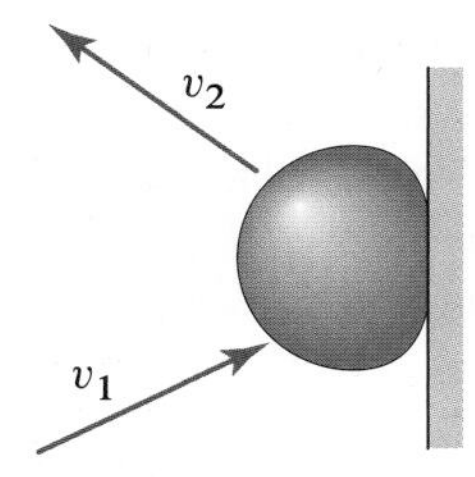

Figure P4.2

Problem 4.2

A soft rubber ball bounces against a wall. Assuming that the wall's deformation due to the ball's impact is negligible, does the contact force due to the wall do positive work, no work, or negative work on the ball?

Problem 4.3

Determine the kinetic energy of the bodies listed below, when modeled as particles. Express all answers using both U.S. Customary units and SI units.

(a) A .30-06 bullet weighing 150 gr (1 lb = 7000 gr) and traveling at 3000 ft/s.

(b) A 25 kg child traveling in a car at 45 km/h.

(c) A 415,000 lb locomotive traveling at 75 mph.

(d) A 20 g metal fragment from a space vehicle traveling at 8000 km/s.

(e) A 3000 lb car traveling at 60 mph.

Problem 4.4

A man whose mass is $m = 80$ kg is in an elevator that accelerates at $a_e = 1.5\,\text{m/s}^2$ over a distance of $d = 2$ m. Determine the work done by the weight force acting on the man and the work done by the normal force the floor of the elevator exerts on him. Explain why the magnitudes of the work of the forces are not the same.

Figure P4.4

Problems 4.5 and 4.6

Consider a 3000 lb car whose speed is increased by 30 mph.

Problem 4.5 Modeling the car as a particle and assuming that the car is traveling on a rectilinear and horizontal stretch of road, determine the amount of work done on the car throughout the acceleration process if the car starts from rest.

Figure P4.5 and P4.6

Problem 4.6 Modeling the car as a particle and assuming that the car is traveling on a rectilinear and horizontal stretch of road, determine the amount of work done on the car throughout the acceleration process if the car has an initial speed of 45 mph.

Figure P4.7

Problem 4.7

A 75 kg skydiver is falling at a speed of 250 km/h when the parachute is deployed, allowing the skydiver to land at a speed of 4 m/s. Modeling the skydiver as a particle, determine the total work done on the skydiver from the moment of parachute deployment until landing.

Problems 4.8 and 4.9

The crate of mass m is pushed to the left until the linear elastic spring of constant k has compressed a distance d from its unstretched length ℓ_0. The crate is released from rest, and friction between the crate and the horizontal surface is negligible.

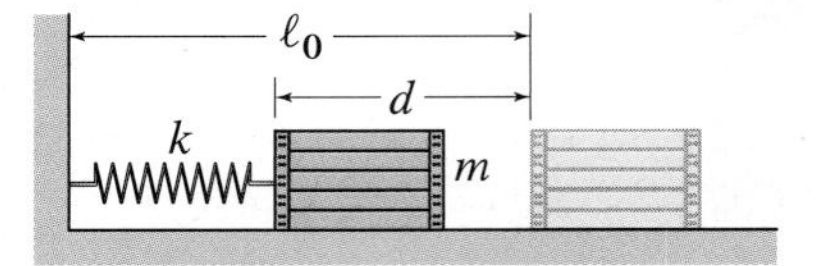

Figure P4.8 and P4.9

Problem 4.8 Using the work-energy principle, determine the speed of the crate at the instant the spring becomes uncompressed.

Problem 4.9 Using the work-energy principle, determine the value of k so that the crate is moving at 3 m/s the instant the spring becomes uncompressed. Use $m = 20$ kg and $d = 0.75$ m.

Problems 4.10 and 4.11

Consider a 1500 kg car whose speed is increased by 45 km/h over a distance of 50 m while traveling up an incline with a 15% grade.

Figure P4.10 and P4.11

Problem 4.10 Modeling the car as a particle, determine the work done on the car if the car starts from rest.

Problem 4.11 Modeling the car as a particle, determine the work done on the car if the car has an initial speed of 60 km/h.

Problems 4.12 and 4.13

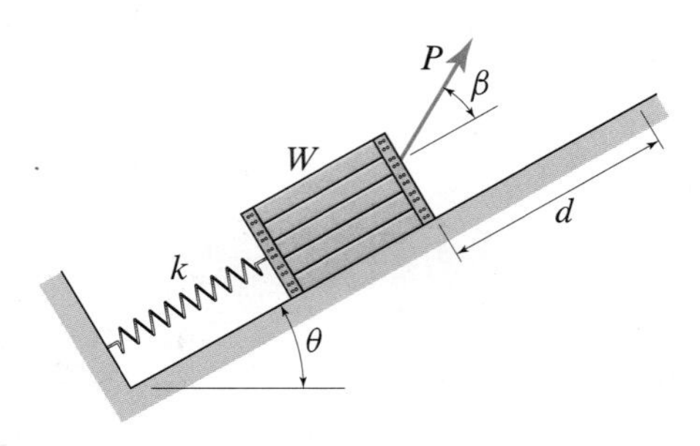

Figure P4.12 and P4.13

The crate moves up the incline a distance $d = 4.5$ ft due to the action of the constant force $P = 100$ lb. The weight of the crate is $W = 65$ lb, and the spring with constant $k = 10$ lb/ft is unstretched before the crate starts moving. Friction between the crate and the incline is negligible, and the angle between the force P and the surface on which the crate slides is $\beta = 30°$.

Problem 4.12 Letting $\theta = 0°$, that is, the surface is horizontal, determine the total work done by all forces after the crate has moved the distance d.

Problem 4.13 Letting $\theta = 30°$, determine the total work done by all forces after the crate has moved the distance d.

Problem 4.14

A 350 kg crate is sliding down a rough incline with a constant speed $v = 7\,\text{m/s}$. Assuming that the angle of the incline is $\theta = 33°$ and that the only forces acting on the crate are gravity, friction, and the normal force between the crate and the incline, determine the work done by friction over every meter slid by the crate.

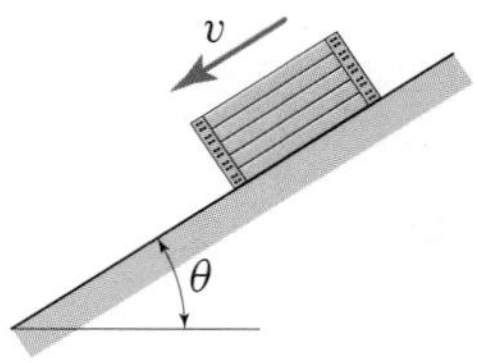

Figure P4.14

Problem 4.15

A vehicle A is stuck on the railroad tracks as a train B approaches with a speed of 120 km/h. As soon as the problem is detected, the train's emergency brakes are activated, locking the wheels and causing the train's wheels to slide relative to the tracks. If the coefficient of kinetic friction between the wheels and the track is 0.2, use the work-energy principle to determine the minimum distance d_{min} at which the brakes must be applied to avoid a collision under the following circumstances:

(a) The train consists of just a 195,000 kg locomotive.

(b) The train consists of a 195,000 kg locomotive and a string of cars whose mass is 10×10^6 kg, all of which can apply brakes and lock their wheels.

(c) The train consists of a 195,000 kg locomotive and a string of cars whose mass is 10×10^6 kg, but only the locomotive can apply its brakes and lock its wheels.

Treat the train as a particle, and assume that the railroad tracks are rectilinear and horizontal.

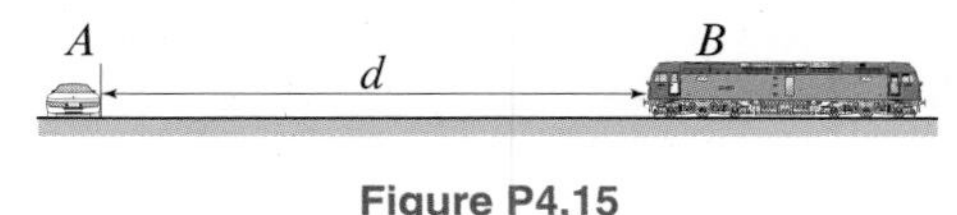

Figure P4.15

Problems 4.16 and 4.17

A classic car is driving down a 20° incline at 45 km/h when the brakes are applied. Treat the car as a particle, and neglect all forces except gravity and friction.

Problem 4.16 Using the work-energy principle, determine the stopping distance if the tires slide and the coefficient of kinetic friction between the tires and the road is 0.7.

Problem 4.17 Using the work-energy principle, determine the minimum stopping distance if the car is retrofitted with antilock brakes and the tires do not slide. Use 0.9 for the coefficient of static friction between the tires and the road.

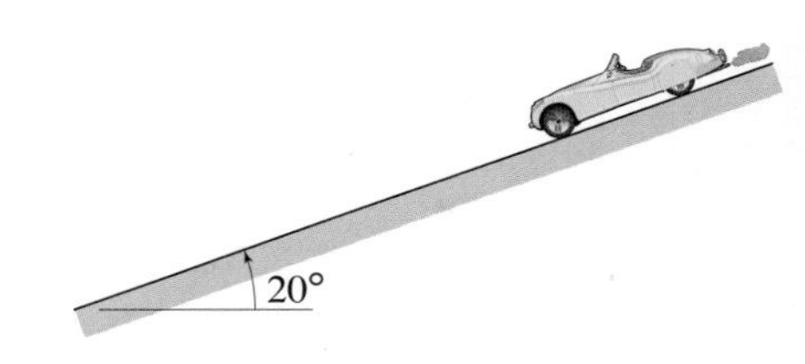

Figure P4.16 and P4.17

Problem 4.18

Two identical cars travel at a speed of 60 mph, one along a newly asphalt-paved straight and horizontal road and the other on a straight and horizontal dirt road. If brakes are applied and if the second car slips during the braking process, what difference will there be in the amount of work done to stop each car?

Problem 4.19

Two identical locomotives A and B are coupled with one and two passenger cars, respectively. Suppose that each passenger car is identical to the others in all respects. If each train (locomotive plus cars) starts from rest, each locomotive exerts the maximum tractive effort (traction force), and assuming that gravity and the tractive effort are the only relevant forces acting on the trains, which of the two trains will have greater kinetic energy after the locomotives have moved 50 m along a horizontal and rectilinear stretch?

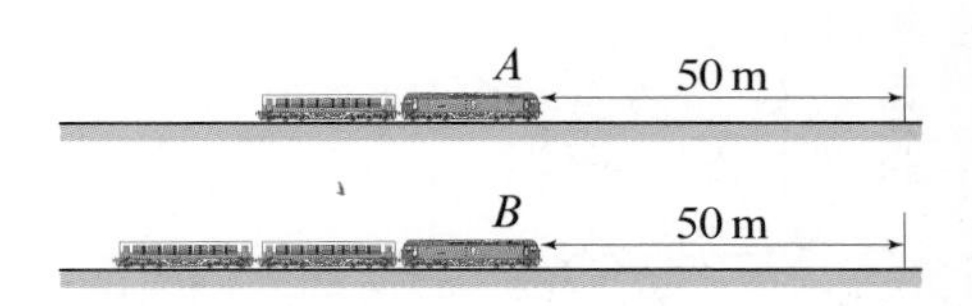

Figure P4.19

Problems 4.20 through 4.22

The crate A of weight $W = 30$ lb is being pulled to the right by the winch at B. The crate starts from rest at $x = 0$ and is pulled a total distance of 15 ft over the rough surface for which the coefficient of kinetic friction is $\mu_k = 0.3$. The force P in the cable due to the winch varies according to the plot, where P is in lb, b is in lb$/\sqrt{\text{ft}}$, and x is in ft. The coefficient of static friction is insufficient to prevent slipping.

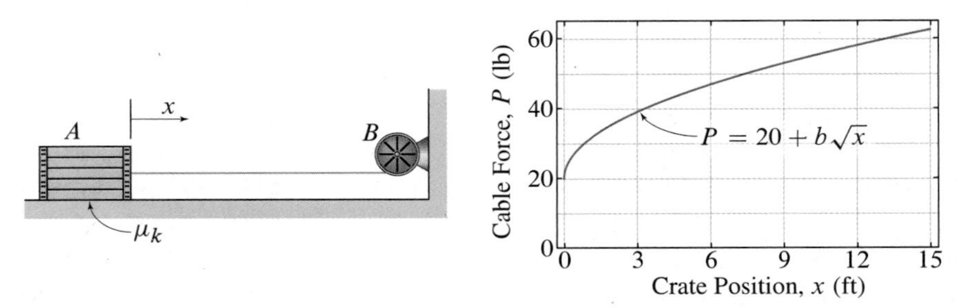

Figure P4.20–P4.22

Problem 4.20 Using the work-energy principle, determine the speed of the block when $b = 11$ lb$/\sqrt{\text{ft}}$ and $x = 15$ ft.

Problem 4.21 Using the work-energy principle, determine the value of b so that the block is moving at 35 ft/s when $x = 10$ ft.

Problem 4.22 Determine how far the crate slides before its speed becomes 20 ft/s with $b = 11$ lb$/\sqrt{\text{ft}}$.

Problem 4.23

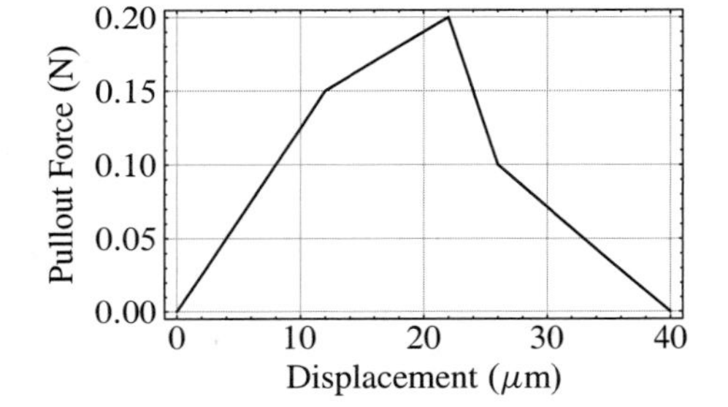

Figure P4.23

Many advanced materials consist of fibers (e.g., made of glass, Kevlar, or carbon) placed within a matrix (such as epoxy, a titanium alloy, etc.). For these materials it is important to assess the bond strength between fibers and matrix, and this is often done with a *pullout test*, in which the tip of a fiber is exposed, the material sampled is properly clamped, and the fiber is pulled out of the matrix. The data collected often consists of a graph like the one shown, in which the force exerted on the fiber is recorded as a function of pullout displacement. With this in mind, the interface toughness assessment process may require a measure of the *energy* expended to pull out the fiber. Use the force-displacement graph shown, which is typical for a glass-fiber reinforced epoxy, to measure the total pullout energy. *Hint:* The work of the pullout force is given by the area under the curve.

Problem 4.24

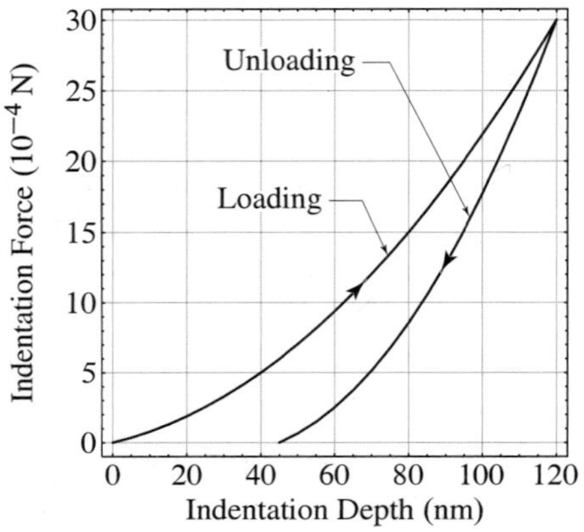

Figure P4.24

Components subjected to large contact forces (e.g., brake disks) are often made of high-grade steel coated with a thin film of a very hard material, such as diamond. A common test to assess the mechanical properties (hardness, elastic moduli, etc.) of the coating is the *nanoindentation test*, which consists of making a controlled dent in the film using a nail-like object, called an *indentor*. During the test, the force applied to the indentor and the indentation depth are measured. The graph shows the curves interpolating the loading and unloading data for a diamond film on steel. Since the unloading curve does not go back to the origin of the plot, the film is permanently deformed during the indentation process. Determine the energy lost to permanent deformation if the indentation force is given by

$$F_I = \begin{cases} \frac{25}{4}x + \frac{5}{32}x^2 & \text{during loading,} \\ 300 - \frac{145}{6}x + \frac{7}{18}x^2 & \text{during unloading,} \end{cases}$$

where F_I is expressed in μN and the indentation depth x is given in nm.

Problem 4.25

An F/A-18 Hornet takes off from an aircraft carrier, using two separate propulsion systems: its two jet engines and a steam-powered catapult. During launch, a fully loaded Hornet weighing about 50,000 lb goes from 0 mph (relative to the aircraft carrier) to 165 mph (measured relative to surface of the Earth) in a distance of 300 ft (measured relative to the aircraft carrier) while each of its two engines is at full power generating about 22,000 lb of thrust. Assuming that the aircraft carrier is traveling in the same direction as the takeoff direction and at a constant speed of 30 knot, determine:

(a) The total work done on the aircraft during launch.

(b) The work done by the catapult on the aircraft during launch.

(c) The force exerted by the catapult on the aircraft.

In solving this problem, model the aircraft as a particle; assume that its trajectory is horizontal and that the catapult assists the aircraft the full 300 ft needed for takeoff; and finally, let all forces be constant and neglect air resistance and friction. Use an inertial reference frame attached to the aircraft carrier.

Figure P4.25

Problem 4.26

Rubber bumpers are commonly used in marine applications to keep boats and ships from getting damaged by docks. Treating the boat C as a particle, neglecting its vertical motion, and neglecting the drag force between the water and the boat C, what is the maximum speed of the boat at impact with the bumper B so that the deflection of the bumper is limited to 6 in.? The weight of the boat is 70,000 lb, and the force compression profile for the rubber bumper is given by $F_B = \beta x^3$, where $\beta = 3.5 \times 10^6$ lb/ft^3 and x is the compression of the bumper.

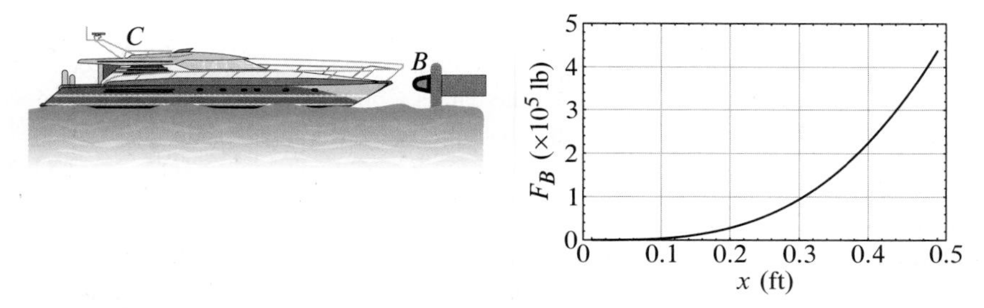

Figure P4.26

Problems 4.27 and 4.28

Packages for transporting delicate items (e.g., a laptop or glass) are designed to "absorb" some of the energy of the impact in order to protect their contents. These energy absorbers can get pretty complicated (e.g., the mechanics of Styrofoam peanuts is not easy), but we can begin to understand how they work by modeling them as a linear elastic spring of constant k that is placed between the contents (an expensive vase) of mass m and the package P. Assume that the vase weighs 6 lb and that the box is dropped from a height of 5 ft. Treat the vase as a particle, and neglect all forces except for gravity and the spring force.

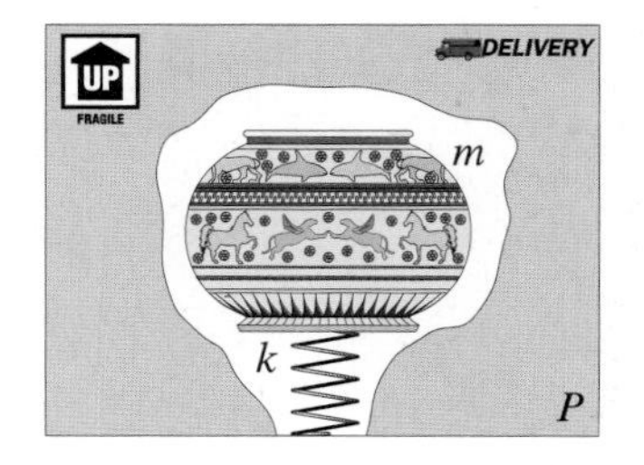

Figure P4.27 and P4.28

Problem 4.27 Determine the maximum displacement of the vase relative to the box and the maximum force on the vase due to the spring if $k = 264$ lb/ft.

Problem 4.28 Plot the maximum displacement of the vase relative to the box and the maximum force on the vase due to the spring as a function of the linear elastic spring constant k. What do these plots tell you about the problem you would encounter in trying to minimize the force on the vase?

Problem 4.29

A block A moves horizontally under the action of a force F whose line of action is parallel to the motion. If the kinetic energy of A as a function of x is that shown (x_1 and x_2 are extrema for T_A), what can you say about the *sign* of F for $x_0 < x < x_1$ and $x_1 < x < x_2$?

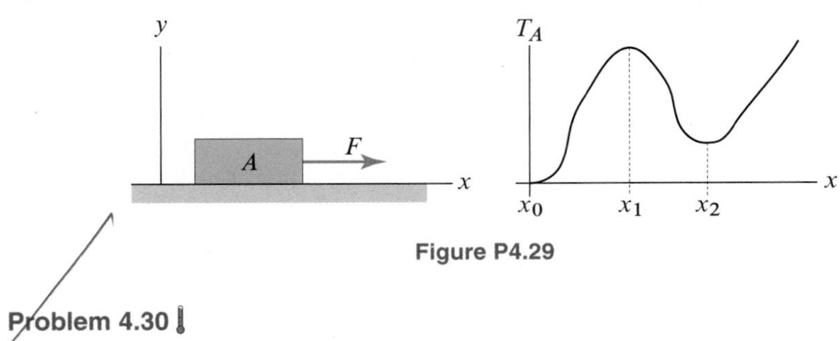

Figure P4.29

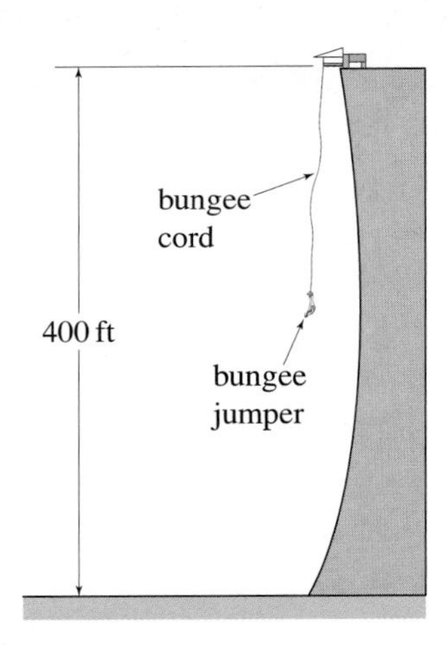

Figure P4.30

Problem 4.30

While the stiffness of an elastic cord can be quite constant (i.e., the force versus displacement curve is a straight line) over a large range of stretch, as a bungee cord is stretched, it softens; that is, the cord tends to get less stiff as it gets longer. Assuming a softening force-displacement relation of the form $k\delta - \beta\delta^3$, where δ (measured in ft) is the displacement of the cord from its unstretched length, considering a bungee cord whose unstretched length is 150 ft, and letting $k = 2.58$ lb/ft, determine the value of the constant β such that a bungee jumper weighing 170 lb and starting from rest gets to the bottom of a 400 ft tower with zero speed.

Problems 4.31 through 4.33

Car bumpers are designed to limit the extent of damage to the car in the case of low-velocity collisions. Consider a 1420 kg passenger car impacting a concrete barrier while traveling at a speed of 5.0 km/h. Model the car as a particle and consider two bumper models: (1) a simple linear spring with constant k and (2) a linear spring of constant k in parallel with a shock-absorbing unit generating a nearly constant force $F_S = 2000$ N over 10 cm.

Problem 4.31 If the bumper is of type 1 and if $k = 9 \times 10^4$ N/m, find the spring compression (distance) necessary to stop the car.

Problem 4.32 If the bumper is of type 1, find the value of k necessary to stop the car when the bumper is compressed 10 cm.

Problem 4.33 If the bumper is of type 2, find the value of k necessary to stop the car when the bumper is compressed 10 cm.

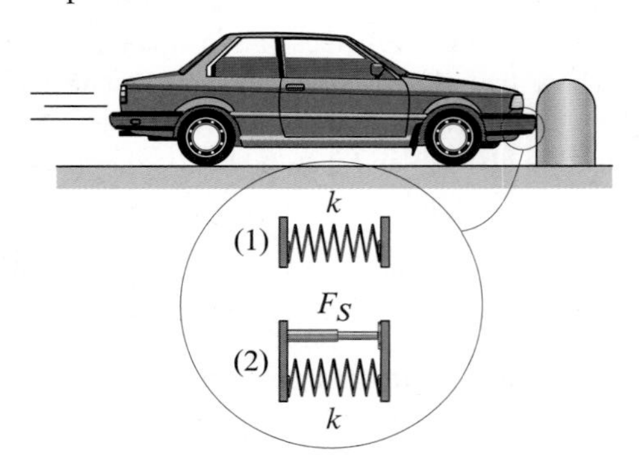

Figure P4.31–P4.33

4.2 Conservative Forces and Potential Energy

In Example 4.1 on p. 245, we saw that the work done by the weight force only depended on the change in height $R(1 - \cos\theta)$ of the ball — the fact that the path of the ball was circular did not matter. In this section, we will see that this is a general property of the weight force. We will also see that there are other forces whose work is independent of the path followed by their points of application in going from an initial to a final position.

Work done by the constant force of gravity

In Fig. 4.6, the particle of mass m moves from ① to ② following the path $\mathcal{L}_{1\text{-}2}$. We wish to determine the work done by the weight force mg during this motion, so it is the only force shown, though there may be other forces acting on the particle. To determine the work done by mg, we apply Eq. (4.7) to obtain

$$(U_{1\text{-}2})_g = \int_{\mathcal{L}_{1\text{-}2}} \vec{F}_g \cdot d\vec{r} = \int_{\mathcal{L}_{1\text{-}2}} -mg\,\hat{\jmath} \cdot (dx\,\hat{\imath} + dy\,\hat{\jmath}) = -mg \int_{y_1}^{y_2} dy, \quad (4.12)$$

where we have used $\vec{F}_g = -mg\,\hat{\jmath}$, $d\vec{r} = dx\,\hat{\imath} + dy\,\hat{\jmath}$, and the fact that mg is constant. Carrying out the final integral, we obtain the work done by mg as

$$\boxed{(U_{1\text{-}2})_g = -mg(y_2 - y_1).} \quad (4.13)$$

Equation (4.13) says that the work done by gravity on a particle moving along an *arbitrary* path is equal to the particle's weight times the particle's change in height $y_1 - y_2$. The path followed between ① and ② is arbitrary because the result in Eq. (4.13) depends only on the endpoints ① and ② rather than the shape of $\mathcal{L}_{1\text{-}2}$.

We now consider another important example of this type of force, namely, a central force, which includes as special cases both the force of a spring and gravity as described by the universal law of gravitation.

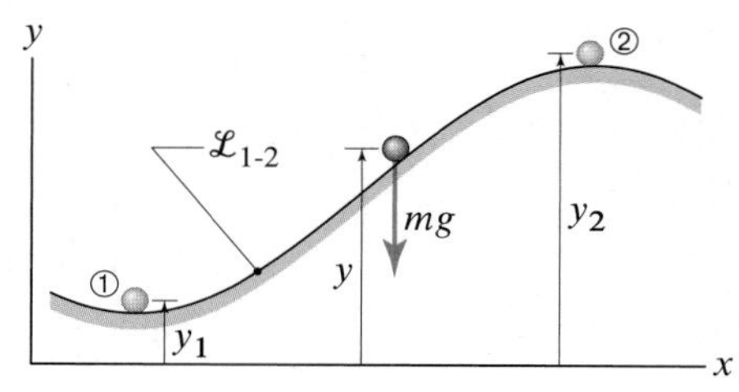

Figure 4.6
A particle of mass m travels between ① and ② along the path $\mathcal{L}_{1\text{-}2}$. The force mg is the only force considered in the work calculation in Eq. (4.12), but it need not be the only force acting on the particle.

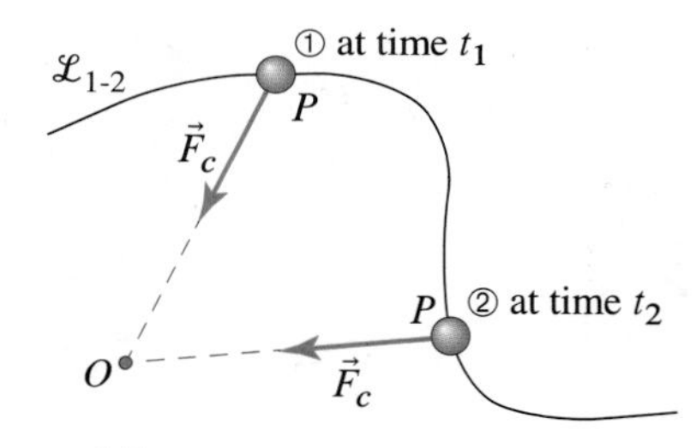

Figure 4.7
Particle P moving under the action of a central force $\vec{F}_c$.

Work of a central force

A *central force* $\vec{F}_c(r)$ acting on a particle P is a force whose line of action always passes through the same point O (see Fig. 4.7) and whose magnitude is a function of the distance r from O to P.

Referring again to Fig. 4.7, we wish to determine the work done by $\vec{F}_c(r)$ on the particle P as it moves along the path $\mathcal{L}_{1\text{-}2}$ from ①, at time t_1, to ②, at time t_2. Figure 4.8 shows the polar coordinate system we will use to describe the motion of P and the central force $\vec{F}_c$. Using the component system shown and applying Eq. (4.10) on p. 243, the work done by $\vec{F}_c$ is given by

$$(U_{1\text{-}2})_c = \int_{t_1}^{t_2} \vec{F} \cdot \vec{v}\,dt = \int_{t_1}^{t_2} F_c(r)\,\hat{u}_r \cdot \left(\dot{r}\,\hat{u}_r + r\dot{\theta}\,\hat{u}_\theta\right) dt, \quad (4.14)$$

where we have used $\vec{F}_c = F_c(r)\,\hat{u}_r$ and Eq. (2.66) on p. 105, which is the expression of the velocity in polar coordinates. Expanding the last dot product in Eq. (4.14) and noting that $\dot{r}\,dt = (dr/dt)\,dt = dr$, we obtain

$$\boxed{(U_{1\text{-}2})_c = \int_{r_1}^{r_2} F_c(r)\,dr.} \quad (4.15)$$

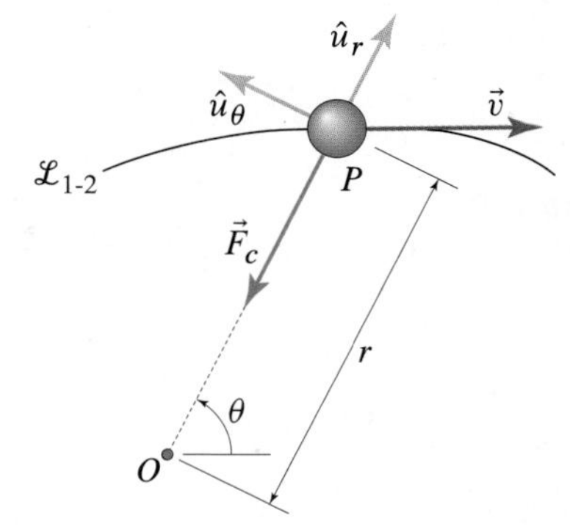

Figure 4.8
A particle P under the action of a central force $\vec{F}_c$. The figure also shows the velocity vector of P, as well as the polar coordinate system used to describe $\vec{F}_c$ and the motion of P.

Work done by a spring force. Figure 4.9(a) shows a linear spring connecting a particle P to the fixed point O while P moves in a plane from point ① to ② along the path $\mathcal{L}_{1\text{-}2}$. Figure 4.9(b) shows that the spring force always acts along the line OP. For a spring whose current length is r and whose unstretched length is L_0, the

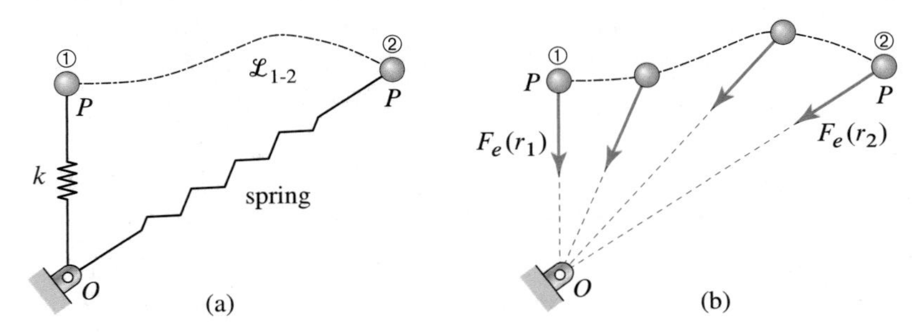

Figure 4.9. (a) A particle P moving along the path $\mathcal{L}_{1\text{-}2}$ from ① to ②. One end of the linear spring is attached to P and the other to the fixed point O. (b) Direction of the spring force as P moves along $\mathcal{L}_{1\text{-}2}$. Note that the figure is showing a situation in which $r > L_0$.

spring's stretch is $\delta = r - L_0$ and the corresponding force is $\vec{F}_e = -k(r - L_0)\,\hat{u}_r$. Thus, using Eq. (4.15), we have

$$(U_{1\text{-}2})_e = \int_{r_1}^{r_2} -k(r - L_0)\,dr = -\tfrac{1}{2}k\Big[(r_2 - L_0)^2 - (r_1 - L_0)^2\Big], \tag{4.16}$$

or

$$\boxed{(U_{1\text{-}2})_e = -\tfrac{1}{2}k\big(\delta_2^2 - \delta_1^2\big),} \tag{4.17}$$

where $\delta_1 = r_1 - L_0$ and $\delta_2 = r_2 - L_0$. As with the work done by the constant force of gravity, the work done by a spring force acting on a particle only depends on the endpoints of the path followed by the particle.

Work done by the force of gravity. Figure 4.10 shows the gravitational force on a particle B of mass m_B exerted by a particle A of mass m_A whose position is fixed. This force is given by Newton's universal law of gravitation (see Eq. (1.5) on p. 3), that is,

$$\vec{F}_{BA} = -\frac{G m_A m_B}{r^2}\,\hat{u}_r. \tag{4.18}$$

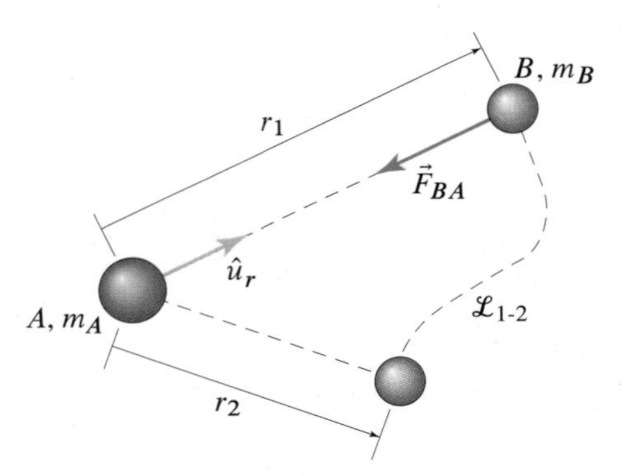

Figure 4.10
The force of gravity $\vec{F}_{BA}$ on B due to its gravitational attraction to A.

Equation (4.15) then gives the work on B as

$$(U_{1\text{-}2})_G = \int_{r_1}^{r_2} -\frac{G m_A m_B}{r^2}\,dr = -G m_A m_B \int_{r_1}^{r_2} \frac{dr}{r^2}, \tag{4.19}$$

which yields

$$\boxed{(U_{1\text{-}2})_G = -G m_A m_B\left(-\frac{1}{r_2} + \frac{1}{r_1}\right).} \tag{4.20}$$

Again, we see that the work done depends only on the initial and final positions of the particle.

Conservative forces and potential energy

In computing the work done on a particle by spring and gravitational forces, we found that this work depended *only* on its initial and final positions, as opposed to depending on the path in some intrinsic way. As we saw in the marginal note "Line integral of an exact differential" on p. 242, if a path or line integral depends on only the limits of integration, then the integrand must be the exact differential of some function. Therefore, for these special forces, we can write the work as

$$U_{1\text{-}2} = \int_{\mathcal{L}_{1\text{-}2}} \vec{F} \cdot d\vec{r} = -\int_{\mathcal{L}_{1\text{-}2}} dV = -\big[V(\vec{r}_2) - V(\vec{r}_1)\big] = -(V_2 - V_1), \tag{4.21}$$

where $\vec{r}_1$ and $\vec{r}_2$ are the position vectors corresponding to ① and ②, respectively, and where V is a scalar function of position called the *potential energy* of the force $\vec{F}$. Forces for which Eq. (4.21) is true are called *conservative forces*.

If all of the forces doing work are conservative, then the statement of the work-energy principle can be given a form reflecting this. Substituting Eq. (4.21) into Eq. (4.9) on p. 242, we obtain the form of the work-energy principle called *conservation of mechanical energy*:

$$\boxed{T_1 + V_1 = T_2 + V_2.} \tag{4.22}$$

Systems for which Eq. (4.22) holds are called *conservative systems*. Equation (4.22) states that the total mechanical energy, that is, the potential plus kinetic energy, is conserved between any two points on the path $\mathcal{L}_{1\text{-}2}$, as long as all forces doing work are conservative. We can now use Eq. (4.21) to determine V for the conservative forces we have encountered.

Potential energy of a constant gravitational force. In Eq. (4.13), we saw that the work done by a constant gravitational force on a particle of mass m is $U_{1\text{-}2} = -mg(y_2 - y_1)$. Comparing this expression with Eq. (4.21) shows that

$$\boxed{V_g = mgy,} \tag{4.23}$$

where V_g is the potential energy corresponding to the constant gravitational force mg and y is the vertical position measured from the $y = 0$ or *datum line*,* with y increasing in the direction *opposite* to gravity. Since it is only the change in height that determines the work done by a constant gravitational force, the vertical position of the datum line is entirely arbitrary.

Potential energy of a spring force. The work done by a spring force was found to be $U_{1\text{-}2} = -\frac{1}{2}k(\delta_2^2 - \delta_1^2)$. Comparing this with Eq. (4.21), we see that

$$\boxed{V_e = \tfrac{1}{2}k\delta^2,} \tag{4.24}$$

where V_e is the elastic potential energy of the linear spring force and δ is *the distance the spring is stretched or compressed from its unstretched length.*

Potential energy of the force of gravity. In Eq. (4.20), we determined the work done by the force of gravity on a particle B due to its interaction with a particle A to be $U_{1\text{-}2} = -Gm_Am_B(-1/r_2 + 1/r_1)$. Comparing this expression with Eq. (4.21), we see that the potential energy associated with a gravitational force between A and B is given by

$$\boxed{V_G = -\frac{Gm_Am_B}{r},} \tag{4.25}$$

where r is the distance between A and B.

Interesting Fact

The minus sign in Eq. (4.21)? The minus sign is there so that *the potential energy V can be viewed as a measure of the potential or capacity to do positive work* in returning to the zero potential energy state. This can be seen by noticing that Eq. (4.21) states that $U_{1\text{-}2} = V_1 - V_2$ so that the work is positive if V decreases and negative if V increases.

Common Pitfall

Elastic potential energy is never negative. It may be tempting to write $V_e = -\frac{1}{2}k\delta^2$ when δ is negative, i.e., when the spring is compressed. However, it is important to remember that the potential energy of a linear elastic spring is never negative because a spring always has the potential to do positive work when it returns to its unstretched length from a stretched or compressed condition. This is reflected in the fact that δ appears in the expression for V as δ^2, which must *always* be positive or zero.

* The word *datum* comes from the Latin verb *dare*, which means to give, and it is used in surveying, mapping, and geology to designate a given point, line, or surface used as a reference.

Work-energy principle for any type of force

The work-energy principle, as given by Eq. (4.9) on p. 242, applies to any system, conservative or nonconservative. When all the forces doing work on a system are conservative, then we can use changes in potential energy to find the work done by those forces (see Eq. (4.21)) instead of computing the work from its integral definition (see Eq. (4.7)). When a problem involves both conservative *and* nonconservative forces, it would still be nice to take advantage of Eq. (4.21) to compute the work of the conservative forces. To do this, we let $(U_{1\text{-}2})_c$ and $(U_{1\text{-}2})_{nc}$ be the work done by conservative and nonconservative forces, respectively. Then, noting that their sum is the total work done, we can write the work-energy principle as

$$T_1 + (U_{1\text{-}2})_c + (U_{1\text{-}2})_{nc} = T_2. \tag{4.26}$$

Using Eq. (4.21), we can write $(U_{1\text{-}2})_c = -(V_2 - V_1)$ so that Eq. (4.26) becomes

$$\boxed{T_1 + V_1 + (U_{1\text{-}2})_{nc} = T_2 + V_2.} \tag{4.27}$$

Equation (4.27) is just as general as Eq. (4.9), but it is easier to apply since the work done by conservative forces, such as spring and gravitational forces, can be included in the V_1 and V_2 terms.

When is a force conservative?

Is there a way to tell whether or not a force is conservative? That is, can we determine whether or not an associated potential energy exists and can we find it? Also, given a potential energy, can we find the associated force?

Equation (4.21) implies that the infinitesimal work of a conservative force $\vec{F}$ can be written as

$$\vec{F} \cdot d\vec{r} = -dV. \tag{4.28}$$

Since $V = V(x, y, z)$ in Cartesian components, we can write the differential of V as

$$dV = \frac{\partial V}{\partial x}\,dx + \frac{\partial V}{\partial y}\,dy + \frac{\partial V}{\partial z}\,dz \tag{4.29}$$

and $\vec{F} \cdot d\vec{r}$ as

$$\begin{aligned}\vec{F} \cdot d\vec{r} &= \left(F_x\,\hat{\imath} + F_y\,\hat{\jmath} + F_z\,\hat{k}\right) \cdot \left(dx\,\hat{\imath} + dy\,\hat{\jmath} + dz\,\hat{k}\right)\\ &= F_x\,dx + F_y\,dy + F_z\,dz.\end{aligned} \tag{4.30}$$

Inserting Eqs. (4.29) and (4.30) into Eq. (4.28), we see that

$$F_x\,dx + F_y\,dy + F_z\,dz = -\frac{\partial V}{\partial x}\,dx - \frac{\partial V}{\partial y}\,dy - \frac{\partial V}{\partial z}\,dz. \tag{4.31}$$

Since dx, dy, and dz are independent of each other, for Eq. (4.31) to be satisfied, it must be true that

$$F_x = -\frac{\partial V}{\partial x}, \quad F_y = -\frac{\partial V}{\partial y}, \quad \text{and} \quad F_z = -\frac{\partial V}{\partial z}. \tag{4.32}$$

You may recognize that, except for the sign, the right-hand sides of Eqs. (4.32) are the components of the gradient of V, so we can say

$$\boxed{\vec{F} = -\left(\frac{\partial V}{\partial x}\,\hat{\imath} + \frac{\partial V}{\partial y}\,\hat{\jmath} + \frac{\partial V}{\partial z}\,\hat{k}\right) = -\vec{\nabla} V,} \tag{4.33}$$

Interesting Fact

The gradient operator. The vector differential operator $\vec{\nabla}$ should be read as *nabla* or *del.* The gradient of the scalar function V is often written grad V. Note that the gradient operator in cylindrical coordinates can be written as

$$\vec{\nabla} = \frac{\partial}{\partial r}\,\hat{u}_r + \frac{1}{r}\frac{\partial}{\partial \theta}\,\hat{u}_\theta + \frac{\partial}{\partial z}\,\hat{u}_z.$$

where $\vec{\nabla} V$ is the gradient of V (also written as 'grad V'). Equation (4.33) answers one of our questions; that is, given a potential V, we can find $\vec{F}$ by computing the negative of the gradient of V.

Now, if we take the curl of $\vec{F}$, which is defined to be $\vec{\nabla} \times \vec{F}$ (also written as curl $\vec{F}$), in Cartesian components, we obtain

$$\begin{aligned}\vec{\nabla} \times \vec{F} &= \left(\frac{\partial}{\partial x}\hat{\imath} + \frac{\partial}{\partial y}\hat{\jmath} + \frac{\partial}{\partial z}\hat{k}\right) \times \left(F_x\,\hat{\imath} + F_y\,\hat{\jmath} + F_z\,\hat{k}\right) \\ &= \left(\frac{\partial F_z}{\partial y} - \frac{\partial F_y}{\partial z}\right)\hat{\imath} + \left(\frac{\partial F_x}{\partial z} - \frac{\partial F_z}{\partial x}\right)\hat{\jmath} + \left(\frac{\partial F_y}{\partial x} - \frac{\partial F_x}{\partial y}\right)\hat{k}. \qquad (4.34)\end{aligned}$$

Substituting Eqs. (4.32) into Eq. (4.34) gives

$$\begin{aligned}\vec{\nabla} \times \vec{F} = \left(-\frac{\partial^2 V}{\partial z \partial y} + \frac{\partial^2 V}{\partial y \partial z}\right)\hat{\imath} + \left(-\frac{\partial^2 V}{\partial x \partial z} + \frac{\partial^2 V}{\partial z \partial x}\right)\hat{\jmath} \\ + \left(-\frac{\partial^2 V}{\partial y \partial x} + \frac{\partial^2 V}{\partial x \partial y}\right)\hat{k} = \vec{0}, \qquad (4.35)\end{aligned}$$

where the equality to zero holds as long as the partial derivatives in Eq. (4.35) are continuous so that we can interchange the order of differentiation of the second derivatives of V, that is,

$$\frac{\partial^2 V}{\partial z\,\partial y} = \frac{\partial^2 V}{\partial y\,\partial z}, \quad \frac{\partial^2 V}{\partial x\,\partial z} = \frac{\partial^2 V}{\partial z\,\partial x}, \quad \text{and} \quad \frac{\partial^2 V}{\partial y\,\partial x} = \frac{\partial^2 V}{\partial x\,\partial y}. \qquad (4.36)$$

Equation (4.35) helps us answer the other question we asked because it shows that *the curl of a conservative force is necessarily equal to zero.* It is possible to prove that the converse of this statement is also true, that is, *if the curl of a force is equal to zero, then the force is conservative.*

End of Section Summary

In this section we discovered that there are forces whose work, in taking a particle from an initial to a final position, does not depend on the path connecting these positions.

Work of a constant gravitational force. For a particle of mass m moving in a constant gravitational field from ① to ②, the work done on the particle is

Eq. (4.13), p. 257

$$(U_{1\text{-}2})_g = -mg(y_2 - y_1),$$

for which gravity must act in the $-y$ direction.

Work done by a spring force. For a particle subject to the force of a linear elastic spring with constant k, the work done on the particle is

Eq. (4.17), p. 258

$$(U_{1\text{-}2})_e = -\tfrac{1}{2}k\left(\delta_2^2 - \delta_1^2\right),$$

where δ_1 and δ_2 are the stretch of the spring at ① and ②, respectively.

Conservative systems. A force is said to be *conservative* if the work done by it depends on only the initial and final position of its point of application, but is otherwise independent of the path connecting these positions. The work of a conservative force can be characterized by a scalar potential energy function V. A *conservative system* is one for which all the forces doing work are conservative. For conservative systems, the work-energy principle becomes the *conservation of mechanical energy*, which is given by

Eq. (4.22), p. 259

$$T_1 + V_1 = T_2 + V_2.$$

Important conservative forces include elastic spring forces and gravitational forces. The *potential energy of a linear elastic spring force* is given by

Eq. (4.24), p. 259

$$V_e = \tfrac{1}{2} k \delta^2,$$

the *potential energy of a constant gravitational force* is given by

Eq. (4.23), p. 259

$$V_g = mgy,$$

and the *potential energy of the force of gravity* is given by

Eq. (4.25), p. 259

$$V_G = -\frac{Gm_A m_B}{r}.$$

Finally, if we have a system in which there are both conservative and nonconservative forces doing work, we can use Eq. (4.9) on p. 242, that is, $T_1 + U_{1\text{-}2} = T_2$, or we can take advantage of the potential energies of the conservative forces by writing the work-energy principle as

Eq. (4.27), p. 260

$$T_1 + V_1 + (U_{1\text{-}2})_{\text{nc}} = T_2 + V_2,$$

where $(U_{1\text{-}2})_{\text{nc}}$ is the work done by nonconservative forces.

EXAMPLE 4.5 *Demolishing a Building: Speed of a Wrecking Ball*

A 2500 lb wrecking ball A is released from rest at 27° with respect to the vertical as shown in Fig. 1. When the ball hits the structure to be demolished, the cable holding the wrecking ball forms an angle of 11°. If the length L of the cable is 30 ft, determine the speed with which the wrecking ball hits the structure.

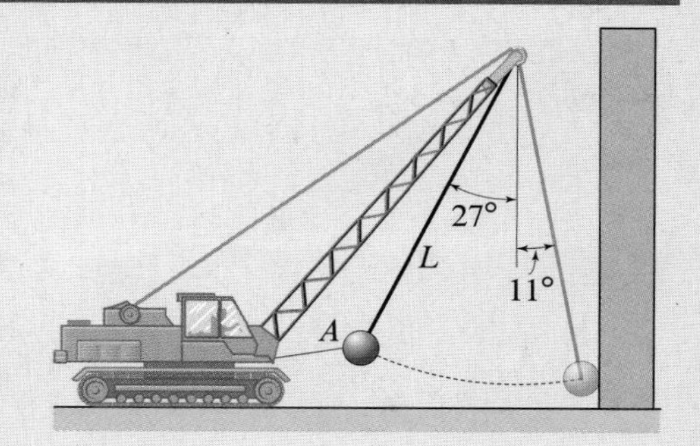

Figure 1

SOLUTION

Road Map & Modeling This problem requires that we relate a change in the position of the wrecking ball to its change in speed, and therefore, the work-energy principle should lead to a solution. Referring to Fig. 2, the wrecking ball is modeled as a particle subject to its own weight and the cable tension. We assume that the cable is inextensible and that the end at O is fixed. Referring to Fig. 3, we denote the release position as ① and the position right before impact with the structure as ②. The only force doing work is gravity, which is a conservative force.

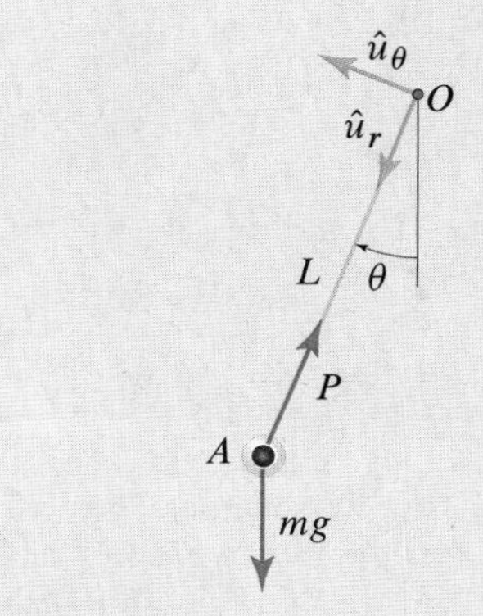

Figure 2
FBD of the wrecking ball.

Governing Equations

Balance Principles We now apply the principle of conservation of mechanical energy between ① and ②, which reads

$$T_1 + V_1 = T_2 + V_2, \tag{1}$$

where

$$T_1 = \tfrac{1}{2}mv_1^2 \quad \text{and} \quad T_2 = \tfrac{1}{2}mv_2^2, \tag{2}$$

and where m is the mass of the wrecking ball and v_1 and v_2 are its speed at ① and ②, respectively.

Force Laws Recalling that only gravity does work, and choosing the datum line as shown in Fig. 3, we have

$$V_1 = -mgL\cos\theta_1 \quad \text{and} \quad V_2 = -mgL\cos\theta_2. \tag{3}$$

Kinematic Equations Since the system is released from rest, we have

$$v_1 = 0. \tag{4}$$

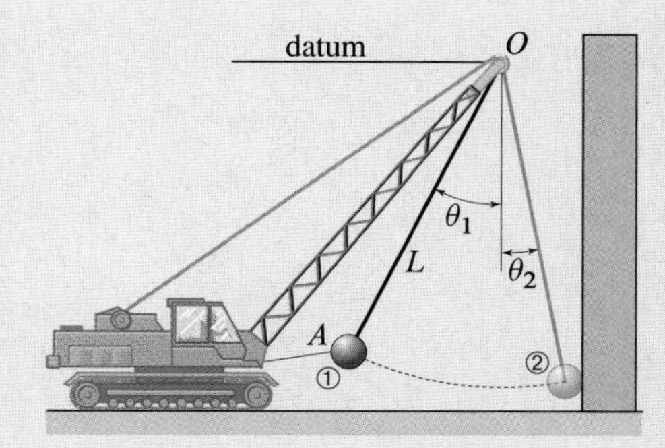

Figure 3
Definition of ① and ②, as well as of the datum line for gravitational potential energy.

Computation Substituting Eqs. (2)–(4) into Eq. (1), we have

$$-mgL\cos\theta_1 = \tfrac{1}{2}mv_2^2 - mgL\cos\theta_2, \tag{5}$$

which, after solving for v_2, yields

$$\boxed{v_2 = \sqrt{2gL(\cos\theta_2 - \cos\theta_1)} = 13.23\,\text{ft/s},} \tag{6}$$

where we have used $\theta_1 = 27°$ and $\theta_2 = -11°$ to obtain the numerical result.

Discussion & Verification The dimensions of the term under the square root sign are length squared over time squared. Therefore, our final result is dimensionally correct, and it has been expressed with correct units.

A Closer Look Notice that the quantity $L(\cos\theta_2 - \cos\theta_1)$ corresponds to the vertical drop of the ball. Hence, the speed achieved by the ball is precisely what the ball would achieve if it were dropped from a height equal to $L(\cos\theta_2 - \cos\theta_1)$.

EXAMPLE 4.6 Bungee Jumping: Conservation of Mechanical Energy

Figure 1
The Verzasca dam in southern Switzerland, which was the location for the bungee jumping scene in the 1995 James Bond film *GoldenEye*.

The mechanics of bungee jumping are rather straightforward, but a miscalculation can have dire consequences. Let's consider the highest bungee jumping location in the world — the Verzasca dam in southern Switzerland. The jump height off of the Verzasca dam is 722 ft. Given a jumper weighing 170 lb, determine the relationship between the stiffness k of the bungee cord and its unstretched length L_0; i.e., find k as a function of L_0, so that the jumper has zero speed at the bottom of the dam. In addition, determine the unstretched length so that the acceleration of the jumper does not exceed $4g$ during the jump.

SOLUTION

Road Map & Modeling Referring to the FBD in Fig. 2, we model the jumper as a particle subject to gravity and the force F_b of the bungee cord, which we model as a linear elastic spring. The jumper begins the jump at ① with zero speed and ends the jump at ② with zero speed (see Fig. 2). Since we know the jumper's speed at ① and ② and we know all the forces doing work on the jumper, we can apply the work-energy principle to the jumper to determine the relationship between the bungee stiffness and its unstretched length. We will then apply Newton's second law to the jumper to determine the maximum acceleration so that we can find k and L_0 for the bungee cord. Note that all the forces acting on the jumper are conservative and that F_b is zero until the jumper falls a distance equal to the unstretched length of the cord.

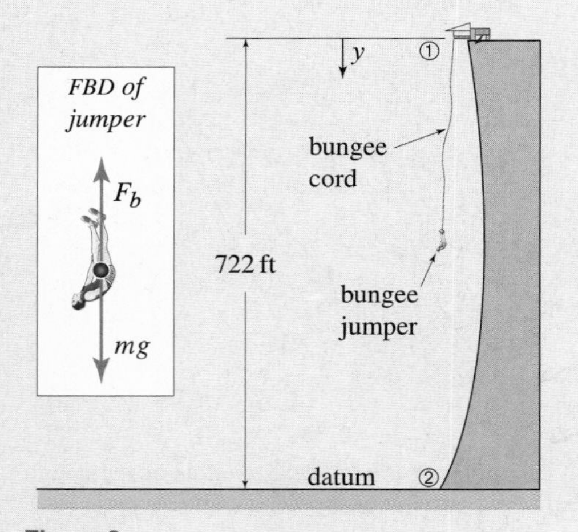

Figure 2
Profile of the Verzasca dam showing the jump platform as well as the bungee jumper. ① and ② are also defined. The blue inset shows the FBD after the jumper has fallen a distance greater than the unstretched length of the cord.

Governing Equations

Balance Principles Applying the work-energy principle between ① and ②, we have

$$T_1 + V_1 = T_2 + V_2, \tag{1}$$

where we have used the fact that all forces doing work are conservative. The kinetic energies can be written as

$$T_1 = \tfrac{1}{2}mv_1^2 \quad \text{and} \quad T_2 = \tfrac{1}{2}mv_2^2, \tag{2}$$

where m is the jumper's mass and v_1 and v_2 are the jumper's speed at ① and ②, respectively. Since we also need to determine the jumper's acceleration, we will write Newton's second law for the jumper in the y direction as

$$\sum F_y\!: \quad mg - F_b = ma_y, \tag{3}$$

where we note that $F_b = 0$ until the bungee cord engages.

Force Laws If we place the datum line for gravitational potential energy at ②, then

$$V_1 = mgh \quad \text{and} \quad V_2 = \tfrac{1}{2}k(h - L_0)^2, \tag{4}$$

where $h = 722$ ft, $mg = 170$ lb, and we have accounted for the potential energy of the bungee cord in V_2. We will also need the force law for the bungee cord, which is given by

$$F_b = k\delta = k(y - L_0), \tag{5}$$

where we note that $F_b = 0$ when $y \le L_0$.

Kinematic Equations Since the jumper starts and ends the jump with zero speed, we have

$$v_1 = v_2 = 0. \tag{6}$$

Computation Substituting Eqs. (2), (4), and (6) into Eq. (1) and solving for k, we obtain

$$k = \frac{2mgh}{(h - L_0)^2}. \tag{7}$$

Equation (7) gives the desired k as a function of L_0, a plot of which is shown in Fig. 3.

We now want to design the bungee system so that the maximum acceleration of a jumper weighing 170 lb does not exceed $a_{max} = 4g$. As with any design problem, there are infinitely many solutions that will satisfy the criteria that the jumper have zero speed at ② and that the jumper's acceleration not exceed $4g$. Referring to the FBD inset in Fig. 2, we know that until the spring engages, the jumper will be in free fall and his acceleration will be g downward. Once the spring engages, the acceleration is determined by solving Eqs. (3) and (5) for a_y, which gives

$$a_y = g - \frac{k}{m}(y - L_0). \tag{8}$$

Recall that Eq. (7) provides a relation between k and L_0 that ensures that the jumper has zero speed at ②. Therefore, using Eq. (7), Eq. (8) becomes

$$a_y = g - \frac{2gh(y - L_0)}{(h - L_0)^2} \quad \text{or} \quad \frac{a_y}{g} = 1 - \frac{2h(y - L_0)}{(h - L_0)^2}, \quad y > L_0. \tag{9}$$

We will choose the following value of the unstretched length of the bungee cord:

$$L_0 = h/2 = 361.0\,\text{ft}. \tag{10}$$

We now need to verify that the design criterion requiring $a_y < 4g$ is always met. To do so, using the chosen L_0 and referring to Fig. 4, we plot a_y/g versus y. The plot shows that the chosen value of L_0 is such that our design goal is met.

Discussion & Verification The right-hand side of Eq. (7) has dimensions of force over length and therefore, has the proper dimensions for k. Equation (7) also implies that the stiffness of the cord must be proportional to the jumper's weight, which is to be expected. In addition, Eq. (7) and the corresponding plot in Fig. 3 demonstrate an interesting aspect of the required stiffness k. If the bungee cord has a very short unstretched length (i.e., the denominator in Eq. (7) will be large), then the stiffness required to stop the jumper after falling 722 ft is very small, and it increases very slowly until L_0 reaches about 500 ft. As L_0 gets closer to the jump height of 722 ft (i.e., the denominator in Eq. (7) will be small), the spring has to get very stiff to be able to stop the jumper in time and, at $L_0 = 722$ ft, the stiffness becomes infinite.

Hence, overall our solution seems to be correct. As far as the value of L_0 is concerned, we have already verified that it satisfies the required criteria.

A Closer Look Figure 4 shows the acceleration of the jumper as a function of distance from the top of the dam for $L_0 = 361.0$ ft. This figure demonstrates that the largest acceleration experienced by the jumper is $3g$, and this occurs at the very end of the jump (that is, at $y = 722$ ft, the acceleration is $3g$ *upward*). As L_0 increases, the maximum acceleration experienced by the jumper increases so that, for example, when $L_0 = 650$ ft, we find that the maximum acceleration of the jumper is over $19g$!

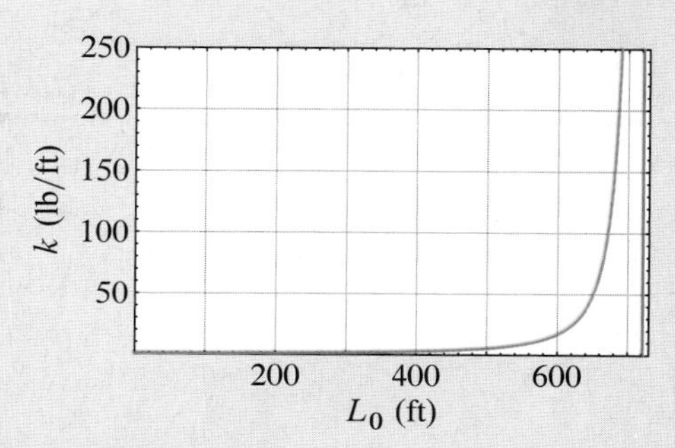

Figure 3
Bungee stiffness k as a function of its unstretched length L_0 required to achieve zero speed at the bottom of the jump (dark green curve). The red vertical line is at $L_0 = 722$ ft.

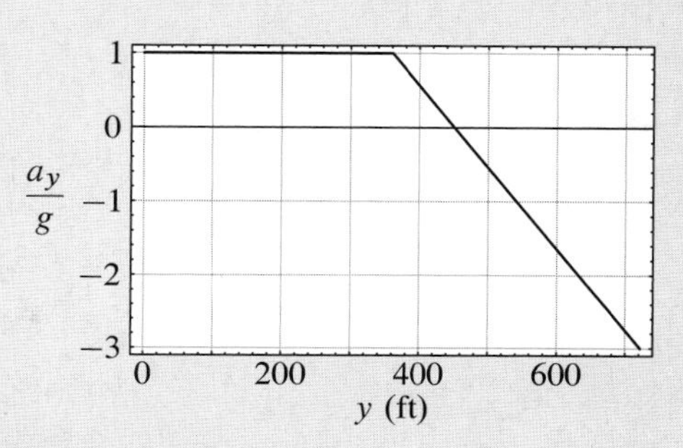

Figure 4
The acceleration of the bungee jumper as a function of y for $L_0 = h/2 = 361.0$ ft.

Interesting Fact

Bungee cord stiffness and unstretched length. We have not taken into account the fact that a real bungee cord has limits to the amount that it can stretch. For example, for $L_0 = 1$ ft, Eq. (7) tells us that $k = 0.4722$ lb/ft for $h = 722$ ft and $mg = 170$ lb. In fact, increasing L_0 to 400 ft only increases k to 2.368 lb/ft. Of course, the 1 ft bungee cord would have to stretch 721 ft or 72,100%, and the 400 ft bungee cord would only have to stretch 300 ft or 80.5%. While a rubber band can be easily stretched to a little less than twice its original length, we are not aware of a rubber band that can stretch to over 700 times its original length!

EXAMPLE 4.7 *Pole Vaulting: Turning Speed into Height*

Figure 1
Sequence of images showing the positions of a vaulter during a vault. The blue pole has been shortened in some of the frames for clarity.

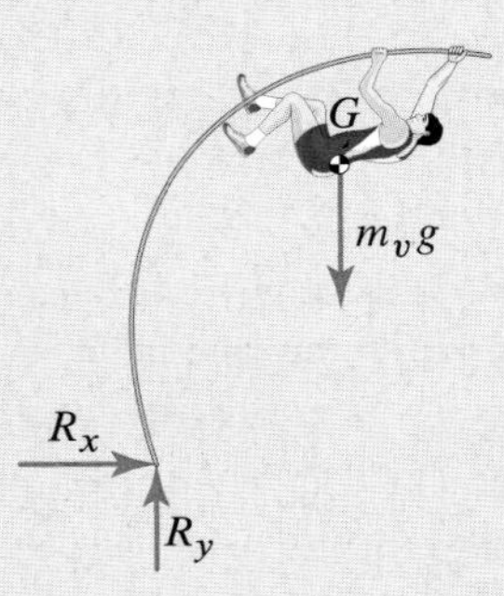

Figure 2
FBD of the vaulter and pole during the vault. The weight force acting on the vaulter acts at his mass center G.

A pole vaulter relies on several skills to achieve maximum height during a vault, but one of the most important is running speed. Speed is important because it is vital that the vaulter turn kinetic energy (running speed) into potential energy (vault height). Figure 1 shows the final moments of a vault during which the forward speed of the vaulter at A is turned into height at B. Modeling the vaulter as a particle, determine the maximum possible height that can be achieved by the "world's fastest pole vaulter," using the following data and assumptions:

1. At the time this book was written, the men's world record in the 100 m dash was 9.72 s, which was established by Usain Bolt of Jamaica on May 31, 2008; the record in the 200 m was 19.32 s, which was set on August 1, 1996, by Michael Johnson of the United States.
2. The pole vaulter is 6 ft tall with mass center at 55% of body height as measured from the ground.
3. *All* of the vaulter's kinetic energy is converted to potential energy during the vault, and no energy is lost from the system consisting of the pole and the vaulter.
4. The vaulter does no work during the vault, and his velocity is zero when he reaches the peak height of the vault.

SOLUTION

Road Map & Modeling We will assume that the vault begins when the vaulter has reached maximum speed. To determine that top speed, we will use the 100 and 200 m dash data as follows: We assume that the vaulter starts from rest and accelerates to top speed with an acceleration that is approximately the same for both races. Since the runner has reached top speed by the time he reaches 100 m, we will assume that he runs at a uniform speed between 100 and 200 m. Therefore, we can approximate his maximum speed as

$$v_{\max} = \frac{200 - 100}{19.32 - 9.72}\ \text{m/s} = 10.42\ \text{m/s}. \tag{1}$$

To complete our model, we need to determine the forces acting on the vaulter. Referring to Fig. 2, we have treated the vaulter as a particle, and we have neglected air resistance as well as any moment that might be applied to the bottom of the pole by the box during the vault.* Note that in the FBD we have not included the weight of the pole. Observe that the point of application of the forces R_x and R_y is fixed so that these forces do no work. Hence, the only force doing work is gravity, which is a conservative force for which we know the potential energy function. Consequently, given that we want to relate changes in speed to changes in position, we can solve this problem by applying the work-energy principle.

Governing Equations

Balance Principles All the forces doing work are conservative, and the work-energy principle is

$$T_1 + V_1 = T_2 + V_2, \tag{2}$$

where, referring to Fig. 3, ① corresponds to the instant the vaulter plants the pole and ② corresponds to the peak height of the vault. The vaulter's kinetic energies at ① and ② are given by

$$T_1 = \tfrac{1}{2} m_v v_1^2 \quad \text{and} \quad T_2 = \tfrac{1}{2} m_v v_2^2, \tag{3}$$

where m_v is the vaulters's mass and v_1 and v_2 are the vaulter's speed at ① and ②, respectively.

* The box is the 20 cm depression in which the vaulter plants the pole to begin the vault.

Interesting Fact

Top human speed. It is believed that the fastest recorded speed achieved by a human was by Donovan Bailey of Canada, who set the world record in the 100 m dash on July 27, 1996, at the Atlanta Olympic Games with a time of 9.84 s. Near the 60 m mark, a radar gun recorded his speed to be 12.1 m/s = 27.1 mph. At the 1997 World Championships in Athens, via more careful measurements, both Maurice Greene and Bailey were clocked at 11.87 m/s (Greene ran 9.86 s to Bailey's 9.91 s). At the same championship, Bailey was clocked at 11.91 m/s in the 4 × 100 m relay. See J. R. Mureika, "A Simple Model for Predicting Sprint-Race Times Accounting for Energy Loss on the Curve," *Canadian Journal of Physics*, **75**(11), 1997, pp. 837–851.

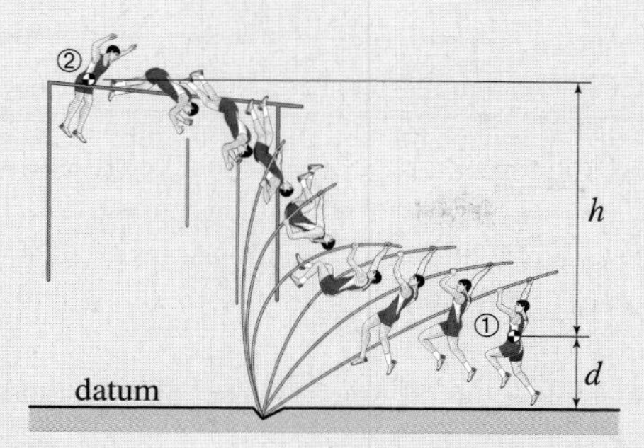

Figure 3. Sequence showing ① and ② within the vault.

Force Laws We will include all potential energy and work terms here. Recalling that the vaulter's weight is the only force doing work, we have (see Fig. 3)

$$V_1 = m_v g d \quad \text{and} \quad V_2 = m_v g(d+h), \tag{4}$$

where we have used the ground as our datum line for gravitational potential energy.

Kinematic Equations Based on the Road Map & Modeling discussion, and ignoring any horizontal component of velocity the vaulter might have at ②, we have

$$v_1 = v_{\text{max}} \quad \text{and} \quad v_2 = 0. \tag{5}$$

Computation Substituting Eqs. (3)–(5) into Eq. (2) and simplifying, we have

$$\tfrac{1}{2}v_{\text{max}}^2 = gh \quad \Rightarrow \quad h = \frac{v_{\text{max}}^2}{2g} = 5.530\,\text{m}, \tag{6}$$

where we have used the value of v_{max} in Eq. (1). Since at ① the vaulter's mass center is a distance $d = 0.55(6)\,\text{ft} = 3.3\,\text{ft} = 1.006\,\text{m}$ above the ground, we can add the distance d to h to find that the maximum height achievable by a pole vaulter is

$$\boxed{h_{\text{max}} = 6.536\,\text{m}.} \tag{7}$$

Discussion & Verification Given that the right-hand side of Eq. (6) has dimensions of length, we can say that our result has the correct dimensions. As far as the correctness of the value we have obtained is concerned, it is larger than the current world record but not far from it (see discussion below). Hence, overall, our solution appears to be correct.

A Closer Look At the time this book was written, the world record in the pole vault was 6.14 m (see marginal note entitled "World records"). The height we obtained is not far from this (it is 8% higher). There are many factors that we have not included in our analysis. Here are a few of these factors, which may contribute to our result being *too high*.

- Vaulters cannot take advantage of all of their speed at the pole plant since they must leap at a takeoff angle of 15–20°.
- While elite male vaulters run faster than 9.0 m/s during the last 5 m of their approach sprint, our estimate of 10.42 m/s was too high.
- Vaulters cannot have zero velocity at the maximum height because they need some horizontal speed to clear the bar.

On the other hand, we have not taken into account the fact that the vaulter can do work during the vault between ① and ②, for example, by pushing with the arms against the pole. This allows the vaulter to go higher than our prediction would suggest, and this additional work can generally add about 0.8 m to the vault.

Interesting Fact

Fiberglass poles. Fiberglass poles were introduced in the early 1960s. Before that, aluminum, steel, or bamboo poles were used. With fiberglass poles, vault heights and records increased dramatically. From 1940 to 1960, the world record increased from 4.7 to 4.8 m, whereas from 1960 to 1963 it increased from 4.8 to 5.1 m. Fiberglass poles were thought to "catapult" the vaulter over the bar, but studies showed otherwise. There are two reasons why fiberglass poles facilitate performances. First, a flexible pole allows for a higher grip height on the pole. A vaulter's grip height is determined by the kinetic energy at takeoff—the higher the kinetic energy, the longer the pole vaulter can rotate to vertical. Second, flexible poles allow vaulters to have a lower takeoff angle since flexible poles can accept and return much more energy than a stiff pole.

Interesting Fact

World records. At the time of writing, the world record in the pole vault was 6.14 m = 20.14 ft, set by Sergey Bubka of the Ukraine in 1994. For comparison, the high jump world record was 2.45 m = 8.04 ft, set by Javier Sotomayor of Cuba in 1993. If all that mattered were kinetic energy, these two records would be the same. However, without a pole this energy cannot be used to gain elevation, and that's why horizontal speed does not play a strong role in the high jump.

EXAMPLE 4.8 *Sliding With and Without Friction*

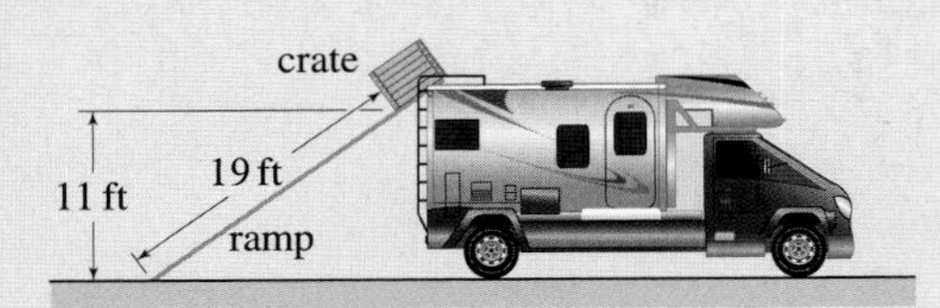

Figure 1
The RV with the ramp and its dimensions.

A crate is unloaded from the top of a recreational vehicle (RV) using the ramp shown in Fig. 1. Determine the speed of the crate as it hits the ground for each of the following two cases if the crate is released from rest in the position shown and if

(a) there is negligible friction between the crate and the ramp; and

(b) the kinetic friction coefficient between the crate and the ramp is $\mu_k = 0.32$.

Assume that static friction is insufficient to prevent slipping.

SOLUTION

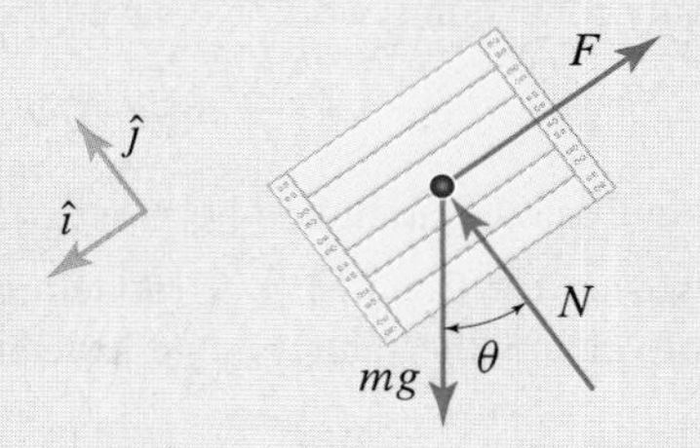

Figure 2. FBD of the crate as it slides down the ramp.

Road Map & Modeling We will model the crate as a particle subject to its own weight, the normal reaction between the crate and the slide, and the friction force F (Fig. 2). The force F will be set to zero when we consider the case of frictionless contact between crate and slide. Since we need to relate a change in the crate's position to a corresponding change in speed, we will apply the work-energy principle. However, since one of the forces doing work is the friction force F, we will also apply Newton's second law in the direction perpendicular to the slide to determine the relation between the normal reaction N and F. We define ① to be the point of release of the crate and ② to be just before the crate hits the ground (see Fig. 3). The length of the ramp l is 19 ft, and the vertical drop h is 11 ft. We will solve Parts (a) and (b) at the same time since the solution for Part (a) is a special case of the solution to case (b). Finally, since it will be used later in the calculations, we note that the angle θ appearing in the FBD of Fig. 2 is such that

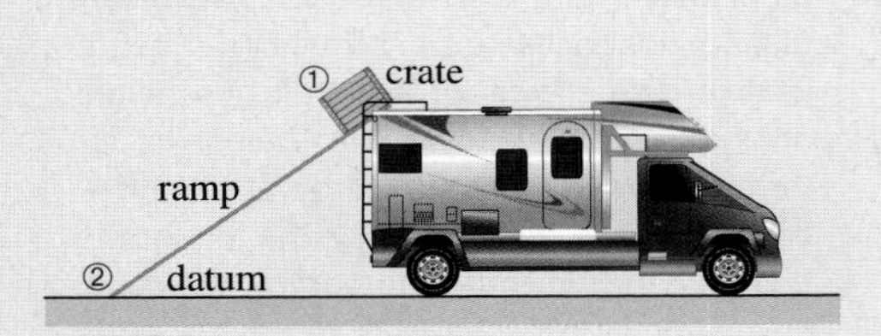

Figure 3
Definition of ① and ② for the work-energy principle.

$$\sin\theta = 11/19 \quad \Rightarrow \quad \theta = 35.38°. \tag{1}$$

Governing Equations

Balance Principles The work-energy principle, applied between ① and ②, reads

$$T_1 + V_1 + (U_{1\text{-}2})_{\text{nc}} = T_2 + V_2. \tag{2}$$

The kinetic energies can be written as

$$T_1 = \tfrac{1}{2}mv_1^2 \quad \text{and} \quad T_2 = \tfrac{1}{2}mv_2^2, \tag{3}$$

where m is the crate's mass and v_1 and v_2 are the crate's speed at ① and ②, respectively. Referring to the FBD in Fig. 2, Newton's second law in the y direction gives

$$\sum F_y\text{:} \quad N - mg\cos\theta = ma_y. \tag{4}$$

Force Laws By setting the datum line at ②, the potential energies in Eq. (2) are

$$V_1 = mgh \quad \text{and} \quad V_2 = 0. \tag{5}$$

The work done by the friction force F is

$$(U_{1\text{-}2})_{\text{nc}} = \int_{\mathcal{L}_{1\text{-}2}} -F\,\hat{\imath}\cdot d\vec{r} = \int_0^l -F\,\hat{\imath}\cdot dx\,\hat{\imath} = \int_0^l -F\,dx, \tag{6}$$

where F is related to the normal force N by the Coulomb friction law for sliding, i.e.,

$$F = \mu_k N. \tag{7}$$

Kinematic Equations Since the crate slides along the ramp, we have

$$a_y = 0. \tag{8}$$

To compute the kinetic energies at ① and ②, we note that

$$v_1 = 0 \tag{9}$$

and that v_2 is the main unknown of the problem.

Computation Using Eqs. (4), (7), and (8), we have

$$F = \mu_k mg\cos\theta. \tag{10}$$

Substituting Eq. (10) into Eq. (6), we then have

$$(U_{1\text{-}2})_{\text{nc}} = -\mu_k mgl\cos\theta. \tag{11}$$

Finally, substituting Eqs. (3), (5), (9), and (11) into Eq. (2), we obtain

$$mgh - \mu_k mgl\cos\theta = \tfrac{1}{2}mv_2^2. \tag{12}$$

For case (a), we have $\mu_k = 0$, so Eq. (12) yields

$$\boxed{v_2 = \sqrt{2gh} = 26.62\ \text{ft/s}.} \tag{13}$$

For case (b), μ_k is different from zero and is equal to the given value of 0.32, so that, from Eq. (12), we have

$$\boxed{v_2 = \sqrt{2g(h - \mu_k l\cos\theta)} = 19.73\ \text{ft/s}.} \tag{14}$$

Discussion & Verification Given that the terms under the square root sign in Eqs. (13) and (14) have dimensions of acceleration times length, our results are dimensionally correct. In addition, the result in Eq. (13) is larger than that in Eq. (14), as expected given that the effect of friction is to oppose the motion of the crate. Therefore, our solution is reasonable.

A Closer Look The result in Eq. (13) corresponds to the speed that the crate would have if the crate had been *dropped* from the top of the RV. While perhaps surprising, this result is indeed correct. In fact, in Part (a), the only two forces acting on the crate are the weight force mg and the normal force N. The normal force does no work, and the work done by the weight force can be easily computed as the force times the component of displacement in the direction of the force (since the weight, in this problem, is a constant force); this component is the vertical drop h. Therefore, the work done by mg is mgh, and that is exactly what it would be if we had dropped the crate from the top of the RV.

EXAMPLE 4.9 Cyclic Work to Promote Swinging Motion

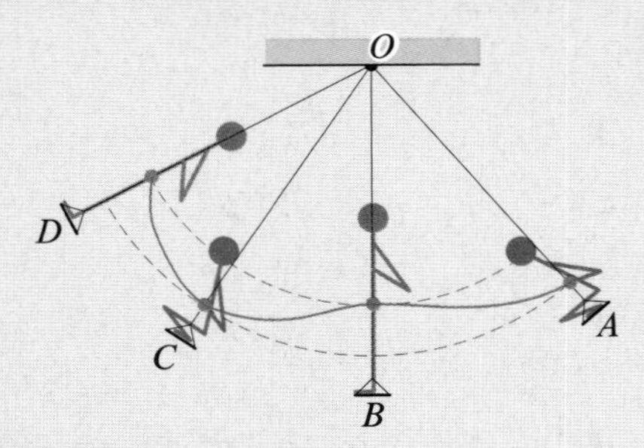

Figure 1
One possible stand-squat sequence during a half swing. You are squatting at A, standing at B, squatting at C, and then standing again at the top of the swing at D. The blue dots indicate the position of your mass center, and the purple line is the trajectory of your mass center. The two dashed lines indicate arcs of nearest and farthest distance of your mass center from point O.

Figure 1 shows you swinging in the standing and squatting positions. As you probably know, if you raise (stand) and lower (squat) yourself as you swing, you can increase the amplitude of swinging with each cycle. While the analysis of the motion shown in Fig. 1 can be very involved,* we can create a simple model to investigate how pumping can increase the amplitude of each swing cycle.

We will analyze the pumping motion depicted in Fig. 2. This motion consists of the following sequence of events:

1. At ①, that is, at the top of a backswing, the speed is zero and you are squatting.
2. At ②, which occurs at the lowest point in the swing, you stand upright to move to ③. This motion is assumed to occur instantaneously.
3. You next swing from ③ to ④ and then back to ⑤ while standing upright.
4. At ⑤, you then instantaneously squat and another cycle begins.

The goal is to find θ_5 as a function of θ_1 due to the pumping motion described above. When you are standing, your mass center is a distance l from the fixed point O; and when you are squatting, your mass center is a distance $l + \delta l$ from O.

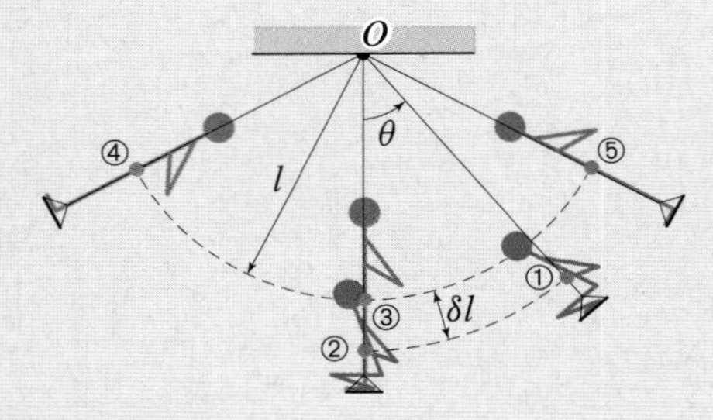

Figure 2
The five key positions of the pumping motion, along with the dimensions of the movement of the mass center.

SOLUTION

Road Map & Modeling As you alternately stand and squat, you will be doing work, and it is this work that will increase the swing angle with each cycle. We will write the work-energy principle to relate the potential energy at the beginning and end of the swing to the work done during the swing. You will be modeled as a particle located at your mass center, which moves as described in Fig. 2. In applying the work-energy principle between ① and ⑤, we will need to calculate the work done as you go from squatting to standing from ② to ③. Your speed at both ① and ⑤ is zero. We will ignore any drag and dissipative forces. Your FBD in an arbitrary position is shown in Fig. 3. This figure shows us that the weight force mg must do work (which we will compute in potential energy terms). Although perhaps not obvious, the force R also does work because it is responsible for lifting your mass center from $l + \delta l$ up to l in going from ② to ③.

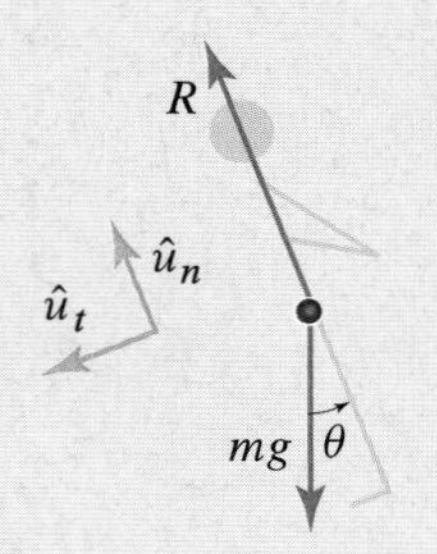

Figure 3
FBD of you on the swing at an arbitrary angle.

Governing Equations

Balance Principles The work-energy principle, applied between ① and ⑤, reads

$$T_1 + V_1 + (U_{1\text{-}5})_{\text{nc}} = T_5 + V_5. \tag{1}$$

The kinetic energies are given by

$$T_1 = \tfrac{1}{2}mv_1^2 \quad \text{and} \quad T_5 = \tfrac{1}{2}mv_5^2, \tag{2}$$

where m is the mass of the person and v_1 and v_5 are the speed of the person at ① and ⑤, respectively. In addition, referring to Fig. 3, Newton's second law applied at ② (i.e., at $\theta = 0$) gives

$$\sum F_n: \quad R - mg = ma_n. \tag{3}$$

* See, for example, W. B. Case and M. A. Swanson, "The Pumping of a Swing from the Seated Position," *American Journal of Physics*, **58**(5), 1990, pp. 463–467, or W. B. Case, "The Pumping of a Swing from the Standing Position," *American Journal of Physics*, **64**(3), 1996, pp. 215–220.

Force Laws Choosing the datum for the gravitational potential energy at ②, we have

$$V_1 = mg(l + \delta l)(1 - \cos\theta_1) \quad \text{and} \quad V_5 = mgl(1 - \cos\theta_5). \tag{4}$$

In addition, recalling that $(U_{1\text{-}5})_{\text{nc}}$ is the work done by R in lifting the mass center from ② to ③, we have

$$(U_{1\text{-}5})_{\text{nc}} = R\,\delta l, \tag{5}$$

where we *assume* that R is constant throughout the lift.

Kinematic Equations At ① and ⑤, we have

$$v_1 = 0 \quad \text{and} \quad v_5 = 0. \tag{6}$$

In addition, the acceleration a_n at ② is given by

$$a_n = \frac{v_2^2}{l + \delta l}. \tag{7}$$

Computation Combining Eqs. (3)–(7) with Eq. (1), we obtain

$$mg(l + \delta l)(1 - \cos\theta_1) + m\left(g + \frac{v_2^2}{l + \delta l}\right)\delta l = mgl(1 - \cos\theta_5). \tag{8}$$

Applying the work-energy principle between ① and ②, during which only the weight force does work, we can relate v_2 to θ_1 as follows:

$$T_1 + V_1 = T_2 + V_2 \quad \Rightarrow \quad mg(l + \delta l)(1 - \cos\theta_1) = \tfrac{1}{2}mv_2^2, \tag{9}$$

which, by solving for $v_2^2/(l + \delta l)$, yields

$$\frac{v_2^2}{l + \delta l} = 2g(1 - \cos\theta_1). \tag{10}$$

Substituting Eq. (10) into Eq. (8) and solving for θ_5, we obtain the final result

$$\boxed{\theta_5 = \cos^{-1}\left[\cos\theta_1 + \frac{\delta l}{l}(3\cos\theta_1 - 4)\right].} \tag{11}$$

Discussion & Verification The argument of the inverse cosine function in Eq. (11) is nondimensional, as it should be. In addition, recalling that the cosine function varies only between -1 and 1, we observe that the term $3\cos\theta_1 - 4$ is always negative. In turn this causes the value of θ_5 to be greater than θ_1 whenever δl is positive. This result confirms that the pumping action will result in an increased swing angle.

A Closer Look Now that we have the angle at the end of a swing cycle θ_5 in terms of the angle at the beginning of a swing cycle θ_1, we can plug in some numbers to see if our pumping scheme works. For example, say that $l = 1.8$ m, $\delta l = 0.2$ m, and $\theta_1 = 40°$. With these numbers, Eq. (11) tells us that $\theta_5 = 54.8°$, thus confirming that we get an increase in swing angle for every cycle (see Figs. 4 and 5). However, notice that Figs. 4 and 5 tell us that if $\theta_1 = 0°$, then by simply standing up (while swinging at zero speed), you will somehow swing to $\theta_5 \approx 27°$! This result is clearly at odds with what happens in reality, and it indicates that our model has some strong limitations. In particular, as discussed in the Helpful Information marginal note, our model cannot be expected to be accurate unless the ratio $\delta l/l$ is sufficiently small. In addition, our model implicitly assumes that $\theta_1 \neq 0$; i.e., that our motion does not start from static equilibrium. The lesson to learn here is that when we construct a physical model, we must be aware of the limitations of the model when interpreting the model's predictions.

Helpful Information

Is it correct to assume that R is constant? Equations (3) and (7) imply that R cannot be constant between ② and ③ because R depends on a_n. In turn, a_n depends on the distance to point O and on v_2, both of which vary as the particle is lifted from ② to ③. The distance from the particle to O changes from $l + \delta l$ to l. In addition, applying the work-energy principle from ② to ③, we know that there must be a change in speed between these two positions because R does work between them. However, it can be argued that the assumption that R is constant between ② and ③ is acceptable as long as δl is small compared to l.

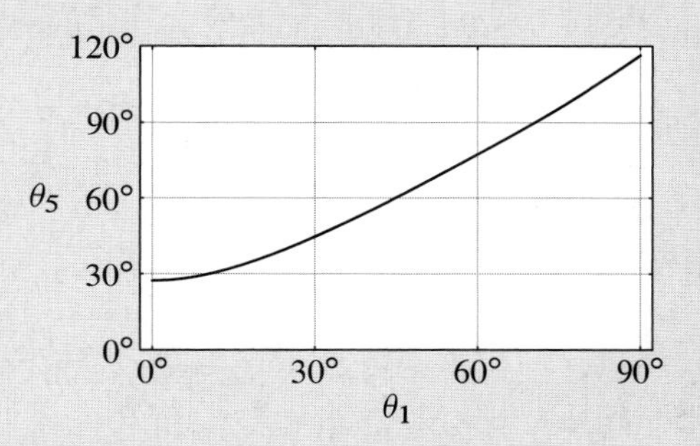

Figure 4
The angle θ_5 as a function of θ_1. This figure, along with Fig. 5, shows that our model gives physically unrealistic results when θ_1 is small.

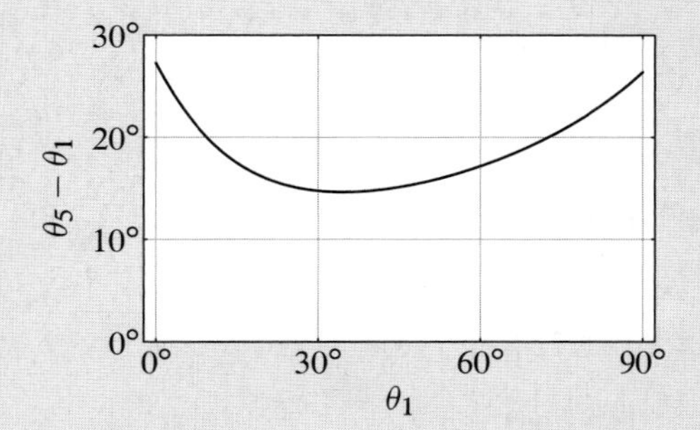

Figure 5
This plot shows the increase in swing amplitude $\theta_5 - \theta_1$ as a function of θ_1. Our model gives physically unrealistic results when θ_1 is small.

EXAMPLE 4.10 A Conservative Force Field

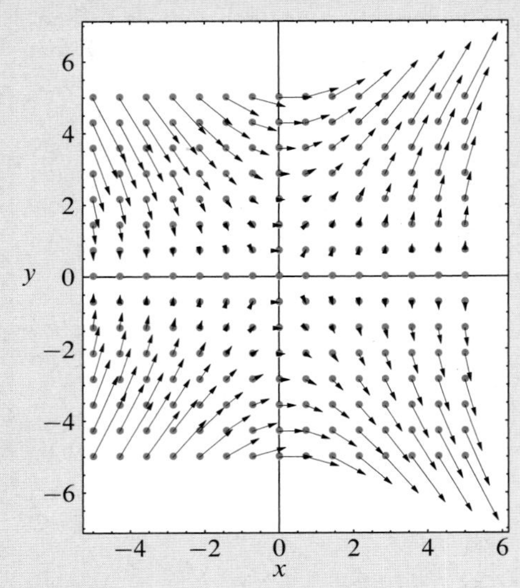

Figure 1
Graphical representation of the force field given in Eq. (1). If each red dot were a particle, then the arrow on each of those dots would represent the force vector on it. In addition, the length of each arrow is proportional to the magnitude of the force.

The force field shown in Fig. 1 is mathematically represented by

$$\vec{F} = y^2\,\hat{\imath} + 2xy\,\hat{\jmath}. \tag{1}$$

Determine the potential energy of $\vec{F}$.

SOLUTION

Road Map & Modeling By computing the curl of the given force field, we can easily verify that curl $\vec{F} = \vec{0}$. Therefore, we know that there is a potential V whose gradient is $-\vec{F}$. With this in mind, our strategy will be to use Eqs. (4.32) on p. 260 since they tell us how the components of $\vec{F}$ relate to V. We can view each of these equations as a simple first-order differential equation that we integrate.

Governing Equations From Eq. (1), the components of $\vec{F}$ are

$$F_x = y^2 \quad \text{and} \quad F_y = 2xy. \tag{2}$$

Using Eqs. (4.32), the relation between the components of $\vec{F}$ and V is

$$F_x = -\frac{\partial V}{\partial x} \quad \text{and} \quad F_y = -\frac{\partial V}{\partial y}. \tag{3}$$

Computation Equating the first of Eqs. (2) with the first of Eqs. (3), we have

$$y^2 = -\frac{\partial V}{\partial x}, \tag{4}$$

which we integrate to obtain

$$V = -xy^2 + f(y), \tag{5}$$

where $f(y)$ is an arbitrary function of y. Next, we equate the second of Eqs. (2) with the second of Eqs. (3) to obtain

$$2xy = -\frac{\partial V}{\partial y}, \tag{6}$$

which we integrate to obtain

$$V = -xy^2 + g(x), \tag{7}$$

where $g(x)$ is an arbitrary function of x. Comparing Eqs. (5) and (7), we see that

$$V = -xy^2 + f(y) \quad \text{and} \quad V = -xy^2 + g(x), \tag{8}$$

and the only way that both of these can be true is if $V = -xy^2 +$ constant. Since only differences in potential energies matter, we can set this constant equal to zero to obtain

$$\boxed{V = -xy^2.} \tag{9}$$

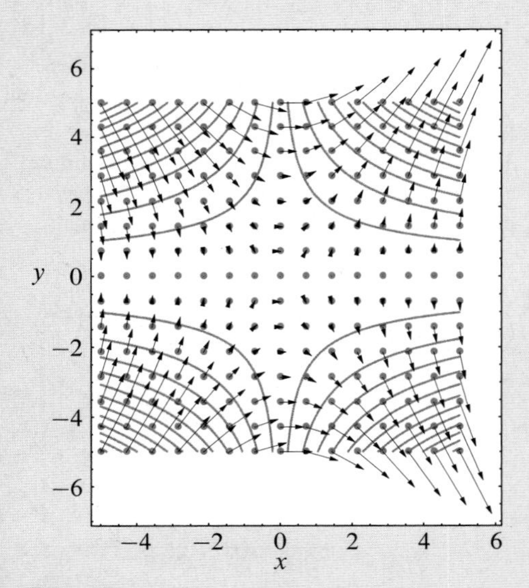

Figure 2
Figure 1 with constant values of V superimposed (i.e., the blue lines).

Discussion & Verification Equation (9) for different values of V has been plotted on top of Fig. 1, and the result is shown in Fig. 2. Notice that the blue lines, that is, lines of constant V, are perpendicular to the force vectors. This is what we should expect since the force field $\vec{F}$ is the negative of the gradient of V. Recall from calculus that the gradient of a scalar function gives, at each point, the direction of greatest change of that function. This agrees with our result since the direction of greatest change of V must be orthogonal to lines of constant V.

Problems

Problem 4.34

The pendulum shown is put in motion with a speed v_0 when $\theta = 0°$. Letting $L = 2$ ft, determine v_0 if the pendulum first comes to a stop at $\theta = 47°$.

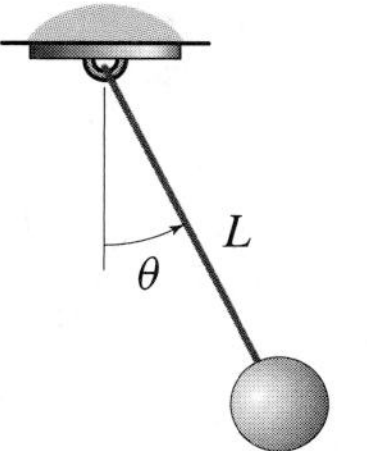

Figure P4.34

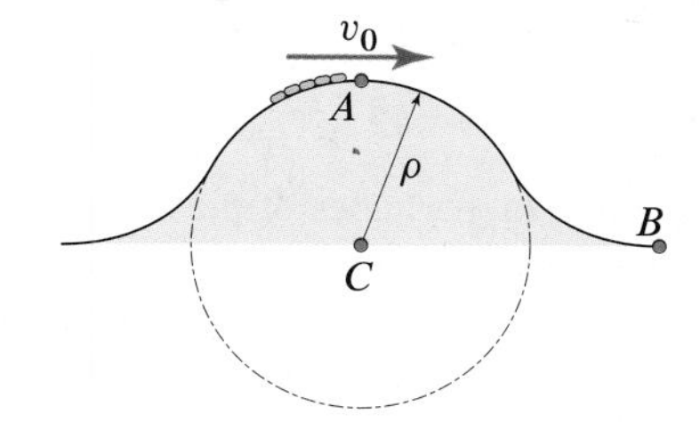

Figure P4.35

Problem 4.35

Point A is the highest point along the roller coaster ride section shown. The inscribed circle at A has radius $\rho = 25$ m and center at C. If point B is on a horizontal line going through C and if the roller coaster has a speed $v_0 = 45$ km/h at A, neglecting friction and air drag and treating the roller coaster as a particle, determine the speed of the roller coaster at B.

Problem 4.36

Assuming that the plunger of a pinball machine has negligible mass and that friction is negligible, determine the spring constant k such that a 2.85 oz ball is released with a speed $v = 15$ ft/s, after pulling back the plunger 2 in. from its rest position, i.e., from the position in which the spring is uncompressed.

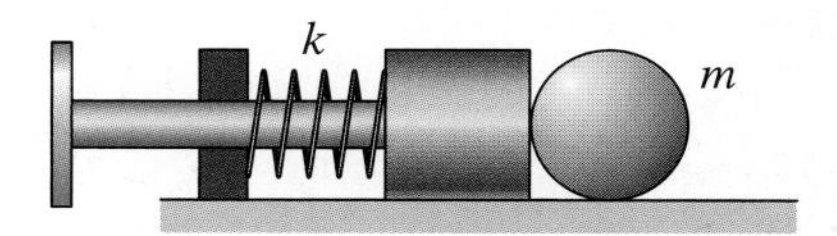

Figure P4.36

Problems 4.37 and 4.38

Consider a 3300 lb car whose speed is increased by 35 mph over a distance of 200 ft while traveling up a rectilinear incline with a 15% grade. Model the car as a particle, assume that the tires do not slip, and neglect *all* sources of frictional losses and drag.

Figure P4.37 and P4.38

Problem 4.37 Determine the work done on the car by the engine if the car starts from rest.

Problem 4.38 Determine the work done on the car by the engine if the car has an initial speed of 30 mph.

Problem 4.39

A classic car is driving down an incline at 60 km/h when its brakes are applied. Treating the car as a particle, neglecting all forces except gravity and friction, and assuming that the tires slip, determine the coefficient of kinetic friction if the car comes to a stop in 55 m and $\theta = 20°$.

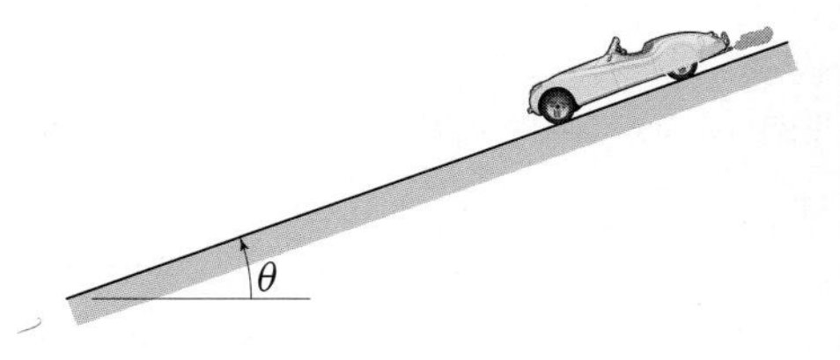

Figure P4.39

Figure P4.40

Problem 4.40

A 75 kg skydiver is falling at a speed of 250 km/h when, at a height of 245 m, the parachute is deployed, allowing the skydiver to land at a speed of 4 m/s. Modeling the skydiver as a particle and assuming that the skydiver follows a perfectly vertical trajectory, determine the average force exerted by the parachute from the moment of deployment until landing.

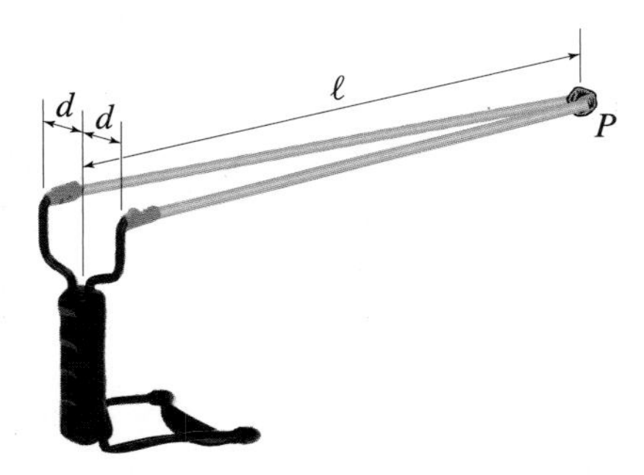

Figure P4.41 and P4.42

Problems 4.41 and 4.42

Each of the two rubber tubes of the wrist rocket has an unstretched length $L_0 = 1$ ft. They are symmetrically pulled back so that $\ell = 3$ ft and then are released from rest. The pellet P that is launched by the wrist rocket weighs 0.145 oz. Neglect the mass of the rubber tubes and any change in height of the pellet while it is in contact with the rubber tubes. Finally, let $d = 1.5$ in.

Problem 4.41 If the desired launch speed of the pellet P is 100 mph, determine the required stiffness of each rubber tube.

Problem 4.42 If the stiffness of each rubber tube is $k = 5$ lb/ft, determine the range of the pellet if it is launched at 45° with respect to the horizontal and if it lands at the same height from which it was launched. Neglect air resistance.

Problems 4.43 through 4.45

The collar C of mass $m = 2$ kg slides freely on the rod EF. At the instant shown, the collar C is moving downward with speed $v_1 = 2.5$ m/s and it is 1/3 of the way between E and F. The dimensions are $L = 2.1$ m, $w = 1.9$ m, $h = 2.3$ m, and $d = 1.2$ m.

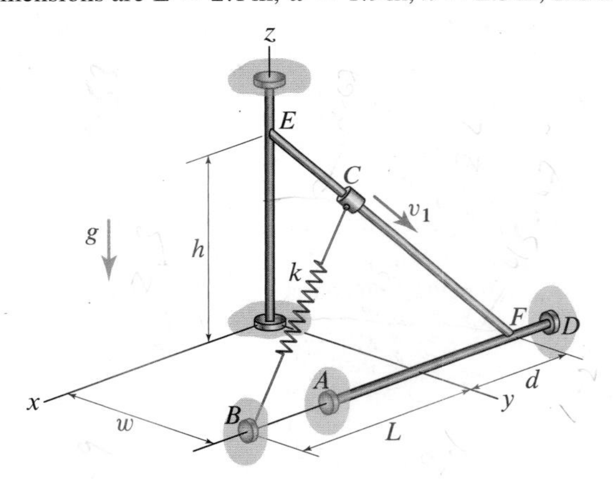

Figure P4.43–P4.45

Problem 4.43 If the unstretched length of the spring is $L_0 = 1$ m, and the spring stiffness is $k = 50$ N/m, determine the speed of the collar when it reaches F. Assume that the attachment point at A does not impede the spring.

Problem 4.44 If the unstretched length of the spring is $L_0 = 4$ m and the spring stiffness is $k = 50$ N/m, determine the speed of the collar when it reaches F. Assume that the attachment point at A does not impede the spring.

Problem 4.45 If the unstretched length of the spring is $L_0 = 1$ m and the spring stiffness is $k = 170$ N/m, show that the collar does not reach F. Assume that the attachment point at A does not impede the spring.

Problems 4.46 and 4.47

The crate of mass $m = 35$ kg is attached to the springs with constants $k_1 = 1$ kN/m and $k_2 = 2$ kN/m. Both springs are unstretched when $\delta = 0$ and the box is centered between the two vertical walls.

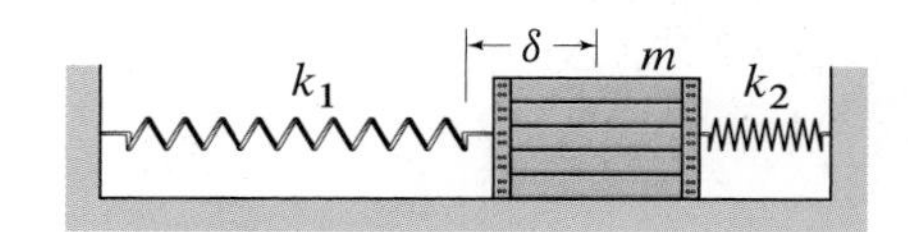

Figure P4.46 and P4.47

Problem 4.46 If the crate is released from rest after being displaced a distance $\delta = 2$ m and friction between the crate and the surface on which it slides is negligible, determine the speed of the crate when it returns to $\delta = 0$.

Problem 4.47 If the crate is released from rest after being displaced a distance $\delta = 2$ m and the coefficient of kinetic friction between the crate and the surface on which it slides is $\mu_k = 0.2$, determine the speed of the crate when $\delta = 0$.

Problem 4.48

Packages for transporting delicate items (e.g., a laptop or glass) are designed to "absorb" some of the energy of the impact in order to protect their contents. These energy absorbers can get very complicated (e.g., the mechanics of Styrofoam peanuts can be complex), but we can begin to understand how they work by modeling them as a linear elastic spring of constant k that is placed between the contents (an expensive vase) of mass m and the package P. Assume that the vase's mass is 3 kg and that the box is dropped from rest from a height of 1.5 m. Treating the vase as a particle and neglecting all forces except for gravity and the spring force, determine the value of the spring constant k so that the maximum displacement of the vase relative to the box is 0.15 m. Assume that the spring relaxes after the box is dropped and that it does not oscillate.

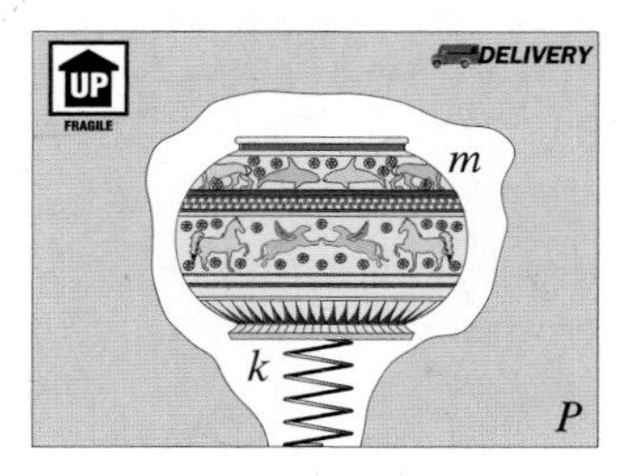

Figure P4.48

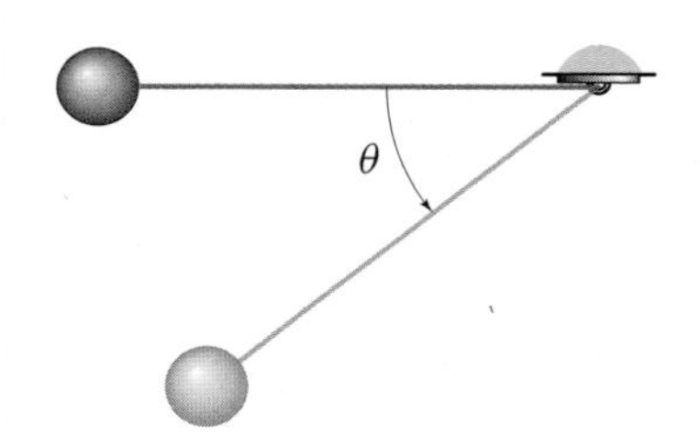

Figure P4.49

Problem 4.49

The pendulum is released from rest when $\theta = 0°$. If the string holding the pendulum bob breaks when the tension is twice the weight of the bob, at what angle does the string break? Treat the pendulum as a particle, ignore air resistance, and let the string be inextensible and massless.

Problem 4.50

The force acting on a stationary electric charge q_A interacting with a charge q_B is described by Coulomb's law and takes the form

$$\vec{F} = k\frac{q_A q_B}{r^2}\hat{u}_r,$$

where $k = 8.9875\times10^9\ \text{N}\cdot\text{m}^2/\text{C}^2$ (C is the symbol for coulomb, the unit used to measure electric charge) is a constant and r is the distance between A and B. This force law is mathematically very similar to Newton's universal gravitation law. With this in mind, determine an expression of the electrostatic potential energy, choosing the datum at infinity, i.e., such that the potential energy is equal to zero when the two charges are separated by an infinite distance.

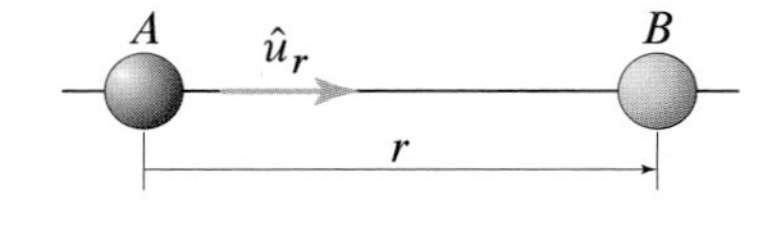

Figure P4.50

Problem 4.51

The force-compression profile of a rubber bumper B is given by $F_B = \beta x^3$, where $\beta = 3.5 \times 10^6\,\text{lb/ft}^3$ and x is the bumper's compression measured in the horizontal direction. Determine the expression for the potential energy of the bumper B. In addition, if the cruiser C weighs 70,000 lb and impacts B with a speed of 5 ft/s, determine the compression required to bring C to a stop. Model C as a particle and neglect C's vertical motion as well as the drag force between the water and the cruiser C.

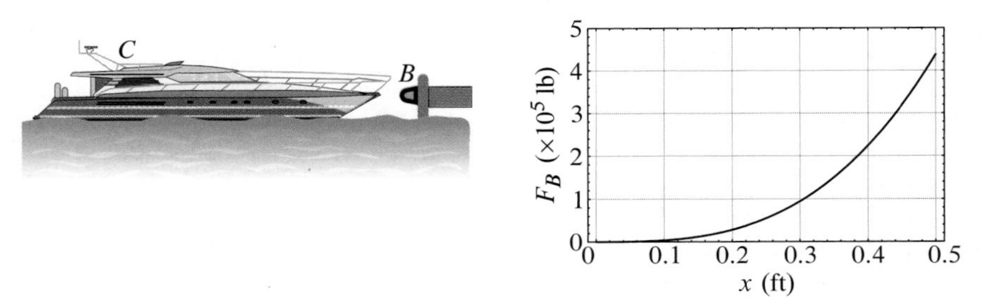

Figure P4.51

Problem 4.52

A satellite orbits the Earth along the orbit shown. The minimum and maximum distances from the center of the Earth are $R_P = 4.5{\times}10^7$ m and $R_A = 6.163{\times}10^7$ m, respectively, where the subscripts P and A stand for *perigee* (the point on the orbit closest to Earth) and *apogee* (the point on the orbit farthest from Earth), respectively. Modeling the satellite as a particle and assuming that the center of the Earth can be chosen as the origin of an inertial frame of reference, if the satellite's speed at P is $|\vec{v}|_P = 3.2{\times}10^3$ m/s, determine the satellite's speed at A.

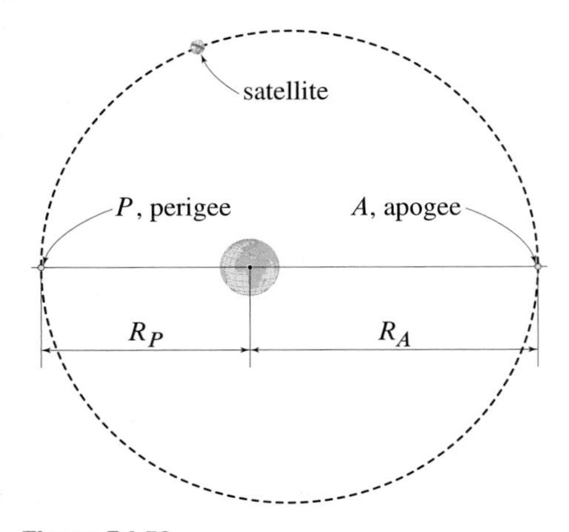

Figure P4.52

Problems 4.53 through 4.55

The arm AB can rotate freely about the pin at A. The spring with stiffness $k = 500$ N/m is designed so that the system is in static equilibrium when $\theta = 0°$. Let $L = 18.2$ cm, $h = 24.6$ cm, and the mass of the ball B be 5 kg. Neglect the mass of arm AB. *Hint:* Sketch an FBD of the ball and the arm together. For Probs. 4.53 and 4.54, let ℓ be the distance between C and D and choose ℓ as the primary unknown.

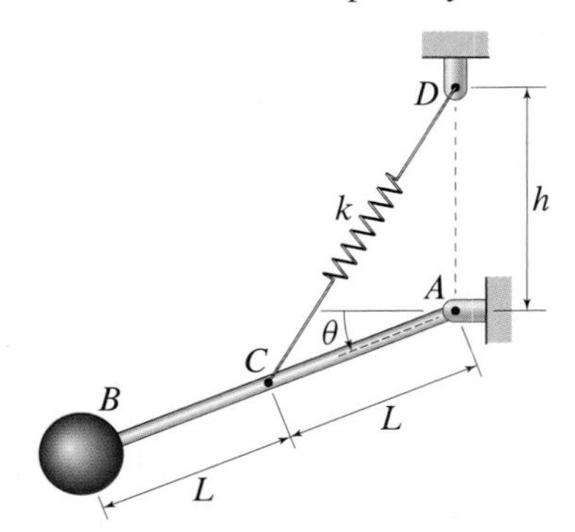

Figure P4.53–P4.55

Problem 4.53 If the system is released from rest when $\theta = -30°$, determine the maximum angle θ reached by the arm AB.

Problem 4.54 If the system is released from rest when $\theta = -30°$, determine the maximum speed achieved by the ball B, and determine the angle at which it occurs.

Problem 4.55 If the arm AB is released from rest when $\theta = 90°$ such that it swings to the left, determine the speed of the ball B when the arm reaches $\theta = 0°$.

Problem 4.56

The package handling system is designed to launch the small package of mass m from A by using a compressed linear spring of constant k. After launch, the package slides along the track until it lands on the conveyor belt at B. The track has small, well-oiled rollers so that you can neglect any energy loss due to the movement of the package along the track. Modeling the package as a particle, determine the minimum initial compression of the spring so that the package gets to B without separating from the track at C. Finally, determine the speed with which the package reaches the conveyor at B.

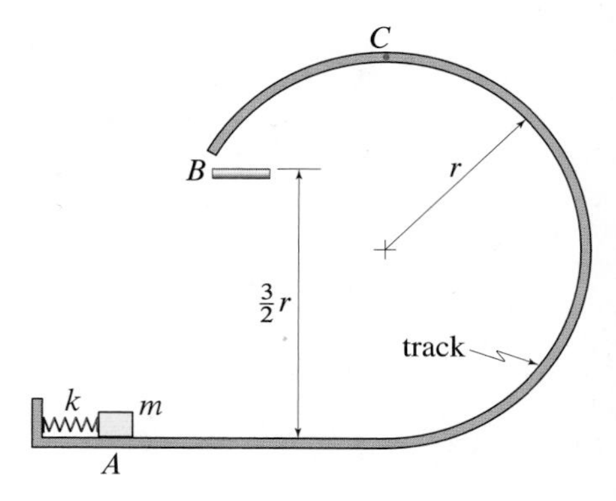

Figure P4.56

Problem 4.57

Compressed gas is used in many circumstances to propel objects within tubes. For example, you can still find pneumatic tubes in use in many banks to receive and send items to drive-through tellers,* and it is compressed gas that propels a bullet out of a gun barrel. Let the cross-sectional area of the tube be given by A and the position of the cylinder by s, and assume that the compressed gas is an ideal gas at constant temperature so that the pressure P times the volume Ω is a constant, i.e., $P\Omega$ = constant. Show that the potential energy of this compressed gas is given by $V = -P_0 s_0 A \ln(s/s_0)$, where P_0 is the initial pressure and s_0 is the initial value of s. Model the cylinder as a particle, and assume that the forces resisting the motion of the cylinder are negligible.

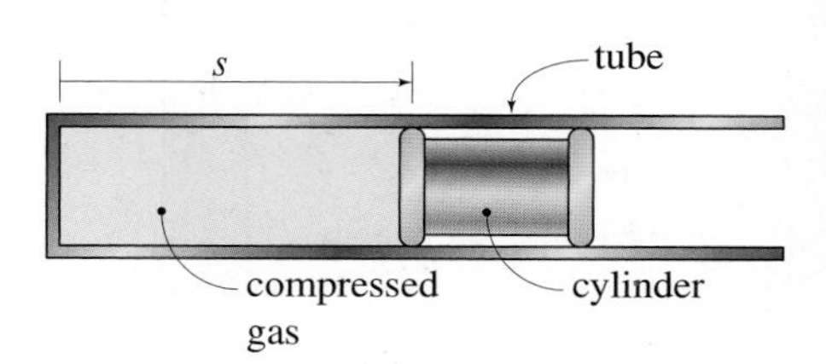

Figure P4.57 and P4.58

Problem 4.58

When a gun fires a bullet, the gun barrel acts as the tube, and the bullet acts as the cylinder in Prob. 4.57. Using the assumptions and result of that problem, determine the velocity of a bullet at the end of a 24 in. gun barrel, given that the bore diameter is 0.458 in., the bullet weight is 300 gr (7000 gr = 1 lb), the initial firing pressure is 27,000 psi, and the initial distance between the back of the bullet and the back wall of the firing chamber (i.e., s_0 in Prob. 4.57) is

(a) 1.855 in. (this distance is realistic and accurate),

(b) 1.5 in., and

(c) explain why the velocity of the bullet at the end of the gun barrel is lower in Part (b) when compared to Part (a).

Problem 4.59

The resistance of a material to fracture is assessed with a fracture test. One such test is the *Charpy impact test*, in which the fracture toughness is assessed by measuring the energy required to break a specimen of a specified geometry. This is done by releasing a heavy pendulum from rest at an angle θ_i and by measuring the maximum swing angle θ_f reached by the pendulum after the specimen is broken. Suppose that in an experiment $\theta_i = 45°$, $\theta_f = 23°$, the weight of the pendulum's bob is 3 lb, and the length of the pendulum is 3 ft. Neglecting the mass of any other component of the testing apparatus, assuming that the pendulum's pivot is frictionless, and treating the pendulum's bob as a particle, determine the fracture energy of the specimen tested. Assume that the fracture energy is the energy required to break the specimen.

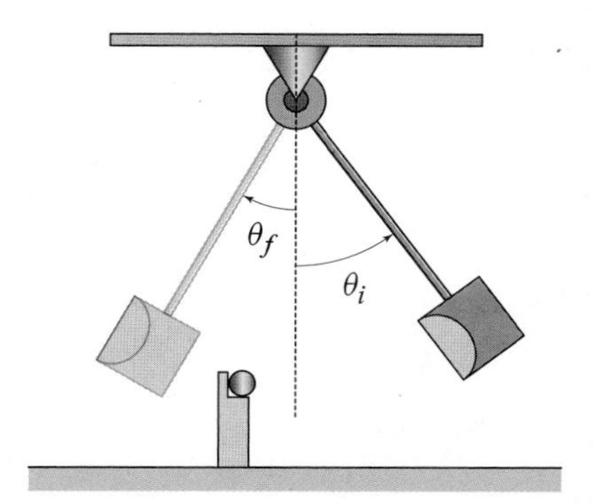

Figure P4.59

---

* Pneumatic tubes were used extensively by postal services in the late 19th and early 20th centuries. By the early 20th century, the cities of Philadelphia, New York, Boston, and Chicago had pneumatic networks, as did Paris, Berlin, and London. Their use for mail did not last past the end of World War I (1918), but almost every department store in the United States had tubes carrying cash and paperwork in the 1920s and 1930s. They still find use in hospitals for the delivery of medication and other items. See Robin Pogrebin, "Underground Mail Road," *New York Times*, May 7, 2001.

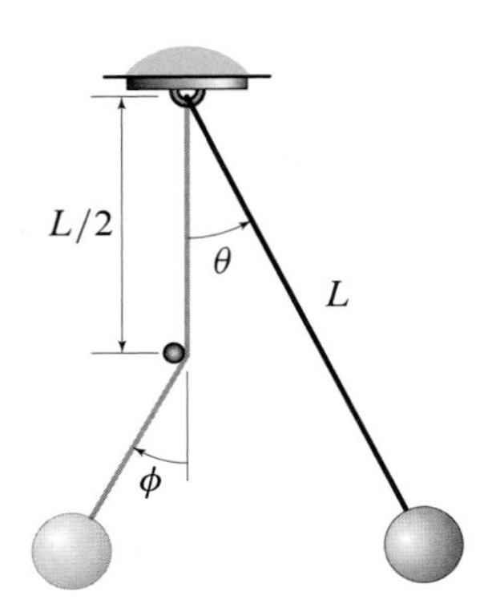

Figure P4.60 and P4.61

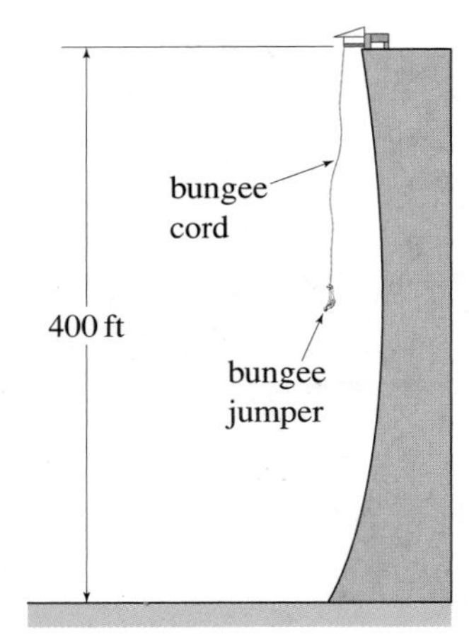

Figure P4.62

Problems 4.60 and 4.61

A pendulum with mass $m = 1.4$ kg and length $L = 1.75$ m is released from rest at an angle θ_i. Once the pendulum has swung to the vertical position (i.e., $\theta = 0$), its cord runs into a small fixed obstacle. In solving this problem, neglect the size of the obstacle, model the pendulum's bob as a particle, model the pendulum's cord as massless and inextensible, and let gravity and the tension in the cord be the only relevant forces.

Problem 4.60 What is the maximum height, measured from its lowest point, reached by the pendulum if $\theta_i = 20°$?

Problem 4.61 If the bob is released from rest at $\theta_i = 90°$, at what angle ϕ does the cord go slack?

Problem 4.62

While the stiffness of an elastic cord can be nearly constant over a large range of deformation, as a bungee cord is stretched, it tends to get less stiff as it gets longer. Assume a softening force-displacement relation of the form $k\delta - \beta\delta^3$, where $k = 2.58$ lb/ft, $\beta = 0.000013$ lb/ft^3, and δ (measured in ft) is the displacement of the cord from its unstretched length. For a bungee cord whose unstretched length is 150 ft, determine

(a) the expression of the cord's potential energy as a function of δ;

(b) the velocity at the bottom of a 400 ft tower of a bungee jumper weighing 170 lb and starting from rest;

(c) the maximum acceleration, expressed in g's, felt by the bungee jumper in question.

Problems 4.63 through 4.65

A crate, initially traveling horizontally with a speed of 18 ft/s, is made to slide down a 14 ft chute inclined at 35°. The surface of the chute has a coefficient of kinetic friction μ_k, and at its lower end, it smoothly lets the crate onto a horizontal trajectory. The horizontal surface at the end of the chute has a coefficient of kinetic friction μ_{k2}. Model the crate as a particle, and assume that gravity and the contact forces between the crate and the sliding surface are the only relevant forces.

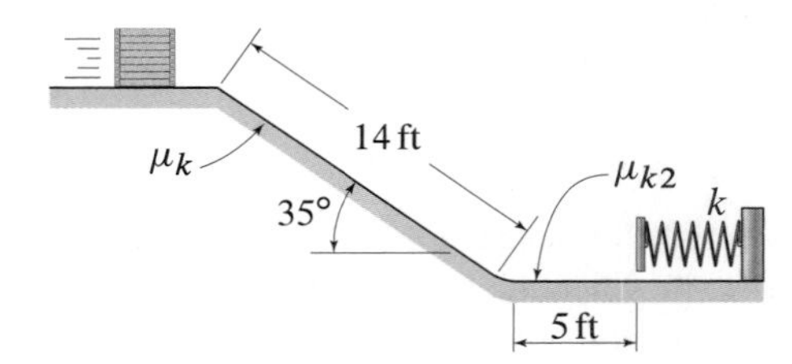

Figure P4.63–P4.65

Problem 4.63 If $\mu_k = 0.35$, what is the speed with which the crate reaches the bottom of the chute (immediately before the crate's trajectory becomes horizontal)?

Problem 4.64 Find μ_k such that the crate's speed at the bottom of the chute (immediately before the crate's trajectory becomes horizontal) is 15 ft/s.

Problem 4.65 Let $\mu_k = 0.5$ and suppose that, once the crate reaches the bottom of the chute and after sliding horizontally for 5 ft, the crate runs into a bumper. If the weight of the crate is $W = 110$ lb, $\mu_{k2} = 0.33$, modeling the bumper as a linear spring with constant k, and neglecting the mass of the bumper, determine the value of k so that the crate comes to a stop 2 ft after impacting with the bumper.

Problem 4.66

The Lennard-Jones force law between two atoms can be represented as

$$f_{ij} = 24\epsilon\left(\frac{\sigma^6}{r_{ij}^7} - \frac{2\sigma^{12}}{r_{ij}^{13}}\right),$$

where r_{ij} is the distance between atoms i and j, and ϵ and σ are material-specific parameters with dimensions of energy and length, respectively. Using this equation, determine the potential between two Ni atoms. Assume the potential between the two atoms is zero when the distance between those two atoms is infinite.

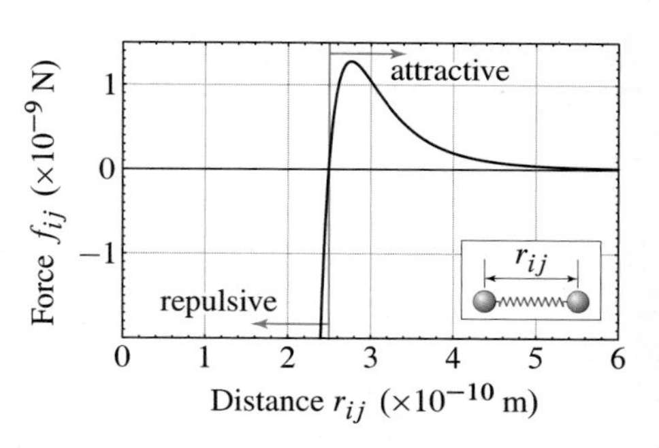

Figure P4.66

Problems 4.67 and 4.68

Spring scales work by measuring the displacement of a spring that supports both the platform and the object, of mass m, whose weight is being measured. Neglect the mass of the platform on which the mass sits, and assume that the spring is uncompressed before the mass is placed on the platform. In addition, assume that the spring is linear elastic with spring constant k. You may have solved these same problems using Newton's second law when doing Prob. 3.27 and 3.28—here use the work-energy principle to solve them.

Problem 4.67 If the mass m is gently placed on the spring scale (i.e., it is dropped from zero height above the scale), determine the *maximum reading* on the scale after the mass is released.

Problem 4.68 If the mass m is gently placed on the spring scale (i.e., it is dropped from zero height above the scale), determine the expression for *maximum velocity* attained by the mass m as the spring compresses.

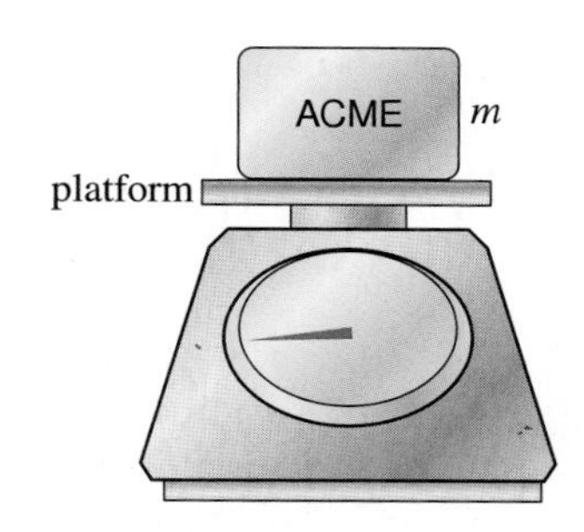

Figure P4.67 and P4.68

Problem 4.69

A 6 lb collar is constrained to travel along a rectilinear and frictionless bar of length $L = 5$ ft. The springs attached to the collar are identical, and they are unstretched when the collar is at B. Treating the collar as a particle, neglecting air resistance, and knowing that at A the collar is moving to the right with a speed of 11 ft/s, determine the linear spring constant k so that the collar reaches D with zero speed. Points E and F are fixed.

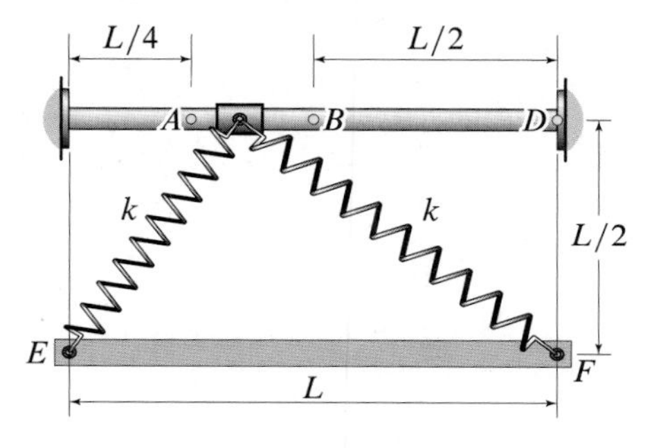

Figure P4.69

Problem 4.70

A 3 kg collar is constrained to travel in the *horizontal plane* along a frictionless ring of radius $R = 0.75$ m. The spring attached to the collar has a spring constant $k = 21$ N/m. Treating the collar as a particle, neglecting air resistance, and knowing that at A the collar is at rest, determine the spring's unstretched length if the collar is to reach point B with a speed of 2 m/s.

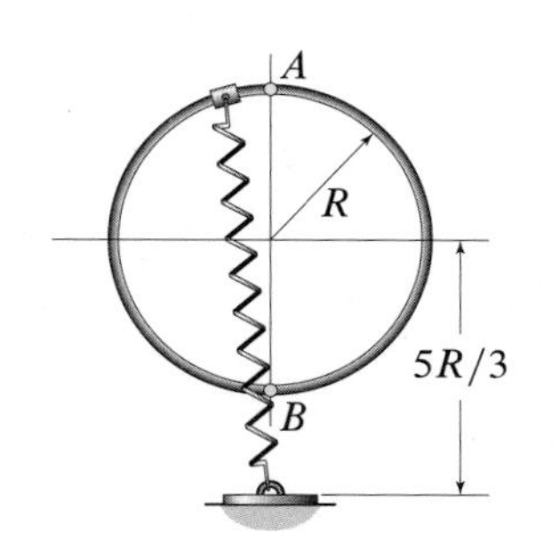

Figure P4.70

Design Problems

Design Problem 4.1

In practice, during a jump, bungee cords stretch to two to four times their unstretched length, and a jumper feels no more than 2.5–3.5g of acceleration. Assume the force in the bungee cord has the mathematical form $k\delta - \beta\delta^3$, where k and β are constants and δ is the amount of stretch in the cord past its unstretched length. Design a cord (that is, design the constants k and β) so that the bungee cord stretches 2.5 times its unstretched length, the acceleration of the jumper does not exceed $3g$, and the bungee cord has zero stiffness at the bottom of a 400 ft drop.

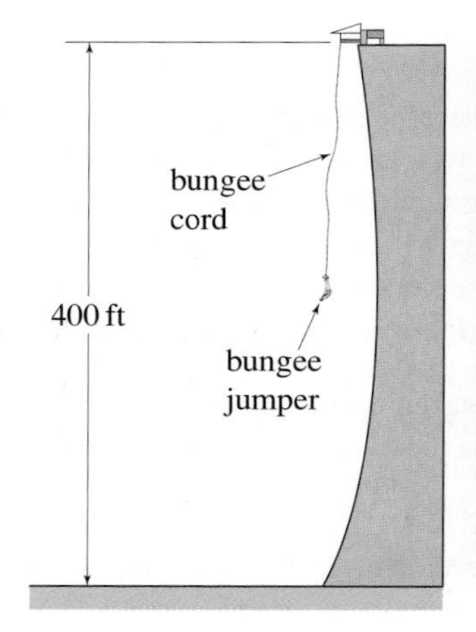

Figure DP4.1

4.3 Work-Energy Principle for Systems of Particles

This section presents the application of the work-energy principle to systems of particles. We will discover a fundamental new concept pertaining to systems of particles that is absent in the single-particle case — internal work.

Internal work and work-energy principle for a system

Referring to Fig. 4.11, consider a system of n particles. The ith particle, whose mass is m_i and position is $\vec{r}_i$, is acted upon by a total *external force* $\vec{F}_i$ and interacts with the other $n - 1$ particles through *internal forces* labeled $\vec{f}_{ij}$ ($j = 1, \ldots, n,\ i \neq j$). Applying Newton's second law to the ith particle, we have

$$\vec{F}_i + \sum_{j=1}^{n} \vec{f}_{ij} = m_i \vec{a}_i, \quad i \neq j. \tag{4.37}$$

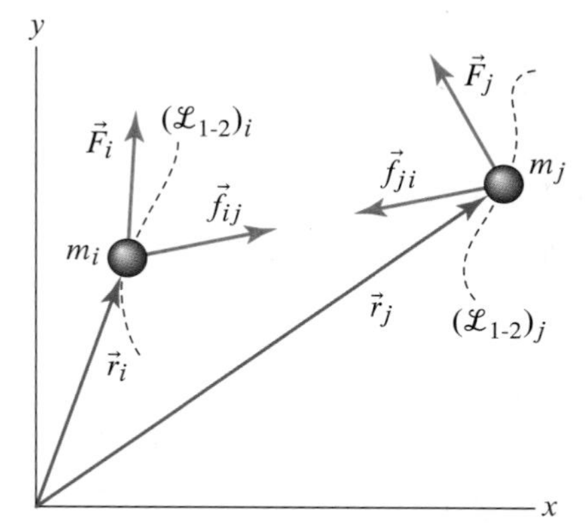

Figure 4.11
Particles i and j of a system of n particles.

As we did for a single particle, we take the dot product of this equation with $d\vec{r}_i$ and integrate along $(\mathscr{L}_{1\text{-}2})_i$, the path traveled by particle i in going from ① to ②. There will be n of these equations (one for each particle). Summing all n of them gives

$$\sum_{i=1}^{n} \int_{(\mathscr{L}_{1\text{-}2})_i} \left(\vec{F}_i + \sum_{j=1}^{n} \vec{f}_{ij} \right) \cdot d\vec{r}_i = \sum_{i=1}^{n} \int_{(\mathscr{L}_{1\text{-}2})_i} m_i \frac{d\dot{\vec{r}}_i}{dt} \cdot d\vec{r}_i, \tag{4.38}$$

where we have written $\vec{a}_i = d\dot{\vec{r}}_i/dt$. Proceeding as we did for a single particle, we can write the right-hand side of Eq. (4.38) as

$$\begin{aligned} \sum_{i=1}^{n} \int m_i \frac{d\dot{\vec{r}}_i}{dt} \cdot d\vec{r}_i &= \sum_{i=1}^{n} \int m_i \vec{v}_i \cdot d\vec{v}_i = \sum_{i=1}^{n} \tfrac{1}{2} m_i \vec{v}_i \cdot \vec{v}_i \Big|_1^2 \\ &= \left(\sum_{i=1}^{n} \tfrac{1}{2} m_i v_i^2 \right)_2 - \left(\sum_{i=1}^{n} \tfrac{1}{2} m_i v_i^2 \right)_1, \end{aligned} \tag{4.39}$$

where each integral is performed over the path $(\mathscr{L}_{1\text{-}2})_i$. We define the *kinetic energy of the particle system* to be the sum of the kinetic energies of each part of the system. Denoting the kinetic energy of the system by T, we have

$$T = \sum_{i=1}^{n} \tfrac{1}{2} m_i v_i^2. \tag{4.40}$$

Using T, Eq. (4.39) can be rewritten as

$$\sum_{i=1}^{n} \int_{(\mathscr{L}_{1\text{-}2})_i} m_i \frac{d\dot{\vec{r}}_i}{dt} \cdot d\vec{r}_i = T_2 - T_1. \tag{4.41}$$

Going back to Eq. (4.38), the left-hand side of this equation can be written as

$$\sum_{i=1}^{n} \int \left(\vec{F}_i + \sum_{j=1}^{n} \vec{f}_{ij} \right) \cdot d\vec{r}_i = \overbrace{\sum_{i=1}^{n} \int \vec{F}_i \cdot d\vec{r}_i}^{(U_{1\text{-}2})_{\text{ext}}} + \overbrace{\sum_{i=1}^{n} \sum_{\substack{j=1 \\ j \neq i}}^{n} \int \vec{f}_{ij} \cdot d\vec{r}_i}^{(U_{1\text{-}2})_{\text{int}}}, \tag{4.42}$$

where $(U_{1\text{-}2})_{\text{ext}}$ and $(U_{1\text{-}2})_{\text{int}}$ represent the work done on the system by *external* and *internal* forces, respectively, and each integral is performed over the path $(\mathcal{L}_{1\text{-}2})_i$. In discussing Newton's second law for systems, we saw that terms similar to the double sum in Eq. (4.42) vanish because of Newton's third law. However, in this case the double sum does not vanish, and it remains an important contribution to the work-energy principle. In fact, for each pair of particles i and j, the double sum that defines the term $(U_{1\text{-}2})_{\text{int}}$ contains the terms

$$\vec{f}_{ij} \cdot d\vec{r}_i + \vec{f}_{ji} \cdot d\vec{r}_j = \vec{f}_{ij} \cdot (d\vec{r}_i - d\vec{r}_j) = \vec{f}_{ij} \cdot d\vec{r}_{i/j}, \tag{4.43}$$

where we have used the fact that $\vec{f}_{ij} = -\vec{f}_{ji}$ by Newton's third law and where, using relative motion kinematics, $d\vec{r}_{i/j} = d\vec{r}_i - d\vec{r}_j$ is the infinitesimal *relative* displacement of particle i with respect to particle j. Equation (4.43) tells us that for internal forces to do no work, either there must be no relative motion or the component of $d\vec{r}_{i/j}$ along $\vec{f}_{ij}$ must be zero for all particle pairs.

Summarizing, we use Eqs. (4.41) and (4.42) to write the work-energy principle for a system of particles as

$$\boxed{T_1 + (U_{1\text{-}2})_{\text{ext}} + (U_{1\text{-}2})_{\text{int}} = T_2,} \tag{4.44}$$

or as

$$\boxed{T_1 + V_1 + (U_{1\text{-}2})^{\text{ext}}_{\text{nc}} + (U_{1\text{-}2})^{\text{int}}_{\text{nc}} = T_2 + V_2,} \tag{4.45}$$

when there are also conservative forces doing work on the system.

We see that in applying the work-energy principle to particle systems, we must account for the work of internal forces. In addition, in view of the discussion following Eq. (4.43), there are situations in which the work of the internal forces vanishes even when there is relative motion, such as in pulley systems with inextensible cables.

Kinetic energy for a system of particles

Here we consider a way of expressing the kinetic energy of a particle system that will help us to understand the expression for the kinetic energy of a rigid body in Chapter 8.

We have defined the kinetic energy for a system of particles as

$$T = \sum_{i=1}^{n} \tfrac{1}{2} m_i \vec{v}_i \cdot \vec{v}_i. \tag{4.46}$$

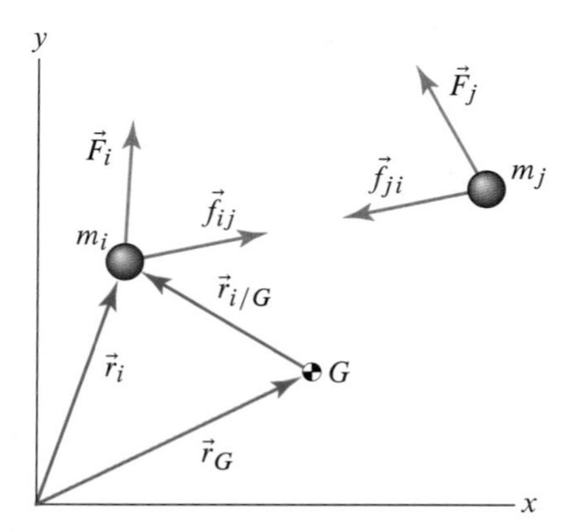

Figure 4.12
Particles i and j of a system of n particles with center of mass at G.

Now, referring to Fig. 4.12, notice that $\vec{r}_i = \vec{r}_G + \vec{r}_{i/G}$, where $\vec{r}_G$ and $\vec{r}_{i/G}$ are the position of the center of mass G and the position of particle i relative to G, respectively. Differentiating this expression for $\vec{r}_i$ with respect to time gives $\vec{v}_i = \vec{v}_G + \vec{v}_{i/G}$, which allows us to write Eq. (4.46) as

$$\begin{aligned} T &= \sum_{i=1}^{n} \tfrac{1}{2} m_i (\vec{v}_G + \vec{v}_{i/G}) \cdot (\vec{v}_G + \vec{v}_{i/G}) \\ &= \sum_{i=1}^{n} \tfrac{1}{2} m_i (\vec{v}_G \cdot \vec{v}_G + 2\vec{v}_G \cdot \vec{v}_{i/G} + \vec{v}_{i/G} \cdot \vec{v}_{i/G}), \end{aligned} \tag{4.47}$$

or

$$T = \tfrac{1}{2}m \underbrace{\vec{v}_G \cdot \vec{v}_G}_{v_G^2} + \underbrace{\left(\sum_{i=1}^{n} m_i \vec{v}_{i/G}\right)}_{\frac{d}{dt}\sum_{i=1}^{n} m_i \vec{r}_{i/G}} \cdot \vec{v}_G + \sum_{i=1}^{n} \tfrac{1}{2} m_i \underbrace{\vec{v}_{i/G} \cdot \vec{v}_{i/G}}_{v_{i/G}^2}, \tag{4.48}$$

where m is the system's total mass. Referring to the second term on the last line of Eq. (4.48), using the definition of center of mass in Eq. (3.35) on p. 218, and recalling that $\vec{r}_{i/G}$ describes the position of each particle *relative to the mass center*, we have

$$\frac{d}{dt}\sum_{i=1}^{n} m_i \vec{r}_{i/G} = \frac{d}{dt}\big(m\, \vec{r}_{G/G}\big). \tag{4.49}$$

Since the term $\vec{r}_{G/G}$ is the position of the mass center relative to the mass center itself, the left-hand side of Eq. (4.49) must always be equal to zero. In conclusion, Eq. (4.48) takes on the form

$$T = \tfrac{1}{2} m v_G^2 + \tfrac{1}{2}\sum_{i=1}^{n} m_i v_{i/G}^2, \tag{4.50}$$

which states that *the kinetic energy of a system of particles depends on the motion of the mass center of the system of particles, as well as the motion of all the particles relative to the mass center.*

Concept Alert

The kinetic energy of a spinning system. Equation (4.50) tells us that a system can have a nonzero kinetic energy even when the center of mass is not moving. A typical example of this idea is the flywheel, which, even with its center of mass remaining stationary, can store a remarkable amount of kinetic energy via its spinning motion.

End of Section Summary

In this section, we developed the work-energy principle for a system of particles, and we discovered that internal forces play an important role in writing this principle. We defined the kinetic energy of a particle system as

Eq. (4.40), p. 281

$$T = \sum_{i=1}^{n} \tfrac{1}{2} m_i v_i^2,$$

where m_i and v_i are the mass and the speed of the ith particle in the system, respectively. Then the work-energy principle for a system of particles was found to be

Eq. (4.45), p. 282

$$T_1 + V_1 + (U_{1\text{-}2})_{\text{nc}}^{\text{ext}} + (U_{1\text{-}2})_{\text{nc}}^{\text{int}} = T_2 + V_2,$$

where V is the potential energy of the system, which may have contributions from both *external* and *internal* forces. The kinetic energy of a system can also be expressed in terms of the speed of the mass center and the relative velocity of the particles in the system with respect to the mass center, that is,

Eq. (4.50), p. 283

$$T = \tfrac{1}{2} m v_G^2 + \tfrac{1}{2}\sum_{i=1}^{n} m_i v_{i/G}^2.$$

EXAMPLE 4.11 A System of Particles with Friction

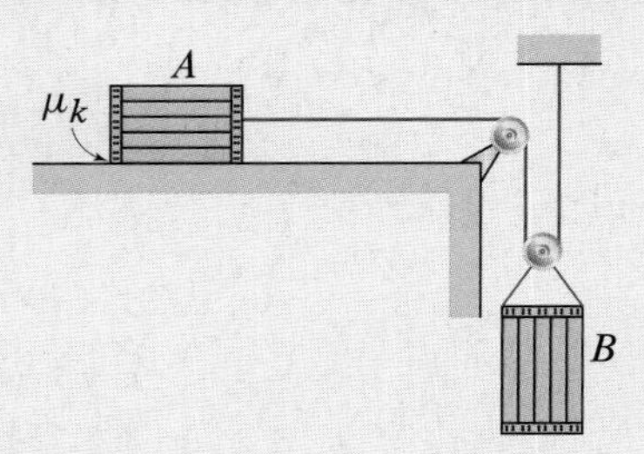

Figure 1
The cables in the pulley system are inextensible, and the coefficient of kinetic friction between the crate A and the horizontal surface is μ_k.

The crates A and B of mass m and $2m$, respectively, are connected by a pulley system (Fig. 1). If the system is released from rest and friction between A and the horizontal surface is insufficient to prevent slipping, determine the speed of crate A after B drops a distance h.

SOLUTION

Road Map & Modeling The FBDs of each of the two crates are shown in Fig. 2. The

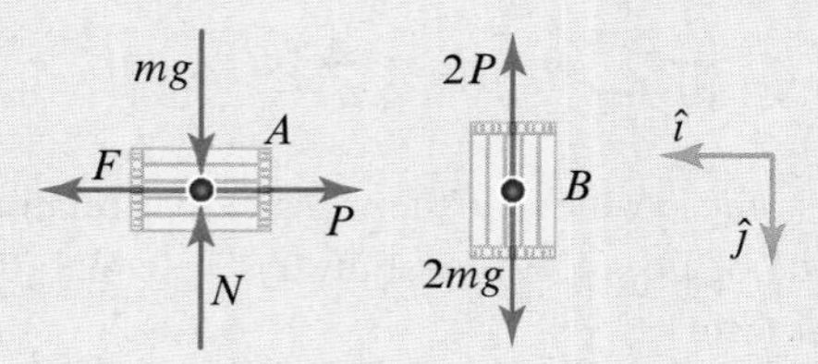

Figure 2. The FBDs of crates A and B.

forces that do work are the friction force F, the cable tension P, and the weight $2mg$. Therefore, we can apply the work-energy principle to A and B together to find the speed of A with only F and $2mg$ doing work.

Governing Equations

Balance Principles Letting ① be when the system is at rest and ② be when B has dropped a distance h, the work-energy principle is

$$T_1 + V_1 + (U_{1\text{-}2})^{\text{ext}}_{\text{nc}} + (U_{1\text{-}2})^{\text{int}}_{\text{nc}} = T_2 + V_2, \tag{1}$$

where the kinetic energies are

$$T_1 = \tfrac{1}{2}mv_{A1}^2 + mv_{B1}^2 \quad \text{and} \quad T_2 = \tfrac{1}{2}mv_{A2}^2 + mv_{B2}^2. \tag{2}$$

Force Laws The weight force on B is conservative, so the potential energies are

$$V_1 = 0 \quad \text{and} \quad V_2 = -2mgh. \tag{3}$$

Since we are treating the cable tensions as external forces, we have

$$(U_{1\text{-}2})^{\text{int}}_{\text{nc}} = 0. \tag{4}$$

The work of friction and the tension P are included in the nonconservative external work term, which gives

$$\begin{aligned}(U_{1\text{-}2})^{\text{ext}}_{\text{nc}} &= \int_{x_{A1}}^{x_{A2}} (F - P)\,dx_A - \int_{y_{B1}}^{y_{B2}} 2P\,dy_B \\ &= \mu_k mg(x_{A2} - x_{A1}) - \int_{x_{A1}}^{x_{A2}} P\,dx_A - 2\int_{y_{B1}}^{y_{B2}} P\,dy_B,\end{aligned} \tag{5}$$

where $x_{A2} - x_{A1}$ is the displacement of crate A and we have used the fact that $F = \mu_k N = \mu_k mg$ is constant ($N = mg$ because A does not move vertically).

Kinematic Equations Since the system starts from rest, so

$$v_{A1} = v_{B1} = 0. \tag{6}$$

Referring to Fig. 3, since the cables are inextensible, we must have

$$x_A + 2y_B = \text{constant}. \tag{7}$$

Taking the differential of Eq. (7) so that we can relate the integrals in Eq. (5), we obtain

$$dx_A + 2dy_B = 0 \quad \Rightarrow \quad dx_A = -2dy_B. \tag{8}$$

In addition, this tells us that if B falls a distance h, then A must move to the right a distance $2h$. Therefore,

$$y_{B2} - y_{B1} = h \quad \text{and} \quad x_{A2} - x_{A1} = -2h. \tag{9}$$

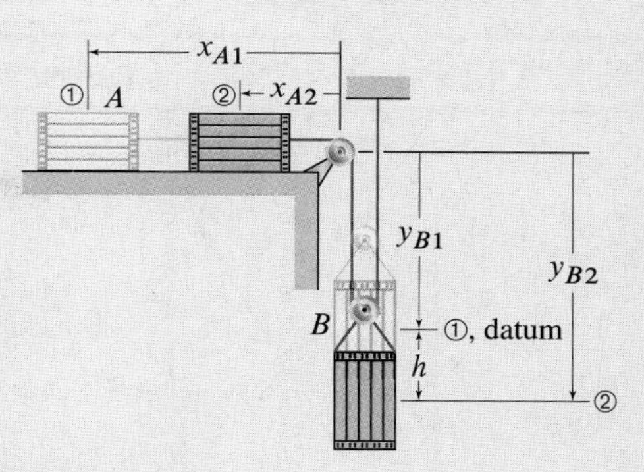

Figure 3
The positions of crates A and B in ① and ②.

Differentiating Eq. (7) with respect to time so that we can relate the velocities of A and B in the expression for kinetic energy, we obtain

$$v_{Ax} + 2v_{By} = 0 \quad \Rightarrow \quad |v_{By}| = \left|-\tfrac{1}{2}v_{Ax}\right| \quad \Rightarrow \quad v_{B2} = \tfrac{1}{2}v_{A2}, \tag{10}$$

where we have evaluated the speed relation in ② to obtain the final result.

Computation Using Eq. (8) to make a change of variables in the first integral in Eq. (5), this equation becomes

$$\begin{aligned}(U_{1\text{-}2})_{\text{nc}}^{\text{ext}} &= \mu_k mg(x_{A2} - x_{A1}) - \int_{y_{B1}}^{y_{B2}} P\,(-2dy_B) - 2\int_{y_{B1}}^{y_{B2}} P\,dy_B \\ &= \mu_k mg(x_{A2} - x_{A1}) + 2\int_{y_{B1}}^{y_{B2}} P\,dy_B - 2\int_{y_{B1}}^{y_{B2}} P\,dy_B \\ &= \mu_k mg(x_{A2} - x_{A1}).\end{aligned} \tag{11}$$

Substituting Eqs. (2), (3), (4), (6), and (9)–(11) into Eq. (1), we obtain

$$\mu_k mg(-2h) = \tfrac{1}{2}mv_{A2}^2 + m\left(\tfrac{1}{2}v_{A2}\right)^2 - 2mgh, \tag{12}$$

where we have used the fact that $(U_{1\text{-}2})_{\text{nc}}^{\text{int}} = 0$. Solving for v_{A2}, we obtain

$$\boxed{v_{A2} = \sqrt{\tfrac{8}{3}gh(1 - \mu_k)}.} \tag{13}$$

Discussion & Verification The right side of Eq. (13) has the dimension of speed, which is correct. Equation (13) also says that v_A increases as h increases, which is reasonable. In addition, notice that the work done by the cables cancels out of Eq. (12), which it should since the cables are inextensible.

EXAMPLE 4.12 Conservation of Energy in a Particle System

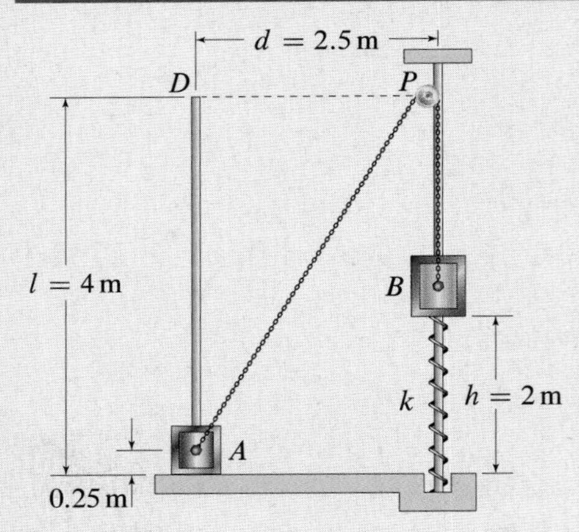

Figure 1
Proposed catapult system with masses A and B connected by the pulley at P. The dimensions shown are those at release.

The simple catapult in Fig. 1 consists of the two blocks A and B connected by the pulley at P. The falling weight B and stretched spring to which it is attached will launch A. Mass B, with $m_B = 8\,\text{kg}$, is released from rest at a distance $h = 2\,\text{m}$ above the ground. At the time of release, A, with $m_A = 1\,\text{kg}$, is in the position shown in Fig. 1. The spring attached to B is linear with constant $k = 40\,\text{N/m}$, and it is unstretched when B reaches the ground. Determine the speeds of A and B immediately before B hits the ground.

SOLUTION

Road Map & Modeling By applying the work-energy principle, we can determine the speed of A at the position of interest. The FBD of the system in Fig. 2 reflects the fact

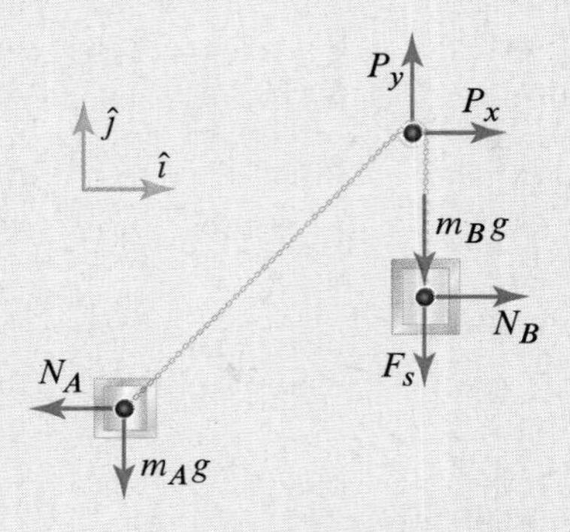

Figure 2. An FBD of the system shown in Fig. 1.

that we are modeling A and B as particles, that we are ignoring any friction, and that we are neglecting the weight of the chain connecting A and B by the pulley P. The only forces doing work as B drops the distance h are the spring force F_s and the two weight forces. The reactions at P do no work since their point of application is fixed. While the chain provides an internal force, for it to do work the two ends of the chain would need to move relative to one another in the direction of the tension (see discussion of Eq. (4.43) on p. 282). That is, the chain would have to stretch. Since the chain is inextensible, its tension does no work. We define ① to be at release (from rest) and ② to be the position at which B hits the ground and the spring becomes unstretched.

Governing Equations

Balance Principles From the discussion above, we see that $(U_{1\text{-}2})^{\text{ext}}_{\text{nc}}$ and $(U_{1\text{-}2})^{\text{int}}_{\text{nc}}$ in Eq. (4.45) are both zero. Therefore, the work-energy principle between ① and ② becomes

$$T_1 + V_1 = T_2 + V_2. \tag{1}$$

The system's kinetic energies can be written as

$$T_1 = \tfrac{1}{2} m_A v_{A1}^2 + \tfrac{1}{2} m_B v_{B1}^2 \quad \text{and} \quad T_2 = \tfrac{1}{2} m_A v_{A2}^2 + \tfrac{1}{2} m_B v_{B2}^2. \tag{2}$$

Force Laws The potential energies in Eq. (1) are made up of the gravitational potential energies of m_A and m_B, as well as the elastic potential energy of the spring. Therefore, referring to Figs. 1 and 3, we have

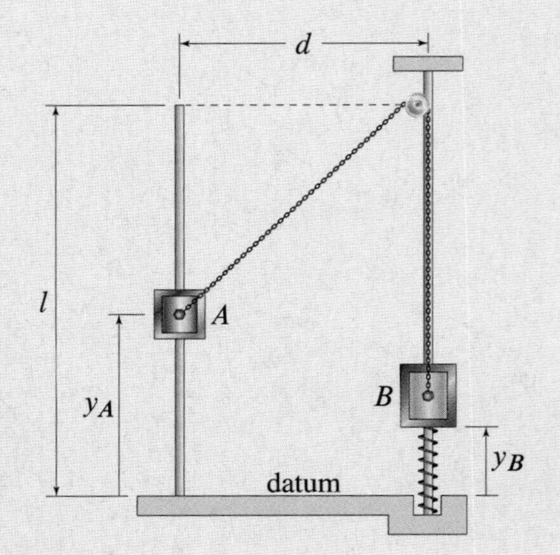

Figure 3
Definition of y_A, y_B, and the datum line for the gravitational potential energy.

$$V_1 = \tfrac{1}{2} k h^2 + m_A g y_{A1} + m_B g y_{B1} \quad \text{and} \quad V_2 = m_A g y_{A2} + m_B g y_{B2}, \tag{3}$$

where we have used the fact that the spring is unstretched at ②.

Kinematic Equations At ① we have

$$v_{A1} = 0 \quad \text{and} \quad v_{B1} = 0. \tag{4}$$

As far as ② is concerned, we note that v_{A2} and v_{B2} are the unknowns of the problem. We also note that both the position and speed of A are related to the position and speed of B. To determine these relations, referring to Fig. 3, we observe that the cord's length is

$$L = \sqrt{(l - y_A)^2 + d^2} + (l - y_B). \tag{5}$$

Since the cord is inextensible, differentiating Eq. (5) with respect to time, we have

$$0 = \frac{-(l - y_A)v_{Ay}}{\sqrt{(l - y_A)^2 + d^2}} - v_{By} \quad \Rightarrow \quad \left| -\frac{l - y_A}{\sqrt{(l - y_A)^2 + d^2}} \right| v_A = v_B, \tag{6}$$

where the introduction of the absolute value in the second equation allows us to replace v_{Ay} and v_{By} with the *speeds* v_A and v_B, respectively. To determine the relationship between the displacement of A and the displacement of B, we can evaluate Eq. (5) at ① and ② as follows:

$$L = \sqrt{(l - y_{A1})^2 + d^2} + (l - y_{B1}) = \sqrt{(l - y_{A2})^2 + d^2} + (l - y_{B2}). \tag{7}$$

The second equality in Eq. (7) implies that

$$y_{B1} - y_{B2} = \sqrt{(l - y_{A1})^2 + d^2} - \sqrt{(l - y_{A2})^2 + d^2}. \tag{8}$$

Recalling that $y_{B1} - y_{B2} = h$ and y_{A1} is known, and solving Eq. (8) for y_{A2}, we have

$$y_{A2} = l - \sqrt{h^2 + (l - y_{A1})^2 - 2h\sqrt{(l - y_{A1})^2 + d^2}}. \tag{9}$$

Computation Substituting Eqs. (2)–(4) into Eq. (1), we obtain

$$\tfrac{1}{2}kh^2 + m_A g y_{A1} + m_B g y_{B1} = \tfrac{1}{2}m_A v_{A2}^2 + \tfrac{1}{2}m_B v_{B2}^2 + m_A g y_{A2} + m_B g y_{B2}. \tag{10}$$

Recalling again that $y_{B1} - y_{B2} = h$, substituting Eqs. (6) into Eq. (10), and solving for v_{A2} yield

$$v_{A2} = \sqrt{\frac{kh^2 + 2m_A g(y_{A1} - y_{A2}) + 2m_B g h}{m_A + m_B (l - y_{A2})^2/[(l - y_{A2})^2 + d^2]}}. \tag{11}$$

Recalling that $l = 4\,\text{m}$, $y_{A1} = 0.25\,\text{m}$, $h = 2\,\text{m}$, and $d = 2.5\,\text{m}$, from Eq. (9) we see that $y_{A2} = 3.814\,\text{m}$. Using this result, along with the given data, from Eqs. (11) and (6) we see that

$$\boxed{v_{A2} = 19.67\,\text{m/s} \quad \text{and} \quad v_{B2} = 1.462\,\text{m/s}.} \tag{12}$$

Discussion & Verification The term multiplying v_A in Eq. (6) is nondimensional, which implies that Eq. (6) is dimensionally correct. The argument of the square root in Eq. (11) has dimensions of energy divided by mass, i.e., speed squared. Therefore, Eq. (11) is also dimensionally correct. To assess whether or not the value for v_{A2} (and therefore, the corresponding value of v_{B2}) in Eq. (12) is reasonable, we can say that v_{A2} should not exceed the speed that we would obtain if *all* of the potential energy of B were turned into the kinetic energy of A. We can find such a value as follows:

$$\tfrac{1}{2}kh^2 + m_B g h = \tfrac{1}{2}m_A v_{A2}^2 \quad \Rightarrow \quad v_{A2} = 21.77\,\text{m/s}. \tag{13}$$

Because the result in Eq. (12) is less than that of Eq. (13), we can say that our answer behaves as expected. Interestingly, even without the spring (i.e., for $k = 0$), we achieve $v_{A2} = 15.29\,\text{m/s}$ with $v_{B2} = 1.137\,\text{m/s}$.

Common Pitfall

Chains can become slack. Having assumed that the chain is inextensible means that the cord's length L must be a constant. By enforcing this constraint via Eq. (5) we are also saying that not only does L remain constant, but also remains *taut*. The solution should be verified by checking whether or not the chain becomes slack. This verification is one element of Prob. 4.96.

EXAMPLE 4.13 Why Does a Snapped Towel Hurt?

Figure 1
A sequence of photos showing a snapped towel. The exposure time of each photo is 0.0008 s. *Top frame:* Preparing for the snap. *Middle frame:* The towel is halfway through the forward part of the snap. The towel is bent like the initial state of our string. *Bottom frame:* The towel has reached its farthest extent. Notice that the last few inches of the towel are blurry due to the high speed at which this part of the towel is moving.

We now analyze a simple model to explain the behavior of a snapped towel, that is, the sting it produces when hitting someone and the loud crack that can be heard at the end of the snap. The model of the towel is shown in Fig. 2(a). Using this string-based model, we attach one end to the ceiling, fold the string back on itself, and then release the unattached end and let it fall freely. For an inextensible string of length L and mass m, determine the speed of the free end of the string as a function of its distance below the ceiling.

SOLUTION

Road Map & Modeling Since we are interested in the speed of the free end of the string after it has fallen a given distance, we apply the work-energy principle. The string is not a system of particles (its mass is distributed in space), but we can apply the concepts for systems of particles by applying the work-energy principle to the equivalent mass center of the system. We assume that the string is inextensible, and we will *view* it as a system of two particles (whose mass sums to that of the string), one particle being the left branch and the other the right (see Fig. 2(b)). These particles will be placed at the midpoint or

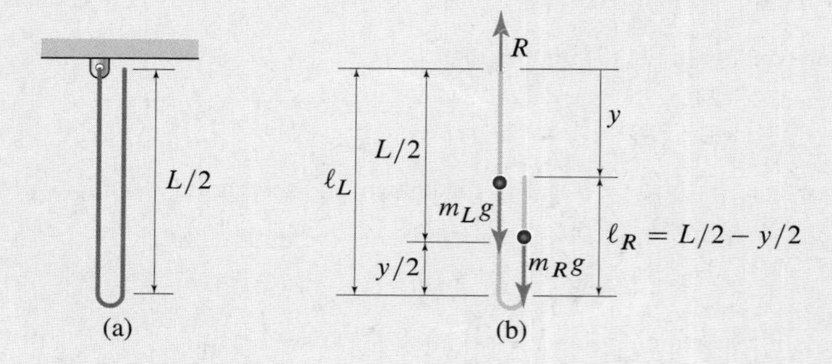

Figure 2. (a) The initial configuration of the string. (b) FBD and kinematics of the string after the free end has fallen a distance y, where $m_L g$ is the weight of the left segment and $m_R g$ is the weight of the right.

mass center of their corresponding branches. Figure 2(b) shows the string in an arbitrary position, which we will call ②, and we will let ① be at release. The weight forces $m_L g$ and $m_R g$ are the only forces doing work on the string, so the system is conservative.

Governing Equations

Balance Principles Since the system is conservative, the work-energy principle is

$$T_1 + V_1 = T_2 + V_2. \tag{1}$$

The kinetic energies are

$$T_1 = \tfrac{1}{2} m_{L1} v_{L1}^2 + \tfrac{1}{2} m_{R1} v_{R1}^2 \quad \text{and} \quad T_2 = \tfrac{1}{2} m_{L2} v_{L2}^2 + \tfrac{1}{2} m_{R2} v_{R2}^2, \tag{2}$$

where m_L and m_R are the masses of the left and right branches of the string, respectively, and where v_L and v_R are the speeds of the mass centers of the left and right branches of the string, respectively. In all cases, the second subscript indicates the position.

Force Laws Assuming that the changing bend in the string does not dissipate any energy, letting ℓ_L and ℓ_R be the lengths of the left and right branches of the string, respectively, and defining the datum for the gravitational potential energy to be at the top of the fixed portion of string, we have

$$V_1 = -m_{L1} g L/4 - m_{R1} g L/4 = -(m_{L1} + m_{R1}) g L/4, \tag{3}$$

$$V_2 = -m_{L2} g \ell_L/2 - m_{R2} g (y + \ell_R/2), \tag{4}$$

where, letting $\rho = m/L$ be the mass per unit length of string, the masses of the left and right branches of the string at ① and ② are given by

$$m_{L1} = \rho L/2, \qquad m_{R1} = \rho L/2, \qquad m_{L2} = \rho \ell_L, \qquad m_{R2} = \rho \ell_R, \tag{5}$$

so that V_1 and V_2 become

$$V_1 = -\tfrac{1}{4} \rho g L^2 \quad \text{and} \quad V_2 = -\rho g \ell_L^2/2 - \rho g \ell_R (y + \ell_R/2). \tag{6}$$

Kinematic Equations Referring to Fig. 2(b), the total length of the string L and the fall distance y are related to the length of the left and right segments as follows:

$$L = \ell_L + \ell_R \quad \text{and} \quad \ell_L = y + \ell_R. \tag{7}$$

Solving Eqs. (7) for ℓ_L and ℓ_R, we obtain

$$\ell_L = L/2 + y/2 \quad \text{and} \quad \ell_R = L/2 - y/2, \tag{8}$$

which are shown in Fig. 2(b).

Since the string is released from rest, at ① we have

$$v_{L1} = 0 \quad \text{and} \quad v_{R1} = 0. \tag{9}$$

Due to the inextensibility of the string, all points on the left segment of string must have a common velocity (as must all points of the right segment). Therefore, since the left branch is attached to the fixed ceiling, we must have

$$v_{L2} = 0. \tag{10}$$

Computation By substituting Eqs. (5), (8), (9), and (10) into Eq. (2) and then simplifying, the kinetic energies become

$$T_1 = 0 \quad \text{and} \quad T_2 = \tfrac{1}{2} \rho \ell_R v_{R2}^2 = \tfrac{1}{4} \rho (L - y) v_{R2}^2. \tag{11}$$

Substituting Eqs. (8) into the second of Eqs. (6) and simplifying, we obtain the final form for V_2 to be

$$V_2 = -\frac{\rho g}{4} (L^2 + 2Ly - y^2). \tag{12}$$

Substituting the first of Eqs. (6), Eqs. (11), and Eq. (12) into Eq. (1) and solving for v_{R2}, we obtain

$$\boxed{v_{R2} = \sqrt{gy \left(\frac{2L - y}{L - y} \right)}.} \tag{13}$$

Discussion & Verification The argument of the square root in Eq. (13) has dimensions of acceleration times length, which is the same as speed squared. Therefore, the answer in Eq. (13) is dimensionally correct. In addition, notice that the value of the speed increases as y increases, as should be expected.

A Closer Look When we let $y \to L$ in Eq. (13), that is, as the string approaches the end of its fall, we see that $v_{R2} \to \infty$ (see Fig. 3). This is an amazing result! Simply take a string, bend it, and drop it as we have done here, and the speed of the tip of the string becomes infinite at the end of the fall. While we know that an infinite speed is not achieved in reality, what the model does imply is that the tip of the string is going *very* fast. In fact, when a real snapped towel is aided by the person snapping it, the loud crack produced is a result of the shock wave generated by the tip of the towel when the speed of the tip becomes greater than the speed of sound.

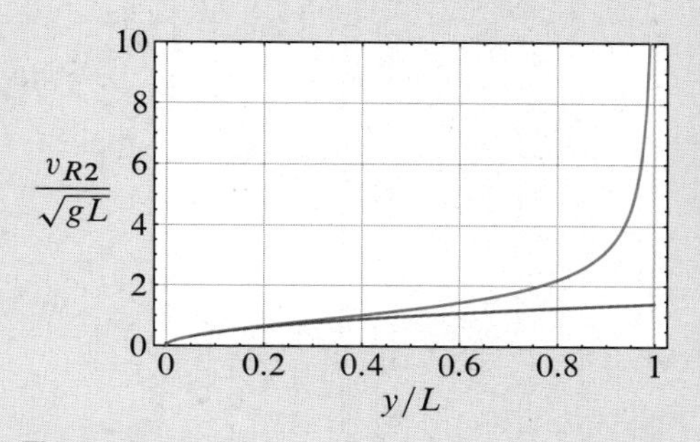

Figure 3
Nondimensional speed of the falling string (red line) and a particle in free fall (blue line) as a function of nondimensional distance. The orange vertical line represents $y = L$.

Interesting Fact

Cracking a whip. When a whip is cracked, the tip reaches a supersonic velocity for a duration of 1.2 ms. In a distance of 45 cm, the tip is accelerated from $M = 1$ to $M = 2.2$ (M is the Mach number, which is the ratio of the object's speed to the speed of sound in the surrounding medium). Assuming a uniform acceleration during this period and the speed of sound in air of 345 m/s, the acceleration of the tip exceeds 51,000 g! See P. Krehl, S. Engemann, and D. Schwenkel, "The Puzzle of Whip Cracking—Uncovered by a Correlation of Whip-Tip Kinematics with Shock Wave Emission," *Shock Waves*, **8**(1), 1998, pp. 1–9.

EXAMPLE 4.14 *Internal Work Due to Friction*

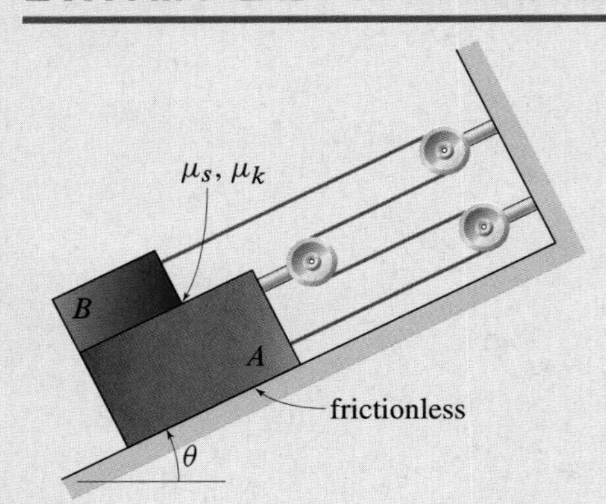

Figure 1
The two-block system connected by pulleys.

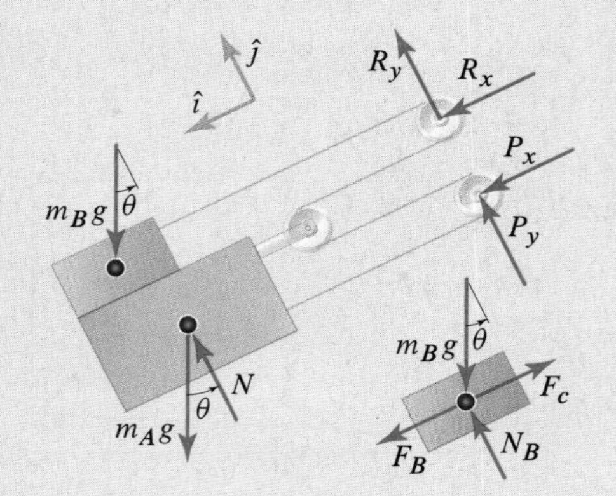

Figure 2
FBD of the system and of just B.

The two blocks A and B of mass $m_A = 4$ kg and $m_B = 1$ kg, respectively, in Fig. 1 are connected by an inextensible cord and the pulley system shown. There is negligible friction between A and the $\theta = 30°$ incline, and the coefficient of kinetic friction between A and B is $\mu_k = 0.1$. Assuming that μ_s is insufficient to prevent slipping and that the system is released from rest, determine the velocity of A and B after B has moved *up the incline* a distance $d = 0.35$ m *relative to A*.

SOLUTION

Road Map & Modeling Since we need to relate changes in position to changes in velocity, we will apply the work-energy principle. In Fig. 2, we have sketched the system's FBD, which includes A and B modeled as particles, the cord, and the pulleys. We ignore friction in the pulleys and assume the cord is inextensible and has negligible mass. The cord tension does no work because the cord is inextensible. However, since A and B slide relative to one another, we need to include the work done by the internal friction force between A and B, so we have also drawn the FBD of B in Fig. 2 so that we can find this friction force. Finally, we will let ① be when the system is released and ② be when B has moved the distance d up the incline relative to A.

Governing Equations

Balance Principles The work-energy principle for the system in Fig. 2 is

$$T_1 + V_1 + (U_{1\text{-}2})_{\text{nc}}^{\text{int}} = T_2 + V_2, \tag{1}$$

where $(U_{1\text{-}2})_{\text{nc}}^{\text{int}}$ is the work done by friction. The kinetic energies are given by

$$T_1 = \tfrac{1}{2}m_A v_{A1}^2 + \tfrac{1}{2}m_B v_{B1}^2 \quad \text{and} \quad T_2 = \tfrac{1}{2}m_A v_{A2}^2 + \tfrac{1}{2}m_B v_{B2}^2, \tag{2}$$

where v_A and v_B are the speed of A and B, respectively, and the second subscript on the speeds indicates position. Again referring to Fig. 2, we sum forces in the y direction on B and obtain

$$\sum F_y: \quad N_B - m_B g \cos\theta = 0, \tag{3}$$

where we have set $a_{By} = 0$ because there is no motion in the y direction.

Force Laws For the system's potential energy we can write

$$V_1 = V_{A1} + V_{B1} = 0, \tag{4}$$

$$V_2 = V_{A2} + V_{B2} = -m_A g\,\Delta x_A \sin\theta - m_B g\,\Delta x_B \sin\theta, \tag{5}$$

where we have placed the datum for the potential energy at ① and where Δx_A and Δx_B are the displacements parallel to the incline of A and B, respectively, between ① and ②. Both Δx_A and Δx_B are unknown, and we will use kinematics to relate them to the given distance d.

The work of the internal friction force F_B is given by

$$(U_{1\text{-}2})_{\text{nc}}^{\text{int}} = -\int_0^d F_B\,dx. \tag{6}$$

Finally, since we know that μ_s is insufficient to prevent slipping, we know that

$$F_B = \mu_k N_B. \tag{7}$$

Kinematic Equations Since the system starts from rest, we have

$$v_{A1} = 0 \quad \text{and} \quad v_{B1} = 0. \tag{8}$$

In addition, we can relate the motions of A and B using pulley kinematics (see Section 2.7 on p. 121). Referring to Fig. 3, we see that for arbitrary x_A and x_B

$$3x_A + x_B = L \quad \Rightarrow \quad 3v_A = -v_B, \tag{9}$$

where L is constant and where v_A and v_B represent the velocity of A and B, respectively, since the motion is one-dimensional. From Eq. (9), at ② we have

$$3v_{A2} = -v_{B2} \quad \text{and} \quad 3\Delta x_A = -\Delta x_B. \tag{10}$$

Finally, we know that A displaces a distance d relative to B. Therefore, we can write

$$\Delta x_{A/B} = d = \Delta x_A - \Delta x_B. \tag{11}$$

Solving Eqs. (10) and (11) for Δx_A and Δx_B, we obtain

$$\Delta x_A = \tfrac{1}{4}d \quad \text{and} \quad \Delta x_B = -\tfrac{3}{4}d. \tag{12}$$

Computation Substituting Eqs. (3) and (7) into Eq. (6), we have

$$(U_{1\text{-}2})_{\text{nc}}^{\text{int}} = -\mu_k m_B g d \cos\theta. \tag{13}$$

Next, we substitute the first of Eqs. (10) into Eq. (2) to obtain

$$T_2 = \tfrac{1}{2}m_A v_{A2}^2 + \tfrac{1}{2}m_B(-3v_{A2})^2 = \tfrac{1}{2}(m_A + 9m_B)v_{A2}^2, \tag{14}$$

and then substitute Eqs. (12) into Eq. (5) to obtain

$$V_2 = \tfrac{1}{4}(3m_B - m_A)gd \sin\theta. \tag{15}$$

We then take Eqs. (4), (8), and (13)–(15) and substitute them into Eq. (1) to obtain one equation for v_{A2}

$$-\mu_k m_B g d \cos\theta = \tfrac{1}{2}(m_A + 9m_B)v_{A2}^2 + \tfrac{1}{4}(3m_B - m_A)gd \sin\theta, \tag{16}$$

which gives

$$v_{A2} = \pm\sqrt{\frac{2gd}{m_A + 9m_B}\left[\tfrac{1}{4}(m_A - 3m_B)\sin\theta - \mu_k m_B \cos\theta\right]}. \tag{17}$$

Observing that A moves down the incline and B moves up the incline, using Eqs. (17) and (9), as well as the given data, we have

$$\boxed{v_{A2} = 0.1424\,\text{m/s} \quad \text{and} \quad v_{B2} = -0.4273\,\text{m/s}.} \tag{18}$$

Discussion & Verification The dimensions of the argument of the square root in Eq. (17) are energy divided by mass, i.e., speed squared. Hence, the expression for v_{A2} in Eq. (17) is dimensionally correct. The expression of v_B in Eqs. (9) is also dimensionally correct. As far as the values in Eqs. (18) are concerned, since B acts as a counterweight for A, a way to check the reasonableness of our result is to compute the speed that A would have if it were disconnected from B and moved a distance $d/4$ down the incline without friction. This speed, given by $(v_{A2})_{\text{no friction}} = \sqrt{2(d/4)g\sin\theta} = 0.9265\,\text{m/s}$, is an upper bound for v_{A2}. That is, v_{A2} must be less than 0.9265 m/s, as indeed it is.

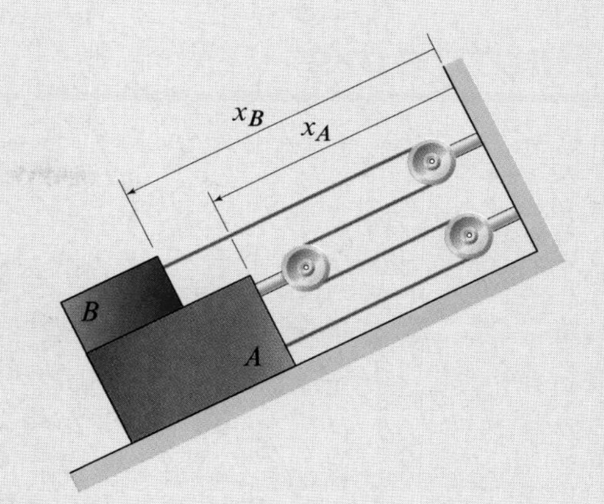

Figure 3
Definition of x_A and x_B for the two blocks.

Helpful Information

What do the ± signs on Eq. (17) mean? In this example we need to determine a *velocity*. Since the motion is one-dimensional, the velocity we are seeking can be equal to the speed or opposite to the speed. The ± sign in Eq. (17) is there to remind us that we need to determine the sign of the velocity.

Problems

Figure P4.71

Problem 4.71

The truck comes to a stop under the action of a constant braking force. During braking, either the crate slides or it does not. Considering the work-energy principle applied to the truck during braking, will the truck stop in a shorter distance (or time) if the crate slides, or will the distance (or time) be shorter if it does not slide? Assume that the truck bed is long enough that you don't have to worry about whether or not the crate hits the truck. Justify your answer.

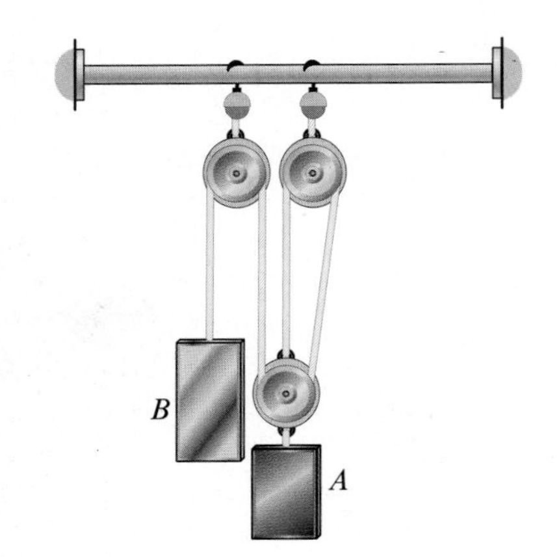

Figure P4.72

Problem 4.72

Consider a pulley system in which bodies A and B have masses $m_A = 2$ kg and $m_B = 10$ kg. If the system is released from rest, neglecting all sources of friction, as well as the inertia of the pulleys, determine the speeds of A and B after B has displaced a distance of 0.6 m downward.

Problem 4.73

A 700 lb floating platform is at rest when a 200 lb crate is thrown onto it with a horizontal speed $v_0 = 12$ ft/s. Once the crate stops sliding relative to the platform, the platform and the crate move with a speed $v = 2.667$ ft/s. Neglecting the vertical motion of the system, as well as any resistance due to the relative motion of the platform with respect to the water, determine the distance that the crate slides relative to the platform if the coefficient of kinetic friction between the platform and the crate is $\mu_k = 0.25$.

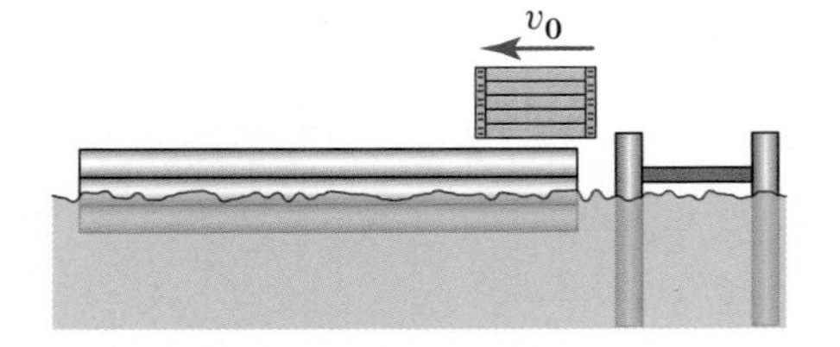

Figure P4.73

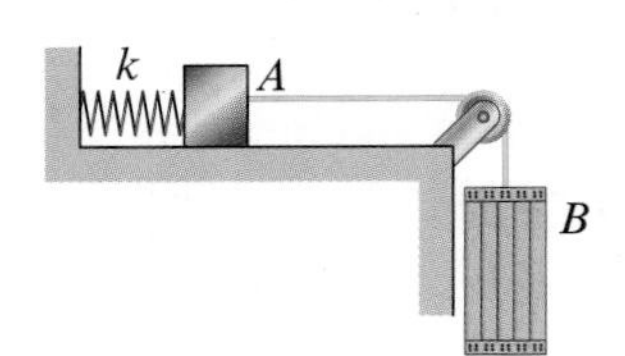

Figure P4.74

Problem 4.74

Blocks A and B, weighing 7 and 15 lb, respectively, are released from rest when the spring is unstretched. If all sources of friction are negligible and $k = 12$ lb/ft, determine the maximum vertical displacement of B from the release position, assuming that A never leaves the horizontal surface shown and the cord connecting A and B is inextensible.

Problems 4.75 through 4.77

Crates A and B of mass 50 kg and 75 kg, respectively, are released from rest. The linear elastic spring has stiffness $k = 500$ N/m. Neglect the mass of the pulleys and cables and neglect friction in the pulley bearings.

Problem 4.75 If $\mu_k = 0$ and the spring is initially unstretched, determine the speed of B after A slides 4 m.

Problem 4.76 If $\mu_k = 0.25$ and the spring is initially unstretched, determine the speed of B after A slides 4 m.

Problem 4.77 If $\mu_k = 0.25$ and the spring is initially stretched 1 m, how far does A slide before the system momentarily comes to rest?

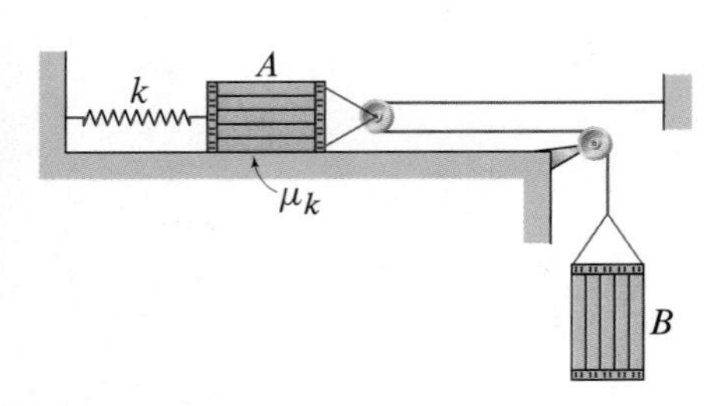

Figure P4.75–P4.77

Problems 4.78 through 4.80

The crates A and B of weight $W_A = 50$ lb and $W_B = 75$ lb, respectively, are connected by a pulley system. The system is released from rest and friction between A and the horizontal surface is insufficient to prevent slipping. The cables in the pulley system are inextensible, and the coefficient of kinetic friction between the crate A and the horizontal surface is μ_k.

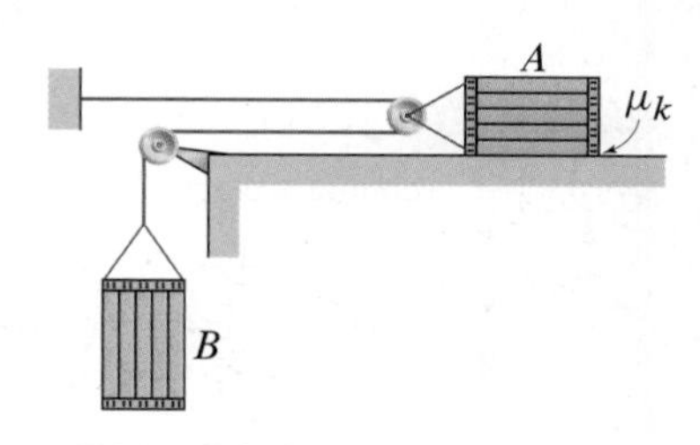

Figure P4.78–P4.80

Problem 4.78 If $\mu_k = 0$, determine the speed of B after A slides 10 ft.

Problem 4.79 Determine the required value of μ_k so that the speed of B is 27 ft/s after it drops 15 ft.

Problem 4.80 If $\mu_k = 0.25$, determine the speed of A after B drops 15 ft.

Problem 4.81

Blocks A and B are released from rest when the spring is unstretched. Block A has a mass $m_A = 2$ kg, and the linear spring has stiffness $k = 7$ N/m. If all sources of friction are negligible, determine the mass of block B such that B has a speed $v_B = 1.5$ m/s after moving 1.2 m downward, assuming that A never leaves the horizontal surface shown and the cord connecting A and B is inextensible.

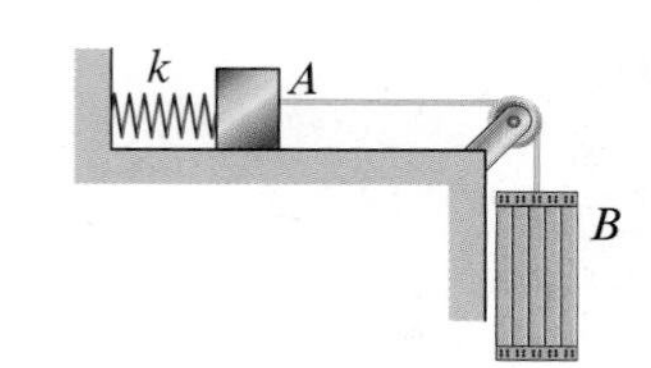

Figure P4.81

Problem 4.82

Spring scales work by measuring the displacement of a spring that supports both the platform of mass m_p and the object of mass m, whose weight is being measured. In your solution, note that most scales read zero when no mass m has been placed on them; that is, they are calibrated so that the weight reading neglects the mass of the platform m_p. Assume that the spring is linear elastic with spring constant k. If the mass m is gently placed on the spring scale (i.e., it is released from zero height above the scale), determine the *maximum reading* on the scale after the mass is released.

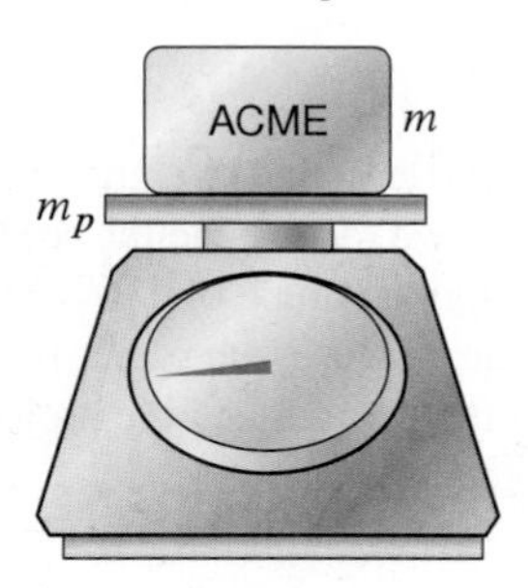

Figure P4.82

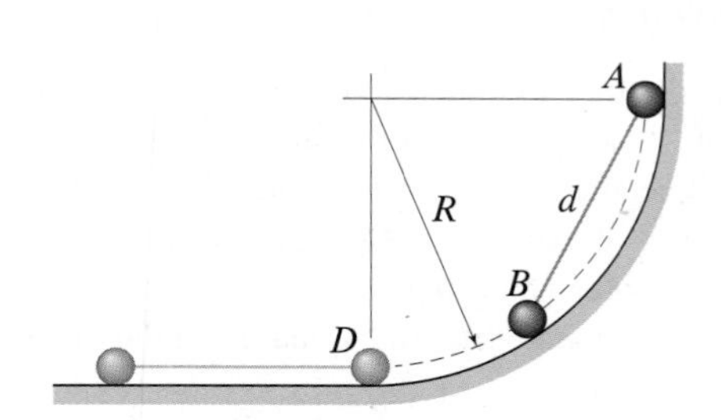

Figure P4.83

Problem 4.83

Two small spheres A and B, each of mass m, are attached at either end of a stiff light rod of length d. The system is released from rest in the position shown. Determine the speed of the spheres when A has reached point D, and determine the normal force between sphere A and the surface on which it is sliding immediately before it reaches point D. Neglect friction, treat the spheres as particles (assume that their diameter is negligible), and neglect the mass of the rod.

Problem 4.84

Solve Example 4.14 by applying the work-energy principle to each block individually, and show that the net work done by the cord on the two blocks is zero.

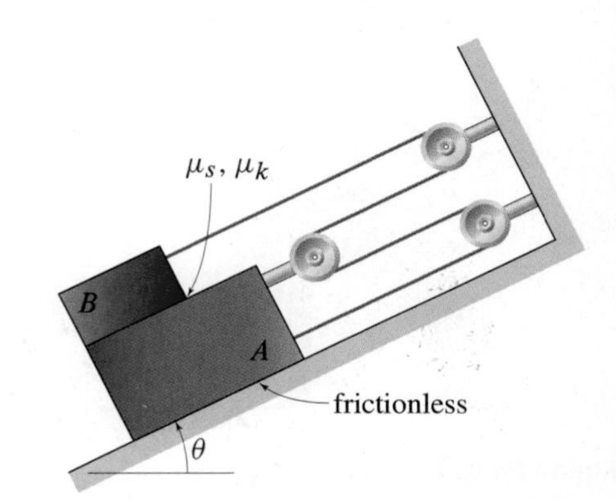

Figure P4.84

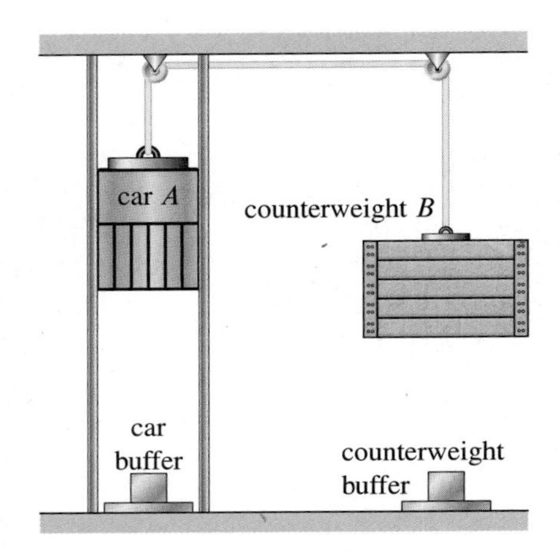

Figure P4.85

Problem 4.85

Consider a simple elevator design in which a 15,000 kg car A is connected to a 12,000 kg counterweight B. Suppose that a failure of the drive system occurs (the failure does not affect the rope connecting A and B) when the car is at rest and 50 m above its buffer, causing the elevator car to fall. Model the car and the counterweight as particles and the cord as massless and inextensible, and model the action of the emergency brakes using a Coulomb friction model with kinetic friction coefficient $\mu_k = 0.5$ and a normal force equal to 35% of the car's weight. Determine the speed with which the car impacts the buffer.

Problem 4.86

Two identical balls, each of mass m, are connected by a string of negligible mass and length $2l$. A short string is attached to the first at its middle and is pulled vertically with a constant force P (exerted by the hand). If the system starts at rest when $\theta = \theta_0$, determine the speed of the two balls as θ approaches 90°. Neglect the size of the balls as well as friction between the balls and the surface on which they slide.

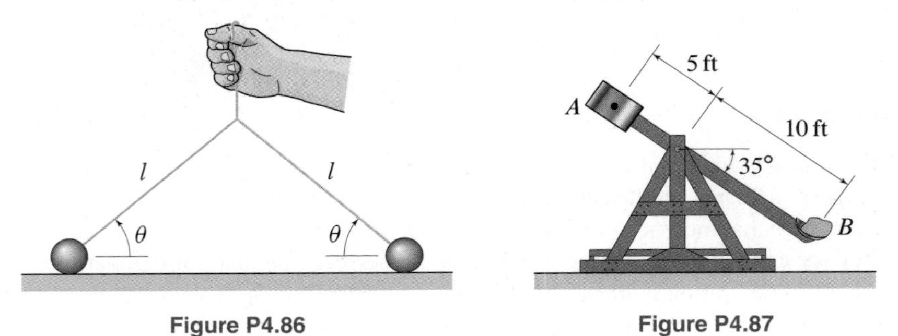

Figure P4.86

Figure P4.87

Problem 4.87

Consider the simple catapult shown in the figure with an 800 lb counterweight A and a 150 lb projectile B. If the system is released from rest as shown, determine the speed of the projectile after the arm rotates (counterclockwise) through an angle of 110°. Model A and B as particles, neglect the mass of the catapult's arm, and assume that friction is negligible. The catapult's frame is fixed with respect to the ground, and the projectile does not separate from the arm during the motion considered.

Problems 4.88 and 4.89

Two blocks A and B weighing 123 and 234 lb, respectively, are released from rest as shown. At the moment of release the spring is unstretched. In solving these problems, model A and B as particles, neglect air resistance, and assume that the cord is inextensible.

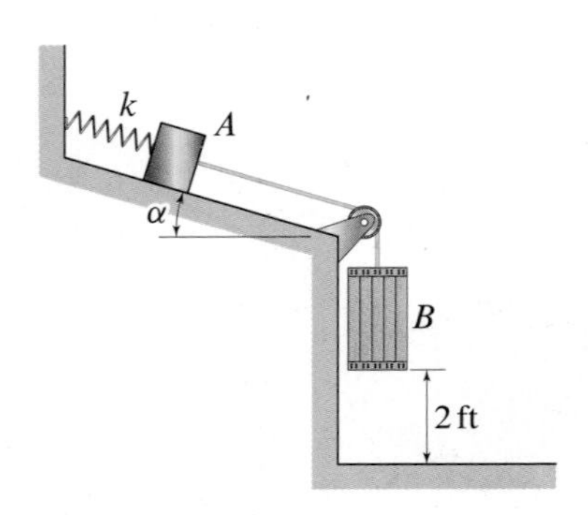

Figure P4.88 and P4.89

Problem 4.88 Determine the maximum speed attained by block B and the distance from the floor where the maximum speed is achieved if $\alpha = 20°$, the contact between A and the incline is frictionless, and the spring constant is $k = 30$ lb/ft.

Problem 4.89 Determine the maximum displacement of block B if $\alpha = 20°$, the contact between A and the incline is frictionless, and the spring constant is $k = 300$ lb/ft.

Problem 4.90

Revisit Example 4.7 on p. 266 and determine the maximum height reached by the pole vaulter, but this time include the mass of the pole in your analysis. To solve the problem, use the following data: the maximum speed achieved by the pole vaulter and pole at the time the vault begins is $v_{max} = 34.19\,\text{ft/s}$; the pole vaulter is 6 ft tall with mass center at 55% of body height (as measured from the ground) and weighs 190 lb; the pole is uniform, has a weight of 5.8 lb, and is 17.1 ft long. In addition, assume that as the pole vaulter sprints before the vault, the pole is carried horizontally at the same height as the vaulter's center of mass relative to the ground. Explain why the vault height you find when the pole is included is higher than the vault height determined in Example 4.7, when the pole was not included.

Figure P4.90

Problems 4.91 through 4.93

The two crates A and B of mass $m_A = 100\,\text{kg}$ and $m_B = 70\,\text{kg}$, respectively, are connected by a system of pulleys. The system is initially at rest, when the man C starts pushing on A with a constant force P. Neglect the mass of the cables and the pulleys and neglect friction in the pulley bearings. Assume that cable segments are long enough so that crate A does not hit any of the pulleys.

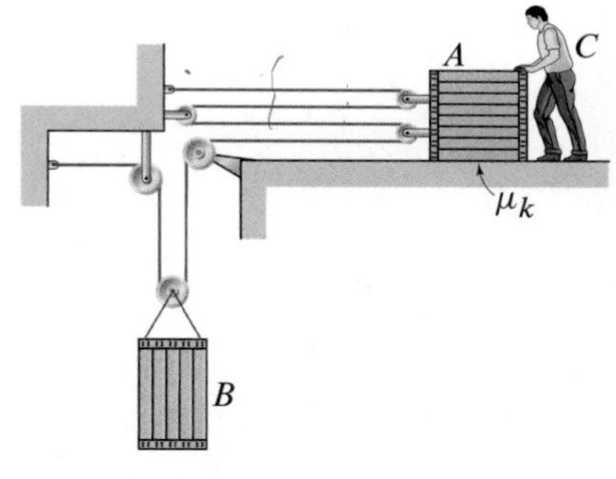

Figure P4.91–P4.93

Problem 4.91 If $\mu_k = 0$ and $P = 375\,\text{N}$, determine the speed of each crate after A slides 6 m.

Problem 4.92 If $\mu_k = 0.2$ and $P = 375\,\text{N}$, determine the speed of each crate after A slides 6 m.

Problem 4.93 If $\mu_k = 0.3$ and $P = 350\,\text{N}$, determine the distance the man needs to push the crate A to achieve a speed of 3 m/s.

Problem 4.94

The rope of mass m and length l is released from rest with a *very* small amount of it hanging over the edge (i.e., $s > 0$, but it is very close to zero). Determine the speed of the rope as a function of s. Assume the surface is smooth.

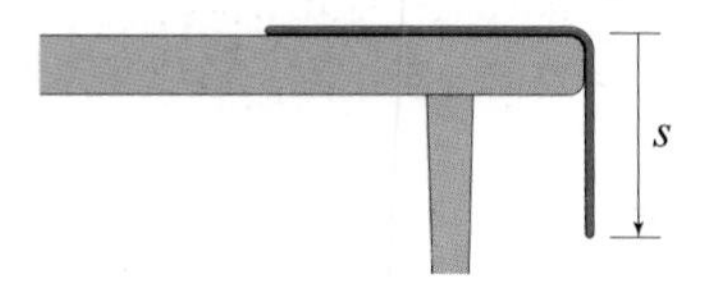

Figure P4.94

Problem 4.95

Consider a system of four identical particles A, B, D, and E, each with a mass of 2.3 kg. These particles are mounted on a wheel of negligible mass whose hub is at O and whose radius is $R = 0.75$ m. The wheel is mounted on a cart H of negligible mass. Suppose that the wheel is spinning counterclockwise with an angular speed $\omega = 3$ rad/s and that the cart is moving to the right with a speed of $v_H = 11$ m/s. Compute the kinetic energy of the system, using

(a) Eq. (4.46) on p. 282,

(b) Eq. (4.50) on p. 283.

Show that the result is the same in both cases.

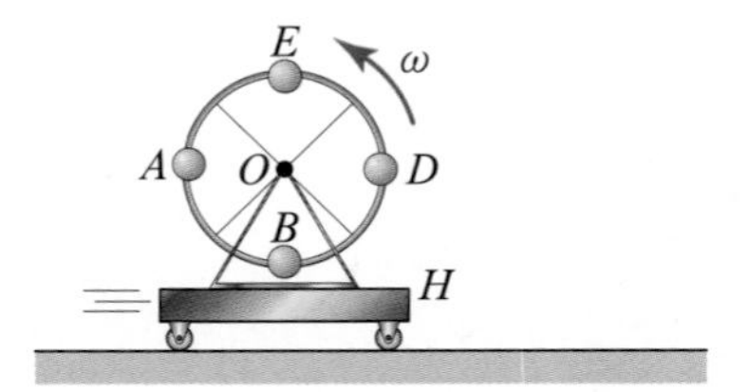

Figure P4.95

Problem 4.96

Reexamine Example 4.12 by choosing ① to be the same as that in the example and ② to be an arbitrary position after release. Determine the speed of A as a function of y_A, and plot the kinetic energy of A as a function of y_A. Use this plot to argue that the cord connecting A and B never goes slack between the instant of release and the impact of B with the ground. *Hint:* To argue that the cord remains taut between the positions of interest, you may want to look at concept Prob. 4.29 on p. 256.

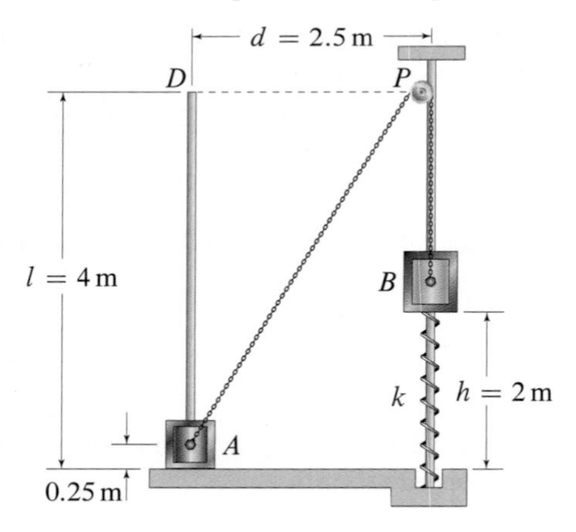

Figure P4.96

Design Problems

Design Problem 4.2

The plunger (the rod and rectangular block attached to the end of it) of a pinball machine has weight $W_p = 5$ oz. The weight of the ball is $W_b = 2.85$ oz, and the machine is inclined at $\theta = 8°$ with the horizontal. Design the springs k_1 and k_2 (i.e., their stiffnesses and unstretched lengths) so that the ball separates from the plunger with a speed $v = 15$ ft/s after pulling back the plunger 2 in. from its rest position and so that the plunger comes no closer than 0.5 in. from the stop. Note that as part of your design, you will also need to specify reasonable values for the dimensions ℓ_1 and ℓ_2.

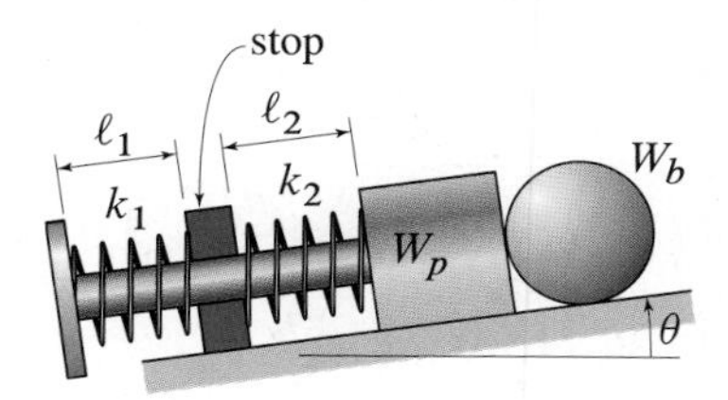

Figure DP4.2

4.4 Power and Efficiency

Power developed by a force

Figure 4.13
The 2009 Ferrari Formula One race car. Machines such as this car produce a tremendous amount of power for their size.

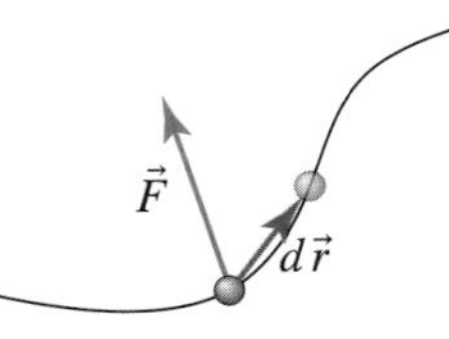

Figure 4.14
A force acting on a point moving along a path.

The concept of *mechanical power* arises from a need to quantify the ability of a motor or other machine to do mechanical work or provide energy in a given amount of time. Referring to Fig. 4.14, we consider a force $\vec{F}$ whose point of application undergoes an infinitesimal displacement $d\vec{r}$ in a corresponding infinitesimal time interval dt. The infinitesimal work done by $\vec{F}$ during the time interval dt is

$$dU = \vec{F} \cdot d\vec{r}. \tag{4.51}$$

The *power* P developed by the force $\vec{F}$ is defined as the time rate at which the $\vec{F}$ does work, that is,

$$\boxed{\text{power} = P = \frac{dU}{dt}.} \tag{4.52}$$

Substituting Eq. (4.51) into Eq. (4.52), we have

$$P = \frac{\vec{F} \cdot d\vec{r}}{dt} = \vec{F} \cdot \frac{d\vec{r}}{dt}. \tag{4.53}$$

Recalling that the velocity of the point of application of $\vec{F}$ is $\vec{v} = d\vec{r}/dt$, we can then rewrite Eq. (4.53) as

$$\boxed{P = \vec{F} \cdot \vec{v}.} \tag{4.54}$$

Dimensions and units of power

Because of its definition, the power developed by a force is a scalar quantity with dimensions of (force) × (length)/(time) or, equivalently, (mass) × (length)^2/(time)^3. In the SI system, the unit of measure for power is called the *watt*, which is abbreviated by the symbol W and is defined as follows:

$$\boxed{1\,\text{W} = 1\,\text{N}\cdot(1\,\text{m/s}) = 1\,\text{J/s}.} \tag{4.55}$$

In the U.S. Customary system, the unit of measure for power is called *horsepower*, which is abbreviated by the symbol hp and is defined as follows:

$$\boxed{1\,\text{hp} = 550\,\text{ft}\cdot\text{lb/s}.} \tag{4.56}$$

The conversion from horsepower to watt is 1 hp = 745.7 W.

Efficiency

Efficiency is a measure of the performance of a motor or a machine. It is the ratio between its power output and the power needed for its operation. *Efficiency*, for which we use the Greek letter ϵ (epsilon), is defined as:

$$\boxed{\epsilon = \frac{\text{power output}}{\text{power input}}.} \tag{4.57}$$

Efficiency is a nondimensional scalar quantity. Since some of the power supplied to a motor or a machine is used to move internal components and to overcome internal frictional resistance, the power output of a motor or machine is always smaller than the power input. Therefore, the efficiency of a motor or a machine is a number that is always smaller than 1.

EXAMPLE 4.15 *Computing the Power of an Aircraft Carrier Catapult*

In Example 4.2 on p. 246, we considered the takeoff of a 50,000 lb F/A-18 Hornet (see Fig. 1) from a stationary aircraft carrier. The F/A-18 goes from 0 to 165 mph in a distance of 300 ft, assisted by a catapult and with each of its two engines generating a constant thrust of 22,000 lb. Neglecting aerodynamic forces, in Example 4.2 we determined that the work done on the aircraft by the catapult was $(U_{1\text{-}2})_{\text{catapult}} = 3.227\times10^7$ ft·lb. Continuing Example 4.2, determine the power developed by the thrust generated by the engines at the beginning and at the end of the takeoff. In addition, determine the average power supplied to the aircraft by the catapult, knowing that the takeoff occurs in roughly 2 s.

Figure 1
An F/A-18 taking off from an aircraft carrier.

SOLUTION

Road Map & Modeling Denoting the beginning and the end of the takeoff by ① and ②, respectively, we know the thrust of the engines and the velocity of the aircraft at ① and ②. Therefore, we can compute the power developed by the engines' thrust with the relation expressing the power developed by a force in terms of the force itself and the velocity vector. To compute the *average* power of the catapult, we will compute the time average of the power developed by the force exerted by the catapult on the aircraft and relate that to the work done by the catapult.

Computation Recall that the quantity $\vec{F}\cdot\vec{v}$ is the power developed by a force $\vec{F}$ whose point of application moves with velocity $\vec{v}$. Since the plane starts from rest, i.e., $\vec{v}_1 = \vec{0}$, letting P_{T1} denote the power developed by the engines' thrust at ①, we must have

$$\boxed{P_{T1} = 0.} \tag{1}$$

Referring to Fig. 2, at ② we will assume that the plane is moving horizontally with a velocity $\vec{v}_2 = v_2\,\hat{\imath}$, where $v_2 = 165\text{ mph} = 242.0\text{ ft/s}$. In addition, assuming that the engines' thrust is also horizontal, the total thrust of the engines is $\vec{F}_T = F_T\,\hat{\imath}$ with $F_T = 44{,}000$ lb. Therefore, letting P_{T2} be the power developed by $\vec{F}_T$ at ②, we have

$$\boxed{P_{T2} = F_T\,\hat{\imath}\cdot v_2\,\hat{\imath} = F_T v_2 = 19{,}360\text{ hp},} \tag{2}$$

where we have used 1 hp = 550 ft·lb/s. Finally, letting P_{cat} and $(P_{\text{avg}})_{\text{cat}}$ be the instantaneous and the average power supplied to the aircraft by the catapult between ① and ②, respectively, we have

$$(P_{\text{avg}})_{\text{cat}} = \frac{1}{t_2 - t_1}\int_{t_1}^{t_2} P_{\text{cat}}\,dt = \frac{1}{t_2 - t_1}\int_{t_1}^{t_2} \frac{dU_{\text{cat}}}{dt}\,dt = \frac{(U_{1\text{-}2})_{\text{cat}}}{t_2 - t_1}. \tag{3}$$

Recalling that $t_2 - t_1 = 2$ s and that $(U_{1\text{-}2})_{\text{cat}} = 3.227\times10^7$ ft·lb, we have

$$\boxed{(P_{\text{avg}})_{\text{cat}} = 29{,}340\text{ hp},} \tag{4}$$

where we have again used 1 hp = 550 ft·lb/s.

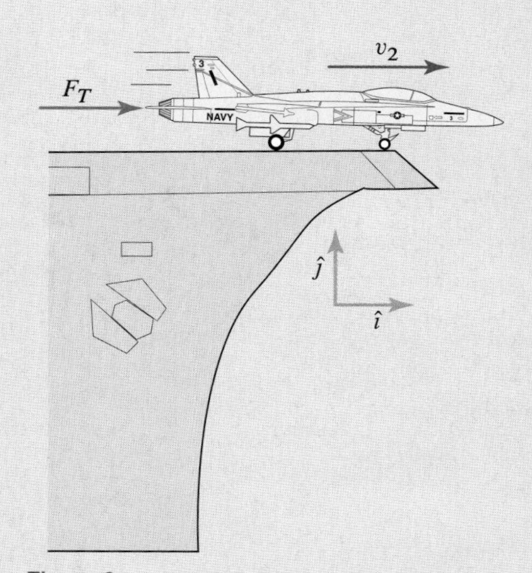

Figure 2
Diagram showing the aircraft's velocity vector and the engines' thrust vector at ②.

Discussion & Verification The results in Eqs. (1) and (2) are obtained as the direct application of a known formula. Hence, we need only verify that we used the correct units, which we have since we expressed our result in horsepower and the data was provided in U.S. Customary units. The result in Eq. (3) was obtained by applying the definition of time average. Keeping this in mind, intuition suggests that the notion of "average power" should correspond to the ratio between the work done by the catapult and the time taken to do such work, which is exactly what our calculation yielded.

EXAMPLE 4.16 Assessing Lift Speed Given the Efficiency

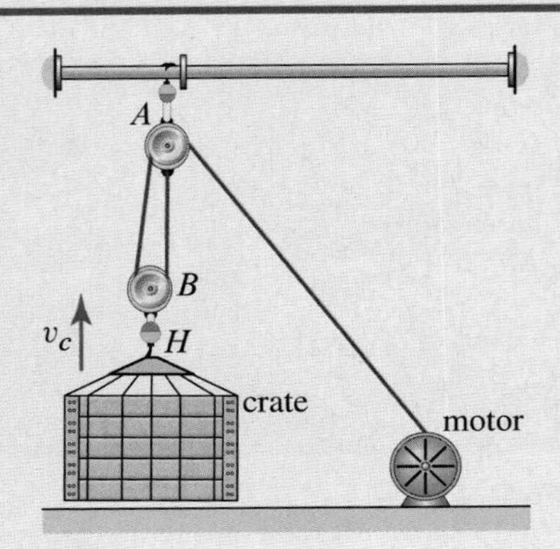

Figure 1
Pulley system lifting a crate with speed v_c.

An electric motor with an efficiency of 85% drives the pulley system shown to lift a 400 kg crate. After the motor is switched on and the system has reached a constant lifting speed, the motor is drawing 15 kW. Assume that the motor is operating at its rated efficiency, that the frictional losses in the pulley system are negligible, and that the weights of the pulleys A and B are negligible. Determine the speed with which the crate is being lifted by considering

(a) an FBD of the system that includes the crate and both pulleys A and B,

(b) an FBD of the crate that has been separated from the system at the hook H.

Neglect the size of the pulleys in your solution.

SOLUTION

Road Map & Modeling We can calculate the output power of the motor since we know its efficiency and input power. This output power is the power used by the pulley system to lift the crate. Since there are no power losses within the pulley system, the output power can be viewed as either *the power of the cable between pulley A and the motor* or *the power developed by the tension in the hook H attached to the crate.* Parts (a) and (b) of the problem reflect these two viewpoints. Hence, we need to use Newton's second law to determine the tension in the cable attached to the motor, and then we will be able to compute the lifting speed by dividing the power developed by this force by the magnitude of the force. We will perform a similar calculation for the hook attached to the crate.

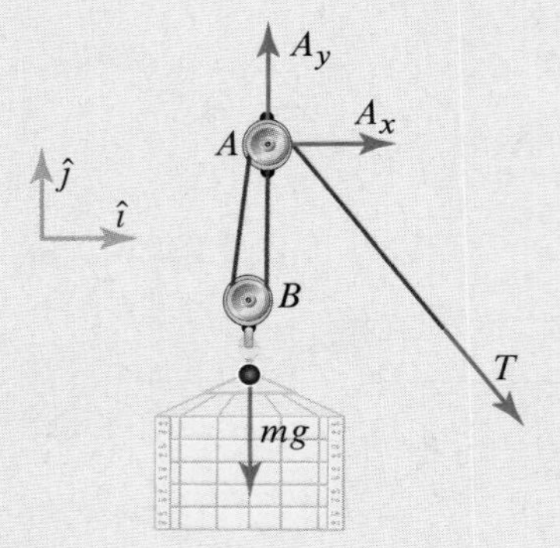

Figure 2
FBD of the pulleys and crate.

Solution for Part (a)

Governing Equations

Balance Principles The FBD of the pulleys and crate is shown in Fig. 2, in which the crate has been modeled as a particle and where T is the tension in the cable being pulled in by the motor. While it is the power developed by the tension T that we want to find, this FBD does not allow us to find T itself. Instead, applying Newton's second law to the FBD in Fig. 3(a), we find

$$\sum F_y:\quad 2T - mg = ma_y. \tag{1}$$

Force Laws All forces are accounted for on the FBD.

Kinematic Equations Since the crate is moving with constant velocity, we have

$$a_y = 0. \tag{2}$$

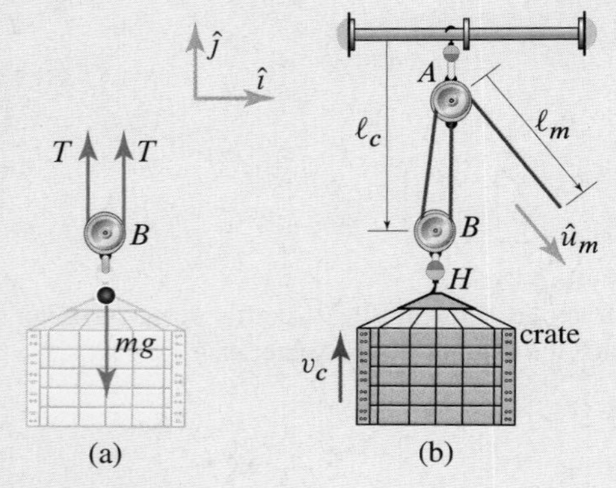

Figure 3
(a) FBD of the crate and the pulley at B. (b) Kinematics of the pulley system connecting the motor to the crate.

To find the speed of the point of application of the tension in the cable between the motor and the pulley at A, we need to refer to the pulley kinematics in Fig. 3(b) and write

$$2\ell_c + \ell_m = \text{constant} \quad \Rightarrow \quad 2\dot{\ell}_c = -\dot{\ell}_m \quad \Rightarrow \quad 2v_c = \dot{\ell}_m, \tag{3}$$

where we have used the fact that $v_c = -\dot{\ell}_c$ and we note that when $\dot{\ell}_m > 0$, the motor is winding in cable.

Computation Substituting Eq. (2) into Eq. (1), we find that

$$T = mg/2. \tag{4}$$

We need to compute the power provided by the motor to the pulley system. Letting P_i and P_o be the motor power input and output, respectively, we have

$$P_i = 15\,\text{kW} \quad \text{and} \quad P_o = \epsilon P_i = 12.75\,\text{kW}. \tag{5}$$

Recalling that P_o is the power developed by the tension T, and that T is in the same direction as the velocity of the cable winding up on the motor, we then must have

$$P_o = T\,\hat{u}_m \cdot \dot{\ell}_m\,\hat{u}_m = T\dot{\ell}_m = \frac{mg}{2}(2v_c) = mgv_c, \tag{6}$$

where we have used Eqs. (3) and (4). Solving Eq. (6) for v_c, we have

$$\boxed{v_c = \frac{P_o}{mg} = 3.249\,\text{m/s},} \tag{7}$$

where we have used P_o from Eqs. (5).

Solution for Part (b)

Governing Equations

Balance Principles The FBD of the crate is shown in Fig. 4, in which the crate has been modeled as a particle subject to its own weight and the tension F_H in the hook. Using the FBD in Fig. 4, Newton's second law yields

$$\sum F_y\text{:}\quad F_H - mg = ma_y. \tag{8}$$

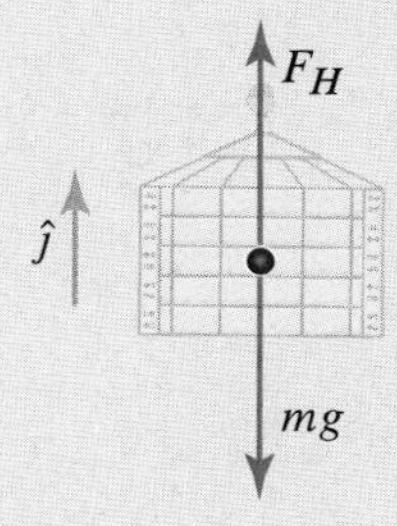

Figure 4
FBD of the crate.

Force Laws All forces are accounted for on the FBD.

Kinematic Equations Since the crate is moving with constant velocity, we have

$$a_y = 0. \tag{9}$$

Computation Substituting Eq. (9) into Eq. (8) and solving for F_H, we have

$$F_H = mg. \tag{10}$$

Again, we need to compute the power provided by the motor to the pulley system. Recalling that P_o is the power developed by the tension F_H, and that F_H is in the same direction as the velocity of the crate, we then must have

$$P_o = F_H\,\hat{\jmath} \cdot v_c\,\hat{\jmath} = F_H v_c, \tag{11}$$

where v_c is the speed of the crate. Solving Eq. (11) for v_c, we have

$$\boxed{v_c = \frac{P_o}{F_H} = \frac{P_o}{mg} = 3.249\,\text{m/s},} \tag{12}$$

where we have used P_o from Eqs. (5).

Discussion & Verification In Parts (a) and (b), P_o has dimensions of force times length divided by time, so the results in Eqs. (7) and (12) have the expected dimensions of length over time. Also, given that the data was provided in SI units, the results are expressed using appropriate units. As far as the value of v_c is concerned, as expected, it is smaller than the maximum possible value of lifting speed given by $P_i/mg = 3.823\,\text{m/s}$. Hence, our solution appears to be correct.

A Closer Look Notice that in Part (a) of this example, we had to analyze the pulley kinematics, but we did not have to do so in Part (b). That is because in pulley systems, the power supplied determines the speed of the payload, regardless of the pulley arrangement. This stems from the fact that power is the product of force and velocity, and the nature of pulley systems is to keep that product constant.

Problems

Figure P4.97

Problem 4.97

Consider a battery that can power a machine whose operation requires 200 W of power. If the battery can keep the machine operating for 8 h and supply energy at a constant rate, determine the amount of energy initially stored in the battery pack.

Problem 4.98

A person takes 16 s to lift a 200 lb crate to a height of 30 ft. Determine the average power supplied by the person.

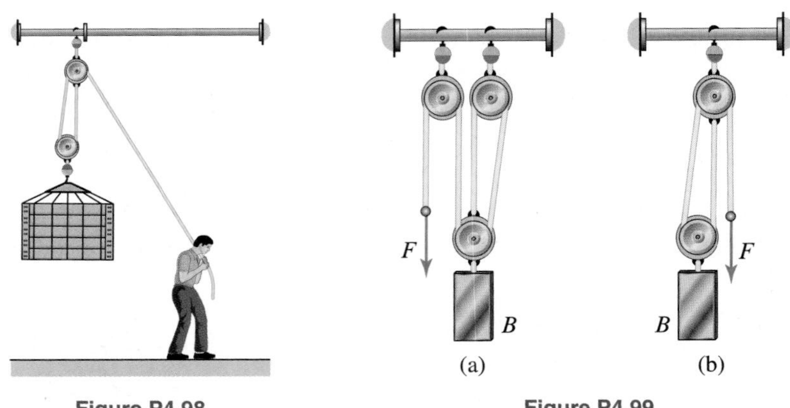

Figure P4.98 **Figure P4.99**

Problem 4.99

The weights B in the pulley systems shown are identical. Assume that the two systems are released from rest, there are no energy losses in the pulleys, and the same constant force F is applied. After 1 s, will the power developed by the force F for system (a) be smaller than, equal to, or greater than that of system (b)?

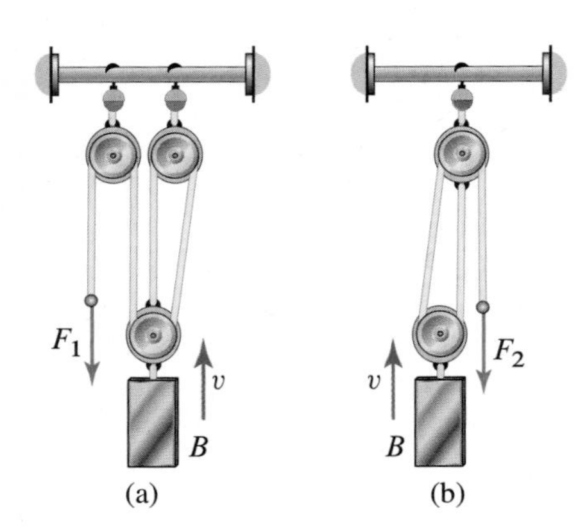

Figure P4.100 and P4.101

Problem 4.100

The weights B in the pulley systems shown are identical and are being lifted at the same constant speed v. Assume there are no energy losses in the pulleys. Is the power developed by force F_1 in (a) smaller than, equal to, or greater than that developed by force F_2 in (b)?

Problem 4.101

For the two pulley systems shown, the block B has a mass $m_B = 254\,\text{kg}$ and is being lifted with a constant speed $v = 3\,\text{m/s}$. Determine the power developed by the forces F_1 and F_2.

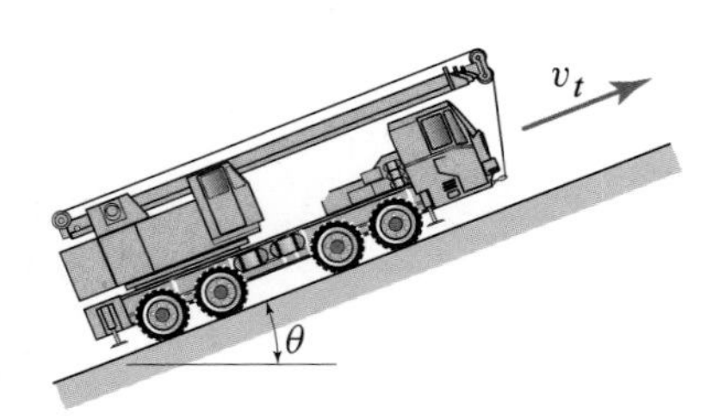

Figure P4.102

Problem 4.102

A truck crane with a mass of 40,000 kg has an engine with a rated power of 300 kW. Assuming that *all* power is transmitted to the wheels and that they do not slip, determine the maximum climb angle θ if the truck is to maintain a constant speed of $v_t = 25\,\text{km/h}$.

Problem 4.103

The 185 lb man climbs a distance $h = 13$ ft to the top of the ladder in 9 s. Determine his average power output.

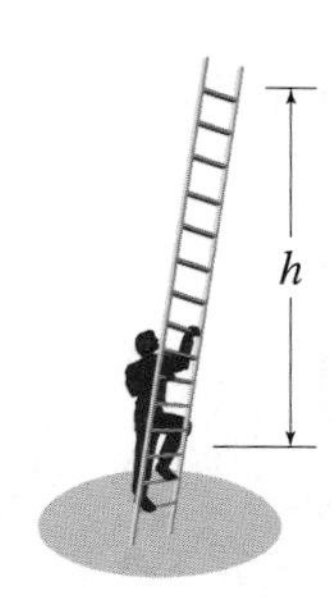

Figure P4.103

Problem 4.104

A cyclist is riding with a speed $v = 18$ mph over an inclined road with $\theta = 15°$. Neglecting aerodynamic drag, if the cyclist were to keep his power output constant, what speed would he attain if θ were equal to 20°?

Figure P4.104

Problem 4.105

A 1500 kg car is traveling up the slope shown at a constant speed. Knowing that the maximum power output of the car is 160 hp, at what speed can the car travel if air resistance is negligible? Also, knowing that 1 L of regular gasoline provides 34.8 MJ of energy, how many liters of gasoline will be required in 1 h if the engine has an efficiency $\epsilon = 0.20$?

Figure P4.105

Problem 4.106

A2600 lb car on a straight horizontal road goes from 0 to 60 mph in 7 s. Neglecting aerodynamic drag, if the force propelling the car is constant during the acceleration phase, determine the power developed by this force 7 s after the beginning of the motion.

Figure P4.106

Problem 4.107

The height h of the aerobic stepper shown is 25 cm. A 65 kg person goes up and down the step once every 2 s. Accounting only for the work done in lifting her body and assuming a muscle efficiency of 25%, determine the calories (1 C = 4.184 kJ) burned by the person in 1 h.

Figure P4.107

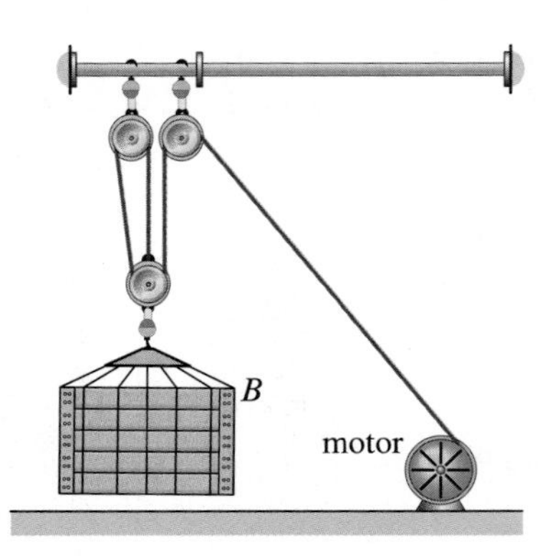

Figure P4.108

Problem 4.108

The motor powering the pulley system shown is drawing 6 kW of power. The crate B has a mass of 250 kg. Determine the speed of the crate if the crate is lifted at a constant speed and the motor's efficiency is $\epsilon = 0.87$.

Problems 4.109 and 4.110

The motor B is used to raise and lower the crate C by a pulley system. At the instant shown, the cable is being retracted by the motor with the constant speed $v_c = 5$ ft/s. The weight of the crate is $W_C = 450$ lb.

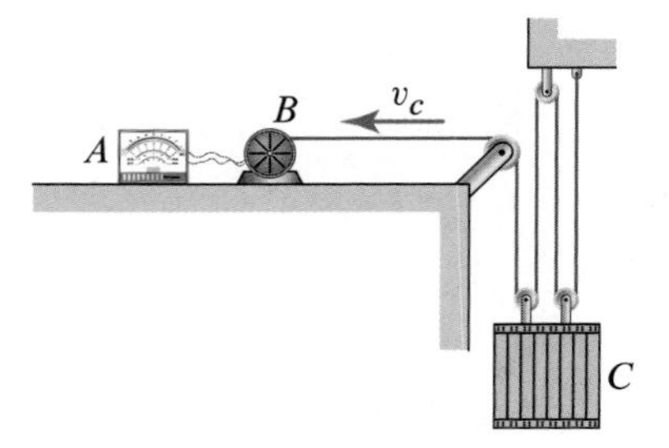

Figure P4.109 and P4.110

Problem 4.109 If the power meter A shows a power input to the motor of 1.36 hp, determine the overall efficiency of the system.

Problem 4.110 If the overall efficiency of the system is 0.82, determine the power input to the motor that the meter at A would read.

Problem 4.111

Assuming that the motor shown has an efficiency $\epsilon = 0.80$, determine the power to be supplied to the motor if it is to pull a 250 lb crate up the incline with a constant speed $v = 7$ ft/s. Assume that the kinetic friction coefficient between the slide and the crate is $\mu_k = 0.25$ and that $\theta = 28°$.

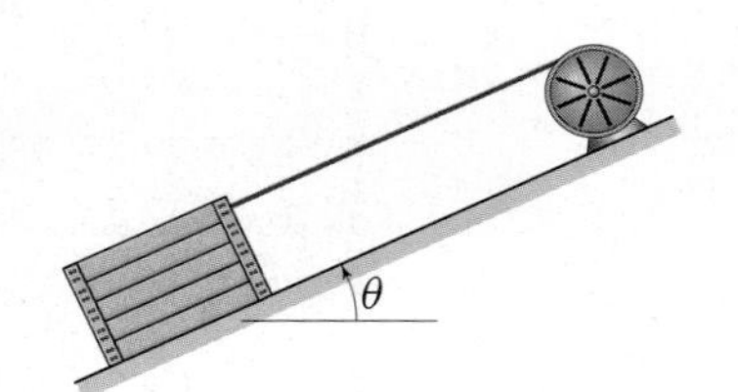

Figure P4.111 and P4.112

Problem 4.112

Assume that the motor shown has an efficiency $\epsilon = 0.82$ and that it draws 6 hp. Determine the constant speed with which the motor can pull a 300 lb crate up the incline if the kinetic friction coefficient between the slide and the crate is $\mu_k = 0.45$ and $\theta = 32°$.

Chapter Review

In this chapter we introduced the concepts of work of a force, kinetic energy of a particle, and kinetic energy of a particle system. We showed that the work done on a system is equal to the change in kinetic energy of the system. This relationship is referred to as the work-energy principle, and it is a direct consequence of Newton's second law. Finally, we introduced the concepts of power developed by a force and efficiency of a motor or a machine.

Work-energy principle for a particle

The *work-energy principle* is expressed by the following equation:

Eq. (4.9), p. 242

$$T_1 + U_{1\text{-}2} = T_2,$$

where $U_{1\text{-}2}$ is the *work* done by the force $\vec{F}$ (see Fig. 4.15) and is defined as

Eq. (4.7), p. 242

$$U_{1\text{-}2} = \int_{\mathcal{L}_{1\text{-}2}} \vec{F} \cdot d\vec{r},$$

and T is the particle's *kinetic energy*, which is defined as

Eq. (4.8), p. 242

$$T = \tfrac{1}{2}mv^2,$$

where m is the mass of the particle and v is its speed. Work and kinetic energy have the following basic properties:

- Work and kinetic energy are *scalar* quantities.
- The work depends on only the component of $\vec{F}$ in the direction of motion, that is, on $\vec{F} \cdot \hat{u}_t$, and its sign is determined by the sign of $\vec{F} \cdot \hat{u}_t$.
- The kinetic energy is *never* negative.

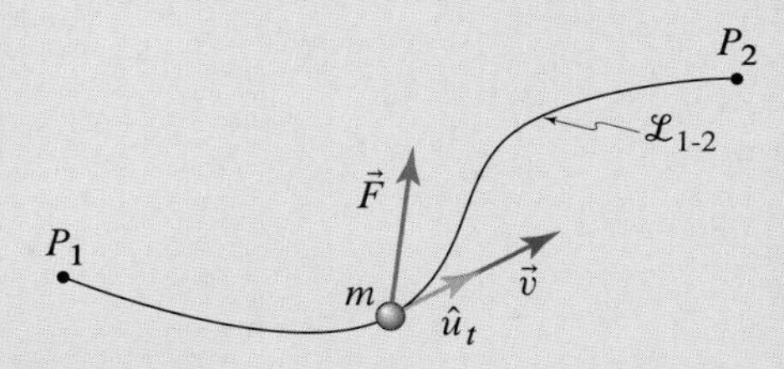

Figure 4.15
Particle of mass m moving along the path $\mathcal{L}_{1\text{-}2}$ while being subjected to the force $\vec{F}$.

Helpful Information

Other forms of the work of a force. By recalling that position and velocity are related by the expression $d\vec{r} = \vec{v}\,dt$ and that, in normal-tangential components, the velocity can be expressed as $\vec{v} = v\,\hat{u}_t$, we can obtain the following two useful forms of the work of a force:

$$U_{1\text{-}2} = \int_{t_1}^{t_2} \vec{F} \cdot \vec{v}\,dt = \int_{s_1}^{s_2} \vec{F} \cdot \hat{u}_t\,ds,$$

where t is time, s is the arc length along the path, and $\hat{u}_t$ is the unit vector tangent to the path and pointing in the direction of motion.

Conservative forces and potential energy

In Section 4.2, we discovered that there are forces whose work, in taking a particle from an initial to a final position, does not depend on the path connecting these positions. Specifically, we derived the following expressions:

Work of a constant gravitational force. For a particle of mass m moving in a constant gravitational field from ① to ②, the work done on the particle is

Eq. (4.13), p. 257

$$(U_{1\text{-}2})_g = -mg(y_2 - y_1),$$

for which gravity must act in the $-y$ direction.

Work done by a spring force. For a particle subject to the force of a linear elastic spring with constant k, the work done on the particle is

Eq. (4.17), p. 258

$$(U_{1\text{-}2})_e = -\tfrac{1}{2}k\left(\delta_2^2 - \delta_1^2\right),$$

where δ_1 and δ_2 are the amount the spring is stretched (or compressed) at ① and ②, respectively.

Conservative systems. A force is said to be *conservative* if its work depends on only the initial and final positions of its point of application, but is otherwise independent of the path connecting these positions. The work of a conservative force can be characterized by a scalar potential energy function. A *conservative system* is one for which all the forces doing work are conservative. For conservative systems, the work-energy principle becomes the *conservation of mechanical energy*, which is given by

Eq. (4.22), p. 259

$$T_1 + V_1 = T_2 + V_2.$$

Important conservative forces include elastic spring forces and gravitational forces (constant or not). The *potential energy of a linear elastic spring force* is given by

Concept Alert

Elastic potential energy is never negative. Whether a spring is elongated or compressed, i.e., whether δ is positive or negative, the potential energy of a spring is never negative!

Eq. (4.24), p. 259

$$V_e = \tfrac{1}{2}k\delta^2.$$

The *potential energy of a constant gravitational force* is given by

Eq. (4.23), p. 259

$$V_g = mgy,$$

where the positive y direction is opposite to gravity. The *potential energy of the force of gravity* is given by

Eq. (4.25), p. 259

$$V_G = -\frac{Gm_A m_B}{r}.$$

In a system for which there are both conservative and nonconservative forces doing work, we can use Eq. (4.9) on p. 242 or we can take advantage of the potential energies of the conservative forces by writing the work-energy principle as

Eq. (4.27), p. 260

$$T_1 + V_1 + (U_{1\text{-}2})_{\text{nc}} = T_2 + V_2,$$

where $(U_{1\text{-}2})_{\text{nc}}$ is the work done by nonconservative forces.

Work-energy principle for systems of particles

In Section 4.3, we developed the work-energy principle for a system of particles, and we discovered that internal forces play an important role in writing this principle. We defined the kinetic energy of a particle system as the sum of the individual kinetic energies of each particle, that is,

Eq. (4.40), p. 281

$$T = \sum_{i=1}^{n} \tfrac{1}{2} m_i v_i^2,$$

where m_i and v_i are the mass and the speed of the ith particle in the system. The work-energy principle for a system of particles was found to be

Eq. (4.45), p. 282

$$T_1 + V_1 + (U_{1\text{-}2})_{\text{nc}}^{\text{ext}} + (U_{1\text{-}2})_{\text{nc}}^{\text{int}} = T_2 + V_2,$$

where V is the potential energy of the system, which may have contributions from both *external* and *internal* forces.

The kinetic energy of a system can also be expressed in terms of the speed of the mass center and the speed of each of the particles relative to the mass center. Specifically, we have

Eq. (4.50), p. 283

$$T = \tfrac{1}{2} m v_G^2 + \tfrac{1}{2} \sum_{i=1}^{n} m_i v_{i/G}^2.$$

Common Pitfall

Internal work due to nonconservative forces. Don't forget that *internal work* due to nonconservative forces needs to be accounted for in the $(U_{1\text{-}2})_{\text{nc}}^{\text{int}}$ term when we write the work-energy principle. The most important example of this type of work is due to internal friction.

Common Pitfall

Internal work due to conservative forces. Don't forget that *internal work* due to conservative forces (e.g., springs) *is not* included in the $(U_{1\text{-}2})_{\text{nc}}^{\text{int}}$ term — it is accounted for by the potential energy V, where, for a system, the potential energy V includes the potential energies of both the external and the internal conservative forces.

Power and efficiency

In Section 4.4, we defined the *power P developed by a force $\vec{F}$* whose point of application moves with a velocity $\vec{v}$ as

Eq. (4.54), p. 298

$$P = \vec{F} \cdot \vec{v},$$

and we observed that power has units of work per unit time.

In addition to power, Section 4.4 introduced the notion of *efficiency*, which is an important concept used to assess the performance of motors and/or machines. Given a machine or a motor whose operation requires a certain power input, efficiency, usually denoted by the Greek letter ϵ (epsilon), is defined as

Eq. (4.57), p. 298

$$\epsilon = \frac{\text{power output}}{\text{power input}},$$

where power output is the work done by the machine or motor per unit time.

Review Problems

Problem 4.113

Rubber bumpers are commonly used in marine applications to keep boats and ships from getting damaged by docks. Consider a boat with a mass of 35,000 kg and a bumper with a force compression profile given by $F_B = \beta x^3$, where β is a constant and x is the compression of the bumper. Treating the boat C as a particle, neglecting its vertical motion, and neglecting the drag force between the water and the boat C, determine β so that if the boat were to impact the bumper with a speed of 4 m/s, the maximum compression of the bumper would be 18 cm.

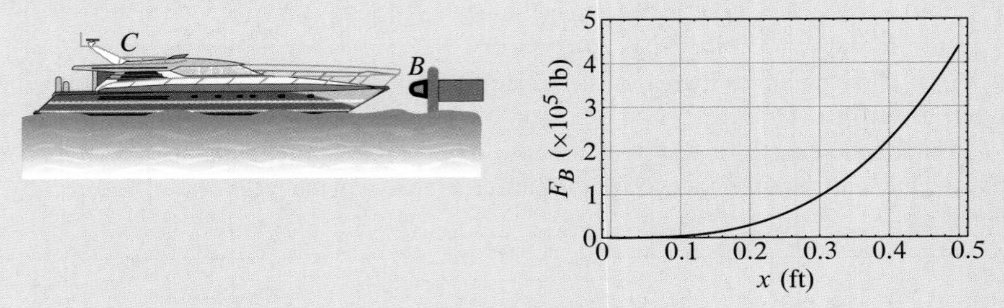

Figure P4.113

Problems 4.114 through 4.116

The crates A and B of mass $m_A = 50$ kg and $m_B = 75$ kg, respectively, are connected by a pulley system. The system is released from rest on the inclined surfaces for which $\theta = 30°$. Friction between A and the inclined surface on which it slides is insufficient to prevent slipping, and friction between B and the incline on which it slides is negligible. The cables in the pulley system are inextensible, and the coefficient of kinetic friction between crate A and the horizontal surface is μ_k.

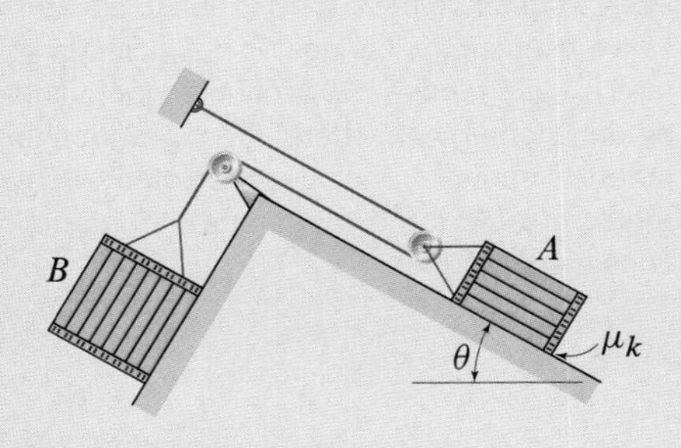

Figure P4.114–P4.116

Problem 4.114 Letting $\mu_k = 0$, determine the speed of B after A slides 5 m up the incline.

Problem 4.115 Determine the required value of μ_k so that the speed of B is 7 m/s after it slides 5 m down the incline.

Problem 4.116 Letting $\mu_k = 0.25$, determine the speed of A after B slides 5 m down the incline.

Problem 4.117

Strength of materials tells us that if a load P is applied at the free end of a cantilevered beam, then the tip displacement δ is given by $\delta = PL^3/(3EI_{cs})$, where L is the length of the beam and E and I_{cs} are constants that depend on the material makeup and the geometry of the cross section, respectively. Determine an expression for the potential energy of a cantilevered beam loaded as shown.

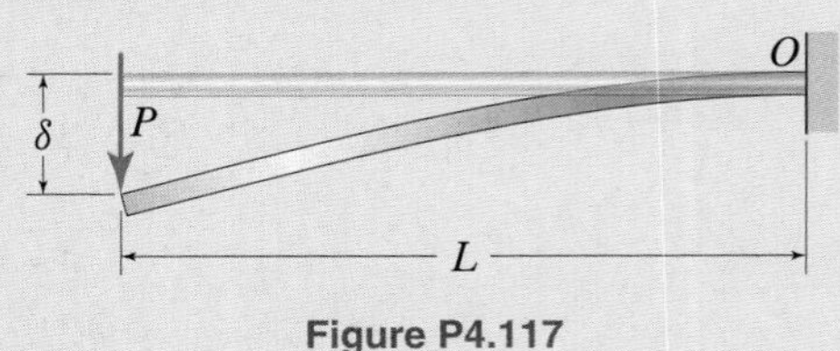

Figure P4.117

Problem 4.118

Starting from the position shown, each horse moves to the right in such a way that the tension in the cord is the same in cases (a) and (b) and remains constant. Knowing that $\beta < \gamma$, and that in both cases (a) and (b), the horse advances by an equal amount L, determine which of the following statements is true: (1) the tension in the cord does more work in (a) than in (b); (2) the tension in the cord does exactly the same amount of work in (a) as in (b); (3) the tension in the cord does less work in (a) than in (b).

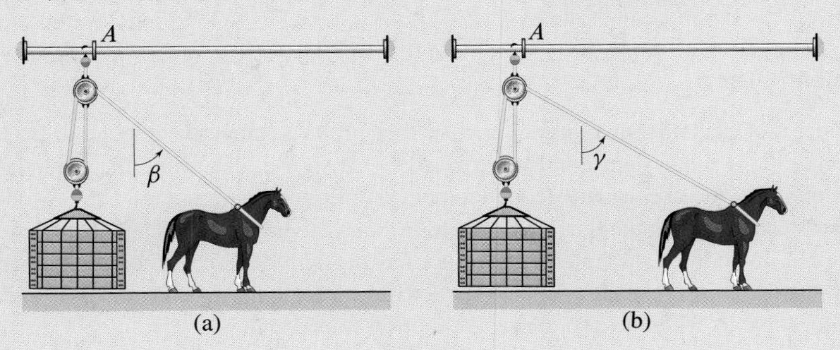

Figure P4.118

Problem 4.119

A metal ball with mass $m = 0.15\,\text{kg}$ is released from rest in a fluid. The magnitude of the resistance due to the fluid is given by $C_d v$, where C_d is a drag coefficient and v is the ball's speed. If $C_d = 2.1\,\text{kg/s}$, determine the total work done on the ball from the moment of release until the ball achieves 99% of terminal velocity.

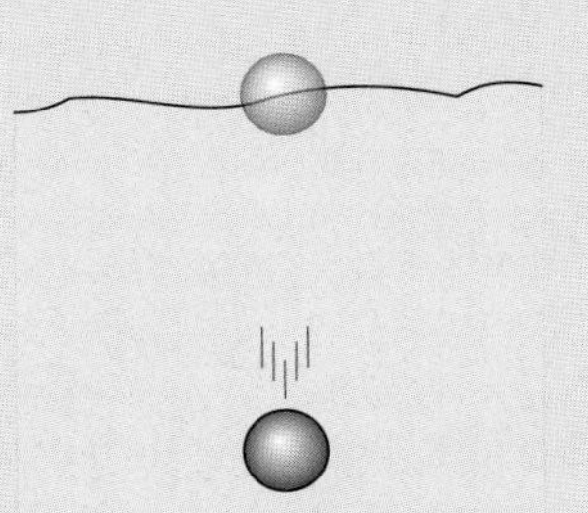

Figure P4.119 and P4.120

Problem 4.120

A metal ball weighing 0.2 lb is released from rest in a fluid. If the magnitude of the resistance due to the fluid is given by $C_d v$, where $C_d = 0.5\,\text{lb·s/ft}$ is a drag coefficient and v is the ball's speed, determine the work done by the drag force during the first 2 s of the ball's motion.

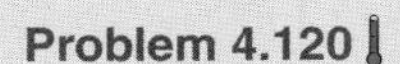

Problem 4.121

A 7 lb collar is constrained to travel along a frictionless vertical ring of radius $R = 1\,\text{ft}$. The spring attached to the collar has a spring constant $k = 20\,\text{lb/ft}$. Treating the collar as a particle, neglecting air resistance, and knowing that, while at rest at A, the collar is displaced gently to the left, determine the spring's unstretched length if the collar is to reach point B with a speed of 15 ft/s.

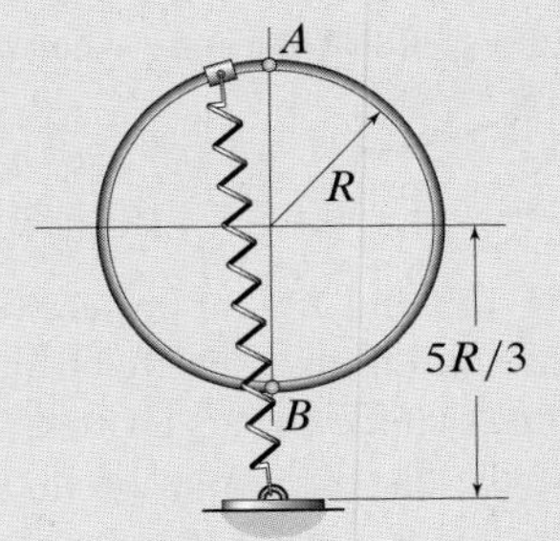

Figure P4.121

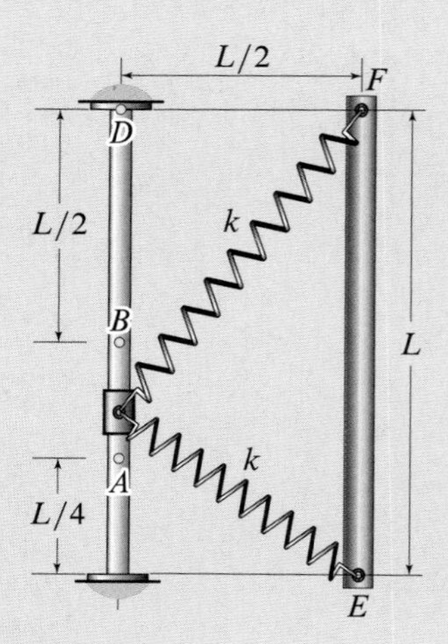

Figure P4.122

Problem 4.122

An 11 kg collar is constrained to travel along a rectilinear and frictionless bar of length $L = 2$ m that lies in the vertical plane. The springs attached to the collar are identical, and they are unstretched when the collar is at B. Treating the collar as a particle, neglecting air resistance, and knowing that at A the collar is moving upward with a speed of 23 m/s, determine the spring constant k so that the collar reaches D with zero speed. Points E and F are fixed.

Problem 4.123

Consider the catapult shown in the figure with a 1200 kg counterweight A and a 330 kg projectile B. If the system is released from rest as shown, determine the speed of the projectile after the arm rotates (counterclockwise) through an angle of 110°. Model A and B as particles; assume that the catapult's arm has negligible mass and that friction is negligible. In addition, assume that the cord has negligible mass, is inextensible, and is always vertical. The catapult's frame is fixed with respect to the ground, and the projectile does not separate from the arm during the motion considered.

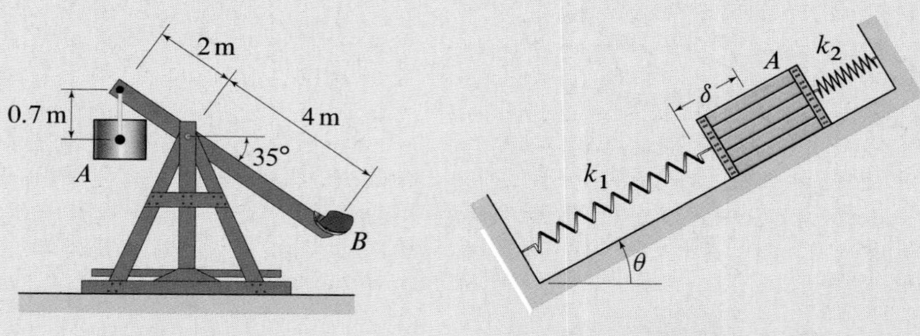

Figure P4.123

Figure P4.124 and P4.125

Problems 4.124 and 4.125

The crate A weighs 100 lb and is attached to the springs with constants $k_1 = 150$ lb/ft and $k_2 = 300$ lb/ft. Both springs are unstretched when $\delta = 0$ and the box is centered between the two walls. The angle of the incline is $\theta = 30°$.

Problem 4.124 If the crate is released from rest after being displaced a distance $\delta = 6$ ft up the incline, and friction between the crate and the inclined surface on which it slides is negligible, determine the speed of the crate when it returns to $\delta = 0$.

Problem 4.125 If the crate is released from rest after being displaced a distance $\delta = 6$ ft and the coefficient of kinetic friction between the crate and the surface on which it slides is $\mu_k = 0.2$, determine the speed of the crate when $\delta = 0$.

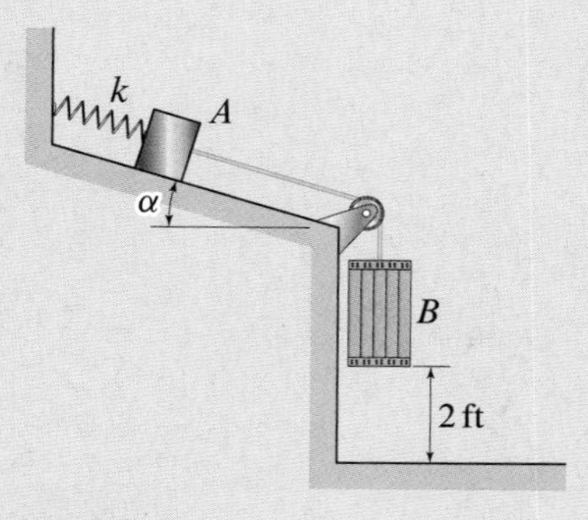

Figure P4.126

Problem 4.126

Two blocks A and B weighing 123 and 234 lb, respectively, are released from rest as shown. At the moment of release the spring is unstretched. Model A and B as particles, neglect air resistance, and assume that the cord is inextensible. Determine the maximum speed attained by block B and the distance from the floor where the maximum speed is achieved if $\alpha = 35°$, the static and kinetic friction coefficients are $\mu_s = 0.25$ and $\mu_k = 0.2$, respectively, and the spring constant is $k = 25$ lb/ft.

Problem 4.127

Spring scales work by measuring the displacement of a spring that supports both the platform of mass m_p and the object, of mass m, whose weight is being measured. In your solution, note that most scales read zero when no mass m has been placed on them; that is, they are calibrated so that the weight reading neglects the mass of the platform m_p. Assume that the spring is linear elastic with spring constant k. If the mass m is gently placed on the spring scale (i.e., it is released from zero height above the scale), determine the *maximum velocity* attained by the mass m as the spring compresses.

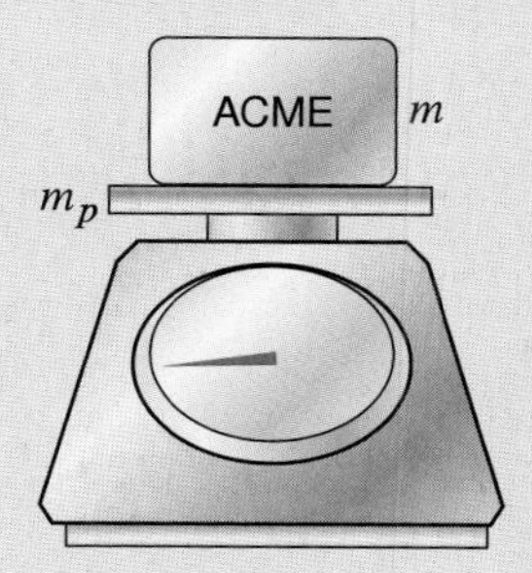

Figure P4.127

Problem 4.128

A cyclist is riding with a constant speed $v = 25\,\text{km/h}$ over an inclined road with $\theta = 15°$. The combined mass of the cyclist and bicycle is 85 kg. Neglecting aerodynamic drag, determine the number of calories (1 C = 4.184 kJ) burned by the cyclist in the course of 15 min if his muscle efficiency is 25%.

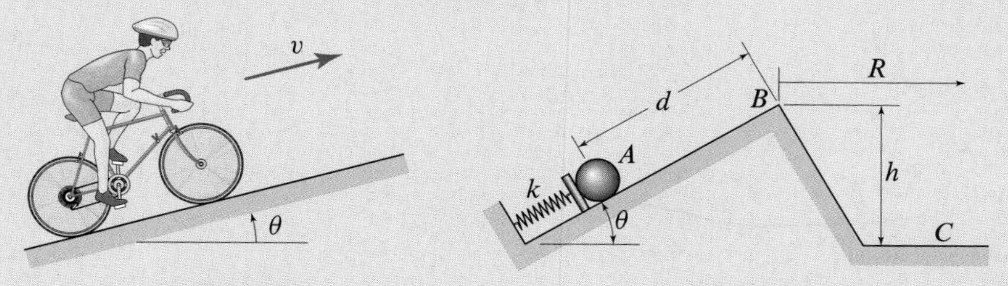

Figure P4.128 Figure P4.129

Problem 4.129

The mass A is pressed against the linear elastic spring with stiffness $k = 500\,\text{N/m}$ such that the spring compresses a distance $\delta = 15\,\text{cm}$. The mass is then released in the position shown and slides a distance $d = 1.2\,\text{m}$ up the incline. The mass leaves the incline at B and becomes a projectile. The angle of the incline is $\theta = 30°$, and the height $h = 0.9\,\text{m}$. Neglecting friction between A and the incline while A is on the incline, determine the range R where the mass first lands on the horizontal surface C. A has mass $m = 0.25\,\text{kg}$.

Problem 4.130

For the two pulley systems shown, the same block B, weighing 150 lb, is being lifted by applying the same constant force $F = 80\,\text{lb}$. If the two systems are released from rest, determine the power developed by the force F in the two cases 1 s after release.

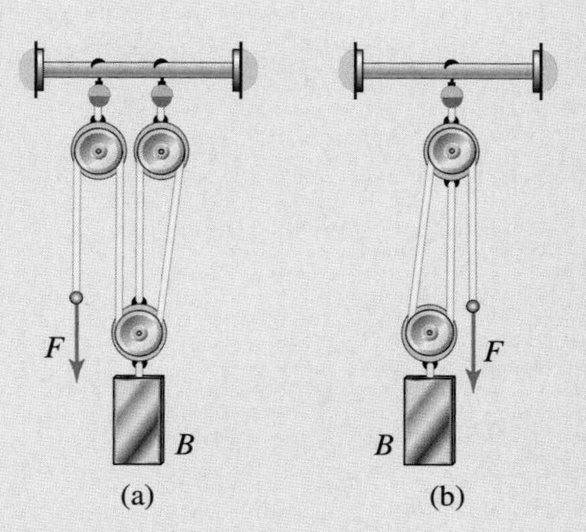

Figure P4.130

Momentum Methods for Particles

The chapter begins by discussing the concepts of momentum and impulse. We will derive a new balance law, the impulse-momentum principle, that *integrates* Newton's second law for a particle with respect to time. This balance law will be essential for the study of impacts.

Next, we will study a new concept that turns out to be fundamental to mechanics — the concept of angular momentum. We will see not only that the concept of angular momentum is useful for solving problems like those found in orbital mechanics, but also that it is fundamental in extending the applicability of Newton's laws to rigid bodies. Finally, we close with the study of mass flows, which is again a topic for which the idea of momentum will be essential.

The Space Shuttle *Atlantis* taking off from launch pad 39A at the Kennedy Space Center. The acceleration of the Shuttle due to the thrust of its rocket engines can be computed starting with the impulse-momentum principle.

5.1 Momentum and Impulse

Impulse-momentum principle

We begin by looking at how a change in velocity is related to the time interval over which that change takes place. To do this, we integrate Newton's second law with respect to time over a time interval $t_1 \leq t \leq t_2$ for a particle of mass m:

$$\int_{t_1}^{t_2} \vec{F}\, dt = \int_{t_1}^{t_2} m\vec{a}\, dt. \tag{5.1}$$

Since $\vec{a}\, dt = d\vec{v}$ (and m, for a particle, is constant), we obtain

$$\int_{t_1}^{t_2} \vec{F}\, dt = m\vec{v}(t_2) - m\vec{v}(t_1). \tag{5.2}$$

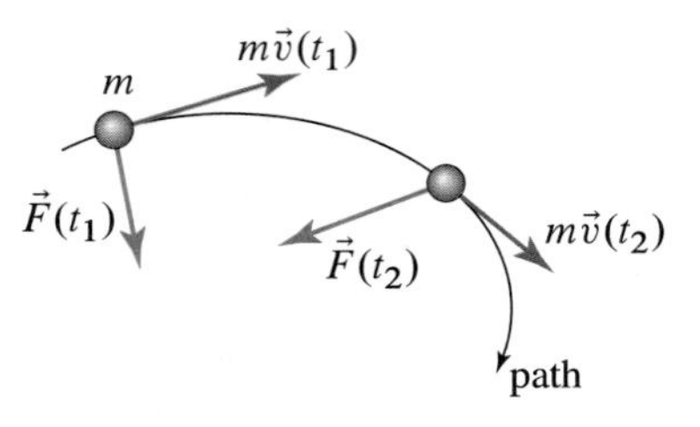

Figure 5.1
A force $\vec{F}(t)$ on the mass m acting over a time interval changes the quantity $m\vec{v}$.

Equation (5.2) states that a change in the quantity $m\vec{v}$ over a certain time interval is related to the *time integral* of the force over that time interval. To better understand the meaning of Eq. (5.2) and to see how it can help us deal with phenomena such as collisions, we will introduce some basic terminology.

Linear momentum of a particle

Since the quantity $m\vec{v}$ plays a prominent role in Eq. (5.2), we give this quantity its own name and symbol. Let the vector quantity $\vec{p}(t)$ be defined as

$$\boxed{\vec{p}(t) = m\vec{v}(t).} \tag{5.3}$$

The quantity $\vec{p}(t)$ is called *linear momentum* or simply *momentum* of the particle of mass m. By definition, the momentum has dimensions of mass times length *divided by* time:

$$[\vec{p}] = [M]\,[L]\,[T]^{-1}. \tag{5.4}$$

Consequently, the units in the SI system are kg·m/s and in the U.S. Customary system are lb·s or slug·ft/s.

Linear impulse of a force

The quantity on the left-hand side of Eq. (5.2), that is,

$$\boxed{\int_{t_1}^{t_2} \vec{F}(t)\,dt,} \tag{5.5}$$

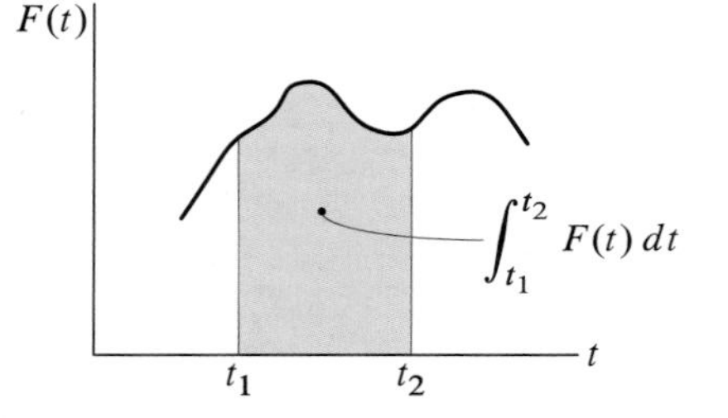

Figure 5.2
Geometric interpretation of the concept of impulse of a force.

is called the *linear impulse* (or simply *impulse*) *of the force* $\vec{F}(t)$ *between times* t_1 *and* t_2. Because it is an integral over time, any component of the impulse can be given a graphical interpretation as the *area under the curve* of a force versus time plot, as shown in Fig. 5.2.

The impulse has dimensions of force times time or mass times length divided by time. Therefore, the units of impulse are the same as for momentum. However, it is common to express the impulse using units of N·s in the SI system and lb·s in the U.S. Customary system.

Impulse-momentum principle for a particle

Using the definition of momentum, Eq. (5.2) can be written as

$$\boxed{\int_{t_1}^{t_2} \vec{F}(t)\,dt = \vec{p}(t_2) - \vec{p}(t_1) \quad \text{or} \quad \vec{p}(t_1) + \int_{t_1}^{t_2} \vec{F}(t)\,dt = \vec{p}(t_2).} \tag{5.6}$$

Equation (5.6) states that *a particle's change in momentum over an interval of time is equal to the impulse imparted to that particle during that same time interval*. The expression in Eq. (5.6) is called the *impulse-momentum principle*.

Equation (5.6) can be written in differential form as

$$\boxed{\vec{F} = \dot{\vec{p}},} \tag{5.7}$$

which can be viewed as a restatement of Newton's second law.

Average force

Knowing the change of momentum of a particle of mass m between two instants t_1 and t_2 is, in general, not enough to allow the reconstruction of the force $\vec{F}(t)$ for every instant in time between t_1 and t_2. However, we can determine the *average force* that caused the change of momentum in question. Figure 5.3 shows a time-varying force and its average over the same time interval. By definition, the average force over the time interval $t_1 \leq t \leq t_2$ is

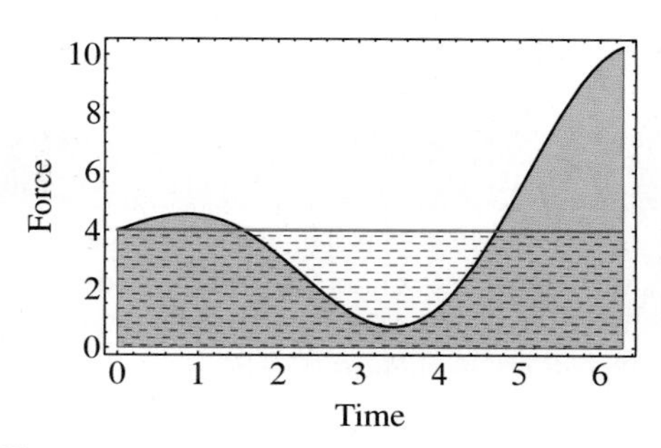

Figure 5.3
A force (the black line) and its average (the red line) over a certain time interval. The area under the force curve (light brown) equals the area under the average force curve (dashed area).

$$\vec{F}_{\text{avg}} = \frac{\int_{t_1}^{t_2} \vec{F}(t)\, dt}{t_2 - t_1}. \tag{5.8}$$

Therefore, using Eqs. (5.6), the average force is

$$\vec{F}_{\text{avg}} = \frac{\vec{p}(t_2) - \vec{p}(t_1)}{t_2 - t_1}. \tag{5.9}$$

The ability to calculate the average force needed to accomplish a change in momentum is useful. Since the magnitude of the actual force must be at least as large as the magnitude of the average force during some part of the time interval, the magnitude of the average force provides a lower bound for the magnitude of the maximum force.

Impulse-momentum principle for systems of particles

To extend the impulse-momentum principle to a system of particles, we need to distinguish between closed and open systems.

Closed systems are systems that do not exchange mass with their surroundings, whereas *open systems* do exchange mass with their surroundings. The mass of a closed system is necessarily constant, whereas an open system can have constant or variable mass. The distinction between these two types of systems becomes important when we talk about systems of particles and, in particular, about *mass flows* and *variable mass systems*, which are open by definition (see Section 5.5).

We consider a closed system of N particles, and we view the total force on each particle as the sum of two parts (see Fig. 5.4):

1. an *external* force $\vec{F}_i$ due to the interaction of particle i with the physical bodies that do not belong to the system,
2. an *internal* force $\sum_{j=1}^{N} \vec{f}_{ij}$ due to the interaction between particle i and all of the other particles in the system ($\vec{f}_{ii} = \vec{0}$ since a particle does not exert a force on itself).

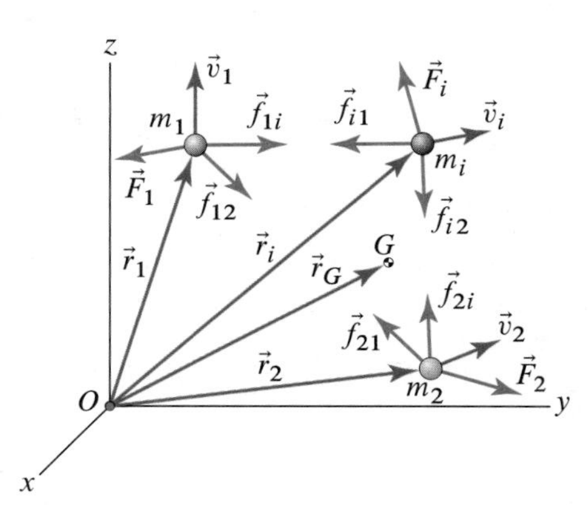

Figure 5.4
A system of particles under the action of internal and external forces. The mass center is at G.

Applying Eq. (5.7), we have

$$\vec{F}_i + \sum_{j=1}^{N} \vec{f}_{ij} = \dot{\vec{p}}_i, \tag{5.10}$$

where $\vec{p}_i = m_i \vec{v}_i$ is the momentum of particle i. Equation (5.10) represents N equations, one for each particle. Let

$$\vec{F} = \sum_{i=1}^{N} \vec{F}_i \quad \text{and} \quad \vec{p} = \sum_{i=1}^{N} \vec{p}_i, \tag{5.11}$$

be the total external force acting on and the total momentum of the system of particles, respectively. Recalling that Newton's third law requires that $\sum_{i=1}^{N}\left(\sum_{j=1}^{N}\vec{f}_{ij}\right)=\vec{0}$, we see that the sum of all N of Eq. (5.10), along with Eqs. (5.11), gives

$$\boxed{\vec{F}=\dot{\vec{p}}.} \tag{5.12}$$

Concept Alert

Internal forces do not change momentum. Internal forces cannot change the momentum of a system. This statement is a fundamental axiom of mechanics.

We encountered this relationship in Eq. (3.37) on p. 218. Equation (5.12) states that internal forces play no role in changing the total momentum of a particle system. In addition, using the definition of center of mass of a closed system of particles, we have

$$\vec{p}=\sum_{i=1}^{N}\vec{p}_i=\sum_{i=1}^{N}m_i\vec{v}_i=m\vec{v}_G, \tag{5.13}$$

where m is the total mass of the system and $\vec{v}_G$ is the velocity of the mass center of the system. Using Eq. (5.13), we can write Eq. (5.12) as

$$\boxed{\vec{F}=\frac{d}{dt}(m\vec{v}_G)=m\vec{a}_G,} \tag{5.14}$$

where we have used the fact that the mass of a closed system of particles is constant. Equation (5.14) states that we can view a system of particles as a single particle, whose mass is equal to the total mass of the system of particles, moving with the mass center of the system of particles.

Newton's second law only applies to a single particle. Leonhard Euler postulated that Eq. (5.12) holds for *any* body whose mass is constant and for any internal force system. For this reason, Eq. (5.12) is generally considered to be a more fundamental axiom governing the motion of bodies than Newton's second law, and it is referred to as *Euler's first law*.

As argued earlier in this section, to directly investigate the relationships between changes in momentum and the applied forces, we can integrate both sides of Eq. (5.12) with respect to time, which yields

$$\boxed{\int_{t_1}^{t_2}\vec{F}(t)\,dt=\vec{p}(t_2)-\vec{p}(t_1),} \tag{5.15}$$

where the left-hand side of Eq. (5.15) is the *total external impulse exerted on the system between t_1 and t_2*. Even though Eqs. (5.6) and (5.15) are visually identical to one another, it is important to keep in mind that the definitions of $\vec{F}$ and $\vec{p}$ are different in the two equations.

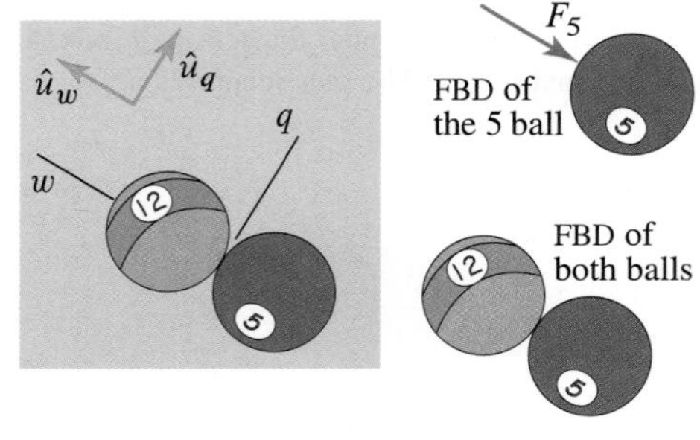

Figure 5.5
Two billiard balls hitting one another, as well as the FBD of the 5 ball and the FBD of both balls during the impact.

Conservation of linear momentum

Figure 5.5 shows two billiard balls hitting one another, as well as the FBD of the 5 ball during impact and the FBD of both balls during impact. Letting q be the line tangent to the impact surface and w the line perpendicular to that surface, if we ignore friction between the two surfaces during impact, then the only force in the qw plane acting on the 5 ball is directed along the w axis, and there is no net external force on the two balls taken together. If the impact begins at time t_1 and ends at time t_2, and if we define our system to consist of both balls, then during the interval $t_1 \le t \le t_2$, the FBD of both balls indicates that

$$\vec{F}_{\text{ext}}(t)=\vec{0}, \tag{5.16}$$

and therefore, by Eq. (5.15), the total momentum of the two-ball system is conserved between t_1 and t_2,

$$\vec{p}(t_1) = \vec{p}(t_2) \quad \Rightarrow \quad m_5\vec{v}_5(t_1) + m_{12}\vec{v}_{12}(t_1) = m_5\vec{v}_5(t_2) + m_{12}\vec{v}_{12}(t_2), \tag{5.17}$$

where we have used Eq. (5.13) and the numeric subscript indicates the ball. In Eq. (5.17), the impulse-momentum principle has become *conservation of momentum* (we saw conservation of energy in Section 4.1). The two billiard balls are an example of an *isolated system*, which is a system that does not exchange mass with its surroundings and which is not acted upon by external forces. *Momentum is always conserved for isolated systems.*

In other applications, including impact problems like the billiard balls, it is common to have a fixed direction in which the force is zero. For example, if we now define the system to consist of *only the* 5 *ball*, for $t_1 \le t \le t_2$, the upper FBD in Fig. 5.5 indicates that

$$\vec{F}_5(t) \cdot \hat{u}_q = 0 \quad \Rightarrow \quad F_{5q} = 0. \tag{5.18}$$

Since the q component of the force on the 5 ball is zero, the q component of Eq. (5.6) indicates that the q component of momentum is conserved for the 5 ball, that is,

$$p_{5q}(t_1) = p_{5q}(t_2) \;\Rightarrow\; m_5 v_{5q}(t_1) = m_5 v_{5q}(t_2) \;\Rightarrow\; v_{5q}(t_1) = v_{5q}(t_2), \tag{5.19}$$

during the impact. Similarly, the momentum and velocity of the 12 ball are conserved in the q direction. The last of Eqs. (5.19) states that, as long as the total external force has a *component* that is zero throughout the time interval, the corresponding *component* of the momentum remains constant. This holds for systems, as well as for individual particles.

Concept Alert

Motion of a system with zero external forces. Since a system's total momentum can be expressed as the product of the system's total mass and the velocity of the system's center of mass, we can conclude that the center of mass of a system with zero external forces is either at rest or moving with a constant velocity.

End of Section Summary

We learned in Chapter 3 that forces lead to changes in velocities since forces cause accelerations. We began this section by learning that forces acting over time change momentum (not just velocity). By integrating Newton's second law, we obtained the *impulse-momentum principle*, which is given by

Eq. (5.6), p. 314

$$\vec{p}(t_1) + \int_{t_1}^{t_2} \vec{F}(t)\,dt = \vec{p}(t_2),$$

where the *linear momentum* (or *momentum*) was defined to be

Eq. (5.3), p. 314

$$\vec{p}(t) = m\vec{v}(t),$$

and a force acting over some time interval was called the *impulse* (or *linear impulse*) and is given by

Eq. (5.5), p. 314

$$\int_{t_1}^{t_2} \vec{F}(t)\,dt.$$

Helpful Information

When should we use the impulse-momentum principle? The impulse-momentum principle provides a natural approach to problems in which we need to relate velocity, force, and time since it relates forces acting over time to changes in momentum.

In addition, we found that without detailed knowledge of the force acting on a particle at every instant in time, we could not determine the change in momentum. On the other hand, knowing just the change in momentum allows us to determine the *average force* acting on a particle during the corresponding time interval, that is,

Eq. (5.9), p. 315

$$\vec{F}_{\text{avg}} = \frac{\vec{p}(t_2) - \vec{p}(t_1)}{t_2 - t_1}.$$

Impulse-momentum principle for systems of particles. When dealing with closed systems of particles, we discovered that we can write the impulse-momentum principle as

Eqs. (5.12) and (5.15), p. 316

$$\vec{F} = \dot{\vec{p}} \quad \text{and} \quad \int_{t_1}^{t_2} \vec{F}(t)\, dt = \vec{p}(t_2) - \vec{p}(t_1),$$

where $\vec{F}$ is the total external force on the particle system and $\vec{p} = \sum_{i=1}^{N} m_i \vec{v}_i$ is the total momentum of the system of particles. Using the definition of the mass center of a system of particles, the impulse-momentum principle can also be written as

Eq. (5.14), p. 316

$$\vec{F} = \frac{d}{dt}\left(m\vec{v}_G\right) = m\vec{a}_G,$$

where m is the total mass of the system of particles, $\vec{v}_G$ is the velocity of its mass center, and $\vec{a}_G$ is the acceleration of its mass center.

Conservation of linear momentum. When there is a direction in which the external force on a system of particles is zero, then the momentum in that direction is constant and is said to be conserved. If the total external force on a system of particles is zero, that is, $\vec{F} = 0$, then the momentum in every direction is constant, and the mass center of the system of particles will move with constant velocity.

EXAMPLE 5.1 *Average Force on an Aircraft Due to the Arresting Cable*

To successfully land on an aircraft carrier, a pilot must use the airplane's tailhook to catch one of four steel arresting cables stretched across the landing deck (see Fig. 1). The hydraulic system activated by the motion of the arresting cable can stop a 54,000 lb aircraft traveling at a speed of 150 mph in a little less than 2 s. Using this information, estimate the average braking force exerted on the plane by the arresting cable.

Figure 1
Military airplane landing on an aircraft carrier with a detailed view of the arresting cable.

SOLUTION

Road Map & Modeling We model the plane as a particle, and we will consider the time between when the tailhook first engages one of the arresting cables t_e and the time at which the plane comes to a stop t_s. Referring to Fig. 2, we only include the force applied by the arresting cable, gravity, and the contact force between the plane and the landing deck. Since we are given a *change of velocity* over a known *time interval* and we need to estimate the value of a force, we can apply the impulse-momentum principle. Since we want to determine the *average* braking force, which is the force that would act on the plane if this force were *constant*, we will apply the impulse-momentum principle under the assumption that the force exerted on the plane by the arresting cable is constant.

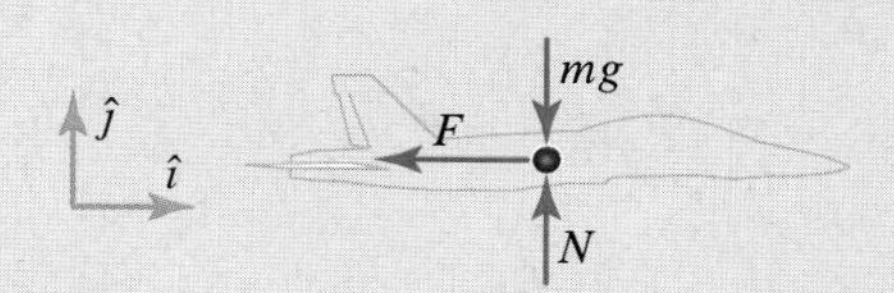

Figure 2
FBD of an airplane landing on a carrier. The force F is the force exerted by the tailhook.

Governing Equations

Balance Principles Applying Eq. (5.6) in component form to the FBD in Fig. 2, we obtain

$$x \text{ momentum:} \quad \int_{t_e}^{t_s} F(t)\,dt = p_x(t_s) - p_x(t_e), \tag{1}$$

$$y \text{ momentum:} \quad \int_{t_e}^{t_s} (N - mg)\,dt = p_y(t_s) - p_y(t_e), \tag{2}$$

where $p_x = mv_x$ and $p_y = mv_y$.

Force Laws All forces that we are modeling are accounted for on the FBD.

Kinematic Equations Since we are assuming that the plane only moves in a straight line parallel to the landing deck, Fig. 2 tells us that the motion is only along the x axis with $v_x(t_e) = 150\,\text{mph} = 220.0\,\text{ft/s}$. Since the plane comes to a stop at time t_s, we have

$$v_x(t_e) = 220.0\,\text{ft/s}, \quad v_x(t_s) = 0\,\text{ft/s}, \quad v_y(t_e) = 0\,\text{ft/s}, \quad v_y(t_s) = 0\,\text{ft/s}. \tag{3}$$

Computation The last two of Eqs. (3) tell us there is no motion in the y direction, so Eq. (2) tells us that the force N and mg cancel each other throughout the time interval of interest. Hence, treating F as a constant, Eqs. (1) and (3) give

$$F(t_s - t_e) = mv_x(t_e) \quad \Rightarrow \quad \boxed{F = \frac{mv_x(t_e)}{t_s - t_e} = 184{,}500\,\text{lb.}} \tag{4}$$

Discussion & Verification The acceleration of the airplane is $a_x = F/m = 110.0\,\text{ft/s}^2 = 3.416g$. Carrier landings are usually considered to be maneuvers with roughly $3g$. Since we have computed an *average* force value, the plane will have accelerations in excess of $3.416g$. However, these higher accelerations would not be sustained for the entire duration of the maneuver (i.e., 2 s).

Interesting Fact

Carrier landing systems. The arresting cables are about 1.5 in. in diameter and are stretched 2–5 in. above the landing deck at 35–40 ft intervals. When one of the cables is caught by an airplane's tailhook, the cable's motion results in the activation of a hydraulic system housed below the deck, whose task is to dissipate the plane's kinetic energy. On carriers, planes land at speeds between 130 and 150 mph, and they are brought to a halt in roughly 300 ft. The limit of human tolerance for sustained acceleration (for more than 2 s) is approximately $8g$.

EXAMPLE 5.2 *Impulse-Momentum Applied to a Racquetball Hitting a Wall*

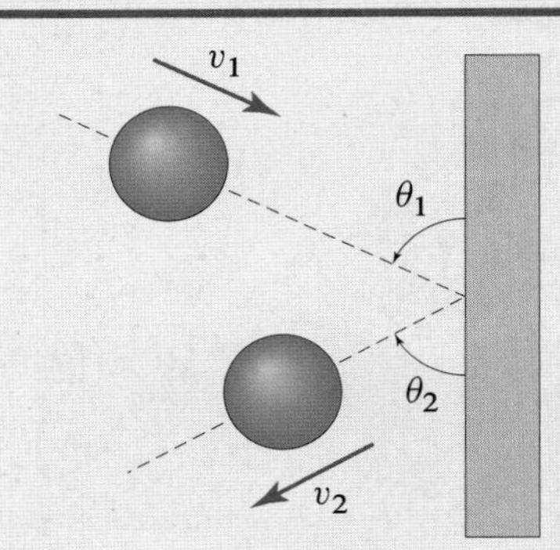

Figure 1
Racquetball approaching a wall at speed v_1 and then rebounding at speed v_2.

Figure 1 shows a 1.4 oz racquetball hitting a wall at 85 mph with an angle $\theta_1 = 65°$. The duration of the impact is 2.5 ms, and the rebound speed is 72.5 mph with the rebound angle $\theta_2 = 61.9°$. Determine the change in momentum, the impulse, and the average force applied to the ball during the impact.

SOLUTION

Road Map & Modeling We are given the mass of the ball and its pre- and postimpact velocities, so we can readily compute its pre- and postimpact momenta, $\vec{p}_1$ and $\vec{p}_2$, respectively. This will allow us to compute the change in momentum and impulse using Eq. (5.6) and the average force using Eq. (5.9). We model the ball as a particle, and we will ignore the effects of gravity during the 2.5 ms impact. The FBD of the ball during the impact in Fig. 2 reflects this model. We have broken the contact force into a normal force N and a friction force F, both of which are a function of time.

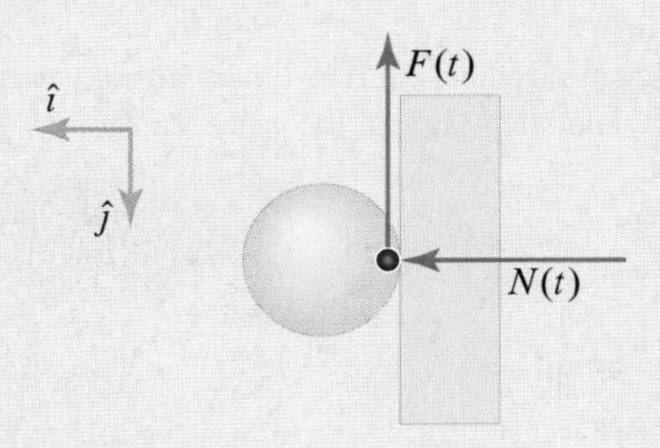

Figure 2
FBD of the racquetball during its impact with the wall.

Governing Equations

Balance Principles Applying Eq. (5.6), the impulse-momentum principle, to the ball during impact, we obtain

$$\Delta\vec{p} = \int_{t_1}^{t_2} \vec{F}(t)\,dt = \vec{p}_2 - \vec{p}_1, \tag{1}$$

where $\Delta\vec{p}$ is the change in momentum, $\int_{t_1}^{t_2} \vec{F}(t)\,dt$ is the impulse applied to the racquetball, $\vec{p}_1 = m\vec{v}_1$ is the momentum just before impact, and $\vec{p}_2 = m\vec{v}_2$ is the momentum just after impact.

Force Laws All forces are accounted for on the FBD.

Kinematic Equations From Fig. 1, we can see the pre- and postimpact velocities of the racquetball are

$$\vec{v}_1 = v_1(-\sin\theta_1\,\hat{\imath} + \cos\theta_1\,\hat{\jmath}), \tag{2}$$

$$\vec{v}_2 = v_2(\sin\theta_2\,\hat{\imath} + \cos\theta_2\,\hat{\jmath}). \tag{3}$$

Computation Substituting Eqs. (2) and (3) into Eq. (1) and evaluating the result using the given quantities, which, when converted, are $m = 0.002717$ slug, $v_1 = 124.7$ ft/s, and $v_2 = 106.3$ ft/s, we obtain the impulse and change in momentum as

$$\begin{aligned}\vec{p}_2 - \vec{p}_1 &= m\big[(v_2\sin\theta_2 + v_1\sin\theta_1)\,\hat{\imath} + (v_2\cos\theta_2 - v_1\cos\theta_1)\,\hat{\jmath}\big]\\ &= (0.5619\,\hat{\imath} - 0.007071\,\hat{\jmath})\ \text{lb·s}.\end{aligned} \tag{4}$$

The average force is found by applying Eq. (5.9) to obtain

$$\vec{F}_{\text{avg}} = \frac{\vec{p}_2 - \vec{p}_1}{t_2 - t_1} = (224.8\,\hat{\imath} - 2.828\,\hat{\jmath})\ \text{lb}. \tag{5}$$

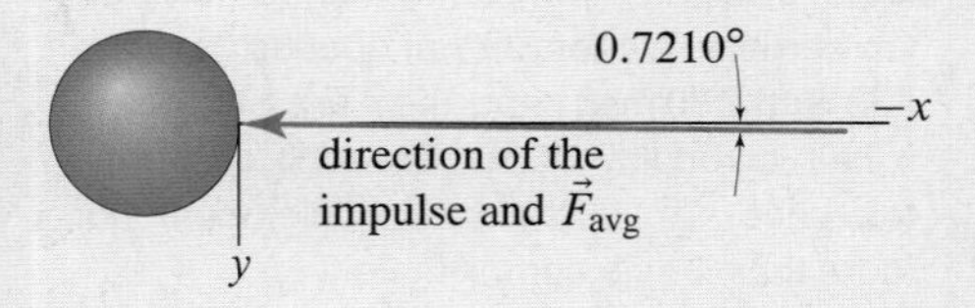

Figure 3
The direction of the change in momentum, impulse, and average force acting on the racquetball during its impact with the wall. This vector deviates from the x axis by less than 1°.

Discussion & Verification Figure 3 shows the direction of the change in momentum of the ball. This is also the direction of the impulse acting on the ball and of the average force on the ball during its impact with the wall. Equation (4) tells us that most of the change in momentum occurs in the x direction. Since the momentum changes very little in the direction parallel to the wall, the frictionless impact assumption we often use for impacts is a good one. The magnitude of the average force is 224.8 lb, which seems about right if you have ever been hit by a racquetball.

EXAMPLE 5.3 *A Person Pulling a Commercial Airliner*

An event at the World's Strongest Man (WSM) competition involves the timed pull of a Boeing 737 airliner over a distance of 82 ft. Given that a Boeing 737 weighs about 75 ton, that the competitor starts pulling with a force of 850 lb, that his strength decreases linearly by 30% during the course of the pull, and that he completes the pull in 40 s, determine how fast he is moving at the end of the pull.

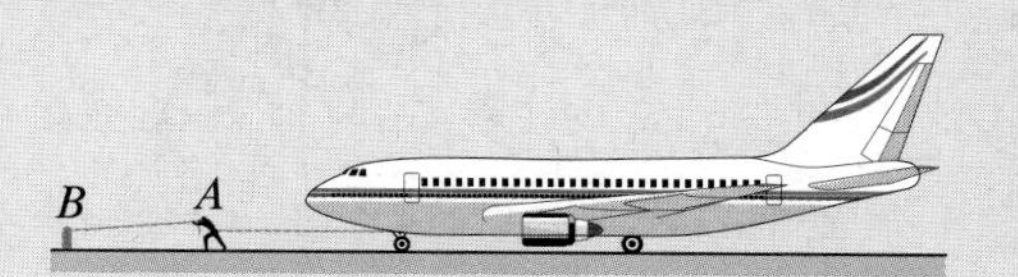

Figure 1
A man at A pulling a Boeing 737 airliner using both his arms (with the rope AB) and his legs.

SOLUTION

Road Map & Modeling We are given enough information to determine the force as a function of time with which the competitor pulls. We know that the system starts from rest and that we want to find its final speed after a given amount of time, so this problem lends itself to the application of the impulse-momentum principle. We neglect the rolling resistance of the airplane and the mass of the competitor, treat the plane as a particle, and assume that the force with which the person pulls the plane is parallel to the motion. Using these assumptions, the FBD of the plane is shown in Fig. 2.

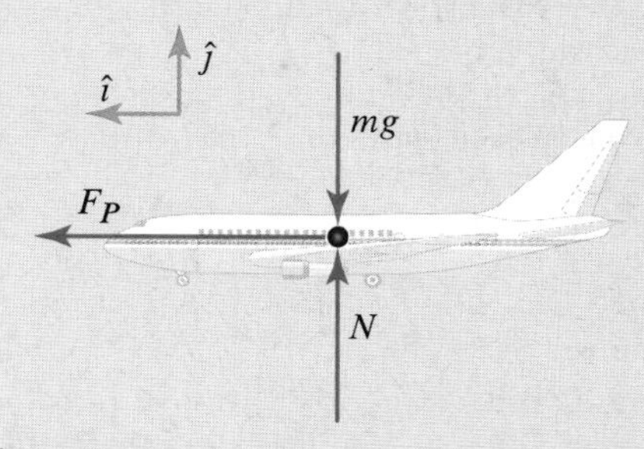

Figure 2
FBD of the 737 as it is being pulled by the WSM competitor.

Governing Equations

Balance Principles Applying the impulse-momentum principle in the x direction to the FBD in Fig. 2 from the time that the person begins pulling until he pulls for 40 s gives

$$x \text{ momentum:} \quad p_1 + \int_{t_1}^{t_2} F_P \, dt = p_2, \tag{1}$$

where $t_1 = 0$ s is the time at which he starts pulling, $t_2 = 40$ s is the time at which he stops pulling, $p_1 = mv_1$ is the momentum of the plane at time t_1, and $p_2 = mv_2$ is the momentum of the plane at time t_2.

Helpful Information

Why have we ignored the mass of the competitor? To include the mass of the competitor in our solution, we would have to realize that the force F_P with which the competitor is pulling the airplane is also pulling *him*. Of course, adding in a 300 lb strongman barely changes the result since he is only 0.2% of the weight of the airplane.

Force Laws To derive the force law $F_P(t)$ for the WSM competitor, we note that he starts pulling with a force of 850 lb and it decreases linearly by 30% over the course of the pull. Therefore, the force versus time curve must be as shown in Fig. 3, and the equation for the curve is

$$F_P = -\frac{0.3(850)}{40} t + 850 = (-6.375t + 850) \text{ lb}. \tag{2}$$

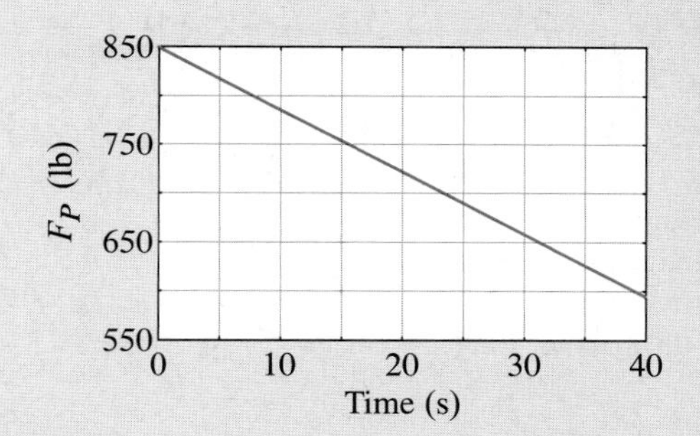

Figure 3
The force F_P versus time curve for the WSM competitor.

Kinematic Equations The kinematic equation for this problem is that the competitor starts from rest, so

$$v_1 = 0. \tag{3}$$

Computation Substituting Eqs. (2) and (3) into Eq. (1), integrating, and using $m = 4658$ slug, we obtain the final speed v_2:

$$\int_0^{40} (-6.375t + 850) \, dt = mv_2 \quad \Rightarrow \quad \boxed{v_2 = 6.204 \text{ ft/s} = 4.230 \text{ mph.}} \tag{4}$$

Discussion & Verification The competitor and the plane are both moving at a little over 4 mph at the end of the 82 ft pull. This is in the range of typical walking speeds, so it is not an unreasonable result. Thus, our choice of F_P in Eq. (2) seems appropriate. Note that a 75 ton Boeing 737 would have substantial rolling resistance, which we have ignored. It is also noted that a person *really* did pull a 75 ton Boeing 737 a distance of 82 ft in 40 s during a WSM competition. Therefore, the initial force generated by the WSM competitor must have actually been larger than the 850 lb that we have estimated.

EXAMPLE 5.4 Walking on a Floating Platform: Conservation of Momentum

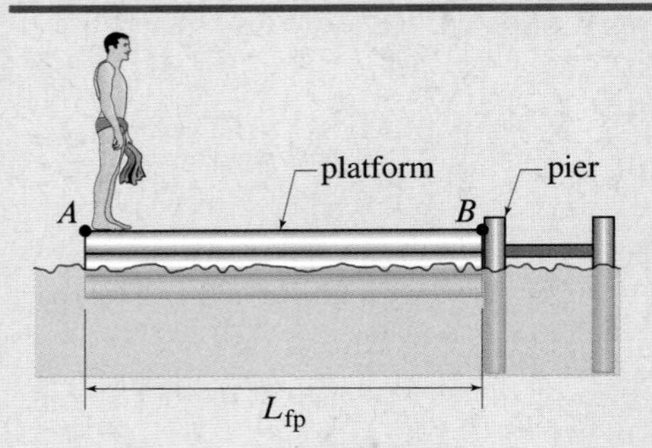

Figure 1
Man on a floating platform.

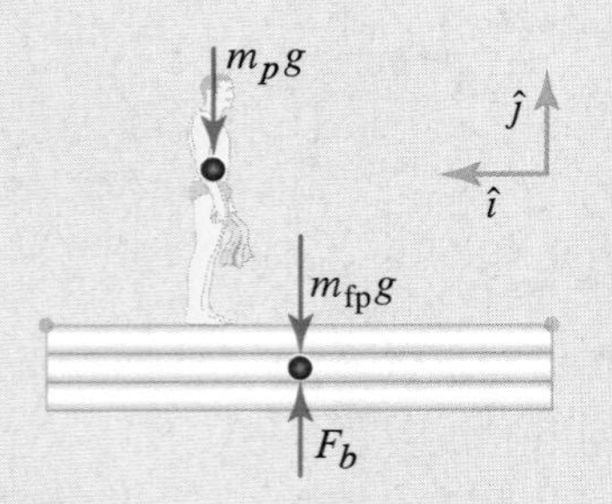

Figure 2
FBD of the system, i.e., the man + platform. The force F_b is the buoyancy force.

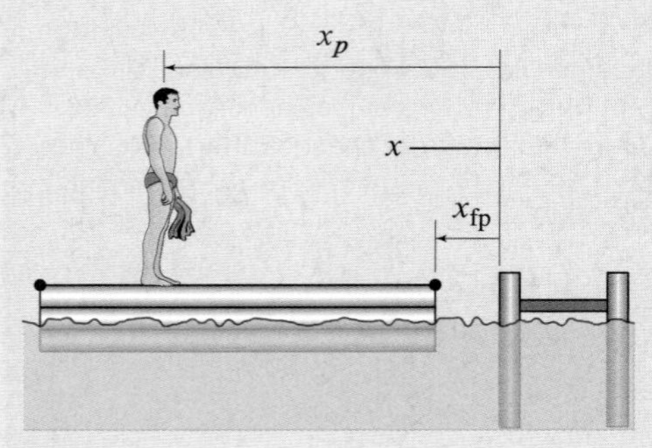

Figure 3
Kinematics diagram showing the origin of the x axis as well as the definition of the positions of the man and the platform.

A man of mass m_p is at end A of a floating platform of mass m_{fp} and length L_{fp}. The man and the platform are initially at rest. The platform is touching the pier (as shown in Fig. 1) when the man starts moving toward B with a constant speed v_0, relative to the platform. Determine the distance between the platform and the pier when the man reaches the other end of the platform at B.

SOLUTION

Road Map & Modeling We model both the man and the platform as particles and ignore any vertical motion and the drag force between the platform and the water. These assumptions imply the FBD shown in Fig. 2, in which F_b is the buoyancy force exerted by the water on the platform. We will let t_A and t_B be the times at which the man is at A and B, respectively, whereas t will denote a generic time instant.

Governing Equations

Balance Principles The FBD in Fig. 2 shows no external forces in the x direction. This implies that the linear momentum in the x direction is conserved (see discussion starting on p. 316). Using Eq. (5.13) on p. 316, we express this fact by writing

$$m v_{Gx}(t_A) = m v_{Gx}(t) \quad \Rightarrow \quad v_{Gx}(t_A) = v_{Gx}(t), \tag{1}$$

where $m = m_p + m_{\text{fp}}$ is the total mass of the system and $t \geq t_A$.

Force Laws All forces are accounted for on the FBD.

Kinematic Equations Both the man and platform start from rest, so

$$v_{Gx}(t_A) = 0. \tag{2}$$

Computation Substituting Eq. (2) into the last of Eqs. (1), we obtain

$$v_{Gx}(t) = 0. \tag{3}$$

Let x_G denote the x coordinate of the system's center of mass. Since $v_{Gx} = \dot{x}_G$, Eq. (3) implies that x_G is constant, so $x_G(t_A) = x_G(t_B)$. Referring to Fig. 3, we then have

$$\frac{1}{m}(m_{\text{fp}} x_{\text{fp}A} + m_p x_{pA}) = \frac{1}{m}(m_{\text{fp}} x_{\text{fp}B} + m_p x_{pB}), \tag{4}$$

where $(m_{\text{fp}} x_{\text{fp}} + m_p x_p)/m = x_G$, and where A and B in the subscripts stand for "at t_A" and "at t_B," respectively. We now observe that x_{fp} and x_p are such that

$$x_{\text{fp}A} = 0, \quad x_{pA} = L_{\text{fp}}, \quad \text{and} \quad x_{pB} = x_{\text{fp}B}. \tag{5}$$

Substituting Eqs. (5) in Eq. (4), we obtain an equation in x_{fp} whose solution is

$$\boxed{x_{\text{fp}B} = \left(\frac{m_p}{m_p + m_{\text{fp}}}\right) L_{\text{fp}}.} \tag{6}$$

Discussion & Verification The final result has dimensions of length, as it should. Equation (6) tells us that the greater m_p, the farther from the pier the platform will end up. In addition, we observe that if something with very little mass (e.g., a flea) walks across the platform, the platform will barely move. Probably both of these observations agree with your experience.

A Closer Look The answer is independent of v_0, that is, regardless of how fast the man walks, the platform will always end up at the same final location.

Problems

Problem 5.1

An airplane performs a turn at constant speed and elevation so as to change its course by 180°. Let A and B designate the beginning and endpoints of the turn. Assuming that the change in mass of the plane due to fuel consumption is negligible, is the airplane's momentum at A different from the airplane's momentum at B? In addition, again neglecting the change in mass between A and B, is the total work done on the plane between A and B positive, negative, or equal to zero?

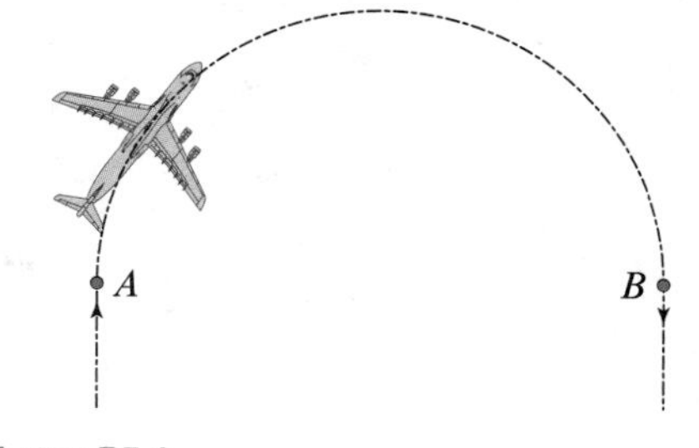

Figure P5.1

Problem 5.2

In a simple force-controlled experiment, two curling stones A and B are made to slide over a sheet of ice. Initially, A and B are at rest on the start line. Then they are acted upon by identical and constant forces $\vec{F}$, which continually push A and B all the way to the finish line. Let $\vec{p}_{A_{FL}}$ and $\vec{p}_{B_{FL}}$ denote the momentum of A and B *at the finish line*, respectively. Assume that the forces $\vec{F}$ are the only nonnegligible forces acting in the plane of motion. If $m_A < m_B$, which of the following statements is true?

(a) $|\vec{p}_{A_{FL}}| < |\vec{p}_{B_{FL}}|$.

(b) $|\vec{p}_{A_{FL}}| = |\vec{p}_{B_{FL}}|$.

(c) $|\vec{p}_{A_{FL}}| > |\vec{p}_{B_{FL}}|$.

(d) There is not enough information given to make a comparison between $|\vec{p}_{A_{FL}}|$ and $|\vec{p}_{B_{FL}}|$.

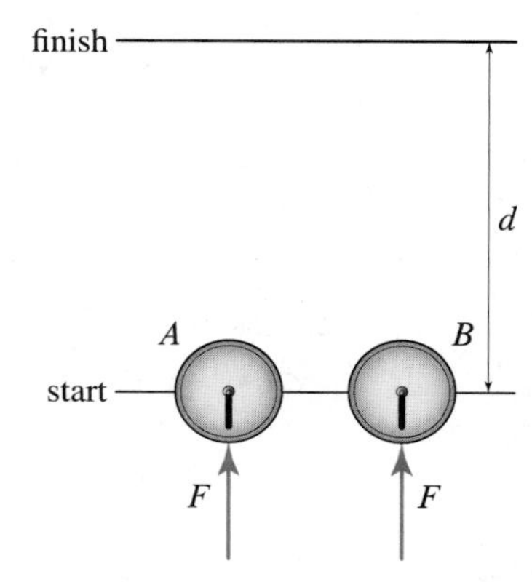

Figure P5.2

Problem 5.3

A train is moving at a constant speed v_t relative to the ground, when a person who is initially at rest (relative to the train) starts running and gains a speed v_0 (relative to the train) after a time interval Δt. Had the person started from rest on the ground (as opposed to on the moving train), would the magnitude of the total impulse exerted on the person during Δt be smaller than, equal to, or larger than the impulse needed to cause the same change in relative velocity in the same amount of time on the moving train? Assume that the person always moves in the direction of motion of the train.

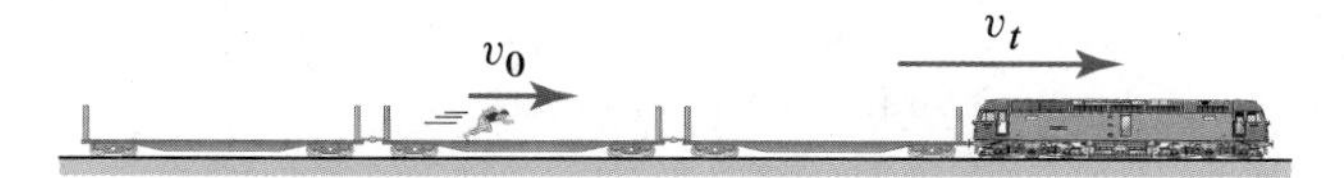

Figure P5.3 and P5.4

Problem 5.4

A train is decelerating at a constant rate, when a person who is initially at rest (relative to the train) starts running and gains a speed v_0 (again relative to the train) after a time interval Δt. Had the person started from rest on the ground (as opposed to on the moving train), would the magnitude of the total impulse exerted on the person during Δt be smaller than, equal to, or larger than the impulse needed to cause the same change in velocity in the same amount of time on the moving train? Assume that the person always moves in the direction of motion of the train and that the train does not reverse its motion during the time interval Δt.

Problem 5.5

A car of mass m collides head-on with a truck of mass $50m$. What is the ratio between the magnitude of the impulse provided by the car to the truck and the magnitude of the impulse provided by the truck to the car during the collision?

Problem 5.6

The spacecraft shown is out in space and is far enough from any other mass (e.g., planets, etc.) so as not to be affected by any gravitational influence (i.e., the net external force on the rocket is approximately zero). The system (i.e., the spacecraft *and all* its fuel) is at rest when it starts at A, and it thrusts all the way to B along the straight line shown using internal chemical rockets (which work by ejecting the fuel mass at very high speeds out the tail of the rocket). We are given that the mass of the system at A is m and that it has ejected half of its mass in thrusting from A to B. What will be the location of the system's mass center when the spacecraft reaches B?

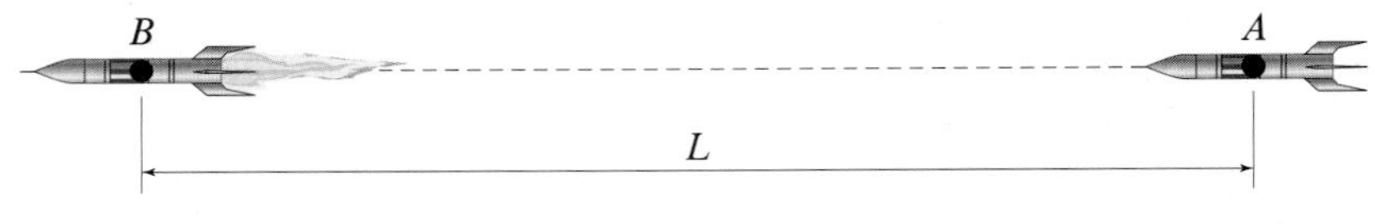

Figure P5.6

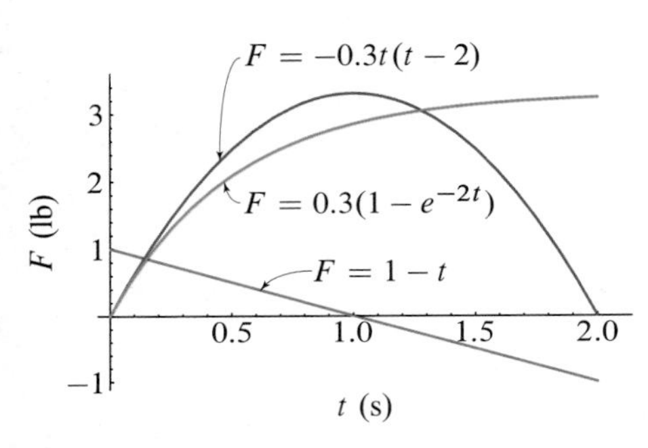

Figure P5.7

Problem 5.7

Use the definition of impulse given in Eq. (5.5) to compute the impulse of the forces shown during the interval $0 \le t \le 2$ s.

Problem 5.8

The mass of the Earth is $m_e = 5.9736 \times 10^{24}$ kg. Modeling the Earth (with everything in and on it) as an isolated system and assuming that the center of the Earth is also the center of mass of the Earth, determine the displacement of the center of the Earth due to

(a) a 2 m jump off the surface by an 85 kg person;

(b) the Space Shuttle, with a mass of 124,000 kg, reaching an orbit of 200 km;

(c) 170,000 km^3 of water being elevated 50 m (these numbers are estimates based on publicly available information about the Aswan Dam at the border between Egypt and Sudan). Use 1 g/cm^3 for the density of water.

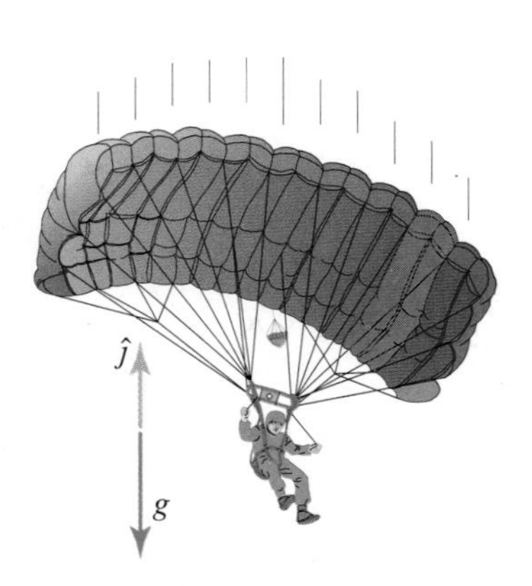

Figure P5.9 and P5.10

Problems 5.9 and 5.10

A 200 lb skydiver deploys the parachute after 10 s of free fall and concludes the jump by touching down with a speed of 15 ft/s. Model the motion of the skydiver as being a vertical drop from rest.

Problem 5.9 Neglecting the change in acceleration due to gravity with elevation, determine the impulse provided by gravity to the skydiver during free fall.

Problem 5.10 Determine the impulse provided by all the forces acting on the skydiver from the beginning of the jump to touchdown.

Figure P5.11

Problem 5.11

Consider an elevator that moves with an operating speed of $2.5\,\text{m/s}$. Suppose that a person who boards the elevator on the ground floor gets off on the fifth floor. Assuming that the elevator has achieved operating speed by the time it reaches the second floor and that it is still moving at its operating speed as it passes the fourth floor, determine the momentum change of a person with a mass of 80 kg between the second and fourth floors if each floor is 4 m high. In addition, determine the impulse of the person's weight during the same time interval.

Problem 5.12

A 180 gr (7000 gr $=$ 1 lb) bullet goes from rest to 3300 ft/s in 0.0011 s. Determine the magnitude of the impulse imparted to the bullet during the given time interval. In addition, determine the magnitude of the average force acting on the bullet.

Figure P5.12

Figure P5.13

Problem 5.13

A 3400 lb car is parked as shown. Determine the impulse of the normal reaction force acting on the car during the span of an hour if $\theta = 15°$.

Problems 5.14 and 5.15

A 3850 lb sports car (driver's weight included), driving along a horizontal rectilinear stretch of road, goes from 0 to 62 mph in 4.2 s.

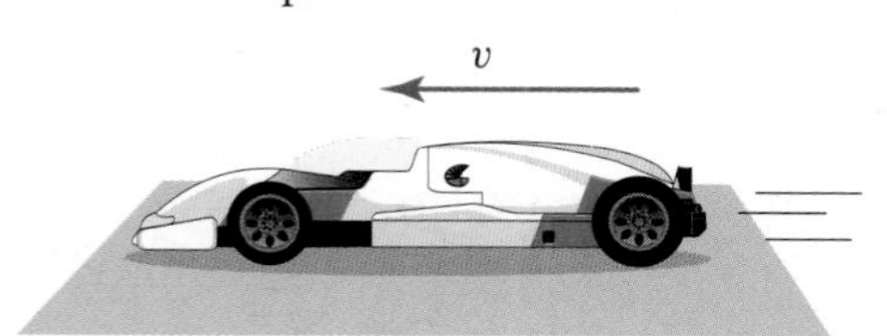

Figure P5.14 and P5.15

Problem 5.14 Determine the magnitude of the average force that needs to be applied to the car for such an acceleration to occur.

Problem 5.15 If the magnitude of the force propelling the car has the form $F_0(1 - e^{-t/\tau})$, with $\tau = 0.5\,\text{s}$, determine F_0.

Problem 5.16

A 75 lb crate is initially at rest when a force $P = 40\,\text{lb}$ is applied to the pulley system as shown. Use the impulse-momentum principle to determine the speed of the crate after 2 s. Neglect the inertia of the rope and of the pulleys, and assume that all the cable segments are purely vertical.

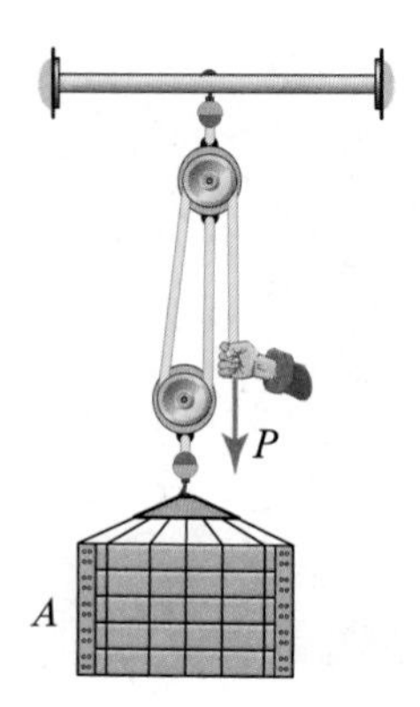

Figure P5.16

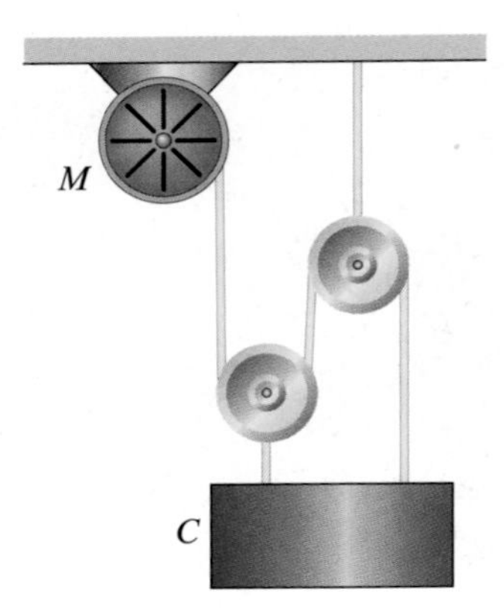

Figure P5.17 and P5.18

Problems 5.17 and 5.18

The pulley system shown is at rest when the motor M starts pulling in rope so as to lift the 120 lb cargo C. For the first second of operation, the motor can produce a tension in the rope of the form $F_0(1 + t/\tau)$, where $F_0 = 40$ lb. Neglect the inertia of the pulleys and of the rope.

Problem 5.17 If $\tau = 0.5$ s, use the impulse-momentum principle to determine the speed of the crate after 1 s.

Problem 5.18 Use the impulse-momentum principle to find the value of τ needed for the cargo C to travel upward with a speed of 10 ft/s after 1 s.

Problem 5.19

A box comes off of a conveyor belt with a speed $v_0 = 3$ m/s and then slides over a low-friction surface. Determine the coefficient of kinetic friction between the crate and floor if the box slides 1.5 s before coming to a stop.

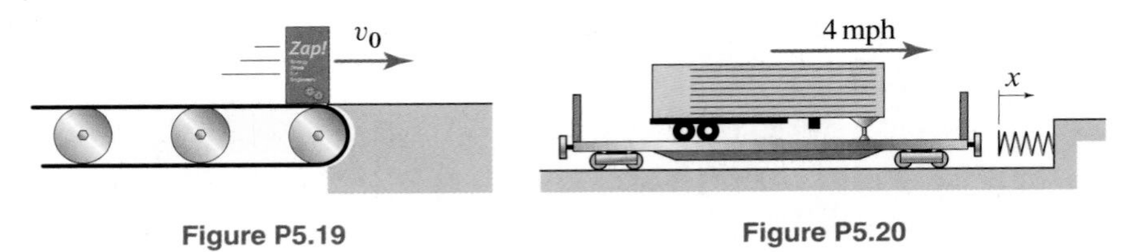

Figure P5.19

Figure P5.20

Problem 5.20

A 60-ton railcar and its cargo, a 27-ton trailer, are moving to the right at 4 mph when they come into contact with a bumper that can bring the system to a stop in 0.78 s. Determine the magnitude of the average force exerted on the railcar by the bumper.

Problem 5.21

A 30,000 lb airplane is flying on a horizontal trajectory with a speed $v_0 = 650$ mph when, at point A, it maneuvers so that at point B it is on a steady climb with $\theta = 40°$ and a speed of 600 mph. If the change in mass of the plane between A and B is negligible, determine the impulse that had to be exerted on the plane in going from A to B.

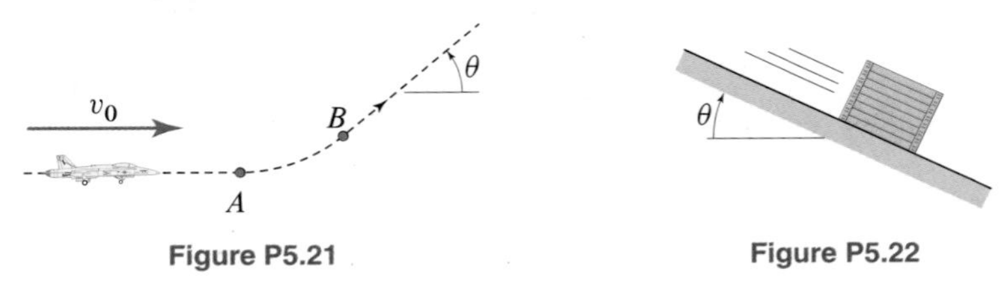

Figure P5.21

Figure P5.22

Problem 5.22

A crate starts sliding from rest down an incline with $\theta = 35°$. Determine the speed of the crate after 2.5 s if the coefficient of kinetic friction between the crate and the incline is $\mu_k = 0.25$. Express the result in SI units.

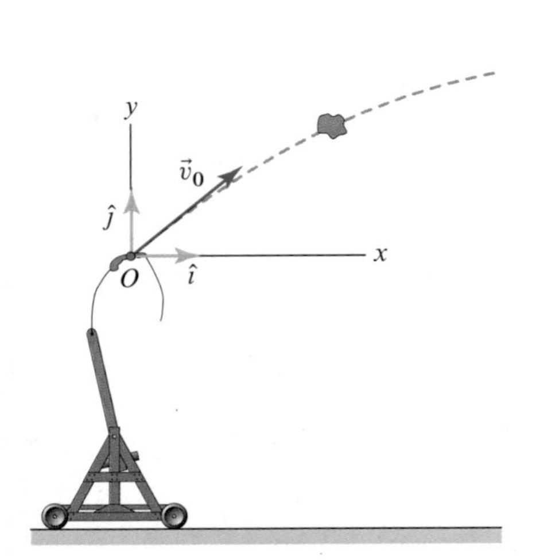

Figure P5.23

Problem 5.23

A trebuchet launches a projectile with an initial velocity $\vec{v}_0$ such that the projectile takes 3 s to achieve its maximum height with a corresponding speed of 145 ft/s. Use the impulse-momentum principle to determine $\vec{v}_0$.

Problem 5.24

A 15 oz football is kicked straight up in the air such that it takes 2.7 s (after separating from the kicker's foot) to achieve its maximum height. Determine the impulse imparted by the kicker to the football, assuming that right before the kick the football is held stationary and that the weight of the football can be neglected while it is in contact with the foot. Also, if the contact between the football and the foot lasts 8×10^{-3} s, determine the average force exerted by the kicker on the football.

Figure P5.24

Problems 5.25 and 5.26

The takeoff runway on carriers is much too short for a modern jetplane to take off on its own. For this reason, the takeoff of carrier planes is assisted by *hydraulic catapults* (Fig. A). The catapult system is housed below the deck except for a relatively small *shuttle* that slides along a rail in the middle of the runway (Fig. B). The front landing gear of carrier planes is equipped with a *tow bar* that, at takeoff, is attached to the catapult shuttle (Fig. C). When the catapult is activated, the shuttle pulls the airplane along the runway and helps the plane reach its takeoff speed. The takeoff runway is approximately 300 ft long, and most modern carriers have three or four catapults.

Figure P5.25 and P5.26

Problem 5.25 In a catapult-assisted takeoff, assume that a 45,000 lb plane goes from 0 to 165 mph in 2 s while traveling along a rectilinear and horizontal trajectory. Also assume that throughout the takeoff the plane's engines are providing 32,000 lb of thrust.

(a) Determine the average force exerted by the catapult on the plane.

(b) Now suppose that the takeoff order is changed so that a small trainer aircraft must take off first. If the trainer's weight and thrust are 13,000 and 5850 lb, respectively, and if the catapult is not reset to match the takeoff specifications for the smaller aircraft, estimate the average acceleration to which the trainer's pilots would be subjected and express the answer in terms of g. What do you think would happen to the trainer's pilot?

Problem 5.26 If the carrier takeoff of a 45,000 lb plane subject to the 32,000 lb thrust of its engines were not assisted by a catapult, estimate how long it would take for a plane to safely take off, i.e., to reach a speed of 165 mph starting from rest. Also, how long a runway would be needed under these conditions?

Problem 5.27

A 10 lb box is released from rest at a distance $d = 5$ ft from the bottom of a smooth chute with $\theta = 30°$. The friction between the chute and the box is negligible, and so is the change in speed of the box in going from the inclined to the horizontal part of the chute. If the duration of the transition between inclined and horizontal motion takes 0.02 s, determine the average force acting on the box during this transition.

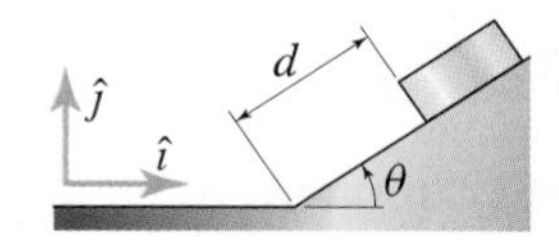

Figure P5.27

Problem 5.28

A ball with a mass $m = 0.25\,\text{kg}$ is dropped from rest from a height $h_1 = 1.5\,\text{m}$ above an incline and rebounds off the incline at a height $h_2 = 0.3\,\text{m}$ above the ground. After the rebound, the ball lands at a horizontal distance $d = 5\,\text{m}$ from the point of contact between the ball and the incline. If the ball takes $0.57\,\text{s}$ to cover the distance d, and if the ball remains in contact with the incline for $0.01\,\text{s}$ during the rebound, determine the average force exerted by the incline onto the ball during the rebound. Express the force using the component system shown. Also, neglect the weight of the ball while it is in contact with the incline.

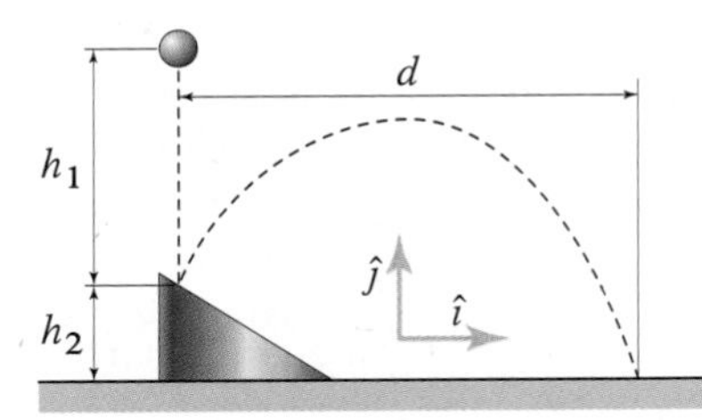

Figure P5.28

Figure P5.29

Problem 5.29

A 1600 kg car, when on a rectilinear and horizontal stretch of road and when the tires do not slip, can go from rest to 100 km/h in 5.5 s. Assuming that the car travels on a straight stretch of road with a 40% slope and that no slip occurs, determine how long it would take to attain a speed of 100 km/h if the car were propelled by the same maximum average force that can be generated on a horizontal road.

Figure P5.30

Problem 5.30

A 1600 kg car, when on a rectilinear and horizontal stretch of road, can go from rest to 100 km/h in 5.5 s.

(a) Assuming that the car travels on such a road, estimate the average value of the force acting on the car for the car to match the expected performance.

(b) Recalling that the force propelling a car is caused by the friction between the driving wheels and the road, and again assuming that the car travels on a rectilinear and horizontal stretch of road, estimate the average value of the friction force acting on the car for the car to match the expected performance. Also estimate the coefficient of friction required to generate such a force.

Figure P5.31

Problem 5.31

A $5\frac{1}{8}$ oz baseball traveling at 80 mph rebounds off a bat with a speed of 160 mph. The ball is in contact with the bat for roughly 10^{-3} s. The incoming velocity of the ball is horizontal, and the outgoing trajectory forms an angle $\alpha = 31^\circ$ with respect to the incoming trajectory.

(a) Determine the impulse provided to the baseball by the bat.

(b) Determine the average force exerted by the bat on the ball.

(c) Determine how much the angle α would change (with respect to 31°) if we were to neglect the effects of the force of gravity on the ball.

Problem 5.32

In an unfortunate incident, a 2.75 kg laptop computer is dropped onto the floor from a height of 1 m. Assuming that the laptop starts from rest, that it rebounds off the floor up to a height of 5 cm, and that the contact with the floor lasts 10^{-3} s, determine the impulse provided by the floor to the laptop and the average acceleration to which the laptop is subjected when in contact with the floor (express this result in terms of g, the acceleration of gravity).

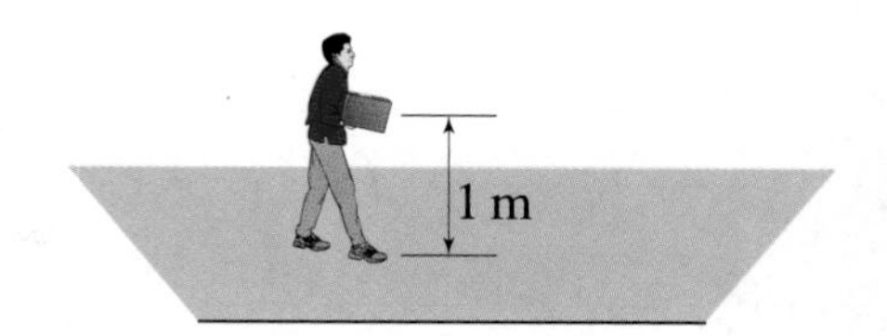

Figure P5.32

Problem 5.33

Three space-junk fragments with masses $m_1 = 7.45$ kg, $m_2 = 3.22$ kg, and $m_3 = 8.45$ kg were the only masses generated from an explosion that split a single body apart. The fragments are traveling as shown with $v_1 = 7701$ m/s, $v_2 = 6996$ m/s, and $v_3 = 6450$ m/s. Assume the velocity vectors of the fragments are coplanar, $\theta = 25°$, and $\phi = 55°$. The angles θ and ϕ are measured with respect to lines that are perpendicular to the velocity $\vec{v}_2$. If the system is *isolated*, determine the mass and velocity of the single body before it exploded.

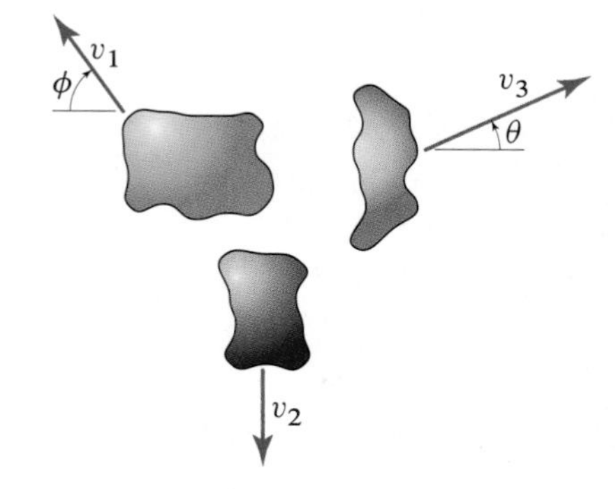

Figure P5.33

Problem 5.34

A 175 lb man, initially at rest at A on a floating platform, starts walking to the right with a constant speed $v_0 = 3$ ft/s relative to the platform until he reaches the right end of the platform at B. The platform is attached to the pier by a rope that is initially slack but that becomes taut before the man reaches B. Neglecting any resistance to the motion of the platform due to the water, determine the impulse provided to the person/platform system by the rope when the rope becomes taut.

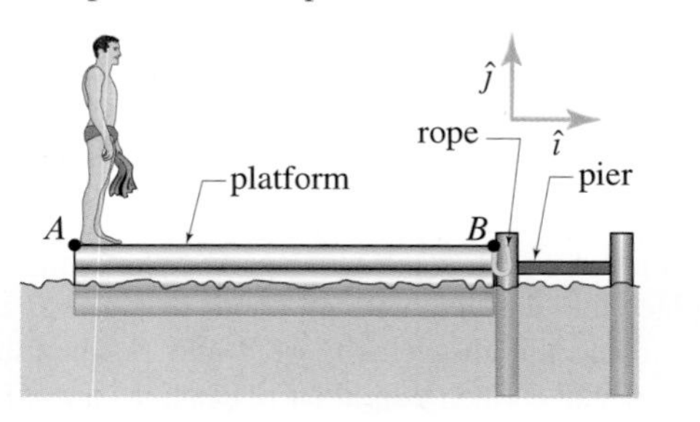

Figure P5.34

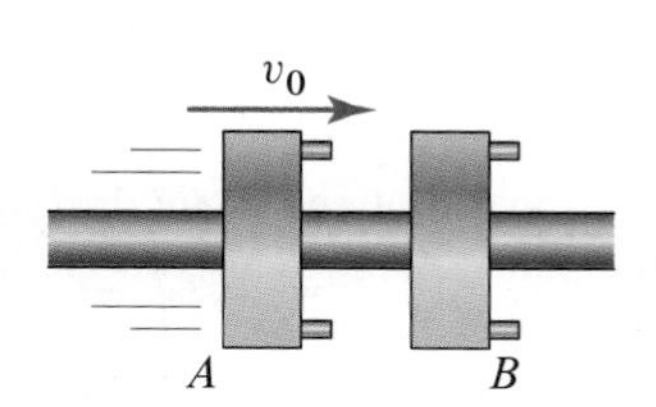

Figure P5.35

Problem 5.35

Collar B is at rest on a smooth horizontal guide when it is impacted by an identical collar A traveling at a speed $v_0 = 12$ ft/s. The collars are designed to lock onto each other on impact, so A and B travel together after impact. Neglecting friction, determine the common velocity of A and B after impact.

Problem 5.36

A 260 gr (1 lb = 7000 gr) bullet B is fired into an 8600 lb SUV A, which is initially moving to the left with a speed $v_A = 15$ mph. If the bullet becomes embedded into the SUV, how fast would the impact speed of the bullet v_B need to be if the SUV is to be stopped cold?

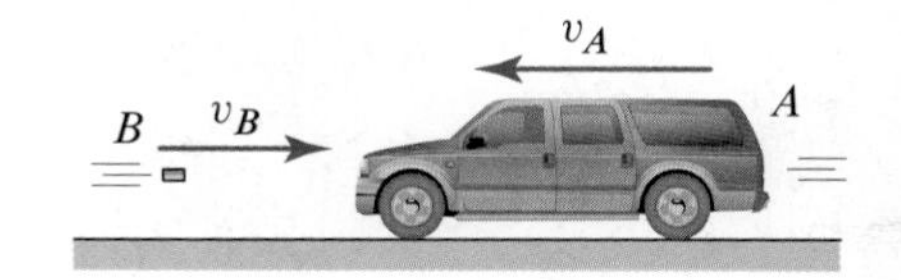

Figure P5.36

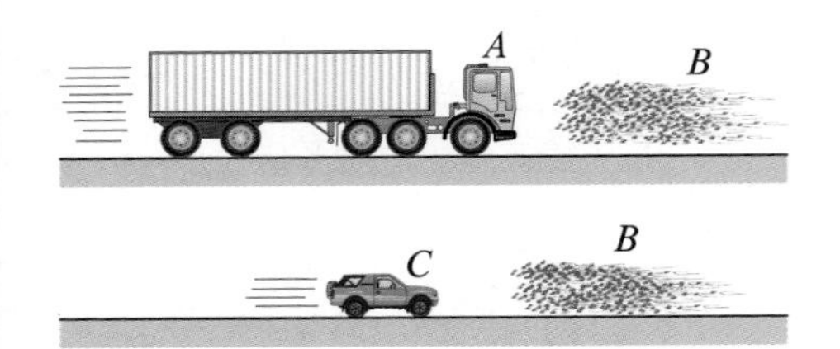

Figure P5.37–P5.39

Problems 5.37 through 5.39

These problems are an introduction to perfectly plastic impact (which we will cover in Section 5.2). In each problem, model the vehicles A and C as particles and treat the swarm of bugs B hitting the vehicles as a single particle. Also assume that the swarm of bugs sticks perfectly to each vehicle (this is what is meant by a *perfectly plastic impact*).

Problem 5.37 An 80,000 lb semitruck A (the maximum weight allowed in many states) is traveling at 70 mph when it encounters a swarm of mosquitoes B. The swarm is traveling at 1 mph in the opposite direction of the truck. Assuming that the entire swarm sticks to the truck, the mass of each mosquito is 2 mg, and that all of these mosquitoes do not significantly damage the truck, how many mosquitoes must have hit the truck if it slows down by 2 mph on impact? If the same number of mosquitoes hit a small SUV C weighing 3000 lb and traveling at 70 mph, by how much would the SUV slow down?

Problem 5.38 An 80,000 lb semitruck A (the maximum weight allowed in many states) is traveling at 70 mph when it encounters a swarm of worker bees B. The swarm is traveling at 12 mph in the opposite direction of the truck. Assuming that the entire swarm sticks to the truck, the mass of each bee is 0.1 g, and that all of these bees do not significantly damage the truck, how many bees must have hit the truck if it slows down by 2 mph on impact? If the same number of bees hit a small SUV C weighing 3000 lb and traveling at 70 mph, by how much would the SUV slow down?

Problem 5.39 An 80,000 lb semitruck A (the maximum weight allowed in many states) is traveling at 70 mph when it encounters a swarm of dragonflies B. The swarm is traveling at 33 mph in the opposite direction of the truck. Assuming that the entire swarm sticks to the truck, the mass of each dragonfly is 0.25 g, and that all of these dragonflies do not significantly damage the truck, how many dragonflies must have hit the truck if it slows down by 2 mph on impact? If the same number of dragonflies hit a small SUV C weighing 3000 lb and traveling at 70 mph, by how much would the SUV slow down?

Problem 5.40

Two canoeists A and B are drifting downstream with a common speed $v_0 = 8\,\text{m/s}$. At some point, A and B use a rope to reduce the distance between them. If A and B can reduce their distance at a rate of 1 m/s, determine the velocity of A and B when they finally come together. Let the masses of A and B (including their respective canoes) be $m_A = 90\,\text{kg}$ and $m_B = 75\,\text{kg}$, respectively. In addition, neglect the drag acting on the canoes due to the water.

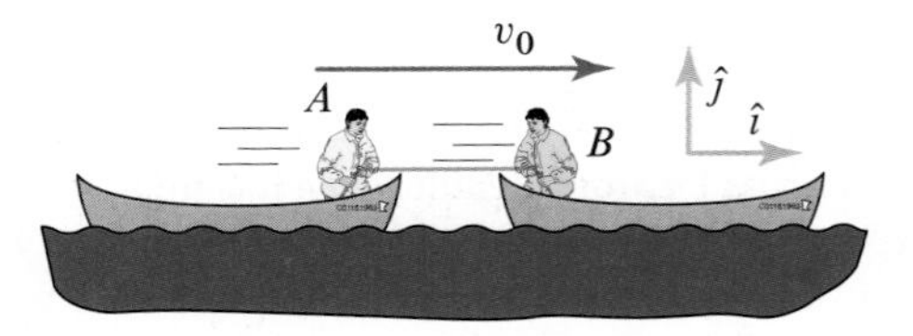

Figure P5.40

Figure P5.41

Problem 5.41

Suppose that a 180 lb person A (the weight includes the the rifle and ammunition before firing) were to fire a 180 gr (7000 gr = 1 lb) bullet B with a muzzle velocity of 3300 ft/s. If the person fires while resting on ice, so that the friction between the person and the ground is negligible, determine the final velocity of both the person and the bullet.

Problems 5.42 and 5.43

Blocks A and B, with masses $m_A = 400\,\text{kg}$ and $m_B = 90\,\text{kg}$, respectively, are initially at rest when block A starts sliding down the incline with $\theta = 30°$.

Problem 5.42 If friction between the two blocks and between block A and the incline is negligible, use the impulse-momentum principle to determine the velocities of A and B 1.5 s after release.

Problem 5.43 Let $\mu_k = 0.15$ be the kinetic coefficient of friction between block A and the incline. If friction between the two blocks is negligible, determine the velocities of A and B 1.5 s after release.

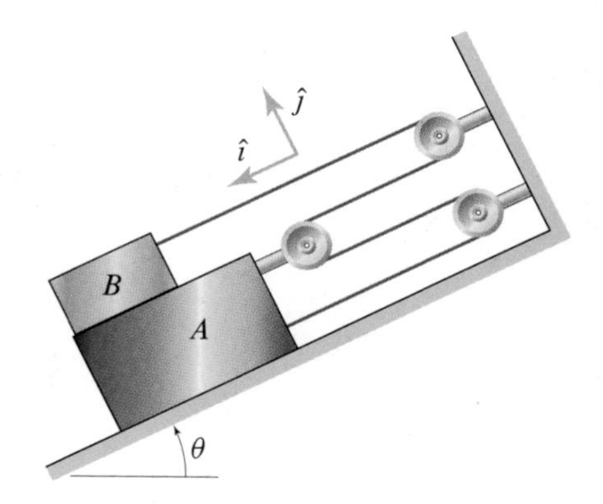

Figure P5.42 and P5.43

Problems 5.44 through 5.46

Two persons A and B weighing 140 and 180 lb, respectively, jump off a floating platform (in the same direction) with a velocity relative to the platform that is completely horizontal and with magnitude $v_0 = 6\,\text{ft/s}$ for both A and B. The floating platform weighs 800 lb. Assume that A, B, and the platform are initially at rest.

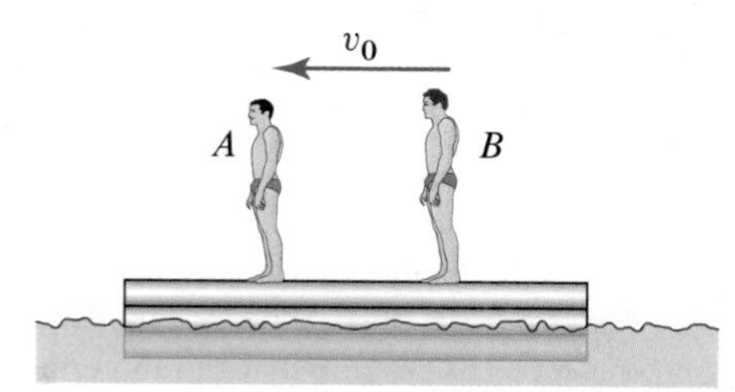

Figure P5.44–P5.46

Problem 5.44 Neglecting the water resistance to the horizontal motion of the platform, determine the speed of the platform after A and B jump at the same time.

Problem 5.45 Neglecting the water resistance to the horizontal motion of the platform, and knowing that B jumps first, determine the speed of the platform after both A and B have jumped.

Problem 5.46 Neglecting the water resistance to the horizontal motion of the platform, and knowing that A jumps first, determine the speed of the platform after both A and B have jumped.

Problem 5.47

A 180 lb man A and a 40 lb child C are at the opposite ends of a 250 lb floating platform P with a length $L_{\text{fp}} = 15\,\text{ft}$. The man, child, and platform are initially at rest at a distance $\delta = 1\,\text{ft}$ from a mooring dock. The child and the man move toward each other with the same speed v_0 relative to the platform. If the drag force due to the water is negligible, determine the distance d from the mooring dock where the child and man will meet.

Problem 5.48

A man A, with a mass $m_A = 85\,\text{kg}$, and a child C, with a mass $m_C = 18\,\text{kg}$, are at the opposite ends of a floating platform P, with a mass $m_P = 150\,\text{kg}$ and a length $L_{\text{fp}} = 6\,\text{m}$. Assume that the man, child, and platform are initially at rest and that the resistance due to the water to the horizontal motion of the platform is negligible. Suppose that the man and child start moving toward each other in such a way that the platform does not move relative to the water. Determine the distance covered by the child to meet the man.

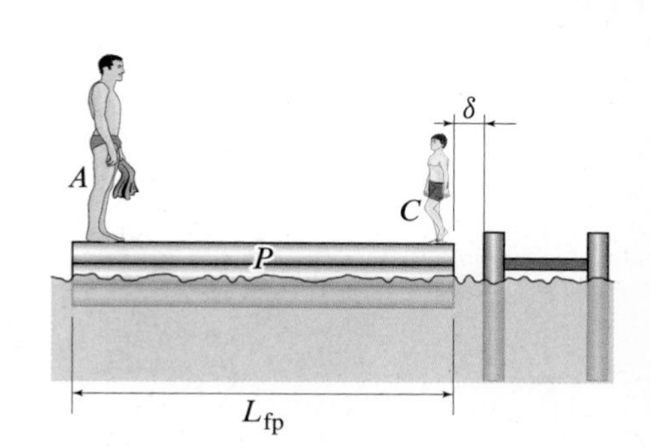

Figure P5.47 and P5.48

Problem 5.49

The 28,000 lb A-10 Thunderbolt is flying at a constant speed of 375 mph when it fires a 4 s burst from its forward-facing seven-barrel Gatling gun. The gun fires 13.2 oz projectiles at a rate of 4200 rounds/min. The muzzle velocity of each projectile is 3250 ft/s. Assuming that each of the plane's two jet engines maintains a constant thrust of 9000 lb, that the plane is subject to a constant air resistance while the gun is firing (equal to that before the burst), and that the plane flies straight and level, determine the plane's change in velocity at the end of the 4 s burst.

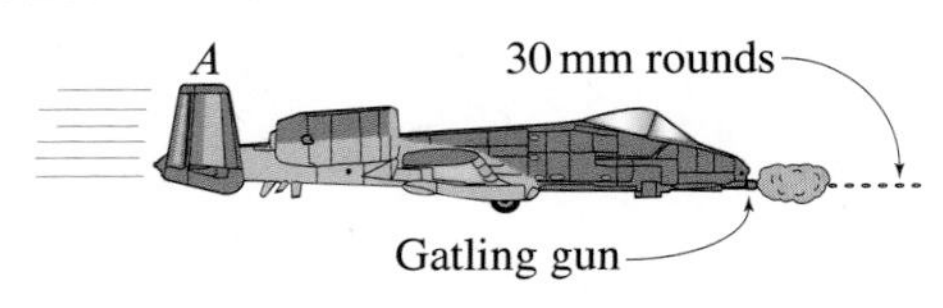

Figure P5.49

Problem 5.50

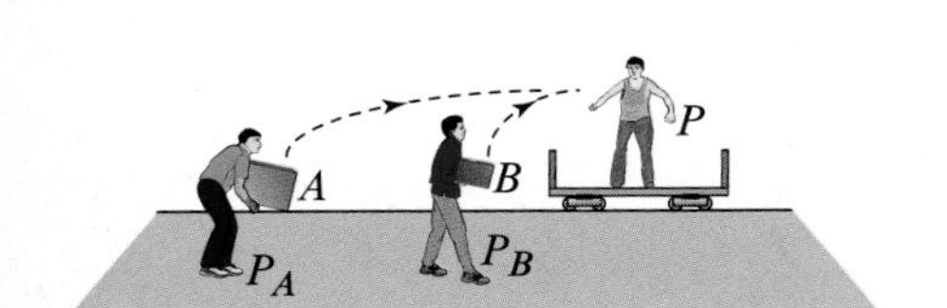

Figure P5.50

A man P on a cart on rails is receiving packages from two men P_A and P_B standing on a stationary platform. Assume that P and the cart have a combined weight of 350 lb and start from rest. In addition, suppose that P_A throws a package A weighing 60 lb, which is received by P with a horizontal speed $v_A = 4.5$ ft/s. After P has received the package from P_A, P_B throws a package B weighing 80 lb, which is received by P with a horizontal speed relative to P and in the same direction as the velocity of P of 5.25 ft/s. Determine the final velocity of P and the cart. Neglect any friction or air resistance acting on P and the cart and assume that the three men are big enough that they can handle throwing around 60 and 80 lb packages.

Problems 5.51 and 5.52

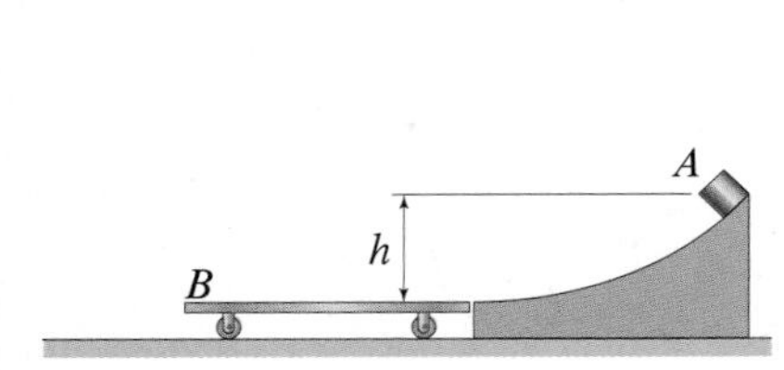

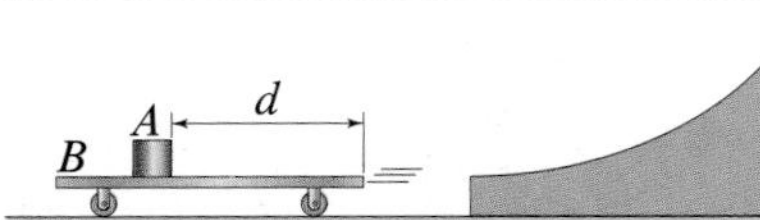

Figure P5.51 and P5.52

Box A, which weighs 47 lb, is released from rest in the position shown in the top figure, where $h = 3$ ft. The box slides down the fixed incline with negligible friction and it lands on a cart B, which is initially at rest and weighs 12 lb. When A reaches the bottom of the incline, the velocity of A is completely horizontal and A starts sliding on the cart. The coefficient of kinetic friction between A and B is $\mu_k = 0.6$.

Problem 5.51 Determine the common speed of A and B after A stops sliding relative to B and the time it takes A to stop sliding relative to B.

Problem 5.52 Determine the distance d from the right end of the cart at which the box A stops relative to the cart.

Problems 5.53 and 5.54

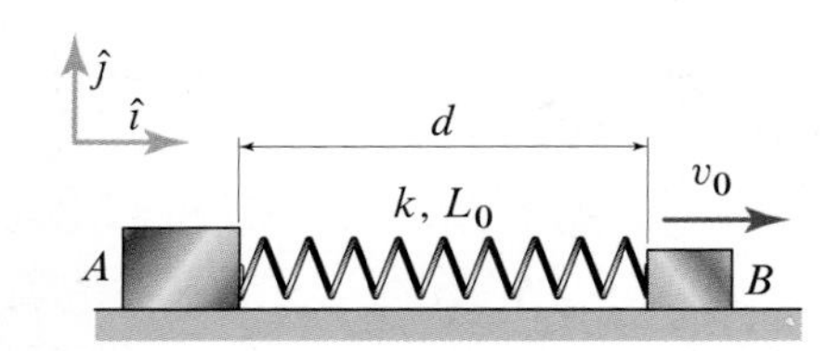

Figure P5.53 and P5.54

Blocks A and B, with masses $m_A = 5$ kg and $m_B = 3$ kg, respectively, are connected by a spring of stiffness $k = 20$ N/m and unstretched length $L_0 = 0.75$ m. The blocks are at rest and separated by the distance L_0 when block B is given a velocity to the right with magnitude $v_0 = 15$ m/s. The friction between the blocks and the horizontal surface on which they rest is negligible.

Problem 5.53 Determine the maximum value of the distance d between A and B that will be achieved during the motion.

Problem 5.54 Assuming that the spring can be squeezed so that $d = 0$, verify that the blocks A and B will collide with one another in the ensuing motion by computing the velocities of the blocks when $d = 0$.

Problem 5.55

Energy storage devices that use spinning flywheels to store energy are starting to become available. To store as much energy as possible, it is important that the flywheel spin as fast as possible. Unfortunately, if it spins too fast, internal stresses in the flywheel cause it to come apart catastrophically. Therefore, it is important to keep the speed at the edge of the flywheel below about 1000 m/s. It is also critical that the flywheel be almost perfectly balanced to avoid the tremendous vibrations that would otherwise result. With this in mind, let the flywheel D, whose diameter is 0.3 m, rotate at $\omega = 60{,}000$ rpm. Assume that the cart B is constrained to move rectilinearly along the smooth guide tracks. Given that the flywheel is not perfectly balanced, that the unbalanced weight A has mass m_A, and that the total mass of the flywheel D, cart B, and electronics package E is m_B, determine the following as a function of θ, the masses, the diameter, and the angular speed of the flywheel:

(a) the amplitude of the motion of the cart,

(b) the maximum speed achieved by the cart.

Neglect the mass of the wheels, assume that initially everything is at rest, and assume that the unbalanced mass is at the edge of the flywheel. Finally, evaluate your answers to Parts (a) and (b) for $m_A = 1$ g (about the mass of a paper clip) and $m_B = 70$ kg (the mass of the flywheel might be about 40 kg).

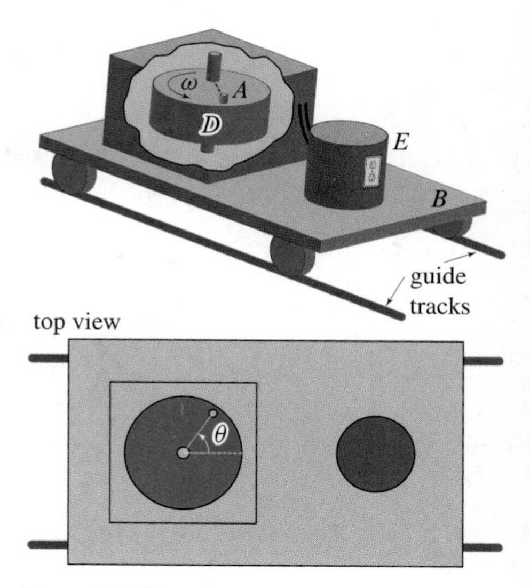

Figure P5.55

Problem 5.56

The 135 lb woman A sits atop the 90 lb cart B, both of which are initially at rest. If the woman slides down the frictionless incline of length $L = 11$ ft, determine the velocity of both the woman and the cart when she reaches the bottom of the incline. Ignore the mass of the wheels on which the cart rolls and any friction in their bearings. The angle $\theta = 26°$.

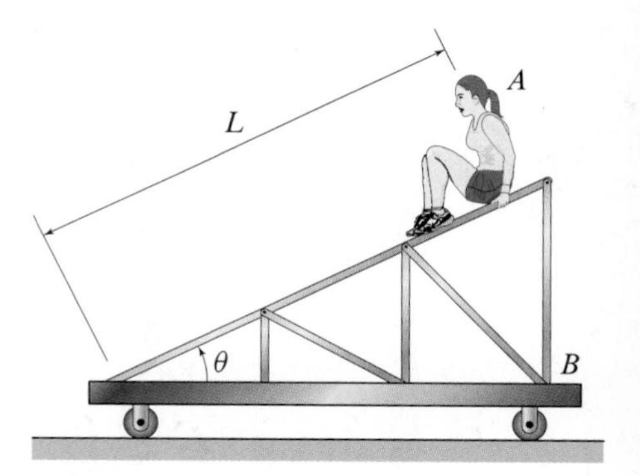

Figure P5.56

Problem 5.57

An Apollo Lunar Module A and Command and Service Module B are moving through space far from any other bodies (so that their gravitational effects can be ignored). When $\theta = 30°$, the two craft are separated using an internal linear elastic spring whose constant is $k = 200{,}000$ N/m and is precompressed 0.5 m. Noting that the mass of the Command and Service Module is about 29,000 kg and that the mass of the Lunar Module is about 15,100 kg, determine their postseparation velocities if their common preseparation velocity is 11,000 m/s.

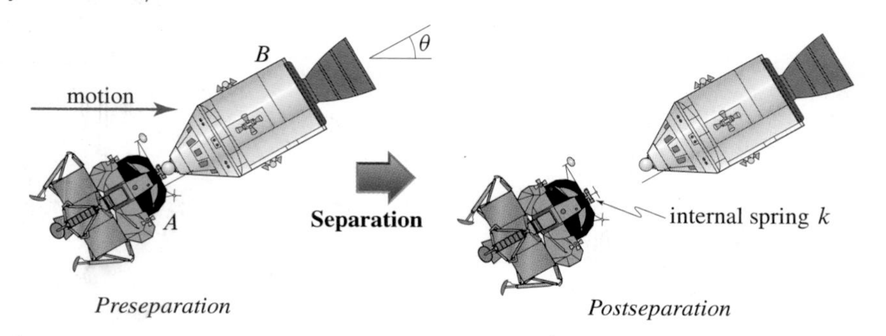

Figure P5.57

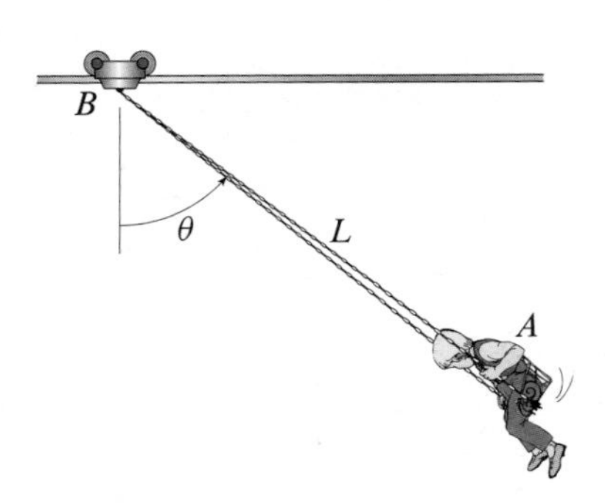

Figure P5.58–P5.61

Problems 5.58 through 5.61

In the ride shown, a person A sits in a seat that is attached by a cable of length L to a freely moving trolley B of mass m_B. The total mass of the person and the seat is m_A. The trolley is constrained by the beam to move only in the horizontal direction. The system is released from rest at the angle $\theta = \theta_0$, and it is allowed to swing in the vertical plane. Neglect the mass of the cable and treat the person and the seat as a single particle.

Problem 5.58 Determine expressions for the velocities of the trolley and the rider the first time that $\theta = 0°$. Evaluate your solution for $W_A = 100$ lb, $W_B = 20$ lb, $L = 15$ ft, and $\theta_0 = 70°$.

Problem 5.59 Determine expressions for the velocities of the trolley and the rider the first time that $\theta = 0°$. After doing so, for given g, L, m_A, and θ_0, determine the maximum speed achievable by the rider at $\theta = 0°$ and the corresponding value of m_B. Evaluate your solution for $W_A = 100$ lb, $L = 15$ ft, and $\theta_0 = 70°$. What would be the motion of B for this value of m_B?

Problem 5.60 Determine the velocity of the trolley and the speed of the rider for any arbitrary value of θ.

Problem 5.61 Determine the equations needed to find the velocity of the trolley and the rider for any arbitrary value of θ. Clearly label all equations and list the corresponding unknowns, showing that you have as many equations as you have unknowns. Solve the equations for the unknowns, and then plot the velocity of the trolley and the speed of the rider as a function of the angle θ for both halves of a full swing of the rider. Use $W_A = 100$ lb, $W_B = 20$ lb, $L = 15$ ft, and $\theta_0 = 70°$.

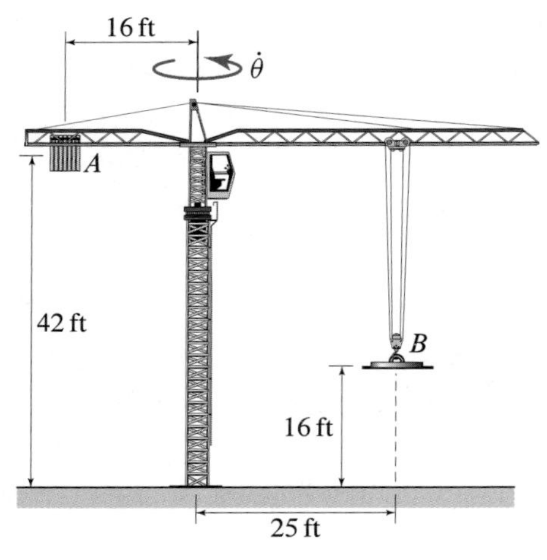

Figure P5.62

Problem 5.62

A tower crane is lifting a 10,000 lb object B at a constant rate of 7 ft/s while rotating at a constant rate of $\dot{\theta} = 0.15$ rad/s. B is also moving outward with a radial velocity of 1.5 ft/s. Assume that the object B does not swing relative to the crane (i.e., it always hangs vertically) and that the crane is fixed to the ground at O.

(a) Determine the radial velocity required of the 20 ton counterweight A to prevent the horizontal motion of the system's center of mass.

(b) Find the total force acting on A and on B.

(c) Determine the velocity and acceleration of the mass center of the system when A moves as determined in Part (a).

5.2 Impact

Impacts are short, dramatic events

Figure 5.6 shows a standard 1.4 oz racquetball hitting a wall. The impact may only

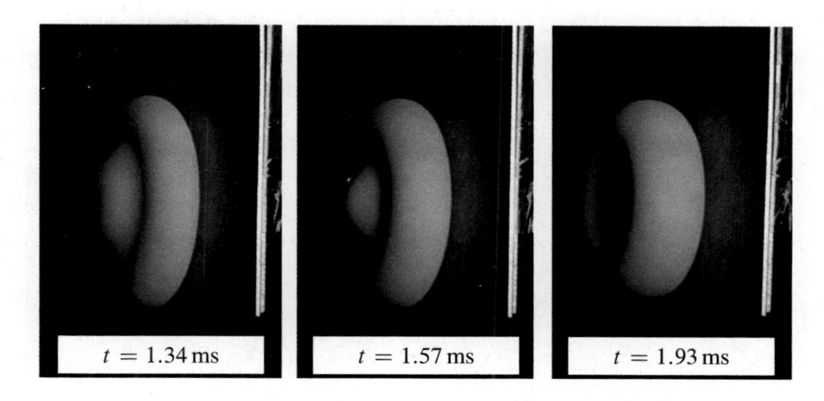

Figure 5.6. High speed photography sequence capturing a racquetball hitting a wall.

last for 2.5 ms, and yet the velocity completely reverses its direction during that time. That is a large change in momentum in a very short time, and there must be a correspondingly large impulse causing that change. The force acting on the ball during the impact, as estimated in Example 5.2, has a component normal to the wall that is almost 100 times larger than the tangent component and more than 1000 times larger than the weight of the ball. Hence, the force *dominating* the impact is not the weight of the ball or the friction with the wall, it is the normal component of the contact force between the ball and the wall.

These observations indicate that impacts occur over time scales so small that they can be modeled as *instantaneous*. In addition, an impact is an event so dominated by constraint forces that all the other forces we normally account for in FBDs can be neglected. We now formalize these ideas into a theory of impacts for particles.

Definition of impact and notation

We will model impacts as events that span *infinitesimal* time intervals. For this reason we adopt the notation that if t is the time of impact, then t^- and t^+ will be the time instants right before and after the impact, respectively. The superscripts $-$ and $+$ will denote the pre- and postimpact values of all quantities in the model. We will also assume that for $t^- \leq t \leq t^+$ the velocities of impacting objects can undergo a finite change, while we neglect any change in position.

Line of impact and contact force between impacting objects

In Fig. 5.7, we assume that contact in an impact occurs at a single point at which we can identify the plane tangent to the contact. The line through the point of contact and normal to the tangent plane is called *line of impact* (LOI). We assume that the LOI is also the line of action of the contact forces between the objects, which is equivalent to neglecting friction at the point of contact. This assumption implies that impacts affect the velocity of impacting objects only along the LOI.

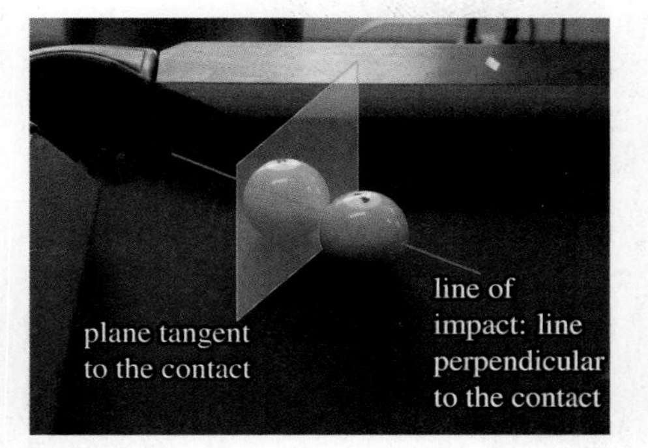

Figure 5.7
Photograph of two billiard balls in contact.

Classification of impacts

Solution strategies for impact problems usually depend on the orientation of the preimpact velocities relative to the LOI. Hence, it is convenient to have a corresponding classification of impacts. Referring to Fig. 5.8, an impact is called *oblique* if one of the preimpact velocities is not parallel to the LOI; otherwise it is called *direct*. An impact is called *central* if the LOI contains the mass centers of the impacting bodies; otherwise it is called *eccentric*. Particles coincide with their own mass centers and can only have central impacts. Eccentric impacts are discussed in Section 8.3.

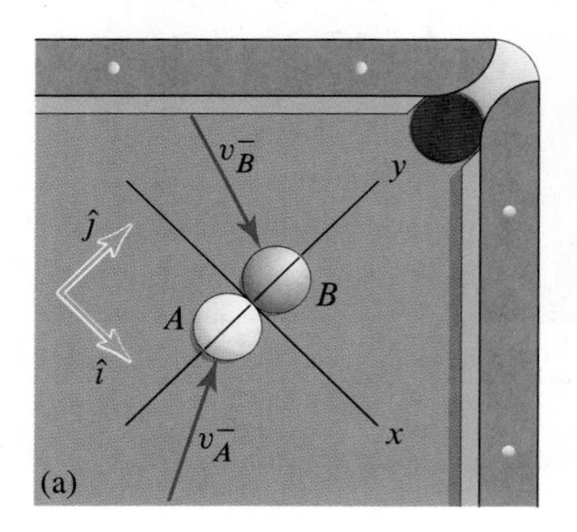

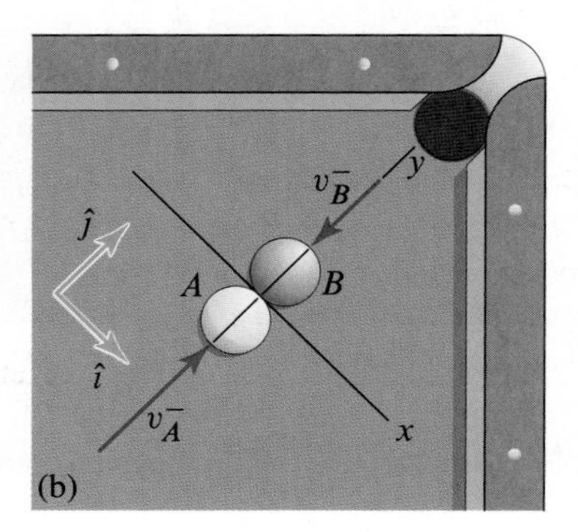

Figure 5.8. (a) Oblique central impact. (b) Direct central impact.

Impulsive forces and impact-relevant FBDs

A force is said to be *impulsive* if it causes a finite change of momentum, and hence velocity, over an infinitesimal time interval. The idea of average force from Section 5.1 then says that the magnitude of an impulsive force must therefore tend to infinity. Because of this, in FBDs of impacting objects, which we call *impact-relevant FBDs*, we neglect any force that is not impulsive.

The concept of an impulsive force is only a modeling tool since we know that forces with infinite magnitude do not exist. *In practice, we model a force as impulsive if its magnitude can grow as large as necessary to guarantee that given constraints are met*. For example, the contact force between two billiard balls becomes as large as necessary to prevent them from passing through one another, and it is therefore modeled as impulsive. By contrast, forces whose magnitude must remain finite, such as weight forces, spring forces, and forces that depend on velocity, cannot be impulsive.

An impact is called *constrained* if the system of impacting objects is subject to *external* impulsive forces; otherwise it is called *unconstrained*. We will present the solution of a constrained impact problem in Example 5.9 on p. 348.

Coefficient of restitution

Figure 5.9
Car A has a speed greater than that of car B and therefore collides with it.

Figure 5.9 shows two cars A and B, with mass m_A and m_B, respectively, with preimpact speeds v_A^- and v_B^-, with $v_B^- < v_A^-$. Car A rear-ends B, and we assume that they separate after the collision. We wish to predict the postimpact velocities of A and B.

For simplicity, we assume the LOI to be parallel to the ground. This implies that neither car will push the other toward the ground and that the reaction forces between the cars and the ground have no reason to grow larger. Thus, we model these reaction forces as nonimpulsive, and we model the overall impact with the impact-relevant FBD in Fig. 5.10, where we have neglected the weights of the cars because they are nonimpulsive. Since there are no external forces in the FBD, the momentum of the system is conserved. Hence, in the x direction, which is the LOI, we have

$$m_A v_{Ax}^- + m_B v_{Bx}^- = m_A v_{Ax}^+ + m_B v_{Bx}^+, \tag{5.20}$$

where $v_{Ax}^- = v_A^-$ and $v_{Bx}^- = v_B^-$ because A and B are moving in the x direction before impact. Equation (5.20), which is the balance law, has two unknowns, v_{Ax}^+ and v_{Bx}^+. We need a second equation in these unknowns to solve the problem. Since we have already written the kinematic equations by relating the speeds to the velocity components of the cars, the second equation will be a *force law*.

Figure 5.10
Impact-relevant FBD of the cars A and B as a system.

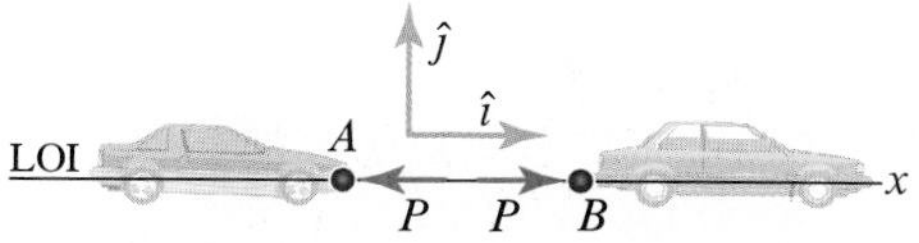

Figure 5.11
Impact-relevant FBD of the cars A and B individually. Recall that we have assumed that the contact force exerted by A and B on one another is only directed along the LOI.

Figure 5.11 shows the impact-relevant FBD of each of the cars individually. The contact force P between A and B is the force that we need to describe in some way. While we cannot describe P as a function of time (impacts span an infinitesimal time interval), we can describe the *effect* of P on A and B by how P measurably affects the pre- and postimpact velocities of A and B along the LOI, which is the line of action of P. With this in mind, experience tells us that:

1. The severity of an impact depends on the *relative* preimpact velocity, also called the *approach velocity*.
2. The "quality" of the *rebound* is described by the postimpact *relative* velocity, also called the *separation velocity*.

The typical plot we obtain by correlating the separation velocity with the approach velocity measured in collision experiments is shown in Fig. 5.12. This plot shows that the separation velocity depends on the approach velocity and that, for higher approach velocities, there is a lack of proportionality between pre- and postimpact relative velocities (this is because more permanent damage is caused by the collision for higher approach velocities). Figure 5.12 also shows that there is a regime (near the origin) in which the approach and separation velocities are *linearly related* to one another. Confining our theory to impacts in their linear range, we therefore adopt the following *impact force law*:

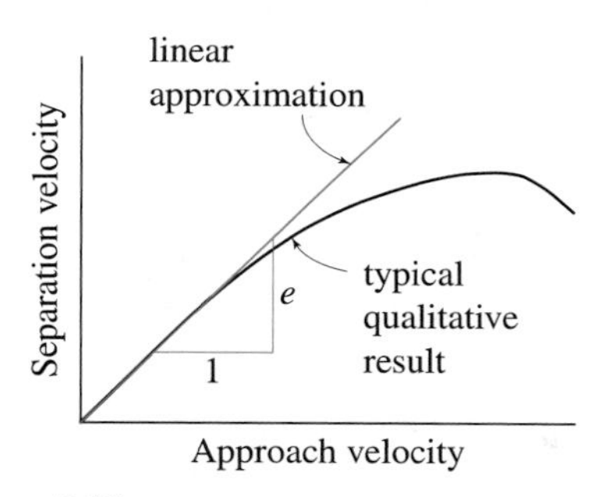

Figure 5.12
Qualitative trend usually found in collision experiments when plotting the post- versus preimpact *relative* velocity (black curve).

$$e = \frac{\text{separation velocity}}{\text{approach velocity}} = \frac{v_{Bx}^+ - v_{Ax}^+}{v_{Ax}^- - v_{Bx}^-}. \tag{5.21}$$

Equation (5.21) is called the *coefficient of restitution equation*, and the constant e, which is determined experimentally, is called the *coefficient of restitution* (COR). The COR equation is the force law we were looking for.

Properties of the COR

Referring to Eq. (5.21), we note the following:

1. The COR e is *dimensionless*.
2. Since $v_{A/Bx}^- = v_{Ax}^- - v_{Bx}^-$ and $v_{B/Ax}^+ = v_{Bx}^+ - v_{Ax}^+$ have the same sign, $e \geq 0$ for any impact.
3. For materials without any energy-producing elements, the separation speed is never greater than the approach speed. This implies that
$$0 \leq e \leq 1. \tag{5.22}$$
4. The COR e depends on the material properties of *both* colliding objects.

Table 5.1
Impact type as a function of COR value.

COR value	Impact type
$e = 0^a$	plastic impact
$0 < e < 1$	elastic impact
$e = 1$	perfectly elastic impact

[a] An impact is perfectly plastic if $\vec{v}_A^+ = \vec{v}_B^+$, that is, the objects stick together after impact.

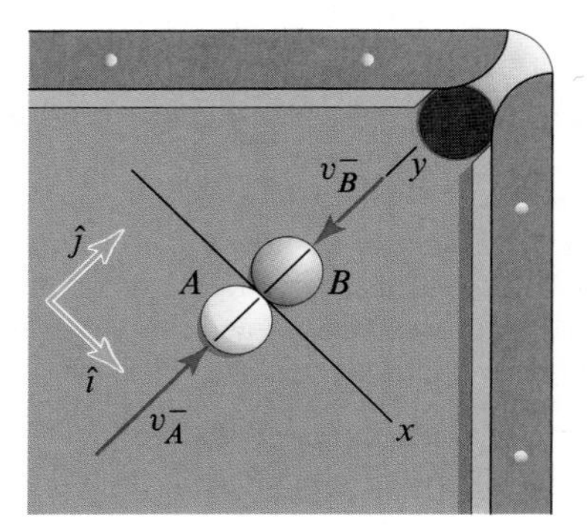

Figure 5.13
Direct central impact of two billiard balls.

An impact is said to be *plastic* if $e = 0$, *elastic* if $0 < e < 1$, and *perfectly elastic* if $e = 1$ (Table 5.1). An impact is called *perfectly plastic* if the colliding objects form a single body after impact. In a perfectly plastic impact the COR equation is replaced by the kinematic condition stating that the colliding bodies share the same postimpact velocity.

Unconstrained direct central impact

Figure 5.13 shows the direct central impact of billiard balls A and B (the preimpact velocities are parallel to the LOI). The impact-relevant FBD in Fig. 5.14(a) shows

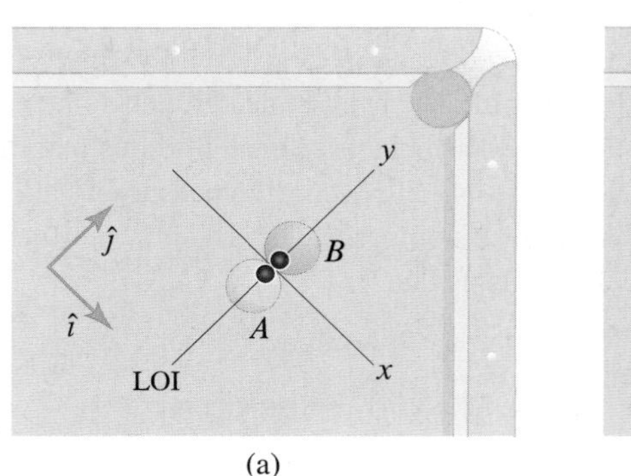

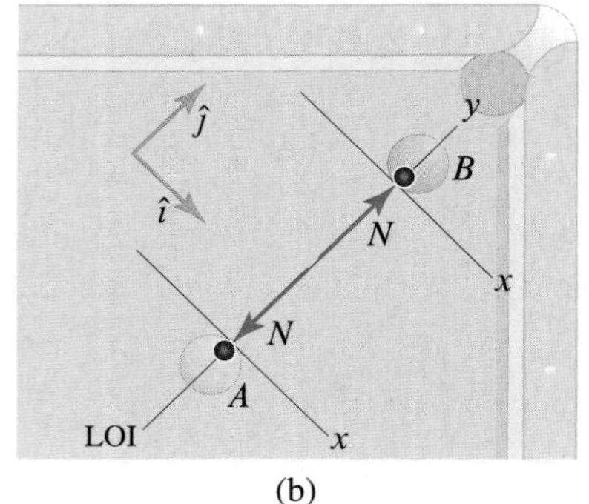

Figure 5.14. (a) FBD of the impact of two billiard balls modeled as particles. (b) FBD of billiard balls A and B individually.

that the impact is unconstrained, i.e., there are no external impulsive forces. Since the motion of A and B is only along the LOI, an unconstrained direct central impact can be solved by applying conservation of momentum and the COR equation along the LOI, which leads to the following two equations:

$$m_A v_{Ay}^- + m_B v_{By}^- = m_A v_{Ay}^+ + m_B v_{By}^+, \tag{5.23}$$

$$v_{By}^+ - v_{Ay}^+ = e\left(v_{Ay}^- - v_{By}^-\right), \tag{5.24}$$

where the y axis coincides with the LOI.

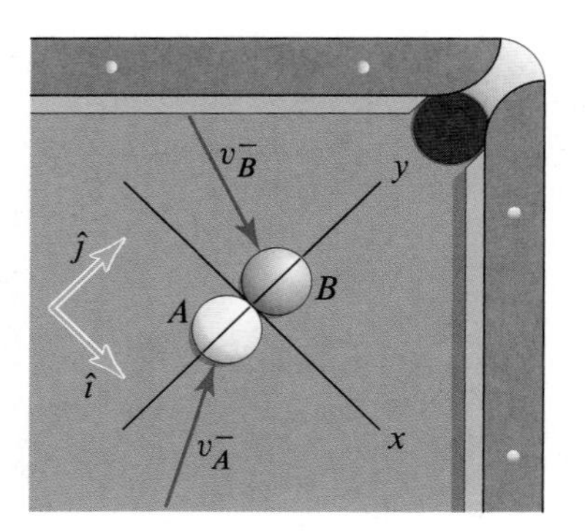

Figure 5.15
Oblique central impact of two billiard balls.

Unconstrained oblique central impact

Figure 5.15 shows an oblique central impact of two billiard balls A and B, in which their masses, preimpact velocities, and the COR are known, and we wish to find the postimpact velocities. The solution requires four equations in the four components of the postimpact velocities. Modeling the impact as unconstrained, the impact-relevant FBDs for A and B as a system and for A and B individually are shown in Figs. 5.14(a) and (b), respectively. Figure 5.14(a) shows that the y component of the momentum of A and B together is conserved, and Fig. 5.14(b) shows that the x component of the momentum is conserved for A and B individually. Therefore, we have

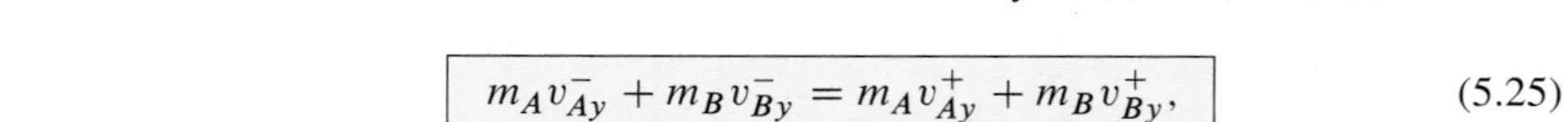

$$m_A v_{Ay}^- + m_B v_{By}^- = m_A v_{Ay}^+ + m_B v_{By}^+, \tag{5.25}$$

$$m_A v_{Ax}^- = m_A v_{Ax}^+ \quad \Rightarrow \quad v_{Ax}^- = v_{Ax}^+, \tag{5.26}$$

$$m_B v_{Bx}^- = m_B v_{Bx}^+ \quad \Rightarrow \quad v_{Bx}^- = v_{Bx}^+. \tag{5.27}$$

The final equation is the force law given by the COR equation applied along the LOI:

$$v_{By}^{+} - v_{Ay}^{+} = e(v_{Ay}^{-} - v_{By}^{-}). \tag{5.28}$$

Equations (5.25)–(5.28) form a system of four equations that can be used to find the postimpact velocity components of A and B.

Impact and energy

Since we have assumed that impacting objects do not move during an impact, the only work done during an impact is done by the impulsive forces, and this work can only be measured by comparing the pre- and postimpact kinetic energies of the colliding objects.

Experience tells us that the *total kinetic energy* of a system of colliding objects can only decrease. The total kinetic energy remains constant only if the collisions are perfectly elastic, which is an idealization. For this reason, it is common to say that energy is lost in a collision. In the case of two particles A and B, the energy loss is usually expressed as a percentage of the preimpact total kinetic energy, i.e.,

$$\text{Percentage of energy loss} = \frac{T^{-} - T^{+}}{T^{-}} \times 100\%, \tag{5.29}$$

where T^{-} and T^{+} are the pre- and postimpact *total* kinetic energies, respectively:

$$T^{-} = \tfrac{1}{2}m_A(v_A^{-})^2 + \tfrac{1}{2}m_B(v_B^{-})^2 \quad \text{and} \quad T^{+} = \tfrac{1}{2}m_A(v_A^{+})^2 + \tfrac{1}{2}m_B(v_B^{+})^2. \tag{5.30}$$

End of Section Summary

In this section we idealized particle impact as *an event spanning an infinitesimal time interval in which objects can experience a finite change in velocity at fixed position.* The model is based on the assumptions summarized in Table 5.2, and we discovered that there are two key elements to *every* impact problem: (1) the application of the impulse-momentum principle and (2) a force law telling us how the colliding objects rebound. When applying the impulse-momentum principle during an impact, only impulsive forces play a role, so they are the only forces included in impact-relevant FBDs.

Table 5.2
Assumptions used in our impact model.

Physical characteristic	**Impact assumption**
duration of impact	infinitesimal
displacement of particle	zero
force on particle	infinite
change in momentum	instantaneous

Problems involving the impact between two particles generally involve four unknowns, so four equations are needed. The geometry of an *unconstrained impact* (for which there are no external impulsive forces) between two particles is shown in Fig. 5.16, and the four equations come from

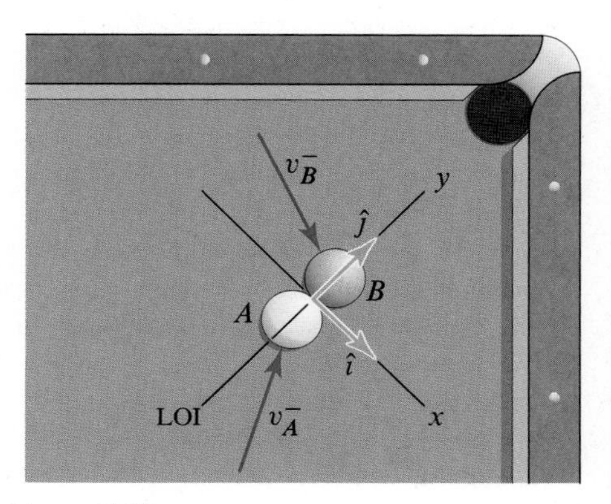

Figure 5.16
Geometry of two impacting particles.

1. Conservation of momentum of the two particles together along the LOI:

Eq. (5.25), p. 338

$$m_A v_{Ay}^{-} + m_B v_{By}^{-} = m_A v_{Ay}^{+} + m_B v_{By}^{+}.$$

2. Conservation of momentum for particle A in the x direction:

Eq. (5.26), p. 338

$$m_A v_{Ax}^{-} = m_A v_{Ax}^{+} \quad \Rightarrow \quad v_{Ax}^{-} = v_{Ax}^{+}.$$

3. Conservation of momentum for particle B in the x direction:

Eq. (5.27), p. 338

$$m_B v_{Bx}^- = m_B v_{Bx}^+ \quad \Rightarrow \quad v_{Bx}^- = v_{Bx}^+.$$

4. COR equation applied along the LOI:

Eq. (5.21), p. 337

$$e = \frac{\text{separation velocity}}{\text{approach velocity}} = \frac{v_{By}^+ - v_{Ay}^+}{v_{Ay}^- - v_{By}^-}.$$

The coefficient of restitution e determines the nature of the rebound between the two particles. When $e = 0$, the impact is called *plastic*; when $0 < e < 1$, the impact is called *elastic*; and when $e = 1$, it is called *perfectly elastic*. In an unconstrained, direct central impact, a plastic collision ($e = 0$) results in the objects sticking together postimpact. This is not necessarily the case in an oblique central impact, as velocity components along the plane of contact may be different.

Impact and energy. If $e < 1$, mechanical energy is lost during the impact. The energy loss is often indicated as a percentage of the preimpact total kinetic energy:

Eq. (5.29), p. 339

$$\text{Percentage of energy loss} = \frac{T^- - T^+}{T^-} \times 100\%,$$

where T^- and T^+ are the pre- and postimpact total kinetic energies, respectively.

EXAMPLE 5.5 *Direct Central Impact of Two Bowling Balls*

Bowling ball A, traveling at 6 ft/s, arrives at a return station and collides with ball B, which is at rest. Balls A and B have weights $W_A = 13$ lb and $W_B = 16$ lb, respectively, they have identical diameters, and the COR for the collision is $e = 0.98$. Determine the postimpact velocities of balls A and B.

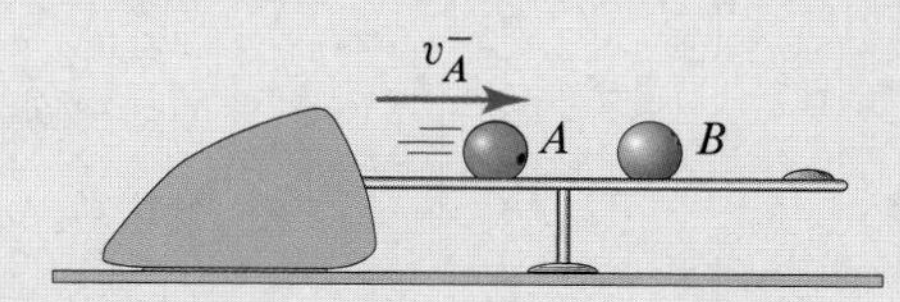

Figure 1
Ball arriving at a bowling ball return station.

SOLUTION

Road Map & Modeling The motion of the two balls takes place along a horizontal line, and so the collision is one-dimensional. Since the diameters of the balls are identical, the LOI is also horizontal, and so the contact force between the balls will be horizontal. This means that the vertical reactions between the balls and the ball return are nonimpulsive. Hence, the impact-relevant FBD for the two balls is shown in Fig. 2. Treating A and B as particles, we model this impact as an unconstrained direct central impact.

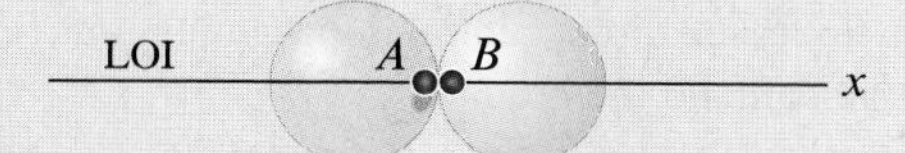

Figure 2. Impact-relevant FBD of the colliding particles.

Governing Equations

Balance Principles Referring to Fig. 2, since there are no impulsive external forces acting on the system, the momentum of the system is conserved in the x direction:

$$m_A v_{Ax}^- + m_B v_{Bx}^- = m_A v_{Ax}^+ + m_B v_{Bx}^+, \tag{1}$$

where m_A and m_B denote the masses of A and B, respectively.

Force Laws The force law characterizing impacts is the COR equation:

$$v_{Bx}^+ - v_{Ax}^+ = e\left(v_{Ax}^- - v_{Bx}^-\right). \tag{2}$$

Kinematic Equations Following the problem statement, the preimpact velocities are

$$v_{Ax}^- = 6\,\text{ft/s} \quad \text{and} \quad v_{Bx}^- = 0\,\text{ft/s}. \tag{3}$$

Computation After substituting Eqs. (3) into Eqs. (1) and (2), we are left with a system of two equations in the two unknowns v_{Ax}^+ and v_{Bx}^+. Solving this system, we obtain

$$v_{Ax}^+ = \frac{m_A v_{Ax}^- + m_B\left[v_{Bx}^- + e\left(v_{Bx}^- - v_{Ax}^-\right)\right]}{m_A + m_B} = -0.5545\,\text{ft/s}, \tag{4}$$

$$v_{Bx}^+ = \frac{m_B v_{Bx}^- + m_A\left[v_{Ax}^- + e\left(v_{Ax}^- - v_{Bx}^-\right)\right]}{m_A + m_B} = 5.326\,\text{ft/s}. \tag{5}$$

Discussion & Verification The solution conforms to common experience in that ball A, being lighter than ball B, rebounds backward (to the left). Also, given that the value of the COR is less than 1, we expect the impact to imply a loss of kinetic energy. A quick calculation based on the given data and the final results tells us that the pre- and postimpact total kinetic energies for the system are $T^- = 7.267$ ft·lb and $T^+ = 7.108$ ft·lb, respectively, with an energy loss of 2.185%. Since T^+ is slightly less than T^-, the solution behaves as expected, given that e is just slightly less than one.

Helpful Information

What about conservation of y momentum? We could, of course, conserve momentum in the y direction, but that would simply be an equation stating that $0 = 0$.

EXAMPLE 5.6 Direct Central Impact of a Ball with the Ground

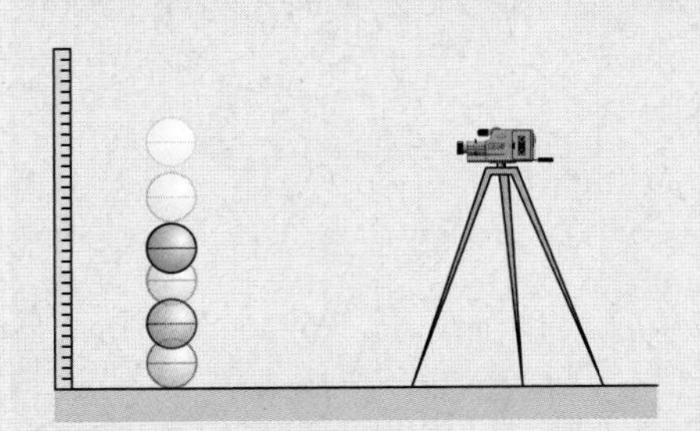

Figure 1
Video camera recording the motion of a ball falling on the ground and rebounding.

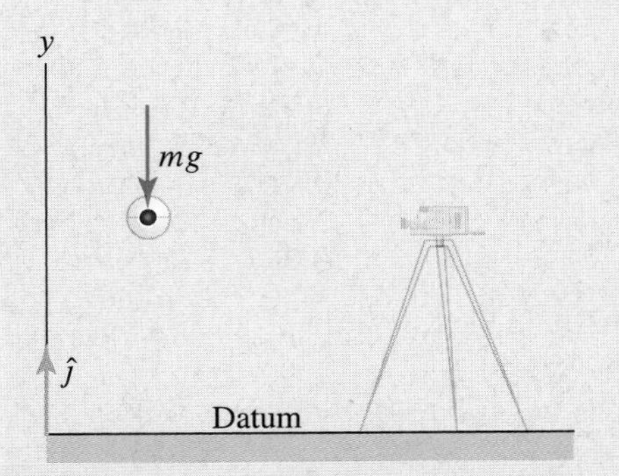

Figure 2
FBD of the ball before and after the impact with the ground.

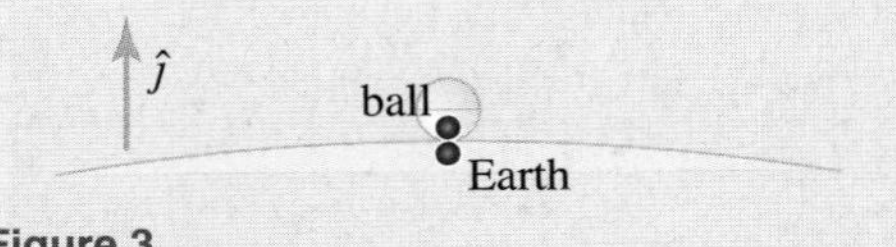

Figure 3
FBD of the ball during the impact with the ground/Earth.

An experiment is carried out to measure the COR for the impact of a ball with the ground (Fig. 1). The experiment consists of video recording the motion of a ball that is initially at rest and that is dropped onto the ground from a known height. By using the video recording we can measure the maximum rebound height of the ball. Use the measurements of the release height and the maximum rebound height to measure the COR for the impact.

SOLUTION

Road Map & Modeling Neglecting air resistance, the FBD before and after impact with the ground is shown in Fig. 2. The FBD of the ball and the ground during the impact is shown in Fig. 3. Referring to Fig. 2, we can use the work-energy principle to relate the release height of the ball to the speed with which it hits the ground. Similarly, we can relate the rebound height to the speed with which the ball rebounds off the ground. Referring to Fig. 3, we can apply the direct central impact equations (Eqs. (5.23) and (5.24)) to the impact between the ball and the ground.

Application of the Work-Energy Principle

Governing Equations

Balance Principles We will let ① denote the state of the ball at the instant of release, ② the state of the ball right before impact with the ground, ③ the state of the ball right after impact, and ④ the state of the ball at its maximum rebound height. Therefore, since all forces are conservative, the application of the work-energy principle between ① and ② and between ③ and ④ takes on the form

$$T_1 + V_1 = T_2 + V_2, \tag{1}$$

$$T_3 + V_3 = T_4 + V_4. \tag{2}$$

The expressions for the kinetic energies are given by

$$T_1 = \tfrac{1}{2}mv_1^2, \quad T_2 = \tfrac{1}{2}mv_2^2, \quad T_3 = \tfrac{1}{2}mv_3^2, \quad T_4 = \tfrac{1}{2}mv_4^2, \tag{3}$$

where v_1 is the speed at release, v_2 is the speed with which the ball collides with the ground, v_3 is the speed with which the ball rebounds off the ground, and v_4 is the speed at the maximum rebound height.

Force Laws Selecting the ground for our datum, we have

$$V_1 = mgh_i, \quad V_2 = 0, \quad V_3 = 0, \quad V_4 = mgh_f, \tag{4}$$

where h_i is its release height and h_f is its rebound height (the subscripts i and f stand for initial and final, respectively).

Kinematic Equations The speeds of the ball at ① and ④ are

$$v_1 = 0 \quad \text{and} \quad v_4 = 0. \tag{5}$$

Combining Eqs. (1)–(5) will allow us to solve for v_2 and v_3.

Application of the Impulse-Momentum Principle

Governing Equations

Balance Principles Referring to the impact-relevant FBD of the ball impacting the ground in Fig. 3, we see that there are no external impulsive forces in the y direction. Therefore, we will solve this problem exactly as we solved Example 5.5, except that now the Earth is one of the particles. Keeping in mind that v_2 and v_3 are speeds, and applying conservation of momentum in the y direction, we obtain

$$-mv_2 + m_e v_{ey}^- = mv_3 + m_e v_{ey}^+, \tag{6}$$

where m_e is the mass of the Earth, v_{ey}^- and v_{ey}^+ are the pre- and postimpact y components of the velocity of the Earth, respectively, and where we have used the fact that the ball moves in the negative and positive y directions before and after impact, respectively.

Force Laws The COR equation for the ball and the Earth is

$$v_3 - v_{ey}^+ = e\left(v_{ey}^- + v_2\right), \tag{7}$$

where, again, we accounted for the direction of the ball's motion before and after impact.

Kinematic Equations The kinematics equations in this situation come from considering the fact that the mass of the Earth is *many* orders of magnitude greater than the mass of the ball (10^{25} times the mass of the ball). Dividing both sides of Eq. (6) by m_e, we obtain

$$-\frac{m}{m_e}v_2 + v_{ey}^- = \frac{m}{m_e}v_3 + v_{ey}^+ \quad \Rightarrow \quad v_{ey}^- = v_{ey}^+ = 0, \tag{8}$$

where we have used $m/m_e \approx 0$ and the fact that the Earth is not moving before the impact to obtain $v_{ey}^- = v_{ey}^+ = 0$.

Computation Combining Eqs. (1)–(5), we have

$$mgh_i = \tfrac{1}{2}mv_2^2 \quad \text{and} \quad \tfrac{1}{2}mv_3^2 = mgh_f, \tag{9}$$

which can be solved for the speeds v_2 and v_3 to obtain

$$v_2 = \sqrt{2gh_i} \quad \text{and} \quad v_3 = \sqrt{2gh_f}. \tag{10}$$

Substituting Eq. (8) into Eq. (7), we obtain

$$v_3 = ev_2 \quad \Rightarrow \quad \boxed{e = \sqrt{\frac{h_f}{h_i}},} \tag{11}$$

where Eqs. (10) have been used.

Discussion & Verification Our final COR formula is consistent with our everyday experience. In fact, if we drop a ball from a height h_i, the maximum rebound height h_f is smaller than h_i, i.e., $h_f < h_i$. Hence, using Eq. (11), we can conclude that $0 \le e \le 1$, which is to say that e is in the expected range. Furthermore, it can be seen that for $e = 1$ we must have $h_f = h_i$, which is to say that the rebound height is equal to the release height only if the collision is perfectly elastic.

A Closer Look Note that we need not write both Eqs. (6) and (7) every time we have to solve a problem in which a particle impacts a nonmoving surface. We can just write

$$v_{\text{particle}}^+ - v_{\text{surface}}^+ = e\left(v_{\text{surface}}^- - v_{\text{particle}}^-\right), \tag{12}$$

along the LOI and then let $v_{\text{surface}}^- = v_{\text{surface}}^+ = 0$, to obtain the result that

$$v_{\text{particle}}^+ = -ev_{\text{particle}}^-, \tag{13}$$

where it is understood that $v_{\text{particle}}^{\pm}$ and $v_{\text{surface}}^{\pm}$ are velocity components along the LOI (i.e., they are not speeds).

Interesting Fact

Motion of the Earth due to impact (and other things). The Earth does move a very small amount when an object impacts it—conservation of momentum tells us that this *must* be so. As Eq. (8) tells us, the mass ratios are so small that the motion of the Earth in a typical impact, such as the one solved here, cannot be measured. Some objects on the Earth that move can cause changes in the Earth's motion. For example, earthquakes, which involve the motion of large portions of the Earth's crust, do produce changes in the Earth's motion that can be detected.

EXAMPLE 5.7 *A Sequence of Particle Collisions*

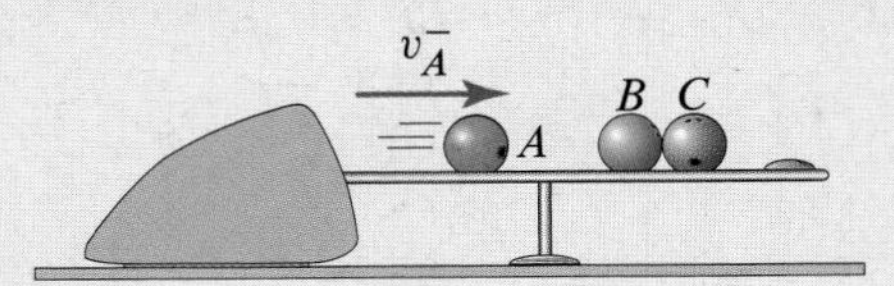

Figure 1
Ball arriving at a bowling ball return station.

Bowling ball A, traveling at 6 ft/s, arrives at a return station and collides with ball B, which is in contact with ball C. Balls B and C are initially at rest. Let $W_A = 16\,\text{lb}$, $W_B = 13\,\text{lb}$, and $W_C = 15\,\text{lb}$ be the weights of A, B, and C, respectively. In addition, let $e_{AB} = 0.98$ and $e_{BC} = 0.94$ be the CORs measured for the individual impacts between balls A and B and balls B and C, respectively. Assuming the balls all have the same diameter, determine the postimpact velocities of all three balls.

SOLUTION

Road Map & Modeling Until now we have only considered the impact of two particles at a time. Since A, B, and C are arranged in series, we will view the problem as a *sequence of two-body impacts*. At a minimum, this sequence consists of a collision between A and B followed by a collision between B and C, though additional collisions may occur. The final motion of the balls must satisfy the inequality

$$v_A^+ \le v_B^+ \le v_C^+, \tag{1}$$

where we assume all motion occurs in the x direction (see Fig. 2) and so the subscript x is omitted.

The balls all have the same diameter, so the LOI is parallel to the guides of the ball return. Therefore, the reaction forces between the balls and the guides are not affected by the impact and can be considered *nonimpulsive*. The resulting impact-relevant FBDs for the A-B and B-C impacts are shown in Fig. 2. As the motion is in the direction parallel to the LOI, all the impacts are unconstrained direct central impacts.

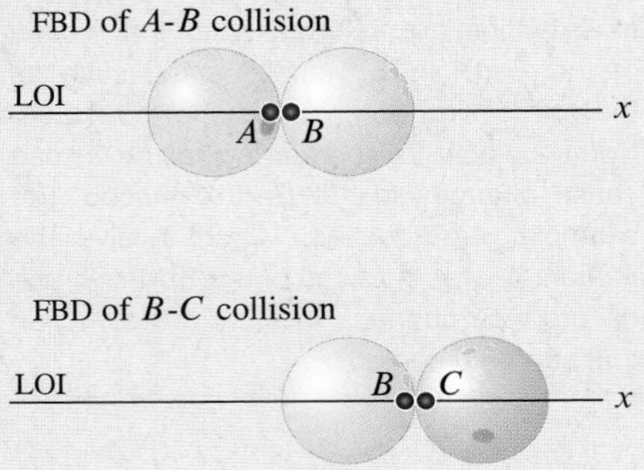

FBD of A-B collision

LOI A B x

FBD of B-C collision

LOI B C x

Figure 2
Impact-relevant FBDs for impacts between A and B and between B and C.

First (A-B) and Second (B-C) Collisions

Governing Equations

Balance Principles The A-B and B-C impacts are *separate but sequential* events. In view of Fig. 2, we can say that momentum is conserved in the x direction for these collisions, that is,

$$m_A v_{A1}^- + m_B v_{B1}^- = m_A v_{A1}^+ + m_B v_{B1}^+, \tag{2}$$
$$m_B v_{B2}^- + m_C v_{C2}^- = m_B v_{B2}^+ + m_C v_{C2}^+, \tag{3}$$

where v_{A1}^+ and v_{B1}^+ are the velocities of A and B right after the first collision; v_{B2}^- and v_{C2}^- are the velocities of B and C right before the second collision; and v_{B2}^+ and v_{C2}^+ are the velocities of B and C right after the second collision. Note that the velocity of B after the first collision must equal the velocity of B before the second, that is,

$$v_{B1}^+ = v_{B2}^-. \tag{4}$$

Force Laws For the first impact (A-B) and second impact (B-C), we have two corresponding COR equations, i.e.,

$$v_{B1}^+ - v_{A1}^+ = e_{AB}\left(v_{A1}^- - v_{B1}^-\right), \tag{5}$$
$$v_{C2}^+ - v_{B2}^+ = e_{BC}\left(v_{B2}^- - v_{C2}^-\right). \tag{6}$$

Kinematic Equations The known values of velocity for the various balls are

$$v_{A1}^- = 6\,\text{ft/s}, \quad v_{B1}^- = 0, \quad \text{and} \quad v_{C2}^- = 0. \tag{7}$$

Computation After we insert Eqs. (4) and (7), Eqs. (2), (3), (5), and (6) form a system of four equations in the four unknowns v_{A1}^+, v_{B1}^+, v_{B2}^+, and v_{C2}^+. For the first collision, we solve Eqs. (2) and (5) for v_{A1}^+ and v_{B1}^+ and then substitute this solution into Eqs. (3) and (6), which can then be solved for v_{B2}^+ and v_{C2}^+. Solving, we obtain

$$v_{A1}^+ = 0.6745\,\text{ft/s}, \quad v_{B1}^+ = 6.554\,\text{ft/s}, \quad v_{B2}^+ = -0.2575\,\text{ft/s}, \quad v_{C2}^+ = 5.904\,\text{ft/s}. \tag{8}$$

Discussion & Verification At this point, our candidate postimpact velocities are v_{A1}^+, v_{B2}^+, and v_{C2}^+. The values given in Eqs. (8) do not satisfy the inequalities in Eq. (1). While the result for C is acceptable, the values for v_{A1}^+ and v_{B2}^+ say that ball A is moving to the right and ball B is moving to the left—therefore, A and B will collide again. Hence, we need to study this third collision and determine if it is the last.

Third Collision (A-B)

Governing Equations

Balance Principles Referring again to the top half of Fig. 2, we can write a conservation of momentum equation for A and B during the third collision as

$$m_A v_{A3}^- + m_B v_{B3}^- = m_A v_{A3}^+ + m_B v_{B3}^+. \tag{9}$$

We need to realize that A is not involved in the second collision, and so the velocity of A after the first collision is the same as the velocity of A before the third collision. In addition, the velocity of B right after the second collision is equal to the velocity of B right before the third collision. Mathematically, these two statements give

$$v_{A1}^+ = v_{A3}^- \quad \text{and} \quad v_{B2}^+ = v_{B3}^-. \tag{10}$$

Force Laws The COR equation for this third impact is given by

$$v_{B3}^+ - v_{A3}^+ = e_{AB}(v_{A3}^- - v_{B3}^-). \tag{11}$$

Kinematic Equations The known velocity components for this collision are

$$v_{A3}^- = v_{A1}^+ = 0.6745\,\text{ft/s} \quad \text{and} \quad v_{B3}^- = v_{B2}^+ = -0.2575\,\text{ft/s}. \tag{12}$$

Computation Substituting Eqs. (12) into Eqs. (9) and (11), we have a system of two equations in the two unknowns v_{A3}^+ and v_{B3}^+ whose solution is

$$v_{A3}^+ = -0.1527\,\text{ft/s} \quad \text{and} \quad v_{B3}^+ = 0.7606\,\text{ft/s}. \tag{13}$$

With these results, we now satisfy Eq. (1), and so the final answers are

$$\boxed{v_A^+ = -0.1527\,\text{ft/s}, \quad v_B^+ = 0.7606\,\text{ft/s}, \quad v_C^+ = 5.904\,\text{ft/s}.} \tag{14}$$

Discussion & Verification At the end of the third collision in the sequence, we know that ball A is traveling to the left while ball B and ball C are traveling to the right. A fourth collision between B and C cannot occur because ball C is moving to the right faster than B. In addition, we should expect that the kinetic energy will decrease in a non-perfectly elastic collision. In fact, the final total kinetic energy of the system is 8.241 ft·lb, whereas the initial kinetic energy was 8.944 ft·lb, with an energy loss of 7.864%.

A Closer Look The reduction in kinetic energy in this example is larger than that we encountered in Example 5.5 (2.185%), but this is to be expected from the accumulated losses in three nonelastic collisions. It should also be noted that while e_{AB} in this example is equal to the COR used in Example 5.5, the COR for the B-C impact has a lower value of 0.94, which further contributes to the energy loss.

EXAMPLE 5.8 *Oblique Central Impact in Air Hockey*

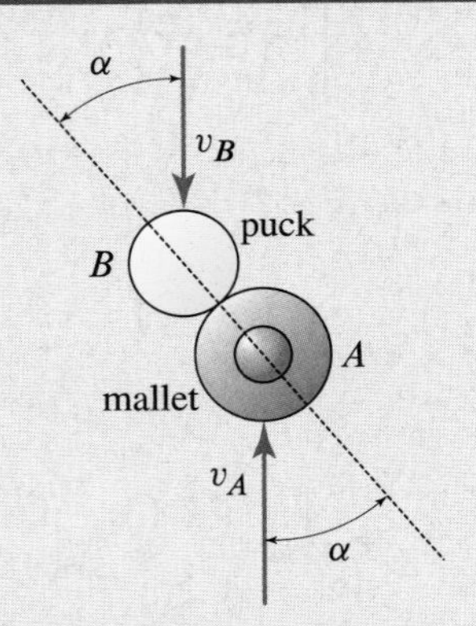

Figure 1
A hockey puck B (yellow) colliding with a mallet A (red) in an air hockey game.

In an air hockey game a 6 oz mallet A is let go and collides with a 1 oz puck B while the puck is in motion (Fig. 1). Assuming that the collision is perfectly elastic, that $\alpha = 40°$, and that the preimpact speeds of the mallet and puck are 9 and 20 ft/s, respectively, determine the postimpact velocities of the mallet and the puck. Finally, verify that the total pre- and postimpact kinetic energies are equal in a perfectly elastic collision.

SOLUTION

Road Map & Modeling Modeling both the puck and the mallet as particles, the LOI is the line connecting their centers, which is the dashed line in Fig. 1. Since the preimpact velocities are not parallel to the LOI, this is an oblique central impact. The impact-relevant FBD for A and B taken together is shown in Fig. 2. The individual FBDs of A and B are

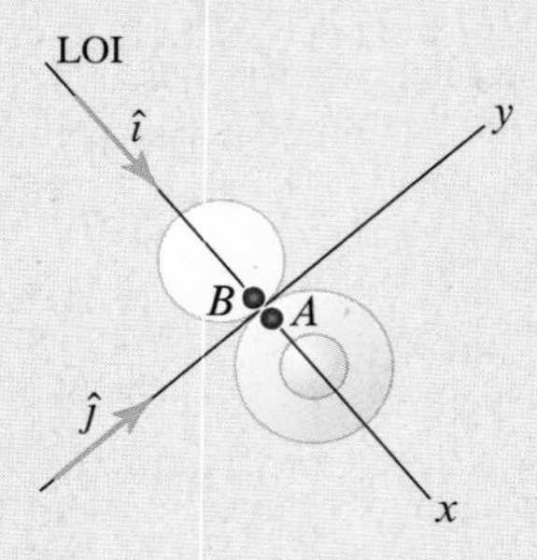

Figure 2. FBD for A and B combined. We have chosen the x axis to coincide with the LOI.

shown in Fig. 3, in which we have assumed that the impact surface is frictionless, and so internal impulsive forces are parallel to the LOI.

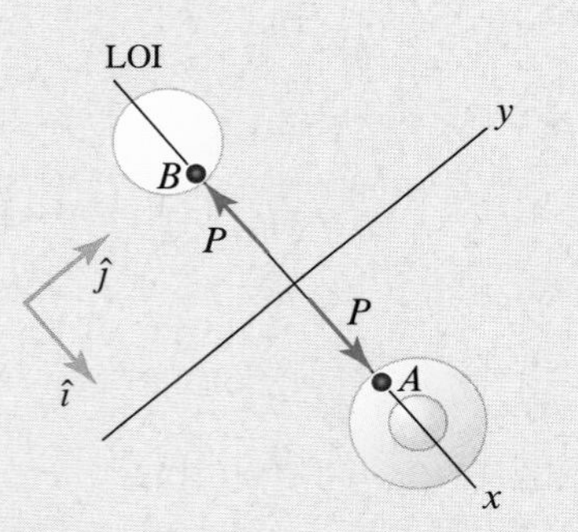

Figure 3
FBDs for A and B taken individually. Force P is the impulsive contact force between A and B during the impact.

Governing Equations

Balance Principles Using the solution strategy developed for unconstrained oblique central impacts, we (1) write a conservation of momentum equation along the LOI for the system, and (2) write two additional conservation of momentum equations, one for each particle, in the direction orthogonal to the LOI:

$$m_A v_{Ax}^- + m_B v_{Bx}^- = m_A v_{Ax}^+ + m_B v_{Bx}^+, \tag{1}$$

$$m_A v_{Ay}^- = m_A v_{Ay}^+ \quad \Rightarrow \quad v_{Ay}^- = v_{Ay}^+, \tag{2}$$

$$m_B v_{By}^- = m_B v_{By}^+ \quad \Rightarrow \quad v_{By}^- = v_{By}^+. \tag{3}$$

Equations (2) and (3) reflect the fact that, for each particle, there are no impulsive forces acting in the direction perpendicular to the LOI (see Fig. 3).

Force Laws Since the collision is perfectly elastic (i.e., $e = 1$), the COR equation for this impact is

$$v_{Bx}^+ - v_{Ax}^+ = v_{Ax}^- - v_{Bx}^-. \tag{4}$$

Kinematic Equations The preimpact velocities are known for both particles, and referring to Fig. 4, they are

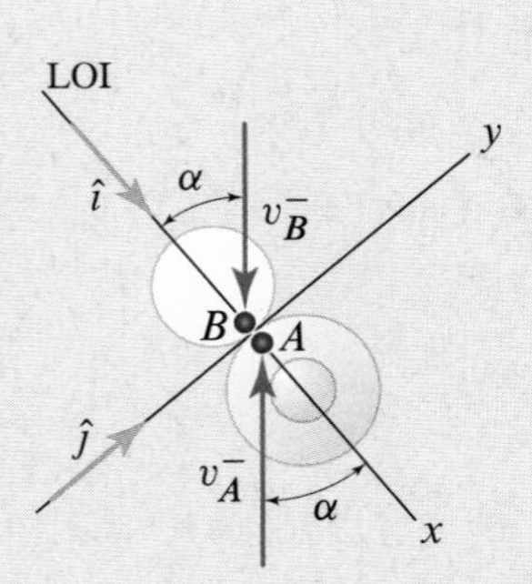

Figure 4
Geometry of the impact between the puck and mallet.

$$v_{Ax}^- = -v_A^- \cos\alpha \quad \Rightarrow \quad v_{Ax}^- = -6.894\,\text{ft/s}, \tag{5}$$

$$v_{Ay}^- = v_A^- \sin\alpha \quad \Rightarrow \quad v_{Ay}^- = 5.785\,\text{ft/s}, \tag{6}$$

$$v_{Bx}^- = v_B^- \cos\alpha \quad \Rightarrow \quad v_{Bx}^- = 15.32\,\text{ft/s}, \tag{7}$$

$$v_{By}^- = -v_B^- \sin\alpha \quad \Rightarrow \quad v_{By}^- = -12.86\,\text{ft/s}. \tag{8}$$

Computation Substituting Eqs. (5)–(8) into Eqs. (1)–(4) gives a system of four equations in the four unknowns v_{Ax}^{+}, v_{Ay}^{+}, v_{Bx}^{+}, and v_{By}^{+}. It is helpful to notice, however, that the equations in the x direction are decoupled from those in the y direction. This means that we can view Eqs. (1) and (4) as forming a system of two equations in the two unknowns v_{Ax}^{+} and v_{Bx}^{+}. The solution of these two equations gives the expressions in Eqs. (4) and (5) in Example 5.5. Substituting in given numerical values yields

$$\boxed{v_{Ax}^{+} = -0.5472\,\text{ft/s} \quad \text{and} \quad v_{Bx}^{+} = -22.76\,\text{ft/s}.} \tag{9}$$

As far as the remaining two unknowns are concerned, these can be obtained directly by substituting Eqs. (6) and (8) into Eqs. (2) and (3), respectively, thus giving

$$\boxed{v_{Ay}^{+} = 5.785\,\text{ft/s} \quad \text{and} \quad v_{By}^{+} = -12.86\,\text{ft/s}.} \tag{10}$$

We can now verify the claim that, for an elastic collision, the pre- and postimpact kinetic energies have the same value. The preimpact total kinetic energy is

$$\boxed{T^{-} = \tfrac{1}{2}m_A(v_A^{-})^2 + \tfrac{1}{2}m_B(v_B^{-})^2 = 0.8599\,\text{ft}\cdot\text{lb},} \tag{11}$$

and the postimpact total kinetic energy is

$$\boxed{T^{+} = \tfrac{1}{2}m_A(v_A^{+})^2 + \tfrac{1}{2}m_B(v_B^{+})^2 = 0.8599\,\text{ft}\cdot\text{lb},} \tag{12}$$

thus verifying that the energy is unchanged during a perfectly elastic collision.

Discussion & Verification The result in Eqs. (11) and (12) is sufficient to give us confidence that our solution is correct. However, it is also helpful to look geometrically at the postimpact velocities. Referring to Fig. 5, we see that

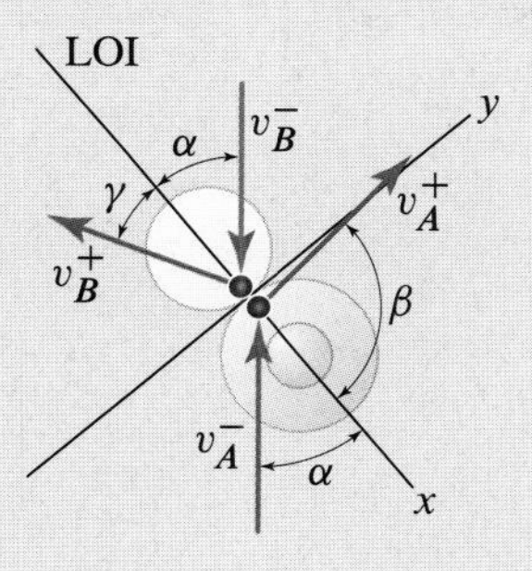

Figure 5. Postimpact geometry of the puck and the mallet.

$$\beta = 180^\circ + \tan^{-1}\left[\frac{v_{Ay}^{+}}{v_{Ax}^{+}}\right] = 95.40^\circ \quad \text{and} \quad \gamma = \tan^{-1}\left[\frac{v_{By}^{+}}{v_{Bx}^{+}}\right] = 29.46^\circ. \tag{13}$$

These angles seem plausible because A and B must have opposite horizontal momenta after impact. In fact, knowing that the total preimpact momentum is vertical and that the total momentum must be conserved, we expect the postimpact horizontal component of the total momentum to be equal to zero. Therefore, we would know immediately that we had made an error if the particles moved both to the left or both to the right after impact.

EXAMPLE 5.9 *Constrained Impact Between a Car and a Truck*

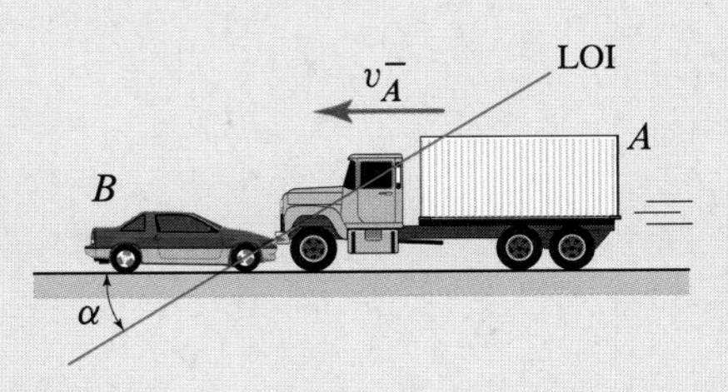

Figure 1
Low-velocity impact between a passenger car and a medium-sized truck. The angle $\alpha = 25°$, the COR for the collision is $e = 0.1$, $m_A = 15{,}000$ kg, $m_B = 1100$ kg, and $v_A^- = 7.3$ km/h.

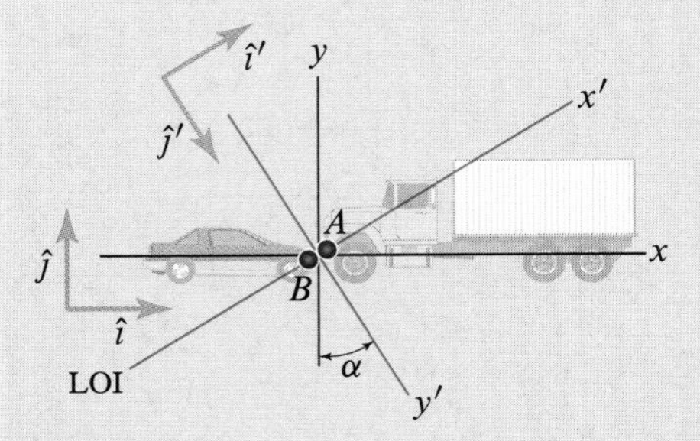

Figure 2
Impact-relevant FBD for the A-B collision under the assumption that the pavement is compliant.

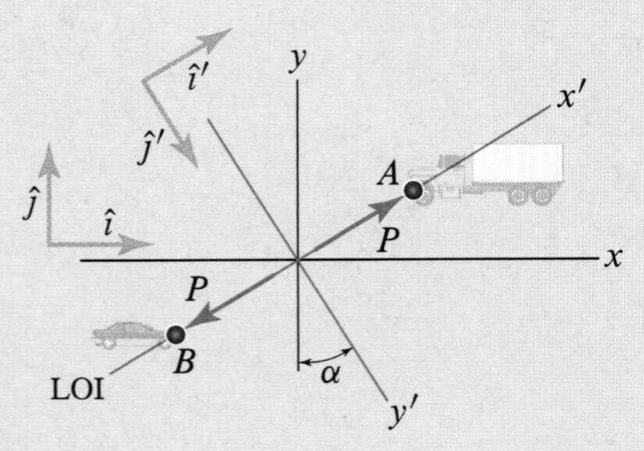

Figure 3
Impact-relevant FBDs for A and B taken individually, under the assumption that the pavement is compliant.

A truck A collides with a parked car B (Fig. 1). Assuming that the friction between A and B and between B and the ground is negligible, compare the postimpact velocities if (a) the pavement is compliant (responds nonimpulsively) and (b) the pavement is rigid (responds impulsively).

SOLUTION

Road Map In (a), the impact is unconstrained since the compliant pavement can't exert an impulsive force. In (b), the impact is modeled as a *constrained* impact. In the latter case, we cannot rely on the general solution strategies developed for unconstrained impact, though we can still solve (b) by analyzing the impact-relevant FBDs.

Part (a): Compliant Pavement

Modeling We model A and B as particles. We neglect friction and gravity because they are nonimpulsive. Due to the orientation of the LOI, A will push B into the pavement, which is *compliant* and thus has a springlike behavior. This means that the reaction between B and the pavement cannot grow large instantaneously; it must, therefore, be nonimpulsive. The impact-relevant FBD for the A-B system under these assumptions is shown in Fig. 2, in which we see that this is an unconstrained oblique central impact. The individual FBDs for A and B are shown in Fig. 3.

Governing Equations

Balance Principles As discussed in Example 5.8, the FBDs in Figs. 2 and 3 allow us to write the following statements of conservation of linear momentum:

$$m_A v_{Ax'}^- + m_B v_{Bx'}^- = m_A v_{Ax'}^+ + m_B v_{Bx'}^+, \tag{1}$$

$$m_A v_{Ay'}^- = m_A v_{Ay'}^+ \quad \Rightarrow \quad v_{Ay'}^- = v_{Ay'}^+, \tag{2}$$

$$m_B v_{By'}^- = m_B v_{By'}^+ \quad \Rightarrow \quad v_{By'}^- = v_{By'}^+. \tag{3}$$

Force Laws The COR equation applied along the LOI gives

$$v_{Bx'}^+ - v_{Ax'}^+ = e\left(v_{Ax'}^- - v_{Bx'}^-\right). \tag{4}$$

Kinematic Equations The preimpact velocities are

$$v_{Ax}^- = -2.028\ \text{m/s}, \quad v_{Ay}^- = 0, \quad v_{Bx}^- = 0, \quad \text{and} \quad v_{By}^- = 0. \tag{5}$$

To transform velocity components from the xy to the $x'y'$ coordinate systems, we use the following relations:

$$\hat{\imath}' = \cos\alpha\,\hat{\imath} + \sin\alpha\,\hat{\jmath} \quad \text{and} \quad \hat{\jmath}' = \sin\alpha\,\hat{\imath} - \cos\alpha\,\hat{\jmath}. \tag{6}$$

Computation The solution of this problem, transformed into the $(\hat{\imath}', \hat{\jmath}')$ component system, is identical to that discussed in Example 5.8. Performing these calculations and transforming back into the xy coordinate system, we obtain

$$v_{Ax}^+ = -1.903\ \text{m/s}, \qquad v_{Ay}^+ = 0.05837\ \text{m/s}, \tag{7}$$

$$v_{Bx}^+ = -1.707\ \text{m/s}, \qquad v_{By}^+ = -0.7960\ \text{m/s}. \tag{8}$$

Part (b): Rigid Pavement

Modeling If the pavement is assumed to be *rigid*, then it is capable of instantaneously providing a reaction force that can grow as large as necessary to prevent B from moving vertically downward. We will, therefore, model this reaction force as impulsive. There is no impulsive reaction force on A since A will tend to be lifted off the ground. Hence, the impact-relevant FBD for the A-B system is that in Fig. 4, whereas the corresponding FBDs for A and B taken individually are given in Fig. 5. This is a *constrained oblique central impact.*

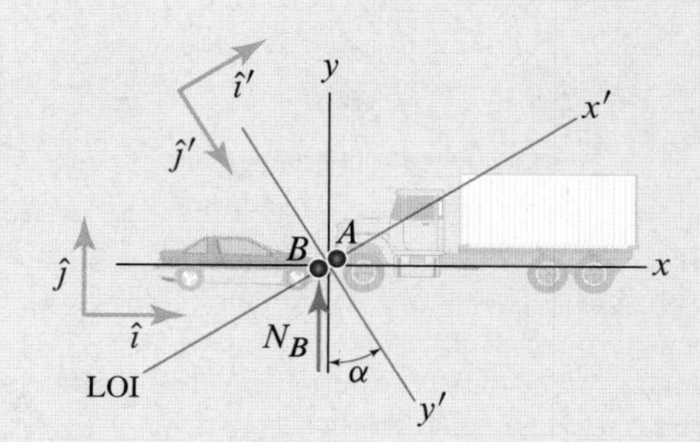

Figure 4
Impact-relevant FBD for the A-B collision under the assumption that the pavement is rigid. Force N_B denotes the impulsive reaction force exerted on B by the pavement.

Governing Equations

Balance Principles In the FBD in Fig. 4, there are no impulsive external forces in the x direction. Hence, the linear momentum for the A-B system is conserved in that direction. Using a similar argument, we know that A's linear momentum is conserved along the y' direction (see Fig. 5). However, the same cannot be said about B. Hence, the application of the impulse-momentum principle yields the following two equations:

$$m_A v_{Ax}^- + m_B v_{Bx}^- = m_A v_{Ax}^+ + m_B v_{Bx}^+, \tag{9}$$

$$m_A v_{Ay'}^- = m_A v_{Ay'}^+ \quad \Rightarrow \quad v_{Ay'}^- = v_{Ay'}^+. \tag{10}$$

Force Laws Referring to Fig. 5, the COR equation is the force law describing the effect of the internal impulsive force P on the components of the velocities of A and B along the LOI:

$$v_{Bx'}^+ - v_{Ax'}^+ = e\big(v_{Ax'}^- - v_{Bx'}^-\big). \tag{11}$$

Kinematic Equations Kinematically, the crucial difference between the current and the previous case is that B is prevented from moving vertically, i.e.,

$$v_{By}^+ = 0. \tag{12}$$

This equation must be added to the rest of the kinematic information we already possess, as expressed by Eqs. (5).

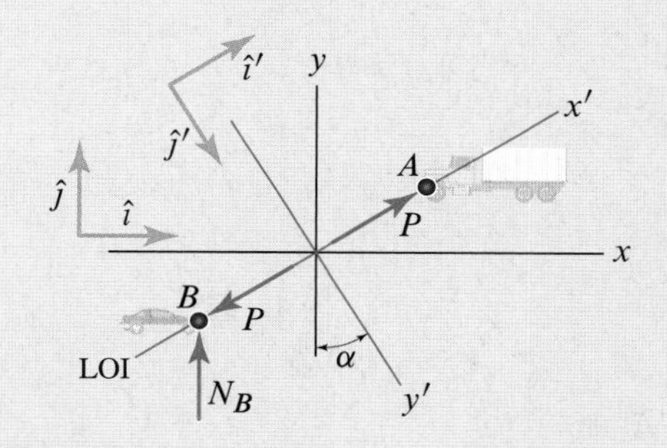

Figure 5
Impact-relevant FBDs for A and B taken individually, under the assumption that the pavement is rigid. Note that the impulsive force P for this case will have, in general, a different intensity than that in Fig. 3.

Computation Since v_{By}^+ is known, expressing the final answer in the xy coordinate system, we have three unknowns: v_{Ax}^+, v_{Ay}^+, and v_{Bx}^+. Hence, we need three corresponding equations. One of these equations is Eq. (9). The other two equations are obtained by rewriting Eqs. (10) and (11) in xy components. Using Eqs. (6) and (12), this change of components gives

$$\big(v_{Bx}^+ - v_{Ax}^+\big)\cos\alpha - v_{Ay}^+ \sin\alpha = e v_{Ax}^- \cos\alpha, \tag{13}$$

$$v_{Ax}^- \sin\alpha = v_{Ax}^+ \sin\alpha - v_{Ay}^+ \cos\alpha. \tag{14}$$

Taking advantage of the fact that $v_{Bx}^- = 0$ and then solving Eqs. (9), (13), and (14) for v_{Ax}^+, v_{Ay}^+, and v_{Bx}^+, we obtain

$$\boxed{v_{Ax}^+ = -1.878\,\text{m/s}, \quad v_{Ay}^+ = 0.07002\,\text{m/s}, \quad v_{Bx}^+ = -2.048\,\text{m/s}.} \tag{15}$$

Discussion & Verification There are several differences between the two solutions. First, we see that if the car B is not allowed to sink into the pavement, then the postimpact velocity of truck A in the y direction for the constrained impact is higher than in the unconstrained impact case. For the constrained impact case, right after impact, the car B moves to the left faster than the truck A, contrary to what happens in the unconstrained impact. Therefore, the solution for the constrained case predicts that B actually moves away from A. By contrast, the unconstrained solution predicts that A is actually going over B. This is not to say that one solution is right and the other is wrong. Both solutions might be correct, depending on the mechanical properties of the vehicles and pavement. All of the considerations mentioned here are limited by the fact that we are modeling A and B as particles, and therefore we are ignoring myriad additional effects of the objects' finite size.

EXAMPLE 5.10 *Obtaining the COR Equation from the Impulse-Momentum Principle*

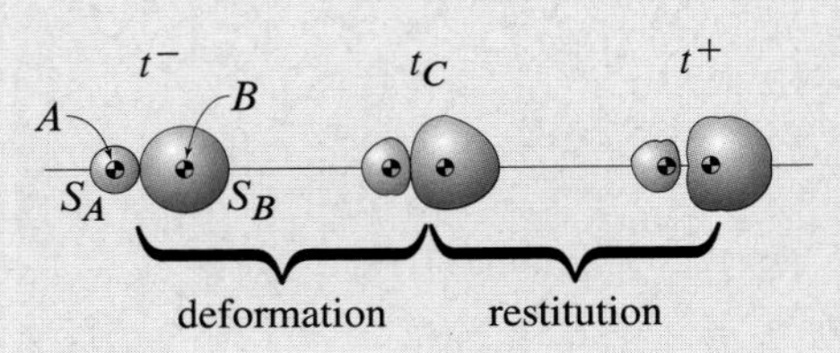

Figure 1
Collision between two deformable bodies with centers of mass at A and B, respectively. Note that at the end of the restitution process there may be some permanent deformation.

In addition to the experimental approach used earlier, Eq. (5.21) can be obtained in a purely *theoretical* way; the key to this alternative derivation is to view colliding objects as *deformable bodies* instead of particles.

Referring to Fig. 1, at time t^- let two *deformable* objects S_A and S_B, with mass centers at points A and B, respectively, begin to impact one another. Since S_A and S_B are deformable, at first A and B move closer to one another until S_A and S_B reach maximum deformation at, say, time t_C. At this time A and B share a common velocity v_C. After t_C, S_A and S_B push away from one another until t^+, at which time they separate. The processes occurring between t^- and t_C are called *deformation* and between t_C and t^+ are called *restitution*. Letting the internal deformation force between S_A and S_B be $D(t)$ and the internal restitution force be $R(t)$ (see Fig. 2), apply the impulse-momentum principle

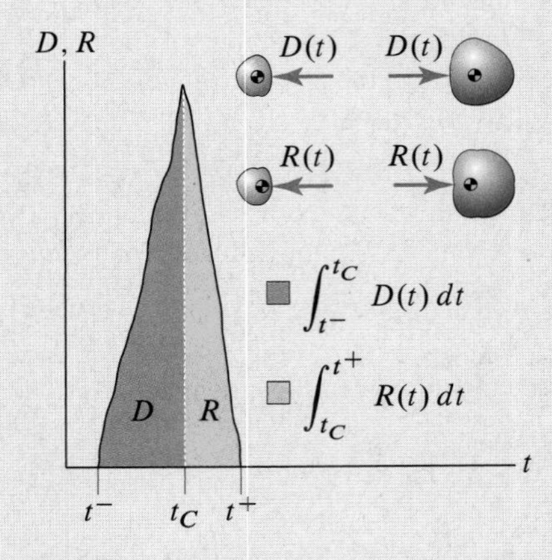

Figure 2. Internal deformation force $D(t)$ and restitution force $R(t)$ between two objects versus time. Since the restitution impulse is always less than the deformation impulse, the orange area is smaller than the purple area. The orange is equal to the purple area only in the ideal case of a perfectly elastic impact.

between t^- and t_C and between t_C and t^+ to each object, and show that the ratio of the restitution impulse to the deformation impulse is equal to the experimental definition of the coefficient of restitution.

SOLUTION

Road Map & Modeling The path to the solution is given to us — we simply need to apply the impulse-momentum principle as instructed in the problem statement and then find the ratio of the restitution to deformation impulses. Since the motion is one-dimensional, we will not use any subscripts to denote velocity components.

Governing Equations

Balance Principles Applying the impulse-momentum principle to S_A and S_B during the deformation process, we obtain

$$m_A v_A^- - \int_{t^-}^{t_C} D(t)\,dt = m_A v_C \quad \text{and} \quad m_B v_B^- + \int_{t^-}^{t_C} D(t)\,dt = m_B v_C, \tag{1}$$

where m_A and m_B are the masses of S_A and S_B, respectively, v_A^- and v_B^- are the preimpact velocities of A and B, respectively, and where $D(t)$ is the deformation force.

Applying the impulse-momentum principle in a similar way for the restitution process, we have

$$m_A v_C - \int_{t_C}^{t^+} R(t)\,dt = m_A v_A^+ \quad \text{and} \quad m_B v_C + \int_{t_C}^{t^+} R(t)\,dt = m_B v_B^+, \tag{2}$$

where v_A^+ and v_B^+ are the postimpact velocities of A and B, respectively, and where $R(t)$ is the restitution force.

Force Laws All forces are accounted for on the FBD.

Kinematic Equations We have determined the kinematic equations by noting that the centers of mass of the two bodies have a common velocity v_C when they reach maximum deformation during the impact.

Computation Eliminating the common velocity v_C from Eqs. (1), we obtain the relative preimpact velocity as

$$v_A^- - v_B^- = \left(\frac{1}{m_A} + \frac{1}{m_B}\right) \int_{t^-}^{t_C} D(t)\,dt. \tag{3}$$

Similarly eliminating the common velocity v_C from Eqs. (2), we obtain the following expression for the relative postimpact velocity:

$$v_B^+ - v_A^+ = \left(\frac{1}{m_A} + \frac{1}{m_B}\right) \int_{t_C}^{t^+} R(t)\,dt. \tag{4}$$

Using Eqs. (3) and (4), we can form the ratio of the restitution impulse to the deformation impulse as

$$\frac{\displaystyle\int_{t_C}^{t^+} R(t)\,dt}{\displaystyle\int_{t^-}^{t_C} D(t)\,dt} = \frac{v_B^+ - v_A^+}{v_A^- - v_B^-} = e, \tag{5}$$

which, as we have indicated, is the same as the experimental definition for the coefficient of restitution e.

Discussion & Verification We have shown that defining the coefficient of restitution to be the ratio of the restitution impulse to the deformation impulse is equivalent to the experimental definition of e.

A Closer Look Equation (5) tells us that the COR can be viewed as the ratio of the restitution impulse to the deformation impulse. This makes it easier to understand why $0 \le e \le 1$: this range of values covers the full spectrum of possibilities ranging from no restitution to full restitution. Another remarkable aspect of Eq. (5) is that the masses of the colliding bodies do not play a role. When we first introduced the COR equation, we implicitly neglected the effect of the masses involved in the impact. Equation (5) *demonstrates* that the masses of the colliding bodies do not play a role in controlling the ratio of separation to approach velocities.

Problems

Problem 5.63

If an impact is an event spanning an infinitesimally small time interval, is the total potential energy of two colliding objects conserved through the impact? What about the potential energy of each individual object?

Problem 5.64

If an impact is an event spanning an infinitesimally small time interval, is the total kinetic energy of two colliding objects conserved through an impact? What about the kinetic energy of each individual object?

Problem 5.65

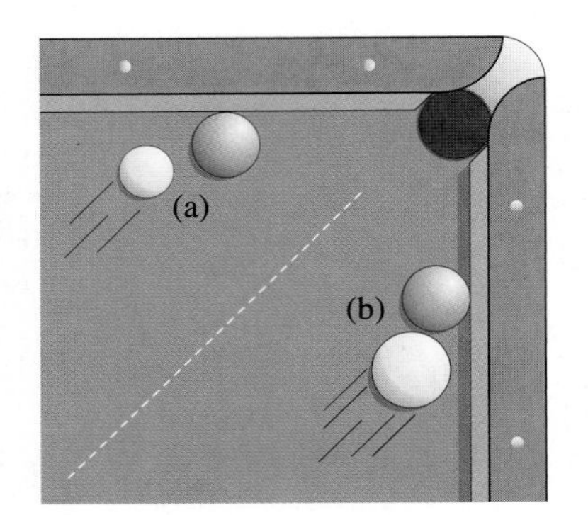

Figure P5.65

Although competition rules prohibit significant difference in size, typical coin-operated pool tables may present players with a significant difference in diameter between the typical object ball (i.e., a colored ball) and the cue ball (i.e., the white ball). In fact, once an object ball goes into a pocket, it is captured by the table, whereas a cue ball must always be returned to the player; and it is not uncommon for the return mechanism to use the difference in ball diameter to separate the cue ball from the rest. Given this, suppose we want to hit a ball resting against the bumper in such a way that, after the collision, it moves along the bumper. Modeling the contact between balls as frictionless, establish whether or not it is possible to execute the shot in question with (a) an undersized cue ball and (b) an oversized cue ball.

Problem 5.66

An 8600 lb Ford Excursion A traveling with a speed $v_A = 55$ mph collides head-on with a 1990 lb Smart Fortwo B traveling in the opposite direction with a speed $v_B = 35$ mph. Determine the postimpact velocity of the two cars if the impact is perfectly plastic.

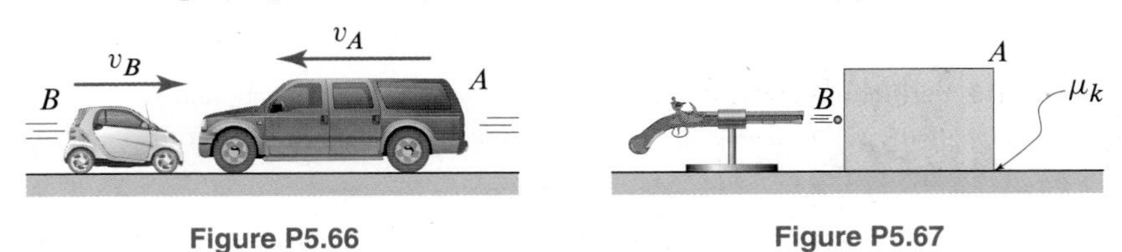

Figure P5.66 Figure P5.67

Problem 5.67

A 323 gr bullet (1 lb = 7000 gr) hits a 2 kg block that is initially at rest. After the collision, the bullet becomes embedded in the block, and they slide a distance of 0.31 m. If the coefficient of friction between the block and the ground is $\mu_k = 0.7$, determine the preimpact speed of the bullet. Although the definition of the unit "grain" is given in terms of pounds, express the answer in SI units.

Problem 5.68

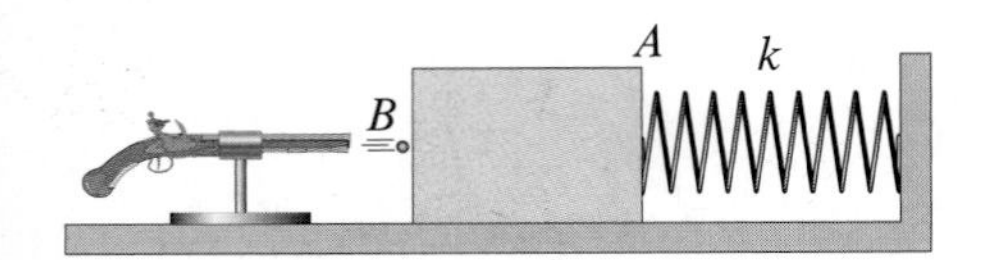

Figure P5.68

A 323 gr bullet B (1 lb = 7000 gr) moving at $v_0 = 800$ ft/s hits a 4 lb block A that is initially at rest. The block is attached to an uncompressed spring of stiffness $k = 6000$ lb/ft. After the collision, the bullet becomes embedded in the block. Neglecting friction between the block and the surface on which it slides, determine the compression of the spring required to bring the system to a stop.

Problems 5.69 through 5.71

The ballistic pendulum used to be a common tool for the determination of the muzzle velocity of bullets as a measure of the performance of firearms and ammunition (nowadays, the ballistic pendulum has been replaced by the ballistic chronograph, an electronic device). The ballistic pendulum is a simple pendulum that allows one to record the maximum swing angle of the pendulum arm caused by the firing of a bullet into the pendulum bob.

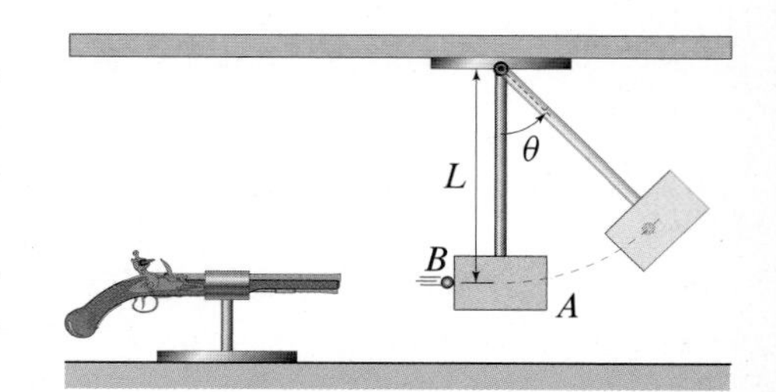

Figure P5.69–P5.71

Problem 5.69 Letting L be the length of the pendulum's arm (whose mass is assumed to be negligible), m_A be the bob's mass, and m_B be the mass of the bullet, and assuming that the pendulum is at rest when the weapon is fired, derive the formula that relates the pendulum's maximum swing angle to the impact velocity of the bullet.

Problem 5.70 Let $L = 1.5\,\text{m}$ and $m_A = 6\,\text{kg}$. For George Washington's 0.58 caliber pistol, which fired a roundball of mass $m_B = 87\,\text{g}$, it is found that the maximum swing angle of the pendulum is $\theta_{\text{max}} = 46°$. Determine the preimpact speed of the bullet B.

Problem 5.71 Suppose we want to build a ballistic pendulum to test rifles using standard NATO 7.62 mm ammunition, i.e., ammunition for which a (single) cartridge weighs roughly 147 gr (1 lb = 7000 gr) and the muzzle speed is typically 2750 ft/s. If the pendulum's length is taken to be 5 ft, and if we are to fire from a short distance so that there is a negligible decrease in speed before the bullet reaches the pendulum, what is the minimum weight we need to give to the pendulum bob to avoid having the pendulum swing to an angle greater than 90°?

Problems 5.72 and 5.73

Two bumper cars A and B collide head-on as shown. The weights of A and B (including the drivers) are $W_A = 635\,\text{lb}$ and $W_B = 650\,\text{lb}$, respectively. In addition, the preimpact speeds of A and B are $v_A^- = 6\,\text{ft/s}$ and $v_B^- = 4\,\text{ft/s}$, respectively.

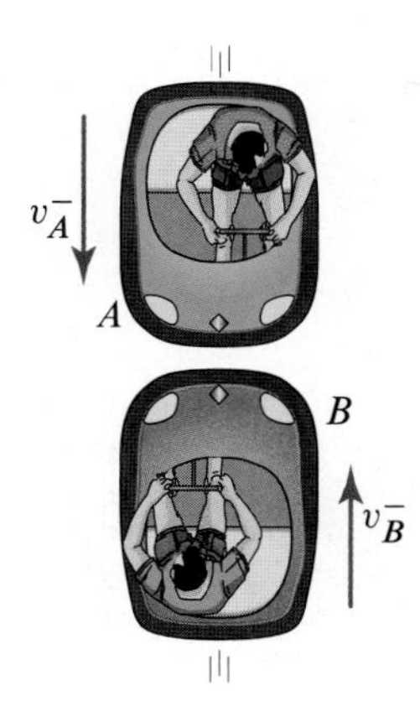

Figure P5.72 and P5.73

Problem 5.72 Determine the postimpact speeds of A and B if the COR for the impact is $e = 0.65$.

Problem 5.73 Determine the COR of the impact and the postimpact speed of B if the postimpact speed of A is equal to zero.

Problem 5.74

Consider a direct central impact for two spheres. Let m_A, m_B, and e denote the mass of sphere A, the mass of sphere B, and the COR, respectively. If sphere B is at rest before the collision, determine the relation that m_A, m_B, and e need to satisfy in order for A to come to a complete stop right after impact.

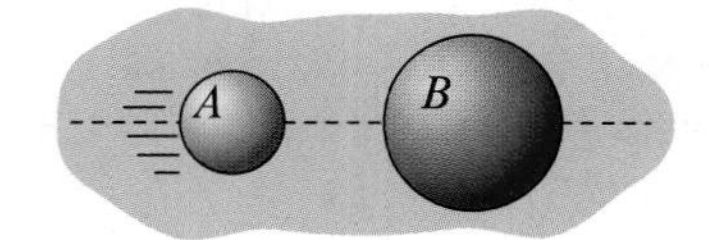

Figure P5.74

Problem 5.75

A bullet B of mass m_B traveling with a speed $v_0 = 1200\,\text{m/s}$ ricochets off a fixed steel plate A of mass m_A. Let $m_A \gg m_B$ so that it can be assumed that $m_B/m_A \to 0$. If the incidence angle of the bullet is $\theta = 15°$ and the COR of the impact is $e = 0.5$, determine the rebound angle ϕ, as well as the bullet's rebound speed.

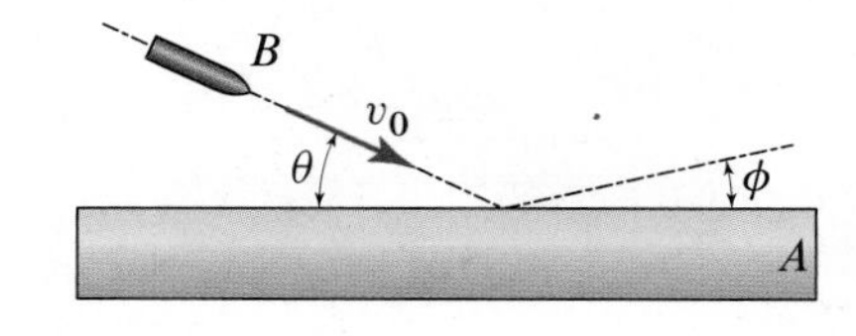

Figure P5.75

Problem 5.76

The official rules of tennis specify that

> The ball shall have a [re]bound of more than 53 in. (134.62 cm) and less than 58 in. (147.32 cm) when dropped 100 in. (254.00 cm) upon a concrete base.

Understanding the expression "when dropped" as "when dropped from rest," determine the range of acceptable CORs for the collision of a tennis ball with concrete.

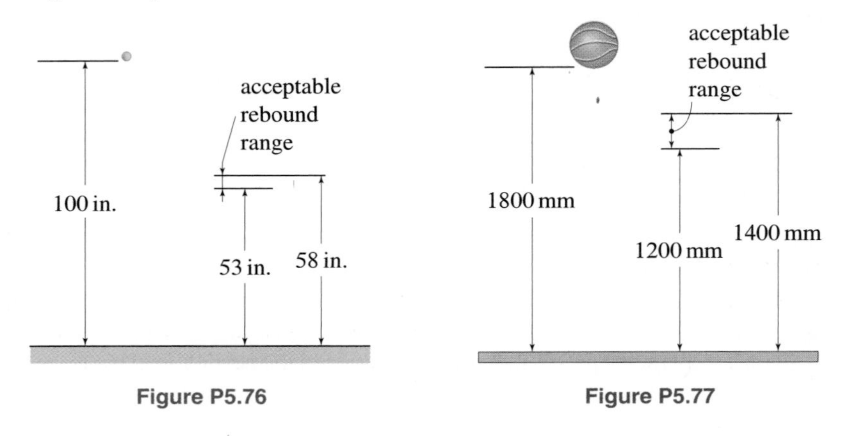

Figure P5.76

Figure P5.77

Problem 5.77

The official rules of basketball specify that a basketball is properly inflated

> such that when it is dropped onto the playing surface from a height of about 1800 mm measured from the bottom of the ball, it will rebound to a height, measured to the top of the ball, of not less than about 1200 mm nor more than about 1400 mm.

Based on this rule, ignoring the diameter of the ball, and understanding the expression "when it is dropped" as "when it is dropped from rest," determine the range of acceptable CORs for the collision between the ball and the court's surface.

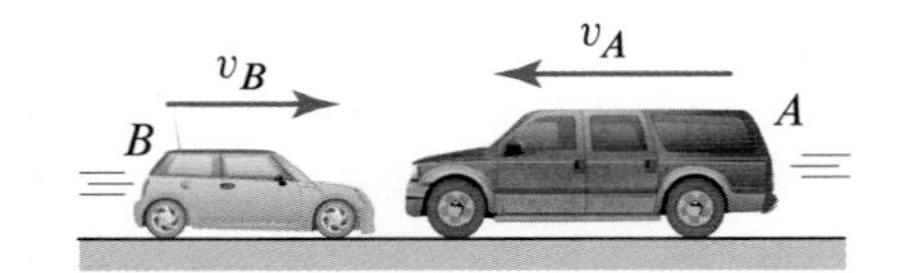

Figure P5.78

Problem 5.78

A Ford Excursion A, with a mass $m_A = 3900\,\text{kg}$ and traveling with a speed $v_A = 25\,\text{km/h}$, collides head-on with a Mini Cooper B, with a mass $m_B = 1200\,\text{kg}$, traveling in the opposite direction with a speed $v_B = 15\,\text{km/h}$. Model the impact of A and B as an unconstrained direct central impact of particles and determine the time needed for each car to stop if the cars slide after impact with $\mu_k = 0.8$ and if the impact's COR is $e = 0.1$.

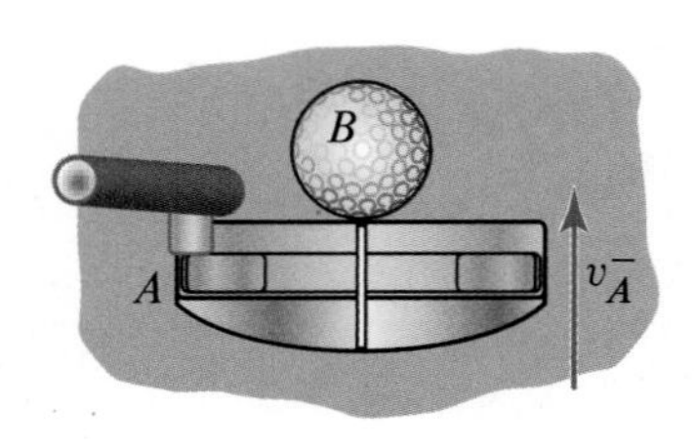

Figure P5.79

Problem 5.79

A golfer strikes a stationary ball B with a putter. At the time of impact, the putter's head A is traveling horizontally with a speed $v_A^- = 1.2\,\text{m/s}$. Model the impact as an unconstrained direct central impact of two particles, and let the masses of A and B be $m_A = 200\,\text{g}$ and $m_B = 46\,\text{g}$, respectively. Determine the COR of the collision if the postimpact speed of B is $v_B^+ = 1.76\,\text{m/s}$.

Problems 5.80 and 5.81

Car A, with $m_A = 1550\,\text{kg}$, is stopped at a red light. Car B, with $m_B = 1865\,\text{kg}$ and a speed of 40 km/h, fails to stop before impacting car A. After impact, cars A and B slide over the pavement with a coefficient of friction $\mu_k = 0.65$.

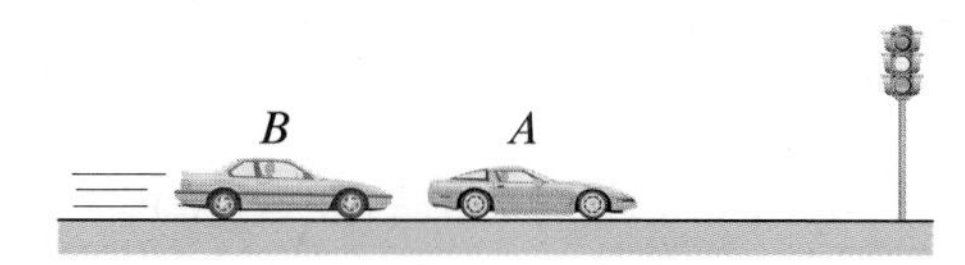

Figure P5.80 and P5.81

Problem 5.80 How far will the cars slide if the cars become entangled?

Problem 5.81 How far will each car slide if the COR for the impact is $e = 0.2$?

Problems 5.82 and 5.83

A platform bench scale consists of a 120 lb plate resting on linear elastic springs whose combined spring constant is $k = 5000\,\text{lb/ft}$. Let $W = k(\delta - \delta_0)$ be the weight measurement actually provided by the scale (that is, it reads zero pounds when nothing is on the plate), where δ_0 is the spring's compression due to the weight of the scale's plate.

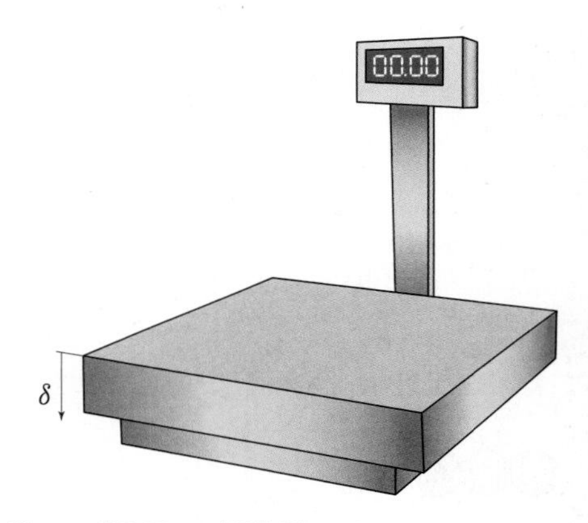

Figure P5.82 and P5.83

Problem 5.82 A 50 lb sack of portland cement is dropped (from rest) onto the scale from a height $h = 4\,\text{ft}$ measured from the scale's plate (there is no rebound of the sack). Determine the maximum weight displayed by the scale.

Problem 5.83 Repeat Prob. 5.82 with $h = 0\,\text{ft}$.

Problem 5.84

Bowling ball A, traveling with a speed $v_A^- = 6\,\text{ft/s}$, arrives at a return station and collides with ball B, which is in contact with ball C. Balls B and C are initially at rest. Let $W_A = W_B = W_C = 16\,\text{lb}$ be the weights of A, B, and C, respectively. Determine the postimpact velocities of A, B, and C if the CORs for all collisions are equal to 1.

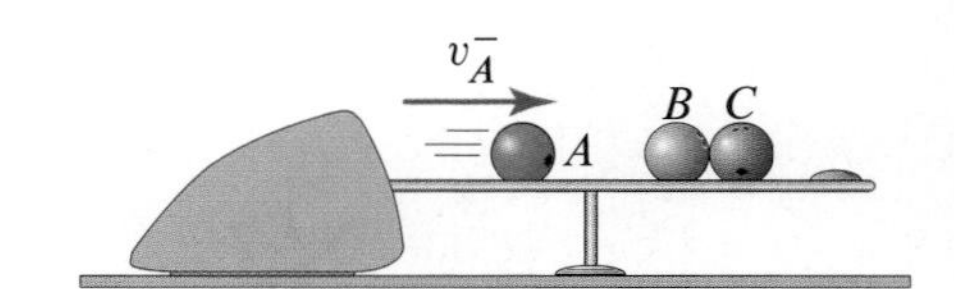

Figure P5.84

Problem 5.85

A steel ball is released from rest at the top of the incline as shown, rolls down the incline, and eventually rebounds off a steel strike plate. The length $L = 0.25\,\text{m}$, $\theta = 25°$, and $h = 0.15\,\text{m}$. Determine the rebound velocity of the ball if the COR between the ball and the strike plate is $e = 0.85$. Neglect friction between the ball and the incline.

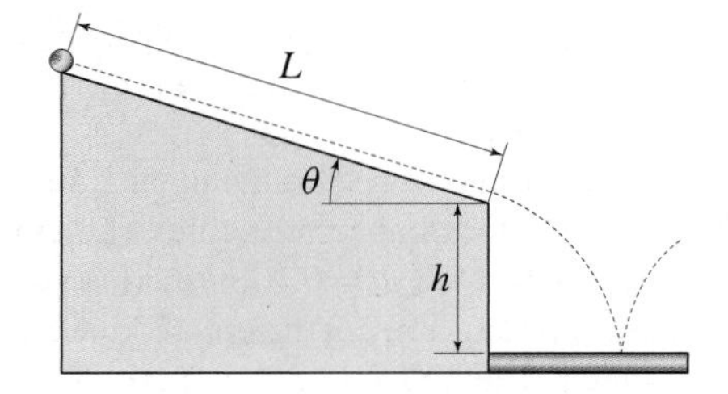

Figure P5.85

Problems 5.86 through 5.88

Newton's cradle is a common desk toy consisting of a number of identical pendulums with steel balls as bobs. These pendulums are arranged in a row in such a way that, when at rest, each ball is tangent to the next and the cords are all vertical. Assume that the COR for the impact of a ball with the next is $e = 1$. *Hint:* Although no gaps are shown between the bobs, assume small gaps are present in all cases so that the interactions can be examined as two-body impacts.

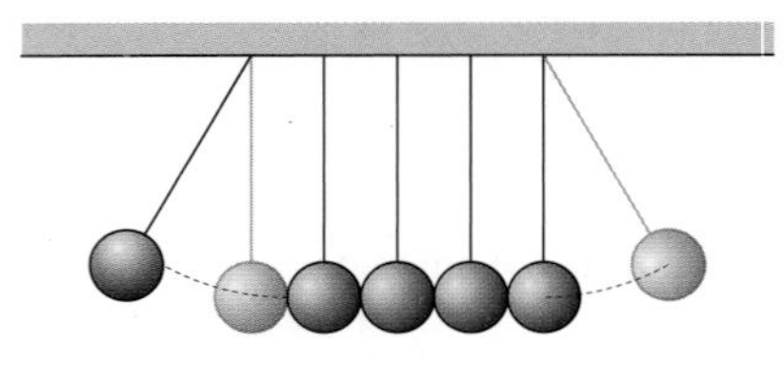

Figure P5.86

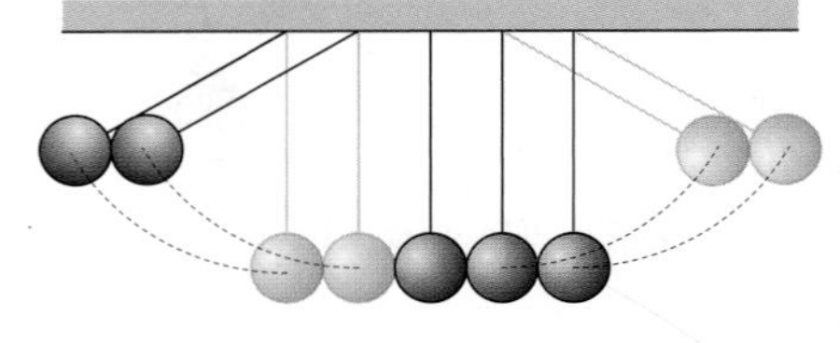

Figure P5.87

Problem 5.86 Explain why if you release the ball to the far left from a certain angle, the ball in question comes to a stop after impact while all the other balls do not seem to move except for the ball to the far right, which swings upward and achieves a maximum swing angle equal to the initial release angle of the ball to the far left.

Problem 5.87 Explain why if you release the two balls to the far left from a certain angle, the balls in question come to a stop right after impact while all the other balls do not seem to move except for the two balls to the far right, which swing upward and achieve a maximum swing angle equal to the initial release angle of the two balls to the far left.

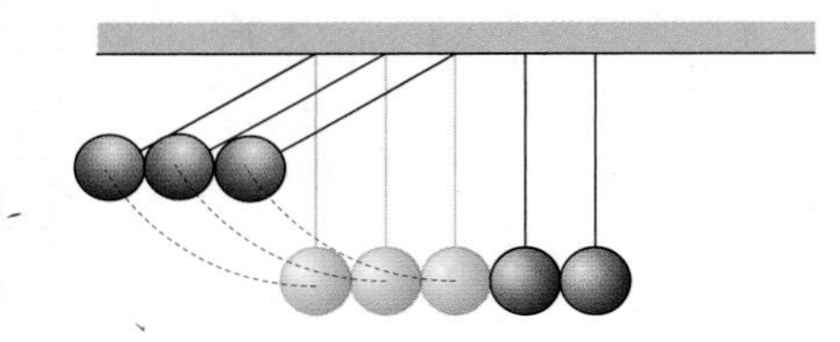

Figure P5.88

Problem 5.88 Predict the swing pattern of the system in the figure if you release from rest, and from a given angle, three of the five balls.

Problems 5.89 through 5.91

A 31,000 lb truck A and a 3970 lb sports car B collide at an intersection. Right before the collision, the truck and the sports car are traveling at $v_A^- = 60$ mph and $v_B^- = 50$ mph. Assume that the entire intersection forms a horizontal surface.

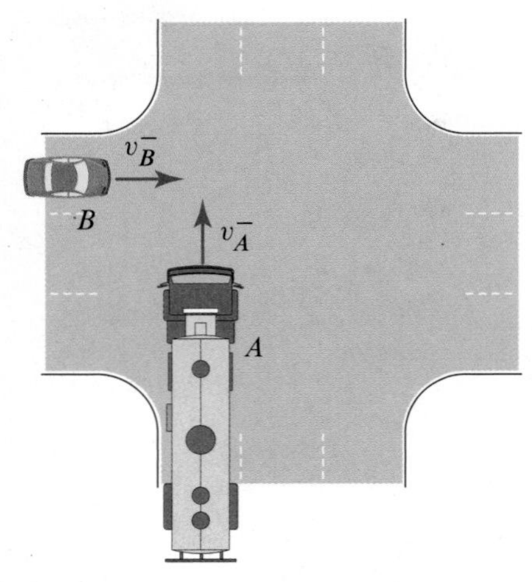

Figure P5.89–P5.91

Problem 5.89 Letting the line of impact be parallel to the ground and to the preimpact velocity of the truck, determine the postimpact velocities of A and B if A and B become entangled. Furthermore, assuming that the truck and the car slide after impact and that the coefficient of kinetic friction is $\mu_k = 0.7$, determine the position at which A and B come to a stop relative to the position they occupied at the instant of impact.

Problem 5.90 Letting the line of impact be parallel to the ground and to the preimpact velocity of the truck, determine the postimpact velocities of A and B if the contact between A and B is frictionless and the COR $e = 0$. Furthermore, assuming that the truck and the car slide after impact and that the coefficient of kinetic friction is $\mu_k = 0.7$, determine the position at which A and B come to a stop relative to the position they occupied at the instant of impact.

Problem 5.91 Letting the line of impact be parallel to the ground and to the preimpact velocity of the truck, determine the postimpact velocities of A and B if the contact between A and B is frictionless and the COR $e = 0.1$. Furthermore, assuming that the truck and the car slide after impact and that the coefficient of kinetic friction is $\mu_k = 0.7$, determine the position at which A and B come to a stop relative to the position they occupied at the instant of impact.

Problems 5.92 and 5.93

Competition billiard balls and tables need to adhere to strict standards (see the Billiard Congress of America for standards in the United States). Specifically, billiard balls must weigh between 5.5 and 6 oz, and they must be 2.25 ± 0.005 in. in diameter.

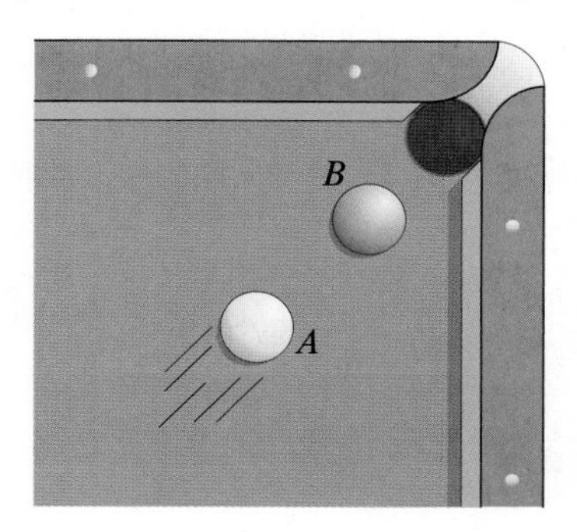

Figure P5.92

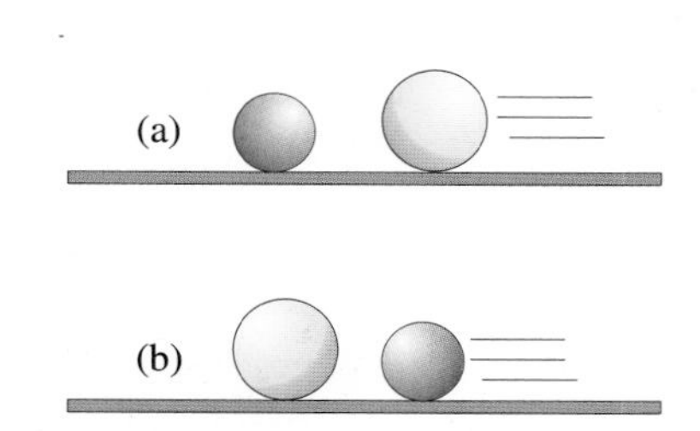

Figure P5.93

Problem 5.92 Treating the objects as particles, determine whether or not it is possible to have a moving ball A hit a stationary ball B so that A stops right after the impact, if A and B have the same diameter but not the same weight (since it is possible to have a weight difference of up to 0.5 oz while staying within regulations). Assume that the COR $e = 1$. *Hint:* Since the balls must be treated as particles and not as finite-size bodies, the fact that one can impart spin to a real ball is not relevant to the solution of this problem.

Problem 5.93 Professional billiard players can easily impart to a ball a speed of 20 mph. Assume the tolerance on the ball diameter to be $1/100$ in. instead of $5/1000$ in. and determine the outcome of the collision between (a) a 2.26 in. diameter ball traveling at 20 mph with a stationary 2.24 in. diameter ball (i.e., each ball is at the extreme limit of tolerance relative to the nominal diameter) and (b) a 2.24 in. diameter ball traveling at 20 mph with a stationary 2.26 in. diameter ball. Assume that the COR $e = 1$ and that the weights of the two balls are identical. Furthermore, assume that the contact between the balls and the table can be treated as essentially frictionless. *Hint:* The difference in size implies that the LOI is *not* parallel to the table.

Problems 5.94 and 5.95

Ball B is stationary when it is hit by an identical ball A as shown, with $\beta = 45°$. The preimpact speed of ball A is $v_0 = 1$ m/s.

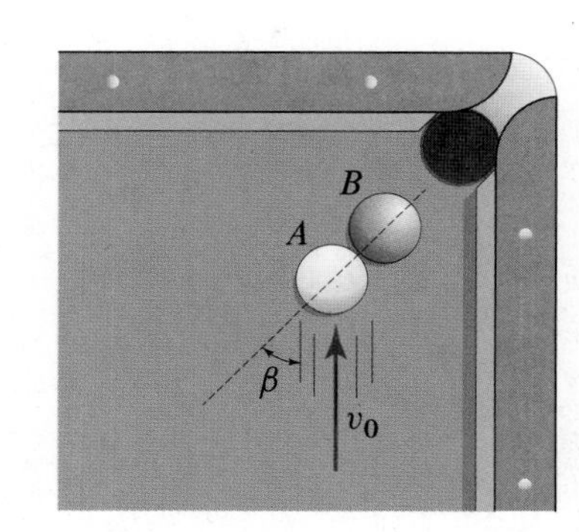

Figure P5.94 and P5.95

Problem 5.94 Determine the postimpact velocity of ball B if the COR of the collision $e = 1$.

Problem 5.95 Determine the postimpact velocity of ball A if the COR of the collision $e = 0.8$.

Problems 5.96 and 5.97

Car B is traveling with a speed v_B^- when it collides with an identical car A moving as shown and with a speed $v_A^- = 2$ ft/s. The angle β is equal to 50°.

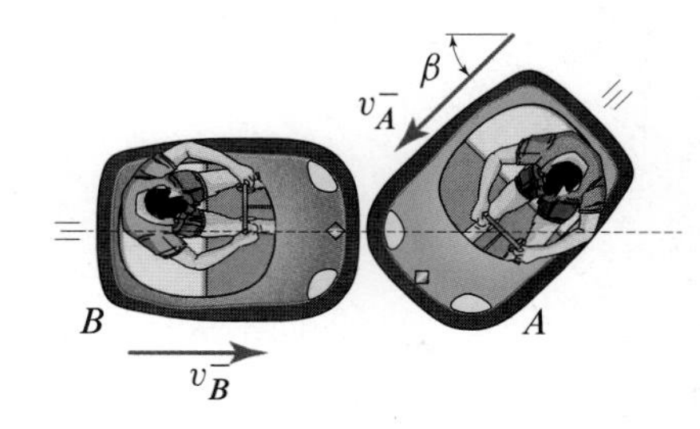

Figure P5.96 and P5.97

Problem 5.96 Determine v_B^- if the COR of the collision is $e = 0.6$ and the postimpact speed of A is equal to 78% of its preimpact value.

Problem 5.97 Determine the COR of the collision e if $v_B^- = 3$ ft/s and the postimpact speed of B is 30% of its preimpact value.

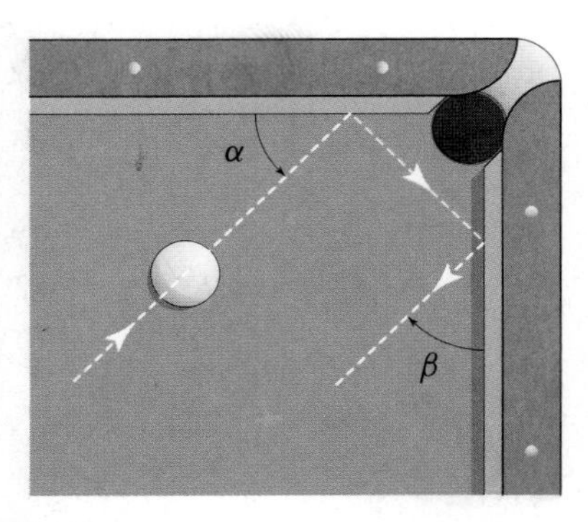

Figure P5.98

Problem 5.98

On a billiard table, the COR e for the impact between a ball and any of the four bumpers should be the same. Assuming that this is the case, determine the angle β after two banks as a function of the initial incidence angle α.

Problems 5.99 and 5.100

Two spheres, A and B, with masses $m_A = 1.35\,\mathrm{kg}$ and $m_B = 2.72\,\mathrm{kg}$, respectively, collide with $v_A^- = 26.2\,\mathrm{m/s}$, and $v_B^- = 22.5\,\mathrm{m/s}$.

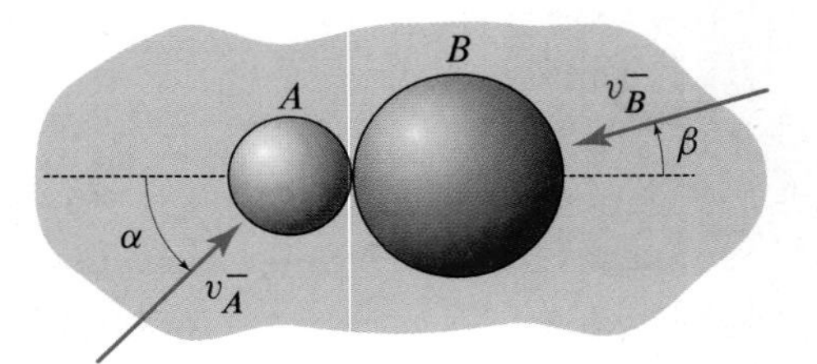

Figure P5.99 and P5.100

Problem 5.99 Compute the postimpact velocities of A and B if $\alpha = 45°$, $\beta = 16°$, the COR is $e = 0.57$, and the contact between A and B is frictionless.

Problem 5.100 Compute the postimpact velocities of A and B if $\alpha = 45°$, $\beta = 16°$, the COR is $e = 0$, and the contact between A and B is frictionless.

Problems 5.101 and 5.102

A 1.34 lb ball is dropped on a 10 lb incline with $\alpha = 33°$. The ball's release height is $h_1 = 5\,\mathrm{ft}$, and the height of the impact point relative to the ground is $h_2 = 0.3\,\mathrm{ft}$. Assume that the contact between the ball and the incline is frictionless, and let the COR for the impact be $e = 0.88$.

Problem 5.101 Compute the distance d at which the ball will hit ground for the first time if the incline cannot move relative to the floor.

Problem 5.102 Compute the distance d at which the ball will hit ground for the first time if the incline can slide without friction relative to the floor.

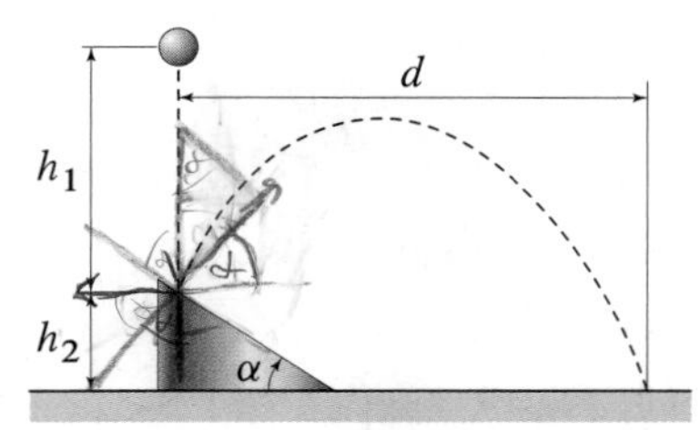

Figure P5.101 and P5.102

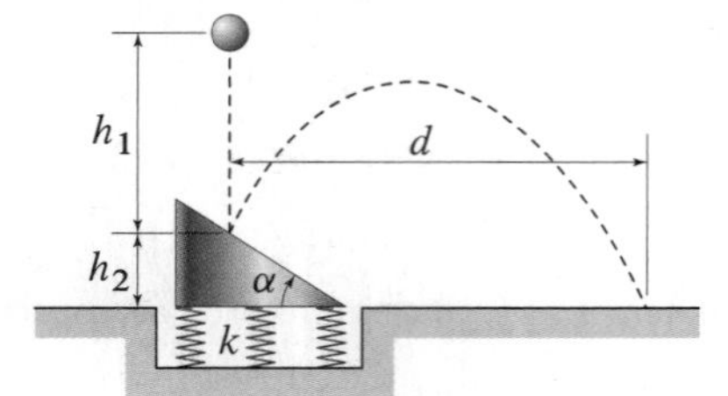

Figure P5.103

Problem 5.103

A 1.34 lb ball is dropped on a 10 lb incline with $\alpha = 33°$. The ball's release height is $h_1 = 5\,\mathrm{ft}$, and the height of the impact point relative to the ground is $h_2 = 0.3\,\mathrm{ft}$. Assume that the contact between the ball and the incline is frictionless, and let the COR for the impact be $e = 0.88$. Compute the distance d at which the ball will hit ground for the first time if the combined stiffness of the supporting springs is $k = 50\,\mathrm{lb/in}$. Assume that the incline can move only vertically.

Problem 5.104

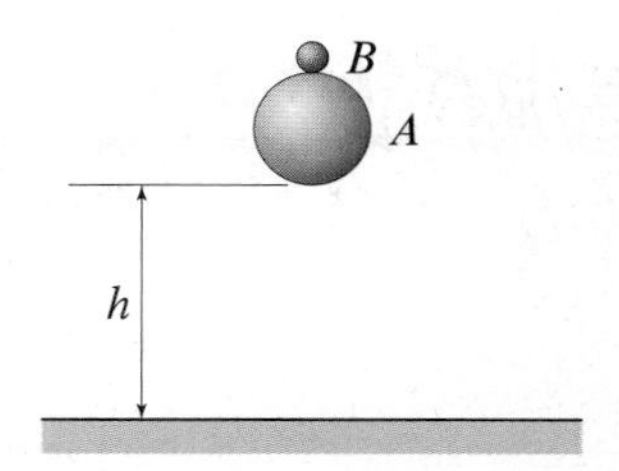

Figure P5.104

Consider two balls A and B that are stacked one on top of the other and dropped from rest from a height h. Let $e_{AG} = 1$ be the COR for the collision of ball A with the ground, and let $e_{AB} = 1$ be the COR for the collision between balls A and B. Finally, assume that the balls can move only vertically and that $m_A \gg m_B$, that is, that $m_B/m_A \approx 0$. Model the combined collision as a sequence of two-body impacts, and predict the rebound speed of ball B as a function of h and g, the acceleration due to gravity. *Hint:* Even though A and B are shown in contact, assume that a small gap is present so that the impact between A and the ground precedes that between A and B.

Problem 5.105

Consider a stack of N balls dropped from rest from a height h. Let all impacts be perfectly elastic, and assume that $m_i \gg m_{i+1}$, that is, that $m_{i+1}/m_i \approx 0$, with $i = 1, \ldots, N-1$ and m_i being the mass of the ith ball. Model the combined collision as a sequence of two-body impacts, and predict the rebound speed of the topmost ball. Assume that the balls can move only vertically. *Hint:* Even though the balls are shown in contact, assume a small gap is present between each pair so that the impact between B_1 and the ground precedes that between B_1 and B_2, etc.

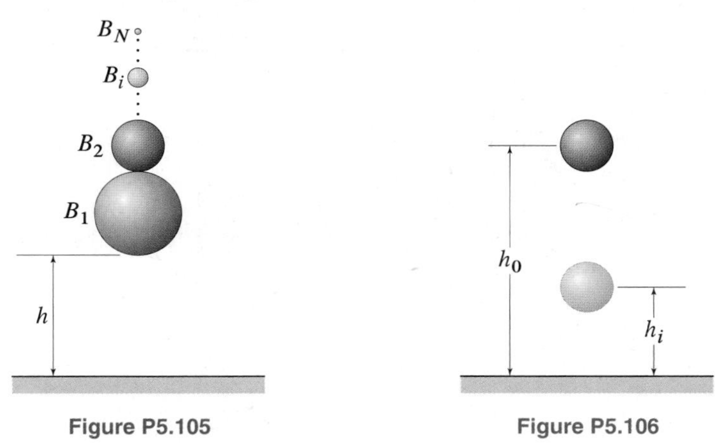

Figure P5.105

Figure P5.106

Problem 5.106

A ball is dropped from rest from a height $h_0 = 1.5$ m. The impact between the ball and the floor has a COR $e = 0.92$. Find the formula that allows you to compute the rebound height h_i of the ith rebound. Furthermore, find the formula that provides the total time required to complete i rebounds. Finally, compute the time t_{stop} that the ball will take to stop bouncing. *Hint:* A formula you may find useful in the solution of this problem is that of the limit value of a geometric series: $\sum_{i=0}^{N-1} e^i = (e^N - 1)/(e - 1)$, with $|e| < 1$.

Design Problems

Design Problem 5.1

In the manufacture of steel balls of the type used for ball bearings, it is important that their material properties be sufficiently uniform. One way to detect gross differences in their material properties is to observe how a ball rebounds when dropped on a hard strike plate. Assuming that each ball has a radius $R = 0.3$ in., design a sorting device to select the balls with $0.900 < \text{COR} < 0.925$. The device consists of an incline defined by the angle θ and length L. The strike plate is placed at the bottom of a well with depth h and width w. Finally, at a distance ℓ from the end of the incline, there is a trap with a diameter d. By releasing a ball from rest at the top of the incline and assuming that the ball slides with negligible friction, the ball will reach the bottom of the incline with a speed v_0, rebound off the strike plate, and then fall directly into the trap (you may want to add a design element that prevents balls from simply rolling into the trap). In your design, choose appropriate values of L, $\theta < 45°$, h, w, d, and ℓ to accomplish the desired task while ensuring that the overall dimensions of the device do not exceed 4 ft in both the horizontal and vertical directions.

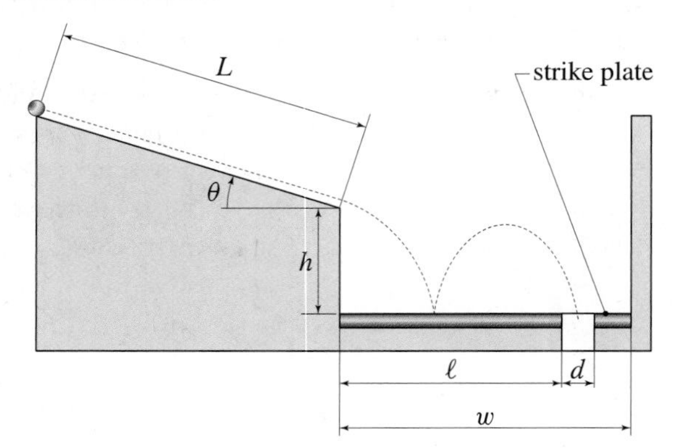

Figure DP5.1

5.3 Angular Momentum

The concept of *angular momentum* not only arises in the study of subjects like the orbital motion of a satellite about a planet, but it is also a key concept in generalizing Newton's laws from particles to more complex models such, as a rigid body. In this generalization, a *fundamental law* is formulated that relates the moment of a force about a point to the time rate of change of angular momentum about that point.

Moment-angular momentum relation for a particle

Angular momentum of a particle

Referring to Fig. 5.17, for a particle Q of mass m and momentum $\vec{p}_Q = m\vec{v}_Q$, the particle's *angular momentum with respect to an arbitrary point* P is denoted by $\vec{h}_P$ and is defined as the moment of $\vec{p}_Q$ about P, i.e.,

$$\vec{h}_P = \vec{r}_{Q/P} \times \vec{p}_Q = \vec{r}_{Q/P} \times m\vec{v}_Q, \tag{5.31}$$

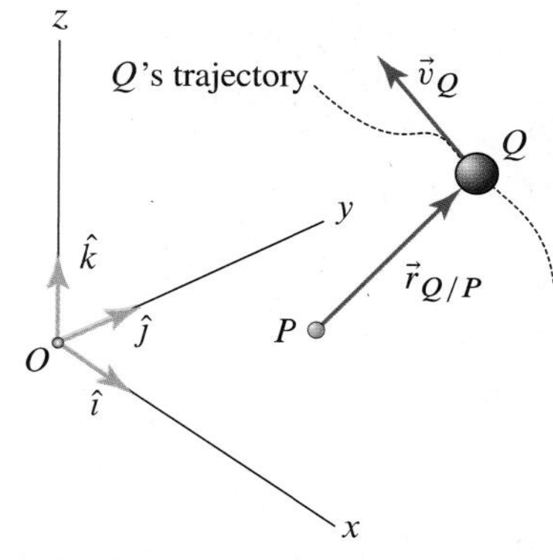

Figure 5.17
A particle Q in motion relative to a point P. Point P need *not* be a stationary point.

where $\vec{r}_{Q/P}$ denotes the position of Q relative to P and where the reference point P is called the *moment center*. Angular momentum is typically expressed using units of kg·m^2/s and slug·ft^2/s in the SI and U.S. Customary systems, respectively.

The relation between the moment of a force and angular momentum can be discovered by taking the time derivative of Eq. (5.31), which yields

$$\dot{\vec{h}}_P = \dot{\vec{r}}_{Q/P} \times m\vec{v}_Q + \vec{r}_{Q/P} \times m\dot{\vec{v}}_Q. \tag{5.32}$$

Recalling that $\dot{\vec{r}}_{Q/P} = \vec{v}_Q - \vec{v}_P$, and that by Newton's second law, $m\dot{\vec{v}}_Q = \vec{F}$, where $\vec{F}$ is the total force on Q, we can rewrite Eq. (5.32) as

$$\dot{\vec{h}}_P = (\vec{v}_Q - \vec{v}_P) \times m\vec{v}_Q + \vec{r}_{Q/P} \times \vec{F}. \tag{5.33}$$

The term

$$\vec{r}_{Q/P} \times \vec{F} = \vec{M}_P, \tag{5.34}$$

is the *moment with respect to* P *of all the forces acting on* Q. Making use of Eq. (5.34) in Eq. (5.33) and recognizing that $\vec{v}_Q \times m\vec{v}_Q = \vec{0}$, we obtain the following important result:

$$\vec{M}_P = \dot{\vec{h}}_P + \vec{v}_P \times m\vec{v}_Q. \tag{5.35}$$

Concept Alert

Angular momentum is the *moment of momentum*. While the term *angular momentum* is most commonly used, conceptually and computationally, it is useful to remember $\vec{h}_P$ as the *moment of momentum*.

Equation (5.35) is the *moment-angular momentum relation for a particle*, and it describes the relation between the moment of the total force acting on Q and the angular momentum of Q, both of which are with respect to the moment center P. Equation (5.35) is not a new law of motion, since it was obtained from Newton's second law, but it is very useful, especially when studying the motion under the action of a central force. Equation (5.35) can be simplified to

$$\vec{M}_P = \dot{\vec{h}}_P, \tag{5.36}$$

in either one of the following cases:

1. the reference point P is fixed, i.e., if $\vec{v}_P = \vec{0}$,
2. $\vec{v}_P$ is parallel to $\vec{v}_Q$, i.e., $\vec{v}_P \times m\vec{v}_Q = \vec{0}$.

Whenever Eq. (5.36) holds, integrating this equation with respect to time over a time interval $t_1 \leq t \leq t_2$, we have

$$\vec{h}_{P1} + \int_{t_1}^{t_2} \vec{M}_P \, dt = \vec{h}_{P2}, \tag{5.37}$$

where $\vec{h}_{P1} = \vec{h}_P(t_1)$, $\vec{h}_{P2} = \vec{h}_P(t_2)$, and the integral term in Eq. (5.37) is called the *angular impulse with respect to* P of all forces acting on Q. Equation (5.37) is called the *angular impulse-momentum principle for a particle*, and it is most useful in those problems in which the moment of all external forces is zero about some fixed point. We can then immediately say that the angular momentum computed with respect to that fixed point is constant. This is especially useful in the solution of problems with a particle moving under the action of a central force.

Angular impulse-momentum for a system of particles

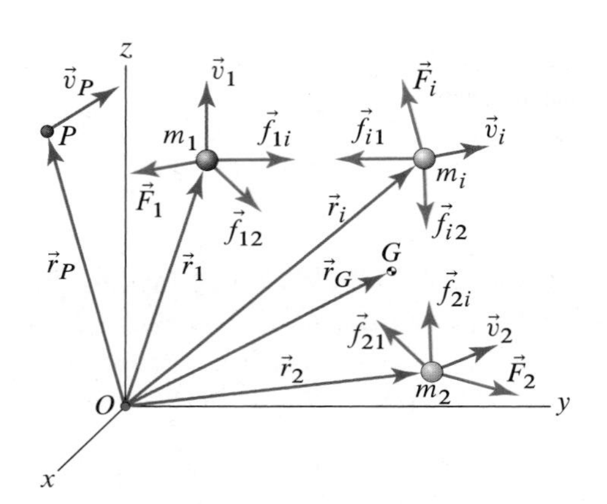

Figure 5.18
A system of particles under the action of internal and external forces. Point P, in general, is a moving point.

We now extend the angular impulse-momentum principle to systems of particles. As was done for the linear impulse-momentum principle for systems, we begin by considering a system of N particles subject to a system of external forces and interacting with one another via pairwise internal forces (see Fig. 5.18). In addition, we assume that the system under consideration is *closed*, that is, it does not exchange mass with its surroundings (see discussion on p. 315 of Section 5.1). Choosing a point P with velocity $\vec{v}_P$ as moment center and applying Eq. (5.35) to the ith particle in the system, we have

$$\vec{M}_{Pi} = \dot{\vec{h}}_{Pi} + \vec{v}_P \times m_i \, \vec{v}_i, \tag{5.38}$$

where, recalling that $\vec{r}_{i/P}$ denotes the position of particle i relative to P,

$$\vec{M}_{Pi} = \vec{r}_{i/P} \times \Bigg(\vec{F}_i + \sum_{\substack{j=1 \\ j \neq i}}^{N} \vec{f}_{ij} \Bigg), \tag{5.39}$$

in which

- $\vec{F}_i$ is the *external* force acting on particle i,
- $\vec{f}_{ij}$ is the *internal* force acting on particle i due to particle j,
- the term $\vec{F}_i + \sum_{\substack{j=1 \\ j \neq i}}^{N} \vec{f}_{ij}$ represents the *total* force acting on particle i.

Summing the contributions in Eq. (5.38) over all the particles in the system, we obtain

$$\sum_{i=1}^{N} \vec{M}_{Pi} = \sum_{i=1}^{N} \dot{\vec{h}}_{Pi} + \sum_{i=1}^{N} \vec{v}_P \times m_i \, \vec{v}_i. \tag{5.40}$$

To simplify Eq. (5.40), we start with the sum on the left-hand side, which can be rewritten as

$$\sum_{i=1}^{N} \vec{M}_{P_i} = \sum_{i=1}^{N} \Big(\vec{r}_{i/P} \times \vec{F}_i \Big) + \sum_{i=1}^{N} \Bigg(\vec{r}_{i/P} \times \sum_{\substack{j=1 \\ j \neq i}}^{N} \vec{f}_{ij} \Bigg). \tag{5.41}$$

The last term in Eq. (5.41) can be eliminated by recalling that Newton's third law is expressed by the following *two* relationships:*

$$\vec{f}_{ij} = -\vec{f}_{ji} \quad \text{and} \quad \vec{f}_{ij} \times (\vec{r}_i - \vec{r}_j) = \vec{0}. \tag{5.42}$$

Next we consider how a given pair of particles d and e contributes to the last term in Eq. (5.41). Referring to Fig. 5.19, consider the terms in the sum

$$\vec{r}_{e/P} \times \vec{f}_{ed} + \vec{r}_{d/P} \times \vec{f}_{de}, \tag{5.43}$$

which, by using the first of Eqs. (5.42), can be rewritten as

$$\vec{r}_{e/P} \times \vec{f}_{ed} + \vec{r}_{d/P} \times \left(-\vec{f}_{ed}\right) = (\vec{r}_{e/P} - \vec{r}_{d/P}) \times \vec{f}_{ed}. \tag{5.44}$$

Using $\vec{r}_{e/P} - \vec{r}_{d/P} = (\vec{r}_e - \vec{r}_P) - (\vec{r}_d - \vec{r}_P) = \vec{r}_e - \vec{r}_d$ and the second of Eqs. (5.42), Eq. (5.44) becomes

$$(\vec{r}_e - \vec{r}_d) \times \vec{f}_{ed} = \vec{0}. \tag{5.45}$$

Equation (5.45) is true for every pair of particles, so the last term in Eq. (5.41) must be zero, and Eq. (5.41) can be simplified to read

$$\vec{M}_P = \sum_{i=1}^{N} \vec{r}_{i/P} \times \vec{F}_i, \tag{5.46}$$

where $\vec{M}_P$ is the total moment with respect to P of *only* the *external* forces.

Now we consider the angular momentum terms in the last summation in Eq. (5.40). Observe that the term $\vec{v}_P$ is not characterized by the index i. This means that we can rewrite the summation containing the term in question as

$$\sum_{i=1}^{N} \vec{v}_P \times m_i\, \vec{v}_i = \vec{v}_P \times \sum_{i=1}^{N} m_i\, \vec{v}_i = \vec{v}_P \times m\vec{v}_G, \tag{5.47}$$

where we have used Eq. (5.13) to replace the sum of all momentum contributions with the term $m\vec{v}_G$ and G is the system's center of mass. We now define the system's *total angular momentum with respect to the moment center P* to be the sum of the angular momenta with respect to P of all the particles in the system, i.e.,

$$\vec{h}_P = \sum_{i=1}^{N} \vec{h}_{Pi}. \tag{5.48}$$

Using Eqs. (5.46)–(5.48), we obtain the *moment-angular momentum relation for a closed system of particles* by rewriting Eq. (5.40) as

$$\vec{M}_P = \dot{\vec{h}}_P + \vec{v}_P \times m\vec{v}_G. \tag{5.49}$$

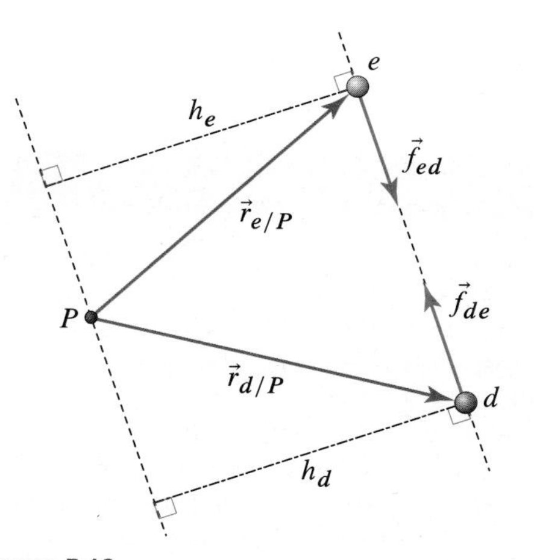

Figure 5.19
Moment of a pair of internal forces that are required to be equal and opposite as well as collinear by Newton's third law. The internal forces $\vec{f}_{ed}$ and $\vec{f}_{de}$ have equal moment arms relative to the line going through P and parallel to $\vec{f}_{ed}$ and $\vec{f}_{de}$, i.e., $h_e = h_d$. Therefore, the overall moment of $\vec{f}_{ed}$ and $\vec{f}_{de}$ with respect to P is equal to zero.

Angular momentum for systems of particles: important special cases

We now consider those cases when Eq. (5.49) can be simplified by making $\vec{v}_P \times m\vec{v}_G = \vec{0}$. This happens when *any* one of the following are true:

* These two relationships were first given in Eqs. (1.3) and (1.4), respectively, and have been repeated here for convenience.

1. P is a fixed point, i.e., when $\vec{v}_P = \vec{0}$,
2. G is a fixed point, i.e., when $\vec{v}_G = \vec{0}$,
3. P coincides with the center of mass so that $\vec{v}_P = \vec{v}_G$, or
4. Vectors $\vec{v}_P$ and $\vec{v}_G$ are parallel to one another.

Common Pitfall

Same appearance, different meaning. Equations (5.50) and (5.51) are identical in appearance to Eqs. (5.36) and (5.37), respectively. However, the quantity $\vec{h}_P$ in Eqs. (5.50) and (5.51) is the total angular momentum of a *system* and therefore is very different from the quantity $\vec{h}_P$ in Eqs. (5.36) and (5.37), which is the angular momentum of a single particle. The reason these equations appear the same is to reinforce the fact that there is a single physical principle at play in both cases.

In all of these cases, Eq. (5.49) takes on the following simplified form:

$$\vec{M}_P = \dot{\vec{h}}_P. \tag{5.50}$$

Furthermore, if any of the conditions listed before hold over a time interval $t_1 \le t \le t_2$, we can integrate Eq. (5.50) with respect to time to obtain

$$\vec{h}_{P1} + \int_{t_1}^{t_2} \vec{M}_P \, dt = \vec{h}_{P2}, \tag{5.51}$$

where $\vec{h}_{P1} = \vec{h}_P(t_1)$ and $\vec{h}_{P2} = \vec{h}_P(t_2)$. Equation (5.51) is the *angular impulse-momentum principle for a system of particles.*

Conservation of angular momentum

Referring to either Eq. (5.37) or Eq. (5.51), if $\vec{M}_P = \vec{0}$ for $t_1 \le t \le t_2$, then we have

$$\vec{h}_{P1} = \vec{h}_{P2}, \tag{5.52}$$

and we say that there is *conservation of angular momentum.*

In some cases it is possible to have conservation of a *component* of the angular momentum instead of the entire angular momentum. This happens when $\vec{M}_P$ is not zero, but a component of $\vec{M}_P$ *along a fixed axis is zero.* For example, consider Fig. 5.20, which shows the FBD of a conical pendulum with bob Q swinging in a horizontal plane. The moment of all forces about point O is

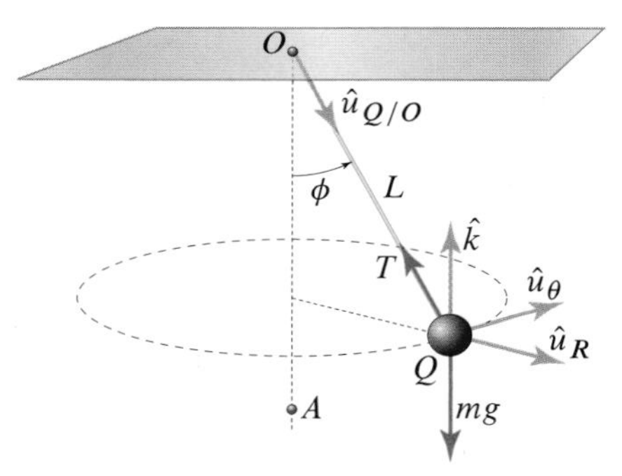

Figure 5.20
FBD of a conical pendulum swinging in a horizontal plane.

$$\begin{aligned}\vec{M}_O &= \vec{r}_{Q/O} \times \vec{F}_Q = L\,\hat{u}_{Q/O} \times \left(-T\,\hat{u}_{Q/O} - mg\,\hat{k}\right) \\ &= L\left(\sin\phi\,\hat{u}_R - \cos\phi\,\hat{k}\right) \times \left(-mg\,\hat{k}\right) \\ &= mgL\sin\phi\,\hat{u}_\theta,\end{aligned} \tag{5.53}$$

where $\vec{F}_Q$ is the total force on Q and $\hat{u}_{Q/O} = \sin\phi\,\hat{u}_R - \cos\phi\,\hat{k}$. Now recall that $\vec{M}_O = \dot{\vec{h}}_O$ by Eq. (5.50). Since we can write $\vec{h}_O = h_{OR}\,\hat{u}_R + h_{O\theta}\,\hat{u}_\theta + h_{Oz}\,\hat{k}$, we have

$$\dot{\vec{h}}_O = \dot{h}_{Or}\,\hat{u}_R + h_{Or}\,\dot{\hat{u}}_R + \dot{h}_{O\theta}\,\hat{u}_\theta + h_{O\theta}\,\dot{\hat{u}}_\theta + \dot{h}_{Oz}\,\hat{k} + h_{Oz}\,\dot{\hat{k}}. \tag{5.54}$$

In cylindrical coordinates, we have $\dot{\hat{k}} = \vec{0}$ because the z axis is fixed, and we have $\dot{\hat{u}}_R = \dot{\theta}\,\hat{u}_\theta$ and $\dot{\hat{u}}_\theta = -\dot{\theta}\,\hat{u}_R$ because the R and θ directions rotate as the pendulum swings. Hence, Eq. (5.54) simplifies to

$$\dot{\vec{h}}_O = (\dot{h}_{Or} - \dot{\theta} h_{O\theta})\,\hat{u}_R + (\dot{h}_{O\theta} + \dot{\theta} h_{OR})\,\hat{u}_\theta + \dot{h}_{Oz}\,\hat{k}. \tag{5.55}$$

Equating Eqs. (5.53) and (5.55) component by component, we can then conclude that $M_{Oz} = 0$ implies $\dot{h}_{Oz} = 0$ and therefore $h_{Oz} =$ constant. By contrast, for example, the fact that $M_{OR} = 0$ only implies that $\dot{h}_{Or} - \dot{\theta} h_{O\theta} = 0$. Again, it is important to keep in mind that $M_{Oz} = 0$ implies $h_{Oz} =$ constant *because the z direction is fixed.*

Euler's first and second laws of motion

Newton's laws of motion regulate the motion of mass points (or particles), where a mass point is an *abstract* entity with *finite* mass and *zero* volume. Using Newton's laws, we derived relations that must be satisfied by the motion of any *closed* system of mass points with pairwise internal forces. These relations, as given in Eq. (5.14) on p. 316 and Eq. (5.49) on p. 363, respectively, are

$$\vec{F} = m\vec{a}_G \quad \text{and} \quad \vec{M}_P = \dot{\vec{h}}_P + \vec{v}_P \times m\vec{v}_G. \tag{5.56}$$

Many common models of the physical world do not view objects as consisting of mass points. For example, the modeling of rigid bodies, which we study later, is based on the notion of *continuous body*, which is not an assemblage of mass points. Therefore, we need to confront the following question:

> What laws of motion do we use in models that are not based on the idea of mass point?

Leonhard Euler was the first to answer this question when he recognized that the force and moment equations given by Eqs. (5.56) were not just by-products of Newton's laws applicable only to particle systems, but equations that could be used as the foundation of the study of motion of any physical system. Euler postulated that, in an inertial frame of reference, the motion of any physical object that does not exchange mass with its surroundings must satisfy the following two laws:

$$\textit{Euler's first law}: \quad \vec{F} = m\vec{a}_G, \tag{5.57}$$

$$\textit{Euler's second law}: \quad \vec{M}_P = \dot{\vec{h}}_P + \vec{v}_P \times m\vec{v}_G, \tag{5.58}$$

where, for a physical system of constant mass m and center of mass G,

- $\vec{F}$ is the resultant of all *external* forces.
- $\vec{v}_G$ and $\vec{a}_G$ are the velocity and acceleration of G, respectively.
- P is an arbitrarily chosen moment center.
- $\vec{M}_P$ is the resultant of the moments of all *external* forces and torques computed with respect to P.
- $\vec{h}_P$ is the angular momentum computed with respect to P.

We will consider Eqs. (5.57) and (5.58) to be the *fundamental laws of motion* of bodies. To recognize Euler's contribution to the laws of mechanics while retaining an awareness of their Newtonian origin, we will refer to these laws as the *Newton-Euler equations*.

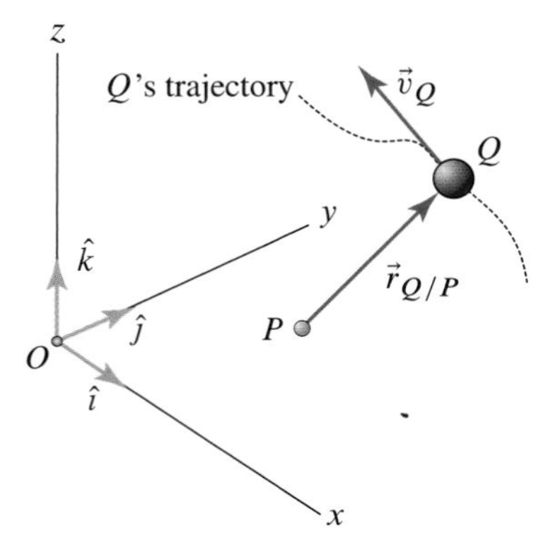

Figure 5.21
A particle Q in motion relative to a point P. Point P need *not* be a stationary point.

End of Section Summary

In this section, we developed the concept of angular momentum for a single particle and for systems of particles. We also derived the *moment-angular momentum relations* for a particle and for a system of particles.

The *angular momentum of a particle Q with respect to the moment center P* (Fig. 5.21) is given by

Eq. (5.31), p. 361

$$\vec{h}_P = \vec{r}_{Q/P} \times \vec{p}_Q = \vec{r}_{Q/P} \times m\vec{v}_Q,$$

where P can be fixed or moving; $\vec{r}_{Q/P}$ is the position of Q relative to P; m and $\vec{v}_Q$ are the mass and the velocity of Q, respectively; and $\vec{p}_Q = m\vec{v}_Q$ is the linear momentum of Q.

The *moment-angular momentum relation* for a single particle is given by

Eq. (5.35), p. 361

$$\vec{M}_P = \dot{\vec{h}}_P + \vec{v}_P \times m\vec{v}_Q,$$

where $\vec{M}_P$ is the moment with respect to P of all the forces acting on Q. If either one of the following conditions is satisfied:

1. The reference point P is fixed, i.e., if $\vec{v}_P = \vec{0}$.

2. $\vec{v}_P$ is parallel to $\vec{v}_Q$, i.e., $\vec{v}_P \times m\vec{v}_Q = \vec{0}$.

the moment-angular momentum relation for a single particle can be simplified to

Eq. (5.36), p. 361

$$\vec{M}_P = \dot{\vec{h}}_P.$$

If either condition (1) or (2) is satisfied for $t_1 \le t \le t_2$, then the moment-angular momentum relation for a single particle can be integrated with respect to time to obtain the angular impulse-momentum principle as

Eq. (5.37), p. 362

$$\vec{h}_{P1} + \int_{t_1}^{t_2} \vec{M}_P \, dt = \vec{h}_{P2},$$

where $\vec{h}_{P1} = \vec{h}_P(t_1)$ and $\vec{h}_{P2} = \vec{h}_P(t_2)$.

For a *closed system of particles*, the moment-angular momentum relation can be given the form

Eq. (5.49), p. 363

$$\vec{M}_P = \dot{\vec{h}}_P + \vec{v}_P \times m\vec{v}_G,$$

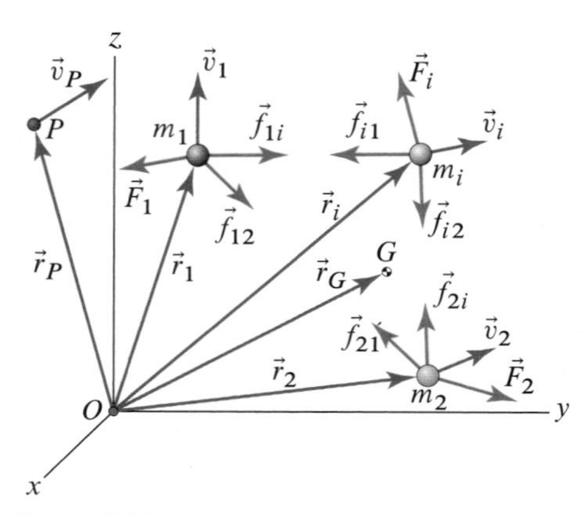

Figure 5.22
A system of particles under the action of internal and external forces. Point P, in general, is a moving point.

where, with reference to Fig. 5.22, $\vec{M}_P$ is the moment with respect to P of *only the external forces* acting on the system, $m = \sum_{i=1}^{N} m_i$ is the total mass of the system,

G is the system's center of mass, $\vec{v}_G$ is the velocity of G, and $\vec{h}_P$ is the total angular momentum, which is defined as

Eq. (5.48), p. 363

$$\vec{h}_P = \sum_{i=1}^{N} \vec{h}_{Pi}.$$

If any one of the following conditions is satisfied:

1. P is a fixed point, i.e., when $\vec{v}_P = \vec{0}$.
2. G is a fixed point, i.e., when $\vec{v}_G = \vec{0}$.
3. P coincides with the center of mass and therefore $\vec{v}_P = \vec{v}_G$.
4. Vectors $\vec{v}_P$ and $\vec{v}_G$ are parallel to one another.

then the moment-angular momentum relation for a closed system of particles simplifies to

Eq. (5.50), p. 364

$$\vec{M}_P = \dot{\vec{h}}_P.$$

Furthermore, if any of the above conditions are satisfied over a time interval $t_1 \leq t \leq t_2$, then the moment-angular momentum relation for a closed system of particles can be integrated with respect to time to obtain the angular impulse-momentum principle as

Eq. (5.51), p. 364

$$\vec{h}_{P1} + \int_{t_1}^{t_2} \vec{M}_P \, dt = \vec{h}_{P2}.$$

EXAMPLE 5.11 *Explaining a Skater's Ability to Change Her Angular Speed*

Figure 1
Three snapshots of a *forward spin*. From left to right, notice how first a leg is retracted and then both arms are retracted. By doing so the skater dramatically increases her spin rate.

In a spin, which is common in ice-skating performances, a skater rotates about a vertical axis that runs through the skater's body. The skater controls the spin rate by extending or retracting the arms and/or legs. When arms and/or legs are brought closer to the spin axis, more of the body mass is placed closer to the spin axis, and the spin rate increases. Conversely, when the mass is distributed away from the spin axis by extending arms and/or legs outward, the spin rate decreases.

To understand this behavior, consider the simple model for the skater in Fig. 2. Disk

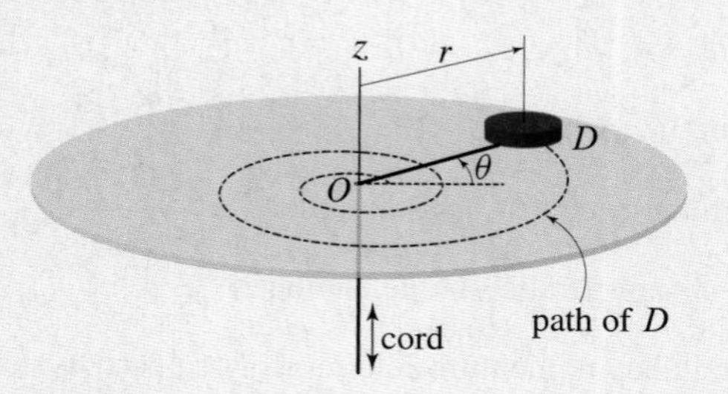

Figure 2. Disk moving in orbit around point O while constrained by a string.

D is connected to O by a string that can be pulled in or let out of the hole in the table at O, which allows for a variable distance between the disk and the spin axis z. Assume that the contact between D and the table is frictionless, that the cord is inextensible and of negligible mass, and that the disk is initially rotating about O at a distance $r_1 = 3\,\mathrm{ft}$ with angular speed $\dot{\theta}_1 = 60\,\mathrm{rpm}$. Using the moment-angular momentum relation, find the angular speed $\dot{\theta}$ of the disk as a function r and show that it behaves similarly to the skater.

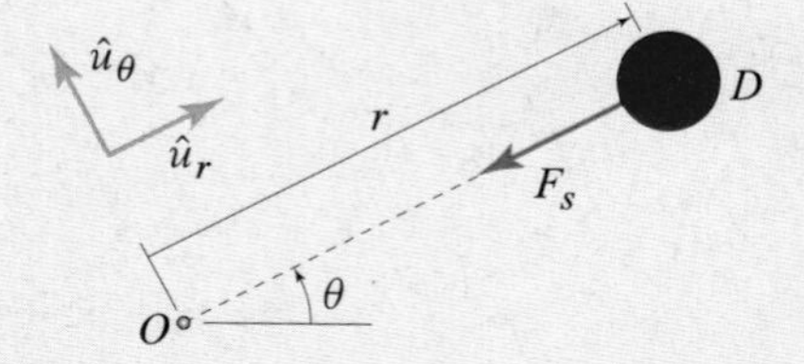

Figure 3
FBD of the disk in Fig. 2.

SOLUTION

Road Map & Modeling The FBD of the disk in Fig. 2 is shown in Fig. 3. Notice that the line of action of the string force F_s always goes through the fixed point O. This will allow us to apply Eq. (5.36) with O as the moment center and with $\vec{M}_O = \vec{0}$. This is a classic example of a central force motion.

Governing Equations

Balance Principles Applying Eq. (5.36) with O as the moment center, we obtain

$$\vec{M}_O = \dot{\vec{h}}_O. \tag{1}$$

Force Laws The force whose moment about O we need to compute is the string force F_s. The moment of this force about O is

$$\vec{M}_O = \vec{F}_s \times \vec{r}_{D/O} = -F_s\,\hat{u}_r \times r\,\hat{u}_r = \vec{0}, \tag{2}$$

since $\hat{u}_r \times \hat{u}_r = \vec{0}$.

Kinematic Equations Computing the angular momentum of the disk D with respect to point O, we obtain

$$\vec{h}_O = \vec{r}_{D/O} \times \vec{p}_D = r\,\hat{u}_r \times m\vec{v}_D = r\,\hat{u}_r \times m\big(\dot{r}\,\hat{u}_r + r\dot{\theta}\,\hat{u}_\theta\big) = mr^2\dot{\theta}\,\hat{u}_z, \tag{3}$$

where we have written $\vec{v}_D$ in polar coordinates.

Computation Substituting Eqs. (2) and (3) into Eq. (1), we obtain

$$\vec{0} = \frac{d}{dt}(mr^2\dot{\theta}\,\hat{u}_z) \quad \Rightarrow \quad mr^2\dot{\theta} = \text{constant}, \tag{4}$$

where we have used the fact that the direction of the unit vector $\hat{u}_z$ is constant. Equation (4) expresses the conservation of the angular momentum of the disk D about point O. Since the quantity $mr^2\dot{\theta}$ is constant, its initial value will not change, so we can say that

$$r_1^2\dot{\theta}_1 = r_2^2\dot{\theta}_2 = 18\pi\ \text{rad}\cdot\text{ft}^2/\text{s}, \tag{5}$$

where we have canceled out the mass, the subscript "2" indicates any later time, and where we have used the fact that $r_1 = 3$ ft and $\dot{\theta}_1 = 60$ rpm $= 2\pi$ rad/s. Solving Eq. (5) for $\dot{\theta}_2$, we obtain

$$\boxed{\dot{\theta}_2 = \frac{18\pi\ \text{rad}\cdot\text{ft}^2/\text{s}}{r_2^2}.} \tag{6}$$

Discussion & Verification As with the skater, Eq. (6) tells us that when r is decreased, $\dot{\theta}$ increases, and when r is increased, $\dot{\theta}$ decreases.

A Closer Look We could have solved this problem just as easily if we had instead applied Eq. (5.37). In doing so, we would have begun with the equation

$$\vec{h}_{O1} + \int_{t_1}^{t_2} \vec{M}_O\,dt = \vec{h}_{O2}. \tag{7}$$

Noting again that $\vec{M}_O$ is zero for all t, we then have that

$$\vec{h}_{O1} = \vec{h}_{O2}, \tag{8}$$

which again expressed the conservation of the angular momentum of the disk D about point O. Noting from Eq. (3) that $\vec{h}_O = mr^2\dot{\theta}\,\hat{u}_z$, we would again recover Eq. (5). In fact, this solution method would probably be considered more straightforward.

EXAMPLE 5.12 *Angular Momentum to Find the Speed of a Satellite*

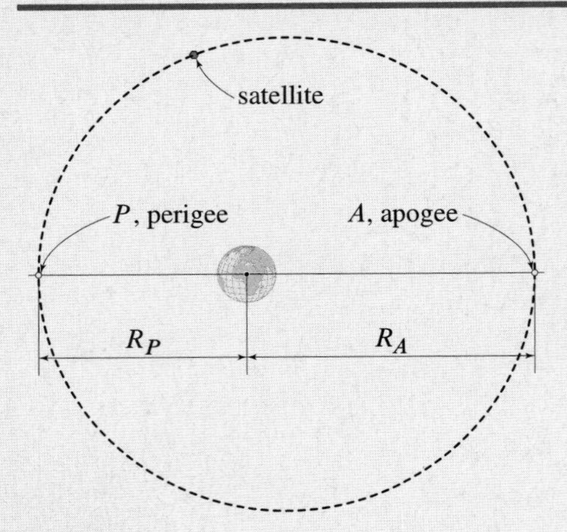

Figure 1
Satellite in elliptical orbit around the Earth. The orbit is drawn to scale with respect to the Earth.

A satellite orbits the Earth as shown in Fig. 1. The minimum and maximum distances from the center of the Earth are $R_P = 4.5\times10^7$ m and $R_A = 6.163\times10^7$ m, respectively, where the subscripts P and A stand for *perigee* (the point on the orbit closest to Earth) and *apogee* (the point on the orbit farthest from Earth), respectively. If the satellite's speed at P is $v_P = 3.2\times10^3$ m/s, determine the satellite's speed at A.

SOLUTION

Road Map & Modeling Referring to Fig. 2, we model both the Earth and the satellite

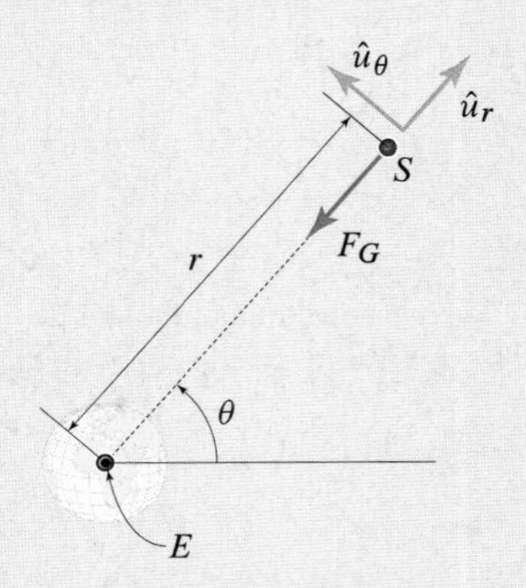

Figure 2. FBD of the satellite along with the chosen coordinate system.

as particles, we assume the Earth to be fixed, and we consider the center of the Earth E as the origin of an inertial frame of reference. We also assume that the only force acting on the satellite is F_G, the gravitational attraction exerted by the Earth. Finally, we assume that the satellite's orbit is planar so that we can study the satellite's motion, using a polar coordinate system with origin at E. These assumptions imply that the force on the satellite is central (it always points toward the fixed point E) and its moment about E is always equal to zero. Therefore, the satellite's angular momentum with respect to E is *conserved* throughout the motion.

Governing Equations

Balance Principles Choosing the fixed point E as our moment center, we can use the moment-angular momentum relation in Eq. (5.35) on p. 361:

$$\vec{M}_E = \dot{\vec{h}}_E, \tag{1}$$

where

$$\vec{M}_E = \vec{r} \times \vec{F}_G \quad \text{and} \quad \vec{h}_E = \vec{r} \times m\vec{v}, \tag{2}$$

and where $\vec{r}$, m, and $\vec{v}$ are the position, mass, and velocity of the satellite, respectively.

Force Laws $\vec{F}_G$ is given by (see Eq. (1.5) on p. 3)

$$\vec{F}_G = -G\,\frac{m_e m}{r^2}\,\hat{u}_r, \tag{3}$$

where G is the universal gravitational constant, m_e is the mass of the Earth, r is the distance between the center of the Earth and the satellite, and $\hat{u}_r$ is the unit vector pointing from the Earth to the satellite.

Kinematic Equations To use Eq. (2), we need $\vec{r}$ and $\vec{v}$ in polar coordinates:

$$\vec{r} = r\,\hat{u}_r \quad \text{and} \quad \vec{v} = v_r\,\hat{u}_r + v_\theta\,\hat{u}_\theta. \tag{4}$$

The velocity components and speed are related as follows:

$$v = \sqrt{v_r^2 + v_\theta^2}. \tag{5}$$

We note that

$$v_{rP} = \dot{r}_P = 0 \quad \text{and} \quad v_{rA} = \dot{r}_A = 0, \tag{6}$$

because, by definition, P and A are the points where the distances between the satellite and the Earth are minimum and maximum, respectively. Therefore, combining Eqs. (5) and (6), we have

$$v_P = \sqrt{v_{\theta P}^2} = |v_{\theta P}| \quad \text{and} \quad v_A = \sqrt{v_{\theta A}^2} = |v_{\theta A}|. \tag{7}$$

Computation Substituting Eqs. (3) and (4) into the first of Eqs. (2), we have

$$\vec{M}_E = r\,\hat{u}_r \times \left(-\frac{Gm_e m}{r^2}\right)\hat{u}_r = \vec{0}, \tag{8}$$

since the line of action of $\vec{F}_G$ goes through the moment center E (we chose E as the moment center to obtain $\vec{M}_E = \vec{0}$). Therefore, from Eq. (1) we have

$$\dot{\vec{h}}_E = \vec{0} \quad \Rightarrow \quad \vec{h}_E = \text{constant throughout the motion}, \tag{9}$$

so that we have

$$\vec{h}_{EP} = \vec{h}_{EA} \quad \Rightarrow \quad mR_P v_{\theta P}\,\hat{k} = mR_A v_{\theta A}\,\hat{k}, \tag{10}$$

where $\vec{h}_{EP}$ and $\vec{h}_{EA}$ are the values of $\vec{h}_E$ at P and A, respectively, and where we have used Eqs. (4) and the second of Eqs. (2). Computing the magnitude of both sides of Eq. (10) and using Eqs. (7), we have

$$mR_P v_P = mR_A v_A, \tag{11}$$

which can be solved for v_A to obtain

$$\boxed{v_A = \frac{R_P}{R_A} v_P = 2337\,\mathrm{m/s}.} \tag{12}$$

Discussion & Verification Since the ratio R_P/R_A is nondimensional, the result in Eq. (12) is dimensionally correct. In addition, our solution confirms what we have learned in the analysis of the skater's spin in Example 5.11. When the angular momentum is conserved, increasing the distance of mass from the spin axis causes a reduction in speed proportional to the ratio between the small and large values of distance from the spin axis. Therefore, the result in Eq. (12) is as expected given that $v_A < v_P$.

A Closer Look This problem was rather simple because we considered the extremes of the orbit (perigee and apogee). Suppose, instead, that we wanted to know the speed of the satellite at the position shown in Fig. 1. Although angular momentum is still conserved at every point in the orbit, it is only at apogee and perigee that there is no radial component of velocity. Therefore, at positions other than apogee and perigee, other principles must be invoked to find the speed of the satellite. We will see how this is done in Section 5.4, although the next example shows some of what is involved.

EXAMPLE 5.13 *Orbit of a Satellite: Angular Momentum and Work-Energy*

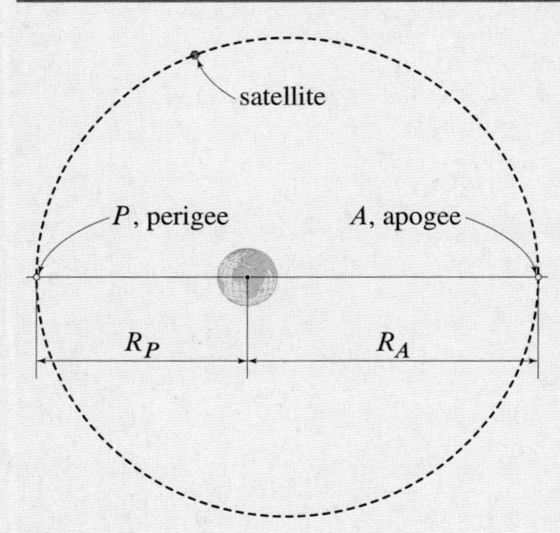

Figure 1
Satellite in orbit around the Earth.

At perigee, a satellite in orbit around the Earth has a speed $v_P = 3.2 \times 10^3$ m/s and a distance from the center of the Earth $R_P = 4.5 \times 10^7$ m. Treating the Earth as stationary, (a) establish whether the satellite's trajectory is circular and (b) determine R_A and the satellite's speed at the *apogee* A, that is, the point on the orbit farthest from Earth.

SOLUTION

Road Map & Modeling We model both the Earth and the satellite as particles, with the Earth being fixed and its center E as the origin of an inertial reference frame (Fig. 2). The satellite is assumed to be subject only to the Earth's gravitational attraction $\vec{F}_G$ and its orbit is assumed to be planar, so we will use a polar coordinate system with origin at E. If the orbit were circular, we would have $R_A = R_P$ and $v_A = v_P$. Since this problem is a central force problem, we can obtain one relation concerning R_A and v_A by the conservation of angular momentum. We can then take advantage of the fact that the force of gravity is conservative to obtain a second relation between R_A and v_A using the work-energy principle.

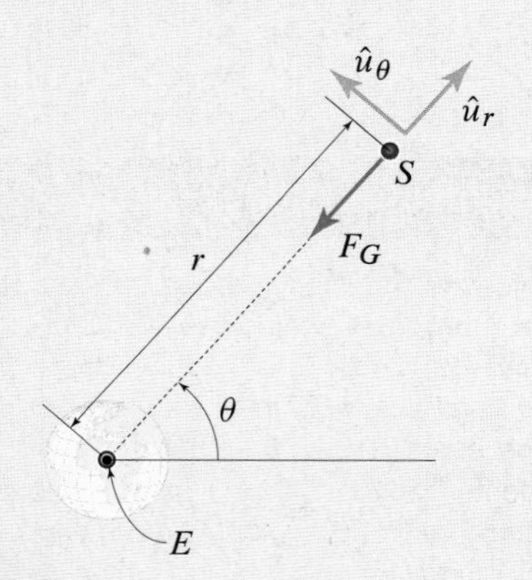

Figure 2
FBD of satellite along with a description of the chosen coordinate and component system.

Governing Equations

Balance Principles Applying Eq. (5.37), we have

$$\vec{h}_{EP} = \text{constant} = \vec{h}_E, \tag{1}$$

where $\vec{h}_{EP}$ is the angular momentum of the satellite about E evaluated at P, where, at a generic position along the satellite's orbit,

$$\vec{h}_E = \vec{r} \times m\vec{v}, \tag{2}$$

and where m, $\vec{r}$, and $\vec{v}$ are the mass, position, and velocity of the satellite, respectively. The work-energy principle, applied between P and a generic point along the orbit, reads

$$T_P + V_P = T + V, \tag{3}$$

where

$$T_P = \tfrac{1}{2} m v_P^2 \quad \text{and} \quad T = \tfrac{1}{2} m v^2, \tag{4}$$

and where v_P and v are the satellite's speed at P and at a generic point on the orbit, respectively.

Force Laws The potential energy of the force $\vec{F}_G$ is (see Eq. (4.25) on p. 259)

$$V = -G\frac{m_e m}{r}, \tag{5}$$

where we recall that $G = 6.674 \times 10^{-11}\ \mathrm{m^3/(kg \cdot s^2)}$ is the universal gravitational constant and $m_e = 5.9736 \times 10^{24}$ kg is the mass of the Earth.

Kinematic Equations In polar coordinates we have

$$\vec{r} = r\,\hat{u}_r, \quad \vec{v} = v_r\,\hat{u}_r + v_\theta\,\hat{u}_\theta, \quad \text{and} \quad v = \sqrt{v_r^2 + v_\theta^2}. \tag{6}$$

Hence, Eq. (2) simplifies to

$$\vec{h}_E = m r v_\theta\,\hat{k}. \tag{7}$$

We also recall (see Example 5.12) that at perigee and apogee we must have

$$v = |v_\theta|. \tag{8}$$

Computation Substituting Eq. (7) into Eq. (1) and recalling that $r_P = R_P$, then anywhere along the orbit we have

$$mR_P v_{\theta P} = mrv_\theta. \tag{9}$$

Computing the absolute value of Eq. (9) and evaluating it at those points along the orbit for which Eq. (8) is true (i.e., for the perigee or apogee), we obtain

$$R_P v_P = rv. \tag{10}$$

Next, substituting Eqs. (4) and (5) into Eq. (3), we obtain

$$\tfrac{1}{2}mv_P^2 - G\frac{m_e m}{R_P} = \tfrac{1}{2}mv^2 - G\frac{m_e m}{r}. \tag{11}$$

Equations (10) and (11) form a system of two equations in the two unknowns r and v, which has the following two solutions:

Solution 1: $\quad v_1 = v_P, \qquad r_1 = R_P, \qquad (12)$

Solution 2: $\quad v_2 = \dfrac{2Gm_e - R_P v_P^2}{R_P v_P}, \qquad r_2 = \dfrac{R_P^2 v_P^2}{2Gm_e - R_P v_P^2}. \qquad (13)$

Solution 1 indicates that P is one of the places on the orbit where Eq. (8) is satisfied. Solution 2 tells us that there is another place on the orbit where Eq. (8) is satisfied.

Part (a): Is the Satellite's Trajectory Circular?

If the trajectory were circular, everywhere along the orbit we would have $r = R_P$ and $v = v_P$, and therefore we would have $v_1 = v_2$ and $r_1 = r_2$, so that we would have

$$v_1 = v_2 \quad \Rightarrow \quad v_P = \sqrt{Gm_e/R_P}. \tag{14}$$

Since the values of v_P, R_P, m_e, and G are given, by substituting these values into Eq. (14), we can verify whether or not the trajectory is circular. Doing so yields

$$(v_P)_{\text{circular orbit}} = \sqrt{\frac{[6.6732\times10^{-11}\,\text{m}^3/(\text{kg}\cdot\text{s}^2)]\,5.9736\times10^{24}\,\text{kg}}{4.5\times10^7\,\text{m}}} = 2976\,\text{m/s} \neq (v_P)_{\text{given}} = 3200\,\text{m/s}. \tag{15}$$

Therefore, the answer to Part (a) is that the satellite's orbit is *not* circular.

Common Pitfall

Proper use of Eq. (14). Equation (14) does not say anything about the *given* values of v_P and R_P. This equation simply states that *if the satellite's orbit were circular*, then the values of v_P and R_P would not be independent—they would be related in the way specified by this equation. Conversely, if the given values of v_P and R_P were not to satisfy Eq. (14), then we would conclude that the satellite's orbit is not a circle.

Part (b): Conditions at Apogee

Since the satellite's trajectory is not circular, there is only one apogee with corresponding values of v_A and R_A given by Solution 2. Substituting the given numerical data in Eqs. (13), we have

$$v_2 = v_A = 2337\,\text{m/s} \quad \text{and} \quad r_2 = R_A = 6.163\times10^7\,\text{m}. \tag{16}$$

Discussion & Verification The solution in Eqs. (16) was obtained from Eqs. (13), which are dimensionally correct. Also, notice that $v_A < v_P$, as we expect, due to the conservation of angular momentum. Hence, our overall solution appears to be correct.

A Closer Look We could have ended this solution with Eqs. (12) and (13); that is, we could have substituted all known quantities into those equations, and we would have obtained the results given in Eqs. (16). Doing so would have also made it clear that the orbit is not circular. On the other hand, by obtaining the condition on the relationship between v_P and R_P for a circular orbit found in Eq. (14), we have created a useful result that can be used in other situations, and we have gained some insight into what it means for an orbit to be circular.

EXAMPLE 5.14 *Controlling the Spin Rate of a Satellite*

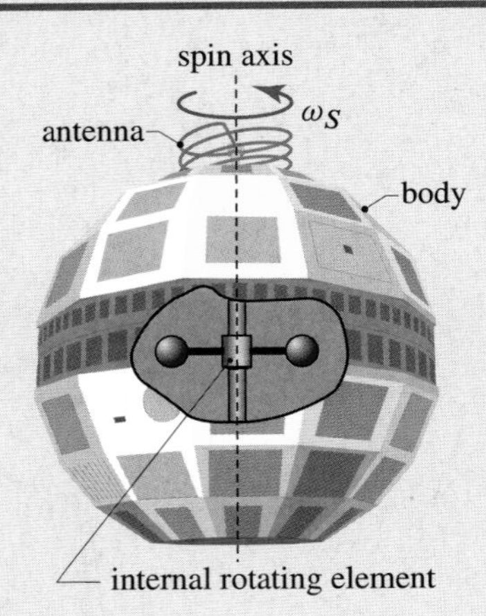

Figure 1
Satellite spinning about an axis. Notice the presence of the *internal* moving element.

> **Interesting Fact**
>
> **Telstar satellite.** The satellite in Fig. 1 is the *Telstar* satellite, the very first operational communications satellite, made by Bell Labs and put into orbit by NASA on July 10, 1962. Before Telstar, transatlantic phone communications relied on cables stretching from the United States to France.

Figure 1 shows a satellite that has been deployed with an angular speed $\omega_S(t_d)$ about the spin axis z, where t_d is the time of deployment. To control the satellite's attitude (its orientation) and overall spin rate, the satellite is equipped with internal motors that can spin up and spin down internal masses. In this problem, we consider a simple setup with just two internal spheres, each of mass m_{int} at a distance R_{int} from the spin axis (where the subscript "int" stands for internal). Assuming that, at deployment, the spheres and the satellite are spinning at the rate ω_S, find the rotational speed of the two internal masses to cause the satellite's body to spin down to one-half of $\omega_S(t_d)$.

SOLUTION

Road Map & Modeling We model the satellite as a *system of mass points*. Referring to Fig. 2, we assume that

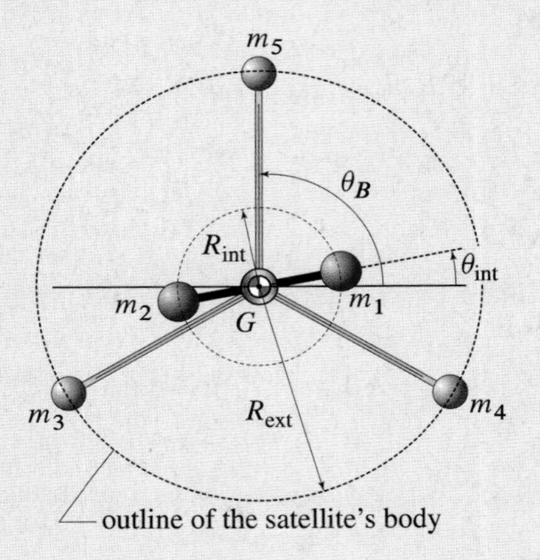

Figure 2. View down the z axis of the *lumped-mass* model of the satellite, which is also the FBD since the system is isolated. Masses m_3, m_4, and m_5 rotate together; masses m_1 and m_2 rotate together, but they can spin at a different rate than the others.

1. The satellite is not acted upon by external forces.
2. The satellite's only motion is a rotation about the z axis, which we will treat as fixed. This allows us to choose an inertial reference frame with origin at the satellite's center of mass G and with one axis coinciding with the spin axis z.
3. The satellite's body has a mass m_B, which is *lumped* into three identical point masses m_3, m_4, and m_5 that are placed at a distance R_{ext} from the spin axis and separated from each other by equal angles of 120° (see Fig. 2).

The position of the internal and external masses can be characterized by the angular coordinates θ_{int} and θ_B, respectively, shown in Fig. 2. Since the system is isolated, its angular momentum about the spin axis is *conserved*.

Governing Equations

Balance Principles Choosing the center of mass G as the moment center and recalling that the system is isolated, we see the system's angular momentum is *constant* and therefore equal to what it was at deployment, i.e.,

$$\vec{h}_G = \text{constant} = \vec{h}_G(t_d), \tag{1}$$

where

$$\vec{h}_G = \sum_{i=1}^{5} \vec{r}_i \times m_i \vec{v}_i, \tag{2}$$

where the subscript i denotes the ith particle in the system.

Force Laws All forces (in this case none) are accounted for on the FBD.

Kinematic Equations Referring to Figs. 3 and 4, for the position vectors we have

$$\vec{r}_i = R_i \hat{u}_{ri}, \tag{3}$$

where $\hat{u}_{ri}$ is the unit vector pointing from G to particle i and where

$$R_i = \begin{cases} R_{\text{int}} & \text{for } i = 1, 2 \\ R_{\text{ext}} & \text{for } i = 3, 4, 5. \end{cases} \tag{4}$$

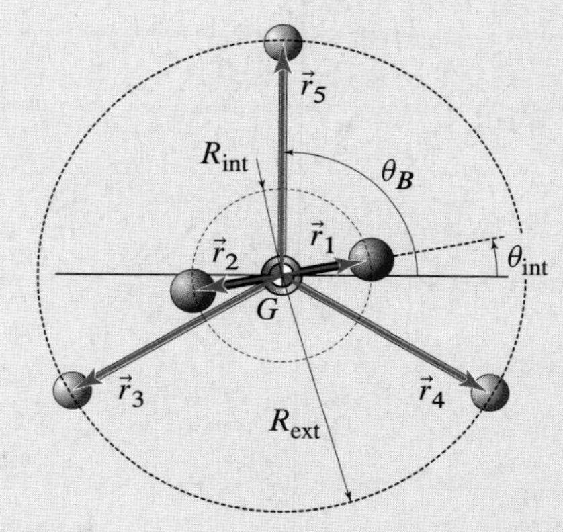

Figure 3
Position vectors for the chosen point masses.

For the velocity vectors, we have

$$\vec{v}_i = R_i \dot{\theta}_i \hat{u}_{\theta i}, \tag{5}$$

where $\hat{u}_{\theta i}$ is the unit vector perpendicular to $\hat{u}_{ri}$ pointing in the direction of growing θ. In addition, we have

$$\dot{\theta}_i = \begin{cases} \dot{\theta}_{\text{int}} & \text{for } i = 1, 2 \\ \dot{\theta}_{\text{ext}} & \text{for } i = 3, 4, 5. \end{cases} \tag{6}$$

Computation Substituting the kinematic relations into Eqs. (1) and (2), we obtain the z component of the angular momentum as

$$3m_{\text{ext}} R_{\text{ext}}^2 \dot{\theta}_{\text{ext}} + 2m_{\text{int}} R_{\text{int}}^2 \dot{\theta}_{\text{int}} = \left(3m_{\text{ext}} R_{\text{ext}}^2 + 2m_{\text{int}} R_{\text{int}}^2\right) \dot{\theta}(t_d), \tag{7}$$

where

$$m_{\text{int}} = m_1 = m_2, \quad \dot{\theta}(t_d) = \omega_S(t_d), \quad m_{\text{ext}} = m_3 = m_4 = m_5 = \frac{m_B}{3}. \tag{8}$$

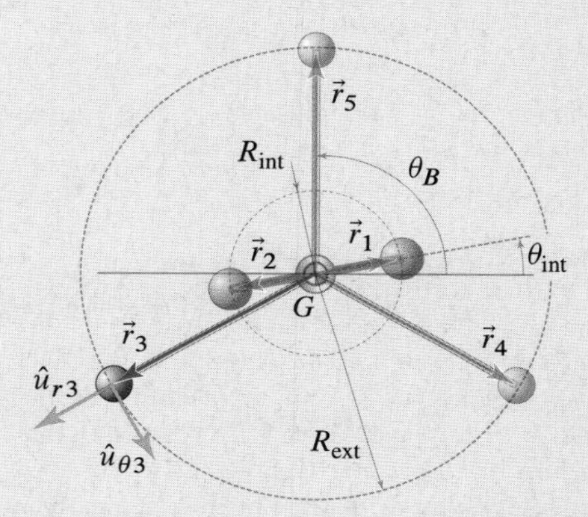

Figure 4
When using a polar coordinate system, there is a set of radial and transverse unit vectors for each point in the system. As an example, this figure shows the radial and transverse unit vectors for m_3 at a particular instant.

In the end, we want to achieve the following result:

$$\dot{\theta}_{\text{ext}} = \tfrac{1}{2}\omega_S(t_d). \tag{9}$$

Hence, substituting the desired value of spin rate in Eq. (7) and solving for $\dot{\theta}_{\text{int}}$, we have

$$\boxed{\dot{\theta}_{\text{int}} = \frac{2m_{\text{int}} R_{\text{int}}^2 + m_B R_{\text{ext}}^2/2}{2m_{\text{int}} R_{\text{int}}^2} \omega_S(t_d).} \tag{10}$$

Discussion & Verification Since the fraction on the right-hand side of Eq. (10) is nondimensional, our final result is dimensionally correct. Notice that the result in Eq. (10) can be written as

$$\dot{\theta}_{\text{int}} = \left(1 + \frac{m_B R_{\text{ext}}^2}{4m_{\text{int}} R_{\text{int}}^2}\right) \omega_S(t_d), \tag{11}$$

which shows that $\dot{\theta}_{\text{int}}$ has the same sign as $\omega_S(t_d)$ and must always be greater than $\omega_S(t_d)$. That is, the internal masses need to rotate in the same direction as the body's initial angular velocity at a rate higher than $\omega_S(t_d)$. This makes physical sense in that if we want to slow down the outer system while keeping the overall angular momentum constant, the internal masses must speed up.

A Closer Look We chose to represent the distributed mass of the satellite as three masses. This choice was entirely arbitrary, and we can see that the final result did not depend on this choice. Had we used four masses, the coefficient preceding m_{ext} in Eq. (7) would have been 4, but now the mass of the body m_B would have been $4m_{\text{ext}}$, so the final result is unchanged.

Interesting Fact

Spinning masses and the Hubble Space Telescope. Many spacecraft, including the Space Shuttle, use small RCS (reaction control system) jets or thrusters to control their attitude. RCS jets cannot be used on the Hubble Space Telescope (HST) because the exhaust gas would damage the delicate mirrors. Therefore, the HST uses four spinning wheels called *reaction wheels* to control its attitude. By orienting the spin axes of these wheels along different directions, engineers can point the HST in any direction with high accuracy.

EXAMPLE 5.15 *Conservation of Angular Momentum: An Advanced Example*

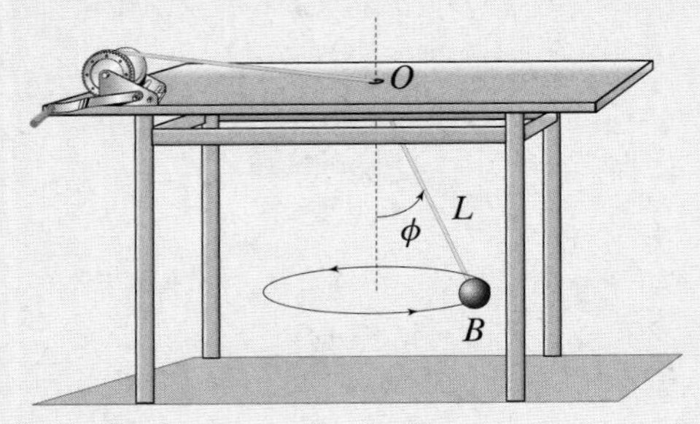

Figure 1
A pendulum with its bob B tracing a horizontal circle and being slowly reeled in.

At time t_0, a pendulum bob B is put in motion along a horizontal circle so that the pendulum describes a right cone with vertex at O, an initial side length $L_0 = 1$ m, and initial opening angle $\phi_0 = 20°$. During a time interval $t_1 \le t \le t_2$, with $t_1 > t_0$, L is made to decrease and then is held constant for $t \ge t_2$. The decrease of L occurs so slowly that, for time $t \ge t_2$, the trajectory of B can be viewed as being a circle lying in a horizontal plane, that is, a circle parallel to the initial trajectory of the pendulum. Use the moment-angular momentum relation to determine the pendulum's initial speed v_0. In addition, at $t = t_2$ determine the speed v_2 and length L_2 if $\phi_2 = 50°$.

SOLUTION

Road Map & Modeling Considering the FBD in Fig. 2, we model B as a particle subject only to its own weight mg and the tension in the cord F_c. We model the cord as inextensible, and we describe the motion of B using a cylindrical coordinate system with origin at O. In obtaining the solution with the moment-angular momentum relation, we need to achieve the following objectives: (1) establish a relation between L, v, and ϕ when the trajectory of B is horizontal and (2) determine how the motions for $t < t_1$ and $t \ge t_2$ are related—that is, determine how the conditions with which the pendulum is initially put in motion affect the motion after the pendulum cord length is decreased.

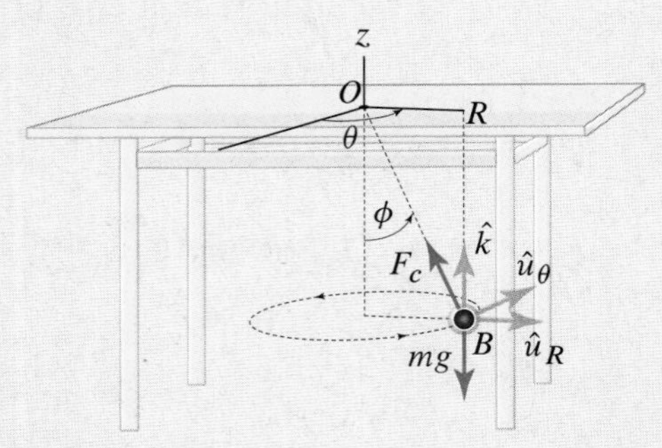

Figure 2
FBD of B with the chosen coordinate and component systems.

Governing Equations

Balance Principles Choosing the fixed point O as moment center, we apply the moment-angular momentum relation as given in Eq. (5.36):

$$\vec{M}_O = \dot{\vec{h}}_O, \tag{1}$$

where

$$\vec{M}_O = \vec{r}_{B/O} \times \left(\vec{F}_c - mg\,\hat{k}\right) \quad \text{and} \quad \vec{h}_O = \vec{r}_{B/O} \times m\vec{v}, \tag{2}$$

and where $\vec{r}_{B/O}$ is the position of B relative to O and $\vec{v}$ is the velocity of B.

Force Laws Since $\vec{F}_c$ is directed from B to O, we must have

$$\vec{F}_c = -F_c \frac{\vec{r}_{B/O}}{|\vec{r}_{B/O}|}. \tag{3}$$

Kinematic Equations Referring to Fig. 3, for $\vec{r}_{B/O}$, we have

$$\vec{r}_{B/O} = L(\sin\phi\,\hat{u}_R - \cos\phi\,\hat{k}). \tag{4}$$

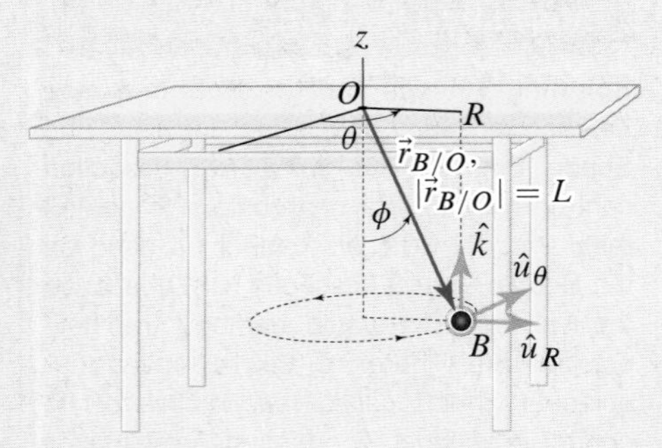

Figure 3
Depiction of the position vector $\vec{r}_{B/O}$. Notice that $\vec{r}_{B/O}$ has no component in the transverse direction and that $|\vec{r}_{B/O}| = L$.

We will leave the discussion of the motion during $t_1 < t < t_2$ for later. For both $t \le t_1$ and $t \ge t_2$, L is constant and the vertical component of the velocity of B is equal to zero, so during these time intervals, we have

$$\vec{v} = v_\theta\,\hat{u}_\theta \quad \text{with} \quad v_\theta = R\dot{\theta} = (L\sin\phi)\dot{\theta} = L\dot{\theta}\sin\phi. \tag{5}$$

To compute $\dot{\vec{h}}_O$ in Eq. (1), we need the time derivatives of the unit vectors $\hat{u}_R$, $\hat{u}_\theta$, and $\hat{k}$, which are (see Sections 2.6 and 2.8)

$$\dot{\hat{u}}_R = \dot{\theta}\,\hat{u}_\theta, \quad \dot{\hat{u}}_\theta = -\dot{\theta}\,\hat{u}_R, \quad \dot{\hat{k}} = \vec{0}. \tag{6}$$

Computation Since O is on the line of action of $\vec{F}_c$, the first of Eqs. (2) yields

$$\vec{M}_O = mgL\sin\phi\,\hat{u}_\theta, \tag{7}$$

where we have used Eq. (4) for $\vec{r}_{B/O}$. Substituting Eqs. (4) and (5) into the second of Eqs. (2), we have

$$\vec{h}_O = mv_\theta L\cos\phi\,\hat{u}_R + mv_\theta L\sin\phi\,\hat{k}. \tag{8}$$

Differentiating Eq. (8) with respect to time and using Eqs. (6), we have

$$\dot{\vec{h}}_O = \tfrac{d}{dt}(mv_\theta L\cos\phi)\,\hat{u}_R + (mv_\theta L\cos\phi)\dot{\theta}\,\hat{u}_\theta + \tfrac{d}{dt}(mv_\theta L\sin\phi)\hat{k}. \tag{9}$$

Enforcing Eq. (1), that is, equating Eqs. (7) and (9) component by component, we have

$$0 = \tfrac{d}{dt}(mv_\theta L\cos\phi) \quad\Rightarrow\quad v_\theta L\cos\phi = K_R, \tag{10}$$

$$mgL\sin\phi = (mv_\theta L\cos\phi)\dot{\theta} \quad\Rightarrow\quad gL\sin\phi = v_\theta^2\frac{\cos\phi}{\sin\phi}, \tag{11}$$

$$0 = \tfrac{d}{dt}(mv_\theta L\sin\phi) \quad\Rightarrow\quad v_\theta L\sin\phi = K_z, \tag{12}$$

where K_R and K_z are constants of integration and where, in Eq. (11), we have used the second of Eqs. (5) to express $\dot{\theta} = v_\theta/(L\sin\phi)$. Equations (10)–(12) describe the state of motion of the system for $t_0 \le t \le t_1$ and for $t \ge t_2$. Recalling that $L_0 = 1$ m and $\phi_0 = 20°$, by using Eq. (11) we can solve for the initial value of v_0

$$\boxed{v_0 = |v_{\theta 0}| = \sqrt{gL_0\frac{\sin^2\phi_0}{\cos\phi_0}} = 1.105\,\text{m/s},} \tag{13}$$

where $|v_\theta| = v$, given that for $t_0 \le t \le t_1$ the motion of B is circular.

For $t \ge t_2$ we only know ϕ_2, and it would seem that there are not enough equations to solve for all the unknowns of the problem, which are L_2, v_2, K_{R2}, and K_{z2}. However, referring to Fig. 2, observe that, *for any time* $t \ge t_0$, none of the forces on B provides a moment about the z axis. Since this axis is fixed, as discussed on p. 364, we must have conservation of angular momentum about the z axis. We now observe that Eq. (12) expresses such a conservation requirement and is valid not just for the time intervals discussed earlier, but for any possible time (this is not true for Eqs. (10) and (11)). *This means that the value of the right-hand side of Eq. (12) must be the same at t_0 as it is at t_2!* Therefore, at $t = t_2$, we can use Eqs. (11) and (12) to write

$$gL_2\sin\phi_2 = v_2^2\frac{\cos\phi_2}{\sin\phi_2} \quad\text{and}\quad v_2L_2\sin\phi_2 = v_0L_0\sin\phi_0, \tag{14}$$

where it is assumed that $v_{\theta 2} > 0$ so that $v_{\theta 2} = v_2$ (see note in the margin). Equations (14) form a system of two equations in the unknowns L_2 and v_2 with solution

$$\boxed{v_2 = \sqrt[3]{g\tan\phi_2 v_0L_0\sin\phi_0} = 1.641\,\text{m/s},} \tag{15}$$

$$\boxed{L_2 = \sqrt[3]{(v_0L_0\sin\phi_0)^2\cos\phi_2/(g\sin^4\phi_2)} = 0.3007\,\text{m}.} \tag{16}$$

Helpful Information

From Eqs. (11) and (12) to Eqs. (14). Equations (14) are perfectly consistent with the discussion leading to them. This is so because the first of Eqs. (14) is given by Eq. (11) evaluated *just at time t_2 when the velocity has no radial component.* By contrast, we see that the left-hand side of the second of Eqs. (14) is Eq. (12) evaluated at time t_2, whereas the right-hand side is Eq. (12) evaluated at time t_0.

Discussion & Verification We leave it to the reader to verify that Eqs. (15) and (16) are dimensionally correct. As the pendulum cord is decreased, the distance between B and the spin axis z is also decreased. Therefore, in compliance with the conservation of angular momentum requirement, we would expect $v_2 > v_0$, as is indeed the case. Hence, overall our solution appears to be correct.

A Closer Look This example was deemed advanced because we had to recognize that part of the solution, Eq. (12), was applicable outside the assumptions underlying the rest of the solution. It was this realization that allowed us to determine how the initial conditions affected the motion at time t_2 *without* having to explicitly compute the motion that takes the system from t_1 to t_2.

Problems

Problem 5.107

Consider the situation depicted in the figure. At the instant shown, how are the angular momenta of particle P with respect to O and Q related?

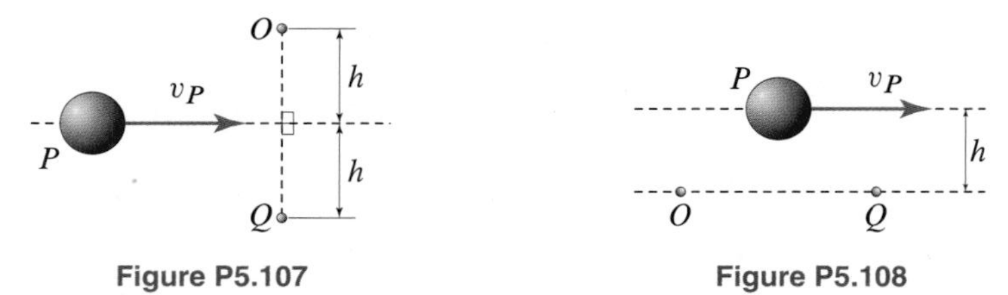

Figure P5.107

Figure P5.108

Problem 5.108

Consider the situation depicted in the figure. At the instant shown, how are the angular momenta of particle P with respect to O and Q related?

Problems 5.109 through 5.111

At the instant shown, a truck A, of weight $W_A = 31{,}000$ lb, and a car B, of weight $W_B = 3970$ lb, are traveling with speeds $v_A = 35$ mph and $v_B = 34$ mph, respectively.

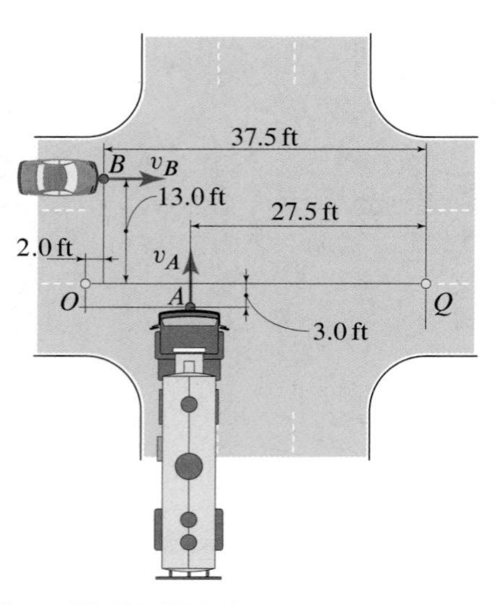

Figure P5.109–P5.111

Problem 5.109 Choosing point O as the moment center, determine the angular momentum (with respect to O) of A and B individually at this instant.

Problem 5.110 Choosing point O as the moment center, determine the angular momentum (with respect to O) of the *particle system* formed by A and B at this instant.

Problem 5.111 Choosing point Q as the moment center, determine the angular momentum (with respect to Q) of the *particle system* formed by A and B at this instant.

Problems 5.112 and 5.113

At time $t_1 = 0$, particle A weighing 5 oz goes through point P of (x, y, z) coordinates $(0, 2, 3)$ ft with a velocity $\vec{v}(t_1) = (3\,\hat{\imath} + 5\,\hat{\jmath} + 7\,\hat{k})$ ft/s. At time t_2, A goes through a point Q of coordinates $(1, 0, 1)$ ft, and with a velocity $\vec{v}(t_2) = (1\,\hat{\imath} + 2\,\hat{\jmath} + 3\,\hat{k})$ ft/s.

Problem 5.112 Determine the angular momentum of A with respect to the origin O at times t_1 and t_2.

Problem 5.113 Determine the time rate of change of the angular momentum of A relative to the origin O at times t_1 and t_2 if the acceleration of A at times t_1 and t_2 is $\vec{a}(t_1) = (1\,\hat{\imath} + 2\,\hat{\jmath} + 3\,\hat{k})$ ft/s^2 and $\vec{a}(t_2) = (9\,\hat{\imath} + 8\,\hat{\jmath} + 7\,\hat{k})$ ft/s^2, respectively.

Problem 5.114

Referring to Problems 5.112 and 5.113, is it possible to compute the velocity and acceleration of A?

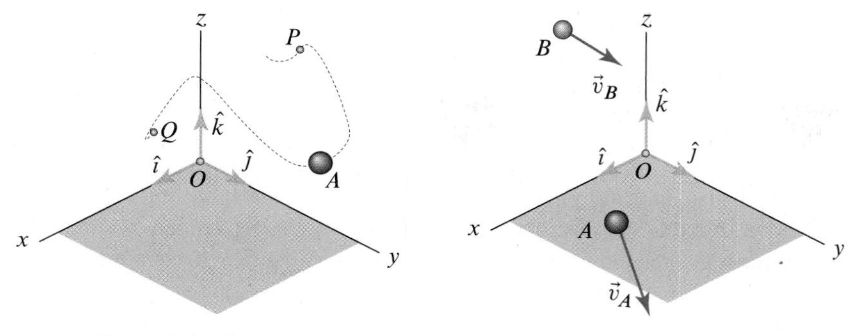

Figure P5.112–P5.114 **Figure P5.115 and P5.116**

Problems 5.115 and 5.116

Particles A and B have masses $m_A = 3\,\text{kg}$ and $m_B = 1.3\,\text{kg}$, respectively. At the instant shown, the (x, y, z) coordinates of A and B are $(3, 2, 0)$ m and $(3, 0, 4)$ m, respectively. In addition, the velocities of A and B are $\vec{v}_A = (7\,\hat{\imath} + 10\,\hat{\jmath})\,\text{m/s}$ and $\vec{v}_B = (-4\,\hat{\imath} - 3\,\hat{k})\,\text{m/s}$, respectively.

Problem 5.115 Determine the angular momentum of the system formed by A and B relative to the origin O at the instant shown.

Problem 5.116 Determine the angular momentum of particle A relative to particle B at the instant shown.

Problem 5.117

A rotor consists of four horizontal blades each of length $L = 4\,\text{m}$ and mass $m = 90\,\text{kg}$ cantilevered from a vertical shaft. The rotor is initially at rest when it is subjected to a moment $M = \beta t$, with $\beta = 60\,\text{N·m/s}$. Modeling each blade as having its mass concentrated at its midpoint, determine the angular speed of the rotor after 10 s.

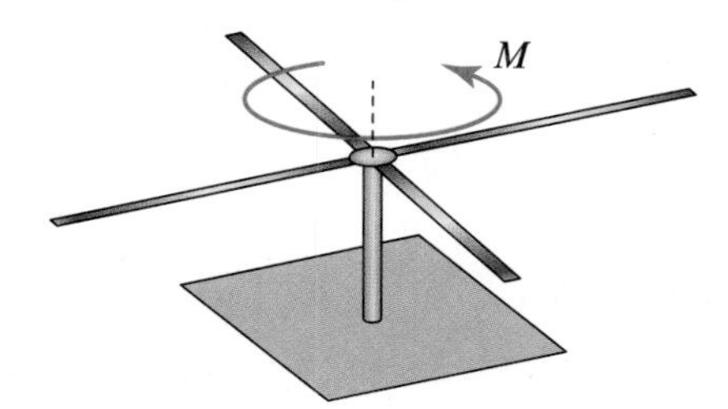

Figure P5.117

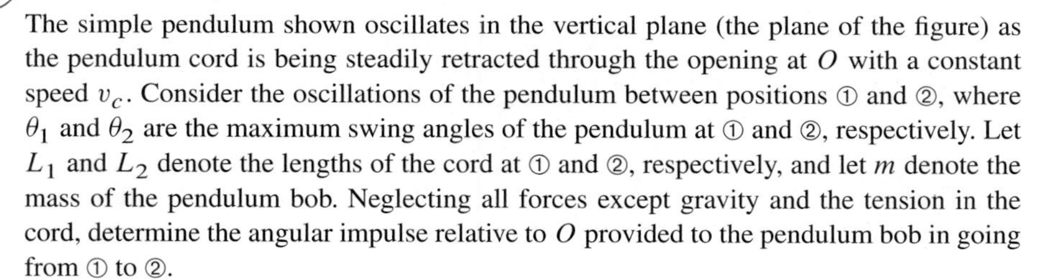

Problem 5.118

The simple pendulum shown oscillates in the vertical plane (the plane of the figure) as the pendulum cord is being steadily retracted through the opening at O with a constant speed v_c. Consider the oscillations of the pendulum between positions ① and ②, where θ_1 and θ_2 are the maximum swing angles of the pendulum at ① and ②, respectively. Let L_1 and L_2 denote the lengths of the cord at ① and ②, respectively, and let m denote the mass of the pendulum bob. Neglecting all forces except gravity and the tension in the cord, determine the angular impulse relative to O provided to the pendulum bob in going from ① to ②.

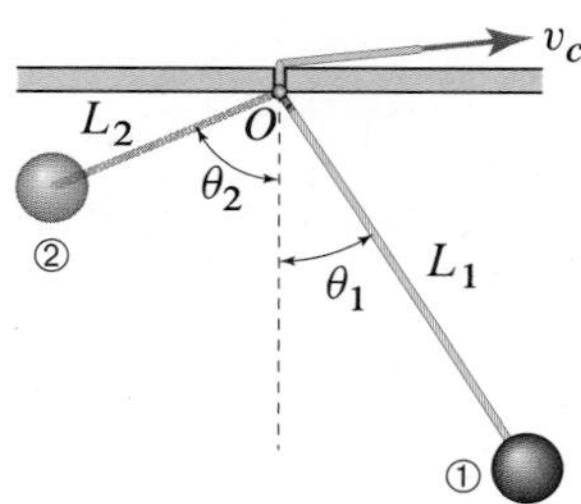

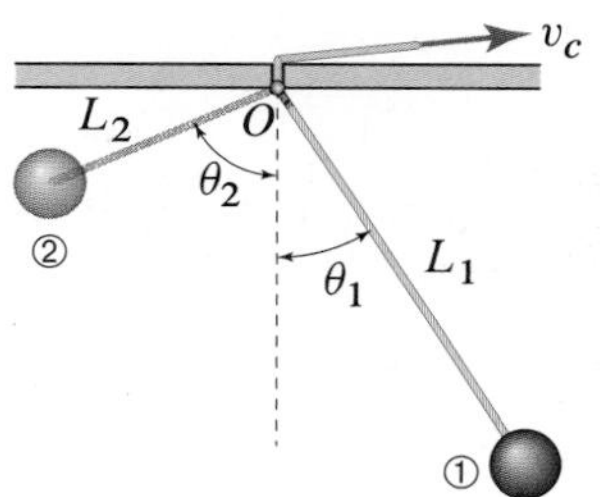

Figure P5.118

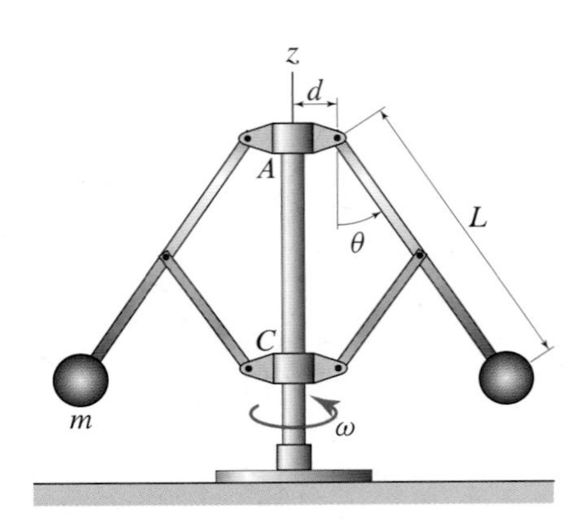

Figure P5.119

Problem 5.119

The object shown is called a *speed governor*, a mechanical device for the regulation and control of the speed of mechanisms. The system consists of two arms of negligible mass at the ends of which are attached two spheres, each of mass m. The upper end of each arm is attached to a fixed collar A. The system is then made to spin with a given angular speed ω_0 at a set opening angle θ_0. Once it is in motion, the opening angle of the governor can be varied by adjusting the position of the collar C (by the application of some force). Let θ represent the generic value of the governor opening angle. If the arms are free to rotate, that is, if no moment is applied to the system about the spin axis after the system is placed in motion, determine the expression of the angular velocity ω of the system as a function of ω_0, θ_0, m, d, and L, where L is the length of each arm and d is the distance of the top hinge point of each arm from the spin axis. Neglect any friction at A and C.

Problems 5.120 and 5.121

Let $\vec{h}_O = 25\,\hat{k}\ \text{kg}\cdot\text{m}^2/\text{s}$ be the angular momentum of a particle A about the origin O of an inertial (x, y, z) reference frame at the instant shown. Let the mass of A be $m = 0.75\,\text{kg}$, and let the coordinates of A at the instant shown be $(1, 2, 0)$ m.

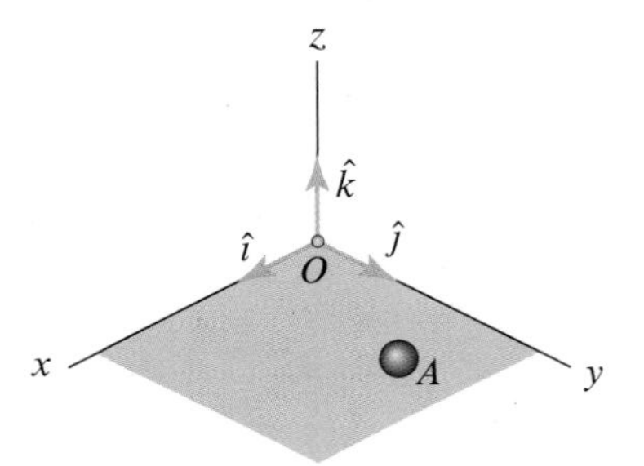

Figure P5.120 and P5.121

Problem 5.120 If A moves only in the xy plane, determine the vector component of the velocity of A perpendicular to the position vector of A at the instant shown.

Problem 5.121 If A moves only in the xy plane, determine the vector component of the acceleration of A perpendicular to the position vector of A at the instant shown if, at this instant, the moment acting on A relative to the origin O is $\vec{M}_O = 2\,\hat{k}\ \text{N}\cdot\text{m}$.

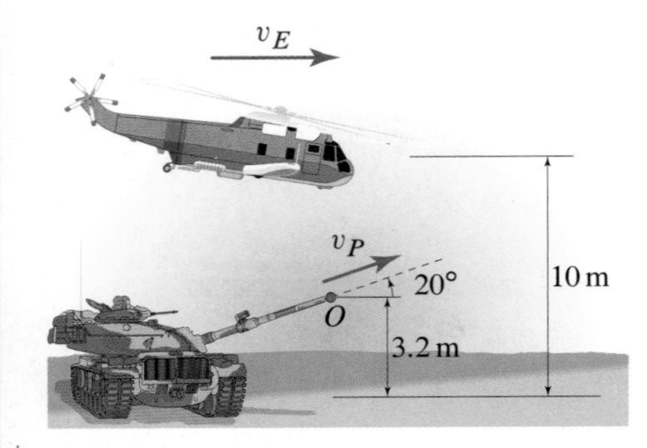

Figure P5.122–P5.124

Problems 5.122 through 5.124

The projectile P of mass $m_P = 18.5\,\text{kg}$ is shot with an initial speed $v_P = 1675\,\text{m/s}$ as shown in the figure. Ignore aerodynamic drag forces on the projectile.

Problem 5.122 Compute the projectile's angular momentum with respect to the point O as a function of time from the time it exits the barrel until the time it hits the ground.

Problem 5.123 Choose point O as moment center. Then verify the validity of the angular impulse-momentum principle as given in Eq. (5.36) by showing that the time derivative of the angular momentum does, in fact, equal the moment.

Problem 5.124 Knowing that the helicopter E happens to have the same horizontal coordinate as the projectile at the instant the projectile leaves the gun and that it moves at a constant speed $v_E = 15\,\text{m/s}$ as shown, and treating E as a moving moment center, verify the angular impulse-momentum principle as given in Eq. (5.35).

Problems 5.125 and 5.126

The simple pendulum in the figure is released from rest when $\theta = 33°$.

Problem 5.125 Knowing that the bob's weight is $W = 2$ lb and that the length is $L = 4$ ft, determine the bob's angular momentum with respect to O as a function of θ.

Problem 5.126 Use the angular impulse-momentum principle in Eq. (5.36) to determine the equations of motion of the pendulum bob.

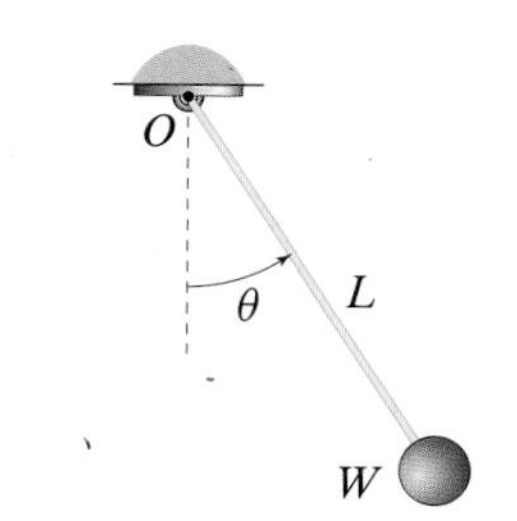

Figure P5.125 and P5.126

Problem 5.127

At the lowest and highest points on its trajectory, the pendulum cord, with a length $L = 2$ ft, forms angles $\phi_1 = 15°$ and $\phi_2 = 50°$ with the vertical direction, respectively. Determine the speed of the pendulum bob corresponding to ϕ_1 and ϕ_2.

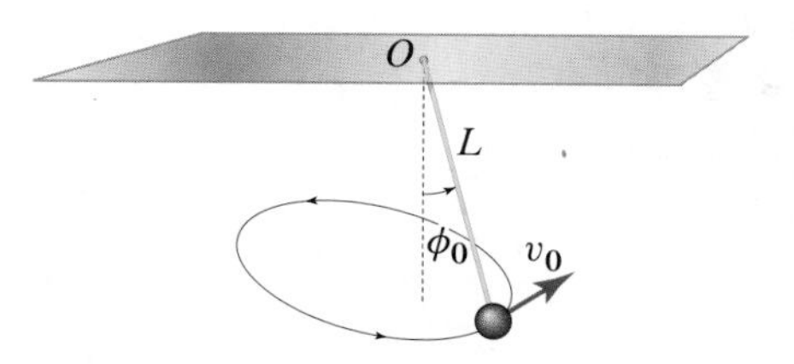

Figure P5.127

Problems 5.128 and 5.129

A collar with mass $m = 2$ kg is mounted on a rotating arm of negligible mass that is initially rotating with an angular velocity $\omega_0 = 1$ rad/s. The collar's initial distance from the z axis is $r_0 = 0.5$ m and $d = 1$ m. At some point, the restraint keeping the collar in place is removed so that the collar is allowed to slide. Assume that the friction between the arm and the collar is negligible.

Problem 5.128 If no external forces and moments are applied to the system, with what speed relative to the arm will the collar impact the end of the arm?

Problem 5.129 Compute the moment that must be applied to the arm, as a function of position along the arm, to keep the arm rotating at a constant angular velocity while the collar travels toward the end of the arm.

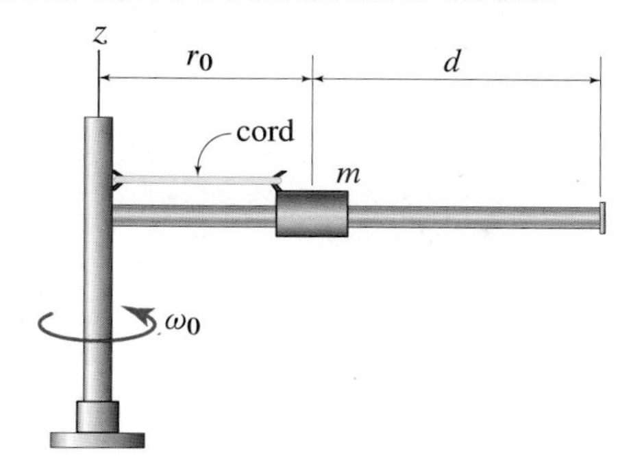

Figure P5.128 and P5.129

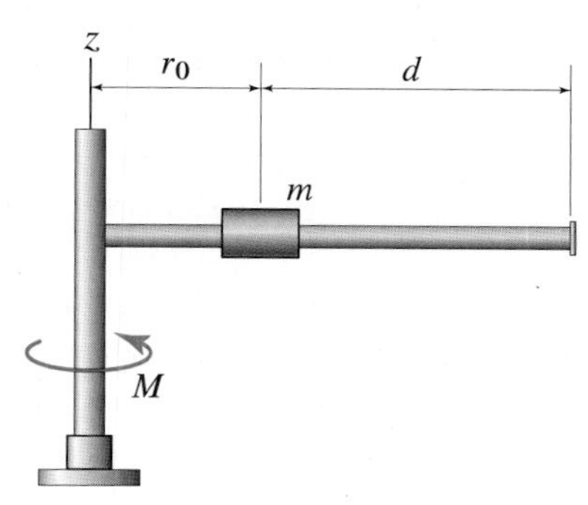

Figure P5.130 and P5.131

Problems 5.130 and 5.131

A collar of mass m is initially at rest on a horizontal arm when a constant moment M is applied to the system to make it rotate. Assume that the mass of the horizontal arm is negligible and that the collar is free to slide without friction.

Problem 5.130 Derive the equations of motion of the system, taking advantage of the angular impulse-momentum principle. *Hint:* Applying the angular impulse momentum principle yields only one of the needed equations of motion.

Problem 5.131 Continue Prob. 5.130 by integrating the collar's equations of motion and determine the time the collar takes to reach the end of the arm. Assume that the collar weighs 1.2 lb and that $M = 20$ ft·lb. Also, at the initial time let $r_0 = 1$ ft and $d = 3$ ft.

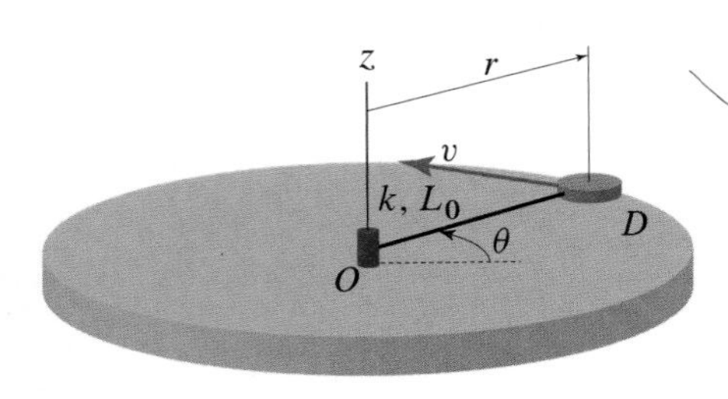

Figure P5.132

Problem 5.132

In a simple model of orbital motion under a central force, a disk D slides with no friction over a horizontal surface while connected to a fixed point O by a linear elastic cord of constant k and unstretched length L_0. Let the mass of D be $m = 0.45\,\text{kg}$ and $L_0 = 1\,\text{m}$. Suppose that when D is at its maximum distance from O, this distance is $r_0 = 1.75\,\text{m}$ and the corresponding speed of D is $v_0 = 4\,\text{m/s}$. Determine the elastic cord constant k such that the minimum distance between D and O is equal to the unstretched length L_0.

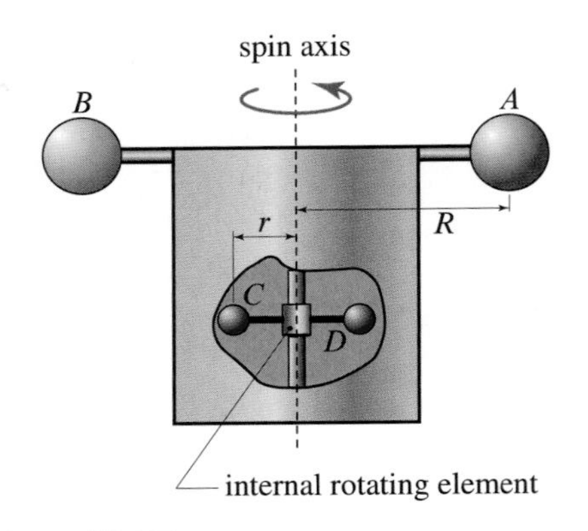

Figure P5.133

Problem 5.133

The body of the satellite shown has a weight that is negligible with respect to the two spheres A and B that are rigidly attached to it, which weigh 150 lb each. The distance between A and B from the spin axis of the satellite is $R = 3.5\,\text{ft}$. Inside the satellite there are two spheres C and D weighing 4 lb mounted on a motor that allows them to spin about the axis of the cylinder at a distance $r = 0.75\,\text{ft}$ from the spin axis. Suppose that the satellite is released from rest and that the internal motor is made to spin up the internal masses at an absolute constant time rate of $5.0\,\text{rad/s}^2$ (measured relative to an inertial observer) for a total of 10 s. Treating the system as isolated, determine the angular speed of the satellite at the end of spin-up.

Problem 5.134

A disk A with mass m moves on a frictionless horizontal surface. The disk is attached to point O with an elastic cord. The disk follows the trajectory shown between ① and ②. The coordinates of A at ① and ② are $(x_1, y_1) = (-1.5, 5.0)\,\text{cm}$ and $(x_2, y_2) = (4.0, -3.0)\,\text{cm}$, respectively. If the velocity of A at ① is parallel to the y axis and has a magnitude $v_1 = 2.0\,\text{m/s}$, determine v_2, the speed of A at ②, knowing that $\theta = 35°$.

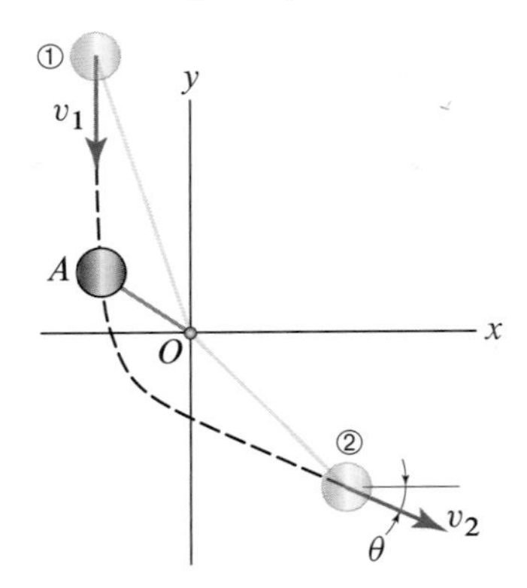

Figure P5.134

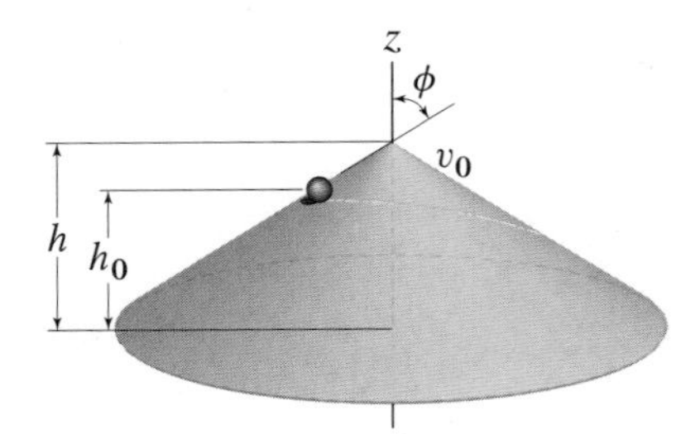

Figure P5.135

Problem 5.135

A sphere of mass m slides over the outer surface of a cone with angle ϕ and height h. The sphere was released at a height h_0 with a velocity of magnitude v_0 and a direction that was completely horizontal. Assume that the opening angle of the cone and the value of v_0 are such that the sphere does not separate from the surface of the cone once put in motion. In addition, assume that the friction between the sphere and the cone is negligible. Determine the vertical component of the sphere's velocity as a function of the vertical position z (measured from the base of the cone), v_0, h, h_0, and ϕ.

Problem 5.136

Consider a planet orbiting the Sun, and let P_1, P_2, P_3, and P_4 be the planet's position at four corresponding time instants t_1, t_2, t_3, and t_4 such that $t_2 - t_1 = t_4 - t_3$. Letting O denote the position of the Sun, determine the ratio between the areas of the orbital sectors $P_1 O P_2$ and $P_3 O P_4$. *Hint:* The area of triangle OAB defined by the two planar vectors $\vec{c}$ and $\vec{d}$ as shown is given by Area$(OAB) = |\vec{c} \times \vec{d}\,|$.

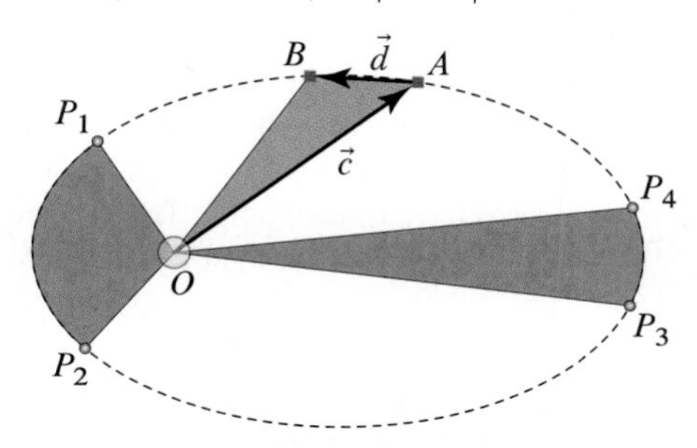

Figure P5.136

Design Problems

Design Problem 5.2

The device shown is designed to remotely rotate a video camera C by counterrotating the two equal masses m with an internal motor that couples the masses and the camera. Assume that the system is mounted on bearings that allow the counterrotating masses and the camera to rotate freely on the vertical mounting post. Model the video camera as two particles, each weighing 3.1 lb and each at a distance of 4.1 in. from the rotation axis. Design the radius of the counterrotating masses, their mass, and their maximum angular velocity so that the angular velocity of the camera never exceeds 10 rpm and so that the camera rotates 90° when the masses have rotated 360°. Treat the counterrotating masses as particles, and neglect the mass of the rod on which they are mounted.

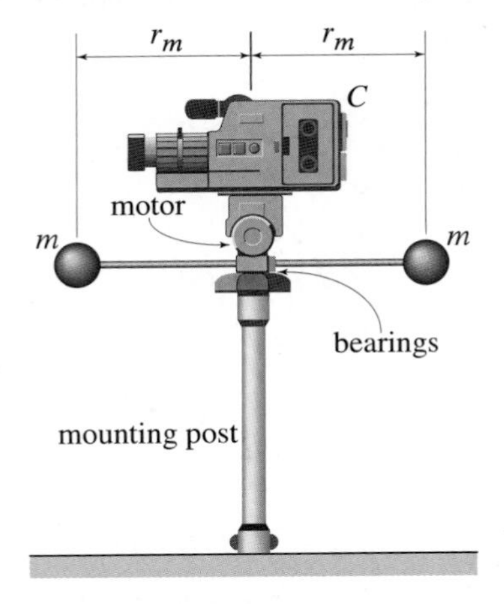

Figure DP5.2

5.4 Orbital Mechanics

Central forces have a special role in dynamics. In Example 3.10 on p. 202, we saw that central force motion is related to an aspect of the motion that is constant, and in Section 5.3 we discovered that this constant is called *angular momentum*. In Section 4.2 on p. 257, we saw that the work done by a central force, such as a spring force or gravity, depends only on the initial and final positions of the point of application of the force, and so those forces are conservative. We will now study one special central force, namely, the central force that arises from Newton's universal law of gravitation. The solutions to the problem of one body orbiting another are provided by combining Newton's second law and this central force. These solutions form the foundation that allows us to predict the motion of planets, Earth satellites (Fig. 5.23), high-altitude rockets, and interplanetary probes.

Figure 5.23
The International Space Station (ISS) as of June 2008. The equations for two-body orbital motion developed in this section form the basis for accurately predicting the orbit of the ISS about the Earth.

Determination of the orbit

We are interested in the motion of a satellite S of mass m that is subject to Newton's universal law of gravitation,

$$F = \frac{Gm_B m}{r^2}, \tag{5.59}$$

where m_B is the mass of the primary, central, or attracting body, G is the universal gravitational constant,* and r is the distance between the centers of mass of the two bodies (see Fig. 5.24). We begin by making three important assumptions:

1. The attracting bodies B and S are treated as particles. This assumption is exactly correct if each body has a spherically symmetric mass distribution. It is a good approximation if the distance between the two bodies is large compared with their dimensions, which is the case for many planets and artificial satellites in high orbits. Because the Earth is oblate and has a nonuniform mass distribution, this assumption starts to break down for artificial satellites in low Earth orbit.
2. The only force acting on the satellite S is the gravitational force F.
3. The primary body B is fixed in space. This assumption works well when m_B is very large compared with m. Note that relaxing this assumption only slightly changes the two-body equations we will derive (see the "Interesting Fact" on the next page).

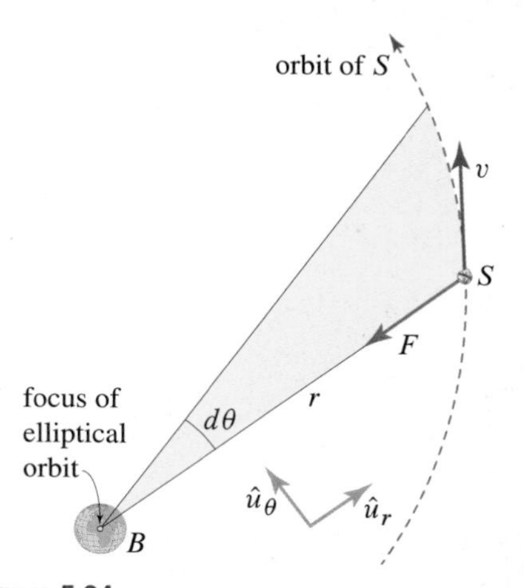

Figure 5.24
Satellite S of mass m orbiting a primary body B of mass m_B under the action of Newton's universal law of gravitation.

Applying Newton's second law to S in the polar component system shown in Fig. 5.24, we obtain

$$\sum F_r: \quad -\frac{Gm_B m}{r^2} = m(\ddot{r} - r\dot{\theta}^2), \tag{5.60}$$

$$\sum F_\theta: \quad 0 = m(r\ddot{\theta} + 2\dot{r}\dot{\theta}), \tag{5.61}$$

where we have substituted in the kinematic equations in polar components and have also used Eq. (5.59). In Section 5.3, we saw that Eq. (5.61) is equivalent to the conservation of angular momentum of S about B, that is,

$$h_B = mr^2\dot{\theta} = \text{constant}. \tag{5.62}$$

Helpful Information

Evaluating Gm_B for the Earth. If the primary body B is the Earth, then from Eq. (1.10) on p. 4 we can write

$$Gm_B = Gm_e = gr_e^2,$$

where m_e is the mass of the Earth, g is the acceleration due to gravity, and r_e is the radius of the Earth, which is 6371 km, or 3959 mi.

* Recall that the generally accepted value of this constant is $G = 6.674\times10^{-11}\ \text{m}^3/(\text{kg}\cdot\text{s}^2) = 3.439\times10^{-8}\ \text{ft}^3/(\text{slug}\cdot\text{s}^2)$.

Interesting Fact

The primary mass really moves. In reality, the satellite S and the primary body B orbit about their common mass center. If this motion of B is taken into account, Eqs (5.60) and (5.61) no longer apply, and the governing equations become

$$-\frac{Gm_B m}{r^2} = \frac{m_B m}{m_B + m}(\ddot{r} - r\dot{\theta}^2),$$

$$0 = \frac{m_B m}{m_B + m}(r\ddot{\theta} + 2\dot{r}\dot{\theta}),$$

where r is the distance between mass centers of B and S. Notice that if $m_B \gg m$, then these two equations are approximated by Eqs. (5.60) and (5.61). See Probs. 2.80 and 2.81 for a one-dimensional version of these two equations.

Equation (5.62) is also a reflection of Kepler's second law of planetary motion. To see this, refer to Fig. 5.24 and note that the area dA swept by the radial line of length r during the time dt (i.e., the yellow area) is equal to $\frac{1}{2}(r\,d\theta)r$. Therefore, the *areal velocity*, which is the time rate at which area is swept by the radial line r, is constant according to Eq. (5.62) and is given by

$$\frac{dA}{dt} = \tfrac{1}{2}r^2\dot{\theta} = \text{constant}. \tag{5.63}$$

This is Kepler's second law, which states that the line joining a planet to the Sun sweeps out equal areas in equal times as the planet travels around the Sun.

Solution of the governing equations

To find the trajectory of S, we need to solve Eqs. (5.60) and (5.61). This is traditionally done by eliminating t and then solving for $r(\theta)$.

To begin, we first rewrite Eqs. (5.60) and (5.61) as

$$\ddot{r} - r\dot{\theta}^2 = -\frac{Gm_B}{r^2}, \tag{5.64}$$

$$r^2\dot{\theta} = \kappa, \tag{5.65}$$

where κ (the Greek letter kappa) is the *angular momentum per unit mass* of the satellite S, which is equal to h_B/m from Eq. (5.62). The chain rule allows us to write $\dot{r}$ and $\ddot{r}$ as

$$\dot{r} = \frac{dr}{d\theta}\frac{d\theta}{dt} = \frac{dr}{d\theta}\frac{\kappa}{r^2} = -\kappa\frac{d}{d\theta}\left(\frac{1}{r}\right), \tag{5.66}$$

$$\ddot{r} = \frac{d\dot{r}}{d\theta}\frac{d\theta}{dt} = \frac{d\dot{r}}{d\theta}\frac{\kappa}{r^2} = \frac{\kappa}{r^2}\frac{d}{d\theta}\left[-\kappa\frac{d}{d\theta}\left(\frac{1}{r}\right)\right] = -\frac{\kappa^2}{r^2}\frac{d^2}{d\theta^2}\left(\frac{1}{r}\right). \tag{5.67}$$

Using Eqs. (5.65) and (5.67), Eq. (5.64) becomes

$$-\frac{\kappa^2}{r^2}\frac{d^2}{d\theta^2}\left(\frac{1}{r}\right) - r\left(\frac{\kappa}{r^2}\right)^2 = -\frac{Gm_B}{r^2}. \tag{5.68}$$

Letting $u = 1/r$, Eq. (5.68) becomes

$$-\kappa^2u^2\frac{d^2u}{d\theta^2} - \kappa^2u^3 = -Gm_Bu^2, \tag{5.69}$$

which, upon canceling u^2 and rearranging, becomes

$$\frac{d^2u}{d\theta^2} + u = \frac{Gm_B}{\kappa^2}. \tag{5.70}$$

As we will see again in Section 9.2, this is a second-order, constant-coefficient, nonhomogeneous, differential equation whose solution can be verified by direct substitution to be

$$u = \frac{1}{r} = C\cos(\theta - \beta) + \frac{Gm_B}{\kappa^2}, \tag{5.71}$$

where C and β are constants of integration.

Equation (5.71) determines the unpowered or free-flight trajectory of the satellite. It is the polar coordinate representation of a *conic section*, which is defined as

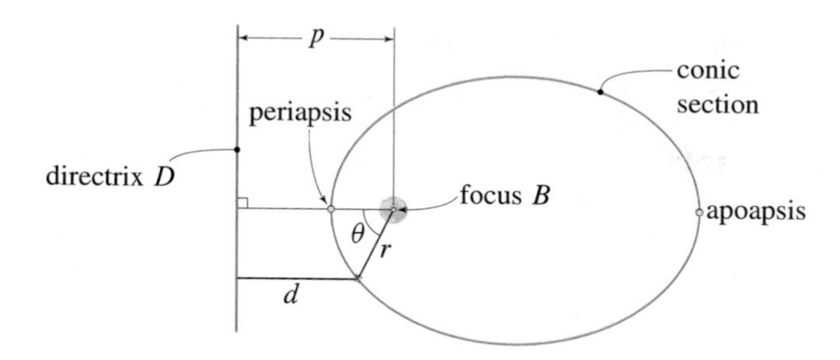

Figure 5.25. Geometry used to define a conic section. In this case, the conic section is an ellipse with eccentricity $e = 0.661$.

follows: Given a point B, called the *focus*, and a line D, called the *directrix*, a conic section is defined as the locus of points for which the ratio (see Fig. 5.25)

$$e = \frac{\text{distance to } B}{\text{distance to } D} = \frac{r}{d} \tag{5.72}$$

is a constant. As we shall soon see, circles, ellipses, parabolas, and hyperbolas are all conic sections. The ratio in Eq. (5.72) is called the *eccentricity* of the conic section. We now need to obtain the constants of integration in Eq. (5.71).

The constant β in Eq. (5.71) can be eliminated by letting the $\theta = 0$ axis be at periapsis, that is, at the point where the orbital radius r is a minimum. At periapsis, $\dot{r} = 0$, which implies that $du/d\theta = 0$ from Eq. (5.66). Applying this condition, we find that $\beta = 0$, which simplifies Eq. (5.71) to

$$\boxed{\frac{1}{r} = C \cos\theta + \frac{Gm_B}{\kappa^2}.} \tag{5.73}$$

Referring to Eq. (5.72) and Fig. 5.25, we can use the definition of eccentricity e to write

$$r = ed = e(p - r\cos\theta), \tag{5.74}$$

where p is the *focal parameter*. Rearranging this equation, we obtain

$$\frac{1}{r} = \frac{1}{p}\cos\theta + \frac{1}{ep}. \tag{5.75}$$

Comparing Eqs. (5.73) and (5.75), we see that

$$p = \frac{1}{C}, \tag{5.76}$$

and that

$$\boxed{e = \frac{C\kappa^2}{Gm_B}.} \tag{5.77}$$

Using Eq. (5.77), we can write (5.73) as

$$\boxed{\frac{1}{r} = \frac{Gm_B}{\kappa^2}(1 + e\cos\theta).} \tag{5.78}$$

Equations (5.73) and (5.78) are equivalent to one another. In the former, we need to determine the constants C and κ, and in the latter we need to determine the constants e and κ (Eq. (5.77) provides the link between these three constants). In either case, these constants are determined by knowing the position and velocity of the satellite at some point on its trajectory.

Helpful Information

Periapsis and apoapsis. An *apsis* is the point of maximum or minimum distance from the focus containing the center of attraction in an elliptical orbit. The point at which the orbital radius is a minimum is called the *periapsis*, and the point at which the orbital radius is a maximum is called the *apoapsis*. For specific celestial bodies, names that apply specifically to orbits about those bodies are generally used rather than periapsis and apoapsis. The following table lists some of these.

Body	Minimum r	Maximum r
Earth	perigee	apogee
Sun	perihelion	aphelion
Mars	periareion	apoareion
Jupiter	perijove	apojove

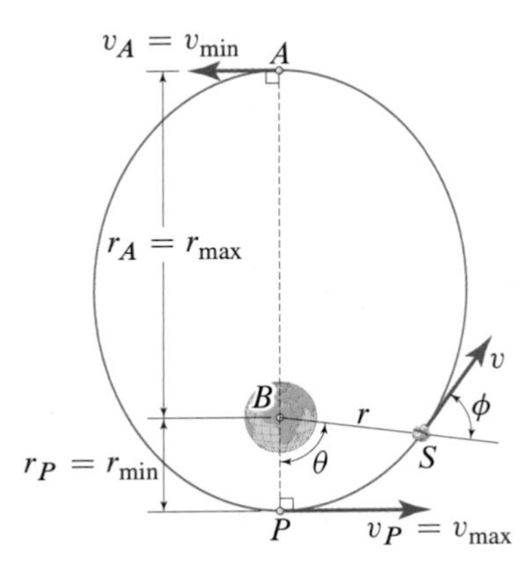

Figure 5.26
A satellite S in orbit about a primary body B. The launch conditions are at periapsis P and are r_P and v_P. The angle ϕ is defined to be the angle between the radial line BS and the velocity vector of the satellite. At P and A, $\phi = 90°$.

Launch conditions at periapsis. Consider a satellite in orbit about a primary body, as shown in Fig. 5.26. If instead of knowing the launch conditions at an arbitrary position S, we know the launch conditions at periapsis (point P), then $\phi_P = 90°$ and $\theta_P = 0°$. Under these conditions, Eq. (5.65) tells us that the angular momentum per unit mass κ becomes

$$\boxed{\kappa = r_P^2 \dot{\theta}_P = r_P v_P.} \tag{5.79}$$

To determine C, we evaluate Eq. (5.73) at P to find that

$$\boxed{C = \frac{1}{r_P}\left(1 - \frac{Gm_B}{r_P v_P^2}\right),} \tag{5.80}$$

where we have used κ from Eq. (5.79). Substituting Eqs. (5.79) and (5.80) into Eq. (5.73), we obtain the following equation for the trajectory of the satellite:

$$\boxed{\frac{1}{r} = \frac{1}{r_P}\left(1 - \frac{Gm_B}{r_P v_P^2}\right)\cos\theta + \frac{Gm_B}{r_P^2 v_P^2}.} \tag{5.81}$$

Conic sections

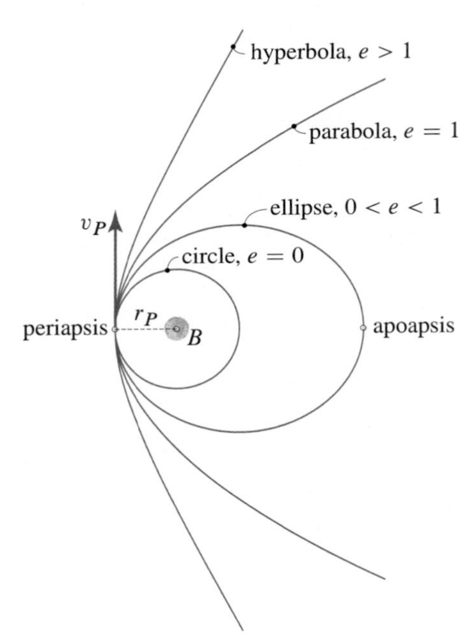

Figure 5.27
The four conic sections, showing r_P and v_P.

We have already mentioned that the solution $r(\theta)$ to the governing orbital equations is a conic section. The type of conic section depends on the eccentricity of the trajectory e (see Fig. 5.27), which, in turn, depends on C and κ via Eq. (5.77). We will now look at each of the four possible cases: $e = 0$, $0 < e < 1$, $e = 1$, and $e > 1$.

Circular orbit ($e = 0$). If r_P and v_P are chosen so that $e = 0$, then Eq. (5.77) tells us that $C\kappa^2 = 0$, which implies that $C = 0$ since κ cannot be zero. If $C = 0$, then Eq. (5.80) tells us that the speed v_c in a circular orbit of radius r_P is given by

$$\frac{1}{r_P}\left(1 - \frac{Gm_B}{r_P v_c^2}\right) = 0 \quad \Rightarrow \quad 1 = \frac{Gm_B}{r_P v_c^2} \quad \Rightarrow \quad \boxed{v_c = \sqrt{\frac{Gm_B}{r_P}}.} \tag{5.82}$$

Elliptical orbit ($0 < e < 1$). For an elliptical orbit, $0 < e < 1$, and when evaluated at periapsis ($\theta = 0°$), Eq. (5.81) gives $r = r_P$ as expected since r_P was chosen at $\theta = 0°$. If we instead evaluate Eq. (5.81) at $\theta = 180°$, we will find r at apoapsis, that is, r_A. Doing so gives

$$\frac{1}{r_A} = \frac{-1}{r_P}\left(1 - \frac{Gm_B}{r_P v_P^2}\right) + \frac{Gm_B}{r_P^2 v_P^2} = \frac{2Gm_B - r_P v_P^2}{r_P^2 v_P^2}, \tag{5.83}$$

or, by simplifying and rearranging,

$$\boxed{r_A = \frac{r_P}{2Gm_B/(r_P v_P^2) - 1},} \tag{5.84}$$

where we recall that the launch conditions are at periapsis.

An additional set of relations for elliptical orbits is obtained by evaluating Eq. (5.75) at periapsis ($\theta = 0°$) and apoapsis ($\theta = 180°$) to obtain

$$\frac{1}{r_P} = \frac{1}{p} + \frac{1}{ep} \quad \Rightarrow \quad r_P = \frac{pe}{1+e}, \tag{5.85}$$

and

$$\frac{1}{r_A} = \frac{-1}{p} + \frac{1}{ep} \quad \Rightarrow \quad r_A = \frac{pe}{1-e}, \tag{5.86}$$

respectively, where p is the focal parameter shown in Fig. 5.25. Taking the ratio of Eqs. (5.85) and (5.86), it is easy to show that r_A can be written in terms of e as

$$r_A = r_P\left(\frac{1+e}{1-e}\right). \tag{5.87}$$

Referring to Fig. 5.28, and using Eqs. (5.85) and (5.86), we obtain

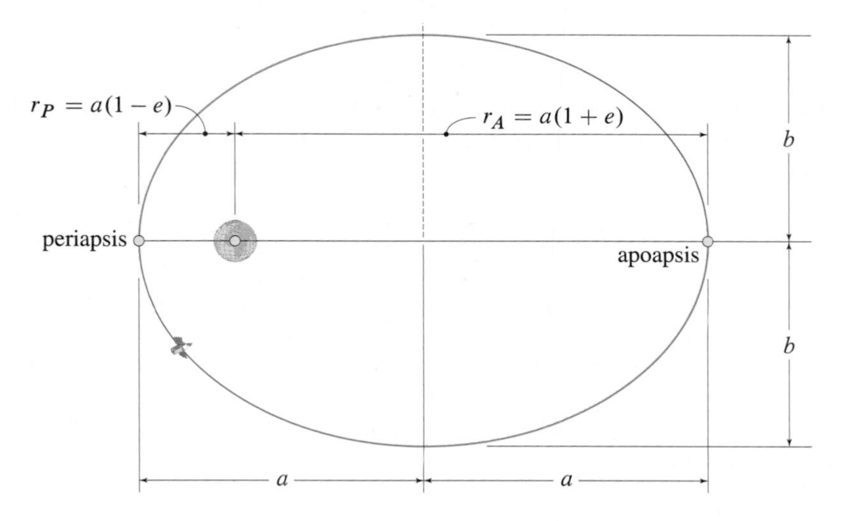

Figure 5.28. The semimajor axis a and semiminor axis b of an ellipse.

$$2a = r_P + r_A = pe\left(\frac{1}{1+e} + \frac{1}{1-e}\right) \quad \Rightarrow \quad a = \frac{pe}{1-e^2}, \tag{5.88}$$

where a is referred to as the *semimajor axis* of the ellipse. Solving Eq. (5.88) for p and substituting the result into Eq. (5.75) give

$$\frac{1}{r} = \frac{1+e\cos\theta}{a(1-e^2)}, \tag{5.89}$$

which, when evaluated at periapsis and apoapsis, gives

$$\boxed{r_P = a(1-e) \quad \text{and} \quad r_A = a(1+e).} \tag{5.90}$$

Equation (5.89) is a statement of Kepler's first law, which states that orbits of planets are ellipses with the Sun at one focus.

To find the period of an elliptical orbit, we refer to the definition of areal velocity in Eq. (5.63) and integrate that equation over the entire ellipse to find that

$$\frac{dA}{dt} = \tfrac{1}{2}r^2\dot{\theta} = \frac{\kappa}{2} \quad \Rightarrow \quad \int_0^A dA = \int_0^\tau \frac{\kappa}{2}\,dt \quad \Rightarrow \quad A = \frac{\kappa}{2}\tau. \tag{5.91}$$

Here τ is the *orbital period*, and we have used Eq. (5.65) to write $\kappa = r^2\dot{\theta}$, which is constant. For an ellipse, analytical geometry tells us that its area is equal to πab,

Figure 5.29
A portrait of Johannes Kepler painted in 1610.

where b is the semiminor axis of the ellipse (see Fig. 5.28). So we can write Eq. (5.91) as

$$\pi ab = \frac{\kappa}{2}\tau \quad \Rightarrow \quad \boxed{\tau = \frac{2\pi ab}{\kappa}.} \tag{5.92}$$

Now, it can be shown that (see Prob. 5.137)

$$b = \sqrt{r_P r_A}, \tag{5.93}$$

and since Eq. (5.88) tells us that $a = (r_P + r_A)/2$, we can write Eq. (5.92) as

$$\boxed{\tau = \frac{\pi}{\kappa}(r_P + r_A)\sqrt{r_P r_A},} \tag{5.94}$$

where κ can be found by using Eq. (5.79).

Kepler's third law states that the square of the orbital period is proportional to the cube of the semimajor axis of that orbit. To show this, we start with Eq. (5.94) and substitute in Eqs. (5.90) to obtain

$$\tau = \frac{\pi}{\kappa}\big[a(1-e) + a(1+e)\big]\sqrt{a(1-e)a(1+e)} = \frac{2\pi}{\kappa}a^2\sqrt{1-e^2}. \tag{5.95}$$

Squaring both sides and noting from Eq. (5.88) that $a(1-e^2) = pe$, we have

$$\tau^2 = \frac{4\pi^2}{\kappa^2}a^3 pe. \tag{5.96}$$

Using Eqs. (5.76) and (5.77), we can write the product pe as $\kappa^2/(Gm_B)$, which means that Eq. (5.96) becomes

$$\boxed{\tau^2 = \frac{4\pi^2}{Gm_B}a^3,} \tag{5.97}$$

which is Kepler's third law.

Parabolic trajectory ($e = 1$). If r_P and v_P are chosen so that the trajectory is parabolic, then $e = 1$. Figure 5.27 indicates that the parabolic trajectory is one that divides those trajectories that are periodic from those that are not. That is, it divides trajectories that return to their initial starting point from those that do not. For a given r_P, the launch velocity v_{par} required to achieve a parabolic trajectory can be found by letting $e = 1$ in Eq. (5.77), which gives $Gm_B = C\kappa^2$, and then substituting C from Eq. (5.80) into that result to obtain

$$Gm_B = \frac{1}{r_P}\left(1 - \frac{Gm_B}{r_P v_{\text{par}}^2}\right)(r_P v_{\text{par}})^2, \tag{5.98}$$

which, when solved for v_{par}, becomes

$$\boxed{v_{\text{par}} = v_{\text{esc}} = \sqrt{\frac{2Gm_B}{r_P}}.} \tag{5.99}$$

The speed given in Eq. (5.99) is often referred to as *escape velocity* since it is the speed required to completely escape the influence of the primary body B.

Interesting Fact

Crashing into the Earth to escape velocity. Comparing Eqs. (5.82) and (5.99), we see that only a factor of $\sqrt{2}$ separates an orbit that crashes into the Earth ($v < v_c$) with one that never returns ($v > v_{\text{esc}}$).

Hyperbolic trajectory ($e > 1$). For a hyperbolic trajectory, the governing equations given by Eqs. (5.77)–(5.81) are used for values of $e > 1$.

Energy considerations

The motion of a satellite S that is governed by Eqs. (5.64) and (5.65) not only conserves angular momentum about the primary body B, but also conserves total mechanical energy. We know this because the only force we are modeling is gravity, and Eq. (4.25) on p. 259 reminds us that this force is conservative. Therefore, the work-energy principle tells us that (see Fig. 5.30)

$$T_P + V_P = \tfrac{1}{2} m v_P^2 - \frac{G m_B m}{r_P} = \text{constant}, \tag{5.100}$$

which, after dividing through by m, gives

$$\tfrac{1}{2} v_P^2 - \frac{G m_B}{r_P} = E \quad \text{or} \quad \frac{\kappa^2}{2 r_P^2} - \frac{G m_B}{r_P} = E, \tag{5.101}$$

where E is the *mechanical energy per unit mass* (sometimes called the *specific energy*) of the satellite. We used Eq. (5.79) to introduce κ. Next, we evaluate Eq. (5.78) at periapsis (i.e., $\theta = 0$) and solve for κ to obtain

$$\kappa^2 = G m_B (1 + e) r_P = G m_B (1 + e) a (1 - e) = G m_B a (1 - e^2), \tag{5.102}$$

where we used the first of Eqs. (5.90) for r_P. Substituting Eq. (5.102) and the first of Eqs. (5.90) into Eq. (5.101), we obtain

$$E = \frac{G m_B a (1 - e^2)}{2 a^2 (1 - e)^2} - \frac{G m_B}{a(1 - e)} \quad \Rightarrow \quad \boxed{E = -\frac{G m_B}{2a},} \tag{5.103}$$

where we see that the total energy of the satellite depends on *only* the semimajor axis a of the orbit.

Now that we have Eq. (5.103), we can apply the work-energy principle at an arbitrary position in the orbit to obtain

$$E = -\frac{G m_B}{2a} = \tfrac{1}{2} v^2 - \frac{G m_B}{r}. \tag{5.104}$$

Solving this equation for v, we have

$$\boxed{v = \sqrt{G m_B \left(\frac{2}{r} - \frac{1}{a} \right)},} \tag{5.105}$$

which is very useful for solving orbital transfer problems and for finding the speed of a satellite at arbitrary positions in its orbit.

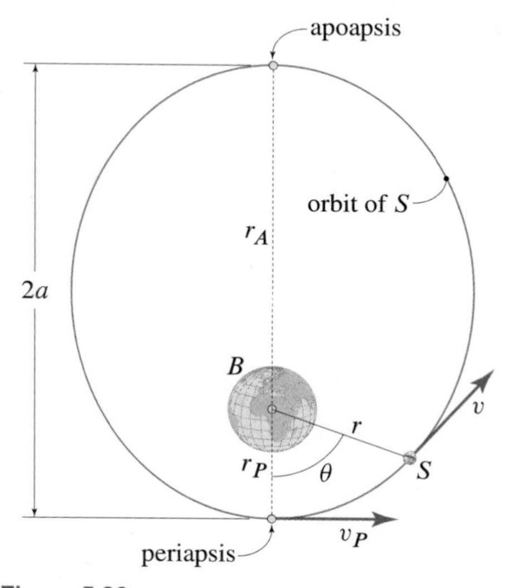

Figure 5.30
A satellite S in an elliptical orbit about a primary body B. The orbital parameters a, r_P, r_A, r, and θ are shown.

End of Section Summary

In this section, we studied the motion of a satellite S of mass m that is subject to Newton's universal law of gravitation due to a body B of mass m_B, which is the primary or attracting body (see Fig. 5.31). We began with these important assumptions:

1. The primary body B and the satellite S are both treated as particles.
2. The only force acting on the satellite S is the force of mutual attraction between B and S.
3. The primary body B is fixed in space.

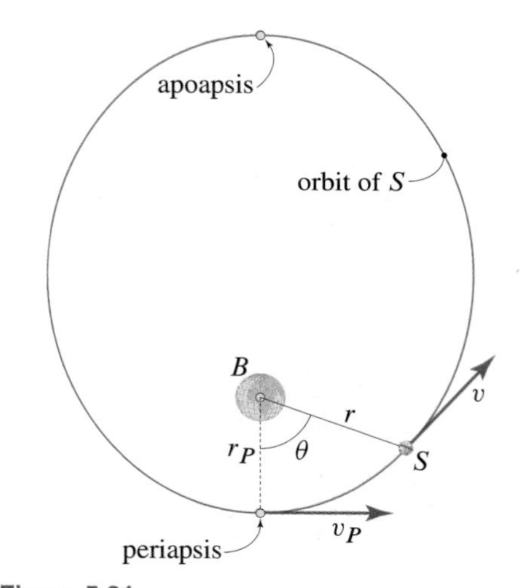

Figure 5.31
A satellite S of mass m orbiting a primary body B of mass m_B on a conic section, which, in this case, is an ellipse. The angle θ and the launch conditions r_P and v_P are defined from periapsis.

Determination of the orbit

Solving the governing equations in polar coordinates, we found that the trajectory of the satellite S under these assumptions is a conic section, whose equation can be written as

Eqs. (5.73) and (5.78), p. 387

$$\frac{1}{r} = C\cos\theta + \frac{Gm_B}{\kappa^2} \quad \text{or}$$

$$\frac{1}{r} = \frac{Gm_B}{\kappa^2}(1 + e\cos\theta),$$

where r is the distance between the centers of mass of S and B; θ is the orbital angle measured relative to periapsis; G is the universal gravitational constant; κ is the angular momentum per unit mass of the satellite S measured about B; C is a constant to be determined; and e is the *eccentricity* of the trajectory, which can be written as

Eq. (5.77), p. 387

$$e = \frac{C\kappa^2}{Gm_B}.$$

If, as is generally the case here, the orbital conditions are known at periapsis, then κ is

Eq. (5.79), p. 388

$$\kappa = r_P v_P,$$

the constant C is

Eq. (5.80), p. 388

$$C = \frac{1}{r_P}\left(1 - \frac{Gm_B}{r_P v_P^2}\right),$$

and the equation describing the trajectory becomes

Eq. (5.81), p. 388

$$\frac{1}{r} = \frac{1}{r_P}\left(1 - \frac{Gm_B}{r_P v_P^2}\right)\cos\theta + \frac{Gm_B}{r_P^2 v_P^2}.$$

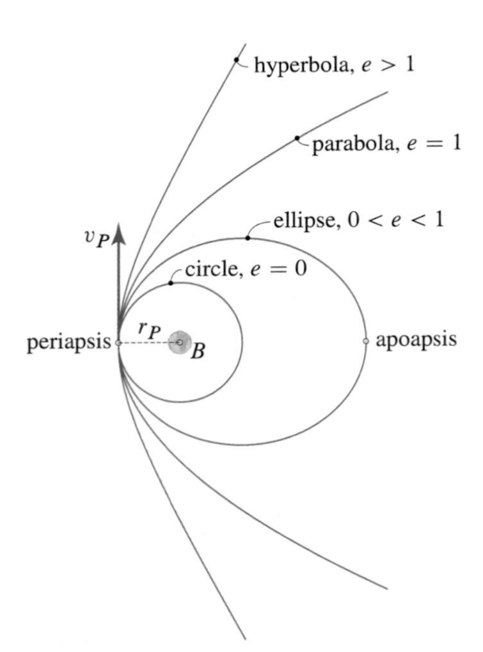

Figure 5.32
The four conic sections, showing r_P and v_P.

Conic sections. Equations (5.73), (5.78), and (5.81) represent equivalent conic sections in polar coordinates. The type of conic section depends on the eccentricity of the trajectory e (see Fig. 5.32), which, in turn, depends on C and κ via Eq. (5.77). There are four types of conic sections, which are determined by the value of the eccentricity e, that is, $e = 0$, $0 < e < 1$, $e = 1$, and $e > 1$.

For a *circular orbit* ($e = 0$), the radius is r_P, and the speed in the orbit is equal to

Eq. (5.82), p. 388

$$v_c = \sqrt{\frac{Gm_B}{r_P}}.$$

For an *elliptical orbit* ($0 < e < 1$), as expected, the radius at periapsis is r_P. The radius at apoapsis is given by

Eq. (5.84), p. 388

$$r_A = \frac{r_P}{2Gm_B/(r_P v_P^2) - 1}.$$

Referring to Fig. 5.33, additional relationships between the semimajor axis a of an

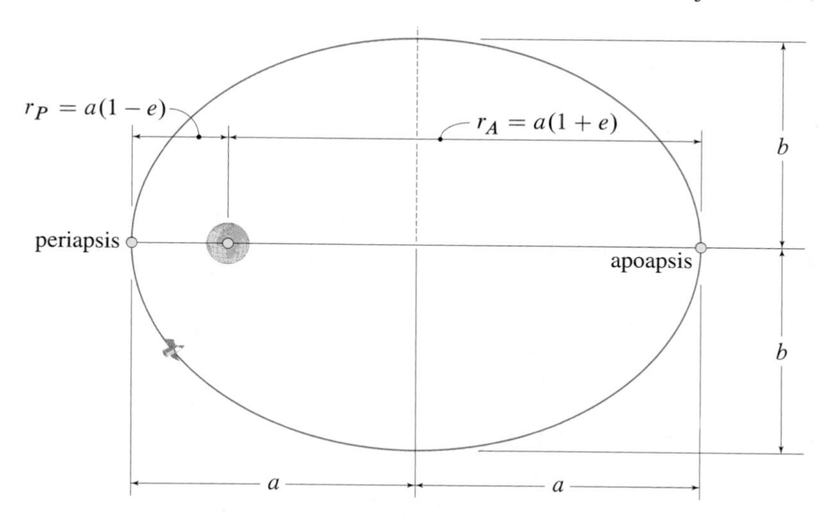

Figure 5.33. The semimajor axis a and semiminor axis b of an ellipse.

elliptical orbit, the eccentricity of the orbit, and the radii at periapsis and apoapsis are, respectively,

Eqs. (5.90), p. 389

$$r_P = a(1 - e) \quad \text{and} \quad r_A = a(1 + e).$$

The period of an elliptical orbit τ can be written in the following two ways:

Eqs. (5.92) and (5.94), p. 390

$$\tau = \frac{2\pi ab}{\kappa} = \frac{\pi}{\kappa}(r_P + r_A)\sqrt{r_P r_A},$$

or, reflecting Kepler's third law, the orbital period can be written as

Eq. (5.97), p. 390

$$\tau^2 = \frac{4\pi^2}{Gm_B} a^3.$$

A *parabolic trajectory* ($e = 1$) is that which divides periodic orbits (which return to their starting location) from trajectories that are not periodic. For a given r_P, the speed v_{par} required to achieve a parabolic trajectory is given by

Eq. (5.99), p. 390

$$v_{\text{par}} = v_{\text{esc}} = \sqrt{\frac{2Gm_B}{r_P}},$$

which is also referred to as the *escape velocity* since it is the speed required to completely escape the influence of the primary body B.

For a *hyperbolic trajectory* ($e > 1$), the governing equations given by Eqs. (5.77)–(5.81) are used for values of $e > 1$.

Energy considerations

Using the work-energy principle, we discovered that the total mechanical energy in an orbit depends on only the semimajor axis a of the orbit and is given by

Eq. (5.103), p. 391

$$E = -\frac{Gm_B}{2a},$$

where E is the *mechanical energy per unit mass* of the satellite. Applying the work-energy principle at an arbitrary location within the orbit, we found that the speed can be written as

Eq. (5.105), p. 391

$$v = \sqrt{Gm_B\left(\frac{2}{r} - \frac{1}{a}\right)}.$$

EXAMPLE 5.16 *Apogee and Perigee Speeds for a Given Orbit*

An artificial satellite is launched from an altitude of 500 km with a velocity that is parallel to the surface of the Earth (Fig. 1). Requiring that the altitude at apogee be 20,000 km and using 6371 km for the radius of the Earth, determine

(a) the required speed at perigee v_P,

(b) the speed at apogee v_A, and

(c) the period of the orbit.

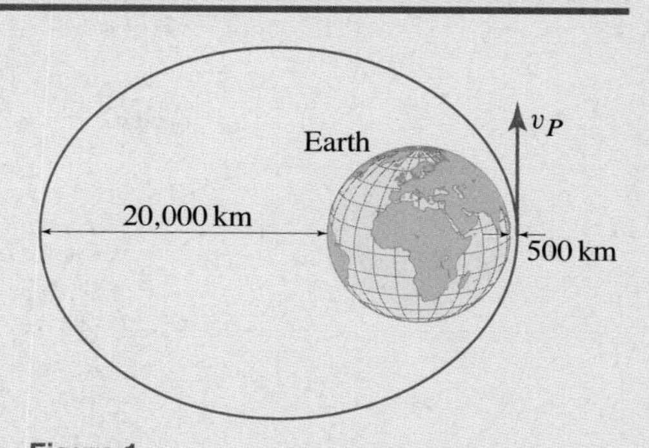

Figure 1
An elliptical orbit about Earth with a launch altitude of 500 km and an apogee altitude of 20,000 km. It can be shown that the orbit has an eccentricity e of 0.5866. The figure is drawn to scale.

SOLUTION

Road Map & Modeling This is an Earth orbit, and we are given r_P and the required r_A, so Eq. (5.84) will allow us to determine the speed at perigee v_P. Knowing r_P and having found v_P, we see that Eq. (5.79) allows us to find κ, which will then allow us to find v_A given that κ is conserved and will allow us to find the orbital period τ using Eq. (5.94). Note that the orbital eccentricity mentioned in the caption of Fig. 1 can be found from either of Eqs. (5.90).

Computation Referring to Fig. 1, we see that an altitude of 500 km at perigee corresponds to

$$r_P = (500 + 6371)\,\text{km} = 6871\times 10^3\,\text{m}. \tag{1}$$

Similarly, the altitude of 20,000 km at apogee means

$$r_A = (20{,}000 + 6371)\,\text{km} = 26{,}370\times 10^3\,\text{m}. \tag{2}$$

Using Eq. (1.10) on p. 4 in Eq. (5.84) and solving for v_P, we find that

$$r_A = \frac{r_P^2 v_P^2}{2gr_e^2 - r_P v_P^2} \quad\Rightarrow\quad v_P = \sqrt{\frac{2gr_A r_e^2}{(r_A + r_P)r_P}} \tag{3}$$

$$\Rightarrow\quad \boxed{v_P = 9589\,\text{m/s} = 34{,}520\,\text{km/h},} \tag{4}$$

where r_e is the radius of the Earth.

To find the speed at apogee, we note from Eq. (5.79) that the angular momentum per unit mass κ is constant and so

$$\kappa = r_P v_P = r_A v_A \quad\Rightarrow\quad v_A = \frac{r_P}{r_A} v_P \tag{5}$$

$$\Rightarrow\quad \boxed{v_A = 2498\,\text{m/s} = 8994\,\text{km/h}.} \tag{6}$$

To determine the period of the orbit, we can use Eq. (5.94) after substituting in κ from Eq. (5) as follows:

$$\boxed{\tau = \frac{\pi}{r_P v_P}(r_P + r_A)\sqrt{r_P r_A} = 21{,}340\,\text{s} = 5.927\,\text{h}.} \tag{7}$$

Discussion & Verification It can be seen that both Eqs. (3) and (5) have dimensions of velocity, as they should. Equation (7) has the dimension of time, as it should. We also note that the speed at apogee is less than the speed at perigee, as it should be, since κ is constant.

Problems

For the problems in this section, use 6371 km or 3959 mi for the radius of the Earth. *Not all orbits or objects are drawn to scale.*

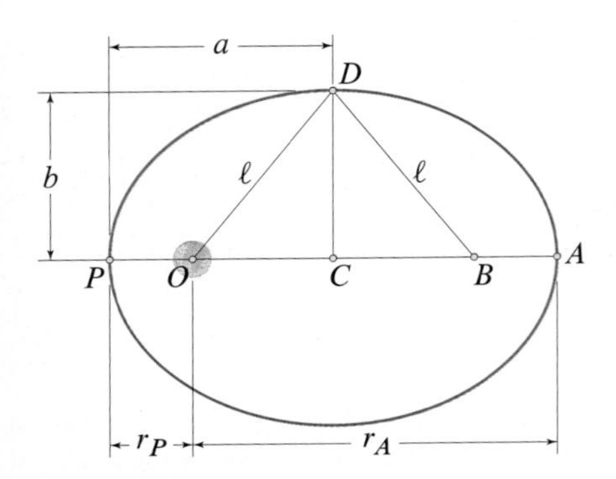

Figure P5.137

Problem 5.137

Using the lengths shown, as well as the property of an ellipse that states that the sum of the distances from each of the foci (i.e., points O and B) to any point on the ellipse is a constant, prove Eq. (5.93), that is, that the length of the semiminor axis can be related to the periapsis and apoapsis radii via $b = \sqrt{r_P r_A}$.

Problem 5.138

Using the last of Eqs. (5.102), along with Eq. (5.103), solve for the eccentricity e as a function of E, κ, and Gm_B.

(a) Using that result, along with fact that $e \geq 0$, show that $E < 0$ corresponds to an elliptical orbit, $E = 0$ corresponds to a parabolic trajectory, and $E > 0$ corresponds to a hyperbolic trajectory.

(b) Show that for $e = 0$, the expression you found for e leads to Eq. (5.82).

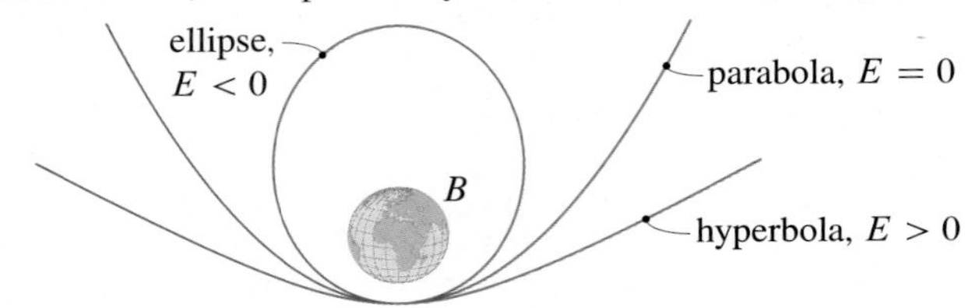

Figure P5.138

Problem 5.139

Assuming that the Sun is the only significant body in the solar system (the mass of the Sun accounts for 99.8% of the mass of the solar system), determine the escape velocity from the Sun as a function of the distance r from its center. What is the value of the escape velocity (expressed in km/h) when r is equal to the radius of Earth's orbit? Use 1.989×10^{30} kg for the mass of the Sun and 150×10^6 km for the radius of Earth's orbit.

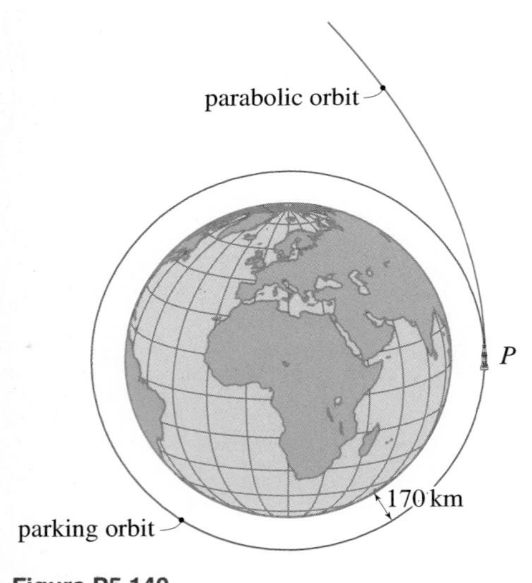

Figure P5.140

Problem 5.140

The S-IVB third stage of the Saturn V rocket, which was used for the Apollo missions, would burn for about 2.5 min to place the spacecraft into a "parking orbit."* Then, after several orbits, it would burn for about 6 min to accelerate the spacecraft to escape velocity to send it to the Moon. Assuming a circular parking orbit with an altitude of 170 km, determine the change in speed needed at P to go from the parking orbit to escape velocity. Assume that the change in speed occurs instantaneously so that you need not worry about changes in orbital position during the engine thrust.

Problem 5.141

In 1705, Edmund Halley (1656–1742), an English astronomer, claimed that the comet sightings of 1531, 1607, and 1682 were all the same comet. He predicted this comet would return again in 1758. Halley did not live to see the comet's return, but it did return

* A *parking orbit* is a temporary orbit of an artificial satellite or spacecraft in preparation for thrusting to another orbit or trajectory.

late in 1758 and reached perihelion in March 1759. In honor of his prediction, this comet was named "Halley." Each elliptical orbit of Halley is slightly different, but the average value of the semimajor axis a is about 17.95 AU.* Using this value, along with the fact that its orbital eccentricity is 0.967 (the orbit is drawn to scale, but the Sun is shown to be 36 times bigger than it should be), determine

(a) the orbital period in years of Halley's comet, and

(b) its distance, in AU, from the Sun at perihelion P and at aphelion A. Look up the orbits of the planets of our solar system on the Web. What planetary orbits is Halley near to at perihelion and aphelion?

Use 1.989×10^{30} kg for the mass of the Sun.

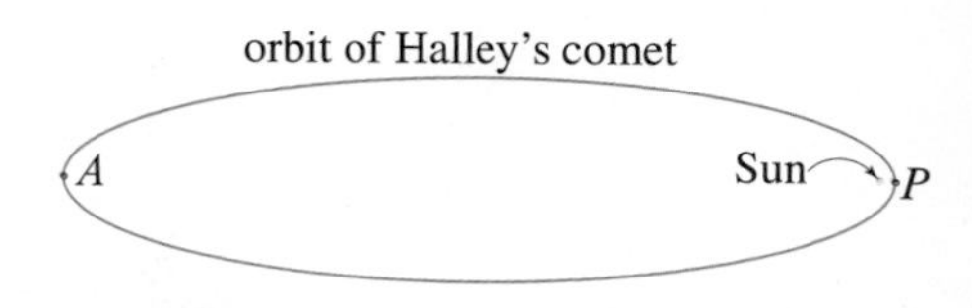

Figure P5.141

Problems 5.142 and 5.143

Explorer 7 was launched on October 13, 1959, with an apogee altitude above the Earth's surface of 1073 km and a perigee altitude of 573 km above the Earth's surface. Its orbital period was 101.4 min.

Problem 5.142 Using this information, calculate Gm_e for the Earth and compare it with gr_e^2.

Problem 5.143 Determine the eccentricity of the Explorer 7's orbit, as well as its speeds at perigee and apogee.

Figure P5.142 and P5.143

Problem 5.144

A *geosynchronous equatorial orbit* is a circular orbit above the Earth's equator that has a period of 1 day (these are sometimes called *geostationary orbits*). These geostationary orbits are of great importance for telecommunications satellites because a satellite orbiting with the same angular rate as the rotation rate of the Earth will appear to hover in the same point in the sky as seen by a person standing on the surface of the Earth. Using this information, determine the altitude h_g and radius r_g of a geostationary orbit (in miles). In addition, determine the speed v_g of a satellite in such an orbit (in miles per hour).

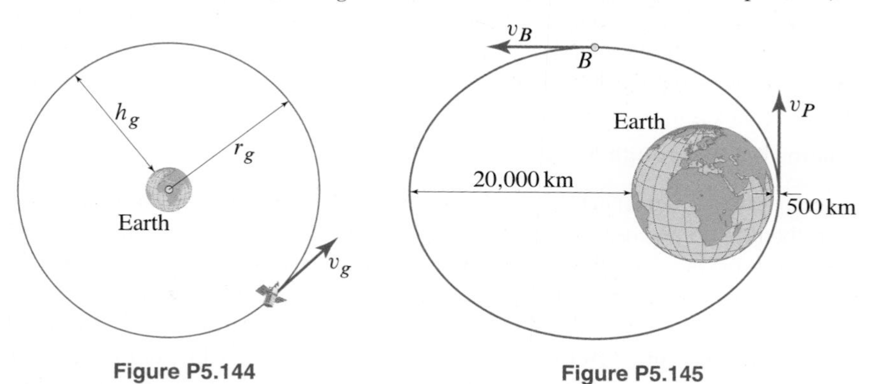

Figure P5.144

Figure P5.145

Problem 5.145

An artificial satellite is launched from an altitude of 500 km with a velocity v_P that is parallel to the surface of the Earth. Requiring that the altitude at apogee be 20,000 km, determine the velocity at B, that is, the position in the orbit when the velocity is first orthogonal to the launch velocity.

* One *astronomical unit* (AU) is the distance between the center of mass of the Earth and that of the Sun and is approximately 1.496×10^8 km $= 9.296 \times 10^7$ mi.

Problem 5.146

The mass of the planet Jupiter is 318 times that of Earth, and its equatorial radius is 71,500 km. If a space probe is in a circular orbit about Jupiter at the altitude of the Galilean moon Callisto (orbital altitude 1.812×10^6 km), determine the change in speed Δv needed in the outer orbit so that the probe reaches a minimum altitude at the orbital radius of the Galilean moon Io (orbital altitude 3.502×10^5 km). Assume that the probe is at the maximum altitude in the transfer orbit when the change in speed occurs and that change in speed is impulsive, that is, it occurs instantaneously.

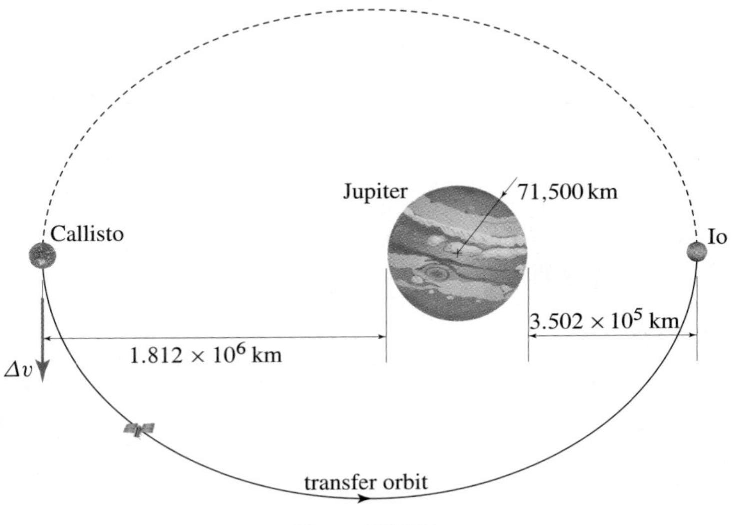

Figure P5.146

Problem 5.147

Figure P5.147

The on-orbit assembly of the International Space Station (ISS) began in 1998 and continues today. The ISS has an apogee altitude above the Earth's surface of 341.9 km and a perigee altitude of 331.0 km above the Earth's surface. Determine its maximum and minimum speeds in orbit, its orbital eccentricity, and its orbital period. Research its actual orbital period and compare it with your calculated value.

Problems 5.148 through 5.150

The optimal way (from an energy standpoint) to transfer from one circular orbit about a primary body B to another circular orbit is via the so-called *Hohmann transfer*, which involves transferring from one circular orbit to another using an elliptical orbit that is tangent to both at the periapsis and apoapsis of the ellipse. The ellipse is uniquely defined because we know r_P (the radius of the inner circular orbit) and r_A (the radius of the outer circular orbit), and therefore we know the semimajor axis a by Eq. (5.88) and the eccentricity e by Eq. (5.87) or Eqs. (5.90). Performing a Hohmann transfer requires two maneuvers, the first to leave the inner (outer) circular orbit and enter the transfer ellipse and the second to leave the transfer ellipse and enter the outer (inner) circular orbit.

Problem 5.148 A spacecraft S_1 needs to transfer from circular low Earth parking orbit with altitude 120 mi above the surface of the Earth to a circular geosynchronous orbit with altitude 22,240 mi. Determine the change in speed Δv_P required at perigee P of the elliptical transfer orbit and the change in speed Δv_A required at apogee A. In addition, compute the time required for the orbital transfer. Assume that the changes in speed are impulsive, that is, they occur instantaneously.

Problem 5.149 A spacecraft S_2 must transfer from a circular Earth orbit whose period is 12 h (i.e., it is overhead twice per day) to a low Earth circular orbit with an altitude of 110 mi. Determine the change in speed Δv_A required at apogee A of the elliptical transfer orbit and the change in speed Δv_P required at perigee P. In addition, compute the time required for the orbital transfer. Assume that the changes in speed are impulsive; that is, they occur instantaneously.

Problem 5.150 A spacecraft S_1 is transferring from circular low Earth parking orbit with altitude 100 mi to a circular orbit with radius r_A. Plot, as a function of r_A for $r_P \le r_A \le 100 r_P$, the change in speed Δv_P required at perigee of the elliptical transfer orbit, as well as the change in speed Δv_A required at apogee. In addition, plot the time as a function of r_A, again for $r_P \le r_A \le 100 r_P$, required for the orbital transfer. Assume that the changes in speed are impulsive; that is, they occur instantaneously.

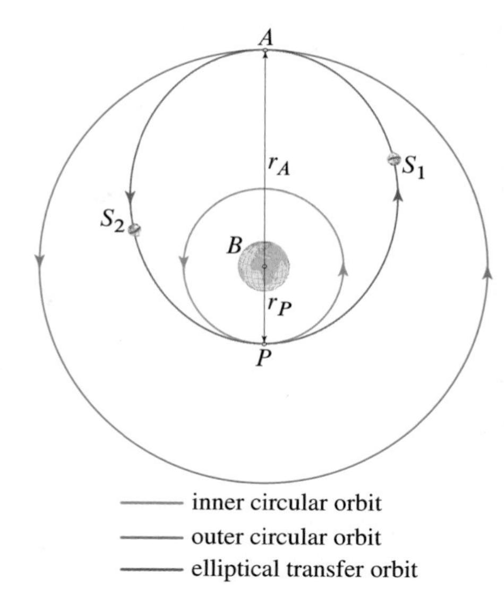

Figure P5.148–P5.152

Problem 5.151

Referring to the description given for Probs. 5.148–5.150, for a Hohmann transfer from an inner circular orbit to an outer circular orbit, what would you expect to be the signs on the change in speed at periapsis and at apoapsis?

Problem 5.152

Referring to the description given for Probs. 5.148–5.150, for a Hohmann transfer from an outer circular orbit to an inner circular orbit, what would you expect to be the signs on the change in speed at periapsis and at apoapsis?

Problem 5.153

During the Apollo missions, while the astronauts were on the Moon with the lunar module (LM), the command module (CM) would fly in a circular orbit around the Moon at an altitude of 60 mi. After the astronauts were done exploring the Moon, the LM would launch from the Moon's surface (at L) and undergo powered flight until burnout at P. This occurred when the LM was approximately 15 mi above the surface of the Moon with its velocity v_{bo} parallel to the surface of the Moon (i.e., at periapsis). It would then fly under the influence of the Moon's gravity until reaching apoapsis A, at which point it would rendezvous with the CM. The radius of the Moon is 1079 mi, and its mass is 0.0123 times that of the Earth. Assume that the changes in speed occur instantaneously.

(a) Determine the required speed v_{bo} at burnout P.

(b) What is the change in speed Δv_{LM} required of the LM at the rendezvous point A?

(c) Determine the time it takes the LM to travel from P to A.

(d) In terms of the angle θ, where should the CM be when the LM reaches P so that they can rendezvous at A?

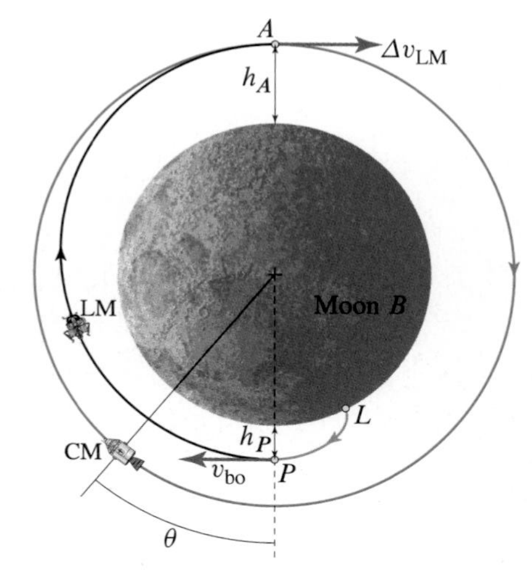

Figure P5.153

Problem 5.154

Use the work-energy principle applied between periapsis P and $r = \infty$, along with the potential energy of the force of gravity given in Eq. (4.25).

(a) Show that a satellite on a hyperbolic trajectory arrives at $r = \infty$ with speed

$$v_\infty = \sqrt{\frac{r_P v_P^2 - 2Gm_B}{r_P}}.$$

(b) In addition, using Eqs. (5.77) and (5.80), show that for a hyperbolic trajectory, $r_P v_P^2 > 2Gm_B$, which means that the square root in the above equation must always yield a real value.

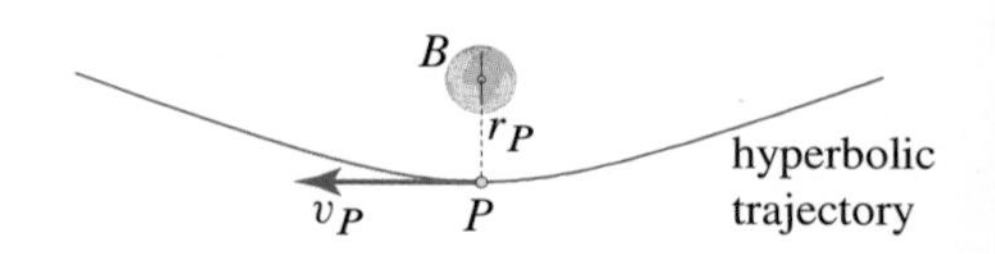

Figure P5.154

Problem 5.155

One option when traveling to Mars from the Earth is to use a Hohmann transfer orbit like that described in Probs. 5.148–5.152. Assuming that the Sun is the primary gravitational influence and ignoring the gravitational influence of Earth and Mars (since the Sun accounts for 99.8% of the mass of the solar system), determine the change in speed required at the Earth Δv_e (perihelion in the transfer orbit) and the required change in speed at Mars Δv_m (aphelion in the transfer orbit) to accomplish the mission to Mars using a Hohmann transfer. In addition, determine the amount of time τ it would take for orbital transfer. Use 1.989×10^{30} kg for the mass of the Sun, assume that the orbits of Earth and Mars are circular, and assume that the changes in speed are impulsive, that is, they occur instantaneously. In addition, use 150×10^6 km for the radius of Earth's orbit and 228×10^6 km for the radius of Mars's orbit.

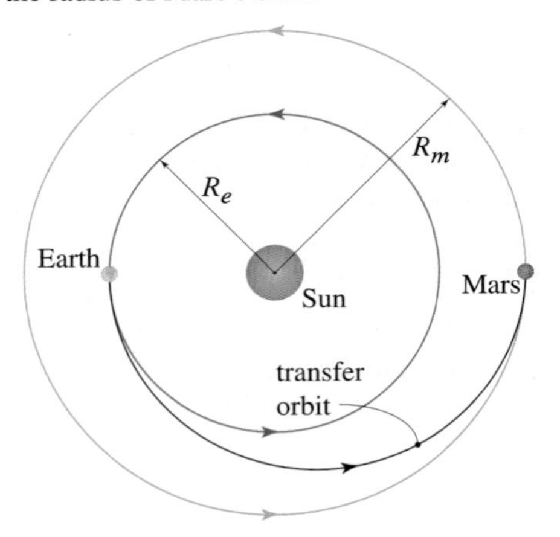

Figure P5.155

5.5 Mass Flows

We now apply the impulse-momentum principles to physical systems that exchange mass with their surrounding environment. While we will focus on problems involving the motion of fluids, the ideas we develop are also applicable to other systems that, while not fluids, move in a fluidlike fashion. For example, in some cases the steady motion of bottles along the bottling line shown in Fig. 5.34 can be modeled as a *continuous* flow of mass. We will see that impulse-momentum principles offer a direct way to compute the forces present in simple fluid motions.

Figure 5.34
Bottles moving along a bottling line. The bottles can be modeled as individual particles/bodies or as mass elements in a continuous mass flow.

Open and closed systems

Before we begin, it is important to recall the distinction between *open* and *closed* systems (which were introduced on p. 315). A *closed system* does not exchange mass with its environment, whereas an *open system* does exchange mass with its environment. The mass of a closed system is constant. An open system can have constant or variable mass. If a physical system is referred to as a *variable mass system*, then such a system is necessarily open. *The impulse-momentum principles presented earlier in the chapter are only applicable to closed systems.*

Steady flows

Figure 5.35 shows part of a pipe filled with a fluid in motion. To simplify our analysis, we assume that the fluid flow is *steady*. By this we mean that the velocity of a fluid particle going through a certain location within the pipe depends only on that location but is otherwise independent of time. On the other hand, the velocity of a particle *can* change as the particle moves from a point to another point within the pipe. In our analysis we also assume that the velocity of the particles at a given cross section is the same.

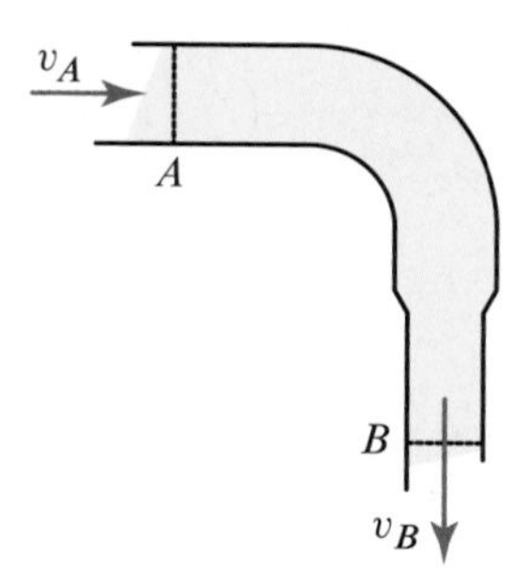

Figure 5.35
A fluid flowing through a curved pipe with variable cross section.

If the velocity of the fluid particles changes, then these particles are accelerating, and by Newton's second law, we conclude that the fluid must be subject to a force, which we now proceed to compute. Because a pipe can be a large structure, we determine only the force exerted on the fluid contained in a given *portion* of the pipe. We will refer to the chosen pipe portion as a *control volume* CV. Referring to Fig. 5.36, we consider a CV defined by the two cross sections A and B. We assume

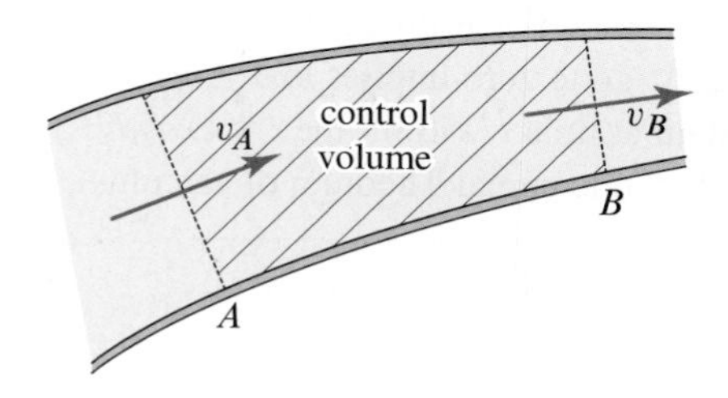

Figure 5.36. A CV corresponding to the portion of a pipe between cross sections A and B.

that the fluid enters the CV at A with velocity $\vec{v}_A$ and exits at B with velocity $\vec{v}_B$. We assume that the flow is such that $\vec{v}_A$ and $\vec{v}_B$ are perpendicular to the cross sections A and B, respectively.

Since mass flows in and out of the CV, *a CV is an open system*. When the flow is steady, then the mass of the fluid in the CV remains constant. This implies that if

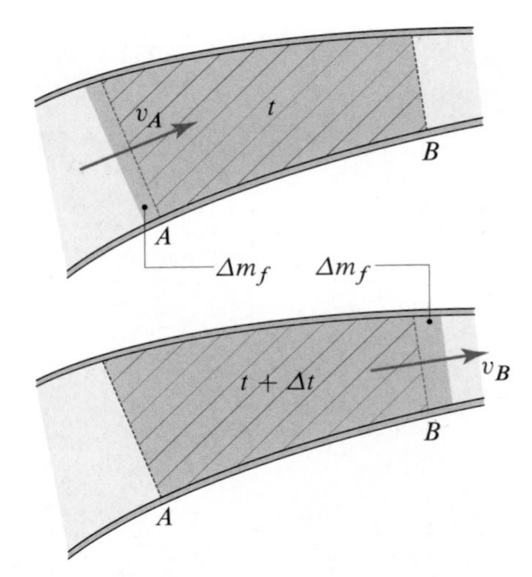

Figure 5.37
A fluid body flowing through a CV defined by the cross sections A and B and behaving as a closed system.

Δm_f (where the subscript f stands for flux) is the mass that flows into the CV at A during any time interval Δt, then Δm_f is also the mass of the fluid that flows out of the CV at B during the same time interval.

The tools we have to relate forces and motion are the impulse-momentum principles, which only apply to closed systems. We will first apply the impulse-momentum principle to a closed system containing the open system of interest, and then we will "shrink" this closed system to make it coincide with the open system contained therein on an instant-by-instant basis. First, referring to Fig. 5.37, we select a body of fluid which at time t fills the chosen CV and has a small element of mass Δm_f about to enter the CV at A. *This will be our closed system.* The fluid element that is about to enter the CV is chosen to be small enough to allow us to assume that, at time t, $\vec{v}_A$ is the velocity of all of its particles. Second, we let Δt be the time taken by the fluid element that is entering at A to flow into the CV. Because the flow is steady, at time $t + \Delta t$, the body will completely fill the CV between A and B and will also include an element of mass Δm_f, which has exited the CV at B (see Fig. 5.37). Since Δm_f is small, we can assume that all of the particles to the right of B have the same velocity $\vec{v}_B$. Since the selected body is a closed system, we can apply to it Eq. (5.15) on p. 316, which yields

$$\int_t^{t+\Delta t} \vec{F}\, dt = \vec{p}(t + \Delta t) - \vec{p}(t), \tag{5.106}$$

where $\vec{F}$ is the total external force acting on the system and, using the stated assumptions, the linear momenta $\vec{p}(t)$ and $\vec{p}(t + \Delta t)$ can be written as

$$\vec{p}(t) = \Delta m_f\, \vec{v}_A + \vec{p}_{\text{cv}}(t), \tag{5.107}$$

$$\vec{p}(t + \Delta t) = \vec{p}_{\text{cv}}(t + \Delta t) + \Delta m_f\, \vec{v}_B. \tag{5.108}$$

The quantity $\vec{p}_{\text{cv}}$ denotes the momentum of the fluid within the CV. Because the flow is steady, we must have

$$\vec{p}_{\text{cv}}(t) = \vec{p}_{\text{cv}}(t + \Delta t). \tag{5.109}$$

Substituting Eqs. (5.107)–(5.109) into Eq. (5.106), simplifying, and dividing all terms by Δt, we have

$$\frac{1}{\Delta t}\int_t^{t+\Delta t} \vec{F}\, dt = \frac{\Delta m_f}{\Delta t}\,(\vec{v}_B - \vec{v}_A). \tag{5.110}$$

Equation (5.110) holds for a closed system that occupies a volume larger than the selected CV. Recalling that Δt is the time it takes Δm_f to flow into and out of the CV, we see that letting Δt go to zero implies that Δm_f must also go to zero and the fluid in the closed system at time t will fill the CV *exactly!* Therefore, by letting Δt go to zero (and using the Fundametal Theorem of calculus), Eq. (5.110) yields

$$\boxed{\vec{F} = \dot{m}_f(\vec{v}_B - \vec{v}_A),} \tag{5.111}$$

where the quantity

$$\boxed{\dot{m}_f = \lim_{\Delta t \to 0} \frac{\Delta m_f}{\Delta t}} \tag{5.112}$$

is called the *mass flow rate* or *mass flux*; it measures the amount of mass flowing into and out of the chosen CV per unit time. The result in Eq. (5.111) applies to open systems and was possible because, in going from Eq. (5.110) to Eq. (5.111), we took a limit that forced the chosen closed system to coincide with the open system we wanted to characterize at time t.

Common Pitfall

The mass in the CV is constant. Often $\dot{m}_f$ is misinterpreted as the time rate of change of the mass contained in the CV. However, the mass of the fluid within the CV is constant because the flow is steady. The quantity $\dot{m}_f$ is simply a measure of the rate at which mass flows into and out of the CV.

Volumetric flow rate

In addition to the mass flux, there is another commonly used measure of the amount of fluid moving through a CV called the *volumetric flow rate*. Referring to Fig. 5.38, we see that, at time t, the volume occupied by the fluid element of mass Δm_f is approximately given by $\Delta \ell_A \, S_A$, where $\Delta \ell_A$ is the pipe length occupied by Δm_f and S_A is the area of the cross section at A. Similarly, at time $t + \Delta t$, the volume occupied by the fluid element of mass Δm_f is $\Delta \ell_B \, S_B$. Because the fluid motion is steady, the quantity Δm_f is the same at A and B. Therefore, letting ρ_A and ρ_B be the fluid density at A and B, respectively, we have

$$\Delta m_f = \rho_A \, \Delta \ell_A \, S_A = \rho_B \, \Delta \ell_B \, S_B. \tag{5.113}$$

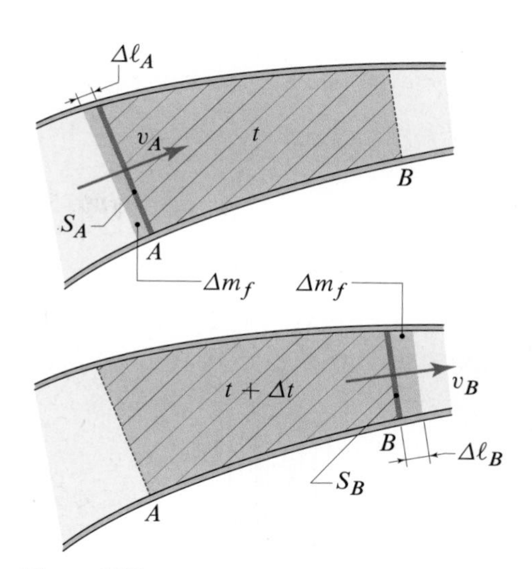

Figure 5.38
Volumes occupied by the fluid element with mass Δm_f upon entering (top) and exiting (bottom) a chosen CV.

Since we have assumed that $\vec{v}_A$ and $\vec{v}_B$ are perpendicular to the cross sections A and B, respectively, we have

$$\lim_{\Delta t \to 0} \frac{\Delta \ell_A}{\Delta t} = v_A \quad \text{and} \quad \lim_{\Delta t \to 0} \frac{\Delta \ell_B}{\Delta t} = v_B, \tag{5.114}$$

where v_A and v_B are the values of the speed of the fluid at A and B. If S is the area of a generic cross section along the pipe and if v is the fluid speed at that cross section, we define the *volumetric flow rate* as the quantity

$$\boxed{Q = vS.} \tag{5.115}$$

Dividing Eq. (5.113) by Δt, letting $\Delta t \to 0$, and using the definition in Eq. (5.115), we have

$$\boxed{\dot{m}_f = \rho_A Q_A = \rho_B Q_B,} \tag{5.116}$$

where Q_A and Q_B are the volume flow rates at A and B.

Moment acting on the fluid

Sometimes it is useful to relate the change in the fluid's angular momentum, computed with respect to a chosen moment center, to the corresponding moment acting on the fluid. Referring to Fig. 5.39, to compute this moment, we choose as moment center the *fixed* point P, we select a body of fluid in the same way as was done for the force calculation, and then we apply the angular impulse-momentum principle given in Eq. (5.51) on p. 364 between times t and $t + \Delta t$. Doing so gives

$$\vec{h}_P(t) + \int_t^{t+\Delta t} \vec{M}_P \, dt = \vec{h}_P(t + \Delta t), \tag{5.117}$$

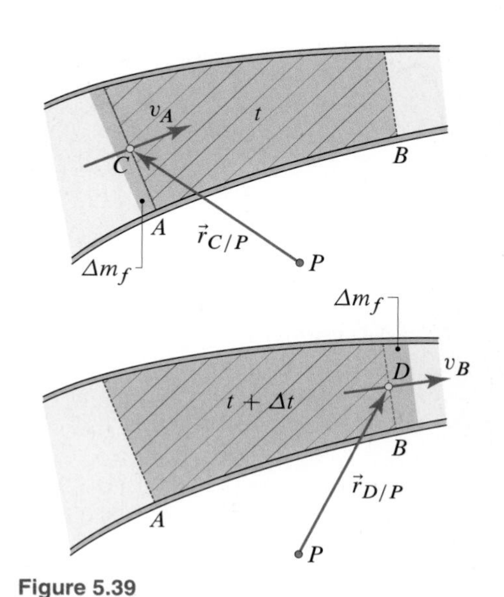

Figure 5.39
Choice of moment center P for the determination of the moment acting on the fluid contained in the CV (shaded area).

where $\vec{h}_P$ is the angular momentum of the selected fluid body and $\vec{M}_P$ is the moment we intend to compute. Because we have assumed that all of the particles in the volume elements of mass Δm_f are moving with velocity $\vec{v}_A$ at time t and $\vec{v}_B$ at time $t + \Delta t$, we have

$$\vec{h}_P(t) = \vec{r}_{C/P} \times \Delta m_f \, \vec{v}_A + \left(\vec{h}_P\right)_{\text{cv}}, \tag{5.118}$$

$$\vec{h}_P(t + \Delta t) = \left(\vec{h}_P\right)_{\text{cv}} + \vec{r}_{D/P} \times \Delta m_f \, \vec{v}_B, \tag{5.119}$$

where C and D are the centers of the cross sections A and B, respectively, $\vec{r}_{C/P}$ and $\vec{r}_{D/P}$ are the positions of C and D with respect to P, respectively, and $\left(\vec{h}_P\right)_{\text{cv}}$ is the angular momentum with respect to P of the fluid contained in the CV. Since the flow is steady, $(\vec{h}_P)_{\text{cv}}$ is a constant. Substituting Eqs. (5.118) and (5.119) into Eq. (5.117), simplifying, and rearranging terms, we have

$$\frac{1}{\Delta t}\int_t^{t+\Delta t} \vec{M}_P\,dt = \frac{\Delta m_f}{\Delta t}(\vec{r}_{D/P}\times\vec{v}_B - \vec{r}_{C/P}\times\vec{v}_A), \tag{5.120}$$

where we have divided all terms by Δt. By proceeding as in the case of the force calculation, i.e., letting $\Delta t \to 0$, Eq. (5.120) yields

$$\boxed{\vec{M}_P = \dot{m}_f(\vec{r}_{D/P}\times\vec{v}_B - \vec{r}_{C/P}\times\vec{v}_A).} \tag{5.121}$$

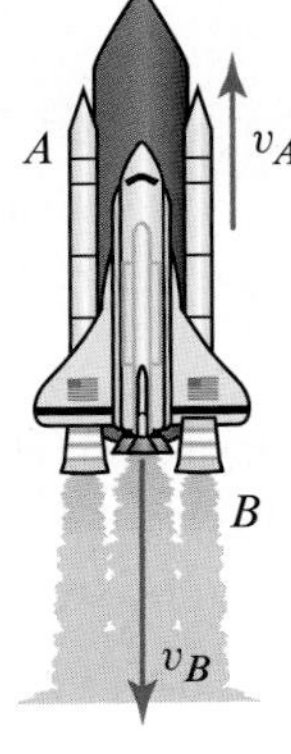

Figure 5.40
A rocket A being propelled by the ejection of combustion gases, which we have labeled B.

Variable mass flows and propulsion

Figure 5.40 shows a body A (the rocket) propelled by the continuous ejection of some material B (combustion gas). Since B used to be part of A before ejection, the mass of A changes with time so that A is a *variable mass system*.* We want to determine the force acting on A, and we will do so using CVs. That is, first we will apply the impulse-momentum principle to a closed system containing the open system of interest, and then we will "shrink" this closed system to make it coincide with the open system contained therein on an instant-by-instant basis.

We now consider a body with mass $m(t)$ at time t such that almost all of its

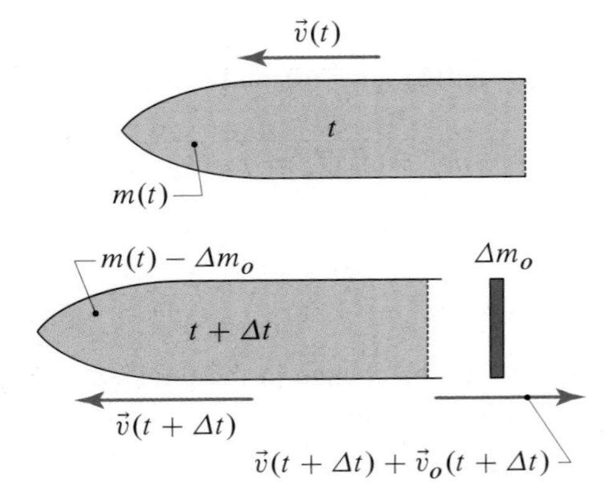

Figure 5.41. A variable mass system that ejects a material element of mass Δm_o over the time interval Δt.

particles travel with a velocity $\vec{v}(t)$ (Fig. 5.41). Some particles, which at time t have a total mass that is negligible with respect to $m(t)$, are being ejected from the body. After an amount of time Δt, the body will have lost an amount of mass Δm_o (the subscript o stands for outflow), and we write

$$m(t + \Delta t) = m(t) - \Delta m_o. \tag{5.122}$$

We assume that all the particles contributing to Δm_o have the same *inertial* velocity $\vec{v} + \vec{v}_o$, where $\vec{v}_o$ is the *relative* velocity of the particles in question with respect to the main body.

As long as the physical system we analyze consists of both the particles of mass Δm_o and the main body of mass m, our system is a closed system. Applying to

* As mentioned at the beginning of this section on p. 401, a variable mass system is *necessarily* open.

this system the impulse-momentum principle given in Eq. (5.15) on p. 316, between times t and $t + \Delta t$, we have

$$\int_t^{t+\Delta t} \vec{F}\, dt = \vec{p}(t + \Delta t) - \vec{p}(t), \tag{5.123}$$

where $F(t)$ is the total external force acting on the system and $\vec{p}(t)$ is the total momentum of the system. Using the stated assumptions, we have

$$\vec{p}(t) = m(t)\vec{v}(t), \tag{5.124}$$

$$\vec{p}(t + \Delta t) = m(t + \Delta t)\vec{v}(t + \Delta t) + \Delta m_o[\vec{v}(t + \Delta t) + \vec{v}_o(t + \Delta t)]. \tag{5.125}$$

Substituting Eq. (5.122) into Eq (5.125), we have

$$\begin{aligned}\vec{p}(t + \Delta t) &= [m(t) - \Delta m_o]\vec{v}(t + \Delta t) + \Delta m_o[\vec{v}(t + \Delta t) + \vec{v}_o(t + \Delta t)] \\ &= m(t)\vec{v}(t + \Delta t) + \Delta m_o \vec{v}_o(t + \Delta t).\end{aligned} \tag{5.126}$$

Substituting Eqs. (5.124) and (5.126) into Eq. (5.123) and collecting the term $m(t)$, we have

$$\int_t^{t+\Delta t} \vec{F}\, dt = m(t)\big[\vec{v}(t + \Delta t) - \vec{v}(t)\big] + \Delta m_o \vec{v}_o(t + \Delta t). \tag{5.127}$$

Dividing Eq. (5.127) by Δt, we obtain

$$\frac{1}{\Delta t}\int_t^{t+\Delta t} \vec{F}\, dt = m(t)\frac{\vec{v}(t + \Delta t) - \vec{v}(t)}{\Delta t} + \frac{\Delta m_o}{\Delta t}\vec{v}_o(t + \Delta t). \tag{5.128}$$

By the definition of time derivative, we have

$$\lim_{\Delta t \to 0} \frac{\vec{v}(t + \Delta t) - \vec{v}(t)}{\Delta t} = \vec{a}(t) \quad \text{and} \quad \lim_{\Delta t \to 0} \frac{\Delta m_o}{\Delta t} = \dot{m}_o(t), \tag{5.129}$$

where $\vec{a}(t)$ is the acceleration of the main body at time t and $\dot{m}_o(t)$ (with $\dot{m}_o \geq 0$) is the rate at which mass flows out of the main body. In addition, by the fundamental theorem of calculus, we have

$$\lim_{\Delta t \to 0} \frac{1}{\Delta t}\int_t^{t+\Delta t} \vec{F}\, dt = \vec{F}. \tag{5.130}$$

Therefore, taking the limit as $\Delta t \to 0$ of the terms in Eq. (5.128) and using Eqs. (5.129) and (5.130), we obtain

$$\vec{F} = m\vec{a} + \dot{m}_o\vec{v}_o, \tag{5.131}$$

where all the terms in Eq. (5.131) are evaluated at time t.

Equation (5.131) applies only to systems that lose mass. But if we follow steps analogous to those that gave us Eq. (5.131) and refer to Fig. 5.42, we can show that if the system also gains mass at the rate $\dot{m}_i$ (the subscript i stands for inflow), with $\dot{m}_i \geq 0$ and with the inflowing mass having a velocity $\vec{v}_i$ *relative* to the main body, Eq. (5.131) can be generalized to

$$\boxed{\vec{F} = m\vec{a} + \dot{m}_o\vec{v}_o - \dot{m}_i\vec{v}_i,} \tag{5.132}$$

where the contribution of the inflowing mass has a sign opposite to that of the outflowing mass.

Common Pitfall

The impulse-momentum principles apply only to closed systems. The impulse-momentum principle can be written as $\vec{F} = \dot{\vec{p}}$, with $\vec{p} = m\vec{v}_G$, where G is the system's center of mass. As we have repeatedly mentioned in this section, this principle applies only to closed systems. Because these systems must have constant mass, $\vec{F} = \dot{\vec{p}}$ implies that $\vec{F} = m\vec{a}_G + \dot{m}\vec{v}_G = m\vec{a}_G$, given that $\dot{m} = 0$.

Unfortunately, sometimes the impulse-momentum principle is erroneously applied to variable mass systems by writing $\vec{F} = m\vec{a}_G + \dot{m}\vec{v}_G$ and claiming that the term $\dot{m}\vec{v}_G$ (which is different from zero for variable mass systems) describes the effect of the mass change. To see that this conclusion is incorrect, consider the case of a rocket engine fired while held fixed in a test rig. In this case, $\vec{a}_G = \vec{0}$ and $\vec{v}_G = \vec{0}$, and if the statement $\vec{F} = m\vec{a}_G + \dot{m}\vec{v}_G$ were applicable to variable mass systems, we would conclude that no force is needed to restrain the rocket motor, thus, contradicting common experience. The correct force balance for a variable mass system is that shown in Eq. (5.132).

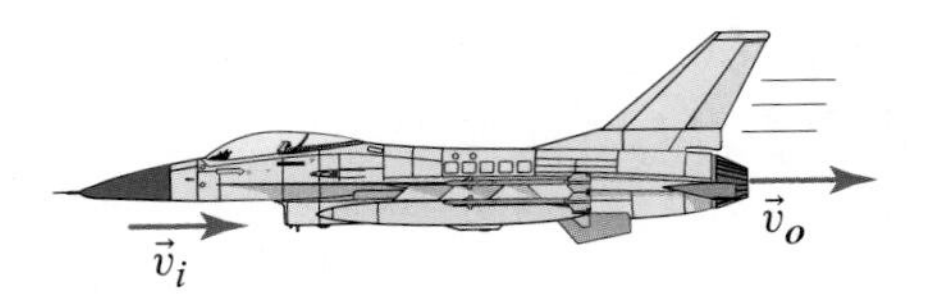

Figure 5.42. A plane with a jet engine. The airplane is taking in air with a mass flow rate $\dot{m}_i$ while combustion gases are ejected with a mass flow rate $\dot{m}_o$. The vector $\vec{v}_i$ is the velocity of the inflowing air *relative* to the plane. The vector $\vec{v}_o$ is the velocity of the outflowing combustion gases *relative* to the plane.

Equation (5.132) is an important result that can be viewed as the generalization of Newton's second law to an open system with variable mass. We arrived at Eq. (5.132) starting from a balance principle applied to a *closed* system, whose mass can only be constant. This was possible because, in going from Eq. (5.128) to Eq. (5.131), we took a limit that forced the chosen closed system to coincide with the variable mass system we wanted to characterize at the time instant t.

In the field of rocket propulsion $\dot{m}_i = 0$. It is often common to move the term $\dot{m}_o \vec{v}_o$ in Eq. (5.132) to the left-hand side of the equation and then to refer to the term $-\dot{m}_o \vec{v}_o$ as the *thrust force* provided by the propulsion system. In jet propulsion, we have both mass outflow and mass inflow so that the thrust is given by the term $\dot{m}_i \vec{v}_i - \dot{m}_o \vec{v}_o$.

End of Section Summary

In this section we have considered mass flows. Specifically, we have considered (1) *steady* mass flows, in which a fluid moves through a conduit with a velocity that depends only on the position within the conduit, and (2) *variable* mass flows, such as the flow of combustion gases out of a rocket.

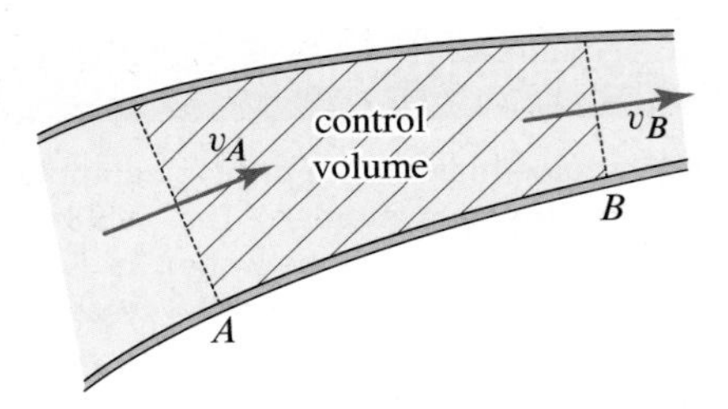

Figure 5.43
A CV corresponding to the portion of a pipe between cross sections A and B.

Steady flows. Given the control volume (CV) shown in Fig. 5.43, where by *control volume* we mean a portion of a conduit delimited by two cross sections, we showed that, in the case of a steady flow, the total external force $\vec{F}$ acting on the fluid in the CV is

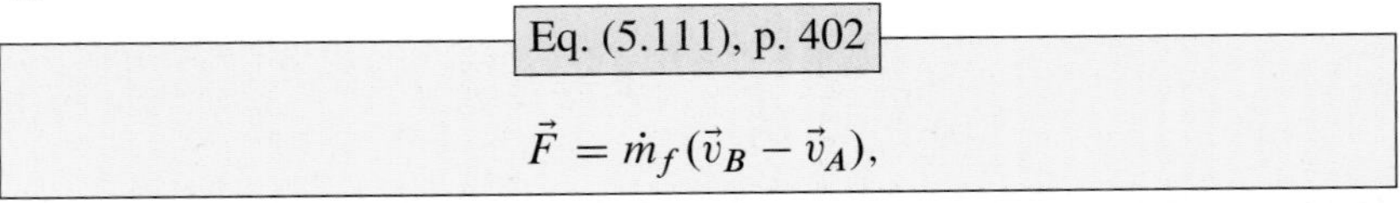

$$\vec{F} = \dot{m}_f(\vec{v}_B - \vec{v}_A),$$

where as long as the cross sections are perpendicular to the flow velocity, $\dot{m}_f$ is the *mass flow rate*, i.e., the amount of mass flowing through a cross section per unit time, and where $\vec{v}_A$ and $\vec{v}_B$ are the flow velocities at the cross sections A and B, respectively. In addition to the mass flow rate, we defined the *volumetric flow rate* as the quantity

Eq. (5.115), p. 403

$$Q = vS,$$

where v is the speed of the fluid at a given cross section and S is the area of the cross section in question. We showed that

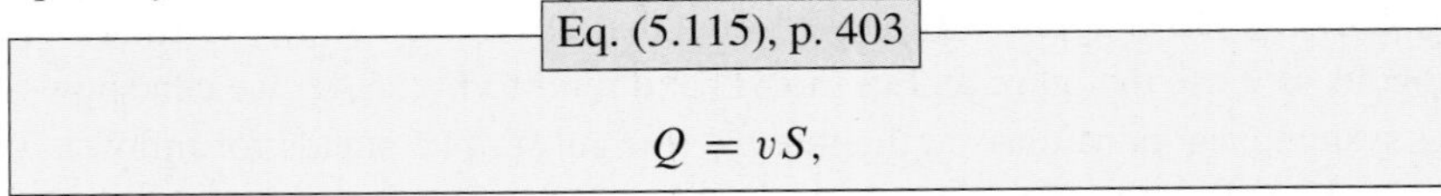

$$\dot{m}_f = \rho_A Q_A = \rho_B Q_B.$$

where ρ_A and ρ_B are the values of the mass density of the fluid at A and B, respectively. Referring to Fig. 5.44, we also showed that, given a fixed point P, the total moment $\vec{M}_P$ acting on the fluid in the CV is

Eq. (5.121), p. 404

$$\vec{M}_P = \dot{m}_f(\vec{r}_{D/P} \times \vec{v}_B - \vec{r}_{C/P} \times \vec{v}_A),$$

where C and D are the centers of the cross sections A and B, respectively.

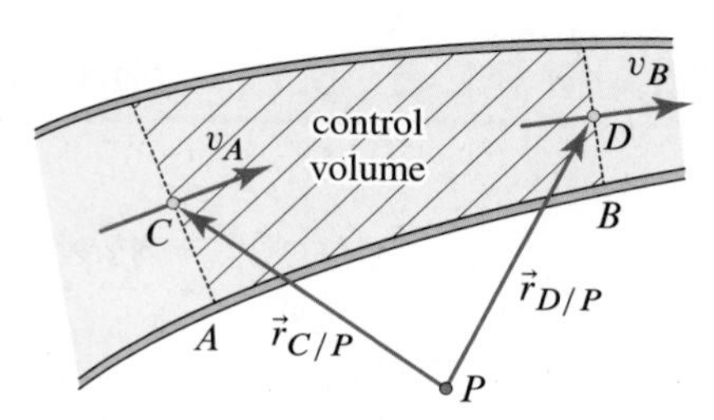

Figure 5.44
A fluid flowing through a CV along with a choice of moment center P for the calculation of angular momenta and moments.

Variable mass flows. With reference to Fig. 5.45, for a body with time varying

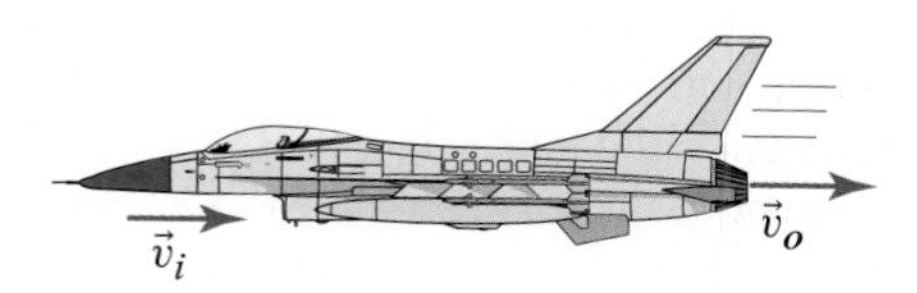

Figure 5.45. A plane with a jet engine. The airplane is taking in air with a mass flow rate $\dot{m}_i$ while combustion gases are ejected with a mass flow rate $\dot{m}_o$. The vector $\vec{v}_i$ is the velocity of the inflowing air *relative* to the plane. The vector $\vec{v}_o$ is the velocity of the outflowing combustion gases *relative* to the plane.

mass $m(t)$ due to an inflow of mass with rate $\dot{m}_i$ and an outflow of mass with the rate $\dot{m}_o$, the total external force acting on the body is given by

Eq. (5.132), p. 405

$$\vec{F} = m\vec{a} + \dot{m}_o\vec{v}_o - \dot{m}_i\vec{v}_i,$$

where $\vec{a}$ is the acceleration of the main body, $\vec{v}_o$ is the *relative* velocity of the outflowing mass with respect to the main body, and $\vec{v}_i$ is the *relative* velocity of the inflowing mass, again relative to the main body.

EXAMPLE 5.17 Force of an Open Water Jet

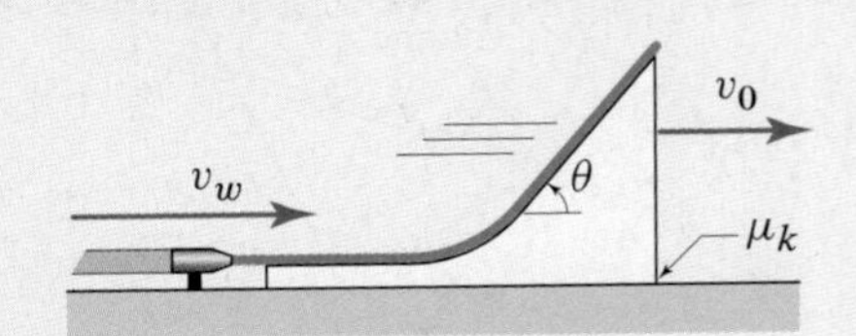

Figure 1

A water jet is let out of a nozzle attached to the ground. The jet has a constant mass flow rate and a speed $v_w = 65$ ft/s relative to the nozzle. The jet strikes a 25 lb incline and causes it to slide at a constant speed $v_0 = 5.5$ ft/s. The kinetic coefficient of friction between the incline and the ground is $\mu_k = 0.43$. Neglecting the effect of gravity and air resistance on the water flow, as well as friction between the water jet and the incline, determine the mass flow rate of the water jet at the nozzle if $\theta = 50°$.

SOLUTION

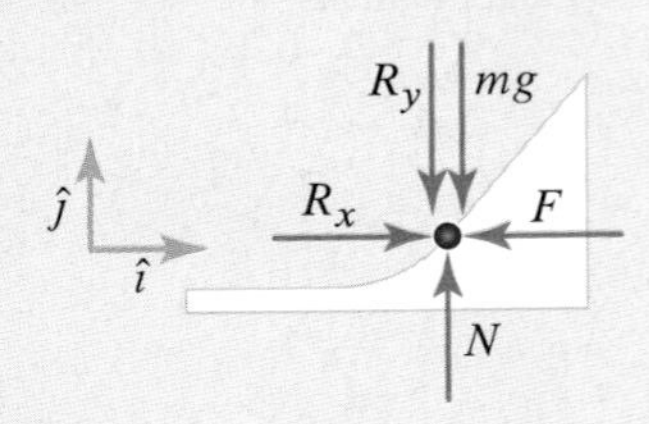

Figure 2
FBD of the incline. The force F is the friction force due to the sliding relative to the ground.

Road Map & Modeling Modeling the incline as a particle and referring to the FBD in Fig. 2, we represent the effect of the water jet on the incline by the reaction forces R_x and R_y. To solve the problem, we need to relate R_x and R_y to the mass flow rate out of the nozzle and then use these relations when applying Newton's second law to the incline. In this way, we will be able to relate the mass flow rate to the friction force opposing the motion of the incline. The key to determining R_x and R_y is to realize that if the friction between the water jet and the incline is negligible, then there is no force that will slow down the flow of water over the incline. This fact, along with the fact that the incline moves at a constant velocity, allows us to model the flow of water over the incline as *steady* and with a constant speed relative to the incline.

Determination of R_x and R_y in Terms of Mass Flow Rate

Governing Equations

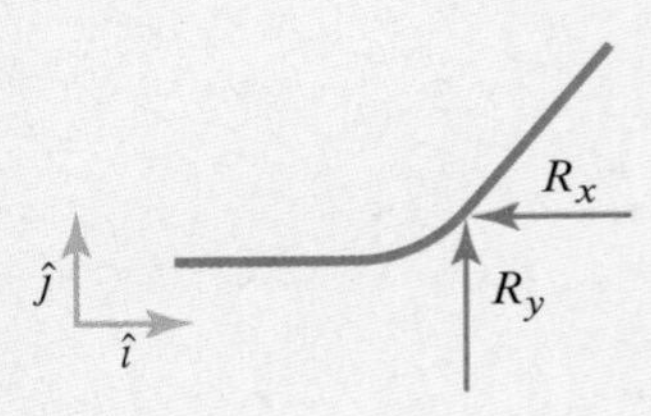

Figure 3
FBD of the water jet. The forces R_x and R_y are equal and opposite to those indicated in Fig. 2 to comply with Newton's third law.

Balance Principles Here we apply the force balance relation for CVs in Eq. (5.111) on p. 402. The terms in this equation must be measured using an *inertial frame of reference*. Because the incline moves at a constant velocity, a reference frame attached to the incline is inertial.* Choosing such a frame and referring to the water jet FBD in Fig. 3, we choose our CV to be the volume occupied by the water flowing over the top surface of the incline. Although this CV is moving with respect to the ground, our choice is acceptable because such a CV is *stationary* with respect to the chosen inertial frame and, as discussed above, the water flow is *steady* over the incline. Then Eq. (5.111) in component form yields (see Fig. 4)

$$\sum F_x: \quad -R_x = \dot{m}_f (v_{Bx} - v_{Ax}), \tag{1}$$

$$\sum F_y: \quad R_y = \dot{m}_f (v_{By} - v_{Ay}), \tag{2}$$

where $\dot{m}_f$ is the mass flow rate that goes past the cross section at A and $\vec{v}_A$ and $\vec{v}_B$ are the velocities of the water flow at A and B, respectively. We note, again, that $\dot{m}_f$, $\vec{v}_A$, and $\vec{v}_B$ are measured by the inertial observer who moves with the incline.

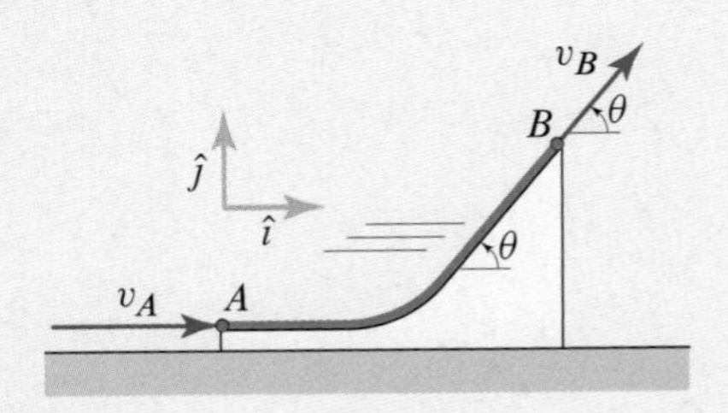

Figure 4
Velocities of the steady flow at A and B as perceived by an observer moving with the incline.

Force Laws All forces are accounted for on the FBD.

Kinematic Equations Using relative kinematics, an observer moving with the incline measures

$$v_{Ax} = v_w - v_0 \quad \text{and} \quad v_{Ay} = 0, \tag{3}$$

where v_w and v_0 are the speeds of the water jet and of the incline measured relative to the ground. Since we are neglecting the friction between the water and the incline, we must have $|\vec{v}_A| = |\vec{v}_B|$. Hence, the components of the velocity of the water at B are

$$v_{Bx} = (v_w - v_0)\cos\theta \quad \text{and} \quad v_{By} = (v_w - v_0)\sin\theta. \tag{4}$$

Let $(\dot{m}_f)_{\text{nz}}$ be the mass flow rate measured at the nozzle. This quantity is the unknown we want to determine. We assume that the water jet has a constant cross section

* This statement is based on the assumption that the ground over which the incline slides can be chosen as an inertial reference frame (see discussion on p. 173).

even when in contact with the incline. Then, letting S denote the flow cross-sectional area, we see from Eq. (5.115), the volumetric flow rate at the nozzle is $Q_{\text{nz}} = v_w S$, and therefore the mass flow rate at the nozzle is

$$(\dot{m}_f)_{\text{nz}} = \rho S v_w, \tag{5}$$

where ρ denotes the mass density of the water. Similarly, the mass flow rate $\dot{m}_f$ measured by an observer moving with the incline is

$$\dot{m}_f = \rho S v_A = \rho S(v_w - v_0) \quad \Rightarrow \quad \dot{m}_f = (\dot{m}_f)_{\text{nz}}(v_w - v_0)/v_w, \tag{6}$$

where we have used Eq. (5) in the first of Eqs. (6).

Computation Substituting Eqs. (3), (4), and the last of Eqs. (6) into Eqs. (1) and (2) gives

$$R_x = \frac{(\dot{m}_f)_{\text{nz}}}{v_w}(1 - \cos\theta)(v_w - v_0)^2 \quad \text{and} \quad R_y = \frac{(\dot{m}_f)_{\text{nz}}}{v_w} \sin\theta\,(v_w - v_0)^2. \tag{7}$$

Discussion & Verification Recalling that the mass flow rate has dimensions of mass over time, we know that Eqs. (7) are dimensionally correct. In addition, given that the right-hand sides of Eqs. (7) have a positive sign under all circumstances, we see that the directions of R_x and R_y are as expected.

$\vec{F} = m\vec{a}$ for the Incline and Determination of $(\dot{m}_f)_{\text{nz}}$

Governing Equations

Balance Principles Using the FBD in Fig. 2, the application of Newton's second law to the incline yields

$$\sum F_x: \quad R_x - F = m a_x, \tag{8}$$

$$\sum F_y: \quad -R_y - mg + N = m a_y. \tag{9}$$

Force Laws Since the incline is sliding, we have

$$F = \mu_k N. \tag{10}$$

Kinematic Equations Because the incline moves with constant velocity, we have

$$a_x = 0 \quad \text{and} \quad a_y = 0. \tag{11}$$

Computation Substituting Eqs. (11) into Eqs. (8) and (9), solving for F and N, and then substituting the result into Eq. (10) yield

$$R_x = \mu_k (R_y + mg). \tag{12}$$

Substituting Eqs. (7) into Eq. (12) and solving for $(\dot{m}_f)_{\text{nz}}$, we have

$$\boxed{(\dot{m}_f)_{\text{nz}} = \frac{\mu_k m g v_w}{(v_w - v_0)^2(1 - \cos\theta - \mu_k \sin\theta)} = 7.096\ \text{slug/s}.} \tag{13}$$

Discussion & Verification Since the terms μ_k and $1 - \cos\theta - \mu_k \sin\theta$ in Eq. (13) are nondimensional, and since the term $g v_w/(v_w - v_0)^2$ has dimensions of 1 over time, our result is dimensionally correct. Our result is directly proportional to the weight of the incline, as well as to the friction coefficient. This is reasonable since, keeping v_w and v_0 fixed, we expect that more water mass per unit time is needed to move a heavier incline over a rougher surface. So our solution appears to be correct.

EXAMPLE 5.18 Geometry of Fluid Motion and Structural Loads

Figure 1
An example of intricate pipeline geometry in a refinement plant.

Pipelines can be quite intricate (see Fig. 1). The fluid going through a bend and/or a change in cross section can exert significant structural loads on the line. Referring to Fig. 2, consider two straight pipes connected by a flanged diverter/reducer of length $\ell = 88$ in., height $h = 62$ in., and internal volume $V = 34\,\text{ft}^3$. Suppose that the flow is steady, the fluid specific weight is $\gamma = 42.5\,\text{lb/ft}^3$ (this is typical of gasoline), and the cross sections at A and B are circular with radii $R_A = 13$ in. and $R_B = 9$ in., respectively. Assume that the center of mass G of the fluid between A and B is located as shown with $d = \ell/2$ and $q = 33$ in. In addition, let $p_A = 1400$ psi and $p_B = 1390$ psi be known measurements of the static pressure at A and B, respectively. If the fluid speed at A is $v_A = 6\,\text{ft/s}$, determine the loads that the fluid exerts on the diverter/reducer. Finally, neglect the weight of the diverter/reducer and sketch its FBD, showing the internal forces at A and B.

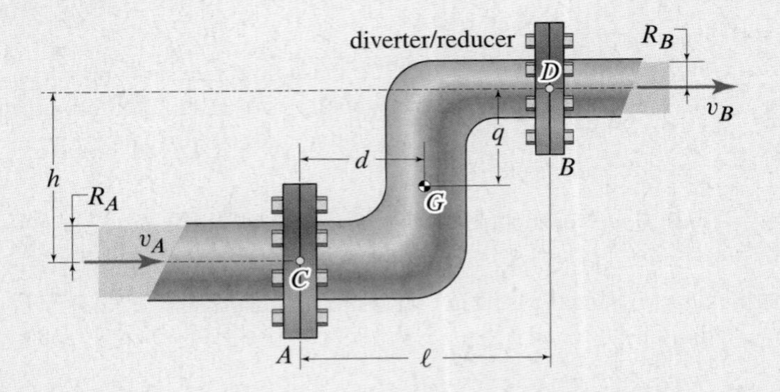

Figure 2. A diverter/reducer connecting two straight pipe segments in the vertical plane. Points C and D indicate the centers of the cross sections A and B, respectively.

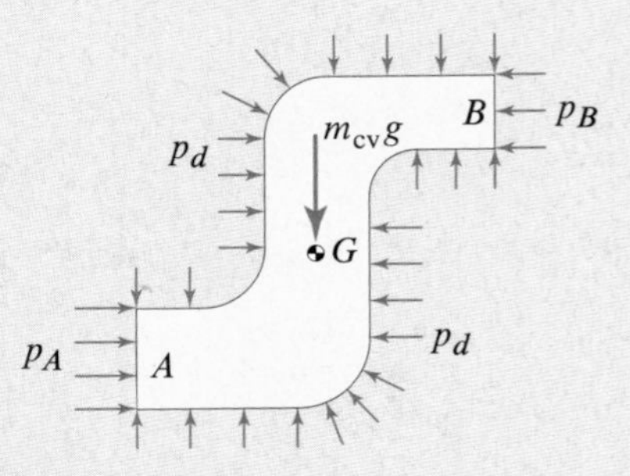

Figure 3
FBD of the fluid moving through the chosen CV, which was taken to be the interior volume of the diverter.

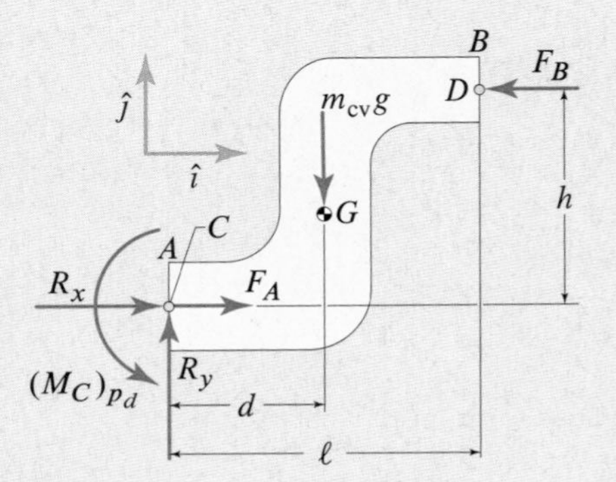

Figure 4
FBD of the fluid within the CV obtained using the concept of equivalent force system.

SOLUTION

Road Map & Modeling The CV is the region occupied by the fluid between the cross sections A and B. Referring to Fig. 3, the fluid in question is subject to the pressure distributions p_A and p_B due to the fluid outside the CV. We assume that p_A and p_B are uniform over A and B, respectively. The fluid in the CV is also subject to the pressure distribution p_d due to the contact with the inner walls of the diverter. Finally, the fluid in the CV is subject to gravity, which is represented by the weight $m_{cv}g$, applied at the fluid's center of mass G. Although we do not have a detailed knowledge of p_d, we can describe the overall effect of p_d using the concept of equivalent force system.* Using this concept, the force system acting on the fluid in the CV can be represented as shown in Fig. 4. The forces F_A and F_B, applied as shown, are equivalent to the pressure distributions p_A and p_B, respectively. The force system equivalent to the pressure distribution p_d consists of the forces R_x and R_y, as well as the moment $(M_C)_{p_d}$, where C has been chosen as the moment center because its location is known (we could have chosen some other convenient reference point such as G or D). It is this force system that we need to compute, and we will do so by applying the force and moment balance for steady flows. Because R_x, R_y, and $(M_C)_{p_d}$ describe the action of the diverter on the fluid in the CV, in sketching the FBD of the diverter we need to include these forces and moment, but with opposite sign to abide by Newton's third law.

* See your statics textbook for the concept of equivalent force system.

Governing Equations

Balance Principles Referring to the FBD in Fig. 4, choosing C as the moment center, and recalling that $m_{vc}g = \gamma V$, the force and impulse-momentum principles in Eqs. (5.111) and (5.121), in component form, give

$$\sum F_x: \quad R_x + F_A - F_B = \dot{m}_f(v_{Bx} - v_{Ax}), \tag{1}$$

$$\sum F_y: \quad R_y - \gamma V = \dot{m}_f(v_{By} - v_{Ay}), \tag{2}$$

$$\sum M_C: \quad (M_C)_{p_d} - \gamma V d + F_B h = \dot{m}_f(v_{By}\ell - v_{Bx}h). \tag{3}$$

Force Laws Since the cross sections at A and B are circular with radii R_A and R_B, respectively, and since we have assumed that the pressure distributions over A and B are uniform, we have

$$F_A = \pi R_A^2 p_A \quad \text{and} \quad F_B = \pi R_B^2 p_B. \tag{4}$$

Kinematic Equations Based on the flow depicted in Fig. 2, we have

$$v_{Ax} = v_A, \quad v_{Ay} = 0, \quad v_{Bx} = v_B, \quad v_{By} = 0. \tag{5}$$

Furthermore, applying Eqs. (5.115) and (5.116), we must have

$$\dot{m}_f = (\gamma/g)\pi R_A^2 v_A = (\gamma/g)\pi R_B^2 v_B \quad \Rightarrow \quad v_B = v_A R_A^2/R_B^2. \tag{6}$$

Computation Substituting Eqs. (4)–(6) into Eqs. (1)–(3), the force system that is equivalent to the pressure distribution p_d is given by

$$R_x = \pi\left(p_B R_B^2 - p_A R_A^2\right) + \frac{\gamma\pi v_A^2 R_A^2(R_A^2 - R_B^2)}{g R_B^2} = -389.4\times 10^3\ \text{lb}, \tag{7}$$

$$R_y = \gamma V = 1445\ \text{lb}, \tag{8}$$

$$(M_C)_{p_d} = \gamma V d - \pi p_B R_B^2 h - \frac{\gamma\pi v_A^2 R_A^4 h}{g R_B^2} = -1.824\times 10^6\ \text{ft}\cdot\text{lb}. \tag{9}$$

Now that we have R_x, R_y, and $(M_C)_{p_d}$, by applying Newton's third law, the FBD for the diverter is that given in Fig. 5, where the internal force system over the cross section A consists of the forces N_A (tension), V_A (shear force), and M_{Ci} (bending moment). Similarly, the internal force system at B is given by N_B, V_B, and M_{Di}.

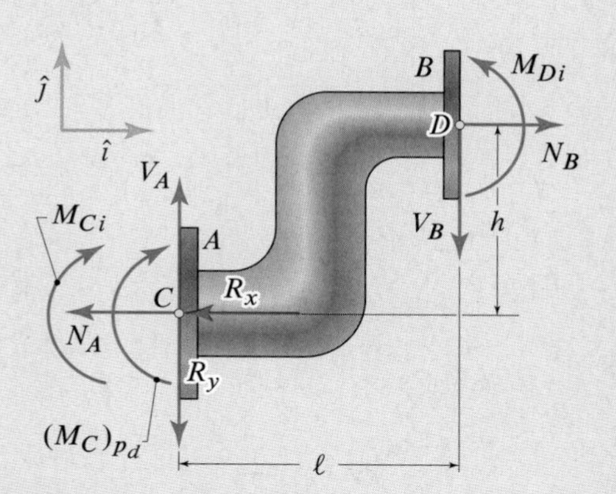

Figure 5
FBD of the diverter/reducer.

Discussion & Verification Since the dimensions of pressure and mass density are force per unit area and mass over length cubed, respectively, our results are dimensionally correct. The fact that R_x is negative in Eq. (7) is consistent with the idea that the fluid motion in the x direction is hindered by the presence of the bend in the line. The result in Eq. (8) also makes sense since it confirms that the diverter is supporting the weight of the fluid in the CV. To explain the result in Eq. (9), recall that if there is no flow (i.e., $v_A = 0$), then the diverter must provide a positive moment to balance the weight. However, if $v_A \neq 0$ and we neglect the weight, then common experience tells us that the flow would cause a counterclockwise rotation of the system, and the diverter must exert a moment in the clockwise direction to prevent the rotation in question. This means that both positive and negative moments are to be expected, and the sign of the result in Eq. (9) tells us that, in our case, the moment due to the fluid motion has the greater effect.

A Closer Look Referring to Fig. 5, if end B of the diverter were free, then the internal forces at B would be equal to zero and an elementary equilibrium calculation would show that we must have $N_A = -R_x$, $V_A = R_y$, and $(M_{C_A})_i = -(M_{C_A})_{p_d}$. That is, if one end is free, we can compute the internal forces at the other end directly in terms of forces computed from the force balance for CVs.

EXAMPLE 5.19 *Hovering Using a Jet Pack*

Figure 1
A pilot with a jet pack.

A jet pack is a rocket propulsion device worn on a person's back that allows the person to become airborne and fly (see Fig. 1). Suppose that the jet pack can hold 75 lb of fuel. Suppose further that when there is no fuel in the pack, the combined weight of the pilot and the jet pack is 180 lb. It is assumed that, in operation, the outflow speed v_o of the ejected material relative to the pack is constant. Neglecting the amount of time it takes for the pilot to start hovering a few feet off the ground, and assuming that the jets are oriented in the direction of gravity, determine v_o so that the fuel will be completely spent after the pilot hovers for 45 s.

SOLUTION

Road Map & Modeling The pilot and the pack form a simple variable mass system. We will, therefore, apply to this system the force balance given in Eq. (5.132) while enforcing the requirement that the mass outflow rate be equal to the time rate of decrease of the system's mass. By doing so, we will be able to relate the speed v_o to the time it takes to exhaust all of the fuel. When we use Eq. (5.132), the thrust due to the ejection of matter from the pack is *not* considered an external force. Hence, given that the pilot is simply hovering, the system's FBD is that shown in Fig. 2, in which we have only included the system's weight. After solving the problem, we will discuss another approach to the solution of propulsion problems according to which the thrust acting on the system is shown on the FBD and, at the same time, the force balance law is made to take on the form $\vec{F} = m\vec{a}$.

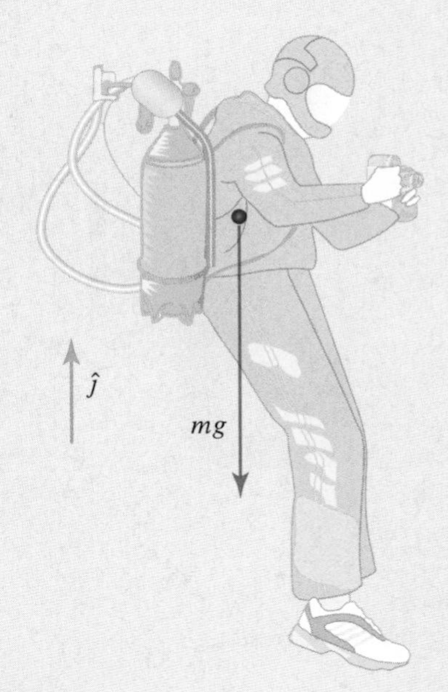

Figure 2
FBD of the system consisting of the pilot and the jet pack.

Governing Equations

Balance Principles Observing that there are no forces acting in the horizontal direction, we see that the only significant component of the force balance law is that in the y direction. Hence, we have

$$-mg = ma_y + \dot{m}_o \vec{v}_o \cdot \hat{\jmath}, \tag{1}$$

where m is the time-varying combined mass of the pilot and of the pack, $\dot{m}_o$ is the time rate at which mass is being ejected from the pack, $\vec{v}_o \cdot \hat{\jmath}$ is the y component of the velocity of the ejected matter relative to the main system, and we have accounted for the fact that the system does not gain mass (i.e., $\dot{m}_i = 0$).

Force Laws All forces are accounted for on the FBD.

Kinematic Equations Since the pilot (with the pack) is hovering, the system is stationary with respect to the ground, which is chosen as our inertial frame. Therefore, we must have

$$a_y = 0 \quad \text{and} \quad \vec{v}_o = -v_o\,\hat{\jmath}. \tag{2}$$

In addition, as already observed, the mass of the system decreases at the rate at which mass is ejected from the pack, so we have

$$\dot{m} = -\dot{m}_o. \tag{3}$$

Computation Substituting Eqs. (2) and (3) into Eq. (1), we have

$$-mg = v_o \dot{m}. \tag{4}$$

Recalling that $\dot{m} = dm/dt$, Eq. (4) can be written as

$$-\frac{g}{v_o}\,dt = \frac{dm}{m}. \tag{5}$$

Integrating this equation from $t = 0$ to the final time $t_f = 45$ s, we have

$$-\int_0^{t_f} \frac{g}{v_o}\,dt = \int_{m(0)}^{m(t_f)} \frac{dm}{m} \quad \Rightarrow \quad -\frac{g}{v_o} t_f = \ln \frac{m(t_f)}{m(0)}. \tag{6}$$

Recalling that $m(0) = (180 + 75)$ lb is the combined mass of the pilot, the pack, and 75 lb of fuel, and that $m(t_f) = 180$ lb is the combined mass of the pilot and the empty pack, we can solve Eq. (6) for v_o to obtain

$$v_o = -\frac{g t_f}{\ln[m(t_f)/m(0)]} = \frac{g t_f}{\ln[m(0)/m(t_f)]} = 4160\,\text{ft/s}. \tag{7}$$

Discussion & Verification Since the argument of the natural logarithm in Eq. (7) is nondimensional, and since the product of acceleration and time has the dimensions of length over time, the result in Eq. (7) has the correct dimensions. As far as the numerical value obtained for v_o is concerned, this result is not far from what is obtained from simple monopropellant rocket engines whose exhaust speeds are typically on the order of 5600 ft/s (although they can get close to 10,000 ft/s).

A Closer Look The problem discussed in this example can be approached by writing the force balance law as $\vec{F} = m\vec{a}$, which is meant to resemble Newton's second law.* If we approach the force balance for a variable mass system using the expression $\vec{F} = m\vec{a}$, then the force $\vec{F}$ includes both those forces that would be considered external according to a strict interpretation of the impulse-momentum principle and those force-like terms that result from the inflow and outflow of mass. Therefore, our FBD would have been that in Fig. 3, where $-m_o\vec{v}_o$ is the thrust force provided by the rocket engine.

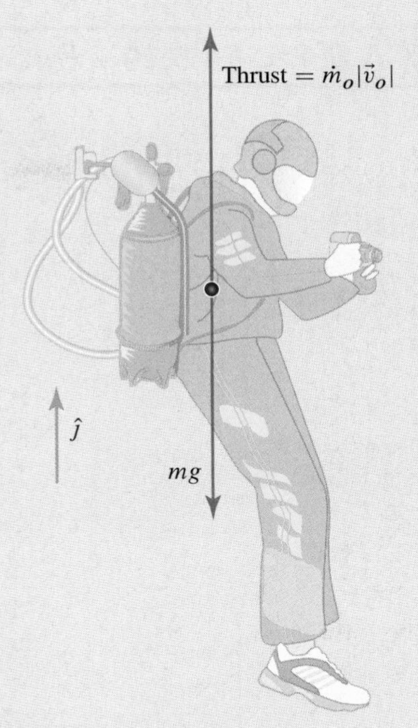

Figure 3
Alternate FBD of the system. The force balance law that must be used with this system is $\vec{F} = m\vec{a}$, where $\vec{F} = -mg\,\hat{\jmath} - \dot{m}_o\vec{v}_o$.

* Writing $\vec{F} = m\vec{a}$ for variable mass systems *cannot* be considered to be the same as applying Newton's second law. This is so because *Newton's second law cannot be applied to variable mass systems.* If $\vec{F} = m\vec{a}$ is applied to a variable mass system, the only correct interpretation that can be given is that what is being applied is actually Eq. (5.132), with $\vec{F} = \vec{F}_{\text{ext}} - \dot{m}_o\vec{v}_o + \dot{m}_i\vec{v}_i$, where $\vec{F}_{\text{ext}}$ is the total external force applied to the system according to a strict interpretation of the impulse-momentum principle.

EXAMPLE 5.20 *Forces in a Falling String*

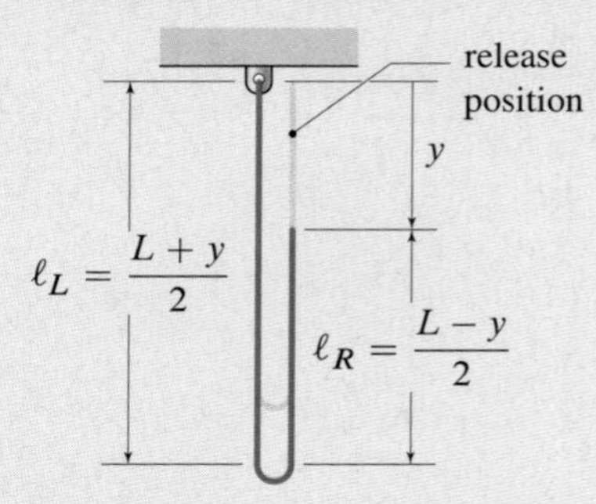

Figure 1
A string falling vertically down.

In Example 4.13 on p. 288, we discovered that the velocity of the free end of a falling inextensible string of length L, released from rest, is given by (see Fig. 1)

$$\dot{y} = \sqrt{gy\frac{2L-y}{L-y}}, \tag{1}$$

where g is the acceleration due to gravity and y is the position of the free end of the string. Letting ρ be the string's mass density per unit length, use Eq. (1) to determine the reaction force R at the ceiling as a function of y by modeling the whole string as a closed system and by modeling the two branches to the right and left of the bend as variable mass systems.

SOLUTION

Modeling the Whole String as a Closed System

Road Map & Modeling If we model the whole string as a closed system, then the string's FBD is that shown in Fig. 2, where the only force other than R is the string's weight (this is consistent with the solution of Example 4.13). Since the system is closed, we can apply the impulse-momentum principle as given in Eq. (5.14) on p. 316.

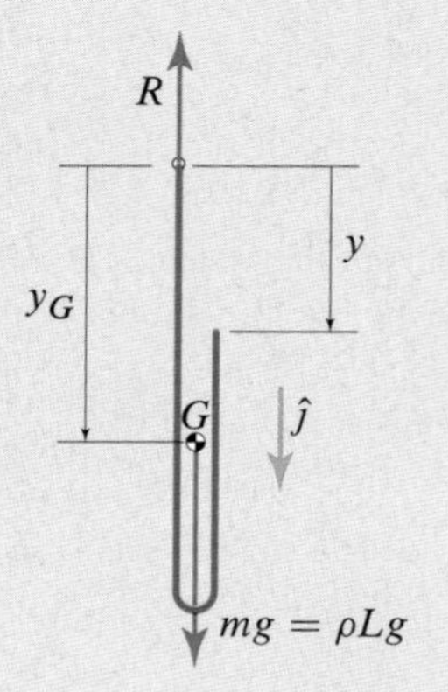

Figure 2
FBD of the string as a whole. The weight of the string has been placed at the string's center of mass, which has been denoted by G.

Governing Equations

Balance Principles Using the FBD in Fig. 2 and Eq. (5.14), we obtain

$$\sum F_y: \quad \rho Lg - R = \rho L a_G, \tag{2}$$

where a_G is the acceleration of the string's mass center and ρL is the string's mass.

Force Laws All forces are accounted for on the FBD.

Kinematic Equations Since $a_G = \ddot{y}_G$, we first find y_G via Eq. (3.35) on p. 218. Recalling that the mass of the string is ρL and referring to Fig. 1, we have

$$\rho L y_G = \rho \ell_L(\ell_L/2) + \rho \ell_R(y + \ell_R/2) \quad \Rightarrow \quad y_G = \frac{1}{4L}(L^2 + 2Ly - y^2), \tag{3}$$

where $\rho\ell_L$ and $\rho\ell_R$ are the masses of the left and right branches of the string, respectively. Differentiating the final result in Eq. (3) twice with respect to time, we obtain

$$a_G = \frac{1}{2L}\big[\ddot{y}(L-y) - \dot{y}^2\big]. \tag{4}$$

Using Eq. (1) along with the chain rule, we have

$$\ddot{y} = \frac{d\dot{y}}{dy}\dot{y} = \frac{\sqrt{g}(2L^2 - 2Ly + y^2)}{2(L-y)^{3/2}\sqrt{y(2L-y)}}\dot{y} \quad \Rightarrow \quad \ddot{y} = \frac{g}{2}\left[1 + \frac{L^2}{(L-y)^2}\right]. \tag{5}$$

Substituting Eq. (1) and the final result in Eq. (5) into Eq. (4), after simplifying we obtain

$$a_G = \frac{g}{4}\left(3 - 3\frac{y}{L} - \frac{L}{L-y}\right). \tag{6}$$

Computation Substituting Eqs. (6) into Eq. (2) and solving for R, we have

$$\boxed{R = \frac{\rho Lg}{4}\left(1 + 3\frac{y}{L} + \frac{L}{L-y}\right).} \tag{7}$$

Variable Mass Systems Modeling

Road Map & Modeling We can model the left and right branches of the string as variable mass systems that exchange mass with each other. Specifically, the left branch gains mass at the expense of the right branch. In this case, separating these systems with a cut at the bend, we have the FBDs in Fig. 3 (see the Helpful Information marginal note for further comments on these FBDs). Then we can apply to each branch the force balance for variable mass systems given in Eq. (5.132).

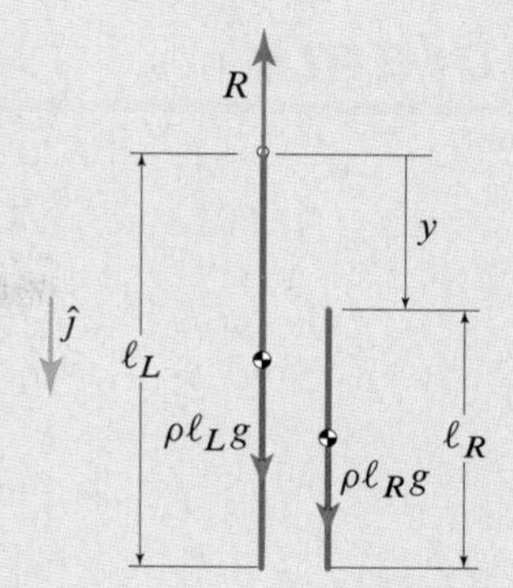

Figure 3
FBDs of the left and right branches of the falling string modeled as variable mass systems (see the Helpful Information marginal note for further comments on these FBDs).

Governing Equations

Balance Principles Using the FBDs in Fig. 3 along with the force balance for variable mass systems, for the left and right branches of the string we have, respectively,

$$\sum F_{yL}: \quad \rho \ell_L g - R = \rho \ell_L a_{yL} - \dot{m}_i \vec{v}_i \cdot \hat{\jmath}, \tag{8}$$

$$\sum F_{yR}: \quad \rho \ell_R g = \rho \ell_R a_{yR} + \dot{m}_o \vec{v}_o \cdot \hat{\jmath}, \tag{9}$$

where $\dot{m}_i$ is the time rate of mass gain of the left branch, $\vec{v}_i$ is the velocity of the mass joining the left branch relative to the velocity of the left branch itself, $\dot{m}_o$ is the time rate of mass loss of the right branch, and $\vec{v}_o$ is the velocity of the mass leaving the right branch relative to the right branch itself.

Force Laws All forces are accounted for on the FBDs.

Kinematic Equations Because of inextensibility, all the mass elements on the left branch must move with the same velocity. The same is true for the mass elements on the right branch. Observing that one point on the left branch is fixed to the ceiling and that the acceleration of the top end of the right branch is $\ddot{y}$, we must have

$$a_{yL} = 0 \quad \text{and} \quad a_{yR} = \ddot{y}. \tag{10}$$

Referring to Fig. 1, the time derivatives of the lengths of the two branches are $\dot{\ell}_L = \dot{y}/2$ and $\dot{\ell}_R = -\dot{y}/2$. Hence, since $m_L = \rho \ell_L$ and $m_R = \rho \ell_R$, we have

$$\dot{m}_i = \dot{m}_L = \rho \dot{\ell}_L = \rho \dot{y}/2 \quad \text{and} \quad \dot{m}_o = -\dot{m}_R = -\rho \dot{\ell}_R = \rho \dot{y}/2. \tag{11}$$

The velocity of the mass elements joining the left branch and leaving the right branch must match the time rate of lengthening and shortening of these branches, i.e.,

$$\vec{v}_i = \dot{\ell}_L \, \hat{\jmath} = \tfrac{1}{2} \dot{y} \, \hat{\jmath} \quad \text{and} \quad \vec{v}_o = \dot{\ell}_R \, \hat{\jmath} = -\tfrac{1}{2} \dot{y} \, \hat{\jmath}. \tag{12}$$

Computation Substituting Eqs. (10)–(12) into Eq. (8) and solving for R, we have

$$R = \rho \ell_L g + \tfrac{1}{4} \rho \dot{y}^2. \tag{13}$$

Recalling that $\dot{y}$ is given in Eq. (1) and $\ell_L = (L + y)/2$, after simplification, we have

$$\boxed{R = \frac{\rho L g}{4} \left(1 + 3\frac{y}{L} + \frac{L}{L - y} \right).} \tag{14}$$

Discussion & Verification Since we have obtained the same result with two very different methods, we can be confident that our final result is correct.

A Closer Look We present a plot of R as a function of y in Fig. 4. Notice that as the string becomes vertical, i.e., as $y \to L$, R goes to infinity. This is so because the free end of the string moves with infinite speed when the string is *almost* completely vertical (i.e., $\dot{y} \to \infty$ as $y \to L$), and therefore R must become *impulsive* to bring the string to a complete stop as soon as the string becomes vertical.

Finally, we note that we did not take advantage of Eq. (9). The reason for this is that substituting Eqs. (10)–(12) into Eq. (9) yields an equation whose solution coincides with Eq. (1) (see Prob. 5.192). If we had not been given Eq. (1), we would have had to use Eq. (9) to obtain the velocity of the free end as a function of y.

Helpful Information

Is something missing from the FBDs in Fig. 3? When we cut some structure and we sketch the FBD of the cut structure, we place on the FBD those forces that act internally to the structure at the location of the cut. However, here we are modeling the two branches as *variable mass systems*, and in this case, we do not include the forces at the cut because of how we derived Eq. (5.132). Specifically, the external forces that appear in Eq. (5.132) do not include any effects due to the exchange of mass.

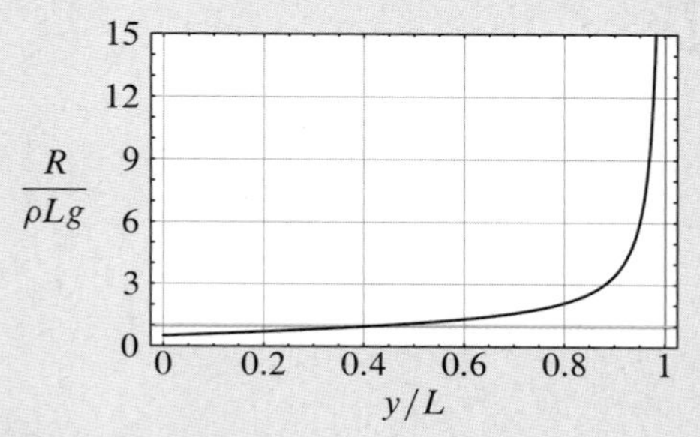

Figure 4
The reaction R (nondimensionalized with respect to the string weight $\rho L g$) at the top of the string as a function of the nondimensional fall distance (y/L). The vertical red line corresponds to the end of the fall, and the horizontal green line corresponds to the weight of the string.

Problems

Problem 5.156

A fluid is in steady motion in the conduit shown. The lines depicted are tangent to the velocity of the fluid particles in the conduit (these lines are called streamlines). Explain whether or not the control volume defined by the cross sections A and B in the figure is consistent with the assumptions laid out in this section.

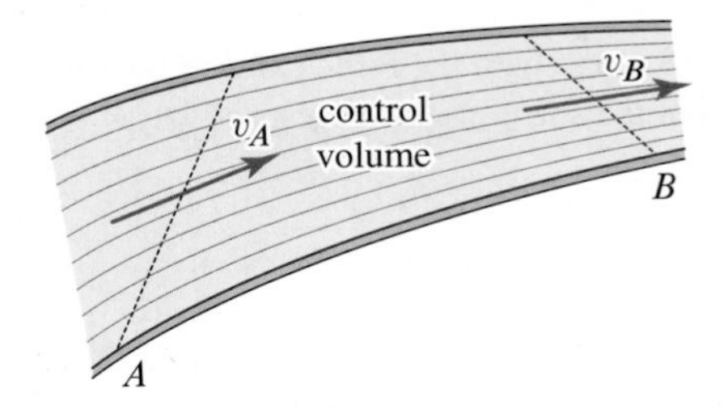

Figure P5.156

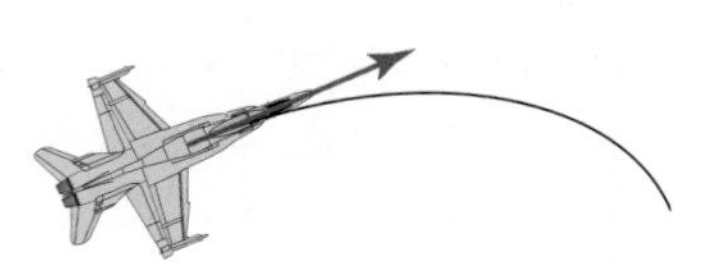

Figure P5.157

Problem 5.157

A hydraulic system is being used to actuate the control surfaces of a plane. Suppose that there is a time interval during which (a) the speed of the hydraulic fluid within a particular line is constant relative to the line itself and (b) the plane is performing a turn. Explain whether or not the force balance for control volumes presented in this section is applicable to the analysis of the hydraulic fluid in question.

Problem 5.158

The cross sections labeled A and B in case (a) are identical to the corresponding cross sections in case (b). Assume that, in both (a) and (b), a fluid in steady motion flows through A with speed v_1 and exits the system at B with a speed v_2. If the pipe sections are to remain stationary and if the mass flow rate is identical in the two cases, determine whether the magnitude of the horizontal force acting on the pipes due to the water flow in case (a) is smaller than, equal to, or larger than that in case (b). In addition, for both (a) and (b), establish the direction of the force.

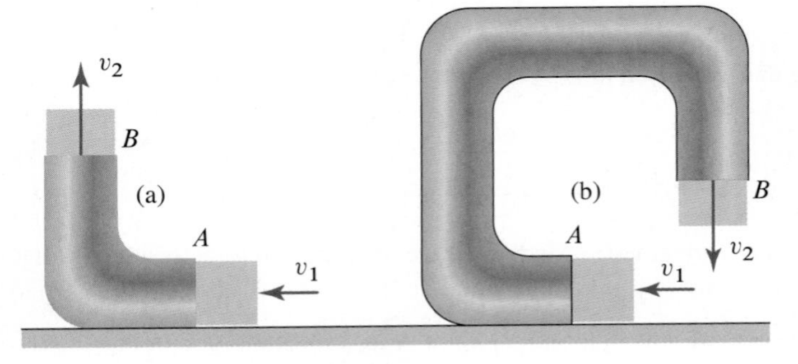

Figure P5.158

Problem 5.159

Experience tells us that when a steady water jet comes out of a nozzle B, the hose line A attached to the nozzle is in tension, that is, the nozzle exerts a force on the hose that is in the direction of the flow. If the end of the nozzle at B were capped to stop the water flow, would the force exerted by the nozzle on the hose decrease, stay the same, or increase?

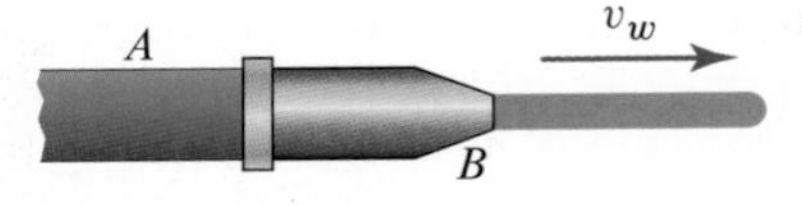

Figure P5.159

Problem 5.160

Revisit Example 5.17 and use the numerical result in Eq. (13) of the example, along with the fact that the specific weight of water is $62.4\,\mathrm{lb/ft^3}$, to determine the volumetric flow rate at the nozzle and the nozzle diameter.

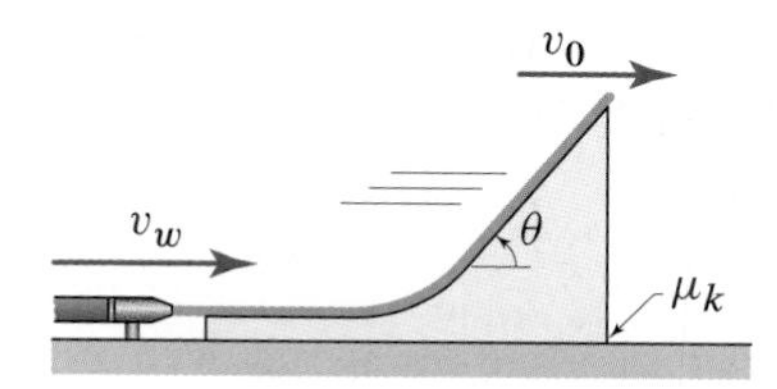

Figure P5.160

Problem 5.161

The rocket shown has 7 lb of propellant with a burnout time (time required to burn all the fuel) of 7 s. Assume that the mass flow rate is constant and that the speed of the exhaust relative to the rocket is also constant and equal to 6500 ft/s. If the rocket is fired from rest, determine the initial weight of the rocket's body if the rocket is to experience an initial acceleration of $6g$.

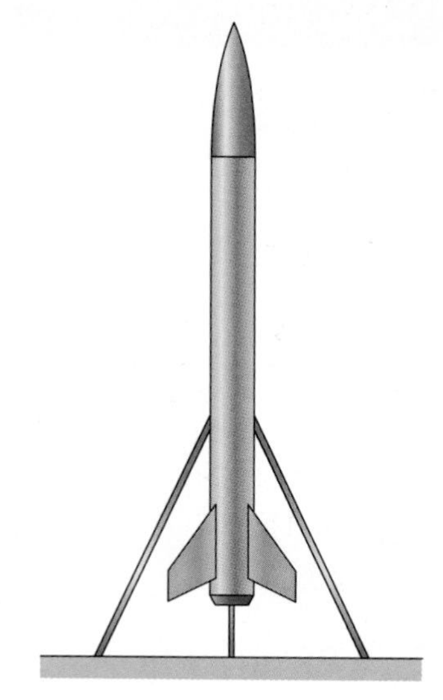

Figure P5.161

Problem 5.162

The tip B of a nozzle is 1.5 in. in diameter, whereas the diameter at A where the hose is attached is 3 in. If water flows through the nozzle at 200 gpm ("gpm" stands for gallons per minute; 1 U.S. gallon is defined as $231\,\mathrm{in.^3}$) and the water static pressure in the line is 300 psi, determine the force necessary to hold the nozzle stationary. Recall that the specific weight of water is $\gamma = 62.4\,\mathrm{lb/ft^3}$, and neglect the atmospheric pressure at B.

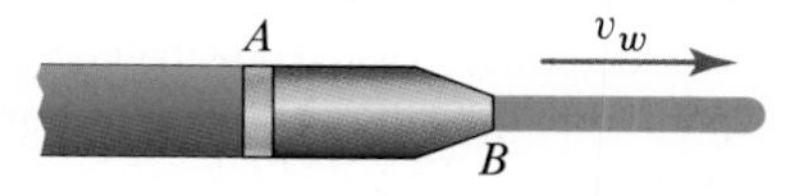

Figure P5.162

Problem 5.163

An intubed fan is mounted on a cart connected to a fixed wall by a linear elastic spring with constant $k = 50\,\mathrm{lb/ft}$. Assume that in a test the fan draws air at A with negligible speed and that the outgoing flow causes the cart to displace to the left so that the spring is stretched by 0.5 ft from its unstretched position. Assuming that the specific weight of the air $\gamma = 7.5\times10^{-2}\,\mathrm{lb/ft^3}$ is constant, and letting the diameter of the tube at B be $d = 4\,\mathrm{ft}$ (the cross section is assumed to be circular), determine the airspeed at B.

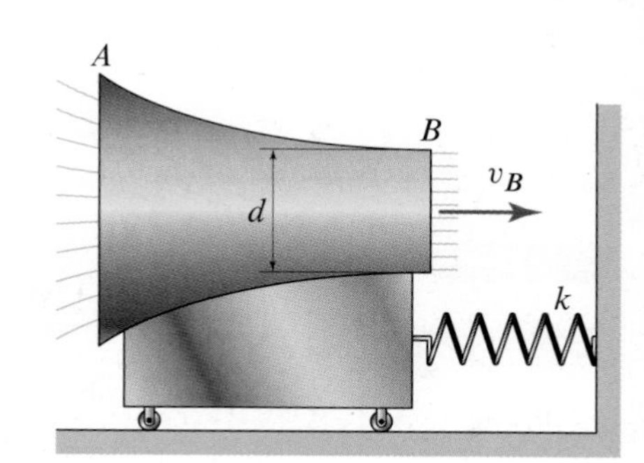

Figure P5.163

Problem 5.164

A test is conducted in which an 80 kg person sitting in a 15 kg cart is propelled by the jets emitted by two household fire extinguishers with a combined initial mass of 18 kg. The cross section of the exhaust nozzles is 3 cm in diameter, and the density of the exhaust is $\rho = 1.98\,\mathrm{kg/m^3}$. The vehicle starts from rest, and it is determined that the initial acceleration of the "jet cart" is $1.8\,\mathrm{m/s^2}$. Recalling that the mass flow rate out of the nozzle is given by $\dot{m}_o = \rho S v_o$, where S is the area of the nozzle cross section and v_o is the exhaust speed, determine v_o at the initial time. Ignore any resistance to the horizontal motion of the cart.

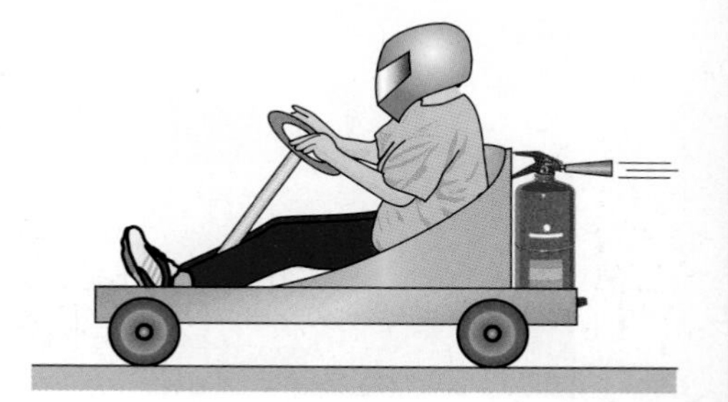

Figure P5.164

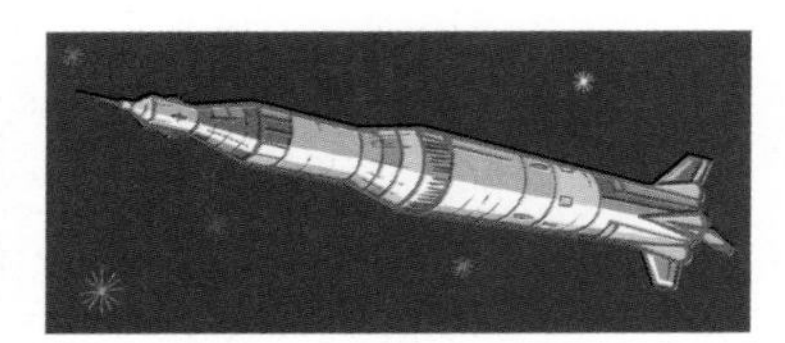

Figure P5.165

Problem 5.165

Consider a rocket in space so that it can be assumed that no external forces act on the rocket. Let v_o be the constant speed of the exhaust gases relative to the rocket. In addition, let $m_b + m_f$ be the total mass of the rocket and its fuel at the initial time, and let m_b be the mass of the body after all the fuel is burned. If the rocket is fired from rest, determine an expression for the maximum speed that the rocket can achieve.

Problem 5.166

A stationary 4 cm diameter nozzle emits a water jet with a speed of 30 m/s. The water jet impinges on a vane with a mass of 15 kg. Recalling that water has a mass density of $1000\,\text{kg/m}^3$, determine the minimum static friction coefficient with the ground such that the vane does not move if $\phi = 20°$ and $\theta = 30°$. Neglect the weight of the water layer in contact with the vane, as well as friction between the water and the vane.

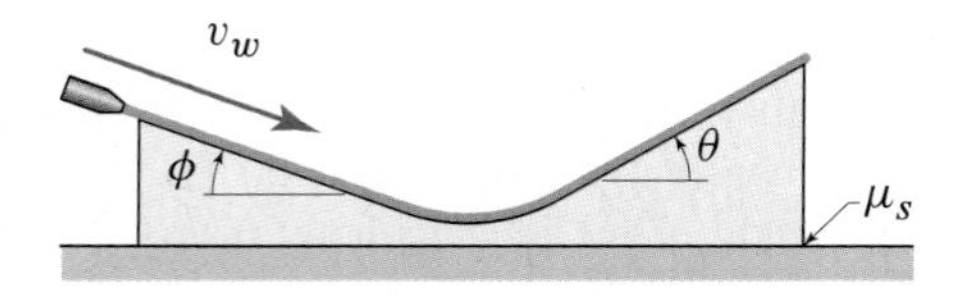

Figure P5.166

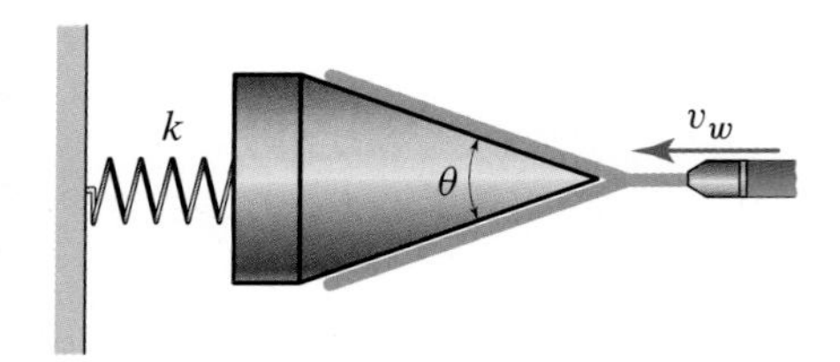

Figure P5.167

Problem 5.167

A diffuser is attached to a structure whose rigidity in the horizontal direction can be modeled by a linear spring with constant k. The diffuser is hit by a water jet issued with a speed $v_w = 55\,\text{ft/s}$ from a 2 in. diameter nozzle. Assume that the friction between the jet and the diffuser is negligible and that the diffuser's motion in the vertical direction can be neglected. Recalling that the specific weight of water is $\gamma = 62.4\,\text{lb/ft}^3$, if the opening angle of the diffuser is $\theta = 40°$, determine k such that the horizontal displacement of the diffuser does not exceed 0.25 in. from the diffuser's rest position. Assume that the water jet splits symmetrically over the diffuser.

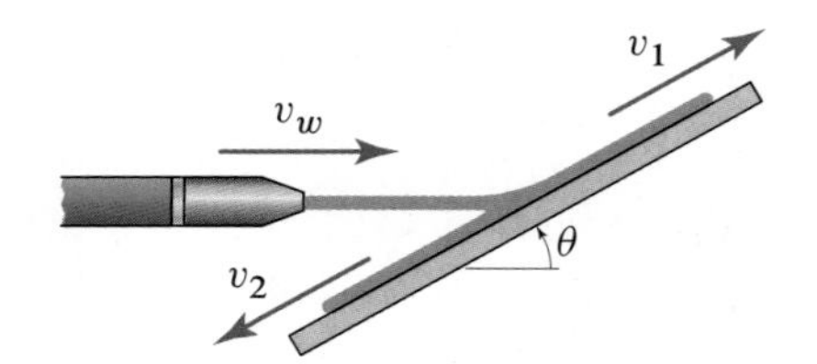

Figure P5.168

Problem 5.168

A water jet with a mass flow rate $\dot{m}_f$ at the nozzle impinges with a speed v_w on a fixed flat vane inclined at an angle θ with respect to the horizontal. Assuming that there is no friction between the water jet and the vane, the jet will split into two flows with mass flow rates $\dot{m}_{f1}$ and $\dot{m}_{f2}$. Neglecting the weight of the water, determine how $\dot{m}_{f1}$ and $\dot{m}_{f2}$ depend on $\dot{m}_f$, v_w, and θ. *Hint:* Due to the no-friction assumption, there is no force that slows down the water in the direction tangent to the vane, and this implies that the momentum in that direction is conserved.

Figure P5.169

Problem 5.169

A person wearing a jet pack lifts off from rest and ascends along a straight vertical trajectory. Let M denote the initial combined mass of the pilot and the equipment, including the fuel in the pack. Assume that the mass flow rate $\dot{m}_o$ and exhaust gas speed v_o are known constants and that the pilot can take off as soon as the rocket engine is started. If the engine exhaust is completely directed in the direction of gravity, determine the expression of the pilot's speed as a function of time, M, $\dot{m}_o$, v_o, and g (the acceleration due to gravity) while the pack is providing a thrust. Neglect air resistance and assume that gravity is constant.

Problem 5.170

A 28,000 lb A-10 Thunderbolt is flying at a constant speed of 375 mph when it fires a 4 s burst from its forward-facing seven-barrel Gatling gun. The gun fires 13.2 oz projectiles at a constant rate of 4200 rounds/min. The muzzle velocity of each projectile is 3250 ft/s. Assume that each of the plane's two jet engines maintains a constant thrust of 9000 lb, that the plane is subject to a constant air resistance while the gun is firing (equal to that before the burst), and that the plane flies straight and level during that time. Determine the plane's change in velocity at the end of the 4 s burst, modeling the airplane's change of mass due to firing as a continuous mass loss.

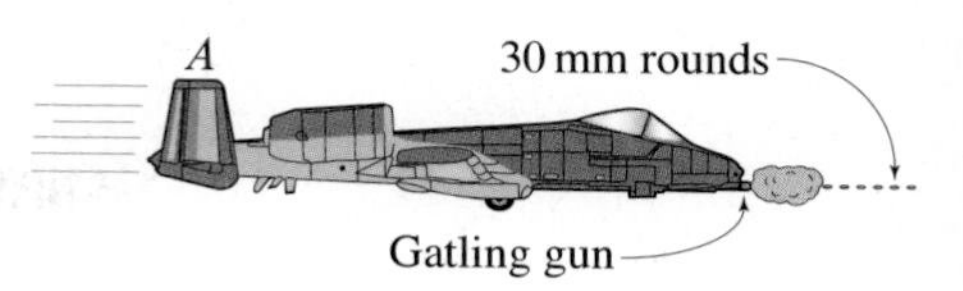

Figure P5.170

Problem 5.171

A faucet is letting out water at a rate of 15 L/min. Assume that the internal diameter d of the faucet is uniform and equal to 1.5 cm, the distance $\ell = 20$ cm, and the static water pressure at the wall is 0.30 MPa. Neglecting the weight of the water inside the faucet, as well as the weight of the faucet itself, determine the forces and the moment that the wall exerts on the faucet. Recall that the density of water is $\rho = 1000\,\mathrm{kg/m^3}$, and neglect the atmospheric pressure at the spout. *Hint:* Define your control volume using a section along the wall.

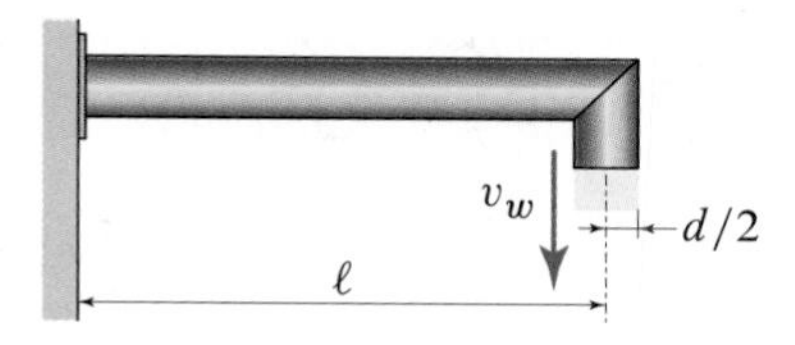

Figure P5.171

Problem 5.172

Consider a wind turbine with a diameter $d = 110$ m and the airflow streamlines shown, which are symmetric relative to the axis of the turbine. Since the airflow is tangent to the streamlines (by definition), these lines can be taken to define the top and bottom surfaces of a control volume. Suppose that pressure measurements indicate that the flow experiences atmospheric pressure at the cross sections A and B (as well as outside the control volume) where the wind speed is $v_A = 7$ m/s and $v_B = 2.5$ m/s, respectively. Furthermore, assume that the average pressure along the streamlines defining the control volume is also atmospheric. Finally, assume that O is on the line of action of the overall weight of the turbine and that the diameter of the flow cross section at A is 85% of the rotor diameter and that the rotor hub is at a distance $h = 75$ m above the ground. If the density of air is constant and equal to $\rho = 1.25\,\mathrm{kg/m^3}$, determine the force exerted by the air on the wind turbine and the reaction moment at the base of the support.

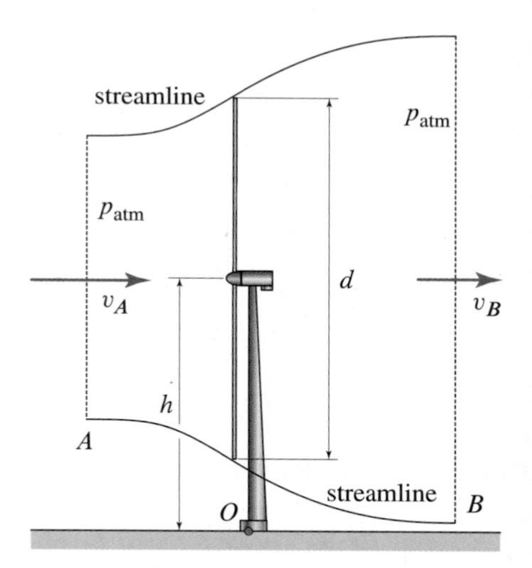

Figure P5.172

Problem 5.173

Let p_A and p_B be given static pressure measurements at the cross sections A and B in the air duct shown. Assume that any cross section between A and B is circular with diameter d. Assume that the flow is steady and that the mass density ρ_A at A is known along with v_A, the speed of the flow at A, and v_B, the speed of the flow at B. Determine the expression of mass density at B and the expression of the force F acting on the fan.

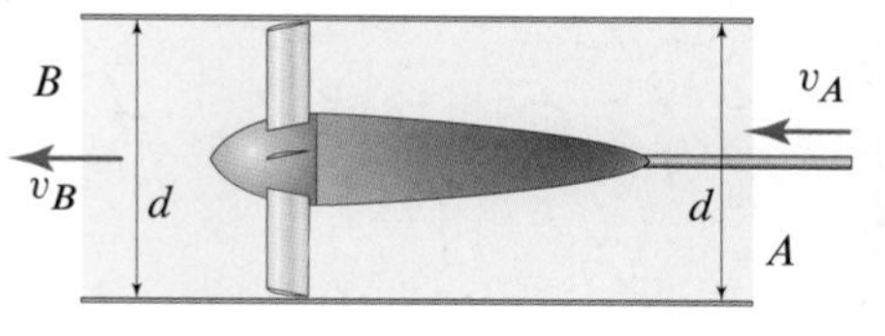

Figure P5.173

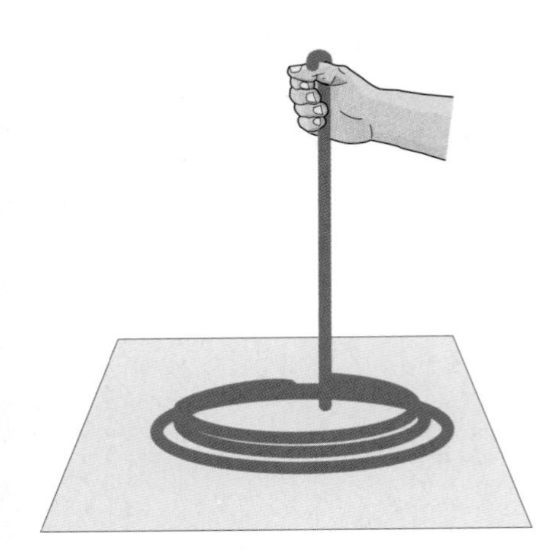
Figure P5.174–P5.176

Problem 5.174

A rope with weight per unit length of 0.1 lb/ft is lifted at a constant upward speed $v_0 = 8\,\mathrm{ft/s}$. Treating the rope as inextensible, determine the force applied to the top end of the rope after it is lifted 9 ft. Assume that the top end of the rope is initially at rest and on the floor. In addition, disregard the horizontal motion associated with the uncoiling of the rope.

Problem 5.175

A rope with mass per unit length of 0.05 kg/m is lifted at a constant upward acceleration $a_0 = 6\,\mathrm{m/s^2}$. Treating the rope as inextensible, determine the force that must be applied at the top end of the rope after it is lifted 3 m. Assume that the top end of the rope is initially at rest and on the floor. In addition, disregard the horizontal motion associated with the uncoiling of the rope.

Problem 5.176

A rope with mass per unit length of 0.05 kg/m is lifted by applying a constant vertical force $F = 10\,\mathrm{N}$. Treating the rope as inextensible, plot the velocity and position of the top end of the string as a function of time for $0 \le t \le 3\,\mathrm{s}$. Assume that the top end of the rope is initially at rest and 1 mm off the floor. In addition, disregard the horizontal motion associated with the uncoiling of the rope.

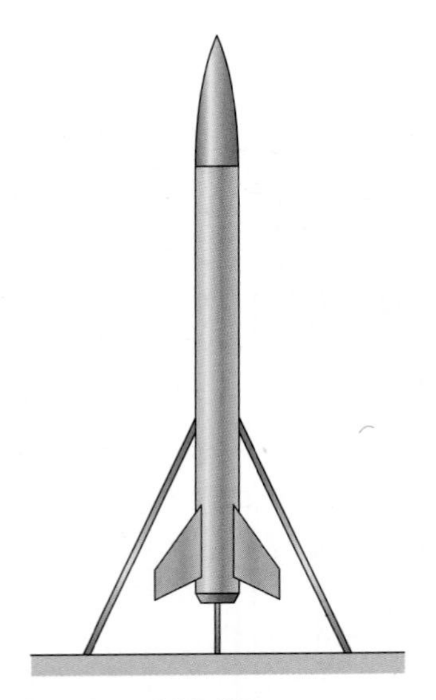
Figure P5.177 and P5.178

Problem 5.177

An amateur rocket with a body weight of 6.5 lb is equipped with a rocket engine holding 2.54 lb of solid propellant with a burnout time (time required to burn all the fuel) of 5.25 s (this is the typical data made available by amateur rocket engine manufacturers). The initial thrust is 68 lb. Assuming that the mass flow rate and the speed of the exhaust relative to the rocket remain constant, determine the exhaust mass flow rate $\dot{m}_o$ and the speed relative to the rocket v_o. In addition, determine the maximum speed achieved by the rocket $v_{\max}$ if the rocket is launched from rest. Neglect air resistance, and assume that gravity does not change with elevation.

Problem 5.178

Continue Prob. 5.177 and determine the maximum height reached by the rocket, again neglecting air resistance and changes of gravity with elevation. *Hint:* For $0 < t < t_0$,

$$\int \ln\left(1 - \frac{t}{t_0}\right) dt = (t_0 - t)\left[1 - \ln\left(1 - \frac{t}{t_0}\right)\right] + C.$$

Problem 5.179

A Pelton impulse wheel, as shown in Fig. P5.136(a), is typically found in hydroelectric power plants and consists of a wheel with a series of buckets attached at the periphery. As shown in Fig. P5.136(b), water jets impinge on the buckets and cause the wheel to spin about its axis (labeled O). Let v_w and $(\dot{m}_f)_{\mathrm{nz}}$ be the speed and the mass flow rate of the water jets at the nozzles (the nozzles are stationary), respectively. As the wheel spins, a given water jet will impinge on a given bucket only for a very small portion of the bucket's trajectory. This fact allows us to model the motion of a bucket relative to a given jet (during the time the bucket interacts with that jet) as essentially rectilinear and with constant relative speed, as was done in Example 5.17. Although each bucket moves away from the jet, the fact that they are arranged in a wheel results in an effective mass flow rate experienced by the vanes is $(\dot{m}_f)_{\mathrm{nz}}$ instead of the reduced mass flow rate computed in Eq. (6) of Example 5.17. With this in mind, consider a bucket, as shown in Fig. P5.136(c), that is moving with a speed v_0 horizontally away from a fixed nozzle, but subject to a mass flow rate $(\dot{m}_f)_{\mathrm{nz}}$. The inside of the bucket is shaped so as to redirect the water jet laterally out (away from the plane of the wheel). The angle θ describes the orientation of the velocity of the fluid relative to the (moving) bucket at B, the point at which the water leaves the bucket. Determine θ and v_0 such that the power transmitted by the water to the wheel is maximum. Express v_0 in terms of v_w.

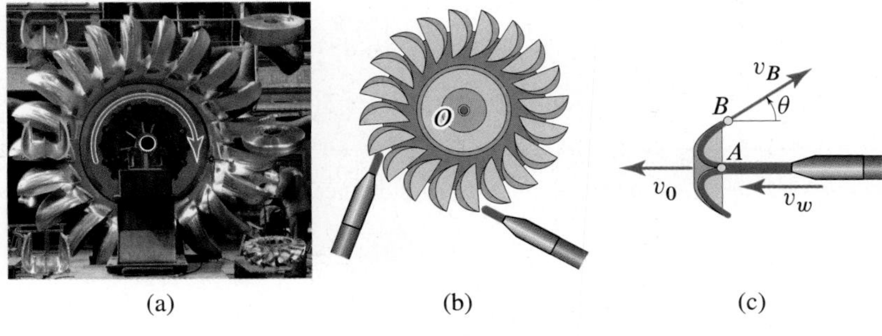

Figure P5.179

Chapter Review

Helpful Information

When should you use the impulse-momentum principle? The impulse-momentum principle provides a natural approach to problems in which you need to relate velocity, force, and time, since it relates forces acting over time to changes in momentum.

Momentum and impulse

We learned in Chapter 3 that forces lead to changes in velocities since forces cause accelerations. We began this section by learning that forces acting over time change momentum (not just velocity). By integrating Newton's second law, we obtained the *impulse-momentum principle*, which is given by

Eq. (5.6), p. 314

$$\vec{p}(t_1) + \int_{t_1}^{t_2} \vec{F}(t)\,dt = \vec{p}(t_2),$$

where the *linear momentum* (or *momentum*) was defined to be

Eq. (5.3), p. 314

$$\vec{p}(t) = m\vec{v}(t),$$

and a force acting over some time interval was called the *impulse* (or *linear impulse*) and is given by

Eq. (5.5), p. 314

$$\int_{t_1}^{t_2} \vec{F}(t)\,dt.$$

We also found that without detailed knowledge of the force acting on a particle at every instant in time, we could not determine the change in momentum. On the other hand, knowing just the change in momentum allows us to determine the *average force* acting on a particle during the corresponding time interval, that is,

Eq. (5.9), p. 315

$$\vec{F}_{\text{avg}} = \frac{\vec{p}(t_2) - \vec{p}(t_1)}{t_2 - t_1}.$$

Impulse-momentum principle for systems of particles. For closed systems of particles, we discovered that we can write the impulse-momentum principle as

Eqs. (5.12) and (5.15), p. 316

$$\vec{F} = \dot{\vec{p}} \quad \text{and} \quad \int_{t_1}^{t_2} \vec{F}(t)\,dt = \vec{p}(t_2) - \vec{p}(t_1),$$

where $\vec{F}$ is the total external force on the particle system and $\vec{p} = \sum_{i=1}^{N} m_i \vec{v}_i$ is the total momentum of the system of particles. Using the definition of the mass center of a system of particles, the impulse-momentum principle can also be written as

Eq. (5.14), p. 316

$$\vec{F} = \frac{d}{dt}(m\vec{v}_G) = m\vec{a}_G,$$

where m is the total mass of the system of particles, $\vec{v}_G$ is the velocity of its mass center, and $\vec{a}_G$ is the acceleration of its mass center.

Conservation of linear momentum. When there is a direction in which the external force on a system of particles is zero, then the momentum in that direction is constant and is said to be conserved. If the total external force on a system of particles is zero, that is, $\vec{F} = 0$, then the momentum in every direction is constant, and the mass center of the system of particles will move with constant velocity.

Impact

In this section we idealized particle impact as *an event spanning an infinitesimal time interval in which objects can experience a finite change in velocity at fixed position*. The model is based on the assumptions summarized in Table 5.3. We discovered that there are two key elements to *every* impact problem: (1) the application of the impulse-momentum principle and (2) a force law telling us how the colliding objects rebound. When applying the impulse-momentum principle during an impact, only impulsive forces play a role, so they are the only forces included in impact-relevant FBDs.

Problems involving the impact between two particles generally involve four unknowns, so four equations are needed. The geometry of an *unconstrained impact* (for which there are no external impulsive forces) between two particles is shown in Fig. 5.46, and the four equations come from

Table 5.3
Assumptions used in our impact model.

Physical characteristic	Impact assumption
duration of impact	infinitesimal
displacement of particle	zero
force on particle	infinite
change in momentum	instantaneous

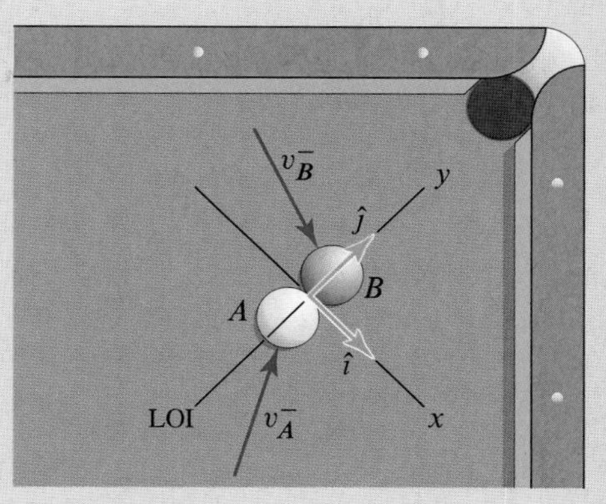

Figure 5.46
Geometry of two impacting particles.

1. Conservation of momentum of the two particles together along the LOI:

Eq. (5.25), p. 338

$$m_A v_{Ay}^- + m_B v_{By}^- = m_A v_{Ay}^+ + m_B v_{By}^+.$$

2. Conservation of momentum for particle A in the x direction:

Eq. (5.26), p. 338

$$m_A v_{Ax}^- = m_A v_{Ax}^+ \quad \Rightarrow \quad v_{Ax}^- = v_{Ax}^+.$$

3. Conservation of momentum for particle B in the x direction:

Eq. (5.27), p. 338

$$m_B v_{Bx}^- = m_B v_{Bx}^+ \quad \Rightarrow \quad v_{Bx}^- = v_{Bx}^+.$$

4. COR equation applied along the LOI:

Eq. (5.21), p. 337

$$e = \frac{\text{separation velocity}}{\text{approach velocity}} = \frac{v_{By}^+ - v_{Ay}^+}{v_{Ay}^- - v_{By}^-}.$$

The coefficient of restitution e determines the nature of the rebound between the two particles. When $e = 0$, the impact is called *plastic*; when $0 < e < 1$, the impact is called *elastic*; and when $e = 1$, it is called *perfectly elastic*. In an unconstrained, direct central impact, a plastic collision ($e = 0$) results in the objects sticking together postimpact. This is not necessarily the case in an oblique central impact as velocity components along the plane of contact may be different.

Impact and energy. If $e < 1$, mechanical energy is lost during the impact. The energy loss is often indicated as a percentage of the preimpact total kinetic energy:

Eq. (5.29), p. 339

$$\text{Percentage of energy loss} = \frac{T^- - T^+}{T^-} \times 100\%,$$

where T^- and T^+ are the pre- and postimpact total kinetic energies, respectively.

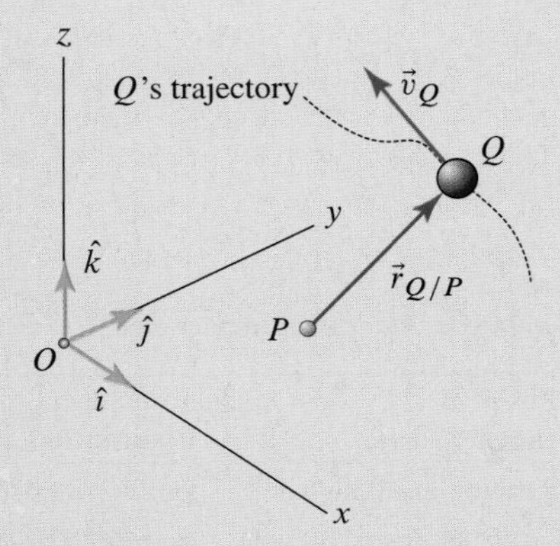

Figure 5.47
A particle Q in motion relative to a point P. Point P need *not* be a stationary point.

Angular momentum

In this section, we developed the concept of angular momentum for a single particle and for systems of particles. We also derived the *moment-angular momentum relations* for a particle and for a system of particles.

Definition of angular momentum of a particle. Referring to Fig. 5.47, the *angular momentum of a particle Q with respect to the moment center P* is given by

Eq. (5.31), p. 361

$$\vec{h}_P = \vec{r}_{Q/P} \times \vec{p}_Q = \vec{r}_{Q/P} \times m\vec{v}_Q,$$

where P can be fixed or moving; $\vec{r}_{Q/P}$ is the position of Q relative to P; m and $\vec{v}_Q$ are the mass and the velocity of Q, respectively; and $\vec{p}_Q = m\vec{v}_Q$ is the linear momentum of Q.

Moment-angular momentum relation for a particle. The *moment-angular momentum relation* for a single particle is given by

Eq. (5.35), p. 361

$$\vec{M}_P = \dot{\vec{h}}_P + \vec{v}_P \times m\vec{v}_Q,$$

where $\vec{M}_P$ is the moment with respect to P of all the forces acting on Q. If either one of the following conditions is satisfied:

1. The reference point P is fixed, i.e., if $\vec{v}_P = \vec{0}$.
2. $\vec{v}_P$ is parallel to $\vec{v}_Q$, i.e., $\vec{v}_P \times m\vec{v}_Q = \vec{0}$.

the moment-angular momentum relation for a single particle can be simplified to

Eq. (5.36), p. 361

$$\vec{M}_P = \dot{\vec{h}}_P.$$

If either condition (1) or (2) is satisfied for $t_1 \leq t \leq t_2$, then the moment-angular momentum relation for a single particle can be integrated with respect to time to obtain the angular impulse-momentum principle as

Eq. (5.37), p. 362

$$\vec{h}_{P1} + \int_{t_1}^{t_2} \vec{M}_P \, dt = \vec{h}_{P2},$$

where $\vec{h}_{P1} = \vec{h}_P(t_1)$ and $\vec{h}_{P2} = \vec{h}_P(t_2)$.

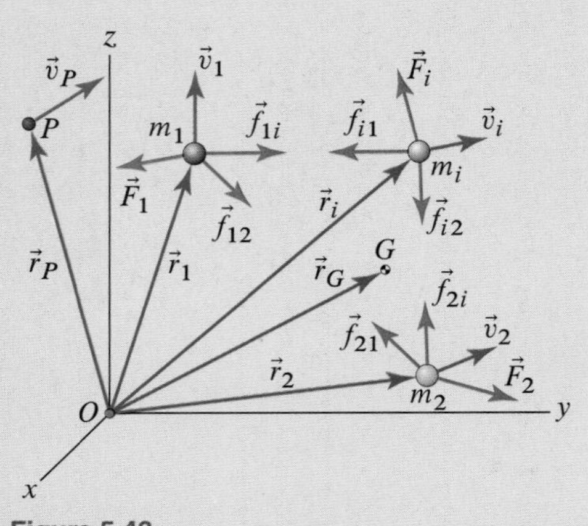

Figure 5.48
A system of particles under the action of internal and external forces. Point P, in general, is a moving point.

Moment-angular momentum relation for a system of particles. For a closed system of particles, the moment-angular momentum relation can be given the form

Eq. (5.49), p. 363

$$\vec{M}_P = \dot{\vec{h}}_P + \vec{v}_P \times m\vec{v}_G.$$

Here, with reference to Fig. 5.48, $\vec{M}_P$ is the moment with respect to P of *only the external forces* acting on the system, $m = \sum_{i=1}^{N} m_i$ is the total mass of the system, G is the system's center of mass, $\vec{v}_G$ is the velocity of G, and $\vec{h}_P$ is the total angular momentum, which is defined as

Eq. (5.48), p. 363

$$\vec{h}_P = \sum_{i=1}^{N} \vec{h}_{Pi}.$$

If any one of the following conditions is satisfied:

1. P is a fixed point, i.e., when $\vec{v}_P = \vec{0}$.
2. G is a fixed point, i.e., when $\vec{v}_G = \vec{0}$.
3. P coincides with the center of mass and therefore $\vec{v}_P = \vec{v}_G$.
4. Vectors $\vec{v}_P$ and $\vec{v}_G$ are parallel.

then the moment-angular momentum relation for a closed system of particles simplifies to

Eq. (5.50), p. 364

$$\vec{M}_P = \dot{\vec{h}}_P.$$

Furthermore, if any of the above conditions are satisfied over a time interval $t_1 \le t \le t_2$, then the moment-angular momentum relation for a closed system of particles can be integrated with respect to time to obtain the angular impulse-momentum principle as

Eq. (5.51), p. 364

$$\vec{h}_{P1} + \int_{t_1}^{t_2} \vec{M}_P \, dt = \vec{h}_{P2}.$$

Orbital mechanics

In orbital mechanics, we studied the motion of a satellite S of mass m that is subject to Newton's universal law of gravitation due to a body B of mass m_B, which is the primary or attracting body (see Fig. 5.49). We began with these important assumptions:

1. The primary body B and the satellite S are both treated as particles.
2. The only force acting on the satellite S is the force of mutual attraction between B and S.
3. The primary body B is fixed in space.

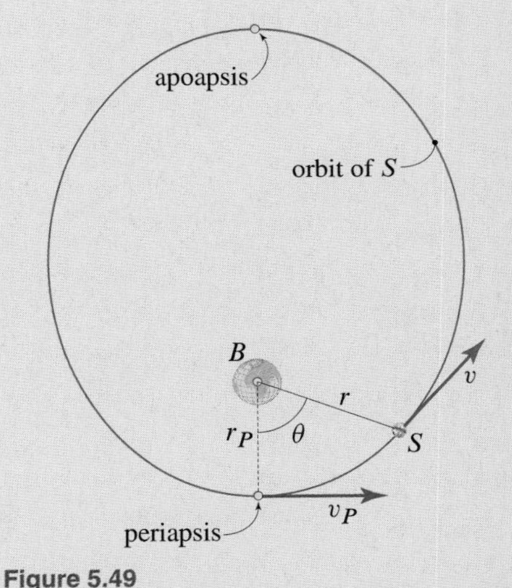

Figure 5.49
A satellite S of mass m orbiting a primary body B of mass m_B on a conic section, which in this case is an ellipse. The angle θ and the launch conditions r_P and v_P are defined from periapsis.

Determination of the orbit. Solving the governing equations in polar coordinates, we found that the trajectory of the satellite S under these assumptions is a conic section, whose equation can be written as

Eqs. (5.73) and (5.78), p. 387

$$\frac{1}{r} = C\cos\theta + \frac{Gm_B}{\kappa^2} \quad \text{or}$$

$$\frac{1}{r} = \frac{Gm_B}{\kappa^2}(1 + e\cos\theta),$$

where r is the distance between the centers of mass of S and B; θ is the orbital angle measured relative to periapsis; G is the universal gravitational constant; κ is the angular momentum per unit mass of the satellite S measured about B; C is a constant to be determined; and e is the *eccentricity* of the trajectory, which can be written as

Eq. (5.77), p. 387

$$e = \frac{C\kappa^2}{Gm_B}.$$

If, as is generally the case here, the orbital conditions are known at periapsis, then κ is

Eq. (5.79), p. 388

$$\kappa = r_P v_P,$$

the constant C is

Eq. (5.80), p. 388

$$C = \frac{1}{r_P}\left(1 - \frac{Gm_B}{r_P v_P^2}\right),$$

and the equation describing the trajectory becomes

Eq. (5.81), p. 388

$$\frac{1}{r} = \frac{1}{r_P}\left(1 - \frac{Gm_B}{r_P v_P^2}\right)\cos\theta + \frac{Gm_B}{r_P^2 v_P^2}.$$

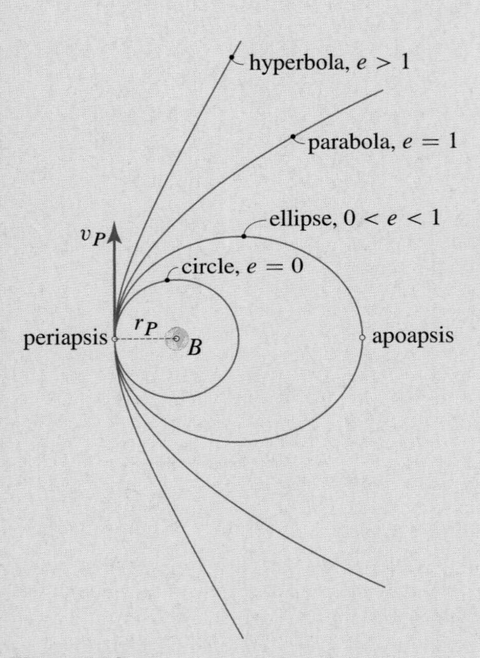

Figure 5.50
The four conic sections, showing r_P and v_P.

Conic sections. Equations (5.73), (5.78), and (5.81) represent equivalent conic sections in polar coordinates. The type of conic sections depends on the eccentricity of the trajectory e (see Fig. 5.50), which, in turn, depends on C and κ via Eq. (5.77). There are four types of conic sections, which are determined by the value of the eccentricity e, that is, $e = 0$, $0 < e < 1$, $e = 1$, and $e > 1$.

For a *circular orbit* ($e = 0$), the radius is r_P and the speed in the orbit is equal to

Eq. (5.82), p. 388

$$v_c = \sqrt{\frac{Gm_B}{r_P}}.$$

For an *elliptical orbit* ($0 < e < 1$), as expected, the radius at periapsis is r_P. The radius at apoapsis is given by

Eq. (5.84), p. 388

$$r_A = \frac{r_P}{2Gm_B/(r_P v_P^2) - 1}.$$

Referring to Fig. 5.51, additional relationships between the semimajor axis a of an elliptical orbit, the eccentricity of the orbit, and the radii at periapsis and apoapsis are, respectively,

Eqs. (5.90), p. 389

$$r_P = a(1 - e) \quad \text{and} \quad r_A = a(1 + e).$$

The period of an elliptical orbit τ can be written in the following two ways:

Eqs. (5.92) and (5.94), p. 390

$$\tau = \frac{2\pi ab}{\kappa} = \frac{\pi}{\kappa}(r_P + r_A)\sqrt{r_P r_A},$$

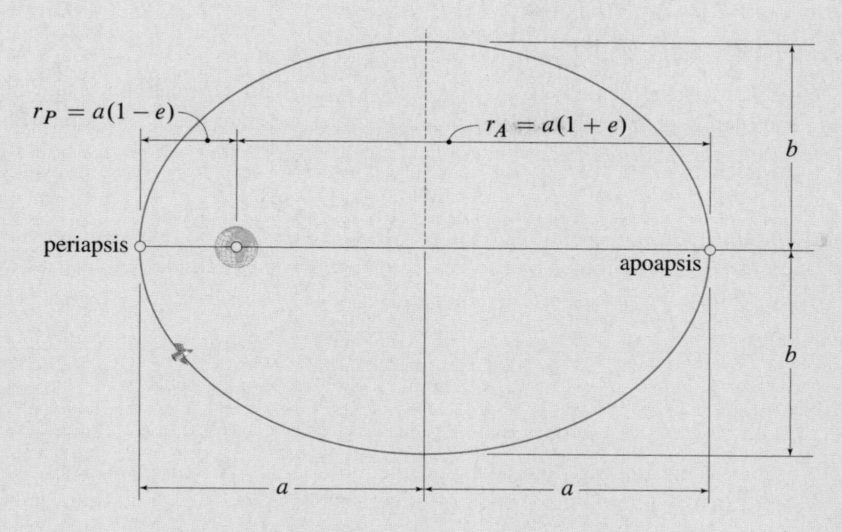

Figure 5.51. The semimajor axis a and semiminor axis b of an ellipse.

or, reflecting Kepler's third law, the orbital period can be written as

Eq. (5.97), p. 390

$$\tau^2 = \frac{4\pi^2}{Gm_B}a^3.$$

A *parabolic trajectory* ($e = 1$) is one that divides periodic orbits (which return to their starting location) from trajectories that are not periodic. For a given r_P, the speed v_{par} required to achieve a parabolic trajectory is given by

Eq. (5.99), p. 390

$$v_{\text{par}} = v_{\text{esc}} = \sqrt{\frac{2Gm_B}{r_P}},$$

which is also referred to as the *escape velocity* since it is the speed required to completely escape the influence of the primary body B.

For a *hyperbolic trajectory* ($e > 1$), the governing equations given by Eqs. (5.77)–(5.81) are used for values of $e > 1$.

Energy considerations. Using the work-energy principle, we discovered that the total mechanical energy in an orbit depends on only the semimajor axis a of the orbit and is given by

Eq. (5.103), p. 391

$$E = -\frac{Gm_B}{2a},$$

where E is the *mechanical energy per unit mass* of the satellite. Applying the work-energy principle at an arbitrary location within the orbit, we found that the speed can be written as

Eq. (5.105), p. 391

$$v = \sqrt{Gm_B\left(\frac{2}{r} - \frac{1}{a}\right)}.$$

Mass flows

In Section 5.5, we considered mass flows. We considered (1) *steady* mass flows, in which a fluid moves through a conduit with a velocity that depends only on the position within the conduit; and (2) *variable* mass flows, such as the flow of combustion gases out of a rocket.

Steady flows. Given the control volume (CV) shown in Fig. 5.52 (a *control volume* is a portion of a conduit delimited by two cross sections), we showed that, in the case of a steady flow, the total external force $\vec{F}$ acting on the fluid in the CV is

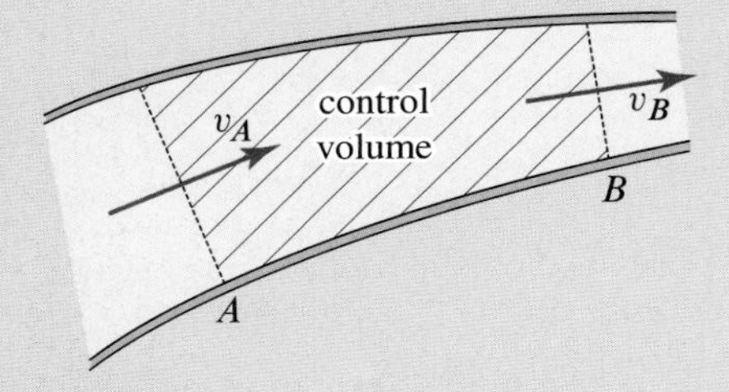

Figure 5.52
A CV corresponding to the portion of a pipe between cross sections A and B.

Eq. (5.111), p. 402

$$\vec{F} = \dot{m}_f(\vec{v}_B - \vec{v}_A),$$

where, as long as the cross sections are perpendicular to the flow velocity, $\dot{m}_f$ is the *mass flow rate*, i.e., the amount of mass flowing through a cross section per unit time, and $\vec{v}_A$ and $\vec{v}_B$ are the flow velocities at the cross sections A and B, respectively. In addition to the mass flow rate, we defined the *volumetric flow rate* as the quantity

Eq. (5.115), p. 403

$$Q = vS,$$

where v is the speed of the fluid at a given cross section and S is the area of the cross section in question. We showed that

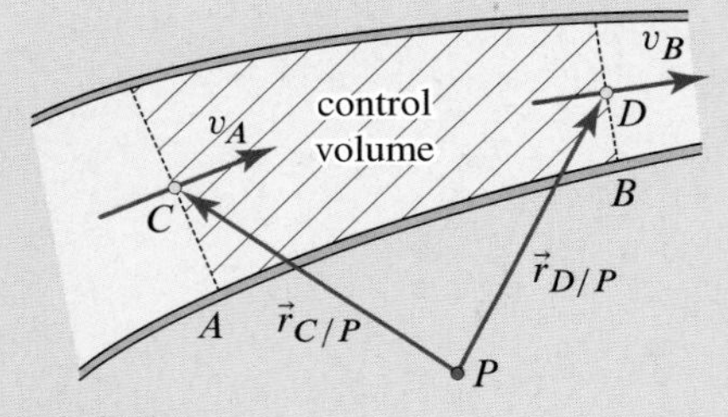

Figure 5.53
A fluid flowing through a CV along with a choice of moment center P for the calculation of angular momenta and moments.

Eq. (5.116), p. 403

$$\dot{m}_f = \rho_A Q_A = \rho_B Q_B,$$

where ρ_A and ρ_B are the values of the mass density of the fluid at A and B, respectively. Referring to Fig. 5.53, we also showed that, given a fixed point P, the total moment $\vec{M}_P$ acting on the fluid in the CV is

Eq. (5.121), p. 404

$$\vec{M}_P = \dot{m}_f(\vec{r}_{D/P} \times \vec{v}_B - \vec{r}_{C/P} \times \vec{v}_A),$$

where C and D are the centers of the cross sections A and B, respectively.

Variable mass flows. With reference to Fig. 5.54, for a body with time varying mass $m(t)$ due to an inflow of mass with rate $\dot{m}_i$ and an outflow of mass with the rate $\dot{m}_o$, the total external force acting on the body is given by

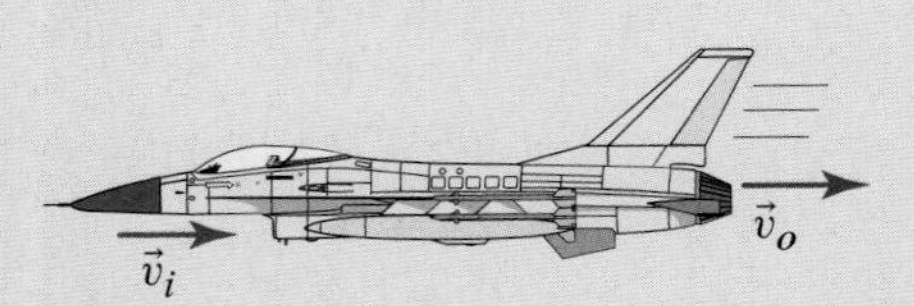

Figure 5.54
A plane with a jet engine. The airplane is taking in air with a mass flow rate $\dot{m}_i$, while combustion gases are ejected with a mass flow rate $\dot{m}_o$. The vector $\vec{v}_i$ is the velocity of the inflowing air *relative* to the plane. The vector $\vec{v}_o$ is the velocity of the outflowing combustion gases *relative* to the plane.

Eq. (5.132), p. 405

$$\vec{F} = m\vec{a} + \dot{m}_o\vec{v}_o - \dot{m}_i\vec{v}_i,$$

where $\vec{a}$ is the acceleration of the main body, $\vec{v}_o$ is the *relative* velocity of the outflowing mass with respect to the main body, and $\vec{v}_i$ is the *relative* velocity of the inflowing mass, again relative to the main body.

Review Problems

Problem 5.180

In Major League Baseball, a pitched ball has been known to hit the head of the batter (sometimes unintentionally and sometimes not). Let the pitcher be, for example, Nolan Ryan who can throw a $5\frac{1}{8}$ oz baseball that crosses the plate at 100 mph.* Studies have shown that the impact of a baseball with a person's head has a duration of about 1 ms. So using Eq. (5.9) on p. 315 and assuming that the rebound speed of the ball after the collision is negligible, determine the magnitude of the average force exerted on the person's head during the impact.

Problem 5.181

A 0.6 kg ball that is initially at rest is dropped on the floor from a height of 1.8 m and has a rebound height of 1.25 m. If the ball spends a total of 0.01 s in contact with the ground, determine the average force applied to the ball by the ground during the rebound. In addition, determine the ratio between the magnitude of the impulse provided to the ball by the ground and the magnitude of the impulse provided to the ball by gravity during the time interval that the ball is in contact with the ground. Neglect air resistance.

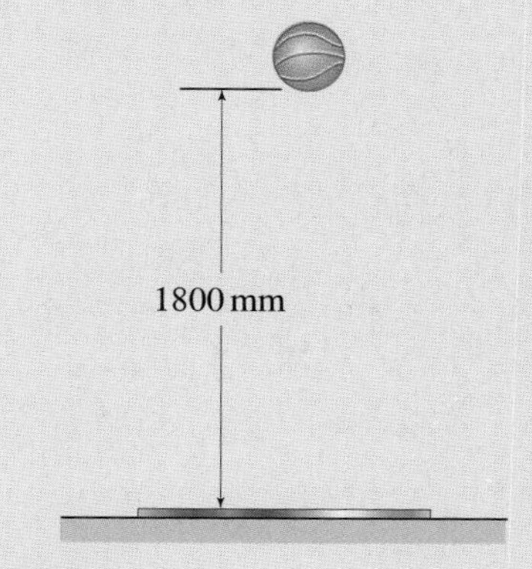

Figure P5.181

Problem 5.182

A person P is initially standing on a cart on rails, which is moving to the right with a speed $v_0 = 2\,\text{m/s}$. The cart is not being propelled by any motor. The combined mass of person P, the cart, and all that is being carried on the cart is 270 kg. At some point a person P_A standing on a stationary platform throws to person P a package A to the right with a mass $m_A = 50\,\text{kg}$. Package A is received by P with a horizontal speed $v_{A/P} = 1.5\,\text{m/s}$. After receiving the package from A, person P throws a package B with a mass $m_B = 45\,\text{kg}$ toward a second person P_B. The package intended for P_B is thrown to the right, i.e., in the direction of the motion of P, and with a horizontal speed $v_{B/P} = 4\,\text{m/s}$ relative to P. Determine the final velocity of the person P. Neglect any friction or air resistance acting on P and the cart.

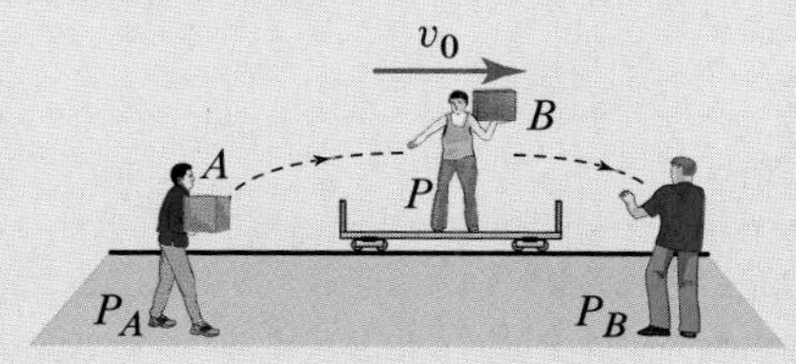

Figure P5.182

Problem 5.183

A Ford Excursion A, with a mass $m_A = 3900\,\text{kg}$, traveling with a speed $v_A = 85\,\text{km/h}$, collides head-on with a Mini Cooper B, with a mass $m_B = 1200\,\text{kg}$, traveling in the opposite direction with a speed $v_B = 40\,\text{km/h}$. Determine the postimpact velocities of the two cars if the impact's coefficient of restitution is $e = 0.22$. In addition, determine the percentage of kinetic energy loss.

Figure P5.183

* This means that he must have thrown the ball at about 108 mph since the ball loses speed at the rate of 1 mph for every 7 ft it travels.

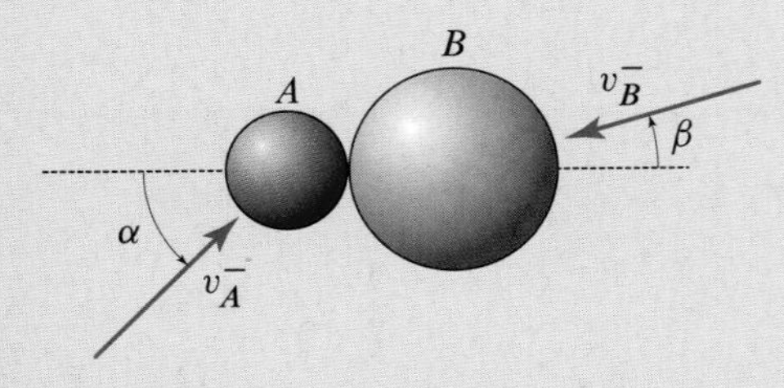

Figure P5.184

Problem 5.184

The two spheres, A and B, with masses $m_A = 1.35\,\text{kg}$ and $m_B = 2.72\,\text{kg}$, respectively, collide with $v_A^- = 26.2\,\text{m/s}$ and $v_B^- = 22.5\,\text{m/s}$. Let $\alpha = 45°$. Compute the value of β if the component of the postimpact velocity of B along the LOI is equal to zero and if the COR is $e = 0.63$.

Problem 5.185

A 31,000 lb truck A and a 3970 lb sports car B collide at an intersection. At the moment of the collision, the truck and the sports car are traveling with speeds $v_A^- = 60\,\text{mph}$ and $v_B^- = 50\,\text{mph}$, respectively. Assume that the entire intersection forms a horizontal surface. Letting the line of impact be parallel to the ground and rotated counterclockwise by $\alpha = 20°$ with respect to the preimpact velocity of the truck, determine the postimpact velocities of A and B if the contact between A and B is frictionless and the COR $e = 0.1$. Furthermore, assuming that the truck and the car slide after impact and that the coefficient of kinetic friction is $\mu_k = 0.7$, determine the position at which A and B come to a stop relative to the position they occupied at the instant of impact.

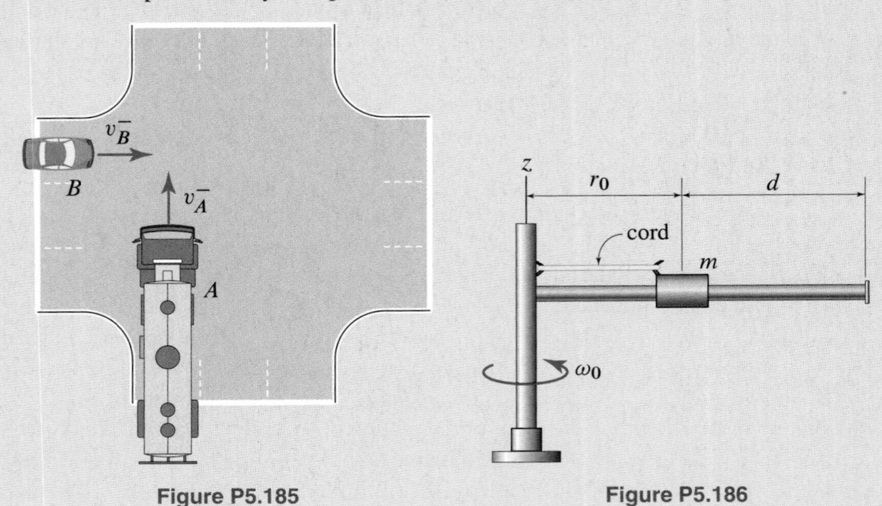

Figure P5.185

Figure P5.186

Problem 5.186

Consider a collar with mass m that is free to slide with no friction along a rotating arm of negligible mass. The system is initially rotating with a constant angular velocity ω_0 while the collar is kept at a distance r_0 from the z axis. At some point, the restraint keeping the collar in place is removed so that the collar is allowed to slide. Determine the expression for the moment that you need to apply to the arm, as a function of time, to keep the arm rotating at a constant angular velocity while the collar travels toward the end of the arm. *Hint:*

$$\int \frac{1}{\sqrt{x^2-1}}\,dx = \ln\left(x + \sqrt{x^2-1}\right) + C.$$

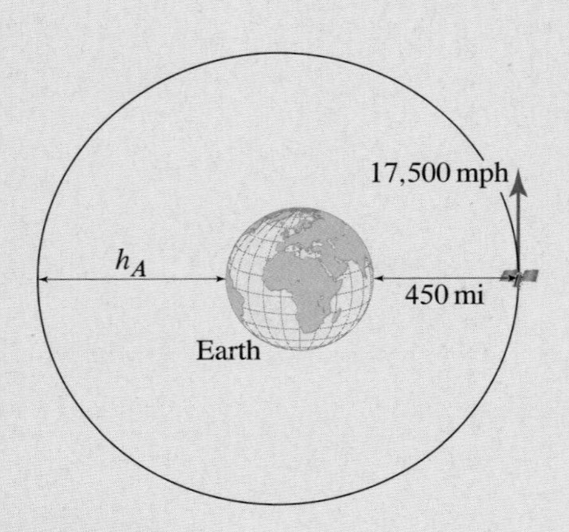

Figure P5.187

Problem 5.187

A satellite is launched parallel to the Earth's surface at an altitude of 450 mi with a speed of 17,500 mph. Determine the apogee altitude h_A above the Earth's surface, as well as the period of the satellite.

Problem 5.188

A spacecraft is traveling at 19,000 mph parallel to the surface of the Earth at an altitude of 250 mi, when it fires a retrorocket to transfer to a different orbit. Determine the change in speed Δv necessary for the spacecraft to reach a minimum altitude of 110 mi during the ensuing orbit. Assume that the change in speed is impulsive; that is, it occurs instantaneously.

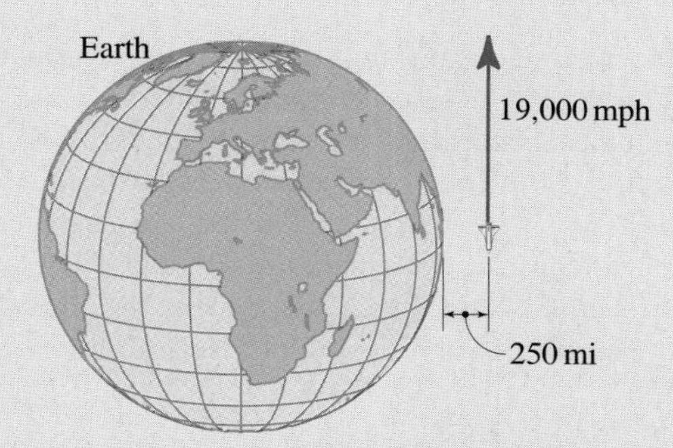

Figure P5.188

Problems 5.189 and 5.190

The optimal way (from an energy standpoint) to transfer from one circular orbit about a primary body (in this case, the Sun) to another circular orbit is via the *Hohmann transfer*, which involves transferring from one circular orbit to another using an elliptical orbit that is tangent to both at the periapsis and apoapsis of the ellipse. This ellipse is uniquely defined because we know the perihelion radius r_e (the radius of the inner circular orbit) and the aphelion radius r_j (the radius of the outer circular orbit), and therefore we know the semimajor axis a via Eq. (5.88) and the eccentricity e via Eq. (5.87) or Eqs. (5.90). Performing a Hohmann transfer requires two maneuvers, the first to leave the inner (outer) circular orbit and enter the transfer ellipse and the second to leave the transfer ellipse and enter the outer (inner) circular orbit. Assume that the orbits of Earth and Jupiter are circular, use 150×10^6 km for the radius of Earth's orbit, use 779×10^6 km for the radius of Jupiter's orbit, and note that the mass of the Sun is 333,000 times that of the Earth.

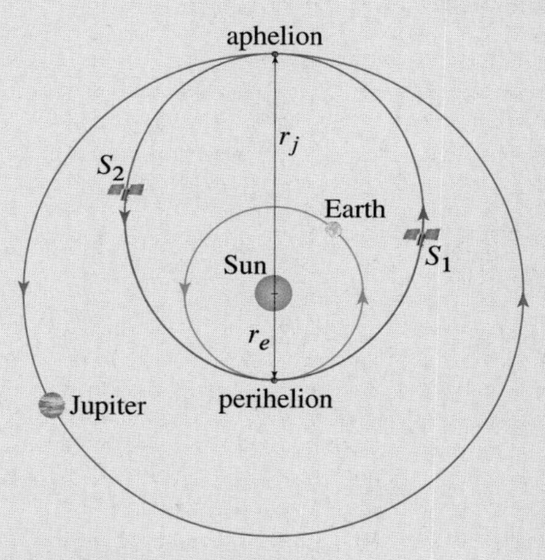

Figure P5.189 and P5.190

Problem 5.189 A space probe S_1 is launched from Earth to Jupiter via a Hohmann transfer orbit. Determine the change in speed Δv_e required at the radius of Earth's orbit of the elliptical transfer orbit (perihelion) and the change in speed Δv_j required at the radius of Jupiter's orbit (aphelion). In addition, compute the time required for the orbital transfer. Assume that the changes in speed are impulsive; that is, they occur instantaneously.

Problem 5.190 A space probe S_2 is at Jupiter and is required to return to the radius of Earth's orbit about the Sun so that it can return samples taken from one of Jupiter's moons. Assuming that the mass of the probe is 722 kg, determine the change in kinetic energy required at Jupiter ΔT_j for the maneuver at aphelion. In addition, determine the change in kinetic energy required at Earth ΔT_e for the perihelion maneuver. Finally, what is the change in potential energy ΔV of the spacecraft in going from Jupiter to the Earth?

Problem 5.191

A water jet is emitted from a nozzle attached to the ground. The jet has a constant mass flow rate $(\dot{m}_f)_{\text{nz}} = 15$ kg/s and a speed v_w relative to the nozzle. The jet strikes a 12 kg incline and causes it to slide at a constant speed $v_0 = 2$ m/s. The kinetic coefficient of friction between the incline and the ground is $\mu_k = 0.25$. Neglecting the effect of gravity and air resistance on the water flow, as well as friction between the water jet and the incline, determine the speed of the water jet at the nozzle if $\theta = 47°$.

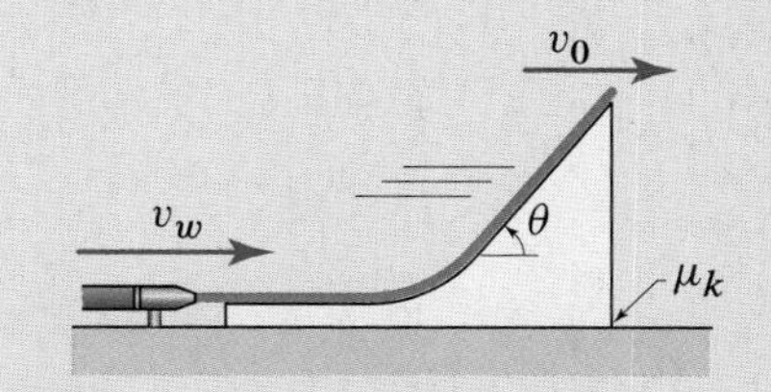

Figure P5.191

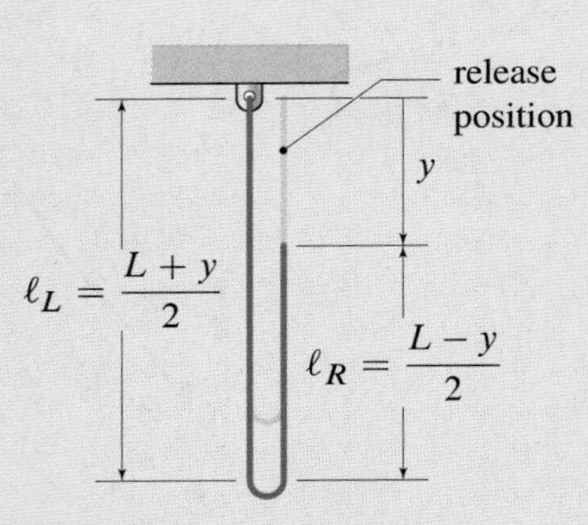

Figure P5.192

Problem 5.192

Revisit Example 5.20 and derive the equation of motion of the free end of the string starting from the force balance for the right branch of the string modeled as a variable mass system.

Problem 5.193

An intubed fan (a fan rotating within a tube or other conduit) is mounted on a cart that is connected to a fixed wall by a linear elastic spring with constant $k = 70\,\mathrm{N/m}$. Assume that in a particular test the fan draws air that enters the tube at A with a speed v_A. The outgoing flow at B has a speed v_B. The flow of air through the tube causes the cart to displace to the left so that the spring is stretched by 0.25 m from its unstretched position. Assume that the density of air is constant throughout the tube and equal to $\rho = 1.25\,\mathrm{kg/m^3}$. In addition, let the tube's cross section be circular, and let the cross-sectional diameters at A and B be $d_A = 3\,\mathrm{m}$ and $d_B = 1.5\,\mathrm{m}$, respectively. Determine the velocities of the airflow at A and B.

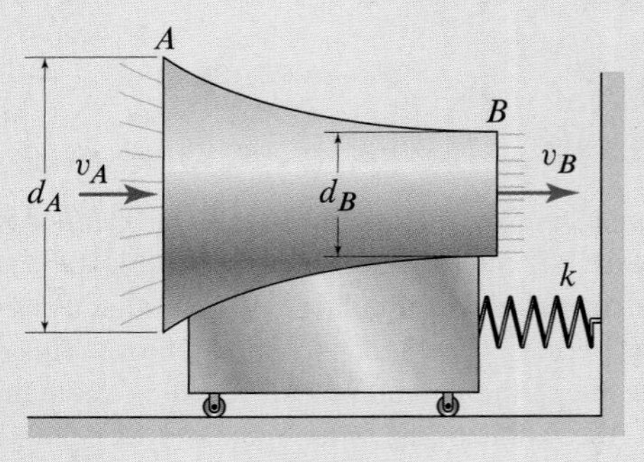

Figure P5.193

Planar Rigid Body Kinematics

6

The Falkirk Wheel in central Scotland. This is a rotating lift system that connects the Union Canal with the Forth and Clyde Canal. Fixed axis rotations are an important special case of rigid body motions.

This chapter begins the study of *rigid body motion* by developing the *planar kinematics of a rigid body*. As we did in Chapter 2, we will describe motion without addressing what causes the motion. We will assume that (1) the body is rigid and its mass is distributed over a *region* of space (see Section 1.2, p. 6) and (2) the velocity of each of the body's points is parallel to a common plane. This may appear daunting because the description of a body's motion requires that we know the motion of each point in the body. However, the rigidity assumption alleviates this difficulty, and we will discover that we can completely characterize the planar motion of a rigid body using only three functions of time. To understand how *rigidity* helps us develop a kinematics that is so efficient, we begin by giving a qualitative description of rigid body motions, and then we proceed with a quantitative analysis. General 3D rigid body motions are examined in Chapter 10.

6.1 Fundamental Equations, Translation, and Rotation About a Fixed Axis

Crank, connecting rod, and piston motion

Figure 6.1 shows the interior of an internal combustion engine. While we have learned to describe the motion of individual points, we have not learned how to characterize the motion of entire bodies. For example, we can find the velocity of a point on one of the connecting rods, but we also want to be able to describe the motion of the connecting rod as a whole. Figure 6.2 shows three sequential views of what we would see in one of the cylinders if we looked down the crankshaft (i.e., the crankshaft axis is perpendicular to the page). Point A represents the axis of the crankshaft, and points B and C represent the pin connections between the connecting rod and the crank and between the connecting rod and the piston, respectively. Because of these connections, the motion of the piston causes a rotation of the crankshaft. As the crankshaft goes through a complete rotation, the piston reverses its motion and slides back up the cylinder so that the overall motion is repeated. This is an

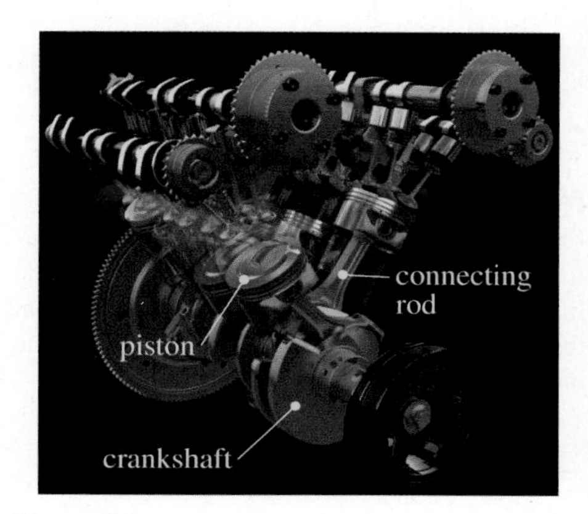

Figure 6.1
Interior view of the EcoBoost engine from Ford, revealing the pistons, connecting rods, and crankshaft.

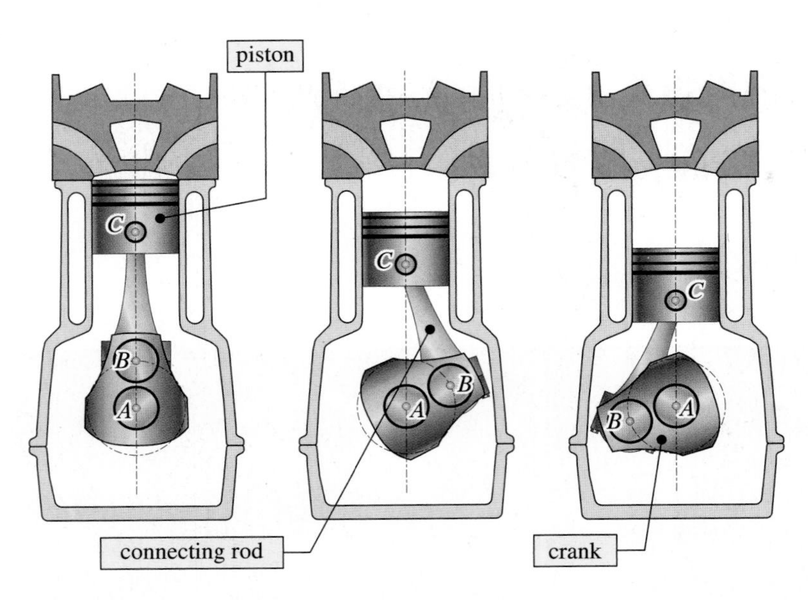

Figure 6.2. Three different positions of the mechanical system formed by a piston, connecting rod, and crank in a typical internal combustion engine. The axis of the crankshaft is perpendicular to the page, and it is represented by the points labeled A.

example of a *slider-crank mechanism.** The dimensions and relative position of the piston, connecting rod, and crankshaft influence the motion of these elements and contribute to the overall engine performance. We will use this system to illustrate most of the kinematics we will present in this chapter.

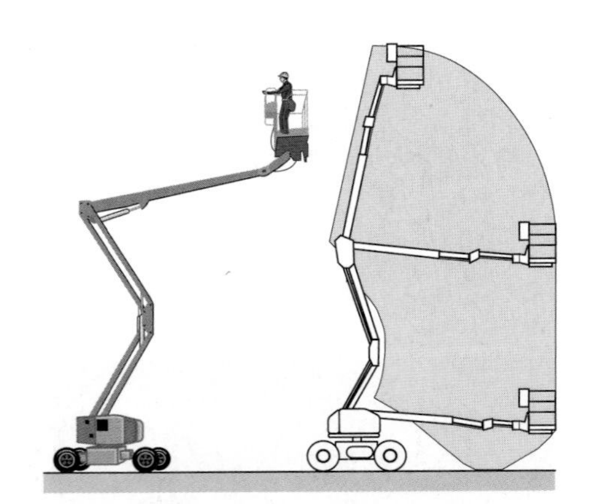

Figure 6.3
Left: a lift with an articulated boom and a platform holding a worker. Right: a schematic of the lift's range of motion (the gray area) showing that the platform remains parallel to the ground while moving along a curved path.

Qualitative description of rigid body motion

Translation

Assume that the cylinder within which the piston moves in Fig. 6.2 is stationary. As the piston moves, we see that the piston rings (appearing as horizontal bands at the top of the piston) always remain horizontal. Therefore, the velocity of point C on the piston is the same as that of *any* other point on the piston. This type of motion is called *translation*, and it is defined by saying that *any line segment connecting two points in the body maintains its original orientation throughout the motion*. Since a translating body does not necessarily move in a straight line (see Fig. 6.3), we classify translations into two categories, *rectilinear translations* or *curvilinear translations*, based on whether the trajectory of each point is a straight line or not. The motion of the piston in Fig. 6.2 is a rectilinear translation, whereas the motion of the platform in Fig. 6.3 is a curvilinear translation.

Rotation about a fixed axis

Figure 6.4 emphasizes the motion of the crank in Fig. 6.2. Point A on the axis of the shaft does not move. Any point on the crank lying on the axis perpendicular to the plane of the figure and going through point A does not move. Because the crank is rigid, off-axis points on the crank (e.g., point B) can only move in a circle centered at A. Any segment connecting points to the shaft axis rotates with the same angular

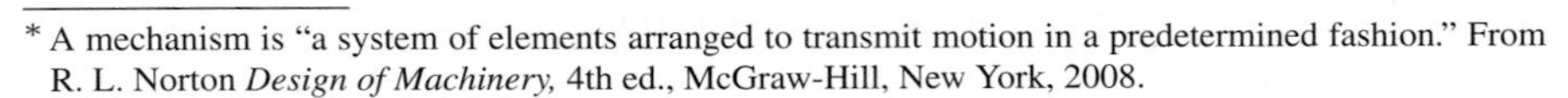

* A mechanism is "a system of elements arranged to transmit motion in a predetermined fashion." From R. L. Norton *Design of Machinery,* 4th ed., McGraw-Hill, New York, 2008.

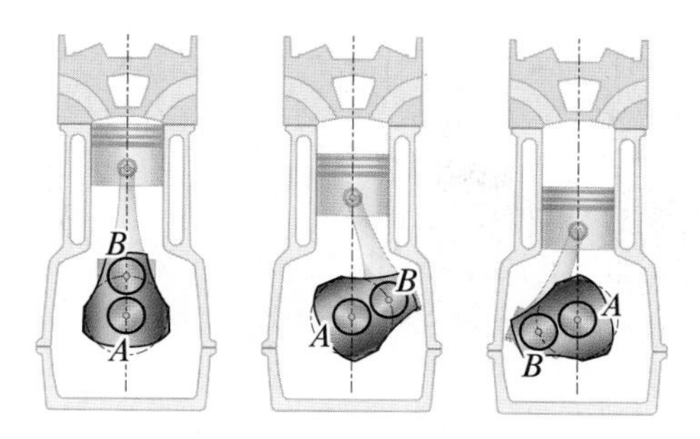

Figure 6.4. A modification of Fig. 6.2 emphasizing the motion of the crank.

velocity as any other segment. This type of motion, where there is a line of points with zero velocity functioning as an axis of rotation, is a *rotation about a fixed axis*.

General planar motion

Looking at the connecting rod, which is emphasized in Fig. 6.5, this motion does not correspond to either of the two cases discussed earlier since (1) we cannot find an axis that is fixed throughout the motion and (2) no two points define a segment that does not change its orientation. The only distinguishing feature of the rod's motion is that none of its points has a component of velocity perpendicular to the plane of the page, which is called the *plane of motion*. This kind of motion is called *general plane motion* or *general planar motion*. This motion can be viewed as the composition of a translation with a rotation about an axis perpendicular to the plane of motion. Note that both the piston's translation and crankshaft's rotation are also planar motions.

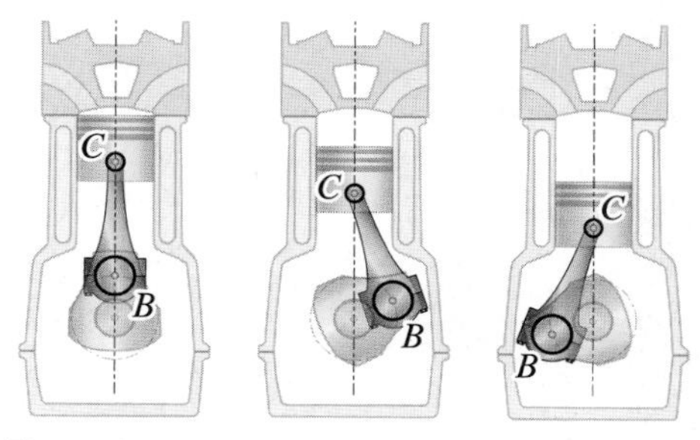

Figure 6.5
A modification of Fig. 6.2 emphasizing the motion of the connecting rod.

General motion of a rigid body

For a rigid body, we can describe the motion of all points by describing (1) the motion of a *single* point and (2) *the rate of change of the body's orientation.*

Figure 6.6 shows the general plane motion of an aircraft carrier from above. We can express the velocity of point B in terms of the velocity of point A using Eq. (2.78) on p. 122 as

$$\vec{v}_B = \vec{v}_A + \vec{v}_{B/A}. \tag{6.1}$$

Noting that $\vec{v}_{B/A} = \dot{\vec{r}}_{B/A}$ and writing $\vec{r}_{B/A}$ as $|\vec{r}_{B/A}|\,\hat{u}_{B/A}$, we can use Eq. (2.48) on p. 81 for the time derivative of a vector, to obtain

$$\vec{v}_{B/A} = \frac{d}{dt}\left(|\vec{r}_{B/A}|\,\hat{u}_{B/A}\right) = \frac{d|\vec{r}_{B/A}|}{dt}\,\hat{u}_{B/A} + |\vec{r}_{B/A}|\,\dot{\hat{u}}_{B/A}$$
$$= |\vec{r}_{B/A}|\vec{\omega}_{AB} \times \hat{u}_{B/A} = \vec{\omega}_{AB} \times \vec{r}_{B/A}, \tag{6.2}$$

where we have used the fact that $d|\vec{r}_{B/A}|/dt = 0$ since *the distance between A and B is constant when they are both on the same rigid body*. Substituting Eq. (6.2) into Eq. (6.1) gives

$$\boxed{\vec{v}_B = \vec{v}_A + \vec{\omega}_{AB} \times \vec{r}_{B/A} = \vec{v}_A + \vec{v}_{B/A},} \tag{6.3}$$

where we note that when A and B are two points on the same rigid body, $\vec{v}_{B/A} = \vec{\omega}_{AB} \times \vec{r}_{B/A}$. Equation (6.3), which relates the motion of points A and B, depends on the angular velocity of the line segment $\overline{AB}$. Figure 6.7 shows that as the aircraft carrier moves from position 1 to position 2, the three line segments $\overline{AB}$, $\overline{BC}$, and

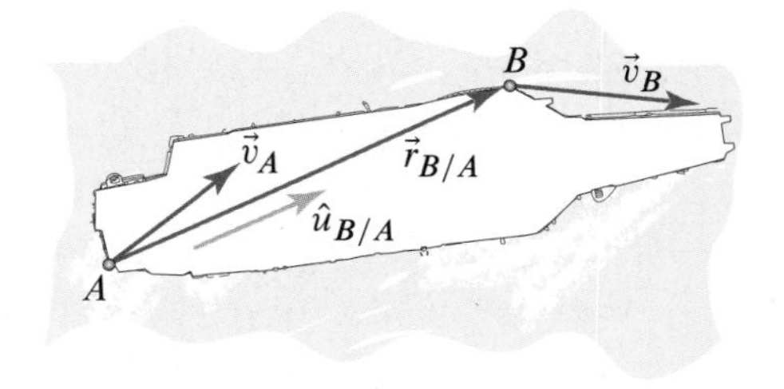

Figure 6.6
Aerial view of an aircraft carrier performing a maneuver. We model the carrier's motion as a planar rigid body motion.

Concept Alert

What $\vec{\omega}$ should we use? The angular velocity is a property of the body, so referring to Fig. 6.7, all of the following equations can be written

$$\vec{v}_A = \vec{v}_B + \vec{\omega}_{\text{body}} \times \vec{r}_{A/B},$$
$$\vec{v}_B = \vec{v}_C + \vec{\omega}_{\text{body}} \times \vec{r}_{B/C},$$
$$\vec{v}_C = \vec{v}_A + \vec{\omega}_{\text{body}} \times \vec{r}_{C/A},$$

where the $\vec{\omega}_{\text{body}}$ in these equations is the angular velocity of the body.

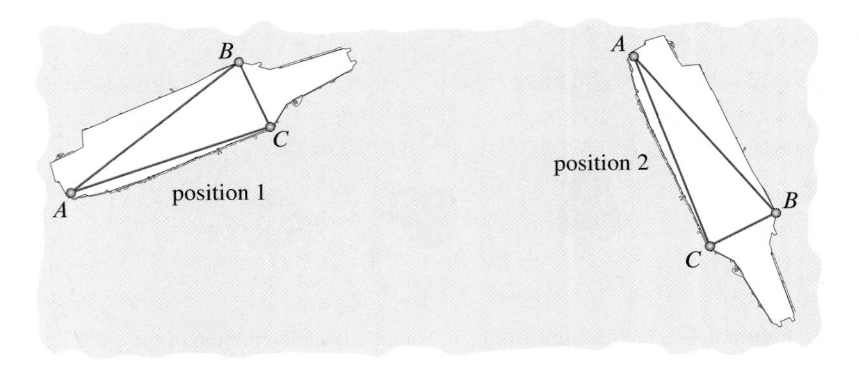

Figure 6.7. Aerial view of an aircraft carrier moving from position 1 to position 2 showing that *all* line segments rotate the same amount as a rigid body moves.

$\overline{CA}$ all rotate the same amount. This implies that *any* line segment on the aircraft carrier will have the same rate of rotation as any other line segment, and so the *angular velocity of a rigid body* $\vec{\omega}_{AB}$ is a property of the body as a whole and not any particular part of it. This means that Eq. (6.3) applies to any two points A and B on the *same* rigid body. The subscripts on angular velocities (and soon, angular accelerations) should be viewed as labels referring to a particular body. For example, in Eq. (6.3), the subscript AB on $\vec{\omega}$ tells us that it is the angular velocity of the body containing points A and B.

The equation relating the acceleration of two points on the same rigid body is found by differentiating Eq. (6.3) with respect to time to obtain

$$\vec{a}_B = \vec{a}_A + \dot{\vec{\omega}}_{AB} \times \vec{r}_{B/A} + \vec{\omega}_{AB} \times \dot{\vec{r}}_{B/A}. \tag{6.4}$$

The quantity $\dot{\vec{\omega}}_{AB}$ is the *angular acceleration of the body* and is denoted by $\vec{\alpha}_{AB}$. Since $\dot{\vec{r}}_{B/A} = \vec{v}_{B/A} = \vec{\omega}_{AB} \times \vec{r}_{B/A}$, Eq. (6.4) can be written as

$$\boxed{\vec{a}_B = \vec{a}_A + \vec{\alpha}_{AB} \times \vec{r}_{B/A} + \vec{\omega}_{AB} \times (\vec{\omega}_{AB} \times \vec{r}_{B/A}) = \vec{a}_A + \vec{a}_{B/A},} \tag{6.5}$$

where we note that when A and B are two points on a rigid body, $\vec{a}_{B/A} = \vec{\alpha}_{AB} \times \vec{r}_{B/A} + \vec{\omega}_{AB} \times (\vec{\omega}_{AB} \times \vec{r}_{B/A})$.

Applying Eqs. (6.3) and (6.5)

Equations (6.3) and (6.5) tell us that we can know the motion of all the points on a rigid body by knowing the motion of a *single* point (e.g., $\vec{v}_A$ and $\vec{a}_A$) and by knowing the rotation of the body (e.g., $\vec{\omega}_{AB}$ and $\vec{\alpha}_{AB}$).

Since we will apply Eqs (6.3) and (6.5) to analyze mechanisms consisting of several rigid bodies, we must always relate points A and B that belong to the same rigid body. For example, in Fig. 6.8, we can use Eqs. (6.3) and (6.5) to relate the velocities and accelerations, respectively, of points A and B because these points are both on the crank. We can do the same for points B and C because they are both on the connecting rod. Since point B is on *both* the crank and the connecting rod, it allows us to relate the motions of the connecting rod and the crank. We cannot use Eqs. (6.3) and (6.5) to relate the motion of points A and C because these points belong to two distinct bodies — the crank and the piston, respectively.

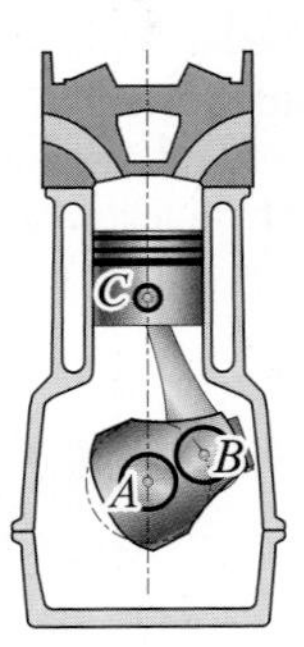

Figure 6.8
Slider-crank mechanism.

Graphical interpretation of Eqs. (6.3) and (6.5)

As we said earlier, a general plane motion can be viewed as the composition of a translation with a rotation about an axis perpendicular to the plane of motion. Figure 6.9 shows this idea for the relative velocity equation given in Eq. (6.3). This

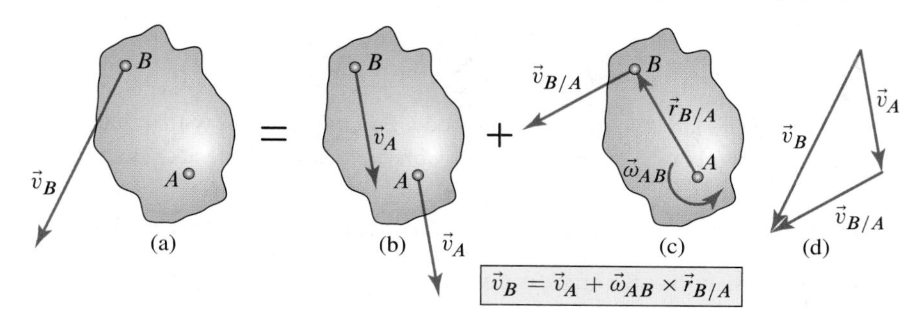

Figure 6.9. A graphical representation of the equation $\vec{v}_B = \vec{v}_A + \vec{\omega}_{AB} \times \vec{r}_{B/A}$.

figure graphically demonstrates that we can add the translation of the body with point A (i.e., $\vec{v}_A$ in Fig. 6.9(b)) to the pure rotation of point B about point A (i.e., $\vec{v}_{B/A} = \vec{\omega}_{AB} \times \vec{r}_{B/A}$ in Fig. 6.9(c)) to obtain the velocity of B (i.e., $\vec{v}_B$ in Figs 6.9(a) and 6.9(d)).

A similar graphical argument can be applied to the relative acceleration equation given in Eq. (6.5). Referring to Fig. 6.10, we again see that we can add the translation

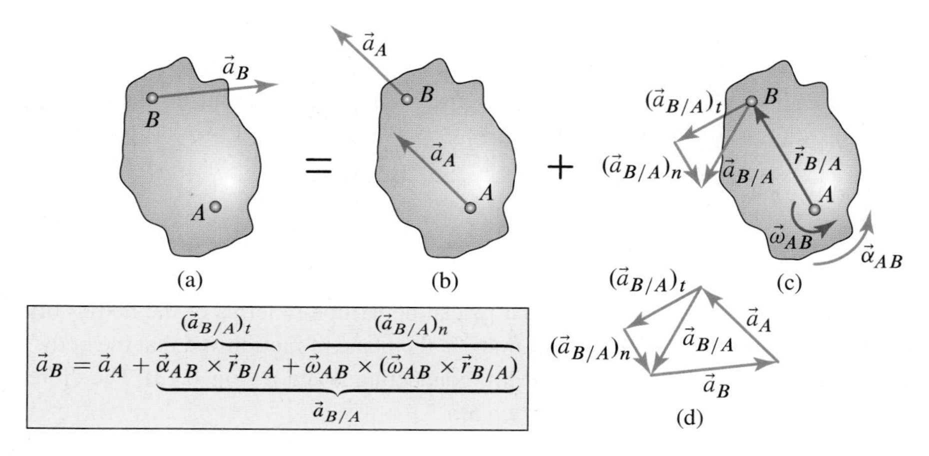

Figure 6.10. A graphical representation of the equation $\vec{a}_B = \vec{a}_A + \vec{\alpha}_{AB} \times \vec{r}_{B/A} + \vec{\omega}_{AB} \times (\vec{\omega}_{AB} \times \vec{r}_{B/A})$. The subscripts t and n stand for the tangential and normal components, respectively, of the relative acceleration $\vec{a}_{B/A}$.

of the body with point A (i.e., $\vec{a}_A$ in Fig. 6.10(b)) to the pure rotation of point B about point A (i.e., $\vec{a}_{B/A} = (\vec{a}_{B/A})_t + (\vec{a}_{B/A})_n = \vec{\alpha}_{AB} \times \vec{r}_{B/A} + \vec{\omega}_{AB} \times (\vec{\omega}_{AB} \times \vec{r}_{B/A})$ in Fig. 6.10(c)) to obtain the acceleration of B (i.e., $\vec{a}_B$ in Figs 6.10(a) and 6.10(d)).

Elementary rigid body motions: translations

Figure 6.11 shows the deployment of a basketball goal. The backboard AB and hoop are attached to an arm, which, in turn, is hinged to two parallel bars CD and EF. The design is meant to ensure that the hoop remains parallel to the floor. Therefore, the backboard, hoop, and supporting arm are in *translation*. This means that the angular

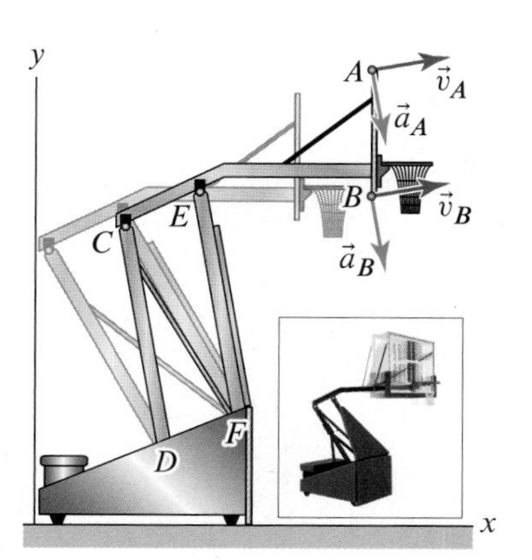

Figure 6.11
The deployment of a portable basketball goal (see inset photo).

velocity and angular acceleration of a translating rigid body are equal to zero, that is,

$$\vec{\omega}_{AB} = \vec{\omega}_{\text{body}} = \vec{0} \quad \text{and} \quad \vec{\alpha}_{AB} = \vec{\alpha}_{\text{body}} = \vec{0}. \tag{6.6}$$

Substituting Eqs. (6.6) into Eqs. (6.3) and (6.5) gives

$$\vec{v}_B = \vec{v}_A \quad \text{and} \quad \vec{a}_B = \vec{a}_A, \tag{6.7}$$

where A and B are any two points on the body. Equations (6.7) say that the motion of a translating rigid body is characterized by one velocity and one acceleration.

Elementary rigid body motions: rotation about a fixed axis

Figure 6.12 shows us that the crank is *rotating about a fixed axis* through point A and

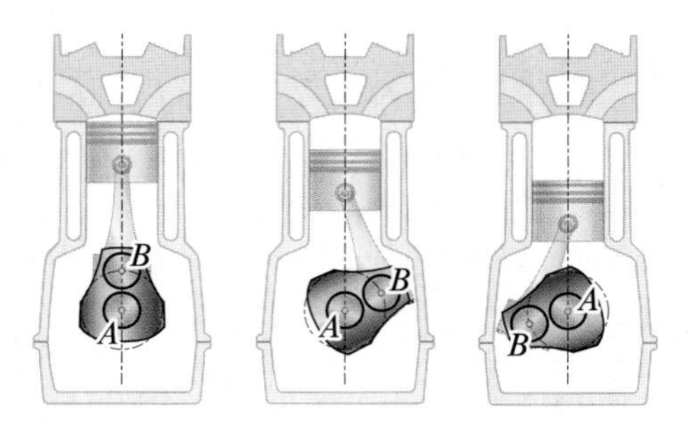

Figure 6.12. Crank motion in a slider-crank mechanism.

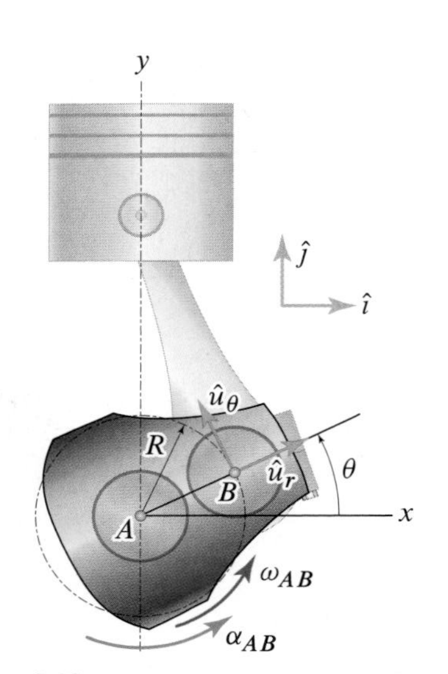

Figure 6.13
Detailed view of the crank of a slider-crank mechanism.

perpendicular to the plane of motion. Referring to Fig. 6.13, this means that $\vec{v}_A = \vec{0}$, $\vec{a}_A = \vec{0}$, and Eqs. (6.3) and (6.5) then give

$$\vec{v}_B = \vec{\omega}_{AB} \times \vec{r}_{B/A}, \tag{6.8}$$

$$\vec{a}_B = \vec{\alpha}_{AB} \times \vec{r}_{B/A} + \vec{\omega}_{AB} \times (\vec{\omega}_{AB} \times \vec{r}_{B/A}). \tag{6.9}$$

Since the motion is planar, $\vec{\omega}_{AB}$ and $\vec{\alpha}_{AB}$ can be written in terms of the body's orientation by noting that points A and B are in the plane of motion and that the body's orientation θ can be defined using the line connecting A and B (Fig. 6.13). Therefore, the vectors $\vec{\omega}_{AB}$ and $\vec{\alpha}_{AB}$ can be written as:

$$\vec{\omega}_{AB} = \omega_{AB}\,\hat{k} = \dot{\theta}\,\hat{k} \quad \text{and} \quad \vec{\alpha}_{AB} = \alpha_{AB}\,\hat{k} = \ddot{\theta}\,\hat{k}, \tag{6.10}$$

where $\hat{k} = \hat{u}_r \times \hat{u}_\theta = \hat{\imath} \times \hat{\jmath}$ is perpendicular to the plane of motion, and ω_{AB} and α_{AB} denote the components of the angular velocity and acceleration, respectively, in the $\hat{k}$ direction. Using the polar component system shown in Fig. 6.13, letting $\vec{r}_{B/A} = R\,\hat{u}_r$, and substituting Eqs. (6.10) into Eqs. (6.8) and (6.9), $\vec{v}_B$ and $\vec{a}_B$ are

$$\vec{v}_B = R\dot{\theta}\,\hat{u}_\theta \quad \text{and} \quad \vec{a}_B = R\ddot{\theta}\,\hat{u}_\theta - R\dot{\theta}^2\,\hat{u}_r, \tag{6.11}$$

which is not surprising since B is in circular motion about A.

The acceleration coming from the term $\vec{\omega}_{AB} \times (\vec{\omega}_{AB} \times \vec{r}_{B/A})$ is equal to $-R\dot{\theta}^2\,\hat{u}_r$. Noting that $\dot{\theta}^2 = \omega_{AB}^2$ and that $-R\,\hat{u}_R = -\vec{r}_{B/A}$, for planar motion, we can write

$$\vec{\omega}_{AB} \times (\vec{\omega}_{AB} \times \vec{r}_{B/A}) = -R\dot{\theta}^2\,\hat{u}_r = -\omega_{AB}^2 \vec{r}_{B/A}, \tag{6.12}$$

$\vec{\omega}_{AB}$

$\vec{r}_{B/A}$

B A

$|\vec{\omega}_{AB} \times \vec{r}_{B/A}| = \omega_{AB} r_{B/A}$

$|\vec{\omega}_{AB} \times (\vec{\omega}_{AB} \times \vec{r}_{B/A})| = \omega_{AB}^2 r_{B/A}$

$\vec{\omega}_{AB} \times (\vec{\omega}_{AB} \times \vec{r}_{B/A}) = \omega_{AB}^2 (-\vec{r}_{B/A})$

plane of motion

$\vec{\omega}_{AB} \times \vec{r}_{B/A}$

Figure 6.14. Geometric demonstration of the equivalence of $\vec{\omega}_{AB} \times (\vec{\omega}_{AB} \times \vec{r}_{B/A})$ and $-\omega_{AB}^2 \vec{r}_{B/A}$ for planar motion.

We can see this geometrically if we note that $\vec{r}_{B/A}$ is always in the plane of motion and that $\vec{\omega}_{AB}$ is always perpendicular to it (see Fig. 6.14). Taking their cross product $\vec{\omega}_{AB} \times \vec{r}_{B/A}$ results in a vector that is in the plane of motion and perpendicular to both $\vec{\omega}_{AB}$ and $\vec{r}_{B/A}$. Finally, taking $\vec{\omega}_{AB} \times (\vec{\omega}_{AB} \times \vec{r}_{B/A})$ (the cross product of the perpendicular purple vectors in Fig. 6.14) results in the vector $-\omega_{AB}^2 \vec{r}_{B/A}$ (the green vector in Fig. 6.14). Using Eq. (6.12), we can write Eq. (6.9) as

$$\vec{a}_B = \vec{\alpha}_{AB} \times \vec{r}_{B/A} - \omega_{AB}^2 \vec{r}_{B/A}. \tag{6.13}$$

This form can save computation in finding accelerations for planar problems.

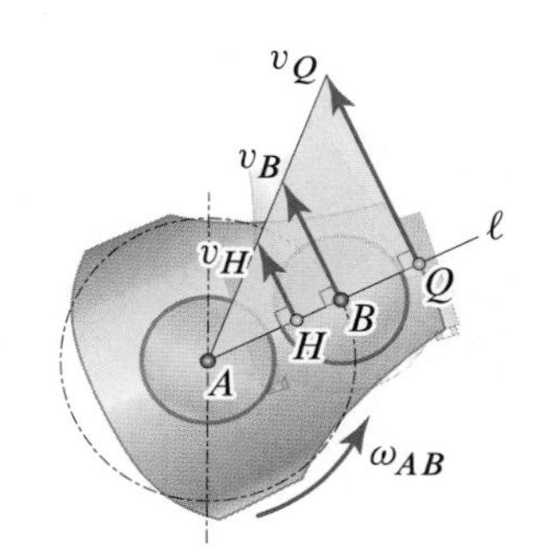

Figure 6.15
Graphical representation of the velocities of points on radial lines originating at the center of rotation.

Graphical interpretation of Eq. (6.8). Referring to the crank in Fig. 6.15, consider the velocity of points H, B, and Q lying on the radial line ℓ with origin at the center of rotation A. Equation (6.8), or the first of Eqs. (6.11), implies that $\vec{v}_H$, $\vec{v}_B$, and $\vec{v}_Q$ are all perpendicular to ℓ (and parallel to one another) and have a magnitude *proportional* to their distance from A. The constant of proportionality is ω_{AB}. Therefore, the distribution of the velocities of points on radial lines can be represented graphically by the triangle as shown.

Planar motion in practice

For general planar motions, we just saw that it is *always* possible to express the term $\vec{\omega}_{AB} \times (\vec{\omega}_{AB} \times \vec{r}_{B/A})$ as $-\omega_{AB}^2 \vec{r}_{B/A}$. Equation (6.5) can then be written as

$$\vec{a}_B = \vec{a}_A + \vec{\alpha}_{AB} \times \vec{r}_{B/A} - \omega_{AB}^2 \vec{r}_{B/A}. \tag{6.14}$$

When relating the motion of two points A and B on the *same* rigid body, we will generally use this version of the acceleration equation.

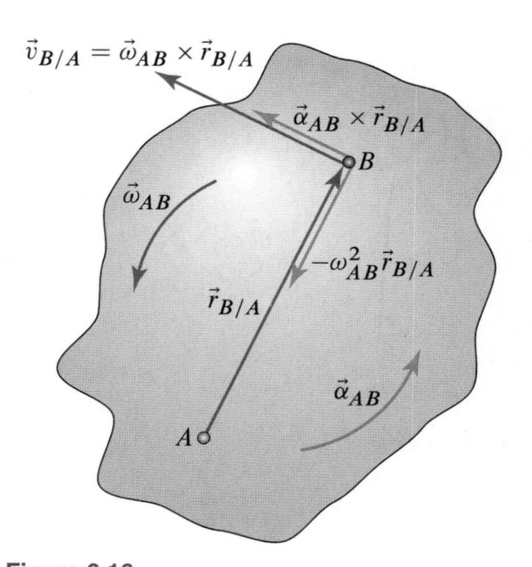

Figure 6.16
A rigid body showing the quantities used in Eqs. (6.3), (6.5), and (6.14).

End of Section Summary

This section began our study of the dynamics of rigid bodies. As with particles, we start with the study of kinematics, which is the focus of this chapter. The key kinematic idea is that a rigid body has only one angular velocity and one angular acceleration; i.e., each is a property of the body as a whole. This allowed us to use the relative velocity equation (Eq. (2.78) on p. 122) and the relation for the time derivative of a vector (Eq. (2.48) on p. 81) to relate the velocities of *two points A and B on the same rigid body* using (see Fig. 6.16)

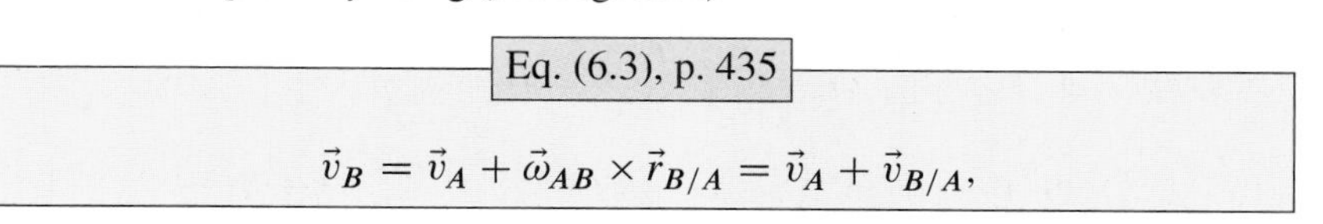

and the accelerations using

Eq. (6.5), p. 436

$$\vec{a}_B = \vec{a}_A + \vec{\alpha}_{AB} \times \vec{r}_{B/A} + \vec{\omega}_{AB} \times (\vec{\omega}_{AB} \times \vec{r}_{B/A}) = \vec{a}_A + \vec{a}_{B/A},$$

which for *planar motion* becomes

Eq. (6.14), p. 439

$$\vec{a}_B = \vec{a}_A + \vec{\alpha}_{AB} \times \vec{r}_{B/A} - \omega_{AB}^2 \vec{r}_{B/A}.$$

Translation. For this motion, the angular velocity and angular acceleration of the body are equal to zero, i.e.,

Eq. (6.6), p. 438

$$\vec{\omega}_{\text{body}} = \vec{0} \quad \text{and} \quad \vec{\alpha}_{\text{body}} = \vec{0},$$

so the velocity and acceleration relations for the body reduce to

Eq. (6.7), p. 438

$$\vec{v}_B = \vec{v}_A \quad \text{and} \quad \vec{a}_B = \vec{a}_A,$$

where A and B are any two points on the body.

Rotation about a fixed axis. For this special motion, there is an axis of rotation perpendicular to the plane of motion that does not move. All points not on the axis of rotation can only move in a circle about that axis. If the axis of rotation is at point A, then the velocity of B is given by

Eq. (6.8), p. 438

$$\vec{v}_B = \vec{\omega}_{AB} \times \vec{r}_{B/A},$$

and its acceleration is

Eq. (6.9), p. 438, and Eq. (6.13), p. 439

$$\vec{a}_B = \vec{\alpha}_{AB} \times \vec{r}_{B/A} + \vec{\omega}_{AB} \times (\vec{\omega}_{AB} \times \vec{r}_{B/A}),$$
$$\vec{a}_B = \vec{\alpha}_{AB} \times \vec{r}_{B/A} - \omega_{AB}^2 \vec{r}_{B/A},$$

where the vectors $\vec{\omega}_{AB}$ and $\vec{\alpha}_{AB}$ are the angular velocity and angular acceleration of the body, respectively, and $\vec{r}_{B/A}$ is the position of B relative to A.

EXAMPLE 6.1 *Engine Pulleys: Fixed Axis Rotation*

Most car engines have a number of belts connecting pulleys on the engine (Fig. 1). The belts are not supposed to slip relative to the pulleys they connect and are used to transmit, as well as synchronize motion between engine parts. For the belt connecting pulley A, which rotates with the crankshaft, with pulley B, which drives the alternator, determine the angular speed of pulley B if the crankshaft is spinning at 2550 rpm and the radii of pulleys A and B are $R_A = 4.25$ in. and $R_B = 2.5$ in., respectively.

Figure 1
Typical car engine with several belts.

SOLUTION

Road Map Referring to Fig. 2, the no-slip condition between the belt and pulleys means that any two points on the belt and pulley that are in contact at a given instant, e.g., C and D or P and Q, must have the same velocity (i.e., $\vec{v}_{C/D} = \vec{v}_{P/Q} = \vec{0}$). Combining this observation with the fact that pulleys A and B rotate about fixed axes and the assumption that the belt is inextensible (all points on the belt must have the same *speed*) will allow us to solve the problem.

Computation From Fig. 2, the inextensibility of the belt implies that

$$|\vec{v}_C| = |\vec{v}_P|. \tag{1}$$

Using this result and recalling the no-slip condition, we find

$$\vec{v}_C = \vec{v}_D \quad \text{and} \quad \vec{v}_P = \vec{v}_Q \quad \Rightarrow \quad |\vec{v}_D| = |\vec{v}_Q|. \tag{2}$$

Referring to Fig. 2, pulleys A and B undergo fixed axis rotation about their respective centers. Applying Eq. (6.8) to each of the pulleys to obtain $\vec{v}_D$ and $\vec{v}_Q$, we find

$$\vec{v}_D = \vec{\omega}_A \times \vec{r}_{D/A} \quad \text{and} \quad \vec{v}_Q = \vec{\omega}_B \times \vec{r}_{Q/B}, \tag{3}$$

and since $\vec{\omega}$ and $\vec{r}$ are perpendicular to each other in each case, the corresponding speeds are

$$|\vec{v}_D| = |\vec{\omega}_A| R_A \quad \text{and} \quad |\vec{v}_Q| = |\vec{\omega}_B| R_B. \tag{4}$$

Substituting Eq. (4) into the last of Eqs. (2), we obtain

$$\boxed{|\vec{\omega}_A| R_A = |\vec{\omega}_B| R_B \quad \Rightarrow \quad |\vec{\omega}_B| = \frac{R_A}{R_B}|\vec{\omega}_A| = 4335\,\text{rpm},} \tag{5}$$

where we have plugged in $|\vec{\omega}_A| = 2550$ rpm, R_A, and R_B.

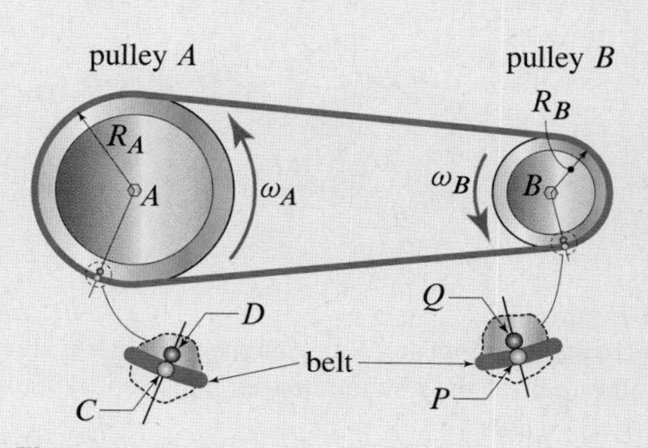

Figure 2
Schematic of the system consisting of pulleys A and B and the belt connecting them. The pulleys are attached to the engine, which is assumed to be stationary.

Discussion & Verification To verify the result in Eq. (5), Fig. 3 shows that the no-slip condition between the belt and pulley A implies that if pulley A rotates through an angle θ_A, then the belt must move an amount $\Delta L = \theta_A R_A$ around pulley A. Since the belt does not slip with respect to pulley B, we must also have $\Delta L = \theta_B R_B$, where θ_B is the angle through which pulley B rotates. Therefore, we must have $\theta_B = (R_A/R_B)\theta_A$. This relation implies that the rates of change of the angle θ_B will be proportional to θ_A via the ratio of the pulley radii R_A/R_B — this is what we obtained in Eq. (5).

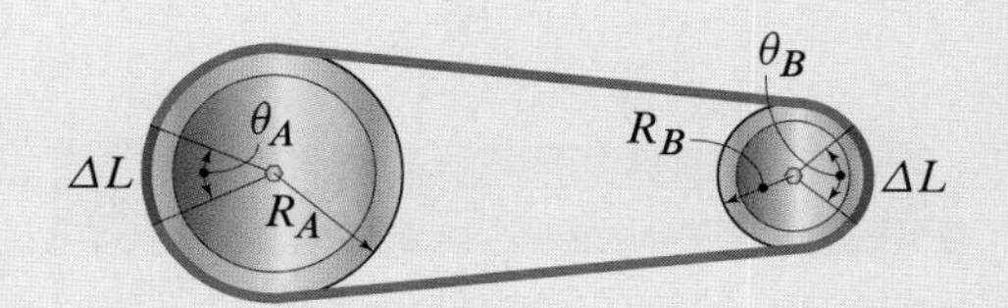

Figure 3
Measure of the linear length of belt going around the pulleys under the no-slip condition.

EXAMPLE 6.2 Gears: Fixed Axis Rotation

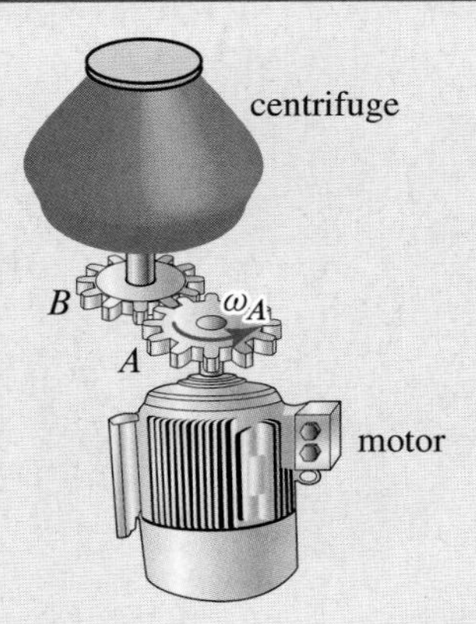

Figure 1
A centrifuge driven by an electric motor via a gear system.

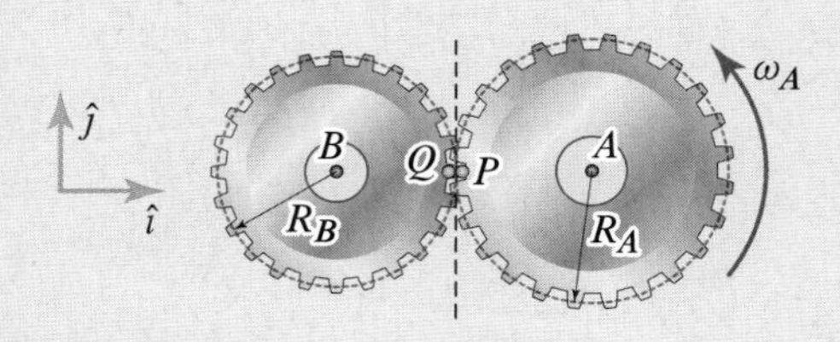

Figure 2
Two meshing gears. Note that the gears' radii are the radii of the circumferences indicated by the dashed lines.

An electric motor with a top angular speed of 3450 rpm is used to spin a centrifuge. The motor's motion is transmitted to the centrifuge by two gears A and B, with radii R_A and R_B, respectively (see Fig. 1). Determine the ratio R_A/R_B if the centrifuge is to achieve 6000 rpm as its top angular speed. In addition, determine the ratio between the magnitude of the angular acceleration of A to that of B during spin-up.

SOLUTION

Road Map Gear teeth are designed so that the gears behave as two wheels rolling without slip on each other. The radius of a gear is the radius of the wheel that the gear is designed to represent (see the dashed red line in Fig. 2), as opposed to the inner or outer radius of the teeth. Therefore, referring to Fig. 2, the no-slip condition between the gears tells us that the velocity of points P and Q must be the same when they are in contact. Furthermore, gears A and B are rotating about the fixed axes of the motor shaft and centrifuge shaft, respectively.

Computation The velocity of points P and Q, which are moving in circular motion about A and B, respectively, can be found by applying Eq. (6.8), which gives

$$\vec{v}_P = \vec{\omega}_A \times \vec{r}_{P/A} = \omega_A\,\hat{k} \times (-R_A)\,\hat{\imath} = -\omega_A R_A\,\hat{\jmath}, \tag{1}$$

$$\vec{v}_Q = \vec{\omega}_B \times \vec{r}_{Q/B} = \omega_B\,\hat{k} \times R_B\,\hat{\imath} = \omega_B R_B\,\hat{\jmath}, \tag{2}$$

where we have assumed that the angular velocities of A and B are both in the positive $\hat{k}$ direction. Enforcing the no-slip condition between the gears, we have

$$\vec{v}_P = \vec{v}_Q \quad\Rightarrow\quad -R_A\omega_A\,\hat{\jmath} = R_B\omega_B\,\hat{\jmath} \quad\Rightarrow\quad -R_A\omega_A = R_B\omega_B, \tag{3}$$

where the minus sign tells us that the two gears spin in opposite directions since R_A and R_B are both positive. Since $|\omega_A| = 3450$ rpm and $|\omega_B| = 6000$ rpm, we can solve Eq. (3) for R_A/R_B to obtain

$$\boxed{\frac{R_A}{R_B} = \left|\frac{\omega_B}{\omega_A}\right| = \frac{6000\text{ rpm}}{3450\text{ rpm}} = 1.739.} \tag{4}$$

To find the ratio of the angular acceleration of gear A to that of gear B, we can obtain the desired information by taking the time derivative of the result in Eq. (3) to obtain

$$-R_A\alpha_A = R_B\alpha_B \quad\Rightarrow\quad \boxed{\left|-\frac{\alpha_A}{\alpha_B}\right| = \frac{R_B}{R_A} = \frac{1}{1.739} = 0.5750.} \tag{5}$$

Discussion & Verification These results are dimensionally correct and are consistent with the result in Example 6.1; that is, Eq. (5) indicates that the angular speed and the magnitude of the angular accelerations of two gears are proportional to each other by the ratio of the gears' radii.

A Closer Look The result in Eq. (4) could be found using ideas developed in Chapter 2, where we learned that the speed of a point in circular motion is equal to the radius of the path times the angular speed. Therefore, from Fig. 2, we have $v_P = R_A|\omega_A|$ and $v_Q = R_B|\omega_B|$. Determining the gear ratio R_A/R_B is then a matter of once again setting $\vec{v}_P = \vec{v}_Q$, which implies $v_P = v_Q$. In turn, this last relation yields Eq. (4).

Interesting Fact

Why use gears? Gears, like belts, are used to transmit rotary motion. Gears are essential when rotary motion *must* be transmitted without slip (e.g., in a watch). Gears are often used to provide *gear reduction*, that is, to connect a shaft with one angular speed with a second shaft with a different angular speed.

EXAMPLE 6.3 *A Carnival Ride: Translation*

Motion platforms, which are used in many of today's amusement parks, are a type of carnival ride in which a platform with seats is made to move always parallel to the ground. Figure 1 shows an elementary type of motion platform, more often found in traveling carnivals than in big amusement parks. Given that the motion platform in the figure (see also Fig. 2) is designed so that the rotating arms AB and CD are of equal length $L = 10\,\text{ft}$ and remain parallel to each other, determine the velocity and acceleration of a person P onboard the ride when ω_{AB} is constant and equal to $1.25\,\text{rad/s}$.

Figure 1
A carnival ride consisting of a motion platform.

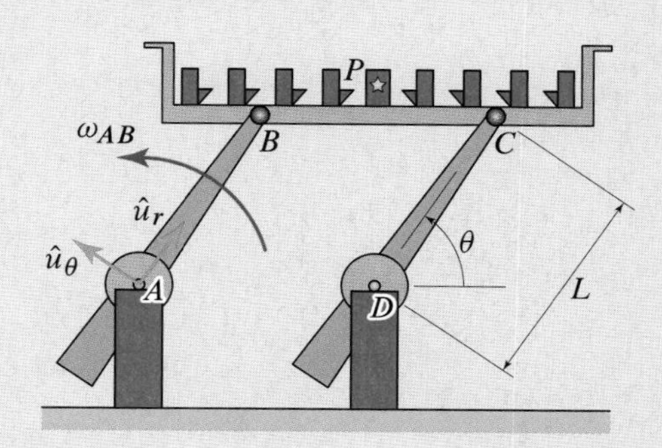

Figure 2
Coordinate definition and geometry for the carnival ride.

SOLUTION

Road Map To simplify the problem, we will assume that the platform and the persons on it form a single rigid body. This means that the two rotating arms and the platform form a *four-bar linkage*.* Because the arms AB and CD are identical in size and are always parallel to each other, the four-bar linkage $ABCD$ always forms a parallelogram, and so the platform BC does not change its orientation and has zero angular velocity. Knowing this and applying the kinematics of fixed axis rotation will allow us to determine $\vec{v}_P$ and $\vec{a}_P$.

Computation Again, from Fig. 2, we see that arms AB and CD are always parallel to each other, and so platform BC does not change its orientation. From the discussion on p. 437, this means the angular velocity of the platform is zero, that is,

$$\omega_{BC} = 0. \tag{1}$$

Also, observe that points B and C move along circles centered at A and D, respectively, where A and D are fixed points. Given the fact that the trajectories of points on BC are not a straight line, BC's motion is a *curvilinear translation*. Consequently, all points on BC, or any rigid extension of it, i.e., any of the passengers, share the same value of velocity, as well as acceleration. Therefore, we have

$$\vec{v}_P = \vec{v}_B = \omega_{AB} L\,\hat{u}_\theta = 12.50\,\text{ft/s}\,\hat{u}_\theta, \tag{2}$$

and

$$\vec{a}_P = \vec{a}_B = -\omega_{AB}^2 L\,\hat{u}_r = -15.62\,\text{ft/s}^2\,\hat{u}_r, \tag{3}$$

where the velocity and acceleration of B were computed using circular motion formulas in Eqs. (6.11).

Discussion & Verification The dimensions and units in Eqs. (2) and (3) are correct, and the magnitudes of the velocity and acceleration of P are reasonable. In particular, the acceleration is not far from those found in general public (as opposed to extreme) carnival rides, and it amounts to a little less than 50% of the acceleration of gravity.

A Closer Look It is important to remember that the direction of $\vec{v}_P$ in Eq. (2) and the direction of $\vec{a}_P$ in Eq. (3) are those shown at point A in Fig. 2. That is, since P is *not* in fixed axis rotation about A, the $\hat{u}_r$ and $\hat{u}_\theta$ in Eqs. (2) and (3) are not those of P relative to A.

* A *four-bar linkage* is a mechanism with four members or links in which one of the links is fixed (link AD in this example) and the other three move in a predetermined fashion (links AB, BC, and CD in this example).

Problems

Problem 6.1

The velocities of points A and B on a disk, which is undergoing planar motion, are such that $|\vec{v}_A| = |\vec{v}_B|$. Is it possible for the disk to be rotating about a fixed axis going through the center of the disk at O? Explain.

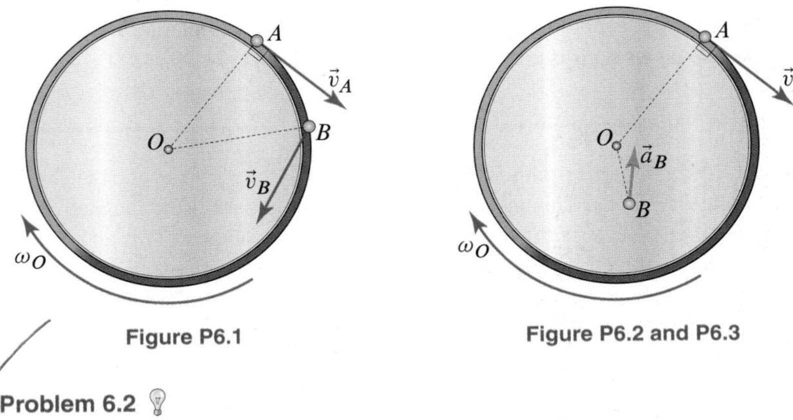

Figure P6.1 Figure P6.2 and P6.3

Problem 6.2

The velocity of a point A and the acceleration of a point B on a disk undergoing planar motion are shown. Is it possible for the disk to be rotating about a fixed axis going through the center of the disk at O? Explain.

Problem 6.3

Assuming that the disk shown is rotating about a fixed axis going through its center at O, determine whether the disk's angular velocity is constant, increasing, or decreasing.

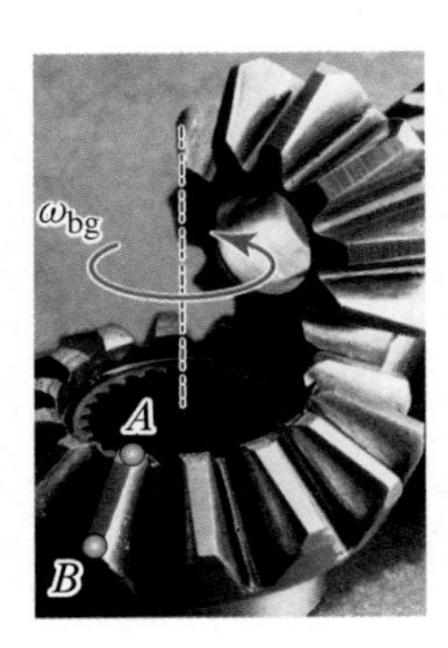

Figure P6.4

Problem 6.4

Do points A and B on the surface of the bevel gear (bg), which rotates with angular velocity ω_{bg}, move relative to each other? At what rate does the distance between A and B change?

Problem 6.5

Letting $R_A = 8$ in., $R_B = 4.2$ in., and $R_C = 6.5$ in., determine the angular velocity of gears B and C when gear A has the angular speed $|\omega_A| = 945$ rpm in the direction shown.

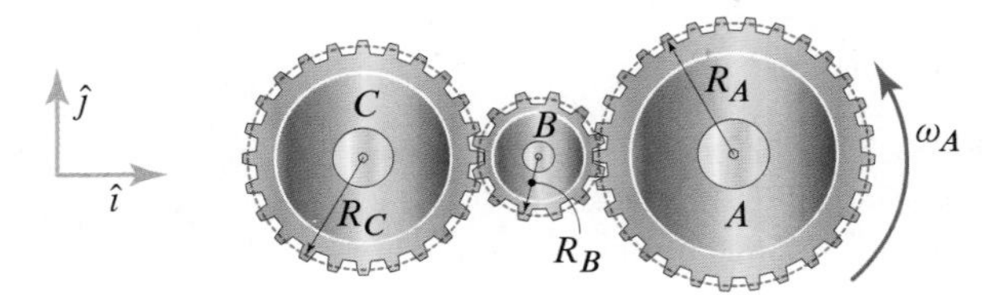

Figure P6.5

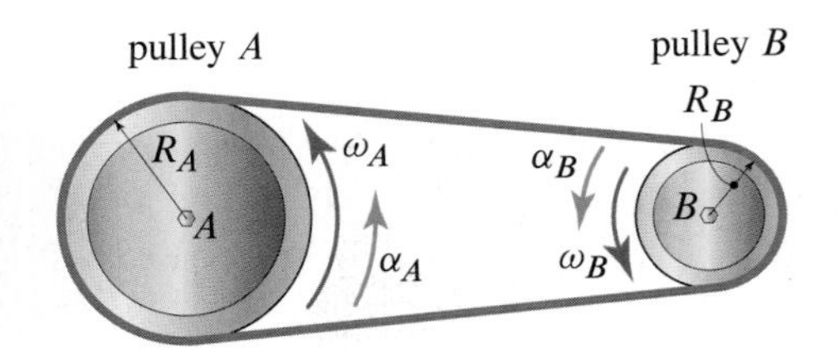

Figure P6.6

Problem 6.6

Letting $R_A = 7.2$ in. and $R_B = 4.6$ in., and assuming that the belt does not slip relative to pulleys A and B, determine the angular velocity and angular acceleration of pulley B when pulley A rotates at 340 rad/s while accelerating at 120 rad/s^2.

Problem 6.7

Letting $R_A = 203$ mm, $R_B = 107$ mm, $R_C = 165$ mm, and $R_D = 140$ mm, determine the angular acceleration of gears B, C, and D when gear A has an angular acceleration with magnitude $|\alpha_A| = 47\,\text{rad/s}^2$ in the direction shown. Note that gears B and C are mounted on the same shaft and they rotate together as a unit.

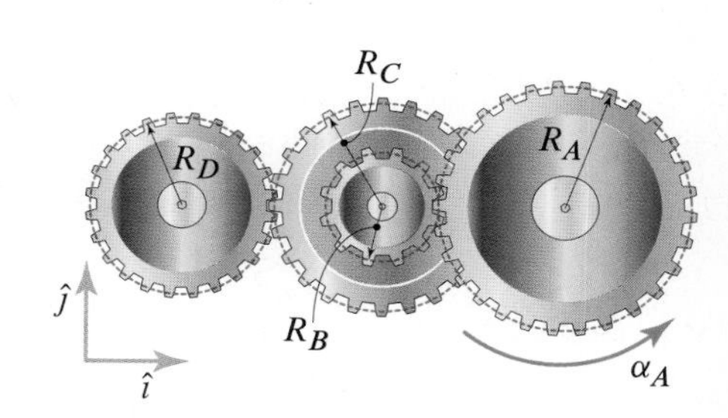

Figure P6.7

Problem 6.8

The bevel gears A and B have nominal radii $R_A = 20$ mm and $R_B = 5$ mm, respectively, and their axes of rotation are mutually perpendicular. If the angular speed of gear A is $\omega_A = 150$ rad/s, determine the angular speed of gear B.

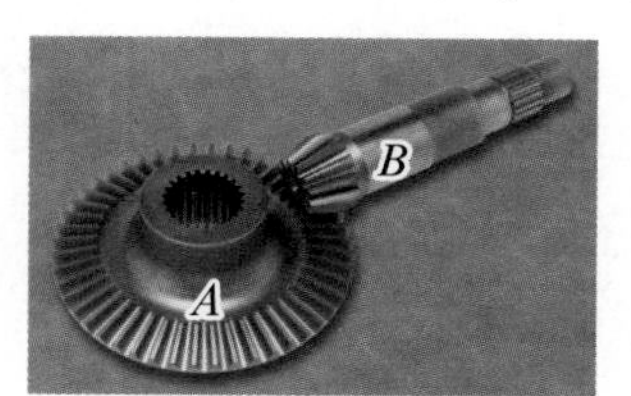

Figure P6.8

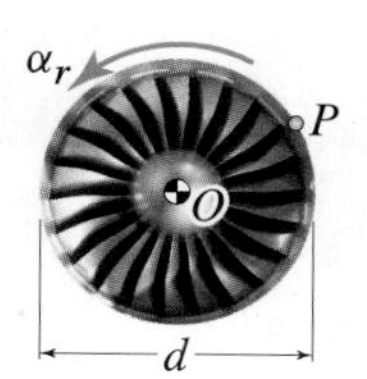

Figure P6.9

Problem 6.9

A rotor with a fixed spin axis through point O is accelerated from rest with a constant angular acceleration $\alpha_r = 0.5\,\text{rad/s}^2$. If the rotor's diameter is $d = 15$ ft, determine the time it takes for the point P on the outer edge of the rotor to reach a speed $v_P = 300$ ft/s and the magnitude of the acceleration of P when the speed v_P is achieved.

Problem 6.10

A Pelton turbine (a type of turbine used in hydroelectric power generation) is spinning at 1100 rpm when the water jets acting on it are shut off, thus causing the turbine to slow down. Assuming that the angular deceleration rate is constant and equal to $1.31\,\text{rad/s}^2$, determine the time it takes for the turbine to stop. In addition, determine the number of revolutions of the turbine during the spin-down.

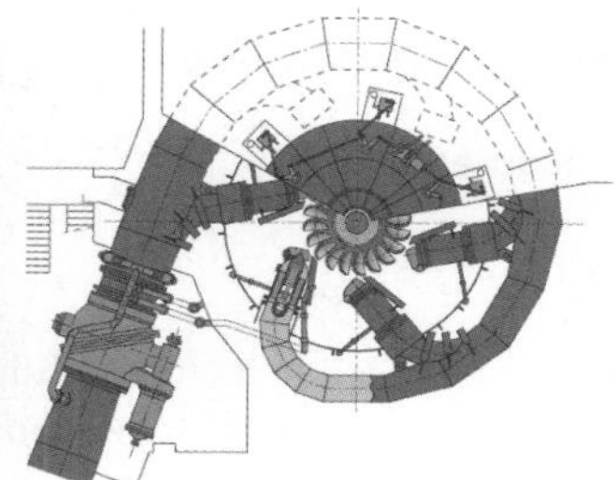

Figure P6.10

Problem 6.11

The sprinkler shown consists of a pipe AB mounted on a hollow vertical shaft. The water comes in the horizontal pipe at O and goes out the nozzles at A and B, causing the pipe to rotate. Letting $d = 7$ in., determine the angular velocity of the sprinkler $\vec{\omega}_s$, and $|\vec{a}_B|$, the magnitude of the acceleration of B, if B is moving with a constant speed $v_B = 20$ ft/s. Assume that the sprinkler does not roll on the ground.

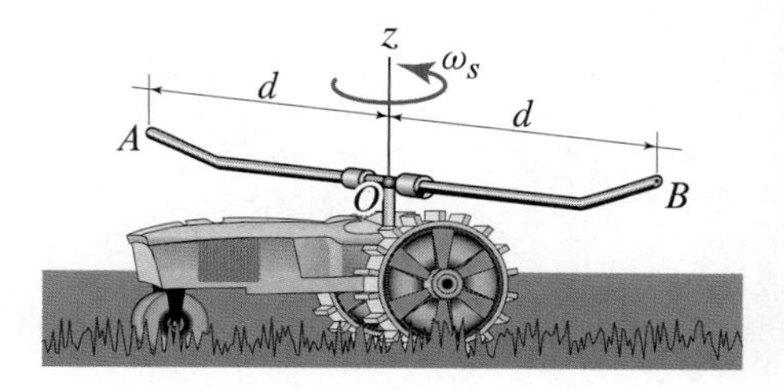

Figure P6.11

Problem 6.12

In a carnival ride, two gondolas spin in opposite directions about a fixed axis. If $\ell = 4$ m, determine the maximum constant angular speed of the gondolas if the magnitude of the acceleration of point A is not to exceed $2.5g$.

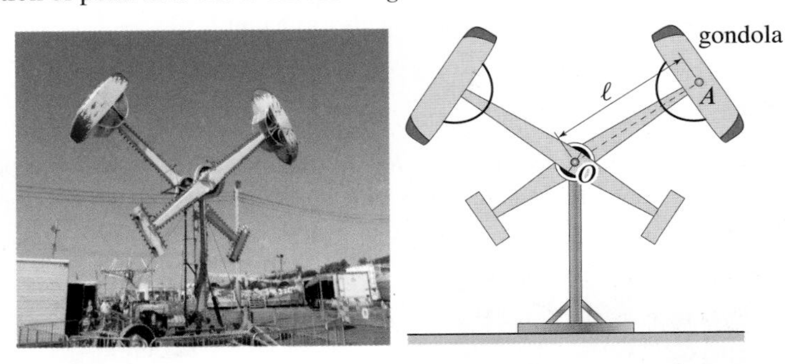

Figure P6.12

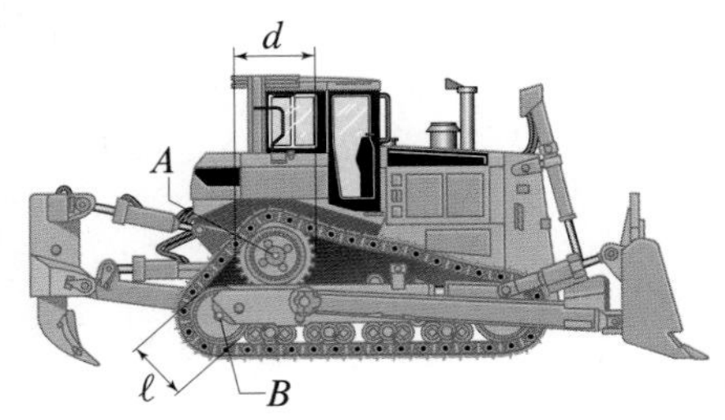

Figure P6.13

Problem 6.13

The tractor shown is stuck with its right track off the ground, and therefore the track is able to move without causing the tractor to move. Letting the diameter of sprocket A be $d = 2.5$ ft and the diameter of sprocket B be $\ell = 2$ ft, determine the angular speed of sprocket B if the sprocket A is rotating at 1 rpm.

Problem 6.14

A battering ram is suspended in its frame with bars AD and BC, which are identical and pinned at their endpoints. At the instant shown, point E on the ram moves with a speed $v_0 = 15$ m/s. Letting $H = 1.75$ m and $\theta = 20°$, determine the magnitude of the angular velocity of the ram at this instant.

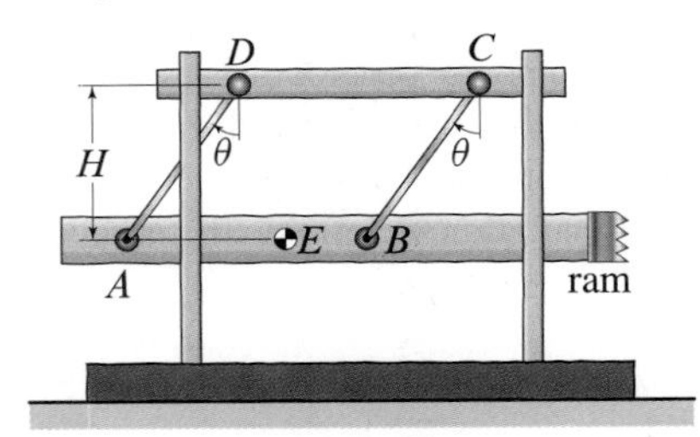

Figure P6.14

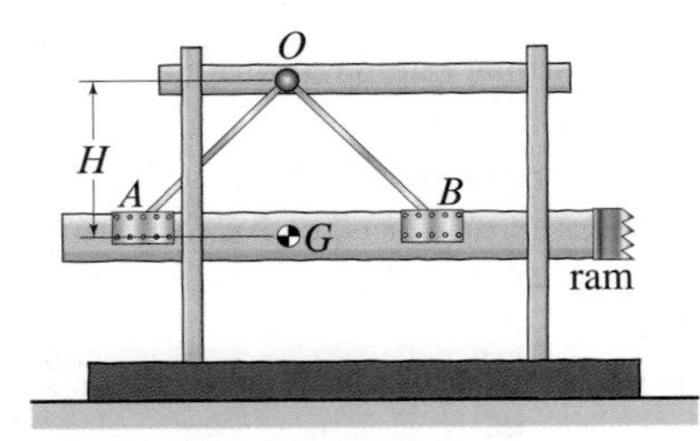

Figure P6.15

Problem 6.15

A battering ram is suspended in its frame with bars OA and OB, which are pinned at O. At the instant shown, point G on the ram is moving to the right with a speed $v_0 = 15$ m/s. Letting $H = 1.75$ m, determine the angular velocity of the ram at this instant.

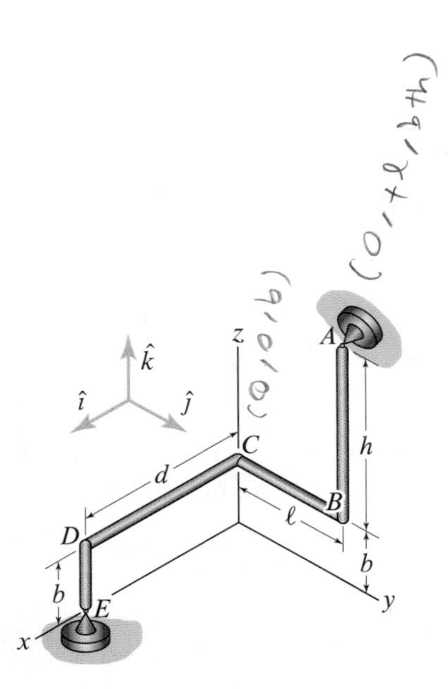

Figure P6.16–P6.18

Problems 6.16 through 6.18

The bent rod rotates about an axis connecting points A and E. All bends in the rod are 90° angles, and the given dimensions are $h = 21$ cm, $\ell = 14.5$ cm, $d = 21$ cm, and $b = 7.6$ cm. Express all your answers using the component system shown.

Problem 6.16 If, at the instant shown, the rod is rotating with a constant angular speed of 25 rad/s and the rotation is counterclockwise as viewed from A looking toward E, determine the velocity and acceleration of point C.

Problem 6.17 If, at the instant shown, the rod is rotating with an angular speed of 20 rad/s, it is decreasing at a rate of $14\,\text{rad/s}^2$, and the rotation is clockwise as viewed from A looking toward E, determine the velocity and acceleration of point B.

Problem 6.18 If, at the instant shown, the rod is rotating with an angular speed of 15 rad/s, it is increasing at a rate of $8\,\text{rad/s}^2$, and the rotation is clockwise as viewed from A looking toward E, determine the velocity and acceleration of point C.

Problems 6.19 through 6.21

The rectangular block is attached to a rod that runs through the block along a diagonal. The rod is mounted in bearings at A and B that allow it to rotate about its own axis. The given dimensions are $h = 8.5\,\text{cm}$, $\ell = 20\,\text{cm}$, and $d = 22\,\text{cm}$. Express all your answers using the component system shown.

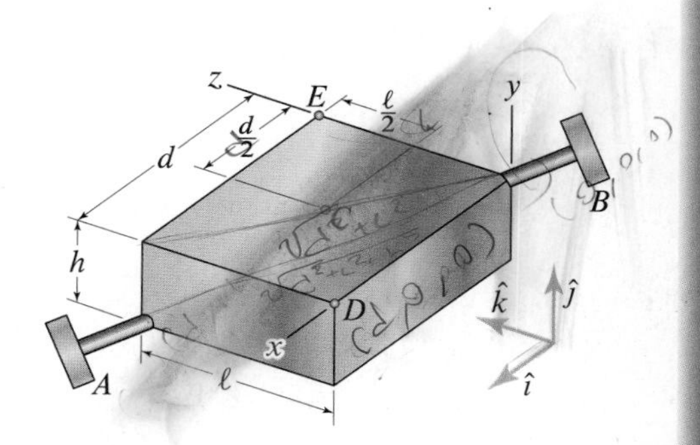

Figure P6.19–P6.21

Problem 6.19 If, at the instant shown, the rod is rotating with an angular speed of 18 rad/s, it is increasing at a rate of $3\,\text{rad/s}^2$, and the rotation is counterclockwise as viewed from A looking toward B, determine the velocity and acceleration of point E.

Problem 6.20 If, at the instant shown, the rod is rotating with an angular speed of 35 rad/s, it is decreasing at a rate of $5\,\text{rad/s}^2$, and the rotation is counterclockwise as viewed from A looking toward B, determine the velocity and acceleration of point D.

Problem 6.21 If, at the instant shown, the rod is rotating with an angular speed of 50 rad/s, it is increasing at a rate of $10\,\text{rad/s}^2$, and the rotation is counterclockwise as viewed from A looking toward B, determine the velocity and acceleration of point C.

Problem 6.22

At the instant shown, the paper is being unrolled with a speed $v_p = 7.5\,\text{m/s}$ and an acceleration $a_p = 1\,\text{m/s}^2$. If, at this instant, the outer radius of the roll is $r = 0.75\,\text{m}$, determine the angular velocity ω_s and acceleration α_s of the roll.

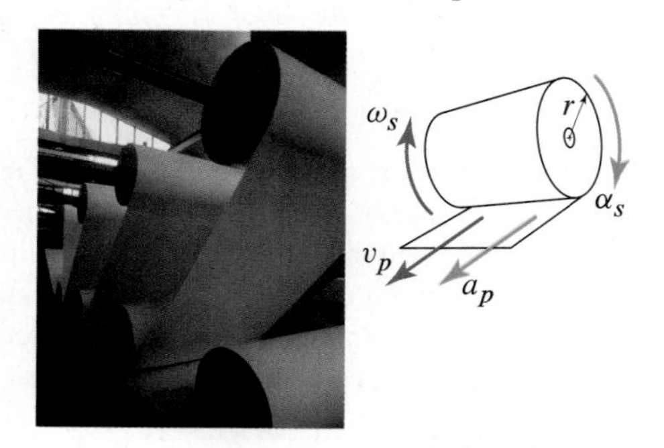

Figure P6.22

Problem 6.23

At the instant shown, the propeller is rotating with an angular velocity $\omega_p = 400\,\text{rpm}$ in the positive z direction and it is slowing down at $2\,\text{rad/s}^2$, where the z axis is also the spin axis of the propeller. Compute the velocity and acceleration of Q, which is 14 ft away from the spin axis. Express your answer in the cylindrical component system shown, which has its origin at O on the z axis and its unit vector $\hat{u}_R$ pointing radially toward point Q.

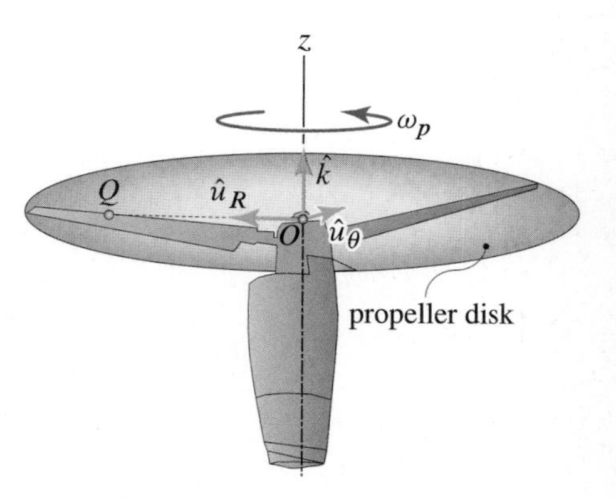

Figure P6.23

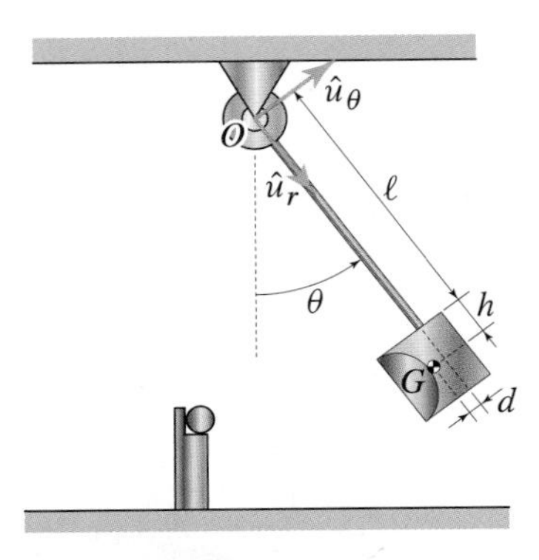

Figure P6.24 and P6.25

Problems 6.24 and 6.25

The hammer of a Charpy impact toughness test machine has the geometry shown, where G is the mass center of the hammer head. Use Eqs. (6.8) and (6.13) and write your answers in terms of the component system shown.

Problem 6.24 Determine the velocity and acceleration of G, assuming $\ell = 500\,\text{mm}$, $h = 65\,\text{mm}$, $d = 25\,\text{mm}$, $\dot{\theta} = -5.98\,\text{rad/s}$, and $\ddot{\theta} = -8.06\,\text{rad/s}^2$.

Problem 6.25 Determine the velocity and acceleration of G as a function of the geometric parameters shown, $\dot{\theta}$ and $\ddot{\theta}$.

Problem 6.26

The bucket of a backhoe is being operated while holding the arm OA fixed. At the instant shown, point B has a horizontal component of velocity $v_0 = 0.25\,\text{ft/s}$ and is vertically aligned with point A. Letting $\ell = 0.9\,\text{ft}$, $w = 2.65\,\text{ft}$, and $h = 1.95\,\text{ft}$, determine the velocity of point C. In addition, assuming that, at the instant shown, point B is not accelerating in the horizontal direction, compute the acceleration of point C. Express your answers using the component system shown.

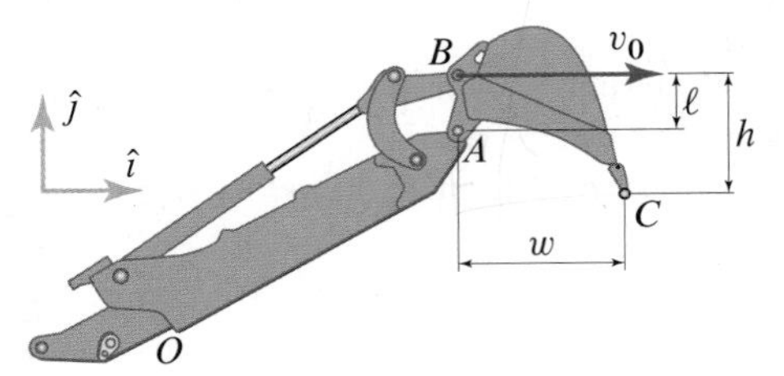

Figure P6.26

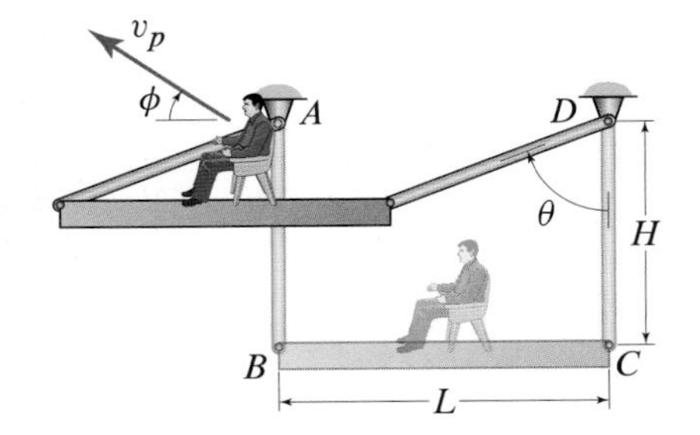

Figure P6.27

Problem 6.27

In a contraption built by a fraternity, a person is sitting at the center of a swinging platform with length $L = 12\,\text{ft}$ that is suspended by two identical arms each of length $H = 10\,\text{ft}$. Determine the angle θ and the angular speed of the arms if the person is moving upward and to the left with a speed $v_p = 25\,\text{ft/s}$ at the angle $\phi = 33°$.

Problem 6.28

The wheel A, with diameter $d = 5\,\text{cm}$, is mounted on the shaft of the motor shown and is rotating with a constant angular speed $\omega_A = 250\,\text{rpm}$. The wheel B, with center at the fixed point O, is connected to A with a belt, which does not slip relative to A or B. The radius of B is $R = 12.5\,\text{cm}$. At point C the wheel B is connected to a saw. If point C is at distance $\ell = 10\,\text{cm}$ from O, determine the velocity and acceleration of C when $\theta = 20°$. Express your answers using the component system shown.

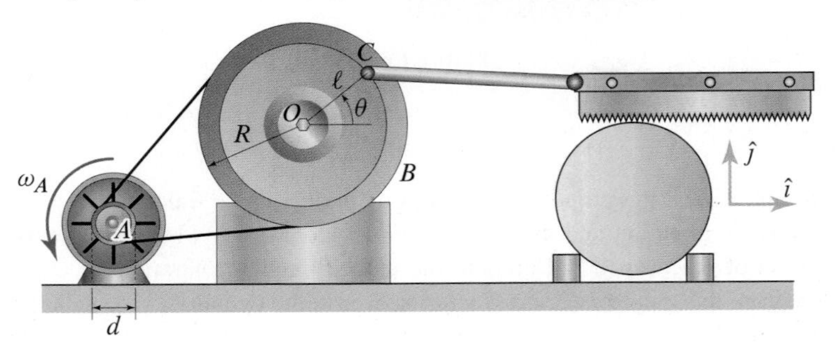

Figure P6.28

Problems 6.29 and 6.30

The mechanism shown is designed to move the tool at H while keeping it oriented vertically. To do so, the rotor in the motor M is attached to the gear A, which drives the gear B. In turn, gear B drives the gear C, which is rigidly attached to the arm EF. Arms EF and DG both have length L and are parallel to one another. The radii of gears A, B, and C are r_A, r_B, and r_C, respectively.

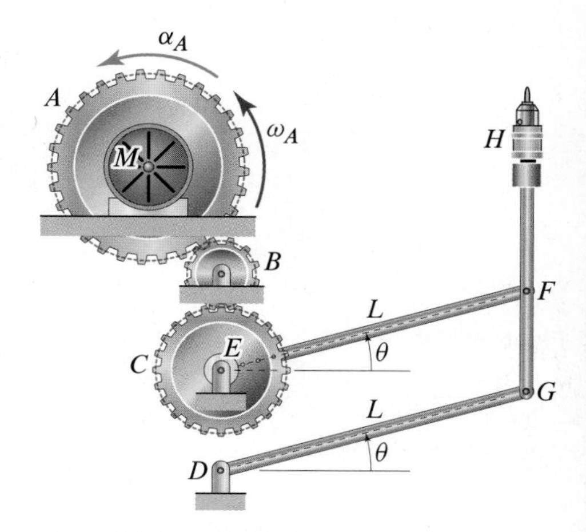

Figure P6.29 and P6.30

Problem 6.29 If the motor A rotates with a constant angular speed ω_A (i.e., $\alpha_A = 0$) in the direction shown, determine the velocity and acceleration of the tool H as functions of the angle θ.

Problem 6.30 If the motor A rotates with angular speed ω_A and angular acceleration in the directions shown, determine the velocity and acceleration of the tool H as functions of the angle θ.

Problem 6.31

An acrobat lands at the end A of a board and, at the instant shown, point A has a downward vertical component of velocity $v_0 = 5.5\,\text{m/s}$. Letting $\theta = 15°$, $\ell = 1\,\text{m}$, and $d = 2.5\,\text{m}$, determine the vertical component of velocity of point B at this instant if the board is modeled as a rigid body.

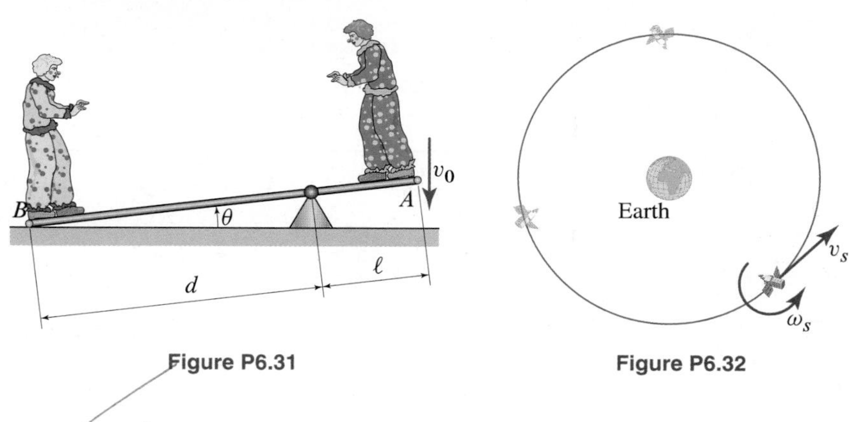

Figure P6.31

Figure P6.32

Problem 6.32

A *geosynchronous equatorial orbit* is a circular orbit above the Earth's equator that has a period of 1 day (these are sometimes called *geostationary orbits*). These geostationary orbits are of great importance for telecommunications satellites because a satellite orbiting with the same angular rate as the rotation rate of the Earth will appear to hover in the same point in the sky as seen by a person standing on the surface of the Earth. Using this information, modeling a geosynchronous satellite as a rigid body, and noting that the satellite has been stabilized so that the same side always faces the Earth, determine the angular speed ω_s of the satellite.

Problem 6.33

Wheels A and C are mounted on the same shaft and rotate together. Wheels A and B are connected by a belt, and so are wheels C and D. The axes of rotation of all the wheels are fixed, and the belts do not slip relative to the wheels they connect. If, at the instant shown, wheel A has an angular velocity $\omega_A = 2\,\text{rad/s}$ and an angular acceleration $\alpha_A = 0.5\,\text{rad/s}^2$, determine the angular velocity and acceleration of wheels B and D. The radii of the wheels are $R_A = 1\,\text{ft}$, $R_B = 0.25\,\text{ft}$, $R_C = 0.6\,\text{ft}$, and $R_D = 0.75\,\text{ft}$.

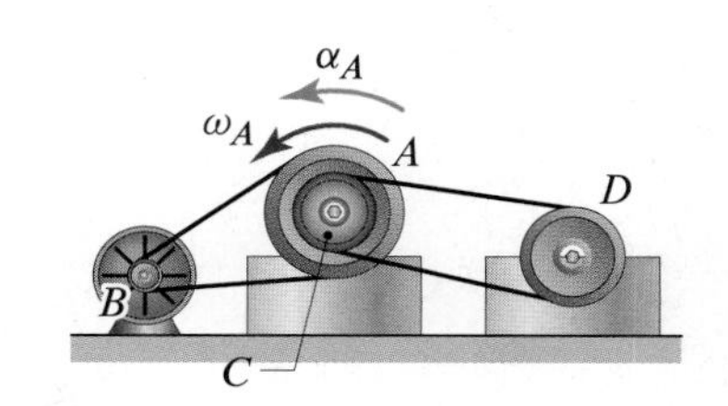

Figure P6.33

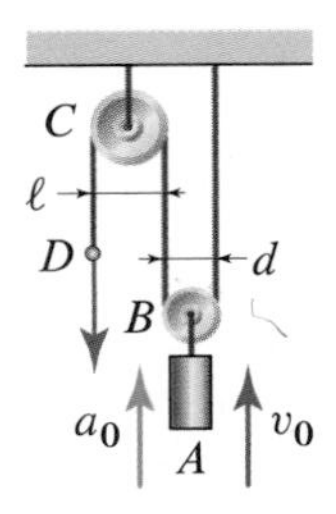

Figure P6.34

Problem 6.34

At the instant shown, A is moving upward with a speed $v_0 = 5\,\text{ft/s}$ and acceleration $a_0 = 0.65\,\text{ft/s}^2$. Assuming that the rope that connects the pulleys does not slip relative to the pulleys, and letting $\ell = 6\,\text{in.}$ and $d = 4\,\text{in.}$, determine the angular velocity and angular acceleration of pulley C.

Problems 6.35 through 6.37

A bicycle has wheels 700 mm in diameter and a gear set with the dimensions given in the table below.

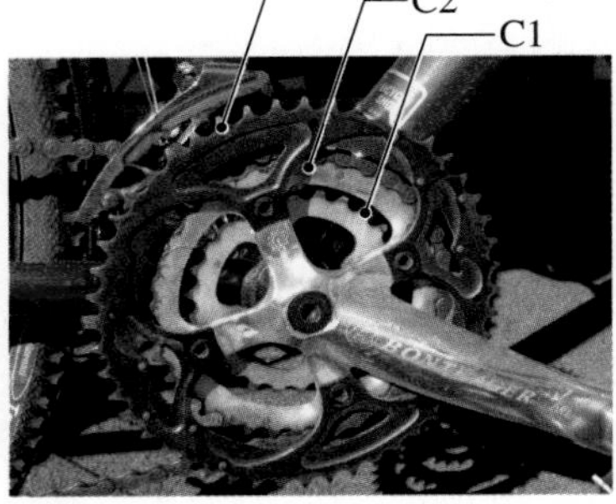

Figure P6.35–P6.37

Crank			
Sprocket	C1	C2	C3
No. of Cogs	26	36	48
Radius (mm)	52.6	72.8	97.0

Cassette (9 speeds)									
Sprocket	S1	S2	S3	S4	S5	S6	S7	S8	S9
No. of Cogs	11	12	14	16	18	21	24	28	34
Radius (mm)	22.2	24.3	28.3	32.3	36.4	42.4	48.5	56.6	68.7

Problem 6.35 If a cyclist has a cadence of 1 Hz, determine the angular speed of the rear wheel in rpm when using the combination of C3 and S2. In addition, knowing that the speed of the cyclist is equal to the radius of the wheel times its angular speed, determine the cyclist's speed in m/s.

Problem 6.36 If a cyclist has a cadence of 68 rpm, determine which combination of chain ring (a sprocket mounted on the crank) and (rear) sprocket would *most closely* make the rear wheel rotate with an angular speed of 127 rpm. Having found a chain ring/sprocket combination, determine the wheel's exact angular speed corresponding to the chosen chain ring/sprocket combination and the given cadence.

Problem 6.37 If a cyclist is pedaling so that the rear wheel rotates with an angular speed of 16 rad/s, determine all possible (rear) sprocket/chain ring (crank-mounted sprocket) combinations that would allow him or her to pedal with a frequency within the range 1.00–1.25 Hz.

Problem 6.38

At the instant shown, the angle $\phi = 30°$, $|\vec{v}_A| = 292\,\text{ft/s}$, and the turbine is rotating clockwise. Letting $\overline{OA} = R$, $\overline{OB} = R/2$, $R = 182\,\text{ft}$, and treating the blades as being equally spaced, determine the velocity of point B at the given instant and express it using the component system shown.

Problem 6.39

At the instant shown, the angle $\phi = 30°$, the turbine is rotating clockwise, and $\vec{a}_B = (70.8\,\hat{\imath} - 12.8\,\hat{\jmath})\,\text{m/s}^2$. Letting $\overline{OA} = R$, $\overline{OB} = R/2$, $R = 55.5\,\text{m}$, and assuming the blades are equally spaced, determine the angular velocity and angular acceleration of the turbine blades, as well as the acceleration of point A at the given instant.

Figure P6.38 and P6.39

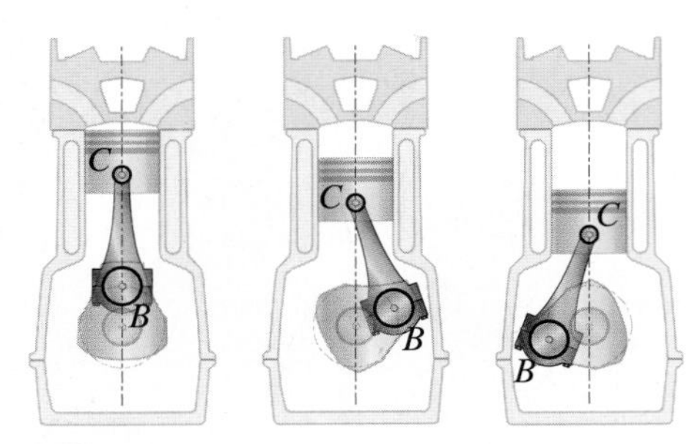

Figure 6.17
Schematic of a slider-crank mechanism emphasizing the motion of the connecting rod.

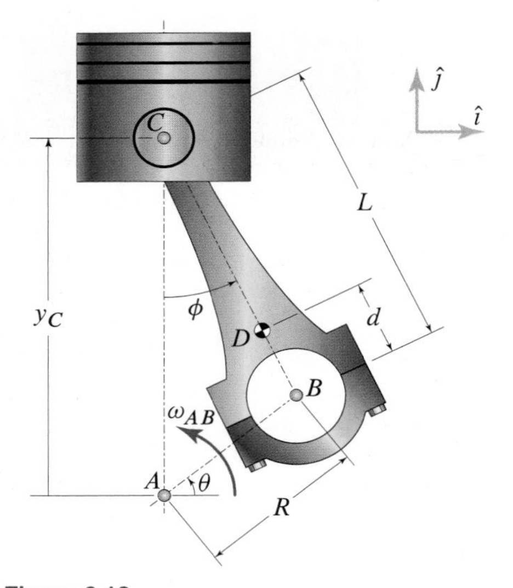

Figure 6.18
Definitions of the geometric parameters used in the slider-crank analysis.

6.2 Planar Motion: Velocity Analysis

In this section, we will develop three different approaches to the velocity analysis of any planar rigid body motion: the vector approach, differentiation of constraints, and instantaneous center of rotation. We will see how each of these approaches applies to the slider-crank mechanism we introduced in Section 6.1 (see Fig. 6.17).

Vector approach

The connecting rod in the slider-crank mechanism in Fig. 6.17 is, in general, planar motion, which means that we can describe the velocity of any of its points by knowing the velocity of one point on the rod and the angular velocity of the rod (see the discussion under *Applying Eqs. (6.3) and (6.5)* on p. 436). On p. 438, we found the velocity of point B, which is in fixed axis rotation about the centerline A of the crank (see Fig. 6.18). Points B and C are on the same rigid body, so we can relate the velocity of C to that of B using Eq. (6.3) on p. 435 in Section 6.1, which gives

$$\boxed{\vec{v}_C = \vec{v}_B + \vec{\omega}_{BC} \times \vec{r}_{C/B}.} \tag{6.15}$$

In planar motion, Eq. (6.15) is a vector equation that represents *two* scalar equations, which allow us to determine $\vec{v}_C$ and $\vec{\omega}_{BC}$. This is so because $\vec{v}_B$ is known, the direction of $\vec{v}_C$ is known (the motion of the piston C is rectilinear along the y axis), $\vec{r}_{C/B}$ can be found in terms of the crank angle θ (see Fig. 6.18), and the axis of rotation for $\vec{\omega}_{BC}$ is known (it is perpendicular to the plane of motion). Therefore, the components v_C and ω_{BC} are the only unknowns in these two scalar equations. We will go through this solution in Example 6.6.

Rolling without slip: velocity analysis

Many applications in engineering involve the motion of disks or wheels rolling over a surface—for example, car wheels rolling on a road, train wheels rolling on tracks, or meshed gears rolling over each other. An important special case of rolling motion is called *rolling without slip* (also called *rolling without slipping* or *rolling without sliding*).

Consider a wheel W rolling over a surface S (S can be moving), as shown in Fig. 6.19. If the wheel and the surface *remain in contact* during the motion and if

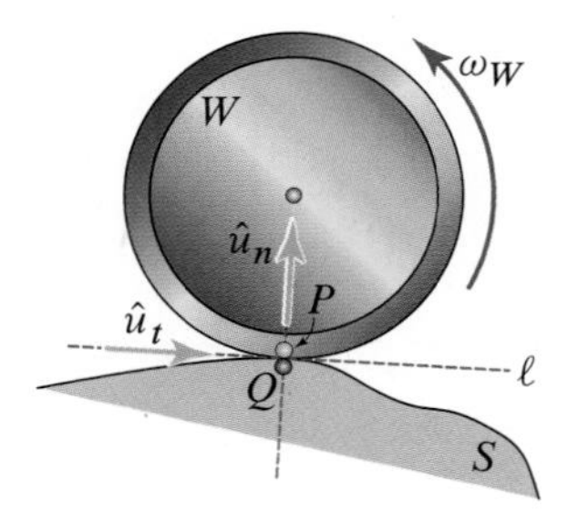

Figure 6.19. A wheel W rolling over a surface S. At the instant shown, the line ℓ is tangent to the two bodies at their point of contact.

the contact points P and Q (on W and S, respectively) move relative to each other, then the only direction in which relative motion can occur at any instant is along the line ℓ that is tangent to both W and S. To say that W is *rolling without slip* over S means that points P and Q *do not move relative to each other*. In terms of velocity, this definition implies that

$$\boxed{\vec{v}_{P/Q} = \vec{0} \quad \Rightarrow \quad \vec{v}_P = \vec{v}_Q.} \tag{6.16}$$

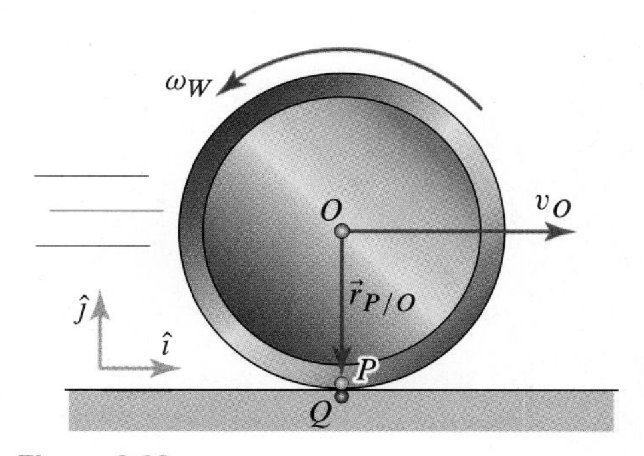

Figure 6.20
A wheel rolling without slip on a horizontal fixed surface.

As an application of Eq. (6.16), consider a wheel of radius R rolling without slipping on a flat stationary surface (Fig. 6.20). To find the wheel's angular velocity when the center of the wheel is moving with speed v_O, we can relate the velocities of points O and P using

$$\vec{v}_P = \vec{v}_O + \vec{\omega}_W \times \vec{r}_{P/O}, \tag{6.17}$$

where, at the instant shown, P is the point on the wheel in contact with the ground, $\vec{v}_O = v_O\,\hat{\imath}$, $\vec{\omega}_W = \omega_W\,\hat{k}$ is the angular velocity of the wheel, and $\vec{r}_{P/O} = -R\,\hat{\jmath}$. Enforcing the no-slip condition given in Eq. (6.16), we have

$$\vec{v}_P = \vec{v}_Q = \vec{0}, \tag{6.18}$$

where Q is the point on the ground in contact with P at this instant and $\vec{v}_Q = \vec{0}$ because the ground is *stationary*. Substituting Eq. (6.18) into Eq. (6.17) and simplifying, we have

$$\vec{0} = v_O\,\hat{\imath} + \omega_W\,\hat{k} \times (-R\,\hat{\jmath}) = v_O\,\hat{\imath} + \omega_W R\,\hat{\imath} \quad \Rightarrow \quad \boxed{\omega_W = -\frac{v_O}{R}.} \tag{6.19}$$

Common Pitfall

***P* is not a fixed point.** Do not interpret Eq. (6.18) as saying that P is fixed because $\vec{v}_P$ is equal to zero. Equation (6.18) *only* holds at the instant when P is touching the ground. In Section 6.3, we will discover that although $\vec{v}_P = \vec{0}$ when P is touching the ground, $\vec{a}_P \neq \vec{0}$ at that instant. That is, P only stops for an instant while it is accelerating away from its current position, so that some other point on the wheel can become the part of the wheel touching the ground.

Differentiation of constraints

Referring to Fig. 6.18, we can find $\vec{\omega}_{BC}$ and $\vec{v}_C$ as functions of θ by writing the appropriate constraint equations and then differentiating them with respect to time. We introduced this idea in Section 2.7 on p. 122, and we now apply it to rigid bodies.

Noting that the piston C only moves in the y direction, $\vec{v}_C = \dot{y}_C\,\hat{\jmath}$. Then, we can write a constraint equation for y_C as

$$y_C = R\sin\theta + L\cos\phi, \tag{6.20}$$

which can be differentiated with respect to time to find

$$\dot{y}_C = R\dot{\theta}\cos\theta - L\dot{\phi}\sin\phi. \tag{6.21}$$

If R, L, and $\theta(t)$ are known, we see that $\dot{\theta} = \omega_{AB}$. To determine ϕ and $\dot{\phi}$, we note that the orientation ϕ of the connecting rod and the crank's orientation θ are related by

$$R\cos\theta = L\sin\phi \quad \Rightarrow \quad \sin\phi = \frac{R}{L}\cos\theta. \tag{6.22}$$

Equation (6.22) can then be differentiated with respect to time to find $\dot{\phi}$, where $\dot{\phi}\,\hat{k} = \vec{\omega}_{BC}$, as a function of θ and $\dot{\theta}$ (see Prob. 6.88), thus concluding our analysis.

Instantaneous center of rotation

When a rigid body rotates about a fixed point Q (where Q is on the body or an extension of the body; see Fig. 6.21), the velocity of any point C on the body is given by

$$\vec{v}_C = \vec{\omega}_{\text{body}} \times \vec{r}_{C/Q}, \tag{6.23}$$

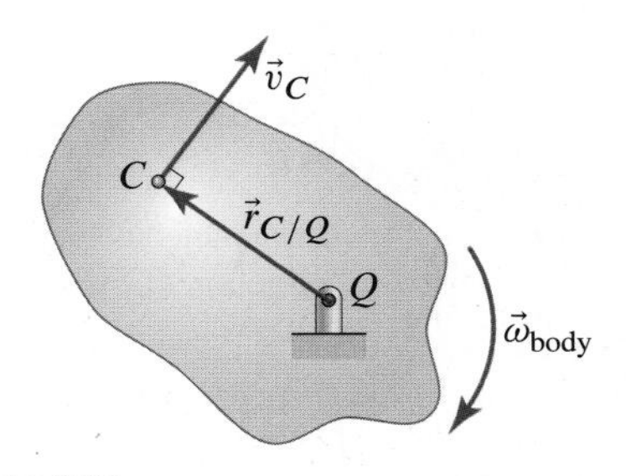

Figure 6.21
A rigid body B rotating about a fixed point Q.

because $\vec{v}_Q = \vec{0}$. If $\vec{v}_Q = \vec{0}$ for all time, we call Q the center of rotation. If $\vec{v}_Q = \vec{0}$ *only at a particular instant in time*, then we call Q the *instantaneous center of rotation* or *instantaneous center* (IC).* If the motion is planar, then $\vec{\omega}_{\text{body}}$ and $\vec{r}_{C/Q}$ are mutually perpendicular, and we can write Eq. (6.23) as

$$|\vec{v}_C| = |\vec{\omega}_{\text{body}}||\vec{r}_{C/Q}| = |\vec{\omega}_{\text{body}}||\vec{r}_{C/\text{IC}}|, \tag{6.24}$$

that is, the speed of C is proportional to its distance from the IC. Because the IC can always be found in planar motion, this formula provides a convenient tool. We now describe the three different possibilities.

Given two nonparallel velocities on a body

Applying the idea of IC to the connecting rod in Fig. 6.22, Eq. (6.23) implies that $\vec{v}_C$

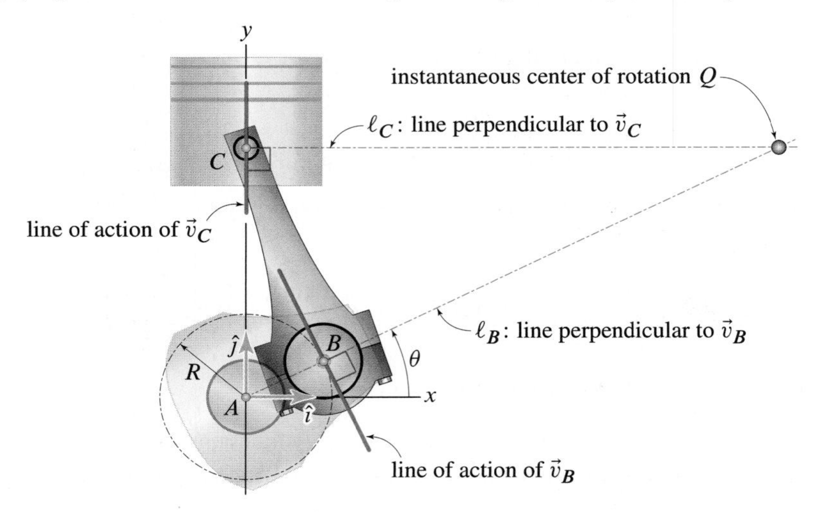

Figure 6.22. Graphical construction for the determination of the IC.

is *perpendicular* to both $\vec{\omega}_{BC}$ and $\vec{r}_{C/Q}$. That means that the IC for the connecting rod must lie in the plane of motion and at the intersection of the line that passes through point C and is perpendicular to $\vec{v}_C$ (line ℓ_C) with the line that passes through B and is perpendicular to $\vec{v}_B$ (line ℓ_B). Thus, at the instant shown, the IC of the connecting rod must be at Q, which is the intersection of ℓ_C and ℓ_B.

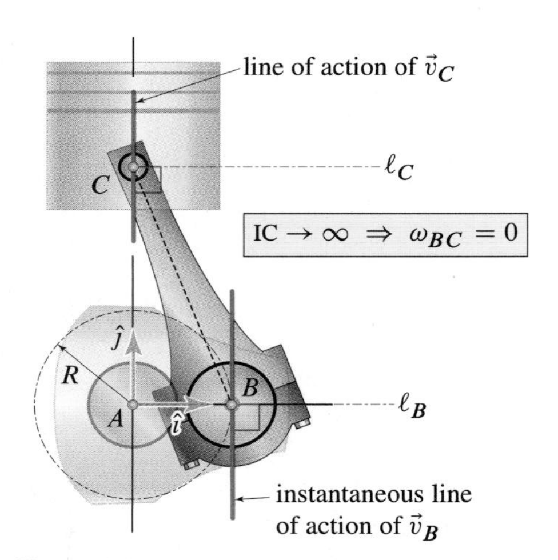

Figure 6.23
At this instant, $\vec{v}_B$ and $\vec{v}_C$ are parallel and ℓ_B and ℓ_C are parallel and distinct. The IC of BC is at infinity, and $\omega_{BC} = 0$.

Given two parallel velocities on a body

The graphical procedure just described does not work if lines ℓ_B and ℓ_C are parallel, of which there are two cases: (1) ℓ_B and ℓ_C are parallel and distinct and (2) ℓ_B and ℓ_C are parallel and coincide.

Case 1: ℓ_B and ℓ_C are parallel and distinct. At the instant when the line AB is perpendicular to AC, $\vec{v}_B$ and $\vec{v}_C$ are parallel, and ℓ_B and ℓ_C are parallel and distinct (see Fig. 6.23). In this case, ℓ_B and ℓ_C intersect at infinity, and the IC is infinitely far away. From Eq. (6.24), the only way that B and C can have finite velocities while being infinitely far from the IC is for the angular velocity of the body to be zero! Thus, if our geometric construction tells us that lines ℓ_B and ℓ_C are *parallel and distinct*, then $\vec{\omega}_{BC} = \vec{0}$ at *that instant* and $\vec{v}_B = \vec{v}_C$.

* The *instantaneous center* is also sometimes called the *instantaneous center of zero velocity*.

Case 2: ℓ_B and ℓ_C coincide. Figure 6.24 shows a wheel rolling without slip while in contact with two parallel surfaces S_1 and S_2 moving with the velocities shown. The no-slip condition at B and C causes $\vec{v}_B$ and $\vec{v}_C$ to be parallel because they must match the velocities of the surfaces S_1 and S_2. The geometrical procedure for the determination of the IC tells us that ℓ_B and ℓ_C coincide. If all we know are the directions of $\vec{v}_B$ and $\vec{v}_C$, then the IC cannot be determined because every point on ℓ_B and ℓ_C is an intersection point for these lines. However, if the values of $v_1 = v_B$ and $v_2 = v_C$ are known, then we *can* find the IC, as well as the body's angular velocity. Referring to Fig. 6.25, recall from Fig. 6.15 on p. 439 that the speed of points on

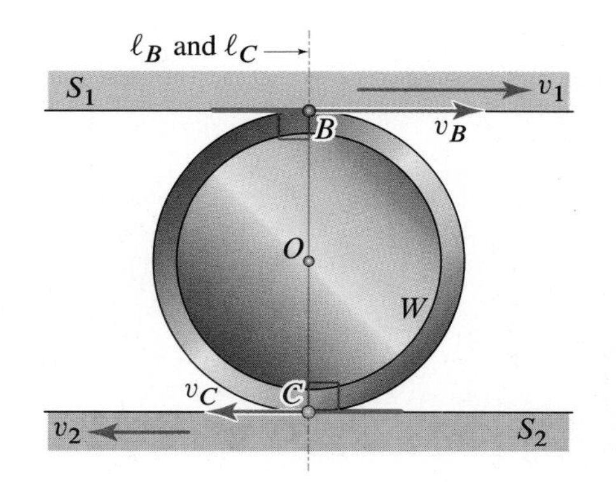

Figure 6.24
Wheel W rolling without slipping over two parallel surfaces.

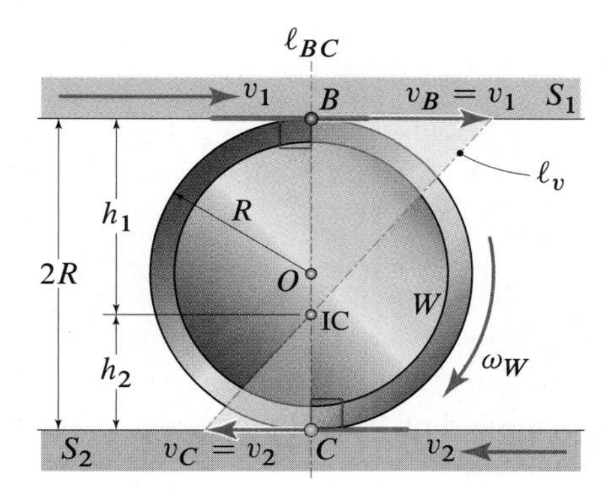

Figure 6.25. Determination of the IC for a wheel W rolling without slipping between two parallel surfaces. The top surface is moving to the right, and the bottom is moving to the left.

line ℓ_{BC} can be graphically represented by right triangles with a vertex at the center of rotation. Therefore, the IC must be at the intersection of ℓ_{BC}, the (radial) line containing B and C, and ℓ_v, the line representing the velocity profile of points on ℓ_{BC}. Since the velocity of points on ℓ_{BC} is proportional to the distance from the IC via the angular speed, we can calculate the angular speed using Eq. (6.11) on p. 438 along with similar triangles as (see Fig. 6.25)

$$\frac{v_B}{h_1} = \frac{v_C}{h_2} = \omega_W. \tag{6.25}$$

Since $h_1 + h_2 = 2R$, Eq. (6.25) becomes

$$\frac{v_B}{\omega_W} + \frac{v_C}{\omega_W} = 2R \quad \Rightarrow \quad \omega_W = \frac{v_B + v_C}{2R}. \tag{6.26}$$

Since the motion is planar, we can assign a direction to ω_W so that our geometrical calculation based on similar triangles allows us to compute the angular velocity of the body.

When both surfaces move in the same direction with known speeds, $\vec{v}_B$ and $\vec{v}_C$ are again parallel and the lines perpendicular to them that run through points B and C coincide in the single line ℓ_{BC} (see Fig. 6.26). The same similar-triangles argument used for the case in Fig. 6.25 tells us that

$$\frac{v_B}{2R + h} = \frac{v_C}{h} = \omega_W. \tag{6.27}$$

Solving the first equality for h and then substituting that into the second, we get

$$h = \frac{2Rv_C}{v_B - v_C} \quad \Rightarrow \quad \omega_W = \frac{v_B - v_C}{2R}. \tag{6.28}$$

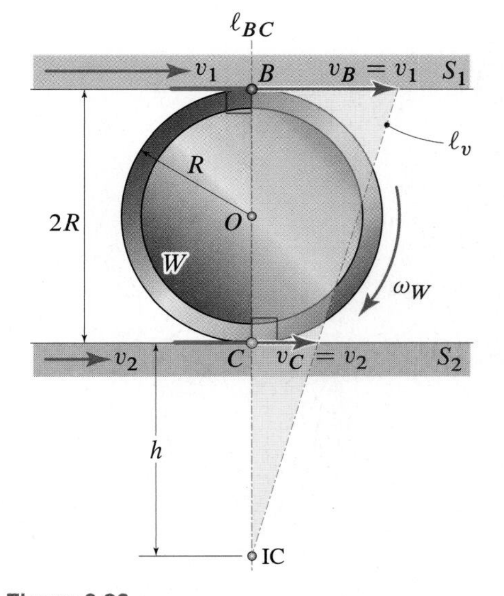

Figure 6.26
Determination of the IC for a wheel W rolling without slipping over two parallel surfaces. Both surfaces are moving to the right, but the top is moving faster than the bottom.

Given a velocity on a body and the body's angular velocity

This case is depicted in Fig. 6.27 for the crank C in an internal combustion engine,

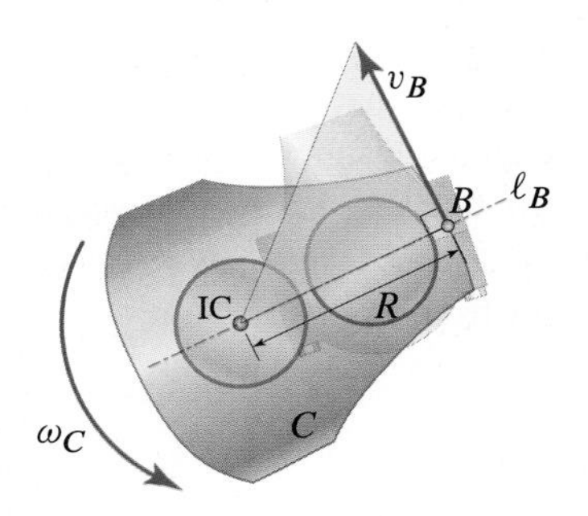

Figure 6.27. Determination of the IC for the case in which the velocity of a point on a body and the body's angular velocity are known.

for which we know the velocity of point B, as well as the angular velocity ω_C of the crank. In this case, the IC is located on line ℓ_B such that the distance from B to the IC is $R = v_B/\omega_C$. We can determine on which side of v_B the IC lies by considering the direction of rotation of the rigid body. In this case, it lies to the left of v_B since the rotation is counterclockwise.

End of Section Summary

This section presents three different ways to analyze the velocities of a rigid body in planar motion: the vector approach, differentiation of constraints, and instantaneous center of rotation.

Vector approach. We saw in Section 6.1 that the equation

Eq. (6.15), p. 452

$$\vec{v}_C = \vec{v}_B + \vec{\omega}_{BC} \times \vec{r}_{C/B},$$

relates the velocity of two points on a rigid body, $\vec{v}_B$ and $\vec{v}_C$, by their relative position $\vec{r}_{C/B}$ and the angular velocity of the body $\vec{\omega}_{BC}$ (see Fig. 6.28).

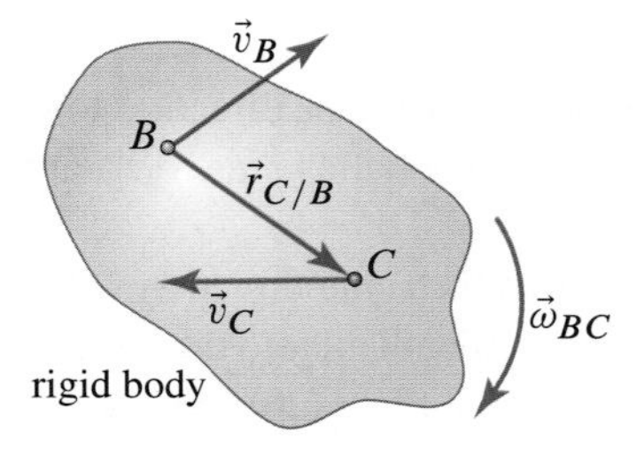

Figure 6.28. A rigid body on which we are relating the velocity of two points B and C.

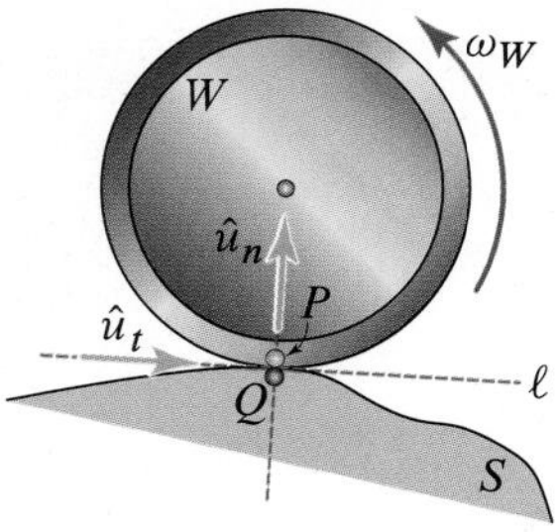

Rolling without slip. When a body rolls without slip over another body, then the two points on the bodies that are in contact at any instant, points P and Q, must have the same velocity (see, for example, the wheel W rolling over the surface S in Fig. 6.29). Mathematically, this means that

Figure 6.29
A wheel W rolling over a surface S. At the instant shown, the line ℓ is tangent to the two bodies at their point of contact.

Eqs. (6.16), p. 453

$$\vec{v}_{P/Q} = \vec{0} \quad \Rightarrow \quad \vec{v}_P = \vec{v}_Q.$$

If a wheel W of radius R is rolling without slipping over a flat, stationary surface, then the point P on the wheel in contact with the surface must have zero velocity (see Fig. 6.30). The consequence is that the angular velocity of the wheel ω_W is related to the velocity of the center v_O and the radius of the wheel R according to

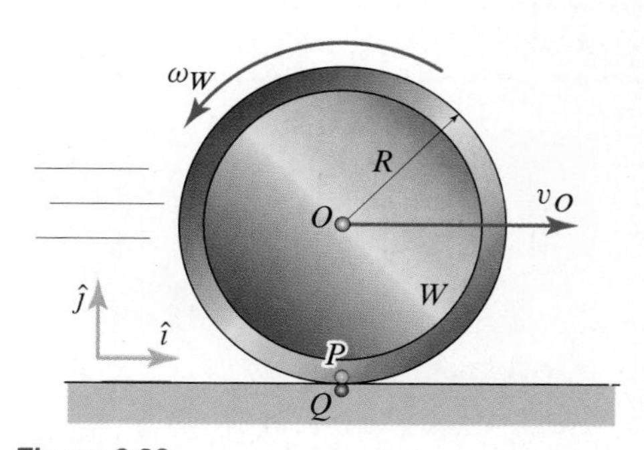

Figure 6.30
A wheel W rolling without slip on a horizontal fixed surface.

Eq. (6.19), p. 453

$$\omega_W = -\frac{v_O}{R},$$

in which the positive direction for ω_W is taken to be the positive z direction using the right-hand rule.

Differentiation of constraints. As we discovered in Section 2.7, it is often convenient to write an equation describing the position of a point of interest, which can then be differentiated with respect to time to find the velocity of that point. For planar motion of rigid bodies, this idea can also apply for describing the position and velocity of a point on a rigid body, as well as for describing the orientation of a rigid body, for which the time derivative provides its angular velocity.

Instantaneous center of rotation. The point on a body (or imaginary extension of the body) whose velocity is zero at a particular instant is called the *instantaneous center of rotation* or *instantaneous center* (IC). The IC can be found geometrically if the velocity is known for two distinct points on a body or if a velocity on the body and the body's angular velocity are known. The three possible geometric constructions are shown in Fig. 6.31.

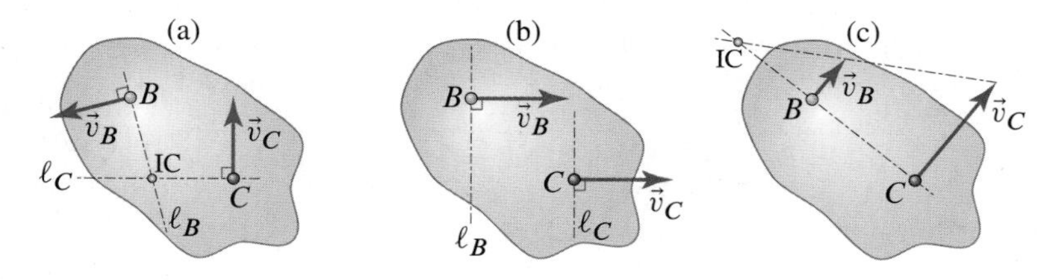

Figure 6.31. The three different possible motions for determining the IC. (a) Two non-parallel velocities are known. (b) The lines of action of two velocities are parallel and distinct; in this case, the IC is at infinity and the body is translating. (c) The lines of action of two parallel velocities coincide.

EXAMPLE 6.4 *A Carnival Ride: Instantaneous Center Analysis*

Figure 1

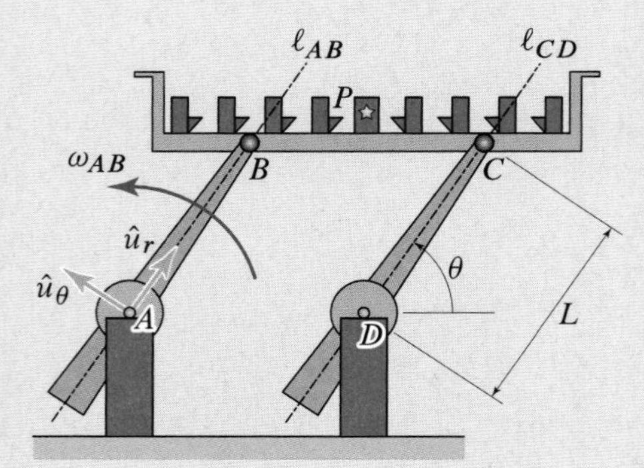

Figure 2
Coordinate definition and geometry for the carnival ride.

Motion platforms, which are used in many of today's amusement parks, are a type of carnival ride in which a platform with seats is made to move always parallel to the ground. Figure 1 shows an elementary type of motion platform, found more often in traveling carnivals than in big amusement parks. Given that the motion platform in the figure (see also Fig. 2) is designed so that the rotating arms AB and CD are of equal length $L = 10\,\text{ft}$ and remain parallel to each other, determine the velocity and acceleration of a person P onboard the ride when ω_{AB} is constant and equal to $1.25\,\text{rad/s}$.

SOLUTION

Road Map To simplify the problem, we will assume that the platform and the persons on it form a single rigid body. This means that the two rotating arms and the platform form a *four-bar linkage*. Because the arms AB and CD are identical in size and are always parallel to each other, the four-bar linkage $ABCD$ always forms a parallelogram. We will use this fact, along with the concept of instantaneous center of rotation, to determine the angular velocity of the platform. Knowing this and applying the kinematics of fixed axis rotation will allow us to determine $\vec{v}_P$ and $\vec{a}_P$.

Computation Referring to Fig. 2, we see that the arms AB and CD are always parallel to each other. Also observe that points B and C move along circles centered at A and D, respectively, where A and D are fixed points. Hence, going through the geometrical procedure to identify the IC of the element BC, we see that lines ℓ_{AB} and ℓ_{CD}, which are perpendicular to $\vec{v}_B$ and $\vec{v}_C$, respectively, are parallel and distinct. This means that the IC of BC is at infinity, and therefore

$$\omega_{BC} = 0. \tag{1}$$

Since the result in Eq. (1) is independent of the value of the angle θ of the arms AB and CD, the motion of element BC is translation. Given the fact that the trajectories of points on BC are not straight lines, BC's motion is a *curvilinear translation*. Consequently, all points on BC, or any rigid extension of it, i.e., any of the passengers, share the same value of velocity, as well as acceleration. In view of this fact, we have

$$\vec{v}_P = \vec{v}_B = \vec{v}_C = \omega_{AB} L\,\hat{u}_\theta = (12.50\,\text{ft/s})\,\hat{u}_\theta, \tag{2}$$

and

$$\vec{a}_P = \vec{a}_B = \vec{a}_C = -\omega_{AB}^2 L\,\hat{u}_r = (-15.62\,\text{ft/s}^2)\,\hat{u}_r, \tag{3}$$

where the velocity and acceleration of B were computed using circular motion formulas (see Eqs. (6.11) on p. 438).

Discussion & Verification The dimensions and units in Eqs. (2) and (3) are correct. The solution of this problem is very elementary; it is performed in a conceptual manner to illustrate the concept of curvilinear translation and its relation to the concept of the IC. As far as the acceleration values are concerned, these are not far from those found in actual general public (as opposed to extreme) carnival rides, and they amount to a little less than 50% of the acceleration of gravity.

EXAMPLE 6.5 *Planetary Gears Rolling Without Slip: Vector Approach*

Planetary gear systems (Fig. 1) are used to transmit power between two shafts (a common application is in car transmissions). The center gear is called the *sun*, the outer gear is called the *ring*, and the inner gears are called *planets*. The planets are mounted on a component called the *planet carrier* (which is not shown in Fig. 1). Referring to Fig. 2, let $R_S = 2\,\text{in.}$, $R_P = 0.67\,\text{in.}$, the ring be fixed, and the angular velocity of the sun gear be $\omega_S = 1500\,\text{rpm}$. Determine $\vec{\omega}_P$ and $\vec{\omega}_{PC}$, the angular velocities of the planet P, and of the planet carrier (*PC*), respectively.

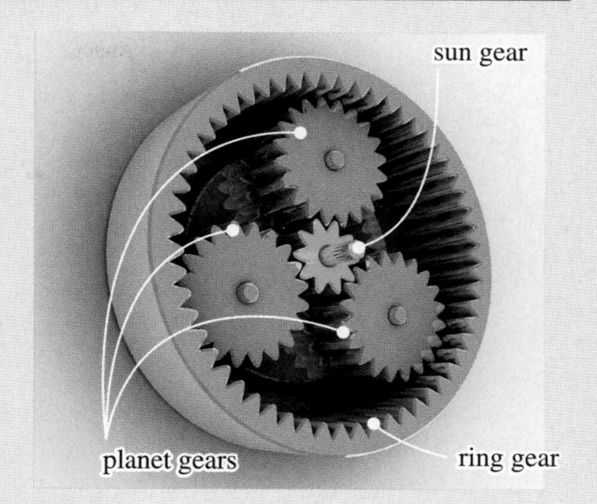

Figure 1
Photo of a planetary gear system.

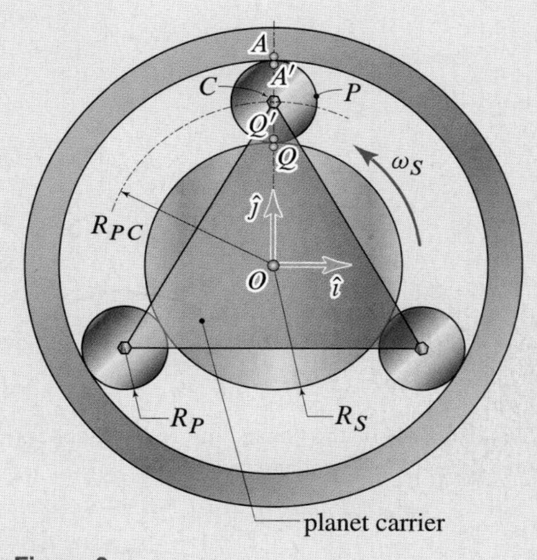

Figure 2
Schematic of a planetary gear system with three planets and a fixed ring.

SOLUTION

Road Map To compute $\vec{\omega}_P$, we need to determine the velocity of two points on P. Two promising candidates are points A' and Q' because their velocities are completely determined by the rolling without slip condition at the A-A' and Q-Q' contacts and because $\vec{v}_A$ and $\vec{v}_Q$ are easily computed. Once $\vec{\omega}_P$ is known, $\vec{\omega}_{PC}$ can be found by finding the velocity of two points on the planet carrier. We will choose O, because its velocity is zero, and C, because it is shared with the planet gear P.

Computation Enforcing the rolling without slip condition at the A-A' and Q-Q' contacts yields

$$\vec{v}_{Q'} = \vec{v}_Q \quad \text{and} \quad \vec{v}_{A'} = \vec{v}_A = \vec{0}, \tag{1}$$

where $\vec{v}_A = \vec{0}$ because the ring is fixed. Next, because the sun gear rotates about the fixed point O, we can write

$$\vec{v}_Q = \vec{\omega}_S \times \vec{r}_{Q/O} = -\omega_S R_S\,\hat{\imath}, \tag{2}$$

where we have used $\vec{\omega}_S = \omega_S\,\hat{k}$ and $\vec{r}_{Q/O} = R_S\,\hat{\jmath}$. In addition, writing $\vec{v}_{Q'}$ using A' as a reference point yields

$$\vec{v}_{Q'} = \vec{v}_{A'} + \vec{\omega}_P \times \vec{r}_{Q'/A'} = 2\omega_P R_P\,\hat{\imath}, \tag{3}$$

where we have used $\vec{\omega}_P = \omega_P\,\hat{k}$ and $\vec{r}_{Q'/A'} = -2R_P\,\hat{\jmath}$, and we have let $v_{A'} = 0$ from Eqs. (1). Substituting Eqs. (2) and (3) into the first of Eqs. (1) gives

$$\boxed{-\omega_S R_S = 2\omega_P R_P \quad \Rightarrow \quad \omega_P = -\frac{R_S}{2R_P}\omega_S = -2239\,\text{rpm}.} \tag{4}$$

Now observe that C is shared by both planet gear P and the planet carrier. This means

$$\vec{v}_C = \vec{v}_{A'} + \omega_P\,\hat{k} \times \vec{r}_{C/A'} = \omega_P R_P\,\hat{\imath}, \tag{5}$$

and

$$\vec{v}_C = \vec{\omega}_{PC} \times \vec{r}_{C/O} = -\omega_{PC} R_{PC}\,\hat{\imath}, \tag{6}$$

where $R_{PC} = R_P + R_S$, $\vec{\omega}_{PC} = \omega_{PC}\,\hat{k}$, and we have enforced the second of Eqs. (1). The two expressions for $\vec{v}_C$ must be equal to each other, that is,

$$\boxed{\omega_{PC} = -\frac{R_P}{R_{PC}}\omega_P = \frac{R_S}{2R_{PC}}\omega_S = 561.8\,\text{rpm},} \tag{7}$$

where we have taken advantage of the solution for ω_P in Eq. (4).

Discussion & Verification To verify the correctness of our results, observe that since $\vec{v}_{A'} = \vec{0}$, point A' is the IC for the planet gear P. Hence, the speeds of point C and Q' are $|\omega_P|R_P$ and $|\omega_P|2R_P$, respectively, which is confirmed by Eqs. (4) and (7).

EXAMPLE 6.6 *Completing the Velocity Analysis of the Connecting Rod*

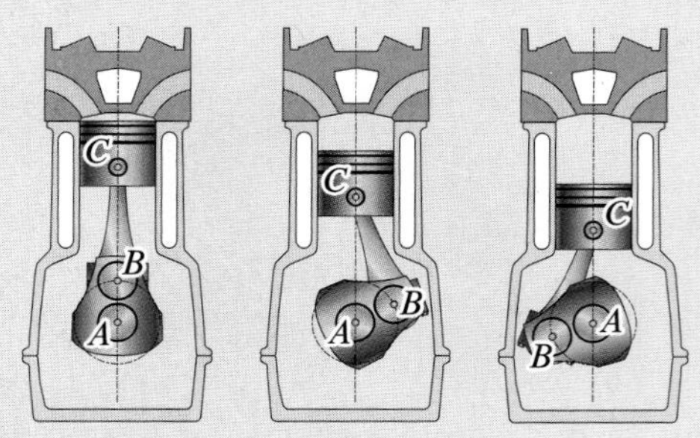

Figure 1

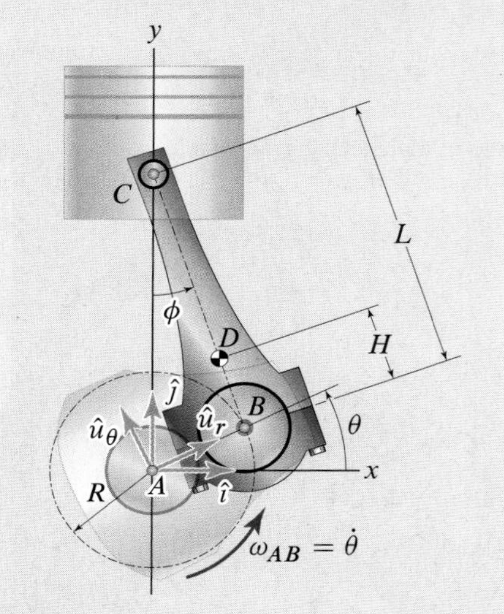

Figure 2
Detailed view of the connecting rod component of a slider-crank mechanism.

On p. 452 we outlined the vector approach for the velocity analysis of the connecting rod (CR) and piston in the slider-crank mechanism shown in Fig. 1. We will complete that analysis here.

Referring to Fig. 2, we are given the radius of the crank R, the length of the CR L, the position of the mass center of the CR H, the angular velocity of the crank ω_{AB}, and the crank angle θ. Determine the angular velocity of the CR $\vec{\omega}_{BC}$, the velocity of the piston $\vec{v}_C$, and the velocity of the mass center of the CR $\vec{v}_D$.

SOLUTION

Road Map The road map for this problem was presented on p. 452.

Computation We begin by recalling that, in Section 6.1, we found $\vec{v}_B$ to be (see Eq. (6.11) on p. 438 and Fig. 2)

$$\vec{v}_B = R\dot{\theta}\,\hat{u}_\theta = R\omega_{AB}(-\sin\theta\,\hat{\imath} + \cos\theta\,\hat{\jmath}), \tag{1}$$

where we have used $\dot{\theta} = \omega_{AB}$ and $\hat{u}_\theta = -\sin\theta\,\hat{\imath} + \cos\theta\,\hat{\jmath}$. Since B and C are both points on the CR, we can relate their velocities using Eq. (6.3) on p. 435, which gives

$$\vec{v}_C = \vec{v}_B + \vec{\omega}_{BC} \times \vec{r}_{C/B}. \tag{2}$$

As discussed on p. 452, Eq. (2) represents two scalar equations in the two unknowns v_C and ω_{BC}. Let's now work out the details to see how.

As for $\vec{\omega}_{BC}$, we can write it as

$$\vec{\omega}_{BC} = \omega_{BC}\,\hat{k}. \tag{3}$$

To complete the right-hand side of Eq. (2), we can write $\vec{r}_{C/B}$ as

$$\vec{r}_{C/B} = L(-\sin\phi\,\hat{\imath} + \cos\phi\,\hat{\jmath}), \tag{4}$$

where we note that the orientation ϕ of the CR and the crank's orientation θ are related by $R\cos\theta = L\sin\phi$, that is,

$$\sin\phi = \frac{R}{L}\cos\theta \quad \text{and} \quad \cos\phi = \frac{\sqrt{L^2 - R^2\cos^2\theta}}{L}. \tag{5}$$

Finally, enforcing the condition $v_{Cx} = 0$, we can write $\vec{v}_C$ as

$$\vec{v}_C = v_{Cy}\,\hat{\jmath}. \tag{6}$$

Substituting Eq. (1) and Eqs. (3)–(6) into Eq. (2) and carrying out the cross product give

$$v_{Cy}\,\hat{\jmath} = -\big(R\omega_{AB}\sin\theta + \omega_{BC}\sqrt{L^2 - R^2\cos^2\theta}\big)\,\hat{\imath} + R\big(\omega_{AB} - \omega_{BC}\big)\cos\theta\,\hat{\jmath}. \tag{7}$$

Equation (7) represents the two scalar equations

$$R\omega_{AB}\sin\theta + \omega_{BC}\sqrt{L^2 - R^2\cos^2\theta} = 0, \tag{8}$$

$$R\big(\omega_{AB} - \omega_{BC}\big)\cos\theta = v_{Cy}, \tag{9}$$

in the unknowns v_{Cy} and ω_{BC}. Solving, we obtain

$$\omega_{BC} = \frac{-\omega_{AB}\sin\theta}{\sqrt{\left(\frac{L}{R}\right)^2 - \cos^2\theta}} \quad \text{and} \quad v_{Cy} = \left[1 + \frac{\sin\theta}{\sqrt{\left(\frac{L}{R}\right)^2 - \cos^2\theta}}\right]R\omega_{AB}\cos\theta, \tag{10}$$

where the solutions have been written to emphasize that, at least for ω_{BC}, the geometry of the mechanism matters only through the ratio L/R. The vectors $\vec{\omega}_{BC}$ and $\vec{v}_C$ are then given by Eqs. (3) and (6), respectively, as

$$\vec{\omega}_{BC} = -\frac{\omega_{AB}\sin\theta}{\sqrt{(L/R)^2 - \cos^2\theta}}\,\hat{k}, \tag{11}$$

$$\vec{v}_C = R\omega_{AB}\cos\theta\left[1 + \frac{\sin\theta}{\sqrt{(L/R)^2 - \cos^2\theta}}\right]\hat{\jmath}. \tag{12}$$

To find $\vec{v}_D$, we note that since $\vec{v}_B$ and $\vec{\omega}_{BC}$ are now both known, $\vec{v}_D$ is readily found by using Eq. (6.3) on p. 435 to relate the motion of D to that of B as

$$\vec{v}_D = \vec{v}_B + \vec{\omega}_{BC} \times \vec{r}_{D/B}. \tag{13}$$

Writing $\vec{r}_{D/B} = H(-\sin\phi\,\hat{\imath} + \cos\phi\,\hat{\jmath})$, substituting Eqs. (1), (3), (5), and (11) into Eq. (13), carrying out the cross product, and simplifying, we obtain

$$\vec{v}_D = R\omega_{AB}\left\{\sin\theta\left(\frac{H}{L} - 1\right)\hat{\imath} + \cos\theta\left[1 + \frac{H\sin\theta}{L\sqrt{(L/R)^2 - \cos^2\theta}}\right]\hat{\jmath}\right\}. \tag{14}$$

Discussion & Verification The answers in Eqs. (11), (12), and (14) are somewhat complicated, but we can see that they have some expected behavior. For example, we expect the piston's velocity to be zero when $\theta = 90°$ and $\theta = 270°$ (when it reaches extreme positions along the y axis), and Eq. (12) tells us that it is. In addition, we expect the angular velocity of the CR to be zero at $\theta = 0°$ and $\theta = 180°$ since its rotation changes direction at those points — Eq. (11) verifies that this is true. Finally, inspection of our three final results tells us that they are all dimensionally correct.

A Closer Look Our results are *general* because they apply for any value of θ and ω_{AB}, as well as for any possible values of R, L, and H, i.e., the geometry of the mechanism. General relations, such as these, are useful because they allow us to know ω_{BC}, v_{Cy}, and $\vec{v}_D$ for *all* values for the parameters θ, ω_{AB}, R, L, and H when designing these machine components. This ability to see how one or more quantities change as parameters are changed is called *parametric analysis*. We now present plots of ω_{BC} and v_{Cy} for the operating conditions that are typical in car engines.

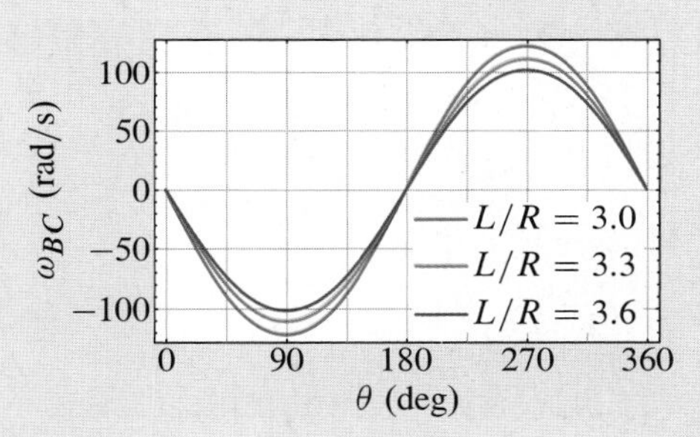

Figure 3
Plot of the angular velocity of the connecting rod for $\omega_{AB} = 3500$ rpm, $L = 150$ mm, and three values of L/R commonly found in practice.

Observe that ω_{BC} and v_{Cy} are periodic functions of θ, so we only need to plot them for one full crank rotation, that is, for $0 \le \theta \le 360°$. In addition, since ω_{BC} and v_{Cy} are directly proportional to ω_{AB}, the plots obtained for one value of ω_{AB} can be rescaled to obtain plots for other ω_{AB} values. Finally, we see that the geometry of the mechanism appears in the equations primarily through the ratio L/R. It is this ratio that is usually found in the analysis of car engine performance. The plots of the functions in Eqs. (11) and (12) are presented in Figs. 3 and 4, respectively, for $\omega_{AB} =$ 3500 rpm (e.g., highway cruising), $L = 150$ mm (values for small block engines typically range between 140 and 155 mm), and three commonly found values of L/R. We see that the smaller L/R, the larger are ω_{BC} and v_{Cy}. In addition, as we discussed above, the piston's velocity is zero when $\theta = 90°$ and $\theta = 270°$. Finally, notice that v_{Cy} looks different for $0° \le \theta < 180°$ when compared with $180° \le \theta \le 360°$. This difference is even more pronounced in the behavior of the piston's acceleration, discussed in Section 6.3. This lack of symmetry tends to disappear for larger values of L/R.

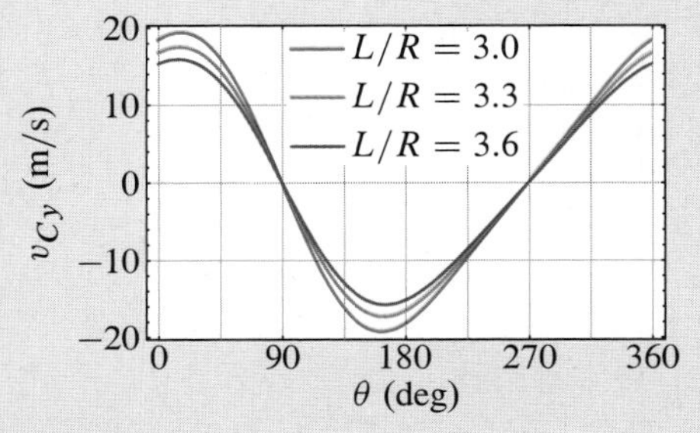

Figure 4
Plot of the piston's velocity for $\omega_{AB} =$ 3500 rpm, $L = 150$ mm, and three values of L/R commonly found in practice.

EXAMPLE 6.7 *A Mechanism with a Slider: Differentiation of Constraints*

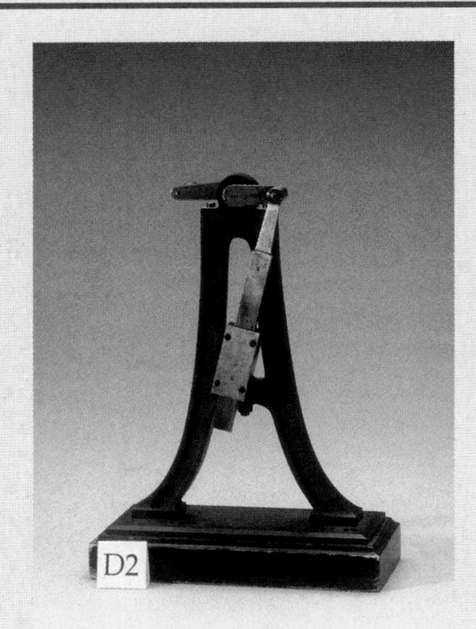

Figure 1
Model illustrating the components of a *swinging block* slider-crank mechanism.

Figure 1 shows a variant of the slider-crank mechanism called a *swinging block* slider crank. First used in various steam locomotive engines in the 1800s, this mechanism is often found in door closing systems (see Fig. 2). Referring to Fig. 3, note that the slider S is directly connected to the crank, and it slides within a swinging block that is free to swing about the pivot at O. For this mechanism we want to derive the relation between the angular velocity of the slider and that of the crank. Also, for $R = 8$ in., $H = 25$ in., $\theta = 20°$, and $\dot{\theta} = 265$ rad/s, we want to determine the velocity of the point P on the slider that is underneath O at this instant.

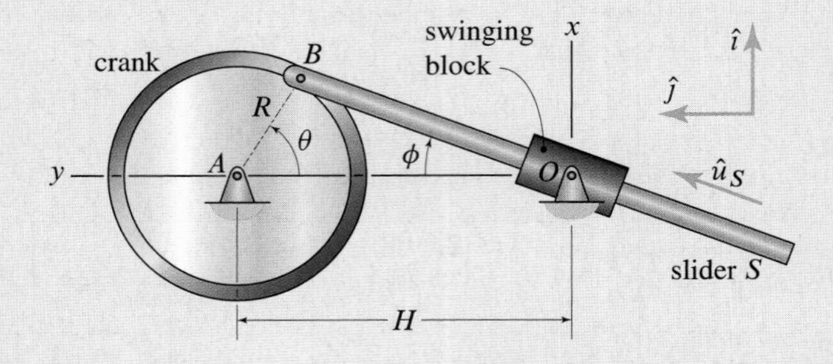

Figure 3. Representation of a *swinging block* slider-crank mechanism with a sliding contact at O.

Figure 2
The pneumatic door closers found on typical storm or screen doors are equivalent to the mechanism shown in Fig. 1.

SOLUTION

Road Map Because θ describes the crank's orientation, the crank's angular velocity is given by $\vec{\omega}_{AB} = \dot{\theta}\,\hat{k}$. Similarly, the slider's angular velocity is $\vec{\omega}_S = -\dot{\phi}\,\hat{k}$, where the minus sign accounts for the fact that if $\dot{\phi} > 0$, the slider rotates clockwise. We can find $\dot{\phi}$ by first relating ϕ to θ and then differentiating the resulting equation with respect to time, which is the differentiation of constraint method of solution. Once $\vec{\omega}_S$ is known, the velocity of any point P on the slider can be found via the relation $\vec{v}_P = \vec{v}_B + \vec{\omega}_S \times \vec{r}_{P/B}$, where point B is on both the crank and the slider.

Computation Focusing on the triangle AOB, throughout the motion we must have

$$R \sin\theta = (H - R\cos\theta)\tan\phi. \tag{1}$$

Differentiating Eq. (1) with respect to time, we obtain

$$R\dot{\theta}\cos\theta = R\dot{\theta}\sin\theta\tan\phi + (H - R\cos\theta)\dot{\phi}\sec^2\phi. \tag{2}$$

Solving Eq. (2) for $\dot{\phi}$ yields

$$\dot{\phi} = \frac{\cos\theta - \sin\theta\tan\phi}{(H - R\cos\theta)\sec^2\phi} R\dot{\theta}, \tag{3}$$

which can be simplified to read

$$\boxed{\dot{\phi} = \frac{R(H\cos\theta - R)\dot{\theta}}{H^2 + R^2 - 2HR\cos\theta} \quad\Rightarrow\quad \vec{\omega}_S = \frac{R(R - H\cos\theta)\dot{\theta}}{H^2 + R^2 - 2HR\cos\theta}\,\hat{k},} \tag{4}$$

where we used Eq. (1) to write $\tan\phi = R\sin\theta/(H - R\cos\theta)$, as well as the identity $\sec^2\phi = 1 + \tan^2\phi$.

For the calculation of $\vec{v}_P$, let t_0 be the instant when $\theta = 20°$. At this instant, we can write

$$\vec{v}_P(t_0) = \vec{v}_B(t_0) + \vec{\omega}_S(t_0) \times \vec{r}_{P/B}(t_0), \tag{5}$$

where P is the point on the slider that, at $t = t_0$, coincides with point O. Since B rotates about the fixed point A, we must have

$$\vec{v}_B = \vec{\omega}_{AB} \times \vec{r}_{B/A} = \dot{\theta}\,\hat{k} \times R(\sin\theta\,\hat{\imath} - \cos\theta\,\hat{\jmath}) = R\dot{\theta}(\cos\theta\,\hat{\imath} + \sin\theta\,\hat{\jmath})$$
$$\Rightarrow \quad \vec{v}_B(t_0) = (166.0\,\hat{\imath} + 60.42\,\hat{\jmath})\ \text{ft/s}, \tag{6}$$

where we have substituted in the given data for θ, $\dot{\theta}$, and R. As for $\vec{r}_{P/B}(t_0)$, since P coincides with the origin, we must have $\vec{r}_P(t_0) = \vec{0}$, so $\vec{r}_{P/B}(t_0) = \vec{r}_P(t_0) - \vec{r}_B(t_0) = -\vec{r}_B(t_0)$. Hence, since $\vec{r}_B = R\sin\theta\,\hat{\imath} + (H - R\cos\theta)\,\hat{\jmath}$, we have

$$\vec{r}_{P/B}(t_0) = -\vec{r}_B(t_0) = (-0.2280\,\hat{\imath} - 1.457\,\hat{\jmath})\ \text{ft}. \tag{7}$$

Next, using Eq. (4), we have

$$\vec{\omega}_S(t_0) = (-104.9\,\hat{k})\ \text{rad/s}. \tag{8}$$

Finally, substituting the results in Eqs. (6), (7), and (8) into Eq. (5), we obtain

$$\boxed{\vec{v}_P(t_0) = (13.20\,\hat{\imath} + 84.34\,\hat{\jmath})\ \text{ft/s}.} \tag{9}$$

Discussion & Verification Intuitively, we would expect $|\omega_S|$ to be smaller than $|\dot{\theta}|$ for all θ. This is the case for the result in Eq. (8). By plotting the function $\omega_S/\dot{\theta}$ (see Fig. 4), we see that $|\omega_S/\dot{\theta}| < 1$ for any θ, that is, $|\omega_S|$ behaves as expected.

A Closer Look Referring to Fig. 3, notice that the axes of the slider and of the swinging block must always coincide; otherwise the mechanism would jam. We can express this condition by saying that, on an instant-by-instant basis, given a point Q on the slider having the same x and y coordinates of a corresponding point Q' on the swinging block, we must have

$$\vec{v}_{Q/Q'} = v_{Q/Q'}\,\hat{u}_S \quad \text{with} \quad \hat{u}_S = \sin\phi\,\hat{\imath} + \cos\phi\,\hat{\jmath}, \tag{10}$$

where $\hat{u}_S$ is a unit vector identifying the orientation of the slider's axis (see Fig. 3). Recall that at time t_0, P has the same x and y coordinates as point O, which is a fixed point. Therefore, rewriting Eq. (10) for points P and O, we have

$$\vec{v}_{P/O}(t_0) = \vec{v}_P(t_0) - \vec{0} = v_P(t_0)\,\hat{u}_S(t_0). \tag{11}$$

Equation (11) says that the vectors $\hat{u}_S(t_0)$ and $\vec{v}_P(t_0)$ must be parallel. This gives the opportunity to check our calculations by comparing the direction of these two vectors. From the second of Eqs. (10), the direction of $\hat{u}_S$ can be expressed as

$$\frac{(\hat{u}_S(t_0))_x}{(\hat{u}_S(t_0))_y} = \frac{\sin\phi(t_0)}{\cos\phi(t_0)} = \tan\phi_0 = 0.1565, \tag{12}$$

where we used Eq. (1) to compute $\tan\phi$ and evaluate it at $t = t_0$. Repeating the calculation for $\vec{v}_P(t_0)$, using Eq. (9), we have

$$\frac{(v_P(t_0))_x}{(v_P(t_0))_y} = \frac{13.20}{84.34} = 0.1565, \tag{13}$$

which implies that $\vec{v}_P$ has the direction we expected.

Helpful Information

Why is P not shown in Fig. 3? The point P cannot be seen in Fig. 3 because it is inside the swinging block pivoted at O and because it happens to coincide with O itself.

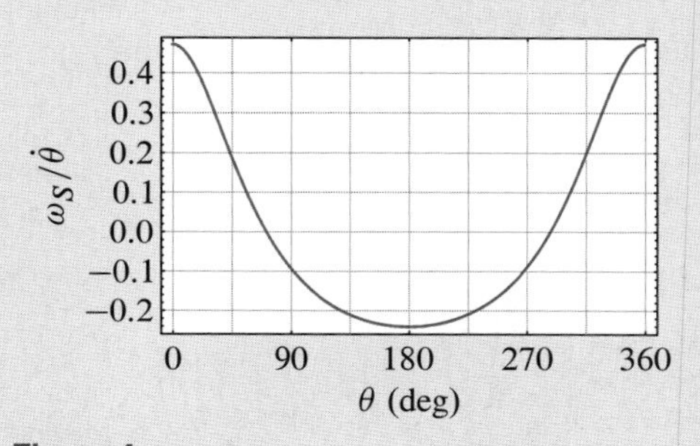

Figure 4
Plot of the function $\omega_S/\dot{\theta}$ as a function of θ over an entire cycle. The expected behavior of ω_S is to always be smaller than $\dot{\theta}$ in absolute value.

EXAMPLE 6.8 *Velocity Analysis of a Four-Bar Linkage: Vector Approach*

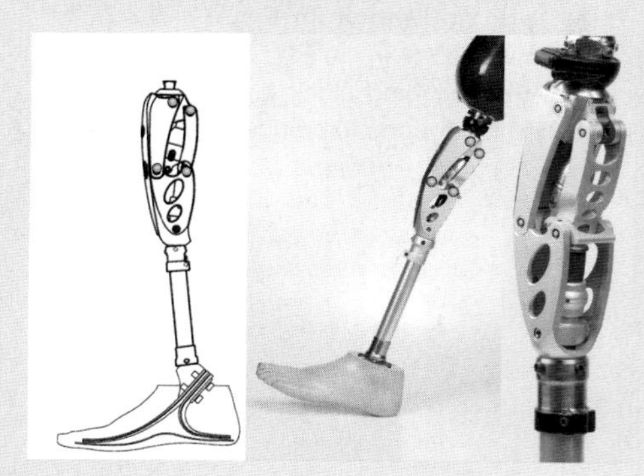

Figure 1

Figure 1 shows three views of a prosthetic leg with an artificial knee joint. The primary kinematic component of this artificial knee joint is a four-bar linkage (see the right panel in Fig. 1 and the system $ABCD$ in Fig. 2). The four-bar linkage consists of the four segments AB, BC, CD, and DA, which are pin-connected and can therefore rotate relative to one another. The mechanism is built in such a way that, given the motion of two of its segments, the motion of the other two is uniquely determined. By varying the relative proportions of its elements, a four-bar linkage system can provide a large variety of controlled motions, and for this reason, four-bar linkages have myriad applications, including engines, sport machines, prosthesis components, drafting tools, and carnival rides. For the mechanism in Fig. 2, assume that the segment AD is fixed, and using the information given in Table 1 for the instant shown, determine the angular velocity of segment BC (the lower leg) if segment AB rotates counterclockwise with a rate $|\vec{\omega}_{AB}| = 1.5\,\text{rad/s}$, i.e., as if walking forward (negative x direction). The acceleration analysis is presented in Example 6.12 on p. 484.

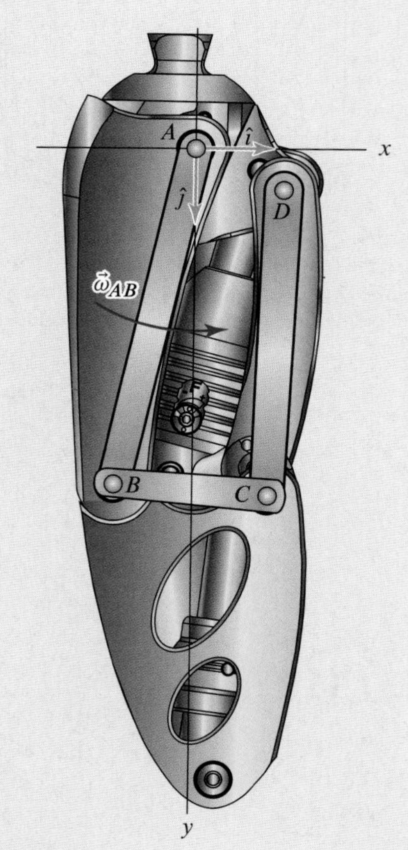

Figure 2
Geometry of the four-bar linkage in the prosthetic leg.

Table 1. Approximate values of the coordinates of the pin centers A, B, C, and D for the system shown in Fig. 2 at the time instant considered.

Points	A	B	C	D
Coordinates (mm)	(0.0, 0.0)	(−27.0, 120)	(26.0, 124)	(30.0, 15.0)

SOLUTION

Road Map This linkage system is a *kinematic chain*, i.e., a system in which motion information is passed from one component to the next along the chain. This means we will start from a point whose motion is known, say, A because it is fixed, and then compute the velocity of point B, which is the next point along the chain $ABCD$, using the equation $\vec{v}_B = \vec{\omega}_{AB} \times \vec{r}_{B/A}$. We repeat this process for segments BC and CD. In doing this, we will generate enough equations to determine the angular velocity of each element along the chain. It is important to keep in mind that the calculations performed in this example hold only at a given instant in time.

Computation For the velocity of B, we have

$$\vec{v}_B = \vec{v}_A + \omega_{AB}\,\hat{k} \times \vec{r}_{B/A}. \tag{1}$$

Letting $\omega_{AB} = -1.5\,\text{rad/s}$, observing that $\vec{r}_{B/A} = \vec{r}_B = (-27\,\hat{\imath} + 120\,\hat{\jmath})\,\text{mm}$, and recalling that $\vec{v}_A = \vec{0}$, Eq. (1) yields

$$\vec{v}_B = (180.0\,\hat{\imath} + 40.50\,\hat{\jmath})\,\text{mm/s}. \tag{2}$$

Since point C is shared by both segments BC and CD, we can express the velocity of C in two independent ways as follows:

$$\vec{v}_C = \vec{v}_B + \omega_{BC}\,\hat{k} \times \vec{r}_{C/B}, \tag{3}$$

and

$$\vec{v}_C = \vec{v}_D + \omega_{CD}\,\hat{k} \times \vec{r}_{C/D}. \tag{4}$$

Observing that

$$\vec{r}_{C/B} = \vec{r}_C - \vec{r}_B = (53.00\,\hat{\imath} + 4.000\,\hat{\jmath})\,\text{mm}, \tag{5}$$

$$\vec{r}_{C/D} = \vec{r}_C - \vec{r}_D = (-4.000\,\hat{\imath} + 109.0\,\hat{\jmath})\,\text{mm}, \tag{6}$$

recalling that $\vec{v}_D = \vec{0}$, and noting that the $\vec{v}_C$ obtained from Eq. (3) must be the same as that obtained from Eq. (4), we obtain

$$[(180.0\,\tfrac{\text{mm}}{\text{s}}) - (4.000\,\text{mm})\omega_{BC}]\,\hat{\imath} + [(40.50\,\tfrac{\text{mm}}{\text{s}}) + (53.00\,\text{mm})\omega_{BC}]\,\hat{\jmath}$$
$$= -(109.0\,\text{mm})\omega_{CD}\,\hat{\imath} - (4.000\,\text{mm})\omega_{CD}\,\hat{\jmath}. \quad (7)$$

Equation (7) is a vector equation equivalent to a linear system of two scalar equations in the unknowns ω_{BC} and ω_{CD}. These equations are

$$(180.0\,\tfrac{\text{mm}}{\text{s}}) - (4.000\,\text{mm})\omega_{BC} = -(109.0\,\text{mm})\omega_{CD}, \quad (8)$$
$$(40.50\,\tfrac{\text{mm}}{\text{s}}) + (53.00\,\text{mm})\omega_{BC} = -(4.000\,\text{mm})\omega_{CD}, \quad (9)$$

which can be solved to obtain

$$\boxed{\omega_{BC} = -0.6378\,\text{rad/s} \quad \text{and} \quad \omega_{CD} = -1.675\,\text{rad/s}.} \quad (10)$$

Discussion & Verification The solution we have obtained seems reasonable in that both segment BC and segment CD rotate counterclockwise, as one would expect, when attempting to walk forward. What is interesting is that the proportions of the segments in the mechanism are such that, in the configuration shown, point B moves down and to the right, that is, in a direction that would cause the foot to move into the ground, which, again, is consistent with what happens when we begin to walk forward from a standing position.

Problems

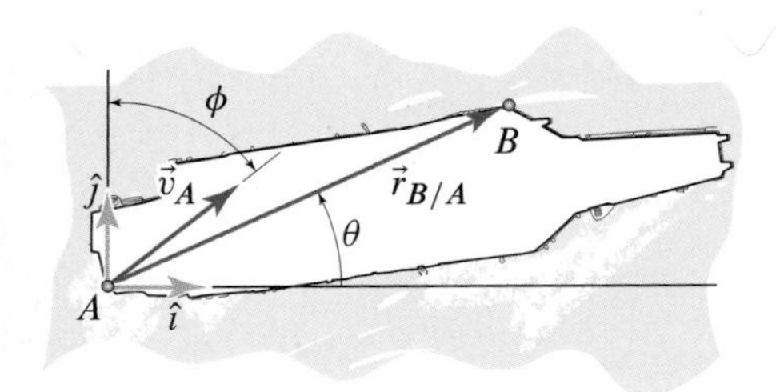

Figure P6.40

Problem 6.40

A carrier is maneuvering so that, at the instant shown, $|\vec{v}_A| = 25$ knots (1 kn is *exactly* equal to 1.852 km/h) and $\phi = 33°$. Letting the distance between A and B be 220 m and $\theta = 22°$, determine $\vec{v}_B$ at the given instant if the ship's turning rate at this instant is $\dot{\theta} = 2°/\text{s}$ clockwise.

Problem 6.41

At the instant shown, the pinion is rotating between two racks with an angular velocity $\omega_P = 55$ rad/s. If the nominal radius of the pinion is $R = 4$ cm and if the lower rack is moving to the right with a speed $v_L = 1.2$ m/s, determine the velocity of the upper rack.

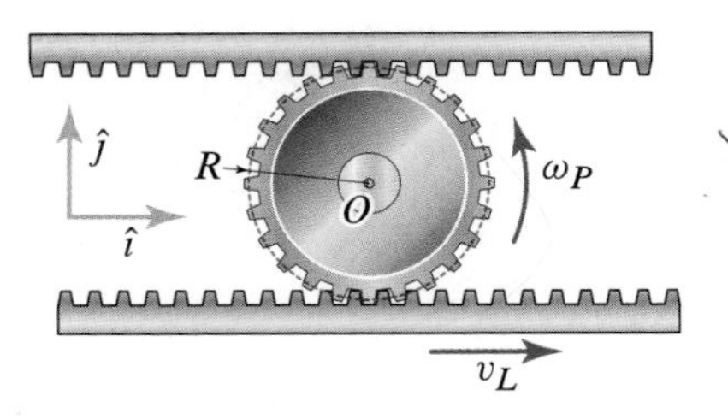

Figure P6.41 and P6.42

Problem 6.42

At the instant shown, the lower rack is moving to the right with a speed of $v_L = 4$ ft/s, while the upper rack is fixed. If the nominal radius of the pinion is $R = 2.5$ in., determine ω_P, the angular velocity of the pinion, as well as the velocity of point O, i.e., the center of the pinion.

Problem 6.43

A bar of length $L = 2.5$ m is pin-connected to a roller at A. The roller is moving along a horizontal rail as shown with $v_A = 5$ m/s. If, at a certain instant, $\theta = 33°$ and $\dot{\theta} = 0.4$ rad/s, compute the velocity of the bar's midpoint C.

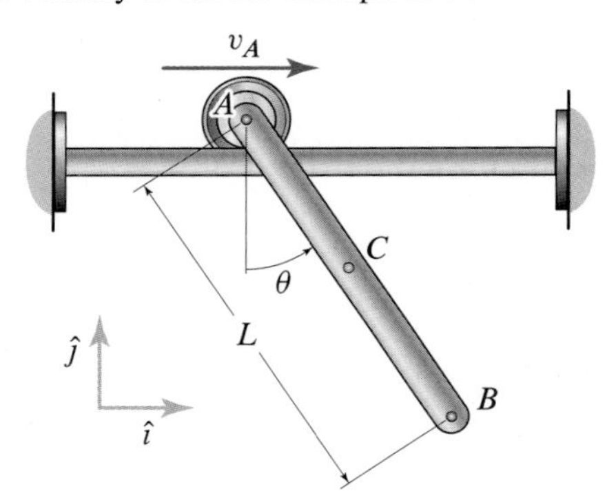

Figure P6.43 and P6.44

Problem 6.44

If the motion of the bar is planar, what would the speed of A need to be for $\vec{v}_C$ to be perpendicular to the bar AB? Why?

Problems 6.45 and 6.46

At the instant shown, the disk of radius r, whose center is at O, is unwinding from the rope, which is attached to the fixed point at D, with the angular speed ω_d.

Problem 6.45 If $r = 6$ in. and $\omega_d = 20$ rad/s, determine the velocities of points A, B, and C.

Problem 6.46 Determine the velocities of points A, B, and C as functions of r and ω_d.

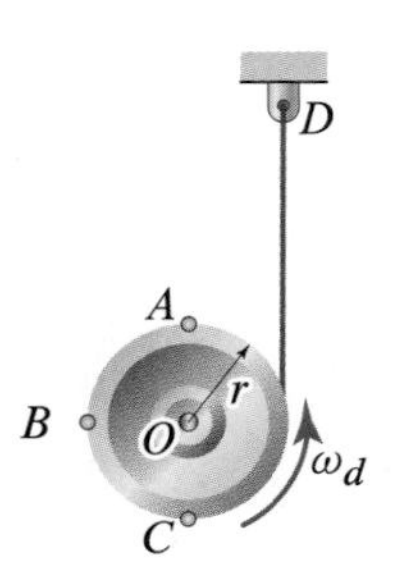

Figure P6.45 and P6.46

Problem 6.47

Points A and B are both on the trailer part of the truck. If the relative velocity of point B with respect to A is as shown, is the body undergoing a planar rigid body motion?

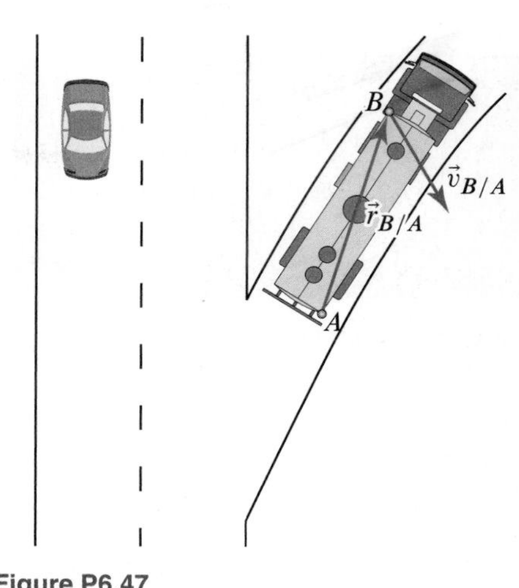

Figure P6.47

Problem 6.48

A truck is moving to the right with a speed $v_0 = 12\,\text{km/h}$, while the pipe section with radius $R = 1.25\,\text{m}$ and center at C rolls without slipping over the truck's bed. The center of the pipe section C is moving to the right at $2\,\text{m/s}$ relative to the truck. Determine the angular velocity of the pipe section and the absolute velocity of C.

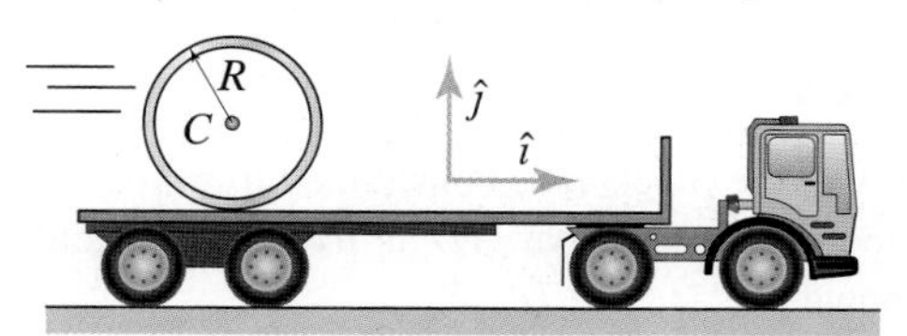

Figure P6.48

Problem 6.49

A wheel W of radius $R_W = 7\,\text{mm}$ is connected to point O via the rotating arm OC, and it rolls without slip over the stationary cylinder S of radius $R_S = 15\,\text{mm}$. If, at the instant shown, $\theta = 47°$ and $\omega_{OC} = 3.5\,\text{rad/s}$, determine the angular velocity of the wheel and the velocity of point Q, where point Q lies on the edge of W and along the extension of the line OC.

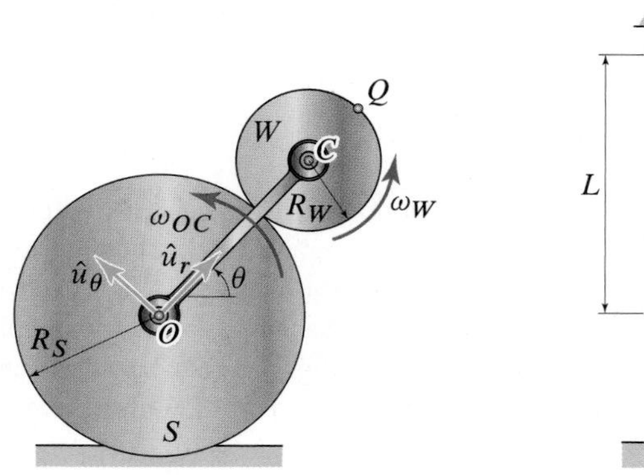

Figure P6.49

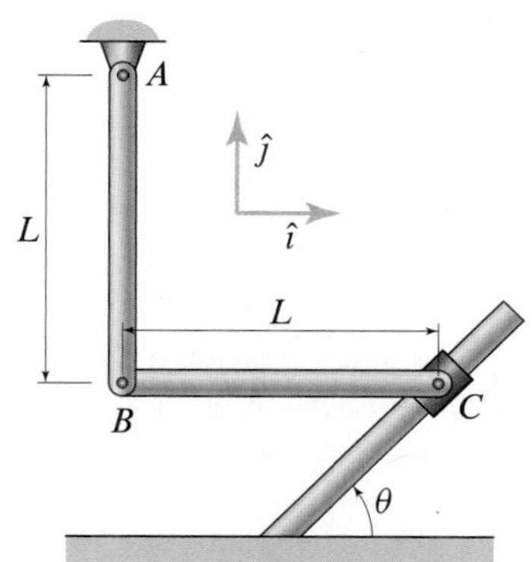

Figure P6.50

Problem 6.50

At the instant shown, bars AB and BC are perpendicular to each other, and bar BC is rotating counterclockwise at $20\,\text{rad/s}$. Letting $L = 2.5\,\text{ft}$ and $\theta = 45°$, determine the angular velocity of bar AB, as well as the velocity of the slider C.

Problem 6.51

For the slider-crank mechanism shown, let $R = 20\,\text{mm}$, $L = 80\,\text{mm}$, and $H = 38\,\text{mm}$. Use the concept of instantaneous center of rotation to determine the values of θ, with $0° \le \theta \le 360°$, for which $v_B = 0$. Also, determine the angular velocity of the connecting rod at these values of θ for $\omega_c = 10\,\text{rad/s}$ in the direction shown.

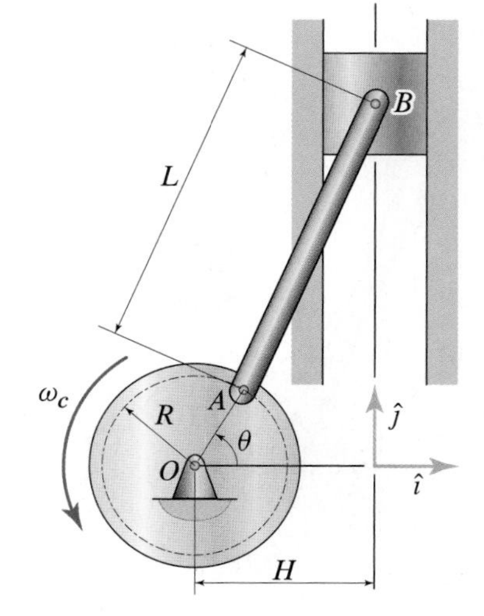

Figure P6.51

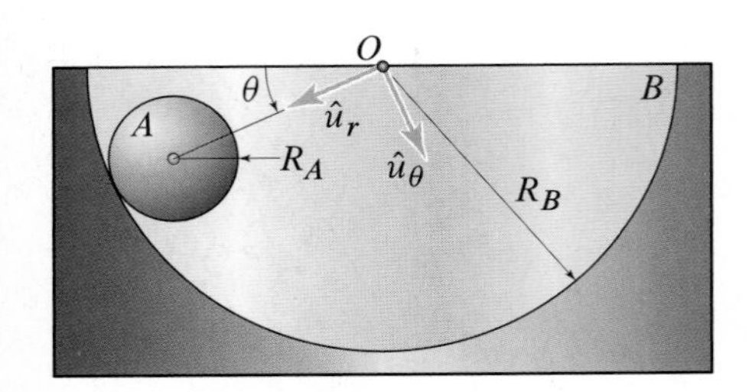

Figure P6.52 and P6.53

Problems 6.52 and 6.53

A ball of radius $R_A = 3$ in. is rolling without slip in a stationary spherical bowl of radius $R_B = 8$ in. Assume that the ball's motion is planar. Express your answers using the component system shown.

Problem 6.52 If the speed of the center of the ball is $v_A = 1.75$ ft/s and if the ball is moving down and to the right, determine the angular velocity of the ball.

Problem 6.53 If the angular speed of the ball $|\omega_A| = 4$ rad/s is counterclockwise, determine the velocity of the center of the ball.

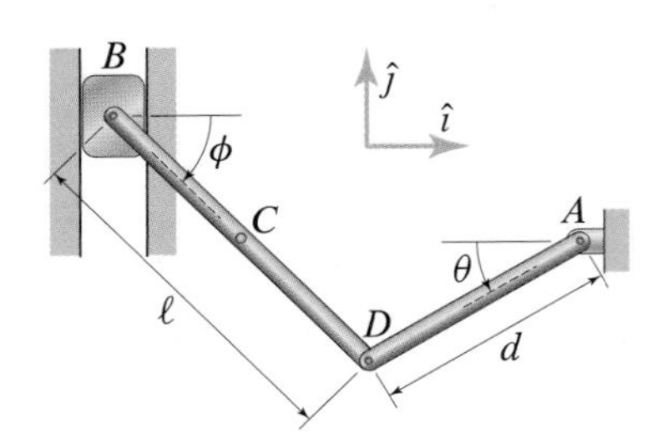

Figure P6.54–P6.56

Problems 6.54 through 6.56

In the mechanism shown, the block B is constrained to move vertically and is attached to the bar BD. The point A on the bar AD is fixed. Express all your answers in the component system shown.

Problem 6.54 At the instant shown, the block B is moving downward at 2.5 ft/s, $\phi = 45°$, and $\theta = 30°$. If $\ell = 12$ in. and $d = 8$ in., determine the angular velocity of bar AD at this instant.

Problem 6.55 At the instant shown, bar AD is rotating counterclockwise at the angular speed $\omega_{AD} = 13$ rad/s, $\phi = 45°$, and $\theta = 30°$. If $\ell = 24$ in. and $d = 16$ in., determine the velocity of the block B at this instant.

Problem 6.56 At the instant shown, the block B is moving downward at 1.5 m/s, $\phi = 45°$, and $\theta = 30°$. If $\ell = 1.2$ m and $d = 0.8$ m, determine the angular velocity of bar AD and the velocity of point C at this instant.

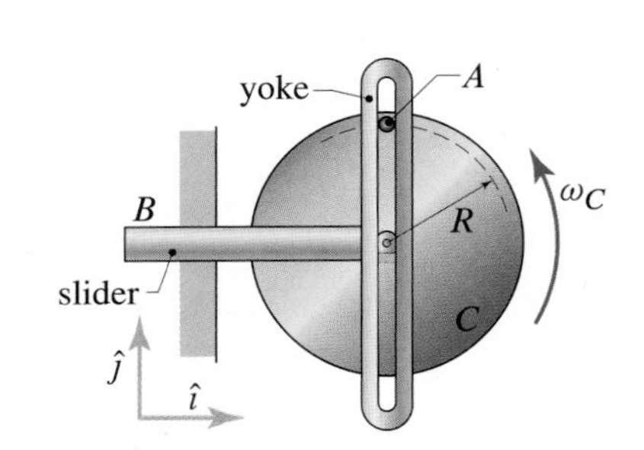

Figure P6.57

Problem 6.57

One way to convert rotational motion into linear motion and vice versa is with the use of a mechanism called a Scotch yoke, which consists of a crank C that is connected to a slider B by a pin A. The pin rotates with the crank while sliding within the yoke, which, in turn, rigidly translates with the slider. This mechanism has been used, for example, to control the opening and closing of valves in pipelines. Letting the radius of the crank be $R = 1.5$ ft, determine the angular velocity ω_C of the crank so that the maximum speed of the slider is $v_B = 90$ ft/s.

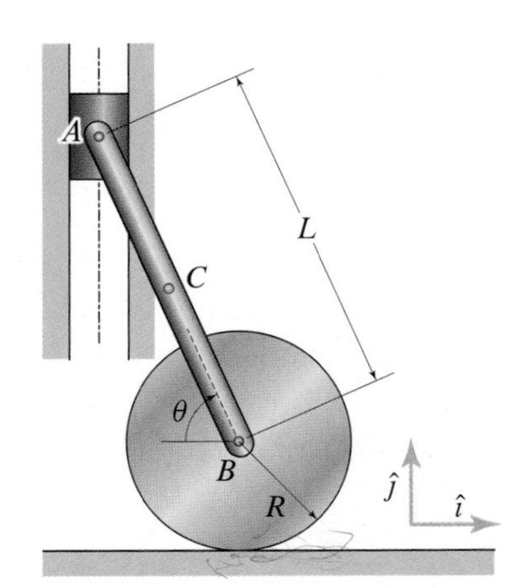

Figure P6.58–P6.60

Problems 6.58 through 6.60

The system shown consists of a wheel of radius $R = 14$ in. rolling on a horizontal surface. A bar AB of length $L = 40$ in. is pin-connected to the center of the wheel and to a slider A that is constrained to move along a vertical guide. Point C is the bar's midpoint.

Problem 6.58 If, when $\theta = 72°$, the wheel is moving to the right so that $v_B = 7$ ft/s, determine the angular velocity of the bar, as well as the velocity of the slider A.

Problem 6.59 If, when $\theta = 53°$, the slider is moving downward with a speed $v_A = 8$ ft/s, determine the velocity of points B and C.

Problem 6.60 If the wheel rolls without slip with a constant counterclockwise angular velocity of 10 rad/s, determine the velocity of the slider A when $\theta = 45°$.

Problems 6.61 and 6.62

As the circular cam, whose center is at A rotates, it causes the follower B to move back and forth. The cam angle is θ, the radius of the cam is R, and the angular speed of the cam is $\dot{\theta} = \omega_{OA}$. The cam is pin-connected to the fixed point O.

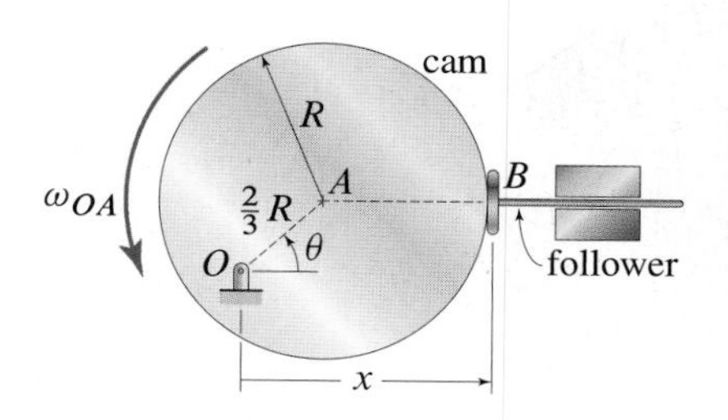

Figure P6.61 and P6.62

Problem 6.61 Using the given x coordinate, determine the velocity of the follower at the instant $\theta = 30°$ if $R = 1.5$ in., and $\dot{\theta} = \omega_{OA} = 1000$ rpm.

Problem 6.62 Using the given x coordinate, determine the velocity of the follower as a function of the cam angle θ, the radius of the cam R, and the given angular speed of the cam ω_{OA}.

Problem 6.63

At the instant shown, the lower rack is moving to the right with a speed of 2.7 m/s while the upper rack is moving to the left with a speed of 1.7 m/s. If the nominal radius of the pinion O is $R = 0.25$ m, determine the angular velocity of the pinion, as well as the position of the pinion's instantaneous center of rotation relative to point O.

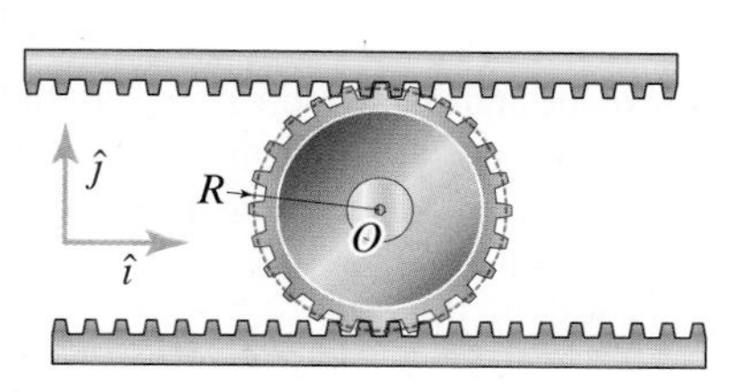

Figure P6.63

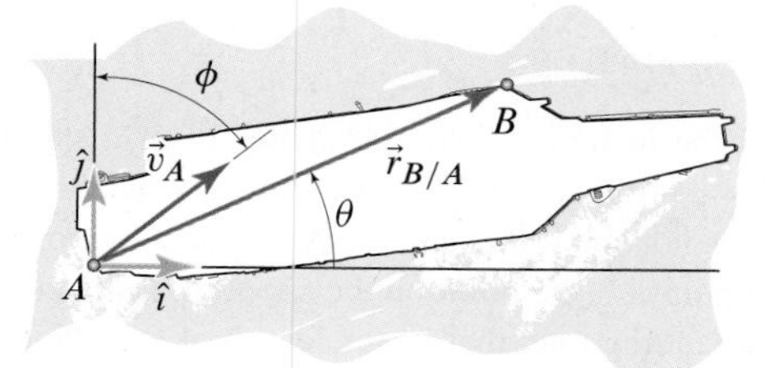

Figure P6.64

Problem 6.64

A carrier is maneuvering so that, at the instant shown, $|\vec{v}_A| = 22$ kn, $\phi = 35°$, $|\vec{v}_B| = 24$ kn (1 kn is equal to 1 nautical mile (nml) per hour or 6076 ft/h). Letting $\theta = 19°$ and the distance between A and B be 720 ft, determine the ship's turning rate at the given instant if the ship is rotating clockwise.

Problems 6.65 and 6.66

For the slider-crank mechanism shown, let $R = 1.9$ in., $L = 6.1$ in., and $H = 1.2$ in. Also, at the instant shown, let $\theta = 27°$ and $\omega_{AB} = 4850$ rpm.

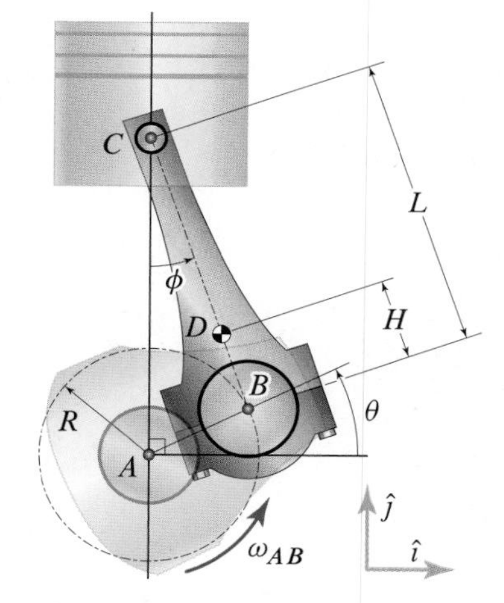

Figure P6.65 and P6.66

Problem 6.65 Determine the velocity of the piston at the instant shown.

Problem 6.66 Determine $\dot{\phi}$ and the velocity of point D at the instant shown.

Problem 6.67

A wheel W of radius $R_W = 7$ mm is connected to point O via the rotating arm OC, and it rolls without slip over the stationary cylinder S of radius $R_S = 15$ mm. If, at the instant shown, $\theta = 63°$ and $\omega_W = 9$ rad/s, determine the angular velocity of the arm OC and the velocity of point P, where point P lies on the edge of W and is vertically aligned with point C.

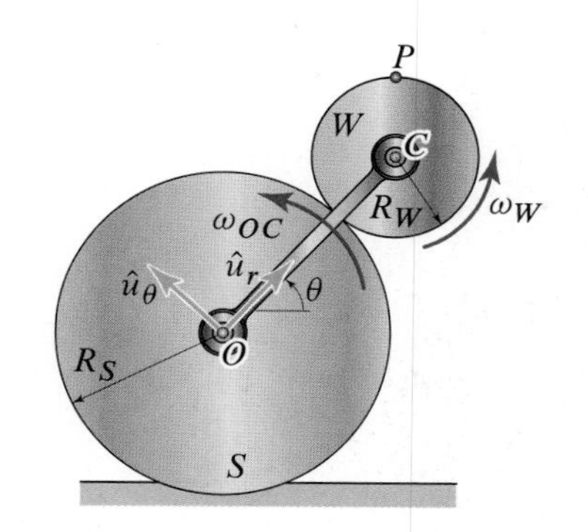

Figure P6.67

Problem 6.68

At the instant shown, the center O of a spool with inner and outer radii $r = 1$ m and $R = 2.2$ m, respectively, is moving up the incline at speed $v_O = 3$ m/s. If the spool does not slip relative to the ground or relative to the cable C, determine the rate at which the cable is wound or unwound, that is, the length of cable being wound or unwound per unit time.

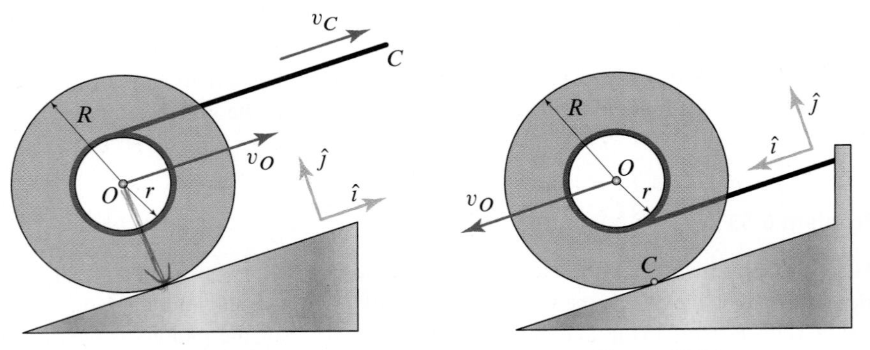

Figure P6.68 Figure P6.69

Problem 6.69

At the instant shown, the center O of a spool with inner and outer radii $r = 3$ ft and $R = 7$ ft, respectively, is moving down the incline at a speed $v_O = 12.2$ ft/s. If the spool does not slip relative to the rope and if the rope is fixed at one end, determine the velocity of point C (the point on the spool that is in contact with the incline), as well as the rope's unwinding rate, that is, the length of rope being unwound per unit time.

Problem 6.70

The bucket of a backhoe is the element AB of the four-bar linkage system $ABCD$. Assume that the points A and D are fixed and that, at the instant shown, point B is vertically aligned with point A, point C is horizontally aligned with point B, and point B is moving to the right with a speed $v_B = 1.2$ ft/s. Determine the velocity of point C at the instant shown, along with the angular velocities of elements BC and CD. Let $h = 0.66$ ft, $e = 0.46$ ft, $\ell = 0.9$ ft, and $w = 1.0$ ft.

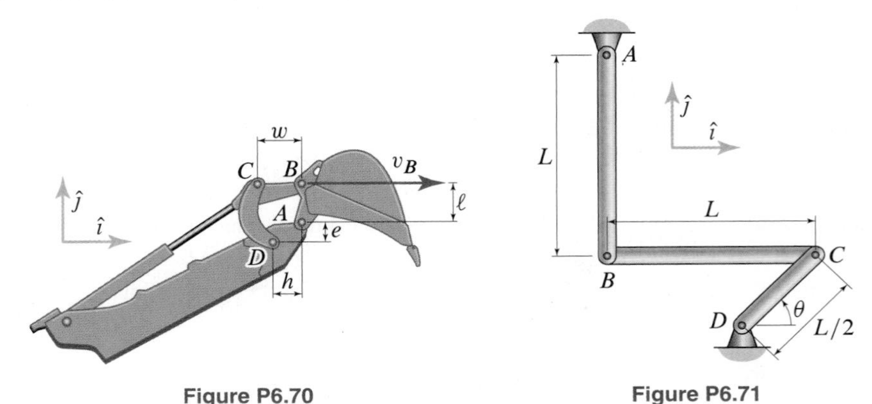

Figure P6.70 Figure P6.71

Problem 6.71

Bar AB is rotating counterclockwise with an angular velocity of 15 rad/s. Letting $L = 1.25$ m, determine the angular velocity of bar CD when $\theta = 45°$.

Problem 6.72

At the instant shown, bars AB and CD are vertical and point C is moving to the left with a speed of 35 ft/s. Letting $L = 1.5$ ft and $H = 0.6$ ft, determine the velocity of point B.

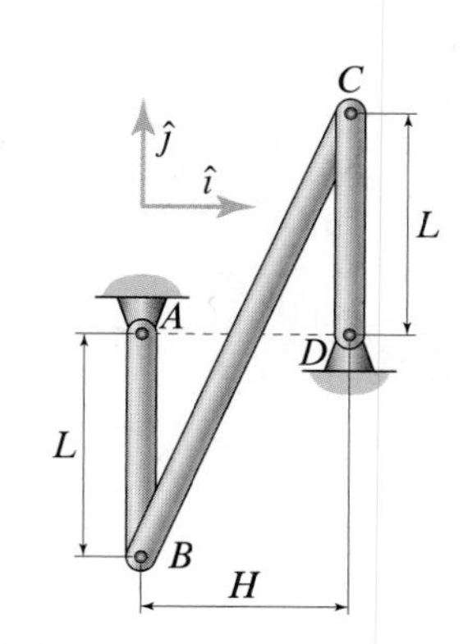

Figure P6.72

Problem 6.73

Collars A and B are constrained to slide along the guides shown and are connected by a bar with length $L = 0.75$ m. Letting $\theta = 45°$, determine the angular velocity of the bar AB at the instant shown if, at this instant, $v_B = 2.7$ m/s.

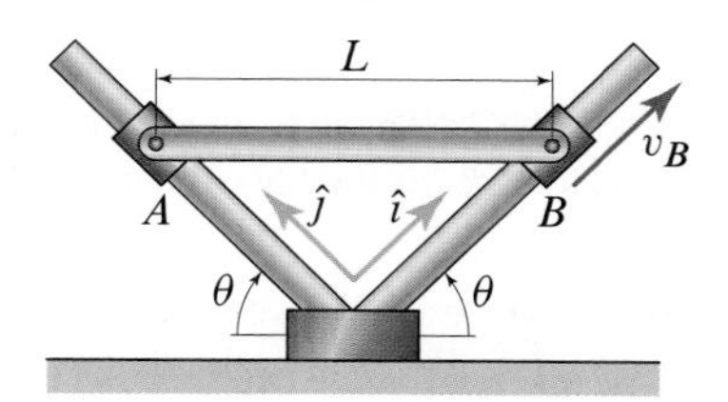

Figure P6.73

Problem 6.74

At the instant shown, an overhead garage door is being shut with point B moving to the left within the horizontal part of the door guide at a speed of 5 ft/s, while point A is moving vertically downward. Determine the angular velocity of the door and the velocity of the counterweight C at this instant if $L = 6$ ft and $d = 1.5$ ft.

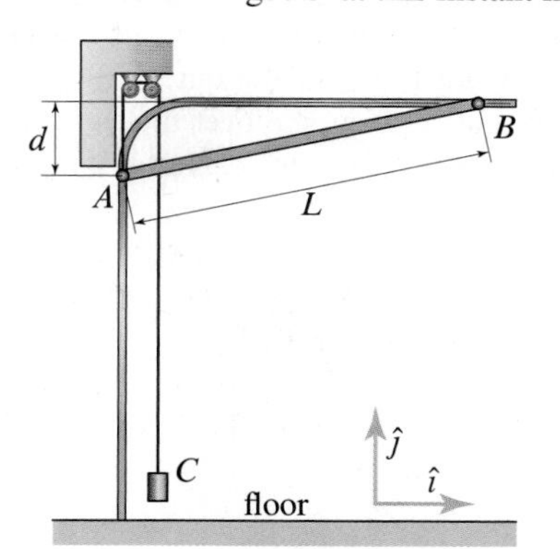

Figure P6.74

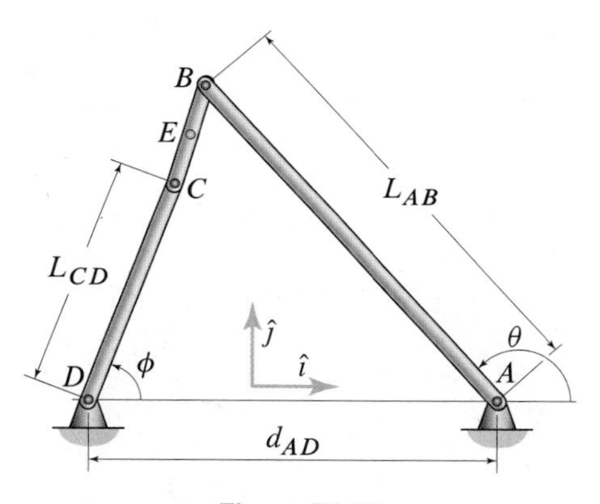

Figure P6.75

Problem 6.75

In the four-bar linkage system shown, the lengths of the bars AB and CD are $L_{AB} =$ 46 mm and $L_{CD} = 25$ mm, respectively. In addition, the distance between points A and D is $d_{AD} = 43$ mm. The dimensions of the mechanism are such that when the angle $\theta = 132°$, the angle $\phi = 69°$. For $\theta = 132°$ and $\dot{\theta} = 27$ rad/s, determine the angular velocity of bars BC and CD, as well as the velocity of the point E, the midpoint of bar BC. Note that the figure is drawn to scale and that bars BC and CD are not collinear.

Problem 6.76

A spool with inner radius $R = 1.5$ m rolls without slip over a horizontal rail as shown. If the cable on the spool is unwound at a rate $v_A = 5$ m/s, in such a way that the unwound cable remains perpendicular to the rail, determine the angular velocity of the spool and the velocity of the spool's center O.

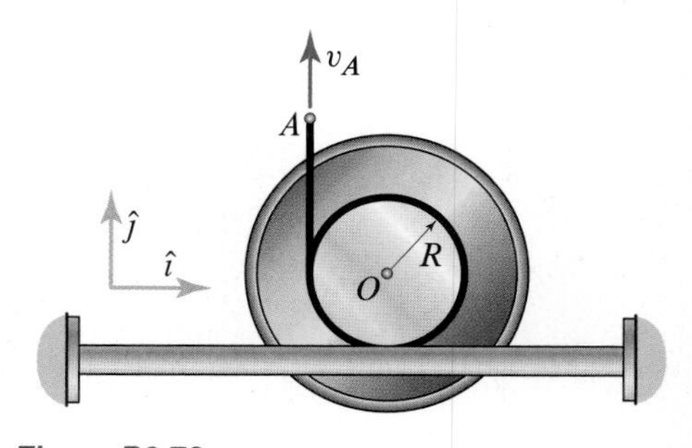

Figure P6.76

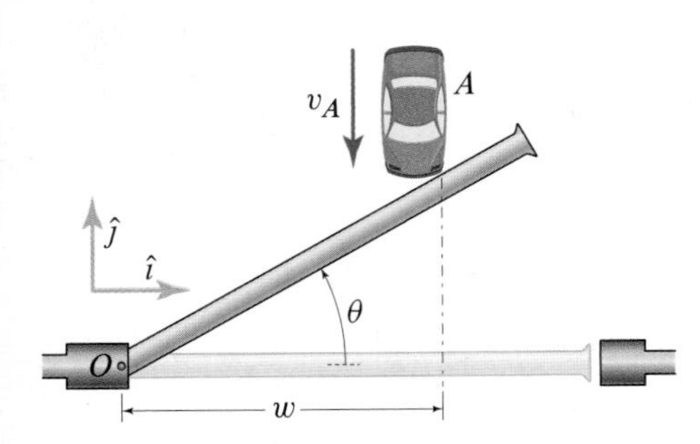

Figure P6.77

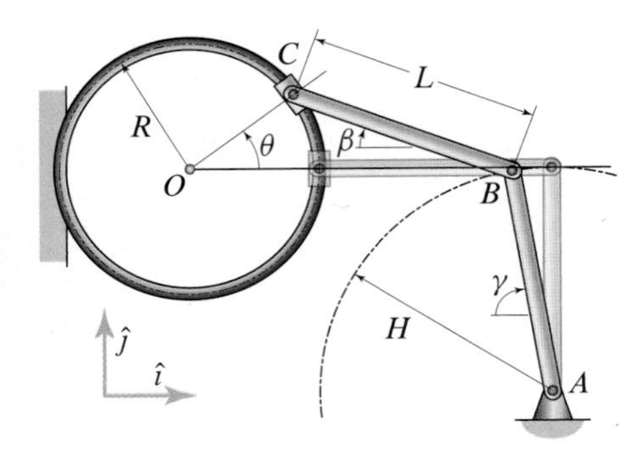

Figure P6.78–P6.80

Problem 6.77

A person is closing a heavy gate with rusty hinges by pushing the gate with car A. If $w = 24$ m and $v_A = 1.2$ m/s, determine the angular velocity of the gate when $\theta = 15°$.

Problems 6.78 through 6.80

In the four-bar linkage system shown, let the circular guide with center at O be fixed and such that, when $\theta = 0°$, the bars AB and BC are vertical and horizontal, respectively. In addition, let $R = 2$ ft, $L = 3$ ft, and $H = 3.5$ ft.

Problem 6.78 When $\theta = 0°$, the collar at C is sliding downward with a speed of 23 ft/s. Determine the angular velocities of the bars AB and BC at this instant.

Problem 6.79 When $\theta = 37°$, $\beta = 25.07°$, $\gamma = 78.71°$, and the collar is sliding clockwise with a speed $v_C = 23$ ft/s. Determine the angular velocities of the bars AB and BC.

Problem 6.80 Determine the general expression for the angular velocities of bars AB and BC as a function of θ, β, γ, R, L, H, and $\dot{\theta}$.

Problem 6.81

At the instant shown, the arm OC rotates counterclockwise with an angular velocity of 35 rpm about the fixed sun gear S of radius $R_S = 3.5$ in. The planet gear P with radius $R_P = 1.2$ in. rolls without slip over both the fixed sun gear and the outer ring gear. Finally, notice that the ring gear is not fixed, and it rolls without slip over the sun gear. Determine the angular velocity of the ring gear and the velocity of the center of the ring gear at the instant shown.

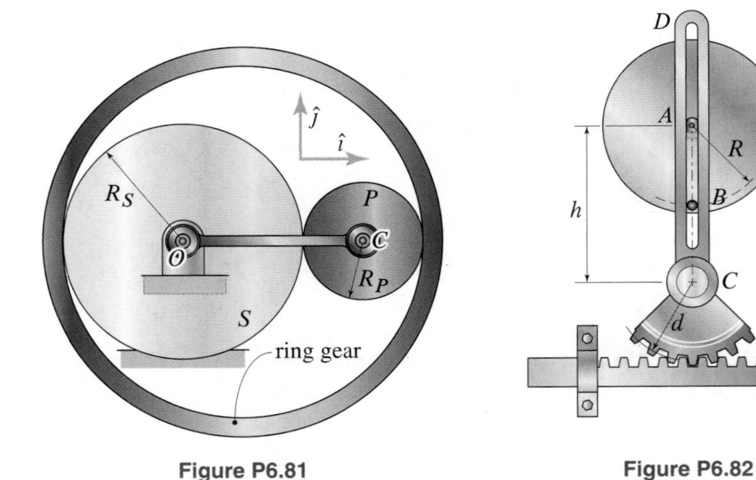

Figure P6.81 Figure P6.82

Problem 6.82

The crank AB is rotating counterclockwise at a constant angular velocity of 12 rad/s while the pin B slides within the slot in the bar CD, which is pinned at C. Letting $R = 0.5$ m, $h = 1$ m, and $d = 0.25$ m, determine the angular velocity of CD at the instant shown (with points A, B, and C vertically aligned), as well as the velocity of the horizontal bar to which bar CD is connected.

Problems 6.83 through 6.87

For the slider-crank mechanism shown, let $R = 20\,\mathrm{mm}$, $L = 80\,\mathrm{mm}$, and $H = 38\,\mathrm{mm}$.

Problem 6.83 If $\dot{\theta} = 1700\,\mathrm{rpm}$, determine the angular velocity of the connecting rod AB and the speed of the slider B for $\theta = 90°$.

Problem 6.84 Determine the angular velocity of the crank OA when $\theta = 20°$ and the slider is moving downward at $15\,\mathrm{m/s}$.

Problem 6.85 Determine the general expression for the velocity of the slider B as a function of θ, $\dot{\theta}$, and the geometrical parameters R, H, and L, using the *vector approach*.

Problem 6.86 Determine the general expression for the velocity of the slider B as a function of θ, $\dot{\theta}$, and the geometrical parameters R, H, and L, using *differentiation of constraints*.

Problem 6.87 Plot the velocity of the slider B as a function of θ, for $0 \le \theta \le 360°$, and for $\dot{\theta} = 1000\,\mathrm{rpm}$, $\dot{\theta} = 3000\,\mathrm{rpm}$, and $\dot{\theta} = 5000\,\mathrm{rpm}$.

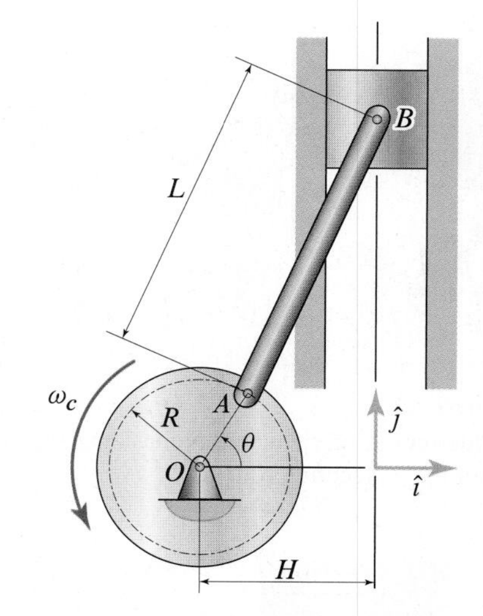

Figure P6.83–P6.87

Problem 6.88

Complete the velocity analysis of the slider-crank mechanism, using differentiation of constraints that was outlined beginning on p. 453. That is, determine the velocity of the piston C and the angular velocity of the connecting rod as a function of the given quantities θ, ω_{AB}, R, and L. Use the component system shown for your answers.

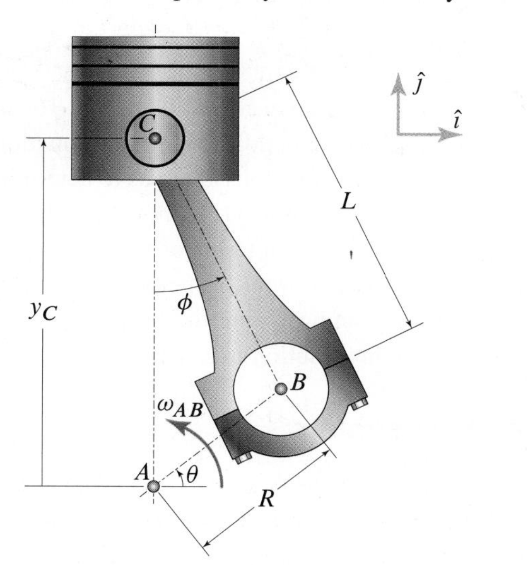

Figure P6.88

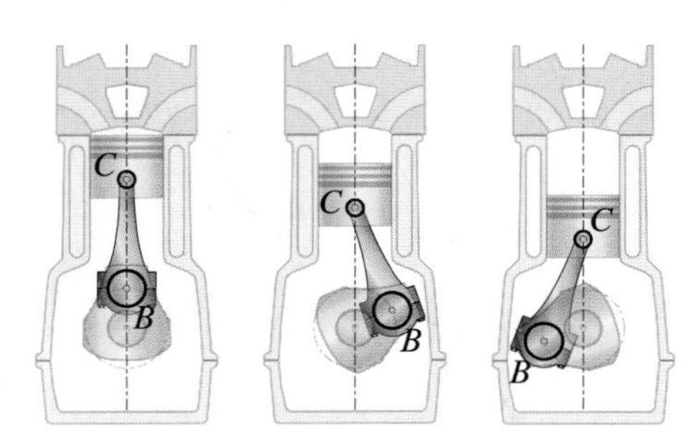

Figure 6.32
A slider-crank mechanism emphasizing the motion of the connecting rod.

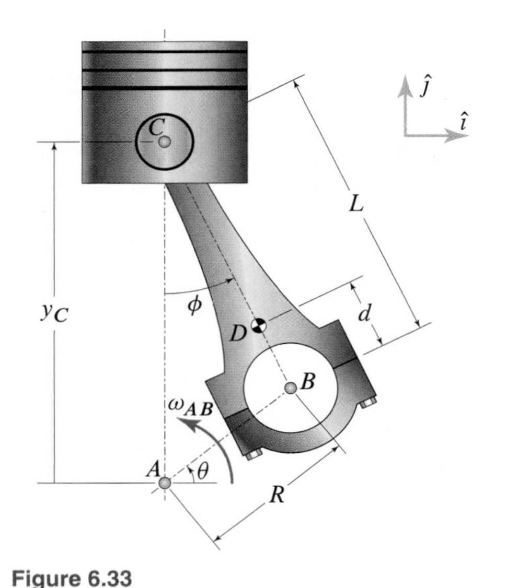

Figure 6.33
The geometric parameters used in the slider-crank analysis. The crank's angular velocity $\dot{\theta} = \omega_{AB}$ is *constant*, though the solution methodology is valid whether or not ω_{AB} is constant.

6.3 Planar Motion: Acceleration Analysis

Figure 6.32 shows the slider-crank mechanism for which we now wish to perform an acceleration analysis. We will develop two different approaches to the problem that are applicable to the acceleration analysis of any planar rigid body motion: the vector approach and differentiation of constraints.

Vector approach

The connecting rod in the slider-crank mechanism (Fig. 6.32) is, in general, planar motion, which means that we can describe the acceleration of any of its points by knowing the acceleration of one point on the rod and the rotational motion of the rod (i.e., its angular velocity and angular acceleration). The velocity analysis was done in Example 6.6 on p. 460, and so we know the angular velocity of the rod $\vec{\omega}_{BC}$. The acceleration analysis proceeds in much the same way.

On p. 438, we found the acceleration of point B, which is in fixed axis rotation about the centerline of the crank A (see Fig. 6.33). Points B and C are on the same rigid body, so we can relate the acceleration of C to that of B using either Eq. (6.5) on p. 436 or Eq. (6.14) on p. 439 in Section 6.1, which give

$$\vec{a}_C = \vec{a}_B + \vec{\alpha}_{BC} \times \vec{r}_{C/B} + \vec{\omega}_{BC} \times (\vec{\omega}_{BC} \times \vec{r}_{C/B}), \tag{6.29}$$

or

$$\vec{a}_C = \vec{a}_B + \vec{\alpha}_{BC} \times \vec{r}_{C/B} - \omega_{BC}^2 \vec{r}_{C/B}, \tag{6.30}$$

respectively, where $\vec{\alpha}_{BC}$ is the angular acceleration of bar BC and $\vec{\omega}_{BC}$ is known. In planar motion, either of the above equations is a vector equation that represents *two* scalar equations. These two scalar equations allow us to determine $\vec{a}_C$ and $\vec{\alpha}_{BC}$ since $\vec{a}_B$ is known, the direction of $\vec{a}_C$ is known (the motion of the piston C is rectilinear along the y axis), $\vec{r}_{C/B}$ can be found in terms of the crank angle θ (see Fig. 6.33), and the axis of rotation for $\vec{\alpha}_{BC}$ is known (it is perpendicular to the plane of motion). Therefore, the components a_C and α_{BC} are the only unknowns in these two scalar equations. We will complete this solution in Example 6.10.

Differentiation of constraints

Referring to Fig. 6.33, we can also determine $\vec{\alpha}_{BC}$ and $\vec{a}_C$ as functions of θ by writing the appropriate constraint equations and then differentiating them with respect to time. We found the slider-crank velocities in Section 6.2 on p. 452; accelerations simply require an additional time derivative.

As was done in Eq. (6.20) on p. 453, we can write the constraint equation for the y coordinate of C as

$$y_C = R \sin\theta + L \cos\phi, \tag{6.31}$$

which can be differentiated twice with respect to time to find the acceleration of C as

$$\ddot{y}_C = a_C = R\ddot{\theta}\cos\theta - R\dot{\theta}^2 \sin\theta - L\ddot{\phi}\sin\phi - L\dot{\phi}^2 \cos\phi. \tag{6.32}$$

As with the velocity analysis, quantities R, L, and $\theta(t)$ are assumed to be known. This implies that we also know $\dot{\theta} = \omega_{AB}$ and $\ddot{\theta} = \alpha_{AB}$, though we see that a_C is also a function of ϕ, $\dot{\phi}$, and $\ddot{\phi}$. In the velocity analysis using differentiation of constraints, we said that we can determine ϕ and $\dot{\phi}$ in terms of known quantities by relating the orientation ϕ of the connecting rod to the crank's orientation θ, using the second constraint equation $\sin\phi = (R/L)\cos\theta$. This equation was differentiated once with respect to time to find $\dot{\phi}$ as a function of θ and $\dot{\theta}$. It can be differentiated twice to obtain $\ddot{\phi}$, where $\ddot{\phi}\,\hat{k} = \vec{\alpha}_{BC}$, as a function of θ, $\dot{\theta}$, and $\ddot{\theta}$ (see Prob. 6.133), which then completes the analysis. We will see this method again in Example 6.11.

Rolling without slip: acceleration analysis

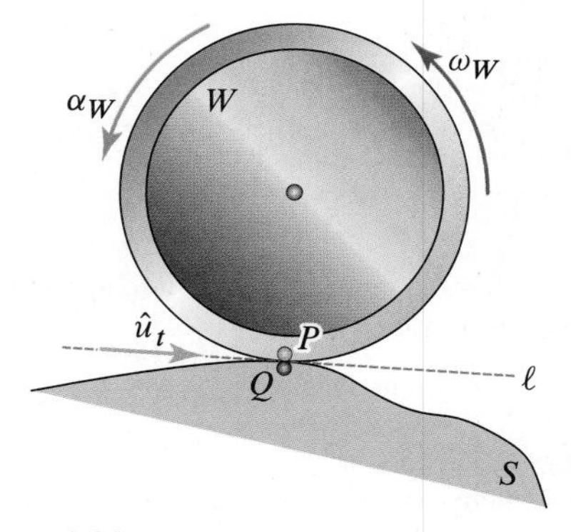

Figure 6.34
A wheel W rolling over a surface S. At the instant shown, the line ℓ is tangent to the path of P and Q.

Referring to Fig. 6.34, the definition of rolling without slip given on p. 452 stated that if the body W is *rolling without slip* over the surface S (S can be moving), then the contact points P and Q (on W and S, respectively) have the same velocity, $\vec{v}_P = \vec{v}_Q$ or $\vec{v}_{P/Q} = \vec{0}$. This condition implies that the component of $\vec{a}_{P/Q}$ tangent to the contact must be equal to zero, that is,

$$\vec{a}_{P/Q} \cdot \hat{u}_t = 0, \tag{6.33}$$

where $\hat{u}_t$ is a unit vector parallel to ℓ, the line tangent to both W and S at their contact. In component form, Eq. (6.33) takes on the form

$$\boxed{(a_{P/Q})_t = 0 \quad\Rightarrow\quad a_{Pt} = a_{Qt}.} \tag{6.34}$$

As an application of Eq. (6.34), we can now determine the acceleration of the point in contact with the ground for a wheel W rolling without slip on a flat stationary surface. If the wheel's center O moves as shown in Fig. 6.35, then

$$\vec{v}_O = v_O\,\hat{\imath} \quad\text{and}\quad \vec{a}_O = a_O\,\hat{\imath}. \tag{6.35}$$

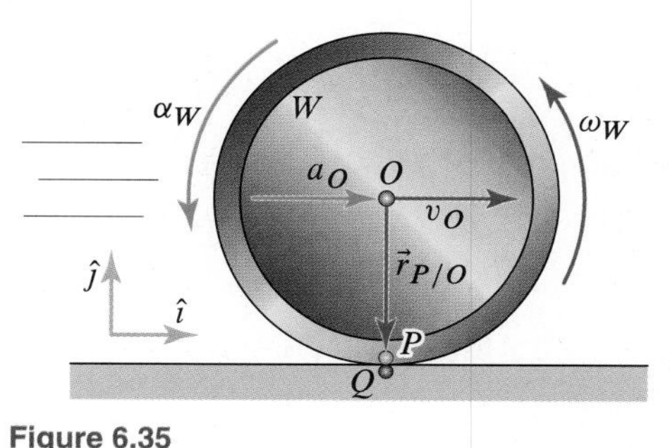

Figure 6.35
A wheel W rolling without slip on a horizontal fixed surface.

We have already discovered that $\vec{v}_P = \vec{0}$ and that $\vec{\omega}_W = -(v_O/R)\,\hat{k}$, so the acceleration of point P is given by

$$\vec{a}_P = \vec{a}_O + \alpha_W\,\hat{k} \times \vec{r}_{P/O} - \omega_W^2\vec{r}_{P/O}, \tag{6.36}$$

which, using $\vec{r}_{P/O} = -R\,\hat{\jmath}$, can be written as

$$\begin{aligned}\vec{a}_P &= a_O\,\hat{\imath} + \alpha_W\,\hat{k} \times (-R)\,\hat{\jmath} - \left(-\frac{v_O}{R}\right)^2(-R\,\hat{\jmath})\\ &= \big(a_O + \alpha_W R\big)\,\hat{\imath} + \left(\frac{v_O^2}{R}\right)\hat{\jmath}.\end{aligned} \tag{6.37}$$

Observe that the tangent to the wheel at the contact point P is in the x direction. Hence, the application of Eq. (6.34) reads

$$a_{Px} = a_{Qx} = 0 \quad\Rightarrow\quad \boxed{\alpha_W = -\frac{a_O}{R},} \tag{6.38}$$

since the point Q on the ground is stationary. This implies that the point P on the wheel is accelerating along the y axis according to

$$\boxed{\vec{a}_P = \frac{v_O^2}{R}\,\hat{\jmath}.} \tag{6.39}$$

This result tells us that although P has zero velocity at the instant considered, it is accelerating, i.e., it is in the process of gaining velocity. This, in turn, will allow P to move away from its current position and thus allow some other point to become the next contact point.

End of Section Summary

This section presents two different ways to analyze the accelerations of a rigid body in planar motion: the vector approach and differentiation of constraints.

Vector approach. We saw in Section 6.1 that, for planar motion, either of the equations

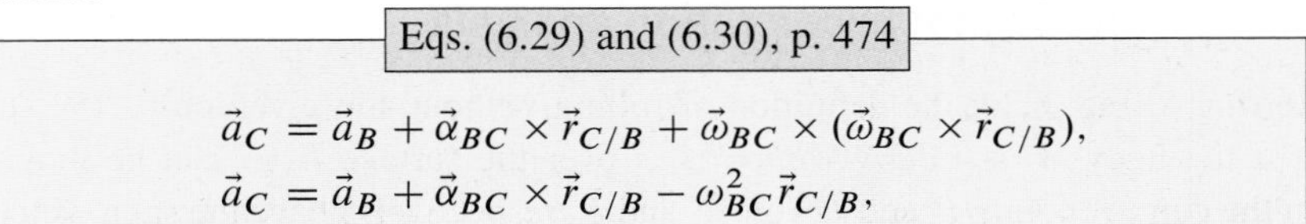

Eqs. (6.29) and (6.30), p. 474

$$\vec{a}_C = \vec{a}_B + \vec{\alpha}_{BC} \times \vec{r}_{C/B} + \vec{\omega}_{BC} \times (\vec{\omega}_{BC} \times \vec{r}_{C/B}),$$
$$\vec{a}_C = \vec{a}_B + \vec{\alpha}_{BC} \times \vec{r}_{C/B} - \omega_{BC}^2 \vec{r}_{C/B},$$

relates the acceleration of two points on a rigid body, $\vec{a}_B$ and $\vec{a}_C$, via their relative position $\vec{r}_{C/B}$, the angular acceleration of the body $\vec{\alpha}_{BC}$, and the angular velocity of the body $\vec{\omega}_{BC}$ (see Fig. 6.36).

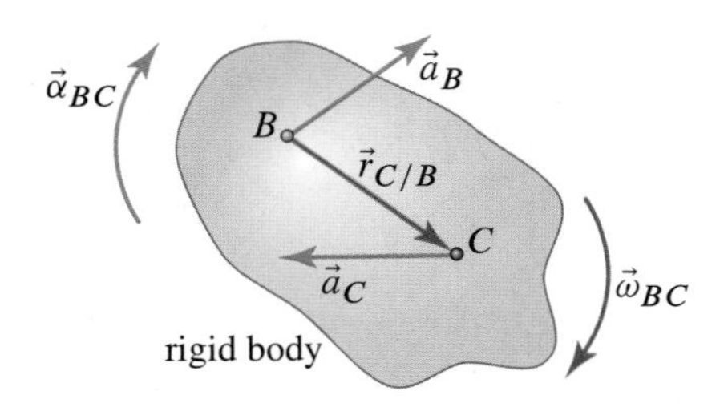

Figure 6.36
A rigid body on which we are relating the acceleration of two points B and C.

Differentiation of constraints. As we discovered in Section 2.7, it is often convenient to write an equation describing the position of a point of interest, which can then be differentiated once with respect to time to find the velocity of that point and twice with respect to time to find the acceleration. For the planar motion of rigid bodies, this idea can also apply for describing the position, velocity, and acceleration of a point on a rigid body, as well as for describing the orientation of a rigid body, for which the first and second time derivatives provide its angular velocity and angular acceleration, respectively (we saw this again in Section 6.2 for velocities).

Rolling without slip. Referring to Fig. 6.37, when a body rolls without slip over another body, the two points on the bodies that are in contact at any instant, points P and Q, must have the same velocity. For accelerations, this means that the component of $\vec{a}_{P/Q}$ tangent to the contact must be equal to zero, that is,

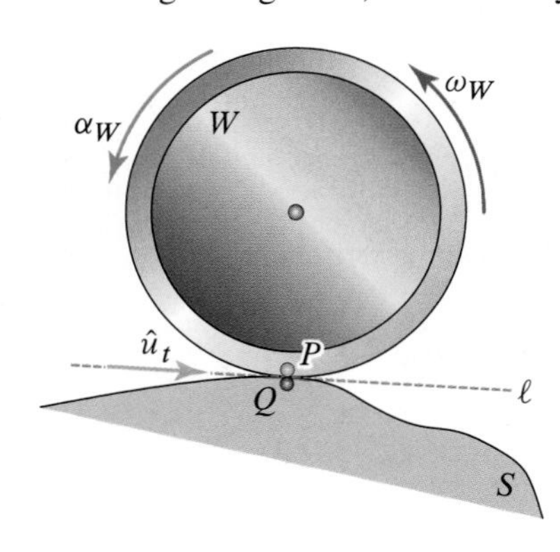

Figure 6.37. A wheel W rolling over a surface S. At the instant shown, the line ℓ is tangent to the path of P and Q.

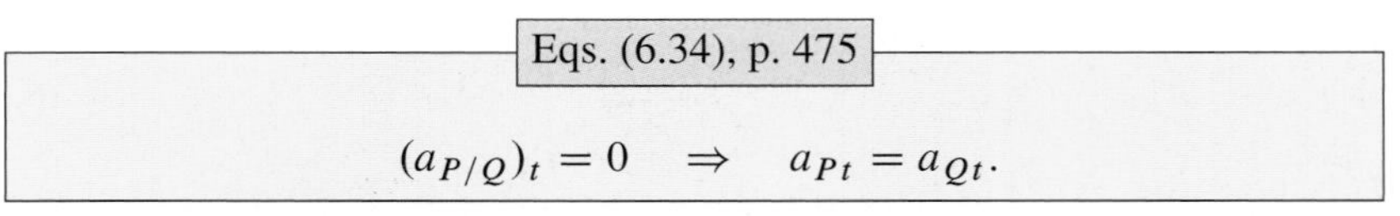

Eqs. (6.34), p. 475

$$(a_{P/Q})_t = 0 \quad \Rightarrow \quad a_{Pt} = a_{Qt}.$$

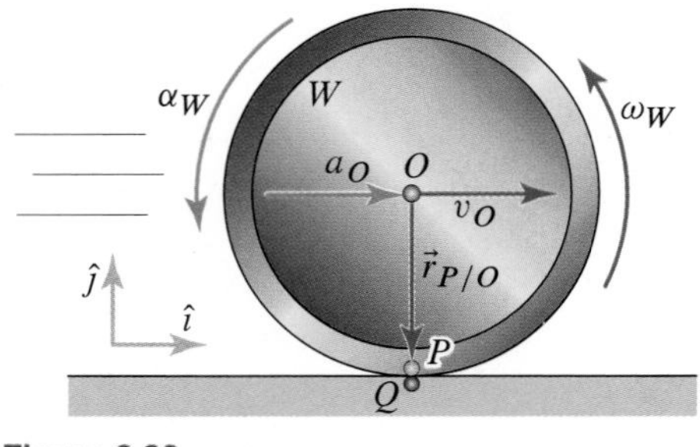

Figure 6.38
A wheel W rolling without slip on a horizontal fixed surface.

If a wheel of radius R is rolling without slip over a flat, stationary surface, then the point P on the wheel in contact with the surface must have zero velocity (see Fig. 6.38). The consequence is that the angular acceleration of the wheel α_W is related to the acceleration of the center a_O and the radius of the wheel R according to

Eq. (6.38), p. 475

$$\alpha_W = -\frac{a_O}{R},$$

in which the positive direction for α_W is taken to be the positive z direction. The point P on the wheel that is in contact with the ground at this instant is accelerating along the y axis according to

Eq. (6.39), p. 475

$$\vec{a}_P = \frac{v_O^2}{R}\,\hat{\jmath}.$$

Even though P has zero velocity at the instant considered, it is accelerating.

EXAMPLE 6.9 Planetary Gears Rolling Without Slip: Vector Approach

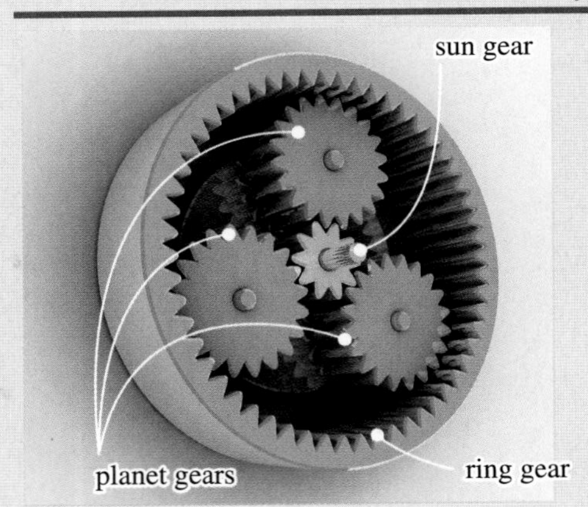

Figure 1
Photo of a planetary gear system.

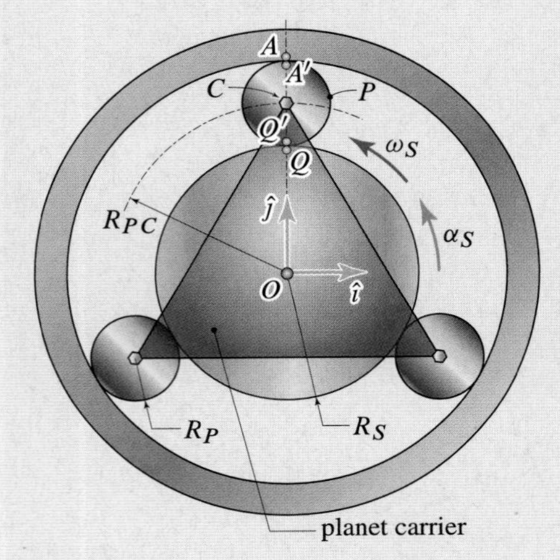

Figure 2
Schematic of a planetary gear system with three planets and a fixed ring. The planet carrier's rotational axis is coincident with that of the sun gear, but they are not directly connected.

Recall from Example 6.5 on p. 459, that planetary gear systems (see Fig. 1) are used to transmit power between two shafts. The center gear is called the *sun*, the outer gear is called the *ring*, and the three inner gears are called *planets*. The planets are mounted on a component called the *planet carrier* (which is not shown in Fig. 1). Referring to Fig. 2, let $R_S = 2$ in., $R_P = 0.67$ in., the ring be fixed, $\omega_S = 1500$ rpm (the same as in Example 6.5), and $\alpha_S = 2.7\,\text{rad/s}^2$. Determine the angular acceleration of the planet gear $\vec{\alpha}_P$, the angular acceleration of the planet carrier $\vec{\alpha}_{PC}$, and the accelerations of points A' and Q'.

SOLUTION

Road Map The solution of this problem has two essential elements: (1) the enforcement of the rolling without slip conditions for the A-A' and the Q-Q' contacts and (2) the enforcement of the constraint that point C on the planet gear P must move along a circle of radius $R_{PC} = R_S + R_P$. We carry out element 1 by first writing the accelerations of points A, A', Q, and Q'. To enforce 2, we write $\vec{a}_C$ twice, the first time viewing C as part of the planet carrier and the second time viewing C as part of P. We then force the two resulting expressions to be equal to each other. In general, before doing the acceleration analysis, we first need to find the angular velocities of all components in the system. This was done in Example 6.5 on p. 459, in which we discovered that $\omega_P = -(R_S/2R_P)\omega_S = -2239$ rpm and $\omega_{PC} = (R_S/2R_{PC})\omega_S = 561.8$ rpm.

Computation The accelerations of A (a fixed point) and A' can be expressed as

$$\vec{a}_A = \vec{0} \quad \text{and} \quad \vec{a}_{A'} = a_{A'x}\,\hat{\imath} + a_{A'y}\,\hat{\jmath}. \tag{1}$$

Observe that the line tangent to the A-A' contact is parallel to the x direction, so that the rolling without slip condition between P and the ring gear implies

$$a_{Ax} = a_{A'x} \quad \Rightarrow \quad a_{A'x} = 0. \tag{2}$$

To find the accelerations of Q and Q', since the sun gear is rotating about the (fixed) z axis, we can express $\vec{a}_Q$ as follows:

$$\vec{a}_Q = \vec{\alpha}_S \times \vec{r}_{Q/O} - \omega_S^2 \vec{r}_{Q/O} = -\alpha_S R_S\,\hat{\imath} - \omega_S^2 R_S\,\hat{\jmath}, \tag{3}$$

where we have set $\vec{\alpha}_S = \alpha_S\,\hat{k}$ and $\vec{r}_{Q/O} = R_S\,\hat{\jmath}$. For Q', using A' as a reference point, we can write

$$\begin{aligned} \vec{a}_{Q'} &= \vec{a}_{A'} + \vec{\alpha}_P \times \vec{r}_{Q'/A'} - \omega_P^2 \vec{r}_{Q'/A'} \\ &= a_{A'y}\,\hat{\jmath} + \alpha_P\,\hat{k} \times (-2R_P)\,\hat{\jmath} - \omega_P^2(-2R_P)\,\hat{\jmath} \\ &= 2R_P\alpha_P\,\hat{\imath} + \left(a_{A'y} + 2R_P\omega_P^2\right)\hat{\jmath}. \end{aligned} \tag{4}$$

We now enforce the rolling without slip condition at the Q-Q' contact by observing that the tangent line to the contact is parallel to the x axis. Therefore, from Eqs. (3) and (4), we have

$$a_{Qx} = a_{Q'x} \quad \Rightarrow \quad -\alpha_S R_S = 2R_P\alpha_P, \tag{5}$$

or

$$\alpha_P = -\frac{R_S}{2R_P}\alpha_S = -4.030\,\text{rad/s}^2 \quad \Rightarrow \quad \boxed{\vec{\alpha}_P = \left(-4.030\,\text{rad/s}^2\right)\hat{k}.} \tag{6}$$

As for the acceleration of point C, when viewed as part of the planet carrier, point C is rotating about point O, so that $\vec{a}_C$ can be given the form

$$\begin{aligned} \vec{a}_C &= \vec{\alpha}_{PC} \times \vec{r}_{C/O} - \omega_{PC}^2 \vec{r}_{C/O} \\ &= -\alpha_{PC} R_{PC}\,\hat{\imath} - \omega_{PC}^2 R_{PC}\,\hat{\jmath}, \end{aligned} \tag{7}$$

where we have set $\vec{r}_{C/O} = R_{PC}\,\hat{\jmath}$ and $\vec{\alpha}_{PC} = \alpha_{PC}\,\hat{k}$, and we note that $R_{PC} = R_S + R_P$. Viewing point C as part of the planet gear P and relating its acceleration to A', we have

$$\begin{aligned}\vec{a}_C &= \vec{a}_{A'} + \vec{\alpha}_P \times \vec{r}_{C/A'} - \omega_P^2 \vec{r}_{C/A'}\\ &= a_{A'y}\,\hat{\jmath} + \alpha_P\,\hat{k} \times \vec{r}_{C/A'} - \omega_P^2 \vec{r}_{C/A'}\\ &= \alpha_P R_P\,\hat{\imath} + \left(a_{A'y} + \omega_P^2 R_P\right)\hat{\jmath}.\end{aligned} \tag{8}$$

The expressions for $\vec{a}_C$ in Eqs. (7) and (8) must be equal to one another. Therefore,

$$-\alpha_{PC} R_{PC} = \alpha_P R_P \quad \text{and} \quad -\omega_{PC}^2 R_{PC} = a_{A'y} + \omega_P^2 R_P. \tag{9}$$

Equations (9) can be solved for the unknowns α_{PC} and $a_{A'y}$ to obtain

$$\alpha_{PC} = -\frac{R_P}{R_{PC}}\alpha_P \quad \text{and} \quad a_{A'y} = -\omega_P^2 R_P - \omega_{PC}^2 R_{PC}. \tag{10}$$

Using Eq. (6) for α_P, the expressions for ω_P and ω_{PC} from the Road Map, and Eq. (2) for $a_{A'x}$, Eqs. (10) become

$$\vec{\alpha}_{PC} = \frac{R_S}{2R_{PC}}\alpha_S\,\hat{k} = \left(1.011\ \text{rad/s}^2\right)\hat{k}, \tag{11}$$

$$\vec{a}_{A'} = -\frac{R_S^2}{4R_P}\left(1 + \frac{R_P}{R_{PC}}\right)\omega_S^2\,\hat{\jmath} = \left(-3839\ \text{ft/s}^2\right)\hat{\jmath}. \tag{12}$$

Finally, substituting the results in Eqs. (6) and (12) and ω_P from the Road Map into Eq. (4), we obtain

$$\vec{a}_{Q'} = -\alpha_S R_S\,\hat{\imath} + \frac{R_S^2}{4R_P}\left(1 - \frac{R_P}{R_{PC}}\right)\omega_S^2\,\hat{\jmath} = (-0.4500\,\hat{\imath} + 2299\,\hat{\jmath})\ \text{ft/s}^2. \tag{13}$$

Discussion & Verification The dimensions of the symbolic answers are all as they should be, and so the units of the numerical answers are also correct.

The results in Eqs. (6) and (11) are not hard to verify by differentiating the corresponding angular velocity equations (we will see this in Example 6.11 on p. 482). While the procedure we have used in our solution is applicable in general, obtaining the component of an angular acceleration by simply differentiating the corresponding component of the angular velocity can be done only under special circumstances, such as when the components in question are with respect to a fixed axis (in our case, the z axis). As far as the acceleration of point A' is concerned, based on our discussion of the rolling without slip condition earlier in this section, because A' was in contact with a stationary surface, we should have expected $\vec{a}_{A'}$ to be completely in the negative y direction and proportional to ω_P^2. This is exactly what we obtained, given that ω_P is proportional to ω_S. As far as $\vec{a}_{Q'}$ is concerned, our expectation was that the x component had to match the motion of the sun gear (and therefore be in the negative x direction) while the y component had to be in the positive y direction and, again, be proportional to ω_P^2, i.e., ω_S^2. Again, these expectations match the obtained results.

EXAMPLE 6.10 *Completing the Acceleration Analysis of the Connecting Rod*

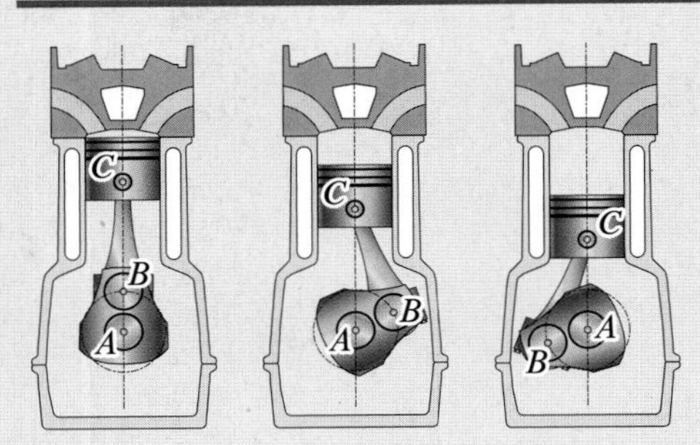

Figure 1

On p. 474 we outlined the vector approach for the acceleration analysis of the connecting rod (CR) and piston in the slider-crank mechanism shown in Fig. 1. We will complete that analysis here.

Referring to Fig. 2, we are given the radius of the crank R, the length of the CR L, the distance from B to the mass center of the CR H, the *constant* angular velocity of the crank $\omega_{AB} = \dot{\theta}$, and the crank angle θ. Determine the angular acceleration of the CR $\vec{\alpha}_{BC}$ and the acceleration of the piston $\vec{a}_C$.

SOLUTION

Road Map The road map for this problem was laid out on p. 474.

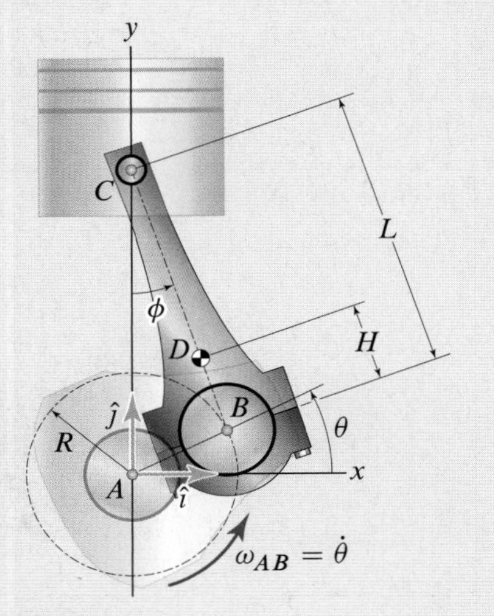

Figure 2
A slider-crank mechanism showing the relevant dimensions.

Computation As with the velocity analysis, we begin by determining the motion of point C, which is constrained to move along the y axis, and so $a_{Cx} = 0$. Applying Eq. (6.14) on p. 439 to the CR and using point B as a reference point for the body, we have

$$\vec{a}_C = \vec{a}_B + \vec{\alpha}_{BC} \times \vec{r}_{C/B} - \omega_{BC}^2 \vec{r}_{C/B}, \tag{1}$$

where $\vec{\alpha}_{BC} = \alpha_{BC}\,\hat{k}$ is the angular acceleration of the CR. Recall that in Example 6.6 on p. 460, we found the angular velocity of the CR to be

$$\omega_{BC} = -\frac{\omega_{AB}\sin\theta}{\sqrt{(L/R)^2 - \cos^2\theta}}, \tag{2}$$

and the position of C relative to B is given by

$$\vec{r}_{C/B} = -L\sin\phi\,\hat{\imath} + L\cos\phi\,\hat{\jmath}, \tag{3}$$

where $\sin\phi$ and $\cos\phi$ are found from the equations

$$\sin\phi = \frac{R}{L}\cos\theta \quad \text{and} \quad \cos\phi = \frac{\sqrt{L^2 - R^2\cos^2\theta}}{L}. \tag{4}$$

Since B is in fixed axis rotation about A with $\omega_{AB} = \dot{\theta} =$ constant, its acceleration is given by (see p. 438)

$$\vec{a}_B = \vec{\alpha}_{AB} \times \vec{r}_{B/A} - \omega_{AB}^2 \vec{r}_{B/A} = -R\omega_{AB}^2(\cos\theta\,\hat{\imath} + \sin\theta\,\hat{\jmath}), \tag{5}$$

where we have used $\alpha_{AB} = \ddot{\theta} = 0$. Substituting Eqs. (3)–(5) into Eq. (1) and simplifying, we obtain

$$\begin{aligned} \vec{a}_C = {} & R\left[\cos\theta(\omega_{BC}^2 - \omega_{AB}^2) - \alpha_{BC}\sqrt{(L/R)^2 - \cos^2\theta}\right]\hat{\imath} \\ & - R\left[\alpha_{BC}\cos\theta + \omega_{BC}^2\sqrt{(L/R)^2 - \cos^2\theta} + \omega_{AB}^2\sin\theta\right]\hat{\jmath}. \end{aligned} \tag{6}$$

Enforcing the condition $a_{Cx} = 0$ in Eq. (6) and solving for α_{BC} yields

$$\alpha_{BC} = \frac{\cos\theta(\omega_{BC}^2 - \omega_{AB}^2)}{\sqrt{(L/R)^2 - \cos^2\theta}}. \tag{7}$$

Substituting Eq. (2) into Eq. (7) and simplifying gives

$$\alpha_{BC} = \frac{\left[1 - (L/R)^2\right]\cos\theta}{\left[(L/R)^2 - \cos^2\theta\right]^{3/2}}\omega_{AB}^2$$

Another way to compute α_{BC}. The method used to obtain Eqs. (8) was laborious, but it only requires a series of algebraic steps. We could have obtained α_{BC} by differentiating ω_{BC} in Eq. (2) with respect to time (this is true here since the motion is planar).

$$\Rightarrow \quad \boxed{\vec{\alpha}_{BC} = \frac{\left[1-(L/R)^2\right]\cos\theta}{\left[(L/R)^2-\cos^2\theta\right]^{3/2}}\omega_{AB}^2\,\hat{k},} \tag{8}$$

where we used $\vec{\alpha}_{BC} = \alpha_{BC}\,\hat{k}$. Substituting Eqs. (2) and (8) into Eq. (6), we obtain

$$\boxed{\vec{a}_C = -R\omega_{AB}^2\left\{\frac{\left[1-(L/R)^2\right]\cos^2\theta}{\left[(L/R)^2-\cos^2\theta\right]^{3/2}} + \frac{\sin^2\theta}{\sqrt{(L/R)^2-\cos^2\theta}} + \sin\theta\right\}\hat{\jmath},} \tag{9}$$

where we have used $a_{Cx} = 0$.

Discussion & Verification The method we illustrated here, i.e., based on the application of the (general) equation $\vec{a}_B = \vec{a}_A + \vec{\alpha}_{AB}\times\vec{r}_{B/A} + \vec{\omega}_{AB}\times(\vec{\omega}_{AB}\times\vec{r}_{B/A})$, while at times laborious, only involves a series of algebraic steps, as opposed to computing accelerations directly through differentiation with respect to time. There are many situations in which it is indeed simpler to differentiate with respect to time than it is to apply the acceleration formula for a rigid body. This strategy for the calculation of accelerations will be demonstrated in Example 6.11 on p. 482.

Common Pitfall

Finding velocities and accelerations by time differentiation. We mentioned that we could have computed Eqs. (8) by time differentiating Eq. (2). In doing this, a common mistake is to take the derivative at a particular instant rather than the general function of time. For example, suppose that we had computed ω_{BC} at a *particular instant* t_0 to be $\omega_{BC}(t_0) = 1234$ rad/s. We shouldn't say that $\alpha_{BC}(t_0) = 0$ because the time derivative of the number 1234 rad/s equals zero. While it is true that the time derivative of a constant is equal to zero, we must first take a time derivative of the *function* $\omega_{BC}(t)$ and then evaluate the result at the time of interest t_0. This concept is illustrated in Example 6.11.

A Closer Look We can now finish the parametric analysis begun in Example 6.6 on p. 460. We plot the angular acceleration of the CR α_{BC}, as well as the acceleration of the piston a_{Cy}, for the same conditions considered in Example 6.11. What is remarkable is the magnitude of the accelerations of the CR. Referring to Figs. 3 and 4, we

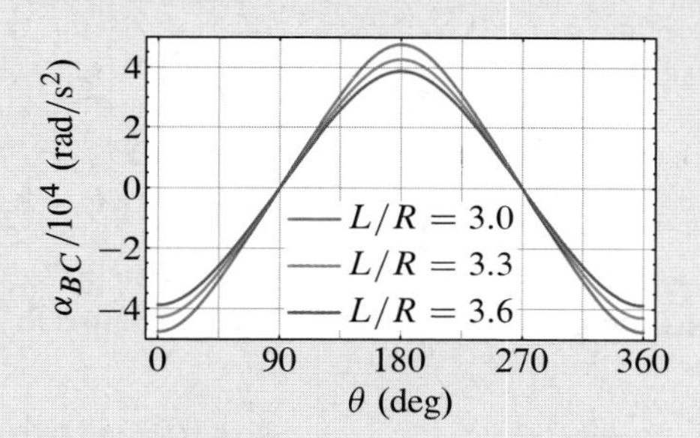

Figure 3. The angular acceleration of the CR for $\dot{\theta} = 3500$ rpm, $L = 150$ mm, and three values of L/R.

see that, for a value of $\dot{\theta} = 3500$ rpm, the angular acceleration reaches values in excess of 40,000 rad/s^2 and the acceleration of the piston reaches values that are 900 times the acceleration of gravity g! Since the accelerations in question are proportional to $\dot{\theta}^2$, if the engine's angular velocity is increased by a factor of, for example, 2, the accelerations we plotted increase by a factor of 4! Thus, for an engine running at 7000 rpm, the piston's accelerations can easily reach $3600g$.

In Chapter 7, we will learn how to use this information to compute the forces and moments that a mechanism, such as the slider-crank, must be able to sustain. We will then understand why connecting rods are typically made of high-grade steel and have a cross-section that looks like an I-beam. As a final remark on Fig. 4, notice that the piston's acceleration has two different behaviors: one for $0° \le \theta < 180°$ and another for $180° \le \theta \le 360°$. We already observed this lack of symmetry during the velocity analysis, and we see now that it is even more pronounced in the acceleration behavior. It is the geometry of the mechanism that generates this lack of symmetry, and the behavior becomes more symmetric as the ratio L/R is increased.

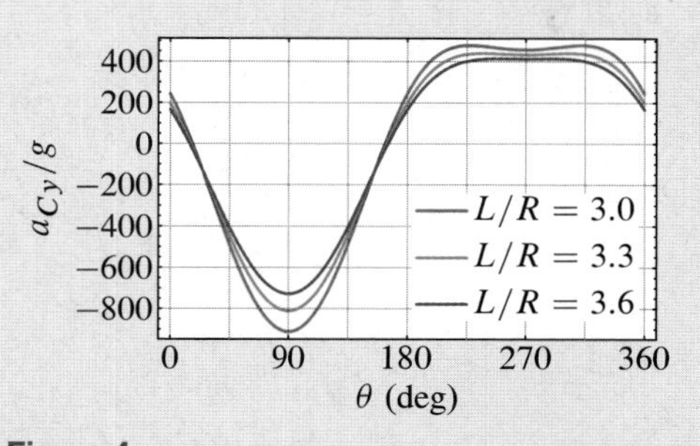

Figure 4
The piston's acceleration for $\dot{\theta} = 3500$ rpm, $L = 150$ mm, and three values of L/R.

EXAMPLE 6.11 *Motion of a Propped Ladder: Differentiation of Constraints*

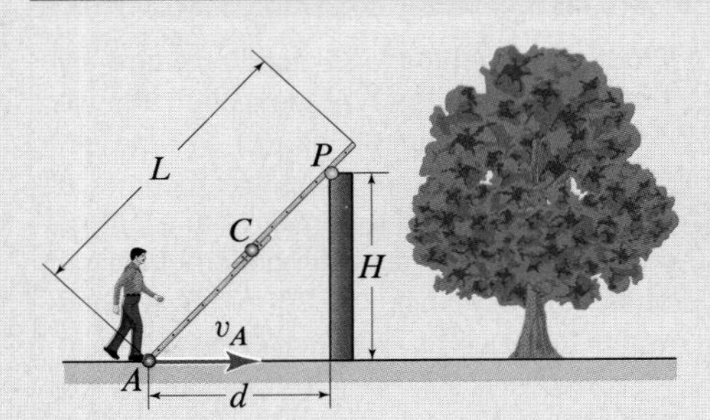

Figure 1

A person is propping up a ladder against a wall by pushing the end A to the right along the ground (see Fig. 1). If, at the instant shown, the speed of A is constant and equal to $v_A = 0.8\,\text{m/s}$, the length $L = 6\,\text{m}$, the height $H = 4\,\text{m}$, and the distance $d = 1.57\,\text{m}$, determine the angular velocity and angular acceleration of the ladder. In addition, determine the acceleration of the midpoint of the ladder C and the acceleration of the point P on the ladder that, at the given instant, is in contact with the wall.

SOLUTION

Road Map Referring to Fig. 2, we can use θ to describe the ladder's orientation so that we can write $\vec{\omega}_L = -\dot{\theta}\,\hat{k}$ and $\vec{\alpha}_L = -\ddot{\theta}\,\hat{k}$ (the minus signs are needed to reconcile the positive direction of θ with the component system used). Thus, we will need to relate $\dot{\theta}$ and $\ddot{\theta}$ to the given data (v_A, H, and d). This can be done by differentiating the constraint relating θ to the position of A with respect to time. We will compute $\vec{a}_C$ using differentiation of constraints and $\vec{a}_P$ using the vector approach to relate $\vec{a}_P$ to the acceleration of a known point on the ladder.

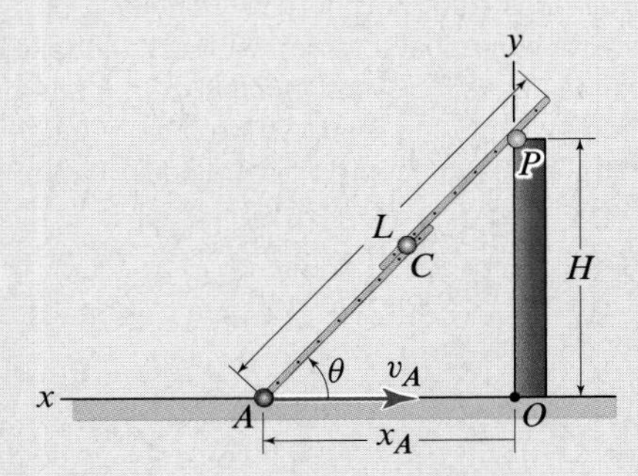

Figure 2
Coordinate system for the ladder kinematics.

Computation Looking at the triangle AOP in Fig. 2, during the ladder's motion we have

$$x_A \tan\theta = H \quad \Rightarrow \quad \dot{x}_A \tan\theta + x_A\dot{\theta}\sec^2\theta = 0, \tag{1}$$

where the second equation was obtained by differentiating the first with respect to time. By using $\sec^2\theta = 1 + \tan^2\theta$, $\tan\theta = H/x_A$, and $\dot{x}_A = -v_A$, the second of Eqs. (1) can be solved for $\dot{\theta}$ to obtain

$$\dot{\theta} = \frac{v_A H}{x_A^2 + H^2} \quad \Rightarrow \quad \dot{\theta}(t_0) = 0.1733\,\text{rad/s}, \tag{2}$$

where t_0 is the time at which $x_A = d = 1.57\,\text{m}$. Having computed $\dot{\theta}$, we can now obtain $\ddot{\theta}$ by time differentiating the first of Eqs. (2) with respect to time to obtain

$$\ddot{\theta} = \frac{2x_A v_A^2 H}{\left(x_A^2 + H^2\right)^2} \quad \Rightarrow \quad \ddot{\theta}(t_0) = 0.02358\,\text{rad/s}^2, \tag{3}$$

where we used the fact that v_A is constant. Since $\vec{\omega}_L = -\dot{\theta}\,\hat{k}$ and $\vec{\alpha}_L = -\ddot{\theta}\,\hat{k}$, Eqs. (2) and (3) imply that

$$\boxed{\vec{\omega}_L(t_0) = -0.1733\,\hat{k}\,\text{rad/s},} \tag{4}$$

$$\boxed{\vec{\alpha}_L(t_0) = -0.02358\,\hat{k}\,\text{rad/s}^2.} \tag{5}$$

We now compute $\vec{a}_C(t_0)$ by differentiation of constraints. This means that we need to take two time derivatives of the general constraint equations for the position of C. Referring to Fig. 3, we can write

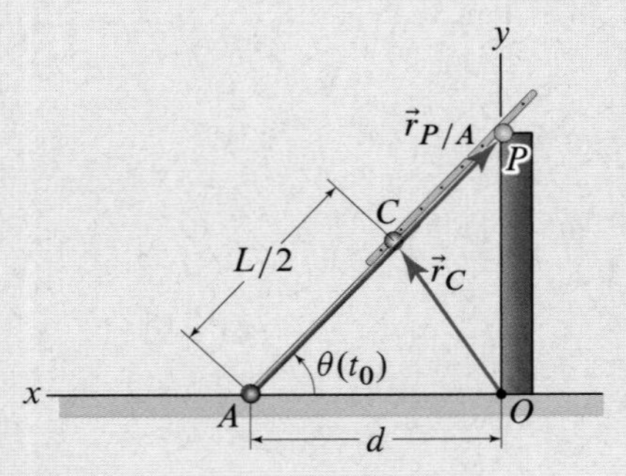

Figure 3
Definition of $\vec{r}_C$ and $\vec{r}_{P/A}$.

$$\vec{r}_C = \left(x_A - \frac{L}{2}\cos\theta\right)\hat{\imath} + \frac{L}{2}\sin\theta\,\hat{\jmath}. \tag{6}$$

Taking one and then two time derivatives of Eq. (6), we obtain

$$\vec{v}_C = \left(-v_A + \frac{L}{2}\dot{\theta}\sin\theta\right)\hat{\imath} + \frac{L}{2}\dot{\theta}\cos\theta\,\hat{\jmath}, \tag{7}$$

and

$$\vec{a}_C = \frac{L}{2}\left[\left(\ddot{\theta}\sin\theta + \dot{\theta}^2\cos\theta\right)\hat{\imath} + \left(\ddot{\theta}\cos\theta - \dot{\theta}^2\sin\theta\right)\hat{\jmath}\right], \tag{8}$$

where we used the fact that v_A is constant. We now find $\theta(t_0)$ by substituting $x_A(t_0) = d = 1.57\,\mathrm{m}$ and $H = 4\,\mathrm{m}$ into the first of Eqs. (1) to obtain

$$\theta(t_0) = \tan^{-1}(4/1.57) = 68.57^\circ. \tag{9}$$

Substituting the results in Eqs. (2), (3), and (9) into Eq. (8), we obtain

$$\boxed{\vec{a}_C(t_0) = \left(0.09876\,\hat{\imath} - 0.05803\,\hat{\jmath}\right)\mathrm{m/s^2}.} \tag{10}$$

Now we compute $\vec{a}_P$ by using the vector method to relate the acceleration of P to that of A. Choosing A as a reference point is convenient because $\vec{a}_A = \vec{0}$ (A is moving at a constant velocity). Since we need to compute $\vec{a}_P$ *at the instant shown*, we can write

$$\vec{a}_P(t_0) = \vec{\alpha}_L(t_0) \times \vec{r}_{P/A}(t_0) - \omega_L^2(t_0)\vec{r}_{P/A}(t_0). \tag{11}$$

Referring to Fig. 3, at time t_0 we have

$$\vec{r}_{P/A}(t_0) = -d\,\hat{\imath} + d\tan\theta(t_0)\,\hat{\jmath}. \tag{12}$$

Substituting Eq. (12), along with the results in Eqs. (4), (5), and (9), into Eq. (11), we have

$$\boxed{\vec{a}_P(t_0) = \left(0.1415\,\hat{\imath} - 0.08312\,\hat{\jmath}\right)\mathrm{m/s^2}.} \tag{13}$$

Discussion & Verification The dimensions and therefore the units of all of our results are as they should be. The signs for the angular velocity and acceleration of the ladder match our expectation, given that the ladder is rotating counterclockwise and the $\hat{k}$ direction is into the page. The values of acceleration are harder to verify without using an alternative solution strategy.

A Closer Look Without actually performing the calculations, if we used the formula $\vec{a}_C = \vec{\alpha}_L \times \vec{r}_{C/A} - \omega_L^2\vec{r}_{C/A}$, we would readily see that the vectors $\vec{\alpha}_L \times \vec{r}_{C/A}$ and $-\omega_L^2\vec{r}_{C/A}$ have positive x components, thus matching the fact that we obtained a positive value for $a_{Cx}(t_0)$. As far as their y components are concerned, we would find that the terms $\vec{\alpha}_L \times \vec{r}_{C/A}$ and $-\omega_L^2\vec{r}_{C/A}$ have positive and negative y components, respectively. However, given that ω_L is larger than α_L in absolute value, we expect that the term with ω_L^2 would dominate with respect to the α_L term. Therefore, overall we expect $a_{Cy}(t_0)$ to be negative, which is exactly what we found. A similar logic can be applied to the discussion of the $\vec{a}_P$ result.

Again, this example is meant to show that the techniques we learned in Chapter 2 are still relevant to the study of rigid bodies and can be used together with the vector method discussed in this chapter.

EXAMPLE 6.12 *Acceleration Analysis of a Four-Bar Linkage: Vector Approach*

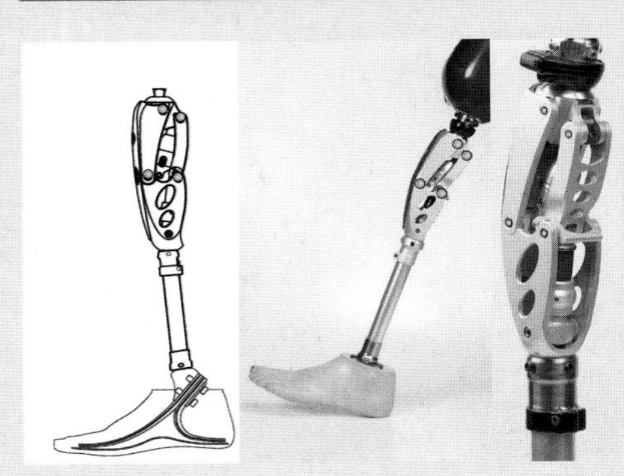

Figure 1

We now continue the kinematic analysis of a prosthetic leg with an artificial knee joint presented in Example 6.8 on p. 464 (see Fig. 1). The primary kinematic component of the artificial knee joint shown is the four-bar linkage system highlighted in Fig. 2. Since the determination of accelerations is crucial in the determination of the forces and moments acting on a mechanism, we will determine the angular accelerations of each link in the system in Fig. 2, at the instant shown, assuming that the AD segment is fixed. We will, also determine the acceleration of point P, which is the midpoint of the link BC. As was done in Example 6.8, we will use the coordinates of points A, B, C, and D given in Table 1. The segment AB is rotating counterclockwise at 1.5 rad/s, and that rate is decreasing at 0.8 rad/s^2.

Table 1. Approximate values of the coordinates of the pin centers A, B, C, and D for the system shown in Fig. 2 at the time instant considered.

Points	A	B	C	D
Coordinates (mm)	(0.0, 0.0)	(−27.0, 120)	(26.0, 124)	(30.0, 15.0)

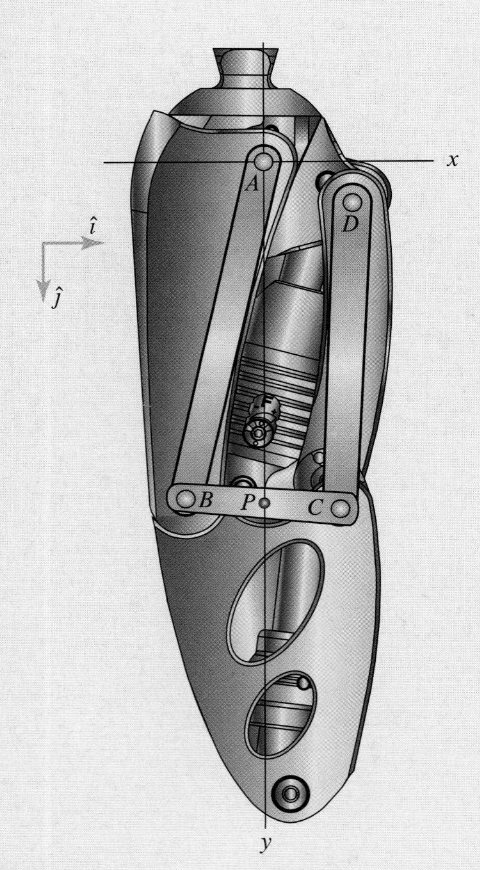

Figure 2
Geometry of the four-bar linkage in the prosthetic leg.

SOLUTION

Road Map We did the velocity analysis for this linkage in Example 6.8 on p. 464, and we will use the angular velocities found there, which were $\omega_{BC} = -0.6378$ rad/s and $\omega_{CD} = -1.675$ rad/s. The calculation of accelerations is similar to the velocity analysis; that is, we will compute the acceleration of B and then the acceleration of C. Because point C is shared by both links BC and CD, we will obtain two independent statements for $\vec{a}_C$, which we will then *require* to be equal. This will allow us to obtain two equations for the angular accelerations of the links BC and CD. Once the angular accelerations are known, then we will compute the acceleration of point P.

Computation Recalling that A is fixed, $\vec{a}_B$ is given by

$$\vec{a}_B = \alpha_{AB}\,\hat{k} \times \vec{r}_{B/A} - \omega_{AB}^2 \vec{r}_{B/A} = -(35.25\,\hat{\imath} + 291.6\,\hat{\jmath})\,\text{mm/s}^2, \tag{1}$$

where we set $\alpha_{AB} = 0.8\,\text{rad/s}^2$ and used $\vec{r}_{B/A} = (-27\,\hat{\imath} + 120\,\hat{\jmath})$ mm from Table 1. Next, since C is shared by both segments BC and CD, we can express $\vec{a}_C$ in the following two independent ways:

$$\vec{a}_C = \vec{a}_B + \alpha_{BC}\,\hat{k} \times \vec{r}_{C/B} - \omega_{BC}^2 \vec{r}_{C/B}, \tag{2}$$

and

$$\vec{a}_C = \vec{a}_D + \alpha_{CD}\,\hat{k} \times \vec{r}_{C/D} - \omega_{CD}^2 \vec{r}_{C/D}, \tag{3}$$

where

$$\vec{r}_{C/B} = \vec{r}_C - \vec{r}_B = (53.00\,\hat{\imath} + 4.000\,\hat{\jmath})\,\text{mm}, \tag{4}$$

$$\vec{r}_{C/D} = \vec{r}_C - \vec{r}_D = (-4.000\,\hat{\imath} + 109.0\,\hat{\jmath})\,\text{mm}, \tag{5}$$

and $\vec{a}_D = \vec{0}$. Substituting Eqs. (1), (4), (5), and the known angular velocities into Eqs. (2) and (3) and setting two expressions for $\vec{a}_C$ equal to one another, we obtain the following vector equation:

$$-\left[56.81\,\tfrac{\text{mm}}{\text{s}^2} + (4.000\,\text{mm})\alpha_{BC}\right]\hat{\imath} + \left[-293.2\,\tfrac{\text{mm}}{\text{s}^2} + (53.00\,\text{mm})\alpha_{BC}\right]\hat{\jmath}$$
$$= \left[11.22\,\tfrac{\text{mm}}{\text{s}^2} - (109.0\,\text{mm})\alpha_{CD}\right]\hat{\imath} - \left[305.7\,\tfrac{\text{mm}}{\text{s}^2} + (4.000\,\text{mm})\alpha_{CD}\right]\hat{\jmath}. \tag{6}$$

Equating $\hat{\imath}$ components and equating $\hat{\jmath}$ components yield the linear system of two equations

$$-56.81\,\tfrac{\text{mm}}{\text{s}^2} - (4.000\,\text{mm})\alpha_{BC} = 11.22\,\tfrac{\text{mm}}{\text{s}^2} - (109.0\,\text{mm})\alpha_{CD}, \tag{7}$$

$$-293.2\,\tfrac{\text{mm}}{\text{s}^2} + (53.00\,\text{mm})\alpha_{BC} = -305.7\,\tfrac{\text{mm}}{\text{s}^2} - (4.000\,\text{mm})\alpha_{CD}, \tag{8}$$

in the two unknowns α_{BC} and α_{CD}, whose solution is

$$\boxed{\alpha_{BC} = -0.2823\,\text{rad/s}^2 \quad \text{and} \quad \alpha_{CD} = 0.6137\,\text{rad/s}^2.} \tag{9}$$

Now that the angular accelerations are known, we can find $\vec{a}_P$ by using

$$\vec{a}_P = \vec{a}_B + \alpha_{BC}\,\hat{k} \times \vec{r}_{P/B} - \omega_{BC}^2 \vec{r}_{P/B}. \tag{10}$$

Since P is the midpoint between B and C, we have

$$\vec{r}_P = \frac{\vec{r}_B + \vec{r}_C}{2}$$

$$\Rightarrow \quad \vec{r}_{P/B} = \vec{r}_P - \vec{r}_B = \frac{\vec{r}_C - \vec{r}_B}{2} = (26.50\,\hat{\imath} + 2.000\,\hat{\jmath})\,\text{mm}. \tag{11}$$

Using Eqs. (1), (9), (11), and the previously computed value of ω_{BC}, from Eq. (10) we obtain

$$\boxed{\vec{a}_P = -(45.46\,\hat{\imath} + 299.9\,\hat{\jmath})\,\text{mm/s}^2.} \tag{12}$$

Discussion & Verification As a first check, we see that the dimensions, and thus the units of all results are as they should be.

Another way to argue that the results we obtained are reasonable is to observe that, *in the position shown*, this four-bar linkage is such that links AB and BC are nearly parallel to each other, while link BC is oriented such that point C has a larger y coordinate than point B. Thus, *in the position shown*, the behavior of this four-bar linkage should not be that different from a similarly sized parallelogram. Therefore, in the position shown, we would expect the angular acceleration of CD to have the same sign as α_{AB}. By the same token, we would expect the angular acceleration of BC to have a sign opposite to that of AB. This is exactly what we obtained. However, obtaining an intuitive understanding of the signs and/or magnitudes of accelerations is not as easy as for velocities. Therefore, when it comes to accelerations, double-checking our calculations is more important than having an intuitive understanding of the mechanism's motion.

Problems

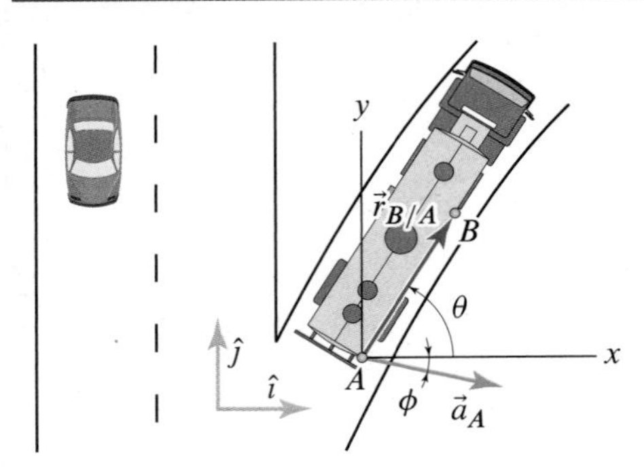

Figure P6.89

Problem 6.89

A truck on an exit ramp is moving in such a way that, at the instant shown, $|\vec{a}_A| = 17\,\text{ft/s}^2$, $\dot{\theta} = -0.3\,\text{rad/s}$, and $\ddot{\theta} = -0.1\,\text{rad/s}^2$. If the distance between points A and B is $d_{AB} = 12\,\text{ft}$, $\theta = 57°$, and $\phi = 13°$, determine $\vec{a}_B$.

Problems 6.90 through 6.92

Let $L = 4\,\text{ft}$, let point A travel parallel to the guide shown, and let C be the midpoint of the bar.

Problem 6.90 If point A is accelerating to the right with $a_A = 27\,\text{ft/s}^2$ and $\dot{\theta} = 7\,\text{rad/s} = \text{constant}$, determine the acceleration of point C when $\theta = 24°$.

Problem 6.91 If point A is accelerating to the right with $a_A = 27\,\text{ft/s}^2$, $\dot{\theta} = 7\,\text{rad/s}$, and $\ddot{\theta} = -0.45\,\text{rad/s}^2$, determine the acceleration of point C when $\theta = 26°$.

Problem 6.92 If, when $\theta = 0°$, A is accelerating to the right with $a_A = 27\,\text{ft/s}^2$ and $\vec{a}_C = \vec{0}$, determine $\dot{\theta}$ and $\ddot{\theta}$.

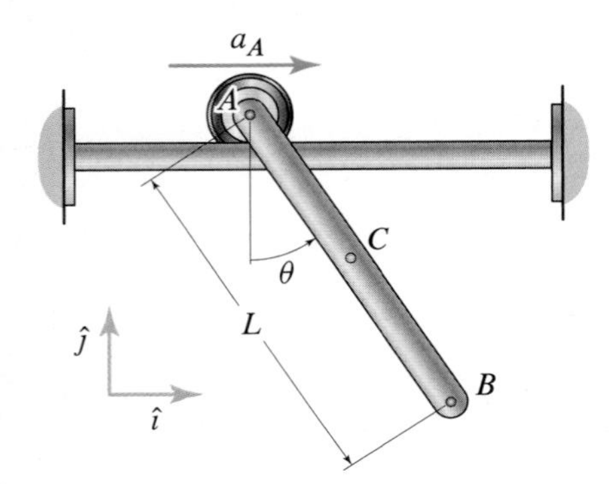

Figure P6.90–P6.92

Problem 6.93

A wheel W of radius $R_W = 5\,\text{cm}$ rolls without slip over the stationary cylinder S of radius $R_S = 12\,\text{cm}$, and the wheel is connected to point O via the arm OC. If $\omega_{OC} = \text{constant} = 3.5\,\text{rad/s}$, determine the acceleration of point Q, which lies on the edge of W and along the extension of the line OC.

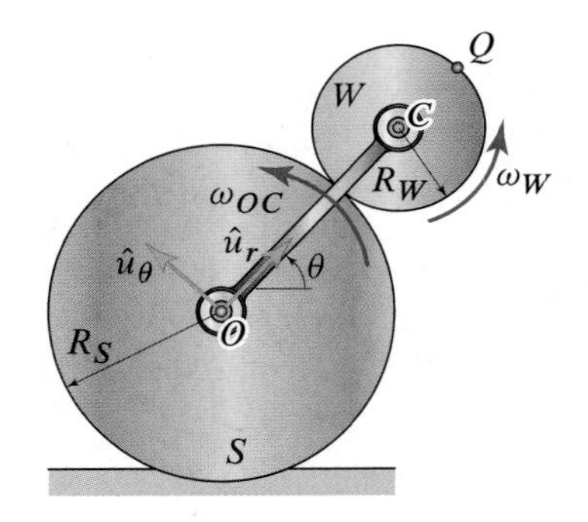

Figure P6.93

Problems 6.94 through 6.97

In the mechanism shown, the block B is constrained to move vertically and is attached to the bar BD. The point A on the bar AD is fixed. Express all your answers in the component system shown.

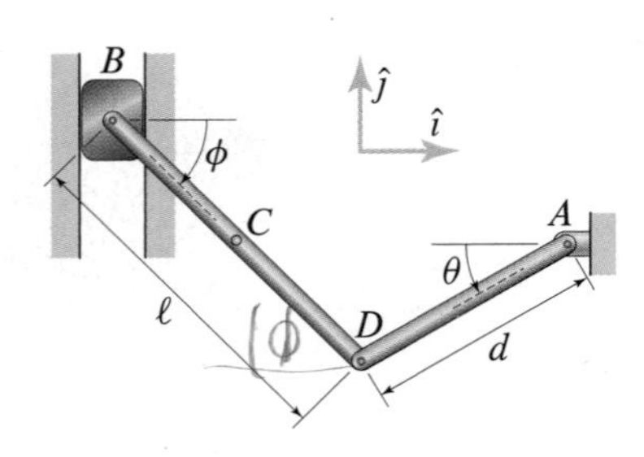

Figure P6.94–P6.97

Problem 6.94 At the instant shown, the block B is moving downward at a constant speed of 2.5 ft/s, $\phi = 45°$, and $\theta = 30°$. If $\ell = 12\,\text{in.}$ and $d = 8\,\text{in.}$, determine the angular acceleration of bar AD at this instant.

Problem 6.95 At the instant shown, bar AD is rotating counterclockwise at a constant angular speed $\omega_{AD} = 13\,\text{rad/s}$, $\phi = 45°$, and $\theta = 30°$. If $\ell = 24\,\text{in.}$ and $d = 16\,\text{in.}$, determine the acceleration of the block B at this instant.

Problem 6.96 At the instant shown, block B is moving downward at $1.5\,\text{m/s}$ and is increasing at $3\,\text{m/s}^2$, $\phi = 45°$, and $\theta = 30°$. If $\ell = 1.2\,\text{m}$ and $d = 0.8\,\text{m}$, determine the angular acceleration of bar AD and the acceleration of point C at this instant.

Problem 6.97 At the instant shown, bar AD is rotating counterclockwise with angular speed $\omega_{AD} = 30\,\text{rad/s}$, which is decreasing at $4\,\text{rad/s}^2$, $\phi = 45°$, and $\theta = 30°$. If $\ell = 24\,\text{cm}$ and $d = 16\,\text{cm}$, determine the acceleration of the block B at this instant.

Problem 6.98

One way to convert rotational motion into linear motion and vice versa is by the use of a mechanism called the Scotch yoke, which consists of a crank C that is connected to a slider B by a pin A. The pin rotates with the crank while sliding within the yoke, which, in turn, rigidly translates with the slider. This mechanism has been used, for example, to control the opening and closing of valves in pipelines. Letting the radius of the crank be $R = 25\,\text{cm}$, determine the angular velocity $\vec{\omega}_C$ and the angular acceleration $\vec{\alpha}_C$ of the crank at the instant shown if $\theta = 25°$ and the slider is moving to the right with a constant speed $v_B = 40\,\text{m/s}$.

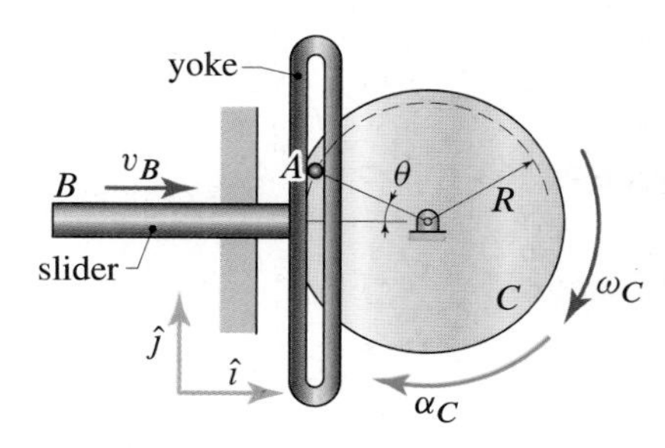

Figure P6.98

Problem 6.99

Collar C moves along a circular guide with radius $R = 2\,\text{ft}$ with a constant speed $v_C = 18\,\text{ft/s}$. At the instant shown, the bars AB and BC are vertical and horizontal, respectively. Letting $L = 4\,\text{ft}$ and $H = 5\,\text{ft}$, determine the angular accelerations of the bars AB and BC at this instant.

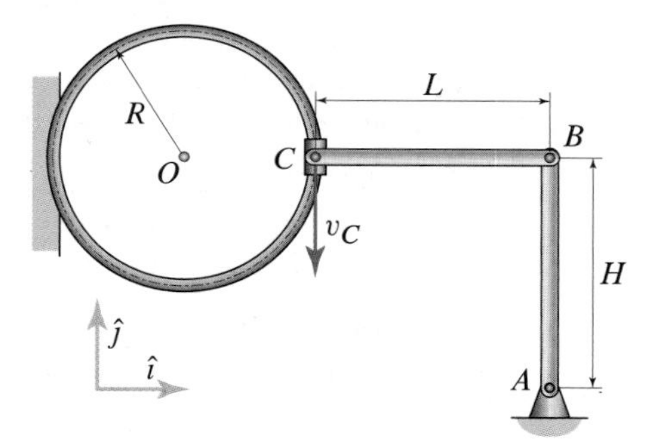

Figure P6.99

Problems 6.100 and 6.101

A sphere S of radius $R_S = 5\,\text{in.}$ is rolling without slip inside a stationary spherical bowl B of radius $R_B = 17\,\text{in.}$ Assume that the motion of the sphere is planar. The center of the sphere is at C and the point of contact between the sphere and the bowl is at P.

Problem 6.100 If, at the instant shown, the center of the sphere C is traveling counterclockwise with a speed $v_C = 32\,\text{ft/s}$ and, such that $\dot{v}_C = 0$, determine the acceleration of C, as well as the acceleration of point P, which is the point on the sphere that is in contact with the bowl at this instant.

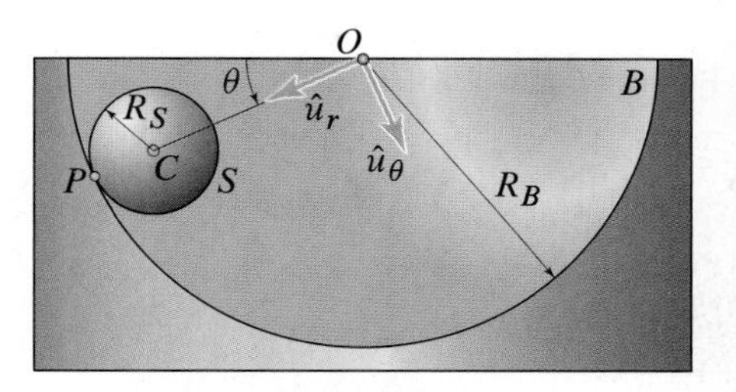

Figure P6.100 and P6.101

Problem 6.101 If, at the instant shown, the center of the sphere C is traveling counterclockwise with a speed $v_C = 32\,\text{ft/s}$ and, such that $\dot{v}_C = 24\,\text{ft/s}^2$, determine the acceleration of C, as well as the acceleration of P, which is the point on the sphere that is in contact with the bowl at this instant.

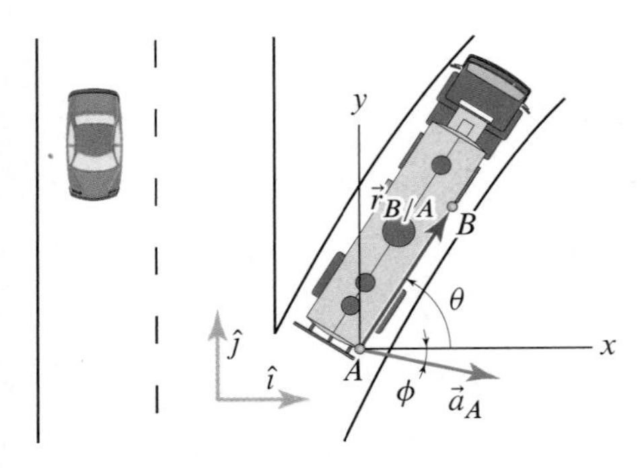

Figure P6.102

Problem 6.102

A truck on an exit ramp is moving in such a way that, at the instant shown, $|\vec{a}_A| = 6\,\text{m/s}^2$ and $\phi = 13°$. Let the distance between points A and B be $d_{AB} = 4\,\text{m}$. If, at this instant, the truck is turning clockwise, $\theta = 59°$, $a_{Bx} = 6.3\,\text{m/s}^2$, and $a_{By} = -2.6\,\text{m/s}^2$, determine the angular velocity and angular acceleration of the truck.

Problems 6.103 and 6.104

As the circular cam whose center is at A rotates, it causes the follower B to move back and forth. The cam angle is θ, the radius of the cam is R, the angular speed of the cam is $\dot{\theta} = \omega_{OA}$, and the angular acceleration of the cam is $\ddot{\theta} = \alpha_{OA}$. The cam is pin-connected to the fixed point O.

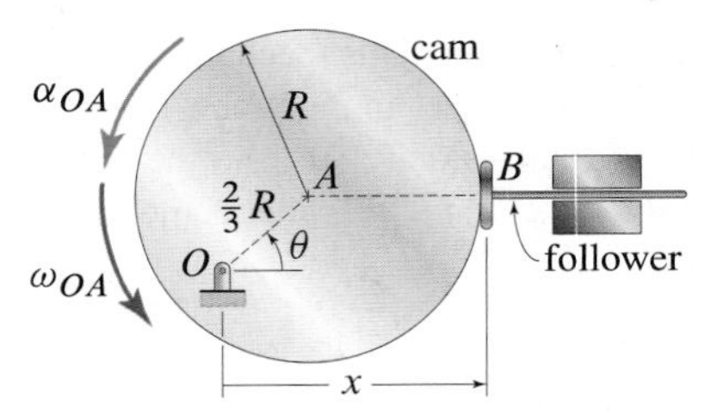

Figure P6.103 and P6.104

Problem 6.103 Using the given x coordinate, determine the acceleration of the follower at the instant $\theta = 30°$ if $R = 1.5\,\text{in.}$, $\dot{\theta} = \omega_{OA} = 1000\,\text{rpm}$, and $\ddot{\theta} = \alpha_{OA} = 25\,\text{rad/s}^2$.

Problem 6.104 Using the given x coordinate, determine the acceleration of the follower as a function of the cam angle θ, the radius of the cam R, the given angular speed of the cam ω_{OA}, and the given angular acceleration of the cam α_{OA}.

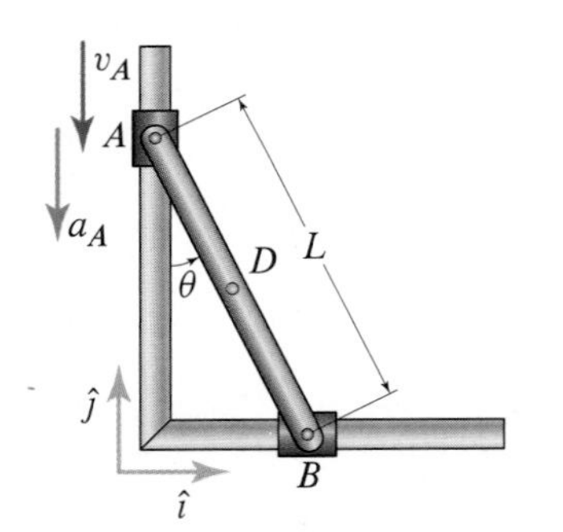

Figure P6.105–P6.107

Problem 6.105

A bar of length $L = 2.5\,\text{m}$ is falling so that, when $\theta = 34°$, $v_A = 3\,\text{m/s}$ and $a_A = 8.7\,\text{m/s}^2$. At this instant, determine the angular acceleration of the bar AB and the acceleration of point D, where D is the midpoint of the bar.

Problem 6.106

A bar of length $L = 8\,\text{ft}$ and midpoint D is falling so that, when $\theta = 27°$, $|\vec{v}_D| = 18\,\text{ft/s}$, and the vertical acceleration of point D is $23\,\text{ft/s}^2$ downward. At this instant, compute the angular acceleration of the bar and the acceleration of point B.

Problem 6.107

Assuming that, for $0° \leq \theta \leq 90°$, v_A is constant, compute the expression for the acceleration of point D, the midpoint of the bar, as a function of θ and v_A.

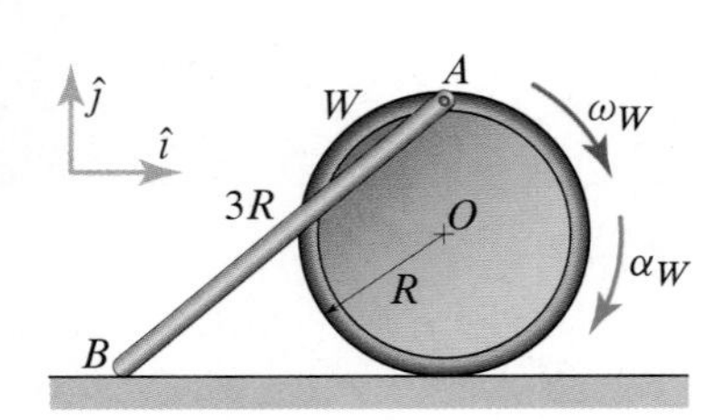

Figure P6.108

Problem 6.108

The wheel W of radius R is rolling without slip with angular speed ω_W and angular acceleration α_W. The bar AB has length $3R$ and is pinned to the outer edge of the wheel at A; the end B remains in contact with the surface. At the instant shown, determine the acceleration of point B. Express your answer in the component system shown.

Problem 6.109

At the instant shown, point B is moving with speed v_B and acceleration a_B. The bar AB has length $3R$; it is pinned to the outer edge of the wheel at A, and the end B remains in contact with the surface. If the wheel rolls without slip, determine the angular velocity $\vec{\omega}_W$ and angular acceleration $\vec{\alpha}_W$ of the wheel at the instant shown. Express all your answers in the component system shown.

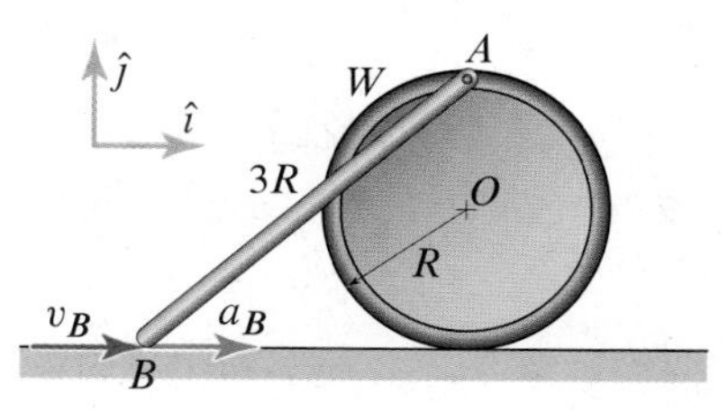

Figure P6.109

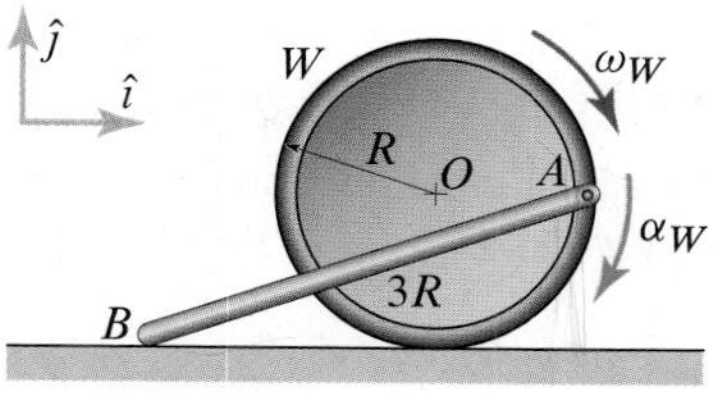

Figure P6.110

Problem 6.110

The wheel W of radius R is rolling without slip with angular speed ω_W and angular acceleration α_W. The bar AB has length $3R$ and is pinned to the outer edge of the wheel at A, and the end B remains in contact with the surface. At the instant shown, determine the acceleration of point B. Express your answer in the component system shown.

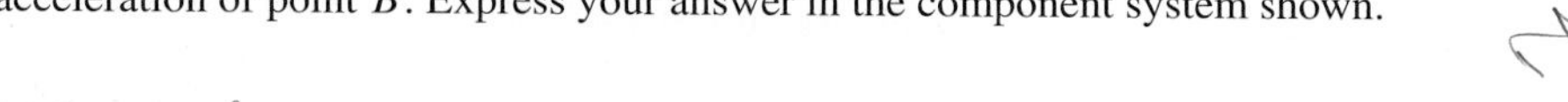

Problem 6.111

At the instant shown, point B is moving with speed v_B and acceleration a_B. The bar AB has length $3R$ and is pinned to the outer edge of the wheel at A, and the end B remains in contact with the surface. If the wheel rolls without slip, determine the angular velocity $\vec{\omega}_W$ and angular acceleration $\vec{\alpha}_W$ of the wheel at the instant shown. Express all your answers in the component system shown.

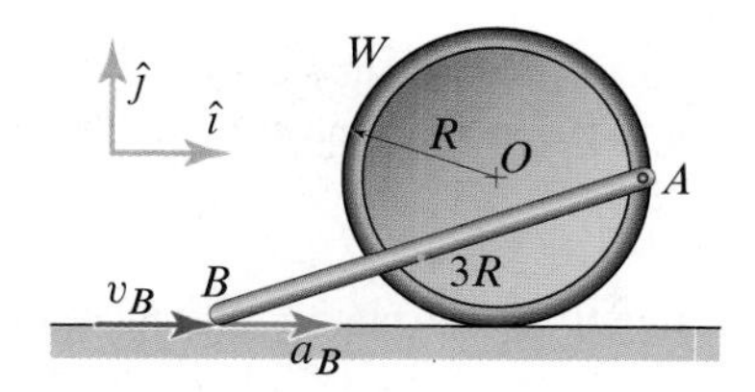

Figure P6.111

Problems 6.112 through 6.114

The wheel W of radius $R = 1.4$ m rolls without slip on a horizontal surface. A bar AB of length $L = 3.7$ m is pin-connected to the center of the wheel and to a slider A constrained to move along a vertical guide. Point C is the bar's midpoint.

Problem 6.112 If the wheel is rolling clockwise with a constant angular speed of 2 rad/s, determine the angular acceleration of the bar when $\theta = 72°$.

Problem 6.113 If the slider A is moving downward with a constant speed 3 m/s, determine the angular acceleration of the wheel when $\theta = 53°$.

Problem 6.114 Determine the general relation expressing the acceleration of the slider A as a function of θ, L, R, the angular velocity of the wheel ω_W, and the angular acceleration of the wheel α_W.

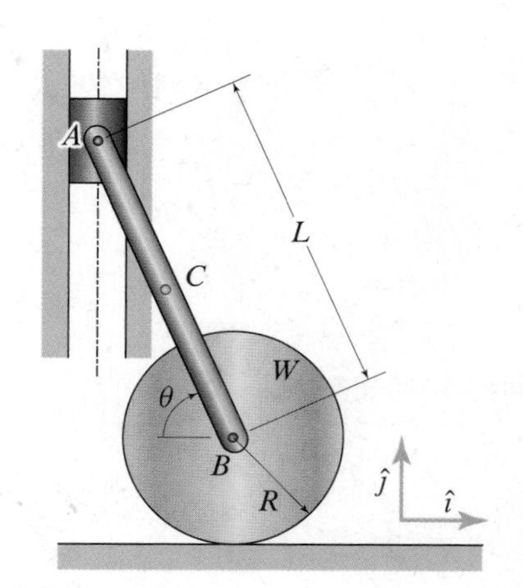

Figure P6.112–P6.114

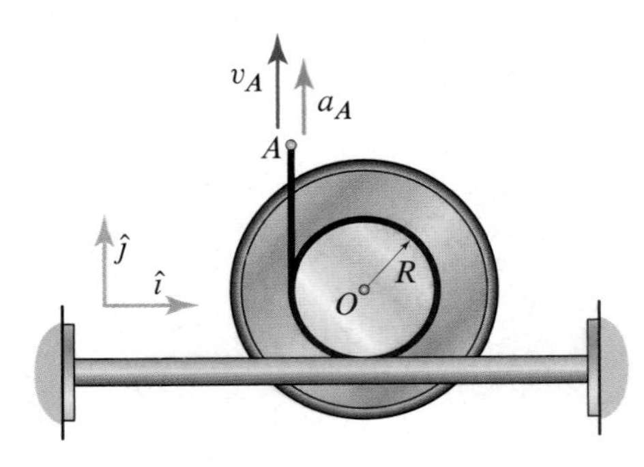

Figure P6.115

Problem 6.115

A spool with inner radius $R = 5\,\text{ft}$ is made to roll without slip over a horizontal rail as shown. If the cable on the spool is unwound in such a way that the free or vertical portion of cable remains perpendicular to the rail, determine the angular acceleration of the spool and the acceleration of the spool's center O. The vertical component of the velocity of point A is $v_A = 12\,\text{ft/s}$, and the vertical component of its acceleration is $a_A = 2\,\text{ft/s}^2$.

Problems 6.116 through 6.119

For the slider-crank mechanism shown, let $R = 0.75\,\text{m}$ and $H = 2\,\text{m}$, and let the length of bar BC be $L_{BC} = 3.25\,\text{m}$.

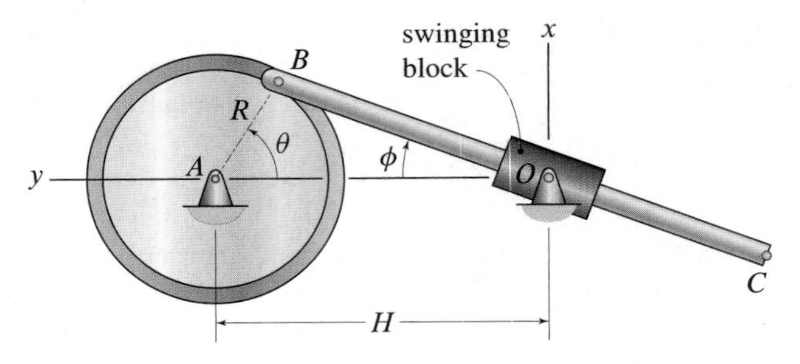

Figure P6.116–P6.119

Problem 6.116 Assume that $\dot{\theta} = 50\,\text{rad/s} = \text{constant}$ and compute the angular acceleration of the slider for $\theta = 27°$.

Problem 6.117 Assume that, at the instant shown, $\theta = 27°$, $\dot{\theta} = 50\,\text{rad/s}$, and $\ddot{\theta} = 15\,\text{rad/s}^2$. Compute the angular acceleration of the slider at this instant, as well as the acceleration of point C.

Problem 6.118 Assuming that $\dot{\theta}$ is constant, determine the expression of the angular acceleration of the slider as a function of θ and $\dot{\theta}$ (and the accompanying geometrical parameters).

Problem 6.119 Letting $\dot{\theta} = 300\,\text{rpm} = \text{constant}$, plot the angular acceleration of the slider as a function of θ for $0° \le \theta \le 360°$. In addition, plot the speed of point C for the same range of θ.

Problems 6.120 through 6.122

A wheel W of radius $R_W = 2\,\text{in.}$ rolls without slip over the stationary cylinder S of radius $R_S = 5\,\text{in.}$, and the wheel is connected to point O via the arm OC.

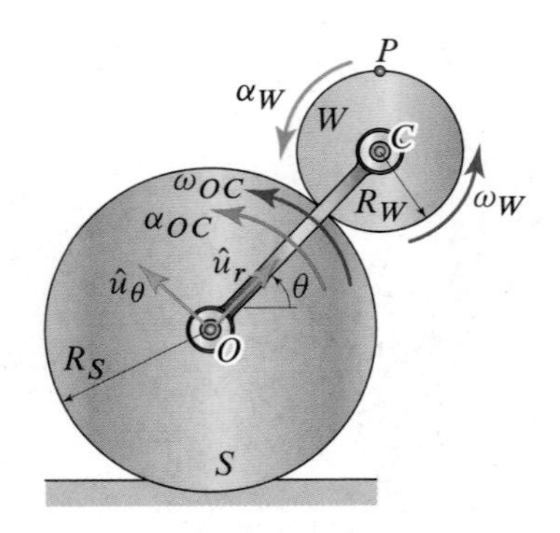

Figure P6.120–P6.122

Problem 6.120 Determine the acceleration of the point on the wheel W that is in contact with S for $\omega_{OC} = 7.5\,\text{rad/s} = \text{constant}$.

Problem 6.121 Determine the acceleration of the point on the wheel W that is in contact with S for $\omega_{OC} = 7.5\,\text{rad/s}$ and $\alpha_{OC} = 2\,\text{rad/s}^2$.

Problem 6.122 If, at the instant shown, $\theta = 63°$, $\omega_W = 9\,\text{rad/s}$, and $\alpha_W = -1.3\,\text{rad/s}^2$, determine the angular acceleration of the arm OC and the acceleration of point P, where P lies on the edge of W and is aligned vertically with point C.

Problem 6.123

At the instant shown, bars AB and BC are perpendicular to each other while the slider C has a velocity $v_C = 24\,\text{m/s}$ and an acceleration $a_C = 2.5\,\text{m/s}^2$ in the directions shown. If $L = 1.75\,\text{m}$ and $\theta = 45°$, determine the angular acceleration of bars AB and BC.

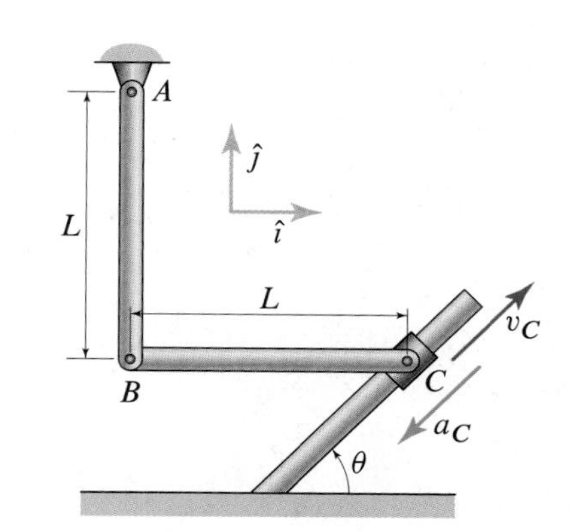

Figure P6.123

Problem 6.124

A flood gate is controlled by the hydraulic cylinder AB. If the length of the cylinder is increased with a constant time rate of 2.5 ft/s, determine the angular acceleration of the gate when $\phi = 0°$. Let $\ell = 10\,\text{ft}$, $h = 2.5\,\text{ft}$, and $d = 5\,\text{ft}$.

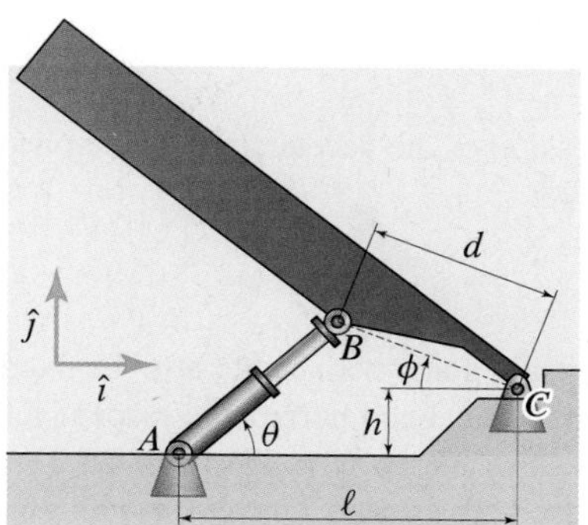

Figure P6.124

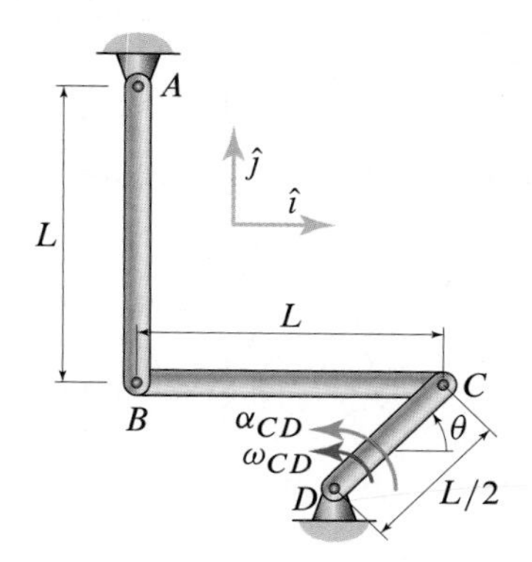

Figure P6.125

Problem 6.125

At the instant shown, bar CD is rotating with an angular velocity 20 rad/s and with angular acceleration $2\,\text{rad/s}^2$ in the directions shown. Furthermore, at this instant $\theta = 45°$. If $L = 2.25\,\text{ft}$, determine the angular accelerations of bars AB and BC.

Problem 6.126

The bucket of a backhoe is the element AB of the four-bar linkage system $ABCD$. Assume that the points A and D are fixed and that the bucket rotates with a constant angular velocity $\omega_{AB} = 0.25\,\text{rad/s}$. In addition, suppose that, at the instant shown, point B is aligned vertically with point A, and C is aligned horizontally with B. Determine the acceleration of point C at the instant shown, along with the angular accelerations of the elements BC and CD. Let $h = 0.66\,\text{ft}$, $e = 0.46\,\text{ft}$, $\ell = 0.9\,\text{ft}$, and $w = 1.0\,\text{ft}$.

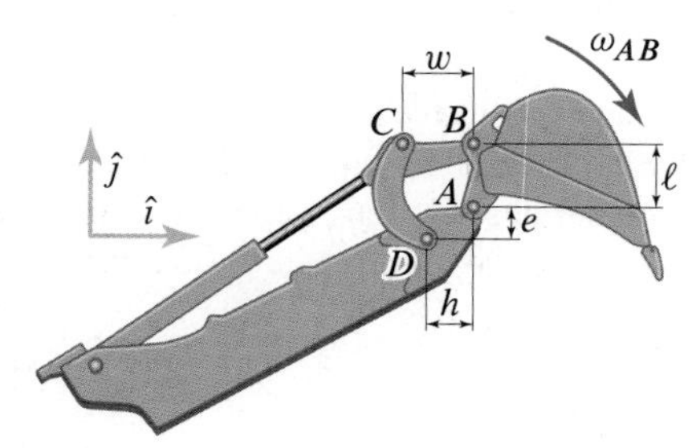

Figure P6.126

Problem 6.127

At the instant shown, bars AB and CD are vertical. In addition, point C is moving to the left with an increasing speed of 4 m/s, and the magnitude of the acceleration of C is $55\,\text{m/s}^2$. If $L = 0.5\,\text{m}$ and $H = 0.2\,\text{m}$, determine the angular accelerations of bars AB and BC.

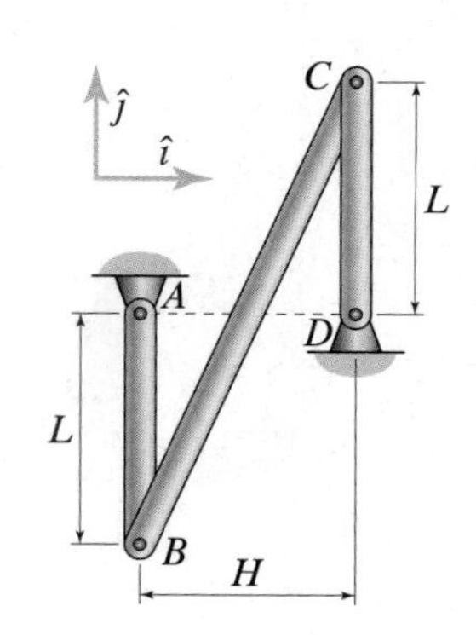

Figure P6.127

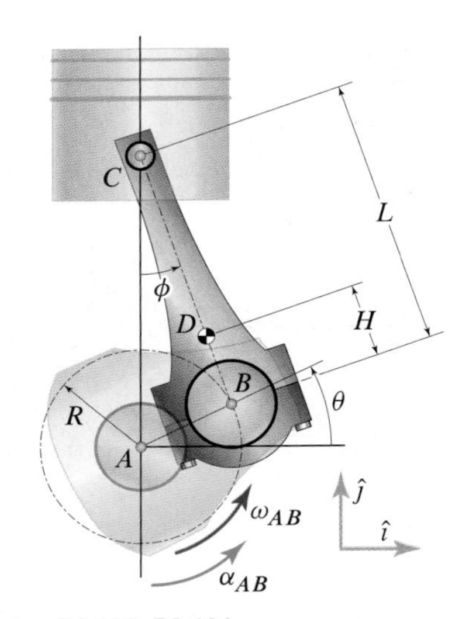

Figure P6.128–P6.131

Problems 6.128 through 6.131

For the slider-crank mechanism shown, let $R = 1.9$ in., $L = 6.1$ in., and $H = 1.2$ in.

Problem 6.128 Assuming that $\omega_{AB} = 4850$ rpm and is constant, determine the angular acceleration of the connecting rod BC and the acceleration of point C at the instant when $\theta = 27°$.

Problem 6.129 Assuming that $\omega_{AB} = 4850$ rpm and is constant, determine the acceleration of point D at the instant when $\phi = 10°$.

Problem 6.130 Assuming that, at the instant shown, $\theta = 31°$, $\omega_{AB} = 4850$ rpm, and $\alpha_{AB} = \dot{\omega}_{AB} = -280\,\text{rad/s}^2$, determine the angular acceleration of the connecting rod and the acceleration of point C.

Problem 6.131 Determine the general expression of the acceleration of the piston C as a function of L, R, θ, $\omega_{AB} = \dot{\theta}$, and $\alpha_{AB} = \ddot{\theta}$.

Problem 6.132

In the four-bar linkage system shown, let the circular guide with center at O be fixed and such that, for $\theta = 0°$, the bars AB and BC are vertical and horizontal, respectively. In addition, let $R = 0.6$ m, $L = 1$ m, and $H = 1.25$ m. When $\theta = 37°$, $\beta = 25.07°$, and $\gamma = 78.71°$, assume collar C is sliding clockwise with a speed 7 m/s. Assuming that, at the instant in question, the speed is increasing and that $|\vec{a}_C| = 93\,\text{m/s}^2$, determine the angular accelerations of the bars AB and BC.

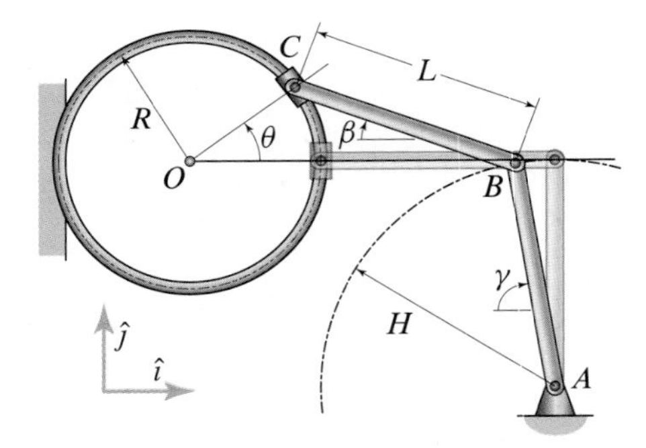

Figure P6.132

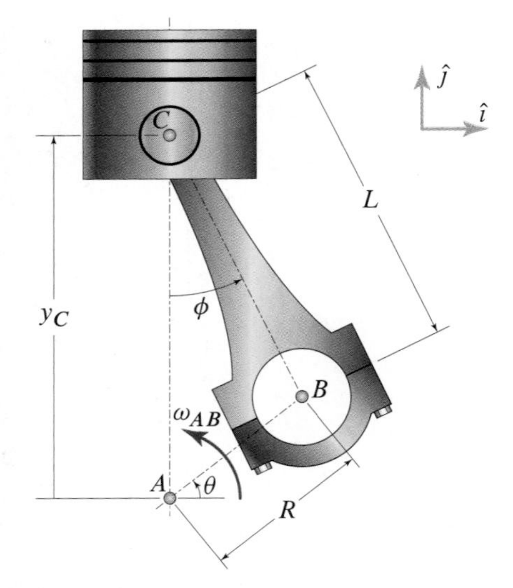

Figure P6.133

Problem 6.133

Complete the acceleration analysis of the slider-crank mechanism using differentiation of constraints that was outlined beginning on p. 474. That is, determine the acceleration of the piston C and the angular acceleration of the connecting rod as a function of the given quantities θ, ω_{AB}, R, and L. Assume that ω_{AB} is constant, and use the component system shown for your answers.

Design Problems

Design Problem 6.1

In some high-performance mountain bikes, one element of the frame is attached to the rest via a four-bar linkage system. Referring to Fig. DP6.1, the four-bar linkage system is defined by the points A, B, C, and D. Notice that this system is also connected to the shock absorber pinned at points E and F. Research the available commercial literature (this information is readily available on the Web), and select a bicycle frame containing a four-bar linkage system, such as that shown below. For the frame you select, obtain the necessary geometric information (again, this information is often made available on the Web by manufacturers), and assume that the points attached to the front part of the frame, which, in Fig. DP6.1 are points A and D, are *fixed*. Then calling ℓ the length of the shock absorber, which in Fig. DP6.1 is the distance between E and F, analyze the kinematics of the linkage system and determine $\dot{\ell}$ and $\ddot{\ell}$ as a function of both the angular velocity and the angular acceleration of the part of the frame to which the wheel is attached, which, in Fig. DP6.1, is the part to the left of points B and C.

Figure DP6.1

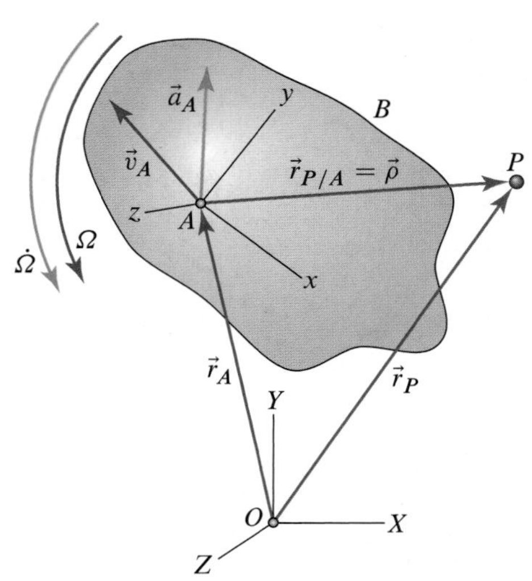

Figure 6.39
Reference frame, position vector, velocity, and acceleration definitions used in describing the motion of a point P relative to a rigid body B.

6.4 Rotating Reference Frames

In problems with multiple bodies or with bodies that have components moving relative to one another, the application of the kinematic relations developed so far is rarely straightforward. For these problems, it is convenient to study the motion of the various parts of the system using more than one frame of reference. For this reason, we now develop a vector-based approach to the study of kinematics that uses both a primary reference frame *and* a secondary reference frame that can rotate and translate relative to the primary frame.

The general kinematic equations for the motion of a point relative to a rotating reference frame

Referring to Fig. 6.39, we will determine the motion of point P using two different reference frames: (1) a reference frame xyz that translates *and* rotates, which we will call the *rotating reference frame*,* and (2) a reference frame XYZ with respect to which we are measuring the motion of P and that we will call the *primary reference frame*. The primary frame will be inertial in kinetics problems. In addition, when the rotating reference frame is attached to a moving rigid body (as it is in Fig. 6.39), we will often call it the *body-fixed frame*. With this as background, we use the following to describe the motion of point P:

1. The motion of the origin (point A) of the rotating reference frame.
2. The angular velocity $\vec{\Omega}$ and angular acceleration $\dot{\vec{\Omega}}$ of the rotating reference frame.
3. How P is seen to be moving by an observer *attached to* the rotating reference frame.

To begin, we note from Fig. 6.39 that the position of P can be written as

$$\vec{r}_P = \vec{r}_A + \vec{r}_{P/A} = \vec{r}_A + \vec{\rho}, \tag{6.40}$$

where $\vec{r}_A$ is the position of the origin of the rotating frame and $\vec{r}_{P/A} = \vec{\rho}$ is the position of P relative to the rotating frame (using $\vec{\rho}$ makes the development more compact). Writing $\vec{\rho}$ in terms of the xyz frame as $\vec{\rho} = \rho_x\,\hat{\imath} + \rho_y\,\hat{\jmath} + \rho_z\,\hat{k}$, Eq. (6.40) becomes†

$$\vec{r}_P = \vec{r}_A + \rho_x\,\hat{\imath} + \rho_y\,\hat{\jmath} + \rho_z\,\hat{k}. \tag{6.41}$$

To find velocities and accelerations, we need to differentiate Eq. (6.41) with respect to time. As we do this, we need to keep in mind that the xyz frame is attached to the body B, and so $\hat{\imath}$ and $\hat{\jmath}$ rotate with B and, therefore, are not constant.

Velocity using a rotating frame

Differentiating Eq. (6.41) with respect to time, we get

$$\vec{v}_P = \vec{v}_A + \dot{\rho}_x\,\hat{\imath} + \rho_x\,\dot{\hat{\imath}} + \dot{\rho}_y\,\hat{\jmath} + \rho_y\,\dot{\hat{\jmath}} + \dot{\rho}_z\,\hat{k} + \rho_z\,\dot{\hat{k}}, \tag{6.42}$$

* Other common names are the *secondary reference frame* or *moving reference frame*.
† For planar motion, we can always consider the plane of motion to be the $z = 0$ plane.

where $\vec{v}_P$ is the velocity of P and $\vec{v}_A$ is the velocity of the origin A of the rotating frame. We can now rewrite this expression, using our knowledge of the time derivative of a unit vector from Eq. (2.46) on p. 81, that is,

$$\dot{\hat{\imath}} = \vec{\omega}_{xyz} \times \hat{\imath}, \quad \dot{\hat{\jmath}} = \vec{\omega}_{xyz} \times \hat{\jmath}, \quad \text{and} \quad \dot{\hat{k}} = \vec{\omega}_{xyz} \times \hat{k}, \tag{6.43}$$

where $\vec{\omega}_{xyz}$ is the angular velocity of $\hat{\imath}$, $\hat{\jmath}$, and $\hat{k}$. Since the xyz frame is body-fixed, we know that all the unit vectors rotate with the body at angular velocity $\vec{\Omega}$, and so

$$\begin{aligned}
\vec{v}_P &= \vec{v}_A + \dot{\rho}_x \hat{\imath} + \dot{\rho}_y \hat{\jmath} + \dot{\rho}_z \hat{k} + \rho_x(\vec{\Omega} \times \hat{\imath}) + \rho_y(\vec{\Omega} \times \hat{\jmath}) + \rho_z(\vec{\Omega} \times \hat{k}) \\
&= \vec{v}_A + \dot{\rho}_x \hat{\imath} + \dot{\rho}_y \hat{\jmath} + \dot{\rho}_z \hat{k} + \vec{\Omega} \times (\rho_x \hat{\imath} + \rho_y \hat{\jmath} + \rho_z \hat{k}) \\
&= \vec{v}_A + \underbrace{\underbrace{\dot{\rho}_x \hat{\imath} + \dot{\rho}_y \hat{\jmath} + \dot{\rho}_z \hat{k}}_{\vec{v}_{P\text{rel}}} + \vec{\Omega} \times \vec{\rho}}_{\dot{\vec{\rho}}},
\end{aligned} \tag{6.44}$$

or

$$\boxed{\vec{v}_P = \vec{v}_A + \vec{v}_{P\text{rel}} + \vec{\Omega} \times \vec{r}_{P/A},} \tag{6.45}$$

where

$$\vec{v}_{P\text{rel}} = \dot{\rho}_x \hat{\imath} + \dot{\rho}_y \hat{\jmath} + \dot{\rho}_z \hat{k}, \tag{6.46}$$

is *the velocity of P relative to the rotating or body-fixed frame* (i.e., *as seen by an observer moving with the body B*) and we have replaced $\vec{\rho}$ with $\vec{r}_{P/A}$. Equation (6.45) is an important development since it allows us to relate the velocities of two points that are *not* on the same rigid body—in this case P and A. In words, and referring to Fig. 6.40, Eq. (6.45) tells us that the velocity of P can be found by using

$\vec{v}_A$ = velocity of the origin of the rotating or body-fixed reference frame (point A in Fig. 6.40)

$\vec{v}_{P\text{rel}}$ = velocity of P as seen by an observer moving with the rotating reference frame xyz

$\vec{\Omega}$ = angular velocity of the rotating reference frame xyz

$\vec{r}_{P/A}$ = vector from the origin of the rotating reference frame to point P

Before looking at accelerations, let's apply Eq. (6.45) in a short example.

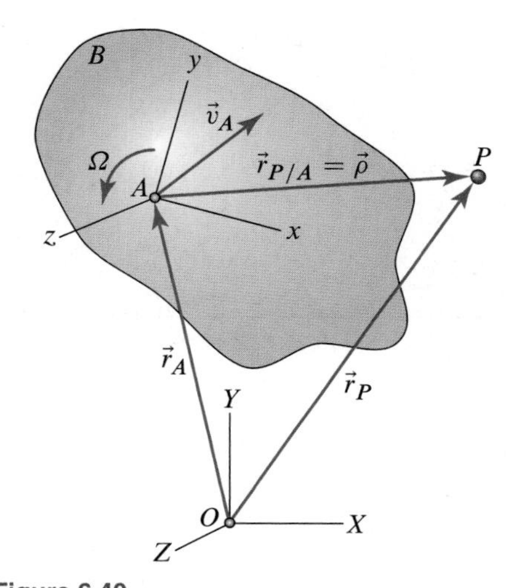

Figure 6.40
The essential ingredients for finding velocities in rotating reference frames.

Mini-Example

Referring to Fig. 6.41, find the velocity of the point P that is moving in the slot in the disk. The angular velocity of the disk ω_0 is constant, and $s(t)$ is known.

Solution

This is a sliding contact (P is not a point on the rotating disk), so we can't apply the kinematic equations developed in Section 6.1. We will apply Eq. (6.45) as

$$\vec{v}_P = \vec{v}_O + \vec{v}_{P\text{rel}} + \vec{\Omega} \times \vec{r}_{P/O}, \tag{6.47}$$

where we have attached the rotating xyz frame to the disk with its origin at the center O, as shown in Fig. 6.41. Interpreting each of these terms, $\vec{v}_O = \vec{0}$ since it is the velocity of the origin of the rotating xyz frame, which is not moving. The term $\vec{v}_{P\text{rel}}$ is the velocity of P as seen by an observer rotating with the disk. If we

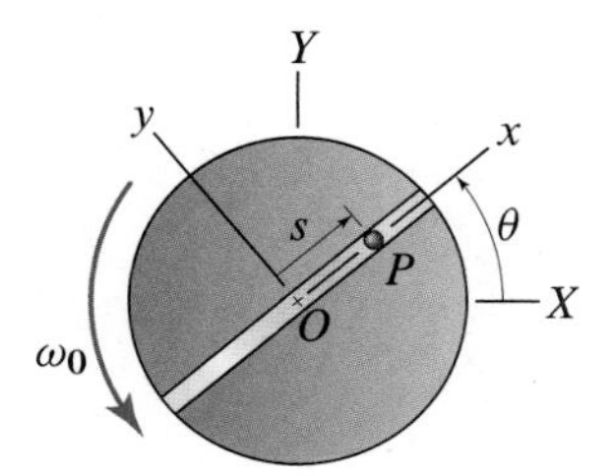

Figure 6.41
A particle moving in the radial slot of a rotating rigid disk.

are sitting on the disk, we see P moving in just the x direction, and so $\vec{v}_{P\text{rel}} = \dot{s}\,\hat{\imath}$. Finally, $\vec{\Omega}$ is the angular velocity of the rotating frame, which is $\omega_0\,\hat{k}$, and $\vec{r}_{P/O}$ is the vector from the origin of the rotating frame to P, which is $\vec{\rho} = s\,\hat{\imath}$. Putting this all in Eq. (6.47), we get

$$\vec{v}_P = \dot{s}\,\hat{\imath} + \omega_0\,\hat{k} \times s\,\hat{\imath} = \dot{s}\,\hat{\imath} + s\omega_0\,\hat{\jmath}. \tag{6.48}$$

Referring to Fig. 6.42, this problem could have been handled using polar coordinates by writing $\vec{v}_P$ as

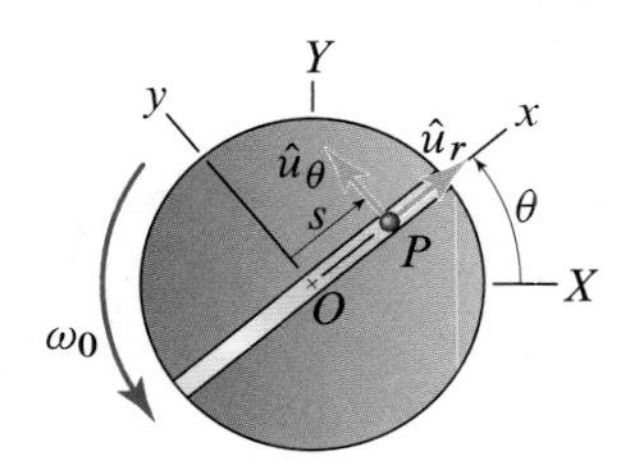

Figure 6.42. The spinning disk of Fig. 6.41 with polar coordinates defined.

$$\vec{v}_P = \dot{r}\,\hat{u}_r + r\dot{\theta}\,\hat{u}_\theta. \tag{6.49}$$

Notice that $r = s$, $\dot{r} = \dot{s}$, $\dot{\theta} = \omega_0$, and the $\hat{\imath}$ and $\hat{\jmath}$ are in the same direction as $\hat{u}_r$ and $\hat{u}_\theta$, respectively. That means that Eqs. (6.48) and (6.49) are giving identical results (as they should!).

Common Pitfall

The velocity and acceleration can be expressed in either the primary or rotating reference frame. We have expressed our final answers in terms of components in the rotating or body-fixed reference frame. Any vector can be expressed in terms of the components of either the rotating frame or the primary frame. For example, in terms of the XYZ frame, we can write the velocity of P in the mini-example as

$$\begin{aligned}\vec{v}_P &= \dot{s}\big(\cos\theta\,\hat{I} + \sin\theta\,\hat{J}\big)\\ &\quad + s\omega_0\big(-\sin\theta\,\hat{I} + \cos\theta\,\hat{J}\big)\\ &= (\dot{s}\cos\theta - s\omega_0\sin\theta)\,\hat{I}\\ &\quad + (\dot{s}\sin\theta + s\omega_0\cos\theta)\,\hat{J}\end{aligned}$$

where $\hat{I}$ and $\hat{J}$ are the unit vectors associated with the XYZ frame.

Acceleration using a rotating frame

To find the acceleration of P, we start by differentiating Eq. (6.44) with respect to time (we use Eq. (6.44) so that we have the component form of $\vec{v}_{P\text{rel}}$) to get

$$\vec{a}_P = \vec{a}_A + \ddot{\rho}_x\,\hat{\imath} + \dot{\rho}_x\,\dot{\hat{\imath}} + \ddot{\rho}_y\,\hat{\jmath} + \dot{\rho}_y\,\dot{\hat{\jmath}} + \ddot{\rho}_z\,\hat{k} + \dot{\rho}_z\,\dot{\hat{k}} + \dot{\vec{\Omega}} \times \vec{\rho} + \vec{\Omega} \times \dot{\vec{\rho}} \tag{6.50}$$

$$\begin{aligned}&= \vec{a}_A + \ddot{\rho}_x\,\hat{\imath} + \ddot{\rho}_y\,\hat{\jmath} + \ddot{\rho}_z\,\hat{k} + \dot{\rho}_x(\vec{\Omega}\times\hat{\imath}) + \dot{\rho}_y(\vec{\Omega}\times\hat{\jmath}) + \dot{\rho}_z(\vec{\Omega}\times\hat{k})\\ &\qquad + \dot{\vec{\Omega}} \times \vec{\rho} + \vec{\Omega} \times \big(\vec{v}_{P\text{rel}} + \vec{\Omega}\times\vec{\rho}\big)\end{aligned} \tag{6.51}$$

$$\begin{aligned}&= \vec{a}_A + \underbrace{\ddot{\rho}_x\,\hat{\imath} + \ddot{\rho}_y\,\hat{\jmath} + \ddot{\rho}_z\,\hat{k}}_{\vec{a}_{P\text{rel}}} + \vec{\Omega} \times \underbrace{\big(\dot{\rho}_x\,\hat{\imath} + \dot{\rho}_y\,\hat{\jmath} + \dot{\rho}_z\,\hat{k}\big)}_{\vec{v}_{P\text{rel}}}\\ &\qquad + \dot{\vec{\Omega}} \times \vec{\rho} + \vec{\Omega} \times \big(\vec{v}_{P\text{rel}} + \vec{\Omega}\times\vec{\rho}\big)\end{aligned}$$

$$= \vec{a}_A + \vec{a}_{P\text{rel}} + 2\vec{\Omega}\times\vec{v}_{P\text{rel}} + \dot{\vec{\Omega}}\times\vec{\rho} + \vec{\Omega}\times\big(\vec{\Omega}\times\vec{\rho}\big), \tag{6.52}$$

where we have used the time derivative of the unit vectors from Eq. (6.43) and $\dot{\vec{\rho}} = \vec{v}_{P\text{rel}} + \vec{\Omega}\times\vec{\rho}$ from Eq. (6.44) to go from Eq. (6.50) to Eq. (6.51). In summary,

$$\boxed{\vec{a}_P = \vec{a}_A + \vec{a}_{P\text{rel}} + 2\vec{\Omega}\times\vec{v}_{P\text{rel}} + \dot{\vec{\Omega}}\times\vec{r}_{P/A} + \vec{\Omega}\times\big(\vec{\Omega}\times\vec{r}_{P/A}\big),} \tag{6.53}$$

where we have replaced $\vec{\rho}$ with $\vec{r}_{P/A}$, and

$$\vec{a}_{P\text{rel}} = \ddot{\rho}_x\,\hat{\imath} + \ddot{\rho}_y\,\hat{\jmath} + \ddot{\rho}_z\,\hat{k} \tag{6.54}$$

is the *acceleration of P relative to the rotating frame* (i.e., *as seen by an observer moving with the body B*). Equation (6.53) looks complicated, but it is readily applied as long as each term is considered individually. In words, the terms in Eq. (6.53) are (see Fig. 6.43):

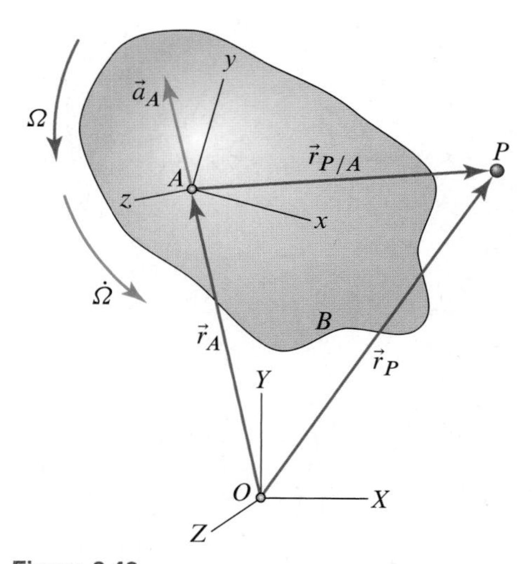

Figure 6.43
The essential quantities needed for finding accelerations in rotating reference frames.

$\vec{a}_A$ = acceleration of the origin of the rotating reference frame

$\vec{a}_{P\text{rel}}$ = acceleration of P as seen by an observer moving with the rotating reference frame

$\vec{\Omega}$ = angular velocity of the rotating reference frame xyz

$\vec{v}_{P\text{rel}}$ = velocity of P as seen by an observer moving with the rotating reference frame xyz

$\dot{\vec{\Omega}}$ = angular acceleration of the rotating reference frame xyz

$\vec{r}_{P/A}$ = vector from the origin of the rotating reference frame to point P

Now, let's revisit the mini-example on p. 495 and find the acceleration of P.

Mini-Example
Referring to Fig. 6.44, find the acceleration of the point P that is moving in the slot in the disk. The disk is rotating with angular velocity ω_0 and angular acceleration α_0, and $s(t)$ is known.

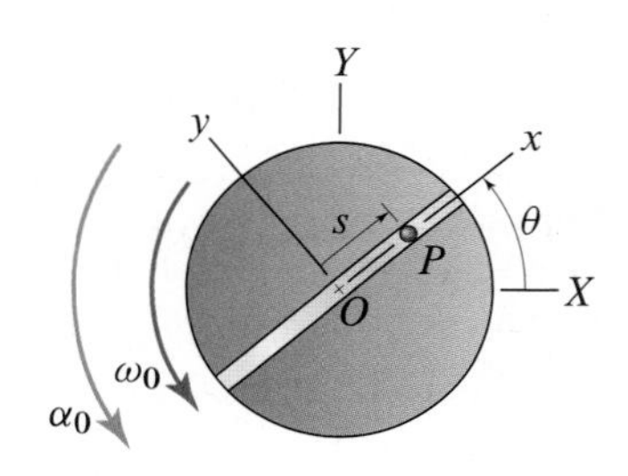

Figure 6.44
A particle moving in the radial slot of a rotating rigid disk.

Solution
We will apply Eq. (6.53) as

$$\vec{a}_P = \vec{a}_O + \vec{a}_{P\text{rel}} + 2\vec{\Omega} \times \vec{v}_{P\text{rel}} + \dot{\vec{\Omega}} \times \vec{r}_{P/O} + \vec{\Omega} \times (\vec{\Omega} \times \vec{r}_{P/O}), \tag{6.55}$$

where we have attached the rotating xyz frame to the disk with its origin at the center O, as shown in Fig. 6.44. First, $\vec{a}_O = \vec{0}$ since the origin of the rotating xyz frame is not moving. Second, $\vec{a}_{P\text{rel}}$ is the acceleration of P as seen by an observer rotating with the disk, which means that $\vec{a}_{P\text{rel}} = \ddot{s}\,\hat{\imath}$. Using similar reasoning, $\vec{v}_{P\text{rel}} = \dot{s}\,\hat{\imath}$. Finally, $\vec{\Omega}$ is the angular velocity of the rotating frame, which is $\omega_0\,\hat{k}$; $\dot{\vec{\Omega}}$ is the angular acceleration of the rotating frame, which is $\alpha_0\,\hat{k}$; and $\vec{r}_{P/O}$ is the vector from the origin of the rotating frame to P, which is $s\,\hat{\imath}$. Putting this all in Eq. (6.55), we get

$$\begin{aligned}\vec{a}_P &= \ddot{s}\,\hat{\imath} + 2\omega_0\,\hat{k} \times \dot{s}\,\hat{\imath} + \alpha_0\,\hat{k} \times s\,\hat{\imath} + \omega_0\,\hat{k} \times (\omega_0\,\hat{k} \times s\,\hat{\imath}) \\ &= (\ddot{s} - s\omega_0^2)\,\hat{\imath} + (s\alpha_0 + 2\dot{s}\omega_0)\,\hat{\jmath}.\end{aligned} \tag{6.56}$$

Using polar coordinates gives us the same result (see Prob. 6.134).

Useful tidbits when using rotating reference frames

- Equations (6.45) and (6.53) apply in a 2D or 3D context. For planar motion, the acceleration given in Eq. (6.53) can be written as

$$\vec{a}_P = \vec{a}_A + \vec{a}_{P\text{rel}} + 2\vec{\Omega} \times \vec{v}_{P\text{rel}} + \dot{\vec{\Omega}} \times \vec{r}_{P/A} - \Omega^2 \vec{r}_{P/A}, \tag{6.57}$$

where $\Omega = |\vec{\Omega}|$.

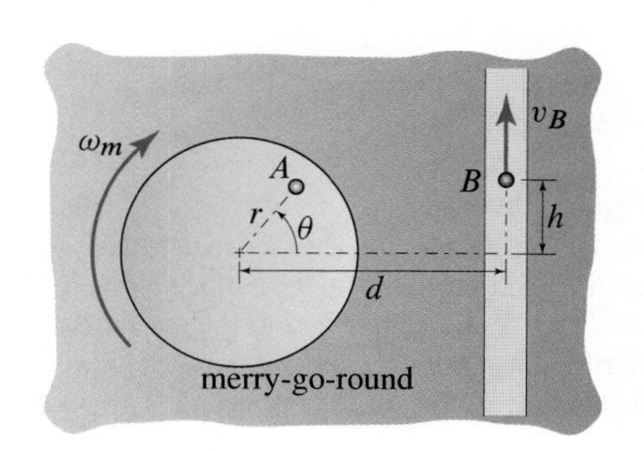

Figure 6.45
A person A standing on a spinning merry-go-round as the person B walks on a sidewalk in the direction shown.

- The terms $\vec{v}_{P\text{rel}}$ and $\vec{a}_{P\text{rel}}$ in Eqs. (6.45) and (6.53) tell us the *apparent* motion of something as seen by a moving observer. Referring to Fig. 6.45, the person at A is not moving relative to the merry-go-round, which is rotating at the constant rate ω_m, and the person at B is walking at a constant speed along the sidewalk in the direction shown. The velocity and acceleration of B as seen by A are *not* $\vec{v}_{B/A}$ and $\vec{a}_{B/A}$, respectively; they are instead

$$\vec{v}_{B\mathrm{rel}} = \vec{v}_B - \vec{v}_A - \vec{\omega}_m \times \vec{r}_{B/A}, \tag{6.58}$$

$$\vec{a}_{B\mathrm{rel}} = \vec{a}_B - \vec{a}_A - 2\vec{\omega}_m \times \vec{v}_{P\mathrm{rel}} - \vec{\alpha}_m \times \vec{r}_{B/A} + \omega_m^2 \vec{r}_{B/A}, \tag{6.59}$$

respectively. In the given situation, $\vec{\alpha}_m$ and $\vec{a}_B$ would both be zero. We will explore this idea in Example 6.15.

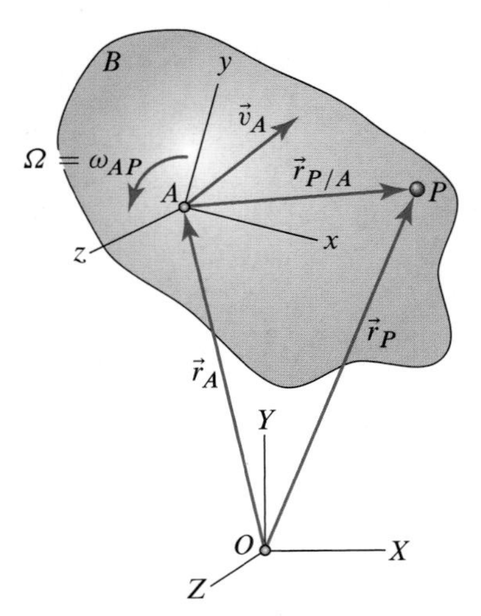

Figure 6.46
Relating the velocity of P to that of B when P is a point *on* the rigid body.

- When deciding where to attach the rotating frame, choose a frame that makes it easy to determine each of the terms in Eqs. (6.45) and (6.53).
- If P is a point on the body B, as shown in Fig. 6.46, then $\vec{v}_{P\mathrm{rel}} = \vec{0}$ in Equation (6.45), and we have

$$\vec{v}_P = \vec{v}_A + \vec{\Omega} \times \vec{r}_{P/A} = \vec{v}_A + \vec{\omega}_{AP} \times \vec{r}_{P/A},$$

which is just Eq. (6.3) on p. 435. A corresponding simplification applies to the acceleration in Eq. (6.53) leading to Eq. (6.5).

Coriolis component of acceleration

The term $2\vec{\Omega} \times \vec{v}_{P\mathrm{rel}}$ in Eq. (6.53) is known as the *Coriolis acceleration* of P; this term results from two equal, but different effects: (1) the change in direction of $\vec{v}_{P\mathrm{rel}}$ due to $\vec{\Omega}$ and (2) the effect of $\vec{\Omega}$ on the change in magnitude of $\vec{r}_{P/A}$ relative to the rotating reference frame. It is named after Gaspard-Gustave Coriolis (1792–1843), who studied mechanics and mathematics in France. He gave the terms *work* and *kinetic energy* their present scientific meaning, and in an 1835 publication, Coriolis showed that Newton's laws may be applied in a rotating reference frame, as long as an "extra" acceleration is added to the equations of motion. The phenomenon in which a person feels as though she or he is being thrown sideways when moving on a rotating platform is a consequence of this acceleration. In fact, this acceleration gives rise to the ccw circulation around low-pressure systems and cw circulation around high-pressure systems in the northern hemisphere. This circulation around high- and low-pressure systems can be understood if we consider the motion of an object on the surface of the Earth as seen by an observer moving with the Earth.

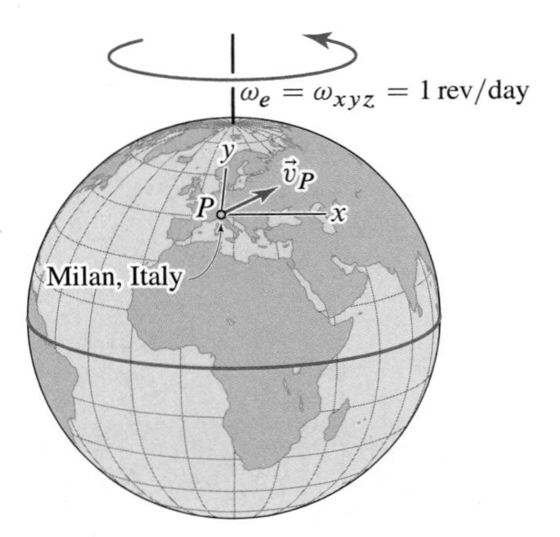

Figure 6.47
An object P moving due east in Milan, Italy, with speed v_P relative to the surface of the Earth. The xyz frame is attached to the surface of the Earth with x pointing east, y north, and z is the local vertical. The latitude of Milan is 45° north.

Figure 6.47 depicts an object P moving both east and north in Milan, Italy, at a constant speed $v_P = \sqrt{v_{Px}^2 + v_{Py}^2}$ relative to the surface of the Earth. The xyz frame is attached to the surface of the Earth with its origin in Milan (which we will call point A) and with x pointing east, y pointing north, and z pointing radially outward from the center of the Earth. The angular velocity of the Earth is $\omega_e = 1$ rev/day (assumed to be constant), but this is also the angular velocity of the rotating frame ω_{xyz} since it is attached to the Earth. Referring to Eq. (6.53) and noting that the origin of the rotating frame, which we are calling point A, is also the location of Milan, we have that

$$\vec{a}_A = \vec{a}_{\mathrm{Milan}}, \qquad \vec{a}_{P\mathrm{rel}} = \vec{0}, \qquad \vec{\Omega} = \vec{\omega}_e, \tag{6.60}$$

$$\vec{v}_{P\mathrm{rel}} = v_{Px}\,\hat{\imath} + v_{Py}\,\hat{\jmath}, \qquad \dot{\vec{\Omega}} = \vec{0}, \qquad \vec{r}_{P/A} = \vec{0}. \tag{6.61}$$

Referring to Fig. 6.48, we can write the nonzero quantities in the xyz frame as

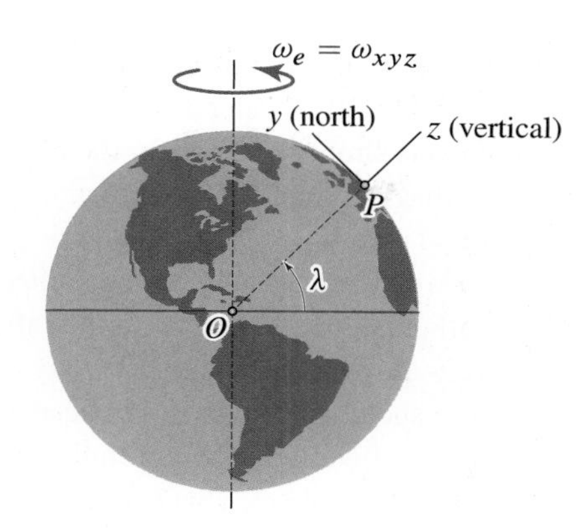

Figure 6.48. A different perspective of Fig. 6.47 showing the latitude λ of the object P in Milan, Italy.

$$\begin{aligned}\vec{a}_A &= \vec{a}_O + \vec{\alpha}_e \times \vec{r}_{A/O} + \vec{\omega}_e \times (\vec{\omega}_e \times \vec{r}_{A/O}) = \vec{\omega}_e \times (\vec{\omega}_e \times \vec{r}_{A/O}) \\ &= \omega_e(\cos\lambda\,\hat{\jmath} + \sin\lambda\,\hat{k}) \times \left[\omega_e(\cos\lambda\,\hat{\jmath} + \sin\lambda\,\hat{k}) \times R_e\,\hat{k}\right] \\ &= R_e\omega_e^2\cos\lambda(\sin\lambda\,\hat{\jmath} - \cos\lambda\,\hat{k}), \qquad (6.62)\end{aligned}$$

$$\vec{\Omega} = \omega_e(\cos\lambda\,\hat{\jmath} + \sin\lambda\,\hat{k}), \qquad (6.63)$$

where R_e is the radius of the Earth, and where we have used the fact that points A and P coincide at this instant. Substituting Eqs. (6.60)–(6.63) into Eq. (6.53), the acceleration of P is

$$\vec{a}_P = R_e\omega_e^2\cos\lambda(\sin\lambda\,\hat{\jmath} - \cos\lambda\,\hat{k}) + 2\omega_e(\cos\lambda\,\hat{\jmath} + \sin\lambda\,\hat{k}) \times (v_{Px}\,\hat{\imath} + v_{Py}\,\hat{\jmath}). \qquad (6.64)$$

The first term in this equation is the normal acceleration due to the rotation of the Earth, and it is zero at the poles and largest on the equator. The second term is the Coriolis acceleration and, in this case, it is given by

$$\vec{a}_{\text{Coriolis}} = 2\omega_e\sin\lambda(-v_{Py}\,\hat{\imath} + v_{Px}\,\hat{\jmath}) - 2\omega_e v_{Px}\cos\lambda\,\hat{k}. \qquad (6.65)$$

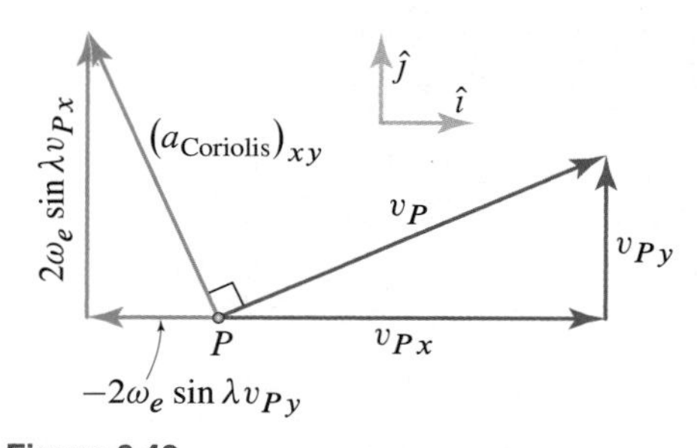

Figure 6.49
The velocity of P relative to the surface of the Earth and the component of the resulting Coriolis acceleration that lies in the xy plane.

Notice that the (vector) component of the Coriolis acceleration in the xy plane has, in turn, a component that is *perpendicular* to $\vec{v}_P$, and it points to the left of $\vec{v}_P$ if one is facing in the direction of the relative motion (see Fig. 6.49). Referring back to Example 3.11 on p. 204, this means that P will be deflected to the right if no external forces prevent it from doing so.*

Now, low- and high-pressure systems are so named because the air pressure in them is lower and higher, respectively, than the surrounding air pressure. Since air flows from areas of higher pressure to areas of lower pressure, we see that air will try to flow *into* the low from its surroundings and *away* from the high to its surroundings (the black arrows in Fig. 6.50). As air flows into the low, the Coriolis component of acceleration causes it to deflect to the right (the red arrows in Fig. 6.50). Since every "particle" of air is deflected to the right, the overall motion becomes that of a circulation *around* the low — a circulation in the ccw direction. A similar argument tells us that the circulation around a high must be in the cw direction. In the southern hemisphere, for an Earth-fixed xyz frame oriented as it was in the northern hemisphere,

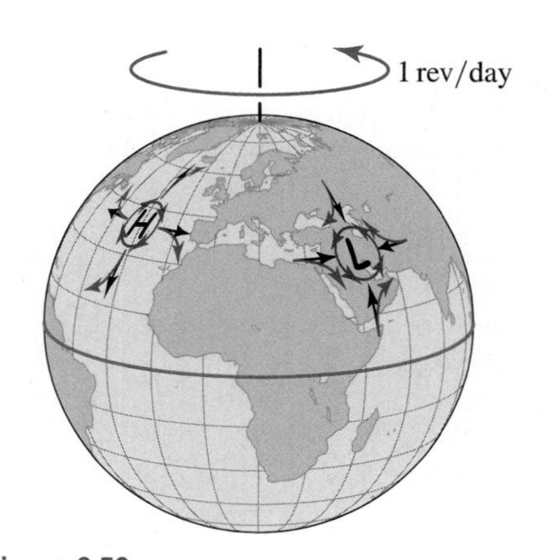

Figure 6.50
Air flow around low- and high-pressure systems in the northern hemisphere.

* Equation (6.65) also tells us that the object will be deflected upward, but that is not a relevant effect for this discussion.

Interesting Fact

Toilets flushing. You may have heard that it is the *Coriolis force* or *Coriolis acceleration* that causes the draining water in a toilet or in a kitchen sink to rotate in one direction in the northern hemisphere and in the opposite direction in the southern hemisphere. As we shall see in Chapter 10 when we do a careful three-dimensional analysis, the effect of the Coriolis component of acceleration is tiny when compared with other factors, such as the shape of the toilet bowl, the angle of the water jets in the toilet, etc. Even in something as large as a swimming pool, it is negligible.

the sign on the $\hat{k}$ component of $\vec{\Omega}$ in Eq. (6.63) will reverse, which means that the sign on the xy component of the Coriolis acceleration will reverse. Therefore, in the southern hemisphere, by arguments similar to those for the northern hemisphere, circulation will reverse direction, that is, around a low-pressure system it is clockwise, and it is counterclockwise around a high.

So, why don't you have to worry about the Coriolis component of acceleration when you play baseball or basketball? It is there, it is just that the angular speed of the Earth is *really* small. It is easy to show that $1\,\text{rev/day} = 0.00007272\,\text{rad/s}$. Although it is present when you shoot a basketball, this acceleration would need to act for a *long* time to have a visible effect — the sort of times (days) experienced by a particle of air moving through the atmosphere. That is not to say that engineers don't need to worry about the Coriolis component of acceleration — they do. However, if the kinematic analysis of a system is carried out using the ideas presented in this section, the Coriolis component of acceleration will *always* be taken into account "automatically."

End of Section Summary

In this section, we developed the kinematic equations that allow us to relate the motion of two points that *are not* on the same rigid body. Referring to Fig. 6.51, this means we can relate the velocity of point P to that of A using the relation

Eq. (6.45), p. 495

$$\vec{v}_P = \vec{v}_A + \vec{v}_{P\text{rel}} + \vec{\Omega} \times \vec{r}_{P/A},$$

where

$\vec{v}_A$ = velocity of the origin of the rotating or body-fixed reference frame (point A in Fig. 6.51)

$\vec{v}_{P\text{rel}}$ = velocity of P as seen by an observer moving with the rotating reference frame

$\vec{\Omega}$ = angular velocity of the rotating reference frame

$\vec{r}_{P/A}$ = position of point P relative to A

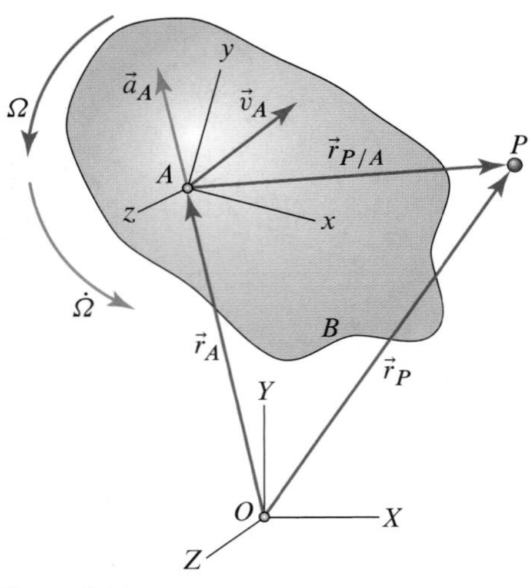

Figure 6.51
The essential ingredients for finding velocities and accelerations using a rotating reference frame.

We also found that we can relate the acceleration of point P to that of A by using

Eq. (6.53), p. 496

$$\vec{a}_P = \vec{a}_A + \vec{a}_{P\text{rel}} + 2\vec{\Omega} \times \vec{v}_{P\text{rel}} + \dot{\vec{\Omega}} \times \vec{r}_{P/A} + \vec{\Omega} \times (\vec{\Omega} \times \vec{r}_{P/A}),$$

where, in addition to the terms defined for $\vec{v}_P$, we have

$\vec{a}_A$ = acceleration of the origin of the rotating or body-fixed reference frame (point A in Fig. 6.51)

$\vec{a}_{P\text{rel}}$ = acceleration of P as seen by an observer moving with the rotating reference frame

$2\vec{\Omega} \times \vec{v}_{P\text{rel}}$ = Coriolis acceleration of P

$\dot{\vec{\Omega}}$ = angular acceleration of the rotating reference frame

EXAMPLE 6.13 A Particle in a Rotating Slot: Equation of Motion and Forces

Figure 1 shows a small object P sliding in a slot relative to a rotating disk D. The disk is rotating in the horizontal plane with angular velocity ω_0 and angular acceleration α_0. The particle P is constrained to move in the slot, which is a distance d from the center of the disk. In addition, a linear elastic spring of constant k is attached to the particle, such that the spring is undeformed when the particle is at $s = 0$. Determine the equation(s) of motion of the particle and the normal force between the particle and the slot.

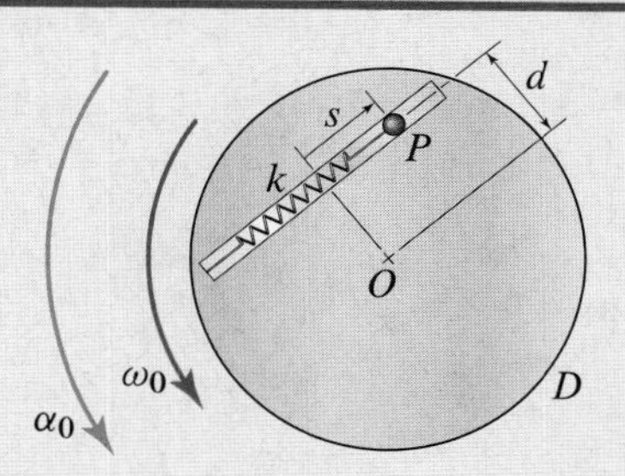

Figure 1

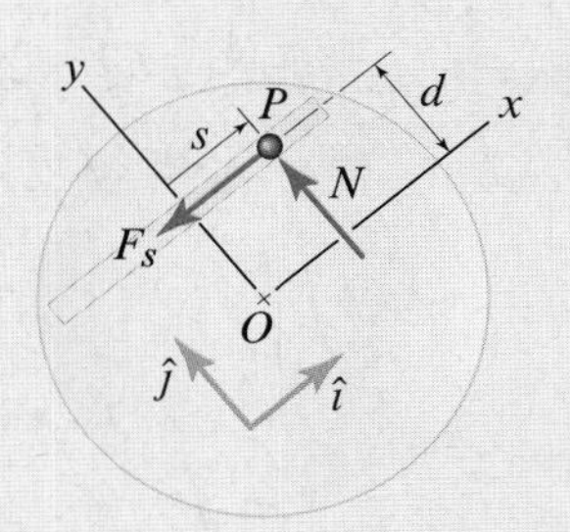

Figure 2
The FBD of the particle, as well as the definition of the rotating reference frame. The unit vectors $\hat{\imath}$ and $\hat{\jmath}$ rotate with the disk.

SOLUTION

Road Map & Modeling Newton's second law will provide the equation of motion for the particle and the normal force between the particle and the slot. Since P is sliding *relative to* the disk, we will attach a rotating reference frame to the disk to find $\vec{a}_P$. The FBD of the particle at an arbitrary position s is shown in Fig. 2, where F_s is the spring force acting on P and N is the normal force acting on P due to the slot. We neglect friction between the particle and the slot. In addition, we attach a rotating xyz frame to the disk with its origin at point O, which is the center of the disk. Since this system has one degree of freedom, we expect to obtain one equation of motion.

Governing Equations

Balance Principles Referring to Fig. 2, Newton's second law applied to the particle gives

$$\sum F_x\colon\ -F_s = ma_{Px}, \quad \text{and} \quad \sum F_y\colon\ N = ma_{Py}, \tag{1}$$

where a_{Px} and a_{Py} are the x and y components, respectively, of the acceleration of P.

Force Laws The spring is undeformed at $s = 0$, so the spring force law is

$$F_s = ks. \tag{2}$$

Kinematic Equations To determine a_{Px} and a_{Py}, we apply Eq. (6.57), which gives

$$\vec{a}_P = a_{Px}\,\hat{\imath} + a_{Py}\,\hat{\jmath} = \vec{a}_O + \vec{a}_{P\text{rel}} + 2\vec{\Omega} \times \vec{v}_{P\text{rel}} + \dot{\vec{\Omega}} \times \vec{r}_{P/O} - \Omega^2 \vec{r}_{P/O}, \tag{3}$$

where $\vec{a}_O = \vec{0}$ since point O is not moving, $\vec{a}_{P\text{rel}} = \ddot{s}\,\hat{\imath}$, $\vec{v}_{P\text{rel}} = \dot{s}\,\hat{\imath}$, $\vec{\Omega} = \omega_0\,\hat{k}$, $\dot{\vec{\Omega}} = \alpha_0\,\hat{k}$, and $\vec{r}_{P/O} = s\,\hat{\imath} + d\,\hat{\jmath}$. Substituting these terms into Eq. (3), we obtain

$$\begin{aligned} \vec{a}_P &= \ddot{s}\,\hat{\imath} + 2\omega_0\,\hat{k} \times \dot{s}\,\hat{\imath} + \alpha_0\,\hat{k} \times (s\,\hat{\imath} + d\,\hat{\jmath}) - \omega_0^2(s\,\hat{\imath} + d\,\hat{\jmath}) \\ &= \left(\ddot{s} - d\alpha_0 - s\omega_0^2\right)\hat{\imath} + \left(2\dot{s}\omega_0 + s\alpha_0 - d\omega_0^2\right)\hat{\jmath}, \qquad (4) \\ &\Rightarrow a_{Px} = \ddot{s} - d\alpha_0 - s\omega_0^2 \quad \text{and} \quad a_{Py} = 2\dot{s}\omega_0 + s\alpha_0 - d\omega_0^2. \qquad (5) \end{aligned}$$

Computation Substituting Eqs. (2) and (5) into Eqs. (1), we obtain

$$-ks = m\left(\ddot{s} - d\alpha_0 - s\omega_0^2\right) \quad \text{and} \quad N = m\left(2\dot{s}\omega_0 + s\alpha_0 - d\omega_0^2\right). \tag{6}$$

Rearranging the first of Eqs. (6), we obtain the equation of motion of the particle as

$$\boxed{\ddot{s} + \left(\frac{k}{m} - \omega_0^2\right)s = d\alpha_0.} \tag{7}$$

The second of Eqs. (6) tells us that once we know $s(t)$ from the solution of Eq. (7), then we can find the normal force between the particle and the slot from

$$\boxed{N = m\left(2\dot{s}\omega_0 + s\alpha_0 - d\omega_0^2\right).} \tag{8}$$

Discussion & Verification Note that the dimension of each term in Eq. (7) is that of acceleration, and the dimension of each term in Eq. (8) is that of force, so both results are dimensionally consistent.

Interesting Fact

Analysis of Eq. (7) if $\alpha_0 = 0$. If you have taken a differential equations course, you may recognize that if $\alpha_0 = 0$ in Eq. (7), it becomes a second-order, constant coefficient, homogeneous linear ordinary differential equation. The solutions are harmonic functions (i.e., sines and cosines) if $k/m > \omega_0^2$, and they are growing exponentials if $k/m < \omega_0^2$. This property would allow us to use a system like this to monitor when the disk starts rotating faster than some critical value, since the particle would then hit the end of the slot.

EXAMPLE 6.14 *Analysis of a Reciprocating Rectilinear Motion Mechanism*

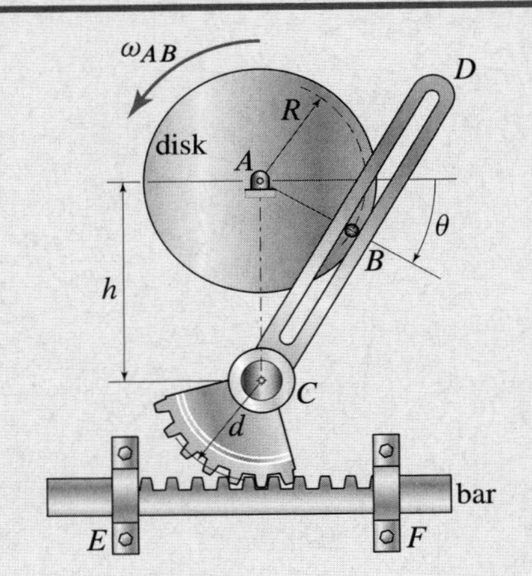

Figure 1

The *reciprocating rectilinear motion* mechanism shown in Fig. 1 consists of a disk pinned at its center at A that rotates with a constant angular velocity ω_{AB}, a slotted arm CD that is pinned at C, and a bar that can oscillate within the guides at E and F. As the disk rotates, the peg at B moves within the slotted arm, causing it to rock back and forth. As the arm rocks, it provides a slow advance and a quick return to the reciprocating bar due to the change in distance between C and B. For the position shown ($\theta = 30°$), determine the

(a) Angular velocity and angular acceleration of the slotted arm CD

(b) Velocity and acceleration of the bar

Evaluate your results for $\omega_{AB} = 60$ rpm, $R = 0.1$ m, $h = 0.2$ m, and $d = 0.12$ m.

SOLUTION

Road Map This is a mechanism with a sliding contact, so we cannot use Eqs. (6.3) and (6.5) to relate the motion of the disk to that of the slotted arm CD. Therefore, we will use a rotating reference frame to relate the motion of the pin at B to the pivot point at C, and we will attach the frame to arm CD, as shown in the schematic of the mechanism in Fig. 2.

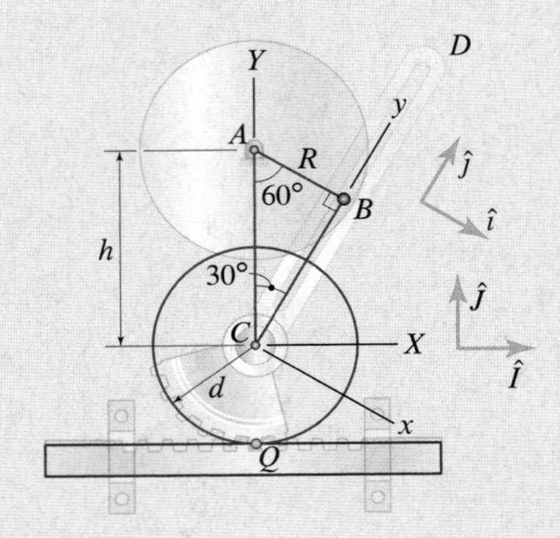

Figure 2
The primary and rotating reference frame for the reciprocating rectilinear motion mechanism. Note that the right angle between AB and BC is a consequence of the given lengths and the angle between AC and BC at this instant. The angle between AB and BC is not a right angle in general.

Computation Since we are relating the motion of B to C, for velocities we can write

$$\vec{v}_B = \vec{v}_C + \vec{v}_{B\text{rel}} + \vec{\Omega}_{CD} \times \vec{r}_{B/C}, \tag{1}$$

where $\vec{v}_{B\text{rel}}$ is the velocity of B as seen by an observer in the rotating frame and $\vec{\Omega}_{CD} = \vec{\omega}_{CD}$ is *both* the angular velocity of the rotating frame and the angular velocity of the arm CD. Now, since B is rotating in a circle about A, $\vec{v}_B$ is found using Eq. (6.8) to be

$$\vec{v}_B = \vec{\omega}_{AB} \times \vec{r}_{B/A} = \omega_{AB}\,\hat{k} \times R\,\hat{\imath} = R\omega_{AB}\,\hat{\jmath}. \tag{2}$$

In addition, we can see that

$$\vec{v}_C = \vec{0}, \quad \vec{v}_{B\text{rel}} = \dot{r}_{B/C}\,\hat{\jmath}, \quad \vec{\Omega}_{CD} = \Omega_{CD}\,\hat{k}, \quad \vec{r}_{B/C} = r_{B/C}\,\hat{\jmath}, \tag{3}$$

where $r_{B/C} = \sqrt{h^2 - R^2} = 0.1732$ m at this instant. Substituting Eqs. (2) and (3) into Eq. (1), we obtain

$$R\omega_{AB}\,\hat{\jmath} = \dot{r}_{B/C}\,\hat{\jmath} + \Omega_{CD}\,\hat{k} \times r_{B/C}\,\hat{\jmath}, \tag{4}$$

which is equivalent to two scalar equations for Ω_{CD} and $\dot{r}_{B/C}$. Solving, we find that $\dot{r}_{B/C} = R\omega_{AB} = 0.6283$ m/s and

$$\Omega_{CD} = 0\,\text{rad/s} \quad \Rightarrow \quad \boxed{\vec{\Omega}_{CD} = \vec{\omega}_{CD} = \vec{0}\,\text{rad/s.}} \tag{5}$$

For the acceleration analysis, we apply Eq. (6.53) on p. 496,

$$\vec{a}_B = \vec{a}_C + \vec{a}_{B\text{rel}} + 2\vec{\Omega}_{CD} \times \vec{v}_{B\text{rel}} + \dot{\vec{\Omega}}_{CD} \times \vec{r}_{B/C} - \Omega_{CD}^2 \vec{r}_{B/C} \tag{6}$$

$$= \vec{a}_C + \vec{a}_{B\text{rel}} + \dot{\vec{\Omega}}_{CD} \times \vec{r}_{B/C}, \tag{7}$$

where to go from Eq. (6) to Eq. (7) we have used Eq. (5). Since B is moving in a circle centered at A, we can use Eq. (6.13) to find $\vec{a}_B$ as

$$\vec{a}_B = -\omega_{AB}^2 \vec{r}_{B/A} = -\omega_{AB}^2 R\,\hat{\imath}. \tag{8}$$

In addition, $\vec{a}_{B\text{rel}}$ and $\dot{\vec{\Omega}}_{CD}$ in Eq. (7) are given by

$$\vec{a}_{B\text{rel}} = \ddot{r}_{B/C}\,\hat{\jmath} \quad \text{and} \quad \dot{\vec{\Omega}}_{CD} = \dot{\Omega}_{CD}\,\hat{k}. \tag{9}$$

Finally, noting that $\vec{a}_C = \vec{0}$ and substituting the last of Eqs. (3) and Eqs. (8) and (9) into Eq. (7), we obtain

$$-\omega_{AB}^2 R\,\hat{\imath} = \ddot{r}_{B/C}\,\hat{\jmath} + \dot{\Omega}_{CD}\,\hat{k} \times r_{B/C}\,\hat{\jmath}, \tag{10}$$

which is equivalent to two scalar equations for $\ddot{r}_{B/C}$ and $\dot{\Omega}_{CD}$. Solving, we find that $\ddot{r}_{B/C} = 0$ and

$$\dot{\Omega}_{CD} = \frac{R}{r_{B/C}}\omega_{AB}^2 = 22.79\,\text{rad/s}^2 \quad \Rightarrow \quad \boxed{\dot{\vec{\Omega}}_{CD} = \vec{\alpha}_{CD} = (22.79\,\text{rad/s}^2)\,\hat{k}.} \tag{11}$$

Referring to Fig. 2, now that we have the angular velocity and angular acceleration of the slotted arm CD, we can find the velocity of the bar using

$$\boxed{\vec{v}_{\text{bar}} = \vec{v}_Q = \vec{\omega}_{CD} \times \vec{r}_{Q/C} = \vec{0},} \tag{12}$$

since $\vec{\omega}_{CD} = \vec{0}$ at this instant. In addition, the acceleration of the bar is given by

$$\vec{a}_{\text{bar}} = (\vec{a}_Q \cdot \hat{I})\,\hat{I} = \vec{\alpha}_{CD} \times \vec{r}_{Q/C} = \alpha_{CD}\,\hat{K} \times (-d\hat{J}) = d\alpha_{CD}\,\hat{I}, \tag{13}$$

that is, $\vec{a}_{\text{bar}}$ is just the tangential component of acceleration of Q. Therefore, the acceleration of the bar at this instant is

$$\boxed{\vec{a}_{\text{bar}} = d\alpha_{CD}\,\hat{I} = (2.735\,\text{m/s}^2)\,\hat{I}.} \tag{14}$$

Discussion & Verification The dimensions and thus the units of the final results are as they should be.

Looking more deeply, we note that the angular velocity of the slotted arm CD is zero at this instant. This makes sense since the line AB on the rotating disk is perpendicular to the slot in the arm at this instant. Therefore, *at this instant*, the velocity of B is parallel to the slot, and so it is not inducing any rotation of the arm containing the slot. Consequently, the velocity of the bar must be zero at this instant since it is the angular velocity of the slotted arm that imparts a velocity to the bar.

On the other hand, we note that the slotted arm does have an angular acceleration. This also makes sense since an instant before the arm is in this position, and an instant after, the slotted arm must have an angular velocity, and therefore there must be an angular acceleration causing this change in angular velocity.

In Probs. 6.154 and 6.155, we will see why this mechanism is sometimes called a *quick-return mechanism*.

EXAMPLE 6.15 Actual Versus Perceived Motion: Finding $\vec{v}_{B\mathrm{rel}}$ and $\vec{a}_{B\mathrm{rel}}$

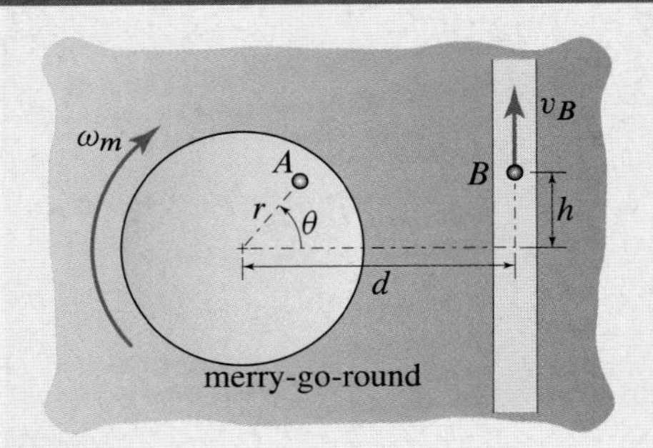

Figure 1

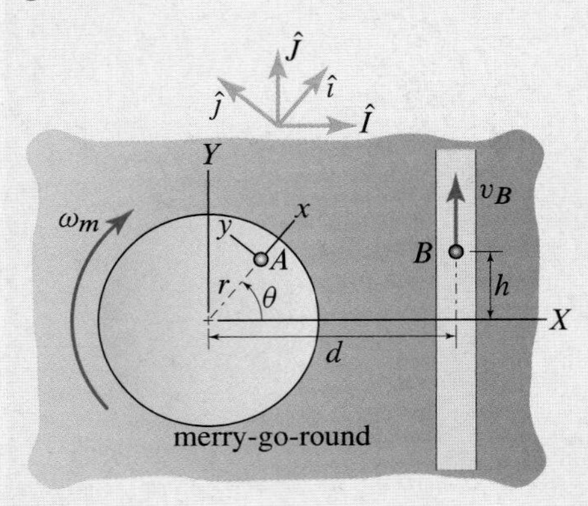

Figure 2
Definition of the primary and rotating frames of reference.

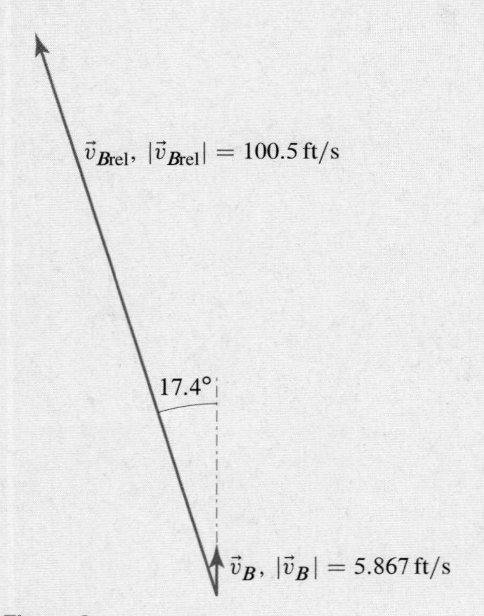

Figure 3
A comparison of the velocity of B in the stationary frame, i.e., $\vec{v}_B$, with the velocity B as seen by A, i.e., $\vec{v}_{B\mathrm{rel}}$.

A person A is standing on a merry-go-round (i.e., not moving relative to it), which is rotating at the constant rate ω_m, and a person B is walking in a straight line at constant speed v_B along a sidewalk in a stationary frame (see Fig. 1). Determine the velocity and acceleration of B *as seen by* A.* Evaluate the results for $v_B = 4$ mph, $\omega_m = 3$ rad/s, $\theta = 45°$, $d = 30$ ft, and $h = 10$ ft.

SOLUTION

Road Map Since A is not moving relative to the merry-go-round, the motion of B as seen by A is equivalent to the motion of B as seen by an observer rotating with the merry-go-round. As we mention on p. 497, the velocity of B as seen by A is *not* equal to $\vec{v}_{B/A}$ since that quantity gives the velocity of B as seen by A only if A is not rotating. Therefore, recall that the term $\vec{v}_{B\mathrm{rel}}$ in Eq. (6.45) is *the velocity of B as seen by an observer moving with the rotating frame*—this is exactly what we want (along with $\vec{a}_{B\mathrm{rel}}$ for the acceleration of B as seen by A).

Computation Referring to Fig. 2, the primary XYZ frame is as shown, and the rotating xyz frame is attached to the merry-go-round with its origin at A. We want to compute $\vec{v}_{B\mathrm{rel}}$, which is given by Eq. (6.45) to be

$$\vec{v}_{B\mathrm{rel}} = \vec{v}_B - \vec{v}_A - \vec{\omega}_m \times \vec{r}_{B/A}, \tag{1}$$

where $\vec{\omega}_m$ is also the angular velocity of the rotating xyz frame since the frame is attached to the merry-go-round. Evaluating the three terms above, we find

$$\vec{v}_B = v_B\,\hat{J} = v_B(\sin\theta\,\hat{\imath} + \cos\theta\,\hat{\jmath}), \tag{2}$$

$$\vec{v}_A = -r\omega_m\,\hat{\jmath}, \tag{3}$$

$$\begin{aligned}\vec{\omega}_m \times \vec{r}_{B/A} &= -\omega_m\,\hat{k} \times \left(-r\,\hat{\imath} + d\,\hat{I} + h\,\hat{J}\right)\\ &= -\omega_m\,\hat{k} \times [-r\,\hat{\imath} + d(\cos\theta\,\hat{\imath} - \sin\theta\,\hat{\jmath}) + h(\sin\theta\,\hat{\imath} + \cos\theta\,\hat{\jmath})]\\ &= \omega_m(h\cos\theta - d\sin\theta)\,\hat{\imath} - \omega_m(d\cos\theta + h\sin\theta - r)\,\hat{\jmath}.\end{aligned} \tag{4}$$

Substituting Eqs. (2)–(4) into Eq. (1), we get

$$\begin{aligned}\vec{v}_{B\mathrm{rel}} &= (v_B\sin\theta - h\omega_m\cos\theta + d\omega_m\sin\theta)\,\hat{\imath}\\ &\quad + (v_B\cos\theta + h\omega_m\sin\theta + d\omega_m\cos\theta)\,\hat{\jmath} \qquad (5)\\ &= (46.57\,\hat{\imath} + 89.00\,\hat{\jmath})\ \text{ft/s}, \qquad (6)\end{aligned}$$

where the numerical result was obtained using $v_B = 4$ mph $= 5.867$ ft/s, $h = 10$ ft, $d = 30$ ft, $\omega_m = 3$ rad/s, and $\theta = 45°$. Figure 3 shows the actual velocity of B, i.e., $\vec{v}_B$, as well as the velocity B as seen by A, i.e., $\vec{v}_{B\mathrm{rel}}$.

Now computing $\vec{a}_{B\mathrm{rel}}$ using Eq. (6.57), we have

$$\vec{a}_{B\mathrm{rel}} = \vec{a}_B - \vec{a}_A - 2\vec{\omega}_m \times \vec{v}_{B\mathrm{rel}} - \vec{\alpha}_m \times \vec{r}_{B/A} + \omega_m^2\vec{r}_{B/A}, \tag{7}$$

where, in the given situation, $\vec{\alpha}_m$ and $\vec{a}_B$ are both zero. Computing the other terms in Eq. (7), we have

$$\vec{a}_A = -r\omega_m^2\,\hat{\imath}, \tag{8}$$

$$\begin{aligned}2\vec{\omega}_m \times \vec{v}_{B\mathrm{rel}} &= 2\omega_m\,[h\omega_m\sin\theta + (v_B + d\omega_m)\cos\theta]\,\hat{\imath}\\ &\quad + 2\omega_m\,[h\omega_m\cos\theta - (v_B + d\omega_m)\sin\theta]\,\hat{\jmath},\end{aligned} \tag{9}$$

* Person A cannot turn his or her head to follow B as the merry-go-round rotates, or else we will have introduced yet another rotation.

$$\omega_m^2 \vec{r}_{B/A} = \omega_m^2 \left[(h \sin\theta + d\cos\theta - r)\,\hat{\imath} + (h\cos\theta - d\sin\theta)\,\hat{\jmath} \right]. \tag{10}$$

Substituting Eqs. (8)–(10), as well as $\vec{\alpha}_m = \vec{0}$ and $\vec{a}_B = \vec{0}$ into Eq. (7), we obtain

$$\begin{aligned} \vec{a}_{B\text{rel}} &= -\omega_m \left[(2v_B + d\omega_m)\cos\theta + h\omega_m \sin\theta \right] \hat{\imath} \\ &\qquad + \omega_m \left[(2v_B + d\omega_m)\sin\theta - h\omega_m \cos\theta \right] \hat{\jmath} \qquad (11) \\ &= (-279.4\,\hat{\imath} + 152.2\,\hat{\jmath})\ \text{ft/s}^2, \qquad (12) \end{aligned}$$

where the numerical result was obtained using $v_B = 4$ mph $= 5.867$ ft/s, $h = 10$ ft, $d = 30$ ft, $\omega_m = 3$ rad/s, and $\theta = 45°$. Figure 4 shows the acceleration of B as perceived by A, i.e., $\vec{a}_{B\text{rel}}$, as well as the actual acceleration of B, which is zero, so only a dot is shown.

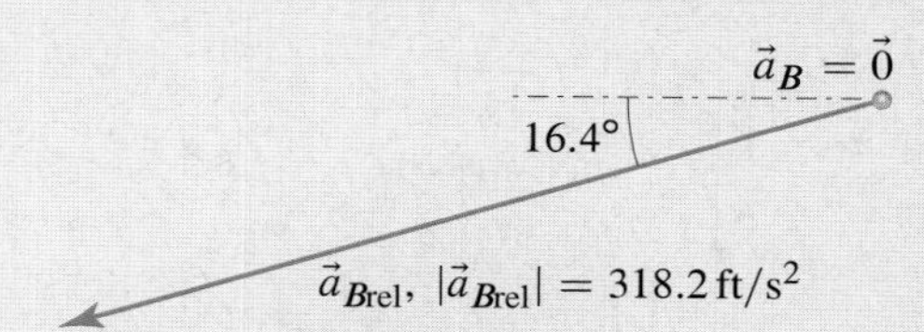

Figure 4
The acceleration of B as perceived by A, i.e., $\vec{a}_{B\text{rel}}$, as well as the actual acceleration of B, which is zero, so only a dot is shown.

Discussion & Verification The dimensions of both $\vec{v}_{B\text{rel}}$ and $\vec{a}_{B\text{rel}}$ are as they should be, that is, velocity and acceleration, respectively. More importantly, this example illustrates an important idea — the motion of an object (i.e., its velocity and acceleration) that one perceives when on a rotating frame is *very* different from the actual motion of the object. Figure 3 shows the vast difference between the actual velocity vector of B and the velocity of B as seen by A who is rotating with the merry-go-round. Person B is walking at 4 mph, but A sees B moving at 68.49 mph. As far as acceleration is concerned, B is not accelerating at all, yet A sees B moving at $318.2\ \text{ft/s}^2 = 9.882g$!

A Closer Look Notice that the relative velocity $\vec{v}_{B\text{rel}}$ and acceleration $\vec{a}_{B\text{rel}}$ of B as seen by A do not depend on r, which is the radius of the circular path of A. This happens in this example because the person A, who is observing the relative quantities $\vec{v}_{B\text{rel}}$ and $\vec{a}_{B\text{rel}}$, is not moving relative to the moving reference frame. That is, A and the moving reference frame are a single rigid body. Because of this, the motion of A always cancels with part of the motion of B as seen by A. For example, referring to the velocities, we see that $\vec{v}_A = -r\omega_m\,\hat{\jmath}$ and that part of $\vec{\omega}_m \times \vec{r}_{B/A}$ is given by $r\omega_m\,\hat{\jmath}$. Therefore, when these terms are combined in Eq. (1), they cancel. A similar cancellation occurs for the relative acceleration.

Problems

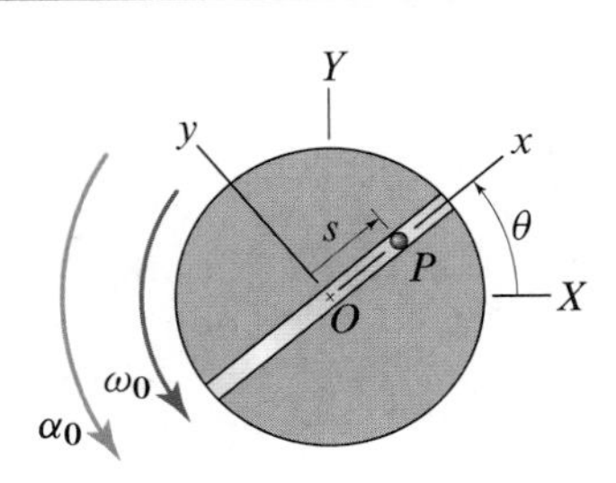

Figure P6.134

Problem 6.134

Using polar coordinates, obtain the acceleration of point P of the the mini-example on p. 497; that is, obtain Eq. (6.56).

Problems 6.135 through 6.137

The disk D is rotating with angular speed ω_D about its central point O in the direction shown. The particle P moves in the slot with known motion $s(t)$, which is measured relative to the disk. Express your answers using the xyz frame that is attached to the disk D with its origin at O as shown.

Problem 6.135 Determine the velocity of P as a function of d, s and its time derivatives, and ω_D.

Problem 6.136 If ω_D is constant, determine the Coriolis acceleration of P and its total acceleration as functions of d, s and its time derivatives, and ω_D.

Problem 6.137 If ω_D is increasing at the rate $\dot{\omega}_D$, determine the Coriolis acceleration of P and its total acceleration as functions of d, s and its time derivatives, ω_D, and $\dot{\omega}_D$.

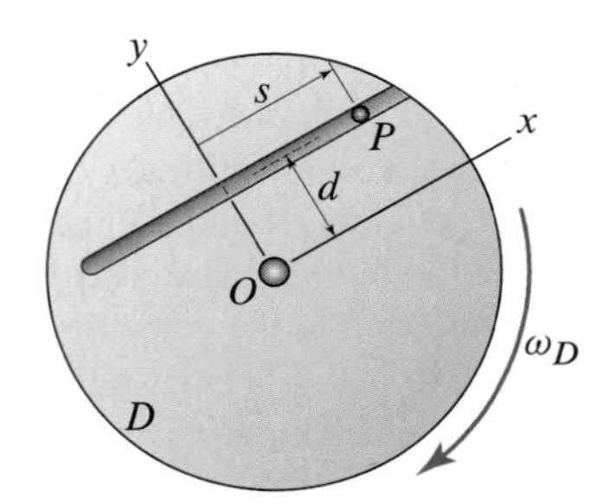

Figure P6.135–P6.137

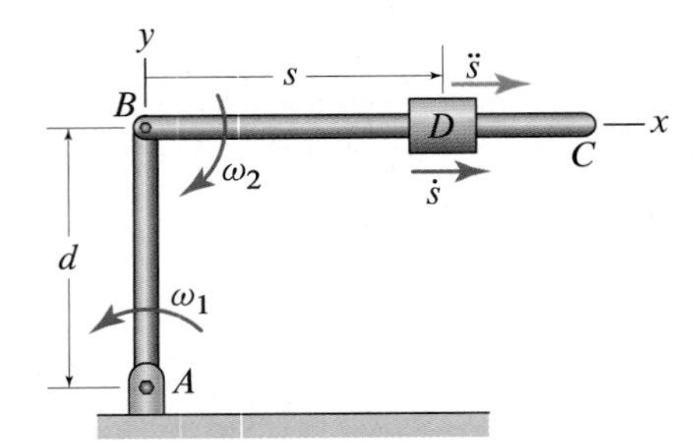

Figure P6.138 and P6.139

Problems 6.138 and 6.139

Bar AB rotates about point A with constant angular speed ω_1. Bar BC is pinned to bar AB at B and rotates with constant angular speed ω_2 *relative to* bar AB about the pin at B. The position of collar D relative to bar BC is given by the coordinate s, which is a known function of time. Express your answers in the xyz frame that is attached to bar BC with its origin at B. *Hint:* As we will see in Eq. (10.14), $\vec{\omega}_{BC} = \vec{\omega}_{AB} + \vec{\omega}_{BC/AB} = \vec{\omega}_1 + \vec{\omega}_2$.

Problem 6.138 Determine the velocity of the collar D in terms of d, ω_1, ω_2, and the position coordinate s and its time derivatives.

Problem 6.139 Determine the acceleration of the collar D in terms of d, ω_1, ω_2, and the position coordinate s and its time derivatives.

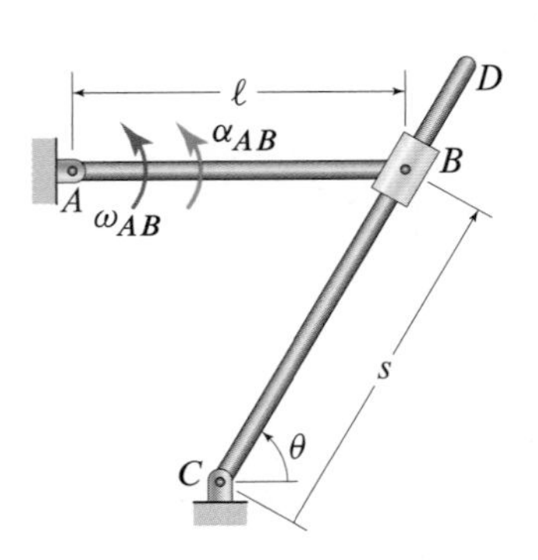

Figure P6.140 and P6.141

Problems 6.140 and 6.141

Bar AB is pinned to the fixed support at A, and the collar B is pinned to the bar at its opposite end. The bar CD can slide freely through the collar at B. At the instant shown, bar AB is horizontal, $\ell = 1.2\,\text{m}$, $s = 1.07\,\text{m}$, $\theta = 60°$, and $\omega_{AB} = 40\,\text{rad/s}$.

Problem 6.140 If α_{AB} is zero at the instant shown, determine the angular velocity and angular acceleration of the bar CD.

Problem 6.141 If $\alpha_{AB} = 15\,\text{rad/s}^2$ at the instant shown, determine the angular velocity and angular acceleration of the bar CD.

Problems 6.142 and 6.143

Bar AB is pinned to the fixed support at A, and the pin P is fixed to the disk at radius R_i. The disk with outer radius R_o rolls without slipping over the horizontal flat surface with angular speed ω_d and angular acceleration α_d in the directions shown. At the instant shown, $s = 35$ in., $R_i = 8.3$ in., $R_o = 11$ in., and $\omega_d = 15\,\text{rad/s}$.

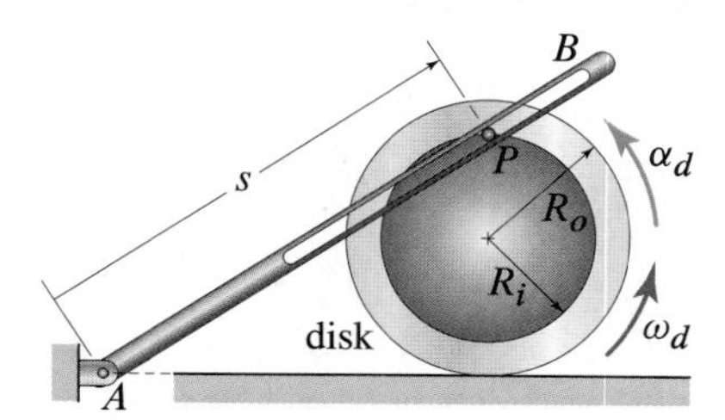

Figure P6.142 and P6.143

Problem 6.142 If α_d is zero at the instant shown, determine the angular velocity and angular acceleration of the bar AB.

Problem 6.143 If $\alpha_d = 30\,\text{rad/s}^2$ at the instant shown, determine the angular velocity and angular acceleration of the bar AB.

Problem 6.144

A vertical shaft has a base B that is stationary relative to an inertial reference frame with vertical axis Z. Arm OA is attached to the vertical shaft and rotates about the Z axis with an angular velocity $\omega_{OA} = 5\,\text{rad/s}$ and an angular acceleration $\alpha_{OA} = 1.5\,\text{rad/s}^2$. The z axis is coincident with the Z axis, but is part of a reference frame that rotates with the arm OA. The x axis of the rotating reference frame coincides with the axis of the arm OA. At the instant shown, the y axis is perpendicular to the page and directed into the page. At this instant, the collar C is sliding along OA with a constant speed $v_C = 5\,\text{ft/s}$ and is at a distance $d = 1.2$ ft from the z axis. Compute the inertial acceleration of the collar, and express it relative to the rotating coordinate system.

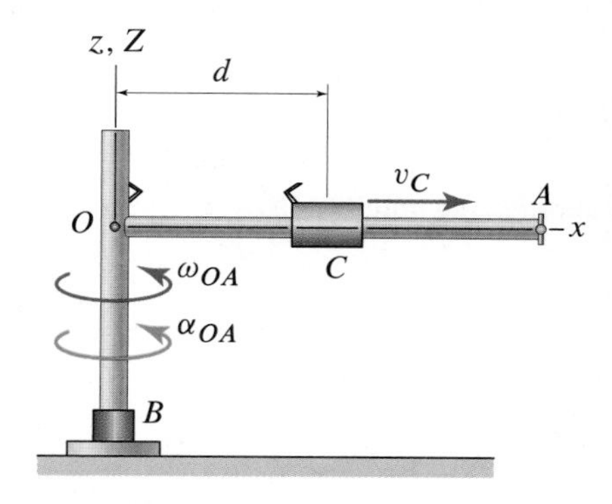

Figure P6.144

Problem 6.145

Let frames A and B be the frames with origins at points A and B, respectively. Point B does not move relative to point A. The velocity and acceleration of point P relative to frame B are

$$\vec{v}_{P\text{rel}} = (-6.14\,\hat{\imath}_B + 23.7\,\hat{\jmath}_B)\,\text{ft/s} \quad \text{and} \quad \vec{a}_{P\text{rel}} = (3.97\,\hat{\imath}_B + 4.79\,\hat{\jmath}_B)\,\text{ft/s}^2.$$

Knowing that, at the instant shown, frame B rotates relative to frame A at a constant angular velocity $\omega_B = 1.2\,\text{rad/s}$, that the position of point P relative to frame B is $\vec{r}_{P/B} = (8\,\hat{\imath}_B + 4.5\,\hat{\jmath}_B)$ ft, and that frame A is fixed, determine the velocity and acceleration of P at the instant shown and express the results using the frame A component system.

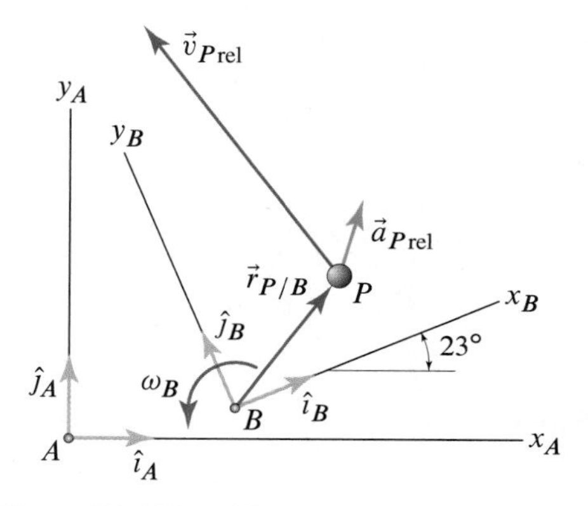

Figure P6.145 and P6.146

Problem 6.146

Repeat Prob. 6.145, but express the results using the component system of frame B.

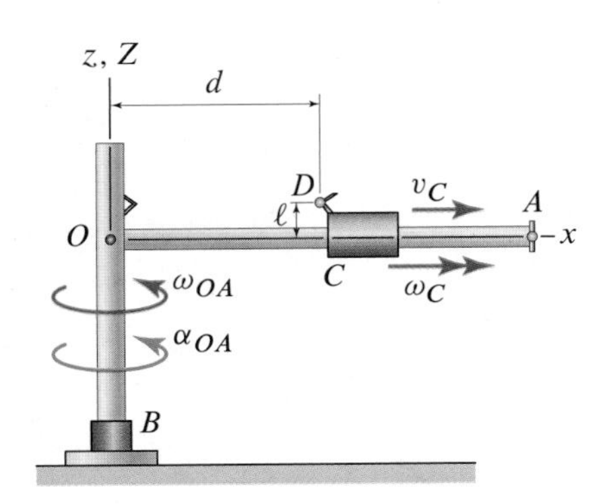

Figure P6.147

Problem 6.147

A vertical shaft has a base B that is stationary relative to an inertial reference frame with vertical axis Z. Arm OA is attached to the vertical shaft and rotates about the Z axis with an angular velocity $\omega_{OA} = 5\,\text{rad/s}$ and an angular acceleration $\alpha_{OA} = 1.5\,\text{rad/s}^2$. The z axis is coincident with the Z axis but is part of a reference frame that rotates with the arm OA. The x axis of the rotating reference frame coincides with the axis of the arm OA. At the instant shown, the y axis is perpendicular to the page and directed into the page. At this instant, the collar C is sliding along OA with a constant speed $v_C = 3.32\,\text{m/s}$ and is rotating with a constant angular velocity $\omega_C = 2.3\,\text{rad/s}$ relative to the arm OA. At the instant shown, point D happens to be in the xz plane and is at a distance $\ell = 0.05\,\text{m}$ from the x axis and at a distance $d = 0.75\,\text{m}$ from the z axis. Compute the inertial acceleration of point D, and express it relative to the rotating reference frame.

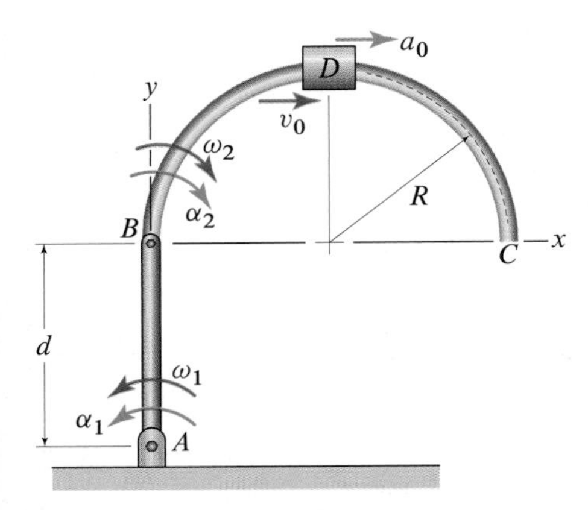

Figure P6.148 and P6.149

Problems 6.148 and 6.149

Bar AB rotates about point A with angular speed ω_1 and angular acceleration α_1, both in the directions shown. The curved bar BC is pinned to bar AB at B and rotates with angular speed ω_2 and angular acceleration α_2 about the pin at B in the directions shown, *relative to* bar AB. At this instant, collar D is at the top of the curved bar BC. The speed and acceleration of D relative to bar BC are given by v_0 and a_0, respectively. Express your answers in the xyz frame that is attached to bar BC with its origin at B. *Hint:* As we will see in Eq. (10.14), $\vec{\omega}_{BC} = \vec{\omega}_{AB} + \vec{\omega}_{BC/AB} = \vec{\omega}_1 + \vec{\omega}_2$.

Problem 6.148 Determine the velocity of the collar D in terms of d, ω_1, ω_2, R, and v_0.

Problem 6.149 Determine the acceleration of the collar D in terms of d, R, ω_1, ω_2, α_1, α_2, v_0, and a_0.

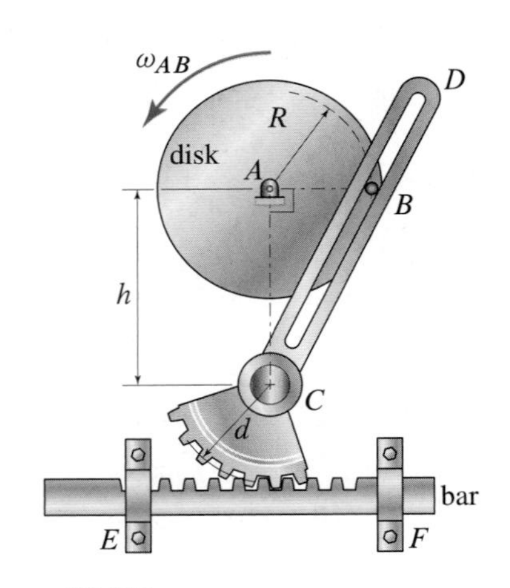

Figure P6.150

Problems 6.150 through 6.154

The *reciprocating rectilinear motion* mechanism consists of a disk pinned at its center at A that rotates with a constant angular velocity ω_{AB}, a slotted arm CD that is pinned at C, and a bar that can oscillate within the guides at E and F. As the disk rotates, the peg at B moves within the slotted arm, causing it to rock back and forth. As the arm rocks, it provides a slow advance and a quick return to the reciprocating bar due to the change in distance between C and B.

Problem 6.150 For the position shown, determine

(a) The angular velocity of the slotted arm CD and the velocity of the bar

(b) The angular acceleration of the slotted arm CD and the acceleration of the bar

Evaluate your results for $\omega_{AB} = 120\,\text{rpm}$, $R/h = 0.5$, and $d = 0.12\,\text{m}$.

Problem 6.151 For the position shown, determine

(a) The angular velocity of the slotted arm CD and the velocity of the bar

(b) The angular acceleration of the slotted arm CD and the acceleration of the bar

Evaluate your results for $\omega_{AB} = 90\,\text{rpm}$, $R/h = 0.5$, and $d = 0.12\,\text{m}$.

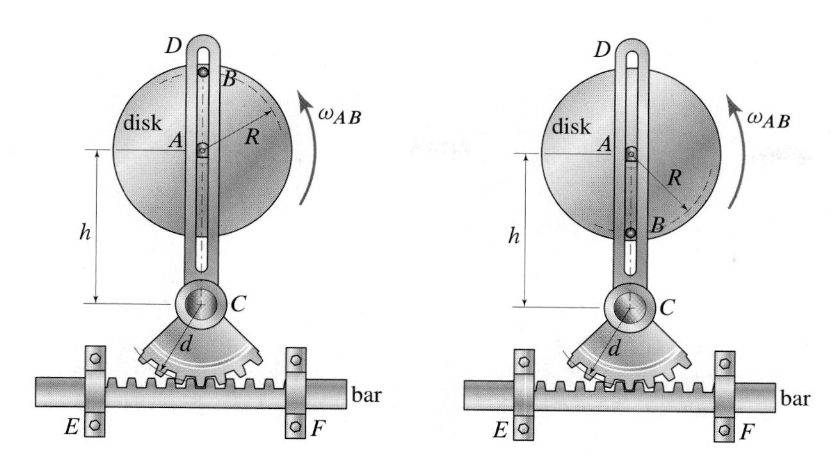

Figure P6.151

Figure P6.152

Problem 6.152 For the position shown, determine

(a) The angular velocity of the slotted arm CD and the velocity of the bar

(b) The angular acceleration of the slotted arm CD and the acceleration of the bar

Evaluate your results for $\omega_{AB} = 60\,\text{rpm}$, $R/h = 0.5$, and $d = 0.12\,\text{m}$.

Problem 6.153 For the arbitrary position shown, determine

(a) The angular velocity of the slotted arm CD and the velocity of the bar

(b) The angular acceleration of the slotted arm CD and the acceleration of the bar

as functions of θ, $\delta = R/h$, d, and ω_{AB}

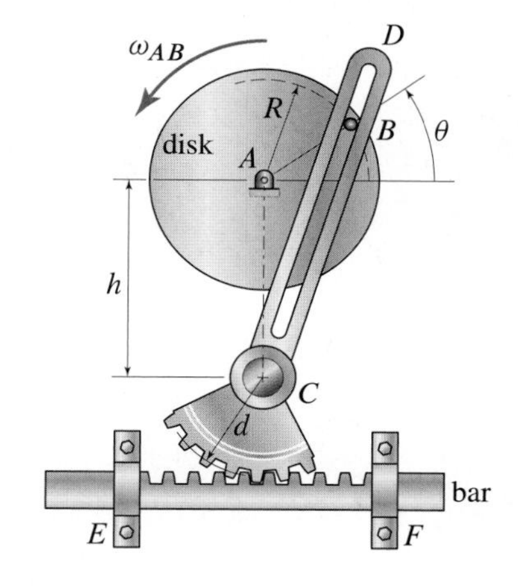

Figure P6.153 and P6.154

Problem 6.154 Determine the angular velocity and angular acceleration of the slotted arm CD as functions of θ, $\delta = R/h$, d, and ω_{AB}. After doing so, plot the velocity and acceleration of the bar as a function of the disk angle θ for one full cycle of the disk's motion and for $\omega_{AB} = 90\,\text{rpm}$, $d = 0.12\,\text{m}$, and

(a) $\delta = R/h = 0.1$

(b) $\delta = R/h = 0.3$

(c) $\delta = R/h = 0.6$

(d) $\delta = R/h = 0.9$

Explain why this mechanism is often referred to as a *quick return mechanism*.

Problem 6.155

The *reciprocating rectilinear motion* mechanism of Probs. 6.150–6.154 is often referred to as a *quick-return mechanism* since it can move much more quickly in one direction than the other. To see this, we will determine the velocity of the bar in each of the two positions shown (position 1 when B is farthest from C and position 2 when B is closest to C) under the assumption that the disk, whose center is at A, rotates with a constant angular velocity $\omega_{AB} = 120\,\text{rpm}$. Find the velocity of the bar in positions 1 and 2 for $d = 0.12\,\text{m}$ and for

(a) $R/h = 0.3$

(b) $R/h = 0.8$

Comment on which of the two R/h values would provide a better quick return and why.

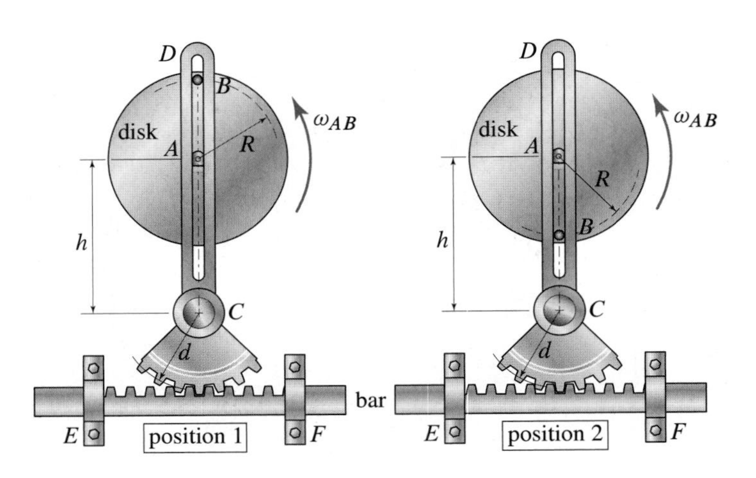

Figure P6.155

Problem 6.156

The wheel D rotates with a constant angular velocity $\omega_D = 14\,\text{rad/s}$ about the fixed point O, which is assumed to be stationary relative to an inertial frame of reference. The xyz frame rotates with the wheel. Collar C slides along the bar AB with a constant velocity $v_C = 4\,\text{ft/s}$ relative to the xyz frame. Letting $\ell = 0.25\,\text{ft}$, determine the inertial velocity and acceleration of C when $\theta = 25°$. Express the result with respect to the xyz frame.

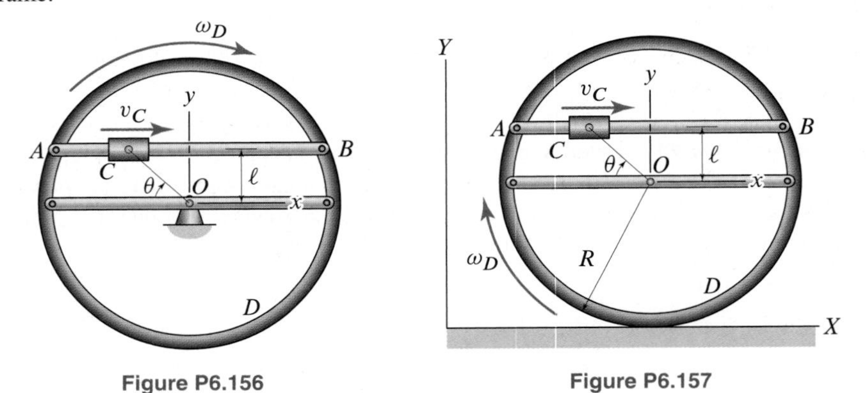

Figure P6.156 Figure P6.157

Problem 6.157

The wheel D rotates without slipping over a flat surface. The XYZ frame shown is inertial, whereas the xyz frame is attached to D at O and rotates with it at a constant angular velocity $\omega_D = 14\,\text{rad/s}$. Collar C slides along the bar AB with a constant velocity $v_C = 4\,\text{ft/s}$ relative to the xyz frame. Letting $\ell = 0.25\,\text{ft}$ and $R = 1\,\text{ft}$, determine the inertial velocity and acceleration of C when $\theta = 25°$ and the xyz frame is parallel to the XYZ frame as shown. Express your result in both the xyz and XYZ frames.

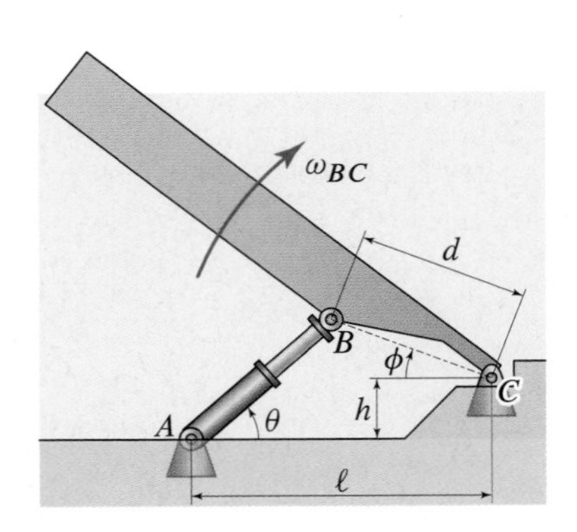

Figure P6.158

Problem 6.158

A floodgate is controlled by the motion of the hydraulic cylinder AB. If the gate BC is to be lifted with a constant angular velocity $\omega_{BC} = 0.5\,\text{rad/s}$, determine $\dot{d}_{AB}$ and $\ddot{d}_{AB}$, where d_{AB} is the distance between points A and B when $\phi = 0$. Let $\ell = 10\,\text{ft}$, $h = 2.5\,\text{ft}$, and $d = 5\,\text{ft}$.

Problem 6.159

The wheel D rotates without slipping over a flat surface. The XYZ frame shown is inertial, whereas the xyz frame is attached to D and rotates with it at a constant angular velocity ω_D. Collar C slides along the bar AB with a velocity v_C relative to the xyz frame. Suppose that ℓ, θ, and R are given and that we want to determine the inertial acceleration of C when the xyz frame is parallel to the XYZ frame as shown. Would the expression of the inertial acceleration of the collar in the two frames be different or the same?

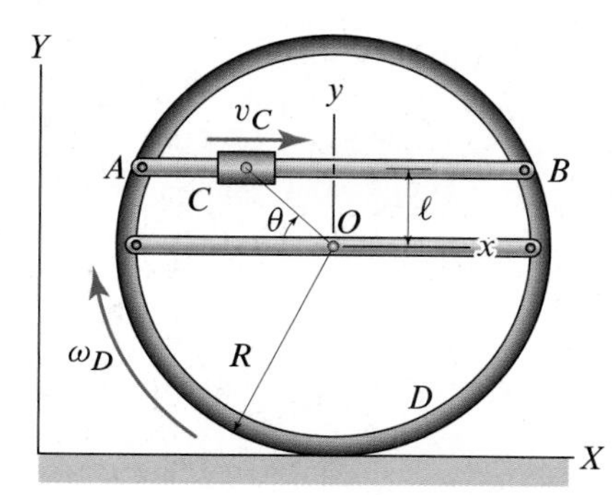

Figure P6.159

Problem 6.160

The Pioneer 3 spacecraft was a spin-stabilized spacecraft launched on December 6, 1958, by the U.S. Army Ballistic Missile agency in conjunction with NASA. It was designed with a despin mechanism consisting of two equal masses A and B that could be spooled out to the end of two wires of variable length $\ell(t)$ when triggered by a hydraulic timer.* As a prelude to Probs. 7.95, 7.96, and 8.114, we will find the velocity and acceleration of each of the two masses. To do this, assume that masses A and B are initially at positions A_0 and B_0, respectively. After the masses are released, they begin to unwind symmetrically, and the length of the cord attaching each mass to the spacecraft of radius R is $\ell(t)$. Given that the angular velocity of the spacecraft at each instant is $\omega_s(t)$, determine the velocity and acceleration of mass A in components expressed in the rotating reference frame whose origin is at Q, as well as R, $\ell(t)$, and $\omega_s(t)$. Note that the rotating frame is always aligned with the unwinding cord, and Q is the point on the cord that is about to unwind at time t. *Hint:* The point Q moves around the periphery of the spacecraft with angular speed $\dot{\theta}$ and angular acceleration $\ddot{\theta}$. It is not fixed to the spacecraft.

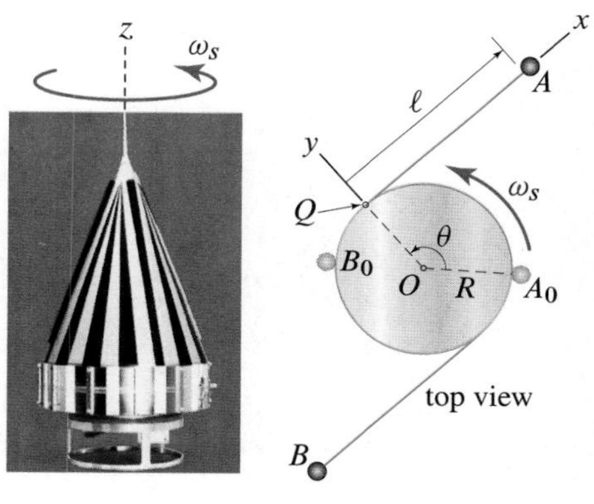

Figure P6.160

Problem 6.161

At the instant shown, the wheel D rolls without slipping over a flat surface with an angular velocity $\omega_D = 14\,\text{rad/s}$ and an angular acceleration $\alpha_D = 1.1\,\text{rad/s}^2$. The XYZ frame shown is inertial, whereas the xyz frame is attached to D. At the instant shown, the collar C is sliding along the bar AB with a velocity $v_C = 4\,\text{ft/s}$ and acceleration $a_C = 7\,\text{ft/s}^2$, both relative to the xyz frame. Letting $\ell = 0.25\,\text{ft}$ and $R = 1\,\text{ft}$, determine the inertial velocity and acceleration of C when $\theta = 25°$ and the xyz frame is parallel to the XYZ frame as shown. Express your result in both the xyz and the XYZ frames.

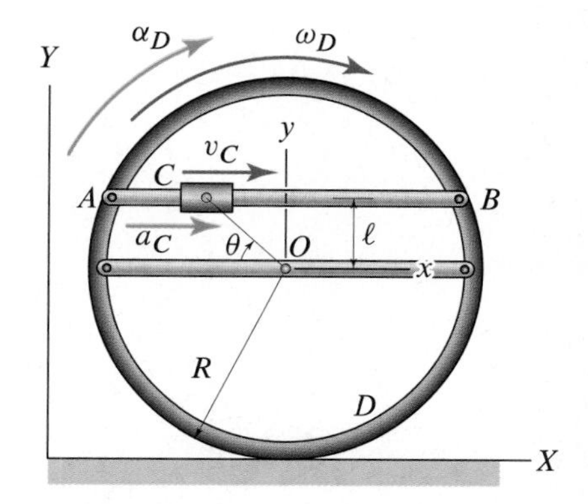

Figure P6.161

* The spacecraft was intended as a lunar probe, but it failed to go past the Moon and into a heliocentric (sun-centered) orbit as planned. It did reach an altitude of 107,400 km before falling back to Earth. It was a cone-shaped probe 58 cm high and 25 cm diameter at its base and was designed with a despin mechanism consisting of two 7 g masses that could be spooled out to the end of two 150 cm wires when triggered by a hydraulic timer 10 h after launch. The masses would slow the spacecraft spin from 400 to 6 rpm, and then the masses and wires would be released.

Chapter Review

Fundamental equations, translation, and rotation about a fixed axis

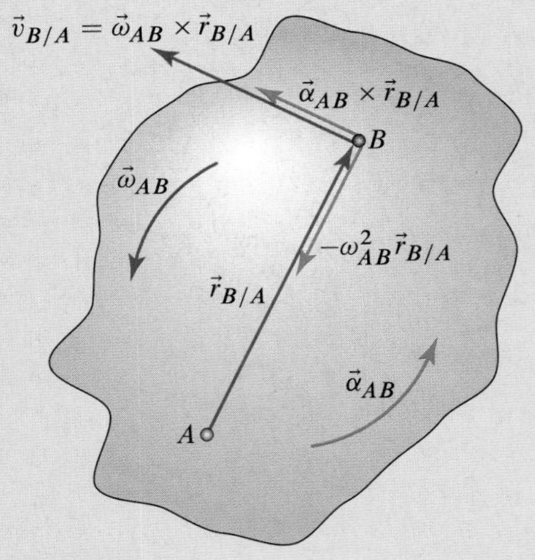

Figure 6.52
A rigid body showing the quantities used in Eqs. (6.3), (6.5), and (6.14).

This section began our study of the dynamics of rigid bodies. As with particles, we started with the study of kinematics, which is the focus of this chapter. The key kinematic idea is that a rigid body has only one angular velocity and one angular acceleration; i.e., each is a property of the body as a whole. This allowed us to use the relative velocity equation (Eq. (2.78) on p. 122) and the relation for the time derivative of a vector (Eq. (2.48) on p. 81) to relate the velocities of *two points A and B on the same rigid body* using (see Fig. 6.52)

Eq. (6.3), p. 435

$$\vec{v}_B = \vec{v}_A + \vec{\omega}_{AB} \times \vec{r}_{B/A} = \vec{v}_A + \vec{v}_{B/A},$$

and the accelerations using

Eq. (6.5), p. 436

$$\vec{a}_B = \vec{a}_A + \vec{\alpha}_{AB} \times \vec{r}_{B/A} + \vec{\omega}_{AB} \times (\vec{\omega}_{AB} \times \vec{r}_{B/A}) = \vec{a}_A + \vec{a}_{B/A},$$

which for *planar motion* becomes

Eq. (6.14), p. 439

$$\vec{a}_B = \vec{a}_A + \vec{\alpha}_{AB} \times \vec{r}_{B/A} - \omega_{AB}^2 \vec{r}_{B/A}.$$

Translation. For this motion, the angular velocity and angular acceleration of the body are equal to zero, i.e.,

Eqs. (6.6), p. 438

$$\vec{\omega}_{\text{body}} = \vec{0} \quad \text{and} \quad \vec{\alpha}_{\text{body}} = \vec{0},$$

so that the velocity and acceleration relations for the body reduce to

Eqs. (6.7), p. 438

$$\vec{v}_B = \vec{v}_A \quad \text{and} \quad \vec{a}_B = \vec{a}_A,$$

where A and B are any two points on the body.

Rotation about a fixed axis. For this special motion, there is an axis of rotation perpendicular to the plane of motion that does not move. All points not on the axis of rotation can only move in a circle about that axis. If the axis of rotation is at point A, then the velocity of B is given by

Eq. (6.8), p. 438

$$\vec{v}_B = \vec{\omega}_{AB} \times \vec{r}_{B/A},$$

and its acceleration is

Eq. (6.9), p. 438, and Eq. (6.13), p. 439

$$\vec{a}_B = \vec{\alpha}_{AB} \times \vec{r}_{B/A} + \vec{\omega}_{AB} \times (\vec{\omega}_{AB} \times \vec{r}_{B/A}),$$
$$\vec{a}_B = \vec{\alpha}_{AB} \times \vec{r}_{B/A} - \omega_{AB}^2 \vec{r}_{B/A},$$

where the vectors $\vec{\omega}_{AB}$ and $\vec{\alpha}_{AB}$ are the angular velocity and angular acceleration of the body, respectively, and $\vec{r}_{B/A}$ is the vector that describes the position of B relative to A.

Planar motion: velocity analysis

This section presented three different ways to analyze the velocities of a rigid body in planar motion: the vector approach, differentiation of constraints, and instantaneous center of rotation.

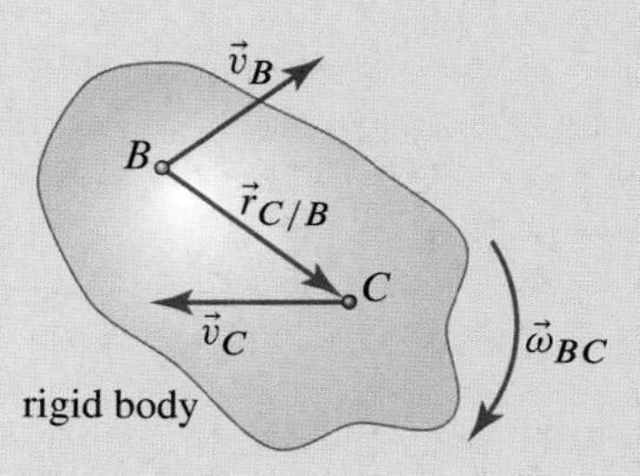

Figure 6.53
A rigid body on which we are relating velocity of two points B and C.

Vector approach. We saw in Section 6.1 that the equation

Eq. (6.15), p. 452

$$\vec{v}_C = \vec{v}_B + \vec{\omega}_{BC} \times \vec{r}_{C/B},$$

relates the velocity of two points on a rigid body $\vec{v}_B$ and $\vec{v}_C$ via their relative position $\vec{r}_{C/B}$ and the angular velocity of the body $\vec{\omega}_{BC}$ (see Fig. 6.53).

Rolling without slip. When a body rolls without slip over another body (see, for example, the wheel W rolling over the surface S in Fig. 6.54), then the two points on the bodies that are in contact at any instant, points P and Q, must have the same velocity. Mathematically, this means that

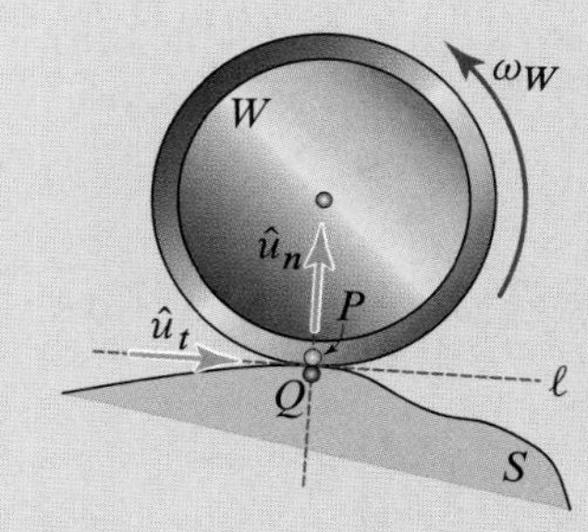

Figure 6.54
A wheel W rolling over a surface S. At the instant shown, the line ℓ is tangent to the path of points P and Q.

Eqs. (6.16), p. 453

$$\vec{v}_{P/Q} = \vec{0} \quad \Rightarrow \quad \vec{v}_P = \vec{v}_Q.$$

If a wheel of radius R is rolling without slip over a flat, stationary surface, then the point P on the wheel in contact with the surface must have zero velocity (see Fig. 6.55). The consequence is that the angular velocity of the wheel ω_W is related to the velocity of the center v_O and the radius of the wheel R according to

Eq. (6.19), p. 453

$$\omega_W = -\frac{v_O}{R},$$

in which the positive direction for ω_W is taken to be the positive z direction using the right-hand rule.

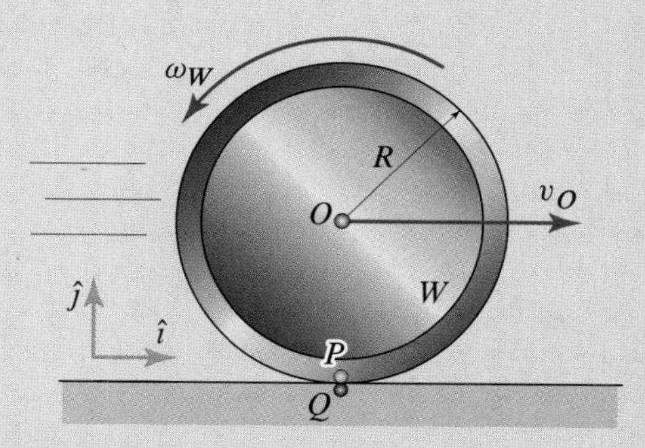

Figure 6.55
A wheel rolling without slip on a horizontal fixed surface.

Differentiation of constraints. As we discovered in Section 2.7, it is often convenient to write an equation describing the position of a point of interest, which can then be differentiated with respect to time to find the velocity of that point. For planar motion of rigid bodies, this idea can also apply for describing the position and velocity of a point on a rigid body, as well as for describing the orientation of a rigid body, for which the time derivative provides its angular velocity.

Instantaneous center of rotation. The point on a body (or imaginary extension of the body) whose velocity is zero at a particular instant is called the *instantaneous center of rotation* or *instantaneous center* (IC). The IC can be found geometrically if the velocity is known for two distinct points on a body or if a velocity on the body and the body's angular velocity are known. The three possible geometric constructions are shown in Fig. 6.56.

Planar motion: acceleration analysis

This section presented two different ways to analyze the accelerations of a rigid body in planar motion: the vector approach and differentiation of constraints.

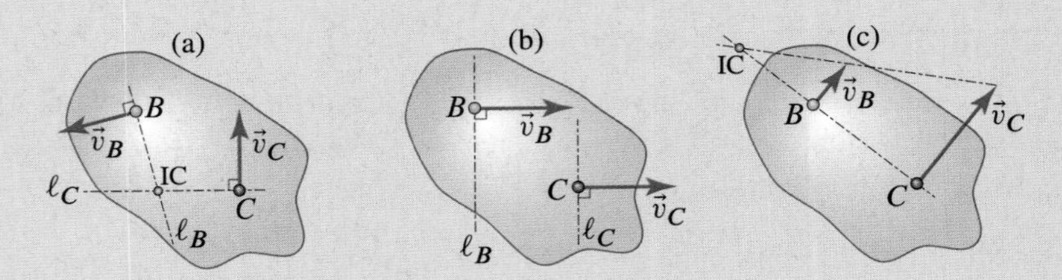

Figure 6.56. The three different possible motions for determining the IC. (a) Two nonparallel velocities are known. (b) The lines of action of two velocities are parallel and distinct; in this case, the IC is at infinity and the body is translating. (c) The lines of action of two parallel velocities coincide.

Vector approach. For planar motion, either of the equations

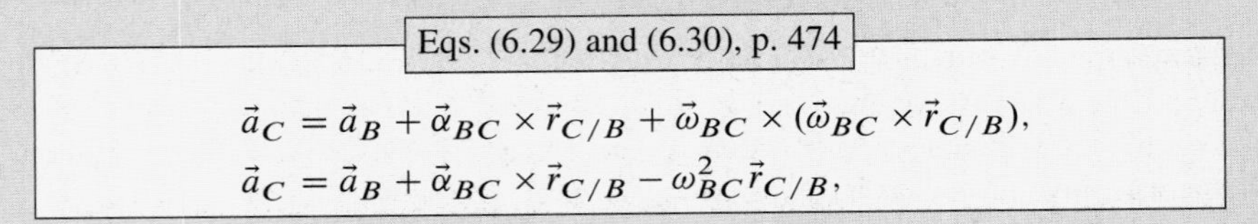

Eqs. (6.29) and (6.30), p. 474

$$\vec{a}_C = \vec{a}_B + \vec{\alpha}_{BC} \times \vec{r}_{C/B} + \vec{\omega}_{BC} \times (\vec{\omega}_{BC} \times \vec{r}_{C/B}),$$

$$\vec{a}_C = \vec{a}_B + \vec{\alpha}_{BC} \times \vec{r}_{C/B} - \omega_{BC}^2 \vec{r}_{C/B},$$

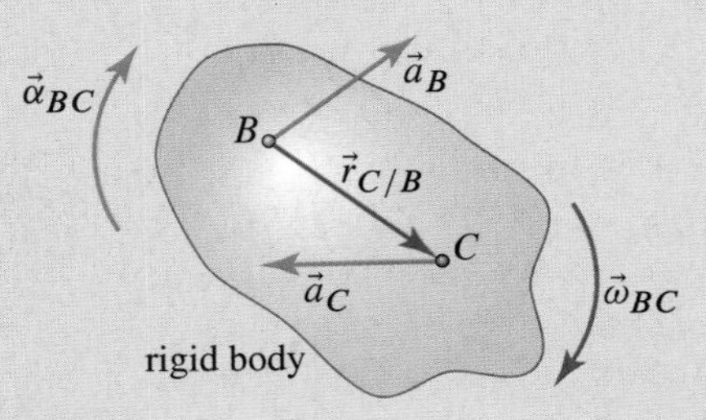

Figure 6.57
A rigid body on which we are relating the acceleration of two points B and C.

relates the acceleration of two points on a rigid body $\vec{a}_B$ and $\vec{a}_C$ via their relative position $\vec{r}_{C/B}$, the angular acceleration of the body $\vec{\alpha}_{BC}$, and the angular velocity of the body $\vec{\omega}_{BC}$ (see Fig. 6.57).

Differentiation of constraints. It is often convenient to write an equation describing the position of a point of interest, which can then be differentiated with respect to time to find the velocity and acceleration of that point. For planar motion of rigid bodies, this idea can also apply for describing the position, velocity, and acceleration of a point on a rigid body, as well as for describing the orientation of a rigid body, for which time derivatives provide its angular velocity and angular acceleration.

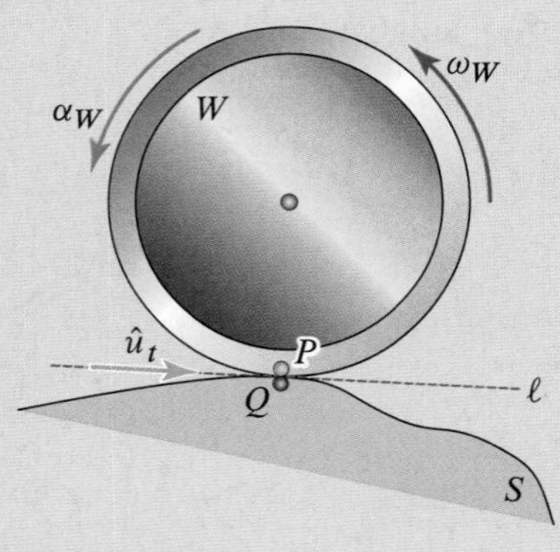

Figure 6.58
A wheel W rolling over a surface S. At the instant shown, the line ℓ is tangent to the path of P and Q.

Rolling without slip. When a body rolls without slip over another body (see Fig. 6.58), then the two points on the bodies that are in contact at any instant, points P and Q, must have the same velocity. For accelerations, this means that the component of $\vec{a}_{P/Q}$ tangent to the contact must be equal to zero, that is,

Eqs. (6.34), p. 475

$$(a_{P/Q})_t = 0 \quad \Rightarrow \quad a_{Pt} = a_{Qt}.$$

If a wheel of radius R is rolling without slip over a flat, stationary surface, then the point P on the wheel in contact with the surface must have zero velocity (see Fig. 6.59). The consequence is that the angular acceleration of the wheel α_W is related to the

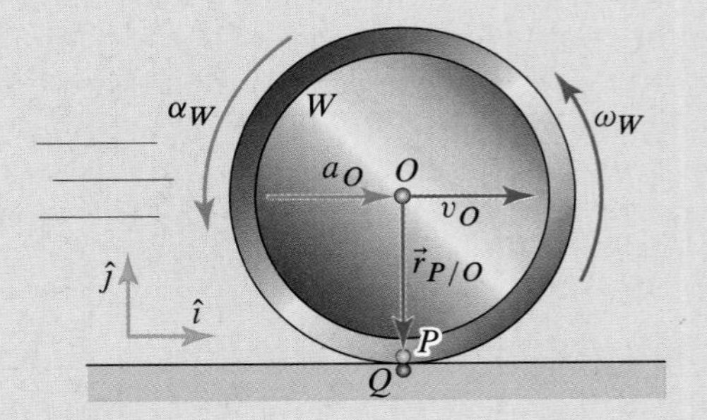

Figure 6.59. A wheel rolling without slip on a horizontal fixed surface.

acceleration of the center a_O and the radius of the wheel R according to

Eq. (6.19), p. 453

$$\alpha_W = -\frac{a_O}{R},$$

in which the positive direction for α_W is taken to be the positive z direction. The point P on the wheel that is in contact with the ground at this instant is accelerating along the y axis according to

Eq. (6.39), p. 475

$$\vec{a}_P = \frac{v_O^2}{R}\,\hat{\jmath}.$$

Even though P has zero velocity at the instant considered, it is accelerating.

Rotating reference frames

In this section, we developed the kinematic equations that allow us to relate the motion of two points that *are not* on the same rigid body. Referring to Fig. 6.60, this means we can relate the velocity of point P to that of A using the relation

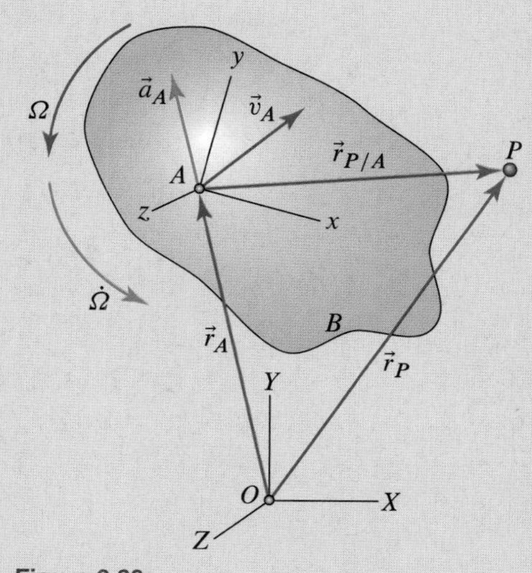

Figure 6.60
The essential ingredients for finding velocities and accelerations using a rotating reference frame.

Eq. (6.45), p. 495

$$\vec{v}_P = \vec{v}_A + \vec{v}_{P\,\mathrm{rel}} + \vec{\Omega} \times \vec{r}_{P/A},$$

where

$\vec{v}_A$ = velocity of the origin of the rotating or body-fixed reference frame (point A in Fig. 6.60)

$\vec{v}_{P\,\mathrm{rel}}$ = velocity of P as seen by an observer moving with the rotating reference frame

$\vec{\Omega}$ = angular velocity of the rotating reference frame

$\vec{r}_{P/A}$ = position of point P relative to A

We also found that we can relate the acceleration of point P to that of A using

Eq. (6.53), p. 496

$$\vec{a}_P = \vec{a}_A + \vec{a}_{P\,\mathrm{rel}} + 2\vec{\Omega} \times \vec{v}_{P\,\mathrm{rel}} + \dot{\vec{\Omega}} \times \vec{r}_{P/A} + \vec{\Omega} \times (\vec{\Omega} \times \vec{r}_{P/A}),$$

where, in addition to the terms defined for $\vec{v}_P$, we have

$\vec{a}_A$ = acceleration of the origin of the rotating or body-fixed reference frame (point A in Fig. 6.60)

$\vec{a}_{P\,\mathrm{rel}}$ = acceleration of P as seen by an observer moving with the rotating reference frame

$2\vec{\Omega} \times \vec{v}_{P\,\mathrm{rel}}$ = Coriolis acceleration of P

$\dot{\vec{\Omega}}$ = angular acceleration of the rotating reference frame

Review Problems

Problem 6.162

The manually operated road barrier shown has a total length $l = 15.7$ ft and has a counterweight C whose position is identified by the distances $d = 2.58$ ft and $\delta = 1.4$ ft. The barrier is pinned at O and can move in the vertical plane. If the barrier is raised and then released so that the end B of the barrier hits the support with a speed $v_B = 1.5$ ft/s, determine the speed of the counterweight C when B hits A.

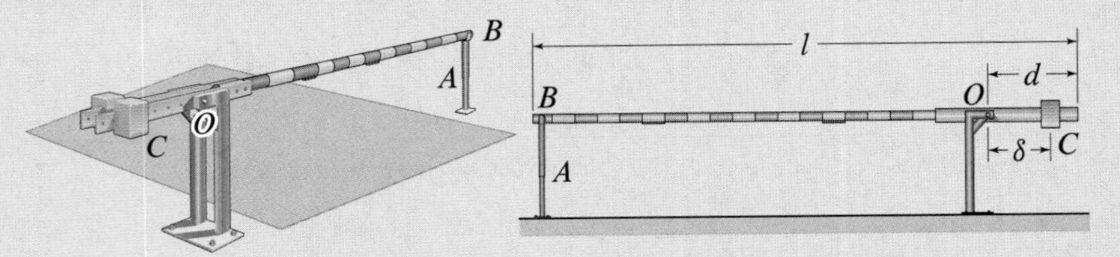

Figure P6.162

Problems 6.163 and 6.164

Bar AB rotates about point A with angular speed ω_1 and angular acceleration α_1, both in the directions shown. Bar BC rotates about the pin at B relative to bar AB with angular speed ω_2 and angular acceleration α_2, both in the directions shown. The position of collar D relative to bar BC is given by the coordinate s, which is a known function of time. Express your answers in the xyz frame that is attached to bar BC with its origin at B. *Hint:* As we will see in Eq. (10.14), $\vec{\omega}_{BC} = \vec{\omega}_{AB} + \vec{\omega}_{BC/AB} = \vec{\omega}_1 + \vec{\omega}_2$.

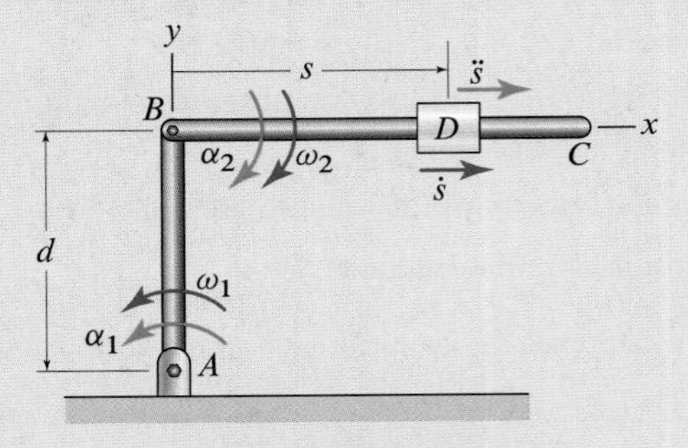

Figure P6.163 and P6.164

Problem 6.163 Determine an expression for the Coriolis acceleration of the collar D.

Problem 6.164 Determine an expression for the acceleration of the collar D in terms of d, ω_1, ω_2, α_1, α_2, and the position coordinate s and its time derivatives.

Problem 6.165

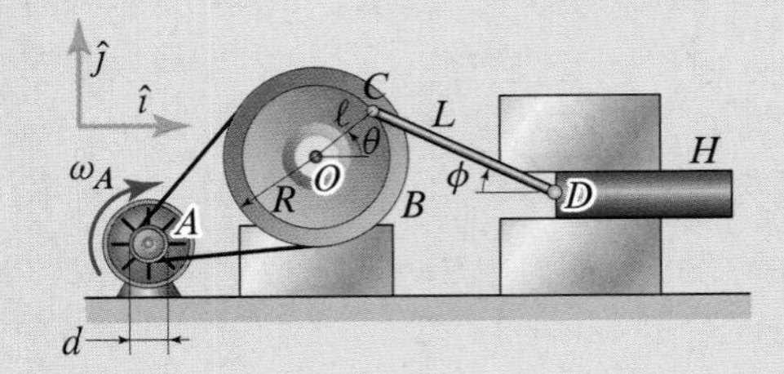

Figure P6.165

At the instant shown, the hammer head H is moving to the right with a speed $v_H = 45$ ft/s and the angle $\theta = 20°$. Assuming that the belt does not slip relative to wheels A and B and assuming that wheel A is mounted on the shaft of the motor shown, determine the angular velocity of the motor at the instant shown. The diameter of wheel A is $d = 0.25$ ft, the radius of wheel B is $R = 0.75$ ft, and point C is at a distance $\ell = 0.72$ ft from O, which is the center of the wheel B. Finally, let CD have a length $L = 2$ ft, and assume that, at the instant shown, $\phi = 25°$.

Problems 6.166 and 6.167

The bar AB is pinned to the ground at A, and its angle θ with respect to the vertical is a known function of time. The bar BC is pinned to the bar AB at B, and its orientation ϕ is measured with respect to the bar AB. The position of the collar D is measured relative to the bar BC by the coordinate s, which is also a known function of time. The length of the bar AB is $d = 2\,\text{ft}$. At the instant shown, $\theta = 30°$, $\phi = 30°$, $\dot{\theta} = 5\,\text{rad/s}$, $\ddot{\theta} = 3\,\text{rad/s}^2$, $\dot{\phi} = -5\,\text{rad/s}$, $\ddot{\phi} = 4\,\text{rad/s}^2$, $s = 2.5\,\text{ft}$, $\dot{s} = 10\,\text{ft/s}$, and $\ddot{s} = -15\,\text{ft/s}^2$. Express all your answers in the xyz frame that is attached to the bar BC and whose origin is at B. *Hint:* As we will see in Eq. (10.14), $\vec{\omega}_{BC} = \vec{\omega}_{AB} + \vec{\omega}_{BC/AB}$.

Problem 6.166 Determine the velocity of the collar D at this instant.

Problem 6.167 Determine the acceleration of the collar D at this instant.

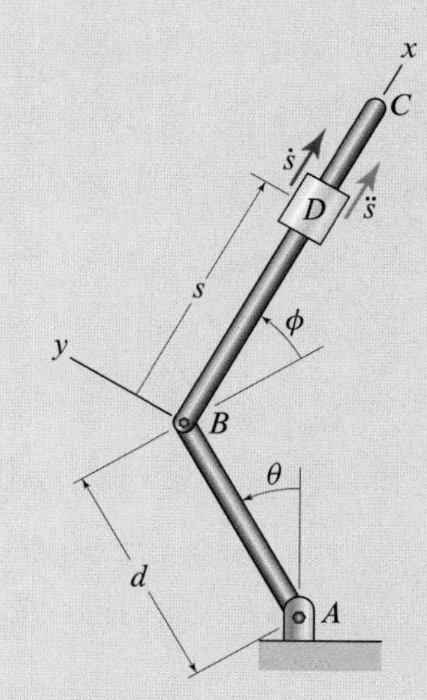

Figure P6.166 and P6.167

Problem 6.168

Assuming that the rope does not slip relative to any of the pulleys in the system, determine the velocity and acceleration of A and D, knowing that the angular velocity and acceleration of pulley B are $\omega_B = 7\,\text{rad/s}$ and $\alpha_B = 3\,\text{rad/s}^2$, respectively. The diameters of pulleys B and C are $d = 25\,\text{cm}$ and $\ell = 34\,\text{cm}$, respectively.

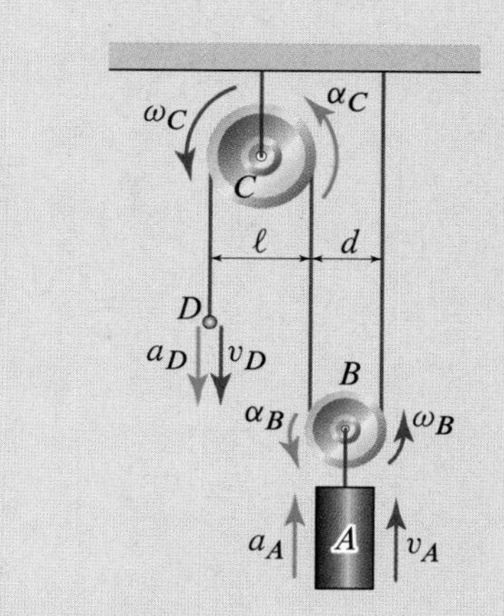

Figure P6.168

Problem 6.169

At the instant shown, bar AB rotates with a constant angular velocity $\omega_{AB} = 24\,\text{rad/s}$. Letting $L = 0.75\,\text{m}$ and $H = 0.85\,\text{m}$, determine the angular acceleration of bar BC when bars AB and CD are as shown, i.e., parallel and horizontal.

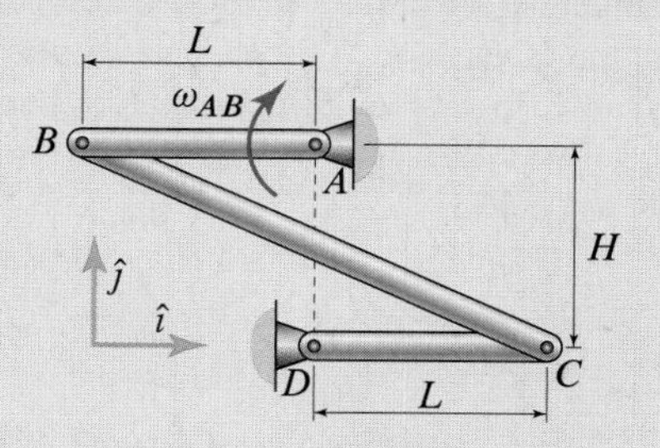

Figure P6.169

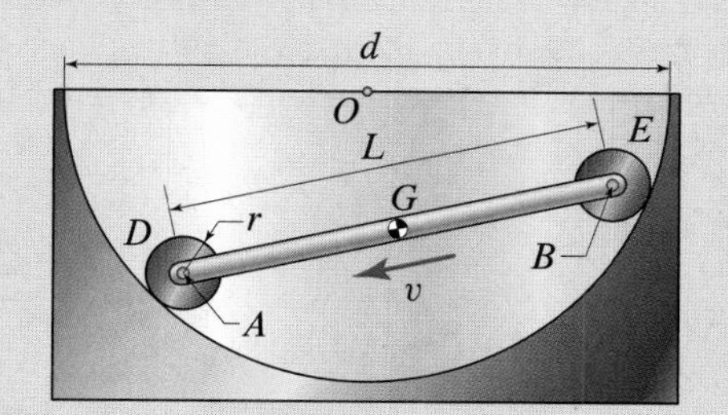

Figure P6.170

Problem 6.170

A slender bar AB of length $L = 1.45\,\text{ft}$ is mounted on two identical disks D and E pinned at A and B, respectively, and of radius $r = 1.5\,\text{in}$. The bar is allowed to move within a cylindrical bowl with center at O and diameter $d = 2\,\text{ft}$. At the instant shown, the center G of the bar is moving with a speed $v = 7\,\text{ft/s}$. Determine the angular velocity of the bar at the instant shown.

Problem 6.171

The bucket of a backhoe is the element AB of the four-bar linkage system $ABCD$. The bucket's motion is controlled by extending or retracting the hydraulic arm EC. Assume that the points A, D, and E are fixed and that the bucket is made to rotate with a constant angular velocity $\omega_{AB} = 0.25\,\text{rad/s}$. In addition, suppose that, at the instant shown, point B is vertically aligned with point A, and point C is horizontally aligned with B. Letting d_{EC} denote the distance between points E and C, determine $\dot{d}_{EC}$ and $\ddot{d}_{EC}$ at the instant shown. Let $h = 0.66\,\text{ft}$, $e = 0.46\,\text{ft}$, $\ell = 0.9\,\text{ft}$, $w = 1.0\,\text{ft}$, $d = 4.6\,\text{ft}$, and $q = 3.2\,\text{ft}$.

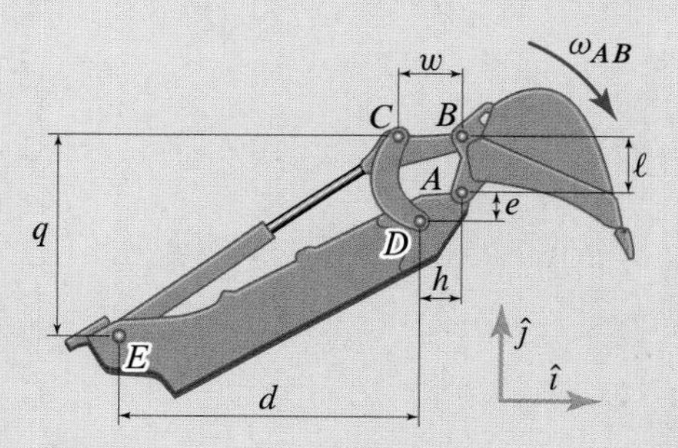

Figure P6.171

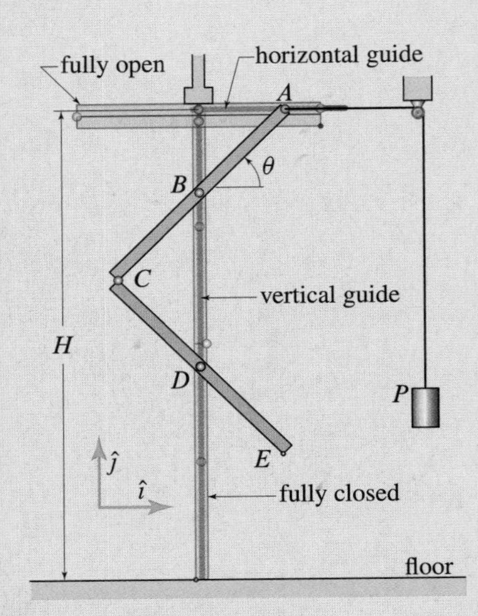

Figure P6.172 and P6.173

Problems 6.172 and 6.173

An overhead fold-up door with height $H = 30\,\text{ft}$ consists of two identical sections hinged at C. The roller at A moves along a horizontal guide, whereas the rollers at B and D, which are the midpoints of sections AC and CE, move along a vertical guide. The door's operation is assisted by a counterweight P. Express your answers using the component system shown.

Problem 6.172 If at the instant shown, the angle $\theta = 55°$ and P is moving upward with a speed $v_P = 15\,\text{ft/s}$, determine the velocity of point E, as well as the angular velocities of sections AC and CE.

Problem 6.173 If at the instant shown, $\theta = 45°$, and A is moving to the right with a speed $v_A = 2\,\text{ft/s}$ while decelerating at a rate of $1.5\,\text{ft/s}^2$, determine the acceleration of point E.

Problem 6.174

A vertical shaft has a base B that is stationary relative to an inertial reference frame with vertical axis Z. Arm OA is attached to the vertical shaft and rotates about the Z axis with an angular velocity ω_{OA} and an angular acceleration α_{OA}. The xyz frame is attached at C, such that Z and z are always parallel and the x axis of the rotating reference frame coincides with the axis of the arm OA. At this instant, the collar C is at a distance d from the Z axis, is sliding along OA with a constant speed v_C (relative to the arm OA), and is rotating with a constant angular velocity ω_C (relative to the xyz frame). Point D is attached to the collar and is at a distance ℓ from the x axis. At the instant shown, point D happens to be in the plane that is rotated by an angle ϕ from the xz plane. Compute expressions for the inertial velocity and acceleration of point D at the instant shown in terms of the parameters given, and express them relative to the rotating coordinate system.

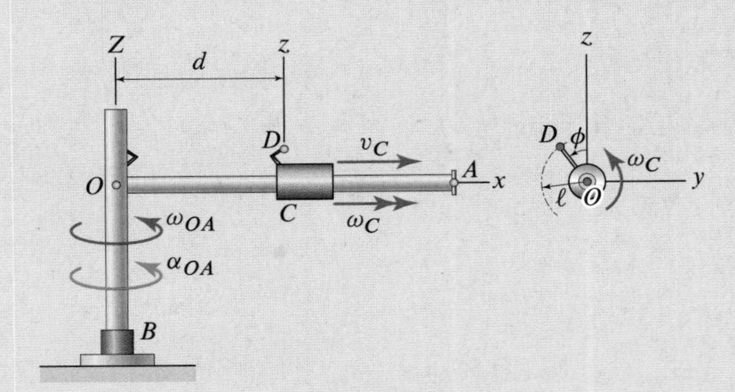

Figure P6.174

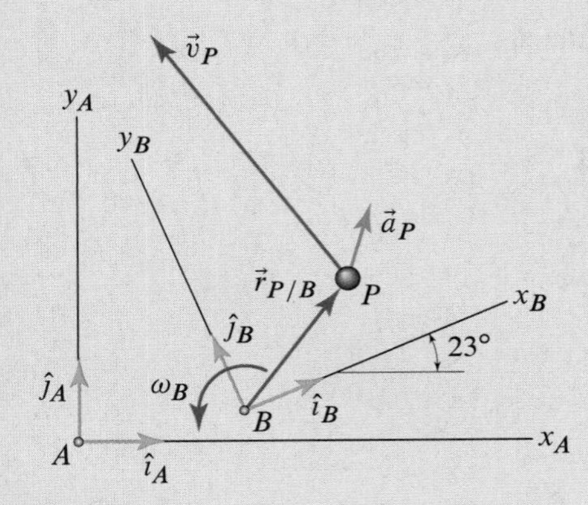

Figure P6.175

Problem 6.175

Let frames A and B have their origins at points A and B, respectively. Point B does not move relative to point A. The velocity and acceleration of point P relative to frame A, which is fixed, are

$$\vec{v}_P = (-14.9\,\hat{\imath}_A + 19.4\,\hat{\jmath}_A)\,\text{ft/s} \quad \text{and} \quad \vec{a}_P = (1.78\,\hat{\imath}_A + 5.96\,\hat{\jmath}_A)\,\text{ft/s}^2.$$

Knowing that frame B rotates relative to frame A at a constant angular velocity $\omega_B = 1.2\,\text{rad/s}$ and that the position of point P relative to frame B is $\vec{r}_{P/B} = (8\,\hat{\imath}_B + 4.5\,\hat{\jmath}_B)\,\text{ft}$, determine the velocity and acceleration of P relative to frame B at the instant shown. Express the results using the frame A component system.

Problem 6.176

The wheel D rotates without slip over a curved cylindrical surface with constant radius of curvature $L = 1.6$ ft and center at the fixed point E. The XY frame is attached to the rolling surface, and it is inertial. The xy frame is attached to D at O and rotates with it at a constant angular velocity $\omega_D = 14$ rad/s. Collar C slides along the bar AB with a constant velocity $v_C = 4$ ft/s relative to the xy frame. At the instant shown, points O and E are vertically aligned. Letting $\ell = 0.25$ ft and $R = 1$ ft, determine the inertial velocity and acceleration of C at the instant shown when $\theta = 25°$ and the xy frame is parallel to the XY frame. Express your result in the xy and XY frames.

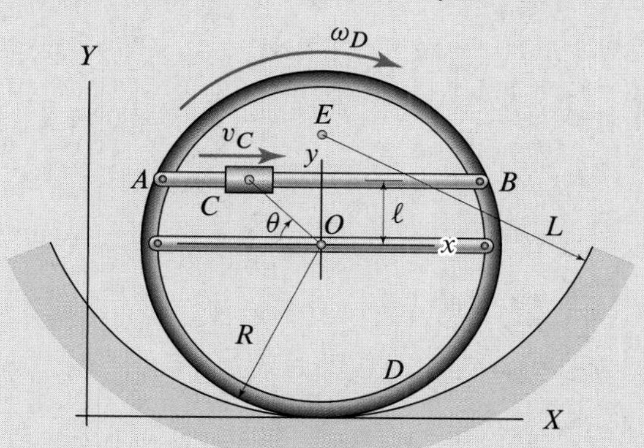

Figure P6.176

Problem 6.177

The wheel D rotates without slip over a curved cylindrical surface with constant radius of curvature $L = 1.6$ ft and center at the fixed point E. The XY frame is attached to the rolling surface, and it is inertial. The xy frame is attached to D at O and rotates with it at an angular velocity $\omega_D = 14$ rad/s and an angular acceleration $\alpha_D = 1.3\,\text{rad/s}^2$. Collar C slides along the bar AB with a constant velocity $v_C = 4$ ft/s relative to the xy frame. Letting $\ell = 0.25$ ft and $R = 1$ ft, determine the inertial velocity and acceleration of C, at the instant shown, when $\theta = 25°$ and the xy frame is parallel to the XY frame. Express your result in the xy and XY frames.

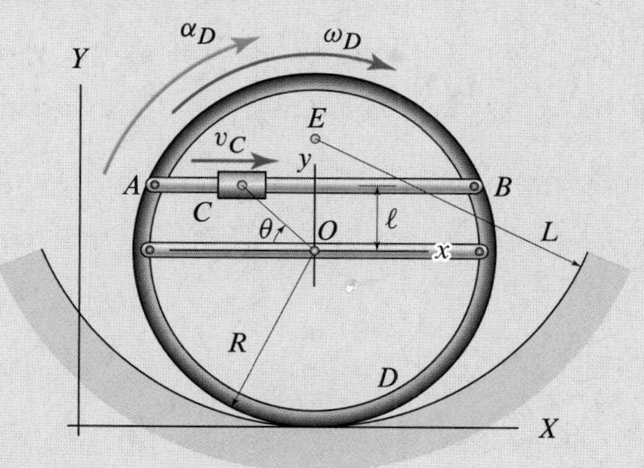

Figure P6.177

Newton-Euler Equations for Planar Rigid Body Motion

Casey Stoner of the Ducati MotoGP Team, performing a turn with his racing motorcycle. By modeling the motorcycle as a rigid body we can determine the maximum acceleration of the motorcycle for which it does not slip or tip.

We now begin the study of rigid body kinetics. In this chapter, we develop the balance principles for forces and moments, this time for a rigid body. In Chapter 8, we will study energy and momentum methods for rigid bodies. By the time we complete this chapter, we will know how to combine the Newton-Euler equations for a rigid body with the force laws and the kinematic equations to obtain the equations governing the body's motion.

In rigid body kinetics we will be writing force *and* moment equations. Therefore, in the spirit of the discussion in Section 5.3, when referring to the force and moment equations for a rigid body, we refer to these equations as *Newton-Euler equations* instead of Newton's second law.

7.1 Newton-Euler Equations: Bodies Symmetric with Respect to the Plane of Motion

We now develop and apply the force and moment balance laws for rigid bodies. The moment balance law accounts for the fact that, in general, we do not have concurrent force systems acting on rigid bodies and we must now account for the rotation or tendency for rotation of objects.

Linear momentum: translational equations

To obtain the translational equations of motion, we use Euler's first law, which is given by Eq. (5.57) on p. 365 and applies to any system with constant mass:

$$\boxed{\vec{F} = m\vec{a}_G,} \tag{7.1}$$

where $\vec{F}$ is the resultant of all *external* forces, m is the total mass of the system, and $\vec{a}_G$ is the inertial acceleration of its mass center, which is given by

$$\vec{a}_G = \frac{d^2\vec{r}_G}{dt^2}. \tag{7.2}$$

Referring to Fig. 7.1, $\vec{r}_G$ is the position of the *mass center* of body B relative to point

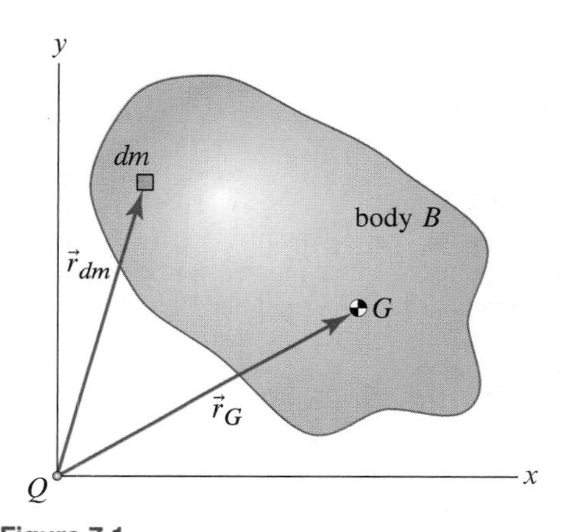

Figure 7.1
Vectors needed for the determination of the position of the mass center $\vec{r}_G$ of a rigid body.

Q and is given by

$$\vec{r}_G = \frac{1}{m} \int_B \vec{r}_{dm} \, dm, \tag{7.3}$$

where the integral is performed over the body B and $\vec{r}_{dm}$ is the position of the infinitesimal element of mass dm.

Angular momentum: rotational equations

Since Euler's second law relates moments to angular momentum, it provides the rotational equations of motion. We begin with the moment–angular momentum relation given by Eq. (5.58) on p. 365, which is

$$\vec{M}_P = \dot{\vec{h}}_P + \vec{v}_P \times m\vec{v}_G, \tag{7.4}$$

Helpful Information

Are there any restrictions on P? No, point P can be *any* point in space. On p. 525 we will show what the equations look like when we restrict P to be on the body (or an arbitrary extension of the body).

where

- $\vec{M}_P$ is the moment of all external forces about point P, *which is an arbitrary point in space*
- $\vec{h}_P$ is the angular momentum of the entire system with respect to point P
- $\vec{v}_P$ is the velocity of the arbitrary point P
- $\vec{v}_G$ is the velocity of the mass center of the system
- m is the total mass of the system

When applying Eq. (7.4) to a rigid body, all terms are easy to interpret except for $\dot{\vec{h}}_P$. To find $\dot{\vec{h}}_P$, recall that, for a system of particles, $\vec{h}_P$ was computed using the definitions in Eqs. (5.31) and (5.48) (on pp. 361 and 363, respectively), which yield

$$\vec{h}_P = \sum_{i=1}^{N} \vec{r}_{i/P} \times m_i \vec{v}_i, \tag{7.5}$$

where N is the number of particles in the system, $\vec{r}_{i/P}$ is the position of particle i relative to the point P, and $m_i \vec{v}_i$ is the linear momentum of particle i. For a rigid body whose mass is distributed in space, we can generalize Eq. (7.5) by replacing the summation over the number of particles with an integral over the body. When we do so, particle i of mass m_i becomes the infinitesimal element of mass dm, and Eq. (7.5) becomes

$$\vec{h}_P = \int_B \vec{r}_{dm/P} \times \vec{v}_{dm} \, dm, \tag{7.6}$$

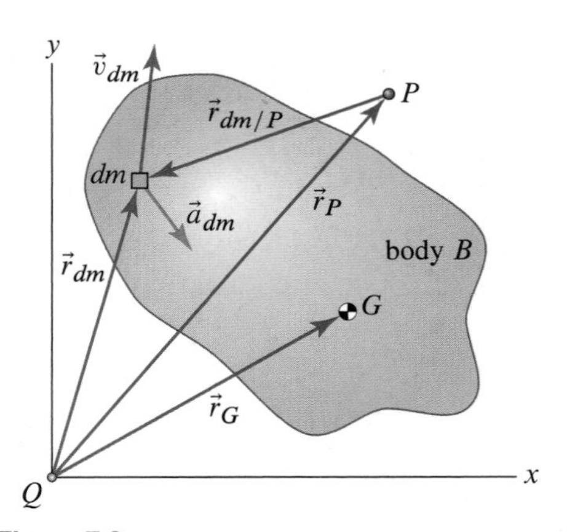

Figure 7.2
Points and corresponding position vectors needed to develop the moment–angular momentum relationship for a rigid body.

where $\vec{v}_{dm}$ is the velocity of the element of mass dm (see Fig. 7.2). Since we need $\dot{\vec{h}}_P$, we differentiate Eq. (7.6) with respect to time to obtain

$$\dot{\vec{h}}_P = \int_B \vec{v}_{dm/P} \times \vec{v}_{dm} \, dm + \int_B \vec{r}_{dm/P} \times \vec{a}_{dm} \, dm. \tag{7.7}$$

Using $\vec{v}_{dm/P} = \vec{v}_{dm} - \vec{v}_P$, Eq. (7.7) becomes

$$\begin{aligned} \dot{\vec{h}}_P &= \int_B (\vec{v}_{dm} - \vec{v}_P) \times \vec{v}_{dm} \, dm + \int_B \vec{r}_{dm/P} \times \vec{a}_{dm} \, dm \\ &= \underbrace{-\int_B \vec{v}_P \times \vec{v}_{dm} \, dm}_{\text{integral } A} + \underbrace{\int_B \vec{r}_{dm/P} \times \vec{a}_{dm} \, dm}_{\text{integral } B}. \end{aligned} \tag{7.8}$$

Since $\vec{v}_P$ does not depend on dm, integral A can be written as

$$-\int_B \vec{v}_P \times \vec{v}_{dm}\, dm = -\vec{v}_P \times \int_B \vec{v}_{dm}\, dm = -\vec{v}_P \times \frac{d}{dt}\int_B \vec{r}_{dm}\, dm$$
$$= -\vec{v}_P \times \frac{d}{dt}(m\vec{r}_G) = -\vec{v}_P \times m\vec{v}_G, \quad (7.9)$$

where we have used the definition of the mass center of a rigid body given in Eq. (7.3). Substituting Eq. (7.9) into Eq. (7.8) gives

$$\dot{\vec{h}}_P = -\vec{v}_P \times m\vec{v}_G + \underbrace{\int_B \vec{r}_{dm/P} \times \vec{a}_{dm}\, dm}_{\text{integral } B}. \quad (7.10)$$

Substituting Eq. (7.10) into Eq. (7.4), we are left with just integral B, that is,

$$\vec{M}_P = \int_B \vec{r}_{dm/P} \times \vec{a}_{dm}\, dm, \quad (7.11)$$

since $-\vec{v}_P \times m\vec{v}_G$ in Eq. (7.10) cancels with $\vec{v}_P \times m\vec{v}_G$ in Eq. (7.4). We will now interpret the integral in Eq. (7.11) for the case in which the rigid body is symmetric with respect to the plane of motion (Chapter 10 covers the case when it is not).

Helpful Information

Which terms can be pulled outside the integral signs? In going from Eq. (7.8) to Eq. (7.9), $\vec{v}_P$ is pulled out of the integral. We can pull $\vec{v}_P$ out since it is a property of a single point and does not vary with dm. We will see this again in Eq. (7.21) where, for example, the same is true for a_{Gy}, which is also a property of a particular point, and for α_B, which is a property of the body as a whole.

Bodies symmetric with respect to the plane of motion

Referring to Fig. 7.3, since the body B is rigid, we can relate $\vec{a}_{dm}$ to $\vec{a}_G$ via Eq. (6.13) on p. 439 as

$$\vec{a}_{dm} = \vec{a}_G + \vec{\alpha}_B \times \vec{q} - \omega_B^2 \vec{q}, \quad (7.12)$$

where $\vec{q}$ is the position of dm relative to G. It is now convenient to write all vectors in Cartesian components as:

$$\vec{M}_P = M_{Px}\,\hat{\imath} + M_{Py}\,\hat{\jmath} + M_{Pz}\,\hat{k}, \quad (7.13)$$
$$\vec{a}_G = a_{Gx}\,\hat{\imath} + a_{Gy}\,\hat{\jmath}, \quad (7.14)$$
$$\vec{q} = q_x\,\hat{\imath} + q_y\,\hat{\jmath}, \quad (7.15)$$
$$\vec{r}_{dm/P} = \vec{r}_{G/P} + \vec{q} = (x_{G/P} + q_x)\,\hat{\imath} + (y_{G/P} + q_y)\,\hat{\jmath}, \quad (7.16)$$
$$\vec{\omega}_B = \omega_B\,\hat{k} \quad \text{and} \quad \vec{\alpha}_B = \alpha_B\,\hat{k}, \quad (7.17)$$

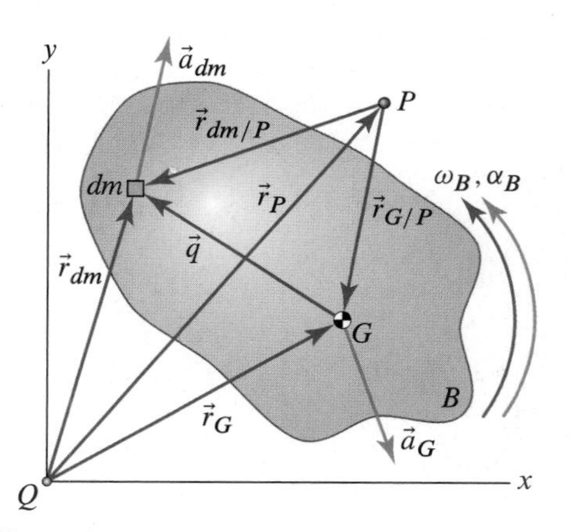

Figure 7.3
The definitions of all needed kinematic and kinetic quantities for a rigid body moving in planar motion.

where Eqs. (7.14) and (7.17) reflect the constraint that the body's motion is planar and Eqs. (7.15) and (7.16) reflect our assumption that the body is symmetric with respect to the plane of motion. Substituting Eqs. (7.14), (7.15), and (7.17) into Eq. (7.12), we obtain $\vec{a}_{dm}$ in component form as

$$\vec{a}_{dm} = \left(a_{Gx} - \alpha_B q_y - \omega_B^2 q_x\right)\hat{\imath} + \left(a_{Gy} + \alpha_B q_x - \omega_B^2 q_y\right)\hat{\jmath}. \quad (7.18)$$

Substituting Eqs. (7.13), (7.16), and (7.18) into Eq. (7.11), expanding the cross product, and equating components, we obtain the following three expressions for the rotational equations of motion:

$$M_{Px} = 0, \quad (7.19)$$

$$M_{Py} = 0, \tag{7.20}$$

$$\begin{aligned} M_{Pz} = \alpha_B \int_B (q_x^2 + q_y^2)\, dm + \big(x_{G/P} a_{Gy} - y_{G/P} a_{Gx}\big) \int_B dm \\ + \big(a_{Gy} + \alpha_B x_{G/P} + \omega_B^2 y_{G/P}\big) \int_B q_x\, dm \\ + \big(-a_{Gx} + \alpha_B y_{G/P} - \omega_B^2 x_{G/P}\big) \int_B q_y\, dm, \end{aligned} \tag{7.21}$$

where we have taken all terms that do not depend on the element of mass dm outside of the integrals. Observe that *there are no moments about the x or y axes when the body is symmetric with respect to the plane of motion*. This is a direct consequence of our assumption that the body is symmetric with respect to the plane of motion — it is not true if that assumption is violated.

The second integral in Eq. (7.21) is the total mass of the body, that is,

$$\int_B dm = m. \tag{7.22}$$

Concept Alert

Developing intuition for mass moments of inertia. Just as mass is a measure of an object's resistance to *linear* acceleration, the mass moment of inertia in Eq. (7.27) is a measure of an object's resistance to *angular* acceleration by accounting for the distribution of mass of a body with respect to the rotation axis. For example, it is much more difficult to spin a long stick perpendicular to its long axis (x-spin) than parallel to its long axis (y-spin).

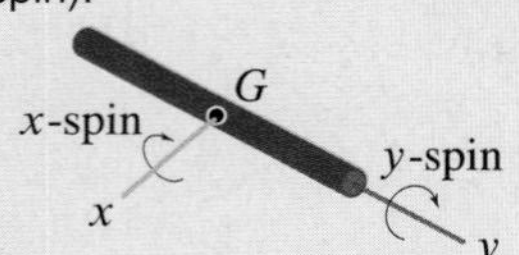

This is reflected in the mass moments of inertia of a thin stick, for which $I_{Gx} = \frac{1}{12}ml^2$ and $I_{Gy} \approx 0$, where m is the mass of the stick and l is its length.

Recalling that Eq. (7.3) defines the position of the mass center relative to point Q (see Fig. 7.3), in component form, Eq. (7.3) becomes

$$\vec{r}_G = x_{G/Q}\,\hat{\imath} + y_{G/Q}\,\hat{\jmath} = \frac{1}{m}\int_B (x_{dm/Q}\,\hat{\imath} + y_{dm/Q}\,\hat{\jmath})\, dm, \tag{7.23}$$

or

$$m x_{G/Q} = \int_B x_{dm/Q}\, dm \quad \text{and} \quad m y_{G/Q} = \int_B y_{dm/Q}\, dm. \tag{7.24}$$

Since $\vec{q} = q_x\,\hat{\imath} + q_y\,\hat{\jmath}$ defines the position of each element of mass dm relative to the mass center G, the component form of the definition of the mass center of a rigid body in Eq. (7.24) tells us that (see Fig. 7.3)

$$\int_B q_x\, dm = \int_B x_{dm/G}\, dm = m x_{G/G} = 0, \tag{7.25}$$

$$\int_B q_y\, dm = \int_B y_{dm/G}\, dm = m y_{G/G} = 0, \tag{7.26}$$

where we see that these integrals just define the position of the mass center relative to the mass center. There is only one integral left to interpret, which is

$$\int_B (q_x^2 + q_y^2)\, dm = I_G. \tag{7.27}$$

This integral defines the *mass moment of inertia of the rigid body about an axis perpendicular to the plane of motion passing through the mass center G*, which we will label as I_G. Because the motion is planar, we will usually refer to the term in question simply as the *mass moment of inertia*, and the subscript on I will uniquely identify the axis with respect to which I is calculated (see Appendix A).

Using Eqs. (7.22)–(7.27), we can see that Eq. (7.21) becomes

$$\boxed{M_P = I_G \alpha_B + m\big(x_{G/P} a_{Gy} - y_{G/P} a_{Gx}\big),} \tag{7.28}$$

where $M_{Pz} = M_P$ to simplify the notation for motion. Equation (7.28) is the most general rotational equation for the planar motion of a rigid body that is symmetric with respect to the plane of motion. Equation (7.28) and the two equations represented by Eq. (7.1) give the three equations needed to describe the planar motion of a rigid body.

Now that we have the rotational equation we need, note that

- If the body is *not* symmetric with respect to the xy plane of motion, then moments in the x and/or y directions will be required to maintain planar motion.
- Equation (7.28) can be written using vector notation as

$$\vec{M}_P = I_G\vec{\alpha}_B + \vec{r}_{G/P} \times m\vec{a}_G, \tag{7.29}$$

where, since the motion is planar, $\vec{M}_P = M_P\,\hat{k}, \vec{\alpha}_B = \alpha_B\,\hat{k}, \vec{r}_{G/P} = x_{G/P}\,\hat{\imath} + y_{G/P}\,\hat{\jmath}$, and $\vec{a}_G = a_{Gx}\,\hat{\imath} + a_{Gy}\,\hat{\jmath}$. Again, *Eqs.* (7.28) *and* (7.29) *apply when* P *is an arbitrary point in space*, that is, it *does not* have to be a point on the rigid body B.

- If *any* one of the following conditions is true:
  1. Point P is the mass center G, so that $\vec{r}_{G/P} = \vec{0}$.
  2. $\vec{a}_G = \vec{0}$ (i.e., G is fixed or moves with constant velocity).
  3. $\vec{r}_{G/P}$ is parallel to $\vec{a}_G$.

  then Eq. (7.28) reduces to

$$M_P = I_G\alpha_B. \tag{7.30}$$

Concept Alert

Why do very different bodies have the same mass moments of inertia? Referring to the table inside the back cover of the book and the figures below, notice that for a thin plate $(I_x)_{\text{plate}} = \frac{1}{12}mb^2$, which is the same as that of a thin rod, $(I_x)_{\text{rod}} = \frac{1}{12}ml^2$ (if they have the same length, i.e., $l = b$).

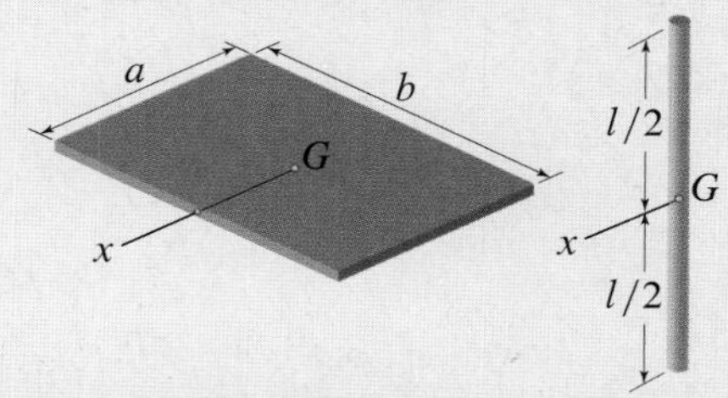

Why is this? Looking down the x axis, the thin plate resembles a thin rod. As long as the thin plate extends uniformly behind its projection (e.g., it doesn't taper in or out) and it has the same mass as a thin rod, it must have the same mass moment of inertia I_G.

What if the moment center is on the rigid body? Equation (7.29) is valid for any possible choice of moment center P. We now derive a version of Eq. (7.29) that is applicable when the moment center is a point O *on* the rigid body or on an arbitrary extension of the rigid body.

Referring to Fig. 7.4, if points O and G are both on the rigid body, we can write the acceleration of G as

$$\vec{a}_G = \vec{a}_O + \vec{\alpha}_B \times \vec{r}_{G/O} - \omega_B^2\vec{r}_{G/O}, \tag{7.31}$$

and use the parallel axis theorem (see Appendix A) to write I_G as

$$I_G = I_O - mr_{G/O}^2, \tag{7.32}$$

where I_O is the mass moment of inertia of the body about an axis perpendicular to the plane of motion passing through point O. Substituting Eqs. (7.31) and (7.32) into Eq. (7.29), we obtain

$$\vec{M}_O = (I_O - mr_{G/O}^2)\vec{\alpha}_B + \vec{r}_{G/O} \times m(\vec{a}_O + \vec{\alpha}_B \times \vec{r}_{G/O} - \omega_B^2\vec{r}_{G/O}) \tag{7.33}$$

$$= (I_O - mr_{G/O}^2)\vec{\alpha}_B + \vec{r}_{G/O} \times m\vec{a}_O + m(\vec{r}_{G/O}\cdot\vec{r}_{G/O})\vec{\alpha}_B - m(\vec{r}_{G/O}\cdot\vec{\alpha}_B)\vec{r}_{G/O} \tag{7.34}$$

$$= (I_O - mr_{G/O}^2)\vec{\alpha}_B + \vec{r}_{G/O} \times m\vec{a}_O + mr_{G/O}^2\vec{\alpha}_B, \tag{7.35}$$

where we have used the vector identity $\vec{a} \times (\vec{e} \times \vec{c}) = (\vec{a}\cdot\vec{c})\vec{e} - (\vec{a}\cdot\vec{e})\vec{c}$, $\vec{r}_{G/O} \times \vec{r}_{G/O} = \vec{0}$, $\vec{r}_{G/O}\cdot\vec{r}_{G/O} = r_{G/O}^2$, and the fact that $\vec{r}_{G/O}$ is orthogonal to $\vec{\alpha}_B$ (so that $\vec{r}_{G/O}\cdot\vec{\alpha}_B = 0$). Making the final simplification by canceling the two terms involving $mr_{G/O}^2\vec{\alpha}_B$, Eq. (7.35) becomes

$$\vec{M}_O = I_O\vec{\alpha}_B + \vec{r}_{G/O} \times m\vec{a}_O. \tag{7.36}$$

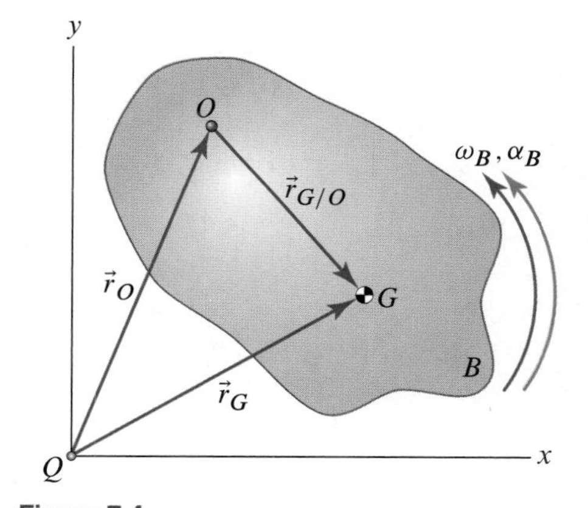

Figure 7.4
An arbitrary rigid body with the point O being a point *on* the rigid body.

Because of the assumptions in going from Eq. (7.29) to Eq. (7.36), Eq. (7.36) is subject to the restriction that point O *must be* a point on the rigid body B or an extension of the rigid body. Finally, we note that if *any* of the following is true:

1. Point O is the mass center G so that $\vec{r}_{G/O} = \vec{0}$.
2. $\vec{a}_O = \vec{0}$ (i.e., O is fixed or moves with constant velocity).
3. $\vec{r}_{G/O}$ is parallel to $\vec{a}_O$.

then Eq. (7.36) becomes

$$\boxed{M_O = I_O \alpha_B,} \tag{7.37}$$

where we have used the scalar form to reflect the fact that the motion is planar. Equation (7.37) is particularly useful when bodies are undergoing fixed axis rotation about an axis through O.

Graphical interpretation of the equations of motion

Equation (7.29) has a graphical interpretation that can help us remember it and understand it physically. Unfortunately, this interpretation is also *very* easy to misapply, and therefore we need to be careful. With this warning in mind, it does provide a convenient way of applying and remembering Eq. (7.29).

We begin by referring to Fig. 7.5. The left side of the figure shows a rigid body

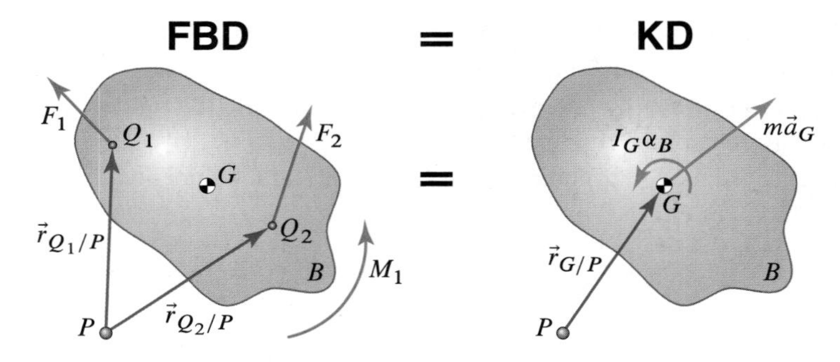

Figure 7.5. The free body diagram and kinetic diagram of a general rigid body. Equating them and writing the associated equations always give the correct equations of motion, i.e., Eqs. (7.1) and (7.29).

that is being acted upon by a number of external forces and moments—it is the FBD of the rigid body. The right side of the figure introduces a new diagram called a *kinetic diagram* (KD). The kinetic diagram *always* contains the $I_G\vec{\alpha}_B$ vector and the $m\vec{a}_G$ vector, which, although it has units of force, we will color green since these vectors originate from accelerations. Now, if we always draw the KD in this way, then by setting the FBD *equal to* the KD, we always recover Eqs. (7.1) and (7.29), which are the Newton-Euler equations for a rigid body. We can see this by noting that the left-hand side of Eq. (7.1) (i.e., $\vec{F}_R$) and the left-hand side of Eq. (7.29) (i.e., $\vec{M}_P$) are readily obtained from the FBD in Fig. 7.5. The right-hand side of Eq. (7.1) is simply $m\vec{a}_G$, which is what one obtains from the KD. The right-hand side of Eq. (7.29) comes from taking moments about P on the KD in Fig. 7.5. Thus, if we always draw the KD by including the $m\vec{a}_G$ vector (written in a convenient component system) and the $I_G\alpha_B$ vector, and we equate forces and moments on the FBD with forces and moments on the KD, we will always end up with the correct Newton-Euler equations for that rigid body.

Common Pitfall

The KD must be consistent with the kinematics. The positive directions of $I_G\alpha_B$ and $m\vec{a}_G$ on the KD must be consistent with the positive directions for α_B and $\vec{a}_G$ in the kinematic equations. If they are not, sign errors will end up polluting the problem solution.

End of Section Summary

In this section, we developed the Newton-Euler equations (equations of motion) for a rigid body. We began by showing that the translational equations are given by *Euler's first law*, which is

Eq. (7.1), p. 521

$$\vec{F} = m\vec{a}_G,$$

where $\vec{F}$ is the resultant of all *external* forces, m is the mass of the rigid body, and $\vec{a}_G$ is the inertial acceleration of its mass center (see Fig. 7.6).

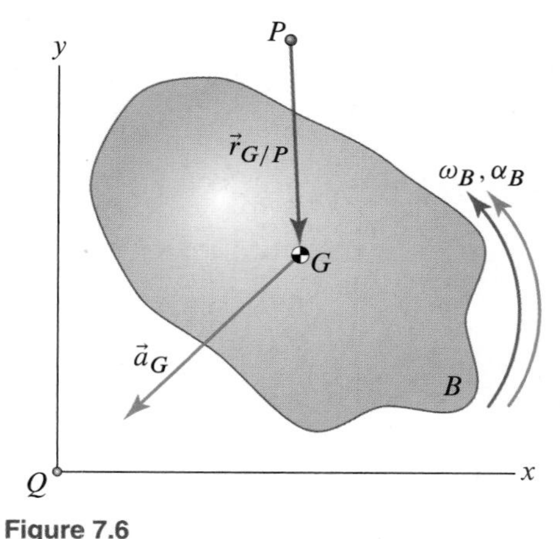

Figure 7.6
The relevant kinematic quantities for the Newton-Euler equations for a rigid body.

Bodies symmetric with respect to the plane of motion. For rigid bodies, we also need rotational equations of motion. Applying *Euler's second law*, i.e., the moment-angular momentum relationship for a system of particles, we were able to show that, *for a rigid body that is symmetric with respect to the plane of motion*, the most general form of the rotational equations of motion is given by (see Fig. 7.6)

Eq. (7.28), p. 524, and Eq. (7.29), p. 525

$$M_P = I_G\alpha_B + m(x_{G/P}a_{Gy} - y_{G/P}a_{Gx}),$$
$$\vec{M}_P = I_G\vec{\alpha}_B + \vec{r}_{G/P} \times m\vec{a}_G,$$

where the second equation is simply the vector form of the first, and

- M_P is the total moment about P in the z direction
- I_G is the mass moment of inertia of the body about its mass center G
- α_B is the angular acceleration of the body
- m is the total mass of the body
- $\vec{a}_G = a_{Gx}\,\hat{\imath} + a_{Gy}\,\hat{\jmath}$ is the acceleration of the mass center
- $\vec{r}_{G/P} = x_{G/P}\,\hat{\imath} + y_{G/P}\,\hat{\jmath}$ is the position of the mass center G relative to the moment center P

Now, in addition, if *any* one of the following conditions is true:

1. Point P is the mass center G, so that $\vec{r}_{G/P} = \vec{0}$.
2. $\vec{a}_G = \vec{0}$ (i.e., G moves with constant velocity).
3. $\vec{r}_{G/P}$ is parallel to $\vec{a}_G$.

then Eqs. (7.28) and (7.29) reduce to

Eq. (7.30), p. 525

$$M_P = I_G\alpha_B.$$

If the moment center is a point O on the rigid body or an arbitrary extension of the rigid body, then an alternate form of Eq. (7.29) is (see Fig. 7.7)

Eq. (7.36), p. 525

$$\vec{M}_O = I_O\vec{\alpha}_B + \vec{r}_{G/O} \times m\vec{a}_O,$$

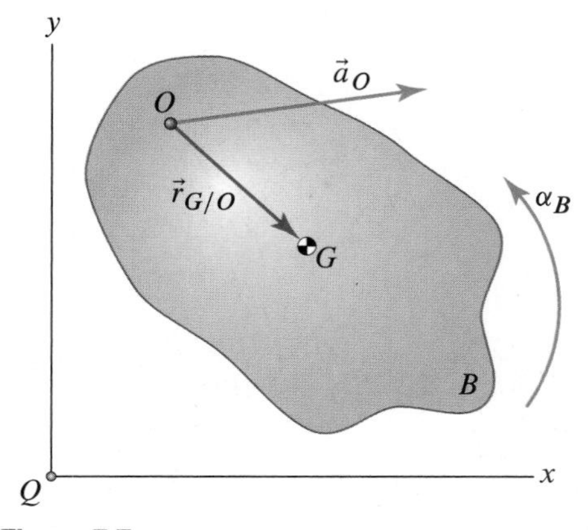

Figure 7.7
The relevant kinematic quantities for the rotational equations of motion of a rigid body when the moment center O is a point on the rigid body.

where I_O is the mass moment of inertia of the body about an axis perpendicular to the plane of motion passing through point O and $\vec{a}_O$ is the acceleration of point O.

If *any* one of the following is true:

1. Point O is the mass center G so that $\vec{r}_{G/O} = \vec{0}$.
2. $\vec{a}_O = \vec{0}$ (i.e., P moves with constant velocity).
3. $\vec{r}_{G/O}$ is parallel to $\vec{a}_O$.

then Eq. (7.36) becomes

Eq. (7.37), p. 526

$$M_O = I_O \alpha_B.$$

Graphical interpretation of the equations of motion. Referring to Fig. 7.8, there is a graphical/visual way of obtaining Eqs. (7.1) and (7.29). We begin by drawing the FBD of the rigid body, including all forces and moments, and then we draw the KD (kinetic diagram) of the rigid body, which includes the vectors $I_G \alpha_B$ and $m\vec{a}_G$. As shown in Fig. 7.8, we graphically equate these two diagrams, and we write the equations generated by that equation. In doing so, we automatically obtain Eqs. (7.1) and (7.29).

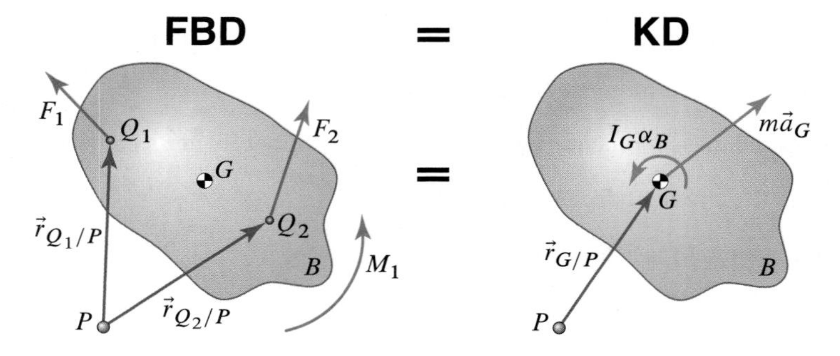

Figure 7.8. The free body diagram and kinetic diagram of a general rigid body. Equating them and writing the associated equations always give the correct equations of motion, i.e., Eqs. (7.1) and (7.29).

7.2 Newton-Euler Equations: Translation

Figure 7.9
A motorcycle and rider undergoing only translation. Only the tires are not translating, though we are neglecting their rotational inertia.

Our first application of the equations developed in Section 7.1 is to rigid bodies that translate in the plane of motion.

Figure 7.9 shows a motorcycle accelerating to the right. If we treat the motorcycle and rider as a single rigid body, and we ignore the inertia of the tires, then we can model the system using the FBD shown in Fig. 7.10, where the mass center of the system (i.e., rider plus motorcycle) is at G, the mass of the system is m, F_A is the friction force between the rear tire and the road, and the relevant dimensions are as shown. Since all points on the rigid body move in parallel straight horizontal lines, the system is only translating, which means that all angular velocities and angular accelerations are zero. Therefore, the first two Newton-Euler equations are the two scalar components of Eq. (7.1), which is (see Fig. 7.11)

$$\boxed{\vec{F} = m\vec{a}_G.} \tag{7.38}$$

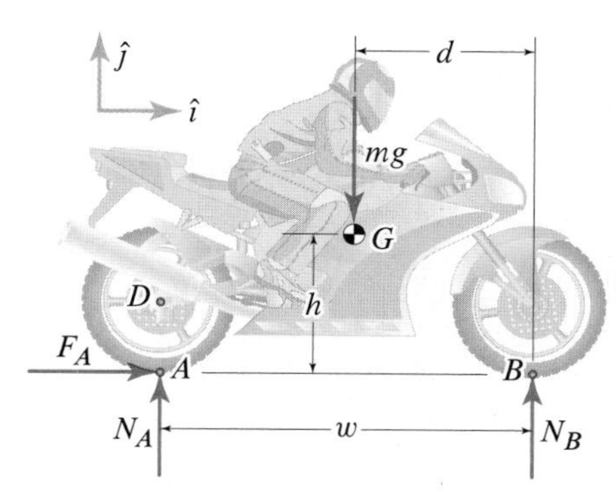

Figure 7.10
The FBD of the rider and motorcycle in Fig. 7.9. There is no friction force at B on the front tire since we are ignoring its rotational inertia.

Referring to Fig. 7.11, the third Newton-Euler equation is given by a rotational equation of motion. If the moment center is an arbitrary point P, then the rotational equation is given by

$$\boxed{\vec{M}_P = \vec{r}_{G/P} \times m\vec{a}_G,} \tag{7.39}$$

where we have used the fact that $\vec{\alpha}_B = \vec{0}$ for pure translation. If the moment center is an arbitrary point P *and* any one of the following conditions is true:

1. Point P is the mass center G, so that $\vec{r}_{G/P} = \vec{0}$.
2. $\vec{a}_G = \vec{0}$ (i.e., G is fixed or moves with constant velocity).
3. $\vec{r}_{G/P}$ is parallel to $\vec{a}_G$.

then, when written in scalar form, Eq. (7.39) reduces to

$$\boxed{M_P = 0.} \tag{7.40}$$

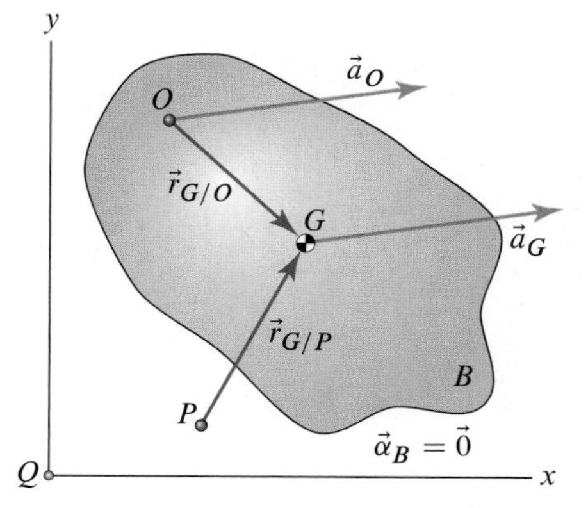

Figure 7.11
Illustration of the important points and kinematic quantities for a rigid body in translation for which $\vec{a}_G = \vec{a}_O$.

Referring to Fig. 7.11, if the moment center is a point O on the translating rigid body, then we can apply Eq. (7.36), which is

$$\boxed{\vec{M}_O = \vec{r}_{G/O} \times m\vec{a}_O,} \tag{7.41}$$

where we have again used $\vec{\alpha}_B = \vec{0}$. If, in addition to the moment center being on the translating rigid body, *any* of the following is true:

1. Point O is the mass center G so that $\vec{r}_{G/O} = \vec{0}$.
2. $\vec{a}_O = \vec{0}$ (i.e., O is fixed or moves with constant velocity).
3. $\vec{r}_{G/O}$ is parallel to $\vec{a}_O$.

then, when written in scalar form, Eq. (7.41) becomes

$$\boxed{M_O = 0.} \tag{7.42}$$

Note that the graphical interpretation of the equations of motion that equates the KD with the FBD is always applicable. See p. 526.

EXAMPLE 7.1 *Maximum Acceleration of a Motorcycle*

Figure 1

For the motorcycle and a rider in Fig. 1 with a combined weight of 650 lb, determine the largest possible acceleration, such that the motorcycle does not pop a wheelie, and find the minimum value of μ_s compatible with this motion. Treat the rider and motorcycle as a single rigid body, assume that the motorcycle is driven by its rear wheel, ignore the rotational inertia of the front wheel, and use the dimensions shown in Fig. 2.

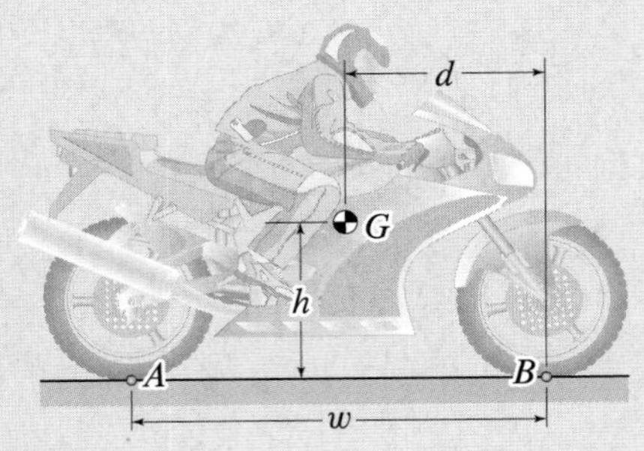

Figure 2
Relevant dimensions for the motorcycle shown in Fig. 1. The mass center of the system is at G, $w = 57.5$ in., $h = 22.5$ in., and $d = 27.8$ in..

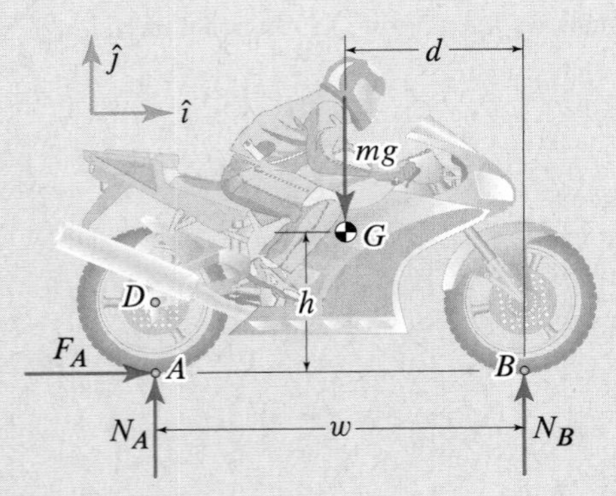

Figure 3
The FBD of the motorcycle in Fig. 2.

SOLUTION

Road Map & Modeling The FBD of the system is shown in Fig. 3. Since we are looking for the largest possible acceleration, we will solve for the condition in which the front tire is just about to leave the ground, and so $N_B = 0$ in our solution.

Governing Equations

Balance Principles Applying Eqs. (7.38) and (7.39), we obtain

$$\sum F_x: \quad F_A = m a_{Gx}, \tag{1}$$

$$\sum F_y: \quad N_A + N_B - mg = m a_{Gy}, \tag{2}$$

$$\sum M_A: \quad N_B w - mg(w-d) = m a_{Gy}(w-d) - m a_{Gx} h. \tag{3}$$

Force Laws Since we want the maximum acceleration and the minimum value of μ_s (i.e., slip must be impending), we have

$$N_B = 0 \quad \text{and} \quad F_A = \mu_s N_A. \tag{4}$$

Kinematic Equations The kinematic equations are

$$a_{Gx} = a_{\text{max}} \quad \text{and} \quad a_{Gy} = 0. \tag{5}$$

Computation Substituting Eqs. (4) and (5) into Eqs. (1)–(3), we obtain

$$\mu_s N_A = m a_{\text{max}}, \tag{6}$$

$$N_A - mg = 0, \tag{7}$$

$$-mg(w-d) = -m a_{\text{max}} h, \tag{8}$$

which are three equations for the unknowns N_A, μ_s, and a_{max}. Solving, we obtain $N_A = mg = 650$ lb and

$$\mu_s = \frac{w-d}{h} = 1.320, \tag{9}$$

$$a_{\text{max}} = \left(\frac{w-d}{h}\right) g = 42.50\ \text{ft/s}^2. \tag{10}$$

Discussion & Verification The dimensions of all three final answers are correct in that N_A has the dimensions of force, μ_s is dimensionless, and a_{max} has the dimensions of acceleration. A coefficient of static friction of 1.320 may seem high. However, it is not unusual to achieve coefficients this high for good "sticky" tires on asphalt.

EXAMPLE 7.2 *Hang Angle of an Accelerating Disk*

The pin A, which is rigidly attached to the uniform disk of radius R, is accelerating to the right in the overhead track at a_0. If the mass of the disk is m, determine the constant angle θ at which the disk will hang with the prescribed motion. In addition, determine the angle that the reaction force at A on the disk forms with the vertical and plot it and θ as a function of a_0.

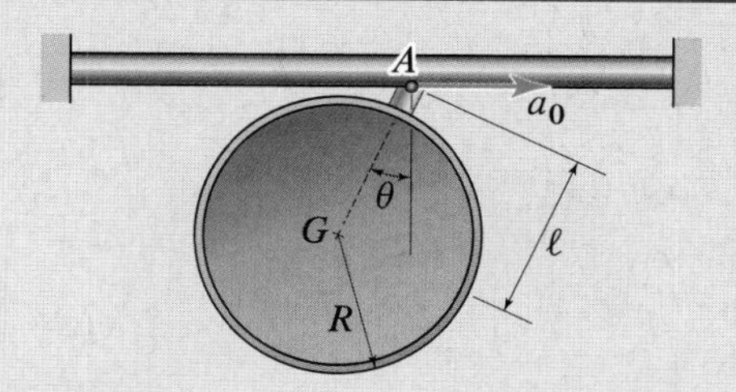

Figure 1

SOLUTION

Road Map & Modeling Since A is moving horizontally and the angle θ is constant, we conclude that the disk is only translating. Referring to Fig. 2, we can use the prescribed motion of the system and the Newton-Euler equations to determine the angles θ and β, the latter of which describes the orientation of the reaction force R_A at A.

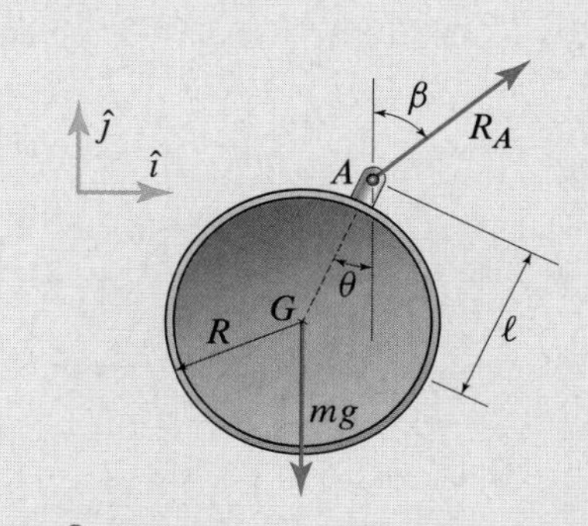

Figure 2
The FBD of the disk in Fig. 1 showing the angle β of the resultant force on the pin at A.

Governing Equations

Balance Principles Applying the Newton-Euler equations given by Eqs. (7.38) and (7.39), we obtain

$$\sum F_x: \quad R_A \sin\beta = m a_{Gx}, \tag{1}$$

$$\sum F_y: \quad R_A \cos\beta - mg = m a_{Gy}, \tag{2}$$

$$\sum M_A: \quad mg\ell \sin\theta = m a_{Gx} \ell \cos\theta - m a_{Gy} \ell \sin\theta. \tag{3}$$

Force Laws All forces are accounted for on the FBD.

Kinematic Equations Since the motion is only horizontal, we have that

$$a_{Gx} = a_0 \quad \text{and} \quad a_{Gy} = 0, \tag{4}$$

where a_0 is the given acceleration of point A.

Computation Substituting Eqs. (4) into Eqs. (1)–(3), we obtain

$$R_A \sin\beta = m a_0, \tag{5}$$

$$R_A \cos\beta - mg = 0, \tag{6}$$

$$g \sin\theta = a_0 \cos\theta. \tag{7}$$

From Eq. (7), we see that the hang angle of the disk is given by

$$\boxed{\theta = \tan^{-1}\left(\frac{a_0}{g}\right).} \tag{8}$$

To determine the angle β, we eliminate R_A from Eqs. (5) and (6) to obtain

$$\left(\frac{mg}{\cos\beta}\right)\sin\beta = m a_0 \quad \Rightarrow \quad \boxed{\beta = \tan^{-1}\left(\frac{a_0}{g}\right).} \tag{9}$$

A plot of θ and β as functions of a_0/g can be seen in Fig. 3.

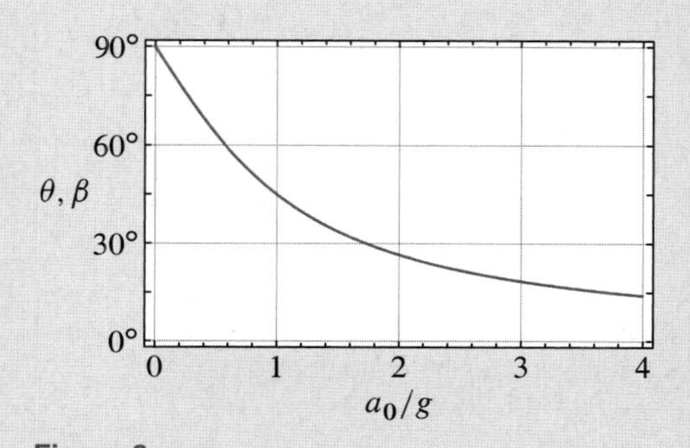

Figure 3
Plot of θ and β as functions of a_0/g. The two curves are identical.

Discussion & Verification We first note that the dimensions of the results in Eqs. (8) and (9) are as they should be, that is, the arguments of the inverse tangent functions are dimensionless, as they should be. The results in Eqs. (8) and (9), which are plotted in Fig. 3, tell us that the reaction force on the disk at the pin A is always aligned with the line AG. This seems reasonable since there should not be a moment about the mass center G if the disk does not have an angular acceleration.

EXAMPLE 7.3 *Translation: Slipping Versus Tipping with Friction*

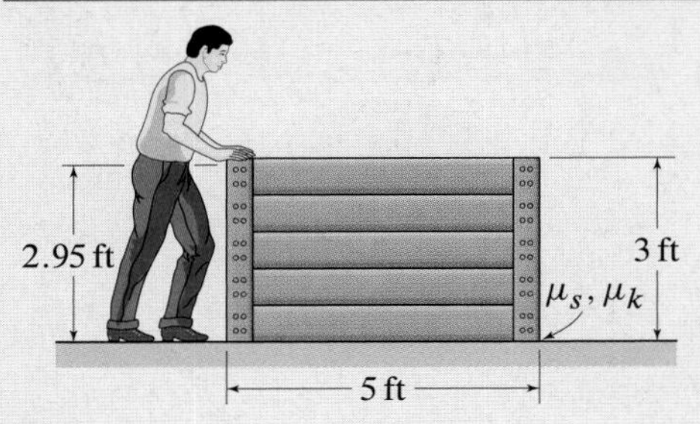

Figure 1
A person pushing a crate across a level surface.

A uniform flat crate is pushed with a constant horizontal force of 95 lb across a rough surface (Fig. 1). The force is applied 2.95 ft above the floor, and the crate is 5 ft long, 3 ft high, and weighs 120 lb. The coefficients of static and kinetic friction between the crate and the surface are $\mu_s = 0.4$ and $\mu_k = 0.35$, respectively. Verify that the crate slips and does not tip, and determine its acceleration.

SOLUTION

Road Map & Modeling The FBD of the crate is shown in Fig. 2, where P is the pushing

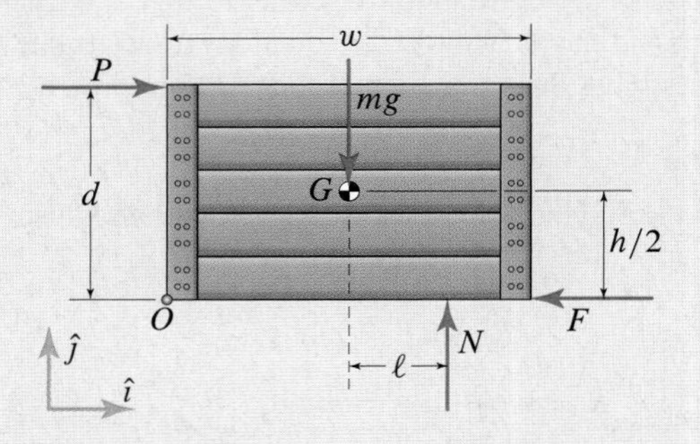

Figure 2. FBD of the crate shown in Fig. 1. The mass center is at G.

force applied at a height d above the floor, F is the friction force, and N is the equivalent normal force on the crate due to the ground. The normal force N is that point force that is equivalent to the actual distributed normal force on the bottom of the crate. Since we don't know the exact form of that distributed force, we place N at an unknown position ℓ, which will be determined as part of the solution.* In addition, w, h, and mg are the crate's width, height, and weight, respectively. We are to verify that the crate does not tip, so we will first solve the problem by assuming just that. During the verification, we will discuss what to look for if the crate were to tip.

> **Helpful Information**
>
> **How do we know the crate slips?** We can answer this question by solving the statics problem associated with the FBD in Fig. 2. Writing the equilibrium equations, we see that Eqs. (1)–(3) are still valid, except that all accelerations are now zero. Solving these modified versions of Eqs. (1)–(3), we obtain
>
> $$N = mg = 120.0\,\text{lb},$$
> $$F = P = 95.00\,\text{lb},$$
> $$\ell = \frac{dP}{mg} = 2.335\,\text{ft}.$$
>
> For the crate to slip, $F \geq \mu_s N$, or
>
> $$F = 95.00\,\text{lb} \overset{?}{\geq} (0.4)(120.0\,\text{lb}) = 48.00\,\text{lb} = \mu_s N.$$
>
> We can see that F is, in fact, greater than $\mu_s N$, and so the crate does slip.

Governing Equations

Balance Principles Applying Eqs. (7.1) and (7.30), the Newton-Euler equations for the FBD in Fig. 2 are

$$\sum F_x: \quad P - F = ma_{Gx}, \tag{1}$$

$$\sum F_y: \quad N - mg = ma_{Gy}, \tag{2}$$

$$\sum M_G: \quad N\ell - P(d - h/2) - Fh/2 = I_G\alpha_c, \tag{3}$$

where a_{Gx} and a_{Gy} are the x and y components of the acceleration of the mass center G, respectively, α_c is the crate's angular acceleration, and I_G is the crate's mass moment of inertia, which is given by

$$I_G = \tfrac{1}{12}m(w^2 + h^2). \tag{4}$$

Force Laws The friction force can be related to N using the Coulomb law for sliding friction, which is

$$F = \mu_k N. \tag{5}$$

* To review this idea, see Sections 5.2 and 9.1 of M. E. Plesha, G. L. Gray, and F. Costanzo, *Engineering Mechanics: Statics*, McGraw-Hill Publishing, Chicago, 2010.

Kinematic Equations Letting ω_c denote the crate's angular velocity, we can relate the acceleration of G to O using $\vec{a}_G = \vec{a}_O + \vec{\alpha}_c \times \vec{r}_{G/O} - \omega_c^2 \vec{r}_{G/O}$, and since we are assuming that the crate slips and does not tip, we have that $\vec{a}_O$ is only in the x direction. Therefore, we can write

$$\omega_c = 0, \quad \alpha_c = 0, \quad \text{and} \quad a_{Gy} = 0. \tag{6}$$

Computation Plugging Eqs. (4)–(6) into Eqs. (1)–(3), we obtain the following three equations for the unknowns N, a_{Gx}, and ℓ:

$$P - \mu_k N = m a_{Gx}, \tag{7}$$

$$N - mg = 0, \tag{8}$$

$$N\ell - P(d - h/2) - h\mu_k N/2 = 0. \tag{9}$$

Solving Eqs. (7)–(9), we obtain

$$N = mg = 120.0\,\text{lb}, \tag{10}$$

$$a_{Gx} = P/m - \mu_k g = 14.22\,\text{ft/s}^2, \tag{11}$$

$$\ell = \tfrac{1}{2}h\mu_k + \frac{P}{mg}(d - h/2) = 1.673\,\text{ft}, \tag{12}$$

where we have used the given data to obtain the numerical results.

We have the acceleration of the crate, but we need to verify that it doesn't tip. The key is that the equivalent normal force N needs to be located *within* the crate; that is, it can't be located outside the right or left edges of the crate. The idea behind this criterion is that we have assumed that the crate *does not* tip, and so our solution must be compatible with that assumption. A normal force outside the boundaries of the crate would mean that a wider base (all other parameters being the same) would be required to prevent tipping. With this said, since $\ell \leq w/2$ (i.e., $1.673\,\text{ft} \leq 2.5\,\text{ft}$), the crate does not tip.

Discussion & Verification

- The dimensions of the solutions in Eqs. (10)–(12) are all as they should be.
- It is reasonable that a crate with the given dimensions would not tip under the given circumstances.
- Although it is hard to know whether $0.44g$ is reasonable for a_{Gx}, it is certainly true that its direction is as expected.

A Closer Look

- Figure 3 shows how ℓ (i.e., the distance of N to the right of the centerline of the crate) varies as P (the applied load) is increased from 0 to 200 lb. As the statics solution tells us, before P reaches $\mu_s N$, the crate does not even move, and so we get the *dark red* preslip curve. Once the crate starts to slip, there is a small sudden drop in ℓ since there is a small sudden drop in the friction force as it goes from $\mu_s N$ to $\mu_k N$, and we get the *dark green* postslip curve. The *red* lines indicate those values of ℓ and P at which the normal force reaches the edge of the crate; at this point, the crate would also start to tip.
- We assumed that the crate slipped, but did not tip. We were then able to verify that assumption. It is important to realize that we can assume any motion we like and that the correctness of our assumption can always be verified using our solution. For example, under the given conditions, if we assumed that the crate tips and slips, we would discover that an impossible motion is obtained, and we would then be able to rule out that possibility. We will explore this possibility and others in the exercises.

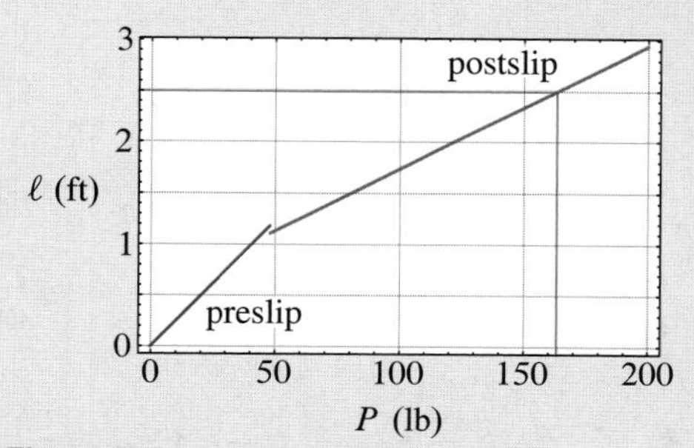

Figure 3
The normal force offset ℓ as a function of the applied load P.

Problems

Figure P7.1

Problem 7.1

The roadster weighs 2750 lb, its mass is evenly distributed between its front and rear wheels, and it accelerates from 0 to 60 mph in 7.0 s. If the acceleration is uniform and if the rear wheels do not slip, determine the forces on each of the front and rear wheels due to the pavement. Also determine the minimum coefficient of static friction compatible with this motion. Assume that the mass is evenly distributed between the right and left sides of the car, neglect the rotational inertia of the front wheels, and assume that the front wheels roll freely.

Problem 7.2

The conveyor is moving the cans at a constant speed $v_0 = 18\,\text{ft/s}$ when, to proceed to the next step in packaging, the cans are transferred onto a stationary surface at A. If each can weighs 0.95 lb, $w = 2.71$ in., $h = 5$ in., and $\mu_k = 0.3$ between the cans and the stationary surface, determine the time and distance it takes for each can to stop. In addition, show that the cans don't tip. Treat each can as a uniform circular cylinder.

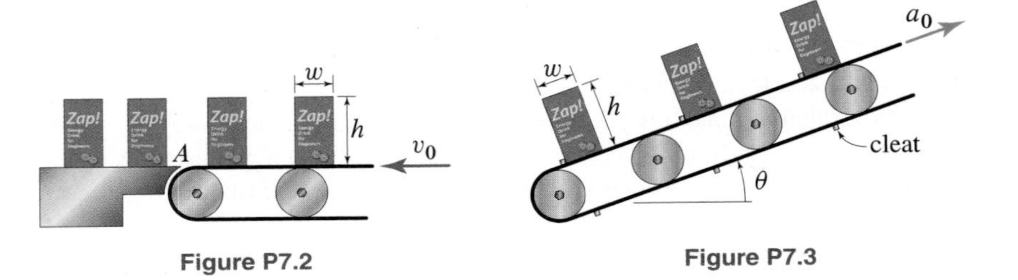

Figure P7.2

Figure P7.3

Problem 7.3

Determine the maximum acceleration a_0 of the conveyor so that the cans do not tip over the cleats. The cleats completely prevent slipping, but are not tall enough to dynamically influence tipping. Treat each can as a uniform circular cylinder of mass m.

Figure P7.4

Problem 7.4

The file cabinet, which weighs 230 lb, is being pushed to the right with a horizontal force of 70 lb, which is applied a distance h from the floor. If the mass center G of the file cabinet is $d = 24$ in. from the floor and the width of the file cabinet is $w = 15$ in., determine the maximum height h at which the file cabinet can be pushed so that it does not tip, and determine the corresponding acceleration of the file cabinet. Assume that friction between the file cabinet and the floor is negligible.

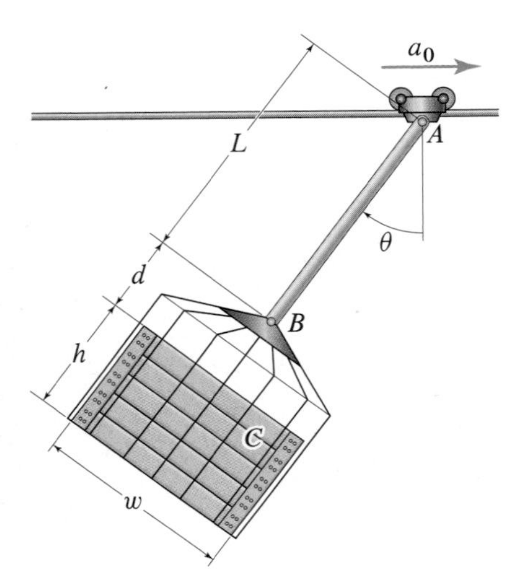

Figure P7.5 and P7.6

Problems 7.5 and 7.6

The uniform slender bar AB has a weight $W_{AB} = 150$ lb while the crate's weight is $W_C = 500$ lb. The bar AB is rigidly attached to the cage containing the crate. Neglect the mass of the cage, and assume that the mass of the crate is uniformly distributed. Furthermore, let $L = 8.5$ ft, $d = 2.5$ ft, $h = 4$ ft, and $w = 6$ ft.

Problem 7.5 If the trolley is accelerating with $a_0 = 11\,\text{ft/s}^2$, determine θ so that the bar-crate system translates with the trolley.

Problem 7.6 If the bar-crate system is translating with the trolley so that $\theta = 26°$, determine the acceleration a_0 of the trolley.

Problem 7.7

A person is pushing a lawnmower of mass $m = 38$ kg and with $h = 0.75$ m, $d = 0.25$ m, $\ell_A = 0.28$ m, and $\ell_B = 0.36$ m. Assuming that the force exerted on the lawnmower by the person is completely horizontal and that the mass center of the lawnmower is at G, and neglecting the rotational inertia of the wheels, determine the minimum value of this force that causes the rear wheels (labeled A) to lift off the ground. In addition, determine the corresponding acceleration of the mower.

Figure P7.7 and P7.8

Problem 7.8

The mower is self-propelled through its rear wheels. If a person were to apply a purely horizontal force, would this force help or hinder the rear wheels' contribution to the forward motion of the mower? That is, would the rear wheels slip less easily or more easily?

Problem 7.9

The stationary wheel loader is lifting the load at B vertically with acceleration $a_0 = 10\,\text{ft/s}^2$. If the weight of the load B is 10,000 lb, the weight of the wheel loader is 33,000 lb, $\ell = 118$ in., $h = 90$ in., and $d = 27$ in., determine the reactions on each of its four wheels if the wheel loader is laterally symmetric. In addition, compute the static load on each of the four wheels (i.e., with $a_0 = 0$) and compare it with the corresponding dynamic load. The center of mass of the wheel loader is at A.

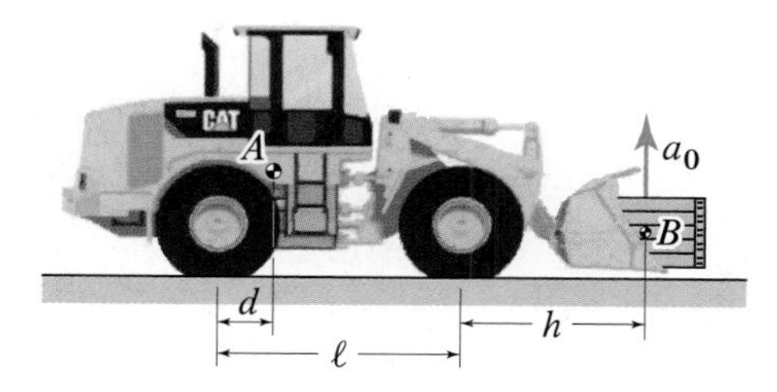

Figure P7.9 and P7.10

Problem 7.10

The stationary wheel loader is lifting the load at B vertically with acceleration a_0. If the weight of the load B is 10,000 lb, the weight of the wheel loader is 33,000 lb, $\ell = 118$ in., $h = 90$ in., and $d = 27$ in., determine the largest value of a_0 for which the reaction on each of the rear wheels does not drop below 1000 lb. The center of mass of the wheel loader is at A.

Problem 7.11

The door is at rest when the man M starts pushing it to the right with a constant horizontal force of 40 lb. If the weight of the door is 300 lb, its width and height are both $\ell = 10$ ft, its center of mass G is at its geometric center, and the man pushes at a height $h = 4$ ft, determine the speed of the door after it has moved 10 ft and determine the reactions at the rollers A and B.

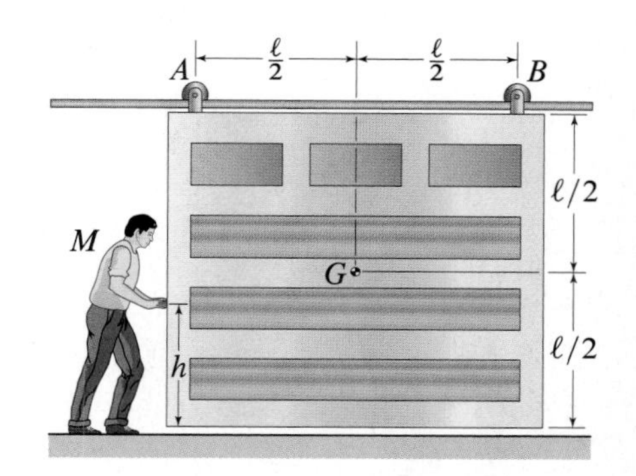

Figure P7.11 and P7.12

Problem 7.12

The 300 lb door is at rest when the man M starts pushing it to the right. If the width and height of the door are both $\ell = 10$ ft, its center of mass G is at its geometric center, and the man pushes at a height $h = 4$ ft, determine the horizontal force with which he must push if he is to move the door 10 ft to the right in 3 s, and determine the reactions at the rollers A and B.

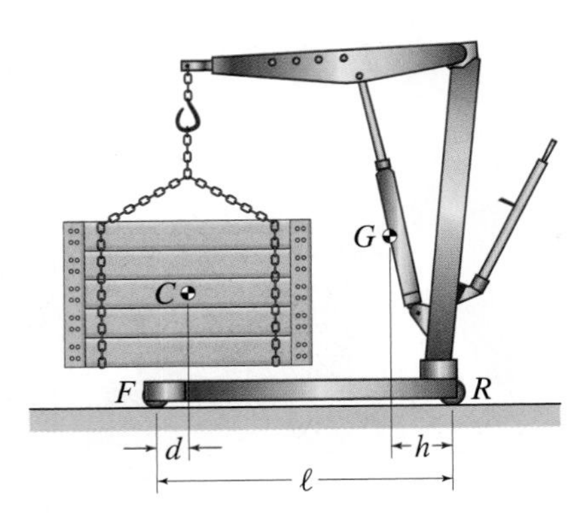

Figure P7.13 and P7.14

Problem 7.13

The portable hydraulic crane with mass $m_G = 60\,\text{kg}$ is stationary as it lifts the crate with mass $m_C = 110\,\text{kg}$ vertically with acceleration $a_C = 2.5\,\text{m/s}^2$. The mass center of the crate is at C and the mass center of the crane is at G. If $d = 0.145\,\text{m}$, $\ell = 1.3\,\text{m}$, and $h = 0.27\,\text{m}$, determine the reaction on each of the two front F and the two rear R wheels.

Problem 7.14

The portable hydraulic crane with mass $m_G = 60\,\text{kg}$ is stationary as it lifts the crate with mass $m_C = 110\,\text{kg}$ vertically with acceleration a_C. The mass center of the crate is at C and the mass center of the crane is at G. If $d = 0.145\,\text{m}$, $\ell = 1.3\,\text{m}$, and $h = 0.27\,\text{m}$, determine the maximum value of the acceleration a_C, such that no wheel reaction exceeds 400 N. Assume that the crane is laterally symmetric so that the loads on the front wheels are equal and the loads on the rear wheels are equal.

Problems 7.15 and 7.16

During takeoff, the thrust from each of the two engines of a Boeing 737 is $T = 95\,\text{kN}$. The mass center of the airplane is located at G, the horizontal distance between the main landing gear and G is $d = 5.15\,\text{m}$, the horizontal distance between the front landing gear and the main landing gear is $\ell = 11.1\,\text{m}$, and the height of the mass center above the ground is $h = 2.36\,\text{m}$. The mass of the plane is $m = 74{,}200\,\text{kg}$, and the thrust from each engine is a distance $\delta = 1.15\,\text{m}$ above the ground.

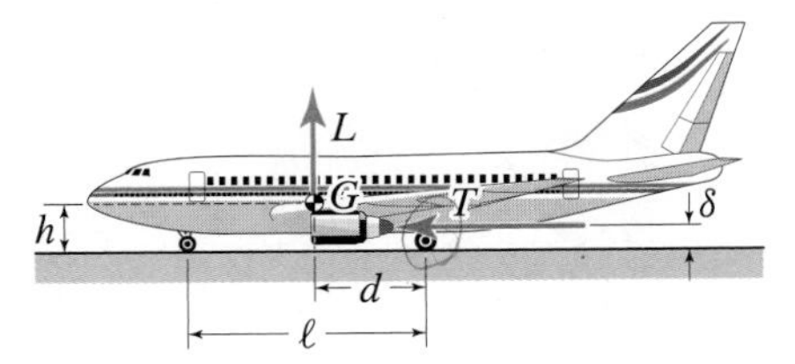

Figure P7.15 and P7.16

Problem 7.15 As the airplane begins its takeoff, the lift L from its wings can be neglected. Determine the normal reaction on the front landing gear and each of the two wheels of the main landing gear at the start of takeoff.

Problem 7.16 Assuming that the wings are the only surfaces generating lift during takeoff, determine the lift L generated by each wing when the normal force on the main landing gear is 25% of its initial takeoff value. The line of action of L goes through G.

Figure P7.17 and P7.18

Problems 7.17 and 7.18

A file cabinet weighing 230 lb is being pushed to the right with a horizontal force P applied a distance h from the floor. The width of the file cabinet is $w = 15\,\text{in.}$, its mass center G is a distance $d = 2\,\text{ft}$ above the floor, and static friction is insufficient to prevent slipping between the cabinet and the floor.

Problem 7.17 If $P = 70\,\text{lb}$ and the coefficient of kinetic friction between the cabinet and the floor is $\mu_k = 0.28$, determine the maximum height h at which the cabinet can be pushed so that it does not tip over, and find the corresponding acceleration of the cabinet.

Problem 7.18 If the coefficient of kinetic friction between the file cabinet and the floor is $\mu_k = 0.15$ and the horizontal force P is applied at a height $h = 42\,\text{in.}$, determine the maximum value of P that can be applied so that the cabinet does not tip over, and find the corresponding acceleration of the cabinet.

Problem 7.19

A conveyor belt must accelerate the cans from rest to $v = 18\,\text{ft/s}$ as quickly as possible. Treating each can as a uniform circular cylinder weighing 1.1 lb, find the minimum possible time to reach v so that the cans do not tip or slip on the conveyer. Assume that the acceleration is uniform, and use $w = 2.71$ in., $h = 5$ in., and $\mu_s = 0.5$.

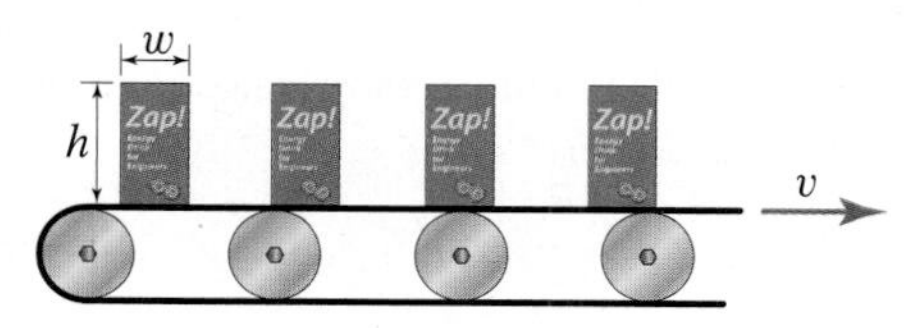

Figure P7.19

Problem 7.20

The uniform slender bar AB, with mass $m_{AB} = 75\,\text{kg}$ and length $L = 4.5\,\text{m}$, is pin-connected at A to a trolley accelerating with $a_0 = 3\,\text{m/s}^2$ along a horizontal rail. A crate with uniformly distributed mass $m_C = 250\,\text{kg}$, height $h = 1.5\,\text{m}$, and width $w = 2\,\text{m}$ is contained in a cage with negligible mass that is pin-connected to AB at B. The distance between B and the top of the crate is $d = 0.75\,\text{m}$. Determine the angles ϕ and θ so that the bar and the crate translate with the trolley.

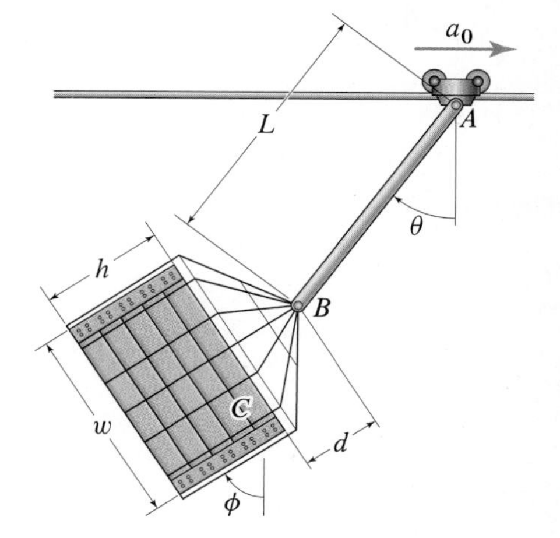

Figure P7.20

Problem 7.21

The system shown lies in the vertical plane. The trolley A is moving to the right with a constant acceleration a_A. Attached to the trolley is a rope AB of negligible mass. Attached to the end of the rope is a thin uniform bar BC of length L and mass m. When the trolley A is accelerating at constant a_A, the angles θ and ϕ are both *constant*. Determine these two constant angles in this steady state as functions of one or more of the given quantities (i.e., L, m, and a_A). Note that $\cos(\frac{\pi}{2} - x) = \sin x$.

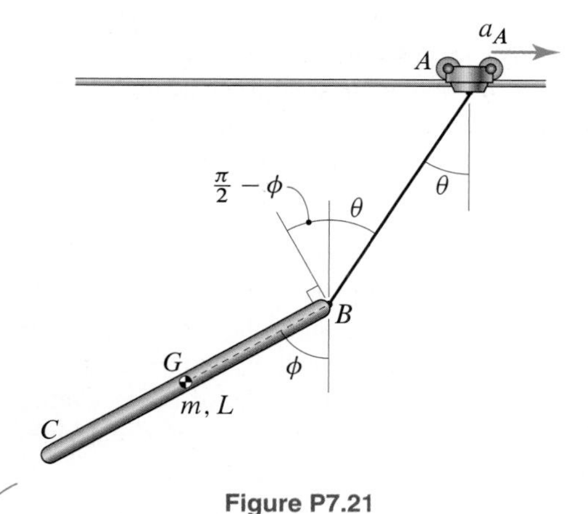

Figure P7.21

Problem 7.22

The stationary excavator is vertically lifting the load at A with acceleration $a_0 = 2.5\,\text{m/s}^2$. If $\ell = 4\,\text{m}$, $w = 3.65\,\text{m}$, $d = 0.82\,\text{m}$, the mass of the load at A is $m_A = 8300\,\text{kg}$, and the total mass of the excavator is $m_G = 20{,}000\,\text{kg}$, determine the equivalent normal force acting on each of the two tracks and its location relative to the rear of the track δ. The center of mass of the excavator is at G.

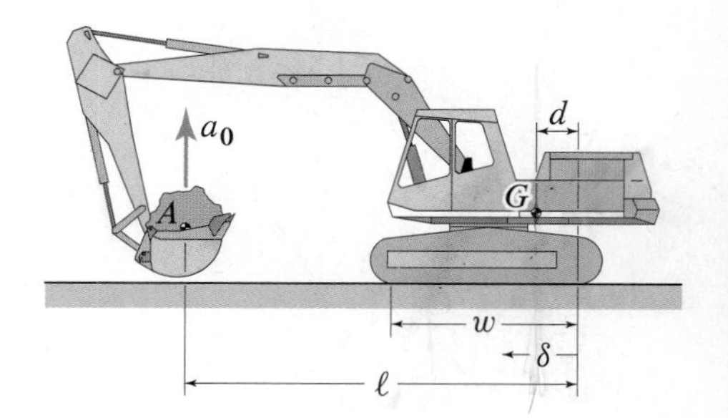

Figure P7.22

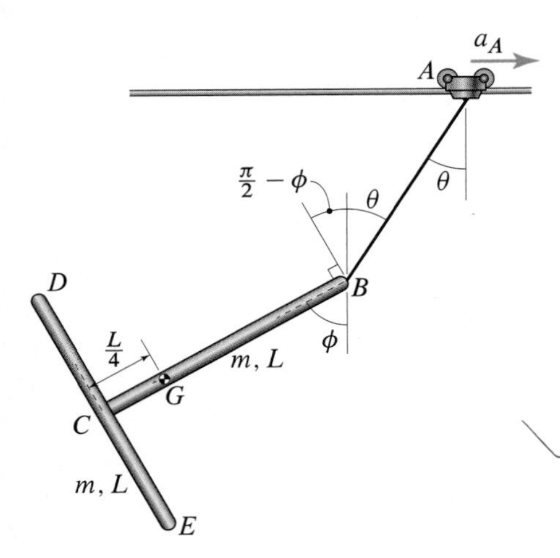

Figure P7.23

Problem 7.23

The system shown lies in the vertical plane. The trolley A is moving to the right with a constant acceleration a_A. Attached to the trolley is a rope AB of negligible mass. Attached to the end of the rope is a T-bar consisting of two thin uniform bars BC and DE, each of which has length L and mass m (bar BC is attached at the midpoint of bar DE). When the trolley A is accelerating at constant a_A, the angles θ and ϕ are both *constant*. Determine these two constant angles in this steady state. Note that $\cos(\frac{\pi}{2} - x) = \sin x$ and the mass center of the T-bar is at G.

Problem 7.24

The 3300 lb front-wheel-drive car whose mass center is at A is pulling a 4300 lb trailer whose mass center is at B. The car and trailer start from rest and accelerate uniformly to 60 mph in 18 s. Determine the forces on all tires, as well as the total force acting on the car due to the trailer. In addition, determine the friction required so that the wheels of the car do not slip. Assume that the car and trailer are laterally symmetric and that the rotational inertia of the wheels is negligible. Note that the mass center of the trailer is directly above the axle of the rear wheel.

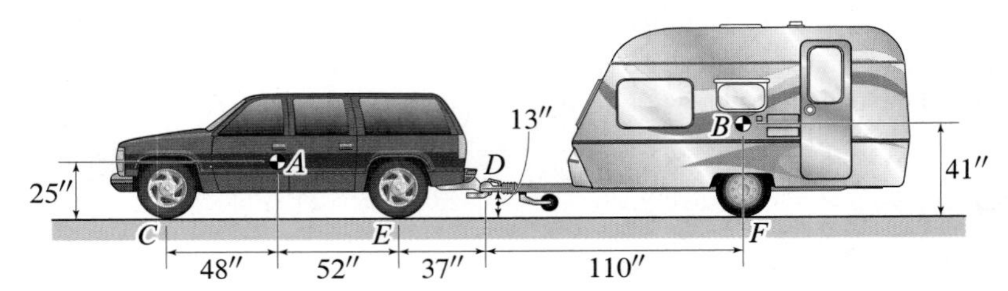

Figure P7.24 and P7.25

Problem 7.25

The 3300 lb front-wheel-drive car, which is pulling a 4300 lb trailer, is traveling 60 mph and applies its brakes to come to a stop. Assuming that all four wheels of the car assist in the braking and that $\mu_s = 0.85$, determine the minimum possible stopping distance, and find the forces on all tires, as well as the total force acting on the car due to the trailer. Assume that the car and trailer are laterally symmetric. Note that the mass center of the trailer is directly above the axle of the rear wheel.

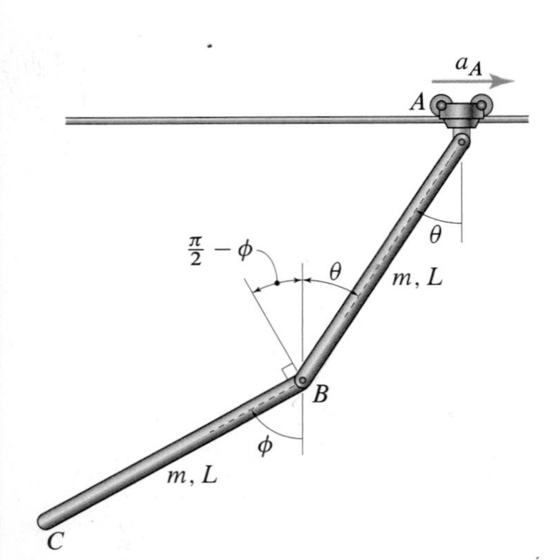

Figure P7.26

Problem 7.26

The system shown lies in the vertical plane. The trolley A is moving to the right with a constant acceleration a_A. Attached to the trolley by a pin is a thin uniform bar AB of mass m and length L. Attached to the end of the bar AB by a pin is a thin uniform bar BC of mass m and length L. When the trolley A is accelerating at constant a_A, the angles θ and ϕ are both *constant*. Determine these two constant angles in this steady state as functions of one or more of the given quantities (i.e., L, m, and a_A). Note that $\cos(\frac{\pi}{2} - x) = \sin x$.

7.3 Newton-Euler Equations: Rotation About a Fixed Axis

We now apply the equations developed in Section 7.1 to the motion of rigid bodies that are rotating about a fixed axis.

Consider, for example, the paper cutter shown in Fig. 7.12. The axis about which the arm rotates is fixed, and the equations we derived in Section 7.1 can be easily specialized to deal with systems like this one. Modeling the arm as a uniform slender bar of mass m and length L, and neglecting all friction and other moments (e.g., torsional springs) at the axis of rotation, we obtain the model and FBD shown in Fig. 7.13. As usual, the first two Newton-Euler equations are the two scalar components of Eq. (7.1), which is

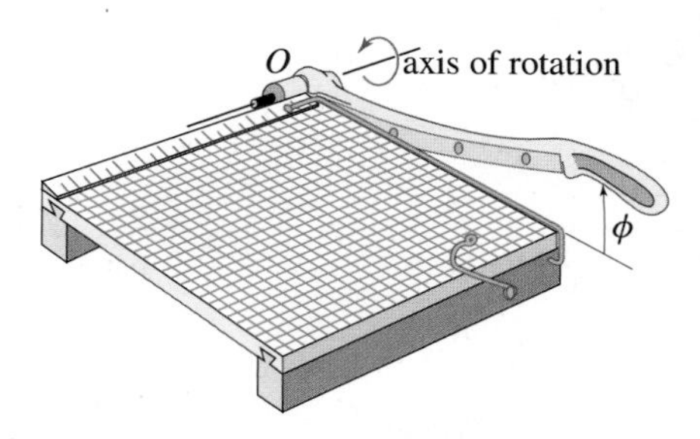

Figure 7.12
A paper cutter. The arm containing the cutting blade rotates about a fixed axis at O. The opening angle of the arm is defined by the angle ϕ.

$$\boxed{\vec{F} = m\vec{a}_G.} \tag{7.43}$$

Referring to Fig. 7.14, for rotation about a fixed axis through O, we can see that,

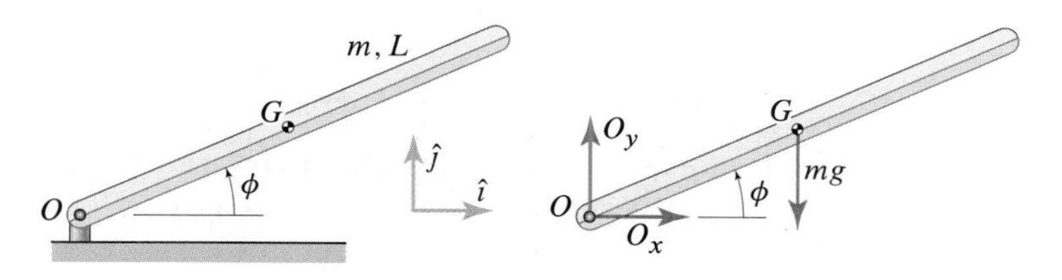

Figure 7.13. A model (left) of the cutting arm of the cutting board in Fig. 7.12, along with the FBD of that model (right).

since G is moving in a circle centered at O, the acceleration of G can be written as

$$\vec{a}_G = \vec{\alpha}_B \times \vec{r}_{G/O} - \omega_B^2 \vec{r}_{G/O}. \tag{7.44}$$

Equation (7.44) implies that, for a rotation about a fixed axis, $\vec{a}_G$ is *completely* determined by the angular velocity and angular acceleration.

Referring again to Fig. 7.14, the third Newton-Euler equation is given by a rotational equation of motion. If the moment center is an arbitrary point P, then the rotational equation is given by

$$\boxed{\vec{M}_P = I_G \vec{\alpha}_B + \vec{r}_{G/P} \times m\vec{a}_G,} \tag{7.45}$$

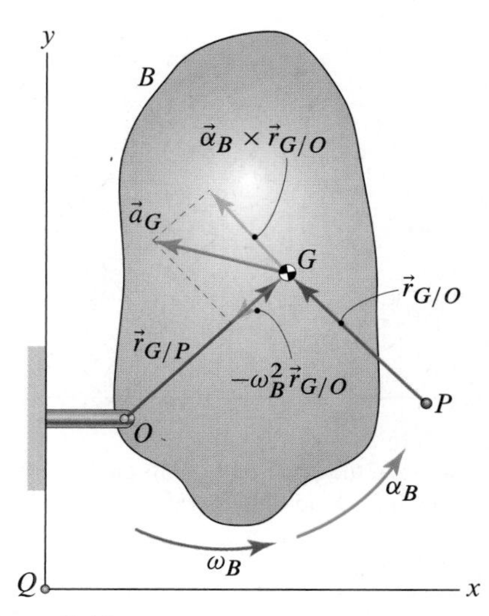

Figure 7.14
Illustration of the important points and kinematic quantities for a rigid body in translation.

where again, the acceleration of G is determined by Eq. (7.44). If the moment center is an arbitrary point P *and* any one of the following conditions is true:

1. Point P is the mass center G, so that $\vec{r}_{G/P} = \vec{0}$.

2. $\vec{a}_G = \vec{0}$ (i.e., G is fixed or moves with constant velocity).

3. $\vec{r}_{G/P}$ is parallel to $\vec{a}_G$.

then, when written in scalar form, Eq. (7.45) reduces to

$$\boxed{M_P = I_G \alpha_B.} \tag{7.46}$$

When a rigid body is rotating about a fixed axis through point O, then we typically want to use that axis as the moment center. In this case, we can apply Eq. (7.37), which is

$$\boxed{M_O = I_O \alpha_B,} \tag{7.47}$$

where we have used the fact that $\vec{a}_O = \vec{0}$ in this case.

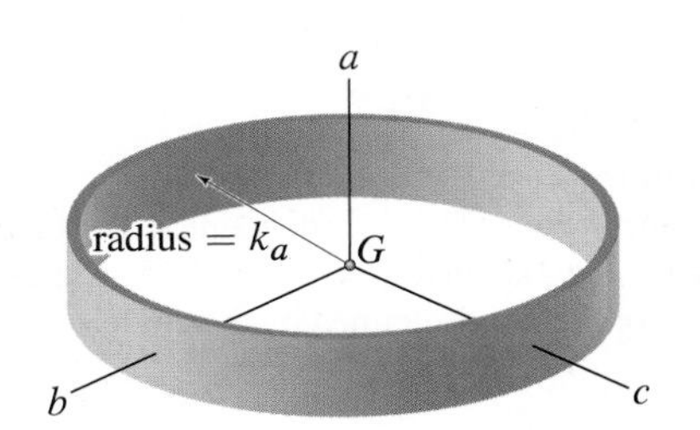

Figure 7.15
A thin ring whose mass is all a distance k_a from the a axis. The radius of gyration of this object with respect to the a axis would be k_a since its mass moment of inertia would be $I_a = mk_a^2$.

Radius of gyration

In some handbooks and other references, the property of mass moment of inertia is described in terms of the *radius of gyration*. The radius of gyration k_a, with respect to an axis a, is defined in terms of the mass moment of inertia about that axis as

$$k_a = \sqrt{\frac{I_a}{m}}, \tag{7.48}$$

where m is the mass of the body in question. Therefore, if a body of mass m had all its mass concentrated at a distance k_a from the axis a, its mass moment of inertia would be I_a (see Fig. 7.15).* The radius of gyration has the dimension of length.

End of Section Summary

Referring to Fig. 7.16, for a body in fixed axis rotation about a point, the three Newton-Euler equations of motion consist of the two scalar components of

Eq. (7.43), p. 539

$$\vec{F} = m\vec{a}_G,$$

along with a moment equation. If the moment center P is arbitrary, then the moment equation is given by

Eqs. (7.45) and (7.46), p. 539

$$\vec{M}_P = I_G \vec{\alpha}_B + \vec{r}_{G/P} \times m\vec{a}_G,$$
$$M_P = I_G \alpha_B,$$

where the second equation applies when $\vec{r}_{G/P} \times m\vec{a}_G = \vec{0}$. If, as is typical with fixed-axis rotation, the moment center O is on the axis of rotation, then the moment equation is

Eq. (7.47), p. 540

$$M_O = I_O \alpha_B.$$

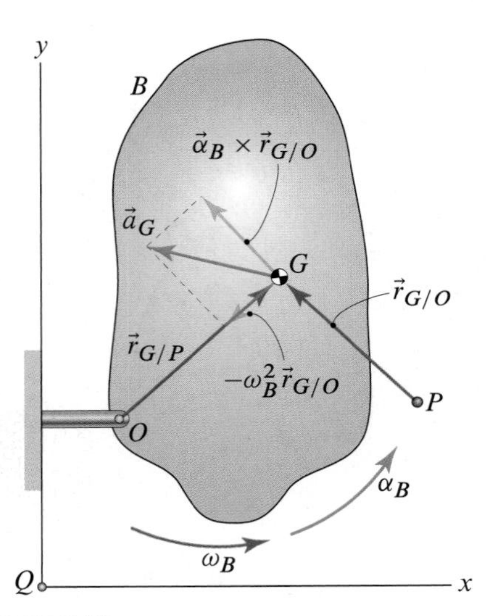

Figure 7.16
Illustration of the important points and kinematic quantities for a rigid body in translation.

*The radius of gyration can also be interpreted in terms of the statistical measure known as *standard deviation*. In fact, the radius of gyration can be thought of as the "standard deviation of the mass distribution."

EXAMPLE 7.4 *Angular Speed of the Paper Cutter Arm When Released from Rest*

For the paper cutter model shown in Fig. 1, determine the angular speed with which the cutting arm will reach the horizontal position if it is released from rest at $\phi_0 = 30°$. As shown in Fig. 1, model the cutting arm as a uniform slender bar of length L and mass m. Neglect friction in the pin at O.

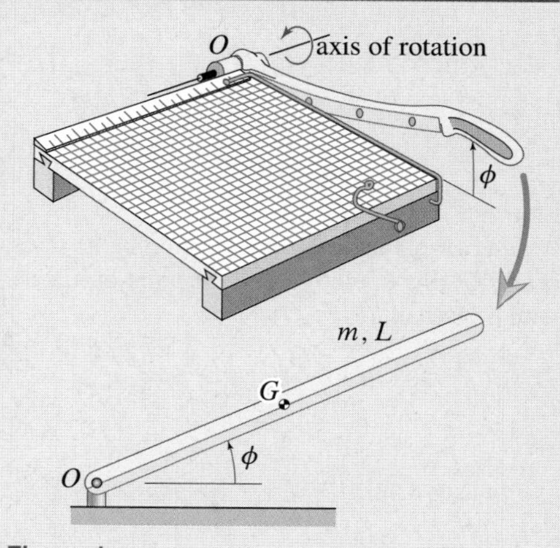

Figure 1
A paper cutter (top) and the slender bar model of the cutting arm of the cutter (bottom).

SOLUTION

Road Map & Modeling The FBD of the cutting arm at an arbitrary angle ϕ is shown in Fig. 2. Applying the Newton-Euler equations, we can find $\ddot{\phi}$ as a function of ϕ and then integrate that to find the angular speed $\dot{\phi}$ when the cutting arm reaches the horizontal position.

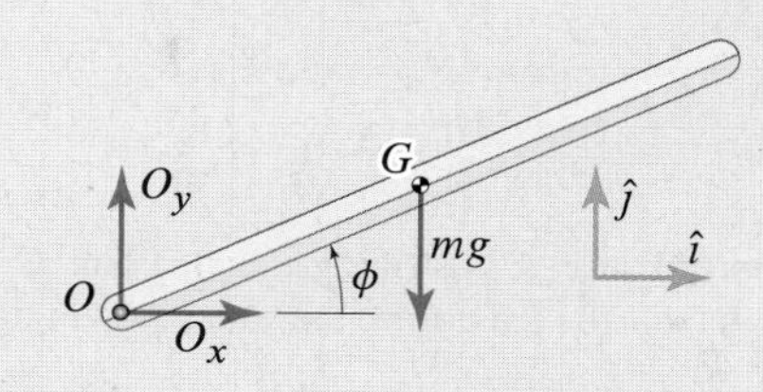

Figure 2
The FBD of the paper cutting arm model in Fig. 1 at an arbitrary angle ϕ.

Governing Equations

Balance Principles Applying Eqs. (7.43) and (7.47) to the FBD in Fig. 2, we obtain

$$\sum F_x: \quad O_x = ma_{Gx}, \tag{1}$$

$$\sum F_y: \quad O_y - mg = ma_{Gy}, \tag{2}$$

$$\sum M_O: \quad -mg\tfrac{L}{2}\cos\phi = I_O\alpha_{\text{arm}}, \tag{3}$$

where, via the parallel axis theorem, $I_O = \frac{1}{3}mL^2$.

Force Laws All forces are accounted for on the FBD.

Kinematic Equations From Eq. (7.44) and from the fact that $\alpha_{\text{arm}} = \ddot{\phi}$, the kinematic equations must be

$$a_{Gx}\,\hat{\imath} + a_{Gy}\,\hat{\jmath} = \ddot{\phi}\,\hat{k} \times \frac{L}{2}(\cos\phi\,\hat{\imath} + \sin\phi\,\hat{\jmath}) - \dot{\phi}^2\frac{L}{2}(\cos\phi\,\hat{\imath} + \sin\phi\,\hat{\jmath}), \tag{4}$$

or

$$a_{Gx} = -\frac{L}{2}(\ddot{\phi}\sin\phi + \dot{\phi}^2\cos\phi) \quad \text{and} \quad a_{Gy} = \frac{L}{2}(\ddot{\phi}\cos\phi - \dot{\phi}^2\sin\phi). \tag{5}$$

Computation Using Eqs. (5), along with $\alpha_{\text{arm}} = \ddot{\phi}$, Eqs. (1)–(3) become

$$O_x = -m\frac{L}{2}(\ddot{\phi}\sin\phi + \dot{\phi}^2\cos\phi), \tag{6}$$

$$O_y - mg = m\frac{L}{2}(\ddot{\phi}\cos\phi - \dot{\phi}^2\sin\phi), \tag{7}$$

$$-mg\tfrac{L}{2}\cos\phi = \tfrac{1}{3}mL^2\ddot{\phi}. \tag{8}$$

Rearranging Eq. (8) and cancelling an mL, we obtain

$$\ddot{\phi} = -\frac{3g}{2L}\cos\phi \quad \Rightarrow \quad \frac{d\dot{\phi}}{d\phi}\dot{\phi} = -\frac{3g}{2L}\cos\phi$$

$$\Rightarrow \quad \int_0^{\dot{\phi}_f} \dot{\phi}\,d\dot{\phi} = -\frac{3g}{2L}\int_{30°}^{0°} \cos\phi\,d\phi$$

$$\Rightarrow \quad \frac{1}{2}\dot{\phi}_f^2 = \frac{3g}{4L} \quad \Rightarrow \quad \boxed{\dot{\phi}_f = -\sqrt{\frac{3g}{2L}},} \tag{9}$$

where we have used the chain rule and integrated from the point of release to the horizontal position of the arm and we have chosen the minus sign when taking the square root since the arm is rotating clockwise.

Discussion & Verification The dimension of the final result in Eq. (9) is 1/time, as it should be.

EXAMPLE 7.5 The Sweet Spot of a Baseball Bat

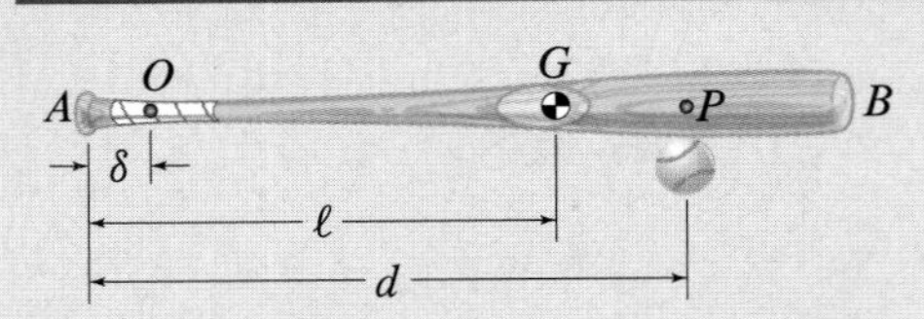

Figure 1
A baseball bat showing the location of its mass center G, pivot point O, and the point of impact with a baseball.

In baseball or softball, you may have experienced that feeling when the ball is hit "just right," that is, you can barely feel the bat hit the ball, but the ball goes *a long way*. This may happen when you hit the ball near the "sweet spot" of the bat. The sweet spot is thought to involve the vibrational modes of the bat, as well as a point called the *center of percussion*.* Figure 1 shows a ball hitting a bat at a distance d from the knob A when the batter has "choked up" a distance δ. Assuming that the batter swings at his or her wrists (i.e., the pivot point is O), determine the distance d at which the ball should be hit so that, no matter how large the force applied at P to the bat by the ball, the lateral force (i.e., perpendicular to the bat) at O is zero. That point P defines the *center of percussion,* and its location depends on the pivot point O. Assume the bat has mass m, its mass center is at G, and I_G is its mass moment of inertia. Use $\delta = 2$ in., and for a typical bat used in Major League Baseball whose stated weight is 32 oz and whose length is 34 in., $\ell = 22.5$ in., $m = 0.0630$ slug, and $I_G = 0.0413\,\text{slug}\cdot\text{ft}^2$.

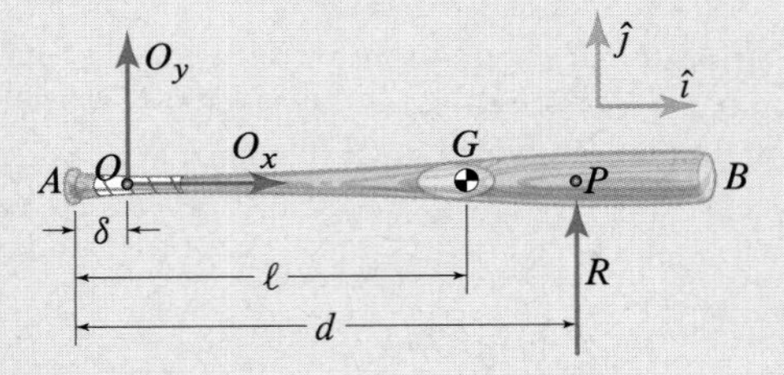

Figure 2
FBD of a bat as it is striking a ball. Point O is considered to be fixed since the batter is swinging at the wrists.

SOLUTION

Road Map & Modeling The FBD of the bat as it is striking the ball is shown in Fig. 2. We have neglected the moment that the batter applies to the bat because its effect is negligible next to that of the forces due to the collision of the ball against the bat. We have neglected the weight of the bat for the same reason. Since the goal is to find the distance d, such that the lateral reaction force O_y is zero, we will want to determine O_y as a function of that distance d.

Governing Equations

Balance Principles Applying Eqs. (7.43) and (7.46) to the FBD in Fig. 2 and summing moments about G, we obtain the following Newton-Euler equations:

$$\sum F_x: \quad O_x = ma_{Gx}, \tag{1}$$

$$\sum F_y: \quad O_y + R = ma_{Gy}, \tag{2}$$

$$\sum M_G: \quad R(d-\ell) - O_y(\ell-\delta) = I_G\alpha_{\text{bat}}, \tag{3}$$

where I_G is given.

Force Laws All forces are accounted for on the FBD.

Kinematic Equations Since the axis of the bat is parallel to the x axis, the x component of acceleration is simply the normal acceleration of G toward O, and the y component of acceleration is the tangential component, that is

$$a_{Gx} = -(\ell-\delta)\omega_{\text{bat}}^2 \quad \text{and} \quad a_{Gy} = (\ell-\delta)\alpha_{\text{bat}}, \tag{4}$$

where we are assuming positive directions for ω_{bat} and α_{bat}.

Computation Substituting Eqs. (4) into Eqs. (1)–(3), we obtain

$$O_x = -m(\ell-\delta)\omega_{\text{bat}}^2, \tag{5}$$

$$O_y + R = m(\ell-\delta)\alpha_{\text{bat}}, \tag{6}$$

$$R(d-\ell) - O_y(\ell-\delta) = I_G\alpha_{\text{bat}}. \tag{7}$$

Eliminating α_{bat} from Eqs. (6) and (7) and solving for O_y, we find that

$$O_y = \frac{mR(d-\ell)(\ell-\delta) - RI_G}{I_G + m(\ell-\delta)^2}. \tag{8}$$

* See R. Cross, "The Sweet Spot of a Baseball Bat," *American Journal of Physics*, **66**(9), 1998, pp. 772–779.

This is the O_y as a function of d that we desired. Since we want the distance d for which $O_y = 0$, we set this result equal to zero and then solve for the distance d to obtain

$$\boxed{d = \frac{I_G + m\ell(\ell - \delta)}{m(\ell - \delta)} = 2.259\,\text{ft} = 27.10\,\text{in.},} \tag{9}$$

where we have used $\delta = 2\,\text{in.} = 0.1667\,\text{ft}$, $\ell = 22.5\,\text{in.} = 1.875\,\text{ft}$, $m = 0.0630\,\text{slug}$, and $I_G = 0.0413\,\text{slug}\cdot\text{ft}^2$.

Discussion & Verification The dimension of the final result in Eq. (9) is length, as it should be. In addition, we see that the location of the sweet spot depends on the distance $\ell - \delta$ from the mass center to the pivot point, that is, it depends on the location of the pivot point and is not an inherent property of the bat.

A Closer Look How is the center of percussion useful? Knowing the location of the center of percussion is important in pendulum-type impact test machines, which are designed to measure the failure resistance of a material to an impulsive force. This is done by measuring the impact energy, which is the energy absorbed by the test specimen prior to failure. If the pendulum arm that strikes the specimen does so at the arm's center of percussion, the force transmitted to the frame of the machine during the test will be minimized.

EXAMPLE 7.6 Rotation About a Fixed Axis: Moments About a Fixed Point

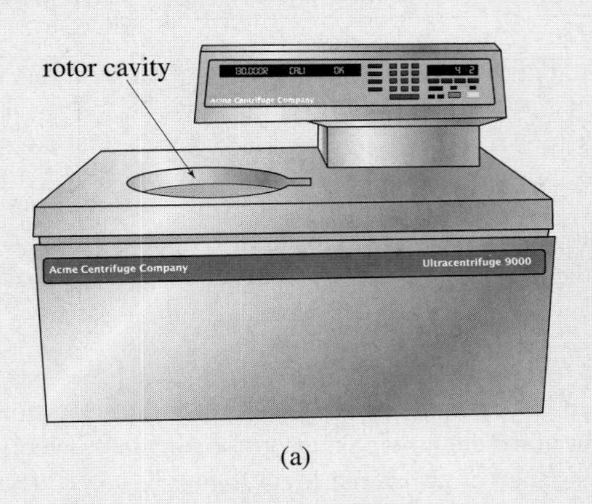

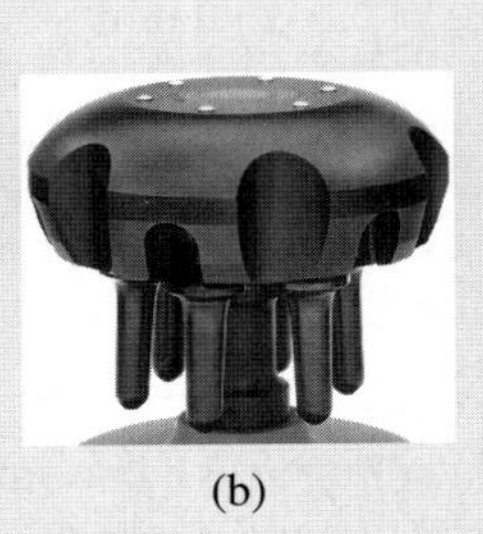

Figure 1. (a) A tabletop ultracentrifuge. (b) Swinging-bucket centrifuge rotor that holds six sample tubes.

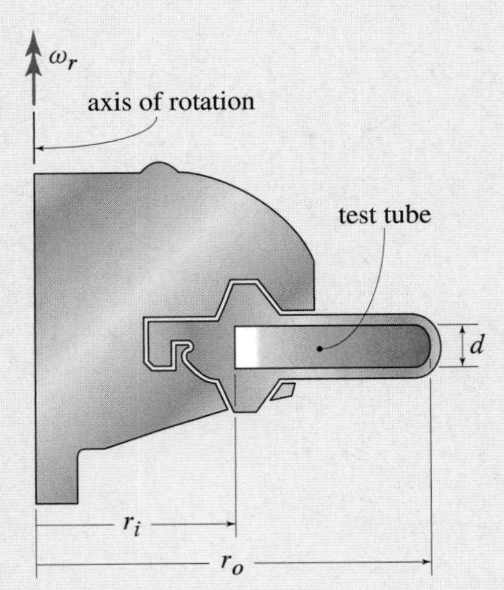

Figure 2
Cross-section of one-half of the centrifuge rotor shown in Fig. 1(b). $r_i = 63.1$ mm, $r_o = 120.5$ mm, $d = 11.0$ mm, and $\omega_r = 60{,}000$ rpm.

Centrifuges like the one shown in Fig. 1(a) can generate accelerations exceeding 1 million g. With the *swinging-bucket* rotor shown in Fig. 1(b), the centrifuge can spin at 60,000 rpm and achieve an acceleration of $485{,}000g$ at the ends of the buckets. As the rotor spins up, the buckets that hang from the bottom of the rotor swing up and eventually assume the nearly horizontal position shown in Fig. 2.*

(a) Determine the radial force and the moment parallel to the axis of rotation required to hold the test tube in place when the rotor is spinning at its maximum rated speed of 60,000 rpm.

(b) What are the implications of these loads on the rotor bearings?

Assume that the test tube and its contents (e.g., blood) can be modeled as a uniform circular cylinder with mass 10 g, and ignore gravity.

SOLUTION

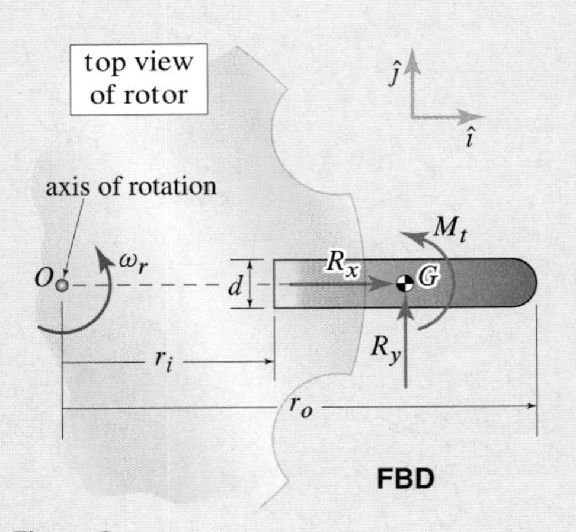

Figure 3
Top view of the FBD of the test tube in Fig. 2. The mass center of the test tube is at G.

Road Map & Modeling The forces shown on the FBD of the test tube in Fig. 3 form the *equivalent force-couple system* to the system of forces that are actually acting on the tube. The FBD could also show a force in the z direction, R_z, as well as moments in both the x and y directions, M_{Ox} and M_{Oy}, respectively. Summing forces in the z direction would simply tell us that $R_z = mg$. Since the body is symmetric with respect to the plane of motion, it follows that $M_{Ox} = M_{Oy} = 0$, and so those moment equations become part of a statics problem. In this case, we are only interested in R_x, R_y, and M_t, so we have chosen the FBD given in Fig. 3. We will approximate the test tube as a uniform circular cylinder, thus ignoring any motion of the fluid within the tube and the nonuniform shape and mass distribution of the tube. Since we *know* the motion of the test tube, we can find all the velocities and accelerations needed to use Eq. (7.43), as well as all the moment equations we have developed. Therefore, the forces and moments will fall right out of the Newton-Euler equations once the kinematics is included.

Governing Equations

Balance Principles Equating the FBD and KD of the test tube shown in Figs. 3 and 4, respectively (this is equivalent to applying Eqs. (7.43) and (7.45) to the FBD in Fig. 3),

* The final orientation cannot be exactly horizontal, but at these high accelerations, it is so close to horizontal that we are treating it as such.

the Newton-Euler equations for the test tube are

$$\sum F_x: \quad R_x = ma_{Gx}, \tag{1}$$

$$\sum F_y: \quad R_y = ma_{Gy}, \tag{2}$$

$$\sum M_O: \quad M_t + R_y\left(\frac{r_o + r_i}{2}\right) = I_G\alpha_r + m\left[x_{G/O}a_{Gy} - y_{G/O}a_{Gx}\right], \tag{3}$$

where I_G is the mass moment of inertia of the test tube and a_{Gx} and a_{Gy} are the x and y components of the acceleration of the mass center G, respectively. Modeling the test tube as a uniform circular cylinder, we calculate its mass moment of inertia as

$$\begin{aligned} I_G &= \tfrac{1}{12}m(3r^2 + h^2) = \tfrac{1}{12}(0.01\,\text{kg})\left[3(0.0055\,\text{m})^2 + (0.0574\,\text{m})^2\right] \\ &= 2.821\times10^{-6}\,\text{kg·m}^2, \end{aligned} \tag{4}$$

where we have used $m = 0.01$ kg, $r = d/2 = 0.0055$ m, and $h = r_o - r_i = 0.0574$ m.

Force Laws All forces are accounted for on the FBD.

Kinematic Equations As for the accelerations on the right-hand side of Eqs. (1)–(3), since we know the motion of the rotor, these are readily found by relating $\vec{a}_G$ to $\vec{a}_O$, where point O is on an arbitrary rigid body extension of the test tube. Doing this, we obtain

$$\vec{a}_G = \vec{a}_O + \vec{\alpha}_r \times \vec{r}_{G/O} - \omega_r^2\vec{r}_{G/O}, \tag{5}$$

in which we note that $\vec{a}_O = \vec{0}$ since it is on the axis of rotation and $\vec{\alpha}_r = \vec{0}$ since the rotor has reached its constant final speed. Substituting in $\omega_r = 60{,}000$ rpm $= 6283$ rad/s and $\vec{r}_{G/O} = (r_o + r_i)/2\,\hat{\imath} = 0.0918\,\text{m}\,\hat{\imath}$, we have

$$\vec{a}_G = -(6283\,\text{rad/s})^2(0.0918\,\text{m}\,\hat{\imath}) = (-3.624\times10^6\,\text{m/s}^2)\,\hat{\imath}. \tag{6}$$

Computation Substituting Eqs. (4), (6), $\vec{r}_{G/O}$, and $\alpha_r = 0$ into Eqs. (1)–(3), we obtain

$$R_x = -(0.01\,\text{kg})(3.624\times10^6\,\text{m/s}^2) = -36{,}240\,\text{N}, \tag{7}$$

$$R_y = 0\,\text{N}, \tag{8}$$

$$M_t + R_y\left(\frac{0.1205\,\text{m} + 0.0631\,\text{m}}{2}\right) = 0 \quad \Rightarrow \quad M_t = 0\,\text{N·m}, \tag{9}$$

where we have substituted $R_y = 0$ from Eq. (8) to obtain the final result in Eq. (9).

Discussion & Verification The force needed to keep the test tube in place is 36,240 N (8147 lb), even though the test tube only has a mass of 10 g (an average paper clip has a mass of about 1 g). Note that an equal and opposite force is acting on the rotor bearing and that force is rotating around the bearing 60,000 times per minute or 1000 times per second! Therefore, not only is the rotor subject to failure due to a huge load imbalance, but also it is subject to failure due to fatigue loading (see marginal note). This means that balancing the rotor is essential to safely operate the centrifuge.

A Closer Look When using kinetic diagrams to obtain the equations of motion for a rigid body, we are *always* applying Eq. (7.29) (which becomes Eq. (7.45) for fixed-axis rotation). So, for example, we could have written the moment equation as $M_O = I_O\alpha_r$ by using Eq. (7.47), which is equivalent to Eq. (7.29), but it is not applying the graphical idea of equating the FBD to the KD.

Also note that even though R_x is very large after the test tube has been spun up, that state may not be the critical one from a design perspective. For example, during spin-up, the angular velocity will not be at its maximum value, but the angular acceleration may be very large, which would make R_y and M_t nonzero. These will induce shear and bending loads that must also be accounted for.

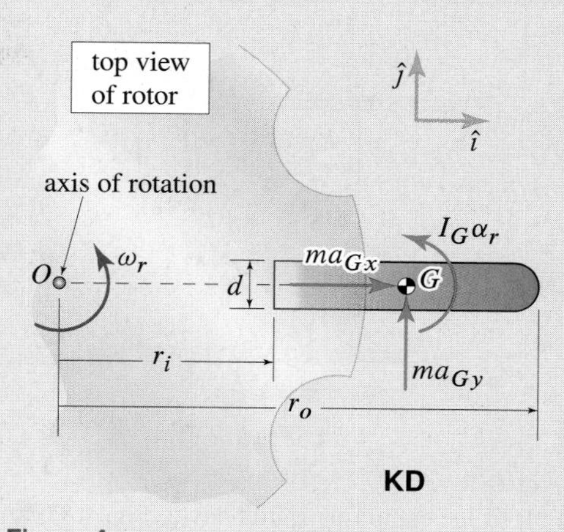

Figure 4
Top view of the KD of the test tube in Fig. 2. Equating this KD to the FBD in Fig. 3, we obtain the test tube's Newton-Euler equations.

Interesting Fact

Cyclic loading and fatigue. The fact that, under the given conditions, the rotor bearing experiences a cyclic load 1000 times per second means that it will quickly experience a large number of load cycles. It turns out that even a rather low stress can cause an object to break after millions of load cycles. The higher the stress, the smaller the number of cycles required. This mechanism of failure is called *fatigue*. Since the number of load cycles on the rotor bearing of a centrifuge grows quickly, even a small imbalance can cause failure due to fatigue. To learn more about fatigue, see W. D. Callister, Jr., *Materials Science and Engineering: An Introduction*, 7th ed., John Wiley & Sons, 2006.

EXAMPLE 7.7 Analysis of a Composite Rigid Body

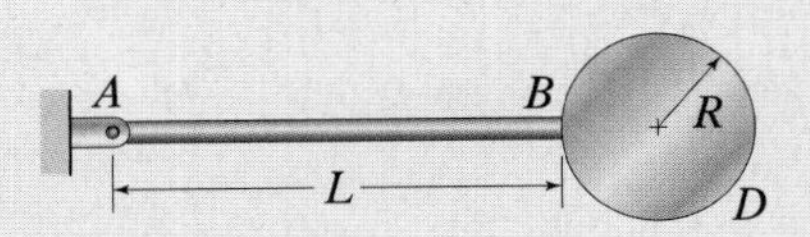

Figure 1

A composite body consisting of a uniform thin bar AB of length L and a uniform disk D of radius $R = L/5$ is pinned at A and lies in the vertical plane. The mass of the bar is $m_{AB} = m$, and the mass of the disk is $m_D = 3m$. If the system is released from rest in the position shown, determine its initial angular acceleration.

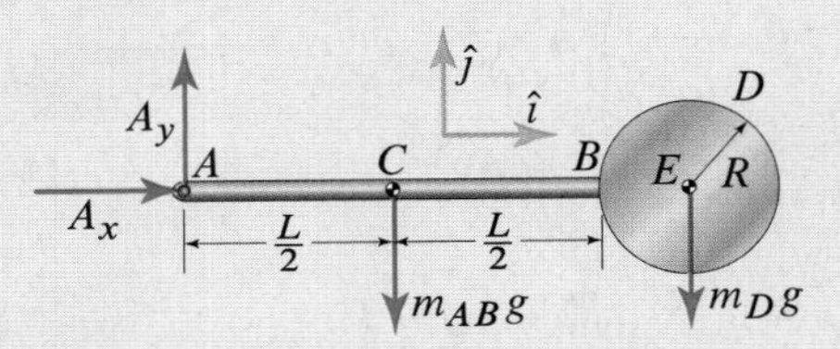

Figure 2
FBD of the composite rigid body in Fig. 1. Point C is the mass center of the thin bar and point E is the mass center of the disk.

SOLUTION

Road Map & Modeling We will demonstrate two ways to solve this problem. For both methods, the FBD of the system is as shown in Fig. 2. Since we are interested in the initial angular acceleration, we will apply the Newton-Euler equations to determine it. In the first solution method, we will sum moments about point A and apply Eq. (7.47). In the second solution method, we will sum moments about point A, but we will make use of the kinetic diagram (KD). We will, of course, obtain the same answer, but it will be informative to compare the methods.

First Solution Method

Governing Equations

Balance Principles Appying Eqs. (7.43) and (7.47) to the FBD in Fig. 2, we obtain

$$\sum F_x: \quad A_x = m_{AB} a_{Cx} + m_D a_{Ex}, \tag{1}$$

$$\sum F_y: \quad A_y - m_{AB} g - m_D g = m_{AB} a_{Cy} + m_D a_{Ey}, \tag{2}$$

$$\sum M_A: \quad -m_{AB} g \tfrac{L}{2} - m_D g (L + R) = I_A \alpha, \tag{3}$$

where α is the angular acceleration of the composite body and where the parallel axis theorem gives I_A for the composite body as

$$\begin{aligned} I_A &= (I_A)_{\text{bar}} + (I_A)_{\text{disk}} \\ &= \tfrac{1}{12} m_{AB} L^2 + m_{AB} \left(\tfrac{L}{2}\right)^2 + \tfrac{1}{2} m_D R^2 + m_D (L + R)^2 \\ &= \tfrac{1}{3} m_{AB} L^2 + m_D \left[\tfrac{1}{2} R^2 + (L + R)^2\right] && (4) \\ &= \tfrac{707}{150} m L^2, && (5) \end{aligned}$$

where we have used $m_{AB} = m$, $m_D = 3m$, and $R = L/5$ to obtain the final result.

Force Laws All forces are accounted for on the FBD.

Kinematic Equations Since Eq. (3) is sufficient to determine the initial angular acceleration of the composite body, there is no need to find $\vec{a}_C$ and $\vec{a}_E$. Therefore, no additional equations are needed for the kinematics.

Computation Substituting Eq. (5) into Eq. (3), we obtain

$$-\tfrac{41}{10} m g L = \tfrac{707}{150} m L^2 \alpha \quad \Rightarrow \quad \boxed{\vec{\alpha} = -\frac{615}{707} \frac{g}{L} \hat{k},} \tag{6}$$

where we have used $m_{AB} = m$, $m_D = 3m$, and $R = L/5$ in Eq. (3).

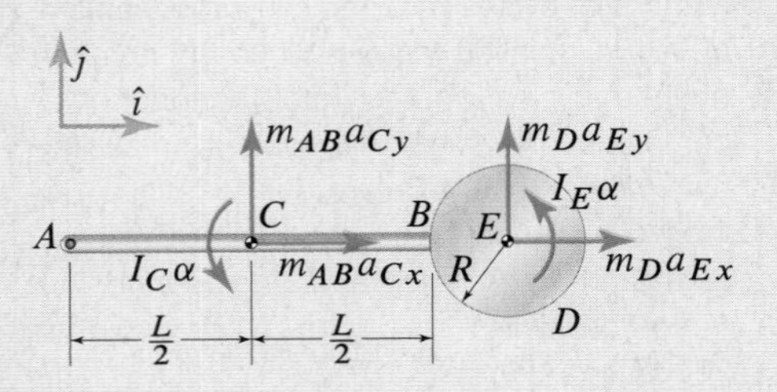

Figure 3
The KD for the composite body in Fig. 1. Note that, as usual, all unknown acceleration components have been drawn in their positive directions, even though we may intuitively see that the initial accelerations of the bar will not be in these directions. Our solution will account for this and will tell us when a component is negative.

Second Solution Method

Governing Equations

Balance Principles Equating the FBD in Fig. 2 to the KD in Fig. 3, we obtain the following Newton-Euler equations for the composite body:

$$\sum F_x: \quad A_x = m_{AB} a_{Cx} + m_D a_{Ex}, \tag{7}$$

$$\sum F_y: \quad A_y - m_{AB}g - m_D g = m_{AB}a_{Cy} + m_D a_{Ey}, \tag{8}$$

$$\sum M_A: \quad -m_{AB}g\tfrac{L}{2} - m_D g(L+R) = I_C\alpha + I_E\alpha + m_{AB}a_{Cy}\tfrac{L}{2} + m_D a_{Ey}(L+R), \tag{9}$$

where the mass moments of inertia are given by

$$I_C = \tfrac{1}{12}m_{AB}L^2 = \tfrac{1}{12}mL^2 \quad \text{and} \quad I_E = \tfrac{1}{2}m_D R^2 = \tfrac{3}{50}mL^2, \tag{10}$$

where we have used $m_{AB} = m$, $m_D = 3m$, and $R = L/5$.

Force Laws All forces are accounted for on the FBD.

Kinematic Equations With this second solution method, we need to find a_{Cy} and a_{Ey} since they appear in the moment equation. Since the composite body is released from rest, all velocities are zero. With this in mind, we can relate the accelerations of points C and E to the angular acceleration of the composite body as follows:

$$\vec{a}_C = \vec{\alpha} \times \vec{r}_{C/A} = \alpha\,\hat{k} \times \tfrac{L}{2}\,\hat{\imath} = \tfrac{L}{2}\alpha\,\hat{\jmath}, \tag{11}$$

$$\vec{a}_E = \vec{\alpha} \times \vec{r}_{E/A} = \alpha\,\hat{k} \times (L+R)\,\hat{\imath} = (L+R)\alpha\,\hat{\jmath} = \tfrac{6}{5}L\alpha\,\hat{\jmath}, \tag{12}$$

where we have used $R = L/5$.

Computation Substituting Eqs. (10)–(12) into Eq. (9) and using $m_{AB} = m$, $m_D = 3m$, and $R = L/5$, we obtain

$$-\tfrac{41}{10}mgL = \tfrac{707}{150}mL^2\alpha \quad \Rightarrow \quad \boxed{\vec{\alpha} = -\frac{615}{707}\frac{g}{L}\,\hat{k},} \tag{13}$$

which is identical to the result obtained in Eq. (6).

Discussion & Verification As expected, both solution methods give the same result. This gives us confidence that our final result is correct. With the first method, we were able to avoid computing the accelerations of points C and E in terms of the angular acceleration of the composite body. With the second method, we were able to avoid the computation of the overall mass moment of inertia using the parallel axis theorem. Which method you choose is really a matter of convenience.

A Closer Look The second solution method used in this example demonstrates the utility of the kinetic diagram when applied to composite rigid bodies. We will continue to see this in the rest of this chapter.

Problems

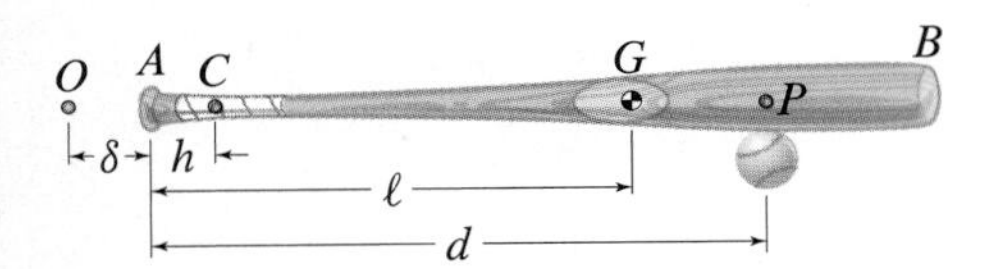

Figure P7.27

Problem 7.27

Following up on Example 7.5 on p. 542, now assume that the pivot point O is close to the center of mass of the batter so that it is a distance $\delta = 3$ in. from the knob of the bat at A. Determine the location of the center of percussion P relative to the knob at A, and show that the position of P is independent of the location of point C at which the batter grips the bat. Recall that P is the point at which the ball should be hit so that, no matter how large the force applied at P to the bat by the ball, the lateral force (i.e., perpendicular to the bat) felt by the batter at the grip C is zero. Assume the bat has mass m, its mass center is at G, and I_G is its mass moment of inertia. Evaluate your answer for a typical bat used in Major League Baseball whose weight is 32 oz and whose length is 34 in., $\ell = 22.5$ in., $m = 0.0630$ slug, and $I_G = 0.0413\,\text{slug}\cdot\text{ft}^2$. Ignore the weight force on the bat.

Problem 7.28

For the centrifuge rotor and test tube given in Example 7.6 on p. 544, assume that all the test tubes are locked into their horizontal position and that the rotor is uniformly accelerated from rest to 60,000 rpm in 9.5 min. Determine, as a function of time, the forces and moments on one of the test tubes during this spin-up phase of motion. Assume that each test tube and its contents can be modeled as a uniform circular cylinder with a mass of 10 g, and ignore gravity.

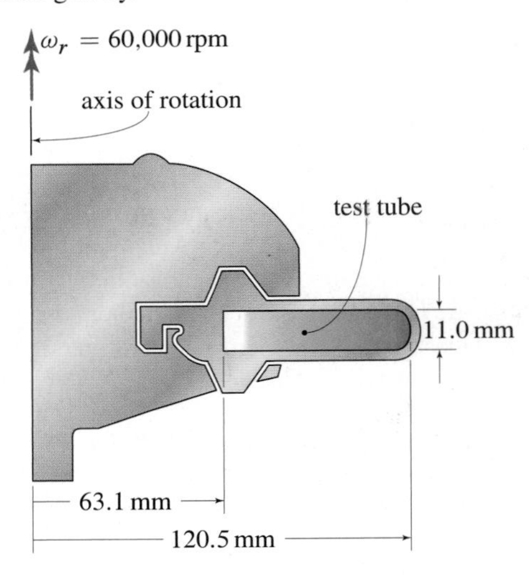

Figure P7.28

Problems 7.29 and 7.30

The spool is pinned at its center at O, about which it can spin freely. The radius of the spool is $R = 0.15$ m, its radius of gyration is $k_O = 0.11$ m, and the mass of the spool is $m_s = 5$ kg. The mass B is suspended from the periphery of the spool by a chain of negligible mass that moves over the spool without slip. The mass of B is $m_B = 7$ kg.

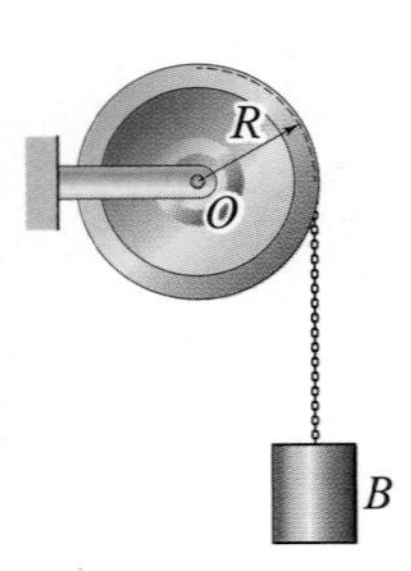

Figure P7.29 and P7.30

Problem 7.29 If the system is released from rest, determine the angular acceleration of the spool and the tension in the chain.

Problem 7.30 If the system is released from rest, determine the number of revolutions of the spool and the time it takes for mass B to achieve a speed of 10 m/s.

Problems 7.31 through 7.33

The uniform disk of radius $R = 0.8\,\text{ft}$ and weight $W = 20\,\text{lb}$ is pin-connected to the link AB and is pulled on its periphery by a force P via a rope that is wrapped around the disk. The coefficient of kinetic friction between the disk and the surface on which it sits is $\mu_k = 0.5$. Neglect the mass of link AB and assume that μ_s is insufficient to prevent slipping.

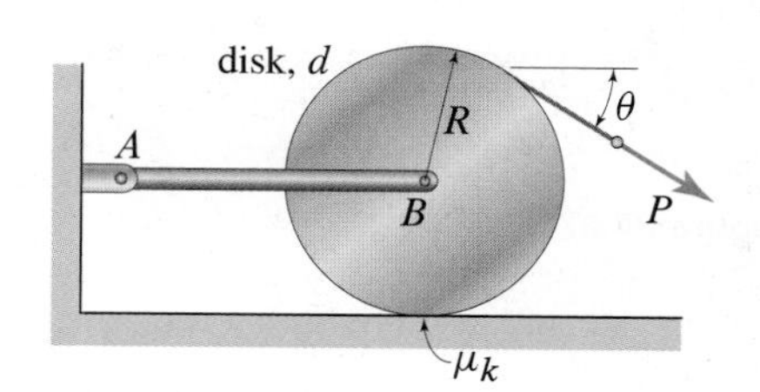

Figure P7.31–P7.33

Problem 7.31 For $P = 15\,\text{lb}$ and $\theta = 0°$, determine the angular acceleration of the disk α_d and the time it takes to achieve an angular velocity of $\omega_d = 35\,\text{rad/s}$, assuming that the disk starts from rest.

Problem 7.32 For $P = 25\,\text{lb}$ and $\theta = 90°$, determine the angular acceleration of the disk α_d and the number of revolutions for it to achieve an angular velocity of $\omega_d = 45\,\text{rad/s}$, assuming that the disk starts from rest.

Problem 7.33 For $P = 25\,\text{lb}$, determine the angular acceleration of the disk α_d as a function of the pull angle θ.

Problem 7.34

A classic balsa wood toy airplane is powered by a rubber band that winds up when its four-bladed propeller is rotated by hand. When released, the rubber band unwinds and the propeller starts spinning, thus propelling the plane around the room. Model each propeller blade as a slender rod of length $L = 6\,\text{cm}$ and mass $m = 2\,\text{g}$, and neglect air resistance. If we model the rubber band as a linear elastic torsional spring with constant $k_t = 6 \times 10^{-7}\,\text{N·m/rad}$, determine the angular speed of the propeller if it is initially wound up 20 revolutions.

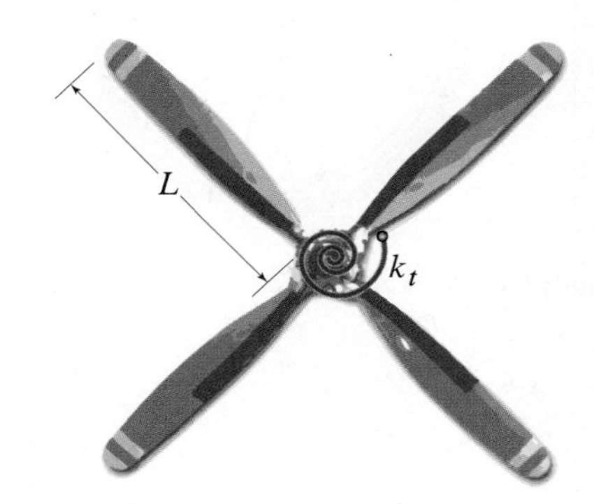

Figure P7.34

Problem 7.35

The driveway gate is hinged at its right end and is pushed open with a force P. Where should the force P be applied (i.e., where should A be located) so that the force acting on the hinge due to the gate always acts along a line parallel to the gate and in the plane of the gate during the entire time the gate is opening? Neglect the weight force acting on the gate, and model the gate as a uniform thin bar as shown below the photo.

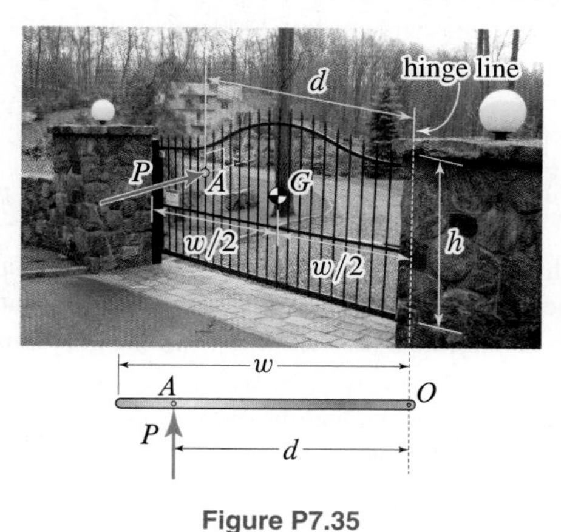

Figure P7.35

Problem 7.36

Assuming the helicopter is on the ground, and modeling each of its four blades as a slender rod with weight $W = 400\,\text{lb}$ and length $L = 24\,\text{ft}$, determine the constant moment that must be applied by the engine to the mast at O to spin the blades from rest to an angular speed of 289 rpm in 90 s.

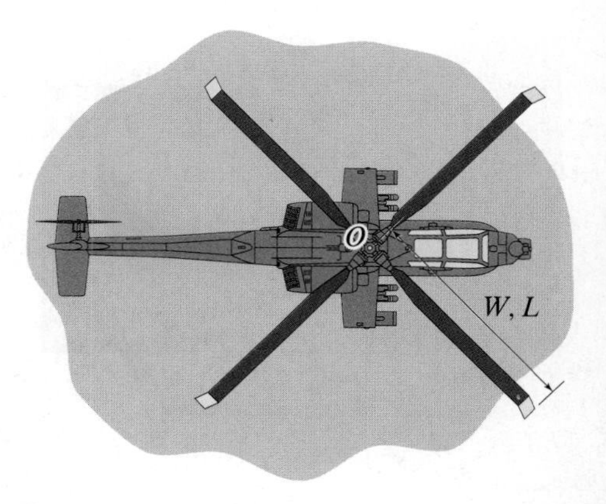

Figure P7.36

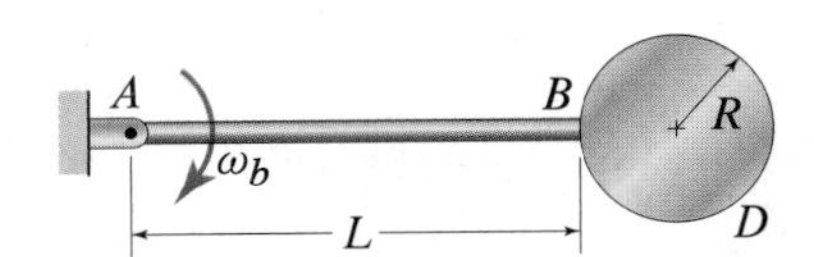

Figure P7.37

Problem 7.37

The composite body lies in the vertical plane and is rotating with angular speed $\omega_b = 11\,\text{rad/s}$ at the instant shown. The mass of the disk D is $m_b = 4\,\text{kg}$, the mass of the bar AB is $m_{AB} = 1.5\,\text{kg}$, the length of the bar AB is $L = 50\,\text{cm}$, and the radius of the disk is $R = 10\,\text{cm}$. At the instant shown, determine the angular acceleration of the composite body and the horizontal and vertical reactions on the body at A.

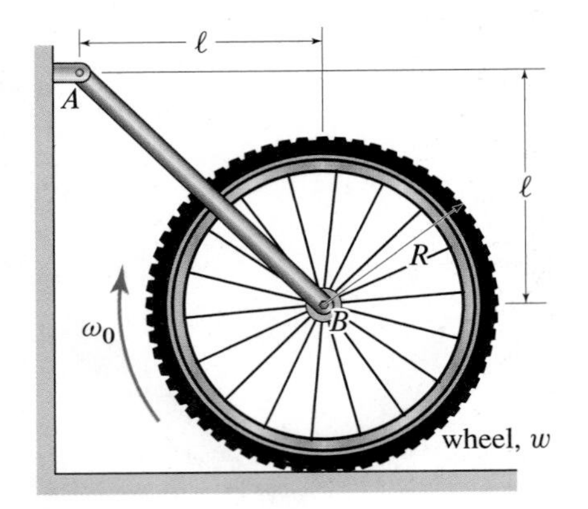

Figure P7.38–P7.40

Problems 7.38 through 7.40

The bar AB is pinned to a fixed support at A at one end and to the center of the bike wheel B at its other end. The bike wheel is spinning with angular speed $\omega_0 = 50\,\text{rad/s}$ when it is gently placed on the horizontal surface. The mass of the bike wheel is $m_w = 2.5\,\text{kg}$, its radius is $R = 33\,\text{cm}$, and its mass moment of inertia is $I_B = 0.2486\,\text{kg·m}^2$.

Problem 7.38 If the mass of bar AB is negligible and the coefficient of kinetic friction between the wheel and the surface is $\mu_k = 0.6$, determine the time it takes for the wheel to come to a complete stop, and find the reactions at A and B on bar AB.

Problem 7.39 If the mass of bar AB is $m_{AB} = 1.5\,\text{kg}$ and the coefficient of kinetic friction between the wheel and the surface is $\mu_k = 0.6$, determine the time it takes for the wheel to come to a complete stop, and find the reactions at A and B on bar AB.

Problem 7.40 If the mass of bar AB is $m_{AB} = 1.5\,\text{kg}$, determine the coefficient of kinetic friction between the wheel and the surface so that it takes 2 s to come to a complete stop. In addition, find the reactions at A and B on bar AB for these conditions.

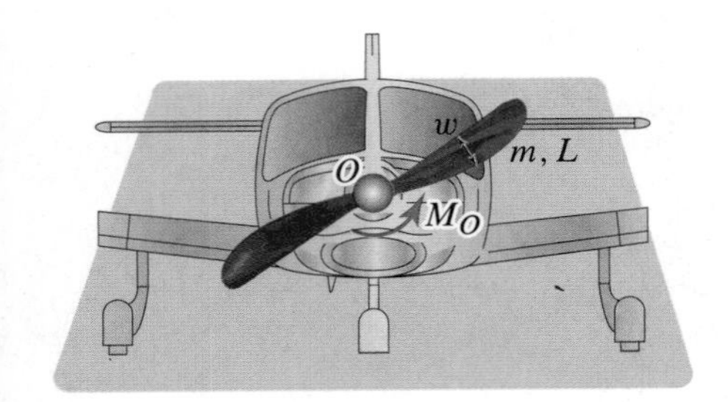

Figure P7.41

Problem 7.41

Assume that the plane is on the ground, and model each of its two propeller blades as a rectangular plate of mass $m = 29\,\text{kg}$, length $L = 137\,\text{cm}$, and width $w = 20\,\text{cm}$. If the engine torque M_O applied to the propeller shaft is $410\,\text{N·m}$, determine the time it takes for the blades to achieve an angular speed of 2000 rpm if the blades start from rest. What would be the required time if, instead, the blades were modeled as thin rods of length L and mass m?

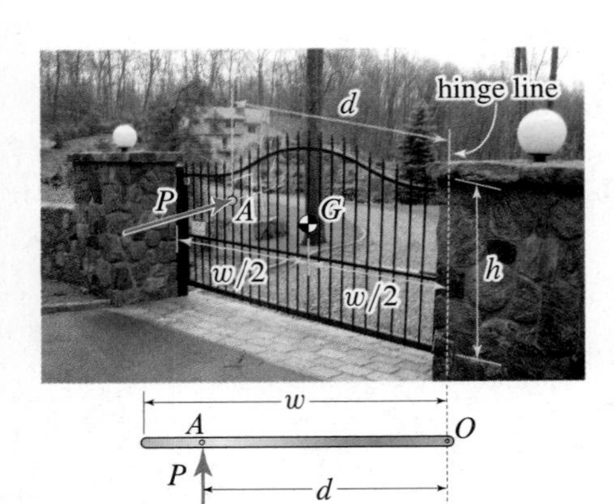

Figure P7.42 and P7.43

Problems 7.42 and 7.43

The driveway gate is hinged at its right end and can swing freely in the horizontal plane. The gate is pushed open by the force P that always acts perpendicular to the plane of the gate at point A, which is a horizontal distance d from the gate hinge. The weight of the gate is $W = 215\,\text{lb}$, and its mass center is at G, which is a distance $w/2$ from each end of the gate, where $w = 16\,\text{ft}$. Assume that the gate is initially at rest, and model the gate as a uniform thin bar as shown below the photo.

Problem 7.42 Given that a force of $P = 20\,\text{lb}$ is applied at the center of mass of the gate (i.e., $d = w/2$), determine the reactions at the hinge O after the force P has been continuously applied for 2 s.

Problem 7.43 Given that a force of $P = 20\,\text{lb}$ is applied at the center of percussion of the gate, determine the reactions at the hinge O after the force P has been continuously applied for 2 s.

Problems 7.44 and 7.45

The uniform thin bar AB of length L and mass m is released from rest in the horizontal position shown.

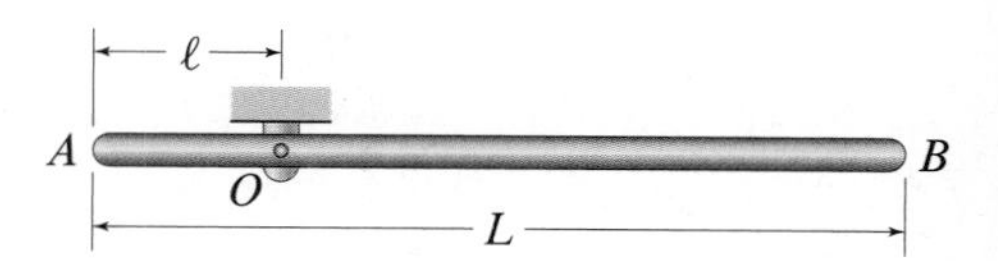

Figure P7.44 and P7.45

Problem 7.44 Determine the distance ℓ at which the pin O should be located from the end of the bar so that it has the maximum possible angular acceleration α_{max}, and find that angular acceleration.

Problem 7.45 Determine the distance ℓ at which the pin should be located from the end of the bar so that it has the maximum possible angular acceleration α_{max}, and find that angular acceleration. In addition, determine the angular acceleration α_0 of the bar when $\ell = 0$, and then find the ratio α_{max}/α_0.

Problems 7.46 and 7.47

The cutting arm of the paper cutter is pinned about a fixed axis at O, and its angle relative to the horizontal is measured by ϕ. A linear elastic torsional spring at O with constant k_t keeps the arm from falling when not in use. Model the cutting arm as a uniform slender bar of length $L = 20$ in. and weight $W = 2.5$ lb. Neglect friction in the pin at O.

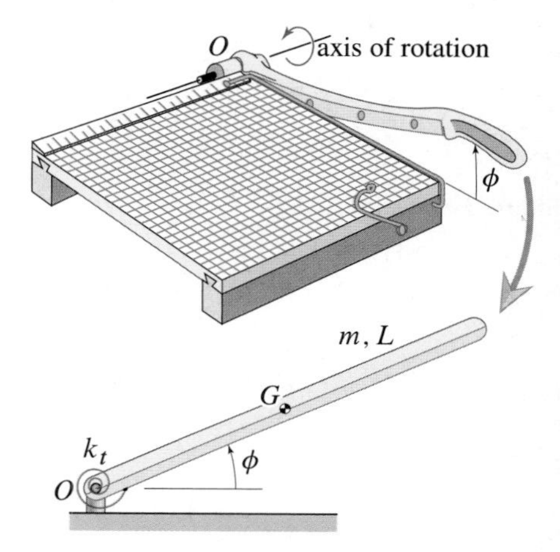

Figure P7.46 and P7.47

Problem 7.46 Determine the angular speed with which the cutting arm will reach the horizontal position if it is released from rest at $\phi_i = 70°$ with $k_t = 1.6$ ft·lb/rad. Assume that the torsional spring is undeformed when $\phi = 90°$.

Problem 7.47 Determine the value of the torsional spring constant k_t so that when the cutting arm is released from rest at $\phi_i = 70°$, it reaches $\phi_f = 15°$ with zero angular speed. Assume that the torsional spring is undeformed when $\phi = 90°$.

Problem 7.48

The uniform thin platform AB of length L and mass m_p is pinned both at A and at D. A uniform crate of height h, width w, and mass m_c is placed at the end of the platform a distance ℓ from the pin at A. The system is at rest when the pin at A breaks. Determine the angular acceleration of the platform and crate, as well as the force on the platform due to the pin at D, immediately after the pin at A breaks. Assume that the crate and the platform do not separate immediately after the pin fails and that friction is sufficient to prevent slipping between the platform and crate.

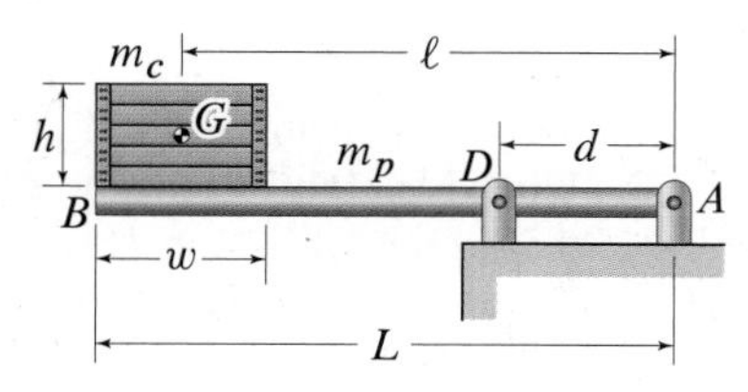

Figure P7.48

Problems 7.49 and 7.50

The ladder of mass m and length L is released from rest at the angle θ_0. Model the ladder as a uniform slender bar.

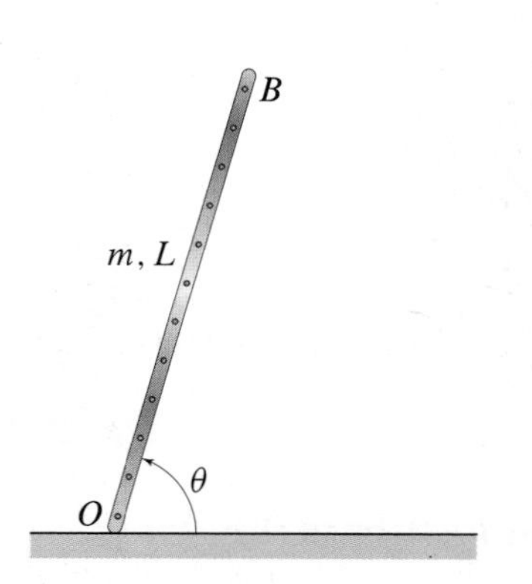

Figure P7.49 and P7.50

Problem 7.49 If the friction at O between the ladder and the ground is sufficient to prevent slipping, and the ladder is given a slight nudge from rest at $\theta = 90°$, determine the angular speed of the ladder when it reaches $\theta = 0$.

Problem 7.50 If the friction at O between the ladder and the ground is sufficient to prevent slipping, and the ladder is given a slight nudge from rest at $\theta = 90°$, determine the normal and frictional forces at O as a function of the angle θ, and find the minimum coefficient of static friction that is compatible with this motion.

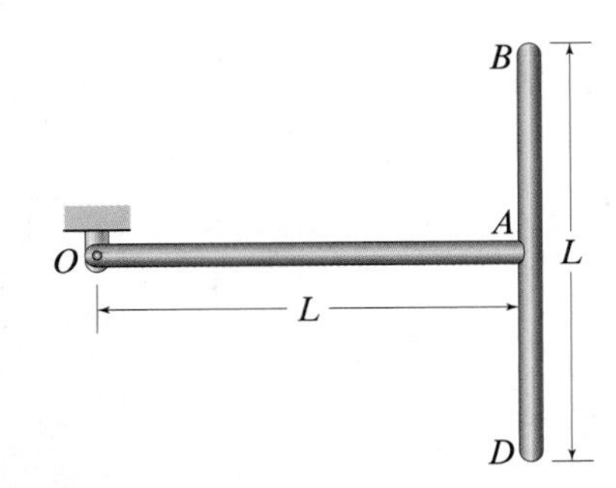

Figure P7.51 and P7.52

Problems 7.51 and 7.52

The T-bar consists of two thin rods, OA and BD, each of length $L = 1.5$ m and mass $m = 12$ kg, that are connected to the frictionless pin at O. The rods are welded together at A and lie in the vertical plane.

Problem 7.51 If the rods are released from rest in the position shown, determine the force on the pin at O, as well as the angular acceleration of the rods immediately after release.

Problem 7.52 If, at the instant shown, the system is rotating clockwise with angular velocity $\omega_0 = 7$ rad/s, determine the force on the pin at O, as well as the angular acceleration of the rods.

Problem 7.53

One way to measure the mass moment of inertia of any body relative to its mass center G is to horizontally suspend it by a string at one end A, and to support it by a scale C at the other end B. The location of points A and B is not important as long as we know the distance d between them and the mass m of the body. When the body is statically supported as described, we take note of the reading on the scale. We then cut the string at A and note the new reading on the scale immediately after the string is cut. Knowing the mass m of the body, the distance d, and the reading on the scale immediately before N_b and after N_a the string is cut, determine I_G for the body.

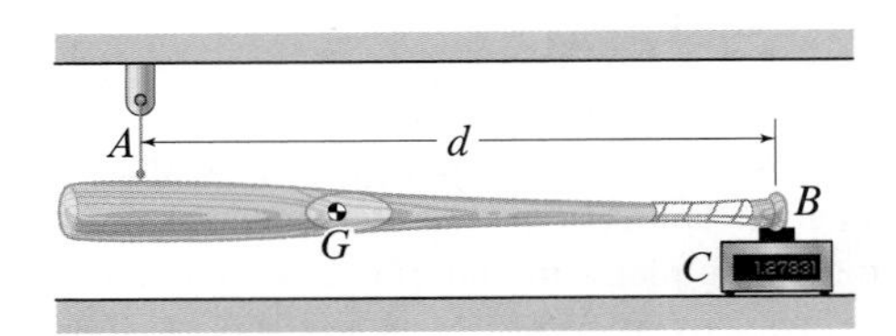

Figure P7.53

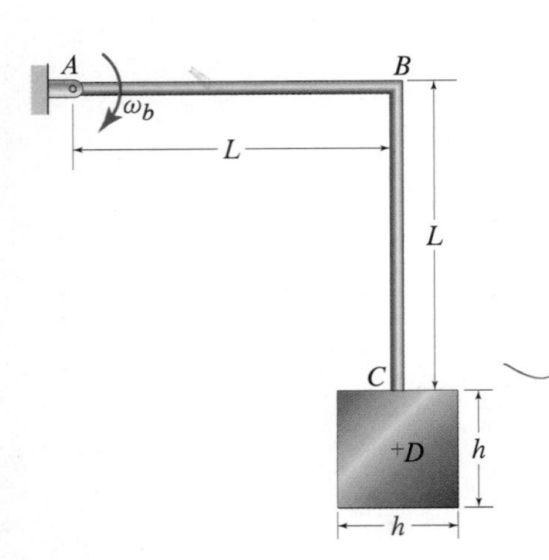

Figure P7.54–P7.56

Problems 7.54 through 7.56

The composite rigid body lies in the vertical plane and consists of the uniform block D and the L-shaped bar ABC. The block is rigidly attached to the L-shaped bar, which is uniform. Each segment of the L-shaped bar has length $L = 0.75$ m and mass $m_{AB} = m_{BC} = 2$ kg, the mass of the block is $m_D = 5$ kg, and the width and height of the block are $h = 0.3$ m.

Problem 7.54 If the system is released from rest in the position shown, determine the initial angular acceleration of the composite body and the reaction at the pin A on the composite body.

Problem 7.55 If the system is rotating clockwise with the angular speed $\omega_b = 10$ rad/s in the position shown, determine the reaction at the pin A on the composite body.

Problem 7.56 If the system is rotating clockwise with the angular speed $\omega_b = 10$ rad/s in the position shown, determine the internal reaction at the rigid joint B on the bar BC of the composite body.

7.4 Newton-Euler Equations: General Plane Motion

We now apply the equations developed in Section 7.1 to the motion of rigid bodies that are in general planar motion, that is, they are both translating and rotating.

Newton-Euler equations for general plane motion

Referring to Fig. 7.17, the first two Newton-Euler equations are always the two scalar components of Euler's first law, Eq. (7.1), which is

$$\boxed{\vec{F} = m\vec{a}_G.} \tag{7.49}$$

Newton-Euler equations for an arbitrary moment center

Again referring to Fig. 7.17, when the moment center is an arbitrary point P, the third Newton-Euler equation is given by Eq. (7.29), which we recall here as

$$\boxed{\vec{M}_P = I_G\vec{\alpha}_B + \vec{r}_{G/P} \times m\vec{a}_G,} \tag{7.50}$$

where, since the motion is planar, $\vec{M}_P = M_P\,\hat{k}$, $\vec{\alpha}_B = \alpha_B\,\hat{k}$, $\vec{r}_{G/P} = x_{G/P}\,\hat{\imath} + y_{G/P}\,\hat{\jmath}$, and $\vec{a}_G = a_{Gx}\,\hat{\imath} + a_{Gy}\,\hat{\jmath}$. If *any* one of the following conditions is true:

1. Point P is the mass center G, so that $\vec{r}_{G/P} = \vec{0}$.
2. $\vec{a}_G = \vec{0}$ (i.e., G is fixed or moves with constant velocity).
3. $\vec{r}_{G/P}$ is parallel to $\vec{a}_G$.

then $\vec{r}_{G/P} \times m\vec{a}_G = \vec{0}$, and Eq. (7.50) reduces to

$$\boxed{M_P = I_G\alpha_B,} \tag{7.51}$$

where we have written the result in scalar form.

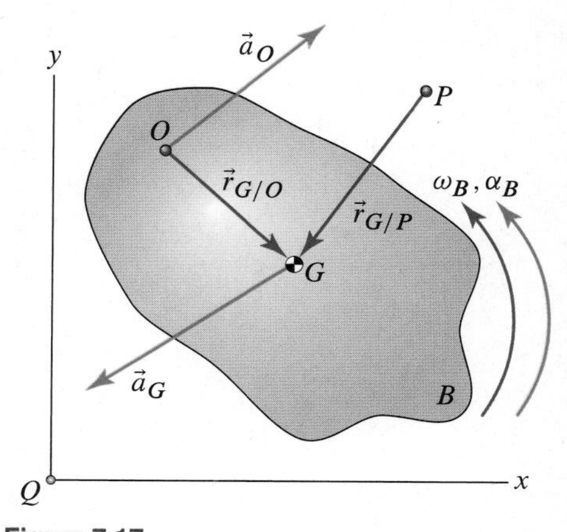

Figure 7.17
A general rigid body defining all relevant quantities needed to write its Newton-Euler equations.

Newton-Euler equations when the moment center is on the rigid body

Referring to Fig. 7.17, when the moment center is a point O on the rigid body, Eq. (7.36) on p. 525 can be used to obtain the third Newton-Euler equation, which is

$$\boxed{\vec{M}_O = I_O\vec{\alpha}_B + \vec{r}_{G/O} \times m\vec{a}_O.} \tag{7.52}$$

The more useful form of this equation is obtained if $\vec{r}_{G/O} \times m\vec{a}_O = \vec{0}$. This occurs if *any* of the following is true:

1. Point O is the mass center G so that $\vec{r}_{G/O} = \vec{0}$.
2. $\vec{a}_O = \vec{0}$ (i.e., O is fixed or moves with constant velocity).
3. $\vec{r}_{G/O}$ is parallel to $\vec{a}_O$.

Then Eq. (7.52) becomes

$$\boxed{M_O = I_O\alpha_B,} \tag{7.53}$$

where we have used the scalar form to reflect the fact that the motion is planar.

EXAMPLE 7.8 Analysis of a Falling Rigid Body

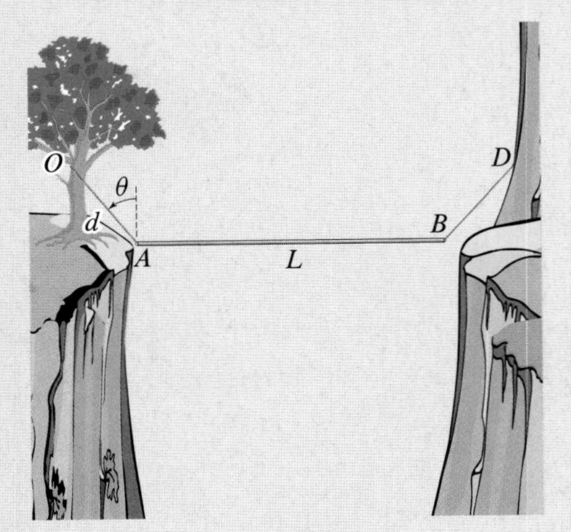

Figure 1

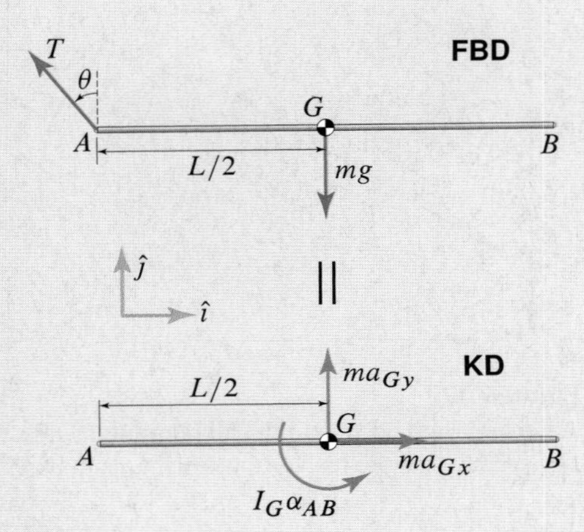

Figure 2
FBD (top) and KD (bottom) of the platform in Fig. 1.

As part of a movie stunt, a long, thin, 388 lb platform whose length is $L = 39$ ft has been rigged across a ravine using two ropes OA and BD. Rope OA of length $d = 13.4$ ft is securely tied to the tree, but rope BD has been tied to a carabiner at D that has not been adequately fastened to the rock face. After everything is set up as shown in Fig. 1, the carabiner at D breaks free, and the platform starts to fall. Determine the angular acceleration of the platform and the tension in the rope OA immediately after the rope BD breaks free. The initial value of θ is $39°$.

SOLUTION

Road Map & Modeling The FBD of the platform immediately after the carabiner breaks is shown in Fig. 2. We model the platform as a slender rod, and our key assumption is that, immediately after the rope BD breaks, all velocities are zero. In addition, we will neglect the mass of each rope and assume that they are inextensible. If the ropes are massless, then neither gravity nor acceleration will affect the behavior of the ropes. Therefore, assuming that rope OA is in tension, it will behave as a straight-line segment with constant length. That is, at the time instant considered, OA can be treated as if it were a massless rigid body. Of course, we will need to verify that it doesn't go slack.

Using this model, applying the Newton-Euler equations to the platform will allow us to determine the forces and accelerations once we have determined the kinematics of the platform immediately after rope BD breaks.

Governing Equations

Balance Principles By equating the FBD and KD of the platform shown in Fig. 2 (this is equivalent to applying Eqs. (7.1) and (7.29) to the FBD in Fig. 2), the Newton-Euler equations for the platfom are

$$\sum F_x: \quad -T\sin\theta = ma_{Gx}, \tag{1}$$

$$\sum F_y: \quad T\cos\theta - mg = ma_{Gy}, \tag{2}$$

$$\sum M_G: \quad -\frac{L}{2}T\cos\theta = I_G\alpha_{AB}, \tag{3}$$

where T is the tension in the rope and I_G is the mass moment of inertia of the platform, which is given by

$$I_G = \tfrac{1}{12}mL^2. \tag{4}$$

Force Laws All forces are accounted for on the FBD, although we must verify that $T > 0$ to make sure that the rope doesn't go slack. If the rope does go slack, we will solve the problem with the knowledge that $T = 0$.

Kinematic Equations We can see from Eqs. (1)–(3) that we need to relate $\vec{a}_G$ to the angular acceleration of the platform, subject to the constraint that point A moves in a circle about O and all velocities are zero immediately after release. Relating A to O, we get

$$\begin{aligned}\vec{a}_A &= \vec{a}_O + \vec{\alpha}_{OA}\times\vec{r}_{A/O} - \omega_{OA}^2\vec{r}_{A/O}\\ &= \alpha_{OA}\,\hat{k}\times d(\sin\theta\,\hat{\imath} - \cos\theta\,\hat{\jmath}) = d\alpha_{OA}\cos\theta\,\hat{\imath} + d\alpha_{OA}\sin\theta\,\hat{\jmath},\end{aligned} \tag{5}$$

since $\vec{a}_O = \vec{0}$ and all velocities are zero. Relating G to A, we get

$$\begin{aligned}\vec{a}_G &= \vec{a}_A + \vec{\alpha}_{AB}\times\vec{r}_{G/A} - \omega_{AB}^2\vec{r}_{G/A}\\ &= d\alpha_{OA}(\cos\theta\,\hat{\imath} + \sin\theta\,\hat{\jmath}) + \alpha_{AB}\,\hat{k}\times(L/2)\,\hat{\imath}\\ &= d\alpha_{OA}\cos\theta\,\hat{\imath} + (d\alpha_{OA}\sin\theta + L\alpha_{AB}/2)\,\hat{\jmath},\end{aligned} \tag{6}$$

where we have used the expression for $\vec{a}_A$ from Eq. (5) and set all velocities to zero.

Computation We can substitute Eqs. (4) and (6) into Eqs. (1)–(3) to obtain the following three equations in the three unknowns T, α_{AB}, and α_{OA}:

$$-T\sin\theta = md\alpha_{OA}\cos\theta, \tag{7}$$

$$T\cos\theta - mg = m\left(d\alpha_{OA}\sin\theta + \frac{L}{2}\alpha_{AB}\right), \tag{8}$$

$$-\frac{L}{2}T\cos\theta = \tfrac{1}{12}mL^2\alpha_{AB}. \tag{9}$$

Solving these three equations for the three unknowns, we obtain

$$T = \frac{2mg\cos\theta}{5+3\cos(2\theta)} = 107.2\,\text{lb}, \tag{10}$$

$$\alpha_{AB} = -\frac{12g\cos^2\theta}{L[5+3\cos(2\theta)]} = -1.064\,\text{rad/s}^2, \tag{11}$$

$$\alpha_{OA} = -\frac{2g\sin\theta}{d[5+3\cos(2\theta)]} = -0.5378\,\text{rad/s}^2, \tag{12}$$

where we have used $L = 39\,\text{ft}$, $d = 13.4\,\text{ft}$, $\theta = 39^\circ$, $m = (388\,\text{lb})/(32.2\,\text{ft/s}^2) = 12.05\,\text{slug}$, and $g = 32.2\,\text{ft/s}^2$ to obtain the final numerical results.

Discussion & Verification

- The dimensions and the units of the final results in Eqs. (10)–(12) are all as they should be.
- The initial angular accelerations of both the rope OA and the bar AB are negative, as expected, since they should both initially rotate clockwise.
- It is difficult to have a sense of the magnitude of the tension in the rope OA, but it should certainly be positive, which it is. The fact that $T > 0$ confirms that our use of Eq. (5) was correct. In Eq. (5) we treated the rope as a massless rigid body, and this is acceptable only as long as the rope does not go slack.
- Referring to Fig. 2, right after the carabiner fails, we expect point G to accelerate downward, i.e., $a_{Gy} < 0$. Since from Eq. (2) we have $T = m(g+a_{Gy})/\cos\theta$, the expectation that $a_{Gy} < 0$ implies the expectation that $T < mg/\cos\theta = 499.3\,\text{lb}$. The result in Eq. (10) is consistent with this expectation.

EXAMPLE 7.9 Rolling Without Slip

Figure 1

When accelerating from 0 to 60 mph, the rear-wheel drive roadster shown in Fig. 1 has the loads shown in Fig. 2 applied to *each* of the two rear wheels from the rear axle of the 2570 lb car. Given that each wheel weighs 47 lb, has a mass moment of inertia I_G of 0.989 slug·ft^2, and has a diameter of 24.3 in., determine

(a) The normal and frictional forces between the wheel and the ground

(b) The minimum coefficient of static friction required for the wheel to roll without slip

(c) The time it takes for the car to reach 60 mph

Assume that the car accelerates uniformly while moving over a flat and level surface.

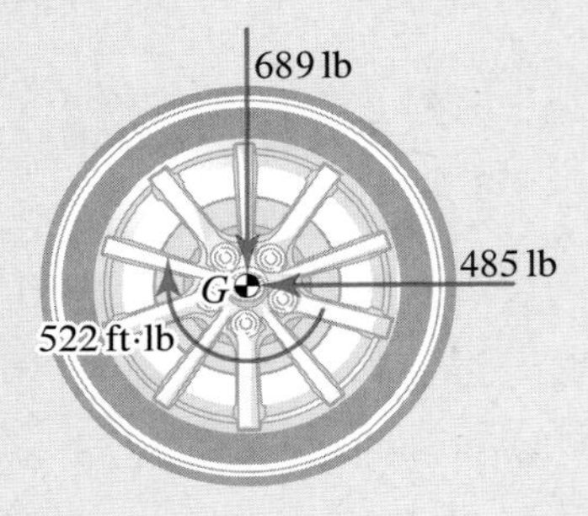

Figure 2
The loads applied by the axle to one of the rear wheels of the roadster shown in Fig. 1.

Interesting Fact

Where do the forces shown in Fig. 2 come from? In Prob. 7.85, one can find the forces on the rear wheels due to the axle by first performing an analysis of the entire car to get the friction and normal forces on the rear wheels, and then isolating one of the rear wheels and analyzing it.

SOLUTION

Road Map & Modeling The FBD of the wheel is shown in Fig. 3, where, from Fig. 2, the vertical force is V, the horizontal force is H, and the moment is M_a. Since we know all the loads causing the wheel to move and the inertia properties of the wheel, we should be able to determine how the wheel moves by solving the Newton-Euler equations. We will use the normal and friction forces that we find to determine the minimum friction coefficient required for rolling without slip.

Governing Equations

Balance Principles Based on the FBD in Fig. 3, the Newton-Euler equations are

$$\sum F_x: \quad F - H = m a_{Gx}, \tag{1}$$

$$\sum F_y: \quad N - V - mg = m a_{Gy}, \tag{2}$$

$$\sum M_G: \quad Fr - M_a = I_G \alpha_w, \tag{3}$$

where a_{Gx} and a_{Gy} are the x and y components of the acceleration of the mass center G, the friction force acting at the bottom of the wheel is F, the normal force between the ground and the wheel is N, the radius of the wheel is $r = (24.3/2)$ in. $= 1.012$ ft, and α_w is the angular acceleration of the wheel. Also, the inertia properties of the wheel are its mass m and its mass moment of inertia I_G, which in this case are

$$m = \frac{47\,\text{lb}}{32.2\,\text{ft/s}^2} = 1.460\,\text{slug} \quad \text{and} \quad I_G = 0.9890\,\text{slug·ft}^2. \tag{4}$$

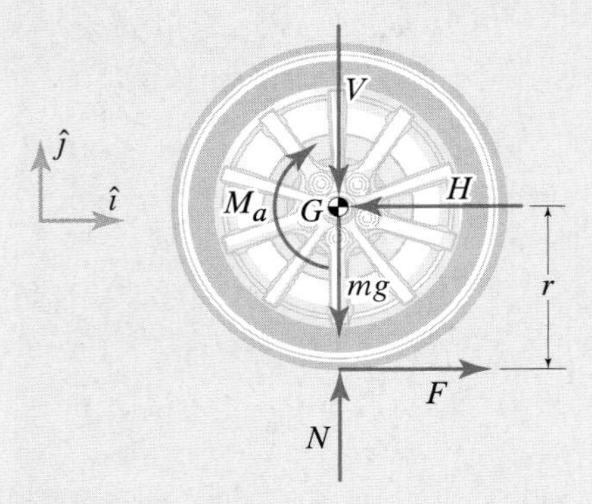

Figure 3
The FBD of one of the rear wheels of the car shown in Fig. 1.

Force Laws The inequality that must be satisfied for the wheel to roll without slip is

$$|F| \le \mu_s |N|, \tag{5}$$

where μ_s is the coefficient of static friction between the wheel and the ground.

Kinematic Equations Since the car is on a flat surface, the center of the wheel cannot undergo any vertical motion, and since we are assuming that the wheel is rolling without slip, we have the following two kinematic constraints:

$$a_{Gy} = 0 \quad \text{and} \quad a_{Gx} = -r\alpha_w, \tag{6}$$

where the minus sign comes from the fact that α_w has been assumed to be positive in the positive z direction.

Computation Substituting Eqs. (6) into Eqs. (1)–(3), we obtain the following three equations:

$$F - H = -mr\alpha_w, \tag{7}$$

$$N - V - mg = 0, \tag{8}$$

$$Fr - M_a = I_G \alpha_w, \tag{9}$$

for the three unknowns N, F, and α_w. Solving, we obtain

$$\boxed{\begin{aligned} N &= V + mg = 736.0\,\text{lb}, \\ F &= \frac{I_G H + mrM_a}{I_G + mr^2} = 503.4\,\text{lb}, \\ \alpha_w &= \frac{rH - M_a}{I_G + mr^2} = -12.45\,\text{rad/s}^2, \end{aligned}} \tag{10–12}$$

where we have used $V = 689\,\text{lb}$, $H = 485\,\text{lb}$, $m = (47\,\text{lb})/(32.2\,\text{ft/s}^2) = 1.460\,\text{slug}$, $g = 32.2\,\text{ft/s}^2$, $I_G = 0.9890\,\text{slug}\cdot\text{ft}^2$, $r = (12.15\,\text{in.})/(12\,\text{in./ft}) = 1.012\,\text{ft}$, and $M_a = 522\,\text{ft}\cdot\text{lb}$, to obtain the final numerical results. Now that we know F and N, we can find the minimum value of μ_s that is compatible with the no-slip assumption by simply using the equality in Eq. (5), that is,

$$\mu_s \geq \left|\frac{F}{N}\right| \quad \Rightarrow \quad \boxed{(\mu_s)_{\text{min}} = 0.6840.} \tag{13}$$

Finally, to determine the time it takes for the car to reach 60 mph, we first find a_{Gx} from the second of Eqs. (6) as

$$a_{Gx} = -(1.012\,\text{ft})(-12.45\,\text{rad/s}^2) = 12.60\,\text{ft/s}^2, \tag{14}$$

and then we apply Eq. (2.32) on p. 49 since we are assuming the acceleration is uniform, that is,

$$v = v_0 + a_{Gx}t \quad \Rightarrow \quad 88\,\text{ft/s} = \left(12.60\,\text{ft/s}^2\right)t \quad \Rightarrow \quad \boxed{t = 6.982\,\text{s.}} \tag{15}$$

Discussion & Verification

- The dimensions of each of the results in Eqs. (10)–(12) are correct.
- The value of static friction found in Eq. (13) is reasonable for a tire on asphalt.
- The "0 to 60" time we found in Eq. (15) is consistent with times found in the product literature for a roadster like the one analyzed here.

A Closer Look We are given all the nonconstraint forces acting on a wheel, which allows us to then determine the motion of the wheel and whether it slips. In this case, since one of the things we were looking for was the minimum μ_s for rolling without slip, we could assume no slip and then find the μ_s compatible with that assumption. Had we not been told whether the wheel slips, then we could have assumed no slip and verified that assumption in the usual way by comparing the needed friction with the available friction (we would need to be given the static friction coefficient). If we discovered that the wheel slips, then Eq. (5) would become $F = \mu_k N$ and the second of Eqs. (6) would no longer be valid.

EXAMPLE 7.10 *Example 3.9 Revisited*

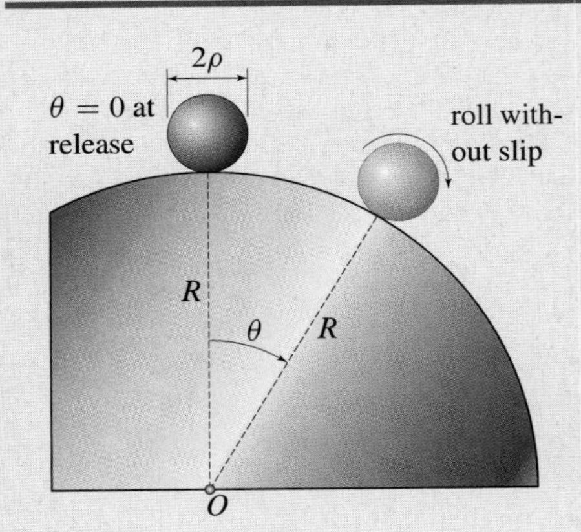

Figure 1

In Example 3.9, we released a small sphere from the top of a semicylinder and, by modeling it as a particle, we determined that it separated from the semicylinder at $\theta = 48.2°$ (see Fig. 1). Here we wish to determine the value of θ at which the small sphere separates if we treat it as a uniform sphere of radius ρ and mass m. We release the sphere from the top of the semicylinder by giving it a *slight* nudge to the right, and we assume that there is sufficient friction between the semicylinder and the sphere for the sphere to roll without slip.

SOLUTION

Road Map & Modeling As with Example 3.9, the key is to find the normal force between the sphere and the semicylinder as a function of θ and then say that the sphere separates at the location where this force becomes zero.

The FBD of the sphere as it slides down the semicylinder is shown in Fig. 2. The FBD has been drawn at an arbitrary angle θ since we need to find that angle at which N becomes zero, and so we need to find N for *any* θ. Since the motion of the mass center of the sphere is along a circular path until it separates from the surface, we will use polar coordinates for the solution. Note that the friction force F has been drawn in the indicated direction since the sphere is passively rolling down the semicylinder, i.e., it is not driven.

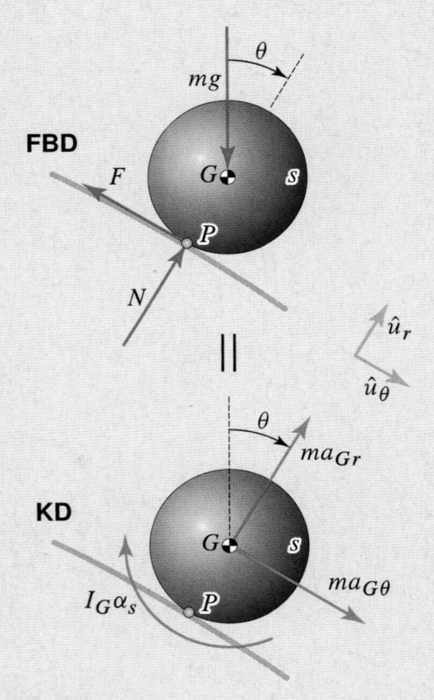

Figure 2
FBD of the sphere s and the polar component system drawn at an arbitrary angle θ. Note that, in this component system, positive z points into the page.

Governing Equations

Balance Principles Equating the FBD and KD of the sphere shown in Fig. 2 (this is equivalent to applying Eqs. (7.1) and (7.29) to the FBD in Fig. 2), the Newton-Euler equations for the sphere are

$$\sum F_r: \quad N - mg\cos\theta = ma_{Gr}, \tag{1}$$

$$\sum F_\theta: \quad -F + mg\sin\theta = ma_{G\theta}, \tag{2}$$

$$\sum M_P: \quad mg\rho\sin\theta = I_G\alpha_s + \rho m a_{G\theta}, \tag{3}$$

where m is the mass of the sphere, F is the friction force between the sphere and semicylinder, and the mass moment of inertia I_G of the sphere is

$$I_G = \tfrac{2}{5}m\rho^2. \tag{4}$$

Force Laws To ensure that the sphere rolls without slip, we must have

$$|F| \le \mu_s |N|. \tag{5}$$

Kinematic Equations If we want N as a function of θ, then the accelerations need to be expressed as functions of θ. We begin by writing $\vec{a}_G$ using polar coordinates as

$$\vec{a}_G = -\dot{\theta}^2(R+\rho)\,\hat{u}_r + \ddot{\theta}(R+\rho)\,\hat{u}_\theta, \tag{6}$$

since G is moving in a circle centered at O. Equation (6) implies that

$$a_{Gr} = -\dot{\theta}^2(R+\rho), \tag{7}$$

$$a_{G\theta} = \ddot{\theta}(R+\rho). \tag{8}$$

Now that we have a_{Gr} and $a_{G\theta}$ as a function of θ, we need $\alpha_s(\theta)$. We can get this by finding $\omega_s(\theta)$, the sphere's angular velocity, and then differentiating with respect to time. Relating $\vec{v}_G$ to $\vec{v}_P$, we obtain

$$\vec{v}_G = \vec{v}_P + \vec{\omega}_s \times \vec{r}_{G/P} \quad \Rightarrow \quad \dot{\theta}(R+\rho)\,\hat{u}_\theta = \vec{v}_P + \vec{\omega}_s \times \vec{r}_{G/P}, \tag{9}$$

where v_G has been written using polar coordinates. Noting that $\vec{v}_P = \vec{0}$ and using components, Eq. (9) becomes

$$\dot{\theta}(R+\rho)\,\hat{u}_\theta = \omega_s\,\hat{u}_z \times \rho\,\hat{u}_r \quad \Rightarrow \quad \omega_s = \left(\frac{R+\rho}{\rho}\right)\dot{\theta}, \tag{10}$$

where we have used $\hat{u}_z \times \hat{u}_r = \hat{u}_\theta$. Differentiating Eq. (10), we obtain

$$\alpha_s = \left(\frac{R+\rho}{\rho}\right)\ddot{\theta}. \tag{11}$$

Computation We get the equations of motion for the sphere by substituting Eqs. (4), (7), (8), and (11) into Eqs. (1)–(3), which gives

$$N - mg\cos\theta = -m\dot{\theta}^2(R+\rho), \tag{12}$$

$$-F + mg\sin\theta = m\ddot{\theta}(R+\rho), \tag{13}$$

$$g\sin\theta = \tfrac{7}{5}(R+\rho)\ddot{\theta}, \tag{14}$$

which are three equations to solve for N, F, and θ (a differential equation must be solved to get θ). However, all we really want is $N(\theta)$. To get $N(\theta)$, we can see from Eq. (12) that we will need to get $\dot{\theta}$ as a function of θ — we can do this by using the chain rule, i.e., $\ddot{\theta} = \dot{\theta}\,d\dot{\theta}/d\theta$, and then integrating Eq. (14) as follows:

$$\int_0^{\dot{\theta}} \dot{\theta}\,d\dot{\theta} = \frac{5g}{7(R+\rho)}\int_0^{\theta} \sin\theta\,d\theta \quad \Rightarrow \quad \dot{\theta}^2 = \frac{10g}{7(R+\rho)}(1-\cos\theta). \tag{15}$$

Substituting Eqs. (15) into Eq. (12), N as a function of θ is

$$N = \tfrac{1}{7}mg(17\cos\theta - 10). \tag{16}$$

Therefore, calling θ_{sep} the separation angle, the sphere separates from the surface, i.e., N becomes zero, when

$$17\cos\theta_{\text{sep}} - 10 = 0 \quad \Rightarrow \quad \theta_{\text{sep}} = \pm 53.97^\circ + n360^\circ,\ n = 0, \pm 1, \ldots, \pm\infty. \tag{17}$$

Since we are only interested in $0^\circ \le \theta \le 90^\circ$, the only acceptable answer is

$$\boxed{\theta_{\text{sep}} = 53.97^\circ.} \tag{18}$$

Discussion & Verification When we compare this example with Example 3.9, we see that when rotary inertia plays a role in the dynamics, as it does in this example, the object separates from the surface almost 6° farther down the cylinder, independent of R, ρ, g, and m. Given that this result is "in the same ballpark" as the 48.19° separation angle for a particle, it helps build some confidence that the result for a sphere is correct.

A Closer Look In Example 3.9, we treated the object sliding down the semicylinder as a particle. A finite-sized sphere would act as a particle if the contact interface were frictionless. Therefore, we should be able to recover the result of Example 3.9 if, in this example, (1) we let the friction go to zero and (2) we account for the fact that the mass center is $R+\rho$ from the center of the semicylinder at O. We will see this in Prob. 7.68.

Common Pitfall

Can we *really* satisfy Eq. (5)? We stated at the beginning that there is enough friction between the surface and the sphere to prevent the sphere from slipping on the surface. Is this possible? It isn't, and let's quickly see why. To obtain $F(\theta)$, we can solve Eq. (14) for $\ddot{\theta}$, substitute the result into Eq. (13), and then solve for F. Now $N(\theta)$ is found in Eq. (16). Taking the ratio of the two as given by Eq. (5), we obtain

$$\mu_s = \left|\frac{F}{N}\right| = \left|\frac{2\sin\theta}{17\cos\theta - 10}\right|.$$

Notice that as the sphere rolls down the semicylinder and θ approaches the separation position, the denominator $17\cos\theta - 10$ goes to zero (see Eq. (17)), and so μ_s goes to ∞. This actually tells us that *it isn't possible for the sphere to roll without slipping until it separates from the semicylinder since infinite friction would be required.*

Helpful Information

Why does the sphere go 6° farther than the particle? The answer lies in the speed of the objects as they fall down the cylinder. We haven't covered the work-energy principle for rigid bodies, but we know that in conservative systems, a decrease in potential energy leads to a corresponding increase in kinetic energy. As the particle and sphere move down the cylinder, for a given height change, they each experience the same increase in kinetic energy. For the particle, all the kinetic energy goes into its speed. For the sphere, some goes into its translational speed, but some also goes into the energy associated with its rotation. Either the sphere or the particle separates from the cylinder when they are moving fast enough that their v^2/ρ acceleration overcomes the normal component of mg. It takes the sphere a little longer to get up to that speed since some of its energy goes into rotation, so it separates at a larger angle.

EXAMPLE 7.11 A System with Multiple Rigid Bodies

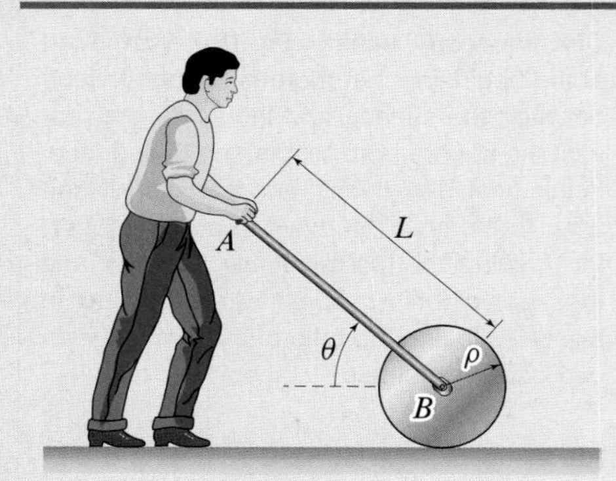

Figure 1

A man starts pushing the lawn roller shown in Fig. 1, such that the angle θ remains at a constant 40° and the center of the lawn roller at B accelerates to the right at a constant $0.45\,\mathrm{m/s^2}$. Given that the mass of the roller is 100 kg, the mass of the handle is 4 kg, $\rho = 25\,\mathrm{cm}$, and $L = 1.1\,\mathrm{m}$, determine the force at A that the man must apply to the handle to achieve this motion and the minimum necessary coefficient of static friction between the roller and ground if the roller is to roll without slip. Treat the roller as a uniform circular cylinder and the handle as a thin rod.

SOLUTION

Road Map & Modeling The FBD of the handle is shown in Fig. 2, and the FBD of the roller is shown in Fig. 3. We have let m_r be the mass of the roller, m_h be the mass of the handle, and F and N be the friction and normal forces, respectively, between the roller and the ground. Since we are treating the bar and roller as uniform, we have placed their mass centers at their geometric centers. These FBDs tell us that this is a problem in which we must analyze *two* rigid bodies. To do so, we will write a set of Newton-Euler equations for each, which will result in a set of *six* equations. Since the kinematics are entirely known, the unknowns will then turn out to be six forces in the system.

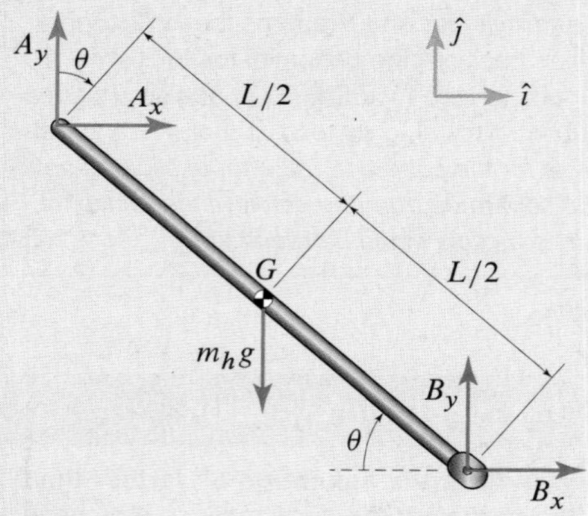

Figure 2
FBD of the handle of the lawn roller shown in Fig. 1.

Governing Equations

Balance Principles The Newton-Euler equations corresponding to the FBD of the handle in Fig. 2 are

$$\sum F_x: \qquad A_x + B_x = m_h a_{Gx}, \tag{1}$$

$$\sum F_y: \qquad A_y + B_y - m_h g = m_h a_{Gy}, \tag{2}$$

$$\sum M_G: \quad B_x \frac{L}{2}\sin\theta + B_y \frac{L}{2}\cos\theta - A_x \frac{L}{2}\sin\theta - A_y \frac{L}{2}\cos\theta = I_G \alpha_{AB}, \tag{3}$$

where $I_G = \frac{1}{12} m_h L^2$. The Newton-Euler equations corresponding to the FBD of the roller in Fig. 3 are

$$\sum F_x: \qquad -B_x - F = m_r a_{Bx}, \tag{4}$$

$$\sum F_y: \quad N - B_y - m_r g = m_r a_{By}, \tag{5}$$

$$\sum M_B: \qquad -F\rho = I_B \alpha_r, \tag{6}$$

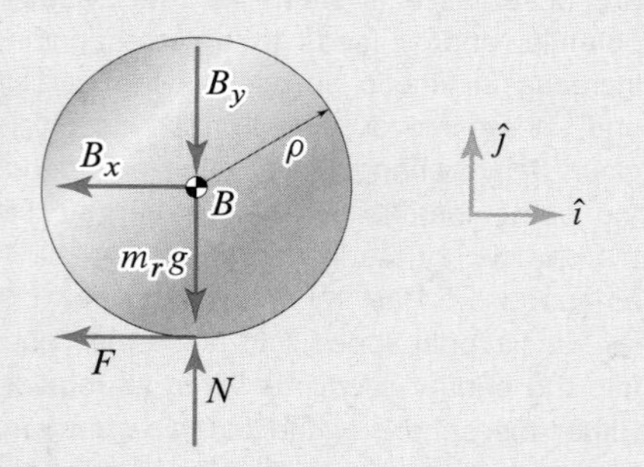

Figure 3
FBD of the roller of the lawn roller shown in Fig. 1.

where $I_B = \frac{1}{2} m_r \rho^2$ and α_r is the angular acceleration of the roller.

Force Laws The force law for this system is the friction inequality that must be satisfied for the roller to roll without slip, that is,

$$|F| \le \mu_s |N|. \tag{7}$$

All other forces are accounted for on the FBD.

Kinematic Equations Kinematically, we know θ is a constant, and so the bar AB is in pure translation. This implies that

$$\omega_{AB} = \alpha_{AB} = 0 \quad \Rightarrow \quad \vec{a}_G = \vec{a}_B. \tag{8}$$

In addition, since the roller is rolling without slip over a flat surface, we can say that a_{Bx} is known and is equal to the given acceleration of $0.45\,\mathrm{m/s^2}$ and that

$$a_{By} = 0 \quad \text{and} \quad \alpha_r = -\frac{a_{Bx}}{\rho}. \tag{9}$$

Computation Substituting Eqs. (7)–(9) into Eqs. (1)–(6), we obtain the six equations

$$A_x + B_x = m_h a_{Bx}, \tag{10}$$

$$A_y + B_y - m_h g = 0, \tag{11}$$

$$\frac{L}{2}\big[(B_x - A_x)\sin\theta + (B_y - A_y)\cos\theta\big] = 0, \tag{12}$$

$$-B_x - F = m_r a_{Bx}, \tag{13}$$

$$N - B_y - m_r g = 0, \tag{14}$$

$$F\rho = \tfrac{1}{2} m_r \rho a_{Bx}, \tag{15}$$

which we can solve for the six unknowns A_x, A_y, B_x, B_y, F, and N. This system is "weakly coupled," which means that each equation involves only a small number of the six unknowns. This makes it easier to solve the equations by hand. For example, since we know a_{Bx}, we can immediately find F using Eq. (15). We can then substitute F and a_{Bx} into Eq. (13) to find B_x. We can then substitute B_x and a_{Bx} into Eq. (10) to find A_x. Continuing similarly, we find all six solutions as

$$A_x = (m_h + \tfrac{3}{2} m_r) a_{Bx} = 69.30\,\text{N}, \tag{16}$$

$$A_y = \tfrac{1}{2}[m_h g - (m_h + 3m_r) a_{Bx} \tan\theta] = -37.77\,\text{N}, \tag{17}$$

$$B_x = -\tfrac{3}{2} m_r a_{Bx} = -67.50\,\text{N}, \tag{18}$$

$$B_y = \tfrac{1}{2}[m_h g + (m_h + 3m_r) a_{Bx} \tan\theta] = 77.01\,\text{N}, \tag{19}$$

$$F = \tfrac{1}{2} m_r a_{Bx} = 22.50\,\text{N}, \tag{20}$$

$$N = \tfrac{1}{2}[(m_h + 2m_r) g + (m_h + 3m_r) a_{Bx} \tan\theta] = 1058\,\text{N}. \tag{21}$$

The force that must be applied at A is given by

$$\boxed{\vec{A} = (69.30\,\hat{\imath} - 37.77\,\hat{\jmath})\,\text{N},} \tag{22}$$

or $|\vec{A}| = \sqrt{A_x^2 + A_y^2} = 78.92\,\text{N}$ at the angle shown in Fig. 4.

Now that we have the friction and normal forces, we can use the friction inequality in Eq. (7) to determine how much friction is needed to ensure that the roller rolls without slipping, that is,

$$\boxed{\mu_s \ge \left|\frac{F}{N}\right| = 0.02127.} \tag{23}$$

Discussion & Verification The dimension of each final result in Eqs. (16)–(21) is correct, and the magnitude of the required force at A is reasonable. Notice that Eq. (23) tells us that not much friction is needed for the roller to roll without slip. This is so because of the (considerable) weight of the roller in relation to the small value of acceleration that is being imparted to the roller by the person pushing it.

A Closer Look The force the man must apply to the handle of the lawn roller is given either by Eqs. (16) and (17) or by Eq. (22). Notice that the force that must be applied at A is not parallel to the handle. The reason is that the handle has mass—if we were to let m_h be zero in our model, we would find that the angle in Eq. (22) would be $-40°$ and the force at A would be directed along the handle.

If we now compute the magnitude of the force at B, we obtain

$$|\vec{B}| = \sqrt{B_x^2 + B_y^2} = 102.4\,\text{N}, \tag{24}$$

where we have also shown this force on the handle at B in Fig. 4. Notice that even though the bar is pin-connected at each end, it is not a two-force member. This is so for two distinct reasons. The first is that the weight of the bar has not been neglected, and the second is that the center of mass of the bar is accelerating.

Helpful Information

Solving six simultaneous equations. While these equations are not difficult to solve by hand, mathematical software packages, such as Mathematica, Maple, MATHCAD, or MATLAB are invaluable for quickly solving systems like this.

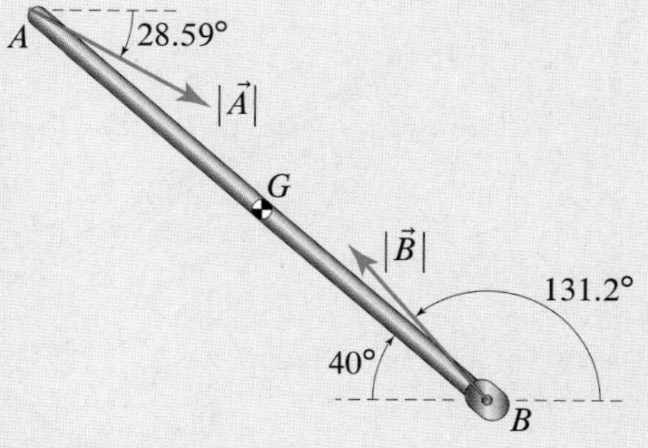

Figure 4
The forces at the ends of the lawn roller handle.

EXAMPLE 7.12 Derivation of Equations of Motion

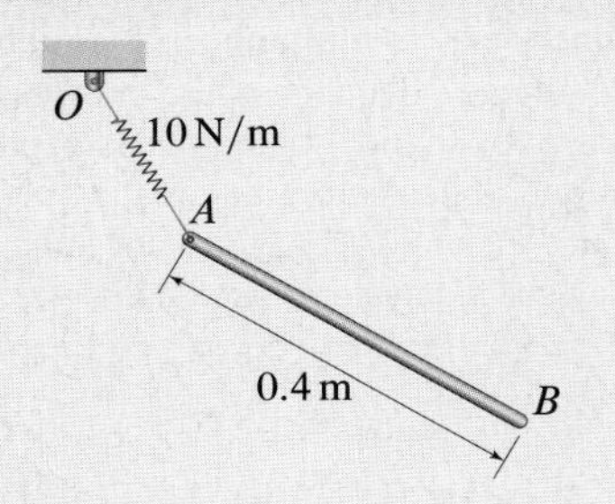

Figure 1

Placing a thin rod at the end of a spring suspended from the ceiling, and then letting the rod swing freely, results in very complicated motions. Write the equations of motion for this system, and then study the motion of the rod for two different sets of initial conditions using computer simulations. Use a rod 0.4 m long with a mass of 0.1 kg. Assume the spring is linear elastic with constant 10 N/m and with an unstretched length of 0.2 m. In the first simulation, use $\theta(0) = 30°$, $\phi(0) = 60°$, $r(0) = 0.2$ m, and $\dot{\theta}(0) = \dot{\phi}(0) = \dot{r}(0) = 0$; and in the second, use the same conditions except let $r(0) = 0.4$ m.

SOLUTION

Road Map & Modeling Figure 2 shows the coordinates r, θ, and ϕ used to define the

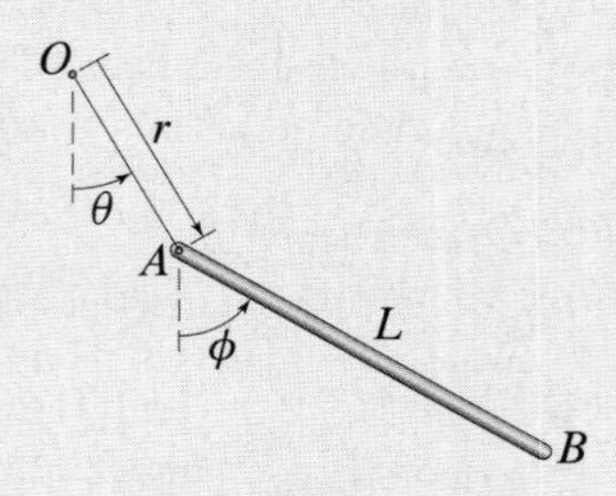

Figure 2. Definition of the coordinates r, θ, and ϕ used to define the position of the rod.

position of the rod, where r is the length of the spring, L is the length of the rod, and θ and ϕ define the angle of the spring and of the rod with respect to the vertical, respectively. Ignoring the mass of the spring, the hanging rod has three degrees of freedom. Therefore, we will need to derive three equations of motion. We will obtain these by writing the Newton-Euler equations for the rod using the FBD shown in Fig. 3, where F_s is the force on the rod due to the spring. Notice that we are using two coordinate systems in the FBD — a global Cartesian system and a polar coordinate system aligned with the spring (and thus the spring force) that we will use to describe the motion of point A.

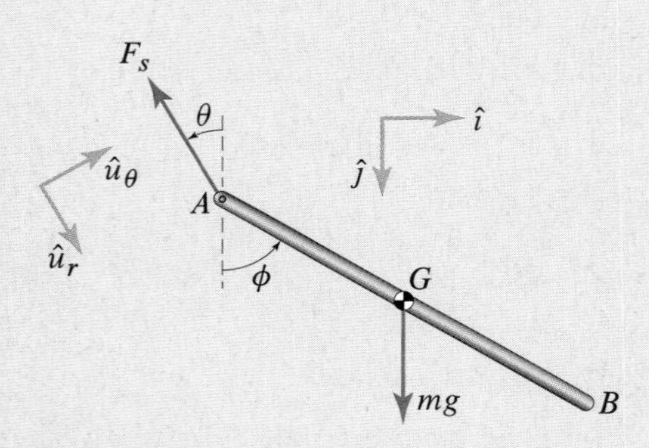

Figure 3
The FBD of the thin rod in Fig. 1.

Governing Equations

Balance Principles The Newton-Euler equations corresponding to the FBD in Fig. 3 are given by

$$\sum F_x: \quad -F_s \sin\theta = ma_{Gx}, \tag{1}$$

$$\sum F_y: \quad mg - F_s \cos\theta = ma_{Gy}, \tag{2}$$

$$\sum M_G: \quad \vec{r}_{A/G} \times \vec{F}_s = I_G \alpha_{AB}\, \hat{k}, \tag{3}$$

where $I_G = \frac{1}{12} m L^2$,

$$\vec{r}_{A/G} = \tfrac{L}{2}(-\sin\phi\,\hat{\imath} - \cos\phi\,\hat{\jmath}) \quad \text{and} \quad \vec{F}_s = F_s(-\sin\theta\,\hat{\imath} - \cos\theta\,\hat{\jmath}), \tag{4}$$

and α_{AB} is the angular acceleration of the rod. Substituting Eqs. (4) into Eq. (3), Eq. (3) becomes

$$\tfrac{1}{2} L F_s \sin(\phi - \theta) = I_G \alpha_{AB}, \tag{5}$$

where we have used the trigonometric identity $\sin\phi\cos\theta - \sin\theta\cos\phi = \sin(\phi - \theta)$.

Force Laws The one force that has not been accounted for in Fig. 3 is that of the linear elastic spring, which is

$$F_s = k(r - r_0), \tag{6}$$

where r_0 is the unstretched length of the spring.

Kinematic Equations Since we are using r, θ, and ϕ as the three coordinates to define the position of the rod, we need to write $\vec{a}_G$ and α_{AB} in terms of those coordinates and their derivatives. We can do this by relating $\vec{a}_G$ to $\vec{a}_A$ as follows:

$$\vec{a}_G = \vec{a}_A + \vec{\alpha}_{AB} \times \vec{r}_{G/A} - \omega_{AB}^2 \vec{r}_{G/A}, \tag{7}$$

where ω_{AB} is the angular velocity of the rod and we can write

$$\vec{\alpha}_{AB} = -\ddot{\phi}\,\hat{k}, \quad \omega_{AB} = -\dot{\phi}, \quad \text{and} \quad \vec{r}_{G/A} = -\vec{r}_{A/G} = \tfrac{L}{2}(\sin\phi\,\hat{\imath} + \cos\phi\,\hat{\jmath}). \tag{8}$$

Using the polar coordinate system defined in Fig. 3, we can write $\vec{a}_A$ as

$$\vec{a}_A = (\ddot{r} - r\dot{\theta}^2)\,\hat{u}_r + (r\ddot{\theta} + 2\dot{r}\dot{\theta})\,\hat{u}_\theta, \tag{9}$$

in which

$$\hat{u}_r = \sin\theta\,\hat{\imath} + \cos\theta\,\hat{\jmath} \quad \text{and} \quad \hat{u}_\theta = \cos\theta\,\hat{\imath} - \sin\theta\,\hat{\jmath}. \tag{10}$$

Substituting Eqs. (8)–(10) into Eq. (7), the components of $\vec{a}_G$ become

$$a_{Gx} = (\ddot{r} - r\dot{\theta}^2)\sin\theta + (r\ddot{\theta} + 2\dot{r}\dot{\theta})\cos\theta + \tfrac{L}{2}\ddot{\phi}\cos\phi - \tfrac{L}{2}\dot{\phi}^2\sin\phi, \tag{11}$$

$$a_{Gy} = (\ddot{r} - r\dot{\theta}^2)\cos\theta - (r\ddot{\theta} + 2\dot{r}\dot{\theta})\sin\theta - \tfrac{L}{2}\ddot{\phi}\sin\phi - \tfrac{L}{2}\dot{\phi}^2\cos\phi. \tag{12}$$

Computation Now that we have assembled all the pieces, the equations of motion are obtained by substituting Eqs. (6), (8), (11), and (12) into Eqs. (1), (2), and (5) to obtain the three equations of motion:

$$(\ddot{r} - r\dot{\theta}^2)\sin\theta + (r\ddot{\theta} + 2\dot{r}\dot{\theta})\cos\theta + \tfrac{L}{2}\ddot{\phi}\cos\phi - \tfrac{L}{2}\dot{\phi}^2\sin\phi + \frac{k}{m}(r - r_0)\sin\theta = 0, \tag{13}$$

$$(\ddot{r} - r\dot{\theta}^2)\cos\theta - (r\ddot{\theta} + 2\dot{r}\dot{\theta})\sin\theta - \tfrac{L}{2}\ddot{\phi}\sin\phi - \tfrac{L}{2}\dot{\phi}^2\cos\phi + \frac{k}{m}(r - r_0)\cos\theta = g, \tag{14}$$

$$\frac{L}{6}\ddot{\phi} + \frac{k}{m}(r - r_0)\sin(\phi - \theta) = 0. \tag{15}$$

Computer simulations of the rod motion are shown in Figs. 4 and 5.

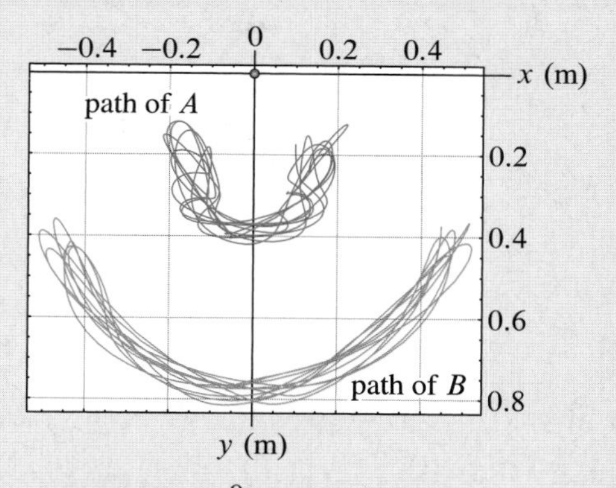

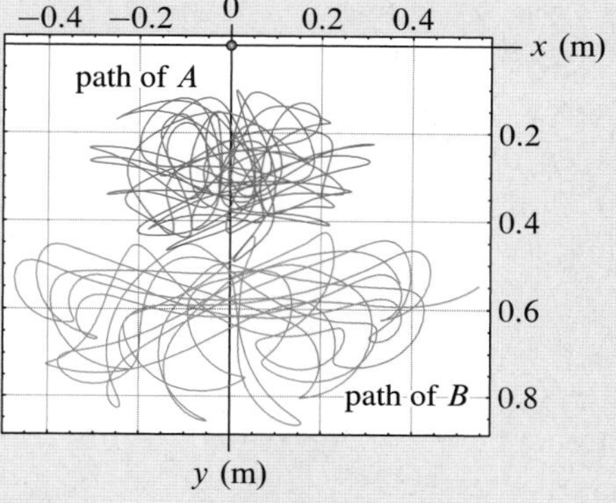

Figure 4
Paths of ends A and B for the first (top) and second (bottom) sets of initial conditions.

Discussion & Verification

- Each term in Eqs. (13) and (14) has been divided by m, so each term should have the units of acceleration, which it does.
- Each term in Eq. (15) has been divided by mL, so each term should have the units of acceleration, which it does.

A Closer Look Figure 4 shows the trajectories of the ends of the bar for the first 10 s after release for the first (top) and second (bottom) sets of initial conditions. The first set of initial conditions releases the bar when the spring is unstretched, and the second set has the bar stretched to twice its unstretched length at release (everything else is equal). In the second case, adding that additional initial energy to the system dramatically changes the ensuing motion; that is, it goes from a fairly regular pattern to one in which the bar is moving very irregularly.

Figure 5 shows stroboscopic images of the movement of the thin rod for the first 10 s of motion. In each case, the earliest image is the one in light gray (labeled "initial"), and each successive image becomes a darker purple (with the last one labeled "final"). The time between successive images is 0.5 s. These figures demonstrate the regular motion associated with the first set of initial conditions and the irregular motion associated with the second set. Systems, such as this one, whose motion is described by a set of nonlinear differential equations can be *very* sensitive to how the system is put in motion. This is one of the subjects of *chaos theory*.*

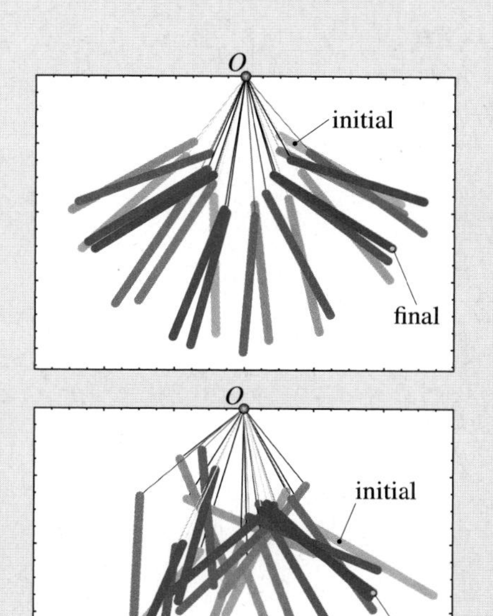

Figure 5
Stroboscopic image sequence of the rod for the first (top) and second (bottom) sets of initial conditions.

* See, for example, S. H. Strogatz, *Nonlinear Dynamics and Chaos: With Applications to Physics, Biology, Chemistry and Engineering*, Perseus Books, Reading, Mass., 1994.

Problems

Problem 7.57

The sphere, cylinder, and thin ring each have mass m and radius r. Each is released from rest on identical inclines. Assuming they all roll without slipping, which will have the largest initial angular acceleration? In addition, which will reach the bottom of the incline first?

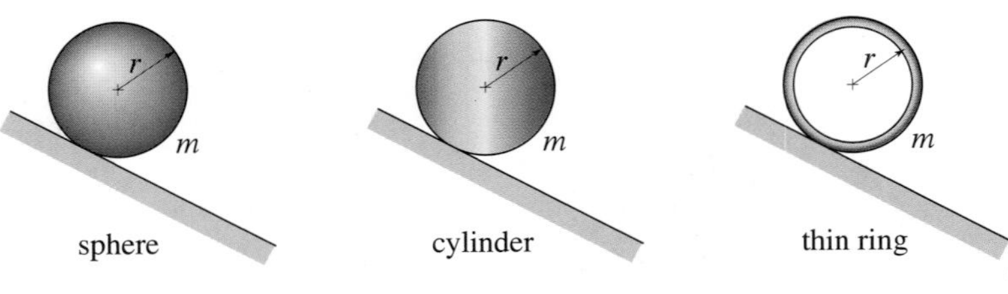

Figure P7.57

Problems 7.58 and 7.59

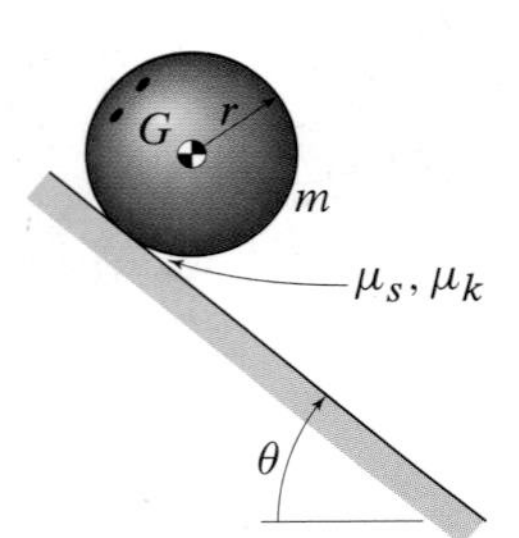

Figure P7.58 and P7.59

A bowling ball of radius r, mass m, and radius of gyration k_G is released from rest on a rough surface that is inclined at the angle θ with respect to the horizontal. The coefficients of static and kinetic friction between the ball and the incline are μ_s and μ_k, respectively. Assume that the mass center G is at the geometric center.

Problem 7.58 Assuming the ball rolls without slip, determine expressions for the angular acceleration of the ball and the friction and normal force between the ball and the incline. In addition, find the minimum value of μ_s that is compatible with this motion.

Problem 7.59 Let the weight of the ball be 14 lb, the radius be 4.25 in., and the radius of gyration be $k_G = 2.6$ in. If the incline is 10 ft long, determine the time it takes the ball to reach the bottom of the incline and the speed of G when it reaches the bottom. Use $\theta = 40°$, $\mu_s = 0.2$, and $\mu_k = 0.15$.

Problems 7.60 and 7.61

A bowling ball is thrown onto a lane with a backspin ω_0 and forward velocity v_0. The mass of the ball is m, its radius is r, its radius of gyration is k_G, and the coefficient of kinetic friction between the ball and the lane is μ_k. Assume the mass center G is at the geometric center.

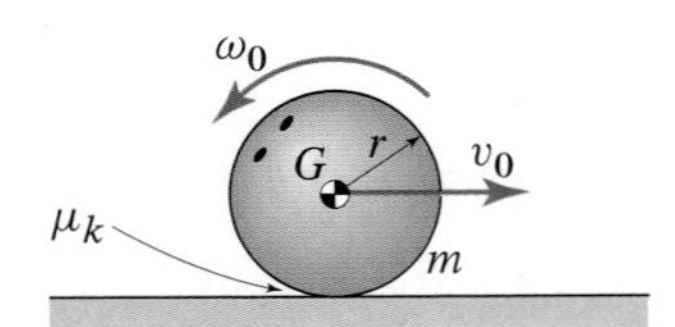

Figure P7.60 and P7.61

Problem 7.60 Find the acceleration of G and the ball's angular acceleration while the ball is slipping.

Problem 7.61 For a 14 lb ball with $r = 4.25$ in., $k_G = 2.6$ in, $\omega_0 = 10$ rad/s, and $v_0 = 17$ mph, determine the time it takes for the ball to start rolling without slip and its speed when it does so. In addition, determine the distance it travels before it starts rolling without slip. Use $\mu_k = 0.10$.

Problem 7.62

Solve Example 7.3 on p. 532 by assuming that the crate slips *and* tips. In doing so, show that this motion is not possible for the given conditions since part of your solution will not be physically admissible.

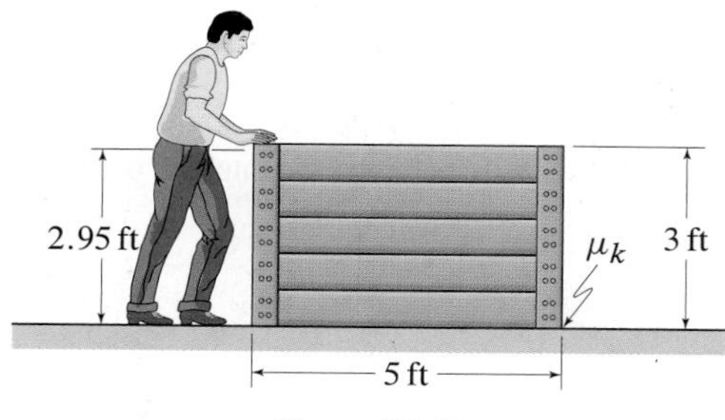

Figure P7.62

Problems 7.63 and 7.64

A shop sign, with a uniformly distributed mass $m = 30\,\text{kg}$, $h = 1.5\,\text{m}$, $w = 2\,\text{m}$, and $d = 0.6\,\text{m}$, is at rest when cord AB suddenly breaks.

Problem 7.63 Modeling AB and CD as inextensible and with negligible mass, determine the tension in cord CD and the acceleration of the sign's center of mass immediately after AB breaks.

Problem 7.64 Modeling AB and CD as elastic cords with negligible mass and stiffness $k = 8000\,\text{N/m}$, determine the tension in cord CD and the acceleration of the sign's center of mass immediately after AB breaks.

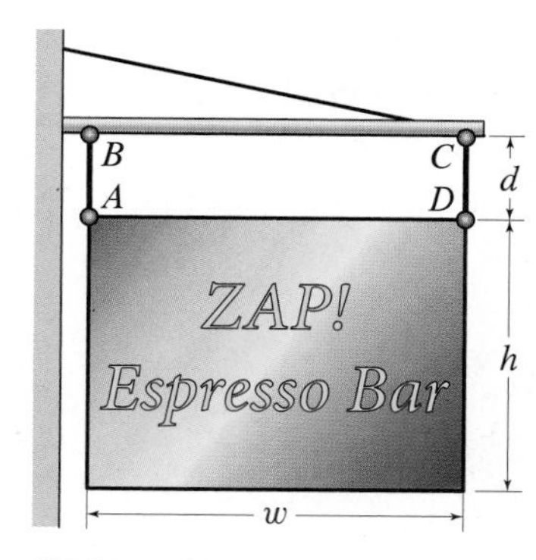

Figure P7.63 and P7.64

Problem 7.65

Solve Example 7.3 on p. 532 by assuming that the crate *just* tips. In doing so, show that this motion is not possible for the given conditions since part of your solution will not be physically admissible.

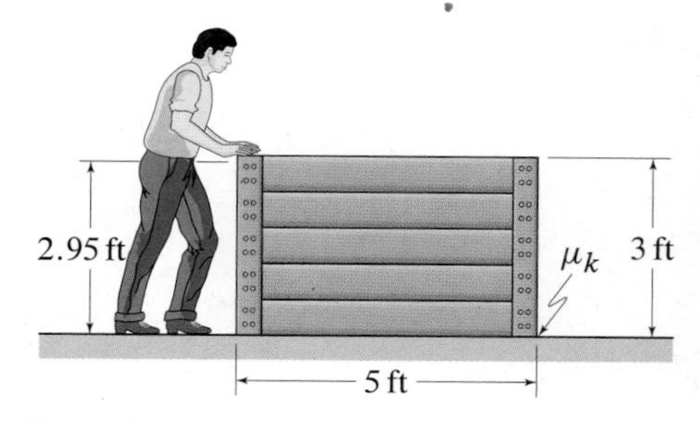

Figure P7.65

Problem 7.66

The cord, which is wrapped around the inner radius of the spool of mass m, is pulled vertically at A by a constant force P, causing the spool to roll over the horizontal bar BD. Assuming that the cord is inextensible and of negligible mass, that the spool rolls without slip, and that its radius of gyration is k_G, determine the angular acceleration of the spool and the total force between the spool and the bar.

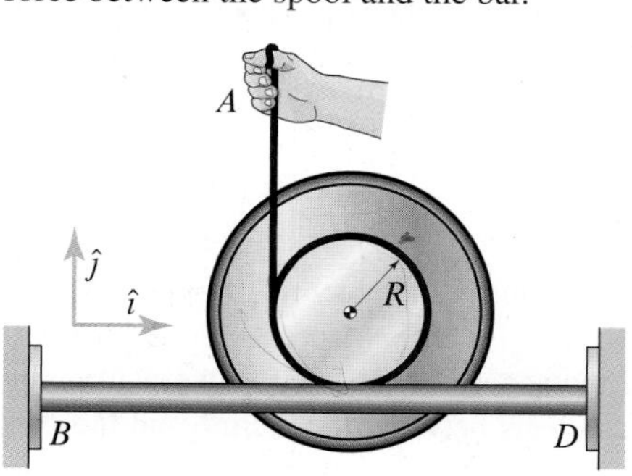

Figure P7.66

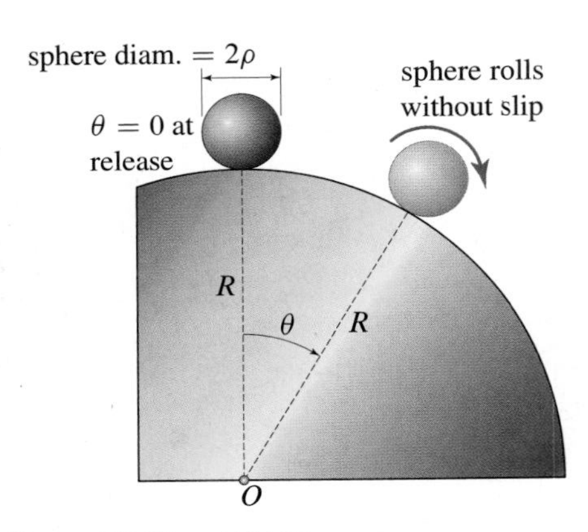

Figure P7.67 and P7.68

Problem 7.67

Refer to the systems in Example 3.9 on p. 200 (particle separating from semicylinder) and Example 7.10 on p. 558 (sphere separating from semicylinder).

(a) Determine the speed of the particle and that of the sphere when each separates from the semicylinder.

(b) Compare their speeds of separation and explain the sources of any difference.

(c) Determine the value of ρ such that the sphere and the particle separate at the same speed.

Problem 7.68

Referring to the systems in Example 3.9 on p. 200 (particle separating from semicylinder) and Example 7.10 on p. 558 (sphere separating from semicylinder), show that the sphere dynamically behaves just as a particle if the interface between the sphere and the semicylinder is frictionless. In this case, that will mean that the sphere separates from the semicylinder at the same location as the particle.

Problem 7.69

Referring to the system in Example 7.10 on p. 558 (and conveniently ignoring the Pitfall on p. 559 so that we can assume that the object rolls without slip), how would the results change if one were to release a uniform cylinder, of mass m and radius ρ, instead of a sphere?

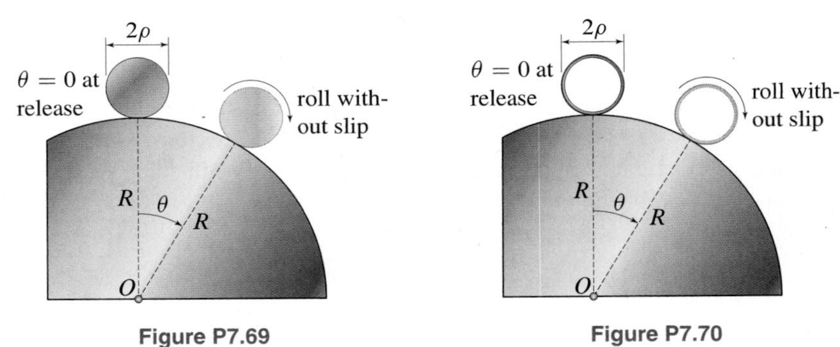

Figure P7.69

Figure P7.70

Problem 7.70

Referring to the system in Example 7.10 on p. 558 (and conveniently ignoring the Pitfall on p. 559 so that we can assume that the object rolls without slip), how would the results change if one were to release a uniform thin ring, of mass m and radius ρ, instead of a sphere?

Problem 7.71

A spool of mass $m = 300\,\text{kg}$, inner and outer radii $\rho = 1.5\,\text{m}$ and $R = 2\,\text{m}$, respectively, and radius of gyration $k_G = 1.8\,\text{m}$, is placed on an incline with $\theta = 43°$. The cable that is wrapped around the spool and attached to the wall is initially taut. If the static and kinetic friction coefficients between the incline and the spool are $\mu_s = 0.35$ and $\mu_k = 0.3$, respectively, determine the acceleration of G, the angular acceleration of the spool, and the tension in the cable once the spool is released from rest.

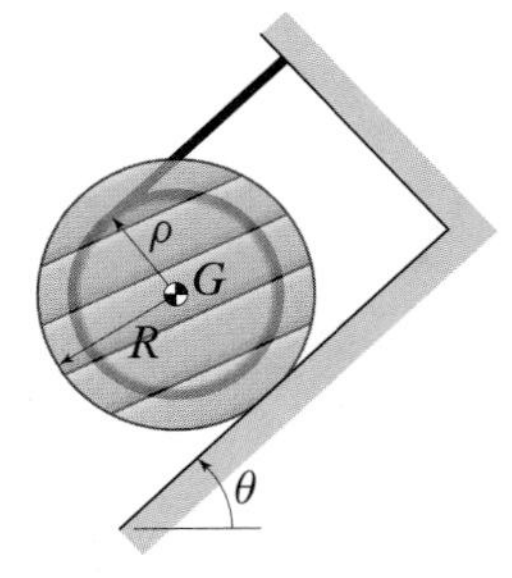

Figure P7.71

Problem 7.72

In Prob. 7.1 you were told to neglect the rotational inertia of the front wheels — would including it really make a difference? Let's see.

A certain roadster can go from 0 to 60 mph in 7.0 s. The weight of the car (including the two front wheels) is 2750 lb, the weight of each of its front wheels is 47 lb, and they each have a mass moment of inertia I_G of 0.989 slug·ft^2. To determine the effect of the rotational inertia of the front wheels, perform the following analysis:

(a) Isolate one of the front wheels and determine the friction force that must be acting on the wheel for it to accelerate as given. *Hint:* The weight of the car on the front wheel is not known, but it is not needed to find the friction force since we are assuming that friction is sufficient to prevent slipping of the front wheels.

(b) Note that it is the friction force that makes the rotational motion of each front wheel possible, and if the mass moment of inertia I_G of the front wheels were zero, then the friction force would be zero. Therefore, by neglecting the rotational inertia of the front wheels, the car would not be slowed by the friction forces found in (a). In other words, when we *do* account for the rotational inertia of the front wheels, we can then conclude that there is a force equal to twice the friction force that is "retarding" the motion of the car. Use this fact, and your result from (a), to determine the 0 to 60 mph time of this same roadster with front wheels that have no rotational inertia.

Figure P7.72

Problems 7.73 through 7.76

The uniform thin rod AB couples the slider A, which moves along a frictionless guide, to the wheel B, which rolls without slip over a horizontal surface. While not required, the use of computer algebra software is recommended for Probs. 7.74–7.76.

Problem 7.73 Assuming that A and B have negligible mass, that the mass of AB is m_{AB}, and that the system is released from rest at the angle θ, determine, immediately after release, the angular acceleration of the rod AB, the acceleration of the center of the wheel at B, and the angular acceleration of the wheel.

Problem 7.74 Assuming that A has negligible mass, B is a uniform disk of mass m_B, the mass of AB is m_{AB}, and the system is released from rest at the angle θ, determine, immediately after release, the angular acceleration of the rod AB, the acceleration of the center of the wheel at B, and the angular acceleration of the wheel.

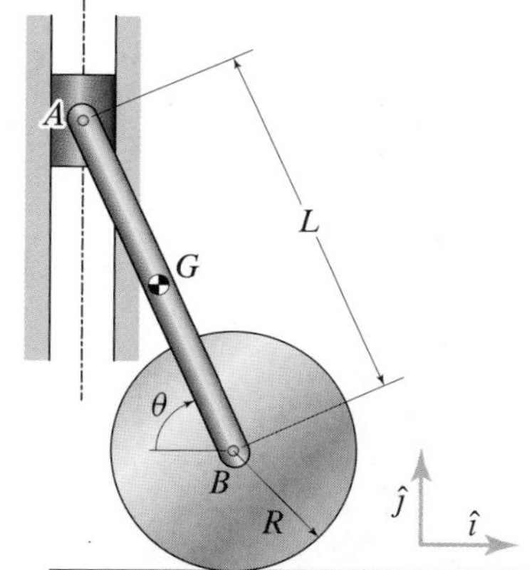

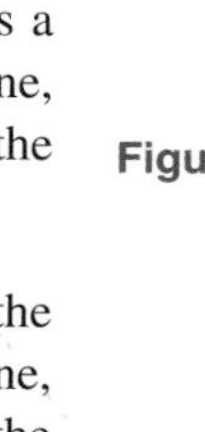

Figure P7.73–P7.76

Problem 7.75 Assuming that AB has negligible mass, the mass of A is m_A, B is a uniform disk of mass m_B, and the system is released from rest at the angle θ, determine, immediately after release, the angular acceleration of the rod AB, the acceleration of the center of the wheel at B, and the angular acceleration of the wheel.

Problem 7.76 Assuming that the mass of A is m_A, B is a uniform disk of mass m_B, the mass of AB is m_{AB}, and that the system is released from rest at the angle θ, determine, immediately after release, the angular acceleration of the rod AB, the acceleration of the center of the wheel at B, and the angular acceleration of the wheel.

Problem 7.77

A spool of mass $m = 220$ kg, inner and outer radii $\rho = 1.75$ m and $R = 2.25$ m, respectively, and radius of gyration $k_G = 1.9$ m, is being lowered down an incline with $\theta = 29°$. There is no slip between the spool and the cable as the spool moves down the incline. If the static and kinetic friction coefficients between the incline and the spool are $\mu_s = 0.4$ and $\mu_k = 0.35$, respectively, determine the acceleration of G, the angular acceleration of the spool, and the tension in the cable if the system is released from rest.

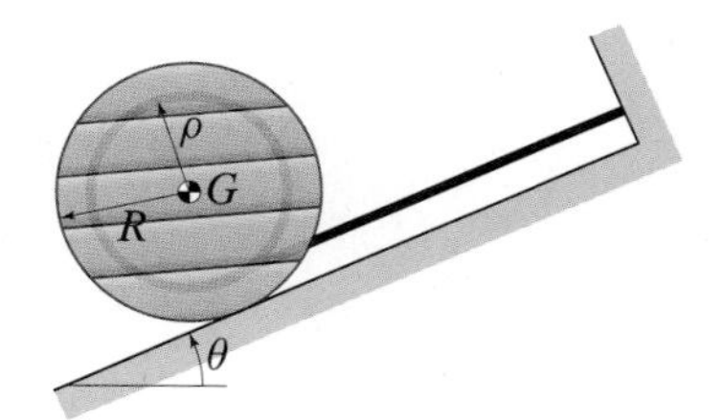

Figure P7.77

Figure P7.78

Problem 7.78

An inextensible cord of negligible mass is wound around a homogeneous circular object. Assume that the cord is pulled to the right while remaining horizontal, and determine the value of the object's mass moment of inertia I_G, such that the object rolls without slip no matter how large the tension in the cord. What is the shape of such an object?

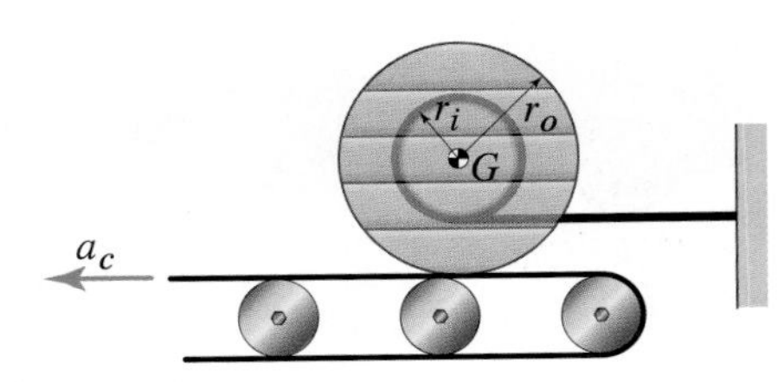

Figure P7.79

Problem 7.79

The spool of mass m, radius of gyration k_G, inner radius r_i, and outer radius r_o is placed on a horizontal conveyer belt. The cable that is wrapped around the spool and attached to the wall is initially taut. Both the spool and the conveyer belt are initially at rest when the conveyer belt starts moving with acceleration a_c. If the coefficient of static friction between the conveyer belt and spool is μ_s, determine

(a) The maximum acceleration of the conveyer belt so that the spool rolls without slipping on the belt

(b) The initial tension in the cable that attaches the spool to the wall

(c) The angular acceleration of the spool

Evaluate your answers for $m = 500\,\text{kg}$, $k_G = 1.3\,\text{m}$, $\mu_s = 0.5$, $r_i = 0.8\,\text{m}$, and $r_o = 1.6\,\text{m}$.

Problems 7.80 through 7.82

The thin uniform bar AB of mass m and length L hangs from a wheel at A, which rolls freely on the horizontal bar DE. In the following problems, neglect the mass of the wheel and assume that the wheel never separates from the horizontal bar.

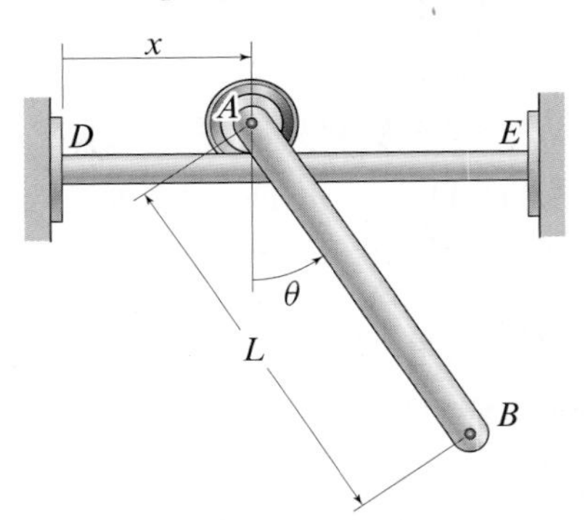

Figure P7.80–P7.82

Problem 7.80 If the bar is released from rest at the angle θ, determine, immediately after release, the angular acceleration of the bar, the force on the bar at A, and the acceleration of end A.

Problem 7.81 Find the equation(s) of motion of the bar, using the coordinates x and θ shown on the figure as the dependent variables.

Problem 7.82 Find the equation(s) of motion of the bar using the coordinates x and θ shown on the figure as the dependent variables and then simulate the system's behavior by numerically solving the equations of motion for 5 s, using $m = 2\,\text{kg}$, $L = 0.6\,\text{m}$, $x(0) = 0\,\text{m}$, $\dot{x}(0) = 0\,\text{m/s}$, $\theta(0) = 60°$, and $\dot{\theta}(0) = 0\,\text{rad/s}$. Plot x and θ for $0 \le t \le 5\,\text{s}$.

Problems 7.83 and 7.84

A uniform thin rod is slightly nudged at B from the $\theta = 0$ position so that it falls to the right. The coefficient of static friction between the rod and the floor is μ_s.

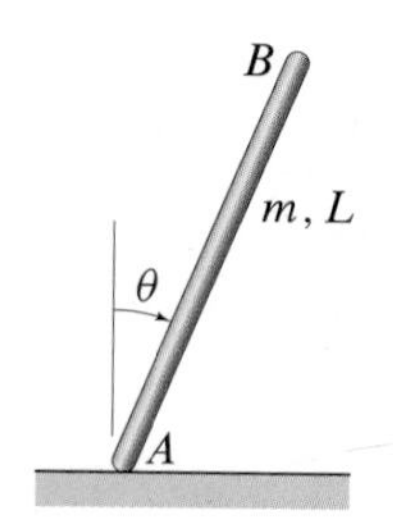

Figure P7.83 and P7.84

Problem 7.83

(a) Determine as a function of θ the normal force (N) and the frictional force (F) exerted by the ground on the rod as the rod falls over.

(b) Knowing that the rod will slip when $|F/N|$ exceeds μ_s, determine whether the rod will slip as it falls.

Problem 7.84

(a) Determine as a function of θ the normal force (N) and the frictional force (F) exerted by the ground on the rod as the rod falls over.

(b) Knowing that the rod will slip when $|F/N|$ exceeds μ_s, determine whether the rod will slip as it falls.

(c) Plot $F/(mg)$, $N/(mg)$, and $|F/N|$ as a function of θ for $0 \leq \theta \leq \pi/2$ rad. Use those plots to show that for smaller values of μ_s, end A of the rod slips to the left, and for larger values of μ_s, it slips to the right.

Problem 7.85

The roadster weighs 2570 lb, and its mass is evenly distributed between its front and rear wheels. It can accelerate from 0 to 60 mph in 6.98 s. The rear wheel, shown in the blowup above the roadster, weighs 47 lb, its mass center is at its geometric center, and its mass moment of inertia I_B is 0.989 slug·ft^2. With this in mind, we want to determine the forces on the rear wheel shown in Fig. 2 of Example 7.9.

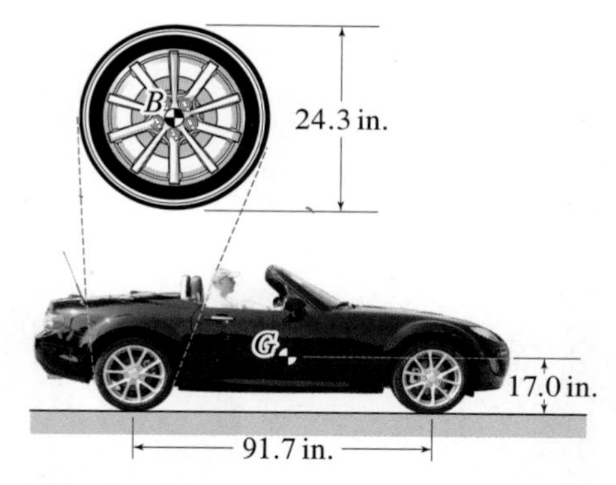

Figure P7.85

(a) Assuming that its acceleration is uniform, determine the forces on the front and rear wheels due to the pavement.

(b) Now that you have the normal and friction forces between the rear wheels and the pavement, isolate one of the rear wheels and determine the forces and moments exerted by the axle on that rear wheel.

Assume that the mass is evenly distributed between the right and left sides of the car and that friction is sufficient to prevent slipping of the wheels.

Problem 7.86

A spool of mass m, radius r, and radius of gyration k_G rolls without slipping on the incline, whose angle with respect to the horizontal is θ. A linear elastic spring with constant k and unstretched length L_0 connects the center of the spool to a fixed wall. Determine the equation(s) of motion of the spool, using the x coordinate shown.

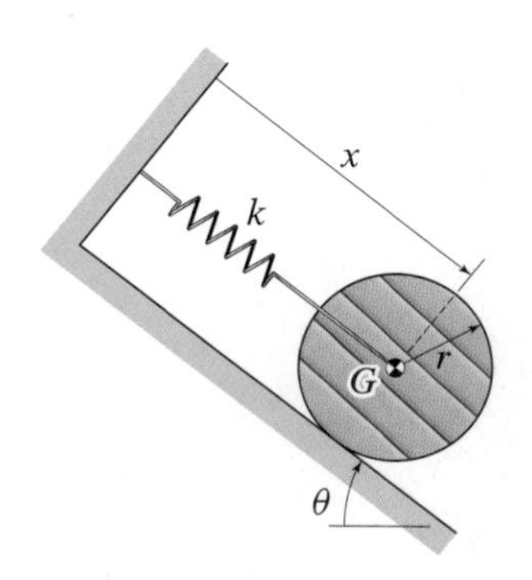

Figure P7.86 and P7.87

Problem 7.87

A spool of mass $m = 200$ kg, radius $r = 0.8$ m, and radius of gyration $k_G = 0.65$ m rolls without slipping on the incline, whose angle with respect to the horizontal is $\theta = 38°$. A linear elastic spring with constant $k = 500$ N/m and unstretched length $L_0 = 1.5$ m connects the center of the spool to a fixed wall. Determine the equation(s) of motion of the spool, using the x coordinate shown; solve them for 15 s, using the initial conditions $x(0) = 2.5$ m and $\dot{x}(0) = 0$ m/s; and then plot x versus t. What is the approximate period of oscillation of the spool?

Problem 7.88

The uniform bar AB of mass m and length L is leaning against the corner with $\theta \approx 0$ when end B is given a slight nudge so that end A starts sliding down the wall as B slides along the floor. Assuming that friction is negligible between the bar and the two surfaces against which it is sliding, determine the angle θ at which end A will lose contact with the vertical wall.

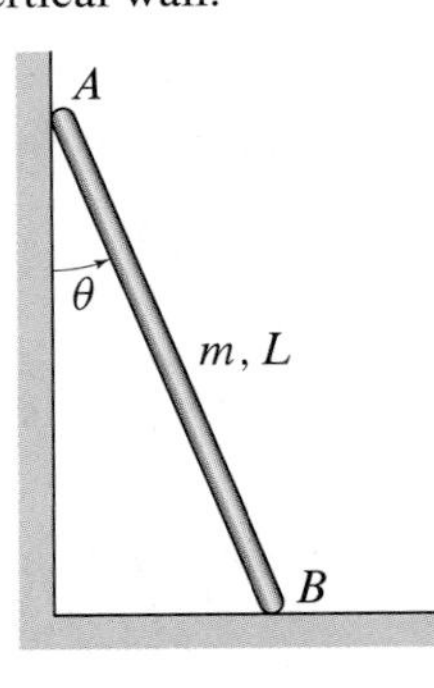

Figure P7.88

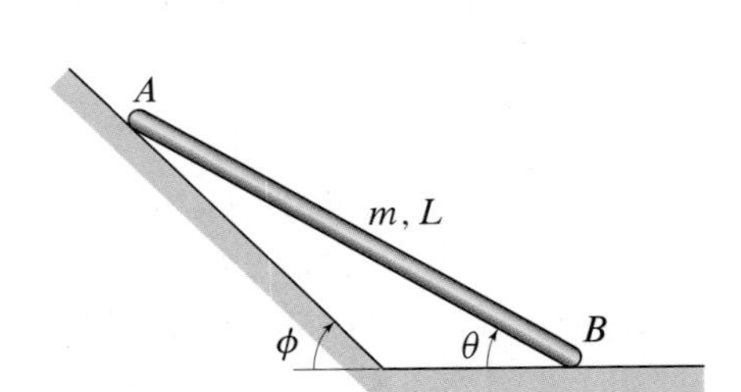

Figure P7.89

Problem 7.89

The uniform thin bar, which is leaning on the incline, is released from rest in the position shown and slides in the vertical plane. The contacts between the bar and the surface at ends A and B have negligible friction. Determine the angular acceleration of the bar immediately after it is released. Evaluate your answer for $m = 3\,\text{kg}$, $L = 0.75\,\text{m}$, $\phi = 45°$, and $\theta = 30°$.

Problems 7.90 through 7.93

The uniform ball of radius ρ and mass m is gently placed in the bowl B with inner radius R and is released. The angle ϕ measures the position of the center of the ball at G with respect to a vertical line through O. Assume that the system lies in the vertical plane. *Hint:* In working the following problems, we recommend using the $r\phi$ coordinate system shown.

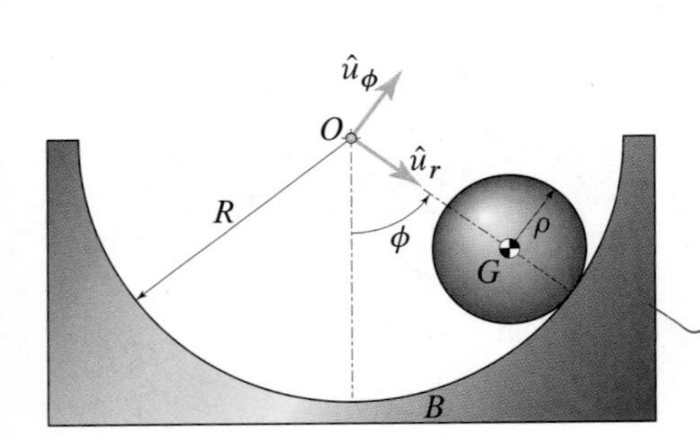

Figure P7.90–P7.93

Problem 7.90 Assuming that the ball rolls without slip, determine the acceleration of the center of the ball at G, the angular acceleration of the ball, and the force on the ball due to the bowl immediately after the ball is released.

Problem 7.91 Assuming that the ball rolls without slip, that it weighs 3 lb, is at the position $\phi = 40°$, and is moving clockwise at 10 ft/s, determine the acceleration of the center of the ball at G and the normal and friction force between the ball and the bowl. Use $R = 4\,\text{ft}$ and $\rho = 1.2\,\text{ft}$.

Problem 7.92 Assuming that friction is sufficient to prevent slipping, derive the equation(s) of motion of the ball in terms of the angle ϕ.

Problem 7.93 Assume that friction is sufficient to prevent slipping.

(a) Derive the equation(s) of motion of the ball in terms of the angle ϕ.

(b) Determine the friction force as a function of ϕ.

(c) Letting $\phi(0) = \phi_0$ and $\dot{\phi}(0) = 0$, with $0° < \phi_0 < 90°$, integrate the equation(s) of motion to determine the normal force as a function of ϕ.

(d) Using the results of Parts (b) and (c), and given a value for μ_s, determine the maximum value of $\phi(0) = \phi_0$ so that the ball does not slip.

Problem 7.94

An important problem in billiards or pool is the determination of the height at which you should hit the cue ball to give it backspin, topspin, or no spin. With this in mind, at what height h should the cue hit the ball so that the ball always rolls without slip, regardless of how hard the ball is hit and how much friction is available? Assume a uniform ball of mass m and radius r. You can determine this position without having to worry about the impact between the cue and the ball by studying an arbitrary horizontal force applied to a ball at height h.

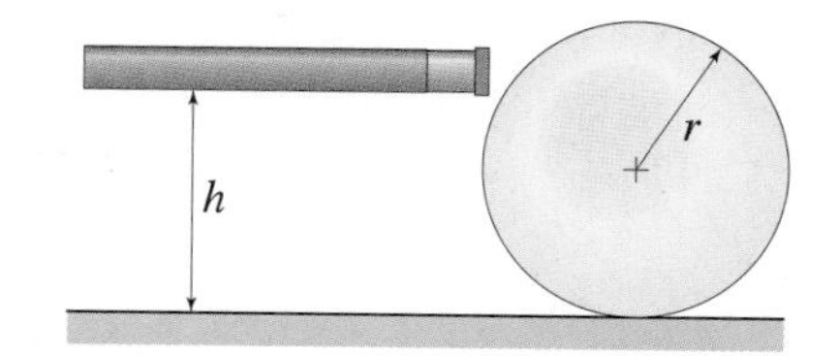

Figure P7.94

Problems 7.95 and 7.96

The Pioneer 3 spacecraft was a spin-stabilized spacecraft launched on December 6, 1958, by the U.S. Army Ballistic Missile agency in conjunction with NASA. It was designed with a despin mechanism consisting of two equal masses A and B, each of mass m, that could be spooled out to the ends of two wires of variable length $\ell(t)$ when triggered by a hydraulic timer. As the masses unwound, they would slow the spacecraft's spin from an initial angular velocity $\omega_s(0)$ to the final angular velocity $(\omega_s)_{\text{final}}$, and then the weights and wires would be released. Assume that masses A and B are initially at positions A_0 and B_0, respectively, before the wire begins to unwind, that the mass moment of inertia of the spacecraft is I_O (this does not include the two masses A and B), and that gravity and the mass of each wire are negligible. *Hint:* Refer to Prob. 6.160 if you need help with the kinematics.

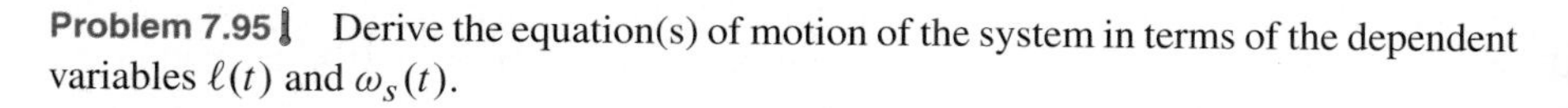

Problem 7.95 Derive the equation(s) of motion of the system in terms of the dependent variables $\ell(t)$ and $\omega_s(t)$.

Problem 7.96 Derive the equation(s) of motion of the system in terms of the dependent variables $\ell(t)$ and $\omega_s(t)$. After doing so:

(a) Use a computer to solve the equations of motion for $0 \le t \le 4$ s, using $R = 12.5$ cm, $m = 7$ g, $I_O = 0.0277\ \text{kg}\cdot\text{m}^2$, and the initial conditions $\omega_s(0) = 400$ rpm, $\ell(0) = 0.01$ m, and $\dot{\ell}(0) = 0$ m/s.

(b) Determine the time at which the angular velocity of the spacecraft becomes zero (this can be done by plotting the solution for ω_s and then estimating the time or by using numerical root finding to determine when $\omega_s = 0$).

(c) Determine the length ℓ of each of the wires at the instant that the angular velocity of the spacecraft becomes zero.

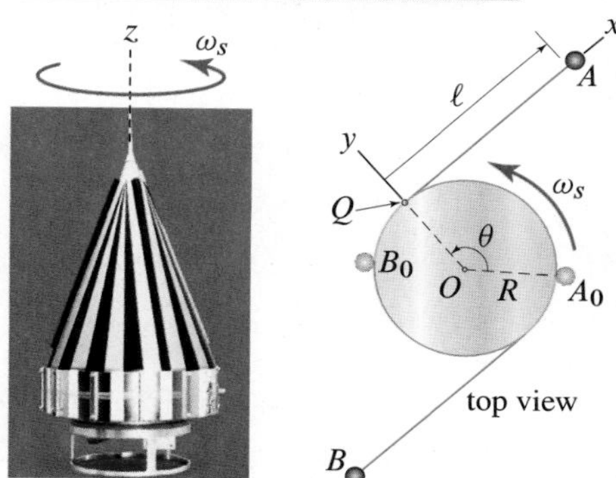

Figure P7.95 and P7.96

Problem 7.97

An SUV is pushing a large drum to the right with force P, using its front bumper. The drum has mass m and radius of gyration k_G. The static and kinetic friction coefficients between the drum and the ground and between the drum and the SUV are μ_s and μ_k, respectively.

(a) Assuming that there is no slipping between the drum and the ground, determine the acceleration of the drum and the minimum value of μ_s that is consistent with this motion.

(b) Determine the acceleration of point G and the angular acceleration of the drum if P is increased so that the drum slips relative to the ground.

Figure P7.97

Problems 7.98 and 7.99

The disk A rolls without slipping on a horizontal surface. End B of bar BC is pinned to the edge of the disk A, and end C of the bar BC can slide freely along the horizontal surface. In addition, bar BC is pushed by the force $P = mg$ at its left end. The mass of bar BC is m_{BC}, and the mass of the disk A is m_A. The system is initially at rest.

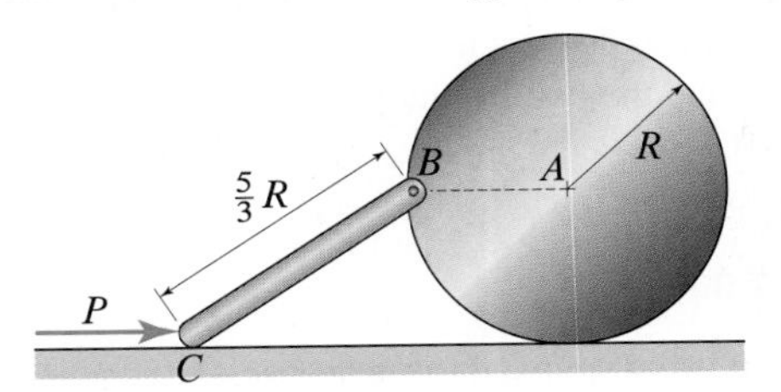

Figure P7.98 and P7.99

Problem 7.98 Determine the acceleration of the center of the disk A and the angular acceleration of bar BC immediately after the force P is applied if $m_A = m_{BC} = m$.

Problem 7.99 Determine the acceleration of the center of disk A and the angular acceleration of bar BC immediately after force P is applied if $m_{BC} = m$ and the mass of the disk m_A is negligible.

Problem 7.100

The crank AB in the slider-crank mechanism is rotating counterclockwise with constant angular velocity ω_{AB}. The crankshaft radius is R, the length of the connecting rod BC is L, and the distance from the mass center of the connecting rod D to the end of the crank at B is d. The mass of the connecting rod is m_D, the mass moment of inertia of the connecting rod is I_D, and the mass of the piston is m_C.

(a) Using the component system shown, determine the x and y components of the forces on the connecting rod at B and C as functions of the crank angle θ.

(b) Using $\omega_{AB} = 5700\,\text{rpm}$, $R = 48.5\,\text{mm}$, $L = 141\,\text{mm}$, $d = 36.4\,\text{mm}$, $m_D = 0.439\,\text{kg}$, $I_D = 0.00144\,\text{kg}\cdot\text{m}^2$, and $m_C = 0.434\,\text{kg}$, plot each of the four force components, the magnitude of the forces at B and C, and the moment acting on the connecting rod about point D, all as a function of θ, for one full rotation of the crank.

Hint: The kinematics of this problem have been considered in Example 6.10 on p. 480.

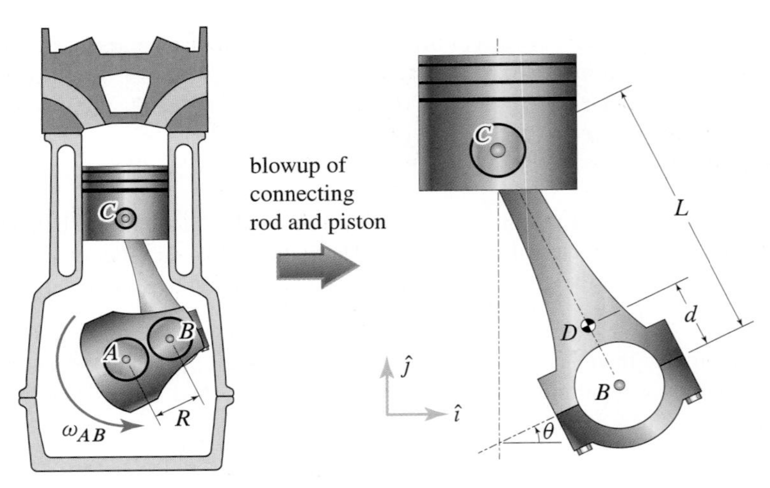

Figure P7.100

Design Problems

Design Problem 7.1

Revisit the calculations done at the beginning of the chapter concerning the determination of the maximum acceleration that can be achieved by a motorcycle without causing the front wheel to lift off the ground. Specifically, construct a new model of the motorcycle by selecting a real-life motorcycle and researching its geometry and inertia properties, including the inertia properties of the wheels. Then analyze your model to determine how the maximum acceleration in question depends on the horizontal and vertical positions of the center of mass with respect to the points of contact between the ground and the wheels. Include in your analysis a comparison of results that account for the inertia of the front wheel with results that neglect the inertia of the front wheel.

Figure DP7.1

Design Problem 7.2

One end of a seat belt on passenger cars is wound around a ratchet wheel that can be locked when the deceleration of the car exceeds a set value. In the sketch shown, consider a ratchet wheel that can rotate about the fixed point A. The ratchet wheel has cogs and is positioned above a weight. The weight can pivot about point B and is rigidly connected to a pawl that, for a sufficiently large rotation, will lock the motion of the ratchet wheel. Research realistic dimensions for the ratchet wheel, and design a locking mechanism, such that the belt's motion will be stopped for horizontal decelerations greater than $0.5g$, where g is the acceleration due to gravity. In your design you may use a spring to limit the motion of the weight.

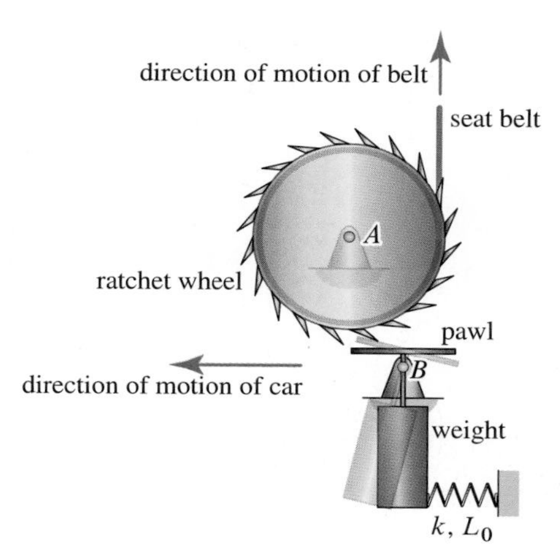

Figure DP7.2

Chapter Review

Newton-Euler Equations: Bodies Symmetric with Respect to the Plane of Motion

For a rigid body, the translational Newton-Euler equations are given by *Euler's first law*, which is

Eq. (7.1), p. 521

$$\vec{F} = m\vec{a}_G,$$

where $\vec{F}$ is the resultant of all *external* forces, m is the mass of the rigid body, and $\vec{a}_G$ is the inertial acceleration of its mass center (see Fig. 7.18).

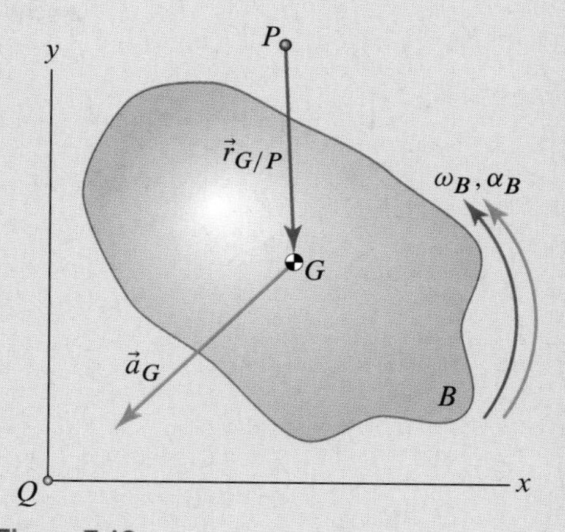

Figure 7.18
The relevant kinematic quantities for the Newton-Euler equations for a rigid body.

The moment center is an arbitrary point P. For *a rigid body that is symmetric with respect to the plane of motion*, the third Newton-Euler equation is the rotational equation of motion given by (see Fig. 7.18)

Eq. (7.28), p. 524, and Eq. (7.29), p. 525

$$M_P = I_G \alpha_B + m\left(x_{G/P} a_{Gy} - y_{G/P} a_{Gx}\right),$$
$$\vec{M}_P = I_G \vec{\alpha}_B + \vec{r}_{G/P} \times m\vec{a}_G,$$

where the second equation is simply the vector form of the first, and

- M_P is the total moment about P in the z direction
- I_G is the mass moment of inertia of the body about its mass center G
- α_B is the angular acceleration of the body
- m is the total mass of the body
- $\vec{a}_G = a_{Gx}\,\hat{\imath} + a_{Gy}\,\hat{\jmath}$ is the acceleration of the mass center
- $\vec{r}_{G/P} = x_{G/P}\,\hat{\imath} + y_{G/P}\,\hat{\jmath}$ is the position of the mass center G relative to the moment center P

If *any* one of the following conditions is true, that is, if $\vec{r}_{G/P} \times m\vec{a}_G = \vec{0}$:

1. Point P is the mass center G, so that $\vec{r}_{G/P} = \vec{0}$.
2. $\vec{a}_G = \vec{0}$ (i.e., G moves with constant velocity).
3. $\vec{r}_{G/P}$ is parallel to $\vec{a}_G$.

then Eqs. (7.28) and (7.29) reduce to

Eq. (7.30), p. 525

$$M_P = I_G \alpha_B.$$

The moment center is a point O on the body. If the moment center is a point O on the rigid body or an arbitrary extension of the rigid body, then an alternate form of Eq. (7.29) is (see Fig. 7.19)

Eq. (7.36), p. 525

$$\vec{M}_O = I_O \vec{\alpha}_B + \vec{r}_{G/O} \times m\vec{a}_O,$$

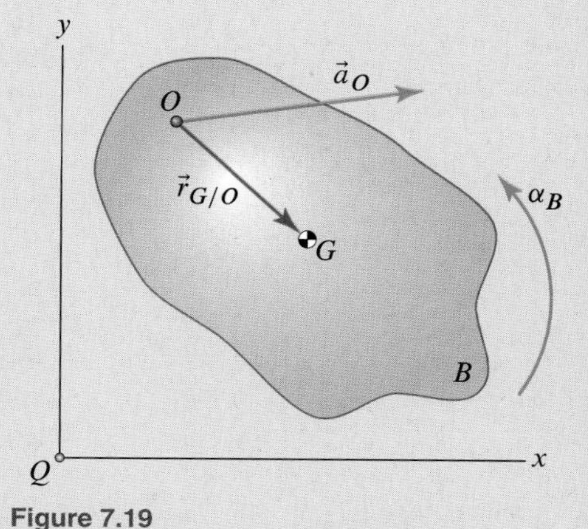

Figure 7.19
The relevant kinematic quantities for the rotational equations of motion of a rigid body when the moment center O is a point on the rigid body.

where I_O is the mass moment of inertia of the body about an axis perpendicular to the plane of motion passing through point O and $\vec{a}_O$ is the acceleration of point O. If *any* one of the following is true, that is, if $\vec{r}_{G/O} \times m\vec{a}_O = \vec{0}$:

1. Point O is the mass center G so that $\vec{r}_{G/O} = \vec{0}$.
2. $\vec{a}_O = \vec{0}$ (i.e., P moves with constant velocity).
3. $\vec{r}_{G/O}$ is parallel to $\vec{a}_O$.

then Eq. (7.36) becomes

Eq. (7.37), p. 526

$$M_O = I_O \alpha_B.$$

Graphical interpretation of the equations of motion. Referring to Fig. 7.20, there is a graphical/visual way of obtaining Eqs. (7.1) and (7.29). We begin by drawing the FBD of the rigid body, including all forces and moments, and then we draw the KD (kinetic diagram) of the rigid body, which includes the vectors $I_G \alpha_B$ and $m\vec{a}_G$. As shown in Fig. 7.20, we graphically equate these two diagrams, and we write the equations generated by that equation. In doing so, we automatically obtain Eqs. (7.1) and (7.29).

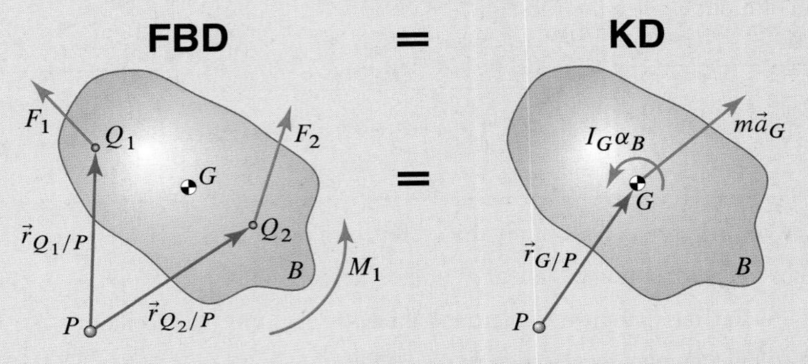

Figure 7.20. The free body diagram and kinetic diagram of a general rigid body. Equating them and writing the associated equations always give the correct equations of motion, i.e., Eqs. (7.1) and (7.29).

Newton-Euler Equations: Translation

For a body that is only translating, the first two Newton-Euler equations are the two scalar components of Euler's first law, which is (see Fig. 7.21)

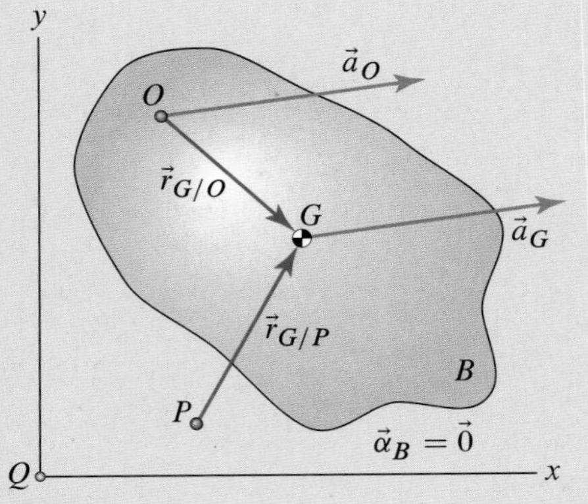

Figure 7.21
Illustration of the important points and kinematic quantities for a rigid body in translation.

Eq. (7.38), p. 529

$$\vec{F} = m\vec{a}_G.$$

If the moment center is an arbitrary point P, since $\vec{\alpha}_B = \vec{0}$ for a body in pure translation, then the rotational equation is given by

Eq. (7.39), p. 529

$$\vec{M}_P = \vec{r}_{G/P} \times m\vec{a}_G.$$

If the moment center is an arbitrary point P *and* any one of the following conditions is true, thus making $\vec{r}_{G/P} \times m\vec{a}_G = \vec{0}$:

1. Point P is the mass center G, so that $\vec{r}_{G/P} = \vec{0}$.
2. $\vec{a}_G = \vec{0}$ (i.e., G is fixed or moves with constant velocity).
3. $\vec{r}_{G/P}$ is parallel to $\vec{a}_G$.

then, when written in scalar form, Eq. (7.39) reduces to

Eq. (7.40), p. 529

$$M_P = 0.$$

Referring to Fig. 7.21, if the moment center is a point O on the translating rigid body, then we can apply Eq. (7.36) with $\vec{\alpha}_B = \vec{0}$, which is

Eq. (7.41), p. 529

$$\vec{M}_O = \vec{r}_{G/O} \times m\vec{a}_O,$$

where we have again used $\vec{\alpha}_B = \vec{0}$. If, in addition to the moment center being on the translating rigid body, *any* of the following is true:

1. Point O is the mass center G so that $\vec{r}_{G/O} = \vec{0}$.
2. $\vec{a}_O = \vec{0}$ (i.e., O is fixed or moves with constant velocity).
3. $\vec{r}_{G/O}$ is parallel to $\vec{a}_O$.

then, when written in scalar form, Eq. (7.41) becomes

Eq. (7.42), p. 529

$$M_O = 0.$$

Newton-Euler Equations: Rotation About a Fixed Axis

Referring to Fig. 7.22, for a body in fixed-axis rotation about a point, the three Newton-Euler equations of motion consist of the two scalar components of

Eq. (7.43), p. 539

$$\vec{F} = m\vec{a}_G,$$

along with a moment equation. If the moment center P is arbitrary, then the moment equation is given by

Eqs. (7.45) and (7.46), p. 539

$$\vec{M}_P = I_G\vec{\alpha}_B + \vec{r}_{G/P} \times m\vec{a}_G,$$
$$M_P = I_G\alpha_B,$$

where the second equation applies when $\vec{r}_{G/P} \times m\vec{a}_G = \vec{0}$. If, as is typical with fixed-axis rotation, the moment center O is on the axis of rotation, then the moment equation is

Eq. (7.47), p. 540

$$M_O = I_O\alpha_B.$$

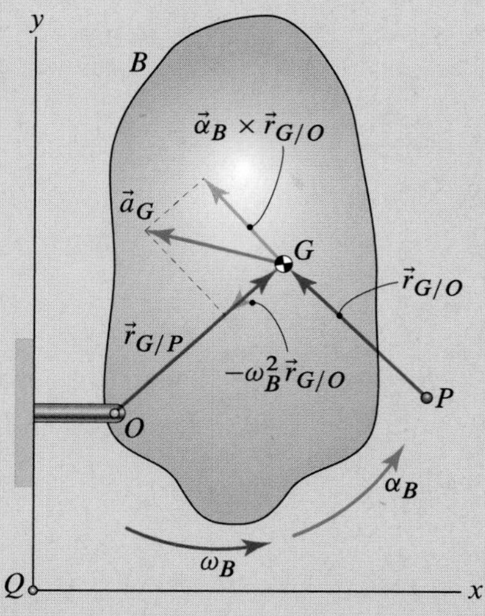

Figure 7.22
Illustration of the important points and kinematic quantities for a rigid body in translation.

Newton-Euler Equations: General Plane Motion

Referring to Fig. 7.23, when a body is translating *and* rotating, the first two Newton-Euler equations are always the two scalar components of Euler's first law, which is

Eq. (7.49), p. 553

$$\vec{F} = m\vec{a}_G.$$

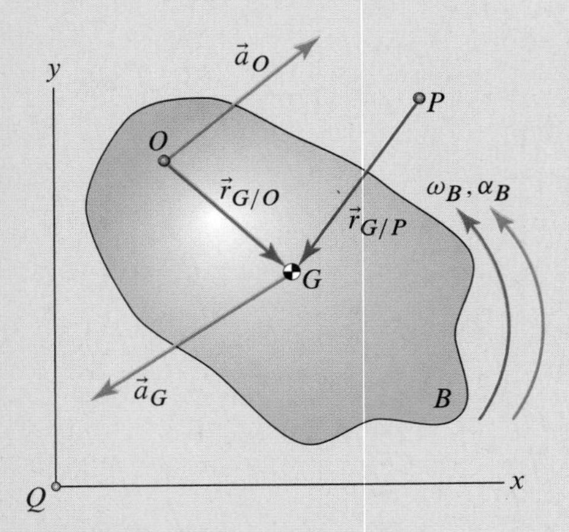

Figure 7.23
A general rigid body defining all relevant quantities needed to write its Newton-Euler equations.

Newton-Euler equations for an arbitrary moment center. Again referring to Fig. 7.23, when the moment center is an arbitrary point P, the third Newton-Euler equation is given by

Eq. (7.50), p. 553

$$\vec{M}_P = I_G \vec{\alpha}_B + \vec{r}_{G/P} \times m\vec{a}_G,$$

where, since the motion is planar, $\vec{M}_P = M_P\,\hat{k}$, $\vec{\alpha}_B = \alpha_B\,\hat{k}$, $\vec{r}_{G/P} = x_{G/P}\,\hat{\imath} + y_{G/P}\,\hat{\jmath}$, and $\vec{a}_G = a_{Gx}\,\hat{\imath} + a_{Gy}\,\hat{\jmath}$. If *any* one of the following conditions is true:

1. Point P is the mass center G, so that $\vec{r}_{G/P} = \vec{0}$.
2. $\vec{a}_G = \vec{0}$ (i.e., G is fixed or moves with constant velocity).
3. $\vec{r}_{G/P}$ is parallel to $\vec{a}_G$.

then $\vec{r}_{G/P} \times m\vec{a}_G = \vec{0}$, and Eq. (7.50) reduces to

Eq. (7.51), p. 553

$$M_P = I_G \alpha_B,$$

where we have written the result in scalar form.

Newton-Euler equations when the moment center is on the rigid body. Referring to Fig. 7.23, when the moment center is a point O on the rigid body, Eq. (7.36) on p. 525 can be used to obtain the third Newton-Euler equation, which is

Eq. (7.52), p. 553

$$\vec{M}_O = I_O \vec{\alpha}_B + \vec{r}_{G/O} \times m\vec{a}_O.$$

The more useful form of this equation is obtained if $\vec{r}_{G/O} \times m\vec{a}_O = \vec{0}$. This occurs if *any* of the following is true:

1. Point O is the mass center G so that $\vec{r}_{G/O} = \vec{0}$.
2. $\vec{a}_O = \vec{0}$ (i.e., O is fixed or moves with constant velocity).
3. $\vec{r}_{G/O}$ is parallel to $\vec{a}_O$.

and then Eq. (7.52) becomes

Eq. (7.53), p. 553

$$M_O = I_O \alpha_B,$$

where we have used the scalar form to reflect the fact that the motion is planar.

Review Problems

Problems 7.101 through 7.106

The figure shows a weapon called a battering ram (modern large battering rams are typically mounted on armored vehicles). The ram has a weight $W_r = 2500$ lb, center of mass at E, and radius of gyration $k_E = 6.5$ ft. Also let the distance between points A and B and between points C and D be 6 ft. In addition, let $h = 4.5$ ft and $d = 3$ ft. Finally, let the connections at points A, B, C, and D be pin connections, and assume that the cart does not move while the ram swings.

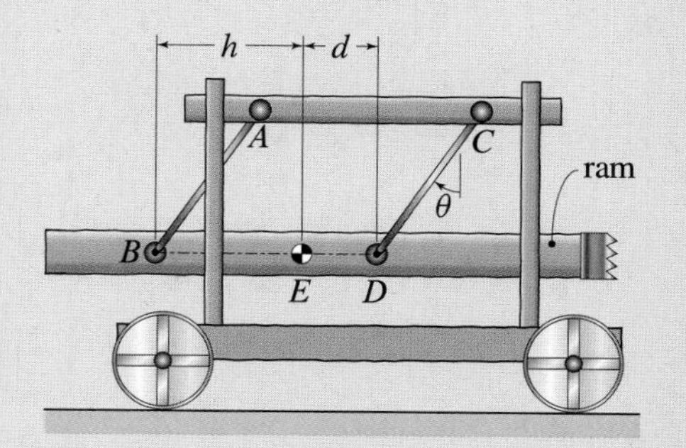

Figure P7.101–P7.106

Problem 7.101 Assuming that the ram is suspended by inextensible cords of negligible mass, determine the tension in the cords and the acceleration of E immediately after the ram is released from rest at $\theta = 75°$.

Problem 7.102 Let the ram be at rest with $\theta = 0°$. Assume that the cords AB and CD are inextensible and of negligible mass. Also assume that cord AB breaks suddenly. Determine the tension in cord CD and the acceleration of E immediately after AB breaks.

Problem 7.103 Assume that AB and CD are inextensible cords with negligible mass. In addition, assume that, at the instant shown, $\theta = 10°$ and the ram is swinging forward with $|\vec{v}_E| = 7$ ft/s. At this instant, determine the acceleration of E, as well as the reaction forces at points A and C.

Problem 7.104 Assume that AB and CD are uniform thin rods weighing 100 lb each. If the ram is released from rest when $\theta = 63°$, determine the acceleration of E, as well as the reaction forces at points A and C immediately after release.

Problem 7.105 Assume that AB and CD are uniform thin rods weighing 100 lb each and that the ram is at rest with $\theta = 0°$. Assume that bar CD breaks suddenly, and determine the acceleration of E and the forces at A immediately after CD breaks.

Problem 7.106 Assume that AB and CD are uniform thin rods weighing 100 lb each. Assume that, at the instant shown, $\theta = 10°$ and the ram is swinging forward with $|\vec{v}_E| = 7$ ft/s. At this instant, determine the acceleration of the ram, as well as the reaction forces at points A and C.

Problems 7.107 through 7.109

A spool has a weight $W = 450$ lb, outer and inner radii $R = 6$ ft and $\rho = 4.5$ ft, respectively, radius of gyration $k_G = 4.0$ ft, and mass center at G. The spool is being pulled to the right as shown, and the cable wrapped around the spool is inextensible and of negligible mass.

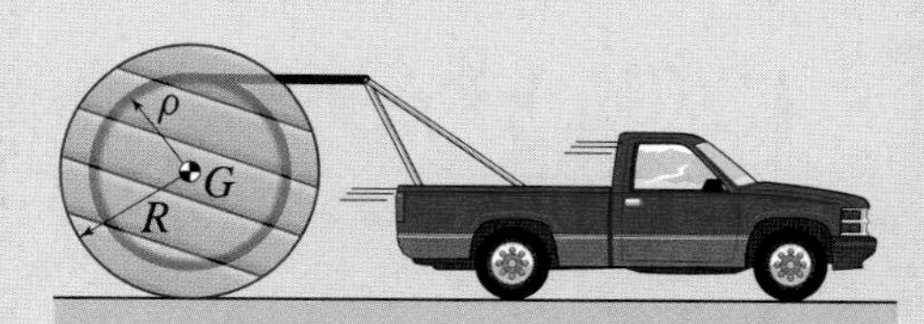

Figure P7.107–P7.109

Problem 7.107 Assume that the spool rolls without slipping with respect to both the cable and the ground. If the pickup truck pulls the cable with a force $P = 125$ lb, determine the acceleration of the center of the spool and the minimum value of the static friction coefficient between the spool and the ground that is compatible with this motion.

Problem 7.108 Assume that the static friction coefficient between the spool and the ground is $\mu_s = 0.75$, and determine the maximum value of the force that the truck could exert on the cable without causing the spool to slip relative to the ground.

Problem 7.109 Assume that the static and kinetic friction coefficients between the spool and the ground are $\mu_s = 0.25$ and $\mu_k = 0.2$. Furthermore, assume that the spool is initially at rest and that the pickup truck pulls the spool with a force of 550 lb. Determine the initial acceleration of the center of the spool.

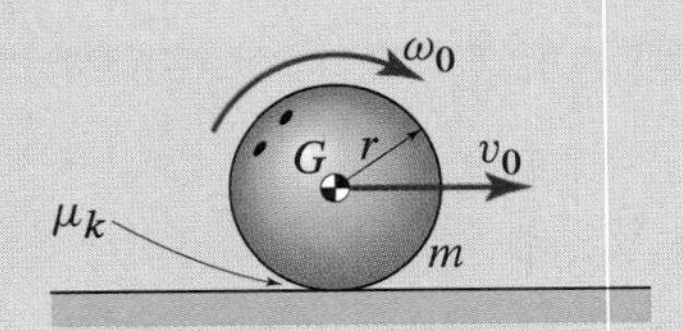

Figure P7.110 and P7.111

Problems 7.110 and 7.111

A bowling ball is thrown onto a lane with a forward spin ω_0 and forward velocity v_0. The mass of the ball is m, its radius is r, its radius of gyration is given by k_G, and the coefficient of kinetic friction between the ball and the lane is μ_k. Assume the mass center G is at the geometric center.

Problem 7.110 Assuming that $v_0 > r\omega_0$, determine the acceleration of the center of the ball and the angular acceleration of the ball until it starts rolling without slip.

Problem 7.111 Assuming that $v_0 < r\omega_0$, determine the acceleration of the center of the ball and the angular acceleration of the ball until it starts rolling without slip.

Problems 7.112 and 7.113

The car, as seen from the front, is traveling at a constant speed v_c on a turn of constant radius R that is banked at an angle θ with respect to the horizontal. The coefficient of static friction between the tires and the road is μ_s. The car's center of mass is at G.

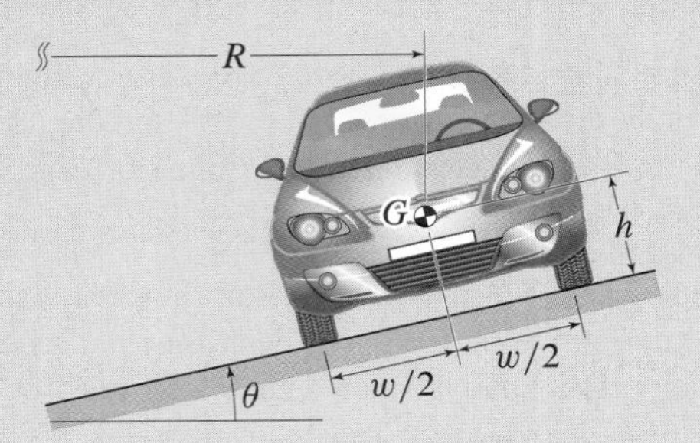

Figure P7.112 and P7.113

Problem 7.112 Determine the bank angle θ so that there is no tendency to slip or tip, i.e., so that no friction is required to keep the car on the road.

Problem 7.113 For a given bank angle θ, and assuming that the car does not tip, find the maximum speed v_m that the car can achieve without slipping.

Problems 7.114 through 7.116

An overhead fold-up door, with height h and mass m, consists of two identical sections hinged at C. The roller at A moves along a horizontal guide, whereas the rollers at B and D, which are the midpoints of sections AC and CE, move along a vertical guide. The door's operation is assisted by two identical springs attached to the horizontally moving rollers (only one of the two springs is shown). The springs are stretched an amount δ_0 when the door is fully open.

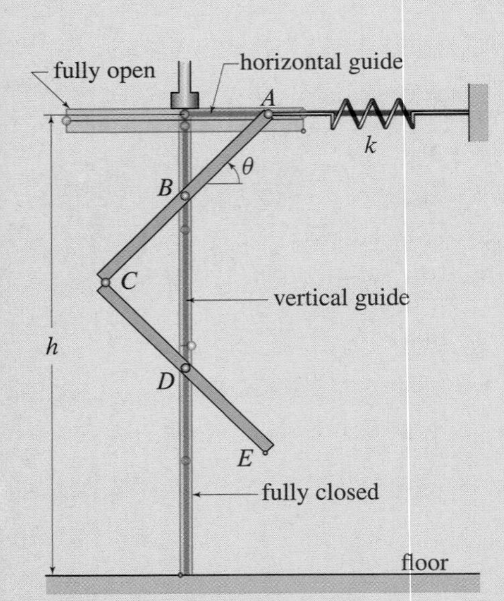

Figure P7.114–P7.116

Problem 7.114 Let $h = 10\,\text{m}$ and $m = 380\,\text{kg}$. In addition, let $k = 2400\,\text{N/m}$ and $\delta_0 = 0.15\,\text{m}$. Assuming that the door is released from rest when $\theta = 10°$ and that all sources of friction are negligible, determine the angular acceleration of each section of the door right after release.

Problem 7.115 Assuming that friction between the rollers and the guide can be neglected, determine the equation(s) of motion of the system.

Problem 7.116 Let $h = 10\,\text{m}$ and $m = 320\,\text{kg}$. In addition, let $k = 2400\,\text{N/m}$ and $\delta_0 = 0.15\,\text{m}$. Assuming that the door is released from rest when $\theta = 5°$, determine the time the door will take to close and the speed of E at closing.

Problems 7.117 through 7.119

A wheel with center O, radius R, weight W, radius of gyration k_G, and center of mass G at a distance ρ from O is released from rest on a rough incline. The angle ϕ is the angle between the segment OG (which rotates with the wheel) and the horizontal.

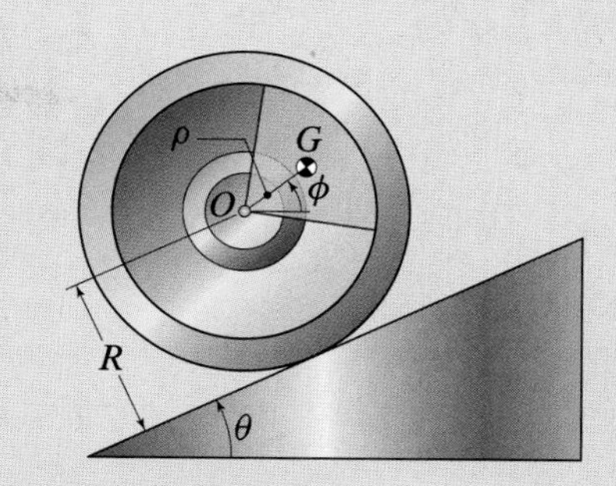

Figure P7.117–P7.119

Problem 7.117 Let $R = 1.5\,\text{ft}$, $\rho = 0.8\,\text{ft}$, $k_G = 0.6\,\text{ft}$, $W = 4\,\text{lb}$, and $\theta = 25°$. In addition, let $\phi = 35°$ at the instant of release. Determine the minimum coefficient of static friction so that the wheel starts moving while rolling without slip. In addition, determine the corresponding angular acceleration right after release.

Problem 7.118 Assuming that there is enough friction for the wheel to roll without slip, determine the equation(s) of motion of the wheel, as well as the constraint force equations, that is, those equations that would allow you to compute the reaction forces at the contact point with the incline if the motion were known.

Problem 7.119 Let $R = 1.5\,\text{ft}$, $\rho = 0.8\,\text{ft}$, $k_G = 0.6\,\text{ft}$, $W = 4\,\text{lb}$, and $\theta = 25°$, and let $\phi = 60°$ at the instant of release. Assuming that there is sufficient friction for the wheel to roll without slip and that the incline is sufficiently long that we need not worry about the wheel reaching the end of the incline, determine the equation(s) of motion of the wheel and expressions for the friction and normal forces at the point of contact between the wheel and the incline. Then, integrate the equation(s) of motion as a function of time for $0 \le t \le 2\,\text{s}$. Plot the normal force as a function of time over the given time interval and determine if and when the wheel loses contact with the incline.

Problems 7.120 and 7.121

The uniform slender bar AB has mass m_{AB} and length L. The crate has a uniformly distributed mass m_C and dimensions h and w. Bar AB is pin-connected to the trolley at A and to the crate at B. The trolley is constrained to move along the horizontal guide shown. Point O on the trolley's guide is a fixed reference point. Neglect the mass of the trolley and friction.

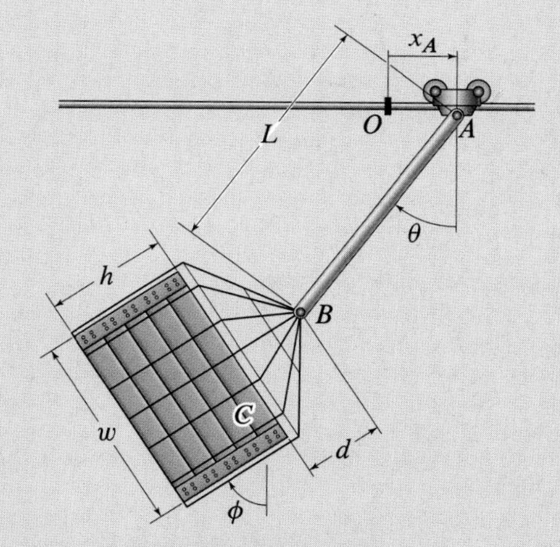

Figure P7.120 and P7.121

Problem 7.120 Derive the equations of motion of the system and express them in terms of the variables x_A, θ, and ϕ along with their time derivatives.

Problem 7.121 Let $m_{AB} = 75\,\text{kg}$, $L = 4.5\,\text{m}$, $m_C = 250\,\text{kg}$, $d = 0.5\,\text{m}$, $h = 1.5\,\text{m}$, and $w = 2\,\text{m}$. Finally, assume that the system is released from rest when $x_A = 0$, $\theta = 30°$, and $\phi = 45°$. Plot x_A, θ, and ϕ as functions of time for $0 \le t \le 15\,\text{s}$.

Problem 7.122

A drum of mass m_d, radius R, radius of gyration k_G, and with center of mass at G is placed on a cart of mass m_c for transport. The system is initially at rest, and the cart is pushed to the right with the force P. The coefficient of static friction between the cart and the drum is μ_s. Neglecting the mass of the wheels, determine the maximum force P that can be applied to the cart so that the drum does not slip on the cart, and find the corresponding acceleration of the cart and of point G.

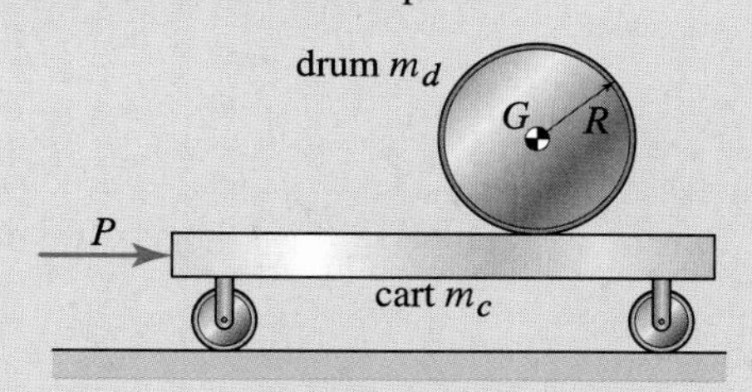

Figure P7.122

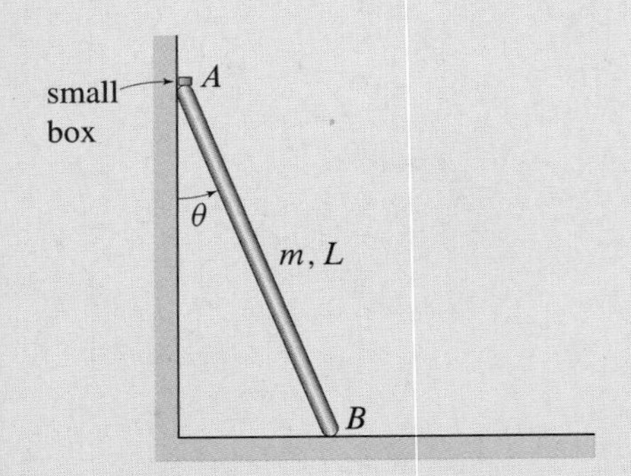

Figure P7.123

Problem 7.123

The uniform bar AB of mass m and length L is leaning against the corner with $\theta \approx 0$. A small box is placed on top of the bar at A. End B of the bar is given a slight nudge so that end A starts sliding down the wall as B slides along the floor. Assuming that friction is negligible between the bar and the two surfaces against which it is sliding, and neglecting the weight of the box, determine the angle θ at which the box will lose contact with the bar.

Problem 7.124

Bars AB and BC are uniform with masses $m_{AB} = 2\,\text{kg}$ and $m_{BC} = 1\,\text{kg}$, respectively. Their lengths are $L = 1.25\,\text{m}$ and $H = 0.75\,\text{m}$. Bar BC is pin-connected to a fixed support at C that is a distance $\delta = 0.2\,\text{m}$ from the ground. Bar AB is pin-connected at A to a uniform wheel with radius $R = 0.60\,\text{m}$ and mass $m_{OA} = 5\,\text{kg}$. Note that A is at a distance ρ from the center of the wheel. At the instant shown, A is vertically aligned with O; in addition, AB and BC are parallel and perpendicular to the ground, respectively. At this instant, bar BC is rotating clockwise with an angular velocity of $2\,\text{rad/s}$ and an angular acceleration of $1.2\,\text{rad/s}^2$. Assuming that the wheel rolls without slip, determine the force P that is applied to the wheel at the instant shown. In addition, determine the minimum static coefficient of friction necessary for the wheel not to slip.

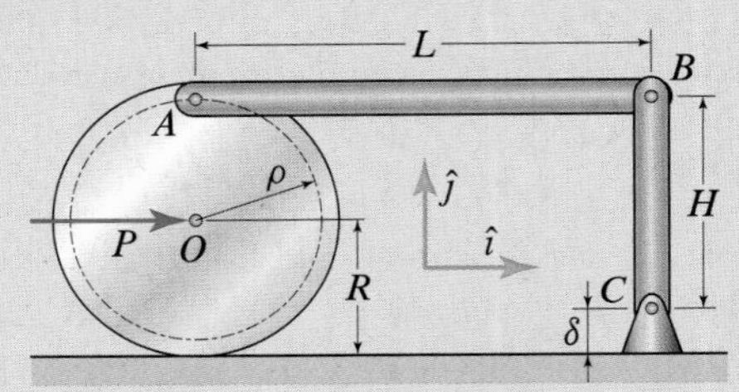

Figure P7.124

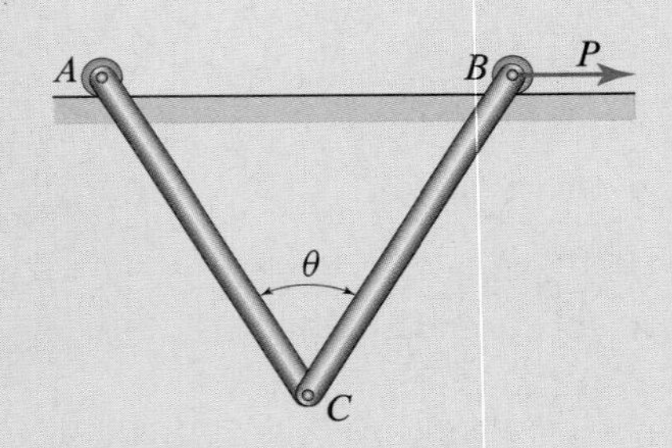

Figure P7.125

Problem 7.125

Two identical uniform bars are pinned together at one end, and each bar has a roller at its other end. The rollers can roll freely along the horizontal surface as shown. Each bar has mass m and length L, and a horizontal force P is applied to bar BC at B. Although the bars can rotate relative to each other, for a given value of P there exists a corresponding value of θ such that the system moves with θ constant.

(a) Find the forces on the bars at A and B and show that they are independent of P.

(b) Determine θ as a function of P and find θ_0 as $P \to 0$ and θ_∞ as $P \to \infty$.

Neglect the mass of the rollers and any friction in their bearings.

Energy and Momentum Methods for Rigid Bodies

In this chapter, we apply three fundamental balance laws to rigid bodies: the work-energy principle, the linear impulse-momentum principle, and the angular impulse-momentum principle. Each of these balance principles stems from the Newton-Euler equations we studied in Chapter 7, and they make it easier for us to solve many problems that would be far more challenging if solved using the Newton-Euler approach as presented in Chapter 7.

Astronaut Dave Williams working at the installation of a new control moment gyroscope on the International Space Station. The concept of angular momentum is essential to understand the working principles of gyroscopes.

8.1 Work-Energy Principle for Rigid Bodies

Kinetic energy of rigid bodies in planar motion

Referring to Fig. 8.1, consider an infinitesimal mass element dm with velocity $\vec{v}_{dm}$ and kinetic energy $dT_{dm} = \frac{1}{2}(dm)v_{dm}^2$ in a body B. The kinetic energy of the body B is therefore

$$T = \int_B \tfrac{1}{2} v_{dm}^2 \, dm. \tag{8.1}$$

Because G and dm are two points on the same rigid body, we can write

$$\vec{v}_{dm} = \vec{v}_G + \vec{\omega}_B \times \vec{q}, \tag{8.2}$$

where G and $\vec{\omega}_B$ are the mass center and angular velocity of B, respectively, and $\vec{q}$ is the position of dm relative to G. In planar motion, $\vec{\omega}_B = \omega_B \, \hat{k}$ and $\vec{q} = q_x \, \hat{\imath} + q_y \, \hat{\jmath}$, so $\vec{\omega}_B \times \vec{q} = \omega_B(-q_y \, \hat{\imath} + q_x \, \hat{\jmath})$. Using Eq. (8.2), we can then write

$$\begin{aligned} v_{dm}^2 &= \vec{v}_{dm} \cdot \vec{v}_{dm} \\ &= \vec{v}_G \cdot \vec{v}_G + 2\vec{v}_G \cdot \omega_B(-q_y \, \hat{\imath} + q_x \, \hat{\jmath}) + \omega_B^2(q_x^2 + q_y^2) \\ &= v_G^2 + 2\omega_B(-v_{Gx}q_y + v_{Gy}q_x) + \omega_B^2(q_x^2 + q_y^2), \end{aligned} \tag{8.3}$$

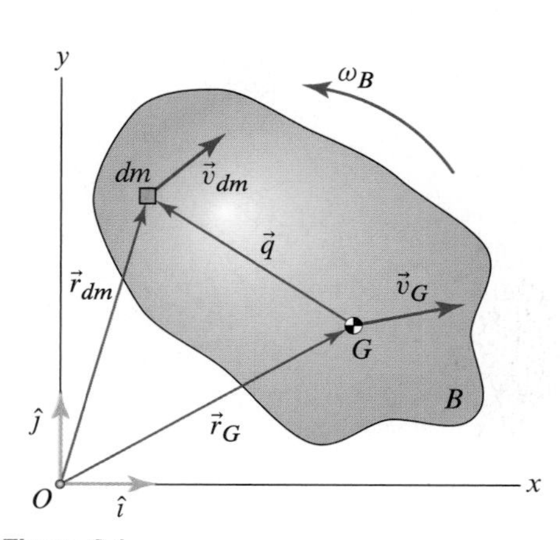

Figure 8.1
A rigid body in planar motion.

where we have used $\vec{v}_G = v_{Gx}\,\hat{\imath} + v_{Gy}\,\hat{\jmath}$. Substituting Eq. (8.3) into Eq. (8.1) gives

$$T = \int_B \left[\tfrac{1}{2}v_G^2 + \omega_B(-v_{Gx}q_y + v_{Gy}q_x) + \tfrac{1}{2}\omega_B^2(q_x^2 + q_y^2)\right] dm. \tag{8.4}$$

We can simplify Eq. (8.4), starting with the term $\frac{1}{2}v_G^2$, as

$$\int_B \tfrac{1}{2}v_G^2\,dm = \tfrac{1}{2}v_G^2 \int_B dm = \tfrac{1}{2}mv_G^2, \tag{8.5}$$

where $m = \int_B dm$ is the total mass of B. The last term in the integral of Eq. (8.4) can be written as

$$\int_B \tfrac{1}{2}\omega_B^2(q_x^2 + q_y^2)\,dm = \tfrac{1}{2}\omega_B^2 \int_B (q_x^2 + q_y^2)\,dm = \tfrac{1}{2}I_{Gz}\omega_B^2, \tag{8.6}$$

where we have used Eq. (7.27) on p. 524 to introduce the moment of inertia. The second term in the integral of Eq. (8.4) can be simplified as

$$\int_B \omega_B(-v_{Gx}q_y + v_{Gy}q_x)\,dm = -\omega_B v_{Gx}\int_B q_y\,dm + \omega_B v_{Gy}\int_B q_x\,dm = 0, \tag{8.7}$$

where we have used the fact that both $\int_B q_x\,dm$ and $\int_B q_y\,dm$ are zero because they measure the position of G with respect to G (see Eqs. (7.25) and (7.26) on p. 524). Substituting Eqs. (8.5)–(8.7) into Eq. (8.4) yields

Concept Alert

Kinetic energy of a rigid body. The kinetic energy of a rigid body consists of two parts. The first part, $\frac{1}{2}mv_G^2$, is due to motion of the center of mass (translational kinetic energy), and the second, $\frac{1}{2}I_G\omega_B^2$, is due to the rotational motion of points *relative* to the center of mass (rotational kinetic energy). A rigid body can have kinetic energy even when the center of mass does not move.

$$\boxed{T = \tfrac{1}{2}mv_G^2 + \tfrac{1}{2}I_G\omega_B^2,} \tag{8.8}$$

where we have written the shorthand I_G for I_{Gz} (as in Chapter 7). From Eq. (8.8), the kinetic energy of a rigid body in planar motion consists of

1. The term $\frac{1}{2}mv_G^2$, often called the *translational kinetic energy*, which is the kinetic energy of the body in translation
2. The term $\frac{1}{2}I_G\omega_B^2$, often called the *rotational kinetic energy*, which is the kinetic energy of the body in fixed-axis rotation about the mass center G and with the axis of rotation perpendicular to the plane of motion

Kinetic energy: rotation about a fixed axis

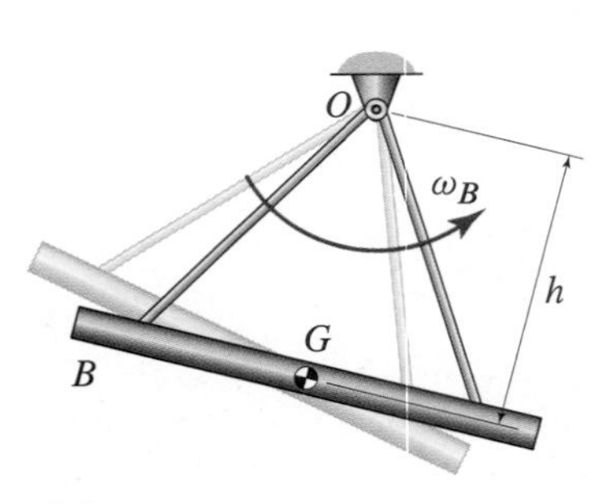

Figure 8.2
A platform B swinging about the pin at O.

Figure 8.2 shows a platform B supported by two bars that are both pinned at O and rigidly connected to B. The platform B rotates about the fixed axis perpendicular to the plane of the figure and going through O. If ω_B is the angular speed of B, rigid body kinematics tells us that the center of mass G of B has speed $v_G = \omega_B h$, where h is the distance between G and O. Equation (8.8) then tells us that the kinetic energy of B is

$$T = \tfrac{1}{2}mv_G^2 + \tfrac{1}{2}I_G\omega_B^2 = \tfrac{1}{2}m\omega_B^2h^2 + \tfrac{1}{2}I_G\omega_B^2, \tag{8.9}$$

which simplifies to

$$T = \tfrac{1}{2}(mh^2 + I_G)\omega_B^2 = \tfrac{1}{2}I_O\omega_B^2, \tag{8.10}$$

where, using the parallel axis theorem, $I_O = mh^2 + I_G$ is the mass moment of inertia of B relative to the axis of rotation (see App. A). This calculation shows that the kinetic energy of a rigid body B in a fixed-axis rotation can always be written as

$$\boxed{T = \tfrac{1}{2}I_O\omega_B^2,} \tag{8.11}$$

where I_O is the mass moment of inertia relative to the axis of rotation. This result is useful because fixed-axis rotations are common in applications.

Using the IC to find kinetic energy. Since the kinetic energy only depends on the velocities, we can also use Eq. (8.11) to determine the kinetic energy of a rigid body using the IC as point O. Referring to Fig. 8.3, which shows a uniform thin bar AB of mass m and length L sliding down a corner, we see that the IC is easy to locate and that it is a distance $\ell = L/2$ from the mass center G of the bar. If we know the bar's orientation θ and either v_A or v_B, then we can find ω_{AB} using either of the following expressions:

$$\omega_{AB} = \frac{v_A}{L \sin\theta} \quad \text{or} \quad \omega_{AB} = \frac{v_B}{L \cos\theta}. \tag{8.12}$$

Knowing ω_{AB}, we can then compute the kinetic energy of bar AB as

$$T = \tfrac{1}{2} I_{\text{IC}} \omega_{AB}^2 = \tfrac{1}{2}\left(I_G + m\ell^2\right)\omega_{AB}^2, \tag{8.13}$$

where I_G is the mass moment of inertia of the bar. Example 8.1 uses the IC to compute the kinetic energy of a bicycle.

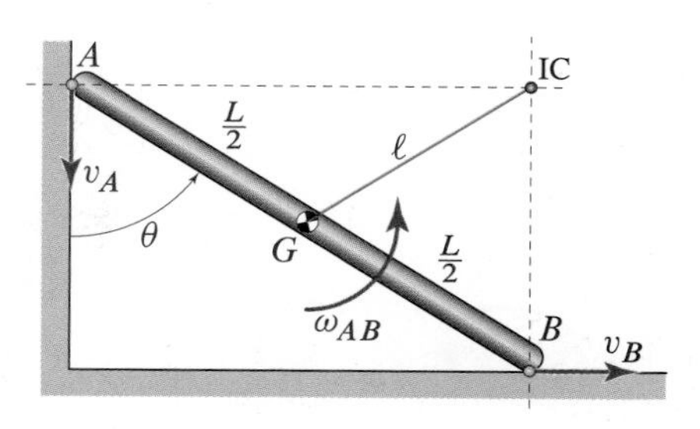

Figure 8.3
A bar sliding down a corner, identifying the kinematic parameters needed to find its kinetic energy using its IC.

Work-energy principle for a rigid body

Since a rigid body can be viewed as a rigid *system* of mass elements, the form of the work-energy principle for a rigid body is the same as that obtained for a *system* of particles, which is given in Eq. (4.44) on p. 282 as

$$T_1 + (U_{1\text{-}2})_{\text{ext}} + (U_{1\text{-}2})_{\text{int}} = T_2, \tag{8.14}$$

where T is the system's kinetic energy and $(U_{1\text{-}2})_{\text{ext}}$ and $(U_{1\text{-}2})_{\text{int}}$ are the work done in going from ① to ② by the forces that are external and internal to the system, respectively. Since we have already discussed the kinetic energy T, here we focus on the term $(U_{1\text{-}2})_{\text{int}}$, leaving the discussion about $(U_{1\text{-}2})_{\text{ext}}$ for later. As discussed in Section 4.3, the internal forces do work only when the system *deforms*. In the case of a *rigid* body, regardless of the internal forces, the *rigidity* of the body prevents deformation, and therefore the internal forces do no work, i.e.,

$$(U_{1\text{-}2})_{\text{int}} = 0. \tag{8.15}$$

Substituting Eq. (8.15) into Eq. (8.14), we get the work-energy principle for a rigid body

$$\boxed{T_1 + U_{1\text{-}2} = T_2,} \tag{8.16}$$

where $U_{1\text{-}2}$ is *only the work of the external forces and couples*. Equation (8.16) shows that the work-energy principle for a rigid body has the same form as that for a single particle. Next, we see how to compute term $U_{1\text{-}2}$ in Eq. (8.16).

Work done on rigid bodies

In statics we learned that a general force system consists of both forces and couples. We will briefly review how to compute the work of forces, and then we will learn how to compute the work of a couple.

Figure 8.4 shows a rigid body acted upon by n forces, labeled $\vec{F}_i$ $(i = 1, \ldots, n)$. The work of this force system is given by

$$U_{1\text{-}2} = \sum_{i=1}^{n} \int_{(\mathcal{L}_{1\text{-}2})_i} \vec{F}_i \cdot d\vec{r}_i, \tag{8.17}$$

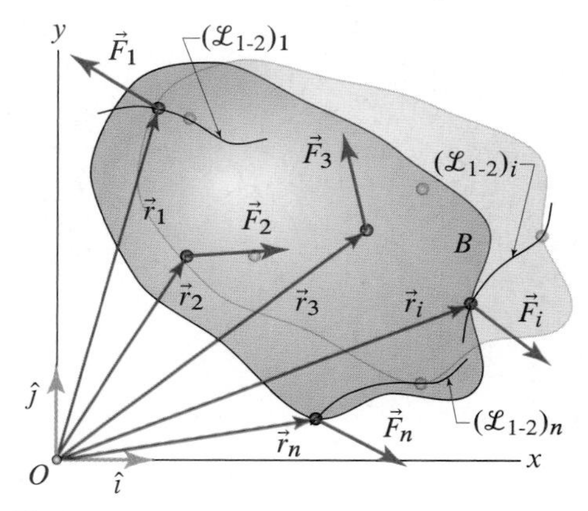

Figure 8.4
A rigid body under the action of a force system.

where $(\mathcal{L}_{1\text{-}2})_i$ is the *path of the point of application of force* $\vec{F}_i$ and $\vec{r}_i$ is the position of the *point of application* force $\vec{F}_i$. Equation (8.17) states that the work done by a force system is the sum of work contributions due to each force, where each contribution is computed by applying what we learned in Chapter 4.

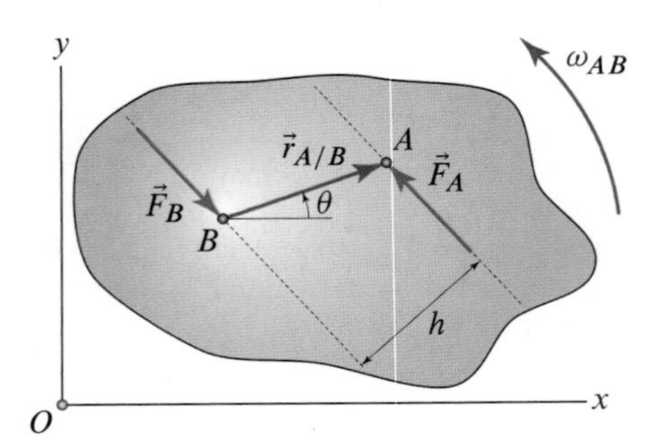

Figure 8.5
Rigid body subject to a couple.

Figure 8.5 shows a rigid body subject to a couple consisting of two equal and opposite forces $\vec{F}_A$ and $\vec{F}_B$ with parallel lines of action separated by the distance h. The moment of this couple is

$$\vec{M} = \vec{r}_{A/B} \times \vec{F}_A = \vec{r}_{B/A} \times \vec{F}_B, \tag{8.18}$$

where A and B are any two arbitrarily chosen points on the lines of action of $\vec{F}_A$ and $\vec{F}_B$, respectively. Applying Eq. (8.17) to the case of $\vec{F}_A$ and $\vec{F}_B$, we have

$$U_{1\text{-}2} = \int_{(\mathcal{L}_{1\text{-}2})_A} \vec{F}_A \cdot d\vec{r}_A + \int_{(\mathcal{L}_{1\text{-}2})_B} \vec{F}_B \cdot d\vec{r}_B. \tag{8.19}$$

Since $d\vec{r}_A = \vec{v}_A\,dt$ and $d\vec{r}_B = \vec{v}_B\,dt$, Eq. (8.19) can be rewritten as

$$U_{1\text{-}2} = \int_{t_1}^{t_2} (\vec{F}_A \cdot \vec{v}_A + \vec{F}_B \cdot \vec{v}_B)\,dt = \int_{t_1}^{t_2} \vec{F}_A \cdot (\vec{v}_A - \vec{v}_B)\,dt, \tag{8.20}$$

where we have used the fact that $\vec{F}_B = -\vec{F}_A$. Because the body is rigid, $\vec{v}_A - \vec{v}_B = \vec{\omega}_{AB} \times \vec{r}_{A/B}$, so Eq. (8.20) becomes

$$U_{1\text{-}2} = \int_{t_1}^{t_2} \vec{F}_A \cdot (\vec{\omega}_{AB} \times \vec{r}_{A/B})\,dt = \int_{t_1}^{t_2} \vec{\omega}_{AB} \cdot (\vec{r}_{A/B} \times \vec{F}_A)\,dt, \tag{8.21}$$

where the identity $\vec{a} \cdot (\vec{b} \times \vec{c}) = \vec{b} \cdot (\vec{c} \times \vec{a})$ has been used. The term $\vec{r}_{A/B} \times \vec{F}_A$ is equal to $\vec{M}$ from Eq. (8.18). Also, $\vec{\omega}_{AB} = \omega_{AB}\,\hat{k}$ and $\vec{M} = M\,\hat{k}$. Hence, Eq. (8.21) becomes

$$\boxed{U_{1\text{-}2} = \int_{t_1}^{t_2} M\omega_{AB}\,dt = \int_{\theta_1}^{\theta_2} M\,d\theta,} \tag{8.22}$$

since in planar rigid body motions $d\theta = \omega_{AB}\,dt$, where $d\theta$ is the infinitesimal angular displacement of the body. If M is constant, Eq. (8.22) simplifies to $U_{1\text{-}2} = M(\theta_2 - \theta_1)$, where $\theta_2 - \theta_1$ is the angular displacement of the body between ① and ②. The angles in Eq. (8.22) *must* be measured in radians.

Potential energy and conservation of energy

If some of the forces acting on a rigid body are conservative, we can account for their work using their associated potential energy.* Therefore, the work-energy principle for a rigid body moving between ① and ② under a general force system can be given the form

$$\boxed{T_1 + V_1 + (U_{1\text{-}2})_{\text{nc}} = T_2 + V_2,} \tag{8.23}$$

where V is the total potential energy of the body and where $(U_{1\text{-}2})_{\text{nc}}$ is the work of nonconservative forces, i.e, forces for which we do not have a potential energy function. Equation (8.23) has the same form as Eq. (4.27) on p. 260, which is the general form of the work-energy principle for a particle. If the entire force system consists of conservative forces, then Eq. (8.23) reduces to the familiar statement of conservation of mechanical energy, which is

$$\boxed{T_1 + V_1 = T_2 + V_2.} \tag{8.24}$$

* See Section 4.2 on p. 257 to review the concepts of conservative forces and potential energy.

Potential energy of a torsional spring

Torsional springs are designed to provide a moment in response to an angular displacement (Fig. 8.6). A typical application of a torsional spring is shown in Fig. 8.7(a), where we see a bar pinned at one end with a spring. If the bar is subject to an angular displacement θ, then the deformation in the torsional spring causes a moment to be applied to the bar in the direction opposite to the angular displacement, as shown in Fig. 8.7(b). The magnitude of the moment generated is a function of the angular

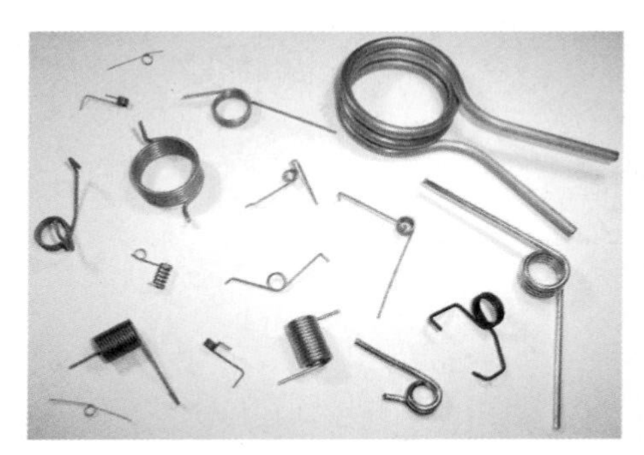

Figure 8.6
A collection of torsional springs.

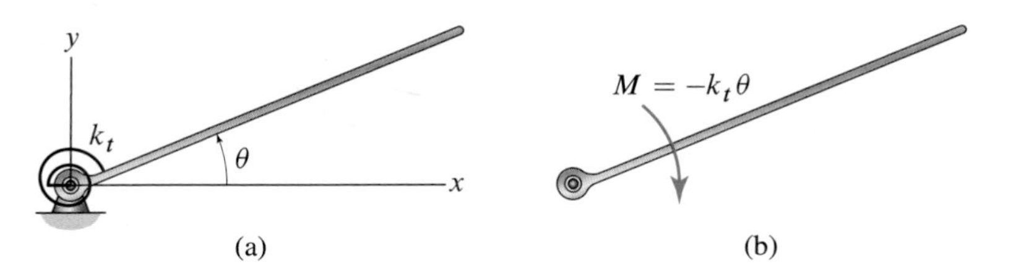

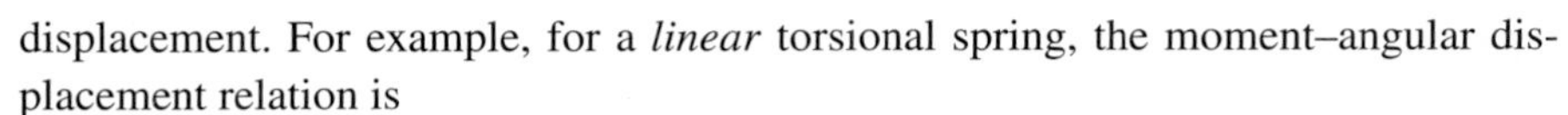

Figure 8.7. Bar with linear torsional spring (a) and spring's reaction moment (b).

displacement. For example, for a *linear* torsional spring, the moment–angular displacement relation is

$$M = -k_t\theta, \tag{8.25}$$

where k_t is the *torsional spring constant*. The dimensions of k_t are those of force × length/angle. Therefore, the SI units of k_t are N·m/rad, and in U.S. Customary units they are ft·lb/rad.

Applying Eq. (8.22) to compute the work done by a torsional spring, we obtain

$$(U_{1\text{-}2})_{\text{torsional spring}} = -\int_{\theta_1}^{\theta_2} k_t\theta\, d\theta = -\left(\tfrac{1}{2}k_t\theta_2^2 - \tfrac{1}{2}k_t\theta_1^2\right). \tag{8.26}$$

Equation (8.26) tells us that the work of a torsional spring depends only on the initial and final positions of the spring, which means that we can account for its work using a potential energy function. Since the relation between work and potential energy is $U_{1\text{-}2} = -(V_2 - V_1)$, the potential energy of a linear torsional spring is

$$\boxed{V_{\text{torsional spring}} = \tfrac{1}{2}k_t\theta^2.} \tag{8.27}$$

Potential energy of a constant gravitational force

A particle occupies a single point, so finding its change in height to compute the work done on it by gravity is straightforward. For rigid bodies, it may not be immediately obvious what point should be used to compute the change in height. Referring to the left side of Fig. 8.8, we see that, for example, point A moves to A' when the rigid body moves from ① to ②, but the net force of gravity does not act at A, it acts at G. Therefore, when computing the change in height of a rigid body for purposes of computing its gravitational potential energy, we still use Eq. (4.23) on p. 259, that is,

$$\boxed{V_g = mgy,} \tag{8.28}$$

where y now *measures the height of the mass center* with respect to the arbitrarily chosen datum line. Referring to the right side of Fig. 8.8, we see that a rigid body that simply rotates about its mass center G has no change in gravitational potential, even though *infinitely many* points on the body change their height (e.g., A to A').

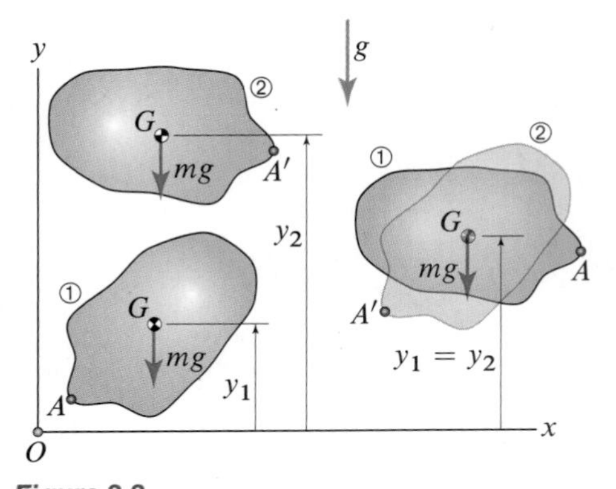

Figure 8.8
Left: A rigid body that rotates and whose mass center G has changed its height. Right: A rigid body that rotates and whose mass center G has *not* changed its height.

Work-energy principle for systems

Whether we view a physical system as consisting of particles, rigid bodies, or a mixture of the two, the statement of the work-energy principle takes on a form that is always the same. Therefore, avoiding unnecessary rederivations, we simply state the work-energy principle for any physical system as

$$T_1 + V_1 + (U_{1\text{-}2})^{\text{ext}}_{\text{nc}} + (U_{1\text{-}2})^{\text{int}}_{\text{nc}} = T_2 + V_2, \tag{8.29}$$

where

- T is the total kinetic energy, given by the sum of the kinetic energy of each individual part;
- V is the total potential energy, consisting of contributions from all conservative forces whether external or internal to the system;
- $(U_{1\text{-}2})^{\text{ext}}_{\text{nc}}$ is the work of all *external* forces without a potential energy; and
- $(U_{1\text{-}2})^{\text{int}}_{\text{nc}}$ is the work of all *internal* forces without a potential energy.

Equation (8.29) has the same form as Eq. (4.45) on p. 282.

Common Pitfall

Work of internal forces for rigid bodies. The internal work term $(U_{1\text{-}2})^{\text{int}}_{\text{nc}}$ in Eq. (8.29) does *not* refer to work internal to a single rigid body—it refers to the work done between two or more rigid bodies in a system of rigid bodies.

Power

We first discussed the power developed by a force in Section 4.4 on p. 298. Whether we model an object as a particle or as a rigid body, the power developed by the force $\vec{F}$ as its point of application moves with a velocity $\vec{v}$ is the work done by the force per unit time and is

$$\text{Power developed by a force } \vec{F} = \frac{dU}{dt} = \vec{F} \cdot \vec{v}. \tag{8.30}$$

As far as the power of a couple is concerned, referring to Fig 8.9, we can derive its form as a direct application of the fundamental theorem of calculus to the expression of the work of a couple given in Eq. (8.21), i.e.,

$$\text{Power developed by a couple} = \frac{dU}{dt} = \vec{\omega}_{AB} \cdot (\vec{r}_{A/B} \times \vec{F}_A) = \vec{M} \cdot \vec{\omega}_{AB}, \tag{8.31}$$

where $\vec{M}$ is the moment of the couple.

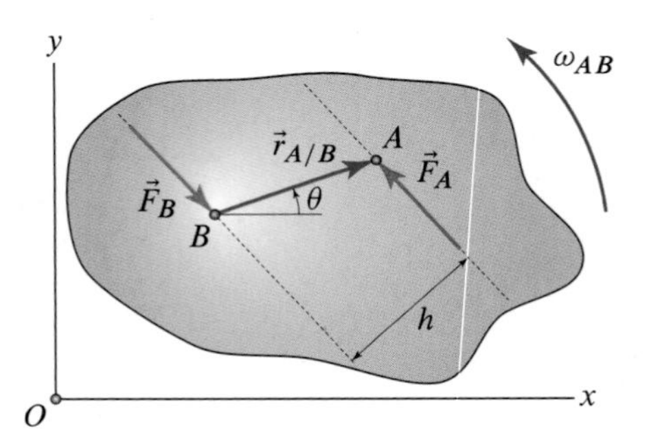

Figure 8.9
Rigid body subject to a couple.

End of Section Summary

The kinetic energy T of a rigid body B in planar motion is given by

Eq. (8.8), p. 584

$$T = \tfrac{1}{2} m v_G^2 + \tfrac{1}{2} I_G \omega_B^2,$$

or by

Eq. (8.11), p. 584

$$T = \tfrac{1}{2} I_O \omega_B^2,$$

where m, G, and I_G are the body's mass, center of mass, and mass moment of inertia, respectively. The second equation applies to fixed-axis rotation, where I_O is the mass moment of inertia relative to the axis of rotation. The work-energy principle for a rigid body is

Eq. (8.16), p. 585

$$T_1 + U_{1\text{-}2} = T_2,$$

where $U_{1\text{-}2}$ is the work done on the body in going from ① to ② by only the external forces. If we make use of the potential energy V of conservative forces, the work-energy principle can also be written as

Eqs. (8.23) and (8.24), p. 586

$$T_1 + V_1 + (U_{1\text{-}2})_{\text{nc}} = T_2 + V_2 \quad \text{(general systems)},$$
$$T_1 + V_1 = T_2 + V_2 \quad \text{(conservative systems)},$$

where the second expression applies only to a conservative system and states that the total mechanical energy of the system is conserved. For systems of rigid bodies or mixed systems of rigid bodies and particles, the work-energy principle is

Eq. (8.29), p. 588

$$T_1 + V_1 + (U_{1\text{-}2})^{\text{ext}}_{\text{nc}} + (U_{1\text{-}2})^{\text{int}}_{\text{nc}} = T_2 + V_2.$$

Here, V is the total potential energy (of both external and internal conservative forces). $(U_{1\text{-}2})^{\text{ext}}_{\text{nc}}$ and $(U_{1\text{-}2})^{\text{int}}_{\text{nc}}$ are the work contributions due to external and internal forces for which we do not have a potential energy, respectively.

Referring to Fig. 8.10, in planar rigid body motion, the work of a couple is

Eq. (8.22), p. 586

$$U_{1\text{-}2} = \int_{t_1}^{t_2} M\omega_{AB}\, dt = \int_{\theta_1}^{\theta_2} M\, d\theta.$$

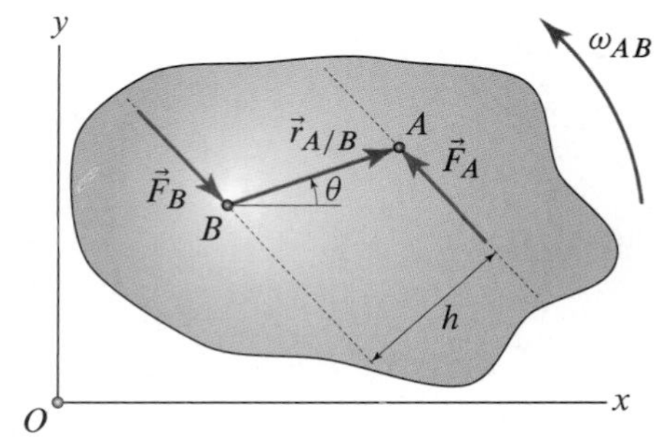

Figure 8.10
Rigid body subject to a couple.

Here, $d\theta$ is the body's infinitesimal angular displacement, and M is the component of the moment of the couple in the direction perpendicular to the plane of motion, taken to be positive in the direction of positive θ, and θ *must* be measured in radians. In addition, the power developed by a couple with moment $\vec{M}$ is computed as

Eq. (8.31), p. 588

$$\text{Power developed by a couple} = \frac{dU}{dt} = \vec{M} \cdot \vec{\omega}_{AB},$$

where $\vec{\omega}_{AB}$ is the body's angular velocity.

EXAMPLE 8.1 *Kinetic Energy Computation with Rolling Without Slip*

Figure 1
A cyclist riding on a horizontal road.

A cyclist rides on a horizontal road at a constant speed $v_0 = 35\,\mathrm{km/h}$. The mass of the frame (the bicycle without the wheels) is $m_f = 2.25\,\mathrm{kg}$. The front and the rear wheels have the same diameter $d = 0.7\,\mathrm{m}$. The masses of the front and rear wheels are $m_{\mathrm{fw}} = 0.737\,\mathrm{kg}$ and $m_{\mathrm{rw}} = 0.933\,\mathrm{kg}$, respectively, and the mass moments of inertia of the front and rear wheels are $I_C = 0.0510\,\mathrm{kg{\cdot}m^2}$ and $I_D = 0.0501\,\mathrm{kg{\cdot}m^2}$, respectively (see Fig. 2). Assuming the wheels roll without slip, compute the kinetic energy of just the bicycle. Neglect the kinetic energy of the pedals, chain, and other components not specifically mentioned.

SOLUTION

Road Map We are given a system's motion, and we are asked to compute the system's kinetic energy, where the system consists of a bicycle frame and two wheels. The kinetic energy of the bicycle is found by computing the kinetic energy of each part and then adding the individual contributions. The solution does not require any knowledge of the forces acting on the system. Therefore, as in kinematics problems, our solution will consist of only the computation step.

Computation Letting T denote the kinetic energy of the system, we have

$$T = T_f + T_{\mathrm{fw}} + T_{\mathrm{rw}}, \tag{1}$$

where T_f, T_{fw}, and T_{rw} are the kinetic energies of the frame, front wheel, and rear wheel, respectively. Because the system is traveling along a horizontal road, the frame is translating in the x direction with speed v_0 (see Fig. 2). Therefore, since the angular velocity

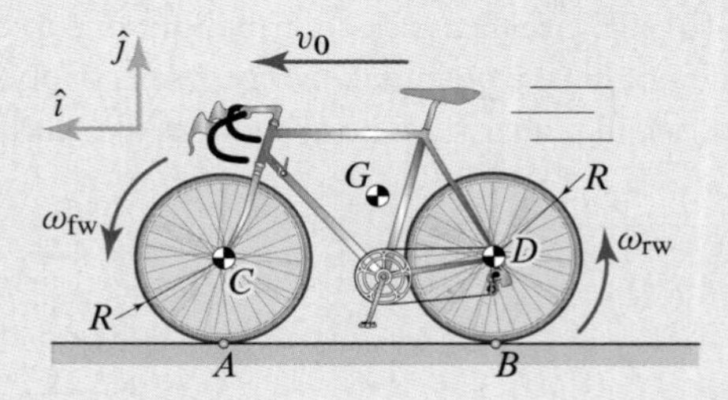

Figure 2. Bicycle moving at constant speed v_0. The points G, C, and D are the centers of mass of the frame, front wheel, and rear wheel, respectively.

of the frame is equal to zero, applying Eq. (8.8) on p. 584, we have

$$T_f = \tfrac{1}{2} m_f v_0^2 = 106.3\,\mathrm{J}. \tag{2}$$

Applying Eq. (8.8) to the front wheel, we have

$$T_{\mathrm{fw}} = \tfrac{1}{2} m_{\mathrm{fw}} v_C^2 + \tfrac{1}{2} I_C \omega_{\mathrm{fw}}^2, \tag{3}$$

where v_C and ω_{fw} are the speed of the center of mass and the angular speed of the front wheel, respectively. Because the wheel rolls without slip, we have

$$v_C = v_0 = R\omega_{\mathrm{fw}} \quad \Rightarrow \quad \omega_{\mathrm{fw}} = \omega_0 = \frac{v_0}{R} = 27.78\,\mathrm{rad/s}, \tag{4}$$

where ω_0 is the angular speed of both wheels since they have the same radius $R = d/2 = 0.35\,\mathrm{m}$. Substituting Eq. (4) into Eq. (3) gives

$$T_{\mathrm{fw}} = \tfrac{1}{2} m_{\mathrm{fw}} R^2 \omega_0^2 + \tfrac{1}{2} I_C \omega_0^2 = \tfrac{1}{2}\left(I_C + m_{\mathrm{fw}} R^2\right)\omega_0^2 = 54.51\,\mathrm{J}. \tag{5}$$

Because the rear wheel is undergoing the same motion as the front wheel, the expression for T_{rw} has the same form as that in Eq. (5). Therefore, we have

$$T_{\text{rw}} = \tfrac{1}{2}\left(I_D + m_{\text{rw}}R^2\right)\omega_0^2 = 63.42\,\text{J}. \tag{6}$$

Substituting the results in Eqs. (2), (5), and (6) into Eq. (1), the total kinetic energy of the bicycle is

$$\boxed{T = 224.3\,\text{J}.} \tag{7}$$

Discussion & Verification Our answer was obtained by combining contributions from Eqs. (2), (5), and (6), each of which is dimensionally correct. The units used in the final answer are appropriate since the problem's data were given in SI units.

One way to check whether or not our result is reasonable is to verify that the computed kinetic energy is larger than what we would compute if the system were just translating. If the wheels did not rotate, their rotational kinetic energy would not contribute to the system's total kinetic energy, and therefore the corresponding system's kinetic energy would have to be smaller than that in Eq. (7). Computing the kinetic energy as if the system were just translating gives $T_{\text{translation}} = \frac{1}{2}(m_f + m_{\text{fw}} + m_{\text{rw}})v_0^2 = 185.3\,\text{J}$, which is less than our computed value, as expected.

A Closer Look An important observation about our calculation is that we solved the problem by applying Eq. (8.8) on p. 584, i.e., the general formula for the kinetic energy of a rigid body in planar motion. In so doing, for the wheels, we ended up with the following two expressions:

$$T_{\text{fw}} = \tfrac{1}{2}\left(I_C + m_{\text{fw}}R^2\right)\omega_0^2 \quad \text{and} \quad T_{\text{rw}} = \tfrac{1}{2}\left(I_D + m_{\text{rw}}R^2\right)\omega_0^2. \tag{8}$$

The first term in parentheses is equivalent to applying the parallel axis theorem to find the mass moment of inertia of the front wheel about A, that is,

$$I_A = I_C + m_{\text{fw}}R^2, \tag{9}$$

where, referring to Fig. 2, point A is the contact point between the wheel and the ground. What is special about point A is that *it is the instantaneous center of rotation of the front wheel*. Similarly, for the rear wheel we have $I_D + m_{\text{rw}}R^2 = I_B$, which is the mass moment of inertia of the rear wheel about B, where point B is the IC of the rear wheel. These observations point to the fact that we could have computed the kinetic energy of the wheels using Eq. (8.11) on p. 584 as

$$T_{\text{fw}} = \tfrac{1}{2}I_A\omega_0^2 \quad \text{and} \quad T_{\text{rw}} = \tfrac{1}{2}I_B\omega_0^2. \tag{10}$$

EXAMPLE 8.2 *The Work-Energy Principle for Rotation About a Fixed Axis*

The manually operated road barrier shown in Fig. 1 is easily opened and closed by hand due to the counterweight. Referring to Fig. 2, model the arm of the barrier as a *uniform* thin bar of mass m_a that is pinned at O and with mass center at A, and model the counterweight as a particle of mass m_c. Determine the angular velocity of the arm and the speed of end B as the arm reaches the horizontal position if it is nudged from rest when it is vertical and falls freely. Evaluate your answers for $l = 15.7$ ft, $d = 2.58$ ft, $\delta = 1.4$ ft, a 45 lb arm, and a 160 lb counterweight.

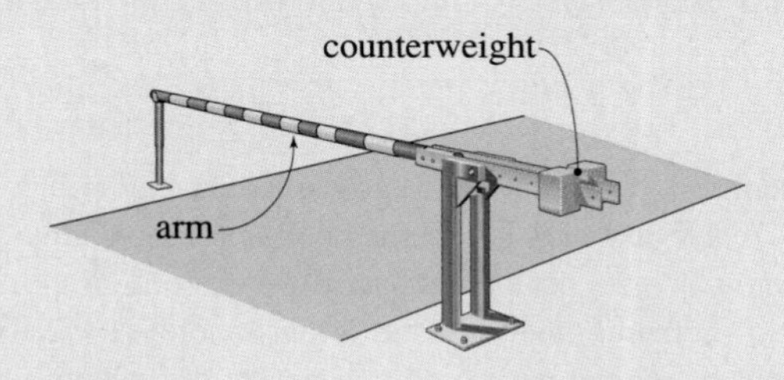

Figure 1. A manually operated road barrier. Note the counterweight on the right end.

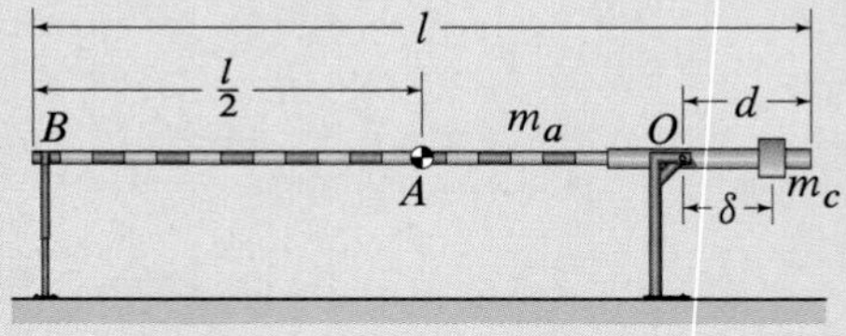

Figure 2
The relevant dimensions of the road barrier. The arm has mass m_a and the counterweight has mass m_c.

SOLUTION

Road Map & Modeling Since we want to relate the change in speed of the arm to its displacement, we will apply the work-energy principle. As can be seen in the FBD of the arm in Fig. 3, only weight forces do work, so this is a conservative system. We will assume that there are no losses due to friction or drag.

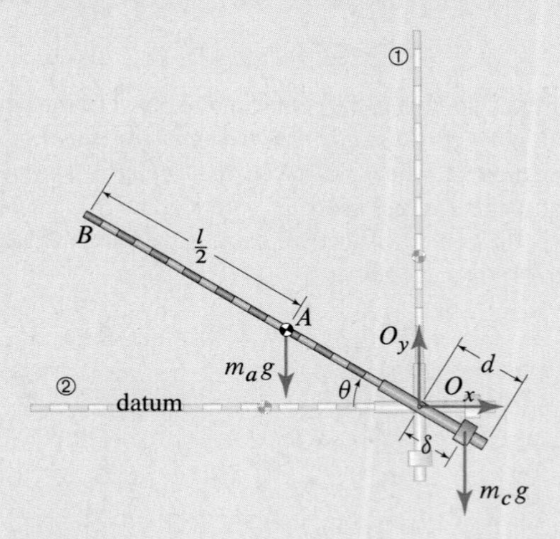

Figure 3. The FBD of the road barrier as it is falling from the vertical position.

Governing Equations

Balance Principles Since the system is conservative, the work-energy principle gives

$$T_1 + V_1 = T_2 + V_2, \tag{1}$$

where ① is when the arm is vertical ($\theta = \pi/2$ rad) and ② is when it is horizontal ($\theta = 0$). At ①, the kinetic energy is zero, and at ②, the kinetic energy must account for the arm and the counterweight, and so

$$T_1 = 0 \quad \text{and} \quad T_2 = \underbrace{\tfrac{1}{2} m_c v_{c2}^2}_{\text{counterweight}} + \underbrace{\tfrac{1}{2} I_O \omega_{a2}^2}_{\text{arm}}, \tag{2}$$

where v_{c2} is the speed of the counterweight at ②, I_O is the mass moment of inertia of the arm with respect to point O, ω_{a2} is the angular velocity of the arm at ②, and we have used Eq. (8.11) to compute the kinetic energy of the arm. Using the parallel axis theorem, the mass moment of inertia of the arm with respect to point O is given by

$$I_O = \tfrac{1}{12} m_a l^2 + m_a \left(\tfrac{1}{2} l - d\right)^2. \tag{3}$$

Force Laws The potential energies of the system at ① and ② are

$$V_1 = m_a g \left(\tfrac{1}{2} l - d\right) - m_c g \delta \quad \text{and} \quad V_2 = 0. \tag{4}$$

Kinematic Equations Since we want to solve for ω_{a2}, we will need to write v_{c2} in terms of ω_{a2}, which is readily done as

$$v_{c2} = \delta \omega_{a2}. \tag{5}$$

Computation Substituting Eqs. (2)–(5) into Eq. (1), we obtain

$$m_a g \left(\tfrac{1}{2} l - d\right) - m_c g \delta = \tfrac{1}{2} m_c (\delta \omega_{a2})^2 + \tfrac{1}{2} \left[\tfrac{1}{12} m_a l^2 + m_a \left(\tfrac{1}{2} l - d\right)^2\right] \omega_{a2}^2, \tag{6}$$

which, solving for ω_{a2}, gives

$$\boxed{\omega_{a2} = \sqrt{\frac{2g\left[m_a(l/2 - d) - m_c \delta\right]}{m_c \delta^2 + m_a\left[l^2/12 + (l/2 - d)^2\right]}} = 0.5835\,\text{rad/s},} \tag{7}$$

and so the speed of the end of the arm at B is

$$\boxed{v_{B2} = (l - d)\omega_{a2} = 7.655\,\text{ft/s}.} \tag{8}$$

Discussion & Verification The dimensions in Eqs. (7) and (8) are correct. If $m_c \delta$ is increased, the argument of the square root in Eq. (7) becomes negative, and the solution is no longer meaningful. This makes sense because we expect that if $m_c \delta$ exceeds a critical value, the barrier will never reach the horizontal position. Different design configurations for the arm can be explored in the exercises and design problems.

A Closer Look Problems 8.55 and 8.56 take a closer look at the problem examined in this example by taking into account the rotational inertia (and thus rotational kinetic energy) of the counterweight. Even without solving those problems, we can predict what the inclusion of rotational inertia will do to our results in Eqs. (7) and (8).

It won't change the potential energy of the counterweight, but it will add a kinetic energy term to the right side of Eq. (6). Once this kinetic energy term is written in terms of ω_{a2}, we see that the denominator in Eq. (7) will get larger and so, all other things being equal, ω_{a2} will decrease.

EXAMPLE 8.3 *Work-Energy Principle and Rolling Without Slip*

Figure 1
A roadster on a horizontal stretch of road.

A rear-wheel-drive car accelerates from rest to a final speed v_f over a distance L along a horizontal stretch. The front wheels do not slip, and we neglect rolling resistance, bearing friction, and air drag. If the geometric center of each wheel is also the wheel's mass center G, determine those forces on each front wheel that do work, and express that work in terms of the given speed v_f, the wheel's radius r, mass m_w, and mass moment of inertia I_G. Can this expression for work be used to find the average friction force acting on the wheel?

SOLUTION

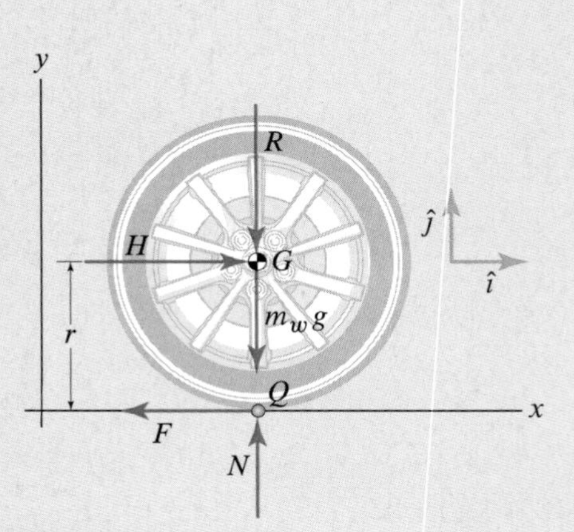

Figure 2
FBD of one of the front wheels. The point G identifies both the geometric center of the wheel and the wheel's center of mass.

Road Map & Modeling The work-energy principle tells us that the work done on a wheel is the difference between the wheel's final and initial kinetic energy. Since we have enough information to determine these kinetic energies, the problem is solved by computing their difference. As for calculating the average friction force at the ground, we will need to see if and how the friction force appears in the expression for the work done on the wheel. Since the work-energy principle accounts for the work of all forces acting on the wheel, it is important to sketch the wheel's FBD, as shown in Fig. 2. The only friction we consider is that at the ground, and no couple is assumed to be acting on the wheel because the front wheels are *not* the driving wheels. The force H comes from the fact that the front axle of the car is pushing the wheel forward, and the force R is due to the weight of the car pushing down on the tire.

Governing Equations

Balance Principles The work-energy principle for each front wheel is

$$T_1 + U_{1\text{-}2} = T_2, \tag{1}$$

where the roadster's speed is zero at ① and v_0 at ②. The kinetic energy T can then be written as

$$T_1 = 0 \quad \text{and} \quad T_2 = \frac{1}{2} m_w v_{G2}^2 + \frac{1}{2} I_G \omega_{w2}^2, \tag{2}$$

where v_{G2} is the speed of G and ω_{w2} is the angular speed of the wheel, both at ②.

Force Laws To express the work of each force appearing on the FBD, recall that if the point of application of a force $\vec{P}$ displaces by $d\vec{r}$, the corresponding work dU is

$$dU = \vec{P} \cdot d\vec{r} \quad \text{with} \quad d\vec{r} = \vec{v}\, dt, \tag{3}$$

where $\vec{v}$ is the velocity of the point of application of $\vec{P}$. Referring to Fig. 2,

1. Forces R and $m_w g$ are oriented vertically and do no work because their points of application move horizontally.
2. F and N do no work because the rolling-without-slip condition demands that $\vec{v}_Q = \vec{0}$, where Q is the point of application of F and N.

Therefore, the only force doing work is the force H, and we can write

$$U_{1\text{-}2} = \int_{(\mathcal{L}_{1\text{-}2})_G} H\,\hat{\imath} \cdot d\vec{r}_G = \int_{x_{G1}}^{x_{G2}} H\, dx_G, \tag{4}$$

where G is the point of application of H and x_G is the horizontal position of G.

Kinematic Equations At ② we have

$$v_{G2} = v_f. \tag{5}$$

To find ω_{w2}, we apply the kinematics of rolling without slip using

$$\vec{v}_G = \vec{v}_Q + \omega_w \hat{k} \times r\,\hat{\jmath} \quad \text{with} \quad \vec{v}_Q = \vec{0}, \tag{6}$$

so that at ② we have

$$\vec{v}_{G2} = v_f\,\hat{\imath} = \omega_{w2}\,\hat{k} \times r\,\hat{\jmath} \quad \Rightarrow \quad \omega_{w2} = -\frac{v_f}{r}. \tag{7}$$

Computation Combining Eqs. (5) and (7) with Eq. (2), we have

$$T_1 = 0 \quad \text{and} \quad T_2 = \frac{1}{2} m_w v_f^2 + \frac{1}{2} I_G \left(-\frac{v_f}{r}\right)^2. \tag{8}$$

Substituting Eqs. (8) into Eq. (1) and solving for $U_{1\text{-}2}$, we have

$$U_{1\text{-}2} = \int_{x_{G1}}^{x_{G2}} H\,dx_G = \frac{1}{2} m_w v_f^2 + \frac{1}{2} I_G \left(-\frac{v_f}{r}\right)^2, \tag{9}$$

where we have also made use of Eq. (4). Equation (9) can be simplified to

$$\boxed{U_{1\text{-}2} = \int_{x_{G1}}^{x_{G2}} H\,dx_G = \frac{1}{2}\left(I_G + m_w r^2\right)\left(\frac{v_f}{r}\right)^2.} \tag{10}$$

Now that we have found the expression for $U_{1\text{-}2}$, we see that the friction force does not appear in it. Therefore, we *cannot* calculate the average value of the friction force acting on the wheel directly from Eq. (10).

Discussion & Verification The term v_f/r in Eq. (10) has dimensions of 1/time. Since I_G is a mass moment of inertia, it has dimensions of mass × length^2. Therefore, overall, the result in Eq. (10) has dimensions of $(\text{mass} \times \text{length}/\text{time}^2)(\text{length})$, i.e., dimensions of work, as it should.

In general, rolling without slip requires a friction force to act on the rolling body at the point of contact between the body in question and the rolling surface. However, this friction force does not dissipate any energy because the absence of slip prevents the friction force from doing work (i.e., the friction is being applied at a point that is not moving at that instant). Therefore, if a *rigid* body is rolling without slip on a *rigid* surface, the mechanical energy of the body will be conserved during the motion. This is an important result because in many problems involving rolling without slip, we can apply conservation of energy even if a friction force does appear on the body's FBD.

Interesting Fact

If friction does no work, then what is "rolling resistance"? Rolling resistance occurs when one object rolls without slip over another object *and* one or both of the objects deform. Not only does the deformation itself dissipate energy, but also the equivalent normal force between the object and the ground impedes the motion.

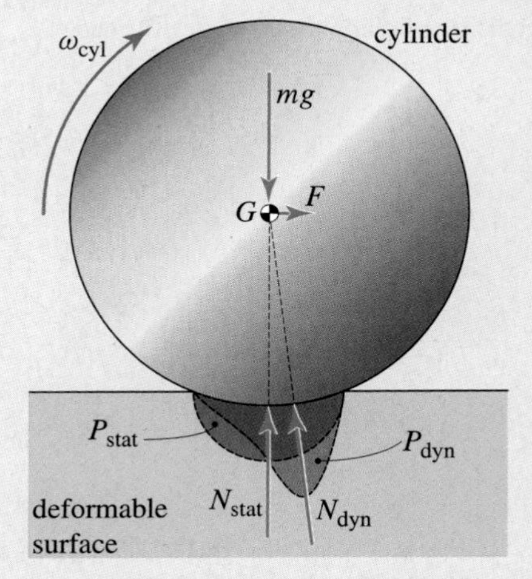

The figure above shows a rigid cylinder on a deformable surface. When $F = 0$ and the object is at rest, the pressure distribution on the object due to the deformation of the surface is given by P_{stat}, and the equivalent normal force is given by the vertical force N_{stat}. On the other hand, when a force F is applied to the cylinder so that it rolls with a constant angular velocity ω_{cyl}, then the pressure distribution is that given by P_{dyn}, and the equivalent normal force is N_{dyn}. Notice that there is a horizontal component of N_{dyn} that is impeding the forward motion of the cylinder. Since the cylinder is moving with a constant velocity, the horizontal component of N_{dyn} must be equal and opposite to F.

EXAMPLE 8.4 Computing Engine Power Output and Wheel Torque

Figure 1
A roadster on a horizontal stretch of road.

The 2600 lb* rear-wheel-drive roadster shown in Fig. 1 accelerates from rest to 60 mph in 7 s. Each of the four wheels has a diameter $d = 24.3$ in., weighs 47 lb, and has a mass moment of inertia $I_G = 0.989\,\text{slug}\cdot\text{ft}^2$ with respect to its mass center. Assuming that the car accelerates uniformly, that we neglect rolling resistance, bearing friction, and air drag, and that the wheels do not slip relative to the ground, estimate the average engine power output, as well as the torque provided to the rear wheels.

SOLUTION

Road Map & Modeling We can use the work-energy principle to compute the total work done on the car as the difference between the final and initial kinetic energies. Referring to the FBD in Fig. 2, we see that none of the external forces does work since the wheels roll without slip and the car's trajectory is horizontal. All the work done on the car is due to internal forces, namely, the torque provided by the engine. The corresponding average power output is then computed by dividing this work by the given time interval. To measure the average torque provided by the engine, we need to consider the forces internal to the car, depicted in Fig. 3. The work done by the forces R_x and R_y applied

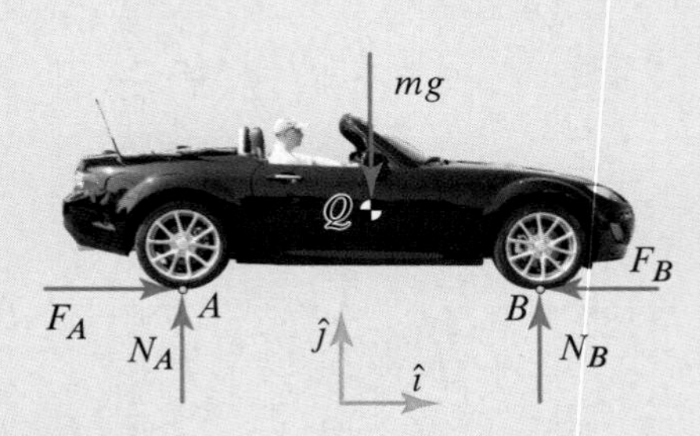

Figure 2
FBD of the car as a whole. Q is the center of mass of the whole car, and A and B are the contact points between the ground and the rear and front wheels, respectively.

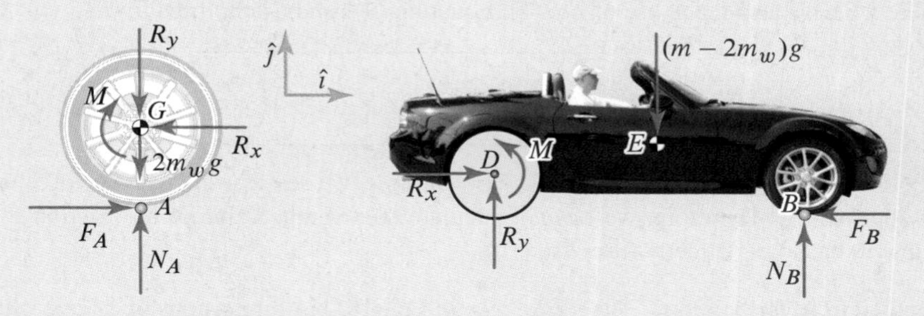

Figure 3. FBDs of the rear wheels and of the rest of car. Point E is the mass center of the car without the rear wheels. Points G and D are coincident: G is part of the wheel, whereas D is part of the axle on which the wheel is mounted.

at G on a rear wheel is equal and opposite to the work done by R_x and R_y applied at D because G and D do not move relative to one another. By contrast, the torque M does work because the rear wheels rotate relative to the rest of the car. Therefore, we can estimate M by dividing the work of the torque by the angle swept by the wheels during the car's motion.

Governing Equations

Balance Principles The general form of the work-energy principle for a system is

$$T_1 + V_1 + (U_{1\text{-}2})^{\text{ext}}_{\text{nc}} + (U_{1\text{-}2})^{\text{int}}_{\text{nc}} = T_2 + V_2, \tag{1}$$

where the car's speed is equal to zero at ① and is equal to 60 mph = 88.00 ft/s at ②. The system's kinetic energy at ① and ② can be written as

$$T_1 = 0 \quad \text{and} \quad T_2 = \underbrace{\tfrac{1}{2}mv_2^2}_{\text{tr KE}} + \underbrace{4\left(\tfrac{1}{2}I_G\omega_{w2}^2\right)}_{\text{rot KE}} = \tfrac{1}{2}mv_2^2 + 2I_G\omega_{w2}^2, \tag{2}$$

where m is the mass of the roadster (including its wheels), tr KE is the translational kinetic energy of the car, rot KE is the rotational kinetic energy of all four wheels, v_2 is the translational speed of the car and the wheels, and ω_{w2} is the angular velocity of the wheels.

* This weight includes the wheels, the fuel, and the passenger.

Force Laws The external forces do no work, so we can write

$$(U_{1\text{-}2})_{\text{nc}}^{\text{ext}} = 0, \quad V_1 = 0, \quad \text{and} \quad V_2 = 0. \tag{3}$$

As for $(U_{1\text{-}2})_{\text{nc}}^{\text{int}}$, referring to Fig. 3, because the points G and D do not move relative to one another, R_x and R_y do no work. Consequently, all the internal work is done by the torque M, which can be written as

$$(U_{1\text{-}2})_{\text{nc}}^{\text{int}} = \int_{\theta_1}^{\theta_2} M\,d\theta = M\,(\theta_2 - \theta_1), \tag{4}$$

where θ measures the rotation of the rear wheels (see Fig. 4), and we have assumed that M is constant between ① and ②.

Kinematic Equations At ②, the translational speed v_2 is given, and the corresponding ω_{w2} can be found by enforcing the rolling-without-slip condition, which gives

$$v_2 = 88.00\,\text{ft/s} \quad \text{and} \quad \omega_{w2} = \frac{v_2}{d/2} = 86.91\,\text{rad/s}. \tag{5}$$

Note that to compute M, we will need the value of $\theta_2 - \theta_1$, which can be calculated if the distance L traveled by the car between ① and ② is known. The distance L can be computed since we have assumed that the car is uniformly accelerating. Letting $\Delta t = 7\,\text{s}$, the constant acceleration of the car is $a = v_2/\Delta t = 12.57\,\text{ft/s}^2$. Therefore, using constant acceleration equations, we have

$$L = \frac{1}{2} a \Delta t^2 = 308.0\,\text{ft} \quad \Rightarrow \quad \theta_2 - \theta_1 = \frac{L}{d/2} = 304.2\,\text{rad}. \tag{6}$$

Computation Combining Eqs. (5) with Eqs. (2) and substituting in known values, we have

$$T_1 = 0 \quad \text{and} \quad T_2 = 327{,}600\,\text{ft}\cdot\text{lb}. \tag{7}$$

Substituting Eqs. (7) into Eq. (1) and using Eqs. (3), for the average power P_{avg} we have

$$\boxed{(U_{1\text{-}2})_{\text{nc}}^{\text{int}} = 327{,}600\,\text{ft}\cdot\text{lb} \;\Rightarrow\; P_{\text{avg}} = \frac{(U_{1\text{-}2})_{\text{nc}}^{\text{int}}}{\Delta t} = 46{,}800\,\frac{\text{ft}\cdot\text{lb}}{\text{s}} = 85.09\,\text{hp}.} \tag{8}$$

Using Eq. (4), along with the first of Eqs. (8) and the result in Eq. (6), we have

$$\boxed{M\,(\theta_2 - \theta_1) = (U_{1\text{-}2})_{\text{nc}}^{\text{int}} \quad \Rightarrow \quad M = \frac{(U_{1\text{-}2})_{\text{nc}}^{\text{int}}}{\theta_2 - \theta_1} = 1077\,\text{ft}\cdot\text{lb}.} \tag{9}$$

Discussion & Verification The answers are all dimensionally correct. As far as the acceptability of each value is concerned, the calculated power may seem low with respect to the power of typical sports car engines (e.g., 170 hp at an engine speed of 6000 rpm). However, the power indicated in ads and product literature is a peak value corresponding to a specific angular velocity of the crankshaft. In our case, we have calculated an *average* power value, and, as such, it is not surprising that it is smaller than the peak values seen in advertisements. As far as the torque is concerned, the value we have obtained is not that unusual if we keep in mind that it measures the torque provided to the driving wheels. That is, the computed torque value should not be confused with the typically much smaller value of torque reported in product literature, which is the torque output at the crankshaft, i.e., before the transmission gets it to the wheels.

Helpful Information

Work of internal forces. The work done by R_x and R_y on the rear wheels is equal and opposite to the work done by these forces on the rest of the car because the relative displacement of points G and D (see Fig. 3) is equal to zero. Therefore, the overall work of the internal forces R_x and R_y is equal to zero. If the displacement of G relative to D were not equal to zero, the car would be broken into two parts. By contrast, the rear wheels rotate relative to the car, so the *relative angular displacement* between these wheels and the rest of the car is not equal to zero. This allows the torque M to do (internal) work.

Figure 4
Definition of the angle θ measuring the rotation of the rear wheels. The red line is an arbitrarily chosen reference line that rotates with the wheels.

EXAMPLE 8.5 An Overhead Fold-Up Door: Conservation of Energy

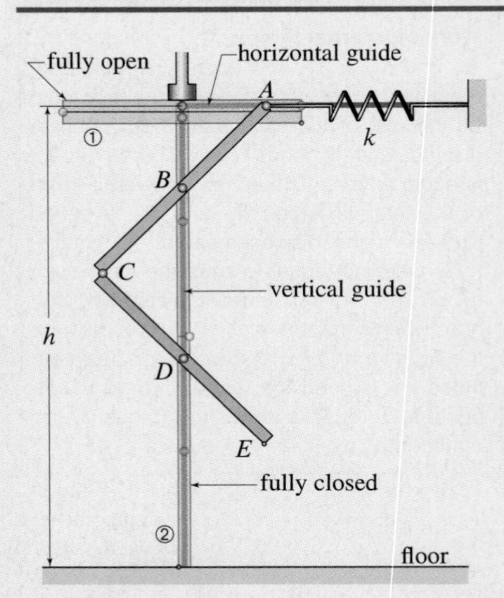

Figure 1
Side view of an overhead door at various positions during its operation.

An overhead door, with height $h = 30\,\text{ft}$ and weight $W = 800\,\text{lb}$, consists of two identical sections hinged at C. The roller at A moves along a horizontal guide, whereas the rollers at B and D, which are the midpoints of sections AC and CE, move along a vertical guide. The door's operation is assisted by two identical springs attached to the horizontally moving rollers (only one of the two springs is shown). The springs are stretched 0.25 ft when the door is fully open. Determine the minimum value of the spring constant k if the roller at A is to strike the left end of the horizontal guide with a maximum speed of 1.5 ft/s after the door is released from rest in the fully open position.

SOLUTION

Road Map & Modeling Since the desired value of k is found by relating changes in speed to changes in position, we will use the work-energy principle as a solution method. We will treat the two identical sections AC and CE as uniform thin rigid bodies, neglect the inertia of the rollers, neglect the mass of the springs, and neglect all friction. Letting ① and ② be the fully open and fully closed positions, respectively, the system's FBD for a generic position between ① and ② is that shown in Fig. 2. The force in each of the two springs is F_s, and N_A, N_B, and N_D are the reactions at each of the rollers A, B, and D, respectively. Note that the reactions at the rollers do no work because each roller moves in a direction perpendicular to the reaction force acting on it. The remaining forces W and F_s are conservative, so we have conservation of mechanical energy. Our solution strategy will be to find the value of k for which the speed of A is $v_{max} = 1.5\,\text{ft/s}$ and then show that larger values of k result in values of the speed of A at ② that are less than v_{max}.

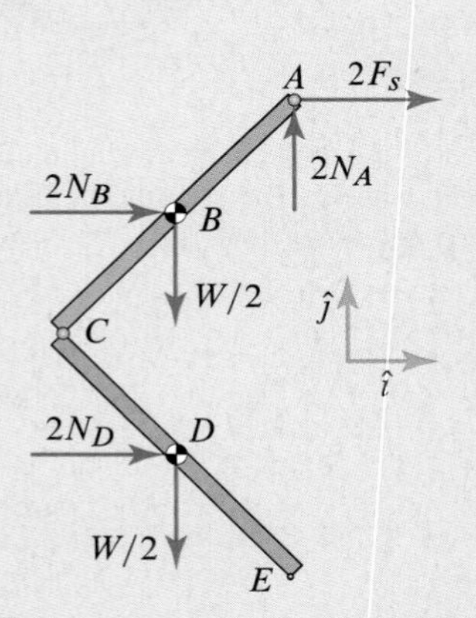

Figure 2
Side view of the system's FBD for a generic position between ① and ②. The factor of 2 in front of F_s, N_A, N_B, and N_D is necessary because there are two springs and two sets of rollers on the door (in the view shown only one set is visible). Note that the centers of mass of sections AC and CE coincide with points B and D, respectively.

Governing Equations

Balance Principles Since mechanical energy is conserved, we can write

$$T_1 + V_1 = T_2 + V_2, \tag{1}$$

where, since the system is at rest at ①, T_1 and T_2 can be expressed as

$$T_1 = 0 \quad \text{and} \quad T_2 = \tfrac{1}{2} m_{AC} v_{B2}^2 + \tfrac{1}{2} I_B \omega_{AC2}^2 + \tfrac{1}{2} m_{CE} v_{D2}^2 + \tfrac{1}{2} I_D \omega_{CE2}^2, \tag{2}$$

where m_{AC} and I_B are the mass and mass moment of inertia of AC, respectively, and m_{CE} and I_D are the mass and mass moment of inertia of CE, respectively. Since AC and CE are identical, we have

$$m_{AC} = m_{CE} = \frac{W/2}{g} \quad \text{and} \quad I_B = I_D = \frac{1}{12}\left(\frac{W}{2g}\right)\left(\frac{h}{2}\right)^2 = \frac{Wh^2}{96g}. \tag{3}$$

Force Laws Choosing the datum for gravitational potential energy as shown in Fig. 3 and recalling that there are two springs, we have

$$V_1 = 2\left(\tfrac{1}{2}k\delta_1^2\right) \quad \text{and} \quad V_2 = \frac{W}{2} y_{B2} + \frac{W}{2} y_{D2} + 2\left(\tfrac{1}{2}k\delta_2^2\right), \tag{4}$$

where $\delta_1 = 0.25\,\text{ft}$, and where

$$y_{B2} = -h/4, \quad y_{D2} = -3h/4, \quad \text{and} \quad \delta_2 = \delta_1 + h/4. \tag{5}$$

Kinematic Equations When at ②, B and D are at the lower limit of their respective motion ranges. In addition, given that B and D cannot move in the x direction, it must therefore be true that $\vec{v}_{B2} = \vec{0} = \vec{v}_{D2}$, and so

$$v_{B2} = 0 \quad \text{and} \quad v_{D2} = 0. \tag{6}$$

Referring to Fig. 3, observe that AC and CE rotate in opposite directions while remaining mirror images of one another relative to the line bisecting the angle ACE. Therefore, we must have

$$\omega_{AC} = -\omega_{CE} \quad \Rightarrow \quad \omega_{AC2} = -\omega_{CE2}. \tag{7}$$

In addition, since $\vec{v}_B = \vec{v}_A + \omega_{AC}\,\hat{k} \times \vec{r}_{B/A}$ and since $\vec{v}_{B2} = \vec{0}$, at ② we can write

$$\vec{v}_{B2} = \vec{0} = (v_{Ax})_2\,\hat{\imath} + \omega_{AC2}\,\hat{k} \times (-h/4)\,\hat{\jmath} = \big[(v_{Ax})_2 + h\omega_{AC2}/4\big]\,\hat{\imath}, \tag{8}$$

where we have used the fact that A can only move in the horizontal direction. Solving Eq. (8) for ω_{AC2} and using Eq. (7), we obtain

$$\omega_{AC2} = -4(v_{Ax})_2/h \quad \text{and} \quad \omega_{CE2} = 4(v_{Ax})_2/h. \tag{9}$$

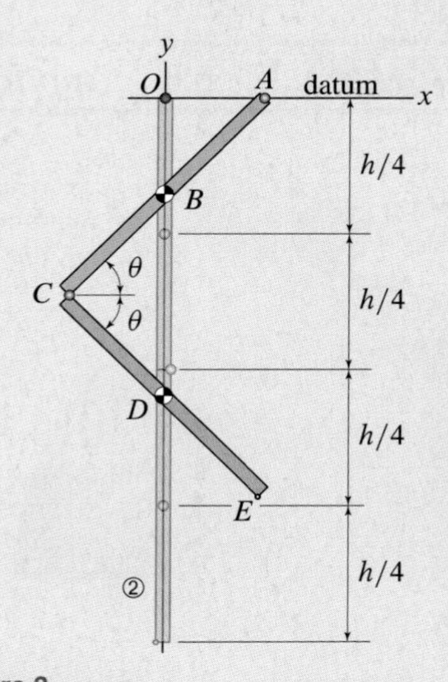

Figure 3
Coordinate system used with the indication of the datum choice. Note that, for convenience, only ② is shown. When the system is at ①, sections AC and CE lie on the x axis since their thickness has been neglected.

Computation Using Eq. (2), along with Eqs. (3), (6), and (9), we have

$$T_1 = 0 \quad \text{and} \quad T_2 = \frac{Wh^2}{96g}\frac{16(v_{Ax})_2^2}{h^2} = \frac{W(v_{Ax})_2^2}{6g}. \tag{10}$$

Substituting Eqs. (5) into Eqs. (4) and simplifying, we have

$$V_1 = k\delta_1^2 \quad \text{and} \quad V_2 = k\left(\delta_1^2 + \frac{\delta_1 h}{2} + \frac{h^2}{16}\right) - \frac{Wh}{2}. \tag{11}$$

Substituting Eqs. (10) and (11) into Eq. (1) and simplifying, we obtain

$$0 = \frac{W(v_{Ax})_2^2}{6g} + k\left(\frac{\delta_1 h}{2} + \frac{h^2}{16}\right) - \frac{Wh}{2}, \tag{12}$$

which can be solved for k to obtain

$$\boxed{k = 8\frac{W}{g}\frac{3hg - (v_{Ax})_2^2}{24\delta_1 h + 3h^2} = 199.8\,\text{lb/ft},} \tag{13}$$

where we have let $(v_{Ax})_2 = v_{\text{max}} = 1.5\,\text{ft/s}$.

Discussion & Verification The result in Eq. (13) is consistent with the fact that k has dimensions of force over length, and the units used to express the numerical result are therefore appropriate. In addition, the result matches our expectation that the slower $(v_{Ax})_2$ is, the stiffer the spring must be. Therefore, the value of $k = 199.8\,\text{lb/ft}$ is the value of k we were looking for.

A Closer Look If we were to use the value of k in Eq. (13), each of the springs would be subject to $F_{s2} = k\delta_2 = 1549\,\text{lb}$ when at ②. If this value of force is judged to be too high, we can design a door with a smaller value of the spring constant (and therefore of force) using a system of counterweights that are lowered when the door opens and that are lifted when the door closes. A properly designed counterweight system can make the use of springs unnecessary.

EXAMPLE 8.6 *Application of the Work-Energy Principle and $\vec{F} = m\vec{a}$*

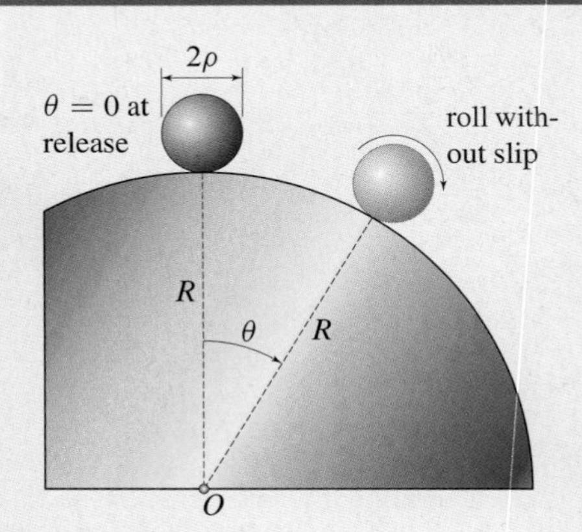

Figure 1

Let's revisit Example 7.10 (on p. 558), in which a solid uniform sphere was released from the top of a semicylinder and we determined where it separated from the semicylinder as it rolled *without slipping*. As was done in Example 7.10, we assume that the sphere is uniform; has mass m and radius ρ; and is released from rest by giving it a *slight* nudge to the right so that it begins to roll along the surface (Fig. 1). Assuming that there is sufficient friction between the surface and the sphere, such that the sphere will not slip on the surface, we want to determine the angle θ at which the sphere separates from the surface. The purpose of this example is to show how the work-energy principle can be combined with the methods studied in Chapter 7 to obtain a solution without direct integration of the system's equations of motion.

SOLUTION

Road Map & Modeling The FBD of the sphere is shown in Fig. 2. The sphere will begin to separate from the cylinder when the reaction N becomes zero. Therefore, we need to write $\vec{F} = m\vec{a}_G$ in the direction of N (which is the radial direction) and solve for the value of θ at which $N = 0$. This equation will involve N, the sphere's weight, and a_{Gr}. As long as the sphere does not separate, G moves in a circle so that a_{Gr} depends on $\dot{\theta}$ but not $\ddot{\theta}$. This observation is important because it tells us that to find θ such that $N = 0$, we must relate $\dot{\theta}$ to θ, i.e., position to velocity, and this problem is ideal for the work-energy principle. Therefore, our solution strategy will be to combine $F_r = ma_{Gr}$ with the work-energy principle.

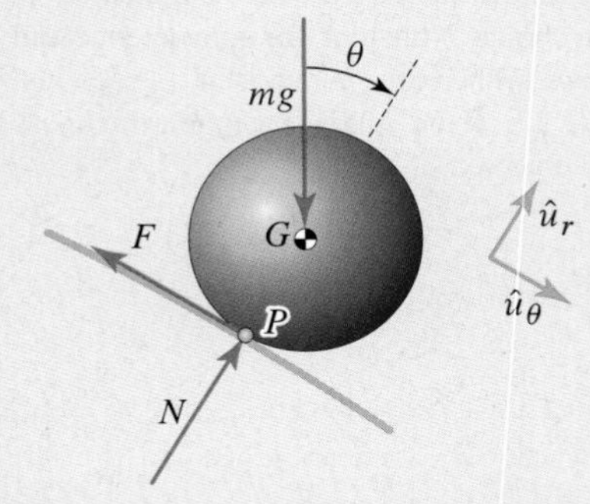

Figure 2
FBD of the sphere drawn at an arbitrary angle $\theta = \theta_2$, which corresponds to ②.

Governing Equations

Balance Principles Letting ① be when $\theta = 0$ and ② be at an arbitrary angle $\theta = \theta_2$, the Newton-Euler equations in the radial direction at ② and the work-energy principle applied to the sphere between ① and ② are, respectively,

$$\sum F_r: \quad N_2 - mg\cos\theta_2 = m\big(a_{Gr}\big)_2, \tag{1}$$

$$T_1 + V_1 + (U_{1\text{-}2})_{\text{nc}} = T_2 + V_2, \tag{2}$$

where T_1 and T_2 can be written as

$$T_1 = 0 \quad \text{and} \quad T_2 = \tfrac{1}{2}mv_{G2}^2 + \tfrac{1}{2}I_G\omega_{s2}^2. \tag{3}$$

Here, v_{G2} and ω_{s2} are the velocity of the mass center of the sphere and the angular velocity of the sphere, respectively, at ②. The sphere's mass moment of inertia is given by

$$I_G = \tfrac{2}{5}m\rho^2. \tag{4}$$

Force Laws Referring to Fig. 2, since the sphere rolls without slip, both F and N do no work, so the only force doing work is the sphere's weight. Referring to the datum line in Fig. 3, we have

$$(U_{1\text{-}2})_{\text{nc}} = 0, \quad V_1 = 0, \quad \text{and} \quad V_2 = -mg(R+\rho)(1-\cos\theta_2). \tag{5}$$

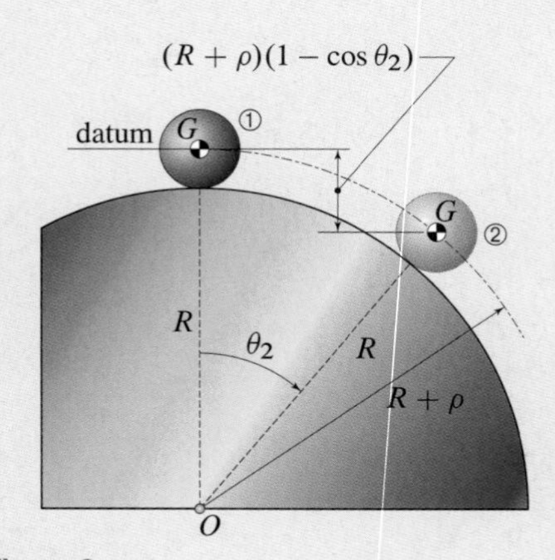

Figure 3
Placement of the datum line coincides with the location of the center of the sphere at ①.

Kinematic Equations Recalling that we are using polar coordinates and that G moves along a circle of radius $R + \rho$, we have

$$\big(a_{Gr}\big)_2 = -(R+\rho)\dot{\theta}_2^2. \tag{6}$$

Next, we focus on the information needed to compute the kinetic energy. We know the kinetic energy at ①, and at ②, since G moves in the θ direction along a circle of radius $R + \rho$, we have

$$v_{G2} = (R+\rho)\dot{\theta}_2. \tag{7}$$

To compute ω_{s2} we need to apply the kinematics of rolling without slip. Rolling without slip implies that $\vec{v}_{P2} = \vec{0}$ (see Fig. 4), and so we can write

$$\vec{v}_{P2} = \vec{0} = \vec{v}_{G2} + \omega_{s2}\,\hat{k} \times (-\rho\,\hat{u}_r). \tag{8}$$

From Eq. (7), we know that $\vec{v}_{G2} = (R+\rho)\dot{\theta}_2\,\hat{u}_\theta$, so Eq. (8) yields

$$(R+\rho)\dot{\theta}_2\,\hat{u}_\theta - \rho\omega_{s2}\,\hat{u}_\theta = \vec{0} \quad\Rightarrow\quad \omega_{s2} = \frac{R+\rho}{\rho}\dot{\theta}_2, \tag{9}$$

where we recall that $\hat{k}$ is defined such that $\hat{u}_r \times \hat{u}_\theta = \hat{k}$.

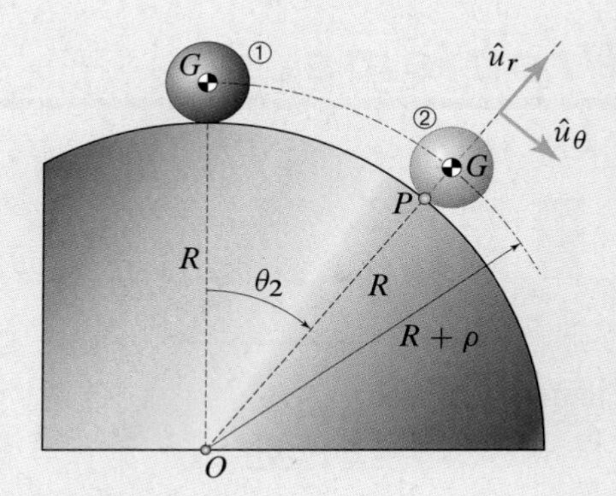

Figure 4
Component system at ②.

Computation Combining Eqs. (7) and (9) with Eq. (3) yields

$$T_1 = 0 \quad\text{and}\quad T_2 = \frac{1}{2}m(R+\rho)^2\dot{\theta}_2^2 + \frac{1}{2}I_G\left(\frac{R+\rho}{\rho}\right)^2\dot{\theta}_2^2, \tag{10}$$

which, using Eq. (4), can be simplified to

$$T_1 = 0 \quad\text{and}\quad T_2 = \tfrac{7}{10}m(R+\rho)^2\dot{\theta}_2^2. \tag{11}$$

Substituting Eqs. (5) and (11) into Eq. (2) gives

$$0 = -mg(R+\rho)(1-\cos\theta_2) + \tfrac{7}{10}m(R+\rho)^2\dot{\theta}_2^2, \tag{12}$$

which can be solved for $\dot{\theta}_2^2$ to obtain

$$\dot{\theta}_2^2 = \frac{10g(1-\cos\theta_2)}{7(R+\rho)}. \tag{13}$$

Substituting Eq. (6) into Eq. (1) and solving for N_2, we obtain

$$N_2 = mg\cos\theta_2 - m(R+\rho)\dot{\theta}_2^2, \tag{14}$$

which, using Eq. (13), yields

$$N_2 = \tfrac{1}{7}mg(17\cos\theta_2 - 10). \tag{15}$$

We want to solve for the value of θ that corresponds to $N = 0$. Therefore, letting $N_2 = 0$ gives

$$\boxed{\theta_2 = \cos^{-1}\left(\tfrac{10}{17}\right) = 53.97^\circ,} \tag{16}$$

where, recalling that $\cos^{-1}(10/17) = \pm 53.97^\circ + n360^\circ$, $n = 0, \pm 1, \ldots, \pm\infty$, we have selected the only physically meaningful solution for θ_2.

Discussion & Verification As expected, our solution matches that found in Example 7.10 on p. 558. What is important here is to notice that we were able to solve the same problem given in Example 7.10 by using a mixed solution strategy that combined both the use of $\vec{F} = m\vec{a}_G$ and the work-energy principle. The use of this principle allowed us to completely bypass the derivation and subsequent integration of the equation of motion in the θ direction.

Problems

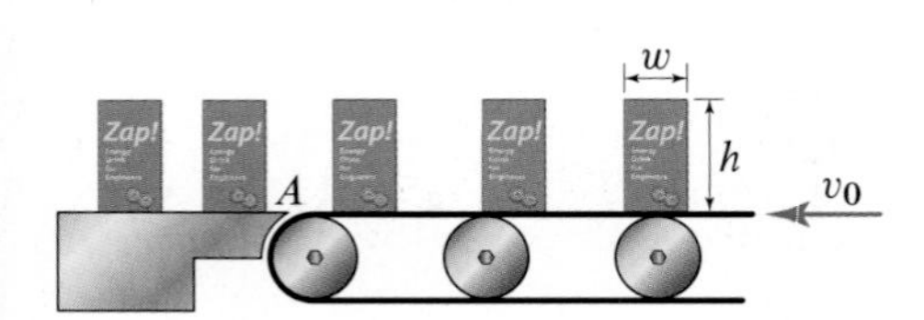

Figure P8.1

Problem 8.1

A conveyor is moving cans at a constant speed v_0 when, to proceed to the next step in packaging, the cans are transferred onto a stationary surface at A. The cans each have mass m, width w, and height h. Assuming that there is friction between each can and the stationary surface, under what conditions would we be able to compute the stopping distance of the cans, using the work-energy principle for a particle?

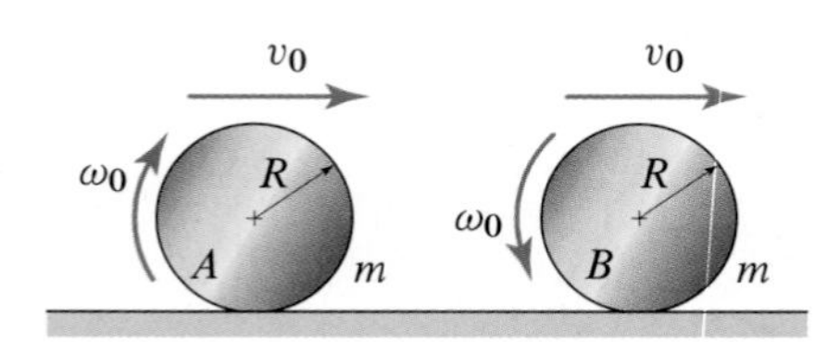

Figure P8.2 and P8.3

Problem 8.2

At the instant shown, the centers of the two identical uniform disks A and B are moving to the right with the same speed v_0. In addition, disk A is rolling clockwise with an angular speed ω_0, while disk B has a backspin with angular speed equal to ω_0. Letting T_A and T_B be the kinetic energies of A and B, respectively, state which of the following statements is true and why: (a) $T_A < T_B$; (b) $T_A = T_B$; (c) $T_A > T_B$.

Problem 8.3

At the instant shown, the centers of the two identical uniform disks A and B, each with mass m and radius R, are moving to the right with the same speed $v_0 = 4\,\text{m/s}$. In addition, disk A is rolling clockwise with an angular speed $\omega_0 = 5\,\text{rad/s}$, while disk B has a backspin with angular speed $\omega_0 = 5\,\text{rad/s}$. Letting $m = 45\,\text{kg}$ and $R = 0.75\,\text{m}$, determine the kinetic energy of each disk.

Problem 8.4

Two identical battering rams are mounted in two different ways on their respective frames as shown. Bars BC and AD are identical and pinned at B and C and at A and D, respectively. Bars FO and HO are rigidly attached to the ram and are pinned at O. At the instant shown, the mass centers of rams 1 and 2, at E and G, respectively, are moving horizontally with speed v_0. Letting T_1 and T_2 be the kinetic energies of rams 1 and 2, respectively, state which of the following statements is true and why: (a) $T_1 < T_2$; (b) $T_1 = T_2$; (c) $T_1 > T_2$.

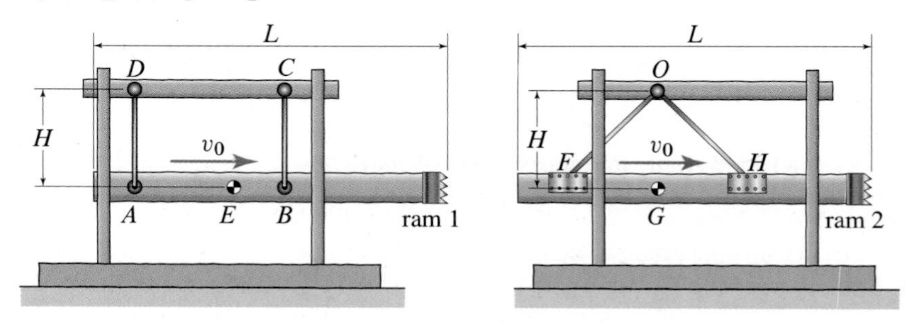

Figure P8.4 and P8.5

Problem 8.5

Two identical battering rams are mounted in two different ways on their respective frames as shown. Bars BC and AD are identical and pinned at B and C and at A and D, respectively. Bars FO and HO are rigidly attached to the ram and are pinned at O. At the instant shown, the centers of mass of rams 1 and 2, at E and G, respectively, are moving horizontally with a speed $v_0 = 20\,\text{ft/s}$. Treating the rams as slender bars with length $L = 10\,\text{ft}$ and weight $W = 1250\,\text{lb}$, and letting $H = 3\,\text{ft}$, compute the kinetic energy of the two rams.

Problem 8.6

A pendulum consists of a uniform disk A of diameter $d = 0.15\,\text{m}$ and mass $m_A = 0.35\,\text{kg}$ attached at the end of a uniform bar B of length $L = 0.75\,\text{m}$ and mass $m_B = 0.8\,\text{kg}$. At the instant shown, the pendulum is swinging with an angular velocity $\omega = 0.24\,\text{rad/s}$ clockwise. Determine the kinetic energy of the pendulum at this instant using Eq. (8.8) on p. 584.

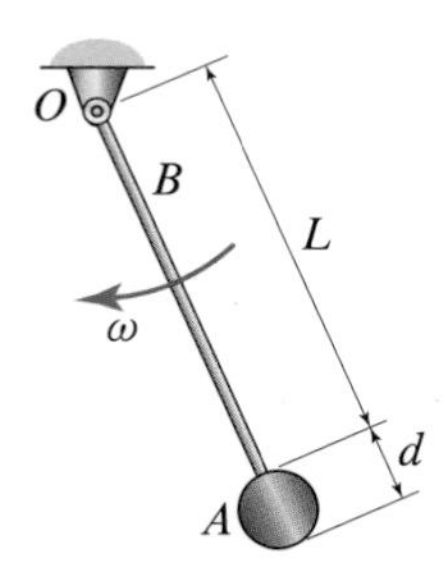

Figure P8.6

Problem 8.7

A 2570 lb car (this includes the weight of the wheels) is traveling on a horizontal flat road at 60 mph. If each wheel has a diameter $d = 24.3\,\text{in.}$ and a mass moment of inertia with respect to its mass center equal to $0.989\,\text{slug}\cdot\text{ft}^2$, determine the kinetic energy of the car. Neglect the rotational energy of all parts except for the wheels, which roll without slip.

Figure P8.7

Problem 8.8

In Example 7.6 on p. 544, we analyzed the forces acting on a test tube in an ultracentrifuge. Recalling that the center of mass G of the test tube was assumed to be at a distance $r = 0.0918\,\text{m}$ from the centrifuge's spin axis, and that the test tube had a mass $m = 0.01\,\text{kg}$ and a mass moment of inertia $I_G = 2.821 \times 10^{-6}\,\text{kg}\cdot\text{m}^2$, determine the kinetic energy of the test tube when it is spun at $\omega = 60{,}000\,\text{rpm}$. If you were to convert the computed kinetic energy to gravitational potential energy, at what height, in meters, relative to the ground could you lift a 10 kg mass?

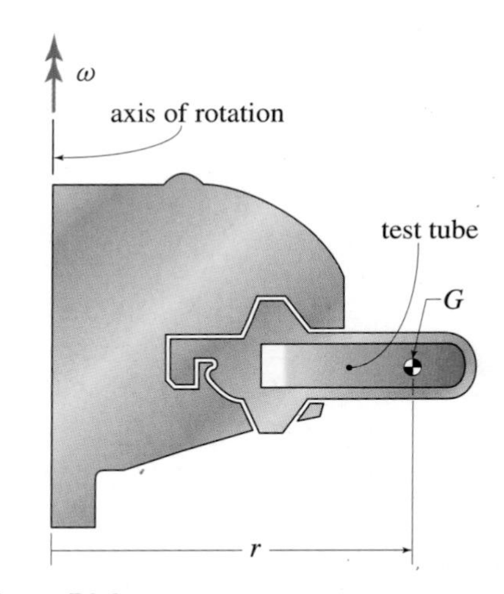

Figure P8.8

Problem 8.9

The uniform thin bars AB, BC, and CD have masses $m_{AB} = 2.3\,\text{kg}$, $m_{BC} = 3.2\,\text{kg}$, and $m_{CD} = 5.0\,\text{kg}$, respectively. The connections at A, B, C, and D are pinned joints. Letting $R = 0.75\,\text{m}$, $L = 1.2\,\text{m}$, and $H = 1.55\,\text{m}$, and $\omega_{AB} = 4\,\text{rad/s}$, compute the kinetic energy T of the system at the instant shown.

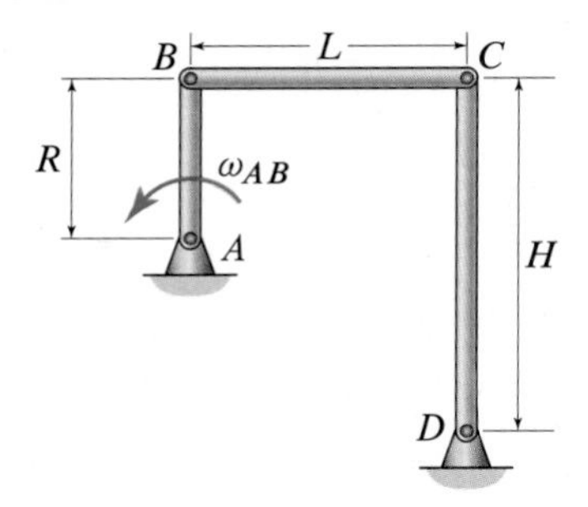

Figure P8.9

Problem 8.10

A T-bar consisting of two uniform bars, each of length $L = 5\,\text{ft}$, is released from rest in the position shown. Neglecting friction, determine the angular speed of the T-bar when point A is directly below point O.

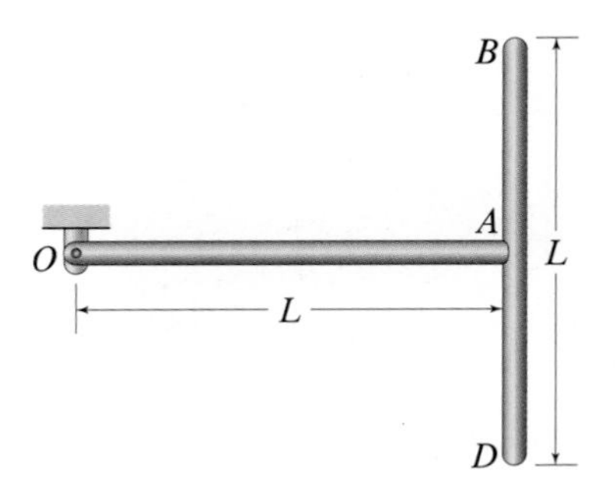

Figure P8.10

Problem 8.11

One of the basement doors is left open in the vertical position when it is given a nudge and allowed to freely fall to the closed position. Given that the door has mass m and that it is modeled as a uniform thin plate of width w and length d, determine its angular velocity when it reaches the closed position. *Hint:* Assume that the door is symmetric with respect to a plane of motion in which the acceleration due to gravity is $g \cos\theta$ rather than g.

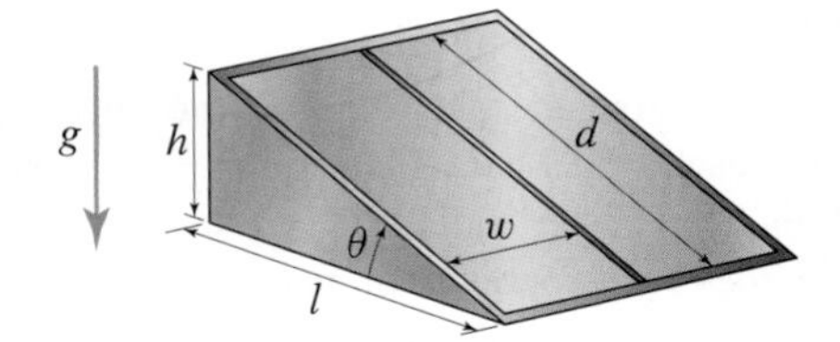

Figure P8.11

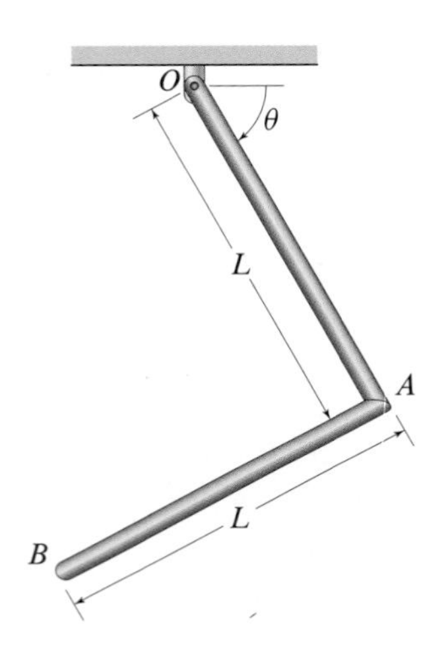

Figure P8.12

Problem 8.12

The L-bar consisting of two uniform bars each of length L is released from rest when $\theta = 90°$. Neglecting friction, determine the smallest value achieved by θ. *Hint:* The equation $\sin\theta + A\cos\theta = B$ admits the solution $\theta = \sin^{-1}(B\cos\phi) - \phi$, with $\phi = \tan^{-1} A$, if $|B\cos\phi| \leq 1$.

Problem 8.13

A turbine rotor with weight $W = 3000\,\text{lb}$, center of mass at the fixed point G, and radius of gyration $k_G = 15\,\text{ft}$ is brought from rest to an angular velocity $\omega = 1500\,\text{rpm}$ in 20 revolutions by applying a constant torque M. Neglecting friction, determine the value of M needed to spin up the rotor as described.

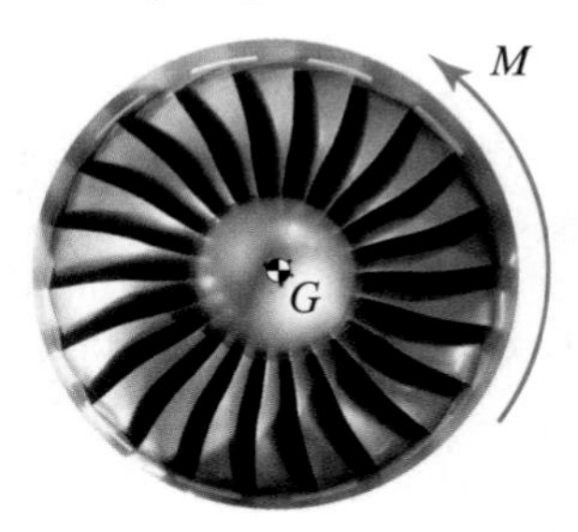

Figure P8.13

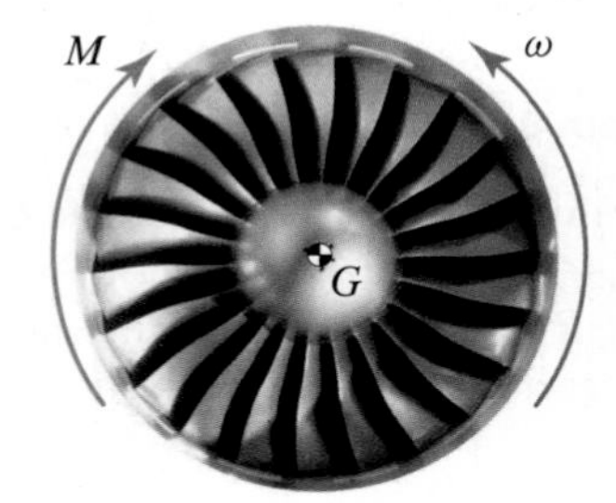

Figure P8.14

Problem 8.14

A turbine rotor with weight $W = 3000\,\text{lb}$, center of mass at the fixed point G, and radius of gyration $k_G = 15\,\text{ft}$ is spinning with an angular speed $\omega = 1200\,\text{rpm}$ when a braking system is engaged that applies a constant torque $M = 3000\,\text{ft}\cdot\text{lb}$. Determine the number of revolutions needed to bring the rotor to a stop.

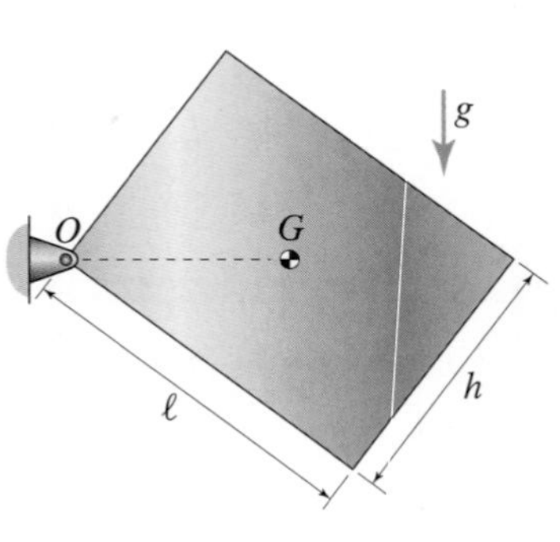

Figure P8.15

Problem 8.15

The uniform rectangular plate of length ℓ, height h, and mass m lies in the vertical plane and is pinned at one corner. If the plate is released from rest in the position shown, determine its angular velocity when the center of mass G is directly below the pivot O. Neglect any friction at the pin at O.

Problems 8.16 and 8.17

A door AB weighing 80 lb is pinned at A and swings in the horizontal plane. The spring CD has stiffness k and is unstretched when $\theta = 0°$. Let $L = 1.5$ ft and $h = 0.5$ ft.

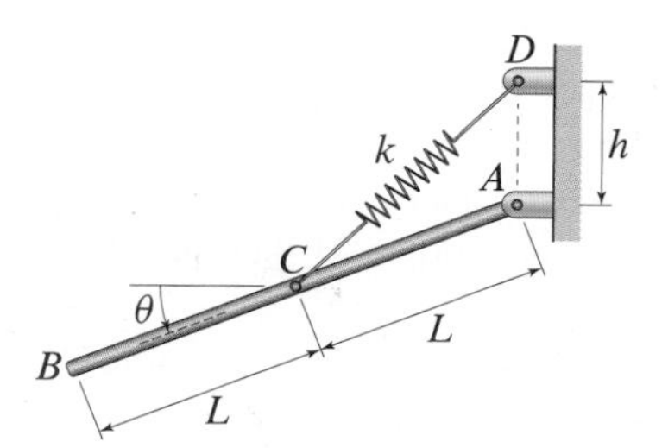

Figure P8.16 and P8.17

Problem 8.16 If the door is rotating counterclockwise and the speed of B is 10 ft/s when $\theta = 0$, determine k, such that the door temporarily stops when $\theta = 90°$. Assume that the spring does not impinge on the mount at A.

Problem 8.17 If the door is released from rest when $\theta = 45°$ and $k = 50$ lb/ft, determine the speed of B when $\theta = 0$.

Problem 8.18

A uniform thin bar AB of length $L = 4$ ft is released from rest at an angle $\theta = \theta_1$. As the bar slides, the ends A and B maintain contact with the surfaces on which they slide. Neglecting friction and knowing that the end A has a speed of 18 ft/s right before hitting the floor, determine θ_1.

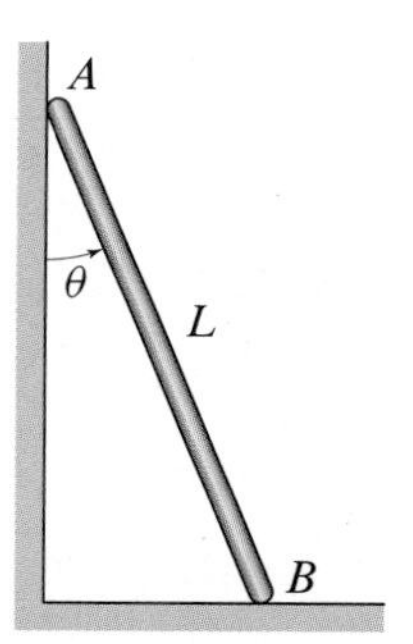

Figure P8.18

Problem 8.19

A uniform thin bar AB of length $L = 3$ ft is released from rest at an angle $\theta = 30°$. As the bar slides, the ends A and B maintain contact with the surfaces on which they slide. The inclination of the wall is $\phi = 50°$. Neglecting friction, determine the angular speed of the bar right before the end A hits the floor.

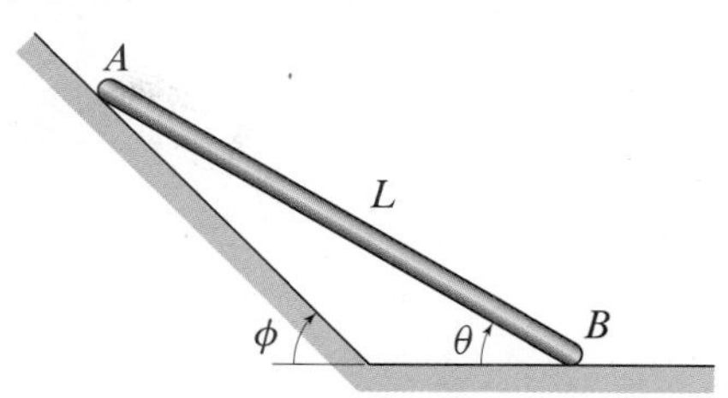

Figure P8.19

Problem 8.20

The disk D, which has mass m, center of mass G, and radius of gyration k_G, is at rest on a flat horizontal surface when the constant moment M is applied to it. The disk is attached at its center to a vertical wall by a linear elastic spring of constant k. The spring is unstretched when the system is at rest. Assuming that the disk rolls without slipping and that it has not yet come to a stop, determine an expression for the angular velocity of the disk after its center G has moved a distance d. After doing so, determine the distance d_s that the disk moves before it comes to a stop.

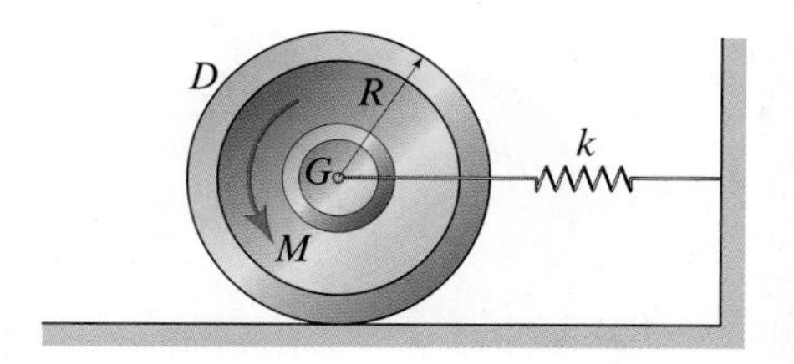

Figure P8.20

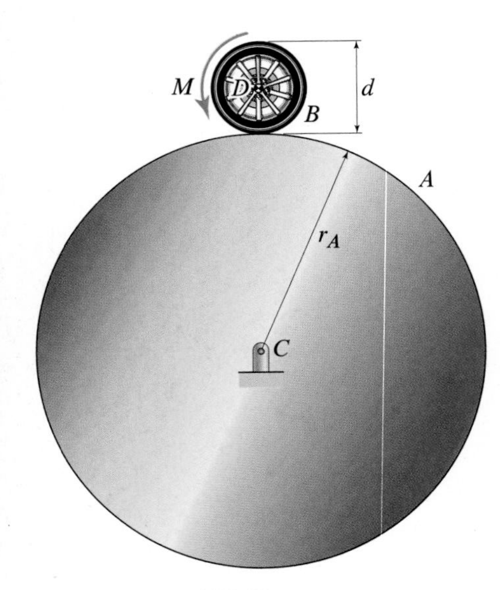

Figure P8.21 and P8.22

Problems 8.21 and 8.22

An automobile wheel test rig consists of a uniform disk A, of mass $m_A = 5000\,\text{kg}$ and radius $r_A = 1.5\,\text{m}$, that can rotate freely about its fixed center C and over which the wheel B of an automobile is made to roll. The wheel B, with center and center of mass at D, is mounted on a shaft (not shown) that holds D fixed while allowing the wheel to rotate about D. The wheel has diameter $d = 0.62\,\text{m}$, mass $m_B = 21.5\,\text{kg}$, and mass moment of inertia about its mass center $I_D = 44\,\text{kg}\cdot\text{m}^2$. Both A and B are initially at rest when B is subject to a constant torque M that causes B to roll without slip over A.

Problem 8.21 If $M = 1500\,\text{N}\cdot\text{m}$, determine the number of revolutions of B needed to reach conditions simulating a car speed of $100\,\text{km/h}$.

Problem 8.22 Determine M if it takes 100 revolutions of the wheel B to achieve conditions simulating a car speed of $60\,\text{km/h}$.

Problems 8.23 and 8.24

An electric motor drawing $15\,\text{kW}$ and with an efficiency of 85% lifts a $400\,\text{kg}$ crate B with a constant speed v_c. Pulley A has radius $r_p = 15\,\text{cm}$, and the center of mass of A is also the center of A. The cord is inextensible and does not slip relative to the pulley. Assume that the friction at the pulley bearings results in a moment about the pulley's center with magnitude $\beta|\omega_p|$ opposing the rotation of the pulley, where β is a constant and $|\omega_p|$ is the angular speed of the pulley. *Hint:* Review Example 4.16 on p. 300. Also note that friction in the pulley bearings causes the tension in the cord to be different on the two sides of the pulley.

Problem 8.23 Determine β if $v_c = 3\,\text{m/s}$.

Problem 8.24 Determine v_c if $\beta = 2\,\text{kg}\cdot\text{m}^2/\text{s}$.

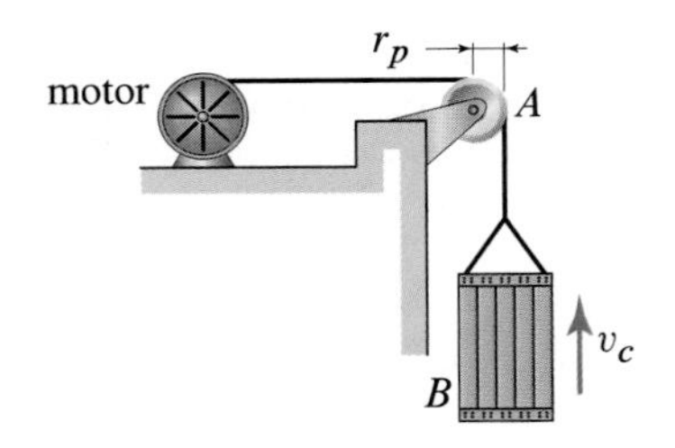

Figure P8.23 and P8.24

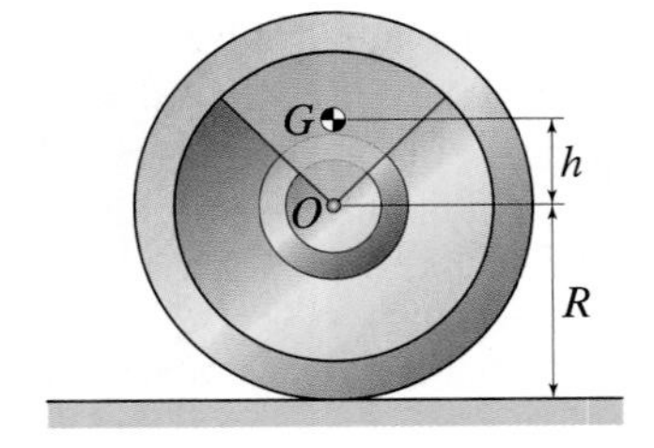

Figure P8.25

Problem 8.25

An eccentric wheel with weight $W = 250\,\text{lb}$, mass center G, and radius of gyration $k_G = 1.32\,\text{ft}$ is initially at rest in the position shown. Letting $R = 1.75\,\text{ft}$ and $h = 0.8\,\text{ft}$, and assuming that the wheel is gently nudged to the right and rolls without slip, determine the speed of O when G is closest to the ground.

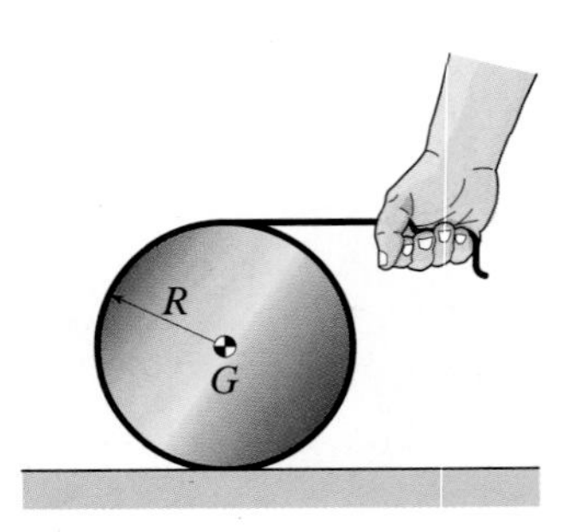

Figure P8.26

Problem 8.26

A cord is wound around a uniform disk of mass $m = 2.5\,\text{kg}$ and radius $R = 10\,\text{cm}$. A person pulls on the cord to the right with a constant horizontal force P. The disk is initially at rest and rolls without slip. Determine P if the center of the disk has a speed of $0.5\,\text{m/s}$ after the hand pulling the cord displaces horizontally to the right by $20\,\text{cm}$.

Problems 8.27 and 8.28

The pendulum consists of a thin bar of length $L = 1.5\,\text{m}$ and mass $m_b = 2\,\text{kg}$, at the end of which is rigidly attached a uniform solid sphere of radius $r = 0.25\,\text{m}$ and mass $m_s = 3\,\text{kg}$. The pendulum is pinned at the top end O of the bar. The pendulum is initially at $\theta = 0$ when the center of the sphere C is given a speed $v_1 = 4\,\text{m/s}$ to the right. The quantity d denotes the distance between O and the center of mass of the bar G. Neglect friction.

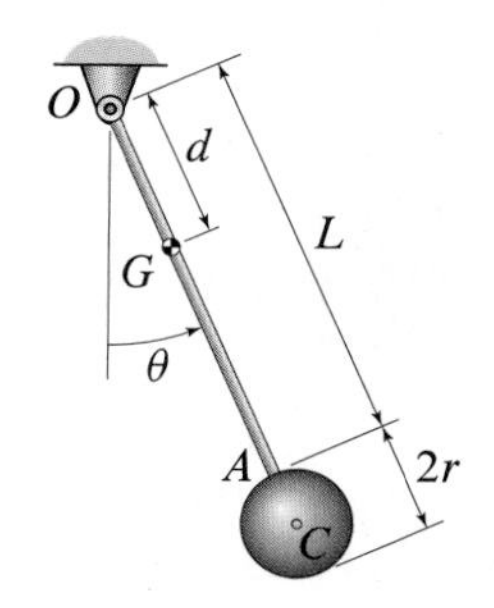

Figure P8.27 and P8.28

Problem 8.27 If the bar is uniform (i.e., $d = L/2$), determine the maximum swing angle of the pendulum.

Problem 8.28 If the radius of gyration of the bar is $k_G = 0.35\,\text{m}$, determine d so that the maximum swing angle is 55°.

Problems 8.29 and 8.30

A 14 lb bowling ball is thrown onto a lane with a backspin angular speed $\omega_0 = 10\,\text{rad/s}$ and forward velocity $v_0 = 17\,\text{mph}$. After a few seconds, the ball starts rolling without slip and moving forward with a speed $v_f = 17.2\,\text{ft/s}$. Let $r = 4.25\,\text{in.}$ be the radius of the ball, and let $k_G = 2.6\,\text{in.}$ be its radius of gyration.

Problem 8.29 Determine the work done by friction on the ball from the initial time until the time that the ball starts rolling without slip.

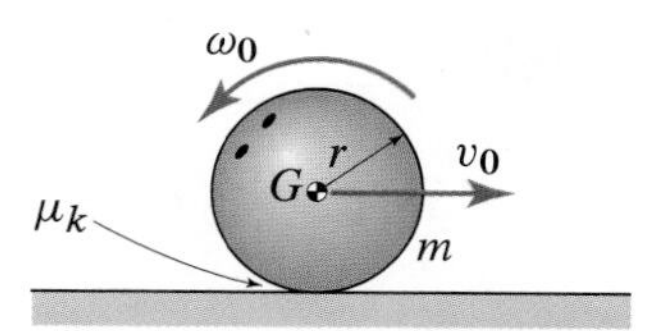

Figure P8.29–P8.31

Problem 8.30 Knowing that the coefficient of kinetic friction between the lane and the ball is $\mu_k = 0.1$, determine the length L_f over which the friction force acts in order to slow down the ball from v_0 to v_f. Does L_f also represent the distance traveled by the center of the ball? Explain.

Problem 8.31

A bowling ball is thrown onto a lane with a forward velocity v_0 and no angular velocity ($\omega_0 = 0$). Because of friction between the lane and the ball, after a short time, the ball starts rolling without slip and moving forward with speed v_f. Let L_G be the distance traveled by the center of the ball while slowing down from v_0 to v_f. In addition, let L_f be the length over which the friction force had to act in order to slow down the ball from v_0 to v_f. State which of the following relations is true and why: (a) $L_G < L_f$; (b) $L_G = L_f$; (c) $L_G > L_f$.

Problems 8.32 and 8.33

The uniform sphere B, of radius $r = 6\,\text{cm}$ and mass $m_B = 5\,\text{kg}$, is rigidly attached to the uniform thin bar AB, which is pinned at A and has a mass $m_{AB} = 8\,\text{kg}$. The system rotates in the vertical plane. The spring CD has stiffness $k = 2000\,\text{N/m}$ and is designed so that the system is in static equilibrium when $\theta = 0°$. Let $L = 18.2\,\text{cm}$ and $h = 24.6\,\text{cm}$.

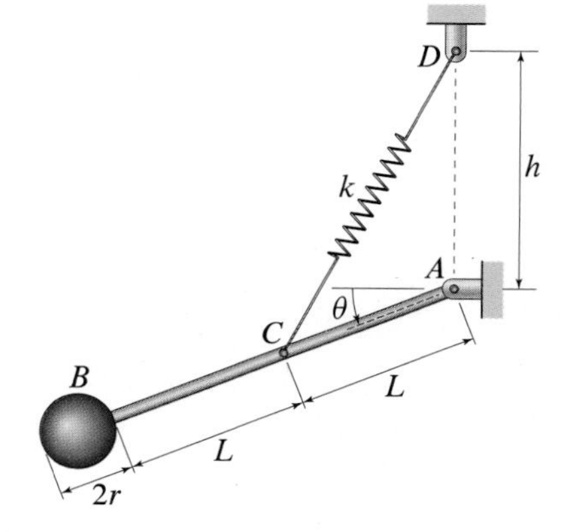

Figure P8.32 and P8.33

Problem 8.32 If the system is released from rest when $\theta = -30°$, determine the angular speed of the arm for $\theta = 0$.

Problem 8.33 If the system is released from rest when $\theta = -30°$, determine the maximum angle θ reached by the arm AB.

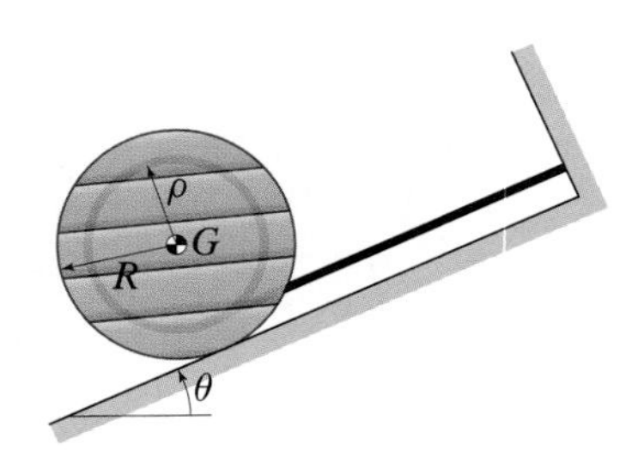

Figure P8.34 and P8.35

Problems 8.34 and 8.35

A 500 lb spool with inner and outer radii $\rho = 4$ ft and $R = 6$ ft, respectively, is released from rest on an incline with $\theta = 30°$. The center of the spool G coincides with its center of mass, and the radius of gyration of the spool is $k_G = 5$ ft. Assume that the only forces acting on the spool after its release are the spool's weight, the tension in the cord, and the contact force between the spool and the incline.

Problem 8.34 If the angular speed of the spool is $\omega_{s2} = 1.2$ rad/s after G has displaced a distance $d = 10$ ft from the release position, determine the work done by friction from the instant of release to when ω_{s2} is achieved.

Problem 8.35 If the spool starts moving immediately after release and the coefficient of kinetic friction between the spool and the incline is $\mu_k = 0.25$, determine the speed of G after G has displaced a distance $d = 10$ ft down the incline.

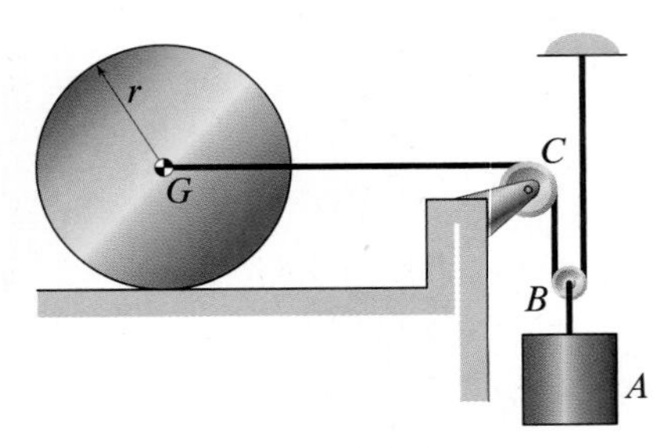

Figure P8.36 and P8.37

Problems 8.36 and 8.37

A 400 lb uniform disk with center G and radius $r = 3$ ft is connected by a pulley system to a counterweight A weighing 75 lb. The system is initially at rest when A is allowed to drop, thus causing the disk to roll without slipping to the right.

Problem 8.36 Neglecting the inertia of the pulley system, determine the speed of G after A has dropped 2 ft.

Problem 8.37 Neglect the inertia of pulley B and the cord, but model pulley C as a uniform disk with radius $r_C = 0.8$ ft and weight $W_C = 50$ lb. Assuming the cord does not slip relative to pulley C, determine the speed of G after A has dropped 2 ft.

Problems 8.38 and 8.39

In a contraption built by a fraternity, a person sits at the center of a swinging platform with mass $m = 400$ kg and length $L = 4$ m suspended by two identical arms of length $H = 3$ m.

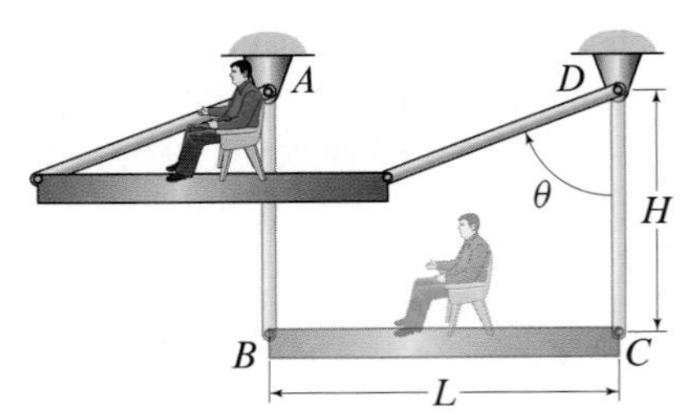

Figure P8.38 and P8.39

Problem 8.38 Neglecting the mass of the arms and of the person, neglecting friction, and assuming that the platform is released from rest when $\theta = 180°$, compute the speed of the person as a function of θ for $0° \leq \theta \leq 180°$. In addition, find the speed of the person for $\theta = 0°$.

Problem 8.39 Neglecting the mass of the person, neglecting friction, letting the mass of each arm be $m_A = 150$ kg, and assuming that the platform is released from rest when $\theta = 180°$, compute the speed of the person as a function of θ for $0° \leq \theta \leq 180°$. In addition, find the speed of the person for $\theta = 0°$.

Problem 8.40

The weights of the uniform thin pin-connected bars AB, BC, and CD are $W_{AB} = 4\,\text{lb}$, $W_{BC} = 6.5\,\text{lb}$, and $W_{CD} = 10\,\text{lb}$, respectively. Letting $\phi = 47°$, $R = 2\,\text{ft}$, $L = 3.5\,\text{ft}$, and $H = 4.5\,\text{ft}$, and knowing that bar AB rotates at an angular velocity $\omega_{AB} = 4\,\text{rad/s}$, compute the kinetic energy T of the system at the instant shown.

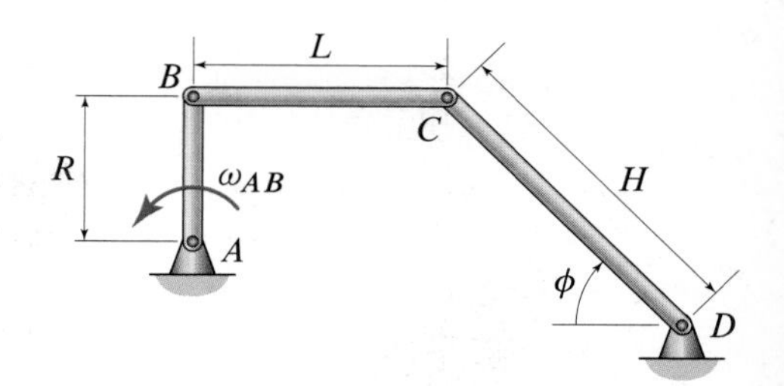

Figure P8.40

Problem 8.41

A payload B of mass $m_B = 50\,\text{kg}$ is lifted via the pulley system shown by the application of a constant force $F = 300\,\text{N}$. The pulleys are identical and can be modeled as uniform disks of radius $r_p = 10\,\text{cm}$ and mass $m_p = 8\,\text{kg}$. The cord does not slip relative to the pulleys. Modeling the cord as inextensible and neglecting friction at the pulley bearings, determine the speed of B after B has been lifted a height $h = 1\,\text{m}$ from its initial rest position. Treat all cable segments as being purely vertical.

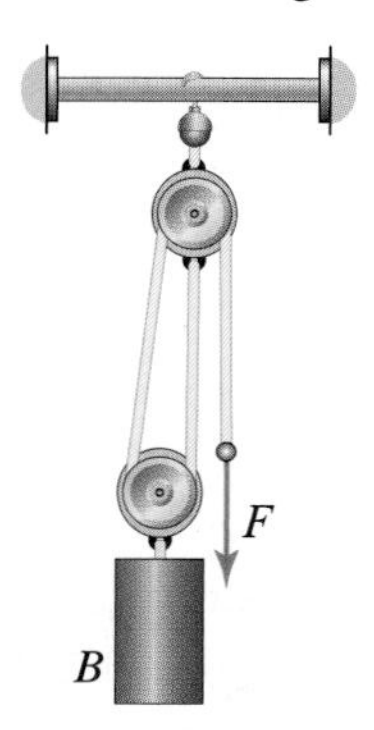

Figure P8.41

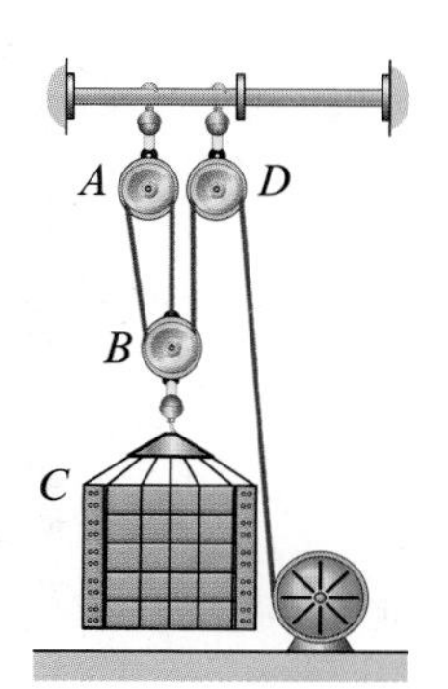

Figure P8.42

Problem 8.42

A winch drawing 9 hp powers a pulley system lifting a 600 lb crate C with a constant speed v_c. The pulleys are all identical and have a radius $r_p = 1.25\,\text{ft}$. The cord is inextensible and does not slip relative to the pulleys. For each pulley, the friction at the pulley bearings produces a moment about the pulley's center with magnitude $\kappa|\omega_p|$ opposing the rotation of the pulley, where $\kappa = 1.5\,\text{lb·ft·s}$ and $|\omega_p|$ is the angular speed of the pulley. Neglecting the inertia of the pulleys and of the cord, and treating segments of cord that do not touch the pulleys as being vertical, determine v_c if the motor's efficiency is $\epsilon = 0.87$. *Hint:* Adapt to this problem the solution in Part (a) of Example 4.16 on p. 300, observing that friction at the pulley bearings causes the tension in the cord on the two sides of a pulley to be different.

Problem 8.43

A 10 lb uniform thin bar BC of length $L = 10\,\text{ft}$ is pinned at B to the edge of a 20 lb uniform disk of radius $R = 3.5\,\text{ft}$. The system is initially at rest in the position shown when a constant horizontal force $P = 60\,\text{lb}$ is applied to the end C. Assume that the disk rolls without slip. In addition, neglect the friction between the end C of the bar and the ground, as well as the friction at the pin. Determine the angular speed of the disk when point B is directly above the center of the disk. *Hint:* To determine the displacement of point C, keep in mind that the overall displacement of point A is $\pi R/2$.

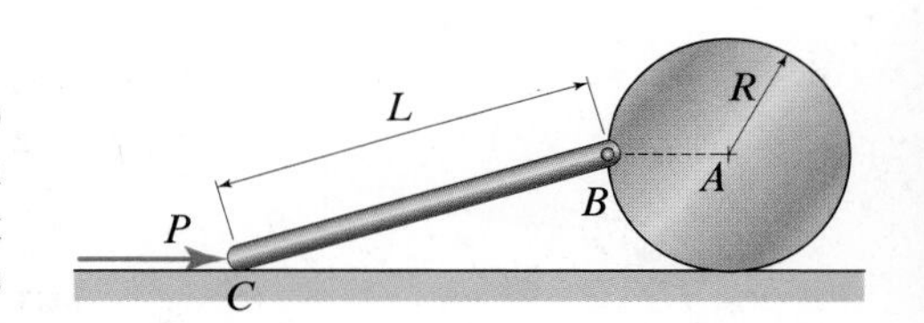

Figure P8.43

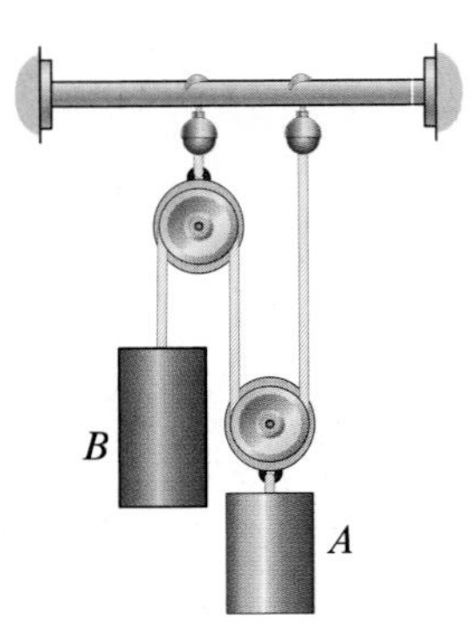

Figure P8.44

Problem 8.44

The two blocks A and B weighing 30 lb and 25 lb, respectively, are released from rest. The pulleys are identical and can be modeled as uniform disks of radius $r_p = 0.75$ ft and weight $W_p = 8$ lb. Modeling the cord as inextensible and neglecting friction at the pulley bearings, determine the speed of A after B has dropped a height $h = 3$ ft.

Problem 8.45

An eccentric wheel with mass $m = 150$ kg, mass center G, and radius of gyration $k_G = 0.4$ m is placed on the incline shown, such that the wheel's center of mass G is vertically aligned with P, which is the point of contact with the incline. If the wheel rolls without slip once it is gently nudged away from its initial placement, letting $R = 0.55$ m, $h = 0.25$ m, $\theta = 25°$, and $d = 0.5$ m, determine whether the wheel arrives at B and, if yes, determine the corresponding speed of the center O. Note that the angle POG is not equal to 90° at release.

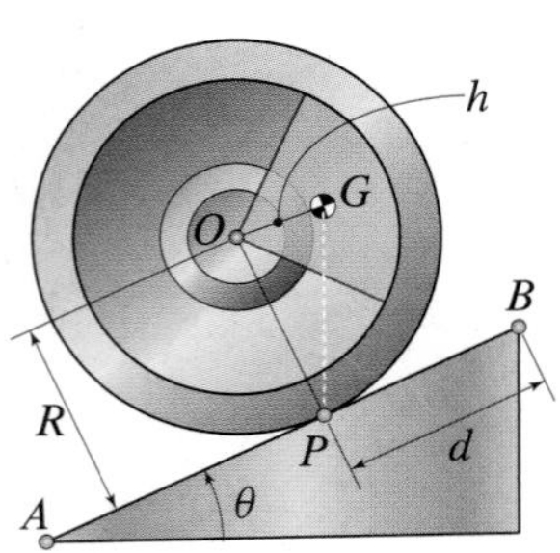

Figure P8.45

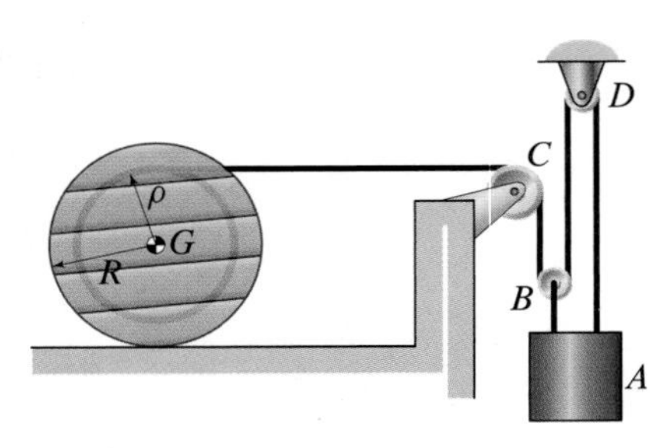

Figure P8.46 and P8.47

Problems 8.46 and 8.47

A spool of mass $m_s = 150$ kg and inner and outer radii $\rho = 0.8$ m and $R = 1.2$ m, respectively, is connected to a counterweight A of mass $m_A = 50$ kg by a pulley system whose cord, at one end, is wound around the inner hub of the spool. The center G of the spool is also the center of mass of the spool, and the radius of gyration of the spool is $k_G = 1$ m. The system is at rest when the counterweight is released, causing the spool to move to the right. Assume that the spool rolls without slip.

Problem 8.46 Neglecting the inertia of the pulley system, determine the angular speed of the spool after the counterweight has dropped 0.5 m.

Problem 8.47 Assume that the inertia of the cord and of pulleys B and D can be neglected, but model pulley C as a uniform disk of mass $m_C = 15$ kg and radius $r_C = 0.3$ m. If the cord does not slip relative to pulley C, determine the angular speed of the spool after A drops 0.5 m.

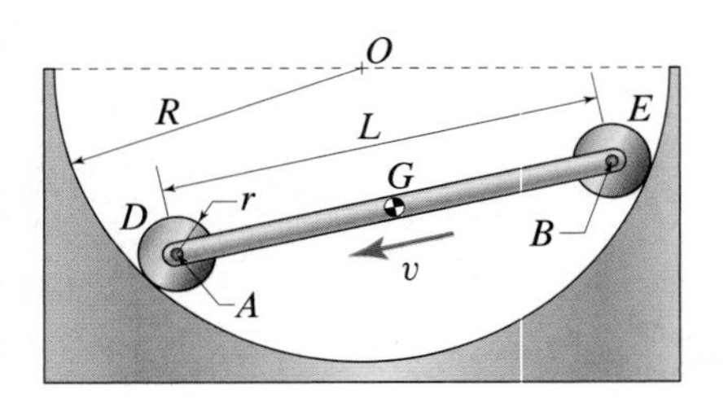

Figure P8.48

Problem 8.48

The uniform slender bar AB has length $L = 1.45$ ft and weight $W_{AB} = 20$ lb. Rollers D and E, which are pinned at A and B, respectively, can be modeled as two identical uniform disks, each with radius $r = 1.5$ in. and weight $W_r = 0.35$ lb. Rollers D and E roll without slip on the surface of a cylindrical bowl with center at O and radius $R = 1$ ft. Determine the system's kinetic energy when G (the center of mass of bar AB) moves with a speed $v = 7$ ft/s.

Problems 8.49 and 8.50

The *Charpy impact test* is one test that measures the resistance of a material to fracture. In this test, the fracture toughness is assessed by measuring the energy required to break a specimen of a given geometry. This is done by releasing a heavy pendulum from rest at an angle θ_i and then measuring the maximum swing angle θ_f reached by the pendulum after the specimen is broken.

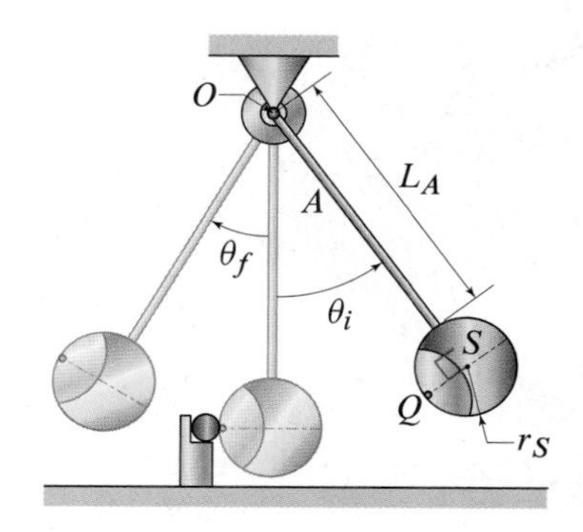

Figure P8.49 and P8.50

Problem 8.49 Consider a test rig in which the striker S (the pendulum's bob) can be modeled as a uniform disk of mass $m_S = 19.5$ kg and radius $r_S = 150$ mm, and the arm can be modeled as a thin rod of mass $m_A = 2.5$ kg and length $L_A = 0.8$ m. Neglecting friction and noting that the striker and the arm are rigidly connected, determine the fracture energy (i.e., the kinetic energy lost in breaking the specimen) in an experiment where $\theta_i = 158°$ and $\theta_f = 43°$.

Problem 8.50 Consider a test rig in which the striker S (the pendulum's bob) can be modeled as a uniform disk of weight $W_S = 40$ lb and radius $r_S = 6$ in., and the arm can be modeled as a thin rod of weight $W_A = 5.5$ lb and length $L_A = 2.75$ ft. If the release angle of the striker is $\theta_i = 158°$ and if the striker impacts the specimen when the pendulum's arm is vertical, determine the speed of the point Q on the striker immediately before the striker impacts with the specimen. Neglect friction, and observe that the striker and the arm are rigidly connected.

Problem 8.51

A crate, with weight $W = 155$ lb and mass center G, is placed on a slide and released from rest as shown. The lower part of the slide is circular, with radius $R = 6$ ft. Model the crate as a uniform body with $b = 3.6$ ft and $h = 2$ ft. Take into account the gap between the crate and the slide when the crate is in its lowest position, and assume that when the crate is in its lowest position on the slide, the crate's center of mass is moving to the left with a speed $v_G = 12$ ft/s. Determine the work done by friction on the crate as the crate moves from the release point to the lowest point on the slide.

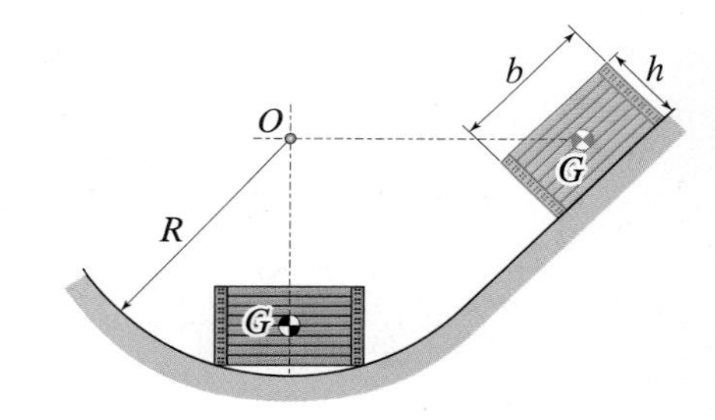

Figure P8.51

Problems 8.52 and 8.53

In a contraption built by a fraternity, a person sits at the center of a swinging platform with weight $W_p = 800$ lb and length $L = 12$ ft suspended by two identical arms each of length $H = 10$ ft and weight $W_a = 200$ lb. The platform, which is at rest when $\theta = 0$, is put in motion by a motor that pumps the ride by exerting a constant moment M in the direction shown, during each upswing, whenever $0 \le \theta \le \theta_p$, while exerting zero moment otherwise.

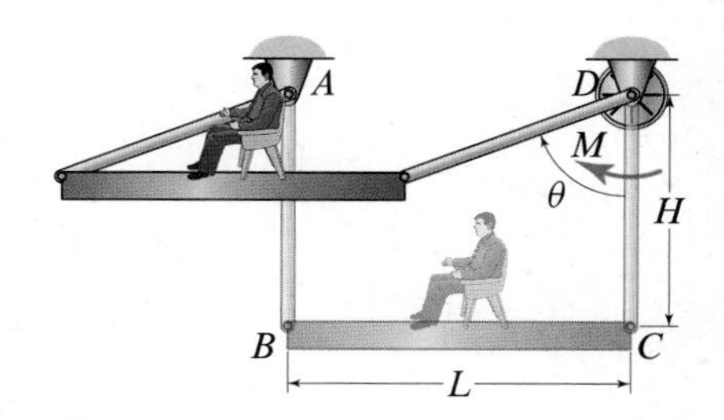

Figure P8.52 and P8.53

Problem 8.52 Neglecting the mass of the person, neglecting friction, letting $M = 900$ ft·lb, and letting $\theta_p = 25°$, find the minimum number of swings necessary to achieve $\theta > 90°$ and the ensuing speed achieved by the person at the lowest point in the swing. Model the arms AB and CD as uniform thin bars.

Problem 8.53 Neglecting the mass of the person, neglecting friction, and letting $\theta_p = 20°$, determine the value of M required to achieve a maximum value of θ equal to 90° in 6 full swings. Model the arms AB and CD as uniform thin bars.

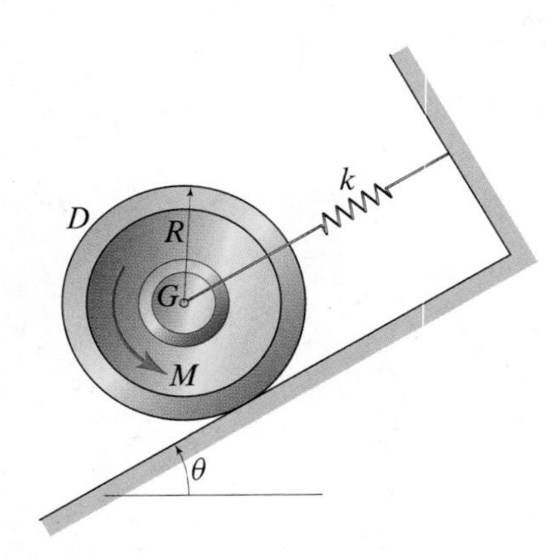

Figure P8.54

Problem 8.54

The disk D, which has weight W, mass center G coinciding with the disk's geometric center, and radius of gyration k_G, is at rest on an incline when the constant moment M is applied to it. The disk is attached at its center to a wall by a linear elastic spring of constant k. The spring is unstretched when the system is at rest. Assuming that the disk rolls without slipping and that it has not yet come to a stop, determine an expression for the angular velocity of the disk after its center G has moved a distance d down the incline. After doing so, using $k = 5\,\text{lb/ft}$, $R = 1.5\,\text{ft}$, $W = 10\,\text{lb}$, and $\theta = 30°$, determine the value of the moment M for the disk to stop after rolling $d_s = 5\,\text{ft}$ down the incline.

Problem 8.55

In Example 8.2 on p. 592, we ignored the rotational inertia of the counterweight. Let's revisit that example and remove that simplifying assumption. Assume that the arm AD is still a uniform thin bar of length $L = 15.7\,\text{ft}$ and weight 45 lb. The hinge O is still $d = 2.58\,\text{ft}$ from the right end of the arm, and the 160 lb counterweight C is still $\delta = 1.4\,\text{ft}$ from the hinge. Now model the counterweight as a uniform block of height $h = 14\,\text{in.}$ and width $w = 9\,\text{in.}$ With this new assumption, solve for the angular velocity of the arm as it reaches the horizontal position after being nudged from the vertical position. Determine the percent change in angular velocity compared with that found in Example 8.2.

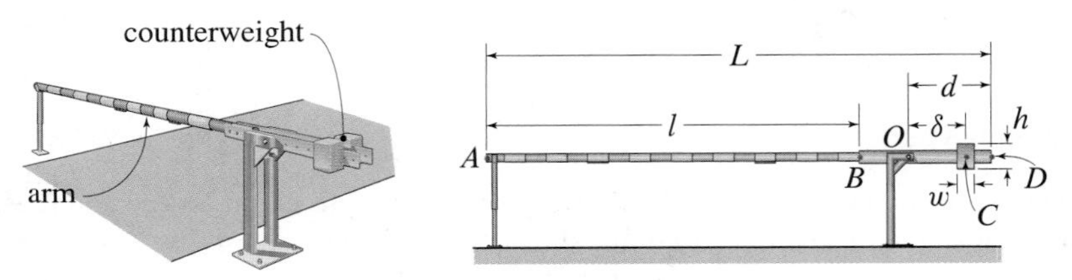

Figure P8.55 and P8.56

Problem 8.56

For the barrier gate shown, assume that the arm consists of a section of aluminum tubing from A to B of length $l = 11.6\,\text{ft}$ and weight 20 lb and a steel support section from B to D of weight 40 lb. The overall length of the arm is $L = 15.7\,\text{ft}$. In addition, the 120 lb counterweight C is placed a distance δ from the hinge at O, and the hinge is $d = 2.58\,\text{ft}$ from the right end of section BD. Model the two sections AB and BD as uniform thin bars, and model the counterweight as a uniform block of height $h = 14\,\text{in.}$ and width $w = 9\,\text{in.}$ Using these new assumptions, determine the distance δ so that the angular velocity of the arm is $0.25\,\text{rad/s}$ as it reaches the horizontal position after being nudged from the vertical position.

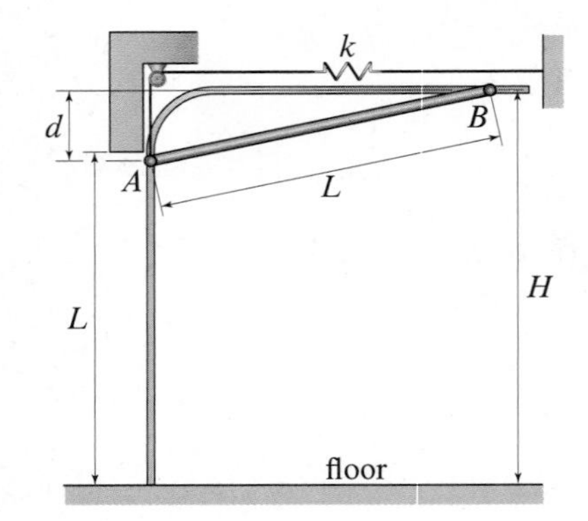

Figure P8.57

Problem 8.57

The figure shows the cross section of a garage door with length $L = 9\,\text{ft}$ and weight $W = 175\,\text{lb}$. At A and B, there are rollers of negligible mass constrained to move in the guide whose horizontal portion is at a distance $H = 11\,\text{ft}$ from the floor. The door's motion is assisted by two springs, each with constant k (only one spring is shown). The door is released from rest when $d = 26\,\text{in.}$ and the spring is stretched 4 in. Neglecting friction, knowing that, when A touches the floor, B is in the vertical portion of the guide, and modeling the door as a uniform thin plate, determine the minimum value of k so that A will strike the ground with a speed no greater than $1\,\text{ft/s}$.

Problem 8.58

The figure shows the cross section of a garage door with length $L = 2.5\,\text{m}$ and mass $m = 90\,\text{kg}$. At the ends A and B, there are rollers of negligible mass constrained to move in the guide whose horizontal portion is at a distance $H = 3\,\text{m}$ from the floor. The door's motion is assisted by two counterweights C, each of mass m_C (only one counterweight is shown). If the door is released from rest when $d = 53\,\text{cm}$, neglecting friction and modeling the door as a uniform thin plate, determine the minimum value of m_C so that A will strike the ground with a speed no greater than $0.25\,\text{m/s}$.

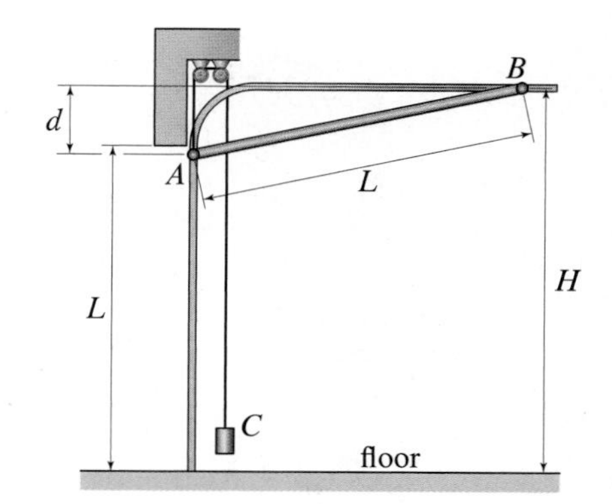

Figure P8.58

Problems 8.59 through 8.61

The uniform thin rod AB is pin-connected to the slider S, which moves along the frictionless guide, and to the disk D, which rolls without slip over the horizontal surface. The pins at A and B are frictionless, and the system is released from rest. Neglect the vertical dimension of S.

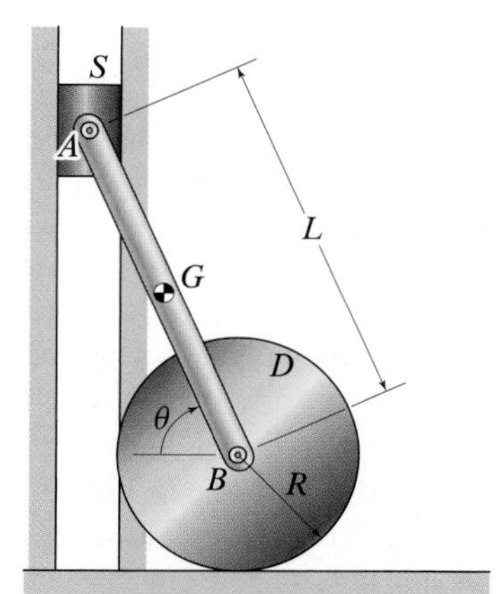

Figure P8.59–P8.61

Problem 8.59 Letting $L = 1.75\,\text{m}$ and $R = 0.6\,\text{m}$, assuming that S and D are of negligible mass, that the mass of rod AB is $m_{AB} = 7\,\text{kg}$, and that the system is released from the angle $\theta_0 = 65°$, determine the speed of the slider S when it strikes the ground.

Problem 8.60 Letting $L = 4.5\,\text{ft}$ and $R = 1.2\,\text{ft}$, assuming that AB is of negligible mass, the weight of S is $W_S = 3\,\text{lb}$, D is a uniform disk of weight $W_D = 9\,\text{lb}$, and the system is released from the angle $\theta_0 = 67°$, determine the speed of the slider S when it strikes the ground.

Problem 8.61 Letting $L = 1.75\,\text{m}$ and $R = 0.6\,\text{m}$, assuming that the mass of S is $m_S = 4.2\,\text{kg}$, D is a uniform disk of mass $m_D = 12\,\text{kg}$, the mass of AB is $m_{AB} = 7\,\text{kg}$, and that the system is released from the angle $\theta = 69°$, determine the speed and the direction of motion of point B when the slider S strikes the ground.

Problem 8.62

Revisit Example 8.5 on p. 598 and replace the two springs with a system of two counterweights P (only one counterweight is shown) each of weight W_P. Recalling that the door's weight is $W = 800\,\text{lb}$ and that the total height of the door is $H = 30\,\text{ft}$, if the door is released from rest in the fully open position and friction is negligible, determine the minimum value of W_P so that A will strike the left end of the horizontal guide with a speed no greater than $0.5\,\text{ft/s}$.

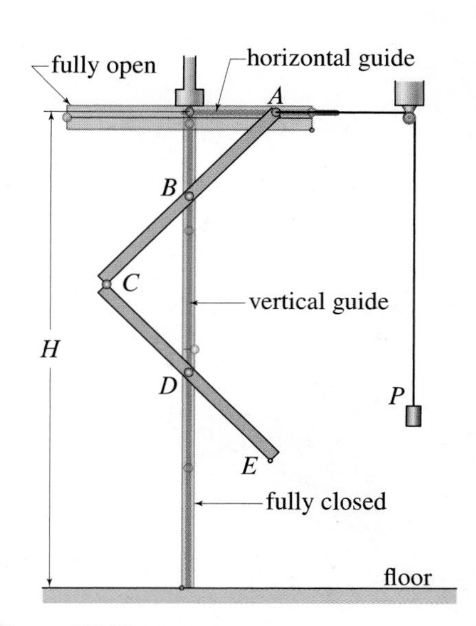

Figure P8.62

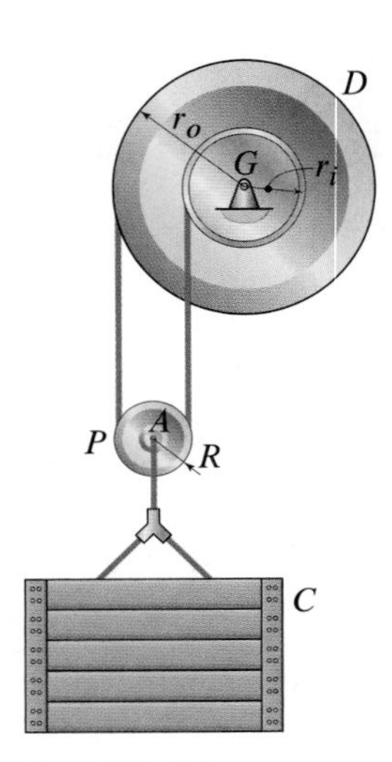

Figure P8.63 and P8.64

Problems 8.63 and 8.64

The double pulley D has mass of 15 kg, center of mass G coinciding with its geometric center, radius of gyration $k_G = 10\,\text{cm}$, outer radius $r_o = 15\,\text{cm}$, and inner radius $r_i = 7.5\,\text{cm}$. It is connected to the pulley P with radius R by a cord of negligible mass that unwinds without slip from the inner and outer spools of the double pulley D. The crate C, which has a mass of 20 kg, is released from rest, and the inner and outer parts of the double pulley rotate together as a single unit.

Problem 8.63 Neglecting the mass of the pulley P, determine the speed of the crate C and the angular velocity of the pulley D after the crate has dropped a distance $h = 2\,\text{m}$.

Problem 8.64 Assuming that the pulley P has a mass of 1.5 kg and a radius of gyration $k_A = 3.5\,\text{cm}$, determine the speed of the crate C and the angular velocity of the pulley D after the crate has dropped a distance $h = 2\,\text{m}$.

Problems 8.65 through 8.67

Torsional springs provide a simple propulsion mechanism for toy cars. When the rear wheels are rotated as if the car were moving backward, they cause a torsional spring (with one end attached to the axle and the other to the body of the car) to wind up and store energy. Therefore, a simple way to charge the spring is to place the car onto a surface and to pull it backward, making sure that the wheels roll without slipping. Note that the torsional spring can only be wound by pulling the car backward; that is, *the forward motion of the car unwinds the spring.*

Figure P8.65 and P8.66

Problem 8.65 Let the weight of the car (body and wheels) be $W = 5\,\text{oz}$, the weight of each of the wheels be $W_w = 0.15\,\text{oz}$, and the radius of the wheels be $r = 0.25\,\text{in.}$, where the wheels roll without slip and can be treated as uniform disks. Neglecting friction internal to the car and letting the car's torsional spring be linear with constant $k_t = 0.0002\,\text{ft}\cdot\text{lb/rad}$, determine the maximum speed achieved by the car if it is released from rest after pulling it back a distance $L = 0.75\,\text{ft}$ from a position in which the spring is unwound.

Problem 8.66 Let the weight of the car (body and wheels) be $W = 5\,\text{oz}$, the weight of each of the wheels be $W_w = 0.15\,\text{oz}$, and the radius of the wheels be $r = 0.25\,\text{in.}$, where the wheels roll without slip and can be treated as uniform disks. In addition, let the torque M provided by the nonlinear torsional spring be given by $M = -\beta\theta^3$, where $\beta = 0.5\times10^{-6}\,\text{ft}\cdot\text{lb/rad}^3$, θ is the angular displacement of the rear axle, and the minus sign in front of β indicates that M acts opposite to the direction of θ. Neglecting any friction internal to the car, determine the maximum speed achieved by the car if it is released from rest after pulling it back a distance $L = 0.75\,\text{ft}$ from a position in which the spring is unwound.

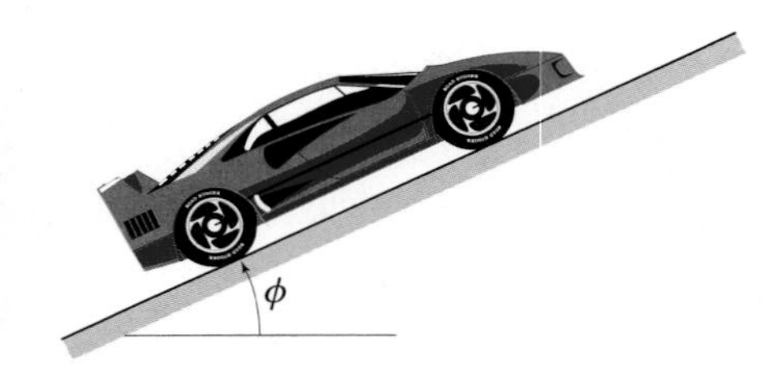

Figure P8.67

Problem 8.67 Let the mass of the car (body and wheels) be $m = 120\,\text{g}$, the mass of each of the wheels be $m_w = 5\,\text{g}$, and the radius of the wheels be $r = 6\,\text{mm}$, where the wheels roll without slip and can be treated as uniform disks. In addition, let the car's torsional spring be linear with constant $k_t = 0.00025\,\text{N}\cdot\text{m/rad}$. Neglecting any friction internal to the car, if the angle of the incline is $\phi = 25^\circ$ and the car is released from rest after pulling it back a distance $L = 25\,\text{cm}$ from a position in which the spring is unwound, determine the maximum distance d_{max} that the car will travel up the incline (from its release point), the maximum speed v_{max} achieved by the car, and the distance $d_{v\text{max}}$ (from the release point) at which v_{max} is achieved.

Problem 8.68

The figure shows the cross section of a garage door with length $L = 2.5\,\text{m}$ and mass $m = 90\,\text{kg}$. At the ends A and B, there are rollers of negligible mass constrained to move in a vertical and a horizontal guide, respectively. The door's motion is assisted by two counterweights (only one counterweight is shown), each of mass $m_C = 22\,\text{kg}$. If the door is released from rest when horizontal, neglecting friction and modeling the door as a uniform thin plate, determine the speed with which B strikes the left end of the horizontal guide.

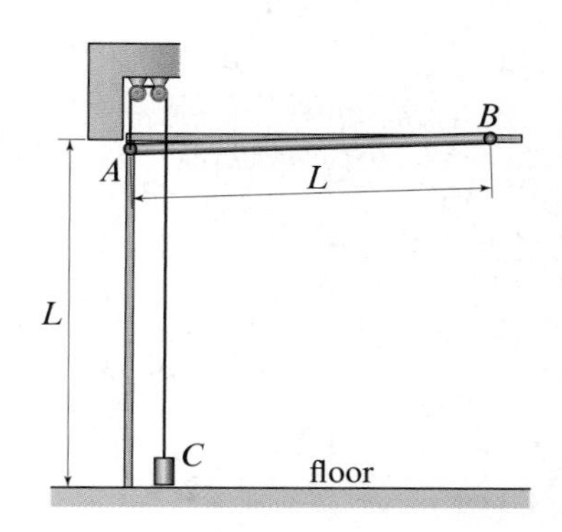

Figure P8.68

Problem 8.69

A stick of length L and mass m is in equilibrium while standing on its end A when end B is gently nudged to the right, causing the stick to fall. Model the stick as a uniform slender bar, and assume that there is friction between the stick and the ground. Under these assumptions, there is a value of θ, let's call it θ_{max}, such that the stick *must* start slipping before reaching θ_{max} for *any* value of the coefficient of static friction μ_s. To find the value of θ_{max}, follow the steps below.

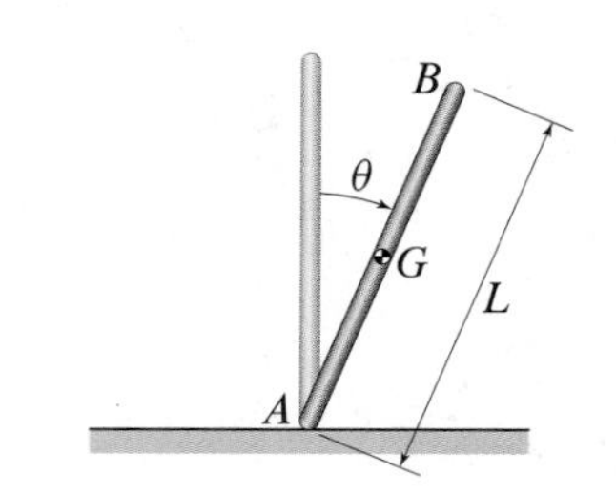

Figure P8.69

(a) Letting F and N be the friction and normal forces, respectively, between the stick and the ground, draw the FBD of the stick as it falls. Then set the sum of forces in the horizontal and vertical directions equal to the corresponding components of $m\vec{a}_G$. Express the components of $\vec{a}_G$ in terms of θ, $\dot{\theta}$, and $\ddot{\theta}$. Finally, express F and N as functions of θ, $\dot{\theta}$, and $\ddot{\theta}$.

(b) Use the work-energy principle to find an expression for $\dot{\theta}^2(\theta)$. Differentiate the expression for $\dot{\theta}^2(\theta)$ with respect to time, and find an expression for $\ddot{\theta}(\theta)$.

(c) Substitute the expressions for $\dot{\theta}^2(\theta)$ and $\ddot{\theta}(\theta)$ into the expressions for F and N to obtain F and N as functions of θ. For impending slip, $|F/N|$ must be equal to the coefficient of static friction. Use this fact to determine θ_{max}.

Problem 8.70

A stick of length L and mass m is in equilibrium while standing on its end A when the end B is gently nudged to the right, causing the stick to fall. Letting μ_s be the coefficient of static friction between the stick and the ground and modeling the stick as a uniform slender bar, find the largest value of μ_s for which the stick slides to the left, as well as the corresponding value of θ at which sliding begins. To solve this problem, follow the steps below.

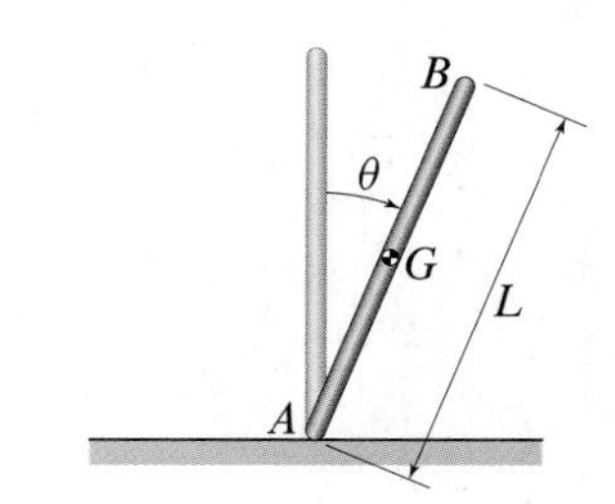

Figure P8.70

(a) Let F and N be the friction and normal forces, respectively, between the stick and the ground, and let F be positive to the right and N positive upward. Draw the FBD of the stick as it falls. Then set the sum of forces in the horizontal and vertical directions equal to the corresponding components of $m\vec{a}_G$. Express the components of $\vec{a}_G$ in terms of θ, $\dot{\theta}$, and $\ddot{\theta}$. Finally, express F and N as functions of θ, $\dot{\theta}$, and $\ddot{\theta}$.

(b) Use the work-energy principle to find an expression for $\dot{\theta}^2(\theta)$. Differentiate the expression for $\dot{\theta}^2(\theta)$ with respect to time, and find an expression for $\ddot{\theta}(\theta)$.

(c) Substitute the expressions for $\dot{\theta}^2(\theta)$ and $\ddot{\theta}(\theta)$ into the expressions for F and N to obtain F and N as functions of θ. When slip is impending (i.e., when $|F| = \mu_s|N|$), $|F/N|$ must be equal to the static coefficient of friction. Therefore, compute the maximum value of $|F/N|$ by differentiating it with respect to θ and setting the resulting derivative equal to zero.

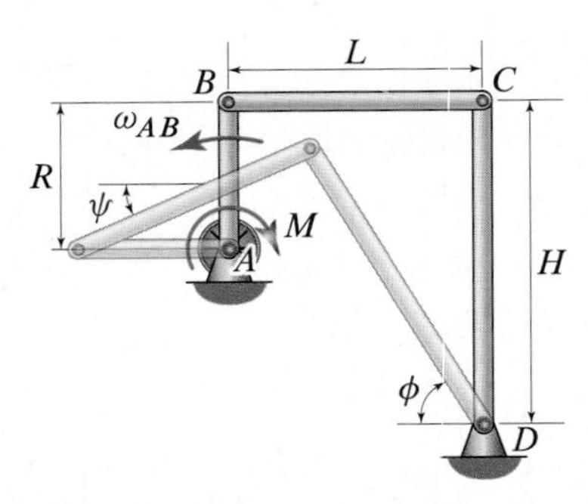

Figure P8.71

Problem 8.71

The uniform thin pin-connected bars AB, BC, and CD have masses $m_{AB} = 2.3$ kg, $m_{BC} = 3.2$ kg, and $m_{CD} = 5.0$ kg, respectively. In addition, $R = 0.75$ m, $L = 1.2$ m, and $H = 1.55$ m. When bars AB and CD are vertical, AB is rotating with angular speed $\omega_{AB} = 4$ rad/s in the direction shown. At this instant, the motor connected to AB starts to exert a constant torque M in the direction opposite to ω_{AB}. If the motor stops AB after AB has rotated 90° counterclockwise, determine M and the maximum power output of the motor during the stopping phase. In the final position, $\phi = 64.36°$ and $\psi = 29.85°$.

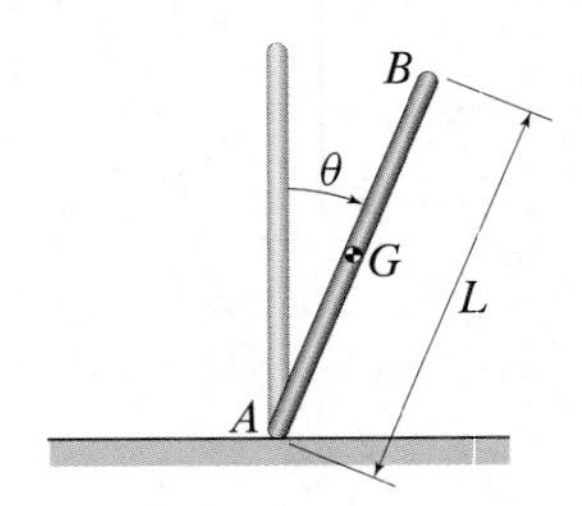

Figure P8.72

Problem 8.72

A stick of length L and mass m is in equilibrium while standing on its end A when end B is gently nudged to the right, causing the stick to fall. Letting the coefficient of static friction between the stick and the ground be $\mu_s = 0.7$ and modeling the stick as a uniform slender bar, find the value of θ at which end A of the stick starts slipping, and determine the corresponding direction of slip. As part of the solution, plot the absolute value of the ratio between the friction and normal force as a function of θ. To solve this problem, follow the steps below.

(a) Letting F and N be the friction and normal forces, respectively, between the stick and the ground, draw the FBD of the stick as it falls. Then set the sum of forces in the horizontal and vertical directions equal to the corresponding components of $m\vec{a}_G$. Express the components of $\vec{a}_G$ in terms of θ, $\dot{\theta}$, and $\ddot{\theta}$. Finally, express F and N as functions of θ, $\dot{\theta}$, and $\ddot{\theta}$.

(b) Use the work-energy principle to find an expression for $\dot{\theta}^2(\theta)$. Differentiate the expression for $\dot{\theta}^2(\theta)$ with respect to time, and find an expression for $\ddot{\theta}(\theta)$.

(c) After substituting the expressions for $\dot{\theta}^2(\theta)$ and $\ddot{\theta}(\theta)$ into the expressions for F and N, plot $|F/N|$ as a function of θ. For impending slip, $|F/N|$ must be equal to μ_s. Therefore, the desired value of θ corresponds to the intersection of the plot of $|F/N|$ with the horizontal line intercepting the vertical axis at the value 0.7. After determining the desired value of θ, the direction of slip can be found by determining the sign of F evaluated at the θ computed.

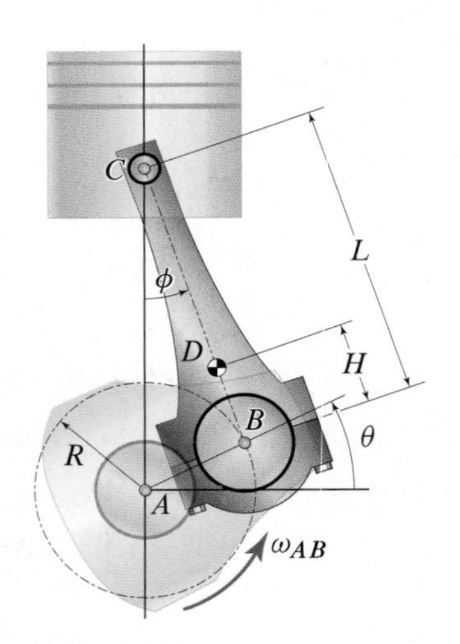

Figure P8.73 and P8.74

Problems 8.73 and 8.74

For the slider-crank mechanism shown, let $L = 141$ mm, $R = 48.5$ mm, and $H = 36.4$ mm. In addition, observing that D is the center of mass of the connecting rod, let the mass moment of inertia of the connecting rod be $I_D = 0.00144\ \text{kg}\cdot\text{m}^2$ and the mass of the connecting rod be $m = 0.439$ kg.

Problem 8.73 Letting $\omega_{AB} = 2500$ rpm, compute the kinetic energy of the connecting rod for $\theta = 90°$ and for $\theta = 180°$.

Problem 8.74 Plot the kinetic energy of the connecting rod as a function of the crank angle θ over one full cycle of the crank for $\omega_{AB} = 2500$ rpm, 5000 rpm, and 7500 rpm.

Design Problems

Design Problem 8.1

The opening and closing of the manually operated road barrier is assisted by the counterweight C and linear elastic torsional spring with constant k_t that is mounted at O. Assume that the length of the arm is $L = 15.7$ ft and that it consists of a section of aluminum tubing from A to B of length $l = 11.6$ ft and weight 20 lb and a steel support section from B to D of weight 40 lb. Model both sections of the arm as uniform thin rods. Model the counterweight as a uniform rectangular rigid body of weight W_c, height h, and width w, and let the hinge O be a distance $d = 2.58$ ft from the right end of section BD.

Using these assumptions, design the unspecified parameters δ, h, w, W_C, and the torsional spring (its stiffness k_t and the position at which it is undeformed), so that a small nudge will close the barrier from the vertical position and so that the arm will reach the closed position with an angular velocity that is less than 0.25 rad/s. In addition, make sure that the barrier is still easy to open.

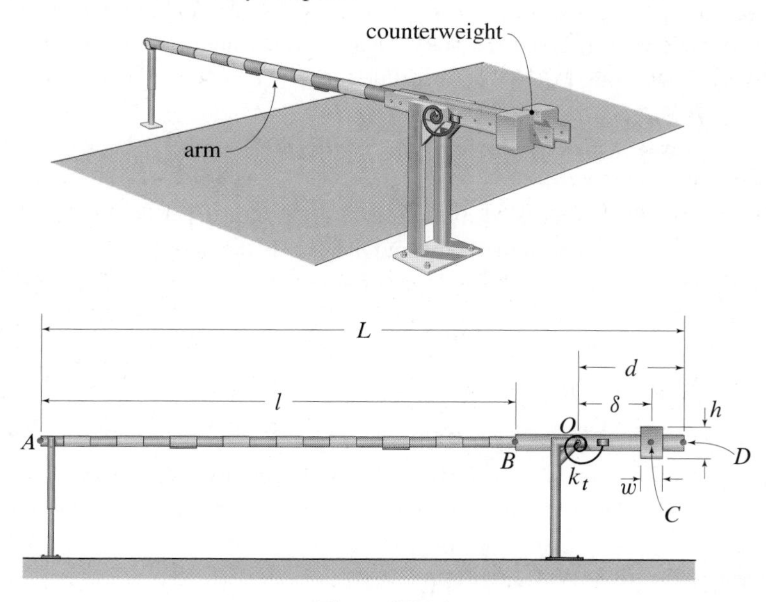

Figure DP8.1

8.2 Momentum Methods for Rigid Bodies

In this section, we develop both the linear and the angular impulse-momentum principles for rigid bodies. This is a departure from what was done in Chapter 5, where we devoted individual sections to each of these principles. The reason for this different approach is that, for rigid bodies, the linear and angular impulse-momentum principles must often be applied together to get a complete picture of a body's motion, as shown in the following example.

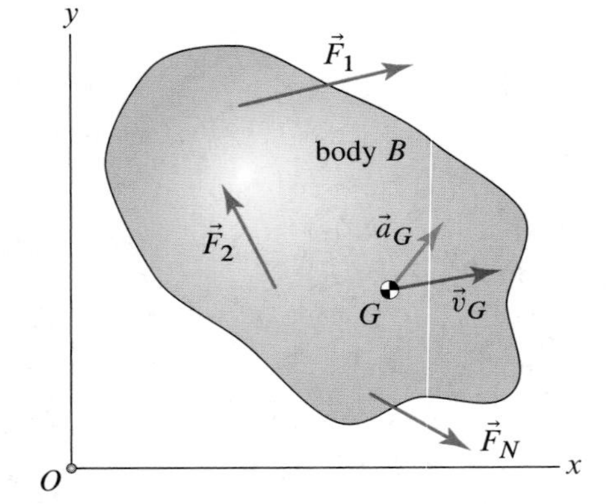

Figure 8.11
A rigid body under the action of a system of forces.

Impulse-momentum principle for a rigid body

A rigid body's mass center moves according to Eq. (7.1) on p. 521, i.e.,

$$\vec{F} = m\vec{a}_G, \tag{8.32}$$

where, referring to Fig. 8.11, $\vec{F} = \vec{F}_1 + \vec{F}_2 + \cdots + \vec{F}_N$ is the total force acting on the body and m and $\vec{a}_G$ are the body's mass and acceleration of the mass center, respectively. Integrating Eq. (8.32) over time for $t_1 \le t \le t_2$, we obtain

$$\int_{t_1}^{t_2} \vec{F}\,dt = \int_{t_1}^{t_2} m\vec{a}_G\,dt = m\vec{v}_G(t_2) - m\vec{v}_G(t_1). \tag{8.33}$$

Here, using concepts introduced in Section 5.1, the first term in Eq. (8.33) is the total *linear impulse* acting on the body, and $m\vec{v}_G(t)$ is the body's *linear momentum*, which we denote by $\vec{p}$. As we have done for a particle (see Eq. (5.6) on p. 314), we can rewrite Eq. (8.33) as

$$m\vec{v}_{G1} + \int_{t_1}^{t_2} \vec{F}\,dt = m\vec{v}_{G2} \quad \text{or} \quad \vec{p}_1 + \int_{t_1}^{t_2} \vec{F}\,dt = \vec{p}_2, \tag{8.34}$$

where the subscripts 1 and 2 indicate the values of a quantity at t_1 and t_2, respectively. Equations (8.34) express the linear impulse-momentum principle for a rigid body.

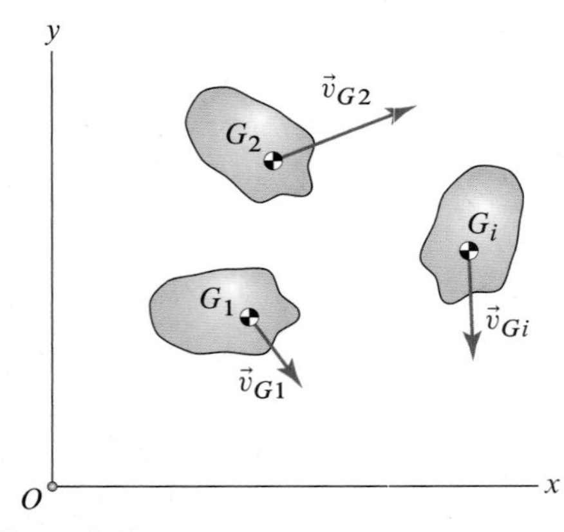

Figure 8.12
A system of rigid bodies.

Extension of Eq. (8.34) for a system

Recalling that $\vec{F}$ is the sum of only the *external forces*, Eqs. (8.34) can be applied to a system of rigid bodies if we properly compute $\vec{p}$. For a system of N rigid bodies (see Fig. 8.12), the system's total momentum is

$$\vec{p} = \sum_{i=1}^{N} m_i \vec{v}_{Gi}(t), \tag{8.35}$$

where m_i and $\vec{v}_{Gi}$ ($i = 1, \ldots, N$) are the mass and the velocity of the mass center, respectively, of body i.

Conservation of linear momentum

If $\vec{F} = \vec{0}$ for $t_1 \le t \le t_2$, Eqs. (8.34) reduce to

$$m\vec{v}_{G1} = m\vec{v}_{G2} \quad \text{or} \quad \vec{p}_1 = \vec{p}_2, \tag{8.36}$$

which states that the system's momentum is conserved for $t_1 \le t \le t_2$. In many applications the total external force $\vec{F}$ is not equal to zero over the time interval considered, but there is a direction, say q, along which the *component* $F_q = 0$ for $t_1 \le t \le t_2$. In this case, we can write

$$m(v_{Gq})_1 = m(v_{Gq})_2 \quad \text{or} \quad p_{q1} = p_{q2}, \tag{8.37}$$

that is, the momentum is conserved in the q direction.

Angular impulse-momentum principle for a rigid body

The moment-angular momentum equation governing the motion of a rigid body is Eq. (7.4) on p. 522, i.e.,

$$\vec{M}_P = \dot{\vec{h}}_P + \vec{v}_P \times m\vec{v}_G, \tag{8.38}$$

where P is an arbitrarily chosen moment center (see Fig. 8.13), $\vec{v}_P$ is the velocity of P, and $\vec{M}_P$ is the total moment relative to P due to the external force system acting on the body. The quantity $\vec{h}_P$ is the body's angular momentum relative to P; it was defined in Eq. (7.6) on p. 522 as

$$\vec{h}_P = \int_B \vec{r}_{dm/P} \times \vec{v}_{dm}\, dm. \tag{8.39}$$

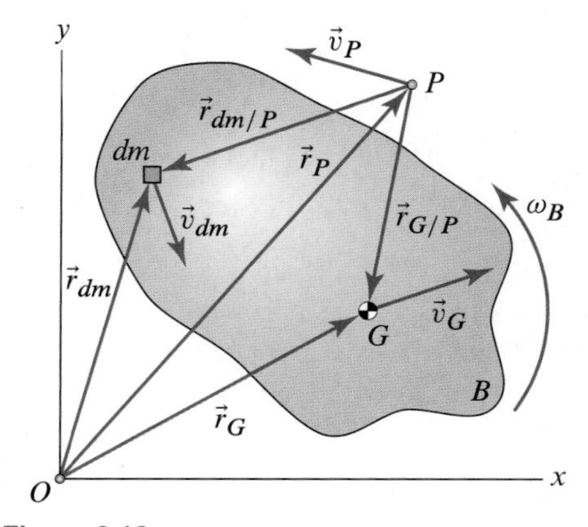

Figure 8.13
The quantities needed to obtain the angular momentum relationships for a rigid body.

While Eqs. (8.38) and (8.39) are valid for any type of body and for any motion, the applications we focus on in this chapter concern the planar motion of rigid bodies that are symmetric with respect to the plane of motion. Therefore, as shown in Eq. (B.25) of App. B, $\vec{h}_P$ can be given the following compact form:

$$\boxed{\vec{h}_P = I_G\vec{\omega}_B + \vec{r}_{G/P} \times m\vec{v}_G,} \tag{8.40}$$

where I_G and $\vec{\omega}_B$ are the body's mass moment of inertia and angular velocity, respectively. If the point P is in the same plane as G and *any* one of the following is true:

1. P is a fixed point (i.e., $\vec{v}_P = \vec{0}$); or
2. P coincides with G (i.e., $\vec{v}_P = \vec{v}_G \Rightarrow \vec{v}_P \times \vec{v}_G = \vec{0}$); or
3. P and G move parallel to one another (i.e., $\vec{v}_P \times \vec{v}_G = \vec{0}$);

Eq. (8.38) simplifies to

$$\boxed{\vec{M}_P = \dot{\vec{h}}_P.} \tag{8.41}$$

If the assumptions underlying Eq. (8.41) hold throughout a time interval $t_1 \le t \le t_2$, then integrating Eq. (8.41) over this time interval, we obtain the traditional form of the angular impulse-momentum principle as

$$\boxed{\vec{h}_{P1} + \int_{t_1}^{t_2} \vec{M}_P\, dt = \vec{h}_{P2},} \tag{8.42}$$

where $\vec{h}_{P1}$ and $\vec{h}_{P2}$ are the values of $\vec{h}_P$ at times t_1 and t_2, respectively. If the moment center is taken to be the mass center G, then $\vec{r}_{G/P} = \vec{0}$, and then using Eq. (8.40) in Eq. (8.42), the angular impulse-momentum principle becomes

$$\boxed{I_{G1}\omega_{B1} + \int_{t_1}^{t_2} M_{Gz}\, dt = I_{G2}\omega_{B2}.} \tag{8.43}$$

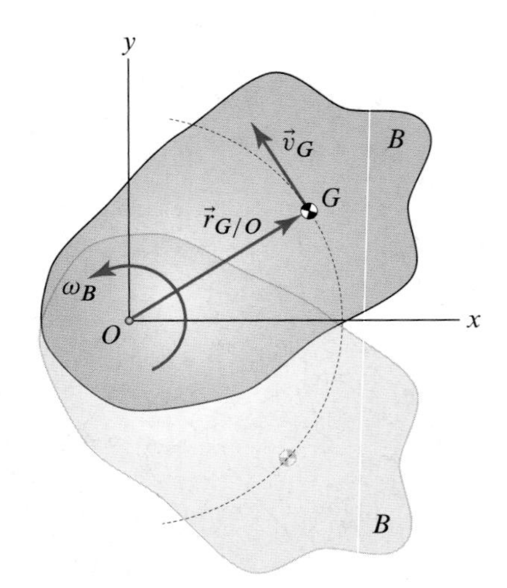

Figure 8.14
A rigid body in a fixed axis rotation.

Here, M_{Gz} is the z component of the moment about G, subscripts 1 and 2 indicate the value of a quantity at t_1 and t_2, respectively, and Eq. (8.43) has been written in scalar form because, under the current assumptions, the only nonzero component of Eq. (8.42) is perpendicular to the plane of motion. Equation (8.43) is valid even if I_G changes with time (see Example 8.9).

If a body B is in fixed-axis rotation about a point O (Fig. 8.14), then $\vec{v}_G = \vec{\omega}_B \times \vec{r}_{G/O}$, and by choosing the center of rotation O as our moment center, Eq. (8.40) becomes

$$\begin{aligned}\vec{h}_O &= I_G\vec{\omega}_B + \vec{r}_{G/O} \times m(\vec{\omega}_B \times \vec{r}_{G/O}) \\ &= \left(I_G + m|\vec{r}_{G/O}|^2\right)\vec{\omega}_B = I_O\vec{\omega}_B,\end{aligned} \tag{8.44}$$

where, by the parallel axis theorem, $I_O = I_G + m|\vec{r}_{G/O}|^2$ is the body's mass moment of inertia about the axis of rotation. Using Eq. (8.44), the angular impulse-momentum principle for a body that is symmetric with respect to the plane of motion and under fixed-axis rotation takes on the form

$$\boxed{I_{O1}\omega_{B1} + \int_{t_1}^{t_2} M_{Oz}\,dt = I_{O2}\omega_{B2},} \tag{8.45}$$

where O is the center of rotation and M_{Oz} is the z component of $\vec{M}_O$. We have written Eq. (8.45) in scalar form because, under the current assumptions, the only nonzero component of Eq. (8.42) is normal to the plane of motion.

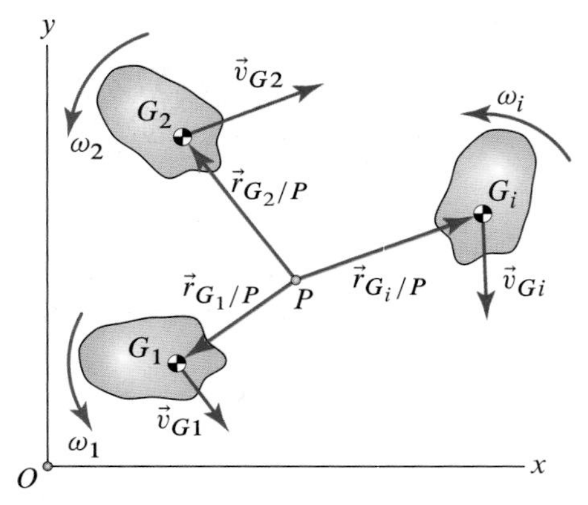

Figure 8.15
A system of rigid bodies.

Angular impulse-momentum principle for a system

Equations (8.42) and (8.43) apply to systems of rigid bodies if the assumptions underlying these equations are satisfied by each element of the system. Referring to Fig. 8.15, for a system of N rigid bodies, in which body i has angular velocity $\vec{\omega}_i$, mass center G_i, mass m_i, and mass moment of inertia I_{Gi}, $\vec{h}_P$ is

$$\vec{h}_P = \sum_{i=1}^{N}\left(I_{Gi}\vec{\omega}_i + \vec{r}_{Gi/P} \times m_i\vec{v}_{Gi}\right), \tag{8.46}$$

where $\vec{v}_{Gi}$ is the velocity of G_i and $\vec{r}_{Gi/P}$ is the position of G_i relative to P.

Conservation of angular momentum

If $\vec{M}_P = \vec{0}$ for $t_1 \leq t \leq t_2$, Eq. (8.42) implies that

$$\vec{h}_{P1} = \vec{h}_{P2} = \text{constant}, \tag{8.47}$$

which states that the body's angular momentum relative to P is conserved. Another useful result is obtained when $\vec{M}_P \neq \vec{0}$, but there is a *fixed* direction, say q, along which $M_{Pq} = 0$. In this case, we can write

$$(h_{Pq})_1 = (h_{Pq})_2 = \text{constant}, \tag{8.48}$$

and apply conservation of angular momentum in the q direction.

End of Section Summary

The linear impulse-momentum principle for a rigid body is (see Fig. 8.16)

Eqs. (8.34), p. 618

$$m\vec{v}_{G1} + \int_{t_1}^{t_2} \vec{F}\,dt = m\vec{v}_{G2} \quad \text{or} \quad \vec{p}_1 + \int_{t_1}^{t_2} \vec{F}\,dt = \vec{p}_2,$$

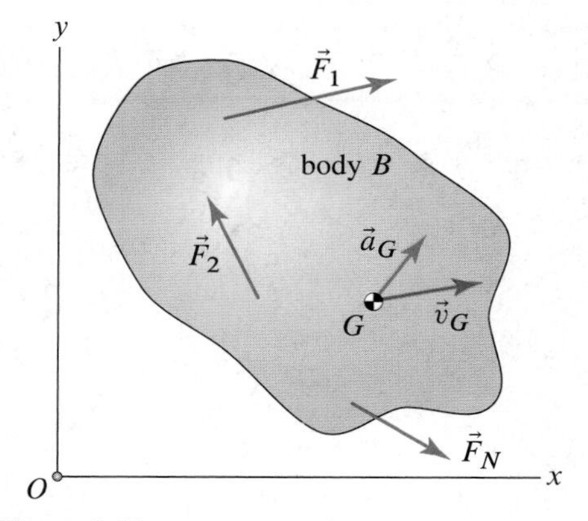

Figure 8.16
A rigid body under the action of a system of force.

where $\vec{F}$ is the total external force on B, $\vec{p} = m\vec{v}_G$ is B's linear momentum, and $\vec{v}_G$ is the velocity of B's center of mass. If $\vec{F} = \vec{0}$, we have

Eqs. (8.36), p. 618

$$m\vec{v}_{G1} = m\vec{v}_{G2} \quad \text{or} \quad \vec{p}_1 = \vec{p}_2,$$

and we say that the body's momentum is conserved. If P in Fig. 8.17 is a moment center coplanar with G and if B is symmetric relative to the plane of motion, the angular momentum of B relative to P is

Eq. (8.40), p. 619

$$\vec{h}_P = I_G\vec{\omega}_B + \vec{r}_{G/P} \times m\vec{v}_G,$$

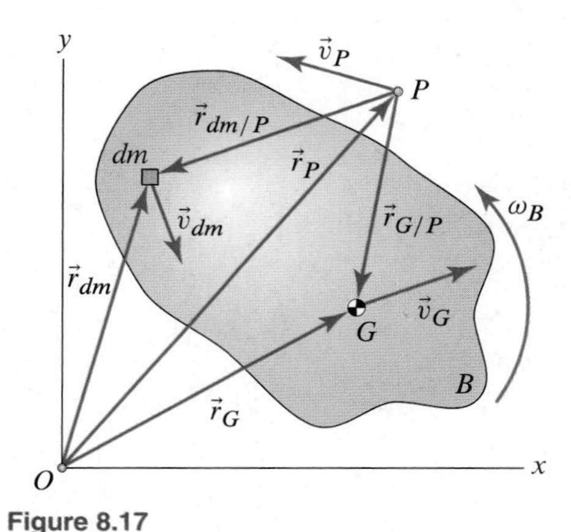

Figure 8.17
The quantities needed to obtain the angular momentum relationships for a rigid body.

where I_G is the mass moment of inertia of B, $\vec{\omega}_B$ is the angular velocity of B, and $\vec{r}_{G/P}$ is the position of G relative to P. If P is chosen so that (1) P is fixed or (2) P coincides with G or (3) P and G move parallel to one another, then $\vec{M}_P = \dot{\vec{h}}_P$, where $\vec{M}_P$ is the moment relative to P of the *external* force system acting on B. When this equation holds, by integrating with respect to time over a time interval $t_1 \le t \le t_2$, we have

Eq. (8.42), p. 619

$$\vec{h}_{P1} + \int_{t_1}^{t_2} \vec{M}_P\,dt = \vec{h}_{P2}.$$

When P coincides with G or if the body undergoes a fixed-axis rotation about a point O as shown in Fig. 8.18, then the above equation becomes

Eq. (8.43), p. 619, and Eq. (8.45), p. 620

$$I_{G1}\omega_{B1} + \int_{t_1}^{t_2} M_{Gz}\,dt = I_{G2}\omega_{B2},$$

$$I_{O1}\omega_{B1} + \int_{t_1}^{t_2} M_{Oz}\,dt = I_{O2}\omega_{B2},$$

respectively, where I_O is the mass moment of inertia about the fixed axis of rotation.

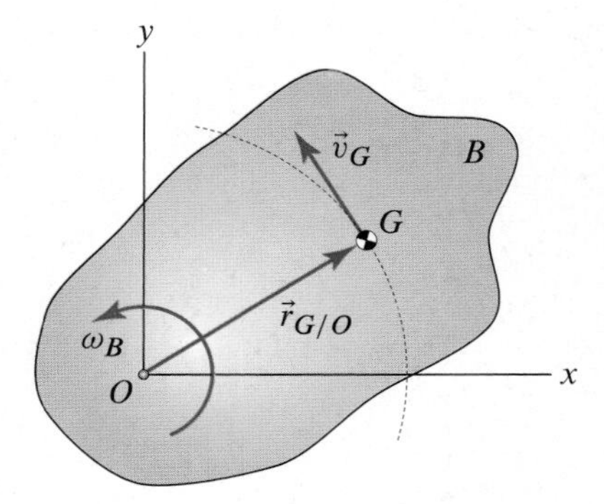

Figure 8.18
A rigid body in a fixed-axis rotation.

EXAMPLE 8.7 *A Rolling Wheel: Computing Angular Momentum*

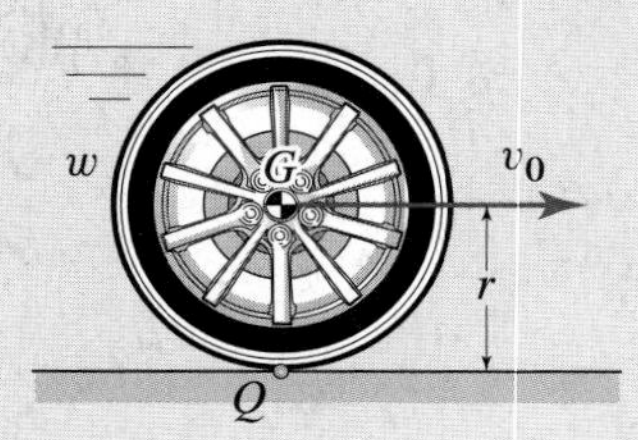

Figure 1

The wheel w shown in Fig. 1 has radius r and mass m. Point G is both the wheel's center of mass and the wheel's geometric center. The radius of gyration is k_G. The wheel is rolling without slip with G moving to the right with a speed v_0. Compute the wheel's angular momentum relative to both G and Q, which is the point on the wheel in contact with the ground.

SOLUTION

Road Map This problem is solved by a direct application of the expression for the angular momentum of a rigid body given in Eq. (8.40) on p. 619.

Computation Referring to Fig. 2 and applying Eq. (8.40), the angular momentum of the wheel relative to its center of mass is

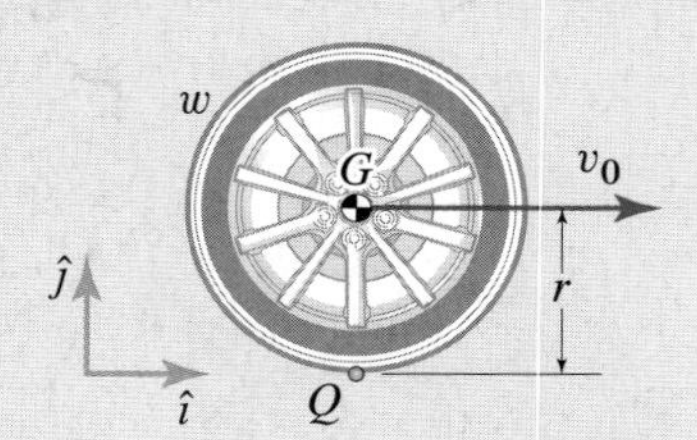

Figure 2
Wheel rolling without slip.

$$\vec{h}_G = I_G\vec{\omega}_w + \vec{r}_{G/G} \times m\vec{v}_G, \tag{1}$$

where $\vec{r}_{G/G} = \vec{0}$ and $I_G = mk_G^2$. Recalling that because of rolling without slip we must have $\vec{\omega}_w = -(v_0/r)\,\hat{k}$, Eq. (1) can be rewritten as

$$\boxed{\vec{h}_G = -mk_G^2\left(\frac{v_0}{r}\right)\hat{k}.} \tag{2}$$

Again referring to Fig. 2 and applying Eq. (8.40), the angular momentum of the wheel relative to Q is

$$\vec{h}_Q = I_G\vec{\omega}_w + \vec{r}_{G/Q} \times m\vec{v}_G. \tag{3}$$

In this case, we have

$$\vec{r}_{G/Q} = r\,\hat{\jmath} \quad \text{and} \quad \vec{v}_G = v_0\,\hat{\imath} \quad \Rightarrow \quad \vec{r}_{G/Q} \times m\vec{v}_G = -mrv_0\,\hat{k}. \tag{4}$$

Substituting the last of Eqs. (4) into Eq. (3) and recalling that $I_G\vec{\omega}_w = -mk_G^2(v_0/r)\,\hat{k}$, we have

$$\boxed{\vec{h}_Q = -mrv_0\,\hat{k} - mk_G^2\frac{v_0}{r}\,\hat{k} = -mv_0\left(r + \frac{k_G^2}{r}\right)\hat{k}.} \tag{5}$$

Discussion & Verification To verify that the results in Eqs. (2) and (5) are dimensionally correct, recall that angular momentum has the dimensions of *moment of the momentum*, i.e., of mass × velocity × length, which is what we see in Eqs. (2) and (5).

A Closer Look Using the parallel axis theorem, we can give the expression in Eq. (5) a much more compact form. Let I_Q denote the wheel's mass moment of inertia relative to Q. Using the parallel axis theorem, we have

$$I_Q = I_G + mr^2 = m\left(k_G^2 + r^2\right). \tag{6}$$

Going back to Eq. (5) and factoring $1/r$ out of the term in parentheses, we can rewrite this equation as

$$\vec{h}_Q = -m\frac{v_0}{r}\left(r^2 + k_G^2\right)\hat{k} = I_Q\vec{\omega}_w, \tag{7}$$

where we have taken advantage of Eq. (6) and the fact that $\vec{\omega}_w = -(v_0/r)\,\hat{k}$. Comparing Eq. (7) with Eq. (8.44) on p. 620, we could interpret the result by saying that the wheel appears as though it were in a fixed-axis rotation about Q. This interpretation is appropriate because Q is the wheel's instantaneous center of rotation.

EXAMPLE 8.8 *A Rolling Pipe: Application of Impulse and Momentum*

A pipe section A of radius r, center G, and mass m is gently placed (i.e., with zero velocity) on a conveyor belt moving with a constant speed v_0 to the right, as shown in Fig. 1. The friction between the belt and pipe will cause the pipe to move to the right, as well as to rotate and, eventually, to roll without slip. Determine the velocity of G and the angular velocity of the pipe when it rolls without slip.

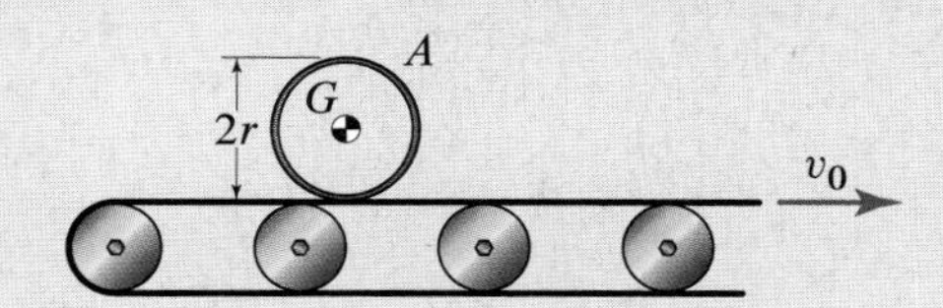

Figure 1
A pipe section of radius r lowered very gently over a conveyor belt.

SOLUTION

Road Map & Modeling Because the pipe A is stationary when it is placed on the conveyor belt, A must slip relative to the belt until rolling without slip begins. Modeling A as a uniform rigid body, until rolling without slip begins, the FBD of A is that shown in Fig. 2. Assuming that G does not move in the vertical direction, the motion of the body is determined by the impulse provided by the friction force, which is the only force acting in the horizontal direction. We can then solve the problem by applying the linear and angular impulse-momentum principles, along with the kinematic relations that describe rolling without slip over a moving surface.

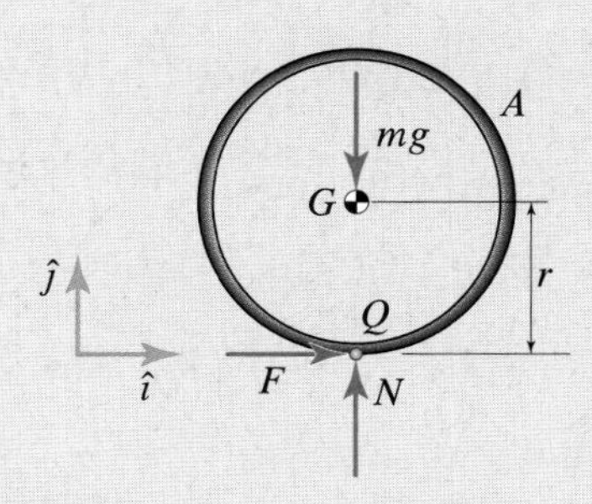

Figure 2
FBD of the pipe section at an instant between the time at which the pipe is placed over the conveyor and the time at which the pipe starts rolling without slip.

Governing Equations

Balance Principles Let t_1 denote the time at which A is placed on the conveyor and t_2 denote the time at which the pipe starts rolling without slip. The impulse-momentum principle in the x direction is

$$m(v_{Gx})_1 + \int_{t_1}^{t_2} F\,dt = m(v_{Gx})_2. \tag{1}$$

Choosing the mass center G as moment center, the angular impulse-momentum principle applied between t_1 and t_2 is

$$\vec{h}_{G1} + \int_{t_1}^{t_2} (\vec{r}_{Q/G} \times F\,\hat{\imath})\,dt = \vec{h}_{G2}, \tag{2}$$

where, using Eq. (8.40) on p. 619 and modeling the pipe section as a thin ring,

$$\vec{h}_{G1} = I_G\vec{\omega}_{A1} = mr^2\omega_{A1}\,\hat{k} \quad \text{and} \quad \vec{h}_{G2} = I_G\vec{\omega}_{A2} = mr^2\omega_{A2}\,\hat{k}. \tag{3}$$

Force Laws All forces are accounted for on the FBD.

Kinematic Equations At time t_1, A is stationary, so we have

$$(v_{Gx})_1 = 0 \quad \text{and} \quad \omega_{A1} = 0. \tag{4}$$

At time t_2, A rolls without slip over the moving belt, which means that $\vec{v}_{Q2} = v_0\,\hat{\imath}$. For G we have

$$\vec{v}_{G2} = \vec{v}_{Q2} + \omega_{A2}\,\hat{k} \times \vec{r}_{G/Q} \quad \Rightarrow \quad \vec{v}_{G2} = (v_0 - r\omega_{A2})\,\hat{\imath}. \tag{5}$$

Computation After expanding the cross-product, the integrand in the second term of Eq. (2) can be written as follows:

$$\vec{r}_{Q/G} \times F\,\hat{\imath} = -r\,\hat{\jmath} \times F\,\hat{\imath} = Fr\,\hat{k}. \tag{6}$$

Substituting Eqs. (3), the second of Eqs. (4), and Eq. (6) into Eq. (2), we have

$$r\int_{t_1}^{t_2} F\,dt = mr^2\omega_{A2}, \tag{7}$$

where we have pulled r outside the integral because it is constant.

Helpful Information

Is the friction force F constant? In the integral in Eq. (7), we left the friction force F inside the integral because we did not know F as a function of time. The important point to understand here is that the final solution to this problem does not require that we know F as a function of time. With this said, starting from the system's FBD and what we learned in Chapter 7, we can show that in this problem F is constant for $t_1 < t < t_2$, and it then becomes equal to zero as soon as the pipe section starts rolling without slip.

Substituting the first of Eqs. (4) and the last of Eqs. (5) into Eq. (1), we have

$$\int_{t_1}^{t_2} F\,dt = m(v_0 - r\omega_{A2}). \tag{8}$$

Substituting Eq. (8) into Eq. (7), we obtain

$$mr(v_0 - r\omega_{A2}) = mr^2\omega_{A2} \quad \Rightarrow \quad \boxed{\omega_{A2} = \frac{v_0}{2r}.} \tag{9}$$

Substituting ω_{A2} from Eq. (9) into the last of Eqs. (5), we have

$$\boxed{\vec{v}_{G2} = \tfrac{1}{2} v_0\,\hat{\imath}.} \tag{10}$$

Discussion & Verification The result we obtained is dimensionally correct and consistent with the FBD in that friction will cause the pipe section to move to the right and rotate counterclockwise.

A Closer Look The problem's solution does not depend on the mass of the object, only on its shape. That is, we would have obtained a different result had we modeled the pipe section as, say, a cylinder.

Note that we could have obtained the solution by enforcing the conservation of angular momentum about point Q without invoking the linear impulse-momentum principle. To see this, referring to the system FBD in Fig. 2, observe that the moment of the external forces about Q is equal to zero. Normally this observation does not help much since Q is neither a fixed point nor the system's center of mass. However, referring to the list preceding Eq. (8.41) on p. 619, point Q does move parallel to the center of mass G. Therefore, since $\vec{M}_Q = \vec{0}$, Eq. (8.41) implies that $\vec{h}_{Q1} = \vec{h}_{Q2}$. Furthermore, since the pipe section was stationary when it was placed on the conveyor belt, we must have $\vec{h}_{Q1} = \vec{0}$. This fact, along with Eq. (8.40) on p. 619, yields

$$I_G\omega_{A2}\,\hat{k} + \vec{r}_{G/Q} \times m\vec{v}_{G2} = \vec{0} \quad \Rightarrow \quad mr^2\omega_{A2} - mr(v_{Gx})_2 = 0. \tag{11}$$

The result in Eq. (11), along with the rolling-without-slip condition in Eq. (5), yields the same solution we derived in Eqs. (9) and (10).

EXAMPLE 8.9 *A Spinning Skater: Conservation of Angular Momentum*

The skater in Fig. 1 begins to spin with her arms completely stretched out and then brings her arms close to her body to increase her spin rate. In Section 5.3 on p. 361, we modeled the skater using a single particle. Here, we revisit the problem by modeling the skater as a system of rigid bodies, as shown in Fig. 2. Except for her arms,* her body is modeled as a cylinder of radius $r_b = 0.55$ ft, weight $W_b = 102$ lb, and radius of gyration $k_G = 0.3$ ft, where G is her body's center of mass. Each arm has weight $W_a = 7.4$ lb and length $\ell = 2.2$ ft and is divided into an upper arm and a forearm. The upper arm and forearm are each modeled as a uniform thin rod weighing $W_a/2$ and with length $\ell/2$. Assuming that the skater starts spinning with a rate $\omega_0 = 60$ rpm, as shown, and that her arms are stretched out, determine her spin rate (a) if her upper arms are kept stretched out and her forearms are folded so as to overlap with her upper arms; and (b) if her arms are placed vertically downward next to her body.

Figure 1
Three snapshots of a *forward spin*. By extending and retracting her arms and leg, the skater controls her spin rate.

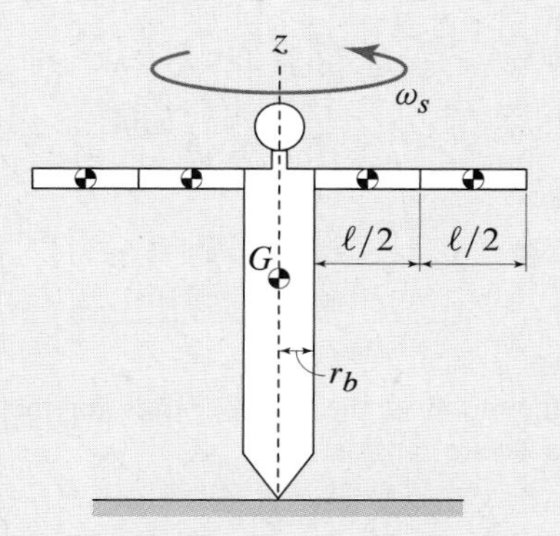

Figure 2
Model of the skater as a system of rigid bodies.

SOLUTION

Road Map & Modeling Neglecting friction between the skater and ice, the skater's FBD for $t_1 \le t \le t_2$ is shown in Fig. 3, where t_1 is the time at which the spin begins and t_2 is the time at which one of the positions corresponding to (a) or (b) is achieved. None of the external forces in the FBD contributes to a moment about the z axis. Assuming that the spin axis coincides with the z axis for $t_1 \le t \le t_2$, the condition $M_z = 0$ causes the angular momentum about this axis to be conserved. Since we need to find only one scalar unknown, namely, the angular velocity at t_2, satisfying this conservation statement will lead us to the solution.

Governing Equations

Balance Principles Referring to Fig. 3, since the skater's body is in a fixed-axis rotation about the z axis with $M_{Oz} = 0$, Eq. (8.45) on p. 620 implies

$$I_{O1}\omega_{s1} = I_{O2}\omega_{s2}, \tag{1}$$

where ω_s is the angular velocity of the skater and I_O is the mass moment of inertia of the skater about O (or any other point along the z axis).

In this problem, *the mass moment of inertia changes as the skater moves her arms!* Referring to Fig. 2, when completely outstretched, the upper arm and forearm can be viewed as forming a single uniform thin rod of mass m_a and length ℓ with mass center $r_b + \ell/2$ away from the z axis. Therefore, applying the parallel axis theorem, we have

$$I_{O1} = \underbrace{m_b k_G^2}_{\text{body}} + \underbrace{2\left[\tfrac{1}{12} m_a \ell^2 + m_a \left(r_b + \tfrac{1}{2}\ell\right)^2\right]}_{\text{each arm}}, \tag{2}$$

where m_b is the mass of her body. Equation (2) can be simplified as

$$I_{O1} = m_b k_G^2 + 2m_a\left(r_b^2 + r_b\ell + \tfrac{1}{3}\ell^2\right). \tag{3}$$

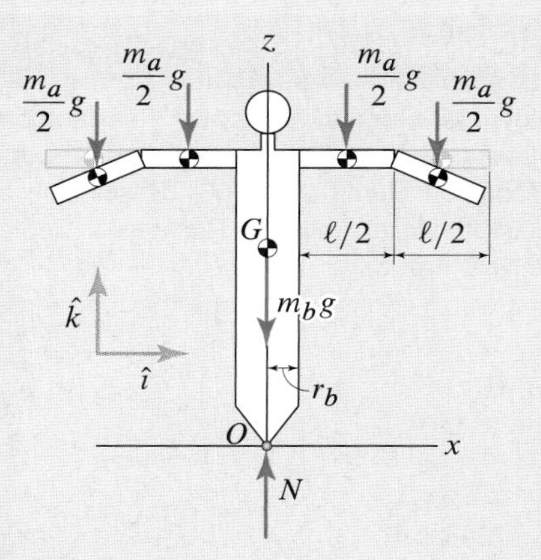

Figure 3
FBD of a spinning skater.

Referring to Fig. 4(a), for case (a), when the skater folds her forearms horizontally, using the parallel axis theorem again, at time t_2 we have

$$(I_{O2})_{\text{out}} = \underbrace{m_b k_G^2}_{\text{body}} + \underbrace{4\left[\frac{1}{12}\frac{m_a}{2}\left(\frac{\ell}{2}\right)^2 + \frac{m_a}{2}\left(r_b + \frac{\ell}{4}\right)^2\right]}_{\text{each upper arm and each forearm}}, \tag{4}$$

* We use *arm* according to its common meaning, i.e., everything from the shoulder to the tip of the fingers. However, in medical anatomy, an arm is only what lies between shoulder and elbow.

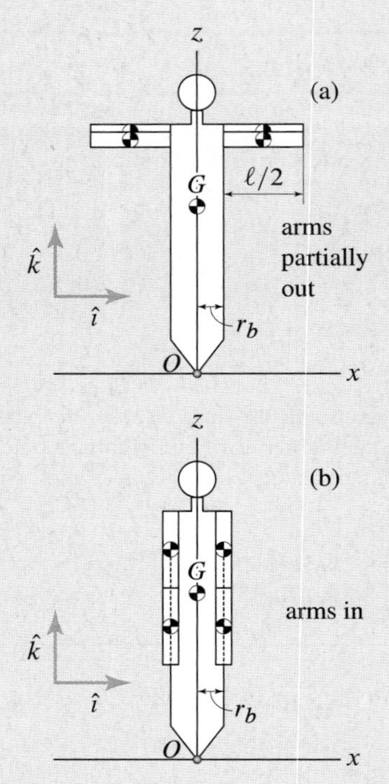

Figure 4
Configuration of the skater's arms for the two cases we are considering.

where the subscript *out* indicates that the arms are partially stretched out. Equation (4) can be simplified to

$$(I_{O2})_{\text{out}} = m_b k_G^2 + m_a\left(2r_b^2 + r_b\ell + \frac{\ell^2}{6}\right). \tag{5}$$

When her arms are folded completely downward, as in Fig. 4(b), we have

$$(I_{O2})_{\text{in}} = \underbrace{m_b k_G^2}_{\text{body}} + \underbrace{2m_a r_b^2}_{\text{each arm}}, \tag{6}$$

where the subscript *in* indicates that the arms are completely downward.

Force Laws All forces are accounted for on the FBD.

Kinematic Equations We know that the skater is initially spinning at ω_0 and so

$$\omega_{s1} = \omega_0. \tag{7}$$

Computation Substituting Eq. (7) into Eq. (1) and solving for ω_{s2}, we have $\omega_{s2} = (I_{O1}/I_{O2})\omega_0$, and therefore for the two cases considered we have

$$(\omega_{s2})_{\text{out}} = \frac{m_b k_G^2 + 2m_a\left(r_b^2 + r_b\ell + \frac{\ell^2}{3}\right)}{m_b k_G^2 + m_a\left(2r_b^2 + r_b\ell + \frac{\ell^2}{6}\right)}\omega_0 = 116.4\,\text{rpm} \tag{8}$$

and

$$(\omega_{s2})_{\text{in}} = \frac{m_b k_G^2 + 2m_a\left(r_b^2 + r_b\ell + \frac{\ell^2}{3}\right)}{m_b k_G^2 + 2m_a r_b^2}\omega_0 = 243.6\,\text{rpm}, \tag{9}$$

where we have used Eqs. (3), (5), and (6) and where we have plugged in the given data to obtain the numerical results.

Discussion & Verification In each of Eqs. (8) and (9) the final spin rate is greater than the initial spin rate ω_0, as expected. In addition, the result in Eq. (9) is larger than that in Eq. (8); i.e., the increase in spin rate for the case where the skater's arms are completely against her body is larger than that when only the forearms are folded, again as expected. Finally, the spin rates that we have obtained are certainly within reach of professional skaters (see the Interesting Fact in the margin).

Interesting Fact

How fast can a skater spin? Russian-born Natalia Kanounnikova established a world record on March 27, 2006, by spinning at 308 rpm while skating at Rockefeller Center in New York City.

A Closer Look The particle solution to this problem was given in Eq. (5) on p. 369 in Example 5.11, which, in terms of the current variables, is

$$\omega_{s2} = (r_1^2/r_2^2)\omega_0, \tag{10}$$

where r_1 and r_2 are the distances between the arms and the spin axis at times t_1 and t_2, respectively. The simplicity of Eq. (10) is due to the fact that we ignored the dimensions of the body and its parts and only considered the mass of the arms. However, it is important to understand that the particle and rigid body solutions are not that different in spirit. We can rewrite Eq. (10) as

$$\omega_{s2} = \frac{2m_a}{2m_a}\frac{r_1^2}{r_2^2}\omega_0 = \frac{2m_a r_1^2}{2m_a r_2^2}\omega_0 = \frac{(I_{O1})_p}{(I_{O2})_p}\omega_0 \Rightarrow (I_{O1})_p\omega_0 = (I_{O2})_p\omega_{s2}, \tag{11}$$

where $(I_{O1})_p$ and $(I_{O2})_p$ represent the mass moments of inertia relative to the spin axis for the particle model. Recalling that $\omega_0 = \omega_{s1}$ and comparing the last of Eqs. (11) with Eq. (1), we see that the only difference between the two models is how the mass moments of inertia are calculated.

EXAMPLE 8.10 *Space Shuttle Docking with ISS: Conservation of Momentum*

Figure 1 shows the Space Shuttle docked with the International Space Station (ISS). To explore what docking entails, we consider the simplified 2D scenario in Fig. 2, in which the Shuttle A docks to the ISS B with a speed $v_0 = 0.03\,\mathrm{m/s}$. We wish to determine the velocities of A and B *immediately after* they dock, assuming that no spacecraft attitude controls are exerted on A or B and assuming that, after docking, A and B *form a single rigid body*. Referring to Fig. 2, we will use the following data: the mass and mass moment of inertia of A are $m_A = 120\times10^3\,\mathrm{kg}$ and $I_C = 14\times10^6\,\mathrm{kg{\cdot}m^2}$, respectively; the mass and mass moment of inertia of B are $m_B = 180\times10^3\,\mathrm{kg}$ and $I_D = 34\times10^6\,\mathrm{kg{\cdot}m^2}$, respectively; the dimensions are $\ell = 24\,\mathrm{m}$ and $h = 8\,\mathrm{m}$. Note that we are *not* assuming that A and B are rectangles in Fig. 2. Since a body's mass and mass moment of inertia completely describe it, these rectangles are used only to describe the relative position of points C and D.

Figure 1
Artist's rendition of the Space Shuttle *Discovery* docked to the International Space Station.

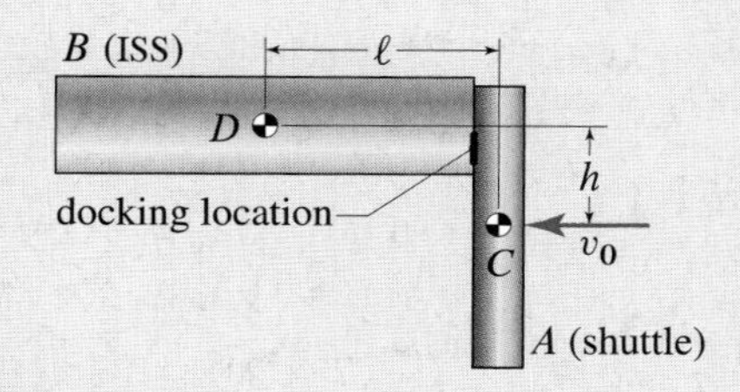

Figure 2
Relative positions of the mass centers C and D of A and B, respectively, at docking. The rectangles shown are not physical models, they are used only to describe the relative position of C and D.

SOLUTION

Road Map & Modeling Since we are assuming that A and B join to form a single rigid body, we can use rigid body kinematics to describe the motion of the A-B-body via the motion of only two points, namely, C and D, provided we know their relative position, which is given in Fig. 2. We will neglect all gravitational effects and assume that B is initially at rest relative to an inertial frame. Since we want the motion immediately after docking, we can assume that the positions of A and B are still the same as those at the time of docking. This allows us to make no distinction between the positions of the system immediately before and after docking. Finally, recalling that no attitude controls are used, the FBD of the system right before *and* right after docking is that in Fig. 3, so the system's linear and angular momenta are conserved. These conservation statements give three scalar equations which, when combined with the assumption that A and B form a single rigid body, are sufficient to solve the problem.

Governing Equations

Balance Principles In components, the conservation of total linear momentum is

$$m_A(v_{Cx})_1 + m_B(v_{Dx})_1 = m_A(v_{Cx})_2 + m_B(v_{Dx})_2, \tag{1}$$

$$m_A(v_{Cy})_1 + m_B(v_{Dy})_1 = m_A(v_{Cy})_2 + m_B(v_{Dy})_2, \tag{2}$$

where the subscripts 1 and 2 indicate right before and right after docking, respectively.

Choosing the fixed point O as the moment center, the conservation of total angular momentum is

$$(\vec{h}_O)_{A1} + (\vec{h}_O)_{B1} = (\vec{h}_O)_{A2} + (\vec{h}_O)_{B2}, \tag{3}$$

where, because A and B do not move significantly between times t_1 and t_2, we have

$$(\vec{h}_O)_{A1} = I_C\vec{\omega}_{A1} + \vec{r}_{C/O}\times m_A\vec{v}_{C1}, \quad (\vec{h}_O)_{B1} = I_D\vec{\omega}_{B1} + \vec{r}_{D/O}\times m_B\vec{v}_{D1}, \tag{4}$$

$$(\vec{h}_O)_{A2} = I_C\vec{\omega}_{A2} + \vec{r}_{C/O}\times m_A\vec{v}_{C2}, \quad (\vec{h}_O)_{B2} = I_D\vec{\omega}_{B2} + \vec{r}_{D/O}\times m_B\vec{v}_{D2}. \tag{5}$$

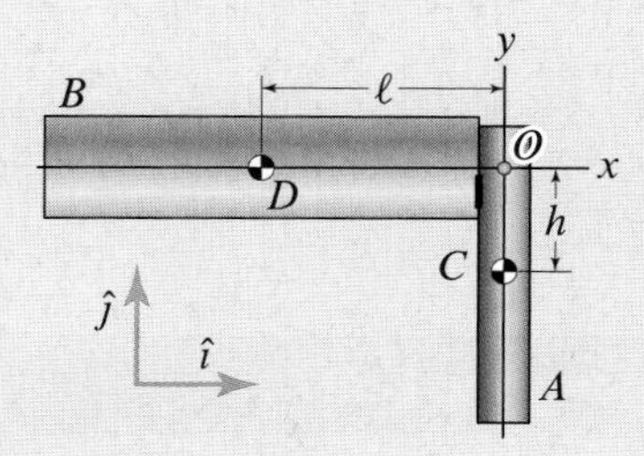

Figure 3
FBD of A and B right before *and* after docking. The coordinate system shown is *fixed in space*, and A and B can move relative to it.

Force Laws All forces are accounted for on the FBD.

Kinematic Equations Before docking,

$$(v_{Cx})_1 = -v_0, \quad (v_{Cy})_1 = 0, \quad \omega_{A1} = 0, \tag{6}$$

$$(v_{Dx})_1 = 0, \quad (v_{Dy})_1 = 0, \quad \omega_{B1} = 0. \tag{7}$$

After docking, A and B form a single rigid body, so we have

$$\omega_{A2} = \omega_{B2} = \omega_{AB} \quad \text{and} \quad \vec{v}_{C2} = \vec{v}_{D2} + \omega_{AB}\,\hat{k}\times\vec{r}_{C/D}, \tag{8}$$

where ω_{AB} is the common angular velocity of A and B immediately after docking. The relative position vectors in Eqs. (4), (5), and (8) are given by

$$\vec{r}_{C/O} = -h\,\hat{\jmath}, \quad \vec{r}_{D/O} = -\ell\,\hat{\imath}, \quad \text{and} \quad \vec{r}_{C/D} = \ell\,\hat{\imath} - h\,\hat{\jmath}. \tag{9}$$

Computation Substituting the first two of Eqs. (6) and (7) into Eqs. (1) and (2), we have

$$-m_A v_0 = m_A (v_{Cx})_2 + m_B (v_{Dx})_2, \tag{10}$$
$$0 = m_A (v_{Cy})_2 + m_B (v_{Dy})_2. \tag{11}$$

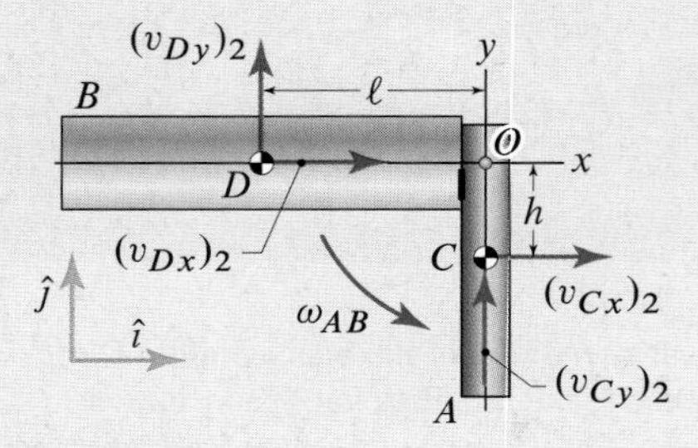

Figure 4
Sketch of the velocity components of the system right after docking.

Referring to Fig. 4 and substituting Eqs. (6), (7), the first of Eqs. (8), and the first two of Eqs. (9) into Eqs. (4) and (5), we have

$$(\vec{h}_O)_{A1} = -m_A h v_0\,\hat{k}, \quad (\vec{h}_O)_{A2} = \big[I_C \omega_{AB} + m_A h (v_{Cx})_2\big]\,\hat{k}, \tag{12}$$
$$(\vec{h}_O)_{B1} = \vec{0}, \quad (\vec{h}_O)_{B2} = \big[I_D \omega_{AB} - m_A \ell (v_{Dy})_2\big]\,\hat{k}. \tag{13}$$

Substituting Eqs. (12) and (13) into Eq. (3), we obtain

$$-m_A h v_0 = (I_C + I_D)\omega_{AB} + m_A h (v_{Cx})_2 - m_B \ell (v_{Dy})_2. \tag{14}$$

Equation (14) is in scalar form because the only nonzero component of Eq. (3) is in the z direction. Finally, substituting the last of Eqs. (9) into the second of Eqs. (8), expanding the cross-product, and expressing the result in components, we have

$$(v_{Cx})_2 = (v_{Dx})_2 + \omega_{AB} h \quad \text{and} \quad (v_{Cy})_2 = (v_{Dy})_2 + \omega_{AB} \ell. \tag{15}$$

Equations (10), (11), (14), and (15) form a system of five equations in the five unknowns $(v_{Cx})_2$, $(v_{Cy})_2$, $(v_{Dx})_2$, $(v_{Dy})_2$, and ω_{AB}. The solution to these five equations is found to be

$$(v_{Cx})_2 = \frac{-m_A\big(I + m_B h^2 + \frac{m_A m_B}{m}\ell^2\big)v_0}{m_A m_B d^2 + m I} = -0.01288\,\text{m/s}, \tag{16}$$
$$(v_{Cy})_2 = \frac{-m_B \frac{m_A m_B}{m} h \ell v_0}{m_A m_B d^2 + m I} = -0.002645\,\text{m/s}, \tag{17}$$
$$(v_{Dx})_2 = \frac{-m_A\big(I + \frac{m_A m_B}{m}\ell^2\big)v_0}{m_A m_B d^2 + m I} = -0.01141\,\text{m/s}, \tag{18}$$
$$(v_{Dy})_2 = \frac{m_A \frac{m_A m_B}{m} h \ell v_0}{m_A m_B d^2 + m I} = 0.001763\,\text{m/s}, \tag{19}$$
$$\omega_{AB} = \frac{-m_A m_B h v_0}{m_A m_B d^2 + m I} = -0.0001837\,\text{rad/s}, \tag{20}$$

where $m = m_A + m_B$, $d = \sqrt{h^2 + \ell^2}$, and $I = I_C + I_D$.

Discussion & Verification The results appear reasonable since the computed velocities are comparable to v_0. In addition, the signs appear correct in that, after docking, we expect both A and B to move to the left and the AB-body to rotate clockwise. This rotation then causes C and D to move slightly downward and upward, respectively.

A Closer Look We assumed that A and B form a rigid body after docking because we did not know the exact position of the docking location. A better assumption is that A and B become pinned to each other after docking. In this way, we better capture the effect of the local flexibility of the docking location. This possibility is considered in Prob. 8.131 on p. 655.

Interesting Fact

Mass of the ISS. The assembly of the ISS began in 1998. Eventually the ISS mass will be about 472,000 kg. The mass and moment of inertia given in the problem are rough estimates based on stage 9A.1 of the ISS assembly process (see J. A. Wojtowicz, "Dynamic Properties of the International Space Station throughout the Assembly Process," Report N. A282843, Air Force Institute of Technology, Wright-Patterson AFB, Ohio, 1998).

Problems

Problem 8.75

Disks A and B have identical masses and mass moments of inertia about their respective mass centers. Point C is both the geometric center and center of mass of disk A. Points O and D are the geometric center and center of mass of disk B, respectively. If, at the instant shown, the two disks are rotating about their centers with the same angular velocity ω_0, determine which of the following statements is true and why: (a) $|(\vec{h}_C)_A| < |(\vec{h}_O)_B|$, (b) $|(\vec{h}_C)_A| = |(\vec{h}_O)_B|$, (c) $|(\vec{h}_C)_A| > |(\vec{h}_O)_B|$.

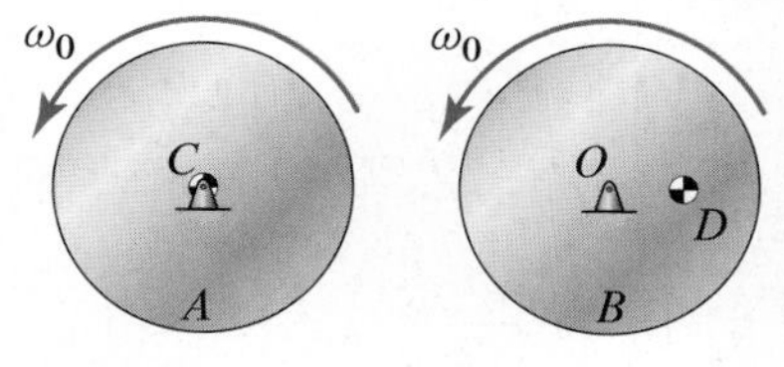

Figure P8.75

Problem 8.76

Body B has mass m and mass moment of inertia I_G, where G is the mass center of B. If B is translating as shown, determine which of the following statements is true and why: (a) $|(\vec{h}_E)_B| < |(\vec{h}_P)_B|$, (b) $|(\vec{h}_E)_B| = |(\vec{h}_P)_B|$, (c) $|(\vec{h}_E)_B| > |(\vec{h}_P)_B|$.

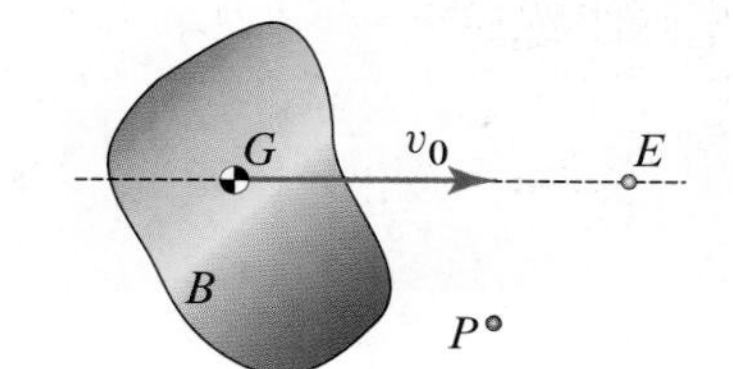

Figure P8.76

Problem 8.77

The rear-wheel-drive car can go from rest to 60 mph in $\Delta t = 8$ s. Assume that the wheels are all identical and that their geometric centers coincide with their mass centers. Let M_{rear} be the average moment applied to one of the rear wheels during Δt and computed relative to the wheel's center. Finally, let M_{front} be the average moment applied to one of the front wheels during Δt and computed relative to the wheel's center. Modeling the wheels as rigid bodies, determine which of the following statements is true and why: (a) $|M_{\text{rear}}| < |M_{\text{front}}|$, (b) $|M_{\text{rear}}| = |M_{\text{front}}|$, (c) $|M_{\text{rear}}| > |M_{\text{front}}|$.

Figure P8.77 and P8.78

Problem 8.78

The rear-wheel-drive car can go from rest to 60 mph in $\Delta t = 8$ s. Assume that its wheels are identical, with their geometric centers coinciding with their mass centers. Let F_{avg} be the average friction force acting on the system during Δt due to contact with the ground. Modeling the car and the wheels as rigid bodies, does the value of F_{avg} change whether or not we account for the rotational inertia of the wheels? Why?

Problem 8.79

A uniform disk of mass m and radius R rolls to the right without slip, such that the speed of the center of mass is v_G. Provide an expression for the linear momentum of the disk in terms of the given quantities. In addition, provide an expression for the angular momentum of the disk relative to O, the point of contact with the ground. Express your answers using the component system shown.

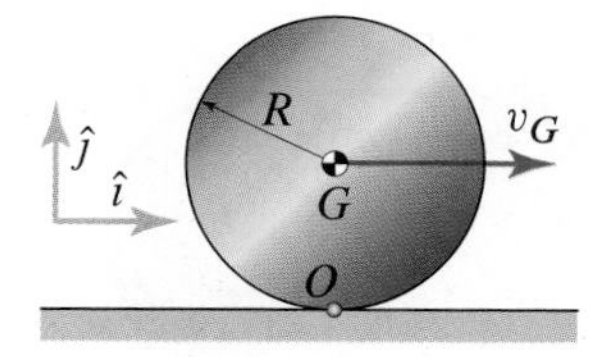

Figure P8.79

Problem 8.80

At the instant shown, the eccentric wheel with center at O and center of mass at G is rotating counterclockwise without slip with an angular speed $\omega_w = 10\,\text{rad/s}$. The weight of the wheel is $W = 90\,\text{lb}$. In addition, let $R = 2\,\text{ft}$, $h = 1\,\text{ft}$, and the radius of gyration $k_G = 1.45\,\text{ft}$. At the instant shown, determine the linear momentum of the disk and the angular momentum about C, the point of contact with the ground. Express your answers in the component system shown.

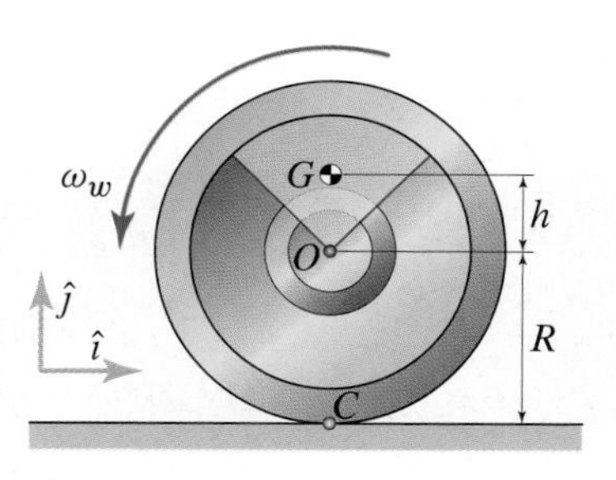

Figure P8.80

Figure P8.81

Problem 8.81

The top of the Space Needle in Seattle, Washington, hosts a revolving restaurant that goes through one full revolution every 47 min under the action of a motor with a power output of 1.5 hp. The portion of the restaurant that rotates is a ring-shaped turntable with internal and external radii $r_i = 33.3$ ft and $r_o = 47.3$ ft, respectively, and approximate weight $W = 125$ tons (1 ton $= 2000$ lb). Use the given values of power output and angular speed to estimate the torque M that the engine provides. Then, assuming that the motor can provide a constant torque equal to M, neglecting all friction, and modeling the turntable as a uniform body, determine the time t_s that it takes to spin up the revolving restaurant from rest to its working angular speed.

Problem 8.82

A rotor, spinning freely about the fixed point O, consists of a thin uniform bar AB that functions as a hub and two identical blades pinned at A and B, respectively. The dimensions of the system are: $d = 0.5$ m, $\ell = 5$ m, and $w = 0.3$ m. The bar AB has a mass $m_{AB} = 30$ kg, and each of the blades has a mass $m_b = 20$ kg. Each blade can be modeled as a uniform thin plate. The angular speed is $\omega_r = 100$ rpm when the angle θ is equal to 90°. At some point, an internal mechanism causes the blades to change orientation relative to AB in such a way that θ becomes constant and equal to 180°. Neglecting aerodynamic forces and friction at the bearings at O, determine the angular speed of the rotor after $\theta = 180°$.

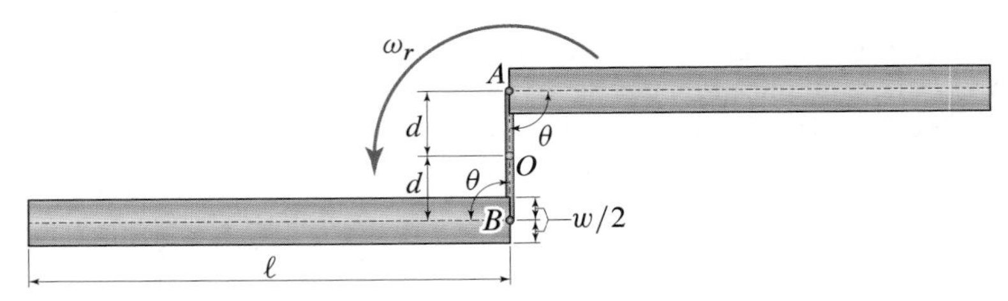

Figure P8.82

Problems 8.83 and 8.84

A uniform disk of mass $m = 20$ kg and radius $R = 0.75$ m is being pulled to the left with a constant horizontal force P by the cord wrapped around it. Assume that the disk starts from rest and that it rolls without slip.

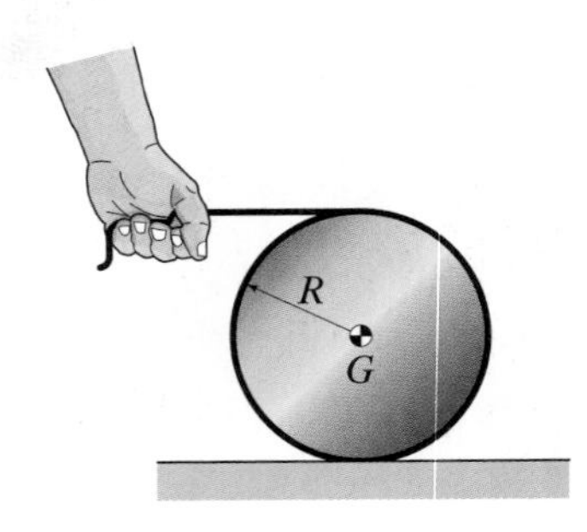

Figure P8.83 and P8.84

Problem 8.83 If $P = 30$ N, apply the impulse-momentum principles to determine the angular speed of the disk after 4 s.

Problem 8.84 Apply the impulse-momentum principles to determine P if the center of the disk achieves a speed $v_G = 5$ m/s after 3 s.

Problem 8.85

Moving on a straight and horizontal stretch of road, the rear-wheel-drive car shown can go from rest to 60 mph in $\Delta t = 8$ s. The car weighs 2570 lb (the weight includes the wheels). Each wheel has diameter $d = 24.3$ in., mass moment of inertia relative to its own center of mass $I_G = 0.989\ \text{slug}\cdot\text{ft}^2$, and the center of mass of each wheel coincides with its geometric center. Determine the average friction force F_{avg} acting on the car during Δt. In addition, if the wheels roll without slip, for each wheel, determine the average moment M_{avg}, computed relative to the wheel's center, that is applied to the wheel during Δt.

Figure P8.85

Problem 8.86

A spool with radius $R = 3\,\text{ft}$ is released from rest on an incline with $\theta = 30°$, and its center, which coincides with its center of mass, is observed to reach a speed of $8\,\text{ft/s}$ two seconds after release. If the spool rolls without slip, use the impulse-momentum principles to determine the radius of gyration of the spool k_G.

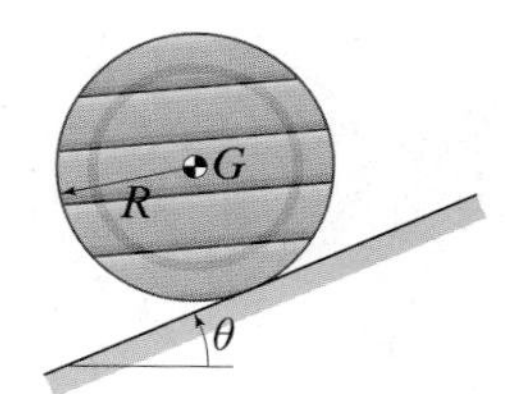

Figure P8.86

Problem 8.87

An eccentric wheel B weighing $150\,\text{lb}$ has its mass center G at a distance $d = 4\,\text{in.}$ from the wheel's center O. The wheel is in the horizontal plane and is spun from rest by applying a constant torque $M = 32\,\text{ft·lb}$. Determine the wheel's radius of gyration k_G if it takes 2 s to spin up the wheel to 140 rpm. Neglect all possible sources of friction.

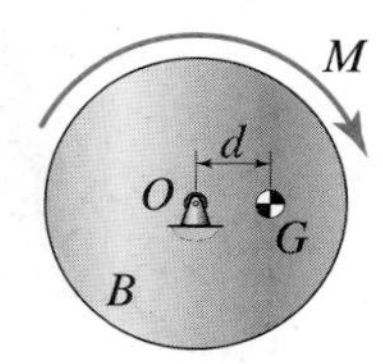

Figure P8.87

Problem 8.88

The uniform thin pin-connected bars AB, BC, and CD have masses $m_{AB} = 2.3\,\text{kg}$, $m_{BC} = 3.2\,\text{kg}$, and $m_{CD} = 5.0\,\text{kg}$, respectively. Letting $R = 0.75\,\text{m}$, $L = 1.2\,\text{m}$, and $H = 1.55\,\text{m}$, and knowing that bar AB rotates at a constant angular velocity $\omega_{AB} = 4\,\text{rad/s}$, compute the angular momentum of bar AB about A, of bar BC about A, and bar CD about D at the instant shown.

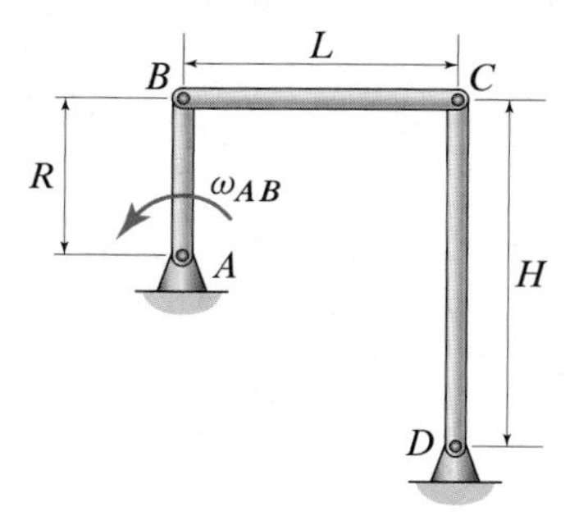

Figure P8.88

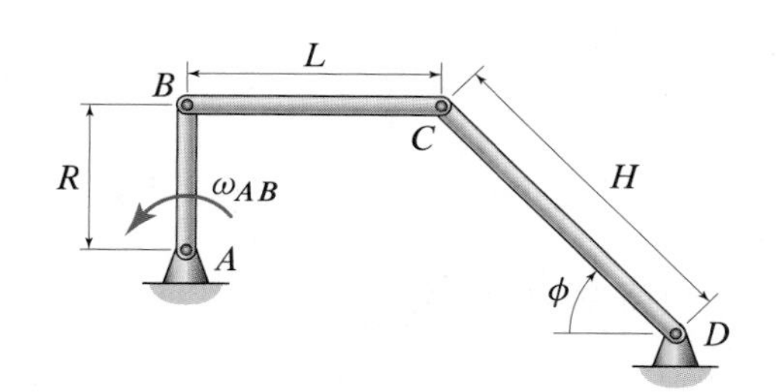

Figure P8.89

Problem 8.89

The weights of the uniform thin pin-connected bars AB, BC, and CD are $W_{AB} = 4\,\text{lb}$, $W_{BC} = 6.5\,\text{lb}$, and $W_{CD} = 10\,\text{lb}$, respectively. Letting $\phi = 47°$, $R = 2\,\text{ft}$, $L = 3.5\,\text{ft}$, and $H = 4.5\,\text{ft}$, and knowing that bar AB rotates at a constant angular velocity $\omega_{AB} = 4\,\text{rad/s}$, compute the magnitude of the linear momentum of the system at the instant shown.

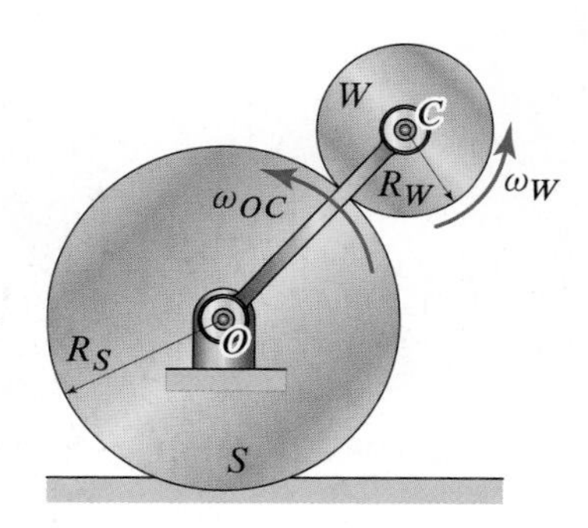

Figure P8.90

Problem 8.90

A uniform disk W of radius $R_W = 7\,\text{mm}$ and mass $m_W = 0.15\,\text{kg}$ is connected to point O via the rotating arm OC. Disk W also rolls without slip over the stationary cylinder S of radius $R_S = 15\,\text{mm}$. Assuming that $\omega_W = 25\,\text{rad/s}$, determine the angular momentum of W about its own center of mass C, as well as about point O.

Problem 8.91

A rotor B with center of mass G, weight $W = 3000\,\text{lb}$, and radius of gyration $k_G = 15\,\text{ft}$ is spinning with an angular speed $\omega_B = 1200\,\text{rpm}$ when a braking system is applied to it, providing a time-dependent torque $M = M_0(1 + ct)$, with $M_0 = 3000\,\text{ft·lb}$ and $c = 0.01\,\text{s}^{-1}$. If G is also the geometric center of the rotor and is a fixed point, determine the time t_s that it takes to stop the rotor.

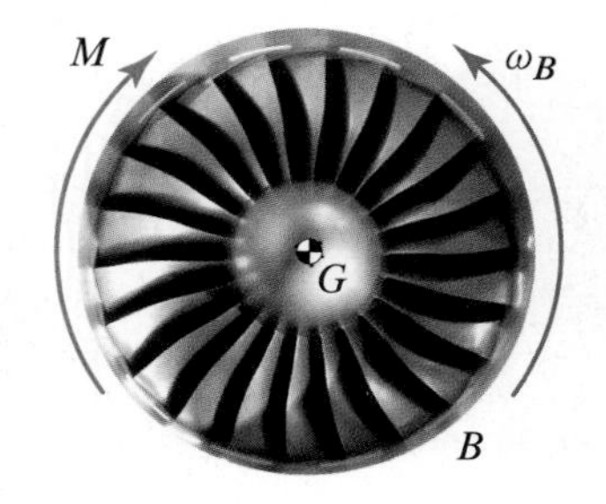

Figure P8.91

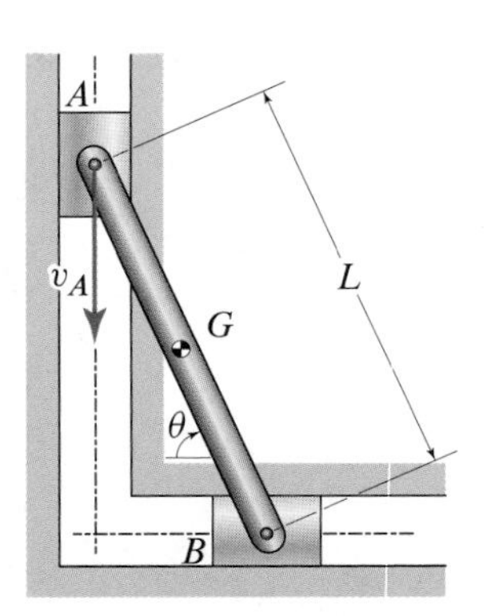

Figure P8.92

Problem 8.92

The uniform bar AB has length $L = 4.5\,\text{ft}$ and weight $W_{AB} = 14\,\text{lb}$. At the instant shown, $\theta = 67°$ and $v_A = 5.8\,\text{ft/s}$. Determine the magnitude of the linear momentum of AB, as well as the angular momentum of AB about its mass center G at the instant shown.

Problem 8.93

A uniform pipe section A of radius r, mass center G, and mass m is gently placed (i.e., with zero velocity) on a conveyor belt moving with a constant speed v_0 to the right. Friction between the belt and pipe causes the pipe to move to the right and eventually to roll without slip. If μ_k is the coefficient of kinetic friction between the pipe and the conveyor belt, find an expression for t_r, the time it takes for A to start rolling without slip. *Hint:* Using the methods of Chapter 7, we can show that the force between the pipe section and the belt is constant.

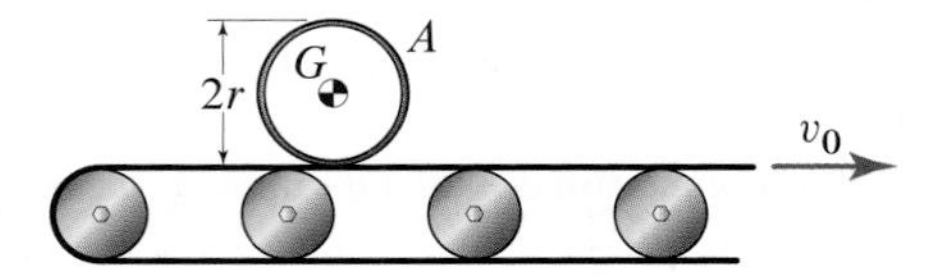

Figure P8.93

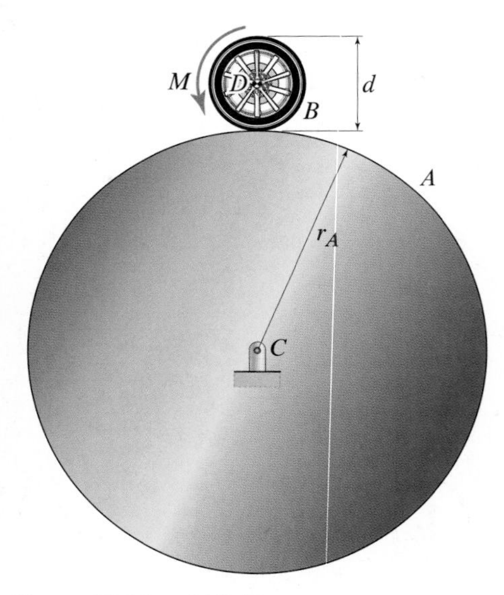

Figure P8.94 and P8.95

Problems 8.94 and 8.95

An automobile wheel test rig consists of a uniform disk A, of mass $m_A = 5000\,\text{kg}$ and radius $r_A = 1.5\,\text{m}$, that can rotate freely about its fixed center C and over which the wheel of an automobile is made to roll. A wheel B, whose center and center of mass coincide at D, is mounted on a shaft (not shown) that holds D fixed while it allows the wheel to rotate about D. The wheel has diameter $d = 0.62\,\text{m}$, mass $m_B = 21.5\,\text{kg}$, and mass moment of inertia about its mass center $I_D = 44\,\text{kg·m}^2$. Both A and B are initially at rest when B is subject to a constant torque M that causes B to roll without slip on A.

Problem 8.94 If $M = 1500\,\text{N·m}$, use the angular impulse-momentum principle to determine how long it takes to reach conditions simulating a car speed of $100\,\text{km/h}$.

Problem 8.95 Use the angular impulse-momentum principle and determine M if it takes 15 seconds to achieve conditions simulating a car speed of $60\,\text{km/h}$.

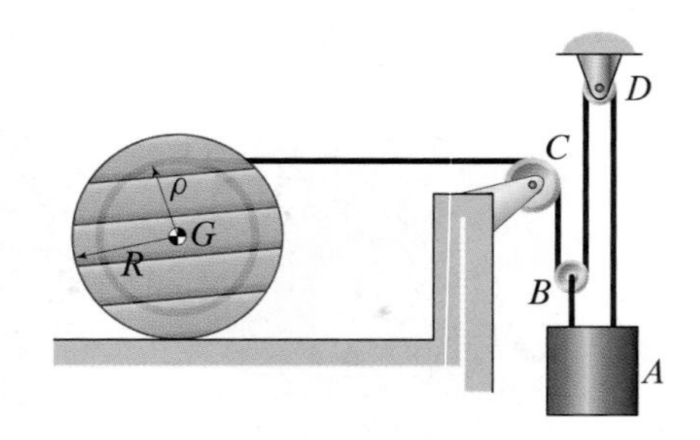

Figure P8.96 and P8.97

Problems 8.96 and 8.97

A spool of mass $m_s = 150\,\text{kg}$ and inner and outer radii $\rho = 0.8\,\text{m}$ and $R = 1.2\,\text{m}$, respectively, is connected to a counterweight A of mass $m_A = 50\,\text{kg}$ by a pulley system whose cord, at one end, is wound around the inner hub of the spool. The center G and the center of mass of the spool coincide, and the radius of gyration of the spool is $k_G = 1\,\text{m}$. The system is at rest when the counterweight is released, causing the spool to move to the right. The spool rolls without slip, and the cord unwinds from the spool without slip.

Problem 8.96 Neglecting the inertia of the pulley system, use the impulse-momentum principles to determine the angular speed of the spool 3 s after release.

Problem 8.97 Assume that the inertia of the cord and of pulleys B and D can be neglected, but model pulley C as a uniform disk mass $m_C = 15\,\text{kg}$ and radius $r_C = 0.3\,\text{m}$. If the cord does not slip relative to pulley C, use the impulse-momentum principles to determine the angular speed of the spool 3 s after release.

Problem 8.98

An 0.8 lb collar with center of mass at G and a uniform cylindrical horizontal arm A of length $L = 1$ ft, radius $r_i = 0.022$ ft, and weight $W_A = 1.5$ lb are rotating as shown with $\omega_0 = 1.5$ rad/s while the collar's mass center is at a distance $d = 0.44$ ft from the z axis. The vertical shaft has radius $e = 0.03$ ft and negligible mass. After the cord restraining the collar is cut, the collar slides with no friction relative to the arm. Assuming that no external forces and moments are applied to the system, determine the collar's impact speed with the end of A if (a) the collar is modeled as a particle coinciding with its own mass center (in this case, neglect the collar's dimensions), and (b) the collar is modeled as a uniform hollow cylinder with length $\ell = 0.15$ ft, inner radius r_i, and outer radius $r_o = 0.048$ ft.

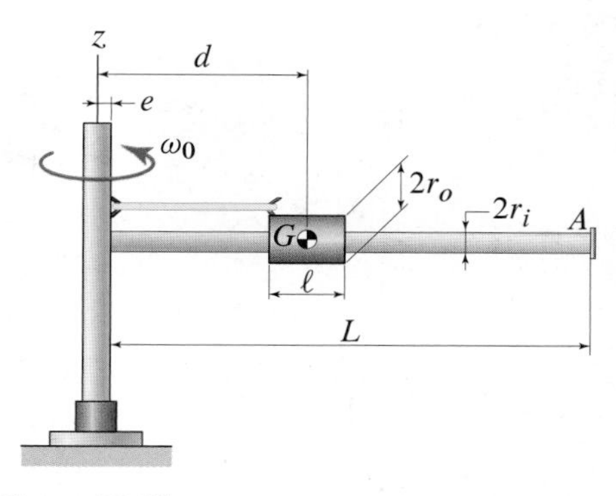

Figure P8.98

Problem 8.99

The uniform disk A, of mass $m_A = 1.2$ kg and radius $r_A = 0.25$ m, is mounted on a vertical shaft that can translate along the horizontal guide C. The uniform disk B, of mass $m_B = 0.85$ kg and radius $r_B = 0.38$ m, is mounted on a fixed vertical shaft. Both disks A and B can rotate about their own axes, namely, ℓ_A and ℓ_B, respectively. Disk A is initially spun with $\omega_A = 1000$ rpm and then brought into contact with B, which is initially stationary. The contact is maintained by a spring, and due to friction between A and B, disk B starts spinning and eventually A and B will stop slipping *relative to one another*. Neglecting any friction except at the contact between the two disks, determine the angular velocities of A and B when slipping stops.

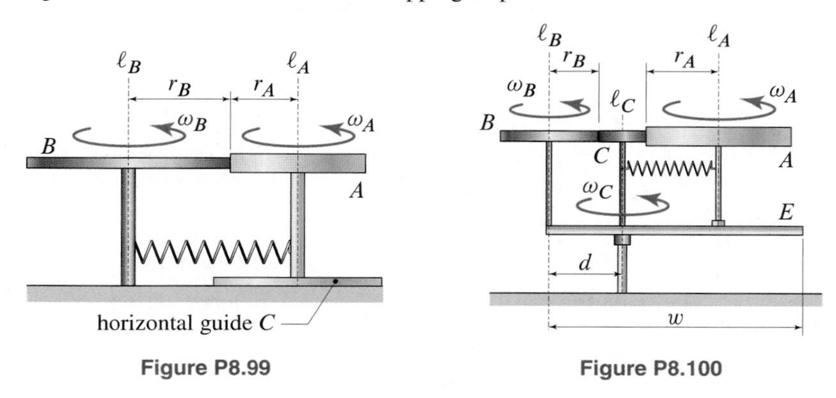

Figure P8.99

Figure P8.100

Problem 8.100

The uniform disk A, of mass $m_A = 1.2$ kg and radius $r_A = 0.25$ m, is mounted on a vertical shaft that can translate along the horizontal arm E. The uniform disk B, of mass $m_B = 0.85$ kg and radius $r_B = 0.18$ m, is mounted on a vertical shaft that is rigidly attached to arm E. Disk A can rotate about axis ℓ_A, disk B can rotate about axis ℓ_B, and the arm E, along with disk C, can rotate about the fixed axis ℓ_C. Disk C has negligible mass and is rigidly attached to E so that they rotate together. While keeping both B and C stationary, disk A is spun to $\omega_A = 1200$ rpm. Disk A is then brought in contact with disk C (contact is maintained by a spring), and B and C (and the arm E) are then allowed to freely rotate. Due to friction between A and C, disks C (and arm E) and B start spinning. Eventually, A and C stop slipping relative to one another. Disk B always rotates without slip over C. Let $d = 0.27$ m and $w = 0.95$ m. If the only elements of the system that have mass are A and B, and if all friction in the system can be neglected except for that between A and C and between C and B, determine the angular speeds of A and C when they stop slipping relative to one another.

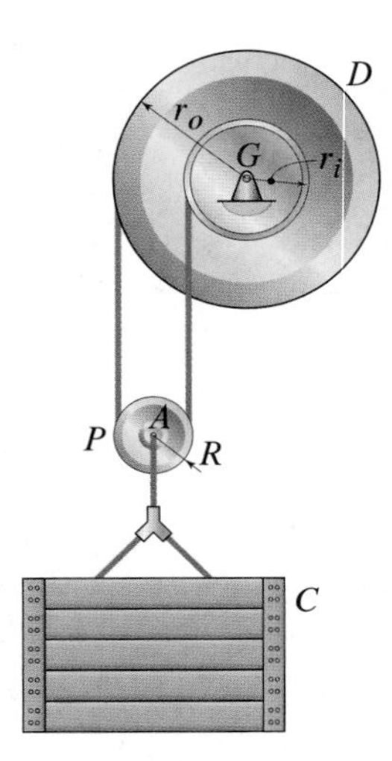

Figure P8.101 and P8.102

Problems 8.101 and 8.102

The double pulley D has mass of $m_D = 15\,\text{kg}$, center of mass G coinciding with its geometric center, radius of gyration $k_G = 10\,\text{cm}$, outer radius $r_o = 15\,\text{cm}$, and inner radius $r_i = 7.5\,\text{cm}$. It is connected to the pulley P with radius $R = (r_o - r_i)/2$ by a cord of negligible mass that unwinds from the inner and outer spools of the double pulley D. The crate C, which has a mass $m_C = 20\,\text{kg}$, is released from rest. The cord does not slip relative to the pulleys, and the inner and outer pulleys rotate as a single unit.

Problem 8.101 Neglecting the mass of the pulley P, use the impulse-momentum principles to determine the speed of the crate 4 s after release.

Problem 8.102 Assuming that the pulley P has a mass of 1.5 kg and a radius of gyration $k_A = 3.5\,\text{cm}$, use the impulse-momentum principles to determine the speed of the crate 4 s after release.

Problem 8.103

Some pipe sections of radius r and mass m are being unloaded and placed in a row against a wall. The first of these pipe sections, A, is made to roll without slipping into a corner with an angular velocity ω_0 as shown. Upon touching the wall, A does not rebound, but slips against the ground and against the wall. Modeling A as a uniform thin ring with center at G and letting μ_g and μ_w be the coefficients of kinetic friction of the contacts between A and the ground and between A and the wall, respectively, determine an expression for the angular velocity of A as a function of time from the moment A touches the wall until it stops. *Hint:* Using the methods learned in Chapter 7, we can show that the friction forces at the ground and at the wall are constant.

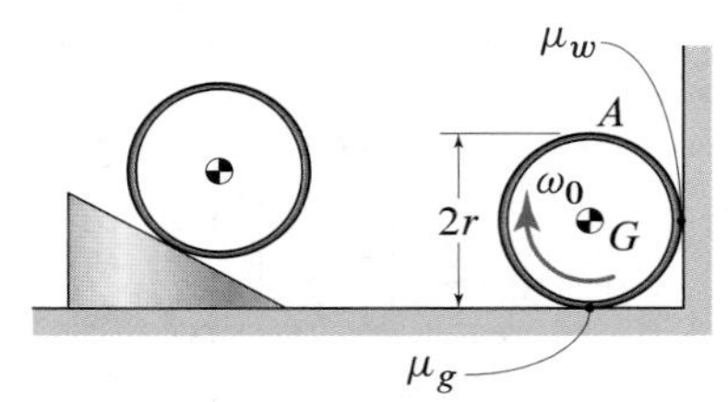

Figure P8.103

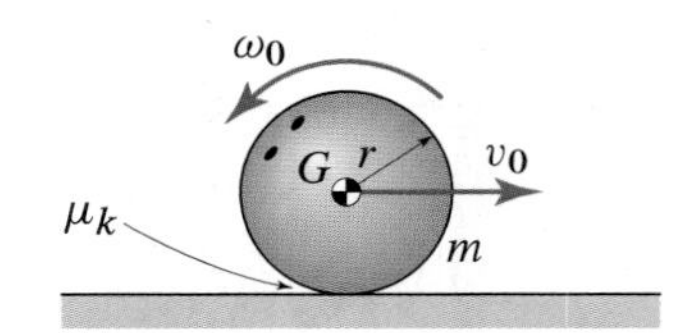

Figure P8.104

Problem 8.104

A 14 lb bowling ball is thrown onto a lane with a backspin $\omega_0 = 9\,\text{rad/s}$ and forward velocity $v_0 = 18\,\text{mph}$. Point G is both the geometric center and the mass center of the ball. After a few seconds, the ball starts rolling without slip. Let $r = 4.25\,\text{in.}$, and let the radius of gyration of the ball be $k_G = 2.6\,\text{in.}$ If the coefficient of kinetic friction between the ball and the floor is $\mu_k = 0.1$, determine the speed v_f that the ball will achieve when it starts rolling without slip. In addition, determine the time t_r the ball takes to achieve v_f. *Hint:* Using the methods of Chapter 7, we can show that the force between the ball and the floor is constant.

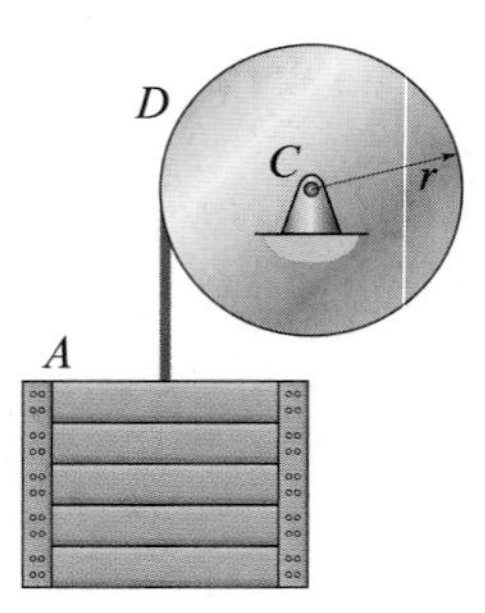

Figure P8.105

Problem 8.105

A crate A with weight $W_A = 250\,\text{lb}$ is hanging from a rope wound around a uniform drum D of radius $r = 1.2\,\text{ft}$, weight $W_D = 125\,\text{lb}$, and center C. The system is initially at rest when the restraining system holding the drum stationary fails, thus causing the drum to rotate, the rope to unwind, and, consequently, the crate to fall. Assuming that the rope does not stretch or slip relative to the drum and neglecting the inertia of the rope, determine the speed of the crate 1.5 s after the system starts to move.

Problem 8.106

A toy helicopter consists of a rotor A, a body B, and a small ballast C. The axis of rotation of the rotor goes through G, which is the center of mass of the body B and ballast C. While holding the body (and ballast) fixed, the rotor is spun as shown with a given angular velocity ω_0. If there is *no friction* between the helicopter's body and the rotor's shaft, will the body of the helicopter start spinning once the toy is released?

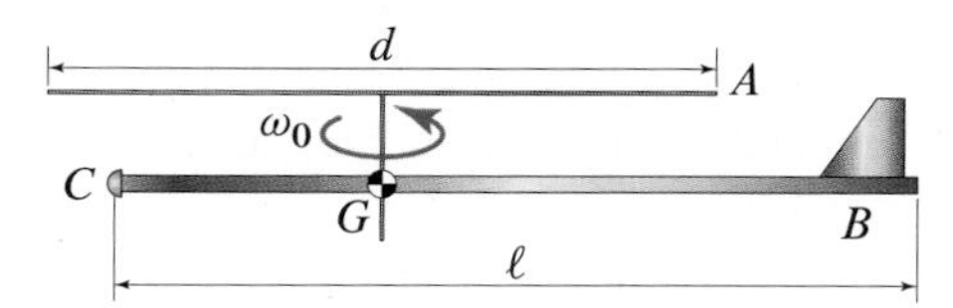

Figure P8.106 and P8.107

Problem 8.107

A toy helicopter consists of a rotor A with diameter $d = 10$ in. and weight $W_r = 0.09 \times 10^{-3}$ oz, a thin body B of length $\ell = 12$ in. and weight $W_B = 0.144 \times 10^{-3}$ oz, and a small ballast C placed at the front end of the body and with weight $W_C = 0.0723 \times 10^{-3}$ oz. The ballast's weight is such that the axis of the rotation of the rotor goes through G, which is the center of mass of the body B and ballast C. While holding the body (and ballast) fixed, the rotor is spun as shown with $\omega_0 = 150$ rpm. Neglecting aerodynamic effects, the weights of the rotor's shaft and the body's tail, and assuming there is friction between the helicopter's body and the rotor's shaft, determine the angular velocity of the body once the toy is released and the angular velocity of the rotor decreases to 120 rpm. Model the body as a uniform thin rod and the ballast as a particle. Assume that the rotor and the body remain horizontal after release.

Problem 8.108

A cord, which is wrapped around the inner radius of the spool of mass $m = 35$ kg, is pulled vertically at A by a constant force $P = 120$ N (the cord is pulled in such a way that it remains vertical), causing the spool to roll over the horizontal bar BD. The inner radius of the spool is $R = 0.3$ m, and the center of mass of the spool is at G, which also coincides with the geometric center of the spool. The spool's radius of gyration is $k_G = 0.18$ m. Assuming that the spool starts from rest, that the cord's inertia and extensibility can be neglected, and that the spool rolls without slip, determine the speed of the spool's center 3 s after the application of the force. In addition, determine the minimum static friction coefficient for rolling without slip to be maintained during the time interval in question.

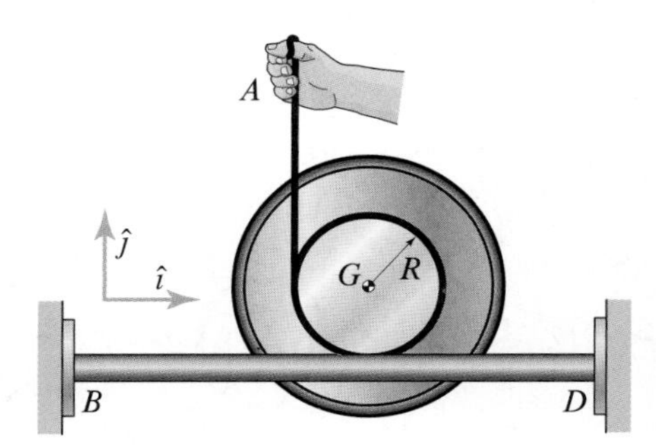

Figure P8.108

Problem 8.109

A spool has weight $W = 450$ lb, outer and inner radii $R = 6$ ft and $\rho = 4.5$ ft, respectively, center of mass G coinciding with its geometric center, and radius of gyration $k_G = 4.0$ ft. The spool is at rest when it is pulled to the right as shown. The cable wrapped around the spool can be modeled as being inextensible and of negligible mass. Assume that the spool rolls without slip relative to both the cable and the ground. If the cable is pulled with a force $P = 125$ lb, determine the speed of the center of the spool after 2 s and the minimum value of the static friction coefficient between the spool and the ground necessary to guarantee rolling without slip.

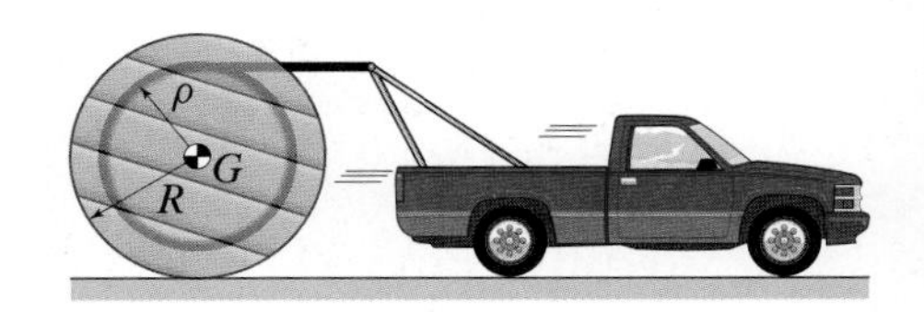

Figure P8.109

Figure P8.110

Problem 8.110

The wind turbine in the figure consists of three equally spaced blades that are rotating as shown about the fixed point O with an angular velocity $\omega_0 = 30$ rpm. Suppose that each 38,000 lb blade can be modeled as a narrow uniform rectangle of length $b = 182$ ft, width $a = 12$ ft, and negligible thickness, with one of its corners coinciding with the center of rotation O. The orientation of each blade can be controlled by rotating the blade about an axis going through the center O that coincides with the blade's leading edge. Neglecting aerodynamic forces and any source of friction, and assuming that the turbine is freely rotating, determine the turbine's angular velocity ω_f after each blade has been rotated 90° about its own leading edge.

Problem 8.111

Cars A and B collide as shown. Neglecting the effect of friction, what would be the angular velocity of A and B immediately after impact if A and B were to form a single rigid body as a result of the collision? In solving the problem, let C and D be the mass centers of A and B, respectively, and use the following data: the weight of A is $W_A = 3130$ lb, the radius of gyration of A is $k_C = 34.5$ in., the speed of A right before impact is $v_A = 12$ mph, the weight of B is $W_B = 3520$ lb, the radius of gyration of B is $k_D = 39.3$ in., the speed of B right before impact is $v_B = 15$ mph, $d = 19$ in., and $\ell = 144$ in. Finally, assume that while A and B form a single rigid body right after impact, the mass center of the rigid body formed by A and B coincides with the mass center of the A-B system right before impact.

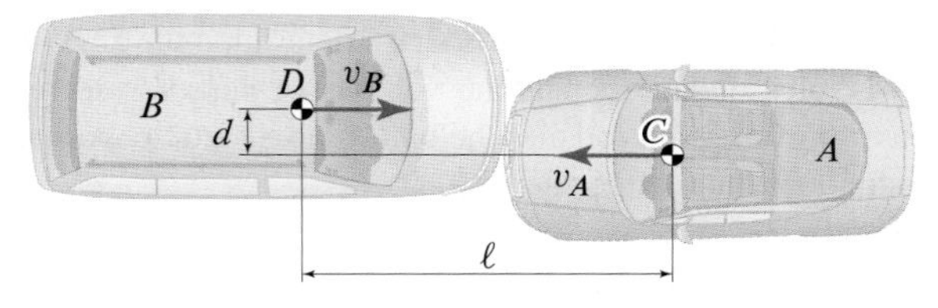

Figure P8.111

Problem 8.112

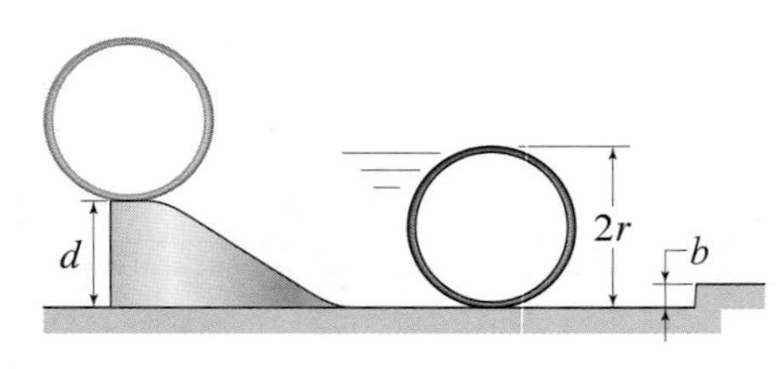

Figure P8.112

Some pipe sections are gently nudged from rest down an incline and roll without slipping all the way to a step of height b. Assume that each pipe section does not slide or rebound against the step, so that the pipes move as if hinged at the corner of the step. Modeling a pipe as a uniform thin ring of mass m and radius r, and letting d be the height from which the pipes are released, determine the minimum value of d so that the pipes can roll over the step. *Hint:* When a pipe hits the corner of the step, its motion changes almost instantaneously from rolling without slip on the ground to a fixed-axis rotation about the corner of the step. Model this transition, using the ideas presented in Section 5.2 on p. 335. That is, assume that there is an infinitesimal time interval right after the impact between a pipe and the corner of the step in which the pipe does not change its position significantly, the pipe loses contact with the ground, and its weight is negligible relative to the contact forces between the pipe and the step.

Problem 8.113

A crane has a boom A of mass m_A and length ℓ that can rotate in the horizontal plane about a fixed point O. A trolley B of mass m_B is mounted on one side of A, such that the mass center of B is always at a distance e from the longitudinal axis of A. The position of B is controlled by a cable and a system of pulleys. Both A and B are initially at rest in the position shown, where d is the initial distance of B from O measured along the longitudinal axis of A. The boom A is free to rotate about O and, for a short time interval $0 \le t \le t_f$, B moves with constant acceleration a_0 without reaching the end of A. Letting I_O be the mass moment of inertia of A, modeling B as a particle, and accounting only for the inertia of A and B, determine the direction of rotation of A and the angle θ swept by A from $t = 0$ to $t = t_f$. Neglect the mass of the cable and of the pulleys.

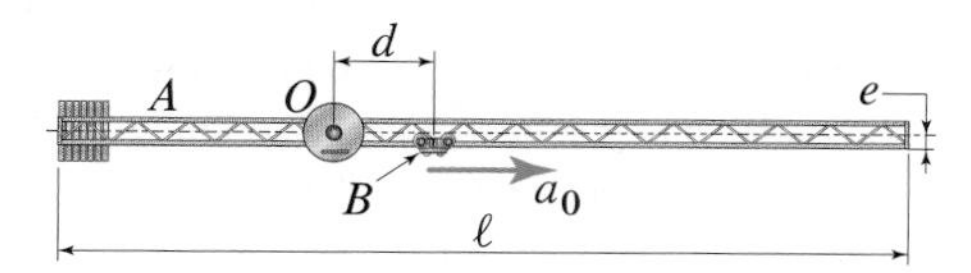

Figure P8.113

Problem 8.114

Following up on parts (b) and (c) of the Pioneer 3 despin in Prob. 7.96, it turns out that we can analytically determine the length of the unwound wire needed to achieve *any* value of ω_s by using conservation of energy and conservation of angular momentum. In doing so, let the masses of A and B each be m, and the mass moment of inertia of the spacecraft body be I_O. Let the initial conditions of the system be $\omega_s(0) = \omega_0$, $\ell(0) = 0$, and $\dot{\ell}(0) = 0$, and neglect gravity and the mass of each wire.

(a) Find the velocity of each of the masses A and B as a function of the wire length $\ell(t)$, the angular velocity of the spacecraft body $\omega_s(t)$, and the radius of the spacecraft R. *Hint:* This part of the problem involves just kinematics—refer to Prob. 6.160 if you need help with the kinematics.

(b) Apply the work-energy principle to the spacecraft system between the time just before the masses start to unwind and any arbitrary later time. You should obtain an expression relating ℓ, $\dot{\ell}$, ω_s, and constants. *Hint:* No external work is done on the system.

(c) Since no external forces act on the system, its total angular momentum must be conserved about point O. Relate the angular momentum for this system between the time just before the masses start unwinding and any arbitrary later time. As with Part (b), you should obtain an expression relating ℓ, $\dot{\ell}$, ω_s, and constants.

(d) Solve the energy and angular momentum equations obtained in Parts (b) and (c), respectively, for $\dot{\ell}$ and ω_s. Now, letting $\omega_s = 0$, show that the length of the unwound wire when the angular velocity of the spacecraft body is zero is given by $\ell_{\omega_s=0} = \sqrt{(I_O + 2mR^2)/(2m)}$.

(e) From your solutions for $\dot{\ell}$ and ω_s in Part (d), find the equations for $\ell(t)$ and $\omega_s(t)$. These are the general solutions to the nonlinear equations of motion found in Prob. 7.95.

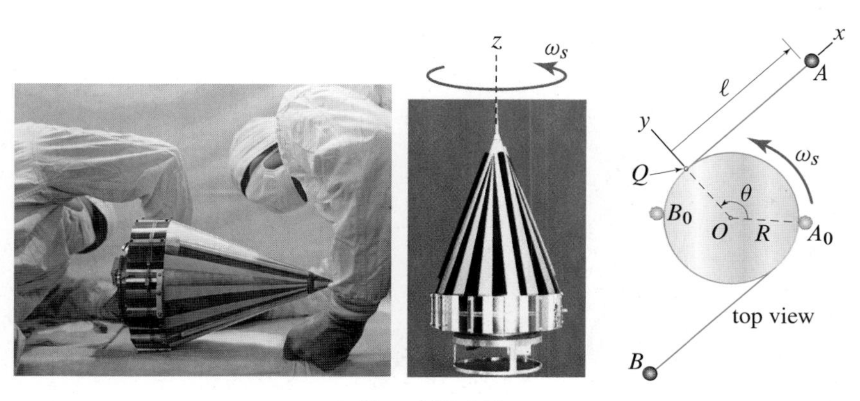

Figure P8.114

Design Problems

Design Problem 8.2

The Pioneer 3 spacecraft was a spin-stabilized spacecraft launched on December 6, 1958, by the U.S. Army Ballistic Missile agency in conjunction with NASA. It was designed with a despin mechanism consisting of two equal masses A and B, each of mass m, that could be spooled out to the ends of two wires of variable length $\ell(t)$ when triggered by a hydraulic timer. As the masses unwound, they would slow the spacecraft's spin from an initial angular velocity $\omega_s(0)$ to the final angular velocity $(\omega_s)_{\text{final}}$, and then the weights and wires would be released. Assume that masses A and B are initially at positions A_0 and B_0, respectively, before the wire begins to unwind, that the mass moment of inertia of the spacecraft is I_O (this does not include the two masses A and B), and neglect gravity and the mass of each wire. With this in mind, you are given the task of despinning the spacecraft body from an angular velocity of 400 rpm to *any* angular velocity in the range $-400\,\text{rpm} < \omega_s < 400\,\text{rpm}$. To do this, a transducer is used that senses the tension in one of the wires at every instant. Design the total length of the wires to achieve the desired range of angular velocities, and determine the tension in the wires as a function of the angular velocity of the spacecraft so that the sensor can know when the spacecraft has achieved the desired velocity and can release the wires and masses.

Use $R = 12.5\,\text{cm}$, $m = 7\,\text{g}$, $I_O = 0.0277\,\text{kg}\cdot\text{m}^2$, and the initial conditions $\ell(0) = 0.01\,\text{m}$ and $\dot{\ell}(0) = 0\,\text{m/s}$, with the despin mechanism shown. *Hint:* Refer to Prob. 6.160 on p. 511 if you need help with the kinematics.

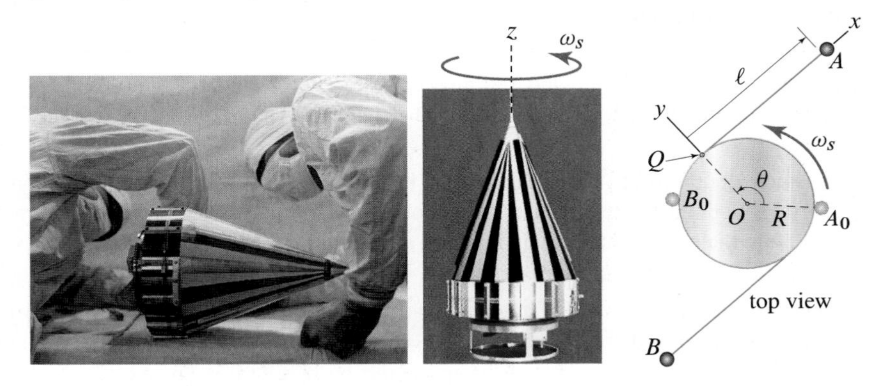

Figure DP8.2

8.3 Impact of Rigid Bodies

In this section we continue the study of impacts, begun in Section 5.2 on p. 335, by considering impacts between rigid bodies. We will use the notation and concepts introduced in Section 5.2. As a reminder,

- We denote the value of a quantity right before and right after a collision with the superscripts − and +, respectively (for example, v_A^+ denotes the speed of a point A right after impact).
- Our modeling of impacts is based on the concept of *impulsive force* (see p. 336). As in Section 5.2, this modeling assumption is reflected in our *impact-relevant* FBDs, which include *only* impulsive forces.

The difference between rigid body and particle impact

Figure 8.19 shows a collision between two bumper cars A and B. Assume that we know the COR (coefficient of restitution) e for the collision, the masses of A and B, the location of their mass centers, their preimpact velocities, and their mass moments of inertia about their respective mass centers. Since the motion is planar, the postimpact velocities of A and B are described by six pieces of information: the velocity components of the mass centers of A and B (four unknowns) and the angular velocities of A and B (two unknowns). Consequently, we will need six scalar equations to solve for the six unknowns. Solving this rigid body impact problem will not require any new theory.

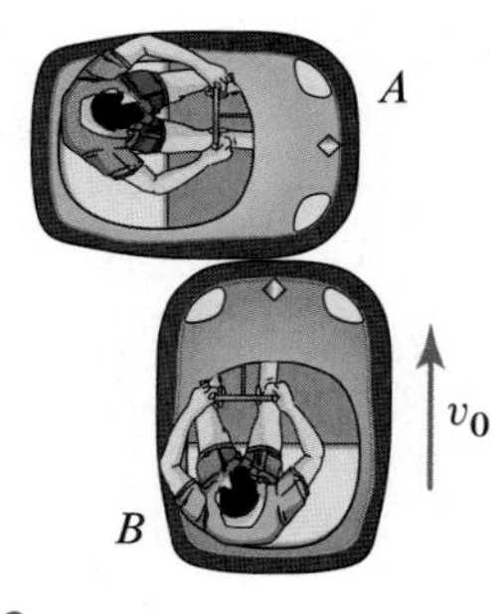

Figure 8.19
A collision between two bumper cars. Before impact, car A is at rest and B is moving with speed v_0 in the direction shown.

Figure 8.20 shows the FBDs of A and B during the impact. Since there are no

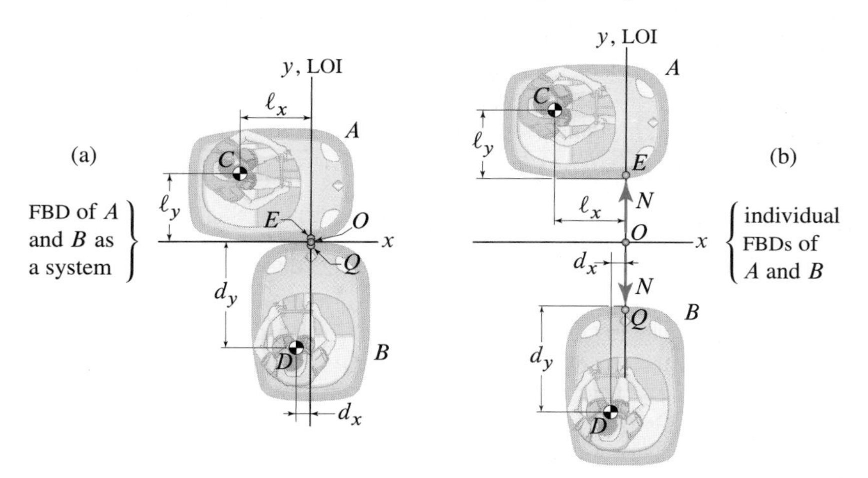

Figure 8.20. (a) Impact-relevant FBD of A and B as a system, and (b) the individual FBDs of A and B. The contact points E and Q belong to A and B, respectively, and coincide with O.

external impulsive forces, the system's total linear and angular momenta are conserved. As with particle impact, conservation of linear momentum will provide three equations from these FBDs, and the application of the COR equation along the LOI will provide a fourth equation. The final two equations are obtained by conserving angular momentum about point O for each of the two bumper cars. Therefore, we see that the analysis of rigid body impact adds two conservation of angular momentum equations to the four equations we used in particle impact.

Rigid body impact: basic nomenclature and assumptions

As in the case of particle impacts (see Section 5.2 on p. 335), a rigid body impact is called *plastic* if the COR $e = 0$. We call a rigid body impact *perfectly plastic* if the colliding bodies form a single rigid body after impact. An impact is *elastic* if $0 < e < 1$, and the ideal case with $e = 1$ is called *perfectly elastic*. In addition, an impact is *unconstrained* if the impacting objects are not acted upon by any external impulsive forces; otherwise the impact is called *constrained*.

Rigid body impacts can be very complex, and we will consider only those cases, such as the bumper car collision problem, in which the following assumptions hold:

1. The impact involves only two bodies in planar motion where no impulsive force has a component perpendicular to the plane of motion.
2. Contact between any two rigid bodies occurs only at one point, and at this point we can clearly define the LOI.
3. The contact between the bodies is frictionless.

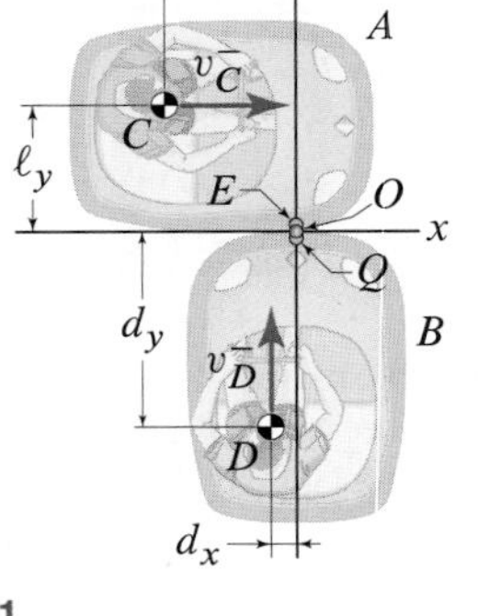

Figure 8.21
Preimpact system configuration of an oblique eccentric impact.

Classification of impacts

In Section 5.2, we classified impacts based on (1) the position of the mass centers of the colliding objects in relation to the LOI and (2) the orientation of the preimpact velocities in relation to the LOI (see Table 8.1). Based on Table 8.1, the bumper car collision in Fig. 8.19 is a *direct eccentric* impact because the cars' mass centers are not on the LOI and because the preimpact velocity of the mass center of B is parallel to the LOI while the velocity of A is zero. If, right before impact, car A were moving as shown in Fig. 8.21, then one of the mass centers would have a preimpact velocity not parallel to the LOI, and the impact would be oblique and eccentric.

Table 8.1. Classification of impacts.

Impact geometry criteria		**Impact type**
Preimpact velocities	*Centers of mass*	
parallel to LOI	on LOI	direct central
parallel to LOI	not on LOI	direct eccentric
not parallel to LOI	on LOI	oblique central
not parallel to LOI	not on LOI	oblique eccentric

Central impact

In a *central impact*, the centers of mass of the two colliding objects lie on the LOI. There are two types of central impacts: direct and oblique (see Table 8.1). No matter the type, under the assumptions introduced above, rigid body central impacts have two important characteristics:

1. The angular velocities are *conserved* through the impact.
2. The COR equation can be written directly in terms of the velocity components along the LOI of the mass centers.

To illustrate these properties, consider the collision of two identical billiard balls A and B, as shown in Fig. 8.22. Ball A is initially stationary, while B is rolling without slip to the left at a speed v_0. For a perfectly elastic collision, we want to determine the postimpact velocities of the two balls. We begin by sketching the system's impact-relevant FBD (Fig. 8.23(a)), and the impact-relevant FBDs of the individual bodies (Fig. 8.23(b)). The FBDs should show the LOI and a coordinate system aligned with the LOI. Points E and Q belong to A and B, respectively. The origin O of the chosen coordinate system coincides with E and Q at the time of impact, but is otherwise understood to be a point *fixed* in space. The FBDs in Fig. 8.23 do not show any reaction force between the balls and the table, which indicates that we have regarded these forces as *nonimpulsive*. This is because we have assumed that the contact between the two balls is frictionless, which implies that no impulsive friction force can be generated along the y direction in response to the sliding of point Q relative to point E. The absence of a vertical impulsive force at the point of contact implies that no corresponding impulsive *reaction* force is generated in the vertical direction at the supporting surface.

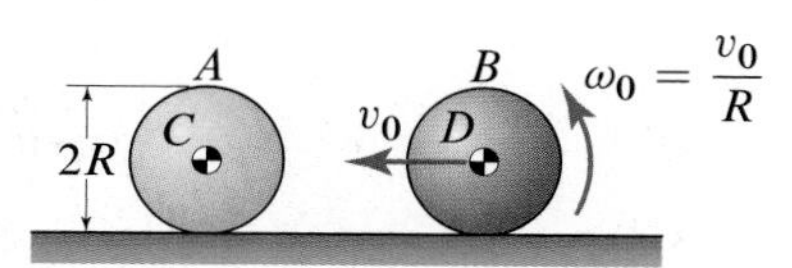

Figure 8.22
Example of a direct central impact.

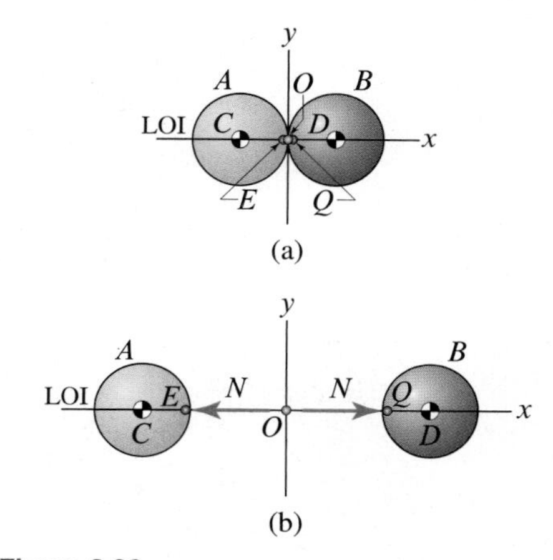

Figure 8.23
Impact-relevant FBDs (a) of the system as a whole and (b) of the two balls individually.

Because the impact is central, the impulsive force N has no moment with respect to C and D. Summing moments about the mass center of each ball using Eq. (8.43) on p. 619, we can then say that $I_C\omega_A^- = I_C\omega_A^+$ and $I_D\omega_B^- = I_D\omega_B^+$, where I_C and I_D are the mass moments of inertia of A and B, respectively. Therefore, we conclude that

$$\boxed{\omega_A^- = \omega_A^+ \quad \text{and} \quad \omega_B^- = \omega_B^+,} \tag{8.49}$$

that is, the angular velocities of A and B are conserved through the impact.

Because there are no impulsive forces in the direction perpendicular to the LOI, the components of velocity in that direction do not change through the impact. Thus, we have $v_{Cy}^- = v_{Cy}^+ = 0$ and $v_{Dy}^- = v_{Dy}^+ = 0$, and so the only remaining unknowns are v_{Cx}^+ and v_{Dx}^+. To find these unknowns, we enforce the conservation of the system's linear momentum along the LOI, i.e.,

$$m_A v_{Cx}^- + m_B v_{Dx}^- = m_A v_{Cx}^+ + m_B v_{Dx}^+, \tag{8.50}$$

and the COR equation *for the contacting points E and Q*, that is,

$$v_{Ex}^+ - v_{Qx}^+ = e\left(v_{Qx}^- - v_{Ex}^-\right). \tag{8.51}$$

Using rigid body kinematics to relate the motion of points E and Q to the mass centers on their respective bodies, we obtain

$$\vec{v}_E^{\pm} = \vec{v}_C^{\pm} + \omega_A^{\pm}\,\hat{k} \times \vec{r}_{E/C} \quad \text{and} \quad \vec{v}_Q^{\pm} = \vec{v}_D^{\pm} + \omega_B^{\pm}\,\hat{k} \times \vec{r}_{Q/D}, \tag{8.52}$$

where we have used the fact that $\vec{r}_{E/C}$ and $\vec{r}_{Q/D}$ do not change during the impact. Figure 8.24 shows that $\vec{r}_{E/C} = R\,\hat{\imath}$ and $\vec{r}_{Q/D} = -R\,\hat{\imath}$, and so

$$\omega_A^{\pm}\,\hat{k} \times \vec{r}_{E/C} = \omega_A^{\pm} R\,\hat{\jmath} \quad \text{and} \quad \omega_B^{\pm}\,\hat{k} \times \vec{r}_{Q/D} = -\omega_B^{\pm} R\,\hat{\jmath}. \tag{8.53}$$

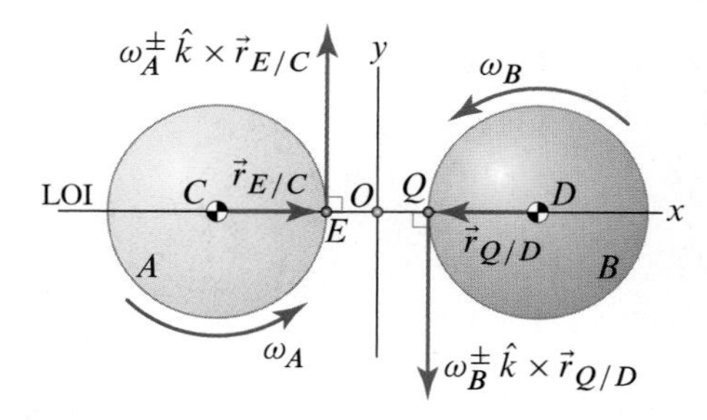

Figure 8.24
Kinematics of a central impact.

Comparing Eqs. (8.52) and (8.53), we see that the x components, i.e., those along the LOI, of $\vec{v}_E^{\pm}$ and $\vec{v}_Q^{\pm}$ must be identical to the x components of $\vec{v}_C^{\pm}$ and $\vec{v}_D^{\pm}$, respectively, i.e.,

$$v_{Ex}^{\pm} = v_{Cx}^{\pm} \quad \text{and} \quad v_{Qx}^{\pm} = v_{Dx}^{\pm}. \tag{8.54}$$

Equations (8.54) allow us to write the COR equation directly in terms of the velocity of the mass centers, i.e.,

$$\boxed{v_{Cx}^+ - v_{Dx}^+ = e\left(v_{Dx}^- - v_{Cx}^-\right).} \tag{8.55}$$

Concept Alert

Central impacts and COR equation. In a central impact the COR equation can be written directly in terms of the velocity components of the mass centers.

Going back to the billiard ball impact problem, recalling the given preimpact conditions and that $e = 1$, we obtain the following solution:

$$v^+_{Cx} = -v_0, \quad \omega^+_A = 0, \quad v^+_{Dx} = 0, \quad \text{and} \quad \omega^+_B = v_0/R. \tag{8.56}$$

Even though $v^+_{Dx} = 0$, because $\omega^+_B = v_0/R$, ball B does *not* stop after impact. The equations $v^+_{Dx} = 0$ and $\omega^+_B = v_0/R$ taken together imply that, right after impact, the mass center of ball B has zero velocity for an instant, while ball B slips against the pool table. With friction between the table and balls, the friction force due to sliding will cause the center of mass of B to start moving again to the left, as if to chase ball A. Had we modeled the balls as particles, we would have concluded that B simply stopped after impact.

Eccentric impact

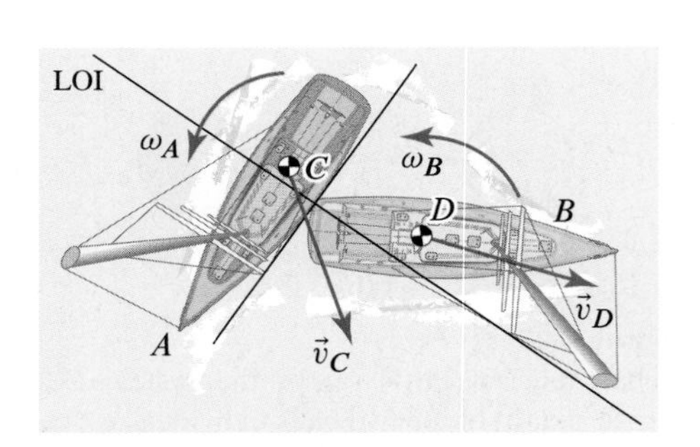

Figure 8.25
Collision between two boats.

Figure 8.25 shows an eccentric impact between two boats. It is *eccentric* because at least one of the mass centers of the colliding bodies is not on the LOI. An eccentric impact not only affects the velocities of the mass centers, but it also affects the bodies' angular velocities.

For the two boats colliding in Fig. 8.25, as usual, we start the solution with the FBD of the system and the FBDs of the individual bodies at the time of impact, which are shown in Fig. 8.26(a) and (b), respectively. On these diagrams we clearly indicate

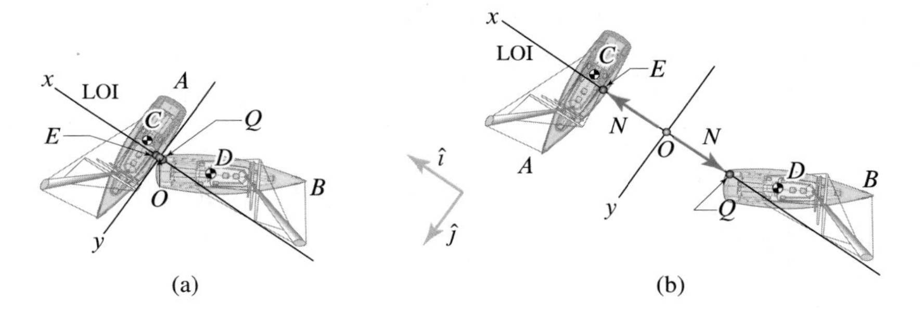

Figure 8.26. FBDs of the colliding bodies (a) as a system and (b) individually.

the LOI and the chosen coordinate system. In addition, we choose the origin O so as to coincide with the points of contact E and Q between the two rigid bodies at the time of impact, and we keep in mind that O is a *fixed* point. The reason for choosing O as stated is that it is a convenient point to use as the moment center when enforcing the angular impulse-momentum principle.

The solution of an unconstrained rigid body eccentric impact problem is governed by six scalar equations. The first of these equations enforces the conservation of linear momentum for the system along the LOI, that is,

$$m_A v^-_{Cx} + m_B v^-_{Dx} = m_A v^+_{Cx} + m_B v^+_{Dx}. \tag{8.57}$$

The second and third of these equations are

$$v^-_{Cy} = v^+_{Cy} \quad \text{and} \quad v^-_{Dy} = v^+_{Dy}, \tag{8.58}$$

which follow from the frictionless contact assumption (assumption 3 on p. 640). This assumption implies that the linear momentum of each colliding body is conserved in the direction perpendicular to the LOI.

The fourth equation is the COR equation, which is first written in terms of the components of the velocities of the contact points along the LOI, i.e.,

$$v_{Ex}^{+} - v_{Qx}^{+} = e\left(v_{Qx}^{-} - v_{Ex}^{-}\right). \tag{8.59}$$

It is then rewritten in terms of the velocity components of the mass centers, making sure to satisfy rigid body kinematics, which requires that

$$\vec{v}_E^{\pm} = \vec{v}_C^{\pm} + \omega_A^{\pm}\,\hat{k} \times \vec{r}_{E/C} \quad \text{and} \quad \vec{v}_Q^{\pm} = \vec{v}_D^{\pm} + \omega_B^{\pm}\,\hat{k} \times \vec{r}_{Q/D}. \tag{8.60}$$

Referring to Fig. 8.27, because E, Q, and O coincide at the time of impact,

$$\vec{r}_{E/C} = -\vec{r}_C = -x_C\,\hat{\imath} - y_C\,\hat{\jmath} \quad \text{and} \quad \vec{r}_{Q/D} = -\vec{r}_D = -x_D\,\hat{\imath} - y_D\,\hat{\jmath}. \tag{8.61}$$

Then, substituting Eqs. (8.61) into Eqs. (8.60), carrying out the cross products, and rearranging terms, we obtain

$$\vec{v}_E^{\pm} = \left(v_{Cx}^{\pm} + \omega_A^{\pm} y_C\right)\hat{\imath} + \left(v_{Cy}^{\pm} - \omega_A^{\pm} x_C\right)\hat{\jmath}, \tag{8.62}$$

$$\vec{v}_Q^{\pm} = \left(v_{Dx}^{\pm} + \omega_B^{\pm} y_D\right)\hat{\imath} + \left(v_{Dy}^{\pm} - \omega_B^{\pm} x_D\right)\hat{\jmath}. \tag{8.63}$$

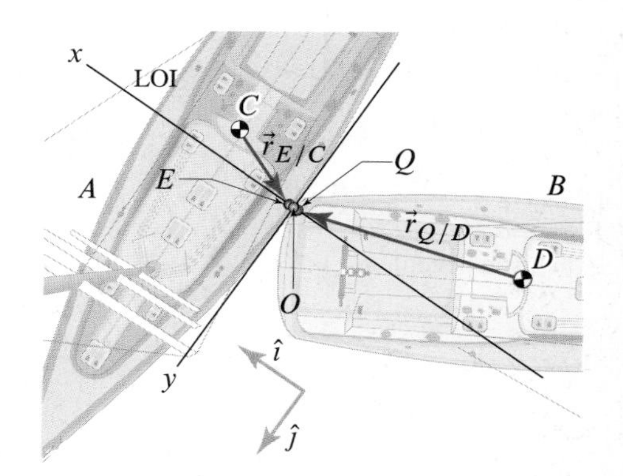

Figure 8.27
Relative position vectors of E and Q with respect to C and D, respectively.

Using Eqs. (8.62) and (8.63), Eq. (8.59) can be written as:

$$v_{Cx}^{+} + \omega_A^{+} y_C - v_{Dx}^{+} - \omega_B^{+} y_D = e\left(v_{Dx}^{-} + \omega_B^{-} y_D - v_{Cx}^{-} - \omega_A^{-} y_C\right). \tag{8.64}$$

Since the line of action of the contact forces between A and B goes through point O, the remaining two equations say that each of the angular momenta of A and B relative to O is conserved, that is,

$$\left(\vec{h}_O^{-}\right)_A = \left(\vec{h}_O^{+}\right)_A \quad \text{and} \quad \left(\vec{h}_O^{-}\right)_B = \left(\vec{h}_O^{+}\right)_B, \tag{8.65}$$

where $\left(\vec{h}_O\right)_A$ and $\left(\vec{h}_O\right)_B$ are the angular momenta of A and B relative to O, respectively. Applying Eq. (8.40) on p. 619, $\left(\vec{h}_O^{\pm}\right)_A$ and $\left(\vec{h}_O^{\pm}\right)_B$ can be written as

$$\left(\vec{h}_O^{\pm}\right)_A = I_C\vec{\omega}_A^{\pm} + \vec{r}_{C/O} \times m_A\vec{v}_C^{\pm}, \tag{8.66}$$

$$\left(\vec{h}_O^{\pm}\right)_B = I_D\vec{\omega}_B^{\pm} + \vec{r}_{D/O} \times m_B\vec{v}_D^{\pm}, \tag{8.67}$$

where I_C and I_D are the mass moments of inertia of A and B, respectively. In Eqs. (8.66) and (8.67), the relative position vectors $\vec{r}_{C/O}$ and $\vec{r}_{D/O}$ are assumed to be constant through the impact, and therefore do not need the superscript $\pm$. Referring to Fig. 8.26(b), the physical justification for Eqs. (8.65) is that the impulsive forces acting on A and B provide no moment about the fixed point O. Observe that Eqs. (8.65) yield only two scalar equations because the only nonzero component of these equations is perpendicular to the plane of motion.

Equations (8.57), (8.58), (8.64), and (8.65) are all that is needed to solve the most general case of unconstrained oblique eccentric impact of two rigid bodies (under the assumptions stated on p. 640).

Concept Alert

Conservation of angular momentum does not mean conservation of angular velocity. Equations (8.66) and (8.67) make clear that conservation of angular momentum does *not* imply that the angular velocity is conserved. Equations (8.65) demand that the left-hand sides of Eqs. (8.66) and (8.67) remain constant (through the impact), and this is possible if the corresponding angular velocities and mass center velocities on the right-hand sides of Eqs. (8.66) and (8.67) change in concert.

Constrained eccentric impact

In a *constrained impact* one or both of the colliding bodies are subject to external impulsive forces (see Example 5.9, p. 348). Modeling these impacts can be challenging, and we consider only a simple case in which one of the impacting bodies is constrained to move in fixed-axis rotation.

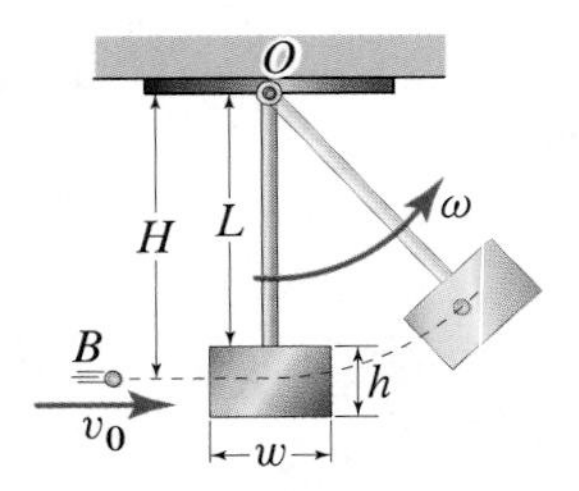

Figure 8.28
A ballistic pendulum.

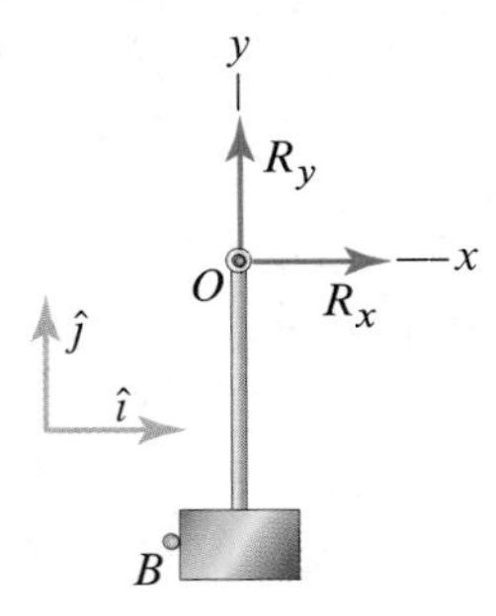

Figure 8.29
FBD of the ballistic pendulum and bullet system at the time of impact.

Referring to Fig. 8.28, consider a ballistic pendulum consisting of a uniform thin rod of mass m_r and length L pinned at O and a target block of mass m_t, width w, and height h. Suppose that a bullet of mass m_b is fired horizontally with a speed v_0 as shown. Assuming that the bullet becomes embedded in the block, we wish to determine the postimpact velocity of the pendulum-bullet system.

Since the only possible motion of the pendulum is a fixed-axis rotation about O, the only piece of information we need to describe the system's postimpact behavior is ω_p^+, the postimpact angular velocity of the pendulum. As usual, we sketch the system's FBD at the time of impact, given in Fig. 8.29, making sure to include only impulsive forces. The pin reactions R_x and R_y appear on the FBD because they will take on whatever value is required of them to keep O from moving, and, as such, they are impulsive. The presence of R_x and R_y makes the impact a *constrained* impact. Observe that R_x and R_y, while impulsive, provide no moment about the *fixed* point O, so the system's total angular momentum relative to O must be conserved through the impact. Calling $(\vec{h}_O)_b$ and $(\vec{h}_O)_p$ the angular momenta relative to O of the bullet and the pendulum, respectively, we have

$$(\vec{h}_O^-)_b + (\vec{h}_O^-)_p = (\vec{h}_O^+)_b + (\vec{h}_O^+)_p, \tag{8.68}$$

where, modeling the bullet as a particle and recalling that the pendulum can move only in a fixed-axis rotation about O (see Eq. (8.44) on p. 620),

$$(\vec{h}_O^\pm)_b = \vec{r}_{B/O} \times m_b \vec{v}_B^\pm \quad \text{and} \quad (\vec{h}_O^\pm)_p = (I_O)_p \vec{\omega}_p^\pm, \tag{8.69}$$

where $(I_O)_p$ is the mass moment of inertia of the pendulum relative to O and where the relative position vector $\vec{r}_{B/O}$ does not have the superscript $\pm$ because it is treated as a constant through the impact. Since the pendulum is initially stationary and the bullet becomes embedded in the block, substituting Eqs. (8.69) into Eq. (8.68), carrying out the cross products, and simplifying gives

$$m_b v_0 H = \left[(I_O)_p + m_b H^2\right]\omega_p^+. \tag{8.70}$$

Solving Eq. (8.70) for the postimpact angular velocity of the pendulum-bullet system, we have

$$\omega_p^+ = \frac{m_b v_0 H}{(I_O)_p + m_b H^2}. \tag{8.71}$$

The problem just discussed illustrates a key element of the solution of most constrained impact problems, namely, the identification of a fixed point about which the system's angular momentum is conserved. Examples 8.12 and 8.13 demonstrate the use of this strategy in the case of more involved physical situations.

End of Section Summary

In this section we studied planar rigid body impacts. We learned that, contrary to particle impacts, rigid bodies can experience *eccentric* impacts. These are collisions in which at least one of the mass centers of the impacting bodies does not lie on the LOI. We also learned that the basic concepts used in particle impacts are applicable to rigid body impacts, and we reviewed solution strategies for a variety of situations.

As with particle impacts, we say that a rigid body impact is *plastic* if the COR $e = 0$. A rigid body impact will be called *perfectly plastic* if the colliding bodies form a single rigid body after impact. An impact is *elastic* if the COR e is such that $0 < e < 1$. Finally, the ideal case with $e = 1$ is referred to as a *perfectly elastic* impact. In addition, an impact is *unconstrained* if the system consisting of the two impacting objects is not subject to external impulsive forces; otherwise, the impact is called *constrained*.

We have considered only impacts satisfying the following assumptions:

1. The impact involves only two bodies in planar motion where no impulsive force has a component perpendicular to the plane of motion.

2. Contact between any two rigid bodies occurs at only one point, and at this point we can clearly define the LOI.

3. The contact between the bodies is frictionless.

We recommend organizing the solution of any impact problem as follows:

- Begin with an FBD of the impacting bodies as a system and FBDs for each of the colliding bodies. Neglect nonimpulsive forces.

- Choose a coordinate system with the origin coincident with the points that come into contact at the time of impact. Recall that the origin of such a coordinate system is a fixed point.

- Enforce the linear and/or the angular impulse-momentum principles for the system and/or for the individual bodies. In applying the angular impulse-momentum principle for the whole system, the moment center should be a fixed point, whereas for an individual body, the moment center should be a fixed point or the body's mass center.

- For plastic, elastic, and perfectly elastic impacts, the COR equation is first written using the velocity components along the LOI of the points that actually come into contact. For example, for the impact shown in Fig. 8.30, the COR equation is first written as

Eq. (8.59), p. 643

$$v_{Ex}^{+} - v_{Qx}^{+} = e\left(v_{Qx}^{-} - v_{Ex}^{-}\right),$$

where the COR e is such that $0 \leq e \leq 1$. The COR equation must then be rewritten in terms of the colliding bodies' angular velocities and velocities of the mass centers using rigid body kinematics. For the situation in Fig. 8.30, this means rewriting the COR equation using the relations $\vec{v}_E^{\pm} = \vec{v}_C^{\pm} + \omega_A^{\pm}\,\hat{k} \times \vec{r}_{E/C}$ and $\vec{v}_Q^{\pm} = \vec{v}_D^{\pm} + \omega_B^{\pm}\,\hat{k} \times \vec{r}_{Q/D}$. Notice that the relative position vectors $\vec{r}_{E/C}$ and $\vec{r}_{Q/D}$ do not have the $\pm$ superscript because they are treated as constants during the impact.

- In perfectly plastic impacts, kinematic constraint equations must be enforced that express the fact that two bodies form a single rigid body after impact.

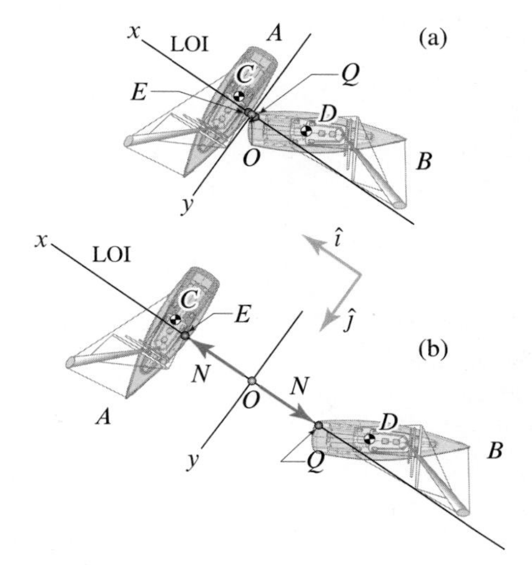

Figure 8.30
FBDs of the colliding bodies (a) as a system and (b) individually.

EXAMPLE 8.11 *Rigid Body Central Impact*

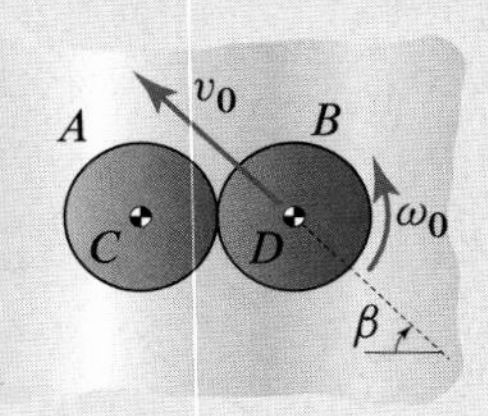

Figure 1
Two colliding hockey pucks.

Two identical hockey pucks sliding over ice collide as shown in Fig. 1. If $\beta = 43°$, the COR is $e = 0.95$, A is initially at rest, and puck B is moving with speed $v_0 = 30\,\text{m/s}$ and angular speed $\omega_0 = 4\,\text{rad/s}$ as shown, determine the postimpact velocities of A and B.

SOLUTION

Road Map & Modeling We start with the FBD of the system and the FBDs of the colliding bodies separately, shown in Figs. 2 and 3, respectively, where we also indicate the LOI and a convenient coordinate system. Because C and D lie on the LOI, the impact is *central*. As discussed earlier in the section, we assume that the contact between the pucks is frictionless. This assumption and the fact that the impact is central allow us to immediately state that the angular velocities of the bodies are unaffected by the impact and that the COR equation can be expressed directly in terms of velocity components of the mass centers. These simplifications make the problem solution very similar to that of a particle impact problem.

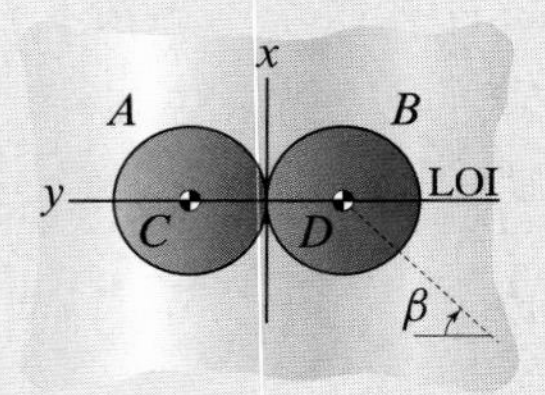

Figure 2
Impact-relevant FBD of the two pucks as a system.

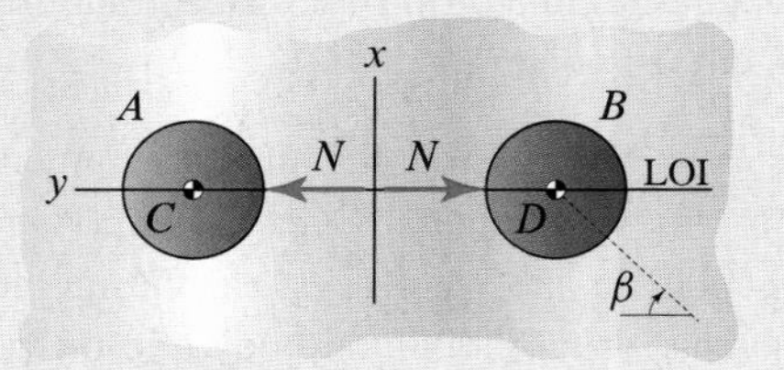

Figure 3. Impact-relevant FBDs of each of two pucks.

Governing Equations

Balance Principles There are no external impulsive forces acting on the system (Fig. 2). Therefore, the system's linear momentum is conserved along the LOI, which gives

$$m_A v_{Cy}^- + m_B v_{Dy}^- = m_A v_{Cy}^+ + m_B v_{Dy}^+. \tag{1}$$

No impulsive force acts on A or B in the direction perpendicular to the LOI (Fig. 3). Therefore, we can write

$$v_{Cx}^- = v_{Cx}^+ \quad \text{and} \quad v_{Dx}^- = v_{Dx}^+. \tag{2}$$

Since the impact is central (and the contact between the pucks is frictionless), the angular impulse-momentum principle for A and B individually yields the following two equations:

$$\omega_A^+ = \omega_A^- \quad \text{and} \quad \omega_B^+ = \omega_B^-. \tag{3}$$

Force Laws The COR equation can be expressed directly in terms of the velocity components along the LOI of the mass centers, so we have

$$v_{Cy}^+ - v_{Dy}^+ = e\big(v_{Dy}^- - v_{Cy}^-\big). \tag{4}$$

Kinematic Equations The preimpact velocities of A and B are

$$\omega_A^- = 0, \qquad v_{Cx}^- = 0, \qquad v_{Cy}^- = 0, \tag{5}$$

$$\omega_B^- = \omega_0, \qquad v_{Dx}^- = v_0 \sin\beta, \qquad v_{Dy}^- = v_0 \cos\beta. \tag{6}$$

Computation Equations (1)–(4) form a system of six equations in the six unknowns ω_A^+, v_{Cx}^+, v_{Cy}^+, ω_B^+, v_{Dx}^+, and v_{Dy}^+. Recalling that $m_A = m_B$, this system of equations can be solved to obtain

$$\omega_A^+ = 0, \qquad v_{Cx}^+ = 0, \qquad v_{Cy}^+ = \frac{1+e}{2} v_0 \cos\beta, \tag{7}$$

$$\omega_B^+ = \omega_0, \qquad v_{Dx}^+ = v_0 \sin\beta, \qquad v_{Dy}^+ = \frac{1-e}{2} v_0 \cos\beta, \tag{8}$$

which, upon substitution of the given data, yields the following numerical answer:

$$\omega_A^+ = 0\,\text{rad/s}, \qquad v_{Cx}^+ = 0\,\text{m/s}, \qquad v_{Cy}^+ = 21.39\,\text{m/s}, \tag{9}$$

$$\omega_B^+ = 4\,\text{rad/s}, \qquad v_{Dx}^+ = 20.46\,\text{m/s}, \qquad v_{Dy}^+ = 0.5485\,\text{m/s}. \tag{10}$$

Discussion & Verification The solution appears to be reasonable in that, as expected, the center of mass of puck A moves only along the LOI: this is the expected behavior of any impact in which there is no friction between the colliding bodies. Since the impact is central (and the contact between the pucks is frictionless), the angular velocities of the colliding bodies are conserved. In addition to these considerations, we should check that the postimpact kinetic energy of the system is smaller than the preimpact kinetic energy since the COR was less than 1. Going through this verification is a bit involved in this problem because neither the masses of A and B nor their respective mass moments of inertia, I_C and I_D, were given. Using Eqs. (5) and (6), the preimpact kinetic energy T^- is

$$\begin{aligned} T^- &= \tfrac{1}{2} m_A \left(v_C^-\right)^2 + \tfrac{1}{2} I_C \left(\omega_A^-\right)^2 + \tfrac{1}{2} m_B \left(v_D^-\right)^2 + \tfrac{1}{2} I_D \left(\omega_B^-\right)^2 \\ &= \tfrac{1}{2} m_B v_0^2 + \tfrac{1}{2} I_D \omega_0^2. \end{aligned} \tag{11}$$

Using Eqs. (7) and (8), the postimpact kinetic energy T^+ is

$$\begin{aligned} T^+ &= \tfrac{1}{2} m_A \left(v_C^+\right)^2 + \tfrac{1}{2} I_C \left(\omega_A^+\right)^2 + \tfrac{1}{2} m_B \left(v_D^+\right)^2 + \tfrac{1}{2} I_D \left(\omega_B^+\right)^2 \\ &= \tfrac{1}{8} m_A (1+e)^2 v_0^2 \cos^2\beta \\ &\quad + \tfrac{1}{8} m_B \left[4\sin^2\beta + (1-e)^2 \cos^2\beta\right] v_0^2 + \tfrac{1}{2} I_D \omega_0^2. \end{aligned} \tag{12}$$

Since the two pucks are identical, $I_C = I_D$. Letting $m = m_A = m_B$, subtracting Eq. (12) from Eq. (11), and simplifying, we have

$$T^- - T^+ = \tfrac{1}{4}\left(1 - e^2\right) m v_0^2 \cos^2\beta, \tag{13}$$

where we have used the fact that $\frac{1}{2} m_B v_0^2 - \frac{1}{2} m_B v_0^2 \sin^2\beta = \frac{1}{2} m v_0^2 \cos^2\beta$. Finally, since $e < 1$, we have $1 - e^2 > 0$, and consequently the right-hand side of Eq. (13) is positive, i.e.,

$$T^- - T^+ > 0 \quad \Rightarrow \quad T^+ < T^-, \tag{14}$$

as expected.

EXAMPLE 8.12 *Constrained Impact of Two Rigid Bodies*

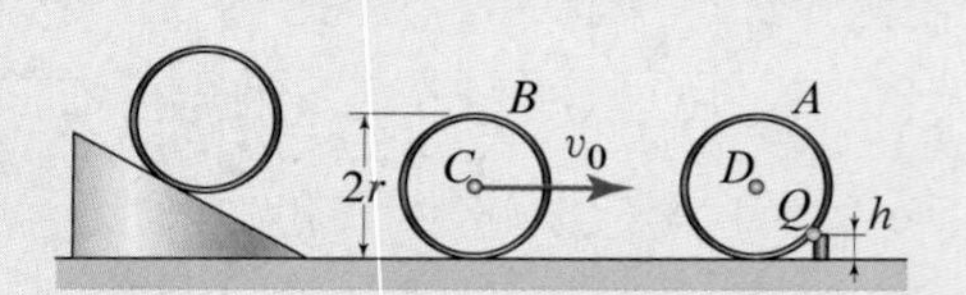

Figure 1
Pipe sections being placed horizontally. Points C and D are the centers of B and A, respectively. Point Q is the point on A in contact with the block of height h.

Identical pipe sections of radius $r = 1.5$ ft and weight $W = 200$ lb are being unloaded and aligned horizontally. A block of height $h = 6$ in. is fixed to the ground and used to hold in place A, the first pipe in the row (see Fig. 1). If the next pipe section B rolls without slip to the right with a speed v_0 and bumps into A, the COR for the A-B collision is $e = 0.85$, and A does not rebound off the block or slide relative to it, then determine the smallest value of v_0 that would make A roll over the block.

SOLUTION

Road Map The solution can be organized in two parts. In the first part, we will study the collision of A and B and determine the postimpact motion of A and B. Once the motion of A is described as a function of v_0, in the second part we will find the *smallest* value of v_0 that makes A roll over the block by relating the postimpact kinetic energy of A to the amount of work needed to move A by a vertical distance h against the direction of gravity.

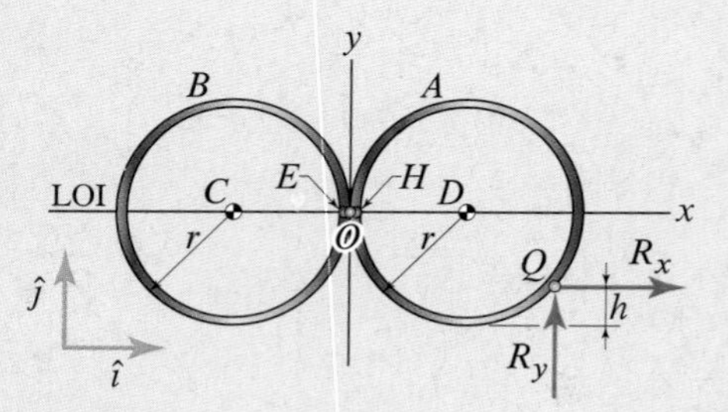

Figure 2
Impact-relevant FBD of A and B combined at the time of impact.

The Impact of A and B

Modeling The FBD of the system is shown in Fig. 2, and the FBDs of each colliding body are shown in Fig. 3. At the time of impact, the origin of the coordinate system O, which is a fixed point, coincides with the contact points E and H, belonging to B and A, respectively. Notice that we are dealing with a constrained impact due to the presence of external impulsive forces at Q. These forces exist because A is initially in contact with a fixed block and does not slide relative to it. This also means that A moves as if in a fixed-axis rotation about Q.

Governing Equations

Balance Principles The external impulsive forces in Fig. 2 provide no moment about the fixed point Q, and therefore the system's angular momentum about Q is conserved,

$$(\vec{h}_Q^-)_A + (\vec{h}_Q^-)_B = (\vec{h}_Q^+)_A + (\vec{h}_Q^+)_B, \tag{1}$$

where, viewing both pipe sections as uniform thin rings of radius r,

$$(\vec{h}_Q^\pm)_A = I_Q \vec{\omega}_A^\pm, \quad \text{with} \quad I_Q = mr^2 + mr^2 = 2mr^2, \tag{2}$$

$$(\vec{h}_Q^\pm)_B = I_C \vec{\omega}_B^\pm + \vec{r}_{C/Q} \times m\vec{v}_C^\pm, \quad \text{with} \quad I_C = mr^2. \tag{3}$$

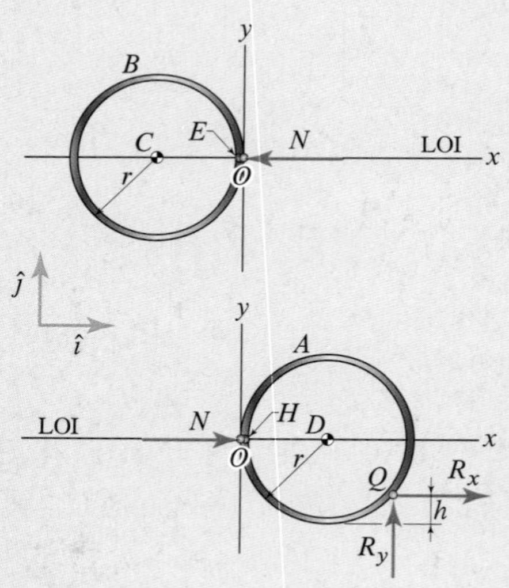

Figure 3
Impact-relevant FBD of A and B individually at the time of impact.

I_Q and I_C are the mass moments of inertia of A relative to Q and of B relative to C, respectively. Note that, in calculating I_Q, we used the parallel axis theorem. The first of Eqs. (2) holds because A moves as if pinned at Q.

The impulsive force N, which acts on B, points toward C. Therefore, since m (the mass of B) and I_C are constant, the y component of the linear momentum of B, as well as the angular momentum of B relative to C are conserved, that is,

$$v_{Cy}^- = v_{Cy}^+ \quad \text{and} \quad \omega_B^- = \omega_B^+. \tag{4}$$

Note that the impulsive force system acting on A does *not* allow us to write relations for A analogous to those in Eqs. (4) for B.

Force Laws Since the contact points between A and B are E and H, the COR equation reads

$$v_{Hx}^+ - v_{Ex}^+ = e(v_{Ex}^- - v_{Hx}^-). \tag{5}$$

Kinematic Equations Before impact, A is stationary and B rolls without slip, so

$$v_{Hx}^- = 0, \quad \omega_A^- = 0, \quad v_{Cy}^- = 0, \quad \text{and} \quad \omega_B^- = -v_0/r. \tag{6}$$

To express $v_{Ex}^{\pm}$ and v_{Hx}^{+} in Eq. (5) in terms of $\vec{v}_C^{+}$, ω_B^{+}, and ω_A^{+}, recall that $\vec{v}_E = \vec{v}_C + \vec{\omega}_B \times \vec{r}_{E/C}$ and $\vec{v}_H = \vec{\omega}_A \times \vec{r}_{H/Q}$ (A rotates about Q). Consequently, we have

$$v_{Ex}^{-} = v_0, \quad v_{Ex}^{+} = v_{Cx}^{+}, \quad \text{and} \quad v_{Hx}^{+} = -\omega_A^{+}(r-h). \tag{7}$$

Computation Substituting Eqs. (2) and (3) into Eq. (1), taking advantage of Eqs. (4) and the last two of Eqs. (6), expanding the cross products, and simplifying, we have

$$-v_0(r-h) = -v_{Cx}^{+}(r-h) + 2r^2\omega_A^{+}, \tag{8}$$

where we have written this equation in scalar form because the only nonzero component of Eq. (1) is in the z direction. Substituting the first of Eqs. (6) and all of Eqs. (7) into Eq. (5), we have

$$-\omega_A^{+}(r-h) - v_{Cx}^{+} = ev_0. \tag{9}$$

Equations (8) and (9) form a system of two equations for v_{Cx}^{+} and ω_A^{+}. Solving these equations, we get the postimpact motion of A and B in terms of v_0:

$$\omega_A^{+} = -\frac{(1+e)(r-h)v_0}{(r-h)^2 + 2r^2} \quad \text{and} \quad v_{Cx}^{+} = \frac{h^2 - 2hr + (1-2e)r^2}{(r-h)^2 + 2r^2}v_0. \tag{10}$$

The Work-Energy Principle Applied to A

Modeling After the impact, A rolls over the block as if pinned at Q, and A's FBD prior to reaching the top of the block is that in Fig. 4. Since Q is fixed, the reactions at Q do no work, so energy is conserved during the postimpact motion of A.

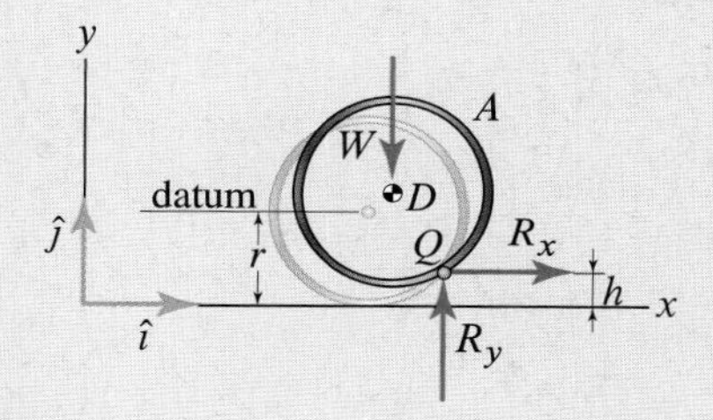

Figure 4
FBD of A following the impact with B.

Governing Equations

Balance Principles Choosing ① to be immediately after impact and ② when A gets to the top of the block with zero velocity (because we want to compute the *minimum* v_0), we have

$$T_{A1} + V_{A1} = T_{A2} + V_{A2}, \quad \text{where} \quad T_{A1} = \tfrac{1}{2}I_Q\omega_{A1}^2 \quad \text{and} \quad T_{A2} = \tfrac{1}{2}I_Q\omega_{A2}^2, \tag{11}$$

and where $I_Q = 2(W/g)r^2$.

Force Laws Choosing the datum line as shown in Fig. 4, we have

$$V_{A1} = 0 \quad \text{and} \quad V_{A2} = Wh. \tag{12}$$

Kinematic Equations Based on our modeling assumptions, we have

$$\omega_{A1} = \omega_A^{+} \quad \text{and} \quad \omega_{A2} = 0. \tag{13}$$

Computation Combining Eqs. (11)–(13), recalling that ω_A^{+} is given by the first of Eqs. (10), and solving for v_0, we have

$$\boxed{v_0 = \frac{\sqrt{gh}\left[(r-h)^2 + 2r^2\right]}{r(1+e)(r-h)} = 7.953\,\text{ft/s}.} \tag{14}$$

Discussion & Verification The solution seems reasonable since the computed v_0, along with Eqs. (10), gives $\omega_A^{+} = -2.675\,\text{rad/s}$ and $v_{Cx}^{+} = -4.085\,\text{ft/s}$, i.e., after impact A rotates clockwise and B moves to the left, as one would expect. In fact, the first of Eqs. (10) indicates that the rotation will always be clockwise, as long as $r > h$. Furthermore, the value of v_0 was expected to be larger than $\sqrt{gh} = 4.012\,\text{ft/s}$, which is the v_0 needed by B to get to a height equal to h off the ground by simply rolling without slip along any path and without any collisions. Finally, using the computed value of v_0 to compute the pre- and postimpact values of the system's kinetic energy, we have $T^{-} = 392.8\,\text{ft·lb} > T^{+} = 348.2\,\text{ft·lb}$, as it should.

EXAMPLE 8.13 *Modeling a Catch as a Rigid Body Impact*

Figure 1
A midair catch. The acrobat hanging at the knee from the trapeze is called the *catcher*, and the other acrobat is called the *flyer*.

Using the model shown in Fig. 2, what is the average force the catcher needs to apply to catch the flyer if the catcher is in an ideal position (i.e., has zero velocity) and the flyer has a free-fall speed $v_0 = 2\,\text{m/s}$ (after dropping about 20 cm)? In this model, the catcher/trapeze system is viewed as the nonuniform slender bar of mass $m_c = 90\,\text{kg}$, length $L = 4\,\text{m}$, mass center at E, mass moment of inertia $I_E = 30\,\text{kg}\cdot\text{m}^2$, and $w = 1\,\text{m}$. The flyer is modeled as a uniform slender bar of length $H = 2\,\text{m}$ and mass $m_f = 70\,\text{kg}$. We model the grasp between catcher and flyer as a pin connection and assume that, at the instant shown, catcher and flyer both have zero angular velocity. Finally, we assume that catcher and flyer can establish a firm grasp in 0.15 s (the best athletes' reaction times are 120–160 ms), and since the catch happens quickly, we model it as an impact.

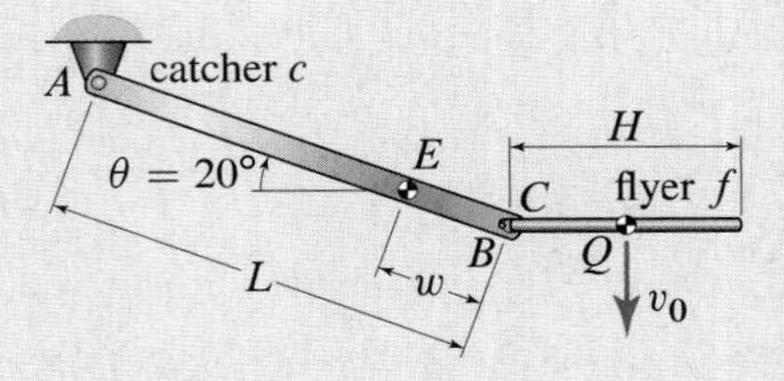

Figure 2. Model of the catch between two trapeze artists. The situation shown is at the time of the catch.

SOLUTION

Road Map & Modeling By modeling the catch as an impact, we can (1) relate the pre- and postcatch velocities using the linear impulse-momentum principle and compute the impulse exerted by c on f, and (2) divide this impulse by the reaction time to compute the desired average force. Note that the impact we are modeling cannot be classified into any of the elastic or plastic categories found on p. 640, since the two bodies attach but can rotate relative to one another. As usual, we consider the system's FBD and the FBDs of c and f individually, shown in Figs. 3 and 4, respectively. At the time of catch, the fixed origin O of the coordinate system coincides with points B and C at which c and f grasp one another. Since c can only move in a fixed-axis rotation about A, the postcatch velocities consist of four quantities: the angular velocities of c and f and the two components of $\vec{v}_Q$.

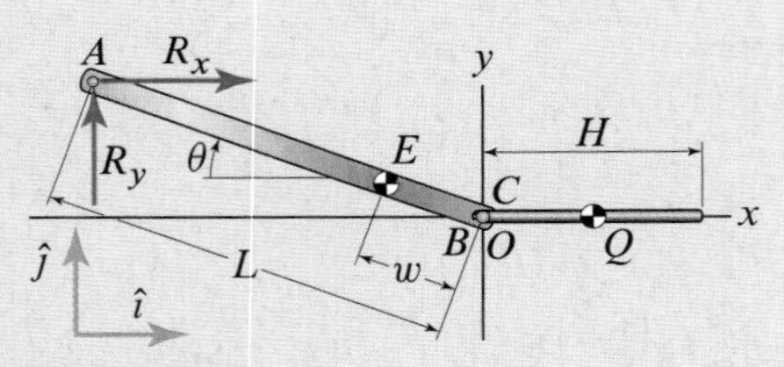

Figure 3
Impact-relevant FBD of the catcher-flyer system. Points B, C, and O are coincident.

Governing Equations

Balance Principles Referring to Fig. 4 and to Eq. (8.34) on p. 618, applying the linear impulse-momentum principle to f gives

$$m_f \vec{v}_Q^- + \int_{t^-}^{t^+} (N_x\,\hat{\imath} + N_y\,\hat{\jmath})\,dt = m_f \vec{v}_Q^+. \tag{1}$$

The forces R_x and R_y have no moment about the fixed point A (see Fig. 3), and so the system's angular momentum about A is conserved, that is,

$$(\vec{h}_A^-)_c + (\vec{h}_A^-)_f = (\vec{h}_A^+)_c + (\vec{h}_A^+)_f, \tag{2}$$

where

$$(\vec{h}_A^\pm)_c = \left[I_E + m_c(L-w)^2\right]\vec{\omega}_c^\pm \quad \text{and} \quad (\vec{h}_A^\pm)_f = I_Q\vec{\omega}_f^\pm + \vec{r}_{Q/A} \times m_f \vec{v}_Q^\pm. \tag{3}$$

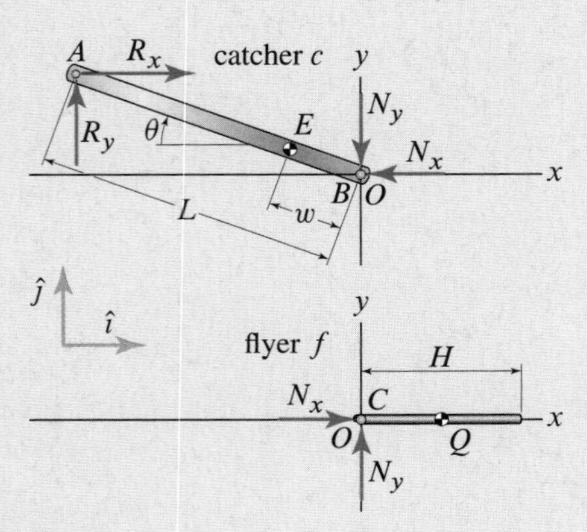

Figure 4
Impact-relevant FBD of c and f individually.

Note that the first of Eqs. (3) accounts for the fixed-axis rotation of c about A. Next, since C and O coincide at the time of catch (see Fig. 4), N_x and N_y provide no moment on f

about O. Thus, the angular momentum of f relative to O, $(\vec{h}_O)_f$, must be conserved, that is,

$$(\vec{h}_O^-)_f = (\vec{h}_O^+)_f \quad \text{where} \quad (\vec{h}_O^\pm)_f = I_Q \vec{\omega}_f^\pm + \vec{r}_{Q/O} \times m_f \vec{v}_Q^\pm. \tag{4}$$

Force Laws All forces are accounted for on the FBD. Also, in this impact we do not have a COR equation.

Kinematic Equations After the catch $\vec{v}_B^+ = \vec{v}_C^+$, so we must have

$$\vec{\omega}_c^+ \times \vec{r}_{B/A} = \vec{v}_Q^+ + \vec{\omega}_f^+ \times \vec{r}_{C/Q}, \tag{5}$$

where Eq. (5) enforces both rigid body kinematics and the fact that A is a fixed point.

Computation Substituting Eqs. (3) into Eq. (2), we have

$$-m_f v_0[L\cos\theta + (H/2)] = \left[I_E + m_c(L-w)^2\right]\omega_c^+ + I_Q\omega_f^+ \\ + m_f v_{Qx}^+ L\sin\theta + m_f v_{Qy}^+[L\cos\theta + (H/2)], \tag{6}$$

where Eq. (6) is in scalar form since the only nonzero component of Eq. (2) is in the z direction. Proceeding similarly with Eqs. (4), we have

$$-m_f v_0(H/2) = I_Q\omega_f^+ + m_f v_{Qy}^+(H/2). \tag{7}$$

Expanding the products in Eq. (5) and expressing the result in components, we have

$$\omega_c^+ L\sin\theta = v_{Qx}^+ \quad \text{and} \quad \omega_c^+ L\cos\theta = v_{Qy}^+ - \omega_f^+(H/2). \tag{8}$$

Equations (6)–(8) form a system of four equations in the four unknowns ω_c^+, ω_f^+, v_{Qx}^+, and v_{Qy}^+. Recalling that $I_Q = m_f H^2/12$, the solution of this system of equations is

$$\omega_c^+ = -\frac{L m_f v_0 \cos\theta}{4[I_E + m_c(L-w)^2] + L^2 m_f(1 + 3\sin^2\theta)} = -0.1080\,\text{rad/s}, \tag{9}$$

$$\omega_f^+ = -\frac{3v_0}{H}\frac{2I_E + 2m_f L^2\sin^2\theta + 2m_c(L-w)^2}{4[I_E + m_c(L-w)^2] + L^2 m_f(1 + 3\sin^2\theta)} = -1.196\,\text{rad/s}, \tag{10}$$

$$v_{Qx}^+ = -\frac{L^2 m_f v_0 \cos\theta\sin\theta}{4[I_E + m_c(L-w)^2] + L^2 m_f(1 + 3\sin^2\theta)} = -0.1477\,\text{m/s}, \tag{11}$$

$$v_{Qy}^+ = -v_0\frac{3I_E + m_f L^2(1 + 2\sin^2\theta) + 3m_c(L-w)^2}{4[I_E + m_c(L-w)^2] + L^2 m_f(1 + 3\sin^2\theta)} = -1.601\,\text{m/s}. \tag{12}$$

Recalling that $t^+ - t^- = 0.15$ s, using the definition of average force over the time interval $t^+ - t^-$, and employing Eq. (1), we have $\vec{N}_{\text{avg}} = \frac{1}{t^+ - t^-}\int_{t^-}^{t^+}(N_x\,\hat{\imath} + N_y\,\hat{\jmath})\,dt = \frac{m_f}{t^+ - t^-}(\vec{v}_Q^+ - \vec{v}_Q^-)$, which gives

$$\boxed{(N_x)_{\text{avg}} = -68.94\,\text{N} \quad \text{and} \quad (N_y)_{\text{avg}} = 186.0\,\text{N} \quad \Rightarrow \quad \left|\vec{N}_{\text{avg}}\right| = 198.3\,\text{N}.} \tag{13}$$

Discussion & Verification Each of the results in Eqs. (9)–(12) has a negative value. This is as expected since it makes physical sense that, after the catch, the angular velocities of both the catcher and the flyer are clockwise, and that the flyer moves down and to the left. In addition, as expected, $v_{Qy}^+ < v_0$, i.e., the catcher slows down the flyer's drop. As far as the forces are concerned, the signs are also as expected, and their value is consistent with the velocity result and the given reaction time.

Problems

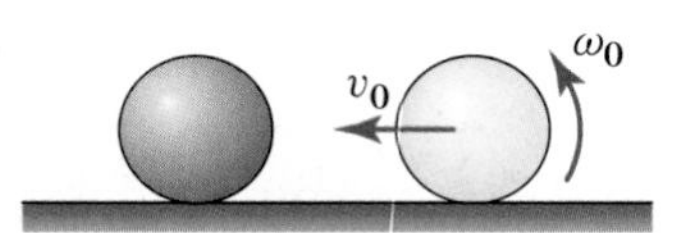

Figure P8.115

Problem 8.115

A *stop shot* is a pool shot in which the cue ball (white) stops upon striking the object ball (aqua). Modeling the collision between the two balls as a perfectly elastic collision of two rigid bodies with frictionless contact, determine which condition must be true for the preimpact angular velocity of the cue ball in order to properly execute a stop shot: (a) $\omega_0 < 0$; (b) $\omega_0 = 0$; (c) $\omega_0 > 0$.

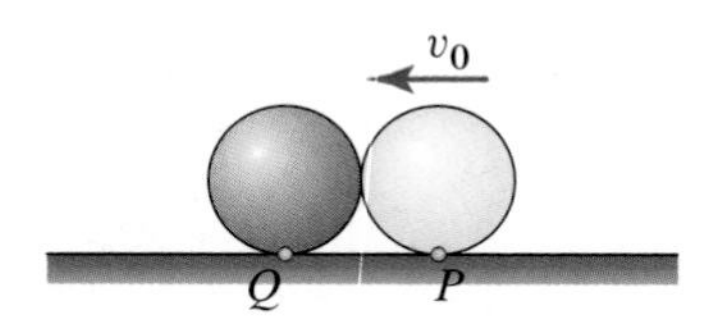

Figure P8.116

Problem 8.116

The cue ball (white) is rolling without slip to the left, and its center is moving with a speed $v_0 = 6\,\text{ft/s}$ while the object ball (aqua) is stationary. The diameter d of the two balls is the same and is equal to 2.25 in. The coefficient of restitution of the impact is $e = 0.98$. Let $W_c = 6\,\text{oz}$ and $W_o = 5.5\,\text{oz}$ be the weights of the cue ball and object ball, respectively. Let P and Q be the points on the cue ball and on the object ball, respectively, that are in contact with the table at the time of impact. Assuming that the contact between the two balls is frictionless and modeling the balls as uniform spheres, determine the postimpact velocities of P and Q.

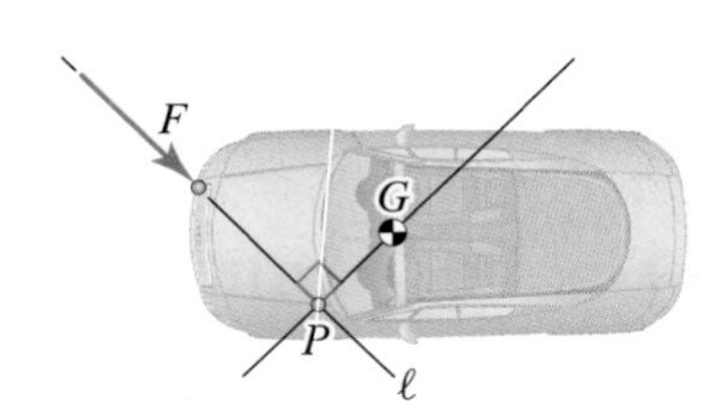

Figure P8.117

Problem 8.117

Consider the impact-relevant FBD of a car involved in a collision. Assume that, at the time of impact, the car was stationary. In addition, assume that the impulsive force F, with line of action ℓ, is the only impulsive force acting on the car at the time of impact. The point P at the intersection of ℓ and the line perpendicular to ℓ and passing through G, the center of mass of the car, is sometimes referred to as the *center of percussion* (for an alternative definition of center of percussion see Example 7.5 on p. 542). Is it true that, at the time of impact, the instantaneous center of rotation of the car lies on the same line as P and G?

Problem 8.118

A basketball with mass $m = 0.6\,\text{kg}$ is rolling without slipping as shown when it hits a small step with $\ell = 7\,\text{cm}$. Letting the ball's diameter be $r = 12.0\,\text{cm}$, modeling the ball as a thin spherical shell (the mass moment of inertia of a spherical shell about its mass center is $\frac{2}{3}mr^2$), and assuming that the ball does not rebound off the step or slip relative to it, determine v_0 such that the ball barely makes it over the step.

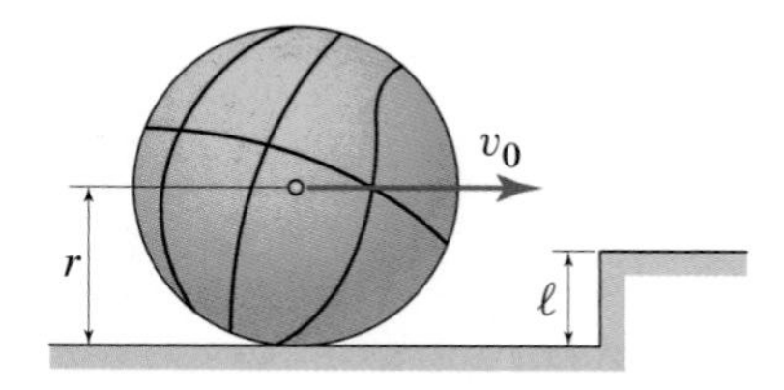

Figure P8.118

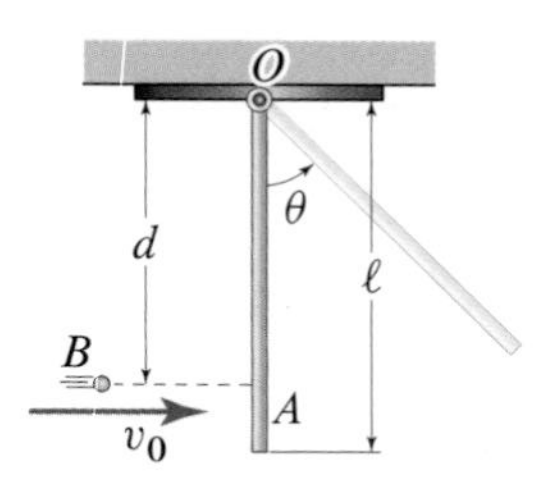

Figure P8.119

Problem 8.119

A bullet B of mass m_b is fired with a speed v_0 as shown against a uniform thin rod A of length ℓ and mass m_r that is pinned at O. Determine the distance d, such that no horizontal reaction is felt at the pin when the bullet strikes the rod.

Problem 8.120

A bullet B weighing 147 gr (1 lb = 7000 gr) is fired with a speed v_0 as shown and becomes embedded in the center of a rubber block of dimensions $h = 4.5$ in. and $w = 6$ in. weighing $W_{rb} = 2$ lb. The rubber block is attached to the end of a uniform thin rod A of length $L = 1.5$ ft and weight $W_r = 5$ lb that is pinned at O. After the impact, the rod (with the block and the bullet embedded in it) swings upward to an angle of 60°. Determine the speed of the bullet right before impact.

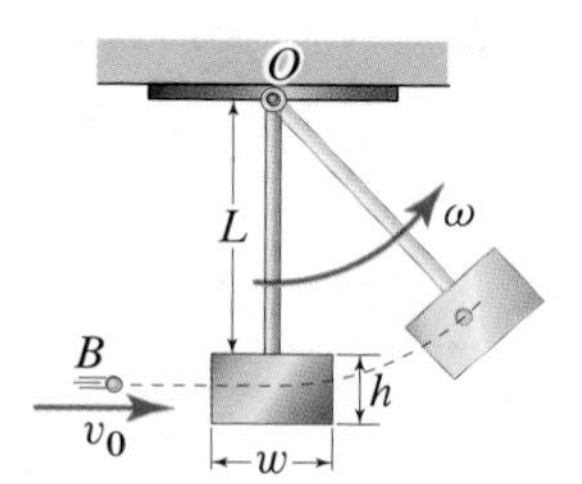

Figure P8.120

Problem 8.121

Solve the problem in Example 7.5 on p. 542 using momentum methods and the concept of impulsive force. Specifically, consider a ball hitting a bat at a distance d from the handle when the batter has "choked up" a distance δ. Find the "sweet spot" P (more properly called the center of percussion*) of the bat B by determining the distance d at which the ball should be hit so that the lateral force (i.e., perpendicular to the bat) at O is zero. Assume that the bat is pinned at O, it has mass m, the mass center is at G, and the mass moment of inertia is I_G.

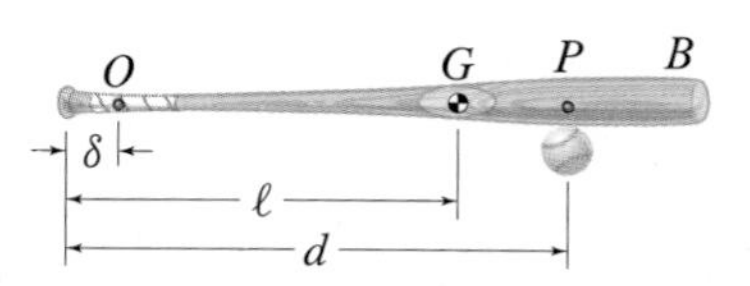

Figure P8.121

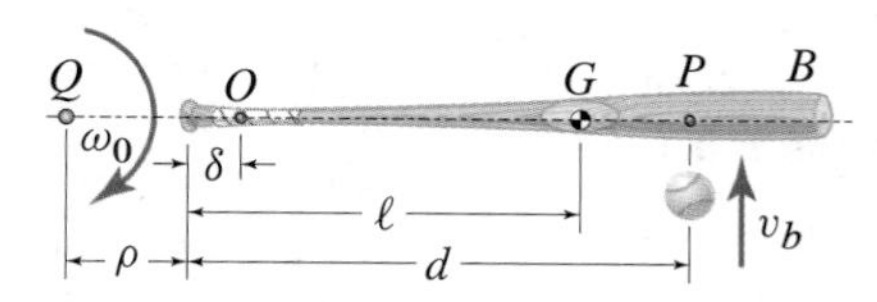

Figure P8.122

Problem 8.122

A batter is swinging a 34 in. long bat with weight $W_B = 32$ oz, mass center G, and mass moment of inertia $I_G = 0.0413\,\text{slug·ft}^2$. The center of rotation of the bat is point Q. Compute the distance d identifying the position of point P, the bat's "sweet spot" or center of percussion, such that the batter will not feel any impulsive forces at O where he is grasping the bat. In addition, knowing that the ball, weighing 5 oz, is traveling at a speed $v_b = 90$ mph and that the batter is swinging the bat with an angular velocity $\omega_0 = 45$ rad/s, determine the speed of the ball and the angular velocity of the bat immediately after impact. To solve the problem, use the following data: $\delta = 6$ in., $\rho = 14$ in., $\ell = 22.5$ in., and COR $e = 0.5$.

Problem 8.123

A thin homogeneous bar A of length $\ell = 1.75$ m and mass $m = 23$ kg is translating as shown with a speed $v_0 = 12$ m/s when it collides with the fixed obstacle B. Modeling the contact between the bar and obstacle as frictionless, letting $\beta = 32°$, and letting the distance $d = 0.46$ m, determine the angular velocity of the bar immediately after the collision, knowing that the COR for the impact is $e = 0.74$.

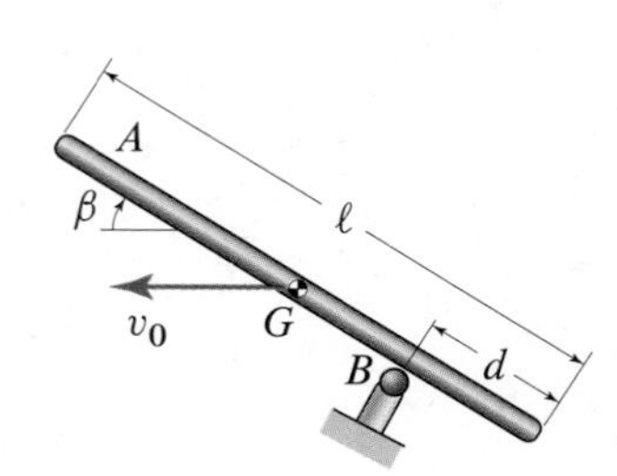

Figure P8.123

Problem 8.124

A uniform bar A with a hook H at the end is dropped from rest as shown from a height $d = 3$ ft over a fixed pin B. Letting the weight and length of A be $W = 100$ lb and $\ell = 7$ ft, respectively, determine the angle θ that the bar will sweep through if the bar becomes hooked with B and does not rebound. Although bar A becomes hooked with B, assume that there is no friction between the hook and the pin.

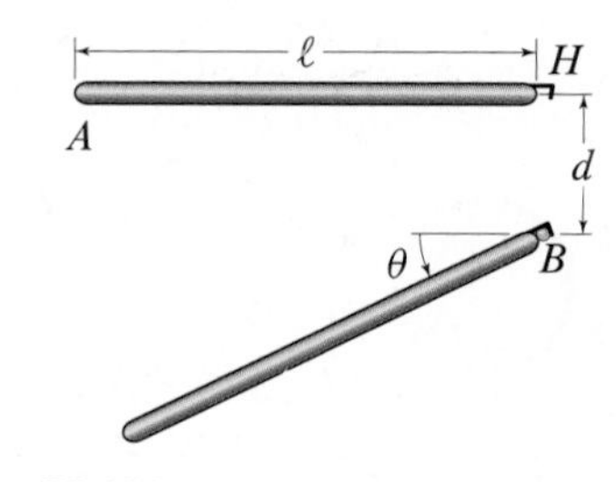

Figure P8.124

* See R. Cross, "The Sweet Spot of a Baseball Bat," *American Journal of Physics*, **66**(9), 1998, pp. 772–779.

Problem 8.125

A drawbridge of length $\ell = 30$ ft and weight $W = 600$ lb is released in the position shown and freely pivots clockwise until it strikes the right end of the moat. If the COR for the collision between the bridge and the ground is $e = 0.45$ and if the contact point between the bridge and the ground is effectively ℓ away from the bridge's pivot point, determine the angle to which the bridge rebounds after the collision. Neglect any possible source of friction.

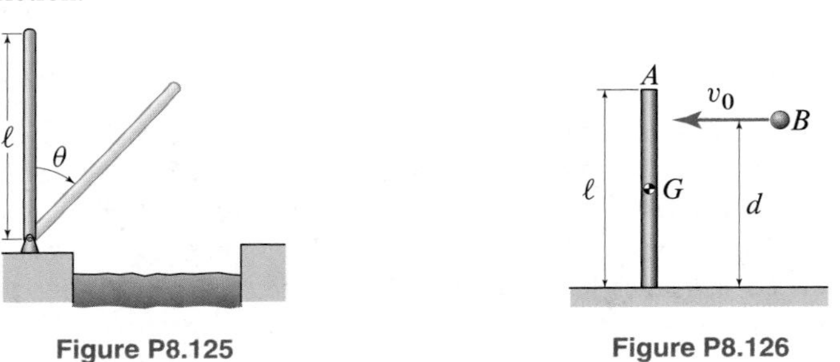

Figure P8.125 **Figure P8.126**

Problem 8.126

A stick A with length $\ell = 1.55$ m and mass $m_A = 6$ kg is in static equilibrium as shown when a ball B with mass $m_B = 0.15$ kg traveling at a speed $v_0 = 30$ m/s strikes the stick at distance $d = 1.3$ m from the lower end of the stick. If the COR for the impact is $e = 0.85$, determine the velocity of the mass center G of the stick, as well as the stick's angular velocity right after the impact.

Problem 8.127

A gymnast on the uneven parallel bars has a vertical speed v_0 and no angular speed when she grasps the upper bar. Model the gymnast as a single uniform rigid bar A of weight $W = 92$ lb and length $\ell = 6$ ft. Neglecting all friction, letting $\beta = 12°$, and assuming that the upper bar B does not move after the gymnast grasps it, determine the minimum speed v_0 for the gymnast to swing (counterclockwise) into the horizontal position on the other side of the bar. Assume that, during the motion, the friction between the gymnast's hands and the upper bar is negligible.

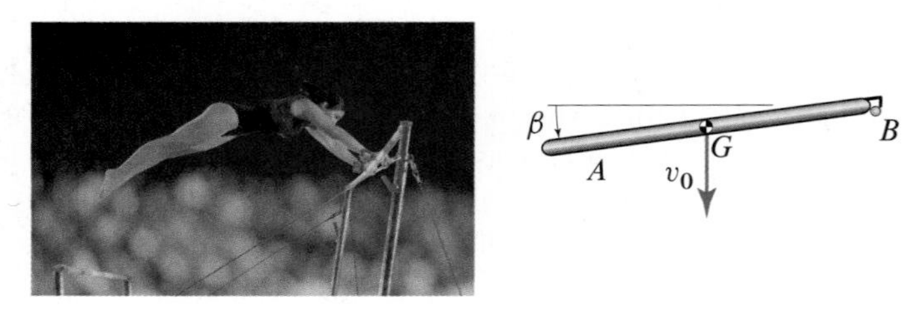

Figure P8.127

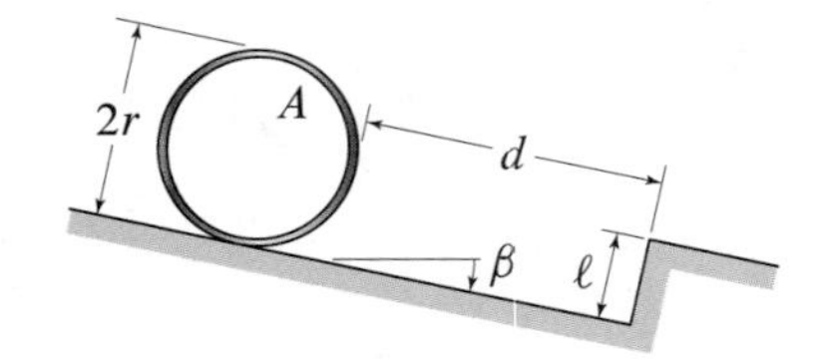

Figure P8.128

Problem 8.128

A uniform thin ring A of mass $m = 7$ kg and radius $r = 0.5$ m is released from rest as shown and rolls without slip until it meets a step of height $\ell = 0.45$ m. Letting $\beta = 12°$ and assuming that the ring does not rebound off the step or slip relative to it, determine the distance d, such that the ring barely makes it over the step.

Problem 8.129

Two identical uniform bars AB and BD are pin-connected at B, and bar BD has a hook at the free end. The two bars are dropped as shown from a height $d = 3$ ft over a fixed pin E (shown in cross section). Letting the weight and length of each bar be $W = 100$ lb and $\ell = 7$ ft, respectively, determine the angular velocities of AB and BD immediately after bar BD becomes hooked on E and does not rebound. *Hint:* The angular momentum of bar AB is conserved about B during impact.

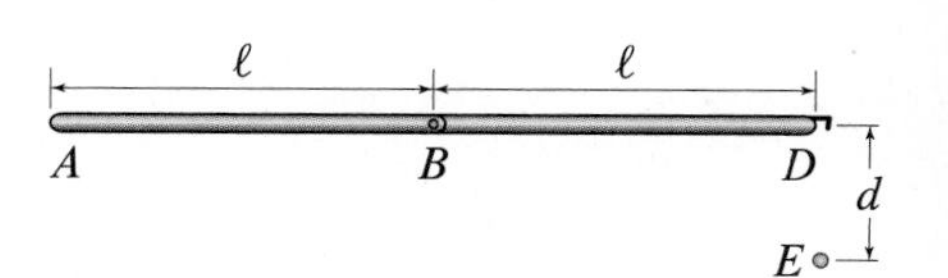

Figure P8.129

Problem 8.130

Cars A and B collide as shown. Determine the angular velocities of A and B immediately after impact if the COR is $e = 0.35$. In solving the problem, let C and D be the mass centers of A and B, respectively. In addition, enforce assumption 3 on p. 640 and use the following data: $W_A = 3130$ lb (weight of A), $k_C = 34.5$ in. (radius of gyration of A), $v_C = 12$ mph (speed of the mass center of A), $W_B = 3520$ lb (weight of B), $k_D = 39.3$ in. (radius of gyration of B), $v_D = 15$ mph (speed of the mass center of B), $d = 19$ in., $\ell = 79$ in., $\delta = 7.1$ in., $\rho = 65$ in., and $\beta = 12°$.

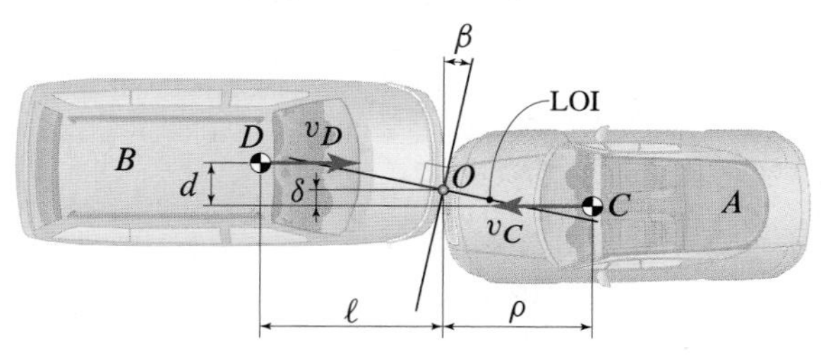

Figure P8.130

Problem 8.131

Consider the collision of two rigid bodies A and B, which, referring to Example 8.10 on p. 627, models the docking of the Space Shuttle (body A) to the International Space Station (body B). As in Example 8.10, we assume that B is stationary relative to an inertial frame of reference and that A translates as shown. In contrast to Example 8.10, here we assume that A and B join at point Q but, due to the flexibility of the docking system, can rotate relative to one another. Determine the angular velocities of A and B right after docking if $v_0 = 0.03$ m/s. In solving the problem, let C and D be the centers of mass of A and B, respectively. In addition, let the mass and mass moment of inertia of A be $m_A = 120 \times 10^3$ kg and $I_C = 14 \times 10^6$ kg·m², respectively, and the mass and mass moment of inertia of B be $m_B = 180 \times 10^3$ kg and $I_D = 34 \times 10^6$ kg·m², respectively. Finally, use the following dimensions: $\ell = 24$ m, $d = 8$ m, $\rho = 2.6$ m, and $\delta = 2.4$ m.

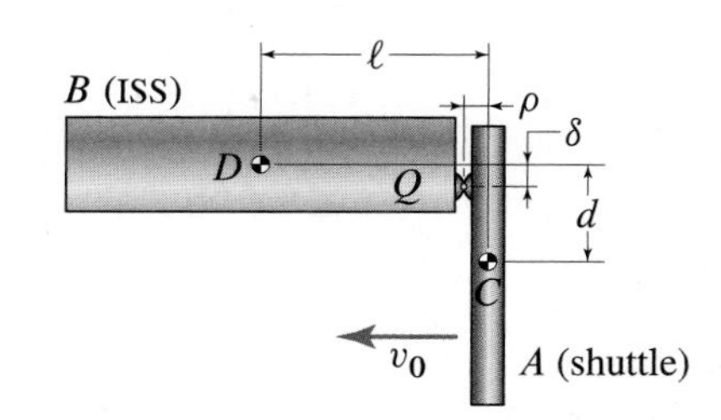

Figure P8.131

Chapter Review

Work-energy principle for rigid bodies

In this section, we discovered that the kinetic energy T of a rigid body B in planar motion is given by

Eq. (8.8), p. 584

$$T = \tfrac{1}{2} m v_G^2 + \tfrac{1}{2} I_G \omega_B^2,$$

where m, G, and I_G are the body's mass, center of mass, and mass moment of inertia, respectively, or by

Eq. (8.11), p. 584

$$T = \tfrac{1}{2} I_O \omega_B^2,$$

which applies to fixed-axis rotation, and where I_O is the mass moment of inertia relative to the axis of rotation. The work-energy principle for a rigid body is

Eq. (8.16), p. 585

$$T_1 + U_{1\text{-}2} = T_2,$$

where $U_{1\text{-}2}$ is the work done on the body in going from ① to ② by only the external forces. If we make use of the potential energy V of conservative forces, the work-energy principle can also be written as

Eqs. (8.23) and (8.24), p. 586

$$T_1 + V_1 + (U_{1\text{-}2})_{\text{nc}} = T_2 + V_2 \qquad \text{(general systems)},$$
$$T_1 + V_1 = T_2 + V_2 \qquad \text{(conservative systems)},$$

where the second expression applies only to a conservative system and states that the total mechanical energy of the system is conserved. For systems of rigid bodies or mixed systems of rigid bodies and particles, the work-energy principle is

Eq. (8.29), p. 588

$$T_1 + V_1 + (U_{1\text{-}2})_{\text{nc}}^{\text{ext}} + (U_{1\text{-}2})_{\text{nc}}^{\text{int}} = T_2 + V_2,$$

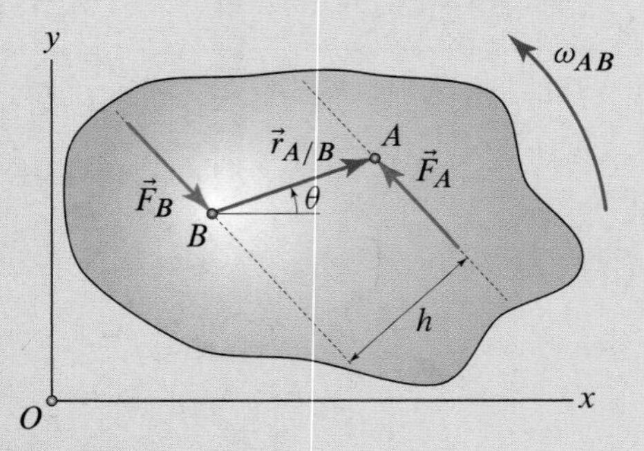

Figure 8.31
Rigid body subject to a couple.

where V is the total potential energy (of both external and internal conservative forces) and $(U_{1\text{-}2})_{\text{nc}}^{\text{ext}}$ and $(U_{1\text{-}2})_{\text{nc}}^{\text{int}}$ are the work contributions due to external and internal forces for which we do not have a potential energy, respectively.

Referring to Fig. 8.31, in planar rigid body motion, the work of a couple is

Eq. (8.22), p. 586

$$U_{1\text{-}2} = \int_{t_1}^{t_2} M \omega_{AB}\, dt = \int_{\theta_1}^{\theta_2} M\, d\theta,$$

where $d\theta$ is the body's infinitesimal angular displacement, M is the component of the moment of the couple in the direction perpendicular to the plane of motion, taken to be positive in the direction of positive θ, and θ *must* be measured in radians. In addition, the power developed by a couple with moment $\vec{M}$ is computed as

Eq. (8.31), p. 588

$$\text{Power developed by a couple} = \frac{dU}{dt} = \vec{M} \cdot \vec{\omega}_{AB}.$$

Momentum methods for rigid bodies

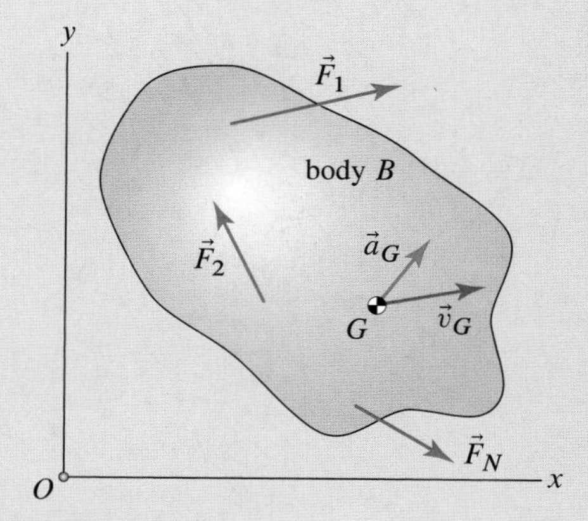

Figure 8.32
A rigid body under the action of a system of forces.

In this section, we derived the linear and the angular impulse-momentum principles for rigid bodies. Referring to Fig. 8.32, the linear impulse-momentum principle for a rigid body reads

Eqs. (8.34), p. 618

$$m\vec{v}_{G1} + \int_{t_1}^{t_2} \vec{F}\, dt = m\vec{v}_{G2} \quad \text{or} \quad \vec{p}_1 + \int_{t_1}^{t_2} \vec{F}\, dt = \vec{p}_2,$$

where $\vec{F}$ is the total external force on B, $\vec{p} = m\vec{v}_G$ is B's linear momentum, and $\vec{v}_G$ is the velocity of B's center of mass. If $\vec{F} = \vec{0}$, we have

Eqs. (8.36), p. 618

$$m\vec{v}_{G1} = m\vec{v}_{G2} \quad \text{or} \quad \vec{p}_1 = \vec{p}_2,$$

and we say that the body's momentum is conserved. If P in Fig. 8.33 is a moment center coplanar with G and if B is symmetric relative to the plane of motion, the angular momentum of B relative to P is

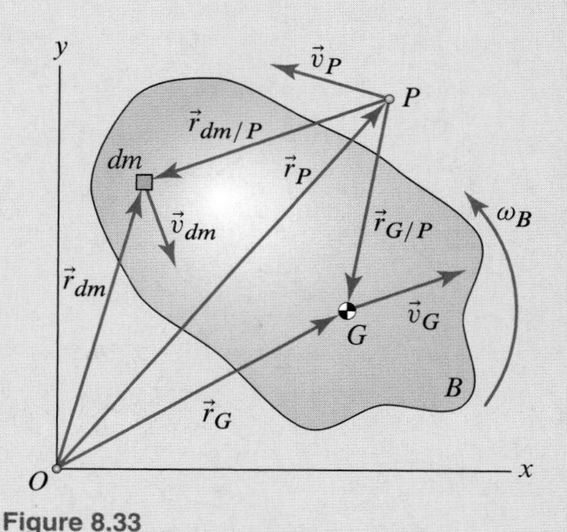

Figure 8.33
The quantities needed to obtain the angular momentum relationships for a rigid body.

Eq. (8.40), p. 619

$$\vec{h}_P = I_G \vec{\omega}_B + \vec{r}_{G/P} \times m\vec{v}_G,$$

where I_G is the mass moment of inertia of B, $\vec{\omega}_B$ is the angular velocity of B, and $\vec{r}_{G/P}$ is the position of G relative to P. If P is chosen so that (1) P is fixed or (2) P coincides with G or (3) P and G move parallel to one another, then $\vec{M}_P = \dot{\vec{h}}_P$, where $\vec{M}_P$ is the moment relative to P of the *external* force system acting on B. When this equation holds, by integrating with respect to time over a time interval $t_1 \le t \le t_2$, we have

Eq. (8.42), p. 619

$$\vec{h}_{P1} + \int_{t_1}^{t_2} \vec{M}_P\, dt = \vec{h}_{P2}.$$

When P coincides with G, or if the body undergoes a fixed-axis rotation about a point O as shown in Fig. 8.34, then the above equation becomes

Eq. (8.43), p. 619, and Eq. (8.45), p. 620

$$I_{G1}\omega_{B1} + \int_{t_1}^{t_2} M_{Gz}\, dt = I_{G2}\omega_{B2},$$

$$I_{O1}\omega_{B1} + \int_{t_1}^{t_2} M_{Oz}\, dt = I_{O2}\omega_{B2},$$

respectively, where I_O is the mass moment of inertia about the fixed-axis of rotation.

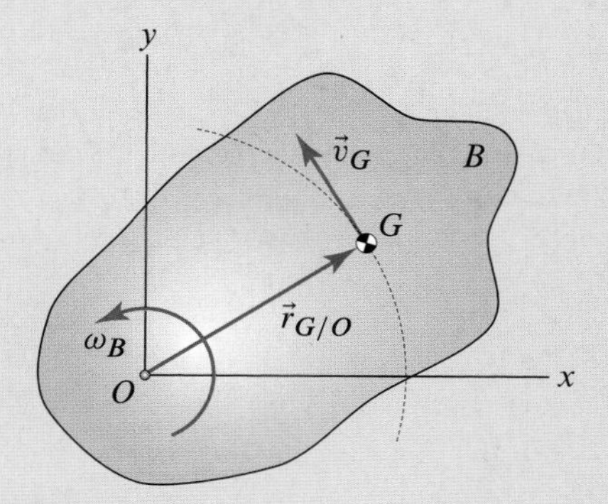

Figure 8.34
A rigid body in a fixed-axis rotation.

Impact of rigid bodies

In this section, we studied planar rigid body impacts. We learned that, contrary to particle impacts, rigid bodies can experience *eccentric* impacts. These are collisions in which at least one of the mass centers of the impacting bodies does not lie on the LOI. We also learned that the basic concepts used in particle impacts are applicable to rigid body impacts, and we reviewed solution strategies for a variety of situations.

As with particle impacts, we say that a rigid body impact is *plastic* if the COR $e = 0$. A rigid body impact will be called *perfectly plastic* if the colliding bodies form a single rigid body after impact. An impact is *elastic* if the COR e is such that $0 < e < 1$. Finally, the ideal case with $e = 1$ is referred to as a *perfectly elastic* impact. In addition, an impact is *unconstrained* if the system consisting of the two impacting objects is not subject to external impulsive forces; otherwise, the impact is called *constrained*.

We have considered only impacts satisfying the following assumptions:

1. The impact involves only two bodies in planar motion where no impulsive force has a component perpendicular to the plane of motion.
2. Contact between any two rigid bodies occurs at only one point, and at this point we can clearly define the LOI.
3. The contact between the bodies is frictionless.

We recommend organizing the solution of any impact problem as follows:

- Begin with an FBD of the impacting bodies as a system and FBDs for each of the colliding bodies. Neglect nonimpulsive forces.
- Choose a coordinate system with the origin coincident with the points that come into contact at the time of impact. Recall that the origin of such a coordinate system is a fixed point.
- Enforce the linear and/or the angular impulse-momentum principles for the system and/or for the individual bodies. In applying the angular impulse-momentum principle for the whole system, the moment center should be a fixed point, whereas for an individual body, the moment center should be a fixed point or the body's mass center.
- For plastic, elastic, and perfectly elastic impacts, the COR equation is first written using the velocity components along the LOI of the points that actually come into contact. For example, for the impact shown in Fig. 8.35, the COR equation is first written as

Eq. (8.59), p. 643

$$v_{Ex}^{+} - v_{Qx}^{+} = e\left(v_{Qx}^{-} - v_{Ex}^{-}\right),$$

where the COR e is such that $0 \le e \le 1$. The COR equation must then be rewritten in terms of the angular velocities and velocities of the mass centers using rigid body kinematics. For the situation in Fig. 8.35, this means rewriting the COR equation using the relations $\vec{v}_E^{\pm} = \vec{v}_C^{\pm} + \omega_A^{\pm}\,\hat{k}\times\vec{r}_{E/C}$ and $\vec{v}_Q^{\pm} = \vec{v}_D^{\pm} + \omega_B^{\pm}\,\hat{k}\times\vec{r}_{Q/D}$. Notice that the relative position vectors $\vec{r}_{E/C}$ and $\vec{r}_{Q/D}$ do not have the $\pm$ superscript because they are treated as constants during the impact.

- In perfectly plastic impacts, kinematic constraint equations must be enforced that express the fact that two bodies form a single rigid body after impact.

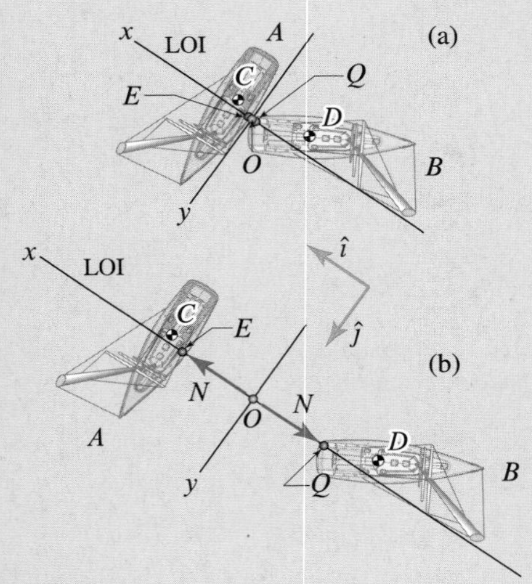

Figure 8.35
FBDs of the colliding bodies (a) as a system and (b) individually.

Review Problems

Problem 8.132

A uniform thin ring A and a uniform disk B roll without slip as shown. Letting T_A and T_B be the kinetic energies of A and B, respectively, if the two objects have the same mass and radius and if their centers are moving with the same speed v_0, state which of the following statements is true and why: (a) $T_A < T_B$; (b) $T_A = T_B$; (c) $T_A > T_B$.

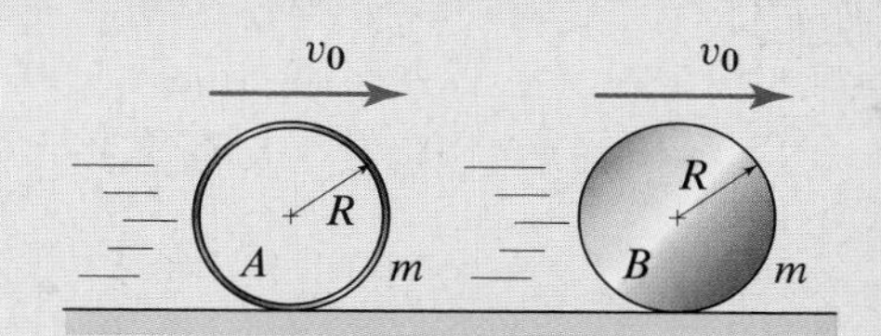

Figure P8.132

Problem 8.133

At the instant shown, the disk D, which has mass m and radius of gyration k_G, is rolling without slip down the flat incline with angular velocity ω_0. The disk is attached at its center to a wall by a linear elastic spring of constant k. If, at the instant shown, the spring is unstretched, determine the distance d down the incline that the disk rolls before coming to a stop. Use $k = 65\,\text{N/m}$, $R = 0.3\,\text{m}$, $m = 10\,\text{kg}$, $k_G = 0.25\,\text{m}$, $\omega_0 = 60\,\text{rpm}$, and $\theta = 30°$.

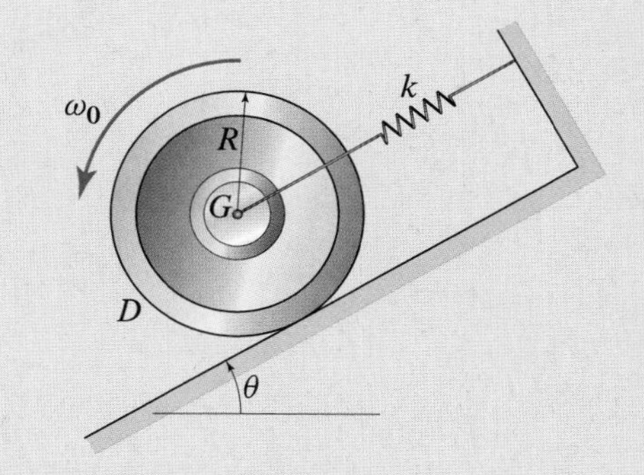

Figure P8.133

Problem 8.134

A pendulum consists of a uniform disk A of diameter $d = 5\,\text{in.}$ and weight $W_A = 0.25\,\text{lb}$ attached at the end of a uniform bar B of length $L = 2.75\,\text{ft}$ and weight $W_B = 1.3\,\text{lb}$. At the instant shown, the pendulum is swinging with an angular velocity $\omega = 0.55\,\text{rad/s}$ clockwise. Determine the kinetic energy of the pendulum at this instant, using Eq. (8.11) on p. 584.

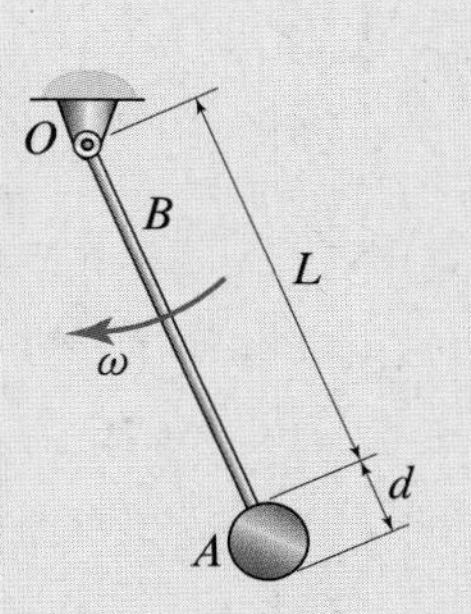

Figure P8.134

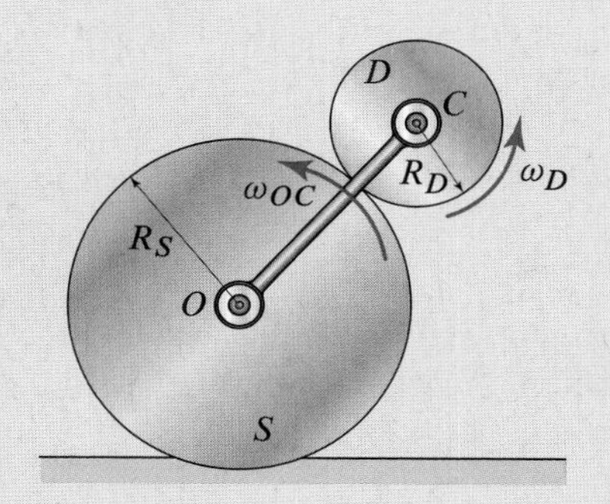

Figure P8.135

Problem 8.135

A uniform disk D of radius $R_D = 7\,\text{mm}$ and mass $m_D = 0.15\,\text{kg}$ is connected to point O via the rotating arm OC and rolls without slip over the stationary cylinder S of radius $R_S = 15\,\text{mm}$. Assuming that $\omega_D = 25\,\text{rad/s}$, and treating the arm OC as a uniform slender bar of length $L = R_D + R_S$ and mass $m_{OC} = 0.08\,\text{kg}$, determine the kinetic energy of the system.

Problem 8.136

The figure shows the cross section of a garage door with length $L = 9\,\text{ft}$ and weight $W = 175\,\text{lb}$. At the ends A and B there are rollers of negligible mass constrained to move in a vertical and a horizontal guide, respectively. The door's motion is assisted by two springs (only one spring is shown), each with constant $k = 9.05\,\text{lb/ft}$. If the door is released from rest when horizontal and the spring is stretched 4 in., neglecting friction, and modeling the door as a uniform thin plate, determine the speed with which B strikes the left end of the horizontal guide.

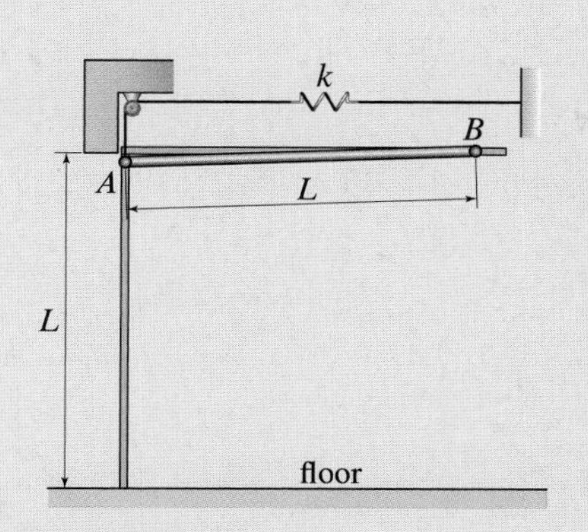

Figure P8.136

Problem 8.137

Body B has mass m and mass moment of inertia I_G, where G is the mass center of B. If B is in fixed-axis rotation about its center of mass G, determine which of the following statements is true and why: (a) $|(\vec{h}_E)_B| < |(\vec{h}_P)_B|$, (b) $|(\vec{h}_E)_B| = |(\vec{h}_P)_B|$, (c) $|(\vec{h}_E)_B| > |(\vec{h}_P)_B|$.

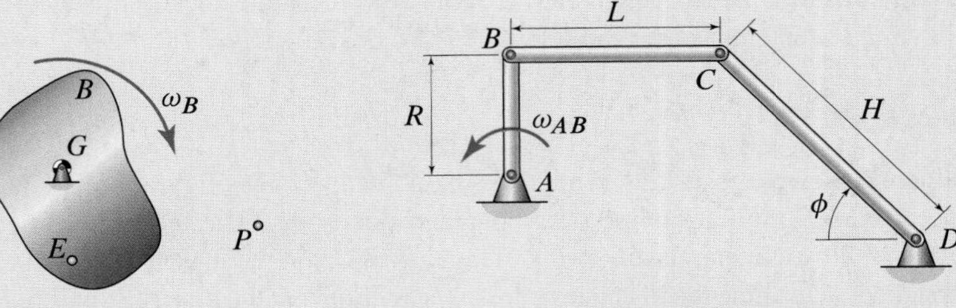

Figure P8.137

Figure P8.138

Problem 8.138

The weights of the uniform thin pin-connected bars AB, BC, and CD are $W_{AB} = 4$ lb, $W_{BC} = 6.5$ lb, and $W_{CD} = 10$ lb, respectively. Letting $\phi = 47°$, $R = 2$ ft, $L = 3.5$ ft, and $H = 4.5$ ft, and knowing that bar AB rotates at a constant angular velocity $\omega_{AB} = 4$ rad/s, compute the angular momentum of the system about D at the instant shown.

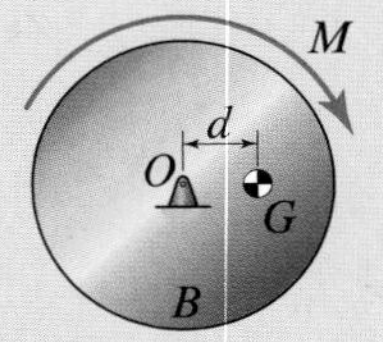

Figure P8.139

Problem 8.139

Consider Prob. 8.87 on p. 631 in which an eccentric wheel B is spun from rest under the action of a known torque M. In that problem, it was said that the wheel was in the *horizontal* plane. Is it possible to solve Prob. 8.87 by just applying Eq. (8.42) on p. 619 if the wheel is in the *vertical* plane? Why?

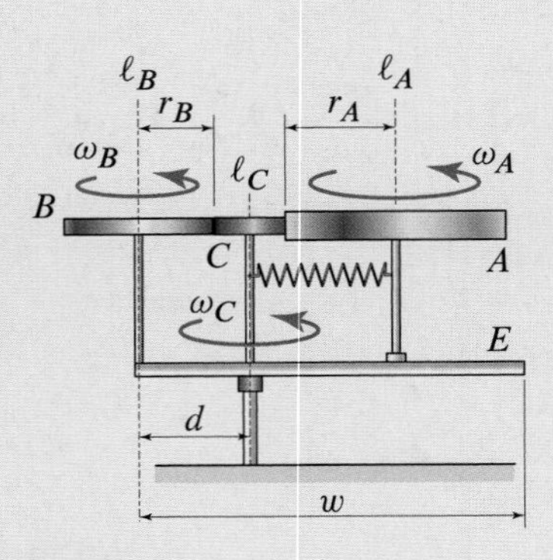

Figure P8.140

Problem 8.140

The uniform disk A of mass $m_A = 1.2$ kg and radius $r_A = 0.25$ m, is mounted on a vertical shaft that can translate along the horizontal rod E. The uniform disk B, of mass $m_B = 0.85$ kg and radius $r_B = 0.18$ m, is mounted on a vertical shaft that is rigidly attached to E. Disk C has a negligible mass and is rigidly attached to E; i.e., C and E form a single rigid body. Disk A can rotate about the axis ℓ_A, disk B can rotate about the axis ℓ_B, and the arm E along with C can rotate about the fixed axis ℓ_C. While keeping both B and C stationary, disk A is initially spun with $\omega_A = 1200$ rpm. Disk A is then brought in contact with C (contact is maintained by a spring), and at the same time, both B and C (and the arm E) are free to rotate. Due to friction between A and C, C along with E and disk B start spinning. Eventually A and C will stop slipping relative to one another. Disk B always rotates without slip over C. Let $d = 0.27$ m and $w = 0.95$ m. Assuming that the only elements of the system that have mass are A, B, and E and that $m_E = 0.3$ kg, and assuming that all friction in the system can be neglected except for that between A and C and between C and B, determine the angular speeds of A, B, and C (the angular velocity of C is the same as that of E since they form a single rigid body), when A and C stop slipping relative to one another.

Problem 8.141

A billiard ball is rolling without slipping with a speed $v_0 = 6\,\mathrm{ft/s}$ as shown when it hits the rail. According to regulations, the nose of the rail is at a height from the table bed of 63.5% of the ball's diameter (i.e., $\ell/(2r) = 0.635$). Model the impact with the rail as perfectly elastic, neglect friction between the ball and the rail, as well as between the ball and the table, and neglect any vertical motion of the ball. Based on the stated assumptions, determine the velocity of the point of contact between the ball and the table right after impact. The diameter of the ball is $2r = 2.25\,\mathrm{in.}$, and the weight of the ball is $W = 5.5\,\mathrm{oz}$.

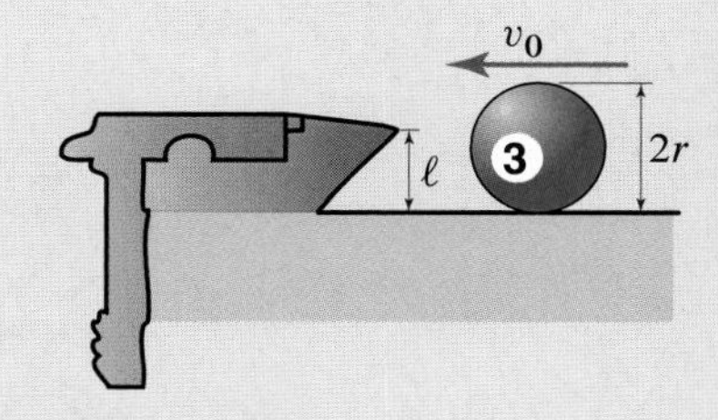

Figure P8.141

Problem 8.142

A basketball with mass $m = 0.6\,\mathrm{kg}$ is rolling without slipping as shown when it hits a small step with $\ell = 7\,\mathrm{cm}$. Letting the ball's diameter be $r = 12.0\,\mathrm{cm}$, modeling the ball as a thin spherical shell (the mass moment of inertia of a spherical shell about its mass center is $\frac{2}{3}mr^2$), and assuming that the ball does not rebound off the step or slip relative to it, determine the maximum value of v_0 for which the ball will roll over the step without losing contact with it.

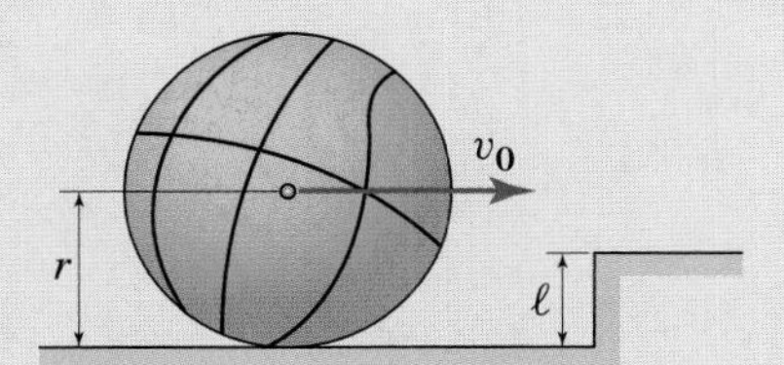

Figure P8.142

Problem 8.143

A bullet B weighing 147 gr (1 lb = 7000 gr) is fired with a speed $v_0 = 2750\,\mathrm{ft/s}$ as shown against a thin uniform rod A of length $\ell = 3\,\mathrm{ft}$, weight $W_r = 35\,\mathrm{lb}$, and pinned at O. If $d = 1.5\,\mathrm{ft}$ and the COR for the impact is $e = 0.25$, determine the bar's angular velocity immediately after the impact. In addition, determine the maximum value of the angle θ to which the bar swings after impact.

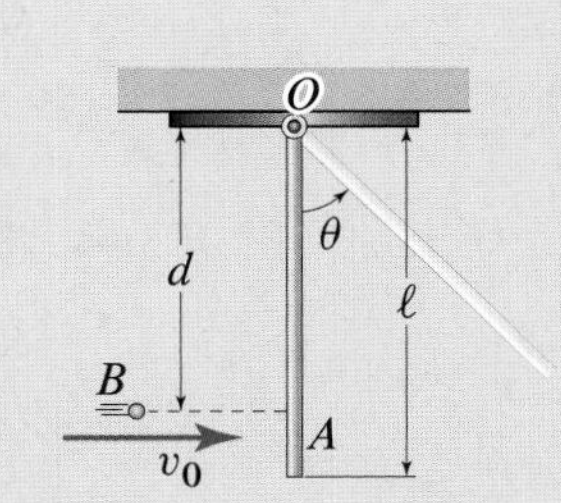

Figure P8.143

Problem 8.144

An airplane is about to crash-land on only one wheel with a vertical component of speed $v_0 = 2\,\mathrm{ft/s}$ and zero roll, pitch, and yaw. Determine the vertical component of velocity of the center of mass of the airplane G, as well as the airplane's angular velocity immediately after touching down, assuming that (1) the only available landing gear is rigid and rigidly attached to the airplane, (2) the coefficient of restitution between the landing gear and the ground is $e = 0.1$, (3) the airplane can be modeled as a rigid body, (4) the mass center G and the point of first contact between the landing gear and the ground are in the same plane perpendicular to the longitudinal axis of the airplane, and (5) friction between the landing gear and the ground is negligible. In solving the problem use the following data: $W = 2500\,\mathrm{lb}$ (weight of the airplane), G is the mass center of the airplane, $k_G = 3\,\mathrm{ft}$ is the radius of gyration of the airplane, and $d = 5.08\,\mathrm{ft}$.

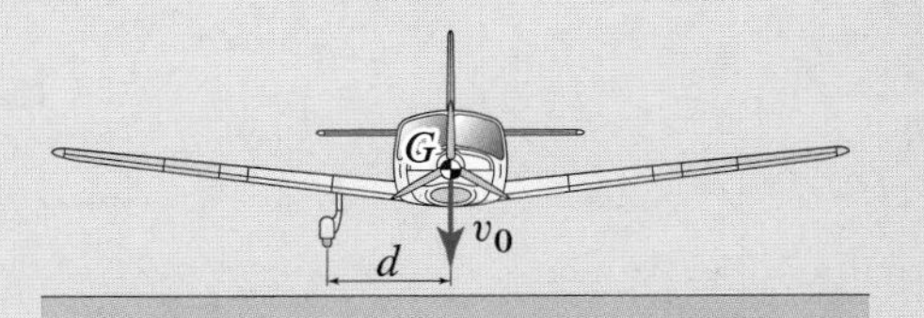

Figure P8.144

Mechanical Vibrations

A *vibration* is a type of dynamic behavior in which a system or part of a system oscillates about an equilibrium position. Vibrations occur in mechanical, electrical, thermal, and fluid systems, but we will consider only those occurring in mechanical systems. Vibrations in mechanical systems can be undesirable (e.g., vibration in structures can lead to failure and can create unwanted noise), or they can be desirable (e.g., vibration in fluid systems can dissipate unwanted energy, and music would be impossible without vibrations).

The Thelonious Monk Quartet in 1957 at the Five Spot Café in New York City (from left to right, John Coltrane, Shadow Wilson, Thelonious Monk, and Ahmed Abdul-Malik). The music we hear is the result of vibrations induced in air by musical instruments.

9.1 Undamped Free Vibration

Oscillation of a railcar after coupling

Recall Example 3.5 on p. 182 in which a railcar ran into a large spring that was designed to stop it (see Fig. 9.1). In that example, we were interested in the maximum compression of the spring and the time it took to stop the railcar. After maximum compression was reached, the spring would push the railcar back. If the railcar were to couple to the spring, the railcar would overshoot the equilibrium position of the spring and the railcar would start oscillating back and forth on the tracks. Let's look at this motion.

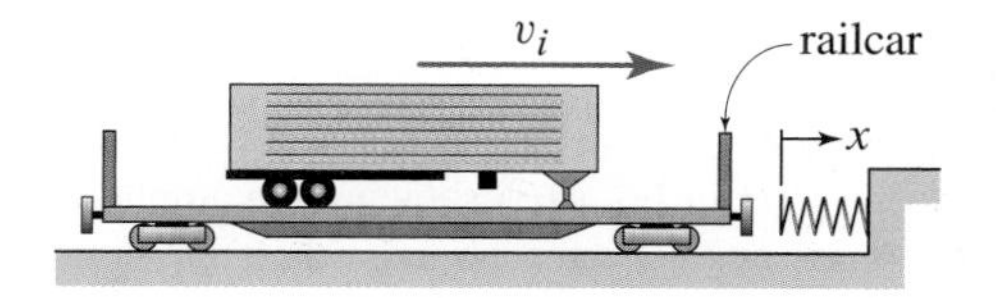

Figure 9.1
A railcar hitting a spring. The coordinate x measures the displacement of the spring from its equilibrium position. Recall from Example 3.5 that the railcar and its load weigh 87 tons and are moving at 4 mph at impact. Also, recall that we found $k = 22{,}270\,\text{lb/ft}$. We assume that the trailer does not move relative to the railcar.

In that example, we found the equation of motion of the railcar to be

$$\ddot{x} + \frac{k}{m}x = 0, \tag{9.1}$$

where k is the spring constant, m is the mass of the railcar and its load, and x is measured from the equilibrium position of the spring (see Fig. 9.1). Using $x(0) = x_i$

and $\dot{x}(0) = v_i$ for initial conditions, we were able to integrate this equation of motion to obtain time as a function of position (see Eq. (11) on p. 183), which can be inverted to obtain

$$x(t) = \frac{\sqrt{v_i^2 + \frac{k}{m}x_i^2}}{\sqrt{k/m}} \sin\left[\sqrt{\frac{k}{m}}\,t + \tan^{-1}\left(\frac{x_i\sqrt{k/m}}{v_i}\right)\right]. \tag{9.2}$$

Here, we have treated both x_i and v_i as positive quantities and then used the trigonometric identity $\sin^{-1}\left(1/\sqrt{z^2+1}\right) = \tan^{-1}(1/z)$. Equation (9.2) looks complicated, but it is really of the form

$$x(t) = C\sin(\omega_n t + \phi), \tag{9.3}$$

where

$$\omega_n = \sqrt{\frac{k}{m}}, \tag{9.4}$$

$$C = \sqrt{\frac{v_i^2}{\omega_n^2} + x_i^2}, \tag{9.5}$$

$$\tan\phi = \frac{x_i\omega_n}{v_i}. \tag{9.6}$$

The quantity ω_n is a constant called the *natural frequency** of vibration, and it is expressed in rad/s in both the SI and U.S. Customary unit systems. The quantities C and ϕ, called the *amplitude* and *phase angle* of vibration, respectively, are constants that depend on the initial conditions and ω_n. Consistent with Eq. (9.5), the *amplitude* of vibration C is understood to be a positive quantity. The angle ϕ can be determined via Eq. (9.6) for $v_i \neq 0$. When $v_i = 0$, ϕ can be chosen to be equal to $-\pi/2$ or $\pi/2$ rad, depending on whether $x_i < 0$ or $x_i > 0$, respectively.

Equation (9.3), which is a solution of Eq. (9.1), describes a *harmonic motion* with natural frequency ω_n. For this reason, the physical system modeled by Eq. (9.1) is called a *harmonic oscillator*. Equation (9.1) represents an *undamped* vibration because there are no terms that depend on $\dot{x}$, which would occur with viscous damping or in some models of aerodynamic drag. Equation (9.1) also represents a *free* vibration since it is homogeneous; that is, there are only terms containing the dependent variable x and no terms that are functions of time or are constant.† The function in Eq. (9.3) is plotted in Fig. 9.2.

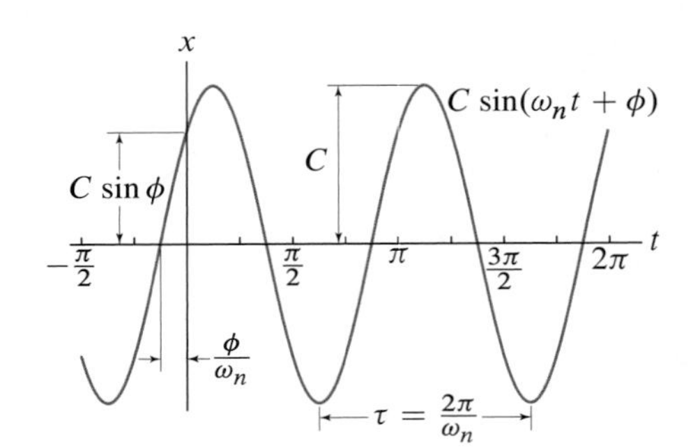

Figure 9.2
Plot of $x(t)$ in Eq. (9.3) showing the amplitude C, phase angle ϕ, and period τ of a harmonic oscillator.

The oscillator represented by Eq. (9.3) completes one cycle in the time

$$\tau = \frac{2\pi}{\omega_n}. \tag{9.7}$$

The quantity τ (the Greek letter tau) is called the *period* of the vibration (see Fig. 9.2). Finally, the number of cycles of vibration per unit of time is called the *frequency*, and it is defined as

$$f = \frac{1}{\tau} = \frac{\omega_n}{2\pi}. \tag{9.8}$$

Generally the frequency f is expressed in cycles per second or *hertz* (Hz).

Applying these ideas to the railcar, we see that if it couples to the spring, its natural frequency is $\omega_n = 2.030$ rad/s, its period is $\tau = (2\pi\text{ rad})/(2.030\text{ rad/s}) =$

* The natural frequency is also sometimes called the *circular frequency*.

† We will see in Section 9.2 that this is equivalent to saying that there is no forcing function in Eq. (9.1).

3.095 s, and its frequency is $f = 0.3231$ Hz, where we have used $k = 22{,}270$ lb/ft and $m = 5404$ slug. The amplitude of vibration is $C = 2.890$ ft and the phase angle is $\phi = 0$, where we have used $x_i = 0$ ft and $v_i = 5.867$ ft/s.

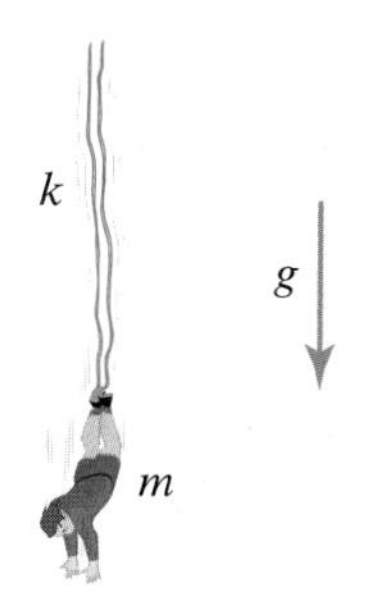

Figure 9.3
A bungee jumper of mass m hanging from a bungee cord of stiffness k.

Mini-Example

A jumper whose mass is $m = 65$ kg is hanging in equilibrium from a linear elastic bungee cord with constant $k = 200$ N/m. The jumper is then pulled down 5 m and released from rest (see Fig. 9.3). Determine the equation governing the ensuing vibration, the period, the amplitude, and the phase angle of the vibration. Treat the jumper as a particle.

Solution

The FBD of the jumper after being pulled a distance y below the $y = 0$ static equilibrium position is shown in Fig. 9.4. Summing forces in the y direction gives

$$\sum F_y: \quad mg - F_s = ma_y, \tag{9.9}$$

where F_s is the force in the bungee cord and $a_y = \ddot{y}$. Since y is measured from the static equilibrium position, the force in the bungee cord must be (see Fig. 9.5)

$$F_s = mg + ky. \tag{9.10}$$

Substituting Eq. (9.10) into Eq. (9.9), we obtain

$$mg - (mg + ky) = m\ddot{y} \quad \Rightarrow \quad \ddot{y} + \frac{k}{m}y = 0. \tag{9.11}$$

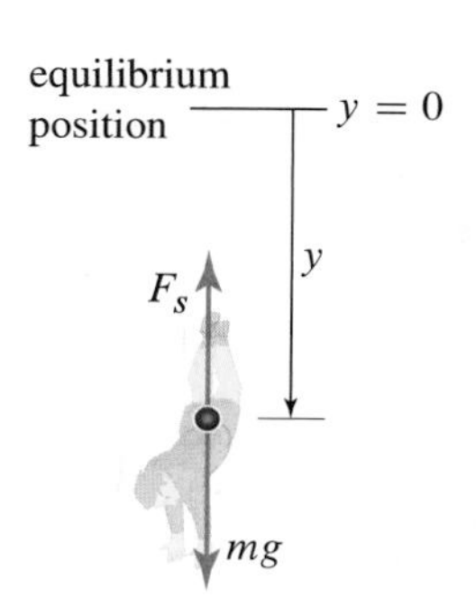

Figure 9.4
FBD of the bungee jumper, along with the coordinate system used in the mini-example.

Equation (9.11) is of the same form as Eq. (9.1), and so the jumper's natural frequency of vibration is $\omega_n = \sqrt{k/m} = 1.754$ rad/s, and the corresponding period of vibration is $\tau = 2\pi/\omega_n = 3.582$ s. Since $v_i = 0$ and $x_i = 5$ m, Eq. (9.5) tells us that the jumper's amplitude of vibration is 5 m, i.e., as expected, the jumper oscillates about the static equilibrium position. Finally, since $v_i = 0$ and $x_i > 0$, we can choose the phase angle to be $\phi = \pi/2$ rad.

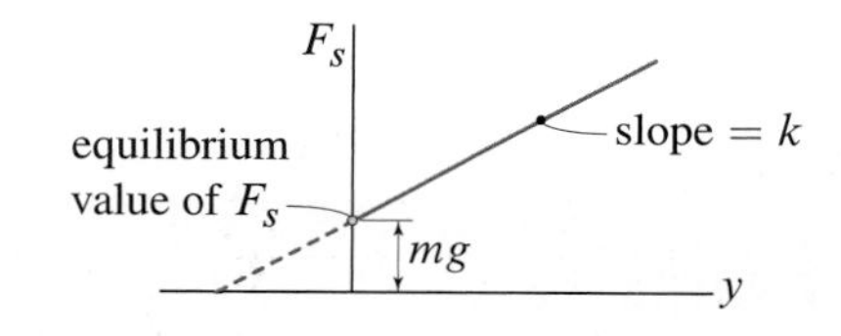

Figure 9.5
The force in the bungee cord as a function of y. The position $y = 0$ corresponds to static equilibrium.

One of the lessons of this mini-example is that *it is convenient to choose the origin of the displacement variable to be at the equilibrium position of the system* rather than at the position of zero spring deflection. Doing this allows us to ignore the equal and opposite forces associated with equilibrium.

Standard form of the harmonic oscillator

Based on the preceding development, we define a *harmonic oscillator* to be any one degree of freedom (DOF, see definition on p. 174) system whose equation of motion can be given the form

$$\ddot{x} + \omega_n^2 x = 0, \tag{9.12}$$

where ω_n is the natural frequency* of the oscillator and x is the coordinate for the one DOF system. Equation (9.12) is called the *standard form* of the harmonic oscillator equation.

As we have seen in Eq. (9.3), *we know the complete vibrational solution for any system whose equation of motion can be put in the form of Eq.* (9.12). We also note

* In the case of a spring-mass system, ω_n takes the form in Eq. (9.4).

that an alternative form of the solution to Eq. (9.12) is

$$x(t) = A\cos\omega_n t + B\sin\omega_n t, \tag{9.13}$$

where the solution in Eq. (9.3) is recovered from Eq. (9.13) if we let

$$C = \sqrt{A^2 + B^2} \quad \text{and} \quad \tan\phi = \frac{A}{B}. \tag{9.14}$$

Noting again that $x(0) = x_i$ and $\dot{x}(0) = v_i$, we find $A = x_i$ and $B = v_i/\omega_n$, and so Eq. (9.13) becomes

$$\boxed{x(t) = x_i\cos\omega_n t + \frac{v_i}{\omega_n}\sin\omega_n t.} \tag{9.15}$$

Linearizing nonlinear systems

Not all vibrating one-DOF systems are harmonic oscillators described by Eq. (9.12). However, many systems can be approximated as harmonic oscillators. As an example, consider the uniform thin bar in Fig. 9.6(a) that is pinned and suspended at one end. Using the FBD in Fig. 9.6(b) and the methods of Chapter 7, the equation of

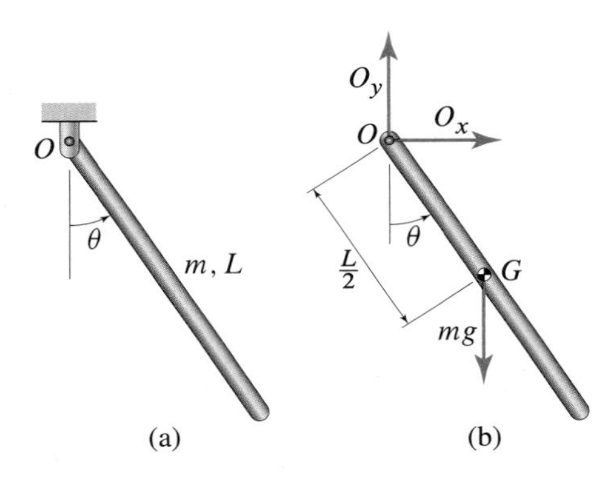

Figure 9.6. (a) A uniform swinging bar. (b) The FBD of the swinging bar.

motion for this bar is

$$\ddot{\theta} + \frac{3g}{2L}\sin\theta = 0. \tag{9.16}$$

This equation is *nonlinear* in θ because of the presence of $\sin\theta$, which is a *nonlinear* function of θ (in general, nonlinear equations are more challenging to solve than linear equations). Equation (9.16) can be written in the standard form $\ddot{\theta} + \omega_n^2\theta = 0$ if we consider only vibrations for small values of θ, although this creates an approximate version of the original equation. As seen in Fig. 9.7, when θ is small, $\sin\theta$ behaves as θ, and Eq. (9.16) becomes

$$\ddot{\theta} + \frac{3g}{2L}\theta = 0, \tag{9.17}$$

which is in standard form with $\omega_n = \sqrt{3g/(2L)}$. This process, in which a nonlinear ordinary differential equation is approximated to be linear, is called *linearization*. We will explore linearization further in the example problems.

Helpful Information

Mathematical justification for small-angle approximations. Figure 9.7 graphically demonstrates that $\sin x$ behaves as x for small x. We can also see this with the Taylor series. The Taylor series expansion of $\sin x$ about $x = 0$ is

$$\sin x = x - \frac{x^3}{3!} + \frac{x^5}{5!} - \cdots.$$

If x is small, then x^3, x^5, and higher-order terms will be so small that $\sin x \approx x$. Similarly, the Taylor series expansion of $\cos x$ about $x = 0$ is

$$\cos x = 1 - \frac{x^2}{2!} + \frac{x^4}{4!} - \cdots.$$

If x is small, then x^2, x^4, and higher-order terms will be so small that $\cos x \approx 1$.

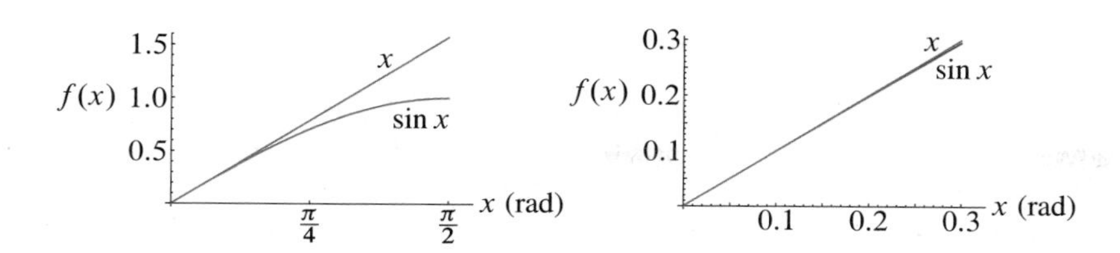

Figure 9.7. Plots of $f(x) = \sin x$ and $f(x) = x$ for two different ranges of x. The plot on the left shows that for large x, the two curves diverge. The plot on the right shows that for $x \lesssim 0.3$ rad the two curves are almost indistinguishable.

Energy method

The equations of motion for all three of the systems considered in this section were derived by applying the Newton-Euler equations to the FBD of the particle or body of interest. In addition, all three of these systems are conservative; that is, all forces doing work are conservative (the spring on the railcar, the bungee cord and gravity on the bungee jumper, and gravity on the swinging bar). For systems like these, the fact that energy is conserved can provide a way to derive the equation of motion.

Finding the equation of motion

To see how to find the equation of motion using conservation of mechanical energy, consider the bungee jumper mini-example on p. 665. Since the bungee jumper is a conservative system, we know that the work-energy principle gives

$$T_1 + V_1 = T_2 + V_2, \tag{9.18}$$

where ① is at release and ② is any subsequent position. This implies that

$$T + V = \text{constant} \quad \Rightarrow \quad \frac{d}{dt}(T + V) = 0, \tag{9.19}$$

where we have dropped the use of the subscript 2 to reinforce the idea that ② is *any* position following ①. If we now compute T and V at an arbitrary position for the bungee jumper, we find that the potential energy is given by (see Fig. 9.8)

$$V = V_e + V_g = \tfrac{1}{2}k(y + \delta_{\text{st}})^2 - mgy, \tag{9.20}$$

where y is measured from the static equilibrium position of the jumper and δ_{st} is the amount of stretch in the bungee cord at the static equilibrium position. The kinetic energy is given by

$$T = \tfrac{1}{2}m\dot{y}^2. \tag{9.21}$$

Substituting Eqs. (9.20) and (9.21) into Eq. (9.19) and taking the time derivative, we find that

$$\frac{d}{dt}(T + V) = k(y + \delta_{\text{st}})\dot{y} - mg\dot{y} + m\dot{y}\ddot{y} = 0. \tag{9.22}$$

Rewriting Eq. (9.22) as $[k(y + \delta_{\text{st}}) - mg + m\ddot{y}]\dot{y} = 0$, we see that this equation is satisfied for any value of $\dot{y}$ if and only if

$$k(y + \delta_{\text{st}}) - mg + m\ddot{y} = 0. \tag{9.23}$$

Since $k\delta_{\text{st}} = mg$, we recover the harmonic oscillator equation

$$m\ddot{y} + ky = 0, \tag{9.24}$$

which is equivalent to Eq. (9.11) for the same bungee jumper.

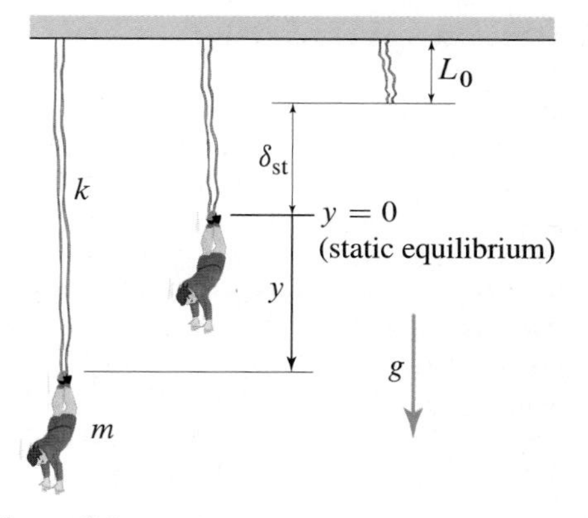

Figure 9.8
The unstretched length of the bungee cord L_0, the static equilibrium position of the jumper ($y = 0$), and the jumper at an arbitrary y position.

The energy method and linearization

When using the energy method to find the linearized equations of motion of a mechanical system, we can begin by approximating the kinetic and potential energies as quadratic functions of position and velocity before taking their time derivative. Then the time derivative of the quadratic approximation of the energy yields equations of motion that are *automatically* linear. The reason for starting with the quadratic approximation of the energy is that, in some cases, linearizing the equations of motion obtained from the nonapproximated form of the energy is more involved than developing the quadratic approximation of the energy.

To see what we mean by "approximating the kinetic and potential energies as quadratic functions of position and velocity," consider again the uniform thin bar in Fig. 9.6. Writing the kinetic and potential energies at an arbitrary angle θ, we find that

$$T = \tfrac{1}{2} I_O \dot{\theta}^2 = \tfrac{1}{6} m L^2 \dot{\theta}^2, \tag{9.25}$$

$$V = -\tfrac{1}{2} mgL \cos\theta \approx -\tfrac{1}{2} mgL\left(1 - \tfrac{1}{2}\theta^2\right). \tag{9.26}$$

Notice that the kinetic energy T in Eq. (9.25) is a quadratic function of the angular velocity $\dot{\theta}$, and therefore it does not need to be approximated in any way. By contrast, notice that the potential energy $V = -\frac{1}{2} mgL \cos\theta$ is not a quadratic function of θ, but we could approximate it as such by using the first *two* terms in the Taylor series expansion of $\cos\theta$ (see the Helpful Information marginal note on p. 666). Using Eqs. (9.25) and (9.26) in $\frac{d}{dt}(T + V) = 0$, we obtain

$$\tfrac{1}{3} m L^2 \dot{\theta}\ddot{\theta} + \tfrac{1}{2} mgL\theta\dot{\theta} = 0 \quad \Rightarrow \quad \ddot{\theta} + \frac{3g}{2L}\theta = 0, \tag{9.27}$$

which is exactly what we obtained in Eq. (9.17). The simple message here is that when we use the energy method to derive the equation of motion, it is important to approximate the kinetic and potential energies as quadratic functions of position and velocities *before* taking their time derivatives.

For future reference, we note that in linearizing the sine function, and in approximating the cosine function as a quadratic function of its argument, we use the following relations:

$$\boxed{\sin\theta \approx \theta \quad \text{and} \quad \cos\theta \approx 1 - \theta^2/2,} \tag{9.28}$$

respectively. In addition, we note that since the quadratic term in the power series expansion of the sine function is equal to zero (see the Helpful Information marginal note on p. 666), the linearized form of the sine function can also be viewed as the quadratic approximation of the sine function.

Helpful Information

Natural frequency, stiffness, and mass. For a simple spring-mass system

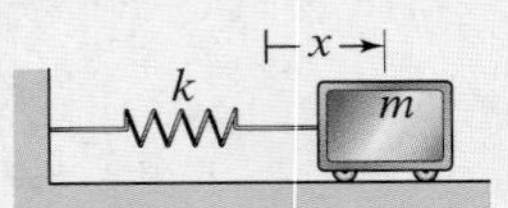

whose equation of motion is $m\ddot{x} + kx = 0$, the natural frequency of vibration is $\sqrt{k/m}$ (x is measured from the position of m when the spring is undeformed). Notice in Eq. (9.27) that the 'm' is $\frac{1}{3}mL^2$ and the 'k' is $\frac{1}{2}mgL$. This is common in harmonic oscillators, so it is useful to introduce the notions of *effective mass* and *effective stiffness*. The effective mass is an inertial quantity that is the coefficient of the second time derivative of the position, but it is not always just m. The effective stiffness is a restoring quantity that is the coefficient of the position, but it is not always just k.

End of Section Summary

Any one DOF system whose equation of motion is of the form

Eq. (9.12), p. 665

$$\ddot{x} + \omega_n^2 x = 0,$$

is called a *harmonic oscillator*, and the above expression is referred to as the *standard form* of the harmonic oscillator equation. The solution of this equation can be written as (see Fig. 9.9)

Eq. (9.3), p. 664

$$x(t) = C \sin(\omega_n t + \phi),$$

where ω_n is the *natural frequency*, C is the *amplitude*, and ϕ is the *phase angle* of vibration.

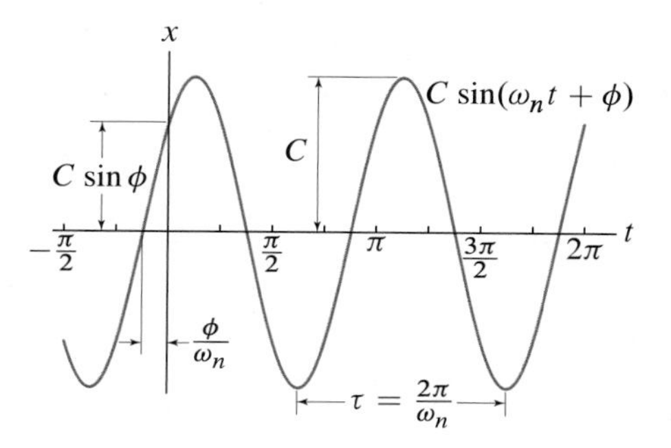

Figure 9.9
Plot of Eq. (9.3) showing the amplitude C, phase angle ϕ, and period τ of a harmonic oscillator.

A simple example of a harmonic oscillator is a system consisting of a block of mass m attached at the free end of a spring with constant k and with the other end fixed (see Fig. 9.10). The natural frequency of such a system is given by

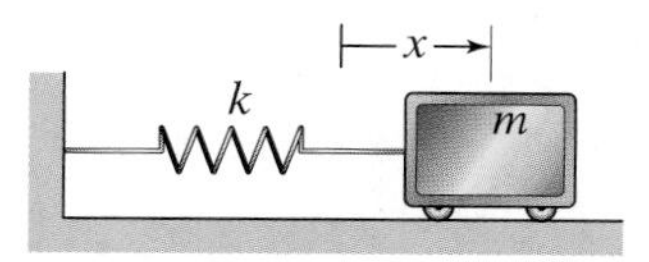

Figure 9.10
A simple spring-mass harmonic oscillator whose equation of motion is given by $m\ddot{x} + kx = 0$ and for which $\omega_n = \sqrt{k/m}$. The position x is measured from the location of m when the spring is undeformed.

Eq. (9.4), p. 664

$$\omega_n = \sqrt{\frac{k}{m}}.$$

In addition, the amplitude C and the phase angle ϕ are given by, respectively,

Eqs. (9.5) and (9.6), p. 664

$$C = \sqrt{\frac{v_i^2}{\omega_n^2} + x_i^2} \quad \text{and} \quad \tan\phi = \frac{x_i \omega_n}{v_i},$$

where, by letting $t = 0$ be the initial time, $x_i = x(0)$ (i.e., x_i is the initial position) and $v_i = \dot{x}(0)$ (i.e., v_i is the initial velocity). If $v_i = 0$, then ϕ can be chosen equal to $-\pi/2$ or $\pi/2$ rad for $x_i < 0$ and $x_i > 0$, respectively. An alternative form of the solution to Eq. (9.12) is given by

Eq. (9.15), p. 666

$$x(t) = x_i \cos \omega_n t + \frac{v_i}{\omega_n} \sin \omega_n t.$$

The *period* of the oscillation is given by

Eq. (9.7), p. 664

$$\text{Period} = \tau = \frac{2\pi}{\omega_n},$$

and the *frequency* of vibration is

Eq. (9.8), p. 664

$$\text{Frequency} = f = \frac{1}{\tau} = \frac{\omega_n}{2\pi}.$$

Energy method. For conservative systems, the work-energy principle tells us that the quantity $T + V$ is constant, and so its time derivative must be zero. This provides a convenient way to obtain the equations of motion via

Eq. (9.19), p. 667

$$\frac{d}{dt}(T + V) = 0 \quad \Rightarrow \quad \text{equations of motion.}$$

When we apply the energy method to determine the linearized equations of motion, it is often convenient to first approximate the kinetic and potential energies as quadratic functions of position and velocity and then take derivatives with respect to time. This process yields equations of motion that are linear.

In approximating the sine and cosine functions as quadratic functions of their arguments, we use the relations

Eq. (9.28), p. 668

$$\sin\theta \approx \theta \qquad \text{and} \qquad \cos\theta \approx 1 - \theta^2/2.$$

Since the quadratic term in the power series expansion of the sine function is equal to zero, the linearized form of the sine function can also be viewed as the quadratic approximation of the sine function.

EXAMPLE 9.1 *Finding the Moment of Inertia of a Rigid Body*

When the connecting rod shown in Fig. 1 is suspended from the knife-edge at point O and displaced slightly so that it oscillates like a pendulum, its period of oscillation is 0.77 s. The mass center G is located a distance $L = 110$ mm from O, and the mass of the connecting rod is 661 g. Using this information, determine the mass moment of inertia of the connecting rod about G.

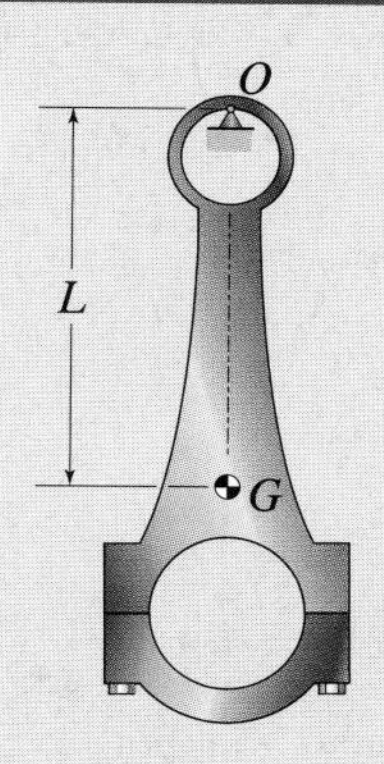

Figure 1
A connecting rod hinged on a knife-edge at O and allowed to oscillate freely in the plane of the page. Point G is the rod's mass center.

SOLUTION

Road Map & Modeling If we write the equation of motion of the connecting rod for small angles, then, as with a pendulum, we should be able to write it in standard form and extract the natural frequency of vibration. The natural frequency will depend on the mass moment of inertia of the connecting rod, which should then allow us to solve for the moment of inertia. The FBD of the connecting rod is shown in Fig. 2.

Governing Equations

Balance Principles Summing moments about the fixed point O (see Fig. 2), we obtain

$$\sum M_O: \quad -mgL\sin\theta = I_O\alpha_{cr}, \tag{1}$$

where I_O is the moment of inertia of the connecting rod with respect to point O, and α_{cr} is the angular acceleration of the connecting rod.

Force Laws All forces are accounted for on the FBD.

Kinematic Equations The only kinematic equation is $\alpha_{cr} = \ddot{\theta}$.

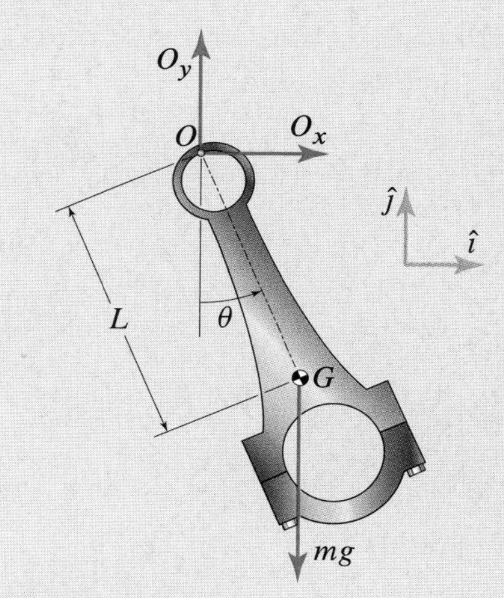

Figure 2
FBD of the connecting rod during oscillation.

Computation Substituting the kinematic equation into Eq. (1) and rearranging, we obtain

$$\ddot{\theta} + \frac{mgL}{I_O}\sin\theta = 0. \tag{2}$$

For small θ, $\sin\theta \approx \theta$ and so Eq. (2) becomes

$$\ddot{\theta} + \frac{mgL}{I_O}\theta = 0. \tag{3}$$

Equation (3) is in standard form, so we know that

$$\omega_n^2 = \frac{mgL}{I_O} \quad \Rightarrow \quad I_O = \frac{mgL}{\omega_n^2} \quad \Rightarrow \quad I_O = \frac{mgL\tau^2}{4\pi^2}, \tag{4}$$

where we have used $\omega_n^2 = 4\pi^2/\tau^2$ from Eq. (9.7). Noting that the parallel axis theorem states that $I_G = I_O - mL^2$, we find

$$I_G = I_O - mL^2 = \frac{mgL\tau^2}{4\pi^2} - mL^2 \quad \Rightarrow \quad I_G = mL^2\left(\frac{g\tau^2}{4\pi^2 L} - 1\right), \tag{5}$$

which, when evaluated for $\tau = 0.77$ s, $L = 0.11$ m, and $m = 0.661$ kg, yields

$$\boxed{I_G = 0.002714\,\text{kg}\cdot\text{m}^2.} \tag{6}$$

Discussion & Verification The dimensions of Eq. (5) are mass times length squared, as they should be. In addition, the first of Eqs. (4) tells us that the natural frequency of vibration is inversely proportional to the square root of the mass moment of inertia (as it is inversely proportional to the mass), which agrees with our intuition.

A Closer Look This simple measurement is tremendously useful. We typically talk about evaluating the mass moment of inertia from either its integral definition or by using the method of composite bodies (see App. A). A complicated shape like this connecting rod does not lend itself to either method, but here we have a simple experiment that allows us to determine the mass moment of inertia.

EXAMPLE 9.2 *Vibration of a Silicon Nanowire Modeled as a Rigid Bar*

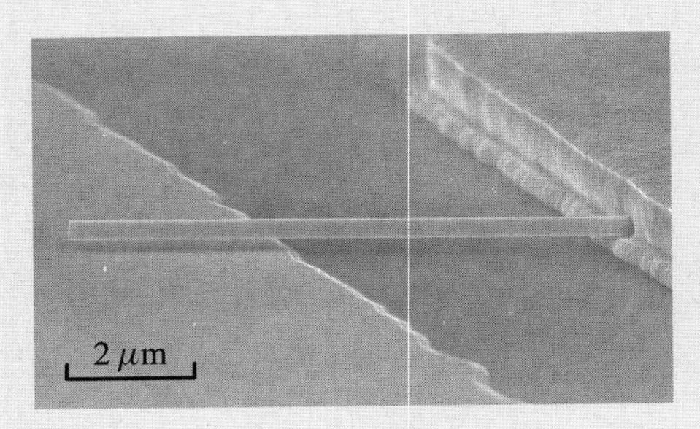

Figure 1
A field-emission scanning electronic microscope image of a silicon (Si) nanowire. From Mingwei Li et al., "Bottom-up Assembly of Large-Area Nanowire Resonator Arrays," *Nature Nanotechnology*, **3**(2), 2008, pp. 88–92.

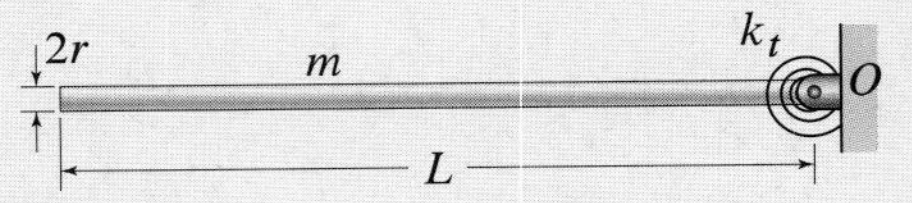

Figure 2
A rigid bar and torsional spring model of a flexible nanowire.

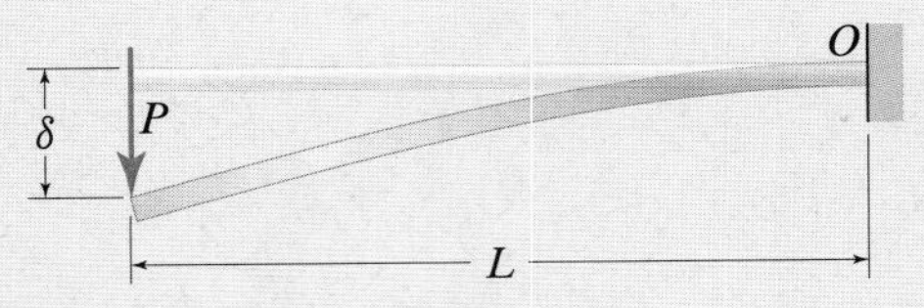

Figure 3
A flexible cantilever beam subject to a load P at its end. The deflection is given by Eq. (1).

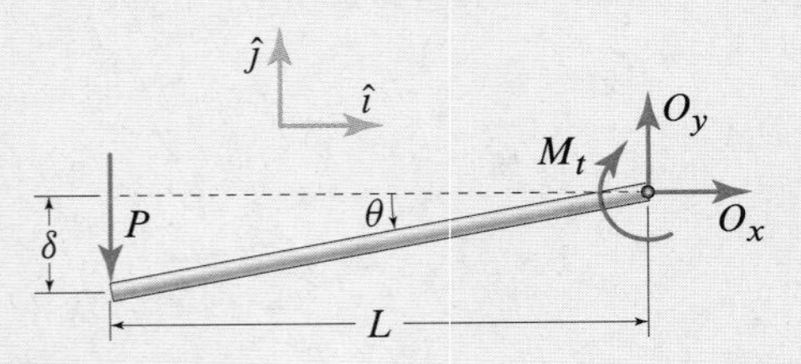

Figure 4
FBD of the rigid bar for finding the equivalent torsional spring constant k_t.

The natural frequency of a vibrating spring-mass system is a function of the equivalent mass and stiffness according to the relation $\omega_n = \sqrt{k_{\text{eq}}/m_{\text{eq}}}$. Therefore, if the cantilevered silicon nanowire (SiNW) shown in Fig. 1 were to vibrate, it would have a different natural frequency as shown than if additional mass were added to the end of the wire. Vibrating nanoelectromechanical systems (NEMS), such as this SiNW, have been proposed for use in chip-based sensor arrays as ultrasensitive mass detectors to detect masses in the zeptogram (zg) range. Such a wire would be capable of detecting small numbers of viruses!

Given a uniform Si nanowire with a circular cross section that is 9.8 μm long and 330 nm in diameter, compute its natural frequency, using a rigid bar model with all the flexibility lumped in a linear torsional spring at the base of the wire (see Fig. 2). Use $\rho = 2330\,\text{kg/m}^3$ for the density of silicon and $E = 152\,\text{GPa}$ for its modulus of elasticity.*

SOLUTION

Road Map & Modeling To obtain the linear torsional spring constant k_t, we will employ a result from mechanics of materials, which states that the deflection of a cantilevered bar subjected to a load P at its end is (see Fig. 3)

$$P = \frac{3EI_{\text{cs}}}{L^3}\delta, \tag{1}$$

where E is its modulus of elasticity and $I_{\text{cs}} = \frac{1}{4}\pi r^4$ is the centroidal *area* moment of inertia of the beam's cross section. Using Eq. (1), we will find the value of the torsional spring constant in Fig. 2 that gives this same deflection for a given load P. Once we have k_t, we can apply the Newton-Euler equations to then obtain the equation of motion in the form of Eq. (9.12).

Finding the Torsional Spring Constant

Governing Equations

Balance Principles Referring to the FBD in Fig. 4, taking moments about point O, and noting that this is a *statics* problem for the purpose of finding k_t, we obain

$$\sum M_O\text{:}\quad PL - M_t = 0, \tag{2}$$

where M_t is the moment due to the torsional spring, and we note that the moment arm for the load P is the distance L for small θ.

Force Laws For a torsional spring, we have $M_t = k_t\theta$.

Kinematic Equations For small θ, we relate δ and θ using $\delta = L\theta$.

Computation Substituting the force law and the kinematic relation into Eq. (2) and solving the resulting equation for k_t, we obtain

$$k_t = \frac{PL^2}{\delta} \quad\Rightarrow\quad k_t = \frac{3EI_{\text{cs}}}{L}, \tag{3}$$

where we have used Eq. (1) in going from the first to the second expression for k_t.

---

* The modulus of elasticity is a material property representing material flexibility. Recall that $1\,\text{Pa} = 1\,\text{N/m}^2$.

The Equation of Motion for the Rigid Bar

Governing Equations

Balance Principles Now that we have k_t, for the bar in Fig. 2, we draw the FBD shown in Fig. 5, and we sum moments about point O to obtain

$$\sum M_O\colon\quad -M_t = I_O \alpha_{\text{bar}}, \tag{4}$$

where the mass moment of inertia of the bar with respect to O is $I_O = \frac{1}{3}mL^2$.

Force Laws The expression for M_t is unchanged and is given by $M_t = k_t\theta$.

Kinematic Equations The kinematic equation relating α_{bar} to θ is

$$\alpha_{\text{bar}} = \ddot{\theta}. \tag{5}$$

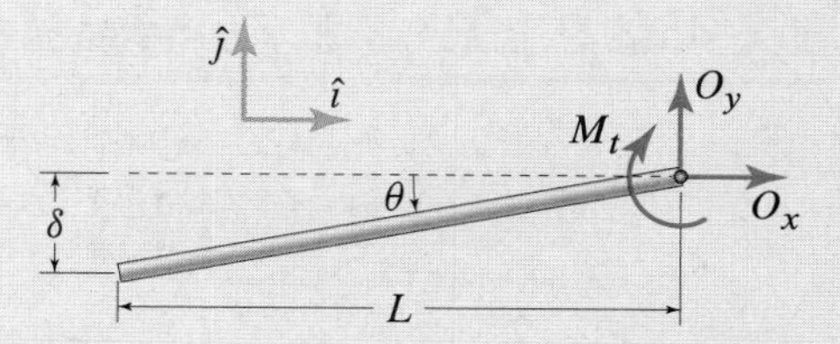

Figure 5
FBD of the bar while vibrating.

Computation Substituting Eq. (3), the force law, and Eq. (5) into Eq. (4), we obtain the equation of motion as

$$-\frac{3EI_{\text{cs}}}{L}\theta = \frac{1}{3}mL^2\ddot{\theta} \quad\Rightarrow\quad \ddot{\theta} + \frac{9EI_{\text{cs}}}{mL^3}\theta = 0, \tag{6}$$

where we recall that $I_{\text{cs}} = \frac{1}{4}\pi r^4$. Comparing Eq. (6) to Eq. (9.12), we see that the natural frequency is

$$\boxed{\omega_n = 3\sqrt{\frac{EI_{\text{cs}}}{mL^3}}.} \tag{7}$$

To obtain a numerical value for ω_n, we find that the volume of the nanowire is $\pi r^2 L$ and that the mass is then $m = \rho\pi r^2 L = 1.953\times 10^{-15}$ kg. The area moment of inertia is $I_{\text{cs}} = \frac{1}{4}\pi r^4 = 5.821\times 10^{-28}\,\text{m}^4$. Using these results, along with $L = 9.8\times 10^{-6}$ m and $E = 152\times 10^9\,\text{N/m}^2$, we find that

$$\boxed{\omega_n = 2.081\times 10^7\,\text{rad/s} \quad\text{and}\quad f = 3.313\,\text{MHz}.} \tag{8}$$

Discussion & Verification The quantity under the square root in Eq. (7) has dimensions of 1 over time squared. Hence, the dimensions of ω_n are 1 over time, as they should be. While it is hard to know what the frequency of vibration of a cantilevered bar of this size should be, in the Closer Look below, we will see that our model actually is quite good.

A Closer Look From the theory of vibration of continuous systems, one can show that a continuous system, like this nanowire, vibrates with infinitely many natural frequencies. The first (i.e., the smallest) of these frequencies provides the relevant comparison and is given by

$$(\omega_n)_{\text{exact}} = 3.516\sqrt{\frac{EI_{\text{cs}}}{mL^3}}, \tag{9}$$

which, when compared with Eq. (7), tells us that our model is only off by about 15%. Evaluating Eq. (9) numerically, we find that

$$(\omega_n)_{\text{exact}} = 2.439\times 10^7\,\text{rad/s} \quad\text{and}\quad f_{\text{exact}} = 3.883\,\text{MHz}. \tag{10}$$

In Prob. 9.17, we have the opportunity to examine another model for a cantilevered wire such as this and see how the natural frequency changes with the addition of a few zeptograms of virus to the end of the wire.

Interesting Fact

Natural frequency of a wooden yardstick. As a comparison, it is interesting to compute the natural frequency of a wooden yardstick using Eq. (9). Using properties typical of a wooden yardstick, that is, $E = 12$ GPa, $m = 0.0614$ kg, a 28 mm × 4 mm cross section (which gives $I_{\text{cs}} = 1.493\times 10^{-10}\,\text{m}^4$), and $L = 0.9144$ m, we find that $\omega_n = 21.72$ rad/s, which corresponds to $f = 3.457$ Hz. The nanowire's natural frequency is 1.1 million times higher!

EXAMPLE 9.3 Energy Method: Equation of Motion of a Diving Board

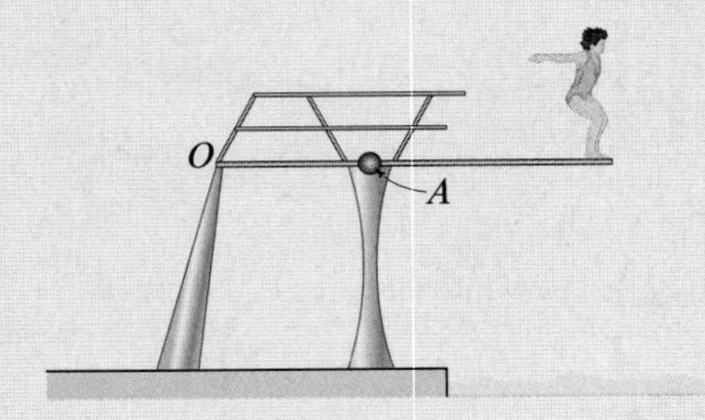

Figure 1

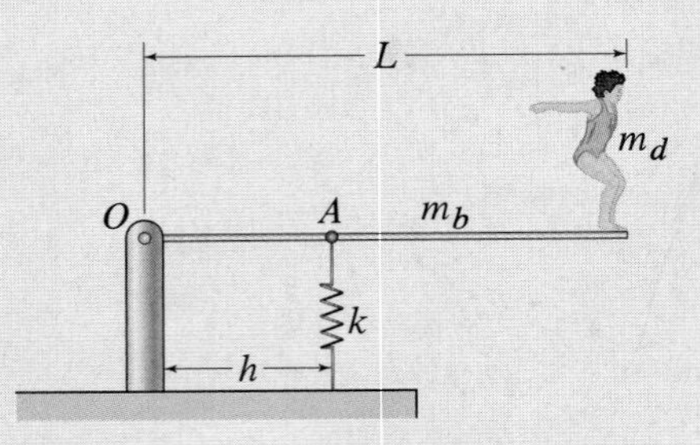

Figure 2
The model used to analyze the oscillation of the diving board and diver.

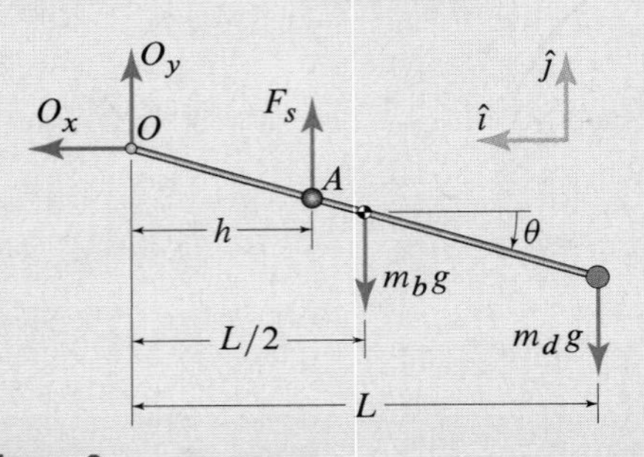

Figure 3
FBD of the board and diver as they oscillate. Note that the dimensions shown apply only when θ is small, so $\cos\theta \approx 1$.

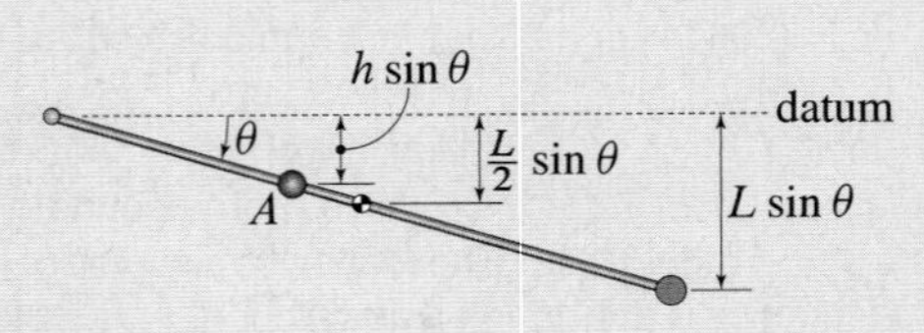

Figure 4
The vertical displacements of the relevant points on the diving board as it rotates.

A diver is causing the end of the diving board shown in Fig. 1 to oscillate. Referring to Fig. 2, we will model the board as a thin, uniform, *rigid* plate of mass m_b and length L and assume that the board is pinned at O. To model the elastic response of the board, we assume that the board oscillates due to a spring of stiffness k attached to the board at what was the fulcrum at A. In addition, we will model the diver as a point mass of mass m_d standing at the end of the board. Use the energy method to find the equation of motion for small rotations of the board.

SOLUTION

Road Map & Modeling The system is conservative since only the spring and the two weight forces do work as the system oscillates. This allows us to write the sum of the kinetic and potential energies of the system at an arbitrary position and then differentiate that sum with respect to time to obtain the equation of motion.

Governing Equations

Balance Principles Since energy is conserved, we can write that the sum of the kinetic and potential energies is constant, so

$$T + V = \text{constant} \quad \Rightarrow \quad \frac{d}{dt}(T+V) = 0. \tag{1}$$

The kinetic energy is given by

$$T = \tfrac{1}{2} I_O \dot{\theta}^2 + \tfrac{1}{2} m_d v_d^2, \tag{2}$$

where $I_O = \frac{1}{3} m_b L^2$ is the mass moment of inertia of the diving board with respect to point O and v_d is the speed of the diver.

Force Laws The potential energy of the system at an arbitrary angle θ is (see Fig. 4)

$$V = \tfrac{1}{2} k(\delta_{\text{st}} + h\sin\theta)^2 - m_b g \tfrac{L}{2}\sin\theta - m_d g L\sin\theta \tag{3}$$

$$\approx \tfrac{1}{2} k(\delta_{\text{st}} + h\theta)^2 - m_b g \tfrac{L}{2}\theta - m_d g L\theta, \tag{4}$$

where δ_{st} is the compression of the spring when the diving board is in static equilibrium with the diver on it, i.e., when $\theta = 0$, and where we have approximated V as a quadratic function of θ (i.e., position) in approximating Eq. (3) as Eq. (4).

Kinematic Equations Since the diver is rotating about the fixed point at O, the diver's speed is $v_d = L\dot{\theta}$.

Computation Substituting the potential energy in Eq. (4), the kinetic energy in Eq. (2), and the kinematic equation into Eq. (1), and then taking the time derivative, we obtain

$$(I_O + m_d L^2)\dot{\theta}\ddot{\theta} + k(\delta_{\text{st}} + h\theta)h\dot{\theta} - \tfrac{1}{2} m_b g L\dot{\theta} - m_d g L\dot{\theta} = 0. \tag{5}$$

Canceling $\dot{\theta}$, we note that the term $kh\delta_{\text{st}}$ is the moment about O due to the spring force needed to hold the system in equilibrium (i.e., at $\theta = 0$). This moment is equal and opposite to the moment about O created by the two weight forces, that is, $-\frac{1}{2} m_b g L - m_d g L$. Therefore, Eq. (5) reduces to the final equation of motion (we have used $I_O = \frac{1}{3} m_b L^2$)

$$\ddot{\theta} + \frac{kh^2}{\left(\frac{1}{3} m_b + m_d\right)L^2}\theta = 0. \tag{6}$$

Discussion & Verification The coefficient of θ in Eq. (6) has dimensions of 1 over time squared, as it should.

Problems

Problem 9.1

Show that Eq. (9.15) is equivalent to Eq. (9.3) if $C = \sqrt{A^2 + B^2}$ and $\tan\phi = A/B$.

Problem 9.2

Derive the formula for the mass moment of inertia of an arbitrarily shaped rigid body about its mass center based on the body's period of oscillation τ when suspended as a pendulum. Assume that the mass of the body m is known and that the location of the mass center G is known relative to the pivot point O.

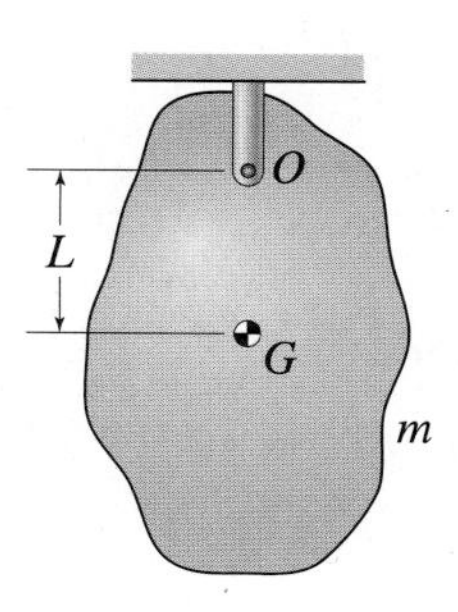

Figure P9.2

Problem 9.3

The thin ring of radius R and mass m is suspended by the pin at O. Determine its period of vibration if it is displaced a small amount and released.

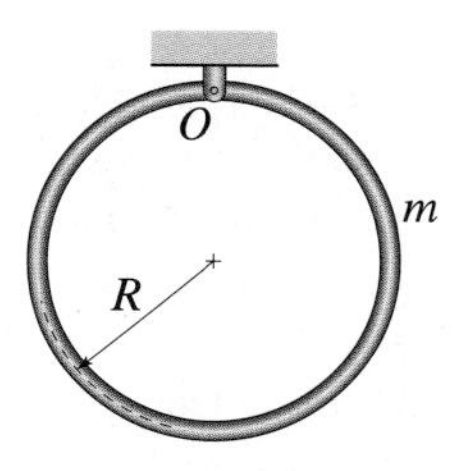

Figure P9.3

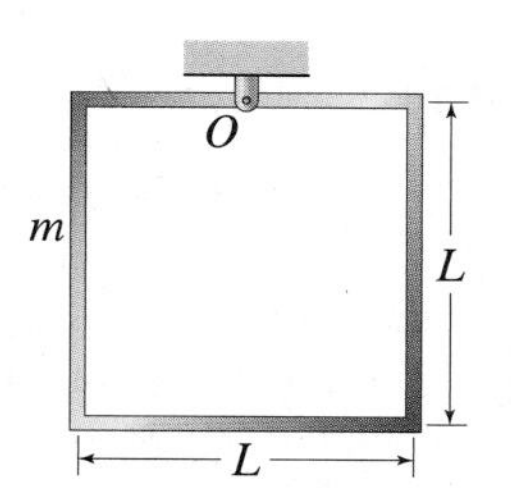

Figure P9.4

Problem 9.4

The thin square hoop has mass m and is suspended by the pin at O. Determine its period of vibration if it is displaced a small amount and released.

Problem 9.5

The swinging bar and the vibrating block of mass m are made to vibrate on Earth, and their respective natural frequencies are measured. The two systems are then taken to the Moon and are again allowed to vibrate at their respective natural frequencies. How will the natural frequency of each system change when compared with that on the Earth, and which of the two systems will experience the larger change in natural frequency?

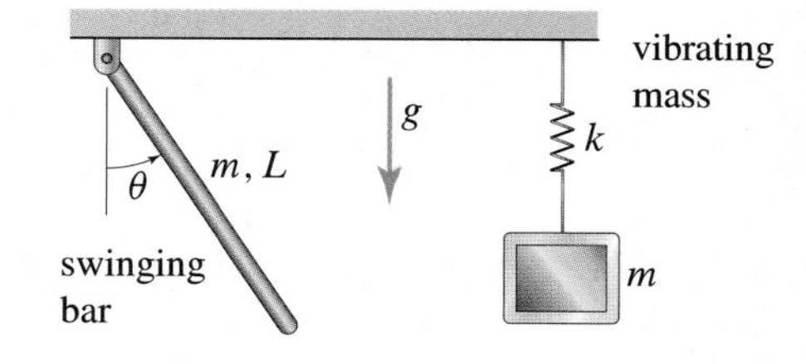

Figure P9.5

Problem 9.6

The uniform disk of radius R and thickness t is attached to the thin shaft of radius r, length L, and negligible mass. The end A of the shaft is fixed. From mechanics of materials, it can be shown that if a torque M_z is applied to the free end of the shaft, then it can be related to the twist angle θ via

$$\theta = \frac{M_z L}{GJ},$$

where G is the shear modulus of elasticity of the shaft and $J = \frac{\pi}{2} r^4$ is the polar moment of inertia of the cross-sectional area of the shaft. Letting ρ be the mass density of the disk. Using the given relationship between M_z and θ, determine the natural frequency of vibration of the disk in terms of the given dimensions and material properties when it is given a small angular displacement θ in the plane of the disk.

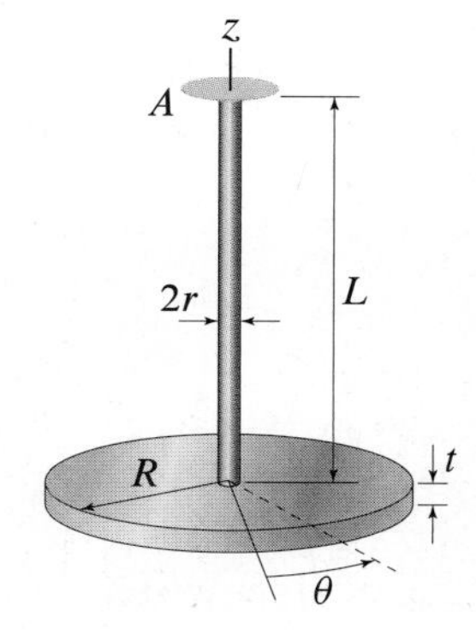

Figure P9.6

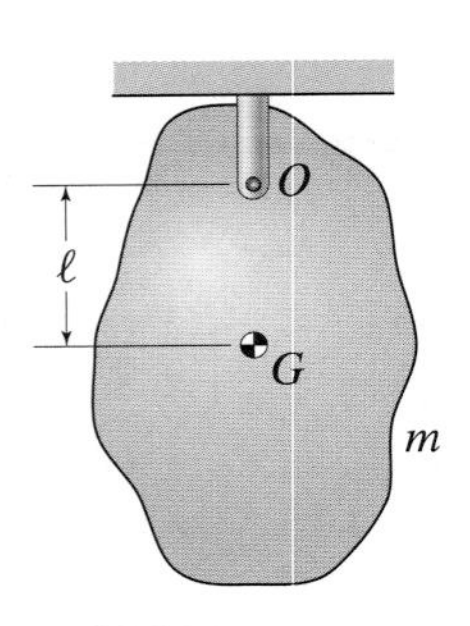

Figure P9.7 and P9.8

Problems 9.7 and 9.8

A rigid body of mass m, mass center at G, and mass moment of inertia I_G is pinned at an arbitrary point O and allowed to oscillate as a pendulum.

Problem 9.7 By writing the Newton-Euler equations, determine the distance ℓ from G to the pivot point O so that the pendulum has the highest possible natural frequency of oscillation.

Problem 9.8 Using the energy method, determine the distance ℓ from G to the pivot point O so that the pendulum has the highest possible natural frequency of oscillation.

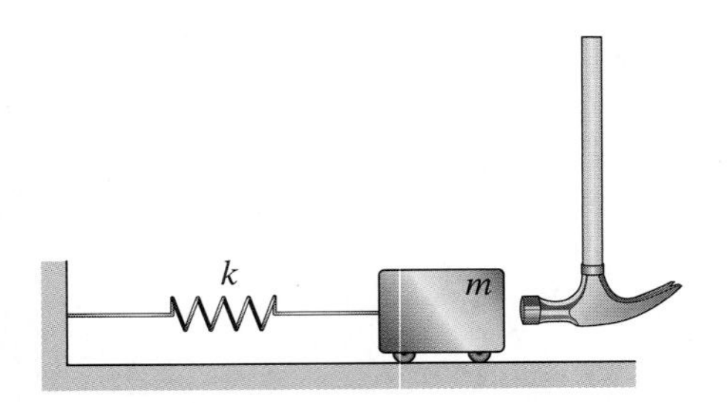

Figure P9.9

Problem 9.9

A block of mass $m = 3\,\mathrm{kg}$ is in equilibrium when a hammer hits it, imparting a velocity v_0 of $2\,\mathrm{m/s}$ to it. If k is $120\,\mathrm{N/m}$, determine the amplitude of the ensuing vibration and find the maximum acceleration experienced by the block.

Problem 9.10

A construction worker C is standing at the midpoint of a 14 ft long pine board that is simply supported. The board is a standard 2×12, so its cross-sectional dimensions are as shown. Assuming the worker weighs 180 lb and he flexes his knees once to get the board oscillating, determine his vibration frequency. Neglect the weight of the beam, and use the fact that a load P applied to a simply supported beam will deflect the center of the beam $PL^3/(48EI_{\mathrm{cs}})$, where L is the length of the beam, E is its modulus of elasticity, and I_{cs} is the area moment of inertia of the cross section of the beam. The elastic modulus of pine is 1.8×10^6 psi.

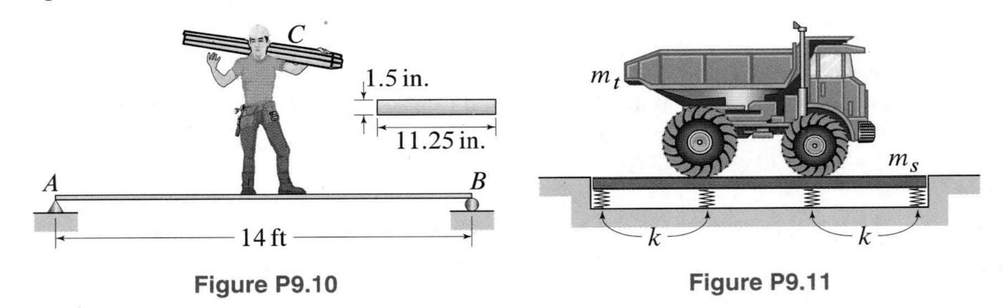

Figure P9.10

Figure P9.11

Problem 9.11

A truck drives onto a deck scale to be weighed, thus causing the truck and scale to vibrate vertically at the natural frequency of the system. The empty truck weighs 74,000 lb, the scale platform weighs 51,000 lb, and the platform is supported by eight identical springs (four of which are shown), each with constant $k = 3.6 \times 10^5\,\mathrm{lb/ft}$. Modeling the truck, its contents, and the concrete deck as a single particle, if a vibration frequency of 3.3 Hz is measured, what is the weight of the payload being carried by the truck?

Figure P9.12

Problem 9.12

The buoy in the photograph can be modeled as a circular cylinder of diameter d and mass m. If the buoy is pushed down in the water, which has density ρ, it will oscillate vertically. Determine the frequency of oscillation. Evaluate your result for $d = 1.2\,\mathrm{m}$, $m = 900\,\mathrm{kg}$, and surface seawater, which has a density of $\rho = 1027\,\mathrm{kg/m^3}$. *Hint:* Use Archimedes' principle, which states that a body wholly or partially submerged in a fluid is buoyed up by a force equal to the weight of the displaced fluid.

Problems 9.13 through 9.16

The L-shaped bar lies in the vertical plane and is pinned at O. One end of the bar has a linear elastic spring with constant k attached to it, and attached at the other end is a sphere of mass m and negligible size. The angle θ is measured from the equilibrium position of the system, and it is assumed to be small.

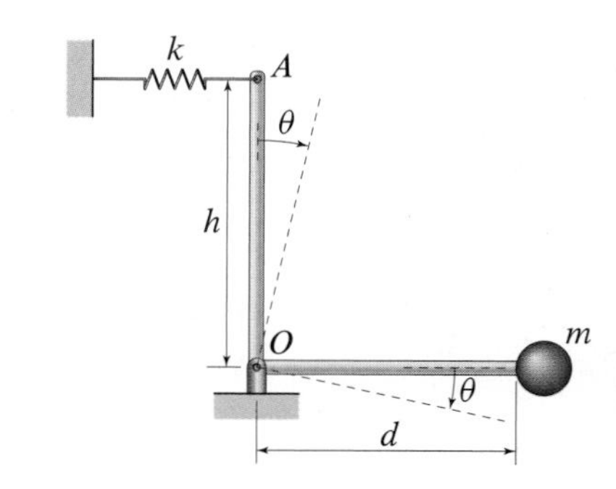

Figure P9.13–P9.16

Problem 9.13 Assuming that the L-shaped bar has negligible mass, determine the natural period of vibration of the system by writing the Newton-Euler equations.

Problem 9.14 Assuming that the L-shaped bar has negligible mass, determine the natural period of vibration of the system using the energy method.

Problem 9.15 Assuming that the L-shaped bar has mass per unit length ρ, determine the natural period of vibration of the system by writing the Newton-Euler equations.

Problem 9.16 Assuming that the L-shaped bar has mass per unit length ρ, determine the natural period of vibration of the system using the energy method.

Problem 9.17

For the silicon nanowire in Example 9.2, use the lumped mass model shown, in which a point mass m is connected to a rod of negligible mass and length L that is pinned at O, to determine the natural frequency ω_n and frequency f of the nanowire. Use the values given in Example 9.2 for the mass of the lumped mass, the length of the massless rod, and the parameters used to determine the spring constant $k = 3EI_{\text{cs}}/L^3$. You may use either δ or θ as the position variable in your solution. Assume that the displacement of m is small so that it moves vertically.

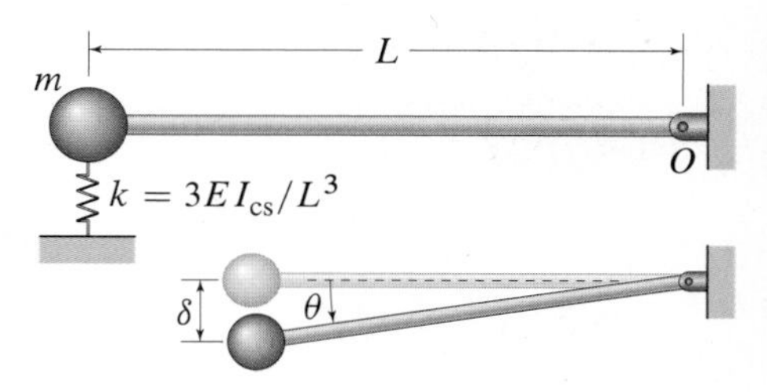

Figure P9.17

Problem 9.18

The small sphere A has mass m and is fixed at the end of the arm OA of negligible mass, which is pinned at O. If the linear elastic spring has stiffness k, determine the equation of motion for small oscillations, using

(a) the vertical position of the mass A as the position coordinate,

(b) the angle formed by the arm OA with the horizontal as the position coordinate.

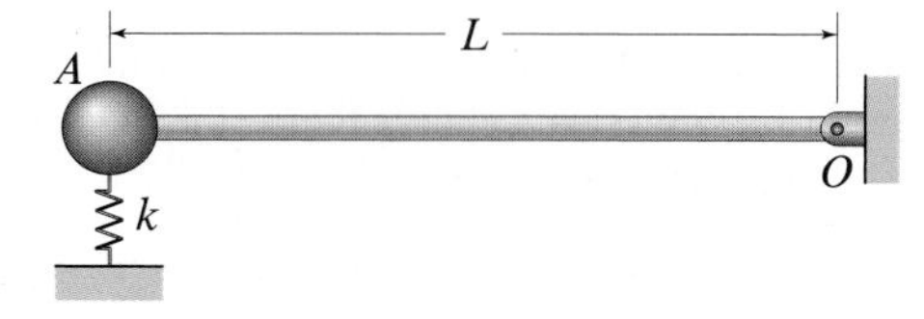

Figure P9.18

Problems 9.19 and 9.20

The uniform cylinder rolls without slipping on a flat surface. Let $k_1 = k_2 = k$ and $r = R/2$. Assume that the horizontal motion of G is small.

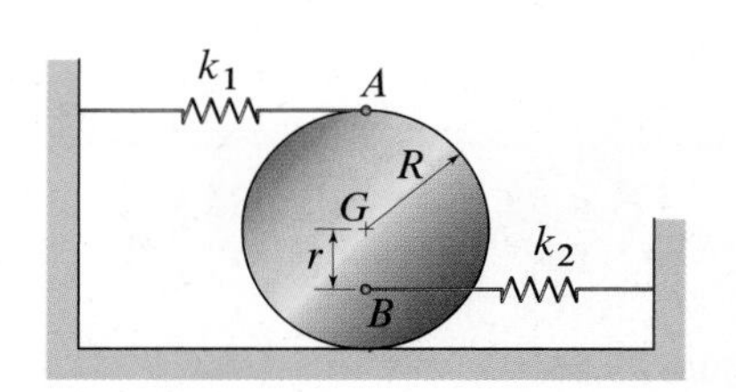

Figure P9.19 and P9.20

Problem 9.19 Determine the equation of motion for the cylinder by writing its Newton-Euler equations. Use the horizontal position of the mass center G as the degree of freedom.

Problem 9.20 Determine the equation of motion for the cylinder using the energy method. Use the horizontal position of the mass center G as the degree of freedom.

Problems 9.21 and 9.22

Grandfather clocks keep time by advancing the hands a set amount per oscillation of the pendulum. Therefore, the pendulum needs to have a very accurate period for the clock to keep time accurately. As a fine adjustment of the pendulum's period, many grandfather clocks have an adjustment nut on a bolt at the bottom of the pendulum disk. Screwing this nut inward or outward changes the mass distribution of the pendulum by moving the pendulum disk closer to or farther from the axis of rotation at O. Model the pendulum as a uniform disk of radius r and mass m_p at the end of a rod of negligible mass and length $L - r$, and assume that the oscillations of θ are small. Let $m_p = 0.7$ kg and $r = 0.1$ m.

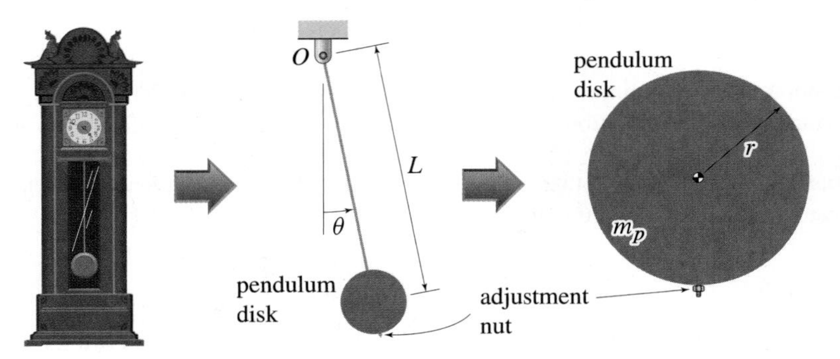

Figure P9.21 and P9.22

Problem 9.21 If the pendulum disk is initially at a distance $L = 0.85$ m from the pin at O, how much would the period of the pendulum change if the adjustment nut with a lead* of 0.5 mm was rotated four complete rotations closer to the disk? In addition, how much time would the clock gain or lose in a 24 h period if this were done?

Problem 9.22 The clock is running slow so that it is losing 2 minutes every 24 hours (i.e., the clock takes 1442 minutes to complete a 1440-minute day). If the pendulum disk is at $L = 0.85$ m, how many turns of the adjustment nut would be needed, and in what direction, to correct the pendulum's period if the screw lead* is 0.5 mm?

Problems 9.23 and 9.24

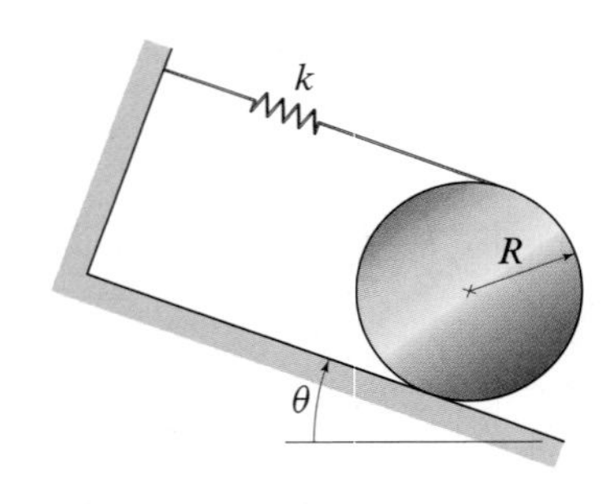

Figure P9.23 and P9.24

The uniform cylinder of mass m and radius R rolls without slipping on the inclined surface. The spring with constant k wraps around the cylinder as it rolls.

Problem 9.23 Determine the equation of motion for the cylinder by writing its Newton-Euler equations. Determine the numerical value of the period of oscillation of the cylinder using $k = 30$ N/m, $m = 10$ kg, $R = 30$ cm, and $\theta = 20°$.

Problem 9.24 Determine the equation of motion for the cylinder using the energy method. Determine the numerical value of the period of oscillation of the cylinder using $k = 30$ N/m, $m = 10$ kg, $R = 30$ cm, and $\theta = 20°$.

Problem 9.25

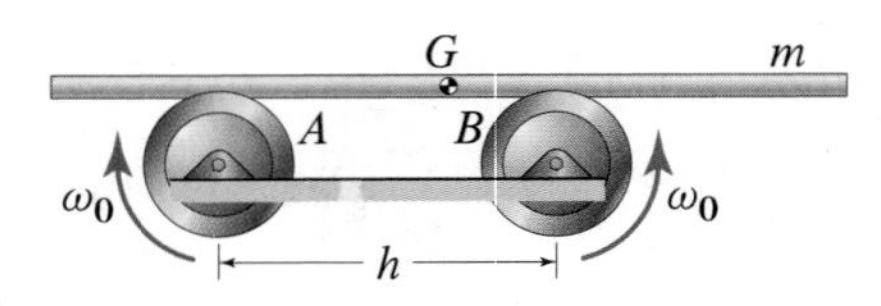

Figure P9.25

A uniform bar of mass m is placed off-center on two counter-rotating drums A and B. Each drum is driven with constant angular speed ω_0, and the coefficient of kinetic friction between the drums and the bar is μ_k. Determine the natural frequency of oscillation of the bar on the rollers. *Hint:* Measure the horizontal position of G relative to the midpoint between the two drums, and assume that the drums rotate sufficiently fast so that the drums are always slipping relative to the bar.

* The *lead of a screw* is the axial distance a nut would advance along the screw during one complete rotation of the nut.

Problem 9.26

The uniform cylinder A of radius r and mass m is released from a small angle θ inside the large cylinder of radius R. Assuming that it rolls without slipping, determine the natural frequency and period of oscillation of A.

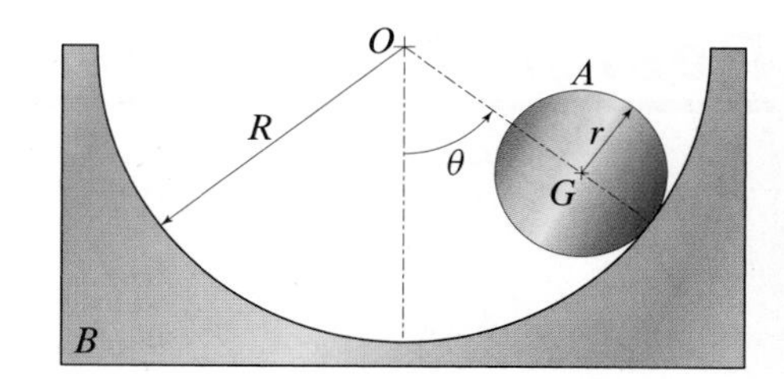

Figure P9.26

Problem 9.27

The uniform sphere A of radius r and mass m is released from a small angle θ inside the large cylinder of radius R. Assuming that it rolls without slipping, determine the natural frequency and period of oscillation of the sphere.

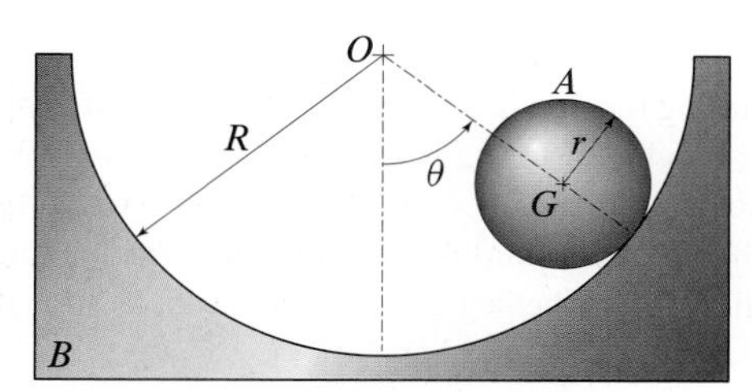

Figure P9.27

Problem 9.28

The U-tube manometer lies in the vertical plane and contains a fluid of density ρ that has been displaced a distance y and oscillates in the tube. If the cross-sectional area of the tube is A and the total length of the fluid in the tube is L, determine the natural period of oscillation of the fluid, using the energy method. *Hint:* As long as the curved portion of the tube is always filled with liquid (i.e., the oscillations do not get large enough to empty part of it), the contribution of the liquid in the curved portion to the potential energy is *constant*.

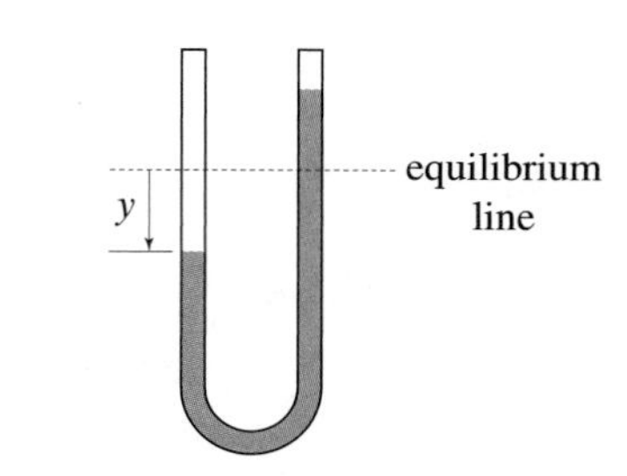

Figure P9.28

Problem 9.29

The uniform semicylinder of radius R and mass m rolls without slip on the horizontal surface. Using the energy method, determine the period of oscillation for small θ.

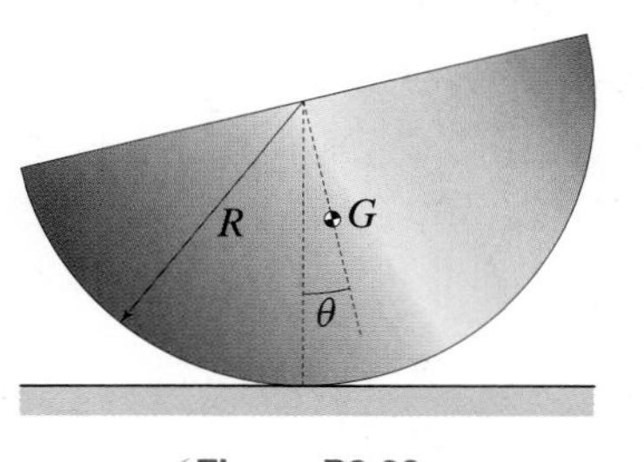

Figure P9.29

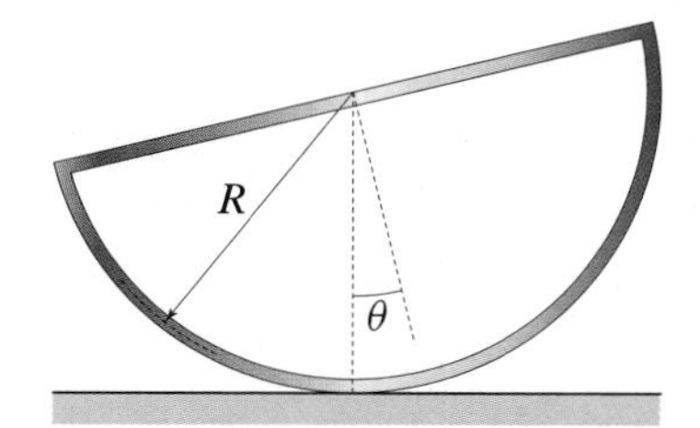

Figure P9.30

Problem 9.30

The thin-shell semicylinder of radius R and mass m rolls without slip on the horizontal surface. Using the energy method, determine the period of oscillation for small θ.

Design Problems

Design Problem 9.1

As part of a manufacturing process, a uniform bar is placed on a centering device consisting of two counter-rotating drums A and B and a frictional slider C that provides light damping. When the bar is placed off-center on the two counter-rotating drums, it oscillates back and forth due to the sliding friction between each drum and the bar, and it eventually settles into the centered position due to friction between the block at C and the bar. Once the bar has been centered, the next step in the manufacturing process can begin.

Assume that the steel bar has length $L = 1.2$ m, radius $r = 22.5$ mm, and density $\rho = 7.85\,\text{g/cm}^3$. Assuming that the initial misalignment (i.e., the initial distance between G and C) of each bar can be up to 0.25 m and that sliding between the drums and the bar never ceases, design the drums (i.e., their radius and what they are made of) and determine their placement so that sliding is maintained for the entire range of motion. Since the damping is light, it can be neglected. In addition, assume that you want to drive each drum with a constant angular speed ω_0.

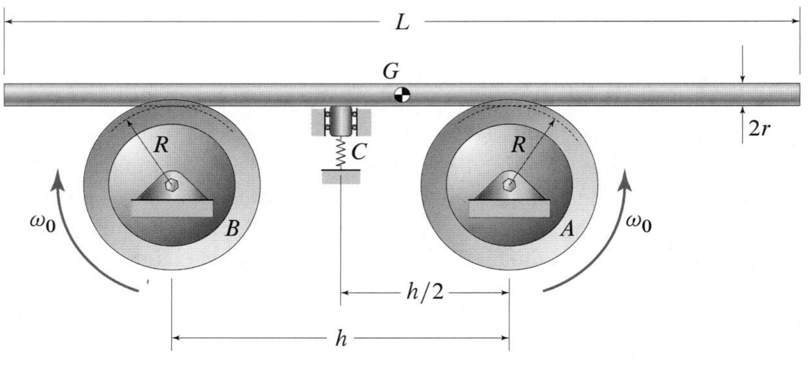

Figure DP9.1

9.2 Undamped Forced Vibration

Many systems are *forced* to vibrate by an external excitation. This section is devoted to the forced vibration of mechanical systems.

Standard form of the forced harmonic oscillator

A standard forced harmonic oscillator is shown in Fig. 9.11, in which the block of mass m is attached to a fixed support by a linear spring of constant k and is also being driven by the time-dependent force $P(t) = F_0 \sin \omega_0 t$. Modeling the block as a particle, its FBD is as shown in Fig. 9.12, where F_s is the spring force acting on the block. Summing forces in the x direction, we obtain

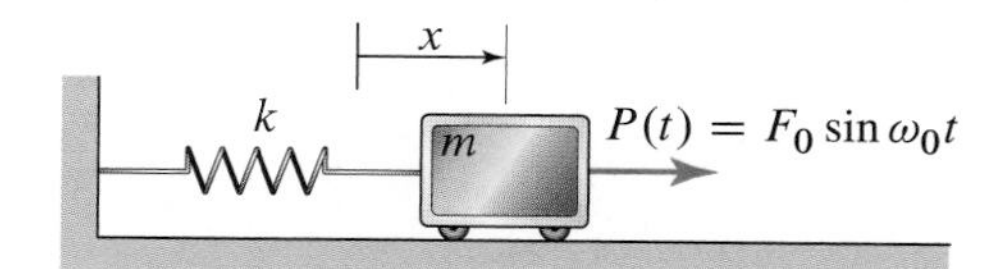

Figure 9.11
A forced harmonic oscillator whose equation of motion is given by Eq. (9.31) with $\omega_n = \sqrt{k/m}$. The position x is measured from the equilibrium position of the system when $F_0 = 0$.

$$\sum F_x: \quad P(t) - F_s = ma_x, \tag{9.29}$$

where the force law is given by $F_s = kx$ and the kinematic equation is $a_x = \ddot{x}$. Substituting these relations, as well as $P(t)$ into Eq. (9.29), we obtain

$$F_0 \sin \omega_0 t - kx = m\ddot{x} \quad \Rightarrow \quad \ddot{x} + \frac{k}{m}x = \frac{F_0}{m} \sin \omega_0 t. \tag{9.30}$$

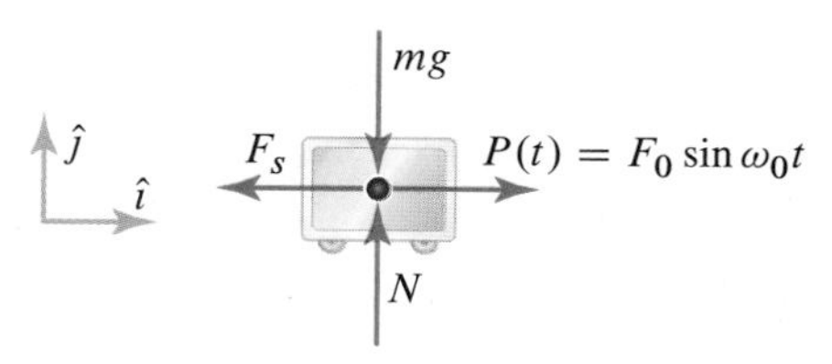

Figure 9.12
FBD of the forced harmonic oscillator in Fig. 9.11.

Noting that $\omega_n^2 = k/m$, this last equation becomes

$$\boxed{\ddot{x} + \omega_n^2 x = \frac{F_0}{m} \sin \omega_0 t,} \tag{9.31}$$

which is the *standard form of the forced harmonic oscillator equation*. It is a *nonhomogeneous* version of Eq. (9.12) on p. 665 as a result of the term $(F_0/m) \sin \omega_0 t$. The term on the right-hand side of Eq. (9.31) is a function of *only* the independent variable t. It is often called a *forcing function* because it forces the system to vibrate. This particular type of forcing is harmonic because it is a harmonic function of time.

The theory of differential equations tells us that the *general solution* of Eq. (9.31) is the sum of the *complementary solution* $x_c(t)$ and a *particular solution* $x_p(t)$. The *complementary solution** is the solution of the associated homogeneous equation (i.e., Eq. (9.12)) given in Eq. (9.3) (or in Eq. (9.13)). The *particular solution* is *any* solution of Eq. (9.31). One way to obtain a particular solution is to guess its form and then verify whether or not the guess is correct. Since it seems reasonable that the response of a forced harmonic oscillator should resemble the forcing, we conjecture that the particular solution x_p is of the form

$$x_p = D \sin \omega_0 t, \tag{9.32}$$

where D is a constant to be determined. We can verify whether our guess is correct by substituting Eq. (9.32) into Eq. (9.31). Doing so yields

$$-D\omega_0^2 \sin \omega_0 t + \omega_n^2 D \sin \omega_0 t = \frac{F_0}{m} \sin \omega_0 t. \tag{9.33}$$

Canceling $\sin \omega_0 t$ and solving for D, we obtain

$$D = \frac{F_0/m}{\omega_n^2 - \omega_0^2} = \frac{F_0/k}{1 - (\omega_0/\omega_n)^2}, \tag{9.34}$$

Interesting Fact

How practical is harmonic forcing? The answer to this question lies in an amazing result due to Jean Baptiste Joseph Fourier (1768–1830) and later contributors. It says that *any* periodic piecewise smooth function can be represented by an infinite series of sines and cosines (called *Fourier series* in honor of Fourier). *This means that any periodic forcing can be regarded as the sum of harmonic functions!* In addition, given the nature of the left side of Eq. (9.31) (i.e., it is linear), it turns out that the overall particular solution for a sum of harmonic forcing terms is simply the sum of the particular solutions for each individual harmonic forcing term. These results taken together allow engineers to easily obtain the solutions to problems with *any* periodic forcing as a sum of simple forced harmonic oscillator solutions. Because periodic forcing is ubiquitous in engineering systems, this is one of the most important results in applied mathematics.

* The complementary solution is sometimes called the *homogeneous solution*.

where we have assumed that $\omega_0 \neq \omega_n$ and have used the fact that $\omega_n^2 = k/m$. Choosing D as in Eq. (9.34), we see that the guess in Eq. (9.32) is correct, and the corresponding particular solution is

$$x_p = \frac{F_0/k}{1 - (\omega_0/\omega_n)^2} \sin \omega_0 t. \tag{9.35}$$

Combining the complementary solution in Eq. (9.13) with the particular solution in Eq. (9.35), the *general solution* to Eq. (9.31) is

$$x = x_c + x_p = A \sin \omega_n t + B \cos \omega_n t + \frac{F_0/k}{1 - (\omega_0/\omega_n)^2} \sin \omega_0 t, \tag{9.36}$$

where, as usual, A and B are constants determined by enforcing the initial conditions.

Equation (9.36) tells us that the vibration of a forced harmonic oscillator is composed of two parts: the complementary solution x_c that describes the *free vibration* of the system and the particular solution x_p that describes the *forced vibration* due to $F_0 \sin \omega_0 t$. As we will see in Section 9.3, the free vibration corresponding to x_c will die out with any amount of damping or energy dissipation, which is always present in real physical systems. For this reason, free vibration is often referred to as *transient vibration*. On the other hand, the forced vibration corresponding to x_p will be there as long as the forcing is there, and so it is often called *steady-state vibration*.*

When a vibration is forced, it is important to know the amplitude of the motion since that will determine the deformation and the deformation-related forces that the system has to endure. Equation (9.35) tells us that the amplitude of the steady-state vibration, which is given by

$$x_{\text{amp}} = \frac{F_0/k}{1 - (\omega_0/\omega_n)^2}, \tag{9.37}$$

depends on the *frequency ratio* ω_0/ω_n. If we now define the *magnification factor* MF for this case to be the ratio of the amplitude x_{amp} of steady-state vibration to the static deflection F_0/k caused by the force F_0, we find it to be

$$\text{MF} = \frac{x_{\text{amp}}}{F_0/k} = \frac{1}{1 - (\omega_0/\omega_n)^2}, \tag{9.38}$$

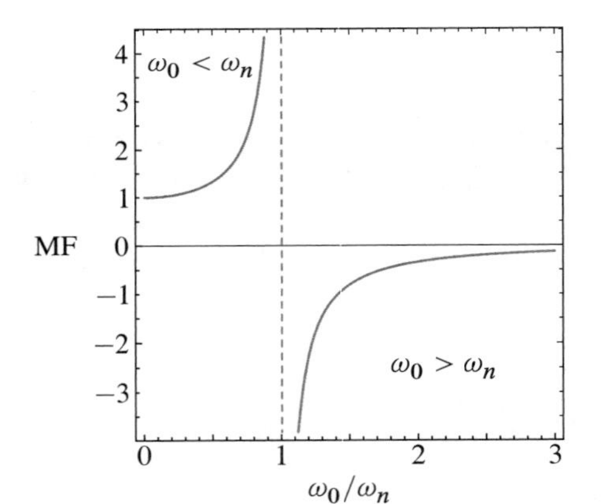

Figure 9.13
MF as a function of the frequency ratio ω_0/ω_n.

a plot of which is shown in Fig. 9.13. Notice that when $\omega_0/\omega_n \ll 1$, MF $\to$ 1. That is, for low-frequency forcing, the motion of the block is in the direction of the forcing (they are said to be *in phase* with one another). We expect this to be the case since, in the limit as $\omega_0/\omega_n \to 0$, we obtain the static deflection. As the forcing frequency ω_0 approaches the natural frequency of the system ω_n, MF increases dramatically and goes to infinity as $\omega_0/\omega_n \to 1$. The situation in which $\omega_0 \approx \omega_n$ is called *resonance*, and it results in very large vibration amplitudes. Resonances in engineering systems are generally undesirable because they result in large displacements and deformations, often leading to premature failure. However, there do exist beneficial resonances in engineering systems, such as in amplifiers or in devices designed to aid sputum clearance of the airways of respiratory patients.† As we will see in

* We do not discuss the solution of Eq. (9.31) for $\omega_0 = \omega_n$ because such a solution does not describe a steady-state vibration.

† See L. C. de Lima et al., "Mechanical Evaluation of a Respiratory Device," *Medical Engineering & Physics*, **27**, 2005, pp. 181–187.

Section 9.3, in a system with even a small amount of energy dissipation or damping, resonance does not result in infinite amplitudes, but the amplitudes can still grow to be *very* large.

Looking back at the system in Fig. 9.11, when $\omega_0 > \omega_n$, the MF is negative and the forcing is out of phase with the motion of the block. Finally, when $\omega_0 \gg \omega_n$, the force changes direction so rapidly compared to the natural frequency of the system that the system remains almost stationary, and the MF $\to 0$.

Harmonic excitation of the support

If the support that couples the spring to the block is excited harmonically rather than the block itself, we obtain a modified form of the forced harmonic oscillator equation. To see this, consider the system shown in Fig. 9.14, which shows a block of mass m that is coupled to a harmonically excited support by a linear spring of constant k. The position of the oscillating support is given by $x_s = X_s \sin \omega_s t$, and the position of the block is given by x. The FBD of this system is shown in Fig. 9.15. Summing forces in the x direction gives

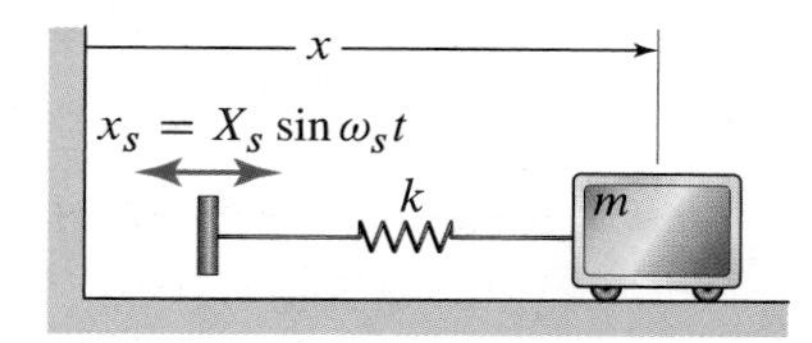

Figure 9.14
A simple harmonic oscillator whose support is being harmonically excited.

$$\sum F_x: \quad -F_s = ma_x, \tag{9.39}$$

where F_s is the force in the spring. Since both ends of the spring are moving, its force law is given by $F_s = k(x - x_s) = k(x - X_s \sin \omega_s t)$, so Eq. (9.39) becomes

$$-k(x - X_s \sin \omega_s t) = ma_x = m\ddot{x}, \tag{9.40}$$

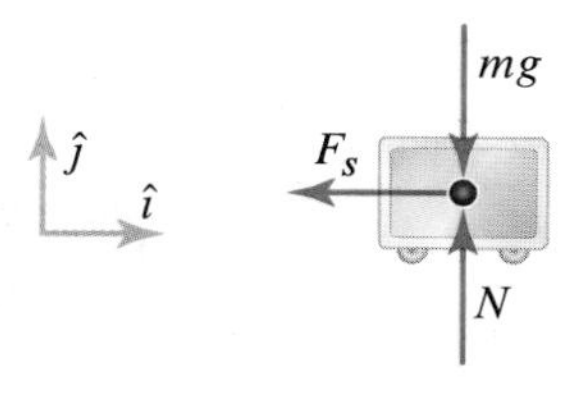

Figure 9.15
FBD of the block for the system shown in Fig. 9.14.

where we have substituted in the kinematic equation $a_x = \ddot{x}$. Upon rearranging, this becomes

$$\ddot{x} + \omega_n^2 x = \frac{kX_s}{m} \sin \omega_s t, \tag{9.41}$$

where $\omega_n^2 = k/m$. Equation (9.41) is of the same form as Eq. (9.31) with the term F_0 replaced by the term kX_s. The results given in Eqs. (9.35)–(9.38) are valid with that same replacement.

End of Section Summary

When a harmonic oscillator is subject to harmonic forcing, the standard form of the equation of motion is

Eq. (9.31), p. 681

$$\ddot{x} + \omega_n^2 x = \frac{F_0}{m} \sin \omega_0 t,$$

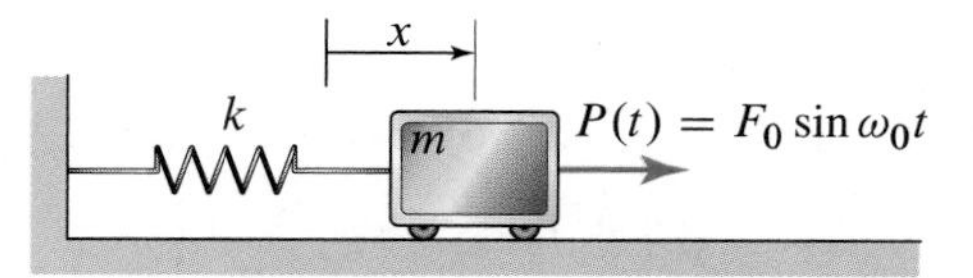

Figure 9.16
A forced harmonic oscillator whose equation of motion is given by Eq. (9.31) with $\omega_n = \sqrt{k/m}$. The position x is measured from the equilibrium position of the block.

where F_0 is the amplitude of the forcing and ω_0 is its frequency (see Fig. 9.16). The general solution to this equation consists of the sum of the complementary solution and a particular solution. The *complementary solution* x_c is the solution of the associated homogeneous equation, which is given by, for example, Eq. (9.13). For $\omega_0 \neq \omega_n$, a particular solution was found to be

Eq. (9.35), p. 682

$$x_p = \frac{F_0/k}{1 - (\omega_0/\omega_n)^2} \sin \omega_0 t,$$

and so the *general solution* is given by

Eq. (9.36), p. 682

$$x = x_c + x_p = A \sin \omega_n t + B \cos \omega_n t + \frac{F_0/k}{1 - (\omega_0/\omega_n)^2} \sin \omega_0 t,$$

where A and B are constants determined by the initial conditions. The amplitude of the steady-state vibration is

Eq. (9.37), p. 682

$$x_{\text{amp}} = \frac{F_0/k}{1 - (\omega_0/\omega_n)^2},$$

which means that the corresponding *magnification factor* MF is

Eq. (9.38), p. 682

$$\text{MF} = \frac{x_{\text{amp}}}{F_0/k} = \frac{1}{1 - (\omega_0/\omega_n)^2}.$$

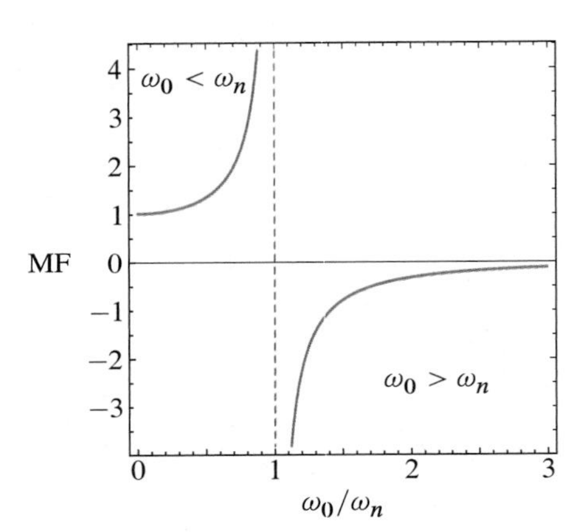

Figure 9.17
MF as a function of the frequency ratio ω_0/ω_n.

The plot of the MF in Fig. 9.17 illustrates the phenomenon of *resonance*, which occurs when $\omega_0 \approx \omega_n$ and which results in very large vibration amplitudes.

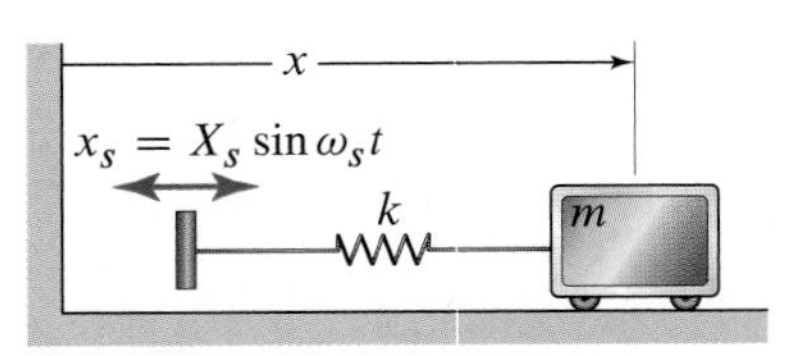

Figure 9.18
A harmonic oscillator whose support is being excited harmonically.

Harmonic excitation of the support. If the support of a structure is excited harmonically rather than the structure itself (see Fig. 9.18), then Eq. (9.31) is still the governing equation, except that F_0 is replaced by the spring constant k times the amplitude of the support vibration X_s. All solutions described previously are then valid with that same replacement.

EXAMPLE 9.4 *Motion of a Weight on a Vibrating Spring Scale*

The block A is resting on the platform P of a spring scale when the lab bench to which it is rigidly attached begins vibrating vertically. The block and the platform are coupled to the base B of the scale by a linear elastic spring that is internal to the scale (Fig. 1). If the combined mass of the block and platform is $m_A = 1.5\,\text{kg}$, the spring constant is $k = 50\,\text{N/m}$, and the vibration is sinusoidal with a frequency of 15 Hz and amplitude of 5 mm, determine the vertical motion of the platform and block as a function of time. Assume that the block A does not separate from the platform P during the vibration.

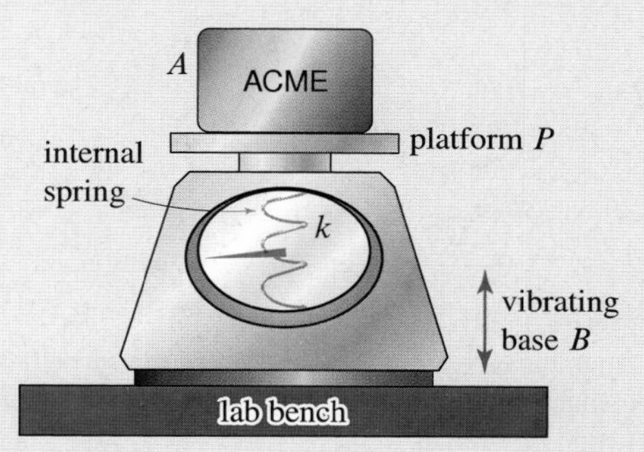

Figure 1
A scale sitting on a lab bench that is vibrating vertically. The base of the scale B is rigidly attached to the lab bench. The platform P and the mass A are coupled to the base of the scale by an internal spring.

SOLUTION

Road Map & Modeling The FBD of the block and platform is shown in Fig. 2, where F_s is the force exerted by the spring on the platform. Since this an example of harmonic excitation of a support, we can apply Newton's second law in the y direction to obtain the equation of motion and then use the results of this section to find the motion.

Governing Equations

Balance Principles Summing forces in the y direction in Fig. 2, we obtain

$$\sum F_y: \quad -F_s - m_A g = m_A a_{Ay}. \tag{1}$$

Force Laws Since both ends of the spring are moving, the force law for the spring is given by

$$F_s = k(y_A - y_B - \delta_s), \tag{2}$$

where y_A is the vertical position of the block measured from the static equilibrium position of the block, y_B is the y position of the base of the scale, and δ_s is the deflection of the spring when the system is in static equilibrium.

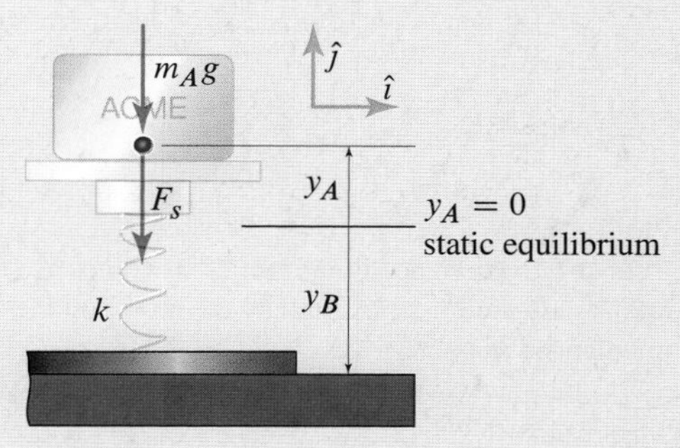

Figure 2
FBD of the block A and platform P in Fig. 1, as well as the definitions of y_A and y_B.

Kinematic Equations For the kinematic equations, we need to express a_{Ay} and y_B in terms of motion of the base B and the block A, which gives

$$a_{Ay} = \ddot{y}_A \quad \text{and} \quad y_B = Y_B \sin\omega_B t, \tag{3}$$

where $\omega_B = 15\,\text{Hz} = 94.25\,\text{rad/s}$ is the frequency of vibration of the base B and Y_B is the amplitude of vibration of the base.

Computation Substituting Eqs. (2) and (3) into Eq. (1), we get the equation of motion

$$-k(y_A - Y_B \sin\omega_B t - \delta_s) - m_A g = m_A \ddot{y}_A \quad \Rightarrow \quad \ddot{y}_A + \frac{k}{m_A} y_A = \frac{kY_B}{m_A}\sin\omega_B t, \tag{4}$$

where we have used the fact that $k\delta_s = m_A g$. Equation (4) is of the same form as Eq. (9.31), so we know that the solution must be given by Eq. (9.36) with $F_0 = kY_B$,

$$y_A = C\sin\omega_n t + D\cos\omega_n t + \frac{Y_B}{1-(\omega_B/\omega_n)^2}\sin\omega_B t, \tag{5}$$

where $\omega_n = \sqrt{k/m_A} = 5.774\,\text{rad/s}$. Since the system starts from rest, the initial conditions are $y_A(0) = 0$ and $\dot{y}_A(0) = 0$. Applying the first initial condition to Eq. (5), we obtain $D = 0$. Applying the second initial condition gives

$$\dot{y}(0) = 0 = C\omega_n + \frac{Y_B\omega_B}{1-(\omega_B/\omega_n)^2} \quad \Rightarrow \quad C = \frac{Y_B\omega_B/\omega_n}{1-(\omega_B/\omega_n)^2}. \tag{6}$$

Substituting Eq. (6), $Y_B = 5\,\text{mm}$, and the numerical values of ω_n and ω_B into Eq. (5), we obtain

$$\boxed{y_A = \big(-0.3075\sin 5.774t - 0.01884\sin 94.25t\big)\,\text{mm}.} \tag{7}$$

Discussion & Verification The plot of Eq. (7) in Fig. 3 shows the high-frequency forcing ω_B superimposed on the much lower-frequency natural vibration of the scale platform.

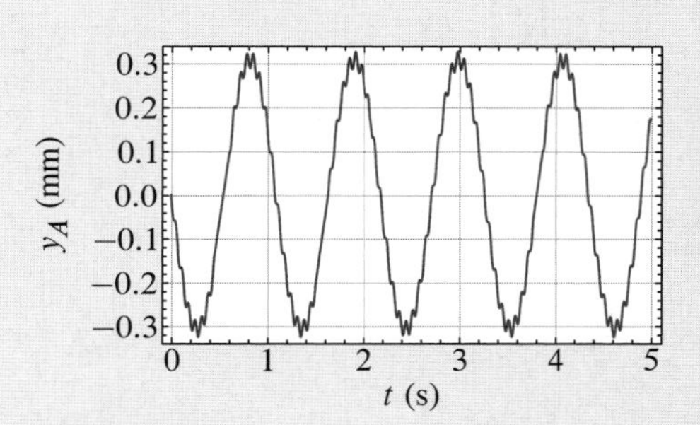

Figure 3
A plot of y_A in Eq. (7).

EXAMPLE 9.5 *Equations of Motion for an Unbalanced Motor*

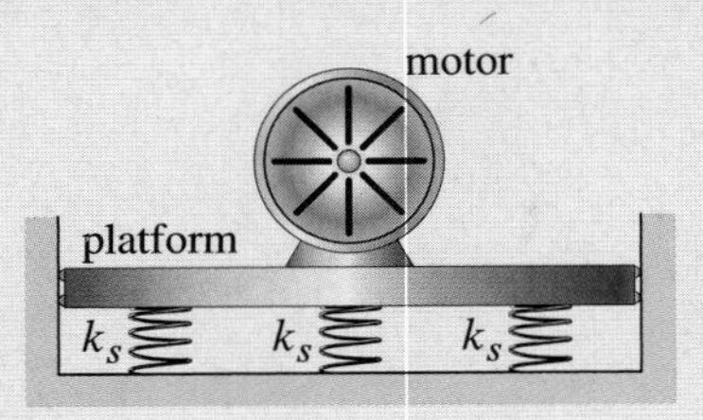

Figure 1
An unbalanced motor mounted on a platform that is elastically suspended on six springs, three of which are shown.

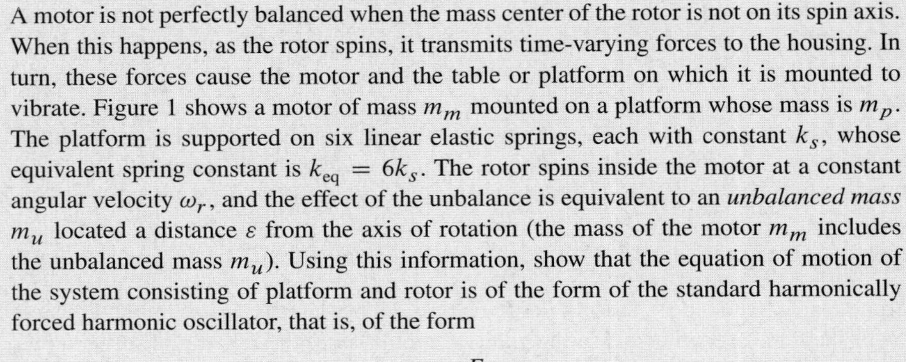

A motor is not perfectly balanced when the mass center of the rotor is not on its spin axis. When this happens, as the rotor spins, it transmits time-varying forces to the housing. In turn, these forces cause the motor and the table or platform on which it is mounted to vibrate. Figure 1 shows a motor of mass m_m mounted on a platform whose mass is m_p. The platform is supported on six linear elastic springs, each with constant k_s, whose equivalent spring constant is $k_{\text{eq}} = 6k_s$. The rotor spins inside the motor at a constant angular velocity ω_r, and the effect of the unbalance is equivalent to an *unbalanced mass* m_u located a distance ε from the axis of rotation (the mass of the motor m_m includes the unbalanced mass m_u). Using this information, show that the equation of motion of the system consisting of platform and rotor is of the form of the standard harmonically forced harmonic oscillator, that is, of the form

$$\ddot{x} + \omega_n^2 x = \frac{F_0}{m} \sin \omega_0 t. \tag{1}$$

SOLUTION

Road Map & Modeling The FBDs of the unbalanced mass and of the motor and platform combination are shown in Figs. 2 and 3, respectively, where we have isolated the unbalanced mass m_u from the motor. By separating the unbalanced mass from the rest of the system, we will be able to determine the forces required to rotate the unbalanced mass, and the equal and opposite forces will be applied to the rest of the system.

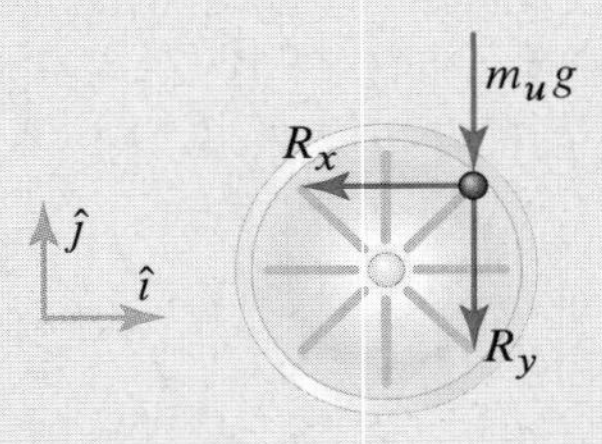

Figure 2
FBD of the unbalanced mass m_u. The forces R_x and R_y are the forces exerted by the motor on the unbalanced mass.

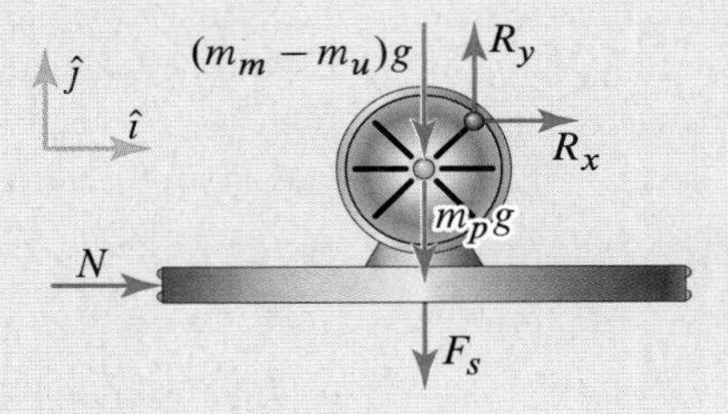

Figure 3. FBD of the motor and platform with the unbalanced mass m_u removed. The forces R_x and R_y are the forces exerted by the particle on the motor and, which, by Newton's third law, are equal and opposite to the corresponding forces in Fig. 2. N is the net normal force on the platform due to the vertical walls.

Governing Equations

Balance Principles Applying Newton's second law in the y direction to the unbalanced mass m_u in Fig. 2, we obtain

$$\left(\sum F_y\right)_{\text{Fig 2}}: \ -R_y - m_u g = m_u a_{uy}, \tag{2}$$

where a_{uy} is the y component of the acceleration of the unbalanced mass. Applying Newton's second law to the FBD of the rotor and platform with the unbalanced mass removed, and observing that the motor and platform can move only in the y direction (see Fig. 3), we obtain

$$\left(\sum F_y\right)_{\text{Fig 3}}: \ R_y - (m_m - m_u)g - m_p g - F_s = (m_m - m_u + m_p)a_{my}, \tag{3}$$

where a_{my} is the y component of acceleration of the motor/platform combination and F_s is the force on the motor/platform due to the springs.

Force Laws The spring force is given by

$$F_s = k_{\text{eq}}(y_m - \delta_s), \tag{4}$$

where δ_s is the deflection of the spring when the system is in static equilibrium.

Kinematic Equations Referring to Fig. 4, we denote the y coordinate of the motor by

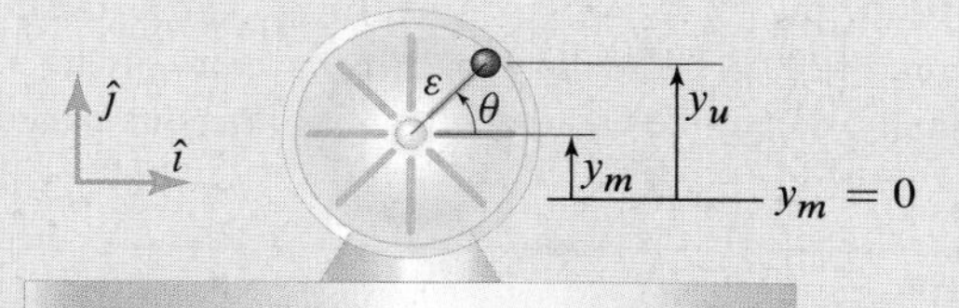

Figure 4. Kinematics of the motor/platform and the unbalanced mass.

y_m, which is measured from the static equilibrium position of the system. Again referring to Fig. 4, the kinematic equation for the unbalanced mass is

$$y_u = y_m + \varepsilon \sin\theta \quad \Rightarrow \quad \ddot{y}_u = \ddot{y}_m - \varepsilon\dot{\theta}^2 \sin\theta \quad \Rightarrow \quad a_{uy} = \ddot{y}_m - \varepsilon\omega_r^2 \sin\theta, \tag{5}$$

where $\dot{\theta} = \omega_r$, $\ddot{y}_u = a_{uy}$, and $\ddot{\theta} = 0$ since ω_r is constant. For the motor/platform, we have

$$a_{my} = \ddot{y}_m. \tag{6}$$

Computation Substituting Eqs. (4)–(6) into Eqs. (2) and (3) and then eliminating R_y from the resulting two equations, we obtain

$$-m_u g - m_u(\ddot{y}_m - \varepsilon\omega_r^2 \sin\theta) - (m_m - m_u)g - m_p g \\ - k_{\text{eq}}(y_m - \delta_s) = (m_m - m_u + m_p)\ddot{y}_m. \tag{7}$$

Canceling terms and rearranging, we obtain

$$(m_m + m_p)\ddot{y}_m + k_{\text{eq}} y_m + (m_m + m_p)g - k_{\text{eq}}\delta_s = m_u \varepsilon\omega_r^2 \sin\theta. \tag{8}$$

Noting that $\theta = \omega_r t$,[†] and that the definition of the static deflection δ_s implies that $(m_m + m_p)g = k_{\text{eq}}\delta_s$, this equation becomes

$$\boxed{\ddot{y}_m + \underbrace{\frac{k_{\text{eq}}}{m_m + m_p}}_{\omega_n^2} y_m = \underbrace{\frac{m_u \varepsilon\omega_r^2}{m_m + m_p}}_{F_0/m} \sin\omega_r t,} \tag{9}$$

which is the equation of motion for the unbalanced motor and platform.

Discussion & Verification As indicated, Eq. 9 is of the same form as Eq. (1), with the correspondence shown in Table 1. We will complete the analysis of this system in Example 9.6.

Table 1. Parameters in Eq. (1) and the corresponding parameters in Eq. (9). In Eq. (1), we are assuming $\omega_n^2 = k/m$.

Eq. (1)	Eq. (9)
k	k_{eq}
m	$m_m + m_p$
F_0	$m_u \varepsilon\omega_r^2$

Helpful Information

Obtaining the equation of motion using the center of mass of the system. Note that Eq. (9) can be obtained more directly by applying Newton's second law to the center of mass of the system shown in Fig. 1. This is how Prob. 9.35 asks you to obtain the equation of motion.

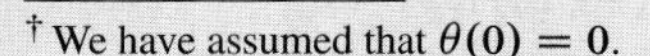

[†] We have assumed that $\theta(0) = 0$.

EXAMPLE 9.6 Response and MF for the Unbalanced Motor

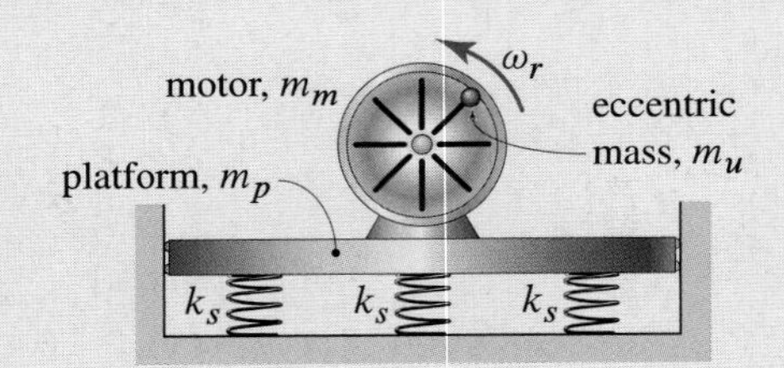

Figure 1

Find the general solution for the displacement, $y_m(t)$, of the unbalanced motor (Fig. 1) subject to the initial conditions $y_m(0) = 0$ and $\dot{y}_m(0) = v_{m0}$. After doing so, plot the solution for $0 < t < 1$ s, using $m_m = 40$ kg, $m_p = 15$ kg, $k_{\text{eq}} = 6k_s = 420{,}000$ N/m, $\varepsilon = 15$ cm, $\omega_r = 1200$ rpm, $y_m(0) = 0$ m, $\dot{y}_m(0) = 0.4$ m/s, and three different values of m_u: 10 g, 100 g, and 1000 g. From the plot, find the approximate maximum amplitude of the vibration. Finally, determine and plot the MF for the unbalanced rotor.

SOLUTION

Road Map & Modeling The equation of motion for our model of the unbalanced motor and platform was derived in Example 9.5 as Eq. (9), so we need only apply the general solution in Eq. (9.36) to determine and plot the response. For the MF, as found in Eq. (9.38), we will need to find a function of ω_r/ω_n where ω_r is the angular velocity of the unbalanced rotor inside the motor.

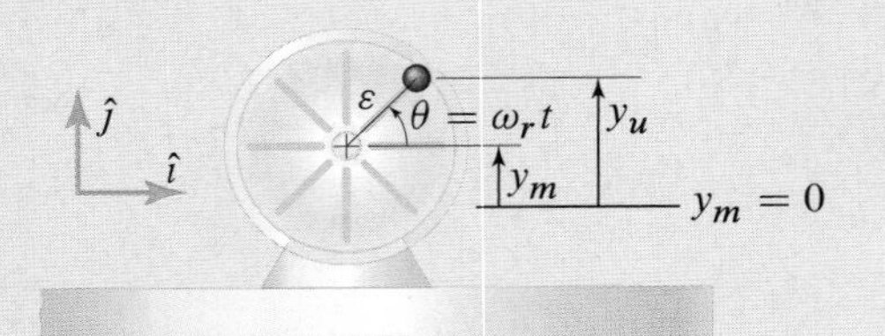

Figure 2
Kinematic definitions for the unbalanced motor and platform.

Governing Equations For convenience, we repeat the equation of motion for the unbalanced motor and platform, which was found in Eq. (9) of Example 9.5 to be

$$\ddot{y}_m + \frac{k_{\text{eq}}}{m_m + m_p} y_m = \frac{m_u \varepsilon \omega_r^2}{m_m + m_p} \sin \omega_r t, \tag{1}$$

where y_m is measured from the static equilibrium position of the system, as shown in Fig. 2.

Computation Equation (1) is of the same form as Eq. (9.31), which is repeated below for convenience

$$\ddot{x} + \omega_n^2 x = \frac{F_0}{m} \sin \omega_0 t, \tag{2}$$

where, comparing Eqs. (1) and (2), we have

$$\omega_n^2 = \frac{k_{\text{eq}}}{m_m + m_p}, \qquad \frac{F_0}{m} = \frac{m_u \varepsilon \omega_r^2}{m_m + m_p}, \qquad \text{and} \qquad \omega_0 = \omega_r. \tag{3}$$

Therefore, the general solution to Eq. (1) can be found using Eq. (9.36), which gives

$$y_m = A \sin \omega_n t + B \cos \omega_n t + \frac{m_u \varepsilon \omega_r^2 / k_{\text{eq}}}{1 - (\omega_r/\omega_n)^2} \sin \omega_r t. \tag{4}$$

For $t = 0$, Eq. (4) gives $y_m(0) = B$. Therefore, recalling that we must have $y_m(0) = 0$, we have

$$B = 0. \tag{5}$$

Differentiating y_m in Eq. (4) with respect to time, we obtain

$$\dot{y}_m = A\omega_n \cos \omega_n t - B\omega_n \sin \omega_n t + \frac{m_u \varepsilon \omega_r^3 / k_{\text{eq}}}{1 - (\omega_r/\omega_n)^2} \cos \omega_r t, \tag{6}$$

and then applying the initial condition $\dot{y}_m(0) = v_{m0}$, we get

$$A\omega_n + \frac{m_u \varepsilon \omega_r^3 / k_{\text{eq}}}{1 - (\omega_r/\omega_n)^2} = v_{m0} \quad \Rightarrow \quad A = \frac{v_{m0}}{\omega_n} - \frac{\omega_r}{\omega_n} \frac{m_u \varepsilon \omega_r^2 / k_{\text{eq}}}{1 - (\omega_r/\omega_n)^2}. \tag{7}$$

Combining Eqs. (4), (5), and (7), the general solution becomes

$$\boxed{y_m = \left[\frac{v_{m0}}{\omega_n} - \frac{\omega_r}{\omega_n} \frac{m_u \varepsilon \omega_r^2 / k_{\text{eq}}}{1 - (\omega_r/\omega_n)^2} \right] \sin \omega_n t + \frac{m_u \varepsilon \omega_r^2 / k_{\text{eq}}}{1 - (\omega_r/\omega_n)^2} \sin \omega_r t,} \tag{8}$$

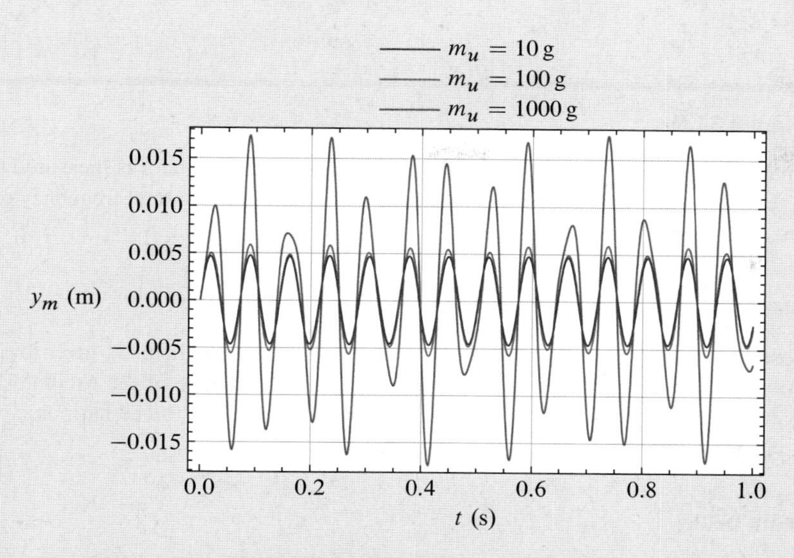

Figure 3. The response y_m of the motor and platform for three values of the unbalanced mass m_u.

a plot of which is shown in Fig. 3. From this figure, we can see that the maximum amplitude of vibration for $m_u = 10\,\mathrm{g}$ is about 5 mm, for $m_u = 100\,\mathrm{g}$ it is about 6 mm, and for $m_u = 1\,\mathrm{kg}$ it is about 17 mm.

To compute MF, we take the amplitude of the particular solution and rearrange it so that ω_r and ω_n always appear as their ratio. Doing this gives

$$|y_{mp}| = \frac{m_u \varepsilon \omega_r^2 / k_{\mathrm{eq}}}{1 - (\omega_r/\omega_n)^2} \quad \Rightarrow \quad |y_{mp}| = \frac{\frac{m_u \varepsilon}{m_m + m_p} (\omega_r/\omega_n)^2}{1 - (\omega_r/\omega_n)^2}. \tag{9}$$

Therefore, MF is given by

$$\boxed{\mathrm{MF} = \frac{|y_{mp}|(m_m + m_p)}{m_u \varepsilon} = \frac{(\omega_r/\omega_n)^2}{1 - (\omega_r/\omega_n)^2},} \tag{10}$$

a plot of which is shown in Fig. 4. In our case, since $\omega_r = 1200\,\mathrm{rpm} = 125.7\,\mathrm{rad/s}$ and $\omega_n = 87.39\,\mathrm{rad/s}$, we have

$$\boxed{\mathrm{MF} = -1.937.} \tag{11}$$

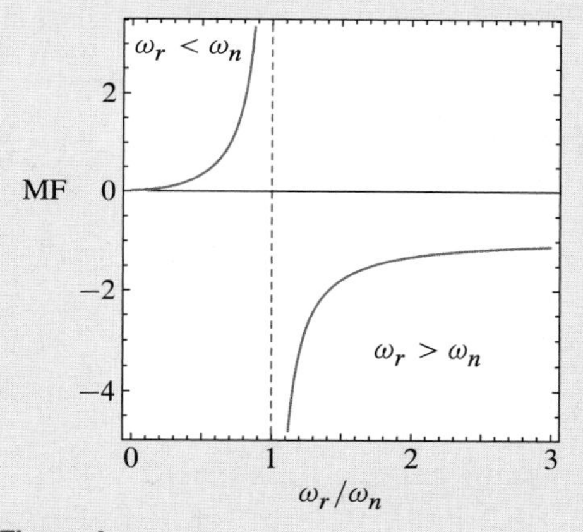

Figure 4
The MF for response of the unbalanced motor as given by Eq. (10).

Discussion & Verification The dimensions of all the terms in the general solution in Eq. (8) are length, as they should be. The amplitude of the vibration found by examining Fig. 3 is a few millimeters in all cases, which seems reasonable given the masses and stiffnesses involved. Finally, MF is dimensionless, again as it should be.

A Closer Look It is interesting to compare the MF of the unbalanced motor plotted in Fig. 4, with the MF in Fig. 9.13, which applies to a particle that is forced directly (see Fig. 9.11). Notice that for small ω_r, the MF in Fig. 4 approaches 0 rather than 1, as it does in Fig. 9.13. This means that when the unbalanced rotor within the motor is spinning *very* slowly, it does not shake the motor and platform by any appreciable amount, which agrees with our intuition. On the other hand, when a particle is harmonically forced directly as in Fig. 9.11 and the forcing frequency is *very* small, the particle moves with the forcing, and so the MF is 1.

Problems

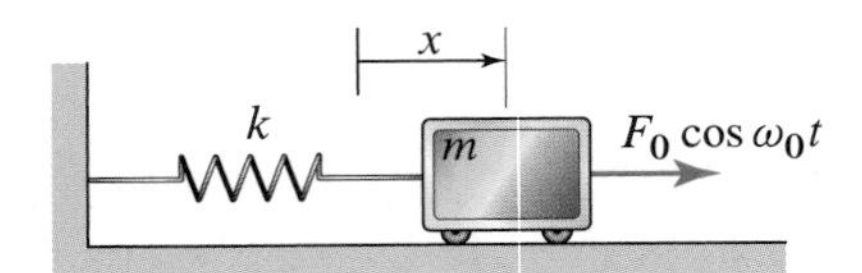

Figure P9.32

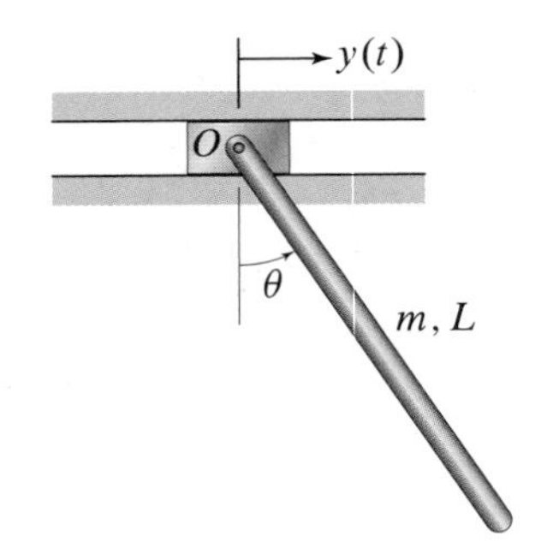

Figure P9.33

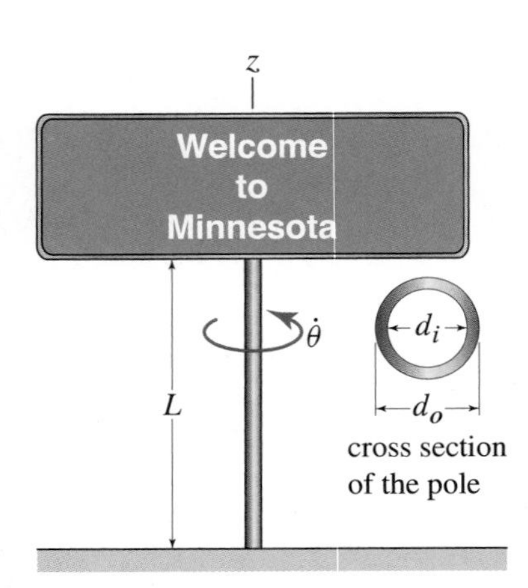

Figure P9.36

Problem 9.31

The magnification factor for a forced (undamped) harmonic oscillator is measured to be equal to 5. Determine the driving frequency of the forcing if the natural frequency of the system is 100 rad/s.

Problem 9.32

Suppose that the equation of motion of a forced harmonic oscillator is given by $\ddot{x} + \omega_n^2 x = (F_0/m)\cos\omega_0 t$. Obtain the expression for the response of the oscillator, and compare it to the response presented in Eq. (9.36) (which is for a forced harmonic oscillator with the equation of motion given in Eq. (9.31)).

Problem 9.33

A uniform bar of mass m and length L is pinned to a slider at O. The slider is forced to oscillate horizontally according to $y(t) = Y \sin\omega_s t$. The system lies in the vertical plane.

(a) Derive the equation of motion of the bar for small angles θ.

(b) Determine the amplitude of steady-state vibration of the bar.

Problem 9.34

Determine the amplitude of vibration of the unbalanced motor we studied in Example 9.6 if the forcing frequency of the motor is $0.95\omega_n$.

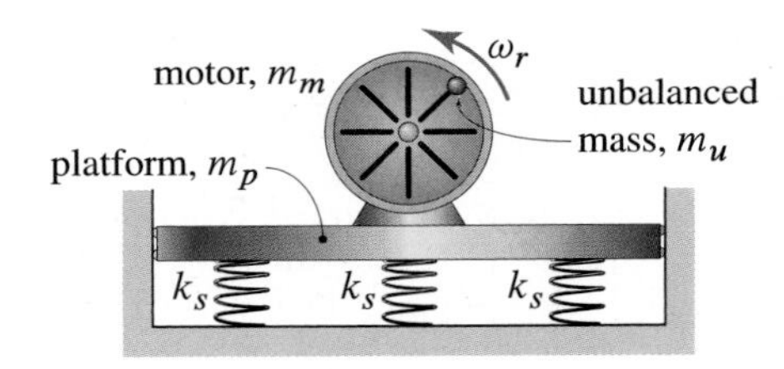

Figure P9.34 and P9.35

Problem 9.35

Derive the equations of motion for the unbalanced motor introduced in this section by applying Newton's second law to the center of mass of the system shown in Fig. 1 of Example 9.5.

Problem 9.36

Consider a sign mounted on a circular hollow steel pole of length $L = 5$ m, outer diameter $d_o = 5$ cm, and inner diameter $d_i = 4$ cm. Aerodynamic forces due to wind provide a harmonic torsional excitation with frequency $f_0 = 3$ Hz and amplitude $M_0 = 10$ N·m about the z axis. The mass center of the sign lies on the central axis z of the pole. The mass moment of inertia of the sign is $I_z = 0.1\ \text{kg·m}^2$. The torsional stiffness of the pole can be estimated as $k_t = \pi G_{\text{st}}(d_o^4 - d_i^4)/(32L)$, where G_{st} is the shear modulus of steel, which is 79 GPa. Neglecting the inertia of the pole, calculate the amplitude of steady-state vibration of the sign.

Problem 9.37

An unbalanced motor is mounted at the tip of a rigid beam of mass m_b and length L. The beam is restrained by a torsional spring of stiffness k_t and an additional support of stiffness k located at the half length of the beam. In the static equilibrium position, the beam is horizontal, and the torsional spring does not exert any moment on the beam. The mass of the motor is m_m, and the unbalance results in a harmonic excitation $F(t) = F_0 \sin \omega_0 t$ in the vertical direction. Derive the equation of motion for the system, assuming that θ is small.

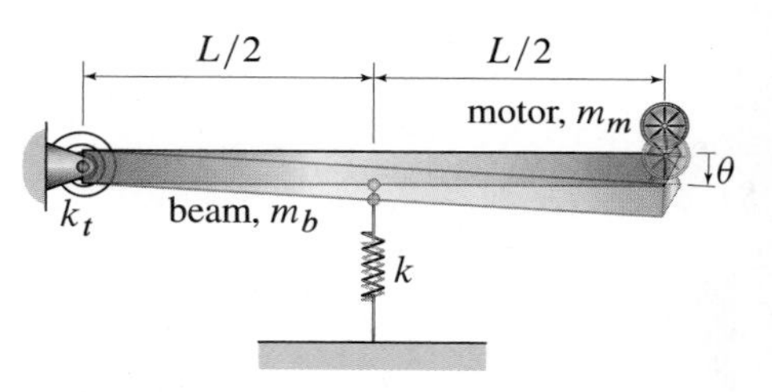

Figure P9.37

Problem 9.38

Revisit Example 9.6 and discuss whether it is possible to obtain the equation of motion of the system via the energy method.

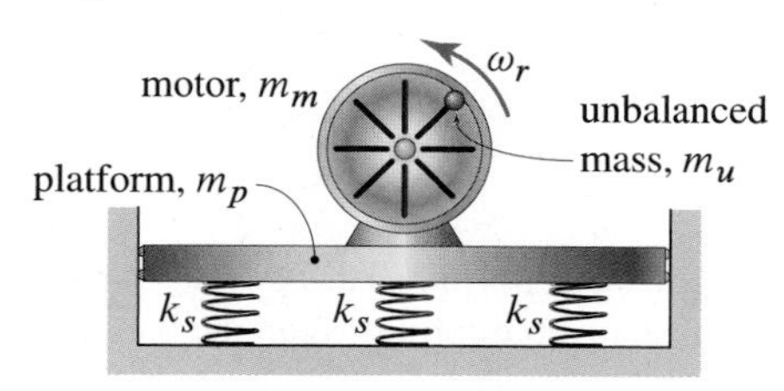

Figure P9.38

Problem 9.39

A fatigue-testing machine for electronic components consists of a platform with an unbalanced motor. Assume that the rotor in the motor spins at $\omega_r = 3000$ rpm, the mass of the platform is $m_p = 20$ kg, the mass of the motor is $m_m = 15$ kg, the unbalanced mass is $m_u = 0.5$ kg, and the equivalent stiffness of the platform suspension is $k_{eq} = nk_s = 5\times10^6$ N/m, where n is the number of springs. For the testing machine, the distance ε between the spin axis of the rotor and the location at which m_u is placed can be varied to obtain the desired vibration level. Calculate the range of values of ε that would provide amplitudes of the particular solution ranging from 0.1 mm to 2 mm.

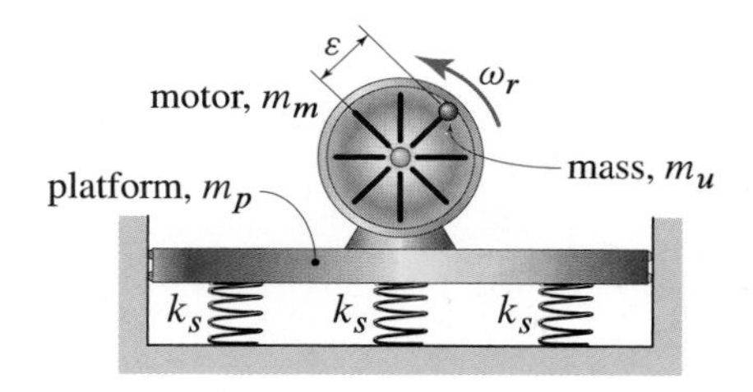

Figure P9.39

Problem 9.40

At time $t = 0$, a forced harmonic oscillator occupies position $x(0) = 0.1$ m and has a velocity $\dot{x}(0) = 0$. The mass of the oscillator is $m = 10$ kg, and the stiffness of the spring is $k = 1000$ N/m. Calculate the motion of the system if the forcing function is $F(t) = F_0 \sin \omega_0 t$, with $F_0 = 10$ N and $\omega_0 = 200$ rad/s.

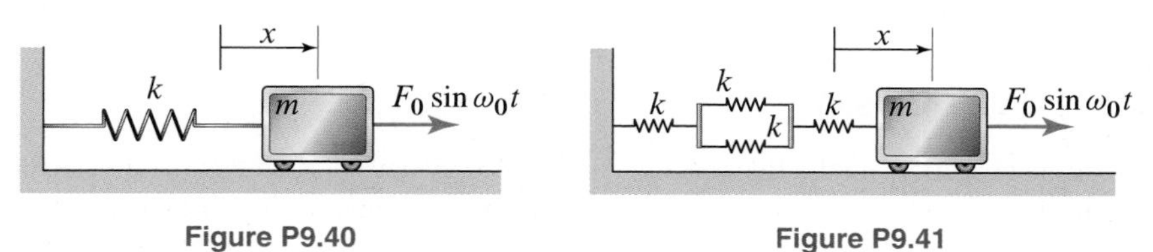

Figure P9.40 Figure P9.41

Problem 9.41

The forced harmonic oscillator shown has a mass $m = 10$ kg. In addition, the harmonic excitation is such that $F_0 = 150$ N and $\omega_0 = 200$ rad/s. If all sources of friction can be neglected, determine the spring constant k such that the magnification factor MF = 5.

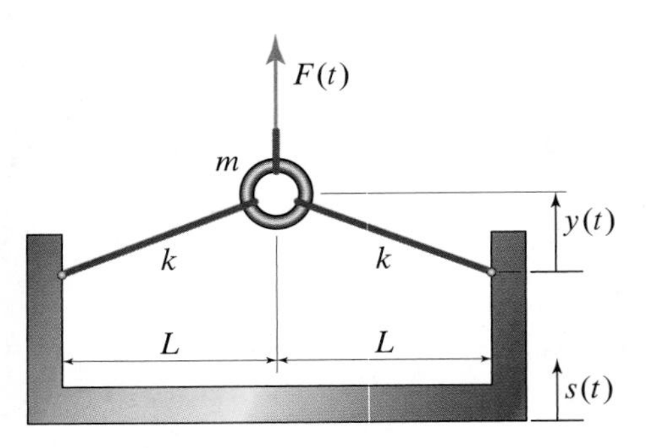

Figure P9.42 and P9.43

Problem 9.42

A ring of mass m is attached by two linear elastic cords with elastic constant k and unstretched length $L_0 < L$ to a support, as shown. Assuming that the pretension in the cords is large, so that the cords' deflection due to the ring's weight can be neglected, find the linearized equation of motion for the case where $F(t) = F_0 \sin \omega_0 t$ and $w(t) = 0$ (i.e., the support is stationary). In addition, find the response of the system for $y(0) = 0$ and $\dot{y}(0) = 0$.

Problem 9.43

A ring of mass m is attached by two linear elastic cords with elastic constant k and unstretched length $L_0 < L$ to a support, as shown. Assuming that the pretension in the cords is large, so that the cords' deflection due to the ring's weight can be neglected, find the linearized equation of motion for the case where $F(t) = 0$ and $s(t) = s_0 \sin \omega t$. In addition, find the response of the system for $y(0) = 0$ and $\dot{y}(0) = 0$.

Problem 9.44

Modeling the beam as a rigid uniform thin bar, ignoring the inertia of the pulleys, assuming that the system is in static equilibrium when the bar is horizontal, and assuming that the cord is inextensible and does not go slack, determine the linearized equation of motion of the system in terms of x, which is the position of A. Finally, determine an expression for the amplitude of the steady-state vibration of block A.

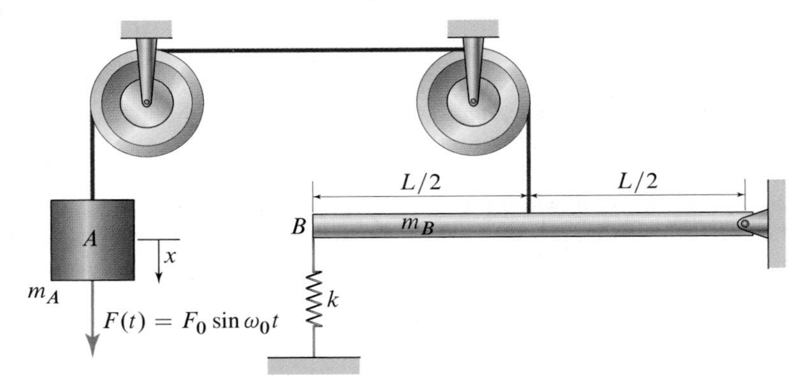

Figure P9.44 and P9.45

Problem 9.45

For the system in Prob. 9.44, determine the maximum forcing frequency ω_0 for steady-state motion, such that the cord does not go slack.

Problems 9.46 and 9.47

One of the propellers on the Beech King Air 200 is unbalanced, such that the unbalanced weight W_u is a distance R from the spin axis of the propeller. The propellers spin at a constant rate ω_p, and the weight of each engine is W_e (this includes the mass of the propeller). Assume that the wing is a uniform beam that is cantilevered at A, has weight W_w and bending stiffness EI, and whose mass center is at G. Treat the engine as a point mass and evaluate your answers for $W_u = 3$ oz, $W_e = 450$ lb, $R = 5.1$ ft, $\omega_p = 2000$ rpm, $EI = 1.13 \times 10^{11}$ lb·in.^{2}, $d = 8.7$ ft, and $h = 10.9$ ft. In addition, ignore the angular motion of the wing in computing the angular velocity and angular acceleration of the propeller, and ignore the time-dependent inertia term in the final equation of motion.

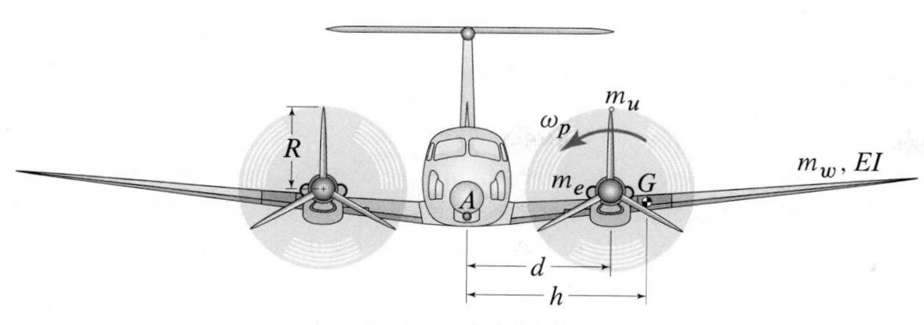

Figure P9.46 and P9.47

Problem 9.46 Neglect the mass of the wing, and model the wing as was done in Example 9.2. Determine the resonance frequency of the system, and find the MF for the given parameters.

Problem 9.47 Let the mass of the wing be $m_w = 350\,\text{lb}/g$, and model the wing as was done in Example 9.2. Determine the resonance frequency of the system, and find the MF for the given parameters.

Design Problems

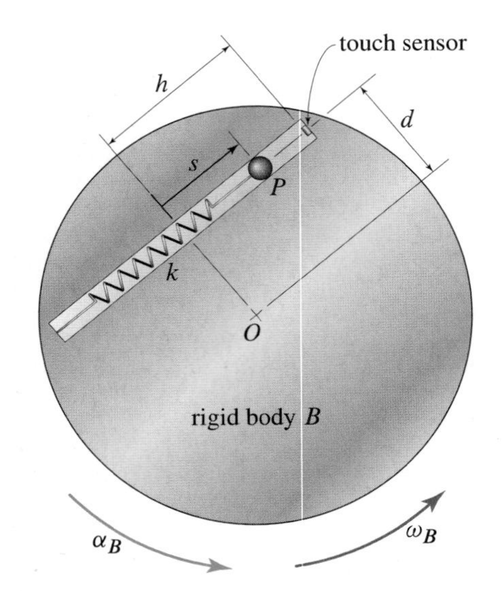

Figure DP9.2

Design Problem 9.2

The device shown can detect when the angular velocity and angular acceleration of a rigid body B achieve a combination of specified values. The device works using the principle that the vibration amplitude of the sphere P depends on both the angular velocity and angular acceleration of the rigid body. When the angular velocity ω_B and angular acceleration α_B reach the appropriate combination, P will contact the touch sensor, thus signaling that the specified values have been reached. With this as background, assume that the rigid body rotates in the horizontal plane with angular velocity ω_B and angular acceleration α_B, and that P is constrained to move in the slot, which is at a distance d from the center of the disk. In addition, a linear elastic spring of constant k is attached to the mass, such that the spring is undeformed when the mass is at $s = 0$.

(a) Derive the equation of motion for P with s as the dependent variable.

(b) Assuming that P is released from rest at $s = 0$, find the solution to the equation of motion found in (a), knowing that the solution to ordinary differential equations of the type

$$\ddot{s} + \omega_n^2 s = D,$$

is given by

$$s(t) = \frac{D}{\omega_n^2} + C_1 \cos \omega_n t + C_2 \sin \omega_n t,$$

where C_1 and C_2 are constants determined from the initial conditions and D is a known constant.

(c) Using the solution for $s(t)$ found above, for given values of d, k, m, ω_B, and α_B, determine the maximum distance from $s = 0$ that P achieves in one cycle.

(d) For a disk-shaped rigid body B whose diameter is 1.5 m, specify the mass of P (treat it as a particle), the spring constant k, the length h, and the distance d so that the touch sensor can detect when the rigid body reaches an angular velocity $\omega_{\text{crit}} = 100$ rpm for a constant angular acceleration $\alpha_B = 1\ \text{rad/s}^2$.

9.3 Viscously Damped Vibration

All mechanical systems exhibit some energy dissipation or damping due to air drag, viscous fluids, friction, and other effects. If the damping is small enough, the undamped solutions obtained in Sections 9.1 and 9.2 will be in close agreement with the damped solution for a short period of time. On the other hand, if we need a solution for a longer period of time or if there is more damping, we need to resort to the solution of the equations that model damped mechanical systems.

In this section, we will consider *linear viscous damping*. This is damping in which the damping force is directly proportional and opposite in sign to the velocity of a body. This type of damping tends to occur when the energy dissipation is due to a fluid (e.g., oil, water, or air), as seen in the shock absorber in Fig. 9.19. In addition, even when the damping is due to other physical mechanisms, linear viscous damping can still be an effective model.

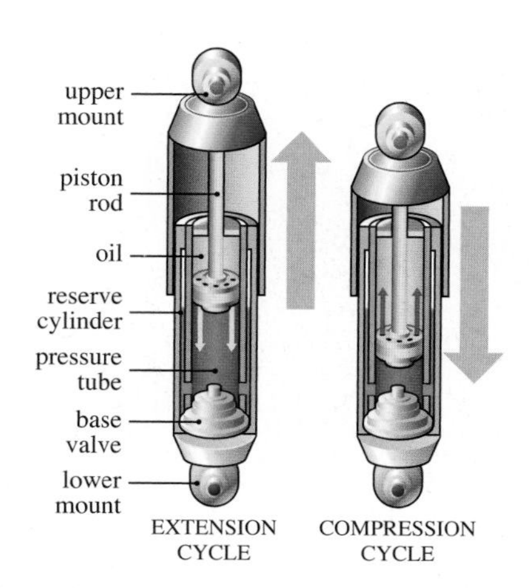

Figure 9.19
A cutaway view of a typical shock absorber, which is used in many suspension systems.

Viscously damped free vibration

The effect of viscous damping is usually modeled by an element called a *dashpot*, a schematic of which is shown in Fig. 9.20. Damping in a dashpot occurs when

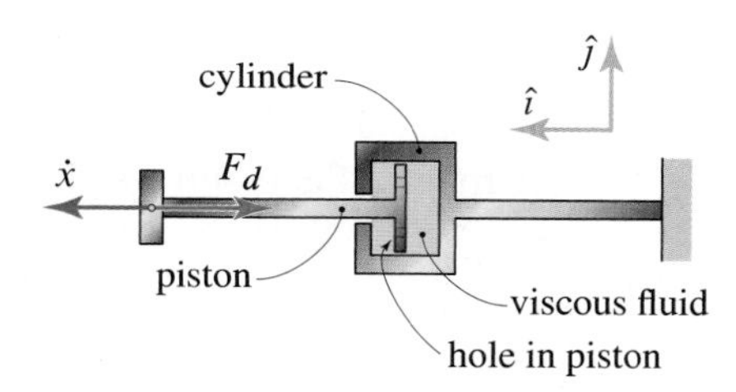

Figure 9.20. Schematic diagram of a dashpot illustrating its basic operation.

the piston moves within the fluid-filled cylinder and forces the fluid to flow either around the piston or through one or more holes in it. Because the fluid within the piston is viscous, this motion results in energy dissipation. Referring to Fig. 9.20, if the relative velocity of the two ends of the piston is $\dot{x}$, then the damping force F_d on the piston is equal to

$$F_d = c\dot{x}, \tag{9.42}$$

where c is a constant called the *coefficient of viscous damping*. This damping coefficient depends on the physical properties of the fluid and the geometry of the dashpot. The coefficient of viscous damping is expressed in lb·s/ft in U.S. Customary units and N·s/m in SI units.

Revisiting Example 3.5 on p. 182 in which a railcar runs into a large spring that is designed to stop it, a dashpot has now been added in parallel with the spring (see Fig. 9.21). Assuming that the railcar couples with the spring and dashpot after it hits them, we obtain the FBD of the railcar shown in Fig. 9.22. Summing forces in the x direction, we find

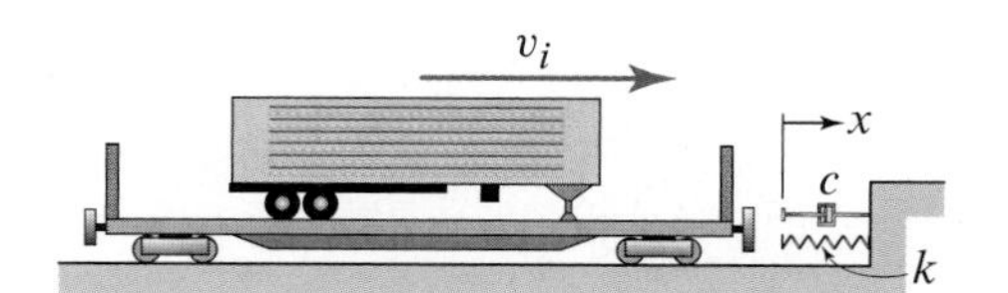

Figure 9.21
Railcar hitting a spring and dashpot. Recall from Example 3.5 that the railcar weighs 87 tons and is moving at 4 mph at impact, and k = 22,270 lb/ft.

$$\sum F_x: \quad -F_d - F_s = m\ddot{x}, \tag{9.43}$$

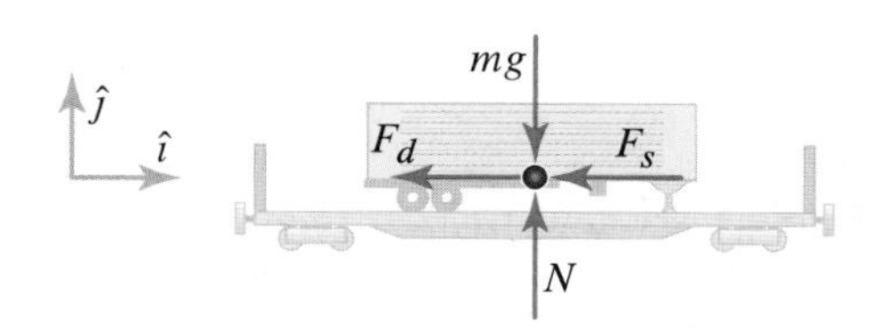

Figure 9.22
FBD of the railcar after it has hit and coupled with the spring and dashpot.

where F_d is the force due to the dashpot, F_s is the force due to the spring, and x measures the displacement of the spring from its equilibrium position. The forces

F_d and F_s can be written as

$$F_d = c\dot{x} \quad \text{and} \quad F_s = kx, \tag{9.44}$$

which allows us to write Eq. (9.43) as

$$\boxed{m\ddot{x} + c\dot{x} + kx = 0.} \tag{9.45}$$

This is the *standard form of the viscously damped harmonic oscillator*. The theory of differential equations tells us that Eq. (9.45) is a linear, second-order, homogeneous, constant coefficient differential equation, and as such, it has solutions of the form

$$x = e^{\lambda t}, \tag{9.46}$$

where λ (the Greek letter lambda) is a constant to be determined. Replacing x in Eq. (9.45) with Eq. (9.46), we find

$$m\lambda^2 e^{\lambda t} + c\lambda e^{\lambda t} + ke^{\lambda t} = 0. \tag{9.47}$$

Factoring out $e^{\lambda t}$ in Eq. (9.47), we have

$$e^{\lambda t}(m\lambda^2 + c\lambda + k) = 0. \tag{9.48}$$

Since $e^{\lambda t}$ never vanishes, to have a solution to Eq. (9.45) we must have

$$m\lambda^2 + c\lambda + k = 0. \tag{9.49}$$

If λ is a root of this quadratic equation, called the *characteristic equation*, then $e^{\lambda t}$ is a solution to Eq. (9.45). The two roots of Eq. (9.49) are given by

$$\lambda_1 = -\frac{c}{2m} + \sqrt{\left(\frac{c}{2m}\right)^2 - \frac{k}{m}} \quad \text{and} \quad \lambda_2 = -\frac{c}{2m} - \sqrt{\left(\frac{c}{2m}\right)^2 - \frac{k}{m}}. \tag{9.50}$$

Referring to Eqs. (9.50), the theory of differential equations tells us that the general solution of Eq. (9.45) takes on one of three possible forms determined by the values of λ_1 and λ_2. Observe that the character of λ_1 and λ_2 depends on whether the term $(c/2m)^2 - k/m$ is positive, zero, or negative. Therefore, we introduce a special value of the damping coefficient called the *critical damping coefficient*, which we denote by c_c and define as the value of c that makes the term $(c/2m)^2 - k/m$ equal to zero, that is,

$$\left(\frac{c_c}{2m}\right)^2 - \frac{k}{m} = 0 \quad \Rightarrow \quad \boxed{c_c = 2m\sqrt{\frac{k}{m}} = 2m\omega_n,} \tag{9.51}$$

where $\omega_n = \sqrt{k/m}$. We now distinguish the three cases mentioned above based on whether $c > c_c$, $c = c_c$, or $c < c_c$.

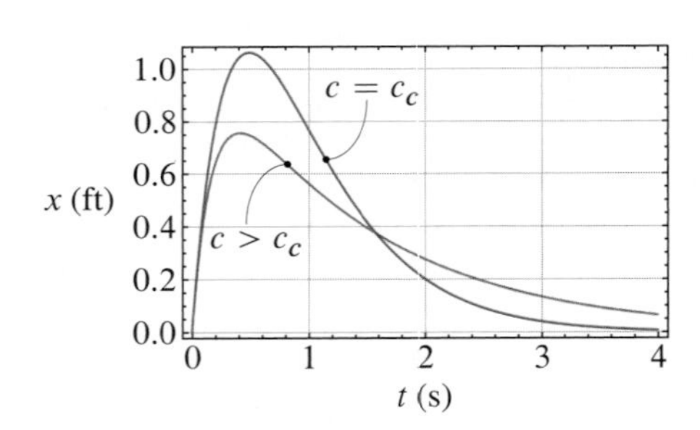

Figure 9.23
The position of the railcar as a function of time after it impacts the spring and dashpot. The blue curve is overdamped with $c = 35{,}000\,\text{lb}{\cdot}\text{s/ft}$, and the red curve is critically damped with $c = c_c \approx 21{,}950\,\text{lb}{\cdot}\text{s/ft}$.

Overdamped system ($c > c_c$)

If $c > c_c$, the term $(c/2m)^2 - k/m$ is positive. Thus, λ_1 and λ_2 are real, distinct, and negative. In this case, the general solution of Eq. (9.45) is

$$\boxed{x = e^{-(c/2m)t}\left(Ae^{t\sqrt{(c/2m)^2 - k/m}} + Be^{-t\sqrt{(c/2m)^2 - k/m}}\right),} \tag{9.52}$$

where A and B are constants that are determined from the initial conditions of the system. The motion represented by Eq. (9.52) is characterized by decaying exponentials, the system does not vibrate, and there is no period associated with the motion (see Fig. 9.23). This type of system is said to be *overdamped*.

Critically damped system ($c = c_c$)

If $c = c_c$, the term $(c/2m)^2 - k/m$ is zero. Thus $\lambda_1 = \lambda_2 = -c_c/2m$. In this case, the general solution of Eq. (9.45) is

$$x = (A + Bt)e^{-\omega_n t}, \tag{9.53}$$

where, again, A and B are constants that are determined from the initial conditions. If $c = c_c$, then c has the smallest value for which no vibration occurs, and the system is said to be *critically damped*. Referring to Fig. 9.23, notice that a critically damped system approaches equilibrium faster than an overdamped system. Critically damped systems are of great interest in engineering applications since they approach equilibrium in the minimum possible time.

Underdamped system ($c < c_c$)

If $c < c_c$, the term $(c/2m)^2 - k/m$ is negative. Thus, λ_1 and λ_2 are complex since they involve the square root of a negative quantity. In this case, the general solution of Eq. (9.45) is

$$x = e^{-(c/2m)t}\big(A \sin \omega_d t + B \cos \omega_d t\big), \tag{9.54}$$

where A and B are determined from the initial conditions, and ω_d is the *damped natural frequency*, which is given by

$$\omega_d = \sqrt{\frac{k}{m} - \left(\frac{c}{2m}\right)^2} = \omega_n \sqrt{1 - (c/c_c)^2}, \tag{9.55}$$

and where we recall that $\omega_n = \sqrt{k/m}$ and $c_c = 2m\omega_n$. A system for which $c < c_c$ is said to be *underdamped*, and for any such system, Eq. (9.55) implies that ω_d is *always* less than ω_n. Using the definition of ω_d given in Eq. (9.55), the *period of damped vibration* is given by

$$\tau_d = 2\pi/\omega_d. \tag{9.56}$$

Note that the solution for x given in Eq. (9.54) can also be written as

$$x = De^{-(c/2m)t} \sin\big(\omega_d t + \phi\big), \tag{9.57}$$

where D and ϕ are constants determined by the initial conditions. The solution in Eq. (9.57) has been plotted in Fig. 9.24 using the data specified in Fig. 9.21 and a value of c that makes the system underdamped.

Damping ratio

In practice, the three cases just discussed are often classified in terms of a nondimensional parameter called the *damping ratio* or *damping factor*, which is usually denoted by the symbol ζ (the Greek letter zeta) and is defined as

$$\zeta = c/c_c. \tag{9.58}$$

Using ζ, the standard form of the viscously damped harmonic oscillator in Eq. (9.45) is rewritten as

$$\ddot{x} + 2\zeta\omega_n x + \omega_n^2 x = 0. \tag{9.59}$$

Interesting Fact

Landing on an aircraft carrier. A carrier-based aircraft has a tailhook, which is attached to an 8 ft bar extending from the rear of the aircraft. When the aircraft lands on the deck of the carrier, the tailhook is supposed to catch one of four steel arresting cables that are stretched across the deck.

Each 1.375 in. thick arresting cable connects to a hydraulic cylinder below deck that acts as a huge dashpot. When the tailhook snags one of the wires, the wire pulls a piston within the fluid-filled hydraulic cylinder. As the piston moves through the cylinder, hydraulic fluid is forced through the small holes at the end of the cylinder, which dissipates the kinetic energy of the aircraft. This system can stop a 54,000 lb aircraft traveling at 150 mph in less than 350 ft. This system can handle a landing every 45 s.

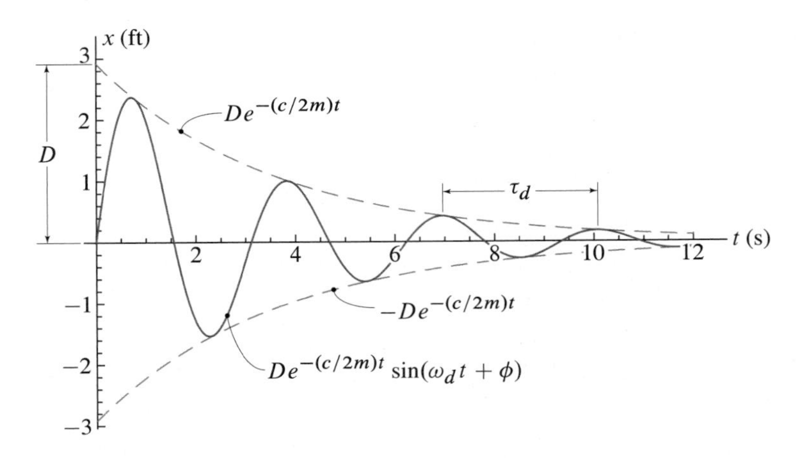

Figure 9.24. The position of the railcar as a function of time after it impacts the spring and dashpot. The red curve is the underdamped solution for $c = 3000\,\text{lb}\cdot\text{s/ft}$.

Overdamped system ($\zeta > 1$). In terms of ζ, an overdamped system is characterized by $\zeta > 1$, and the general solution in Eq. (9.52) is rewritten as

$$x = e^{-\zeta\omega_n t}\left(Ae^{\sqrt{\zeta^2-1}\,\omega_n t} + Be^{-\sqrt{\zeta^2-1}\,\omega_n t}\right). \tag{9.60}$$

Critically damped system ($\zeta = 1$). In terms of ζ, a critically damped system is characterized by $\zeta = 1$, and the general solution in Eq. (9.53) is unchanged.

Underdamped system ($\zeta < 1$). In terms of ζ, an underdamped system is characterized by $\zeta < 1$, and the solutions in Eqs. (9.54) and (9.57) are rewritten as

$$x = e^{-\zeta\omega_n t}\big(A\sin\omega_d t + B\cos\omega_d t\big) = De^{-\zeta\omega_n t}\sin\big(\omega_d t + \phi\big), \tag{9.61}$$

respectively, where, referring to Eq. (9.55), ω_d is expressed in terms of ζ as

$$\boxed{\omega_d = \omega_n\sqrt{1-\zeta^2}.} \tag{9.62}$$

Viscously damped forced vibration

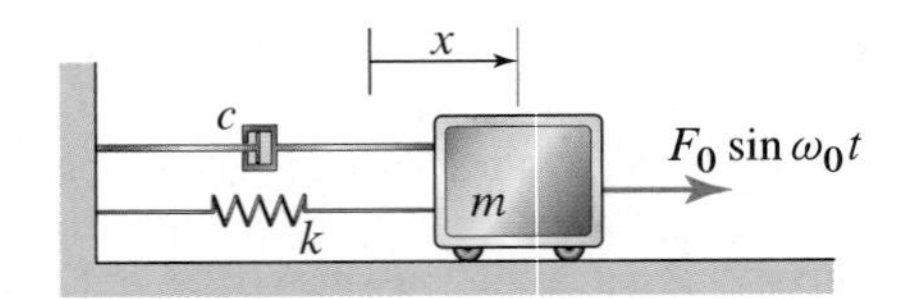

Figure 9.25
A simple damped harmonic oscillator that is harmonically forced.

Here, we consider the case of the vibration of a one degree of freedom system that is both damped *and* forced, for example, the simple system shown in Fig. 9.25, in which we have simply added a dashpot to the system in Fig. 9.11 on p. 681. The equation of motion for this system is given by

$$\boxed{m\ddot{x} + c\dot{x} + kx = F_0\sin\omega_0 t,} \tag{9.63}$$

where x is measured from the equilibrium position of the block. Using the damping ratio ζ, Eq. (9.63) can be written as

$$\boxed{\ddot{x} + 2\zeta\omega_n\dot{x} + \omega_n^2 x = \frac{F_0}{m}\sin\omega_0 t.} \tag{9.64}$$

If, rather than the harmonic forcing being applied to m, the support in Fig. 9.25 is displaced harmonically according to $Y\sin\omega_0 t$, we cannot simply replace F_0 with kY, as we did in Eq. (9.41) in Section 9.2 (see Example 9.7 and Prob. 9.63 for ways

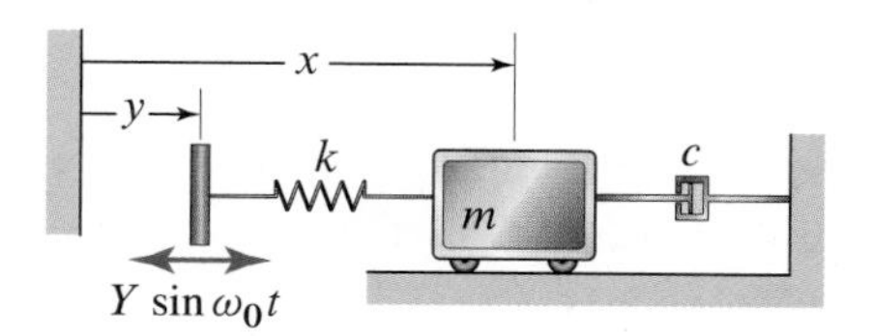

Figure 9.26
Harmonic support displacement of a viscously damped harmonic oscillator.

to handle this situation). On the other hand, if the dashpot remains attached to a fixed support and we harmonically displace the support to which the spring is attached (see Fig. 9.26), then the equation of motion is of the form

$$m\ddot{x} + c\dot{x} + kx = kY \sin\omega_0 t, \tag{9.65}$$

so we *can* replace F_0 with kY in all the corresponding solutions.

As was the case with undamped forced vibration in Section 9.2, the solution to Eq. (9.63) or Eq. (9.64) is the sum of the complementary solution and a particular solution. As a reminder, the complementary solution x_c is the solution to the associated homogeneous equation, which is Eq. (9.45) and for which we found the solution to depend on the level of damping (i.e., whether c is greater than, equal to, or less than the critical damping c_c). Regardless of the level of damping, this complementary solution will die out in time, and thus its contribution is *transient* (see p. 682). As discussed in Section 9.2 on p. 682, the forced vibration associated with a particular solution x_p will exist as long as the forcing does, and it is thus a *steady-state vibration*. Since we have already found the complementary solution, it will be the particular solution that we focus on here.

As with undamped forced vibration, the response of a damped forced system should also resemble the forcing. Therefore, we will assume a particular solution of either of the following forms:

$$\boxed{x_p = A\sin\omega_0 t + B\cos\omega_0 t = D\sin(\omega_0 t - \phi),} \tag{9.66}$$

where, in the first expression, A and B are constants to be determined and, in the second expression, D and ϕ are constants to be determined and D is assumed to be a positive quantity.* While either of these expressions can be used, the latter gives a more easily interpreted result since the amplitude D and phase ϕ are immediately apparent, and so we will use it and proceed to determine D and ϕ. Substituting the second expression for x_p from Eq. (9.66) into Eq. (9.63), we obtain

$$-Dm\omega_0^2 \sin(\omega_0 t - \phi) + Dc\omega_0 \cos(\omega_0 t - \phi) + Dk\sin(\omega_0 t - \phi) = F_0 \sin\omega_0 t. \tag{9.67}$$

Using the trigonometric identities $\sin(\alpha - \beta) = \sin\alpha\cos\beta - \cos\alpha\sin\beta$ and $\cos(\alpha - \beta) = \cos\alpha\cos\beta + \sin\alpha\cos\beta$ and then collecting coefficients of $\sin(\omega_0 t - \phi)$ and $\cos(\omega_0 t - \phi)$, we obtain

$$D(-m\omega_0^2\cos\phi + c\omega_0\sin\phi + k\cos\phi)\sin\omega_0 t \\ + D(m\omega_0^2\sin\phi + c\omega_0\cos\phi - k\sin\phi)\cos\omega_0 t = F_0\sin\omega_0 t. \tag{9.68}$$

Since this equation must be true for all time, we can equate the coefficients of $\sin\omega_0 t$ and $\cos\omega_0 t$ to obtain two equations for the unknowns D and ϕ. Doing this for $\cos\omega_0 t$ allows us to solve for $\tan\phi$ as

$$\boxed{\tan\phi = \frac{c\omega_0}{k - m\omega_0^2} = \frac{2(c/c_c)(\omega_0/\omega_n)}{1 - (\omega_0/\omega_n)^2} = \frac{2\zeta\omega_0/\omega_n}{1 - (\omega_0/\omega_n)^2},} \tag{9.69}$$

where the definitions of ω_n, c_c, and ζ, in Eqs. (9.4), Eq. (9.51), and Eq. (9.58), respectively, have been used. The phase angle ϕ, which is plotted in Fig. 9.27 as a

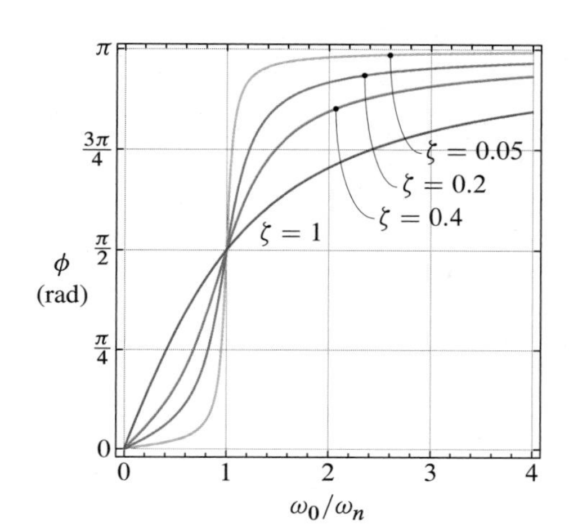

Figure 9.27
The phase angle ϕ as a function of the frequency ratio ω_0/ω_n.

---

* In Eq. (9.66) we have used a sine function of the form $\sin(\omega_0 t - \phi)$ instead of $\sin(\omega_0 t + \phi)$ because it results in a more convenient expression for $\tan\phi$.

function of the frequency ratio ω_0/ω_n for different values of ζ, represents the amount of a cycle by which the response of the system lags the forcing applied to it. Equating the coefficients of $\sin\omega_0 t$, we obtain

$$D = \frac{F_0/\cos\phi}{k - m\omega_0^2 + c\omega_0\tan\phi} = \frac{F_0/\cos\phi}{c\omega_0/\tan\phi + c\omega_0\tan\phi}, \tag{9.70}$$

where, to get the second expression for D, we have used Eq. (9.69). Now multiplying the numerator and denominator by $\tan\phi$ and noting that $1 + \tan^2\phi = 1/\cos^2\phi$, we obtain

$$D = \frac{F_0\sin\phi}{c\omega_0} = \frac{F_0}{c\omega_0}\left\{\frac{2\zeta\omega_0/\omega_n}{\sqrt{\left[1-(\omega_0/\omega_n)^2\right]^2 + \left(2\zeta\omega_0/\omega_n\right)^2}}\right\}, \tag{9.71}$$

where we have used the trigonometric identity that if $\alpha = \tan^{-1}x$, then $\sin\alpha = x/\sqrt{1+x^2}$. Again using the definitions of ω_n, c_c, and ζ as above, D simplifies to

$$D = \frac{F_0/k}{\sqrt{\left[1-(\omega_0/\omega_n)^2\right]^2 + \left(2\zeta\omega_0/\omega_n\right)^2}}. \tag{9.72}$$

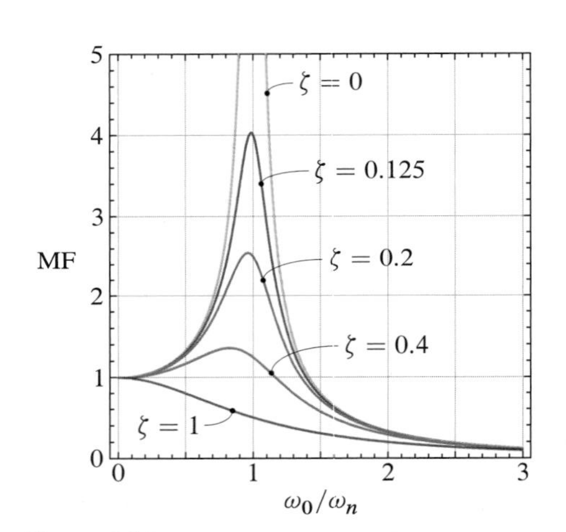

Figure 9.28
The MF as a function of the frequency ratio ω_0/ω_n for various values of the damping ratio ζ.

Now that we have the particular solution, we will, as was done in Section 9.2, define a *magnification factor* for a damped forced harmonic oscillator as the ratio of the amplitude of steady-state vibration D to the static deflection F_0/k. Thus, we obtain

$$\text{MF} = \frac{D}{F_0/k} = \frac{1}{\sqrt{\left[1-(\omega_0/\omega_n)^2\right]^2 + \left(2\zeta\omega_0/\omega_n\right)^2}}. \tag{9.73}$$

A plot of the MF is shown in Fig. 9.28 for various values of the damping ratio ζ. Figure 9.28 illustrates some important features of the behavior of a damped forced harmonic oscillator:

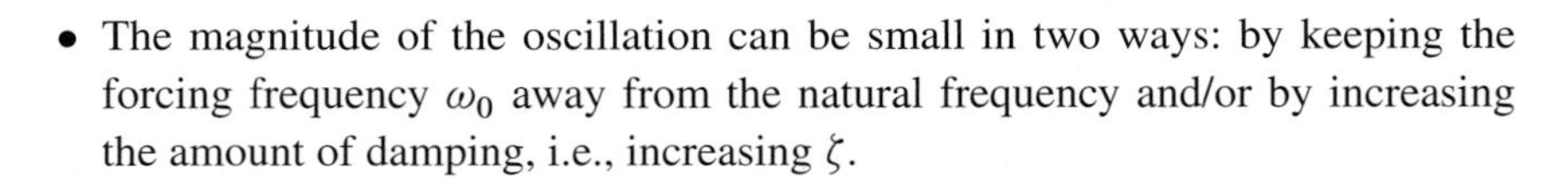

- The magnitude of the oscillation can be small in two ways: by keeping the forcing frequency ω_0 away from the natural frequency and/or by increasing the amount of damping, i.e., increasing ζ.
- As the amount of damping is increased, the peak in the MF moves farther to the left, away from $\omega_0/\omega_n = 1$. The peak for any given value of ζ can be found by using the usual technique from calculus to find the maximum value of a function (see Prob. 9.52).
- Comparing Fig. 9.28 with Fig. 9.13, observe that the effect of damping, while quite noticeable near the resonance frequency, becomes very small for values of ω_0/ω_n away from 1.

End of Section Summary

Viscously damped free vibration. The standard form of the equation of motion for a one degree of freedom viscously damped harmonic oscillator is

Eq. (9.45), p. 696

$$m\ddot{x} + c\dot{x} + kx = 0,$$

where m is the mass, c is the *coefficient of viscous damping*, and k is the linear spring constant. The character of the solution to this equation depends on the amount of damping relative to a specific amount of damping called the *critical damping coefficient*, which is defined as

Eq. (9.51), p. 696

$$c_c = 2m\sqrt{\frac{k}{m}} = 2m\omega_n,$$

where $\omega_n = \sqrt{k/m}$. In particular, if $c \geq c_c$, then the motion is nonoscillatory, whereas if $c < c_c$, then the motion is oscillatory. If $c > c_c$, the system is said to be *overdamped*, and the solution is given by

Eq. (9.52), p. 696

$$x = e^{-(c/2m)t}\left(Ae^{t\sqrt{(c/2m)^2-k/m}} + Be^{-t\sqrt{(c/2m)^2-k/m}}\right),$$

where A and B are constants to be determined from the initial conditions. If $c = c_c$, the system is said to be *critically damped*, and the solution is given by

Eq. (9.53), p. 697

$$x = (A + Bt)e^{-\omega_n t},$$

where, again, A and B are constants to be determined from the initial conditions. Finally, if $c < c_c$, the system is said to be *underdamped*, and the solution is given by

Eq. (9.54), p. 697

$$x = e^{-(c/2m)t}\big(A\sin\omega_d t + B\cos\omega_d t\big),$$

or, equivalently, by

Eq. (9.57), p. 697

$$x = De^{-(c/2m)t}\sin(\omega_d t + \phi),$$

where A and B are constants to be determined from the initial conditions in the first solution, D and ϕ are analogous constants in the second solution, and ω_d is the *damped natural frequency*, which is given by

Eq. (9.55), p. 697, and Eq. (9.62), p. 698

$$\omega_d = \sqrt{\frac{k}{m} - \left(\frac{c}{2m}\right)^2} = \omega_n\sqrt{1-(c/c_c)^2} = \omega_n\sqrt{1-\zeta^2},$$

where $\zeta = c/c_c$ is the *damping ratio*.

Viscously damped forced vibration. The standard form of a viscously damped forced harmonic oscillator is

Eq. (9.63), p. 698

$$m\ddot{x} + c\dot{x} + kx = F_0 \sin\omega_0 t,$$

where F_0 is the amplitude of the forcing function and ω_0 is the frequency of the forcing function. When expressed using the damping ratio ζ, the equation above takes on the form

Eq. (9.64), p. 698

$$\ddot{x} + 2\zeta\omega_n\dot{x} + \omega_n^2 x = \frac{F_0}{m}\sin\omega_0 t.$$

The general solution of either of these equations is the sum of the complementary solution and a particular solution. The complementary solution is transient; i.e., it vanishes as time increases. The particular or steady-state solution is of the form

Eq. (9.66), p. 699

$$x_p = D\sin(\omega_0 t - \phi),$$

where ϕ and D are given by

Eq. (9.69), p. 699, and Eq. (9.72), p. 700

$$\tan\phi = \frac{c\omega_0}{k - m\omega_0^2} = \frac{2(c/c_c)(\omega_0/\omega_n)}{1 - (\omega_0/\omega_n)^2} = \frac{2\zeta\omega_0/\omega_n}{1 - (\omega_0/\omega_n)^2},$$

$$D = \frac{F_0/k}{\sqrt{[1 - (\omega_0/\omega_n)^2]^2 + (2\zeta\omega_0/\omega_n)^2}}.$$

The *magnification factor* for a damped, forced harmonic oscillator is

Eq. (9.73), p. 700

$$\text{MF} = \frac{D}{F_0/k} = \frac{1}{\sqrt{[1 - (\omega_0/\omega_n)^2]^2 + (2\zeta\omega_0/\omega_n)^2}},$$

a plot of which is shown in Fig. 9.29 for various values of the damping ratio ζ.

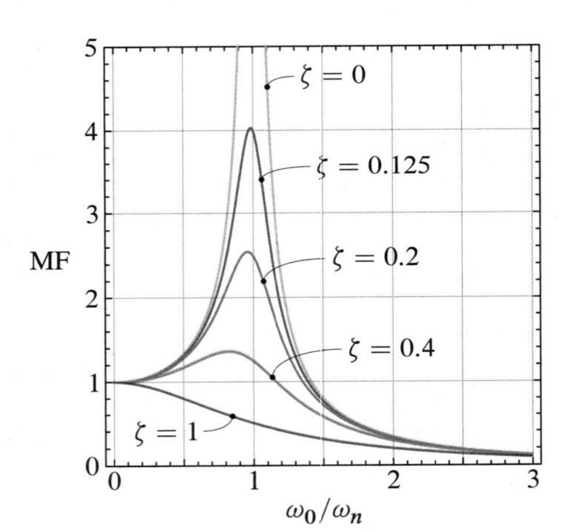

Figure 9.29
The MF as a function of the frequency ratio ω_0/ω_n for various values of the damping ratio ζ.

EXAMPLE 9.7 *Motion of a Weight on a Viscously Damped Vibrating Spring Scale*

The scale shown in Fig. 1 uses an internal spring that couples the platform P to the base B of the scale to determine the weight of an object placed on the platform. So that oscillations of the platform die out quickly after an object is placed on the scale, it also includes an internal dashpot in parallel with the spring to dissipate energy. If the combined mass of the block and platform is $m_P = 1.5\,\text{kg}$ and the spring constant is $k = 50\,\text{N/m}$, determine the critical viscous damping coefficient c_c for the dashpot so that oscillations dissipate quickly.

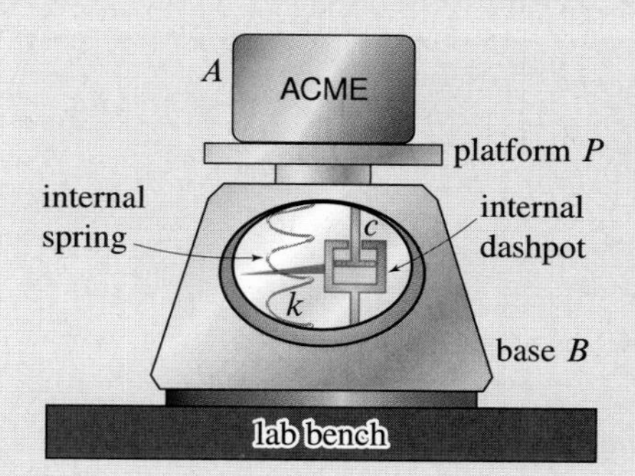

Figure 1
A scale used to measure the mass of an object A that has been placed on it. The platform P and the mass A are coupled to the base of the scale via an internal spring and dashpot. The block A does not separate from the platform P during the vibration.

SOLUTION

Road Map & Modeling The FBD of the block and platform during oscillation of the platform is shown in Fig. 2. We will apply Newton's second law in the y direction to obtain the equation of motion, from which we should be able to extract c_c.

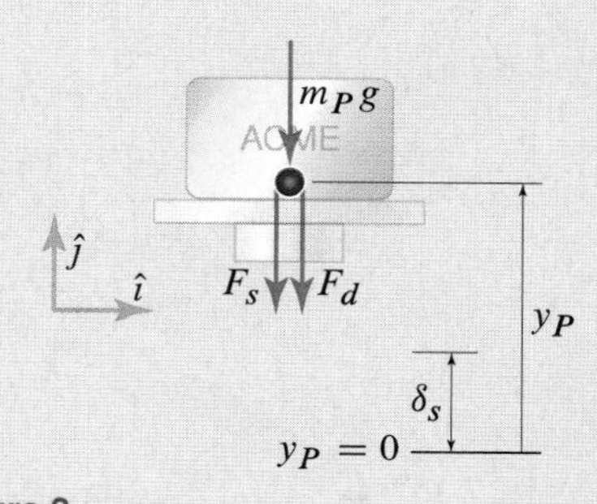

Figure 2
FBD of the block A and platform P in Fig. 1. F_s and F_d are the force exerted by the spring and the force exerted by the dashpot, respectively, on the platform.

Governing Equations

Balance Principles Summing forces in the y direction in Fig. 2, we obtain

$$\sum F_y: \quad -F_s - F_d - m_P g = m_P a_{Py}. \tag{1}$$

Force Laws The force laws for the spring and dashpot are determined by the motion of the platform P, so we obtain

$$F_s = k(y_P - \delta_s) \quad \text{and} \quad F_d = c\dot{y}_P, \tag{2}$$

where y_P is the vertical position of the platform measured from the static equilibrium position of the platform, and δ_s is the deflection of the spring when the system is in static equilibrium (see Fig. 2).

Kinematic Equations For the kinematic equations, we need to express a_{Py} in terms of motion of the platform, which gives

$$a_{Py} = \ddot{y}_P. \tag{3}$$

Computation Substituting Eqs. (2) and (3) into Eq. (1), we get the equation of motion as

$$-k(y_P - \delta_s) - c\dot{y}_P - m_P g = m_P \ddot{y}_P \quad \Rightarrow \quad m_P \ddot{y}_P + c\dot{y}_P + k y_P = 0, \tag{4}$$

where we have used the fact that $k\delta_s = m_P g$. Since Eq. (4) is in the standard form of the viscously damped harmonic oscillator in Eq. (9.45), we can find the critical damping coefficient by using Eq. (9.51), which gives

$$c_c = 2m_P \sqrt{\frac{k}{m_P}} = 2m_P \omega_n \quad \Rightarrow \quad \boxed{c_c = 17.32\,\text{kg/s},} \tag{5}$$

where $m_P = 1.5\,\text{kg}$ and $\omega_n = \sqrt{k/m_P} = 5.774\,\text{rad/s}$.

Discussion & Verification The units of c_c are kg/s, which is equivalent to the SI units for the coefficient of viscous damping stated earlier in this section, that is, N·s/m.

EXAMPLE 9.8 *Critically Damped Free Vibration of a Gate*

Figure 1
A carnival ride with a gate that counts riders.

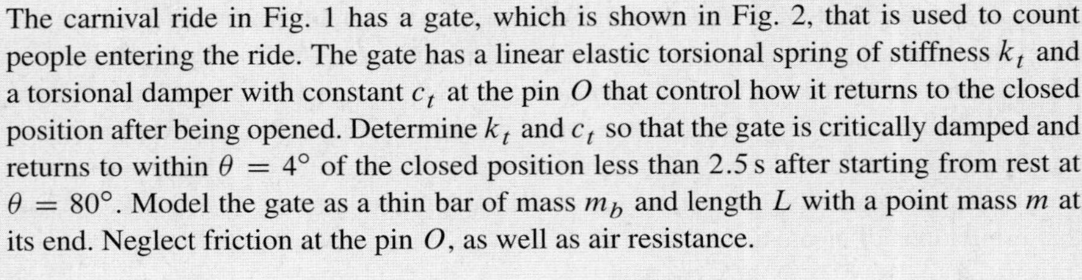

The carnival ride in Fig. 1 has a gate, which is shown in Fig. 2, that is used to count people entering the ride. The gate has a linear elastic torsional spring of stiffness k_t and a torsional damper with constant c_t at the pin O that control how it returns to the closed position after being opened. Determine k_t and c_t so that the gate is critically damped and returns to within $\theta = 4°$ of the closed position less than 2.5 s after starting from rest at $\theta = 80°$. Model the gate as a thin bar of mass m_b and length L with a point mass m at its end. Neglect friction at the pin O, as well as air resistance.

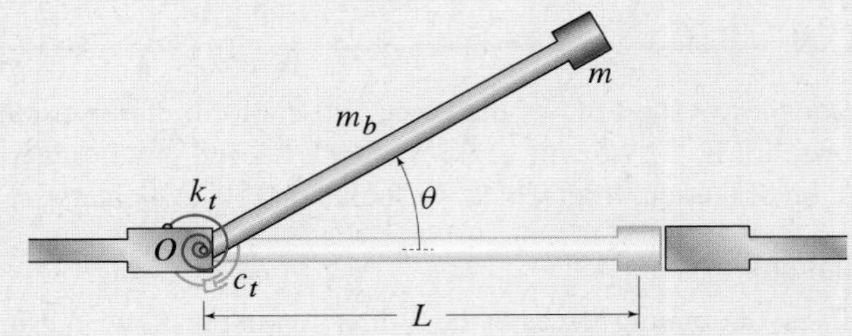

Figure 2. A carnival ride gate for which $L = 42$ in., the weight of the thin bar is 4 lb, and the weight of the point mass is 3 lb. The gate lies in the horizontal plane.

SOLUTION

Road Map & Modeling We will first need to derive the equation of motion for the gate in the standard form of Eq. (9.45), using the FBD of the arm shown in Fig. 3. Once we have the equation of motion, we can find its response using Eq. (9.53). From the response, we can determine the value of k_t needed to get the arm closed in the required time and then use that to find the c_t needed to critically damp the arm.

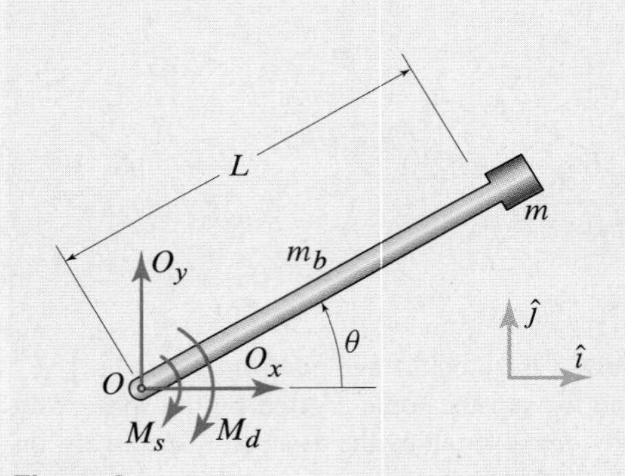

Figure 3
FBD of the carnival ride gate in Fig. 2.

Governing Equations

Balance Principles Summing moments about point O, we obtain

$$\sum M_O\colon \quad -M_s - M_d = I_O \alpha_{\text{gate}}, \tag{1}$$

where M_s is the restoring moment due to the spring, M_d is the damping moment due to the dashpot, and α_{gate} is the angular acceleration of the gate. The mass moment of inertia of the gate with respect to O is computed as

$$I_O = mL^2 + \tfrac{1}{12} m_b L^2 + m_b (L/2)^2 = \left(\tfrac{1}{3} m_b + m\right) L^2. \tag{2}$$

Force Laws The moment laws for the spring and dashpot are

$$M_s = k_t \theta \quad \text{and} \quad M_d = c_t \dot{\theta}. \tag{3}$$

Kinematic Equations The angular acceleration of the gate can be written in terms of θ as $\alpha_{\text{gate}} = \ddot{\theta}$.

Computation Substituting the kinematic relation, as well as Eqs. (2) and (3) into Eq. (1) and rearranging, we obtain the equation of motion of the gate as

$$\left(\tfrac{1}{3} m_b + m\right) L^2 \ddot{\theta} + c_t \dot{\theta} + k_t \theta = 0. \tag{4}$$

This implies that the natural frequency ω_n is given by

$$\omega_n = \sqrt{\frac{k_t}{\left(\frac{1}{3} m_b + m\right) L^2}} = 0.7788 \sqrt{k_t} \text{ rad/s}. \tag{5}$$

To determine k, we will enforce the condition requiring the gate to be within 4° of the closed position in 2.5 s or less to the solution of Eq. (4), which is given by $\theta = (A + Bt)e^{-\omega_n t}$ for critical damping. But first we need to find A and B. Enforcing the condition that $\theta(0) = 80° = 1.396$ rad, we obtain

$$\theta(0) = A = 1.396\,\text{rad}. \tag{6}$$

We now enforce the condition that $\dot{\theta}(0) = 0$ rad/s, which gives

$$\dot{\theta} = Be^{-\omega_n t} + (A + Bt)\big(-\omega_n e^{-\omega_n t}\big) \quad \Rightarrow \quad \dot{\theta}(0) = B - A\omega_n = 0$$
$$\Rightarrow \quad B = 1.396\omega_n. \tag{7}$$

Substituting Eqs. (5)–(7) into the critically damped solution for θ, we obtain

$$\theta = \big(1.396 + 1.087t\sqrt{k_t}\big)e^{-0.7788t\sqrt{k_t}}. \tag{8}$$

To obtain k_t, we will say that we want $\theta(2.5) = 4° = 0.06981$ rad, which gives

$$0.06981 = \big(1.396 + 2.719\sqrt{k_t}\big)e^{-1.947\sqrt{k_t}}, \tag{9}$$

which is a transcendental equation for k_t. This equation can be solved iteratively or by using a mathematical software package (e.g., Mathematica or Matlab). Solving Eq. (9) gives

$$\boxed{k_t = 5.936\,\text{ft·lb/rad.}} \tag{10}$$

Now that we have k_t, the condition for critical damping in Eq. (9.51) tells us that the damping coefficient must be given by

$$c_{tc} = 2\big(\tfrac{1}{3}m_b + m\big)L^2\omega_n \quad \Rightarrow \quad \boxed{c_{tc} = 6.256\,\text{ft·lb·s.}} \tag{11}$$

A plot of the solution in Eq. (8) (using k_t from Eq. (10)) can be found in Fig. 4.

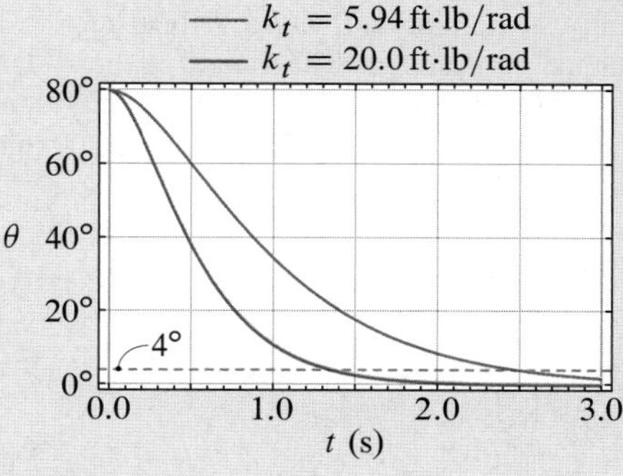

Figure 4
Plot of response of the gate for two different values of the torsional spring constant.

Discussion & Verification The dimensions of each of the two torsional constants found in Eqs. (10) and (11) are as they should be.

A Closer Look Note that the gate could be returned to the closed position even more quickly by increasing the value of k_t (compare the curves for two different values of k_t in Fig. 4). Unfortunately, this has potentially undesirable consequences. Referring to Fig. 5, we see that larger values of k_t result in the angular velocity of the gate being larger for *every* value of θ. This means that someone going through the gate before it closes will get hit harder by the gate (compare $\dot{\theta}$ at points A and B in Fig. 5). In addition, referring to Eq. (11), we see that larger values of k_t mean that the damping coefficient must be larger to achieve critical damping. This would likely imply that the damping mechanism must be more substantial and more expensive.

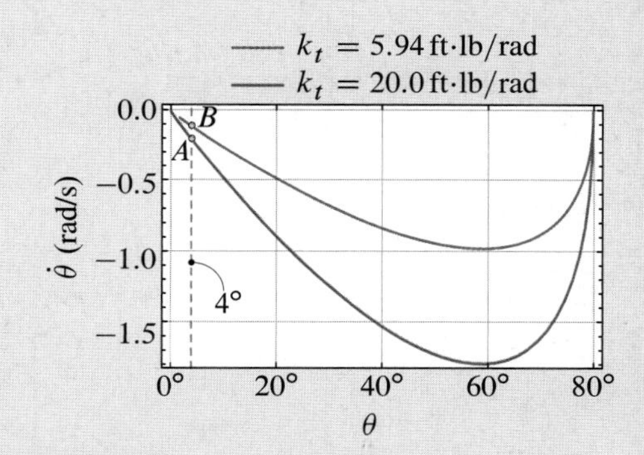

Figure 5
A plot of the angular velocity of the gate $\dot{\theta}$ as a function of its position θ for two different values of the torsional spring constant.

EXAMPLE 9.9 *Response and MF for a Rotating Unbalanced Mass*

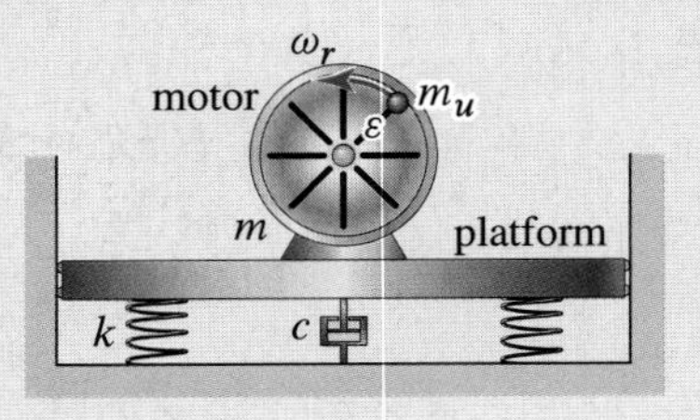

Figure 1
An unbalanced motor on a platform that is elastically supported, and whose vertical motion is damped by dashpots. The values of the system parameters are $m = 55\,\text{kg}$, $k = 420{,}000\,\text{N/m}$, $c = 4000\,\text{N}\cdot\text{s/m}$, $\varepsilon = 15\,\text{cm}$, $\omega_r = 1200\,\text{rpm}$, and m_u is 10 g, 100 g, or 1000 g.

If we add dashpots in parallel with the springs that are supporting the platform to the unbalanced motor we studied in Examples 9.5 and 9.6, we obtain the system shown in Fig. 1. We will assume that all the springs supply a total spring constant k, the dashpots provide a total damping coefficient c, and the combined mass of the motor and of the platform is m. From Example 9.5, we can show that the equation of motion for the unbalanced motor is (see Prob. 9.54)

$$\ddot{y} + 2\zeta\omega_n\dot{y} + \omega_n^2 y = \frac{m_u\varepsilon\omega_r^2}{m}\sin\omega_r t, \tag{1}$$

where y is the vertical position of the motor measured from its static equilibrium position, m_u is the eccentric mass (m includes m_u), ε is the distance from the unbalanced mass to the rotor axis, and ω_r is the angular velocity of the rotor. Using $c = 4000\,\text{N}\cdot\text{s/m}$, determine and plot the steady-state solution, using the parameters given in Example 9.6. In addition, determine and plot the MF for the unbalanced motor and compare it with the MF shown in Fig. 9.28, which applies to Eq. (9.64).

SOLUTION

Road Map & Modeling We know that the steady-state solution to an equation of the form in Eq. (1) is given by Eqs. (9.66), (9.69), and (9.72). Therefore, we need to interpret the forcing amplitude on the right-hand side of Eq. (1) in that context to obtain the steady-state solution. The MF is found from the amplitude of the steady-state solution, so we will look at the expression for D that we obtain from Eq. (9.72) after interpreting the right-hand side of Eq. (1). Once we write the amplitude as a function of ω_r/ω_n and ζ, we will have the desired MF.

Governing Equations As discussed above, the steady-state solution is given by Eq. (9.66), i.e.,

$$y_{\text{ss}} = D\sin(\omega_r t - \phi), \tag{2}$$

where D is given by Eq. (9.72), i.e.,

$$D = \frac{F_0/k}{\sqrt{[1-(\omega_0/\omega_n)^2]^2 + (2\zeta\omega_0/\omega_n)^2}}, \tag{3}$$

and ϕ is given by Eq. (9.69), i.e.,

$$\tan\phi = \frac{c\omega_0}{k - m\omega_0^2} = \frac{2\zeta\omega_0/\omega_n}{1-(\omega_0/\omega_n)^2}, \tag{4}$$

in which $\omega_n = \sqrt{k/m} = 87.39\,\text{rad/s}$ and $\zeta = c/(2m\omega_n) = 0.4161$. Comparing Eq. (1) with Eq. (9.64) on p. 698, we see that

$$\omega_0 = \omega_r \quad \text{and} \quad F_0 = m_u\varepsilon\omega_r^2. \tag{5}$$

Computation The steady-state solution is now found by substituting Eqs. (3)–(5) into Eq. (2). Doing this and then substituting in all given parameters, we find

$$\boxed{y_{\text{ss}} = 0.003519 m_u \sin(125.7t + 0.8423)\ \text{m},} \tag{6}$$

which is plotted in Fig. 2 for the three given values of m_u.

To find the MF, we substitute Eqs. (5) into Eq. (3) to obtain

$$D = \frac{m_u\varepsilon\omega_r^2/k}{\sqrt{[1-(\omega_r/\omega_n)^2]^2 + (2\zeta\omega_r/\omega_n)^2}}. \tag{7}$$

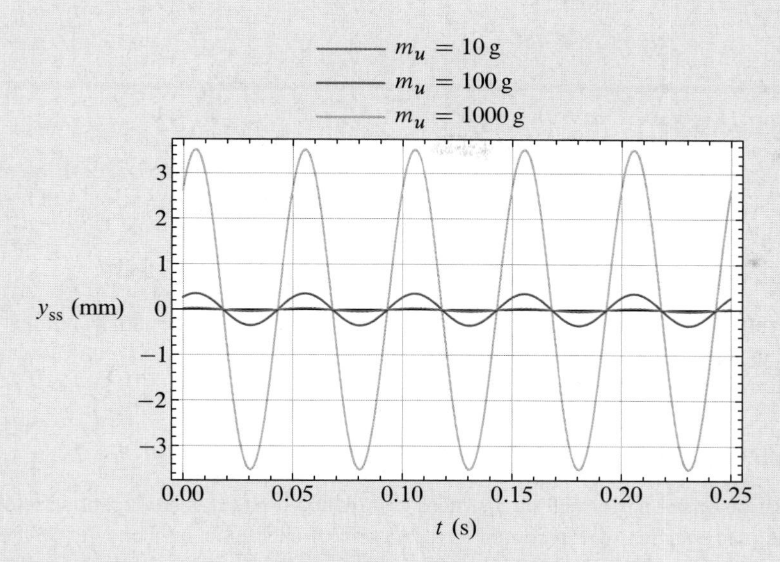

Figure 2. The steady-state solution of the unbalanced motor for three different values of the unbalanced mass m_u.

Focusing on the numerator, we see that we can write it as

$$\frac{m_u \varepsilon \omega_r^2}{k} = \frac{m_u \varepsilon \omega_r^2}{k} \frac{m}{m} = \frac{m_u \varepsilon}{m} \frac{m}{k} \omega_r^2 = \frac{m_u \varepsilon}{m} \frac{\omega_r^2}{\omega_n^2}, \tag{8}$$

where, to obtain the last equality, we have used the fact that $m/k = 1/\omega_n^2$. Substituting Eq. (8) into Eq. (7) and moving $m_u \varepsilon/m$ to the left-hand side, we obtain the MF as

$$\boxed{\text{MF} = \frac{mD}{m_u \varepsilon} = \frac{(\omega_r/\omega_n)^2}{\sqrt{[1-(\omega_r/\omega_n)^2]^2 + (2\zeta\omega_r/\omega_n)^2}},} \tag{9}$$

a plot of which is shown in Fig. 3 for various values of ζ.

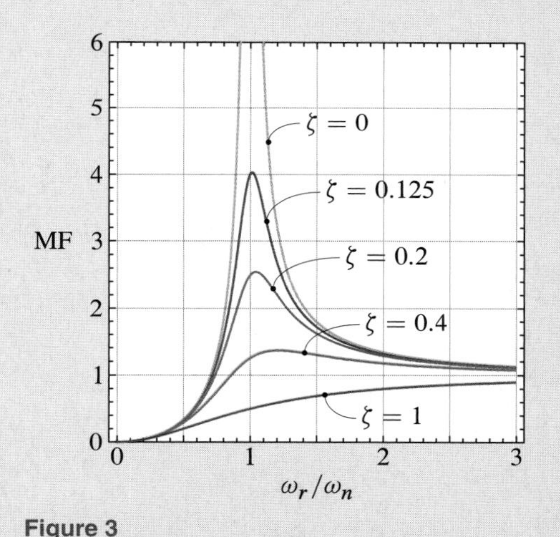

Figure 3
The unbalanced motor MF as a function of the frequency ratio ω_r/ω_n for various values of the damping ratio ζ.

Discussion & Verification Careful examination of the steady-state response given in Eq. (2), with Eqs. (3)–(5) substituted in, reveals that it has the dimension of length, as should be expected. In addition, we see in Fig. 2 that as we increase the amount of the unbalanced mass m_u, the oscillation amplitude increases as expected. The MF in Eq. (9) is dimensionless, as it should be.

A Closer Look Comparing Fig. 3 with Fig. 9.28 on p. 700, we see that for low forcing frequencies:

- When a harmonic oscillator is forced by applying the forcing directly to the oscillator, the oscillator follows the forcing (MF → 1 in Fig. 9.28).
- When a harmonic oscillator is forced by an internal rotating unbalanced mass, the oscillator barely moves (MF → 0 in Fig. 3).

For high forcing frequencies:

- When a harmonic oscillator is forced by applying the forcing directly to the oscillator, the oscillator barely moves (MF → 0 in Fig. 9.28).
- When a harmonic oscillator is forced by an internal rotating unbalanced mass, the oscillator moves with the forcing (MF → 1 in Fig. 3).

Problems

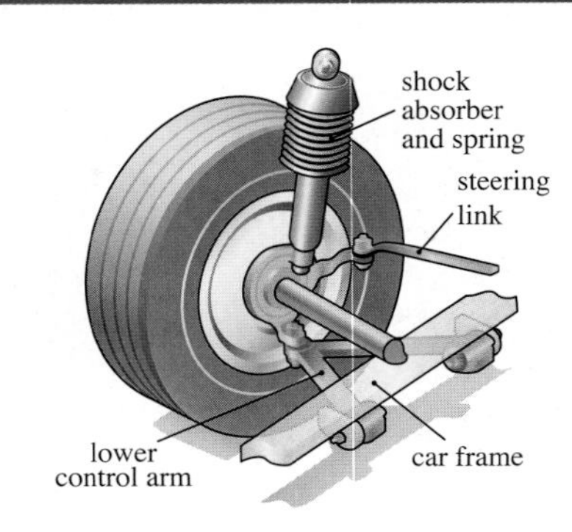

Figure P9.48

Problem 9.48

In the design of a MacPherson strut suspension, what would you choose for the damping ratio ζ? Explain your answer in terms of automotive ride and comfort.

Problem 9.49

For identical systems, one with damping and the other without, would you expect the period of damped vibration to be greater, less than, or equal to the period of undamped vibration? Explain your answer.

Problem 9.50

A vibration test is performed on a structure, in which both the magnification factor MF and the phase angle ϕ are recorded as a function of excitation frequency ω_0. After the test, it is discovered that, for some unfortunate reason, the recording of the magnification factor data is corrupted so that only the phase angle data is available for analysis. Is it possible to determine the resonant frequency from the available data? What can be inferred about the amount of damping in the system from the phase data?

Problem 9.51

Suppose that the equation of motion of a damped forced harmonic oscillator is given by $\ddot{x}+2\zeta\omega_n\dot{x}+\omega_n^2 x = (F_0/m)\cos\omega_0 t$, where x is measured from the equilibrium position of the system. Obtain the expression for the amplitude of the steady-state response of the oscillator, and compare it with the expression presented in Eq. (9.72) (which is for a system with equation of motion $\ddot{x} + 2\zeta\omega_n\dot{x} + \omega_n^2 x = (F_0/m)\sin\omega_0 t$).

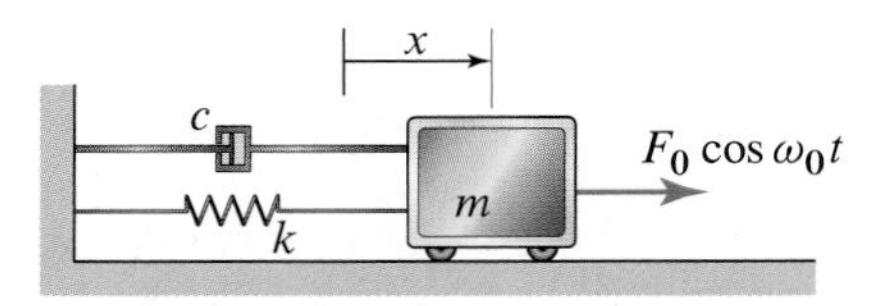

Figure P9.51

Problem 9.52

Differentiate Eq. (9.73) with respect to ω_0/ω_n, and set the result equal to zero to determine the frequency ω_0 at which peaks in the MF curve occur as a function of ζ and ω_n. Use this result to show that the peak always occurs at $\omega_0/\omega_n \le 1$. Finally, determine the value of ζ for which the MF has no peak.

Problem 9.53

Calculate the response described by the equations listed below, in which x is measured in feet and time is measured in seconds.

(a) $5\ddot{x} + 10\dot{x} + 100x = 0$, with $x(0) = 0.1$ and $\dot{x}(0) = -0.1$

(b) $3\ddot{x} + 15\dot{x} + 12x = 0$, with $x(0) = 0$ and $\dot{x}(0) = 0.5$

(c) $\ddot{x} + 10\dot{x} + 25x = 0$, with $x(0) = 0.15$ and $\dot{x}(0) = 0$

(d) $25\ddot{x} + 200\dot{x} + 1500x = 0$, with $x(0) = 0.01$ and $\dot{x}(0) = 0$

Problem 9.54

Derive the equation of motion given in Eq. (1) of Example 9.9 for the system in that example. The independent variable y is measured from the equilibrium position of the system, m is the mass of the motor and platform, c is the total damping coefficient of the dashpots, k is the total constant of the linear elastic springs, ω_r is the angular velocity of the unbalanced rotor, ε is the distance of the eccentric mass from the rotor axis, and m_u is the eccentric mass. Note that m *includes* the eccentric mass so that the nonrotating mass is equal to $m - m_u$.

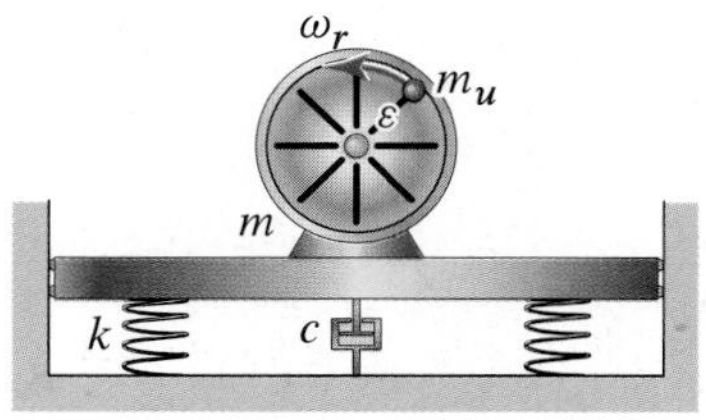

Figure P9.54

Problem 9.55

A module with sensitive electronics is mounted on a panel that vibrates due to excitation from a nearby diesel generator. To prevent fatigue failure, the module is placed on vibration-absorbing mounts. The displacement of the panel is measured to be $y_p(t) = y_0 \sin \omega_0 t$, where $y_0 = 0.001$ m, $\omega_0 = 300$ rad/s, and the time t is measured in seconds. Letting the mass of the electronic module be $m = 0.5$ kg, calculate the amplitude of the vibration of the module if the equivalent stiffness and damping coefficients for all the mounts combined are $k = 10{,}000$ N/m and $c = 40$ N·s/m, respectively.

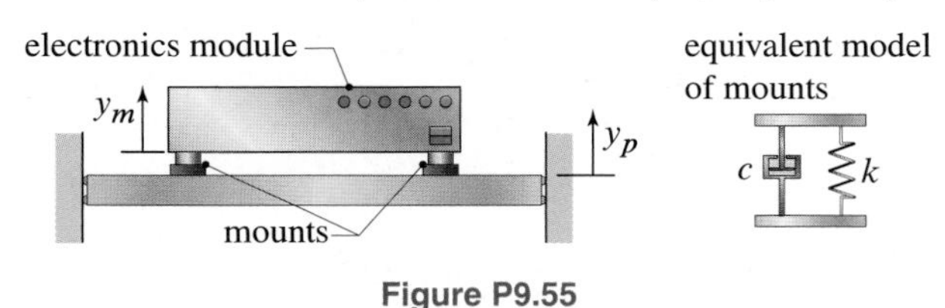

Figure P9.55

Problem 9.56

A hard drive arm undergoes flow-induced vibration caused by the vortices of air produced by a platter that rotates at $\omega_0 = 10{,}000$ rpm. The arm has length $L = 0.037$ m and mass $m = 0.00075$ kg, and it is made from aluminum with a modulus of elasticity $E = 70$ GPa. In addition, assume that the cross section of the arm has an area moment of inertia $I_{cs} = 8.5 \times 10^{-14}$ m^4. Following the steps in Example 9.2 on p. 672, the arm can be modeled as a rigid rod that is pinned at one end and is restrained by a torsional spring with equivalent spring constant $k_t = 3EI_{cs}/L$. In addition to the torsional spring, assume that the arm's motion is affected by a torsional damper with torsional damping coefficient c_t. Assuming that the damping ratio is $\zeta = 0.02$ and that the vortices produce an aerodynamic force with the same frequency as the rotation of the platter, determine the amplitude of the aerodynamic force needed to cause a steady-state vibration amplitude of 0.0001 m at the tip of the arm. Assume that the aerodynamic force is applied at the midpoint B of the hard drive. What vibration amplitude will result if the same excitation is applied to a hard drive arm assembly with the damping ratio of 0.05?

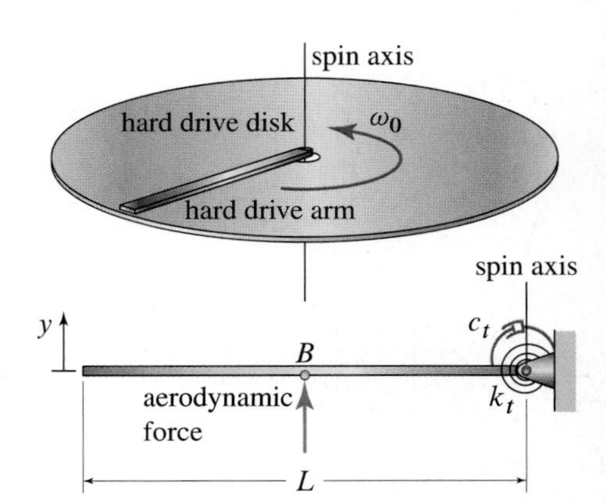

Figure P9.56

Problem 9.57

The mechanism consists of a disk D pinned at G, which is both the geometric center of the disk and its mass center. The outer circumference of the disk has radius $r_o = 0.1$ m and is connected to an element consisting of a linear spring with stiffness $k_1 = 100$ N/m in parallel with a dashpot with damping coefficient $c = 50$ N·s/m. The disk has a hub of radius $r_i = 0.05$ m that is connected to a linear spring with constant $k_2 = 350$ N/m. Knowing that for $\theta = 0$ the disk is in static equilibrium and that the mass moment of inertia of the disk is $I_G = 0.001$ kg·m^2, derive the linearized equation of motion of the disk in terms of θ. In addition, calculate the resulting vibrational motion if the system is released from rest with an initial angular displacement $\theta_i = 0.05$ rad.

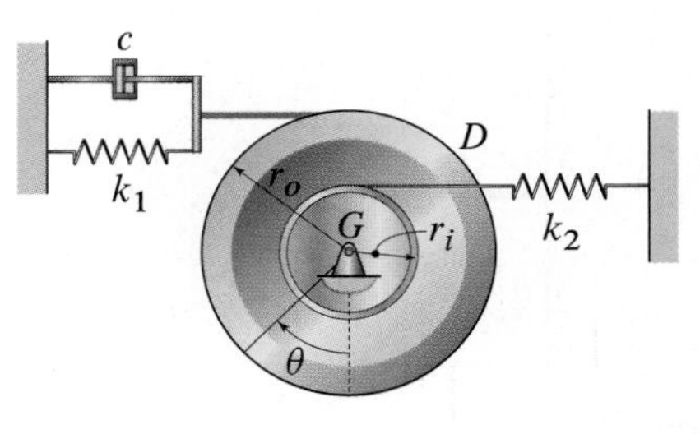

Figure P9.57

Problem 9.58

A box of mass 0.75 kg is thrown on a scale, causing both the scale and the box to move vertically downward with an initial speed of 0.5 m/s. Before the box lands on the scale, the scale is in equilibrium. The total mass of the scale's moving platform and the box is $m = 1.25$ kg. Modeling the platform's support as a spring and dashpot with stiffness $k = 1000$ N/m and damping coefficient $c = 70.7$ N·s/m, find the response of the scale. *Hint:* Place the origin of the y axis at the position of the platform corresponding to the equilibrium configuration of the platform and box together.

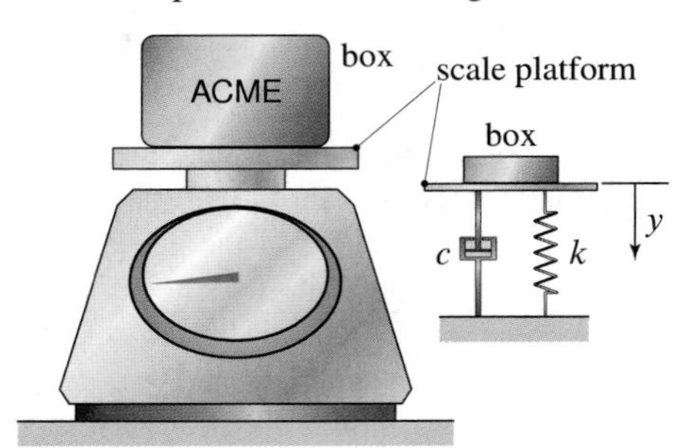

Figure P9.58

Problem 9.59

Consider a simple viscously damped harmonic oscillator governed by Eq. (9.45), and analyze the case in which the damping coefficient c is negative. Calculate the general expression for the response (without taking into account specific initial conditions), using $m = 1$ kg, $c = -1$ N·s/m, and $k = 10$ N/m. Comment on the system's response.

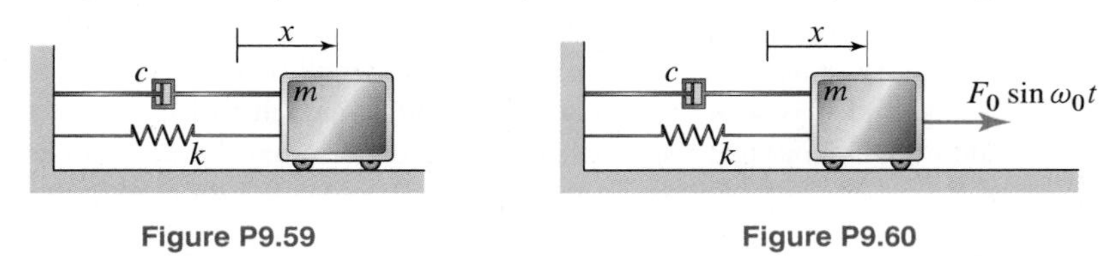

Figure P9.59 **Figure P9.60**

Problem 9.60

The MF for a harmonically excited spring-mass-damper system at $\omega_0/\omega_n \approx 1$ is equal to 5. Calculate the damping ratio of the system. What would the damping ratio be if the MF were equal to 10? Sketch the magnification factor at $\omega_0/\omega_n \approx 1$ as a function of the damping ratio.

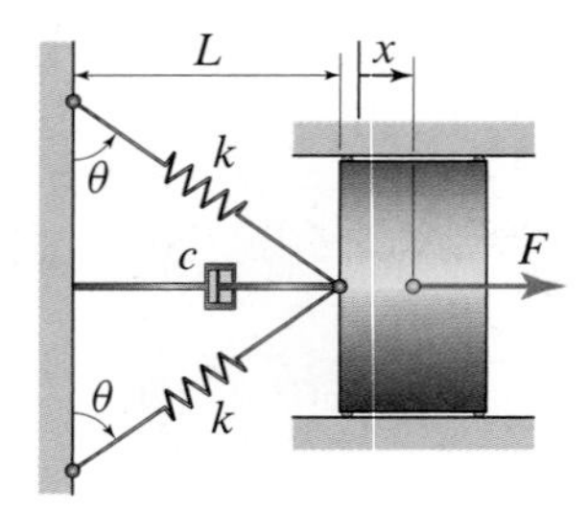

Figure P9.61

Problem 9.61

A slider moves in the horizontal plane under the action of the harmonic forcing $F(t) = F_0 \sin \omega_0 t$. The slider is connected to two identical linear springs, each of which has constant k. When $t = 0$, $x(0) = 0$, the springs are unstretched, $\theta = 45°$, and $L = L_0$. The slider is also connected to a damper with damping coefficient c. Treating F_0, k, c, and L_0 as known quantities, neglecting friction, and letting $\dot{x}(0) = v_i$, (a) derive the equations of motion of the system, (b) derive the linearized equations of motion about the initial position, and (c) determine the amplitude of the steady-state vibrations for the linearized equations of motion.

Problem 9.62

The mechanism shown is a pendulum consisting of a pendulum bob B with mass m and a T-bar, which is pinned at O and has negligible mass. The horizontal portion of the T-bar is connected to two supports, each of which has an identical spring and dashpot system, each with spring constant k and damping coefficient c. The springs are unstretched when B is vertically aligned with the pin at O. Modeling B as a particle, derive the linearized equations of motion of the system. In addition, assuming that the system is underdamped, derive the expression for the damped natural frequency of vibration of the system.

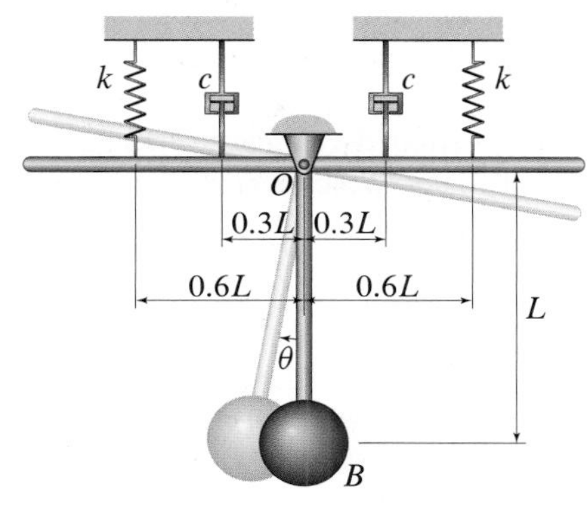

Figure P9.62

Problem 9.63

The block of mass m is coupled to the support A, which is displacing harmonically according to $y = Y \sin \omega_0 t$, by the linear elastic spring with constant k and the dashpot with constant c.

(a) Derive its equation of motion, using x as the independent variable, and explain in what way the resulting equation of motion is not in the form of Eq. (9.65).

(b) Next, let $z = x - y$ and substitute it into the equation of motion found in Part (a). After doing so, show that you obtain an equation of motion in z that is of the same form as Eq. (9.65).

(c) Find the steady-state solution to the equation of motion found in Part (b) and then using that, determine the steady-state solution for x.

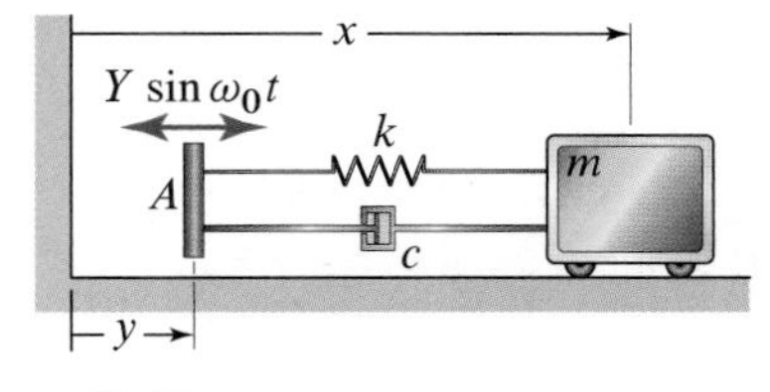

Figure P9.63

Problems 9.64 and 9.65

The block A and the platform P of a spring scale are at rest when the lab bench to which the scale is rigidly attached begins vibrating sinusoidally with a frequency of 15 Hz and amplitude of 5 mm. The block and the platform are coupled to the base B of the scale by a linear elastic spring and a viscous damper that are internal to the scale. The combined mass of the block and platform is $m_A = 1.5$ kg, the spring constant is $k = 50$ N/m, and the viscous damping coefficient is $c = 7.5$ N·s/m.

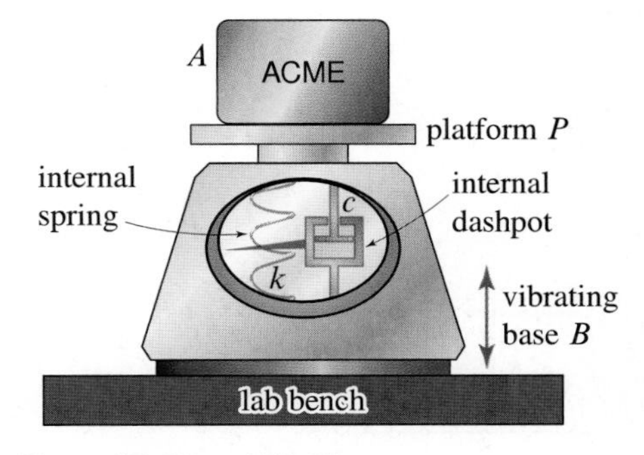

Figure P9.64 and P9.65

Problem 9.64 Determine the vertical motion of the platform and block as a function of time. The base of the scale B is rigidly attached to the lab bench, and the block A does not separate from the platform P during the vibration. *Hint:* Parts (a)–(c) of Prob. 9.63 will be helpful.

Problem 9.65 Determine and plot for 10 s the vertical motion of the platform and block as a function of time. The base of the scale B is rigidly attached to the lab bench, and the block A does not separate from the platform P during the vibration. *Hint:* Parts (a)–(c) of Prob. 9.63 will be helpful.

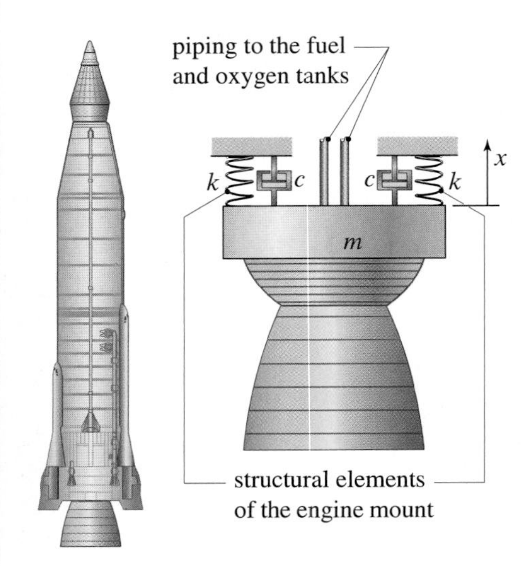

Figure P9.66

Problem 9.66

The engine in the rocket shown is supposed to provide a constant thrust of 5000 kN. The turbopump unit in the engine nominally operates at 7000 rpm, and as a result of a design issue, the actual thrust provided by the engine oscillates harmonically with an amplitude of 10 kN at the same rotational frequency of the turbopump unit. The mass of the engine is $m = 5000\,\text{kg}$. The rest of the rocket is much heavier than the engine and can be treated as being fixed. The engine is mounted to the rocket via two structural members, each of which can be modeled as consisting of a linear spring of stiffness k in parallel with a dashpot with linear viscous damping coefficient c. If $k = 2\times 10^8\,\text{N/m}$ and $c = 100\,\text{N}\cdot\text{s/m}$, determine the amplitude of vibration in the nominal operating regime. Ignore the stiffness and damping due to the piping. *Hint:* If x is measured from the equilibrium position of the engine that results from the combined effect of the thrust and gravity, then the engine is subject to an externally applied forcing equal to $(10\,\text{kN})\sin\omega_0 t$, where ω_0 is the rotational frequency of the turbopump.

Problem 9.67

A simple model for a ship rolling on waves* treats the waves as *sinusoids*. Using this model, it can be shown that a linear model for the roll angle θ is given by

$$I_G\ddot{\theta} + c\dot{\theta} + mg\theta = -\frac{I_G A\omega_0^2}{\lambda}\sin\omega_0 t,$$

where G denotes the mass center of the ship, I_G is the ship's mass moment of inertia, c is a rotational viscous damping constant coefficient, m is the mass of the ship, A is the wave amplitude, and λ is the wavelength of the waves.

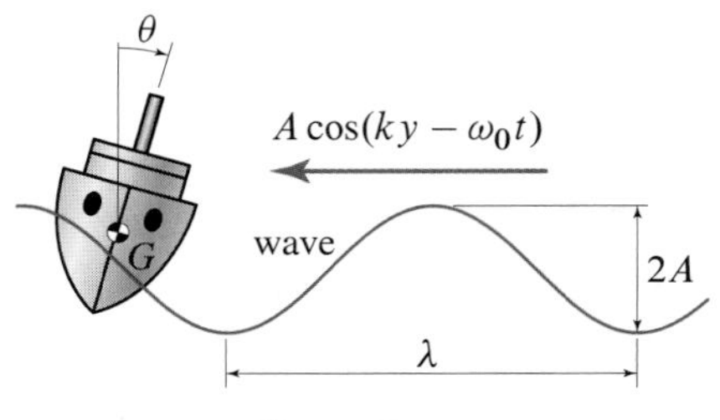

Figure P9.67

(a) What is the natural frequency of the system?

(b) Find the magnification factor for the system.

(c) Assuming that the damping is negligible (i.e., $c \approx 0$), if the maximum amplitude of oscillation that the ship can undergo without capsizing is $\theta_{max} = 1$ rad, find the maximum A so that the crew remains safe.

Problem 9.68

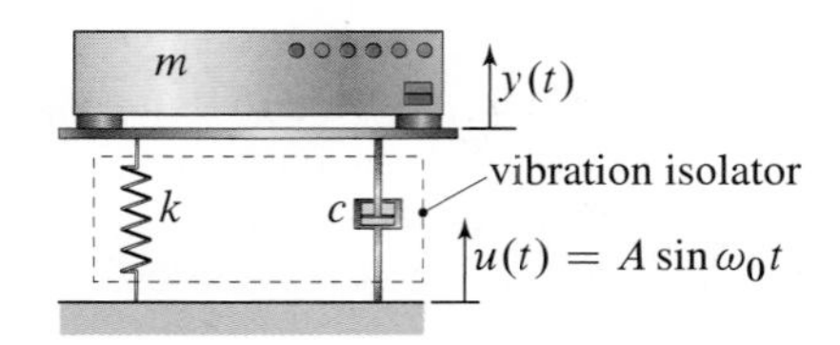

Figure P9.68

A delicate instrument of mass m must be isolated from excessive vibration of the ground, which is described by the function $u(t) = A\sin\omega_0 t$. To do so, we need to design a *vibration isolating mount*, modeled by the spring and dashpot system shown.

(a) Find the equation of motion of the instrument and reduce it to standard form.

(b) Find the steady-state response $y(t)$.

(c) Find the *displacement transmissibility*, i.e., the response amplitude D divided by A, where A is the amplitude of the ground's vibration.

* See J. M. T. Thompson, R. C. T. Rainey, and M. S. Soliman, "Mechanics of Ship Capsize under Direct and Parametric Wave Excitation," *Philosophical Transactions of the Royal Society of London A*, **338**(1651), 1992, pp. 471–490.

Design Problems

Design Problem 9.3

As a result of firing a projectile, a 300 kg naval gun assembly gains momentum in the x direction. We can consider the motion of the assembly to start from the equilibrium position $x(0) = 0$ with an initial velocity $\dot{x}(0) = 50\,\mathrm{m/s}$. Choose values of spring stiffness k and damping coefficient c to provide the fastest return of the assembly to its equilibrium position without oscillation. In addition, make sure that the maximum displacement of the assembly does not exceed 0.1 m. Finally, estimate the time it takes for the gun assembly to return to within 1% of its maximum displacement.

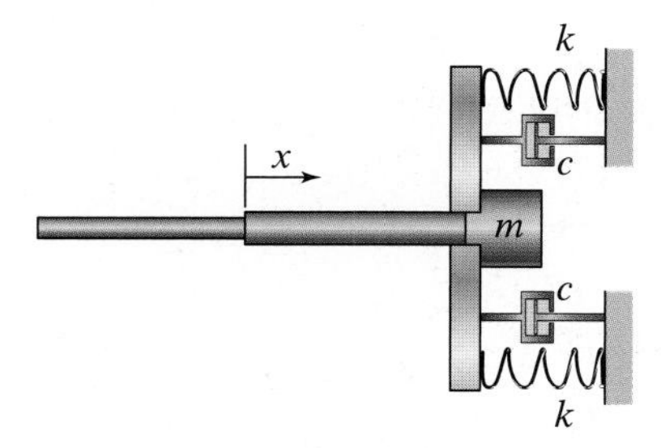

Figure DP9.3

Chapter Review

In this chapter, we studied the vibration or oscillation of mechanical systems about their equilibrium position. We considered only harmonic oscillators, although we did study the effects of viscous damping and harmonic forcing on the response of a harmonic oscillator.

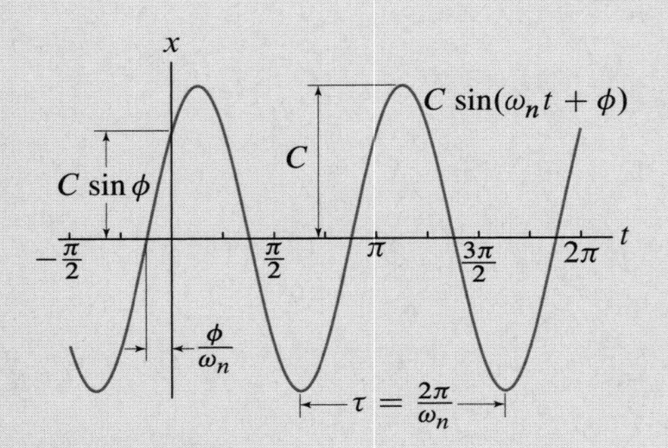

Figure 9.30
Plot of Eq. (9.3) showing the amplitude C, phase angle ϕ, and period τ of a harmonic oscillator.

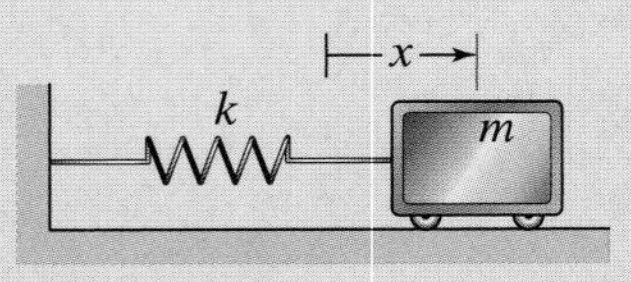

Figure 9.31
A simple spring-mass harmonic oscillator whose equation of motion is given by $m\ddot{x}+kx=0$ and for which $\omega_n=\sqrt{k/m}$.

Undamped free vibration

Any one DOF system whose equation of motion is of the form

Eq. (9.12), p. 665

$$\ddot{x}+\omega_n^2 x=0,$$

is called a *harmonic oscillator*, and the above expression is referred to as the *standard form* of the harmonic oscillator equation. The solution of this equation can be written as (see Fig. 9.30)

Eq. (9.3), p. 664

$$x(t)=C\sin(\omega_n t+\phi),$$

where ω_n is the *natural frequency*, C is the *amplitude*, and ϕ is the *phase angle* of vibration.

A simple example of a harmonic oscillator is a system consisting of a mass m attached at the free end of a spring with constant k and with the other end fixed (see Fig. 9.31). The natural frequency of such a system is given by

Eq. (9.4), p. 664

$$\omega_n=\sqrt{\frac{k}{m}}.$$

In addition, the amplitude C and the phase angle ϕ are given by, respectively,

Eqs. (9.5) and (9.6), p. 664

$$C=\sqrt{\frac{v_i^2}{\omega_n^2}+x_i^2} \quad \text{and} \quad \tan\phi=\frac{x_i\omega_n}{v_i},$$

where we let $t=0$ be the initial time, $x_i=x(0)$ (i.e., x_i is the initial position), and $v_i=\dot{x}(0)$ (i.e., v_i is the initial velocity). If $v_i=0$, then ϕ can be chosen equal to $-\pi/2$ or $\pi/2$ rad for $x_i<0$ and $x_i>0$, respectively. An alternative form of the solution to Eq. (9.12) is given by

Eq. (9.15), p. 666

$$x(t)=x_i\cos\omega_n t+\frac{v_i}{\omega_n}\sin\omega_n t.$$

The *period* of the oscillation is given by

Eq. (9.7), p. 664

$$\text{Period}=\tau=\frac{2\pi}{\omega_n},$$

and the *frequency* of vibration is

Eq. (9.8), p. 664

$$\text{Frequency}=f=\frac{1}{\tau}=\frac{\omega_n}{2\pi}.$$

Energy method. For conservative systems, the work-energy principle tells us that the quantity $T + V$ is constant, and so its time derivative must be zero. This provides a convenient way to obtain the equations of motion via

Eq. (9.19), p. 667

$$\frac{d}{dt}(T + V) = 0 \quad \Rightarrow \quad \text{equations of motion.}$$

When we apply the energy method to determine the linearized equations of motion, it is often convenient to first approximate the kinetic and potential energies as quadratic functions of position and velocity, and then take derivatives with respect to time. This process yields equations of motion that are linear.

In approximating the sine and cosine functions as quadratic functions of their arguments, we use the relations:

Eqs. (9.28), p. 668

$$\sin\theta \approx \theta \quad \text{and} \quad \cos\theta \approx 1 - \theta^2/2.$$

Since the quadratic term in the power series expansion of the sine function is identically equal to zero, the linearized form of the sine function can also be viewed as the quadratic approximation of the sine function.

Undamped forced vibration

When a harmonic oscillator is subject to harmonic forcing, the standard form of the equation of motion is

Eq. (9.31), p. 681

$$\ddot{x} + \omega_n^2 x = \frac{F_0}{m}\sin\omega_0 t,$$

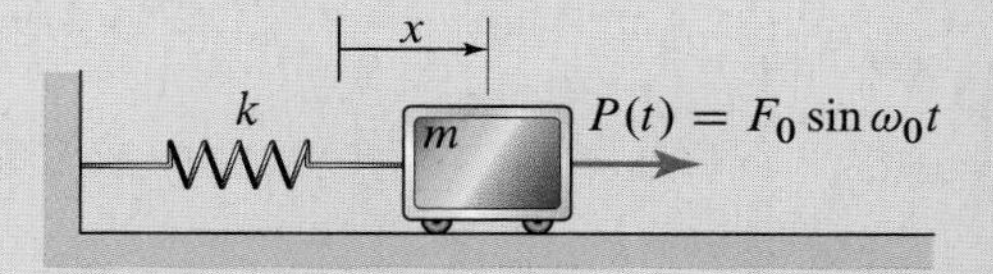

Figure 9.32
A forced harmonic oscillator whose equation of motion is given by Eq. (9.31) with $\omega_n = \sqrt{k/m}$. The position x is measured from the equilibrium position of the mass.

where F_0 is the amplitude of the forcing and ω_0 is its frequency (see Fig. 9.32). The general solution to this equation consists of the sum of the complementary solution and a particular solution. The *complementary solution* x_c is the solution of the associated homogeneous equation, which is given by, for example, Eq. (9.13). For $\omega_0 \neq \omega_n$, a particular solution is

Eq. (9.35), p. 682

$$x_p = \frac{F_0/k}{1 - (\omega_0/\omega_n)^2}\sin\omega_0 t,$$

and so the *general solution* is given by

Eq. (9.36), p. 682

$$x = x_c + x_p = A\sin\omega_n t + B\cos\omega_n t + \frac{F_0/k}{1 - (\omega_0/\omega_n)^2}\sin\omega_0 t,$$

where A and B are constants determined by the initial conditions. The amplitude of the steady-state vibration is

Eq. (9.37), p. 682

$$x_{\text{amp}} = \frac{F_0/k}{1 - (\omega_0/\omega_n)^2},$$

which means that the corresponding *magnification factor* MF is

Eq. (9.38), p. 682

$$\text{MF} = \frac{x_{\text{amp}}}{F_0/k} = \frac{1}{1 - (\omega_0/\omega_n)^2}.$$

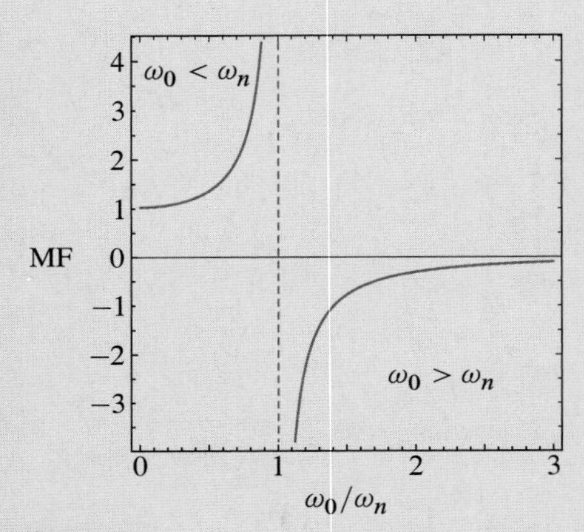

Figure 9.33
MF as a function of the frequency ratio ω_0/ω_n.

The plot of the MF in Fig. 9.33 illustrates the phenomenon of *resonance*, which occurs when $\omega_0 \approx \omega_n$ and results in very large vibration amplitudes.

Harmonic excitation of the support. If the support of a structure is excited harmonically rather than the structure itself (see Fig. 9.34), then Eq. (9.31) is still the governing equation, except that F_0 is replaced by the spring constant k times the amplitude of the support vibration X_s. All solutions previously described are then valid with that same replacement.

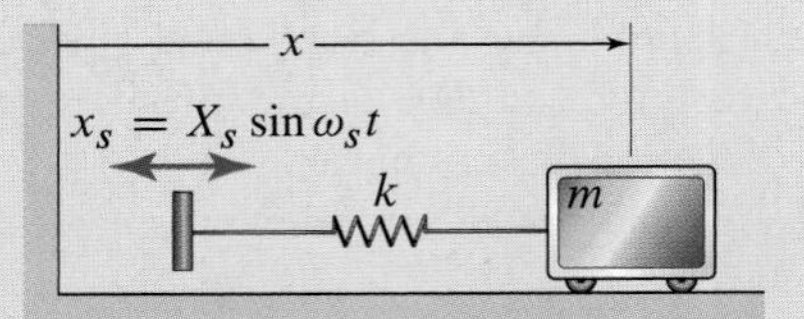

Figure 9.34. A harmonic oscillator whose support is being excited harmonically.

Viscously damped vibration

Viscously damped free vibration. The standard form of the equation of motion for a one degree of freedom viscously damped harmonic oscillator is

Eq. (9.45), p. 696

$$m\ddot{x} + c\dot{x} + kx = 0,$$

where m is the mass, c is the *coefficient of viscous damping*, and k is the linear spring constant. The character of the solution to this equation depends on the amount of damping relative to a specific amount of damping called the *critical damping coefficient*, which is defined as

Eq. (9.51), p. 696

$$c_c = 2m\sqrt{\frac{k}{m}} = 2m\omega_n,$$

where $\omega_n = \sqrt{k/m}$. In particular, if $c \geq c_c$, then the motion is nonoscillatory, whereas if $c < c_c$, then the motion is oscillatory. If $c > c_c$, the system is said to be *overdamped*, and the solution is given by

Eq. (9.52), p. 696

$$x = e^{-(c/2m)t}\left[Ae^{t\sqrt{(c/2m)^2 - k/m}} + Be^{-t\sqrt{(c/2m)^2 - k/m}}\right],$$

where A and B are constants to be determined from the initial conditions. If $c = c_c$, the system is said to be *critically damped*, and the solution is given by

Eq. (9.53), p. 697

$$x = (A + Bt)e^{-\omega_n t},$$

where, again, A and B are constants to be determined from the initial conditions. Finally, if $c < c_c$, the system is said to be *underdamped*, and the solution is given by

Eq. (9.54), p. 697

$$x = e^{-(c/2m)t}(A \sin \omega_d t + B \cos \omega_d t),$$

or, equivalently, by

Eq. (9.57), p. 697

$$x = De^{-(c/2m)t}\sin(\omega_d t + \phi),$$

where A and B are constants to be determined from the initial conditions in the first solution, D and ϕ are analogous constants in the second solution, and ω_d is the *damped natural frequency*, which is given by

Eq. (9.55), p. 697, and Eq. (9.62), p. 698

$$\omega_d = \sqrt{\frac{k}{m} - \left(\frac{c}{2m}\right)^2} = \omega_n\sqrt{1-(c/c_c)^2} = \omega_n\sqrt{1-\zeta^2},$$

where $\zeta = c/c_c$ is the *damping ratio*.

Viscously damped forced vibration. The standard form of a viscously damped forced harmonic oscillator is

Eq. (9.63), p. 698

$$m\ddot{x} + c\dot{x} + kx = F_0 \sin\omega_0 t,$$

where F_0 is the amplitude of the forcing function and ω_0 is the frequency of the forcing function. When expressed using the damping ratio ζ, the equation above takes on the form

Eq. (9.64), p. 698

$$\ddot{x} + 2\zeta\omega_n\dot{x} + \omega_n^2 x = \frac{F_0}{m}\sin\omega_0 t.$$

The general solution to either of these equations is the sum of the complementary solution and a particular solution. The complementary solution is transient; i.e., it vanishes as time increases. The particular or steady-state solution is of the form

Eq. (9.66), p. 699

$$x_p = D\sin(\omega_0 t - \phi),$$

where ϕ and D are given by

Eq. (9.69), p. 699, and Eq. (9.72), p. 700

$$\tan\phi = \frac{c\omega_0}{k - m\omega_0^2} = \frac{2(c/c_c)(\omega_0/\omega_n)}{1-(\omega_0/\omega_n)^2} = \frac{2\zeta\omega_0/\omega_n}{1-(\omega_0/\omega_n)^2},$$

$$D = \frac{F_0/k}{\sqrt{\left[1-(\omega_0/\omega_n)^2\right]^2 + (2\zeta\omega_0/\omega_n)^2}}.$$

The *magnification factor* for a damped, forced harmonic oscillator is

Eq. (9.73), p. 700

$$\text{MF} = \frac{D}{F_0/k} = \frac{1}{\sqrt{\left[1-(\omega_0/\omega_n)^2\right]^2 + (2\zeta\omega_0/\omega_n)^2}},$$

a plot of which is shown in Fig. 9.35 for various values of the damping ratio ζ.

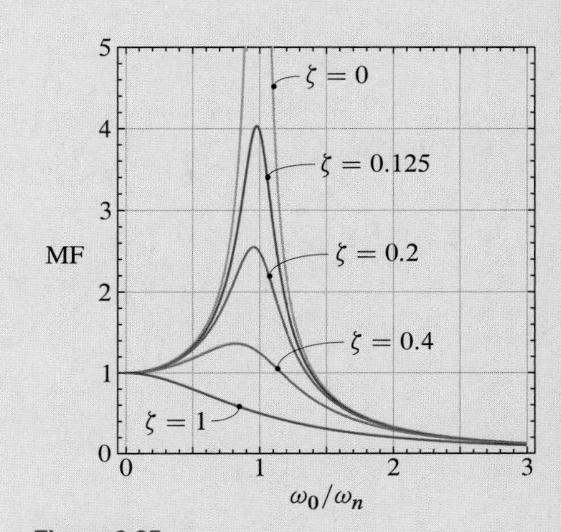

Figure 9.35
The MF as a function of the frequency ratio ω_0/ω_n for various values of the damping ratio ζ.

Review Problems

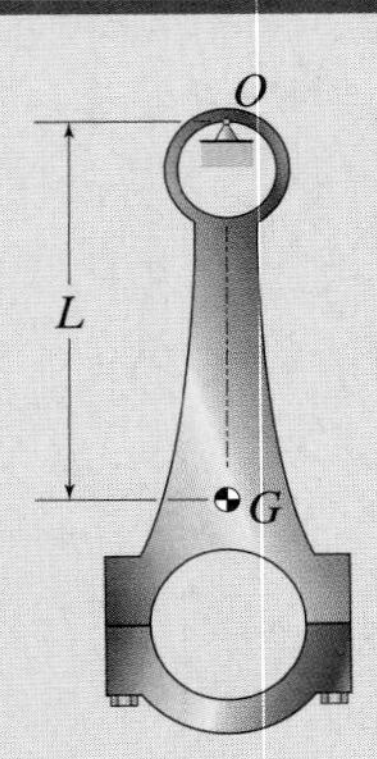

Figure P9.69

Problem 9.69

When the connecting rod shown is suspended from the knife-edge at point O and displaced slightly so that it oscillates as a pendulum, its period of oscillation is 0.77 s. In addition, it is known that the mass center G is located a distance $L = 110\,\text{mm}$ from O and that the mass of the connecting rod is 661 g. Using the energy method, determine the mass moment of inertia of the connecting rod I_G.

Problem 9.70

Derive the equation of motion for the system, in which the springs with constants k_1 and k_2 connecting m to the wall are joined in series. Neglect the mass of the small wheels, and assume that the attachment point A between the two springs has negligible mass. *Hint:* The force in the two springs must be the same; use this fact, along with the fact that the total deflection of the mass must equal the sum of the deflections of the springs, to find an equivalent spring constant k_{eq}.

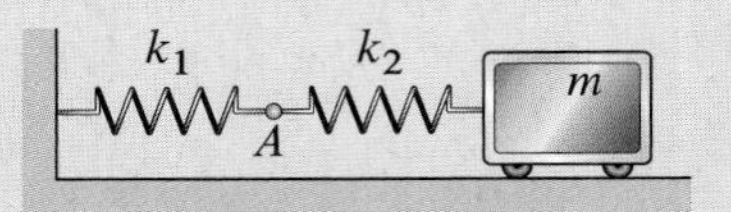

Figure P9.70

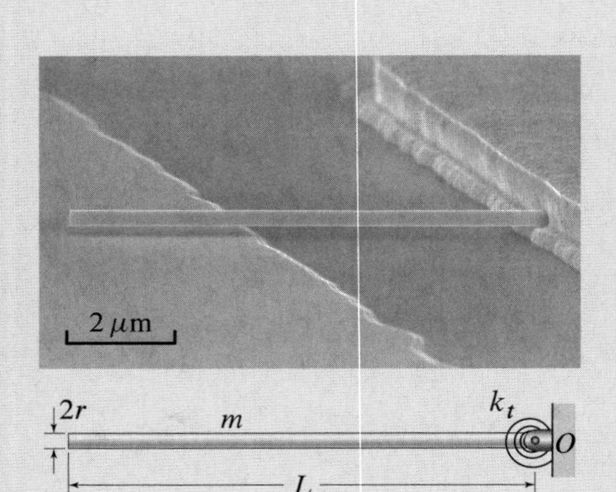

Figure P9.71

Problem 9.71

Revisit Example 9.2 and compute the natural frequency of the silicon nanowire, using the energy method. Use a uniform Si nanowire with a circular cross section that is 9.8 μm long and 330 nm in diameter and with all its flexibility lumped in a torsional spring at the base of the wire. In addition, use $\rho = 2330\,\text{kg/m}^3$ for the density of silicon and $E = 152\,\text{GPa}$ for its modulus of elasticity.

Problem 9.72

Structural health monitoring technology detects damage in civil, aerospace, and other structures. Structural damage is usually comprised of cracking, delaminations, or loose fasteners, which result in the reduction of stiffness. Many structural health monitoring methods are based on tracking changes in natural frequencies. Modeling a structure as a one DOF harmonic oscillator, calculate the change in stiffness needed to cause a 3% reduction in the natural frequency of the structure being monitored.

Problems 9.73 through 9.75

When the electric motor is resting on the beam, the static deflection of the beam is $\delta_s = 15\,\text{mm}$. The motor is not perfectly balanced, so when it is operating the unbalanced mass is equivalent to a mass $m_u = 200\,\text{g}$ at a distance of $\varepsilon = 150\,\text{mm}$ from the axis of the rotor. The combined mass of the motor and sprung mass of the beam is $m_c = 40\,\text{kg}$.

Problem 9.73 Determine the angular speed of the rotor for resonance to occur. Express your answer in revolutions per minute.

Problem 9.74 Determine the amplitude of steady-state vibration of the motor if the rotor is spinning at $\omega_r = 150\,\text{rpm}$.

Problem 9.75 It is determined that the largest allowable vibration amplitude of the motor is 10 mm. Determine the angular speed of the rotor at which this will occur.

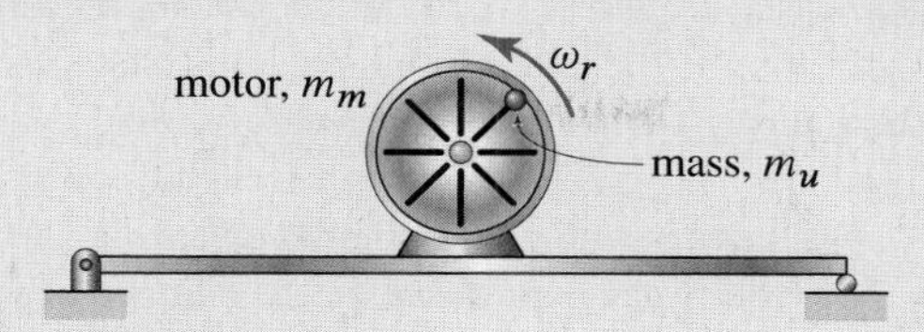

Figure P9.73–P9.75

Problem 9.76

The harmonic oscillator shown has a mass $m = 5\,\text{kg}$, a spring with constant $k = 4000\,\text{N/m}$, and a dashpot with a damping coefficient $c = 20\,\text{N}\cdot\text{s/m}$. Calculate the amplitude F_0 of the sinusoidal excitation force that is necessary to produce a steady-state vibration with a velocity amplitude of $10\,\text{m/s}$ at resonance. What is the corresponding amplitude of the acceleration?

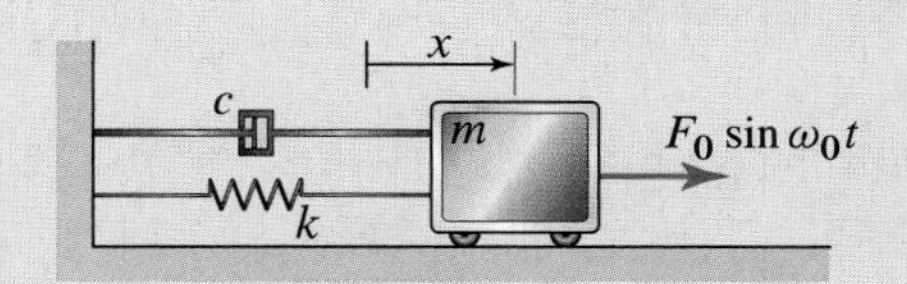

Figure P9.76

Problem 9.77

Modeling the beam as a uniform thin bar, ignoring the inertia of the pulleys, assuming that the system is in static equilibrium when the bar is horizontal, and assuming that the cord is inextensible and does not go slack, determine the linearized equation of motion of the system. In addition, determine the system's natural frequency of vibration. Treat the parameters shown in the figure as known.

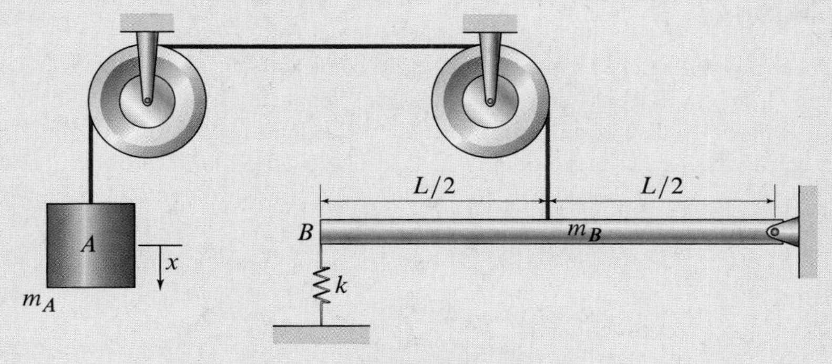

Figure P9.77

Problem 9.78

Revisit Example 9.9 and obtain the expression for the force transmitted to the floor, using the expression for the steady-state response of the unbalanced motor.

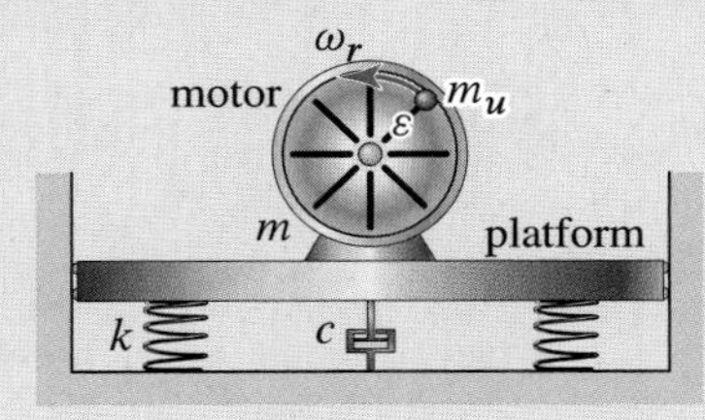

Figure P9.78

Problem 9.79

The system shown is released from rest when both springs are unstretched and $x = 0$. Neglecting the inertia of the pulley P and assuming that the disk rolls without slip, derive the equation of motion of the system in terms of x. Assume that point G is both the mass center of the disk and its geometric center. Treat the quantities k_1, k_2, c, m_1, m_2, and I_G as known, where I_G is the mass moment of inertia of the disk. Finally, assuming that the system is underdamped, derive an expression for the damped natural frequency of the system.

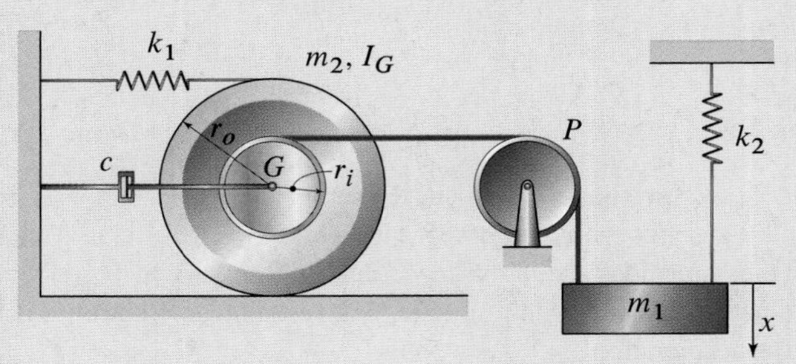

Figure P9.79

Problem 9.80

A ring of mass m is attached by two linear elastic cords to the vertical supports as shown. The cords have elastic constant k and unstretched length $L_0 < L$. Assuming that the pretension in the cords is large enough that the deflection of the cords due to the ring's weight can be neglected, find the nonlinear equation of motion for the mass m.

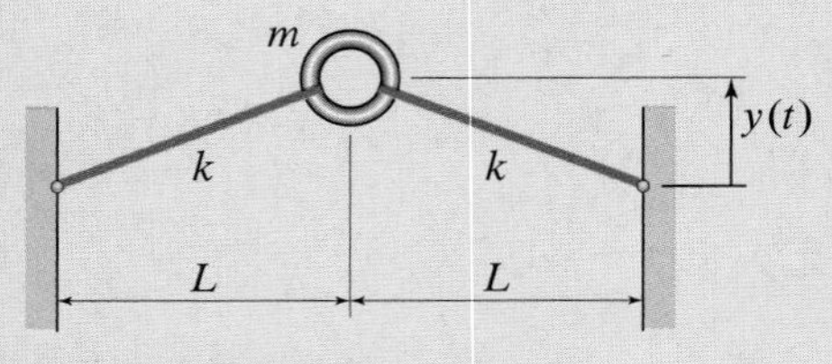

Figure P9.80–P9.82

Problem 9.81

A ring of mass m is attached by two linear elastic cords to the vertical supports as shown. The cords have elastic constant k and unstretched length $L_0 < L$. Assuming that the pretension in the cords is large enough that the deflection of the cords due to the ring's weight can be neglected, use Newton's second law to find the linearized equation of motion about $y = 0$ for the mass m. In addition, determine the natural frequency of the ring's vibration.

Problem 9.82

Solve Prob. 9.81 by finding the linearized equations of motion using the energy method.

Three-Dimensional Dynamics of Rigid Bodies

Photo © Corbis RF

The HH/MH-65 Multi-Mission Cutter Helicopter used by the United States Coast Guard for search and rescue and law enforcement missions. The understanding of three-dimensional dynamics is essential for modeling its rotor blades.

Until now, we have only considered rigid bodies in planar motion that are symmetric with respect to the plane of motion. In this chapter, we will relax these assumptions and allow for not only nonsymmetric bodies, but also for bodies moving in all three dimensions. In Chapter 7, we had three Newton-Euler equations (two force and one moment equation) for a rigid body. For three-dimensional motion, we will find that we need *six* Newton-Euler equations—three moment equations and three force equations. Before we derive these equations, as with planar motion, we need to begin by understanding the kinematics of rigid bodies moving in three dimensions. We do that in Section 10.1. In Section 10.2, we derive the Newton-Euler equations for three-dimensional motion.

10.1 Three-Dimensional Kinematics of Rigid Bodies

In Chapter 6, we considered the planar kinematics of rigid bodies, but most of the equations we derived apply equally well to three-dimensional motion. Equations (6.3) and (6.5) on p. 435, which are given by

$$\vec{v}_C = \vec{v}_A + \vec{\omega}_B \times \vec{r}_{C/A}, \tag{10.1}$$

$$\vec{a}_C = \vec{a}_A + \vec{\alpha}_B \times \vec{r}_{C/A} + \vec{\omega}_B \times \left(\vec{\omega}_B \times \vec{r}_{C/A}\right), \tag{10.2}$$

where $\vec{\omega}_B$ and $\vec{\alpha}_B$ are the angular velocity and angular acceleration of the body B, respectively, both apply for the three-dimensional motion of rigid bodies, as long as A and C are two points on the *same* rigid body (see Fig. 10.1).

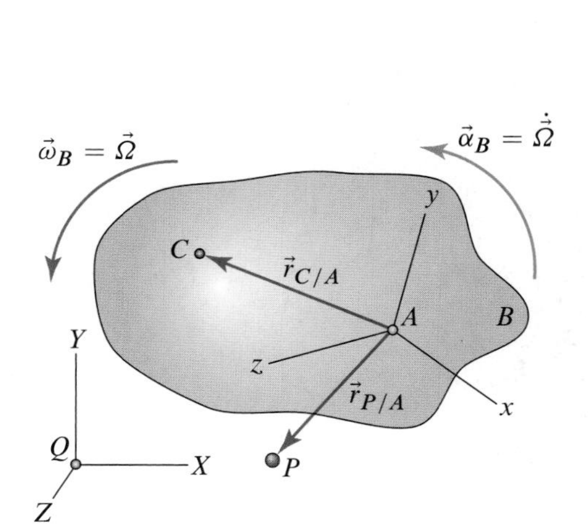

Figure 10.1
Rigid body B displaying the points and kinematic quantities needed to apply Eqs. (10.1)–(10.4). Frame XYZ is the primary reference frame and xyz is the secondary, body-fixed, or rotating reference frame.

The equations we derived for rotating reference frames, which were useful for sliding contacts in planar motion, that is, Eqs. (6.45) on p. 495, and (6.53) on p. 496, also apply in three-dimensional motion, and we repeat them here for convenience:

$$\vec{v}_P = \vec{v}_A + \vec{v}_{P\mathrm{rel}} + \vec{\Omega} \times \vec{r}_{P/A}, \tag{10.3}$$

$$\vec{a}_P = \vec{a}_A + \vec{a}_{P\mathrm{rel}} + 2\vec{\Omega} \times \vec{v}_{P\mathrm{rel}} + \dot{\vec{\Omega}} \times \vec{r}_{P/A} + \vec{\Omega} \times (\vec{\Omega} \times \vec{r}_{P/A}), \tag{10.4}$$

where $\vec{v}_{P\mathrm{rel}}$ and $\vec{a}_{P\mathrm{rel}}$ are the velocity and acceleration, respectively, of the point P as seen by an observer in the rotating xyz reference frame, $\vec{\Omega}$ is the angular velocity of the rotating frame, and $\dot{\vec{\Omega}}$ is the angular acceleration of the rotating frame. We will see that, even though the kinematic equations are familiar, the added difficulty in three-dimensional motion concerns angular velocities and angular accelerations. One angular velocity can cause another angular velocity to change direction, and this adds a dimension to the angular acceleration that we haven't seen before.

Computation of angular accelerations

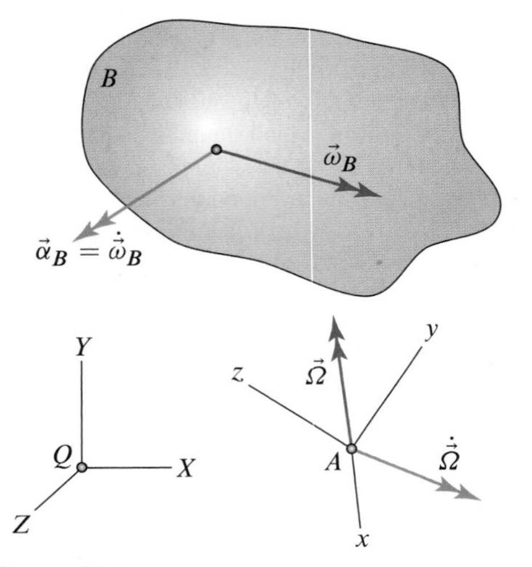

Figure 10.2
The primary reference frame, rotating reference frame, and rigid body B.

When analyzing the kinematics of rigid bodies in three dimensions, we will often need to use a second reference frame that rotates relative to the primary frame. The reference frame is chosen in such a way as to make the analysis of the motion as easy as possible. Sometimes the rotating frame will be attached to a rigid body and will rotate with it. Alternatively, the rotating reference frame may rotate relative to *both* the primary frame *and* the rigid body. Referring to Fig. 10.2, the primary reference frame is XYZ and the rotating frame is xyz. The angular velocity and angular acceleration of the rotating reference frame xyz relative to the primary reference frame are given by $\vec{\Omega}$ and $\dot{\vec{\Omega}}$, respectively. The angular velocity and angular acceleration of the rigid body B relative to the primary reference frame are given by $\vec{\omega}_B$ and $\dot{\vec{\omega}}_B$, respectively.

We now express the angular velocity of the rigid body B in terms of its components in the rotating reference frame as

$$\vec{\omega}_B = \omega_{Bx}\,\hat{\imath} + \omega_{By}\,\hat{\jmath} + \omega_{Bz}\,\hat{k}. \tag{10.5}$$

Differentiating $\vec{\omega}_B$ to find the angular acceleration $\dot{\vec{\omega}}_B = \vec{\alpha}_B$, we obtain

$$\dot{\vec{\omega}}_B = \vec{\alpha}_B = \dot{\omega}_{Bx}\,\hat{\imath} + \dot{\omega}_{By}\,\hat{\jmath} + \dot{\omega}_{Bz}\,\hat{k} + \vec{\Omega} \times \vec{\omega}_B, \tag{10.6}$$

where we have applied Eq. (2.48) on p. 81 to take the time derivative of $\vec{\omega}_B$. Equation (10.6) is *invaluable* for computing the angular acceleration of rigid bodies in three-dimensional motion. Notice that if the rotating frame is body-fixed, that is, if $\vec{\Omega} = \vec{\omega}_B$, then

$$\dot{\vec{\omega}}_B = \dot{\omega}_{Bx}\,\hat{\imath} + \dot{\omega}_{By}\,\hat{\jmath} + \dot{\omega}_{Bz}\,\hat{k}. \tag{10.7}$$

Summing angular velocities

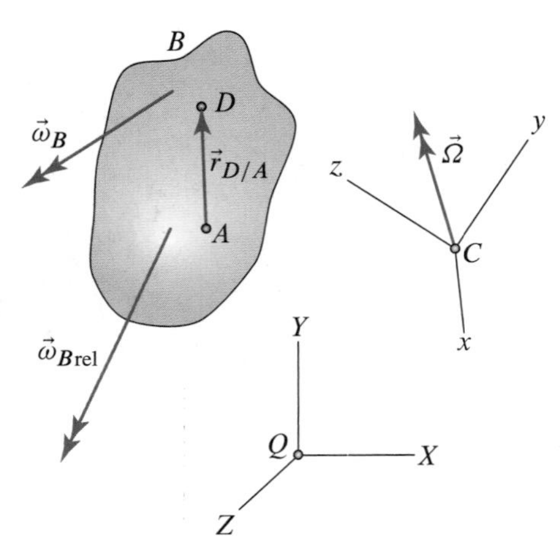

Figure 10.3
A body with angular velocity $\vec{\omega}_B$ and a moving reference frame with angular velocity $\vec{\Omega}$. The angular velocity of the body relative to the moving reference frame is $\vec{\omega}_{B\mathrm{rel}}$.

Referring to Fig. 10.3, to understand how to *sum* angular velocities with one another, we will relate the velocity of point D to point A *and* point C using

$$\vec{v}_D = \vec{v}_A + \vec{\omega}_B \times \vec{r}_{D/A} = \vec{v}_C + \vec{v}_{D\mathrm{rel}} + \vec{\Omega} \times \vec{r}_{D/C}, \tag{10.8}$$

where $\vec{v}_{D\mathrm{rel}}$ is the velocity of D relative to the xyz frame. Since points A and D are two points on the same rigid body, an observer in the moving reference frame would see the velocity of D as

$$\vec{v}_{D\mathrm{rel}} = \vec{v}_{A\mathrm{rel}} + \vec{\omega}_{B\mathrm{rel}} \times \vec{r}_{D/A}, \tag{10.9}$$

where $\vec{\omega}_{B\text{rel}}$ is the angular velocity of the body as seen by an observer in the moving reference frame. Since $\vec{v}_A = \vec{v}_C + \vec{v}_{A\text{rel}} + \vec{\Omega} \times \vec{r}_{A/C}$, we can rearrange it as

$$\vec{v}_{A\text{rel}} = \vec{v}_A - \vec{v}_C - \vec{\Omega} \times \vec{r}_{A/C}. \tag{10.10}$$

Substituting Eq. (10.10) into Eq. (10.9), we obtain

$$\vec{v}_{D\text{rel}} = \vec{v}_A - \vec{v}_C - \vec{\Omega} \times \vec{r}_{A/C} + \vec{\omega}_{B\text{rel}} \times \vec{r}_{D/A}, \tag{10.11}$$

which can be substituted into Eq. (10.8) to obtain

$$\vec{\omega}_B \times \vec{r}_{D/A} = -\vec{\Omega} \times \vec{r}_{A/C} + \vec{\omega}_{B\text{rel}} \times \vec{r}_{D/A} + \vec{\Omega} \times \vec{r}_{D/C}, \tag{10.12}$$

where we have cancelled out $\vec{v}_A$ and $\vec{v}_C$. This equation can be rewritten as

$$\begin{aligned} \vec{\omega}_B \times \vec{r}_{D/A} &= \vec{\Omega} \times \left(\vec{r}_{D/C} - \vec{r}_{A/C}\right) + \vec{\omega}_{B\text{rel}} \times \vec{r}_{D/A} \\ &= \vec{\Omega} \times \vec{r}_{D/A} + \vec{\omega}_{B\text{rel}} \times \vec{r}_{D/A} \\ &= \left(\vec{\Omega} + \vec{\omega}_{B\text{rel}}\right) \times \vec{r}_{D/A}. \end{aligned} \tag{10.13}$$

Since Eq. (10.13) must hold for every pair of points A and D in the body B, we obtain the following important result,

$$\boxed{\vec{\omega}_B = \vec{\Omega} + \vec{\omega}_{B\text{rel}},} \tag{10.14}$$

which states that if the angular velocity $\vec{\Omega}$ of a rotating (usually body-fixed) frame is known and the angular velocity of a body B relative to that frame $\vec{\omega}_{B\text{rel}}$ is known, then the angular velocity of the body $\vec{\omega}_B$ is the sum of those two angular velocities.

Mini-Example

The platform shown in Fig. 10.4 is rotating with constant angular speed ω_p in the direction shown. On the platform is mounted a motor that is spinning with constant angular speed ω_s *relative to the platform* in the direction shown. Given the dimensions shown, determine the acceleration of point Q, which is on the periphery of the motor.

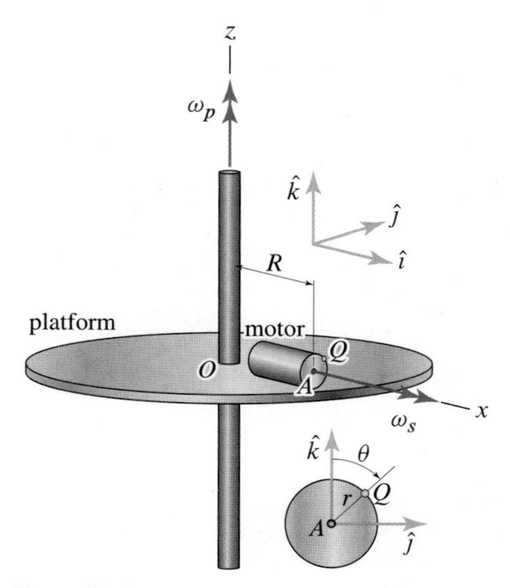

Figure 10.4
A motor that is spinning with angular speed ω_s, mounted on a platform that is spinning with angular speed ω_p.

Solution

We have attached a moving reference frame xyz to the platform with origin at point O. To find the acceleration of point Q on the periphery of the motor, we first make use of Eq. (10.2) to relate the acceleration of Q to the acceleration of A as

$$\vec{a}_Q = \vec{a}_A + \vec{\alpha}_m \times \vec{r}_{Q/A} + \vec{\omega}_m \times \left(\vec{\omega}_m \times \vec{r}_{Q/A}\right), \tag{10.15}$$

where $\vec{\alpha}_m$ and $\vec{\omega}_m$ are the angular acceleration and angular velocity of the motor, respectively. Since point A is moving with constant angular velocity in a circle of radius R, we can write its acceleration as

$$\vec{a}_A = -R\omega_p^2\,\hat{\imath}. \tag{10.16}$$

To determine $\vec{\omega}_m$, we can apply Eq. (10.14) by saying that the angular velocity of the motor is the angular velocity of the rotating xyz frame (which rotates with the platform) plus the angular velocity of the motor as seen by an observer moving with the platform, that is,

$$\vec{\omega}_m = \vec{\omega}_p + \vec{\omega}_{m\text{rel}} \quad \Rightarrow \quad \vec{\omega}_m = \omega_p\,\hat{k} + \omega_s\,\hat{\imath}, \tag{10.17}$$

where $\vec{\omega}_{m\text{rel}}$ is the angular velocity of the motor as seen by an observer on the platform. To find the angular acceleration of the motor, we apply Eq. (10.6) to obtain

$$\vec{\alpha}_m = \dot{\omega}_p \,\hat{k} + \dot{\omega}_m \,\hat{\imath} + \vec{\omega}_p \times \vec{\omega}_m = \omega_p \,\hat{k} \times \left(\omega_p \,\hat{k} + \omega_s \,\hat{\imath}\right) = \omega_p \omega_s \,\hat{\jmath}, \tag{10.18}$$

where we have used the fact that ω_p and ω_m are constant and that $\hat{k} \times \hat{k} = \vec{0}$. Substituting Eqs. (10.16)–(10.18) and $\vec{r}_{Q/A} = r\left(\sin\theta \,\hat{\jmath} + \cos\theta \,\hat{k}\right)$ into Eq. (10.15), we obtain

$$\begin{aligned}\vec{a}_Q &= -R\omega_p^2 \,\hat{\imath} + \omega_p \omega_s \,\hat{\jmath} \times r\left(\sin\theta \,\hat{\jmath} + \cos\theta \,\hat{k}\right) \\ &\qquad + \left(\omega_s \,\hat{\imath} + \omega_p \,\hat{k}\right) \times \left[\left(\omega_s \,\hat{\imath} + \omega_p \,\hat{k}\right) \times r\left(\sin\theta \,\hat{\jmath} + \cos\theta \,\hat{k}\right)\right] \\ &= \left(2r\omega_s\omega_p \cos\theta - R\omega_p^2\right)\hat{\imath} - r\sin\theta\left(\omega_s^2 + \omega_p^2\right)\hat{\jmath} - r\omega_s^2 \cos\theta \,\hat{k}.\end{aligned} \tag{10.19}$$

The expression for the acceleration in Eq. (10.19) is written in terms of the components in the rotating frame, but it represents the acceleration of Q in the primary frame. Therefore, it is inertial if the primary frame is inertial. In addition, because the angular velocity of the platform causes the angular velocity of the motor to change direction, we again see a gyroscopic term in this acceleration, i.e., the term $2r\omega_s\omega_p \cos\theta$.

Equation (10.18) can also be interpreted using Eq. (2.48) on p. 81. That is, the time derivative of $\vec{\omega}_m$ is the sum of its change in magnitude, that is, $\dot{\omega}_m \,\hat{\imath}$, and its change in direction, that is, $\vec{\omega}_p \times \vec{\omega}_m$. Notice that $\vec{\omega}_p$ *is* the angular velocity of $\vec{\omega}_m$.

End of Section Summary

Referring to Fig. 10.5, for two points A and C on the same rigid body B, we can relate their motion using

Eqs. (10.1) and (10.2), p. 721

$$\vec{v}_C = \vec{v}_A + \vec{\omega}_B \times \vec{r}_{C/A},$$
$$\vec{a}_C = \vec{a}_A + \vec{\alpha}_B \times \vec{r}_{C/A} + \vec{\omega}_B \times (\vec{\omega}_B \times \vec{r}_{C/A}),$$

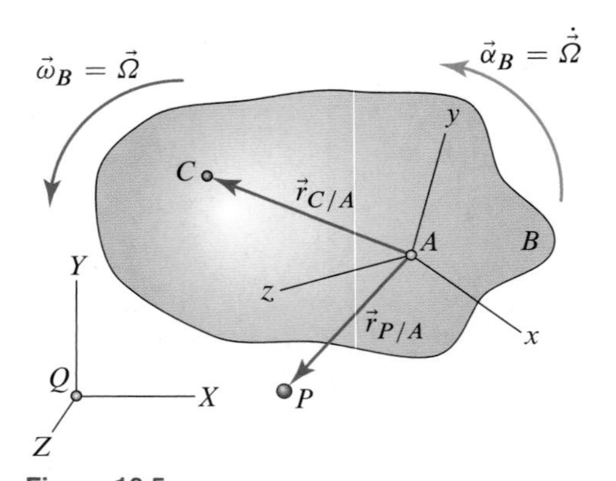

Figure 10.5
Rigid body B displaying the points and kinematic quantities needed to apply Eqs. (10.1)–(10.4). Frame XYZ is the primary reference frame and xyz is the secondary, body-fixed, or rotating reference frame.

where $\vec{\omega}_B$ and $\vec{\alpha}_B$ are the angular velocity and angular acceleration, respectively, of the rigid body. Note that for three-dimensional motion, we *cannot* write $\vec{\omega}_{AB} \times (\vec{\omega}_{AB} \times \vec{r}_{B/A})$ as $-\omega_{AB}^2 \vec{r}_{B/A}$. Referring to Fig. 10.5, if we use a rotating or secondary reference frame with its origin at A, then we can relate the motion of points P and A, which are not necessarily on the same body, using

Eqs. (10.3) and (10.4), p. 722

$$\vec{v}_P = \vec{v}_A + \vec{v}_{P\text{rel}} + \vec{\Omega} \times \vec{r}_{P/A},$$
$$\vec{a}_P = \vec{a}_A + \vec{a}_{P\text{rel}} + 2\vec{\Omega} \times \vec{v}_{P\text{rel}} + \dot{\vec{\Omega}} \times \vec{r}_{P/A} + \vec{\Omega} \times \left(\vec{\Omega} \times \vec{r}_{P/A}\right),$$

where $\vec{v}_{P\text{rel}}$ and $\vec{a}_{P\text{rel}}$ are the velocity and acceleration, respectively, of P as seen by an observer in the rotating frame; and $\vec{\Omega}$ and $\dot{\vec{\Omega}}$ are the angular velocity and angular acceleration, respectively, of the rotating reference frame. The rotating frame is usually attached to a rigid body, though sometimes the rotating reference frame will rotate relative to both the primary frame *and* the rigid body.

Referring to Fig. 10.6, if the rotating frame is *not* attached to the rigid body B, then the angular acceleration of B is given by

Eq. (10.6), p. 722

$$\dot{\vec{\omega}}_B = \vec{\alpha}_B = \dot{\omega}_{Bx}\,\hat{\imath} + \dot{\omega}_{By}\,\hat{\jmath} + \dot{\omega}_{Bz}\,\hat{k} + \vec{\Omega} \times \vec{\omega}_B,$$

where $\vec{\Omega}$ is the angular velocity of the rotating frame relative to the primary frame and $\vec{\omega}_B$ is the angular velocity of B relative to the primary frame.

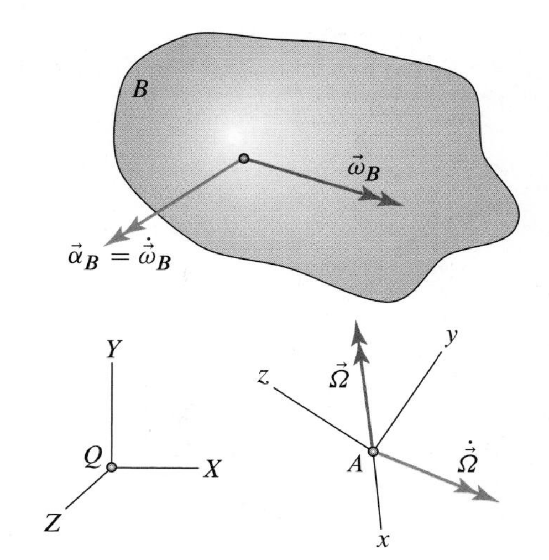

Figure 10.6
The primary reference frame XYZ, rotating reference frame xyz, and rigid body B.

Referring to Fig. 10.7, if the angular velocity of the rigid body B is expressed *relative* to the rotating frame as $\vec{\omega}_{B\mathrm{rel}}$ and the angular velocity of the rotating frame relative to the primary frame is $\vec{\Omega}$, then the angular velocity of B relative to the primary frame, $\vec{\omega}_B$, is

Eq. (10.14), p. 723

$$\vec{\omega}_B = \vec{\Omega} + \vec{\omega}_{B\mathrm{rel}}.$$

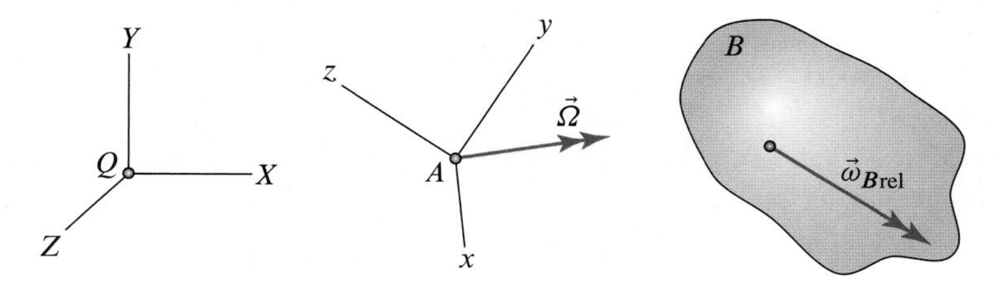

Figure 10.7. A secondary frame rotating with $\vec{\Omega}$ and a rigid body rotating with $\vec{\omega}_{B/A}$ *relative to* the rotating frame.

EXAMPLE 10.1 *Motion of a Disk Rolling Without Slip*

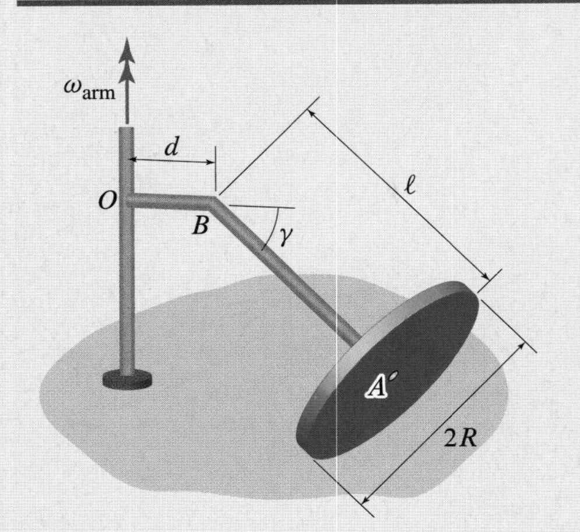

Figure 1

The bent arm in Fig. 1 rotates with constant angular speed ω_{arm} in the direction shown. The wheel of radius R with center at A rotates relative to the bent arm as it rolls without slipping over the horizontal surface. Given the dimensions shown, determine expressions for the angular velocity and angular acceleration of the wheel relative to a primary frame attached to the horizontal surface. Express the results in a frame attached to the bent arm.

SOLUTION

Road Map Referring to Fig. 2, we begin by defining a rotating reference frame that is attached to the arm OBA, with origin at B, and that is aligned as shown. In addition, we let $\vec{\omega}_{w\text{rel}}$ be the angular velocity of the wheel relative to the bent arm. The key kinematic constraints in this problem are that:

(i) Point A moves in a circle centered on the vertical part of the bent arm with constant angular velocity $\vec{\omega}_{\text{arm}}$.

(ii) The wheel rolls without slipping at point C.

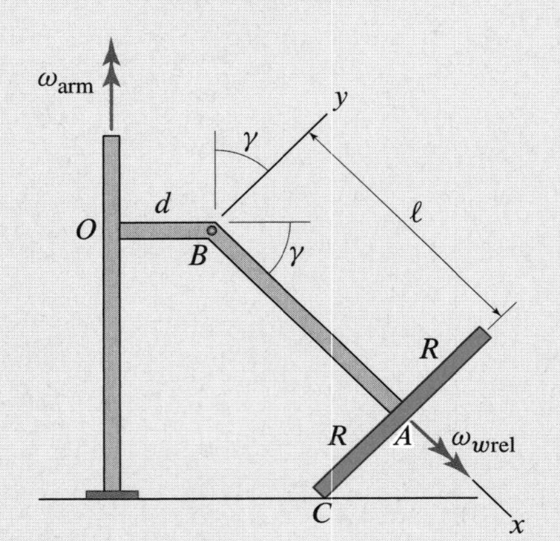

Figure 2
The definition of the rotating reference frame xyz, as well as $\vec{\omega}_{w\text{rel}}$, which is the angular velocity of the wheel relative to the bent arm.

Computation The angular velocity of the wheel can be written as

$$\vec{\omega}_w = \vec{\omega}_{\text{arm}} + \vec{\omega}_{w\text{rel}}, \tag{1}$$

where $\vec{\omega}_{w\text{rel}} = \omega_{w\text{rel}}\,\hat{\imath}$. We can write $\vec{\omega}_{\text{arm}}$ in the rotating frame as

$$\vec{\omega}_{\text{arm}} = \omega_{\text{arm}}(-\sin\gamma\,\hat{\imath} + \cos\gamma\,\hat{\jmath}). \tag{2}$$

To determine $\vec{\omega}_{w\text{rel}}$, we note that the wheel rolls without slipping, and we relate the velocity of A to that of C,

$$\begin{aligned}\vec{v}_A &= \vec{v}_C + \vec{\omega}_w \times \vec{r}_{A/C} = \vec{\omega}_w \times \vec{r}_{A/C}\\ &= \left[\omega_{\text{arm}}(-\sin\gamma\,\hat{\imath} + \cos\gamma\,\hat{\jmath}) + \omega_{w\text{rel}}\,\hat{\imath}\right] \times R\,\hat{\jmath}\\ &= R\left(\omega_{w\text{rel}} - \omega_{\text{arm}}\sin\gamma\right)\hat{k},\end{aligned} \tag{3}$$

where, in the first line, we have taken advantage of the fact that $\vec{v}_C = \vec{0}$ since the wheel rolls without slipping over a stationary surface. Since the velocity of point A is easily seen to be given by

$$\vec{v}_A = -(d + \ell\cos\gamma)\omega_{\text{arm}}\,\hat{k}, \tag{4}$$

we can equate Eqs. (3) and (4) to obtain

$$-(d + \ell\cos\gamma)\omega_{\text{arm}} = R\left(\omega_{w\text{rel}} - \omega_{\text{arm}}\sin\gamma\right)$$

$$\Rightarrow \quad \omega_{w\text{rel}} = \left(\sin\gamma - \frac{d + \ell\cos\gamma}{R}\right)\omega_{\text{arm}}. \tag{5}$$

Substituting Eqs. (2) and (5) into Eq. (1) and noting that $\omega_{w\text{rel}}$ is only in the x direction, we obtain

$$\boxed{\vec{\omega}_w = \left(-\frac{d + \ell\cos\gamma}{R}\,\hat{\imath} + \cos\gamma\,\hat{\jmath}\right)\omega_{\text{arm}}.} \tag{6}$$

To find the angular acceleration of the wheel, we apply Eq. (10.6) in the form

$$\vec{\alpha}_w = \dot{\omega}_{wx}\,\hat{\imath} + \dot{\omega}_{wy}\,\hat{\jmath} + \dot{\omega}_{wz}\,\hat{k} + \vec{\Omega} \times \vec{\omega}_w, \tag{7}$$

where $\vec{\Omega} = \vec{\omega}_{\text{arm}}$ is the angular velocity of the rotating frame. Since all dimensions, as well as ω_{arm} are constant, we can say that $\dot{\omega}_{wx} = \dot{\omega}_{wy} = \dot{\omega}_{wz} = 0$, and so

$$\vec{\alpha}_w = \vec{\omega}_{\text{arm}} \times \left(\vec{\omega}_{\text{arm}} + \vec{\omega}_{w\text{rel}}\right) = \vec{\omega}_{\text{arm}} \times \vec{\omega}_{w\text{rel}} \tag{8}$$

$$= \omega_{\text{arm}}(-\sin\gamma\,\hat{\imath} + \cos\gamma\,\hat{\jmath}) \times \left(\sin\gamma - \frac{d + \ell\cos\gamma}{R}\right)\omega_{\text{arm}}\,\hat{\imath}, \tag{9}$$

or, carrying out the cross products,

$$\vec{\alpha}_w = \left(\frac{d + \ell\cos\gamma}{R} - \sin\gamma\right)\omega_{\text{arm}}^2\cos\gamma\,\hat{k}. \tag{10}$$

Discussion & Verification The dimensions of the results in Eqs. (6) and (10) are one over time and one over time squared, respectively, as they should be.

A Closer Look Referring to Eq. (10), notice that the angular acceleration of the wheel is always parallel to the z direction, which means that it is always perpendicular to the plane formed by the bent arm. We can see why this must be so if, referring to Fig. 2, we note that $\vec{\omega}_{\text{arm}}$ causes the tip of $\vec{\omega}_{w\text{rel}}$ to move away from us, which is in the $-z$ direction. We will see that this angular acceleration plays an important role when we write the Newton-Euler equations for a rigid body moving in three dimensions.

EXAMPLE 10.2 Motion of the Tip of an Airplane Propeller Blade

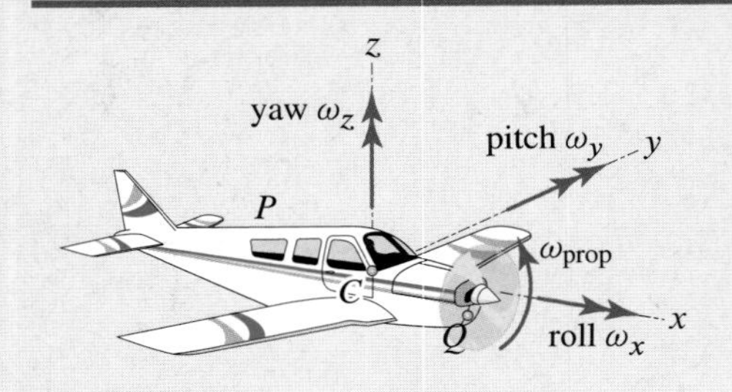

Figure 1
A secondary or moving xyz reference frame attached to an airplane. The primary XYZ reference frame (not shown) is attached to the Earth.

Referring to Fig. 1, assume that the velocity and acceleration of point C are known as $\vec{v}_C = v_{Cx}\,\hat{\imath} + v_{Cy}\,\hat{\jmath} + v_{Cz}\,\hat{k}$ and $\vec{a}_C = a_{Cx}\,\hat{\imath} + a_{Cy}\,\hat{\jmath} + a_{Cz}\,\hat{k}$, respectively, and that the airplane is pitching at a known rate $\omega_y(t)$, but that the roll and yaw rates are both zero. Given the constant angular speed of the propeller in the direction shown, determine expressions for the velocity and acceleration of a point Q, which is a radial distance R from the hub of the propeller at B, as a function of the angle θ (see Fig. 2).

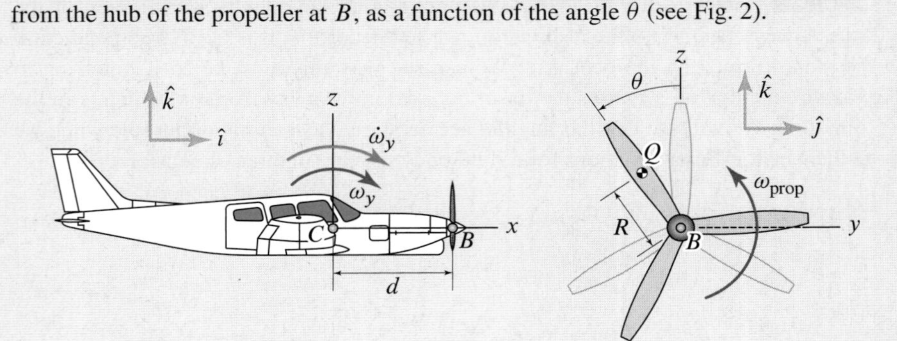

Figure 2. The geometry needed to determine the velocity and acceleration of point Q.

SOLUTION

Road Map & Modeling The secondary frame xyz is attached to the airplane and has its origin at C. We observe that the motion of point C is known and that the axis of the propeller lies on the x axis. This makes the motion of point Q, as seen by an observer in the xyz frame, straightforward to describe because the motion of Q relative to the xyz frame is circular.

Computation Since the xyz frame with origin at C is a body-fixed frame, we can apply Eq. (6.45) to obtain

$$\vec{v}_Q = \vec{v}_C + \vec{v}_{Q\text{rel}} + \vec{\Omega} \times \vec{r}_{Q/C}, \tag{1}$$

where $\vec{\Omega}$ is the angular velocity of the body-fixed reference frame and $\vec{v}_{Q\text{rel}}$ is the velocity of point Q as seen by an observer in the body-fixed reference frame. Since we know that $\omega_x = \omega_z = 0$, then the angular velocity of the rotating frame, expressed in the rotating frame, is

$$\vec{\Omega} = \omega_y\,\hat{\jmath}. \tag{2}$$

As seen by the xyz frame, the point Q is just moving in a circle centered on the x axis and so, using the polar frame shown in Fig. 3, we obtain

$$\vec{v}_{Q\text{rel}} = R\omega_{\text{prop}}\,\hat{u}_\theta. \tag{3}$$

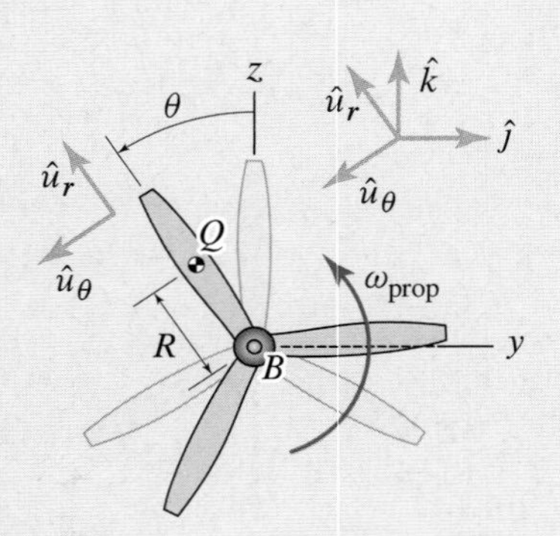

Figure 3
The definition of an additional polar component system to allow us to easily find the velocity and acceleration of Q relative to the rotating frame.

Substituting Eqs. (2) and (3), along with $\vec{v}_C = v_{Cx}\,\hat{\imath} + v_{Cy}\,\hat{\jmath} + v_{Cz}\,\hat{k}$ and $\vec{r}_{Q/C} = d\,\hat{\imath} + R\,\hat{u}_r$, into Eq. (1), we obtain

$$\begin{aligned}\vec{v}_Q &= v_{Cx}\,\hat{\imath} + v_{Cy}\,\hat{\jmath} + v_{Cz}\,\hat{k} + R\omega_{\text{prop}}\,\hat{u}_\theta + \omega_y\,\hat{\jmath} \times (d\,\hat{\imath} + R\,\hat{u}_r)\\ &= v_{Cx}\,\hat{\imath} + v_{Cy}\,\hat{\jmath} + v_{Cz}\,\hat{k} + R\omega_{\text{prop}}(-\cos\theta\,\hat{\jmath} - \sin\theta\,\hat{k})\\ &\qquad + \omega_y\,\hat{\jmath} \times \big[d\,\hat{\imath} + R(\cos\theta\,\hat{k} - \sin\theta\,\hat{\jmath})\big],\end{aligned} \tag{4}$$

or

$$\boxed{\begin{aligned}\vec{v}_{Q\text{rel}} = (v_{Cx} + R\omega_y\cos\theta)\,\hat{\imath} + (v_{Cy} - R\omega_{\text{prop}}\cos\theta)\,\hat{\jmath}\\ + (v_{Cz} - R\omega_{\text{prop}}\sin\theta - d\omega_y)\,\hat{k},\end{aligned}} \tag{5}$$

where we have written $\hat{u}_r$ and $\hat{u}_\theta$ in terms of $\hat{\jmath}$ and $\hat{k}$, carried out the cross products, and collected terms.

To determine the acceleration of Q, we now apply

$$\vec{a}_Q = \vec{a}_C + \vec{a}_{Q\text{rel}} + 2\vec{\Omega} \times \vec{v}_{Q\text{rel}} + \dot{\vec{\Omega}} \times \vec{r}_{Q/C} + \vec{\Omega} \times (\vec{\Omega} \times \vec{r}_{Q/C}), \tag{6}$$

where $\dot{\vec{\Omega}}$ is the angular acceleration of the body-fixed reference frame and $\vec{a}_{Q\text{rel}}$ is the acceleration of point Q as seen by an observer in the body-fixed reference frame. We know or have determined everything in Eq. (6) except for $\dot{\vec{\Omega}}$ and $\vec{a}_{Q\text{rel}}$. For the former, we simply express the angular acceleration of the xyz frame as

$$\dot{\vec{\Omega}} = \dot{\omega}_y \, \hat{\jmath}. \tag{7}$$

For the relative acceleration, since the angular speed of the propeller is constant, point Q only has a normal component of relative acceleration, which gives

$$\vec{a}_{Q\text{rel}} = -R\omega_{\text{prop}}^2 \, \hat{u}_r. \tag{8}$$

Substituting Eqs. (2), (3), (7), and (8), along with $\vec{a}_C = a_{Cx}\,\hat{\imath} + a_{Cy}\,\hat{\jmath} + a_{Cz}\,\hat{k}$, into Eq. (6), we obtain

$$\begin{aligned}
\vec{a}_Q &= a_{Cx}\,\hat{\imath} + a_{Cy}\,\hat{\jmath} + a_{Cz}\,\hat{k} - R\omega_{\text{prop}}^2\,\hat{u}_r + 2\omega_y\,\hat{\jmath} \times R\omega_{\text{prop}}\,\hat{u}_\theta \\
&\qquad + \dot{\omega}_y\,\hat{\jmath} \times (d\,\hat{\imath} + R\,\hat{u}_r) + \omega_y\,\hat{\jmath} \times \big[\omega_y\,\hat{\jmath} \times (d\,\hat{\imath} + R\,\hat{u}_r)\big] \\
&= a_{Cx}\,\hat{\imath} + a_{Cy}\,\hat{\jmath} + a_{Cz}\,\hat{k} - R\omega_{\text{prop}}^2(\cos\theta\,\hat{k} - \sin\theta\,\hat{\jmath}) \\
&\quad + 2\omega_y\,\hat{\jmath} \times R\omega_{\text{prop}}(-\cos\theta\,\hat{\jmath} - \sin\theta\,\hat{k}) + \dot{\omega}_y\,\hat{\jmath} \times \big[d\,\hat{\imath} + R(\cos\theta\,\hat{k} - \sin\theta\,\hat{\jmath})\big] \\
&\qquad + \omega_y\,\hat{\jmath} \times \underbrace{\Big\{\omega_y\,\hat{\jmath} \times \big[d\,\hat{\imath} + R(\cos\theta\,\hat{k} - \sin\theta\,\hat{\jmath})\big]\Big\}}_{R\omega_y \cos\theta\,\hat{\imath} - d\omega_y\,\hat{k}},
\end{aligned} \tag{9}$$

or

$$\boxed{\begin{aligned}
\vec{a}_Q = \big(a_{Cx} &- 2R\omega_y\omega_{\text{prop}}\sin\theta + R\dot{\omega}_y\cos\theta - d\omega_y^2\big)\,\hat{\imath} \\
&+ \big(a_{Cy} + R\omega_{\text{prop}}^2\sin\theta\big)\,\hat{\jmath} \\
&+ \big[a_{Cz} - R\big(\omega_{\text{prop}}^2 + \omega_y^2\big)\cos\theta - d\,\dot{\omega}_y\big]\,\hat{k},
\end{aligned}} \tag{10}$$

where we have again written $\hat{u}_r$ and $\hat{u}_\theta$ in terms of $\hat{\jmath}$ and $\hat{k}$, carried out the cross products, and collected terms.

Discussion & Verification The final results for velocity and acceleration are dimensionally correct. We also see that, even though the motion of point Q on the propeller is *very* complicated as seen by an inertial observer, it is straightforward to determine using a rotating reference frame.

We also note that the velocity and acceleration given in Eqs. (5) and (10), respectively, are measured with respect to some primary XYZ frame (not shown in Figures 1 or 2), and are therefore inertial if XYZ is inertial, *but they are expressed in the body-fixed xyz frame*.

Interesting Fact

Gyroscopic terms in the acceleration. Terms in the acceleration that contain the product of two different angular velocities (e.g., the term containing $\omega_y\omega_{\text{prop}}$ in Eq. (10)) are often called *gyroscopic terms*. They result from the changing direction of angular velocities. This is really the only new result encountered in this section. In this example, the pitch angular velocity of the plane ω_y causes the angular velocity of the propeller to change direction. Thus, the time derivative of $\vec{\omega}_{\text{prop}}$ is not equal to zero because it changes direction due to ω_y.

EXAMPLE 10.3 *Angular Velocity and Angular Acceleration of a Bar in 3D*

The mechanism in Fig. 1 consists of a disk of radius R that rotates with constant angular speed ω_d about the x axis in the direction shown. Attached by a ball joint to the disk at B is the bar AB. End A of bar AB is attached by a ball joint to a collar that slides along the bar CD. Bar CD lies in the yz plane and is inclined at the angle θ with respect to the y axis. At the instant shown, the point B lies in the xy plane. At this instant, determine expressions for the x, y, and z components of the angular velocity of the bar AB, as well as the velocity of the slider A. Treat r and L as given.

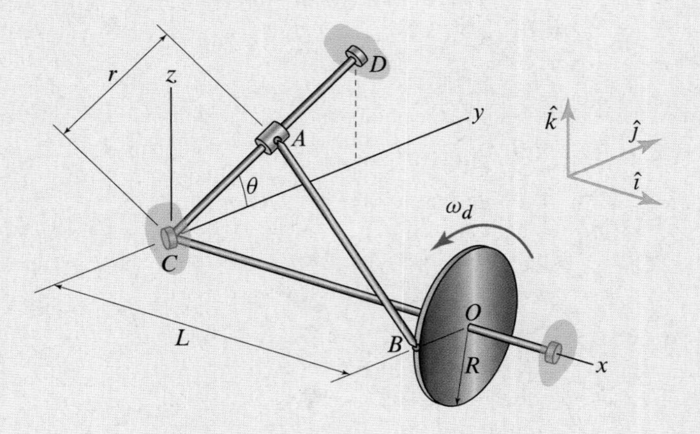

Figure 1.

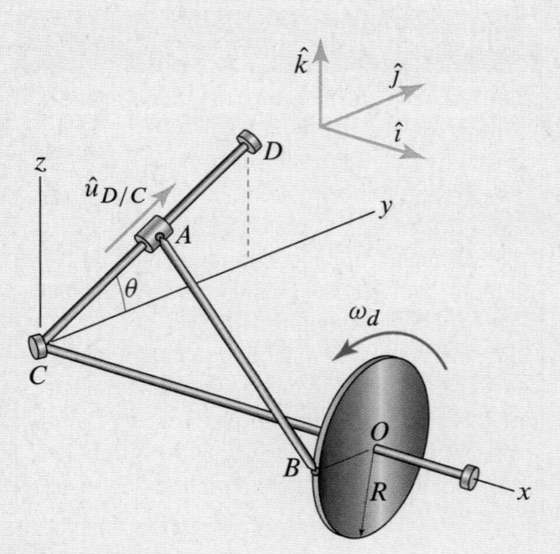

Figure 2
The mechanism showing the unit vector $\hat{u}_{D/C}$.

SOLUTION

Road Map & Modeling The velocity and acceleration of point B are easily determined since it is moving in a circle centered at O. Given that point A is constrained to move in a known direction, we can use Eq. (10.1) to relate the velocity of point A to that of point B, which should lead us to the angular velocity of the bar AB.

Computation At this instant, in terms of the given component system, the velocity of point B can be written as

$$\vec{v}_B = -R\omega_d\,\hat{k}. \tag{1}$$

Applying Eq. (10.1) to points A and B, we obtain

$$\vec{v}_A = \vec{v}_B + \vec{\omega}_{AB} \times \vec{r}_{A/B}, \tag{2}$$

where $\vec{\omega}_{AB}$ is the angular velocity of the bar AB. Since point A is constrained by the collar to move along bar CD, we can write $\vec{v}_A$ as (see Fig. 2)

$$\vec{v}_A = v_A\,\hat{u}_{D/C} = v_A\big(\cos\theta\,\hat{\jmath} + \sin\theta\,\hat{k}\big), \tag{3}$$

where v_A is the component of $\vec{v}_A$ in the direction of $\hat{u}_{D/C}$. The vector $\vec{r}_{A/B}$ can be formed using

$$\begin{aligned}\vec{r}_{A/B} = \vec{r}_A - \vec{r}_B &= r\big(\cos\theta\,\hat{\jmath} + \sin\theta\,\hat{k}\big) - (L\,\hat{\imath} - R\,\hat{\jmath})\\ &= -L\,\hat{\imath} + (R + r\cos\theta)\,\hat{\jmath} + r\sin\theta\,\hat{k}.\end{aligned} \tag{4}$$

Noting that we can write the angular velocity of bar AB as

$$\vec{\omega}_{AB} = \omega_{ABx}\,\hat{\imath} + \omega_{ABy}\,\hat{\jmath} + \omega_{ABz}\,\hat{k}, \tag{5}$$

we can substitute Eqs. (1), (3), (4), and (5) into Eq. (2) to obtain

$$v_A(\cos\theta\,\hat{\jmath} + \sin\theta\,\hat{k}) = -R\omega_d\,\hat{k} + (\omega_{ABx}\,\hat{\imath} + \omega_{ABy}\,\hat{\jmath} + \omega_{ABz}\,\hat{k}) \times \left[-L\,\hat{\imath} + (R + r\cos\theta)\,\hat{\jmath} + r\sin\theta\,\hat{k}\right]. \tag{6}$$

Carrying out the cross product and equating components, we obtain the following three equations:

$$0 = r\omega_{ABy}\sin\theta - (R + r\cos\theta)\omega_{ABz}, \tag{7}$$

$$v_A\cos\theta = -L\omega_{ABz} - r\omega_{ABx}\sin\theta, \tag{8}$$

$$v_A\sin\theta = (R + r\cos\theta)\omega_{ABx} + L\omega_{ABy} - R\omega_d. \tag{9}$$

These three equations have four unknowns, that is, ω_{ABx}, ω_{ABy}, ω_{ABz}, and v_A. With this said, there is one thing that we have not taken into account—any rotation of the bar AB about its own axis does not affect the motion of either point A or point B, so that component of angular velocity is arbitrary as far as Eq. (2) is concerned. Therefore, we can set the component of the angular velocity along the bar AB to zero to obtain the fourth equation (this is equivalent to saying that $\vec{\omega}_{AB}$ is orthogonal to the bar AB), that is,

$$\vec{\omega}_{AB}\cdot\hat{u}_{A/B} = 0 \quad\Rightarrow\quad (\omega_{ABx}\,\hat{\imath} + \omega_{ABy}\,\hat{\jmath} + \omega_{ABz}\,\hat{k})\cdot\frac{\vec{r}_A - \vec{r}_B}{|\vec{r}_A - \vec{r}_B|} = 0, \tag{10}$$

or, substituting in for $\vec{r}_A - \vec{r}_B$ from Eq. (4), we obtain

$$(\omega_{ABx}\,\hat{\imath} + \omega_{ABy}\,\hat{\jmath} + \omega_{ABz}\,\hat{k})\cdot\frac{-L\,\hat{\imath} + (R + r\cos\theta)\,\hat{\jmath} + r\sin\theta\,\hat{k}}{\sqrt{L^2 + (R + r\cos\theta)^2 + r^2\sin^2\theta}} = 0. \tag{11}$$

Expanding the dot product and simplifying, this equation becomes

$$-L\omega_{ABx} + (R + r\cos\theta)\omega_{ABy} + r\sin\theta\omega_{ABz} = 0. \tag{12}$$

Solving Eqs. (7)–(9) and (12) for ω_{ABx}, ω_{ABy}, ω_{ABz}, and v_A, we obtain

$$\omega_{ABx} = \frac{R\omega_d\cos\theta(r^2 + R^2 + 2rR\cos\theta)}{(r + R\cos\theta)(L^2 + r^2 + R^2 + 2rR\cos\theta)}, \tag{13}$$

$$\omega_{ABy} = \frac{LR\omega_d\cos\theta(R + r\cos\theta)}{(r + R\cos\theta)(L^2 + r^2 + R^2 + 2rR\cos\theta)}, \tag{14}$$

$$\omega_{ABz} = \frac{LrR\omega_d\cos\theta\sin\theta}{(r + R\cos\theta)(L^2 + r^2 + R^2 + 2rR\cos\theta)}, \tag{15}$$

and

$$v_A = -\frac{rR\omega_d\sin\theta}{r + R\cos\theta} \quad\Rightarrow\quad \vec{v}_A = -\frac{rR\omega_d\sin\theta}{r + R\cos\theta}(\cos\theta\,\hat{\jmath} + \sin\theta\,\hat{k}). \tag{16}$$

Discussion & Verification The dimension of each of the four results is correct. In addition, given the direction of the angular velocity of the disk, we would expect the collar at A to be moving down the bar CD, and it is.

A Closer Look If we wanted to find the angular acceleration of the bar AB, we could apply

$$\vec{a}_A = \vec{a}_B + \vec{\alpha}_{AB}\times\vec{r}_{A/B} + \vec{\omega}_{AB}\times(\vec{\omega}_{AB}\times\vec{r}_{A/B}), \tag{17}$$

along with a condition similar to Eq. (10) on the angular acceleration $\vec{\alpha}_{AB}$. This is the objective of Problems 10.17 and 10.18.

Problems

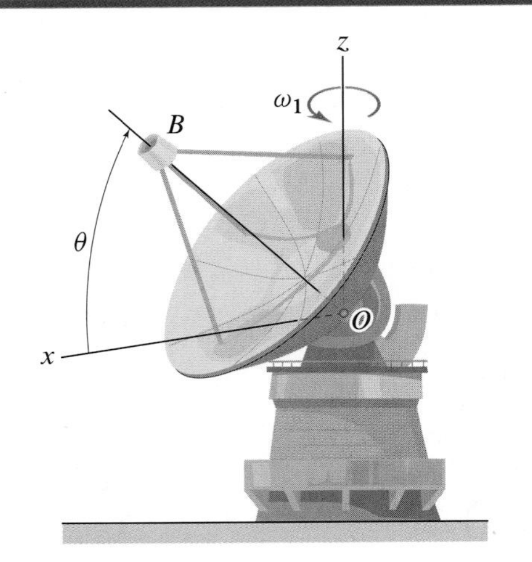

Figure P10.1 and P10.2

Problems 10.1 and 10.2

The radar dish can rotate about the vertical z axis at rate ω_1 and about the horizontal y axis (not shown in the figure) at rate $\dot{\theta}$. The distance between the center of rotation at O and the subreflector at B is ℓ.

Problem 10.1 If ω_1 and $\dot{\theta}$ are both constant, determine the velocity and acceleration of the subreflector B in terms of the elevation angle θ.

Problem 10.2 If $\omega_1(t)$ and $\dot{\theta}(t)$ are known functions of time, determine the velocity and acceleration of the subreflector B in terms of the elevation angle θ.

Problems 10.3 and 10.4

The truncated cone rolls without slipping on the xy plane. At the instant shown, the angular speed about the z axis is ω_1, and it is changing at $\dot{\omega}_1$.

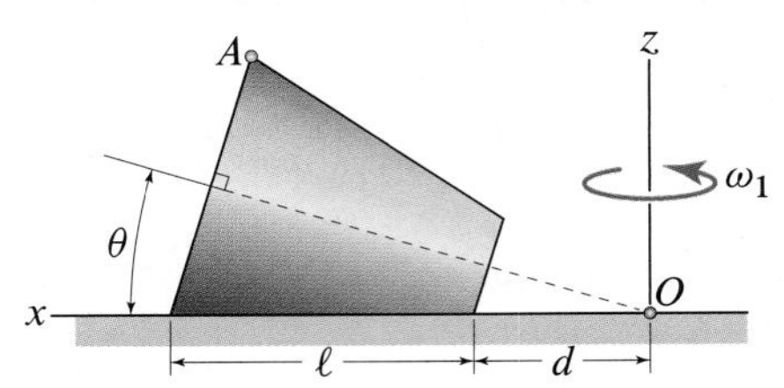

Figure P10.3 and P10.4

Problem 10.3 Determine expressions for the angular velocity and angular acceleration of the cone in terms of ℓ, d, θ, ω_1, and $\dot{\omega}$. Express your answers in the rotating component system shown.

Problem 10.4 Determine expressions for the velocity and acceleration of the point A, which is at the highest point on the cone at this instant, in terms of ℓ, d, θ, ω_1, and $\dot{\omega}$. Express your answers in the rotating component system shown.

Problems 10.5 through 10.8

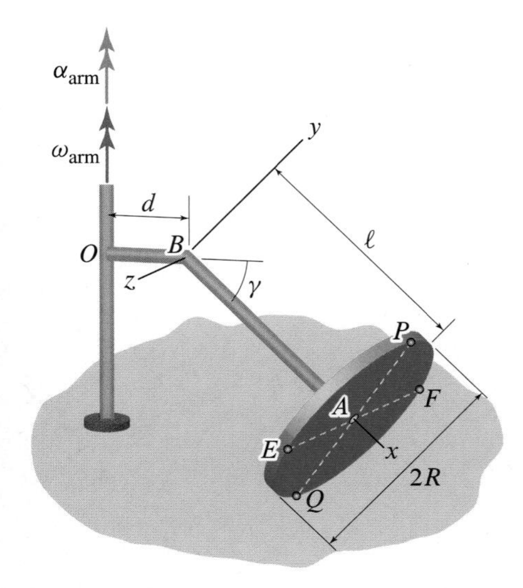

Figure P10.5–P10.8

The bent arm rotates with angular speed ω_{arm} and angular acceleration α_{arm} in the directions shown. The wheel of radius R with center at A rotates relative to the bent arm as it rolls without slipping over the stationary horizontal surface. At the instant shown, the line PQ is perpendicular to the line EF, which is parallel to the horizontal surface (i.e., the line PQ lies in the xy plane). Express your answers using the xyz reference frame that is attached to the arm OBA. Treat d, ℓ, R, and γ as known.

Problem 10.5 Assuming that $\alpha_{arm} = 0$ at the instant shown, determine expressions for the velocity and acceleration of point P.

Problem 10.6 Assuming that $\alpha_{arm} = 0$ at the instant shown, determine expressions for the velocity and acceleration of point E.

Problem 10.7 Assuming that $\alpha_{arm} \neq 0$ at the instant shown, determine expressions for the velocity and acceleration of point P.

Problem 10.8 Assuming that $\alpha_{arm} \neq 0$ at the instant shown, determine expressions for the velocity and acceleration of point E.

Problems 10.9 through 10.11

The fire truck ladder can rotate about the vertical z axis at known rate $\omega_1(t)$, elevate about the horizontal x axis at known rate $\dot{\theta}(t)$, and the ladder can change its length (moving the bucket at B outward or inward) with known rate $\dot{\ell}(t)$.

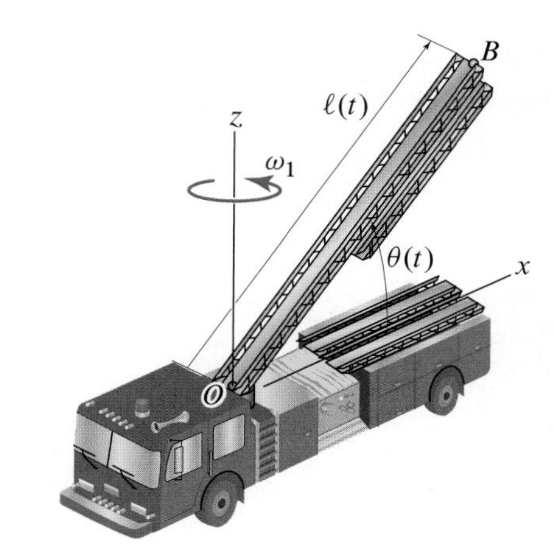

Figure P10.9–P10.11

Problem 10.9 If ω_1 and $\dot{\theta}$ are constant and $\dot{\ell} = 0$, determine expressions for the velocity and acceleration of the bucket at B as functions of the elevation angle θ. Express your answer in the given rotating xyz frame.

Problem 10.10 If ω_1, $\dot{\theta}$, and $\dot{\ell}$ are each constant, determine expressions for the velocity and acceleration of the bucket at B as functions of the elevation angle θ. Express your answer in the given rotating xyz frame.

Problem 10.11 Given $\omega_1(t)$, $\dot{\theta}(t)$, and $\dot{\ell}(t)$ as functions of time, determine expressions for the velocity and acceleration of the bucket at B as functions of the elevation angle θ. Express your answer in the given rotating xyz frame.

Problems 10.12 and 10.13

The bar AB rotates at the rate ω_b about the fixed x axis as shown. The bar CD is attached perpendicularly to AB to form a T-bar. The disk of radius R centered at C rotates at the rate ω_d relative to the arm CD in the direction shown. Assume that the xyz reference frame is attached to the T-bar and that its origin is at D. In addition, assume that the angular rates ω_b and ω_d are *not* constant. Treat h and R as known.

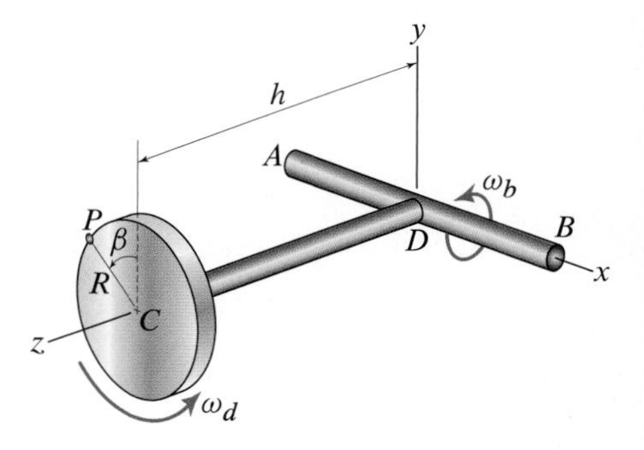

Figure P10.12 and P10.13

Problem 10.12 Determine expressions for the angular velocity and angular acceleration of the disk at C. Express your answer in the rotating xyz reference frame.

Problem 10.13 Determine expressions for the velocity and acceleration of the point P, which lies at the edge of the disk at an arbitrary angle β with respect to the y axis. Express your answer in the rotating xyz reference frame.

Problems 10.14 through 10.16

The cone rolls without slipping over the xz plane and around the y axis with angular speed ω_0 and angular acceleration α_0 in the directions shown. At the instant shown, the line BC is parallel to the surface on which the cone is rolling and the line AD lies on the base of the cone and is perpendicular to the line BC. Treat L and β as known.

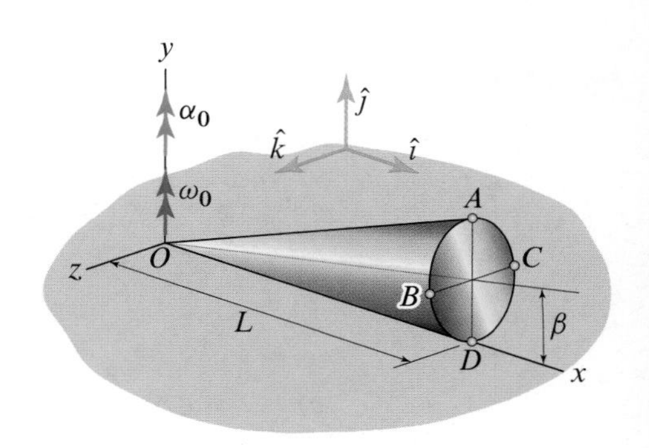

Figure P10.14–P10.16

Problem 10.14 Using the component system shown, determine expressions for the angular velocity $\vec{\omega}_c$ and angular acceleration $\vec{\alpha}_c$ of the cone.

Problem 10.15 Using the component system shown, determine expressions for the velocity and acceleration of point A at this instant.

Problem 10.16 Using the component system shown, determine expressions for the velocity and acceleration of point B at this instant.

Problems 10.17 and 10.18

The mechanism consists of a disk of radius R that rotates with angular speed ω_d and angular acceleration α_d about the x axis in the directions shown. Attached by a ball joint to the disk at B is the bar AB. End A of bar AB is attached by a ball joint to a collar that slides along the bar CD. Bar CD lies in the yz plane and is inclined at the angle θ with respect to the y axis. At the instant shown, the point B lies in the xy plane. Use $R = 1$ ft, $L = 3$ ft, $r = 1.4$ ft, $\theta = 25°$, and $\omega_d = 20$ rad/s.

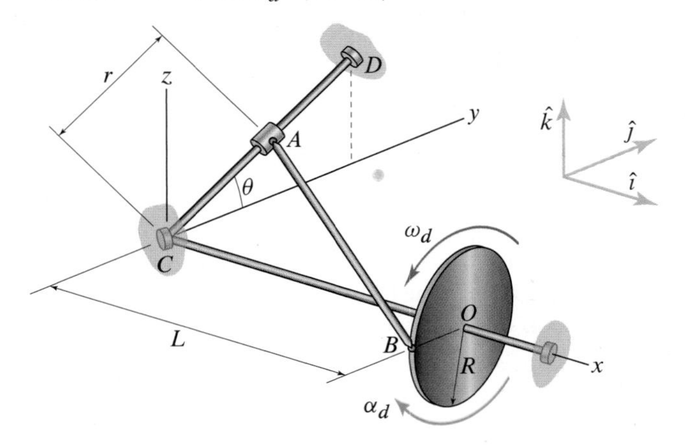

Figure P10.17 and P10.18

Problem 10.17 Assuming that $\alpha_d = 0$ at this instant, determine the angular acceleration of the bar AB, as well as the acceleration of the slider A.

Problem 10.18 Assuming that $\alpha_d = 35\text{ rad/s}^2$ at this instant, determine the angular acceleration of the bar AB, as well as the acceleration of the slider A.

Problems 10.19 and 10.20

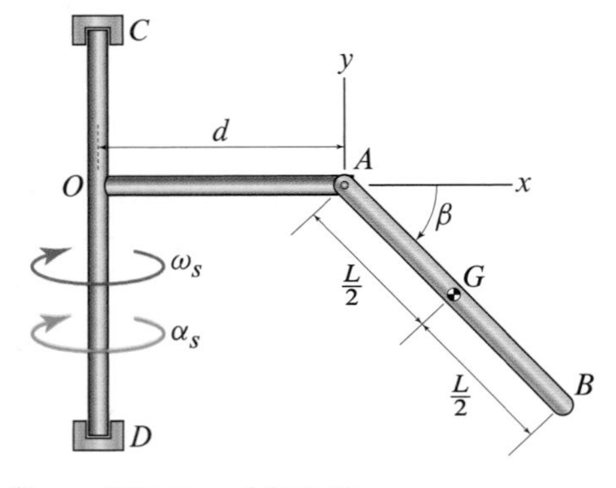

Figure P10.19 and P10.20

The T-bar support structure is mounted in bearings at C and D and spins with angular velocity $\vec{\omega}_s$ and angular acceleration $\vec{\alpha}_s$ in the directions shown. Bar AB of length L is pinned at A to the T-bar support, and its position in the vertical plane is defined by the variable angle β. As we will see in the next section, the Newton-Euler equations for the bar will require that we know the acceleration of G, the angular velocity of the bar AB, and the angular acceleration of the bar AB relative to the primary frame XYZ, which is assumed to be inertial. Express your answers in the xyz frame shown, which is attached to the T-bar support at A. Treat d and L as known.

Problem 10.19 Find expressions for $\vec{a}_G$, $\vec{\omega}_{AB}$, and $\vec{\alpha}_{AB}$ assuming that $\alpha_s = 0$.

Problem 10.20 Find expressions for $\vec{a}_G$, $\vec{\omega}_{AB}$, and $\vec{\alpha}_{AB}$ assuming that $\alpha_s \neq 0$.

Problems 10.21 through 10.24

The mechanism consists of a disk of radius R that rotates with angular speed ω_d and angular acceleration α_d about the x axis in the directions shown. Attached by a ball joint to the disk at B is the bar AB. End A of bar AB is attached by a clevis joint to a collar that slides along the bar CD. Bar CD lies in the yz plane and is inclined at the angle θ with respect to the y axis. At the instant shown, the point B lies in the xy plane. Use $R = 0.2$ m, $L = 0.5$ m, $r = 0.3$ m, $\theta = 25°$, and $\omega_d = 30$ rad/s. *Hint:* The clevis joint constrains the rotation of arm AB *relative to* the collar at A to be perpendicular to the plane formed by bar CD and arm AB. Therefore, the angular velocity of arm AB is the sum of the angular velocity of the collar at A and the angular velocity associated with the change in the angle β.

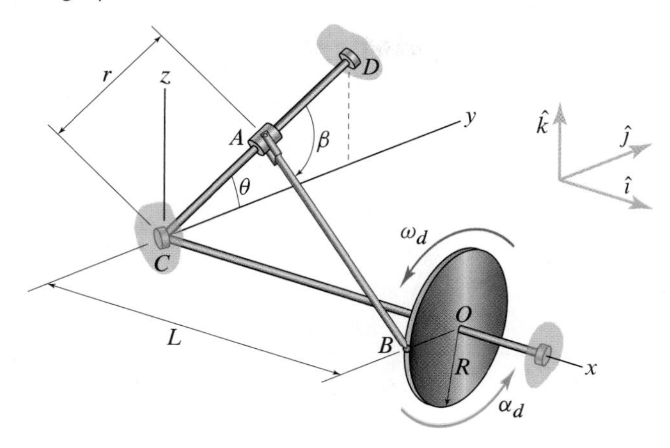

Figure P10.21–P10.24

Problem 10.21 Using the component system shown, determine the velocity of the collar at A.

Problem 10.22 Assuming that $\alpha_d = 15\,\mathrm{rad/s^2}$ at this instant, determine the angular velocity and angular acceleration of the arm AB. Use the component system shown.

Problem 10.23 Assuming that $\alpha_d = 0$ at this instant, determine the acceleration of the collar at A. Use the component system shown.

Problem 10.24 Assuming that $\alpha_d = 15\,\mathrm{rad/s^2}$ at this instant, determine the acceleration of the collar at A. Use the component system shown.

Problems 10.25 and 10.26

Rod AB is attached to the collar at B by a ball joint and to the slider at A by a clevis joint. Collar A slides along the fixed bar OD, which lies in the yz plane and is inclined at the angle θ with respect to the y axis. At the instant shown, collar B lies in the xy plane, and it is moving with the velocity and acceleration shown. Bar CE is fixed and parallel to the z axis. Use $r = 0.3\,\mathrm{m}$, $\theta = 25°$, $L = 0.8\,\mathrm{m}$, $d = 0.1\,\mathrm{m}$, $v_B = 2.5\,\mathrm{m/s}$, and $a_B = 1.5\,\mathrm{m/s^2}$. *Hint:* The clevis joint constrains the rotation of arm AB *relative to* the collar at A to be perpendicular to the plane formed by bar OD and arm AB. Therefore, the angular velocity of arm AB is the sum of the angular velocity of the collar at A and the angular velocity associated with the change in the angle β.

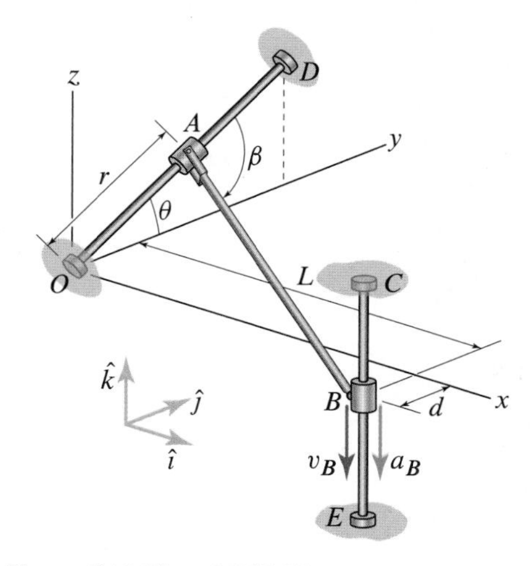

Figure P10.25 and P10.26

Problem 10.25 Using the component system shown, determine the angular velocity and angular acceleration of the bar AB.

Problem 10.26 Using the component system shown, determine the velocity and acceleration of the collar at A.

Problems 10.27 through 10.32

Bar AB of length $L_{AB} = 2.5\,\mathrm{m}$ is attached by ball joints to a collar at A and to a disk at B. The disk lies in the xy plane and its center at E lies on the y axis in the yz plane. The disk rotates about a vertical axis at the constant angular rate $\omega_d = 100\,\mathrm{rpm}$. The dimensions $d = 1.2\,\mathrm{m}$, $h = 0.9\,\mathrm{m}$, and $R = 0.75\,\mathrm{m}$ are given.

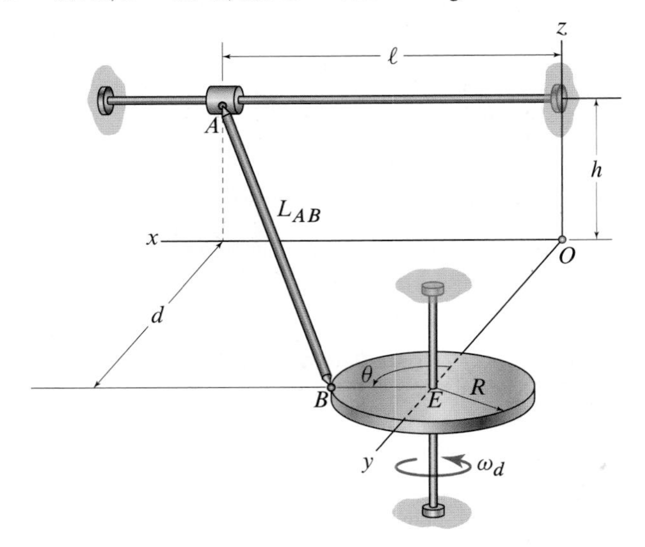

Figure P10.27–P10.32

Problem 10.27 For the disk position shown, that is, $\theta = 90°$, determine the angular velocity of the bar. Express your answer in the given component system, and assume that the angular velocity of the bar is orthogonal to it.

Problem 10.28 For the disk position shown, that is, $\theta = 90°$, determine the angular acceleration of the bar. Express your answer in the given component system, and assume that the angular velocity and angular acceleration of the bar are orthogonal to it.

Problem 10.29 For the disk position $\theta = 0°$, determine the velocity of the collar at A. Express your answer in the given component system, and assume that the angular velocity of the bar is orthogonal to it.

Problem 10.30 For the disk position $\theta = 0°$, determine the acceleration of the collar at A. Express your answer in the given component system, and assume that the angular velocity and angular acceleration of the bar are orthogonal to it.

Problem 10.31 Determine the angular velocity of the bar AB and the velocity of the collar at A for any position θ of the disk. Express your answers in the given component system, and assume that the angular velocity of the bar is orthogonal to it.

Problem 10.32 Determine the angular acceleration of the bar AB and the acceleration of the collar at A for any position θ of the disk. Express your answers in the given component system, and assume that the angular velocity and angular acceleration of the bar are orthogonal to it.

Problems 10.33 through 10.35

The robotic arm shown is used to drill holes during an assembly line manufacturing process. The base of the arm rotates about the vertical axis relative to the platform on which it is mounted at the rate ω_b. The bent arm rotates about the y axis at the rate ω_a relative to the base. To extend and retract the drill bit, the telescoping shaft moves in and out of the bent arm with the speed v_t and acceleration a_t relative to the arm. The angular speed of the drill bit relative to the telescoping arm is ω_d. At this instant, the z axis is perpendicular to the platform. Express all answers using the xyz reference frame that is attached to the bent arm at point O, and *do not* assume that the angular rates ω_b, ω_a, and ω_d are constant. Treat ℓ, r_t, and h as known.

Problem 10.33 Determine expressions for the angular velocity and angular acceleration of the drill bit.

Problem 10.34 Determine expressions for the velocity of point B at the end of the drill bit.

Problem 10.35 Determine expressions for the acceleration of point B at the end of the drill bit.

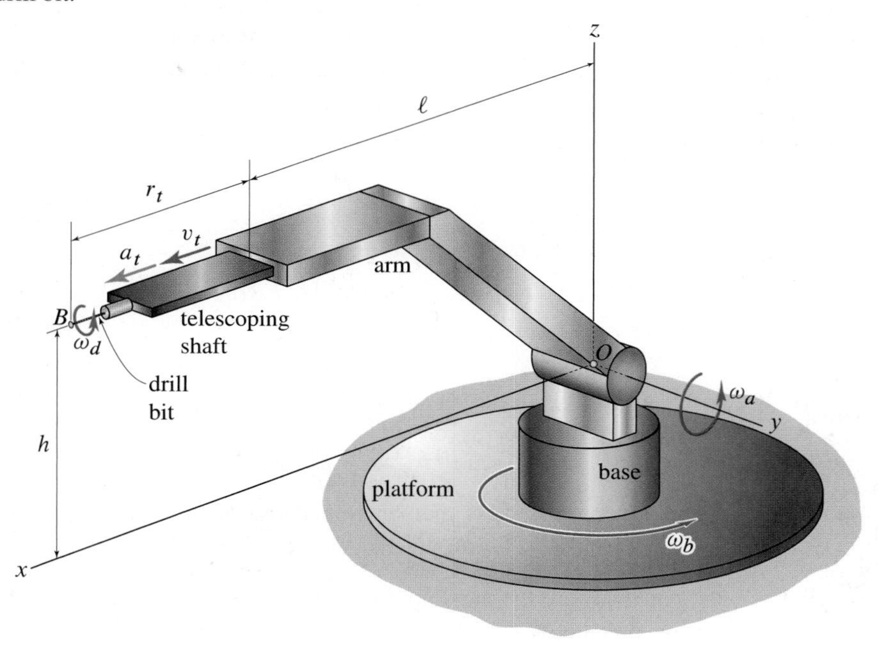

Figure P10.33–P10.35

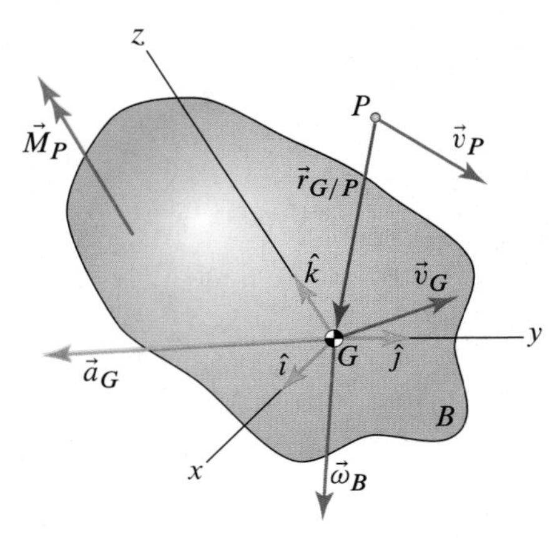

Figure 10.8
Rigid body B showing quantities used in obtaining the Newton-Euler rotational equations of motion for three-dimensional motion.

10.2 Three-Dimensional Kinetics of Rigid Bodies

Newton-Euler equations for three-dimensional motion

Referring to Fig. 10.8, to obtain the Newton-Euler equations for a rigid body, we begin with Euler's first law as given by Eq. (5.57) on p. 365, which is

$$\vec{F} = m\vec{a}_G, \tag{10.20}$$

where $\vec{F}$ is the total external force acting on the rigid body, m is the mass of the rigid body, and $\vec{a}_G$ is the acceleration of the mass center of the rigid body. Referring to Fig. 10.8, we next apply Euler's second law as given by Eq. (5.58) on p. 365, that is,

$$\vec{M}_P = \dot{\vec{h}}_P + \vec{v}_P \times m\vec{v}_G, \tag{10.21}$$

where $\vec{M}_P$ is the total external moment acting on the rigid body about an arbitrary moment center P, $\vec{h}_P$ is the angular momentum of the rigid body about P, $\vec{v}_P$ is the velocity of the point P, and $\vec{v}_G$ is the velocity of the mass center of the rigid body. To make this equation useful, we need an expression for the angular momentum of a rigid body. Equations (B.15) on p. A-13 and (B.18) on p. A-14 in Appendix B provide two such expressions, both of which we repeat below for convenience.

$$\begin{aligned}\vec{h}_P = {} & \big(I_{Gx}\omega_{Bx} - I_{Gxy}\omega_{By} - I_{Gxz}\omega_{Bz}\big)\,\hat{\imath} \\ & + \big(-I_{Gxy}\omega_{Bx} + I_{Gy}\omega_{By} - I_{Gyz}\omega_{Bz}\big)\,\hat{\jmath} \\ & + \big(-I_{Gxz}\omega_{Bx} - I_{Gyz}\omega_{By} + I_{Gz}\omega_{Bz}\big)\,\hat{k} + \vec{r}_{G/P} \times m\vec{v}_G,\end{aligned} \tag{10.22}$$

$$\{h_P\} = [I_G]\{\omega_B\} + \vec{r}_{G/P} \times m\vec{v}_G, \tag{10.23}$$

where:

- The column vector $\{h_P\} = h_{Px}\,\hat{\imath} + h_{Py}\,\hat{\jmath} + h_{Pz}\,\hat{k}$.
- I_{Gx}, I_{Gy}, and I_{Gz} are the mass moments of inertia of the rigid body with respect to its mass center G (see Appendix A).
- I_{Gxy}, I_{Gxz}, and I_{Gyz} are the mass products of inertia of the rigid body with respect to its mass center G (see Appendix A).
- $[I_G]$ is the inertia matrix of B with respect to the xyz axes, and its entries consist of the moments and products of inertia as follows:

$$[I_G] = \begin{bmatrix} I_{Gx} & -I_{Gxy} & -I_{Gxz} \\ -I_{Gxy} & I_{Gy} & -I_{Gyz} \\ -I_{Gxz} & -I_{Gyz} & I_{Gz} \end{bmatrix}. \tag{10.24}$$

- ω_{Bx}, ω_{By}, and ω_{Bz} are the angular velocity components of the rigid body written in the xyz frame and $\{\omega_B\} = \vec{\omega}_B = \omega_{Bx}\,\hat{\imath} + \omega_{By}\,\hat{\jmath} + \omega_{Bz}\,\hat{k}$.
- $\vec{r}_{G/P} = x_{G/P}\,\hat{\imath} + y_{G/P}\,\hat{\jmath} + z_{G/P}\,\hat{k}$.
- $\vec{v}_G = v_{Gx}\,\hat{\imath} + v_{Gy}\,\hat{\jmath} + v_{Gz}\,\hat{k}$.

Helpful Information

The xyz reference frame is not, in general, attached to B. The angular velocity of the body $\vec{\omega}_B$ and the moments and products of inertia are all written in the components of the xyz frame, which is rotating with angular velocity $\vec{\Omega}$. Note that, at this point, the body can be rotating relative to the xyz frame so that the angular velocity of the body and the angular velocity of the frame are, in general, not the same.

To obtain the rotational Newton-Euler equations represented by Eq. (10.21), we now differentiate $\vec{h}_P$ with respect to time. Starting with the time derivative of the last term on the right-hand side of Eq. (10.22), we obtain

$$\begin{aligned}\frac{d}{dt}(\vec{r}_{G/P} \times m\vec{v}_G) &= \dot{\vec{r}}_{G/P} \times m\vec{v}_G + \vec{r}_{G/P} \times m\vec{a}_G \\ &= (\vec{v}_G - \vec{v}_P) \times m\vec{v}_G + \vec{r}_{G/P} \times m\vec{a}_G \\ &= -\vec{v}_P \times m\vec{v}_G + \vec{r}_{G/P} \times m\vec{a}_G.\end{aligned} \tag{10.25}$$

When taking the time derivative of the first three terms in Eq. (10.22), we need to differentiate all terms that are time varying. This includes not only the angular velocity components and Cartesian unit vectors, but also moments and products of inertia if the xyz frame is allowed to rotate relative to the rigid body B.* To greatly simplify the equations, while still allowing us to write the rotational equations for almost any rigid body, we will assume that the xyz frame is *attached* to the rigid body B so that the moment and products of inertia are *constant* with respect to that reference frame. Therefore, the time derivative of the first three terms of Eq. (10.22) is

$$\begin{aligned}\frac{d}{dt}([I_G]\{\omega_B\}) = {} & (I_{Gx}\dot{\omega}_{Bx} - I_{Gxy}\dot{\omega}_{By} - I_{Gxz}\dot{\omega}_{Bz})\,\hat{\imath} \\ & + (-I_{Gxy}\dot{\omega}_{Bx} + I_{Gy}\dot{\omega}_{By} - I_{Gyz}\dot{\omega}_{Bz})\,\hat{\jmath} \\ & + (-I_{Gxz}\dot{\omega}_{Bx} - I_{Gyz}\dot{\omega}_{By} + I_{Gz}\dot{\omega}_{Bz})\,\hat{k} \\ & + (I_{Gx}\omega_{Bx} - I_{Gxy}\omega_{By} - I_{Gxz}\omega_{Bz})(\vec{\Omega} \times \hat{\imath}) \\ & + (-I_{Gxy}\omega_{Bx} + I_{Gy}\omega_{By} - I_{Gyz}\omega_{Bz})(\vec{\Omega} \times \hat{\jmath}) \\ & + (-I_{Gxz}\omega_{Bx} - I_{Gyz}\omega_{By} + I_{Gz}\omega_{Bz})(\vec{\Omega} \times \hat{k}),\end{aligned} \tag{10.26}$$

where $\vec{\Omega}$ is the angular velocity of the xyz reference frame. Since the xyz frame is now attached to the rigid body, we have that $\vec{\Omega} = \vec{\omega}_B = \omega_{Bx}\,\hat{\imath} + \omega_{By}\,\hat{\jmath} + \omega_{Bz}\,\hat{k}$, so that, after expanding the cross products and collecting terms, Eq. (10.26) becomes

$$\begin{aligned}\frac{d}{dt}([I_G]\{\omega_B\}) = {} & \big[I_{Gx}\dot{\omega}_{Bx} + (I_{Gz} - I_{Gy})\,\omega_{By}\omega_{Bz} + I_{Gxy}(\omega_{Bx}\omega_{Bz} - \dot{\omega}_{By}) \\ & \quad - I_{Gxz}(\omega_{Bx}\omega_{By} + \dot{\omega}_{Bz}) - I_{Gyz}(\omega_{By}^2 - \omega_{Bz}^2)\big]\,\hat{\imath} \\ & + \big[I_{Gy}\dot{\omega}_{By} + (I_{Gx} - I_{Gz})\,\omega_{Bx}\omega_{Bz} + I_{Gyz}(\omega_{Bx}\omega_{By} - \dot{\omega}_{Bz}) \\ & \quad - I_{Gxz}(\omega_{Bz}^2 - \omega_{Bx}^2) - I_{Gxy}(\dot{\omega}_{Bx} + \omega_{By}\omega_{Bz})\big]\,\hat{\jmath} \\ & + \big[I_{Gz}\dot{\omega}_{Bz} + (I_{Gy} - I_{Gx})\,\omega_{Bx}\omega_{By} - I_{Gyz}(\omega_{Bx}\omega_{Bz} + \dot{\omega}_{By}) \\ & \quad - I_{Gxz}(\dot{\omega}_{Bx} - \omega_{By}\omega_{Bz}) - I_{Gxy}(\omega_{Bx}^2 - \omega_{By}^2)\big]\,\hat{k}.\end{aligned} \tag{10.27}$$

Substituting Eqs. (10.25) and (10.27) into Eq. (10.21), and writing

$$\vec{M}_P = M_{Px}\,\hat{\imath} + M_{Py}\,\hat{\jmath} + M_{Pz}\,\hat{k}, \tag{10.28}$$

$$\vec{r}_{G/P} = x_{G/P}\,\hat{\imath} + y_{G/P}\,\hat{\jmath} + z_{G/P}\,\hat{k}, \tag{10.29}$$

$$\vec{a}_G = a_{Gx}\,\hat{\imath} + a_{Gy}\,\hat{\jmath} + a_{Gz}\,\hat{k}, \tag{10.30}$$

* See the discussion under *Practical use of Eqs. (B.15) and (B.16)* in Appendix B on p. A-13.

then the three components of Eq. (10.21) become

$$M_{Px} = I_{Gx}\dot{\omega}_{Bx} + (I_{Gz} - I_{Gy})\,\omega_{Bz}\omega_{By} + I_{Gxy}(\omega_{Bx}\omega_{Bz} - \dot{\omega}_{By}) - I_{Gxz}(\omega_{Bx}\omega_{By} + \dot{\omega}_{Bz}) - I_{Gyz}(\omega_{By}^2 - \omega_{Bz}^2) + m(y_{G/P}a_{Gz} - z_{G/P}a_{Gy}), \tag{10.31}$$

$$M_{Py} = I_{Gy}\dot{\omega}_{By} + (I_{Gx} - I_{Gz})\,\omega_{Bx}\omega_{Bz} + I_{Gyz}(\omega_{Bx}\omega_{By} - \dot{\omega}_{Bz}) - I_{Gxz}(\omega_{Bz}^2 - \omega_{Bx}^2) - I_{Gxy}(\dot{\omega}_{Bx} + \omega_{By}\omega_{Bz}) + m(z_{G/P}a_{Gx} - x_{G/P}a_{Gz}), \tag{10.32}$$

$$M_{Pz} = I_{Gz}\dot{\omega}_{Bz} + (I_{Gy} - I_{Gx})\,\omega_{By}\omega_{Bx} - I_{Gyz}(\omega_{Bx}\omega_{Bz} + \dot{\omega}_{By}) - I_{Gxz}(\dot{\omega}_{Bx} - \omega_{By}\omega_{Bz}) - I_{Gxy}(\omega_{Bx}^2 - \omega_{By}^2) + m(x_{G/P}a_{Gy} - y_{G/P}a_{Gx}), \tag{10.33}$$

where we have canceled $-\vec{v}_P \times m\vec{v}_G$ in Eq. (10.25) with $\vec{v}_P \times m\vec{v}_G$ in Eq. (10.21). These moment equations are rather complicated due to the nonzero products of inertia, but they can be greatly simplified if the xyz axes that are attached to the body B are *principal body axes* (see Appendix A). In that case, all the products of inertia become zero, and Eqs. (10.31)–(10.33) become

$$M_{Px} = I_{Gx}\dot{\omega}_{Bx} + (I_{Gz} - I_{Gy})\,\omega_{Bz}\omega_{By} + m(y_{G/P}a_{Gz} - z_{G/P}a_{Gy}), \tag{10.34}$$

$$M_{Py} = I_{Gy}\dot{\omega}_{By} + (I_{Gx} - I_{Gz})\,\omega_{Bx}\omega_{Bz} + m(z_{G/P}a_{Gx} - x_{G/P}a_{Gz}), \tag{10.35}$$

$$M_{Pz} = I_{Gz}\dot{\omega}_{Bz} + (I_{Gy} - I_{Gx})\,\omega_{By}\omega_{Bx} + m(x_{G/P}a_{Gy} - y_{G/P}a_{Gx}). \tag{10.36}$$

Concept Alert

Recognizing principal body axes. While mass moments and products of inertia are discussed in detail in Appendix A, for all the rigid bodies we will consider here, the principal axes of inertia can be easily recognized. In particular, all products of inertia containing a coordinate whose corresponding axis is perpendicular to a plane of symmetry for a body must be zero, as long as the origin of the coordinate system lies in that plane of symmetry.

If *any* one of the following special cases can be applied:

(1) P is the mass center G of the body B so that $\vec{r}_{P/G} = \vec{0}$.

(2) $\vec{a}_G = \vec{0}$, that is, the mass center of the body moves with constant velocity.

(3) $\vec{r}_{G/P}$ is parallel to $\vec{a}_G$ so that the cross product $\vec{r}_{G/P} \times m\vec{a}_G = \vec{0}$.

then Eqs. (10.34)–(10.36) become

$$M_{Px} = I_{Gx}\dot{\omega}_{Bx} + (I_{Gz} - I_{Gy})\,\omega_{Bz}\omega_{By}, \tag{10.37}$$

$$M_{Py} = I_{Gy}\dot{\omega}_{By} + (I_{Gx} - I_{Gz})\,\omega_{Bx}\omega_{Bz}, \tag{10.38}$$

$$M_{Pz} = I_{Gz}\dot{\omega}_{Bz} + (I_{Gy} - I_{Gx})\,\omega_{By}\omega_{Bx}. \tag{10.39}$$

Equations (10.37)–(10.39) are the famous *Euler's equations of rotational motion for a rigid body.*

Equations (10.31)–(10.33), or Eqs. (10.34)–(10.36), or Eqs. (10.37)–(10.39), along with Eq. (10.20), give us the six Newton-Euler equations of motion that we need for a rigid body in three-dimensional motion. Before moving on, let's summarize the assumptions and conditions for each of the three sets of rotational equations (see Fig. 10.9).

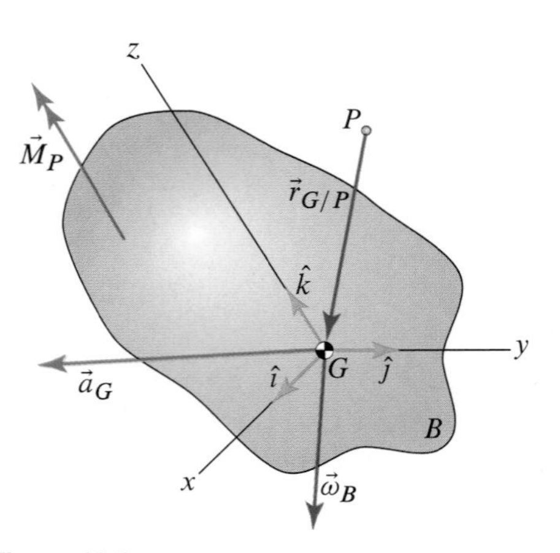

Figure 10.9
Illustration of the kinetic and kinematic quantities needed to understand and apply Eqs. (10.31)–(10.39).

Conditions for applying Eqs. (10.31)–(10.33)

- Point P is arbitrary, and the xyz reference frame is attached to the body B.
- The moments of inertia are expressed relative to a set of axes that are parallel to xyz and whose origin is at G, the center of mass of the body.
- The angular velocities and their time derivatives are inertial, but are expressed in the rotating xyz frame.

Conditions for applying Eqs. (10.34)–(10.36) All the conditions for applying Eqs. (10.31)–(10.33) plus

- The xyz axes are principal axes of inertia so that all products of inertia are zero.

Conditions for applying Eqs. (10.37)–(10.39) All the conditions for applying Eqs. (10.34)–(10.36) plus

- $\vec{r}_{G/P} \times m\vec{a}_G = \vec{0}$, which can happen in any of the following three ways:
  - ◇ P is the mass center G of the body B so that $\vec{r}_{G/P} = \vec{0}$,
  - ◇ $\vec{a}_G = \vec{0}$, that is, the mass center of the body moves with constant velocity,
  - ◇ $\vec{r}_{G/P}$ is parallel to $\vec{a}_G$ so that the cross product $\vec{r}_{G/P} \times m\vec{a}_G = \vec{0}$.

Rotational equations of motion for a body with an axis of radial symmetry

Many mechanical systems have axisymmetric parts that spin about their axis of symmetry. For these systems, any reference frame that has an axis coinciding with the axis of symmetry of that body will be a principal one (i.e., its products of inertia will be zero; see Appendix A). For example, referring to Fig. 10.10, we introduce

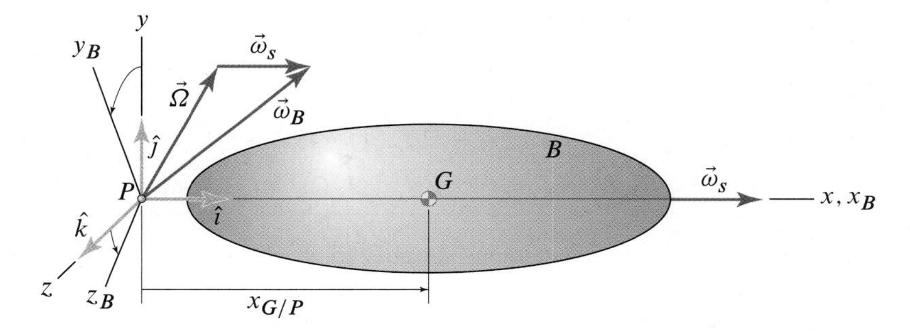

Figure 10.10. A rigid body B spinning about its axis of radial symmetry x_B relative to the xyz frame with angular velocity $\vec{\omega}_s$. The angular velocity of the xyz frame is $\vec{\Omega}$, relative to an underlying inertial frame.

the principal body axes $x_B y_B z_B$ with the x_B axis coinciding with the axis of radial symmetry of the body B. The xyz frame rotates with angular velocity $\vec{\Omega}$, the angular velocity of the body is $\vec{\omega}_B$, and the body spins relative to the xyz frame about its axis of symmetry with angular velocity $\vec{\omega}_s$. With these definitions, we can write that angular velocity of the body as

$$\vec{\omega}_B = \vec{\Omega} + \vec{\omega}_s, \tag{10.40}$$

or, in component form relative to the xyz frame as

$$\omega_{Bx} = \Omega_x + \omega_s, \quad \omega_{By} = \Omega_y, \quad \text{and} \quad \omega_{Bz} = \Omega_z. \tag{10.41}$$

We will again apply Euler's second law as given by Eq. (10.21). Referring back to Eq. (10.22), $[I_G]\{\omega_B\}$ for the rigid body in this case is

$$\begin{aligned} [I_G]\{\omega_B\} &= I_{Gx}\omega_{Bx}\,\hat{\imath} + I_{Gy}\omega_{By}\,\hat{\jmath} + I_{Gy}\omega_{Bz}\,\hat{k} \\ &= I_{Gx}(\Omega_x + \omega_s)\,\hat{\imath} + I_{Gy}\Omega_y\,\hat{\jmath} + I_{Gy}\Omega_z\,\hat{k}, \end{aligned} \tag{10.42}$$

where we have used the fact that all products of inertia are zero, have substituted in Eq. (10.41), and have used the fact that $I_{Gz} = I_{Gy}$. Differentiating Eq. (10.42) with respect to time, we obtain

$$\frac{d}{dt}([I_G]\{\omega_B\}) = I_{Gx}(\dot{\Omega}_x + \dot{\omega}_s)\,\hat{\imath} + I_{Gy}\dot{\Omega}_y\,\hat{\jmath} + I_{Gy}\dot{\Omega}_z\,\hat{k} + I_{Gx}(\Omega_x + \omega_s)(\vec{\Omega}\times\hat{\imath}) + I_{Gy}\Omega_y(\vec{\Omega}\times\hat{\jmath}) + I_{Gy}\Omega_z(\vec{\Omega}\times\hat{k}). \tag{10.43}$$

Expanding the cross products and collecting terms, this equation becomes

$$\frac{d}{dt}([I_G]\{\omega_B\}) = I_{Gx}(\dot{\Omega}_x + \dot{\omega}_s)\,\hat{\imath} + \left[I_{Gy}\dot{\Omega}_y + I_{Gx}(\Omega_x + \omega_s)\Omega_z - I_{Gy}\Omega_x\Omega_z\right]\hat{\jmath} + \left[I_{Gy}\dot{\Omega}_z - I_{Gx}(\Omega_x + \omega_s)\Omega_y + I_{Gy}\Omega_x\Omega_y\right]\hat{k}. \tag{10.44}$$

Noting that $\dot{\vec{h}}_P$ is the sum of Eqs. (10.25) and (10.44), and writing

$$\vec{M}_P = M_{Px}\,\hat{\imath} + M_{Py}\,\hat{\jmath} + M_{Pz}\,\hat{k}, \tag{10.45}$$

$$\vec{r}_{G/P} = x_{G/P}\,\hat{\imath}, \tag{10.46}$$

$$\vec{a}_G = a_{Gx}\,\hat{\imath} + a_{Gy}\,\hat{\jmath} + a_{Gz}\,\hat{k}, \tag{10.47}$$

then the three components of Eq. (10.21) become

$$M_{Px} = I_{Gx}(\dot{\Omega}_x + \dot{\omega}_s), \tag{10.48}$$

$$M_{Py} = I_{Gy}\dot{\Omega}_y + I_{Gx}(\Omega_x + \omega_s)\Omega_z - I_{Gy}\Omega_x\Omega_z - m x_{G/P} a_{Gz}, \tag{10.49}$$

$$M_{Pz} = I_{Gy}\dot{\Omega}_z - I_{Gx}(\Omega_x + \omega_s)\Omega_y + I_{Gy}\Omega_x\Omega_y + m x_{G/P} a_{Gy}. \tag{10.50}$$

If $\vec{r}_{G/P} \times m\vec{a}_G = \vec{0}$, then these equations simplify to

$$M_{Px} = I_{Gx}(\dot{\Omega}_x + \dot{\omega}_s), \tag{10.51}$$

$$M_{Py} = I_{Gy}\dot{\Omega}_y + I_{Gx}(\Omega_x + \omega_s)\Omega_z - I_{Gy}\Omega_x\Omega_z, \tag{10.52}$$

$$M_{Pz} = I_{Gy}\dot{\Omega}_z - I_{Gx}(\Omega_x + \omega_s)\Omega_y + I_{Gy}\Omega_x\Omega_y. \tag{10.53}$$

Planar motion of bodies *not* symmetric with respect to the plane of motion

In Chapter 7, we emphasized that the Newton-Euler rotational equations we derived there were for the planar motion of bodies symmetric with respect to the plane of motion. What if we have a rigid body undergoing planar motion, but the body is not symmetric with respect to the plane of motion? The translational equation, i.e., Eq. (10.20), doesn't change in this case. The rotational or moment equations are given by Eqs. (10.31)–(10.33), but a number of the terms can be eliminated. To see which terms we can eliminate, refer to the system in Fig. 10.11 in which the bar AB is *rigidly* attached to the T-bar support (i.e., the angle β is constant), which is, in turn, spinning about the vertical axis with angular velocity ω_s and angular acceleration α_s. The xyz frame has its origin at G, and it rotates with the arm AB and the support.

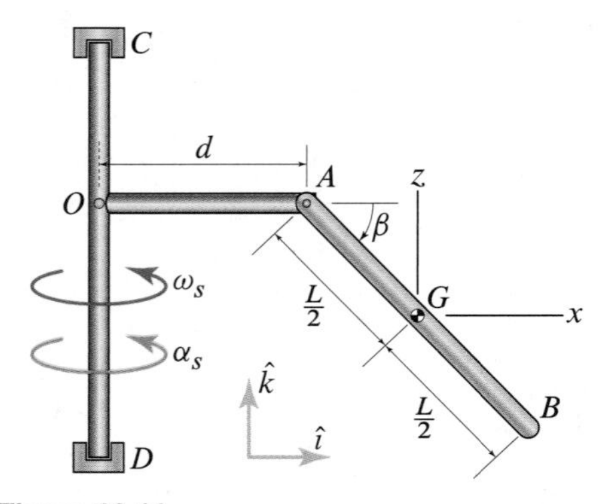

Figure 10.11
An example of a rigid body undergoing planar motion in which the rigid body is *not* symmetric with respect to the plane of motion.

- ω_{Bx}, ω_{By}, $\dot{\omega}_{Bx}$, and $\dot{\omega}_{By}$ are zero since the motion is planar in the xy plane.
- a_{Gz} is zero since G does not move in the z direction.
- $z_{G/P}$ is zero since the mass center lies at the origin of the xyz frame.

With these simplifications, Eqs. (10.31)–(10.33) become

$$M_{Px} = -I_{Gxz}\dot{\omega}_{Bz} + I_{Gyz}\omega_{Bz}^2, \tag{10.54}$$

$$M_{Py} = -I_{Gyz}\dot{\omega}_{Bz} - I_{Gxz}\omega_{Bz}^2, \tag{10.55}$$

$$M_{Pz} = I_{Gz}\dot{\omega}_{Bz} + m\big(x_{G/P}a_{Gy} - y_{G/P}a_{Gx}\big). \tag{10.56}$$

Equations (10.54)–(10.56) are the most general rotational equations for the planar motion of a rigid body. As is discussed in Appendix A, I_{Gz}, I_{Gxz}, and I_{Gyz} are measures of the distribution of mass of a body with respect to the indicated axes, with I_{Gxz} and I_{Gyz} being measures of the symmetry of the body with respect to those axes.

- We now see that if a body is *not* symmetric about the xy plane of motion, then moments in the x and/or y directions will be required to maintain planar motion.
- If the body is symmetric with respect to the xy plane of motion, then $I_{Gxz} = I_{Gyz} = 0$, and we are left with

$$M_{Pz} = I_G\alpha_B + m\big(x_{G/P}a_{Gy} - y_{G/P}a_{Gx}\big), \tag{10.57}$$

where we have again let $I_{Gz} = I_G$ for the symmetric case and have let $\dot{\omega}_{Bz} = \alpha_G$. This equation is identical to Eq. (7.28) on p. 524 in Chapter 7.

In Example 10.6, we will analyze the bar AB in Fig. 10.11 to determine the required forces and moments at A for the arm to move as described.

Helpful Information

The utility of Eqs. (10.54)–(10.56). While Eqs. (10.54)–(10.56) apply to systems like the one shown in Fig. 10.11, they are usually not the most convenient set of rotational equations to use. The reason is that one needs to compute products of inertia to apply Eqs. (10.54)–(10.56), which is usually a tedious task. We will see in Example 10.6 that using Eqs. (10.34)–(10.36) or Eqs. (10.37)–(10.39) is not conceptually difficult and eliminates the need to calculate products of inertia.

Kinetic energy of a rigid body in three-dimensional motion

The work-energy principle for a rigid body is the same whether its motion is two-dimensional or three-dimensional. Therefore, the work-energy principle as given by any of Eqs. (8.16) on p. 585, (8.23) on p. 586, (8.24) on p. 586, or (8.29) on p. 588 is valid for three-dimensional motion. However, the expression for the kinetic energy of a rigid body changes dramatically when we consider three-dimensional motion.

Referring to Fig. 10.12, to derive the kinetic energy for a rigid body B moving in three-dimensional motion, we begin as we did in Section 8.1 and write the kinetic energy as

$$T = \int_B \tfrac{1}{2} v_{dm}^2\, dm, \tag{10.58}$$

where dm is an element of mass in the body B, v_{dm} is the speed of the mass element dm, and the integral is over the entire body B. Because the mass center G and dm are two points on the same rigid body, we can write

$$\vec{v}_{dm} = \vec{v}_G + \vec{\omega}_B \times \vec{q}, \tag{10.59}$$

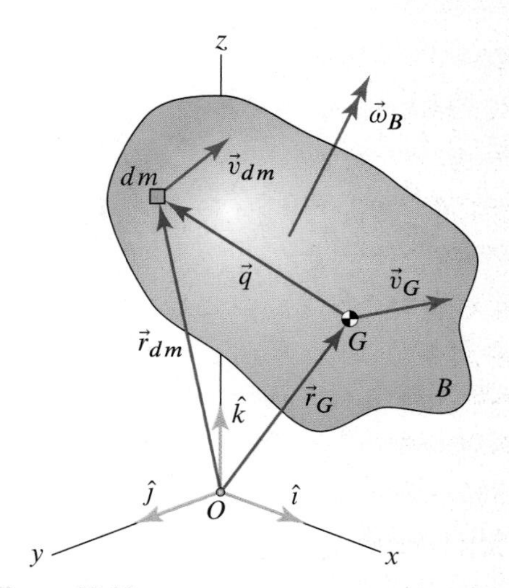

Figure 10.12
The quantities needed for deriving the kinetic energy of a rigid body in three-dimensional motion.

where $\vec{\omega}_B$ is the angular velocity of B and $\vec{q}$ is the position of dm relative to G. To evaluate the integral in Eq. (10.58), we write the vectors in component form as

$$\vec{v}_G = v_{Gx}\,\hat{\imath} + v_{Gy}\,\hat{\jmath} + v_{Gz}\,\hat{k}, \tag{10.60}$$

$$\vec{\omega}_B = \omega_{Bx}\,\hat{\imath} + \omega_{By}\,\hat{\jmath} + \omega_{Bz}\,\hat{k}, \tag{10.61}$$

$$\vec{q} = q_x\,\hat{\imath} + q_y\,\hat{\jmath} + q_z\,\hat{k}. \tag{10.62}$$

Substituting Eqs. (10.60)–(10.62) into Eq. (10.59) and noting that $v_{dm}^2 = \vec{v}_{dm} \cdot \vec{v}_{dm}$, we find that

$$\begin{aligned} v_{dm}^2 &= \vec{v}_{dm} \cdot \vec{v}_{dm} \quad (10.63) \\ &= v_{Gx}^2 + v_{Gy}^2 + v_{Gz}^2 + \omega_{Bx}^2(q_y^2 + q_z^2) + \omega_{By}^2(q_x^2 + q_z^2) \\ &\quad + \omega_{Bz}^2(q_x^2 + q_y^2) - 2q_xq_y\omega_{Bx}\omega_{By} - 2q_xq_z\omega_{Bx}\omega_{Bz} \\ &\quad - 2q_yq_z\omega_{By}\omega_{Bz} + 2(v_{Gy}\omega_{Bz} - v_{Gz}\omega_{By})q_x \\ &\quad + 2(v_{Gz}\omega_{Bx} - v_{Gx}\omega_{Bz})q_y + 2(v_{Gx}\omega_{By} - v_{Gy}\omega_{Bx})q_z. \quad (10.64) \end{aligned}$$

Noting that $v_{Gx}^2 + v_{Gy}^2 + v_{Gz}^2 = v_G^2$, Eq. (10.58) becomes

$$\begin{aligned} T = \frac{1}{2}\bigg[& v_G^2 \int_B dm + \omega_{Bx}^2 \int_B (q_y^2 + q_z^2)\, dm + \omega_{By}^2 \int_B (q_x^2 + q_z^2)\, dm \\ & + \omega_{Bz}^2 \int_B (q_x^2 + q_y^2)\, dm - 2\omega_{Bx}\omega_{By} \int_B q_xq_y\, dm \\ & - 2\omega_{Bx}\omega_{Bz} \int_B q_xq_z\, dm - 2\omega_{By}\omega_{Bz} \int_B q_yq_z\, dm \\ & + 2(v_{Gy}\omega_{Bz} - v_{Gz}\omega_{By}) \int_B q_x\, dm \\ & + 2(v_{Gz}\omega_{Bx} - v_{Gx}\omega_{Bz}) \int_B q_y\, dm \\ & + 2(v_{Gx}\omega_{By} - v_{Gy}\omega_{Bx}) \int_B q_z\, dm \bigg]. \end{aligned} \quad (10.65)$$

The first integral defines the mass of the rigid body. The last three integrals are zero since they measure the position of the mass center relative to the mass center. From Eqs. (A.1)–(A.3) in Appendix A on pp. A-1–A-2, we see that the second, third, and fourth integrals define the mass moments of inertia of the rigid body with respect to the xyz axes whose origin is at the mass center of the body, that is, they define I_{Gx}, I_{Gy}, and I_{Gz}, respectively. From Eqs. (A.4)–(A.6) on p. A-2 of Appendix A, the fifth, sixth, and seventh integrals define the mass products of inertia of the rigid body with respect to the indicated axes whose origin is at the mass center of the body, that is, they define I_{Gxy}, I_{Gxz}, and I_{Gyz}, respectively. Substituting these definitions for the mass properties into Eq. (10.65), we obtain the final form of the kinetic energy of a rigid body for three-dimensional motion,

$$\begin{aligned} T = \tfrac{1}{2}mv_G^2 + \tfrac{1}{2}I_{Gx}\omega_{Bx}^2 + \tfrac{1}{2}I_{Gy}\omega_{By}^2 + \tfrac{1}{2}I_{Gz}\omega_{Bz}^2 \\ - I_{Gxy}\omega_{Bx}\omega_{By} - I_{Gxz}\omega_{Bx}\omega_{Bz} - I_{Gyz}\omega_{By}\omega_{Bz}. \end{aligned} \quad (10.66)$$

End of Section Summary

Newton-Euler equations in 3D. There are six Newton-Euler equations for a rigid body in three-dimensional motion. Three are given by the component form of Euler's first law, which is

Eq. (10.20), p. 738

$$\vec{F} = m\vec{a}_G,$$

where $\vec{F}$ is the total external force acting on the rigid body, m is the mass of the rigid body, and $\vec{a}_G$ is the acceleration of the mass center of the rigid body (see Fig. 10.13).

The other three equations are given by Euler's second law. We derived three different versions of these equations, each successive version with more restrictions on their application. The last two versions are the most useful in practice, the first of which is given by

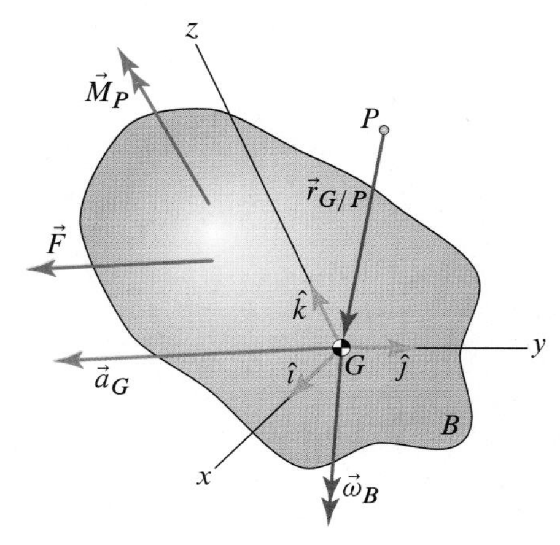

Figure 10.13
Illustration of the kinetic and kinematic quantities needed to understand and apply Eqs. (10.20) and (10.31)–(10.39).

Eqs. (10.34)–(10.36), p. 740

$$M_{Px} = I_{Gx}\dot{\omega}_{Bx} + (I_{Gz} - I_{Gy})\,\omega_{Bz}\omega_{By} + m(y_{G/P}a_{Gz} - z_{G/P}a_{Gy}),$$
$$M_{Py} = I_{Gy}\dot{\omega}_{By} + (I_{Gx} - I_{Gz})\,\omega_{Bx}\omega_{Bz} + m(z_{G/P}a_{Gx} - x_{G/P}a_{Gz}),$$
$$M_{Pz} = I_{Gz}\dot{\omega}_{Bz} + (I_{Gy} - I_{Gx})\,\omega_{By}\omega_{Bx} + m(x_{G/P}a_{Gy} - y_{G/P}a_{Gx}),$$

where (see Fig. 10.13):

- Point P is arbitrary, the xyz reference frame is attached to the body B, and xyz are principal axes of inertia so that all products of inertia are zero.
- The moments of inertia are expressed relative to a set of axes that are parallel to xyz and whose origin is at G, the center of mass of the body.
- The angular velocities and their time derivatives are inertial, but are expressed in the rotating xyz frame.

The final set of equations, also known as *Euler's equations of rotational motion for a rigid body*, are given by

Eqs. (10.37)–(10.39), p. 740

$$M_{Px} = I_{Gx}\dot{\omega}_{Bx} + (I_{Gz} - I_{Gy})\,\omega_{Bz}\omega_{By},$$
$$M_{Py} = I_{Gy}\dot{\omega}_{By} + (I_{Gx} - I_{Gz})\,\omega_{Bx}\omega_{Bz},$$
$$M_{Pz} = I_{Gz}\dot{\omega}_{Bz} + (I_{Gy} - I_{Gx})\,\omega_{By}\omega_{Bx},$$

where, in addition to the conditions for Eqs. (10.34)–(10.36), we also have that:

- $\vec{r}_{G/P} \times m\vec{a}_G = \vec{0}$, which can happen in any of the following three ways:
  - $\Rightarrow$ P is the mass center G of the body B so that $\vec{r}_{G/P} = \vec{0}$,
  - $\Rightarrow$ $\vec{a}_G = \vec{0}$, that is, the mass center of the body moves with constant velocity,
  - $\Rightarrow$ $\vec{r}_{G/P}$ is parallel to $\vec{a}_G$ so that the cross product $\vec{r}_{G/P} \times m\vec{a}_G = \vec{0}$.

Rotational equations of motion for a body with an axis of radial symmetry. For a rigid body with an axis of radial symmetry, any coordinate system that has an axis coinciding with the axis of symmetry of that body will be a principal coordinate system. Referring to Fig. 10.14, $x_B y_B z_B$ are principal body axes with the x_B axis coinciding with the axis of radial symmetry of the body B. The xyz frame rotates with angular velocity $\vec{\Omega}$, the angular velocity of the body is $\vec{\omega}_B$, and the body spins relative to the xyz frame about its axis of symmetry with angular velocity $\vec{\omega}_s$. With these definitions, the rotational equations of motion of the body are

Eqs. (10.48)–(10.50), p. 742

$$M_{Px} = I_{Gx}(\dot{\Omega}_x + \dot{\omega}_s),$$
$$M_{Py} = I_{Gy}\dot{\Omega}_y + I_{Gx}(\Omega_x + \omega_s)\Omega_z - I_{Gy}\Omega_x\Omega_z - m x_{G/P}a_{Gz},$$
$$M_{Pz} = I_{Gy}\dot{\Omega}_z - I_{Gx}(\Omega_x + \omega_s)\Omega_y + I_{Gy}\Omega_x\Omega_y + m x_{G/P}a_{Gy}.$$

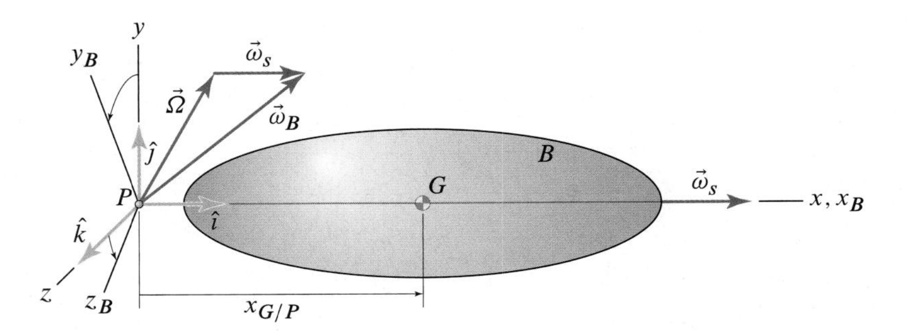

Figure 10.14. A rigid body B spinning about its axis of radial symmetry x_B relative to the xyz frame with angular velocity $\vec{\omega}_s$. The angular velocity of the xyz frame is $\vec{\Omega}$, relative to an underlying inertial frame.

If $\vec{r}_{G/P} \times m\vec{a}_G = \vec{0}$, then these equations become

Eqs. (10.51)–(10.53), p. 742

$$M_{Px} = I_{Gx}(\dot{\Omega}_x + \dot{\omega}_s),$$
$$M_{Py} = I_{Gy}\dot{\Omega}_y + I_{Gx}(\Omega_x + \omega_s)\Omega_z - I_{Gy}\Omega_x\Omega_z,$$
$$M_{Pz} = I_{Gy}\dot{\Omega}_z - I_{Gx}(\Omega_x + \omega_s)\Omega_y + I_{Gy}\Omega_x\Omega_y.$$

Planar motion of bodies *not* symmetric with respect to the plane of motion. In Chapter 7, we emphasized that the Newton-Euler rotational equations derived there were for the planar motion of bodies symmetric with respect to the plane of motion. If the rigid body is undergoing planar motion, but the body is not symmetric with respect to the plane of motion, the governing rotational equations are:

Eqs. (10.54)–(10.56), p. 743

$$M_{Px} = -I_{Gxz}\dot{\omega}_{Bz} + I_{Gyz}\omega_{Bz}^2,$$
$$M_{Py} = -I_{Gyz}\dot{\omega}_{Bz} - I_{Gxz}\omega_{Bz}^2,$$
$$M_{Pz} = I_{Gz}\dot{\omega}_{Bz} + m\big(x_{G/P}a_{Gy} - y_{G/P}a_{Gx}\big),$$

where the plane of motion is the xy plane and the moments and products of inertia are measured with respect to a set of axes whose origin is at the mass center G.

Kinetic energy of a rigid body in 3D. The work-energy principle for a rigid body is the same whether its motion is two-dimensional or three-dimensional. Therefore, the work-energy principle, as given by any of Eqs. (8.16) on p. 585, (8.23) on p. 586, (8.24) on p. 586, or (8.29) on p. 588, is valid for the three-dimensional motion of this chapter. The expression for the kinetic energy of a rigid body in three-dimensional motion is:

Eq. (10.66), p. 744

$$T = \tfrac{1}{2}mv_G^2 + \tfrac{1}{2}I_{Gx}\omega_{Bx}^2 + \tfrac{1}{2}I_{Gy}\omega_{By}^2 + \tfrac{1}{2}I_{Gz}\omega_{Bz}^2$$
$$- I_{Gxy}\omega_{Bx}\omega_{By} - I_{Gxz}\omega_{Bx}\omega_{Bz} - I_{Gyz}\omega_{By}\omega_{Bz},$$

where the moments and products of inertia are measured with respect to axes whose origin is at the mass center of the rigid body.

EXAMPLE 10.4 *Gyroscopic Moment on a Spinning Disk*

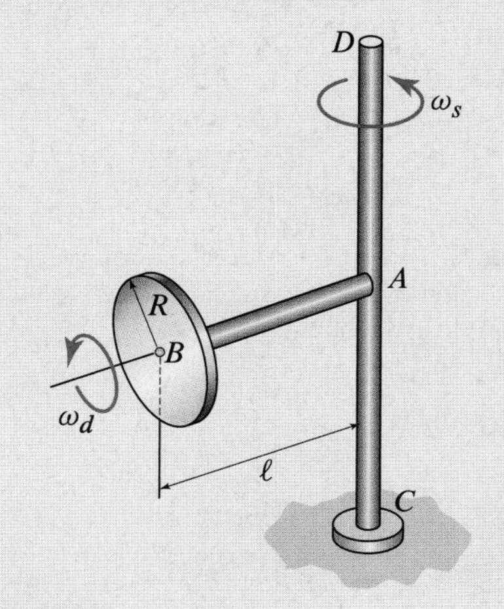

Figure 1

Referring to Fig. 1, the T-bar support spins about the line CD with constant angular speed ω_s. The disk at B spins with constant angular speed ω_d relative to the T-bar. If the mass of the disk is m_d and the mass of the T-bar is negligible, determine the reactions at the support A required for this motion.

SOLUTION

Road Map & Modeling The FBD of the disk and the arm AB is shown in Fig. 2. If we attach a rotating xyz frame to the T-bar as shown in Fig. 2, then the disk will spin about an axis of symmetry relative to that frame. Therefore, to find reactions at A, we can apply Eqs. (10.20) and (10.51)–(10.53) for the force and moment equations. By taking A as the moment center, we will obtain the reactions at A.

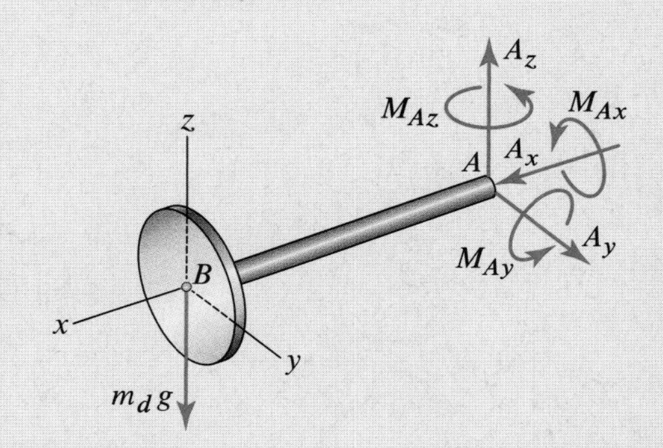

Figure 2
The FBD of the disk and bar AB for the system in Fig. 1. The mass of the bar AB is negligible.

Governing Equations

Balance Principles Applying Eqs. (10.20) and (10.51)–(10.53) to the FBD in Fig. 2, we obtain

$$\sum F_x: \quad A_x = m_d a_{Bx}, \tag{1}$$

$$\sum F_y: \quad A_y = m_d a_{By}, \tag{2}$$

$$\sum F_z: \quad A_z - m_d g = m_d a_{Bz}, \tag{3}$$

$$\sum M_{Ax}: \quad M_{Ax} = I_{Bx}(\dot{\Omega}_x + \dot{\omega}_d), \tag{4}$$

$$\sum M_{Ay}: \quad M_{Ay} - m_d g\ell = I_{By}\dot{\Omega}_y + I_{Bx}(\Omega_x + \omega_d)\Omega_z - I_{By}\Omega_x\Omega_z, \tag{5}$$

$$\sum M_{Az}: \quad M_{Az} = I_{By}\dot{\Omega}_z - I_{Bx}(\Omega_x + \omega_d)\Omega_y + I_{By}\Omega_x\Omega_y. \tag{6}$$

Force Laws All forces are accounted for on the FBD.

Kinematic Equations Determining the acceleration of B, we obtain

$$\vec{a}_B = \vec{\alpha}_{AB} \times \vec{r}_{B/A} - \omega_{AB}^2 \vec{r}_{B/A} = -\omega_s^2 \ell\,\hat{\imath}, \tag{7}$$

where we have used the fact that $\vec{\alpha}_{AB} = \vec{0}$. For the angular velocity and angular acceleration components, noting that $\vec{\Omega}$ is the angular velocity of the rotating xyz frame, we have

$$\Omega_x = \Omega_y = 0, \quad \Omega_z = \omega_s, \quad \dot{\Omega}_x = \dot{\Omega}_y = \dot{\Omega}_z = 0, \quad \dot{\omega}_d = 0. \tag{8}$$

Computation Substituting Eqs. (7) and (8) into the right sides of Eqs. (1)–(6), we obtain

$$A_x = -m_d \ell \omega_s^2, \qquad A_y = 0, \qquad A_z = m_d g, \tag{9}$$

$$M_{Ax} = 0, \qquad M_{Ay} = m_d\left(\tfrac{1}{2}R^2\omega_d\omega_s - g\ell\right), \qquad M_{Az} = 0, \tag{10}$$

where we have used $I_{Bx} = \frac{1}{2} m_d R^2$.

Discussion & Verification The dimensions of each result in Eqs. (9) and (10) are as they should be. The force reactions given in Eq. (9) are all as expected, that is, we should expect a reaction opposing the weight of the disk, and we should expect a tension in the bar due to the normal acceleration resulting from the circular path of the mass center of the disk B centered at A. We also see that no moments are required in the x and z directions to maintain the given motion. Part of the moment in the y direction is due to the weight of the disk, but the other part is not intuitive since it is a gyroscopic moment due to the fact that the rotation of the support ω_s is causing the rotation of the disk ω_d to change direction.

EXAMPLE 10.5 *Forces and Moments at the Base of a Rotating Propeller*

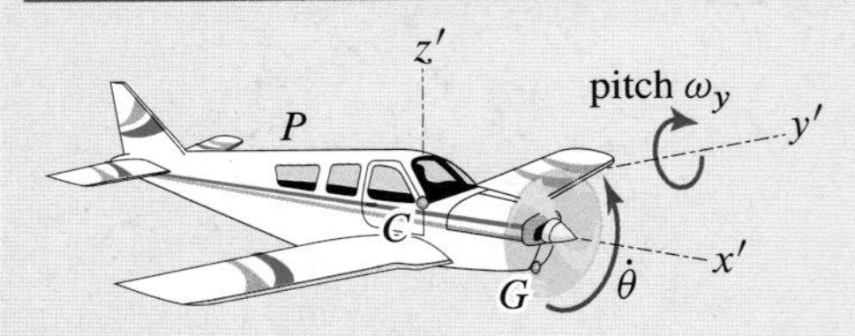

Figure 1
The secondary or moving $x'y'z'$ reference is frame attached to an airplane at C.

The airplane shown in Fig. 1 is pitching about the body-fixed y' axis. The acceleration of point C on the plane is $\vec{a}_C = a_{Cx'}\,\hat{\imath}' + a_{Cz'}\,\hat{k}'$, and the constant angular speed of the propeller $\dot{\theta}$ is measured relative to the airplane. Referring to Fig. 2, consider the propeller blade with center of mass at G. If m is the mass of the propeller blade, determine the forces and moments applied to its base B required for the blade to undergo the prescribed motion. Assume that the blade is a uniform thin rod of length L. The $x'y'z'$ frame is *attached to the airplane* at C with the x' axis aligned with the axis of rotation of the propeller. Ignore gravity.

SOLUTION

Road Map & Modeling The xyz frame is attached to the propeller blade and rotates relative to the $x'y'z'$ frame as shown in Fig. 2. The FBD of the propeller blade is as shown in Fig. 3, where B_x, B_y, B_z are the resultant forces acting on the blade and M_{Bx}, M_{By}, M_{Bz} are the resultant moments measured relative to point B. Since the motion of the propeller is specified, we can write the six Newton-Euler equations to solve for the three forces and three moments applied at B necessary to sustain the given motion.

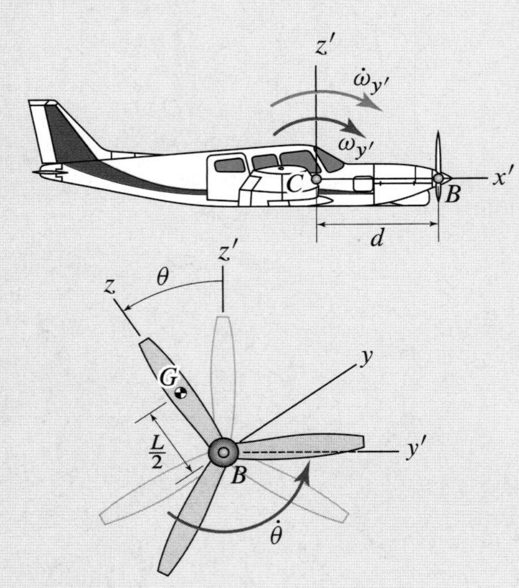

Figure 2
A side view of the airplane and a front view of the propeller showing the two moving reference frames $x'y'z'$ and xyz, as well as the relevant dimensions. The xyz frame is *attached to the propeller* with the z axis aligned with the axis of the blade in question and with the x and x' axes always parallel to one another.

Governing Equations

Balance Principles Summing forces on the propeller blade, we obtain

$$B_x = ma_{Gx}, \quad B_y = ma_{Gy}, \quad \text{and} \quad B_z = ma_{Gz}. \tag{1}$$

Since the xyz frame is attached to the propeller blade and aligned with its principle axes, we can apply either Eqs. (10.34)–(10.36) or Eqs. (10.37)–(10.39). If we sum moments about G, we can apply Eqs. (10.37)–(10.39), but then we will have to include the moments due to B_x and B_y. If we sum moments about B, then we need to apply Eqs. (10.34)–(10.36) because the moment center is accelerating. Since we will have to compute $\vec{a}_G$ as part of the solution anyway, let's use Eqs. (10.34)–(10.36), which give

$$\sum M_{Bx}\colon\ M_{Bx} = I_{Gx}\dot{\omega}_{px} + (I_{Gz} - I_{Gy})\omega_{pz}\omega_{py} + m(y_G a_{Gz} - z_G a_{Gy}), \tag{2}$$

$$\sum M_{By}\colon\ M_{By} = I_{Gy}\dot{\omega}_{py} + (I_{Gx} - I_{Gz})\omega_{px}\omega_{pz} + m(z_G a_{Gx} - x_G a_{Gz}), \tag{3}$$

$$\sum M_{Bz}\colon\ M_{Bz} = I_{Gz}\dot{\omega}_{pz} + (I_{Gy} - I_{Gx})\omega_{py}\omega_{px} + m(x_G a_{Gy} - y_G a_{Gx}), \tag{4}$$

where the mass moments of inertia $I_{Gx} = I_{Gy} = \frac{1}{12}mL^2$ and $I_{Gz} = 0$ are computed relative to the mass center, $\vec{r}_{G/B} = x_G\,\hat{\imath} + y_G\,\hat{\jmath} + z_G\,\hat{k}$, and where $\vec{\omega}_p$ and $\dot{\vec{\omega}}_p$ are the angular velocity and angular acceleration of the propeller, respectively.

Force Laws All forces are accounted for on the FBD.

Kinematic Equations We need to determine the acceleration of G, the angular velocity and angular acceleration of the propeller, $\vec{\omega}_p$ and $\dot{\vec{\omega}}_p$, respectively, and the vector $\vec{r}_{G/B}$. Beginning with the acceleration of G, we first write

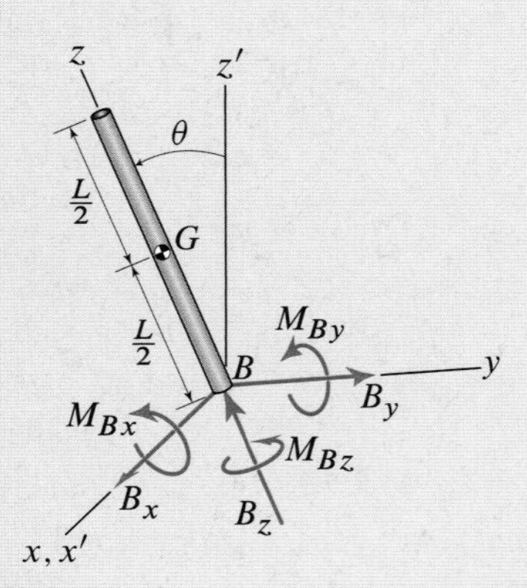

Figure 3
FBD of one of the propeller blades modeled as a uniform thin rod when it is at an arbitrary angle θ with respect to the z' axis.

$$\begin{aligned}\vec{a}_B &= \vec{a}_C + \vec{\alpha}_{\text{plane}} \times \vec{r}_{B/C} + \vec{\omega}_{\text{plane}} \times (\vec{\omega}_{\text{plane}} \times \vec{r}_{B/C})\\ &= a_{Cx'}\,\hat{\imath}' + a_{Cz'}\,\hat{k}' + \dot{\omega}_{y'}\,\hat{\jmath}' \times d\,\hat{\imath}' + \omega_{y'}\,\hat{\jmath}' \times (\omega_{y'}\,\hat{\jmath}' \times d\,\hat{\imath}')\\ &= (a_{Cx'} - d\omega_{y'}^2)\,\hat{\imath} + (a_{Cz'} - d\dot{\omega}_{y'})\,\hat{k}',\end{aligned} \tag{5}$$

where $\vec{\alpha}_{\text{plane}} = \dot{\omega}_{y'}\,\hat{\jmath}'$, the acceleration of C was given as $a_{Cx'}\,\hat{\imath}' + a_{Cz'}\,\hat{k}'$, and we have used the fact that $\hat{\imath} = \hat{\imath}'$ since the x and x' axes coincide. We can now relate the acceleration of G to that of B using

$$\vec{a}_G = \vec{a}_B + \dot{\vec{\omega}}_p \times \vec{r}_{G/B} + \vec{\omega}_p \times (\vec{\omega}_p \times \vec{r}_{G/B}), \tag{6}$$

where $\vec{a}_B$ is given by Eq. (5) and the position of G relative to B is

$$\vec{r}_{G/B} = x_G\,\hat{\imath} + y_G\,\hat{\jmath} + z_G\,\hat{k} = \tfrac{L}{2}\,\hat{k}. \tag{7}$$

To find $\vec{\omega}_p$ and $\dot{\vec{\omega}}_p$, we note that the angular velocity of the propeller is equal to the sum of the angular velocity of the plane and the angular velocity of the propeller relative to the plane. Therefore,

$$\begin{aligned}\vec{\omega}_p &= \vec{\omega}_{\text{plane}} + \vec{\omega}_{p/\text{plane}} = \omega_{y'}\,\hat{\jmath}' + \dot{\theta}\,\hat{\imath} = \omega_{y'}(\cos\theta\,\hat{\jmath} - \sin\theta\,\hat{k}) + \dot{\theta}\,\hat{\imath}\\ &= \dot{\theta}\,\hat{\imath} + \omega_{y'}\cos\theta\,\hat{\jmath} - \omega_{y'}\sin\theta\,\hat{k},\end{aligned} \tag{8}$$

where we have used $\hat{\jmath}' = \cos\theta\,\hat{\jmath} - \sin\theta\,\hat{k}$. To find $\dot{\vec{\omega}}_p$, we differentiate $\vec{\omega}_p$ to obtain

$$\begin{aligned}\dot{\vec{\omega}}_p &= (\dot{\omega}_{y'}\cos\theta - \omega_{y'}\dot{\theta}\sin\theta)\,\hat{\jmath} - (\dot{\omega}_{y'}\sin\theta + \omega_{y'}\dot{\theta}\cos\theta)\,\hat{k}\\ &\qquad + \dot{\theta}(\vec{\omega}_p \times \hat{\imath}) + \omega_{y'}\cos\theta(\vec{\omega}_p \times \hat{\jmath}) - \omega_{y'}\sin\theta(\vec{\omega}_p \times \hat{k}) && (9)\\ &= (\dot{\omega}_{y'}\cos\theta - \omega_{y'}\dot{\theta}\sin\theta)\,\hat{\jmath} - (\dot{\omega}_{y'}\sin\theta + \omega_{y'}\dot{\theta}\cos\theta)\,\hat{k}\\ &\qquad + (\vec{\omega}_p \times \dot{\theta}\,\hat{\imath}) + (\vec{\omega}_p \times \omega_{y'}\cos\theta\,\hat{\jmath}) - (\vec{\omega}_p \times \omega_{y'}\sin\theta\,\hat{k}) && (10)\\ &= (\dot{\omega}_{y'}\cos\theta - \omega_{y'}\dot{\theta}\sin\theta)\,\hat{\jmath} - (\dot{\omega}_{y'}\sin\theta + \omega_{y'}\dot{\theta}\cos\theta)\,\hat{k}\\ &\qquad + \underbrace{\vec{\omega}_p \times \underbrace{(\dot{\theta}\,\hat{\imath} + \omega_{y'}\cos\theta\,\hat{\jmath} - \omega_{y'}\sin\theta\,\hat{k})}_{\vec{\omega}_p}}_{\vec{\omega}_p\times\vec{\omega}_p=\vec{0}} && (11)\\ &= (\dot{\omega}_{y'}\cos\theta - \omega_{y'}\dot{\theta}\sin\theta)\,\hat{\jmath} - (\dot{\omega}_{y'}\sin\theta + \omega_{y'}\dot{\theta}\cos\theta)\,\hat{k}, && (12)\end{aligned}$$

where we have used the fact that $\dot{\theta}$ is constant and that the angular velocity of the xyz frame is $\vec{\omega}_p$. Substituting Eqs. (5), (7), (8), and (12) into Eq. (6), we obtain

$$\begin{aligned}\vec{a}_G = \Big(a_{Cx'} - d\omega_{y'}^2 + \tfrac{L}{2}\dot{\omega}_{y'}\cos\theta - L\dot{\theta}\omega_{y'}\sin\theta\Big)\,\hat{\imath}&\\ + \Big(a_{Cz'} - d\dot{\omega}_{y'} - \tfrac{L}{2}\omega_{y'}^2\cos\theta\Big)\sin\theta\,\hat{\jmath}&\\ + \Big[(a_{Cz'} - d\dot{\omega}_{y'})\cos\theta - \tfrac{L}{2}\dot{\theta}^2 - \tfrac{L}{2}\omega_{y'}^2\cos^2\theta\Big]\,\hat{k}.&\end{aligned} \tag{13}$$

Computation Substituting Eq. (13) into Eq. (1) gives the forces at B, which are

$$B_x = m\Big(a_{Cx'} - d\omega_{y'}^2 + \tfrac{L}{2}\dot{\omega}_{y'}\cos\theta - L\dot{\theta}\omega_{y'}\sin\theta\Big), \tag{14}$$

$$B_y = m\Big(a_{Cz'} - d\dot{\omega}_{y'} - \tfrac{L}{2}\omega_{y'}^2\cos\theta\Big)\sin\theta, \tag{15}$$

$$B_z = m\Big[(a_{Cz'} - d\dot{\omega}_{y'})\cos\theta - \tfrac{L}{2}\dot{\theta}^2 - \tfrac{L}{2}\omega_{y'}^2\cos^2\theta\Big]. \tag{16}$$

Substituting Eqs. (8), (12), and (13) into Eqs. (2)–(4), we obtain the following three expressions for the moments at the base of the propeller

$$M_{Bx} = \frac{mL}{6}\Big(-3a_{Cz'} + 3d\dot{\omega}_{y'} + 2L\omega_{y'}^2\cos\theta\Big)\sin\theta, \tag{17}$$

$$M_{By} = \frac{mL}{6}\Big(3a_{Cx'} - 3d\omega_{y'}^2 + 2L\dot{\omega}_{y'}\cos\theta - 4L\dot{\theta}\omega_{y'}\sin\theta\Big), \tag{18}$$

$$M_{Bz} = 0. \tag{19}$$

Discussion & Verification The dimensions of each term in Eqs. (14)–(19) are as they should be. In addition, we would not expect that a moment about the z axis on the propeller would be required, and indeed that moment is zero.

EXAMPLE 10.6 Planar Motion of a Rigid Body That is Not Symmetric With Respect to the Plane of Motion

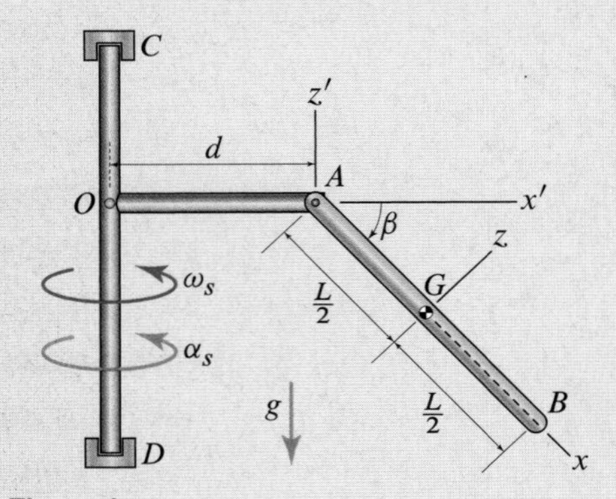

Figure 1
A rigid body undergoing planar motion in which the rigid body is *not* symmetric with respect to the plane of motion.

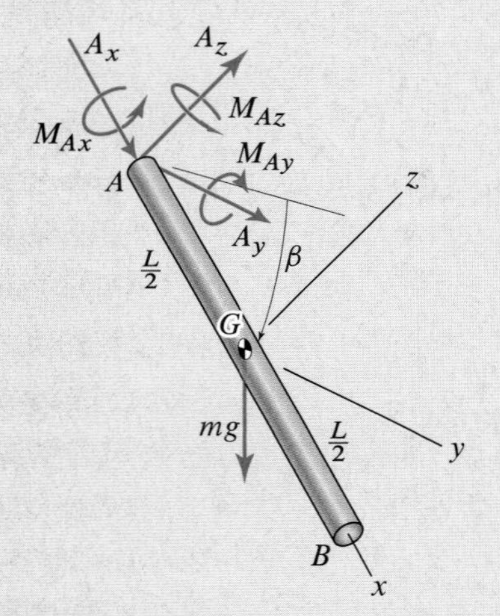

Figure 2
FBD of the bar AB in Fig. 1.

The bar AB is *rigidly* attached to the T-bar support (i.e., the angle β is constant), which is, in turn, spinning about the vertical axis with angular velocity $\vec{\omega}_s$ and angular acceleration $\vec{\alpha}_s$ as shown. The xyz frame has its origin at G, and it is attached to the arm AB. Assuming that the arm AB lies in the vertical plane, determine the forces and moments exerted by the T-bar onto the bar AB at A required for this motion, and express them in the $x'y'z'$ frame that is attached to the T-bar with its origin at A.

SOLUTION

Road Map & Modeling An FBD of the arm AB is shown in Fig. 2. The motion of the bar AB is specified, so we can apply the six Newton-Euler equations to the bar to determine the six unknown forces and moments at A required to sustain this motion.

Governing Equations

Balance Principles Summing forces in the three coordinate directions in Fig. 2, we obtain

$$\sum F_x: \quad A_x + mg\sin\beta = ma_{Gx}, \tag{1}$$

$$\sum F_y: \quad A_y = ma_{Gy}, \tag{2}$$

$$\sum F_z: \quad A_z - mg\cos\beta = ma_{Gz}, \tag{3}$$

where a_{Gx}, a_{Gy}, and a_{Gz} are the components of $\vec{a}_G$ in the indicated coordinate directions. Applying Eqs. (10.34)–(10.36), we can sum moments about A to obtain

$$\sum M_{Ax}: \quad M_{Ax} = I_{Gx}\dot{\omega}_{ABx} + (I_{Gz} - I_{Gy})\,\omega_{ABz}\omega_{ABy} + m(y_{G/A}a_{Gz} - z_{G/A}a_{Gy}), \tag{4}$$

$$\sum M_{Ay}: \quad M_{Ay} + mg\tfrac{L}{2}\cos\beta = I_{Gy}\dot{\omega}_{ABy} + (I_{Gx} - I_{Gz})\,\omega_{ABx}\omega_{ABz} + m(z_{G/A}a_{Gx} - x_{G/A}a_{Gz}), \tag{5}$$

$$\sum M_{Az}: \quad M_{Az} = I_{Gz}\dot{\omega}_{ABz} + (I_{Gy} - I_{Gx})\,\omega_{ABy}\omega_{ABx} + m(x_{G/A}a_{Gy} - y_{G/A}a_{Gx}), \tag{6}$$

where all moments of inertia are computed with respect to the mass center G, $\vec{\omega}_{AB}$ is the angular velocity of the bar AB, and $\dot{\vec{\omega}}_{AB}$ is the angular acceleration of bar AB. The mass moments of inertia are given by

$$I_{Gx} = 0 \quad \text{and} \quad I_{Gy} = I_{Gz} = \tfrac{1}{12}mL^2. \tag{7}$$

Force Laws All forces are accounted for on the FBD.

Kinematic Equations The components of the position of G relative to A are

$$x_{G/A} = L/2 \quad \text{and} \quad y_{G/A} = z_{G/A} = 0. \tag{8}$$

Since G is moving in a circle centered on the vertical line CD, its acceleration is readily found to be

$$\vec{a}_G = -(d + \tfrac{L}{2}\cos\beta)\omega_s^2\,\hat{\imath}' + (d + \tfrac{L}{2}\cos\beta)\alpha_s\,\hat{\jmath}' \tag{9}$$

$$= -(d + \tfrac{L}{2}\cos\beta)\omega_s^2(\cos\beta\,\hat{\imath} + \sin\beta\,\hat{k}) + (d + \tfrac{L}{2}\cos\beta)\alpha_s\,\hat{\jmath} \tag{10}$$

$$= -(d + \tfrac{L}{2}\cos\beta)(\omega_s^2\cos\beta\,\hat{\imath} - \alpha_s\,\hat{\jmath} + \omega_s^2\sin\beta\,\hat{k}). \tag{11}$$

The angular velocity of the bar is given to be

$$\vec{\omega}_{AB} = \omega_s\,\hat{k}' = \omega_s(-\sin\beta\,\hat{\imath} + \cos\beta\,\hat{k}), \tag{12}$$

and the angular acceleration of the bar is

$$\dot{\vec{\omega}}_{AB} = \alpha_s\,\hat{k}' = \alpha_s(-\sin\beta\,\hat{\imath} + \cos\beta\,\hat{k}). \tag{13}$$

Computation Substituting Eq. (11) into Eqs. (1)–(3) and solving for the forces at A, we obtain

$$A_x = -mg\sin\beta - m\left(d + \tfrac{L}{2}\cos\beta\right)\omega_s^2\cos\beta, \tag{14}$$

$$A_y = m\left(d + \tfrac{L}{2}\cos\beta\right)\alpha_s, \tag{15}$$

$$A_z = mg\cos\beta - m\left(d + \tfrac{L}{2}\cos\beta\right)\omega_s^2\sin\beta. \tag{16}$$

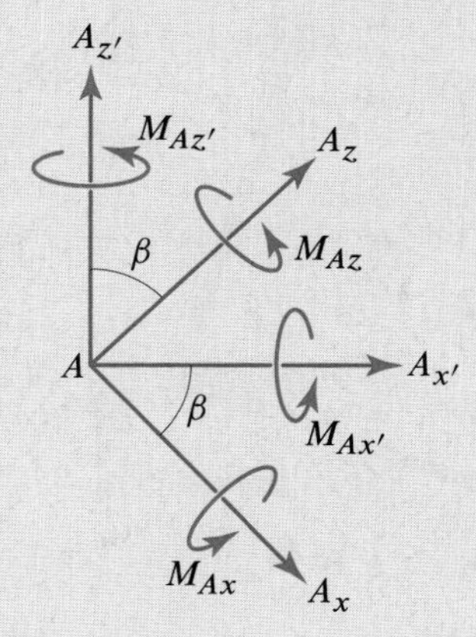

Figure 3

Referring to Fig. 3, these forces can be written in terms of the $x'y'z'$ directions as

$$A_{x'} = A_x\cos\beta + A_z\sin\beta \tag{17}$$

$$= \left[-mg\sin\beta - m\left(d + \tfrac{L}{2}\cos\beta\right)\omega_s^2\cos\beta\right]\cos\beta$$

$$+ \left[mg\cos\beta - m\left(d + \tfrac{L}{2}\cos\beta\right)\omega_s^2\sin\beta\right]\sin\beta \tag{18}$$

$$= -m\left(d + \tfrac{L}{2}\cos\beta\right)\omega_s^2 \quad\Rightarrow\quad \boxed{A_{x'} = -m\left(d + \tfrac{L}{2}\cos\beta\right)\omega_s^2,} \tag{19}$$

$$A_{y'} = A_y = m\left(d + \tfrac{L}{2}\cos\beta\right)\alpha_s \quad\Rightarrow\quad \boxed{A_{y'} = m\left(d + \tfrac{L}{2}\cos\beta\right)\alpha_s,} \tag{20}$$

$$A_{z'} = -A_x\sin\beta + A_z\cos\beta \tag{21}$$

$$= -\left[-mg\sin\beta - m\left(d + \tfrac{L}{2}\cos\beta\right)\omega_s^2\cos\beta\right]\sin\beta$$

$$+ \left[mg\cos\beta - m\left(d + \tfrac{L}{2}\cos\beta\right)\omega_s^2\sin\beta\right]\cos\beta \tag{22}$$

$$= mg \quad\Rightarrow\quad \boxed{A_{z'} = mg.} \tag{23}$$

Substituting Eqs. (7), (8), and (11)–(13) into Eqs. (4)–(6), and then solving for M_{Ax}, M_{Ay}, and M_{Az}, we obtain

$$M_{Ax} = 0, \tag{24}$$

$$M_{Ay} = \tfrac{1}{6}mL\left[-3g\cos\beta + \omega_s^2(3d + 2L\cos\beta)\sin\beta\right], \tag{25}$$

$$M_{Az} = \tfrac{1}{6}mL(3d + 2L\cos\beta)\,\alpha_s. \tag{26}$$

Substituting Eqs. (24)–(26) into the relationships relating the moments in the two different reference frames, $M_{Ax'} = M_{Ax}\cos\beta + M_{Az}\sin\beta$, $M_{Ay'} = M_{Ay}$, and $M_{Az'} = -M_{Az}\sin\beta + M_{Az}\cos\beta$, gives

$$\boxed{M_{Ax'} = \tfrac{1}{6}mL(3d + 2L\cos\beta)\,\alpha_s\sin\beta,} \tag{27}$$

$$\boxed{M_{Ay'} = \tfrac{1}{6}mL\left[-3g\cos\beta + \omega_s^2(3d + 2L\cos\beta)\sin\beta\right],} \tag{28}$$

$$\boxed{M_{Az'} = \tfrac{1}{6}mL(3d + 2L\cos\beta)\,\alpha_s\cos\beta.} \tag{29}$$

Discussion & Verification The final results for the forces and moments are all dimensionally as we expect them to be. In addition, given the motion of the bar AB, the directions of the forces are as expected. As for the moments, we would expect a negative moment in the y' direction due to the weight of the bar, and that is what we found. In addition, notice that there is a moment in the positive y' direction that is proportional to the square of the angular velocity. This moment results from the tendency of the bar AB to rotate upward when it is spinning. To keep the angle β constant, a moment needs to be applied in the positive y' direction.

EXAMPLE 10.7 *Computing the Kinetic Energy of a Bar Moving in 3D*

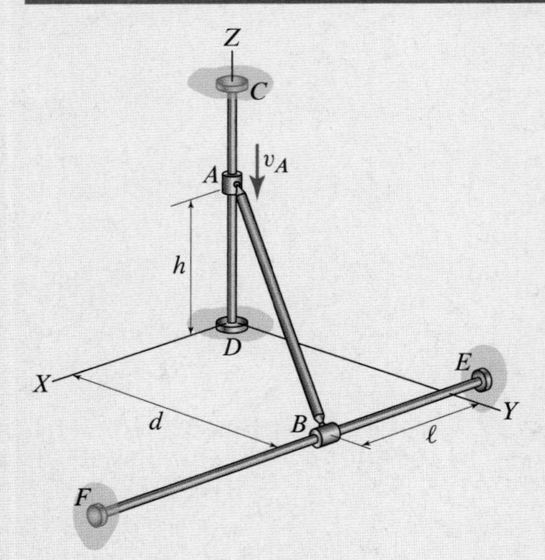

Figure 1

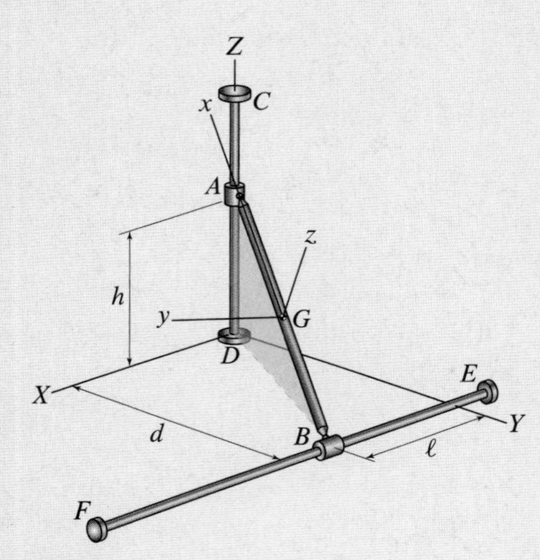

Figure 2
The system in Fig. 1 showing the body-fixed xyz frame and the plane (in light blue) to which the y axis is perpendicular.

At the instant shown, collar A is moving downward with speed v_A. The uniform thin bar AB has mass m and is attached to collars A and B with ball-and-socket joints. Neglecting the dimensions of the collars, determine the kinetic energy of bar AB at this instant. Assume that the angular velocity of the rod AB is orthogonal to the line AB.

SOLUTION

Road Map & Modeling The kinetic energy of the bar can be computed using Eq. (10.66). Since we would like to avoid computing products of inertia, we will attach a set of principal body axes to the bar AB as shown in Fig. 2, with the x axis aligned with the bar and the y axis perpendicular to the plane defined by the bar AB and the bar CD.

Governing Equations

Balance Principles Using the principal axes shown in Fig. 2, the kinetic energy of the bar is given by

$$T = \tfrac{1}{2}mv_G^2 + \tfrac{1}{2}I_{Gx}\omega_{ABx}^2 + \tfrac{1}{2}I_{Gy}\omega_{ABy}^2 + \tfrac{1}{2}I_{Gz}\omega_{ABz}^2, \tag{1}$$

where $I_{Gx} = 0$, $I_{Gy} = I_{Gz} = \frac{1}{12}mL^2$, and where L is the length of the bar AB, which is given by

$$L = \sqrt{d^2 + h^2 + \ell^2}. \tag{2}$$

Force Laws We are not applying the work-energy principle here, so there are no force laws.

Kinematic Equations Referring to Eq. (1), we need to determine the speed of the mass center of the bar AB, as well as the components of its angular velocity expressed in the xyz frame. Starting with the angular velocity, we can relate the velocity of B to that of A using

$$\vec{v}_B = \vec{v}_A + \vec{\omega}_{AB} \times \vec{r}_{B/A}, \tag{3}$$

$$v_B\,\hat{\imath} = -v_A\,\hat{K} + \big(\omega_{ABX}\,\hat{I} + \omega_{ABY}\,\hat{J} + \omega_{ABZ}\,\hat{K}\big) \times \big(\ell\,\hat{I} + d\,\hat{J} - h\,\hat{K}\big). \tag{4}$$

Expanding and equating coefficients, Eq. (4) becomes

$$v_B = -h\omega_{ABY} - d\omega_{ABZ}, \tag{5}$$

$$0 = h\omega_{ABX} + \ell\omega_{ABZ}, \tag{6}$$

$$0 = -v_A + d\omega_{ABX} - \ell\omega_{ABY}. \tag{7}$$

Implementing the constraint that the angular velocity of the rod AB be perpendicular to AB, we obtain

$$\vec{\omega}_{AB} \cdot \hat{u}_{B/A} = 0 \quad \Rightarrow \quad \big(\omega_{ABX}\,\hat{I} + \omega_{ABY}\,\hat{J} + \omega_{ABZ}\,\hat{K}\big) \cdot \frac{\vec{r}_{B/A}}{|\vec{r}_{B/A}|} = 0$$

$$\Rightarrow \quad \big(\omega_{ABX}\,\hat{I} + \omega_{ABY}\,\hat{J} + \omega_{ABZ}\,\hat{K}\big) \cdot \frac{\big(\ell\,\hat{I} + d\,\hat{J} - h\,\hat{K}\big)}{L} = 0, \tag{8}$$

which, upon expanding and multiplying through by L, becomes

$$\ell\omega_{ABX} + d\omega_{ABY} - h\omega_{ABZ} = 0. \tag{9}$$

Solving Eqs. (5)–(7) and Eq. (9) for ω_{ABX}, ω_{ABY}, ω_{ABZ}, and v_B, we obtain

$$\omega_{ABX} = \frac{d}{L^2}\,v_A, \quad \omega_{ABY} = -\frac{h^2 + \ell^2}{L^2\ell}\,v_A, \quad \omega_{ABZ} = -\frac{dh}{L^2\ell}\,v_A, \tag{10}$$

and $v_B = hv_A/\ell$. We now need to transform the components of the angular velocity vector from the XYZ frame to the xyz frame. Referring to Fig. 2, we see that

$$\hat{\imath} = -\hat{u}_{B/A} = \hat{u}_{A/B} = \frac{-1}{L}(\ell\,\hat{I} + d\,\hat{J} - h\,\hat{K}), \tag{11}$$

$$\hat{\jmath} = \hat{K} \times \hat{u}_{A/B} = \frac{d\,\hat{I} - \ell\,\hat{J}}{\sqrt{d^2+\ell^2}}, \tag{12}$$

$$\hat{k} = \hat{\imath} \times \hat{\jmath} = \frac{1}{L\sqrt{d^2+\ell^2}}\Big[h\ell\,\hat{I} + dh\,\hat{J} + (d^2+\ell^2)\,\hat{K}\Big]. \tag{13}$$

We can now find each of the three components of $\vec{\omega}_{AB}$ expressed in the xyz frame by dotting it with each of the three unit vectors in the xyz frame, that is,

$$\omega_{ABx} = \vec{\omega}_{AB}\cdot\hat{\imath} \tag{14}$$

$$= (\omega_{ABX}\,\hat{I} + \omega_{ABY}\,\hat{J} + \omega_{ABZ}\,\hat{K})\cdot\frac{-1}{L}(\ell\,\hat{I} + d\,\hat{J} - h\,\hat{K}) = 0, \tag{15}$$

$$\omega_{ABy} = \vec{\omega}_{AB}\cdot\hat{\jmath} \tag{16}$$

$$= (\omega_{ABX}\,\hat{I} + \omega_{ABY}\,\hat{J} + \omega_{ABZ}\,\hat{K})\cdot\frac{d\,\hat{I} - \ell\,\hat{J}}{\sqrt{d^2+\ell^2}} = \frac{v_A}{\sqrt{L^2-h^2}}, \tag{17}$$

$$\omega_{ABz} = \vec{\omega}_{AB}\cdot\hat{k} \tag{18}$$

$$= (\omega_{ABX}\,\hat{I} + \omega_{ABY}\,\hat{J} + \omega_{ABZ}\,\hat{K})\cdot\frac{1}{L\sqrt{d^2+\ell^2}}\Big[h\ell\,\hat{I} + dh\,\hat{J} + (d^2+\ell^2)\,\hat{K}\Big] \tag{19}$$

$$= -\frac{dhv_A}{L\ell\sqrt{d^2+\ell^2}}, \tag{20}$$

where we have substituted in Eq. (10) to get the final results. Now that we have the angular velocity components in the xyz frame, all that is left is to find v_G. This can now easily be done using

$$\vec{v}_G = \vec{v}_A + \vec{\omega}_{AB} \times \vec{r}_{G/A} \tag{21}$$

$$= -v_A\,\hat{K} + \frac{v_A}{L^2}\left[d\,\hat{I} - \frac{(h^2+\ell^2)}{\ell}\,\hat{J} - \frac{dh}{\ell}\,\hat{K}\right] \times \tfrac{1}{2}(\ell\,\hat{I} + d\,\hat{J} - h\,\hat{K}) \tag{22}$$

$$= \frac{h}{2\ell}\,v_A\,\hat{I} - \frac{1}{2}v_A\,\hat{K}, \tag{23}$$

and so

$$v_G^2 = \vec{v}_G\cdot\vec{v}_G = \frac{h^2+\ell^2}{4\ell^2}\,v_A^2. \tag{24}$$

Computation Substituting the moments of inertia, Eqs. (15), (17), (20), and (24) into Eq. (1) and simplifying, we obtain

$$\boxed{T = \frac{(h^2+\ell^2)}{6\ell^2}\,mv_A^2,}$$

where we have used the $L = \sqrt{d^2+h^2+\ell^2}$.

Discussion & Verification The dimensions of the kinetic energy are as expected. We see that the kinetic energy depends on the square of the only speed in the problem v_A, so that seems reasonable.

Common Pitfall

An easier way to compute T? The special condition we applied to the angular velocity of the bar is what led to the result that $\omega_{ABx} = 0$, as it should since x is aligned with the axis of the bar. This means that the angular velocity of the bar must be perpendicular to it at every instant, and so it is like the bar is in planar motion at every instant. This means that we *could* have computed the kinetic energy of the bar using the equation

$$T = \tfrac{1}{2}mv_G^2 + \tfrac{1}{2}I_G\omega_{AB}^2,$$

where I_G is the moment of inertia of the bar about any axis perpendicular to the bar through G and ω_{AB} is the angular speed of the bar. While that would have been much easier, it is a very special case, and it is better to learn how to compute T the way we did here so that we can do it in more general situations.

Problems

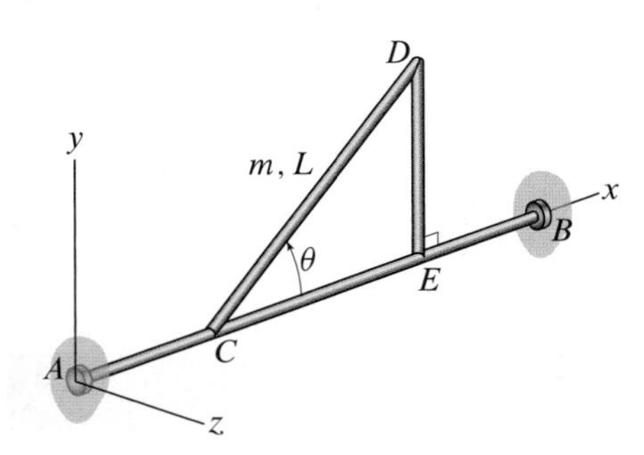

Figure P10.36 and P10.37

Problems 10.36 and 10.37

The angled bar CDE is rigidly attached to the horizontal shaft AB, which can rotate freely in the bearings at A and B. The system is released from rest when the segment DE is vertical. Segment CD has mass m and length L. Segments CD and DE have the same linear density.

Problem 10.36 Determine expressions for the angular velocity of the system when it has rotated 180°.

Problem 10.37 Determine expressions for the reactions at the bearings A and B when the system has rotated 90°.

Problem 10.38

The system is at rest when a time-dependent moment $M(t) = 3t^{3/2}$ N·m is applied to the shaft AB starting at $t = 0$. If the mass of the plate is $m = 10$ kg, its width $w = 0.5$ m, and its height $h = 0.25$ m, determine the angular speed of the system after 10 s. Neglect the mass of the horizontal shaft AB.

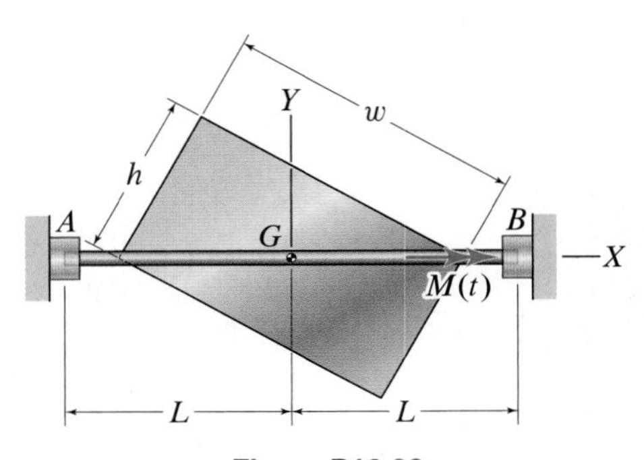

Figure P10.38

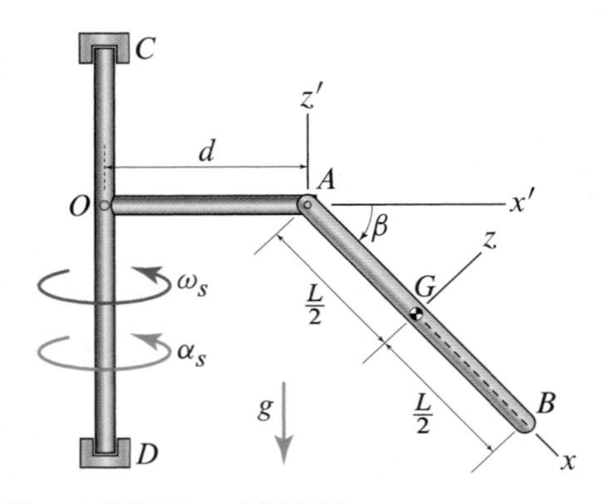

Figure P10.39 and P10.40

Problems 10.39 and 10.40

The uniform bar AB of length L and mass m is attached to the T-bar support by a frictionless pin at A. The $x'y'z'$ frame is attached to the T-bar at A and is aligned as shown. The support rotates with angular speed ω_s, and the angle between the bar AB and the x' axis is β.

Problem 10.39 Find the equation of motion of the bar AB in terms of the angle β for the case where ω_s is constant.

Problem 10.40 Find the equation of motion of the bar AB in terms of the angle β for the case where ω_s is not constant, that is, $\dot{\omega}_s = \alpha_s$.

Problem 10.41

The uniform bar AB of length L and mass m is attached to the T-bar support by a frictionless pin at A. The mass of the T-bar is negligible, and it rotates freely in the bearings at C and D. The $x'y'z'$ frame is attached to the T-bar at A and is aligned as shown. The angle between the bar AB and the x' axis is β, and the angle measuring the orientation of the T-bar is ϕ. Determine the equations of motion of the system in terms of the angles β and ϕ.

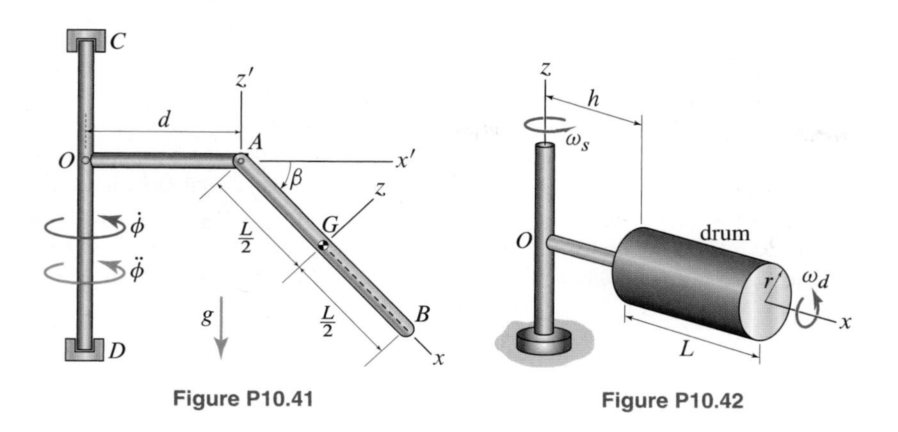

Figure P10.41

Figure P10.42

Problem 10.42

The uniform drum of length L, radius r, and mass m is spinning about its axis of symmetry with constant angular speed ω_d relative to the T-bar support. The T-bar support is rotating with constant angular speed ω_s about the vertical z axis. Neglecting the mass of the horizontal support of length h, determine the forces and moments at O acting on the horizontal bar required to sustain this motion.

Problem 10.43

The uniform bar AB of length L and mass m is attached to the T-bar support by a pin at A. The $x'y'z'$ frame is attached to the T-bar at A and is aligned as shown. The support rotates with angular speed ω_s, and the angle between the bar AB and the x' axis is a known function of time $\beta(t)$. Determine the kinetic energy of the bar AB.

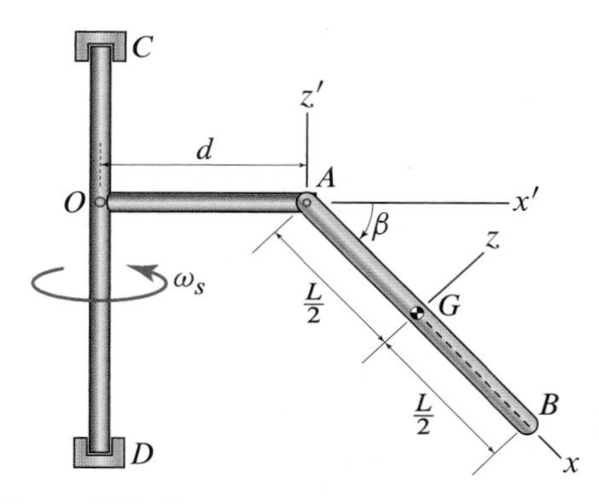

Figure P10.43

Problems 10.44 through 10.47

When $h_1 = 0.6\,\text{m}$, $\ell_1 = 0.5\,\text{m}$, and $d = 0.9\,\text{m}$, collar A is moving downward at speed $v_{A1} = 3\,\text{m/s}$. The uniform thin bar AB has mass $m_{AB} = 4\,\text{kg}$ and length L and is attached to collars A and B with ball-and-socket joints. Collars A and B slide smoothly on rods CD and EF, respectively, and collar B is attached to the stop at E by a linear elastic spring with constant $k = 200\,\text{N/m}$ and unstretched length $2\ell_1$. Neglect the dimensions of the collars and assume that the angular velocity of the rod AB is such that AB does not spin about its axis.

Problem 10.44 Assuming that the spring is absent and that the masses of the collars at A and B are negligible, determine the speed of the collar A when it reaches D.

Problem 10.45 Assuming that the masses of the collars at A and B are negligible, determine the speed of the collar A when it reaches D.

Problem 10.46 Assuming that the spring is absent and that the mass of collar A is $m_A = 0.5\,\text{kg}$ and the mass of collar B is $m_B = 0.5\,\text{kg}$, determine the speed of the collar A when it reaches D.

Problem 10.47 Assuming that the mass of collar A is $m_A = 0.5\,\text{kg}$ and the mass of collar B is $m_B = 0.5\,\text{kg}$, determine the speed of the collar A when it reaches D.

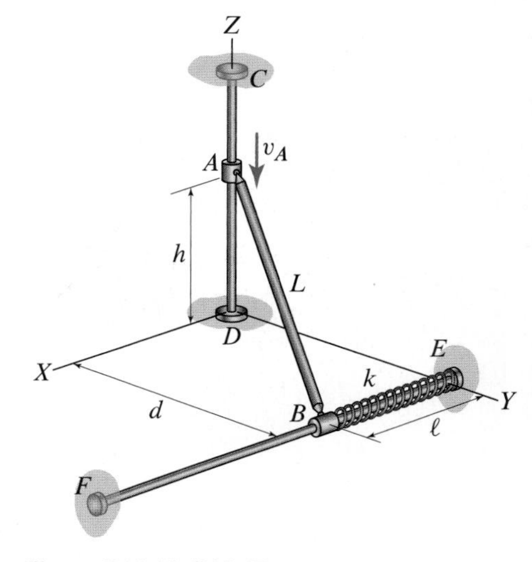

Figure P10.44–P10.47

Problem 10.48

The thin disk of mass m and radius R rotates with constant angular speed $\dot{\psi}$ about the pin at A. The pin itself twists with the light shaft OA of length L at a constant rate $\dot{\theta}$ about the horizontal y axis. In turn, the mechanism at O allows the horizontal bar to be driven about the vertical axis BC at a constant precession rate $\dot{\phi}$. All motion occurs with negligible friction. Determine the total moment exerted on the disk at A, and express it in the given rotating reference frame whose origin is at A and that is attached to the bar OA. The position shown (when the disk is vertically aligned) corresponds to $\theta = 0$.

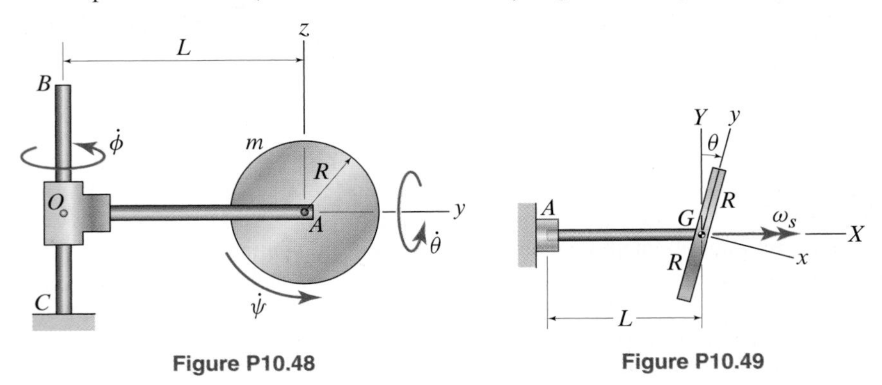

Figure P10.48

Figure P10.49

Problem 10.49

The thin uniform disk of radius R and mass m is mounted on the horizontal shaft, such that the mass center of the disk is on the axis of rotation. Due to an error in manufacturing, the disk has a misalignment angle θ relative to the shaft, such that the position shown occurs only once for each revolution of the shaft. If the shaft is rotating with a constant angular speed ω_s, determine the reactions at the bearing A on the shaft.

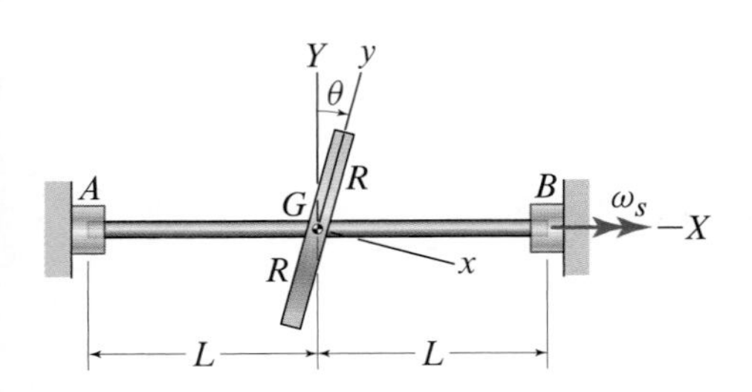

Figure P10.50

Problem 10.50

The thin uniform disk of radius R and mass m is mounted on the horizontal shaft, such that the mass center of the disk is on the axis of rotation. Due to an error in manufacturing, the disk has a misalignment angle θ relative to the shaft, such that the position shown occurs only once for each revolution of the shaft. If the shaft is rotating with a constant angular speed ω_s, determine the reactions at the bearings A and B on the shaft.

Problems 10.51 and 10.52

Unlike the rotor shown in Example 7.6, centrifuge rotors that spin at *very* high speeds (> 100,000 rpm) don't have swinging buckets to hold the test tubes. For these high-speed rotors, the buckets are fixed at an oblique angle. With this in mind, even if we model the test tube as a uniform circular cylinder, the tube itself is no longer symmetric with respect to the plane of motion, and so the products of inertia in Eqs. (10.54) and (10.55) are not all zero. For this fixed-angle rotor, assume a maximum angular velocity of 130,000 rpm, and assume that the angle of each test tube is fixed at $\theta = 35°$ relative to the spin axis of the rotor. Model the test tube as a uniform circular cylinder of diameter $d = 11$ mm, so that the mass center is at the midpoint between $r_o = 120.5$ mm and $r_i = 63.1$ mm. In addition, use the following inertia properties: $m = 10$ g, $(I_{zz})_C = 2.87\times10^{-6}$ kg·m², $(I_{Gxz})_C = -3.89\times10^{-6}$ kg·m², and $(I_{Gyz})_C = 0.00$ kg·m².

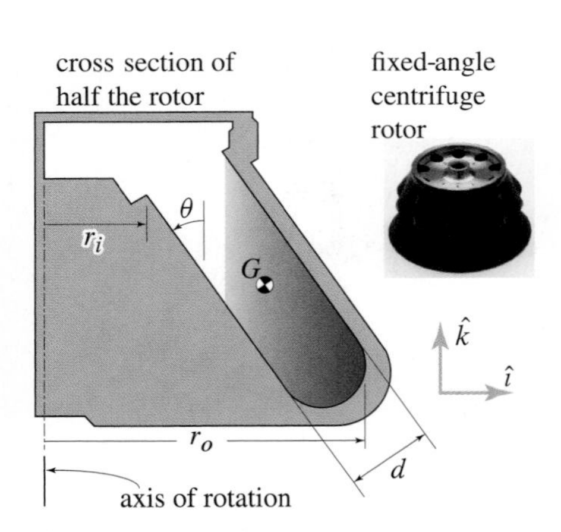

Figure P10.51 and P10.52

Problem 10.51 Assuming that the rotor has reached its terminal speed of 130,000 rpm, use Eqs. (10.54)–(10.56) to determine the net forces and moments acting at the mass center of the test tube to sustain this motion.

Problem 10.52 Assume that the rotor accelerates uniformly from rest and takes 9.5 min to reach its terminal speed of 130,000 rpm. Determine, as a function of time, the net forces and moments acting at the mass center of the test tube to sustain this motion. Use Eqs. (10.54)–(10.56) for your solution.

Problem 10.53

The horizontal shaft AB is spinning with a constant angular speed ω_s in the direction shown. If the mass of the uniform rectangular plate is m and it has the dimensions shown, determine the reactions on the shaft AB due to the bearings at A and B. Neglect the mass of the shaft AB.

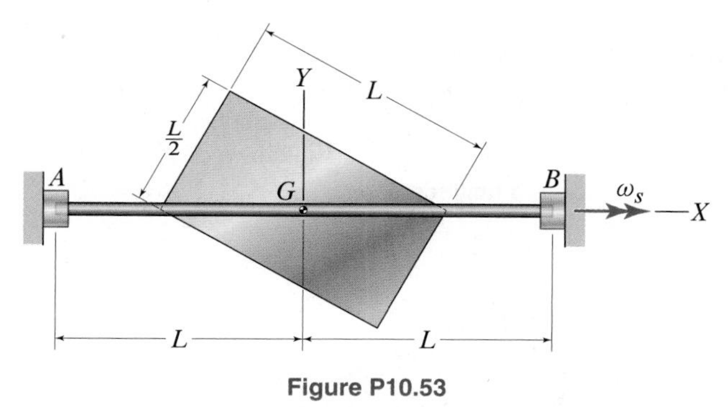

Figure P10.53

Problem 10.54

The L-shaped bar OCD is pin-connected at O to the vertical bar AB. The segments OC and CD are uniform, and each has mass m and length L. The bar AB rotates about its own axis at the constant speed ω_s. Determine the angular speed ω_s required to keep the L-shaped bar in the position shown.

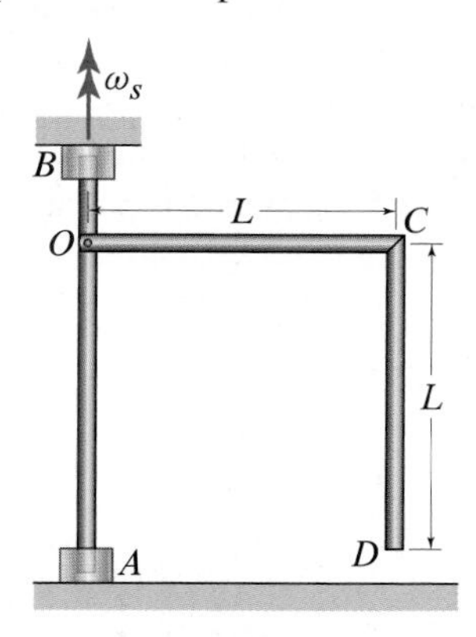

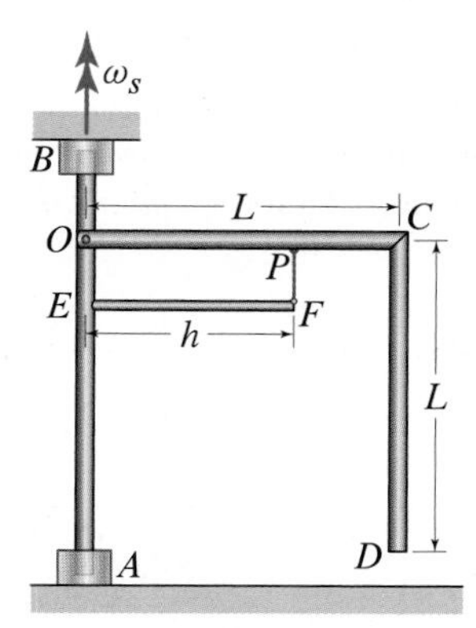

Figure P10.54 **Figure P10.55**

Problem 10.55

The L-shaped bar OCD is pin-connected at O to the vertical bar AB. The segments OC and CD are uniform, and each has mass m and length L. The bar AB rotates about its own axis at the constant speed ω_s. The horizontal bar EF is attached to the bar AB at E, and there is a string attaching the bar EF to the bar OC at a distance h from the spin axis.

(a) Determine the angular speed ω_s required to keep the L-shaped bar in the position shown, such that there is zero tension in the string FP.

(b) For angular speeds ω_t greater than ω_s found in Part (a), determine the tension in the string FP as a function of ω_t.

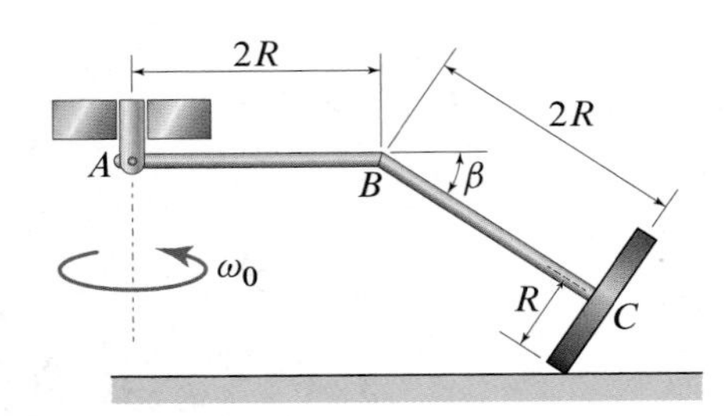

Figure P10.56

Problem 10.56

A thin uniform disk of radius R and mass m rolls without slipping over the horizontal surface. The disk, whose center is at C, can rotate freely relative to the bent shaft ABC, which precesses with constant angular speed ω_0 about the vertical axis. Friction in the pin connecting the bent shaft to the vertical bar at A is negligible. Neglecting the mass of the bent shaft, determine an expression for the magnitude of the normal force exerted on the disk by the horizontal surface.

Problem 10.57

The uniform cone of mass m, length L, and apex angle 2β rolls without slipping over the horizontal surface. The cone rotates at constant ω_0 about a fixed vertical axis intersecting the apex A. Determine an expression for the maximum value of ω_0 for which the cone will not tip over the rim at B. *Hint:* Model the distributed normal force acting on the cone as a concentrated normal force acting at an unknown distance d from A and find an expression for that distance as a function of ω_0.

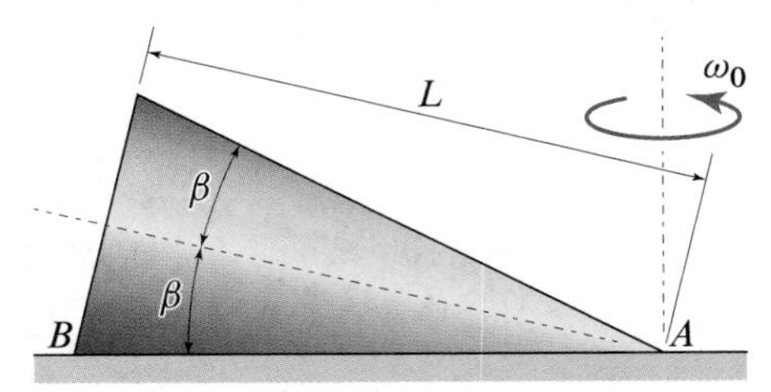

Figure P10.57

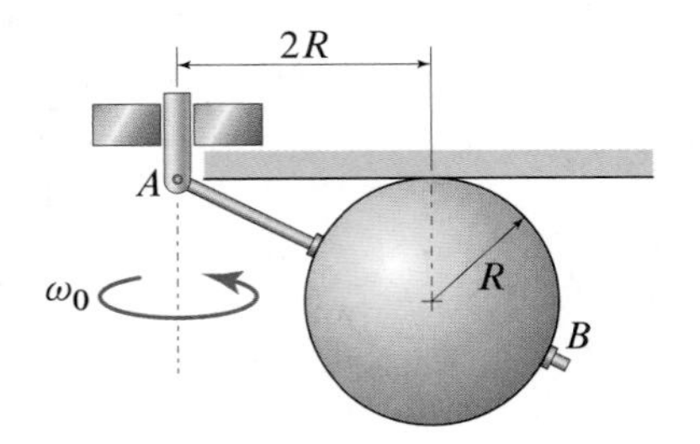

Figure P10.58

Problem 10.58

The uniform sphere of mass m and radius R can spin freely relative to the shaft AB, whose mass is negligible. The shaft AB precesses about the vertical axis with constant angular speed ω_0. Assuming that the friction between the sphere and the horizontal surface above is sufficient to prevent slipping and neglecting friction in the pin at A, determine an expression for the *minimum* value of ω_0 for which this motion is possible.

Chapter Review

Three-dimensional kinematics of rigid bodies

Referring to Fig. 10.15, for two points A and C on the same rigid body B, we can relate their motion using

Eqs. (10.1) and (10.2), p. 721

$$\vec{v}_C = \vec{v}_A + \vec{\omega}_B \times \vec{r}_{C/A},$$
$$\vec{a}_C = \vec{a}_A + \vec{\alpha}_B \times \vec{r}_{C/A} + \vec{\omega}_B \times (\vec{\omega}_B \times \vec{r}_{C/A}),$$

where $\vec{\omega}_B$ and $\vec{\alpha}_B$ are the angular velocity and angular acceleration, respectively, of the rigid body. Note that for three-dimensional motion, we *cannot* write $\vec{\omega}_{AB} \times (\vec{\omega}_{AB} \times \vec{r}_{B/A})$ as $-\omega_{AB}^2 \vec{r}_{B/A}$. Referring to Fig. 10.15, if we use a rotating or secondary reference frame with its origin at A, then we can relate the motion of points P and A, which are not necessarily on the same body, using

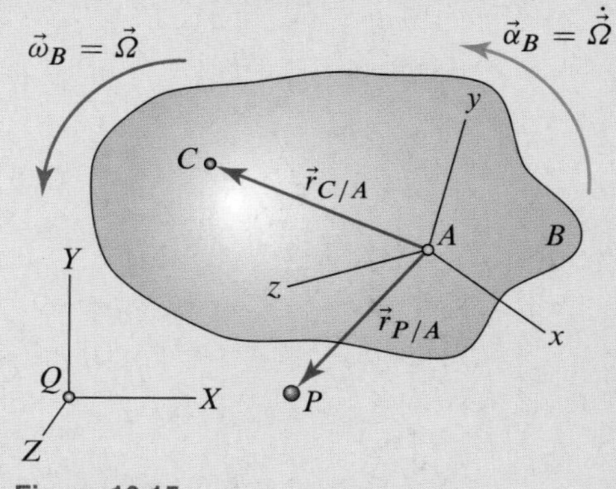

Figure 10.15
Rigid body B displaying the points and kinematic quantities needed to apply Eqs. (10.1)–(10.4). Frame XYZ is the primary reference frame and xyz is the secondary, body-fixed, or rotating reference frame.

Eqs. (10.3) and (10.4), p. 722

$$\vec{v}_P = \vec{v}_A + \vec{v}_{P\text{rel}} + \vec{\Omega} \times \vec{r}_{P/A},$$
$$\vec{a}_P = \vec{a}_A + \vec{a}_{P\text{rel}} + 2\vec{\Omega} \times \vec{v}_{P\text{rel}} + \dot{\vec{\Omega}} \times \vec{r}_{P/A} + \vec{\Omega} \times \big(\vec{\Omega} \times \vec{r}_{P/A}\big),$$

where $\vec{v}_{P\text{rel}}$ and $\vec{a}_{P\text{rel}}$ are the velocity and acceleration, respectively, of P as seen by an observer in the rotating frame; and $\vec{\Omega}$ and $\dot{\vec{\Omega}}$ are the angular velocity and angular acceleration, respectively, of the rotating reference frame. The rotating frame is usually attached to a rigid body, though sometimes the rotating reference frame will rotate relative to both the primary frame *and* the rigid body.

Referring to Fig. 10.16, if the rotating frame is *not* attached to the rigid body B, then the angular acceleration of B is given by

Eq. (10.6), p. 722

$$\dot{\vec{\omega}}_B = \vec{\alpha}_B = \dot{\omega}_{Bx}\,\hat{\imath} + \dot{\omega}_{By}\,\hat{\jmath} + \dot{\omega}_{Bz}\,\hat{k} + \vec{\Omega} \times \vec{\omega}_B,$$

where $\vec{\Omega}$ is the angular velocity of the rotating frame relative to the primary frame and $\vec{\omega}_B$ is the angular velocity of B relative to the primary frame.

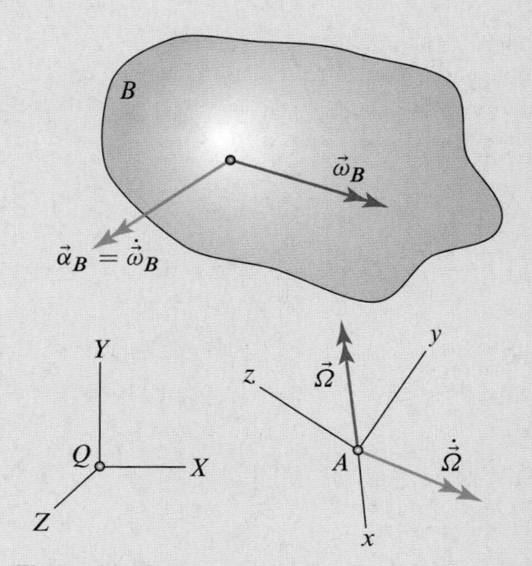

Figure 10.16
The primary reference frame XYZ, rotating reference frame xyz, and rigid body B.

Referring to Fig. 10.17, if the angular velocity of the rigid body B is expressed *relative* to the rotating frame as $\vec{\omega}_{B\text{rel}}$ and the angular velocity of the rotating frame relative to the primary frame is $\vec{\Omega}$, then the angular velocity of B relative to the primary frame, $\vec{\omega}_B$, is

Eq. (10.14), p. 723

$$\vec{\omega}_B = \vec{\Omega} + \vec{\omega}_{B\text{rel}}.$$

Kinetics: equations of motion and kinetic energy

Newton-Euler equations in 3D. There are six Newton-Euler equations for a rigid body in three-dimensional motion. Three are given by the component form of Euler's first law, which is

Eq. (10.20), p. 738

$$\vec{F} = m\vec{a}_G,$$

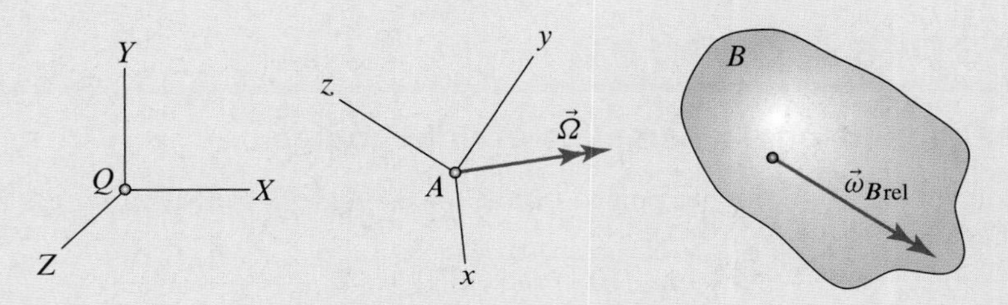

Figure 10.17. A secondary frame rotating with $\vec{\Omega}$ and a rigid body rotating with $\vec{\omega}_{B/A}$ *relative to* the rotating frame.

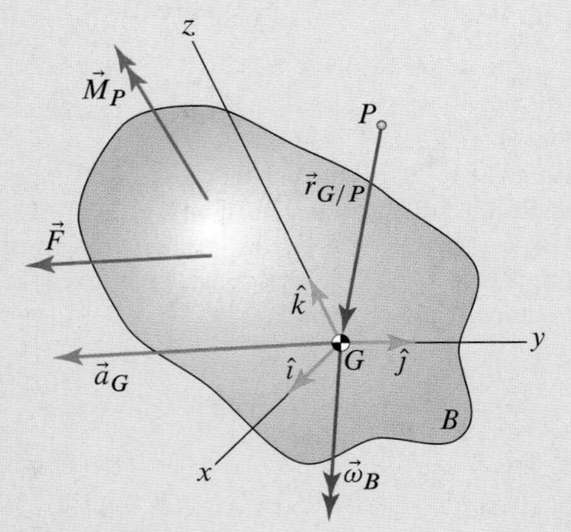

Figure 10.18
Illustration of the kinetic and kinematic quantities needed to understand and apply Eqs. (10.20) and (10.31)–(10.39).

where $\vec{F}$ is the total external force acting on the rigid body, m is the mass of the rigid body, and $\vec{a}_G$ is the acceleration of the mass center of the rigid body (see Fig. 10.18). The other three equations are given by Euler's second law. We derived three different versions of these equations under three corresponding sets of assumptions, each set more restrictive than the next. The last two versions are the most useful in practice, the first of which is given by

Eqs. (10.34)–(10.36), p. 740

$$
\begin{aligned}
M_{Px} &= I_{Gx}\dot{\omega}_{Bx} + \left(I_{Gz} - I_{Gy}\right)\omega_{Bz}\omega_{By} + m\left(y_{G/P}a_{Gz} - z_{G/P}a_{Gy}\right),\\
M_{Py} &= I_{Gy}\dot{\omega}_{By} + \left(I_{Gx} - I_{Gz}\right)\omega_{Bx}\omega_{Bz} + m\left(z_{G/P}a_{Gx} - x_{G/P}a_{Gz}\right),\\
M_{Pz} &= I_{Gz}\dot{\omega}_{Bz} + \left(I_{Gy} - I_{Gx}\right)\omega_{By}\omega_{Bx} + m\left(x_{G/P}a_{Gy} - y_{G/P}a_{Gx}\right),
\end{aligned}
$$

where (see Fig. 10.13):

- Point P is arbitrary, the xyz reference frame is attached to the body B, and xyz are principal axes of inertia, so all products of inertia are zero.
- The moments of inertia are expressed relative to a set of axes that are parallel to xyz and whose origin is at G, the center of mass of the body.
- The angular velocities and their time derivatives are inertial, but are expressed in the rotating xyz frame.

The final set of equations, also known as *Euler's equations of rotational motion for a rigid body*, are given by

Eqs. (10.37)–(10.39), p. 740

$$
\begin{aligned}
M_{Px} &= I_{Gx}\dot{\omega}_{Bx} + \left(I_{Gz} - I_{Gy}\right)\omega_{Bz}\omega_{By},\\
M_{Py} &= I_{Gy}\dot{\omega}_{By} + \left(I_{Gx} - I_{Gz}\right)\omega_{Bx}\omega_{Bz},\\
M_{Pz} &= I_{Gz}\dot{\omega}_{Bz} + \left(I_{Gy} - I_{Gx}\right)\omega_{By}\omega_{Bx},
\end{aligned}
$$

where, in addition to the conditions for Eqs. (10.34)–(10.36), we also have that:

- $\vec{r}_{G/P} \times m\vec{a}_G = \vec{0}$, which can happen in any of the following three ways:
  - $\Rightarrow$ P is the mass center G of the body B so that $\vec{r}_{G/P} = \vec{0}$,
  - $\Rightarrow$ $\vec{a}_G = \vec{0}$, that is, the mass center of the body moves with constant velocity,
  - $\Rightarrow$ $\vec{r}_{G/P}$ is parallel to $\vec{a}_G$ so that the cross product $\vec{r}_{G/P} \times m\vec{a}_G = \vec{0}$.

Rotational equations of motion for a body with an axis of radial symmetry. For a rigid body with an axis of radial symmetry, any coordinate system that has an axis coinciding with the axis of symmetry of that body will be a principal coordinate system. Referring to Fig. 10.19, $x_B y_B z_B$ are principal body axes with the x_B axis coinciding with the axis of radial symmetry of the body B. The xyz frame rotates with angular

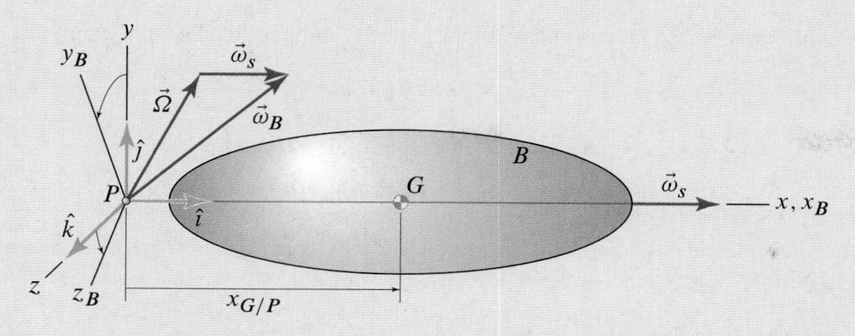

Figure 10.19. A rigid body B spinning about its axis of radial symmetry x_B relative to the xyz frame with angular velocity $\vec{\omega}_s$. The angular velocity of the xyz frame is $\vec{\Omega}$, relative to an underlying inertial frame.

velocity $\vec{\Omega}$, the angular velocity of the body is $\vec{\omega}_B$, and the body spins relative to the xyz frame about its axis of symmetry with angular velocity $\vec{\omega}_s$. With these definitions, the rotational equations of motion of the body are

Eqs. (10.48)–(10.50), p. 742

$$M_{Px} = I_{Gx}(\dot{\Omega}_x + \dot{\omega}_s),$$
$$M_{Py} = I_{Gy}\dot{\Omega}_y + I_{Gx}(\Omega_x + \omega_s)\Omega_z - I_{Gy}\Omega_x\Omega_z - m x_{G/P} a_{Gz},$$
$$M_{Pz} = I_{Gy}\dot{\Omega}_z - I_{Gx}(\Omega_x + \omega_s)\Omega_y + I_{Gy}\Omega_x\Omega_y + m x_{G/P} a_{Gy}.$$

If $\vec{r}_{G/P} \times m\vec{a}_G = \vec{0}$, then these equations become

Eqs. (10.51)–(10.53), p. 742

$$M_{Px} = I_{Gx}(\dot{\Omega}_x + \dot{\omega}_s),$$
$$M_{Py} = I_{Gy}\dot{\Omega}_y + I_{Gx}(\Omega_x + \omega_s)\Omega_z - I_{Gy}\Omega_x\Omega_z,$$
$$M_{Pz} = I_{Gy}\dot{\Omega}_z - I_{Gx}(\Omega_x + \omega_s)\Omega_y + I_{Gy}\Omega_x\Omega_y.$$

Planar motion of bodies *not* symmetric with respect to the plane of motion. In Chapter 7, we emphasized that the Newton-Euler rotational equations derived there were for the planar motion of bodies symmetric with respect to the plane of motion. If the rigid body is undergoing planar motion, but the body is not symmetric with respect to the plane of motion, the governing rotational equations are:

Eqs. (10.54)–(10.56), p. 743

$$M_{Px} = -I_{Gxz}\dot{\omega}_{Bz} + I_{Gyz}\omega_{Bz}^2,$$
$$M_{Py} = -I_{Gyz}\dot{\omega}_{Bz} - I_{Gxz}\omega_{Bz}^2,$$
$$M_{Pz} = I_{Gz}\dot{\omega}_{Bz} + m(x_{G/P}a_{Gy} - y_{G/P}a_{Gx}),$$

where the plane of motion is the xy plane and the moments and products of inertia are measured with respect to a set of axes whose origin is at the mass center G.

Kinetic energy of a rigid body in 3D. The work-energy principle for a rigid body is the same whether its motion is two-dimensional or three-dimensional. Therefore, the work-energy principle as given by any of Eqs. (8.16) on p. 585, (8.23) on p. 586, (8.24) on p. 586, or (8.29) on p. 588 is valid for the three-dimensional motion of this chapter.

The expression for the kinetic energy of a rigid body in three-dimensional motion is:

Eq. (10.66), p. 744

$$T = \tfrac{1}{2} m v_G^2 + \tfrac{1}{2} I_{Gx} \omega_{Bx}^2 + \tfrac{1}{2} I_{Gy} \omega_{By}^2 + \tfrac{1}{2} I_{Gz} \omega_{Bz}^2 - I_{Gxy} \omega_{Bx} \omega_{By} - I_{Gxz} \omega_{Bx} \omega_{Bz} - I_{Gyz} \omega_{By} \omega_{Bz},$$

where the moments and products of inertia are measured with respect to axes whose origin is at the mass center of the rigid body.

Review Problems

Problems 10.59 through 10.62

The bar AB rotates about the vertical y axis with angular speed $\omega_b(t)$. The disk, whose center is at C, can rotate freely relative to the arm CD as it rolls without slipping on the horizontal surface.

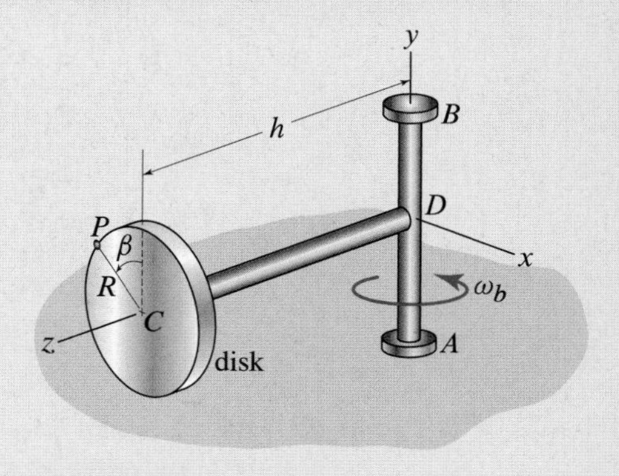

Figure P10.59–P10.62

Problem 10.59 If ω_b is constant, determine the angular velocity $\vec{\omega}_d$ of the disk. Express your answer in the rotating xyz component system shown.

Problem 10.60 If $\omega_b(t)$ is a known function of time, determine the angular acceleration $\vec{\alpha}_d$ of the disk. Express your answer in the rotating xyz component system shown.

Problem 10.61 If ω_b is constant, determine the velocity $\vec{v}_P$ of the point P on the periphery of the disk as a function of the angle β. Express your answer in the rotating xyz component system shown.

Problem 10.62 If $\omega_b(t)$ is a known function of time, determine the acceleration $\vec{a}_P$ of the point P on the periphery of the disk as a function of the angle β. Express your answer in the rotating xyz component system shown.

Problems 10.63 and 10.64

The shaft AB rotates with angular speed $\omega_0(t)$ about the vertical axis. The uniform thin rod CD of length L is rigidly attached to the end of the horizontal arm OG, which is rigidly attached to AB. The rod CD is tilted from the vertical position through the angle θ in the XY plane about the $-Z$ axis. Express your answers in the XYZ frame.

Problem 10.63 If the angular speed ω_0 of the vertical shaft is constant, determine the reaction at O on the horizontal shaft required for this motion.

Problem 10.64 If the angular speed $\omega_0(t)$ of the vertical shaft is not constant, determine the reaction at O on the horizontal shaft required for this motion.

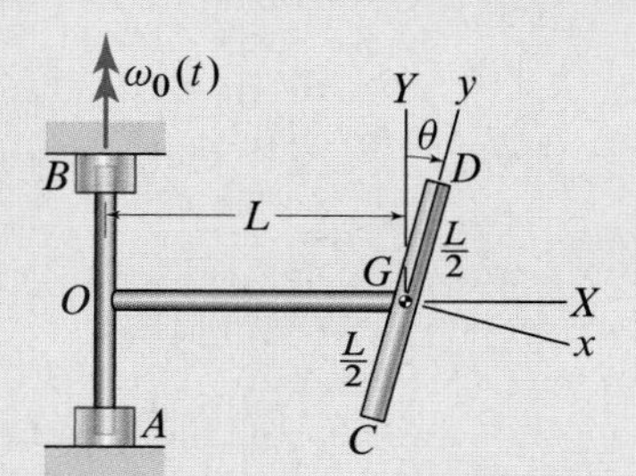

Figure P10.63 and P10.64

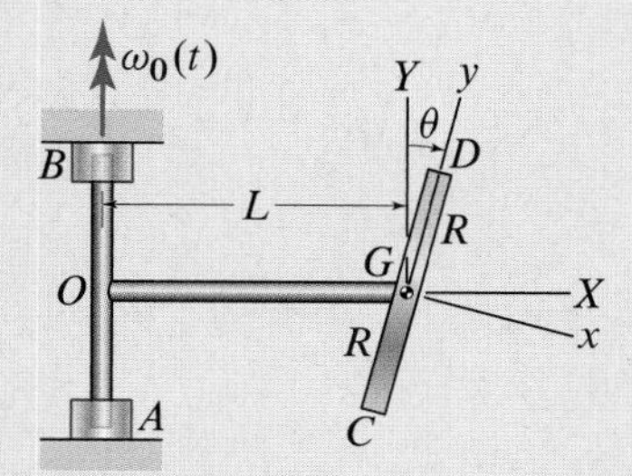

Figure P10.65 and P10.66

Problems 10.65 and 10.66

The shaft AB rotates with angular speed $\omega_0(t)$ about the vertical axis. The uniform thin disk CD of radius R is rigidly attached to the end of the horizontal arm OG, which is rigidly attached to AB. The disk CD is tilted from the vertical position through the angle θ in the XY plane about the $-Z$ axis. Express your answers in the XYZ frame.

Problem 10.65 If the angular speed ω_0 of the vertical shaft is constant, determine the reaction at O on the horizontal shaft required for this motion.

Problem 10.66 If the angular speed $\omega_0(t)$ of the vertical shaft is not constant, determine the reaction at O on the horizontal shaft required for this motion.

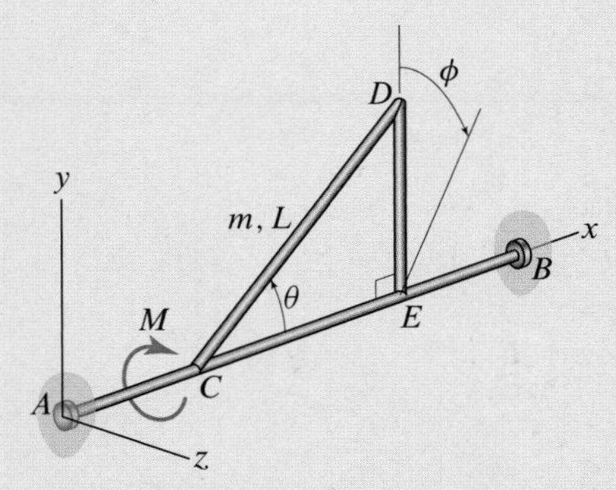

Figure P10.67 and P10.68

Problems 10.67 and 10.68

The moment M is applied to the shaft AB at $t = 0$ and when $\phi = 0$. The angled bar CDE is rigidly attached to the shaft AB. Segment CD has mass m and length L. Segments CD and DE have the same linear density. The rotational inertia of the shaft is negligible as is friction in the bearings at A and B.

Problem 10.67 If $M = M(\phi) = 3\phi^{1/2}$, determine an expression for the angular velocity of the system after it has undergone three revolutions.

Problem 10.68 If $M = M(t) = 5t^{1/3}$, determine an expression for the angular velocity of the system when $t = 15$ s.

Problems 10.69 through 10.74

Bar AB of length $L_{AB} = 2.5$ m is attached by a fork and clevis joint to the collar at A and by a ball joint to the disk at B. The disk lies in the xy plane, and its center at E lies on the y axis in the yz plane. The disk rotates about a vertical axis at the constant angular rate $\omega_d = 100$ rpm. The dimensions $d = 1.2$ m, $h = 0.9$ m, and $R = 0.75$ m are given. *Hint:* The clevis joint constrains the rotation of arm AB *relative to* the collar at A to be perpendicular to the plane formed by bar CD and arm AB. Therefore, the angular velocity of arm AB is the sum of the angular velocity of the collar at A and the angular velocity associated with the change in the angle β, which lies in the plane formed by bars CD and AB.

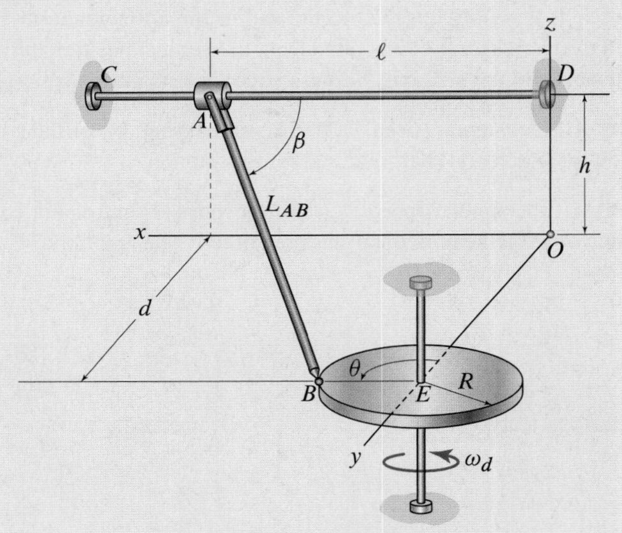

Figure P10.69–P10.74

Problem 10.69 For the disk position shown, that is, $\theta = 90°$, determine the angular velocity of the bar. Express your answer in the given component system, and assume that the angular velocity of the bar is orthogonal to it.

Problem 10.70 For the disk position shown, that is, $\theta = 90°$, determine the angular acceleration of the bar. Express your answer in the given component system, and assume that the angular velocity and angular acceleration of the bar are orthogonal to it.

Problem 10.71 For the disk position $\theta = 0°$, determine the velocity of the collar at A. Express your answer in the given component system, and assume that the angular velocity of the bar is orthogonal to it.

Problem 10.72 For the disk position $\theta = 0°$, determine the acceleration of the collar at A. Express your answer in the given component system, and assume that the angular velocity and angular acceleration of the bar are orthogonal to it.

Problem 10.73 Determine an expression for the angular velocity of the bar AB and the velocity of the collar at A for any position θ of the disk. Express your answers in the given component system, and assume that the angular velocity of the bar is orthogonal to it.

Problem 10.74 Determine the angular acceleration of the bar AB and the acceleration of the collar at A for any position θ of the disk. Express your answers in the given component system, and assume that the angular velocity and angular acceleration of the bar are orthogonal to it.

Mass Moments of Inertia

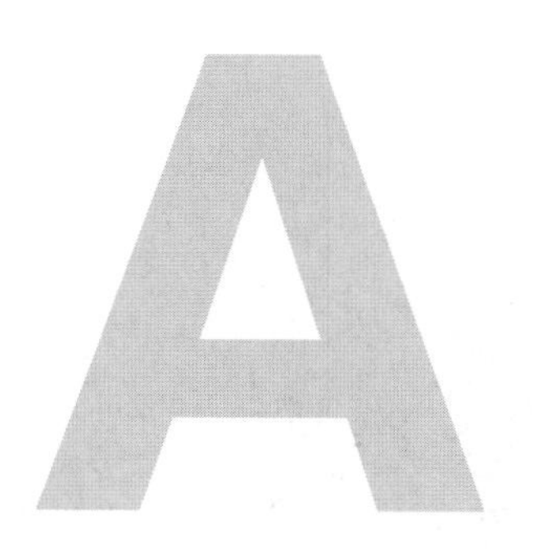

The International Space Station as it existed in June 2008. Accurately determining its mass moments and products of inertia requires sophisticated computer codes, but the principles used in those codes are no different than what we study in this book.

Mass moments and products of inertia are measures of how the mass is distributed within a body. Mass moments and products of inertia of a body depend on its geometry (size and shape), the density of the material at each point in the body, and the axes selected for measuring them.

Mass moments of inertia and *mass products of inertia* are measures of how the mass is distributed within a body. Mass moments and products of inertia appear in the rotational equations of motion for a rigid body (Chapter 7), the kinetic energy of a rigid body (Chapter 8), the angular momentum of a rigid body (Chapter 8 and Appendix B), and the three-dimensional dynamics of rigid bodies.

Definition of mass moments and products of inertia

The *mass moment of inertia of the element of mass dm* within the body B about any of the three coordinate axes x, y, or z, is defined as the product of the mass of the element and the square of the shortest distance from the element to the axis in question. For example, referring to Fig. A.1, the mass moment of inertia of dm about the y axis is $dI_{Py} = r_y^2\,dm$, where we include the origin P of the coordinate system in the subscript so that different sets of axes can be identified. The mass moments of inertia of the entire body B with respect to the three axes are found by integrating the differential moments of inertia over the entire body. Therefore, the *mass moments of inertia for the body B* shown in Fig. A.1 are defined as

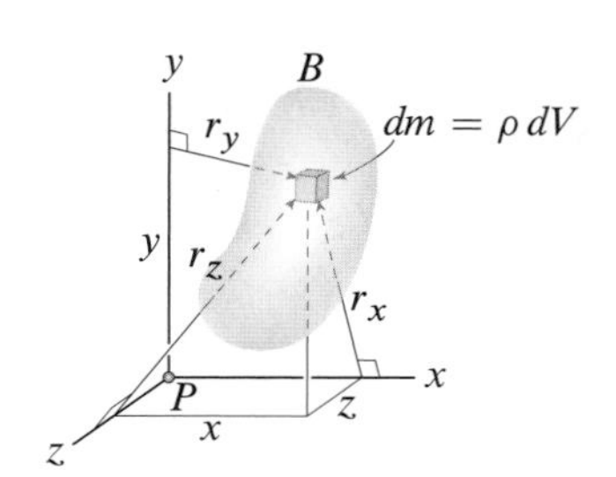

Figure A.1
A body B with mass m, density ρ, and volume V. The scalar quantities r_x, r_y, and r_z are radial distances from the x, y, and z axes, respectively, to the mass element dm.

$$I_{Px} = \int_B r_x^2\,dm = \int_B (y^2 + z^2)\,dm, \tag{A.1}$$

$$I_{Py} = \int_B r_y^2 \, dm = \int_B (x^2 + z^2) \, dm, \tag{A.2}$$

$$I_{Pz} = \int_B r_z^2 \, dm = \int_B (x^2 + y^2) \, dm, \tag{A.3}$$

where:

r_x, r_y, and r_z are shown in Fig. A.1 and are the radial distances (i.e., moment arms) from the x, y, and z axes, respectively, to the mass element dm.

x, y, and z are shown in Fig. A.1 and are the coordinates of the mass element dm.

I_{Px}, I_{Py}, and I_{Pz} are the *mass moments of inertia of the body B about the x, y, and z axes*, respectively.

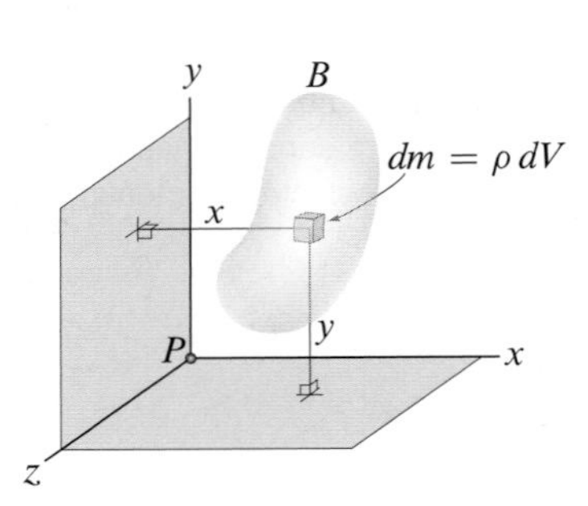

Figure A.2
A body B with mass m demonstrating the mass product of inertia $dI_{Pxy} = xy\,dm$ of the element of mass dm.

The three *mass products of inertia for the element of mass dm* are defined with respect to the three possible pairs of orthogonal planes as the perpendicular distance from each pair of planes to the mass element. For example, referring to Fig. A.2, the mass product of inertia of the element dm with respect to the two planes xz and yz is $dI_{Pxy} = xy\,dm$. The mass products of inertia of the entire body B with respect to the three pairs of planes are found by integrating the differential products of inertia over the entire body. The *mass products of inertia for the body B* shown in Fig. A.2 are defined as

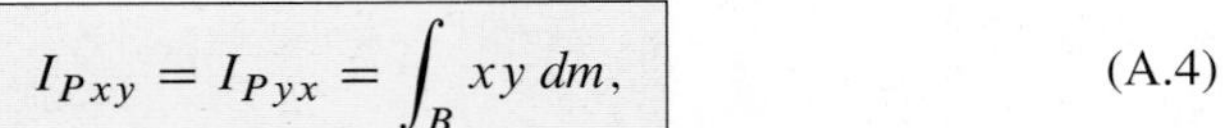

$$I_{Pxy} = I_{Pyx} = \int_B xy \, dm, \tag{A.4}$$

$$I_{Pyz} = I_{Pzy} = \int_B yz \, dm, \tag{A.5}$$

$$I_{Pxz} = I_{Pzx} = \int_B xz \, dm, \tag{A.6}$$

where:

I_{Pxy}, I_{Pyz}, and I_{Pxz} are the *products of inertia of the mass with respect the xz-yz, xy-xz, and xy-yz plane pairs*, respectively.

Remarks

- When referring to mass moments of inertia, we often omit the word "mass" when it is obvious from the context that we are dealing with mass moments of inertia as opposed to area moments of inertia.

- In each of Eqs. (A.1)–(A.3), two equivalent integral expressions are provided, and each is useful depending on the geometry of the object under consideration.

- The moments of inertia in Eqs. (A.1)–(A.6) measure the *second moment* of the mass distribution. That is, to determine I_{Px}, I_{Py}, and I_{Pz} in Eqs. (A.1)–(A.3), the moment arms r_x, r_y, and r_z are *squared*. The second integral in each of these expressions is obtained by noting that $r_x^2 = y^2 + z^2$, and similarly for r_y^2 and r_z^2. For the products of inertia I_{Pxy}, I_{Pyz}, and I_{Pxz} in Eqs. (A.4)–(A.6), the product of two different moment arms is used.

- In Eqs. (A.1)–(A.6), x, y, and z have dimensions of length, and dm has the dimension of mass. Hence, all mass moments of inertia have dimensions of $(mass)(length)^2$ and are expressed in slug·ft² and kg·m² in the U.S. Customary and SI unit systems, respectively.

- When the x, y, and z axes pass through the center of mass of an object, we denote these axes as x', y', and z', and we refer to the moments of inertia associated with these axes as *mass center moments of inertia* with the designations I_{Gx}, I_{Gy}, etc.
- The quantities I_{Px}, I_{Py}, and I_{Pz} are never negative. The products of inertia I_{Pxy}, I_{Pyz}, and I_{Pxz} may be positive, zero, or negative, as discussed below.
- Evaluation of moments of inertia using composite shapes is possible using the parallel axis theorem, as discussed later in this appendix.

How are mass moments of inertia used?

It is useful to discuss why there are six mass moments of inertia, how they differ from each other, and how they are used.

Figure A.3
The International Space Station with a Space Shuttle docked to it.

Moments of inertia I_{Px}, I_{Py}, and I_{Pz}. In Fig. A.3, the International Space Station with a docked Space Shuttle is shown. Modeling the system as rigid, if a moment M_x about the x axis is applied to it, the system will begin to undergo an angular acceleration about the x axis. The value of the angular acceleration is proportional to the mass moment of inertia about the x axis I_{Px}. Furthermore, the larger I_{Px} is, the lower the angular acceleration will be for a given value of M_x. Similar remarks apply to moments applied about the y and z axes and the influence that moments of inertia I_{Py} and I_{Pz} have on angular accelerations about these axes.

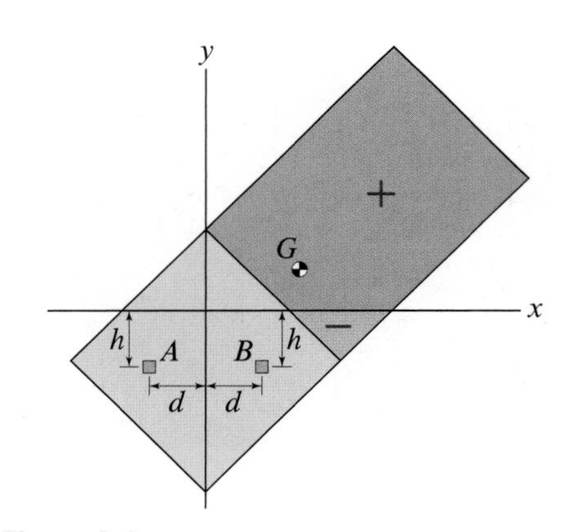

Figure A.4
A uniform body and a coordinate system used to understand products of inertia.

Products of inertia I_{Pxy}, I_{Pyz}, and I_{Pxz}. Products of inertia measure the asymmetry of a body's mass distribution with respect to the xy, yz, and xz planes. Products of inertia can have a positive, zero, or negative value, depending on the shape and mass distribution of an object, the selection of the x, y, and z axes, and the location of the origin of the xyz coordinate system. Figure A.4 shows the cross section of a uniform body at some arbitrary z coordinate. The orange shaded region shows that part of the cross section that is symmetric about the y axis. The two mass elements A and B (shown in blue) are an equal distance d from the yz plane, and both have the same y coordinate, i.e., $y = -h$. Referring to the integrals for I_{Pxy} and I_{Pxz} in Eqs. (A.4) and (A.6), $xy\, dm$ and $xz\, dm$ for the left element A have the opposite sign of the analogous quantities for the right element B. Therefore, the mass in the orange region does not contribute to the products of inertia I_{Pxy} and I_{Pxz}. By contrast, the mass in the yellow shaded region has no corresponding region that is symmetric with respect to the yz plane, and since the product xy for all points in that region is positive, it contributes positively to I_{Pxy}. Using a similar argument, the green shaded region contributes negatively to I_{Pxy}. If the body in Fig. A.4 had a uniform mass distribution, it would have $I_{Pxy} > 0$ since the yellow region is larger than the green. Note that nothing can be said about the sign of I_{Pxz} since we don't know if the cross section in Fig. A.4 is at a positive or negative z coordinate.

The preceding arguments lead us to the conclusion that if the body in Fig. A.4 consisted of only the area shaded in orange, then both I_{Pxy} and I_{Pxz} would be zero. This argument implies that:

> *All products of inertia containing a coordinate that is perpendicular to a plane of symmetry for a body must be zero, as long as the origin of the coordinate system lies in that plane of symmetry.*

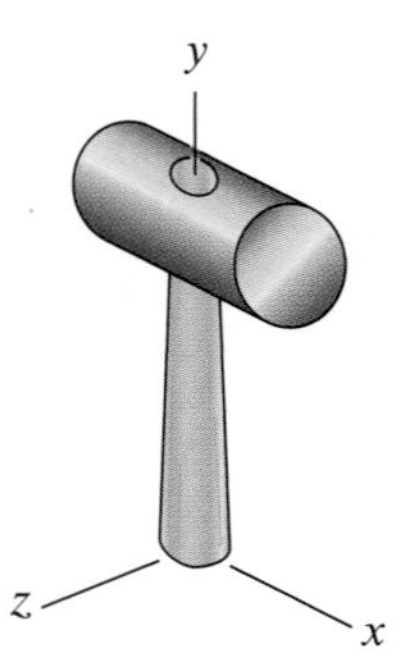

Figure A.5
If the mallet's geometry and mass distribution are both symmetric about the xy plane, then the mallet is said to be a *symmetric object*. If the mallet is also symmetric about the yz plane, it may be called a *doubly symmetric object*.

For example, if an object is symmetric about the xy plane, such as the mallet shown in Fig. A.5, then $I_{Pxz} = I_{Pyz} = 0$. If an object is symmetric about at least two of the xy, yz, and xz planes, then all of the products of inertia are zero. For example, the products of inertia are zero for a uniform solid of revolution if one of the coordinate directions coincides with the axis of revolution.

Objects that have one or more nonzero products of inertia may display complicated behavior in three-dimensional motions. Forcing such a body to undergo planar motion generally requires the application of moments along directions in that plane of motion. For example, the Space Station in Fig. A.3 is nonsymmetric about the xyz axes shown, and thus it has nonzero products of inertia. If a moment M_x about the x axis is applied, the Space Station, in addition to rotating about the x axis, will also rotate about the y and/or z axes (or moments about those axes will be required to prevent it from doing so).

Radius of gyration

Rather than using mass moments of inertia to quantify the distribution of mass within a body, the *radii of gyration* are often used. For a body of mass m, the *radii of gyration* are directly related to the mass moments of inertia, and are defined as

$$k_{Px} = \sqrt{\frac{I_{Px}}{m}}, \qquad k_{Py} = \sqrt{\frac{I_{Py}}{m}}, \qquad k_{Pz} = \sqrt{\frac{I_{Pz}}{m}}, \tag{A.7}$$

where k_{Px}, k_{Py}, and k_{Pz} are called the *radii of gyration of the body about the x, y, and z axes*, respectively. The radii of gyration have units of *length*.

Parallel axis theorem

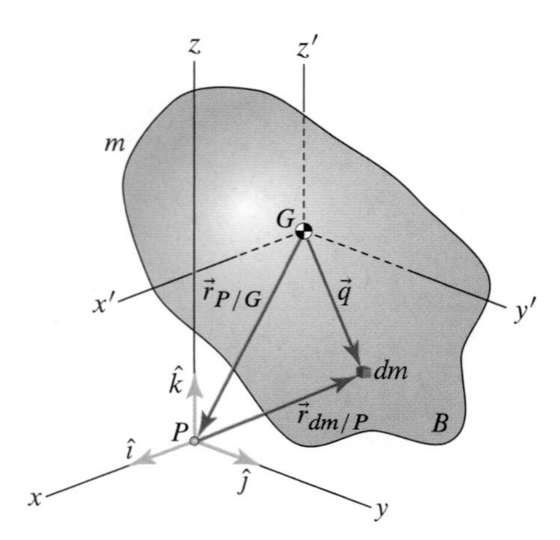

Figure A.6
An object with mass m and center of mass at point G. The x and x' axes, the y and y' axes, and the z and z' axes are parallel to one another, respectively.

The parallel axis theorem relates mass moments and products of inertia I_{Px}, I_{Py}, I_{Pz}, I_{Pxy}, I_{Pyz}, and I_{Pxz} to the mass center moments and products of inertia I_{Gx}, I_{Gy}, I_{Gz}, I_{Gxy}, I_{Gyz}, and I_{Gxz}. Chapters 7–10 demonstrate how important the parallel axis theorem is for the dynamics of rigid bodies.

Referring to the body B and coordinate systems in Fig. A.6, the xyz axes are parallel to the $x'y'z'$ axes, respectively, the origin of the xyz axes is at an arbitrary point P, and the origin of the $x'y'z'$ system is at the mass center G of B.

Parallel axis theorem for moments of inertia

Referring to Fig. A.6, we begin by noting that the mass moment of inertia of the body B about the x axis is given by Eq. (A.1), which is repeated here for convenience as

$$I_{Px} = \int_B (y^2 + z^2)\, dm. \tag{A.8}$$

In addition, we can write the position of a mass element dm relative to G as

$$\vec{q} = \vec{r}_{P/G} + \vec{r}_{dm/P}. \tag{A.9}$$

Writing Eq. (A.9) in component form, we obtain

$$x'\,\hat{\imath} + y'\,\hat{\jmath} + z'\,\hat{k} = \left[(r_{P/G})_x\,\hat{\imath} + (r_{P/G})_y\,\hat{\jmath} + (r_{P/G})_z\,\hat{k}\right] + \left(x\,\hat{\imath} + y\,\hat{\jmath} + z\,\hat{k}\right). \tag{A.10}$$

Substituting Eq. (A.10) into Eq. (A.8) results in

$$I_{Px} = \int_B \left[(y' - (r_{P/G})_y)^2 + (z' - (r_{P/G})_z)^2\right] dm, \tag{A.11}$$

$$= \int_B (y'^2 + z'^2)\, dm + \left[(r_{P/G})_y^2 + (r_{P/G})_z^2\right] \int_B dm$$
$$- 2(r_{P/G})_y \int_B y'\, dm - 2(r_{P/G})_z \int_B z'\, dm. \tag{A.12}$$

The first term in Eq. (A.12) is the mass center moment of inertia about the x' axis $I_{Gx'} = I_{Gx}$ (see Eq. (A.1)). The second integral in Eq. (A.12) is the mass m of B. Finally, since x', y', and z' measure the position of dm relative to G, the last two terms in Eq. (A.12) measure the position of the mass center of B relative to G, and so they must both be zero. Therefore, Eq. (A.12) becomes

$$I_{Px} = I_{Gx} + m\left[(r_{P/G})_y^2 + (r_{P/G})_z^2\right]. \tag{A.13}$$

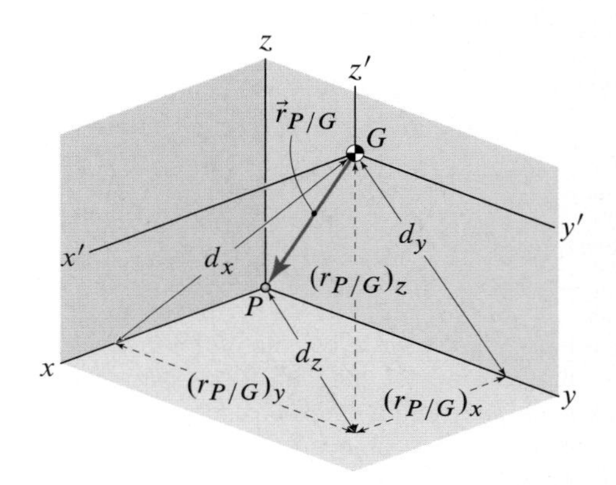

Figure A.7
The xyz and $x'y'z'$ axes are parallel with separation distances d_x, d_y, and d_z, respectively.

Referring to Fig. A.7, we see that $(r_{P/G})_y^2 + (r_{P/G})_z^2$ is the square of the perpendicular distance d_x between the x and x' axes, i.e.,

$$d_x^2 = (r_{P/G})_y^2 + (r_{P/G})_z^2, \tag{A.14}$$

so that Eq. (A.13) becomes

$$I_{Px} = I_{Gx} + md_x^2. \tag{A.15}$$

Similarly, substituting Eq. (A.10) into Eqs. (A.2) and (A.3), we obtain the following:

$$I_{Py} = I_{Gy} + m\left[(r_{P/G})_x^2 + (r_{P/G})_z^2\right] = I_{Gy} + md_y^2, \tag{A.16}$$
$$I_{Pz} = I_{Gz} + m\left[(r_{P/G})_x^2 + (r_{P/G})_y^2\right] = I_{Gz} + md_z^2, \tag{A.17}$$

where, referring to Fig. A.7, we have used the fact that

$$d_y^2 = (r_{P/G})_x^2 + (r_{P/G})_z^2 \quad \text{and} \quad d_z^2 = (r_{P/G})_x^2 + (r_{P/G})_y^2. \tag{A.18}$$

Summarizing these results, we have the *parallel axis theorem for moments of inertia*, which relates the mass moments of inertia with respect to the x, y, and z axes to the mass center moments of inertia as follows:

$$I_{Px} = I_{Gx} + md_x^2 = I_{Gx} + m\left[(r_{P/G})_y^2 + (r_{P/G})_z^2\right], \tag{A.19}$$
$$I_{Py} = I_{Gy} + md_y^2 = I_{Gy} + m\left[(r_{P/G})_x^2 + (r_{P/G})_z^2\right], \tag{A.20}$$
$$I_{Pz} = I_{Gz} + md_z^2 = I_{Gz} + m\left[(r_{P/G})_x^2 + (r_{P/G})_y^2\right]. \tag{A.21}$$

Parallel axis theorem for products of inertia

Referring to Fig. A.8, an $x'y'z'$ coordinate system is defined, with origin at the center of mass G of the object B, and the x, y, and z axes are parallel to the x', y', and z' axes, respectively. The product of inertia of the body B about the x and y axes is

$$I_{Pxy} = \int_B xy\, dm. \tag{A.22}$$

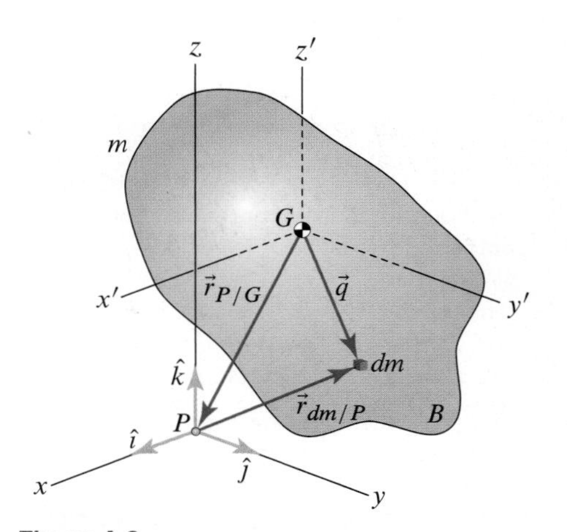

Figure A.8
An object with mass m and center of mass at point G. The x and x' axes, the y and y' axes, and the z and z' axes are parallel to one another, respectively.

Substituting in Eq. (A.10) for x and y, Eq. (A.22) becomes

$$I_{Pxy} = \int_B [x' - (r_{P/G})_x][y' - (r_{P/G})_y]\, dm \tag{A.23}$$

$$= \int_B x'y'\, dm + (r_{P/G})_x (r_{P/G})_y \int_B dm - (r_{P/G})_y \int_B x'\, dm - (r_{P/G})_x \int_B y'\, dm. \tag{A.24}$$

The last two integrals are zero by definition of center of mass. The first integral is the mass center product of inertia about the $x'y'$ axes, and the second integral defines the mass of the body B. Therefore, Eq. (A.24) becomes

$$I_{Pxy} = I_{Gxy} + m(r_{P/G})_x (r_{P/G})_y. \tag{A.25}$$

Using a similar approach, we can find I_{Pyz} and I_{Pxz} in terms of the mass center products of inertia. In summary, the *parallel axis theorem for products of inertia* is

$$I_{Pxy} = I_{Gxy} + m(r_{P/G})_x (r_{P/G})_y, \tag{A.26}$$

$$I_{Pxz} = I_{Gxz} + m(r_{P/G})_x (r_{P/G})_z, \tag{A.27}$$

$$I_{Pyz} = I_{Gyz} + m(r_{P/G})_y (r_{P/G})_z, \tag{A.28}$$

where we recall that I_{Gxy}, I_{Gxz}, and I_{Gyz} are the mass center products of inertia and $(r_{P/G})_x$, $(r_{P/G})_y$, and $(r_{P/G})_z$ are defined in Eq. (A.10) and can be seen in Fig. A.7.

Common Pitfall

Parallel axis theorem. Consider the body whose center of mass is at point G, as well as the parallel axes x_1, x_2, and mass center axis x'.

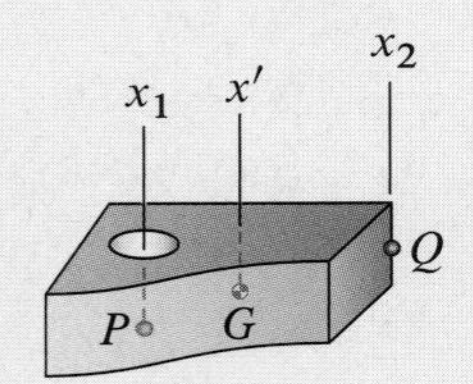

If I_{Px} is known, a common error is to use the parallel axis theorem to *directly* determine I_{Qx}. In the parallel axis theorem, one of the axes is always a mass center axis. Thus, with I_{Px} known, the parallel axis theorem must be employed *twice*, the first time to determine I_{Gx} from I_{Px}, and the second time to use I_{Gx} to determine I_{Qx}.

Principal moments of inertia

The moments and products of inertia for a rigid body B depend on both the origin and orientation of the coordinate axes used to compute them. For a given origin, we can find a unique orientation of the coordinate axes, such that all the products of inertia are zero when computed with respect to those axes. These special axes are called the *principal axes of inertia*, and the corresponding moments of inertia are called the *principal moments of inertia* $\bar{I}_x$, $\bar{I}_y$, and $\bar{I}_z$. Let's see how we find these axes and the corresponding moments of inertia.

Equation (B.16) in Appendix B states that the angular momentum of a rigid body with respect to its mass center G, that is $\vec{h}_G$, can be written as

$$\vec{h}_G = \left(I_{Gx}\omega_{Bx} - I_{Gxy}\omega_{By} - I_{Gxz}\omega_{Bz}\right)\hat{\imath} + \left(-I_{Gxy}\omega_{Bx} + I_{Gy}\omega_{By} - I_{Gyz}\omega_{Bz}\right)\hat{\jmath} + \left(-I_{Gxz}\omega_{Bx} - I_{Gyz}\omega_{By} + I_{Gz}\omega_{Bz}\right)\hat{k}, \tag{A.29}$$

$$= \{h_G\} = [I_G]\{\omega_B\}, \tag{A.30}$$

where $\{\omega_B\} = \vec{\omega}_B$ is the angular velocity of the rigid body B, $\{h_G\} = \vec{h}_G$, and $[I_G]$ is the *inertia tensor* or *inertia matrix*, which is written as

$$[I_G] = \begin{bmatrix} I_{Gx} & -I_{Gxy} & -I_{Gxz} \\ -I_{Gxy} & I_{Gy} & -I_{Gyz} \\ -I_{Gxz} & -I_{Gyz} & I_{Gz} \end{bmatrix}. \tag{A.31}$$

Helpful Information

The column matrix representation of a vector. In representing the vector $\vec{\omega}_B$ as

$$\{\omega_B\} = \begin{Bmatrix} \omega_{Bx} \\ \omega_{By} \\ \omega_{Bz} \end{Bmatrix}$$

the first row contains the x component, the second row the y component, and the third row the z component of the vector. The same interpretation applies to $\{h_G\}$. When we multiply $[I_G]$ by $\{\omega_B\}$, we obtain the vector $\{h_G\}$ whose first, second, and third rows are the x, y, and z components of the vector, respectively.

We now assume that the rigid body is spinning about just *one* of its principal axes of inertia whose moment of inertia is $\bar{I}$. If this is the case, then the angular momentum can be written as

$$\vec{h}_G = \bar{I}\vec{\omega}_B = \bar{I}\omega_{Bx}\,\hat{\imath} + \bar{I}\omega_{By}\,\hat{\jmath} + \bar{I}\omega_{Bz}\,\hat{k}. \tag{A.32}$$

The expressions for $\vec{h}_G$ given in Eq. (A.29) and Eq. (A.32) must be equal to one another. Since the Cartesian basis vectors are linearly independent, we can equate the coefficients of $\hat{\imath}$, $\hat{\jmath}$, and $\hat{k}$ in these two equations to obtain a linear system of equations in the unknowns ω_{Bx}, ω_{By}, and ω_{Bz}, which can be expressed in matrix notation as

$$\begin{bmatrix} I_{Gx}-\bar{I} & -I_{Gxy} & -I_{Gxz} \\ -I_{Gxy} & I_{Gy}-\bar{I} & -I_{Gyz} \\ -I_{Gxz} & -I_{Gyz} & I_{Gz}-\bar{I} \end{bmatrix} \begin{Bmatrix} \omega_{Bx} \\ \omega_{By} \\ \omega_{Bz} \end{Bmatrix} = \begin{Bmatrix} 0 \\ 0 \\ 0 \end{Bmatrix}. \tag{A.33}$$

This system of equations has a nonzero solution if and only if the determinant of the coefficient matrix is zero, which implies that

$$\begin{aligned} \bar{I}^3 - (I_{Gx} + I_{Gy} + I_{Gz})\bar{I}^2 - (I_{Gxy}^2 + I_{Gxz}^2 + I_{Gyz}^2 - I_{Gx}I_{Gy} \\ - I_{Gx}I_{Gz} - I_{Gy}I_{Gz})\bar{I} + (I_{Gx}I_{Gyz}^2 + I_{Gy}I_{Gxz}^2 + I_{Gz}I_{Gxy}^2 \\ - I_{Gx}I_{Gy}I_{Gz} + 2I_{Gxy}I_{Gyz}I_{Gxz}) = 0. \end{aligned} \tag{A.34}$$

Equation (A.34) is called the *characteristic equation*. The three roots of this equation represent the three principal moments of inertia $\bar{I}_{G1}$, $\bar{I}_{G2}$, and $\bar{I}_{G3}$. The corresponding directions of those principal moments of inertia are found by substituting each in turn into

$$\begin{bmatrix} I_{Gx}-\bar{I}_{Gi} & -I_{Gxy} & -I_{Gxz} \\ -I_{Gxy} & I_{Gy}-\bar{I}_{Gi} & -I_{Gyz} \\ -I_{Gxz} & -I_{Gyz} & I_{Gz}-\bar{I}_{Gi} \end{bmatrix} \begin{Bmatrix} x_i \\ y_i \\ z_i \end{Bmatrix} = \begin{Bmatrix} 0 \\ 0 \\ 0 \end{Bmatrix}, \quad i = 1, 2, 3 \tag{A.35}$$

which then represents three equations in terms of the unknowns x_i, y_i, and z_i. Each of the three solutions are then the components of a vector in the principal direction corresponding to the principal moment of inertia used to obtain it.

Mini-Example

Given the system of particles lying in the $z = 0$ plane in Fig. A.9, determine the principal moments of inertia and their corresponding directions. Since the system is discrete, the moments and products of inertia are computed using sums instead of integrals as

$$\begin{aligned} I_{Ox} &= \sum_{i=1}^{4} m_i(y_i^2 + z_i^2) \\ &= m(-2\ell)^2 + 2m(-\ell)^2 + 3m(4\ell)^2 + 4m(\ell)^2 = 58m\ell^2, \end{aligned} \tag{A.36}$$

$$\begin{aligned} I_{Oy} &= \sum_{i=1}^{4} m_i(x_i^2 + z_i^2) \\ &= m(\ell)^2 + 2m(-2\ell)^2 + 3m(-2\ell)^2 + 4m(2\ell)^2 = 37m\ell^2, \end{aligned} \tag{A.37}$$

$$\begin{aligned} I_{Oz} &= \sum_{i=1}^{4} m_i(x_i^2 + y_i^2) \\ &= m\big[(\ell)^2 + (-2\ell)^2\big] + 2m\big[(-2\ell)^2 + (-\ell)^2\big] \\ &\quad + 3m\big[(-2\ell)^2 + (4\ell)^2\big] + 4m\big[(2\ell)^2 + (\ell)^2\big] = 95m\ell^2, \end{aligned} \tag{A.38}$$

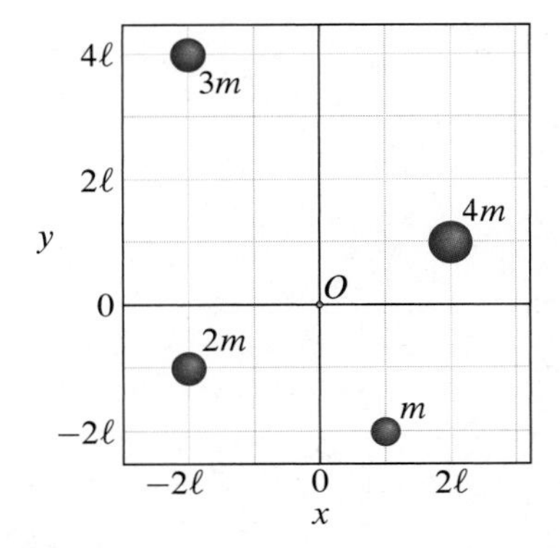

Figure A.9
Four point masses in the $z = 0$ plane. The coordinates are dimensionless, and the masses are in terms of an arbitrary m.

$$I_{Oxy} = \sum_{i=1}^{4} m_i x_i y_i$$
$$= m(\ell)(-2\ell) + 2m(-2\ell)(-\ell)$$
$$+ 3m(-2\ell)(4\ell) + 4m(2\ell)(\ell) = -14m\ell^2, \tag{A.39}$$

$$I_{Oxz} = \sum_{i=1}^{4} m_i x_i z_i = 0, \tag{A.40}$$

$$I_{Oyz} = \sum_{i=1}^{4} m_i y_i z_i = 0. \tag{A.41}$$

Substituting the results from Eqs. (A.36)–(A.41) into Eq. (A.31) as

$$[I_O] = \begin{bmatrix} 58m\ell^2 & 14m\ell^2 & 0 \\ 14m\ell^2 & 37m\ell^2 & 0 \\ 0 & 0 & 95m\ell^2 \end{bmatrix}, \tag{A.42}$$

Eq. (A.34) becomes

$$\bar{I}^3 - 190m\ell^2\bar{I}^2 + 10{,}975m^2\ell^4\bar{I} - 185{,}250m^3\ell^6 = 0. \tag{A.43}$$

The three roots of this equation are

$$\bar{I}_{O1} = 95m\ell^2, \quad \bar{I}_{O2} = 65m\ell^2, \quad \text{and} \quad \bar{I}_{O3} = 30m\ell^2, \tag{A.44}$$

which are the three principal moments of inertia. Substituting $\bar{I}_{O1}$ and Eq. (A.42) into Eq. (A.35), we obtain

$$-37m\ell^2 x_1 + 14m\ell^2 y_1 = 0, \tag{A.45}$$
$$14m\ell^2 x_1 - 58m\ell^2 y_1 = 0, \tag{A.46}$$
$$0z_1 = 0. \tag{A.47}$$

the only solution of which is $x_1 = y_1 = 0$ and z_1 is arbitrary. Therefore, the principal direction corresponding to the principal moment of inertia $\bar{I}_{O1}$ is the z direction (the blue axis in Fig. A.10). Similarly substituting $\bar{I}_{O2}$, we obtain

$$-7m\ell^2 x_2 + 14m\ell^2 y_2 = 0, \tag{A.48}$$
$$14m\ell^2 x_2 - 28m\ell^2 y_2 = 0, \tag{A.49}$$
$$30m\ell^2 z_2 = 0, \tag{A.50}$$

the solution of which is $y_2 = \frac{1}{2}x_2$ and $z_2 = 0$. This direction is shown in Fig. A.10 as the line labeled $\bar{I}_{O2}$. Finally, similarly substituting $\bar{I}_{O3}$, we obtain

$$28m\ell^2 x_3 + 14m\ell^2 y_3 = 0, \tag{A.51}$$
$$14m\ell^2 x_3 + 7m\ell^2 y_3 = 0, \tag{A.52}$$
$$65m\ell^2 z_3 = 0, \tag{A.53}$$

the solution of which is $y_3 = -2x_3$ and $z_3 = 0$. This direction is shown in Fig. A.10 as the line labeled $\bar{I}_{O3}$.

Notice that the principal directions are mutually orthogonal and, in this case, they happen to go through the point masses themselves.

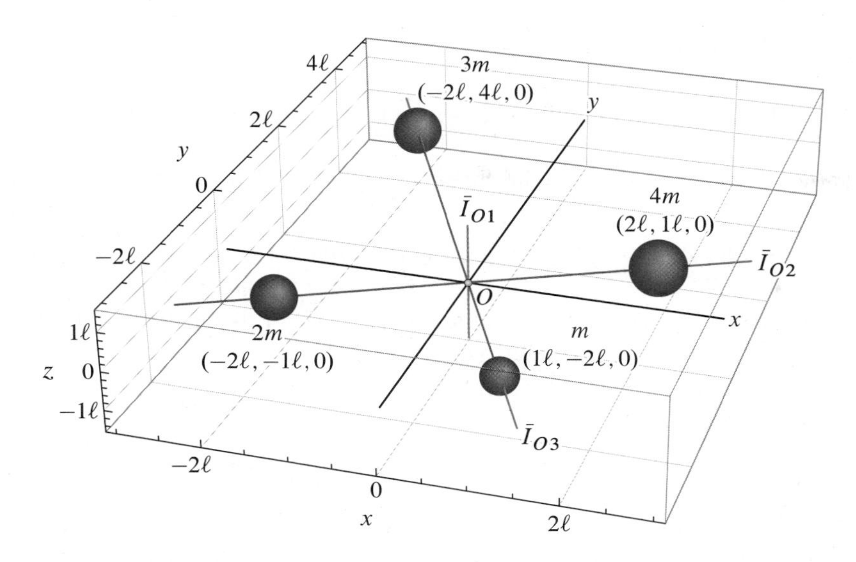

Figure A.10. The locations of the four masses showing the three principal directions.

Moment of inertia about an arbitrary axis

Referring to Fig. A.11, suppose we know all the components of the inertia tensor relative to the xyz axes shown and that we wish to find the moment of inertia of the body about the arbitrarily oriented axis ℓ. The direction of ℓ is defined by the unit vector $\hat{u}_\ell$. We know that the definition of mass moment of inertia states the moment of inertia about the ℓ axis is given by

$$I_{P\ell} = \int_B \left(|\vec{r}_{dm/P}| \sin\theta\right)^2 dm, \tag{A.54}$$

where θ is the angle between the positive directions of the $\vec{r}_{dm/P}$ and $\hat{u}_\ell$ vectors and $|\vec{r}_{dm/P} \sin\theta|$ is the perpendicular distance from ℓ to dm. If we now notice that

$$|\vec{r}_{dm/P} \times \hat{u}_\ell| = |\vec{r}_{dm/P}||\hat{u}_\ell| \sin\theta = |\vec{r}_{dm/P}| \sin\theta, \tag{A.55}$$

and therefore that

$$\left(|\vec{r}_{dm/P}| \sin\theta\right)^2 = (\vec{r}_{dm/P} \times \hat{u}_\ell) \cdot (\vec{r}_{dm/P} \times \hat{u}_\ell), \tag{A.56}$$

then we can write the integral in Eq. (A.54) as

$$I_{P\ell} = \int_B (\vec{r}_{dm/P} \times \hat{u}_\ell) \cdot (\vec{r}_{dm/P} \times \hat{u}_\ell)\, dm. \tag{A.57}$$

If we now write $\vec{r}_{dm/P}$ and $\hat{u}_\ell$ in component form as

$$\vec{r}_{dm/P} = x\,\hat{\imath} + y\,\hat{\jmath} + z\,\hat{k}, \tag{A.58}$$

$$\hat{u}_\ell = u_{\ell x}\,\hat{\imath} + u_{\ell y}\,\hat{\jmath} + u_{\ell z}\,\hat{k}, \tag{A.59}$$

so that

$$\vec{r}_{dm/P} \times \hat{u}_\ell = (u_{\ell z} y - u_{\ell y} z)\,\hat{\imath} + (u_{\ell x} z - u_{\ell z} x)\,\hat{\jmath} + (u_{\ell y} x - u_{\ell x} y)\,\hat{k}, \tag{A.60}$$

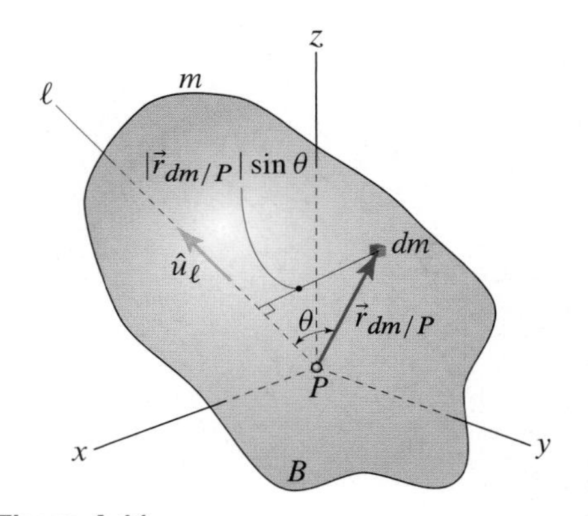

Figure A.11
The quantities needed to find the moment of inertia of a body B about an arbitrarily oriented axis ℓ.

then Eq. (A.57) becomes

$$I_{P\ell} = \int_B \left[(u_{\ell z} y - u_{\ell y} z)^2 + (u_{\ell x} z - u_{\ell z} x)^2 + (u_{\ell y} x - u_{\ell x} y)^2\right] dm \tag{A.61}$$

$$\begin{aligned} &= u_{\ell x}^2 \int_B (y^2 + z^2)\, dm + u_{\ell y}^2 \int_B (x^2 + z^2)\, dm + u_{\ell z}^2 \int_B (x^2 + y^2)\, dm \\ &\quad - 2u_{\ell x} u_{\ell y} \int_B xy\, dm - 2u_{\ell x} u_{\ell z} \int_B xz\, dm - 2u_{\ell y} u_{\ell z} \int_B yz\, dm. \end{aligned} \tag{A.62}$$

The integrals in Eq. (A.62) are the *known* components of the inertia tensor relative to the xyz axes, so we can write Eq. (A.62) as

$$\begin{aligned} I_{P\ell} = {} & u_{\ell x}^2 I_{Px} + u_{\ell y}^2 I_{Py} + u_{\ell z}^2 I_{Pz} \\ & - 2u_{\ell x} u_{\ell y} I_{Pxy} - 2u_{\ell x} u_{\ell z} I_{Pxz} - 2u_{\ell y} u_{\ell z} I_{Pyz}. \end{aligned} \tag{A.63}$$

Thus, if we know the moments and products of inertia for a body B about a given set of axes xyz (i.e., the inertia tensor relative to these axes), we can find the moment of inertia of that body about an arbitrarily oriented axis ℓ using Eq. (A.63), where we note that the components of the unit vector $\hat{u}_\ell$, namely, $u_{\ell x}$, $u_{\ell y}$, and $u_{\ell z}$, are also called the *direction cosines* of $\hat{u}_\ell$ relative to the x, y, and z axes, respectively.

Evaluation of moments of inertia using composite shapes

The parallel axis theorem written for composite shapes is

$$I_x = \sum_{i=1}^{n} \left(I_{Gx} + d_x^2\, m\right)_i, \tag{A.64}$$

where n is the number of shapes, I_{Gx} is the mass moment of inertia for shape i about its mass center x' axis, d_x is the *shift distance* for shape i (i.e., the distance between the x axis and the x' axis for shape i), and m is the mass for shape i. Similar expressions may be written for I_y and I_z. To use Eq. (A.64), it is necessary to know the mass moment of inertia for each of the composite shapes about its mass center axis, and this generally must be obtained by integration or, when possible, by consulting a table of moments of inertia for common shapes, such as the Table of Properties of Solids on the inside back cover. A common error is to use the parallel axis theorem to relate moments of inertia between two parallel axes where neither of them is a mass center axis.

Angular Momentum of a Rigid Body

Hurricane Isabelle on September 15, 2003, as seen from the International Space Station. Even though the air particles near the eye of a hurricane have a small moment arm relative to the center of the eye, they contribute significantly to the total angular momentum because they have the highest speed.

In Section 8.2, we used the angular momentum of a rigid body to derive the angular impulse-momentum principle for a rigid body. We will now show how we obtained Eq. (8.40) on p. 619 for the angular momentum of a rigid body. Along the way, we will derive the angular momentum for a rigid body in three-dimensional motion since that result will be useful in Chapter 10.

We now obtain the angular momentum of a rigid body $\vec{h}_P$, as given by Eq. (8.40) on p. 619, which we repeat here as

$$\vec{h}_P = I_G \vec{\omega}_B + \vec{r}_{G/P} \times m\vec{v}_G, \tag{B.1}$$

where I_G and $\vec{\omega}_B$ are the body's mass moment of inertia and angular velocity, respectively (see Fig. B.1). We could derive Eq. (B.1) directly, but it will be useful to derive the general three-dimensional form of the angular momentum of a rigid body and then simplify the result to Eq. (B.1) since the general result will also be useful in Chapter 10.

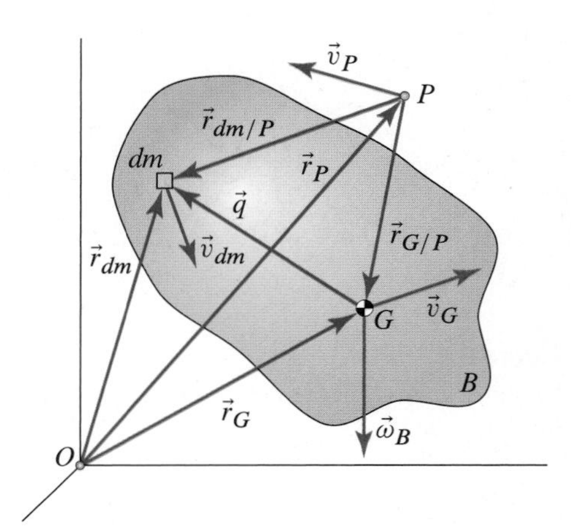

Figure B.1
A rigid body in motion with the definitions of the vectors needed to characterize the body's angular momentum.

Angular momentum of a rigid body undergoing three-dimensional motion

Referring to Fig. B.1, recall that the angular momentum of a rigid body B about the arbitrary point P was defined in Eq. (7.6) on p. 522 to be

$$\vec{h}_P = \int_B \vec{r}_{dm/P} \times \vec{v}_{dm}\, dm. \tag{B.2}$$

The vectors $\vec{r}_{dm/P}$ and $\vec{v}_{dm}$ indicate the position of the infinitesimal mass element dm relative to P and the velocity of dm, respectively. Letting $\vec{q}$ indicate the position of dm relative to the mass center G, and noting that G and dm are two points on the same rigid body, we can write

$$\vec{r}_{dm/P} = \vec{q} + \vec{r}_{G/P} \quad \text{and} \quad \vec{v}_{dm} = \vec{v}_G + \vec{\omega}_B \times \vec{q}. \tag{B.3}$$

Substituting Eqs. (B.3) into Eq. (B.2) and expanding the cross products, we have

$$\vec{h}_P = \int_B \vec{q} \times (\vec{\omega}_B \times \vec{q})\, dm + \int_B \vec{q} \times \vec{v}_G\, dm + \int_B \vec{r}_{G/P} \times (\vec{\omega}_B \times \vec{q})\, dm + \int_B \vec{r}_{G/P} \times \vec{v}_G\, dm. \tag{B.4}$$

To simplify the right-hand side of Eq. (B.4), recall that $\vec{q}$ measures the position of each mass element dm relative to the mass center G of the body. Therefore, the definition of center of mass requires that

$$\int_B \vec{q}\, dm = m\vec{r}_{G/G} = \vec{0}. \tag{B.5}$$

Since the quantities $\vec{v}_G$, $\vec{r}_{G/P}$, and $\vec{\omega}_B$ are not a function of position within B, the last three terms on the right-hand side of Eq. (B.4) become

$$\int_B \vec{q} \times \vec{v}_G\, dm = \left(\int_B \vec{q}\, dm\right) \times \vec{v}_G = \vec{0}, \tag{B.6}$$

$$\int_B \vec{r}_{G/P} \times (\vec{\omega}_B \times \vec{q})\, dm = \vec{r}_{G/P} \times \left[\vec{\omega}_B \times \left(\int_B \vec{q}\, dm\right)\right] = \vec{0}, \tag{B.7}$$

and

$$\int_B \vec{r}_{G/P} \times \vec{v}_G\, dm = (\vec{r}_{G/P} \times \vec{v}_G) \int_B dm = \vec{r}_{G/P} \times m\vec{v}_G. \tag{B.8}$$

Substituting Eqs. (B.6)–(B.8) into Eq. (B.4) then gives

$$\vec{h}_P = \int_B \vec{q} \times (\vec{\omega}_B \times \vec{q})\, dm + \vec{r}_{G/P} \times m\vec{v}_G. \tag{B.9}$$

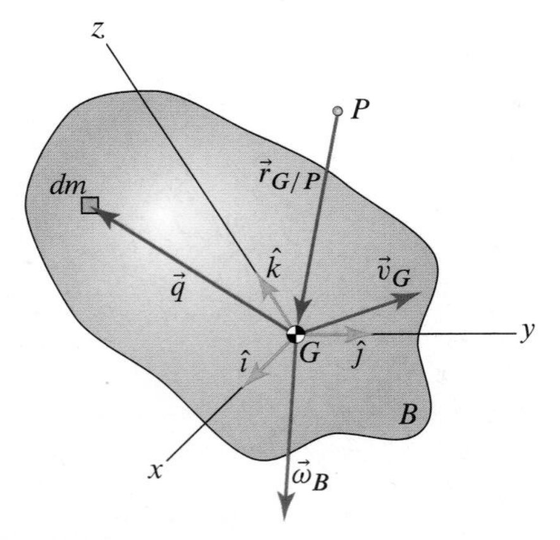

Figure B.2
Vector definitions needed in the derivation of $\vec{h}_P$ for a rigid body.

The second term on the right side of Eq. (B.9) needs no further interpretation since $\vec{r}_{G/P}$ is the position of the mass center of the rigid body relative to the reference point P, and $\vec{v}_G$ is the velocity of the mass center of the rigid body. As for the integral on the right side of Eq. (B.9), writing $\vec{q}$ and $\vec{\omega}_B$ in Cartesian components as (see Fig. B.2)

$$\vec{q} = q_x\,\hat{\imath} + q_y\,\hat{\jmath} + q_z\,\hat{k} \quad \text{and} \quad \vec{\omega}_B = \omega_{Bx}\,\hat{\imath} + \omega_{By}\,\hat{\jmath} + \omega_{Bz}\,\hat{k}, \tag{B.10}$$

and then substituting them into the integral on the right side of Eq. (B.9), we obtain

$$\begin{aligned}\int_B \vec{q} \times (\vec{\omega}_B \times \vec{q})\, dm = &\int_B \left[\left(q_y^2 + q_z^2\right)\omega_{Bx} - q_x q_y \omega_{By} - q_x q_z \omega_{Bz}\right] dm\,\hat{\imath} \\ &+ \int_B \left[-q_x q_y \omega_{Bx} + \left(q_x^2 + q_z^2\right)\omega_{By} - q_y q_z \omega_{Bz}\right] dm\,\hat{\jmath} \\ &+ \int_B \left[-q_x q_z \omega_{Bx} - q_y q_z \omega_{By} + \left(q_x^2 + q_y^2\right)\omega_{Bz}\right] dm\,\hat{k}.\end{aligned} \tag{B.11}$$

Bringing the components of $\vec{\omega}_B$ outside of the integrals since they are not a function of position within B, and then using the definitions of moment and product of inertia given in Eqs. (A.1)–(A.6) on p. A-1, we obtain

$$\int_B \vec{q} \times (\vec{\omega}_B \times \vec{q})\, dm = \left(I_{Gx}\omega_{Bx} - I_{Gxy}\omega_{By} - I_{Gxz}\omega_{Bz}\right)\hat{\imath} + \left(-I_{Gxy}\omega_{Bx} + I_{Gy}\omega_{By} - I_{Gyz}\omega_{Bz}\right)\hat{\jmath} + \left(-I_{Gxz}\omega_{Bx} - I_{Gyz}\omega_{By} + I_{Gz}\omega_{Bz}\right)\hat{k}, \tag{B.12}$$

where the moments and products of inertia *are with respect to the mass center G* since $\vec{q}$ measures the position of dm with respect to G. Writing $\vec{r}_{G/P}$ and $\vec{v}_G$ in Cartesian components as

$$\vec{r}_{G/P} = x_{G/P}\,\hat{\imath} + y_{G/P}\,\hat{\jmath} + z_{G/P}\,\hat{k}, \tag{B.13}$$

$$\vec{v}_G = v_{Gx}\,\hat{\imath} + v_{Gy}\,\hat{\jmath} + v_{Gz}\,\hat{k}, \tag{B.14}$$

and then substituting Eqs. (B.12)–(B.14) into Eq. (B.9), the *angular momentum of a rigid body* is given by

$$\vec{h}_P = \left(I_{Gx}\omega_{Bx} - I_{Gxy}\omega_{By} - I_{Gxz}\omega_{Bz}\right)\hat{\imath} + \left(-I_{Gxy}\omega_{Bx} + I_{Gy}\omega_{By} - I_{Gyz}\omega_{Bz}\right)\hat{\jmath} + \left(-I_{Gxz}\omega_{Bx} - I_{Gyz}\omega_{By} + I_{Gz}\omega_{Bz}\right)\hat{k} + \vec{r}_{G/P} \times m\vec{v}_G. \tag{B.15}$$

If any of the following conditions is satisfied:

1. The reference point P is the mass center G of the rigid body so that $\vec{r}_{G/P} = \vec{0}$.
2. The mass center G of the rigid body is a fixed point so that $\vec{v}_G = \vec{0}$.
3. The velocity of the mass center G, $\vec{v}_G$, is parallel to the position of G relative to P, $\vec{r}_{G/P}$.

then Eq. (B.15) simplifies to

$$\vec{h}_P = \left(I_{Gx}\omega_{Bx} - I_{Gxy}\omega_{By} - I_{Gxz}\omega_{Bz}\right)\hat{\imath} + \left(-I_{Gxy}\omega_{Bx} + I_{Gy}\omega_{By} - I_{Gyz}\omega_{Bz}\right)\hat{\jmath} + \left(-I_{Gxz}\omega_{Bx} - I_{Gyz}\omega_{By} + I_{Gz}\omega_{Bz}\right)\hat{k}, \tag{B.16}$$

$$= h_{Px}\,\hat{\imath} + h_{Py}\,\hat{\jmath} + h_{Pz}\,\hat{k}. \tag{B.17}$$

Conditions 1 and 2 are very common in practice, so Eq. (B.16) is frequently used.

Practical use of Eqs. (B.15) and (B.16)

The xyz reference frame referred to in Eqs. (B.15) and (B.16) can be any frame (inertial or not) as long as velocity and angular velocity components are computed with respect to an inertial frame. With that in mind, if a body rotates relative to the xyz frame, the moments and products of inertia in Eqs. (B.15) and (B.16) will be *time dependent*. Referring to Fig. B.3, notice that as the rectangular body moves in the xy plane, its distribution of mass changes relative to the nonmoving xyz reference frame, and therefore its moments and products of inertia will be time dependent relative to that frame. On the other hand, the moments and products of inertia relative to the $x'y'z'$ frame that is attached to the body will be *constant*. We will see in applications that attaching a reference frame to the body will be the preferred strategy for solving problems. We will refer to these frames as *body fixed*.

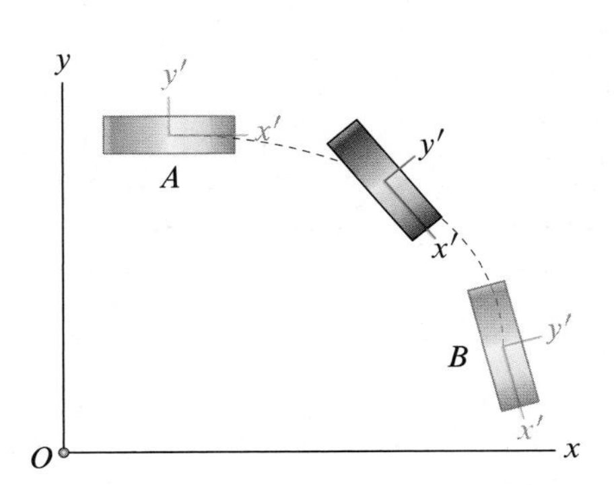

Figure B.3
A rectangular rigid body rotating in planar motion relative to the *fixed* xy frame. The $x'y'$ is *attached* to the body.

A compact way to write Eqs. (B.15) and (B.16)

Using a combination of matrix and vector notation, Eq. (B.15) can be written as

$$\boxed{\{h_P\} = [I_G]\{\omega_B\} + \vec{r}_{G/P} \times m\vec{v}_G,} \tag{B.18}$$

where $\{h_P\}$ is the angular momentum vector of the body B with respect to P, $\{\omega_B\}$ is the angular velocity vector of B, and $[I_G]$ is the *inertia matrix* or *inertia tensor* of B with respect to the xyz axes. In matrix form *and* vector form, these quantities are given by

$$\{h_P\} = \begin{Bmatrix} h_{Px} \\ h_{Py} \\ h_{Pz} \end{Bmatrix} = \vec{h}_P = h_{Px}\,\hat{\imath} + h_{Py}\,\hat{\jmath} + h_{Pz}\,\hat{k}, \tag{B.19}$$

$$\{\omega_B\} = \begin{Bmatrix} \omega_{Bx} \\ \omega_{By} \\ \omega_{Bz} \end{Bmatrix} = \vec{\omega}_B = \omega_{Bx}\,\hat{\imath} + \omega_{By}\,\hat{\jmath} + \omega_{Bz}\,\hat{k}, \tag{B.20}$$

and

$$[I_G] = \begin{bmatrix} I_{Gx} & -I_{Gxy} & -I_{Gxz} \\ -I_{Gxy} & I_{Gy} & -I_{Gyz} \\ -I_{Gxz} & -I_{Gyz} & I_{Gz} \end{bmatrix}.$$

Standard matrix multiplication of $[I_G]$ and $\{\omega_B\}$, with the addition of the vector cross product $\vec{r}_{G/P} \times m\vec{v}_G$, then gives $\{h_P\}$ as given in either Eq. (B.15) or (B.18). With this as background, we see that Eq. (B.16) can be written much more compactly as

$$\boxed{\{h_P\} = [I_G]\{\omega_B\}.} \tag{B.21}$$

> **Helpful Information**
>
> **The column matrix representation of a vector.** In representing the vector $\vec{h}_P$ as
>
> $$\{h_P\} = \begin{Bmatrix} h_{Px} \\ h_{Py} \\ h_{Pz} \end{Bmatrix}$$
>
> the first row contains the x component, the second row the y component, and the third row the z component of the vector. The same interpretation applies to $\{\omega_B\}$. Similarly, when we multiply $[I_G]$ by $\{\omega_B\}$, we obtain a vector whose first, second, and third rows are the x, y, and z components of the vector, respectively.

Angular momentum of a rigid body in planar motion

We are now in a position to develop Eq. (B.1), which we recall is the angular momentum of a rigid body in planar motion that is symmetric with respect to the plane of motion.

If the rigid body B is in planar motion and the motion is occurring in the xy plane, then

$$\omega_{Bx} = \omega_{By} = 0, \quad v_{Gz} = 0, \quad \text{and} \quad z_{G/P} = 0, \tag{B.22}$$

which means that Eq. (B.15) becomes

$$\boxed{\begin{aligned} \vec{h}_P = {} & -I_{Gxz}\omega_{Bz}\,\hat{\imath} - I_{Gyz}\omega_{Bz}\,\hat{\jmath} \\ & + \big[I_{Gz}\omega_{Bz} + m\big(x_{G/P}v_{Gy} - y_{G/P}v_{Gx}\big)\big]\,\hat{k}, \end{aligned}} \tag{B.23}$$

where, again, the mass moments and products of inertia are computed with respect to the mass center G of the rigid body.

Finally, if the xy plane is also a plane of symmetry for the body, then $I_{Gxz} = 0$ and $I_{Gyz} = 0$, and so Eq. (B.23) can be written as

$$\vec{h}_P = \big[I_{Gz}\omega_{Bz} + m\big(x_{G/P}v_{Gy} - y_{G/P}v_{Gx}\big)\big]\,\hat{k}. \tag{B.24}$$

Noting that for planar motion, $I_{Gz} = I_G$, $\omega_{Bz}\,\hat{k} = \vec{\omega}_B$, and that $\vec{r}_{G/P} \times m\vec{v}_G = m\big(x_{G/P}v_{Gy} - y_{G/P}v_{Gx}\big)$, we can write Eq. (B.24) as

$$\boxed{\vec{h}_P = I_G\vec{\omega}_B + \vec{r}_{G/P} \times m\vec{v}_G,} \tag{B.25}$$

which is the form of $\vec{h}_P$ used in Eq. (8.40).

Answers to Even-Numbered Problems

The answers to all even-numbered problems, except for Concept Questions, are provided in a freely downloadable PDF file at:

`http://www.mhhe.com/pgc2e`

Providing answers in this manner allows for the inclusion of plots for Computer Problems and gives students the convenience of having the answers to even-numbered problems on their smart phone or tablet computer.

PHOTO CREDITS

Chapter 1

Opener: © SuperStock, Inc./SuperStock; 1.2(right-left): NASA; 1.17: © University of Washington Libraries, Special Collections UW21422.

Chapter 2

Opener: © Paul Slaughter/www.slaughterphoto.com; p. 46: NASA; 2.10: © NICOLAS ASFOURI/AFP/Getty Images; p. 56: © Terrance Klassen/Alamy; p. 58: U.S. Navy photo by Seaman Daniel A. Barker; p. 61: © Universal Studios; p. 66: © David Lees/CORBIS; 2.13: © Thomas Barwick/Getty Images; p. 76: U.S. Army Photo; p. 90: "Design and Analysis of a Surface Micromachined Spiral-Channel Viscous Pump," by M. I. Kilani, P. C. Galambos, Y. S. Haik, C-H. Chen, *Journal of Fluids Engineering*, Vol. 125, pp. 339-344, 2003. Originally published by ASME; p. 97: NASA; p. 102: Courtesy of the Department of Energy; p. 108: © Charles O'Rear/CORBIS; p. 115: "Design and Analysis of a Surface Micromachined Spiral-Channel Viscous Pump," by M. I. Kilani, P. C. Galambos, Y. S. Haik, C-H. Chen, *Journal of Fluids Engineering*, Vol. 125, pp. 339-344, 2003. Originally published by ASME; 2.34: © Joe Jennings/Jennings Productions; pp. 148-167: © Gary L. Gray.

Chapter 3

Opener: © Robert King; 3.4: © Gary L. Gray; p. 176: © David A. Northcott/CORBIS; 3.8: © Jonathan Ferrey/Getty Images; p. 196: © Andrew Redington/Getty Images; 3.16: U.S. Navy.

Chapter 4

Opener: © Eric Gaillard/Reuters/Corbis; p. 246: U.S. Navy photo by Mass Communication Specialist 3rd Class Torrey W. Lee; p. 251: NASA; p. 255: U.S. Navy photo by Mass Communication Specialist 3rd Class Torrey W. Lee; p. 264: © GFC Collection/Alamy RF; p. 280: © tbkmedia.de/Alamy; p. 288: © Gary L. Gray; 4.13: © Ferrari Press Office, HO/AP Photo; p. 299: U.S. Navy photo by Mass Communication Specialist 3rd Class Torrey W. Lee; p. 303: © Gary L. Gray.

Chapter 5

Opener: NASA; p. 319: PH3 Christopher Mobley/U.S. Navy; p. 325: © Rooney, Irving & Associates, Ltd.; p. 327a: U.S. Navy photo by Photographer's Mate 2nd Class H. Dwain Willis; p. 327b: PHAN James Farrally II, U.S. Navy; p. 327c: U.S. Navy photo by Photographer's Mate 3rd Class (AW) J. Scott Campbell; 5.6: © Loren M. Winters, Durham, NC; 5.7: © Gary L. Gray; p. 368: © Jill Braaten; 5.23-5.29, pp. 397-398: NASA; 5.34: © Sean Gallup/Getty Images; p. 410: © Royalty-Free/CORBIS; pp. 412-418: Courtesy of JetPack International, LLC; p. 421: Courtesy of Andritz Hydro, Austria.

Chapter 6

Opener: © John Peter Photography/Alamy; 6.1, p. 441: © Fordimages.com; p. 443: © Majdi Mohammed/AP Photo; p. 444: © Lawrence Manning/Corbis RF; p. 445(top left): © Arrow Gear Company; p. 445(top right): NASA; p. 445(bottom left): Courtesy of Andritz Hydro, Austria; p. 445(bottom right): Courtesy of Voith Hydro, Germany; p. 446: © Gary L. Gray; p. 447: © David Lees/CORBIS; p. 450: © Gary L. Gray; p. 451: © Martin Child/Getty Images RF; p. 458: © Majdi Mohammed/AP Photo; p. 459: © Vladimir Vakhrin/iStockphoto.com; p. 462(top): © Jon Reis; p. 462(bottom): © The McGraw-Hill Companies, Inc./Photo by Lucinda Dowell; p. 464: Courtesy of Otto Bock HealthCare, Germany; p. 478: © Vladimir Vakhrin/iStockphoto.com; p. 484: Courtesy of Otto Bock HealthCare, Germany; p. 493: Courtesy of Specialized Bicycle Components; p. 511: NASA.

Chapter 7

Opener: © Quinn Rooney/Getty Images; p. 534: © Gary L. Gray; p. 544: Courtesy of Beckman Coulter, Inc.; pp. 549-550: Courtesy of Amazing Gates of America; pp. 556-569: © Gary L. Gray; p. 571: NASA; p. 573: Courtesy of Ducati Motor Holding.

Chapter 8

Opener: NASA; 8.6: © Joe Kilmer/Penninsula Spring Corp.; p. 590: © Polka Dot Images/Jupiterimages RF; pp. 594-603: © Gary L. Gray; p. 604(top): © The McGraw-Hill Companies, Inc./Photo by Lucinda Dowell; p. 604(bottom): NASA; p. 625(left-right): © Jill Braaten; p. 627: European Space Agency; p. 629: © Gary L. Gray; p. 630(top): © Golden Gate Images/Alamy RF; p. 630(bottom): © Gary L. Gray; p. 631: NASA; p. 636: © Martin Child/Getty Images RF; pp. 637-638: NASA; p. 650: © John Lund/Getty Images; p. 654: © 2009 Jupiterimages Corporation RF.

Chapter 9

Opener: Photo by Don Schlitten; p. 672: Reprinted by permission from Macmillan Publishers Ltd: *Nature Nanotechnology*, M. Li, R.B. Bhiladvala, T.J. Morrow, J.A. Sioss, K.K. Lew et al., "Bottom-up assembly of large-area nanowire resonator arrays," 3(2):88-92, copyright 2008; p. 676: © Jeppe Wikstrom/Getty Images; p. 697: PH3 Christopher Mobley/U.S. Navy; p. 704: © Gary L. Gray; p. 718: Reprinted by permission from Macmillan Publishers Ltd: *Nature Nanotechnology*, M. Li, R.B. Bhiladvala, T.J. Morrow, J.A. Sioss, K.K. Lew et al., "Bottom-up assembly of large-area nanowire resonator arrays," 3(2):88-92, copyright 2008.

Chapter 10

Opener: © Corbis RF; p. 756: Photo courtesy of Beckman Coulter, Inc.

Appendix A

Opener: NASA; A.3: European Space Agency.

Appendix B

Opener: NASA.

INDEX

A

Abdul-Malik, Ahmed, 663
Acceleration analysis
  Coriolis component, 498–500
  example problems, 478–85, 530, 544–47, 554–57
  rigid bodies in three dimensions, 722
  rigid bodies in two dimensions, 474–75, 513–15
  rotating reference frames, 496–97, 515
Acceleration as function of position, 47–49, 53, 155–56
Acceleration as function of time, 51
Acceleration as function of velocity, 52, 56–57
Acceleration due to gravity. *See also* Gravitational force
  to calculate well depth, 54–55
  defined, 3–4
  effects on projectile motions, 68
Acceleration vectors
  components in circular motion, 84
  form in Cartesian component systems, 33
  normal-tangential components, 92, 93, 95–98
  notation, 6
  polar coordinates, 105
  properties, 32, 155
Accuracy of numbers, 12
Acrobatic catch example, 650–51
Active guidance systems, 71
Addition of vectors, 8
Aerodynamic drag, 55, 57, 196–97
Aircraft, person pulling, 321
Aircraft carrier landings, 319, 697
Aircraft carrier launch forces, 246–47, 299, 327
Aircraft tracking examples, 85–86, 106, 150
Air hockey, 346–47
Altitude, acceleration due to gravity and, 4
Ames Research Center, 46, 97
Amplitude, 664, 714
Andretti, Marco, 192
Angular acceleration. *See also* Acceleration analysis
  particles in planar motion, 49, 58–59
  rigid bodies in planar motion, 436, 546–47
  rigid bodies in three dimensions, 722, 724, 726–31, 759
Angular impulse-momentum principle
  for impacts between rigid bodies, 639–44
  for particles, 362, 424
  in place of Newton's third law, 2
  for rigid bodies, 619–20, 623–24
  for systems of particles, 362–64
Angular impulse with respect to *P*, 362
Angular momentum. *See also* Conservation of angular momentum
  computing for rigid bodies, 622, A-11–A-14
  conservation for particles, 107, 364
  conservation for rigid bodies, 620, 625–28, 643, 644
  defined for particles, 361–62, 424
  example problems, 368–77
  rigid bodies in planar motion, A-14
  rigid bodies in three dimensions, 738, A-6–A-7, A-11–A-14
  rotational equations of motion, 522–26
Angular momentum per unit mass, 386
Angular speed, 58, 455, 541
Angular velocity
  defined, 49
  rigid bodies in planar motion, 435, 436
  rigid bodies in three dimensions, 722–24, 726–31, 759
  in rigid body collisions, 641
  of rotations, 81, 97–98
Anticommutative property of vector cross product, 8
Apoapses, 387, 388–89
Apogee, 212, 276, 370–73, 395
Apollo missions, 399
Apparent motions, 497–98, 504–5
Approach velocity, 337
Apses, 387
Arbitrary axes, moments of inertia about, A-9–A-10
Arbitrary moment center, 553, 578
Arc length, 92
Areal velocity, 386
Argument, 10
Arresting cables, 319, 697
Artificial knee joint examples, 464–65, 484–85
Artillery, 76
Atlantis shuttle, 313
Atoms, forces between, 279
Authalic mean radius of Earth, 3
Automobile engines, 433–34, 441
Average braking force, 319
Average force
  defined, 315, 422
  relating to kinetic energy, 246–47
Average speed, 31
Average velocity vectors, 30
Axes of rotation, 80–81
Axioms, 169

B

Bailey, Donovan, 266
Balance principles, 170
Balancing centrifuges, 545
Ball bearings, 79, 360
Ballistic pendulums, 353, 644
Baseball bats, 542–43
Baseball trajectory example, 72–73
Base dimensions, 10
Base units, 10
Basketballs, 354
Belts on pulleys, 441
Billiard ball standards, 357
Binormal unit vectors, 93
Body-fixed frames, 494, A-13
Boeing 737 airliner, 321
Bolt, Usain, 47, 266
Braking test example, 53
Bubka, Sergey, 267
Bumper cars, 639, 640
Bungee jumping examples, 264–65, 665, 667

C

Cable tension, 195
Canoe examples, 220–21
Cantilevered bar deflection, 672
Car engines, 433–34, 441
Carets, 6
Carnival ride examples, 443, 458, 704–5
Cartesian components, 6, 9–10. *See also* Components

Cartesian coordinate system
basic concepts, 5, 155
Newton's second law applied, 192, 193
radii of curvature, 93
relation to components of vectors, 32–33
representing vectors in, 6, 7
three dimensions in, 144, 160
Catapult systems
aircraft carrier, 246–47, 299, 327
conservation of energy example, 286–87
Catch example, 650–51
Center for Gravitational Biology Research, 46, 97
Center of percussion, 542–43, 652
Centers of mass (particle systems)
calculating motions, 218–19, 235
in closed versus open systems, 363
defining position, 218
use in determining kinetic energy, 282–83
Centers of mass (rigid bodies), 640, 641
Central forces
example problems, 202–3
gravitational, 385
work on particles, 257–58
Central impacts
between particles, 336
between rigid bodies, 640–42, 646–47
Centrifuges
Center for Gravitational Biology Research, 46, 97–98
electric motors for, 442
fixed angle, 756
with swinging-bucket rotors, 544–45
for uranium enrichment, 102
Chain rule, 48
Chameleons, 176
Chaos theory, 563
Characteristic equation, 696, A-7
Charpy impact test, 277, 611
Circular frequency, 664
Circular motion
example problems, 97–98, 200–201
one-dimensional relations, 49, 156
time derivatives of vectors and, 83–84
Circular orbits, 373, 388, 426
Closed systems
defined, 315, 401
impulse-momentum principle application, 401, 404, 405
modeling control volumes as, 402, 404
Coefficient of kinetic friction, 171, 235
Coefficient of restitution
in particle collisions, 79, 336–39, 350–51
in rigid body collisions, 639, 640, 641, 643, 658
Coefficient of static friction, 171, 234
Coefficient of viscous damping, 695, 716
Collector plates, 148–49
Collisions. *See* Impacts
Coltrane, John, 663
Comet Halley, 396–97
Complementary solutions, 681, 682, 715, 717
Component form of Newton's second law, 170
Components of vectors, 9–10, 15, 32–33
Component systems, 192–94
Composite rigid bodies, 546–47
Composite shapes method, A-10
Compressed gas, 277
Conic sections, 386–87, 388–90, 426–27
Connecting rods. *See also* Slider-crank mechanisms
acceleration analysis, 474–75, 480–81
basic functions, 433–34
construction, 481
general planar motions, 435
oscillation example, 671
velocity analysis examples, 460–61
Conservation of angular momentum
for particles, 107, 364, 369–77
for rigid bodies, 620, 625–28
in rigid body collisions, 639, 643, 644
Conservation of linear momentum
example problems, 220–21, 322, 627–28
for rigid bodies, 618–19
in rigid body collisions, 639, 642
for systems of particles, 219, 316–17, 423
Conservation of mechanical energy
defined, 259, 306
example problems, 264–65, 286–87, 598–99
harmonic oscillation, 667
rigid bodies in planar motion, 586
Conservative forces
defined, 259, 306
example problems, 272
identifying, 260–61
in potential energy, 307
Conservative systems, 259, 306
Constant acceleration
in one-dimensional problems, 49
of projectile motion, 68
propeller example, 58–59
Constant forces, work of, 243–44
Constant of universal gravitation, 3
Constant velocity, 106, 150
Constrained eccentric impacts, 643–44
Constrained impacts
between particles, 336, 348–49, 423
between rigid bodies, 640, 643–44, 648–49, 658
Constrained systems
example problems, 129–32
relative motion in, 122–24, 158
Constraint force equations, 453
Contact forces, 5
Continuous bodies, 365
Control volumes, 401–3
Conversion factors, 11
Coordinate lines, 33
Coordinate systems
basic concepts, 5
Cartesian system, 5, 6, 144, 148–49, 155
cylindrical, 142–43, 147, 159
polar, 105, 157–58
spherical, 143–44, 150, 159–60
Coriolis, Gaspard-Gustave, 498
Coriolis acceleration, 498–500
Coulomb, Charles Augustin de, 171
Coulomb friction model, 171–73, 179, 181, 234–35
Couples, 586, 588
Cranes, 147
Crankshafts
basic functions, 433–34
in Ford EcoBoost engine, 433
planar motion, 435
Critical damping coefficient, 696, 716
Critically damped systems, 697, 698, 704–5, 716
Cross products
finding for multiple vectors, 83, 84
finding for two vectors, 8, 10
Curvature of a path, 92
Curve geometry, 93
Curvilinear motion, 192–94
Curvilinear translations, 434, 443, 458
Cyclic loading, 545

Cylindrical coordinates
basic concepts, 142–43, 159
example problems, 147
Newton's second law applied, 193–94

D

Damped natural frequency, 697, 717
Damped vibrations, 695–700, 703–7, 716–17
Damping factor, 697
Damping ratio, 697, 717
Dashpots, 695, 698–99, 704
Datum lines, 259
Deformation, 350, 595
Degrees of freedom
defined, 174, 234
in harmonic oscillators, 665
rotating reference frame examples, 501
Del, 260
Derivatives, time. *See* Time derivatives of vectors
Derived dimensions, 10
Derived units, 10
Design, role of dynamics in, 26–27
Despin mechanisms, 511, 571, 638
Determinants, finding cross products with, 10
Differentiation of constraints
acceleration analysis by, 474–75, 482–83, 514
in relating speed to position, 249
relative motion analysis by, 122–24, 158
velocity analysis by, 453, 462–63, 513
Dimensional analysis, 18–20
Dimensional homogeneity, 18
Direct central impacts, 338, 341–45
Direct eccentric impacts, 640
Direct impacts, 336, 640
Direction cosines, A-10
Direction of rotation, 81
Directrices, 387
Displacement vectors, 30, 155
Diverters, 410–11
Diving boards, 674
Docking example, 627–28
Door closers, 462
Dot products, 8
Double-headed arrows, 7
Doubly symmetric objects, A-4
Drag
projectile motion with, 196–97
role in terminal velocity, 57
when to disregard, 55
Drag coefficients, 196
Duffing's equation, 191
Dynamics
basic concepts, 1–2, 5–12
basic role in design, 26–27

E

Earth
acceleration due to gravity, 3–4
Coriolis acceleration, 498–500
motion due to impacts, 343
as particle, 370
radius, 3
Eccentric impacts, 336, 642–44, 657
Eccentricity of conic sections, 387, 426
EcoBoost engine, 433
Effective mass and stiffness, 668
Efficiency, 298, 300–301, 307
Elastic impacts, 338, 423, 640, 641, 658
Elastic potential energy, 259
Electric motor examples, 300–301
Electrostatic precipitators, 148–49
Elementary motions in particle kinematics, 47–49, 155–56
Elevation angles in projectile motion, 70–73
Elliptical orbits, 388–90, 426–27
Energy method, 667–68, 674, 715
Engines, 433–34, 441, 596–97. *See also* Slider-crank mechanisms
Equals sign, 11
Equations of motion, defined, 173, 234
Equivalent force systems, 410
Escape velocity, 390, 427
Euler, Leonhard, 2, 316
Euler's first law, 316, 365, 521, 575, 738, 759–60
Euler's rotational equations for rigid bodies, 740, 760
Euler's second law, 365, 522, 738, 760
Exact differentials, 242
Explorer 7, 397
External forces in particle systems
defined, 217
effects on system motion, 218–19
impulse-momentum principle, 315–16
work-energy principle, 281–82

F

F/A-18 Hornet aircraft, 246, 247, 299
Falkirk Wheel, 433
Falling rigid body example, 554–55
Falling string examples, 288–89, 414–15
Fatigue, 545
Fiberglass poles, 267
First law of motion (Newton), 2
Fixed-angle centrifuge rotors, 756
Fixed axis rotation. *See* Rotation about a fixed axis
Flight time, 69
Floating platform examples, 322
Fluid flows, 401–6, 408–11
Flywheels, 231, 283, 333
Focal parameters, 387
Foci of ellipses, 387
Foot-pounds, 243
Force concepts, 5
Forced vibrations
damped, 697–99, 717
undamped, 681–83, 715–16
Force fields, 272
Force laws, 170, 171–73
Forces, power and efficiency, 298
Force systems, work on rigid bodies, 585–86
Force vectors in free body diagrams, 170
Forcing functions, 681
Ford Excursion, 354, 429
Formula 1 racing, 95
Forward spin, 625
Four-bar linkages, 443, 458, 464–65, 484–85
Fourier, Jean Baptiste Joseph, 681
Fracture toughness tests, 277
Frames of reference. *See* Reference frames
Free body diagrams, impact-relevant, 336
Free body diagrams in dynamics, 169–70
Free vibrations
damped, 695–98, 716–17
example problems, 671–74, 704–5
undamped, 663–68, 714–15
Frequency, 664, 714
Frequency ratio, 682
Friction forces
Coulomb model, 171–73, 179, 181, 234–35
as function of time, 623
impending slip examples, 178–81, 198–99
internal work examples, 290–91
rolling without slip examples, 556–57, 594–95
sliding examples, 224–25, 268–69

G

Gear reduction, 442
Gears, 442, 459
General planar motion
  defined, 435, 437
  Newton-Euler equations, 553, 554–63, 577–78
  rigid body kinetics, 577–78
General solution, equations of motion for harmonic forcing, 681, 682, 715
Geometrical constraints, 122–24, 158
Geometry of curves, 93
Geostationary orbits, 397, 449
Geosynchronous equatorial orbits, 397, 449
Golf balls, 196–97
Governing equations, types, 170, 234
Gradient operator, 260
Grandfather clocks, 678
Gravitation, Newton's law, 2–4, 107, 385
Gravitational force
  orbital mechanics, 385–91, 425–27
  potential energy, 259, 587
  work on particles, 257, 258, 305
Greene, Maurice, 266
Gun tackles, 137–38
Gyration, radii of, 540
Gyroscopic moment on spinning disk, 747
Gyroscopic terms, 729

H

Halley, Edmund, 396–97
Hang angle example, 531
Harmonic motion, 664
Harmonic oscillators
  defined, 664, 714
  forced, 681–83, 707, 715, 717
  natural frequency, 668
  standard form, 665–66, 714
  viscously damped, 696, 697, 717
Hertz, 664
High-pressure systems, 499–500
Hockenheim racing track, 100
Hockey pucks, 646–47
Hohmann transfers, 398, 431
Homogeneity, dimensional, 18
Horsepower, 298
Howitzers, 76
Hubble Space Telescope, 375
Hyperbolic trajectories, 390, 427

I

Impact force law, 337
Impact-relevant FBDs, 336
Impacts
  classifying, 335–36, 640–44
  coefficient of restitution in, 336–39
  constrained, 336
  energy of, 339, 423
  example problems, 341–51, 646–51
  modeling between particles, 335, 423
  of rigid bodies, 639–44, 657–58
Impact tests, 277, 543
Impending slip, 172, 178–81, 198–99
Impulse, 314
Impulse-momentum principle
  angular, for impacts between rigid bodies, 639–44
  angular, for particles, 362
  angular, for rigid bodies, 619–20, 623–24, 657
  collisions between particles, 423
  as conservation law, 316–17
  example problems, 319–22, 342–43, 349, 350–51
  linear, for particles, 313–16, 422
  linear, for rigid bodies, 618–19, 623–24, 657
  for mass flows, 401–6, 428
Impulsive forces, 336, 639
Indentors, 254
Inertia concepts, 5
Inertial reference frames
  basic properties, 173–74
  example problems, 178
  rigid bodies in planar motion, A-13
  in water jet example, 408
Inertia matrix, A-6, A-14
Inertia properties, 5
Inertia tensor, A-6, A-14
Instantaneous center of rotation, 453–56, 513, 585
Integration in elementary motion calculations, 47–48
Internal combustion engines, 433–34, 596–97. *See also* Slider-crank mechanisms
Internal forces in particle systems
  in closed systems, 315–16
  defined, 217, 218
  work-energy principle, 281–82
Internal work
  example problems, 290–91
  in systems of particles, 281–82, 307
International Space Station
  altitude above Earth, 398
  control moment gyroscope installation, 583
  June 2008 photo, 385, A-1
  mass, 628
  Space Shuttle docking, 627, A-3
Isolated systems, 317

J

Jerk, 116
Jet packs, 412–13
Jet propulsion, 59, 406
Johnson, Michael, 266
Joules, 243
Jupiter, 3

K

Kanounnikova, Natalia, 626
Kepler, Johannes, 2, 389
Kepler's first law of motion, 389
Kepler's second law of motion, 386
Kepler's third law of motion, 390
Kinematic chains, 464
Kinematic equations, 170
Kinematics, 29–33. *See also* Particle kinematics; Rigid body kinematics
Kinetic diagrams, 526
Kinetic energy
  approximating as quadratic function of position and velocity, 668
  bicycle example, 590–91
  converting to potential energy, 266–67
  of impacts, 339, 423
  relating to average force, 246–47
  rigid bodies in planar motion, 583–85
  rigid bodies in three-dimensional motion, 743–44, 752–53, 761–62
  for systems of particles, 281, 282–83, 307
  work and, 242–43, 305
Kinetic friction coefficient, 171, 235
Kinetics, 169. *See also* Rigid body kinetics
King Kong, 61

L

Ladders, 131–32, 482–83
Lateral G-force, 95
Lawn roller example, 560–61
Lead of screw, 678
Lennard-Jones force law, 279

Lift coefficient, 24
Lift speed example, 300–301
Linear elastic springs, 173, 235
Linear impulse, 314, 422, 618
Linear impulse-momentum principle
for particles, 313–16, 422
for rigid bodies, 618–19, 623–24
Linearization, 666, 668
Linear momentum. *See also* Conservation of linear momentum
conservation law, 316–17
defined, 422
of particles, 314
of rigid bodies, 618–19
translational equations of motion, 521–22
Linear torsional spring constant, 672
Linear viscous damping, 695
Line integrals, 242
Lines of action, in free body diagrams, 170
Lines of impact, 335–36
Low-pressure systems, 499–500
Lumped-mass models, 374

M

M777 lightweight 155mm howitzer, 76
Magnification factors
damped, forced harmonic oscillators, 700, 717
determining for unbalanced motor, 688–89
forced, undamped vibrations, 682–83, 715–16
Magnitude of vectors, 7, 80, 81
Mars orbiter, 11
Mass, basic concepts, 5, 10
Mass, effective, 668
Mass center moments of inertia, A-3, A-5
Mass center motions. *See* Centers of mass (particle systems)
Mass centers of rigid bodies, 521–22
Mass detectors, 672
Mass flow rate, 402
Mass flows
example problems, 408–15
impulse-momentum principle, 401–6, 428
Mass flux, 402
Mass moments of inertia
about arbitrary axis, A-9–A-10
in angular impulse-momentum principle for rigid bodies, 620
applications, A-3–A-4
in centrifuge example, 545
defining, 524, A-1–A-3
determining from oscillations, 671
parallel axis theorem, A-4–A-6
principal moments, A-6–A-8
as radii of gyration, 540, A-4
for spinning skater, 625
Mass points, 365
Mass products of inertia, A-2, A-3–A-4, A-5–A-6
Material frame indifference principle, 2
McIlroy, Rory, 196
Mechanical energy per unit mass, 391, 427
Mechanical power, 298
Mechanical vibrations. *See* Vibrations
Mechanicians, 26
Meridional Earth radius, 3
Merry-go-round examples, 204–5
Micro spiral pumps, 90, 115
Midair catch example, 650–51
Mini Cooper, 354, 429
Minnesota cities example, 16–17
Modeling, 27
Molecular dynamics, 217
Moment-angular momentum relation
Euler's second law, 365, 522
for particles, 361–62, 424
for systems of particles, 363, 424–25
Moment center, 361, 540, 575–76
Moment of inertia about arbitrary axis, A-9–A-10
Moments acting on fluids, 403–4
Moments of couples, 586
Moments of inertia. *See* Mass moments of inertia
Momentum of particles, 314, 316–17. *See also* Impulse-momentum principle
Monaco Grand Prix, 95
Monk, Thelonius, 663
Motion laws, 2
Motion platforms, 443, 458
Motorcycles, 530
Motors, unbalanced, 686–89, 706–7
Moving frames, 121
Moving targets, 127–28

N

Nabla, 260
Nanoelectromechanical systems, 672
Nanoindentation tests, 254
Nanotechnology, 217, 672–73
NATO 7:62 mm ammunition, 353
Natural frequency, 664, 668, 714
Neptune, 3
Newton, Isaac, 1, 107
Newton-Euler equations
application to rigid bodies in planar motion, 521–26, 575–76
example problems using, 530–33, 554–63
fundamental importance, 365
general planar motion, 553, 577–78
from kinetic diagrams, 526
rotation about a fixed axis, 539–40, 577
three-dimensional motions, 738–43, 759–60
translation in plane of motion, 529, 576–77
Newtons (SI units), 11
Newton's cradle, 356
Newton's laws, 2. *See also* Second law of motion (Newton); Third law of motion (Newton)
Nonconservative forces, 260, 307
Nonlinear systems, 666
Normal-tangential components, particle paths in planar motion, 92–93, 95–98, 157–58
No-slip conditions, 171, 441, 442. *See also* Rolling without slip
Nuclear reactors, 102
Number of degrees of freedom, 174

O

Oblique central impacts, 338–39, 346–49
Oblique impacts, 336, 338–39, 346–49, 640
Octopus ride, 153
One-dimensional motions, 47–49
Open systems, 315, 401–2
Optimal angle in projectile motion, 73
Orbital mechanics
example problems, 395
major concepts, 385–91, 425–26
Orbital periods, 389–90
Orbits of satellites
acceleration analysis, 107
determining, 385–90, 425–26
energy considerations, 391
equations of motion, 212
geostationary, 397, 449
speed, 276, 370–73, 395

Origins, 5
Oscillators. *See also* Vibrations
example problems, 671–74
forced harmonic, 681–83
railcar illustration, 663–65
simple, 24
Osculating circles, 93
Output power, electric motor, 300–301
Overdamped systems, 696, 698, 716
Overhead door example, 598–99

P

Packaging, energy absorbing, 189, 255, 275
Paper cutters, 539, 541
Parabolic trajectories, 39, 390, 427
Parallel axis theorem, A-4–A-6, A-10
Parametric analysis, 461, 481
Parking orbits, 396
Particle kinematics
basic concepts, 29–33
differentiation of geometrical constraints, 122–24
elementary motions, 47–49, 155–56
major concepts, 155
normal-tangential components of planar motion, 92–93, 157–58
polar coordinates in planar motion, 105
projectile motion, 68–73, 156
rectilinear motions, 169–74
relative motion concepts, 121–22, 158
three-dimensional coordinate systems, 142–44
time derivative of a vector, 80–81, 156–57
Particles
angular momentum, 361–62
basic concepts, 6
conservative forces and potential energy, 257–61
linear impulse-momentum principle, 313–16
systems of, 217–19, 281–83
treating Earth as, 370
work-energy principle, 241–44, 305
Particular solutions, 681–82, 717
Path component system, 192, 193
Paths. *See also* Trajectories
defined, 30
normal-tangential components, 92–93, 95–98
relation to work, 257, 258–59
Pelton impulse wheels, 420–21
Pelton turbines, 445
Pendulums
ballistic, 353, 644
Charpy impact test, 277
clock, 678
conservation of angular momentum example, 376–77
Newton's cradle, 356
spherical, 152
Perceived motions, 504–5
Perfectly elastic impacts
between particles, 338, 346–47, 423
between rigid bodies, 640, 641, 658
Perfectly plastic impacts
between particles, 330, 338
between rigid bodies, 640, 657–58
Periapses, 387, 388–89
Perigee, 212, 276, 370–73, 395
Periodic forcing, 681
Periods of elliptical orbits, 389–90
Periods of vibration, 664, 697, 714
Phase angle, 664, 714
Pioneer 3 spacecraft, 511, 571, 638
Pipeline example, 410–11
Pistons
basic functions, 433–34
characterizing motions, 434, 435, 452
maximum accelerations, 481
Planar motion. *See also* Rigid body kinematics; Rigid body kinetics
bodies not symmetric with respect to plane of motion, 742–43, 750–51, 761
general, 435, 437
normal-tangential components, 92–93, 157–58
polar coordinates, 105
Plane angles, 11
Planes of motion, 435
Planetary gear system examples, 459, 478–79
Planet carriers, 478
Plastic impacts, 338, 423, 640, 657–58
Platform bench scales, 355
Pneumatic door closers, 462
Pneumatic tubes, 277
Polar coordinates
basic application to planar motion, 105, 157–58
example problems in planar motion, 106–11
Newton's second law applied, 192–93
Polar slope, 90
Pole vaulting examples, 266–67
Position
acceleration as function, 47–49, 53, 155–56
defined, 155
relating to speed, 248–50
Position vectors
example problems, 16–17
form in Cartesian component systems, 33
notation, 6
polar coordinates, 105
properties, 30
Potential energy
approximating as quadratic function of position and velocity, 668
converting kinetic energy to, 266–67
defined, 259
rigid bodies in planar motion, 586–87
of various forces, 259, 306
Power
basic concepts, 298, 307, 588
computing engine output, 596–97
of couples, 588
example problems, 299–301
Prefixes, 11–12
Preimpact velocities, 640
Primary reference frames, 494
Primitive concepts, 5
Principal axes of inertia, A-6
Principal body axes, 740, 741
Principal moments of inertia, 24, A-6–A-8
Principal unit normals, 92–93
Principle of material frame indifference, 2
Products of inertia, A-2, A-3–A-4, A-5–A-6
Projectile motion
example problems, 69–73, 96, 110–11, 196–97
modeling, 68, 156
Propellers, 58–59, 728–29, 748–49
Propulsion, mass flows in, 404–6
Prosthetic leg examples, 464–65, 484–85
Pulleys
fixed axis rotation, 441
geometrical constraints illustrated by, 123–24, 129–30
system examples, 222–23
in winch and block systems, 248–50
Pullout tests, 254
Pumping while swinging, 270–71
"Punkin Chunkin" competition, 96

Q

Quadratic mean radius of Earth, 3
Quick-return mechanisms, 503, 509

R

Racquetballs, 320, 335
Radar gun examples, 125
Radial components of acceleration, 105
Radial components of velocity, 105
Radial symmetry, 741–42, 760–61
Radians, 11
Radii of curvature, 93, 95, 96
Radii of gyration, 540, A-4
Radius of Earth, 3
Railcar oscillation, 663–65, 695–96
Rascasse, 95
Rate of separation, 133, 134
RCS jets, 375
Reaction wheels, 375
Reciprocating rectilinear motions, 117, 502–3, 508
Rectilinear motion
  force laws, 171–73
  Newton's second law applied, 169–71
  polar coordinates example, 108–9
  reciprocating, 117, 502–3, 508
  relations in particle kinematics, 47–49
Rectilinear translations, 434
Red Arrows aerobatic team, 169
Reference frames
  inertial, 173–74, 178, 408, A-13
  rotating, 494–500, 501–5, 515, 721–24, 759
Reference points for position vectors, 30
Relative motion
  basic concepts, 121–22, 158
  in constrained systems, 122–24, 158
  example problems, 125–32
Resonances, 682–83, 716
Restitution coefficient. *See* Coefficient of restitution
Restitution in particle collisions, 350
Right-hand Cartesian coordinate system, 5
Right-hand rule, 5, 81
Rigid bodies, 6
Rigid body kinematics
  acceleration analysis in planar motion, 474–75
  basic planar motion concepts, 433–39, 512
  rotating reference frames, 494–500
  in three dimensions, 721–24, 759
  velocity analysis in planar motion, 452–56
Rigid body kinetics
  angular impulse-momentum principle, 619–20, 657
  angular momentum equations, 522–26, 575–76, A-11–A-14
  general planar motion, 553
  impacts between bodies, 639–44, 657–58
  linear impulse-momentum principle, 618–19, 657
  linear momentum equations, 521–22
  in planar motion, 583–85
  rotation about a fixed axis, 539–40, 577
  in three dimensions, 738–44, 759–62, A-11–A-14
  translation in plane of motion, 529
  work-energy principle, 583–88, 656
Ring gears, 459, 478
Road barrier example, 592–93
Robotic arm examples, 108–9
Rocket propulsion, 404–6, 412–13
Rocket sled, 51
Rolling resistance, 595
Rolling without slip
  acceleration analysis for, 475, 514–15
  disk motion in three dimensions example, 726–27
  friction examples, 556–57
  impulse-momentum principle examples, 623–24
  planetary gear examples, 459, 478–79
  velocity analysis for, 452–53, 455, 513
  work-energy principle examples, 590–91, 594–95, 600–601
Rotating reference frames, 494–500, 501–5, 515, 721–24, 759
Rotation about a fixed axis
  angular impulse-momentum principle, 620
  example problems, 441–42, 544–45, 592–93
  kinetic energy of rigid bodies, 584–85
  Newton-Euler equations, 539–40, 577
  qualitative description, 434–35
  velocity and acceleration with, 438–39, 512
Rotational kinetic energy, 584
Rotational motions
  angular momentum equations for rigid bodies, 522–26
  Euler's equations for rigid bodies, 740, 760
  relation to time derivative of a vector, 80–81
Ryan, Nolan, 429

S

Satellites
  acceleration during orbital motion, 107
  angular momentum examples, 370–75
  apogee and perigee, 212, 276, 370–73
  orbital mechanics concepts, 385–91, 425–27
  speed and orbital period example, 395
Saturn V rocket, 396
Scalar components. *See* Components
Scalar product, 8
Scalars, 6, 7, 8
School of Athens, 1
Scotch yokes, 468, 487
Second law of motion (Newton)
  applied to curvilinear motion, 107, 192–94
  applied to rectilinear motion of particles, 169–74
  as balance principle, 170
  Euler's first law compared, 365
  framework for kinetics applications, 234
  friction and spring forces, 171–73
  impulse-momentum principle and, 313–14
  inertial reference frame requirement, 173–74
  scalar form, 10
  statement of, 2, 5
  in systems of particles, 217–19
  work-energy principle and, 241–43
Semimajor axes, 389
Separation velocity, 337
Sequences of collisions, 344–45
Shift distance, A-10
Shock absorbers, 126, 695
Shuttles on aircraft carrier catapults, 327
Significant digits, 12
Silicon fracture, 217
Silicon nanowire example, 672–73
Sine function, linearizing, 668
Sinusoids, 712
SI system, 10–11, 12, 242–43
Skater's spin, 368–69, 625–26

Ski jumpers, 76
Skydiver descent example, 56–57
Skysurfing, 142
Slackness in chains, 287
Slider-crank mechanisms
  acceleration analysis, 46, 474–75
  types of rigid-body motions, 433–35, 438–39
  velocity analysis, 46, 452–56, 460–63
Sliding motions, 224–25, 268–69
Slip
  example problems, 532–33
  friction forces in, 171–72, 224–25, 268–69
Slugs, 10
Snapped towels, 288, 289
Sotomayor, Javier, 267
Sound speed, 54–55, 58, 59
Space, 5
Space curves, 30
Space Needle, 630
Space Shuttle, 313, 375, 627
Specific energy of satellites, 391
Speed
  converting to height, 266–67
  defined, 31, 155
  relating to position, 248–50
  relating to velocity and acceleration, 38–39
Speed governors, 380
Spheres, particles versus, 559
Spherical coordinates
  basic concepts, 143–44, 159–60
  example problems, 150
  Newton's second law applied, 194
Spherical pendulums, 152
Spinning motions, 283, 368–69
Spin rate of satellite, 374–75
Spiral pumps, 90, 115
Spring and rod example, 562–63
Spring constant, 173, 235
Springs
  crate-on-incline example, 177
  forces on, 173, 235
  linear spring and disk example, 202–3
  modeling packaging as, 189, 255, 275
  in nanowire example, 672–73
  potential energy, 259, 587
  railcar oscillation, 663–65, 695–96
  railcar stopping example, 182–83
  work on particles, 258, 306
Spring scales, 188, 230, 685, 703
Standard form of forced harmonic oscillator equation, 681–83
Standard form of harmonic oscillator equation, 665–66, 714
Static friction coefficient, 171, 234
Static-kinetic friction transitions, 178–81
Stationary frames, 121
Steady flows, 401–4, 428
Steady-state vibration, 682
Stiffness of springs, 668
Stomp rockets, 78
Stoner, Casey, 521
Stop shots, 652
Stretch, 173
Strike plates, 79, 360
Strings as models, 20, 288–89, 414–15
Strong form of Newton's third law, 2
Sun, 3, 396
Sun gears, 459, 478
Supersonic effects on propellers, 58, 59
Supports, harmonic excitation, 716
Support structures, harmonic excitation, 683
Sweet spots, 542–43
Swinging block slider cranks, 91, 118, 462–63
Swinging-bucket rotors, 544
Swinging motion examples, 270–71
Symbols for vectors, 2, 6
Symmetric geometry, 741–42, 760–61
Symmetric objects, A-4
System modeling, 27
Systems (physical), work-energy principle, 588
Systems of particles
  angular impulse-momentum principle, 362–64, 425
  closed versus open, 315, 401
  linear impulse-momentum principle, 315–16, 422
  Newton's second law applied, 217–19, 235
  work-energy principle, 281–83, 306–7, 585
Systems of rigid bodies, 620

T

Tacoma Narrows bridge, 26
Tailhooks, 319, 697
Talladega Superspeedway, 198
Tangential components of acceleration, 92, 93, 96, 98
Taylor series expansions, 666
Telstar satellite, 374
Tennis balls, 354
Terminal velocity, 57
Thelonious Monk Quartet, 663
Thin spaces, 12
Third law of motion (Newton)
  effects on systems of particles, 218, 363
  statement of, 2
Three-dimensional component systems, 193–94
Three-dimensional dynamics
  coordinate systems, 142–44
  example problems, 726–31, 747–53
  rigid body kinematics, 721–24, 759
  rigid body kinetics, 738–44, 759–62
Thrust force, 406
Time, acceleration as function, 51
Time concepts, 5
Time derivatives of vectors
  example problems, 83–86
  major concepts, 80–81, 156–57
  with normal-tangential components, 93
  notation, 29–30
  velocity and acceleration with, 33
Time differentiation, 481
Time of impact in projectile motion, 69
Time-varying processes, 1–2
Tipping motions, 532–33
Toilet flushing, 500
Top-slewing cranes, 147
Torque, 596–97
Torsional spring constant, 587, 672
Torsional spring potential energy, 587
Tow bars, 327
Towel snapping, 288, 289
Tracking, 85–86, 106, 150
Trajectories. *See also* Paths; Projectile motion
  defined, 30, 155
  normal-tangential components, 92–93, 95–98
  parabolic, 39, 390, 427
  relation to acceleration vectors, 32, 37, 38–39
  relation to velocity vectors, 31
Transient vibration, 682
Translation
  defined, 434
  example problems, 443, 532–33
  rigid bodies in planar motion, 529, 576–77
  velocity and acceleration with, 437–38, 512
Translational kinetic energy, 584

Trapeze catch example, 650–51
Trebuchets, 96
Trigonometric identities, 249
Turbines, 445
Two-dimensional component systems, 192–93

U

U.S. Customary system, 10–11, 242–43, 298
Unbalanced motors, 686–89, 706–7
Unconstrained eccentric impacts, 642–43
Unconstrained impacts
  between particles, 336, 338–39, 348–49, 423
  between rigid bodies, 640, 642–43, 658
Undamped vibrations
  defined, 664
  forced, 681–83, 715–16
  free, 663–68, 714–15
Underdamped systems, 697, 698, 716–17
Unit conversions, 11, 20
Units
  of power, 298
  systems of, 10–12
  work and energy, 242–43
Unit tangent vectors, 93
Unit vectors
  acceleration and, 15
  defined, 6
  as functions of time, 110
  time derivatives, 80–81
Universal gravitation, 2–4, 107, 385
Universal gravitational constant, 3, 385
Unstretched length of a spring, 173
Uranium enrichment, 102

V

V-22 Osprey aircraft, 58
Variable mass flows, 404–6, 412–15, 428
Variable mass systems, 315, 401
Vector addition, 8
Vector approaches
  acceleration analysis, 474, 478–81, 484–85, 514
  velocity analysis, 452–53, 513
Vector components, 7
Vector differential operator, 260
Vectors
  basic concepts in kinematics, 29–33
  Cartesian representation, 6
  finding components, 9–10, 15
  notation, 2, 6
  operations defined, 8
Velocity, acceleration as function, 52
Velocity, determining from acceleration, 56–57
Velocity analysis
  example problems, 458–65
  in rigid body planar motion, 452–56, 513
  rotating reference frames, 494–96, 515
Velocity vectors
  example problems, 35–39
  form in Cartesian component systems, 33
  normal-tangential components, 92, 93, 96
  notation, 6
  polar coordinates, 105
  properties, 30–31, 155
Verzasca dam, 264
Vibrations
  defined, 663
  example problems, 671–74, 685–89, 703–7
  undamped and forced, 681–83, 715–16
  undamped and free, 663–68, 714–15
  viscously damped, 695–700, 716–17
Vlašić, Blanka, 241
Volumetric flow rate, 403, 428
Volumetric radius of Earth, 3

W

Water jet example, 408–9
Watts, Naomi, 61
Watts (power), 298
Weight, 11
Well depth measurement, 54–55
Wheel torque, 596–97
Whip cracking, 289
Whip-upon-whip purchases, 138
Williams, Dave, 583
Wilson, Shadow, 663
Winch-and-block systems, 248–50
Wings
  lift coefficients, 24
  modeling, 27
  propellers as, 59
Wooding, Jim, 29
Woods, Tiger, 197
Work
  calculating for a force, 243–44
  of central forces on particles, 257–58
  of force systems on rigid bodies, 585–86
  relation to kinetic energy, 242–43, 305
Work-energy principle
  conservative forces and potential energy, 257–61
  example problems, 245–50, 263–72, 284–91, 342, 590–601
  for impacts, 342, 649
  particle applications, 241–44, 305
  in rigid body kinetics, 583–88, 656, 743–44, 761–62
  for systems of particles, 281–83, 306–7
World's Strongest Man competition, 321
World Championship "Punkin Chunkin" competition, 96
World records, 266, 267
Wrecking balls, 195, 263

Y

Yardstick natural frequency, 673

Properties of lines and areas. Length L, area A, centroid C, and area moments of inertia.

Quarter & Semicircular Arcs

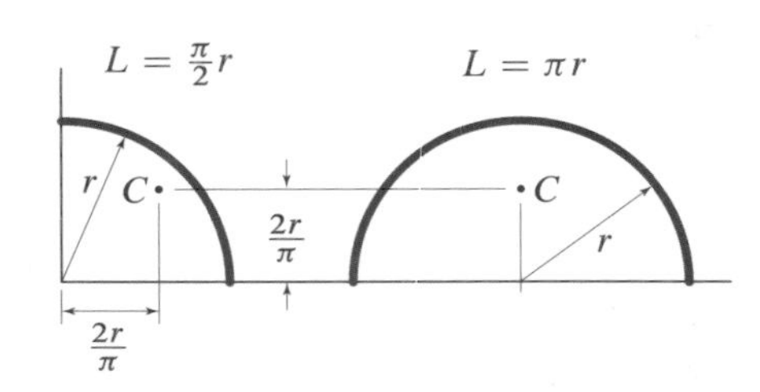

$L = \frac{\pi}{2}r$ $\quad$ $L = \pi r$

Segment of Circular Arc

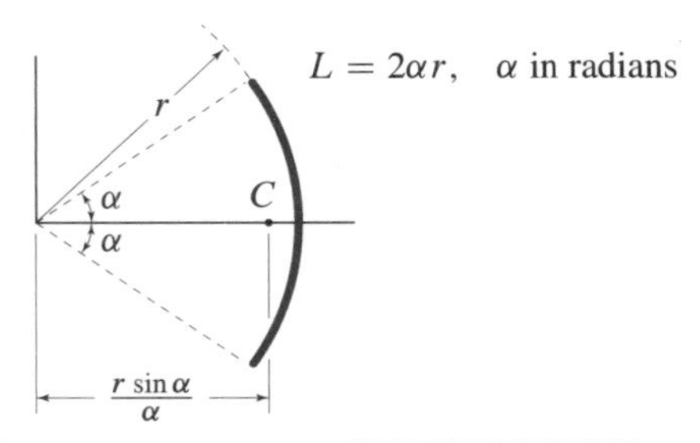

$L = 2\alpha r$, $\quad \alpha$ in radians

Rectangular Area

$A = bh$

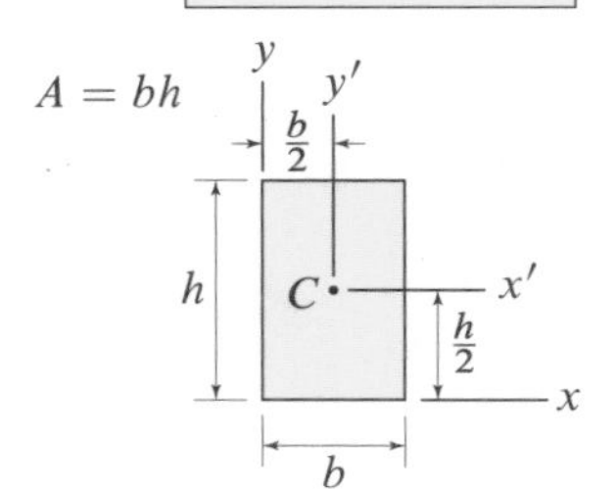

$I_{x'} = \frac{1}{12}bh^3$, $I_{y'} = \frac{1}{12}hb^3$,
$I_x = \frac{1}{3}bh^3$, $I_y = \frac{1}{3}hb^3$

Right Triangular Area

$A = \frac{1}{2}bh$

$I_{x'} = \frac{1}{36}bh^3$, $I_{y'} = \frac{1}{36}hb^3$,
$I_x = \frac{1}{12}bh^3$, $I_y = \frac{1}{4}hb^3$

Triangular Area

$A = \frac{1}{2}bh$

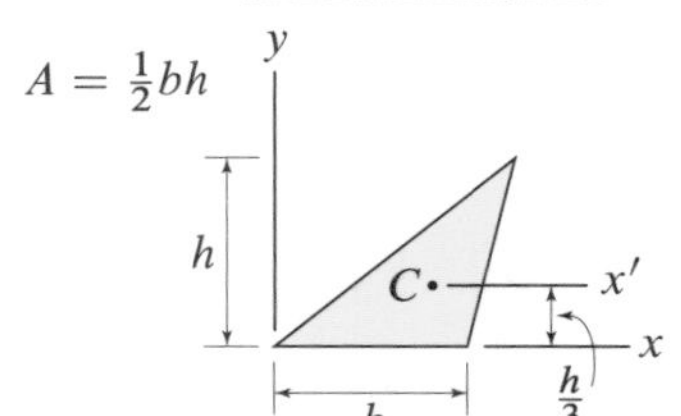

$I_{x'} = \frac{1}{36}bh^3$,
$I_x = \frac{1}{12}bh^3$

Circular Area

$A = \pi r^2$

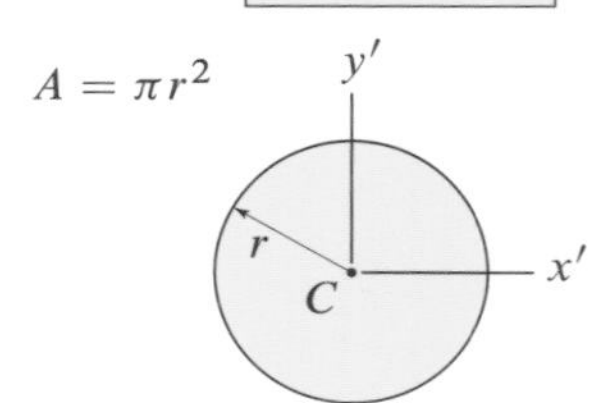

$I_{x'} = I_{y'} = \frac{\pi}{4}r^4$
$J_C = \frac{\pi}{2}r^4$

Semicircular Area

$A = \frac{1}{2}\pi r^2$

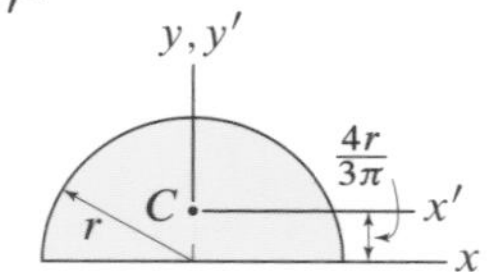

$I_{x'} = (\frac{\pi}{8} - \frac{8}{9\pi})r^4$, $I_{y'} = \frac{1}{8}\pi r^4$,
$I_x = I_y = \frac{1}{8}\pi r^4$

Quarter Circular Area

$A = \frac{1}{4}\pi r^2$

$I_{x'} = I_{y'} = (\frac{\pi}{16} - \frac{4}{9\pi})r^4$,
$I_x = I_y = \frac{1}{16}\pi r^4$